Atomic Masses of the Elements
Based on the 2005 IUPAC Table of Atomic Masses

Name	Symbol	Atomic Number	Atomic Mass
Actinium*	Ac	89	227
Aluminum	Al	13	26.981538
Americium*	Am	95	243
Antimony	Sb	51	121.760
Argon	Ar	18	39.948
Arsenic	As	33	74.92160
Astatine*	At	85	210
Barium	Ba	56	137.327
Berkelium*	Bk	97	247
Beryllium	Be	4	9.012182
Bismuth	Bi	83	208.98038
Bohrium*	Bh	107	264
Boron	B	5	10.811
Bromine	Br	35	79.904
Cadmium	Cd	48	112.411
Calcium	Ca	20	40.078
Californium*	Cf	98	251
Carbon	C	6	12.0107
Cerium	Ce	58	140.116
Cesium	Cs	55	132.90545
Chlorine	Cl	17	35.4527
Chromium	Cr	24	51.9961
Cobalt	Co	27	58.933200
Copper	Cu	29	63.546
Curium*	Cm	96	247
Darmstadtium*	Ds	110	271
Dubnium*	Db	105	262
Dysprosium	Dy	66	162.500
Einsteinium*	Es	99	252
Erbium	Er	68	167.26
Europium	Eu	63	151.964
Fermium*	Fm	100	257
Fluorine	F	9	18.9984032
Francium*	Fr	87	233
Gadolinium	Gd	64	157.25
Gallium	Ga	31	69.723
Germanium	Ge	32	72.61
Gold	Au	79	196.96655
Hafnium	Hf	72	178.49
Hassium*	Hs	108	277
Helium	He	2	4.002602
Holmium	Ho	67	164.93032
Hydrogen	H	1	1.00794
Indium	In	49	114.818
Iodine	I	53	126.90447
Iridium	Ir	77	192.217
Iron	Fe	26	55.845
Krypton	Kr	36	83.80
Lanthanum	La	57	138.9055
Lawrencium*	Lr	103	262
Lead	Pb	82	207.2
Lithium	Li	3	6.941
Lutetium	Lu	71	174.967
Magnesium	Mg	12	24.3050
Manganese	Mn	25	54.938049
Meitnerium*	Mt	109	268
Mendelevium*	Md	101	258
Mercury	Hg	80	200.59
Molybdenum	Mo	42	95.94
Neodymium	Nd	60	144.24
Neon	Ne	10	20.1797
Neptunium*	Np	93	237
Nickel	Ni	28	58.6934
Niobium	Nb	41	92.90638
Nitrogen	N	7	14.00674
Nobelium*	No	102	259
Osmium	Os	76	190.23
Oxygen	O	8	15.9994
Palladium	Pd	46	106.42
Phosphorus	P	15	30.973762
Platinum	Pt	78	195.078
Plutonium*	Pu	94	244
Polonium*	Po	84	209
Potassium	K	19	39.0983
Praseodymium	Pr	59	140.90765
Promethium*	Pm	61	145
Protactinium	Pa	91	231.03588
Radium*	Ra	88	226
Radon*	Rn	86	222
Rhenium	Re	75	186.207
Rhodium	Rh	45	102.90550
Roentgenium*	Rg	111	272
Rubidium	Rb	37	85.4678
Ruthenium	Ru	44	101.07
Rutherfordium*	Rf	104	261
Samarium	Sm	62	150.36
Scandium	Sc	21	44.955910
Seaborgium*	Sg	106	266
Selenium	Se	34	78.96
Silicon	Si	14	28.0855
Silver	Ag	47	107.8682
Sodium	Na	11	22.989770
Strontium	Sr	38	87.62
Sulfur	S	16	32.066
Tantalum	Ta	73	180.9479
Technetium*	Tc	43	98
Tellurium	Te	52	127.60
Terbium	Tb	65	158.92534
Thallium	Tl	81	204.3833
Thorium	Th	90	232.0381
Thulium	Tm	69	168.93421
Tin	Sn	50	118.710
Titanium	Ti	22	47.867
Tungsten	W	74	183.94
Uranium	U	92	238.0289
Vanadium	V	23	50.9415
Xenon	Xe	54	131.29
Ytterbium	Yb	70	173.04
Yttrium	Y	39	88.90585
Zinc	Zn	30	65.39
Zirconium	Zr	40	91.224

* This element has no stable isotopes. The atomic mass given is that of the isotope with the longest known half-life.

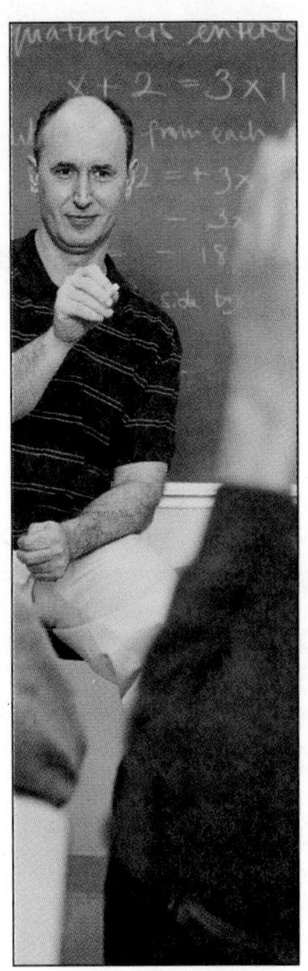

Every one of your students has the potential to make a difference. And realizing that potential starts right here, in your course.

WILEY PLUS

When students succeed in your course—when they stay on-task and make the breakthrough that turns confusion into confidence—they are empowered to realize the possibilities for greatness that lie within each of them. We know your goal is to create an environment where students reach their full potential and experience the exhilaration of academic success that will last them a lifetime. *WileyPLUS* can help you reach that goal.

Wiley**PLUS** is a suite of resources—including the complete online text—that will help your students:

- come to class better prepared for your lectures
- get immediate feedback and context-sensitive help on assignments and quizzes
- track their progress throughout the course

www.wileyplus.com

88% of students surveyed said it improved their understanding of the material.*

FOR INSTRUCTORS

Wiley**PLUS** is built around the activities you perform in your class each day. With Wiley**PLUS** you can:

Create Assignments

Automate the assigning and grading of homework or quizzes by using the provided question banks, or by writing your own.

Prepare & Present

Create outstanding class presentations using a wealth of resources such as PowerPoint™ slides, image galleries, Interactive LearningWare, and more. You can even add materials you have created yourself.

Track Student Progress

Keep track of your students' progress and analyze individual and overall class results.

Now Available with WebCT!

> "It has been a great help, and I believe it has helped me to achieve a better grade."
>
> Michael Morris, *Columbia Basin College*

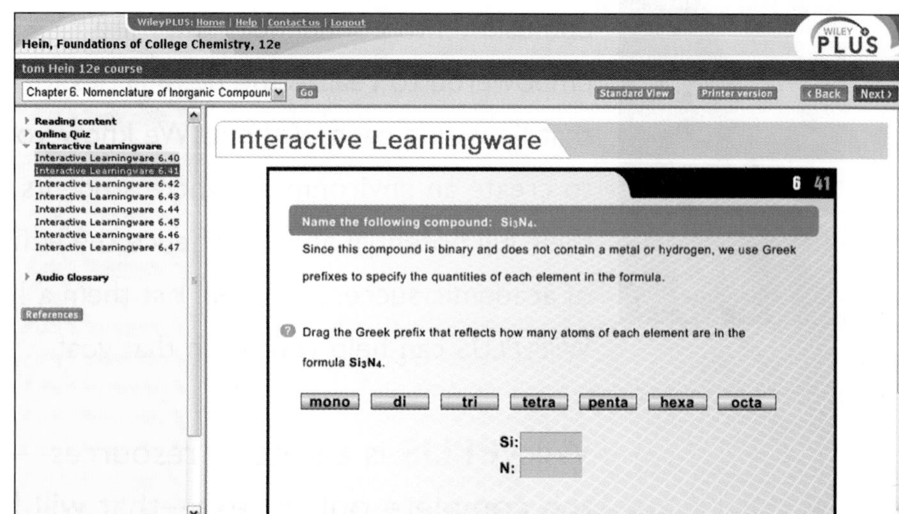

FOR STUDENTS

You have the potential to make a difference!

WileyPLUS is a powerful online system packed with features to help you make the most of your potential and get the best grade you can!

With Wiley**PLUS** you get:

- A complete online version of your text and other study resources.
- Problem-solving help, instant grading, and feedback on your homework and quizzes.
- The ability to track your progress and grades throughout the term.

For more information on what *WileyPLUS* can do to help you and your students reach their potential, please visit www.wileyplus.com.

76% of students surveyed said it made them better prepared for tests. *

*Based on a survey of 972 student users of *WileyPLUS*

Introduction to General, Organic, and Biochemistry

NINTH EDITION

Introduction to General, Organic, and Biochemistry

NINTH EDITION

Morris Hein
Mount San Antonio College

Scott Pattison
Ball State University

Susan Arena
University of Illinois, Urbana–Champaign

Leo R. Best
Mount San Antonio College

John Wiley & Sons, Inc

We dedicate this edition to the memory of our esteemed colleague Leo R. Best, who was one of the original authors of this textbook.

Vice-President & Executive Publisher: *Kaye Pace*
Associate Publisher: *Petra Recter*
Marketing Manager: *Amanda Wygal Wainer*
Editorial Project Coordinator: *Catherine Donovan*
Production Manager: *Dorothy Sinclair*
Senior Production Editor: *Sandra Dumas*
Outside Project Management: *Ingrao Associates*
Director of Creative Services: *Harry Nolan*
Senior Designer: *Hope Miller*
Photo Department Manager: *Hilary Newman*
Photo Editor: *Tara Sanford*
Senior Media Editor: *Thomas Kulesa*
Cover Photo: *Courtesy Dr. D.L. Fernandes, Ludger Ltd., UK.*
Cover: α-Subunit of human chorionic gonadotropin is decorated with a biantennary glycan (space-filling structure).

This book was set in 10/12 Minister Light by Preparé Inc. and printed and bound by QuebecorWorld. The cover was printed by QuebecorWorld.

This book is printed on acid free paper. ∞

ISBN-13 978-0-470-12925-8

Printed in the United States of America

10 9 8 7 6 5 4 3

ABOUT THE AUTHORS

Morris Hein is professor emeritus of chemistry at Mt. San Antonio College, where he regularly taught the preparatory chemistry course and organic chemistry. He is the original author of *Foundations of College Chemistry* and his name has become synonymous with clarity, meticulous accuracy, and a step-by-step approach that students can follow. Over the years, more than three million students have learned chemistry using a text by Morris Hein. In addition to *Foundations of College Chemistry, 12E*, he is co-author of *Introduction to General, Organic and Biochemistry, 9E*. He is also the co-author of *Foundations of Chemistry in the Laboratory, 12E* and *Introduction to General, Organic and Biochemistry in the Laboratory, 9E*.

Scott E. Pattison lives in Muncie, Indiana, where he is a professor of chemistry at Ball State University. He maintains active, current research involving zinc metabolism that provides laboratory experience for both undergraduate and graduate students. A dedicated teacher at the university level for twenty-seven years, his primary area of instruction is biochemistry; however, he greatly enjoys teaching medical/nursing chemistry and general chemistry. Scott became a co-author with Morris Hein on the third edition of *Introduction to General, Organic, and Biochemistry*, and he brings his knowledge of students, subject matter, and current research to the text. In addition to his professional career, he volunteers in local schools from preschool through high school, providing extra science experiences for tomorrow's university students.

Susan Arena has taught chemistry to students at many levels including middle school, high school, community college and most recently at University of Illinois, Urbana-Champaign. She especially focuses on using active learning techniques to improve the understanding of concepts in chemistry. Susan currently authors chemistry texts and electronic media, and presents workshops for teachers in using active learning and electronic media to teach chemistry. She collaborated with Morris Hein on the seventh edition of *Foundations of College Chemistry* and became a co-author on the eighth and subsequent editions.

Leo R. Best taught chemistry for twenty-four years at Mt. San Antonio College. He had been a collaborator and co-author with Morris Hein since the first edition of *Introduction to General, Organic, and Biochemistry*. He was co-author of *Foundations of Chemistry in the Laboratory* and *Introduction to General, Organic, and Biochemistry in the Laboratory*.

PREFACE

This new Ninth Edition of *Introduction to General, Organic, and Biochemistry* presents chemistry as a modern, vital subject and is designed to make introductory chemistry accessible to all beginning students. The goal for this edition is to continue to present chemistry in a clear, engaging manner that will stimulate students to further their scientific knowledge as they prepare for health sciences, nursing, and other careers. *Introduction to General, Organic, and Biochemistry* was originally intended for students who had never taken a chemistry class but have a limited mathematical background, or those who had a significant interruption in their studies and are now returning to school to pursue various career objectives. The central focus is the same as it has been from the first edition: making the wide variety of chemical processes, occurring both within our bodies and in our surroundings, accessible to students and teaching them the problem-solving skills they will need in their future studies.

In preparing this new edition we considered the comments and suggestions of students and instructors to design a revision that builds on the strengths of previous editions. We have especially tried to relate chemistry to the real lives of our students as we develop the principles that form the foundation for the futher study of general, organic, and biochemistry.

Development of Problem-Solving Skills

We all want our students to develop real skills in solving problems. We believe that a key to the success of this text is the fact that our problem-solving approach works for students. It is a step-by-step process that teaches the use of units and shows the change from one unit to the next. Students learn concepts most easily in a step-by-step process. In this edition we continue to use examples to incorporate fundamental mathematical skills, scientific notation, and significant figures. Painstaking care has been taken to show each step in the problem-solving process and to give *alternative methods for solutions* where appropriate. These alternative methods give students flexibility in choosing the one that works best for them. We continue to use four significant figures for atomic and molar masses for consistency and for rounding off answers appropriately. We have been meticulous in providing answers, correctly rounded, for students who have difficulty with mathematics.

Fostering Student Skills *Attitude* plays a critical role in problem solving. We encourage students to learn that a systematic approach to solving problems is better than simple memorization. Throughout the book we encourage students to begin by writing down the facts or data given in a problem and to think their way through the problem to an answer, which is then checked to see if it makes sense. Once we have laid the foundations of concepts, we highlight the steps so students can locate them easily. Important rules and equations are highlighted for emphasis and ready reference.

Student Practice Practice problems follow the examples in the text, with answers provided at the end of the chapter, and in this edition a number of new practice problems have been added throughout the text. The end of each chapter begins with *Review Questions,* which help students review key terms and concepts, as well as material presented in tables and figures. This is followed by *Paired Exercises,* covering concepts and numerical exercises, where two similar exercises are presented side by side. *Additional Exercises,* includes further practice problems presented in a more random order. Finally, *Challenge Exercises* present problems designed to take the student beyond the basic chapter material. This new edition includes many new exercises and problems. Answers for selected questions and exercises appear in Appendix VI and answers for *Putting It Together* review exercises appear in Appendix VII.

Emphasis on Real-World Aspects

We continue to emphasize the less theoretical aspects of general chemistry early in the book, leaving the more abstract theory for later. This sequence seems especially appropriate in a course where students are encountering chemistry for the very first time. Atoms, molecules, and reactions are all an integral part of the chemical nature of matter. A sound understanding of these topics allows the student to develop a basic understanding of chemical properties and vocabulary.

We build toward a basic knowledge of organic and biochemistry for the health science student. Thus, we stress the nomenclature, structure, and reactivity of major organic functional groups. In turn, the basic biochemical concepts rest on this foundation. We encourage the students to apply their understanding to examples drawn from medicine, nutrition, agriculture, and so on.

Chapters 1 through 3 present the basic mathematics and the language of chemistry, including an explanation of the metric system and significant figures. In Chapter 4 we present chemical properties—the ability of a substance to form new substances. Then, in Chapter 5, students encounter the history and language of basic atomic theory.

We continue to present new material at a level appropriate for the beginning student by emphasizing nomenclature, composition of compounds, and reactions in Chapters 6 through 9 before moving into the details of modern atomic theory. Some applications of the Periodic Table are shown in early chapters and discussed in detail in Chapters 10 and 11. Students gain confidence in their own ability to identify and work with chemicals in the laboratory before tackling the molecular models of matter. As practicing chemists we have little difficulty connecting molecular models and chemical properties. Students, especially those with no prior chemistry background, may not share this ability to connect the molecular models and the macroscopic properties of matter. Those instructors who feel it is essential to teach atomic theory and bonding early in the course can cover Chapters 10 and 11 immediately following Chapter 5.

In Chapters 19 through 26 we introduce organic chemistry. We have reviewed and carefully selected organic reactions to illustrate the reactivities of important functional groups. Three general categories of organic reactions—substitution, elimination, and addition—are introduced and, where possible, subsequent chapters present reactions within this conceptual framework.

IUPAC nomenclature is emphasized in this edition, but we have also considered how organics are named in everyday usage. Thus, we present a common name if it continues to be widely used.

Finally, we examine the principles of biochemistry in Chapters 27 through 35 and again strive to illustrate current, relevant applications such as the Human Genome Project, gene therapy, and sources for, and uses of, industrial enzymes. Because biochemistry is increasingly a visual science, many new molecular models have been incorporated to help students better comprehend biochemical functions. As such, Chapter 29, "Amino Acids, Polypeptides, and Proteins," features a three-dimensional structure-to-function approach and Chapter 30, "Enzymes," stresses a qualitative and visual approach to enzymes.

Realistic molecular pictures are used throughout both the organic and biochemistry sections. Important individual molecules are shown in the margins and molecular pictures are incorporated into selected chemical reactions. It is our intent that students be able to see the real changes in molecules that are occurring constantly in nature.

New to This Edition

In this Ninth Edition we have tried to build on the strengths of the previous editions. We have worked to update the language and to provide students with clear explanations for concepts and useful ways to solve chemistry problems. We continually strive to keep the material at the same level so students can easily read and use the text to learn chemistry. Some specific changes in the text are highlighted below:

- **Chapter 1** has been rewritten to include a discussion of matter and physical and chemical changes. This new material engages the students immediately into the vocabulary of chemistry and the particulate nature of matter. We have added new end-of-chapter review material and exercises so that students immediately begin to review and practice what they are learning.
- **Chapter 3** has been revised for a greater emphasis on classification of matter and to include an introduction to the periodic table. This introductory material supports laboratory work and presents an overview of how the periodic table provides the big picture for how the elements are related.
- **Chapter 13** has been reorganized to move the discussion of hydrogen bonding and hydrates forward to complete the discussion of liquids. The information on water has been collected in a new section called Water, a Unique Liquid and serves to discuss water and its properties as an example of a unique liquid.
- **Chapter 19** starts the organic/biochemistry sections by considering the various methods for representing molecules. This includes an introduction to line structures of organic compounds. More examples of realistic molecular pictures are included throughout the remaining chapters, often in parallel

with molecular formulas or ball-and-stick molecular depictions. Students have the opportunity to see actual shapes and relative sizes of many common molecules.

- **Chapter 20** continues a theme started in Chapter 19: Some simple organic reactions are easiest to understand when considered as a step-wise process—an organic mechanism. Common organic mechanisms help students to predict products of simple organic reactions and help students to start dealing with the general question—why do these molecules react the way they do?
- **Chapter 27** now includes a section covering complex carbohydrates including the blood group substances as examples of glycoproteins.
- **Chapter 32** has been significantly updated to include the latest nutrition information. Among other changes, new nutrition measures, the Dietary Reference Intake (DRI) and the Estimated Energy Requirement (EER) are carefully defined and the new USDA food pyramid is described.
- **Molecular art**: Learning chemistry requires the ability to connect the macroscopic world of everyday life to the microscopic world of atoms and molecules. In this edition we have added molecular art to macroscopic pictures to emphasize this connection.
- **Chapter Reviews** have been reformatted to include key terms and to review each section in a bulleted format similar to what a student might use to review the material. We have included summary art from the chapter in this section to aid visual learners.
- **Design and Illustration Program:** This edition has a fresh new design, enhanced by an updated art and photo program. New photos appear throughout.
- **Problem Solving:** Because the development of problem-solving skills is essential to learning chemistry, in this new design the sections on problem solving are highlighted throughout for easy student reference.
- **New Exercises:** Many *new* end of chapter exercises and questions have been added. *Challenge Exercises* have also been added to most chapters.
- **Math Skills Learning Aids: A Review of Mathematics** is provided as Appendix I for students who need assistance.
- **New Book Companion Web Site:** This site supplies a number of helpful resources for students and instructors, including **Interactive Learningware Problems**, **Online Quizzing**, and **Power Point Lecture Slides.**
- **Chemistry in Action** sections, a number new to this edition, show the impact of chemistry in a variety of practical applications. Given the rapid growth of knowledge in organic and biochemistry over 90% of the Chemistry in Action sections are new to this edition. These essays cover a range of relevant topics and introduce experimental information on new chemical discoveries and applications.
- **Important statements**, equations, and laws are highlighted for emphasis. A **Glossary** is provided to help students review key terms, with section numbers given for each term to guide the student to the contextual definition. The margins of the glossary pages are color tinted to provide ready access.
- **End of chapter questions and exercises**, provide practice and review of the chapter material. *Paired exercises* present two parallel exercises, side by side, so the student can solve one problem, check the answer in Appendix VI,

and use the same problem-solving skills with the second exercise. **Additional exercises** are provided at the end of most chapters, arranged in random order, to encourage students to review the chapter material. **Answers to selected questions and exercises** are given in Appendix VI.

- **Putting It Together** review sections appear after every 2 to 4 chapters and include additional conceptually oriented exercises for effective self-review. These review sections are printed on colored pages for easy reference and provide students with a helpful summary of the preceding material. **Answers to Putting It Together** review sections are given in Appendix VII.
- Directions on using a calculator to solve problems are given in Appendix II, **Using a Scientific Calculator**.
- **Units of measurement** are shown in table format in Appendix IV and in the endpapers.

Learning Aids

To help the beginning student gain the confidence necessary to master technical material we have refined and enhanced a series of learning aids:

- Important **terms** are set off in bold type where they are defined, and are printed in blue in the margin. Most glossary terms are also defined in the glossary.
- Worked **examples** show students the how of problem solving before they are asked to tackle problems on their own.
- **Practice problems** permit immediate reinforcement of a skill shown in the example problems. Answers are provided at the end of the chapter to encourage students to check their problem solving immediately.
- **Marginal notations** help students understand basic concepts and problem-solving techniques. These are printed in green ink to clearly distinguish them from text and vocabulary terms.

Learning Aids: Math Skills For students who may need help with the mathematical aspects of chemistry, the following learning aids are available:

- A **Review of Mathematics**, covering the basic functions, is provided in Appendix I.
- **Math Survival Guide: Tips and Tricks for Science Students**, 2nd Edition, by Jeffrey R. Appling and Jean C. Richardson, a brief paperback summary of basic skills that can be packaged with the text, provides an excellent resource for students who need help with the mathematical aspects of chemistry.

Student and Instructor Supplements

For the Student

Student Solutions Manual by Morris Hein, Scott Pattison, Susan Arena and Kathy Mitchell, ISBN 978-0-470-24765-5: includes answers and solutions to all end-of-chapter questions and exercises.

Introduction to General, Organic and Biochemistry in the Laboratory, 9th Edition by Morris Hein, Judith Peisen and James M. Ritchey, ISBN 978-0-470-23965-0: includes 42 experiments for a laboratory program that may accompany the lecture course. It has been completely updated and revised to reflect the most current terminology and environmental standards, and features up-to-date information on waste disposal and safe laboratory procedures. The Manual also includes study aides and exercises.

Math Survival Guide: Tips and Tricks for Science Students, 2nd Edition, Jeffrey R. Appling and Jean C. Richardson, ISBN 978-0-471-27054-6 a brief paperback summary of basic skills with practice exercises in every chapter. This guide provides an excellent resource for students who need help with the mathematical aspects of chemistry.

Student Companion Web Site at www.wiley.com/college/hein The Web Site includes a number of helpful tools and materials to enhance student learning:

- **Interactive Learningware:** a step-by-step problem-solving tutorial program that guides students through over 40 interactive problems, prepared by Bette Kruez, University of Michigan-Dearborn and Margaret Kimble, Indiana-Purdue University, Fort Wayne.
- The website also features **Online Quizzes**, organized by chapter, to provide students with additional practice and immediate reinforcement. These questions have been prepared by Bette Kreuz, University of Michigan-Dearborn and Igor Alabugin, Florida State University, Tallahassee.

For the Instructor

WileyPLUS combines the complete, dynamic online text with all the teaching and learning resources you need, in an easy-to-use system. ***WileyPLUS*** allows you to deliver all or a portion of your course online. With ***WileyPLUS*** you can:

- Create and assign online homework that is automatically graded and closely correlated to the text. The questions, organized by chapter and topic, offer students practice with instant feedback that explains why an answer choice is right or wrong.
- Manage your students' results in the online gradebook.
- Build media-rich class presentations.
- Customize your course to meet your course objectives.
- All instructor supplements (test bank, PPTs, etc.) are available for download from within ***WileyPLUS*** for convenience.

Provided at **no charge** when packaged with a new textbook or students can purchase the access code stand alone.

Text and WileyPLUS bundle 978-0-470-28682-1
Stand alone WileyPLUS: 978-0-470-11729-3

Instructor's Companion Web Site at www.wiley.com/college/hein Instructors have access to all student Web Site features and, also, the following useful resources:

Test Bank, prepared by Anthony Stellato, Suffolk County Community College and Paul D. Root, Henry Ford Community College. Includes more than 3,400 true-false, multiple-choice and open-ended questions and answers.

Computerized Test Bank, the test bank contains true-false, multiple-choice, open-ended questions, and answers to all test questions. Available in two formats (MAC and PC).

Instructor's Manual for Introduction to General, Organic, and Biochemistry in the Laboratory, 9E, by Morris Hein and Judith Peisen. Fully updated and revised. Includes information on the management of the lab, evaluation of experiments, notes for individual experiments, a list of reagents and amounts needed for each experiment, and answer keys to each experiment report form and all exercises.

Digital Image Library: images from the text are available online in JPEG format. Instructors may use these to customize their presentations and to provide additional visual support for quizzes and exams.

Power Point Lecture Slides: created by Eugene Passer, Bronx Community College and Jerry Poteat, Georgia Perimeter College, Dunwoody. These slides contain lecture outlines and key topics from each chapter of the text, along with supporting artwork and figures from the text.

Acknowledgments

No textbook can be completed without the untiring efforts of many publication professionals. We thank the talented staff at John Wiley & Sons, especially Editorial Project Coordinator Catherine Donovan who worked very hard to control the many aspects of this project. Much credit also goes to the Project Manager Suzanne Ingrao of Ingrao Associates for her unfailing attention to detail and persistence in moving this book through the numerous stages of production.

We are grateful for the many helpful comments from colleagues and students who, over the years, have made this book possible. We hope they will continue to share their ideas for change with us, either directly or through our publisher.

We are especially thankful for the support, friendship, and constant encouragement of our spouses, Edna, Joan and Steve.

Our sincere appreciation goes to the following reviewers who were kind enough to read and give their professional comments:

Rebecca Barlag, *Ohio University*
Tara S. Carpenter, *University of Maryland, Baltimore County*
James D. Carr, *University of Nebraska*
Eugene F. Douglass, *University of North Carolina-Pembroke*
Larry Emme, *Chemeketa Community College*
Gerald Ittenbach, *Fayetteville Technical Community College*
Michael L. Kirby, *Des Moines Area Community College*
Stephen Milczanowski, *Florida Community College-Jacksonville*
Suzanne Williams, *Northern Michigan University*

Morris Hein, Scott Pattison, and Susan Arena

BRIEF CONTENTS

CONTENTS

Introduction to General, Organic, and Biochemistry

NINTH EDITION

CHAPTER 1

An Introduction to Chemistry

The colors, fragrances, and textures of a tulip garden are all the results of chemistry.

Chapter Outline

Do you know how the battery in your car works to start your engine? Have you ever wondered how a tiny seedling can grow into a corn stalk taller than you in just one season? Perhaps you have been mesmerized by the flames in your fireplace on a romantic evening as they change color and form. And think of your relief when you dropped a container and found that it was plastic, not glass. These phenomena are the result of chemistry that occurs all around us, all the time. Chemical changes bring us beautiful colors, warmth, light, and products to make our lives function more smoothly. Understanding, explaining, and using the diversity of materials we find around us is what chemistry is all about.

1.1 Why Study Chemistry?

A knowledge of chemistry is useful to virtually everyone—we see chemistry occurring around us every day. An understanding of chemistry is useful to doctors, health care professionals, attorneys, homemakers, businesspeople, firefighters, and environmentalists just to name a few. Even if you're not planning to work in any of these fields, chemistry is important and is used by us every day. Learning about the benefits and risks associated with chemicals will help you to be an informed citizen, able to make intelligent choices concerning the world around you. Studying chemistry teaches you to solve problems and communicate with others in an organized and logical manner. These skills will be helpful in college and throughout your career.

1.2 The Nature of Chemistry

chemistry

Key words are highlighted in bold and color in the margin to alert you to new terms defined in the text.

What is chemistry? One dictionary gives this definition: "**Chemistry** is the science of the composition, structure, properties, and reactions of matter, especially of atomic and molecular systems." Another, somewhat simpler definition is "Chemistry is the science dealing with the composition of *matter* and the changes in composition that matter undergoes." Neither of these definitions is entirely adequate. Chemistry and physics form a fundamental branch of knowledge. Chemistry is also closely related to biology, not only because living organisms are made of material substances but also because life itself is essentially a complicated system of interrelated chemical processes.

The scope of chemistry is extremely broad. It includes the whole universe and everything, animate and inanimate, in it. Chemistry is concerned with the composition and changes in the composition of matter and also with the energy and energy changes associated with matter. Through chemistry we seek to learn and to understand the general principles that govern the behavior of all matter.

The chemist, like other scientists, observes nature and attempts to understand its secrets: What makes a tulip red? Why is sugar sweet? What is occurring when iron rusts? Why is carbon monoxide poisonous? Problems such as these—some of which have been solved, some of which are still to be solved—are all part of what we call chemistry.

A chemist may interpret natural phenomena, devise experiments that reveal the composition and structure of complex substances, study methods for improving natural processes, or synthesize substances. Ultimately, the efforts of successful chemists advance the frontiers of knowledge and at the same time contribute to the well-being of humanity.

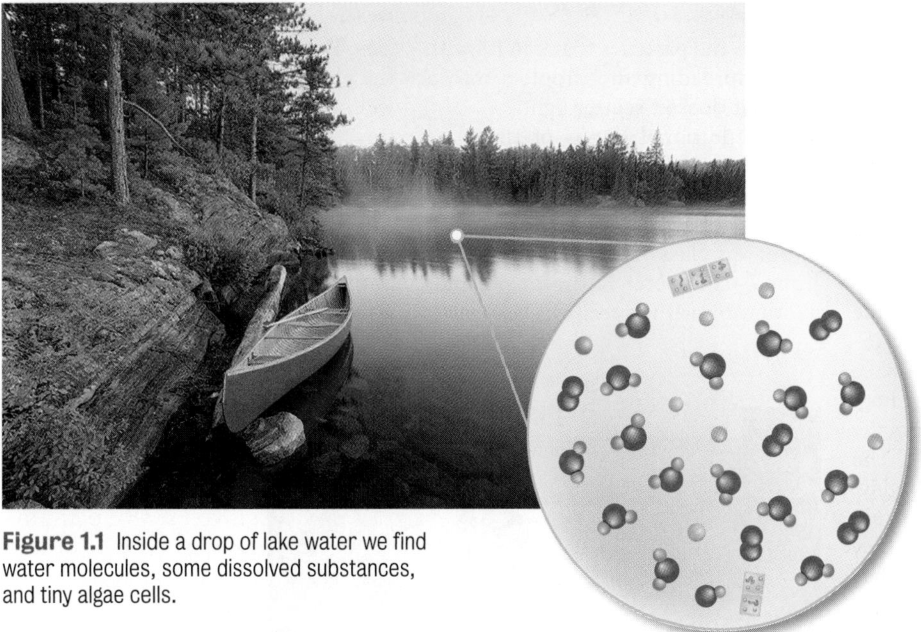

Figure 1.1 Inside a drop of lake water we find water molecules, some dissolved substances, and tiny algae cells.

1.3 Thinking Like a Chemist

Chemists take a special view of things in order to understand the nature of the chemical changes taking place. Chemists "look inside" everyday objects to see how the basic components are behaving. To understand this approach, let's consider a lake. When we view the lake from a distance, we get an overall picture of the water and shoreline. This overall view is called the *macroscopic* picture.

As we approach the lake we begin to see more details—rocks, sandy beach, plants submerged in the water, and aquatic life. We get more and more curious. What makes the rocks and sand? What kind of organisms live in the water? How do plants survive underwater? What lies hidden in the water? We can use a microscope to learn the answers to some of these questions. Within the water and the plants, we can see single cells and inside them organelles working to keep the organisms alive. For answers to other questions, we need to go even further inside the lake. A drop of lake water can itself become a mysterious and fascinating *microscopic* picture full of molecules and motion. (See Figure 1.1). A chemist looks into the world of atoms and molecules and their motions. Chemistry makes the connection between the *microscopic* world of molecules and the *macroscopic* world of everyday objects.

Think about the water in the lake. On the surface it has beauty and colors, and it gently laps the shore of the lake. What is the microscopic nature of water? It is composed of tiny molecules represented as

In this case H represents a hydrogen atom and O an oxygen atom. The water molecule is represented by H_2O since it is made up of two hydrogen atoms and one oxygen atom.

CHEMISTRY IN ACTION • Clearing the Fog

If only someone could find a way to keep that annoying fog off mirrors in the bathroom on a more permanent basis. Michael Rabner, a materials scientist at MIT, just might have figured out how to clear the fog from glass surfaces. He found his inspiration in the beautiful lotus plant. It turns out that lotus leaves repel water so well that when raindrops hit the leaf, the drops remain spherical. He began exploring the possibility of trying to make a coating that did the same thing. Fogging occurs when water drops from the air condense on a cool surface. The drops scatter light, creating the fogging. The majority of antifogging sprays are polymers that flatten the drops so they do not scatter light. Unfortunately, they wear off the surface quickly.

Rubner and his colleagues found a nanocoating (a very thin coating) that had the opposite effect of the lotus leaf. Their coating is extremely water loving, or superhydrophilic. It is composed of tiny, hydrophilic glass particles that are packed together very irregularly. This leaves tiny pockets between the particles that can fill with water, spreading the droplets into a film that doesn't scatter light.

An additional bonus of the glass and air coating is that it prevents glare. Nearly 100% of light travels through this nanocoat, compared to 92% on untreated glass. The coating is made from 7-nanometer layers of polymer alternating with layers of glass particles. The layers are constructed so that the glass particles don't pack together well, producing a Swiss cheese structure. To increase the durability of the coating, it is heated to 500°C to fuse the glass particles and disintegrate the polymer. This makes the coating scratch-resistant and also makes the antifog property last for more than a year.

A lotus flower seen through nanocoated glass.

1.4 A Scientific Approach to Problem Solving

One of the most common and important things we do every day is to solve problems. For example,

- You have two exams and a laboratory report due on Monday. How should you divide your time?
- You leave for school and learn from the radio that there is a big accident on the freeway. What is your fastest alternate route to avoid the traffic problem?
- You need to buy groceries, mail some packages, attend your child's soccer game, and pick up the dry cleaning. What is the most efficient sequence of events?

We all face these kind of problems and decisions. A logical approach can be useful for solving daily problems:

1. Define the problem. We first need to recognize we have a problem and state it clearly, including all the known information. When we do this in science, we call it *making an observation*.
2. Propose possible solutions to the problem. In science this is called *making a hypothesis*.

3. Decide which is the best way to proceed or solve the problem. In daily life we use our memory of past experiences to help us. In the world of science we *perform an experiment*.

Using a scientific approach to problem solving is worthwhile. It helps in all parts of your life whether you plan to be a scientist, doctor, businessperson, or writer.

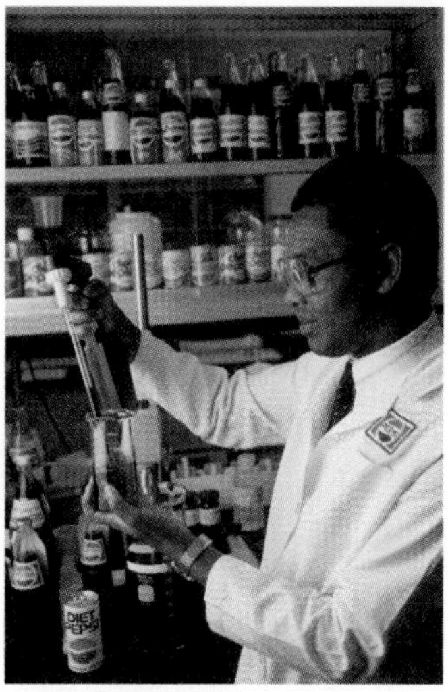

Scientists employ the scientific method every day in their laboratory work.

1.5 The Scientific Method

Chemists work together and also with other scientists to solve problems. As scientists conduct studies they ask many questions, and their questions often lead in directions that are not part of the original problem. The amazing developments from chemistry and technology usually involve what we call the **scientific method**, which can generally be described as follows:

scientific method

1. **Collect the facts or data** that are relevant to the problem or question at hand. This is usually done by planned experimentation. The data are then analyzed to find trends or regularities that are pertinent to the problem.
2. **Formulate a hypothesis** that will account for the data and that can be tested by further experimentation.
3. **Plan and do additional experiments to test the hypothesis**.
4. **Modify the hypothesis** as necessary so that it is compatible with all the pertinent data.

Confusion sometimes arises regarding the exact meanings of the words *hypothesis*, *theory*, and *law*. A **hypothesis** is a tentative explanation of certain facts that provides a basis for further experimentation. A well-established hypothesis is often called a **theory** or model. Thus a theory is an explanation of the general principles of certain phenomena with considerable evidence or facts to support it. Hypotheses and theories explain natural phenomena, whereas **scientific laws** are simple statements of natural phenomena to which no exceptions are known under the given conditions.

hypothesis

theory

scientific laws

These four steps are a broad outline of the general procedure that is followed in most scientific work, but they are not a "recipe" for doing chemistry or any other science (Figure 1.2). Chemistry is an experimental science, however, and much of its progress has been due to application of the scientific method through systematic research.

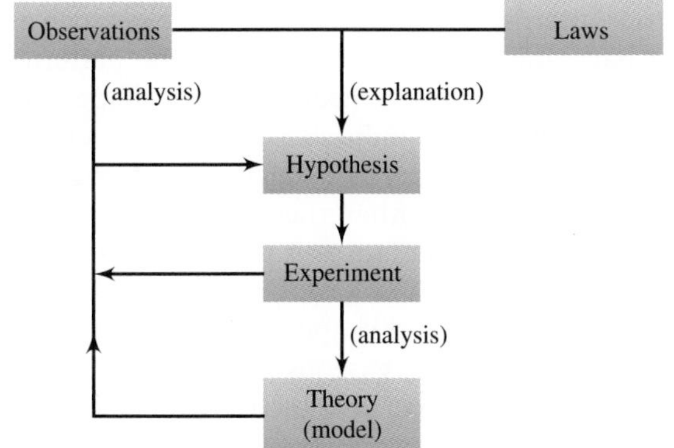

Figure 1.2
The scientific method.

Chemists continue to explore the effects of nicotine on the smoker in order to help those who want to quit smoking.

We study many theories and laws in chemistry; this makes our task as students easier because theories and laws summarize important aspects of the sciences. Certain theories and models advanced by great scientists in the past have since been substantially altered and modified. Such changes do not mean that the discoveries of the past are any less significant. Modification of existing theories and models in the light of new experimental evidence is essential to the growth and evolution of scientific knowledge. Science is dynamic.

1.6 The Particulate Nature of Matter

The entire universe consists of matter and energy. Every day we come into contact with countless kinds of matter. Air, food, water, rocks, soil, glass, and this book are all different types of matter. Broadly defined, **matter** is *anything* that has mass and occupies space.

matter

Matter may be quite invisible. For example, if an apparently empty test tube is submerged mouth downward in a beaker of water, the water rises only slightly into the tube. The water cannot rise further because the tube is filled with invisible matter: air (see Figure 1.3).

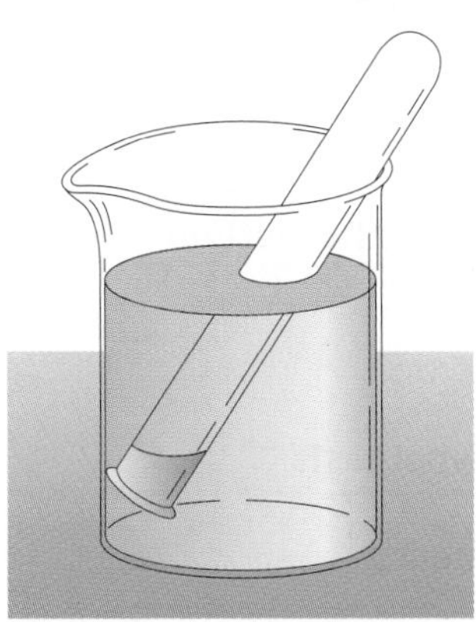

Figure 1.3 An apparently empty test tube is submerged, mouth downward, in water. Only a small volume of water rises into the tube, which is actually filled with invisible matter—air.

To the macroscopic eye, matter appears to be continuous and unbroken. We are impressed by the great diversity of matter. Given its many forms, it is difficult to believe that on a microscopic level all of matter is composed of discrete, tiny, fundamental particles called *atoms*. It is truly amazing to understand that the fundamental particles in ice cream are very similar to the particles in air that we breathe. Matter is actually discontinuous and is composed of discrete, tiny particles called *atoms*.

1.7 Physical States of Matter

Matter exists in three physical states: solid, liquid, and gas (see Figure 1.4). A **solid** has a definite shape and volume, with particles that cling rigidly to one another. The shape of a solid can be independent of its container. In Figure 1.4a we see water in its solid form. Another example, a crystal of sulfur, has the same shape and volume whether it is placed in a beaker or simply laid on a glass plate.

solid

Most commonly occurring solids, such as salt, sugar, quartz, and metals, are *crystalline*. The particles that form crystalline materials exist in regular, repeating, three-dimensional, geometric patterns (see Figure 1.5). Some solids such as plastics, glass, and gels do not have any regular, internal geometric pattern. Such solids are called **amorphous** solids. (*Amorphous* means "without shape or form").

amorphous

liquid

A **liquid** has a definite volume but not a definite shape, with particles that stick firmly but not rigidly. Although the particles are held together by strong attractive forces and are in close contact with one another, they are able to move freely. Particle mobility gives a liquid fluidity and causes it to take the shape of the container in which it is stored. Note how water looks as a liquid in Figure 1.4b.

gas

A **gas** has indefinite volume and no fixed shape, with particles that move independently of one another. Particles in the gaseous state have gained enough energy to overcome the attractive forces that held them together as liquids or solids. A gas presses continuously in all directions on the walls of any container.

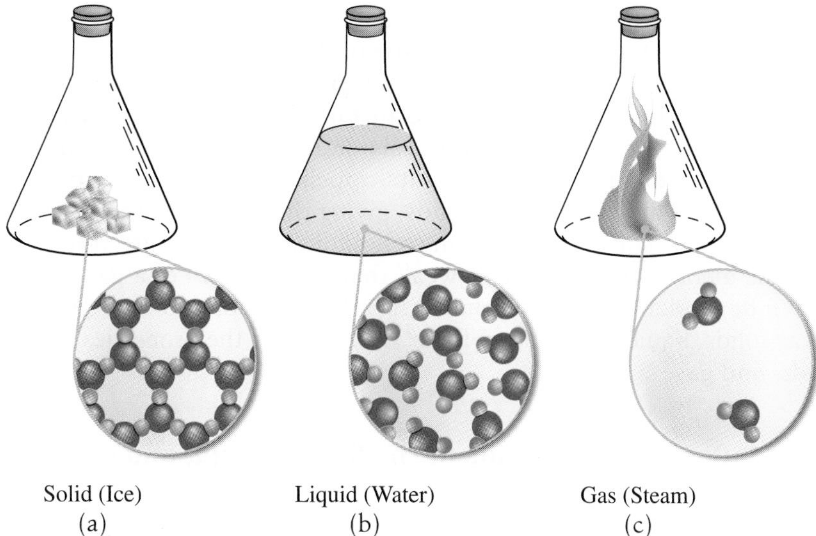

Figure 1.4 The three states of matter. (a) Solid—water molecules are held together rigidly and are very close to each other. (b) Liquid—water molecules are close together but are free to move around and slide over each other. (c) Gas—water molecules are far apart and move freely and randomly.

Because of this quality, a gas completely fills a container. The particles of a gas are relatively far apart compared with those of solids and liquids. The actual volume of the gas particles is very small compared with the volume of the space occupied by the gas. Observe the large space between the water molecules in Figure 1.4c compared to ice and liquid water. A gas therefore may be compressed into a very small volume or expanded almost indefinitely. Liquids cannot be compressed to any great extent, and solids are even less compressible than liquids.

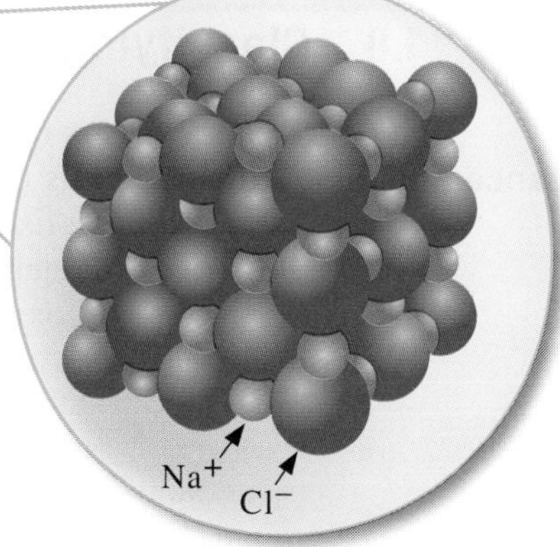

Figure 1.5 A large crystal of table salt. A salt crystal is composed of a three-dimensional array of particles.

If a bottle of ammonia solution is opened in one corner of the laboratory, we can soon smell its familiar odor in all parts of the room. The ammonia gas escaping from the solution demonstrates that gaseous particles move freely and rapidly and tend to permeate the entire area into which they are released.

Although matter is discontinuous, attractive forces exist that hold the particles together and give matter its appearance of continuity. These attractive forces are strongest in solids, giving them rigidity; they are weaker in liquids, but still strong enough to hold liquids to definite volumes. In gases, the attractive forces are so weak that the particles of a gas are practically independent of one another. Table 1.1 lists common materials that exist as solids, liquids, and gases. Table 1.2 compares the properties of solids, liquids, and gases.

Table 1.1 Common Materials in the Solid, Liquid, and Gaseous States of Matter

Solids	Liquids	Gases
Aluminum	Alcohol	Acetylene
Copper	Blood	Air
Gold	Gasoline	Butane
Polyethylene	Honey	Carbon dioxide
Salt	Mercury	Chlorine
Sand	Oil	Helium
Steel	Vinegar	Methane
Sulfur	Water	Oxygen

Table 1.2 Physical Properties of Solids, Liquids, and Gases

State	Shape	Volume	Particles	Compressibility
Solid	Definite	Definite	Rigidly clinging; tightly packed	Very slight
Liquid	Indefinite	Definite	Mobile; adhering	Slight
Gas	Indefinite	Indefinite	Independent of each other and relatively far apart	High

1.8 Classifying Matter

The term *matter* refers to all materials that make up the universe. Many thousands of distinct kinds of matter exist. A **substance** is a particular kind of matter with a definite, fixed composition. Sometimes known as *pure substances*, substances are either elements or compounds. Familiar examples of elements are copper, gold, and oxygen. Familiar compounds are salt, sugar, and water. We'll discuss elements and compounds in more detail in Chapter 3.

substance

(a)

(b)

(a) Water is the liquid in the beaker, and the white solid in the spoon is sugar.
(b) Sugar can be dissolved in the water to produce a solution.

We classify a sample of matter as either *homogeneous* or *heterogeneous* by examining it. **Homogeneous** matter is uniform in appearance and has the same properties throughout. Matter consisting of two or more physically distinct phases is **heterogeneous**. A **phase** is a homogeneous part of a system separated from other parts by physical boundaries. A **system** is simply the body of matter under consideration. Whenever we have a system in which visible boundaries exist between the parts or components, that system has more than one phase and is heterogeneous. It does not matter whether these components are in the solid, liquid, or gaseous states.

homogeneous

heterogeneous
phase
system

A pure substance may exist as different phases in a heterogeneous system. Ice floating in water, for example, is a two-phase system made up of solid water and liquid water. The water in each phase is homogeneous in composition, but because two phases are present, the system is heterogeneous.

A **mixture** is a material containing two or more substances and can be either heterogeneous or homogeneous. Mixtures are variable in composition. If we add a spoonful of sugar to a glass of water, a heterogeneous mixture is formed immediately. The two phases are a solid (sugar) and a liquid (water). But upon stirring, the sugar dissolves to form a homogeneous mixture or solution. Both substances are still present: All parts of the solution are sweet and wet. The proportions of sugar and water can be varied simply by adding more sugar and stirring to dissolve. Solutions do not have to be liquid. For example, air is a homogeneous mixture of gases. Solid solutions also exist. Brass is a homogeneous solution of copper and zinc.

mixture

Many substances do not form homogeneous mixtures. If we mix sugar and fine white sand, a heterogeneous mixture is formed. Careful examination may be needed to decide that the mixture is heterogeneous because the two phases (sugar and sand) are both white solids. Ordinary matter exists mostly as mixtures.

Flowcharts can help you to visualize the connections between concepts.

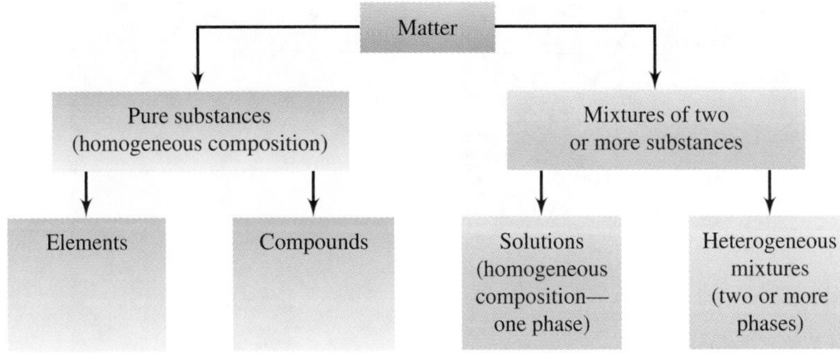

Figure 1.6 Classification of matter. A pure substance is always homogeneous in composition, whereas a mixture always contains two or more substances and may be either homogeneous or heterogeneous.

Granite is a heterogeneous mixture.

If we examine soil, granite, iron ore, or other naturally occurring mineral deposits, we find them to be heterogeneous mixtures. Figure 1.6 illustrates the relationships of substances and mixtures.

Distinguishing Mixtures from Pure Substances

Single substances—elements or compounds—seldom occur naturally in a pure state. Air is a mixture of gases; seawater is a mixture of a variety of dissolved minerals; ordinary soil is a complex mixture of minerals and various organic materials.

How is a mixture distinguished from a pure substance? A mixture always contains two or more substances that can be present in varying concentrations. Let's consider two examples.

Homogeneous Mixture Homogeneous mixtures (solutions) containing either 5% or 10% salt in water can be prepared simply by mixing the correct amounts of salt and water. These mixtures can be separated by boiling away the water, leaving the salt as a residue.

Heterogeneous Mixture The composition of a heterogeneous mixture of sulfur crystals and iron filings can be varied by merely blending in either more sulfur or more iron filings. This mixture can be separated physically by using a magnet to attract the iron.

(a)

(b)

(a) When iron and sulfur exist as pure substances, only the iron is attracted to a magnet.
(b) A mixture of iron and sulfur can be separated by using the difference in magnetic attraction.

Chapter 1 Review

1.1 Why Study Chemistry?

- Chemistry is important to everyone because chemistry occurs all around us in our daily lives.

1.2 The Nature of Chemistry

KEY TERM
Chemistry

- Chemistry is the science dealing with matter and the changes in composition that matter undergoes.
- Chemists seek to understand the general principles governing the behavior of all matter.

1.3 Thinking Like a Chemist

- Chemistry "looks inside" ordinary objects to study how their components behave.
- Chemistry connects the macroscopic and microscopic worlds.

1.4 A Scientific Approach to Problem Solving

- Scientific thinking helps us solve problems in our daily lives.
- General steps for solving problems include:
  - Defining the problem
  - Proposing possible solutions
  - Solving the problem

1.5 The Scientific Method

KEY TERMS
Scientific method
Hypothesis
Theory
Scientific laws

- The scientific method is a procedure for processing information in which we:
  - Collect the facts
  - Formulate a hypothesis
  - Plan and do experiments
  - Modify the hypothesis if necessary

1.6 The Particulate Nature of Matter

KEY TERM
Matter

- Matter is anything with the following two characteristics:
  - Has mass
  - Occupies space
- On the macroscopic level matter is continuous.
- On the microscopic level matter is discontinuous and composed of atoms.

1.7 Physical States of Matter

KEY TERMS
Solid
Amorphous
Liquid
Gas

- Solid—rigid substance with a definite shape
- Liquid—fluid substance with a definite volume that takes the shape of its container
- Gas—takes the shape and volume of its container

1.8 Classifying Matter

KEY TERMS
Substance
Homogeneous
Heterogeneous
Phase
System
Mixture

- Matter can be classified as a pure substance or a mixture.
- A mixture has variable composition:
  - Homogeneous mixtures have the same properties throughout.
  - Heterogeneous mixtures have different properties in different parts of the system.
- A pure substance always has the same composition.
  - There are two types of pure substances
    - Elements
    - Compounds

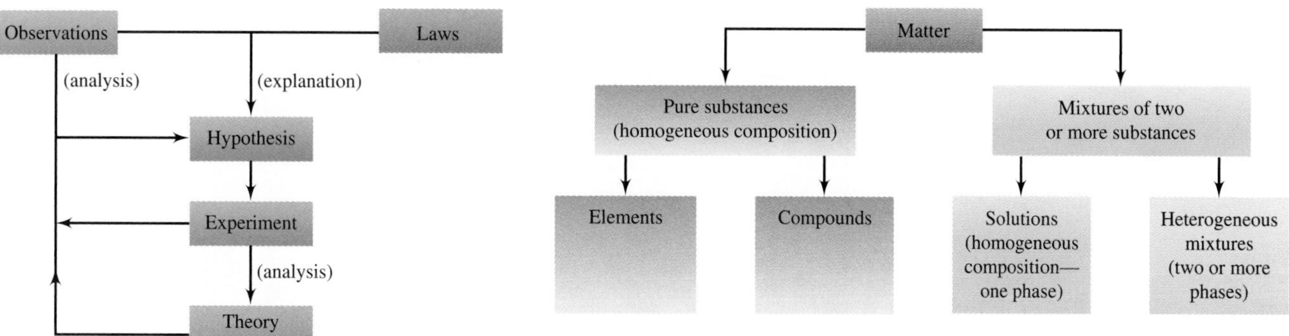

Review Questions

All questions with blue numbers have answers in the appendix of the text.

1. Explain the difference between
   (a) a hypothesis and a theory
   (b) a theory and a scientific law
2. Define a phase.
3. How many phases are present in the graduated cylinder?

4. Consider each of the following statements and determine whether it represents an observation, a hypothesis, a theory, or a scientific law:
   (a) The battery in my watch must be dead since it is no longer keeping time.
   (b) My computer must have a virus since it is not working properly.
   (c) The air feels cool.
   (d) The candle burns more brightly in pure oxygen than in air because oxygen supports combustion.
   (e) My sister wears red quite often.
   (f) A pure substance has a definite, fixed composition.
5. List four different substances in each of the three states of matter.
6. In terms of the properties of the ultimate particles of a substance, explain
   (a) why a solid has a definite shape but a liquid does not
   (b) why a liquid has a definite volume but a gas does not
   (c) why a gas can be compressed rather easily but a solid cannot be compressed appreciably
7. What evidence can you find in Figure 1.3 that gases occupy space?
8. Which liquids listed in Table 1.1 are not mixtures?
9. Which of the gases listed in Table 1.1 are not pure substances?
10. When the stopper is removed from a partly filled bottle containing solid and liquid acetic acid at 16.7°C, a strong vinegarlike odor is noticeable immediately. How many acetic acid phases must be present in the bottle? Explain.
11. Is the system enclosed in the bottle in Question 10 homogeneous or heterogeneous? Explain.
12. Is a system that contains only one substance necessarily homogeneous? Explain.
13. Is a system that contains two or more substances necessarily heterogeneous? Explain.
14. Distinguish between homogeneous and heterogeneous mixtures.
15. Which of the following are pure substances?
    (a) sugar
    (b) sand
    (c) gold
    (d) maple syrup
    (e) eggs

Paired Exercises

All exercises with blue numbers have answers in the appendix of the text.

1. Refer to the illustration and determine which state(s) of matter are present.

2. Refer to the illustration and determine which states(s) of matter are present.

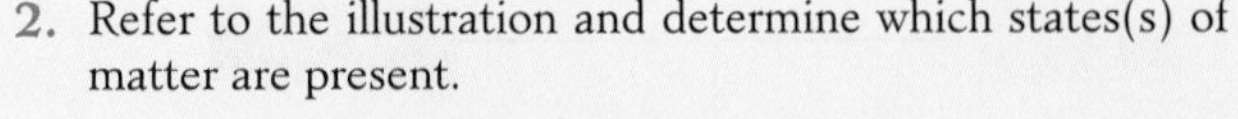

3. Look at the photo and determine whether it represents a homogeneous or heterogeneous mixture.

4. Look at the maple leaf below and determine whether it represents a homogeneous or heterogeneous mixture.

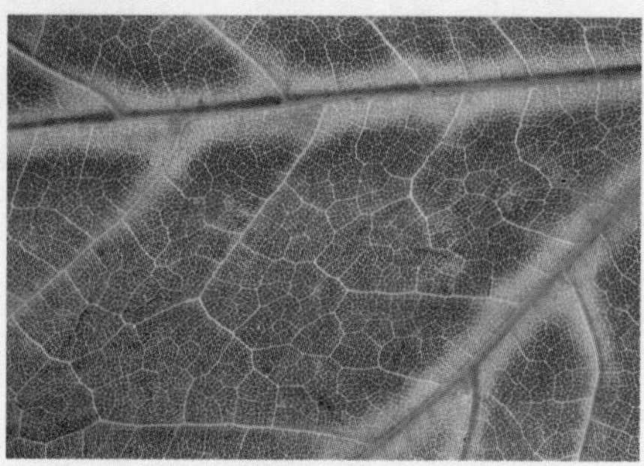

5. For each of the following mixtures, state whether it is homogeneous or heterogeneous:
   (a) tap water
   (b) carbonated beverage
   (c) oil and vinegar salad dressing
   (d) people in a football stadium

6. For each of the following mixtures, state whether it is homogeneous or heterogeneous:
   (a) stainless steel
   (b) motor oil
   (c) soil
   (d) a tree

Additional Exercises

All exercises with blue numbers have answers in the appendix of the text.

7. At home, check your kitchen and bathroom cabinets for five different substances, then read the labels and list the first ingredient of each.

8. During the first week of a new semester, consider that you have enrolled in five different classes, each of which meets for 3 hours per week. For every 1 hour that is spent in class, a minimum of 1 hour is required outside of class to complete assignments and study for exams. You also work 20 hours per week, and it takes you 1 hour to drive to the job site and back home. On Friday nights, you socialize with your friends. You are fairly certain that you will be able to successfully complete the semester with good grades. Show how the steps in the scientific method can help you predict the outcome of the semester.

CHAPTER 2

Standards for Measurement

Different-sized and -calibrated containers are needed to measure quantities of liquids accurately.

Chapter Outline

Doing an experiment in chemistry is very much like cooking a meal in the kitchen. It's important to know the ingredients and the amounts of each in order to have a tasty product. Working on your car requires specific tools, in exact sizes. Buying new carpeting or draperies is an exercise in precise and accurate measurement for a good fit. A small difference in the concentration or amount of medication a pharmacist gives you may have significant effects on your well-being. As we saw in Chapter 1, observation is an important part of the scientific process. Observations can be *qualitative* (the substance is a blue solid) or *quantitative* (the mass of the substance is 4.7 grams). A quantitative observation is called a **measurement**. Both a number and a unit are required for a measurement. For example, at home in your kitchen measuring 3 flour is not possible. We need to know the unit on the 3. It is cups, grams, or tablespoons? A measurement of 3 cups tells us both the amount and the size of the measurement. In this chapter we will discuss measurements and the rules for calculations using these measurements.

measurement

2.1 Scientific Notation

Scientists often use numbers that are very large or very small in measurements. For example, the Earth's age is estimated to be about 4,500,000 (4.5 billion) years. Numbers like these are bulky to write, so to make them more compact scientists use powers of 10. Writing a number as the product of a number between 1 and 10 multiplied by 10 raised to some power is called **scientific notation.**

scientific notation

To learn how to write a number in scientific notation, let's consider the number 2468. To write this number in scientific notation:

1. Move the decimal point in the original number so that it is located after the first nonzero digit.

   2468 → 2.468 (decimal moves three places to the left)

2. Multiply this new number by 10 raised to the proper exponent (power). The proper exponent is equal to the number of places that the decimal point was moved.

   2.468×10^3

3. The sign on the exponent indicates the direction the decimal was moved.

   moved right → negative exponent
   moved left → positive exponent

Examples show you problem-solving techniques in a step-by-step form. Study each one and then try the Practice Exercises.

Example 2.1

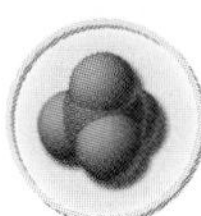

Write 5283 in scientific notation.

SOLUTION

5283.
3

Place the decimal between the 5 and the 2. Since the decimal was moved three places to the left, the power of 10 will be 3 and the number 5.283 is multiplied by 10^3.

5.283×10^3

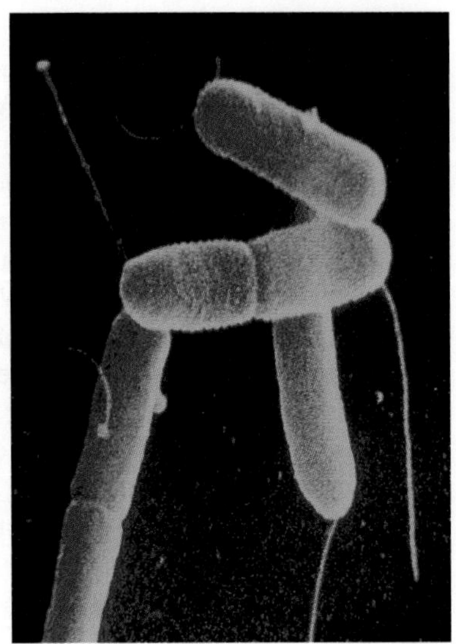

Scientific notation is a useful way to write very large numbers such as the distance between Earth and the moon, or very small numbers such as the length of these *E. coli* bacteria (shown here as a colored scanning electron micrograph ×30,000).

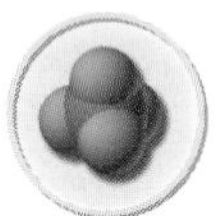

Example 2.2 Write 4,500,000,000 in scientific notation (two significant figures).

SOLUTION 4 500 000 000. (9) Place the decimal between the 4 and the 5. Since the decimal was moved nine places to the left, the power of 10 will be 9 and the number 4.5 is multiplied by 10^9.

4.5×10^9

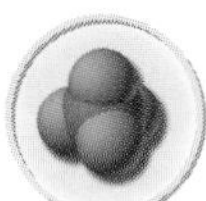

Example 2.3 Write 0.000123 in scientific notation.

SOLUTION 0.000123 (4) Place the decimal between the 1 and the 2. Since the decimal was moved four places to the right, the power of 10 will be −4 and the number 1.23 is multiplied by 10^{-4}.

1.23×10^{-4}

Practice 2.1

Write the following numbers in scientific notation:

(a) 1200 (four digits)
(b) 6,600,000 (two digits)
(c) 0.0468
(d) 0.00003

2.2 Measurement and Uncertainty

To understand chemistry, it is necessary to set up and solve problems. Problem solving requires an understanding of the mathematical operations used to manipulate numbers. Measurements are made in an experiment, and chemists use these data to calculate the extent of the physical and chemical changes occurring in the substances that are being studied. By appropriate calculations an experiment's results can be compared with those of other experiments and summarized in ways that are meaningful.

A measurement is expressed by a numerical value together with a unit of that measurement. For example,

numerical value

70.0 kilograms = 154 pounds

unit

Whenever a measurement is made with an instrument such as a thermometer or ruler, an estimate is required. We can illustrate this by measuring temperature. Suppose we measure temperature on a thermometer calibrated in degrees and observe that the mercury stops between 21 and 22 (see Figure 2.1a). We then know that the temperature is at least 21 degrees and is less than 22 degrees. To express the temperature with greater precision, we estimate that the mercury is about two-tenths the distance between 21 and 22. The temperature is therefore 21.2 degrees. The last digit (2) has some uncertainty because it is an estimated value. Because the last digit has some uncertainty (we made a visual estimate), it may be different when another person makes the same measurement. If three more people make this same reading, the results might be

Person	Measurement
1	21.2
2	21.3
3	21.1
4	21.2

Notice that the first two digits of the measurements did not change (they are *certain*). The last digit in these measurements is *uncertain*. The custom in science is to record all of the certain digits and the first uncertain number.

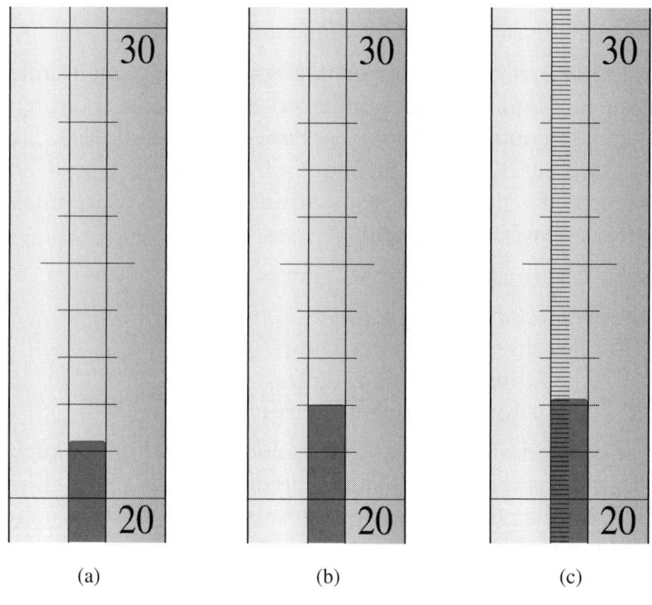

Figure 2.1
Measuring temperature (°C) with various degrees of precision.

Numbers obtained from a measurement are never exact values. They always have some degree of uncertainty due to the limitations of the measuring instrument and the skill of the individual making the measurement. The numerical value recorded for a measurement should give some indication of its reliability (precision). To express maximum precision, this number should contain all the digits that are known plus one digit that is estimated. This last estimated digit introduces some uncertainty. Because of this uncertainty, every number that expresses a measurement can have only a limited number of digits. These digits, used to express a measured quantity, are known as **significant figures.**

significant figures

Now let's return to our temperature measurements. In Figure 2.1a the temperature is recorded as 21.2 degrees and is said to have three significant figures. If the mercury stopped exactly on the 22 (Figure 2.1b), the temperature would be recorded as 22.0 degrees. The zero is used to indicate that the temperature was estimated to a precision of one-tenth degree. Finally, look at Figure 2.1c. On this thermometer, the temperature is recorded as 22.11°C (four significant figures). Since the thermometer is calibrated to tenths of a degree, the first estimated digit is the hundredths.

2.3 Significant Figures

Since all measurements involve uncertainty, we must use the proper number of significant figures in each measurement. In chemistry we frequently do calculations involving measurements, so we must understand what happens when we do arithmetic on numbers containing uncertainties. We'll learn several rules for doing these calculations and figuring out how many significant figures to have in the result. You will need to follow these rules throughout the calculations in this text.

The first thing we need to learn is how to determine how many significant figures are in a number.

Rules for Counting Significant Figures

Rule 1 *Nonzero digits.* All nonzero digits are significant.

Rule 2 *Exact numbers.* Some numbers are exact and have an infinite number of significant figures. Exact numbers occur in simple counting operations; when you count 25 dollars, you have exactly 25 dollars. Defined numbers, such as 12 inches in 1 foot, 60 minutes in 1 hour, and 100 centimeters in 1 meter, are also considered to be exact numbers. Exact numbers have no uncertainty.

Rules for significant figures should be memorized for use throughout the text.

Rule 3 *Zeros.* A zero is *significant* when it is

- between nonzero digits:
  205 has three significant figures
  2.05 has three significant figures
  61.09 has four significant figures
- at the end of a number that includes a decimal point:
  0.500 has three significant figures (5, 0, 0)
  25.160 has five significant figures (2, 5, 1, 6, 0)
  3.00 has three significant figures (3, 0, 0)
  20. has two significant figures (2, 0)

A zero is *not significant* when it is

- before the first nonzero digit. These zeros are used to locate a decimal point:
  0.0025 has two significant figures (2, 5)
  0.0108 has three significant figures (1, 0, 8)
- at the end of a number without a decimal point:
  1000 has one significant figure (1)
  590 has two significant figures (5, 9)

One way of indicating that these zeros are significant is to write the number using scientific notation. Thus if the value 1000 has been determined to four significant figures, it is written as 1.000×10^3. If 590 has only two significant figures, it is written as 5.9×10^2.

Practice 2.2

How many significant figures are in each of these measurements?

(a) 4.5 inches	(e) 25.0 grams
(b) 3.025 feet	(f) 12.20 liters
(c) 125.0 meters	(g) 100,000 people
(d) 0.001 mile	(h) 205 birds

Answers to Practice Exercises are found at the end of each chapter.

Rounding Off Numbers

When we do calculations on a calculator, we often obtain answers that have more digits than are justified. It is therefore necessary to drop the excess digits in order to express the answer with the proper number of significant figures. When digits are dropped from a number, the value of the last digit retained is determined by a process known as **rounding off numbers**. Two rules will be used in this book for rounding off numbers:

rounding off numbers

Rules for Rounding Off

Rule 1 When the first digit after those you want to retain is 4 or less, that digit and all others to its right are dropped. The last digit retained is not changed. The following examples are rounded off to four digits:

Not all schools use the same rules for rounding. Check with your instructor for variations in these rules.

74.693 = 74.69 (This digit is dropped.: 3)

1.00629 = 1.006 (These two digits are dropped.: 29)

Rule 2 When the first digit after those you want to retain is 5 or greater, that digit and all others to the right are dropped and the last digit retained is increased by one. These examples are rounded off to four digits:

1.026868 = 1.027 (These three digits are dropped.: 868; This digit is changed to 7.)

18.02500 = 18.03 (These three digits are dropped.: 500; This digit is changed to 3.)

12.899 = 12.90 (This digit is dropped.: 9; These two digits are changed to 90.)

Practice 2.3

Round off these numbers to the number of significant figures indicated:

(a) 42.246 (four)
(b) 88.015 (three)
(c) 0.08965 (three)
(d) 0.08965 (two)
(e) 225.3 (three)
(f) 14.150 (three)

2.4 Significant Figures in Calculations

The results of a calculation based on measurements cannot be more precise than the least precise measurement.

Use your calculator to check your work in the examples. Compare your results to be sure you understand the mathematics.

Multiplication or Division

In calculations involving multiplication or division, the answer must contain the same number of significant figures as in the measurement that has the least number of significant figures. Consider the following examples:

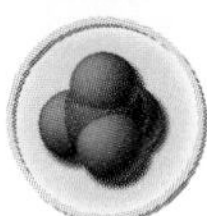

Example 2.4 (190.6)(2.3) = 438.38

SOLUTION The value 438.38 was obtained with a calculator. The answer should have two significant figures because 2.3, the number with the fewest significant figures, has only two significant figures.

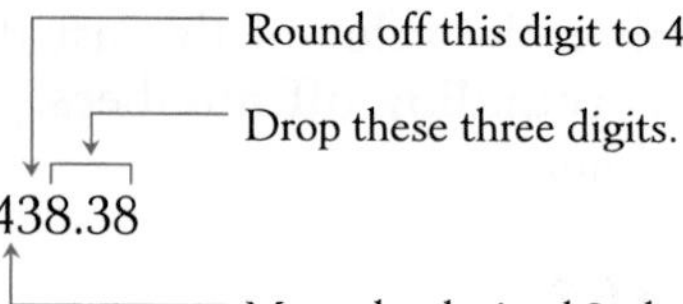
Round off this digit to 4.
Drop these three digits.
438.38
Move the decimal 2 places to the left to express in scientific notation.

The correct answer is 440 or 4.4×10^2.

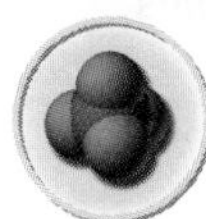

Example 2.5 $\frac{(13.59)(6.3)}{12} = 7.13475$

SOLUTION The value 7.13475 was obtained with a calculator. The answer should contain two significant figures because 6.3 and 12 each have only two significant figures.

Drop these four digits.

7.13475

This digit remains the same.

The correct answer is 7.1.

Practice 2.4

(a) (134 in.)(25 in.) = ?

(b) $\dfrac{213 \text{ miles}}{4.20 \text{ hours}} = ?$

(c) $\dfrac{(2.2)(273)}{760} = ?$

(d) $0.0321 \times 42 = ?$

(e) $\dfrac{0.0450}{0.00220} = ?$

(f) $\dfrac{1.280}{0.345} = ?$

Addition or Subtraction

The results of an addition or a subtraction must be expressed to the same precision as the least precise measurement. This means the result must be rounded to the same number of decimal places as the value with the fewest decimal places (blue line in examples).

Example 2.6

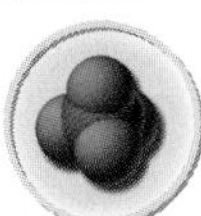

Add 125.17, 129, and 52.2.

SOLUTION

```
125.|17
129.|
 52.|2
----------
306.|37
```

The number with the least precision is 129. Therefore the answer is rounded off to the nearest unit: 306.

Example 2.7

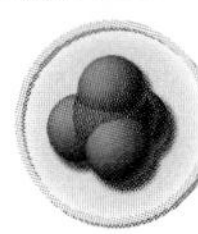

Subtract 14.1 from 132.56.

SOLUTION

```
  132.5|6
-  14.1|
----------
  118.4|6
```

14.1 is the number with the least precision. Therefore the answer is rounded off to the nearest tenth: 118.5.

Example 2.8

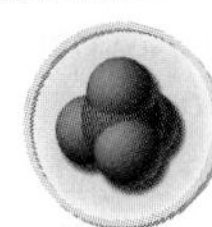

Subtract 120 from 1587.

SOLUTION

```
  158|7
-  12|0
--------
  146|7
```

120 is the number with least precision. The zero is not considered significant; therefore the answer must be rounded to the nearest ten: 1470 or 1.47×10^3.

Example 2.9 Add 5672 and 0.00063.

SOLUTION

$$\begin{array}{r|l} 5672 & \\ +\quad 0. & 00063 \\ \hline 5672. & 00063 \end{array}$$

Note: When a very small number is added to a large number, the result is simply the original number.

The number with least precision is 5672. So the answer is rounded off to the nearest unit: 5672.

Example 2.10 $\dfrac{1.039 - 1.020}{1.039} = ?$

SOLUTION $\dfrac{1.039 - 1.020}{1.039} = 0.018286814$

The value 0.018286814 was obtained with a calculator. When the subtraction in the numerator is done,

$$1.039 - 1.020 = 0.019$$

the number of significant figures changes from four to two. Therefore the answer should contain two significant figures after the division is carried out:

0.018286814 — Drop these six digits (286814). This digit (8) remains the same.

The correct answer is 0.018, or 1.8×10^{-2}.

Practice 2.5

How many significant figures should the answer in each of these calculations contain?

(a) (14.0)(5.2)

(b) (0.1682)(8.2)

(c) $\dfrac{(160)(33)}{4}$

(d) 8.2 + 0.125

(e) 119.1 − 3.44

(f) $\dfrac{94.5}{1.2}$

(g) 1200 + 6.34

(h) 1.6 + 23 − 0.005

If you need to brush up on your math skills, refer to the "Mathematical Review" in Appendix I.

Additional material on mathematical operations is given in Appendix I, "Mathematical Review." Study any portions that are not familiar to you. You may need to do this at various times during the course when additional knowledge of mathematical operations is required.

2.5 The Metric System

The **metric system**, or **International System** (**SI**, from *Système International*), is a decimal system of units for measurements of mass, length, time, and other physical quantities. Built around a set of standard units, the metric system uses factors of 10 to express larger or smaller numbers of these units. To express quantities that are larger or smaller than the standard units, prefixes are added to the names of the units. These prefixes represent multiples of 10, making the metric system a decimal system of measurements. Table 2.1 shows the names, symbols, and numerical values of the common prefixes. Some examples of the more commonly used prefixes are

metric system or SI

1 *kilo*meter = 1000 meters
1 *kilo*gram = 1000 grams
1 *milli*meter = 0.001 meter
1 *micro*second = 0.000001 second

The prefixes most commonly used in chemistry are shown in bold.

Table 2.1 Common Prefixes and Numerical Values for SI Units

Prefix	Symbol	Numerical value	Power of 10 equivalent
giga	G	1,000,000,000	10^9
mega	M	1,000,000	10^6
kilo	k	1,000	10^3
hecto	h	100	10^2
deka	da	10	10^1
—	—	1	10^0
deci	d	0.1	10^{-1}
centi	c	0.01	10^{-2}
milli	m	0.001	10^{-3}
micro	μ	0.000001	10^{-6}
nano	n	0.000000001	10^{-9}
pico	p	0.000000000001	10^{-12}
femto	f	0.000000000000001	10^{-15}

Most products today list both systems of measurement on their labels.

The common standard units in the International System, their abbreviations, and the quantities they measure are given in Table 2.2. Other units are derived from these units. The metric system, or International System, is currently used by most of the countries in the world, not only in scientific and technical work, but also in commerce and industry.

Table 2.2 International System's Standard Units of Measurement

Quantity	Name of unit	Abbreviation
Length	meter	m
Mass	kilogram	kg
Temperature	kelvin	K
Time	second	s
Amount of substance	mole	mol
Electric current	ampere	A
Luminous intensity	candela	cd

Measurement of Length

meter (m)

Common length relationships:
$1\,m = 10^6\,\mu m = 10^{10}\,Å$
$= 100\,cm = 1000\,mm$
$1\,cm = 10\,mm = 0.01\,m$
$1\,in. = 2.54\,cm$
$1\,mile = 1.609\,km$

See inside back cover for a table of conversions.

The standard unit of length in the metric system is the **meter (m)**. When the metric system was first introduced in the 1790s, the meter was defined as one ten-millionth of the distance from the equator to the North Pole, measured along the meridian passing through Dunkirk, France. The latest definition describes a meter as the distance that light travels in a vacuum during 1/299,792,458 of a second.

A meter is 39.37 inches, a little longer than 1 yard. One meter equals 10 decimeters, 100 centimeters, or 1000 millimeters (see Figure 2.2). A kilometer contains 1000 meters. Table 2.3 shows the relationships of these units.

The nanometer (10^{-9} m) is used extensively in expressing the wavelength of light, as well as in atomic dimensions. See the inside back cover for a complete table of common conversions.

Table 2.3 Metric Units of Length

Unit	Abbreviation	Meter equivalent	Exponential equivalent
kilometer	km	1000 m	10^3 m
meter	m	1 m	10^0 m
decimeter	dm	0.1 m	10^{-1} m
centimeter	cm	0.01 m	10^{-2} m
millimeter	mm	0.001 m	10^{-3} m
micrometer	μm	0.000001 m	10^{-6} m
nanometer	nm	0.000000001 m	10^{-9} m
angstrom	Å	0.0000000001 m	10^{-10} m

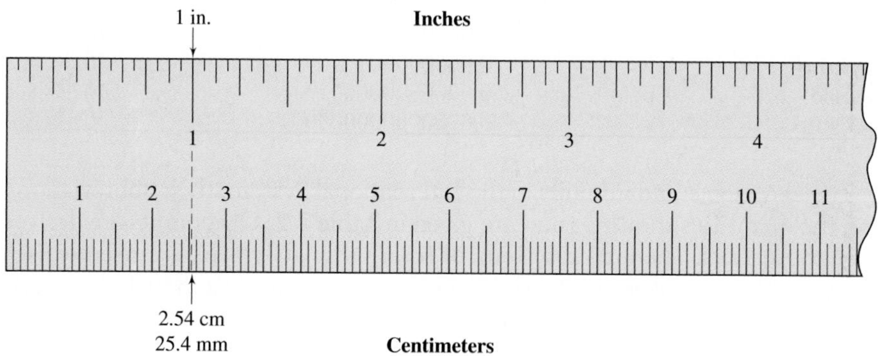

Figure 2.2
Comparison of the metric and American systems of length measurement: 2.54 cm = 1 in.

2.6 Problem Solving

See Appendix II for help in using a scientific calculator.

Many chemical principles can be illustrated mathematically. Learning how to set up and solve numerical problems in a systematic fashion is *essential* in the study of chemistry.

Usually a problem can be solved by several methods. But in all methods it is best to use a systematic, orderly approach. The *dimensional analysis method* is emphasized in this book because it

- provides a systematic, straightforward way to set up problems.
- gives a clear understanding of the principles involved.
- trains you to organize and evaluate data.
- helps to identify errors, since unwanted units are not eliminated if the setup of the problem is incorrect.

Steps for solving problems are highlighted in color for easy reference.

The basic steps for solving problems are

Step 1 **Read.** Read the problem carefully. Determine what is known and what is to be solved for, and write them down. It is important to label *all* factors and measurements with the proper units.

Step 2 **Plan.** Determine which principles are involved and which unit relationships are needed to solve the problem. You may need to refer to tables for needed data.

Step 3 **Set up.** Set up the problem in a neat, organized, and logical fashion, making sure that unwanted units cancel. Use sample problems in the text as guides for setting up the problem.

Step 4 **Calculate.** Proceed with the necessary mathematical operations. Make certain that your answer contains the proper number of significant figures.

Step 5 **Check.** Check the answer to see if it is reasonable.

A few more words about problem solving: Don't allow any formal method of problem solving to limit your use of common sense and intuition. If a problem is clear to you and its solution seems simpler by another method, by all means use it. But in the long run you should be able to solve many otherwise difficult problems by using dimensional analysis.

Dimensional analysis converts one unit to another unit by using conversion factors.

$$\textbf{unit}_1 \times \textbf{conversion factor} = \textbf{unit}_2$$

Important equations and statements are boxed or highlighted in color.

If you want to know how many millimeters are in 2.5 meters, you need to convert meters (m) to millimeters (mm). Start by writing

$$\text{m} \times \text{conversion factor} = \text{mm}$$

This conversion factor must accomplish two things: It must cancel (or eliminate) meters *and* it must introduce millimeters—the unit wanted in the answer. Such a conversion factor will be in fractional form and have meters in the denominator and millimeters in the numerator:

$$\cancel{\text{m}} \times \frac{\text{mm}}{\cancel{\text{m}}} = \text{mm}$$

We know that multiplying a measurement by 1 does not change its value. Since our conversion factors both equal 1, we can multiply the measurement by the appropriate one to convert units.

We know that 1 m = 1000 mm. From this relationship we can write two conversion factors—by dividing both sides of the equation by the same quantity:

$$\frac{1\text{ m}}{1000\text{ mm}} = \frac{1000\text{ mm}}{1000\text{ mm}} \longrightarrow \frac{1\text{ m}}{1000\text{ mm}} = 1$$

or

$$\frac{1\text{ m}}{1\text{ m}} = \frac{1000\text{ mm}}{1\text{ m}} \longrightarrow 1 = \frac{1000\text{ mm}}{1\text{ m}}$$

Therefore the two conversion factors are

$$\frac{1\text{ m}}{1000\text{ mm}} \quad \text{and} \quad \frac{1000\text{ mm}}{1\text{ m}}$$

Choosing the conversion factor $\frac{1000\text{ mm}}{1\text{ m}}$, we can set up the calculation for the conversion of 2.5 m to millimeters:

$$(2.5\,\cancel{\text{m}})\left(\frac{1000\text{ mm}}{1\,\cancel{\text{m}}}\right) = 2500\text{ mm} \quad \text{or} \quad 2.5 \times 10^3\text{ mm}$$

(two significant figures)

Note that, in making this calculation, units are treated as numbers; meters in the numerator are canceled by meters in the denominator.

Now suppose you need to change 215 centimeters to meters. We start with

$$\text{cm} \times \text{conversion factor} = \text{m}$$

The conversion factor must have centimeters in the denominator and meters in the numerator:

$$\cancel{\text{cm}} \times \frac{\text{m}}{\cancel{\text{cm}}} = \text{m}$$

From the relationship 100 cm = 1 m, we can write a factor that will accomplish this conversion:

$$\frac{1\text{ m}}{100\text{ cm}}$$

Now set up the calculation using all the data given:

$$(215\,\cancel{\text{cm}})\left(\frac{1\text{ m}}{100\,\cancel{\text{cm}}}\right) = 2.15\text{ m}$$

Units are emphasized in problems by using color and flow diagrams to help visualize the steps in the process.

Some problems require a series of conversions to reach the correct units in the answer. For example, suppose we want to know the number of seconds in 1 day. We need to convert from the unit of days to seconds in this manner:

day → hours → minutes → seconds

This sequence requires three conversion factors, one for each step. We convert days to hours (hr), hours to minutes (min), and minutes to seconds (s). The conversions can be done individually or in a continuous sequence:

$$\cancel{\text{day}} \times \frac{\text{hr}}{\cancel{\text{day}}} \longrightarrow \cancel{\text{hr}} \times \frac{\text{min}}{\cancel{\text{hr}}} \longrightarrow \cancel{\text{min}} \times \frac{\text{s}}{\cancel{\text{min}}} = \text{s}$$

$$\cancel{\text{day}} \times \frac{\cancel{\text{hr}}}{\cancel{\text{day}}} \times \frac{\cancel{\text{min}}}{\cancel{\text{hr}}} \times \frac{\text{s}}{\cancel{\text{min}}} = \text{s}$$

Inserting the proper factors, we calculate the number of seconds in 1 day to be

$$(1\ \cancel{\text{day}})\left(\frac{24\ \cancel{\text{hr}}}{1\ \cancel{\text{day}}}\right)\left(\frac{60\ \cancel{\text{min}}}{1\ \cancel{\text{hr}}}\right)\left(\frac{60\ \text{s}}{1\ \cancel{\text{min}}}\right) = 86{,}400.\ \text{s}$$

All five digits in 86,400 are significant, since all the factors in the calculation are exact numbers.

Label all factors with the proper units.

The dimensional analysis used in the preceding examples shows how unit conversion factors are derived and used in calculations. As you become more proficient, you can save steps by writing the factors directly in the calculation. The following examples show the conversion from American units to metric units.

Example 2.11

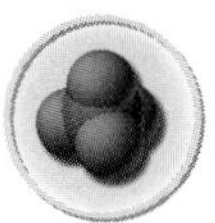

How many centimeters are in 2.00 ft?

SOLUTION

The stepwise conversion of units from feet to centimeters may be done in this sequence. Convert feet to inches; then convert inches to centimeters:

ft → in. → cm

The conversion factors needed are

$$\frac{12\ \text{in.}}{1\ \text{ft}} \quad \text{and} \quad \frac{2.54\ \text{cm}}{1\ \text{in.}}$$

$$(2.00\ \cancel{\text{ft}})\left(\frac{12\ \text{in.}}{1\ \cancel{\text{ft}}}\right) = 24.0\ \text{in.}$$

$$(24.0\ \cancel{\text{in.}})\left(\frac{2.54\ \text{cm}}{1\ \cancel{\text{in.}}}\right) = 61.0\ \text{cm}$$

Since 1 ft and 12 in. are exact numbers, the number of significant figures allowed in the answer is three, based on the number 2.00.

Example 2.12

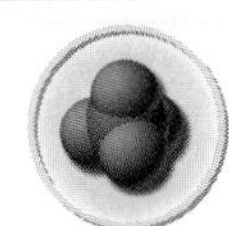

How many meters are in a 100.-yd football field?

SOLUTION

The stepwise conversion of units from yards to meters may be done by this sequence, using the proper conversion factors:

yd → ft → in. → cm → m

$$(100.\ \cancel{\text{yd}})\left(\frac{3\ \text{ft}}{1\ \cancel{\text{yd}}}\right) = 300.\ \text{ft}$$

$$(300.\ \cancel{\text{ft}})\left(\frac{12\ \text{in.}}{1\ \cancel{\text{ft}}}\right) = 3600\ \text{in.}$$

$$(3600\ \cancel{\text{in.}})\left(\frac{2.54\ \text{cm}}{1\ \cancel{\text{in.}}}\right) = 9144\ \text{cm}$$

$$(9144\ \cancel{\text{cm}})\left(\frac{1\ \text{m}}{100\ \cancel{\text{cm}}}\right) = 91.4\ \text{m} \quad \text{(three significant figures)}$$

If you need help in doing the calculation all at once on your calculator, see Appendix II.

Examples 2.11 and 2.12 may be solved using a linear expression and writing down conversion factors in succession. This method often saves one or two calculation steps and allows numerical values to be reduced to simpler terms, leading to simpler calculations. The single linear expressions for Examples 2.11 and 2.12 are

$$(2.00\ \cancel{\text{ft}})\left(\frac{12\ \cancel{\text{in.}}}{1\ \cancel{\text{ft}}}\right)\left(\frac{2.54\ \text{cm}}{1\ \cancel{\text{in.}}}\right) = 61.0\ \text{cm}$$

$$(100.\ \cancel{\text{yd}})\left(\frac{3\ \cancel{\text{ft}}}{1\ \cancel{\text{yd}}}\right)\left(\frac{12\ \cancel{\text{in.}}}{1\ \cancel{\text{ft}}}\right)\left(\frac{2.54\ \cancel{\text{cm}}}{1\ \cancel{\text{in.}}}\right)\left(\frac{1\ \text{m}}{100\ \cancel{\text{cm}}}\right) = 91.4\ \text{m}$$

Using the units alone (Example 2.12), we see that the stepwise cancellation proceeds in succession until the desired unit is reached:

$$\cancel{\text{yd}} \times \frac{\cancel{\text{ft}}}{\cancel{\text{yd}}} \times \frac{\cancel{\text{in.}}}{\cancel{\text{ft}}} \times \frac{\cancel{\text{cm}}}{\cancel{\text{in.}}} \times \frac{\text{m}}{\cancel{\text{cm}}} = \text{m}$$

Practice 2.6

(a) How many meters are in 10.5 miles?
(b) What is the area of a 6.0-in. × 9.0-in. rectangle in square meters?

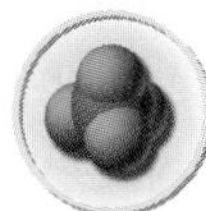

Example 2.13 How many cubic centimeters (cm^3) are in a box that measures 2.20 in. by 4.00 in. by 6.00 in.?

SOLUTION First we need to determine the volume of the box in cubic inches ($in.^3$) by multiplying the length × width × height:

$$(2.20\ \text{in.})(4.00\ \text{in.})(6.00\ \text{in.}) = 52.8\ \text{in.}^3$$

Now we convert $in.^3$ to cm^3 by using the inches and centimeters relationship (1 in. = 2.54 cm) three times:

$$\cancel{\text{in.}^3} \times \frac{\text{cm}}{\cancel{\text{in.}}} \times \frac{\text{cm}}{\cancel{\text{in.}}} \times \frac{\text{cm}}{\cancel{\text{in.}}} = \text{cm}^3$$

$$(52.8\ \cancel{\text{in.}^3})\left(\frac{2.54\ \text{cm}}{1\ \cancel{\text{in.}}}\right)\left(\frac{2.54\ \text{cm}}{1\ \cancel{\text{in.}}}\right)\left(\frac{2.54\ \text{cm}}{1\ \cancel{\text{in.}}}\right) = 865\ \text{cm}^3$$

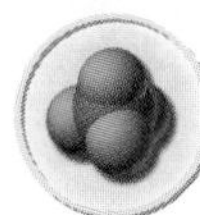

Example 2.14 A driver of a car is obeying the speed limit of 55 miles per hour. How fast is the car traveling in kilometers per second?

SOLUTION Three conversions are needed to solve this problem:

mi → km

hr → min → s

To convert mi → km,

$$\left(\frac{55\ \cancel{\text{mi}}}{\text{hr}}\right)\left(\frac{1.609\ \text{km}}{1\ \cancel{\text{mi}}}\right) = 88\frac{\text{km}}{\text{hr}}$$

Next we must convert hr → min → s. Notice that hours is in the denominator of our quantity, so the conversion factor must have hours in the numerator:

$$\left(\frac{88\ \text{km}}{\cancel{\text{hr}}}\right)\left(\frac{1\ \cancel{\text{hr}}}{60\ \cancel{\text{min}}}\right)\left(\frac{1\ \cancel{\text{min}}}{60\ \text{s}}\right) = 0.024\frac{\text{km}}{\text{s}}$$

CHEMISTRY IN ACTION • Keeping Track of Units

Why do scientists worry about units? The National Aeronautics and Space Administration (NASA) recently was reminded of just why keeping track of units is so important. In 1999 a $125 million satellite was lost in the atmosphere of Mars because scientists made some improper assumptions about units. NASA's scientists at the Jet Propulsion Lab (JPL) in Pasadena, California, received thrust data from the satellite's manufacturer, Lockheed Martin Aeronautics in Denver, Colorado. Unfortunately, the Denver scientists used American units in their measurements and the JPL scientists assumed the units were metric. This mistake caused the satellite to fall 100 km lower into the Mars atmosphere than planned. The spacecraft burned up from the friction with the atmosphere.

Measuring and using units correctly is very important. In fact, it can be critical, as we have just seen. For example, a Canadian jet almost crashed when the tanks were filled with 22,300 pounds (instead of kilograms) of fuel. Calculations for distance were based on kg, and the jet almost ran out of fuel before landing at the destination. Correct units are also important to ensure perfect fits for household purchases such as drapes, carpet, or appliances. Be sure to pay attention to units in both your chemistry problems and in everyday life!

Mars climate orbiter.

Practice 2.7

How many cubic meters of air are in a room measuring 8 ft × 10 ft × 12 ft?

2.7 Measuring Mass and Volume

Although we often use mass and weight interchangeably in our everyday lives, they have quite different meanings in chemistry. In science we define the **mass** of an object as the amount of matter in the object. Mass is measured on an instrument called a balance. The **weight** of an object is a measure of the effect of gravity on the object. Weight is determined by using an instrument called a scale, which measures force against a spring. This means mass is independent of the location of an object, but weight is not. In this text we will use the term mass for all of our metric mass measurements. The gram is a unit of mass measurement, but it is a tiny amount of mass; for instance, a U.S. nickel has a mass of about 5 grams.

mass

weight

kilogram

Therefore the *standard unit* of mass in the SI system is the **kilogram** (equal to 1000 g). The amount of mass in a kilogram is defined by international agreement as exactly equal to the mass of a platinum-iridium cylinder (international prototype kilogram) kept in a vault at Sèvres, France. Comparing this unit of mass to 1 lb (16 oz), we find that 1 kg is equal to 2.205 lb. A pound is equal to 453.6 g (0.4536 kg). The same prefixes used in length measurement are used to indicate larger and smaller gram units (see Table 2.4). A balance is used to measure mass. Two examples of balances are shown in Figure 2.3.

Table 2.4 Metric Units of Mass

Unit	Abbreviation	Gram equivalent	Exponential equivalent
kilogram	kg	1000 g	10^3 g
gram	g	1 g	10^0 g
decigram	dg	0.1 g	10^{-1} g
centigram	cg	0.01 g	10^{-2} g
milligram	mg	0.001 g	10^{-3} g
microgram	μg	0.000001 g	10^{-6} g

Common mass relationships:
1 g = 1000 mg
1 kg = 1000 g
1 kg = 2.205 lb
1 lb = 453.6 g

To change grams to milligrams, we use the conversion factor 1000 mg/g. The setup for converting 25 g to milligrams is

$$(25\ \cancel{g})\left(\frac{1000\ \text{mg}}{1\ \cancel{g}}\right) = 25{,}000\ \text{mg} \qquad (2.5 \times 10^4\ \text{mg})$$

(a)

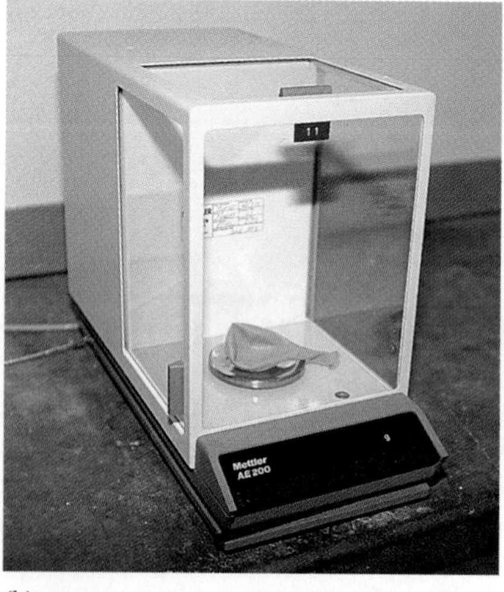

(b)

Figure 2.3
(a) Digital electronic, top-loading balance with a precision of 0.001 g (1 mg); and (b) a digital electronic analytical balance with a precision of 0.0001 g (0.1 mg).

Note that multiplying a number by 1000 is the same as multiplying the number by 10^3 and can be done simply by moving the decimal point three places to the right:

$$(6.428)(1000) = 6428 \qquad (6.\underbrace{428}_{3})$$

To change milligrams to grams, we use the conversion factor 1 g/1000 mg. For example, we convert 155 mg to grams as follows:

$$(155 \cancel{\text{mg}})\left(\frac{1\text{ g}}{1000 \cancel{\text{mg}}}\right) = 0.155\text{ g}$$

Mass conversions from American to metric units are shown in Examples 2.15 and 2.16.

Example 2.15

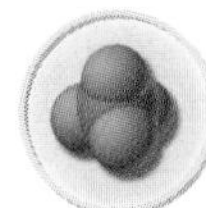

A 1.50-lb package of baking soda contains how many grams?

SOLUTION

We are solving for the number of grams equivalent to 1.50 lb. Since 1 lb = 453.6 g, the conversion factor is 453.6 g/lb:

$$(1.50 \cancel{\text{lb}})\left(\frac{453.6\text{ g}}{1 \cancel{\text{lb}}}\right) = 680.\text{ g}$$

Example 2.16

Suppose four ostrich feathers weigh 1.00 lb. Assuming that each feather is equal in mass, how many milligrams does a single feather weigh?

SOLUTION

The unit conversion in this problem is from 1.00 lb/4 feathers to milligrams per feather. Since the unit *feathers* occurs in the denominator of both the starting unit and the desired unit, the unit conversions are

lb → g → mg

$$\left(\frac{1.00 \cancel{\text{lb}}}{4\text{ feathers}}\right)\left(\frac{453.6 \cancel{\text{g}}}{1 \cancel{\text{lb}}}\right)\left(\frac{1000\text{ mg}}{1 \cancel{\text{g}}}\right) = \frac{113{,}400\text{ mg}}{\text{feather}} (1.13 \times 10^5\text{ mg/feather})$$

Practice 2.8

You are traveling in Europe and wake up one morning to find your mass is 75.0 kg. Determine the American equivalent (in pounds) to see whether you need to go on a diet before you return home.

Practice 2.9

A tennis ball has a mass of 65 g. Determine the American equivalent in pounds.

volume

Volume, as used here, is the amount of space occupied by matter. The SI unit of volume is the *cubic meter* (m^3). However, the liter (pronounced *lee-ter* and abbreviated L) and the milliliter (abbreviated mL) are the standard units of volume used in most chemical laboratories. A **liter** is usually defined as 1 cubic decimeter (1 kg) of water at 4°C.

liter

Common volume relationships:
1 L = 1000 mL = 1000 cm^3
1 mL = 1 cm^3
1 L = 1.057 qt
946.1 mL = 1 qt

The most common instruments or equipment for measuring liquids are the graduated cylinder, volumetric flask, buret, pipet, and syringe, which are illustrated in Figure 2.4. These pieces are usually made of glass and are available in various sizes.

The volume of a cubic or rectangular container can be determined by multiplying its length × width × height. Thus a box 10 cm on each side has a volume of (10 cm)(10 cm)(10 cm) = 1000 cm^3. Let's try some examples.

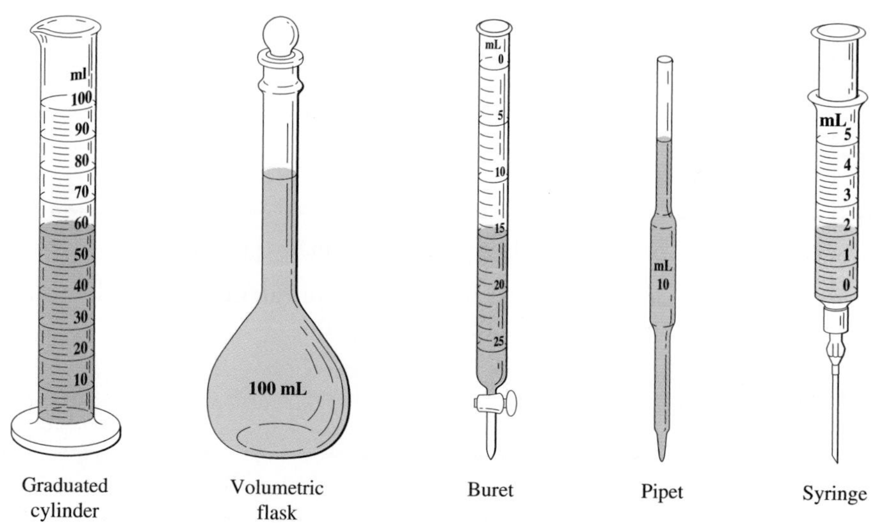

Figure 2.4
Calibrated glassware for measuring the volume of liquids.

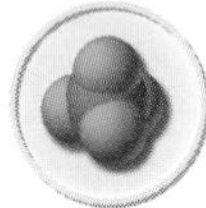

Example 2.17 How many milliliters are contained in 3.5 liters?

SOLUTION The conversion factor to change liters to milliliters is 1000 mL/L:

$$(3.5\ \cancel{L})\left(\frac{1000\ \text{mL}}{\cancel{L}}\right) = 3500\ \text{mL} \qquad (3.5 \times 10^3\ \text{mL})$$

Liters may be changed to milliliters by moving the decimal point three places to the right and changing the units to milliliters:

1.500 L = 1500. mL

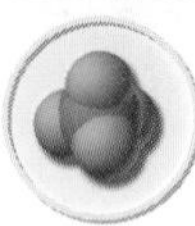

Example 2.18 How many cubic centimeters are in a cube that is 11.1 inches on a side?

SOLUTION First we change inches to centimeters; our conversion factor is 2.54 cm/in.:

$$(11.1\ \cancel{\text{in.}})\left(\frac{2.54\ \text{cm}}{1\ \cancel{\text{in.}}}\right) = 28.2\ \text{cm on a side}$$

Then determine volume (length × width × height):

$$(28.2\ \text{cm})(28.2\ \text{cm})(28.2\ \text{cm}) = 22{,}426\ \text{cm}^3 \qquad (2.24 \times 10^4\ \text{cm}^3)$$

CHEMISTRY IN ACTION • Setting Standards

The kilogram is the standard base unit for mass in the SI system. But who decides just what a kilogram is? Before 1880, a kilogram was defined as the mass of a cubic decimeter of water. But this is difficult to reproduce very accurately since impurities and air are dissolved in water. The density of water also changes with temperature, leading to inaccuracy. So in 1885 a Pt-Ir cylinder was made with a mass of exactly 1 kilogram. This cylinder is kept in a vault at Sèvres, France, outside Paris, along with six copies. In 1889 the kilogram was determined to be the base unit of mass for the metric system and has been weighed against its copies three times in the past 100 years (1890, 1948, 1992) to make sure that it is accurate. The trouble with all this is that the cylinder itself can vary. Experts in the science of measurement want to change the definition of the kilogram to link it to a property of matter that does not change.

The kilogram is the only one of seven SI base units defined in terms of a physical object instead of a property of matter. The meter (redefined in 1983) is defined in terms of the speed of light (1 m = distance light travels in $\frac{1}{299,792,458}$ s). The second is defined in terms of the natural vibration of the cesium atom. Unfortunately, three of the SI base units (mole, ampere, and candela) depend on the definition of the kilogram. If the mass of the kilogram is uncertain, then all of these units are uncertain.

How else could we define a kilogram? Scientists have several ideas. They propose to fix the kilogram to a set value of a physical constant (which does not change). These constants connect experiment to theory. Right now they are continuing to struggle to find a new way to determine the mass of the kilogram. By 2007 they hope to have a new definition of the kilogram based on constants and accurately reproducible anywhere in the world.

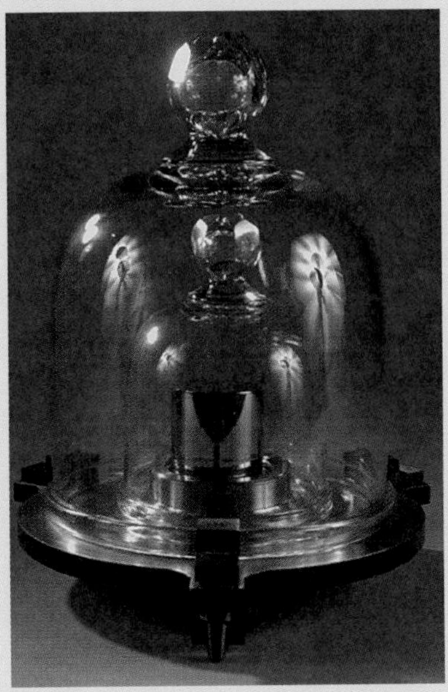

Kilogram artifact in Sevres, France.

Practice 2.10

A bottle of excellent chianti holds 750. mL. What is its volume in quarts?

Practice 2.11

Milk is often purchased by the half gallon. Determine the number of liters equal to this amount.

When doing problems with multiple steps, you should round only at the end of the problem. We are rounding at the end of each step in example problems to illustrate the proper significant figures.

2.8 Measurement of Temperature

heat

Heat is a form of energy associated with the motion of small particles of matter. The term *heat* refers to the quantity of energy within a system or to a quantity of energy added to or taken away from a system. *System*, as used here, simply refers to the entity that is being heated or cooled. Depending on the amount of heat energy present, a given system is said to be hot or cold. **Temperature** is a measure of the intensity of heat, or how hot a system is, regardless of its size. Heat always flows from a region of higher temperature to one of lower temperature. The SI unit of temperature is the kelvin. The common laboratory instrument for measuring temperature is a thermometer (see Figure 2.5).

temperature

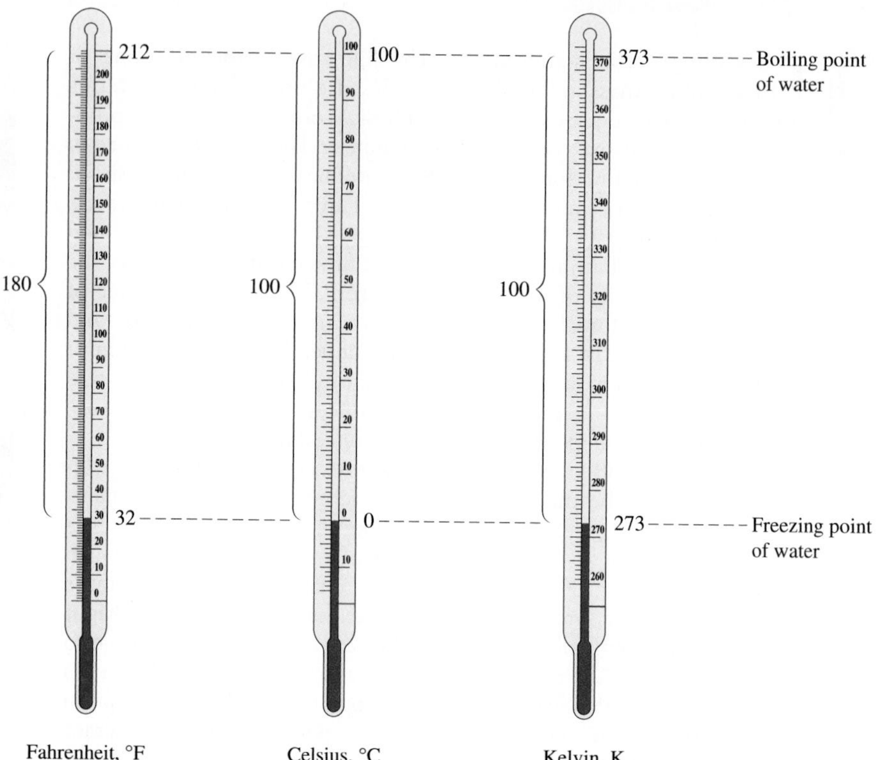

Figure 2.5
Comparison of Celsius, Kelvin, and Fahrenheit temperature scales.

The temperature of a system can be expressed by several different scales. Three commonly used temperature scales are Celsius (pronounced *sell-see-us*), Kelvin (also called absolute), and Fahrenheit. The unit of temperature on the Celsius and Fahrenheit scales is called a *degree*, but the size of the Celsius and the Fahrenheit degree is not the same. The symbol for the Celsius and Fahrenheit degrees is °, and it is placed as a superscript after the number and before the symbol for the scale. Thus, 10.0°C means 10.0 *degrees Celsius*. The degree sign is not used with Kelvin temperatures:

degrees Celsius = °C

Kelvin (absolute) = K

degrees Fahrenheit = °F

On the Celsius scale the interval between the freezing and boiling temperatures of water is divided into 100 equal parts, or degrees. The freezing point of water is assigned a temperature of 0°C and the boiling point of water a temperature of 100°C. The Kelvin temperature scale is known as the absolute temperature scale, because 0 K is the lowest temperature theoretically attainable. The Kelvin zero is 273.15 kelvins below the Celsius zero. A kelvin is equal in size to a Celsius degree. The freezing point of water on the Kelvin scale is 273.15 K. The Fahrenheit scale has 180 degrees between the freezing and boiling temperatures of water. On this scale the freezing point of water is 32°F and the boiling point is 212°F:

$$0°\text{C} \cong 273\ \text{K} \cong 32°\text{F} \qquad 100°\text{C} \cong 373\ \text{K} \cong 212°\text{F}$$

The three scales are compared in Figure 2.5. Although absolute zero (0 K) is the lower limit of temperature on these scales, temperature has no upper limit. Temperatures of several million degrees are known to exist in the sun and in other stars.

By examining Figure 2.5, we can see that there are 100 Celsius degrees and 100 kelvins between the freezing and boiling points of water, but there are 180 Fahrenheit degrees between these two temperatures. Hence, the size of the Celsius degree and the kelvin are the same, but 1 Celsius degree is equal to 1.8 Fahrenheit degrees.

$$\frac{180}{100} = 1.8$$

From these data, mathematical formulas have been derived to convert a temperature on one scale to the corresponding temperature on another scale:

$$K = {}^\circ C + 273.15$$

$${}^\circ F = (1.8 \times {}^\circ C) + 32$$

$${}^\circ C = \frac{{}^\circ F - 32}{1.8}$$

Example 2.19

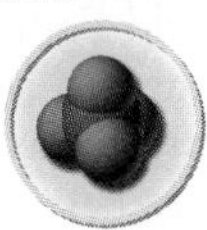

The temperature at which table salt (sodium chloride) melts is 800.°C. What is this temperature on the Kelvin and Fahrenheit scales?

SOLUTION

To calculate K from °C, we use the formula

$$K = {}^\circ C + 273.15$$

$$K = 800.{}^\circ C + 273.15 = 1073\ K$$

To calculate °F from °C, we use the formula

$${}^\circ F = (1.8 \times {}^\circ C) + 32$$

$${}^\circ F = (1.8)(800.{}^\circ C) + 32$$

$${}^\circ F = 1440 + 32 = 1472{}^\circ F$$

Summarizing our calculations, we see that

$$800.{}^\circ C = 1073\ K = 1472{}^\circ F$$

Remember that the original measurement of 800.°C was to the units place, so the converted temperature is also to the units place.

Example 2.20

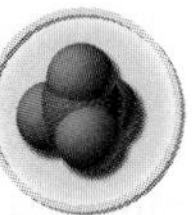

The temperature on December 1 in Honolulu, Hawaii, was 110.°F, a new record. Convert this temperature to °C.

SOLUTION

We use the formula

$${}^\circ C = \frac{{}^\circ F - 32}{1.8}$$

$${}^\circ C = \frac{110. - 32}{1.8} = \frac{78}{1.8} = 43{}^\circ C$$

Example 2.21

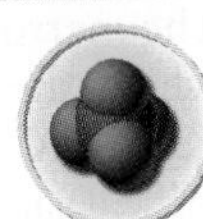

What temperature on the Fahrenheit scale corresponds to −8.0°C? (Notice the negative sign in this problem.)

SOLUTION

$${}^\circ F = (1.8 \times {}^\circ C) + 32$$

$${}^\circ F = (1.8)(-8.0) + 32 = -14 + 32$$

$${}^\circ F = 18{}^\circ F$$

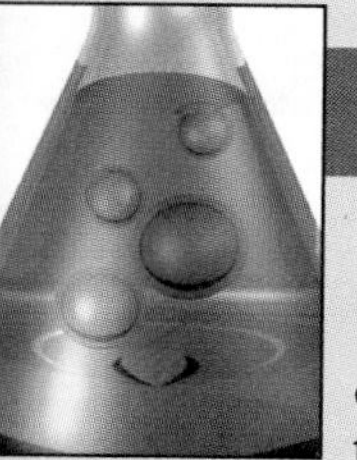

CHEMISTRY IN ACTION • Taking the Temperature of Old Faithful

If you have ever struggled to take the temperature of a sick child, imagine the difficulty in taking the temperature of a geyser. Such are the tasks scientists set for themselves! In 1984, James A. Westphal and Susan W. Keiffer, geologists from the California Institute of Technology, measured the temperature and pressure inside Old Faithful during several eruptions in order to learn more about how a geyser functions. The measurements, taken at eight depths along the upper part of the geyser, were so varied and complicated that the researchers returned to Yellowstone in 1992 to further investigate Old Faithful's structure and functioning. To see what happened between eruptions, Westphal and Keiffer lowered an insulated 2-inch video camera into the geyser. Keiffer had assumed that the vent (opening in the ground) was a uniform vertical tube, but this is not the case. Instead, the geyser appears to be an east–west crack in the Earth that extends downward at least 14 m. In some places it is over 1.8 m wide, and in other places it narrows to less than 15 cm. The walls of the vent contain many cracks, allowing water to enter at several depths. The complicated nature of the temperature data is explained by these cracks. Cool water enters the vent at depths of 5.5 m and 7.5 m. Superheated water and steam blast into the vent 14 m underground. According to Westphal, temperature increases of up to 130°C at the beginning of an eruption suggest that water and steam also surge into the vent from deeper geothermal sources.

During the first 20–30 seconds of an eruption, steam and boiling water shoot through the narrowest part of the vent at near the speed of sound. The narrow tube limits the rate at which the water can shoot from the geyser. When the pressure falls below a critical value, the process slows and Old Faithful begins to quiet down again.

The frequency of Old Faithful's eruptions are not on a precise schedule, but vary from 45 to 105 minutes—the average is about 79 minutes. The variations in time between eruptions depend on the amount of boiling water left in the fissure. Westphal says, "There's no real pattern except that a short eruption is always followed by a long one." Measurements of temperatures inside Old Faithful have given scientists a better understanding of what causes a geyser to erupt.

Old Faithful eruptions shoot into the air an average of 130 feet.

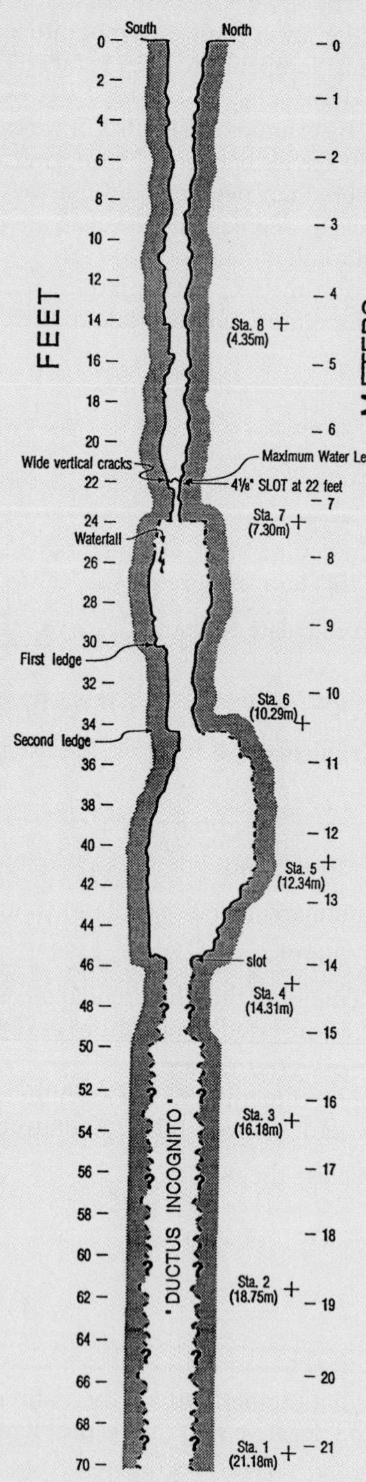

Temperatures used throughout this book are in degrees Celsius (°C) unless specified otherwise. The temperature after conversion should be expressed to the same precision as the original measurement.

Practice 2.12

Helium boils at 4 K. Convert this temperature to °C and then to °F.

Practice 2.13

"Normal" human body temperature is 98.6°F. Convert this to °C and K.

2.9 Density

Density (d) is the ratio of the mass of a substance to the volume occupied by that mass; it is the mass per unit of volume and is given by the equation

density

$$d = \frac{\text{mass}}{\text{volume}}$$

Density is a physical characteristic of a substance and may be used as an aid to its identification. When the density of a solid or a liquid is given, the mass is usually expressed in grams and the volume in milliliters or cubic centimeters.

$$d = \frac{\text{mass}}{\text{volume}} = \frac{\text{g}}{\text{mL}} \quad \text{or} \quad d = \frac{\text{g}}{\text{cm}^3}$$

Since the volume of a substance (especially liquids and gases) varies with temperature, it is important to state the temperature along with the density. For example, the volume of 1.0000 g of water at 4°C is 1.0000 mL; at 20°C, it is 1.0018 mL; and at 80°C, it is 1.0290 mL. Density therefore also varies with temperature.

The density of water at 4°C is 1.0000 g/mL, but at 80°C the density of water is 0.9718 g/mL:

$$d^{4^\circ\text{C}} = \frac{1.0000\text{ g}}{1.0000\text{ mL}} = 1.0000\text{ g/mL}$$

$$d^{80^\circ\text{C}} = \frac{1.0000\text{ g}}{1.0290\text{ mL}} = 0.97182\text{ g/mL}$$

The density of iron at 20°C is 7.86 g/mL.

$$d^{20^\circ\text{C}} = \frac{7.86\text{ g}}{1.00\text{ mL}} = 7.86\text{ g/mL}$$

The densities of a variety of materials are compared in Figure 2.6.

Densities for liquids and solids are usually represented in terms of grams per milliliter (g/mL) or grams per cubic centimeter (g/cm^3). The density of gases, however, is expressed in terms of grams per liter (g/L). Unless otherwise stated, gas densities are given for 0°C and 1 atmosphere pressure (discussed further in Chapter 13). Table 2.5 lists the densities of some common materials.

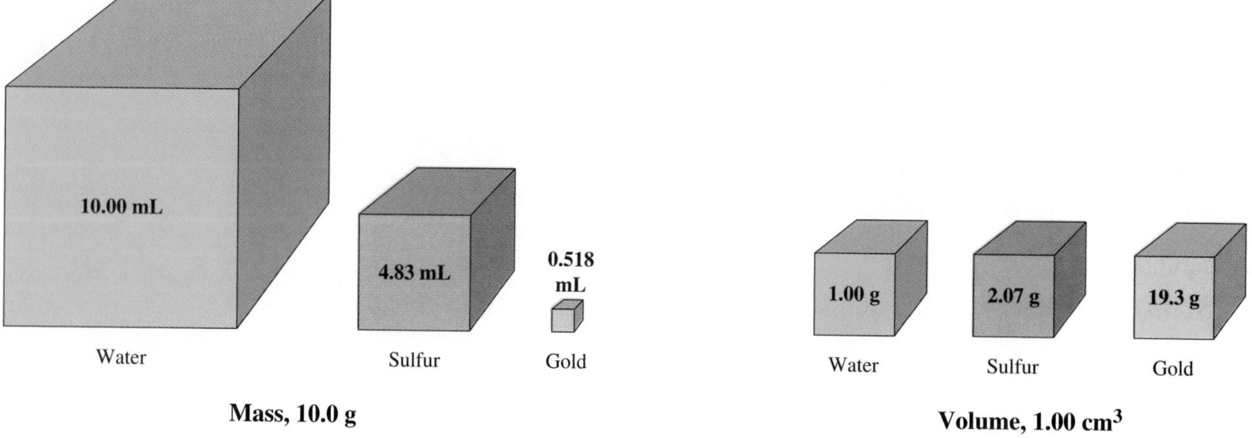

Figure 2.6
(a) Comparison of the volumes of equal masses (10.0 g) of water, sulfur, and gold.
(b) Comparison of the masses of equal volumes (1.00 cm^3) of water, sulfur, and gold.
(Water is at 4°C; the two solids, at 20°C.)

Table 2.5 Densities of Some Selected Materials

Liquids and solids		Gases	
Substance	Density (g/mL at 20°C)	Substance	Density (g/L at 0°C)
Wood (Douglas fir)	0.512	Hydrogen	0.090
Ethyl alcohol	0.789	Helium	0.178
Vegetable oil	0.91	Methane	0.714
Water (4°C)	**1.000***	Ammonia	0.771
Sugar	1.59	Neon	0.90
Glycerin	1.26	Carbon monoxide	1.25
Karo syrup	1.37	Nitrogen	1.251
Magnesium	1.74	**Air**	**1.293***
Sulfuric acid	1.84	Oxygen	1.429
Sulfur	2.07	Hydrogen chloride	1.63
Salt	2.16	Argon	1.78
Aluminum	2.70	Carbon dioxide	1.963
Silver	10.5	Chlorine	3.17
Lead	11.34		
Mercury	13.55		
Gold	19.3		

*For comparing densities, the density of water is the reference for solids and liquids; air is the reference for gases.

When an insoluble solid object is dropped into water, it will sink or float, depending on its density. If the object is less dense than water, it will float, displacing a *mass* of water equal to the mass of the object. If the object is more dense than water, it will sink, displacing a *volume* of water equal to the volume of the object. This information can be used to determine the volume (and density) of irregularly shaped objects.

specific gravity

The **specific gravity** (sp gr) of a substance is the ratio of the density of that substance to the density of another substance, usually water at 4°C. Specific gravity has no units because the density units cancel. The specific gravity tells us how many times as heavy a liquid, a solid, or a gas is as compared to the reference material. Since the density of water at 4°C is 1.00 g/mL, the specific gravity of a solid or liquid is the same as its density in g/mL without the units.

$$\text{sp gr} = \frac{\text{density of a liquid or solid}}{\text{density of water}}$$

Sample calculations of density problems follow.

Example 2.22

What is the density of a mineral if 427 g of the mineral occupy a volume of 35.0 mL?

SOLUTION

We need to solve for density, so we start by writing the formula for calculating density:

$$d = \frac{\text{mass}}{\text{volume}}$$

Then we substitute the data given in the problem into the equation and solve:

$$\text{mass} = 427\text{ g} \qquad \text{volume} = 35.0\text{ mL}$$

$$d = \frac{\text{mass}}{\text{volume}} = \frac{427\text{ g}}{35.0\text{ mL}} = 12.2\text{ g/mL}$$

Example 2.23

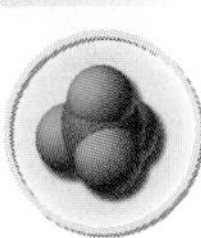

The density of gold is 19.3 g/mL. What is the mass of 25.0 mL of gold?

SOLUTION

There are two ways to solve this problem: (1) Solve the density equation for mass, and then substitute the density and volume data into the new equation and calculate. (2) Solve by dimensional analysis.

When alternative methods of solution are available, more than one is shown in the example. Choose the method you are most comfortable with.

Method 1 (a) Solve the density equation for mass:

$$d = \frac{\text{mass}}{\text{volume}} \qquad d \times \text{volume} = \text{mass}$$

(b) Substitute the data and calculate.

$$\text{mass} = \left(\frac{19.3\text{ g}}{\cancel{\text{mL}}}\right)(25.0\ \cancel{\text{mL}}) = 483\text{ g}$$

Method 2 Dimensional analysis: Use density as a conversion factor, converting

$$\text{mL} \rightarrow \text{g}$$

The conversion of units is

$$\text{mL} \times \frac{\text{g}}{\text{mL}} = \text{g}$$

$$(25.0\ \cancel{\text{mL}})\left(\frac{19.3\text{ g}}{\cancel{\text{mL}}}\right) = 483\text{ g}$$

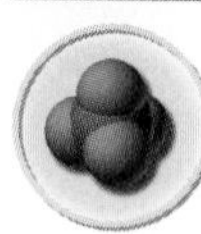

Example 2.24 Calculate the volume (in mL) of 100. g of ethyl alcohol.

SOLUTION From Table 2.5 we see that the density of ethyl alcohol is 0.789 g/mL. This density also means that 1 mL of the alcohol has a mass of 0.789 g (1 mL/0.789 g).

Method 1 Solve the density equation for volume and then substitute the data in the new equation:

$$d = \frac{\text{mass}}{\text{volume}}$$

$$\text{volume} = \frac{\text{mass}}{d}$$

$$\text{volume} = \frac{100.\ \cancel{g}}{0.789\ \cancel{g}/\text{mL}} = 127\ \text{mL}$$

Method 2 Dimensional analysis. For a conversion factor, we can use either

$$\frac{\text{g}}{\text{mL}} \quad \text{or} \quad \frac{\text{mL}}{\text{g}}$$

In this case the conversion is from g → mL, so we use mL/g. Substituting the data,

$$100.\ \cancel{g} \times \frac{1\ \text{mL}}{0.789\ \cancel{g}} = 127\ \text{mL of ethyl alcohol}$$

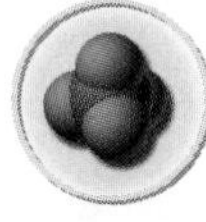

Example 2.25 The water level in a graduated cylinder stands at 20.0 mL before and at 26.2 mL after a 16.74-g metal bolt is submerged in the water. (a) What is the volume of the bolt? (b) What is the density of the bolt?

SOLUTION (a) The bolt will displace a volume of water equal to the volume of the bolt. Thus, the increase in volume is the volume of the bolt:

$$\begin{aligned} 26.2\ \text{mL} &= \text{volume of water plus bolt} \\ -20.0\ \text{mL} &= \text{volume of water} \\ \hline 6.2\ \text{mL} &= \text{volume of bolt} \end{aligned}$$

(b) $d = \dfrac{\text{mass of bolt}}{\text{volume of bolt}} = \dfrac{16.74\ \text{g}}{6.2\ \text{mL}} = 2.7\ \text{g/mL}$

Practice 2.14

Pure silver has a density of 10.5 g/mL. A ring sold as pure silver has a mass of 18.7 g. When it is placed in a graduated cylinder, the water level rises 2.0 mL. Determine whether the ring is actually pure silver or whether the customer should contact the Better Business Bureau.

Practice 2.15

The water level in a metric measuring cup is 0.75 L before the addition of 150. g of shortening. The water level after submerging the shortening is 0.92 L. Determine the density of the shortening.

Chapter 2 Review

2.1 Scientific Notation

KEY TERMS

Measurement
Scientific notation

- Quantitative observations consist of a number and a unit and are called measurements.
- Very large and very small numbers can be represented compactly by using scientific notation:
  - The number is represented as a decimal between 1 and 10 and is multiplied by 10 raised to the appropriate exponent.
  - The sign on the exponent is determined by the direction the decimal point is moved in the original number.

2.2 Measurement and Uncertainty

KEY TERM

Significant figures

- All measurements reflect some amount of uncertainty, which is indicated by the number of significant figures in the measurement.
- The significant figures include all those known with certainty plus one estimated digit.

2.3 Significant Figures

KEY TERM

Rounding off numbers

- Rules exist for counting significant figures in a measurement:
  - Nonzero numbers are always significant.
  - Exact numbers have an infinite number of significant figures.
  - The significance of a zero in a measurement is determined by its position within the number.
- Rules exist for rounding off the result of a calculation to the correct number of significant figures.
  - If the first number after the one you want to retain is 4 or less, that digit and all those after it are dropped.
  - If the first number after the one you want to retain is 5 or greater, that digit and all those after it are dropped and the last digit retained is increased by one.

2.4 Significant Figures in Calculations

- The results of a calculation cannot be more precise than the least precise measurement.
- Rules exist for determining the correct number of significant figures in the result of a calculation.

2.5 The Metric System

KEY TERMS

Metric system (SI)
Meter (m)

- The metric system uses factors of 10 and a set of standard units for measurements.
- Length in the metric system is measured by the standard unit of the meter.

2.6 Problem Solving

- Dimensional analysis is a common method used to solve problems:
  - Conversion factors are used to convert one unit into another:

    $\text{unit}_1 \times \text{conversion factor} = \text{unit}_2$
- The basic steps for solving problems are:
  - Read the problem and write down the known information.
  - Plan how you will solve the problem by deciding what principles or equations are needed.
  - Set up the calculation using dimensional analysis, making sure all the units cancel except the unit required in the answer.
  - Calculate the result and round to the correct number of significant figures.
  - Check the answer to determine whether it is reasonable.

2.7 Measuring Mass and Volume

KEY TERMS

Mass
Weight
Kilogram (kg)
Volume
Liter (L)

- The standard unit for mass in the metric system is the kilogram. In chemistry we often use the gram instead, as we tend to work in smaller quantities.
- Volume is the amount of space occupied by matter.
- The standard unit for volume is the cubic meter. In chemistry we usually use the volume unit of the liter or the milliliter.

2.8 Measurement of Temperature

KEY TERMS

Heat
Temperature

- There are three commonly used temperature scales: Fahrenheit, Celsius, and Kelvin.
- We can convert among the temperature scales by using mathematical formulas:
  - $K = °C + 273.15$
  - $°F = (1.8 \times °C) + 32$
  - $°C = \dfrac{°F - 32}{1.8}$

2.9 Density

KEY TERMS

Density
Specific gravity

- The density of a substance is the amount of matter (mass) in a given volume of the substance:

$$d = \frac{\text{mass}}{\text{volume}}$$

- Specific gravity is the ratio of the density of a substance to the density of another reference substance (usually water).

Review Questions

All questions with blue numbers have answers in the appendix of the text.

1. How many centimeters make up 1 km? (Table 2.3)
2. What is the metric equivalent of 3 in.? (Figure 2.3)
3. Why is the neck of a 100-mL volumetric flask narrower than the top of a 100-mL graduated cylinder? (Figure 2.4)
4. Describe the order of the following substances (top to bottom) if these three substances were placed in a 100-mL graduated cylinder: 25 mL glycerin, 25 mL mercury, and a cube of magnesium 2.0 cm on an edge. (Table 2.5)
5. Arrange these materials in order of increasing density: salt, vegetable oil, lead, and ethyl alcohol. (Table 2.5)
6. Ice floats in vegetable oil and sinks in ethyl alcohol. The density of ice must lie between what numerical values? (Table 2.5)
7. Distinguish between heat and temperature.
8. Distinguish between density and specific gravity.
9. State the rules used in this text for rounding off numbers.
10. Compare the number of degrees between the freezing point of water and its boiling point on the Fahrenheit, Kelvin, and Celsius temperature scales. (Figure 2.5)
11. Why does an astronaut weigh more on Earth than in space when his or her mass remains the same in both places?
12. Ice floats in water, yet ice is simply frozen water. If the density of water is 1.0 g/mL, how is this possible?

Paired Exercises

All exercises with blue numbers have answers in the appendix of the text.

1. What is the numerical meaning of each of the following?
(a) kilogram
(b) centimeter
(c) microliter
(d) millimeter
(e) deciliter

2. What is the correct unit of measurement for each of the following?
(a) 1000 meters
(b) 0.1 gram
(c) 0.000001 liter
(d) 0.01 meter
(e) 0.001 liter

3. State the abbreviation for each of the following units:
(a) gram
(b) microgram
(c) centimeter
(d) micrometer
(e) milliliter
(f) deciliter

4. State the abbreviation for each of the following units:
(a) milligram
(b) kilogram
(c) meter
(d) nanometer
(e) angstrom
(f) microliter

Significant Figures, Rounding, Exponential Notation

5. Are the zeros in these numbers significant?
(a) 503
(b) 0.007
(c) 4200
(d) 3.0030
(e) 100.00
(f) 8.00×10^2

6. Are the zeros significant in these numbers?
(a) 63,000
(b) 6.004
(c) 0.00543
(d) 8.3090
(e) 60.
(f) 5.0×10^{-4}

7. How many significant figures are in each of the following numbers?
(a) 0.025
(b) 22.4
(c) 0.0404
(d) 5.50×10^3

8. State the number of significant figures in each of the following numbers:
(a) 40.0
(b) 0.081
(c) 129,042
(d) 4.090×10^{-3}

9. Round each of the following numbers to three significant figures:
(a) 93.246
(b) 0.02857
(c) 4.644
(d) 34.250

10. Round each of the following numbers to three significant figures:
(a) 8.8726
(b) 21.25
(c) 129.509
(d) 1.995×10^6

11. Express each of the following numbers in exponential notation:
(a) 2,900,000 (c) 0.00840
(b) 0.587 (d) 0.0000055

12. Write each of the following numbers in exponential notation:
(a) 0.0456 (c) 40.30
(b) 4082.2 (d) 12,000,000

13. Solve the following problems, stating answers to the proper number of significant figures:
(a) $12.62 + 1.5 + 0.25 = ?$
(b) $(2.25 \times 10^3)(4.80 \times 10^4) = ?$
(c) $\frac{(452)(6.2)}{14.3} = ?$
(d) $(0.0394)(12.8) = ?$
(e) $\frac{0.4278}{59.6} = ?$
(f) $10.4 + 3.75(1.5 \times 10^4) = ?$

14. Evaluate each of the following expressions. State the answer to the proper number of significant figures:
(a) $15.2 - 2.75 + 15.67$
(b) $(4.68)(12.5)$
(c) $\frac{182.6}{4.6}$
(d) $1986 + 23.84 + 0.012$
(e) $\frac{29.3}{(284)(415)}$
(f) $(2.92 \times 10^{-3})(6.14 \times 10^5)$

15. Change these fractions into decimals. Express each answer to three significant figures:
(a) $\frac{5}{6}$ (c) $\frac{12}{16}$
(b) $\frac{3}{7}$ (d) $\frac{9}{18}$

16. Change each of the following decimals to fractions in lowest terms:
(a) 0.25 (c) 1.67
(b) 0.625 (d) 0.8888

17. Solve each of these equations for x:
(a) $3.42x = 6.5$
(b) $\frac{x}{12.3} = 7.05$
(c) $\frac{0.525}{x} = 0.25$

18. Solve each equation for the variable:
(a) $x = \frac{212 - 32}{1.8}$
(b) $8.9\frac{\text{g}}{\text{mL}} = \frac{40.90\ \text{g}}{x}$
(c) $72°\text{F} = 1.8x + 32$

Unit Conversions

19. Complete the following metric conversions using the correct number of significant figures:
(a) 28.0 cm to m (e) 6.8×10^4 mg to kg
(b) 1000. m to km (f) 8.54 g to kg
(c) 9.28 cm to mm (g) 25.0 mL to L
(d) 10.68 g to mg (h) 22.4 L to μL

20. Complete the following metric conversions using the correct number of significant figures:
(a) 4.5 cm to Å (e) 0.65 kg to mg
(b) 12 nm to cm (f) 5.5 kg to g
(c) 8.0 km to mm (g) 0.468 L to mL
(d) 164 mg to g (h) 9.0 μL to mL

21. Complete the following American/metric conversions using the correct number of significant figures:
(a) 42.2 in. to cm (d) 42.8 kg to lb
(b) 0.64 m to in. (e) 3.5 qt to mL
(c) 2.00 in.^{2} to cm^2 (f) 20.0 L to gal

22. Make the following conversions using the correct number of significant figures:
(a) 35.6 m to ft (d) 95 lb to g
(b) 16.5 km to mi (e) 20.0 gal to L
(c) 4.5 in.^{3} to mm^3 (f) 4.5×10^4 ft^3 to m^3

23. After you have worked out at the gym on a stationary bike for 45 minutes, the distance gauge indicates that you have traveled 15.2 miles. What was your rate in km/hr?

24. A competitive college runner ran a 5-K (5.0-km) race in 15 minutes, 23 seconds. What was her pace in miles per hour?

25. A pharmacy technician is asked to prepare an antibiotic IV that will contain 500. mg of cephalosporin for every 100 mL of normal saline solution. The total volume of saline solution will be 1 L. How many grams of the cephalosporin will be needed for this IV?

26. An extra-strength aspirin tablet contains 0.500 gram of the active ingredient, acetylsalicylic acid. Aspirin strength used to be measured in grains. If 1 grain = 60 mg, how many grains of the active ingredient are in 1 tablet? (Report your answer to 3 significant figures.)

27. The velocity of light is 1.86×10^8 mi/hr. The distance of Mercury from the sun is approximately 5.78×10^7 km. How many minutes will it take for light from the sun to travel to Mercury?

28. How many days would it take to lose 25 lb if the average weight loss is 1.0 kg per week?

29. A personal trainer uses calipers on a client to determine his percent body fat. After taking the necessary measurements, the personal trainer determines that the client's body contains 11.2% fat by mass (11.2 lb of fat per 100 lb of body mass). If the client weighs 225 lb, how many kg of fat does he have?

30. The weight of a diamond is measured in carats. How many pounds does a 5.75-carat diamond weigh? 1 carat = 200. mg

31. A competitive high school swimmer takes 52 seconds to swim 100. yards. What is his rate in m/min?

32. In 2005, Jarno Trulli was the Pole Winner of the US Grand Prix Race with a speed of 133 miles per hour. What was his speed in cm/s?

33. If gasoline is $2.50 per gallon for regular grade, how much will it cost a commuter to drive to and from work each day if the round-trip distance is 45.5 miles and his car averages 11 km/L?

34. The price of gold varies greatly and has been as high as $875 per ounce. What is the value of 250 g of gold at $559 per ounce? Gold is priced by troy ounces (14.58 troy ounces = 1 lb).

35. An adult ruby-throated hummingbird has an average mass of 3.2 g, while an adult California condor may attain a mass of 21 lb. How many hummingbirds would it take to equal the mass of one condor?

36. A very strong camel can carry 990 lb. If one straw weighs 1.5 grams, how many straws can the camel carry without breaking his back?

37. Assuming that there are 20. drops in 1.0 mL, how many drops are in 1.0 gallon?

38. How many liters of oil are in a 42-gal barrel of oil?

39. Calculate the number of milliliters of water in a cubic foot of water.

40. Oil spreads in a thin layer on water called an "oil slick." How much area in m^2 will 200 cm^3 of oil cover if it forms a layer 0.5 nm thick?

41. A textbook is 27 cm long, 21 cm wide, and 4.4 cm thick. What is the volume in:
(a) cubic centimeters?
(b) liters?
(c) cubic inches?

42. An aquarium measures 16 in. × 8 in. × 10 in. How many liters of water does it hold? How many gallons?

43. A toddler in Italy visits the family doctor. The nurse takes the child's temperature, which reads 38.8°C. (a) Convert this temperature to °F. (b) If 98.6°F is considered normal, does the child have a fever?

44. Driving to the grocery store, you notice the temperature is 45°C. Determine what this temperature is on the Fahrenheit scale and what season of the year it might be.

45. Make the following conversions and include an equation for each one:
(a) 162°F to °C
(b) 0.0°F to K
(c) −18°C to °F
(d) 212 K to °C

46. Make the following conversions and include an equation for each one:
(a) 32°C to °F
(b) −8.6°F to °C
(c) 273°C to K
(d) 100 K to °F

*47. At what temperature are the Fahrenheit and Celsius temperatures exactly equal?

*48. At what temperature are Fahrenheit and Celsius temperatures the same in value but opposite in sign?

49. The normal boiling point of ethyl alcohol is 173.3°F. What is this temperature in °C?

50. The normal boiling point of oxygen, O_2, is −183°C. What is this temperature in °F?

51. Calculate the density of a liquid if 50.00 mL of the liquid have a mass of 78.26 g.

52. A 12.8-mL sample of bromine has a mass of 39.9 g. What is the density of bromine?

53. When a 32.7-g piece of chromium metal was placed into a graduated cylinder containing 25.0 mL of water, the water level rose to 29.6 mL. Calculate the density of the chromium.

54. An empty graduated cylinder has a mass of 42.817 g. When filled with 50.0 mL of an unknown liquid, it has a mass of 106.773 g. What is the density of the liquid?

55. Concentrated hydrochloric acid has a density of 1.19 g/mL. Calculate the mass of 250.0 mL of this acid.

56. What mass of mercury (density 13.6 g/mL) will occupy a volume of 25.0 mL?

Additional Exercises

All exercises with blue numbers have answers in the appendix of the text.

57. You have measured out 10.0123576 grams of NaCl. What amount should you report if the precision of the balance you used was
(a) + or − 0.01 g?
(b) + or − 0.001 g?
(c) + or − 0.0001 g?

58. Suppose you want to add 100 mL of solvent to a reaction flask. Which piece of glassware shown in Figure 2.4 would be the **best** choice for accomplishing this task and why?

59. A reaction requires 21.5 g of $CHCl_3$. No balance is available, so it will have to be measured by volume. How many mL of $CHCl_3$ need to be taken? (Density of $CHCl_3$ is 1.484 g/mL.)

60. A 25.27-g sample of pure sodium was prepared for an experiment. How many mL of sodium is this? (Density of sodium is 0.97 g/mL.)

61. One liter of homogenized whole milk has a mass of 1032 g. What is the density of the milk in grams per milliliter? in kilograms per liter?

62. The volume of blood plasma in adults is 3.1 L. Its density is 1.03 g/cm^3. Approximately how many pounds of blood plasma are there in your body?

63. The dashed lane markers on an interstate highway are 2.5 ft long and 4.0 in. wide. One (1.0) qt of paint covers 43 ft^2. How many dashed lane markers can be painted with 15 gal of paint?

64. Will a hollow cube with sides of length 0.50 m hold 8.5 L of solution? Depending on your answer, how much additional solution would be required to fill the container or how many times would the container need to be filled to measure the 8.5 L?

65. The accepted toxic dose of mercury is 300 μg/day. Dental offices sometimes contain as much as 180 μg of mercury per cubic meter of air. If a nurse working in the office ingests 2×10^4 L of air per day, is he or she at risk for mercury poisoning?

66. Which is the higher temperature, 4.5°F or −15°C? Show your calculation.

67. According to the National Heart, Lung, and Blood Institute, LDL-cholesterol levels of less than 130 mg of LDL-cholesterol per deciliter of blood are desirable for heart health in humans. On the average, a human has 4.7 L of whole blood. What is the maximum number of grams of LDL-cholesterol that a human should have?

68. Pure water at 4.0°C has a density of 1.00 g/mL. How many pounds does 2.50 gallons of water weigh?

69. The height of a horse is measured in hands (1 hand = exactly 4 inches). How many meters is a horse that measures 14.2 hands?

70. Camels have been reported to drink as much as 22.5 gallons of water in 12 hours. How many liters can they drink in 30. days?

71. You are given three cubes, A, B, and C; one is magnesium, one is aluminum, and the third is silver. All three cubes have the same mass, but cube A has a volume of 25.9 mL, cube B has a volume of 16.7 mL, and cube C has a volume of 4.29 mL. Identify cubes A, B, and C.

72. How many cubic centimeters are in 1.00 cubic inch?

* **73.** A cube of aluminum has a mass of 500. g. What will be the mass of a cube of gold of the same dimensions?

74. A 25.0-mL sample of water at 90°C has a mass of 24.12 g. Calculate the density of water at this temperature.

75. The mass of an empty container is 88.25 g. The mass of the container when filled with a liquid (d = 1.25 g/mL) is 150.50 g. What is the volume of the container?

76. Which liquid will occupy the greater volume, 50 g of water or 50 g of ethyl alcohol? Explain.

77. The Sacagawea gold-colored dollar coin has a mass of 8.1 g and contains 3.5% manganese. What is its mass in ounces (1 lb = 16 oz) and how many ounces of Mn are in this coin?

78. The density of sulfuric acid is 1.84 g/mL. What volume of this acid will weigh 100. g?

79. The density of palladium at 20°C is 12.0 g/mL and at 1550°C the density is 11.0 g/mL. What is the change in volume (in mL) of 1.00 kg Pd in going from 20°C to 1550°C?

* **80.** As a solid substance is heated, its volume increases but its mass remains the same. Sketch a graph of density versus temperature showing the trend you expect. Briefly explain.

81. A gold bullion dealer advertised a bar of pure gold for sale. The gold bar had a mass of 3300 g and measured 2.00 cm × 15.0 cm × 6.00 cm. Was the bar pure gold? Show evidence for your answer.

82. A 35.0-mL sample of ethyl alcohol (density 0.789 g/mL) is added to a graduated cylinder that has a mass of 49.28 g. What will be the mass of the cylinder plus the alcohol?

83. The largest nugget of gold on record was found in 1872 in New South Wales, Australia, and had a mass of 93.3 kg. Assuming the nugget is pure gold, what is its volume in cubic centimeters? What is it worth by today's standards if gold is \$559/oz? (14.58 troy oz = 1 lb)

Challenge Exercises

All exercises with blue numbers have answers in the appendix of the text.

***84.** Your boss found a piece of metal in the lab and wants you to determine what the metal is. She's pretty sure that the metal is either of lead, aluminum, or silver. The lab bench has a balance and a 100-mL graduated cylinder with 50 mL of water in it. You decide to weigh the metal and find that it has a mass of 20.25 g. After dropping the piece of metal into the graduated cylinder containing water, you notice that the volume increased to 57.5 mL. Identify the metal.

***85.** Forgetful Freddie placed 25.0 mL of a liquid in a graduated cylinder with a mass of 89.450 g when empty. When Freddie placed a metal slug with a mass of 15.454 g into the cylinder, the volume rose to 30.7 mL. Freddie was asked to calculate the density of the liquid and of the metal slug from his data, but he forgot to obtain the mass of the liquid. He was told that if he found the mass of the cylinder containing the liquid and the slug, he would have enough data for the calculations. He did so and found its mass to be 125.934 g. Calculate the density of the liquid and of the metal slug.

Answers to Practice Exercises

2.1 (a) $1200 = 1.200 \times 10^3$ (left means positive exponent)
(b) $6{,}600{,}000 = 6.6 \times 10^6$ (left means positive exponent)
(c) $0.0468 = 4.68 \times 10^{-2}$ (right means negative exponent)
(d) $0.00003 = 3 \times 10^{-5}$ (right means negative exponent)

2.2 (a) 2; (b) 4; (c) 4; (d) 1; (e) 3; (f) 4; (g) 1; (h) 3

2.3 (a) 42.25 (Rule 2); (b) 88.0 (Rule 1); (c) 0.0897 (Rule 2); (d) 0.090 (Rule 2); (e) 225 (Rule 1); (f) 14.2 (Rule 2)

2.4 (a) 3350 in.^{2} = 3.4×10^3 in.^{2}; (b) 50.7 mi/hr; (c) 0.79; (d) 1.3; (e) 20.5; (f) 3.71

2.5 (a) 2; (b) 2; (c) 1; (d) 2; (e) 4; (f) 2; (g) 2; (h) 2

2.6 (a) 1.69×10^4 m; (b) 3.5×10^{-2} m^2

2.7 30 m^3 or 3×10^1 m^3

2.8 165 lb

2.9 0.14 lb

2.10 0.793 qt

2.11 1.89 L (the number of significant figures is arbitrary)

2.12 −269°C, −452°F

2.13 37.0°C, 310.2 K

2.14 The density is 9.35 g/mL; therefore the ring is not pure silver. The density of silver is 10.5 g/mL.

2.15 0.88 g/mL

CHAPTER 3

Elements and Compounds

Flower varieties are often classified by color in the nursery.

Chapter Outline

In Chapter 1 we learned that matter can be divided into the broad categories of pure substances and mixtures. We further learned that pure substances are either elements or compounds. The chemical elements are very important to us in our daily lives. In tiny amounts they play a large role in our health and metabolism. Metallic elements are used for the skin of airplanes, buildings, and sculpture. In this chapter we explore the nature of the chemical elements and begin to learn how chemists classify them.

3.1 Elements

element

All words in English are formed from an alphabet consisting of only 26 letters. All known substances on Earth—and most probably in the universe, too—are formed from a sort of "chemical alphabet" consisting of over 100 known elements. An **element** is a fundamental or elementary substance that cannot be broken down by chemical means to simpler substances. Elements are the building blocks of all substances. The elements are numbered in order of increasing complexity beginning with hydrogen, number 1. Of the first 92 elements, 88 are known to occur in nature. The other four—technetium (43), promethium (61), astatine (85), and francium (87)—either do not occur in nature or have only transitory existences during radioactive decay. With the exception of number 94, plutonium, elements above number 92 are not known to occur naturally but have been synthesized, usually in very small quantities, in laboratories. The discovery of trace amounts of element 94 (plutonium) in nature has been reported. No elements other than those on Earth have been detected on other bodies in the universe.

See the periodic table on the inside front cover.

Most substances can be decomposed into two or more simpler substances. Water can be decomposed into hydrogen and oxygen. Sugar can be decomposed into carbon, hydrogen, and oxygen. Table salt is easily decomposed into sodium and chlorine. An element, however, cannot be decomposed into simpler substances by ordinary chemical changes.

If we could take a small piece of an element, say copper, and divide it and subdivide it into smaller and smaller particles, we would finally come to a single unit of copper that we could no longer divide and still have copper (see Figure 3.1).

Figure 3.1
The surface of a penny is made up of tiny identical copper atoms packed tightly together.

This smallest particle of an element that can exist is called an **atom**, which is also the smallest unit of an element that can enter into a chemical reaction. Atoms are made up of still smaller subatomic particles. However, these subatomic particles (described in Chapter 5) do not have the properties of elements.

atom

3.2 Distribution of Elements

Elements are distributed unequally in nature, as shown in Figure 3.2. At normal room temperature two of the elements, bromine and mercury, are liquids. Eleven elements—hydrogen, nitrogen, oxygen, fluorine, chlorine, helium, neon, argon, krypton, xenon, and radon—are gases. All the other elements are solids.

Ten elements make up about 99% of the mass of the Earth's crust, seawater, and atmosphere. Oxygen, the most abundant of these, constitutes about 50% of this mass. The distribution of the elements shown in Figure 3.2 includes the Earth's crust to a depth of about 10 miles, the oceans, fresh water, and the atmosphere but does not include the mantle and core of the Earth, which are believed to consist of metallic iron and nickel. Because the atmosphere contains relatively little matter, its inclusion has almost no effect on the distribution. But the inclusion of fresh and salt water does have an appreciable effect since water contains about 11.2% hydrogen. Nearly all of the 0.9% hydrogen shown in the second graph in Figure 3.2 is from water.

The average distribution of the elements in the human body is shown in Figure 3.2c. Note again the high percentage of oxygen.

3.3 Names of the Elements

The names of the elements come to us from various sources. Many are derived from early Greek, Latin, or German words that describe some property of the element. For example, iodine is taken from the Greek word *iodes,* meaning

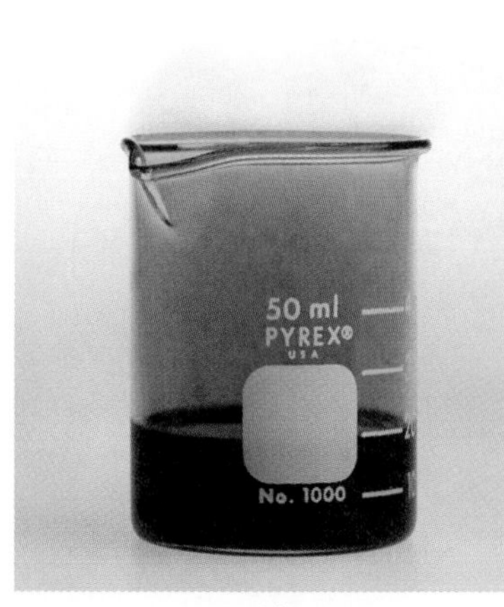

(a)

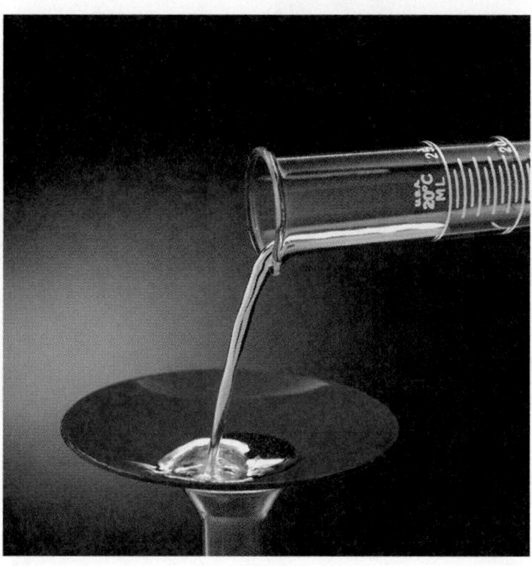

(b)

(a) Bromine (Br_2) and (b) mercury (Hg) are elements that are liquid at room temperature.

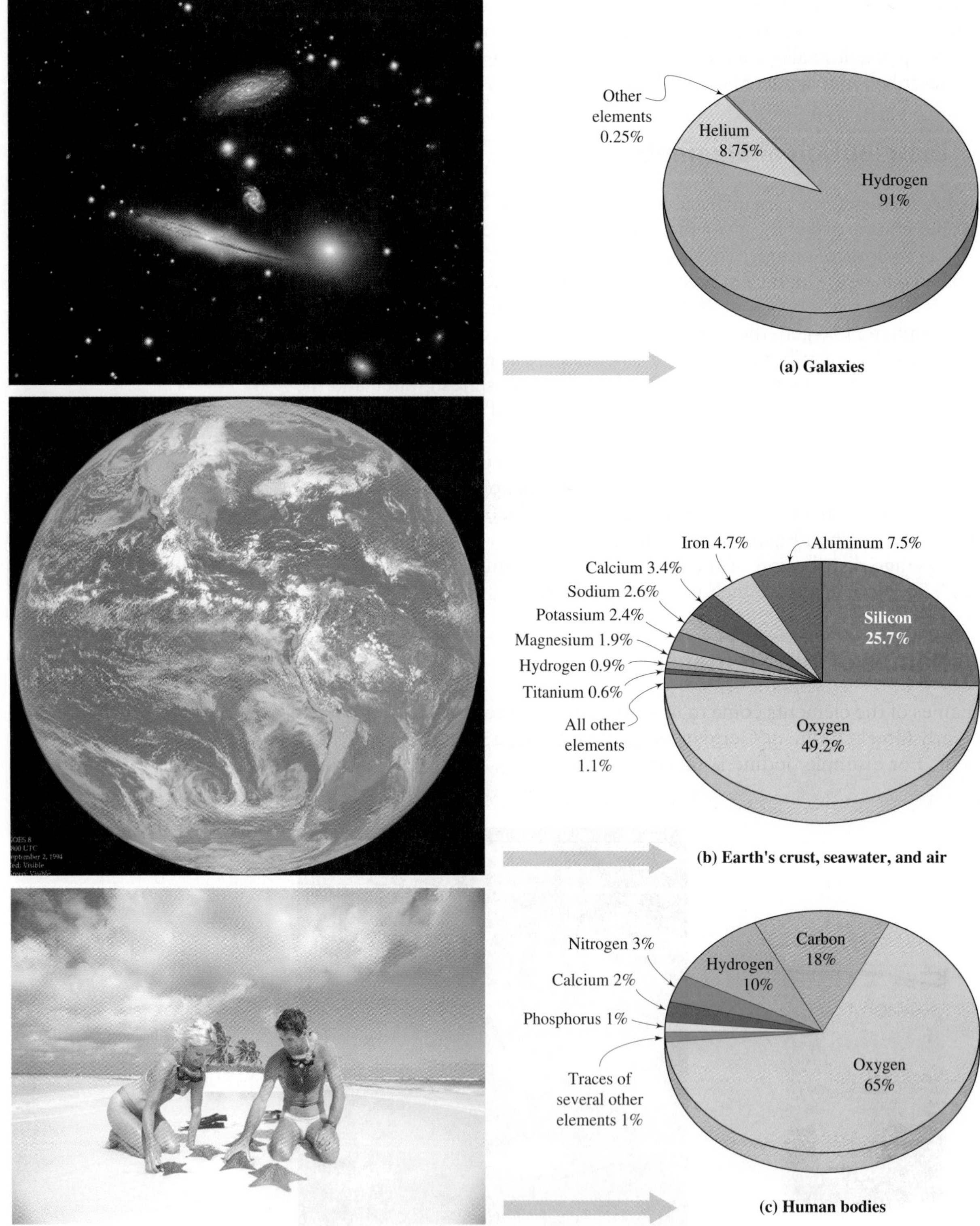

Figure 3.2
Distribution of the common elements in nature.

"violetlike," and iodine is certainly violet in the vapor state. The name of the metal bismuth originates from the German words *weisse masse,* which means "white mass." Miners called it *wismat*; it was later changed to *bismat,* and finally to bismuth. Some elements are named for the location of their discovery—for example, germanium, discovered in 1886 by a German chemist. Others are named in commemoration of famous scientists, such as einsteinium and curium, named for Albert Einstein and Marie Curie, respectively.

3.4 Symbols of the Elements

We all recognize Mr., N.Y., and Ave. as abbreviations for mister, New York, and avenue. In a like manner, each element also has an abbreviation; these are called **symbols** of the elements. Fourteen elements have a single letter as their symbol, and the rest have two letters. A symbol stands for the element itself, for one atom of the element, and (as we shall see later) for a particular quantity of the element.

symbol

Rules governing symbols of elements are as follows:

1. Symbols have either one or two letters.
2. If one letter is used, it is capitalized.
3. If two letters are used, only the first is capitalized.

Examples: Iodine I Barium Ba

The symbols and names of all the elements are given in the table on the inside front cover of this book. Table 3.1 lists the more commonly used elements and their symbols. Examine Table 3.1 carefully and you will note that most of

Iodine in its elemental form is a bluish-black crystal. The symbol for iodine is I.

The metal sodium is soft enough to cut with a knife. The symbol for sodium is Na.

Table 3.1 Symbols of the Most Common Elements

Element	Symbol	Element	Symbol	Element	Symbol
Aluminum	Al	Gold	Au	Platinum	Pt
Antimony	Sb	Helium	He	Plutonium	Pu
Argon	Ar	Hydrogen	H	Potassium	K
Arsenic	As	Iodine	I	Radium	Ra
Barium	Ba	Iron	Fe	Silicon	Si
Bismuth	Bi	Lead	Pb	Silver	Ag
Boron	B	Lithium	Li	Sodium	Na
Bromine	Br	Magnesium	Mg	Strontium	Sr
Cadmium	Cd	Manganese	Mn	Sulfur	S
Calcium	Ca	Mercury	Hg	Tin	Sn
Carbon	C	Neon	Ne	Titanium	Ti
Chlorine	Cl	Nickel	Ni	Tungsten	W
Chromium	Cr	Nitrogen	N	Uranium	U
Cobalt	Co	Oxygen	O	Xenon	Xe
Copper	Cu	Palladium	Pd	Zinc	Zn
Fluorine	F	Phosphorus	P		

The colors of these varieties of quartz are the result of the presence of different metallic elements in the samples.

the symbols start with the same letter as the name of the element that is represented. A number of symbols, however, appear to have no connection with the names of the elements they represent (see Table 3.2). These symbols have been carried over from earlier names (usually in Latin) of the elements and are so firmly implanted in the literature that their use is continued today.

Special care must be taken in writing symbols. Capitalize only the first letter, and use a lowercase second letter if needed. This is important. For example, consider Co, the symbol for the element cobalt. If you write CO (capital C and capital O), you will have written the two elements carbon and oxygen (the *formula* for carbon monoxide), *not* the single element cobalt. Also, make sure that you write the letters distinctly; otherwise, Co (for cobalt) may be misread as Ca (for calcium).

Table 3.2 Symbols of the Elements Derived from Early Names*

Present name	Symbol	Former name
Antimony	Sb	Stibium
Copper	Cu	Cuprum
Gold	Au	Aurum
Iron	Fe	Ferrum
Lead	Pb	Plumbum
Mercury	Hg	Hydrargyrum
Potassium	K	Kalium
Silver	Ag	Argentum
Sodium	Na	Natrium
Tin	Sn	Stannum
Tungsten	W	Wolfram

*These symbols are in use today even though they do not correspond to the current name of the element.

Knowledge of symbols is essential for writing chemical formulas and equations, and will be needed in the remainder of this book and in any future chemistry courses you may take. One way to learn the symbols is to practice a few minutes a day by making flash cards of names and symbols and then practicing daily. Initially it is a good plan to learn the symbols of the most common elements shown in Table 3.1.

3.5 Introduction to the Periodic Table

Almost all chemistry classrooms have a chart called the *periodic table* hanging on the wall. It shows all the chemical elements and contains a great deal of useful information about them. As we continue our study of chemistry, we will learn much more about the periodic table. For now let's begin with the basics.

A simple version of the periodic table is shown in Table 3.3. Notice that in each box there is the symbol for the element and, above it, a number called the *atomic number*. For example nitrogen is [7 **N**] and gold is [79 **Au**].

The elements are placed in the table in order of increasing atomic number in a particular arrangement designed by Dimitri Mendeleev in 1869. His arrangement organizes the elements with similar chemical properties in columns called **families or groups**. An example of this is the column

families or groups of elements

2 **He**
10 **Ne**
18 **Ar**
36 **Kr**
54 **Xe**
86 **Rn**

These elements are all gases and nonreactive. The group is called the **noble gases**. Other groups with special names are the **alkali metals** (under the 1A on the table), **alkaline earth metals** (Group 2A), and **halogens** (Group 7A).

noble gases
alkali metals
alkaline earth metals
halogens
representative elements
transition elements

The tall columns of the periodic table (1A–7A and the noble gases) are known as the **representative elements**. Those elements in the center section of the periodic table and called **transition elements**.

Metals, Nonmetals, and Metalloids

metal

The elements can be classified as metals, nonmetals, and metalloids. Most of the elements are metals. We are familiar with them because of their widespread use in tools, construction materials, automobiles, and so on. But nonmetals are equally useful in our everyday life as major components of clothing, food, fuel, glass, plastics, and wood. Metalloids are often used in the electronics industry.

Tiny computer chips contain silicon, a metalloid.

The **metals** are solids at room temperature (mercury is an exception). They have high luster, are good conductors of heat and electricity, are *malleable* (can be rolled or hammered into sheets), and are *ductile* (can be drawn into wires). Most metals have a high melting point and high density. Familiar metals are aluminum, chromium, copper, gold, iron, lead, magnesium, mercury, nickel, platinum, silver, tin, and zinc. Less familiar but still important metals are calcium, cobalt, potassium, sodium, uranium, and titanium.

Metals have little tendency to combine with each other to form compounds. But many metals readily combine with nonmetals such as chlorine, oxygen, and sulfur to form compounds such as metallic chlorides, oxides, and sulfides. In nature, minerals are composed of the more reactive metals combined with other elements. A few of the less reactive metals such as copper, gold, and silver are sometimes found in a native, or free, state. Metals are often mixed with one another to form homogeneous mixtures of solids called alloys. Some examples are brass, bronze, steel, and coinage metals.

nonmetal

Nonmetals, unlike metals, are not lustrous, have relatively low melting points and densities, and are generally poor conductors of heat and electricity.

Carbon, phosphorus, sulfur, selenium, and iodine are solids; bromine is a liquid; the rest of the nonmetals are gases. Common nonmetals found uncombined in nature are carbon (graphite and diamond), nitrogen, oxygen, sulfur, and the noble gases (helium, neon, argon, krypton, xenon, and radon).

Nonmetals combine with one another to form molecular compounds such as carbon dioxide (CO_2), methane (CH_4), butane (C_4H_{10}), and sulfur dioxide (SO_2). Fluorine, the most reactive nonmetal, combines readily with almost all other elements.

metalloid Several elements (boron, silicon, germanium, arsenic, antimony, tellurium, and polonium) are classified as **metalloids** and have properties that are intermediate between those of metals and those of nonmetals. The intermediate position of these elements is shown in Table 3.3. Certain metalloids, such as boron, silicon, and germanium, are the raw materials for the semiconductor devices that make the electronics industry possible.

Table 3.3 The Periodic Table

- Metals
- Metalloids
- Nonmetals

1A	2A											3A	4A	5A	6A	7A	Noble gases
1 H																	2 He
3 Li	4 Be											5 B	6 C	7 N	8 O	9 F	10 Ne
11 Na	12 Mg											13 Al	14 Si	15 P	16 S	17 Cl	18 Ar
19 K	20 Ca	21 Sc	22 Ti	23 V	24 Cr	25 Mn	26 Fe	27 Co	28 Ni	29 Cu	30 Zn	31 Ga	32 Ge	33 As	34 Se	35 Br	36 Kr
37 Rb	38 Sr	39 Y	40 Zr	41 Nb	42 Mo	43 Tc	44 Ru	45 Rh	46 Pd	47 Ag	48 Cd	49 In	50 Sn	51 Sb	52 Te	53 I	54 Xe
55 Cs	56 Ba	57 La*	72 Hf	73 Ta	74 W	75 Re	76 Os	77 Ir	78 Pt	79 Au	80 Hg	81 Tl	82 Pb	83 Bi	84 Po	85 At	86 Rn
87 Fr	88 Ra	89 Ac†	104 Rf	105 Db	106 Sg	107 Bh	108 Hs	109 Mt	110 Ds	111 Rg							

*	58 Ce	59 Pr	60 Nd	61 Pm	62 Sm	63 Eu	64 Gd	65 Tb	66 Dy	67 Ho	68 Er	69 Tm	70 Yb	71 Lu
†	90 Th	91 Pa	92 U	93 Np	94 Pu	95 Am	96 Cm	97 Bk	98 Cf	99 Es	100 Fm	101 Md	102 No	103 Lr

3.6 Elements in Their Natural States

Most substances around us are mixtures or compounds. Elements tend to be reactive, and they combine with other elements to form compounds. It is rare to find elements in nature in pure form. There are some exceptions, however. Gold, for example, can be found as nuggets. Silver and platinum can also be found in nature in pure form. In fact, these metals are sometimes called the

CHEMISTRY IN ACTION • Using Oxygen to Revive Damaged Art

Treating damaged works of art is a tricky and time-consuming process that often produces mixed results. Now NASA has found a way to help bring damaged paintings back to life. NASA was breaking down oxygen molecules into oxygen atoms (a process that happens in nature in the upper atmosphere). The researchers were using the atomic oxygen to test the durability of satellite materials when they discovered that atomic oxygen could remove organic materials from the surface of objects without damaging the object. Bad news for satellite parts (which needed their coatings) but great news for the art world.

The chief conservator at the Cleveland Museum of Art tried the atomic oxygen treatment on two paintings damaged in a church fire in Cleveland. Although the paintings were not extremely valuable (and so were good subjects for an experiment), all other attempts to restore them had failed. Atomic oxygen proved to work wonders, and the soot and char came off to reveal the image below it. Since the treatment is a gas, the underlying layers were not harmed. The treatment doesn't work on everything and won't replace other techniques altogether, but the conservator was impressed enough to continue to work with NASA on the process.

A small, portable atomic oxygen unit has also been built to treat artworks that have been damaged in a small area (by grafitti, etc.). A beam of atomic oxygen is applied directly to the damaged area to "spot clean" a piece of art. This technique was used to clean a lipstick mark from Andy Warhol's painting *Bathtub* (1961). A cosmetic party had been held at the Andy Warhol Museum in 1997. Free lipstick samples had been distributed at the party, and one reveler decided to kiss the painting! The painting was not varnished, so the lipstick stuck fast to the paint. After a day of treatment with the atomic oxygen beam, the lipstick mark was gone. Space technology can now be used to revive and restore art!

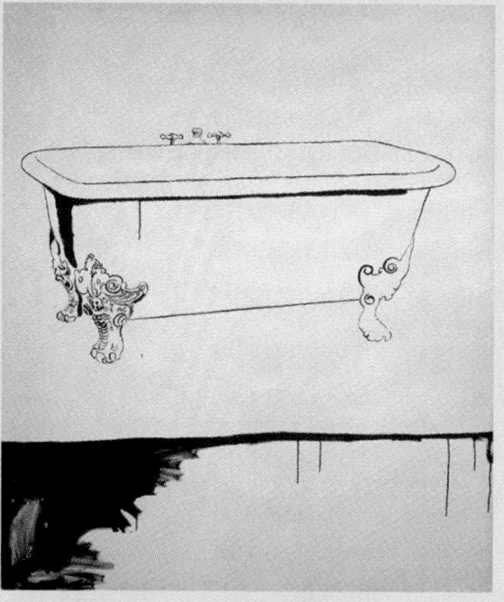

This famous Andy Warhol painting was spot cleaned by using a beam of atomic oxygen.

noble metals since they have a low reactivity. The noble gases are also not reactive and can be found in nature in uncombined form. Helium gas, for example, consists of tiny helium atoms moving independently.

Air can also be divided into its component gases. It is mainly composed of nitrogen and oxygen gases. But when we "look inside" these gases, we find tiny molecules (N_2 and O_2) instead of independent atoms like we see in the noble gases.

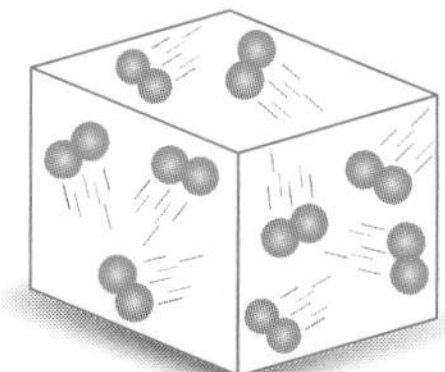

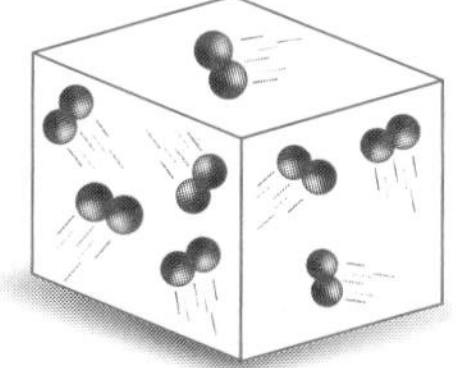

Nitrogen and oxygen gases are composed of molecules (N_2, ●●) and (O_2, ●●).

3.7 Elements That Exist as Diatomic Molecules

Diatomic molecules each contain exactly two atoms (alike or different). Seven elements in their uncombined state are **diatomic molecules**. Their symbols, formulas, and brief descriptions are listed in Table 3.4. Whether found free in nature or prepared in the laboratory, the molecules of these elements always contain two atoms. The formulas of the free elements are therefore always written to show this molecular composition: H_2, N_2, O_2, F_2, Cl_2, Br_2, and I_2.

diatomic molecules

Table 3.4 Elements That Exist as Diatomic Molecules

Element	Symbol	Molecular formula	Normal state
Hydrogen	H	H_2	Colorless gas
Nitrogen	N	N_2	Colorless gas
Oxygen	O	O_2	Colorless gas
Fluorine	F	F_2	Pale yellow gas
Chlorine	Cl	Cl_2	Yellow-green gas
Bromine	Br	Br_2	Reddish-brown liquid
Iodine	I	I_2	Bluish-black solid

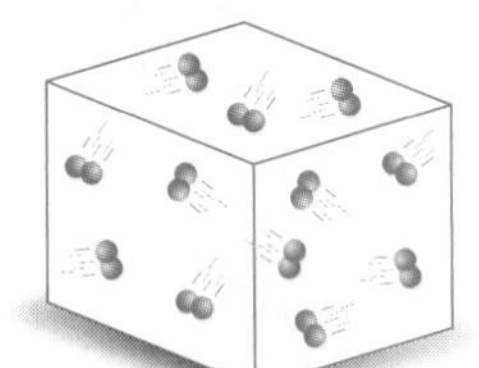

Container of hydrogen molecules.

It is important to see that symbols can designate either an atom or a molecule of an element. Consider hydrogen and oxygen. Hydrogen gas is present in volcanic gases and can be prepared by many chemical reactions. Regardless of their source, all samples of free hydrogen gas consist of diatomic molecules.

Free hydrogen is designated by the formula H_2, which also expresses its composition. Oxygen makes up about 21% by volume of the air that we breathe. This free oxygen is constantly being replenished by photosynthesis; it can also be prepared in the laboratory by several reactions. The majority of free oxygen is diatomic and is designated by the formula O_2. Now consider water, a compound designated by the formula H_2O (sometimes HOH). Water contains neither free hydrogen (H_2) nor free oxygen (O_2). The H_2 part of the formula H_2O simply indicates that two atoms of hydrogen are combined with one atom of oxygen to form water.

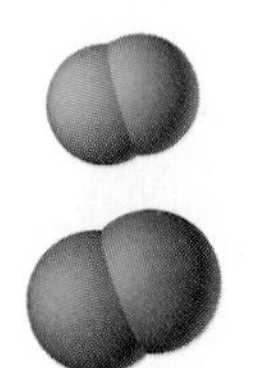

H_2 (blue) and O_2 (red) molecule.

Symbols are used to designate elements, show the composition of molecules of elements, and give the elemental composition of compounds.

Practice 3.1

Identify the physical state of each of the following elements at room temperature (20°C):

H, Na, Ca, N, S, Fe, Cl, Br, Ne, Hg

Hint: You may need to use a resource (such as the Internet or a chemical handbook) to assist you.

Practice 3.2

Identify each of the following elements as a nonmetal, metal, or metalloid:

Na, F, Cr, Mo, Kr, Si, Cu, Sb, I, S

3.8 Compounds

compound

A **compound** is a distinct substance that contains two or more elements chemically combined in a definite proportion by mass. Compounds, unlike elements, can be decomposed chemically into simpler substances—that is, into simpler compounds and/or elements. Atoms of the elements in a compound are combined in whole-number ratios, never as fractional parts. Compounds fall into two general types, *molecular* and *ionic*. Figure 3.3 illustrates the classification of compounds.

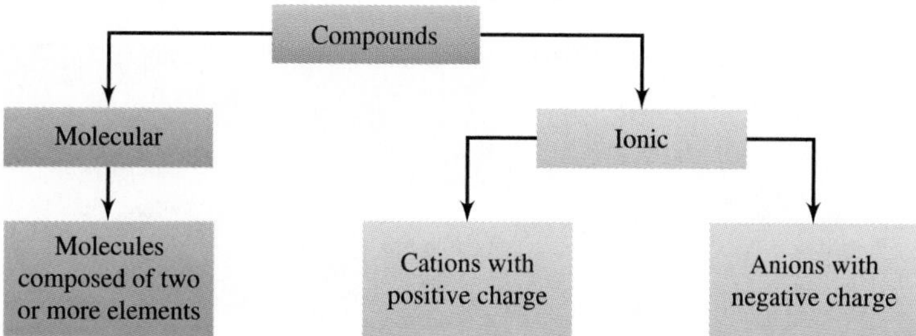

Figure 3.3
Compounds can be classified as molecular or ionic. Ionic compounds are held together by attractive forces between their positive and negative charges. Molecular compounds are held together by covalent bonds.

A **molecule** is the smallest uncharged individual unit of a compound formed by the union of two or more atoms. Water is a typical molecular compound. If we divide a drop of water into smaller and smaller particles, we finally obtain a single molecule of water consisting of two hydrogen atoms bonded to one oxygen atom, as shown in Figure 3.4a. This molecule is the ultimate particle of water; it cannot be further subdivided without destroying the water molecule and forming hydrogen and oxygen.

molecule

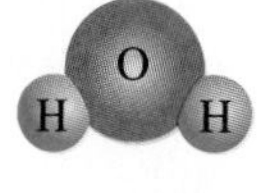

(a) H_2O

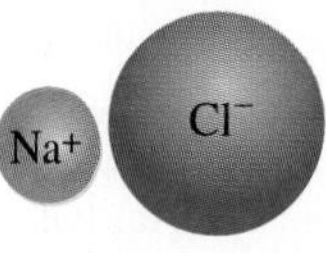

(b) NaCl

Figure 3.4
Representation of molecular and ionic (nonmolecular) compounds. (a) Two hydrogen atoms combined with an oxygen atom to form a molecule of water. (b) A positively charged sodium ion and a negatively charged chloride ion form the compound sodium chloride.

An **ion** is a positively or negatively charged atom or group of atoms. An ionic compound is held together by attractive forces that exist between positively and negatively charged ions. A positively charged ion is called a **cation** (pronounced *cat-eye-on*); a negatively charged ion is called an **anion** (pronounced *an-eye-on*).

ion
cation
anion

Sodium chloride is a typical ionic compound. The ultimate particles of sodium chloride are positively charged sodium ions and negatively charged chloride ions, shown in Figure 3.4b. Sodium chloride is held together in a crystalline structure by the attractive forces existing between these oppositely charged ions. Although ionic compounds consist of large aggregates of cations and anions, their formulas are normally represented by the simplest possible ratio of the atoms in the compound. For example, in sodium chloride the ratio is one sodium ion to one chlorine ion, so the formula is NaCl.

There are more than 11 million known registered compounds, with no end in sight as to the number that will be prepared in the future. Each compound is unique and has characteristic properties. Let's consider two compounds, water and sodium chloride, in some detail. Water is a colorless, odorless, tasteless liquid that can be changed to a solid (ice) at 0°C and to a gas (steam) at 100°C. Composed of two atoms of hydrogen and one atom of oxygen per molecule, water is 11.2% hydrogen and 88.8% oxygen by mass. Water reacts chemically with sodium to produce hydrogen gas and sodium hydroxide, with lime to produce calcium hydroxide, and with sulfur trioxide to produce sulfuric acid. When water is decomposed, it forms hydrogen and oxygen molecules (see Figure 3.5). No other compound has all these exact physical and chemical properties; they are characteristic of water alone.

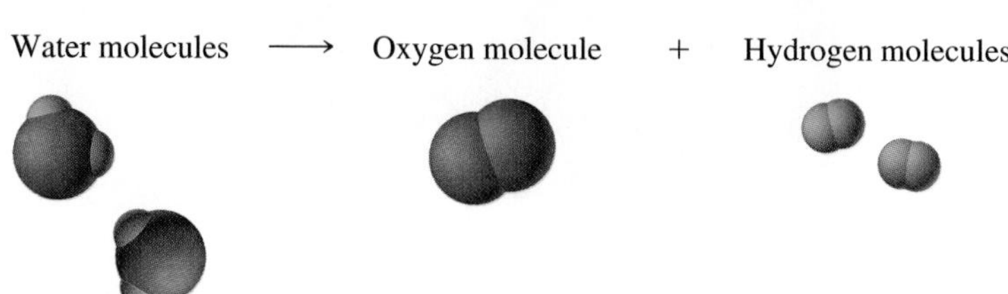

Figure 3.5
A representation of the decomposition of water into oxygen and hydrogen molecules.

CHEMISTRY IN ACTION • Cars: Is Hydrogen the Answer?

The increasing cost of petroleum and the increased pollution in our cities have sparked a race to find alternative fuels for our cars. Manufacturers agree that the best choice for automobiles of the future is hydrogen fuel cells. The greatest controversy now is how to obtain and deliver the hydrogen required for the fuel cells.

General Motors scientists are working to get hydrogen by "cracking," or re-forming, gasoline. Basically, larger hydrocarbon molecules are cracked, or broken down, to produce hydrogen and smaller hydrocarbons. This could be accomplished right onboard the vehicle. In fact, a Chevrolet S-10 truck has been designed as a fuel cell/battery hybrid. The fuel cell gets hydrogen from an onboard re-former, but the re-former and fuel cell equipment take up half the truck's bed. After adding batteries (for acceleration), the total weight of the truck is about 6300 pounds! The good news is the fuel economy is 40 miles per gallon with a range of 525 miles. This combined with the possibility of re-forming the gasoline or natural gas into hydrogen at gas stations (once enough hybrid vehicles are on the road) give General Motors optimism to continue development.

Chrysler Town and Country Natrium fuel cell vehicle.

DaimlerChrysler is also working on fuel cells using another process. The company's engineers plan to bypass fossil fuels altogether. In March 2002, the firm unveiled the Natrium fuel cell Town and Country minivan prototype. This electric van is fueled by hydrogen obtained from a noncombustible aqueous solution of sodium borohydride. The liquid is carried in a large fuel tank and is re-formed onboard to release hydrogen to the fuel cells. Since Natrium uses a nonfossil fuel and has a range of 300 miles without taking up passenger and cargo space, DaimlerChrysler plans to continue development in this direction. The biggest hurdle to overcome is the lack of an economic means to get the fuel from a borax mine to a filling station. Hopefully, one or both of these methods will produce economical alternatives to our present-day automobiles.

Sodium chloride is a colorless crystalline substance with a ratio of one atom of sodium to one atom of chlorine. Its composition by mass is 39.3% sodium and 60.7% chlorine. It does not conduct electricity in its solid state; it dissolves in water to produce a solution that conducts electricity. When a current is passed through molten sodium chloride, solid sodium and gaseous chlorine are produced (see Figure 3.6). These specific properties belong to sodium

Figure 3.6
When sodium chloride (a) is decomposed, it forms sodium metal (b) and chlorine gas (c).

chloride and to no other substance. Thus, a compound may be identified and distinguished from all other compounds by its characteristic properties. We consider these chemical properties further in Chapter 4.

3.9 Chemical Formulas

chemical formula

Chemical formulas are used as abbreviations for compounds. A **chemical formula** shows the symbols and the ratio of the atoms of the elements in a compound. Sodium chloride contains one atom of sodium per atom of chlorine; its formula is NaCl. The formula for water is H_2O; it shows that a molecule of water contains two atoms of hydrogen and one atom of oxygen.

subscript

The formula of a compound tells us which elements it is composed of and how many atoms of each element are present in a formula unit. For example, a unit of sulfuric acid is composed of two atoms of hydrogen, one atom of sulfur, and four atoms of oxygen. We could express this compound as HHSOOOO, but this is cumbersome, so we write H_2SO_4 instead. The formula may be expressed verbally as "H-two-S-O-four." Numbers that appear partially below the line and to the right of a symbol of an element are called **subscripts**. Thus, the 2 and the 4 in H_2SO_4 are subscripts (See Figure 3.7). Characteristics of chemical formulas are as follows:

1. The formula of a compound contains the symbols of all the elements in the compound.
2. When the formula contains one atom of an element, the symbol of that element represents that one atom. The number 1 is not used as a subscript to indicate one atom of an element.

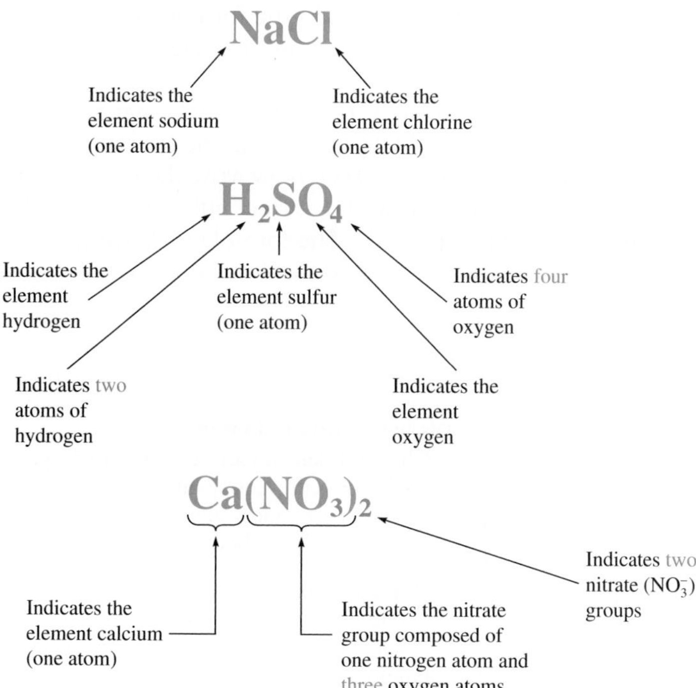

Figure 3.7
Explanation of the formulas NaCl, H_2SO_4, and $Ca(NO_3)_2$.

3. When the formula contains more than one atom of an element, the number of atoms is indicated by a subscript written to the right of the symbol of that atom. For example, the 2 in H_2O indicates two atoms of H in the formula.
4. When the formula contains more than one of a group of atoms that occurs as a unit, parentheses are placed around the group, and the number of units of the group is indicated by a subscript placed to the right of the parentheses. Consider the nitrate group, NO_3^-. The formula for sodium nitrate, $NaNO_3$, has only one nitrate group, so no parentheses are needed. Calcium nitrate, $Ca(NO_3)_2$, has two nitrate groups, as indicated by the use of parentheses and the subscript 2. $Ca(NO_3)_2$ has a total of nine atoms: one Ca, two N, and six O atoms. The formula $Ca(NO_3)_2$ is read as "C-A [pause] N-O-three taken twice."
5. Formulas written as H_2O, H_2SO_4, $Ca(NO_3)_2$, and $C_{12}H_{22}O_{11}$ show only the number and kind of each atom contained in the compound; they do not show the arrangement of the atoms in the compound or how they are chemically bonded to one another.

Practice 3.3

How would you read these formulas aloud?
(a) KBr (b) $PbCl_2$ (c) $CaCO_3$ (d) $Mg(OH)_2$

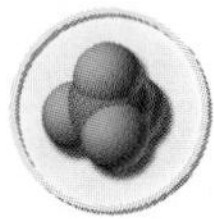

Example 3.1 Write formulas for the following compounds; the atomic composition is given. (a) hydrogen chloride: 1 atom hydrogen + 1 atom chlorine; (b) methane: 1 atom carbon + 4 atoms hydrogen; (c) glucose: 6 atoms carbon + 12 atoms hydrogen + 6 atoms oxygen.

SOLUTION

(a) First write the symbols of the atoms in the formula: H Cl. Since the ratio of atoms is one to one, we bring the symbols together to give the formula for hydrogen chloride as HCl.
(b) Write the symbols of the atoms: C H. Now bring the symbols together and place a subscript 4 after the hydrogen atom. The formula is CH_4.
(c) Write the symbols of the atoms: C H O. Now write the formula, bringing together the symbols followed by the correct subscripts according to the data given (six C, twelve H, six O). The formula is $C_6H_{12}O_6$.

Chapter 3 Review

3.1 Elements

KEY TERMS

Element
Atom

- All matter consists of about 100 elements.
- An element is a fundamental chemical substance.
- The smallest particle of an element is an atom.
- Elements cannot be broken down by chemical means to a simpler substance.

3.2 Distribution of Elements

- Chemical elements are not distributed equally in nature.
- Hydrogen is the most abundant element in the universe.
- Oxygen is the most abundant element on the Earth and in the human body.

3.3 Names of the Elements

- Names for the chemical elements come from a variety of sources, including Latin, location of discovery, and famous scientists.

3.4 Symbols of the Elements

KEY TERM

Symbols

- Rules for writing symbols for the elements are:
  - One or two letters
  - If one letter, use a capital
  - If two letters, only the first is a capital

3.5 Introduction to the Periodic Table

KEY TERMS

Group	Representative elements
Noble gases	Transition elements
Alkali metals	Metals
Alkaline earth metal	Nonmetals
Halogens	Metalloids

- The periodic table was designed by Dimitri Mendeleev and arranges the elements according to their atomic numbers and in groups by their chemical properties.
- Elements can be classified as representative or as transition elements.
- Elements can also be classified by special groups with similar chemical properties. Such groups include the noble gases, alkali metals, alkaline earth metals, and halogens.
- Elements can be classified as metals, nonmetals, or metalloids:
  - Most elements are metals.
  - Metals have the following properties:
    - High luster
    - Good conductor of heat and electricity
    - Malleable
  - Nonmetals have the following properties:
    - Not lustrous
    - Poor conductors of heat and electricity

3.6 Elements in Their Natural States

- Most elements are found in nature combined with other elements.
- Elements that are found in uncombined form in nature include gold, silver, copper, and platinum as well as the noble gases (He, Ne, Ar, Kr, Xe, Rn).

3.7 Elements That Exist as Diatomic Molecules

KEY TERM

Diatomic molecules

- Diatomic molecules contain exactly two atoms (alike or different).
- Seven elements exist as diatomic molecules—H_2, N_2, O_2, F_2, Cl_2, Br_2, and I_2.

3.8 Compounds

KEY TERMS

Compound
Molecule
Ion
Cation
Anion

- A compound is a substance that contains two or more elements chemically combined in a definite proportion by mass.
- There are two general types of compounds:
  - Molecular—formed of individual molecules composed of atoms
  - Ionic—formed from ions that are either positive or negative
    - Cation—positively charged ion
    - Anion—negatively charged ion

3.9 Chemical Formulas

KEY TERMS

Chemical formula
Subscript

- A chemical formula shows the symbols and the ratios of atoms for the elements in a chemical compound.
- Characteristics of chemical formulas include:
  - It contains symbols of all elements in the compound.
  - The symbol represents one atom of the element.
  - If more than one atom of an element is present, the number of atoms is indicated by a subscript.
  - Parentheses are used to show multiple groups of atoms occurring as a unit in the compound.
  - A formula does not show the arrangement of the atoms in the compound.

Review Questions

All questions with blue numbers have answers in the appendix of the text.

1. Are there more atoms of silicon or hydrogen in the Earth's crust, seawater, and atmosphere? Use Figure 3.2 and the fact that the mass of a silicon atom is about 28 times that of a hydrogen atom.
2. Give the symbols for each of the following elements:
   (a) silver (b) oxygen (c) hydrogen (d) carbon
   (e) iron (f) nitrogen (g) magnesium (h) potassium
3. Give the names for each of these elements:
   (a) Na (b) F (c) Ni (d) Zn
   (e) Ne (f) He (g) Ca (h) Cl
4. What does the symbol of an element stand for?
5. Write down what you believe to be the symbols for the elements phosphorus, aluminum, hydrogen, potassium, magnesium, sodium, nitrogen, nickel, and silver. Check yourself by looking up the correct symbols in Table 3.1.

6. Interpret the difference in meanings for each of these pairs: (a) Si and SI (b) Pb and PB (c) 4 P and P_4
7. List six elements and their symbols in which the first letter of the symbol is different from that of the name. (Table 3.2)
8. Write the names and symbols for the 14 elements that have only one letter as their symbol. (See periodic table on inside front cover.)
9. Distinguish between an element and a compound.
10. How many metals are there? nonmetals? metalloids? (Table 3.3)
11. Of the ten most abundant elements in the Earth's crust, seawater, and atmosphere, how many are metals? nonmetals? metalloids? (Figure 3.2)
12. Of the six most abundant elements in the human body, how many are metals? nonmetals? metalloids? (Figure 3.2)
13. Why is the symbol for gold Au rather than G or Go?
14. Give the names of (a) the solid diatomic nonmetal and (b) the liquid diatomic nonmetal. (Table 3.4)
15. Distinguish between a compound and a mixture.
16. What are the two general types of compounds? How do they differ from each other?
17. What is the basis for distinguishing one compound from another?
18. What is the major difference between a cation and an anion?
19. Write the names and formulas of the elements that exist as diatomic molecules. (Table 3.4)

Paired Exercises

All exercises with blue numbers have answers in the appendix of the text.

1. Which of the following are diatomic molecules?
   (a) Cl_2 (b) CO (c) CCl_4 (d) NO_2 (e) N_2 (f) HBr (g) ClO_2 (h) H_2O

2. Which of the following are diatomic molecules?
   (a) CO_2 (b) H_2S (c) H_2 (d) P_4 (e) CS_2 (f) NO (g) ClF

3. What elements are present in each compound?
   (a) potassium iodide KI
   (b) sodium carbonate Na_2CO_3
   (c) aluminum oxide Al_2O_3
   (d) calcium bromide $CaBr_2$
   (e) acetic acid $HC_2H_3O_2$

4. What elements are present in each compound?
   (a) magnesium bromide $MgBr_2$
   (b) carbon tetrachloride CCl_4
   (c) nitric acid HNO_3
   (d) barium sulfate $BaSO_4$
   (e) aluminum phosphate $AlPO_4$

5. Write the formula for each compound (the composition is given after each name):
   (a) zinc oxide — 1 atom Zn, 1 atom O
   (b) potassium chlorate — 1 atom K, 1 atom Cl, 3 atoms O
   (c) sodium hydroxide — 1 atom Na, 1 atom O, 1 atom H
   (d) ethyl alcohol — 2 atoms C, 6 atoms H, 1 atom O

6. Write the formula for each compound (the composition is given after each name):
   (a) aluminum bromide — 1 atom Al, 3 atoms Br
   (b) calcium fluoride — 1 atom Ca, 2 atoms F
   (c) lead(II) chromate — 1 atom Pb, 1 atom Cr, 4 atoms O
   (d) benzene — 6 atoms C, 6 atoms H

7. Explain the meaning of each symbol and number in these formulas:
   (a) H_2O
   (b) Na_2SO_4
   (c) $HC_2H_3O_2$

8. Explain the meaning of each symbol and number in these formulas:
   (a) $AlBr_3$
   (b) $Ni(NO_3)_2$
   (c) $C_{12}H_{22}O_{11}$ (sucrose)

9. How many atoms are represented in each formula?
   (a) KF (b) $CaCO_3$ (c) $K_2Cr_2O_7$ (d) $NaC_2H_3O_2$ (e) $(NH_4)_2C_2O_4$

10. How many atoms are represented in each formula?
    (a) NaCl (b) N_2 (c) $Ba(ClO_3)_2$ (d) CCl_2F_2 (Freon) (e) $Al_2(SO_4)_3$

11. How many atoms of oxygen are represented in each formula?
    (a) H_2O (b) $CuSO_4$ (c) H_2O_2 (d) $Fe(OH)_3$ (e) $Al(ClO_3)_3$

12. How many atoms of hydrogen are represented in each formula?
    (a) H_2 (b) $Ba(C_2H_3O_2)_2$ (c) $C_6H_{12}O_6$ (d) $HC_2H_3O_2$ (e) $(NH_4)_2Cr_2O_7$

13. Identify each of the following as a pure substance or a mixture:
(a) a bottle of Gatorade (d) coffee
(b) aluminum (e) fluorine gas
(c) a piece of bread (f) potassium carbonate

14. Identify each of the following as a pure substance or a mixture:
(a) the human body (d) copper
(b) bottled spring water (e) Mountain Dew
(c) aluminum oxide (f) oak wood

15. For Question 13, state whether each pure substance is an element or a compound.

16. For Question 14, state whether each pure substance is an element or a compound.

17. Classify each of the following as an element, compound, or mixture:

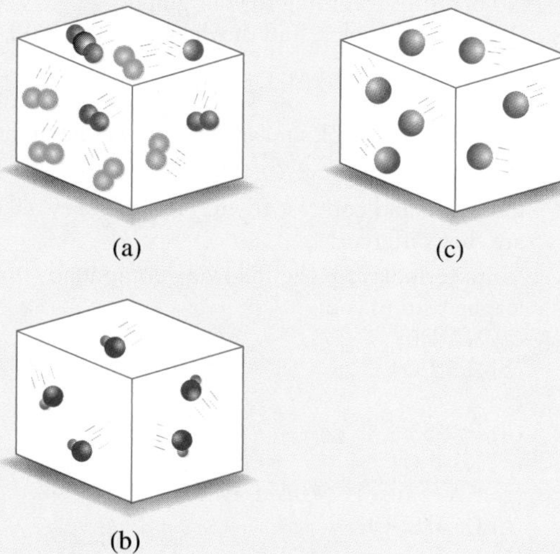

18. Classify each of the following as an element, compound, or mixture:

(a)

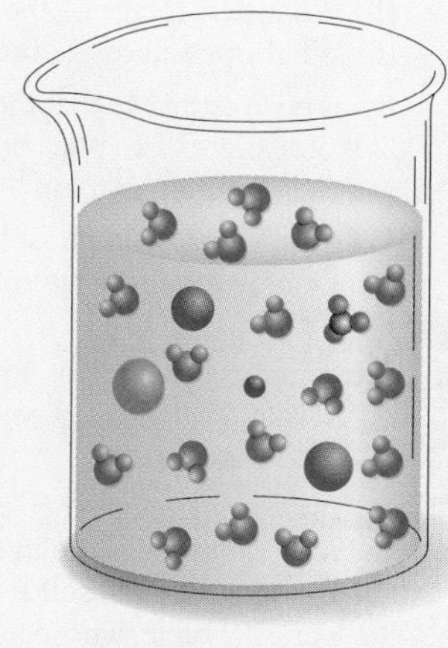

(b) (c)

19. Classify each material as an element, compound, or mixture:
(a) xenon (c) crude oil
(b) sugar (d) nitric acid

20. Classify each material as an element, compound, or mixture:
(a) carbon monoxide (c) mouthwash
(b) iced tea (d) nickel

21. Is there a pattern to the location of the gaseous elements on the periodic table? If so, describe it.

22. Is there a pattern to the location of the liquid elements on the periodic table? If so, describe it.

23. What percent of the first 36 elements on the periodic table are metals?

24. What percent of the first 36 elements on the periodic table are solids at room temperature?

Additional Exercises

All exercises with blue numbers have answers in the appendix of the text.

25. Consider a homogeneous mixture of salt dissolved in water. What method could you use to separate the two components of the mixture?

26. Consider a heterogeneous mixture of golf balls, tennis balls, and footballs. What method could you use to separate the three components of the mixture?

27. You accidentally poured salt into your large grind pepper shaker. This is the only pepper that you have left and you need it to cook a meal, but you don't want the salt to be mixed in. How can you successfully separate these two components?

28. On the periodic table at the front of this book, do you notice anything about the atoms that make up the following ionic compounds: NaCl, KI, and $MgBr_2$? (*Hint:* Look at the position of the atoms in the given compounds on the periodic table.)

29. How many total atoms are present in each of the following compounds?
(a) CO
(b) BF_3
(c) HNO_3
(d) $KMnO_4$
(e) $Ca(NO_3)_2$
(f) $Fe_3(PO_4)_2$

30. The formula for vitamin B_{12} is $C_{63}H_{88}CoN_{14}O_{14}P$.
(a) How many atoms make up one molecule of vitamin B_{12}?
(b) What percentage of the total atoms are carbon?
(c) What fraction of the total atoms are metallic?

***31.** It has been estimated that there is 4×10^{-4} mg of gold per liter of seawater. At a price of \$19.40/g, what would be the value of the gold in 1 km^3(1×10^{15} cm^3) of the ocean?

32. Calcium dihydrogen phosphate is an important fertilizer. How many atoms of hydrogen are there in ten formula units of $Ca(H_2PO_4)_2$?

33. How many total atoms are there in one molecule of $C_{145}H_{293}O_{168}$?

34. Name the following:
(a) three elements, all metals, beginning with the letter M
(b) four elements, all solid nonmetals
(c) five elements, all solids in the first five rows of the periodic table, whose symbols start with letters different from the element name.

35. How would you separate a mixture of sugar and sand and isolate each in its pure form?

36. How many total atoms are there in seven dozen molecules of nitric acid, HNO_3?

37. Make a graph using the data below. Plot the density of air in grams per liter along the *x*-axis and temperature along the *y*-axis.

Temperature (°C)	Density (g/L)
0	1.29
10	1.25
20	1.20
40	1.14
80	1.07

(a) What is the relationship between density and temperature according to your graph?
(b) From your plot, find the density of air at these temperatures:
5°C 25°C 70°C

38. These formulas look similar but represent different things.
8 S S_8
Compare and contrast them. How are they alike? How are they different?

39. Write formulas for the following compounds that a colleague read to you:
(a) NA-CL
(b) H2-S-O4
(c) K2-O
(d) Fe2-S3
(e) K3-P-O4
(f) CA (pause) CN taken twice
(g) C6-H12-O6
(h) C2-H5 (pause) OH
(i) CR (pause) NO3 taken three times

40. The abundance of iodine in seawater is 5.0×10^{-8} % by mass. How many kilograms of seawater must be treated to obtain 1.0 g iodine?

Challenge Exercises

All exercises with blue numbers have answers in the appendix of the text.

***41.** Write the chemical formulas of the neutral compounds that would result if the following ions were combined. Use the Charges of Common Ions table on the inside back cover of the book to help you.
(a) ammonium ion and chloride ion
(b) hydrogen ion and hydrogen sulfate ion
(c) magnesium ion and iodide ion
(d) iron(II) ion and fluoride ion
(e) lead(II) ion and phosphate ion
(f) aluminum ion and oxide ion

***42.** Write formulas of all the compounds that will form between the first five of the Group 1A and 2A metals and the oxide ion.

Answers to Practice Exercises

3.1

gases	H, N, Cl, Ne
liquids	Br, Hg
solids	Na, Cu, S, Fe

3.2

nonmetal	F, Kr, I, S
metal	Na, Cr, Mo, Cu
metalloid	Si, Sb

3.3 (a) K–BR
(b) P–B–CL–2
(c) CA (pause) CO–3
(d) MG (pause) OH taken twice

CHAPTER 4

Properties of Matter

Iron can be melted at very high temperatures and then cast into a variety of shapes.

Chapter Outline

The world we live in is a kaleidoscope of sights, sounds, smells, and tastes. Our senses help us to describe these objects in our lives. For example, the smell of freshly baked cinnamon rolls creates a mouthwatering desire to gobble down a sample. Just as sights, sounds, smells, and tastes form the properties of the objects around us, each substance in chemistry has its own unique properties that allow us to identify it and predict its interactions.

These interactions produce both physical and chemical changes. When you eat an apple, the ultimate metabolic result is carbon dioxide and water. These same products are achieved by burning logs. Not only does a chemical change occur in these cases, but an energy change occurs as well. Some reactions release energy (as does the apple or the log) whereas others require energy, such as the production of steel or the melting of ice. Over 90% of our current energy comes from chemical reactions.

4.1 Properties of Substances

properties

physical properties

chemical properties

How do we recognize substances? Each substance has a set of **properties** that are characteristic of that substance and give it a unique identity. Properties—the "personality traits" of substances—are classified as either physical or chemical. **Physical properties** are the inherent characteristics of a substance that can be determined without altering its composition; they are associated with its physical existence. Common physical properties include color, taste, odor, state of matter (solid, liquid, or gas), density, melting point, and boiling point. **Chemical properties** describe the ability of a substance to form new substances, either by reaction with other substances or by decomposition.

Let's consider a few of the physical and chemical properties of chlorine. Physically, chlorine is a gas at room temperature about 2.4 times heavier than air. It is yellowish-green in color and has a disagreeable odor. Chemically, chlorine will not burn but will support the combustion of certain other substances. It can be used as a bleaching agent, as a disinfectant for water, and in many chlorinated substances such as refrigerants and insecticides. When chlorine combines with the metal sodium, it forms a salt called sodium chloride. These properties, among many others, help us characterize and identify chlorine.

Substances, then, are recognized and differentiated by their properties. Table 4.1 lists four substances and several of their common physical properties. Information about physical properties, such as that given in Table 4.1,

Table 4.1 Physical Properties of Chlorine, Water, Sugar, and Acetic Acid

Substance	Color	Odor	Taste	Physical state	Melting point (°C)	Boiling point (°C)
Chlorine	Yellowish-green	Sharp, suffocating	Sharp, sour	Gas (20°C)	−101.6	−34.6
Water	Colorless	Odorless	Tasteless	Liquid	0.0	100.0
Sugar	White	Odorless	Sweet	Solid	—	Decomposes 170–186
Acetic acid	Colorless	Like vinegar	Sour	Liquid	16.7	118.0

CHEMISTRY IN ACTION • Making Money

Chemists are heavily involved in the manufacture of our currency. In fact, in a very real way, the money industry depends on chemistry and finding substances with the correct properties. The most common paper currency in the United States is the dollar bill. Chemistry is used to form the ink and paper and in processes used to defeat counterfeiters. The ink used on currency has to do a variety of things. It must be just the right consistency to fill the fine lines of the printing plate and release onto the paper without smearing. The ink must dry almost immediately, since the sheets of currency fly out of the press and into stacks 10,000 sheets tall. The pressure at the bottom of the stack is large, and the ink must not stick to the back of the sheet above it.

The security of our currency also depends on substances in the ink, which is optically variable. This color-changing ink shifts from green to black depending on how the bill is tilted. The ink used for the numbers in the lower right corner on the front of $10, $20, $50, and $100 bills shows this color change.

Once the currency is printed, it has to undergo durability tests. The bills are soaked in different solvents for 24 hours to make sure that they can stand dry cleaning and chemical exposure to household items such as gasoline. Then the bills must pass a washing-machine test to be sure that, for example, your $20 bill is still intact in your pocket after the washer is through with it. Paper currency is a blend of 75% cotton and 25% linen. Last, the bills must pass the crumple test, in which they are rolled up, put in a metal tube, crushed by a plunger, removed, and flattened. This process is repeated up to 50 times.

Once a bill is in circulation, Federal Reserve banks screen it using light to measure wear. If the bill gets too dirty, it is shredded and sent to a landfill. The average $1 bill stays in circulation for only 18 months. No wonder Congress decided in 1997 to revive the dollar coin by creating a new dollar coin to succeed the Susan B. Anthony dollar.

Dollar bill.

Currency in stacks.

is available in handbooks of chemistry and physics. Scientists don't pretend to know all the answers or to remember voluminous amounts of data, but it is important for them to know where to look for data in the literature and on the Internet.

Many chemists have reference books such as *Handbook of Chemistry and Physics* to use as a resource.

No two substances have identical physical and chemical properties.

4.2 Physical Changes

physical change

Matter can undergo two types of changes, physical and chemical. **Physical changes** are changes in physical properties (such as size, shape, and density) or changes in the state of matter without an accompanying change in composition. The changing of ice into water and water into steam are physical changes from one state of matter into another. No new substances are formed in these physical changes.

When a clean platinum wire is heated in a burner flame, the appearance of the platinum changes from silvery metallic to glowing red. This change is physical because the platinum can be restored to its original metallic appearance by cooling and, more importantly, because the composition of the platinum is not changed by heating and cooling.

4.3 Chemical Changes

chemical change

In a **chemical change**, new substances are formed that have different properties and composition from the original material. The new substances need not resemble the original material in any way.

When a clean copper wire is heated in a burner flame, the appearance of the copper changes from coppery metallic to glowing red. Unlike the platinum wire, the copper wire is not restored to its original appearance by cooling but instead becomes a black material. This black material is a new substance called copper(II) oxide. It was formed by chemical change when copper combined with oxygen in the air during the heating process. The unheated wire was essentially 100% copper, but the copper(II) oxide is 79.9% copper and 20.1% oxygen. One gram of copper will yield 1.251 g of copper(II) oxide (see Figure 4.1). The platinum is changed only physically when heated, but the copper is changed both physically and chemically when heated.

When 1.000 g of copper reacts with oxygen to yield 1.251 g of copper(II) oxide, the copper must have combined with 0.251 g of oxygen. The percentage of copper and oxygen can be calculated from these data [the copper and oxygen each being a percent of the total mass of copper(II) oxide]:

$$1.000 \text{ g copper} + 0.251 \text{ g oxygen} \rightarrow 1.251 \text{ g copper(II) oxide}$$

$$\frac{1.000 \text{ g copper}}{1.251 \text{ g copper(II) oxide}} \times 100 = 79.94\% \text{ copper}$$

$$\frac{0.251 \text{ g oxygen}}{1.251 \text{ g copper(II) oxide}} \times 100 = 20.1\% \text{ oxygen}$$

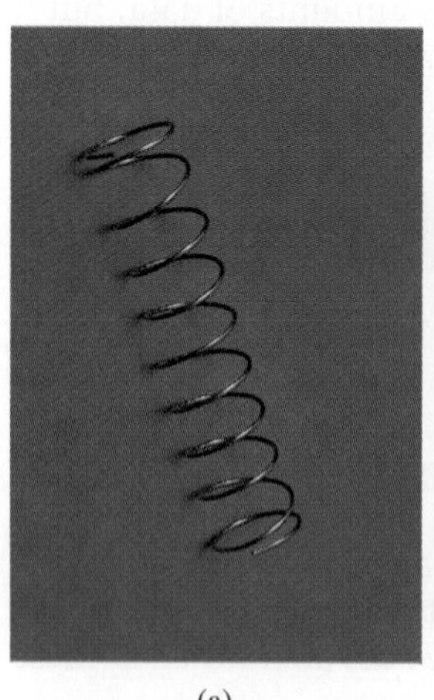
(a)

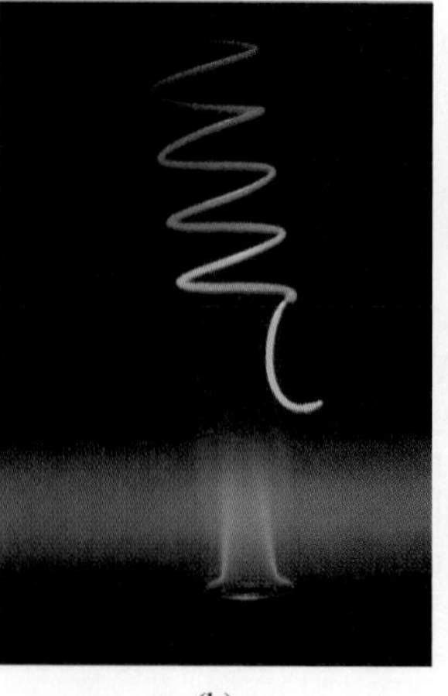
(b)

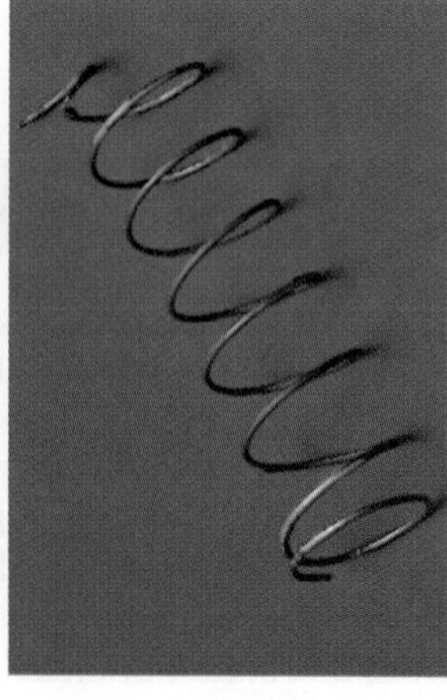
(c)

Figure 4.1
Chemical change: Forming of copper(II) oxide from copper and oxygen:
(a) Before heating, the wire is 100% copper (1.000 g)
(b) Copper and oxygen from the air combine chemically when the wire is heated.
(c) After heating, the wire is black copper(II) oxide (79.9% copper, 20.1% oxygen) (1.251 g).

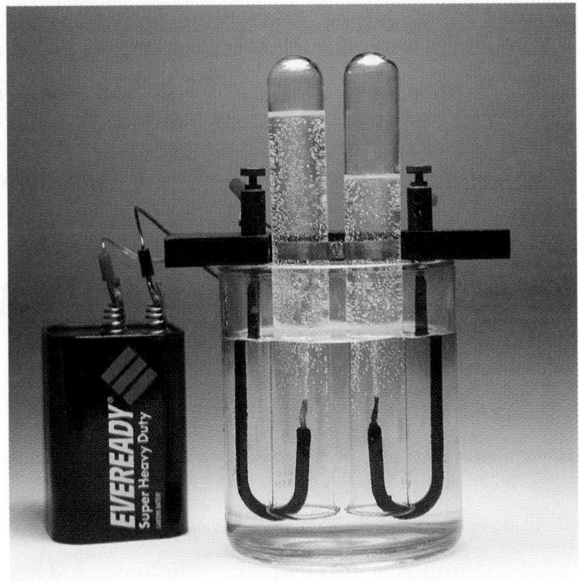

Figure 4.2
Electrolysis of water produces hydrogen gas (on the right) and oxygen gas (on the left). Note the ratio of the gases is 2:1.

Water can be decomposed chemically into hydrogen and oxygen. This is usually accomplished by passing electricity through the water in a process called *electrolysis*. Hydrogen collects at one electrode while oxygen collects at the other (see Figure 4.2). The composition and the physical appearance of the hydrogen and the oxygen are quite different from that of water. They are both colorless gases, but each behaves differently when a burning splint is placed into the sample: The hydrogen explodes with a pop while the flame brightens considerably in the oxygen (oxygen supports and intensifies the combustion of the wood). From these observations, we conclude that a chemical change has taken place.

Chemists have devised a shorthand method for expressing chemical changes in the form of **chemical equations**. The two previous examples of chemical changes can be represented by the following molecular representations, word and symbol equations.

chemical equations

Reaction 1					
Type of equation	**Reactants**		**Products**		
Word	**water**	$\xrightarrow[\text{energy}]{\text{electrical}}$	**hydrogen**	+	**oxygen**
Molecular		$\xrightarrow[\text{energy}]{\text{electrical}}$		+	
Symbol (formula)	$2\ H_2O$	$\xrightarrow[\text{energy}]{\text{electrical}}$	$2\ H_2$	+	O_2

Reaction 2						
Type of equation	Reactants					Products
Word	copper	+	oxygen	$\xrightarrow{\Delta}$		copper(II) oxide
Molecular		+		$\xrightarrow{\Delta}$		
Symbol (formula)	2 Cu	+	O_2	$\xrightarrow{\Delta}$		2 CuO

In reaction 1 water decomposes into hydrogen and oxygen when electrolyzed. In reaction 2 copper plus oxygen when heated produce copper(II) oxide. The arrow means "produces," and it points to the products. The Greek letter delta (Δ) represents heat. The starting substances (water, copper, and oxygen) are called the **reactants** and the substances produced [hydrogen, oxygen, and copper(II) oxide] are called the **products**. We will learn more about writing chemical equations in later chapters.

reactants
products

Physical change usually accompanies a chemical change. Table 4.2 lists some common physical and chemical changes; note that wherever a chemical change occurs, a physical change also occurs. However, wherever a physical change is listed, only a physical change occurs.

4.4 Conservation of Mass

law of conservation of mass

The **law of conservation of mass** states that no change is observed in the total mass of the substances involved in a chemical change. This law, tested by extensive laboratory experimentation, is the basis for the quantitative mass relationships among reactants and products.

The decomposition of water into hydrogen and oxygen illustrates this law. Thus, 100.0 g of water decompose into 11.2 g of hydrogen and 88.8 g of oxygen:

$$\begin{array}{lll} \text{water} \longrightarrow & \text{hydrogen} + & \text{oxygen} \\ 100.0\text{ g} & 11.2\text{ g} & 88.8\text{ g} \end{array}$$

$$\underset{\text{reactant}}{100.0\text{ g}} \longrightarrow \underset{\text{products}}{100.0\text{ g}}$$

In a chemical reaction

mass of reactants = mass of products

Table 4.2 Physical or Chemical Changes of Some Common Processes

Process taking place	Type of change	Accompanying observations
Rusting of iron	Chemical	Shiny, bright metal changes to reddish-brown rust.
Boiling of water	Physical	Liquid changes to vapor.
Burning of sulfur in air	Chemical	Yellow, solid sulfur changes to gaseous, choking sulfur dioxide.
Boiling an egg	Chemical	Liquid white and yolk change to solids.
Combustion of gasoline	Chemical	Liquid gasoline burns to gaseous carbon monoxide, carbon dioxide, and water.
Digestion of food	Chemical	Food changes to liquid nutrients and partially solid wastes.
Sawing of wood	Physical	Smaller pieces of wood and sawdust are made from a larger piece of wood.
Burning of wood	Chemical	Wood burns to ashes, gaseous carbon dioxide, and water.
Heating of glass	Physical	Solid becomes pliable during heating, and the glass may change its shape.

Sawing wood is a physical charge.

Combustion of gasoline is a chemical change.

4.5 Energy

From the early discovery that fire can warm us and cook our food to our discovery that nuclear reactors can be used to produce vast amounts of controlled energy, our technical progress has been directed by our ability to produce, harness, and utilize energy. **Energy** is the capacity of matter to do work. Energy exists in many forms; some of the more familiar forms are mechanical, chemical, electrical, heat, nuclear, and radiant or light energy. Matter can have both potential and kinetic energy.

energy

potential energy

Potential energy (PE) is stored energy, or energy that an object possesses due to its relative position. For example, a ball located 20 ft above the ground has more potential energy than when located 10 ft above the ground and will bounce higher when allowed to fall. Water backed up behind a dam represents potential energy that can be converted into useful work in the form of electrical or mechanical energy. Gasoline is a source of chemical potential energy. When gasoline burns (combines with oxygen), the heat released is associated with a decrease in potential energy. The new substances formed by burning have less chemical potential energy than the gasoline and oxygen.

kinetic energy

Kinetic energy (KE) is energy that matter possesses due to its motion. When the water behind the dam is released and allowed to flow, its potential energy is changed into kinetic energy, which can be used to drive generators and produce electricity. Moving bodies possess kinetic energy. We all know the results when two moving vehicles collide: Their kinetic energy is expended in the crash that occurs. The pressure exerted by a confined gas is due to the kinetic energy of rapidly moving gas particles.

The mechanical energy of falling water is converted to electrical energy at the hydroelectric plant at Niagara Falls.

Energy can be converted from one form to another form. Some kinds of energy can be converted to other forms easily and efficiently. For example, mechanical energy can be converted to electrical energy with an electric generator at better than 90% efficiency. On the other hand, solar energy has thus far been directly converted to electrical energy at an efficiency of only about 15%. In chemistry, energy is most frequently expressed as heat.

4.6 Heat: Quantitative Measurement

The SI-derived unit for energy is the joule (pronounced *jool*, rhyming with *tool*, and abbreviated J). Another unit for heat energy, which has been used for many years, is the calorie (abbreviated cal). The relationship between joules and calories is

$$4.184\ \text{J} = 1\ \text{cal} \quad \text{(exactly)}$$

To give you some idea of the magnitude of these heat units, 4.184 **joules**, or 1 **calorie**, is the quantity of heat energy required to change the temperature of 1 gram of water by 1°C, usually measured from 14.5°C to 15.5°C.

joule
calorie

Since joules and calories are rather small units, kilojoules (kJ) and kilocalories (kcal) are used to express heat energy in many chemical processes. The kilocalorie is also known as the nutritional, or large, Calorie (spelled with a capital C and abbreviated Cal). In this book, heat energy will be expressed in joules.

1 kJ = 1000 J

1 kcal = 1000 cal = 1 Cal

The difference in the meanings of the terms *heat* and *temperature* can be seen by this example: Visualize two beakers, A and B. Beaker A contains 100 g of water at 20°C, and beaker B contains 200 g of water also at 20°C. The beakers are heated until the temperature of the water in each reaches 30°C. The temperature of the water in the beakers was raised by exactly the same amount, 10°C. But twice as much heat (8368 J) was required to raise the temperature of the water in beaker B as was required in beaker A (4184 J).

In the middle of the 18th century, Joseph Black, a Scottish chemist, was experimenting with the heating of elements. He heated and cooled equal masses of iron and lead through the same temperature range. Black noted that much more heat was needed for the iron than for the lead. He had discovered a fundamental property of matter—namely, that every substance has a characteristic heat capacity. Heat capacities may be compared in terms of specific heats. The **specific heat** of a substance is the quantity of heat (lost or gained) required to change the temperature of 1 g of that substance by 1°C. It follows then that the specific heat of liquid water is 4.184 J/g°C (1.000 cal/g°C). The specific heat of water is high compared with that of most substances. Aluminum and copper, for example, have specific heats of 0.900 J/g°C and 0.385 J/g°C, respectively (see Table 4.3). The relation of mass, specific heat, temperature change (Δt), and quantity of heat lost or gained by a system is expressed by this general equation:

specific heat

$$\left(\begin{array}{c}\textbf{mass of}\\ \textbf{substance}\end{array}\right)\left(\begin{array}{c}\textbf{specific heat}\\ \textbf{of substance}\end{array}\right)(\Delta t) = \textbf{heat}$$

Thus the amount of heat needed to raise the temperature of 200. g of water by 10.0°C can be calculated as follows:

$$(200.\ \cancel{\text{g}})\left(\frac{4.184\ \text{J}}{\cancel{\text{g}}\cancel{^\circ\text{C}}}\right)(10.0\ \cancel{^\circ\text{C}}) = 8.37 \times 10^3\ \text{J}$$

Table 4.3 Specific Heat of Selected Substances

Substance	Specific heat (J/g°C)	Specific heat (cal/g°C)
Water	4.184	1.000
Ethyl alcohol	2.138	0.511
Ice	2.059	0.492
Aluminum	0.900	0.215
Iron	0.473	0.113
Copper	0.385	0.0921
Gold	0.131	0.0312
Lead	0.128	0.0305

Example 4.1

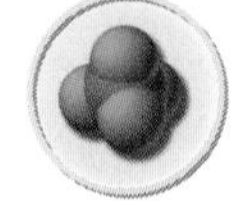

Calculate the specific heat of a solid in J/g°C and cal/g°C if 1638 J raise the temperature of 125 g of the solid from 25.0°C to 52.6°C.

SOLUTION

Substitute the data into the equation (mass)(specific heat)(Δt) = heat and solve for specific heat:

$$\text{specific heat} = \frac{\text{heat}}{\text{g} \times \Delta t}$$

$$\text{heat} = 1638\,\text{J} \qquad \text{mass} = 125\,\text{g} \qquad \Delta t = 52.6°\text{C} - 25.0°\text{C} = 27.6°\text{C}$$

$$\text{specific heat} = \frac{1638\,\text{J}}{(125\,\text{g})(27.6°\text{C})} = 0.475\,\text{J/g°C}$$

Now convert joules to calories using 1.000 cal/4.184 J:

$$\text{specific heat} = \left(\frac{0.475\,\cancel{\text{J}}}{\text{g°C}}\right)\left(\frac{1.000\,\text{cal}}{4.184\,\cancel{\text{J}}}\right) = 0.114\,\text{cal/g°C}$$

Example 4.2

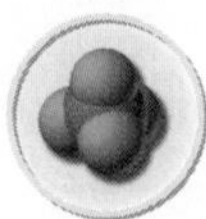

A sample of a metal with a mass of 212 g is heated to 125.0°C and then dropped into 375 g water at 24.0°C. If the final temperature of the water is 34.2°C, what is the specific heat of the metal? (Assume no heat losses to the surroundings.)

SOLUTION

When the metal enters the water, it begins to cool, losing heat to the water. At the same time, the temperature of the water rises. This process continues until the temperature of the metal and the temperature of the water are equal, at which point (34.2°C) no net flow of heat occurs.

The heat lost or gained by a system is given by (mass)(specific heat)(Δt) = energy change. First calculate the heat gained by the water, and then calculate the specific heat of the metal:

$$\text{temperature rise of the water} \qquad (\Delta t) = 34.2°\text{C} - 24.0°\text{C} = 10.2°\text{C}$$

$$\text{heat gained by the water} = (375\,\cancel{\text{g}})\left(\frac{4.184\,\text{J}}{\cancel{\text{g°C}}}\right)(10.2\,\cancel{°\text{C}}) = 1.60 \times 10^4\,\text{J}$$

The metal dropped into the water must have a final temperature the same as the water (34.2°C):

$$\text{temperature drop by the metal} \qquad (\Delta t) = 125.0°\text{C} - 34.2°\text{C} = 90.8°\text{C}$$

$$\text{heat lost by the metal} = \text{heat gained by the water} = 1.60 \times 10^4\,\text{J}$$

To determine the specific heat of the metal, we rearrange the equation (mass)(specific heat)(Δt) = heat to obtain

$$\text{specific heat} = \frac{\text{heat}}{(\text{mass})(\Delta t)}$$

$$\text{specific heat of the metal} = \frac{1.60 \times 10^4\,\text{J}}{(212\,\text{g})(90.8^\circ\text{C})} = 0.831\,\text{J/g}^\circ\text{C}$$

Practice 4.1

Calculate the quantity of energy needed to heat 8.0 g of water from 42.0°C to 45.0°C.

Practice 4.2

A 110.0-g sample of metal at 55.5°C raises the temperature of 150.0 g of water from 23.0°C to 25.5°C. Determine the specific heat of the metal in J/g°C.

A rain forest in Marajo Island, Amazon estuary, Brazil.

4.7 Energy in Chemical Changes

In all chemical changes, matter either absorbs or releases energy. Chemical changes can produce different forms of energy. For example, electrical energy to start automobiles is produced by chemical changes in the lead storage battery. Light energy is produced by the chemical change that occurs in a light stick. Heat and light are released from the combustion of fuels. All the energy needed for our life processes—breathing, muscle contraction, blood circulation, and so on—is produced by chemical changes occurring within the cells of our bodies.

Rapid combustion provides the energy for lifting the space shuttle.

CHEMISTRY IN ACTION • Popping Popcorn

What's the secret to getting all those kernels in a bag of microwave popcorn to pop? Food scientists have looked at all sorts of things to improve the poppability of popcorn. In every kernel of popcorn there is starch and a little moisture. As the kernel heats up, the water vaporizes and the pressure builds up in the kernel. Finally, the kernel bursts and the starch expands into the white foam of popped corn.

Researchers in Indiana tested 14 types of popcorn to determine the most poppable. They found a range of 4–47% of unpopped kernels and determined that the kernels that held moisture the best produced the fewest unpopped kernels. Further analysis led them to the conclusion that the determining factor was the pericarp.

The pericarp is the hard casing on the popcorn kernel. It is made of long polymers of cellulose. When the kernel is heated up, the long chains of cellulose line up and make a strong crystalline structure. This strong structure keeps the water vapor inside the kernel, allowing the pressure to rise high enough to pop the kernel.

Getting the most pops comes from the molecular changes resulting from heating the kernel. Energy and chemistry combine to produce a tasty snack!

Popcorn.

Conversely, energy is used to cause chemical changes. For example, a chemical change occurs in the electroplating of metals when electrical energy is passed through a salt solution in which the metal is submerged. A chemical change occurs when radiant energy from the sun is used by green plants in the process of photosynthesis. And as we saw, a chemical change occurs when electricity is used to decompose water into hydrogen and oxygen. Chemical changes are often used primarily to produce energy rather than to produce new substances. The heat or thrust generated by the combustion of fuels is more important than the new substances formed.

4.8 Conservation of Energy

An energy transformation occurs whenever a chemical change occurs (see Figure 4.3). If energy is absorbed during the change, the products will have more chemical potential energy than the reactants. Conversely, if energy is given off in

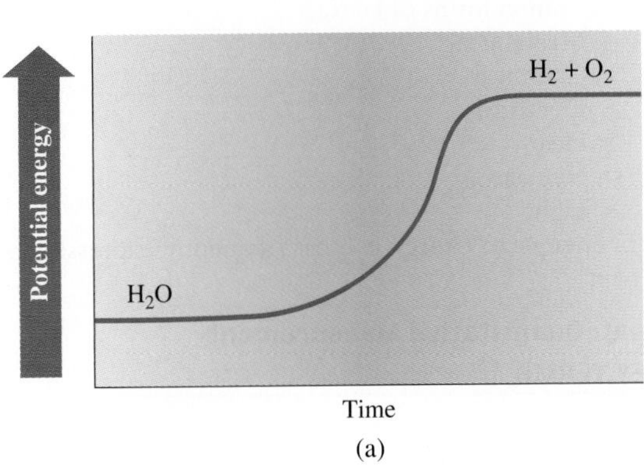

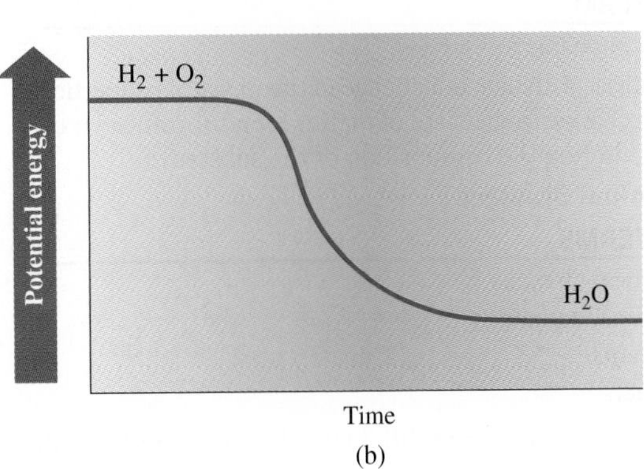

Figure 4.3
(a) In electrolysis of water, energy is absorbed by the system, so the products H_2 and O_2 have a higher potential energy.
(b) When hydrogen is burned (in O_2), energy is released and the product (H_2O) has lower potential energy.

a chemical change, the products will have less chemical potential energy than the reactants. Water can be decomposed in an electrolytic cell by absorbing electrical energy. The products, hydrogen and oxygen, have a greater chemical potential energy level than that of water (see Figure 4.3a). This potential energy is released in the form of heat and light when the hydrogen and oxygen are burned to form water again (see Figure 4.3b). Thus, energy can be changed from one form to another or from one substance to another, and therefore is not lost.

law of conservation of energy

The energy changes occurring in many systems have been thoroughly studied. No system has ever been found to acquire energy except at the expense of energy possessed by another system. This is the **law of conservation of energy**: Energy can be neither created nor destroyed, though it can be transformed from one form to another.

Chapter 4 Review

4.1 Properties of Substances

KEY TERMS

Properties
Physical properties
Chemical properties

- Each substance has a set of properties that are characteristic of the substance and give it a unique identity.
- Physical properties are inherent in the substance and can be determined without altering the composition of the substance:
  - Color
  - Taste
  - Odor
  - State of matter
  - Density
  - Melting point
  - Boiling point
- Chemical properties describe the ability of the substance to interact with other substances to form different substances.

4.2 Physical Changes

KEY TERM

Physical changes

- A physical change is a change in the physical properties or a change in the state of matter for a substance without altering the composition of the substance.

4.3 Chemical Changes

KEY TERMS

Chemical change
Chemical equations
Reactants
Products

- In a chemical change different substances are formed that have different properties and composition from the original material.
- Chemical changes can be represented by chemical equations:
  - Word equation
  - Molecular equation
  - Symbol (formula) equation

4.4 Conservation of Mass

KEY TERM

Law of conservation of mass

- The law of conservation of mass states that matter is neither gained nor lost during a chemical reaction.

4.5 Energy

KEY TERMS

Energy
Potential energy
Kinetic energy

- Energy is the capacity to do work.
- Potential energy is energy that results from position or is stored within a substance.
- Kinetic energy is the energy matter possesses as a result of its motion.
- Energy can be converted from one form to another:
  - Common forms of energy:
    - Mechanical
    - Chemical
    - Electrical
    - Heat
    - Nuclear
    - Light
- In chemistry, energy is most frequently expressed as heat.

4.6 Heat: Quantitative Measurement

KEY TERMS

Joule
Calorie
Specific heat

- The SI unit for heat is the joule
  $4.184\,\text{J} = 1\,\text{cal}$

- The calorie is defined as the amount of heat required to change the temperature of 1 gram of water 1°C.
- Every substance has a characteristic heat capacity:
  - The specific heat of a substance is a measure of its heat capacity.
  - The specific heat is the quantity of heat required to change the temperature of 1 gram of the substance by 1°C.
- Heat lost or gained by a system can be calculated by

heat = (mass of substance) (specific heat of substance) (Δt).

4.7 Energy in Chemical Changes

- In all chemical changes matter either absorbs or releases energy.
- Energy can be used to cause chemical change.
- Chemical changes are often used to produce energy rather than to produce new substances.

4.8 Conservation of Energy

KEY TERM

Law of conservation of energy

- Energy can neither be created or destroyed. It can be transformed from one form to another.
- The energy released or absorbed in a chemical reaction can be summarized in graphical form.

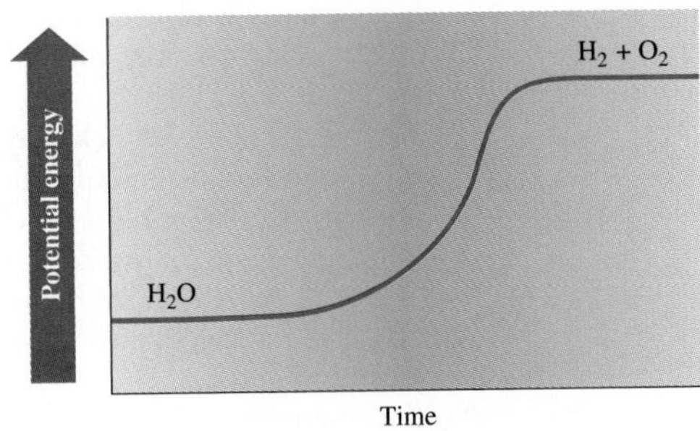

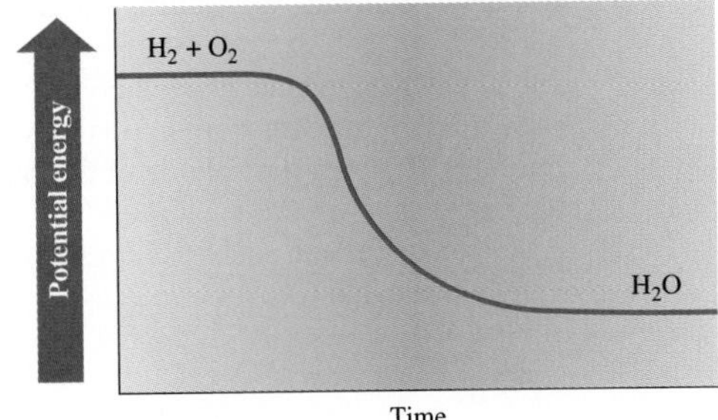

Review Questions

All questions with blue numbers have answers in the appendix of the text.

1. In what physical state does acetic acid exist at 393 K? (Table 4.1)
2. In what physical state does chlorine exist at −65°C? (Table 4.1)
3. What evidence of chemical change is visible when electricity is run through water? (Figure 4.2)
4. What is the fundamental difference between a chemical change and a physical change?
5. Distinguish between potential and kinetic energy.
6. Calculate the boiling point of acetic acid in
   (a) kelvins
   (b) degrees Fahrenheit (Table 4.1)
7. Calculate the melting point of acetic acid in (Table 4.1)
   (a) degrees Fahrenheit
   (b) kelvins
8. When water boils, bubbles form. What is inside those bubbles?
9. Is dissolving salt in water a physical or chemical change and why? How could you verify your answer?

Paired Exercises

All exercises with blue numbers have answers in the appendix of the text.

1. Determine whether each of the following represents a physical property or a chemical property:
   (a) Vinegar has a pungent odor.
   (b) Carbon cannot be decomposed.
   (c) Sulfur is a bright yellow.
   (d) Sodium chloride is a crystalline solid.
   (e) Water does not burn.
   (f) Mercury is a liquid at 25°C.
   (g) Oxygen is not combustible.
   (h) Aluminum combines with oxygen to form a protective oxide coating.
2. Determine whether each of the following represents a physical property or a chemical property:
   (a) Fluorine gas has a greenish-yellow tint.
   (b) The density of water at 4°C is 1.000 g/mL.
   (c) Hydrogen gas is very flammable.
   (d) Aluminum is a solid at 25°C.
   (e) Water is colorless and odorless.
   (f) Lemon juice tastes sour.
   (g) Gold does not tarnish.
   (h) Copper cannot be decomposed.

3. Cite the evidence that indicates that only physical changes occur when a platinum wire is heated in a Bunsen burner flame.

4. Cite the evidence that indicates that both physical and chemical changes occur when a copper wire is heated in a Bunsen burner flame.

5. Identify the reactants and products when a copper wire is heated in air in a Bunsen burner flame.

6. Identify the reactants and products for the electrolysis of water.

7. State whether each of the following represents a chemical change or merely a physical change:
(a) A steak is cooked on a grill until well done.
(b) In the lab, students firepolish the end of a glass rod. The jagged edge of the glass has become smooth.
(c) Chlorine bleach is used to remove a coffee stain on a white lab coat.
(d) When two clear and colorless aqueous salt solutions are mixed together, the solution turns cloudy and yellow.
(e) One gram of an orange crystalline solid is heated in a test tube, producing a green powdery solid whose volume is 10 times the volume of the original substance.
(f) In the lab, a student cuts a 20-cm strip of magnesium metal into 1-cm pieces.

8. State whether each of the following represents a chemical change or merely a physical change:
(a) A few grams of sucrose (table sugar) are placed in a small beaker of deionized water; the sugar crystals "disappear," and the liquid in the beaker remains clear and colorless.
(b) A copper statue, over time, turns green.
(c) When a teaspoon of baking soda (sodium bicarbonate) is placed into a few ounces of vinegar (acetic acid), volumes of bubbles (effervescence) are produced.
(d) When a few grams of a blue crystalline solid are placed into a beaker of deionized water, the crystals "disappear" and the liquid becomes clear and blue in color.
(e) In the lab, a student mixes 2 mL of sodium hydroxide with 2 mL of hydrochloric acid in a test tube. He notices that the test tube has become very warm to the touch.
(f) A woman visits a hairdresser and has her hair colored a darker shade of brown. After several weeks the hair, even though washed several times, has not changed back to the original color.

9. Are the following examples of potential energy or kinetic energy?
(a) a ball sitting on the top of a hill
(b) a stretched rubber band
(c) a lit match
(d) diving into a pool
(e) a person rolling down a hill

10. Are the following examples of potential energy or kinetic energy?
(a) runners poised at a starting line
(b) a stretched arrow bow
(c) running water
(d) rubbing your hands together
(e) a swing when it is poised at its highest position

11. What happens to the kinetic energy of a speeding car when the car is braked to a stop?

12. What energy transformation is responsible for the fiery reentry of the space shuttle into Earth's atmosphere?

13. Indicate with a plus sign (+) any of these processes that require energy, and a negative sign (−) any that release energy:
(a) melting ice
(b) relaxing a taut rubber band
(c) a rocket launching
(d) striking a match
(e) a Slinky toy (spring) "walking" down stairs

14. Indicate with a plus sign (+) any of these processes that require energy, and a negative sign (−) any that release energy:
(a) boiling water
(b) releasing a balloon full of air with the neck open
(c) a race car crashing into a wall
(d) cooking a potato in a microwave oven
(e) ice cream freezing in an ice cream maker

15. How many joules of energy are required to raise the temperature of 75 g of water from 20.0°C to 70.0°C?

16. How many joules of energy are required to raise the temperature of 65 g of iron from 25°C to 95°C?

17. How many joules of heat are required to heat 25.0 g of ethyl alcohol from the prevailing room temperature, 22.5°C, to its boiling point, 78.5°C?

18. How many joules of heat are required to heat 35.0 g of isopropyl alcohol from the prevailing room temperature, 21.2°C, to its boiling point, 82.4°C? (The specific heat of isopropyl alcohol is 2.604 J/g°C.)

19. A 250.0-g metal bar requires 5.866 kJ to change its temperature from 22°C to 100.0°C. What is the specific heat of the metal?

20. A 1.00-kg sample of antimony absorbed 30.7 kJ, thus raising the temperature of the antimony from 20.0°C to its melting point (630.0°C). Calculate the specific heat of antimony.

21. A 325.0-g piece of gold at 427°C is dropped into 200.0 g of water at 22.0°C. (The specific heat of gold is 0.131 J/g°C.) Calculate the final temperature of the mixture. (Assume no heat loss to the surroundings.)

22. A 500.0-g iron bar at 212°C is placed in 2.0 L of water at 24.0°C. What will be the change in temperature of the water? (Assume no heat is lost to the surroundings.)

Additional Exercises

All exercises with blue numbers have answers in the appendix of the text.

23. A 110.0-g sample of a gray-colored, unknown, pure metal was heated to 92.0°C and put into a coffee-cup calorimeter containing 75.0 g of water at 21.0°C. When the heated metal was put into the water, the temperature of the water rose to a final temperature of 24.2°C. The specific heat of water is 4.184 J. (a) What is the specific heat of the metal. (b) Is it possible that the metal is either iron or lead? Explain.

24. A 40.0-g sample of an unknown, yellowish-brown, pure metal was heated to 62.0°C and put into a coffee-cup calorimeter containing 85.0 g of water at 19.2°C. The water was heated by the hot metal to a temperature of 21.0°C. (a) What is the specific heat of the metal? (b) Is it possible that the metal is gold? Explain.

*25. You learned about the law of conservation of mass in this chapter. Suppose you burn a piece of wood that has a mass of 20 kg. You then determine the mass of the ashes remaining after the wood has been burned. Its mass is less than 20 kg. According to the law of conservation of mass, the mass before reaction should equal the mass after reaction (in this case burning) is over. What happened?

26. A sample of metal with a mass 250. g is heated to 130.°C and dropped into 425 g of water at 26.0°C. The final temperature of the water is 38.4°C. What is the specific heat capacity of the metal if 1.2×10^3 J are lost to the environment?

27. 18 g of oxygen are reacted with sufficient nitrogen to make 56 g of nitrogen dioxide according the following reaction:

$$2\,O_2(g) + N_2(g) \longrightarrow 2\,NO_2(g)$$

If the reaction goes to completion, how much nitrogen reacted?

28. The specific heat of zinc is 0.096 cal/g°C. Determine the energy required to raise the temperature of 250.0 g of zinc from room temperature (24°C) to 150.0°C.

29. If 40.0 kJ of energy are absorbed by 500.0 g of water at 10.0°C, what is the final temperature of the water?

*30. The heat of combustion of a sample of coal is 5500 cal/g. What quantity of this coal must be burned to heat 500.0 g of water from 20.0°C to 90.0°C?

31. One gram of anthracite coal gives off 7000. cal when burned. How many joules is this? How many grams of anthracite are required to raise the temperature of 4.0 L of water from 20.0°C to 100.0°C?

(For Questions 32 and 33, you may need to refer to Table 4.3)

32. A 100.0-g sample of copper is heated from 10.0°C to 100.0°C.
   (a) Determine the number of calories needed. (The specific heat of copper is 0.0921 cal/g°C.)
   (b) The same amount of heat is added to 100.0 g of Al at 10.0°C. (The specific heat of Al is 0.215 cal/g°C.) Which metal gets hotter, the copper or the aluminum?

33. A 500.0-g piece of iron is heated in a flame and dropped into 400.0 g of water at 10.0°C. The temperature of the water rises to 90.0°C. How hot was the iron when it was first removed from the flame? (The specific heat of iron is 0.473 J/g°C.)

*34. A 20.0-g piece of metal at 203°C is dropped into 100.0 g of water at 25.0°C. The water temperature rises to 29.0°C. Calculate the specific heat of the metal (J/g°C). Assume that all of the heat lost by the metal is transferred to the water and no heat is lost to the surroundings.

*35. Assuming no heat loss by the system, what will be the final temperature when 50.0 g of water at 10.0°C are mixed with 10.0 g of water at 50.0°C?

36. Three 500.0-g pans of iron, aluminum, and copper are each used to fry an egg. Which pan fries the egg (105°C) the quickest? Explain.

*37. At 6:00 P.M., you put a 300.0-g copper pan containing 800.0 g of water (all at room temperature, which is 25°C) on the stove. The stove supplies 628 J/s. When will the water reach the boiling point? (Assume no heat loss.)

38. Why does blowing gently across the surface of a cup of hot coffee help to cool it? Why does inserting a spoon into the coffee do the same thing?

39. If you are boiling some potatoes in a pot of water, will they cook faster if the water is boiling vigorously than if the water is only gently boiling? Explain your reasoning.

40. Homogenized whole milk contains 4% butterfat by volume. How many milliliters of fat are there in a glass (250 mL) of milk? How many grams of butterfat (d = 0.8 g/mL) are in this glass of milk?

41. A 100.0-mL volume of mercury (density = 13.6 g/mL) is put into a container along with 100.0 g of sulfur. The two substances react when heated and result in 1460 g of a dark, solid matter. Is that material an element or a compound? Explain. How many grams of mercury were in the container? How does this support the law of conservation of matter?

Challenge Exercises

All exercises with blue numbers have answers in the appendix of the text.

*42. Suppose a ball is sitting at the top of a hill. At this point, the ball has potential energy. The ball rolls down the hill, so the potential energy is converted into kinetic energy. When the ball reaches the bottom of the hill, it goes halfway up the hill on the other side and stops. If energy is supposed to be conserved, then why doesn't the ball go up the other hill to the same level as it started from?

43. From the following molecular picture determine whether a physical or a chemical change has occurred. Justify your answer.

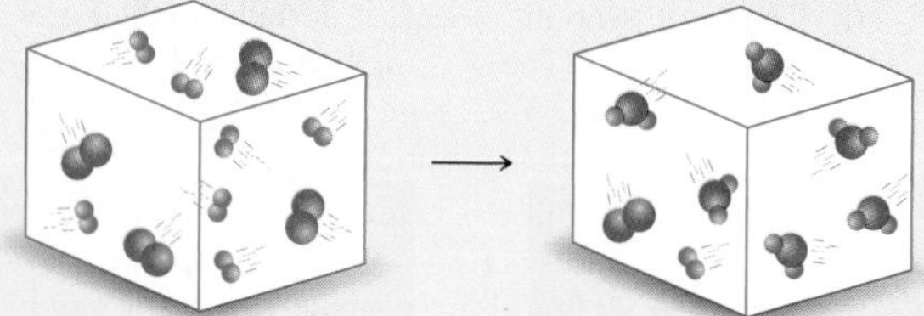

44. From the following illustration (a) describe the change that has occurred and (b) determine whether the change was a physical change or a chemical change. Justify your answer.

45. (a) Describe the change(s) that you see in the following illustration. (b) Was this a physical or a chemical change?

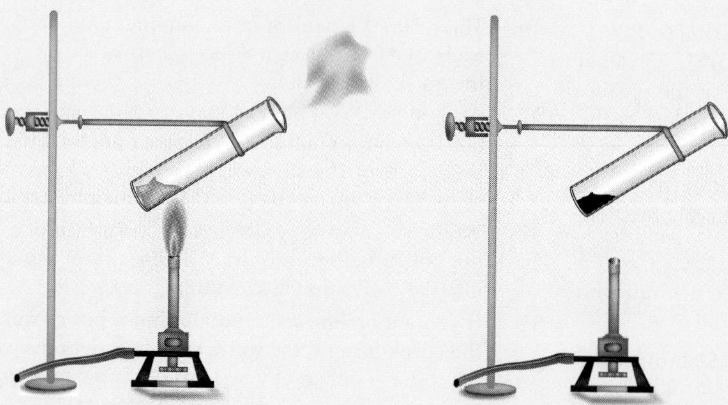

Answers to Practice Exercises

4.1 1.0×10^2 J = 24 cal

4.2 0.477 J/g°C

PUTTING IT TOGETHER Review for Chapters 1–4

Answers for Putting It Together Reviews are found in Appendix VII.

Multiple Choice:

Choose the correct answer to each of the following.

1. 1.00 cm is equal to how many meters?
   (a) 2.54 (b) 100. (c) 10.0 (d) 0.0100
2. 1.00 cm is equal to how many inches?
   (a) 0.394 (b) 0.10 (c) 12 (d) 2.54
3. 4.50 ft is how many centimeters?
   (a) 11.4 (b) 21.3 (c) 454 (d) 137
4. The number 0.0048 contains how many significant figures?
   (a) 1 (b) 2 (c) 3 (d) 4
5. Express 0.00382 in scientific notation.
   (a) 3.82×10^{3} (c) 3.82×10^{-2}
   (b) 3.8×10^{-3} (d) 3.82×10^{-3}
6. 42.0°C is equivalent to
   (a) 273 K (c) 108°F
   (b) 5.55°F (d) 53.3°F
7. 267°F is equivalent to
   (a) 404 K (c) 540 K
   (b) 116°C (d) 389 K
8. An object has a mass of 62 g and a volume of 4.6 mL. Its density is
   (a) 0.074 mL/g (c) 7.4 g/mL
   (b) 285 g/mL (d) 13 g/mL
9. The mass of a block is 9.43 g and its density is 2.35 g/mL. The volume of the block is
   (a) 4.01 mL (c) 22.2 mL
   (b) 0.249 mL (d) 2.49 mL
10. The density of copper is 8.92 g/mL. The mass of a piece of copper that has a volume of 9.5 mL is
    (a) 2.6 g (c) 0.94 g
    (b) 85 g (d) 1.1 g
11. An empty graduated cylinder has a mass of 54.772 g. When filled with 50.0 mL of an unknown liquid, it has a mass of 101.074 g. The density of the liquid is
    (a) 0.926 g/mL (c) 2.02 g/mL
    (b) 1.08 g/mL (d) 1.85 g/mL
12. The conversion factor to change grams to milligrams is
    (a) $\frac{100 \text{ mg}}{1 \text{ g}}$ (c) $\frac{1 \text{ g}}{1000 \text{ mg}}$
    (b) $\frac{1 \text{ g}}{100 \text{ mg}}$ (d) $\frac{1000 \text{ mg}}{1 \text{ g}}$
13. What Fahrenheit temperature is twice the Celsius temperature?
    (a) 64°F (c) 200°F
    (b) 320°F (d) 546°F
14. A gold alloy has a density of 12.41 g/mL and contains 75.0% gold by mass. The volume of this alloy that can be made from 255 g of pure gold is
    (a) 4.22×10^{3} mL (c) 27.4 mL
    (b) 2.37×10^{3} mL (d) 20.5 mL
15. A lead cylinder ($V = \pi r^2 h$) has radius 12.0 cm and length 44.0 cm and a density of 11.4 g/mL. The mass of the cylinder is
    (a) 2.27×10^{5} g (c) 1.78×10^{3} g
    (b) 1.89×10^{5} g (d) 3.50×10^{5} g
16. The following units can all be used for density *except*
    (a) g/cm^3 (b) kg/m^3 (c) g/L (d) kg/m^2
17. 37.4 cm × 2.2 cm equals
    (a) 82.28 cm^2 (c) 82 cm^2
    (b) 82.3 cm^2 (d) 82.2 cm^2
18. The following elements are among the five most abundant by mass in the Earth's crust, seawater, and atmosphere *except*
    (a) oxygen (c) silicon
    (b) hydrogen (d) aluminum
19. Which of the following is a compound?
    (a) lead (c) potassium
    (b) wood (d) water
20. Which of the following is a mixture?
    (a) water (c) wood
    (b) chromium (d) sulfur
21. How many atoms are represented in the formula Na_2CrO_4?
    (a) 3 (b) (5) (c) (7) (d) 8
22. Which of the following is a characteristic of metals?
    (a) ductile (c) extremely strong
    (b) easily shattered (d) dull
23. Which of the following is a characteristic of nonmetals?
    (a) always a gas
    (b) poor conductor of electricity
    (c) shiny
    (d) combines only with metals
24. When a pure substance was analyzed, it was found to contain carbon and chlorine. This substance must be classified as
    (a) an element
    (b) a mixture
    (c) a compound
    (d) both a mixture and a compound
25. Chromium, fluorine, and magnesium have the symbols
    (a) Ch, F, Ma (c) Cr, F, Mg
    (b) Cr, Fl, Mg (d) Cr, F, Ma
26. Sodium, carbon, and sulfur have the symbols
    (a) Na, C, S (c) Na, Ca, Su
    (b) So, C, Su (d) So, Ca, Su
27. Coffee is an example of
    (a) an element (c) a homogeneous mixture
    (b) a compound (d) a heterogeneous mixture
28. The number of oxygen atoms in $Al(C_2H_3O_2)_3$ is
    (a) 2 (b) 3 (c) 5 (d) 6

29. Which of the following is a mixture?
(a) water (c) sugar solution
(b) iron(II) oxide (d) iodine

30. Which is the most compact state of matter?
(a) solid (c) gas
(b) liquid (d) amorphous

31. Which is not characteristic of a solution?
(a) a homogeneous mixture
(b) a heterogeneous mixture
(c) one that has two or more substances
(d) one that has a variable composition

32. A chemical formula is a combination of
(a) symbols (c) elements
(b) atoms (d) compounds

33. The number of nonmetal atoms in $Al_2(SO_3)_3$ is
(a) 5 (b) 7 (c) 12 (d) 14

34. Which of the following is not a physical property?
(a) boiling point (c) bleaching action
(b) physical state (d) color

35. Which of the following is a physical change?
(a) A piece of sulfur is burned.
(b) A firecracker explodes.
(c) A rubber band is stretched.
(d) A nail rusts.

36. Which of the following is a chemical change?
(a) Water evaporates.
(b) Ice melts.
(c) Rocks are ground to sand.
(d) A penny tarnishes.

37. When 9.44 g of calcium are heated in air, 13.22 g of calcium oxide are formed. The percent by mass of oxygen in the compound is
(a) 28.6% (b) 40.0% (c) 71.4% (d) 13.2%

38. Barium iodide, BaI_2, contains 35.1% barium by mass. An 8.50-g sample of barium iodide contains what mass of iodine?
(a) 5.52 g (b) 2.98 g (c) 3.51 g (d) 6.49 g

39. Mercury(II) sulfide, HgS, contains 86.2% mercury by mass. The mass of HgS that can be made from 30.0 g of mercury is
(a) 2586 g (b) 2.87 g (c) 25.9 g (d) 34.8 g

40. The changing of liquid water to ice is known as a
(a) chemical change
(b) heterogeneous change
(c) homogeneous change
(d) physical change

41. Which of the following does not represent a chemical change?
(a) heating of copper in air
(b) combustion of gasoline
(c) cooling of red-hot iron
(d) digestion of food

42. Heating 30. g of water from 20. °C to 50. °C requires
(a) 30. cal (c) 3.8×10^3 J
(b) 50. cal (d) 6.3×10^3 J

43. The specific heat of aluminum is 0.900 J/g°C. How many joules of energy are required to raise the temperature of 20.0 g of Al from 10.0°C to 15.0°C?
(a) 79 J (b) 90. J (c) 100. J (d) 112 J

44. A100.-g iron ball (specific heat = 0.473 J/g°C) is heated to 125°C and is placed in a calorimeter holding 200. g of water at 25.0°C. What will be the highest temperature reached by the water?
(a) 43.7°C (c) 65.3°C
(b) 30.4°C (d) 35.4°C

45. Which has the highest specific heat?
(a) ice (b) lead (c) water (d) aluminum

46. When 20.0 g of mercury are heated from 10.0°C to 20.0°C, 27.6 J of energy are absorbed. What is the specific heat of mercury?
(a) 0.726 J/ g°C (c) 2.76 J/g°C
(b) 0.138 J/ g°C (d) no correct answer given

47. Changing hydrogen and oxygen into water is a
(a) physical change
(b) chemical change
(c) conservation reaction
(d) no correct answer given

Free Response Questions:

Answer each of the following. Be sure to include your work and explanations in a clear, logical form.

1. You decide to go sailing in the tropics with some friends. Once there, you listen to the marine forecast, which predicts in-shore wave heights of 1.5 meters, offshore wave heights of 4 meters, and temperature of 27°C. Your friend the captain is unfamiliar with the metric system and he needs to know whether it is safe for your small boat and if it will be warm. He asks you to convert the measurements to feet and Fahrenheit, respectively.

2. Jane is melting butter in a copper pot on the stove. If she knows how much heat her stove releases per minute, what other measurements does she need to determine how much heat the butter absorbed? She assumes that the stove does not lose any heat to the surroundings.

3. Julius decided to heat 75 g $CaCO_3$ to determine how much carbon dioxide is produced. (*Note*: When $CaCO_3$ is heated, it produces CaO and carbon dioxide.) He collected the carbon dioxide in a balloon. Julius found the mass of the CaO remaining was 42 g. If 44 g of carbon dioxide take up 24 dm^3 of space, how many *liters* of gas were trapped in the balloon?

Use these pictures to answer Question 4.

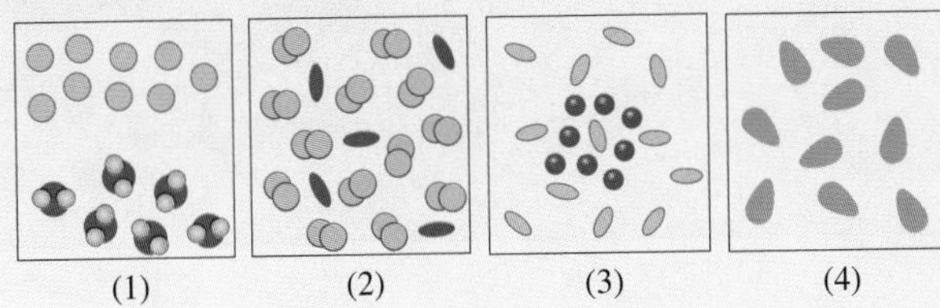

(1) (2) (3) (4)

4. (a) Which picture best describes a homogeneous mixture?
   (b) How would you classify the contents of the other containers?
   (c) Which picture contains a compound? Explain how you made your choice.

Use these pictures to answer Question 5.

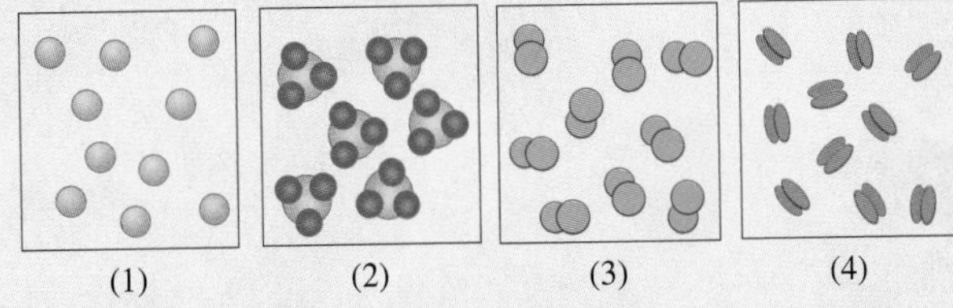

(1) (2) (3) (4)

5. (a) Which picture best represents fluorine gas? Why?
   (b) Which other elements could that picture also represent?
   (c) Which of the pictures could represent SO_3 gas?
6. Sue and Tim each left a one-quart bowl outside one night. Sue's bowl was full of water and covered while Tim's was empty and open. The next day there was a huge snowstorm that filled Tim's bowl with snow. The temperature that night went down to 12°F.
   (a) Which bowl would require less energy to bring its contents to room temperature (25°C)? Why?
   (b) What temperature change (°C) is required to warm the bowls to 25°C?
   (c) How much heat (in kJ) is required to raise the temperature of the contents of Sue's bowl to 0°C (without converting the ice to water)?
   (d) Did the water in Sue's bowl undergo chemical or physical changes or both?
7. One cup of Raisin Bran provides 60.% of the U.S. recommended daily allowance (RDA) of iron.
   (a) If the cereal provides 11 mg of iron, what is the U.S. RDA for Fe?
   (b) When the iron in the cereal is extracted, it is found to be the pure element. What is the volume of iron in a cup of the cereal?
8. 16% of the U.S. RDA for Ca is 162 mg.
   (a) What mass of calcium phosphate, $Ca_3(PO_4)_2$, provides 162 mg of calcium?
   (b) Is $Ca_3(PO_4)_2$ an element, mixture, or compound?
   (c) Milk is a good source of calcium in the diet. If 120 mL of skim milk provide 13% of the U.S. RDA for Ca, how many cups of milk should you drink per day if that is your only source of calcium?
9. Absent-minded Alfred put down a bottle containing silver on the table. When he went to retrieve it, he realized he had forgotten to label the bottle. Unfortunately there were two full bottles of the same size side-by-side. Alfred realized he had placed a bottle of mercury on the same table last week. State two ways Alfred can determine which bottle contains silver without opening the bottles.
10. Suppose 25 g of solid sulfur and 35 g of oxygen gas are placed in a sealed container.
    (a) Does the container hold a mixture or a compound?
    (b) After heating, the container was weighed. From a comparison of the total mass before heating to the total mass after heating, can you tell whether a reaction took place? Explain.
    (c) After the container is heated, all the contents are gaseous. Has the density of the container including its contents changed? Explain.

CHAPTER 5

Early Atomic Theory and Structure

Multiple lightning strikes over the desert in Arizona. Lightning occurs when electrons move to neutralize a charge difference between the clouds and the Earth.

Chapter Outline

Pure substances are classified as elements or compounds, but just what makes a substance possess its unique properties? How small a piece of salt will still taste salty? Carbon dioxide puts out fires, is used by plants to produce oxygen, and forms dry ice when solidified. But how small a mass of this material still behaves like carbon dioxide? Substances are in their simplest identifiable form at the atomic, ionic, or molecular level. Further division produces a loss of characteristic properties.

What particles lie within an atom or ion? How are these tiny particles alike? How do they differ? How far can we continue to divide them? Alchemists began the quest, early chemists laid the foundation, and modern chemists continue to build and expand on models of the atom.

5.1 Early Thoughts

The structure of matter has long intrigued and engaged us. The earliest models of the atom were developed by the ancient Greek philosophers. About 440 B.C. Empedocles stated that all matter was composed of four "elements"—earth, air, water, and fire. Democritus (about 470–370 B.C.) thought that all forms of matter were composed of tiny indivisible particles, which he called atoms, derived from the Greek word *atomos*, meaning "indivisible." He held that atoms were in constant motion and that they combined with one another in various ways. This hypothesis was not based on scientific observations. Shortly thereafter, Aristotle (384–322 B.C.) opposed the theory of Democritus and instead endorsed and advanced the Empedoclean theory. So strong was the influence of Aristotle that his theory dominated the thinking of scientists and philosophers until the beginning of the 17th century.

5.2 Dalton's Model of the Atom

More than 2000 years after Democritus, the English schoolmaster John Dalton (1766–1844) revived the concept of atoms and proposed an atomic model based on facts and experimental evidence (Figure 5.1). His theory, described in a series of papers published from 1803 to 1810, rested on the idea of a different kind of atom for each element. The essence of **Dalton's atomic model** may be summed up as follows:

Dalton's atomic model

1. Elements are composed of minute, indivisible particles called atoms.
2. Atoms of the same element are alike in mass and size.
3. Atoms of different elements have different masses and sizes.
4. Chemical compounds are formed by the union of two or more atoms of different elements.
5. Atoms combine to form compounds in simple numerical ratios, such as one to one, one to two, two to three, and so on.
6. Atoms of two elements may combine in different ratios to form more than one compound.

Dalton's atomic model stands as a landmark in the development of chemistry. The major premises of his model are still valid, but some of his statements must be modified or qualified because later investigations have shown that (1) atoms are composed of subatomic particles; (2) not all the atoms of a specific element have the same mass; and (3) atoms, under special circumstances, can be decomposed.

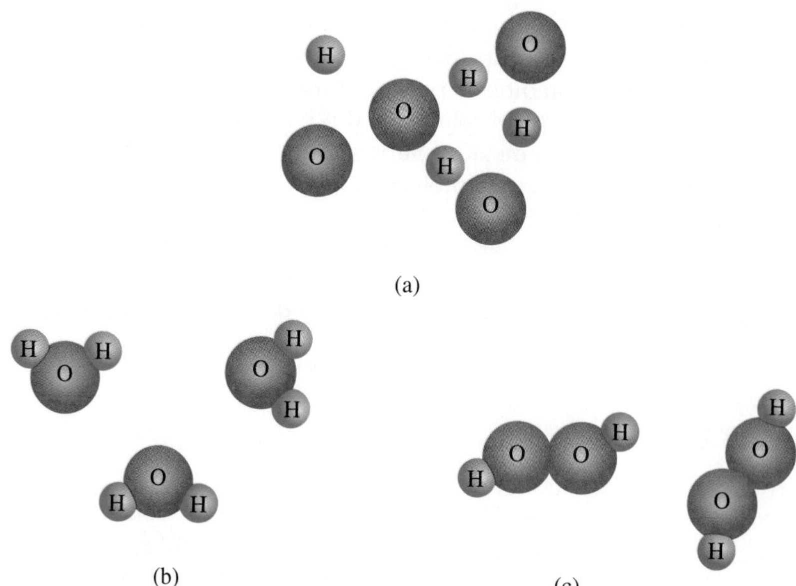

Figure 5.1
(a) Dalton's atoms were individual particles, the atoms of each element being alike in mass and size but different in mass and size from other elements. (b) and (c) Dalton's atoms combine in specific ratios to form compounds.

In chemistry we use models (theories) such as Dalton's atomic model to explain the behavior of atoms, molecules, and compounds. Models are modified to explain new information. We frequently learn the most about a system when our models (theories) fail. That is the time when we must rethink our explanation and determine whether we need to modify our model or propose a new or different model to explain the behavior.

5.3 Composition of Compounds

A large number of experiments extending over a long period have established the fact that a particular compound always contains the same elements in the same proportions by mass. For example, water always contains 11.2% hydrogen and 88.8% oxygen by mass (see Figure 5.1b). The fact that water contains hydrogen and oxygen in this particular ratio does not mean that hydrogen and oxygen cannot combine in some other ratio but rather that a compound with a different ratio would not be water. In fact, hydrogen peroxide is made up of two atoms of hydrogen and two atoms of oxygen per molecule and contains 5.9% hydrogen and 94.1% oxygen by mass; its properties are markedly different from those of water (see Figure 5.1c).

	Water	Hydrogen peroxide
	11.2% H	5.9% H
	88.8% O	94.1% O
Atomic composition	2 H + 1 O	2 H + 2 O

natural law
law of definite composition

We often summarize our general observations regarding nature into a statement called a **natural law**. In the case of the composition of a compound, we use the **law of definite composition**, which states that a compound always contains two or more elements chemically combined in a definite proportion by mass.

Let's consider two elements, oxygen and hydrogen, that form more than one compound. In water, 8.0 g of oxygen are present for each gram of hydrogen. In hydrogen peroxide, 16.0 g of oxygen are present for each gram of hydrogen. The masses of oxygen are in the ratio of small whole numbers, 16:8 or 2:1. Hydrogen peroxide has twice as much oxygen (by mass) as does water. Using Dalton's atomic model, we deduce that hydrogen peroxide has twice as many oxygen atoms per hydrogen atom as water. In fact, we now write the formulas for water as H_2O and for hydrogen peroxide as H_2O_2. See Figure 5.1b and c.

The **law of multiple proportions** states atoms of two or more elements may combine in different ratios to produce more than one compound.

law of multiple proportions

Some examples of the law of multiple proportions are given in Table 5.1. The reliability of this law and the law of definite composition is the cornerstone of the science of chemistry. In essence, these laws state that (1) the composition of a particular substance will always be the same no matter what its origin or how it is formed, and (2) the composition of different compounds formed from the same elements will always be unique.

Table 5.1 Selected Compounds Showing Elements That Combine to Give More Than One Compound

Compound	Formula	Percent composition
Copper(I) chloride	$CuCl$	64.2% Cu, 35.8% Cl
Copper(II) chloride	$CuCl_2$	47.3% Cu, 52.7% Cl
Methane	CH_4	74.9% C, 25.1% H
Octane	C_8H_{18}	85.6% C, 14.4% H
Methyl alcohol	CH_4O	37.5% C, 12.6% H, 49.9% O
Ethyl alcohol	C_2H_6O	52.1% C, 13.1% H, 34.7% O
Glucose	$C_6H_{12}O_6$	40.0% C, 6.7% H, 53.3% O

You need to recognize the difference between a *law* and a *model (theory)*. A law is a summary of observed behavior. A model (theory) is an attempt to explain the observed behavior. This means that laws remain constant—that is, they do not undergo modification—while theories (models) sometimes fail and are modified or discarded over time.

5.4 The Nature of Electric Charge

You've probably received a shock after walking across a carpeted area on a dry day. You may have also experienced the static associated with combing your hair and have had your clothing cling to you. These phenomena result from an accumulation of *electric charge*. This charge may be transferred from one object to another. The properties of electric charge are as follows:

1. Charge may be of two types, positive and negative.
2. Unlike charges attract (positive attracts negative), and like charges repel (negative repels negative and positive repels positive).

3. Charge may be transferred from one object to another, by contact or induction.
4. The less the distance between two charges, the greater the force of attraction between unlike charges (or repulsion between identical charges). The force of attraction (F) can be expressed using the following equation:

$$F = \frac{kq_1q_2}{r^2}$$

where q_1 and q_2 are the charges, r is the distance between the charges, and k is a constant.

5.5 Discovery of Ions

When ions are present in a solution of salt water and an electric current is passed through the solution, the light bulb glows.

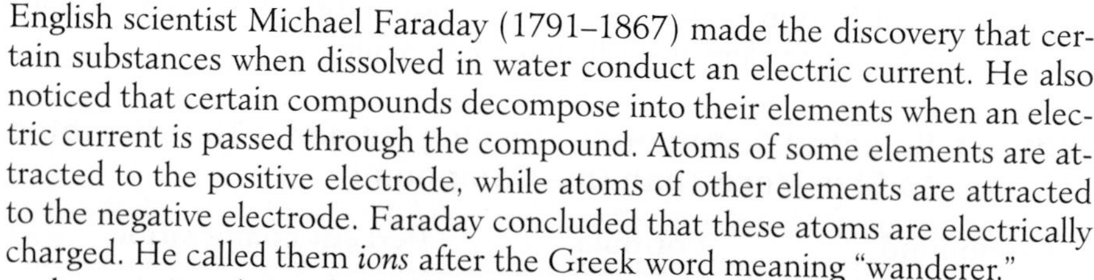

English scientist Michael Faraday (1791–1867) made the discovery that certain substances when dissolved in water conduct an electric current. He also noticed that certain compounds decompose into their elements when an electric current is passed through the compound. Atoms of some elements are attracted to the positive electrode, while atoms of other elements are attracted to the negative electrode. Faraday concluded that these atoms are electrically charged. He called them *ions* after the Greek word meaning "wanderer."

Any moving charge is an electric current. The electrical charge must travel through a substance known as a conducting medium. The most familiar conducting media are metals formed into wires.

The Swedish scientist Svante Arrhenius (1859–1927) extended Faraday's work. Arrhenius reasoned that an ion is an atom (or a group of atoms) carrying a positive or negative charge. When a compound such as sodium chloride (NaCl) is melted, it conducts electricity. Water is unnecessary. Arrhenius's explanation of this conductivity was that upon melting, the sodium chloride dissociates, or breaks up, into charged ions, Na^+ and Cl^-. The Na^+ ions move toward the negative electrode (cathode), whereas the Cl^- ions migrate toward the positive electrode (anode). Thus positive ions are called *cations*, and negative ions are called *anions*.

From Faraday's and Arrhenius's work with ions, Irish physicist G. J. Stoney (1826–1911) realized there must be some fundamental unit of electricity associated with atoms. He named this unit the electron in 1891. Unfortunately, he had no means of supporting his idea with experimental proof. Evidence remained elusive until 1897, when English physicist J. J. Thomson (1856–1940) was able to show experimentally the existence of the electron.

5.6 Subatomic Parts of the Atom

The concept of the atom—a particle so small that until recently it could not be seen even with the most powerful microscope—and the subsequent determination of its structure stand among the greatest creative intellectual human achievements.

Any visible quantity of an element contains a vast number of identical atoms. But when we refer to an atom of an element, we isolate a single atom from the multitude in order to present the element in its simplest form. Figure 5.2 shows individual atoms as we can see them today.

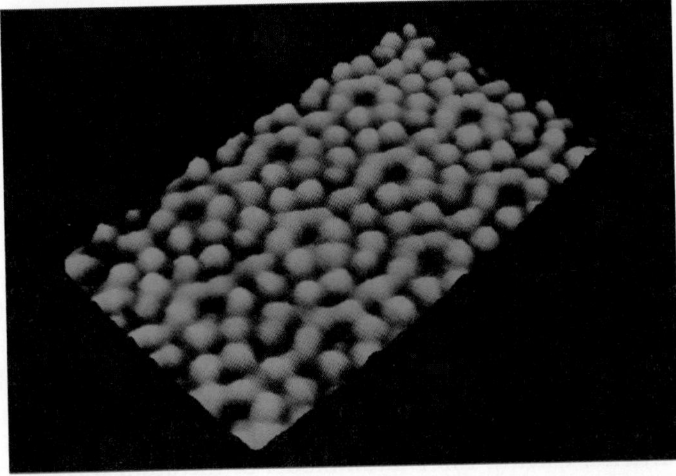

Figure 5.2
A scanning tunneling microscope shows an array of copper atoms.

What is this tiny particle we call the atom? The diameter of a single atom ranges from 0.1 to 0.5 nanometer ($1\,\text{nm} = 1 \times 10^{-9}\,\text{m}$). Hydrogen, the smallest atom, has a diameter of about 0.1 nm. To arrive at some idea of how small an atom is, consider this dot (•), which has a diameter of about 1 mm, or $1 \times 10^{6}\,\text{nm}$. It would take 10 million hydrogen atoms to form a line of atoms across this dot. As inconceivably small as atoms are, they contain even smaller particles, the **subatomic particles**, including electrons, protons, and neutrons.

Subatomic particles

The development of atomic theory was helped in large part by the invention of new instruments. For example, the Crookes tube, developed by Sir William Crookes (1832–1919) in 1875, opened the door to the subatomic structure of the atom (Figure 5.3). The emissions generated in a Crookes tube are called *cathode rays*. J. J. Thomson demonstrated in 1897 that cathode rays (1) travel in straight lines, (2) are negative in charge, (3) are deflected by electric and magnetic fields, (4) produce sharp shadows, and (5) are capable of moving a small paddle wheel. This was the experimental discovery of the fundamental unit of charge—the electron.

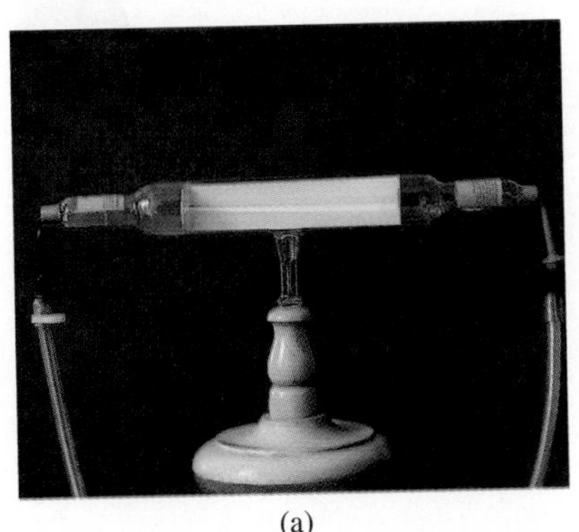

(a)

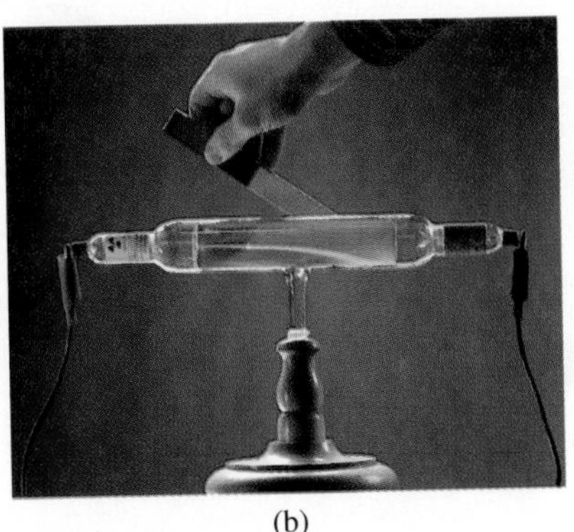

(b)

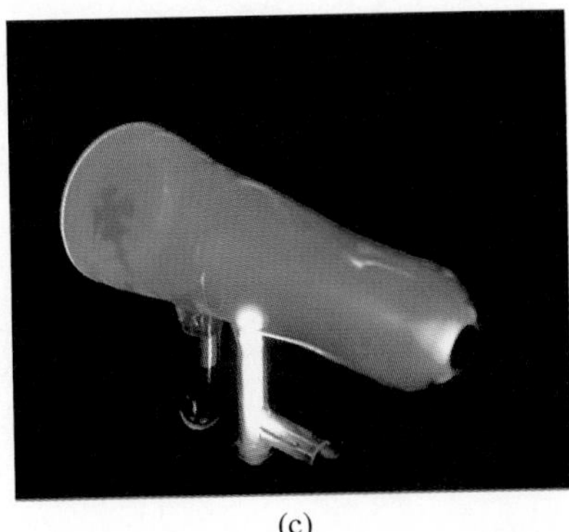

(c)

Figure 5.3
Crookes tube. Emissions generated in Crookes tube (a) travel in straight lines and are negative in charge, (b) are deflected by a magnetic field, and (c) provide a sharp shadow of the cross in the center of the tube.

electron

The **electron** (e^-) is a particle with a negative electrical charge and a mass of 9.110×10^{-28} g. This mass is 1/1837 the mass of a hydrogen atom. Although the actual charge of an electron is known, its value is too cumbersome for practical use and has therefore been assigned a relative electrical charge of -1. The size of an electron has not been determined exactly, but its diameter is believed to be less than 10^{-12} cm.

Protons were first observed by German physicist Eugen Goldstein (1850–1930) in 1886. However, it was Thomson who discovered the nature of the proton. He showed that the proton is a particle, and he calculated its mass to be about 1837 times that of an electron. The **proton** (p) is a particle with actual mass of 1.673×10^{-24} g. Its relative charge ($+1$) is equal in magnitude, but opposite in sign, to the charge on the electron. The mass of a proton is only very slightly less than that of a hydrogen atom.

proton

Thomson had shown that atoms contain both negatively and positively charged particles. Clearly, the Dalton model of the atom was no longer acceptable. Atoms are not indivisible but are instead composed of smaller parts. Thomson proposed a new model of the atom.

Thomson model of the atom

In the **Thomson model of the atom**, the electrons are negatively charged particles embedded in the atomic sphere. Since atoms are electrically neutral, the sphere also contains an equal number of protons, or positive charges. A neutral atom could become an ion by gaining or losing electrons.

Positive ions were explained by assuming that the neutral atom loses electrons. An atom with a net charge of $+1$ (for example, Na^+ or Li^+) has lost one electron. An atom with a net charge of $+3$ (for example, Al^{3+}) has lost three electrons (Figure 5.4a).

Negative ions were explained by assuming that additional electrons can be added to atoms. A net charge of -1 (for example, Cl^- or F^-) is produced by the addition of one electron. A net charge of -2 (for example, O^{2-} or S^{2-}) requires the addition of two electrons (Figure 5.4b).

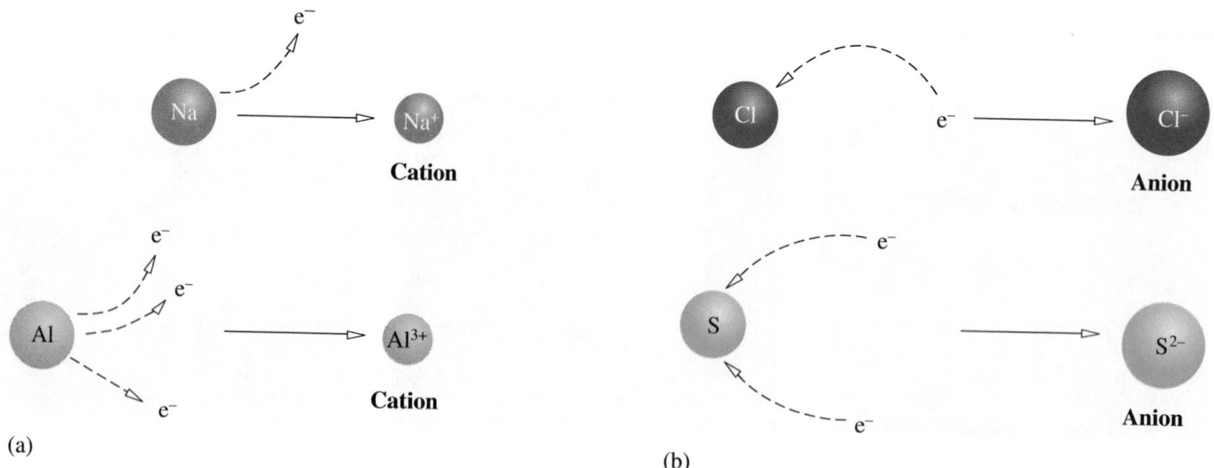

Figure 5.4
(a) When one or more electrons are lost from an atom, a cation is formed. (b) When one or more electrons are added to a neutral atom, an anion is formed.

The third major subatomic particle was discovered in 1932 by James Chadwick (1891–1974). This particle, the **neutron** (n), has neither a positive nor a negative charge and has an actual mass (1.675×10^{-24} g) which is only very slightly greater than that of a proton. The properties of these three subatomic particles are summarized in Table 5.2.

neutron

Nearly all the ordinary chemical properties of matter can be explained in terms of atoms consisting of electrons, protons, and neutrons. The discussion of atomic structure that follows is based on the assumption that atoms contain only these principal subatomic particles. Many other subatomic particles, such as mesons, positrons, neutrinos, and antiprotons, have been discovered, but it is not yet clear whether all these particles are actually present in the atom or whether they are produced by reactions occurring within the nucleus. The fields of atomic and high-energy physics have produced a long list of subatomic particles. Descriptions of the properties of many of these particles are to be found in physics textbooks.

Table 5.2 Electrical Charge and Relative Mass of Electrons, Protons, and Neutrons

Particle	Symbol	Relative electrical charge	Actual mass (g)
Electron	e^-	−1	9.110×10^{-28}
Proton	p	+1	1.673×10^{-24}
Neutron	n	0	1.675×10^{-24}

Example 5.1

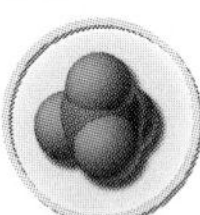

The mass of a helium atom is 6.65×10^{-24} g. How many atoms are in a 4.0-g sample of helium?

SOLUTION

$$(4.0\text{ g})\left(\frac{1\text{ atom He}}{6.65 \times 10^{-24}\text{ g}}\right) = 6.0 \times 10^{23}\text{ atoms He}$$

Practice 5.1

The mass of an atom of hydrogen is 1.673×10^{-24} g. How many atoms are in a 10.0-g sample of hydrogen?

5.7 The Nuclear Atom

The discovery that positively charged particles are present in atoms came soon after the discovery of radioactivity by Henri Becquerel (1852–1908) in 1896. Radioactive elements spontaneously emit alpha particles, beta particles, and gamma rays from their nuclei (see Chapter 18).

By 1907 Ernest Rutherford (1871–1937) had established that the positively charged alpha particles emitted by certain radioactive elements are ions of the element helium. Rutherford used these alpha particles to establish the

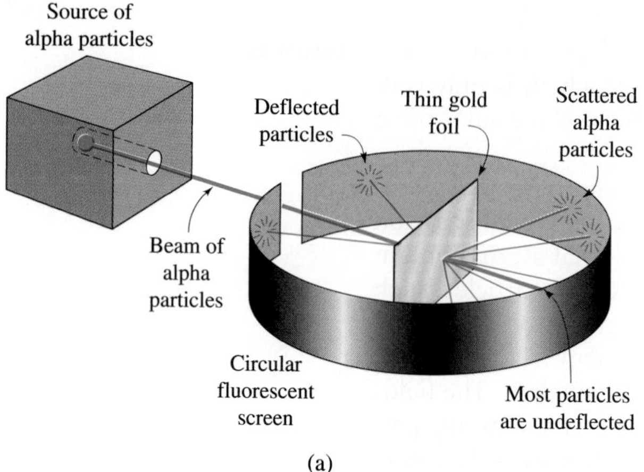

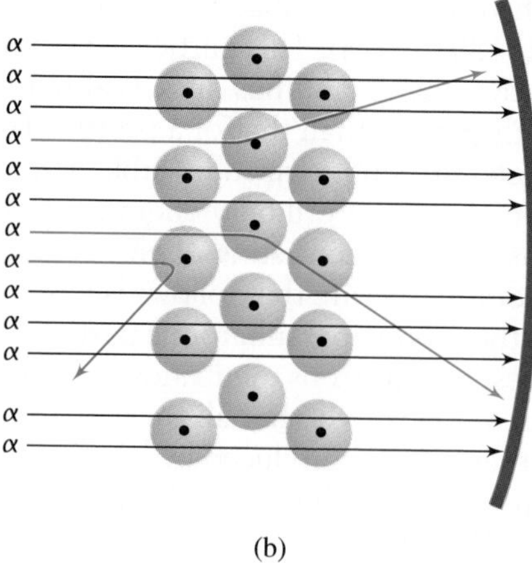

Figure 5.5
(a) Rutherford's experiment on alpha-particle scattering, where positive alpha particles (α), emanating from a radioactive source, were directed at a thin gold foil. (b) Deflection (red) and scattering (blue) of the positive alpha particles by the positive nuclei of the gold atoms.

nuclear nature of atoms. In experiments performed in 1911, he directed a stream of positively charged helium ions (alpha particles) at a very thin sheet of gold foil (about 1000 atoms thick). See Figure 5.5a. He observed that most of the alpha particles passed through the foil with little or no deflection; but a few of the particles were deflected at large angles, and occasionally one even bounced back from the foil (Figure 5.5b). It was known that like charges repel each other and that an electron with a mass of 1/1837 amu could not possibly have an appreciable effect on the path of a 4-amu alpha particle, which is about 7350 times more massive than an electron. Rutherford therefore reasoned that each gold atom must contain a positively charged mass occupying a relatively tiny volume and that, when an alpha particle approaches close enough to this positive mass, it is deflected. Rutherford spoke of this positively charged mass as the *nucleus* of the atom. Because alpha particles have relatively high masses, the extent of the deflections (some actually bounced back) indicated to Rutherford that the nucleus is very heavy and dense. (The density of the nucleus of a hydrogen atom is about 10^{12} g/cm^3—about 1 trillion times the density of water.) Because most of the alpha particles passed through the thousand or so gold atoms without any apparent deflection, he further concluded that most of an atom consists of empty space.

When we speak of the mass of an atom, we are referring primarily to the mass of the nucleus. The nucleus contains all the protons and neutrons, which represent more than 99.9% of the total mass of any atom (see Table 5.1). By way of illustration, the largest number of electrons known to exist in an atom is 111. The mass of even 111 electrons is only about 1/17 of the mass of a single proton or neutron. The mass of an atom therefore is primarily determined by the combined masses of its protons and neutrons.

General Arrangement of Subatomic Particles

The alpha-particle scattering experiments of Rutherford established that the atom contains a dense, positively charged nucleus. The later work of Chadwick demonstrated that the atom contains neutrons, which are particles with mass,

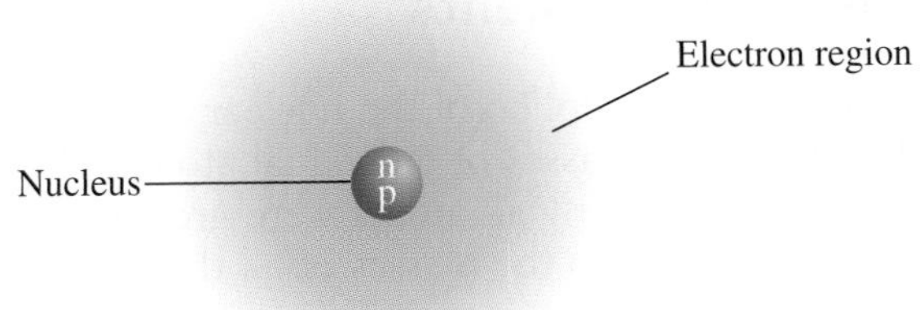

Figure 5.6
In the nuclear model of the atom, protons (p) and neutrons (n) are located in the nucleus. The electrons are found in the remainder of the atom (which is mostly empty space because electrons are very tiny).

but no charge. Rutherford also noted that light, negatively charged electrons are present and offset the positive charges in the nucleus. Based on this experimental evidence, a model of the atom and the location of its subatomic particles was devised in which each atom consists of a **nucleus** surrounded by electrons (see Figure 5.6). The nucleus contains protons and neutrons but does not contain electrons. In a neutral atom the positive charge of the nucleus (due to protons) is exactly offset by the negative electrons. Because the charge of an electron is equal to, but of opposite sign than, the charge of a proton, a neutral atom must contain exactly the same number of electrons as protons. However, this model of atomic structure provides no information on the arrangement of electrons within the atom.

nucleus

A neutral atom contains the same number of protons and electrons.

Atomic Numbers of the Elements

The **atomic number** of an element is the number of protons in the nucleus of an atom of that element. The atomic number determines the identity of an atom. For example, every atom with an atomic number of 1 is a hydrogen atom; it contains one proton in its nucleus. Every atom with an atomic number of 6 is a carbon atom; it contains 6 protons in its nucleus. Every atom with an atomic number of 92 is a uranium atom; it contains 92 protons in its nucleus. The atomic number tells us not only the number of positive charges in the nucleus but also the number of electrons in the neutral atom, since a neutral atom contains the same number of electrons and protons.

atomic number

atomic number = number of protons in the nucleus

You don't need to memorize the atomic numbers of the elements because a periodic table is usually provided in texts, in laboratories, and on examinations. The atomic numbers of all elements are shown in the periodic table on the inside front cover of this book and are also listed in the table of atomic masses on the inside front endpapers.

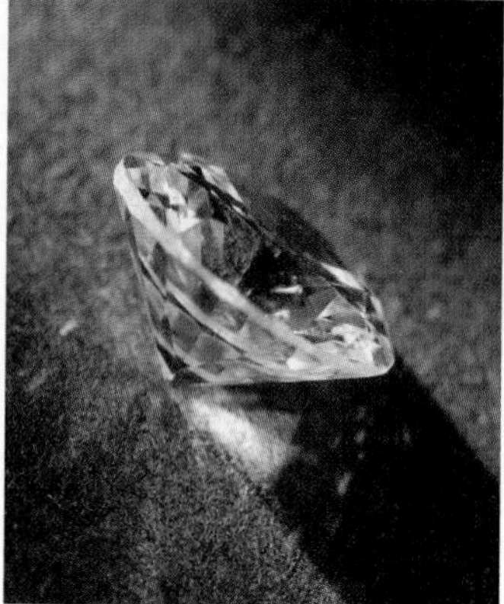

Carbon, shown here as a beautiful diamond, has six protons and six electrons in each atom.

5.8 Isotopes of the Elements

Shortly after Rutherford's conception of the nuclear atom, experiments were performed to determine the masses of individual atoms. These experiments showed that the masses of nearly all atoms were greater than could be accounted for by simply adding up the masses of all the protons and electrons that were known to be present in an atom. This fact led to the concept of the neutron, a particle with no charge but with a mass about the same as that of a proton. Because this particle has no charge, it was very difficult to detect, and the existence of the neutron was not proven experimentally until 1932. All atomic nuclei except that of the simplest hydrogen atom contain neutrons.

All atoms of a given element have the same number of protons. Experimental evidence has shown that, in most cases, all atoms of a given element do not have identical masses. This is because atoms of the same element may have different numbers of neutrons in their nuclei.

isotopes

Atoms of an element having the same atomic number but different atomic masses are called **isotopes** of that element. Atoms of the various isotopes of an element therefore have the same number of protons and electrons but different numbers of neutrons.

Three isotopes of hydrogen (atomic number 1) are known. Each has one proton in the nucleus and one electron. The first isotope (protium), without a neutron, has a mass number of 1; the second isotope (deuterium), with one neutron in the nucleus, has a mass number of 2; the third isotope (tritium), with two neutrons, has a mass number of 3 (see Figure 5.7).

mass number

The three isotopes of hydrogen may be represented by the symbols $^{1}_{1}H$, $^{2}_{1}H$, $^{3}_{1}H$, indicating an atomic number of 1 and mass numbers of 1, 2, and 3, respectively. This method of representing atoms is called *isotopic notation*. The subscript (Z) is the atomic number; the superscript (A) is the **mass number**, which is the sum of the number of protons and the number of neutrons in the nucleus. The hydrogen isotopes may also be referred to as hydrogen-1, hydrogen-2, and hydrogen-3.

The mass number of an element is the sum of the protons and neutrons in the nucleus.

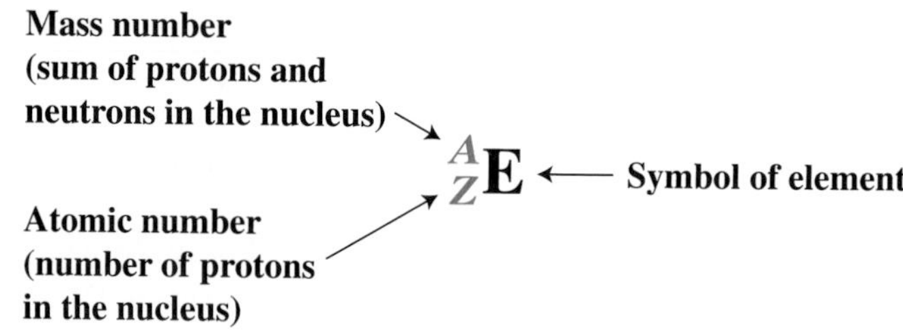

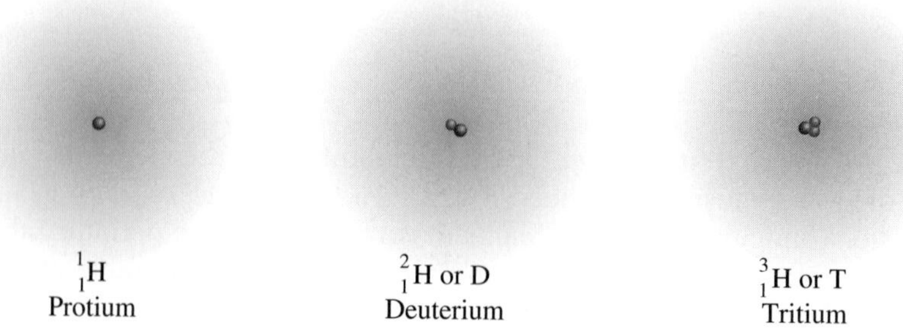

Figure 5.7
The isotopes of hydrogen. The number of protons (purple) and neutrons (blue) are shown within the nucleus. The electron (e^{-}) exists outside the nucleus.

CHEMISTRY IN ACTION • Isotope Detectives

Scientists are learning to use isotopes to determine the origin of drugs and gems. It turns out that isotope ratios similar to those used in carbon dating can also identify the source of cocaine or the birthplace of emeralds.

Researchers with the Drug Enforcement Agency (DEA) have created a database of the origin of coca leaves that pinpoints the origin of the leaves with a 90% accuracy. Cocaine keeps a chemical signature of the environment where it grew. Isotopes of carbon and nitrogen are found in a particular ratio based on climatic conditions in the growing region. These ratios correctly identified the source of 90% of the samples tested, according to James Ehleringer of the University of Utah, Salt Lake City. This new method can trace drugs a step further back than current techniques, which mainly look at chemicals introduced by processing practices in different locations. This could aid in tracking the original exporters and stopping production at the source.

An uncut (left) and cut (right) emerald from Brazil.

It turns out that a similar isotopic analysis of oxygen has led researchers in France to be able to track the birthplace of emeralds. Very high quality emeralds have few inclusions (microscopic cavities). Gemologists use these inclusions and the material trapped in them to identify the source of the gem. High-quality gems can now also be identified by using an oxygen isotope ratio. These tests use an ion microscope that blasts a few atoms from the gems' surface (with virtually undetectable damage). The tiny sample is analyzed for its oxygen isotope ratio and then compared to a database from emerald mines around the world. Using the information, gemologists can determine the mine from which the emerald was born. Since emeralds from Colombian mines are valued much more highly than those from other countries, this technique can be used to help collectors know just what they are paying for, as well as to identify the history of treasured emeralds.

Most of the elements occur in nature as mixtures of isotopes. However, not all isotopes are stable; some are radioactive and are continuously decomposing to form other elements. For example, of the seven known isotopes of carbon, only two, carbon-12 and carbon-13, are stable. Of the seven known isotopes of oxygen, only three—$^{16}_{8}O$, $^{17}_{8}O$, and $^{18}_{8}O$—are stable. Of the fifteen known isotopes of arsenic, $^{75}_{33}As$ is the only one that is stable.

Practice 5.2

How many protons, neutrons, and electrons are in each of the three stable isotopes of oxygen?

5.9 Atomic Mass

The mass of a single atom is far too small to measure on a balance, but fairly precise determinations of the masses of individual atoms can be made with an instrument called a *mass spectrometer* (see Figure 5.8). The mass of a single

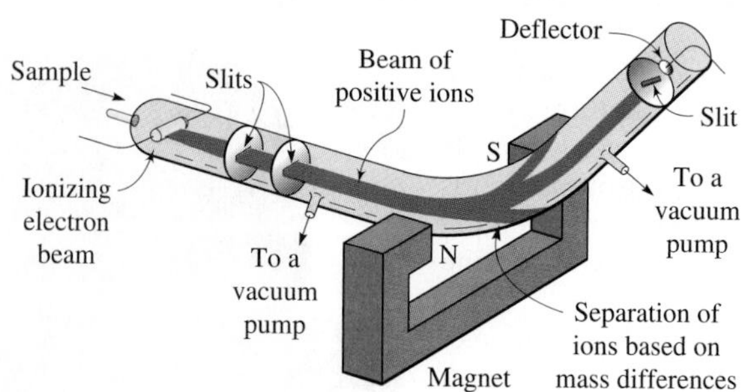

Figure 5.8
A modern mass spectrometer. A beam of positive ions is produced from the sample as it enters the chamber. The positive ions are then accelerated as they pass through slits in an electric field. When the ions enter the magnetic field, they are deflected differently, depending on mass and charge. The ions are then detected at the end of the tube. From intensity and position of the lines on the mass spectrogram, the different isotopes of the elements and their relative amounts can be determined.

hydrogen atom is 1.673×10^{-24} g. However, it is neither convenient nor practical to compare the actual masses of atoms expressed in grams; therefore, a table of relative atomic masses using *atomic mass units* was devised. (The term *atomic weight* is sometimes used instead of *atomic mass*.) The carbon isotope having six protons and six neutrons and designated carbon-12, or ${}^{12}_{6}C$, was chosen as the standard for atomic masses. This reference isotope was assigned a value of exactly 12 atomic mass units (amu). Thus, 1 **atomic mass unit** is defined as equal to exactly 1/12 of the mass of a carbon-12 atom. The actual mass of a carbon-12 atom is 1.9927×10^{-23} g, and that of one atomic mass unit is 1.6606×10^{-24} g. In the table of atomic masses, all elements then have values that are relative to the mass assigned to the reference isotope, carbon-12.

atomic mass unit

$1 \text{ amu} = 1.6606 \times 10^{-24}$ g

A table of atomic masses is given on the inside front cover of this book. Hydrogen atoms, with a mass of about 1/12 that of a carbon atom, have an average atomic mass of 1.00797 amu on this relative scale. Magnesium atoms, which are about twice as heavy as carbon, have an average mass of 24.305 amu. The average atomic mass of oxygen is 15.9994 amu.

Since most elements occur as mixtures of isotopes with different masses, the atomic mass determined for an element represents the average relative mass of all the naturally occurring isotopes of that element. The atomic masses of the individual isotopes are approximately whole numbers, because the relative masses of the protons and neutrons are approximately 1.0 amu each. Yet we find that the atomic masses given for many of the elements deviate considerably from whole numbers.

For example, the atomic mass of rubidium is 85.4678 amu, that of copper is 63.546 amu, and that of magnesium is 24.305 amu. The deviation of an atomic mass from a whole number is due mainly to the unequal occurrence of the various isotopes of an element. The two principal isotopes of copper are ${}^{63}_{29}Cu$ and ${}^{65}_{29}Cu$. It is apparent that copper-63 atoms are the more abundant isotope, since the atomic mass of copper, 63.546 amu, is closer to 63 than to 65 amu (see Figure 5.9). The actual values of the copper isotopes observed by mass spectra determination are shown in the following table:

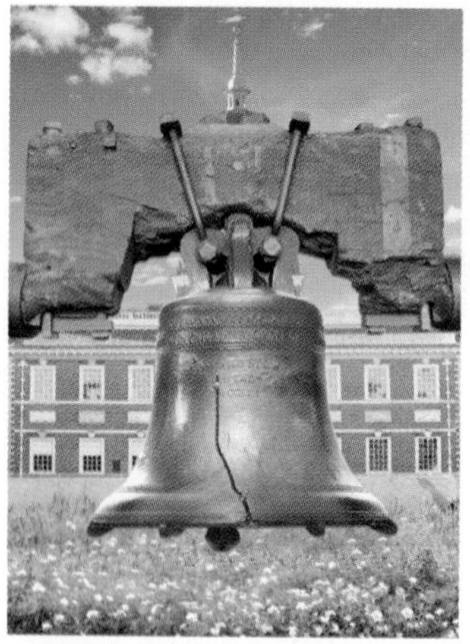

The copper used in casting the Liberty bell contains a mixture of the isotopes of copper.

Isotope	Isotopic mass (amu)	Abundance (%)	Average atomic mass (amu)
${}^{63}_{29}Cu$	62.9298	69.09	63.55
${}^{65}_{29}Cu$	64.9278	30.91	

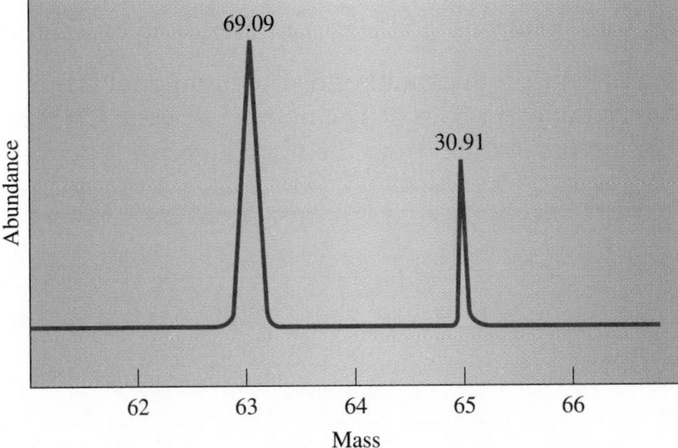

Figure 5.9
A typical reading from a mass spectrometer. The two principal isotopes of copper are shown with the abundance (%) given.

The average atomic mass can be calculated by multiplying the atomic mass of each isotope by the fraction of each isotope present and adding the results. The calculation for copper is

The average atomic mass is a weighted average of the masses of all the isotopes present in the sample.

$$\begin{aligned}(62.9298\ \text{amu})(0.6909) &= 43.48\ \text{amu}\\(64.9278\ \text{amu})(0.3091) &= \underline{20.07\ \text{amu}}\\ &\ 63.55\ \text{amu}\end{aligned}$$

The **atomic mass** of an element is the average relative mass of the isotopes of that element compared to the atomic mass of carbon-12 (exactly 12.0000 . . . amu).

atomic mass

The relationship between mass number and atomic number is such that if we subtract the atomic number from the mass number of a given isotope, we obtain the number of neutrons in the nucleus of an atom of that isotope. Table 5.3 shows this method of determining the number of neutrons. For example, the fluorine atom ($^{19}_{9}F$), atomic number 9, having a mass of 19 amu, contains 10 neutrons:

mass number	−	atomic number	=	number of neutrons
19	−	9	=	10

The atomic masses given in the table on the front endpapers of this book are values accepted by international agreement. You need not memorize atomic masses. In the calculations in this book, the use of atomic masses to four significant figures will give results of sufficient accuracy. (See periodic table.)

Use four significant figures for atomic masses in this text.

Table 5.3 Determination of the Number of Neutrons in an Atom by Subtracting Atomic Number from Mass Number

	Hydrogen ($^{1}_{1}H$)	Oxygen ($^{16}_{8}O$)	Sulfur ($^{32}_{16}S$)	Fluorine ($^{19}_{9}F$)	Iron ($^{56}_{26}Fe$)
Mass number	1	16	32	19	56
Atomic number	(−)1	(−)8	(−)16	(−)9	(−)26
Number of neutrons	0	8	16	10	30

Example 5.2 How many protons, neutrons, and electrons are found in an atom of $^{14}_{6}C$?

SOLUTION The element is carbon, atomic number 6. The number of protons or electrons equals the atomic number and is 6. The number of neutrons is determined by subtracting the atomic number from the mass number: $14 - 6 = 8$.

Practice 5.3

How many protons, neutrons, and electrons are in each of these isotopes?
(a) $^{16}_{8}O$, (b) $^{80}_{35}Br$, (c) $^{235}_{92}U$, (d) $^{64}_{29}Cu$

Practice 5.4

What is the atomic number and the mass number of the elements that contain
(a) 9 electrons (b) 24 protons and 28 neutrons (c) $^{197}_{79}X$

What are the names of these elements?

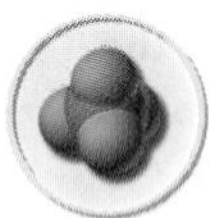

Example 5.3 Chlorine is found in nature as two isotopes, $^{37}_{17}Cl$ (24.47%) and $^{35}_{17}Cl$ (75.53%). The atomic masses are 36.96590 and 34.96885 amu, respectively. Determine the average atomic mass of chlorine.

SOLUTION Multiply each mass by its percentage and add the results to find the average:

$$(0.2447)(36.96590\text{ amu}) + (0.7553)(34.96885\text{ amu})$$
$$= 35.4575\text{ amu}$$
$$= 35.46\text{ amu (4 significant figures)}$$

Practice 5.5

Silver occurs as two isotopes with atomic masses 106.9041 and 108.9047 amu, respectively. The first isotope represents 51.82% and the second 48.18%. Determine the average atomic mass of silver.

Chapter 5 Review

5.1 Early Thoughts

- Greek model of matter:
  - Four elements—earth, air, water, fire
  - Democritus—atoms (indivisible particles) make up matter
  - Aristotle—opposed atomic ideas

5.2 Dalton's Model of the Atom

KEY TERM

Dalton's atomic model

- Summary of Dalton's model of the atom:
  - Elements are composed of atoms.
  - Atoms of the same element are alike (mass and size).
  - Atoms of different elements are different in mass and size.
  - Compounds form by the union of two or more atoms of different elements.
  - Atoms form compounds in simple numerical ratios.
  - Atoms of 2 elements may combine in different ratios to form different compounds.

5.3 Composition of Compounds

KEY TERMS

Natural law
Law of definite composition
Law of multiple proportions

- The law of definite composition states that a compound always contains two or more elements combined in a definite proportion by mass.
- The law of multiple proportions states that atoms of two or more elements may combine in different ratios to form more than one compound.

5.4 The Nature of Electric Charge

- The properties of electric charge:
  - Charges are one of two types—positive or negative.
  - Unlike charges attract and like charges repel.
  - Charge is transferred from one object to another by contact or by induction.
  - The force of attraction between charges is expressed by

$$F = \frac{kq_1q_2}{r^2}.$$

5.5 Discovery of Ions

- Michael Faraday discovered electrically charged ions.
- Svante Arrhenius explained that conductivity results from the dissociation of compounds into ions:
  - Cation—positive charge—attracted to negative electrode (cathode)
  - Anion—negative charge—attracted to positive electrode (anode)

5.6 Subatomic Parts of the Atom

KEY TERMS
Subatomic particles
Electron
Proton
Thomson model of the atom
Neutron

- Atoms contain smaller subatomic particles:
  - Electron—negative charge, 1/1837 mass of proton
  - Proton—positive charge, 1.673×10^{-24} g
  - Neutron—no charge, 1.675×10^{-24} g
- Thomson model of the atom:
  - Negative electrons are embedded in a positive atomic sphere.
  - Number of protons = number of electrons in a neutral atom.
  - Ions are formed by losing or gaining electrons.

5.7 The Nuclear Atom

KEY TERMS
Nucleus
Atomic number

- Rutherford gold foil experiment modified the Thomson model to a nuclear model of the atom:
  - Atoms are composed of a nucleus containing protons and/or neutrons surrounded by electrons, which occupy mostly empty space.
  - Neutral atoms contain equal numbers of protons and electrons.

5.8 Isotopes of the Elements

KEY TERMS
Isotopes
Mass number

- The mass number of an element is the sum of the protons and neutrons in the nucleus.

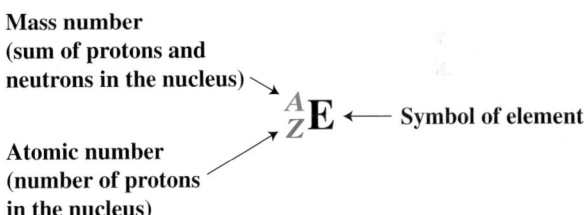

$$^{A}_{Z}\mathrm{E}$$

5.9 Atomic Mass

KEY TERMS
Atomic mass unit
Atomic mass

- The average atomic mass is a weighted average of the masses of all the isotopes present in the sample.
- The number of neutrons in an atom is determined by

mass number − atomic number = number of neutrons

$A - Z =$ neutrons

Review Questions

All questions with blue numbers have answers in the appendix of the text.

1. What are the atomic numbers of (a) copper, (b) nitrogen, (c) phosphorus, (d) radium, and (e) zinc?
2. A neutron is approximately how many times heavier than an electron?
3. From the chemist's point of view, what are the essential differences among a proton, a neutron, and an electron?
4. Distinguish between an atom and an ion.
5. What letters are used to designate atomic number and mass number in isotopic notation of atoms?
6. In what ways are isotopes alike? In what ways are they different?

Paired Exercises

All exercises with blue numbers have answers in the appendix of the text.

1. In Section 5.3, there is a statement about the composition of water. It says that water (H_2O) contains 8 grams of oxygen for every 1 gram of hydrogen. Show why this statement is true.

2. In Section 5.3, there is a statement about the composition of hydrogen peroxide. It says that hydrogen peroxide (H_2O_2) contains 16 grams of oxygen for every 1 gram of hydrogen. Show why this statement is true.

3. Explain why, in Rutherford's experiments, some alpha particles were scattered at large angles by the gold foil or even bounced back.

4. What experimental evidence led Rutherford to conclude the following?
   (a) The nucleus of the atom contains most of the atomic mass.
   (b) The nucleus of the atom is positively charged.
   (c) The atom consists of mostly empty space.

5. Describe the general arrangement of subatomic particles in the atom.

6. What part of the atom contains practically all its mass?

7. What contribution did these scientists make to atomic models of the atom?
   (a) Dalton
   (b) Thomson
   (c) Rutherford

8. Consider the following models of the atom: (a) Dalton, (b) Thomson, (c) Rutherford. How does the location of the electrons in an atom vary? How does the location of the atom's positive matter compare?

9. Explain why the atomic masses of elements are not whole numbers.

10. Is the isotopic mass of a given isotope ever an exact whole number? Is it always? (Consider the masses of $^{12}_{6}C$ and $^{63}_{29}Cu$.)

11. What special names are given to the isotopes of hydrogen?

12. List the similarities and differences in the three isotopes of hydrogen.

13. What is the nuclear composition of the five naturally occurring isotopes of germanium having mass numbers 70, 72, 73, 74, and 76?

14. What is the nuclear composition of the five naturally occurring isotopes of zinc having mass numbers 64, 66, 67, 68, and 70?

15. Write isotopic notation symbols for each of the following:
   (a) $Z = 29,\quad A = 65$
   (b) $Z = 20,\quad A = 45$
   (c) $Z = 36,\quad A = 84$

16. Write isotopic notation symbols for each of the following:
   (a) $Z = 47,\quad A = 109$
   (b) $Z = 8,\quad A = 18$
   (c) $Z = 26,\quad A = 57$

17. Give the nuclear composition and isotopic notation for
   (a) an atom containing 27 protons, 32 neutrons, and 27 electrons
   (b) an atom containing 15 protons, 16 neutrons, and 15 electrons
   (c) an atom containing 110 neutrons, 74 electrons, and 74 protons
   (d) an atom containing 92 electrons, 143 neutrons, and 92 protons

18. Give the nuclear composition and isotopic notation for
   (a) an atom containing 12 protons, 13 neutrons, and 12 electrons
   (b) an atom containing 51 neutrons, 40 electrons, and 40 protons
   (c) an atom containing 50 protons, 50 electrons, and 72 neutrons
   (d) an atom containing 122 neutrons, 80 protons, and 80 electrons

19. An unknown element contains 24 protons, 21 electrons, and has mass number 54. Answer the following questions:
   (a) What is the atomic number of this element?
   (b) What is the symbol of the element?
   (c) How many neutrons does it contain?

20. An unknown element contains 35 protons, 36 electrons, and has mass number 80. Answer the following questions:
   (a) What is the atomic number of this element?
   (b) What is the symbol of the element?
   (c) How many neutrons does it contain?

21. Naturally occurring lead exists as four stable isotopes: ^{204}Pb with a mass of 203.973 amu (1.480%); ^{206}Pb, 205.974 amu (23.60%); ^{207}Pb, 206.9759 amu (22.60%); and ^{208}Pb, 207.9766 amu (52.30%). Calculate the average atomic mass of lead.

22. Naturally occurring magnesium consists of three stable isotopes: ^{24}Mg, 23.985 amu (78.99%); ^{25}Mg, 24.986 amu (10.00%); and ^{26}Mg, 25.983 amu (11.01%). Calculate the average atomic mass of magnesium.

23. 68.9257 amu is the mass of 60.4% of the atoms of an element with only two naturally occurring isotopes. The atomic mass of the other isotope is 70.9249 amu. Determine the average atomic mass of the element. Identify the element.

24. A sample of enriched lithium contains 30.00% ^{6}Li (6.015 amu) and 70.00% ^{7}Li (7.016 amu). What is the average atomic mass of the sample?

25. An average dimension for the radius of an atom is 1.0×10^{-8} cm, and the average radius of the nucleus is 1.0×10^{-13} cm. Determine the ratio of atomic volume to nuclear volume. Assume that the atom is spherical [$V = (4/3)\pi r^3$ for a sphere].

*26. An aluminum atom has an average diameter of about 3.0×10^{-8} cm. The nucleus has a diameter of about 2.0×10^{-13} cm. Calculate the ratio of the atom's diameter to its nucleus.

Additional Exercises

All exercises with blue numbers have answers in the appendix of the text.

27. What experimental evidence supports these statements?
   (a) The nucleus of an atom is small.
   (b) The atom consists of both positive and negative charges.
   (c) The nucleus of the atom is positive.

28. What is the relationship between the following two atoms:
   (a) one atom with 10 protons, 10 neutrons, and 10 electrons; and another atom with 10 protons, 11 neutrons, and 10 electrons
   (b) one atom with 10 protons, 11 neutrons, and 10 electrons; and another atom with 11 protons, 10 neutrons, and 11 electrons

29. The radius of a carbon atom in many compounds is 0.77×10^{-8} cm. If the radius of a Styrofoam ball used to represent the carbon atom in a molecular model is 1.5 cm, how much of an enlargement is this?

30. How is it possible for there to be more than one kind of atom of the same element?

31. Which element contains the largest number of neutrons per atom: ^{210}Bi, ^{210}Po, ^{210}At, or ^{211}At?

*32. An unknown element Q has two known isotopes: ^{60}Q and ^{63}Q. If the average atomic mass is 61.5 amu, what are the relative percentages of the isotopes?

*33. The actual mass of one atom of an unknown isotope is 2.18×10^{-22} g. Calculate the atomic mass of this isotope.

34. The mass of an atom of argon is 6.63×10^{-24} g. How many atoms are in a 40.0-g sample of argon?

35. Using the periodic table inside the front cover of the book, determine which of the first 20 elements have isotopes that you would expect to have the same number of protons, neutrons, and electrons.

36. Complete the following table with the appropriate data for each isotope given (all are neutral atoms):

Element	Symbol	Atomic number	Mass number	Number of protons	Number of neutrons	Number of electrons
	^{36}Cl					
GOLD			197			
		56			79	
					20	18
			58	28		

37. Complete the following table with the appropriate data for each isotope given (all are neutral atoms):

Element	Symbol	Atomic number	Mass number	Number of protons	Number of neutrons	Number of electrons
	^{134}Xe					
SILVER			107			
				9		
		92			143	92
			41	19		

38. Draw diagrams similar to those shown in Figure 5.4 for the following ions:
(a) F^- (b) Pb^{2+} (c) S^{2-} (d) Al^{3+}

39. Draw pictures similar to those shown in Figure 5.7 for the following isotopes of oxygen:
(a) $^{16}_{8}O$ (b) $^{17}_{8}O$ (c) $^{18}_{8}O$

40. What percent of the total mass of one atom of each of the following elements comes from electrons?
(a) iron (mass of one atom = 9.274×10^{-23} g)
(b) nitrogen (mass of one atom = 2.326×10^{-23} g)
(c) carbon (mass of one atom = 1.994×10^{-23} g)
(d) potassium (mass of one atom = 6.493×10^{-23} g)

41. What percent of the total mass of one atom of each of the following elements comes from protons?
(a) sodium (mass of one atom = 3.818×10^{-23} g)
(b) oxygen (mass of one atom = 2.657×10^{-23} g)
(c) mercury (mass of one atom = 3.331×10^{-22} g)
(d) fluorine (mass of one atom = 3.155×10^{-23} g)

42. Figure 5.6 is a representation of the nuclear model of the atom. The area surrounding the nucleus is labeled as the electron region. What is the electron region?

Challenge Exercise

All exercises with blue numbers have answers in the appendix of the text.

***43.** You have discovered a new element and are trying to determine where on the periodic table it would fit. You decide to do a mass spectrometer analysis of the sample and discover that it contains three isotopes with masses of 270.51 amu, 271.23 amu, and 269.14 amu and relative abundances of 34.07%, 55.12%, and 10.81%, respectively. Sketch the mass spectrometer reading, determine the average atomic mass of the element, estimate its atomic number, and determine its approximate location on the periodic table.

Answers to Practice Exercises

5.1 5.98×10^{24} atoms

5.2
$^{16}_{8}O$ 8p, 8e$^-$, 8n
$^{17}_{8}O$ 8p, 8e$^-$, 9n
$^{18}_{8}O$ 8p, 8e$^-$, 10n

5.3

	protons	neutrons	electrons
(a)	8	8	8
(b)	35	45	35
(c)	92	143	92
(d)	29	35	29

5.4

(a)	atomic number	9
	mass number	19
	name	potassium
(b)	atomic number	24
	mass number	52
	name	chromium
(c)	atomic number	79
	mass number	179
	name	gold

5.5 107.9 amu

CHAPTER 6

Nomenclature of Inorganic Compounds

This exquisite chambered nautilus shell is formed from calcium carbonate, commonly called limestone.

Chapter Outline

6.1 Common and Systematic Names
6.2 Elements and Ions
6.3 Writing Formulas from Names of Ionic Compounds
6.4 Naming Binary Compounds
6.5 Naming Compounds Containing Polyatomic Ions
6.6 Acids

As children, we begin to communicate with other people in our lives by learning the names of objects around us. As we continue to develop, we learn to speak and use language to complete a wide variety of tasks. As we enter school, we begin to learn of other languages—the languages of mathematics, of other cultures, of computers. In each case, we begin by learning the names of the building blocks and then proceed to more abstract concepts. Chemistry has a language all its own—a whole new way of describing the objects so familiar to us in our daily lives. Once we learn the language, we can begin to understand the macroscopic and microscopic world of chemistry.

6.1 Common and Systematic Names

Water (H_2O) and ammonia (NH_3) are almost always referred to by their common names.

Chemical nomenclature is the system of names that chemists use to identify compounds. When a new substance is formulated, it must be named in order to distinguish it from all other substances (see Figure 6.1). In this chapter, we will restrict our discussion to the nomenclature of inorganic compounds—compounds that do not generally contain carbon.

Common names are arbitrary names that are not based on the chemical composition of compounds. Before chemistry was systematized, a substance was given a name that generally associated it with one of its outstanding physical or chemical properties. For example, *quicksilver* is a common name for mercury, and nitrous oxide (N_2O), used as an anesthetic in dentistry, has been called *laughing gas* because it induces laughter when inhaled. Water and ammonia are also common names because neither provides any information about the chemical composition of the compounds. If every substance were assigned a common name, the amount of memorization required to learn over 12 million names would be astronomical.

Common names have distinct limitations, but they remain in frequent use. Common names continue to be used because the systematic name is too long or too technical for everyday use. For example, calcium oxide (CaO) is called *lime* by plasterers; photographers refer to *hypo* rather than sodium thiosulfate ($Na_2S_2O_3$); and nutritionists use the name *vitamin* D_3, instead of 9,10-secocholesta-5,7,10(19)-trien-3-β-ol ($C_{27}H_{44}O$). Table 6.1 lists the common names, formulas, and systematic names of some familiar substances.

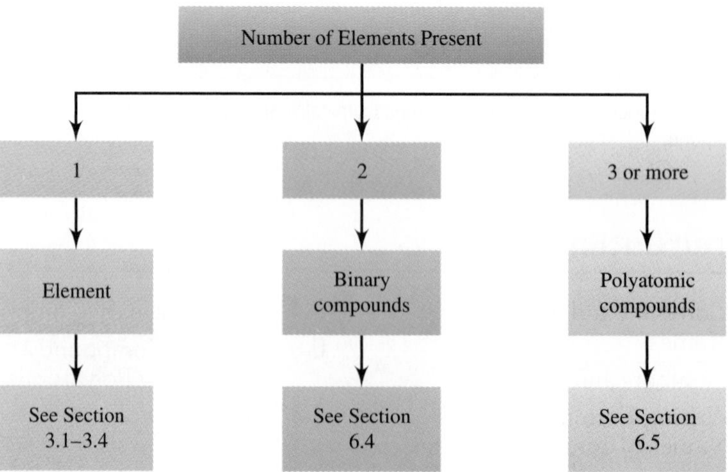

Figure 6.1
Where to find rules for naming inorganic substances in this book.

Table 6.1 Common Names, Formulas, and Chemical Names of Familiar Substances

Common names	Formula	Chemical name
Acetylene	C_2H_2	ethyne
Lime	CaO	calcium oxide
Slaked lime	$Ca(OH)_2$	calcium hydroxide
Water	H_2O	water
Galena	PbS	lead(II) sulfide
Alumina	Al_2O_3	aluminum oxide
Baking soda	$NaHCO_3$	sodium hydrogen carbonate
Cane or beet sugar	$C_{12}H_{22}O_{11}$	sucrose
Borax	$Na_2B_4O_7 \cdot 10\,H_2O$	sodium tetraborate decahydrate
Brimstone	S	sulfur
Calcite, marble, limestone	$CaCO_3$	calcium carbonate
Cream of tartar	$KHC_4H_4O_6$	potassium hydrogen tartrate
Epsom salts	$MgSO_4 \cdot 7\,H_2O$	magnesium sulfate heptahydrate
Gypsum	$CaSO_4 \cdot 2\,H_2O$	calcium sulfate dihydrate
Grain alcohol	C_2H_5OH	ethanol, ethyl alcohol
Hypo	$Na_2S_2O_3$	sodium thiosulfate
Laughing gas	N_2O	dinitrogen monoxide
Lye, caustic soda	$NaOH$	sodium hydroxide
Milk of magnesia	$Mg(OH)_2$	magnesium hydroxide
Muriatic acid	HCl	hydrochloric acid
Plaster of paris	$CaSO_4 \cdot \frac{1}{2}\,H_2O$	calcium sulfate hemihydrate
Potash	K_2CO_3	potassium carbonate
Pyrite (fool's gold)	FeS_2	iron disulfide
Quicksilver	Hg	mercury
Saltpeter (chile)	$NaNO_3$	sodium nitrate
Table salt	$NaCl$	sodium chloride
Vinegar	$HC_2H_3O_2$	acetic acid
Washing soda	$Na_2CO_3 \cdot 10\,H_2O$	sodium carbonate decahydrate
Wood alcohol	CH_3OH	methanol, methyl alcohol

Chemists prefer systematic names that precisely identify the chemical composition of chemical compounds. The system for inorganic nomenclature was devised by the International Union of Pure and Applied Chemistry (IUPAC), which was founded in 1921. The IUPAC meets regularly and constantly reviews and updates the system.

6.2 Elements and Ions

In Chapter 3, we studied the names and symbols for the elements as well as their location on the periodic table. In Chapter 5, we investigated the composition of the atom and learned that all atoms are composed of protons, electrons, and neutrons; that a particular element is defined by the number of protons it contains; and that atoms are uncharged because they contain equal numbers of protons and electrons.

CHEMISTRY IN ACTION • What's in a Name?

When a scientist discovered a new element in the early days of chemistry, he or she had the honor of naming it. Now researchers must submit their choices for a name to an international committee called the International Union of Pure and Applied Chemistry before they can be placed on the periodic table. In 1997, the IUPAC decided on names for the elements from 104 through 111. These eight elements are now called rutherfordium (Rf), dubnium (Db), seaborgium (Sg), bohrium (Bh), hassium (Hs), meitnerium (Mt), darmstadtium (Ds), and roentgenium (Rg).

The new names are a compromise among choices presented by different research teams. The Russians gained recognition for work done at a laboratory in Dubna. Americans gained recognition for Glenn Seaborg, the first living scientist to have an element named after him. The British recognized Ernest Rutherford, who discovered the atomic nucleus. The Germans won recognition both for Lise Meitner, who co-discovered atomic fission, and for one of their labs in the German state of Hesse. Both the Germans and the Russians won recognition for Niels Bohr, whose model of the atom led the way toward modern ideas about atomic structure. The German group that discovered elements 107–111 is recognized in the name darmstadtium, for their lab. Element 111 was named in honor of Wilhelm Roentgen, who discovered X-rays and won the first Nobel Prize in physics.

Niels Bohr

Lise Meitner

Ernest Rutherford

104	105	106	107	108	109	110	111
Rf	**Db**	**Sg**	**Bh**	**Hs**	**Mt**	**Ds**	**Rg**

Wilhelm Roentgen

Glenn Seaborg

The formula for most elements is simply the symbol of the element. In chemical reactions or mixtures, an element behaves as though it were a collection of individual particles. A small number of elements have formulas that are not single atoms at normal temperatures. Seven of the elements are *diatomic* molecules—that is, two atoms bonded together to form a molecule. These diatomic elements are hydrogen, H_2; oxygen, O_2; nitrogen, N_2; fluorine, F_2; chlorine, Cl_2; bromine, Br_2; and iodine, I_2. Two other elements that are commonly polyatomic are S_8, sulfur; and P_4, phosphorus.

Elements Occurring as Polyatomic Molecules

Hydrogen	H_2	Chlorine	Cl_2	Sulfur	S_8
Oxygen	O_2	Fluorine	F_2	Phosphorus	P_4
Nitrogen	N_2	Bromine	Br_2		
		Iodine	I_2		

We have learned that a charged particle, known as an *ion*, can be produced by adding or removing one or more electrons from a neutral atom. For example, potassium atoms contain 19 protons and 19 electrons. To make a potassium ion, we remove one electron, leaving 19 protons and only 18 electrons. This gives an ion with a positive one (+1) charge:

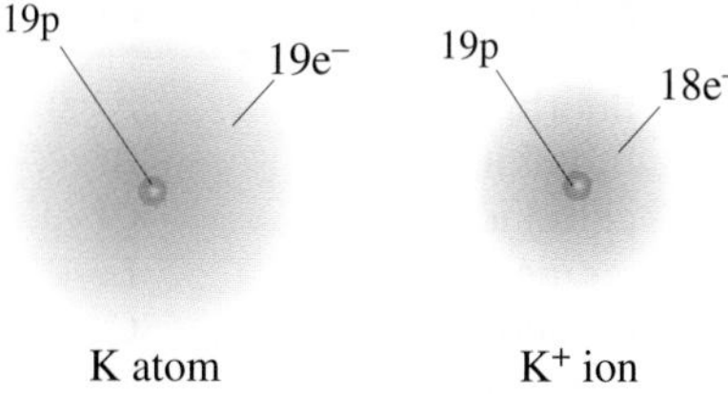

Written in the form of an equation, $K \rightarrow K^+ + e^-$. A positive ion is called a *cation*. Any neutral atom that *loses* an electron will form a cation. Sometimes an atom may lose one electron, as in the potassium example. Other atoms may lose more than one electron:

$$Mg \rightarrow Mg^{2+} + 2e^-$$

or

$$Al \rightarrow Al^{3+} + 3e^-$$

Cations are named the same as their parent atoms, as shown here:

Atom		Ion	
K	potassium	K^+	potassium ion
Mg	magnesium	Mg^{2+}	magnesium ion
Al	aluminum	Al^{3+}	aluminum ion

Ions can also be formed by adding electrons to a neutral atom. For example, the chlorine atom contains 17 protons and 17 electrons. The equal number of positive charges and negative charges results in a net charge of zero for the atom.

If one electron is added to the chlorine atom, it now contains 17 protons and 18 electrons, resulting in a net charge of negative one (−1) on the ion:

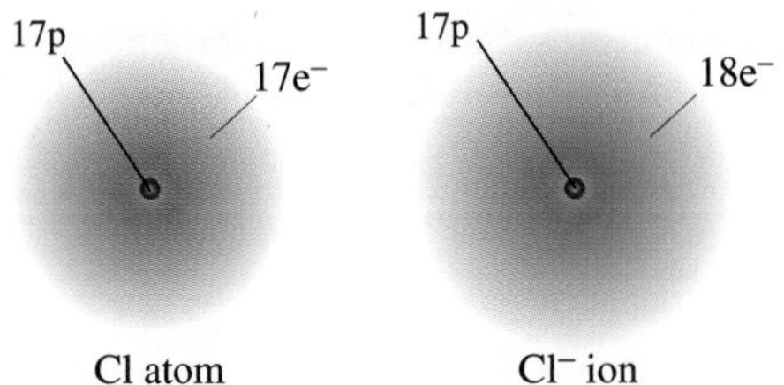

In a chemical equation, this process is summarized as $Cl + e^- \rightarrow Cl^-$. A negative ion is called an *anion*. Any neutral atom that *gains* an electron will form an anion. Atoms may gain more than one electron to form anions with different charges:

$$O + 2e^- \rightarrow O^{2-}$$

$$N + 3e^- \rightarrow N^{3-}$$

Anions are named differently from cations. To name an anion consisting of only one element, use the stem of the parent element name and change the ending to *-ide*. For example, the Cl^- ion is named by using the stem *chlor-* from chlorine and adding *-ide* to form chloride ion. Here are some examples:

Symbol	Name of atom	Ion	Name of ion
F	fluorine	F^-	fluoride ion
Br	bromine	Br^-	bromide ion
Cl	chlorine	Cl^-	chloride ion
I	iodine	I^-	iodide ion
O	oxygen	O^{2-}	oxide ion
N	nitrogen	N^{3-}	nitride ion

Ions are always formed by adding or removing electrons from an atom. Atoms do not form ions on their own. Most often ions are formed when metals combine with nonmetals.

The charge on an ion can often be predicted from the position of the element on the periodic table. Figure 6.2 shows the charges of selected ions from several groups on the periodic table. Notice that all the metals and hydrogen in the far left column (Group 1A) are (1+), all those in the next column (Group 2A) are (2+), and the metals in the next tall column (Group 3A) form (3+) ions. The elements in the lower center part of the table are called *transition metals*. These elements tend to form cations with various positive charges. There is no easy way to predict the charges on these cations. All metals lose electrons to form positive ions.

In contrast, the nonmetals form anions by gaining electrons. On the right side of the periodic table in Figure 6.2, you can see that the atoms in Group 7A form (1−) ions. The nonmetals in Group 6A form (2−) ions. It's important to learn the charges on the ions shown in Figure 6.2 and their relationship to the group number at the top of the column. For nontransition metals the charge is equal to the group number. For nonmetals, the charge is equal to the group number minus 8. We will learn more about why these ions carry their particular charges later in the course.

1A	2A											3A	4A	5A	6A	7A	
H^+																	
Li^+	Be^{2+}													N^{3-}	O^{2-}	F^-	
Na^+	Mg^{2+}											Al^{3+}		P^{3-}	S^{2-}	Cl^-	
K^+	Ca^{2+}				Cr^{2+} Cr^{3+}		Fe^{2+} Fe^{3+}			Cu^+ Cu^{2+}	Zn^{2+}					Br^-	
Rb^+	Sr^{2+}			Transition metals						Ag^+	Cd^{2+}					I^-	
Cs^+	Ba^{2+}																

Figure 6.2
Charges of selected ions in the periodic table.

6.3 Writing Formulas from Names of Ionic Compounds

In Chapters 3 and 5, we learned that compounds can be composed of ions. These substances are called *ionic compounds* and will conduct electricity when dissolved in water. An excellent example of an ionic compound is ordinary table salt. It's composed of crystals of sodium chloride. When dissolved in water, sodium chloride conducts electricity very well, as shown in Figure 6.3.

A chemical compound must have a net charge of zero. If it contains ions, the charges on the ions must add up to zero in the formula for the compound. This is relatively easy in the case of sodium chloride. The sodium ion (1+) and the chloride ion (1−) add to zero, resulting in the formula NaCl. Now consider an ionic compound containing calcium (Ca^{2+}) and fluoride (F^-) ions. How can we write a formula with a net charge of zero? To do this we need one Ca^{2+} and two F^- ions. The correct formula is CaF_2. The subscript 2 indicates that two fluoride ions are needed for each calcium ion. Aluminum oxide is a bit more complicated because it consists of Al^{3+} and O^{2-} ions. Since 6 is the least common multiple of 3 and 2, we have $2(3+) + 3(2-) = 0$, or a formula containing 2 Al^{3+} ions and 3 O^{2-} ions for Al_2O_3. Here are a few more examples of formula writing for ionic compounds:

Name of compound	Ions	Least common multiple	Sum of charges on ions	Formula
Sodium bromide	Na^+, Br^-	1	$(+1) + (-1) = 0$	NaBr
Potassium sulfide	K^+, S^{2-}	2	$2(+1) + (-2) = 0$	K_2S
Zinc sulfate	Zn^{2+}, SO_4^{2-}	2	$(+2) + (-2) = 0$	$ZnSO_4$
Ammonium phosphate	NH_4^+, PO_4^{3-}	3	$3(+1) + (-3) = 0$	$(NH_4)_3PO_4$
Aluminum chromate	Al^{3+}, CrO_4^{2-}	6	$2(+3) + 3(-2) = 0$	$Al_2(CrO_4)_3$

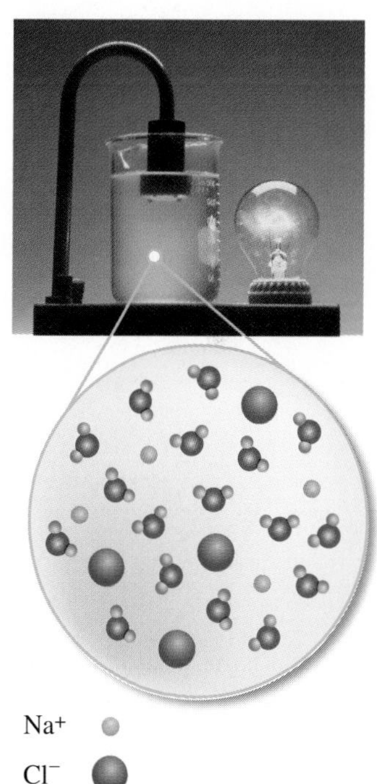

Figure 6.3
A solution of salt water contains Na^+ and Cl^- ions in addition to water molecules. The ions cause the solution to conduct electricity, lighting the bulb.

Example 6.1 Write formulas for (a) calcium chloride, (b) magnesium oxide, and (c) barium phosphide.

SOLUTION (a) Use the following steps for calcium chloride.

Step 1 From the name, we know that calcium chloride is composed of calcium and chloride ions. First write down the formulas of these ions:

Ca^{2+} Cl^{-}

Step 2 To write the formula of the compound, combine the smallest numbers of Ca^{2+} and Cl^{-} ions to give the charge sum equal to zero. In this case the lowest common multiple of the charges is 2:

$$(Ca^{2+}) + 2(Cl^{-}) = 0$$
$$(2+) \quad + 2(1-) = 0$$

Therefore the formula is $CaCl_2$.

(b) Use the same procedure for magnesium oxide:

Step 1 From the name, we know that magnesium oxide is composed of magnesium and oxide ions. First write down the formulas of these ions:

Mg^{2+} O^{2-}

Step 2 To write the formula of the compound, combine the smallest numbers of Mg^{2+} and O^{2-} ions to give the charge sum equal to zero:

$$(Mg^{2+}) + (O^{2-}) = 0$$
$$(2+) \quad + (2-) = 0$$

The formula is MgO.

(c) Use the same procedure for barium phosphide:

Step 1 From the name, we know that barium phosphide is composed of barium and phosphide ions. First write down the formulas of these ions:

Ba^{2+} P^{3-}

Step 2 To write the formula of this compound, combine the smallest numbers of Ba^{2+} and P^{3-} ions to give the charge sum equal to zero. In this case the lowest common multiple of the charges is 6:

$$3(Ba^{2+}) + 2(P^{3-}) = 0$$
$$3(2+) \quad + 2(3-) = 0$$

The formula is Ba_3P_2.

Practice 6.1

Write formulas for compounds containing the following ions:

(a) K^{+} and F^{-}
(b) Ca^{2+} and Br^{-}
(c) Mg^{2+} and N^{3-}
(d) Na^{+} and S^{2-}
(e) Ba^{2+} and O^{2-}
(f) As^{5+} and CO_3^{2-}

6.4 Naming Binary Compounds

Binary compounds contain only two different elements. Many binary compounds are formed when a metal combines with a nonmetal to form a *binary ionic compound*. The metal loses one or more electrons to become a cation while the nonmetal gains one or more electrons to become an anion. The cation is written first in the formula, followed by the anion.

Binary Ionic Compounds Containing a Metal Forming Only One Type of Cation

The chemical name is composed of the name of the metal followed by the name of the nonmetal, which has been modified to an identifying stem plus the suffix *-ide*.

For example, sodium chloride, NaCl, is composed of one atom each of sodium and chlorine. The name of the metal, sodium, is written first and is not modified. The second part of the name is derived from the nonmetal, chlorine, by using the stem *chlor-* and adding the ending *-ide*; it is named *chloride*. The compound name is sodium chloride.

To name these compounds:

1. **Write the name of the cation.**
2. **Write the stem for the anion with the suffix *-ide*.**

NaCl

Elements: Sodium (metal)
Chlorine (nonmetal)
name modified to the stem *chlor-* + *-ide*

Name of compound: Sodium chloride

Stems of the more common negative-ion-forming elements are shown in Table 6.2.

Compounds may contain more than one atom of the same element, but as long as they contain only two different elements and only one compound of these two elements exists, the name follows the rules for binary compounds:

Examples:

$CaBr_2$	Mg_3N_2	Li_2O
calcium bromide	magnesium nitride	lithium oxide

Table 6.2 Examples of Elements Forming Anions

Symbol	Element	Stem	Anion name
Br	bromine	brom	bromide
Cl	chlorine	chlor	chloride
F	fluorine	fluor	fluoride
H	hydrogen	hydr	hydride
I	iodine	iod	iodide
N	nitrogen	nitr	nitride
O	oxygen	ox	oxide
P	phosphorus	phosph	phosphide
S	sulfur	sulf	sulfide

Table 6.3 lists some compounds with names ending in *-ide*.

Table 6.3 Examples of Compounds with Names Ending in *-ide*

Formula	Name	Formula	Name
$AlCl_3$	aluminum chloride	BaS	barium sulfide
Al_2O_3	aluminum oxide	LiI	lithium iodide
CaC_2	calcium carbide	$MgBr_2$	magnesium bromide
HCl	hydrogen chloride	NaH	sodium hydride
HI	hydrogen iodide	Na_2O	sodium oxide

Example 6.2 Name the compound CaF_2.

SOLUTION From the formula it is a two-element compound and follows the rules for binary compounds.

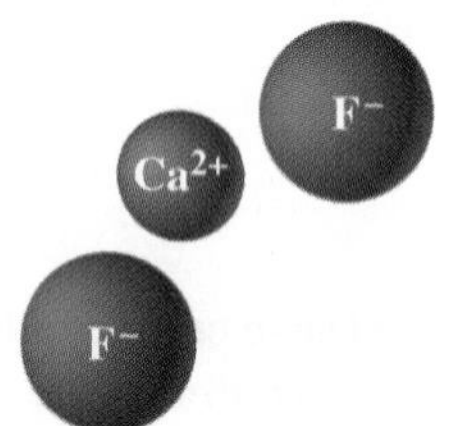

Step 1 The compound is composed of Ca, a metal, and F, a nonmetal. Elements in the 2A column of the periodic table form only one type of cation. Thus, we name the positive part of the compound *calcium*.

Step 2 Modify the name of the second element to the stem *fluor-* and add the binary ending *-ide* to form the name of the negative part, *fluoride*.

The name of the compound is therefore *calcium fluoride*.

Practice 6.2

Write formulas for these compounds:

(a) strontium chloride
(b) potassium iodide
(c) aluminum nitride
(d) calcium sulfide
(e) sodium oxide

Binary Ionic Compounds Containing a Metal That Can Form Two or More Types of Cations

The metals in the center of the periodic table (including the transition metals) often form more than one type of cation. For example, iron can form Fe^{2+} and Fe^{3+} ions, and copper can form Cu^{+} and Cu^{2+} ions. This can be confusing when you are naming compounds. For example, copper chloride could be $CuCl_2$ or CuCl. To resolve this difficulty the IUPAC devised a system, known as the **Stock System**, to name these compounds. This system is currently recognized as the official system to name these compounds, although another older system is sometimes used. In the Stock System, when a compound contains a metal that can form more than one type of cation, the charge on the cation of the metal is designated by a Roman numeral placed in parentheses immediately following the name of the metal. The negative element is treated in the usual manner for binary compounds:

Stock System

Cation charge	+1	+2	+3	+4	+5
Roman numeral	(I)	(II)	(III)	(IV)	(V)

Examples:

$FeCl_2$	iron(II) chloride	Fe^{2+}
$FeCl_3$	iron(III) chloride	Fe^{3+}
CuCl	copper(I) chloride	Cu^{+}
$CuCl_2$	copper(II) chloride	Cu^{2+}

To name compounds using the Stock System:

1. **Write the name of the cation.**
2. **Write the charge on the cation as a Roman numeral in parentheses.**
3. **Write the stem of the anion with the suffix *-ide*.**

The fact that $FeCl_2$ has two chloride ions, each with a −1 charge, establishes that the charge on Fe is +2. To distinguish between the two iron chlorides, $FeCl_2$ is named iron(II) chloride and $FeCl_3$ is named iron(III) chloride:

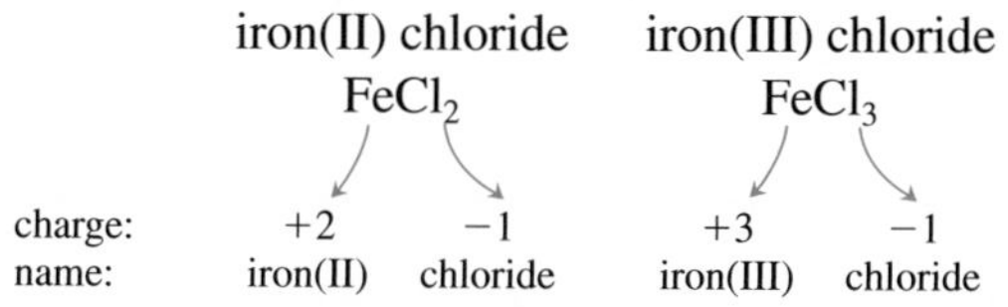

When a metal forms only one possible cation, we need not distinguish one cation from another, so Roman numerals are not needed. Thus we do not say - calcium(II) chloride for $CaCl_2$, but rather calcium chloride, since the charge of calcium is understood to be +2.

To name compounds using the Stock System, do the following:

1. Write the name of the cation.
2. Write the charge on the cation as a Roman numeral in parentheses.
3. Write the stem of the anion with the suffix *-ide*.

In classical nomenclature, when the metallic ion has only two cation types, the name of the metal (usually the Latin name) is modified with the suffixes *-ous* and *-ic* to distinguish between the two. The lower-charge cation is given the *-ous* ending, and the higher one, the *-ic* ending.

Examples:

$FeCl_2$	ferrous chloride	Fe^{2+}	(lower charge cation)
$FeCl_3$	ferric chloride	Fe^{3+}	(higher charge cation)
CuCl	cuprous chloride	Cu^{+}	(lower charge cation)
$CuCl_2$	cupric chloride	Cu^{2+}	(higher charge cation)

Table 6.4 lists some common metals that have more than one type of cation.

Notice that the *ous–ic* naming system does not give the charge of the cation of an element but merely indicates that at least two types of cations exist. The Stock System avoids any possible uncertainty by clearly stating the charge on the cation.

In this book we will use mainly the Stock System.

Table 6.4 Names and Charges of Some Common Metal Ions That Have More Than One Type of Cation

Formula	Stock System name	Classical name	Formula	Stock System name	Classical name
Cu^{1+}	copper(I)	cuprous	Sn^{4+}	tin(IV)	stannic
Cu^{2+}	copper(II)	cupric	Pb^{2+}	lead(II)	plumbous
$Hg^{1+} (Hg_2)^{2+}$	mercury(I)	mercurous	Pb^{4+}	lead(IV)	plumbic
Hg^{2+}	mercury(II)	mercuric	As^{3+}	arsenic(III)	arsenous
Fe^{2+}	iron(II)	ferrous	As^{5+}	arsenic(V)	arsenic
Fe^{3+}	iron(III)	ferric	Ti^{3+}	titanium(III)	titanous
Sn^{2+}	tin(II)	stannous	Ti^{4+}	titanium(IV)	titanic

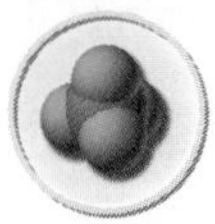

Example 6.3 Name the compound FeS.

SOLUTION This compound follows the rules for a binary compound.

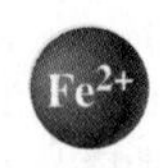

Steps 1,2 It is a compound of Fe, a metal, and S, a nonmetal, and Fe is a transition metal that has more than one type of cation. In sulfides, the charge on the S is −2. Therefore, the charge on Fe must be +2, and the name of the positive part of the compound is *iron(II)*.

Step 3 We have already determined that the name of the negative part of the compound will be *sulfide*.

The name of FeS is *iron(II) sulfide*.

Practice 6.3

Write the name for each of the following compounds using the Stock System:
(a) PbI_2 (b) SnF_4 (c) Fe_2O_3 (d) CuO

Practice 6.4

Write formulas for the following compounds:
(a) tin(IV) chromate
(b) chromium(III) bromide
(c) tin(II) fluoride
(d) copper(I) oxide

Binary Compounds Containing Two Nonmetals

Prefix	Number
mono	1
di	2
tri	3
tetra	4
penta	5
hexa	6
hepta	7
octa	8
nona	9
deca	10

Compounds between nonmetals are molecular, not ionic. Therefore, a different system for naming them is used. In a compound formed between two nonmetals, the element that occurs first in this series is written and named first:

Si, B, P, H, C, S, I, Br, N, Cl, O, F

The name of the second element retains the *-ide* ending as though it were an anion. A Latin or Greek prefix (*mono-*, *di-*, *tri-*, and so on) is attached to the name of each element to indicate the number of atoms of that element in the molecule. The prefix *mono-* is rarely used for naming the first element. Some common prefixes and their numerical equivalences are as shown in the margin table.

Here are some examples of compounds that illustrate this system:

N_2O_3

(di)nitrogen — Indicates two nitrogen atoms

(tri)oxide — Indicates three oxygen atoms

To name these compounds:

1. **Write the name for the first element using a prefix if there is more than one atom of this element.**
2. **Write the stem for the second element followed by the suffix *-ide*. Use a prefix to indicate the number of atoms for the second element.**

CO	carbon monoxide
CO_2	carbon dioxide
PCl_3	phosphorus trichloride
SO_2	sulfur dioxide
P_2O_5	diphosphorus pentoxide
CCl_4	carbon tetrachloride
N_2O	dinitrogen monoxide
N_2O_4	dinitrogen tetroxide
NO	nitrogen monoxide
N_2O_3	dinitrogen trioxide
S_2Cl_2	disulfur dichloride
S_2F_{10}	disulfur decafluoride

These examples illustrate that we sometimes drop the final *o* (mono) or *a* (penta) of the prefix when the second element is oxygen. This avoids creating a name that is awkward to pronounce. For example, CO is carbon monoxide instead of carbon monooxide.

Example 6.4

Name the compound PCl_5.

SOLUTION

Step 1 Phosphorus and chlorine are nonmetals, so the rules for naming binary compounds containing two nonmetals apply. Phosphorus is named first. Therefore the compound is a chloride.

Step 2 No prefix is needed for phosphorus because each molecule has only one atom of phosphorus. The prefix *penta-* is used with chloride to indicate the five chlorine atoms. (PCl_3 is also a known compound.)

Step 3 The name for PCl_5 is *phosphorus pentachloride.*

Phosphorus pentachloride

Practice 6.5

Name these compounds:
(a) Cl_2O (b) SO_2 (c) CBr_4 (d) N_2O_5 (e) NH_3 (f) ICl_3

Acids Derived from Binary Compounds

Certain binary hydrogen compounds, when dissolved in water, form solutions that have *acid* properties. Because of this property, these compounds are given acid names in addition to their regular *-ide* names. For example, HCl is a gas and is called *hydrogen chloride*, but its water solution is known as *hydrochloric acid*. Binary acids are composed of hydrogen and one other nonmetallic element. However, not all binary hydrogen compounds are acids. To express the formula of a binary acid, it's customary to write the symbol of hydrogen first, followed by the symbol of the second element (e.g., HCl, HBr, or H_2S). When we see formulas such as CH_4 or NH_3, we understand that these compounds are not normally considered to be acids.

To name a binary acid, place the prefix *hydro-* in front of, and the suffix *-ic* after, the stem of the nonmetal name. Then add the word *acid*:

To name binary acids:
1. **Write the prefix *hydro-* followed by the stem of the second element and add the suffix *-ic*.**
2. **Write the word *acid*.**

	HCl	H_2S
Examples:	*Hydro*-chlor-*ic acid*	*Hydro*-sulfur-*ic acid*
	(hydrochloric acid)	(hydrosulfuric acid)

Acids are hydrogen-containing substances that liberate hydrogen ions when dissolved in water. The same formula is often used to express binary hydrogen compounds, such as HCl, regardless of whether or not they are dissolved in water. Table 6.5 shows several examples of binary acids.

Table 6.5 Names and Formulas of Selected Binary Acids

Formula	Acid name	Formula	Acid name
HF	Hydrofluoric acid	HI	Hydroiodic acid
HCl	Hydrochloric acid	H_2S	Hydrosulfuric acid
HBr	Hydrobromic acid	H_2Se	Hydroselenic acid

Naming binary compounds is summarized in Figure 6.4.

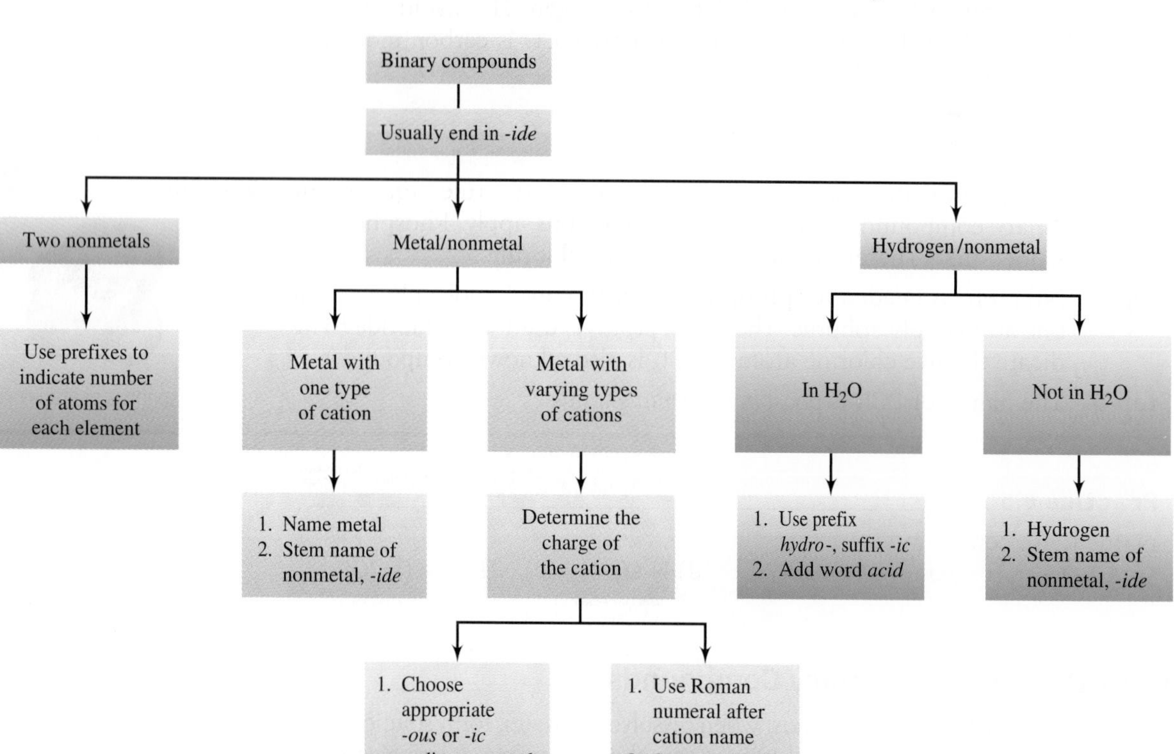

Figure 6.4
Flow diagram for naming binary compounds.

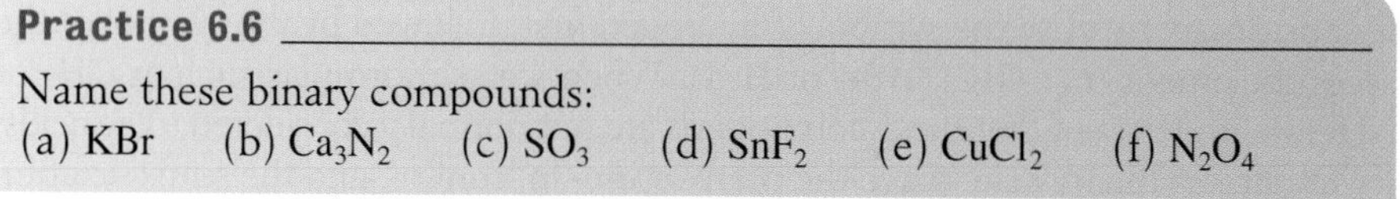

Practice 6.6

Name these binary compounds:
(a) KBr (b) Ca_3N_2 (c) SO_3 (d) SnF_2 (e) $CuCl_2$ (f) N_2O_4

6.5 Naming Compounds Containing Polyatomic Ions

polyatomic ion

A **polyatomic ion** is an ion that contains two or more elements. Compounds containing polyatomic ions are composed of three or more elements and usually consist of one or more cations combined with a negative polyatomic ion. In general, naming compounds containing polyatomic ions is similar to naming binary compounds. The cation is named first, followed by the name for the negative polyatomic ion.

rmanganate crystals

To name these compounds, you must learn to recognize the common polyatomic ions (Table 6.6) and know their charges. Consider the formula $KMnO_4$. You must be able to recognize that it consists of two parts $KMnO_4$. These parts are composed of a K^+ ion and a MnO_4^- ion. The correct name for this compound is potassium permanganate. Many polyatomic ions that contain oxygen are called *oxy-anions*, and generally have the suffix *-ate* or *-ite*. Unfortunately, the suffix doesn't indicate the number of oxygen atoms present. The *-ate* form

Table 6.6 Names, Formulas, and Charges of Some Common Polyatomic Ions

Name	Formula	Charge	Name	Formula	Charge
Acetate	$C_2H_3O_2^-$	−1	Cyanide	CN^-	−1
Ammonium	NH_4^+	+1	Dichromate	$Cr_2O_7^{2-}$	−2
Arsenate	AsO_4^{3-}	−3	Hydroxide	OH^-	−1
Hydrogen carbonate	HCO_3^-	−1	Nitrate	NO_3^-	−1
Hydrogen sulfate	HSO_4^-	−1	Nitrite	NO_2^-	−1
Bromate	BrO_3^-	−1	Permanganate	MnO_4^-	−1
Carbonate	CO_3^{2-}	−2	Phosphate	PO_4^{3-}	−3
Chlorate	ClO_3^-	−1	Sulfate	SO_4^{2-}	−2
Chromate	CrO_4^{2-}	−2	Sulfite	SO_3^{2-}	−2

More ions are listed on the back endpapers.

contains more oxygen atoms than the *-ite* form. Examples include sulfate (SO_4^{2-}), sulfite (SO_3^{2-}), nitrate (NO_3^-), and nitrite (NO_2^-).

Some elements form more than two different polyatomic ions containing oxygen. To name these ions, prefixes are used in addition to the suffix. To indicate more oxygen than in the *-ate* form, we add the prefix *per-*, which is a short form of *hyper-*, meaning "more." The prefix *hypo-*, meaning "less" (oxygen in this case), is used for the ion containing less oxygen than the *-ite* form. An example of this system is shown for the polyatomic ions containing chlorine and oxygen in Table 6.7. The prefixes are also used with other similar ions, such as iodate (IO_3^-), bromate (BrO_3^-), and phosphate (PO_4^{3-}).

To name these compounds:
1. **Write the name of the cation.**
2. **Write the name of the anion.**

Table 6.7 Oxy-Anions and Oxy-Acids of Chlorine

Anion formula	Anion name	Acid formula	Acid name
ClO^-	*hypo*chlor*ite*	HClO	*hypo*chlor*ous* acid
ClO_2^-	chlor*ite*	$HClO_2$	chlor*ous* acid
ClO_3^-	chlor*ate*	$HClO_3$	chlor*ic* acid
ClO_4^-	*per*chlor*ate*	$HClO_4$	*per*chlor*ic* acid

Only four of the common negatively charged polyatomic ions do not use the *ate–ite* system. These exceptions are hydroxide (OH^-), hydrogen sulfide (HS^-), peroxide (O_2^{2-}), and cyanide (CN^-). Care must be taken with these, as their endings can easily be confused with the *-ide* ending for binary compounds (Section 6.4).

There are three common positively charged polyatomic ions as well—the ammonium, the mercury(I) (Hg_2^{2+}), and the hydronium(H_3O^+) ions. The ammonium ion (NH_4^+) is frequently found in polyatomic compounds (Section 6.5), whereas the hydronium ion (H_3O^+) is usually associated with aqueous solutions of acids (Chapter 15).

Practice 6.7

Name these compounds:

(a) $NaNO_3$ (b) $Ca_3(PO_4)_2$ (c) KOH (d) Li_2CO_3 (e) $NaClO_3$

Inorganic compounds are also formed from more than three elements (see Table 6.8). In these cases, one or more of the ions is often a polyatomic ion.

Table 6.8 Names of Selected Compounds That Contain More Than One Kind of Positive Ion

Formula	Name of compound
$KHSO_4$	potassium hydrogen sulfate
$Ca(HSO_3)_2$	calcium hydrogen sulfite
NH_4HS	ammonium hydrogen sulfide
$MgNH_4PO_4$	magnesium ammonium phosphate
NaH_2PO_4	sodium dihydrogen phosphate
Na_2HPO_4	sodium hydrogen phosphate
KHC_2O_4	potassium hydrogen oxalate
$KAl(SO_4)_2$	potassium aluminum sulfate
$Al(HCO_3)_3$	aluminum hydrogen carbonate

Once you have learned to recognize the polyatomic ions, naming these compounds follows the patterns we have already learned. First identify the ions. Name the cations in the order given, and follow them with the names of the anions. Study the following examples:

Compound	Ions	Name
$NaHCO_3$	Na^+; HCO_3^-	sodium hydrogen carbonate
$NaHS$	Na^+; HS^-	sodium hydrogen sulfide
$MgNH_4PO_4$	Mg^{2+}; NH_4^+; PO_4^{3-}	magnesium ammonium phosphate
$NaKSO_4$	Na^+; K^+; SO_4^{2-}	sodium potassium sulfate

6.6 Acids

While we will learn much more about acids later (see Chapter 15), it is helpful to be able to recognize and name common acids both in the laboratory and in class. The simplest way to recognize many acids is to know that acid formulas often begin with hydrogen. Naming binary acids was covered in Section 6.4. Inorganic compounds containing hydrogen, oxygen, and one other element are called *oxy-acids*.

The first step in naming these acids is to determine that the compound in question is really an oxy-acid. The keys to identification are (1) hydrogen is the first element in the compound's formula and (2) the second part of the formula consists of a polyatomic ion containing oxygen.

Hydrogen in an oxy-acid is not specifically designated in the acid name. The presence of hydrogen in the compound is indicated by the use of the word *acid* in the name of the substance. To determine the particular type of acid, the polyatomic ion following hydrogen must be examined. The name of the polyatomic ion is modified in the following manner: (1) *-ate* changes to an *-ic* ending; (2) *-ite* changes to an *-ous* ending. (See Table 6.9.) The compound with the *-ic* ending contains more oxygen than the one with the *-ous* ending.

Table 6.9 Comparison of Acid and Anion Names for Selected Oxy-Acids

Acid	Anion	Acid	Anion	Acid	Anion
H_2SO_4 Sulfuric acid	SO_4^{2-} Sulfate ion	H_2CO_3 Carbonic acid	CO_3^{2-} Carbonate ion	HIO_3 Iodic acid	IO_3^- Iodate ion
H_2SO_3 Sulfurous acid	SO_3^{2-} Sulfite ion	H_3BO_3 Boric acid	BO_3^{3-} Borate ion	$HC_2H_3O_2$ Acetic acid	$C_2H_3O_2^-$ Acetate ion
HNO_3 Nitric acid	NO_3^- Nitrate ion	H_3PO_4 Phosphoric acid	PO_4^{3-} Phosphate ion	$H_2C_2O_4$ Oxalic acid	$C_2O_4^{2-}$ Oxalate ion
HNO_2 Nitrous acid	NO_2^- Nitrite ion	H_3PO_3 Phosphorous acid	PO_3^{3-} Phosphite ion	$HBrO_3$ Bromic acid	BrO_3^- Bromate ion

Consider these examples:

SO_4^{2-} sulf*ate* $\longrightarrow$ H_2SO_4 sulfur*ic* acid (contains four oxygens)
SO_3^{2-} sulf*ite* $\longrightarrow$ H_2SO_3 sulfur*ous* acid (three oxygens)
NO_3^- nitr*ate* $\longrightarrow$ HNO_3 nitr*ic* acid (three oxygens)
NO_2^- nitr*ite* $\longrightarrow$ HNO_2 nitr*ous* acid (two oxygens)

The complete system for naming oxy-acids is shown in Table 6.7 for the various oxy-acids containing chlorine.

Naming polyatomic compounds is summarized in Figure 6.5. We have now looked at ways of naming a variety of inorganic compounds—binary compounds consisting of a metal and a nonmetal and of two nonmetals, binary acids, and polyatomic compounds. These compounds are just a small part of the classified chemical compounds. Most of the remaining classes are in the broad field of organic chemistry under such categories as hydrocarbons, alcohols, ethers, aldehydes, ketones, phenols, and carboxylic acids.

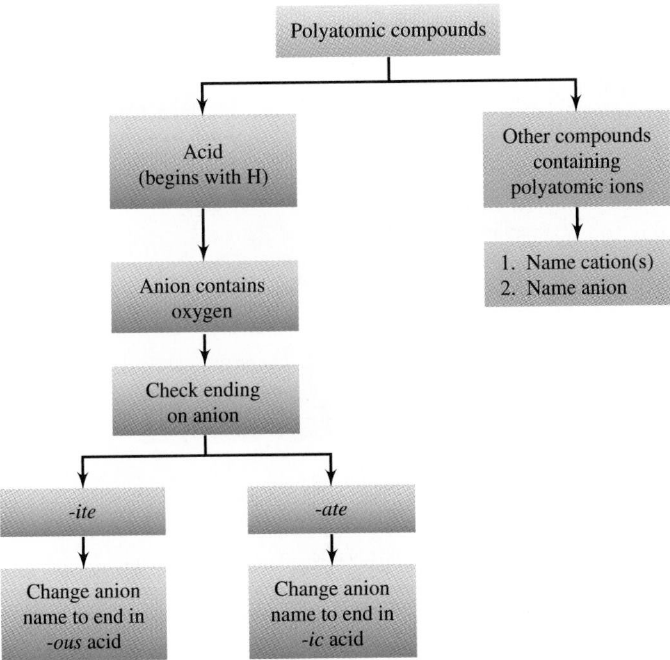

Figure 6.5
Flow diagram for naming polyatomic compounds.

Practice 6.8

Name these compounds:

(a) Cu_2CO_3 (b) $Fe(ClO)_3$ (c) $Sn(C_2H_3O_2)_2$ (d) $HBrO_3$ (e) HBr

Practice 6.9

Write formulas for

(a) lead(II) nitrate
(b) potassium phosphate
(c) mercury(II) cyanide
(d) ammonium chromate

Chapter 6 Review

6.1 Common and Systematic Names

- Common names are arbitrary and do not describe the chemical composition of compounds.
  - Examples—ammonia, water
- A standard system of nomenclature was devised and is maintained by the IUPAC.

6.2 Elements and Ions

- Diatomic elements—H_2, O_2, N_2, F_2, Cl_2, Br_2, I_2.
- Polyatomic elements—P_4, S_8.
- Ions form through the addition or loss of electrons from neutral atoms.
- The charge on ions can often be predicted from the periodic table.

6.3 Writing Formulas from the Names of Ionic Compounds

- Compounds must have a net charge of zero.
- To write a formula from a name:
  - Identify and write the symbols for the elements in the compound.
  - Combine the smallest number of ions required to produce a net charge of zero.

6.4 Naming Binary Compounds

KEY TERM

Stock System

- The rules for naming binary compounds can be summarized in the chart below:

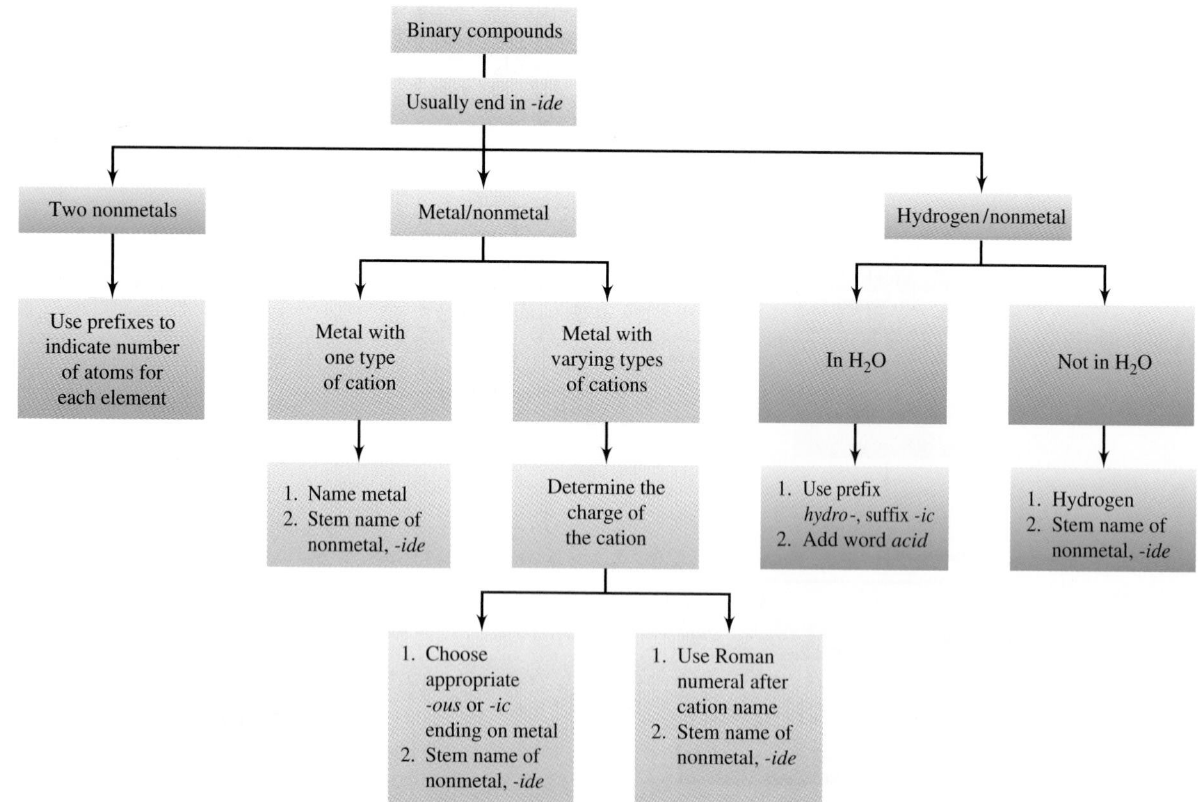

6.5 Naming Compounds Containing Polyatomic Ions

KEY TERM

Polyatomic ion

- The rules for naming these compounds are included in the chart for naming acids below.

6.6 Acids

- The rules for naming acids can be summarized in the chart below:

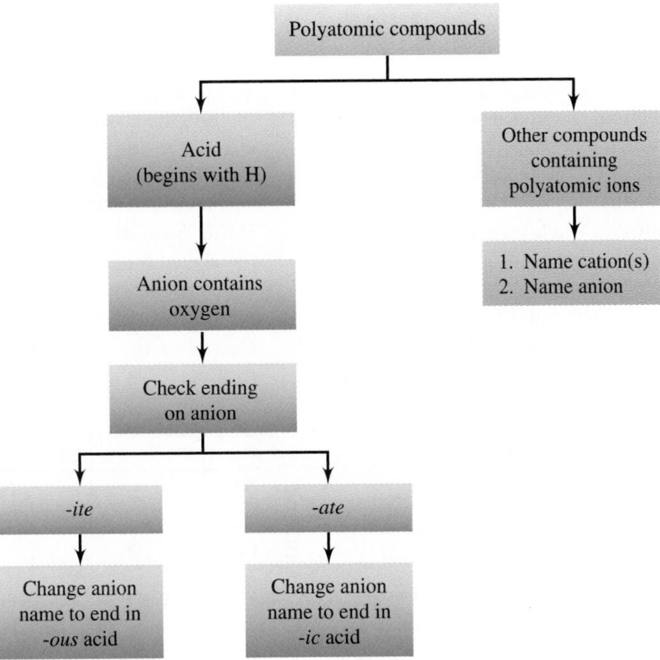

Review Questions

All questions with blue numbers have answers in the appendix of the text.

1. Use the common ion table on the back endpapers of your text to determine the formulas for compounds composed of the following ions:
   (a) sodium and chlorate
   (b) hydrogen and sulfate
   (c) tin(II) and acetate
   (d) copper(I) and oxide
   (e) zinc and hydrogen carbonate
   (f) iron(III) and carbonate
2. Does the fact that two elements combine in a one-to-one atomic ratio mean that the charges on their ions are both 1? Explain.
3. Write the names and formulas for the four oxy-acids containing (a) bromine, (b) iodine. (Table 6.7)
4. Explain why N_2O_5 is called dinitrogen pentoxide.
5. Write the formulas for the compounds formed when a chromium(III) ion is combined with the following anions:
   (a) hydroxide
   (b) nitrate
   (c) nitrite
   (d) hydrogen carbonate
   (e) carbonate
   (f) dichromate
   (g) phosphate
   (h) oxalate
   (i) oxide
   (j) fluoride
   (See Table of Names, Formulas, and Charges of Common Ions on the back endpapers.)
6. Explain why the name for $MgCl_2$ is magnesium chloride but the name for $CuCl_2$ is copper(II) chloride.
7. Where are each of the following located in the periodic table?
   (a) metals
   (b) nonmetals
   (c) transition metals
8. Name two compounds almost always referred to by their common names. What would be their systematic names?

Paired Exercises

All exercises with blue numbers have answers in the appendix of the text.

1. Write the formula of the compound that would be formed between these elements:
(a) Na and I (d) K and S
(b) Ba and F (e) Cs and Cl
(c) Al and O (f) Sr and Br

2. Write the formula of the compound that would be formed between these elements:
(a) Ba and O (d) Be and Br
(b) H and S (e) Li and Se
(c) Al and Cl (f) Mg and P

3. Write formulas for the following cations (don't forget to include the charges): sodium, magnesium, aluminum, copper(II), iron(II), iron(III), lead(II), silver, cobalt(II), barium, hydrogen, mercury(II), tin(II), chromium(III), tin(IV), manganese(II), bismuth(III).

4. Write formulas for the following anions (don't forget to include the charges): chloride, bromide, fluoride, iodide, cyanide, oxide, hydroxide, sulfide, sulfate, hydrogen sulfate, hydrogen sulfite, chromate, carbonate, hydrogen carbonate, acetate, chlorate, permanganate, oxalate.

5. Write the systematic names for the following:
(a) laughing gas (N_2O) (d) galena (PbS)
(b) calcite ($CaCO_3$) (e) muriatic acid (HCl)
(c) alumina (Al_2O_3) (f) table salt (NaCl)

6. Write the systematic names for the following:
(a) slaked lime ($Ca(OH)_2$) (d) baking soda ($NaHCO_3$)
(b) saltpeter ($NaNO_3$) (e) pyrite (FeS_2)
(c) brimstone (S) (f) potash (K_2CO_3)

7. Complete the table, filling in each box with the proper formula.

Cations \ Anions	Br^-	O^{2-}	NO_3^-	PO_4^{3-}	CO_3^{2-}
K^+	KBr				
Mg^{2+}					
Al^{3+}					
Zn^{2+}				$Zn_3(PO_4)_2$	
H^+					

8. Complete the table, filling in each box with the proper formula.

Cations \ Anions	SO_4^{2-}	OH^-	AsO_4^{3-}	$C_2H_3O_2^-$	CrO_4^{2-}
NH_4^+			$(NH_4)_3AsO_4$		
Ca^{2+}					
Fe^{3+}	$Fe_2(SO_4)_3$				
Ag^+					
Cu^{2+}					

9. Write the names of each of the compounds formed in Question 7.

10. Write the names of each of the compounds formed in Question 8.

11. Write formulas for these binary compounds, all of which are composed of nonmetals:
(a) carbon monoxide (f) dinitrogen pentoxide
(b) sulfur trioxide (g) iodine monobromide
(c) carbon tetrabromide (h) silicon tetrachloride
(d) phosphorus trichloride (i) phosphorus pentaiodide
(e) nitrogen dioxide (j) diboron trioxide

12. Write formulas for these compounds:
(a) sodium nitrate (e) silver carbonate
(b) magnesium fluoride (f) calcium phosphate
(c) barium hydroxide (g) potassium nitrite
(d) ammonium sulfate (h) strontium oxide

13. Name these binary compounds, all of which are composed of nonmetals:
(a) CO_2 (f) N_2O_4
(b) N_2O (g) P_2O_5
(c) PCl_5 (h) OF_2
(d) CCl_4 (i) NF_3
(e) SO_2 (j) CS_2

14. Name these compounds:
(a) K_2O (e) Na_3PO_4
(b) NH_4Br (f) Al_2O_3
(c) CaI_2 (g) $Zn(NO_3)_2$
(d) $BaCO_3$ (h) Ag_2SO_4

15. Name these compounds by the Stock (IUPAC) System:
(a) $CuCl_2$ (d) $FeCl_3$
(b) CuBr (e) SnF_2
(c) $Fe(NO_3)_2$ (f) $HgCO_3$

16. Write formulas for these compounds:
(a) tin(IV) bromide (d) mercury(II) nitrite
(b) copper(I) sulfate (e) titanium(IV) sulfide
(c) iron(III) carbonate (f) iron(II) acetate

17. Write formulas for these acids:
(a) hydrochloric acid (d) carbonic acid
(b) chloric acid (e) sulfurous acid
(c) nitric acid (f) phosphoric acid

18. Write formulas for these acids:
(a) acetic acid (d) boric acid
(b) hydrofluoric acid (e) nitrous acid
(c) hydrosulfuric acid (f) hypochlorous acid

19. Name these acids:
(a) HNO_2 (d) HBr (g) HF
(b) H_2SO_4 (e) H_3PO_3 (h) $HBrO_3$
(c) $H_2C_2O_4$ (f) $HC_2H_3O_2$

20. Name these acids:
(a) H_3PO_4 (d) HCl (g) HI
(b) H_2CO_3 (e) HClO (h) $HClO_4$
(c) HIO_3 (f) HNO_3

21. Write formulas for these compounds:
(a) silver sulfite
(b) cobalt(II) bromide
(c) tin(II) hydroxide
(d) aluminum sulfate
(e) manganese(II) fluoride
(f) ammonium carbonate
(g) chromium(III) oxide
(h) copper(II) chloride
(i) potassium permanganate
(j) barium nitrite
(k) sodium peroxide
(l) iron(II) sulfate
(m) potassium dichromate
(n) bismuth(III) chromate

22. Write formulas for these compounds:
(a) sodium chromate
(b) magnesium hydride
(c) nickel(II) acetate
(d) calcium chlorate
(e) lead(II) nitrate
(f) potassium dihydrogen phosphate
(g) manganese(II) hydroxide
(h) cobalt(II) hydrogen carbonate
(i) sodium hypochlorite
(j) arsenic(V) carbonate
(k) chromium(III) sulfite
(l) antimony(III) sulfate
(m) sodium oxalate
(n) potassium thiocyanate

23. Write the name of each compound:
(a) $ZnSO_4$
(b) $HgCl_2$
(c) $CuCO_3$
(d) $Cd(NO_3)_2$
(e) $Al(C_2H_3O_2)_3$
(f) CoF_2
(g) $Cr(ClO_3)_3$
(h) Ag_3PO_4
(i) NiS
(j) $BaCrO_4$

24. Write the name of each compound:
(a) $Ca(HSO_4)_2$
(b) $As_2(SO_3)_3$
(c) $Sn(NO_2)_2$
(d) $FeBr_3$
(e) $KHCO_3$
(f) $BiAsO_4$
(g) $Fe(BrO_3)_2$
(h) $(NH_4)_2HPO_4$
(i) NaClO
(j) $KMnO_4$

25. Write the chemical formula for these substances:
(a) baking soda
(b) lime
(c) Epsom salts
(d) muriatic acid
(e) vinegar
(f) potash
(g) lye

26. Write the chemical formula for these substances:
(a) fool's gold
(b) saltpeter
(c) limestone
(d) cane sugar
(e) milk of magnesia
(f) washing soda
(g) grain alcohol

Additional Exercises

All exercises with blue numbers have answers in the appendix of the text.

27. Write equations similar to those found in Section 6.2 for the formation of:
(a) potassium ion
(b) iodide ion
(c) bromide ion
(d) iron(II) ion
(e) calcium ion
(f) oxide ion

28. Name each of the following polyatomic ions:

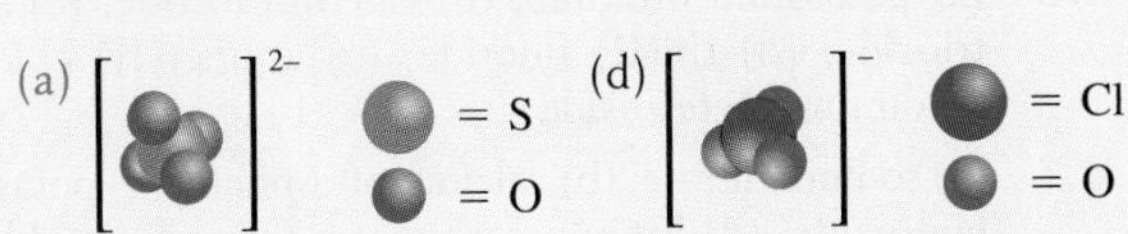

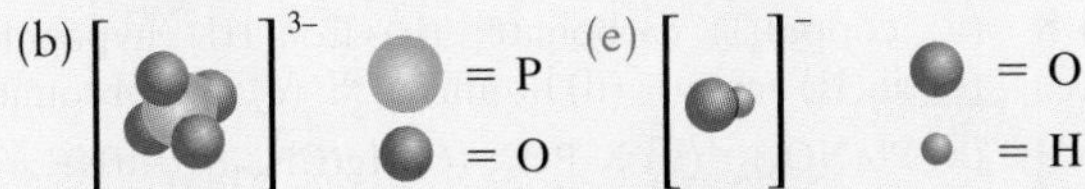

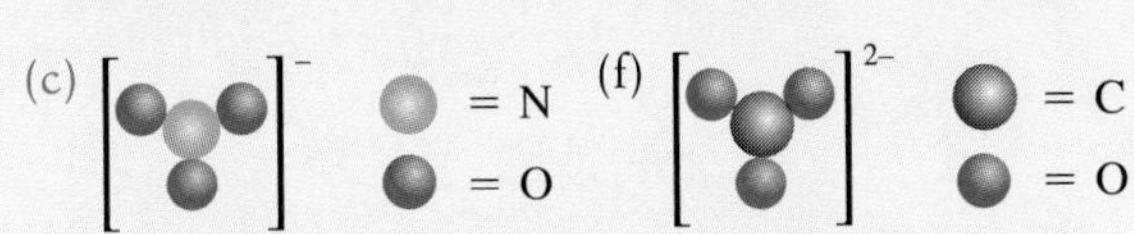

29. Write formulas for all possible compounds formed between the calcium ion and the anions shown in Question 28.

30. Write formulas for all possible compounds formed between the potassium ion and the anions shown in Question 28.

31. Write the formula and name for each of the following compounds:

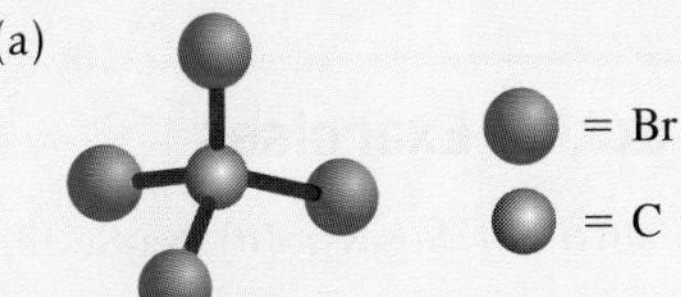

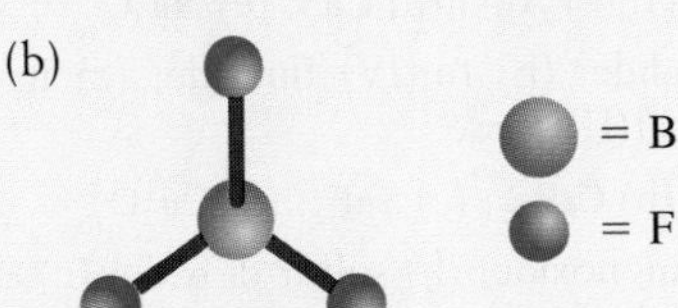

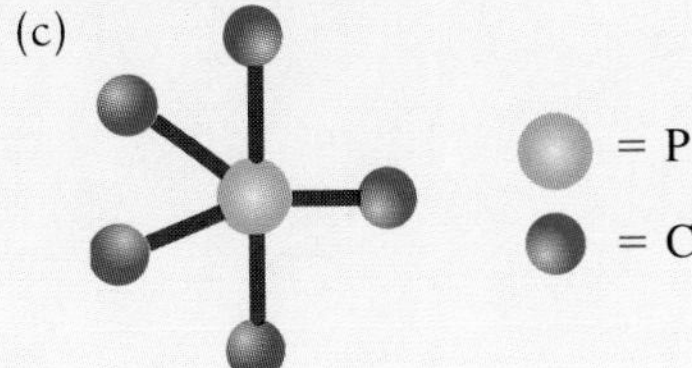

32. Write the formula and name for the following compound:

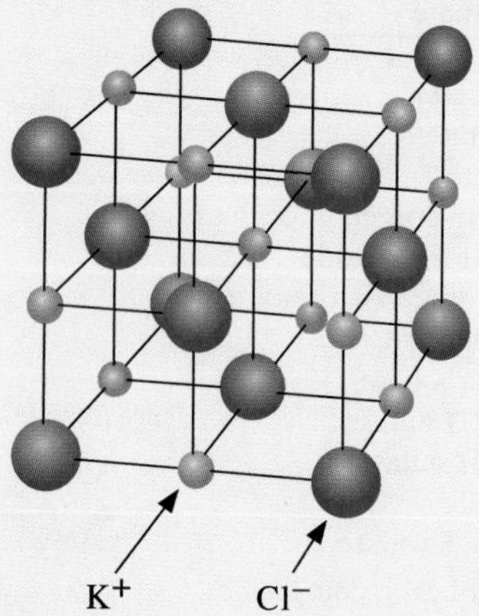

33. State how each of the following is used in naming inorganic compounds: *ide, ous, ic, hypo, per, ite, ate,* Roman numerals.

34. Translate the following sentences into unbalanced formula chemical equations:
(a) Silver nitrate and sodium chloride react to form silver chloride and sodium nitrate.
(b) Iron(III) sulfate and calcium hydroxide react to form iron(III) hydroxide and calcium sulfate.
(c) Potassium hydroxide and sulfuric acid react to form potassium sulfate and water.

35. How many of each type of subatomic particle (protons and electrons) is in
(a) an atom of tin?
(b) a Sn^{2+} ion?
(c) a Sn^{4+} ion?

36. The compound X_2Y_3 is a stable solid. What ionic charge do you expect for X and Y? Explain.

37. The ferricyanide ion has the formula $Fe(CN)_6^{3-}$. Write the formula for the compounds that ferricyanide would form with the cations of elements 3, 13, and 30.

38. Compare and contrast the formulas of
(a) nitride with nitrite
(b) nitrite with nitrate
(c) nitrous acid with nitric acid

Challenge Exercise

All exercises with blue numbers have answers in the appendix of the text.

*39. The compound X_2Y has a mass of 110.27 g. The ratio of X:Y (by mass) in the compound is 2.44:1. What is the compound?

Answers to Practice Exercises

6.1 (a) KF; (b) $CaBr_2$; (c) Mg_3N_2; (d) Na_2S; (e) BaO; (f) $As_2(CO_3)_5$

6.2 (a) $SrCl_2$; (b) KI; (c) AlN; (d) CaS; (e) Na_2O

6.3 (a) lead(II) iodide; (b) tin(IV) fluoride; (c) iron(III) oxide; (d) copper(II) oxide

6.4 (a) $Sn(CrO_4)_2$; (b) $CrBr_3$; (c) SnF_2; (d) Cu_2O

6.5 (a) dichlorine monoxide; (b) sulfur dioxide; (c) carbon tetrabromide; (d) dinitrogen pentoxide; (e) ammonia; (f) iodine trichloride

6.6 (a) potassium bromide; (b) calcium nitride; (c) sulfur trioxide; (d) tin(II) fluoride; (e) copper(II) chloride; (f) dinitrogen tetroxide

6.7 (a) sodium nitrate; (b) calcium phosphate; (c) potassium hydroxide; (d) lithium carbonate; (e) sodium chlorate

6.8 (a) copper(I) carbonate; (b) iron(III) hypochlorite; (c) tin(II) acetate; (d) bromic acid; (e) hydrobromic acid

6.9 (a) $Pb(NO_3)_2$; (b) K_3PO_4; (c) $Hg(CN)_2$; (d) $(NH_4)_2CrO_4$

PUTTING IT TOGETHER Review for Chapters 5–6

Multiple Choice:

Choose the correct answer to each of the following.

1. The concept of positive charge and a small, "heavy" nucleus surrounded by electrons was the contribution of
(a) Dalton (c) Thomson
(b) Rutherford (d) Chadwick
2. The neutron was discovered in 1932 by
(a) Dalton (c) Thomson
(b) Rutherford (d) Chadwick
3. An atom of atomic number 53 and mass number 127 contains how many neutrons?
(a) 53 (c) 127
(b) 74 (d) 180
4. How many electrons are in an atom of $^{40}_{18}Ar$?
(a) 20 (c) 40
(b) 22 (d) no correct answer given
5. The number of neutrons in an atom of $^{139}_{56}Ba$ is
(a) 56 (c) 139
(b) 83 (d) no correct answer given
6. The name of the isotope containing one proton and two neutrons is
(a) protium (c) deuterium
(b) tritium (d) helium
7. Each atom of a specific element has the same
(a) number of protons (c) number of neutrons
(b) atomic mass (d) no correct answer given
8. Which pair of symbols represents isotopes?
(a) $^{23}_{11}Na$ and $^{23}_{12}Na$ (c) $^{63}_{29}Cu$ and $^{29}_{64}Cu$
(b) $^{7}_{3}Li$ and $^{6}_{3}Li$ (d) $^{12}_{24}Mg$ and $^{12}_{26}Mg$
9. Two naturally occurring isotopes of an element have masses and abundance as follows: 54.00 amu (20.00%) and 56.00 amu (80.00%). What is the relative atomic mass of the element?
(a) 54.20 (c) 54.80
(b) 54.40 (d) 55.60
10. Substance X has 13 protons, 14 neutrons, and 10 electrons. Determine its identity.
(a) ^{27}Mg (c) $^{27}Al^{3+}$
(b) ^{27}Ne (d) ^{27}Al
11. The mass of a chlorine atom is 5.90×10^{-23} g. How many atoms are in a 42.0-g sample of chlorine?
(a) 2.48×10^{-21} (c) 1.40×10^{-24}
(b) 7.12×10^{23} (d) no correct answer given
12. The number of neutrons in an atom of $^{108}_{47}Ag$ is
(a) 47 (c) 155
(b) 108 (d) no correct answer given
13. The number of electrons in an atom of $^{27}_{13}Al$ is
(a) 13 (c) 27
(b) 14 (d) 40
14. The number of protons in an atom of $^{65}_{30}Zn$ is
(a) 65 (c) 30
(b) 35 (d) 95
15. The number of electrons in the nucleus of an atom of $^{24}_{12}Mg$ is
(a) 12 (c) 36
(b) 24 (d) no correct answer given

Names and Formulas:

In which of the following is the formula correct for the name given?

1. copper(II) sulfate, $CuSO_4$
2. ammonium hydroxide, NH_4OH
3. mercury(I) carbonate, $HgCO_3$
4. phosphorus triiodide, PI_3
5. calcium acetate, $Ca(C_2H_3O_2)_2$
6. hypochlorous acid, HClO
7. dichlorine heptoxide, Cl_2O_7
8. magnesium iodide, MgI
9. sulfurous acid, H_2SO_3
10. potassium manganate, $KMnO_4$
11. lead(II) chromate, $PbCrO_4$
12. ammonium hydrogen carbonate, NH_4HCO_3
13. iron(II) phosphate, $FePO_4$
14. calcium hydrogen sulfate, $CaHSO_4$
15. mercury(II) sulfate, $HgSO_4$
16. dinitrogen pentoxide, N_2O_5
17. sodium hypochlorite, NaClO
18. sodium dichromate, $Na_2Cr_2O_7$
19. cadmium cyanide, $Cd(CN)_2$
20. bismuth(III) oxide, Bi_3O_2
21. carbonic acid, H_2CO_3
22. silver oxide, Ag_2O
23. ferric iodide, FeI_2
24. tin(II) fluoride, TiF_2
25. carbon monoxide, CO
26. phosphoric acid, H_3PO_3
27. sodium bromate, Na_2BrO_3
28. hydrosulfuric acid, H_2S
29. potassium hydroxide, POH
30. sodium carbonate, Na_2CO_3
31. zinc sulfate, $ZnSO_3$
32. sulfur trioxide, SO_3

33. tin(IV) nitrate, $Sn(NO_3)_4$
34. ferrous sulfate, $FeSO_4$
35. chloric acid, HCl
36. aluminum sulfide, Al_2S_3
37. cobalt(II) chloride, $CoCl_2$
38. acetic acid, $HC_2H_3O_2$
39. zinc oxide, ZnO_2
40. stannous fluoride, SnF_2

Free Response Questions:

Answer each of the following. Be sure to include your work and explanations in a clear, logical form.

1. (a) What is an ion?
   (b) The average mass of a calcium atom is 40.08 amu. Why do we also use 40.08 amu as the average mass of a calcium ion (Ca^{2+})?
2. Congratulations! You discover a new element you name wyzzlebium (Wz). The average atomic mass of Wz was found to be 303.001 amu and its atomic number is 120.
   (a) If the masses of the two isotopes of wyzzlebium are 300.9326 amu and 303.9303 amu, what is the relative abundance of each isotope?
   (b) What are the isotopic notations of the two isotopes? (e.g., ${}^{A}_{Z}Wz$)
   (c) How many neutrons are in one atom of the more abundant isotope?
3. How many protons are in one molecule of dichlorine heptoxide? Is it possible to determine precisely how many electrons and neutrons are in a molecule of dichlorine heptoxide? Why or why not?
4. An unidentified metal forms an ionic compound with phosphate. The metal forms a 2+ cation. If the minimum ratio of protons in the metal to the phosphorus is 6:5, what metal is it? (*Hint:* First write the formula for the ionic compound formed with phosphate anion.)
5. For each of the following compounds, indicate what is wrong with the name and why. If possible, fix the name.
   (a) iron hydroxide
   (b) dipotassium dichromium heptoxide
   (c) sulfur oxide
6. Sulfur dioxide is a gas formed as a by-product of burning coal. Sulfur trioxide is a significant contributor to acid rain. Does the existence of these two substances violate the law of multiple proportions? Explain.
7. (a) Which subatomic particles are not in the nucleus?
   (b) What happens to the size of an atom when it becomes an anion?
   (c) What do an ion of Ca and an atom of Ar have in common?
8. An unidentified atom is found to have an atomic mass 7.18 times that of the carbon-12 isotope.
   (a) What is the mass of the unidentified atom?
   (b) What are the possible identities of this atom?
   (c) Why are you unable to positively identify the element based on the atomic mass and the periodic table?
   (d) If the element formed a compound M_2O, where M is the unidentified atom, identify M by writing the isotopic notation for the atom.
9. Scientists such as Dalton, Thomson, and Rutherford proposed important models, which were ultimately challenged by later technology. What do we know to be false in Dalton's atomic model? What was missing in Thomson's model of the atom? What was Rutherford's experiment that led to the current model of the atom?

CHAPTER 7

Quantitative Composition of Compounds

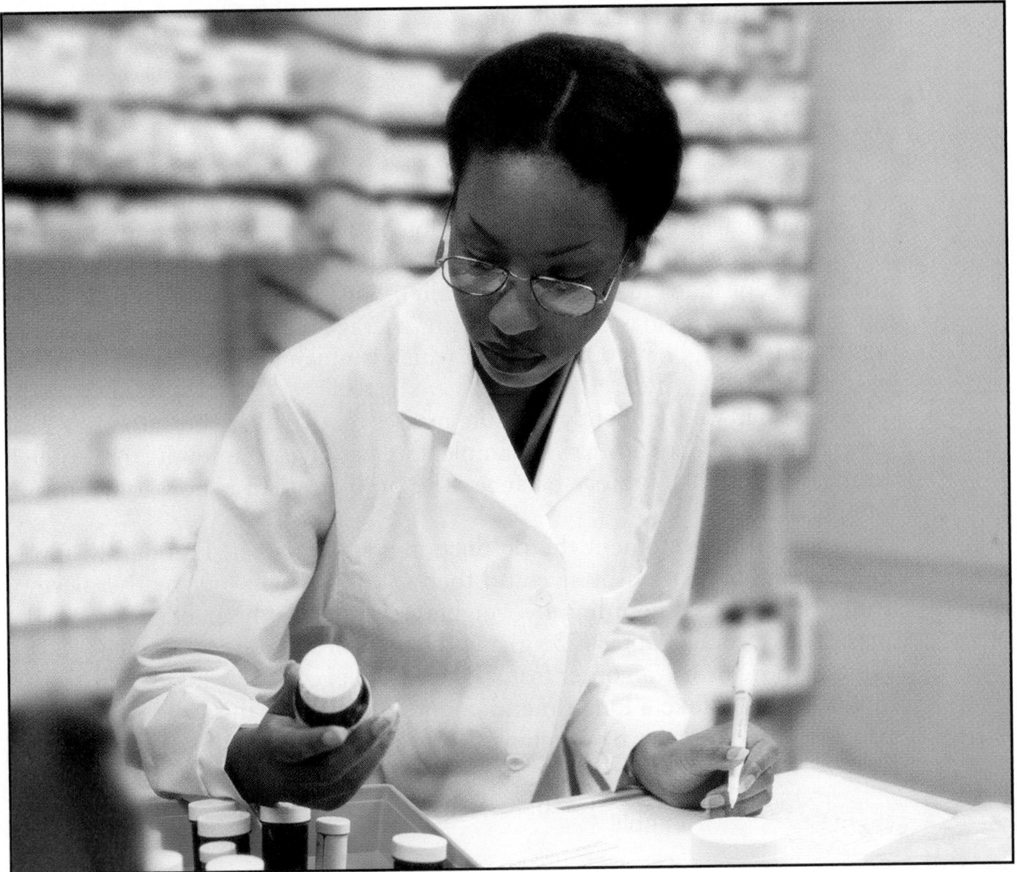

A compounding pharmacist adjusts the quantities of the components for a prescription to precisely meet the needs of his patients.

Chapter Outline

Cereals, cleaning products, and pain remedies all list their ingredients on the package label. The ingredients are listed in order from most to least, but the amounts are rarely given. However, it is precisely these amounts that give products their desired properties and distinguish them from the competition. Understandably, manufacturers carefully regulate the amounts of ingredients to maintain quality and hopefully their customers' loyalty. In the medicines we purchase, these quantities are especially important for safety reasons—for example, they determine whether a medicine is given to children or is safe only for adults.

The composition of compounds is an important concept in chemistry. Determining numerical relationships among the elements in compounds and measuring exact quantities of particles are fundamental tasks that chemists routinely perform in their daily work.

7.1 The Mole

The atom is an incredibly tiny object. Its mass is far too small to measure on an ordinary balance. In Chapter 5 (Section 5.9), we learned to compare atoms using a table of atomic mass units. These units are valuable when we compare the masses of individual atoms (mentally), but they have no practical use in the laboratory. The mass in grams for an "average" carbon atom (atomic mass 12.01 amu) would be 2.00×10^{-23} g, which is much too tiny for the best laboratory balance.

So how can we confidently measure masses for these very tiny atoms? We increase the number of atoms in a sample until we have an amount large enough to measure on a laboratory balance. The problem then is how to count our sample of atoms.

Consider for a moment the produce in a supermarket. Frequently apples and oranges are sorted by size and then sold by weight, not by the piece of fruit. The grocer is counting by weighing. To do this, he needs to know the mass of an "average" apple (235 g) and the mass of an "average" orange (186 g). Now suppose he has an order from the local college for 275 apples and 350 oranges. It would take a long time to count and package this order. The grocer can quickly count fruit by weighing. For example,

$$(275 \text{ apples})\left(\frac{235 \text{ g}}{\text{apple}}\right) = 6.46 \times 10^4 \text{ g} = 64.6 \text{ kg}$$

$$(350 \text{ oranges})\left(\frac{186 \text{ g}}{\text{orange}}\right) = 6.51 \times 10^4 \text{ g} = 65.1 \text{ kg}$$

Oranges can be "counted" by weighing them in the store.

He can now weigh 64.6 kg of apples and 65.1 kg of oranges and pack them without actually counting them. Manufacturers and suppliers often count by weighing. Other examples of counting by weighing include nuts, bolts, and candy.

Chemists also count atoms by weighing. We know the "average" masses of atoms, so we can count atoms by defining a unit to represent a larger number of atoms. Chemists have chosen the mole (mol) as the unit for counting atoms. The mole is a unit for counting just as a dozen or a ream or a gross is used to count:

$$1 \text{ dozen} = 12 \text{ objects}$$
$$1 \text{ ream} = 500 \text{ objects}$$
$$1 \text{ gross} = 144 \text{ objects}$$
$$1 \text{ mole} = 6.022 \times 10^{23} \text{ objects}$$

Note that we use a unit only when it is appropriate. A dozen eggs is practical in our kitchen, a gross might be practical for a restaurant, but a ream of eggs would not be very practical. Chemists can't use dozens, grosses, or reams because atoms are so tiny that a dozen, gross, or ream of atoms still couldn't be measured in the laboratory.

Avogadro's number

The number represented by 1 mol, 6.022×10^{23}, is called **Avogadro's number**, in honor of Amadeo Avogadro (1776–1856), who investigated several quantitative aspects in chemistry. It's difficult to imagine how large Avogadro's number really is, but this example may help: If 10,000 people started to count Avogadro's number, and each counted at the rate of 100 numbers per minute each minute of the day, it would take them over 1 trillion (10^{12}) years to count the total number. So even the tiniest amount of matter contains extremely large numbers of atoms.

Remember that ^{12}C is the reference isotope for atomic masses.

Avogadro's number has been experimentally determined by several methods. How does it relate to atomic mass units? Remember that the atomic mass for an element is the average relative mass of all the isotopes for the element. The atomic mass (expressed in grams) of 1 mole of any element contains the same number of particles (Avogadro's number) as there are in exactly 12 g of ^{12}C. Thus, 1 **mole** of anything is the amount of the substance that contains the same number of items as there are atoms in exactly 12 g of ^{12}C.

mole

$$1 \text{ mole} = 6.022 \times 10^{23} \text{ items}$$

From the definition of mole, we can see that the atomic mass in grams of any element contains 1 mol of atoms. The term *mole* is so commonplace in chemistry that chemists use it as freely as the words *atom* or *molecule*. A mole of atoms, molecules, ions, or electrons represents Avogadro's number of these particles. If we can speak of a mole of atoms, we can also speak of a mole of

(a)

(b)

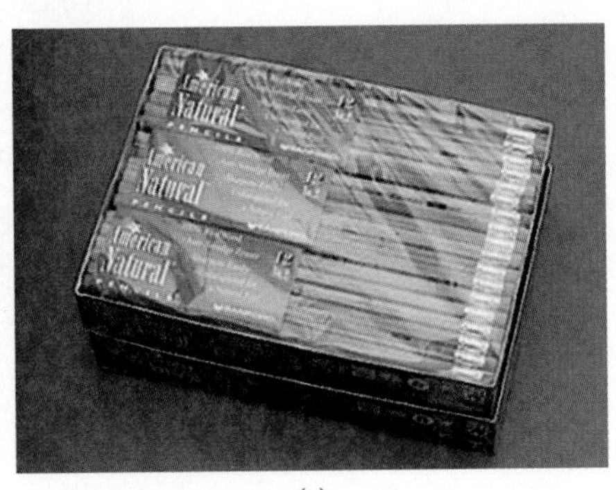

(c)

Units of measurement need to be appropriate for the object being measured. (a) Eggs are measured by the dozen, (b) paper is measured by the ream (500 sheets), and (c) pencils are measured by the gross (144).

molecules, a mole of electrons, or a mole of ions, understanding that in each case we mean 6.022×10^{23} particles:

1 mol of atoms = 6.022×10^{23} atoms
1 mol of molecules = 6.022×10^{23} molecules
1 mol of ions = 6.022×10^{23} ions

Practice 7.1

How many atoms are in 1.000 mole of
(a) Fe (b) H_2 (c) H_2SO_4

molar mass

The atomic mass of an element in grams contains Avogadro's number of atoms and is defined as the **molar mass** of that element. To determine molar mass, we change the units of the atomic mass (found in the periodic table) from atomic mass units to grams. For example, sulfur has an atomic mass of 32.07 amu, so 1 mol of sulfur has a molar mass of 32.07 g and contains 6.022×10^{23} atoms of sulfur. Here are some other examples:

Element	Atomic mass	Molar mass	Number of atoms
H	1.008 amu	1.008 g	6.022×10^{23}
Mg	24.31 amu	24.31 g	6.022×10^{23}
Na	22.99 amu	22.99 g	6.022×10^{23}

To summarize:

In this text molar masses of elements are given to 4 significant figures.

1. The atomic mass expressed in grams is the *molar mass* of an element. It is different for each element. In this text, molar masses are expressed to four significant figures.

   1 molar mass = atomic mass of an element in grams

2. One mole of any element contains Avogadro's number of atoms.

   1 mol of atoms = 6.022×10^{23} atoms

We can use these relationships to make conversions between number of atoms, mass, and moles, as shown in the following examples.

Example 7.1 How many moles of iron does 25.0 g of iron (Fe) represent?

SOLUTION We need to change grams of Fe to moles of Fe. The atomic mass of Fe (from either the periodic table or the table of atomic masses) is 55.85. Use the proper conversion factor to obtain moles:

$$\text{grams Fe} \longrightarrow \text{moles Fe} \qquad (\text{grams Fe})\left(\frac{1 \text{ mol Fe}}{55.85 \text{ g Fe}}\right)$$

$$(25.0 \cancel{\text{g Fe}})\left(\frac{1 \text{ mol Fe}}{55.85 \cancel{\text{g Fe}}}\right) = 0.448 \text{ mol Fe}$$

Example 7.2

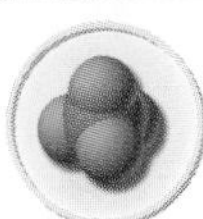

How many magnesium atoms are contained in 5.00 g of Mg?

SOLUTION

We need to change grams of Mg to atoms of Mg:

grams Mg $\longrightarrow$ atoms Mg

We find the atomic mass of magnesium to be 24.31 and set up the calculation using a conversion factor between atoms and grams:

$$(\text{grams Mg})\left(\frac{6.022 \times 10^{23}\ \text{atoms Mg}}{24.31\ \text{g Mg}}\right)$$

$$(5.00\ \cancel{\text{g Mg}})\left(\frac{6.022 \times 10^{23}\ \text{atoms Mg}}{24.31\ \cancel{\text{g Mg}}}\right) = 1.24 \times 10^{23}\ \text{atoms Mg}$$

Alternatively, we could first convert grams of Mg to moles of Mg, which are then changed to atoms of Mg:

grams Mg $\longrightarrow$ moles Mg $\longrightarrow$ atoms Mg

Use conversion factors for each step. The calculation setup is

$$(5.00\ \cancel{\text{g Mg}})\left(\frac{1\ \cancel{\text{mol Mg}}}{24.31\ \cancel{\text{g Mg}}}\right)\left(\frac{6.022 \times 10^{23}\ \text{atoms Mg}}{1\ \cancel{\text{mol Mg}}}\right) = 1.24 \times 10^{23}\ \text{atoms Mg}$$

Thus, 1.24×10^{23} atoms of Mg are contained in 5.00 g of Mg.

Example 7.3

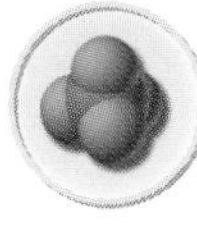

What is the mass, in grams, of one atom of carbon (C)?

SOLUTION

The molar mass of C is 12.01 g. We need to change atoms of C to grams of C:

atoms C $\longrightarrow$ grams C

Our conversion will be

$$(\text{atoms C})\left(\frac{12.01\ \text{g C}}{6.022 \times 10^{23}\ \text{atoms C}}\right)$$

$$(1\ \cancel{\text{atom C}})\left(\frac{12.01\ \text{g C}}{6.022 \times 10^{23}\ \cancel{\text{atoms C}}}\right) = 1.994 \times 10^{-23}\ \text{g C}$$

Example 7.4

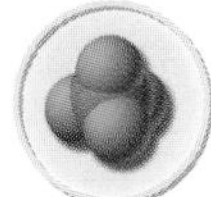

What is the mass of 3.01×10^{23} atoms of sodium (Na)?

SOLUTION

We need the molar mass of Na (22.99 g) and a conversion factor between molar mass and atoms:

atoms Na $\longrightarrow$ grams Na

$$(\text{atoms Na})\left(\frac{22.99\ \text{g Na}}{6.022 \times 10^{23}\ \text{atoms Na}}\right)$$

$$(3.01 \times 10^{23}\ \cancel{\text{atoms Na}})\left(\frac{22.99\ \text{g Na}}{6.022 \times 10^{23}\ \cancel{\text{atoms Na}}}\right) = 11.5\ \text{g Na}$$

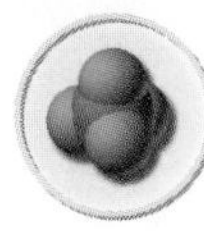

Example 7.5 What is the mass of 0.252 mol of copper (Cu)?

SOLUTION We need the molar mass of Cu (63.55 g) and a conversion factor between molar mass and moles:

moles Cu $\longrightarrow$ grams Cu

$$(\text{moles Cu})\left(\frac{1 \text{ molar mass Cu}}{1 \text{ mol Cu}}\right)$$

$$(0.252 \cancel{\text{mol Cu}})\left(\frac{63.55 \text{ g Cu}}{1 \cancel{\text{mol Cu}}}\right) = 16.0 \text{ g Cu}$$

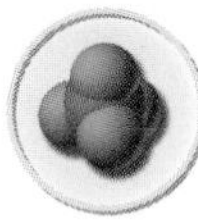

Example 7.6 How many oxygen atoms are present in 1.00 mol of oxygen molecules?

SOLUTION Oxygen is a diatomic molecule with the formula O_2. Therefore, a molecule of oxygen contains two oxygen atoms:

$$\frac{2 \text{ atoms O}}{1 \text{ molecule } O_2}$$

The sequence of conversions is

moles O_2 $\longrightarrow$ molecules O_2 $\longrightarrow$ atoms O

The two conversion factors needed are

$$\frac{6.022 \times 10^{23} \text{ molecules } O_2}{1 \text{ mol } O_2} \quad \text{and} \quad \frac{2 \text{ atoms O}}{1 \text{ molecule } O_2}$$

The calculation is

$$(1.00 \cancel{\text{mol } O_2})\left(\frac{6.022 \times 10^{23} \cancel{\text{molecules } O_2}}{1 \cancel{\text{mol } O_2}}\right)\left(\frac{2 \text{ atoms O}}{1 \cancel{\text{molecules } O_2}}\right)$$
$$= 1.20 \times 10^{24} \text{ atoms O}$$

Practice 7.2

What is the mass of 2.50 mol of helium (He)?

Practice 7.3

How many atoms are present in 0.025 mol of iron?

7.2 Molar Mass of Compounds

A formula unit is indicated by the formula, e.g., Mg, MgS, H_2O, NaCl.

One mole of a compound contains Avogadro's number of *formula units* of that compound. The terms *molecular weight*, *molecular mass*, *formula weight*, and *formula mass* have been used in the past to refer to the mass of 1 mol of a compound. However, the term *molar mass* is more inclusive, because it can be used for all types of compounds.

If the formula of a compound is known, its molar mass can be determined by adding the molar masses of all the atoms in the formula. If more than one atom of any element is present, its mass must be added as many times as it appears in the compound.

Example 7.7

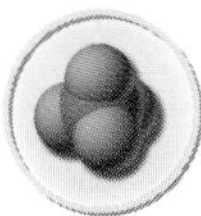

The formula for water is H_2O. What is its molar mass?

SOLUTION

First we look up the molar masses of H (1.008 g) and O (16.00 g); then we add the masses of all the atoms in the formula unit. Water contains two H atoms and one O atom. Thus

$$\begin{aligned} 2\,H &= 2(1.008\,g) = 2.016\,g \\ 1\,O &= 1(16.00\,g) = \underline{16.00\ \ g} \\ & \qquad\qquad\quad 18.02\ \ g = \text{molar mass of } H_2O \end{aligned}$$

Example 7.8

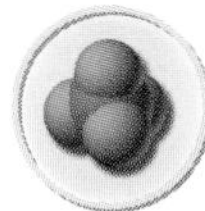

Calculate the molar mass of calcium hydroxide ($Ca(OH)_2$).

SOLUTION

The formula of this substance contains one atom of Ca and two atoms each of O and H. We proceed as in Example 7.7:

$$\begin{aligned} 1\,Ca &= 1(40.08\,g) = 40.08\ \ g \\ 2\,O &= 2(16.00\,g) = 32.00\ \ g \\ 2\,H &= 2(1.008\,g) = \underline{\ 2.016\,g} \\ & \qquad\qquad\quad 74.10\ \ g = \text{molar mass of } Ca(OH)_2 \end{aligned}$$

Practice 7.4

Calculate the molar mass of KNO_3.

In this text we round all molar masses to four significant figures, although you may need to use a different number of significant figures for other work (in the lab).

The mass of 1 mol of a compound contains Avogadro's number of formula units. Consider the compound hydrogen chloride (HCl). One atom of H combines with one atom of Cl to form HCl. When 1 mol of H (1.008 g of H or 6.022×10^{23} H atoms) combines with 1 mol of Cl (35.45 g of Cl or 6.022×10^{23} Cl atoms), 1 mol of HCl (36.46 g of HCl or 6.022×10^{23} HCl

A mole of table salt (in front of a salt shaker) and a mole of water (in the film container) have different sizes but both contain Avogadro's number of formula units.

molecules) is produced. These relationships are summarized in the following table:

H	Cl	HCl
6.022×10^{23} H *atoms*	6.022×10^{23} Cl *atoms*	6.022×10^{23} HCl *molecules*
1 mol H *atoms*	1 mol Cl *atoms*	1 mol HCl *molecules*
1.008 g H	35.45 g Cl	36.46 g HCl
1 molar mass H *atoms*	1 molar mass Cl *atoms*	1 molar mass HCl

In dealing with diatomic elements (H_2, O_2, N_2, F_2, Cl_2, Br_2, and I_2), we must take special care to distinguish between a mole of atoms and a mole of molecules. For example, consider 1 mol of oxygen molecules, which has a mass of 32.00 g. This quantity is equal to 2 mol of oxygen atoms. Remember that 1 mol represents Avogadro's number of the particular chemical entity that is under consideration:

Formula units are often used in place of molecules for substances that contain ions.

$$
\begin{aligned}
1\ \text{mol}\ H_2O &= 18.02\ \text{g}\ H_2O &&= 6.022 \times 10^{23}\ \text{molecules} \\
1\ \text{mol NaCl} &= 58.44\ \text{g NaCl} &&= 6.022 \times 10^{23}\ \text{formula units} \\
1\ \text{mol}\ H_2 &= 2.016\ \text{g}\ H_2 &&= 6.022 \times 10^{23}\ \text{molecules} \\
1\ \text{mol}\ HNO_3 &= 63.02\ \text{g}\ HNO_3 &&= 6.022 \times 10^{23}\ \text{molecules} \\
1\ \text{mol}\ K_2SO_4 &= 174.3\ \text{g}\ K_2SO_4 &&= 6.022 \times 10^{23}\ \text{formula units}
\end{aligned}
$$

Create the appropriate conversion factor by placing the unit desired in the numerator and the unit to be eliminated in the denominator.

$$
\begin{aligned}
1\ \text{mol} &= 6.022 \times 10^{23}\ \text{formula units or molecules} \\
&= 1\ \text{molar mass of a compound}
\end{aligned}
$$

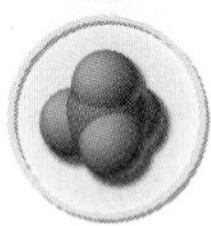

Example 7.9 What is the mass of 1 mol of sulfuric acid (H_2SO_4)?

SOLUTION We look up the masses of hydrogen, sulfur, and oxygen and solve in a manner similar to Examples 7.7 and 7.8:

$$
\begin{aligned}
2\,\text{H} &= 2(1.008\ \text{g}) = 2.016\ \text{g} \\
1\,\text{S} &= 1(32.07\ \text{g}) = 32.07\ \text{g} \\
4\,\text{O} &= 4(16.00\ \text{g}) = \underline{64.00\ \text{g}} \\
& \qquad\qquad\quad 98.09\ \text{g} = \text{mass of 1 mol of } H_2SO_4
\end{aligned}
$$

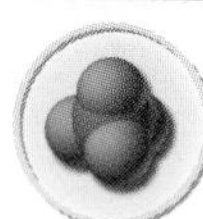

Example 7.10 How many moles of sodium hydroxide (NaOH) are there in 1.00 kg of sodium hydroxide?

SOLUTION First we know that

$$\text{molar mass} = (\overset{\text{Na}}{22.99\ \text{g}} + \overset{\text{O}}{16.00\ \text{g}} + \overset{\text{H}}{1.008\ \text{g}})\ \text{or}\ 40.00\ \text{g NaOH}$$

To convert grams to moles we use the conversion factor $\frac{1\ \text{mol NaOH}}{40.00\ \text{g NaOH}}$ and this conversion sequence:

$$\text{kg NaOH} \longrightarrow \text{g NaOH} \longrightarrow \text{mol NaOH}$$

The calculation is

$$(1.00\ \cancel{kg\ NaOH})\left(\frac{1000\ \cancel{g\ NaOH}}{\cancel{kg\ NaOH}}\right)\left(\frac{1\ mol\ NaOH}{40.00\ \cancel{g\ NaOH}}\right) = 25.0\ mol\ NaOH$$

$$1.00\ kg\ NaOH = 25.0\ mol\ NaOH$$

Example 7.11

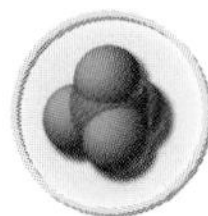

What is the mass of 5.00 mol of water?

SOLUTION

First we know that

$$1\ mol\ H_2O = 18.02\ g\ (Example\ 7.7)$$

The conversion is

$$mol\ H_2O \longrightarrow g\ H_2O$$

To convert moles to grams, we use the conversion factor $\frac{18.02\ g\ H_2O}{1\ mol\ H_2O}$

The calculation is

$$(5.00\ \cancel{mol\ H_2O})\left(\frac{18.02\ g\ H_2O}{1\ \cancel{mol\ H_2O}}\right) = 90.1\ g\ H_2O$$

Example 7.12

How many molecules of hydrogen chloride (HCl) are there in 25.0 g of hydrogen chloride?

SOLUTION

From the formula, we find that the molar mass of HCl is 36.46 g $(\overset{H}{1.008\ g} + \overset{Cl}{35.45\ g})$. The sequence of conversions is

$$g\ HCl \longrightarrow mol\ HCl \longrightarrow molecules\ HCl$$

Using the conversion factors

$$\frac{1\ mol\ HCl}{36.46\ g\ HCl} \quad and \quad \frac{6.022 \times 10^{23}\ molecules\ HCl}{1\ mol\ HCl}$$

we get

$$(25.0\ \cancel{g\ HCl})\left(\frac{1\ \cancel{mol\ HCl}}{36.46\ \cancel{g\ HCl}}\right)\left(\frac{6.022 \times 10^{23}\ molecules\ HCl}{1\ \cancel{mol\ HCl}}\right) = 4.13 \times 10^{23}\ molecules\ HCl$$

Practice 7.5

What is the mass of 0.150 mol of Na_2SO_4?

Practice 7.6

How many moles and molecules are there in 500.0 g of $HC_2H_3O_2$?

7.3 Percent Composition of Compounds

percent composition of a compound

Percent means parts per 100 parts. Just as each piece of pie is a percent of the whole pie, each element in a compound is a percent of the whole compound. The **percent composition of a compound** is the *mass percent* of each element in the compound. The molar mass represents the total mass, or 100%, of the compound. Thus, the percent composition of water, H_2O, is 11.19% H and 88.79% O by mass. According to the law of definite composition, the percent composition must be the same no matter what size sample is taken.

H
11.19%
O
88.79%

The percent composition of a compound can be determined (1) from knowing its formula or (2) from experimental data.

Percent Composition from Formula

If the formula of a compound is known, a two-step process is needed to determine the percent composition:

Step 1 Calculate the molar mass (Section 7.2).
Step 2 Divide the total mass of each element in the formula by the molar mass and multiply by 100. This gives the percent composition:

$$\frac{\text{total mass of the element}}{\text{molar mass}} \times 100 = \text{percent of the element}$$

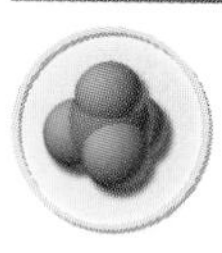

Example 7.13 Calculate the percent composition of sodium chloride (NaCl).

SOLUTION **Step 1** Calculate the molar mass of NaCl:

$$\begin{aligned} 1\,\text{Na} &= 1(22.99\text{ g}) = 22.99\text{ g} \\ 1\,\text{Cl} &= 1(35.45\text{ g}) = \underline{35.45\text{ g}} \\ & \qquad\qquad\qquad 58.44\text{ g (molar mass)} \end{aligned}$$

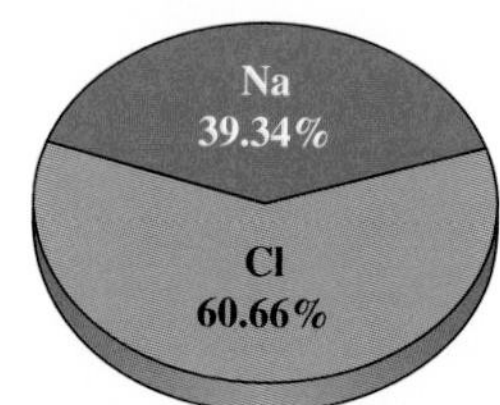

Step 2 Now calculate the percent composition. We know there are 22.99 g Na and 35.45 g Cl in 58.44 g NaCl:

$$\text{Na: } \left(\frac{22.99\,\cancel{\text{g}}\text{ Na}}{58.44\,\cancel{\text{g}}}\right)(100) = 39.34\%\text{ Na}$$

$$\text{Cl: } \left(\frac{35.45\,\cancel{\text{g}}\text{ Cl}}{58.44\,\cancel{\text{g}}}\right)(100) = 60.66\%\text{ Cl}$$

$$100.00\%\text{ total}$$

In any two-component system, if the percent of one component is known, the other is automatically defined by the difference; that is, if Na is 39.34%, then Cl is 100% − 39.34% = 60.66%. However, the calculation of the percent of each component should be carried out, since this provides a check against possible error. The percent composition data should add up to 100 ± 0.2%.

Example 7.14

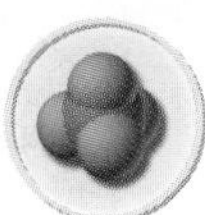

Calculate the percent composition of potassium chloride (KCl).

SOLUTION

Step 1 Calculate the molar mass of KCl:

$$\begin{aligned} 1\,\text{K} &= 1(39.10\text{ g}) = 39.10\text{ g} \\ 1\,\text{Cl} &= 1(35.45\text{ g}) = \underline{35.45\text{ g}} \\ & \qquad\qquad\quad\; 74.55\text{ g (molar mass)} \end{aligned}$$

Step 2 Now calculate the percent composition. We know there are 39.10 g K and 35.45 g Cl in 74.55 g KCl:

$$\text{K:}\ \left(\frac{39.10\,\cancel{\text{g}}\,\text{K}}{74.55\,\cancel{\text{g}}}\right)(100) = 52.45\%\ \text{K}$$

$$\text{Cl:}\ \left(\frac{35.45\,\cancel{\text{g}}\,\text{Cl}}{74.55\,\cancel{\text{g}}}\right)(100) = \underline{47.55\%\ \text{Cl}}$$

$$100.00\%\ \text{total}$$

Comparing the results calculated for NaCl and for KCl, we see that NaCl contains a higher percentage of Cl by mass, although each compound has a one-to-one atom ratio of Cl to Na and Cl to K. The reason for this mass percent difference is that Na and K do not have the same atomic masses.

It is important to realize that when we compare *1 mol of NaCl with 1 mol of KCl,* each quantity contains the same number of Cl atoms—namely, 1 mol of Cl atoms. However, if we compare *equal masses* of NaCl and KCl, there will be more Cl atoms in the mass of NaCl, since NaCl has a higher mass percent of Cl.

1 mol NaCl contains	100.00 g NaCl contains	1 mol KCl contains	100.00 g KCl contains
1 mol Na	39.34 g Na	1 mol K	52.45 g K
1 mol Cl	60.66 g Cl	1 mol Cl	47.55 g Cl
	60.66% Cl		47.55% Cl

Example 7.15

Calculate the percent composition of potassium sulfate (K_2SO_4):

SOLUTION

Step 1 Calculate the molar mass of K_2SO_4:

$$\begin{aligned} 2\,\text{K} &= 2(39.10\text{ g}) = 78.20\text{ g} \\ 1\,\text{S} &= 1(32.07\text{ g}) = 32.07\text{ g} \\ 4\,\text{O} &= 4(16.00\text{ g}) = \underline{64.00\text{ g}} \\ & \qquad\qquad\quad\; 174.3\ \ \text{g (molar mass)} \end{aligned}$$

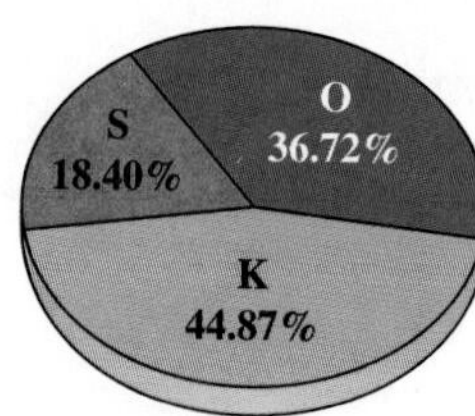

Step 2 Now calculate the percent composition. We know there are 78.20 g of K, 32.07 g of S, and 64.00 g of O in 174.3 g of K_2SO_4:

$$\text{K:}\left(\frac{78.20\ \cancel{g}\ \text{K}}{174.3\ \cancel{g}}\right)(100) = 44.87\%\ \text{K}$$

$$\text{S:}\left(\frac{32.07\ \cancel{g}\ \text{S}}{174.3\ \cancel{g}}\right)(100) = 18.40\%\ \text{S}$$

$$\text{O:}\left(\frac{64.00\ \cancel{g}\ \text{O}}{174.3\ \cancel{g}}\right)(100) = 36.72\%\ \text{O}$$

99.99% total

Practice 7.7

Calculate the percent composition of $Ca(NO_3)_2$.

Practice 7.8

Calculate the percent composition of K_2CrO_4.

Percent Composition from Experimental Data

The percent composition can be determined from experimental data without knowing the formula of a compound:

Step 1 Calculate the mass of the compound formed.
Step 2 Divide the mass of each element by the total mass of the compound and multiply by 100.

Example 7.16 Zinc oxide is a compound with many uses from preventing sunburn to a pigment in white paint. When heated in the air, 1.63 g of zinc (Zn) combines with 0.40 g of oxygen (O_2) to form zinc oxide. Calculate the percent composition of the compound formed:

SOLUTION **Step 1** First calculate the total mass of the compound formed:

1.63 g Zn
0.40 g O_2
2.03 g = total mass of product

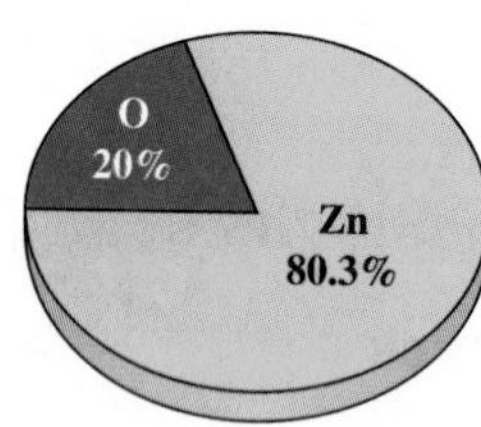

Step 2 Divide the mass of each element by the total mass (2.03 g) and multiply by 100:

$$\left(\frac{1.63\ \cancel{g}\ \text{Zn}}{2.03\ \cancel{g}}\right)(100) = 80.3\%\ \text{Zn} \qquad \left(\frac{0.40\ \cancel{g}\ \text{O}}{2.03\ \cancel{g}}\right)(100) = 20.\%\ \text{O}$$

The compound formed contains 80.3% Zn and 20.% O.

CHEMISTRY IN ACTION • Vanishing Coins?

Modern technology is changing our coins in some pretty interesting ways. The first U.S. coins were produced during the late 1700s from "coin silver," an alloy of 90% silver and 10% copper. Coins were made following a very simple rule: The mass of the coin reflects is relative value. Therefore a half dollar weighed half as much as a dollar coin and a quarter weighed $^1/_4$ as much as a dollar coin.

In the twentieth century our society began using machines to collect coins. We use vending machines, parking meters, toll baskets, Laundromats, slot machines, and video games to make our lives more convenient and fun. This change produced a huge demand for coins. At the same time, the price of silver rose rapidly.

The solution? Make coins from a different alloy that would still work in all of our machines. This was a tricky process since the machines require a specific mass and electric "resistivity." In the mid-1960s new coins which sandwiched a layer of copper between layers of an alloy of copper and nickel began to replace quarters and dimes. And a new Susan B. Anthony dollar appeared briefly in 1979. Finally, in 1997 Congress decided to revive the dollar coin once more.

Congress specified that the new dollar coin had to look and feel different to consumers, be golden in color, and have a distinctive edge. At the same time, it needed to fool vending machines so that both the Susan B. Anthony dollar and the new coin would work. Designers decided on the portrait of Sacagawea, a Shoshone woman and her child. She was a guide on the Lewis and Clark expedition. On the back is an eagle. The chemists struggled to find a golden alloy for the coin. It is slightly larger than a quarter and weighs 8.1 grams. Vending machines determine a coin's value by its size, weight and its electromagnetic signature. This electromagnetic signature is hard to duplicate. All the golden coins tested had three times the conductivity of the silver Anthony coin. Metallurgists finally tried adding manganese and zinc to the copper core and found a "golden" alloy that fooled the vending machines! Sacagawea dollars are made of an alloy which is 77% copper, 12% zinc, 7% manganese and 4% nickel. And on top of it all, the new coins only cost 12 cents each to make, leaving an 88 cent profit for the mint on each coin!

The future of coins is becoming less certain, however, as our machines are converted to electronic devices which use magnetic strips and a swipe of a card in place of "coins" and scan our cars for toll pass information. Some day soon, coins may vanish altogether from our pockets into collector's albums.

Sacagawea coin

Practice 7.9

Aluminum chloride is formed by reacting 13.43 g aluminum with 53.18 g chlorine. What is the percent composition of the compound?

7.4 Empirical Formula versus Molecular Formula

empirical formula

The **empirical formula,** or *simplest formula*, gives the smallest whole-number ratio of atoms present in a compound. This formula gives the relative number of atoms of each element in the compound.

molecular formula

The **molecular formula** is the true formula, representing the total number of atoms of each element present in one molecule of a compound. It is entirely possible that two or more substances will have the same percent composition yet be distinctly different compounds. For example, acetylene (C_2H_2) is a common gas used in welding; benzene (C_6H_6) is an important solvent obtained from coal tar and is used in the synthesis of styrene and nylon. Both acetylene and benzene contain 92.3% C and 7.7% H. The smallest ratio of C and H corresponding to these percentages is CH (1 : 1). Therefore, the *empirical* formula for both acetylene and benzene is CH, even though the *molecular* formulas are C_2H_2 and C_6H_6, respectively. Often the molecular formula is the same as the empirical formula. If the molecular formula is not the same, it will be an integral (whole number) multiple of the empirical formula. For example,

$$\text{CH} = \text{empirical formula}$$
$$(\text{CH})_2 = C_2H_2 = \text{acetylene} \quad (\text{molecular formula})$$
$$(\text{CH})_6 = C_6H_6 = \text{benzene} \quad (\text{molecular formula})$$

Table 7.1 compares the formulas of these substances. Table 7.2 shows empirical and molecular formula relationships of other compounds.

Table 7.1 Molecular Formulas of Two Compounds Having an Empirical Formula with a 1:1 Ratio of Carbon and Hydrogen Atoms

Formula	Composition		Molar mass
	% C	% H	
CH (empirical)	92.3	7.7	13.02 (empirical)
C_2H_2 (acetylene)	92.3	7.7	26.04 (2 × 13.02)
C_6H_6 (benzene)	92.3	7.7	78.12 (6 × 13.02)

Table 7.2 Some Empirical and Molecular Formulas

Substance	Empirical formula	Molecular formula	Substance	Empirical formula	Molecular formula
Acetylene	CH	C_2H_2	Diborane	BH_3	B_2H_6
Benzene	CH	C_6H_6	Hydrazine	NH_2	N_2H_4
Ethylene	CH_2	C_2H_4	Hydrogen	H	H_2
Formaldehyde	CH_2O	CH_2O	Chlorine	Cl	Cl_2
Acetic acid	CH_2O	$C_2H_4O_2$	Bromine	Br	Br_2
Glucose	CH_2O	$C_6H_{12}O_6$	Oxygen	O	O_2
Hydrogen chloride	HCl	HCl	Nitrogen	N	N_2
Carbon dioxide	CO_2	CO_2	Iodine	I	I_2

7.5 Calculating Empirical Formulas

We can establish empirical formulas because (1) individual atoms in a compound are combined in whole-number ratios, and (2) each element has a specific atomic mass.

To calculate an empirical formula, we need to know (1) the elements that are combined, (2) their atomic masses, and (3) the ratio by mass or percentage in which they are combined. If elements A and B form a compound, we may represent the empirical formula as A_xB_y, where x and y are small whole numbers that represent the atoms of A and B. To write the empirical formula, we must determine x and y:

Step 1 Assume a definite starting quantity (usually 100.0 g) of the compound, if not given, and express the mass of each element in grams.

Step 2 Convert the grams of each element into moles using each element's molar mass. This conversion gives the number of moles of atoms of each element in the quantity assumed in Step 1. At this point, these numbers will usually not be whole numbers.

Step 3 Divide each value obtained in Step 2 by the smallest of these values. If the numbers obtained are whole numbers, use them as subscripts and write the empirical formula. If the numbers obtained are not whole numbers, go on to Step 4.

Step 4 Multiply the values obtained in Step 3 by the smallest number that will convert them to whole numbers. Use these whole numbers as the subscripts in the empirical formula. For example, if the ratio of A to B is 1.0 : 1.5, multiply both numbers by 2 to obtain a ratio of 2 : 3. The empirical formula then is A_2B_3.

Some common fractions and their decimal equivalents are

$\frac{1}{4} = 0.25$
$\frac{1}{3} = 0.333\ldots$
$\frac{2}{3} = 0.666\ldots$
$\frac{1}{2} = 0.5$
$\frac{3}{4} = 0.75$

Multiply the decimal equivalent by the number in the denominator of the fraction to get a whole number: 4(0.75) = 3.

In many of these calculations, results will vary somewhat from an exact whole number; this can be due to experimental errors in obtaining the data or from rounding off numbers. Calculations that vary by no more than ±0.1 from a whole number usually are rounded off to the nearest whole number. Deviations greater than about 0.1 unit usually mean that the calculated ratios need to be multiplied by a factor to make them all whole numbers. For example, an atom ratio of 1 : 1.33 should be multiplied by 3 to make the ratio 3 : 4.

Example 7.17

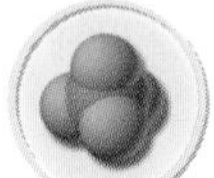

Calculate the empirical formula of a compound containing 11.19% hydrogen (H) and 88.79% oxygen (O):

SOLUTION

Step 1 Express each element in grams. Assuming 100.00 g of material, we know that the percent of each element equals the grams of each element and that the percent sign can be omitted:

$$H = 11.19\,g$$
$$O = 88.79\,g$$

Step 2 Convert the grams of each element to moles:

$$H: (11.19\,\cancel{g\,H})\left(\frac{1\text{ mol H atoms}}{1.008\,\cancel{g\,H}}\right) = 11.10\text{ mol H atoms}$$

$$O: (88.79\,\cancel{g\,O})\left(\frac{1\text{ mol O atoms}}{16.00\,\cancel{g\,O}}\right) = 5.549\text{ mol O atoms}$$

The formula could be expressed as $H_{11.10}O_{5.549}$. However, it's customary to use the smallest whole-number ratio of atoms. This ratio is calculated in Step 3.

Step 3 Change these numbers to whole numbers by dividing them by the smaller number:

$$H = \frac{11.10\ \text{mol}}{5.549\ \text{mol}} = 2.000 \qquad O = \frac{5.549\ \text{mol}}{5.549\ \text{mol}} = 1.000$$

In this step, the ratio of atoms has not changed, because we divided the number of moles of each element by the same number:

The simplest ratio of H to O is 2:1.

Empirical formula = H_2O

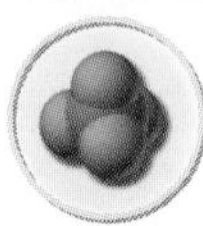

Example 7.18 The analysis of a salt shows that it contains 56.58% potassium (K); 8.68% carbon (C); and 34.73% oxygen (O). Calculate the empirical formula for this substance.

SOLUTION **Steps 1 and 2** After changing the percentage of each element to grams, we find the relative number of moles of each element by multiplying by the proper conversion factor:

$$\text{K: } (56.58\ \cancel{\text{g K}})\left(\frac{1\ \text{mol K atoms}}{39.10\ \cancel{\text{g K}}}\right) = 1.447\ \text{mol K atoms}$$

$$\text{C: } (8.68\ \cancel{\text{g C}})\left(\frac{1\ \text{mol C atoms}}{12.01\ \cancel{\text{g C}}}\right) = 0.723\ \text{mol C atoms}$$

$$\text{O: } (34.73\ \cancel{\text{g O}})\left(\frac{1\ \text{mol O atoms}}{16.00\ \cancel{\text{g O}}}\right) = 2.171\ \text{mol O atoms}$$

Step 3 Divide each number of moles by the smallest value:

$$K = \frac{1.447\ \text{mol}}{0.723\ \text{mol}} = 2.00$$

$$C = \frac{0.723\ \text{mol}}{0.723\ \text{mol}} = 1.00$$

$$O = \frac{2.171\ \text{mol}}{0.723\ \text{mol}} = 3.00$$

The simplest ratio of K:C:O is 2:1:3.

Empirical formula = K_2CO_3

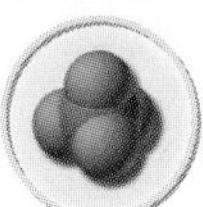

Example 7.19 A sulfide of iron was formed by combining 2.233 g of iron (Fe) with 1.926 g of sulfur (S). What is the empirical formula of the compound?

SOLUTION **Steps 1 and 2** The grams of each element are given, so we use them directly in our calculations. Calculate the relative number of moles of each element by multiplying the grams of each element by the proper conversion factor:

$$\text{Fe: } (2.233\ \cancel{\text{g Fe}})\left(\frac{1\text{ mol Fe atoms}}{55.85\ \cancel{\text{g Fe}}}\right) = 0.03998\text{ mol Fe atoms}$$

$$\text{S: } (1.926\ \cancel{\text{g S}})\left(\frac{1\text{ mol S atoms}}{32.07\ \cancel{\text{g S}}}\right) = 0.06006\text{ mol S atoms}$$

Step 3 Divide each number of moles by the smaller of the two numbers:

$$\text{Fe} = \frac{0.03998\text{ mol}}{0.03998\text{ mol}} = 1.000 \qquad \text{S} = \frac{0.06006\text{ mol}}{0.03998\text{ mol}} = 1.502$$

Step 4 We still have not reached a ratio that gives whole numbers in the formula, so we double each value to obtain a ratio of 2.000 Fe atoms to 3.000 S atoms. Doubling both values does not change the ratio of Fe and S atoms:

$$\text{Fe: } (1.000)2 = 2.000$$

$$\text{S: } (1.502)2 = 3.004$$

$$\text{Empirical formula} = Fe_2S_3$$

Practice 7.10

Calculate the empirical formula of a compound containing 52.14% C, 13.12% H, and 34.73% O.

Practice 7.11

Calculate the empirical formula of a compound that contains 43.7% phosphorus and 56.3% O by mass.

7.6 Calculating the Molecular Formula from the Empirical Formula

The molecular formula can be calculated from the empirical formula if the molar mass is known. The molecular formula, as stated in Section 7.4, will be equal either to the empirical formula or some multiple of it. For example, if the empirical formula of a compound of hydrogen and fluorine is HF, the molecular formula can be expressed as $(HF)_n$, where $n = 1, 2, 3, 4, \ldots$. This n means that the molecular formula could be HF, H_2F_2, H_3F_3, H_4F_4, and so on. To determine the molecular formula, we must evaluate n:

$$n = \frac{\text{molar mass}}{\text{mass of empirical formula}} = \text{number of empirical formula units}$$

What we actually calculate is the number of units of the empirical formula contained in the molecular formula.

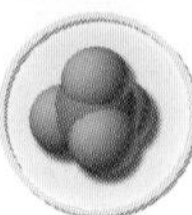

Example 7.20 A compound of nitrogen and oxygen with a molar mass of 92.00 g was found to have an empirical formula of NO_2. What is its molecular formula?

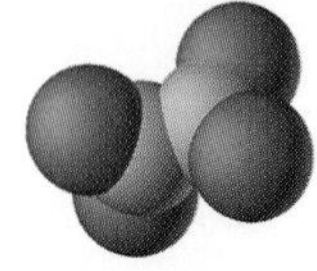

N_2O_4

SOLUTION Let n be the number of NO_2 units in a molecule; then the molecular formula is $(NO_2)_n$. Each NO_2 unit has a mass of [14.01 g + 2(16.00 g)] or 46.01 g. The molar mass of $(NO_2)_n$ is 92.00 g, and the number of 46.01 units in 92.00 is 2:

$$n = \frac{92.00\text{ g}}{46.01\text{ g}} = 2 \quad \text{(empirical formula units)}$$

The molecular formula is $(NO_2)_2$, or N_2O_4.

Example 7.21 Propylene is a compound frequently polymerized to make polypropylene which is used for rope, laundry bags, blankets, carpets and textile fibers. Propylene has a molar mass of 42.08 g and contains 14.3% H and 85.7% C. What is its molecular formula?

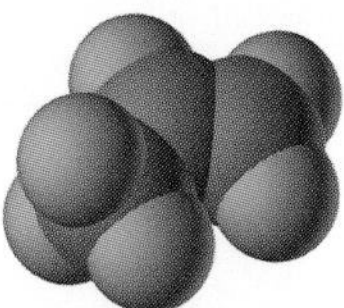

C_3H_6

SOLUTION First find the empirical formula:

$$\text{C:} \quad (85.7\,\cancel{\text{g C}})\left(\frac{1\text{ mol C atoms}}{12.01\,\cancel{\text{g C}}}\right) = 7.14\text{ mol C atoms}$$

$$\text{H:} \quad (14.3\,\cancel{\text{g H}})\left(\frac{1\text{ mol H atoms}}{1.008\,\cancel{\text{g H}}}\right) = 14.2\text{ mol H atoms}$$

Divide each value by the smaller number of moles:

$$\text{C} = \frac{7.14\text{ mol}}{7.14\text{ mol}} = 1.00$$

$$\text{H} = \frac{14.2\text{ mol}}{7.14\text{ mol}} = 1.99$$

Empirical formula = CH_2

We determine the molecular formula from the empirical formula and the molar mass:

Molecular formula = $(CH_2)_n$

Molar mass = 42.08 g

Each CH_2 unit has a mass of (12.01 g + 2.016 g) or 14.03 g. The number of CH_2 units in 42.08 g is 3:

$$n = \frac{42.08\text{ g}}{14.03\text{ g}} = 3 \quad \text{(empirical formula units)}$$

The molecular formula is $(CH_2)_3$, or C_3H_6.

When propylene is polymerized into polypropylene it can be used for rope and other recreational products.

Practice 7.12

Calculate the empirical and molecular formulas of a compound that contains 80.0% C, 20.0% H, and has a molar mass of 30.00 g.

Chapter 7 Review

7.1 The Mole

KEY TERMS

Avogadro's number
Mole

- We count atoms by weighing them since they are so tiny.
- 1 mole = 6.022×10^{23} items.
- Avogadro's number is 6.022×10^{23}.

7.2 Molar Mass of Compounds

- One mole of a compound contains Avogadro's number of formula units of that compound.
- The mass (grams) of one mole of a compound is the molar mass
- Molar mass is determined by adding the molar masses of all the atoms in a formula.
- Molar masses are given to four significant figures in this text.

7.3 Percent Composition of Compounds

KEY TERM

Percent composition of a compound

- To determine the percent composition from a formula:
  - Calculate the molar mass
  - For each element in the formula

$$\frac{\text{total mass of the element}}{\text{molar mass of the compound}} \times 100 = \text{percent of the element}$$

- To determine percent composition from experimental data:
  - Calculate the mass of the compound formed
  - For each element in the formula

$$\frac{\text{mass of the element}}{\text{mass of the compound formed}} \times 100 = \text{percent of the element}$$

7.4 Empirical Formula versus Molecular Formula

KEY TERMS

Empirical formula
Molecular formula

- The empirical formula is the simplest formula giving the smallest whole number ratio of atoms present in a compound
- The molecular formula is the true formula representing the total number of atoms of each element present in one molecule of the compound
- Two or more substances may have the same empirical formulas but different molecular formulas

7.5 Calculating Empirical Formulas

- To determine the empirical formula for a compound you need to know:
  - The elements that are combined
  - Their atomic masses
  - The ratio of masses or percentage in which they are combined
- Empirical formulas are represented in the form A_xB_y. To determine the empirical formula of this compound:
  - Assume a starting quantity (100.0 g is a good choice)
  - Convert mass (g) to moles for each element
  - Divide each element's moles by the smallest number of moles.
    - If the ratios are whole numbers, use them as subscripts and write the empirical formula.
  - If the ratios are not whole numbers, multiply them all by the smallest number which will convert them to whole number.
    - Use the whole numbers to write the empirical formula

7.6 Calculating the Molecular Formula from the Empirical Formula

- The molecular formula is calculated from the empirical formula when the molar mass is known:
  - $n = \dfrac{\text{molar mass}}{\text{mass of empirical formula}}$ = number of empirical formula units
  - The molecular formula is $(A_xB_y)_n$

Review Questions

All questions with blue numbers have answers in the appendix of the text.

1. What is a mole?
2. Which would have a higher mass: a mole of K atoms or a mole of Au atoms?
3. Which would contain more atoms: a mole of K atoms or a mole of Au atoms?
4. Which would contain more electrons: a mole of K atoms or a mole of Au atoms?
5. What is molar mass?
6. If the atomic mass scale had been defined differently, with an atom of $^{12}_{6}C$ defined as a mass of 50 amu, would this have any effect on the value of Avogadro's number? Explain.
7. What is the numerical value of Avogadro's number?
8. What is the relationship between Avogadro's number and the mole?

9. Complete these statements, supplying the proper quantity.
(a) A mole of O atoms contains _____ atoms.
(b) A mole of O_2 molecules contains _____ molecules.
(c) A mole of O_2 molecules contains _____ atoms.
(d) A mole of O atoms has a mass of _____ grams.
(e) A mole of O_2 molecules has a mass of _____ grams.

10. How many molecules are present in 1 molar mass of sulfuric acid (H_2SO_4)? How many atoms are present?

11. What is the difference between an empirical formula and a molecular formula?

12. In calculating the empirical formula of a compound from its percent composition, why do we choose to start with 100.0 g of the compound?

Paired Exercises

All exercises with blue numbers have answers in the appendix of the text.

Molar Masses

1. Determine the molar masses of these compounds:
(a) KBr
(b) Na_2SO_4
(c) $Pb(NO_3)_2$
(d) C_2H_5OH
(e) $HC_2H_3O_2$
(f) Fe_3O_4
(g) $C_{12}H_{22}O_{11}$
(h) $Al_2(SO_4)_3$
(i) $(NH_4)_2HPO_4$

2. Determine the molar masses of these compounds:
(a) NaOH
(b) Ag_2CO_3
(c) Cr_2O_3
(d) $(NH_4)_2CO_3$
(e) $Mg(HCO_3)_2$
(f) C_6H_5COOH
(g) $C_6H_{12}O_6$
(h) $K_4Fe(CN)_6$
(i) $BaCl_2 \cdot 2\,H_2O$

Moles and Avogadro's Number

3. How many moles of atoms are contained in the following?
(a) 22.5 g Zn
(b) 0.688 g Mg
(c) 4.5×10^{22} atoms Cu
(d) 382 g Co
(e) 0.055 g Sn
(f) 8.5×10^{24} molecules N_2

4. How many moles are contained in the following?
(a) 25.0 g NaOH
(b) 44.0 g Br_2
(c) 0.684 g $MgCl_2$
(d) 14.8 g CH_3OH
(e) 2.88 g Na_2SO_4
(f) 4.20 lb ZnI_2

5. Calculate the number of grams in each of the following:
(a) 0.550 mol Au
(b) 15.8 mol H_2O
(c) 12.5 mol Cl_2
(d) 3.15 mol NH_4NO_3

6. Calculate the number of grams in each of the following:
(a) 4.25×10^{-4} mol H_2SO_4
(b) 4.5×10^{22} molecules CCl_4
(c) 0.00255 mol Ti
(d) 1.5×10^{16} atoms S

7. How many molecules are contained in each of the following:
(a) 1.26 mol O_2
(b) 0.56 mol C_6H_6
(c) 16.0 g CH_4
(d) 1000. g HCl

8. How many molecules are contained in each of the following:
(a) 1.75 mol Cl_2
(b) 0.27 mol C_2H_6O
(c) 12.0 g CO_2
(d) 1000. g CH_4

9. How many atoms are contained in each of the following:
(a) 11 molecules C_2H_5OH
(b) 25.0 mol Ag
(c) 0.0986 g Xe
(d) 72.5 g $CHCl_3$

10. How many atoms are contained in each of the following:
(a) 18 molecules N_2O_5
(b) 10.0 mol Au
(c) 75.2 g BF_3
(d) 15.2 g U

11. Calculate the mass in grams of each of the following:
(a) 1 atom Pb
(b) 1 atom Ag
(c) 1 molecule H_2O
(d) 1 molecule $C_3H_5(NO_3)_3$

12. Calculate the mass in grams of each of the following:
(a) 1 atom Au
(b) 1 atom U
(c) 1 molecule NH_3
(d) 1 molecule $C_6H_4(NH_2)_2$

13. Make the following conversions:
(a) 8.66 mol Cu to grams Cu
(b) 125 mol Au to kilograms Au
(c) 10. atoms C to moles C
(d) 5000 molecules CO_2 to moles CO_2

14. Make the following conversions:
(a) 28.4 g S to moles S
(b) 2.50 kg NaCl to moles NaCl
(c) 42.4 g Mg to atoms Mg
(d) 485 mL Br_2 ($d = 3.12$ g/mL) to moles Br_2

15. Exactly 1 mol of carbon disulfide contains
(a) how many carbon disulfide molecules?
(b) how many carbon atoms?
(c) how many sulfur atoms?
(d) how many total atoms of all kinds?

16. One (1.000) mole of ammonia contains
(a) how many ammonia molecules?
(b) how many nitrogen atoms?
(c) how many hydrogen atoms?
(d) how many total atoms of all kinds?

17. How many atoms of oxygen are contained in each of the following?
(a) 16.0 g O_2
(b) 0.622 mol MgO
(c) 6.00×10^{22} molecules $C_6H_{12}O_6$

18. How many atoms of oxygen are contained in each of the following?
(a) 5.0 mol MnO_2
(b) 255 g $MgCO_3$
(c) 5.0×10^{18} molecules H_2O

19. Calculate the number of
(a) grams of silver in 25.0 g AgBr
(b) grams of nitrogen in 6.34 mol $(NH_4)_3PO_4$
(c) grams of oxygen in 8.45×10^{22} molecules SO_3

20. Calculate the number of
(a) grams of chlorine in 5.0 g $PbCl_2$
(b) grams of hydrogen in 4.50 mol H_2SO_4
(c) grams of hydrogen in 5.45×10^{22} molecules NH_3

Percent Composition

21. Calculate the percent composition by mass of these compounds:
(a) NaBr
(b) $KHCO_3$
(c) $FeCl_3$
(d) $SiCl_4$
(e) $Al_2(SO_4)_3$
(f) $AgNO_3$

22. Calculate the percent composition by mass of these compounds:
(a) $ZnCl_2$
(b) $NH_4C_2H_3O_2$
(c) MgP_2O_7
(d) $(NH_4)_2SO_4$
(e) $Fe(NO_3)_3$
(f) ICl_3

23. Calculate the percent of iron in the following compounds:
(a) FeO
(b) Fe_2O_3
(c) Fe_3O_4
(d) $K_4Fe(CN)_6$

24. Which of the following chlorides has the highest and which has the lowest percentage of chlorine, by mass, in its formula?
(a) KCl
(b) $BaCl_2$
(c) $SiCl_4$
(d) LiCl

25. A 6.20-g sample of phosphorus was reacted with oxygen to form an oxide with a mass of 14.20 g. Calculate the percent composition of the compound.

26. A sample of ethylene chloride was analyzed to contain 6.00 g of C, 1.00 g of H, and 17.75 g of Cl. Calculate the percent composition of ethylene chloride.

27. Examine the following formulas. Which compound has the
(a) higher percent by mass of hydrogen: H_2O or H_2O_2?
(b) lower percent by mass of nitrogen: NO or N_2O_3?
(c) higher percent by mass of oxygen: NO_2 or N_2O_4?
Check your answers by calculation if you wish.

28. Examine the following formulas. Which compound has the
(a) lower percent by mass of chlorine: $NaClO_3$ or $KClO_3$?
(b) higher percent by mass of sulfur: $KHSO_4$ or K_2SO_4?
(c) lower percent by mass of chromium: Na_2CrO_4 or $Na_2Cr_2O_7$?
Check your answers by calculation if you wish.

Empirical and Molecular Formulas

29. Determine the empirical formula for the following:
(a) a compound made from 5.94% hydrogen and 94.06% oxygen with a total molar mass of 34.02 g
(b) a compound made from 80.34% zinc and 19.66% oxygen with a total molar mass of 81.39 g
(c) a compound made from 35.18% iron, 44.66% chlorine and 20.16% oxygen with a total molar mass of 158.75 g
(d) a compound made from 26.19% nitrogen, 7.55% hydrogen and 66.26% chlorine with a total molar mass of 53.50 g

30. Determine the empirical formula for the following:
(a) a compound made from 32.86% potassium and 67.14% bromine with a total molar mass of 119.00 g
(b) a compound made from 63.50% silver, 8.25% nitrogen and 28.25% oxygen with a total molar mass of 169.91 g
(c) a compound made from 54.09% calcium, 2.72% hydrogen and 43.18% oxygen with a total molar mass of 74.10 g
(d) a compound made from 2.06% hydrogen, 32.69% sulfur and 65.25% oxygen with a total molar mass of 98.09 g

31. Calculate the empirical formula of each compound from the percent compositions given:
(a) 63.6% N, 36.4% O
(b) 46.7% N, 53.3% O
(c) 25.9% N, 74.1% O
(d) 43.4% Na, 11.3% C, 45.3% O
(e) 18.8% Na, 29.0% Cl, 52.3% O
(f) 72.02% Mn, 27.98% O

32. Calculate the empirical formula of each compound from the percent compositions given:
(a) 64.1% Cu, 35.9% Cl
(b) 47.2% Cu, 52.8% Cl
(c) 51.9% Cr, 48.1% S
(d) 55.3% K, 14.6% P, 30.1% O
(e) 38.9% Ba, 29.4% Cr, 31.7% O
(f) 3.99% P, 82.3% Br, 13.7% Cl

33. A sample of tin having a mass of 3.996 g was oxidized and found to have combined with 1.077 g of oxygen. Calculate the empirical formula of this oxide of tin.

34. A 3.054-g sample of vanadium (V) combined with oxygen to form 5.454 g of product. Calculate the empirical formula for this compound.

35. In a laboratory experiment a student burned a 2.465-g sample of copper, which reacted with oxygen from the air, to form an oxide of copper. The oxide had a mass of 2.775 g. Calculate the empirical formula for this compound.

36. A 5.276-g sample of an unknown compound was found to contain 3.989 g of mercury and the rest, chlorine. Calculate the empirical formula of this chloride of mercury.

37. Hydroquinone is an organic compound commonly used as a photographic developer. It has a molar mass of 110.1 g/mol and a composition of 65.45% C, 5.45% H, and 29.09% O. Calculate the molecular formula of hydroquinone.

38. Fructose is a very sweet natural sugar that is present in honey, fruits, and fruit juices. It has a molar mass of 180.1 g/mol and a composition of 40.0% C, 6.7% H, and 53.3% O. Calculate the molecular formula of fructose.

39. Ethanedioic acid, a compound that is present in many vegetables, has a molar mass of 90.04 g/mol and a composition of 26.7% C, 2.2% H, and 71.1% O. What is the molecular formula for this substance?

40. Butyric acid is a compound that is present in butter. It has a molar mass of 88.11 g/mol and is composed of 54.5% C, 9.2% H, and 36.3% O. What is the molecular formula for this subtance?

41. Calculate the percent composition and determine the molecular formula and the empirical formula for the nitrogen-oxygen compound that results when 12.04 g of nitrogen is reacted with enough oxygen to produce 39.54 g of product. The molar mass of the product is 92.02 g.

42. Calculate the percent composition and determine the molecular formula and the empirical formula for the carbon-hydrogen-oxygen compound that results when 30.21 g of carbon, 40.24 g of oxygen, and 5.08 g of hydrogen are reacted to produce a product with a molar mass of 180.18 g.

43. The compound XYZ_3 has a molar mass of 100.09 g and a percent composition (by mass) of 40.04% X, 12.00% Y, and 47.96% Z. What is the compound?

44. The compound $X_2(YZ_3)_3$ has a molar mass of 282.23 g and a percent composition (by mass) of 19.12% X, 29.86% Y, and 51.02% Z. What is the compound?

Additional Exercises

All exercises with blue numbers have answers in the appendix of the text.

45. White phosphorus is one of several forms of phosphorus and exists as a waxy solid consisting of P_4 molecules. How many atoms are present in 0.350 mol of P_4?

46. How many grams of sodium contain the same number of atoms as 10.0 g of potassium?

47. One atom of an unknown element is found to have a mass of 1.79×10^{-23} g. What is the molar mass of this element?

48. How many molecules of sugar are in a 5-lb bag of sugar? The formula for table sugar or sucrose is $C_{12}H_{22}O_{11}$.

49. If a stack of 500 sheets of paper is 4.60 cm high, what will be the height, in meters, of a stack of Avogadro's number of sheets of paper?

50. There are about 6.1 billion (6.1×10^9) people on earth. If exactly 1 mol of dollars were distributed equally among these people, how many dollars would each person receive?

51. If 20. drops of water equal 1.0 mL (1.0 cm^3),
(a) how many drops of water are there in a cubic mile of water?
(b) what would be the volume in cubic miles of a mole of drops of water?

52. Silver has a density of 10.5 g/cm^3. If 1.00 mol of silver were shaped into a cube,
(a) what would be the volume of the cube?
(b) what would be the length of one side of the cube?

53. Given 1.00-g samples of each of the compounds, CO_2, O_2, H_2O, and CH_3OH,
(a) which sample will contain the largest number of molecules?
(b) which sample will contain the largest number of atoms?
Show proof for your answers.

54. How many grams of Fe_2S_3 will contain a total number of atoms equal to Avogadro's number?

55. How many grams of calcium must be combined with phosphorus to form the compound Ca_3P_2?

56. An iron ore contains 5% $FeSO_4$ by mass. How many grams of iron could be obtained from 1.0 ton of this ore?

57. How many grams of lithium will combine with 20.0 g of sulfur to form the compound Li_2S?

58. Calculate the percentage of
(a) mercury in $HgCO_3$
(b) oxygen in $Ca(ClO_3)_2$
(c) nitrogen in $C_{10}H_{14}N_2$ (nicotine)
(d) Mg in $C_{55}H_{72}MgN_4O_5$ (chlorophyll)

59. Zinc and sulfur react to form zinc sulfide, ZnS. If we mix 19.5 g of zinc and 9.40 g of sulfur, have we added sufficient sulfur to fully react all the zinc? Show evidence for your answer.

60. Pyrethrins are a group of compounds that are isolated from flowers and used as insecticidal agents. One form has the formula $C_{21}H_{28}O_3$. What is the percent composition of this compound?

61. Diphenhydramine hydrochloride, a drug used commonly as an antihistamine, has the formula $C_{17}H_{21}NO \cdot HCl$. What is the percent composition of each element in this compound?

62. What is the percent composition of each element in sucrose, $C_{12}H_{22}O_{11}$?

63. Aspirin is well known as a pain reliever (analgesic) and as a fever reducer (antipyretic). It has a molar mass of 180.2 g/mol and a composition of 60.0% C, 4.48% H, and 35.5% O. Calculate the molecular formula of aspirin.

64. How many grams of oxygen are contained in 8.50 g of $Al_2(SO_4)_3$?

65. Gallium arsenide is one of the newer materials used to make semiconductor chips for use in supercomputers. Its composition is 48.2% Ga and 51.8% As. What is the empirical formula?

66. Calcium tartrate is used as a preservative for certain foods and as an antacid. It contains 25.5% C, 2.1% H, 21.3% Ca, and 51.0% O. What is the empirical formula for calcium tartrate?

67. The compositions of four different compounds of carbon and chlorine follow. Determine the empirical formula and the molecular formula for each compound.

Percent C	Percent Cl	Molar mass (g)
(a) 7.79	92.21	153.8
(b) 10.13	89.87	236.7
(c) 25.26	74.74	284.8
(d) 11.25	88.75	319.6

68. How many years is a mole of seconds?

69. A normal penny has a mass of about 2.5 g. If we assume the penny to be pure copper (which means the penny is very old since newer pennies are a mixture of copper and zinc), how many atoms of copper does it contain?

70. What would be the mass (in grams) of one thousand trillion molecules of glycerin ($C_3H_8O_3$)?

71. If we assume that there are 6.1 billion people on the Earth, how many moles of people is this?

72. An experimental catalyst used to make polymers has the following composition: Co, 23.3%; Mo, 25.3%; and Cl, 51.4%. What is the empirical formula for this compound?

73. If a student weighs 18 g of aluminum and needs twice as many atoms of magnesium as she has of aluminum, how many grams of Mg does she need?

74. If 10.0 g of an unknown compound composed of carbon, hydrogen, and nitrogen contains 17.7% N and 3.8×10^{23} atoms of hydrogen, what is its empirical formula?

75. A substance whose formula is A_2O (A is a mystery element) is 60.0% A and 40.0% O. Identify the element A.

76. For the following compounds whose molecular formulas are given, indicate the empirical formula:

(a) $C_6H_{12}O_6$	glucose
(b) C_8H_{18}	octane
(c) $C_3H_6O_3$	lactic acid
(d) $C_{25}H_{52}$	paraffin
(e) $C_{12}H_4Cl_4O_2$	dioxin (a powerful poison)

Challenge Exercises

All exercises with blue numbers have answers in the appendix of the text.

77. The compound $A(BC)_3$ has a molar mass of 78.01 grams and a percent composition of 34.59% A, 61.53% B, and 3.88% C. Determine the identity of the elements A, B, and C. What is the percent composition by mass of the compound A_2B_3?

78. A 2.500 g sample of an unknown compound containing only C, H, and O was burned in O_2. The products were 4.776 g of CO_2 and 2.934 g of H_2O.
(a) What is the percent composition of the original compound?
(b) What is the empirical formula of the compound?

Answers to Practice Exercises

7.1 (a) 6.022×10^{23} atoms Fe
(b) 1.204×10^{24} atoms H
(c) 4.215×10^{24} atoms

7.2 10.0 g helium

7.3 1.5×10^{22} atoms

7.4 101.1 g KNO_3

7.5 21.3 g Na_2SO_4

7.6 8.326 mol and 5.014×10^{24} molecules $HC_2H_3O_2$

7.7 24.42% Ca; 17.07% N; 58.50% O

7.8 40.27% K; 26.78% Cr; 32.96% O

7.9 20.16% Al; 79.84% Cl

7.10 C_2H_6O

7.11 P_2O_5

7.12 The empirical formula is CH_3; the molecular formula is C_2H_6.

CHAPTER 8

Chemical Equations

Chapter Outline

The thermite reaction is a reaction between elemental aluminum and iron oxide. This reaction produces so much energy the iron becomes molten. One use of the thermite reaction is to weld railroad rails.

In the world today, we continually strive to express information in a concise, useful manner. From early childhood, we are taught to translate our ideas and desires into sentences. In mathematics, we learn to describe numerical relationships and situations through mathematical expressions and equations. Historians describe thousands of years of history in 500-page textbooks. Filmmakers translate entire events, such as the Olympics, into a few hours of entertainment.

Chemists use chemical equations to describe reactions they observe in the laboratory or in nature. Chemical equations provide us with the means to (1) summarize the reaction, (2) display the substances that are reacting, (3) show the products, and (4) indicate the amounts of all component substances in a reaction.

8.1 The Chemical Equation

Chemical reactions always involve change. Atoms, molecules, or ions rearrange to form different substances, sometimes in a spectacular manner. For example, the thermite reaction (shown in the chapter opening photo) is a reaction between aluminum metal and iron(III) oxide, which produces molten iron and aluminum oxide. The substances entering the reaction are called the **reactants** and the substances formed are called the **products**. In our example,

reactants
products

reactants	aluminum iron(III) oxide
products	iron aluminum oxide

During reactions chemical bonds are broken and new bonds are formed. The reactants and products may be present as solids, liquids, gases, or in solution.

In a chemical reaction atoms are neither created nor destroyed.
All atoms present in the reactants must also be present in the products.

A **chemical equation** is a shorthand expression for a chemical change or reaction. A chemical equation uses the chemical symbols and formulas of the reactants and products and other symbolic terms to represent a chemical reaction. The equations are written according to this general format:

chemical equation

1. Reactants are separated from products by an arrow ($\longrightarrow$) that indicates the direction of the reaction. The reactants are placed to the left and the products to the right of the arrow. A plus sign (+) is placed between reactants and between products when needed.

   $$\underbrace{Al + Fe_2O_3}_{\text{reactants}} \longrightarrow \underbrace{Fe + Al_2O_3}_{\text{products}}$$

2. Coefficients (whole numbers) are placed in front of substances to balance the equation and to indicate the number of units (atoms, molecules, moles, ions) of each substance reacting or being produced. When no number is shown, it is understood that one unit of the substance is indicated.

   $$2\,Al + Fe_2O_3 \longrightarrow 2\,Fe + Al_2O_3$$

3. Conditions required to carry out the reaction may, if desired, be placed above or below the arrow or equality sign. For example, a delta sign placed over the arrow ($\xrightarrow{\Delta}$) indicates that heat is supplied to the reaction.

$$2\,Al + Fe_2O_3 \xrightarrow{\Delta} 2\,Fe + Al_2O_3$$

4. The physical state of a substance is indicated by the following symbols: (*s*) for solid state; (*l*) for liquid state; (*g*) for gaseous state; and (*aq*) for substances in aqueous solution. States are not always given in chemical equations.

$$2\,Al(s) + Fe_2O_3(s) \xrightarrow{\Delta} Fe(l) + Al_2O_3(s)$$

Commonly used symbols are given in Table 8.1.

Table 8.1 Symbols Commonly Used in Chemical Equations

Symbol	Meaning
$+$	Plus or added to (placed between substances)
$\longrightarrow$	Yields; produces (points to products)
(*s*)	Solid state (written after a substance)
(*l*)	Liquid state (written after a substance)
(*g*)	Gaseous state (written after a substance)
(*aq*)	Aqueous solution (substance dissolved in water)
Δ	Heat is added (when written above or below arrow)

8.2 Writing and Balancing Chemical Equations

balanced equation

To represent the quantitative relationships of a reaction, the chemical equation must be balanced. A **balanced equation** contains the same number of each kind of atom on each side of the equation. The balanced equation therefore obeys the law of conservation of mass.

Correct formulas are not changed to balance an equation.

Every chemistry student must learn to *balance* equations. Many equations are balanced by trial and error, but care and attention to detail are still required. The way to balance an equation is to adjust the number of atoms of each element so that they are the same on each side of the equation, but a correct formula is never changed in order to balance an equation. The general procedure for balancing equations is as follows:

Study this procedure carefully and refer to it when you work examples.

Step 1 **Identify the reaction.** Write a description or word equation for the reaction. For example, let's consider mercury(II) oxide decomposing into mercury and oxygen.

$$\text{mercury(II) oxide} \xrightarrow{\Delta} \text{mercury} + \text{oxygen}$$

Step 2 **Write the unbalanced (skeleton) equation.** Make sure that the formula for each substance is correct and that reactants are written to the left and products to the right of the arrow. For our example,

$$HgO \xrightarrow{\Delta} Hg + O_2$$

The correct formulas must be known or determined from the periodic table, lists of ions, or experimental data.

Step 3 **Balance the equation.** Use the following process as necessary:

(a) Count and compare the number of atoms of each element on each side of the equation and determine those that must be balanced:

Hg is balanced (1 on each side)

O needs to be balanced

(1 on reactant side, 2 on product side)

(b) Balance each element, one at a time, by placing whole numbers (coefficients) in front of the formulas containing the unbalanced element. It is usually best to balance metals first, then nonmetals, then hydrogen and oxygen. Select the smallest coefficients that will give the same number of atoms of the element on each side. A coefficient placed before a formula multiplies every atom in the formula by that number (e.g., 2 H_2SO_4 means two molecules of sulfuric acid and also means four H atoms, two S atoms, and eight O atoms). Place a 2 in front of HgO to balance O:

$$2\,HgO \xrightarrow{\Delta} Hg + O_2$$

(c) Check all other elements after each individual element is balanced to see whether, in balancing one element, other elements have become unbalanced. Make adjustments as needed. Now Hg is not balanced. To adjust this, we write a 2 in front of Hg:

$$2\,HgO \xrightarrow{\Delta} 2\,Hg + O_2 \quad \text{(balanced)}$$

(d) Do a final check, making sure that each element and/or polyatomic ion is balanced and that the smallest possible set of whole-number coefficients has been used:

$$2\,HgO \xrightarrow{\Delta} 2\,Hg + O_2 \quad \text{(correct form)}$$

$$4\,HgO \xrightarrow{\Delta} 4\,Hg + 2\,O_2 \quad \text{(incorrect form)}$$

Leave elements that are in two or more formulas (on the same side of the equation) unbalanced until just before balancing hydrogen and oxygen.

Not all chemical equations can be balanced by the simple method of inspection just described. The following examples show *stepwise* sequences leading to balanced equations. Study each one carefully.

Example 8.1

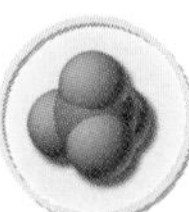

Write the balanced equation for the reaction that takes place when magnesium metal is burned in air to produce magnesium oxide.

SOLUTION

Step 1 *Word equation:*

magnesium + oxygen ⟶ magnesium oxide
reactants (R) product (P)

Step 2 *Skeleton equation:*

$$Mg + O_2 \longrightarrow MgO \quad \text{(unbalanced)}$$

R = reactant
P = product

Step 3 *Balance:*

(a) Mg is balanced.
Oxygen is not balanced. Two O atoms appear on the left side and one on the right side.

R	1 Mg	2 O
P	1 Mg	1 O

R	2 Mg	2 O
P	2 Mg	2 O

(b) Place the coefficient 2 in front of MgO:

$$Mg + O_2 \longrightarrow 2\,MgO \qquad \text{(unbalanced)}$$

(c) Now Mg is not balanced. One Mg atom appears on the left side and two on the right side. Place a 2 in front of Mg:

R	2 Mg	2 O
P	2 Mg	2 O

$$2\,Mg + O_2 \longrightarrow 2\,MgO \qquad \text{(balanced)}$$

(d) *Check:* Each side has two Mg and two O atoms.

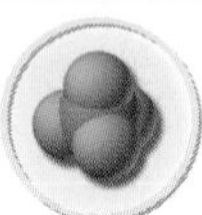

Example 8.2 When methane, CH_4, undergoes complete combustion, it reacts with oxygen to produce carbon dioxide and water. Write the balanced equation for this reaction.

SOLUTION

Step 1 *Word equation:*

$$\text{methane} + \text{oxygen} \longrightarrow \text{carbon dioxide} + \text{water}$$

Step 2 *Skeleton equation:*

R	1 C	4 H	2 O
P	1 C	2 H	3 O

$$CH_4 + O_2 \longrightarrow CO_2 + H_2O \qquad \text{(unbalanced)}$$

Step 3 *Balance:*

(a) Carbon is balanced.
Hydrogen and oxygen are not balanced.

(b) Balance H atoms by placing a 2 in front of H_2O:

R	1 C	4 H	2 O
P	1 C	4 H	4 O

$$CH_4 + O_2 \longrightarrow CO_2 + 2\,H_2O \qquad \text{(unbalanced)}$$

Each side of the equation has four H atoms; oxygen is still not balanced. Place a 2 in front of O_2 to balance the oxygen atoms:

$$CH_4 + 2\,O_2 \longrightarrow CO_2 + 2\,H_2O \qquad \text{(balanced)}$$

R	1 C	4 H	4 O
P	1 C	4 H	4 O

(c) The other atoms remain balanced.

(d) *Check*: The equation is correctly balanced; it has one C, four O, and four H atoms on each side.

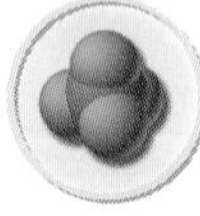

Example 8.3 Oxygen and potassium chloride are formed by heating potassium chlorate. Write a balanced equation for this reaction:

SOLUTION

Step 1 *Word equation:*

$$\text{potassium chlorate} \xrightarrow{\Delta} \text{potassium chloride} + \text{oxygen}$$

Step 2 *Skeleton equation:*

$$KClO_3 \xrightarrow{\Delta} KCl + O_2 \qquad \text{(unbalanced)}$$

Step 3 *Balance:*

R	1 K	1 Cl	3 O
P	1 K	1 Cl	2 O

(a) Potassium and chlorine are balanced.
Oxygen is unbalanced (three O atoms on the left and two on the right side).

(b) How many oxygen atoms are needed? The subscripts of oxygen (3 and 2) in $KClO_3$ and O_2 have a least common multiple of 6. Therefore, coefficients for $KClO_3$ and O_2 are needed to get six O atoms on each side. Place a 2 in front of $KClO_3$ and a 3 in front of O_2:

R	2 K	2 Cl	6 O
P	1 K	1 Cl	6 O

$$2\,KClO_3 \xrightarrow{\Delta} KCl + 3\,O_2 \qquad \text{(unbalanced)}$$

(c) Now K and Cl are not balanced. Place a 2 in front of KCl, which balances both K and Cl at the same time:

$$2\,KClO_3 \xrightarrow{\Delta} 2\,KCl + 3\,O_2 \qquad \text{(balanced)}$$

(d) *Check:* Each side now contains two K, two Cl, and six O atoms.

R	2 K	2 Cl	6 O
P	2 K	2 Cl	6 O

Example 8.4

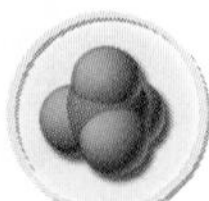

Silver nitrate reacts with hydrogen sulfide to produce silver sulfide and nitric acid. Write a balanced equation for this reaction.

SOLUTION

Step 1 *Word equation:*

silver nitrate + hydrogen sulfide ⟶ silver sulfide + nitric acid

Step 2 *Skeleton equation:*

$$AgNO_3 + H_2S \longrightarrow Ag_2S + HNO_3 \qquad \text{(unbalanced)}$$

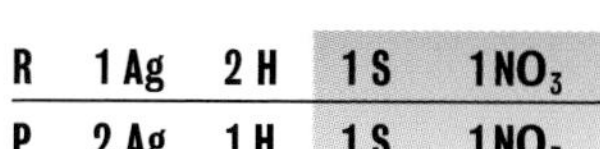

R	1 Ag	2 H	1 S	1 NO_3
P	2 Ag	1 H	1 S	1 NO_3

Step 3 *Balance:*

(a) Ag and H are unbalanced.

(b) Place a 2 in front of $AgNO_3$ to balance Ag:

$$2\,AgNO_3 + H_2S \longrightarrow Ag_2S + HNO_3 \qquad \text{(unbalanced)}$$

R	2 Ag	2 H	1 S	2 NO_3
P	2 Ag	1 H	1 S	1 NO_3

(c) H and NO_3^- are still unbalanced. Balance by placing a 2 in front of HNO_3:

$$2\,AgNO_3 + H_2S \longrightarrow Ag_2S + 2\,HNO_3 \qquad \text{(balanced)}$$

R	2 Ag	2 H	1 S	2 NO_3
P	2 Ag	2 H	1 S	2 NO_3

In this example, N and O atoms are balanced by balancing the NO_3^- ion as a unit.

(d) The other atoms remain balanced.

(e) *Check:* Each side has two Ag, two H, and one S atom. Also, each side has two NO_3^- ions.

Example 8.5

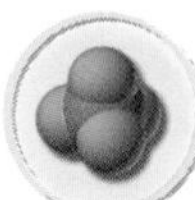

When aluminum hydroxide is mixed with sulfuric acid, the products are aluminum sulfate and water. Write a balanced equation for this reaction.

SOLUTION

Step 1 *Word equation:*

aluminum hydroxide + sulfuric acid ⟶ aluminum sulfate + water

Step 2 *Skeleton equation:*

$$Al(OH)_3 + H_2SO_4 \longrightarrow Al_2(SO_4)_3 + H_2O \qquad \text{(unbalanced)}$$

R	1 Al	1 SO_4	3 O	5 H
P	2 Al	3 SO_4	1 O	2 H

Step 3 *Balance:*

(a) All elements are unbalanced.

(b) Balance Al by placing a 2 in front of $Al(OH)_3$. Treat the unbalanced SO_4^{2-} ion as a unit and balance by placing a 3 in front of H_2SO_4:

$$2\,Al(OH)_3 + 3\,H_2SO_4 \longrightarrow Al_2(SO_4)_3 + H_2O \qquad \text{(unbalanced)}$$

R	2 Al	3 SO_4	6 O	12 H
P	2 Al	3 SO_4	1 O	2 H

Balance the unbalanced H and O by placing a 6 in front of H_2O:

$$2\,Al(OH)_3 + 3\,H_2SO_4 \longrightarrow Al_2(SO_4)_3 + 6\,H_2O \qquad \text{(balanced)}$$

(c) The other atoms remain balanced.

(d) *Check:* Each side has 2 Al, 12 H, 3 S, and 18 O atoms.

R	2 Al	3 SO_4	6 O	12 H
P	2 Al	3 SO_4	6 O	12 H

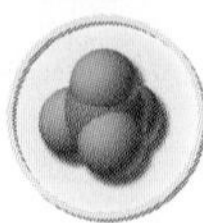

Example 8.6 When the fuel in a butane gas stove undergoes complete combustion, it reacts with oxygen to form carbon dioxide and water. Write the balanced equation for this reaction.

SOLUTION

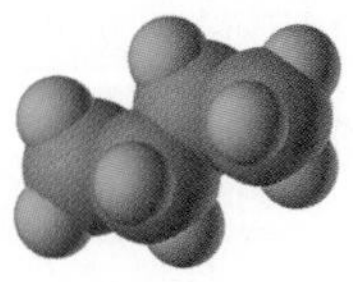

Butane

Step 1 *Word equation:*

butane + oxygen ⟶ carbon dioxide + water

Step 2 *Skeleton equation:*

$C_4H_{10} + O_2 \longrightarrow CO_2 + H_2O$ (unbalanced)

Step 3 *Balance:*

R	4 C	10 H	2 O
P	1 C	2 H	3 O

(a) All elements are unbalanced.

(b) Balance C by placing a 4 in front of CO_2:

$C_4H_{10} + O_2 \longrightarrow 4\,CO_2 + H_2O$ (unbalanced)

Balance H by placing a 5 in front of H_2O:

$C_4H_{10} + O_2 \longrightarrow 4\,CO_2 + 5\,H_2O$ (unbalanced)

R	4 C	10 H	2 O
P	4 C	10 H	13 O

Oxygen remains unbalanced. The oxygen atoms on the right side are fixed because $4\,CO_2$ and $5\,H_2O$ are derived from the single C_4H_{10} molecule on the left. When we try to balance the O atoms, we find that there is no whole number that can be placed in front of O_2 to bring about a balance, so we double the coefficients of each substance, and then balance the oxygen:

$2\,C_4H_{10} + 13\,O_2 \longrightarrow 8\,CO_2 + 10\,H_2O$ (balanced)

R	8 C	20 H	26 O
P	8 C	20 H	26 O

(c) The other atoms remain balanced.

(d) *Check:* Each side now has 8 C, 20 H, and 26 O atoms.

Practice 8.1

Write a balanced formula equation:

aluminum + oxygen ⟶ aluminum oxide

Practice 8.2

Write a balanced formula equation:

magnesium hydroxide + phosphoric acid ⟶ magnesium phosphate + water

8.3 Information in a Chemical Equation

Depending on the particular context in which it is used, a formula can have different meanings. A formula can refer to an individual chemical entity (atom, ion, molecule, or formula unit) or to a mole of that chemical entity. For example, the formula H_2O can mean any of the following:

1. 2 H atoms and 1 O atom
2. 1 molecule of water
3. 1 mol of water
4. 6.022×10^{23} molecules of water
5. 18.02 g of water

CHEMISTRY IN ACTION • CO Poisoning—A Silent Killer

Carbon monoxide is one possible product when carbon-containing fuels such as natural gas, propane, heating oil, or gasoline are burned. For example, when methane, the main component of natural gas, burns, the reaction is

$$CH_4(g) + 2\,O_2(g) \longrightarrow 2\,H_2O(g) + CO_2(g)$$

However, if not enough oxygen is available, carbon monoxide will be produced instead of carbon dioxide:

$$2\,CH_4(g) + 3\,O_2(g) \longrightarrow 4\,H_2O(g) + 2\,CO(g)$$

Carbon monoxide is colorless and has no flavor or smell. It often can poison several people at once, since it can build up undetected in enclosed spaces. The symptoms of carbon monoxide poisoning are so easily missed that people often don't realize that they have been poisoned. Common symptoms are headache, ringing in ears, nausea, dizziness, weakness, confusion, and drowsiness. A low concentration of CO in the air may go completely unnoticed while the toxicity level increases in the blood.

Carbon monoxide in the blood is poisonous because of its ability to stick to a hemoglobin molecule. Both CO and O_2 bind to the same place on the hemoglobin molecule. The CO sticks more tightly than O_2. We use hemoglobin to transport oxygen to the various tissues from the lungs. Unfortunately, CO also binds to hemoglobin, forming a molecule called carboxyhemoglobin. In fact, hemoglobin prefers CO to O_2. Once a CO is bound to a hemoglobin molecule, it cannot carry an oxygen molecule. So, until the CO is released, that molecule of hemoglobin is lost as an O_2 carrier. If you breathe normal air (about 20% oxygen) containing 0.1% CO, in just one hour the CO binds to 50% of the hemoglobin molecules. To keep CO from being a silent killer, there are three possible actions:

1. Immediate treatment of victims to restore the hemoglobin
2. Detection of CO levels before poisoning occurs
3. Conversion of CO to CO_2 to eliminate the threat.

If a person already has CO poisoning, the treatment must focus on getting CO released from the hemoglobin. This can be done by giving the victim oxygen or by placing the victim in a hyperbaric chamber (at about 3 atm pressure). These treatments reduce the time required for carboxyhemoglobin to release CO from between 4 and 6 h to 30 to 90 min. This treatment also supplies more oxygen to the system to help keep the victim's brain functioning.

A CO detector can sound an alarm before the CO level becomes toxic in a home. Current CO detectors sound an alarm when levels increase rapidly or when lower concentrations are present over a long time. Inside a CO detector is a beam of infrared light that shines on a chromophore (a substance that turns darker with increasing CO levels). If the infrared light passing through the chromophore is too low, the alarm sounds. Typically, these alarms ring when CO levels reach 70 ppm, or greater than 30 ppm for 30 days.

The final option is to convert CO to CO_2 before toxicity occurs. David Schreyer from NASA's Langley Research Center has devised a catalyst (tin hydroxide with small amounts of Pd) that speeds up the conversion of CO to CO_2 at room temperature. He and his colleagues have adapted their catalyst to home ventilation systems to eliminate even low levels of CO before it has a chance to build up. Using all of these approaches can help to eliminate 500–10,000 cases of CO poisoning each year.

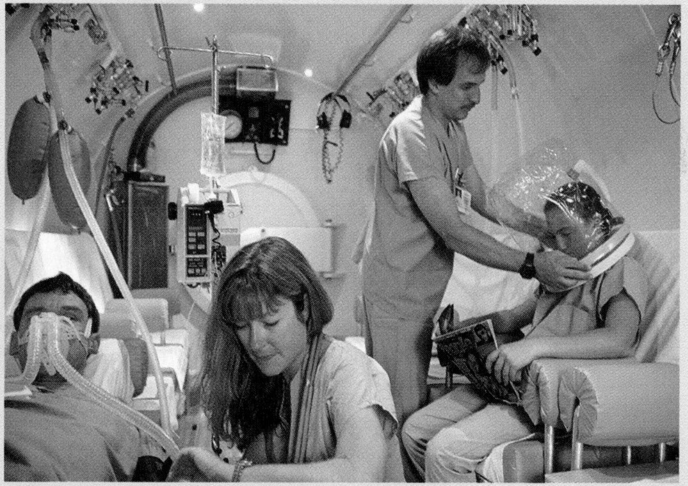
People in a hyperbaric chamber.

Formulas used in equations can represent units of individual chemical entities or moles, the latter being more commonly used. For example, in the reaction of hydrogen and oxygen to form water,

$2\,H_2$	+	O_2	$\longrightarrow$	$2\,H_2O$
2 molecules hydrogen		1 molecule oxygen		2 molecules water
2 mol hydrogen		1 mol oxygen		2 mol water

We generally use moles in equations because molecules are so small.

As indicated earlier, a chemical equation is a shorthand description of a chemical reaction. Interpretation of a balanced equation gives us the following information:

1. What the reactants are and what the products are
2. The formulas of the reactants and products
3. The number of molecules or formula units of reactants and products in the reaction
4. The number of atoms of each element involved in the reaction
5. The number of moles of each substance

Consider the reaction that occurs when propane gas (C_3H_8) is burned in air; the products are carbon dioxide (CO_2) and water (H_2O). The balanced equation and its interpretation are as follows:

Propane		Oxygen		Carbon dioxide		Water
$C_3H_8(g)$	+	$5\ O_2(g)$	→	$3\ CO_2(g)$	+	$4\ H_2O(g)$
1 molecule		5 molecules		3 molecules		4 molecules
3 atoms C 8 atoms H		10 atoms O		3 atoms C 6 atoms O		8 atoms H 4 atoms O
1 mol		5 mol		3 mol		4 mol
44.09 g		5(32.00 g) = (160.0 g)		3(44.01 g) = (132.0 g)		4(18.02 g) = (72.08 g)

Practice 8.3

Consider the reaction that occurs when hydrogen gas reacts with chlorine gas to produce gaseous hydrogen chloride:

(a) Write a word equation for this reaction.
(b) Write a balanced formula equation including the state for each substance.
(c) Label each reactant and product to show the relative amounts of each substance.
  (1) number of molecules
  (2) number of atoms
  (3) number of moles
  (4) mass
(d) What mass of HCl would be produced if you reacted 2 mol hydrogen gas with 2 mol chlorine gas?

The quantities involved in chemical reactions are important when working in industry or the laboratory. We will study the relationship among quantities of reactants and products in the next chapter.

8.4 Types of Chemical Equations

Chemical equations represent chemical changes or reactions. Reactions are classified into types to assist in writing equations and in predicting other reactions. Many chemical reactions fit one or another of the four principal reaction types that we discuss in the following paragraphs. Reactions are also classified as oxidation–reduction. Special methods are used to balance complex oxidation–reduction equations. (See Chapter 17.)

Combination Reaction

combination reaction

In a **combination reaction**, two reactants combine to give one product. The general form of the equation is

$$A + B \longrightarrow AB$$

in which A and B are either elements or compounds and AB is a compound. The formula of the compound in many cases can be determined from a knowledge of the ionic charges of the reactants in their combined states. Some reactions that fall into this category are given here:

1. metal + oxygen $\longrightarrow$ metal oxide:

$$2\,Mg(s) + O_2(g) \xrightarrow{\Delta} 2\,MgO(s)$$
$$4\,Al(s) + 3\,O_2(g) \xrightarrow{\Delta} 2\,Al_2O_3(s)$$

2. nonmetal + oxygen $\longrightarrow$ nonmetal oxide:

$$S(s) + O_2(g) \xrightarrow{\Delta} SO_2(g)$$
$$N_2(g) + O_2(g) \xrightarrow{\Delta} 2\,NO(g)$$

3. metal + nonmetal $\longrightarrow$ salt:

$$2\,Na(s) + Cl_2(g) \longrightarrow 2\,NaCl(s)$$
$$2\,Al(s) + 3\,Br_2(l) \longrightarrow 2\,AlBr_3(s)$$

4. metal oxide + water $\longrightarrow$ metal hydroxide:

$$Na_2O(s) + H_2O(l) \longrightarrow 2\,NaOH(aq)$$
$$CaO(s) + H_2O(l) \longrightarrow Ca(OH)_2(aq)$$

5. nonmetal oxide + water $\longrightarrow$ oxy-acid:

$$SO_3(g) + H_2O(l) \longrightarrow H_2SO_4(aq)$$
$$N_2O_5(s) + H_2O(l) \longrightarrow 2\,HNO_3(aq)$$

Flames and sparks result when aluminum foil is dropped into liquid bromine.

Decomposition Reaction

decomposition reaction

In a **decomposition reaction**, a single substance is decomposed, or broken down, to give two or more different substances. This reaction may be considered the reverse of combination. The starting material must be a compound, and the products may be elements or compounds. The general form of the equation is

$$AB \longrightarrow A + B$$

Hydrogen peroxide decomposes to steam ($H_2O(g)$) and oxygen.

Predicting the products of a decomposition reaction can be difficult and requires an understanding of each individual reaction. Heating oxygen-containing compounds often results in decomposition. Some reactions that fall into this category are

1. Metal oxides. Some metal oxides decompose to yield the free metal plus oxygen; others give another oxide, and some are very stable, resisting decomposition by heating:

$$2\,HgO(s) \xrightarrow{\Delta} 2\,Hg(l) + O_2(g)$$

$$2\,PbO_2(s) \xrightarrow{\Delta} 2\,PbO(s) + O_2(g)$$

2. Carbonates and hydrogen carbonates decompose to yield CO_2 when heated:

$$CaCO_3(s) \xrightarrow{\Delta} CaO(s) + CO_2(g)$$

$$2\,NaHCO_3(s) \xrightarrow{\Delta} Na_2CO_3(s) + H_2O(g) + CO_2(g)$$

3. Miscellaneous reactions in this category:

$$2\,KClO_3(s) \xrightarrow{\Delta} 2\,KCl(s) + 3\,O_2(g)$$

$$2\,NaNO_3(s) \xrightarrow{\Delta} 2\,NaNO_2(g) + O_2(g)$$

$$2\,H_2O_2(l) \xrightarrow{\Delta} 2\,H_2O(l) + O_2(g)$$

Single-Displacement Reaction

single-displacement reaction

In a **single-displacement reaction**, one element reacts with a compound to replace one of the elements of that compound, yielding a different element and a different compound. The general forms of the equation follow.

If A is a *metal*, A will replace B to form AC, provided that A is a more reactive metal than B.

$$A + BC \longrightarrow B + AC$$

If A is a *halogen*, it will replace C to form BA, provided that A is a more reactive halogen than C.

$$A + BC \longrightarrow C + BA$$

A brief activity series of selected metals (and hydrogen) and halogens is shown in Table 8.2. This series is listed in descending order of chemical activity, with the most active metals and halogens at the top. Many chemical reactions can be predicted from an activity series because the atoms of any element in the series will replace the atoms of those elements below it. For example, zinc metal will replace hydrogen from a hydrochloric acid solution. But copper metal, which is below hydrogen on the list and thus less reactive than hydrogen, will not replace hydrogen from a hydrochloric acid solution. Here are some reactions that fall into this category:

When pieces of zinc metal are placed in hydrochloric acid, hydrogen bubbles form immediately.

1. metal + acid $\longrightarrow$ hydrogen + salt

$$Zn(s) + 2\,HCl(aq) \longrightarrow H_2(g) + ZnCl_2(aq)$$

$$2\,Al(s) + 3\,H_2SO_4(aq) \longrightarrow 3\,H_2(g) + Al_2(SO_4)_3(aq)$$

2. metal + water $\longrightarrow$ hydrogen + metal hydroxide or metal oxide

$2\,Na(s) + 2\,H_2O \longrightarrow H_2(g) + 2\,NaOH(aq)$

$Ca(s) + 2\,H_2O \longrightarrow H_2(g) + Ca(OH)_2(aq)$

$3\,Fe(s) + \underset{\text{steam}}{4\,H_2O(g)} \longrightarrow 4\,H_2(g) + Fe_3O_4(s)$

3. metal + salt $\longrightarrow$ metal + salt

$Fe(s) + CuSO_4(aq) \longrightarrow Cu(s) + FeSO_4(aq)$

$Cu(s) + 2\,AgNO_3(aq) \longrightarrow 2\,Ag(s) + Cu(NO_3)_2(aq)$

4. halogen + halide salt $\longrightarrow$ halogen + halide salt

$Cl_2(g) + 2\,NaBr(aq) \longrightarrow Br_2(l) + 2\,NaCl(aq)$

$Cl_2(g) + 2\,KI(aq) \longrightarrow I_2(s) + 2\,KCl(aq)$

Table 8.2 Activity Series

Metals	Halogens
K	F_2
Ca	Cl_2
Na	Br_2
Mg	I_2
Al	
Zn	
Fe	
Ni	
Sn	
Pb	
H	
Cu	
Ag	
Hg	
Au	

(↑ increasing activity)

A common chemical reaction is the displacement of hydrogen from water or acids (shown in 1 and 2 above). This reaction is a good illustration of the relative reactivity of metals and the use of the activity series. For example,

- K, Ca, and Na displace hydrogen from cold water, steam (H_2O), and acids.
- Mg, Al, Zn, and Fe displace hydrogen from steam and acids.
- Ni, Sn, and Pb displace hydrogen only from acids.
- Cu, Ag, Hg, and Au do not displace hydrogen.

Example 8.7

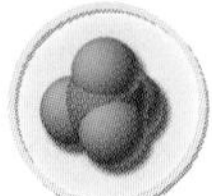

Will a reaction occur between (a) nickel metal and hydrochloric acid and (b) tin metal and a solution of aluminum chloride? Write balanced equations for the reactions.

SOLUTION

(a) Nickel is more reactive than hydrogen, so it will displace hydrogen from hydrochloric acid. The products are hydrogen gas and a salt of Ni^{2+} and Cl^- ions:

$Ni(s) + 2\,HCl(aq) \longrightarrow H_2(g) + NiCl_2(aq)$

(b) According to the activity series, tin is less reactive than aluminum, so no reaction will occur:

$Sn(s) + AlCl_3(aq) \longrightarrow$ no reaction

Practice 8.4

Write balanced equations for these reactions:
(a) iron metal and a solution of magnesium chloride
(b) zinc metal and a solution of lead(II) nitrate

Double-Displacement Reaction

double-displacement reaction

In a **double-displacement reaction**, two compounds exchange partners with each other to produce two different compounds. The general form of the equation is

$$AB + CD \longrightarrow AD + CB$$

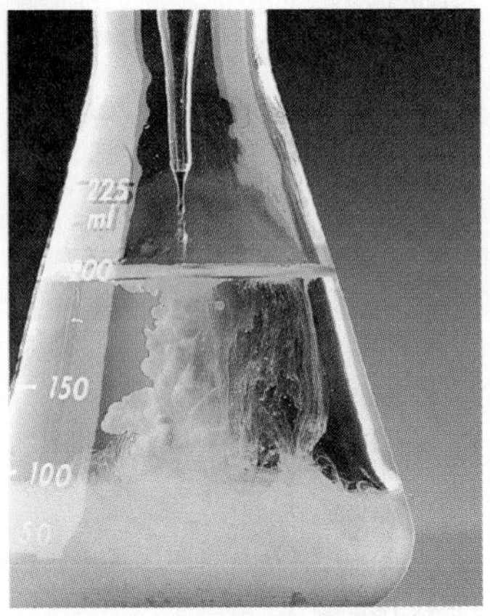

A double-displacement reaction results from pouring a clear, colorless solution of $Pb(NO_3)_2$ into a clear, colorless solution of KI, forming a yellow precipitate of PbI_2.

This reaction can be thought of as an exchange of positive and negative groups, in which A combines with D and C combines with B. In writing formulas for the products, we must account for the charges of the combining groups.

It's also possible to write an equation in the form of a double-displacement reaction when a reaction has not occurred. For example, when solutions of sodium chloride and potassium nitrate are mixed, the following equation can be written:

$$NaCl(aq) + KNO_3(aq) \longrightarrow NaNO_3(aq) + KCl(aq)$$

When the procedure is carried out, no physical changes are observed, indicating that no chemical reaction has taken place.

A double-displacement reaction is accompanied by evidence of such reactions as

1. The evolution of heat
2. The formation of an insoluble precipitate
3. The production of gas bubbles

Let's look at some of these reactions more closely:

Neutralization of an Acid and a Base The production of a molecule of water from an H^+ and an OH^- ion is accompanied by a *release of heat*, which can be detected by touching the reaction container. For neutralization reactions, $H^+ + OH^- \longrightarrow H_2O$:

$$\text{acid} + \text{base} \longrightarrow \text{salt} + \text{water}$$
$$HCl(aq) + NaOH(aq) \longrightarrow NaCl(aq) + H_2O(l)$$
$$H_2SO_4(aq) + Ba(OH)_2(aq) \longrightarrow BaSO_4(s) + 2\,H_2O(l)$$

Metal Oxide + Acid *Heat is released* by the production of a molecule of water.

$$\text{metal oxide} + \text{acid} \longrightarrow \text{salt} + \text{water}$$
$$CuO(s) + 2\,HNO_3(aq) \longrightarrow Cu(NO_3)_2(aq) + H_2O(l)$$
$$CaO(s) + 2\,HCl(aq) \longrightarrow CaCl_2(aq) + H_2O(l)$$

Formation of an Insoluble Precipitate The solubilities of the products can be determined by consulting the solubility table in Appendix V to see whether one or both of the products are insoluble in water. An insoluble product (precipitate) is indicated by placing (*s*) after its formula in the equation.

$$BaCl_2(aq) + 2\,AgNO_3(aq) \longrightarrow 2\,AgCl(s) + Ba(NO_3)_2(aq)$$
$$Pb(NO_3)_2(aq) + 2\,KI(aq) \longrightarrow PbI_2(s) + 2\,KNO_3(aq)$$

Formation of a Gas A gas such as HCl or H_2S may be produced directly, as in these two examples:

$$H_2SO_4(l) + NaCl(s) \longrightarrow NaHSO_4(s) + HCl(g)$$
$$2\,HCl(aq) + ZnS(s) \longrightarrow ZnCl_2(aq) + H_2S(g)$$

A gas can also be produced indirectly. Some unstable compounds formed in a double-displacement reaction, such as H_2CO_3, H_2SO_3, and NH_4OH, will decompose to form water and a gas:

$$2\,HCl(aq) + Na_2CO_3(aq) \longrightarrow 2\,NaCl(aq) + H_2CO_3(aq) \longrightarrow 2\,NaCl(aq) + H_2O(l) + CO_2(g)$$
$$2\,HNO_3(aq) + K_2SO_3(aq) \longrightarrow 2\,KNO_3(aq) + H_2SO_3(aq) \longrightarrow 2\,KNO_3(aq) + H_2O(l) + SO_2(g)$$
$$NH_4Cl(aq) + NaOH(aq) \longrightarrow NaCl(aq) + NH_4OH(aq) \longrightarrow NaCl(aq) + H_2O(l) + NH_3(g)$$

Example 8.8

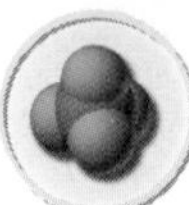

Write the equation for the reaction between aqueous solutions of hydrobromic acid and potassium hydroxide.

SOLUTION

First we write the formulas for the reactants. They are HBr and KOH. Then we classify the type of reaction that would occur between them. Because the reactants are compounds, one an acid and the other a base, the reaction will be of the neutralization type:

$$\text{acid} + \text{base} \longrightarrow \text{salt} + \text{water}$$

Now rewrite the equation using the formulas for the known substances:

$$HBr(aq) + KOH(aq) \longrightarrow \text{salt} + H_2O$$

In this reaction, which is a double-displacement type, the H^+ from the acid combines with the OH^- from the base to form water. The ionic compound must be composed of the other two ions, K^+ and Br^-. We determine the formula of the ionic compound to be KBr from the fact that K is a +1 cation and Br is a −1 anion. The final balanced equation is

$$HBr(aq) + KOH(aq) \longrightarrow KBr(aq) + H_2O(l)$$

Example 8.9

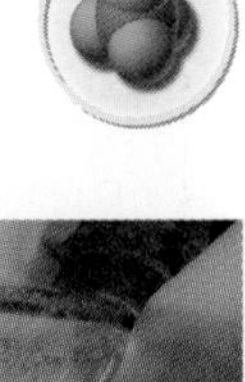

Complete and balance the equation for the reaction between aqueous solutions of barium chloride and sodium sulfate.

SOLUTION

First determine the formulas for the reactants. They are $BaCl_2$ and Na_2SO_4. Then classify these substances as acids, bases, or ionic compounds. Both substances are ionic compounds. Since both substances are compounds, the reaction will be of the double-displacement type. Start writing the equation with the reactants:

$$BaCl_2(aq) + Na_2SO_4(aq) \longrightarrow$$

If the reaction is double-displacement, Ba^{2+} will be written combined with SO_4^{2-}, and Na^+ with Cl^- as the products. The balanced equation is

$$BaCl_2(aq) + Na_2SO_4(aq) \longrightarrow BaSO_4 + 2\,NaCl$$

The final step is to determine the nature of the products, which controls whether or not the reaction will take place. If both products are soluble, we have a mixture of all the ions in solution. But if an insoluble precipitate is formed, the reaction will definitely occur. We know from experience that NaCl is fairly soluble in water, but what about $BaSO_4$? Consulting the solubility table in Appendix V, we see that $BaSO_4$ is insoluble in water, so it will be a precipitate in the reaction. Thus, the reaction will occur, forming a precipitate. The equation is

$$BaCl_2(aq) + Na_2SO_4(aq) \longrightarrow BaSO_4(s) + 2\,NaCl(aq)$$

When barium chloride is poured into a solution of sodium sulfate, a white precipitate of barium sulfate forms.

Practice 8.5

Complete and balance the equations for these reactions:
(a) potassium phosphate + barium chloride
(b) hydrochloric acid + nickel carbonate
(c) ammonium chloride + sodium nitrate

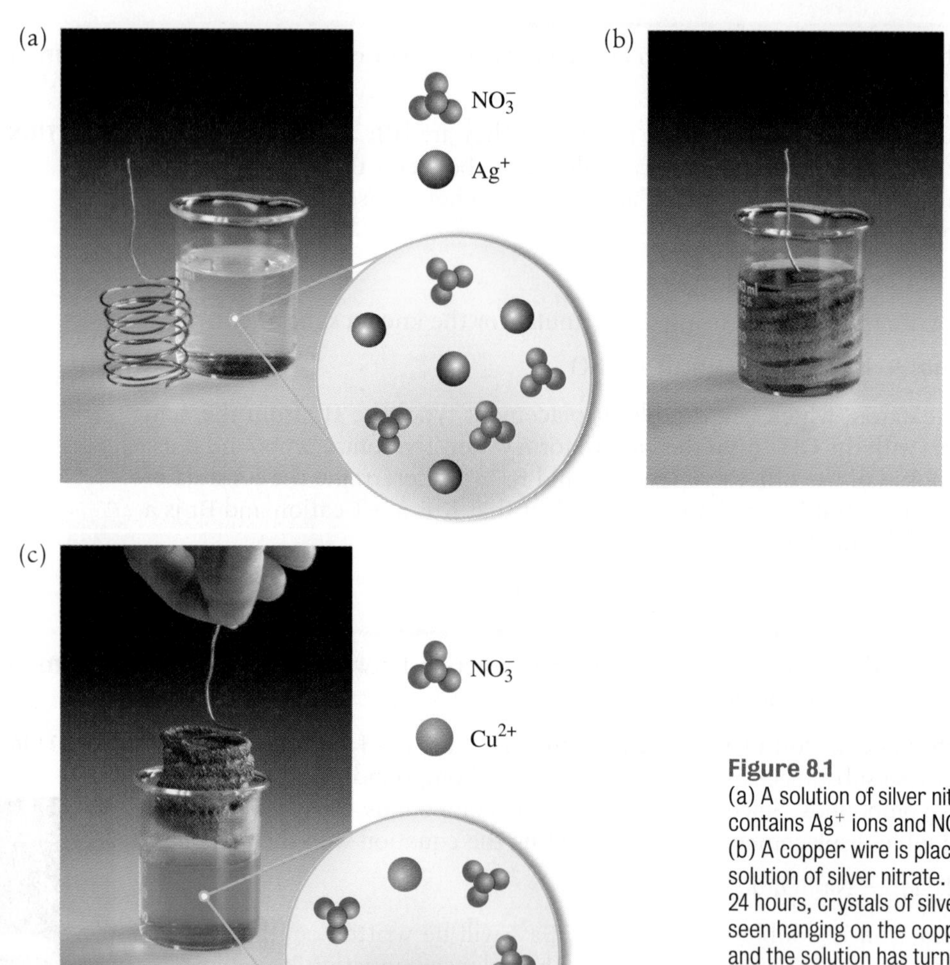

Figure 8.1
(a) A solution of silver nitrate contains Ag^+ ions and NO_3^- ions. (b) A copper wire is placed in a solution of silver nitrate. After 24 hours, crystals of silver are seen hanging on the copper wire and the solution has turned blue, indicating copper ions are present there. (c) The silver metal clings to the copper wire and the blue solution contains Cu^{2+} and NO_3^- ions.

Some of the reactions you attempt may fail because the substances are not reactive or because the proper conditions for reaction are not present. For example, mercury(II) oxide does not decompose until it is heated; magnesium does not burn in air or oxygen until the temperature reaches a certain point. When silver is placed in a solution of copper(II) sulfate, no reaction occurs. When copper wire is placed in a solution of silver nitrate, a single-displacement reaction takes place because copper is a more reactive metal than silver. (See Figure 8.1.)

The successful prediction of the products of a reaction is not always easy. The ability to predict products correctly comes with knowledge and experience. Although you may not be able to predict many reactions at this point, as you continue to experiment you will find that reactions can be categorized and that prediction of the products becomes easier, if not always certain.

8.5 Heat in Chemical Reactions

Energy changes always accompany chemical reactions. One reason reactions occur is that the products attain a lower, more stable energy state than the reactants. When the reaction leads to a more stable state, energy is released to the surroundings as heat and/or work. When a solution of a base is neutralized by the addition of an acid, the liberation of heat energy is signaled by an immediate rise in the temperature of the solution. When an automobile engine burns gasoline, heat is certainly liberated; at the same time, part of the liberated energy does the work of moving the automobile.

Reactions are either exothermic or endothermic. **Exothermic reactions** liberate heat; **endothermic reactions** absorb heat. In an exothermic reaction, heat is a product and may be written on the right side of the equation for the reaction. In an endothermic reaction, heat can be regarded as a reactant and is written on the left side of the equation. Here are two examples:

exothermic reaction
endothermic reaction

$$H_2(g) + Cl_2(g) \longrightarrow 2\,HCl(g) + 185\text{ kJ} \quad (\textit{exothermic})$$

$$N_2(g) + O_2(g) + 181\text{ kJ} \longrightarrow 2\,NO(g) \quad (\textit{endothermic})$$

The quantity of heat produced by a reaction is known as the **heat of reaction**. The units used can be kilojoules or kilocalories. Consider the reaction represented by this equation:

heat of reaction

$$C(s) + O_2(g) \longrightarrow CO_2(g) + 393\text{ kJ}$$

When the heat released is expressed as part of the equation, the substances are expressed in units of moles. Thus, when 1 mol (12.01 g) of C combines with 1 mol (32.00 g) of O_2, 1 mol (44.01 g) of CO_2 is formed and 393 kJ of heat are released. In this reaction, as in many others, the heat energy is more useful than the chemical products.

This cornfield is a good example of the endothermic reactions happening through photosynthesis in plants.

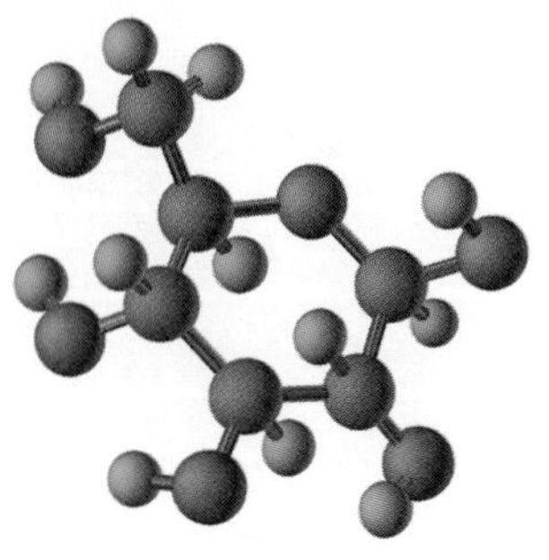
Glucose

Aside from relatively small amounts of energy from nuclear processes, the sun is the major provider of energy for life on Earth. The sun maintains the temperature necessary for life and also supplies light energy for the endothermic photosynthetic reactions of green plants. In photosynthesis, carbon dioxide and water are converted to free oxygen and glucose:

$$6\,CO_2 + 6\,H_2O + 2519\,\text{kJ} \longrightarrow \underset{\text{glucose}}{C_6H_{12}O_6} + 6\,O_2$$

Nearly all of the chemical energy used by living organisms is obtained from glucose or compounds derived from glucose.

The major source of energy for modern technology is fossil fuel—coal, petroleum, and natural gas. The energy is obtained from the combustion (burning) of these fuels, which are converted to carbon dioxide and water. Fossil fuels are mixtures of **hydrocarbons**, compounds containing only hydrogen and carbon.

hydrocarbon

Natural gas is primarily methane, CH_4. Petroleum is a mixture of hydrocarbons (compounds of carbon and hydrogen). Liquefied petroleum gas (LPG) is a mixture of propane (C_3H_8) and butane (C_4H_{10}).

The combustion of these fuels releases a tremendous amount of energy, but reactions won't occur to a significant extent at ordinary temperatures. A spark or a flame must be present before methane will ignite. The amount of energy that must be supplied to start a chemical reaction is called the **activation energy**. In an exothermic reaction, once this activation energy is provided, enough energy is then generated to keep the reaction going.

activation energy

Here are some examples:

$$CH_4(g) + 2\,O_2(g) \longrightarrow CO_2(g) + 2\,H_2O(g) + 890\,\text{kJ}$$

$$C_3H_8(g) + 5\,O_2(g) \longrightarrow 3\,CO_2(g) + 4\,H_2O(g) + 2200\,\text{kJ}$$

$$2\,C_8H_{18}(l) + 25\,O_2(g) \longrightarrow 16\,CO_2(g) + 18\,H_2O(g) + 10{,}900\,\text{kJ}$$

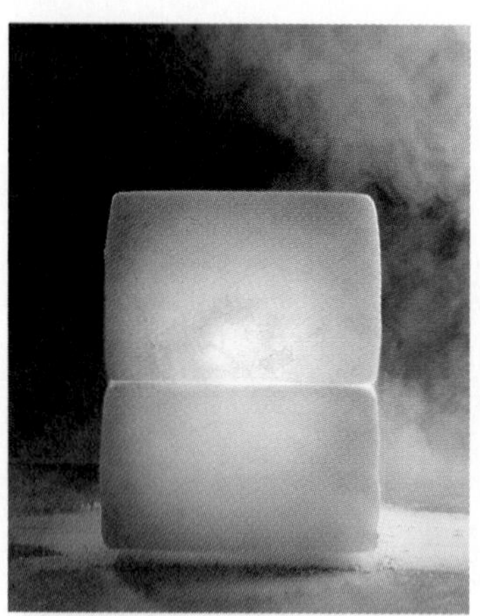
Combustion of Mg inside a block of dry ice (CO_2) makes a glowing lantern.

Be careful not to confuse an exothermic reaction that requires heat (activation energy) to get it *started* with an endothermic process that requires energy to keep it going. The combustion of magnesium, for example, is highly exothermic, yet magnesium must be heated to a fairly high temperature in air before combustion begins. Once started, however, the combustion reaction goes very vigorously until either the magnesium or the available supply of oxygen is exhausted. The electrolytic decomposition of water to hydrogen and oxygen is highly endothermic. If the electric current is shut off when this process is going on, the reaction stops instantly. The relative energy levels of reactants and products in exothermic and in endothermic processes are presented graphically in Figures 8.2 and 8.3.

Examples of endothermic and exothermic processes can be easily demonstrated. In Figure 8.2, the products are at a higher potential energy than the reactants. Energy has therefore been absorbed, and the reaction is endothermic. An endothermic reaction takes place when you apply a cold pack to an injury. When a cold pack is activated, ammonium chloride (NH_4Cl) dissolves in water. For example, temperature changes from 24.5°C to 18.1°C result

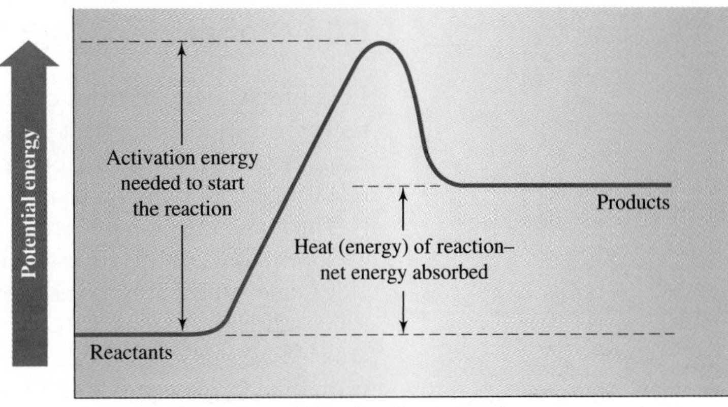

Figure 8.2
An endothermic process occurs when a cold pack containing an ampule of solid NH_4Cl is broken, releasing the crystals into water in the pack. The dissolving of NH_4Cl in water is endothermic, producing a salt solution that is cooler than the surroundings. This process is represented graphically to show the energy changes between the reactants and the products.

when 10 g of NH_4Cl are added to 100 mL of water. Energy, in the form of heat, is taken from the immediate surroundings (water), causing the salt solution to become cooler.

In Figure 8.3, the products are at a lower potential energy than the reactants. Energy (heat) is given off, producing an exothermic reaction. Here, potassium chlorate ($KClO_3$) and sugar are mixed and placed into a pile. A drop of concentrated sulfuric acid is added, creating a spectacular exothermic reaction.

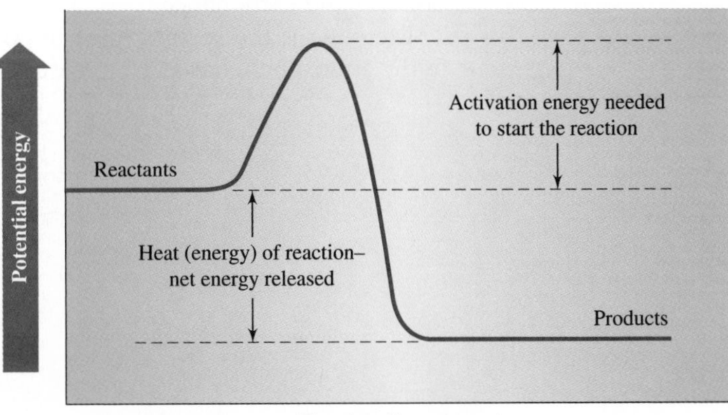

Figure 8.3
Exothermic reaction between $KClO_3$ and sugar. A sample of $KClO_3$ and sugar are well mixed and placed on a fireproof pad. Several drops of concentrated H_2SO_4 are used to ignite the mixture.

8.6 Global Warming: The Greenhouse Effect

Fossil fuels, derived from coal and petroleum, provide the energy we use to power our industries, heat and light our homes and workplaces, and run our cars. As these fuels are burned, they produce carbon dioxide and water, releasing over 50 billion tons of carbon dioxide into our atmosphere each year.

The concentration of carbon dioxide has been monitored by scientists since 1958. Analysis of the air trapped in a core sample of snow from Antartica provides data on carbon dioxide levels for the past 160,000 years. The results of this study show that as the carbon dioxide increased, the global temperature increased as well. The levels of carbon dioxide remained reasonably constant from the last ice age, 100,000 years ago, until the Industrial Revolution. Since then, the concentration of carbon dioxide in our atmosphere has risen 15% to an all-time high. (See Figure 8.4.)

Carbon dioxide is a minor component in our atmosphere and is not usually considered to be a pollutant. The concern expressed by scientists arises from the dramatic increase occurring in the Earth's atmosphere. Without the influence of humans in the environment, the exchange of carbon dioxide between plants and animals would be relatively balanced. Our continued use of fossil fuels led to an increase of 7.4% in carbon dioxide between 1900 and 1970 and an additional 3.5% increase during the 1980s. Continued increases were observed in the 1990s.

Besides our growing consumption of fossil fuels, other factors contribute to increased carbon dioxide levels in the atmosphere: Rain forests are being destroyed by cutting and burning to make room for increased population and agricultural needs. Carbon dioxide is added to the atmosphere during the burning, and the loss of trees diminishes the uptake of carbon dioxide by plants.

About half of all the carbon dioxide released into our atmosphere each year remains there, thus increasing its concentration. The other half is absorbed by plants during photosynthesis or is dissolved in the ocean to form hydrogen carbonates and carbonates.

Methane is the second most important greenhouse gas. Its concentration in the atmosphere has also increased significantly since the 1850s, as shown in

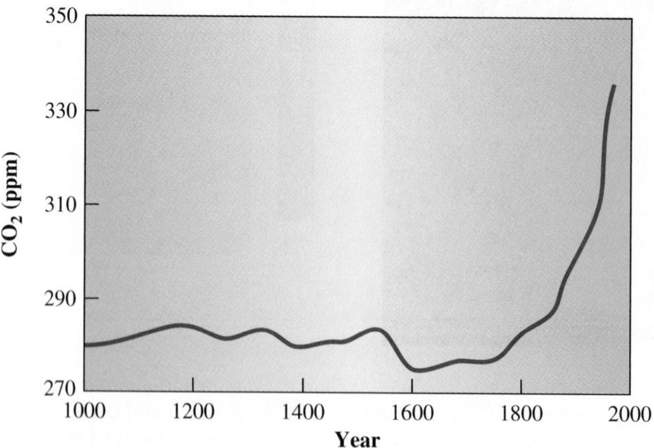

Figure 8.4
Concentration of CO_2 in the atmosphere.

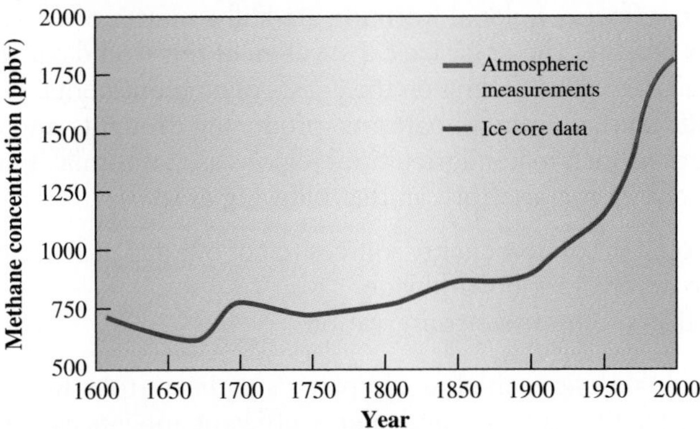

Figure 8.5
Concentration of methane in the atmosphere.

Figure 8.5. Methane is produced by cows, termites, agriculture, and anaerobic bacteria. Coal mining and oil wells also release methane into the atmosphere. The greenhouse effect of methane is 20 times more than that of CO_2, but there is less methane in the atmosphere. It presently contributes about 25% of the global warming.

Carbon dioxide and other greenhouse gases, such as methane and water, act to warm our atmosphere by trapping heat near the surface of the Earth. Solar radiation strikes the Earth and warms the surface. The warmed surface then reradiates this energy as heat. The greenhouse gases absorb some of this heat energy from the surface, which then warms our atmosphere. (See Figure 8.6.) A similar principle is illustrated in a greenhouse, where sunlight comes through the glass yet heat cannot escape. The air in the greenhouse warms, producing a climate considerably different than that in nature. In the atmosphere, these greenhouse gases are producing dramatic changes in our climate.

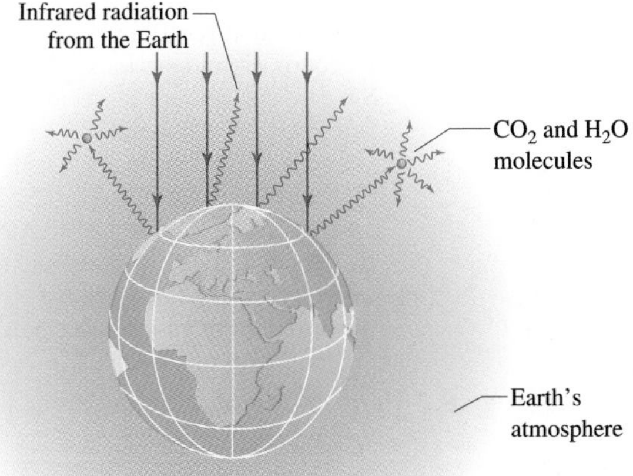

Figure 8.6
Visible light from the sun (red arrows) penetrates the atmosphere and strikes the Earth. Part of this light is changed to infrared radiation. Molecules of CO_2 and H_2O in the atmosphere absorb infrared radiation, acting like the glass of a greenhouse by trapping the energy.

Long-term effects of global warming are still a matter of debate. One consideration is whether the polar ice caps will melt; this would cause a rise in sea level and lead to major flooding on the coasts of our continents. Further effects could include shifts in rainfall patterns, producing droughts and extreme seasonal changes in such major agricultural regions as California. To reverse these trends will require major efforts in the following areas:

- The development of new energy sources to cut our dependence on fossil fuels
- An end to deforestation worldwide
- Intense efforts to improve conservation

On an individual basis each of us can play a significant role. Recycling, switching to more fuel-efficient cars, and energy-efficient appliances, heaters, and air conditioners all would result in decreased energy consumption and less carbon dioxide being released into our atmosphere.

Chapter 8 Review

8.1 The Chemical Equation

KEY TERMS

Reactants
Products
Chemical equation

- A chemical equation is shorthand for expressing a chemical change or reaction.
- In a chemical reaction atoms are neither created nor destroyed.
- All atoms in the reactants must be present in the products.

8.2 Writing and Balancing Chemical Equations

KEY TERM

Balanced equation

- To balance a chemical equation:
  - Identify the reaction.
  - Write the unbalanced (skeleton) equation.
  - Balance the equation:
    - Count the number of atoms of each element on each side and determine which need to be balanced.
    - Balance each element (one at a time) by placing whole numbers (coefficients) in front of the formulas containing the unbalanced element:
      - Begin with metals, then nonmetals, then H and O.
    - Check the other elements to see if they have become unbalanced in the process of balancing the chosen element. If so, rebalance as needed.
    - Do a final check to make sure all elements are balanced.

8.3 Information in a Chemical Equation

- The following information can be found in a chemical equation:
  - Identity of reactants and products
  - Formulas for reactants and products
  - Number of formula units for reactants and products
  - Number of atoms of each element in the reaction
  - Number of moles of each substance

8.4 Types of Chemical Equations

KEY TERMS

Combination reaction
Decomposition reaction
Single-displacement reaction
Double-displacement reaction

- Combination reactions $A + B \rightarrow AB$
- Decomposition reactions $AB \rightarrow A + B$
- Single-displacement reactions:
  - In which A is a metal $A + BC \rightarrow B + AC$
  - In which A is a halogen $A + BC \rightarrow C + BA$
  - Double-displacement reactions $AB + CD \rightarrow AD + CB$
- Evidence for a chemical reaction:
  - Evolution of heat
  - Formation of an insoluble precipitate
  - Production of a gas

8.5 Heat in Chemical Reactions

KEY TERMS

Exothermic reactions
Endothermic reactions
Heat of reaction
Hydrocarbons
Activation energy

- Exothermic reactions release heat.

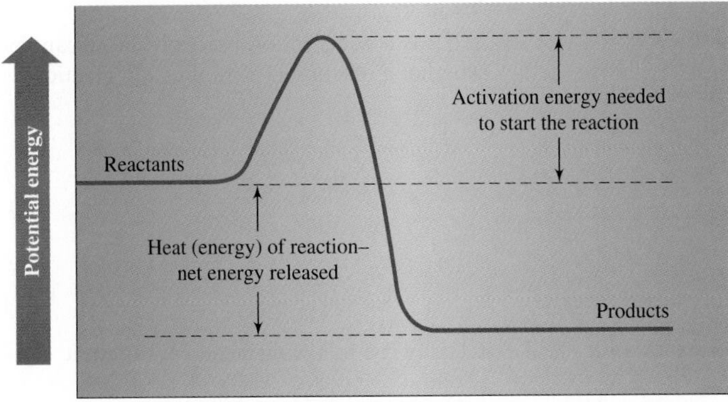

- Endothermic reactions absorb heat.

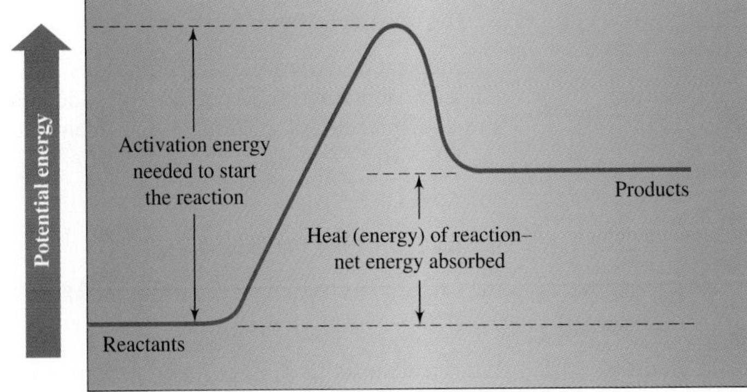

- The amount of heat released or absorbed in a chemical reaction is called the heat of reaction:
  - It can be written as a reactant or product in the chemical equation.
  - Units are joules (J) or kilojoules (kJ).
- The major source of energy for modern technology is hydrocarbon combustion.
- To initiate a chemical reaction, activation energy is required:
  - In exothermic reactions this energy is returned and more is released, which allows the reaction to continue on its own.
  - In endothermic reactions energy must be added to start and continue to be added to sustain the reaction.

8.6 Global Warming: The Greenhouse Effect

- Carbon dioxide levels are now increasing on our planet every year.
- Carbon dioxide and other greenhouse gases warm the atmosphere by trapping heat near the surface of the Earth.

Review Questions

All questions with blue numbers have answers in the appendix of the text.

1. What is the purpose of balancing equations?
2. What is represented by the numbers (coefficients) that are placed in front of the formulas in a balanced equation?
3. In a balanced chemical equation:
   (a) are atoms conserved?
   (b) are molecules conserved?
   (c) are moles conserved?
   Explain yours answers briefly.

4. Explain how endothermic reactions differ from exothermic reactions.
5. In Section 8.2, there are small charts in color along the margins. What information is represented in those charts?
6. What is meant by the physical state of a substance? What symbols are used to represent these physical states and what does each symbol mean?
7. What information does the activity series shown in Table 8.2 give?
8. What is a combustion reaction? How can you identify whether a reaction is a combustion reaction?

Paired Exercises

All exercises with blue numbers have answers in the appendix of the text.

1. Classify the following as an endothermic or exothermic reaction.
   (a) freezing water
   (b) the reaction inside an ice pack
   (c) burning wood
   (d) combustion of Mg in dry ice
   (e) melting ice

2. Classify the following as an endothermic or exothermic reaction.
   (a) making popcorn in a microwave
   (b) a burning match
   (c) boiling water
   (d) burning rocket fuel
   (e) the reaction inside a heat pack

3. Balance each of the following equations. Classify each reaction as combination, decomposition, single-displacement, or double-displacement.
   (a) $H_2 + O_2 \longrightarrow H_2O$
   (b) $C + Fe_2O_3 \longrightarrow Fe + CO$
   (c) $H_2SO_4 + NaOH \longrightarrow H_2O + Na_2SO_4$
   (d) $Al_2(CO_3)_3 \xrightarrow{\Delta} Al_2O_3 + CO_2$
   (e) $NH_4I + Cl_2 \longrightarrow NH_4Cl + I_2$

4. Balance each of the following equations. Classify each reaction as combination, decomposition, single-displacement, or double-displacement.
   (a) $H_2 + Br_2 \longrightarrow HBr$
   (b) $Al + C \xrightarrow{\Delta} Al_4C_3$
   (c) $Ba(ClO_3)_2 \xrightarrow{\Delta} BaCl_2 + O_2$
   (d) $CrCl_3 + AgNO_3 \longrightarrow Cr(NO_3)_3 + AgCl$
   (e) $H_2O_2 \longrightarrow H_2O + O_2$

5. What reactant(s) is (are) required to form an oxide product?

6. What reactant(s) is (are) required to form a salt?

7. Balance the following equations:
   (a) $MnO_2 + CO \longrightarrow Mn_2O_3 + CO_2$
   (b) $Mg_3N_2 + H_2O \longrightarrow Mg(OH)_2 + NH_3$
   (c) $C_3H_5(NO_3)_3 \longrightarrow CO_2 + H_2O + N_2 + O_2$
   (d) $FeS + O_2 \longrightarrow Fe_2O_3 + SO_2$
   (e) $Cu(NO_3)_2 \longrightarrow CuO + NO_2 + O_2$
   (f) $NO_2 + H_2O \longrightarrow HNO_3 + NO$
   (g) $Al + H_2SO_4 \longrightarrow Al_2(SO_4)_3 + H_2$
   (h) $HCN + O_2 \longrightarrow N_2 + CO_2 + H_2O$
   (i) $B_5H_9 + O_2 \longrightarrow B_2O_3 + H_2O$

8. Balance the following equations:
   (a) $SO_2 + O_2 \longrightarrow SO_3$
   (b) $Al + MnO_2 \xrightarrow{\Delta} Mn + Al_2O_3$
   (c) $Na + H_2O \longrightarrow NaOH + H_2$
   (d) $AgNO_3 + Ni \longrightarrow Ni(NO_3)_2 + Ag$
   (e) $Bi_2S_3 + HCl \longrightarrow BiCl_3 + H_2S$
   (f) $PbO_2 \xrightarrow{\Delta} PbO + O_2$
   (g) $LiAlH_4 \xrightarrow{\Delta} LiH + Al + H_2$
   (h) $KI + Br_2 \longrightarrow KBr + I_2$
   (i) $K_3PO_4 + BaCl_2 \longrightarrow KCl + Ba_3(PO_4)_2$

9. Change these word equations into formula equations and balance them:
   (a) water $\longrightarrow$ hydrogen + oxygen
   (b) acetic acid + potassium hydroxide $\longrightarrow$ potassium acetate + water
   (c) phosphorus + iodine $\longrightarrow$ phosphorus triiodide
   (d) aluminum + copper(II) sulfate $\longrightarrow$ copper + aluminum sulfate
   (e) ammonium sulfate + barium chloride $\longrightarrow$ ammonium chloride + barium sulfate
   (f) sulfur tetrafluoride + water $\longrightarrow$ sulfur dioxide + hydrogen fluoride
   (g) chromium(III) carbonate $\xrightarrow{\Delta}$ chromium(III) oxide + carbon dioxide

10. Change these word equations into formula equations and balance them:
   (a) copper + sulfur $\xrightarrow{\Delta}$ copper(I) sulfide
   (b) phosphoric acid + calcium hydroxide $\longrightarrow$ calcium phosphate + water
   (c) silver oxide $\xrightarrow{\Delta}$ silver + oxygen
   (d) iron(III) chloride + sodium hydroxide $\longrightarrow$ iron(III) hydroxide + sodium chloride
   (e) nickel(II) phosphate + sulfuric acid $\longrightarrow$ nickel(II) sulfate + phosphoric acid
   (f) zinc carbonate + hydrochloric acid $\longrightarrow$ zinc chloride + water + carbon dioxide
   (g) silver nitrate + aluminum chloride $\longrightarrow$ silver chloride + aluminum nitrate

11. For each of the following reactions, predict the products, converting each to a balanced formula equation:
(a) Aqueous solutions of sulfuric acid and sodium hydroxide are mixed together. (Heat is released during the reaction.)
(b) Aqueous solutions of lead(II) nitrate and potassium bromide are mixed together. (The solution turns cloudy white during the reaction.)
(c) Aqueous solutions of ammonium chloride and silver nitrate are mixed together. (The solution turns cloudy white during the reaction.)
(d) Solid calcium carbonate is mixed with acetic acid. (Bubbles of gas are formed during the reaction.)

12. For each of the following reactions, predict the products, converting each to a balanced formula equation:
(a) Aqueous solutions of copper(II) sulfate and potassium hydroxide are mixed together. (The solution turns cloudy and light blue during the reaction.)
(b) Aqueous solutions of phosphoric acid and sodium hydroxide are mixed together. (Heat is produced during the reaction.)
(c) Solid sodium bicarbonate is mixed with phosphoric acid. (Bubbles of gas are formed during the reaction.)
(d) Aqueous solutions of aluminum chloride and lead(II) nitrate are mixed together. (The solution turns cloudy white during the reaction.)

13. Use the activity series to predict which of the following reactions will occur. Complete and balance the equations. Where no reaction will occur, write "no reaction" as the product.
(a) $Ag(s) + H_2SO_4(aq) \longrightarrow$
(b) $Cl_2(g) + NaBr(aq) \longrightarrow$
(c) $Mg(s) + ZnCl_2(aq) \longrightarrow$
(d) $Pb(s) + AgNO_3(aq) \longrightarrow$

14. Use the activity series to predict which of the following reactions will occur. Complete and balance the equations. Where no reaction will occur, write "no reaction" as the product.
(a) $Cu(s) + FeCl_3(aq) \longrightarrow$
(b) $H_2(g) + Al_2O_3(aq) \longrightarrow$
(c) $Al(s) + HBr(aq) \longrightarrow$
(d) $I_2(s) + HCl(aq)$

15. Complete and balance the equations for these reactions. All reactions yield products.
(a) $H_2 + I_2 \longrightarrow$
(b) $CaCO_3 \xrightarrow{\Delta}$
(c) $Mg + H_2SO_4 \longrightarrow$
(d) $FeCl_2 + NaOH \longrightarrow$

16. Complete and balance the equations for these reactions. All reactions yield products.
(a) $SO_2 + H_2O \longrightarrow$
(b) $SO_3 + H_2O \longrightarrow$
(c) $Ca + H_2O \longrightarrow$
(d) $Bi(NO_3)_3 + H_2S \longrightarrow$

17. Complete and balance the equations for these reactions. All reactions yield products.
(a) $Ba + O_2 \longrightarrow$
(b) $NaHCO_3 \xrightarrow{\Delta} Na_2CO_3 +$
(c) $Ni + CuSO_4 \longrightarrow$
(d) $MgO + HCl \longrightarrow$
(e) $H_3PO_4 + KOH \longrightarrow$

18. Complete and balance the equations for these reactions. All reactions yield products.
(a) $C + O_2 \longrightarrow$
(b) $Al(ClO_3)_3 \xrightarrow{\Delta} O_2 +$
(c) $CuBr_2 + Cl_2 \longrightarrow$
(d) $SbCl_3 + (NH_4)_2S \longrightarrow$
(e) $NaNO_3 \xrightarrow{\Delta} NaNO_2 +$

19. Interpret these chemical reactions in terms of the number of moles of each reactant and product:
(a) $MgBr_2 + 2\,AgNO_3 \longrightarrow Mg(NO_3)_2 + 2\,AgBr$
(b) $N_2 + 3\,H_2 \longrightarrow 2\,NH_3$
(c) $2\,C_3H_7OH + 9\,O_2 \longrightarrow 6\,CO_2 + 8\,H_2O$

20. Interpret these equations in terms of the relative number of moles of each substance involved and indicate whether the reaction is exothermic or endothermic:
(a) $2\,Na + Cl_2 \longrightarrow 2\,NaCl + 822\ kJ$
(b) $PCl_5 + 92.9\ kJ \longrightarrow PCl_3 + Cl_2$

21. Write balanced equations for each of these reactions, including the heat term:
(a) Lime, CaO, is converted to slaked lime $Ca(OH)_2$ by reaction with water. The reaction liberates 65.3 kJ of heat for each mole of lime reacted.
(b) The industrial production of aluminum metal from aluminum oxide is an endothermic electrolytic process requiring 1630 kJ per mole of Al_2O_3. Oxygen is also a product.

22. Write a balanced equation for these reactions. Include a heat term on the appropriate side of the equation.
(a) Powdered aluminum will react with crystals of iodine when moistened with dishwashing detergent. The reaction produces violet sparks and flaming aluminum. The major product is aluminum iodide (AlI_3). The detergent is not a reactant.
(b) Copper(II) oxide (CuO), a black powder, can be decomposed to produce pure copper by heating the powder in the presence of methane gas (CH_4). The products are copper, carbon dioxide, and water vapor.

23. Determine what reactants would form the given products. Give a balanced equation for each and classify the reaction as a combination, decomposition, single-displacement, or double-displacement reaction.
(a) $AgCl(s) + O_2(g)$
(b) $H_2(g) + FeSO_4(aq)$
(c) $ZnCl_2(s)$
(d) $KBr(aq) + H_2O(l)$

24. Determine what reactants would form the given products. Give a balanced equation for each and classify the reaction as a combination, decomposition, single-displacement, or double-displacement reaction.
(a) $Pb(s) + Ni(NO_3)_2(aq)$
(b) $Mg(OH)_2(s)$
(c) $Hg(l) + O_2(g)$
(d) $PbCO_3(s) + NH_4Cl(aq)$

Additional Exercises

All exercises with blue numbers have answers in the appendix of the text.

25. Name one piece of evidence that a chemical reaction is actually taking place in each of these situations:
(a) making a piece of toast
(b) frying an egg
(c) striking a match

26. Balance this equation, using the smallest possible whole numbers. Then determine how many atoms of oxygen appear on each side of the equation:

$$P_4O_{10} + HClO_4 \longrightarrow Cl_2O_7 + H_3PO_4$$

27. Suppose that in a balanced equation the term $7\ Al_2(SO_4)_3$ appears.
(a) How many atoms of aluminum are represented?
(b) How many atoms of sulfur are represented?
(c) How many atoms of oxygen are represented?
(d) How many atoms of all kinds are represented?

28. Name two pieces of information that can be obtained from a balanced chemical equation. Name two pieces of information that the equation does not provide.

29. Make a drawing to show six molecules of ammonia gas decomposing to form hydrogen and nitrogen gases.

30. Explain briefly why this single-displacement reaction will not take place:

$$Zn + Mg(NO_3)_2 \longrightarrow \text{no reaction}$$

31. A student does an experiment to determine where titanium metal should be placed on the activity series chart. He places newly cleaned pieces of titanium into solutions of nickel(II) nitrate, lead(II) nitrate, and magnesium nitrate. He finds that the titanium reacts with the nickel(II) nitrate and lead(II) nitrate solutions, but not with the magnesium nitrate solution. From this information, place titanium in the activity series in a position relative to these ions.

32. Complete and balance the equations for these combination reactions:
(a) $K + O_2 \longrightarrow$
(b) $Al + Cl_2 \longrightarrow$
(c) $CO_2 + H_2O \longrightarrow$
(d) $CaO + H_2O \longrightarrow$

33. Complete and balance the equations for these decomposition reactions:
(a) $HgO \xrightarrow{\Delta}$
(b) $NaClO_3 \xrightarrow{\Delta}$
(c) $MgCO_3 \xrightarrow{\Delta}$
(d) $PbO_2 \xrightarrow{\Delta} PbO +$

34. Complete and balance the equations for these single-displacement reactions:
(a) $Zn + H_2SO_4 \longrightarrow$
(b) $AlI_3 + Cl_2 \longrightarrow$
(c) $Mg + AgNO_3 \longrightarrow$
(d) $Al + CoSO_4 \longrightarrow$

35. Complete and balance the equations for these double-displacement reactions:
(a) $ZnCl_2 + KOH \longrightarrow$
(b) $CuSO_4 + H_2S \longrightarrow$
(c) $Ca(OH)_2 + H_3PO_4 \longrightarrow$
(d) $(NH_4)_3PO_4 + Ni(NO_3)_2 \longrightarrow$
(e) $Ba(OH)_2 + HNO_3 \longrightarrow$
(f) $(NH_4)_2S + HCl \longrightarrow$

36. Predict which of the following double-displacement reactions will occur. Complete and balance the equations. Where no reaction will occur, write "no reaction" as the product.
(a) $AgNO_3(aq) + KCl(aq) \longrightarrow$
(b) $Ba(NO_3)_2(aq) + MgSO_4(aq) \longrightarrow$
(c) $H_2SO_4(aq) + Mg(OH)_2(aq) \longrightarrow$
(d) $MgO(s) + H_2SO_4(aq) \longrightarrow$
(e) $Na_2CO_3(aq) + NH_4Cl(aq) \longrightarrow$

37. Write balanced equations for each of the following combustion reactions:

(a)  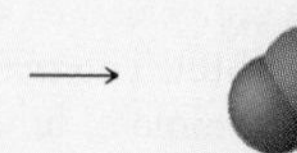 + 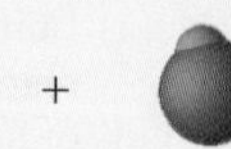

(b)   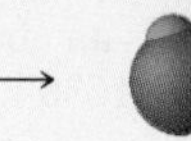

38. Write balanced equations for the complete combustion of these hydrocarbons:
(a) ethane, C_2H_6
(b) benzene, C_6H_6
(c) heptane, C_7H_{16}

39. List the factors that contribute to an increase in carbon dioxide in our atmosphere.

40. List three gases considered to be greenhouse gases. Explain why they are given this name.

41. How can the effects of global warming be reduced?

42. What happens to carbon dioxide released into our atmosphere?

Challenge Exercise

All exercises with blue numbers have answers in the appendix of the text.

*43. You are given a solution of the following cations in water: Ag^+, Co^{2+}, Ba^{2+}, Zn^{2+}, and Sn^{2+}. You want to separate the ions out *one at a time* by precipitation using the following reagents: NaF, NaI, Na_2SO_4, and NaCl.
(a) In what order would you add the reagents to ensure that *only one cation* is precipitating out at a time? Use the solubility table in Appendix V to help you. (*Note:* A precipitate that is slightly soluble in water is still considered to precipitate out of the solution.)
(b) Why are the anionic reagents listed above all sodium salts?

Answers to Practice Exercises

8.1 $4\,Al + 3\,O_2 \longrightarrow 2\,Al_2O_3$

8.2 $3\,Mg(OH)_2 + 2\,H_3PO_4 \longrightarrow Mg_3(PO_4)_2 + 6\,H_2O$

8.3 (a) hydrogen gas + chlorine gas $\longrightarrow$ hydrogen chloride gas

(b) $H_2(g) + Cl_2(g) \longrightarrow 2\,HCl(g)$

(c)

$H_2(g)$	+	$Cl_2(g)$	$\longrightarrow$	$2\,HCl(g)$
1 molecule		1 molecule		2 molecules
2 atoms H		2 atoms Cl		2 atoms H 2 atoms Cl
1 mol 2.016 g		1 mol 70.90 g		2 mol 2 (36.46 g) (72.92 g)

(d) $2\text{ mol } H_2 + 2\text{ mol } Cl_2 \longrightarrow 4\text{ mol } HCl\,(145.8\text{ g})$

8.4 (a) $Fe + MgCl_2 \longrightarrow$ no reaction

(b) $Zn(s) + Pb(NO_3)_2(aq) \longrightarrow Pb(s) + Zn(NO_3)_2(aq)$

8.5 (a) $2\,K_3PO_4(aq) + 3\,BaCl_2(aq) \longrightarrow Ba_3(PO_4)_2(s) + 6\,KCl(aq)$

(b) $2\,HCl(aq) + NiCO_3(aq) \longrightarrow NiCl_2(aq) + H_2O(l) + CO_2(g)$

(c) $NH_4Cl(aq) + NaNO_3(aq) \longrightarrow$ no reaction

CHAPTER 9

Calculations from Chemical Equations

We quickly learn as children that accurate measurement and calculation are required to bake a masterpiece. Although we don't see the chemical reactions occurring, we must scale up or down correctly to increase or decrease a recipe.

Chapter Outline

The old adage "waste not, want not" is appropriate in our daily life and in the laboratory. Determining correct amounts comes into play in almost all professions. A seamstress determines the amount of material, lining, and trim necessary to produce a gown for her client by relying on a pattern or her own experience to guide the selection. A carpet layer determines the correct amount of carpet and padding necessary to recarpet a customer's house by calculating the floor area. The IRS determines the correct deduction for federal income taxes from your paycheck based on your expected annual income.

The chemist also finds it necessary to calculate amounts of products or reactants by using a balanced chemical equation. With these calculations, the chemist can control the amount of product by scaling the reaction up or down to fit the needs of the laboratory, and can thereby minimize waste or excess materials formed during the reaction.

9.1 A Short Review

Molar Mass The sum of the atomic masses of all the atoms in an element or compound is **molar mass**. The term *molar mass* also applies to the mass of a mole of any formula unit—atoms, molecules, or ions; it is the atomic mass of an atom or the sum of the atomic masses in a molecule or an ion (in grams).

molar mass

Relationship between Molecule and Mole A molecule is the smallest unit of a molecular substance (e.g., Br_2), and a mole is Avogadro's number (6.022×10^{23}) of molecules of that substance. A mole of bromine (Br_2) has the same number of molecules as a mole of carbon dioxide, a mole of water, or a mole of any other molecular substance. When we relate molecules to molar mass, 1 molar mass is equivalent to 1 mol, or 6.022×10^{23} molecules.

The term *mole* also refers to any chemical species. It represents a quantity (6.022×10^{23} particles) and may be applied to atoms, ions, electrons, and formula units of nonmolecular substances. In other words,

$$1 \text{ mole} = \begin{cases} 6.022 \times 10^{23} \text{ molecules} \\ 6.022 \times 10^{23} \text{ formula units} \\ 6.022 \times 10^{23} \text{ atoms} \\ 6.022 \times 10^{23} \text{ ions} \end{cases}$$

Other useful mole relationships are

$$\text{molar mass} = \frac{\text{grams of a substance}}{\text{number of moles of the substance}}$$

$$\text{molar mass} = \frac{\text{grams of a monatomic element}}{\text{number of moles of the element}}$$

$$\text{number of moles} = \frac{\text{number of molecules}}{6.022 \times 10^{23} \text{ molecules/mole}}$$

A mole of water, salt, and any gas all have the same number of particles (6.022×10^{23}).

Balanced Equations When using chemical equations for calculations of mole–mass–volume relationships between reactants and products, the equations must be balanced. *Remember:* The number in front of a formula in a balanced chemical equation represents the number of moles of that substance in the chemical reaction.

9.2 Introduction to Stoichiometry

We often need to calculate the amount of a substance that is either produced from, or needed to react with, a given quantity of another substance. The area of chemistry that deals with quantitative relationships among reactants and products is known as **stoichiometry** (*stoy-key-ah-meh-tree*). Solving problems in stoichiometry requires the use of *moles* in the form of *mole ratios*.

stoichiometry

mole ratio

A **mole ratio** is a ratio between the number of moles of any two species involved in a chemical reaction. For example, in the reaction

$$\underset{2\text{ mol}}{2\,H_2} + \underset{1\text{ mol}}{O_2} \longrightarrow \underset{2\text{ mol}}{2\,H_2O}$$

six mole ratios can be written:

$$\frac{2\text{ mol }H_2}{1\text{ mol }O_2} \quad \frac{2\text{ mol }H_2}{2\text{ mol }H_2O} \quad \frac{1\text{ mol }O_2}{2\text{ mol }H_2}$$

$$\frac{1\text{ mol }O_2}{2\text{ mol }H_2O} \quad \frac{2\text{ mol }H_2O}{2\text{ mol }H_2} \quad \frac{2\text{ mol }H_2O}{1\text{ mol }O_2}$$

We use the mole ratio to convert the number of moles of one substance to the corresponding number of moles of another substance in a chemical reaction. For example, if we want to calculate the number of moles of H_2O that can be obtained from 4.0 mol of O_2, we use the mole ratio 2 mol H_2O/1 mol O_2:

$$(4.0\,\cancel{\text{mol }O_2})\left(\frac{2\text{ mol }H_2O}{1\,\cancel{\text{mol }O_2}}\right) = 8.0\text{ mol }H_2O$$

There are three basic steps to use in solving stoichiometry problems:

Write and balance the equation before you begin the problem.

1. Convert the quantity of starting substance to moles (if it is not given in moles).
2. Convert the moles of starting substance to moles of desired substance.
3. Convert the moles of desired substance to the units specified in the problem.

Like balancing chemical equations, making stoichiometric calculations requires practice. Several worked examples follow. Study this material and practice on the problems at the end of this chapter.

Use a balanced equation.

Step 1 **Determine the number of moles of starting substance.** Identify the starting substance from the data given in the problem statement. Convert the quantity of the starting substance to moles, if it is not already done:

$$\text{moles} = (\text{grams})\left(\frac{1\ \text{mole}}{\text{molar mass}}\right)$$

Step 2 **Determine the mole ratio of the desired substance to the starting substance.** The number of moles of each substance in the balanced equation is indicated by the coefficient in front of each substance. Use these coefficients to set up the mole ratio:

$$\text{mole ratio} = \frac{\text{moles of desired substance in the equation}}{\text{moles of starting substance in the equation}}$$

As in all problems with units, the desired quantity is in the numerator, and the quantity to be eliminated is in the denominator.

Multiply the number of moles of starting substance (from Step 1) by the mole ratio to obtain the number of moles of desired substance:

$$\text{moles of desired substance} = \underbrace{\left(\text{moles of starting substance}\right)}_{\text{From Step 1}}\left(\frac{\text{moles of desired substance in the balanced equation}}{\text{moles of starting substance in the balanced equation}}\right)$$

Units of moles of starting substance cancel in the numerator and denominator.

Step 3 **Calculate the desired substance in the units specified in the problem.** If the answer is to be in moles, the calculation is complete. If units other than moles are wanted, multiply the moles of the desired substance (from Step 2) by the appropriate factor to convert moles to the units required. For example, if grams of the desired substance are wanted,

$$\text{grams} = \underbrace{(\text{moles})}_{\text{From Step 2}}\left(\frac{\text{molar mass}}{1\ \text{mol}}\right)$$

If moles $\longrightarrow$ atoms, use $\dfrac{6.022 \times 10^{23}\ \text{atoms}}{1\ \text{mol}}$

If moles $\longrightarrow$ molecules, use $\dfrac{6.022 \times 10^{23}\ \text{molecules}}{1\ \text{mol}}$

The steps for converting the mass of a starting substance A to either the mass, atoms, or molecules of desired substance B are summarized in Figure 9.1.

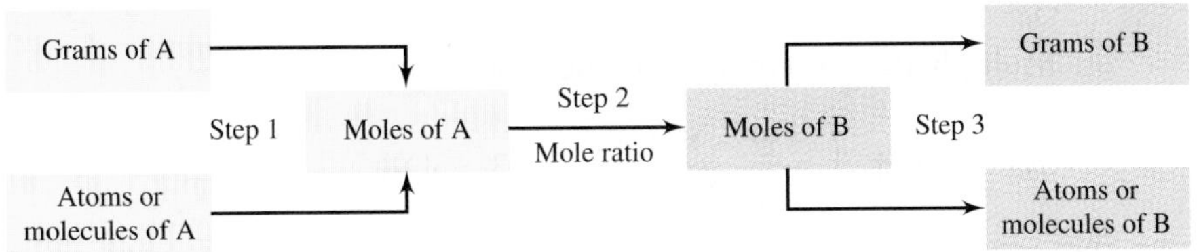

Figure 9.1
Steps for converting starting substance A to mass, atoms, or molecules of desired substance B.

9.3 Mole–Mole Calculations

Let's solve stoichiometric problems for mole–mole calculations. The quantity of starting substance is given in moles and the quantity of desired substance is requested in moles.

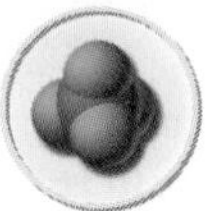

Example 9.1 How many moles of carbon dioxide will be produced by the complete reaction of 2.0 mol of glucose ($C_6H_{12}O_6$), according to the following equation?

$$\underset{1\ \text{mol}}{C_6H_{12}O_6} + \underset{6\ \text{mol}}{6\,O_2} \longrightarrow \underset{6\ \text{mol}}{6\,CO_2} + \underset{6\ \text{mol}}{6\,H_2O}$$

SOLUTION The balanced equation states that 6 mol of CO_2 will be produced from 1 mol of $C_6H_{12}O_6$. Even though we can readily see that 12 mol of CO_2 will be formed from 2.0 mol of $C_6H_{12}O_6$, let's use the mole-ratio method to solve the problem.

Step 1 The number of moles of starting substance is 2.0 mol $C_6H_{12}O_6$.

Step 2 The conversion needed is

$$\text{moles } C_6H_{12}O_6 \longrightarrow \text{moles } CO_2$$

Multiply 2.0 mol of glucose (given in the problem) by this mole ratio:

The mole ratio (shown in color) is exact and does not affect the number of significant figures in the answer.

$$(2.0\ \cancel{\text{mol } C_6H_{12}O_6})\left(\frac{6\ \text{mol } CO_2}{1\ \cancel{\text{mol } C_6H_{12}O_6}}\right) = 12\ \text{mol } CO_2$$

Note how the units work; the moles of $C_6H_{12}O_6$ cancel, leaving the answer in units of moles of CO_2.

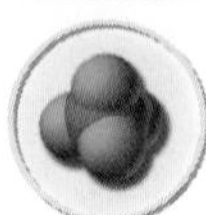

Example 9.2 How many moles of ammonia can be produced from 8.00 mol of hydrogen reacting with nitrogen? The balanced equation is

$$3\,H_2 + N_2 \longrightarrow 2\,NH_3$$

SOLUTION **Step 1** The starting substance is 8.00 mol of H_2.

Step 2 The conversion needed is

$$\text{moles } H_2 \longrightarrow \text{moles } NH_3$$

The balanced equation states that we get 2 mol of NH_3 for every 3 mol of H_2 that react. Set up the mole ratio of desired substance (NH_3) to starting substance (H_2):

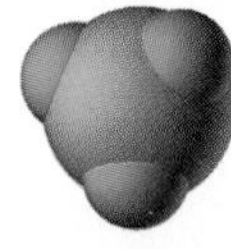

Ammonia

$$\text{mole ratio} = \frac{2\ \text{mol } NH_3}{3\ \text{mol } H_2} \quad \text{(from balanced equation)}$$

Multiply the 8.00 mol H_2 by the mole ratio:

$$(8.00\ \cancel{\text{mol } H_2})\left(\frac{2\ \text{mol } NH_3}{3\ \cancel{\text{mol } H_2}}\right) = 5.33\ \text{mol } NH_3$$

Example 9.3

Given the balanced equation

$$\underset{1\text{ mol}}{K_2Cr_2O_7} + \underset{6\text{ mol}}{6\,KI} + 7\,H_2SO_4 \longrightarrow Cr_2(SO_4)_3 + 4\,K_2SO_4 + \underset{3\text{ mol}}{3\,I_2} + 7\,H_2O$$

calculate (a) the number of moles of potassium dichromate ($K_2Cr_2O_7$) that will react with 2.0 mol of potassium iodide (KI) and (b) the number of moles of iodine (I_2) that will be produced from 2.0 mol of potassium iodide.

SOLUTION

Since the equation is balanced, we are concerned only with $K_2Cr_2O_7$, KI, and I_2, and we can ignore all the other substances. The equation states that 1 mol of $K_2Cr_2O_7$ will react with 6 mol of KI to produce 3 mol of I_2.

(a) Calculate the number of moles of $K_2Cr_2O_7$.

Step 1 The starting substance is 2.0 mol of KI.
Step 2 The conversion needed is

$$\text{moles KI} \longrightarrow \text{moles } K_2Cr_2O_7$$

Set up the mole ratio of desired substance to starting substance:

$$\text{mole ratio} = \frac{1\text{ mol } K_2Cr_2O_7}{6\text{ mol KI}} \quad \text{(from balanced equation)}$$

Multiply the moles of starting material by this ratio:

$$(2.0\,\cancel{\text{mol KI}})\left(\frac{1\text{ mol } K_2Cr_2O_7}{6\,\cancel{\text{mol KI}}}\right) = 0.33\text{ mol } K_2Cr_2O_7$$

(b) Calculate the number of moles of I_2:

Step 1 The moles of starting substance are 2.0 mol KI as in part (a).
Step 2 The conversion needed is

$$\text{moles KI} \longrightarrow \text{moles } I_2$$

Set up the mole ratio of desired substance to starting substance:

$$\text{mole ratio} = \frac{3\text{ mol } I_2}{6\text{ mol KI}} \quad \text{(from balanced equation)}$$

Multiply the moles of starting material by this ratio:

$$(2.0\,\cancel{\text{mol KI}})\left(\frac{3\text{ mol } I_2}{6\,\cancel{\text{mol KI}}}\right) = 1.0\text{ mol } I_2$$

Example 9.4

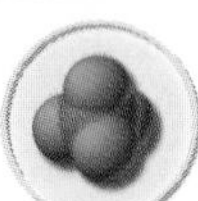

How many molecules of water can be produced by reacting 0.010 mol of oxygen with hydrogen?

SOLUTION

The balanced equation is $2\,H_2 + O_2 \longrightarrow 2\,H_2O$.

The sequence of conversions needed in the calculation is

$$\text{moles } O_2 \longrightarrow \text{moles } H_2O \longrightarrow \text{molecules } H_2O$$

Step 1 The starting substance is 0.010 mol O_2.

Step 2 The conversion needed is moles $O_2 \longrightarrow$ moles H_2O. Set up the mole ratio of desired substance to starting substance:

$$\text{mole ratio} = \frac{2 \text{ mol } H_2O}{1 \text{ mol } O_2} \quad \text{(from balanced equation)}$$

Multiply 0.010 mol O_2 by the mole ratio:

$$(0.010 \cancel{\text{mol } O_2})\left(\frac{2 \text{ mol } H_2O}{1 \cancel{\text{mol } O_2}}\right) = 0.020 \text{ mol } H_2O$$

Step 3 Since the problem asks for molecules instead of moles of H_2O, we must convert moles to molecules. Use the conversion factor $(6.022 \times 10^{23}$ molecules)/mole:

$$(0.020 \cancel{\text{mol}} \; H_2O)\left(\frac{6.022 \times 10^{23} \text{ molecules}}{1 \cancel{\text{mol}}}\right) = 1.2 \times 10^{22} \text{ molecules } H_2O$$

Note that 0.020 mol is still quite a large number of water molecules.

Practice 9.1

How many moles of aluminum oxide will be produced from 0.50 mol of oxygen?

$$4\,Al + 3\,O_2 \longrightarrow 2\,Al_2O_3$$

Practice 9.2

How many moles of aluminum hydroxide are required to produce 22.0 mol of water?

$$2\,Al(OH)_3 + 3\,H_2SO_4 \longrightarrow Al_2(SO_4)_3 + 6\,H_2O$$

9.4 Mole–Mass Calculations

The object of this type of problem is to calculate the mass of one substance that reacts with or is produced from a given number of moles of another substance in a chemical reaction. If the mass of the starting substance is given, we need to convert it to moles. We use the mole ratio to convert moles of starting substance to moles of desired substance. We can then change moles of desired substance to mass. Each example is solved in two ways.

Method A. **Step by step**
Method B. **Continuous calculation**, where the individual steps are combined in a single line.

Select the method that is easier for you and use it to practice solving problems.

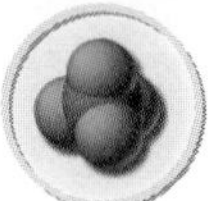

Example 9.5 What mass of hydrogen can be produced by reacting 6.0 mol of aluminum with hydrochloric acid?

SOLUTION The balanced equation is $2\,Al(s) + 6\,HCl(aq) \longrightarrow 2\,AlCl_3(aq) + 3\,H_2(g)$.

Method A. Step by Step

Step 1 The starting substance is 6.0 mol of aluminum.

Step 2 Calculate moles of H_2

moles Al $\longrightarrow$ moles H_2

$$(6.0\ \cancel{\text{mol Al}})\left(\frac{3\ \text{mol H}_2}{2\ \cancel{\text{mol Al}}}\right) = 9.0\ \text{mol H}_2$$

Step 3 Convert moles of H_2 to grams:

moles H_2 $\longrightarrow$ grams H_2

$$(9.0\ \cancel{\text{mol H}_2})\left(\frac{2.016\ \text{g H}_2}{1\ \cancel{\text{mol H}_2}}\right) = 18\ \text{g H}_2$$

We see that 18 g of H_2 can be produced by reacting 6.0 mol of Al with HCl.

Method B. Continuous Calculation

moles Al $\longrightarrow$ moles H_2 $\longrightarrow$ grams H_2

$$(6.0\ \cancel{\text{mol Al}})\left(\frac{3\ \cancel{\text{mol H}_2}}{2\ \cancel{\text{mol Al}}}\right)\left(\frac{2.016\ \text{g H}_2}{1\ \cancel{\text{mol H}_2}}\right) = 18\ \text{g H}_2$$

Example 9.6

How many moles of water can be produced by burning 325 g of octane (C_8H_{18})? The balanced equation is $2\,C_8H_{18}(l) + 25\,O_2(g) \longrightarrow 16\,CO_2(g) + 18\,H_2O(g)$.

SOLUTION

Method A. Step by Step

Step 1 The starting substance is 325 g C_8H_{18}. Convert 325 g of C_8H_{18} to moles:

grams C_8H_{18} $\longrightarrow$ moles C_8H_{18}

$$(325\ \cancel{\text{g C}_8\text{H}_{18}})\left(\frac{1\ \text{mol C}_8\text{H}_{18}}{114.2\ \cancel{\text{g C}_8\text{H}_{18}}}\right) = 2.85\ \text{mol C}_8\text{H}_{18}$$

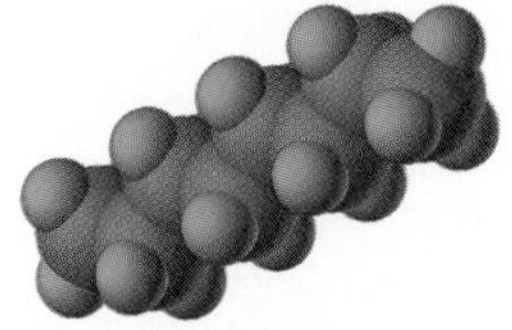

Octane

Step 2 Calculate the moles of water:

moles C_8H_{18} $\longrightarrow$ moles H_2O

$$(2.85\ \cancel{\text{mol C}_8\text{H}_{18}})\left(\frac{18\ \text{mol H}_2\text{O}}{2\ \cancel{\text{mol C}_8\text{H}_{18}}}\right) = 25.7\ \text{mol H}_2\text{O}$$

Method B. Continuous Calculation

grams C_8H_{18} $\longrightarrow$ moles C_8H_{18} $\longrightarrow$ moles H_2O

$$(325\ \cancel{\text{g C}_8\text{H}_{18}})\left(\frac{1\ \cancel{\text{mol C}_8\text{H}_{18}}}{114.2\ \cancel{\text{g C}_8\text{H}_{18}}}\right)\left(\frac{18\ \text{mol H}_2\text{O}}{2\ \cancel{\text{mol C}_8\text{H}_{18}}}\right) = 25.6\ \text{mol H}_2\text{O}$$

The answers for the different methods vary in the last digit. This results from rounding off at different times in the calculation. Check with your instructor to find the appropriate rules for your course.

Practice 9.3

How many moles of potassium chloride and oxygen can be produced from 100.0 g of potassium chlorate?

$$2\,KClO_3 \longrightarrow 2\,KCl + 3\,O_2$$

Practice 9.4

How many grams of silver nitrate are required to produce 0.25 mol of silver sulfide?

$$2\,AgNO_3 + H_2S \longrightarrow Ag_2S + 2\,HNO_3$$

9.5 Mass–Mass Calculations

Use either the step-by-step method or the continuous-calculation method to solve these problems.

Solving mass–mass stoichiometry problems requires all the steps of the mole-ratio method. The mass of starting substance is converted to moles. The mole ratio is then used to determine moles of desired substance, which, in turn, is converted to mass.

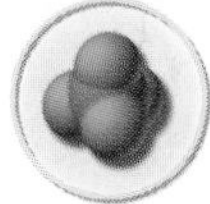

Example 9.7 What mass of carbon dioxide is produced by the complete combustion of 100. g of the hydrocarbon pentane, C_5H_{12}? The balanced equation is

$$C_5H_{12} + 8\,O_2 \longrightarrow 5\,CO_2 + 6\,H_2O$$

SOLUTION Method A. Step by Step

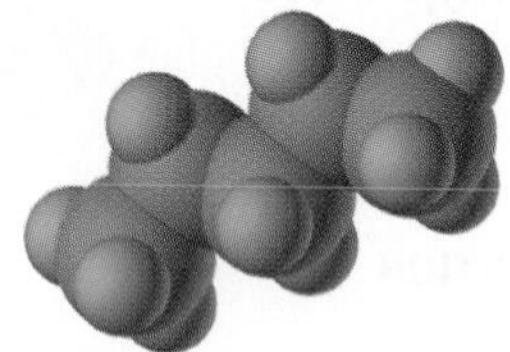

Pentane

Step 1 The starting substance is 100. g of C_5H_{12}. Convert 100. g of C_5H_{12} to moles:

grams $\longrightarrow$ moles

$$(100.\ \cancel{g\,C_5H_{12}})\left(\frac{1\ mol\ C_5H_{12}}{72.15\ \cancel{g\,C_5H_{12}}}\right) = 1.39\ mol\ C_5H_{12}$$

Step 2 Calculate the moles of CO_2:

moles C_5H_{12} $\longrightarrow$ moles CO_2

$$(1.39\ \cancel{mol\,C_5H_{12}})\left(\frac{5\ mol\ CO_2}{1\ \cancel{mol\,C_5H_{12}}}\right) = 6.95\ mol\ CO_2$$

Step 3 Convert moles of CO_2 to grams:

moles CO_2 $\longrightarrow$ grams CO_2

$$(6.95\ \cancel{mol\,CO_2})\left(\frac{44.01\ g\ CO_2}{1\ \cancel{mol\,CO_2}}\right) = 306\ g\ CO_2$$

Remember: Round off as appropriate for your particular course.

Method B. Continuous Calculation

grams C_5H_{12} $\longrightarrow$ moles C_5H_{12} $\longrightarrow$ moles CO_2 $\longrightarrow$ grams CO_2

$$(100.\ \cancel{g\,C_5H_{12}})\left(\frac{1\ \cancel{mol\,C_5H_{12}}}{72.15\ \cancel{g\,C_5H_{12}}}\right)\left(\frac{5\ \cancel{mol\,CO_2}}{1\ \cancel{mol\,C_5H_{12}}}\right)\left(\frac{44.01\ g\ CO_2}{1\ \cancel{mol\,CO_2}}\right) = 305\ g\ CO_2$$

Example 9.8

How many grams of nitric acid (HNO_3) are required to produce 8.75 g of dinitrogen monoxide (N_2O) according to the following equation?

$$\underset{10\text{ mol}}{4\,\text{Zn}(s) + 10\,\text{HNO}_3(aq)} \longrightarrow 4\,\text{Zn(NO}_3)_2(aq) + \underset{1\text{ mol}}{\text{N}_2\text{O}(g)} + 5\,\text{H}_2\text{O}(l)$$

SOLUTION

Method A. Step by Step

Step 1 The starting substance for this calculation is 8.75 g of N_2O:

grams $N_2O \longrightarrow$ moles N_2O

$$(8.75\,\cancel{\text{g N}_2\text{O}})\left(\frac{1\text{ mol N}_2\text{O}}{44.02\,\cancel{\text{g N}_2\text{O}}}\right) = 0.199\text{ mol N}_2\text{O}$$

Step 2 Calculate the moles of HNO_3 by the mole-ratio method:

moles $N_2O \longrightarrow$ moles HNO_3

$$(0.199\,\cancel{\text{mol N}_2\text{O}})\left(\frac{10\text{ mol HNO}_3}{1\,\cancel{\text{mol N}_2\text{O}}}\right) = 1.99\text{ mol HNO}_3$$

Step 3 Convert moles of HNO_3 to grams:

moles $HNO_3 \longrightarrow$ grams HNO_3

$$(1.99\,\cancel{\text{mol HNO}_3})\left(\frac{63.02\text{ g HNO}_3}{1\,\cancel{\text{mol HNO}_3}}\right) = 125\text{ g HNO}_3$$

Method B. Continuous Calculation

grams $N_2O \longrightarrow$ moles $N_2O \longrightarrow$ moles $HNO_3 \longrightarrow$ grams HNO_3

$$(8.75\,\cancel{\text{g N}_2\text{O}})\left(\frac{1\,\cancel{\text{mol N}_2\text{O}}}{44.02\,\cancel{\text{g N}_2\text{O}}}\right)\left(\frac{10\,\cancel{\text{mol HNO}_3}}{1\,\cancel{\text{mol N}_2\text{O}}}\right)\left(\frac{63.02\text{ g HNO}_3}{1\,\cancel{\text{mol HNO}_3}}\right) = 125\text{ g HNO}_3$$

Practice 9.5

How many grams of chromium(III) chloride are required to produce 75.0 g of silver chloride?

$$\text{CrCl}_3 + 3\,\text{AgNO}_3 \longrightarrow \text{Cr(NO}_3)_3 + 3\,\text{AgCl}$$

Practice 9.6

What mass of water is produced by the complete combustion of 225.0 g of butane (C_4H_{10})?

$$2\,\text{C}_4\text{H}_{10} + 13\,\text{O}_2 \longrightarrow 8\,\text{CO}_2 + 10\,\text{H}_2\text{O}$$

CHEMISTRY IN ACTION • A Shrinking Technology

The microchip has revolutionized the field of electronics. Engineers at Bell Laboratories, Massachusetts Institute of Technology, the University of California, and Stanford University are racing to produce parts for tiny machines and robots. New techniques now produce gears smaller than a grain of sand and motors lighter than a speck of dust.

To produce ever smaller computers, calculators, and even microbots (micro-sized robots), precise quantities of chemicals in exact proportions are required. The secret to producing minute circuits is to print the entire circuit or blueprint at one time. Computers are used to draw a chip. This image is then transferred onto a pattern, or mask, with details finer than a human hair. In a process similar to photography, light is then shined through the mask onto a silicon-coated surface. The areas created on the silicon exhibit high or low resistance to chemical etching. Chemicals then etch away the silicon.

Micromachinery is produced in the same way. First a thin layer of silicon dioxide is applied (sacrificial material), then a layer of polysilicon is carefully applied (structural material). A mask is then applied and the whole structure is covered with plasma (excited gas). The plasma acts as a tiny sandblaster, removing everything the mask doesn't protect. This process is repeated as the entire machine is constructed. When the entire assembly is complete, the whole machine is placed in hydrofluoric acid, which dissolves all the sacrificial material and permits the various parts of the machine to move.

To turn these micromachines into true microbots, current research is focusing on locomotion and sensing imaging systems. Possible uses for these microbots include "smart" pills, which could contain sensors or drug reservoirs (currently used in birth control). Tiny pumps, once inside the body, will dispense the proper amount of medication at precisely the correct site. These microbots are currently in production for the treatment of diabetes (to release insulin).

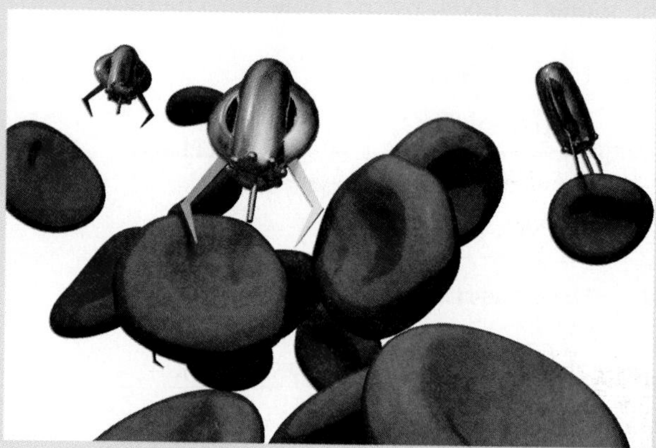

These nanorobots attach themselves to red blood cells.

9.6 Limiting Reactant and Yield Calculations

limiting reactant

In many chemical processes, the quantities of the reactants used are such that one reactant is in excess. The amount of the product(s) formed in such a case depends on the reactant that is not in excess. This reactant is called the **limiting reactant**—it limits the amount of product that can be formed.

Consider the case illustrated in Figure 9.2. How many bicycles can be assembled from the parts shown? The limiting part in this case is the number of pedal assemblies; only three bicycles can be built because there are only three pedal assemblies. The wheels and frames are parts in excess.

Let's consider a chemical example at the molecular level in which seven molecules of H_2 are combined with four molecules of Cl_2 (Figure 9.3a). How many molecules of HCl can be produced according to this reaction?

$$H_2 + Cl_2 \longrightarrow 2\,HCl$$

If the molecules of H_2 and Cl_2 are taken apart and recombined as HCl (Figure 9.3b), we see that eight molecules of HCl can be formed before we run out of Cl_2. Therefore the Cl_2 is the limiting reactant and H_2 is in excess—three molecules of H_2 remain unreacted.

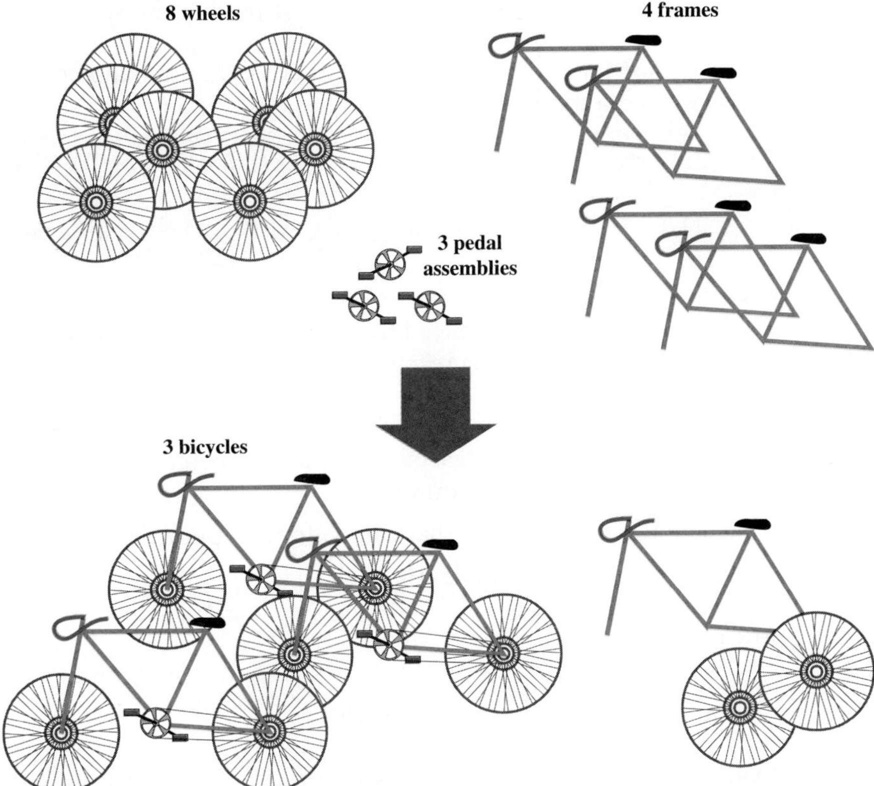

Figure 9.2
The number of bicycles that can be built from these parts is determined by the "limiting reactant" (the pedal assemblies).

When problem statements give the amounts of two reactants, one of them is usually a limiting reactant. We can identify the limiting reactant using the following method:

1. Calculate the amount of product (moles or grams, as needed) formed from each reactant.
2. Determine which reactant is limiting. (The reactant that gives the least amount of product is the limiting reactant; the other reactant is in excess.)

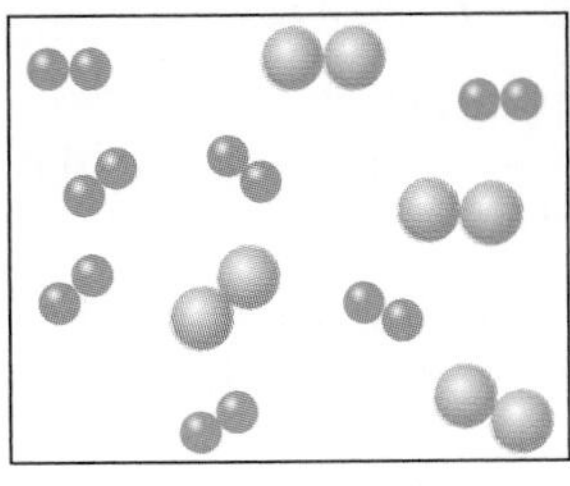

(a) Before reaction

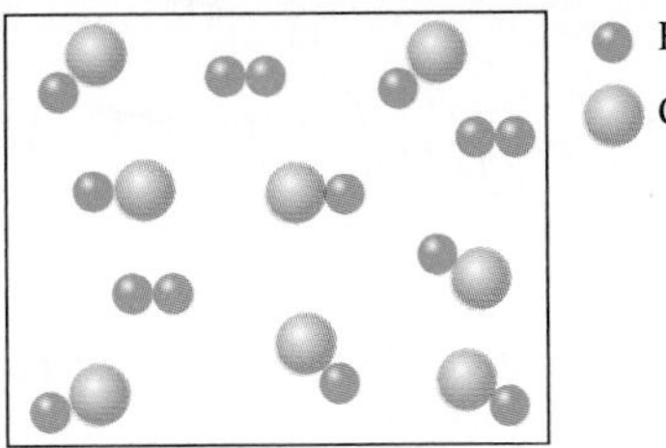

(b) After reaction

Figure 9.3
Reaction of H_2 with Cl_2. Eight molecules of HCl are formed and three molecules of H_2 remain. The limiting reactant is Cl_2.

Once we know the limiting reactant, the amount of product formed can be determined.

Sometimes we must find the amount of the other reactant remaining after the reaction, so we add a third step:

> **3.** Calculate the amount of the other reactant required to react with the limiting reactant, then subtract this amount from the starting quantity of the reactant. This gives the amount of that substance that remains unreacted.

Example 9.9 How many moles of HCl can be produced by reacting 4.0 mol H_2 and 3.5 mol Cl_2? Which compound is the limiting reactant?

SOLUTION

$$H_2(g) + Cl_2(g) \longrightarrow 2\,HCl(g)$$

Since we have the amounts of both reactants, we need to determine which reactant limits the amount of product formed.

Step 1 Calculate the moles of HCl that can be formed from each reactant:

$$\text{Moles of } H_2 \text{ or } Cl_2 \longrightarrow \text{moles HCl}$$

$$(4.0\text{ mol } H_2)\frac{(2\text{ mol HCl})}{(1\text{ mol } H_2)} = 8.0\text{ mol HCl}$$

$$(3.5\text{ mol } Cl_2)\frac{(2\text{ mol HCl})}{(1\text{ mol } Cl_2)} = 7.0\text{ mol HCl}$$

Step 2 Determine the limiting reactant. The limiting reactant is Cl_2 because it produces less HCl than H_2. The H_2 is in excess. The yield of product is 7.0 mol HCl. (*Note*: In this reaction, the limiting reactant can be determined by inspection. From the equation, you can see that 1 mol H_2 reacts with 1 mol Cl_2. Therefore, when reacting 4.0 mol H_2 and 3.5 mol Cl_2, Cl_2 is the reactant that limits the amount of HCl produced, since it is present in smaller amount.)

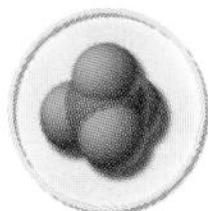

Example 9.10 How many moles of Fe_3O_4 can be obtained by reacting 16.8 g Fe with 10.0 g H_2O? Which substance is the limiting reactant? Which substance is in excess?

SOLUTION

$$3\,Fe(s) + 4\,H_2O(g) \xrightarrow{\Delta} Fe_3O_4(s) + 4\,H_2(g)$$

Step 1 Calculate the moles of Fe_3O_4 that can be formed from each reactant:

The continuous-calculation method is shown here. You can also use the step-by-step method to determine the requested substance.

$$\text{g reactant} \longrightarrow \text{mol reactant} \longrightarrow \text{mol } Fe_3O_4$$

$$(16.8\,\cancel{\text{g Fe}})\left(\frac{1\,\cancel{\text{mol Fe}}}{55.85\,\cancel{\text{g Fe}}}\right)\left(\frac{1\text{ mol } Fe_3O_4}{3\,\cancel{\text{mol Fe}}}\right) = 0.100\text{ mol } Fe_3O_4$$

$$(10.0\,\cancel{\text{g } H_2O})\left(\frac{1\,\cancel{\text{mol } H_2O}}{18.02\,\cancel{\text{g } H_2O}}\right)\left(\frac{1\text{ mol } Fe_3O_4}{4\,\cancel{\text{mol } H_2O}}\right) = 0.139\text{ mol } Fe_3O_4$$

Step 2 Determine the limiting reactant. The limiting reactant is Fe because it produces less Fe_3O_4; the H_2O is in excess. The yield of product is 0.100 mol of Fe_3O_4.

Example 9.11

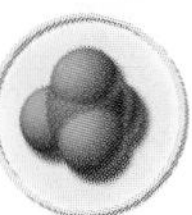

How many grams of silver bromide (AgBr) can be formed when solutions containing 50.0 g of $MgBr_2$ and 100.0 g of $AgNO_3$ are mixed together? How many grams of the excess reactant remain unreacted?

$$MgBr_2(aq) + 2\,AgNO_3(aq) \longrightarrow 2\,AgBr(s) + Mg(NO_3)_2(aq)$$

SOLUTION

Step 1 Calculate the grams of AgBr that can be formed from each reactant.

g reactant ⟶ mol reactant ⟶ mol AgBr ⟶ g AgBr

$$(50.0\,\cancel{g\,MgBr_2})\left(\frac{1\,\cancel{mol\,MgBr_2}}{184.1\,\cancel{g\,MgBr_2}}\right)\left(\frac{2\,\cancel{mol\,AgBr}}{1\,\cancel{mol\,MgBr_2}}\right)\left(\frac{187.8\,g\,AgBr}{1\,\cancel{mol\,AgBr}}\right) = 102\,g\,AgBr$$

$$(100.0\,\cancel{g\,AgNO_3})\left(\frac{1\,\cancel{mol\,AgNO_3}}{169.9\,\cancel{g\,AgNO_3}}\right)\left(\frac{2\,\cancel{mol\,AgBr}}{2\,\cancel{mol\,AgNO_3}}\right)\left(\frac{187.8\,g\,AgBr}{1\,\cancel{mol\,AgBr}}\right) = 110.5\,g\,AgBr$$

Step 2 Determine the limiting reactant. The limiting reactant is $MgBr_2$ because it gives less AgBr; $AgNO_3$ is in excess. The yield is 102 g AgBr.

Step 3 Calculate the grams of unreacted $AgNO_3$. Calculate the grams of $AgNO_3$ that will react with 50.0 g of $MgBr_2$:

g $MgBr_2$ ⟶ mol $MgBr_2$ ⟶ mol $AgNO_3$ ⟶ g $AgNO_3$

$$(50.0\,\cancel{g\,MgBr_2})\left(\frac{1\,\cancel{mol\,MgBr_2}}{184.1\,\cancel{g\,MgBr_2}}\right)\left(\frac{2\,\cancel{mol\,AgNO_3}}{1\,\cancel{mol\,MgBr_2}}\right)\left(\frac{169.9\,g\,AgNO_3}{1\,\cancel{mol\,AgNO_3}}\right) = 92.3\,g\,AgNO_3$$

Thus 92.3 g of $AgNO_3$ reacts with 50.0 g of $MgBr_2$. The amount of $AgNO_3$ that remains unreacted is

$$100.0\,g\,AgNO_3 - 92.3\,g\,AgNO_3 = 7.7\,g\,AgNO_3\ \text{unreacted}$$

The final mixture will contain 102 g AgBr(*s*), 7.7 g $AgNO_3$, and an undetermined amount of $Mg(NO_3)_2$ in solution.

Practice 9.7

How many grams of hydrogen chloride can be produced from 0.490 g of hydrogen and 50.0 g of chlorine?

$$H_2(g) + Cl_2(g) \longrightarrow 2\,HCl(g)$$

Practice 9.8

How many grams of barium sulfate will be formed from 200.0 g of barium nitrate and 100.0 g of sodium sulfate?

$$Ba(NO_3)_2(aq) + Na_2SO_4(aq) \longrightarrow BaSO_4(s) + 2\,NaNO_3(aq)$$

The quantities of the products we have been calculating from chemical equations represent the maximum yield (100%) of product according to the reaction represented by the equation. Many reactions, especially those involving organic substances, fail to give a 100% yield of product. The main reasons for this failure are the side reactions that give products other than the main product and the fact that many reactions are reversible. In addition, some product may be lost in handling and transferring from one vessel to another. The **theoretical yield** of a reaction is the calculated amount of product that can be obtained from a given amount of reactant, according to the chemical equation. The **actual yield** is the amount of product that we finally obtain.

theoretical yield

actual yield

percent yield

The **percent yield** is the ratio of the actual yield to the theoretical yield multiplied by 100. Both the theoretical and the actual yields must have the same units to obtain a percent:

$$\frac{\text{actual yield}}{\text{theoretical yield}} \times 100 = \text{percent yield}$$

For example, if the theoretical yield calculated for a reaction is 14.8 g, and the amount of product obtained is 9.25 g, the percent yield is

Round off as appropriate for your particular course.

$$\text{percent yield} = \left(\frac{9.25\,\cancel{g}}{14.8\,\cancel{g}}\right)(100) = 62.5\%$$

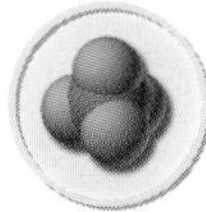

Example 9.12 Carbon tetrachloride (CCl_4) was prepared by reacting 100. g of carbon disulfide and 100. g of chlorine. Calculate the percent yield if 65.0 g of CCl_4 was obtained from the reaction:

$$CS_2 + 3\,Cl_2 \longrightarrow CCl_4 + S_2Cl_2$$

SOLUTION We need to determine the limiting reactant to calculate the quantity of CCl_4 (theoretical yield) that can be formed. Then we can compare that amount with the 65.0 g CCl_4 (actual yield) to calculate the percent yield.

Step 1 Determine the theoretical yield. Calculate the grams of CCl_4 that can be formed from each reactant:

$$\text{g reactant} \longrightarrow \text{mol reactant} \longrightarrow \text{mol } CCl_4 \longrightarrow \text{g } CCl_4$$

$$(100.\,\cancel{g\,CS_2})\left(\frac{1\,\cancel{mol\,CS_2}}{76.15\,\cancel{g\,CS_2}}\right)\left(\frac{1\,\cancel{mol\,CCl_4}}{1\,\cancel{mol\,CS_2}}\right)\left(\frac{153.8\,g\,CCl_4}{1\,\cancel{mol\,CCl_4}}\right) = 202\,g\,CCl_4$$

$$(100.\,\cancel{g\,Cl_2})\left(\frac{1\,\cancel{mol\,Cl_2}}{70.90\,\cancel{g\,Cl_2}}\right)\left(\frac{1\,mol\,CCl_4}{3\,\cancel{mol\,Cl_2}}\right)\left(\frac{153.8\,g\,CCl_4}{1\,\cancel{mol\,CCl_4}}\right) = 72.3\,g\,CCl_4$$

Step 2 Determine the limiting reactant. The limiting reactant is Cl_2 because it gives less CCl_4. The CS_2 is in excess. The theoretical yield is 72.3 g CCl_4.

Step 3 Calculate the percent yield. According to the equation, 72.3 g of CCl_4 are the maximum amount or theoretical yield of CCl_4 possible from 100. g of Cl_2. Actual yield is 65.0 g of CCl_4:

$$\text{percent yield} = \left(\frac{65.0\,\cancel{g}}{72.3\,\cancel{g}}\right)(100) = 89.9\%$$

Example 9.13

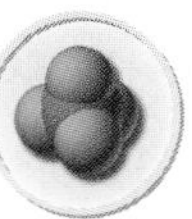

Silver bromide was prepared by reacting 200.0 g of magnesium bromide and an adequate amount of silver nitrate. Calculate the percent yield if 375.0 g of silver bromide were obtained from the reaction:

$$MgBr_2 + 2\,AgNO_3 \longrightarrow Mg(NO_3)_2 + 2\,AgBr$$

SOLUTION

Step 1 Determine the theoretical yield. Calculate the grams of AgBr that can be formed:

$$\text{g MgBr}_2 \longrightarrow \text{mol MgBr}_2 \longrightarrow \text{mol AgBr} \longrightarrow \text{g AgBr}$$

$$(200.0\ \text{g}\ \cancel{\text{MgBr}_2})\left(\frac{1\ \text{mol}\ \cancel{\text{MgBr}_2}}{184.1\ \text{g}\ \cancel{\text{MgBr}_2}}\right)\left(\frac{2\ \text{mol}\ \cancel{\text{AgBr}}}{1\ \text{mol}\ \cancel{\text{MgBr}_2}}\right)\left(\frac{187.8\ \text{g AgBr}}{1\ \text{mol}\ \cancel{\text{AgBr}}}\right) = 408.0\ \text{g AgBr}$$

The theoretical yield is 408.0 g AgBr.

Step 2 Calculate the percent yield. According to the equation, 408.0 g AgBr are the maximum amount of AgBr possible from 200.0 g $MgBr_2$. Actual yield is 375.0 g AgBr:

$$\text{percent yield} = \frac{375.0\ \cancel{\text{g AgBr}}}{408.0\ \cancel{\text{g AgBr}}} \times 100 = 91.91\%$$

Practice 9.9

Aluminum oxide was prepared by heating 225 g of chromium(II) oxide with 125 g of aluminum. Calculate the percent yield if 100.0 g of aluminum oxide were obtained:

$$2\,Al + 3\,CrO \longrightarrow Al_2O_3 + 3\,Cr$$

When solving problems, you will achieve better results if you work in an organized manner.

1. Write data and numbers in a logical, orderly manner.
2. Make certain that the equations are balanced and that the computations are accurate and expressed to the correct number of significant figures.
3. Check the units. Units are very important; a number without units has little or no meaning.

Chapter 9 Review

9.1 A Short Review

KEY TERM

Molar mass

- The molar mass is the sum of the atomic masses of all the atoms in an element or compound.
- $1 \text{ mole} = \begin{Bmatrix} 6.022 \times 10^{23} \text{ molecules} \\ 6.022 \times 10^{23} \text{ formula units} \\ 6.022 \times 10^{23} \text{ atoms} \\ 6.022 \times 10^{23} \text{ ions} \end{Bmatrix}$
- When you use chemical equations for calculations, the chemical equations must be balanced first.

9.2 Introduction to Stoichiometry

KEY TERMS

Stoichiometry
Mole ratio

- Solving stoichiometry problems requires the use of moles and mole ratios.
- To solve a stoichiometry problem:
  - Convert the quantity of starting substance to moles.
  - Convert the moles of starting substance to moles of desired substance.
  - Convert the moles of desired substance to the appropriate unit.

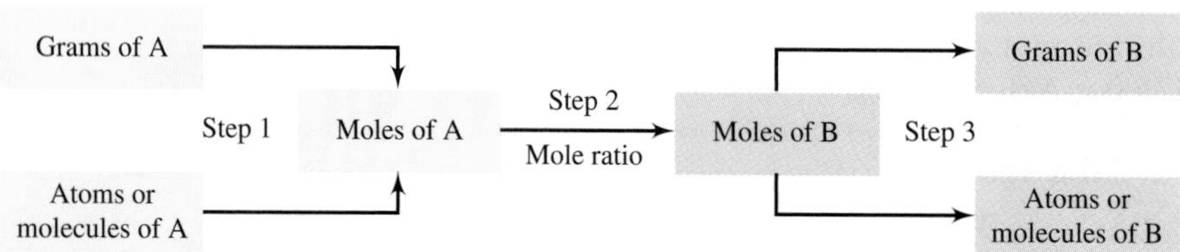

9.3 Mole–Mole Calculations

- If the starting substance is given in moles and the unit specified for the answer is moles:
  - Convert the moles of starting substance to moles of desired substance.

9.4 Mole–Mass Calculations

- If the starting substance is given in moles and the unit specified for the answer is grams:
  - Convert the moles of starting substance to moles of desired substance.
  - Convert the moles of desired substance to grams.

9.5 Mass–Mass Calculations

- If the starting substance is given in grams and the unit specified for the answer is grams:
  - Convert the quantity of starting substance to moles.
  - Convert the moles of starting substance to moles of desired substance.
  - Convert the moles of desired substance to grams.

9.6 Limiting Reactant and Yield Calculations

KEY TERMS

Limiting reactant
Theoretical yield
Actual yield
Percent yield

- To identify the limiting reactant in a reaction:
  - Calculate the amount of product formed from each reactant.
  - Determine the limiting reactant by selecting the one that gives the least amount of product.
- To determine the actual amount of product formed in a limiting reactant situation:
  - Use the result calculated that is the least amount of product.
- To determine the amount of the other reactants required to react with the limiting reactant:
  - Calculate the amount of the other reactant needed to react with the limiting reactant.
  - Subtract this amount from the original amount of the other reactant to find the amount of excess reactant (unreacted).
- The theoretical yield of a chemical reaction is the calculated amount of product from a given amount of reactant (the limiting reactant).
- The actual yield is the amount of product actually obtained experimentally.
- The percent yield is

$$\frac{\text{actual yield}}{\text{theoretical yield}} \times 100 = \text{percent yield}$$

Review Questions

All questions with blue numbers have answers in the appendix of the text.

1. What is a mole ratio?
2. What piece of information is needed to convert grams of a compound to moles of the same compound?
3. Phosphine (PH_3) can be prepared by the hydrolysis of calcium phosphide, Ca_3P_2:

   $$Ca_3P_2 + 6\,H_2O \longrightarrow 3\,Ca(OH)_2 + 2\,PH_3$$

   Based on this equation, which of the following statements are correct? Show evidence to support your answer.
   (a) One mole of Ca_3P_2 produces 2 mol of PH_3.
   (b) One gram of Ca_3P_2 produces 2 g of PH_3.
   (c) Three moles of $Ca(OH)_2$ are produced for each 2 mol of PH_3 produced.
   (d) The mole ratio between phosphine and calcium phosphide is

   $$\frac{2 \text{ mol } PH_3}{1 \text{ mol } Ca_3P_2}.$$

   (e) When 2.0 mol of Ca_3P_2 and 3.0 mol of H_2O react, 4.0 mol of PH_3 can be formed.
   (f) When 2.0 mol of Ca_3P_2 and 15.0 mol of H_2O react, 6.0 mol of $Ca(OH)_2$ can be formed.
   (g) When 200. g of Ca_3P_2 and 100. g of H_2O react, Ca_3P_2 is the limiting reactant.
   (h) When 200. g of Ca_3P_2 and 100. g of H_2O react, the theoretical yield of PH_3 is 57.4 g.
4. The equation representing the reaction used for the commercial preparation of hydrogen cyanide is

   $$2\,CH_4 + 3\,O_2 + 2\,NH_3 \longrightarrow 2\,HCN + 6\,H_2O$$

   Based on this equation, which of the following statements are correct? Rewrite incorrect statements to make them correct.
   (a) Three moles of O_2 are required for 2 mol of NH_3.
   (b) Twelve moles of HCN are produced for every 16 mol of O_2 that react.
   (c) The mole ratio between H_2O and CH_4 is

   $$\frac{6 \text{ mol } H_2O}{2 \text{ mol } CH_4}$$

   (d) When 12 mol of HCN are produced, 4 mol of H_2O will be formed.
   (e) When 10 mol of CH_4, 10 mol of O_2, and 10 mol of NH_3 are mixed and reacted, O_2 is the limiting reactant.
   (f) When 3 mol each of CH_4, O_2, and NH_3 are mixed and reacted, 3 mol of HCN will be produced.
5. What is the difference between the theoretical and the actual yield of a chemical reaction?
6. How is the percent yield of a reaction calculated?

Paired Exercises

All exercises with blue numbers have answers in the appendix of the text.

Mole Review

1. Calculate the number of moles in these quantities:
   (a) 25.0 g KNO_3
   (b) 56 millimol NaOH
   (c) 5.4×10^2 g $(NH_4)_2C_2O_4$
   (d) 16.8 mL H_2SO_4 solution (d = 1.727 g/mL, 80.0% H_2SO_4 by mass)
2. Calculate the number of moles in these quantities:
   (a) 2.10 kg $NaHCO_3$
   (b) 525 mg $ZnCl_2$
   (c) 9.8×10^{24} molecules CO_2
   (d) 250 mL ethyl alcohol, C_2H_5OH (d = 0.789 g/mL)
3. Calculate the number of grams in these quantities:
   (a) 2.55 mol $Fe(OH)_3$
   (b) 125 kg $CaCO_3$
   (c) 10.5 mol NH_3
   (d) 72 millimol HCl
   (e) 500.0 mL of liquid Br_2 (d = 3.119 g/mL)
4. Calculate the number of grams in these quantities:
   (a) 0.00844 mol $NiSO_4$
   (b) 0.0600 mol $HC_2H_3O_2$
   (c) 0.725 mol Bi_2S_3
   (d) 4.50×10^{21} molecules glucose, $C_6H_{12}O_6$
   (e) 75 mL K_2CrO_4 solution (d = 1.175 g/mL, 20.0% K_2CrO_4 by mass)
5. Which contains the larger number of molecules, 10.0 g H_2O or 10.0 g H_2O_2? Show evidence for your answer.
6. Which contains the larger numbers of molecules, 25.0 g HCl or 85.0 g $C_6H_{12}O_6$? Show evidence for your answer.

Mole Ratio

7. Given the equation for the combustion of isopropyl alcohol

$$2\,C_3H_7OH + 9\,O_2 \longrightarrow 6\,CO_2 + 8\,H_2O$$

set up the mole ratio of

(a) CO_2 to C_3H_7OH
(b) C_3H_7OH to O_2
(c) O_2 to CO_2
(d) H_2O to C_3H_7OH
(e) CO_2 to H_2O
(f) H_2O to O_2

8. For the reaction

$$3\,CaCl_2 + 2\,H_3PO_4 \longrightarrow Ca_3(PO_4)_2 + 6\,HCl$$

set up the mole ratio of

(a) $CaCl_2$ to $Ca_3(PO_4)_2$
(b) HCl to H_3PO_4
(c) $CaCl_2$ to H_3PO_4
(d) $Ca_3(PO_4)_2$ to H_3PO_4
(e) HCl to $Ca_3(PO_4)_2$
(f) H_3PO_4 to HCl

9. How many moles of CO_2 can be produced from 7.75 mol C_2H_5OH? (Balance the equation first.)

$$C_2H_5OH + O_2 \longrightarrow CO_2 + H_2O$$

10. How many moles of Cl_2 can be produced from 5.60 mol HCl?

$$HCl + O_2 \longrightarrow Cl_2 + H_2O$$

11. Given the equation

$$MnO_2(s) + HCl(aq) \longrightarrow Cl_2(g) + MnCl_2(aq) + H_2O(l) \text{ (unbalanced)}$$

(a) How many moles of HCl will react with 1.05 mol of MnO_2?
(b) How many moles of $MnCl_2$ will be produced when 1.25 mol of H_2O are formed?

12. Given the equation

$$Al_4C_3 + 12\,H_2O \longrightarrow 4\,Al(OH)_3 + 3\,CH_4$$

(a) How many moles of water are needed to react with 100. g of Al_4C_3?
(b) How many moles of $Al(OH)_3$ will be produced when 0.600 mol of CH_4 is formed?

13. How many grams of sodium hydroxide can be produced from 500 g of calcium hydroxide according to this equation?

$$Ca(OH)_2 + Na_2CO_3 \longrightarrow 2\,NaOH + CaCO_3$$

14. How many grams of zinc phosphate, $Zn_3(PO_4)_2$, are formed when 10.0 g of Zn are reacted with phosphoric acid?

$$3\,Zn + 2\,H_3PO_4 \longrightarrow Zn_3(PO_4)_2 + 3\,H_2$$

15. In a blast furnace, iron(III) oxide reacts with coke (carbon) to produce molten iron and carbon monoxide:

$$Fe_2O_3 + 3\,C \xrightarrow{\Delta} 2\,Fe + 3\,CO$$

How many kilograms of iron would be formed from 125 kg of Fe_2O_3?

16. How many grams of steam and iron must react to produce 375 g of magnetic iron oxide, Fe_3O_4?

$$3\,Fe(s) + 4\,H_2O(g) \xrightarrow{\Delta} Fe_3O_4(s) + 4\,H_2(g)$$

17. Ethane gas, C_2H_6, burns in air (i.e., reacts with the oxygen in air) to form carbon dioxide and water:

$$2\,C_2H_6 + 7\,O_2 \longrightarrow 4\,CO_2 + 6\,H_2O$$

(a) How many moles of O_2 are needed for the complete combustion of 15.0 mol of ethane?
(b) How many grams of CO_2 are produced for each 8.00 g of H_2O produced?
(c) How many grams of CO_2 will be produced by the combustion of 75.0 g of C_2H_6?
(d) If the reaction produces 2.75 mol of H_2O, how many grams of CO_2 are also produced?
(e) How many grams of O_2 are need to completely react with 25.0 mol C_2H_6?
(f) How many grams of C_2H_6 will be needed to produce 125 g of water?

18. Given the equation

$$4\,FeS_2 + 11\,O_2 \longrightarrow 2\,Fe_2O_3 + 8\,SO_2$$

(a) How many moles of Fe_2O_3 can be made from 1.00 mol of FeS_2?
(b) How many moles of O_2 are required to react with 4.50 mol of FeS_2?
(c) If the reaction produces 1.55 mol of Fe_2O_3, how many moles of SO_2 are produced?
(d) How many grams of SO_2 can be formed from 0.512 mol of FeS_2?
(e) If the reaction produces 40.6 g of SO_2, how many moles of O_2 were reacted?
(f) How many grams of FeS_2 are needed to produce 221 g of Fe_2O_3?

Limiting Reactant and Percent Yield

19. Draw the molecules of product(s) formed for each of the following reactions; then determine which substance is the limiting reactant:

(a)

H_2
O_2

(b)

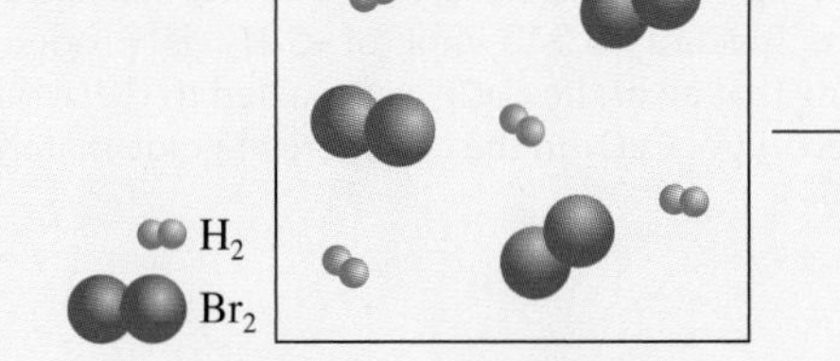

20. Draw the molecules of product(s) formed for the following reactions; then determine which substance is the limiting reactant:

(a)

(b)

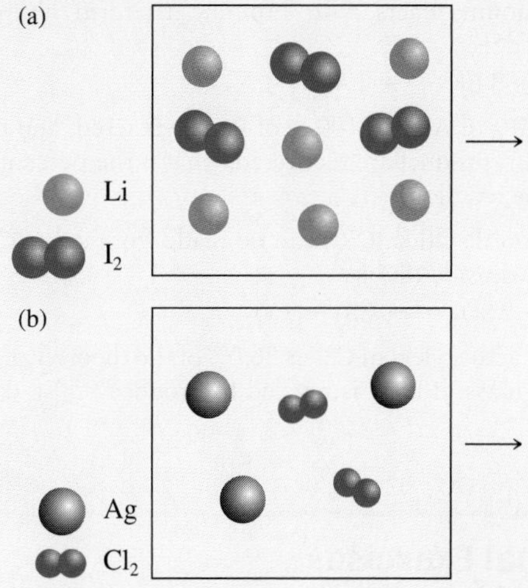

21. Draw pictures similar to those in Question 19 and determine which is the limiting reactant in the following reactions:

(a) Eight atoms of potassium react with five molecules of chlorine to produce potassium chloride.
(b) Ten atoms of aluminum react with three molecules of oxygen to produce aluminum oxide.

22. Draw pictures similar to those in Question 20 and determine which is the limiting reactant in the following reactions:

(a) Eight molecules of nitrogen react with six molecules of oxygen to produce nitrogen dioxide.
(b) Fifteen atoms of iron react with twelve molecules of water to produce magnetic iron oxide (Fe_3O_4) and hydrogen.

23. In the following equations, determine which reactant is the limiting reactant and which reactant is in excess. The amounts used are given below each reactant. Show evidence for your answers.

(a) $KOH + HNO_3 \longrightarrow KNO_3 + H_2O$
16.0 g 12.0 g

(b) $2\,NaOH + H_2SO_4 \longrightarrow Na_2SO_4 + H_2O$
10.0 g 10.0 g

24. In the following equations, determine which reactant is the limiting reactant and which reactant is in excess. The amounts used are given below each reactant. Show evidence for your answers.

(a) $2\,Bi(NO_3)_3 + 3\,H_2S \longrightarrow Bi_2S_3 + 6\,HNO_3$
50.0 g 6.00 g

(b) $3\,Fe + 4\,H_2O \longrightarrow Fe_3O_4 + 4\,H_2$
40.0 g 16.0 g

25. The reaction for the combustion of propane is

$$C_3H_8 + 5\,O_2 \longrightarrow 3\,CO_2 + 4\,H_2O$$

(a) If 20.0 g of C_3H_8 and 20.0 g of O_2 are reacted, how many moles of CO_2 can be produced?
(b) If 20.0 g of C_3H_8 and 80.0 g of O_2 are reacted, how many moles of CO_2 can be produced?
(c) If 2.0 mol of C_3H_8 and 14.0 mol of O_2 are placed in a closed container and they react to completion (until one reactant is completely used up), what compounds will be present in the container after the reaction, and how many moles of each compound are present?

26. The reaction for the combustion of propene is

$$2\,C_3H_6 + 9\,O_2 \longrightarrow 6\,CO_2 + 6\,H_2O$$

(a) If 15.0 g of C_3H_6 and 15 g of O_2 are reacted, how many moles of H_2Ocan be produced?
(b) If 12.0 g of C_3H_6 and 25.0 g of O_2 are reacted, how many moles of H_2O can be produced?
(c) If 5.0 mol of C_3H_6 and 15.0 mol of O_2 are reacted to completion (until one reactant is completely used up), how many moles of CO_2 can be produced? Which reactant is left over when the reaction has gone to completion in a closed container?

***27.** When a certain nonmetal whose formula is X_8 burns in air, XO_3 forms. Write a balanced equation for this reaction. If 120.0 g of oxygen gas are consumed completely, along with 80.0 g of X_8, identify element X.

29. Aluminum reacts with bromine to form aluminum bromide:

$$2\,Al + 3\,Br_2 \longrightarrow 2\,AlBr_3$$

If 25.0 g of Al and 100. g of Br_2 are reacted, and 64.2 g of $AlBr_3$ product are recovered, what is the percent yield for the reaction?

***31.** Carbon disulfide, CS_2, can be made from coke (C) and sulfur dioxide (SO_2):

$$3\,C + 2\,SO_2 \longrightarrow CS_2 + CO_2$$

If the actual yield of CS_2 is 86.0% of the theoretical yield, what mass of coke is needed to produce 950 g of CS_2?

*28. When a particular metal X reacts with HCl, the resulting products are XCl_2 and H_2. Write and balance the equation. When 78.5 g of the metal react completely, 2.42 g of hydrogen gas result. Identify the element X.

30. Iron was reacted with a solution containing 400. g of copper(II) sulfate. The reaction was stopped after 1 hour, and 151 g of copper were obtained. Calculate the percent yield of copper obtained:

$$Fe(s) + CuSO_4(aq) \longrightarrow Cu(s) + FeSO_4(aq)$$

*32. Acetylene (C_2H_2) can be manufactured by the reaction of water and calcium carbide, CaC_2:

$$CaC_2(s) + 2\,H_2O(l) \longrightarrow C_2H_2(g) + Ca(OH)_2(aq)$$

When 44.5 g of commercial-grade (impure) calcium carbide are reacted, 0.540 mol of C_2H_2 is produced. Assuming that all of the CaC_2 was reacted to C_2H_2, what is the percent of CaC_2 in the commercial-grade material?

Additional Exercises

All exercises with blue numbers have answers in the appendix of the text.

33. A tool set contains 6 wrenches, 4 screwdrivers, and 2 pliers. The manufacturer has 1000 pliers, 2000 screwdrivers, and 3000 wrenches in stock. Can an order for 600 tool sets be filled? Explain briefly.

34. What is the difference between using a number as a subscript and using a number as a coefficient in a chemical equation?

35. For mass–mass calculations, why is it necessary to convert the grams of starting material to moles of starting material, then determine the moles of product from the moles of starting material, and convert the moles of product to grams of product? Why can't you calculate the grams of product directly from the grams of starting material?

*36. Oxygen masks for producing O_2 in emergency situations contain potassium superoxide (KO_2). It reacts according to this equation:

$$4\,KO_2 + 2\,H_2O + 4\,CO_2 \longrightarrow 4\,KHCO_3 + 3\,O_2$$

(a) If a person wearing such a mask exhales 0.85 g of CO_2 every minute, how many moles of KO_2 are consumed in 10.0 minutes?

(b) How many grams of oxygen are produced in 1.0 hour?

37. Ethyl alcohol is the result of fermentation of sugar, $C_6H_{12}O_6$:

$$C_6H_{12}O_6 \longrightarrow 2\,C_2H_5OH + 2\,CO_2$$

(a) How many grams of ethyl alcohol and how many grams of carbon dioxide can be produced from 750 g of sugar?

(b) How many milliliters of alcohol ($d = 0.79$ g/mL) can be produced from 750 g of sugar?

38. The methyl alcohol (CH_3OH) used in alcohol burners combines with oxygen gas to form carbon dioxide and water. How many grams of oxygen are required to burn 60.0 mL of methyl alcohol ($d = 0.72$ g/mL)?

39. Hydrazine (N_2H_4) and hydrogen peroxide (H_2O_2) have been used as rocket propellants. They react according to the following equation:

$$7\,H_2O_2 + N_2H_4 \longrightarrow 2\,HNO_3 + 8\,H_2O$$

(a) When 75 kg of hydrazine react, how many grams of nitric acid can be formed?

(b) When 250 L of hydrogen peroxide ($d = 1.41$ g/mL) react, how many grams of water can be formed?

(c) How many grams of hydrazine will be required to react with 725 g hydrogen peroxide?

(d) How many grams of water can be produced when 750 g hydrazine combine with 125 g hydrogen peroxide?

(e) How many grams of the excess reactant in part (d) are left unreacted?

40. Chlorine gas can be prepared according to the reaction

$$16\,HCl + 2\,KMnO_4 \longrightarrow 5\,Cl_2 + 2\,KCl + 2\,MnCl_2 + 8\,H_2O$$

(a) How many moles of $MnCl_2$ can be produced when 25 g $KMnO_4$ is mixed with 85 g HCl?

(b) How many grams of water will be produced when 75 g KCl are produced?

(c) What is the percent yield of Cl_2 if 150 g HCl is reacted, producing 75 g of Cl_2?

(d) When 25 g HCl react with 25 g $KMnO_4$, how many grams of Cl_2 can be produced?

(e) How many grams of the excess reactant in part (d) are left unreacted?

41. Silver tarnishes in the presence of hydrogen sulfide (which smells like rotten eggs) and oxygen because of the reaction

$4\,Ag + 2\,H_2S + O_2 \longrightarrow 2\,Ag_2S + 2\,H_2O$

(a) How many grams of silver sulfide can be formed from a mixture of 1.1 g Ag, 0.14 g H_2S, and 0.080 g O_2?
(b) How many more grams of H_2S would be needed to completely react all of the Ag?

42. Solid CaO picks up water from the atmosphere to form $Ca(OH)_2$.

$CaO(s) + H_2O(g) \longrightarrow Ca(OH)_2(s)$

A beaker containing CaO weighs 26.095 g. When allowed to sit open to the atmosphere, it absorbs water to a final mass of 26.500 g. What is the mass of the beaker?

43. When a solution containing 15 g of lead(II) nitrate and a solution containing 15 g of potassium iodide are mixed, a yellow precipitate is formed.
(a) What is the name and formula for the solid product formed?
(b) What type of reaction occurred?
(c) After filtration and drying, the solid product has a mass of 6.68 g. What is the percent yield for the reaction?

44. A 10.00-g mixture of KNO_3 and KCl is reacted with $AgNO_3$ to give 4.33 g AgCl. What is the percent composition for the mixture?

45. After 180.0 g of zinc were dropped into a beaker of hydrochloric acid and the reaction ceased, 35 g of unreacted zinc remained in the beaker:

$Zn + HCl \longrightarrow ZnCl_2 + H_2$

(a) How many grams of hydrogen gas were produced?
(b) How many grams of HCl were reacted?
(c) How many more grams of HCl would be required to completely react with the original sample of zinc?

46. Use this equation to answer (a) and (b):

$Fe(s) + CuSO_4(aq) \longrightarrow Cu(s) + FeSO_4(aq)$

(a) When 2.0 mol of Fe and 3.0 mol of $CuSO_4$ are reacted, what substances will be present when the reaction is over? How many moles of each substance are present?
(b) When 20.0 g of Fe and 40.0 g of $CuSO_4$ are reacted, what substances will be present when the reaction is over? How many grams of each substance are present?

47. Methyl alcohol (CH_3OH) is made by reacting carbon monoxide and hydrogen in the presence of certain metal oxide catalysts. How much alcohol can be obtained by reacting 40.0 g of CO and 10.0 g of H_2? How many grams of excess reactant remain unreacted?

$CO(g) + 2\,H_2(g) \longrightarrow CH_3OH(l)$

*48. Ethyl alcohol (C_2H_5OH), also called grain alcohol, can be made by the fermentation of sugar:

$C_6H_{12}O_6 \longrightarrow 2\,C_2H_5OH + 2\,CO_2$
glucose — ethyl alcohol

If an 84.6% yield of ethyl alcohol is obtained,
(a) what mass of ethyl alcohol will be produced from 750 g of glucose?
(b) what mass of glucose should be used to produce 475 g of C_2H_5OH?

***49.** Both $CaCl_2$ and $MgCl_2$ react with $AgNO_3$ to precipitate AgCl. When solutions containing equal masses of $CaCl_2$ and $MgCl_2$ are reacted with $AgNO_3$, which salt solution will produce the larger amount of AgCl? Show proof.

*50. An astronaut excretes about 2500 g of water a day. If lithium oxide (Li_2O) is used in the spaceship to absorb this water, how many kilograms of Li_2O must be carried for a 30-day space trip for three astronauts?

$Li_2O + H_2O \longrightarrow 2\,LiOH$

***51.** Much commercial hydrochloric acid is prepared by the reaction of concentrated sulfuric acid with sodium chloride:

$H_2SO_4 + 2\,NaCl \longrightarrow Na_2SO_4 + 2\,HCl$

How many kilograms of concentrated H_2SO_4, 96% H_2SO_4 by mass, are required to produce 20.0 L of concentrated hydrochloric acid (d = 1.20 g/mL, 42.0% HCl by mass)?

52. Three chemical reactions that lead to the formation of sulfuric acid are

$S + O_2 \longrightarrow SO_2$
$2\,SO_2 + O_2 \longrightarrow 2\,SO_3$
$SO_3 + H_2O \longrightarrow H_2SO_4$

Starting with 100.0 g of sulfur, how many grams of sulfuric acid will be formed, assuming a 10% loss in each step? What is the percent yield of H_2SO_4?

53. A 10.00-g mixture of $NaHCO_3$ and Na_2CO_3 was heated and yielded 0.0357 mol H_2O and 0.1091 mol CO_2. Calculate the percent composition of the mixture:

$NaHCO_3 \longrightarrow Na_2O + CO_2 + H_2O$
$Na_2CO_3 \longrightarrow Na_2O + CO_2$

Challenge Exercises

All exercises with blue numbers have answers in the appendix of the text.

*54. When 12.82 g of a mixture of $KClO_3$ and NaCl are heated strongly, the $KClO_3$ reacts according to this equation:

$2\,KClO_3(s) \longrightarrow 2\,KCl(s) + 3\,O_2(g)$

The NaCl does not undergo any reaction. After the heating, the mass of the residue (KCl and NaCl) is 9.45 g. Assuming that all the loss of mass represents loss of oxygen gas, calculate the percent of $KClO_3$ in the original mixture.

*55. Gastric juice contains about 3.0 g HCl per liter. If a person produces about 2.5 L of gastric juice per day, how many antacid tablets, each containing 400. mg of $Al(OH)_3$, are needed to neutralize all the HCl produced in 1 day?

$$Al(OH)_3(s) + 3\,HCl(aq) \longrightarrow AlCl_3(aq) + 3\,H_2O(l)$$

*56. Phosphoric acid, H_3PO_4, can be synthesized from phosphorus, oxygen, and water according to these two reactions:

$$4\,P + 5\,O_2 \longrightarrow P_4O_{10}$$
$$P_4O_{10} + 6\,H_2O \longrightarrow 4\,H_3PO_4$$

Starting with 20.0 g P, 30.0 g O_2, and 15.0 g H_2O, what is the mass of phosphoric acid that can be formed?

Answers to Practice Exercises

9.1 0.33 mol Al_2O_3
9.2 7.33 mol $Al(OH)_3$
9.3 0.8157 mol KCl
9.4 85 g $AgNO_3$
9.5 27.6 g $CrCl_3$
9.6 348.8 g H_2O
9.7 17.7 g HCl
9.8 164.3 g $BaSO_4$
9.9 88.5% yield

PUTTING IT TOGETHER: Review for Chapters 7–9

Multiple Choice:

Choose the correct answer to each of the following.

1. 4.0 g of oxygen contain
(a) 1.5×10^{23} atoms of oxygen
(b) 4.0 molar masses of oxygen
(c) 0.50 mol of oxygen
(d) 6.022×10^{23} atoms of oxygen

2. One mole of hydrogen atoms contains
(a) 2.0 g of hydrogen
(b) 6.022×10^{23} atoms of hydrogen
(c) 1 atom of hydrogen
(d) 12 g of carbon-12

3. The mass of one atom of magnesium is
(a) 24.31 g (c) 12.00 g
(b) 54.94 g (d) 4.037×10^{-23} g

4. Avogadro's number of magnesium atoms
(a) has a mass of 1.0 g
(b) has the same mass as Avogadro's number of sulfur atoms
(c) has a mass of 12.0 g
(d) is 1 mol of magnesium atoms

5. Which of the following contains the largest number of moles?
(a) 1.0 g Li (c) 1.0 g Al
(b) 1.0 g Na (d) 1.0 g Ag

6. The number of moles in 112 g of acetylsalicylic acid (aspirin), $C_9H_8O_4$, is
(a) 1.61 (c) 112
(b) 0.622 (d) 0.161

7. How many moles of aluminum hydroxide are in one antacid tablet containing 400 mg of $Al(OH)_3$?
(a) 5.13×10^{-3} (c) 5.13
(b) 0.400 (d) 9.09×10^{-3}

8. How many grams of Au_2S can be obtained from 1.17 mol of Au?
(a) 182 g (c) 364 g
(b) 249 g (d) 499 g

9. The molar mass of $Ba(NO_3)_2$ is
(a) 199.3 (c) 247.3
(b) 261.3 (d) 167.3

10. A 16-g sample of oxygen
(a) is 1 mol of O_2
(b) contains 6.022×10^{23} molecules of O_2
(c) is 0.50 molecule of O_2
(d) is 0.50 molar mass of O_2

11. What is the percent composition for a compound formed from 8.15 g of zinc and 2.00 g of oxygen?
(a) 80.3% Zn, 19.7% O (c) 70.3% Zn, 29.7% O
(b) 80.3% O, 19.7% Zn (d) 65.3% Zn, 34.7% O

12. Which of these compounds contains the largest percentage of oxygen?
(a) SO_2 (c) N_2O_3
(b) SO_3 (d) N_2O_5

13. 2.00 mol of CO_2
(a) have a mass of 56.0 g
(b) contain 1.20×10^{24} molecules
(c) have a mass of 44.0 g
(d) contain 6.00 molar masses of CO_2

14. In Ag_2CO_3, the percent by mass of
(a) carbon is 43.5% (c) oxygen is 17.4%
(b) silver is 64.2% (d) oxygen is 21.9%

15. The empirical formula of the compound containing 31.0% Ti and 69.0% Cl is
(a) TiCl (c) $TiCl_3$
(b) $TiCl_2$ (d) $TiCl_4$

16. A compound contains 54.3% C, 5.6% H, and 40.1% Cl. The empirical formula is
(a) CH_3Cl (c) $C_2H_4Cl_2$
(b) C_2H_5Cl (d) C_4H_5Cl

17. A compound contains 40.0% C, 6.7% H, and 53.3% O. The molar mass is 60.0 g/mol. The molecular formula is
(a) $C_2H_3O_2$ (c) C_2HO
(b) C_3H_8O (d) $C_2H_4O_2$

18. How many chlorine atoms are in 4.0 mol of PCl_3?
(a) 3 (c) 12
(b) 7.2×10^{24} (d) 2.4×10^{24}

19. What is the mass of 4.53 mol of Na_2SO_4?
(a) 142.1 g (c) 31.4 g
(b) 644 g (d) 3.19×10^{-2} g

20. The percent composition of Mg_3N_2 is
(a) 72.2% Mg, 27.8% N (c) 83.9% Mg, 16.1% N
(b) 63.4% Mg, 36.6% N (d) no correct answer given

21. How many grams of oxygen are contained in 0.500 mol of Na_2SO_4?
(a) 16.0 g (c) 64.0 g
(b) 32.0 g (d) no correct answer given

22. The empirical formula of a compound is CH. If the molar mass of this compound is 78.11, then the molecular formula is
(a) C_2H_2 (c) C_6H_6
(b) C_5H_{18} (d) no correct answer given

23. The reaction

$$BaCl_2 + (NH_4)_2CO_3 \longrightarrow BaCO_3 + 2\,NH_4Cl$$

is an example of
(a) combination (c) single displacement
(b) decomposition (d) double displacement

24. When the equation

$Al + O_2 \longrightarrow Al_2O_3$

is properly balanced, which of the following terms appears?
(a) 2 Al (c) 3 Al
(b) 2 Al_2O_3 (d) 2 O_2

25. Which equation is *incorrectly* balanced?

(a) $2\,KNO_3 \xrightarrow{\Delta} 2\,KNO_2 + O_2$

(b) $H_2O_2 \longrightarrow H_2O + O_2$

(c) $2\,Na_2O_2 + 2\,H_2O \longrightarrow 4\,NaOH + O_2$

(d) $2\,H_2O \xrightarrow[H_2SO_4]{\text{electrical energy}} 2\,H_2 + O_2$

26. The reaction

$2\,Al + 3\,Br_2 \longrightarrow 2\,AlBr_3$

is an example of
(a) combination (c) decomposition
(b) single displacement (d) double displacement

27. When the equation

$PbO_2 \xrightarrow{\Delta} PbO + O_2$

is balanced, one term in the balanced equation is
(a) PbO_2 (c) 3 PbO
(b) 3 O_2 (d) O_2

28. When the equation

$Cr_2S_3 + HCl \longrightarrow CrCl_3 + H_2S$

is balanced, one term in the balanced equation is
(a) 3 HCl (c) 3 H_2S
(b) $CrCl_3$ (d) 2 Cr_2S_3

29. When the equation

$F_2 + H_2O \longrightarrow HF + O_2$

is balanced, one term in the balanced equation is
(a) 2 HF (c) 4 HF
(b) 3 O_2 (d) 4 H_2O

30. When the equation

$NH_4OH + H_2SO_4 \longrightarrow$

is completed and balanced, one term in the balanced equation is
(a) NH_4SO_4 (c) H_2OH
(b) 2 H_2O (d) 2 $(NH_4)_2SO_4$

31. When the equation

$H_2 + V_2O_5 \longrightarrow V +$

is completed and balanced, one term in the balanced equation is
(a) 2 V_2O_5 (c) 2 V
(b) 3 H_2O (d) 8 H_2

32. When the equation

$Al(OH)_3 + H_2SO_4 \longrightarrow Al_2(SO_4)_3 + H_2O$

is balanced, the sum of the coefficients will be
(a) 9 (c) 12
(b) 11 (d) 15

33. When the equation

$H_3PO_4 + Ca(OH)_2 \longrightarrow H_2O + Ca_3(PO_4)_2$

is balanced, the proper sequence of coefficients is
(a) 3, 2, 1, 6 (c) 2, 3, 1, 6
(b) 2, 3, 6, 1 (d) 2, 3, 3, 1

34. When the equation

$Fe_2(SO_4)_3 + Ba(OH)_2 \longrightarrow$

is completed and balanced, one term in the balanced equation is
(a) $Ba_2(SO_4)_3$ (c) 2 $Fe_2(SO_4)_3$
(b) 2 $Fe(OH)_2$ (d) 2 $Fe(OH)_3$

35. For the reaction

$2\,H_2 + O_2 \longrightarrow 2\,H_2O + 572.4\text{ kJ}$

which of the following is not true?
(a) The reaction is exothermic.
(b) 572.4 kJ of heat are liberated for each mole of water formed.
(c) 2 mol of hydrogen react with 1 mol of oxygen.
(d) 572.4 kJ of heat are liberated for each 2 mol of hydrogen reacted.

36. How many moles is 20.0 g of Na_2CO_3?
(a) 1.89 mol (c) 212 mol
(b) 2.12×10^3 mol (d) 0.189 mol

37. What is the mass in grams of 0.30 mol of $BaSO_4$?
(a) 7.0×10^3 g (c) 70. g
(b) 0.13 g (d) 700.2 g

38. How many molecules are in 5.8 g of acetone, C_3H_6O?
(a) 0.10 molecule
(b) 6.0×10^{22} molecules
(c) 3.5×10^{24} molecules
(d) 6.0×10^{23} molecules

Problems 39–45 refer to the reaction

$2\,C_2H_4 + 6\,O_2 \longrightarrow 4\,CO_2 + 4\,H_2O$

39. If 6.0 mol of CO_2 are produced, how many moles of O_2 were reacted?
(a) 4.0 mol (c) 9.0 mol
(b) 7.5 mol (d) 15.0 mol

40. How many moles of O_2 are required for the complete reaction of 45 g of C_2H_4?
(a) 1.3×10^2 mol (c) 112.5 mol
(b) 0.64 mol (d) 4.8 mol

41. If 18.0 g of CO_2 are produced, how many grams of H_2O are produced?
(a) 7.37 g (c) 9.00 g
(b) 3.68 g (d) 14.7 g

42. How many moles of CO_2 can be produced by the reaction of 5.0 mol of C_2H_4 and 12.0 mol of O_2?
(a) 4.0 mol (c) 8.0 mol
(b) 5.0 mol (d) 10. mol

43. How many moles of CO_2 can be produced by the reaction of 0.480 mol of C_2H_4 and 1.08 mol of O_2?
(a) 0.240 mol
(b) 0.960 mol
(c) 0.720 mol
(d) 0.864 mol

44. How many grams of CO_2 can be produced from 2.0 g of C_2H_4 and 5.0 g of O_2?
(a) 5.5 g
(b) 4.6 g
(c) 7.6 g
(d) 6.3 g

45. If 14.0 g of C_2H_4 are reacted and the actual yield of H_2O is 7.84 g, the percent yield in the reaction is
(a) 0.56%
(b) 43.6%
(c) 87.1%
(d) 56.0%

Problems 46–48 refer to the equation

$$H_3PO_4 + MgCO_3 \longrightarrow Mg_3(PO_4)_2 + CO_2 + H_2O$$

46. The sequence of coefficients for the balanced equation is
(a) 2, 3, 1, 3, 3
(b) 3, 1, 3, 2, 3
(c) 2, 2, 1, 2, 3
(d) 2, 3, 1, 3, 2

47. If 20.0 g of carbon dioxide are produced, the number of moles of magnesium carbonate used is
(a) 0.228 mol
(b) 1.37 mol
(c) 0.910 mol
(d) 0.454 mol

48. If 50.0 g of magnesium carbonate react completely with H_3PO_4, the number of grams of carbon dioxide produced is
(a) 52.2 g
(b) 26.1 g
(c) 13.1 g
(d) 50.0 g

49. When 10.0 g of $MgCl_2$ and 10.0 g of Na_2CO_3 are reacted in

$$MgCl_2 + Na_2CO_3 \longrightarrow MgCO_3 + 2\,NaCl$$

the limiting reactant is
(a) $MgCl_2$
(b) Na_2CO_3
(c) $MgCO_3$
(d) NaCl

50. When 50.0 g of copper are reacted with silver nitrate solution in

$$Cu + 2\,AgNO_3 \longrightarrow Cu(NO_3)_2 + 2\,Ag$$

148 g of silver are obtained. What is the percent yield of silver obtained?
(a) 87.1%
(b) 84.9%
(c) 55.2%
(d) no correct answer given

Free Response Questions:

Answer each of the following. Be sure to include your work and explanations in a clear, logical form.

1. Compound X requires 104 g of O_2 to produce 2 moles of CO_2 and 2.5 moles of H_2O.
(a) What is the empirical formula for X?
(b) What additional information would you need to determine the molecular formula for X?

2. Consider the reaction of sulfur dioxide with oxygen to form sulfur trioxide taking place in a closed container.

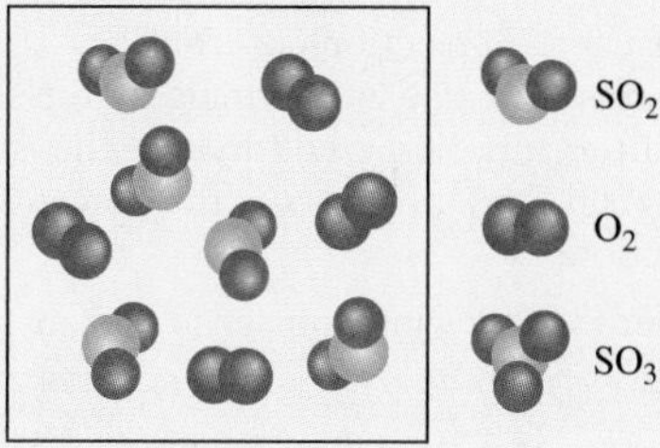

(a) Draw what you would expect to see in the box at the completion of the reaction.
(b) If you begin with 25 g of SO_2 and 5 g of oxygen gas, which is the limiting reagent?
(c) Is the following statement true or false? "When SO_2 is converted into SO_3, the percent composition of S in the compounds changes from 33% to 25%." Explain.

3. The percent composition of compound Z is 63.16% C and 8.77% H. When compound Z burns in air, the only products are carbon dioxide and water. The molar mass for Z is 114.
(a) What is the molecular formula for compound Z?
(b) What is the balanced reaction for Z burning in air?

4. Compound A decomposes at room temperature in an exothermic reaction, while compound B requires heating before it will decompose in an endothermic reaction.
(a) Draw reaction profiles (potential energy vs. reaction progress) for both reactions.
(b) The decomposition of 0.500 mol of $NaHCO_3$ to form sodium carbonate, water, and carbon dioxide requires 85.5 kJ of heat. How many grams of water could be collected and how many kJ of heat were absorbed when 24.0 g of CO_2 are produced?
(c) Could $NaHCO_3$ be compound A or compound B? Explain your reasoning.

5. Aqueous ammonium hydroxide reacts with aqueous cobalt(II) sulfate to produce aqueous ammonium sulfate and solid cobalt(II) hydroxide. When 38.0 g of one of the reactants were fully reacted with enough of the other reactant, 8.09 g of ammonium sulfate were obtained, which corresponded to a 25.0% yield.
(a) What type of reaction took place?
(b) Write the balanced chemical equation.
(c) What is the theoretical yield of ammonium sulfate?
(d) Which reactant was the limiting reagent?

6. $C_6H_{12}O_6 \longrightarrow 2\,C_2H_5OH + 2\,CO_2(g)$
(a) If 25.0 g of $C_6H_{12}O_6$ were used, and only 11.2 g of C_2H_5OH were produced, how much reactant was left over and what volume of gas was produced? (*Note*: One mole of gas occupies 24.0 L of space at room temperature.)
(b) What was the yield of the reaction?
(c) What type of reaction took place?

7. When solutions containing 25 g each of lead(II) nitrate and potassium iodide are mixed, a yellow precipitate results.
   (a) What type of reaction occurred?
   (b) What is the name and formula for the solid product?
   (c) If after filtration and drying, the solid product weighed 7.66 g, what was the percent yield for the reaction?

8. Consider the following unbalanced reaction:

   $XNO_3 + CaCl_2 \longrightarrow XCl + Ca(NO_3)_2$

   (a) If 30.8 g of $CaCl_2$ produced 79.6 g of XCl, what is X?
   (b) Would X be able to displace hydrogen from an acid?

9. Consider the following reaction: $H_2O_2 \longrightarrow H_2O + O_2$
   (a) If at the end of the reaction there are eight water molecules and eight oxygen molecules, what was in the flask at the start of the reaction?
   (b) Does the following reaction profile indicate the reaction is exothermic or endothermic?

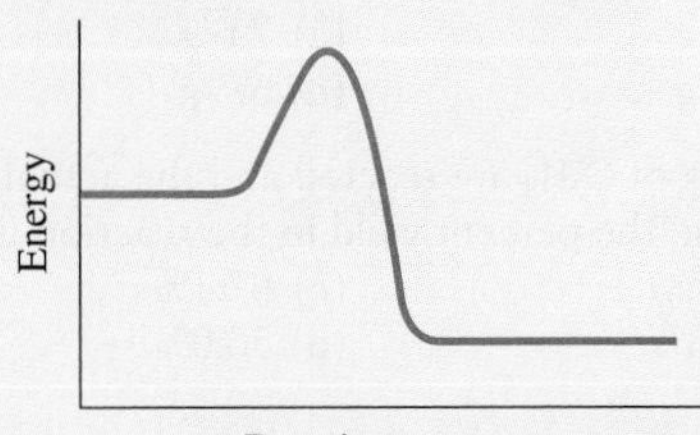

   (c) What type of reaction is given above?
   (d) What is the empirical formula for hydrogen peroxide?

CHAPTER 10

Modern Atomic Theory and the Periodic Table

Electrons moving between energy levels in neon produce the colors in this neon walkway at Epcot Center.

Chapter Outline

How do we go about studying an object that is too small to see? Think back to that birthday present you could look at, but not yet open. Judging from the wrapping and size of the box was not very useful, but shaking, turning, and lifting the package all gave indirect clues to its contents. After all your experiments were done, you could make a fairly good guess about the contents. But was your guess correct? The only way to know for sure would be to open the package.

Chemists have the same dilemma when they study the atom. Atoms are so very small that it isn't possible to use the normal senses to describe them. We are essentially working in the dark with this package we call the atom. However, our improvements in instruments (X-ray machines and scanning tunneling microscopes) and measuring devices (spectrophotometers and magnetic resonance imaging, MRI) as well as in our mathematical skills are bringing us closer to revealing the secrets of the atom.

10.1 A Brief History

In the last 200 years, vast amounts of data have been accumulated to support atomic theory. When atoms were originally suggested by the early Greeks, no physical evidence existed to support their ideas. Early chemists did a variety of experiments, which culminated in Dalton's model of the atom. Because of the limitations of Dalton's model, modifications were proposed first by Thomson and then by Rutherford, which eventually led to our modern concept of the nuclear atom. These early models of the atom work reasonably well—in fact, we continue to use them to visualize a variety of chemical concepts. There remain questions that these models cannot answer, including an explanation of how atomic structure relates to the periodic table. In this chapter, we will present our modern model of the atom; we will see how it varies from and improves upon the earlier atomic models.

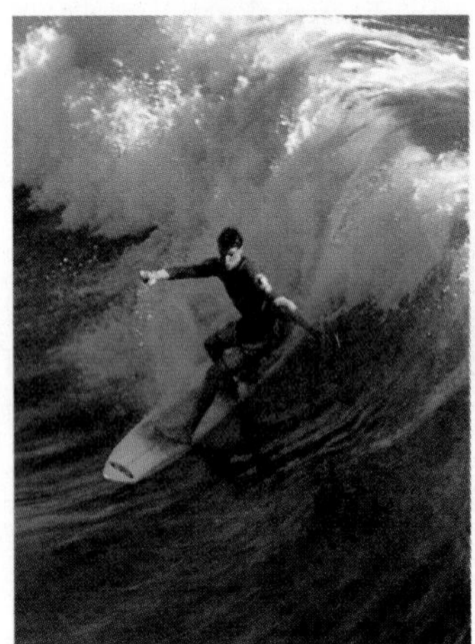

Surfers judge the wavelength, frequency, and speed of waves to get the best ride.

10.2 Electromagnetic Radiation

Scientists have studied energy and light for centuries, and several models have been proposed to explain how energy is transferred from place to place. One way energy travels through space is by *electromagnetic radiation*. Examples of electromagnetic radiation include light from the sun, X-rays in your dentist's office, microwaves from your microwave oven, radio and television waves, and radiant heat from your fireplace. While these examples seem quite different, they are all similar in some important ways. Each shows wavelike behavior, and all travel at the same speed in a vacuum (3.00×10^8 m/s).

The study of wave behavior is a topic for another course, but we need some basic terminology to understand atoms. Waves have three basic characteristics: wavelength, frequency, and speed. **Wavelength** (lambda, λ) is the distance between consecutive peaks (or troughs) in a wave, as shown in Figure 10.1. **Frequency** (nu, ν) tells how many waves pass a particular point per second. **Speed** (v) tells how fast a wave moves through space.

wavelength

frequency

speed

Light is one form of electromagnetic radiation and is usually classified by its wavelength, as shown in Figure 10.2. Visible light, as you can see, is only a tiny part of the electromagnetic spectrum. Some examples of electromagnetic radiation involved in energy transfer outside the visible region are hot coals in

CHEMISTRY IN ACTION • You Light Up My Life

Birds in the parrot family have an unusual way to attract their mates—their feathers glow in the dark! This phenomenon is called fluorescence. It results from the absorption of ultraviolet (UV) light, which is then reemitted at longer wavelengths that both birds and people can see. In everyday life this happens in a fluorescent bulb or in the many glow-in-the-dark products such as light sticks.

Kathleen Arnold from the University of Glasgow, Scotland, discovered that the feathers of parrots that fluoresced were only those used in display or those shown off during courtship. She decided to experiment using budgerigars, with their natural colors. The researchers offered birds a choice of two companion birds, which were smeared with petroleum jelly. One of the potential companions also had a UV blocker in the petroleum jelly. The birds clearly preferred companions without the UV blocker, leading the researchers to conclude that the parrots prefer to court radiant partners. The researchers also tested same-sex companions and discovered they did not prefer radiant companions.

Perhaps we may find that the glow of candlelight really does add to the radiance of romance. Ultraviolet light certainly does for parrots!

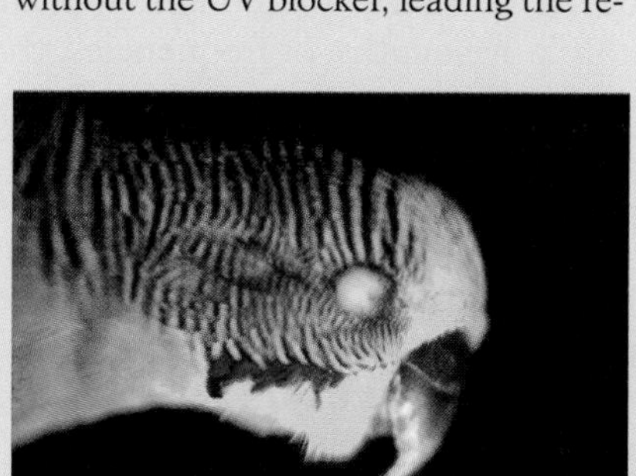

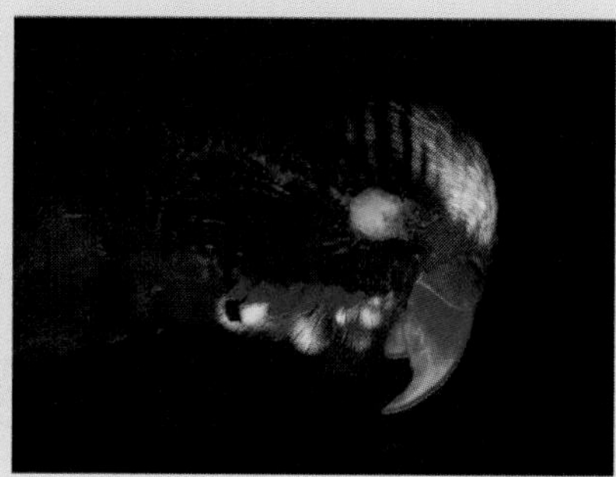

Budgerigars under normal light and under UV light showing the glow used to attract mates.

your backyard grill, which transfer infrared radiation to cook your food, and microwaves, which transfer energy to water molecules in the food, causing them to move more quickly and thus raise the temperature of your food.

We have evidence for the wavelike nature of light. We also know that a beam of light behaves like a stream of tiny packets of energy called **photons**. So what is light exactly? Is it a particle? Is it a wave? Scientists have agreed to explain the properties of electromagnetic radiation by using both wave and particle properties. Neither explanation is ideal, but currently these are our best models.

photons

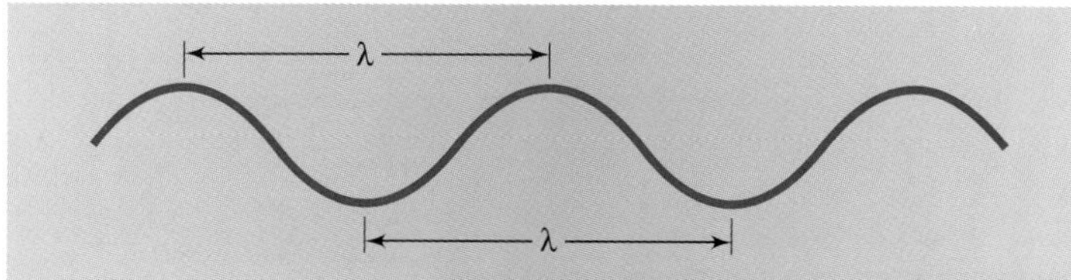

Figure 10.1
The wavelength of this wave is shown by λ. It can be measured from peak to peak or trough to trough.

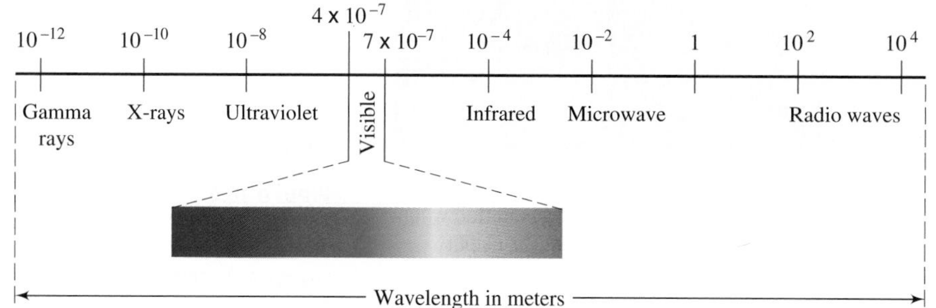

Figure 10.2
The electromagnetic spectrum.

10.3 The Bohr Atom

As scientists struggled to understand the properties of electromagnetic radiation, evidence began to accumulate that atoms could radiate light. At high temperatures, or when subjected to high voltages, elements in the gaseous state give off colored light. Brightly colored neon signs illustrate this property of matter very well. When the light emitted by a gas is passed through a prism or diffraction grating, a set of brightly colored lines called a **line spectrum** results (Figure 10.3). These colored lines indicate that the light is being emitted only at certain wavelengths, or frequencies, that correspond to specific colors. Each element possesses a unique set of these spectral lines that is different from the sets of all the other elements.

line spectrum

In 1912–1913, while studying the line spectrum of hydrogen, Niels Bohr (1885–1962), a Danish physicist, made a significant contribution to the rapidly growing knowledge of atomic structure. His research led him to believe that electrons exist in specific regions at various distances from the nucleus. He also visualized the electrons as revolving in orbits around the nucleus, like planets rotating around the sun, as shown in Figure 10.4.

Bohr's first paper in this field dealt with the hydrogen atom, which he described as a single electron revolving in an orbit about a relatively heavy nucleus. He applied the concept of energy quanta, proposed in 1900 by the German physicist Max Planck (1858–1947), to the observed line spectrum of hydrogen. Planck stated that energy is never emitted in a continuous stream but only in small, discrete packets called **quanta** (Latin, *quantus*, "how much"). From this, Bohr theorized that electrons have several possible energies corresponding to several possible orbits at different distances from the nucleus. Therefore an electron has to be in one specific energy level; it cannot exist between energy levels. In other words, the energy of the electron is said to be quantized. Bohr also stated that when a hydrogen atom absorbed one or more quanta of energy, its electron would "jump" to a higher energy level.

quanta

Bohr was able to account for spectral lines of hydrogen this way. A number of energy levels are available, the lowest of which is called the **ground state**.

ground state

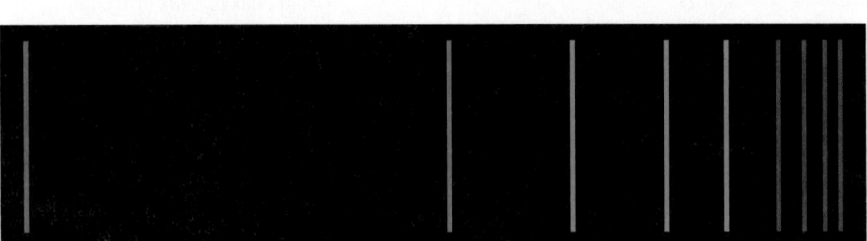

Figure 10.3
Line spectrum of hydrogen. Each line corresponds to the wavelength of the energy emitted when the electron of a hydrogen atom, which has absorbed energy, falls back to a lower principal energy level.

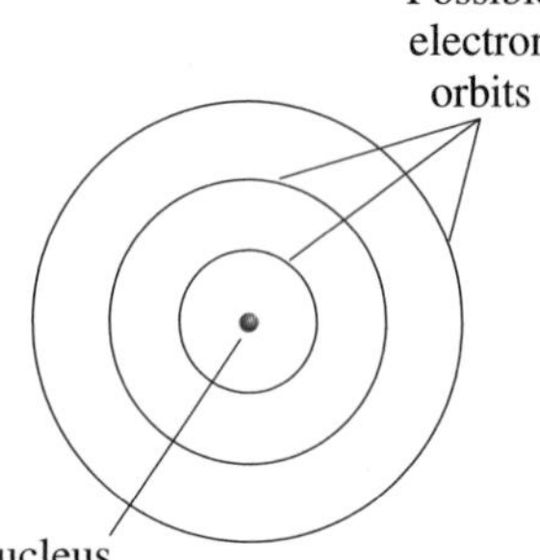

Figure 10.4
The Bohr model of the hydrogen atom described the electron revolving in certain allowed circular orbits around the nucleus.

When an electron falls from a high energy level to a lower one (say, from the fourth to the second), a quantum of energy is emitted as light at a specific frequency, or wavelength (Figure 10.5). This light corresponds to one of the lines visible in the hydrogen spectrum (Figure 10.3). Several lines are visible in this spectrum, each one corresponding to a specific electron energy-level shift within the hydrogen atom.

The chemical properties of an element and its position in the periodic table depend on electron behavior within the atoms. In turn, much of our knowledge of the behavior of electrons within atoms is based on spectroscopy. Niels Bohr contributed a great deal to our knowledge of atomic structure by (1) suggesting quantized energy levels for electrons and (2) showing that spectral lines result from the radiation of small increments of energy (Planck's quanta) when electrons shift from one energy level to another. Bohr's calculations succeeded very well in correlating the experimentally observed spectral lines with electron energy levels for the hydrogen atom. However, Bohr's methods of calculation did not succeed for heavier atoms. More theoretical work on atomic structure was needed.

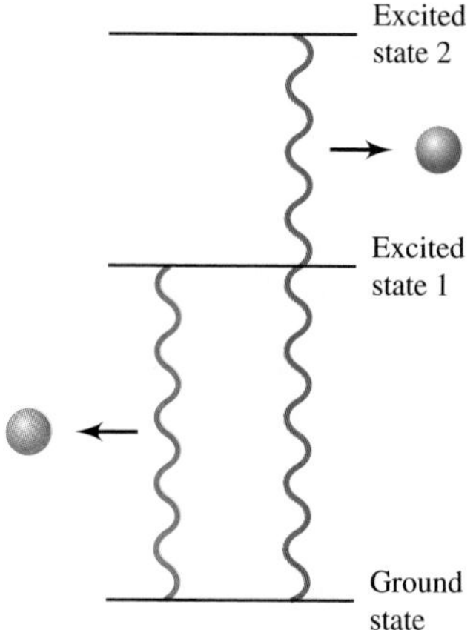

Figure 10.5
When an excited electron returns to the ground state, energy is emitted as a photon is released. The color (wavelength) of the light is determined by the difference in energy between the two states (excited and ground).

In 1924, the French physicist Louis de Broglie suggested a surprising hypothesis: All objects have wave properties. De Broglie used sophisticated mathematics to show that the wave properties for an object of ordinary size, such as a baseball, are too small to be observed. But for smaller objects, such as an electron, the wave properties become significant. Other scientists confirmed de Broglie's hypothesis, showing that electrons do exhibit wave properties. In 1926, Erwin Schrödinger, an Austrian physicist, created a mathematical model that described electrons as waves. Using Schrödinger's wave mechanics, we can determine the *probability* of finding an electron in a certain region around the nucleus the atom.

orbital

This treatment of the atom led to a new branch of physics called *wave mechanics* or *quantum mechanics*, which forms the basis for our modern understanding of atomic structure. Although the wave-mechanical description of the atom is mathematical, it can be translated, at least in part, into a visual model. It's important to recognize that we cannot locate an electron precisely within an atom; however, it is clear that electrons are not revolving around the nucleus in orbits as Bohr postulated. The electrons are instead found in *orbitals*. An **orbital** is pictured in Figure 10.6 as a region in space around the nucleus where there is a high probability of finding a given electron.

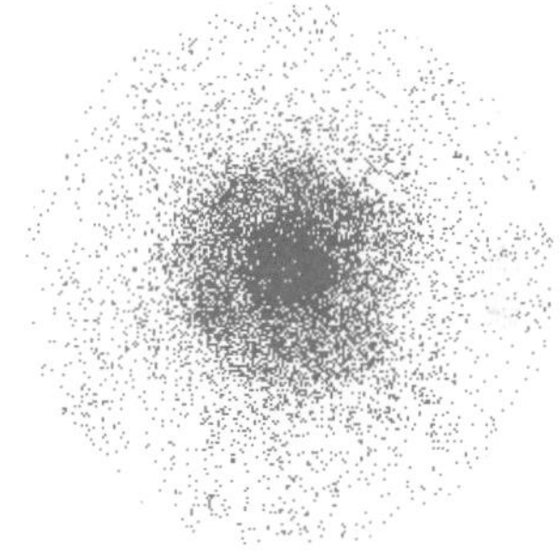

Figure 10.6
An orbital for a hydrogen atom. The intensity of the dots shows that the electron spends more time closer to the nucleus.

10.4 Energy Levels of Electrons

One of the ideas Bohr contributed to the modern concept of the atom was that the energy of the electron is quantized—that is, the electron is restricted to only certain allowed energies. The wave-mechanical model of the atom also predicts discrete **principal energy levels** within the atom. These energy levels are designated by the letter n, where n is a positive integer (Figure 10.7). The lowest principal energy level corresponds to $n = 1$, the next to $n = 2$, and so on. As n increases, the energy of the electron increases, and the electron is found on average farther from the nucleus.

principal energy levels

Each principal energy level is divided into **sublevels**, which are illustrated in Figure 10.8. The first principal energy level has one sublevel. The second principal energy level has two sublevels, the third energy level has three sublevels, and so on. Each of these sublevels contains spaces for electrons called orbitals.

sublevel

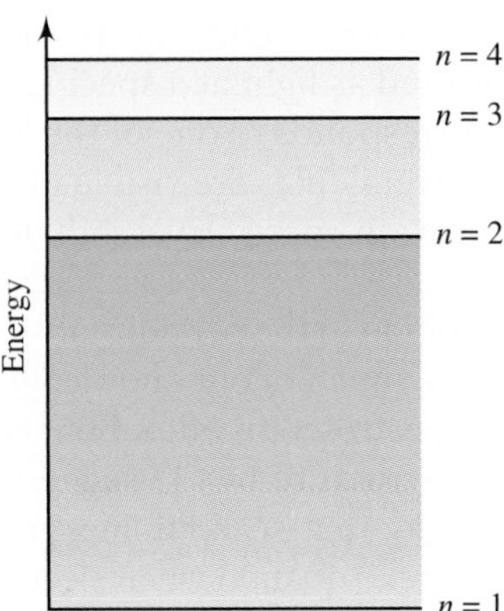

Figure 10.7
The first four principal energy levels in the hydrogen atom. Each level is assigned a principal quantum number n.

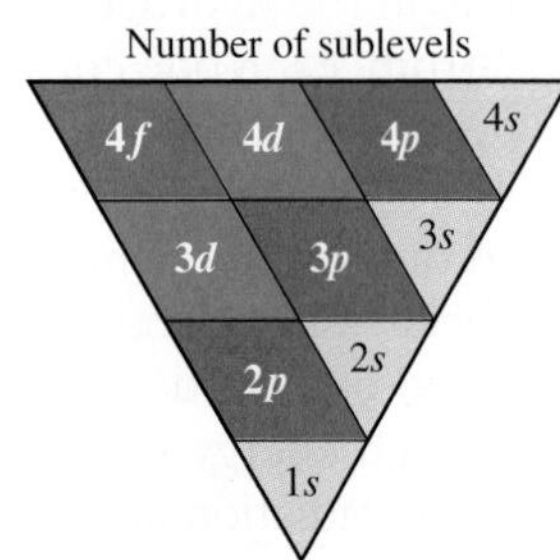

Figure 10.8
The types of orbitals on each of the first four principal energy levels.

In each sublevel the electrons are found within specified orbitals (s, p, d, f). Let's consider each principal energy level in turn. The first principal energy level ($n = 1$) has one sublevel or type of orbital. It is spherical in shape and is designated as 1*s*. It is important to understand what the spherical shape of the 1*s* orbital means. The electron does *not* move around on the surface of the sphere, but rather the surface encloses a space where there is a 90% probability that the electron may be found. It might help to consider these orbital shapes in the same way we consider the atmosphere. There is no distinct dividing line between the atmosphere and "space." The boundary is quite fuzzy. The same is true for atomic orbitals. Each has a region of highest density roughly corresponding to its shape. The probability of finding the electron outside this region drops rapidly but never quite reaches zero. Scientists often speak of orbitals as electron "clouds" to emphasize the fuzzy nature of their boundaries.

spin

How many electrons can fit into a 1*s* orbital? To answer this question, we need to consider one more property of electrons. This property is called **spin**. Each electron appears to be spinning on an axis, like a globe. It can only spin in two directions. We represent this spin with an arrow: ↑ or ↓. In order to occupy the same orbital, electrons must have *opposite* spins. That is, two electrons with the same spin cannot occupy the same orbital. This gives us the answer to our question: An atomic orbital can hold a maximum of two electrons, which must have opposite spins. This rule is called the **Pauli exclusion principle**. The first principal energy level contains one type of orbital (1*s*) that holds a maximum of two electrons.

Pauli exclusion principle

What happens with the second principal energy level ($n = 2$)? Here we find two sublevels, 2*s* and 2*p*. Like 1*s* in the first principal energy level, the 2*s* orbital is spherical in shape but is larger in size and higher in energy. It also holds a maximum of two electrons. The second type of orbital is designated by 2*p*. The 2*p* sublevel consists of three orbitals: $2p_x$, $2p_y$, and $2p_z$. The shape of *p* orbitals is quite different from the *s* orbitals, as shown in Figure 10.9.

Each *p* orbital has two "lobes." Remember, the space enclosed by these surfaces represents the regions of probability for finding the electrons 90% of the time. There are three separate *p* orbitals, each oriented in a different direction,

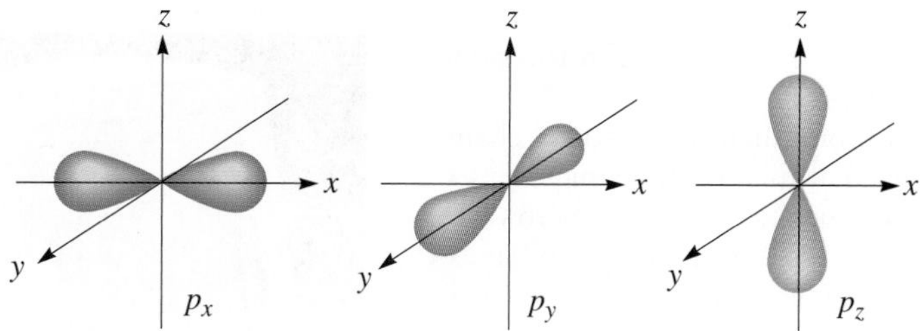

Figure 10.9
Perspective representation of the p_x, p_y, and p_z atomic orbitals.

and each p orbital can hold a maximum of two electrons. Thus the total number of electrons that can reside in all three p orbitals is six. To summarize our model, the first principal energy level of an atom has a 1s orbital. The second principal energy level has a 2s and three 2p orbitals labeled $2p_x$, $2p_y$, and $2p_z$, as shown in Figure 10.10.

Notice that there is a correspondence between the energy level and number of sublevels.

The third principal energy level has three sublevels labeled 3s, 3p, and 3d. The 3s orbital is spherical and larger than the 1s and 2s orbitals. The $3p_x$, $3p_y$, $3p_z$ orbitals are shaped like those of the second level, only larger. The five 3d orbitals have the shapes shown in Figure 10.11. You don't need to memorize these shapes, but notice that they look different from the s or p orbitals.

Each time a new principal energy level is added, we also add a new sublevel. This makes sense because each energy level corresponds to a larger average distance from the nucleus, which provides more room on each level for new sublevels containing more orbitals.

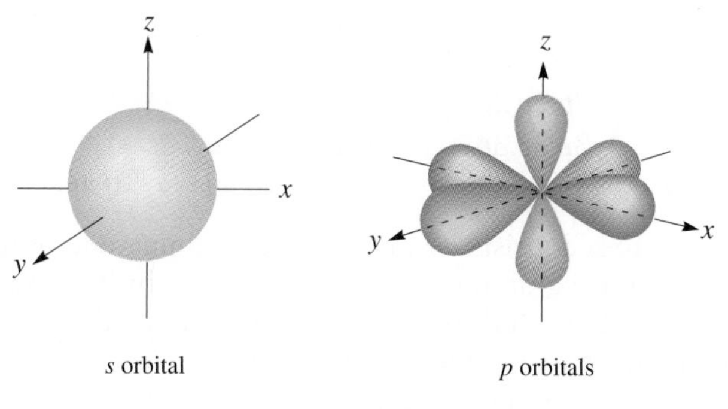

Figure 10.10
Orbitals on the second principal energy level are one 2s and three 2p orbitals.

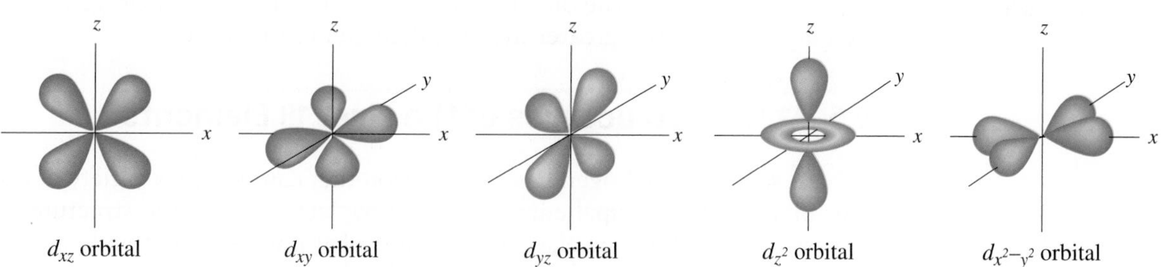

Figure 10.11
The five d orbitals are found in the third principal energy level along with one 3s orbital and three 3p orbitals.

CHEMISTRY IN ACTION • Atomic Clocks

Imagine a clock that keeps time to within 1 second over a million years. The National Institute of Standards and Technology in Boulder, Colorado, has an atomic clock that does just that—a little better than your average alarm, grandfather, or cuckoo clock! This atomic clock serves as the international standard for time and frequency. How does it work?

Within the glistening case are several layers of magnetic shielding. In the heart of the clock is a small oven that heats cesium metal to release cesium atoms, which are collected into a narrow beam (1 mm wide). The beam of atoms passes down a long evacuated tube while being excited by a laser until all the cesium atoms are in the same electron state.

The atoms then enter another chamber filled with reflecting microwaves. The frequency of the microwaves (9,192, 631,770 cycles per second) is exactly the same frequency required to excite a cesium atom from its ground state to the next higher energy level. These excited cesium atoms then release electromagnetic radiation in a process known as fluorescence. Electronic circuits maintain the microwave frequency at precisely the right level to keep the cesium atoms moving from one level to the next. One second is equal to 9,192,631,770 of these vibrations. The clock is set to this frequency and can keep accurate time for over a million years.

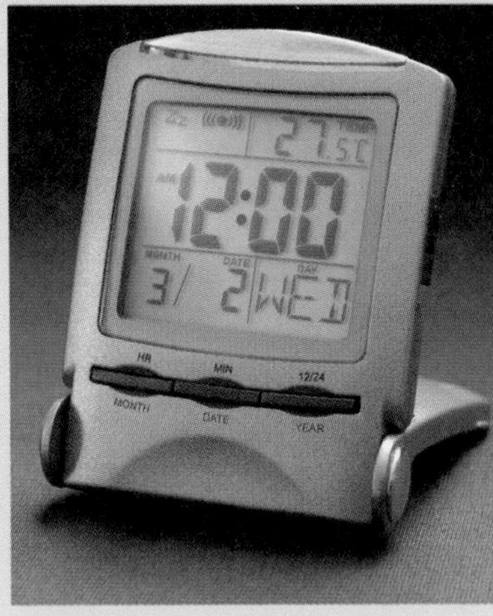

This clock automatically updates itself by comparing time with an atomic clock by radio signal.

The pattern continues with the fourth principal energy level. It has 4*s*, 4*p*, 4*d*, and 4*f* orbitals. There are one 4*s*, three 4*p*, five 4*d*, and seven 4*f* orbitals. The shapes of the *s*, *p*, and *d* orbitals are the same as those for lower levels, only larger. We will not consider the shapes of the *f* orbitals. Remember that for all *s*, *p*, *d*, and *f* orbitals, the maximum number of electrons per orbital is two. We summarize each principal energy level:

$n = 1$	1*s*			
$n = 2$	2*s*	2*p* 2*p* 2*p*		
$n = 3$	3*s*	3*p* 3*p* 3*p*	3*d* 3*d* 3*d* 3*d* 3*d*	
$n = 4$	4*s*	4*p* 4*p* 4*p*	4*d* 4*d* 4*d* 4*d* 4*d*	4*f* 4*f* 4*f* 4*f* 4*f* 4*f* 4*f*

The hydrogen atom consists of a nucleus (containing one proton) and one electron occupying a region outside of the nucleus. In its ground state, the electron occupies a 1*s* orbital, but by absorbing energy the electron can become *excited* and move to a higher energy level.

The hydrogen atom can be represented as shown in Figure 10.12. The diameter of the nucleus is about 10^{-13} cm, and the diameter of the electron orbital is about 10^{-8} cm. The diameter of the electron cloud of a hydrogen atom is about 100,000 times greater than the diameter of the nucleus.

Figure 10.12
The modern concept of a hydrogen atom consists of a proton and an electron in an *s* orbital. The shaded area represents a region where the electron may be found with 90% probability.

10.5 Atomic Structures of the First 18 Elements

We have seen that hydrogen has one electron that can occupy a variety of orbitals in different principal energy levels. Now let's consider the structure of atoms with more than one electron. Because all atoms contain orbitals similar to those found in hydrogen, we can describe the structures of atoms beyond hydrogen by systematically placing electrons in these hydrogenlike orbitals. We use the following guidelines:

1. No more than two electrons can occupy one orbital.
2. Electrons occupy the lowest energy orbitals available. They enter a higher energy orbital only when the lower orbitals are filled. For the atoms beyond hydrogen, orbital energies vary as $s < p < d < f$ for a given value of n.
3. Each orbital in a sublevel is occupied by a single electron before a second electron enters. For example, all three p orbitals must contain one electron before a second electron enters a p orbital.

We can use several methods to represent the atomic structures of atoms, depending on what we are trying to illustrate. When we want to show both the nuclear makeup and the electron structure of each principal energy level (without orbital detail), we can use a diagram such as Figure 10.13.

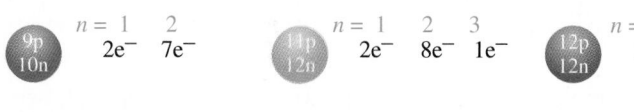

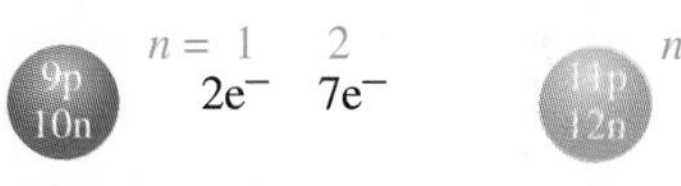

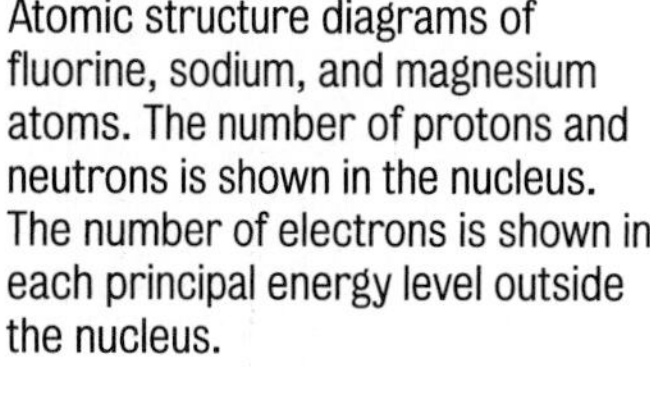

Figure 10.13
Atomic structure diagrams of fluorine, sodium, and magnesium atoms. The number of protons and neutrons is shown in the nucleus. The number of electrons is shown in each principal energy level outside the nucleus.

Often we are interested in showing the arrangement of the electrons in an atom in their orbitals. There are two ways to do this. The first method is called the **electron configuration**. In this method, we list each type of orbital, showing the number of electrons in it as an exponent. An electron configuration is read as follows:

electron configuration

$$2p^6$$

Principal energy level → 2; Type of orbital → p; Number of electrons in sublevel orbitals → 6

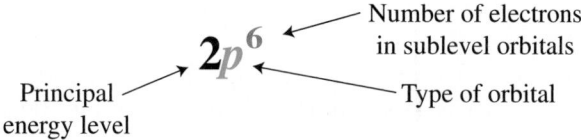

We can also represent this configuration with an **orbital diagram** in which boxes represent the orbitals (containing small arrows indicating the electrons). When the orbital contains one electron, an arrow, pointing upward (↑), is placed in the box. A second arrow, pointing downward (↓), indicates the second electron in that orbital.

orbital diagram

Let's consider each of the first 18 elements on the periodic table in turn. The order of filling for the orbitals in these elements is 1s, 2s, 2p, 3s, 3p, and 4s. Hydrogen, the first element, has only one electron. The electron will be in the 1s orbital because this is the most favorable position (where it will have the greatest attraction for the nucleus). Both representations are shown here:

	Orbital diagram	Electron configuration
H	[↑]	$1s^1$

Helium, with two electrons, can be shown as

	Orbital diagram	Electron configuration
He	[↑↓]	$1s^2$

The first energy level, which can hold a maximum of two electrons, is now full. An atom with three electrons will have its third electron in the second energy level. Thus, in lithium (atomic number 3), the first two electrons are in the 1s orbital, and the third electron is in the 2s orbital of the second energy level.

Lithium has the following structure:

	1s	2s	
Li	↑↓	↑	$1s^22s^1$

All four electrons of beryllium are *s* electrons:

	1s	2s	
Be	↑↓	↑↓	$1s^22s^2$

The next six elements illustrate the filling of the *p* orbitals. Boron has the first *p* electron. Because *p* orbitals all have the same energy, it doesn't matter which of these orbitals fills first:

	1s	2s	2p			
B	↑↓	↑↓	↑			$1s^22s^22p^1$

Carbon is the sixth element. It has two electrons in the 1*s* orbital, two electrons in the 2*s* orbital, and two electrons to place in the 2*p* orbitals. Because it is more difficult for the *p* electrons to pair up than to occupy a second *p* orbital, the second *p* electron is located in a different *p* orbital. We could show this by writing $2p_x^12p_y^1$, but we usually write it as $2p^2$; it is *understood* that the electrons are in different *p* orbitals. The spins on these electrons are alike, for reasons we will not explain here.

	1s	2s	2p			
C	↑↓	↑↓	↑	↑		$1s^22s^22p^2$

Nitrogen has seven electrons. They occupy the 1*s*, 2*s*, and 2*p* orbitals. The third *p* electron in nitrogen is still unpaired and is found in the $2p_z$ orbital:

	1s	2s	2p			
N	↑↓	↑↓	↑	↑	↑	$1s^22s^22p^3$

Oxygen is the eighth element. It has two electrons in both the 1*s* and 2*s* orbitals and four electrons in the 2*p* orbitals. One of the 2*p* orbitals is now occupied by a second electron, which has a spin opposite the electron already in that orbital:

	1s	2s	2p			
O	↑↓	↑↓	↑↓	↑	↑	$1s^22s^22p^4$

The next two elements are fluorine with 9 electrons and neon with 10 electrons:

	1s	2s	2p			
F	↑↓	↑↓	↑↓	↑↓	↑	$1s^22s^22p^5$
Ne	↑↓	↑↓	↑↓	↑↓	↑↓	$1s^22s^22p^6$

With neon, the first and second energy levels are filled as shown in Table 10.1. The second energy level can hold a maximum of eight electrons, $2s^22p^6$.

Sodium, element 11, has two electrons in the first energy level and eight electrons in the second energy level, with the remaining electron occupying the 3*s* orbital in the third energy level:

	1s	2s	2p			3s	
Na	↑↓	↑↓	↑↓	↑↓	↑↓	↑	$1s^22s^22p^63s^1$

Table 10.1 Orbital Filling for the First Ten Elements*

Number	Element	Orbitals 1s	2s	2p	Electron configuration
1	H	[↑]			$1s^1$
2	He	[↑↓]			$1s^2$
3	Li	[↑↓]	[↑]		$1s^22s^1$
4	Be	[↑↓]	[↑↓]		$1s^22s^2$
5	B	[↑↓]	[↑↓]	[↑][ ][ ]	$1s^22s^22p^1$
6	C	[↑↓]	[↑↓]	[↑][↑][ ]	$1s^22s^22p^2$
7	N	[↑↓]	[↑↓]	[↑][↑][↑]	$1s^22s^22p^3$
8	O	[↑↓]	[↑↓]	[↑↓][↑][↑]	$1s^22s^22p^4$
9	F	[↑↓]	[↑↓]	[↑↓][↑↓][↑]	$1s^22s^22p^5$
10	Ne	[↑↓]	[↑↓]	[↑↓][↑↓][↑↓]	$1s^22s^22p^6$

*Boxes represent the orbitals grouped by sublevel. Electrons are shown by arrows.

Magnesium (12), aluminum (13), silicon (14), phosphorus (15), sulfur (16), chlorine (17), and argon (18) follow in order. Table 10.2 summarizes the filling of the orbitals for elements 11–18.

The electrons in the outermost (highest) energy level of an atom are called the **valence electrons**. For example, oxygen, which has the electron configuration of $1s^22s^22p^4$, has electrons in the first and second energy levels. Therefore the second principal energy level is the valence level for oxygen. The 2*s* and 2*p* electrons are the valence electrons. In the case of magnesium ($1s^22s^22p^63s^2$), the valence electrons are in the 3*s* orbital, since these are outermost electrons. Valence electrons are involved in bonding atoms together to form compounds and are of particular interest to chemists, as we will see in Chapter 11.

valence electrons

Table 10.2 Orbital Diagrams and Electron Configurations for Elements 11–18

Number	Element	Orbitals 1s	2s	2p	3s	3p	Electron configuration
11	Na	[↑↓]	[↑↓]	[↑↓][↑↓][↑↓]	[↑]		$1s^22s^22p^63s^1$
12	Mg	[↑↓]	[↑↓]	[↑↓][↑↓][↑↓]	[↑↓]		$1s^22s^22p^63s^2$
13	Al	[↑↓]	[↑↓]	[↑↓][↑↓][↑↓]	[↑↓]	[↑][ ][ ]	$1s^22s^22p^63s^23p^1$
14	Si	[↑↓]	[↑↓]	[↑↓][↑↓][↑↓]	[↑↓]	[↑][↑][ ]	$1s^22s^22p^63s^23p^2$
15	P	[↑↓]	[↑↓]	[↑↓][↑↓][↑↓]	[↑↓]	[↑][↑][↑]	$1s^22s^22p^63s^23p^3$
16	S	[↑↓]	[↑↓]	[↑↓][↑↓][↑↓]	[↑↓]	[↑↓][↑][↑]	$1s^22s^22p^63s^23p^4$
17	Cl	[↑↓]	[↑↓]	[↑↓][↑↓][↑↓]	[↑↓]	[↑↓][↑↓][↑]	$1s^22s^22p^63s^23p^5$
18	Ar	[↑↓]	[↑↓]	[↑↓][↑↓][↑↓]	[↑↓]	[↑↓][↑↓][↑↓]	$1s^22s^22p^63s^23p^6$

Practice 10.1

Give the electron configuration for the valence electrons in these elements.
(a) B (b) N (c) Na (d) Cl

10.6 Electron Structures and the Periodic Table

We have seen how the electrons are assigned for the atoms of elements 1–18. How do the electron structures of these atoms relate to their position on the periodic table? To answer this question, we need to look at the periodic table more closely.

The periodic table represents the efforts of chemists to organize the elements logically. Chemists of the early 19th century had sufficient knowledge of the properties of elements to recognize similarities among groups of elements. In 1869, Dimitri Mendeleev (1834–1907) of Russia and Lothar Meyer (1830–1895) of Germany independently published periodic arrangements of the elements based on increasing atomic masses. Mendeleev's arrangement is the precursor to the modern periodic table and his name is associated with it. The modern periodic table is shown in Figure 10.14 and on the inside front cover of the book.

period Each horizontal row in the periodic table is called a **period**, as shown in Figure 10.14. There are seven periods of elements. The number of each period corresponds to the outermost energy level that contains electrons for elements in that period. Those in Period 1 contain electrons only in energy level 1 while those in Period 2 contain electrons in levels 1 and 2. In Period 3, electrons are found in levels 1, 2, and 3, and so on.

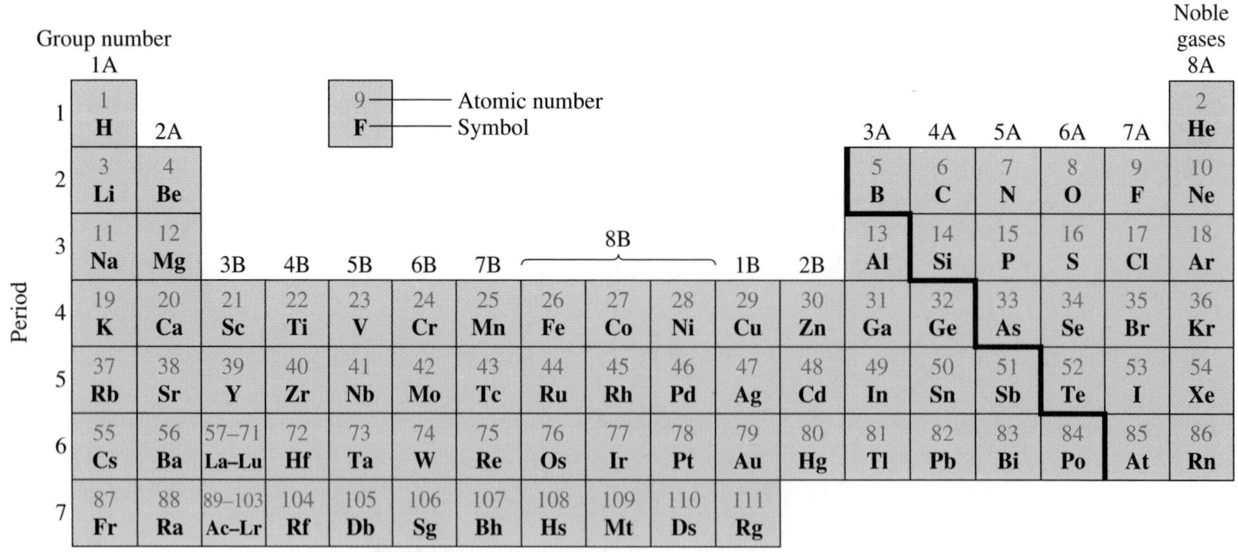

Group number

Key: 9 — Atomic number; F — Symbol

Period	1A	2A	3B	4B	5B	6B	7B	8B	8B	8B	1B	2B	3A	4A	5A	6A	7A	Noble gases 8A
1	1 H																	2 He
2	3 Li	4 Be											5 B	6 C	7 N	8 O	9 F	10 Ne
3	11 Na	12 Mg											13 Al	14 Si	15 P	16 S	17 Cl	18 Ar
4	19 K	20 Ca	21 Sc	22 Ti	23 V	24 Cr	25 Mn	26 Fe	27 Co	28 Ni	29 Cu	30 Zn	31 Ga	32 Ge	33 As	34 Se	35 Br	36 Kr
5	37 Rb	38 Sr	39 Y	40 Zr	41 Nb	42 Mo	43 Tc	44 Ru	45 Rh	46 Pd	47 Ag	48 Cd	49 In	50 Sn	51 Sb	52 Te	53 I	54 Xe
6	55 Cs	56 Ba	57–71 La–Lu	72 Hf	73 Ta	74 W	75 Re	76 Os	77 Ir	78 Pt	79 Au	80 Hg	81 Tl	82 Pb	83 Bi	84 Po	85 At	86 Rn
7	87 Fr	88 Ra	89–103 Ac–Lr	104 Rf	105 Db	106 Sg	107 Bh	108 Hs	109 Mt	110 Ds	111 Rg							

Figure 10.14
The periodic table of the elements.

Elements that behave in a similar manner are found in **groups** or **families**. These form the vertical columns on the periodic table. Several systems exist for numbering the groups. In one system, the columns are numbered from left to right using the numbers 1–18. However, we use a system that numbers the columns with numbers and the letters A and B, as shown in Figure 10.14. The A groups are known as the **representative elements**. The B groups and Group 8 are called the **transition elements**. In this book we will focus on the representative elements. The groups (columns) of the periodic table often have family names. For example, the group on the far right side of the periodic table (He, Ne, Ar, Kr, Xe, and Rn) is called the *noble gases*. Group 1A is called the *alkali metals*, Group 2A the *alkaline earth metals*, and Group 7A the *halogens*.

groups, families

representative elements
transition elements

Both numbering systems are shown on the inside front cover.

How is the structure of the periodic table related to the atomic structures of the elements? We've just seen that the periods of the periodic table are associated with the energy level of the outermost electrons of the atoms in that period. Look at the valence electron configurations of the elements we have just examined (Figure 10.15). Do you see a pattern? The valence electron configuration for the elements in each column is the same. The chemical behavior and properties of elements in a particular family must therefore be associated with the electron configuration of the elements. The number for the principal energy level is different. This is expected since each new period is associated with a different energy level for the valence electrons.

The electron configurations for elements beyond these first 18 become long and tedious to write. We often abbreviate the electron configuration using the following notation:

Na $[Ne]3s^1$

Look carefully at Figure 10.15 and you will see that the *p* orbitals are full at the noble gases. By placing the symbol for the noble gas in square brackets, we can abbreviate the complete electron configuration and focus our attention on the valence electrons (the electrons we will be interested in when we discuss bonding in Chapter 11). To write the abbreviated electron configuration for any element, go back to the previous noble gas and place its symbol in square brackets. Then list the valence electrons. Here are some examples:

B	$1s^22s^22p^1$	$[He]2s^22p^1$
Cl	$1s^22s^22p^63s^23p^5$	$[Ne]3s^23p^5$
Na	$1s^22s^22p^63s^1$	$[Ne]3s^1$

1A							Noble gases
1 **H** $1s^1$	2A	3A	4A	5A	6A	7A	2 **He** $1s^2$
3 **Li** $2s^1$	4 **Be** $2s^2$	5 **B** $2s^22p^1$	6 **C** $2s^22p^2$	7 **N** $2s^22p^3$	8 **O** $2s^22p^4$	9 **F** $2s^22p^5$	10 **Ne** $2s^22p^6$
11 **Na** $3s^1$	12 **Mg** $3s^2$	13 **Al** $3s^23p^1$	14 **Si** $3s^23p^2$	15 **P** $3s^23p^3$	16 **S** $3s^23p^4$	17 **Cl** $3s^23p^5$	18 **Ar** $3s^23p^6$

Figure 10.15
Valence electron configurations for the first 18 elements.

The sequence for filling the orbitals is exactly as we would expect up through the $3p$ orbitals. The third energy level might be expected to fill with $3d$ electrons before electrons enter the $4s$ orbital, but this is not the case. The behavior and properties of the next two elements, potassium (19) and calcium (20), are very similar to the elements in Groups 1A and 2A respectively. They clearly belong in these groups. The other elements in Group 1A and Group 2A have electron configurations that indicate valence electrons in the s orbitals. For example, since the electron configuration is connected to the element's properties, we should place the last electrons for potassium and calcium in the $4s$ orbital. Their electron configurations are

K	$1s^22s^22p^63s^23p^64s^1$	or	$[Ar]4s^1$
Ca	$1s^22s^22p^63s^23p^64s^2$	or	$[Ar]4s^2$

Practice 10.2

Write the abbreviated electron configuration for the following elements:
(a) Br (b) Sr (c) Ba (d) Te

Elements 21–30 belong to the elements known as *transition elements*. Electrons are placed in the $3d$ orbitals for each of these elements. When the $3d$ orbitals are full, the electrons fill the $4p$ orbitals to complete the fourth period. Let's consider the overall relationship between orbital filling and the periodic table. Figure 10.16 illustrates the type of orbital filling and its location on the periodic table. The tall columns on the table (labeled 1A–7A, and noble gases) are often called the *representative elements*. Valence electrons in these elements

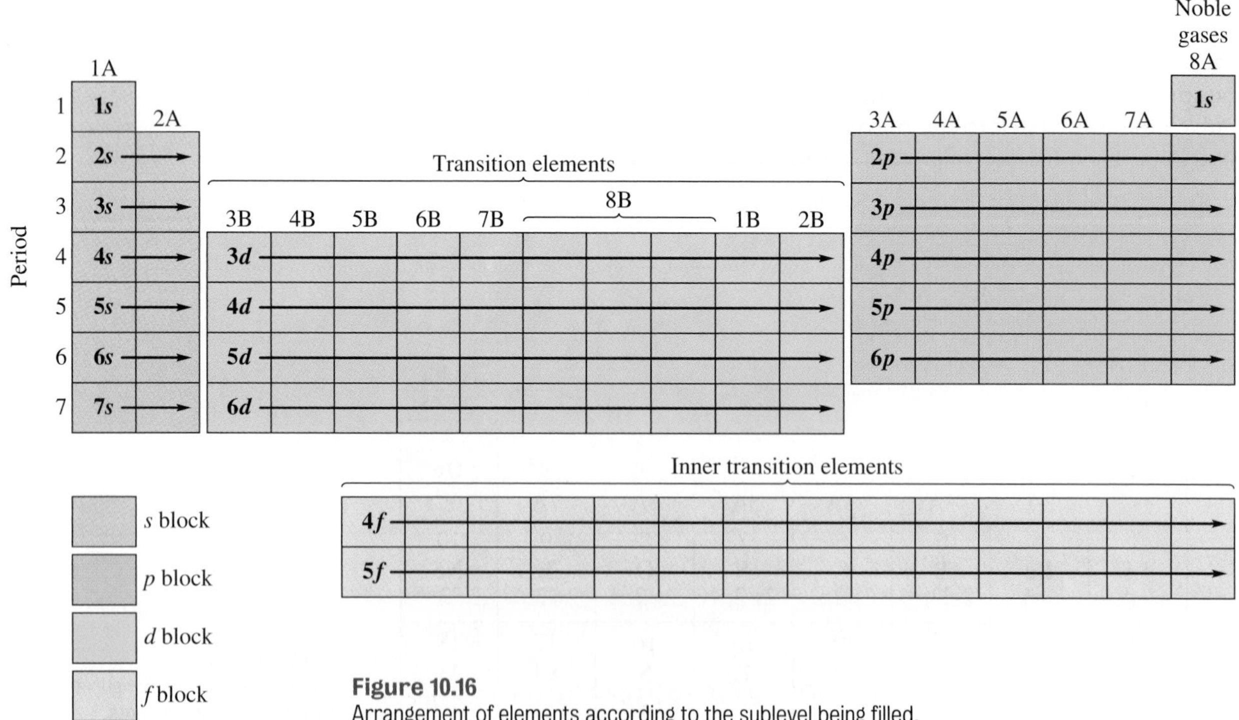

Figure 10.16
Arrangement of elements according to the sublevel being filled.

occupy *s* and *p* orbitals. The period number corresponds to the energy level of the valence electrons. The elements in the center of the periodic table (shown in) are the transition elements where the *d* orbitals are being filled. Notice that the number for the *d* orbitals is one less than the period number. The two rows shown at the bottom of the table in Figure 10.16 are called the *inner transition elements* or the *lanthanide* and *actinide series*. The last electrons in these elements are placed in the *f* orbitals. The number for the *f* orbitals is always two less than that of the *s* and *p* orbitals. A periodic table is almost always available to you, so if you understand the relationship between the orbitals and the periodic table, you can write the electron configuration for any element. There are several minor variations to these rules, but we won't concern ourselves with them in this course.

Example 10.1

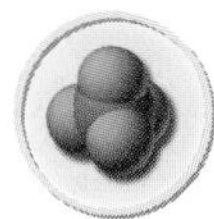

Use the periodic table to write the electron configuration for phosphorus and tin.

SOLUTION

Phosphorus is element 15 and is located in Period 3, Group 5A. The electron configuration must have a full first and second energy level:

P $1s^22s^22p^63s^23p^3$ or $[Ne]3s^23p^3$

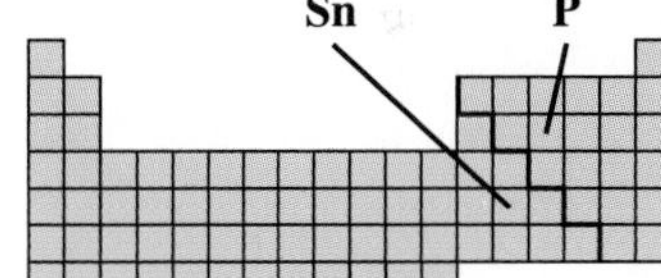

You can determine the electron configuration by looking across the period and counting the element blocks.

Tin is element 50 in Period 5, Group 4A, two places after the transition metals. It must have two electrons in the $5p$ series. Its electron configuration is

Sn $1s^22s^22p^63s^23p^64s^23d^{10}4p^65s^24d^{10}5p^2$ or $[Kr]5s^24d^{10}5p^2$

Notice that the *d* series of electrons always has a principal energy level number one less than its period.

Practice 10.3

Use the periodic table to write the complete electron configuration for (a) O, (b) Ca, and (c) Ti.

The early chemists classified the elements based only on their observed properties, but modern atomic theory gives us an explanation for why the properties of elements vary periodically. For example, as we "build" atoms by filling orbitals with electrons, the same orbitals occur on each energy level. This means that the same electron configuration reappears regularly for each level. Groups of elements show similar chemical properties because of the similarity of these outermost electron configurations.

In Figure 10.17, only the electron configuration of the outermost electrons is given. This periodic table illustrates these important points:

1. The number of the period corresponds with the highest energy level occupied by electrons in that period.
2. The group numbers for the representative elements are equal to the total number of outermost electrons in the atoms of the group. For example, elements in Group 7A always have the electron configuration ns^2np^5. The *d* and *f* electrons are always in a lower energy level than the highest energy level and so are not considered as outermost electrons.

Period	Group number 1A	2A	3B	4B	5B	6B	7B	8B	8B	8B	1B	2B	3A	4A	5A	6A	7A	Noble gases 8A
1	1 **H** $1s^1$																	2 **He** $1s^2$
2	3 **Li** $2s^1$	4 **Be** $2s^2$											5 **B** $2s^22p^1$	6 **C** $2s^22p^2$	7 **N** $2s^22p^3$	8 **O** $2s^22p^4$	9 **F** $2s^22p^5$	10 **Ne** $2s^22p^6$
3	11 **Na** $3s^1$	12 **Mg** $3s^2$											13 **Al** $3s^23p^1$	14 **Si** $3s^23p^2$	15 **P** $3s^23p^3$	16 **S** $3s^23p^4$	17 **Cl** $3s^23p^5$	18 **Ar** $3s^23p^6$
4	19 **K** $4s^1$	20 **Ca** $4s^2$	21 **Sc** $4s^23d^1$	22 **Ti** $4s^23d^2$	23 **V** $4s^23d^3$	24 **Cr** $4s^13d^5$	25 **Mn** $4s^23d^5$	26 **Fe** $4s^23d^6$	27 **Co** $4s^23d^7$	28 **Ni** $4s^23d^8$	29 **Cu** $4s^13d^{10}$	30 **Zn** $4s^23d^{10}$	31 **Ga** $4s^24p^1$	32 **Ge** $4s^24p^2$	33 **As** $4s^24p^3$	34 **Se** $4s^24p^4$	35 **Br** $4s^24p^5$	36 **Kr** $4s^24p^6$
5	37 **Rb** $5s^1$	38 **Sr** $5s^2$	39 **Y** $5s^24d^1$	40 **Zr** $5s^24d^2$	41 **Nb** $5s^14d^4$	42 **Mo** $5s^14d^5$	43 **Tc** $5s^14d^6$	44 **Ru** $5s^14d^7$	45 **Rh** $5s^14d^8$	46 **Pd** $5s^04d^{10}$	47 **Ag** $5s^14d^{10}$	48 **Cd** $5s^24d^{10}$	49 **In** $5s^25p^1$	50 **Sn** $5s^25p^2$	51 **Sb** $5s^25p^3$	52 **Te** $5s^25p^4$	53 **I** $5s^25p^5$	54 **Xe** $5s^25p^6$
6	55 **Cs** $6s^1$	56 **Ba** $6s^2$	57 **La** $6s^25d^1$	72 **Hf** $6s^25d^2$	73 **Ta** $6s^25d^3$	74 **W** $6s^25d^4$	75 **Re** $6s^25d^5$	76 **Os** $6s^25d^6$	77 **Ir** $6s^25d^7$	78 **Pt** $6s^15d^9$	79 **Au** $6s^15d^{10}$	80 **Hg** $6s^25d^{10}$	81 **Tl** $6s^26p^1$	82 **Pb** $6s^26p^2$	83 **Bi** $6s^26p^3$	84 **Po** $6s^26p^4$	85 **At** $6s^26p^5$	86 **Rn** $6s^26p^6$
7	87 **Fr** $7s^1$	88 **Ra** $7s^2$	89 **Ac** $7s^26d^1$	104 **Rf** $7s^26d^2$	105 **Db** $7s^26d^3$	106 **Sg** $7s^26d^4$	107 **Bh** $7s^26d^5$	108 **Hs** $7s^26d^6$	109 **Mt** $7s^26d^7$	110 **Ds** $7s^16d^9$	111 **Rg** $7s^16d^{10}$							

Figure 10.17
Outermost electron configurations.

3. The elements of a family have the same outermost electron configuration except that the electrons are in different energy levels.
4. The elements within each of the s, p, d, f blocks are filling the s, p, d, f orbitals, as shown in Figure 10.16.
5. Within the transition elements, some discrepancies in the order of filling occur. (Explanation of these discrepancies and similar ones in the inner transition elements are beyond the scope of this book.)

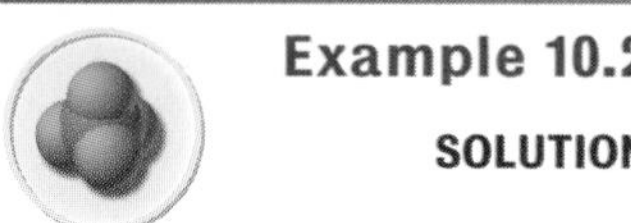

Example 10.2 Write the complete electron configuration of a zinc atom and a rubidium atom.

SOLUTION The atomic number of zinc is 30; it therefore has 30 protons and 30 electrons in a neutral atom. Using Figure 10.14, we see that the electron configuration of a zinc atom is $1s^22s^22p^63s^23p^64s^23d^{10}$. Check by adding the superscripts, which should equal 30.

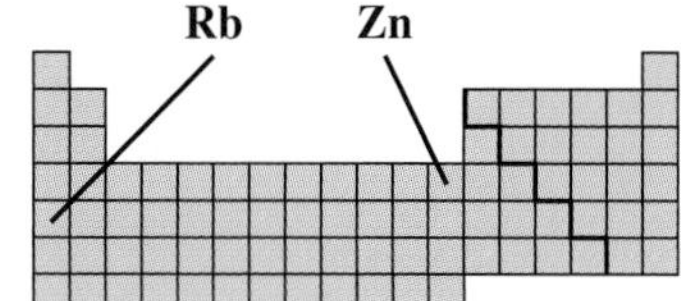

The atomic number of rubidium is 37; therefore it has 37 protons and 37 electrons in a neutral atom. With a little practice using a periodic table, you can write the electron configuration directly. The electron configuration of a rubidium atom is $1s^22s^22p^63s^23p^64s^23d^{10}4p^65s^1$. Check by adding the superscripts, which should equal 37.

Practice 10.4

Write the complete electron configuration for a gallium atom and a lead atom.

Chapter 10 Review

10.1 A Brief History

- Atomic theory has changed over the past 200 years:
  - Early Greek ideas
  - Dalton's model
  - Thomson's model
  - Rutherford's model
  - Modern atomic theory

10.2 Electromagnetic Radiation

KEY TERMS

Wavelength
Frequency
Speed
Photons

- Basic wave characteristics include:
  - Wavelength (λ)
  - Frequency (ν)
  - Speed (v)
- Light is a form of electromagnetic radiation.
- Light can also be considered to be composed of energy packets called photons, which behave like particles.

10.3 The Bohr Atom

KEY TERMS

Line spectrum
Quanta
Ground state
Orbital
Bohr atom

- The study of atomic spectra led Bohr to propose that:
  - Electrons are found in quantized energy levels in an atom.
  - Spectral lines result from the radiation of quanta of energy when the electron moves from a higher level to a lower level.
- The chemical properties of an element and its position on the periodic table depend on electrons.
- De Broglie suggested that all objects have wave properties.
- Schrödinger created a mathematical model to describe electrons as waves.
  - Electrons are located in orbitals, or regions of probability, around the nucleus of an atom.

10.4 Energy Levels of Electrons

KEY TERMS

Principal energy levels
Sublevels
Spin
Pauli exclusion principle

- Modern atomic theory predicts that
  - Electrons are found in discrete principal energy levels (n = 1, 2, 3 . . .):
    - Energy levels contain sublevels.

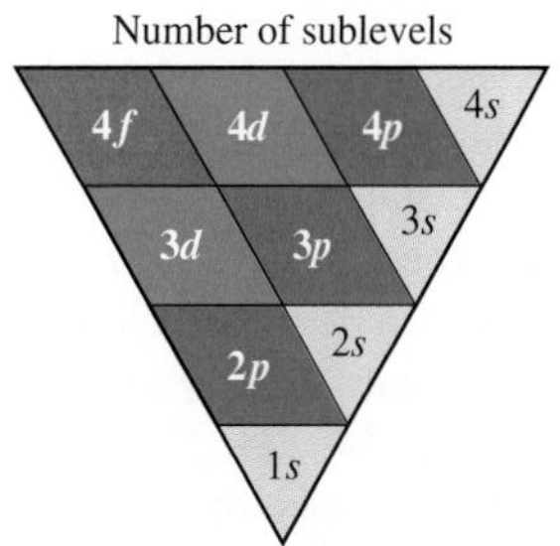

    - Two electrons fit into each orbital but must have opposite spin to do so.

10.5 Atomic Structures of the First 18 Elements

KEY TERMS

Electron configuration
Orbital diagram
Valence electrons

- Guidelines for writing electron configurations:
  - Not more than two electrons per orbital
  - Electrons fill lowest-energy levels first:
    - $s < p < d < f$ for a given value of n
  - Orbitals on a given sublevel are each filled with a single electron before pairing of electrons begins to occur.
- For the representative elements, only electrons in the outermost energy level (valence electrons) are involved in bonding.

10.6 Electron Structures and the Periodic Table

KEY TERMS

Period
Group (family)
Representative element
Transition element

- Elements in horizontal rows on the periodic table contain elements whose valence electrons (s and p) are generally on the same energy level as the number of the row.
- Elements that are chemically similar are arranged in columns (groups) on the periodic table.
- The valence electron configuration of elements in a group or family are the same but they are located in different principal energy levels.

Review Questions

All questions with blue numbers have answers in the appendix of the text.

1. What is an orbital?
2. Under what conditions can a second electron enter an orbital already containing one electron?
3. What is a valence shell?
4. What are valence electrons and why are they important?
5. What is meant when we say the electron structure of an atom is in its ground state?
6. How do 1*s* and 2*s* orbitals differ? How are they alike?
7. What letters are used to designate the types of orbitals?
8. List the following orbitals in order of increasing energy: 2*s*, 2*p*, 4*s*, 1*s*, 3*d*, 3*p*, 4*p*, 3*s*.
9. How many *s* electrons, *p* electrons, and *d* electrons are possible in any energy level?
10. What is the major difference between an orbital and a Bohr orbit?
11. Explain how and why Bohr's model of the atom was modified to include the cloud model of the atom.
12. Sketch the *s*, p_x, p_y, and p_z orbitals.
13. In the designation $3d^7$, give the significance of 3, d, and 7.
14. Describe the difference between transition and representative elements.
15. From the standpoint of electron structure, what do the elements in the *s* block have in common?
16. Write symbols for elements with atomic numbers 8, 16, 34, 52, and 84. What do these elements have in common?
17. Write the symbols of the family of elements that have seven electrons in their outermost energy level.
18. What is the greatest number of elements to be found in any period? Which periods have this number?
19. From the standpoint of energy level, how does the placement of the last electron in the Group A elements differ from that of the Group B elements?
20. Find the places in the periodic table where elements are not in proper sequence according to atomic mass. (See inside of front cover for periodic table.)

Paired Exercises

All exercises with blue numbers have answers in the appendix of the text.

1. How many protons are in the nucleus of an atom of these elements?
   (a) H (b) B (c) Sc (d) U

2. How many protons are in the nucleus of an atom of these elements?
   (a) F (b) Ag (c) Br (d) Sb

3. Give the electron configuration for these elements:
   (a) B (b) Ti (c) Zn (d) Br (e) Sr

4. Give the electron configuration for these elements:
   (a) chlorine (b) silver (c) lithium (d) iron (e) iodine

5. Explain how the spectral lines of hydrogen occur.

6. Explain how Bohr used the data from the hydrogen spectrum to support his model of the atom.

7. How many orbitals exist in the third principal energy level? What are they?

8. How many electrons can be present in the fourth principal energy level?

9. Write orbital diagrams for these elements:
   (a) N (b) Cl (c) Zn (d) Zr (e) I

10. Write orbital diagrams for these elements:
   (a) Si (b) S (c) Ar (d) V (e) P

11. For each of the orbital diagrams given, write out the corresponding electron configurations.

(a) O [↑↓] [↑↓] [↑↓|↑|↑]

(b) Ca [↑↓] [↑↓] [↑↓|↑↓|↑↓] [↑↓] [↑↓|↑↓|↑↓] [↑↓]

(c) Ar [↑↓] [↑↓] [↑↓|↑↓|↑↓] [↑↓] [↑↓|↑↓|↑↓]

(d) Br [↑↓] [↑↓] [↑↓|↑↓|↑↓] [↑↓] [↑↓|↑↓|↑↓] [↑↓] [↑↓|↑↓|↑↓|↑↓|↑↓] [↑↓|↑↓|↑]

(e) Fe [↑↓] [↑↓] [↑↓|↑↓|↑↓] [↑↓] [↑↓|↑↓|↑↓] [↑↓] [↑↓|↑|↑|↑|↑]

12. For each of the orbital diagrams given, write out the corresponding electron configurations.

(a) Li [↑↓] [↑]

(b) P [↑↓] [↑↓] [↑↓|↑↓|↑↓] [↑↓] [↑|↑|↑]

(c) Zn [↑↓] [↑↓] [↑↓|↑↓|↑↓] [↑↓] [↑↓|↑↓|↑↓] [↑↓] [↑↓|↑↓|↑↓|↑↓|↑↓]

(d) Na [↑↓] [↑↓] [↑↓|↑↓|↑↓] [↑]

(e) K [↑↓] [↑↓] [↑↓|↑↓|↑↓] [↑↓] [↑↓|↑↓|↑↓] [↑]

13. Which elements have these electron configurations?
(a) $1s^22s^22p^63s^2$
(b) $1s^22s^22p^63s^23p^1$
(c) $1s^22s^22p^63s^23p^64s^23d^8$
(d) $1s^22s^22p^63s^23p^64s^23d^5$

14. Which elements have these electron configurations?
(a) $[Ar]4s^23d^1$
(b) $[Ar]4s^23d^{10}$
(c) $[Kr]5s^24d^{10}5p^2$
(d) $[Xe]6s^1$

15. Show the electron configurations for elements with these atomic numbers:
(a) 8 (d) 23
(b) 11 (e) 28
(c) 17 (f) 34

16. Show the electron configurations for elements with these atomic numbers:
(a) 9 (d) 39
(b) 26 (e) 52
(c) 31 (f) 10

17. Identify the element and draw orbital diagrams for elements with these atomic numbers:
(a) 22 (d) 35
(b) 18 (e) 25
(c) 33

18. Identify the element and draw orbital diagrams for elements with these atomic numbers:
(a) 15 (d) 34
(b) 30 (e) 19
(c) 20

19. For each of the electron configurations given, write the corresponding orbital diagrams.
(a) F $1s^22s^22p^5$
(b) S $1s^22s^22p^63s^23p^4$
(c) Co $1s^22s^22p^63s^23p^64s^23d^7$
(d) Kr $1s^22s^22p^63s^23p^64s^23d^{10}4p^6$
(e) Ru $1s^22s^22p^63s^23p^64s^23d^{10}4p^65s^24d^6$

20. For each of the electron configurations given, write the corresponding orbital diagrams.
(a) Cl $1s^22s^22p^63s^23p^5$
(b) Mg $1s^22s^22p^63s^2$
(c) Ni $1s^22s^22p^63s^23p^64s^23d^8$
(d) Cu $1s^22s^22p^63s^23p^64s^23d^9$
(e) Ba $1s^22s^22p^63s^23p^64s^23d^{10}4p^65s^24d^{10}5p^66s^2$

21. Identify these elements from their atomic structure diagrams:
(a) (16p 16n) $2e^-$ $8e^-$ $6e^-$
(b) (28p 32n) $2e^-$ $8e^-$ $16e^-$ $2e^-$

22. Diagram the atomic structures (as you see in Exercise 21) for these elements:
(a) $^{27}_{13}Al$ (b) $^{48}_{22}Ti$

23. Why is the 11th electron of the sodium atom located in the third energy level rather than in the second energy level?

24. Why is the last electron in potassium located in the fourth energy level rather than in the third energy level?

25. What electron structure do the noble gases have in common?

26. What is unique about the noble gases, from an electron point of view?

27. How are elements in a period related to one another?

28. How are elements in a group related to one another?

29. How many valence electrons do each of the following elements have?
(a) C (d) I
(b) S (e) B
(c) K

30. How many valence electrons do each of the following elements have?
(a) N (d) Ba
(b) P (e) Al
(c) O

31. What do the electron structures of the alkali metals have in common?

32. Why would you expect the elements zinc, cadmium, and mercury to be in the same chemical family?

33. Pick the electron structures that represent elements in the same chemical family:
(a) $1s^22s^1$
(b) $1s^22s^22p^4$
(c) $1s^22s^22p^2$
(d) $1s^22s^22p^63s^23p^4$
(e) $1s^22s^22p^63s^23p^6$
(f) $1s^22s^22p^63s^23p^64s^2$
(g) $1s^22s^22p^63s^23p^64s^1$
(h) $1s^22s^22p^63s^23p^64s^23d^1$

34. Pick the electron structures that represent elements in the same chemical family:
(a) $[He]2s^22p^6$
(b) $[Ne]3s^1$
(c) $[Ne]3s^2$
(d) $[Ne]3s^23p^3$
(e) $[Ar]4s^23d^{10}$
(f) $[Ar]4s^23d^{10}4p^6$
(g) $[Ar]4s^23d^5$
(h) $[Kr]5s^24d^{10}$

35. In the periodic table, calcium, element 20, is surrounded by elements 12, 19, 21, and 38. Which of these have physical and chemical properties most resembling calcium?

36. In the periodic table, phosphorus, element 15, is surrounded by elements 14, 7, 16, and 33. Which of these have physical and chemical properties most resembling phosphorus?

37. Classify the following elements as metals, nonmetals, or metalloids (review Chapter 3 if you need help):
(a) potassium
(b) plutonium
(c) sulfur
(d) antimony

38. Classify the following elements as metals, nonmetals, or metalloids (review Chapter 3 if you need help):
(a) iodine
(b) tungsten
(c) molybdenum
(d) germanium

39. In which period and group does an electron first appear in an *f* orbital?

40. In which period and group does an electron first appear in a *d* orbital?

41. How many electrons occur in the valence level of Group 7A and 7B elements? Why are they different?

42. How many electrons occur in the valence level of Group 3A and 3B elements? Why are they different?

Additional Exercises

All exercises with blue numbers have answers in the appendix of the text.

43. Using only the periodic table, explain how the valence energy level and the number of valence electrons could be determined.

44. Using the periodic table only, identify the valence energy level and the number of valence electrons for each of the following elements:
(a) Li
(b) Cl
(c) Si
(d) S
(e) Be

45. Which of the following would have the same number of valence electrons?
(a) Na^+
(b) O
(c) Li
(d) F^-
(e) Ne

46. Name the group in which each of the following elements appear:
(a) $1s^22s^22p^5$
(b) $1s^22s^22p^63s^23p^64s^2$
(c) $1s^22s^22p^63s^1$
(d) $1s^22s^22p^63s^23p^64s^23d^{10}4p^6$
(e) $1s^2$
(f) $1s^22s^22p^63s^23p^64s^23d^{10}4p^65s^1$

47. Write the electron configuration of each of the following neutral atoms. (You may need to refer to Figure 3.2.)
(a) the four most abundant elements in the Earth's crust, seawater, and air
(b) the five most abundant elements in the human body

48. What is the maximum number of electrons that can reside in the following:
(a) a *p* orbital
(b) a *d* sublevel
(c) the third principal energy level
(d) an *s* orbital
(e) an *f* sublevel

49. Give the names of each of the following elements based on the information given:
(a) the second element in Period 3
(b) [Ne]$3s^2 3p^3$
(c) 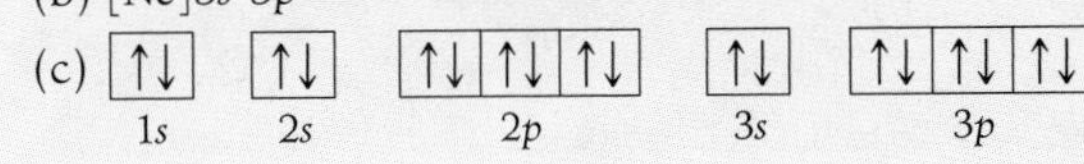

50. Why does the emission spectrum for nitrogen reveal many more spectral lines than that for hydrogen?

51. Suppose we use a foam ball to represent a typical atom. If the radius of the ball is 1.5 cm and the radius of a typical atom is 1.0×10^{-8} cm, how much of an enlargement is this? Use a ratio to express your answer.

52. List the first element on the periodic table that satisfies each of these conditions:
(a) a completed set of p orbitals
(b) two $4p$ electrons
(c) seven valence electrons
(d) three unpaired electrons

53. Oxygen is a gas. Sulfur is a solid. What is it about their electron structures that causes them to be grouped in the same chemical family?

54. In which groups are the transition elements located?

55. How do the electron structures of the transition elements differ from those of the representative elements?

56. The atomic numbers of the noble gases are 2, 10, 18, 36, 54, and 86. What are the atomic numbers for the elements with six electrons in their outermost electron configuration?

57. What is the family name for
(a) Group 1A?
(b) Group 2A?
(c) Group 7A?

58. What sublevel is being filled in
(a) Period 3, Group 3A to 7A?
(b) Period 5, transition elements?
(c) the lanthanide series?

59. Classify each of the following as a noble gas, a representative element, or a transition metal. Also indicate whether the element is a metal, nonmetal, or metalloid.
(a) Na
(b) N
(c) Mo
(d) Ra
(e) As
(f) Ne

60. If element 36 is a noble gas, in which groups would you expect elements 35 and 37 to occur?

61. Write a paragraph describing the general features of the periodic table.

62. Some scientists have proposed the existence of element 117. If it were to exist,
(a) what would its electron configuration be?
(b) how many valence electrons would it have?
(c) what element would it likely resemble?
(d) to what family and period would it belong?

63. What is the relationship between two elements if
(a) one of them has 10 electrons, 10 protons, and 10 neutrons and the other has 10 electrons, 10 protons, and 12 neutrons?
(b) one of them has 23 electrons, 23 protons, and 27 neutrons and the other has 24 electrons, 24 protons, and 26 neutrons?

64. Is there any pattern for the location of gases on the periodic table? for the location of liquids? for the location of solids?

Challenge Exercises

All exercises with blue numbers have answers in the appendix of the text.

***65.** A valence electron in an atom of sulfur is excited by heating a sample. The electron jumps from the s orbital to the p orbital. What is the electron configuration of the excited sulfur atom, and what would the orbital diagram look like?

***66.** Element 87 is in Group 1A, Period 7. In how many principal energy levels are electrons located? Describe its outermost energy level.

***67.** Use the periodic table to explain why metals tend to lose electrons and nonmetals tend to gain electrons.

***68.** Show how the periodic table helps determine the expected electron configuration for any element.

Answers to Practice Exercises

10.1 (a) $2s^2 2p^1$ (b) $2s^2 2p^3$ (c) $3s^1$ (d) $3s^2 3p^5$

10.2 (a) [Ar]$4s^2 4p^5$ (b) [Kr]$5s^2$ (c) [Xe]$6s^2$ (d) [Kr]$5s^2 5p^4$

10.3 (a) O $1s^2 2s^2 2p^4$
(b) Ca $1s^2 2s^2 2p^6 3s^2 3p^6 4s^2$
(c) Ti $1s^2 2s^2 2p^6 3s^2 3p^6 4s^2 3d^2$

10.4 Ga, $1s^2 2s^2 2p^6 3s^2 3p^6 4s^2 3d^{10} 4p^1$
Pb, $1s^2 2s^2 2p^6 3s^2 3p^6 4s^2 3d^{10} 4p^6 5s^2 4d^{10} 5p^6 6s^2 4f^{14} 5d^{10} 6p^2$

CHAPTER 11

Chemical Bonds: The Formation of Compounds from Atoms

This colorfully lighted limestone cave reveals dazzling stalactites and stalagmites, formed from calcium carbonate.

Chapter Outline

For centuries we've been aware that certain metals cling to a magnet. We've seen balloons sticking to walls. Why? High-speed levitation trains are heralded to be the wave of the future. How do they function? In each case, forces of attraction and repulsion are at work.

Human interactions also suggest that "opposites attract" and "likes repel." Attractions draw us into friendships and significant relationships, whereas repulsive forces may produce debate and antagonism. We form and break apart interpersonal bonds throughout our lives.

In chemistry, we also see this phenomenon. Substances form chemical bonds as a result of electrical attractions. These bonds provide the tremendous diversity of compounds found in nature.

11.1 Periodic Trends in Atomic Properties

Although atomic theory and electron configuration help us understand the arrangement and behavior of the elements, it's important to remember that the design of the periodic table is based on observing properties of the elements. Before we use the concept of atomic structure to explain how and why atoms combine to form compounds, we need to understand the characteristic properties of the elements and the trends that occur in these properties on the periodic table. These trends allow us to use the periodic table to accurately predict properties and reactions of a wide variety of substances.

Metals and Nonmetals

In Section 3.5, we classified elements as metals, nonmetals, or metalloids. The heavy stair-step line beginning at boron and running diagonally down the periodic table separates the elements into metals and nonmetals. Metals are usually lustrous, malleable, and good conductors of heat and electricity. Nonmetals are just the opposite—nonlustrous, brittle, and poor conductors. Metalloids are found bordering the heavy diagonal line and may have properties of both metals and nonmetals.

Most elements are classified as metals (see Figure 11.1). Metals are found on the left side of the stair-step line, while the nonmetals are located toward the upper right of the table. Note that hydrogen does not fit into the division of metals and nonmetals. It displays nonmetallic properties under normal conditions, even though it has only one outermost electron like the alkali metals. Hydrogen is considered to be a unique element.

It is the chemical properties of metals and nonmetals that interest us most. Metals tend to lose electrons and form positive ions, while nonmetals tend to gain electrons and form negative ions. When a metal reacts with a nonmetal, electrons are often transferred from the metal to the nonmetal.

Atomic Radius

The relative radii of the representative elements are shown in Figure 11.2. Notice that the radii of the atoms tend to increase down each group and that they tend to decrease from left to right across a period.

The increase in radius down a group can be understood if we consider the electron structure of the atoms. For each step down a group, an additional

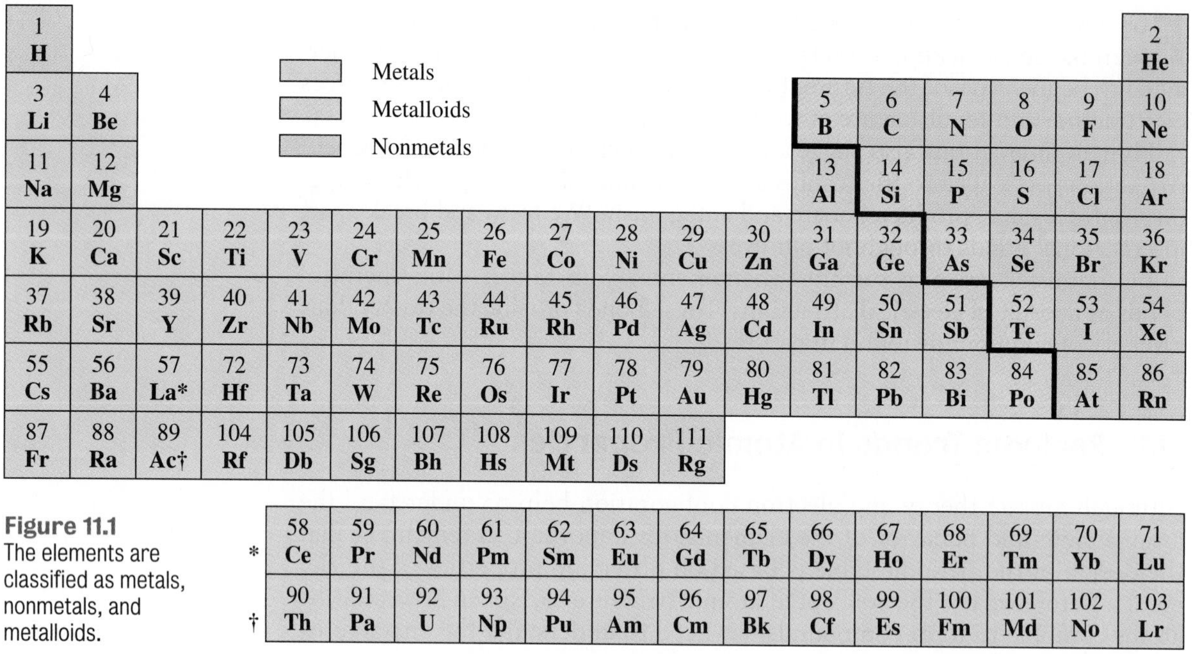

Metals
Metalloids
Nonmetals

1 **H**																	2 **He**
3 **Li**	4 **Be**											5 **B**	6 **C**	7 **N**	8 **O**	9 **F**	10 **Ne**
11 **Na**	12 **Mg**											13 **Al**	14 **Si**	15 **P**	16 **S**	17 **Cl**	18 **Ar**
19 **K**	20 **Ca**	21 **Sc**	22 **Ti**	23 **V**	24 **Cr**	25 **Mn**	26 **Fe**	27 **Co**	28 **Ni**	29 **Cu**	30 **Zn**	31 **Ga**	32 **Ge**	33 **As**	34 **Se**	35 **Br**	36 **Kr**
37 **Rb**	38 **Sr**	39 **Y**	40 **Zr**	41 **Nb**	42 **Mo**	43 **Tc**	44 **Ru**	45 **Rh**	46 **Pd**	47 **Ag**	48 **Cd**	49 **In**	50 **Sn**	51 **Sb**	52 **Te**	53 **I**	54 **Xe**
55 **Cs**	56 **Ba**	57 **La***	72 **Hf**	73 **Ta**	74 **W**	75 **Re**	76 **Os**	77 **Ir**	78 **Pt**	79 **Au**	80 **Hg**	81 **Tl**	82 **Pb**	83 **Bi**	84 **Po**	85 **At**	86 **Rn**
87 **Fr**	88 **Ra**	89 **Ac†**	104 **Rf**	105 **Db**	106 **Sg**	107 **Bh**	108 **Hs**	109 **Mt**	110 **Ds**	111 **Rg**							

*	58 **Ce**	59 **Pr**	60 **Nd**	61 **Pm**	62 **Sm**	63 **Eu**	64 **Gd**	65 **Tb**	66 **Dy**	67 **Ho**	68 **Er**	69 **Tm**	70 **Yb**	71 **Lu**
†	90 **Th**	91 **Pa**	92 **U**	93 **Np**	94 **Pu**	95 **Am**	96 **Cm**	97 **Bk**	98 **Cf**	99 **Es**	100 **Fm**	101 **Md**	102 **No**	103 **Lr**

Figure 11.1
The elements are classified as metals, nonmetals, and metalloids.

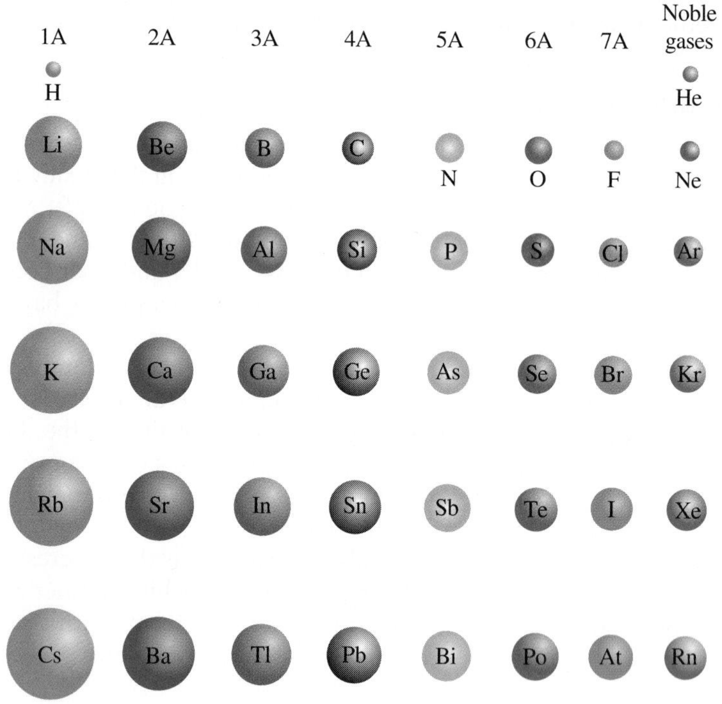

1A	2A	3A	4A	5A	6A	7A	Noble gases
H							He
Li	Be	B	C	N	O	F	Ne
Na	Mg	Al	Si	P	S	Cl	Ar
K	Ca	Ga	Ge	As	Se	Br	Kr
Rb	Sr	In	Sn	Sb	Te	I	Xe
Cs	Ba	Tl	Pb	Bi	Po	At	Rn

Figure 11.2
Relative atomic radii for the representative elements. Atomic radius decreases across a period and increases down a group in the periodic table.

energy level is added to the atom. The average distance from the nucleus to the outside edge of the atom must increase as each new energy level is added. The atoms get bigger as electrons are placed in these new higher-energy levels.

Understanding the decrease in atomic radius across a period requires more thought, however. As we move from left to right across a period, electrons within the same block are being added to the same principal energy level. Within a given energy level, we expect the orbitals to have about the same size. We would then expect the atoms to be about the same size across the period. But each time an electron is added, a proton is added to the nucleus as well. The increase in positive charge (in the nucleus) pulls the electrons closer to the nucleus, which results in a gradual decrease in atomic radius across a period.

Ionization Energy

ionization energy

The **ionization energy** of an atom is the energy required to remove an electron from the atom. For example,

$$\text{Na} + \text{ionization energy} \longrightarrow \text{Na}^+ + e^-$$

The first ionization energy is the amount of energy required to remove the first electron from an atom, the second is the amount required to remove the second electron from that atom, and so on.

Table 11.1 gives the ionization energies for the removal of one to five electrons from several elements. The table shows that even higher amounts of energy are needed to remove the second, third, fourth, and fifth electrons. This makes sense because removing electrons leaves fewer electrons attracted to the same positive charge in the nucleus. The data in Table 11.1 also show that an extra-large ionization energy (blue) is needed when an electron is removed from a noble gas–like structure, clearly showing the stability of the electron structure of the noble gases.

First ionization energies have been experimentally determined for most elements. Figure 11.3 plots these energies for representative elements in the first four periods. Note these important points:

1. Ionization energy in Group A elements decreases from top to bottom in a group. For example, in Group 1A the ionization energy changes from 520 kJ/mol for Li to 419 kJ/mol for K.
2. Ionization energy gradually increases from left to right across a period. Noble gases have a relatively high value, confirming the nonreactive nature of these elements.

Table 11.1 Ionization Energies for Selected Elements*

	Required amounts of energy (kJ/mol)				
Element	1st e^-	2nd e^-	3rd e^-	4th e^-	5th e^-
H	1,314				
He	2,372	5,247			
Li	520	7,297	11,810		
Be	900	1,757	14,845	21,000	
B	800	2,430	3,659	25,020	32,810
C	1,088	2,352	4,619	6,222	37,800
Ne	2,080	3,962	6,276	9,376	12,190
Na	496	4,565	6,912	9,540	13,355

*Values are expressed in kilojoules per mole, showing energies required to remove 1 to 5 electrons per atom. Blue type indicates the energy needed to remove an electron from a noble gas electron structure.

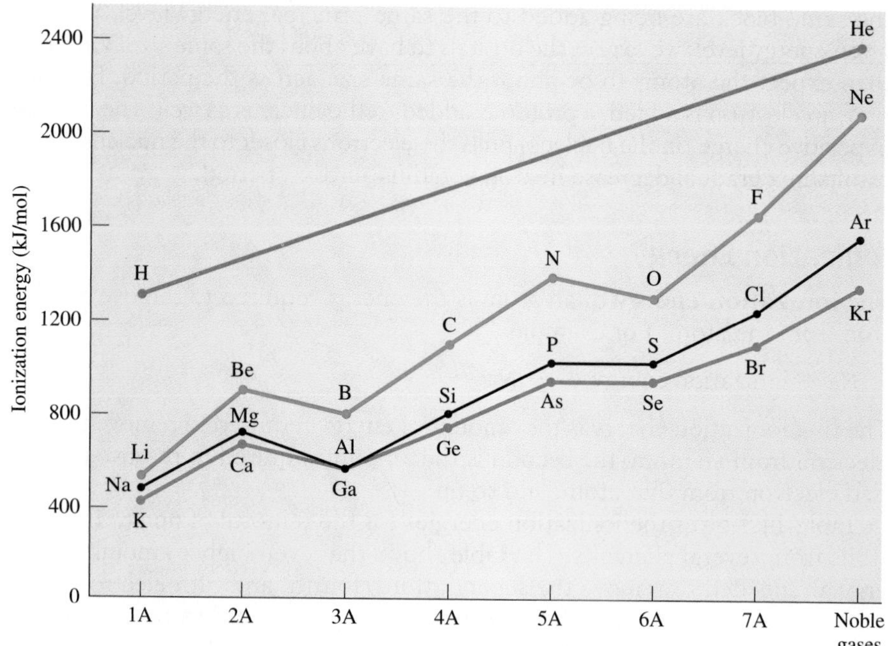

Figure 11.3
Periodic relationship of the first ionization energy for representative elements in the first four periods.

Metals don't behave in exactly the same manner. Some metals give up electrons much more easily than others. In the alkali metal family, cesium gives up its 6*s* electron much more easily than the metal lithium gives up its 2*s* electron. This makes sense when we consider that the size of the atoms increases down the group. The distance between the nucleus and the outer electrons increases and the ionization energy decreases. The most chemically active metals are located at the lower left of the periodic table.

Nonmetals have relatively large ionization energies compared to metals. Nonmetals tend to gain electrons and form anions. Since the nonmetals are located at the right side of the periodic table, it is not surprising that ionization energies tend to increase from left to right across a period. The most active nonmetals are found in the *upper* right corner of the periodic table (excluding the noble gases).

11.2 Lewis Structures of Atoms

Metals tend to form cations (positively charged ions) and nonmetals form anions (negatively charged ions) in order to attain a stable valence electron structure. For many elements this stable valence level contains eight electrons (two *s* and six *p*), identical to the valence electron configuration of the noble gases. Atoms undergo rearrangements of electron structure to lower their chemical potential energy (or to become more stable). These rearrangements are accomplished by losing, gaining, or sharing electrons with other atoms. For example, a hydrogen atom could accept a second electron and attain an electron structure the same as the noble gas helium. A fluorine atom could gain an electron and attain an electron structure like neon. A sodium atom could lose one electron to attain an electron structure like neon.

1A	2A	3A	4A	5A	6A	7A	Noble Gases
H·							He:
Li·	Be:	:Ḃ	:Ċ·	:Ṇ̇·	·Ö̤:	:F̤̈:	:N̤̈e:
Na·	Mg:	:Ȧl	:Ṡi·	:Ṗ̣·	·S̤̈:	:C̤̈l:	:Ä̤r:
K·	Ca:						

Figure 11.4
Lewis structures of the first 20 elements. Dots represent electrons in the outermost s and p energy levels only.

Lewis structure

The valence electrons in the outermost energy level of an atom are responsible for the electron activity that occurs to form chemical bonds. The **Lewis structure** of an atom is a representation that shows the valence electrons for that atom. American chemist Gilbert N. Lewis (1875–1946) proposed using the symbol for the element and dots for electrons. The number of dots placed around the symbol equals the number of s and p electrons in the outermost energy level of the atom. Paired dots represent paired electrons; unpaired dots represent unpaired electrons. For example, **H·** is the Lewis symbol for a hydrogen atom, $1s^1$; **:Ḃ** is the Lewis symbol for a boron atom, with valence electrons $2s^22p^1$. In the case of boron, the symbol B represents the boron nucleus and the $1s^2$ electrons; the dots represent only the $2s^22p^1$ electrons.

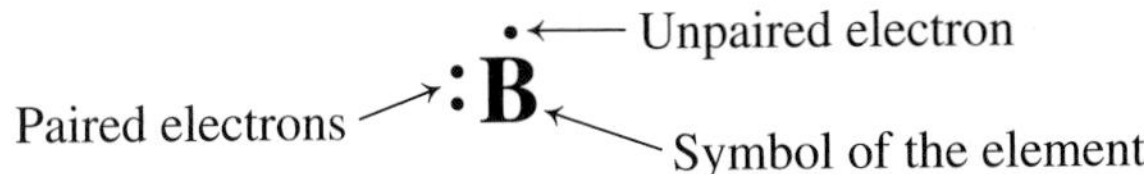

The Lewis method is used not only because of its simplicity of expression but also because much of the chemistry of the atom is directly associated with the electrons in the outermost energy level. Figure 11.4 shows Lewis structures for the elements hydrogen through calcium.

Example 11.1

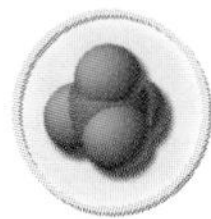

Write the Lewis structure for a phosphorus atom.

SOLUTION

First establish the electron structure for a phosphorus atom, which is $1s^22s^22p^63s^23p^3$. Note that there are five electrons in the outermost energy level; they are $3s^23p^3$. Write the symbol for phosphorus and place the five electrons as dots around it.

:Ṗ̣·

The $3s^2$ electrons are paired and are represented by the paired dots. The $3p^3$ electrons, which are unpaired, are represented by the single dots.

A quick way to determine the correct number of dots (electrons) for a Lewis structure is to use the Group number. For the A groups on the periodic table, the Group number is the same as the number of electrons in the Lewis structure.

Practice 11.1

Write the Lewis structure for the following elements:
(a) N (b) Al (c) Sr (d) Br

11.3 The Ionic Bond: Transfer of Electrons from One Atom to Another

The chemistry of many elements, especially the representative ones, is to attain an outer electron structure like that of the chemically stable noble gases. With the exception of helium, this stable structure consists of eight electrons in the outermost energy level (see Table 11.2).

Table 11.2 Arrangement of Electrons in the Noble Gases*

Noble gas	Symbol	Electron structure $n = 1$	2	3	4	5	6
Helium	He	$\mathbf{1s^2}$					
Neon	Ne	$1s^2$	$\mathbf{2s^22p^6}$				
Argon	Ar	$1s^2$	$2s^22p^6$	$\mathbf{3s^23p^6}$			
Krypton	Kr	$1s^2$	$2s^22p^6$	$3s^23p^63d^{10}$	$\mathbf{4s^24p^6}$		
Xenon	Xe	$1s^2$	$2s^22p^6$	$3s^23p^63d^{10}$	$4s^24p^64d^{10}$	$\mathbf{5s^25p^6}$	
Radon	Rn	$1s^2$	$2s^22p^6$	$3s^23p^63d^{10}$	$4s^24p^64d^{10}4f^{14}$	$5s^25p^65d^{10}$	$\mathbf{6s^26p^6}$

*Each gas except helium has eight electrons in its outermost energy level.

Let's look at the electron structures of sodium and chlorine to see how each element can attain a structure of 8 electrons in its outermost energy level. A sodium atom has 11 electrons: 2 in the first energy level, 8 in the second energy level, and 1 in the third energy level. A chlorine atom has 17 electrons: 2 in the first energy level, 8 in the second energy level, and 7 in the third energy level. If a sodium atom transfers or loses its 3*s* electron, its third energy level becomes vacant, and it becomes a sodium ion with an electron configuration identical to that of the noble gas neon. This process requires energy:

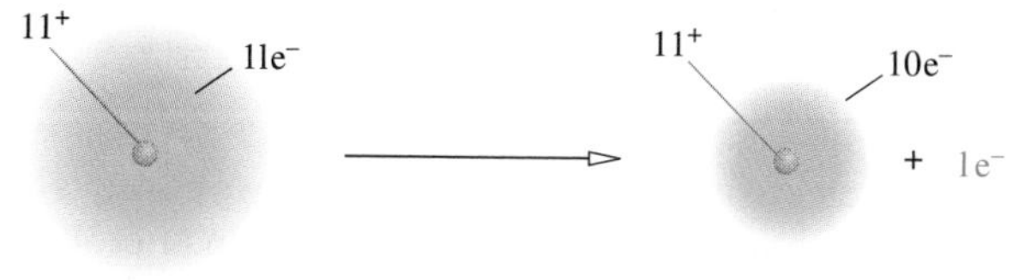

11+ 11e⁻ → 11+ 10e⁻ + 1e⁻

Na atom ($1s^22s^22p^63s^1$) → Na^+ ion ($1s^22s^22p^6$)

An atom that has lost or gained electrons will have a positive or negative charge, depending on which particles (protons or electrons) are in excess. Remember that a charged particle or group of particles is called an *ion*.

By losing a negatively charged electron, the sodium atom becomes a positively charged particle known as a sodium ion. The charge, +1, results because the nucleus still contains 11 positively charged protons, and the electron orbitals contain only 10 negatively charged electrons. The charge is indicated by a plus sign (+) and is written as a superscript after the symbol of the element (Na^+).

A chlorine atom with seven electrons in the third energy level needs one electron to pair up with its one unpaired 3*p* electron to attain the stable outer electron structure of argon. By gaining one electron, the chlorine atom becomes a chloride ion (Cl^-), a negatively charged particle containing 17 protons and 18 electrons. This process releases energy:

17+ 17e⁻ + 1e⁻ ⟶ 17+ 18e⁻

Cl atom ($1s^22s^22p^63s^23p^5$) → Cl^- ion ($1s^22s^22p^63s^23p^6$)

Consider sodium and chlorine atoms reacting with each other. The 3*s* electron from the sodium atom transfers to the half-filled 3*p* orbital in the chlorine atom to form a positive sodium ion and a negative chloride ion. The compound sodium chloride results because the Na^+ and Cl^- ions are strongly attracted to each other by their opposite electrostatic charges. The force holding the oppositely charged ions together is called an ionic bond:

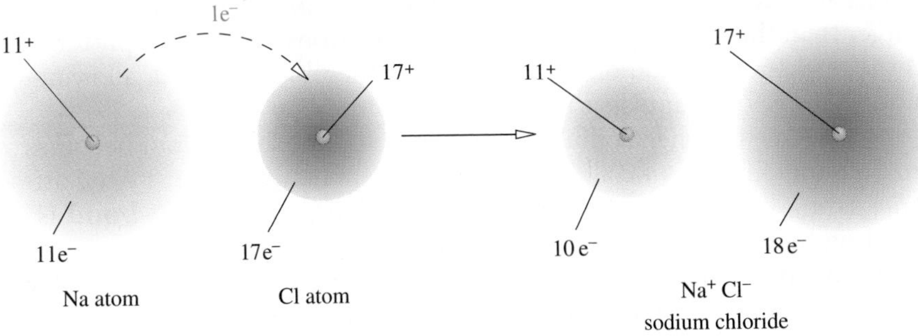

The Lewis representation of sodium chloride formation is

$$\text{Na}\cdot + \cdot\ddot{\underset{..}{\text{Cl}}}: \longrightarrow [\text{Na}]^+[:\ddot{\underset{..}{\text{Cl}}}:]^-$$

The chemical reaction between sodium and chlorine is a very vigorous one, producing considerable heat in addition to the salt formed. When energy is released in a chemical reaction, the products are more stable than the reactants. Note that in NaCl both atoms attain a noble gas electron structure.

Sodium chloride is made up of cubic crystals in which each sodium ion is surrounded by six chloride ions and each chloride ion by six sodium ions, except at the crystal surface. A visible crystal is a regularly arranged aggregate of millions of these ions, but the ratio of sodium to chloride ions is 1 : 1, hence the formula NaCl. The cubic crystalline lattice arrangement of sodium chloride is shown in Figure 11.5.

Figure 11.6 contrasts the relative sizes of sodium and chlorine atoms with those of their ions. The sodium ion is smaller than the atom due primarily to two factors: (1) The sodium atom has lost its outermost electron, thereby

These tiny NaCl crystals (on a penny) show the cubic structure illustrated in Figure 11.5.

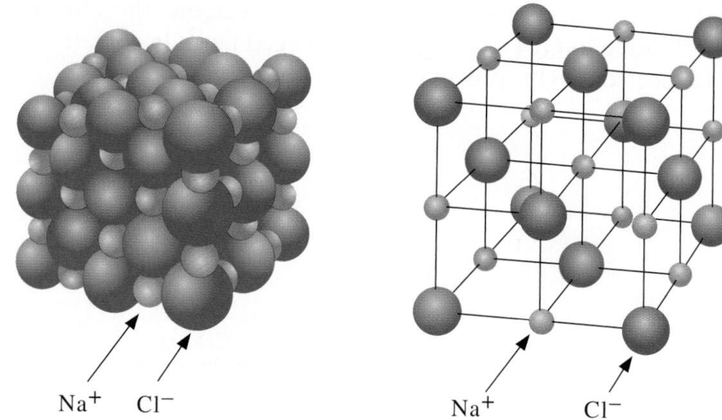

Figure 11.5
Sodium chloride crystal. Diagram represents a small fragment of sodium chloride, which forms cubic crystals. Each sodium ion is surrounded by six chloride ions, and each chloride ion is surrounded by six sodium ions. The tiny NaCl crystals show the cubic crystal structure of salt.

Remember: A cation is always smaller than its parent atom whereas an anion is always larger than its parent atom.

reducing its size; and (2) the 10 remaining electrons are now attracted by 11 protons and are thus drawn closer to the nucleus. Conversely, the chloride ion is larger than the atom because (1) it has 18 electrons but only 17 protons and (2) the nuclear attraction on each electron is thereby decreased, allowing the chlorine atom to expand as it forms an ion.

ionic bond

We've seen that when sodium reacts with chlorine, each atom becomes an ion. Sodium chloride, like all ionic substances, is held together by the attraction existing between positive and negative charges. An **ionic bond** is the attraction between oppositely charged ions.

Ionic bonds are formed whenever one or more electrons are transferred from one atom to another. Metals, which have relatively little attraction for their valence electrons, tend to form ionic bonds when they combine with nonmetals.

It's important to recognize that substances with ionic bonds do not exist as molecules. In sodium chloride, for example, the bond does not exist solely between a single sodium ion and a single chloride ion. Each sodium ion in the crystal attracts six near-neighbor negative chloride ions; in turn, each negative chloride ion attracts six near-neighbor positive sodium ions (see Figure 11.5).

A metal will usually have one, two, or three electrons in its outer energy level. In reacting, metal atoms characteristically lose these electrons, attain

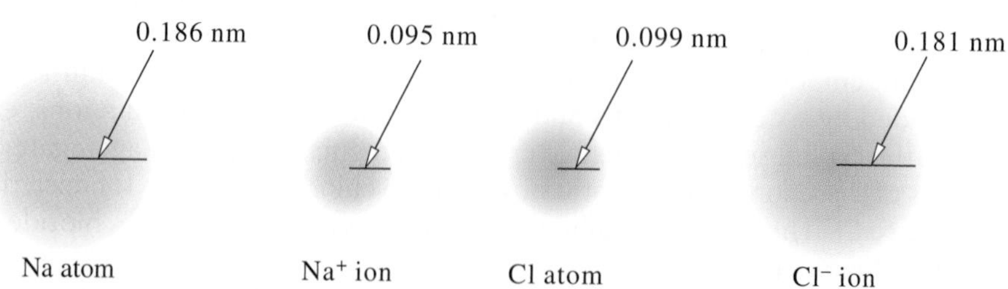

Figure 11.6
Relative radii of sodium and chlorine atoms and their ions.

Table 11.3 Change in Atomic Radii (nm) of Selected Metals and Nonmetals*

Atomic radius		Ionic radius		Atomic radius		Ionic radius	
Li	0.152	Li^+	0.060	F	0.071	F^-	0.136
Na	0.186	Na^+	0.095	Cl	0.099	Cl^-	0.181
K	0.227	K^+	0.133	Br	0.114	Br^-	0.195
Mg	0.160	Mg^{2+}	0.065	O	0.074	O^{2-}	0.140
Al	0.143	Al^{3+}	0.050	S	0.103	S^{2-}	0.184

*The metals shown lose electrons to become positive ions. The nonmetals gain electrons to become negative ions.

the electron structure of a noble gas, and become positive ions. A nonmetal, on the other hand, is only a few electrons short of having a noble gas electron structure in its outer energy level and thus has a tendency to gain electrons. In reacting with metals, nonmetal atoms characteristically gain one, two, or three electrons; attain the electron structure of a noble gas; and become negative ions. The ions formed by loss of electrons are much smaller than the corresponding metal atoms; the ions formed by gaining electrons are larger than the corresponding nonmetal atoms. The dimensions of the atomic and ionic radii of several metals and nonmetals are given in Table 11.3.

Practice 11.2

What noble gas structure is formed when an atom of each of these metals loses all its valence electrons? Write the formula for the metal ion formed.
(a) K (b) Mg (c) Al (d) Ba

Study the following examples. Note the loss and gain of electrons between atoms; also note that the ions in each compound have a noble gas electron structure.

Example 11.2

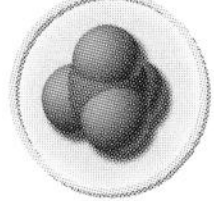

Explain how magnesium and chlorine combine to form magnesium chloride, $MgCl_2$.

SOLUTION

A magnesium atom of electron structure $1s^22s^22p^63s^2$ must lose two electrons or gain six to reach a stable electron structure. If magnesium reacts with chlorine and each chlorine atom can accept only one electron, two chlorine atoms will be needed for the two electrons from each magnesium atom. The compound formed will contain one magnesium ion and two chloride ions. The magnesium atom, having lost two electrons, becomes a magnesium ion with a +2 charge. Each chloride ion will have a −1 charge. The transfer of electrons from a magnesium atom to two chlorine atoms is shown in the following illustration:

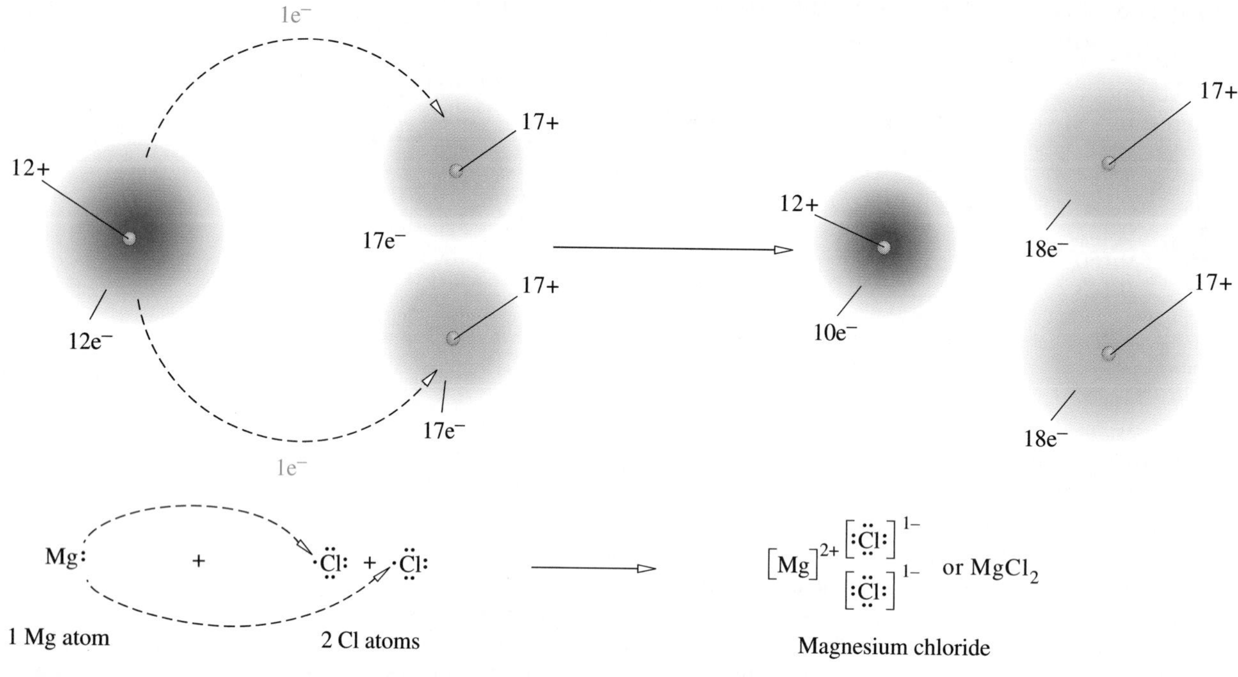

Example 11.3 Explain the formation of sodium fluoride (NaF) from its elements.

SOLUTION

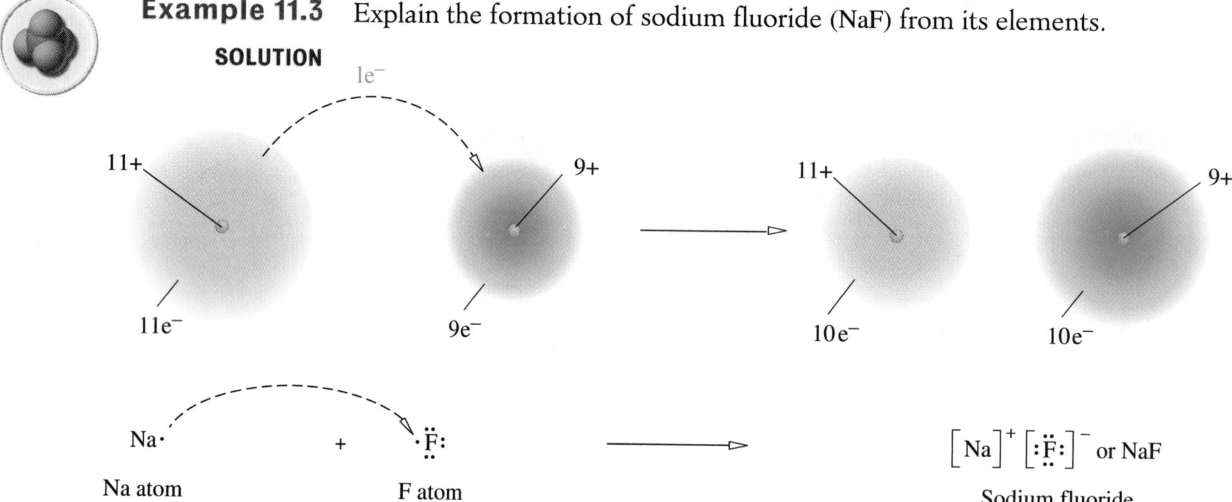

The fluorine atom, with seven electrons in its outer energy level, behaves similarly to the chlorine atom.

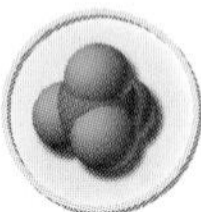

Example 11.4 Explain the formation of aluminum fluoride (AlF_3) from its elements.

SOLUTION

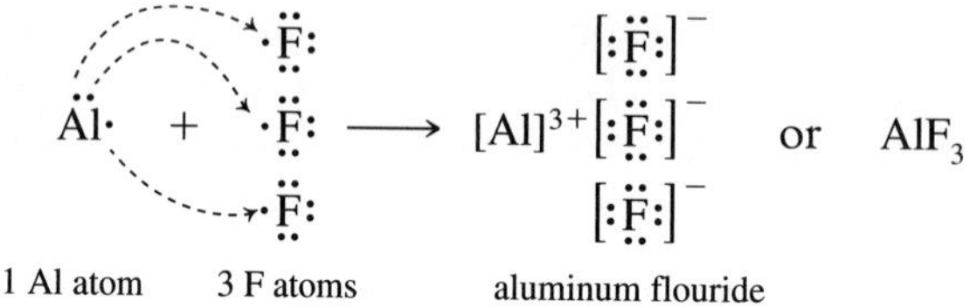

Each fluorine atom can accept only one electron. Therefore, three fluorine atoms are needed to combine with the three valence electrons of one aluminum atom. The aluminum atom has lost three electrons to become an aluminum ion (Al^{3+}) with a +3 charge.

Example 11.5

Explain the formation of magnesium oxide (MgO) from its elements.

SOLUTION

1e⁻ 12+ 8+ 12+ 8+ 12e⁻ 8e⁻ 10e⁻ 10e⁻ 1e⁻

Mg: + ·Ö: ⟶ $[Mg]^{2+}[:\ddot{O}:]^{2-}$ or MgO

Mg atom O atom Magnesium oxide

The magnesium atom, with two electrons in the outer energy level, exactly fills the need of one oxygen atom for two electrons. The resulting compound has a ratio of one magnesium atom to one oxygen atom. The oxygen (oxide) ion has a −2 charge, having gained two electrons. In combining with oxygen, magnesium behaves the same way as when it combines with chlorine—it loses two electrons.

Example 11.6

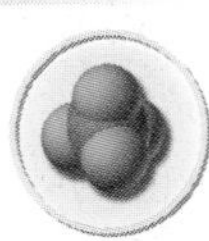

Explain the formation of sodium sulfide (Na_2S) from its elements.

SOLUTION

Na· + Na· + ·S̈: ⟶ $[Na]^+ [Na]^+ [:\ddot{S}:]^{2-}$ or Na_2S

2 Na atoms 1 S atom sodium sulfide

Two sodium atoms supply the electrons that one sulfur atom needs to make eight in its outer energy level.

Example 11.7

Explain the formation of aluminum oxide (Al_2O_3) from its elements.

SOLUTION

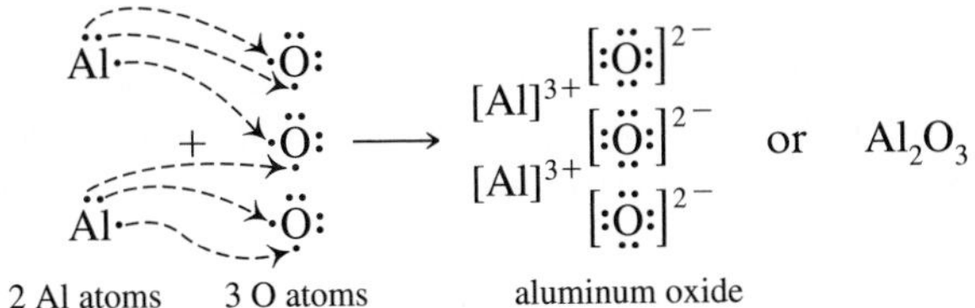

One oxygen atom, needing two electrons, cannot accommodate the three electrons from one aluminum atom. One aluminum atom falls one electron short of the four electrons needed by two oxygen atoms. A ratio of two atoms of aluminum to three atoms of oxygen, involving the transfer of six electrons (two to each oxygen atom), gives each atom a stable electron configuration.

Note that in each of these examples, outer energy levels containing eight electrons were formed in all the negative ions. This formation resulted from the pairing of all the *s* and *p* electrons in these outer energy levels.

11.4 Predicting Formulas of Ionic Compounds

In previous examples, we learned that when a metal and a nonmetal react to form an ionic compound, the metal loses one or more electrons to the nonmetal. In Chapter 6, where we learned to name compounds and write formulas, we saw that Group 1A metals always form +1 cations, whereas Group 2A form +2 cations. Group 7A elements form −1 anions and Group 6A form −2 anions.

It stands to reason, then, that this pattern is directly related to the stability of the noble gas configuration. Metals lose electrons to attain the electron configuration of a noble gas (the previous one on the periodic table). A nonmetal forms an ion by gaining enough electrons to achieve the electron configuration of the noble gas following it on the periodic table. These observations lead us to an important chemical principle:

In almost all stable chemical compounds of representative elements, each atom attains a noble gas electron configuration. This concept forms the basis for our understanding of chemical bonding.

We can apply this principle in predicting the formulas of ionic compounds. To predict the formula of an ionic compound, we must recognize that chemical compounds are always electrically neutral. In addition, the metal will lose electrons to achieve noble gas configuration and the nonmetal will gain electrons to achieve noble gas configuration. Consider the compound formed between barium and sulfur. Barium has two valence electrons, whereas sulfur has six valence electrons:

Ba $[Xe]6s^2$ S $[Ne]3s^23p^4$

If barium loses two electrons, it will achieve the configuration of xenon. By gaining two electrons, sulfur achieves the configuration of argon. Consequently, a pair of electrons is transferred between atoms. Now we have Ba^{2+} and S^{2-}. Since compounds are electrically neutral, there must be a ratio of one Ba to one S, giving the formula BaS.

The same principle works for many other cases. Since the key lies in the electron configuration, the periodic table can be used to extend the prediction even further. Because of similar electron structures, the elements in a family

Table 11.4 Formulas of Compounds Formed by Alkali Metals

Lewis structure	Oxides	Chlorides	Bromides	Sulfates
Li·	Li_2O	LiCl	LiBr	Li_2SO_4
Na·	Na_2O	NaCl	NaBr	Na_2SO_4
K·	K_2O	KCl	KBr	K_2SO_4
Rb·	Rb_2O	RbCl	RbBr	Rb_2SO_4
Cs·	Cs_2O	CsCl	CsBr	Cs_2SO_4

generally form compounds with the same atomic ratios. In general, if we know the atomic ratio of a particular compound—say, NaCl—we can predict the atomic ratios and formulas of the other alkali metal chlorides. These formulas are LiCl, KCl, RbCl, CsCl, and FrCl (see Table 11.4).

Similarly, if we know that the formula of the oxide of hydrogen is H_2O, we can predict that the formula of the sulfide will be H_2S, because sulfur has the same valence electron structure as oxygen. Recognize, however, that these are only predictions; it doesn't necessarily follow that every element in a group will behave like the others or even that a predicted compound will actually exist. For example, knowing the formulas for potassium chlorate, bromate, and iodate to be $KClO_3$, $KBrO_3$, and KIO_3, we can correctly predict the corresponding sodium compounds to have the formulas $NaClO_3$, $NaBrO_3$, and $NaIO_3$. Fluorine belongs to the same family of elements (Group 7A) as chlorine, bromine, and iodine, so it would appear that fluorine should combine with potassium and sodium to give fluorates with the formulas KFO_3 and $NaFO_3$. However, potassium and sodium fluorates are not known to exist.

In the discussion in this section, we refer only to representative metals (Groups 1A, 2A, and 3A). The transition metals (Group B) show more complicated behavior (they form multiple ions), and their formulas are not as easily predicted.

Example 11.8

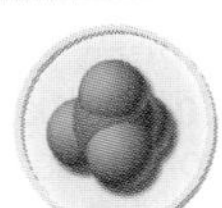

The formula for calcium sulfide is CaS and that for lithium phosphide is Li_3P. Predict formulas for (a) magnesium sulfide, (b) potassium phosphide, and (c) magnesium selenide.

SOLUTION

(a) Look up calcium and magnesium in the periodic table; they are both in Group 2A. The formula for calcium sulfide is CaS, so it's reasonable to predict that the formula for magnesium sulfide is MgS.

(b) Find lithium and potassium in the periodic table; they are in Group 1A. Since the formula for lithium phosphide is Li_3P, it's reasonable to predict that K_3P is the formula for potassium phosphide.

(c) Find selenium in the periodic table; it is in Group 6A just below sulfur. Therefore it's reasonable to assume that selenium forms selenide in the same way that sulfur forms sulfide. Since MgS was the predicted formula for magnesium sulfide in part (a), we can reasonably assume that the formula for magnesium selenide is MgSe.

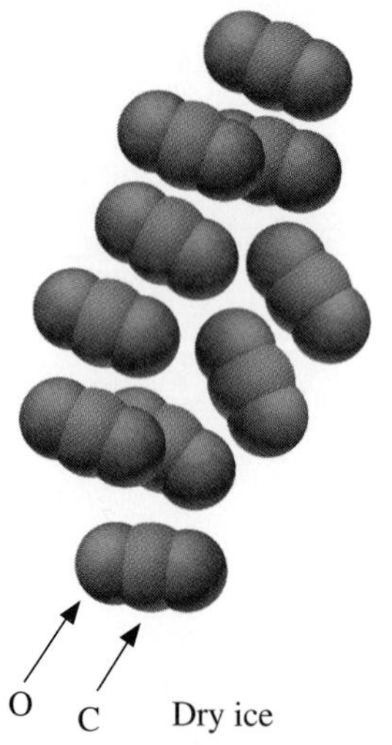

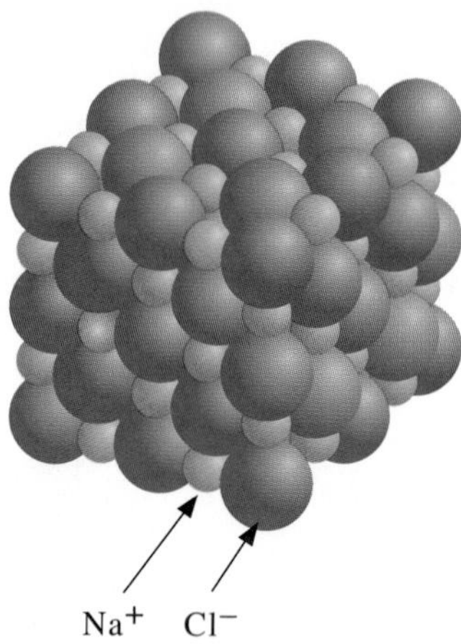

Figure 11.7
Solid carbon dioxide (dry ice) is composed of individual covalently bonded molecules of CO_2 closely packed together. Table salt is a large aggregate of Na^+ and Cl^- ions instead of molecules.

Practice 11.3

The formula for sodium oxide is Na_2O. Predict the formula for
(a) sodium sulfide
(b) rubidium oxide

Practice 11.4

The formula for barium phosphide is Ba_3P_2. Predict the formula for
(a) magnesium nitride
(b) barium arsenide

11.5 The Covalent Bond: Sharing Electrons

Some atoms do not transfer electrons from one atom to another to form ions. Instead they form a chemical bond by sharing pairs of electrons between them. A **covalent bond** consists of a pair of electrons shared between two atoms. This bonding concept was introduced in 1916 by G. N. Lewis. In the millions of known compounds, the covalent bond is the predominant chemical bond.

True molecules exist in substances in which the atoms are covalently bonded. It is proper to refer to molecules of such substances as hydrogen, chlorine, hydrogen chloride, carbon dioxide, water, or sugar (Figure 11.7). These substances contain only covalent bonds and exist as aggregates of molecules. We don't use the term *molecule* when talking about ionically bonded compounds such as sodium chloride, because such substances exist as large aggregates of positive and negative ions, not as molecules (Figure 11.7).

A study of the hydrogen molecule gives us an insight into the nature of the covalent bond and its formation. The formation of a hydrogen molecule (H_2) involves the overlapping and pairing of 1*s* electron orbitals from two hydrogen atoms, shown in Figure 11.8. Each atom contributes one electron of the pair that is shared jointly by two hydrogen nuclei. The orbital of the electrons now includes both hydrogen nuclei, but probability factors show that the most likely place to find the electrons (the point of highest electron density) is between the two nuclei. The two nuclei are shielded from each other by the pair of electrons, allowing the two nuclei to be drawn very close to each other.

The formula for chlorine gas is Cl_2. When the two atoms of chlorine combine to form this molecule, the electrons must interact in a manner similar to that shown in the hydrogen example. Each chlorine atom would be more stable with eight electrons in its outer energy level. But chlorine atoms are identical, and neither is able to pull an electron away from the other. What happens

covalent bond

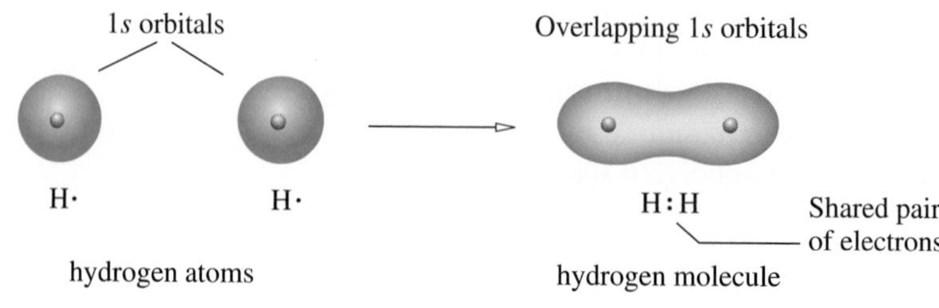

Figure 11.8
The formation of a hydrogen molecule from two hydrogen atoms. The two 1*s* orbitals overlap, forming the H_2 molecule. In this molecule the two electrons are shared between the atoms, forming a covalent bond.

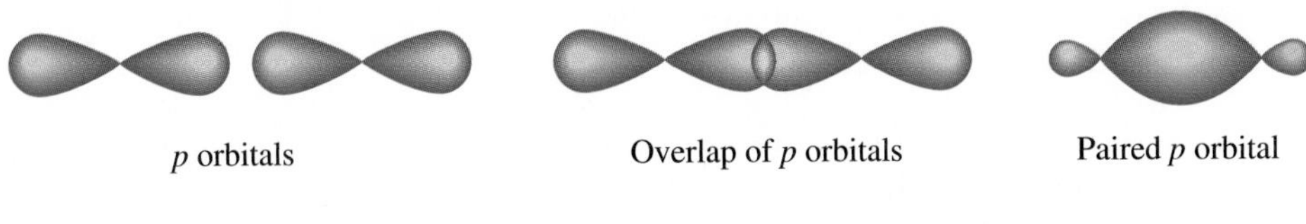

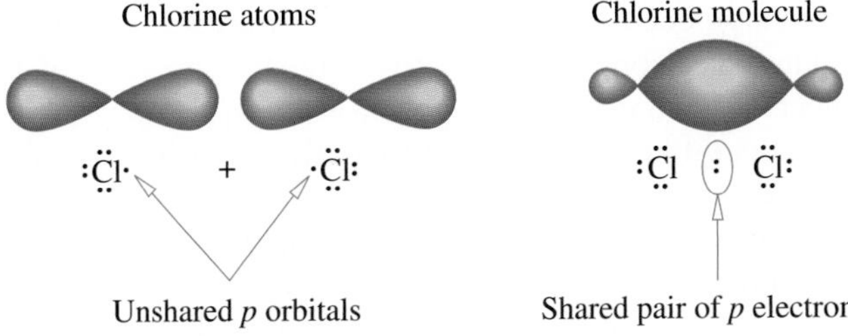

Figure 11.9
Pairing of *p* electrons in the formation of a chlorine molecule.

is this: The unpaired $3p$ electron orbital of one chlorine atom overlaps the unpaired $3p$ electron orbital of the other atom, resulting in a pair of electrons that are mutually shared between the two atoms. Each atom furnishes one of the pair of shared electrons. Thus, each atom attains a stable structure of eight electrons by sharing an electron pair with the other atom. The pairing of the p electrons and the formation of a chlorine molecule are illustrated in Figure 11.9. Neither chlorine atom has a positive or negative charge, because both contain the same number of protons and have equal attraction for the pair of electrons being shared. Other examples of molecules in which electrons are equally shared between two atoms are hydrogen (H_2), oxygen (O_2), nitrogen (N_2), fluorine (F_2), bromine (Br_2), and iodine (I_2). Note that more than one pair of electrons may be shared between atoms:

H:H	:F:F:	:Br:Br:	:I:I:	:O::O:	:N:::N:
hydrogen	fluorine	bromine	iodine	oxygen	nitrogen

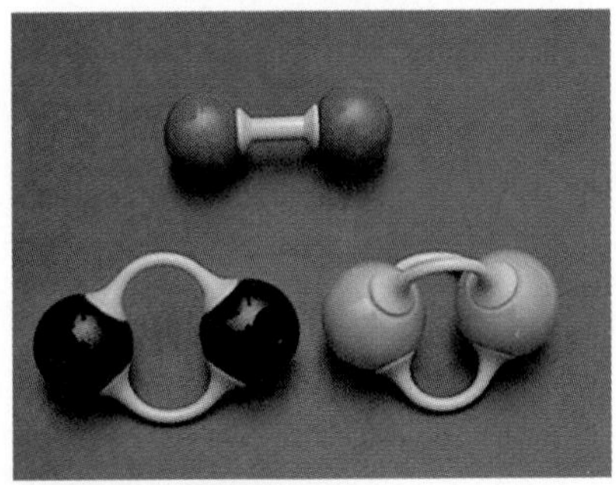

Molecular models for F_2 (green, single bond), O_2 (black, double bond), and N_2 (blue, triple bond).

The Lewis structure given for oxygen does not adequately account for all the properties of the oxygen molecule. Other theories explaining the bonding in oxygen molecules have been advanced, but they are complex and beyond the scope of this book.

In writing structures, we commonly replace the pair of dots used to represent a shared pair of electrons with a dash (—). One dash represents a single bond; two dashes, a double bond; and three dashes, a triple bond. The six structures just shown may be written thus:

H—H :F—F: :Br—Br: :I—I: :O=O: :N≡N:

Remember: A dash represents a shared pair of electrons.

The ionic bond and the covalent bond represent two extremes. In ionic bonding the atoms are so different that electrons are transferred between them, forming a charged pair of ions. In covalent bonding, two identical atoms share electrons equally. The bond is the mutual attraction of the two nuclei for the shared electrons. Between these extremes lie many cases in which the atoms are not different enough for a transfer of electrons but are different enough that the electron pair cannot be shared equally. This unequal sharing of electrons results in the formation of a **polar covalent bond**.

polar covalent bond

11.6 Electronegativity

When two *different* kinds of atoms share a pair of electrons, a bond forms in which electrons are shared unequally. One atom assumes a partial positive charge and the other a partial negative charge with respect to each other. This difference in charge occurs because the two atoms exert unequal attraction for the pair of shared electrons. The attractive force that an atom of an element has for shared electrons in a molecule or polyatomic ion is known as its **electronegativity**. Elements differ in their electronegativities. For example, both hydrogen and chlorine need one electron to form stable electron configurations. They share a pair of electrons in hydrogen chloride (HCl). Chlorine is more electronegative and therefore has a greater attraction for the shared electrons than does hydrogen. As a result, the pair of electrons is displaced toward the chlorine atom, giving it a partial negative charge and leaving the hydrogen atom with a partial positive charge. Note that the electron is not transferred entirely to the chlorine atom (as in the case of sodium chloride) and that no ions are formed. The entire molecule, HCl, is electrically neutral. A partial charge is usually indicated by the Greek letter delta, δ. Thus, a partial positive charge is represented by $\delta+$ and a partial negative charge by $\delta-$.

electronegativity

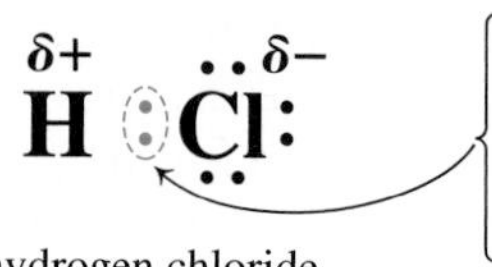

hydrogen chloride

The pair of shared electrons in HCl is closer to the more electronegative chlorine atom than to the hydrogen atom, giving chlorine a partial negative charge with respect to the hydrogen atom.

A scale of relative electronegativities, in which the most electronegative element, fluorine, is assigned a value of 4.0, was developed by the Nobel laureate (1954 and 1962) Linus Pauling (1901–1994). Table 11.5 shows that the relative electronegativity of the nonmetals is high and that of the metals is low. These electronegativities indicate that atoms of metals have a greater tendency to lose electrons than do atoms of nonmetals and that nonmetals have a greater tendency to gain electrons than do metals. The higher the electronegativity value, the greater the attraction for electrons. Note that electronegativity generally increases from left to right across a period and decreases down a group for the representative elements. The highest electronegativity is 4.0 for fluorine, and the lowest is 0.7 for francium and cesium. It's important to remember that the higher the electronegativity, the stronger an atom attracts electrons.

The polarity of a bond is determined by the difference in electronegativity values of the atoms forming the bond (see Figure 11.10). If the electronegativities are the same, the bond is **nonpolar covalent** and the electrons are shared equally. If the atoms have greatly different electronegativities, the bond is very *polar*. At the extreme, one or more electrons are actually transferred and an ionic bond results.

nonpolar covalent bond

dipole

A **dipole** is a molecule that is electrically asymmetrical, causing it to be oppositely charged at two points. A dipole is often written as (+ −). A hydrogen chloride molecule is polar and behaves as a small dipole. The HCl dipole may be written as H+⟶Cl. The arrow points toward the negative end of the dipole. Molecules of H_2O, HBr, and ICl are polar:

H+⟶Cl H+⟶Br I+⟶Cl H+⟶O⟵+H

Table 11.5 Three-Dimensional Representation of Electronegativity

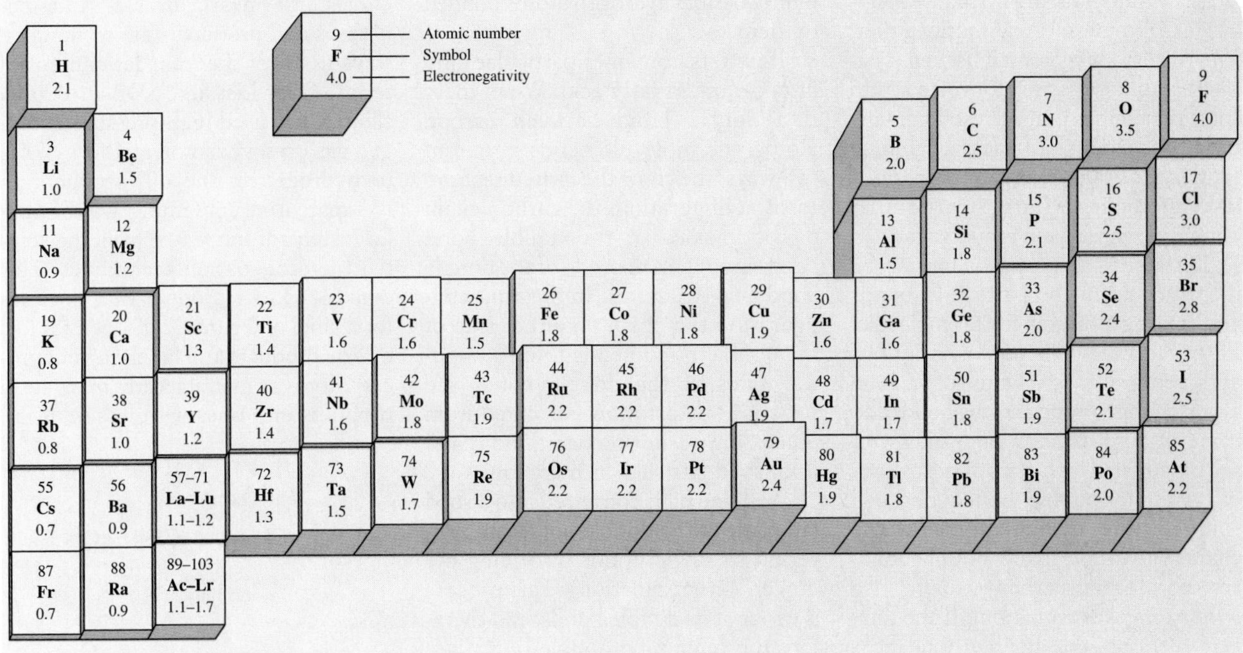

1 H 2.1																
3 Li 1.0	4 Be 1.5											5 B 2.0	6 C 2.5	7 N 3.0	8 O 3.5	9 F 4.0
11 Na 0.9	12 Mg 1.2											13 Al 1.5	14 Si 1.8	15 P 2.1	16 S 2.5	17 Cl 3.0
19 K 0.8	20 Ca 1.0	21 Sc 1.3	22 Ti 1.4	23 V 1.6	24 Cr 1.6	25 Mn 1.5	26 Fe 1.8	27 Co 1.8	28 Ni 1.8	29 Cu 1.9	30 Zn 1.6	31 Ga 1.6	32 Ge 1.8	33 As 2.0	34 Se 2.4	35 Br 2.8
37 Rb 0.8	38 Sr 1.0	39 Y 1.2	40 Zr 1.4	41 Nb 1.6	42 Mo 1.8	43 Tc 1.9	44 Ru 2.2	45 Rh 2.2	46 Pd 2.2	47 Ag 1.9	48 Cd 1.7	49 In 1.7	50 Sn 1.8	51 Sb 1.9	52 Te 2.1	53 I 2.5
55 Cs 0.7	56 Ba 0.9	57–71 La–Lu 1.1–1.2	72 Hf 1.3	73 Ta 1.5	74 W 1.7	75 Re 1.9	76 Os 2.2	77 Ir 2.2	78 Pt 2.2	79 Au 2.4	80 Hg 1.9	81 Tl 1.8	82 Pb 1.8	83 Bi 1.9	84 Po 2.0	85 At 2.2
87 Fr 0.7	88 Ra 0.9	89–103 Ac–Lr 1.1–1.7														

How do we know whether a bond between two atoms is ionic or covalent? The difference in electronegativity between the two atoms determines the character of the bond formed between them. As the difference in electronegativity increases, the polarity of the bond (or percent ionic character) increases.

If the electronegativity difference between two bonded atoms is greater than 1.7–1.9, the bond will be more ionic than covalent.

If the electronegativity difference is greater than 2.0, the bond is strongly ionic. If the electronegativity difference is less than 1.5, the bond is strongly covalent.

Practice 11.5

Which of these compounds would you predict to be ionic and which would be covalent?

(a) $SrCl_2$ (d) RbBr
(b) PCl_3 (e) LiCl
(c) NH_3 (f) CS_2

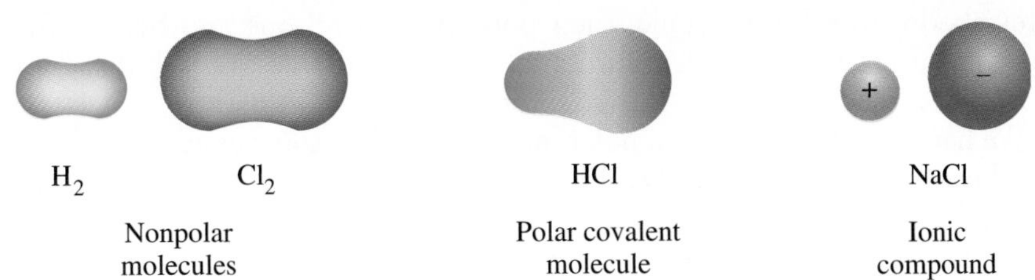

Figure 11.10
Nonpolar, polar covalent, and ionic compounds.

CHEMISTRY IN ACTION • *Trans*-forming Fats

Trans fats are virtually everywhere in the American diet. These fats remain when vegetable oils are converted into solid substances that are used in many processed foods. In fact, beginning January 1, 2006, manufacturers must label products to show their *trans* fat content. So just what is a *trans* fat?

Different categories of fat can be identified from the pattern of bonds and hydrogen atoms in the molecule. Fatty acids are one component of fats and contain long chains of carbon atoms with hydrogen atoms bonded to some or all of the carbon atoms. Unsaturated fats (such as corn or soybean oil) contain double bonds between some of the carbon atoms in the chains. A carbon that is doubly bonded to another carbon usually also bonds to a hydrogen atom. If the fatty acid has only one double bond, it is monounsaturated (such as olive oil). If the fatty acid contains more than one double bond, it is polyunsaturated. All fat with no double bonds is saturated—all the carbon atoms have the maximum possible hydrogen atoms bonded to them.

Trans fats contain a particular kind of unsaturated fatty acid. When there is a double bond between carbon atoms, the molecule bends in one of two ways: the *cis* or the *trans* direction. In *cis* configuration, the carbon chain on both sides of the double bond bends in to the same side of the double bond (see structure). In the *trans* configuration the chain on either side of the double bond bends toward opposite sides of the double bond (see structure). Most *trans* fats come from processing oils for prepared foods and from solid fats such as margarine.

When an oil is converted into a solid fat, some of the double bonds are converted to single bonds by adding hydrogen (hydrogenation). This process is easier at *cis* double bonds, and therefore the remaining double bonds are mainly in the *trans* configuration. These *trans* fatty acids tend to stack together, making a solid easier than the *cis* forms. Studies have linked diets high in *trans* fats to poor health, high cholesterol, heart disease, and diabetes. Food producers are working on ways to lower the *trans* fat content of foods. Gary List at USDA in Peoria, Illinois, has used high-pressure hydrogen gas on soybean oil at 140°–170°C to hydrogenate the oil, producing a soft margarine containing 5%–6% *trans* fat instead of the ~40% from the standard hydrogenation techniques. This could lead to a product that qualifies for a label of 0 g *trans* fat. Look for lots of new products and labels in your grocery stores as manufactures *trans* -form products into ones with less *trans* fat.

$CH_3-(CH_2)_7-CH=CH-(CH_2)_7-C(=O)OH$ (H atoms on the same side)

cis fatty acid

$CH_3-(CH_2)_7-CH=CH-(CH_2)_7-C(=O)OH$ (H atoms on opposite sides)

trans fatty acid

Care must be taken to distinguish between polar bonds and polar molecules. A covalent bond between different kinds of atoms is always polar. But a molecule containing different kinds of atoms may or may not be polar, depending on its shape or geometry. Molecules of HF, HCl, HBr, HI, and ICl are all polar because each contains a single polar bond. However, CO_2, CH_4, and CCl_4 are nonpolar molecules despite the fact that all three contain polar bonds. The carbon dioxide molecule O=C=O is nonpolar because the carbon–oxygen dipoles cancel each other by acting in opposite directions.

$$\overset{\longleftarrow\!+}{O=}C\overset{+\!\longrightarrow}{=O}$$

dipoles in opposite directions

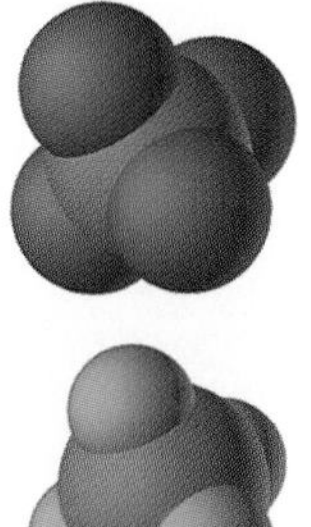

Spacefilling molecular model CCl_4 (top) and methane (CH_4).

Carbon tetrachloride (CCl_4) is nonpolar because the four C—Cl polar bonds are identical, and since these bonds emanate from the center to the corners of a tetrahedron in the molecule, their polarities cancel one another. Methane has the same molecular structure and is also nonpolar. We will discuss the shapes of molecules later in this chapter.

We have said that water is a polar molecule. If the atoms in water were linear like those in carbon dioxide, the two O—H dipoles would cancel each other, and the molecule would be nonpolar. However, water is definitely polar and has a nonlinear (bent) structure with an angle of 105° between the two O—H bonds.

CHEMISTRY IN ACTION • Goal! A Spherical Molecule

One of the most diverse elements in the periodic table is carbon. Graphite and diamond, two well-known forms of elemental carbon, both contain extended arrays of carbon atoms. In graphite the carbon atoms are arranged in sheets, and the bonding between the sheets is very weak. This property makes graphite useful as a lubricant and as a writing material. Diamond consists of transparent octahedral crystals in which each carbon atom is bonded to four other carbon atoms. This three-dimensional network of bonds gives diamond the property of hardness for which it is noted. In the 1980s, a new form of carbon was discovered in which the atoms are arranged in relatively small clusters.

When Harold Kroto of the University of Sussex, England, and Richard Smalley of Rice University, Texas, discovered a strange carbon molecule of formula C_{60}, they deduced that the most stable arrangement for the atoms would be in the shape of a soccer ball. In thinking about possible arrangements, the scientists considered the geodesic domes designed by R. Buckminster Fuller in the 1960s. This cluster form of carbon was thus named *buckminsterfullerene* and is commonly called buckyballs.

Buckyballs have captured the imagination of a variety of chemists. Research on buckyballs has led to a host of possible applications for these molecules. If metals are bound to the carbon atoms, the fullerenes become superconducting; that is, they conduct electrical current without resistance, at very low temperature. Scientists are now able to make buckyball compounds that superconduct at temperatures of 45 K. Other fullerenes are being used in lubricants and optical materials.

Chemists at Yale University have managed to trap helium and neon inside buckyballs. This is the first time chemists have ever observed helium or neon in a compound of any kind. They found that at temperatures from 1000°F to 1500°F one of the covalent bonds linking neighboring carbon atoms in the buckyball breaks. This opens a window in the fullerene molecule through which a helium or neon atom can enter the buckyball. When the fullerene is allowed to cool, the broken bond between carbon atoms re-forms, shutting the window and trapping the helium or neon atom inside the buckyball. Since the trapped helium or neon cannot react or share electrons with its host, the resulting compound has forced scientists to invent a new kind of chemical formula to describe the compound. The relationship between the "prisoner" helium or neon and the host buckyball is shown with an @ sign. A helium fullerene containing 60 carbon atoms would thus be $He@C_{60}$.

Buckyballs can be tailored to fit a particular size requirement. Raymond Schinazi of the Emory University School of Medicine, Georgia, made a buckyball to fit the active site of a key HIV enzyme that paralyzes the virus, making it noninfectious in human cells. The key to making this compound was preparing a water-soluble buckyball that would fit in the active site of the enzyme. Eventually, scientists created a water-soluble fullerene molecule that has two charged arms to grasp the binding site of the enzyme. It is toxic to the virus but doesn't appear to harm the host cells.

Scandium atoms trapped in a buckyball.

"Raspberries" of fullerene lubricant.

The relationships among types of bonds are summarized in Figure 11.11. It is important to realize that bonding is a continuum; that is, the difference between ionic and covalent is a gradual change.

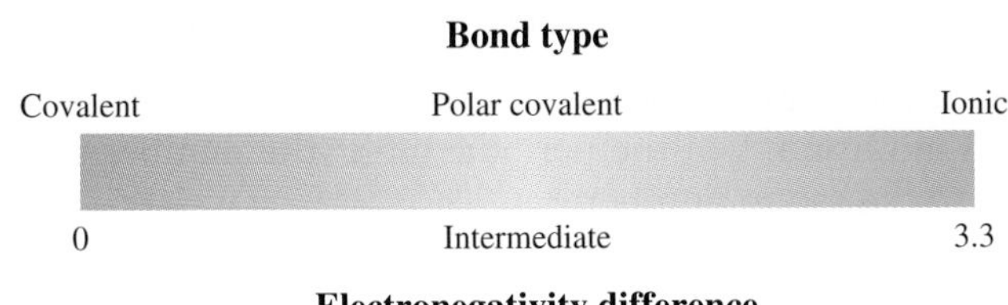

Figure 11.11
Relating bond type to electronegativity difference between atoms.

11.7 Lewis Structures of Compounds

As we have seen, Lewis structures are a convenient way of showing the covalent bonds in many molecules or ions of the representative elements. In writing Lewis structures, the most important consideration for forming a stable compound is that the atoms attain a noble gas configuration.

The most difficult part of writing Lewis structures is determining the arrangement of the atoms in a molecule or an ion. In simple molecules with more than two atoms, one atom will be the central atom surrounded by the other atoms. Thus, Cl_2O has two possible arrangements, Cl—Cl—O or Cl—O—Cl. Usually, but not always, the single atom in the formula (except H) will be the central atom.

Although Lewis structures for many molecules and ions can be written by inspection, the following procedure is helpful for learning to write them:

Remember: The number of valence electrons of Group A elements is the same as their group number in the periodic table.

Step 1 Obtain the total number of valence electrons to be used in the structure by adding the number of valence electrons in all the atoms in the molecule or ion. If you are writing the structure of an ion, add one electron for each negative charge or subtract one electron for each positive charge on the ion.

Step 2 Write the skeletal arrangement of the atoms and connect them with a single covalent bond (two dots or one dash). Hydrogen, which contains only one bonding electron, can form only one covalent bond. Oxygen atoms are not normally bonded to each other, except in compounds known to be peroxides. Oxygen atoms normally have a maximum of two covalent bonds (two single bonds or one double bond).

Step 3 Subtract two electrons for each single bond you used in Step 2 from the total number of electrons calculated in Step 1. This gives you the net number of electrons available for completing the structure.

Step 4 Distribute pairs of electrons (pairs of dots) around each atom (except hydrogen) to give each atom a noble gas structure.

Step 5 If there are not enough electrons to give these atoms eight electrons, change single bonds between atoms to double or triple bonds by shifting unbonded pairs of electrons as needed. Check to see that each atom has a noble gas electron structure (two electrons for hydrogen and eight for the others). A double bond counts as four electrons for each atom to which it is bonded.

Example 11.9 How many valence electrons are in each of these atoms: Cl, H, C, O, N, S, P, I?

SOLUTION You can look at the periodic table to determine the electron structure, or, if the element is in Group A of the periodic table, the number of valence electrons is equal to the group number:

Atom	Group	Valence electrons
Cl	7A	7
H	1A	1
C	4A	4
O	6A	6
N	5A	5
S	6A	6
P	5A	5
I	7A	7

Example 11.10

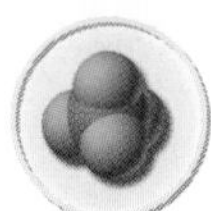

Write the Lewis structure for water (H_2O).

SOLUTION

Step 1 The total number of valence electrons is eight, two from the two hydrogen atoms and six from the oxygen atom.

Step 2 The two hydrogen atoms are connected to the oxygen atom. Write the skeletal structure:

```
H O     or     H O H
 H
```

Place two dots between the hydrogen and oxygen atoms to form the covalent bonds:

```
H:O     or     H:O:H
  ..
  H
```

Step 3 Subtract the four electrons used in Step 2 from eight to obtain four electrons yet to be used.

Step 4 Distribute the four electrons in pairs around the oxygen atom. Hydrogen atoms cannot accommodate any more electrons:

```
   ..              ..
H—O:     or     H—O—H
  |               ..
  H
```

These arrangements are Lewis structures because each atom has a noble gas electron structure. Note that the shape of the molecule is not shown by the Lewis structure.

Example 11.11

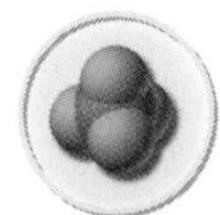

Write Lewis structures for a molecule of methane (CH_4).

SOLUTION

Step 1 The total number of valence electrons is eight, one from each hydrogen atom and four from the carbon atom.

Step 2 The skeletal structure contains four H atoms around a central C atom. Place two electrons between the C and each H.

```
  H         H
                ..
H C H     H:C:H
                ..
  H         H
```

Step 3 Subtract the eight electrons used in Step 2 from eight (obtained in Step 1) to obtain zero electrons yet to be placed. Therefore the Lewis structure must be as written in Step 2:

```
                      H
   H                  |
 H:C:H      or    H — C — H
   H                  |
                      H
```

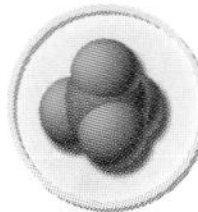

Example 11.12 Write the Lewis structure for a molecule of carbon tetrachloride (CCl_4).

SOLUTION **Step 1** The total number of valence electrons to be used is 32, 4 from the carbon atom and 7 from each of the four chlorine atoms.

Step 2 The skeletal structure contains the four Cl atoms around a central C atom. Place 2 electrons between the C and each Cl:

```
    Cl          Cl
 Cl C Cl     Cl:C:Cl
    Cl          Cl
```

Step 3 Subtract the 8 electrons used in Step 2 from 32 (obtained in Step 1) to obtain 24 electrons yet to be placed.

Step 4 Distribute the 24 electrons (12 pairs) around the Cl atoms so that each Cl atom has 8 electrons around it:

```
                             ..
     ..                     :Cl:
    :Cl:                     |
 .. .. ..                ..      ..
:Cl:C:Cl:      or       :Cl — C — Cl:
 .. .. ..                ..      ..
    :Cl:                     |
     ..                     :Cl:
                             ..
```

This arrangement is the Lewis structure; CCl_4 contains four covalent bonds.

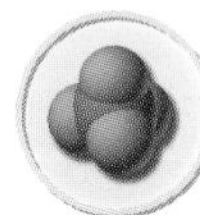

Example 11.13 Write a Lewis structure for CO_2.

SOLUTION **Step 1** The total number of valence electrons is 16, 4 from the C atom and 6 from each O atom.

Step 2 The two O atoms are bonded to a central C atom. Write the skeletal structure and place 2 electrons between the C and each O atom.

O:C:O

Step 3 Subtract the 4 electrons used in Step 2 from 16 (found in Step 1) to obtain 12 electrons yet to be placed.

Step 4 Distribute the 12 electrons (six pairs) around the C and O atoms. Several possibilities exist:

```
 ..   ..      ..  ..      .. ..
:O:C:O:      :O:C:O:     :O:C:O:
 ..   ..        .. ..       .. ..
   I            II          III
```

Step 5 Not all the atoms have 8 electrons around them (noble gas structure). Remove one pair of unbonded electrons from each O atom in structure I and place one pair between each O and the C atom, forming two double bonds:

$$:\ddot{O}::C::\ddot{O}: \quad \text{or} \quad :\ddot{O}=C=\ddot{O}:$$

Each atom now has 8 electrons around it. The carbon is sharing four pairs of electrons, and each oxygen is sharing two pairs. These bonds are known as double bonds because each involves sharing two pairs of electrons.

Practice 11.6

Write the Lewis structures for the following:

(a) PBr_3 (b) $CHCl_3$ (c) HF (d) H_2CO (e) N_2

Although many compounds attain a noble gas structure in covalent bonding, there are numerous exceptions. Sometimes it's impossible to write a structure in which each atom has 8 electrons around it. For example, in BF_3 the boron atom has only 6 electrons around it, and in SF_6 the sulfur atom has 12 electrons around it.

Although there are exceptions, many molecules can be described using Lewis structures where each atom has a noble gas electron configuration. This is a useful model for understanding chemistry.

11.8 Complex Lewis Structures

Most Lewis structures give bonding pictures that are consistent with experimental information on bond strength and length. There are some molecules and polyatomic ions for which no single Lewis structure consistent with all characteristics and bonding information can be written. For example, consider the nitrate ion, NO_3^-. To write a Lewis structure for this polyatomic ion, we use the following steps.

Step 1 The total number of valence electrons is 24, 5 from the nitrogen atom, 6 from each oxygen atom, and 1 from the −1 charge.

Step 2 The three O atoms are bonded to a central N atom. Write the skeletal structure and place two electrons between each pair of atoms. Since we have an extra electron in this ion, resulting in a −1 charge, we enclose the group of atoms in square brackets and add a − charge as shown.

$$\left[\begin{array}{c} O \\ \vdots \\ O:N:O \end{array}\right]^-$$

Step 3 Subtract the 6 electrons used in Step 2 from 24 (found in Step 1) to obtain 18 electrons yet to be placed.

CHEMISTRY IN ACTION • Cleaner Showers through Chemistry

Keeping the shower area sparkling clean and free of mildew is a job none of us enjoy. Now thanks to chemist Bob Black of Jacksonville, Florida, there is a product that cleans the shower without any scrubbing! Black struggled with his home shower until he finally decided he really needed a new product that would solve the mildew and scrubbing problem.

His search was based on the following needs:

1. A molecule to lift deposits off the walls of the shower
2. A way to prevent hard-water deposits from forming
3. A wetting agent to wet the walls and rinse off deposits

In all cases, he limited his search to substances nontoxic and environmentally safe.

Black used a molecule called a glycol ether to lift deposits off the shower wall. This molecule is a long chain with a polar end and a nonpolar end. Substances that are nonpolar (such as grease, oils, and organic material) are attracted to the nonpolar end of the molecules. The molecules cluster together to form micelles (with the polar end pointed out). The polar sphere dissolves in the polar water from the shower, washing off the organic deposits.

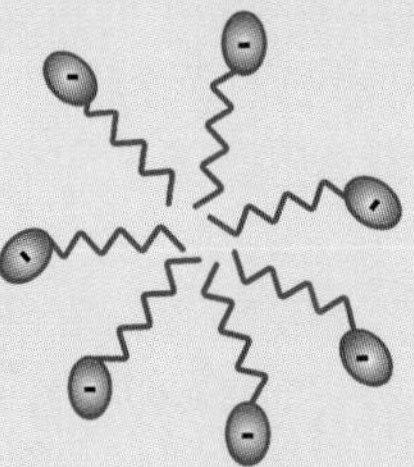

A micelle

Preventing the hard-water deposits from forming on the shower walls required use of a molecule called EDTA. This molecule bonds to ions (like Ca^{2+}, Mg^{2+}, or Fe^{3+}) and prevents the formation of soap scum or hard-water deposits on the walls.

Lastly, Black added isopropyl alcohol to wet the shower wall and also to disturb the mildew fungus. He mixed all the ingredients in just the right proportions and found he no longer needed to work so hard to clean the shower.

Fortunately for all of us, Black shared his solution with friends, who also liked it. Black patented his new product and began mass production. You can now find Clean Shower in your local grocery store!

Simply spraying your shower regularly with Clean Shower will keep it free of deposits and mildew.

Step 4 Distribute the 18 electrons around the N and O atoms:

:Ö ← electron deficient
:Ö:N̈:Ö:

Step 5 One pair of electrons is still needed to give all the N and O atoms a noble gas structure. Move the unbonded pair of electrons from the N atom and place it between the N and the electron-deficient O atom, making a double bond.

[:Ö=N(–:Ö:)–:Ö:]⁻ (O double-bonded above N) or [:Ö:–N(–:Ö:)=Ö:]⁻ or [Ö=N(–:Ö:)–:Ö:]⁻

Are these all valid Lewis structures? Yes, so there really are three possible Lewis structures for NO_3^-.

A molecule or ion that has multiple correct Lewis structures shows *resonance*.

resonance structure

Each of these Lewis structures is called a **resonance structure**. In this book, however, we will not be concerned with how to choose the correct resonance structure for a molecule or ion. Therefore any of the possible resonance structures may be used to represent the ion or molecule.

Example 11.14

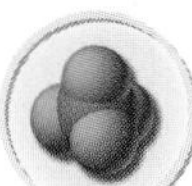

Write the Lewis structure for a carbonate ion (CO_3^{2-}).

SOLUTION

Step 1 These four atoms have 22 valence electrons plus 2 electrons from the −2 charge, which makes 24 electrons to be placed.

Step 2 In the carbonate ion, the carbon is the central atom surrounded by the three oxygen atoms. Write the skeletal structure and place 2 electrons (or a single line) between each C and O:

```
O
|
C—O
|
O
```

Step 3 Subtract the 6 electrons used in Step 2 from 24 (from Step 1) to give 18 electrons yet to be placed.

Step 4 Distribute the 18 electrons around the three oxygen atoms and indicate that the carbonate ion has a −2 charge:

```
[    :Ö:    ]2−
[     |     ]
[     C     ]
[   /   \   ]
[ :O:   :O: ]
```

The difficulty with this structure is that the carbon atom has only six electrons around it instead of a noble gas octet.

Step 5 Move one of the nonbonding pairs of electrons from one of the oxygens and place them between the carbon and the oxygen. Three Lewis structures are possible:

```
[    :Ö:    ]2−        [    :Ö:    ]2−        [    :O:    ]2−
[     |     ]          [     |     ]          [     ||    ]
[     C     ]    or    [     C     ]    or    [     C     ]
[   //  \   ]          [   /  \\   ]          [   /   \   ]
[ O:    :O: ]          [ :O:    O  ]          [ :O:   :O: ]
```

Practice 11.7

Write the Lewis structure for each of the following:

(a) NH_3 (b) H_3O^+ (c) NH_4^+ (d) HCO_3^-

11.9 Compounds Containing Polyatomic Ions

A polyatomic ion is a stable group of atoms that has either a positive or a negative charge and behaves as a single unit in many chemical reactions. Sodium carbonate, Na_2CO_3, contains two sodium ions and a carbonate ion. The carbonate ion (CO_3^{2-}) is a polyatomic ion composed of one carbon atom and three oxygen atoms and has a charge of −2. One carbon and three oxygen atoms have a total of 22 electrons in their outer energy levels. The carbonate ion contains 24 outer electrons and therefore has a charge of −2. In this case, the 2 additional electrons come from the two sodium atoms, which are now sodium ions:

$$[\mathrm{Na}]^+ \left[\begin{array}{c} :\ddot{\mathrm{O}}: \\ | \\ \mathrm{C} \\ :\ddot{\mathrm{O}}\diagup \quad \diagdown\!\!=\ddot{\mathrm{O}} \end{array} \right]^{2-} [\mathrm{Na}]^+ \qquad \left[\begin{array}{c} :\ddot{\mathrm{O}}: \\ | \\ \mathrm{C} \\ :\ddot{\mathrm{O}}\diagup \quad \diagdown\!\!=\ddot{\mathrm{O}} \end{array} \right]^{2-}$$

sodium carbonate — carbonate ion

Sodium carbonate has both ionic and covalent bonds. Ionic bonds exist between each of the sodium ions and the carbonate ion. Covalent bonds are present between the carbon and oxygen atoms within the carbonate ion. One important difference between the ionic and covalent bonds in this compound can be demonstrated by dissolving sodium carbonate in water. It dissolves in water, forming three charged particles—two sodium ions and one carbonate ion—per formula unit of Na_2CO_3:

$$\underset{\text{sodium carbonate}}{\mathrm{Na_2CO_3}(s)} \xrightarrow{\text{water}} \underset{\text{sodium ions}}{2\,\mathrm{Na}^+(aq)} + \underset{\text{carbonate ion}}{\mathrm{CO_3^{2-}}(aq)}$$

The CO_3^{2-} ion remains as a unit, held together by covalent bonds; but where the bonds are ionic, dissociation of the ions takes place. Do not think, however, that polyatomic ions are so stable that they cannot be altered. Chemical reactions by which polyatomic ions can be changed to other substances do exist.

11.10 Molecular Shape

So far in our discussion of bonding we have used Lewis structures to represent valence electrons in molecules and ions, but they don't indicate anything regarding the molecular or geometric shape of a molecule. The three-dimensional arrangement of the atoms within a molecule is a significant feature in understanding molecular interactions. Let's consider several examples illustrated in Figure 11.12.

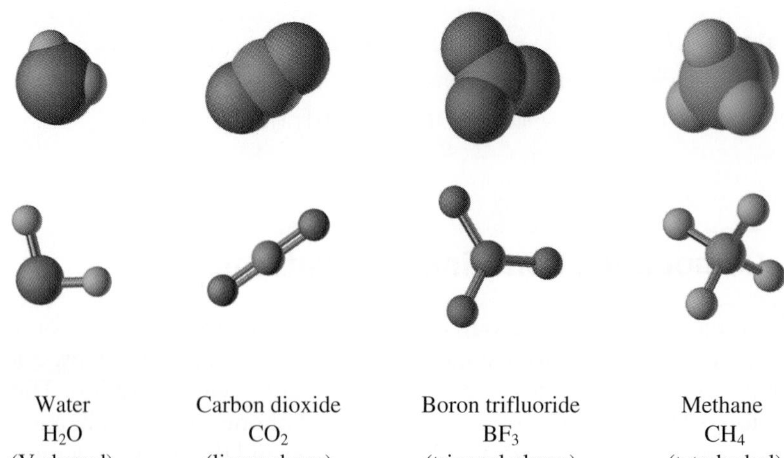

Figure 11.12
Geometric shapes of common molecules. Each molecule is shown as a ball-and-stick model (showing the bonds) and as a spacefilling model (showing the shape).

CHEMISTRY IN ACTION • Strong Enough to Stop a Bullet?

What do color-changing pens, bullet-resistant vests, and calculators have in common? The chemicals that make each of them work are liquid crystals. These chemicals find numerous applications; you are probably most familiar with liquid crystal displays (LCDs) and color-changing products, but these chemicals are also used to make superstrong synthetic fibers.

Molecules in a normal crystal remain in an orderly arrangement, but in a liquid crystal the molecules can flow *and* maintain an orderly arrangement at the same time. Liquid crystal molecules are linear and polar. Since the atoms tend to lie in a relatively straight line, the molecules are generally much longer than they are wide. These polar molecules are attracted to each other and are able to line up in an orderly fashion, without solidifying.

Liquid crystals with twisted arrangements of molecules give us novelty color-changing products. In these liquid crystals the molecules lie side by side in a nearly flat layer. The next layer is similar, but at an angle to the one below. The closely packed flat layers have a special effect on light. As the light strikes the surface, some of it is reflected from the top layer and some from lower layers. When the same wavelength is reflected from many layers, we see a color. (This is similar to the rainbow of colors formed by oil in a puddle on the street or the film of a soap bubble.) As the temperature is increased, the molecules move faster, causing a change in the angle and the space between the layers. This results in a color change in the reflected light. Different compounds change color within different temperature ranges, allowing a variety of practical and amusing applications.

Liquid crystal (nematic) molecules that lie parallel to one another are used to manufacture very strong synthetic fibers. Perhaps the best example of these liquid crystals is Kevlar, a synthetic fiber used in bullet-resistant vests, canoes, and parts of the space shuttle. Kevlar is a synthetic polymer, like nylon or polyester, that gains strength by passing through a liquid crystal state during its manufacture.

In a typical polymer, the long molecular chains are jumbled together, somewhat like spaghetti. The strength of the material is limited by the disorderly arrangement. The trick is to get the molecules to line up parallel to each other. Once the giant molecules have been synthesized, they are dissolved in sulfuric acid. At the proper concentration the molecules align, and the solution is forced through tiny holes in a nozzle and further aligned. The sulfuric acid is removed in a water bath, thereby forming solid fibers in near-perfect alignment. One strand of Kevlar is stronger than an equal-sized strand of steel. It has a much lower density as well, making it a material of choice in bullet-resistant vests.

Kevlar is used to make protective vests for police.

Water is known to have the geometric shape known as "bent" or "V-shaped." Carbon dioxide exhibits a linear shape. BF_3 forms a third molecular shape called *trigonal planar* since all the atoms lie in one plane in a triangular arrangement. One of the more common molecular shapes is the tetrahedron, illustrated by the molecule methane (CH_4).

How do we predict the geometric shape of a molecule? We will now study a model developed to assist in making predictions from the Lewis structure.

11.11 The Valence Shell Electron Pair Repulsion (VSEPR) Model

The chemical properties of a substance are closely related to the structure of its molecules. A change in a single site on a large biomolecule can make a difference in whether or not a particular reaction occurs.

Instrumental analysis can be used to determine exact spatial arrangements of atoms. Quite often, though, we only need to be able to predict the approximate structure of a molecule. A relatively simple model has been developed to allow us to make predictions of shape from Lewis structures.

Nonbonding pairs of electrons are not shown here, so you can focus your attention on the shapes, not electron arrangement.

The VSEPR model is based on the idea that electron pairs will repel each other electrically and will seek to minimize this repulsion. To accomplish this minimization, the electron pairs will be arranged around a central atom as far apart as possible. Consider $BeCl_2$, a molecule with only two pairs of electrons surrounding the central atom. These electrons are arranged 180° apart for maximum separation:

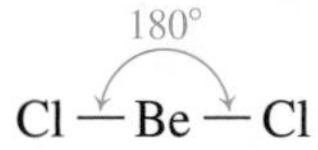

180°

Cl — Be — Cl

linear structure

This molecular structure can now be labeled as a **linear structure**. When only two pairs of electrons surround a central atom, they should be placed 180° apart to give a linear structure.

What occurs when there are only three pairs of electrons around the central atom? Consider the BF_3 molecule. The greatest separation of electron pairs occurs when the angles between atoms are 120°:

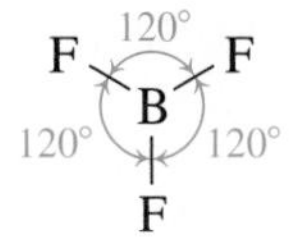

120°
F F
B
120° 120°
F

trigonal planar structure

This arrangement of atoms is flat (planar) and, as noted earlier, is called **trigonal planar structure**. When three pairs of electrons surround an atom, they should be placed 120° apart to show the trigonal planar structure.

Now consider the most common situation (CH_4), with four pairs of electrons on the central carbon atom. In this case the central atom exhibits a noble gas electron structure. What arrangement best minimizes the electron pair repulsions? At first it seems that an obvious choice is a 90° angle with all the atoms in a single plane:

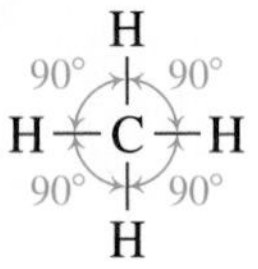

H
90° 90°
H — C — H
90° 90°
H

However, we must consider that molecules are three-dimensional. This concept results in a structure in which the electron pairs are actually 109.5° apart:

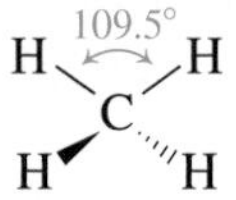

109.5°
H H
C
H H

tetrahedral structure

In this diagram the wedged line seems to protrude from the page whereas the dashed line recedes. Two representations of this arrangement, known as **tetrahedral structure**, are illustrated in Figure 11.13. When four pairs of electrons surround a central atom, they should be placed 109.5° apart to give them a tetrahedral structure.

The VSEPR model is based on the premise that we are counting electron pairs. It's quite possible that one or more of these electron pairs may be nonbonding (lone) pairs. What happens to the molecular structure in these cases? Consider the ammonia molecule. First we draw the Lewis structure to determine the number of electron pairs around the central atom:

$$\begin{array}{c} H:\ddot{\underset{\cdot\cdot}{N}}:H \\ H \end{array}$$

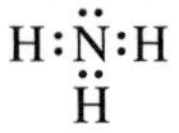

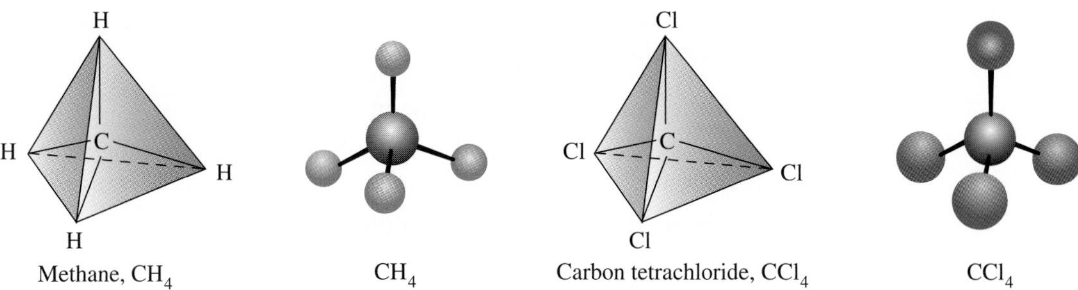

Figure 11.13
Ball-and-stick models of methane and carbon tetrachloride. Methane and carbon tetrachloride are nonpolar molecules because their polar bonds cancel each other in the tetrahedral arrangement of their atoms. The carbon atoms are located in the centers of the tetrahedrons.

Since there are four pairs of electrons, the arrangement of electrons around the central atom will be tetrahedral (Figure 11.14a). However, only three of the pairs are bonded to another atom, so the molecule itself is pyramidal. It is important to understand that the placement of the electron pairs determines the shape but the name for the molecule is determined by the position of the atoms themselves. Therefore, ammonia is pyramidal. See Figure 11.14c.

Now consider the effect of two unbonded pairs of electrons in the water molecule. The Lewis structure for water is

$$\begin{array}{l} \text{H}-\ddot{\text{O}}\text{:} \\ \quad\;\; | \\ \quad\;\; \text{H} \end{array}$$

The four electron pairs indicate a tetrahedral electron arrangement is necessary (see Figure 11.15a). The molecule is not called tetrahedral because two of the electron pairs are unbonded pairs. The water molecule is "bent," as shown in Figure 11.15c.

The arrangement of electron pairs around an atom determines its shape, but we name the shape of molecules by the position of the atoms.

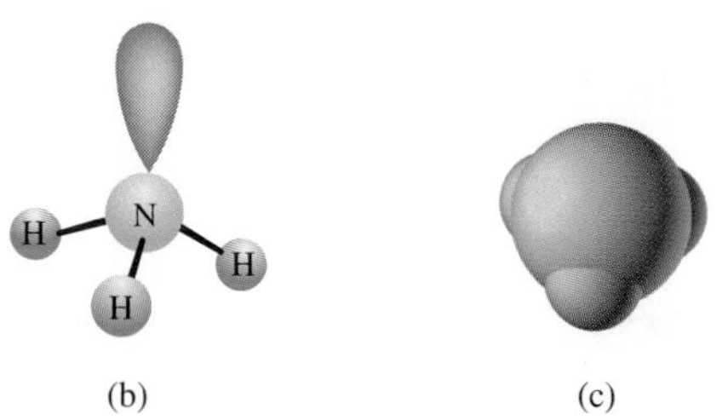

Figure 11.14
(a) The tetrahedral arrangement of electron pairs around the N atom in the NH_3 molecule. (b) Three pairs are shared and one is unshared. (c) The NH_3 molecule is pyramidal.

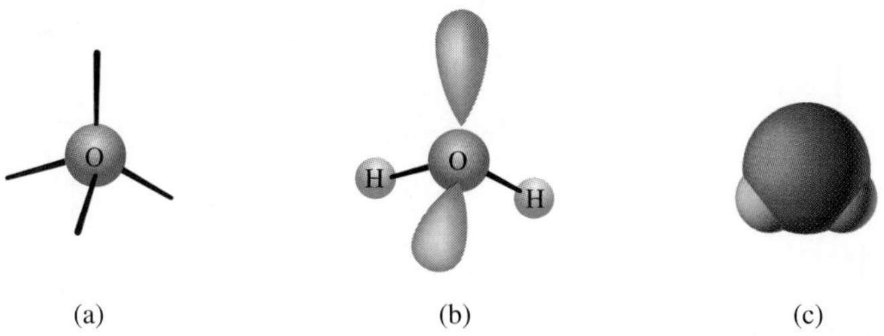

Figure 11.15
(a) The tetrahedral arrangement of the four electron pairs around oxygen in the H_2O molecule. (b) Two of the pairs are shared and two are unshared. (c) The H_2O molecule is bent.

Let's summarize the VSEPR model. To determine the molecular shape for a substance, follow these steps:

Step 1 Draw the Lewis structure for the molecule.

Step 2 Count the electron pairs and arrange them to minimize repulsions (as far apart as possible). This determines the electron pair arrangement.

Step 3 Determine the positions of the atoms.

Step 4 Name the molecular structure from the position of the atoms.

It is important to recognize that the placement of the electron pairs determines the structure but the name of the molecular structure is determined by the position of the atoms. Table 11.6 shows the results of this process. Note that when the number of electron pairs is the same as the number of atoms, the electron pair arrangement and the molecular structure are the same. But when the number of atoms and the number of electron pairs are not the same, the molecular structure is different from the electron pair arrangement. This is illustrated when the number of electron pairs is four (a tetrahedral arrangement) in Table 11.6.

Table 11.6 Arrangement of Electron Pairs and Molecular Structure

Number of electron pairs	Electron pair arrangement	Ball-and-stick model	Bonds	Molecular structure	Molecular structure model
2	Linear	180°	2	Linear	
3	Trigonal planar	120°	3	Trigonal planar	
4	Tetrahedral	109.5°	4	Tetrahedral	
4	Tetrahedral	109.5°	3	Trigonal pyramidal	
4	Tetrahedral	109.5°	2	Bent	

Practice 11.8

Predict the shape for CF_4, NF_3, and BeI_2.

Example 11.15

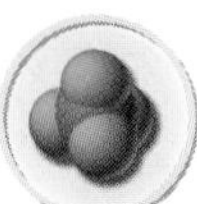

SOLUTION

Predict the molecular shape for these molecules: H_2S, CCl_4, $AlCl_3$.

1. Draw the Lewis structure.
2. Count the electron pairs around the central atom and determine the electron arrangement that will minimize repulsions.
3. Determine the positions of the atoms and name the shape of the molecule.

Molecule	Lewis structure	Number of electron pairs	Electron pair arrangement	Molecular shape
H_2S	H:S̤̈:H	4	tetrahedral	bent
CCl_4	:C̤̈l: :C̤̈l:C̤̈:C̤̈l: :C̤̈l:	4	tetrahedral	tetrahedral
$AlCl_3$	:C̤̈l: :C̤̈l:Äl:C̤̈l:	3	trigonal planar	trigonal planar

Chapter 11 Review

11.1 Periodic Trends in Atomic Properties

KEY TERM

Ionization energy

- Metals and nonmetals
- Atomic radius:
  - Increases down a group
  - Decreases across a row
- Ionization energy:
  - Energy required to remove an electron from an atom
  - Decreases down a group
  - Increases across a row

11.2 Lewis Structures of Atoms

KEY TERM

Lewis structure

- A Lewis structure is a representation of the atom where the symbol represents the element and dots around the symbol represent the valence electrons.
- To determine a Lewis structure for representative elements, use the Group number as the number of electrons to place around the symbol for the element.

11.3 The Ionic Bond: Transfer of Electrons from One Atoms to Another

KEY TERM

Ionic bond

- The goal of bonding is to achieve stability:
  - For representative elements, this stability can be achieved by attaining a valence electron structure of a noble gas.
- In an ionic bond stability is attained by transferring an electron from one atom to another:
  - The atom that loses an electron becomes a cation:
    - Positive ions are smaller than their parent atoms.
    - Metals tend to form cations.

- The atom gaining an electron becomes an anion:
  - Negative ions are larger than their parent atoms.
  - Nonmetals tend to form anions.
- Ionic compounds do not exist as molecules:
  - Ions are attracted by multiple ions of the opposite charge to form a crystalline structure.

11.4 Predicting Formulas of Ionic Compounds

- Chemical compounds are always electrically neutral.
- Metals lose electrons and nonmetals gain electrons to form compounds.
- Stability is achieved (for representative elements) by attaining a noble gas electron configuration.

11.5 The Covalent Bond: Sharing Electrons

KEY TERMS

Covalent bond
Polar covalent bond

- Covalent bonds are formed when two atoms share a pair of electrons between them:
  - This is the predominant type of bonding in compounds.
  - True molecules exist in covalent compounds.
  - Overlap of orbitals forms a covalent bond.
- Unequal sharing of electrons results in a polar covalent bond.

11.6 Electronegativity

KEY TERMS

Electronegativity
Nonpolar covalent bond
Dipole

- Electronegativity is the attractive force an atom has for a shared electrons in a molecule or polyatomic ion.
- Electrons spend more time closer to the more electronegative atom in a bond forming a polar bond.
- The polarity of a bond is determined by the electronegativity difference between the atoms involved in the bond:
  - The greater the difference, the more polar the bond is.
  - At the extremes:
    - Large differences result in ionic bonds.
    - Tiny differences (or no difference) result(s) in a nonpolar covalent bond.
  - A molecule that is electrically asymmetrical has a dipole, resulting in charged areas within the molecule.

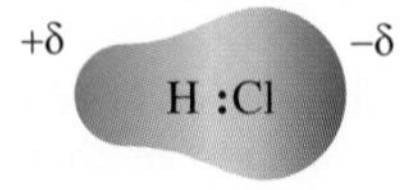

hydrogen chloride

- If the electronegativity difference between two bonded atoms is greater than 1.7–1.9, the bond will be more ionic than covalent.
- Polar bonds do not always result in polar molecules.

11.7 Lewis Structures of Compounds

- To write a Lewis structure for a compound:
  - Add all the valence electrons in all the atoms in the compound. For ions, adjust the number accordingly.
  - Write the skeletal arrangement for the atoms and connect them with single bonds (2 electrons per bond).
  - Subtract the electrons used in the bonds from the total valence electrons.
  - Distribute pairs of electrons around each atom to give each atom a noble gas structure.
  - If there are not enough electrons, convert single bonds between atoms to multiple bonds to attain noble gas structures.

11.8 Complex Lewis Structures

KEY TERM

Resonance structure

- When a single unique Lewis structure cannot be drawn for a molecule, resonance structures (multiple Lewis structures) are used to represent the molecule

11.9 Compounds Containing Polyatomic Ions

- Polyatomic ions behave like a single unit in many chemical reactions.
- The bonds within a polyatomic ion are covalent.

11.10 Molecular Shape

- Lewis structures do not indicate the shape of a molecule.

11.11 The Valence Shell Electron Pair Repulsion (VSEPR) Model

KEY TERMS

Linear structure
Trigonal planar structure
Tetrahedral structure
Bent structure

- VSEPR model:
  - Electron pairs around an atom tend to orient themselves in space as far apart as possible to minimize repulsive forces.
  - The arrangement of electron pairs around an atom determines its structure, but the molecular shape is determined by the position of the atoms in the molecule

Review Questions

All questions with blue numbers have answers in the appendix of the text.

1. Rank these elements according to the radii of their atoms, from smallest to largest: Na, Mg, Cl, K, and Rb. (Figure 11.2)
2. Explain why much more ionization energy is required to remove the first electron from neon than from sodium. (Table 11.1)
3. Explain the large increase in ionization energy needed to remove the third electron from beryllium compared with that needed for the second electron. (Table 11.1)
4. Does the first ionization energy increase or decrease from top to bottom in the periodic table for the alkali metal family? Explain. (Figure 11.3)
5. Does the first ionization energy increase or decrease from top to bottom in the periodic table for the noble gas family? Explain. (Figure 11.3)
6. Why does barium (Ba) have a lower ionization energy than beryllium (Be)? (Figure 11.3)
7. Why is there such a large increase in the ionization energy required to remove the second electron from a sodium atom as opposed to the first? (Table 11.1)
8. Which element in the pair has the larger atomic radius? (Figure 11.2)
   (a) Na or K
   (b) Na or Mg
   (c) O or F
   (d) Br or I
   (e) Ti or Zr
9. In Groups 1A–7A, which element in each group has the smallest atomic radius? (Figure 11.2)
10. Why does the atomic size increase in going down any family of the period table?
11. Why are only valence electrons represented in a Lewis structure?
12. Why do metals tend to lose electrons and nonmetals tend to gain electrons when forming ionic bonds?
13. State whether the elements in each group gain or lose electrons in order to achieve a noble gas configuration. Explain.
    (a) Group 1A
    (b) Group 2A
    (c) Group 6A
    (d) Group 7A
14. What is the difference between a polar bond and a polar molecule?
15. What is the difference between electron pair arrangement and molecular shape?
16. What is the purpose of a Lewis structure?
17. In a polar covalent bond, how do you determine which atom has a partial negative charge ($\delta-$) and which has a partial positive charge ($\delta+$)?
18. In a Lewis structure, what do the dots represent and what do the lines represent?
19. All the atoms within each Group A family of elements can be represented by the same Lewis structure. Complete the following table, expressing the Lewis structure for each group. (Use E to represent the elements.) (Figure 11.4)

Group	1A	2A	3A	4A	5A	6A	7A
	E·						

20. Draw the Lewis structure for Cs, Ba, Tl, Pb, Po, At, and Rn. How do these structures correlate with the group in which each element occurs?
21. In which general areas of the periodic table are the elements with (a) the highest and (b) the lowest electronegativities located?
22. What are valence electrons?
23. Explain why potassium usually forms a K^+ ion, but not a K^{2+} ion.
24. Why does an aluminum ion have a +3 charge?

Paired Exercises

All exercises with blue numbers have answers in the appendix of the text.

1. Which one in each pair has the larger radius? Explain.
   (a) a calcium atom or a calcium ion
   (b) a chlorine atom or a chloride ion
   (c) a magnesium ion or an aluminum ion
   (d) a sodium atom or a silicon atom
   (e) a potassium ion or a bromide ion
2. Which one in each pair has the larger radius? Explain.
   (a) Fe^{2+} or Fe^{3+}
   (b) a potassium atom or a potassium ion
   (c) a sodium ion or a chloride ion
   (d) a strontium atom or an iodine atom
   (e) a rubidium ion or a strontium ion

3. Using the table of electronegativity values (Table 11.5), indicate which element is more positive and which is more negative in these compounds:
(a) H_2O
(b) NaF
(c) NH_3
(d) PbS
(e) NO
(f) CH_4

4. Using the table of electronegativity values (Table 11.5), indicate which element is more positive and which is more negative in these compounds:
(a) HCl
(b) LiH
(c) CCl_4
(d) IBr
(e) MgH_2
(f) OF_2

5. Classify the bond between these pairs of elements as principally ionic or principally covalent (use Table 11.5):
(a) sodium and chlorine
(b) carbon and hydrogen
(c) chlorine and carbon
(d) calcium and oxygen

6. Classify the bond between these pairs of elements as principally ionic or principally covalent (use Table 11.5):
(a) hydrogen and sulfur
(b) barium and oxygen
(c) fluorine and fluorine
(d) potassium and fluorine

7. Explain what happens to the electron structures of Mg and Cl atoms when they react to form $MgCl_2$.

8. Write an equation representing each of the following:
(a) the change of a fluorine atom to a fluoride ion
(b) the change of a calcium atom to a calcium ion

9. Use Lewis structures to show the electron transfer that enables these ionic compounds to be formed:
(a) MgF_2
(b) K_2O

10. Use Lewis structures to show the electron transfer that enables these ionic compounds to be formed:
(a) CaO
(b) NaBr

11. How many valence electrons are in each of these atoms? H, K, Mg, He, Al

12. How many valence electrons are in each of these atoms? Si, N, P, O, Cl

13. How many electrons must be gained or lost for the following to achieve a noble gas electron structure?
(a) a calcium atom
(b) a sulfur atom
(c) a helium atom

14. How many electrons must be gained or lost for the following to achieve a noble gas electron structure?
(a) a chloride ion
(b) a nitrogen atom
(c) a potassium atom

15. Determine whether the following atoms will form an ionic compound or a molecular compound and give the formula of the compound.
(a) sodium and chlorine
(b) carbon and 4 hydrogen
(c) magnesium and bromine
(d) 2 bromine
(e) carbon and 2 oxygen

16. Determine whether each of the following atoms will form a nonpolar covalent compound or a polar covalent compound, and give the formula of the compound.
(a) 2 oxygen
(b) hydrogen and bromine
(c) oxygen and 2 hydrogen
(d) 2 iodine

17. Let E be any representative element. Following the pattern in the table, write formulas for the hydrogen and oxygen compounds of the following:
(a) Na
(b) Ca
(c) Al
(d) Sn

Group 1A	2A	3A	4A	5A	6A	7A
EH E_2O	EH_2 EO	EH_3 E_2O_3	EH_4 EO_2	EH_3 E_2O_5	H_2E EO_3	HE E_2O_7

18. Let E be any representative element. Following the pattern in the table, write formulas for the hydrogen and oxygen compounds of the following:
(a) Sb
(b) Se
(c) Cl
(d) C

Group 1A	2A	3A	4A	5A	6A	7A
EH E_2O	EH_2 EO	EH_3 E_2O_3	EH_4 EO_2	EH_3 E_2O_5	H_2E EO_3	HE E_2O_7

19. The formula for sodium sulfate is Na_2SO_4. Write the names and formulas for the other alkali metal sulfates.

20. The formula for calcium bromide is $CaBr_2$. Write the names and formulas for the other alkaline earth metal bromides.

21. Write Lewis structures for the following:
(a) Na
(b) Br^-
(c) O^{2-}

22. Write Lewis structures for the following:
(a) Ga
(b) Ga^{3+}
(c) Ca^{2+}

23. Classify the bonding in each compound as ionic or covalent:
(a) H_2O
(b) NaCl
(c) MgO
(d) Br_2

24. Classify the bonding in each compound as ionic or covalent:
(a) HCl
(b) $BaCl_2$
(c) NH_3
(d) SO_2

25. Predict the type of bond that would be formed between the following pairs of atoms:
(a) Na and N
(b) N and S
(c) Br and I

26. Predict the type of bond that would be formed between the following pairs of atoms:
(a) H and Si
(b) O and F
(c) Ca and I

27. Draw Lewis structures for the following:
(a) H_2
(b) N_2
(c) Cl_2

28. Draw Lewis structures for the following:
(a) O_2
(b) Br_2
(c) I_2

29. Draw Lewis structures for the following:
(a) NCl_3
(b) H_2CO_3
(c) C_2H_6
(d) $NaNO_3$

30. Draw Lewis structures for the following:
(a) H_2S
(b) CS_2
(c) NH_3
(d) NH_4Cl

31. Draw Lewis structures for the following:
(a) Ba^{2+}
(b) Al^{3+}
(c) SO_3^{2-}
(d) CN^-
(e) HCO_3^-

32. Draw Lewis structures for the following:
(a) I^-
(b) S^{2-}
(c) CO_3^{2-}
(d) ClO_3^-
(e) NO_3^-

33. Classify these molecules as polar or nonpolar:
(a) H_2O
(b) HBr
(c) CF_4

34. Classify these molecules as polar or nonpolar:
(a) F_2
(b) CO_2
(c) NH_3

35. Give the number and arrangement of the electron pairs around the central atom:
(a) C in CCl_4
(b) S in H_2S
(c) Al in AlH_3

36. Give the number and arrangement of the electron pairs around the central atom:
(a) Ga in $GaCl_3$
(b) N in NF_3
(c) Cl in ClO_3^-

37. Use VSEPR theory to predict the structure of these polyatomic ions:
(a) sulfate ion
(b) chlorate ion
(c) periodate ion

38. Use VSEPR theory to predict the structure of these polyatomic ions:
(a) ammonium ion
(b) sulfite ion
(c) phosphate ion

39. Use VSEPR theory to predict the shape of these molecules:
(a) SiH_4
(b) PH_3
(c) SeF_2

40. Use VSEPR theory to predict the shape of these molecules:
(a) SiF_4
(b) OF_2
(c) Cl_2O

41. Element X reacts with sodium to form the compound Na_2X and is in the second period on the periodic table. Identify this element.

42. Element Y reacts with oxygen to form the compound Y_2O and has the lowest ionization energy of any Period 4 element on the periodic table. Identify this element.

Additional Exercises

All exercises with blue numbers have answers in the appendix of the text.

43. Write Lewis structures for hydrazine (N_2H_4) and hydrazoic acid (HN_3).

44. Draw Lewis structures and give the molecular or ionic shape of each of the following compounds:
(a) NO_2^-
(b) SO_4^{2-}
(c) $SOCl_2$
(d) Cl_2O

45. Draw Lewis structures for each of the following compounds:
(a) ethane (C_2H_6)
(b) ethylene (C_2H_4)
(c) acetylene (C_2H_2)

46. Identify the element on the periodic table that satisfies each description:
(a) transition metal with the largest atomic radius
(b) alkaline earth metal with the greatest first ionization energy
(c) least dense member of the nitrogen family
(d) alkali metal with the greatest ratio of neutrons to protons
(e) most electronegative transition metal

47. Choose the element that fits each description:
(a) the lower electronegativity: As or Zn
(b) the lower chemical reactivity: Ba or Be
(c) the fewer valence electrons: N or Ne

48. Identify two reasons why fluorine has a much higher electronegativity than neon.

49. When one electron is removed from an atom of Li, it has two left. Helium atoms also have two electrons. Why is more energy required to remove the second electron from Li than to remove the first from He?

50. Group 1B elements (see the periodic table on the inside cover of your book) have one electron in their outer energy level, as do Group 1A elements. Would you expect them to form compounds such as CuCl, AgCl, and AuCl? Explain.

51. The formula for lead(II) bromide is $PbBr_2$: predict formulas for tin(II) and germanium(II) bromides.

52. Why is it not proper to speak of sodium chloride molecules?

53. What is a covalent bond? How does it differ from an ionic bond?

54. Briefly comment on the structure $Na:\ddot{\underset{..}{O}}:Na$ for the compound Na_2O.

55. What are the four most electronegative elements?

56. Rank these elements from highest electronegativity to lowest: Mg, S, F, H, O, Cs.

57. Is it possible for a molecule to be nonpolar even though it contains polar covalent bonds? Explain.

58. Why is CO_2 a nonpolar molecule, whereas CO is a polar molecule?

59. Estimate the bond angle between atoms in these molecules:
(a) H_2S
(b) NH_3
(c) NH_4^+
(d) $SiCl_4$

60. Consider the two molecules BF_3 and NF_3. Compare and contrast them in terms of the following:
(a) valence-level orbitals on the central atom that are used for bonding
(b) shape of the molecule
(c) number of lone electron pairs on the central atom
(d) type and number of bonds found in the molecule

61. With respect to electronegativity, why is fluorine such an important atom? What combination of atoms on the periodic table results in the most ionic bond?

62. Why does the Lewis structure of each element in a given group of representative elements on the periodic table have the same number of dots?

63. A sample of an air pollutant composed of sulfur and oxygen was found to contain 1.40 g sulfur and 2.10 g oxygen. What is the empirical formula for this compound? Draw a Lewis structure to represent it.

64. A dry-cleaning fluid composed of carbon and chlorine was found to have the composition 14.5% carbon and 85.5% chlorine. Its known molar mass is 166 g/mol. Draw a Lewis structure to represent the compound.

Challenge Exercises

All exercises with blue numbers have answers in the appendix of the text.

*65. Determine whether the following Lewis structures are correct. If they are incorrect, state why and provide the correct Lewis structure.

(a) CO_2 $\quad \ddot{O}-\ddot{C}=\ddot{O}$

(b) ClO_2^- $\quad [:\ddot{O}-\ddot{Cl}-\ddot{O}:]^-$

(c) SF_6 $\quad$ S bonded to six F atoms, each F with three lone pairs

(d) NO_3^- $\quad \left[\ddot{O}=N=\ddot{O} \text{ with } :\ddot{O}: \text{ singly bonded to N}\right]^-$

(e) HCN $\quad H-\ddot{C}=N:$

(f) SO_4^{2-} $\quad \left[:\ddot{O}-S-\ddot{O}: \text{ with two } :O: \text{ doubly bonded to S}\right]^{2-}$

*66. The first ionization energy for lithium is 520 kJ/mol. How much energy would be required to change 25 g of lithium atoms to lithium ions? (Refer to Table 11.1.)

*67. What is the total amount of energy required to remove the first two electrons from 15 moles of sodium atoms? (Refer to Table 11.1.)

Answers to Practice Exercises

11.1 (a) :Ṅ·
(b) :Ȧl
(c) Sr:
(d) :B̈r·

11.2 (a) Ar; K^+
(b) Ar; Mg^{2+}
(c) Ne; Al^{3+}
(d) Xe; Ba^{2+}

11.3 (a) Na_2S
(b) Rb_2O

11.4 (a) Mg_3N_2
(b) Ba_3As_2

11.5 ionic: (a), (d), (e)
covalent: (b), (c), (f)

11.6 (a)
```
 :Br:
 :P:Br:
 :Br:
```
(b)
```
     H
 :Cl:C:Cl:
     :Cl:
```
(c) H:F:

(d)
```
  O:
H:C:H
```
(e) :N:::N:

11.7 (a)
```
H:N:H
  H
```
(b)
```
[H:O:H]+
   H
```
(c)
```
[  H  ]+
[H:N:H]
[  H  ]
```
(d)
```
[     :O:   ]-
[H:O:C::O:  ]
```

11.8 CF_4, tetrahedral; NF_3, pyramidal; BeI_2, linear

PUTTING IT TOGETHER: Review for Chapters 10–11

Multiple Choice:

Choose the correct answer to each of the following.

1. The concept of electrons existing in specific orbits around the nucleus was the contribution of

(a) Thomson (c) Bohr
(b) Rutherford (d) Schrödinger

2. The correct electron structure for a fluorine atom (F) is

(a) $1s^22s^22p^5$ (c) $1s^22s^22p^43s^1$
(b) $1s^22s^22p^23s^23p^1$ (d) $1s^22s^22p^3$

3. The correct electron structure for $_{48}Cd$ is

(a) $1s^22s^22p^63s^23p^64s^23d^{10}$
(b) $1s^22s^22p^63s^23p^64s^23d^{10}4p^65s^24d^{10}$
(c) $1s^22s^22p^63s^23p^64s^23d^{10}4p^64d^4$
(d) $1s^22s^22p^63s^23p^64s^24p^64d^{10}5s^25d^{10}$

4. The correct electron structure of $_{23}V$ is

(a) $[Ar]4s^23d^3$ (c) $[Ar]4s^24d^3$
(b) $[Ar]4s^24p^3$ (d) $[Kr]4s^23d^3$

5. Which of the following is the correct atomic structure for $^{48}_{22}Ti$?

(a)

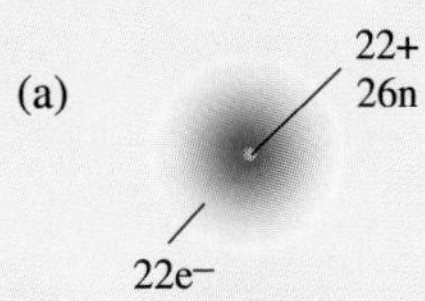

(b)

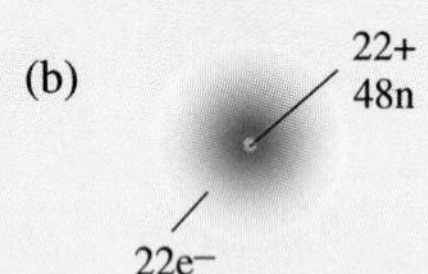

(c)

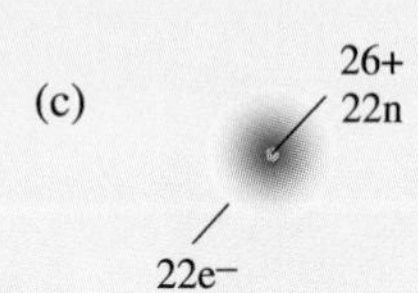

(d)

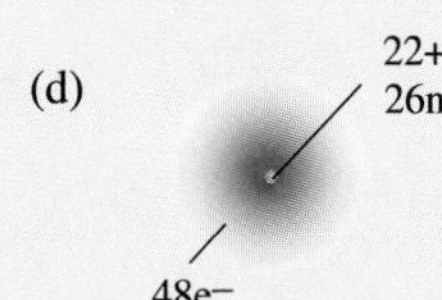

6. The number of orbitals in a *d* sublevel is

(a) 3 (c) 7
(b) 5 (d) no correct answer given

7. The number of electrons in the third principal energy level in an atom having the electron structure $1s^22s^22p^63s^23p^2$ is

(a) 2 (c) 6
(b) 4 (d) 8

8. The total number of orbitals that contain at least one electron in an atom having the structure $1s^22s^22p^63s^23p^2$ is

(a) 5 (c) 14
(b) 8 (d) no correct answer given

9. Which of these elements has two *s* and six *p* electrons in its outer energy level?

(a) He (c) Ar
(b) O (d) no correct answer given

10. Which element is not a noble gas?

(a) Ra (c) He
(b) Xe (d) Ar

11. Which element has the largest number of unpaired electrons?

(a) F (c) Cu
(b) S (d) N

12. How many unpaired electrons are in the electron structure of $_{24}Cr$, $[Ar]4s^13d^5$?

(a) 2 (c) 5
(b) 4 (d) 6

13. Groups 3A–7A plus the noble gases form the area of the periodic table where the electron sublevels being filled are

(a) *p* sublevels (c) *d* sublevels
(b) *s* and *p* sublevels (d) *f* sublevels

14. In moving down an A group on the periodic table, the number of electrons in the outermost energy level

(a) increases regularly
(b) remains constant
(c) decreases regularly
(d) changes in an unpredictable manner

15. Which of the following is an incorrect formula?

(a) NaCl (c) AlO
(b) K_2O (d) BaO

16. Elements of the noble gas family

(a) form no compounds at all
(b) have no valence electrons
(c) have an outer electron structure of ns^2np^6 (helium excepted), where *n* is the period number
(d) no correct answer given

17. The lanthanide and actinide series of elements are

(a) representative elements
(b) transition elements
(c) filling in *d*-level electrons
(d) no correct answer given

18. The element having the structure $1s^22s^22p^63s^23p^2$ is in Group

(a) 2A (c) 4A
(b) 2B (d) 4B

19. In Group 5A, the element having the smallest atomic radius is

(a) Bi (c) As
(b) P (d) N

20. In Group 4A, the most metallic element is

(a) C (c) Ge
(b) Si (d) Sn

21. Which group in the periodic table contains the least reactive elements?

(a) 1A (c) 3A
(b) 2A (d) noble gases

22. Which group in the periodic table contains the alkali metals?

(a) 1A (c) 3A
(b) 2A (d) 4A

23. An atom of fluorine is smaller than an atom of oxygen. One possible explanation is that, compared to oxygen, fluorine has

(a) a larger mass number
(b) a smaller atomic number
(c) a greater nuclear charge
(d) more unpaired electrons

24. If the size of the fluorine atom is compared to the size of the fluoride ion,

(a) they would both be the same size.
(b) the atom is larger than the ion.
(c) the ion is larger than the atom.
(d) the size difference depends on the reaction.

25. Sodium is a very active metal because

(a) it has a low ionization energy.
(b) it has only one outermost electron.
(c) it has a relatively small atomic mass.
(d) all of the above

26. Which of the following formulas is not correct?

(a) Na^+ (c) Al^{3+}
(b) S^- (d) F^-

27. Which of the following molecules does not have a polar covalent bond?

(a) CH_4 (c) CH_3OH
(b) H_2O (d) Cl_2

28. Which of the following molecules is a dipole?

(a) HBr (c) H_2
(b) CH_4 (d) CO_2

29. Which of the following has bonding that is ionic?

(a) H_2 (c) H_2O
(b) MgF_2 (d) CH_4

30. Which of the following is a correct Lewis structure?

(a) :Ö:C:Ö: (with lone pairs below each O) (c) :Cl::Cl: (with lone pairs above and below each Cl)

(b)
```
      :Cl:
  :Cl:C:Cl:
      :Cl:
```
(d) :N:N: (with lone pairs above each N)

31. Which of the following is an incorrect Lewis structure?

(a) H:N:H (with lone pairs above and below N)

(c)
```
   H
 H:C:H
   H
```

(b)
```
 :O:H
  H
```
(d) :N:::N:

32. The correct Lewis structure for SO_2 is

(a) :Ö:S:Ö: (c) :Ö::S::Ö:
(b) :Ö:S::Ö: (d) :Ö:S:Ö:

33. Carbon dioxide (CO_2) is a nonpolar molecule because

(a) oxygen is more electronegative than carbon
(b) the two oxygen atoms are bonded to the carbon atom
(c) the molecule has a linear structure with the carbon atom in the middle
(d) the carbon-oxygen bonds are polar covalent

34. When a magnesium atom participates in a chemical reaction, it is most likely to

(a) lose 1 electron
(b) gain 1 electron
(c) lose 2 electrons
(d) gain 2 electrons

35. If X represents an element of Group 3A, what is the general formula for its oxide?

(a) X_3O_4 (c) XO
(b) X_3O_2 (d) X_2O_3

36. Which of the following has the same electron structure as an argon atom?

(a) Ca^{2+} (c) Na^+
(b) Cl^0 (d) K^0

37. As the difference in electronegativity between two elements decreases, the tendency for the elements to form a covalent bond

(a) increases
(b) decreases
(c) remains the same
(d) sometimes increases and sometimes decreases

38. Which compound forms a tetrahedral molecule?

(a) NaCl (c) CH_4
(b) CO_2 (d) $MgCl_2$

39. Which compound has a bent (V-shaped) molecular structure?

(a) NaCl (c) CH_4
(b) CO_2 (d) H_2O

40. Which compound has double bonds within its molecular structure?

(a) NaCl (c) CH_4
(b) CO_2 (d) H_2O

41. The total number of valence electrons in a nitrate ion, NO_3^- is

(a) 12 (c) 23
(b) 18 (d) 24

42. The number of electrons in a triple bond is

(a) 3 (c) 6
(b) 4 (d) 8

43. The number of unbonded pairs of electrons in H_2O is

(a) 0 (c) 2
(b) 1 (d) 4

44. Which of the following does not have a noble gas electron structure?

(a) Na (c) Ar
(b) Sc^{3+} (d) O^{2-}

Free Response Questions:

Answer each of the following. Be sure to include your work and explanations in a clear, logical form.

1. An alkaline earth metal, M, combines with a halide, X. Will the resulting compound be ionic or covalent? Why? What is the Lewis structure for the compound?

2. "All electrons in atoms with even atomic numbers are paired." Is this statement true or false? Explain your answer using an example.

3. Discuss whether the following statement is true or false: "All nonmetals have two valence electrons in an *s* sublevel with the exception of the noble gases, which have at least one unpaired electron in a *p* sublevel."

4. The first ionization energy (IE) of potassium is lower than the first IE for calcium, but the second IE of calcium is lower than the second IE of potassium. Use an electron configuration or size argument to explain this trend in ionization energies.

5. Chlorine has a very large first ionization energy, yet it forms a chloride ion relatively easily. Explain.

6. Three particles have the same electron configuration. One is a cation of an alkali metal, one is an anion of the halide in the third period, and the third particle is an atom of a noble gas. What are the identities of the three particles (including charges)? Which particle should have the smallest atomic/ionic radius, which should have the largest, and why?

7. Why is the Lewis structure of AlF_3 not written as

```
       ..
      :F:
      /
 ..  /
:F—Al
 ..  \
      \
      :F:
       ..
```

What is the correct Lewis structure and which electrons are shown in a Lewis structure?

8. Why does carbon have a maximum of four covalent bonds?

9. Both NCl_3 and BF_3 have a central atom bonded to three other atoms, yet one is pyramidal and the other is trigonal planar. Explain.

10. Draw the Lewis structure of the atom whose electron configuration is $1s^22s^22p^63s^23p^64s^23d^{10}4p^5$. Would you expect this atom to form an ionic, nonpolar covalent, or polar covalent bond with sulfur?

CHAPTER 12

The Gaseous State of Matter

Properties of gases are transformed to art and sport at the hot-air balloon festival in Albuquerque, New Mexico.

Chapter Outline

Our atmosphere is composed of a mixture of gases, including nitrogen, oxygen, argon, carbon dioxide, ozone, and trace amounts of others. These gases are essential to life, yet they can also create hazards to us. For example, carbon dioxide is valuable when it is taken in by plants and converted to carbohydrates, but it is also associated with the potentially hazardous greenhouse effect. Ozone surrounds the Earth at high altitudes and protects us from harmful ultraviolet rays, but it also destroys rubber and plastics. We require air to live, yet scuba divers must be concerned about oxygen poisoning and the "bends."

In chemistry, the study of the behavior of gases allows us to understand our atmosphere and the effects that gases have on our lives.

12.1 General Properties

A mole of water occupies 18 mL as a liquid but would fill this box (22.4 L) as a gas at the same temperature.

Gases are the least dense and most mobile of the three states of matter. A solid has a rigid structure, and its particles remain in essentially fixed positions. When a solid absorbs sufficient heat, it melts and changes into a liquid. Melting occurs because the molecules (or ions) have absorbed enough energy to break out of the rigid structure of the solid. The molecules or ions in the liquid are more energetic than they were in the solid, as indicated by their increased mobility. Molecules in the liquid state cling to one another. When the liquid absorbs additional heat, the more energetic molecules break away from the liquid surface and go into the gaseous state—the most mobile state of matter. Gas molecules move at very high velocities and have high kinetic energy. The average velocity of hydrogen molecules at 0°C is over 1600 meters (1 mile) per second. Mixtures of gases are uniformly distributed within the container in which they are confined.

The same quantity of a substance occupies a much greater volume as a gas than it does as a liquid or a solid. For example, 1 mol of water (18.02 g) has a volume of 18 mL at 4°C. This same amount of liquid water would occupy about 22,400 mL in the gaseous state—more than a 1200-fold increase in volume. We may assume from this difference in volume that (1) gas molecules are relatively far apart, (2) gases can be greatly compressed, and (3) the volume occupied by a gas is mostly empty space.

12.2 The Kinetic-Molecular Theory

Careful scientific studies of the behavior and properties of gases were begun in the 17th century by Robert Boyle (1627–1691). His work was carried forward by many investigators, and the accumulated data were used in the second half of the 19th century to formulate a general theory to explain the behavior and properties of gases. This theory is called the **kinetic-molecular theory (KMT)**. The KMT has since been extended to cover, in part, the behavior of liquids and solids.

kinetic-molecular theory (KMT)

The KMT is based on the motion of particles, particularly gas molecules. A gas that behaves exactly as outlined by the theory is known as an **ideal gas**. No ideal gases exist, but under certain conditions of temperature and pressure, real gases approach ideal behavior or at least show only small deviations from it.

ideal gas

Under extreme conditions, such as very high pressure and low temperature, real gases deviate greatly from ideal behavior. For example, at low temperature and high pressure many gases become liquids.

The principal assumptions of the KMT are as follows:

1. Gases consist of tiny particles.
2. The distance between particles is large compared with the size of the particles themselves. The volume occupied by a gas consists mostly of empty space.
3. Gas particles have no attraction for one another.
4. Gas particles move in straight lines in all directions, colliding frequently with one another and with the walls of the container.
5. No energy is lost by the collision of a gas particle with another gas particle or with the walls of the container. All collisions are perfectly elastic.
6. The average kinetic energy for particles is the same for all gases at the same temperature, and its value is directly proportional to the Kelvin temperature.

The kinetic energy (KE) of a particle is expressed by the equation

$$\text{KE} = \frac{1}{2}mv^2$$

where m is the mass and v is the velocity of the particle.

All gases have the same kinetic energy at the same temperature. Therefore, from the kinetic energy equation we can see that, if we compare the velocities of the molecules of two gases, the lighter molecules will have a greater velocity than the heavier ones. For example, calculations show that the velocity of a hydrogen molecule is four times the velocity of an oxygen molecule.

Due to their molecular motion, gases have the property of **diffusion**, the ability of two or more gases to mix spontaneously until they form a uniform mixture. The diffusion of gases may be illustrated by the use of the apparatus shown in Figure 12.1. Two large flasks, one containing reddish-brown bromine vapors and the other dry air, are connected by a side tube. When the stopcock between the flasks is opened, the bromine and air will diffuse into each other. After standing awhile, both flasks will contain bromine and air.

diffusion

If we put a pinhole in a balloon, the gas inside will effuse or flow out of the balloon. **Effusion** is a process by which gas molecules pass through a very small orifice (opening) from a container at higher pressure to one at lower pressure.

effusion

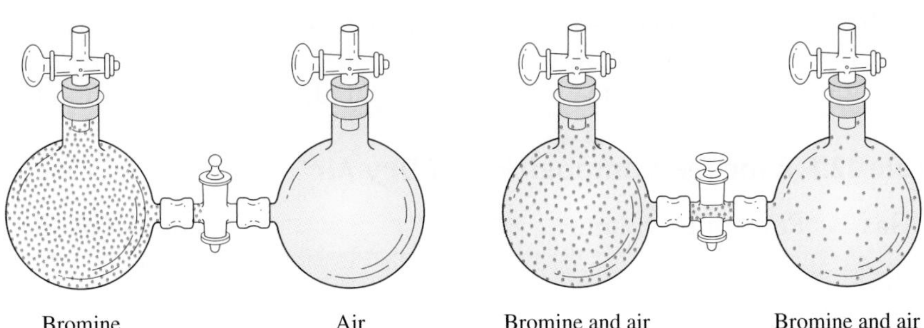

Figure 12.1
Diffusion of gases. When the stopcock between the two flasks is opened, colored bromine molecules can be seen diffusing into the flask containing air.

Thomas Graham (1805–1869), a Scottish chemist, observed that the rate of effusion was dependent on the density of a gas. This observation led to **Graham's law of effusion**.

Graham's law of effusion

The rates of effusion of two gases at the same temperature and pressure are inversely proportional to the square roots of their densities, or molar masses:

$$\frac{\text{rate of effusion of gas A}}{\text{rate of effusion of gas B}} = \sqrt{\frac{dB}{dA}} = \sqrt{\frac{\text{molar mass B}}{\text{molar mass A}}}$$

12.3 Measurement of Pressure of Gases

pressure

Pressure is defined as force per unit area. When a rubber balloon is inflated with air, it stretches and maintains its larger size because the pressure on the inside is greater than that on the outside. Pressure results from the collisions of gas molecules with the walls of the balloon (see Figure 12.2). When the gas is released, the force or pressure of the air escaping from the small neck propels the balloon in a rapid, irregular flight. If the balloon is inflated until it bursts, the gas escaping all at once causes an explosive noise.

The effects of pressure are also observed in the mixture of gases in the atmosphere, which is composed of about 78% nitrogen, 21% oxygen, 1% argon, and other minor constituents by volume (see Table 12.1). The outer boundary of the atmosphere is not known precisely, but more than 99% of the atmosphere is below an altitude of 20 miles (32 km). Thus, the concentration of gas molecules in the atmosphere decreases with altitude, and at about 4 miles the amount of oxygen is insufficient to sustain human life. The gases in the atmosphere exert a pressure known as **atmospheric pressure**. The pressure exerted by a gas depends on the number of molecules of gas present, the temperature, and the volume in which the gas is confined. Gravitational forces hold the atmosphere relatively close to Earth and prevent air molecules from flying off into space. Thus, the atmospheric pressure at any point is due to the mass of the atmosphere pressing downward at that point.

atmospheric pressure

barometer

The pressure of the gases in the atmosphere can be measured with a **barometer**. A mercury barometer may be prepared by completely filling a long tube with pure, dry mercury and inverting the open end into an open dish of mercury. If the tube is longer than 760 mm, the mercury level will drop to a point at which the column of mercury in the tube is just supported by the pressure of the atmosphere. If the tube is properly prepared, a vacuum will exist above

Figure 12.2
The pressure resulting from the collisions of gas molecules with the walls of the balloon keeps the balloon inflated.

Table 12.1 Average Composition of Dry Air

Gas	Percent by volume	Gas	Percent by volume
N_2	78.08	He	0.0005
O_2	20.95	CH_4	0.0002
Ar	0.93	Kr	0.0001
CO_2	0.033	Xe, H_2, and N_2O	Trace
Ne	0.0018		

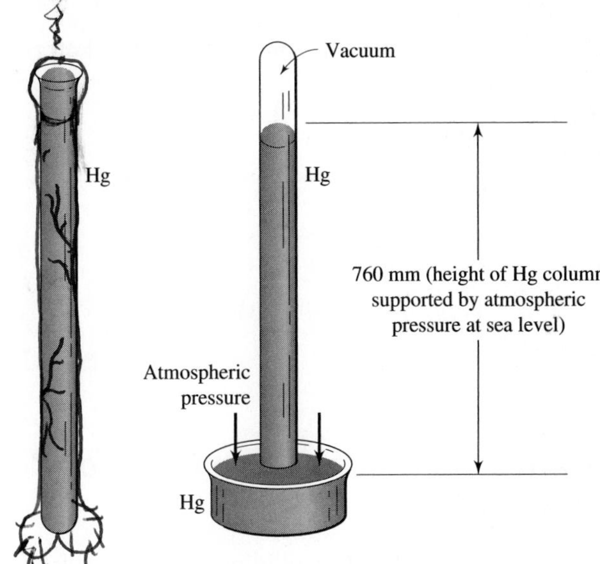

Figure 12.3
Preparation of a mercury barometer. The full tube of mercury at the left is inverted and placed in a dish of mercury.

the mercury column. The weight of mercury, per unit area, is equal to the pressure of the atmosphere. The column of mercury is supported by the pressure of the atmosphere, and the height of the column is a measure of this pressure (see Figure 12.3). The mercury barometer was invented in 1643 by the Italian physicist E. Torricelli (1608–1647), for whom the unit of pressure *torr* was named.

1 atmosphere

Air pressure is measured and expressed in many units. The standard atmospheric pressure, or simply **1 atmosphere** (atm), is the pressure exerted by a column of mercury 760 mm high at a temperature of 0°C. The normal pressure of the atmosphere at sea level is 1 atm, or 760 torr, or 760 mm Hg. The SI unit for pressure is the pascal (Pa), where 1 atm = 101,325 Pa, or 101.3 kPa. Other units for expressing pressure are inches of mercury, centimeters of mercury, the millibar (mbar), and pounds per square inch (lb/in.^{2} or psi). The values of these units equivalent to 1 atm are summarized in Table 12.2.

Atmospheric pressure varies with altitude. The average pressure at Denver, Colorado, 1.61 km (1 mile) above sea level, is 630 torr (0.83 atm). Atmospheric pressure is 0.5 atm at about 5.5 km (3.4 miles) altitude.

Pressure is often measured by reading the heights of mercury columns in millimeters on a barometer. Thus pressure may be recorded as mm Hg (torr). In problems dealing with gases, it is necessary to make interconversions among the various pressure units. Since atm, torr, and mm Hg are common pressure units, we give examples involving all three of these units:

$$1 \text{ atm} = 760 \text{ torr} = 760 \text{ mm Hg}$$

Table 12.2 Pressure Units Equivalent to 1 Atmosphere

1 atm
760 torr
760 mm Hg
76 cm Hg
101.325 kPa
1013 mbar
29.9 in. Hg
14.7 lb/in.^{2}

Example 12.1

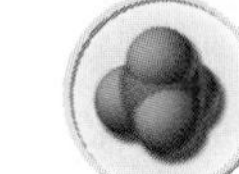

The average atmospheric pressure at Walnut, California, is 740. mm Hg. Calculate this pressure in (a) torr and (b) atmospheres.

SOLUTION

Let's use conversion factors that relate one unit of pressure to another.

(a) To convert mm Hg to torr, use the conversion factor 760 torr/760 mm Hg (1 torr/1 mm Hg):

$$(740. \cancel{\text{mm Hg}})\left(\frac{1 \text{ torr}}{1 \cancel{\text{mm Hg}}}\right) = 740. \text{ torr}$$

(b) To convert mm Hg to atm, use the conversion factor 1 atm/760. mm Hg:

$$(740.\ \cancel{\text{mm Hg}})\left(\frac{1\ \text{atm}}{760.\ \cancel{\text{mm Hg}}}\right) = 0.974\ \text{atm}$$

Practice 12.1

A barometer reads 1.12 atm. Calculate the corresponding pressure in (a) torr and (b) mm Hg.

12.4 Dependence of Pressure on Number of Molecules and Temperature

Pressure is produced by gas molecules colliding with the walls of a container. At a specific temperature and volume the number of collisions depends on the number of gas molecules present. The number of collisions can be increased by increasing the number of gas molecules present. If we double the number of molecules, the frequency of collisions and the pressure should double. We find, for an ideal gas, that this doubling is actually what happens. When the temperature and mass are kept constant, the pressure is directly proportional to the number of moles or molecules of gas present. Figure 12.4 illustrates this concept.

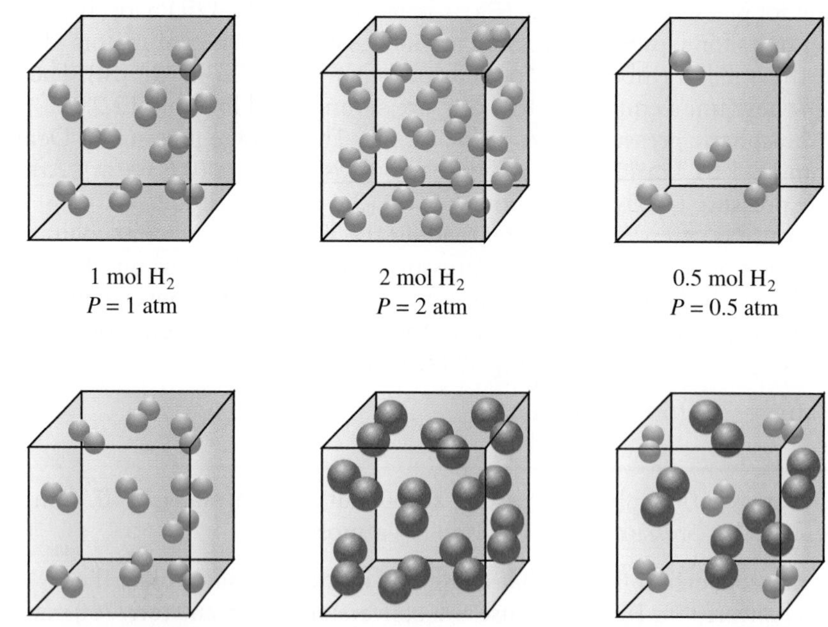

Figure 12.4
The pressure exerted by a gas is directly proportional to the number of molecules present. In each case shown, the volume is 22.4 L and the temperature is 0°C.

A good example of this molecule–pressure relationship may be observed in an ordinary cylinder of compressed gas equipped with a pressure gauge. When the valve is opened, gas escapes from the cylinder. The volume of the cylinder is constant, and the decrease in quantity (moles) of gas is registered by a drop in pressure indicated on the gauge.

The pressure of a gas in a fixed volume also varies with temperature. When the temperature is increased, the kinetic energy of the molecules increases, causing more frequent and more energetic collisions of the molecules with the walls of the container. This increase in collision frequency and energy results in a pressure increase (see Figure 12.5).

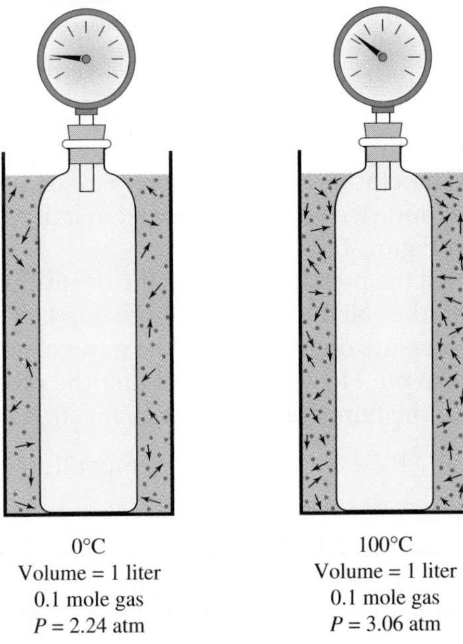

Figure 12.5
The pressure of a gas in a fixed volume increases with increasing temperature. The increased pressure is due to more frequent and more energetic collisions of the gas molecules with the walls of the container at the higher temperature.

12.5 Boyle's Law

Through a series of experiments, Robert Boyle (1627–1691) determined the relationship between the pressure (P) and volume (V) of a particular quantity of a gas. This relationship of P and V is known as **Boyle's law**:

Boyle's law

At constant temperature (T), the volume (V) of a fixed mass of a gas is inversely proportional to the pressure (P), which may be expressed as

$$V \propto \frac{1}{P} \quad \text{or} \quad P_1V_1 = P_2V_2$$

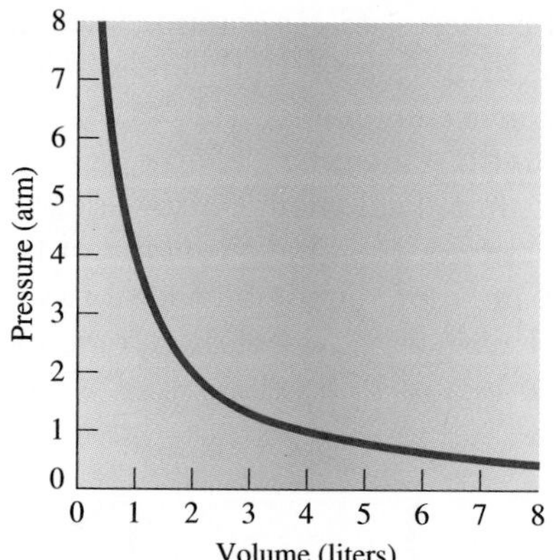

Figure 12.6
Graph of pressure versus volume showing the inverse PV relationship of an ideal gas.

This equation says that the volume varies ($\propto$) inversely with the pressure at constant mass and temperature. When the pressure on a gas is increased, its volume will decrease, and vice versa. The inverse relationship of pressure and volume is graphed in Figure 12.6.

When Boyle doubled the pressure on a specific quantity of a gas, keeping the temperature constant, the volume was reduced to one-half the original volume; when he tripled the pressure on the system, the new volume was one-third the original volume; and so on. His work showed that the product of volume and pressure is constant if the temperature is not changed:

$$PV = \text{constant} \quad \text{or} \quad PV = k \quad \text{(mass and temperature are constant)}$$

Let's demonstrate this law using a cylinder with a movable piston so that the volume of gas inside the cylinder may be varied by changing the external pressure (see Figure 12.7). Assume that the temperature and the number of gas molecules do not change. We start with a volume of 1000 mL and a pressure of 1 atm. When we change the pressure to 2 atm, the gas molecules are crowded closer together, and the volume is reduced to 500 mL. When we increase the pressure to 4 atm, the volume becomes 250 mL.

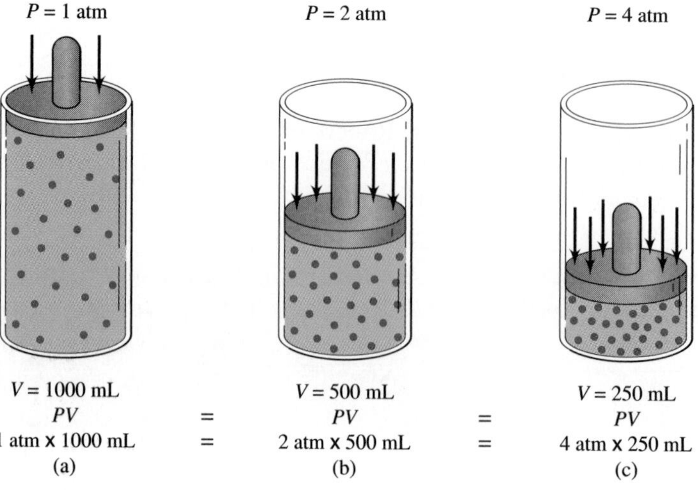

Figure 12.7
The effect of pressure on the volume of a gas.

Note that the product of the pressure times the volume is the same number in each case, substantiating Boyle's law. We may then say that

$$P_1V_1 = P_2V_2$$

where P_1V_1 is the pressure–volume product at one set of conditions, and P_2V_2 is the product at another set of conditions. In each case, the new volume may be calculated by multiplying the starting volume by a ratio of the two pressures involved. Of course, the ratio of pressures used must reflect the direction in which the volume should change. When the pressure is changed from 1 atm to 2 atm, the ratio to be used is 1 atm/2 atm. Now we can verify the results given in Figure 12.7:

1. Starting volume, 1000 mL; pressure change, 1 atm $\longrightarrow$ 2 atm

$$(1000 \text{ mL})\left(\frac{1 \cancel{\text{atm}}}{2 \cancel{\text{atm}}}\right) = 500 \text{ mL}$$

2. Starting volume, 1000 mL; pressure change, 1 atm $\longrightarrow$ 4 atm

$$(1000 \text{ mL})\left(\frac{1 \cancel{\text{atm}}}{4 \cancel{\text{atm}}}\right) = 250 \text{ mL}$$

3. Starting volume, 500 mL; pressure change, 2 atm $\longrightarrow$ 4 atm

$$(500 \text{ mL})\left(\frac{2 \cancel{\text{atm}}}{4 \cancel{\text{atm}}}\right) = 250 \text{ mL}$$

In summary, a change in the volume of a gas due to a change in pressure can be calculated by multiplying the original volume by a ratio of the two pressures. If the pressure is increased, the ratio should have the smaller pressure in the numerator and the larger pressure in the denominator. If the pressure is decreased, the larger pressure should be in the numerator and the smaller pressure in the denominator:

new volume = original volume × ratio of pressures

We use Boyle's law in the following examples. If no mention is made of temperature, assume that it remains constant.

Example 12.2

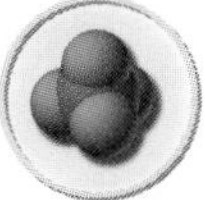

What volume will 2.50 L of a gas occupy if the pressure is changed from 760. mm Hg to 630. mm Hg?

SOLUTION

Method A. Conversion Factors

Step 1 Determine whether pressure is being increased or decreased:

pressure decreases $\longrightarrow$ volume increases

Step 2 Multiply the original volume by a ratio of pressures that will result in an increase in volume:

$$V = (2.50 \text{ L})\left(\frac{760. \cancel{\text{mm Hg}}}{630. \cancel{\text{mm Hg}}}\right) = 3.02 \text{ L (new volume)}$$

Decide which method is the best for you and stick with it.

Method B. Algebraic Equation

Step 1 Organize the given information:

$P_1 = 760.\text{ mm Hg} \qquad V_1 = 2.50\text{ L}$

$P_2 = 630.\text{ mm Hg} \qquad V_2 = ?$

Step 2 Write and solve this equation for the unknown:

$$P_1V_1 = P_2V_2 \qquad V_2 = \frac{P_1V_1}{P_2}$$

Step 3 Put the given information into this equation and calculate:

$$V_2 = \frac{(760.\ \cancel{\text{mm Hg}})(2.50\text{ L})}{630.\ \cancel{\text{mm Hg}}} = 3.02\text{ L}$$

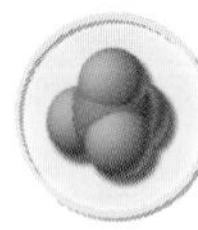

Example 12.3 A given mass of hydrogen occupies 40.0 L at 700. torr. What volume will it occupy at 5.00 atm pressure?

SOLUTION **Method A. Conversion Factors**

Step 1 Determine whether the pressure is being increased or decreased. Note that in order to compare the values, the units must be the same. We'll convert 700. torr to atm:

$$(700.\ \cancel{\text{torr}})\left(\frac{1\text{ atm}}{760\ \cancel{\text{torr}}}\right) = 0.921\text{ atm}$$

The pressure is going from 0.921 atm to 5.00 atm:

pressure increases $\longrightarrow$ volume decreases

Step 2 Multiply the original volume by a ratio of pressures that will result in a decrease in volume:

$$V = (40.0\text{ L})\left(\frac{0.921\ \cancel{\text{atm}}}{5.00\ \cancel{\text{atm}}}\right) = 7.37\text{ L}$$

Method B. Algebraic Equation

Step 1 Organize the given information. Remember to make the pressure units the same:

$P_1 = 700.\text{ torr} = 0.921\text{ atm} \qquad V_1 = 40.0\text{ L}$

$P_2 = 5.00\text{ atm} \qquad V_2 = ?$

Step 2 Write and solve this equation for the unknown:

$$P_1V_1 = P_2V_2 \qquad V_2 = \frac{P_1V_1}{P_2}$$

Step 3 Put the given information into this equation and calculate:

$$V_2 = \frac{(0.921\ \cancel{\text{atm}})(40.0\text{ L})}{5.00\ \cancel{\text{atm}}} = 7.37\text{ L}$$

Example 12.4

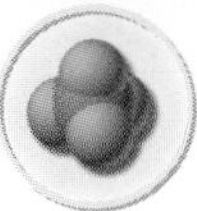

A gas occupies a volume of 200. mL at 400. torr pressure. To what pressure must the gas be subjected in order to change the volume to 75.0 mL?

SOLUTION

Method A. Conversion Factors

Step 1 Determine whether volume is being increased or decreased:

volume decreases ⟶ pressure increases

Step 2 Multiply the original pressure by a ratio of volumes that will result in an increase in pressure:

new pressure = original pressure × ratio of volumes

$$P = (400.\ \text{torr})\left(\frac{200.\ \cancel{\text{mL}}}{75.0\ \cancel{\text{mL}}}\right) = 1067\ \text{torr} \quad \text{or} \quad 1.07 \times 10^3\ \text{torr (new pressure)}$$

Method B. Algebraic Equation

Step 1 Organize the given information (remember to make units the same):

$P_1 = 400.\ \text{torr}$ $V_1 = 200.\ \text{mL}$

$P_2 = ?$ $V_2 = 75.0\ \text{mL}$

Step 2 Write and solve this equation for the unknown:

$$P_1V_1 = P_2V_2 \qquad P_2 = \frac{P_1V_1}{V_2}$$

Step 3 Put the given information into the equation and calculate:

$$P_2 = \frac{P_1V_1}{V_2} = \frac{(400.\ \text{torr})(200.\ \cancel{\text{mL}})}{75.0\ \cancel{\text{mL}}} = 1.07 \times 10^3\ \text{torr}$$

Practice 12.2

A gas occupies a volume of 3.86 L at 0.750 atm. At what pressure will the volume be 4.86 L?

12.6 Charles' Law

The effect of temperature on the volume of a gas was observed in about 1787 by the French physicist J. A. C. Charles (1746–1823). Charles found that various gases expanded by the same fractional amount when they underwent the same change in temperature. Later it was found that if a given volume of any gas initially at 0°C was cooled by 1°C, the volume decreased by $\frac{1}{273}$; if cooled by 2°C, it decreased by $\frac{2}{273}$; if cooled by 20°C, by $\frac{20}{273}$; and so on. Since each degree of cooling reduced the volume by $\frac{1}{273}$, it was apparent that any quantity of any gas would have zero volume if it could be cooled to −273°C. Of course, no real gas can be cooled to −273°C for the simple reason that it would liquefy before that temperature is reached. However, −273°C (more precisely, −273.15°C) is referred to as **absolute zero**; this temperature is the zero point on the Kelvin (absolute) temperature scale—the temperature at which the volume of an ideal, or perfect, gas would become zero.

absolute zero

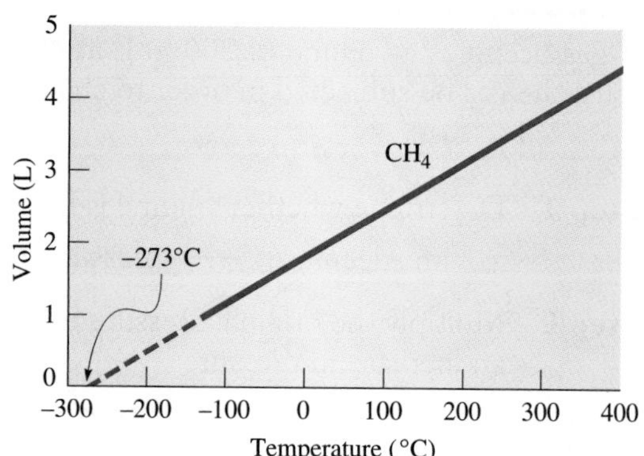

Figure 12.8
Volume–temperature relationship of methane (CH_4). Extrapolated portion of the graph is shown by the broken line.

The volume–temperature relationship for methane is shown graphically in Figure 12.8. Experimental data show the graph to be a straight line that, when extrapolated, crosses the temperature axis at -273.15°C, or absolute zero. This is characteristic for all gases.

Charles' law

In modern form, **Charles' law** is as follows:

> At *constant pressure* the volume of a fixed mass of any gas is directly proportional to the absolute temperature, which may be expressed as
>
> $$V \propto T \quad \text{or} \quad \frac{V_1}{T_1} = \frac{V_2}{T_2}$$

A capital T is usually used for absolute temperature (K) and a small t for °C.

Mathematically this states that the volume of a gas varies directly with the absolute temperature when the pressure remains constant. In equation form, Charles' law may be written as

$$V = kT \quad \text{or} \quad \frac{V}{T} = k \quad \text{(at constant pressure)}$$

where k is a constant for a fixed mass of the gas. If the absolute temperature of a gas is doubled, the volume will double.

To illustrate, let's return to the gas cylinder with the movable or free-floating piston (see Figure 12.9). Assume that the cylinder labeled (a) contains a quantity of gas and the pressure on it is 1 atm. When the gas is heated, the molecules move faster, and their kinetic energy increases. This action should increase the number of collisions per unit of time and therefore increase the pressure. However, the increased internal pressure will cause the piston to rise to a level at which the internal and external pressures again equal 1 atm, as we see in cylinder (b). The net result is an increase in volume due to an increase in temperature.

Another equation relating the volume of a gas at two different temperatures is

$$\frac{V_1}{T_1} = \frac{V_2}{T_2} \quad \text{(constant } P\text{)}$$

where V_1 and T_1 are one set of conditions and V_2 and T_2 are another set of conditions.

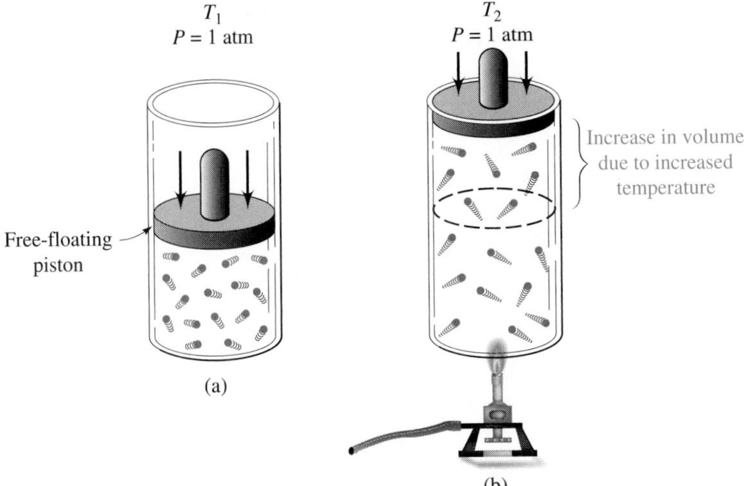

Figure 12.9
The effect of temperature on the volume of a gas. The gas in cylinder (a) is heated from T_1 to T_2. With the external pressure constant at 1 atm, the free-floating piston rises, resulting in an increased volume, shown in cylinder (b).

A simple experiment showing the variation of the volume of a gas with temperature is illustrated in Figure 12.10. A balloon is placed in a beaker and liquid N_2 is poured over it. The volume is reduced, as shown by the collapse of the balloon; when the balloon is removed from the liquid N_2, the gas expands as it warms back to room temperature and the balloon increases in size.

(a)

(b)

(c)

Figure 12.10
The air-filled balloons in (a) are placed in liquid nitrogen (b). The volume of the air decreases tremendously at this temperature. In (c) the balloons are removed from the beaker and are beginning to return to their original volume as they warm back to room temperature.

Example 12.5

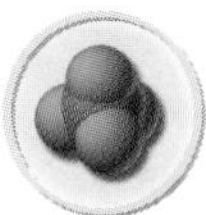

Three liters of hydrogen at $-20.°C$ are allowed to warm to a room temperature of 27°C. What is the volume at room temperature if the pressure remains constant?

SOLUTION

Method A. Conversion Factors

Step 1 Determine whether temperature is being increased or decreased:

$$-20.°C + 273 = 253 \text{ K}$$

$$27°C + 273 = 300. \text{ K}$$

temperature increases $\longrightarrow$ volume increases

Remember: Temperature must be changed to Kelvin in gas law problems. Note that we use 273 to convert instead of 273.15 since our original measurements are to the nearest degree.

Step 2 Multiply the original volume by a ratio of temperatures that will result in an increase in volume:

$$V = (3.00\ \text{L})\left(\frac{300.\ \not{K}}{253\ \not{K}}\right) = 3.56\ \text{L} \quad \text{(new volume)}$$

Method B. Algebraic Equation

Step 1 Organize the given information (remember to make units the same):

$V_1 = 3.00\ \text{L}$ $\quad T_1 = -20.°\text{C} = 253\ \text{K}$

$V_2 = ?$ $\quad T_2 = 27°\text{C} = 300.\ \text{K}$

Step 2 Write and solve the equation for the unknown:

$$\frac{V_1}{T_1} = \frac{V_2}{T_2} \qquad V_2 = \frac{V_1T_2}{T_1}$$

Step 3 Put the given information into the equation and calculate:

$$V_2 = \frac{V_1T_2}{T_1} = \frac{(3.00\ \text{L})(300.\ \not{K})}{253\ \not{K}} = 3.56\ \text{L}$$

Example 12.6 If 20.0 L of oxygen are cooled from 100.°C to 0.°C, what is the new volume?

SOLUTION Since no mention is made of pressure, assume that pressure does not change.

Method A. Conversion Factors

Step 1 Change °C to K:

$100.°\text{C} + 273 = 373\ \text{K}$

$0.°\text{C} + 273 = 273\ \text{K}$

Step 2 The ratio of temperature to be used is 273 K/373 K, because the final volume should be smaller than the original volume. The calculation is

$$V = (20.0\ \text{L})\left(\frac{273\ \not{K}}{373\ \not{K}}\right) = 14.6\ \text{L} \quad \text{(new volume)}$$

Method B. Algebraic Equation

Step 1 Organize the given information (remember to make units coincide):

$V_1 = 20.0\ \text{L}$ $\quad T_1 = 100.°\text{C} = 373\ \text{K}$

$V_2 = ?$ $\quad T_2 = 0.°\text{C} = 273\ \text{K}$

Step 2 Write and solve the equation for the unknown:

$$\frac{V_1}{T_1} = \frac{V_2}{T_2} \qquad V_2 = \frac{V_1T_2}{T_1}$$

Step 3 Put the given information into the equation and calculate:

$$V_2 = \frac{V_1T_2}{T_1} = \frac{(20.0\ \text{L})(273\ \not{K})}{373\ \not{K}} = 14.6\ \text{L}$$

Practice 12.3

A 4.50-L container of nitrogen gas at 28.0°C is heated to 56.0°C. Assuming that the volume of the container can vary, what is the new volume of the gas?

12.7 Gay-Lussac's Law

J. L. Gay-Lussac (1778–1850) was a French chemist involved in the study of volume relationships of gases. The three variables [pressure (P), volume (V), and temperature (T)] are needed to describe a fixed amount of a gas. Boyle's law ($PV = k$) relates pressure and volume at constant temperature; Charles' law ($V = kT$) relates volume and temperature at constant pressure. A third relationship involving pressure and temperature at constant volume is a modification of Charles' law and is sometimes called **Gay-Lussac's law** (Figure 12.11).

Gay-Lussac's law

The pressure of a fixed mass of a gas, at constant volume, is directly proportional to the Kelvin temperature:

$$P = kT \quad \text{or} \quad \frac{P_1}{T_1} = \frac{P_2}{T_2}$$

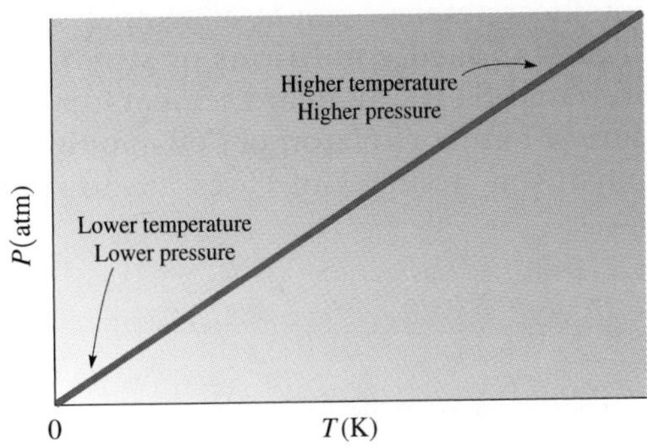

Figure 12.11
Gay-Lussac's law shows that pressure and temperature are directly proportional.

Example 12.7

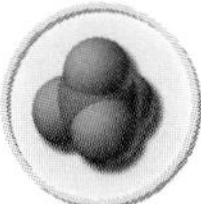

The pressure of a container of helium is 650. torr at 25°C. If the sealed container is cooled to 0.°C, what will the pressure be?

SOLUTION

Method A. Conversion Factors

Step 1 Determine whether temperature is being increased or decreased:

temperature decreases ⟶ pressure decreases

Step 2 Multiply the original pressure by a ratio of Kelvin temperatures that will result in a decrease in pressure:

$$(650.\ \text{torr})\left(\frac{273\ \cancel{K}}{298\ \cancel{K}}\right) = 595\ \text{torr}$$

Method B. Algebraic Equation

Step 1 Organize the given information (remember to make units the same):

$$P_1 = 650.\ \text{torr} \qquad T_1 = 25°\text{C} = 298\ \text{K}$$
$$P_2 = ? \qquad T_2 = 0.°\text{C} = 273\ \text{K}$$

Step 2 Write and solve equation for the unknown:

$$\frac{P_1}{T_1} = \frac{P_2}{T_2} \qquad P_2 = \frac{P_1 T_2}{T_1}$$

Step 3 Put given information into equation and calculate:

$$P_2 = \frac{(650.\ \text{torr})(273\ \cancel{\text{K}})}{298\ \cancel{\text{K}}} = 595\ \text{torr}$$

Practice 12.4

A gas cylinder contains 40.0 L of gas at 45.0°C and has a pressure of 650. torr. What will the pressure be if the temperature is changed to 100.°C?

12.8 Combined Gas Laws

standard conditions
standard temperature and pressure (STP)

To compare volumes of gases, common reference points of temperature and pressure were selected and called **standard conditions** or **standard temperature and pressure** (abbreviated **STP**). Standard temperature is 273.15 K (0°C), and standard pressure is 1 atm or 760 torr or 760 mm Hg or 101.325 kPa. For purposes of comparison, volumes of gases are usually changed to STP conditions:

In this text we'll use 273 K for temperature conversions and calculations. Check with your instructor for rules in your class.

standard temperature = 273.15 K or 0.00°C
standard pressure = 1 atm or 760 torr or 760 mm Hg or 101.325 kPa

When temperature and pressure change at the same time, the new volume may be calculated by multiplying the initial volume by the correct ratios of both pressure and temperature as follows:

$$\text{final volume} = (\text{initial volume})\begin{pmatrix}\text{ratio of}\\ \text{pressures}\end{pmatrix}\begin{pmatrix}\text{ratio of}\\ \text{temperatures}\end{pmatrix}$$

This equation combines Boyle's and Charles' laws, and the same considerations for the pressure and temperature ratios should be used in the calculation. The four possible variations are as follows:

1. Both T and P cause an increase in volume.
2. Both T and P cause a decrease in volume.
3. T causes an increase and P causes a decrease in volume.
4. T causes a decrease and P causes an increase in volume.

The P, V, and T relationships for a given mass of any gas, in fact, may be expressed as a single equation, $PV/T = k$. For problem solving, this equation is usually written

$$\frac{P_1V_1}{T_1} = \frac{P_2V_2}{T_2}$$

where P_1, V_1, and T_1 are the initial conditions and P_2, V_2, and T_2 are the final conditions.

This equation can be solved for any one of the six variables and is useful in dealing with the pressure–volume–temperature relationships of gases. Note what happens to the combined gas law when one of the variables is constant:

- T constant $\rightarrow P_1V_1 = P_2V_2$ Boyle's law
- P constant $\rightarrow \frac{V_1}{T_1} = \frac{V_2}{T_2}$ Charles' law
- V constant $\rightarrow \frac{P_1}{T_1} = \frac{P_2}{T_2}$ Gay-Lussac's law

Note: In the examples below, the use of 273 K does not change the number of significant figures in the temperature. The converted temperature is expressed to the same precision as the original measurement.

Example 12.8

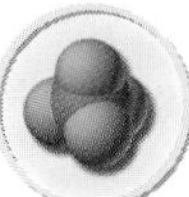

Given 20.0 L of ammonia gas at 5°C and 730. torr, calculate the volume at 50.°C and 800. torr.

SOLUTION

Step 1 Organize the given information, putting temperatures in kelvins:

$P_1 = 730.\ \text{torr}$ $P_2 = 800.\ \text{torr}$

$V_1 = 20.0\ \text{L}$ $V_2 = ?$

$T_1 = 5°\text{C} = 278\ \text{K}$ $T_2 = 50.°\text{C} = 323\ \text{K}$

Method A. Conversion Factors

Step 2 Set up ratios of T and P:

$$T\ \text{ratio} = \frac{323\ \text{K}}{278\ \text{K}} \quad \text{(increase in } T \text{ should increase } V\text{)}$$

$$P\ \text{ratio} = \frac{730.\ \text{torr}}{800.\ \text{torr}} \quad \text{(increase in } P \text{ should decrease } V\text{)}$$

Step 3 Multiply the original volume by the ratios:

$$V_2 = (20.0\ \text{L})\left(\frac{730.\ \cancel{\text{torr}}}{800.\ \cancel{\text{torr}}}\right)\left(\frac{323\ \cancel{\text{K}}}{278\ \cancel{\text{K}}}\right) = 21.2\ \text{L}$$

Method B. Algebraic Equation

Step 2 Write and solve the equation for the unknown. Solve

$$\frac{P_1V_1}{T_1} = \frac{P_2V_2}{T_2}$$

for V_2 by multiplying both sides of the equation by T_2/P_2 and rearranging to obtain

$$V_2 = \frac{V_1P_1T_2}{P_2T_1}$$

Step 3 Put the given information into the equation and calculate:

$$V_2 = \frac{(20.0\ \text{L})(730.\ \cancel{\text{torr}})(323\ \cancel{\text{K}})}{(800.\ \cancel{\text{torr}})(278\ \cancel{\text{K}})} = 21.2\ \text{L}$$

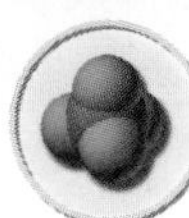

Example 12.9 To what temperature (°C) must 10.0 L of nitrogen at 25°C and 700. torr be heated in order to have a volume of 15.0 L and a pressure of 760. torr?

SOLUTION **Step 1** Organize the given information, putting temperatures in kelvins:

$P_1 = 700.\text{ torr}$ $\quad P_2 = 760.\text{ torr}$

$V_1 = 10.0\text{ L}$ $\quad V_2 = 15.0\text{ L}$

$T_1 = 25°\text{C} = 298\text{ K}$ $\quad T_2 = ?$

Method A. Conversion Factors

Step 2 Set up ratios of V and P:

$$P\text{ ratio} = \frac{760.\text{ torr}}{700.\text{ torr}} \quad \text{(increase in } P \text{ should increase } T\text{)}$$

$$V\text{ ratio} = \frac{15.0\text{ L}}{10.0\text{ L}} \quad \text{(increase in } V \text{ should increase } T\text{)}$$

Step 3 Multiply the original temperature by the ratios:

$$T_2 = (298\text{ K})\left(\frac{760.\text{ torr}}{700.\text{ torr}}\right)\left(\frac{15.0\text{ L}}{10.0\text{ L}}\right) = 485\text{ K}$$

Method B. Algebraic Equation

Step 2 Write and solve the equation for the unknown:

$$\frac{P_1V_1}{T_1} = \frac{P_2V_2}{T_2} \qquad T_2 = \frac{T_1P_2V_2}{P_1V_1}$$

Step 3 Put the given information into the equation and calculate:

$$T_2 = \frac{(298\text{ K})(760.\text{ torr})(15.0\text{ L})}{(700.\text{ torr})(10.0\text{ L})} = 485\text{ K}$$

In either method, since the problem asks for °C, we must subtract 273 from the Kelvin answer:

$$485\text{ K} - 273 = 212°\text{C}$$

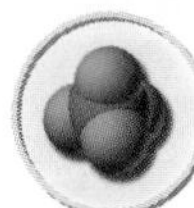

Example 12.10 The volume of a gas-filled balloon is 50.0 L at 20.°C and 742 torr. What volume will it occupy at standard temperature and pressure (STP)?

SOLUTION **Step 1** Organize the given information, putting temperatures in kelvins:

$P_1 = 742\text{ torr}$ $\quad P_2 = 760.\text{ torr}$ (standard pressure)

$V_1 = 50.0\text{ L}$ $\quad V_2 = ?$

$T_1 = 20.°\text{C} = 293\text{ K}$ $\quad T_2 = 273\text{ K}$ (standard temperature)

Method A. Conversion Factors

Step 2 Set up ratios of T and P:

$$T\text{ ratio} = \frac{273\text{ K}}{293\text{ K}} \quad \text{(decrease in } T \text{ should decrease } V\text{)}$$

$$P\text{ ratio} = \frac{742\text{ torr}}{760.\text{ torr}} \quad \text{(increase in } P \text{ should decrease } V\text{)}$$

Step 3 Multiply the original volume by the ratios:

$$V_2 = (50.0\ \text{L})\left(\frac{273\ \cancel{\text{K}}}{293\ \cancel{\text{K}}}\right)\left(\frac{742\ \cancel{\text{torr}}}{760.\ \cancel{\text{torr}}}\right) = 45.5\ \text{L}$$

Method B. Algebraic Equation

Step 2 Write and solve the equation for the unknown:

$$\frac{P_1V_1}{T_1} = \frac{P_2V_2}{T_2} \qquad V_2 = \frac{P_1V_1T_2}{P_2T_1}$$

Step 3 Put the given information into the equation and calculate:

$$V_2 = \frac{(742\ \cancel{\text{torr}})(50.0\ \text{L})(273\ \cancel{\text{K}})}{(760.\ \cancel{\text{torr}})(293\ \cancel{\text{K}})} = 45.5\ \text{L}$$

Practice 12.5

15.00 L of gas at 45.0°C and 800. torr are heated to 400.°C and the pressure changed to 300. torr. What is the new volume?

Practice 12.6

To what temperature must 5.00 L of oxygen at 50.°C and 600. torr be heated in order to have a volume of 10.0 L and a pressure of 800. torr?

12.9 Dalton's Law of Partial Pressures

If gases behave according to the kinetic-molecular theory, there should be no difference in the pressure–volume–temperature relationships whether the gas molecules are all the same or different. This similarity in the behavior of gases is the basis for an understanding of **Dalton's law of partial pressures**:

Dalton's law of partial pressures

The total pressure of a mixture of gases is the sum of the partial pressures exerted by each of the gases in the mixture.

Each gas in the mixture exerts a pressure that is independent of the other gases present. These pressures are called **partial pressures**. Thus, if we have a mixture of three gases, *A*, *B*, and *C*, exerting partial pressures of 50. torr, 150. torr, and 400. torr, respectively, the total pressure will be 600. torr:

partial pressure

$$P_{\text{Total}} = P_A + P_B + P_C$$

$$P_{\text{Total}} = 50.\ \text{torr} + 150.\ \text{torr} + 400.\ \text{torr} = 600.\ \text{torr}$$

We can see an application of Dalton's law in the collection of insoluble gases over water. When prepared in the laboratory, oxygen is commonly collected by the downward displacement of water. Thus the oxygen is not pure but is mixed with water vapor (see Figure 12.12). When the water levels are adjusted to the same height inside and outside the bottle, the pressure of the oxygen plus water vapor inside the bottle is equal to the atmospheric pressure:

$$P_{\text{atm}} = P_{O_2} + P_{H_2O}$$

CHEMISTRY IN ACTION • Messenger Molecules

Traditional "messenger" molecules are amino acids (also known as the building blocks for proteins). You've probably heard of endorphins, which are the messenger molecules associated with "runner's high." Until recently, these molecules (called *neurotransmitters*) were thought to be specific; that is, each neurotransmitter fits into the target cell like a key in a lock. It was also thought that neurotransmitters were stored in tiny pouches where they are manufactured and then released when needed. But then two gases, nitrogen monoxide (NO) and carbon monoxide (CO), were found to act as neurotransmitters. These gases break all the "rules" for neurotransmission because, as gases, they are nonspecific. Gases freely diffuse into nearby cells, so NO and CO neurotransmitters must be made "on demand," since they cannot be stored. Gas neurotransmitters cannot use the lock-and-key model to act on cells, so they must use their chemical properties.

Nitrogen monoxide has been used for nearly a century to dilate blood vessels and increase blood flow, lowering blood pressure. In the late 1980s, scientists discovered that biologically produced NO is an important signaling molecule for nerve cells. Nitrogen monoxide mediates certain neurons that do not respond to traditional neurotransmitters. These NO-sensitive neurons are found in the cardiovascular, respiratory, digestive, and urogenital systems. Nitrogen monoxide also appears to play a role in regulating blood pressure, blood clotting, and neurotransmission. The 1998 Nobel Prize in medicine was awarded to three Americans for their work on nitrogen monoxide's role in the body.

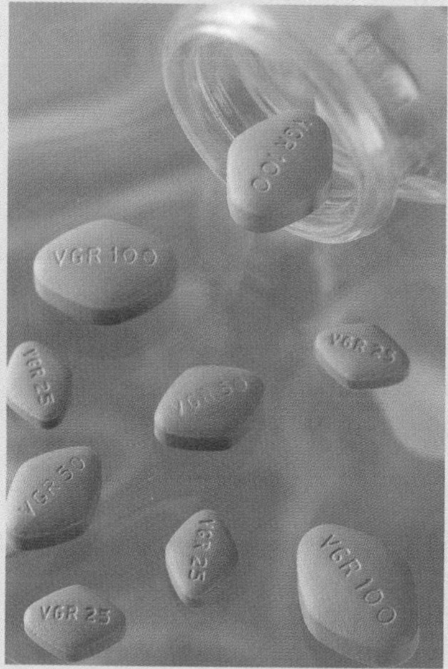

Viagra, the medication for male impotence, is the latest application of our understanding of the role of NO in the body.

Nitrogen monoxide is a key molecule in producing erections in men. When a man is sexually stimulated, NO is released into the penis, where it activates the release of an enzyme that increases the level of a molecule called cGMP and ultimately produces relaxation of smooth muscles, allowing blood to flow into the penis. Viagra, an impotence pill, enhances the effect of NO, inhibiting the enzyme that breaks down the cGMP molecule. This results in more cGMP molecules, and smooth muscle relaxation begins producing an erection.

After NO transmitters were discovered, researchers at Johns Hopkins Medical School reasoned that if one gas acted as a neurotransmitter, so might others. They proposed CO as a possible transmitter because the enzyme used to make it is localized in specific parts of the brain (those responsible for smell and long-term memory). The enzyme used to make messengers was found in exactly the same locations! Researchers then showed that nerve cells made the messenger molecule when stimulated with CO. An inhibitor for CO blocked the messenger molecule production. Researchers are exploring the possibility of still other roles for these gaseous messenger molecules.

To determine the amount of O_2 or any other gas collected over water, we subtract the pressure of the water vapor (vapor pressure) from the total pressure of the gases:

$$P_{O_2} = P_{atm} - P_{H_2O}$$

The vapor pressure of water at various temperatures is tabulated in Appendix IV.

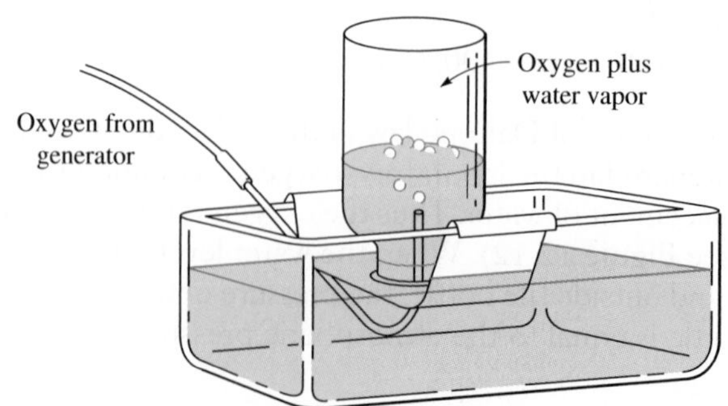

Figure 12.12
Oxygen collected over water.

Example 12.11 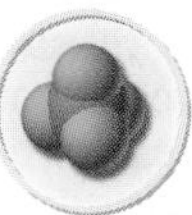

A 500.-mL sample of oxygen was collected over water at 23°C and 760. torr. What volume will the dry O_2 occupy at 23°C and 760. torr? The vapor pressure of water at 23°C is 21.2 torr.

SOLUTION

To solve this problem, we must first determine the pressure of the oxygen alone, by subtracting the pressure of the water vapor (given in Appendix IV) present:

Step 1 Determine the pressure of dry O_2:

$$P_{\text{Total}} = 760.\ \text{torr} = P_{O_2} + P_{H_2O}$$
$$P_{O_2} = 760.\ \text{torr} - 21.2\ \text{torr} = 739\ \text{torr} \qquad (\text{dry } O_2)$$

Step 2 Organize the given information:

$P_1 = 739$ torr $\quad P_2 = 760.$ torr

$V_1 = 500.$ mL $\quad V_2 = ?$

T is constant

Step 3 Solve as a Boyle's law problem:

$$V = \frac{(500.\ \text{mL})(739\ \cancel{\text{torr}})}{760.\ \cancel{\text{torr}}} = 486\ \text{mL dry } O_2$$

Practice 12.7

Hydrogen gas was collected by downward displacement of water. A volume of 600.0 mL of gas was collected at 25.0°C and 740.0 torr. What volume will the dry hydrogen occupy at STP?

12.10 Avogadro's Law

Early in the 19th century, Gay-Lussac studied the volume relationships of reacting gases. His results, published in 1809, were summarized in a statement known as **Gay-Lussac's law of combining volumes**:

Gay-Lussac's law of combining volumes

When measured at the same temperature and pressure, the ratios of the volumes of reacting gases are small whole numbers.

Thus, H_2 and O_2 combine to form water vapor in a volume ratio of 2:1 (Figure 12.13); H_2 and Cl_2 react to form HCl in a volume ratio of 1:1; and H_2 and N_2 react to form NH_3 in a volume ratio of 3:1.

Two years later, in 1811, Amedeo Avogadro (1776–1856) used the law of combining volumes of gases to make a simple but significant and far-reaching generalization concerning gases. **Avogadro's law** states:

Avogadro's law

Equal volumes of different gases at the same temperature and pressure contain the same number of molecules.

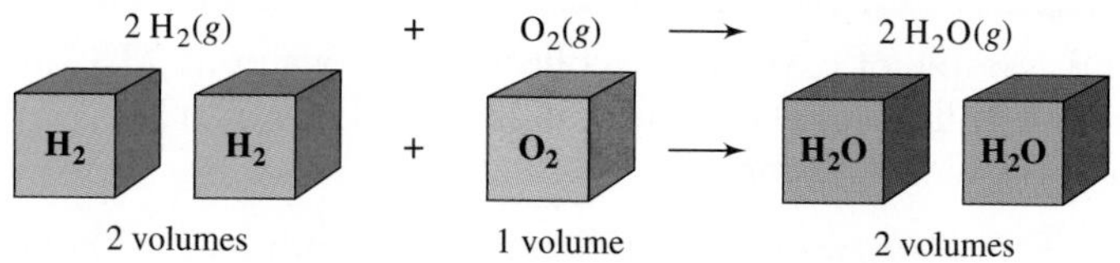

Figure 12.13
Gay-Lussac's law of combining volumes of gases applied to the reaction of hydrogen and oxygen. When measured at the same temperature and pressure, hydrogen and oxygen react in a volume ratio of 2:1.

This law was a real breakthrough in understanding the nature of gases.

1. It offered a rational explanation of Gay-Lussac's law of combining volumes of gases and indicated the diatomic nature of such elemental gases as hydrogen, chlorine, and oxygen.
2. It provided a method for determining the molar masses of gases and for comparing the densities of gases of known molar mass (see Sections 12.11 and 12.12).
3. It afforded a firm foundation for the development of the kinetic-molecular theory.

By Avogadro's law, equal volumes of hydrogen and chlorine at the same temperature and pressure contain the same number of molecules. On a volume basis, hydrogen and chlorine react thus (see Figure 12.14):

hydrogen	+	chlorine	⟶	hydrogen chloride
1 volume		1 volume		2 volumes

Therefore, hydrogen molecules react with chlorine molecules in a 1:1 ratio. Since two volumes of hydrogen chloride are produced, one molecule of hydrogen and one molecule of chlorine must produce two molecules of hydrogen chloride. Therefore, each hydrogen molecule and each chlorine molecule must consist of two atoms. The coefficients of the balanced equation for the reaction give the correct ratios for volumes, molecules, and moles of reactants and products:

H_2	+	Cl_2	$\longrightarrow$	$2\,HCl$
1 volume		1 volume		2 volumes
1 molecule		1 molecule		2 molecules
1 mol		1 mol		2 mol

By like reasoning, oxygen molecules also must contain at least two atoms because one volume of oxygen reacts with two volumes of hydrogen to produce two volumes of water vapor.

The volume of a gas depends on the temperature, the pressure, and the number of gas molecules. Different gases at the same temperature have the same average kinetic energy. Hence, if two different gases are at the same temperature, occupy equal volumes, and exhibit equal pressures, each gas must contain the same number of molecules. This statement is true because systems with identical *PVT* properties can be produced only by equal numbers of molecules having the same average kinetic energy.

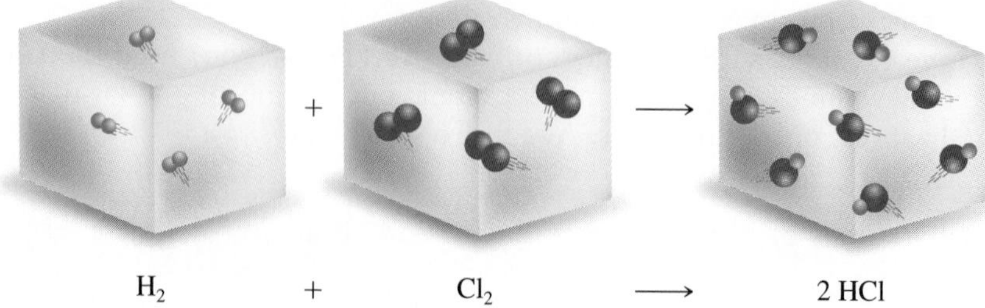

Figure 12.14
Avogadro's law proved the concept of diatomic molecules for hydrogen and chlorine.

12.11 Mole–Mass–Volume Relationships of Gases

Because a mole contains 6.022×10^{23} molecules (Avogadro's number), a mole of any gas will have the same volume as a mole of any other gas at the same temperature and pressure. It has been experimentally determined that the volume occupied by a mole of any gas is 22.4 L at STP. This volume, 22.4 L, is known as the **molar volume** of a gas. The molar volume is a cube about 28.2 cm (11.1 in.) on a side. The molar masses of several gases, each occupying 22.4 L at STP, are shown in Figure 12.15:

molar volume

As with many constants, the molar volume is known more exactly to be 22.414 L. We use 22.4 L in our calculations, since the extra figures don't often affect the result, given the other measurements in the calculation.

> One mole of a gas occupies 22.4 L at STP.

The molar volume is useful for determining the molar mass of a gas or of substances that can be easily vaporized. If the mass and the volume of a gas at STP are known, we can calculate its molar mass. For example, 1.00 L of pure oxygen at STP has a mass of 1.429 g. The molar mass of oxygen may be calculated by multiplying the mass of 1.00 L by 22.4 L/mol:

$$\left(\frac{1.429\text{ g}}{1.00\text{ L}}\right)\left(\frac{22.4\text{ L}}{1\text{ mol}}\right) = 32.0\text{ g/mol} \qquad \text{(molar mass)}$$

If the mass and volume are at other than standard conditions, we change the volume to STP and then calculate the molar mass.

Standard conditions apply only to pressure, temperature, and volume. Mass is not affected.

The molar volume, 22.4 L/mol, is used as a conversion factor to convert grams per liter to grams per mole (molar mass) and also to convert liters to moles. The two conversion factors are

$$\frac{22.4\text{ L}}{1\text{ mol}} \quad \text{and} \quad \frac{1\text{ mol}}{22.4\text{ L}}$$

These conversions must be done at STP except under certain special circumstances.

131.3 g Xe
71.90 g Cl_2
16.04 g CH_4
16.00 g O_2
36.46 g HCl
44.01 g CO_2
22.4 L
28.2 cm
28.2 cm
28.2 cm
64.07 g SO_2
28.02 g N_2
2.016 g H_2
17.03 g NH_3
39.95 g Ar
34.09 g H_2S

Figure 12.15
One mole of a gas occupies 22.4 L at STP. The mass given for each gas is the mass of 1 mol.

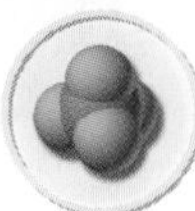

Example 12.12 If 2.00 L of a gas measured at STP have a mass of 3.23 g, what is the molar mass of the gas?

SOLUTION The unit of molar mass is g/mol; the conversion is from

$$\frac{\text{g}}{\text{L}} \longrightarrow \frac{\text{g}}{\text{mol}}$$

The starting amount is $\frac{3.23\text{ g}}{2.00\text{ L}}$. The conversion factor is $\frac{22.4\text{ L}}{1\text{ mol}}$. The calculation is

$$\left(\frac{3.23\text{ g}}{2.00\text{ L}}\right)\left(\frac{22.4\text{ L}}{1\text{ mol}}\right) = 36.2\text{ g/mol} \qquad \text{(molar mass)}$$

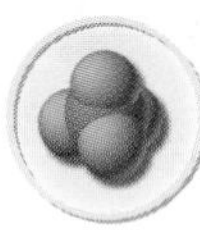

Example 12.13 Measured at 40.°C and 630. torr, the mass of 691 mL of diethyl ether is 1.65 g. Calculate the molar mass of diethyl ether.

Diethyl ether
$(C_2H_5)_2O$

SOLUTION **Step 1** Organize the given information, converting temperatures to kelvin. Note that we must change to STP in order to determine molar mass.

$P_1 = 630.\text{ torr}$ $\qquad P_2 = 760.\text{ torr}$
$V_1 = 691\text{ mL}$ $\qquad V_2 = ?$
$T_1 = 313\text{ K } (40.°\text{C})$ $\qquad T_2 = 273\text{ K}$

Step 2 Use either the conversion factor method or the algebraic method and the combined gas law to correct the volume (V_2) to STP:

$$V_2 = \frac{(691\text{ mL})(273\text{ K})(630.\text{ torr})}{(313\text{ K})(760.\text{ torr})} = 500.\text{ mL} = 0.500\text{ L} \quad \text{(at STP)}$$

Step 3 In the example, V_2 is the volume for 1.65 g of the gas, so we can now find the molar mass by converting g/L to g/mol:

$$\left(\frac{1.65\text{ g}}{0.500\text{ L}}\right)\left(\frac{22.4\text{ L}}{\text{mol}}\right) = 73.9\text{ g/mol}$$

Practice 12.8

A gas with a mass of 86 g occupies 5.00 L at 25°C and 3.00 atm pressure. What is the molar mass of the gas?

12.12 Density of Gases

The density, d, of a gas is its mass per unit volume, which is generally expressed in grams per liter as follows:

$$d = \frac{\text{mass}}{\text{volume}} = \frac{\text{g}}{\text{L}}$$

Because the volume of a gas depends on temperature and pressure, both should be given when stating the density of a gas. The volume of a solid or liquid is hardly affected by changes in pressure and is changed only slightly when the temperature is varied. Increasing the temperature from 0°C to 50°C will reduce the density of a gas by about 18% if the gas is allowed to expand, whereas a 50°C rise in the temperature of water (0°C $\longrightarrow$ 50°C) will change its density by less than 0.2%.

The density of a gas at any temperature and pressure can be determined by calculating the mass of gas present in 1 L. At STP, in particular, the density can be calculated by multiplying the molar mass of the gas by 1 mol/22.4 L:

$$d_{\text{STP}} = \text{molar mass}\left(\frac{1\text{ mol}}{22.4\text{ L}}\right)$$

$$\text{molar mass} = d_{\text{STP}}\left(\frac{22.4\text{ L}}{1\text{ mol}}\right)$$

Table 12.3 lists the densities of some common gases.

Table 12.3 Density of Common Gases at STP

Gas	Molar mass (g/mol)	Density (g/L at STP)	Gas	Molar mass (g/mol)	Density (g/L at STP)
H_2	2.016	0.0900	H_2S	34.09	1.52
CH_4	16.04	0.716	HCl	36.46	1.63
NH_3	17.03	0.760	F_2	38.00	1.70
C_2H_2	26.04	1.16	CO_2	44.01	1.96
HCN	27.03	1.21	C_3H_8	44.09	1.97
CO	28.01	1.25	O_3	48.00	2.14
N_2	28.02	1.25	SO_2	64.07	2.86
air	**(28.9)**	**(1.29)**	Cl_2	70.90	3.17
O_2	32.00	1.43			

Example 12.14

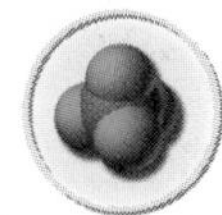

Calculate the density of Cl_2 at STP.

SOLUTION

First calculate the molar mass of Cl_2. It is 70.90 g/mol. Since d = g/L, the conversion is

$$\frac{\text{g}}{\text{mol}} \longrightarrow \frac{\text{g}}{\text{L}}$$

The conversion factor is $\dfrac{1\text{ mol}}{22.4\text{ L}}$:

$$d = \left(\frac{70.90\text{ g}}{1\cancel{\text{ mol}}}\right)\left(\frac{1\cancel{\text{ mol}}}{22.4\text{ L}}\right) = 3.17\text{ g/L}$$

Practice 12.9

The molar mass of a gas is 20. g/mol. Calculate the density of the gas at STP.

12.13 Ideal Gas Law

We've used four variables in calculations involving gases: the volume (V), the pressure (P), the absolute temperature (T), and the number of molecules or moles (n). Combining these variables into a single expression, we obtain

$$V \propto \frac{nT}{P} \qquad \text{or} \qquad V = \frac{nRT}{P}$$

where R is a proportionality constant known as the *ideal gas constant*. The equation is commonly written as

$$PV = nRT$$

ideal gas law

and is known as the **ideal gas law**. This law summarizes in a single expression what we have considered in our earlier discussions. The value and units of R depend on the units of P, V, and T. We can calculate one value of R by taking 1 mol of a gas at STP conditions. Solve the equation for R:

$$R = \frac{PV}{nT} = \frac{(1\text{ atm})(22.4\text{ L})}{(1\text{ mol})(273\text{ K})} = 0.0821\frac{\text{L-atm}}{\text{mol-K}}$$

The units of R in this case are liter-atmospheres (L-atm) per mole kelvin (mol-K). When the value of $R = 0.0821$ L-atm/mol-K, P is in atmospheres, n is in moles, V is in liters, and T is in kelvins.

The ideal gas equation can be used to calculate any one of the four variables when the other three are known.

Example 12.15 What pressure will be exerted by 0.400 mol of a gas in a 5.00-L container at 17°C?

SOLUTION

Step 1 Organize the given information, converting temperature to kelvins:

$P = ?$

$V = 5.00\text{ L}$

$T = 290.\text{ K}$

$n = 0.400\text{ mol}$

Step 2 Write and solve the ideal gas equation for the unknown:

$$PV = nRT \qquad \text{or} \qquad P = \frac{nRT}{V}$$

Step 3 Substitute the given data into the equation and calculate:

$$P = \frac{(0.400\,\cancel{\text{mol}})(0.0821\,\cancel{\text{L}}\cdot\text{atm}/\cancel{\text{mol}\cdot\text{K}})(290.\,\cancel{\text{K}})}{5.00\,\cancel{\text{L}}} = 1.90\text{ atm}$$

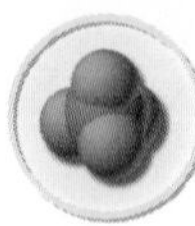

Example 12.16 How many moles of oxygen gas are in a 50.0-L tank at 22.0°C if the pressure gauge reads 2000. lb/in.^{2}?

SOLUTION

Step 1 Organize the given information, converting temperature to kelvins and pressure to atmospheres:

$$P = \left(\frac{2000.\ \cancel{\text{lb}}}{\cancel{\text{in.}^2}}\right)\left(\frac{1\ \text{atm}}{14.7\ \cancel{\text{lb/in.}^2}}\right) = 136.1\ \text{atm}$$

$V = 50.0\ \text{L}$

$T = 295\ \text{K}$

$n = ?$

Step 2 Write and solve the ideal gas equation for the unknown:

$$PV = nRT \qquad \text{or} \qquad n = \frac{PV}{RT}$$

Step 3 Substitute the given information into the equation and calculate:

$$n = \frac{(136.1\ \cancel{\text{atm}})(50.0\ \cancel{\text{L}})}{(0.0821\ \cancel{\text{L}\cdot\text{atm}}/\text{mol}\cdot\cancel{\text{K}})(295\ \cancel{\text{K}})} = 281\ \text{mol}\ O_2$$

Practice 12.10

A 23.8-L cylinder contains oxygen gas at 20.0°C and 732 torr. How many moles of oxygen are in the cylinder?

The molar mass of a gaseous substance can be determined using the ideal gas law. Since molar mass = g/mol, it follows that mol = g/molar mass. Using M for molar mass and g for grams, we can substitute g/M for n (moles) in the ideal gas law to get

This form of the ideal gas law is most useful in problems containing mass instead of moles.

$$PV = \frac{g}{M}RT \qquad \text{or} \qquad M = \frac{gRT}{PV} \qquad \text{(modified ideal gas law)}$$

which allows us to calculate the molar mass, M, for any substance in the gaseous state.

Example 12.17

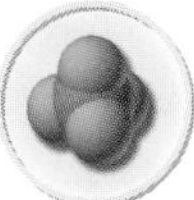

Calculate the molar mass of butane gas, if 3.69 g occupy 1.53 L at 20.0°C and 1.00 atm.

SOLUTION

Change 20.0°C to 293 K and substitute the data into the modified ideal gas law:

$$M = \frac{gRT}{PV} = \frac{(3.69\ \text{g})(0.0821\ \cancel{\text{L}\cdot\text{atm}}/\text{mol}\cdot\cancel{\text{K}})(293\ \cancel{\text{K}})}{(1.00\ \cancel{\text{atm}})(1.53\ \cancel{\text{L}})} = 58.0\ \text{g/mol}$$

Practice 12.11

A 0.286-g sample of a certain gas occupies 50.0 mL at standard temperature and 76.0 cm Hg. Determine the molar mass of the gas.

12.14 Gas Stoichiometry

Mole–Volume and Mass–Volume Calculations

Stoichiometric problems involving gas volumes can be solved by the general mole-ratio method outlined in Chapter 9. The factors 1 mol/22.4 L and 22.4 L/1 mol are used for converting volume to moles and moles to volume, respectively. (See Figure 12.16.) These conversion factors are used under the assumption that the gases are at STP and that they behave as ideal gases. In actual practice, gases are measured at other than STP conditions, and the volumes are converted to STP for stoichiometric calculations.

In a balanced equation, the number preceding the formula of a gaseous substance represents the number of moles or molar volumes (22.4 L at STP) of that substance.

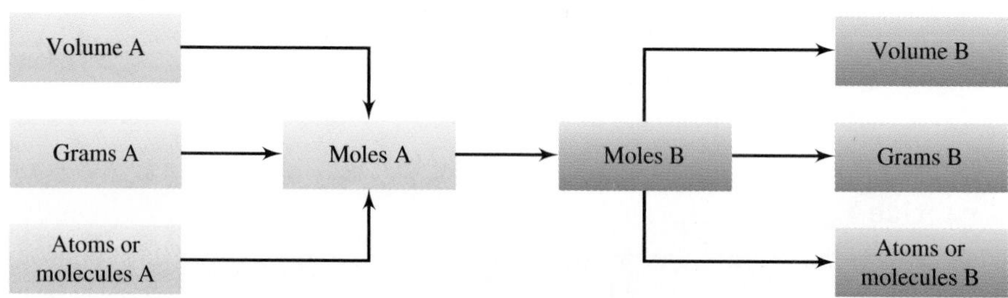

Figure 12.16
Summary of the primary conversions involved in stoichiometry. The conversion for volumes of gases is included.

The following are examples of typical problems involving gases and chemical equations.

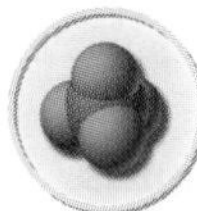

Example 12.18 What volume of oxygen (at STP) can be formed from 0.500 mol of potassium chlorate?

SOLUTION

Step 1 Write the balanced equation:

$$2\,KClO_3(s) \longrightarrow 2\,KCl(s) + 3\,O_2(g)$$

Step 2 The starting amount is 0.500 mol $KClO_3$. The conversion is from

$$\text{mol } KClO_3 \longrightarrow \text{mol } O_2 \longrightarrow \text{L } O_2$$

Step 3 Calculate the moles of O_2, using the mole-ratio method:

$$(0.500\ \cancel{\text{mol KClO}_3})\left(\frac{3\ \text{mol O}_2}{2\ \cancel{\text{mol KClO}_3}}\right) = 0.750\ \text{mol O}_2$$

Step 4 Convert moles of O_2 to liters of O_2. The moles of a gas at STP are converted to liters by multiplying by the molar volume, 22.4 L/mol:

$$(0.750\ \cancel{\text{mol}}\ \text{O}_2)\left(\frac{22.4\ \text{L}}{1\ \cancel{\text{mol}}}\right) = 16.8\ \text{L O}_2$$

Alternatively, setting up a continuous calculation, we obtain

$$(0.500\ \cancel{\text{mol KClO}_3})\left(\frac{3\ \text{mol O}_2}{2\ \cancel{\text{mol KClO}_3}}\right)\left(\frac{22.4\ \text{L}}{1\ \cancel{\text{mol}}}\right) = 16.8\ \text{L O}_2$$

Example 12.19

How many grams of aluminum must react with sulfuric acid to produce 1.25 L of hydrogen gas at STP?

SOLUTION

Step 1 The balanced equation is

$$2\,Al(s) + 3\,H_2SO_4(aq) \longrightarrow Al_2(SO_4)_3(aq) + 3\,H_2(g)$$

Step 2 We first convert liters of H_2 to moles of H_2. Then the familiar stoichiometric calculation from the equation is used. The conversion is

$$L\,H_2 \longrightarrow mol\,H_2 \longrightarrow mol\,Al \longrightarrow g\,Al$$

$$1.25\,\cancel{L\,H_2}\left(\frac{1\,\cancel{mol}}{22.4\,\cancel{L}}\right)\left(\frac{2\,\cancel{mol\,Al}}{3\,\cancel{mol\,H_2}}\right)\left(\frac{26.98\,g\,Al}{1\,\cancel{mol\,Al}}\right) = 1.00\,g\,Al$$

Example 12.20

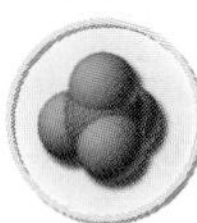

What volume of hydrogen, collected at 30.°C and 700. torr, will be formed by reacting 50.0 g of aluminum with hydrochloric acid?

$$2\,Al(s) + 6\,HCl(aq) \longrightarrow 2\,AlCl_3(aq) + 3\,H_2(g)$$

SOLUTION

In this problem, the conditions are not at STP, so we cannot use the method shown in Example 12.18. Either we need to calculate the volume at STP from the equation and then convert this volume to the conditions given in the problem, or we can use the ideal gas law. Let's use the ideal gas law.

First calculate the moles of H_2 obtained from 50.0 g of Al. Then, using the ideal gas law, calculate the volume of H_2 at the conditions given in the problem:

Step 1 Moles of H_2: The conversion is

$$\text{grams Al} \longrightarrow \text{moles Al} \longrightarrow \text{moles } H_2$$

$$50.0\,\cancel{g\,Al}\left(\frac{1\,\cancel{mol\,Al}}{26.98\,\cancel{g\,Al}}\right)\left(\frac{3\,mol\,H_2}{2\,\cancel{mol\,Al}}\right) = 2.78\,mol\,H_2$$

Step 2 Liters of H_2: Solve $PV = nRT$ for V and substitute the data into the equation:

Convert °C to K: 30.°C + 273 = 303 K.

Convert torr to atm: $(700.\,\cancel{torr})(1\,atm/760.\,\cancel{torr}) = 0.921\,atm$.

$$V = \frac{nRT}{P} = \frac{(2.78\,\cancel{mol}\,H_2)(0.0821\,L\text{-}\cancel{atm})(303\,\cancel{K})}{(0.921\,\cancel{atm})(\cancel{mol\text{-}K})} = 75.1\,L\,H_2$$

(*Note:* The volume at STP is 62.3 L H_2.)

Practice 12.12

If 10.0 g of sodium peroxide (Na_2O_2) react with water to produce sodium hydroxide and oxygen, how many liters of oxygen will be produced at 20.°C and 750. torr?

$$2\,Na_2O_2(s) + 2\,H_2O(l) \longrightarrow 4\,NaOH(aq) + O_2(g)$$

Volume–Volume Calculations

When all substances in a reaction are in the gaseous state, simplifications in the calculation can be made. These are based on Avogadro's law, which states that gases under identical conditions of temperature and pressure contain the same number of molecules and occupy the same volume. Under the standard conditions of temperature and pressure, the volumes of gases reacting are proportional to the numbers of moles of the gases in the balanced equation. Consider the reaction:

$H_2(g)$	$+$	$Cl_2(g)$	$\longrightarrow$	$2\,HCl(g)$
1 mol		1 mol		2 mol
22.4 L		22.4 L		2 × 22.4 L
1 volume		1 volume		2 volumes
Y volume		*Y* volume		2 *Y* volumes

Remember: For gases at the same *T* and *P*, equal volumes contain equal numbers of particles.

In this reaction, 22.4 L of hydrogen will react with 22.4 L of chlorine to give 2 (22.4) = 44.8 L of hydrogen chloride gas. This statement is true because these volumes are equivalent to the number of reacting moles in the equation. Therefore, *Y* volume of H_2 will combine with *Y* volume of Cl_2 to give 2 *Y* volumes of HCl. For example, 100 L of H_2 react with 100 L of Cl_2 to give 200 L of HCl; if the 100 L of H_2 and of Cl_2 are at 50°C, they will give 200 L of HCl at 50°C. When the temperature and pressure before and after a reaction are the same, volumes can be calculated without changing the volumes to STP.

For reacting gases at constant temperature and pressure, volume–volume relationships are the same as mole–mole relationships.

Example 12.21 What volume of oxygen will react with 150. L of hydrogen to form water vapor? What volume of water vapor will be formed?

SOLUTION Assume that both reactants and products are measured at standard conditions. Calculate by using reacting volumes:

$2\,H_2(g)$	$+$	$O_2(g)$	$\longrightarrow$	$2\,H_2O(g)$
2 mol		1 mol		2 mol
2 × 22.4 L		22.4 L		2 × 22.4 L
2 volumes		1 volume		2 volumes
150. L		75 L		150. L

For every two volumes of H_2 that react, one volume of O_2 reacts and two volumes of $H_2O(g)$ are produced:

$$(150.\ \text{L}\ H_2)\left(\frac{1\ \text{volume}\ O_2}{2\ \text{volumes}\ H_2}\right) = 75\ \text{L}\ O_2$$

$$(150.\ \text{L}\ H_2)\left(\frac{2\ \text{volumes}\ H_2O}{2\ \text{volumes}\ H_2}\right) = 150.\ \text{L}\ H_2O$$

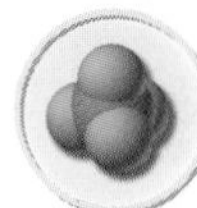

Example 12.22 The equation for the preparation of ammonia is

$$3\,H_2(g) + N_2(g) \xrightarrow{400°\text{C}} 2\,NH_3(g)$$

Assuming that the reaction goes to completion, determine the following:

(a) What volume of H_2 will react with 50.0 L of N_2?
(b) What volume of NH_3 will be formed from 50.0 L of N_2?
(c) What volume of N_2 will react with 100. mL of H_2?
(d) What volume of NH_3 will be produced from 100. mL of H_2?
(e) If 600. mL of H_2 and 400. mL of N_2 are sealed in a flask and allowed to react, what amounts of H_2, N_2, and NH_3 are in the flask at the end of the reaction?

SOLUTION

The answers to parts (a)–(d) are shown in the boxes and can be determined from the equation by inspection, using the principle of reacting volumes:

	$3\,H_2(g)$	+	$N_2(g)$	$\longrightarrow$	$2\,NH_3(g)$
	3 volumes		1 volume		2 volumes
(a)	**[150. L]**		50.0 L		
(b)			50.0 L		**[100. L]**
(c)	100. mL		**[33.3 mL]**		
(d)	100. mL				**[66.7 mL]**

(e) Volume ratio from the equation $= \dfrac{3 \text{ volumes } H_2}{1 \text{ volume } N_2}$

$$\text{Volume ratio used} = \frac{600.\ \text{mL } H_2}{400.\ \text{mL } N_2} = \frac{3 \text{ volumes } H_2}{2 \text{ volumes } N_2}$$

Comparing these two ratios, we see that an excess of N_2 is present in the gas mixture. Therefore, the reactant limiting the amount of NH_3 that can be formed is H_2:

$$\underset{600.\ \text{mL}}{3\,H_2(g)} + \underset{200.\ \text{mL}}{N_2(g)} \longrightarrow \underset{400.\ \text{mL}}{2\,NH_3(g)}$$

To have a 3 : 1 ratio of volumes reacting, 600. mL of H_2 will react with 200. mL of N_2 to produce 400. mL of NH_3, leaving 200. mL of N_2 unreacted. At the end of the reaction the flask will contain 400. mL of NH_3 and 200. mL of N_2.

Practice 12.13

What volume of oxygen will react with 15.0 L of propane (C_3H_8) to form carbon dioxide and water? What volume of carbon dioxide will be formed? What volume of water vapor will be formed?

$$C_3H_8(g) + 5\,O_2(g) \longrightarrow 3\,CO_2(g) + 4\,H_2O(g)$$

12.15 Real Gases

All the gas laws are based on the behavior of an ideal gas—that is, a gas with a behavior that is described exactly by the gas laws for all possible values of P, V, and T. Most real gases actually do behave very nearly as predicted by the gas laws over a fairly wide range of temperatures and pressures. However, when conditions are such that the gas molecules are crowded closely together (high pressure and/or low temperature), they show marked deviations from ideal behavior. Deviations occur because molecules have finite volumes and also have intermolecular attractions, which result in less compressibility at high pressures and greater compressibility at low temperatures than predicted by the gas laws. Many gases become liquids at high pressure and low temperature.

Air Pollution

Chemical reactions occur among the gases that are emitted into our atmosphere. In recent years, there has been growing concern over the effects these reactions have on our environment and our lives.

The outer portion (stratosphere) of the atmosphere plays a significant role in determining the conditions for life at the surface of the Earth. This stratosphere protects the surface from the intense radiation and particles bombarding our planet. Some of the high-energy radiation from the sun acts upon oxygen molecules, O_2, in the stratosphere, converting them into ozone, O_3.

allotrope

Different molecular forms of an element are called **allotropes** of that element. Thus, oxygen and ozone are allotropic forms of oxygen:

$$O_2 \xrightarrow{\text{sunlight}} \underbrace{O + O}_{\text{oxygen atoms}}$$

$$O_2 + O \longrightarrow \underset{\text{ozone}}{O_3}$$

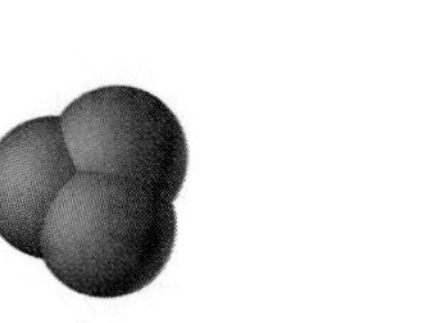

Ozone (O_3)

Ultraviolet radiation from the sun is highly damaging to living tissues of plants and animals. The ozone layer, however, shields the Earth by absorbing ultraviolet radiation and thus prevents most of this lethal radiation from reaching the Earth's surface. The reaction that occurs is the reverse of the preceding one:

$$O_3 \xrightarrow[\text{radiation}]{\text{ultraviolet}} O_2 + O + \text{heat}$$

Scientists have become concerned about a growing hazard to the ozone layer. Chlorofluorocarbon propellants, such as the Freons, CCl_3F and CCl_2F_2, which were used in aerosol spray cans and are used in refrigeration and air-conditioning units, are stable compounds and remain unchanged in the lower atmosphere. But when these chlorofluorocarbons are carried by convection currents to the stratosphere, they absorb ultraviolet radiation and produce chlorine atoms (chlorine free radicals), which in turn react with ozone. The following reaction sequence involving free radicals has been proposed to explain the partial destruction of the ozone layer by chlorofluorocarbons.

A free radical is a species containing an odd number of electrons. Free radicals are highly reactive.

$$\underset{\text{fluorocarbon molecule}}{CCl_3F} \xrightarrow[\text{radiation}]{\text{ultraviolet}} \underset{\text{fluorocarbon free radical}}{\cdot CCl_2F} + \underset{\text{chlorine free radical (atom)}}{Cl\cdot} \quad (1)$$

$$Cl\cdot + O_3 \longrightarrow ClO\cdot + O_2 \quad (2)$$

$$ClO\cdot + O \longrightarrow O_2 + Cl\cdot \quad (3)$$

Because a chlorine atom is generated for each ozone molecule that is destroyed (reactions 2 and 3 can proceed repeatedly), a single chlorofluorocarbon molecule can be responsible for the destruction of many ozone molecules.

CHEMISTRY IN ACTION • Wetland Blanket for Global Warming

Volcanic eruptions send sulfur dioxide high into the atmosphere. The SO_2 gas forms tiny droplets that reflect sunlight back into space, cooling the ground temperatures for up to two years following an eruption. Eventually the sulfur dioxide combines with water vapor and oxygen to form sulfate compounds, which fall back to Earth as acid rain. Researchers in England suggest these sulfates may also cause climate change in the ground by suppressing the release of methane (a greenhouse gas) from wetland areas. The scientists spread powdered sodium sulfate on an area of peat bog in northern Scotland. They added an amount equal to their calculation of how much SO_2 was released during a volcanic eruption (and about 10 times the amount of SO_2 currently released by industrial sources.) The scientists then measured atmospheric concentrations of methane over the bog and found a 42% reduction over the treated areas. Two years later (after halting the sulfate addition) the methane release was still 40% reduced. In fact, the data now suggest a five- to seven-year reduction in methane results from sulfate addition.

How does sulfate reduce methane emission? The methane comes primarily from soil bacteria. The additional sulfates increase the growth of a type of bacteria that uses sulfates to produce energy. This increase, called a bloom, causes the methane-producing bacteria to decrease. These complicated biological interactions can end up influencing atmospheric chemistry. Scientists think more evidence for these effects might be found in polar and glacial ice formed following volcanic eruptions.

Scientists have discovered an annual thinning in the ozone layer over Antarctica. This is what we call the "hole" in the ozone layer. If this hole were to occur over populated regions of the world, severe effects would result, including a rise in the cancer rate, increased climatic temperatures, and vision problems (which have been observed in the penguins of Antarctica). See Figure 12.17.

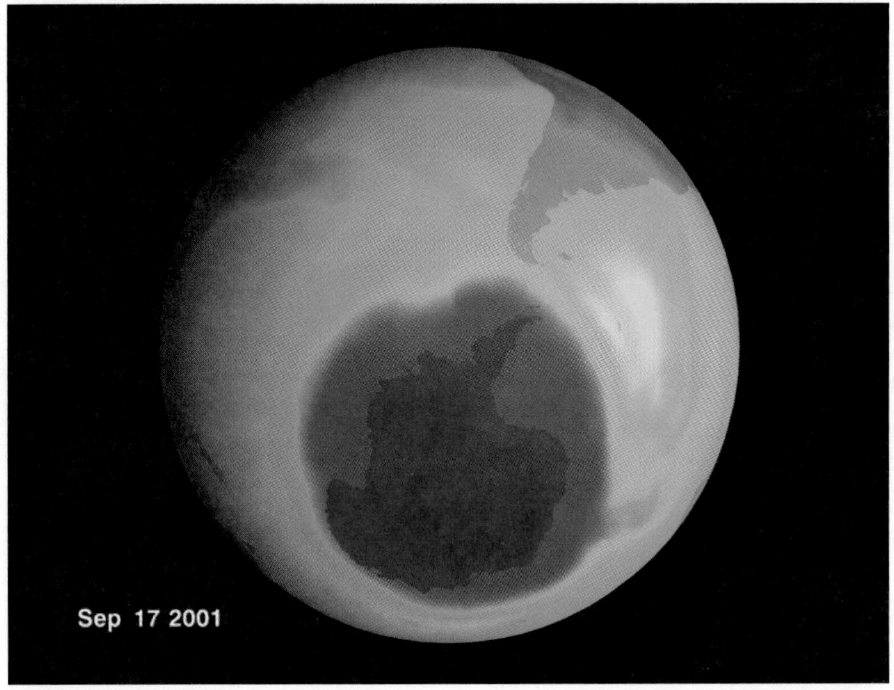

Figure 12.17
Antarctica ozone hole.

Ozone can be prepared by passing air or oxygen through an electrical discharge:

$$3\,O_2(g) + 286\text{ kJ} \xrightarrow[\text{discharge}]{\text{electrical}} 2\,O_3(g)$$

The characteristic pungent odor of ozone is noticeable in the vicinity of electrical machines and power transmission lines. Ozone is formed in the atmosphere during electrical storms and by the photochemical action of ultraviolet radiation on a mixture of nitrogen dioxide and oxygen. Areas with high air pollution are subject to high atmospheric ozone concentrations.

Ozone is not a desirable low-altitude constituent of the atmosphere because it is known to cause extensive plant damage, cracking of rubber, and the formation of eye-irritating substances. Concentrations of ozone greater than 0.1 part per million (ppm) of air cause coughing, choking, headache, fatigue, and reduced resistance to respiratory infection. Concentrations between 10 and 20 ppm are fatal to humans.

In addition to ozone, the air in urban areas contains nitrogen oxides, which are components of smog. The term *smog* refers to air pollution in urban environments. Often the chemical reactions occur as part of a *photochemical process*. Nitrogen monoxide (NO) is oxidized in the air or in automobile engines to produce nitrogen dioxide (NO_2). In the presence of light,

$$NO_2 \xrightarrow{\text{light}} NO + O$$

In addition to nitrogen oxides, combustion of fossil fuels releases CO_2, CO, and sulfur oxides. Incomplete combustion releases unburned and partially burned hydrocarbons.

Society is continually attempting to discover, understand, and control emissions that contribute to this sort of atmospheric chemistry. It is a problem that each one of us faces as we look to the future if we want to continue to support life as we know it on our planet.

Chapter 12 Review

12.1 General Properties

- Gases:
  - Particles are relatively far apart.
  - Particles are very mobile.
  - Gases take the shape and volume of the container.
  - Gases are easily compressible.

12.2 The Kinetic-Molecular Theory

KEY TERMS

Kinetic-molecular theory (KMT)
Ideal gas
Diffusion
Effusion
Graham's law of effusion

- Kinetic-molecular theory assumptions:
  - Gases are tiny particles with no attraction for each other.
  - The distance between particles is great compared to the size of the particles.
  - Gas particles move in straight lines.
  - No energy is lost in particle collisions (perfectly elastic collisions).
  - The average kinetic energy for particles is the same for all gases at the same temperature and pressure.
- A gas that follows the KMT is an ideal gas
- The kinetic energy of a particle is expressed as $KE = \frac{1}{2}mv^2$.
- Gases will diffuse to mix spontaneously.
- Gases can effuse through a small opening.
- Graham's law of effusion states

$$\frac{\text{rate of effusion for gas A}}{\text{rate of effusion for gas B}} = \sqrt{\frac{dB}{dA}} = \sqrt{\frac{\text{molar mass B}}{\text{molar mass A}}}$$

12.3 Measurement of Pressure of Gases

KEY TERMS

Pressure
Atmospheric pressure
Barometer
1 atmosphere

- Pressure is force per unit area.
- Pressure of the atmosphere is measured by using a barometer:
  - Units of pressure include:
    - Atmosphere (atm) = 760 mm Hg.
    - Pascal, 1 atm = 101,325 Pa = 101.3 kPa.
    - Torr, 1 atm = 760 torr.

12.4 Dependence of Pressure on Number of Molecules and Temperature

- Pressure is directly related to the number of molecules in the sample.
- Pressure is directly related to the Kelvin temperature of the sample.

12.5 Boyle's Law

KEY TERM

Boyle's law

- At constant temperature the volume of a gas is inversely proportional to the pressure of the gas:

$$V \propto \frac{1}{P} \quad \text{or} \quad P_1V_1 = P_2V_2$$

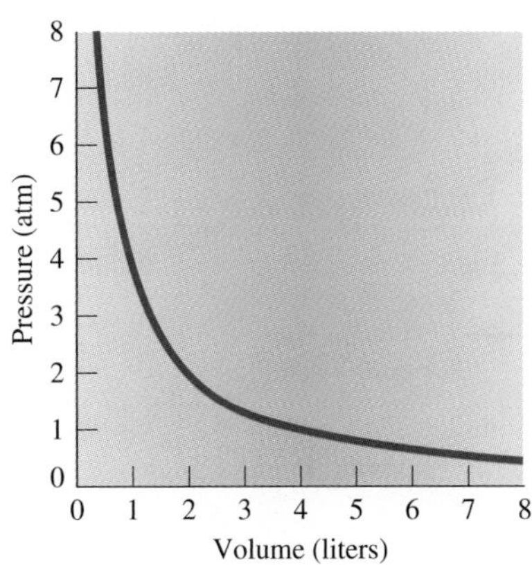

12.6 Charles' Law

KEY TERMS

Absolute zero
Charles' law

- At constant pressure the volume of a gas is directly proportional to the absolute temperature of the gas.

$$V \propto T \quad \text{or} \quad \frac{V_1}{T_1} = \frac{V_2}{T_2}$$

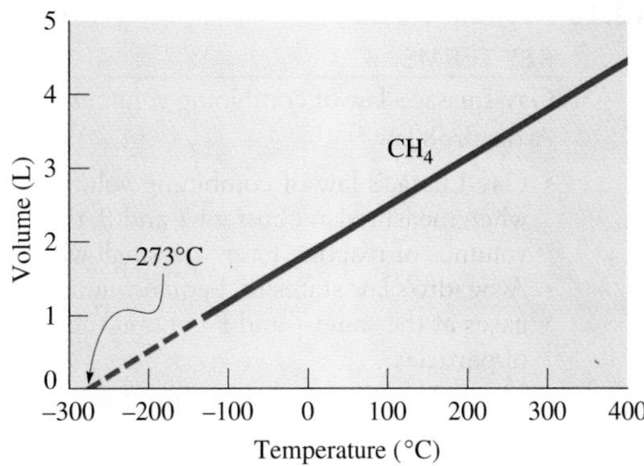

12.7 Gay-Lussac's Law

KEY TERM

Gay-Lussac's law

- At constant volume the pressure of a gas is directly proportional to the absolute temperature of the gas.

$$P \propto T \quad \text{or} \quad \frac{P_1}{T_1} = \frac{P_2}{T_2}$$

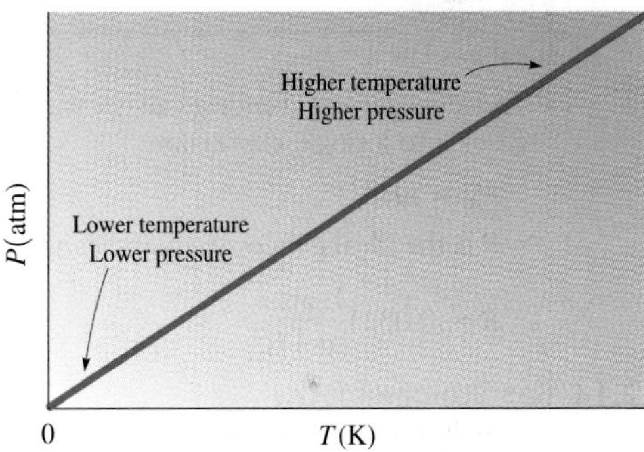

12.8 Combined Gas Law

KEY TERMS

Standard conditions
Standard temperature and pressure (STP)

- The $P\,V\,T$ relationship for gases can be expressed in a single equation known as the combined gas law:

$$\frac{P_1V_1}{T_1} = \frac{P_2V_2}{T_2}$$

12.9 Dalton's Law of Partial Pressures

KEY TERMS

Dalton's law of partial pressures
Partial pressure

- The total pressure of a mixture of gases is the sum of the partial pressures of the component gases in the mixture.
- When a gas is collected over water, the pressure of the collected gas is the difference between the atmospheric pressure and the vapor pressure of water at that temperature.

12.10 Avogadro's Law

KEY TERMS

Gay-Lussac's law of combining volumes
Avogadro's law

- Gay-Lussac's law of combining volumes states that when measured at constant *T* and *P*, the ratios of the volumes of reacting gases are small whole numbers.
- Avogadro's law states that equal volumes of different gases at the same *T* and *P* contain the same number of particles.

12.11 Mole–Mass–Volume Relationships of Gases

KEY TERM

Molar volume

- One mole of any gas occupies 22.4 L at STP.

12.12 Density of Gases

- The density of a gas is usually expressed in units of g/L.
- Temperature and pressure are given for a density of a gas since the volume of the gas depends on these conditions.

12.13 Ideal Gas Law

KEY TERM

Ideal gas law

- The ideal gas law combines all the variables involving gases into a single expression:

$$PV = nRT$$

  - *R* is the ideal gas constant and can be expressed as

$$R = 0.0821 \frac{\text{L-atm}}{\text{mol-K}}$$

12.14 Gas Stoichiometry

- Stoichiometry problems involving gases are solved the same way as other stoichiometry problems. See graphic below.
- For reacting gases (contant *T* and *P*) volume–volume relationships are the same as mole–mole relationships.

12.15 Real Gases

KEY TERM

Allotrope

- Real gases show deviation from the ideal gas law:
  - Deviations occur at:
    - High pressure
    - Low temperature
- These deviations occur because:
  - Molecules have finite volumes.
  - Molecules have intermolecular attractions.
- Ozone in our upper atmosphere protects us from harmful ultraviolet radiation.
- The ozone layer is being destroyed by pollutants such as Freons and other chlorinated hydrocarbons:
  - These molecules produce free radicals, which interact with ozone to convert it to oxygen.
- In the lower atmosphere ozone causes damage to plants, cracking of rubber, and irritations to humans.
- Other air pollutants such as nitrogen oxides and sulfur oxides are found in urban areas, creating smog, which can impact our daily lives.

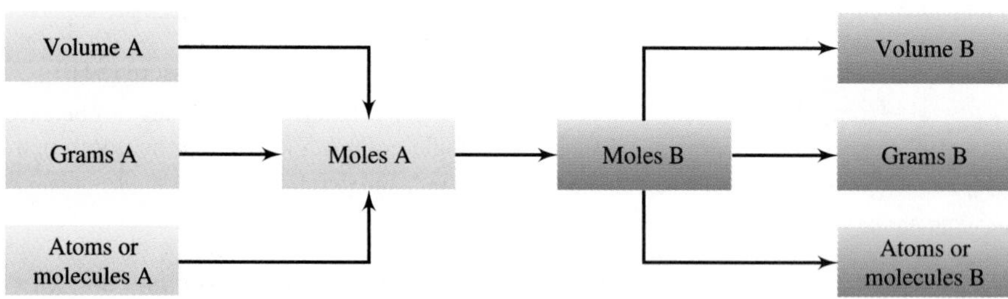

Review Questions

All questions with blue numbers have answers in the appendix of the text.

1. What evidence is used to show diffusion in Figure 12.1? If H_2 and O_2 were in the two flasks, how could you prove that diffusion had taken place?
2. What is meant by pressure (for a gas)?
3. How does the air pressure inside the balloon shown in Figure 12.2 compare with the air pressure outside the balloon? Explain.
4. According to Table 12.1, what two gases are the major constituents of dry air?

5. How does the pressure represented by 1 torr compare in magnitude to the pressure represented by 1 mm Hg? See Table 12.2.
6. In which container illustrated in Figure 12.5 are the molecules of gas moving faster? Assume both gases to be hydrogen.
7. In Figure 12.6, what gas pressure corresponds to a volume of 4 L?
8. How do the data illustrated in Figure 12.6 substantiate Boyle's law?
9. What effect would you observe in Figure 12.9 if T_2 were lower than T_1?
10. In the diagram shown in Figure 12.12, is the pressure of the oxygen plus water vapor inside the bottle equal to, greater than, or less than the atmospheric pressure outside the bottle? Explain.
11. List five gases in Table 12.3 that are more dense than air. Explain the basis for your selection.
12. What are the basic assumptions of the kinetic-molecular theory?
13. Arrange the following gases, all at standard temperature, in order of increasing relative molecular velocities: H_2, CH_4, Rn, N_2, F_2, and He. What is your basis for determining the order?
14. List, in descending order, the average kinetic energies of the molecules in Question 13.
15. What are the four parameters used to describe the behavior of a gas?
16. What are the characteristics of an ideal gas?
17. How is Boyle's law related to the ideal gas law?
18. How is Charles' law related to the ideal gas law?
19. Under what condition of temperature, high or low, is a gas least likely to exhibit ideal behavior? Explain.
20. Under what condition of pressure, high or low, is a gas least likely to exhibit ideal behavior? Explain.
21. Compare, at the same temperature and pressure, equal volumes of H_2 and O_2 as to the following:
(a) number of molecules
(b) mass
(c) number of moles
(d) average kinetic energy of the molecules
(e) rate of effusion
(f) density
22. How does the kinetic-molecular theory account for the behavior of gases as described by
(a) Boyle's law?
(b) Charles' law?
(c) Dalton's law of partial pressures?
23. Explain how the reaction

$$N_2(g) + O_2(g) \xrightarrow{\Delta} 2\,NO(g)$$

proves that nitrogen and oxygen are diatomic molecules.
24. What is the reason for comparing gases to STP?
25. Is the conversion of oxygen to ozone an exothermic or endothermic reaction? How do you know?
26. What compounds are responsible for damaging the ozone layer and why are they hazardous?
27. When constant pressure is maintained, what effect does heating a mole of N_2 gas have on
(a) its density?
(b) its mass?
(c) the average kinetic energy of its molecules?
(d) the average velocity of its molecules?
(e) the number of N_2 molecules in the sample?
28. Write formulas for an oxygen atom, an oxygen molecule, and an ozone molecule. How many electrons are in an oxygen molecule?

Paired Exercises

All exercises with blue numbers have answers in the appendix of the text.

Pressure Units

1. The barometer reads 715 mm Hg. Calculate the corresponding pressure in the following:
(a) atmospheres
(b) inches of Hg
(c) lb/in.^{2}

2. The barometer reads 715 mm Hg. Calculate the corresponding pressure in the following:
(a) torrs
(b) millibars
(c) kilopascals

3. Express the following pressures in atmospheres:
(a) 28 mm Hg
(b) 6000. cm Hg
(c) 795 torr
(d) 5.00 kPa

4. Express the following pressures in atmospheres:
(a) 62 mm Hg
(b) 4250. cm Hg
(c) 225 torr
(d) 0.67 kPa

Boyle's and Charles' Laws

5. A sample of a gas occupies a volume of 525 mL at 625 torr. At constant temperature, what will be the new volume when the pressure changes to the following:
(a) 1.5 atm
(b) 455 mm Hg

6. A sample of a gas occupies a volume of 635 mL at 722 torr. At constant temperature, what will be the new volume when the pressure changes to the following:
(a) 2.5 atm
(b) 795 mm Hg

7. A sample of a gas at 0.75 atm occupies a volume of 521 mL. If the temperature remains constant, what will be the new pressure if the volume increases to 776 mL?

8. A sample of a gas at 1.7 atm occupies a volume of 225 mL. If the temperature remains constant, what will be the new pressure if the volume decreases to 115 mL?

9. Given 6.00 L of N_2 gas at −25°C, what volume will the nitrogen occupy at (a) 0.0°C? (b) 100. K? (Assume constant pressure.)

10. Given 6.00 L of N_2 gas at −25°C, what volume will the nitrogen occupy at (a) 0.0°F? (b) 345. K? (Assume constant pressure.)

Combined Gas Laws

11. A gas occupies a volume of 410 mL at 27°C and 740 mm Hg pressure. Calculate the volume the gas would occupy at STP.

12. A gas occupies a volume of 410 mL at 27°C and 740 mm Hg pressure. Calculate the volume the gas would occupy at 250.°C and 680 mm Hg pressure.

13. An expandable balloon contains 1400. L of He at 0.950 atm pressure and 18°C. At an altitude of 22 miles (temperature 2.0°C and pressure 4.0 torr), what will be the volume of the balloon?

14. A gas occupies 22.4 L at 2.50 atm and 27°C. What will be its volume at 1.50 atm and −5.00°C?

15. A 775-mL sample of NO_2 gas is at STP. If the volume changes to 615 mL and the temperature changes to 25°C, what will be the new pressure?

16. A 2.5-L sample of N_2 is at 19°C and 1.5 atm. What will be the new temperature in °C when the volume changes to 1.5 L and the pressure to 765 torr?

Dalton's Law of Partial Pressures

17. What would be the partial pressure of N_2 gas collected over water at 22°C and 721 torr pressure? (Check Appendix IV for the vapor pressure of water.)

18. What would be the partial pressure of N_2 gas collected over water at 25°C and 705 torr pressure? (Check Appendix IV for the vapor pressure of water.)

19. A mixture contains H_2 at 600. torr pressure, N_2 at 200. torr pressure, and O_2 at 300. torr pressure. What is the total pressure of the gases in the system?

20. A mixture contains H_2 at 325 torr pressure, N_2 at 475 torr pressure, and O_2 at 650. torr pressure. What is the total pressure of the gases in the system?

21. A sample of methane gas, CH_4, was collected over water at 25.0°C and 720. torr. The volume of the wet gas is 2.50 L. What will be the volume of the dry methane at standard pressure?

22. A sample of propane gas, C_3H_8, was collected over water at 22.5°C and 745 torr. The volume of the wet gas is 1.25 L. What will be the volume of the dry propane at standard pressure?

Mole–Mass–Volume Relationships

23. How many moles of O_2 will occupy a volume of 1.75 L at STP?

24. How many moles of N_2 will occupy a volume of 3.50 L at STP?

25. What volume will each of the following occupy at STP?
(a) 6.02×10^{23} molecules of CO_2
(b) 2.5 mol CH_4
(c) 12.5 g oxygen

26. What volume will each of the following occupy at STP?
(a) 1.80×10^{24} molecules of SO_3
(b) 7.5 mol C_2H_6
(c) 25.2 g chlorine

27. How many grams of NH_3 are present in 725 mL of the gas at STP?

28. How many grams of C_3H_6 are present in 945 mL of the gas at STP?

29. How many molecules of NH_3 gas are present in a 1.00-L flask of NH_3 gas at STP?

30. How many molecules of CH_4 gas are present in a 1.00-L flask of CH_4 gas at STP?

31. What volume would result if a balloon were filled with 10.0 grams of chlorine gas at STP?

32. How many grams of methane gas were used to fill a balloon to a volume of 3.0 L at STP?

Density of Gases

33. Calculate the density of the following gases at STP:
(a) Xe
(b) F_2
(c) C_2H_6

34. Calculate the density of the following gases at STP:
(a) Rn
(b) CO_2
(c) C_4H_8

35. Calculate the density of the following:
(a) F_2 gas at STP
(b) F_2 gas at 27°C and 1.00 atm pressure

36. Calculate the density of the following:
(a) Cl_2 gas at STP
(b) Cl_2 gas at 22°C and 0.500 atm pressure

Ideal Gas Law and Stoichiometry

37. At 27°C and 750 torr pressure, what will be the volume of 2.3 mol of Ne?

38. At 25°C and 725 torr pressure, what will be the volume of 0.75 mol of Kr?

39. What volume will a mixture of 5.00 mol of H_2 and 0.500 mol of CO_2 occupy at STP?

40. What volume will a mixture of 2.50 mol of N_2 and 0.750 mol of HCl occupy at STP?

41. What is the Kelvin temperature of a system in which 4.50 mol of a gas occupy 0.250 L at 4.15 atm?

42. How many moles of N_2 gas occupy 5.20 L at 250 K and 0.500 atm?

43. In the lab, students generated and collected hydrogen gas according to the following equation:

$$Zn(s) + H_2SO_4(aq) \rightarrow H_2(g) + ZnSO_4(aq)$$

(a) How many mL of hydrogen gas at STP were generated from 52.7 g of zinc metal?
(b) If 525 mL of hydrogen gas at STP were needed, how many moles of H_2SO_4 would be required?

44. In the lab, students generated and collected oxygen gas according to the following equation:

$$2\,H_2O_2(aq) \rightarrow 2\,H_2O(l) + O_2(g)$$

(a) How many mL of oxygen gas at STP were generated from 50.0 g of H_2O_2?
(b) If 225 mL of oxygen gas at STP were needed for an experiment, how many moles of H_2O_2 would be needed?

45. Consider the following equation:

$$4\,NH_3(g) + 5\,O_2(g) \rightarrow 4\,NO(g) + 6\,H_2O(g)$$

(a) How many liters of oxygen are required to react with 2.5 L NH_3? Both gases are at STP.
(b) How many grams of water vapor can be produced from 25 L NH_3 if both gases are at STP?
(c) How many liters of NO can be produced when 25 L O_2 are reacted with 25 L NH_3? All gases are at the same temperature and pressure.

46. Consider the following equation:

$$C_3H_8(g) + 5\,O_2(g) \rightarrow 3\,CO_2(g) + 4\,H_2O(g)$$

(a) How many liters of oxygen are required to react with 7.2 L C_3H_8? Both gases are at STP.
(b) How many grams of CO_2 will be produced from 35 L C_3H_8 if both gases are at STP?
(c) How many liters of water vapor can be produced when 15 L C_3H_8 are reacted with 15 L O_2? All gases are at the same temperature and pressure.

47. Oxygen gas can be generated by the decomposition of potassium chlorate according to the following equation:

$$2\,KClO_3(s) \rightarrow 2\,KCl(s) + 3\,O_2(g)$$

How many liters of oxygen at STP will be produced when 0.525 kg of KCl is also produced?

48. When glucose is burned in a closed container, carbon dioxide gas and water are produced according to the following equation:

$$C_6H_{12}O_6(s) + 6\,O_2(g) \rightarrow 6\,CO_2(g) + 6\,H_2O(l)$$

How many liters of CO_2 at STP can be produced when 1.50 kg of glucose are burned?

Additional Exercises

All exercises with blue numbers have answers in the appendix of the text.

49. How is it possible that 1 mole of liquid water occupies a volume of 18 mL, but 1 mole of gaseous water occupies a volume of 22.4 L? (See the photo in Section 12.1.)

50. Explain why it is necessary to add air to a car's tires during the winter.

51. You have a 10-L container filled with 0.5 mol of O_2 gas at a temperature of 30.°C with a pressure of 945 torr.

(a) What will happen to the pressure if the container size is doubled while keeping the temperature and number of moles constant?

(b) What will happen to the pressure when the temperature is doubled while keeping the size of the container and the number of moles constant?

(c) What will happen to the pressure when the amount of O_2 gas is cut in half while keeping the size of the container and the temperature constant?

(d) What will happen to the pressure if 1 mole of N_2 gas is added to the container while keeping the temperature and size of the container the same?

52. Sketch a graph to show each of the following relationships:

(a) P vs. V at constant temperature and number of moles

(b) T vs. V at constant pressure and number of moles

(c) T vs. P at constant volume and number of moles

(d) n vs. V at constant temperature and pressure

53. Why is it dangerous to incinerate an aerosol can?

54. What volume does 1 mol of an ideal gas occupy at standard conditions?

55. Which of these occupies the greatest volume?

(a) 0.2 mol of chlorine gas at 48°C and 80 cm Hg

(b) 4.2 g of ammonia at 0.65 atm and −11°C

(c) 21 g of sulfur trioxide at 55°C and 110 kPa

56. Which of these has the greatest density?

(a) SF_6 at STP

(b) C_2H_6 at room conditions

(c) He at −80°C and 2.15 atm

57. A chemist carried out a chemical reaction that produced a gas. It was found that the gas contained 80.0% carbon and 20.0% hydrogen. It was also noticed that 1500 mL of the gas at STP had a mass of 2.01 g.

(a) What is the empirical formula of the compound?

(b) What is the molecular formula of the compound?

(c) What Lewis structure fits this compound?

58. Three gases were added to the same 2.0-L container. The total pressure of the gases was 790 torr at room temperature (25.0°C). If the mixture contained 0.65 g of oxygen gas, 0.58 g of carbon dioxide, and an unknown amount of nitrogen gas, determine the following:

(a) the total number of moles of gas in the container

(b) the number of grams of nitrogen in the container

(c) the partial pressure of each gas in the mixture

***59.** When carbon monoxide and oxygen gas react, carbon dioxide results. If 500. mL of O_2 at 1.8 atm and 15°C are mixed with 500. mL of CO at 800 mm Hg and 60°C, how many milliliters of CO_2 at STP could possibly result?

***60.** One of the methods for estimating the temperature at the center of the sun is based on the ideal gas law. If the center is assumed to be a mixture of gases whose average molar mass is 2.0 g/mol, and if the density and pressure are 1.4 g/cm^3 and 1.3×10^9 atm, respectively, calculate the temperature.

***61.** A soccer ball of constant volume 2.24 L is pumped up with air to a gauge pressure of 13 lb/in.^{2} at 20.0°C. The molar mass of air is about 29 g/mol.

(a) How many moles of air are in the ball?

(b) What mass of air is in the ball?

(c) During the game, the temperature rises to 30.0°C. What mass of air must be allowed to escape to bring the gauge pressure back to its original value?

***62.** A balloon will burst at a volume of 2.00 L. If it is partially filled at 20.0°C and 65 cm Hg to occupy 1.75 L, at what temperature will it burst if the pressure is exactly 1 atm at the time that it breaks?

63. Given a sample of a gas at 27°C, at what temperature would the volume of the gas sample be doubled, the pressure remaining constant?

64. A gas sample at 22°C and 740 torr pressure is heated until its volume is doubled. What pressure would restore the sample to its original volume?

65. A gas occupies 250. mL at 700. torr and 22°C. When the pressure is changed to 500. torr, what temperature (°C) is needed to maintain the same volume?

66. The tires on an automobile were filled with air to 30. psi at 71.0°F. When driving at high speeds, the tires become hot. If the tires have a bursting pressure of 44 psi, at what temperature (°F) will the tires "blow out"?

67. What pressure will 800. mL of a gas at STP exert when its volume is 250. mL at 30°C?

68. How many gas molecules are present in 600. mL of N_2O at 40°C and 400. torr pressure? How many atoms are present? What would be the volume of the sample at STP?

69. An automobile tire has a bursting pressure of 60 lb/in.^{2}. The normal pressure inside the tire is 32 lb/in.^{2}. When traveling at high speeds, a tire can get quite hot. Assuming that the temperature of the tire is 25°C before running it, determine whether the tire will burst when the inside temperature gets to 212°F. Show your calculations.

70. If you prepared a barometer using water instead of mercury, how high would the column of water be at one atmosphere pressure? (Neglect the vapor pressure of water.)

71. How many moles of oxygen are in a 55-L cylinder at 27°C and 2.20×10^3 lb/in.^{2}?

***72.** How many moles of Cl_2 are in one cubic meter (1.00 m^3) of Cl_2 gas at STP?

73. At STP, 560. mL of a gas have a mass of 1.08 g. What is the molar mass of the gas?

74. A gas has a density at STP of 1.78 g/L. What is its molar mass?

75. Using the ideal gas law, $PV = nRT$, calculate the following:

(a) the volume of 0.510 mol of H_2 at 47°C and 1.6 atm pressure

(b) the number of grams in 16.0 L of CH_4 at 27°C and 600. torr pressure

*(c) the density of CO_2 at 4.00 atm pressure and −20.0°C

(d) the molar mass of a gas having a density of 2.58 g/L at 27°C and 1.00 atm pressure.

***76.** Acetylene (C_2H_2) and hydrogen fluoride (HF) react to give difluoroethane:

$$C_2H_2(g) + 2\,HF(g) \longrightarrow C_2H_4F_2(g)$$

When 1.0 mol of C_2H_2 and 5.0 mol of HF are reacted in a 10.0-L flask, what will be the pressure in the flask at 0°C when the reaction is complete?

77. What volume of hydrogen at STP can be produced by reacting 8.30 mol of Al with sulfuric acid? The equation is

$$2\,Al(s) + 3\,H_2SO_4(aq) \longrightarrow Al_2(SO_4)_3(aq) + 3\,H_2(g)$$

78. What are the relative rates of effusion of N_2 and He?

***79.** (a) What are the relative rates of effusion of CH_4 and He?

(b) If these two gases are simultaneously introduced into opposite ends of a 100.-cm tube and allowed to diffuse toward each other, at what distance from the helium end will molecules of the two gases meet?

***80.** A gas has a percent composition by mass of 85.7% carbon and 14.3% hydrogen. At STP the density of the gas is 2.50 g/L. What is the molecular formula of the gas?

***81.** Assume that the reaction

$$2\,CO(g) + O_2(g) \longrightarrow 2\,CO_2(g)$$

goes to completion. When 10. mol of CO and 8.0 mol of O_2 react in a closed 10.-L vessel,

(a) how many moles of CO, O_2, and CO_2 are present at the end of the reaction?

(b) what will be the total pressure in the flask at 0°C?

***82.** If 250 mL of O_2, measured at STP, are obtained by the decomposition of the $KClO_3$ in a 1.20-g mixture of KCl and $KClO_3$,

$$2\,KClO_3(s) \longrightarrow 2\,KCl(s) + 3\,O_2(g)$$

what is the percent by mass of $KClO_3$ in the mixture?

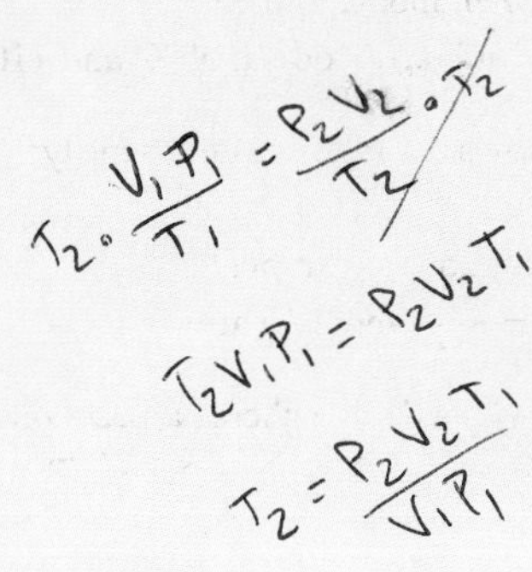

Challenge Exercises

All exercises with blue numbers have answers in the appendix of the text.

***83.** Air has a density of 1.29 g/L at STP. Calculate the density of air on Pikes Peak, where the pressure is 450 torr and the temperature is 17°C.

***84.** Consider the arrangement of gases shown below. If the valve between the gases is opened and the temperature is held constant, determine the following:
(a) the pressure of each gas
(b) the total pressure in the system

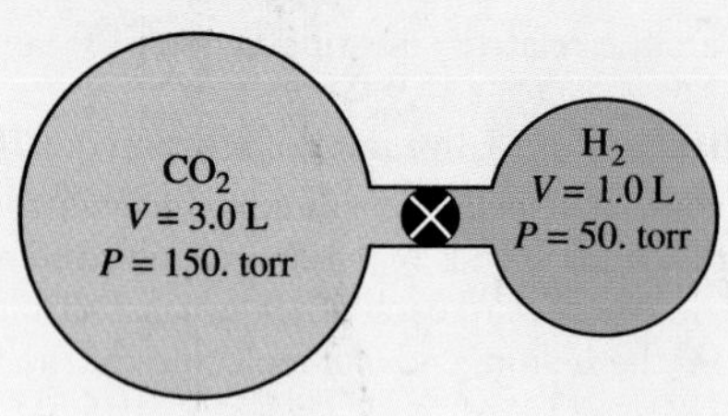

***85.** A steel cylinder contained 50.0 L of oxygen gas under a pressure of 40.0 atm and at a temperature of 25°C. What was the pressure in the cylinder during a storeroom fire that caused the temperature to rise 152°C? (Be careful!)

***86.** A balloon has a mass of 0.5 g when completely deflated. When it is filled with an unknown gas, the mass increases to 1.7 g. You notice on the canister of the unknown gas that it occupies a volume of 0.4478 L at a temperature of 50°C. You note the temperature in the room is 25°C. Identify the gas.

Answers to Practice Exercises

12.1 (a) 851 torr, (b) 851 mm Hg
12.2 0.596 atm
12.3 4.92 L
12.4 762 torr
12.5 84.7 L
12.6 861 K (588°C)
12.7 518 mL
12.8 1.4×10^2 g/mol
12.9 0.89 g/L
12.10 0.953 mol
12.11 128 g/mol
12.12 1.56 L O_2
12.13 75.0 L O_2, 45.0 L CO_2, 60.0 L H_2O

CHAPTER 13

Properties of Liquids

Bushkill Falls and Bridesmaid Falls (lower level) at Delaware Water Gap, Pennsylvania.

Chapter Outline

Planet Earth, that magnificient blue sphere we enjoy viewing from space, is spectacular. Over 75% of Earth is covered with water. We are born from it, drink it, bathe in it, cook with it, enjoy its beauty in waterfalls and rainbows, and stand in awe of the majesty of icebergs. Water supports and enhances life.

In chemistry, water provides the medium for numerous reactions. The shape of the water molecule is the basis for hydrogen bonds. These bonds determine the unique properties and reactions of water. The tiny water molecule holds the answers to many of the mysteries of chemical reactions.

13.1 What Is a Liquid?

In the last chapter, we found that gases contain particles that are far apart, in rapid random motion, and essentially independent of each other. The kinetic-molecular theory, along with the ideal gas law, summarizes the behavior of most gases at relatively high temperatures and low pressures.

Solids are obviously very different from gases. Solids contain particles that are very close together; solids have a high density, compress negligibly, and maintain their shape regardless of container. These characteristics indicate large attractive forces between particles. The model for solids is very different from the one for gases.

Liquids, on the other hand, lie somewhere between the extremes of gases and solids. Liquids contain particles that are close together; liquids are essentially incompressible, and have definite volume. These properties are very similar to those of solids. But liquids also take the shape of their containers; this is closer to the model of a gas.

Although liquids and solids show similar properties, they differ tremendously from gases (see Figure 13.1). No simple mathematical relationship, like the ideal gas law, works well for liquids or solids. Instead, these models are directly related to the forces of attraction between molecules. With these general statements in mind, let's consider some specific properties of liquids.

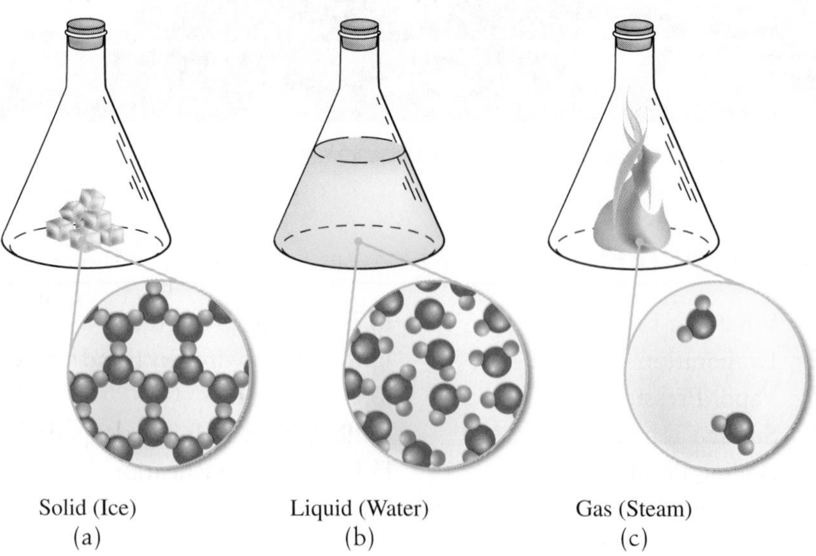

Figure 13.1
The three states of matter. (a) Solid—water molecules are held together rigidly and are very close to each other. (b) Liquid—water molecules are close together but are free to move around and slide over each other. (c) Gas—water molecules are far apart and move freely and randomly.

13.2 Evaporation

When beakers of water, ethyl ether, and ethyl alcohol (all liquids at room temperature) are allowed to stand uncovered, their volumes gradually decrease. The process by which this change takes place is called *evaporation*.

Attractive forces exist between molecules in the liquid state. Not all of these molecules, however, have the same kinetic energy. Molecules that have greater-than-average kinetic energy can overcome the attractive forces and break away from the surface of the liquid to become a gas (see Figure 13.2). **Evaporation,** or **vaporization,** is the escape of molecules from the liquid state to the gas or vapor state.

evaporation
vaporization

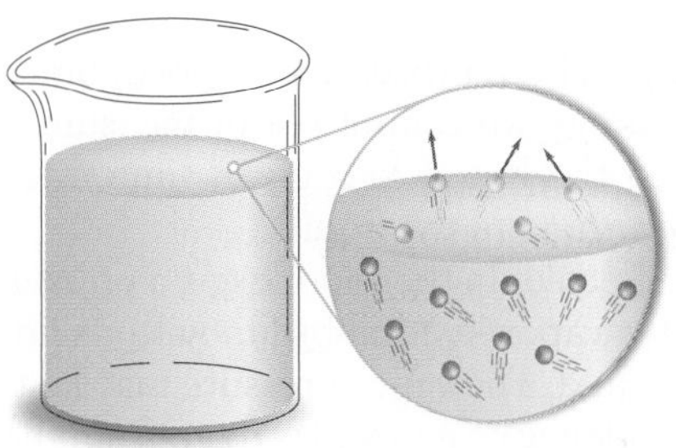

Figure 13.2
The high-energy molecules escape from the surface of a liquid in a process known as evaporation.

In evaporation, molecules of greater-than-average kinetic energy escape from a liquid, leaving it cooler than it was before they escaped. For this reason, evaporation of perspiration is one way the human body cools itself and keeps its temperature constant. When volatile liquids such as ethyl chloride (C_2H_5Cl) are sprayed on the skin, they evaporate rapidly, cooling the area by removing heat. The numbing effect of the low temperature produced by evaporation of ethyl chloride allows it to be used as a local anesthetic for minor surgery.

Solids such as iodine, camphor, naphthalene (moth balls), and, to a small extent, even ice will go directly from the solid to the gaseous state, bypassing the liquid state. This change is a form of evaporation and is called **sublimation:**

sublimation

$$\text{liquid} \xrightarrow{\text{evaporation}} \text{vapor}$$

$$\text{solid} \xrightarrow{\text{sublimation}} \text{vapor}$$

13.3 Vapor Pressure

When a liquid vaporizes in a closed system like that shown in Figure 13.3b, some of the molecules in the vapor or gaseous state strike the surface and return to the liquid state by the process of **condensation.** The rate of condensation increases until it's equal to the rate of vaporization. At this point, the space above the liquid is said to be saturated with vapor, and an equilibrium, or steady state, exists between the liquid and the vapor. The equilibrium equation is

condensation

$$\text{liquid} \underset{\text{condensation}}{\overset{\text{evaporation}}{\rightleftharpoons}} \text{vapor}$$

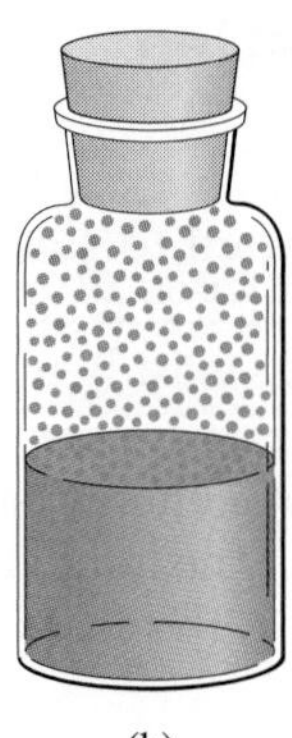

Figure 13.3
(a) Molecules in an open beaker evaporate from the liquid and disperse into the atmosphere. Evaporation will continue until all the liquid is gone. (b) Molecules leaving the liquid are confined to a limited space. With time, the concentration in the vapor phase will increase until an equilibrium between liquid and vapor is established.

This equilibrium is dynamic; both processes—vaporization and condensation—are taking place, even though we cannot see or measure a change. The number of molecules leaving the liquid in a given time interval is equal to the number of molecules returning to the liquid.

vapor pressure

At equilibrium, the molecules in the vapor exert a pressure like any other gas. The pressure exerted by a vapor in equilibrium with its liquid is known as the **vapor pressure** of the liquid. The vapor pressure may be thought of as a measure of the "escaping" tendency of molecules to go from the liquid to the vapor state. The vapor pressure of a liquid is independent of the amount of liquid and vapor present, but it increases as the temperature rises. Figure 13.4 illustrates a liquid–vapor equilibrium and the measurement of vapor pressure.

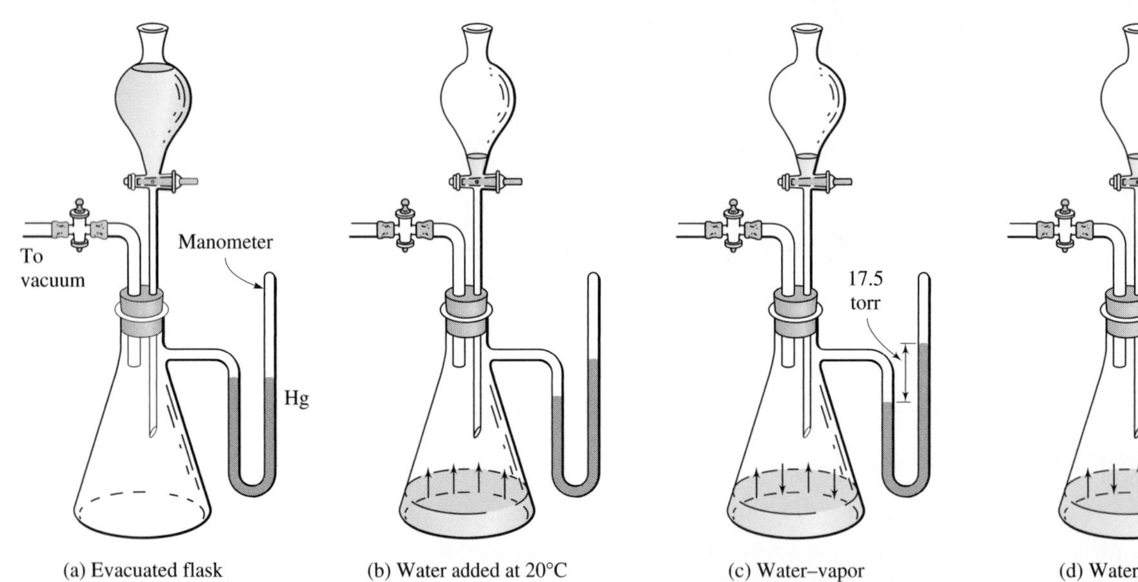

Figure 13.4
Measurement of the vapor pressure of water at 20°C and 30°C. (a) The system is evacuated. The mercury manometer attached to the flask shows equal pressure in both legs. (b) Water has been added to the flask and begins to evaporate, exerting pressure as indicated by the manometer. (c) When equilibrium is established, the pressure inside the flask remains constant at 17.5 torr. (d) The temperature is changed to 30°C, and equilibrium is reestablished with the vapor pressure at 31.8 torr.

Table 13.1 Vapor Pressure of Water, Ethyl Alcohol, and Ethyl Ether at Various Temperatures

	Vapor pressure (torr)		
Temperature (°C)	Water	Ethyl alcohol	Ethyl ether*
0	4.6	12.2	185.3
10	9.2	23.6	291.7
20	17.5	43.9	442.2
30	31.8	78.8	647.3
40	55.3	135.3	921.3
50	92.5	222.2	1276.8
60	152.9	352.7	1729.0
70	233.7	542.5	2296.0
80	355.1	812.6	2993.6
90	525.8	1187.1	3841.0
100	760.0	1693.3	4859.4
110	1074.6	2361.3	6070.1

*Note that the vapor pressure of ethyl ether at temperatures of 40°C and higher exceeds standard pressure, 760 torr, which indicates that the substance has a low boiling point and should therefore be stored in a cool place in a tightly sealed container.

When equal volumes of water, ethyl ether, and ethyl alcohol are placed in beakers and allowed to evaporate at the same temperature, we observe that the ether evaporates faster than the alcohol, which evaporates faster than the water. This order of evaporation is consistent with the fact that ether has a higher vapor pressure at any particular temperature than ethyl alcohol or water. One reason for this higher vapor pressure is that the attraction is less between ether molecules than between alcohol or water molecules. The vapor pressures of these three compounds at various temperatures are compared in Table 13.1.

volatile

Substances that evaporate readily are said to be **volatile.** A volatile liquid has a relatively high vapor pressure at room temperature. Ethyl ether is a very volatile liquid; water is not too volatile; and mercury, which has a vapor pressure of 0.0012 torr at 20°C, is essentially a nonvolatile liquid. Most substances that are normally in a solid state are nonvolatile (solids that sublime are exceptions).

13.4 Surface Tension

Mercury droplets.

surface tension

Have you ever observed water and mercury in the form of small drops? These liquids form drops because liquids have *surface tension.* A droplet of liquid that is not falling or under the influence of gravity (as on the space shuttle) will form a sphere. Spheres minimize the ratio of surface area to volume. The molecules within the liquid are attracted to the surrounding liquid molecules, but at the liquid's surface, the attraction is nearly all inward. This pulls the surface into a spherical shape. The resistance of a liquid to an increase in its surface area is called the **surface tension** of the liquid. Substances with large attractive forces between molecules have high surface tensions. The effect of surface tension in water is illustrated by floating a needle on the surface of still water.

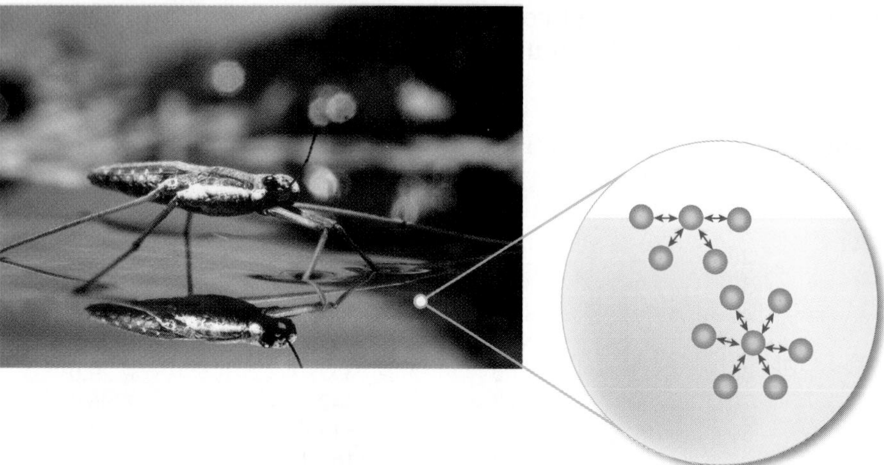

A water strider skims the surface of the water as a result of surface tension. At the molecular level the surface tension results from the net attraction of the water molecules toward the liquid below. In the interior of the water, the forces are balanced in all directions.

Other examples include a water strider walking across a calm pond and water beading on a freshly waxed car. Surface tension is temperature dependent, decreasing with increasing temperature.

capillary action

Liquids also exhibit a phenomenon called **capillary action,** the spontaneous rising of a liquid in a narrow tube. This action results from the *cohesive forces* within the liquid and the *adhesive forces* between the liquid and the walls of the container. If the forces between the liquid and the container are greater than those within the liquid itself, the liquid will climb the walls of the container. For example, consider the California sequoia, a tree that reaches over 200 feet in height. Although water rises only 33 feet in a glass tube (under atmospheric pressure), capillary action causes water to rise from the sequoia's roots to all its parts.

meniscus

The meniscus in liquids is further evidence of cohesive and adhesive forces. When a liquid is placed in a glass cylinder, the surface of the liquid shows a curve called the **meniscus** (see Figure 13.5). The concave shape of water's meniscus shows that the adhesive forces between the glass and water are stronger than the cohesive forces within the water. In a nonpolar substance such as mercury, the meniscus is convex, indicating that the cohesive forces within mercury are greater than the adhesive forces between the glass wall and the mercury.

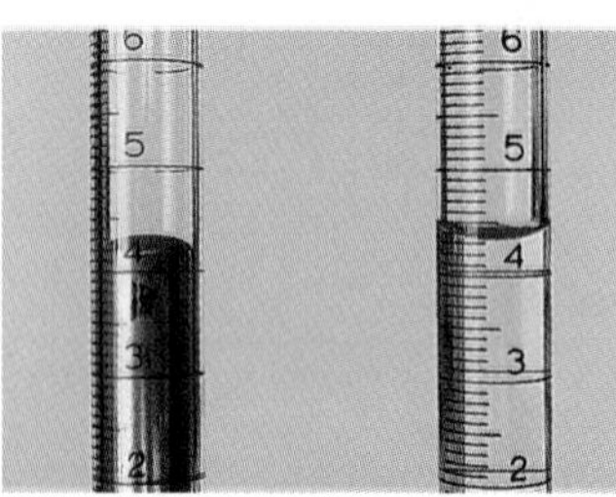

Figure 13.5
The meniscus of mercury (left) and water (right). The meniscus is the characteristic curve of the surface of a liquid in a narrow tube.

13.5 Boiling Point

The boiling temperature of a liquid is related to its vapor pressure. We've seen that vapor pressure increases as temperature increases. When the internal or vapor pressure of a liquid becomes equal to the external pressure, the liquid boils. (By external pressure we mean the pressure of the atmosphere above the liquid.) The boiling temperature of a pure liquid remains constant as long as the external pressure does not vary.

The boiling point (bp) of water is 100°C at 1 atm pressure. Table 13.1 shows that the vapor pressure of water at 100°C is 760 torr. The significant fact here is that the boiling point is the temperature at which the vapor pressure of the water or other liquid is equal to standard, or atmospheric, pressure at sea level.

boiling point

Boiling point is the temperature at which the vapor pressure of a liquid is equal to the external pressure above the liquid.

We can readily see that a liquid has an infinite number of boiling points. When we give the boiling point of a liquid, we should also state the pressure. When we express the boiling point without stating the pressure, we mean it to be the **normal boiling point** at standard pressure (760 torr). Using Table 13.1 again, we see that the normal boiling point of ethyl ether is between 30°C and 40°C, and for ethyl alcohol it is between 70°C and 80°C because for each compound 760 torr lies within these stated temperature ranges. At the normal boiling point, 1 g of a liquid changing to a vapor (gas) absorbs an amount of energy equal to its heat of vaporization (see Table 13.2).

normal boiling point

The boiling point at various pressures can be evaluated by plotting the data of Table 13.1 on the graph in Figure 13.6, where temperature is plotted horizontally along the *x*-axis and vapor pressure is plotted vertically along the *y*-axis.

Table 13.2 Physical Properties of Ethyl Chloride, Ethyl Ether, Ethyl Alcohol, and Water

Substance	Boiling point (°C)	Melting point (°C)	Heat of vaporization J/g (cal/g)	Heat of fusion J/g (cal/g)
Ethyl chloride	12.3	−139	387 (92.5)	—
Ethyl ether	34.6	−116	351 (83.9)	—
Ethyl alcohol	78.4	−112	855 (204.3)	104 (24.9)
Water	100.0	0	2259 (540)	335 (80)

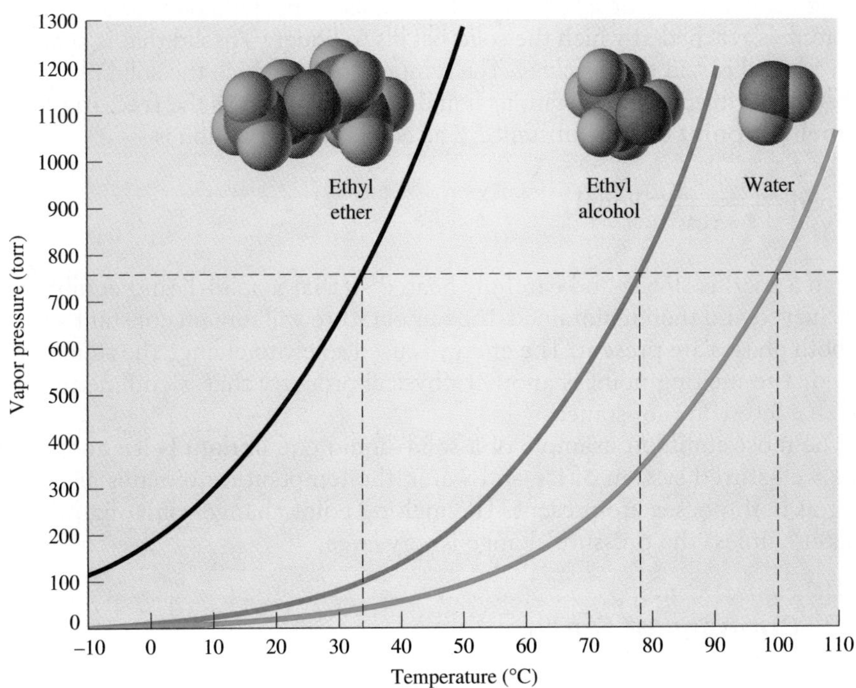

Figure 13.6
Vapor pressure–temperature curves for ethyl ether, ethyl alcohol, and water.

vapor pressure curves

The resulting curves are known as **vapor pressure curves.** Any point on these curves represents a vapor–liquid equilibrium at a particular temperature and pressure. We can find the boiling point at any pressure by tracing a horizontal line from the designated pressure to a point on the vapor pressure curve. From this point, we draw a vertical line to obtain the boiling point on the temperature axis. Three such points are shown in Figure 13.6; they represent the normal boiling points of the three compounds at 760 torr. By reversing this process, you can ascertain at what pressure a substance will boil at a specific temperature. The boiling point is one of the most commonly used physical properties for characterizing and identifying substances.

Practice 13.1

Use the graph in Figure 13.6 to determine the boiling points of ethyl ether, ethyl alcohol, and water at 600 torr.

Practice 13.2

The average atmospheric pressure in Denver is 0.83 atm. What is the boiling point of water in Denver?

13.6 Freezing Point or Melting Point

As heat is removed from a liquid, the liquid becomes colder and colder, until a temperature is reached at which it begins to solidify. A liquid changing into a solid is said to be *freezing,* or *solidifying*. When a solid is heated continuously, a temperature is reached at which the solid begins to liquefy. A solid that is changing into a liquid is said to be *melting*. The temperature at which the solid phase of a substance is in equilibrium with its liquid phase is known as the **freezing point** or **melting point** of that substance. The equilibrium equation is

freezing or melting point

$$\text{solid} \underset{\text{freezing}}{\overset{\text{melting}}{\rightleftharpoons}} \text{liquid}$$

When a solid is slowly and carefully heated so that a solid–liquid equilibrium is achieved and then maintained, the temperature will remain constant as long as both phases are present. The energy is used solely to change the solid to the liquid. The melting point is another physical property that is commonly used for characterizing substances.

The most common example of a solid–liquid equilibrium is ice and water. In a well-stirred system of ice and water, the temperature remains at 0°C as long as both phases are present. The melting point changes only slightly with pressure unless the pressure change is very large.

13.7 Changes of State

The majority of solids undergo two changes of state upon heating. A solid changes to a liquid at its melting point, and a liquid changes to a gas at its boiling point. This warming process can be represented by a graph called a *heating*

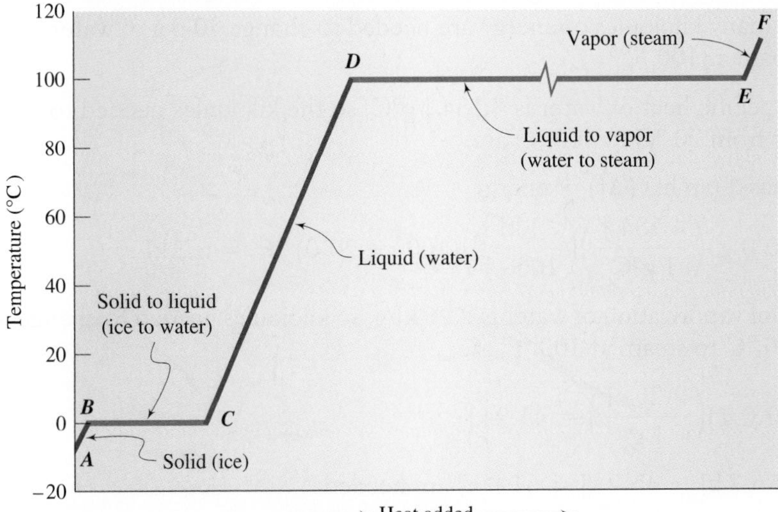

Figure 13.7
Heating curve for a pure substance—the absorption of heat by a substance from the solid state to the vapor state. Using water as an example, the AB interval represents the ice phase; BC interval, the melting of ice to water; CD interval, the elevation of the temperature of water from 0°C to 100°C; DE interval, the boiling of water to steam; and EF interval, the heating of steam.

curve (Figure 13.7). This figure shows ice being heated at a constant rate. As energy flows into the ice, the vibrations within the crystal increase and the temperature rises ($A \longrightarrow B$). Eventually, the molecules begin to break free from the crystal and melting occurs ($B \longrightarrow C$). During the melting process all energy goes into breaking down the crystal structure; the temperature remains constant.

The energy required to change exactly one gram of a solid at its melting point into a liquid is called the **heat of fusion.** When the solid has completely melted, the temperature once again rises ($C \longrightarrow D$); the energy input is increasing the molecular motion within the water. At 100°C, the water reaches its boiling point; the temperature remains constant while the added energy is used to vaporize the water to steam ($D \longrightarrow E$). The **heat of vaporization** is the energy required to change exactly one gram of liquid to vapor at its normal boiling point. The attractive forces between the liquid molecules are overcome during vaporization. Beyond this temperature all the water exists as steam and is being heated further ($E \longrightarrow F$).

heat of fusion

heat of vaporization

Example 13.1

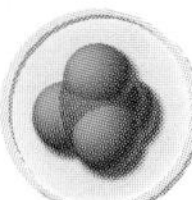

How many joules of energy are needed to change 10.0 g of ice at 0.00°C to water at 20.0°C?

SOLUTION

Ice will absorb 335 J/g (heat of fusion) in going from a solid at 0°C to a liquid at 0°C. An additional 4.184 J/g°C (specific heat of water) is needed to raise the temperature of the water by 1°C.

Joules needed to melt the ice:

$$(10.0\,\cancel{g})\left(\frac{335\,\text{J}}{1\,\cancel{g}}\right) = 3.35 \times 10^3\,\text{J}$$

Joules needed to heat the water from 0.00°C to 20.0°C [(mass) (sp ht)(Δt)]:

$$(10.0\,\cancel{g})\left(\frac{4.184\,\text{J}}{1\,\cancel{g\,°C}}\right)(20.0\,\cancel{°C}) = 837\,\text{J}$$

Thus, $3350\,\text{J} + 837\,\text{J} = 4.19 \times 10^3\,\text{J}$ are needed.

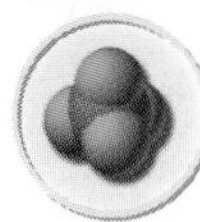

Example 13.2 How many kilojoules of energy are needed to change 20.0 g of water at 20.°C to steam at 100.°C?

SOLUTION The specific heat of water is 4.184 J/g°C, so the kilojoules needed to heat the water from 20.°C to 100.°C are:

$$(\text{mass})\ (\text{sp ht})\ (\Delta t) = \text{energy}$$

$$(20.0\ \cancel{g})\left(\frac{4.184\ \cancel{J}}{1\ \cancel{g°C}}\right)\left(\frac{1\ kJ}{1000\ \cancel{J}}\right)(100. - 20.0)\ \cancel{°C} = 6.7\ kJ$$

Heat of vaporization of water is 2.26 kJ/g, so kilojoules needed to change water at 100.°C to steam at 100.°C are:

$$(20.0\ \cancel{g})\left(\frac{2.26\ kJ}{1\ \cancel{g}}\right) = 45.2\ kJ$$

Thus, 6.7 kJ + 45.2 kJ = 51.9 kJ are needed.

Practice 13.3

How many kilojoules of energy are required to change 50.0 g of ethyl alcohol from 60.0°C to vapor at 78.4°C? The specific heat of ethyl alcohol is 2.138 J/g°C.

13.8 The Hydrogen Bond

Table 13.3 compares the physical properties of H_2O, H_2S, H_2Se, and H_2Te. From this comparison, it is apparent that four physical properties of water—melting point, boiling point, heat of fusion, and heat of vaporization—are extremely high and do not fit the trend relative to the molar masses of the four compounds. For example, if the properties of water followed the progression shown by the other three compounds, we would expect the melting point of water to be below −85°C and the boiling point to be below −60°C.

Why does water exhibit these anomalies? Because liquid water molecules are held together more strongly than other molecules in the same family. The intermolecular force acting between water molecules is called a **hydrogen bond**, which acts like a very weak bond between two polar molecules.

hydrogen bond

Table 13.3 Physical Properties of Water and Other Hydrogen Compounds of Group 6A Elements

Formula	Color	Molar mass (g/mol)	Melting point (°C)	Boiling point, 1 atm (°C)	Heat of fusion J/g (cal/g)	Heat of vapor ization J/g (cal/g)
H_2O	Colorless	18.02	0.00	100.0	335 (80.0)	2.26×10^3(540)
H_2S	Colorless	34.09	−85.5	−60.3	69.9 (16.7)	548 (131)
H_2Se	Colorless	80.98	−65.7	−41.3	31 (7.4)	238 (57.0)
H_2Te	Colorless	129.6	−49	−2	—	179 (42.8)

CHEMISTRY IN ACTION • How Sweet It Is!

Did you think artificial sweeteners were a product of the post–World War II chemical industry? Not so—many of them have been around a long time, and several of the important ones were discovered quite by accident. In 1878, Ira Remsen was working late in his laboratory and realized he was about to miss a dinner with friends. In his haste to leave the lab, he forgot to wash his hands. Later at dinner he broke a piece of bread and tasted it only to discover that it was very sweet. The sweet taste had to be the chemical he had been working with in the lab. Back at the lab, he isolated saccharin—the first of the artificial sweeteners.

In 1937, Michael Sveda was smoking a cigarette in his laboratory (a very dangerous practice to say the least!). He touched the cigarette to his lips and was surprised by the exceedingly sweet taste. The chemical on his hands turned out to be cyclamate, which soon became a staple of the artificial sweetener industry.

In 1965, James Schlatter was researching antiulcer drugs for the pharmaceutical firm G. D. Searle. In the course of his work, he accidentally ingested a small amount of a preparation and found to his surprise that it had an extremely sweet taste. He had discovered aspartame, a molecule consisting of two amino acids joined together. Since only very small quantities of aspartame are necessary to produce sweetness, it proved to be an excellent low-calorie artificial sweetener. More than 50 different molecules have a sweet taste, and it is difficult a find a single binding site that could interact with all of them.

Our taste receptors are composed of proteins that can form hydrogen bonds with other molecules. The proteins contain —NH and —OH groups (with hydrogen available to bond) as well as C=O groups (providing oxygen for hydrogen bonding). "Sweet molecules" also contain H-bonding groups including —OH, $-NH_2$, and O or N. These molecules not only must have the proper atoms to form hydrogen bonds, they must also contain a hydrophobic region (repels H_2O). A new model for binding to a sweetness receptor has been developed at Senomyx in La Jolla, California. The model shows four binding sites that can act independently. Small molecules bind to a pocket on a subunit as shown in the model. Large molecules (such as proteins) bind to a different site above one of the pockets.

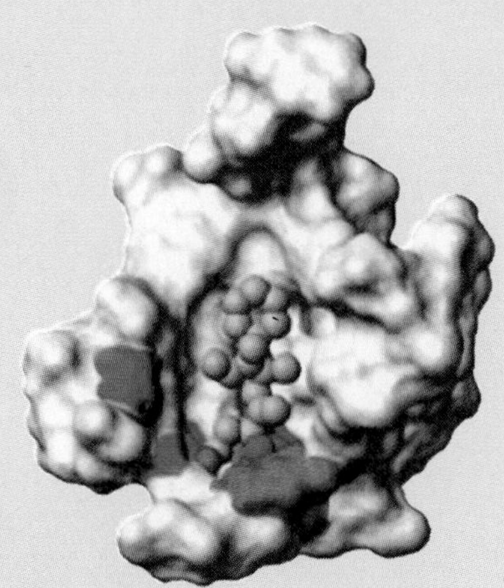

Sweet bondage. Model shows how the sweetener aspartame binds to a site on the sweetness receptor's T1R3 subunit. Red and blue are hydrogen-bond donor and acceptor residues, respectively; aspartame is in gold, except for its carboyxlate (red) and ammonium (blue) groups. Model prepared with MOLMOL (*J. Mol. Graphics* 1996, 14, 51). Adapted from *J. Med. Chem.*

A hydrogen bond is formed between polar molecules that contain hydrogen covalently bonded to a small, highly electronegative atom such as fluorine, oxygen, or nitrogen (F—H, O—H, N—H). A hydrogen bond is actually the dipole–dipole attraction between polar molecules containing these three types of polar bonds.

Compounds that have significant hydrogen-bonding ability are those that contain H covalently bonded to F, O, or N.

Because a hydrogen atom has only one electron, it forms only one covalent bond. When it is bonded to a strong electronegative atom such as oxygen, a hydrogen atom will also be attracted to an oxygen atom of another molecule, forming a dipole–dipole attraction (H bond) between the two molecules. Water has two types of bonds: covalent bonds that exist between hydrogen and oxygen atoms within a molecule and hydrogen bonds that exist between hydrogen and oxygen atoms in *different* water molecules.

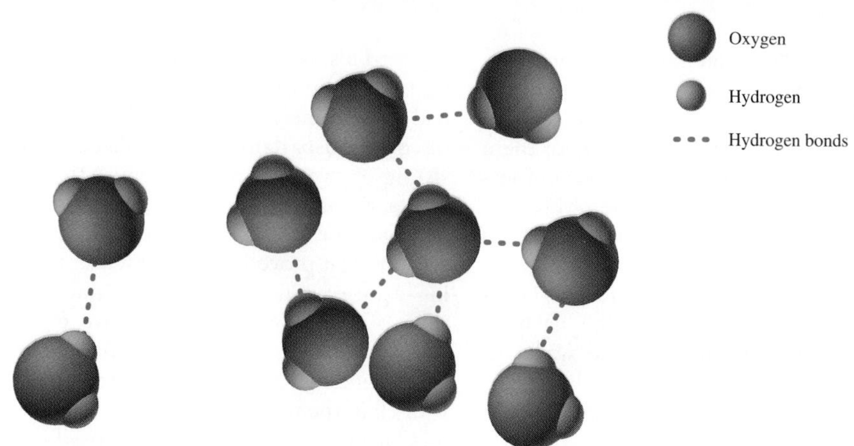

Figure 13.8
Hydrogen bonding. Water in the liquid and solid states exists as aggregates in which the water molecules are linked together by hydrogen bonds.

Hydrogen bonds are *intermolecular* bonds; that is, they are formed between atoms in different molecules. They are somewhat ionic in character because they are formed by electrostatic attraction. Hydrogen bonds are much weaker than the ionic or covalent bonds that unite atoms to form compounds. Despite their weakness, they are of great chemical importance.

The oxygen atom in water can form two hydrogen bonds—one through each of the unbonded pairs of electrons. Figure 13.8 shows two water molecules linked by a hydrogen bond and eight water molecules linked by hydrogen bonds. A dash (—) is used for the covalent bond and a dotted line (····) for the hydrogen bond. In water each molecule is linked to others through hydrogen bonds to form a three-dimensional aggregate of water molecules. This intermolecular hydrogen bonding effectively gives water the properties of a much larger, heavier molecule, explaining in part its relatively high melting point, boiling point, heat of fusion, and heat of vaporization. As water is heated and energy absorbed, hydrogen bonds are continually being broken until at 100°C, with the absorption of an additional 2.26 kJ/g, water separates into individual molecules, going into the gaseous state. Sulfur, selenium, and tellurium are not sufficiently electronegative for their hydrogen compounds to behave like water. The lack of hydrogen bonding is one reason H_2S is a gas, not a liquid, at room temperature.

Fluorine, the most electronegative element, forms the strongest hydrogen bonds. This bonding is strong enough to link hydrogen fluoride molecules together as *dimers,* H_2F_2, or as larger $(HF)_n$ molecular units. The dimer structure may be represented in this way:

```
               ..
              /F:
  H       H  /  ..
   \    .·
    F:··  <—— H bond
    ..
```

Hydrogen bonding can occur between two different atoms that are capable of forming H bonds. Thus, we may have an O····H—N or O—H····N linkage in which the hydrogen atom forming the H bond is between an oxygen and a nitrogen atom. This form of H bond exists in certain types of protein molecules and many biologically active substances.

Example 13.3

Would you expect hydrogen bonding to occur between molecules of these substances?

(a)
```
     H   H
     |   |   ..
 H — C — C — O — H
     |   |   ..
     H   H
```
ethyl alcohol

(b)
```
     H       H
     |  ..   |
 H — C — O — C — H
     |  ..   |
     H       H
```
dimethyl ether

SOLUTION

(a) Hydrogen bonding should occur in ethyl alcohol because one hydrogen atom is bonded to an oxygen atom:

```
     H   H            H   H   H
     |   |   ..       |   |   |
 H — C — C — O — H ···:O — C — C — H
     |   |   ..   ↑   ..  |   |
     H   H      H bond    H   H
```

(b) There is no hydrogen bonding in dimethyl ether because all the hydrogen atoms are bonded only to carbon atoms.

Both ethyl alcohol and dimethyl ether have the same molar mass (46.07 g/mol). Although both compounds have the same molecular formula, C_2H_6O, ethyl alcohol has a much higher boiling point (78.4°C) than dimethyl ether (−23.7°C) because of hydrogen bonding between the alcohol molecules.

Practice 13.4

Would you expect hydrogen bonding to occur between molecules of these substances?

(a)
```
     H   H   H
     |   |   |
 H — C — C — N — H
     |   |   ..
     H   H
```

(b)
```
     H   H   H
     |   |   |
 H — C — N — C — H
     |   ..  |
     H       H
```

(c)
```
                  H
                  |
     H   H    H — C — H
     |   |        |
 H — C — C ——————— N:
     |   |        |
     H   H    H — C — H
                  |
                  H
```

13.9 Hydrates

hydrate
water of hydration
water of crystallization

When certain solutions containing ionic compounds are allowed to evaporate, some water molecules remain as part of the crystalline compound that is left after evaporation is complete. Solids that contain water molecules as part of their crystalline structure are known as **hydrates.** Water in a hydrate is known as **water of hydration** or **water of crystallization.**

CHEMISTRY IN ACTION • Hot Ice—It's a Gas!

Ice could well become the most abundant source of energy in the 21st century. This particular type of ice has unusual properties. Even its existence has been the subject of debate since 1810, when Humphrey Davy synthesized the first hydrate of chlorine. Chemists have long debated whether water could crystallize around a gas to form an unusual state of matter known as a gas hydrate. The material looks just like a chunk of ice left from a winter storm, gray and ugly. But unlike ice, this gas hydrate ice pops and sizzles and a lit match causes it to burst into flames. Left alone, it melts quickly into a puddle of water.

Just what is gas hydrate ice and how does it form? It turns out that gas hydrates form when water and some gases (like propane, ethane, or methane) are present at high pressure and low temperature. These conditions occur in nature at the bottom of the ocean and under permafrost in the Arctic north. Oil companies struggle with hydrate ice since it plugs pipelines and damages oil rigs in oil fields in the Arctic. A gas hydrate is a gas molecule trapped inside the crystal lattice of water molecules. The most common gas is methane produced from the decomposition of organic matter. Sonar mapping has revealed large gas hydrate deposits in polar regions such as Alaska and Siberia and off the southeastern coast of the United States.

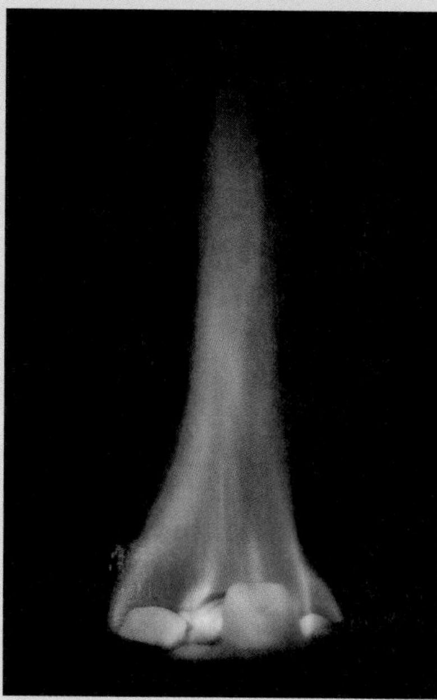

Nuggets of gas hydrate burn as they revert to water (or ice) and gas.

Scientists think that gas hydrates trap huge amounts of natural gas. One cubic meter of methane hydrate can contain the amount of gas that would fill 164 m^3 at room temperature and pressure. If this gas could be brought to the surface economically, it could supply 1000 years of fuel for the United States. As prices increase for energy sources, the costs for liberating gas hydrates become more feasible. At this point, most American research is focused on figuring out how to recover the gas cost-effectively from the hydrates and on transporting the methane to customers. But one day in the not too distant future we could be heating our homes and running our cars by mining gas hydrates from ice!

Formulas for hydrates are expressed by first writing the usual anhydrous (without water) formula for the compound and then adding a dot followed by the number of water molecules present. An example is $BaCl_2 \cdot 2\,H_2O$. This formula tells us that each formula unit of this compound contains one barium ion, two chloride ions, and two water molecules. A crystal of the compound contains many of these units in its crystalline lattice.

In naming hydrates, we first name the compound exclusive of the water and then add the term *hydrate,* with the proper prefix representing the number of water molecules in the formula. For example, $BaCl_2 \cdot 2\,H_2O$ is called *barium chloride dihydrate.* Hydrates are true compounds and follow the law of definite composition. The molar mass of $BaCl_2 \cdot 2\,H_2O$ is 244.2 g/mol; it contains 56.22% barium, 29.03% chlorine, and 14.76% water.

Water molecules in hydrates are bonded by electrostatic forces between polar water molecules and the positive or negative ions of the compound. These forces are not as strong as covalent or ionic chemical bonds. As a result, water of crystallization can be removed by moderate heating of the compound. A partially dehydrated or completely anhydrous compound may result. When $BaCl_2 \cdot 2\,H_2O$ is heated, it loses its water at about 100°C:

$$BaCl_2 \cdot 2\,H_2O(s) \xrightarrow{100^\circ C} BaCl_2(s) + 2\,H_2O(g)$$

(a)

(b)

Figure 13.9
(a) When these blue crystals of $CuSO_4 \cdot 5\,H_2O$ are dissolved in water, a blue solution forms. (b) The anhydrous crystals of $CuSO_4$ are pale green. When water is added, they immediately change color to blue $CuSO_4 \cdot 5\,H_2O$ crystals.

When a solution of copper(II) sulfate ($CuSO_4$) is allowed to evaporate, beautiful blue crystals containing 5 moles water per 1 mole $CuSO_4$ are formed (Figure 13.9a). The formula for this hydrate is $CuSO_4 \cdot 5\,H_2O$; it is called copper(II) sulfate pentahydrate. When $CuSO_4 \cdot 5\,H_2O$ is heated, water is lost, and a pale green-white powder, anhydrous $CuSO_4$, is formed:

$$CuSO_4 \cdot 5\,H_2O(s) \xrightarrow{250^\circ C} CuSO_4(s) + 5\,H_2O(g)$$

When water is added to anhydrous copper(II) sulfate, the foregoing reaction is reversed, and the compound turns blue again (Figure 13.9b). Because of this outstanding color change, anhydrous copper(II) sulfate has been used as an indicator to detect small amounts of water. The formation of the hydrate is noticeably exothermic.

The formula for plaster of paris is $(CaSO_4)_2 \cdot H_2O$. When mixed with the proper quantity of water, plaster of paris forms a dihydrate and sets to a hard mass. It is therefore useful for making patterns for the production of art objects, molds, and surgical casts. The chemical reaction is

$$(CaSO_4)_2 \cdot H_2O(s) + 3\,H_2O(l) \longrightarrow 2\,CaSO_4 \cdot 2\,H_2O(s)$$

Table 13.4 lists a number of common hydrates.

Table 13.4 Selected Hydrates

Hydrate	Name	Hydrate	Name
$CaCl_2 \cdot 2\,H_2O$	calcium chloride dihydrate	$Na_2CO_3 \cdot 10\,H_2O$	sodium carbonate decahydrate
$Ba(OH)_2 \cdot 8\,H_2O$	barium hydroxide octahydrate	$(NH_4)_2C_2O_4 \cdot H_2O$	ammonium oxalate monohydrate
$MgSO_4 \cdot 7\,H_2O$	magnesium sulfate heptahydrate	$NaC_2H_3O_2 \cdot 3\,H_2O$	sodium acetate trihydrate
$SnCl_2 \cdot 2\,H_2O$	tin(II) chloride dihydrate	$Na_2B_4O_7 \cdot 10\,H_2O$	sodium tetraborate decahydrate
$CoCl_2 \cdot 6\,H_2O$	cobalt(II) chloride hexahydrate	$Na_2S_2O_3 \cdot 5\,H_2O$	sodium thiosulfate pentahydrate

Practice 13.5

Write formulas for

(a) beryllium carbonate tetrahydrate
(b) cadmium permanganate hexahydrate
(c) chromium(III) nitrate nonahydrate
(d) platinum(IV) oxide trihydrate

Practice 13.6

Calculate the percent water in Epsom salts, $MgSO_4 \cdot 7\,H_2O$.

13.10 Water, a Unique Liquid

Water is our most common natural resource. It covers about 75% of Earth's surface. Not only is it found in the oceans and seas, in lakes, rivers, streams, and glacial ice deposits, it is always present in the atmosphere and in cloud formations.

About 97% of Earth's water is in the oceans. This *saline* water contains vast amounts of dissolved minerals. More than 70 elements have been detected in the mineral content of seawater. Only four of these—chlorine, sodium, magnesium, and bromine—are now commercially obtained from the sea. The world's *fresh* water comprises the other 3%, of which about two-thirds is locked up in polar ice caps and glaciers. The remaining fresh water is found in groundwater, lakes, rivers and the atmosphere.

Water is an essential constituent of all living matter. It is the most abundant compound in the human body, making up about 70% of total body mass. About 92% of blood plasma is water; about 80% of muscle tissue is water; and about 60% of a red blood cell is water. Water is more important than food in the sense that we can survive much longer without food than without water.

Physical Properties of Water

Water is a colorless, odorless, tasteless liquid with a melting point of 0°C and a boiling point of 100°C at 1 atm. The heat of fusion of water is 335 J/g (80 cal/g). The heat of vaporization of water is 2.26 kJ/g (540 cal/g). The values for water for both the heat of fusion and the heat of vaporization are high compared with those for other substances; this indicates strong attractive forces between the molecules.

Ice and water exist together in equilibrium at 0°C, as shown in Figure 13.10. When ice at 0°C melts, it absorbs 335 J/g in changing into a liquid; the temperature remains at 0°C. To refreeze the water, 335 J/g must be removed from the liquid at 0°C.

In Figure 13.10, both boiling water and steam are shown to have a temperature of 100°C. It takes 418 J to heat 1 g of water from 0°C to 100°C, but water at its boiling point absorbs 2.26 kJ/g in changing to steam. Although boiling water and steam are both at the same temperature, steam contains considerably more heat per gram and can cause more severe burns than hot water.

The maximum density of water is 1.000 g/mL at 4°C. Water has the unusual property of contracting in volume as it is cooled to 4°C and then expanding when cooled from 4°C to 0°C. Therefore, 1 g of water occupies a volume greater than 1 mL at all temperatures except 4°C. Although most liquids

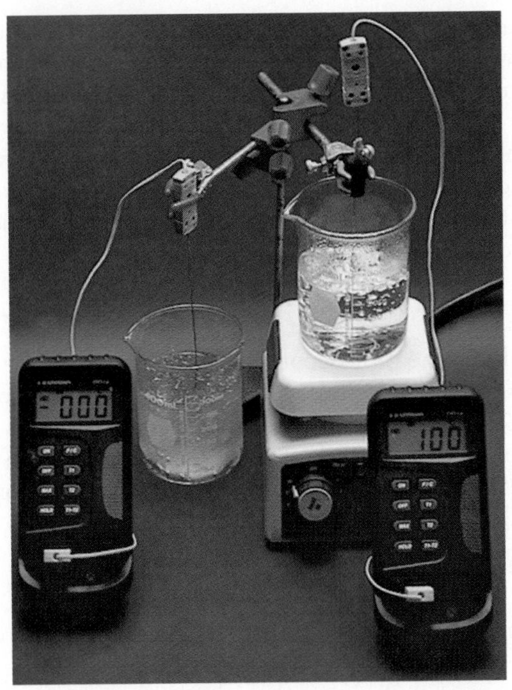

Figure 13.10
Water equilibrium systems.
In the beaker on the left, ice and water are in equilibrium at 0°C; in the beaker on the right, boiling water and steam are in equilibrium at 100°C.

contract in volume all the way down to the point at which they solidify, a large increase (about 9%) in volume occurs when water changes from a liquid at 0°C to a solid (ice) at 0°C. The density of ice at 0°C is 0.917 g/mL, which means that ice, being less dense than water, will float in water.

Structure of the Water Molecule

A single water molecule consists of two hydrogen atoms and one oxygen atom. Each hydrogen atom is bonded to the oxygen atom by a single covalent bond. This bond is formed by the overlap of the 1*s* orbital of hydrogen with an unpaired 2*p* orbital of oxygen. The average distance between the two nuclei is known as the *bond length*. The O—H bond length in water is 0.096 nm. The water molecule is nonlinear and has a bent structure with an angle of about 105° between the two bonds (see Figure 13.11).

Oxygen is the second most electronegative element. As a result, the two covalent OH bonds in water are polar. If the three atoms in a water molecule were aligned in a linear structure, such as $H \mapsto O \leftarrow\!\!\!+ H$, the two polar

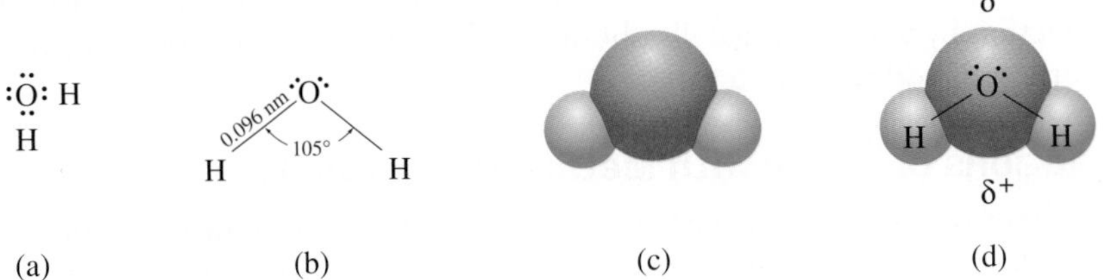

Figure 13.11
Diagrams of a water molecule: (a) electron distribution, (b) bond angle and O—H bond length, (c) molecular orbital structure, and (d) dipole representation.

bonds would be acting in equal and opposite directions and the molecule would be nonpolar. However, water is a highly polar molecule. It therefore does not have a linear structure. When atoms are bonded in a nonlinear fashion, the angle formed by the bonds is called the *bond angle*. In water the HOH bond angle is 105°. The two polar covalent bonds and the bent structure result in a partial negative charge on the oxygen atom and a partial positive charge on each hydrogen atom. The polar nature of water is responsible for many of its properties, including its behavior as a solvent.

Formation of Water

Water is very stable to heat; it decomposes to the extent of only about 1% at temperatures up to 2000°C. Pure water is a nonconductor of electricity. But when a small amount of sulfuric acid or sodium hydroxide is added, the solution is readily decomposed into hydrogen and oxygen by an electric current. Two volumes of hydrogen are produced for each volume of oxygen:

$$2\,H_2O(l) \xrightarrow[H_2SO_4 \text{ or NaOH}]{\text{electrical energy}} 2\,H_2(g) + O_2(g)$$

Water is formed when hydrogen burns in air. Pure hydrogen burns very smoothly in air, but mixtures of hydrogen and air or oxygen explode when ignited. The reaction is strongly exothermic:

$$2\,H_2(g) + O_2(g) \longrightarrow 2\,H_2O(g) + 484\text{ kJ}$$

Water is produced by a variety of other reactions, especially by (1) acid–base neutralizations, (2) combustion of hydrogen-containing materials, and (3) metabolic oxidation in living cells:

1. $HCl(aq) + NaOH(aq) \longrightarrow NaCl(aq) + H_2O(l)$

2. $\underset{\text{acetylene}}{2\,C_2H_2(g)} + 5\,O_2(g) \longrightarrow 4\,CO_2(g) + 2\,H_2O(g) + 1212\text{ kJ}$

$\underset{\text{methane}}{CH_4(g)} + 2\,O_2(g) \longrightarrow CO_2(g) + 2\,H_2O(g) + 803\text{ kJ}$

3. $\underset{\text{glucose}}{C_6H_{12}O_6(aq)} + 6\,O_2(g) \xrightarrow{\text{enzymes}} 6\,CO_2(g) + 6\,H_2O(l) + 2519\text{ kJ}$

Sodium reacts vigorously with water to produce hydrogen gas and sodium hydroxide.

The combustion of acetylene shown in (2) is strongly exothermic and is capable of producing very high temperatures. It is used in oxygen–acetylene torches to cut and weld steel and other metals. Methane is known as natural gas and is commonly used as fuel for heating and cooking. The reaction of glucose with oxygen shown in (3) is the reverse of photosynthesis. It is the overall reaction by which living cells obtain needed energy by metabolizing glucose to carbon dioxide and water.

Reactions of Water with Metals and Nonmetals

The reactions of metals with water at different temperatures show that these elements vary greatly in their reactivity. Metals such as sodium, potassium, and calcium react with cold water to produce hydrogen and a metal hydroxide. A small piece of sodium added to water melts from the heat produced by the reaction, forming a silvery metal ball, which rapidly flits back and forth on the surface of the water. Caution must be used when experimenting with this reaction, because the hydrogen produced is frequently ignited by the sparking of the sodium, and it will explode, spattering sodium. Potassium reacts even more vigorously than sodium. Calcium sinks in water and liberates a gentle stream of hydrogen. The equations for these reactions are

$$2\,Na(s) + 2\,H_2O(l) \longrightarrow H_2(g) + 2\,NaOH(aq)$$
$$2\,K(s) + 2\,H_2O(l) \longrightarrow H_2(g) + 2\,KOH(aq)$$
$$Ca(s) + 2\,H_2O(l) \longrightarrow H_2(g) + Ca(OH)_2(aq)$$

Zinc, aluminum, and iron do not react with cold water but will react with steam at high temperatures, forming hydrogen and a metallic oxide. The equations are

$$Zn(s) + H_2O(g) \longrightarrow H_2(g) + ZnO(s)$$
$$2\,Al(s) + 3\,H_2O(g) \longrightarrow 3\,H_2(g) + Al_2O_3(s)$$
$$3\,Fe(s) + 4\,H_2O(g) \longrightarrow 4\,H_2(g) + Fe_3O_4(s)$$

Copper, silver, and mercury are examples of metals that do not react with cold water or steam to produce hydrogen. We conclude that sodium, potassium, and calcium are chemically more reactive than zinc, aluminum, and iron, which are more reactive than copper, silver, and mercury.

Certain nonmetals react with water under various conditions. For example, fluorine reacts violently with cold water, producing hydrogen fluoride and free oxygen. The reactions of chlorine and bromine are much milder, producing what is commonly known as "chlorine water" and "bromine water," respectively. Chlorine water contains HCl, $HOCl$, and dissolved Cl_2; the free chlorine gives it a yellow-green color. Bromine water contains HBr, $HOBr$, and dissolved Br_2; the free bromine gives it a reddish-brown color. The equations for these reactions are

$$2\,F_2(g) + 2\,H_2O(l) \longrightarrow 4\,HF(aq) + O_2(g)$$
$$Cl_2(g) + H_2O(l) \longrightarrow HCl(aq) + HOCl(aq)$$
$$Br_2(l) + H_2O(l) \longrightarrow HBr(aq) + HOBr(aq)$$

Reactions of Water with Metal and Nonmetal Oxides

Metal oxides that react with water to form hydroxides are known as **basic anhydrides.** Examples are

basic anhydride

$$CaO(s) + H_2O(l) \longrightarrow \underset{\text{calcium hydroxide}}{Ca(OH)_2(aq)}$$

$$Na_2O(s) + H_2O(l) \longrightarrow \underset{\text{sodium hydroxide}}{2\,NaOH(aq)}$$

Certain metal oxides, such as CuO and Al_2O_3, do not form solutions containing OH^- ions because the oxides are insoluble in water.

Practice 13.7

Copper does not react with water to produce hydrogen gas. However, hydrogen reacts with copper(II) oxide to produce water. Write a balanced equation for this reaction.

acid anhydride Nonmetal oxides that react with water to form acids are known as **acid anhydrides.** Examples are

$$CO_2(g) + H_2O(l) \rightleftharpoons \underset{\text{carbonic acid}}{H_2CO_3(aq)}$$

$$SO_2(g) + H_2O(l) \rightleftharpoons \underset{\text{sulfurous acid}}{H_2SO_3(aq)}$$

$$N_2O_5(s) + H_2O(l) \longrightarrow \underset{\text{nitric acid}}{2\,HNO_3(aq)}$$

The word *anhydrous* means "without water." An anhydride is a metal oxide or a nonmetal oxide derived from a base or an oxy-acid by the removal of water. To determine the formula of an anhydride, the elements of water are removed from an acid or base formula until all the hydrogen is removed. Sometimes more than one formula unit is needed to remove all the hydrogen as water. The formula of the anhydride then consists of the remaining metal or nonmetal and the remaining oxygen atoms. In calcium hydroxide, removal of water as indicated leaves CaO as the anhydride:

$$Ca\begin{matrix} O\boxed{H} \\ \boxed{OH} \end{matrix} \xrightarrow{\Delta} CaO + H_2O$$

In sodium hydroxide, H_2O cannot be removed from one formula unit, so two formula units of NaOH must be used, leaving Na_2O as the formula of the anhydride:

$$\begin{matrix} NaO\boxed{H} \\ Na\boxed{OH} \end{matrix} \xrightarrow{\Delta} Na_2O + H_2O$$

The removal of H_2O from H_2SO_4 gives the acid anhydride SO_3:

$$H_2SO_4 \xrightarrow{\Delta} SO_3 + H_2O$$

The foregoing are examples of typical reactions of water but are by no means a complete list of the known reactions of water.

13.11 Water Purification

Natural fresh waters are not pure, but contain dissolved minerals, suspended matter, and sometimes harmful bacteria. The water supplies of large cities are usually drawn from rivers or lakes. Such water is generally unsafe to drink without treatment. To make such water safe to drink, it is treated by some or all of the following processes (see Figure 13.12):

1. *Screening.* Removal of relatively large objects, such as trash, fish, and so on.
2. *Flocculation and sedimentation.* Chemicals, usually lime, CaO, and alum (aluminum sulfate), $Al_2(SO_4)_3$, are added to form a flocculent jellylike precipitate of aluminum hydroxide. This precipitate traps most of the fine suspended matter in the water and carries it to the bottom of the sedimentation basin.
3. *Sand filtration.* Water is drawn from the top of the sedimentation basin and passed downward through fine sand filters. Nearly all the remaining suspended matter and bacteria are removed by the sand filters.

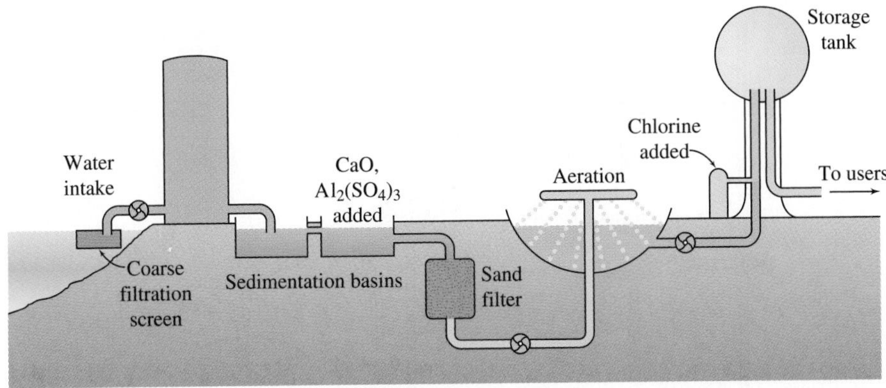

Figure 13.12
Typical municipal water-treatment plant.

4. *Aeration.* Water is drawn from the bottom of the sand filters and is aerated by spraying. The purpose of this process is to remove objectionable odors and tastes.
5. *Disinfection.* In the final stage chlorine gas is injected into the water to kill harmful bacteria before the water is distributed to the public. Ozone is also used to disinfect water. In emergencies, water may be disinfected by simply boiling it for a few minutes.

If the drinking water of children contains an optimum amount of fluoride ion, their teeth will be more resistant to decay.

Water that contains dissolved calcium and magnesium salts is called *hard water.* Ordinary soap does not lather well in hard water; the soap reacts with the calcium and magnesium ions to form an insoluble greasy scum. However, synthetic soaps, known as detergents, have excellent cleaning qualities and do not form precipitates with hard water. Hard water is also undesirable because it causes "scale" to form on the walls of water heaters, tea kettles, coffee pots, and steam irons, which greatly reduces their efficiency.

Many people today find that they want to further purify their drinking and cooking water. Often this is because of objectionable tastes in the municipal drinking water. People add filtration systems to appliances such as refrigerators and coffee pots or use home water filtration systems. People also buy bottled water for drinking and cooking.

Four techniques are used to "soften" hard water:

1. **Distillation** The water is boiled, and the steam formed is condensed into a liquid again, leaving the minerals behind in the distilling vessel. Figure 13.13 illustrates a simple laboratory distillation apparatus. Commercial stills are capable of producing hundreds of liters of distilled water per hour.
2. **Calcium and magnesium precipitation** Calcium and magnesium ions are precipitated from hard water by adding sodium carbonate and lime. Insoluble calcium carbonate and magnesium hydroxide are precipitated and are removed by filtration or sedimentation.
3. **Ion exchange** Hard water is effectively softened as it is passed through a bed or tank of zeolite—a complex sodium aluminum silicate. In this process, sodium ions replace objectionable calcium and magnesium ions, and the water is thereby softened:

$$Na_2(\text{zeolite})(s) + Ca^{2+}(aq) \longrightarrow Ca(\text{zeolite})(s) + 2\,Na^+(aq)$$

The zeolite is regenerated by back-flushing with concentrated sodium chloride solution, reversing the foregoing reaction.

Bottled water provides a convenient source of hydration for athletes.

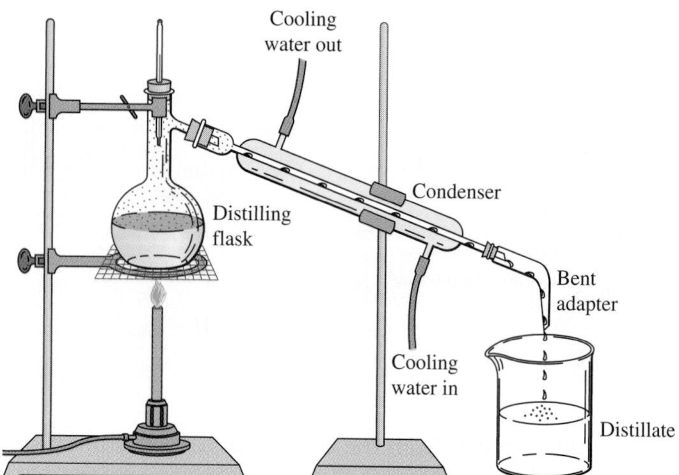

Figure 13.13
Simple laboratory setup for distilling liquids.

4. **Demineralization** Both cations and anions are removed by a two-stage ion-exchange system. Special synthetic organic resins are used in the ion-exchange beds. In the first stage metal cations are replaced by hydrogen ions. In the second stage anions are replaced by hydroxide ions. The hydrogen and hydroxide ions react, and essentially pure, mineral-free water leaves the second stage.

Our oceans are an enormous source of water, but seawater contains about 3.5 lb of salts per 100 lb of water. This 35,000 ppm of dissolved salts makes seawater unfit for agricultural and domestic uses. Water that contains less than 1000 ppm of salts is considered reasonably good for drinking, and safe drinking water is already being obtained from the sea in many parts of the world. Continuous research is being done in an effort to make usable water from our oceans more abundant and economical. See Figure 13.14.

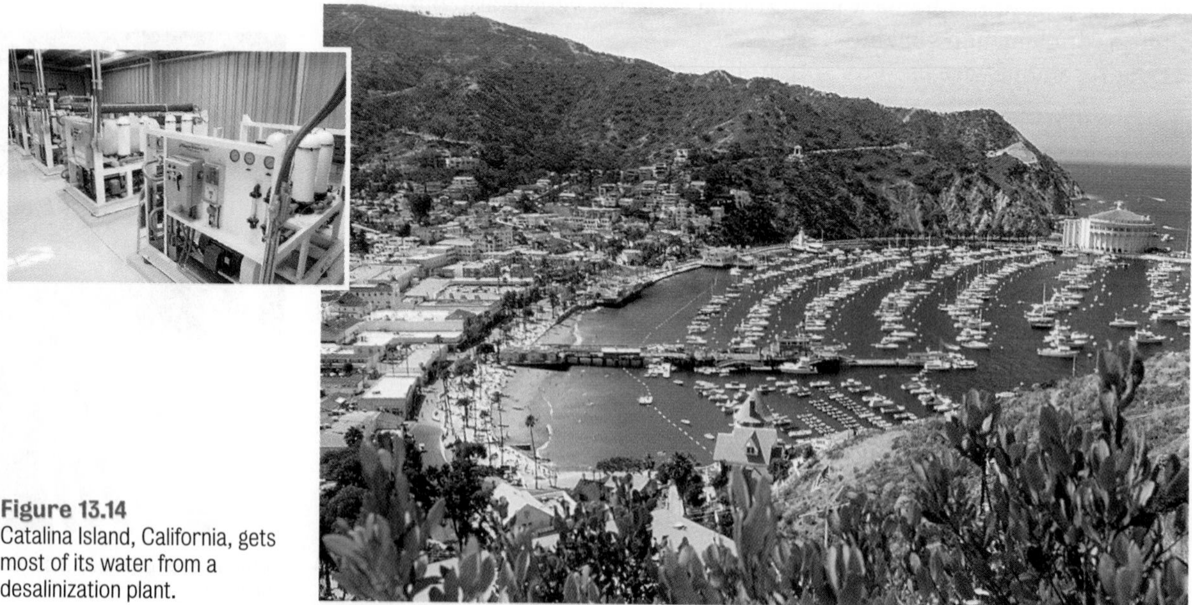

Figure 13.14
Catalina Island, California, gets most of its water from a desalinization plant.

Table 13.5 Classification of Water Pollutants

Type of pollutant	Examples
Oxygen-demanding wastes	Decomposable organic wastes from domestic sewage and industrial wastes of plant and animal origin
Infectious agents	Bacteria, viruses, and other organisms from domestic sewage, animal wastes, and animal process wastes
Plant nutrients	Principally compounds of nitrogen and phosphorus
Organic chemicals	Large numbers of chemicals synthesized by industry, pesticides, chlorinated organic compounds
Other minerals and chemicals	Inorganic chemicals from industrial operations, mining, oil field operations, and agriculture
Radioactive substances	Waste products from mining and processing radioactive materials, airborne radioactive fallout, increased use of radioactive materials in hospitals and research
Heat from industry	Large quantities of heated water returned to water bodies from power plants and manufacturing facilities after use for cooling
Sediment from land erosion	Solid matter washed into streams and oceans by erosion, rain, and water runoff

Our highly technical, industrial society has produced many new chemicals that add to our need for water purification. Many of the newer chemicals are not removed or destroyed by the usual water-treatment processes. For example, among the 66 organic compounds found in the drinking water of a major city on the Mississippi River, 3 are labeled slightly toxic, 17 moderately toxic, 15 very toxic, 1 extremely toxic, and 1 supertoxic. Two are known carcinogens (cancer-producing agents), 11 are suspect, and 3 are metabolized to carcinogens. The U.S. Public Health Service classifies water pollutants under eight broad categories. These categories are shown in Table 13.5.

The disposal of hazardous waste products adds to the water purification problem. These substances are unavoidable in the manufacture of many products we consider indispensable today. One common way to dispose of these wastes is to place them in toxic waste dumps. What has been found after many years of disposing of wastes in this manner is that toxic substances have seeped into the groundwater deposits. As a result, many people have become ill, and water wells have been closed until satisfactory methods of detoxifying this water are found. This problem is serious, because one-half the U.S. population gets its drinking water from groundwater. Cleaning up the thousands of industrial dumps and finding and implementing new and safe methods of disposing of wastes is ongoing and costly.

Chapter 13 Review

13.1 What Is a Liquid?

- Properties:
  - Particles are close together yet free to move.
  - Liquids are incompressible.
  - Liquids have definite volume but take the shape of the container.

13.2 Evaporation

KEY TERMS

Evaporation, or vaporization
Sublimation

- During evaporation, molecules of greater-than-average kinetic energy escape from the liquid.
- Sublimation is the evaporation of a solid directly to a gas.

13.3 Vapor Pressure

KEY TERMS

Condensation
Vapor pressure
Volatile

- Gaseous molecules return to the liquid state through condensation.
- At equilibrium the rate of evaporation equals the rate of condensation.
- The pressure exerted by the vapor in equilibrium with its liquid in a closed container is the vapor pressure of the liquid.
- A volatile substance evaporates readily.

13.4 Surface Tension

KEY TERMS

Surface tension
Capillary action
Meniscus

- The resistance of a liquid to an increase in its surface area is the surface tension of the liquid.
- Capillary action is caused by the cohesive forces within the liquid and adhesive forces between the liquid and the walls of the container:
  - If the forces between the liquid and the container are greater than those within the liquid, the liquid will climb the walls of the container.
- A meniscus is evidence of cohesive and adhesive forces:
  - It is concave if the adhesive forces are stronger than the cohesive forces.
  - It is convex if the cohesive forces are stronger than the adhesive forces.

13.5 Boiling Point

KEY TERMS

Boiling point
Normal boiling point
Vapor pressure curves

- The boiling point of a liquid is the temperature at which its vapor pressure equals the atmospheric pressure:
  - At 1 atm the boiling point is called the normal boiling point for a liquid.

13.6 Freezing Point or Melting Point

KEY TERM

Freezing point or melting point

- The temperature at which a solid is in equilibrium with its liquid phase is the freezing point or melting point.

13.7 Changes of State

KEY TERMS

Heat of fusion
Heat of vaporization

- A graph of the warming of a liquid is called a heating curve:
  - Horizontal lines on the heating curve represent changes of state:
    - The energy required to change 1 g of solid to liquid at its melting point is the heat of fusion.
    - The energy required to change 1 g of liquid to gas at its normal boiling point is the heat of vaporization.
    - The energy required to change the phase of a sample (at its melting or boiling point) is

      energy = (mass)(heat of fusion (or vaporization))
  - The energy required to heat molecules without a phase change is determined by

    energy = (mass)(sp ht)(Δt)

13.8 The Hydrogen Bond

KEY TERM

Hydrogen bond

- A hydrogen bond is the dipole–dipole attraction between polar molecules containing any of these types of bonds: F—H, O—H, N—H.

13.9 Hydrates

KEY TERMS

Hydrate
Water of hydration or water crystallization

- Solids that contain water molecules as part of their crystalline structure are called hydrates.
- Formulas of hydrates are given by writing the formula for the anhydrous compound followed by a dot and then the number of water molecules present.
- Water molecules in hydrates are bonded by electrostatic forces, which are weaker than covalent or ionic bonds:
  - Water of hydration can be removed by heating the hydrate to form a partially dehydrated compound or the anhydrous compound.

13.10 Water, a Unique Liquid

KEY TERMS

Basic anhydride
Acid anhydride

- Water is our most common resource, covering 75% of the Earth's surface.
- Water is a colorless, odorless, tasteless liquid with a melting point of 0°C and boiling point of 100°C at 1 atm
- The water molecule consists of 2 H atoms and 1 O atom bonded together at a 105° bond angle, making the molecule polar.
- Water can be formed in a variety of ways including:
  - Burning hydrogen in air
  - Acid–base neutralizations
  - Combustion of hydrogen-containing materials
  - Metabolic oxidation in living cells
- Water undergoes reactions with many metals and nonmetals, depending on their reactivity.
- Water reacts with metal oxides (basic anhydrides) to form bases (hydroxides).
- Water reacts with nonmetal oxides (acid anhydrides) to form acids.

13.11 Water Purification

- Water can be purified at the commercial level by using some or all of the following processes: screening, flocculation and sedimentation, sand filtration, aeration, and disinfection.
- Water can be softened by one of the following techniques: distillation, calcium and magnesium precipitation, ion exchange, or demineralization.

Review Questions

All questions with blue numbers have answers in the appendix of the text.

1. Compare the potential energy of the three states of water shown in Figure 13.10.
2. In what state (solid, liquid, or gas) would H_2S, H_2Se, and H_2Te be at 0°C? (Table 13.3)
3. What property or properties of liquids are similar to solids?
4. What property or properties of liquids are similar to gases?
5. The temperature of the water in the beaker on the hotplate (Figure 13.10) reads 100°C. What is the pressure of the atmosphere?
6. Diagram a water molecule and point out the negative and positive ends of the dipole.
7. If the water molecule were linear, with all three atoms in a straight line rather than in the shape of a V, as shown in Figure 13.11, what effect would this have on the physical properties of water?
8. How do we specify 1, 2, 3, 4, 5, 6, 7, and 8 molecules of water in the formulas of hydrates? (Table 13.4)
9. Would the distillation setup in Figure 13.13 be satisfactory for separating salt and water? for separating ethyl alcohol and water? Explain.
10. If the liquid in the flask in Figure 13.13 is ethyl alcohol and the atmospheric pressure is 543 torr, what temperature will show on the thermometer? (Use Figure 13.6.)
11. If water were placed in both containers in Figure 13.3, would both have the same vapor pressure at the same temperature? Explain.
12. Why doesn't the vapor pressure of a liquid depend on the amount of liquid and vapor present?
13. In Figure 13.3, in which case, (a) or (b), will the atmosphere above the liquid reach a point of saturation?
14. Suppose a solution of ethyl ether and ethyl alcohol is placed in the closed bottle in Figure 13.3. (Use Figure 13.6 for information on the substances.)
    (a) Are both substances present in the vapor?
    (b) If the answer to part (a) is yes, which has more molecules in the vapor?
15. In Figure 13.4, if 50% more water is added in part (b), what equilibrium vapor pressure will be observed in (c)?
16. At approximately what temperature would each of the substances shown in Figure 13.6 boil when the pressure is 30 torr?
17. Use the graph in Figure 13.6 to find the following:
    (a) boiling point of water at 500 torr
    (b) normal boiling point of ethyl alcohol
    (c) boiling point of ethyl ether at 0.50 atm
18. Consider Figure 13.7.
    (a) Why is line *BC* horizontal? What is happening in this interval?
    (b) What phases are present in the interval *BC*?
    (c) When heating is continued after point *C*, another horizontal line, *DE*, is reached at a higher temperature. What does this line represent?
19. List six physical properties of water.
20. What condition is necessary for water to have its maximum density? What is its maximum density?
21. Account for the fact that an ice–water mixture remains at 0°C until all the ice is melted, even though heat is applied to it.

22. Which contains less energy: ice at 0°C or water at 0°C? Explain.

23. Why does ice float in water? Would ice float in ethyl alcohol ($d = 0.789$ g/mL)? Explain.

24. If water molecules were linear instead of bent, would the heat of vaporization be higher or lower? Explain.

25. The heat of vaporization for ethyl ether is 351 J/g and that for ethyl alcohol is 855 J/g. From this data, which of these compounds has hydrogen bonding? Explain.

26. Would there be more or less H bonding if water molecules were linear instead of bent? Explain.

27. In which condition are there fewer hydrogen bonds between molecules: water at 40°C or water at 80°C? Explain.

28. Which compound

$H_2NCH_2CH_2NH_2$ or $CH_3CH_2CH_2NH_2$

would you expect to have the higher boiling point? Explain. (Both compounds have similar molar masses.)

29. Explain why rubbing alcohol warmed to body temperature still feels cold when applied to your skin.

30. The vapor pressure at 20°C for the following substances is

methyl alcohol	96 torr
acetic acid	11.7 torr
benzene	74.7 torr
bromine	173 torr
water	17.5 torr
carbon tetrachloride	91 torr
mercury	0.0012 torr
toluene	23 torr

(a) Arrange these substances in order of increasing rate of evaporation.
(b) Which substance listed has the highest boiling point? the lowest?

31. Suggest a method whereby water could be made to boil at 50°C.

32. Explain why a higher temperature is obtained in a pressure cooker than in an ordinary cooking pot.

33. What is the relationship between vapor pressure and boiling point?

34. On the basis of the kinetic-molecular theory, explain why vapor pressure increases with temperature.

35. Why does water have such a relatively high boiling point?

36. The boiling point of ammonia, NH_3, is −33.4°C and that of sulfur dioxide, SO_2, is −10.0°C. Which has the higher vapor pressure at −40°C?

37. Explain what is occurring physically when a substance is boiling.

38. Explain why HF (bp = 19.4°C) has a higher boiling point than HCl (bp = −85°C), whereas F_2 (bp = −188°C) has a lower boiling point than Cl_2 (bp = −34°C).

39. Why does a boiling liquid maintain a constant temperature when heat is continuously being added?

40. At what specific temperature will ethyl ether have a vapor pressure of 760 torr?

41. What water temperature would you theoretically expect to find at the bottom of a very deep lake? Explain.

42. Is the formation of hydrogen and oxygen from water an exothermic or an endothermic reaction? How do you know?

Paired Exercises

All exercises with blue numbers have answers in the appendix of the text.

1. Write the formulas for the anhydrides of these acids: $HClO_4$, H_2CO_3, H_3PO_4.

2. Write the formulas for the anhydrides of these acids: H_2SO_3, H_2SO_4, HNO_3.

3. Write the formulas for the anhydrides of these bases: LiOH, NaOH, $Mg(OH)_2$.

4. Write the formulas for the anhydrides of these bases: KOH, $Ba(OH)_2$, $Ca(OH)_2$.

5. Complete and balance these equations:

(a) $Ba(OH)_2 \xrightarrow{\Delta}$

(b) $CH_3OH + O_2 \xrightarrow{\Delta}$
methyl alcohol

(c) $Rb + H_2O \longrightarrow$

(d) $SnCl_2 \cdot 2\,H_2O \xrightarrow{\Delta}$

(e) $HNO_3 + NaOH \longrightarrow$

(f) $CO_2 + H_2O \longrightarrow$

6. Complete and balance these equations:

(a) $Li_2O + H_2O \longrightarrow$

(b) $KOH \xrightarrow{\Delta}$

(c) $Ba + H_2O \longrightarrow$

(d) $Cl_2 + H_2O \longrightarrow$

(e) $SO_3 + H_2O \longrightarrow$

(f) $H_2SO_3 + KOH \longrightarrow$

7. Name these hydrates:
(a) $BaBr_2 \cdot 2\,H_2O$
(b) $AlCl_3 \cdot 6\,H_2O$
(c) $FePO_4 \cdot 4\,H_2O$

8. Name these hydrates:
(a) $MgNH_4PO_4 \cdot 6\,H_2O$
(b) $FeSO_4 \cdot 7\,H_2O$
(c) $SnCl_4 \cdot 5\,H_2O$

9. Distinguish between deionized water and
(a) hard water
(b) soft water

10. Distinguish between deionized water and
(a) distilled water
(b) natural water

11. In which of the following substances would you expect to find hydrogen bonding? Explain your reasons.
(a) C_2H_6
(b) NH_3
(c) H_2O
(d) HI
(e) C_2H_5OH

12. In which of the following substances would you expect to find hydrogen bonding? Explain your reasons.
(a) HF
(b) CH_4
(c) H_2O_2
(d) CH_3OH
(e) H_2

13. For each of the compounds in Question 11 that forms hydrogen bonds, draw a diagram of the two molecules using a dotted line to indicate where the hydrogen bonding will occur.

14. For each of the compounds in Question 12 that forms hydrogen bonds, draw a diagram of the two molecules using a dotted line to indicate where the hydrogen bonding will occur.

15. Water spreads out on a glass plate. In this example, are adhesive forces or cohesive forces strongest? Explain.

16. Rainex is a product that causes water to "bead" instead of spread when sprayed on car windshields. When the water forms these droplets, are adhesive forces or cohesive forces strongest? Explain.

17. How many moles of compound are in 100. g of $CoCl_2 \cdot 6\,H_2O$?

18. How many moles of compound are in 100. g of $FeI_2 \cdot 4\,H_2O$?

19. How many moles of water can be obtained from 100. g of $CoCl_2 \cdot 6\,H_2O$?

20. How many moles of water can be obtained from 100. g of $FeI_2 \cdot 4\,H_2O$?

21. When a person purchases Epsom salts, $MgSO_4 \cdot 7\,H_2O$, what percent of the compound is water?

22. Calculate the mass percent of water in the hydrate $Al_2(SO_4)_3 \cdot 18\,H_2O$.

23. Sugar of lead, a hydrate of lead(II) acetate, $Pb(C_2H_3O_2)_2$, contains 14.2% H_2O. What is the empirical formula for the hydrate?

24. A 25.0-g sample of a hydrate of $FePO_4$ was heated until no more water was driven off. The mass of anhydrous sample is 16.9 g. What is the empirical formula of the hydrate?

25. How many joules are needed to change 120. g of water at 20.°C to steam at 100.°C?

26. How many joules of energy must be removed from 126 g of water at 24°C to form ice at 0°C?

*27. Suppose 100. g of ice at 0°C are added to 300. g of water at 25°C. Is this sufficient ice to lower the temperature of the system to 0°C and still have ice remaining? Show evidence for your answer.

*28. Suppose 35.0 g of steam at 100.°C are added to 300. g of water at 25°C. Is this sufficient steam to heat all the water to 100.°C and still have steam remaining? Show evidence for your answer.

*29. If 75 g of ice at 0.0°C were added to 1.5 L of water at 75°C, what would be the final temperature of the mixture?

*30. If 9560 J of energy were absorbed by 500. g of ice at 0.0°C, what would be the final temperature?

31. How many grams of water will react with the following?
(a) 1.00 g Na
(b) 1.00 g MgO
(c) 1.00 g N_2O_5

32. How many grams of water will react with the following?
(a) 1.00 mol K
(b) 1.00 mol Ca
(c) 1.00 mol SO_3

Additional Exercises

All exercises with blue numbers have answers in the appendix of the text.

33. You walk out to your car just after a rain shower and notice that small droplets of water are scattered on the hood. How was it possible for the water to form those droplets?

34. Which causes a more severe burn: liquid water at 100°C or steam at 100°C? Why?

35. You have a shallow dish of alcohol set into a tray of water. If you blow across the tray, the alcohol evaporates, while the water cools significantly and eventually freezes. Explain why.

36. Regardless of how warm the outside temperature may be, we always feel cool when stepping out of a swimming pool, the ocean, or a shower. Why is this so?

37. Sketch a heating curve for a substance *X* whose melting point is 40°C and whose boiling point is 65°C.
(a) Describe what you will observe as a 60.-g sample of *X* is warmed from 0°C to 100°C.
(b) If the heat of fusion of *X* is 80. J/g, the heat of vaporization is 190. J/g, and if 3.5 J are required to warm 1 g of *X* each degree, how much energy will be needed to accomplish the change in (a)?

38. For the heating curve of water (see Figure 13.7), why doesn't the temperature increase when ice is melting into liquid water and liquid water is changing into steam?

39. Why does the vapor pressure of a liquid increase as the temperature is increased?

40. At the top of Mount Everest, which is just about 29,000 feet above sea level, the atmospheric pressure is about 270 torr. Use Figure 13.6 to determine the approximate boiling temperature of water on Mount Everest.

41. Explain how anhydrous copper(II) sulfate ($CuSO_4$) can act as an indicator for moisture.

42. Write formulas for magnesium sulfate heptahydrate and disodium hydrogen phosphate dodecahydrate.

43. How can soap make soft water from hard water? What objections are there to using soap for this purpose?

44. What substance is commonly used to destroy bacteria in water?

45. What chemical, other than chlorine or chlorine compounds, can be used to disinfect water for domestic use?

46. Some organic pollutants in water can be oxidized by dissolved molecular oxygen. What harmful effect can result from this depletion of oxygen in the water?

47. Why should you not drink liquids that are stored in ceramic containers, especially unglazed ones?

48. Write the chemical equation showing how magnesium ions are removed by a zeolite water softener.

49. Write an equation to show how hard water containing calcium chloride ($CaCl_2$) is softened by using sodium carbonate (Na_2CO_3).

50. Draw a heating curve and label the different phases and phase changes on the curve for solid H_2S at −95 °C and a pressure of 1 atm going to gaseous H_2S. (*Note:* For H_2S the melting point is −82.9°C and the boiling point is −59.6°C.)

51. How many joules are required to change 225 g of ice at 0°C to steam at 100.°C?

52. The molar heat of vaporization is the number of joules required to change 1 mol of a substance from liquid to vapor at its boiling point. What is the molar heat of vaporization of water?

***53.** The specific heat of zinc is 0.096 cal/g°C. Determine the energy required to raise the temperature of 250. g of zinc from room temperature (20.0°C) to 150.°C.

54. How many joules of energy would be liberated by condensing 50.0 mol of steam at 100.0°C and allowing the liquid to cool to 30.0°C?

55. How many kilojoules of energy are needed to convert 100. g of ice at −10.0°C to water at 20.0°C? (The specific heat of ice at −10.0°C is 2.01 J/g°C.)

56. What mass of water must be decomposed to produce 25.0 L of oxygen at STP?

***57.** Suppose 1.00 mol of water evaporates in 1.00 day. How many water molecules, on the average, leave the liquid each second?

58. Compare the volume occupied by 1.00 mol of liquid water at 0°C and 1.00 mol of water vapor at STP.

59. A mixture of 80.0 mL of hydrogen and 60.0 mL of oxygen is ignited by a spark to form water.
(a) Does any gas remain unreacted? Which one, H_2 or O_2?
(b) What volume of which gas (if any) remains unreacted? (Assume the same conditions before and after the reaction.)

60. A student (with slow reflexes) puts his hand in a stream of steam at 100.°C until 1.5 g of water has condensed. If the water then cools to room temperature (20.0°C), how many joules have been absorbed by the student's hand?

61. Determine which of the following molecules would hydrogen bond with other molecules like it. For those that do not hydrogen bond explain why. For those that hydrogen bond draw a diagram of two molecules using a dotted line to indicate where the hydrogen bonding will occur.
(a) Br_2
(b) $CH_3—O—H$
(c) $CH_3—O—CH_3$
(d) H_2O
(e) H_2S

62. The heat of fusion of a substance is given in units of J/g. The specific heat of a substance is given in units of J/g °C. Why is a temperature factor not needed in the units for heat of fusion?

63. How many joules of energy are required to change 50.0 g Cu from 25.0°C to a liquid at its melting point, 1083 °C?
Specific heat of Cu = 0.385 J/g °C
Heat of fusion for Cu = 134 J/g

Challenge Exercises

All exercises with blue numbers have answers in the appendix of the text.

64. Why does a lake freeze from the top down? What significance does this have for life on Earth?

***65.** Suppose 150. g of ice at 0.0°C are added to 0.120 L of water at 45°C. If the mixture is stirred and allowed to cool to 0.0°C, how many grams of ice remain?

***66.** A quantity of sulfuric acid is added to 100. mL of water. The final volume of the solution is 122 mL and has a density of 1.26 g/mL. What mass of acid was added? Assume the density of the water is 1.00 g/mL.

Answers to Practice Exercises

13.1 28°C, 73°C, 93°C

13.2 approximately 95°C

13.3 44.8 kJ

13.4 (a) yes, (b) yes, (c) no

13.5 (a) $BeCO_3 \cdot 4\,H_2O$
(b) $Cd(MnO_4)_2 \cdot 6\,H_2O$
(c) $Cr(NO_3)_3 \cdot 9\,H_2O$
(d) $PtO_2 \cdot 3\,H_2O$

13.6 51.17% H_2O

13.7 $CuO + H_2 \longrightarrow Cu + H_2O$

CHAPTER 14

Solutions

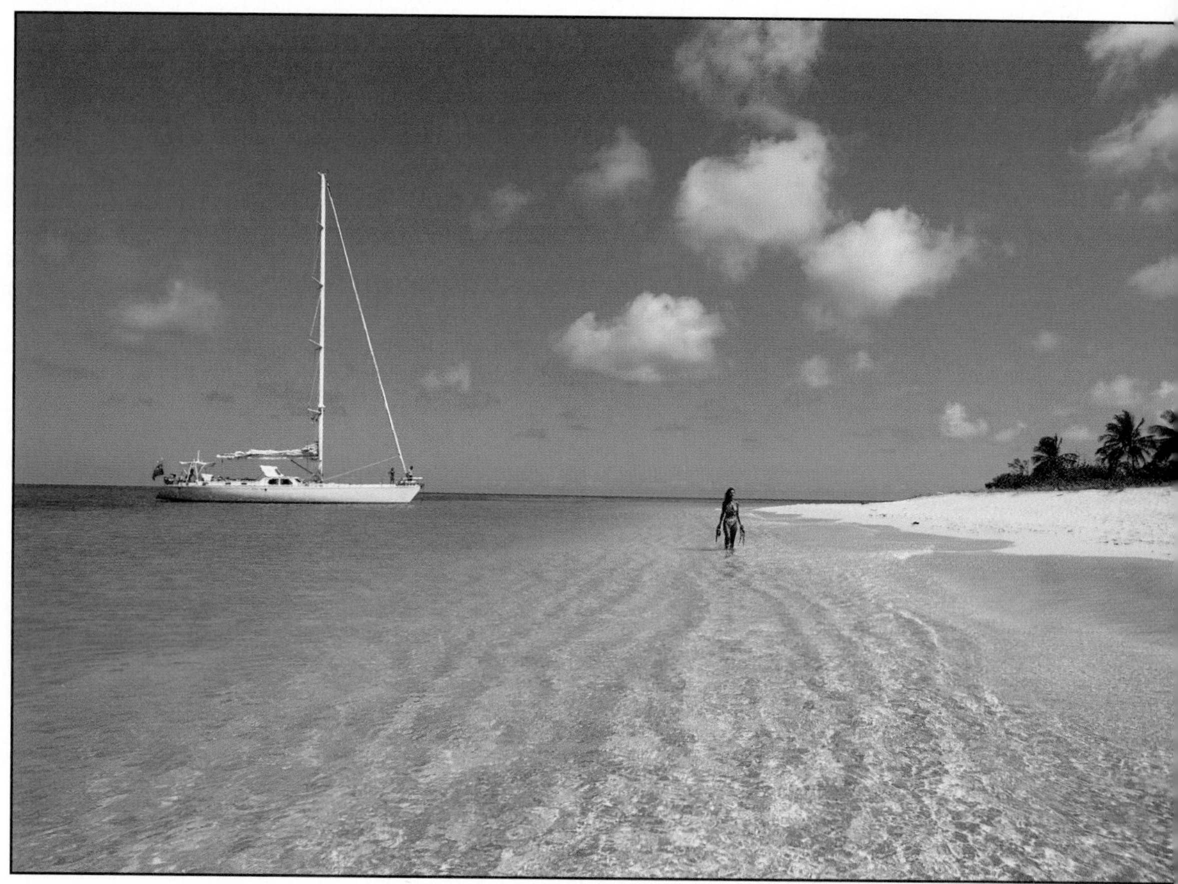

The ocean is a salt solution covering the majority of the Earth's surface.

Chapter Outline

Most substances we encounter in our daily lives are mixtures. Often they are homogeneous mixtures, which are called *solutions*. Some solutions we commonly encounter are shampoo, soft drinks, or wine. Blood plasma is a complex mixture composed of compounds and ions dissolved in water and proteins suspended in the solution. These solutions all have water as a main component, but many common items, such as air, gasoline, and steel, are also solutions that do not contain water. What are the necessary components of a solution? Why do some substances dissolve, while others do not? What effect does a dissolved substance have on the properties of the solution? Answering these questions is the first step in understanding the solutions we encounter in our daily lives.

14.1 General Properties of Solutions

The term **solution** is used in chemistry to describe a system in which one or more substances are homogeneously mixed or dissolved in another substance. A simple solution has two components: a solute and a solvent. The **solute** is the component that is dissolved or is the least abundant component in the solution. The **solvent** is the dissolving agent or the most abundant component in the solution. For example, when salt is dissolved in water to form a solution, salt is the solute and water is the solvent. Complex solutions containing more than one solute and/or more than one solvent are common.

solution

solute

solvent

The three states of matter—solid, liquid, and gas—give us nine different types of solutions: solid dissolved in solid, solid dissolved in liquid, solid dissolved in gas, liquid dissolved in liquid, and so on. Of these, the most common solutions are solid dissolved in liquid, liquid dissolved in liquid, gas dissolved in liquid, and gas dissolved in gas. Some common types of solutions are listed in Table 14.1.

A true solution is one in which the particles of dissolved solute are molecular or ionic in size, generally in the range of 0.1 to 1 nm (10^{-8} to 10^{-7} cm). The properties of a true solution are as follows:

1. A mixture of two or more components—solute and solvent—is homogeneous and has a variable composition; that is, the ratio of solute to solvent can be varied.
2. The dissolved solute is molecular or ionic in size.
3. It is either colored or colorless and is usually transparent.
4. The solute remains uniformly distributed throughout the solution and will not settle out with time.
5. The solute can generally be separated from the solvent by purely physical means (for example, by evaporation).

Table 14.1 Common Types of Solutions

Phase of solution	Solute	Solvent	Example
Gas	gas	gas	air
Liquid	gas	liquid	soft drinks
Liquid	liquid	liquid	antifreeze
Liquid	solid	liquid	salt water
Solid	gas	solid	H_2 in Pt
Solid	solid	solid	brass

CHEMISTRY IN ACTION • Exploding Silicon

We usually consider silicon, a main ingredient in sand and computer chips, to be reasonably inert. So imagine researcher Frederic Mikulec's surprise when he cut a silicon wafer with a diamond scribe and it blew up in his face. Michael Sailor, professor of chemistry and biochemistry at the University of California, San Diego, recalls, "It was just a small explosion, like a cap going off in a cap gun." The event inspired the research team to investigate the process further.

The silicon wafers being used were specially processed to have many tiny holes. Gadolinium nitrate was dissolved in ethanol and added to the porous silicon wafers. After the ethanol evaporated, the gadolinium nitrate remained inside the pores in the silicon chip. Sailor's team discovered that they could explode these chips using a 9-volt battery or by scratching them.

Next the team began to look for ways to use these exploding silicon chips. The scientists added metal ions to the wafers and exploded them. Their results indicated that the elements could be identified by their emission spectra in the flame from the explosion. Currently, groundwater samples must be taken to the lab and analyzed in an emission spectrometer by being mixed with other chemicals and burned. Sailor thinks that these wafers could be used as a sensor for analyzing water contaminants in the field. Imagine being able to test water for heavy metals using an instrument you could hold in the palm of your hand.

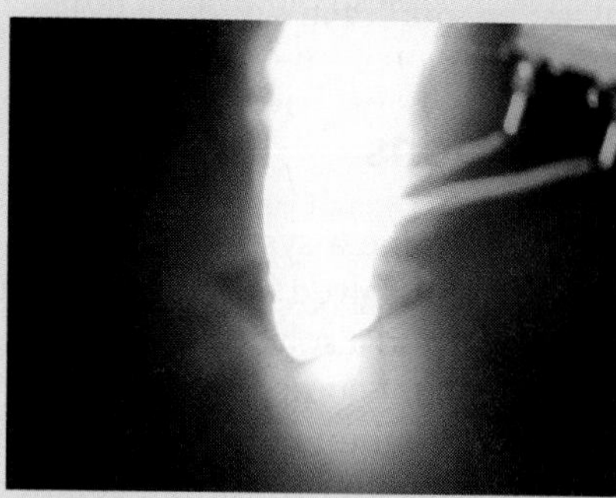

Since the silicon wafers explode like gunpowder, these tiny wafers could also be used to operate tiny micromotors or microvehicles. Sailor also thinks that these silicon wafers could replace blasting caps in explosives.

Let's illustrate these properties using water solutions of sugar and of potassium permanganate. We prepare two sugar solutions, the first containing 10 g of sugar added to 100 mL of water and the second containing 20 g of sugar added to 100 mL of water. Each solution is stirred until all the solute dissolves, demonstrating that we can vary the composition of a solution. Every portion of the solution has the same sweet taste because the sugar molecules are uniformly distributed throughout. If confined so that no solvent is lost, the solution will taste and appear the same a week or a month later. A solution cannot be separated into its components by filtering it. But by carefully evaporating the water, we can recover the sugar from the solution.

Note the beautiful purple trails of $KMnO_4$ as the crystals dissolve.

To observe the dissolving of potassium permanganate ($KMnO_4$), we drop a few crystals of it in a beaker of water. Almost at once, the beautiful purple color of dissolved permanganate ions (MnO_4^-) appears and streams to the bottom of the beaker as the crystals dissolve. After a while, the purple color disperses until it's evenly distributed throughout the solution. This dispersal demonstrates that molecules and ions move about freely and spontaneously (diffuse) in a liquid or solution.

Solution permanency is explained in terms of the kinetic-molecular theory (see Section 12.2). According to the KMT, both the solute and solvent particles (molecules or ions) are in constant random motion. This motion is energetic enough to prevent the solute particles from settling out under the influence of gravity.

14.2 Solubility

solubility

The term **solubility** describes the amount of one substance (solute) that will dissolve in a specified amount of another substance (solvent) under stated conditions. For example, 36.0 g of sodium chloride will dissolve in 100 g of water at 20°C. We say then that the solubility of NaCl in water is 36.0 g/100 g H_2O at 20°C. Solubility is often used in a relative way. For instance, we say that a substance is very soluble, moderately soluble, slightly soluble, or insoluble. Although these terms do not accurately indicate how much solute will dissolve, they are frequently used to describe the solubility of a substance qualitatively.

Two other terms often used to describe solubility are *miscible* and *immiscible*. Liquids that are capable of mixing and forming a homogeneous solution are **miscible**; those that do not form solutions or are generally insoluble in each other are **immiscible**. Methyl alcohol and water are miscible in each other in all proportions. Oil and water are immiscible, forming two separate layers when they are mixed, as shown in Figure 14.1.

miscible
immiscible

The general guidelines for the solubility of common ionic compounds (salts) are given in Figure 14.2. These guidelines have some exceptions, but they provide a solid foundation for the compounds considered in this course. The solubilities of over 200 compounds are given in the Solubility Table in Appendix V. Solubility data for thousands of compounds can be found by consulting standard reference sources.*

The term *salt* is used interchangeably with *ionic compound* by many chemists.

The quantitative expression of the amount of dissolved solute in a particular quantity of solvent is known as the **concentration of a solution**. Several methods of expressing concentration are described in Section 14.6.

concentration of a solution

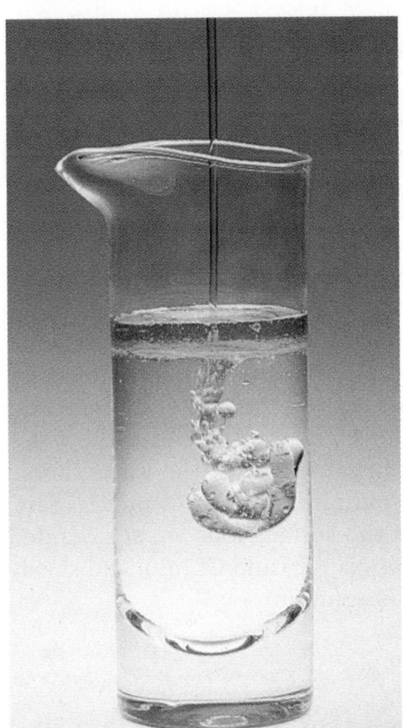

Figure 14.1
An immiscible mixture of oil and water.

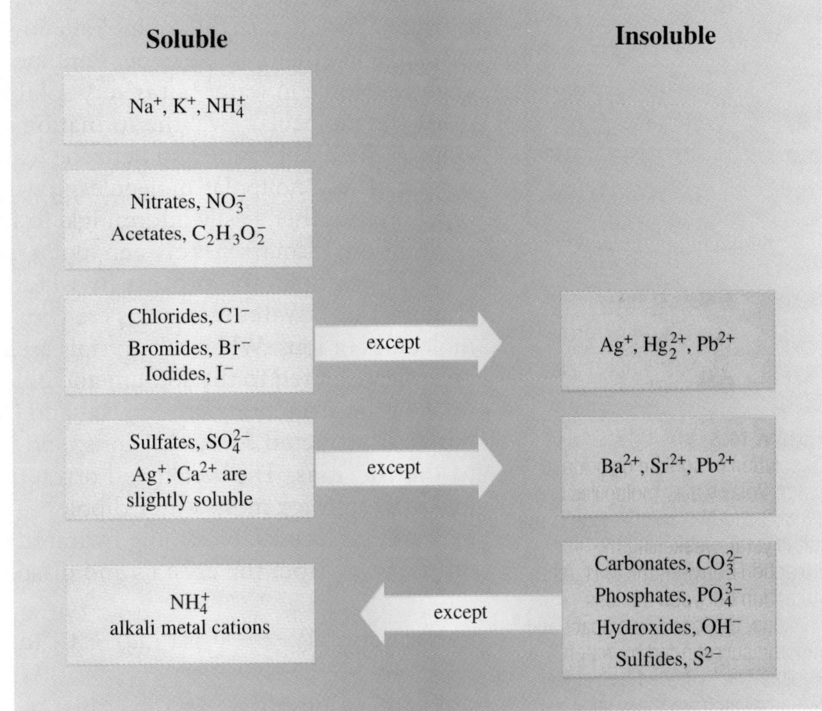

Figure 14.2
The solubility of various common ions. Substances containing the ions on the left are generally soluble in cold water, while those substances containing the ions on the right are insoluble in cold water. The arrows point to the exceptions.

*Two commonly used handbooks are *Lange's Handbook of Chemistry*, 15th ed. (New York: McGrawHill, 1998), and the *Handbook of Chemistry and Physics*, 85th ed. (Cleveland: Chemical Rubber Co., 2004).

14.3 Factors Related to Solubility

Predicting solubilities is complex and difficult. Many variables, such as size of ions, charge on ions, interaction between ions, interaction between solute and solvent, and temperature, complicate the problem. Because of the factors involved, the general rules of solubility given in Figure 14.2 have many exceptions. However, these rules are useful because they do apply to many of the more common compounds that we encounter in the study of chemistry. Keep in mind that these are rules, not laws, and are therefore subject to exceptions. Fortunately, the solubility of a solute is relatively easy to determine experimentally. Now let's examine the factors related to solubility.

The Nature of the Solute and Solvent

= Water

= Na^+

= Cl^-

Figure 14.3
Dissolution of sodium chloride in water. Polar water molecules are attracted to Na^+ and Cl^- ions in the salt crystal, weakening the attraction between the ions. As the attraction between the ions weakens, the ions move apart and become surrounded by water dipoles. The hydrated ions slowly diffuse away from the crystal to become dissolved in solution.

The old adage "like dissolves like" has merit, in a general way. Polar or ionic substances tend to be more compatible with other polar substances. Nonpolar substances tend to be compatible with other nonpolar substances and less miscible with polar substances. Thus ionic compounds, which are polar, tend to be much more soluble in water, which is polar, than in solvents such as ether, hexane, or benzene, which are essentially nonpolar. Sodium chloride, an ionic substance, is soluble in water, slightly soluble in ethyl alcohol (less polar than water), and insoluble in ether and benzene. Pentane, C_5H_{12}, a nonpolar substance, is only slightly soluble in water but is very soluble in benzene and ether.

At the molecular level the formation of a solution from two nonpolar substances, such as hexane and benzene, can be visualized as a process of simple mixing. These nonpolar molecules, having little tendency to either attract or repel one another, easily intermingle to form a homogeneous solution.

Solution formation between polar substances is much more complex. See, for example, the process by which sodium chloride dissolves in water (Figure 14.3). Water molecules are very polar and are attracted to other polar molecules or ions. When salt crystals are put into water, polar water molecules become attracted to the sodium and chloride ions on the crystal surfaces and weaken the attraction between Na^+ and Cl^- ions. The positive end of the water dipole is attracted to the Cl^- ions, and the negative end of the water dipole to the Na^+ ions. The weakened attraction permits the ions to move apart, making room for more water dipoles. Thus the surface ions are surrounded by water molecules, becoming hydrated ions, $Na^+(aq)$ and $Cl^-(aq)$, and slowly diffuse away from the crystals and dissolve in solution:

$$\text{NaCl(crystal)} \xrightarrow{H_2O} Na^+(aq) + Cl^-(aq)$$

Examination of the data in Table 14.2 reveals some of the complex questions relating to solubility.

The Effect of Temperature on Solubility

Temperature affects the solubility of most substances, as shown by the data in Table 14.2. Most solutes have a limited solubility in a specific solvent at a fixed temperature. For most solids dissolved in a liquid, an increase in temperature

Table 14.2 Solubility of Alkali Metal Halides in Water

Salt	Solubility (g salt/100 g H_2O) 0°C	100°C
LiF	0.12	0.14 (at 35°C)
LiCl	67	127.5
LiBr	143	266
LiI	151	481
NaF	4	5
NaCl	35.7	39.8
NaBr	79.5	121
NaI	158.7	302
KF	92.3 (at 18°C)	Very soluble
KCl	27.6	57.6
KBr	53.5	104
KI	127.5	208

results in increased solubility (see Figure 14.4). However, no single rule governs the solubility of solids in liquids with change in temperature. Some solids increase in solubility only slightly with increasing temperature (see NaCl in Figure 14.4); other solids decrease in solubility with increasing temperature (see Li_2SO_4 in Figure 14.4).

On the other hand, the solubility of a gas in water usually decreases with increasing temperature (see HCl and SO_2 in Figure 14.4). The tiny bubbles that form when water is heated are due to the decreased solubility of air at higher temperatures. The decreased solubility of gases at higher temperatures is explained in terms of the KMT by assuming that, in order to dissolve, the gas

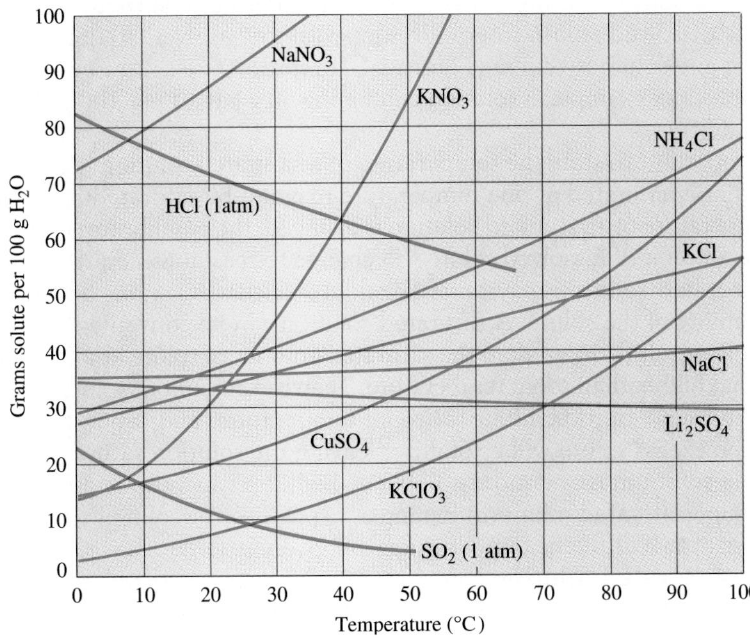

Figure 14.4
Solubility of various compounds in water. Solids are shown in red and gases are shown in blue.

molecules must form bonds of some sort with the molecules of the liquid. An increase in temperature decreases the solubility of the gas because it increases the kinetic energy (speed) of the gas molecules and thereby decreases their ability to form "bonds" with the liquid molecules.

The Effect of Pressure on Solubility

Pouring root beer into a glass illustrates the effect of pressure on solubility. The escaping CO_2 produces the foam.

Small changes in pressure have little effect on the solubility of solids in liquids or liquids in liquids but have a marked effect on the solubility of gases in liquids. The solubility of a gas in a liquid is directly proportional to the pressure of that gas above the solution. Thus the amount of a gas dissolved in solution will double if the pressure of that gas over the solution is doubled. For example, carbonated beverages contain dissolved carbon dioxide under pressures greater than atmospheric pressure. When a can of carbonated soda is opened, the pressure is immediately reduced to the atmospheric pressure, and the excess dissolved carbon dioxide bubbles out of the solution.

Saturated, Unsaturated, and Supersaturated Solutions

At a specific temperature there is a limit to the amount of solute that will dissolve in a given amount of solvent. When this limit is reached, the resulting solution is said to be *saturated*. For example, when we put 40.0 g of KCl into 100 g of H_2O at 20°C, we find that 34.0 g of KCl dissolve and 6.0 g of KCl remain undissolved. The solution formed is a saturated solution of KCl.

Two processes are occurring simultaneously in a saturated solution. The solid is dissolving into solution and, at the same time, the dissolved solute is crystallizing out of solution. This may be expressed as

$$\text{solute (undissolved)} \rightleftharpoons \text{solute (dissolved)}$$

saturated solution

When these two opposing processes are occurring at the same rate, the amount of solute in solution is constant, and a condition of equilibrium is established between dissolved and undissolved solute. Therefore, a **saturated solution** contains dissolved solute in equilibrium with undissolved solute. Thus, any point on any solubility curve (Figure 14.4) represents a saturated solution of that solute. For example, a solution containing 60 g NH_4Cl per 100 g H_2O is saturated at 70°C.

It's important to state the temperature of a saturated solution, because a solution that is saturated at one temperature may not be saturated at another. If the temperature of a saturated solution is changed, the equilibrium is disturbed, and the amount of dissolved solute will change to reestablish equilibrium.

A saturated solution may be either dilute or concentrated, depending on the solubility of the solute. A saturated solution can be conveniently prepared by dissolving a little more than the saturated amount of solute at a temperature somewhat higher than room temperature. Then the amount of solute in solution will be in excess of its solubility at room temperature, and, when the solution cools, the excess solute will crystallize, leaving the solution saturated. (In this case, the solute must be more soluble at higher temperatures and must not form a supersaturated solution.) Examples expressing the solubility of saturated solutions at two different temperatures are given in Table 14.3.

unsaturated solution

An **unsaturated solution** contains less solute per unit of volume than does its corresponding saturated solution. In other words, additional solute can be dissolved in an unsaturated solution without altering any other conditions.

Table 14.3 Saturated Solutions at 20°C and 50°C

Solute	Solubility (g solute/100 g H_2O) 20°C	50°C
NaCl	36.0	37.0
KCl	34.0	42.6
$NaNO_3$	88.0	114.0
$KClO_3$	7.4	19.3
$AgNO_3$	222.0	455.0
$C_{12}H_{22}O_{11}$	203.9	260.4

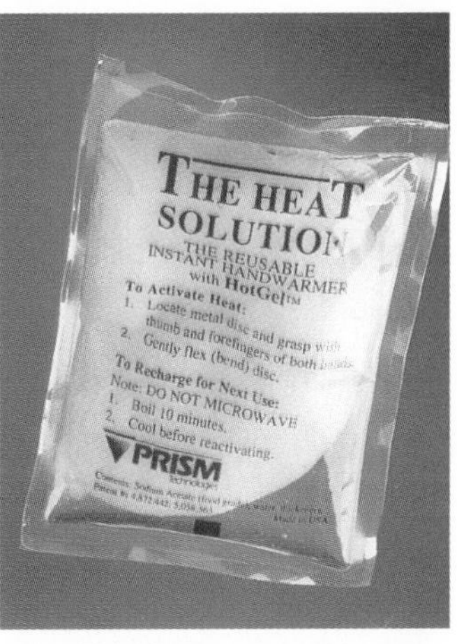

The heat released in this hot pack results from the crystallization of a supersaturated solution of sodium acetate.

Consider a solution made by adding 40 g of KCl to 100 g of H_2O at 20°C (see Table 14.3). The solution formed will be saturated and will contain about 6 g of undissolved salt, because the maximum amount of KCl that can dissolve in 100 g of H_2O at 20°C is 34 g. If the solution is now heated and maintained at 50°C, all the salt will dissolve and even more can be dissolved. The solution at 50°C is unsaturated.

supersaturated solution

In some circumstances, solutions can be prepared that contain more solute than needed for a saturated solution at a particular temperature. These solutions are said to be **supersaturated.** However, we must qualify this definition by noting that a supersaturated solution is unstable. Disturbances, such as jarring, stirring, scratching the walls of the container, or dropping in a "seed" crystal, cause the supersaturation to return to saturation, releasing heat. When a supersaturated solution is disturbed, the excess solute crystallizes out rapidly, returning the solution to a saturated state.

Supersaturated solutions are not easy to prepare but may be made from certain substances by dissolving, in warm solvent, an amount of solute greater than that needed for a saturated solution at room temperature. The warm solution is then allowed to cool very slowly. With the proper solute and careful work, a supersaturated solution will result.

Example 14.1

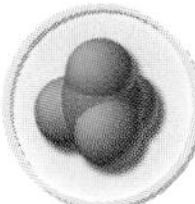

Will a solution made by adding 2.5 g of $CuSO_4$ to 10 g of H_2O be saturated or unsaturated at 20°C?

SOLUTION

We first need to know the solubility of $CuSO_4$ at 20°C. From Figure 14.4, we see that the solubility of $CuSO_4$ at 20°C is about 21 g per 100 g of H_2O. This amount is equivalent to 2.1 g of $CuSO_4$ per 10 g of H_2O.

Since 2.5 g per 10 g of H_2O is greater than 2.1 g per 10 g of H_2O, the solution will be saturated and 0.4 g of $CuSO_4$ will be undissolved.

Practice 14.1

Will a solution made by adding 9.0 g NH_4Cl to 20 g of H_2O be saturated or unsaturated at 50°C?

14.4 Rate of Dissolving Solids

The rate at which a solid dissolves is governed by (1) the size of the solute particles, (2) the temperature, (3) the concentration of the solution, and (4) agitation or stirring. Let's look at each of these conditions:

1. *Particle size.* A solid can dissolve only at the surface that is in contact with the solvent. Because the surface-to-volume ratio increases as size decreases, smaller crystals dissolve faster than large ones. For example, if a salt crystal 1 cm on a side (a surface area of 6 cm^2) is divided into 1000 cubes, each 0.1 cm on a side, the total surface of the smaller cubes is 60 cm^2—a 10-fold increase in surface area (see Figure 14.5).
2. *Temperature.* In most cases, the rate of dissolving of a solid increases with temperature. This increase is due to kinetic effects. The solvent molecules move more rapidly at higher temperatures and strike the solid surfaces more often, causing the rate of dissolving to increase.
3. *Concentration of the solution.* When the solute and solvent are first mixed, the rate of dissolving is at its maximum. As the concentration of the solution increases and the solution becomes more nearly saturated with the solute, the rate of dissolving decreases greatly. The rate of dissolving is graphed in Figure 14.6. Note that about 17 g dissolve in the first five-minute interval, but only about 1 g dissolves in the fourth five-minute interval. Although different solutes show different rates, the rate of dissolving always becomes very slow as the concentration approaches the saturation point.
4. *Agitation or stirring.* The effect of agitation or stirring is kinetic. When a solid is first put into water, it comes in contact only with solvent in its immediate vicinity. As the solid dissolves, the amount of dissolved solute around the solid becomes more and more concentrated, and the rate of dissolving slows down. If the mixture is not stirred, the dissolved solute

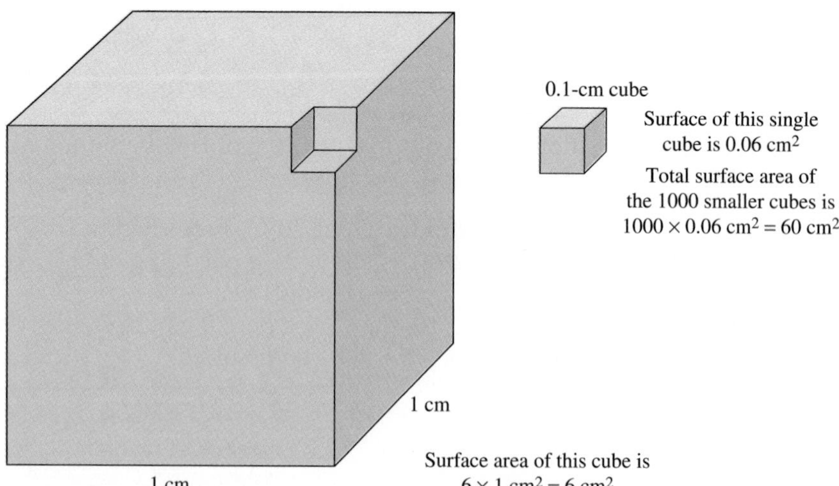
0.1-cm cube
Surface of this single cube is 0.06 cm²
Total surface area of the 1000 smaller cubes is
1000 × 0.06 cm² = 60 cm²
1 cm
1 cm
Surface area of this cube is
6 × 1 cm² = 6 cm²

Figure 14.5
Surface area of crystals. A crystal 1 cm on a side has a surface area of 6 cm^2. Subdivided into 1000 smaller crystals, each 0.1 cm on a side, the total surface area is increased to 60 cm^2.

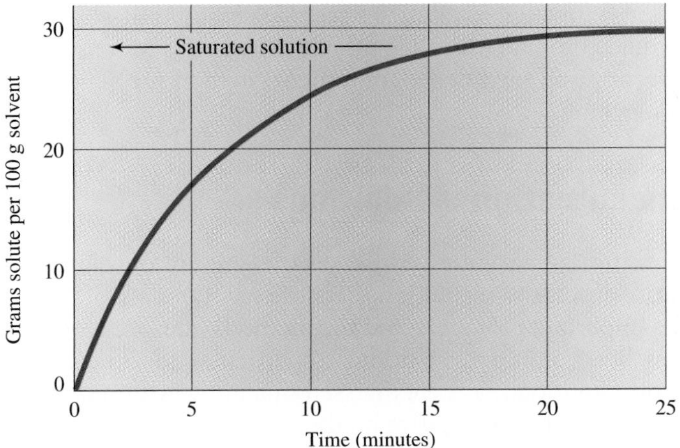

Figure 14.6
Rate of dissolution of a solid solute in a solvent. The rate is maximum at the beginning and decreases as the concentration approaches saturation.

diffuses very slowly through the solution; weeks may pass before the solid is entirely dissolved. Stirring distributes the dissolved solute rapidly through the solution, and more solvent is brought into contact with the solid, causing it to dissolve more rapidly.

14.5 Solutions: A Reaction Medium

Many solids must be put into solution to undergo appreciable chemical reaction. We can write the equation for the double-displacement reaction between sodium chloride and silver nitrate:

$$NaCl + AgNO_3 \longrightarrow AgCl + NaNO_3$$

But suppose we mix solid NaCl and solid $AgNO_3$ and look for a chemical change. If any reaction occurs, it is slow and virtually undetectable. In fact, the crystalline structures of NaCl and $AgNO_3$ are so different that we could separate them by tediously picking out each kind of crystal from the mixture. But if we dissolve the NaCl and $AgNO_3$ separately in water and mix the two solutions, we observe the immediate formation of a white, curdlike precipitate of silver chloride.

Molecules or ions must collide with one another in order to react. In the foregoing example, the two solids did not react because the ions were securely locked within their crystal structures. But when the NaCl and $AgNO_3$ are dissolved, their crystal lattices are broken down and the ions become mobile. When the two solutions are mixed, the mobile Ag^+ and Cl^- ions come into contact and react to form insoluble AgCl, which precipitates out of solution. The soluble Na^+ and NO_3^- ions remain mobile in solution but form the crystalline salt $NaNO_3$ when the water is evaporated:

$$NaCl(aq) + AgNO_3(aq) \longrightarrow AgCl(s) + NaNO_3(aq)$$

$$\underbrace{Na^+(aq) + Cl^-(aq)}_{\text{sodium chloride solution}} + \underbrace{Ag^+(aq) + NO_3^-(aq)}_{\text{silver nitrate solution}} \longrightarrow \underbrace{AgCl(s)}_{\text{silver chloride}} + \underbrace{Na^+(aq) + NO_3^-(aq)}_{\text{sodium nitrate in solution}}$$

The mixture of the two solutions provides a medium or space in which the Ag^+ and Cl^- ions can react. (See Chapter 15 for further discussion of ionic reactions.)

Solutions also function as diluting agents in reactions in which the undiluted reactants would combine with each other too violently. Moreover, a solution of known concentration provides a convenient method for delivering a specific amount of reactant.

14.6 Concentration of Solutions

The concentration of a solution expresses the amount of solute dissolved in a given quantity of solvent or solution. Because reactions are often conducted in solution, it's important to understand the methods of expressing concentration and to know how to prepare solutions of particular concentrations. The concentration of a solution may be expressed qualitatively or quantitatively. Let's begin with a look at the qualitative methods of expressing concentration.

Dilute and Concentrated Solutions

When we say that a solution is *dilute* or *concentrated*, we are expressing, in a relative way, the amount of solute present. One gram of a compound and 2 g of a compound in solution are both dilute solutions when compared with the same volume of a solution containing 20 g of a compound. Ordinary concentrated hydrochloric acid contains 12 mol of HCl per liter of solution. In some laboratories, the dilute acid is made by mixing equal volumes of water and the concentrated acid. In other laboratories the concentrated acid is diluted with two or three volumes of water, depending on its use. The term **dilute solution**, then, describes a solution that contains a relatively small amount of dissolved solute. Conversely, a **concentrated solution** contains a relatively large amount of dissolved solute.

dilute solution

concentrated solution

Mass Percent Solution

The mass percent method expresses the concentration of the solution as the percent of solute in a given mass of solution. It says that for a given mass of solution, a certain percent of that mass is solute. Suppose we take a bottle from the reagent shelf that reads "sodium hydroxide, NaOH, 10%." This statement means that for every 100 g of this solution, 10 g will be NaOH and 90 g will be water. (Note that this amount of solution is 100 g, not 100 mL.) We could also make this same concentration of solution by dissolving 2.0 g of NaOH in 18 g of water. Mass percent concentrations are most generally used for solids dissolved in liquids:

Note that mass percent is independent of the formula for the solute.

$$\text{mass percent} = \frac{\text{g solute}}{\text{g solute} + \text{g solvent}} \times 100 = \frac{\text{g solute}}{\text{g solution}} \times 100$$

As instrumentation advances are made in chemistry, our ability to measure the concentration of dilute solutions is increasing as well. In addition to mass percent, chemists now commonly use **parts per million (ppm):**

parts per million (ppm)

$$\text{parts per million} = \frac{\text{g solute}}{\text{g solute} + \text{g solvent}} \times 1{,}000{,}000$$

Currently, air and water contaminants, drugs in the human body, and pesticide residues are measured in parts per million.

Example 14.2

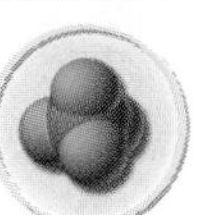

What is the mass percent of sodium hydroxide in a solution that is made by dissolving 8.00 g NaOH in 50.0 g H_2O?

SOLUTION

grams of solute(NaOH) = 8.00 g

grams of solvent (H_2O) = 50.0 g

$$\left(\frac{8.00\text{ g NaOH}}{8.00\text{ g NaOH} + 50.0\text{ g }H_2O}\right)100 = 13.8\%\text{ NaOH solution}$$

Example 14.3

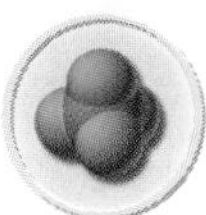

What masses of potassium chloride and water are needed to make 250. g of 5.00% solution?

SOLUTION

The percent expresses the mass of the solute:

250. g = total mass of solution

5.00% of 250. g = (0.0500)(250. g) = 12.5 g KCl (solute)

250. g − 12.5 g = 238 g H_2O

Dissolving 12.5 g KCl in 238 g H_2O gives a 5.00% KCl solution.

Example 14.4

A 34.0% sulfuric acid solution has a density of 1.25 g/mL. How many grams of H_2SO_4 are contained in 1.00 L of this solution?

SOLUTION

Since H_2SO_4 is the solute, we first solve the mass percent equation for grams of solute:

$$\text{mass percent} = \left(\frac{\text{g solute}}{\text{g solution}}\right)100$$

$$\text{g solute} = \frac{(\text{mass percent})(\text{g solution})}{100}$$

The mass percent is given, so we need to determine the grams of solution. The mass of the solution can be calculated from the density data. Convert density (g/mL) to grams:

$$1.00\text{ L} = 1.00 \times 10^3\text{ mL}$$

$$\left(\frac{1.25\text{ g}}{\text{mL}}\right)(1.00 \times 10^3\text{ mL}) = 1250\text{ g (mass of solution)}$$

Now we have all the figures to calculate the grams of solute:

$$\text{g solute} = \frac{(34.0\text{ g }H_2SO_4)(1250\text{ g})}{100\text{ g}} = 425\text{ g }H_2SO_4$$

Thus, 1.00 L of 34.0% H_2SO_4 solution contains 425 g H_2SO_4.

Practice 14.2

What is the mass percent of Na_2SO_4 in a solution made by dissolving 25.0 g Na_2SO_4 in 225.0 g H_2O?

Mass/Volume Percent (m/v)

This method expresses concentration as grams of solute per 100 mL of solution. With this system, a 10.0%-m/v-glucose solution is made by dissolving 10.0 g of glucose in water, diluting to 100 mL, and mixing. The 10.0%-m/v solution could also be made by diluting 20.0 g to 200 mL, 50.0 g to 500 mL, and so on. Of course, any other appropriate dilution ratio may be used:

$$\text{mass/volume percent} = \frac{\text{g solute}}{\text{mL solution}} \times 100$$

Example 14.5 A 3.0% H_2O_2 solution is commonly used as a topical antiseptic to prevent infection. What volume of this solution will contain 10. g of H_2O_2?

SOLUTION First solve the mass/volume percent equation for mL of solution:

$$\text{mL solution} = \frac{(\text{g solute})}{(\text{m/v percent})}(100)$$

$$\text{mL solution} = \frac{(10.\ \text{g solute})}{(3.0\ \text{m/v percent})}(100) = 330\ \text{mL}$$

Volume Percent

Solutions that are formulated from two liquids are often expressed as *volume percent* with respect to the solute. The volume percent is the volume of a liquid in 100 mL of solution. The label on a bottle of ordinary rubbing alcohol reads "isopropyl alcohol, 70% by volume." Such a solution could be made by mixing 70 mL of alcohol with water to make a total volume of 100 mL, but we cannot use 30 mL of water, because the two volumes are not necessarily additive:

$$\text{volume percent} = \frac{\text{volume of liquid in question}}{\text{total volume of solution}} \times 100$$

Volume percent is used to express the concentration of alcohol in beverages. Wines generally contain 12% alcohol by volume. This translates into 12 mL of alcohol in each 100 mL of wine. The beverage industry also uses the concentration unit of *proof* (twice the volume percent). Pure ethyl alcohol is 100% and therefore 200 proof. Scotch whiskey is 86 proof, or 43% alcohol.

Molarity

Mass percent solutions do not equate or express the molar masses of the solute in solution. For example, 1000. g of 10.0% NaOH solution contains 100. g NaOH; 1000. g of 10.0% KOH solution contains 100. g KOH. In terms of moles of NaOH and KOH, these solutions contain

$$\text{mol NaOH} = (100.\ \text{g NaOH})\left(\frac{1\ \text{mol NaOH}}{40.00\ \text{g NaOH}}\right) = 2.50\ \text{mol NaOH}$$

$$\text{mol KOH} = (100.\ \text{g KOH})\left(\frac{1\ \text{mol KOH}}{56.11\ \text{g KOH}}\right) = 1.78\ \text{mol KOH}$$

From these figures, we see that the two 10.0% solutions do not contain the same number of moles of NaOH and KOH. As a result, we find that a 10.0% NaOH solution contains more reactive base than a 10.0% KOH solution.

We need a method of expressing concentration that will easily indicate how many moles of solute are present per unit volume of solution. For this purpose, the concentration known as molarity is used in calculations involving chemical reactions.

A 1-molar solution contains 1 mol of solute per liter of solution. For example, to make a 1-molar solution of sodium hydroxide (NaOH) we dissolve 40.00 g NaOH (1 mol) in water and dilute the solution with more water to a volume of 1 L. The solution contains 1 mol of the solute in 1 L of solution and is said to be 1 molar in concentration. Figure 14.7 illustrates the preparation of a 1-molar solution. Note that the volume of the solute and the solvent together is 1 L.

The concentration of a solution can, of course, be varied by using more or less solute or solvent; but in any case the **molarity** of a solution is the number of moles of solute per liter of solution. The abbreviation for molarity is M. The units of molarity are moles per liter. The expression "2.0 M NaOH" means a 2.0-molar solution of NaOH (2.0 mol, or 80. g, of NaOH dissolved in water to make 1.0 L of solution):

molarity (M)

$$\text{molarity} = M = \frac{\text{number of moles of solute}}{\text{liter of solution}} = \frac{\text{moles}}{\text{liter}}$$

Flasks that are calibrated to contain specific volumes at a particular temperature are used to prepare solutions of a desired concentration. These *volumetric flasks* have a calibration mark on the neck that accurately indicates the measured volume. Molarity is based on a specific volume of solution and therefore will vary slightly with temperature because volume varies with temperature (1000 mL H_2O at 20°C = 1001 mL at 25°C).

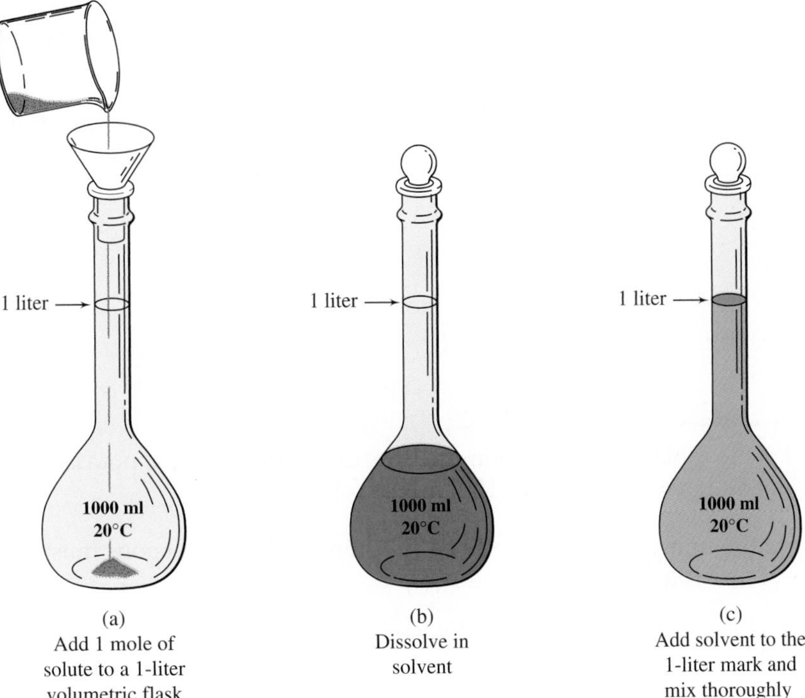
1 liter
1000 ml
20°C
(a)
Add 1 mole of solute to a 1-liter volumetric flask
1 liter
1000 ml
20°C
(b)
Dissolve in solvent
1 liter
1000 ml
20°C
(c)
Add solvent to the 1-liter mark and mix thoroughly

Figure 14.7
Preparation of a 1 M solution.

Suppose we want to make 500 mL of 1 M solution. This solution can be prepared by determining the mass of 0.5 mol of the solute and diluting with water in a 500-mL (0.5-L) volumetric flask. The molarity will be

$$M = \frac{0.5 \text{ mol solute}}{0.5 \text{ L solution}} = 1 \text{ molar}$$

You can see that it isn't necessary to have a liter of solution to express molarity. All we need to know is the number of moles of dissolved solute and the volume of solution. Thus, 0.001 mol NaOH in 10 mL of solution is 0.1 M:

$$\left(\frac{0.001 \text{ mol}}{10 \cancel{\text{mL}}}\right)\left(\frac{1000 \cancel{\text{mL}}}{1 \text{ L}}\right) = 0.1\,M$$

When we stop to think that a balance is not calibrated in moles, but in grams, we can incorporate grams into the molarity formula. We do so by using the relationship

Molarities of concentrated acids commonly used in the laboratory:

HCl	12 M
$HC_2H_3O_2$	17 M
HNO_3	16 M
H_2SO_4	18 M

$$\text{moles} = \frac{\text{grams of solute}}{\text{molar mass}}$$

Substituting this relationship into our expression for molarity, we get

$$M = \frac{\text{mol}}{\text{L}} = \frac{\text{g solute}}{\text{molar mass solute} \times \text{L solution}}$$

We can now determine the mass of any amount of a solute that has a known formula, dilute it to any volume, and calculate the molarity of the solution using this formula.

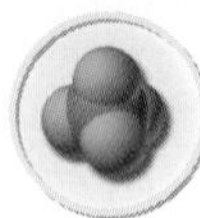

Example 14.6 What is the molarity of a solution containing 1.4 mol of acetic acid ($HC_2H_3O_2$) in 250. mL of solution?

SOLUTION By the unit-conversion method, we note that the concentration given in the problem statement is 1.4 mol per 250. mL (mol/mL). Since molarity = mol/L, the needed conversion is

$$\frac{\text{mol}}{\text{mL}} \longrightarrow \frac{\text{mol}}{\text{L}} = M$$

$$\left(\frac{1.4 \text{ mol}}{250. \cancel{\text{mL}}}\right)\left(\frac{1000 \cancel{\text{mL}}}{\text{L}}\right) = \frac{5.6 \text{ mol}}{\text{L}} = 5.6\,M$$

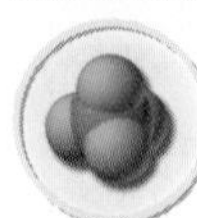

Example 14.7 What is the molarity of a solution made by dissolving 2.00 g of potassium chlorate in enough water to make 150. mL of solution?

SOLUTION We use the unit-conversion method. The steps in the conversions must lead to units of moles/liter:

$$\frac{\text{g } KClO_3}{\text{mL}} \longrightarrow \frac{\text{g } KClO_3}{\text{L}} \longrightarrow \frac{\text{mol } KClO_3}{\text{L}} = M$$

The data are

mass $KClO_3$ = 2.00 g molar mass $KClO_3$ = 122.6 g/mol volume = 150. mL

$$\left(\frac{2.00\ \text{g KClO}_3}{150.\ \text{mL}}\right)\left(\frac{1000\ \text{mL}}{\text{L}}\right)\left(\frac{1\ \text{mol KClO}_3}{122.6\ \text{g KClO}_3}\right) = \frac{0.109\ \text{mol}}{\text{L}} = 0.109\ M\ \text{KClO}_3$$

Example 14.8

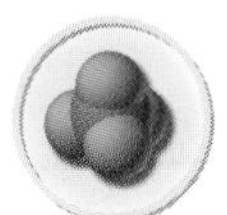

How many grams of potassium hydroxide are required to prepare 600. mL of 0.450 *M* KOH solution?

SOLUTION

The conversion is

milliliters ⟶ liters ⟶ moles ⟶ grams

The data are

$$\text{volume} = 600.\ \text{mL} \qquad M = \frac{0.450\ \text{mol}}{\text{L}} \qquad \text{molar mass KOH} = \frac{56.11\ \text{g}}{\text{mol}}$$

The calculation is

$$(600.\ \text{mL})\left(\frac{1\ \text{L}}{1000\ \text{mL}}\right)\left(\frac{0.450\ \text{mol}}{\text{L}}\right)\left(\frac{56.11\ \text{g KOH}}{\text{mol}}\right) = 15.1\ \text{g KOH}$$

Practice 14.3

What is the molarity of a solution made by dissolving 7.50 g of magnesium nitrate [$Mg(NO_3)_2$] in enough water to make 25.0 mL of solution?

Practice 14.4

How many grams of sodium chloride are needed to prepare 125 mL of a 0.037 *M* NaCl solution?

Example 14.9

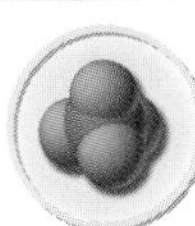

Calculate the number of moles of nitric acid in 325 mL of 16 *M* HNO_3 solution.

SOLUTION

Use the equation

$$\text{moles} = \text{liters} \times M$$

Substitute the data given in the problem and solve:

$$\text{moles} = (0.325\ \text{L})\left(\frac{16\ \text{mol HNO}_3}{1\ \text{L}}\right) = 5.2\ \text{mol HNO}_3$$

Example 14.10

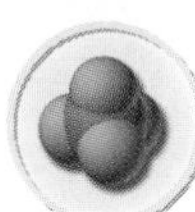

What volume of 0.250 *M* solution can be prepared from 16.0 g of potassium carbonate?

SOLUTION

We start with 16.0 g K_2CO_3; we need to find the volume of 0.250 *M* solution that can be prepared from this amount of K_2CO_3. The conversion therefore is

g K_2CO_3 ⟶ mol K_2CO_3 ⟶ L solution

The data are

$$\text{mass } K_2CO_3 = 16.0\text{ g} \quad M = \frac{0.250\text{ mol}}{1\text{ L}} \quad \text{molar mass } K_2CO_3 = \frac{138.2\text{ g}}{1\text{ mol}}$$

$$(16.0\ \cancel{\text{g } K_2CO_3})\left(\frac{1\ \cancel{\text{mol } K_2CO_3}}{138.2\ \cancel{\text{g } K_2CO_3}}\right)\left(\frac{1\text{ L}}{0.250\ \cancel{\text{mol } K_2CO_3}}\right) = 0.463\text{ L } (463\text{ mL})$$

Thus, 463 mL of 0.250 *M* solution can be made from 16.0 g K_2CO_3.

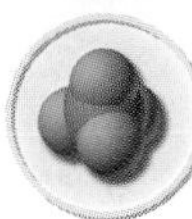

Example 14.11 How many milliliters of 2.00 *M* HCl will react with 28.0 g NaOH?

SOLUTION

Step 1 Write and balance the equation for the reaction:

$$HCl(aq) + NaOH(aq) \longrightarrow NaCl(aq) + H_2O(aq)$$

The equation states that 1 mol of HCl reacts with 1 mol of NaOH.

Step 2 Find the number of moles NaOH in 28.0 g NaOH:

$$\text{g NaOH} \longrightarrow \text{mol NaOH}$$

$$(28.0\text{ g NaOH})\left(\frac{1\text{ mol}}{40.00\text{ g}}\right) = 0.700\text{ mol NaOH}$$

$$28.0\text{ g NaOH} = 0.700\text{ mol NaOH}$$

Step 3 Solve for moles and volume of HCl needed. From Steps 1 and 2, we see that 0.700 mol HCl will react with 0.700 mol NaOH, because the ratio of moles reacting is 1 : 1. We know that 2.00 *M* HCl contain 2.00 mol HCl per liter, and so the volume that contains 0.700 mol HCl will be less than 1 L:

$$\text{mol NaOH} \longrightarrow \text{mol HCl} \longrightarrow \text{L HCl} \longrightarrow \text{mL HCl}$$

$$(0.700\ \cancel{\text{mol NaOH}})\left(\frac{1\ \cancel{\text{mol HCl}}}{1\ \cancel{\text{mol NaOH}}}\right)\left(\frac{1\text{ L HCl}}{2.00\ \cancel{\text{mol HCl}}}\right) = 0.350\text{ L HCl}$$

$$(0.350\ \cancel{\text{L}}\text{ HCl})\left(\frac{1000\text{ mL}}{1\ \cancel{\text{L}}}\right) = 350.\text{ mL HCl}$$

Therefore, 350. mL of 2.00 *M* HCl contain 0.700 mol HCl and will react with 0.700 mol, or 28.0 g, of NaOH.

Practice 14.5

What volume of 0.035 *M* $AgNO_3$ can be made from 5.0 g of $AgNO_3$?

Practice 14.6

How many milliliters of 0.50 *M* NaOH are required to react completely with 25.00 mL of 1.5 *M* HCl?

We've now examined several ways to measure concentration of solutions quantitatively. A summary of these concentration units is found in Table 14.4.

Table 14.4 Concentration Units for Solutions

Units	Symbol	Definition
Mass percent	% m/m	$\frac{\text{mass solute}}{\text{mass solution}} \times 100$
Parts per million	ppm	$\frac{\text{mass solute}}{\text{mass solution}} \times 1{,}000{,}000$
Mass/volume percent	% m/v	$\frac{\text{mass solute}}{\text{mL solution}} \times 100$
Volume percent	% v/v	$\frac{\text{mL solute}}{\text{mL solution}} \times 100$
Molarity	M	$\frac{\text{moles solute}}{\text{L solution}}$
Molality	m	$\frac{\text{moles solute}}{\text{kg solvent}}$

Molality is covered in Section 14.7

Dilution Problems

Chemists often find it necessary to dilute solutions from one concentration to another by adding more solvent to the solution. If a solution is diluted by adding pure solvent, the volume of the solution increases, but the number of moles of solute in the solution remains the same. Thus the moles/liter (molarity) of the solution decreases. Always read a problem carefully to distinguish between (1) how much solvent must be added to dilute a solution to a particular concentration and (2) to what volume a solution must be diluted to prepare a solution of a particular concentration.

Example 14.12

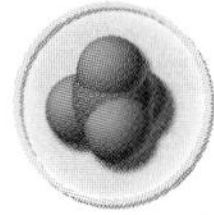

Calculate the molarity of a sodium hydroxide solution that is prepared by mixing 100. mL of 0.20 M NaOH with 150. mL of water. Assume that the volumes are additive.

SOLUTION

This problem is a dilution problem. If we double the volume of a solution by adding water, we cut the concentration in half. Therefore, the concentration of the above solution should be less than 0.10 M. In the dilution, the moles of NaOH remain constant; the molarity and volume change. The final volume is (100. mL + 150. mL), or 250. mL.

To solve this problem, (1) calculate the moles of NaOH in the original solution and (2) divide the moles of NaOH by the final volume of the solution to obtain the new molarity:

Step 1 Calculate the moles of NaOH in the original solution:

$$M = \frac{\text{mol}}{\text{L}} \qquad \text{mol} = \text{L} \times M$$

$$(0.100\ \cancel{\text{L}})\left(\frac{0.20\ \text{mol NaOH}}{1\ \cancel{\text{L}}}\right) = 0.020\ \text{mol NaOH}$$

Step 2 Solve for the new molarity, taking into account that the total volume of the solution after dilution is 250. mL (0.250 L):

$$M = \frac{0.020 \text{ mol NaOH}}{0.250 \text{ L}} = 0.080\ M \text{ NaOH}$$

ALTERNATIVE SOLUTION When the moles of solute in a solution before and after dilution are the same, the moles before and after dilution may be set equal to each other. That is,

$$\text{mol}_1 = \text{mol}_2$$

where mol_1 = moles before dilution and mol_2 = moles after dilution. Then

$$\text{mol}_1 = \text{L}_1 \times M_1 \qquad \text{mol}_2 = \text{L}_2 \times M_2$$
$$\text{L}_1 \times M_1 = \text{L}_2 \times M_2$$

When both volumes are in the same units, a more general statement can be made:

$$V_1 \times M_1 = V_2 \times M_2$$

For this problem,

$$V_1 = 100.\text{ mL} \qquad M_1 = 0.20\ M$$
$$V_2 = 150.\text{ mL} + 100.\text{ mL} \qquad M_2 = \text{(unknown)}$$

Then

$$(100.\text{ mL})(0.20\ M) = (250.\text{ mL})M_2$$

Solving for M_2, we get

$$M_2 = \frac{(100.\ \cancel{\text{mL}})(0.20\ M)}{250.\ \cancel{\text{mL}}} = 0.080\ M \text{ NaOH}$$

Practice 14.7

Calculate the molarity of a solution prepared by diluting 125 mL of 0.400 M $K_2Cr_2O_7$ with 875 mL of water.

Example 14.13 How many grams of silver chloride will be precipitated by adding sufficient silver nitrate to react with 1500. mL of 0.400 M barium chloride solution?

$$2\,AgNO_3(aq) + BaCl_2(aq) \longrightarrow 2\,AgCl(s) + Ba(NO_3)_2(aq)$$

SOLUTION This problem is a stoichiometry problem. The fact that $BaCl_2$ is in solution means that we need to consider the volume and concentration of the solution in order to determine the number of moles of $BaCl_2$ reacting.

Step 1 Determine the number of moles of $BaCl_2$ in 1500. mL of 0.400 M solution:

$$M = \frac{\text{mol}}{\text{L}} \qquad \text{mol} = \text{L} \times M \qquad 1500.\text{ mL} = 1.500\text{ L}$$

$$(1.500\ \cancel{\text{L}})\left(\frac{0.400 \text{ mol } BaCl_2}{\cancel{\text{L}}}\right) = 0.600 \text{ mol } BaCl_2$$

Step 2 Calculate the moles and grams of AgCl:

$$\text{mol BaCl}_2 \longrightarrow \text{mol AgCl} \longrightarrow \text{g AgCl}$$

$$(0.600 \cancel{\text{mol BaCl}_2})\left(\frac{2 \cancel{\text{mol AgCl}}}{1 \cancel{\text{mol BaCl}_2}}\right)\left(\frac{143.4 \text{ g AgCl}}{\cancel{\text{mol AgCl}}}\right) = 172 \text{ g AgCl}$$

Practice 14.8

How many grams of lead(II) iodide will be precipitated by adding sufficient $Pb(NO_3)_2$ to react with 750 mL of 0.250 *M* KI solution?

$$2\,KI(aq) + Pb(NO_3)_2(aq) \longrightarrow PbI_2(s) + 2\,KNO_3(aq)$$

14.7 Colligative Properties of Solutions

Two solutions—one containing 1 mol (60.06 g) of urea (NH_2CONH_2) and the other containing 1 mol (342.3 g) of sucrose ($C_{12}H_{22}O_{11}$) each in 1 kg of water—both have a freezing point of $-1.86°C$, not 0°C as for pure water. Urea and sucrose are distinct substances, yet they lower the freezing point of the water by the same amount. The only thing apparently common to these two solutions is that each contains 1 mol (6.022×10^{23} molecules) of solute and 1 kg of solvent. In fact, when we dissolve 1 mol of any nonionizable solute in 1 kg of water, the freezing point of the resulting solution is $-1.86°C$.

These results lead us to conclude that the freezing point depression for a solution containing 6.022×10^{23} solute molecules (particles) and 1 kg of water is a constant, namely, 1.86°C. Freezing point depression is a general property of solutions. Furthermore, the amount by which the freezing point is depressed is the same for all solutions made with a given solvent; that is, each solvent shows a characteristic *freezing point depression constant*. Freezing point depression constants for several solvents are given in Table 14.5.

The solution formed by the addition of a nonvolatile solute to a solvent has a lower freezing point, a higher boiling point, and a lower vapor pressure than that of the pure solvent. These effects are related and are known as colligative

Table 14.5 Freezing Point Depression and Boiling Point Elevation Constants of Selected Solvents

Solvent	Freezing point of pure solvent (°C)	Freezing point depression constant, K_f $\left(\frac{°\text{C kg solvent}}{\text{mol solute}}\right)$	Boiling point of pure solvent (°C)	Boiling point elevation constant, K_b $\left(\frac{°\text{C kg solvent}}{\text{mol solute}}\right)$
Water	0.00	1.86	100.0	0.512
Acetic acid	16.6	3.90	118.5	3.07
Benzene	5.5	5.1	80.1	2.53
Camphor	178	40	208.2	5.95

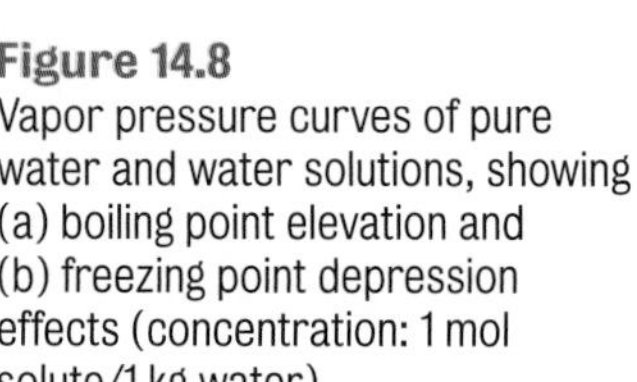

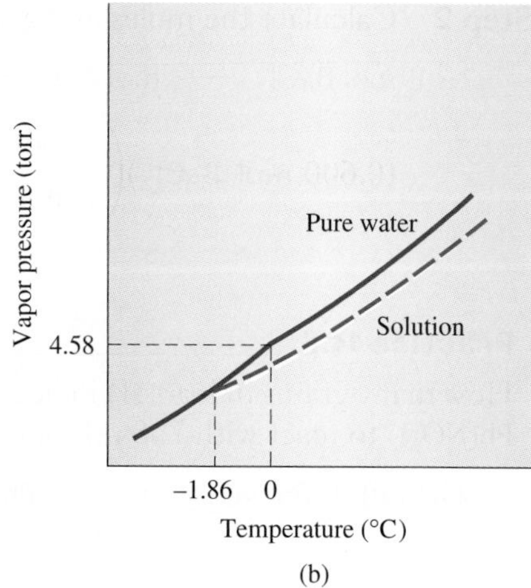

Figure 14.8
Vapor pressure curves of pure water and water solutions, showing (a) boiling point elevation and (b) freezing point depression effects (concentration: 1 mol solute/1 kg water).

colligative properties

properties. The **colligative properties** are properties that depend only on the number of solute particles in a solution, not on the nature of those particles. Freezing point depression, boiling point elevation, and vapor pressure lowering are colligative properties of solutions.

The colligative properties of a solution can be considered in terms of vapor pressure. The vapor pressure of a pure liquid depends on the tendency of molecules to escape from its surface. If 10% of the molecules in a solution are nonvolatile solute molecules, the vapor pressure of the solution is 10% lower than that of the pure solvent. The vapor pressure is lower because the surface of the solution contains 10% nonvolatile molecules and 90% of the volatile solvent molecules. A liquid boils when its vapor pressure equals the pressure of the atmosphere. We can thus see that the solution just described as having a lower vapor pressure will have a higher boiling point than the pure solvent. The solution with a lowered vapor pressure doesn't boil until it has been heated above the boiling point of the solvent (see Figure 14.8a). Each solvent has its own characteristic boiling point elevation constant (Table 14.5). The boiling point elevation constant is based on a solution that contains 1 mol of solute particles per kilogram of solvent. For example, the boiling point elevation constant for a solution containing 1 mol of solute particles per kilogram of water is 0.512°C, which means that this water solution will boil at 100.512°C.

The freezing behavior of a solution can also be considered in terms of lowered vapor pressure. Figure 14.8b shows the vapor pressure relationships of ice, water, and a solution containing 1 mol of solute per kilogram of water. The freezing point of water is at the intersection of the liquid and solid vapor pressure curves (i.e., at the point where water and ice have the same vapor pressure). Because the vapor pressure of the liquid is lowered by the solute, the vapor pressure curve of the solution does not intersect the vapor pressure curve of the solid until the solution has been cooled below the freezing point of pure water. So the solution must be cooled below 0°C in order for it to freeze.

Engine coolant is one application of colligative properties. The addition of coolant to the water in a radiator raises its boiling point and lowers its freezing point.

The foregoing discussion dealing with freezing point depressions is restricted to *un-ionized* substances. The discussion of boiling point elevations is restricted to *nonvolatile* and un-ionized substances. The colligative properties of ionized substances are not under consideration at this point; we will discuss them in Chapter 15.

Sodium chloride or calcium chloride is used to melt ice on snowy streets and highways.

Some practical applications involving colligative properties are (1) use of salt–ice mixtures to provide low freezing temperatures for homemade ice cream, (2) use of sodium chloride or calcium chloride to melt ice from streets, and (3) use of ethylene glycol–water mixtures as antifreeze in automobile radiators (ethylene glycol also raises the boiling point of radiator fluid, thus allowing the engine to operate at a higher temperature).

Both the freezing point depression and the boiling point elevation are directly proportional to the number of moles of solute per kilogram of solvent. When we deal with the colligative properties of solutions, another concentration expression, *molality*, is used. The **molality (m)** of a solute is the number of moles of solute per kilogram of solvent:

molality (m)

$$m = \frac{\text{mol solute}}{\text{kg solvent}}$$

Note that a lowercase m is used for molality concentrations and a capital M for molarity. The difference between molality and molarity is that molality refers to moles of solute *per kilogram of solvent*, whereas molarity refers to moles of solute *per liter of solution*. For un-ionized substances, the colligative properties of a solution are directly proportional to its molality.

Molality is independent of volume. It is a mass-to-mass relationship of solute to solvent and allows for experiments, such as freezing point depression and boiling point elevation, to be conducted at variable temperatures.

The following equations are used in calculations involving colligative properties and molality:

$$\Delta t_f = mK_f \qquad \Delta t_b = mK_b \qquad m = \frac{\text{mol solute}}{\text{kg solvent}}$$

m = molality; mol solute/kg solvent
Δt_f = freezing point depression; °C
Δt_b = boiling point elevation; °C
K_f = freezing point depression constant; °C kg solvent/mol solute
K_b = boiling point elevation constant; °C kg solvent/mol solute

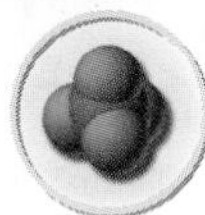

Example 14.14 What is the molality (m) of a solution prepared by dissolving 2.70 g CH_3OH in 25.0 g H_2O?

SOLUTION Since $m = \dfrac{\text{mol solute}}{\text{kg solvent}}$, the conversion is

$$\frac{2.70\text{ g CH}_3\text{OH}}{25.0\text{ g H}_2\text{O}} \longrightarrow \frac{\text{mol CH}_3\text{OH}}{25.0\text{ g H}_2\text{O}} \longrightarrow \frac{\text{mol CH}_3\text{OH}}{1\text{ kg H}_2\text{O}}$$

The molar mass of CH_3OH is (12.01 + 4.032 + 16.00), or 32.04 g/mol:

$$\left(\frac{2.70\ \cancel{\text{g CH}_3\text{OH}}}{25.0\ \cancel{\text{g H}_2\text{O}}}\right)\left(\frac{1\text{ mol CH}_3\text{OH}}{32.04\ \cancel{\text{g CH}_3\text{OH}}}\right)\left(\frac{1000\ \cancel{\text{g H}_2\text{O}}}{1\text{ kg H}_2\text{O}}\right) = \frac{3.37\text{ mol CH}_3\text{OH}}{1\text{ kg H}_2\text{O}}$$

The molality is 3.37 m.

Practice 14.9

What is the molality of a solution prepared by dissolving 150.0 g $C_6H_{12}O_6$ in 600.0 g H_2O?

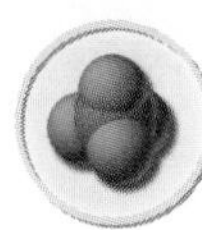

Example 14.15 A solution is made by dissolving 100. g of ethylene glycol ($C_2H_6O_2$) in 200. g of water. What is the freezing point of this solution?

SOLUTION To calculate the freezing point of the solution, we first need to calculate Δt_f, the change in freezing point. Use the equation

$$\Delta t_f = mK_f = \frac{\text{mol solute}}{\text{kg solvent}} \times K_f$$

K_f (for water): $\dfrac{1.86°\text{C kg solvent}}{\text{mol solute}}$ (from Table 14.5)

mol solute: $$(100.\ \cancel{\text{g C}_2\text{H}_6\text{O}_2})\left(\frac{1\text{ mol C}_2\text{H}_6\text{O}_2}{62.07\ \cancel{\text{g C}_2\text{H}_6\text{O}_2}}\right) = 1.61\text{ mol C}_2\text{H}_6\text{O}_2$$

kg solvent: $$(200.\ \cancel{\text{g}}\text{ H}_2\text{O})\left(\frac{1\text{ kg}}{1000\ \cancel{\text{g}}}\right) = 0.200\text{ kg H}_2\text{O}$$

$$\Delta t_f = \left(\frac{1.61\ \cancel{\text{mol C}_2\text{H}_6\text{O}_2}}{0.200\ \cancel{\text{kg H}_2\text{O}}}\right)\left(\frac{1.86°\text{C}\ \cancel{\text{kg H}_2\text{O}}}{1\ \cancel{\text{mol C}_2\text{H}_6\text{O}_2}}\right) = 15.0°\text{C}$$

The freezing point depression, 15.0°C, must be subtracted from 0°C, the freezing point of the pure solvent (water):

$$\text{freezing point of solution} = \text{freezing point of solvent} - \Delta t_f$$
$$= 0.0°\text{C} - 15.0°\text{C} = -15.0°\text{C}$$

Therefore, the freezing point of the solution is −15.0°C. This solution will protect an automobile radiator down to −15.0°C (5°F).

CHEMISTRY IN ACTION • The Scoop on Ice Cream

Ice cream is mainly composed of water (from milk and cream), milk solids, milk fats, and, frequently, various sweeteners (corn syrup or sugar), flavorings, emulsifiers, and stabilizers. But that smooth, creamy, rich flavor and texture are the result of the chemistry of the mixing and freezing process. The rich, smooth texture of great ice cream results from the milk fat. By law, if the carton is labeled "ice cream," it must contain a minimum of 10% milk fat. That carton of ice cream also contains 20%–50% air whipped into the ingredients during the initial mixing process.

H. Douglas Goff, a professor of food science and ice cream expert from Ontario, Canada, says, "There are no real chemical reactions that take place when you make ice cream, but that doesn't mean that there isn't plenty of chemistry." The structure of ice cream contributes greatly to its taste. Tiny air bubbles are formed in the initial whipping process. These bubbles are distributed through a network of fat globules and liquid water. The milk fat has surface proteins on the globules in the milk to keep the fat dissolved in solution. Ice cream manufacturers destabilize these globules using emulsifiers (such as egg yolks, mono- or diglycerides) and let them come together in larger networks sort of like grape clusters.

Once the mixture is fully whipped, it is cooled to begin the freezing process. But ice cream does not freeze at 0°C even though it is 55–64% water. The freezing point is depressed as a colligative property of the ice cream solution. Once the freezing of the water begins, the concentration of the solution increases, which continues to depress the freezing point. Goff tells us that even at a typical serving temperature (−16°C for most ice cream), only about 72% of the water in the ice cream is frozen. The unfrozen solution keeps the ice cream "scoopable."

One last factor that affects the quality of ice cream is the size of the ice crystals. For very smooth ice cream, tiny crystals are needed. To produce these, the ice cream must freeze very slowly. Large crystals give a coarse, grainy texture. Now, as you savor that premium ice cream cone, you'll know just how colligative properties and the chemistry of freezing helped make it so delicious!

Example 14.16

A solution made by dissolving 4.71 g of a compound of unknown molar mass in 100.0 g of water has a freezing point of −1.46°C. What is the molar mass of the compound?

SOLUTION

First substitute the data in $\Delta t_f = mK_f$ and solve for m:

$$\Delta t_f = +1.46 \text{ (since the solvent, water, freezes at 0°C)}$$

$$K_f = \frac{1.86°\text{C kg H}_2\text{O}}{\text{mol solute}}$$

$$1.46°\text{C} = mK_f = m \times \frac{1.86°\text{C kg H}_2\text{O}}{\text{mol solute}}$$

$$m = \frac{1.46\cancel{°\text{C}} \times \text{mol solute}}{1.86\cancel{°\text{C}} \times \text{kg H}_2\text{O}} = \frac{0.785 \text{ mol solute}}{\text{kg H}_2\text{O}}$$

Now convert the data, 4.71 g solute/100.0 g H_2O, to g/mol:

$$\left(\frac{4.71 \text{ g } \cancel{\text{solute}}}{100.0 \cancel{\text{ g H}_2\text{O}}}\right)\left(\frac{1000 \cancel{\text{ g H}_2\text{O}}}{1 \cancel{\text{ kg H}_2\text{O}}}\right)\left(\frac{1 \cancel{\text{ kg H}_2\text{O}}}{0.785 \text{ mol } \cancel{\text{solute}}}\right) = 60.0 \text{ g/mol}$$

The molar mass of the compound is 60.0 g/mol.

Practice 14.10

What is the freezing point of the solution in Practice Exercise 14.9? What is the boiling point?

14.8 Osmosis and Osmotic Pressure

When red blood cells are put into distilled water, they gradually swell and in time may burst. If red blood cells are put in a 5%-urea (or a 5%-salt) solution, they gradually shrink and take on a wrinkled appearance. The cells behave in this fashion because they are enclosed in semipermeable membranes. A **semipermeable membrane** allows the passage of water (solvent) molecules through it in either direction but prevents the passage of larger solute molecules or ions. When two solutions of different concentrations (or water and a water solution) are separated by a semipermeable membrane, water diffuses through the membrane from the solution of lower concentration into the solution of higher concentration. The diffusion of water, either from a dilute solution or from pure water, through a semipermeable membrane into a solution of higher concentration is called **osmosis**.

semipermeable membrane

osmosis

A 0.90% (0.15 *M*) sodium chloride solution is known as a *physiological saline solution* because it is *isotonic* with blood plasma; that is, it has the same concentration of NaCl as blood plasma. Because each mole of NaCl yields about 2 mol of ions when in solution, the solute particle concentration in physiological saline solution is nearly 0.30 *M*. Five-percent-glucose solution (0.28 *M*) is also approximately isotonic with blood plasma. Blood cells neither swell nor shrink in an isotonic solution. The cells described in the preceding paragraph swell in water because water is *hypotonic* to cell plasma. The cells shrink in 5%-urea solution because the urea solution is *hypertonic* to the cell plasma. To prevent possible injury to blood cells by osmosis, fluids for intravenous use are usually made up at approximately isotonic concentration.

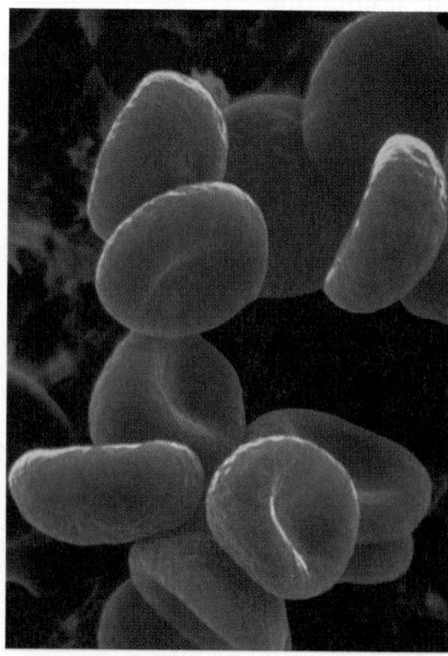

Human red blood cells. Left: In an isotonic solution the concentration is the same inside and outside the cell (0.9% saline). Center: In a hypertonic solution (1.6% saline) water leaves the cells, causing them to crenate (shrink). Right: In a hypotonic solution (0.2% saline) the cells swell as water moves into the cell center. Magnification is 260,000×.

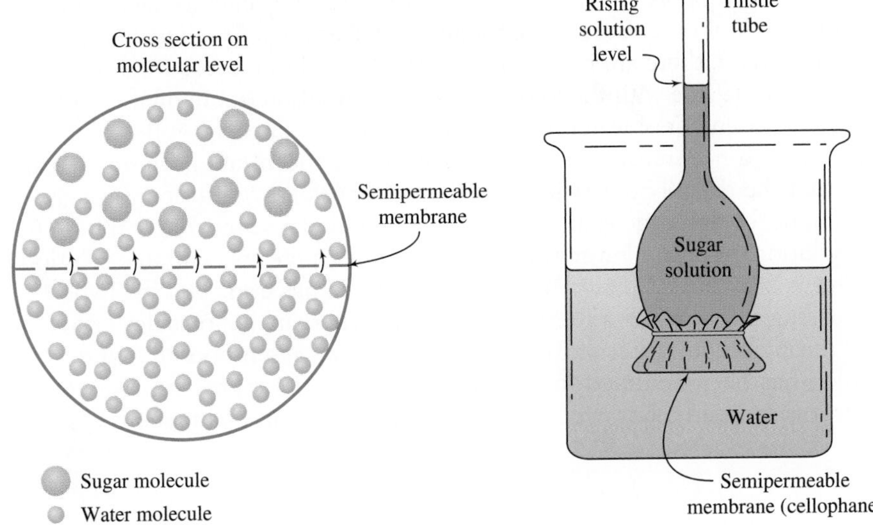

Figure 14.9
Laboratory demonstration of osmosis: As a result of osmosis, water passes through the membrane, causing the solution to rise in the thistle tube.

All solutions exhibit *osmotic pressure,* which is another colligative property. Osmotic pressure is a pressure difference between the system and atmospheric pressure. The osmotic pressure of a system can be measured by applying enough pressure to stop the flow of water due to osmosis in the system. The difference between the applied pressure and atmospheric pressure is the **osmotic pressure**. When pressure greater than the osmotic pressure is applied to a system, the flow of water can be reversed from that of osmosis. This process can be used to obtain useful drinking water from seawater and is known as *reverse osmosis.* Osmotic pressure is dependent only on the concentration of the solute particles and is independent of their nature. The osmotic pressure of a solution can be measured by determining the amount of counterpressure needed to prevent osmosis; this pressure can be very large. The osmotic pressure of a solution containing 1 mol of solute particles in 1 kg of water is about 22.4 atm, which is about the same as the pressure exerted by 1 mol of a gas confined in a volume of 1 L at 0°C.

osmotic pressure

Osmosis has a role in many biological processes, and semipermeable membranes occur commonly in living organisms. An example is the roots of plants, which are covered with tiny structures called root hairs; soil water enters the plant by osmosis, passing through the semipermeable membranes covering the root hairs. Artificial or synthetic membranes can also be made.

Osmosis can be demonstrated with the simple laboratory setup shown in Figure 14.9. As a result of osmotic pressure, water passes through the cellophane membrane into the thistle tube, causing the solution level to rise. In osmosis, the net transfer of water is always from a less concentrated to a more concentrated solution; that is, the effect is toward equalization of the concentration on both sides of the membrane. Note that the effective movement of water in osmosis is always from the region of *higher water concentration* to the region of *lower water concentration.*

Osmosis can be explained by assuming that a semipermeable membrane has passages that permit water molecules and other small molecules to pass in either direction. Both sides of the membrane are constantly being struck by water molecules in random motion. The number of water molecules crossing the membrane is proportional to the number of water molecule-to-membrane impacts per unit of time. Because the solute molecules or ions reduce the concentration of water, there are more water molecules and thus more water molecule impacts on the side with the lower solute concentration (more dilute solution). The greater number of water molecule-to-membrane impacts on the dilute side thus causes a net transfer of water to the more concentrated solution. Again, note that the overall process involves the net transfer, by diffusion through the membrane, of water molecules from a region of higher water concentration (dilute solution) to one of lower water concentration (more concentrated solution).

This is a simplified picture of osmosis. No one has ever seen the hypothetical passages that allow water molecules and other small molecules or ions to pass through them. Alternative explanations have been proposed, but our discussion has been confined to water solutions. Osmotic pressure is a general colligative property, however, and is known to occur in nonaqueous systems.

Chapter 14 Review

14.1 General Properties of Solutions

KEY TERMS

Solution
Solute
Solvent

- A solution is a homogeneous mixture of two or more substances:
  - Consists of solvent—the dissolving agent—and solute—the component(s) dissolved in the solvent
  - Is a homogeneous mixture
  - Contains molecular or ionic particles
  - Can be colored or colorless
  - Can be separated into solute and solvent by a physical separation process

14.2 Solubility

KEY TERMS

Solubility
Miscible
Immiscible
Concentration of a solution

- Solubility describes the amount of solute that will dissolve in a specified amount of solvent.
- Solubility can also be qualitative.
- General guideline for ionic solubility are:

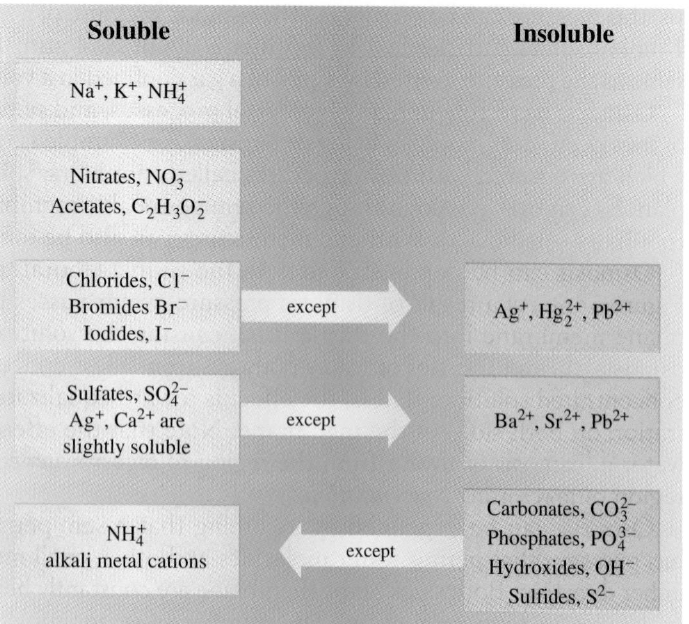

- Liquids can also be classified as miscible (soluble in each other) or immiscible (not soluble in each other).
- The concentration of a solution is the quantitative measurement of the amount of solute that is dissolved in a solution.

14.3 Factors Related to Solubility

KEY TERMS

Saturated solution
Unsaturated solution
Supersaturated solution

- Like tends to dissolve like is a general rule for solvents and solutes.
- As temperature increases:
  - Solubility of a solid in a liquid tends to increase.
  - Solubility of a gas in a liquid tends to decrease.
- As pressure increases:
  - Solubility of a solid in a liquid remains constant.
  - Solubility of a gas in a liquid tends to increase.
- At a specific temperature, the amount of solute that can dissolve in a solvent has a limit:
  - Unsaturated solutions contain less solute than the limit.
  - Saturated solutions contain dissolved solute at the limit.
  - Supersaturated solutions contain more solute than the limit and are therefore unstable:
    - If disturbed, the excess solute will precipitate out of solution.

14.4 Rate of Dissolving Solids

- The rate at which a solute dissolves is determined by these factors:
  - Particle size
  - Temperature
  - Concentration of solution
  - Agitation

14.5 Solutions: A Reaction Medium

- Molecules or ions must collide in order to react.
- Solutions provide a medium for the molecules or ions to collide.

14.6 Concentration of Solutions

KEY TERMS

Dilute solution
Concentrated solution
Parts per million (ppm)
Molarity (M)

- Concentrations can be measured in many ways:

Mass percent	% m/m	$\frac{\text{mass solute}}{\text{mass solution}} \times 100$
Parts per million	ppm	$\frac{\text{mass solute}}{\text{mass solution}} \times 1{,}000{,}000$
Mass/volume percent	% m/v	$\frac{\text{mass solute}}{\text{mL solution}} \times 100$
Volume percent	% v/v	$\frac{\text{mL solute}}{\text{mL solution}} \times 100$
Molarity	M	$\frac{\text{moles solute}}{\text{L solution}}$
Molality	m	$\frac{\text{moles solute}}{\text{kg solvent}}$

- Dilution of solutions requires the addition of more solvent to an existing solution:
  - The number of moles in the diluted solution is the same as that in the original solution.
  - $M_1V_1 = M_2V_2$.

14.7 Colligative Properties of Solutions

KEY TERMS

Colligative properties
Molality *(m)*

- Properties of a solution that depend only on the number of solute particles in solution are called colligative properties:
  - Freezing point depression
    $\Delta t_f = mK_f$
  - Boiling point elevation
    $\Delta t_b = mK_b$
  - Osmotic pressure
- Molality is used in working with colligative properties.

14.8 Osmosis and Osmotic Pressure

KEY TERMS

Semipermeable membrane
Osmosis
Osmotic pressure

- Osmosis is the diffusion of water through a semipermeable membrane:
  - Occurs from dilute solution to a solution of higher concentration
- Osmosis results in osmotic pressure, which is a colligative property of a solution.

Review Questions

All questions with blue numbers have answers in the appendix of the text.

1. Sketch the orientation of water molecules (a) about a single sodium ion and (b) about a single chloride ion in solution.
2. Estimate the number of grams of sodium fluoride that would dissolve in 100 g of water at 50°C. (Table 14.2)
3. What is the solubility at 25°C of these substances (Figure 14.4)?
   (a) potassium chloride
   (b) potassium chlorate
   (c) potassium nitrate

4. What is different in the solubility trend of the potassium halides compared with that of the lithium halides and the sodium halides? (Table 14.2)

5. What is the solubility, in grams of solute per 100 g of H_2O, of (a) $KClO_3$ at 60°C; (b) HCl at 20°C; (c) Li_2SO_4 at 80°C; and (d) KNO_3 at 0°C? (Figure 14.4)

6. Which substance, KNO_3 or NH_4Cl, shows the greater increase in solubility with increased temperature? (Figure 14.4)

7. Does a 2-molal solution in benzene or a 1-molal solution in camphor show the greater freezing point depression? (Table 14.5)

8. What would be the total surface area if the 1-cm cube in Figure 14.5 were cut into cubes 0.01 cm on a side?

9. At which temperatures—10°C, 20°C, 30°C, 40°C, or 50°C—would you expect a solution made from 63 g of ammonium chloride and 150 g of water to be unsaturated? (Figure 14.4)

10. Explain why the rate of dissolving decreases. (Figure 14.6)

11. Explain how a supersaturated solution of $NaC_2H_3O_2$ can be prepared and proven to be supersaturated.

12. Assume that the thistle tube in Figure 14.9 contains 1.0 *M* sugar solution and that the water in the beaker has just been replaced by a 2.0 *M* solution of urea. Would the solution level in the thistle tube continue to rise, remain constant, or fall? Explain.

13. What is a true solution?

14. Name and distinguish between the two components of a solution.

15. Is it always apparent in a solution which component is the solute, for example, in a solution of a liquid in a liquid?

16. Explain why the solute does not settle out of a solution.

17. Is it possible to have one solid dissolved in another? Explain.

18. An aqueous solution of KCl is colorless, $KMnO_4$ is purple, and $K_2Cr_2O_7$ is orange. What color would you expect of an aqueous solution of $Na_2Cr_2O_7$? Explain.

19. Explain why hexane will dissolve benzene but will not dissolve sodium chloride.

20. Some drinks like tea are consumed either hot or cold, whereas others like Coca-Cola are drunk only cold. Why?

21. Why is air considered to be a solution?

22. In which will a teaspoonful of sugar dissolve more rapidly, 200 mL of iced tea or 200 mL of hot coffee? Explain in terms of the KMT.

23. What is the effect of pressure on the solubility of gases in liquids? of solids in liquids?

24. Why do smaller particles dissolve faster than large ones?

25. In a saturated solution containing undissolved solute, solute is continuously dissolving, but the concentration of the solution remains unchanged. Explain.

26. Explain why there is no apparent reaction when crystals of $AgNO_3$ and NaCl are mixed, but a reaction is apparent immediately when solutions of $AgNO_3$ and NaCl are mixed.

27. Why do salt trucks distribute salt over icy roads in the winter?

28. What do we mean when we say that concentrated nitric acid (HNO_3) is 16 molar?

29. Will 1 L of 1 *M* NaCl contain more chloride ions than 0.5 L of 1 *M* $MgCl_2$? Explain.

30. Champagne is usually cooled in a refrigerator prior to opening. It's also opened very carefully. What would happen if a warm bottle of champagne were shaken and opened quickly and forcefully?

31. Describe how you would prepare 750 mL of 5.0 *M* NaCl solution.

32. Explain in terms of the KMT how a semipermeable membrane functions when placed between pure water and a 10% sugar solution.

33. Which has the higher osmotic pressure, a solution containing 100 g of urea (NH_2CONH_2) in 1 kg H_2O or a solution containing 150 g of glucose, $C_6H_{12}O_6$, in 1 kg H_2O?

34. Explain why a lettuce leaf in contact with salad dressing containing salt and vinegar soon becomes wilted and limp whereas another lettuce leaf in contact with plain water remains crisp.

35. A group of shipwreck survivors floated for several days on a life raft before being rescued. Those who had drunk some seawater were found to be suffering the most from dehydration. Explain.

36. Arrange the following bases (in descending order) according to the volume of each that will react with 1 L of 1 *M* HCl:
(a) 1 *M* NaOH
(b) 1.5 *M* $Ca(OH)_2$
(c) 2 *M* KOH
(d) 0.6 *M* $Ba(OH)_2$

37. Explain in terms of vapor pressure why the boiling point of a solution containing a nonvolatile solute is higher than that of the pure solvent.

38. Explain why the freezing point of a solution is lower than the freezing point of the pure solvent.

39. When water and ice are mixed, the temperature of the mixture is 0°C. But if methyl alcohol and ice are mixed, a temperature of −10°C is readily attained. Explain why the two mixtures show such different temperature behavior.

40. Which would be more effective in lowering the freezing point of 500. g of water?
(a) 100. g of sucrose, $C_{12}H_{22}O_{11}$, or 100. g of ethyl alcohol, C_2H_5OH
(b) 100. g of sucrose or 20.0 g of ethyl alcohol
(c) 20.0 g of ethyl alcohol or 20.0 g of methyl alcohol, CH_3OH.

41. What is the difference between molarity and molality?

42. Is the molarity of a 5 *m* aqueous solution of NaCl greater or less than 5 *M*? Explain.

Paired Exercises

All exercises with blue numbers have answers in the appendix of the text.

1. Which of the substances listed below are reasonably soluble and which are insoluble in water? (See Figure 14.2 or Appendix V.)
(a) KOH
(b) $NiCl_2$
(c) ZnS
(d) $AgC_2H_3O_2$
(e) Na_2CrO_4

2. Which of the substances listed below are reasonably soluble and which are insoluble in water? (See Figure 14.2 or Appendix V.)
(a) PbI_2
(b) $MgCO_3$
(c) $CaCl_2$
(d) $Fe(NO_3)_3$
(e) $BaSO_4$

Percent Solution

3. Calculate the mass percent of the following solutions:
(a) 15.0 g KCl + 100.0 g H_2O
(b) 2.50 g Na_3PO_4 + 10.0 g H_2O
(c) 0.20 mol $NH_4C_2H_3O_2$ + 125 g H_2O
(d) 1.50 mol NaOH in 33.0 mol H_2O

4. Calculate the mass percent of the following solutions:
(a) 25.0 g $NaNO_3$ in 125.0 g H_2O
(b) 1.25 g $CaCl_2$ in 35.0 g H_2O
(c) 0.75 mol K_2CrO_4 in 225 g H_2O
(d) 1.20 mol H_2SO_4 in 72.5 mol H_2O

5. A chemistry lab experiment requires 25.2 g of silver nitrate. How many grams of a 15.5% by mass solution of silver nitrate should be used?

6. A reaction requires 25.0 g of sodium chloride. How many grams of a 10.0% by mass solution would provide this amount of solute?

7. In 25 g of a 7.5% by mass solution of $CaSO_4$
(a) how many grams of solute are present?
(b) how many grams of solvent are present?

8. In 75 g of a 12.0% by mass solution of $BaCl_2$
(a) how many grams of solute are present?
(b) how many grams of solvent are present?

9. Calculate the mass/volume percent of a solution made by dissolving 22.0 g of CH_3OH (methanol) in C_2H_5OH (ethanol) to make 100. mL of solution.

10. Calculate the mass/volume percent of a solution made by dissolving 4.20 g of NaCl in H_2O to make 12.5 mL of solution.

11. What is the volume percent of 10.0 mL of CH_3OH (methanol) dissolved in water to a volume of 40.0 mL?

12. What is the volume percent of 2.0 mL of hexane, C_6H_{14}, dissolved in benzene, C_6H_6, to a volume of 9.0 mL?

Molarity

13. Calculate the molarity of the following solutions:
(a) 0.25 mol of solute in 75.0 mL of solution
(b) 1.75 mol of KBr in 0.75 L of solution
(c) 35.0 g of $NaC_2H_3O_2$ in 1.25 L of solution
(d) 75 g of $CuSO_4 \cdot 5\,H_2O$ in 1.0 L of solution

14. Calculate the molarity of the following solutions:
(a) 0.50 mol of solute in 125 mL of solution
(b) 2.25 mol of $CaCl_2$ in 1.50 L of solution
(c) 275 g $C_6H_{12}O_6$ in 775 mL of solution
(d) 125 g $MgSO_4 \cdot 7\,H_2O$ in 2.50 L of solution

15. Calculate the number of moles of solute in each of the following solutions:
(a) 1.5 L of 1.20 M H_2SO_4
(b) 25.0 mL of 0.0015 M $BaCl_2$
(c) 125 mL of 0.35 M K_3PO_4

16. Calculate the number of moles of solute in each of the following solutions:
(a) 0.75 L of 1.50 M HNO_3
(b) 10.0 mL of 0.75 M $NaClO_3$
(c) 175 mL of 0.50 M LiBr

17. Calculate the grams of solute in each of the following solutions:
(a) 2.5 L of 0.75 M K_2CrO_4
(b) 75.2 mL of 0.050 M $HC_2H_3O_2$
(c) 250 mL of 16 M HNO_3

18. Calculate the grams of solute in each of the following solutions:
(a) 1.20 L of 18 M H_2SO_4
(b) 27.5 mL of 1.50 M $KMnO_4$
(c) 120 mL of 0.025 M $Fe_2(SO_4)_3$

19. How many milliliters of 0.750 M H_3PO_4 will contain the following?
(a) 0.15 mol H_3PO_4
(b) 35.5 g H_3PO_4

20. How many milliliters of 0.250 M NH_4Cl will contain the following?
(a) 0.85 mol NH_4Cl
(b) 25.2 g NH_4Cl

Dilution

21. What will be the molarity of the resulting solutions made by mixing the following? Assume that volumes are additive.
(a) 125 mL of 5.0 *M* H_3PO_4 with 775 mL of H_2O
(b) 250 mL of 0.25 *M* Na_2SO_4 with 750 mL of H_2O
(c) 75 mL of 0.50 *M* HNO_3 with 75 mL of 1.5 *M* HNO_3

22. What will be the molarity of the resulting solutions made by mixing the following? Assume that volumes are additive.
(a) 175 mL of 3.0 *M* H_2SO_4 with 275 mL of H_2O
(b) 350 mL of 0.10 *M* $CuSO_4$ with 150 mL of H_2O
(c) 50.0 mL of 0.250 *M* HCl with 25.0 mL of 0.500 *M* HCl

23. Calculate the volume of concentrated reagent required to prepare the diluted solutions indicated:
(a) 15 *M* H_3PO_4 to prepare 750 mL of 3.0 *M* H_3PO_4
(b) 16 *M* HNO_3 to prepare 250 mL of 0.50 *M* HNO_3

24. Calculate the volume of concentrated reagent required to prepare the diluted solutions indicated:
(a) 18 *M* H_2SO_4 to prepare 225 mL of 2.0 *M* H_2SO_4
(b) 15 *M* NH_3 to prepare 75 mL of 1.0 *M* NH_3

25. Calculate the molarity of the solutions made by mixing 125 mL of 6.0 *M* $HC_2H_3O_2$ with the following:
(a) 525 mL of H_2O
(b) 175 mL of 1.5 *M* $HC_2H_3O_2$

26. Calculate the molarity of the solutions made by mixing 175 mL of 3.0 *M* HCl with the following:
(a) 250 mL of H_2O
(b) 115 mL of 6.0 *M* HCl

Stoichiometry

27. Use the equation to calculate the following:

$$3\,Ca(NO_3)_2(aq) + 2\,Na_3PO_4(aq) \rightarrow Ca_3(PO_4)_2(s) + 6\,NaNO_3(aq)$$

(a) the moles $Ca_3(PO_4)_2$ produced from 2.7 mol Na_3PO_4
(b) the moles $NaNO_3$ produced from 0.75 mol $Ca(NO_3)_2$
(c) the moles Na_3PO_4 required to react with 1.45 L of 0.225 *M* $Ca(NO_3)_2$
(d) the grams of $Ca_3(PO_4)_2$ that can be obtained from 125 mL of 0.500 *M* $Ca(NO_3)_2$
(e) the volume of 0.25 *M* Na_3PO_4 needed to react with 15.0 mL of 0.50 *M* $Ca(NO_3)_2$
(f) the molarity (*M*) of the $Ca(NO_3)_2$ solution when 50.0 mL react with 50.0 mL of 2.0 *M* Na_3PO_4

28. Use the equation to calculate the following:

$$2\,NaOH(aq) + H_2SO_4(aq) \rightarrow Na_2SO_4(aq) + 2\,H_2O(l),$$

(a) the moles Na_2SO_4 produced from 3.6 mol H_2SO_4
(b) the moles H_2O produced from 0.025 mol NaOH
(c) the moles NaOH required to react with 2.50 L of 0.125 *M* H_2SO_4
(d) the grams of Na_2SO_4 that can be obtained from 25 mL of 0.050 *M* NaOH
(e) the volume of 0.250 *M* H_2SO_4 needed to react with 25.5 mL of 0.750 *M* NaOH
(f) the molarity (*M*) of the NaOH solution when 48.20 mL react with 35.72 mL of 0.125 *M* H_2SO_4

29. Use the equation to calculate the following:

$$2\,KMnO_4(aq) + 16\,HCl(aq) \rightarrow 2\,MnCl_2(aq) + 5\,Cl_2(g) + 8\,H_2O(l) + 2\,KCl(aq)$$

(a) the moles of H_2O that can be obtained from 15.0 mL of 0.250 *M* HCl
(b) the volume of 0.150 *M* $KMnO_4$ needed to produce 1.85 mol $MnCl_2$
(c) the volume of 2.50 *M* HCl needed to produce 125 mL of 0.525 *M* KCl
(d) the molarity (*M*) of the HCl solution when 22.20 mL react with 15.60 mL of 0.250 *M* $KMnO_4$
(e) the liters of Cl_2 gas at STP produced by the reaction of 125 mL of 2.5 *M* HCl
(f) the liters of Cl_2 gas at STP produced by the reaction of 15.0 mL of 0.750 *M* HCl and 12.0 mL of 0.550 *M* $KMnO_4$

30. Use the equation to calculate the following:

$$K_2CO_3(aq) + 2\,HC_2H_3O_2(aq) \rightarrow 2\,KC_2H_3O_2(aq) + H_2O(l) + CO_2(g)$$

(a) the moles of H_2O that can be obtained from 25.0 mL of 0.150 *M* $HC_2H_3O_2$
(b) the volume of 0.210 *M* K_2CO_3 needed to produce 17.5 mol $KC_2H_3O_2$
(c) the volume of 1.25 *M* $HC_2H_3O_2$ needed to react with 75.2 mL 0.750 *M* K_2CO_3
(d) the molarity (*M*) of the $HC_2H_3O_2$ solution when 10.15 mL react with 18.50 mL of 0.250 *M* K_2CO_3
(e) the liters of CO_2 gas at STP produced by the reaction of 105 mL of 1.5 *M* $HC_2H_3O_2$
(f) the liters of CO_2 gas at STP produced by the reaction of 25.0 mL of 0.350 *M* K_2CO_3 and 25.0 mL of 0.250 *M* $HC_2H_3O_2$

Molality and Colligative Properties

31. Calculate the molality of these solutions:
(a) 14.0 g CH_3OH in 100.0 g of H_2O
(b) 2.50 mol of benzene (C_6H_6) in 250 g of hexane (C_6H_{14})

32. Calculate the molality of these solutions:
(a) 1.0 g $C_6H_{12}O_6$ in 1.0 g H_2O
(b) 0.250 g iodine in 1.0 kg H_2O

33. What is the (a) molality, (b) freezing point, and (c) boiling point of a solution containing 2.68 g of naphthalene ($C_{10}H_8$) in 38.4 g of benzene (C_6H_6)?

*35. The freezing point of a solution of 8.00 g of an unknown compound dissolved in 60.0 g of acetic acid is 13.2°C. Calculate the molar mass of the compound.

34. What is the (a) molality, (b) freezing point, and (c) boiling point of a solution containing 100.0 g of ethylene glycol ($C_2H_6O_2$) in 150.0 g of water?

*36. What is the molar mass of a compound if 4.80 g of the compound dissolved in 22.0 g of H_2O give a solution that freezes at −2.50°C?

Additional Exercises

All exercises with blue numbers have answers in the appendix of the text.

37. What happens to salt (NaCl) crystals when they are dissolved in water?

38. What happens to sugar molecules ($C_{12}H_{22}O_{11}$) when they are dissolved in water?

39. Why do sugar and salt behave differently when dissolved in water?

40. Why don't blood cells shrink or swell in an isotonic sodium chloride solution (0.9% saline)?

41. In the picture of dissolving $KMnO_4$ found in Section 14.1, the compound is forming purple streaks as it dissolves. Why?

42. In Figure 14.4, observe the line for KNO_3. Explain why it slopes up from left to right. How does the slope compare to the slopes of the other substances? What does this mean?

43. How many grams of solution, 10.0% NaOH by mass, are required to neutralize 150 mL of a 1.0 *M* HCl solution?

*44. How many grams of solution, 10.0% NaOH by mass, are required to neutralize 250.0 g of a 1.0 *m* solution of HCl?

*45. A sugar syrup solution contains 15.0% sugar, $C_{12}H_{22}O_{11}$, by mass and has a density of 1.06 g/mL.
(a) How many grams of sugar are in 1.0 L of this syrup?
(b) What is the molarity of this solution?
(c) What is the molality of this solution?

*46. A solution of 3.84 g C_4H_2N (empirical formula) in 250.0 g of benzene depresses the freezing point of benzene 0.614°C. What is the molecular formula for the compound?

47. Hydrochloric acid (HCl) is sold as a concentrated aqueous solution (12.0 mol/L). If the density of the solution is 1.18 g/mL, determine the molality of the solution.

*48. How many grams of KNO_3 are needed to make 450 mL of a solution that is to contain 5.5 mg/mL of potassium ion? Calculate the molarity of the solution.

49. What mass of 5.50% solution can be prepared from 25.0 g KCl?

50. Given a solution containing 16.10 g $C_2H_6O_2$ in 82.0 g H_2O that has a boiling point of 101.62°C, verify that the boiling point elevation constant K_f for water is 0.512°C kg H_2O/mole solute.

51. Physiological saline (NaCl) solutions used in intravenous injections have a concentration of 0.90% NaCl (mass/volume).
(a) How many grams of NaCl are needed to prepare 500.0 mL of this solution?
(b) How much water must evaporate from this solution to give a solution that is 9.0% NaCl (mass/volume)?

*52. A solution is made from 50.0 g KNO_3 and 175 g H_2O. How many grams of water must evaporate to give a saturated solution of KNO_3 in water at 20°C? (See Figure 14.4.)

53. What volume of 70.0% rubbing alcohol can you prepare if you have only 150 mL of pure isopropyl alcohol on hand?

54. At 20°C, an aqueous solution of HNO_3 that is 35.0% HNO_3 by mass has a density of 1.21 g/mL.
(a) How many grams of HNO_3 are present in 1.00 L of this solution?
(b) What volume of this solution will contain 500. g HNO_3?

*55. What is the molarity of a nitric acid solution if the solution is 35.0% HNO_3 by mass and has a density of 1.21 g/mL?

56. To what volume must a solution of 80.0 g H_2SO_4 in 500.0 mL of solution be diluted to give a 0.10 *M* solution?

57. A 10.0-mL sample of 16 *M* HNO_3 is diluted to 500.0 mL. What is the molarity of the final solution?

58. (a) How many moles of hydrogen will be liberated from 200.0 mL of 3.00 *M* HCl reacting with an excess of magnesium? The equation is

$$Mg(s) + 2\,HCl(aq) \longrightarrow MgCl_2(aq) + H_2(g)$$

(b) How many liters of hydrogen gas (H_2) measured at 27°C and 720 torr will be obtained? (*Hint*: Use the ideal gas law.)

59. Which will be more effective in neutralizing stomach acid, HCl: a tablet containing 1.20 g $Mg(OH)_2$ or a tablet containing 1.00g $Al(OH)_3$? Show evidence for your answer.

60. Which would be more effective as an antifreeze in an automobile radiator? A solution containing
(a) 10 kg of methyl alcohol (CH_3OH) or 10 kg of ethyl alcohol (C_2H_5OH)?
(b) 10 *m* solution of methyl alcohol or 10 *m* solution of ethyl alcohol?

*61. Automobile battery acid is 38% H_2SO_4 and has a density of 1.29 g/mL. Calculate the molality and the molarity of this solution.

*62. What is the (a) molality and (b) boiling point of an aqueous sugar, $C_{12}H_{22}O_{11}$, solution that freezes at −5.4°C?

63. A solution of 6.20 g $C_2H_6O_2$ in water has a freezing point of −0.372°C. How many grams of H_2O are in the solution?

64. What (a) mass and (b) volume of ethylene glycol ($C_2H_6O_2$, density = 1.11 g/mL) should be added to 12.0 L of water in an automobile radiator to protect it from freezing at −20°C? (c) To what temperature Fahrenheit will the radiator be protected?

*65. If 150 mL of 0.055 *M* HNO_3 are needed to completely neutralize 1.48 g of an *impure* sample of sodium hydrogen carbonate (baking soda), what percent of the sample is baking soda?

66. (a) How much water must be added to concentrated sulfuric acid (H_2SO_4) (17.8 *M*) to prepare 8.4 L of 1.5 *M* sulfuric acid solution?
(b) How many moles of H_2SO_4 are in each milliliter of the original concentrate?
(c) How many moles are in each milliliter of the diluted solution?

*67. How would you prepare a 6.00 *M* HNO_3 solution if only 3.00 *M* and 12.0 *M* solutions of the acid are available for mixing?

*68. A 20.0-mL portion of an HBr solution of unknown strength is diluted to exactly 240 mL. If 100.0 mL of this diluted solution requires 88.4 mL of 0.37 *M* NaOH to achieve complete neutralization, what was the strength of the original HBr solution?

69. When 80.5 mL of 0.642 *M* $Ba(NO_3)_2$ are mixed with 44.5 mL of 0.743 *M* KOH, a precipitate of $Ba(OH)_2$ forms. How many grams of $Ba(OH)_2$ do you expect?

70. A 0.25 *M* solution of lithium carbonate (Li_2CO_3), a drug used to treat manic depression, is prepared.
(a) How many moles of Li_2CO_3 are present in 45.8 mL of the solution?
(b) How many grams of Li_2CO_3 are in 750 mL of the same solution?
(c) How many milliliters of the solution would be needed to supply 6.0 g of the solute?
(d) If the solution has a density of 1.22 g/mL, what is its mass percent?

Challenge Exercises

All exercises with blue numbers have answers in the appendix of the text.

71. When solutions of hydrochloric acid and sodium sulfite react, a salt, water, and sulfur dioxide gas are produced. How many liters of sulfur dioxide gas at 775 torr and 22°C can be produced when 125 mL of 2.50 *M* hydrochloric acid react with 75.0 mL of 1.75 *M* sodium sulfite?

*72. Consider a saturated solution at 20°C made from 5.549 moles of water and an unknown solute. You determine the mass of the container containing the solution to be 563 g. The mass of the empty container is 375 g. Identify the solute.

Answers to Practice Exercises

14.1 unsaturated

14.2 10.0% Na_2SO_4 solution

14.3 2.02 *M*

14.4 0.27 g NaCl

14.5 0.84 L (840 mL)

14.6 75 mL NaOH solution

14.7 5.00×10^{-2} *M*

14.8 43 g

14.9 1.387 *m*

14.10 freezing point = −2.58°C, boiling point = 100.71°C

PUTTING IT TOGETHER: Review for Chapters 12–14

Multiple Choice:

Choose the correct answer to each of the following.

1. Which of these statements is *not* one of the principal assumptions of the kinetic-molecular theory for an ideal gas?
   (a) All collisions of gaseous molecules are perfectly elastic.
   (b) A mole of any gas occupies 22.4 L at STP.
   (c) Gas molecules have no attraction for one another.
   (d) The average kinetic energy for molecules is the same for all gases at the same temperature.

2. Which of the following is not equal to 1.00 atm?
   (a) 760. cm Hg (c) 760. mm Hg
   (b) 29.9 in. Hg (d) 760. torr

3. If the pressure on 45 mL of gas is changed from 600. torr to 800. torr, the new volume will be
   (a) 60 mL (c) 0.045 L
   (b) 34 mL (d) 22.4 L

4. The volume of a gas is 300. mL at 740. torr and 25°C. If the pressure remains constant and the temperature is raised to 100.°C, the new volume will be
   (a) 240. mL (c) 376 mL
   (b) 1.20 L (d) 75.0 mL

5. The volume of a dry gas is 4.00 L at 15.0°C and 745 torr. What volume will the gas occupy at 40.0°C and 700. torr?
   (a) 4.63 L (c) 3.92 L
   (b) 3.46 L (d) 4.08 L

6. A sample of Cl_2 occupies 8.50 L at 80.0°C and 740. mm Hg. What volume will the Cl_2 occupy at STP?
   (a) 10.7 L (c) 11.3 L
   (b) 6.75 L (d) 6.40 L

7. What volume will 8.00 g O_2 occupy at 45°C and 2.00 atm?
   (a) 0.462 L (c) 9.62 L
   (b) 104 L (d) 3.26 L

8. The density of NH_3 gas at STP is
   (a) 0.760 g/mL (c) 1.32 g/mL
   (b) 0.760 g/L (d) 1.32 g/L

9. The ratio of the relative rate of effusion of methane (CH_4) to sulfur dioxide (SO_2) is
   (a) $^{64}/_{16}$ (c) $^{1}/_{4}$
   (b) $^{16}/_{64}$ (d) $^{2}/_{1}$

10. Measured at 65°C and 500. torr, the mass of 3.21 L of a gas is 3.5 g. The molar mass of this gas is
    (a) 21 g/mole (c) 24 g/mole
    (b) 46 g/mole (d) 130 g/mole

11. Box A contains O_2 (molar mass = 32.0) at a pressure of 200 torr. Box B, which is identical to box A in volume, contains twice as many molecules of CH_4 (molar mass = 16.0) as the molecules of O_2 in box A. The temperatures of the gases are identical. The pressure in box B is
    (a) 100 torr (c) 400 torr
    (b) 200 torr (d) 800 torr

12. A 300.-mL sample of oxygen (O_2) is collected over water at 23°C and 725 torr. If the vapor pressure of water at 23°C is 21.0 torr, the volume of dry O_2 at STP is
    (a) 256 mL (c) 341 mL
    (b) 351 mL (d) 264 mL

13. A tank containing 0.01 mol of neon and 0.04 mol of helium shows a pressure of 1 atm. What is the partial pressure of neon in the tank?
    (a) 0.8 atm (c) 0.2 atm
    (b) 0.01 atm (d) 0.5 atm

14. How many liters of NO_2 (at STP) can be produced from 25.0 g Cu reacting with concentrated nitric acid?

    $$Cu(s) + 4\,HNO_3(aq) \longrightarrow Cu(NO_3)_2(aq) + 2\,H_2O(l) + 2\,NO_2(g)$$

    (a) 4.41 L (c) 17.6 L
    (b) 8.82 L (d) 44.8 L

15. How many liters of butane vapor are required to produce 2.0 L CO_2 at STP?

    $$\underset{\text{butane}}{2\,C_4H_{10}(g)} + 13\,O_2(g) \longrightarrow 8\,CO_2(g) + 10\,H_2O(g)$$

    (a) 2.0 L (c) 0.80 L
    (b) 4.0 L (d) 0.50 L

16. What volume of CO_2 (at STP) can be produced when 15.0 g C_2H_6 and 50.0 g O_2 are reacted?

    $$2\,C_2H_6(g) + 7\,O_2(g) \longrightarrow 4\,CO_2(g) + 6\,H_2O(g)$$

    (a) 20.0 L (c) 35.0 L
    (b) 22.4 L (d) 5.6 L

17. Which of these gases has the highest density at STP?
    (a) N_2O (c) Cl_2
    (b) NO_2 (d) SO_2

18. What is the density of CO_2 at 25°C and 0.954 atm?
    (a) 1.72 g/L (c) 0.985 g/L
    (b) 2.04 g/L (d) 1.52 g/L

19. How many molecules are present in 0.025 mol of H_2 gas?
    (a) 1.5×10^{22} molecules
    (b) 3.37×10^{23} molecules
    (c) 2.40×10^{25} molecules
    (d) 1.50×10^{22} molecules

20. 5.60 L of a gas at STP have a mass of 13.0 g. What is the molar mass of the gas?
    (a) 33.2 g/mol (c) 66.4 g/mol
    (b) 26.0 g/mol (d) 52.0 g/mol

21. The heat of fusion of water is
    (a) 4.184 J/g (c) 2.26 kJ/g
    (b) 335 J/g (d) 2.26 kJ/mol

22. The heat of vaporization of water is
    (a) 4.184 J/g (c) 2.26 kJ/g
    (b) 335 J/g (d) 2.26 kJ/mol

23. The specific heat of water is
    (a) 4.184 J/g°C (c) 2.26 kJ/g°C
    (b) 335 J/g°C (d) 18 J/g°C

24. The density of water at 4°C is
(a) 1.0 g/mL
(b) 80 g/mL
(c) 18.0 g/mL
(d) 14.7 lb/in.^{3}

25. SO_2 can be properly classified as a(n)
(a) basic anhydride
(b) hydrate
(c) anhydrous salt
(d) acid anhydride

26. When compared to H_2S, H_2Se, and H_2Te, water is found to have the highest boiling point because it
(a) has the lowest molar mass
(b) is the smallest molecule
(c) has the highest bonding
(d) forms hydrogen bonds better than the others

27. In which of the following molecules will hydrogen bonding be important?
(a) 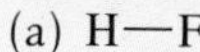
(b) 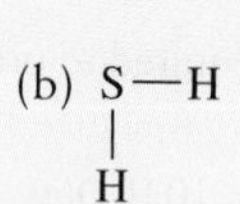
(c) 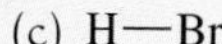
(d) 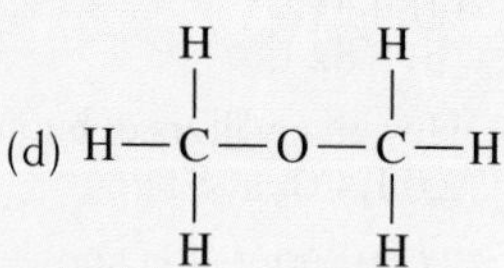

28. Which of the following is an incorrect equation?
(a) $H_2SO_4 + 2\,NaOH \longrightarrow Na_2SO_4 + 2\,H_2O$
(b) $C_2H_6 + O_2 \longrightarrow 2\,CO_2 + 3\,H_2$
(c) $2\,H_2O \xrightarrow[H_2SO_4]{\text{electrolysis}} 2\,H_2 + O_2$
(d) $Ca + 2\,H_2O \longrightarrow H_2 + Ca(OH)_2$

29. Which of the following is an incorrect equation?
(a) $C + H_2O(g) \xrightarrow{1000°C} CO(g) + H_2(g)$
(b) $CaO + H_2O \longrightarrow Ca(OH)_2$
(c) $2\,NO_2 + H_2O \longrightarrow 2\,HNO_3$
(d) $Cl_2 + H_2O \longrightarrow HCl + HOCl$

30. How many kilojoules are required to change 85 g of water at 25°C to steam at 100.°C?
(a) 219 kJ
(b) 27 kJ
(c) 590 kJ
(d) 192 kJ

31. A chunk of 0°C ice, mass 145 g, is dropped into 75 g of water at 62°C. The heat of fusion of water is 335 J/g. The result, after thermal equilibrium is attained, will be
(a) 87 g ice and 133 g liquid water, all at 0°C
(b) 58 g ice and 162 g liquid water, all at 0°C
(c) 220 g water at 7°C
(d) 220 g water at 17°C

32. The formula for iron(II) sulfate heptahydrate is
(a) $Fe_2SO_4 \cdot 7\,H_2O$
(b) $Fe(SO_4)_2 \cdot 6\,H_2O$
(c) $FeSO_4 \cdot 7\,H_2O$
(d) $Fe_2(SO_4)_3 \cdot 7\,H_2O$

33. The process by which a solid changes directly to a vapor is called
(a) vaporization
(b) evaporation
(c) sublimation
(d) condensation

34. Hydrogen bonding
(a) occurs only between water molecules
(b) is stronger than covalent bonding
(c) can occur between NH_3 and H_2O
(d) results from strong attractive forces in ionic compounds

35. A liquid boils when
(a) the vapor pressure of the liquid equals the external pressure above the liquid
(b) the heat of vaporization exceeds the vapor pressure
(c) the vapor pressure equals 1 atm
(d) the normal boiling temperature is reached

36. Consider two beakers, one containing 50 mL of liquid A and the other 50 mL of liquid B. The boiling point of A is 90°C and that of B is 72°C. Which of these statements is correct?
(a) A will evaporate faster than B.
(b) B will evaporate faster than A.
(c) Both A and B evaporate at the same rate.
(d) Insufficient data to answer the question.

37. 95.0 g of 0.0°C ice is added to exactly 100. g of water at 60.0°C. When the temperature of the mixture first reaches 0.0°C, the mass of ice still present is
(a) 0.0 g
(b) 20.0 g
(c) 10.0 g
(d) 75.0 g

38. Which of the following is not a general property of solutions?
(a) a homogeneous mixture of two or more substances
(b) variable composition
(c) dissolved solute breaks down to individual molecules
(d) the same chemical composition, the same chemical properties, and the same physical properties in every part

39. If NaCl is soluble in water to the extent of 36.0 g NaCl/100 g H_2O at 20°C, then a solution at 20°C containing 45 g NaCl/150 g H_2O would be
(a) dilute
(b) saturated
(c) supersaturated
(d) unsaturated

40. If 5.00 g NaCl are dissolved in 25.0 g of water, the percent of NaCl by mass is
(a) 16.7%
(b) 20.0%
(c) 0.20%
(d) no correct answer given

41. How many grams of 9.0% $AgNO_3$ solution will contain 5.3 g $AgNO_3$?
(a) 47.7 g
(b) 0.58 g
(c) 59 g
(d) no correct answer given

42. The molarity of a solution containing 2.5 mol of acetic acid ($HC_2H_3O_2$) in 400. mL of solution is
(a) 0.063 *M*
(b) 1.0 *M*
(c) 0.103 *M*
(d) 6.3 *M*

43. What volume of 0.300 *M* KCl will contain 15.3 g KCl?
(a) 1.46 L
(b) 683 mL
(c) 61.5 mL
(d) 4.60 L

44. What mass of $BaCl_2$ will be required to prepare 200. mL of 0.150 *M* solution?
(a) 0.750 g
(b) 156 g
(c) 6.25 g
(d) 31.2 g

Problems 45–47 relate to the reaction

$$CaCO_3 + 2\,HCl \longrightarrow CaCl_2 + H_2O + CO_2$$

45. What volume of 6.0 M HCl will be needed to react with 0.350 mol of $CaCO_3$?
(a) 42.0 mL (c) 117 mL
(b) 1.17 L (d) 583 mL

46. If 400. mL of 2.0 M HCl react with excess $CaCO_3$, the volume of CO_2 produced, measured at STP, is
(a) 18 L (c) 9.0 L
(b) 5.6 L (d) 56 L

47. If 5.3 g $CaCl_2$ are produced in the reaction, what is the molarity of the HCl used if 25 mL of it reacted with excess $CaCO_3$?
(a) 3.8 M (c) 0.38 M
(b) 0.19 M (d) 0.42 M

48. If 20.0 g of the nonelectrolyte urea ($CO(NH_2)_2$) is dissolved in 25.0 g of water, the freezing point of the solution will be
(a) −2.47°C (c) −24.7°C
(b) −1.40°C (d) −3.72°C

49. When 256 g of a nonvolatile, nonelectrolyte unknown were dissolved in 500. g H_2O, the freezing point was found to be −2.79°C. The molar mass of the unknown solute is
(a) 357 (c) 768
(b) 62.0 (d) 341

50. How many milliliters of 6.0 M H_2SO_4 must you use to prepare 500. mL of 0.20 M sulfuric acid solution?
(a) 30 (c) 12
(b) 17 (d) 100

51. How many milliliters of water must be added to 200. mL of 1.40 M HCl to make a solution that is 0.500 M HCl?
(a) 360. mL (c) 140. mL
(b) 560. mL (d) 280. mL

52. Which procedure is most likely to increase the solubility of most solids in liquids?
(a) stirring
(b) pulverizing the solid
(c) heating the solution
(d) increasing the pressure

53. The addition of a crystal of $NaClO_3$ to a solution of $NaClO_3$ causes additional crystals to precipitate. The original solution was
(a) unsaturated (c) saturated
(b) dilute (d) supersaturated

54. Which of these anions will not form a precipitate with silver ions, Ag^+?
(a) Cl^- (c) Br^-
(b) NO_3^- (d) CO_3^{2-}

55. Which of these salts are considered to be soluble in water?
(a) $BaSO_4$ (c) AgI
(b) NH_4Cl (d) PbS

56. A solution of ethyl alcohol and benzene is 40% alcohol by volume. Which statement is correct?
(a) The solution contains 40 mL of alcohol in 100 mL of solution.
(b) The solution contains 60 mL of benzene in 100 mL of solution.
(c) The solution contains 40 mL of alcohol in 100 g of solution.
(d) The solution is made by dissolving 40 mL of alcohol in 60 mL of benzene.

57. Which of the following is not a colligative property?
(a) boiling point elevation
(b) freezing point depression
(c) osmotic pressure
(d) surface tension

58. When a solute is dissolved in a solvent
(a) the freezing point of the solution increases.
(b) the vapor pressure of the solution increases.
(c) the boiling point of the solution increases.
(d) the concentration of the solvent increases.

59. Which of the following solutions will have the lowest freezing point where X is any element or nonelectrolytic compound?
(a) 1.0 mol X in 1 kg H_2O
(b) 2.0 mol X in 1 kg H_2O
(c) 1.2 mol X in 1 kg H_2O
(d) 0.80 mol X in 1 kg H_2O

60. In the process of osmosis, water passes through a semipermeable membrane
(a) from a more concentrated solution to a dilute solution
(b) from a dilute solution to a more concentrated solution
(c) in order to remove a solute from a solution
(d) so that a sugar solution can become sweeter

Free Response Questions

Answer each of the following. Be sure to include your work and explanations in a clear, logical form.

1. Which solution should have a higher boiling point: 215 mL of a 10.0% (m/v) aqueous KCl solution or 224 mL of a 1.10 M aqueous NaCl solution?

2. A glass containing 345 mL of a soft drink (a carbonated beverage) was left sitting out on a kitchen counter. If the CO_2 released at room temperature (25°C) and pressure (1 atm) occupies 1.40 L, at a minimum, what is the concentration (in ppm) of the CO_2 in the original soft drink (assume the density of the original soft drink is 0.965 g/mL).

3. Dina and Murphy were trying to react 100. mL of a 0.10 M HCl solution with KOH. The procedure called for a 10% KOH solution. Dina made a 10% mass/volume solution while Murphy made a 10% by mass solution. (Assume there is no volume change upon dissolving KOH.) Which solution required less volume to fully react with 100. mL of the HCl solution?

4. A flask containing 825 mL of a solution containing 0.355 mole of $CuSO_4$ was left open overnight. The next morning the flask only contained 755 mL of solution.

(a) What is the concentration (molarity) of the $CuSO_4$ solution remaining in the flask?

(b) Which of the pathways shown below best represents the evaporation of water and why are the others wrong?

(1) $\rightarrow$ H—H(*g*) + O=O(*g*)

(2) $\rightarrow$ $H^+(g)$ + $OH^-(g)$

(3) $\rightarrow$ $H_2O(g)$ + $H_2O(g)$

5. Three students at Jamston High—Zack, Gaye, and Lamont—each had the opportunity to travel over spring break. As part of a project, each of them measured the boiling point of pure water at their vacation spot. Zack found a boiling point of 93.9°C, Gaye measured 101.1°C, and Lamont read 100.°C. Which student most likely went snow skiing near Ely, Nevada, and which student most likely went water skiing in Honolulu? From the boiling point information, what can you surmise about the Dead Sea region, the location of the third student's vacation? Explain.

6. Why does a change in pressure of a gas significantly affect its volume whereas a change in pressure on a solid or liquid has negligible effect on their respective volumes? If the accompanying picture represents a liquid at the molecular level, draw what you might expect a solid and a gas to look like.

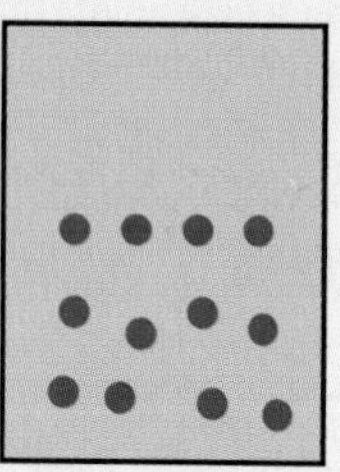

7. (a) If you filled up three balloons with equal volumes of hydrogen, argon, and carbon dioxide gas, all at the same temperature and pressure, which balloon would weigh the most? the least? Explain.

(b) If you filled up three balloons with equal masses of nitrogen, oxygen, and neon, all to the same volume at the same temperature, which would have the lowest pressure?

8. Ray ran a double-displacement reaction using 0.050 mol $CuCl_2$ and 0.10 mol $AgNO_3$. The resulting white precipitate was removed by filtration. The filtrate was accidentally left open on the lab bench for over a week, and when Ray returned the flask contained solid, blue crystals. Ray weighed the crystals and found they had a mass of 14.775 g. Was Ray expecting this number? If not, what did he expect?

9. Why is it often advantageous or even necessary to run reactions in solution rather than mixing two solids? Would you expect reactions run in the gas phase to be more similar to solutions or to solids? Why?

10. A solution of 5.36 g of a molecular compound dissolved in 76.8 g benzene (C_6H_6) and has a boiling point of 81.48°C. What is the molar mass of the compound?

Boiling point for benzene = 80.1°C

K_b for benzene = 2.53°C kg solvent/mol solute

CHAPTER 15

Acids, Bases, and Salts

Citrus fruits are all acidic substances.

Chapter Outline

Acids are important chemicals. They are used in cooking to produce the surprise of tartness (from lemons) and to release CO_2 bubbles from leavening agents in baking. Vitamin C is an acid that is an essential nutrient in our diet. Our stomachs release acid to aid in digestion. Excess stomach acid can produce heartburn and indigestion. Bacteria in our mouths produce acids that can dissolve tooth enamel to form cavities. In our recreational activities we are concerned about acidity levels in swimming pools and spas. Acids are essential in the manufacture of detergents, plastics, and storage batteries. The acid–base properties of substances are found in all areas of our lives. In this chapter we consider the properties of acids, bases, and salts.

15.1 Acids and Bases

The word *acid* is derived from the Latin *acidus,* meaning "sour" or "tart," and is also related to the Latin word *acetum,* meaning "vinegar." Vinegar has been around since antiquity as a product of the fermentation of wine and apple cider. The sour constituent of vinegar is acetic acid ($HC_2H_3O_2$). Characteristic properties commonly associated with acids include the following:

1. sour taste
2. the ability to change the color of litmus, a vegetable dye, from blue to red
3. the ability to react with
   - metals such as zinc and magnesium to produce hydrogen gas
   - hydroxide bases to produce water and an ionic compound (salt)
   - carbonates to produce carbon dioxide

These properties are due to the hydrogen ions, H^+, released by acids in a water solution.

Classically, a *base* is a substance capable of liberating hydroxide ions, OH^-, in water solution. Hydroxides of the alkali metals (Group 1A) and alkaline earth metals (Group 2A), such as LiOH, NaOH, KOH, $Ca(OH)_2$, and $Ba(OH)_2$, are the most common inorganic bases. Water solutions of bases are called *alkaline solutions* or *basic solutions.* Some of the characteristic properties commonly associated with bases include the following:

1. bitter or caustic taste
2. a slippery, soapy feeling
3. the ability to change litmus from red to blue
4. the ability to interact with acids

Several theories have been proposed to answer the question "What is an acid and what is a base?" One of the earliest, most significant of these theories was advanced in 1884 by Svante Arrhenius (1859–1927), a Swedish scientist, who stated that "an acid is a hydrogen-containing substance that dissociates to produce hydrogen ions, and a base is a hydroxide-containing substance that dissociates to produce hydroxide ions in aqueous solutions." Arrhenius postulated that the hydrogen ions are produced by the dissociation of acids in water and that the hydroxide ions are produced by the dissociation of bases in water:

$$\underset{\text{acid}}{HA} \xrightarrow{H_2O} H^+(aq) + A^-(aq)$$

$$\underset{\text{base}}{MOH} \xrightarrow{H_2O} M^+(aq) + OH^-(aq)$$

An Arrhenius acid solution contains an excess of H^+ ions.
An Arrhenius base solution contains an excess of OH^- ions.

In 1923, the Brønsted–Lowry proton transfer theory was introduced by J. N. Brønsted (1897–1947), a Danish chemist, and T. M. Lowry (1847–1936), an English chemist. This theory states that an acid is a proton donor and a base is a proton acceptor.

A Brønsted–Lowry acid is a proton (H^+) donor.
A Brønsted–Lowry base is a proton (H^+) acceptor.

Consider the reaction of hydrogen chloride gas with water to form hydrochloric acid:

$$HCl(g) + H_2O(l) \longrightarrow H_3O^+(aq) + Cl^-(aq) \qquad (1)$$

In the course of the reaction, HCl donates, or gives up, a proton to form a Cl^- ion, and H_2O accepts a proton to form the H_3O^+ ion. Thus, HCl is an acid and H_2O is a base, according to the Brønsted–Lowry theory.

A hydrogen ion (H^+) is nothing more than a bare proton and does not exist by itself in an aqueous solution. In water H^+ combines with a polar water molecule to form a hydrated hydrogen ion (H_3O^+) commonly called a **hydronium ion.** The H^+ is attracted to a polar water molecule, forming a bond with one of the two pairs of unshared electrons:

hydronium ion

$$H^+ + H{:}\ddot{\underset{\cdot\cdot}{O}}{:}\underset{H}{} \longrightarrow \left[H{:}\ddot{\underset{\cdot\cdot}{O}}{:}H\right]^+$$

hydronium ion

Note the electron structure of the hydronium ion. For simplicity we often use H^+ instead of H_3O^+ in equations, with the explicit understanding that H^+ is always hydrated in solution.

When a Brønsted–Lowry acid donates a proton, as illustrated in the equation below, it forms the conjugate base of that acid. When a base accepts a proton, it forms the conjugate acid of that base. A conjugate acid and base are produced as products. The formulas of a conjugate acid–base pair differ by one proton (H^+). Consider what happens when HCl(g) is bubbled through water, as shown by this equation:

conjugate acid–base pair

$$HCl(g) + H_2O(l) \longrightarrow Cl^-(aq) + H_3O^+(aq)$$

conjugate acid–base pair

acid base base acid

The conjugate acid–base pairs are $HCl—Cl^-$ and $H_3O^+—H_2O$. The conjugate base of HCl is Cl^-, and the conjugate acid of Cl^- is HCl. The conjugate base of H_3O^+ is H_2O, and the conjugate acid of H_2O is H_3O^+.

Another example of conjugate acid–base pairs can be seen in this equation:

$$NH_4^+ + H_2O \longrightarrow H_3O^+ + NH_3$$

acid base acid base

Here the conjugate acid–base pairs are $NH_4^+—NH_3$ and $H_3O^+—H_2O$.

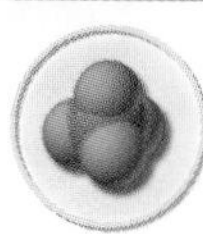

Example 15.1 Write the formula for (a) the conjugate base of H_2O and of HNO_3, and (b) the conjugate acid of SO_4^{2-} and of $C_2H_3O_2^-$.

SOLUTION (a) To write the conjugate base of an acid, remove one proton from the acid formula:

Remember: The difference between an acid or a base and its conjugate is one proton, H^+.

$$H_2O \xrightarrow{-H^+} OH^- \quad \text{(conjugate base)}$$

$$HNO_3 \xrightarrow{-H^+} NO_3^- \quad \text{(conjugate base)}$$

Note that, by removing an H^+, the conjugate base becomes more negative than the acid by one minus charge.

(b) To write the conjugate acid of a base, add one proton to the formula of the base:

$$SO_4^{2-} \xrightarrow{+H^+} HSO_4^- \quad \text{(conjugate acid)}$$

$$C_2H_3O_2^- \xrightarrow{+H^+} HC_2H_3O_2 \quad \text{(conjugate acid)}$$

In each case the conjugate acid becomes more positive than the base by a +1 charge due to the addition of H^+.

Practice 15.1

Indicate the conjugate base for these acids:

(a) H_2CO_3 (b) HNO_2 (c) $HC_2H_3O_2$

Practice 15.2

Indicate the conjugate acid for these bases:

(a) HSO_4^- (b) NH_3 (c) OH^-

A more general concept of acids and bases was introduced by Gilbert N. Lewis. The Lewis theory deals with the way in which a substance with an unshared pair of electrons reacts in an acid–base type of reaction. According to this theory a base is any substance that has an unshared pair of electrons (electron pair donor), and an acid is any substance that will attach itself to or accept a pair of electrons.

A Lewis acid is an electron pair acceptor.
A Lewis base is an electron pair donor.

In the reaction

$$\underset{\text{acid}}{H^+} + \underset{\text{base}}{:\overset{\displaystyle H}{\underset{\displaystyle H}{\ddot{\underset{..}{N}}}}:H} \longrightarrow \left[H:\overset{\displaystyle H}{\underset{\displaystyle H}{\ddot{\underset{..}{N}}}}:H \right]^+$$

Table 15.1 Summary of Acid–Base Definitions

Theory	Acid	Base
Arrhenius	A hydrogen-containing substance that produces hydrogen ions in aqueous solution	A hydroxide-containing substance that produces hydroxide ions in aqueous solution
Brønsted–Lowry	A proton (H^+) donor	A proton (H^+) acceptor
Lewis	Any species that will bond to an unshared pair of electrons (electron pair acceptor)	Any species that has an unshared pair of electrons (electron pair donor)

The H^+ is a Lewis acid and $:NH_3$ is a Lewis base. According to the Lewis theory, substances other than proton donors (e.g., BF_3) behave as acids:

$$\underset{\text{acid}}{F:\overset{F}{\underset{F}{\ddot{B}}}} + \underset{\text{base}}{:\overset{H}{\underset{H}{\ddot{N}}}:H} \longrightarrow F:\overset{F}{\underset{F}{\ddot{B}}}:\overset{H}{\underset{H}{\ddot{N}}}:H$$

These three theories, which explain how acid–base reactions occur, are summarized in Table 15.1. We will generally use the theory that best explains the reaction under consideration. Most of our examples will refer to aqueous solutions. Note that in an aqueous acidic solution the H^+ ion concentration is always greater than OH^- ion concentration. And vice versa—in an aqueous basic solution the OH^- ion concentration is always greater than the H^+ ion concentration. When the H^+ and OH^- ion concentrations in a solution are equal, the solution is *neutral*; that is, it is neither acidic nor basic.

15.2 Reactions of Acids

In aqueous solutions, the H^+ or H_3O^+ ions are responsible for the characteristic reactions of acids. The following reactions are in an aqueous medium:

Reaction with Metals Acids react with metals that lie above hydrogen in the activity series of elements to produce hydrogen and an ionic compound (salt) (see Section 17.5):

$$\text{acid} + \text{metal} \longrightarrow \text{hydrogen} + \text{ionic compound}$$

$$2\,HCl(aq) + Ca(s) \longrightarrow H_2(g) + CaCl_2(aq)$$

$$H_2SO_4(aq) + Mg(s) \longrightarrow H_2(g) + MgSO_4(aq)$$

$$6\,HC_2H_3O_2(aq) + 2\,Al(s) \longrightarrow 3\,H_2(g) + 2\,Al(C_2H_3O_2)_3(aq)$$

Acids such as nitric acid (HNO_3) are oxidizing substances (see Chapter 17) and react with metals to produce water instead of hydrogen. For example,

$$3\,Zn(s) + 8\,HNO_3(\text{dilute}) \longrightarrow 3\,Zn(NO_3)_2(aq) + 2\,NO(g) + 4\,H_2O(l)$$

Reaction with Bases The interaction of an acid and a base is called a *neutralization reaction.* In aqueous solutions, the products of this reaction are a salt and water:

acid + base ⟶ salt + water

$HBr(aq) + KOH(aq) \longrightarrow KBr(aq) + H_2O(l)$

$2\,HNO_3(aq) + Ca(OH)_2(aq) \longrightarrow Ca(NO_3)_2(aq) + 2\,H_2O(l)$

$2\,H_3PO_4(aq) + 3\,Ba(OH)_2(aq) \longrightarrow Ba_3(PO_4)_2(s) + 6\,H_2O(l)$

Reaction with Metal Oxides This reaction is closely related to that of an acid with a base. With an aqueous acid solution, the products are a salt and water:

acid + metal oxide ⟶ salt + water

$2\,HCl(aq) + Na_2O(s) \longrightarrow 2\,NaCl(aq) + H_2O(l)$

$H_2SO_4(aq) + MgO(s) \longrightarrow MgSO_4(aq) + H_2O(l)$

$6\,HCl(aq) + Fe_2O_3(s) \longrightarrow 2\,FeCl_3(aq) + 3\,H_2O(l)$

Reaction with Carbonates Many acids react with carbonates to produce carbon dioxide, water, and an ionic compound:

Carbonic acid (H_2CO_3) is not the product because it is unstable and spontaneously decomposes into water and carbon dioxide.

$H_2CO_3(aq) \longrightarrow CO_2(g) + H_2O(l)$

acid + carbonate ⟶ salt + water + carbon dioxide

$2\,HCl(aq) + Na_2CO_3(aq) \longrightarrow 2\,NaCl(aq) + H_2O(l) + CO_2(g)$

$H_2SO_4(aq) + MgCO_3(s) \longrightarrow MgSO_4(aq) + H_2O(l) + CO_2(g)$

$HCl(aq) + NaHCO_3(aq) \longrightarrow NaCl(aq) + H_2O(l) + CO_2(g)$

15.3 Reactions of Bases

The OH^- ions are responsible for the characteristic reactions of bases. The following reactions are in an aqueous medium:

Reaction with Acids Bases react with acids to produce a salt and water. See reaction of acids with bases in Section 15.2.

amphoteric

Amphoteric Hydroxides Hydroxides of certain metals, such as zinc, aluminum, and chromium, are **amphoteric;** that is, they are capable of reacting as either an acid or a base. When treated with a strong acid, they behave like bases; when reacted with a strong base, they behave like acids:

Strong acids and bases are discussed in Section 15.7.

$Zn(OH)_2(s) + 2\,HCl(aq) \longrightarrow ZnCl_2(aq) + 2\,H_2O(l)$

$Zn(OH)_2(s) + 2\,NaOH(aq) \longrightarrow Na_2Zn(OH)_4(aq)$

Reaction of NaOH and KOH with Certain Metals Some amphoteric metals react directly with the strong bases sodium hydroxide and potassium hydroxide to produce hydrogen:

base + metal + water ⟶ salt + hydrogen

$2\,NaOH(aq) + Zn(s) + 2\,H_2O(l) \longrightarrow Na_2Zn(OH)_4(aq) + H_2(g)$

$2\,KOH(aq) + 2\,Al(s) + 6\,H_2O(l) \longrightarrow 2\,KAl(OH)_4(aq) + 3\,H_2(g)$

Practice 15.3

Write the formulas and names of the acid and base from which these salts are formed.

(a) K_3PO_4 (b) $MgBr_2$ (c) LiCl (d) $FeCO_3$

CHEMISTRY IN ACTION • A Cool Fizz

Try this experiment with your friends: Chill some soft drinks well. Next, pour the soft drink into a glass. Stick your tongue into the liquid and time how long you can keep it there.

What causes the tingling we feel in our mouth (or on our tongue)? Many people believe it is the bubbles of carbon dioxide, but scientists have found that is not the answer. The tingling is caused by "chemisthesis." Bruce Bryant at Monell Chemical Senses Center in Philadelphia says that chemisthesis is a chemically induced sensation that does not involve taste or odor receptors. The tongue-tingling response to the soft drink is caused by production of protons (H^+ ions) released when an enzyme (carbonic anhydrase) acts on CO_2. The H^+ ions acidify nerve endings, producing the sensation of tingling.

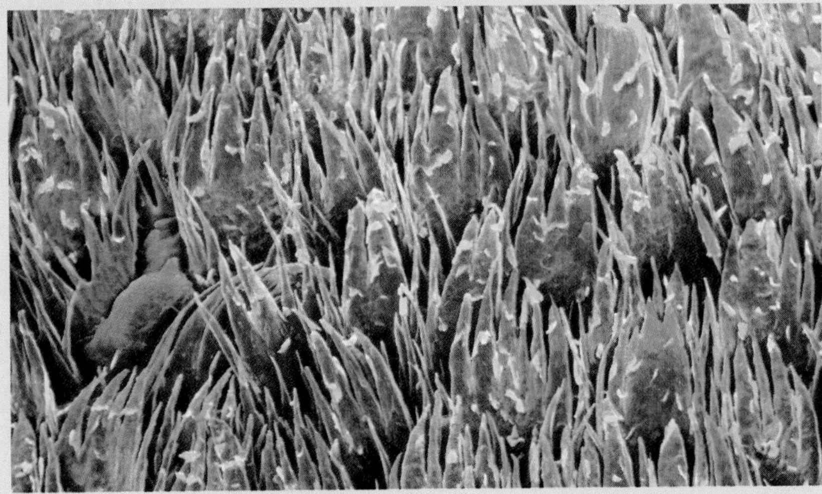

Tongue papillae contain nerve endings that send temperature and tactile information to the brain.

Carbon dioxide also stimulates other neurons when the drink is cold. This means that at a constant pressure of CO_2, a cold drink will produce more tingling than a room-temperature drink. If the drink is at room temperature, CO_2 will increase the cool feeling. Chilling a soda increases the effect of the protons on the nerve endings. At the same time, the high concentration of CO_2 in a freshly opened soda makes the cold soda feel even colder. Together these chemical effects make it really painful to keep your tongue in the chilled soda.

Chemisthesis is important for survival, as well as for having fun with friends.

Bryant says, "It tells you that something is chemically impinging on your body, that tissue is in imminent danger. Burrowing animals can sense toxic levels of carbon dioxide and even feel a sting when exposed to those levels."

15.4 Salts

Salts are very abundant in nature. Most of the rocks and minerals of Earth's mantle are salts of one kind or another. Huge quantities of dissolved salts also exist in the oceans. Salts can be considered compounds derived from acids and bases. They consist of positive metal or ammonium ions combined with negative nonmetal ions (OH^- and O^{2-} excluded). The positive ion is the base counterpart and the nonmetal ion is the acid counterpart:

Chemists use the terms *ionic compound* and *salt* interchangeably.

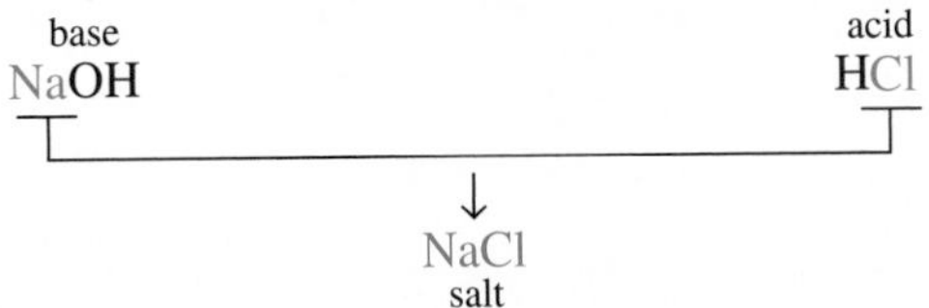

Salts are usually crystalline and have high melting and boiling points.

From a single acid such as hydrochloric acid (HCl), we can produce many chloride compounds by replacing the hydrogen with metal ions (e.g., NaCl, KCl, RbCl, $CaCl_2$, $NiCl_2$). Hence the number of known salts greatly exceeds the number of known acids and bases. If the hydrogen atoms of a binary acid are replaced by a nonmetal, the resulting compound has covalent bonding and is therefore not considered to be ionic (e.g., PCl_3, S_2Cl_2, Cl_2O, NCl_3, ICl).

You may want to review Chapter 6 for nomenclature of acids, bases, and salts.

15.5 Electrolytes and Nonelectrolytes

We can show that solutions of certain substances are conductors of electricity with a simple conductivity apparatus, which consists of a pair of electrodes connected to a voltage source through a light bulb and switch (see Figure 15.1). If the medium between the electrodes is a conductor of electricity, the light bulb will glow when the switch is closed. When chemically pure water is placed in the beaker and the switch is closed, the light does not glow, indicating that water is a virtual nonconductor. When we dissolve a small amount of sugar in the water and test the solution, the light still does not glow, showing that a sugar solution is also a nonconductor. But when a small amount of salt, NaCl, is dissolved in water and this solution is tested, the light glows brightly. Thus the salt solution conducts electricity. A fundamental difference exists between the chemical bonding in sugar and that in salt. Sugar is a covalently bonded (molecular) substance; common salt is a substance with ionic bonds.

electrolyte
nonelectrolyte

Substances whose aqueous solutions are conductors of electricity are called **electrolytes.** Substances whose solutions are nonconductors are known as **nonelectrolytes.** The classes of compounds that are electrolytes are acids, bases, and other ionic compounds (salts). Solutions of certain oxides also are conductors because the oxides form an acid or a base when dissolved in water. One major difference between electrolytes and nonelectrolytes is that electrolytes are capable of producing ions in solution, whereas nonelectrolytes do not have this property. Solutions that contain a sufficient number of ions will conduct an electric current. Although pure water is essentially a nonconductor, many city water supplies contain enough dissolved ionic matter to cause the light to glow dimly when the water is tested in a conductivity apparatus. Table 15.2 lists some common electrolytes and nonelectrolytes.

Acids, bases, and salts are electrolytes.

(a)

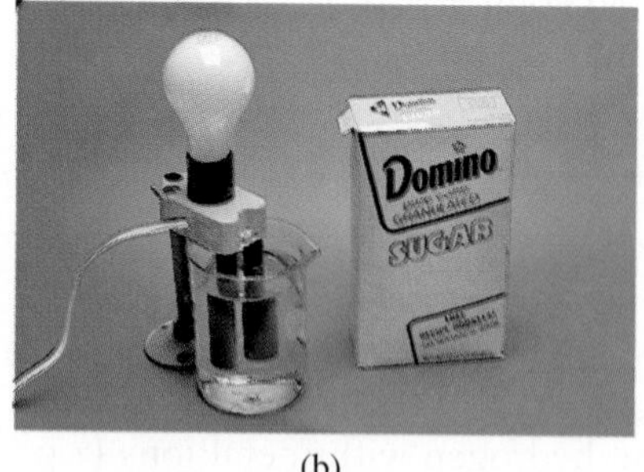

(b)

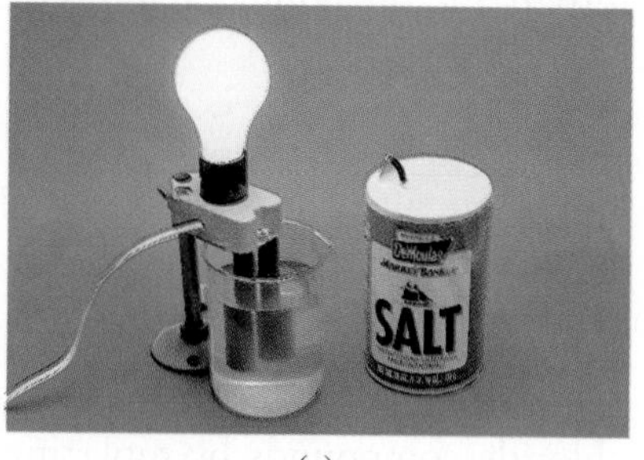

(c)

Figure 15.1
A conductivity apparatus shows the difference in conductivity of solutions. (a) Distilled water does not conduct electricity. (b) Sugar water is a nonelectrolyte. (c) Salt water is a strong electrolyte and conducts electricity.

These strange mineral formations called "tufa" exist at Mono Lake, California. Tufa is formed by water bubbling through sand saturated with NaCl, Na_2CO_3, and Na_2SO_4.

Table 15.2 Representative Electrolytes and Nonelectrolytes

Electrolytes		Nonelectrolytes	
H_2SO_4	$HC_2H_3O_2$	$C_{12}H_{22}O_{11}$ (sugar)	CH_3OH (methyl alcohol)
HCl	NH_3	C_2H_5OH (ethyl alcohol)	$CO(NH_2)_2$ (urea)
HNO_3	K_2SO_4	$C_2H_4(OH)_2$ (ethylene glycol)	O_2
NaOH	$NaNO_3$	$C_3H_5(OH)_3$ (glycerol)	H_2O

15.6 Dissociation and Ionization of Electrolytes

Arrhenius received the 1903 Nobel Prize in chemistry for his work on electrolytes. He found that a solution conducts electricity because the solute dissociates immediately upon dissolving into electrically charged particles (ions). The movement of these ions toward oppositely charged electrodes causes the solution to be a conductor. According to his theory, solutions that are relatively poor conductors contain electrolytes that are only partly dissociated. Arrhenius also believed that ions exist in solution whether or not an electric current is present. In other words, the electric current does not cause the formation of ions. Remember that positive ions are cations; negative ions are anions.

We have seen that sodium chloride crystals consist of sodium and chloride ions held together by ionic bonds. **Dissociation** is the process by which the ions of a salt separate as the salt dissolves. When placed in water, the sodium and chloride ions are attracted by the polar water molecules, which surround each ion as it dissolves. In water, the salt dissociates, forming hydrated sodium

dissociation

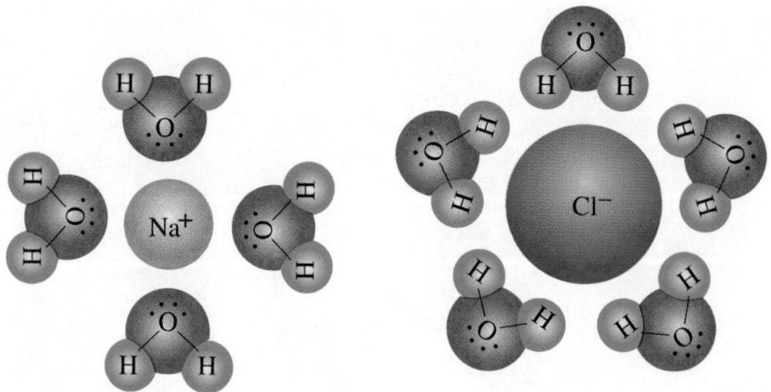

Figure 15.2
Hydrated sodium and chloride ions. When sodium chloride dissolves in water, each Na^+ and Cl^- ion becomes surrounded by water molecules. The negative end of the water dipole is attracted to the Na^+ ion, and the positive end is attracted to the Cl^- ion.

and chloride ions (see Figure 15.2). The sodium and chloride ions in solution are surrounded by a specific number of water molecules and have less attraction for each other than they had in the crystalline state. The equation representing this dissociation is

$$NaCl(s) + (x + y)\,H_2O \longrightarrow Na^+(H_2O)_x + Cl^-(H_2O)_y$$

A simplified dissociation equation in which the water is omitted but understood to be present is

$$NaCl(s) \longrightarrow Na^+(aq) + Cl^-(aq)$$

Remember that sodium chloride exists in an aqueous solution as hydrated ions, not as NaCl units, even though the formula NaCl (or $Na^+ + Cl^-$) is often used in equations.

The chemical reactions of salts in solution are the reactions of their ions. For example, when sodium chloride and silver nitrate react and form a precipitate of silver chloride, only the Ag^+ and Cl^- ions participate in the reaction. The Na^+ and NO_3^- remain as ions in solution:

$$Ag^+(aq) + Cl^-(aq) \longrightarrow AgCl(s)$$

ionization

Ionization is the formation of ions; it occurs as a result of a chemical reaction of certain substances with water. Glacial acetic acid (100% $HC_2H_3O_2$) is a liquid that behaves as a nonelectrolyte when tested by the method described in Section 15.5. But a water solution of acetic acid conducts an electric current (as indicated by the dull-glowing light of the conductivity apparatus). The equation for the reaction with water, which forms hydronium and acetate ions, is

$$\underset{\text{acid}}{HC_2H_3O_2} + \underset{\text{base}}{H_2O} \rightleftharpoons \underset{\text{acid}}{H_3O^+} + \underset{\text{base}}{C_2H_3O_2^-}$$

or, in the simplified equation,

$$HC_2H_3O_2 \rightleftharpoons H^+ + C_2H_3O_2^-$$

In this ionization reaction, water serves not only as a solvent but also as a base according to the Brønsted–Lowry theory.

Hydrogen chloride is predominantly covalently bonded, but when dissolved in water, it reacts to form hydronium and chloride ions:

$$HCl(g) + H_2O(l) \longrightarrow H_3O^+(aq) + Cl^-(aq)$$

When a hydrogen chloride solution is tested for conductivity, the light glows brilliantly, indicating many ions in the solution.

Ionization occurs in each of the preceding two reactions with water, producing ions in solution. The necessity for water in the ionization process can be demonstrated by dissolving hydrogen chloride in a nonpolar solvent such as hexane and testing the solution for conductivity. The solution fails to conduct electricity, indicating that no ions are produced.

The terms *dissociation* and *ionization* are often used interchangeably to describe processes taking place in water. But, strictly speaking, the two are different. In the dissociation of a salt, the salt already exists as ions; when it dissolves in water, the ions separate, or dissociate, and increase in mobility. In the ionization process, ions are produced by the reaction of a compound with water.

15.7 Strong and Weak Electrolytes

strong electrolyte
weak electrolyte

Electrolytes are classified as strong or weak depending on the degree, or extent, of dissociation or ionization. **Strong electrolytes** are essentially 100% ionized in solution; **weak electrolytes** are much less ionized (based on comparing 0.1 *M* solutions). Most electrolytes are either strong or weak, with a few classified as moderately strong or weak. Most salts are strong electrolytes. Acids and bases that are strong electrolytes (highly ionized) are called *strong acids* and *strong bases.* Acids and bases that are weak electrolytes (slightly ionized) are called *weak acids* and *weak bases.*

For equivalent concentrations, solutions of strong electrolytes contain many more ions than do solutions of weak electrolytes. As a result, solutions of strong electrolytes are better conductors of electricity. Consider two solutions, 1 *M* HCl and 1 *M* $HC_2H_3O_2$. Hydrochloric acid is almost 100% ionized; acetic acid is about 1% ionized. (See Figure 15.3.) Thus, HCl is a strong acid and $HC_2H_3O_2$ is a weak acid. Hydrochloric acid has about 100 times as many hydronium ions in solution as acetic acid, making the HCl solution much more acidic.

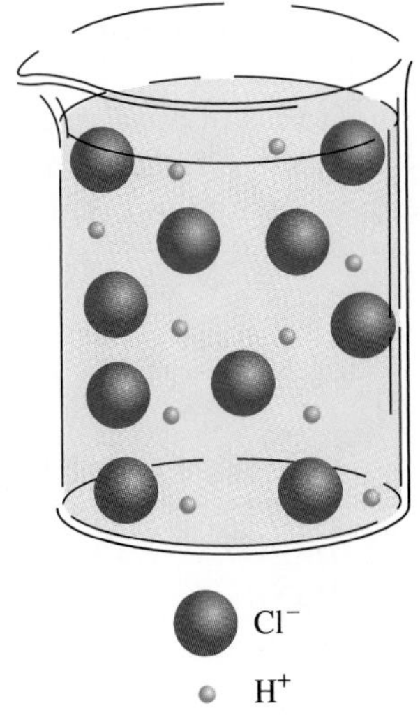

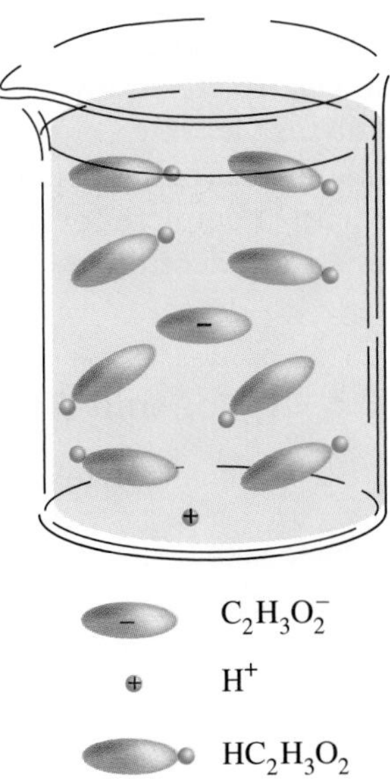

Figure 15.3
HCl solution (left) is 100% ionized, while in $HC_2H_3O_2$ solution (right) almost all of the solute is in molecular form. HCl is a strong acid, while $HC_2H_3O_2$ is a weak acid. Note: The water molecules in the solution are not shown in this figure.

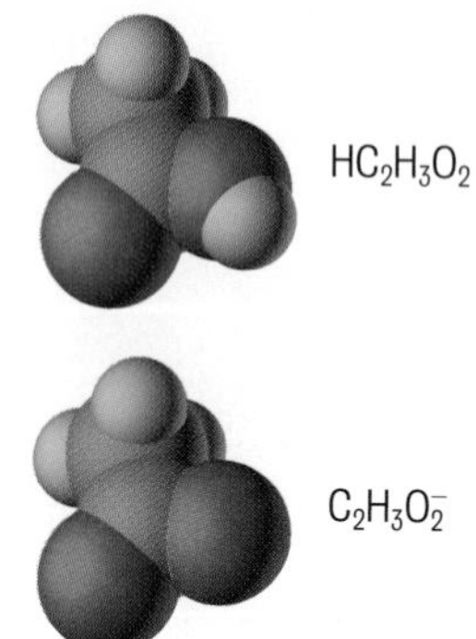

We can distinguish between strong and weak electrolytes experimentally using the apparatus described in Section 15.5. A 1 *M* HCl solution causes the light to glow brilliantly, but a 1 *M* $HC_2H_3O_2$ solution causes only a dim glow. The strong base sodium hydroxide (NaOH) can be distinguished in a similar fashion from the weak base ammonia (NH_3). The ionization of a weak electrolyte in water is represented by an equilibrium equation showing that both the un-ionized and ionized forms are present in solution. In the equilibrium equation of $HC_2H_3O_2$ and its ions, we say that the equilibrium lies "far to the left" because relatively few hydrogen and acetate ions are present in solution:

$$HC_2H_3O_2(aq) \rightleftharpoons H^+(aq) + C_2H_3O_2^-(aq)$$

We have previously used a double arrow in an equation to represent reversible processes in the equilibrium between dissolved and undissolved solute in a saturated solution. A double arrow ($\rightleftharpoons$) is also used in the ionization equation of soluble weak electrolytes to indicate that the solution contains a considerable amount of the un-ionized compound in equilibrium with its ions in solution. (See Section 16.1 for a discussion of reversible reactions.) A single arrow is used to indicate that the electrolyte is essentially all in the ionic form in the solution. For example, nitric acid is a strong acid; nitrous acid is a weak acid. Their ionization equations in water may be indicated as

$$HNO_3(aq) \xrightarrow{H_2O} H^+(aq) + NO_3^-(aq)$$

$$HNO_2(aq) \overset{H_2O}{\rightleftharpoons} H^+(aq) + NO_2^-(aq)$$

Practically all soluble salts, acids (such as sulfuric, nitric, and hydrochloric acids), and bases (such as sodium, potassium, calcium, and barium hydroxides) are strong electrolytes. Weak electrolytes include numerous other acids and bases such as acetic acid, nitrous acid, carbonic acid, and ammonia. The terms *strong acid, strong base, weak acid,* and *weak base* refer to whether an acid or base is a strong or weak electrolyte. A brief list of strong and weak electrolytes is given in Table 15.3.

Electrolytes yield two or more ions per formula unit upon dissociation—the actual number being dependent on the compound. Dissociation is complete or nearly complete for nearly all soluble ionic compounds and for certain other strong electrolytes, such as those given in Table 15.3. The following

Table 15.3 Strong and Weak Electrolytes

Strong electrolytes		Weak electrolytes	
Most soluble salts	$HClO_4$	$HC_2H_3O_2$	$H_2C_2O_4$
H_2SO_4	NaOH	H_2CO_3	H_3BO_3
HNO_3	KOH	HNO_2	HClO
HCl	$Ca(OH)_2$	H_2SO_3	NH_3
HBr	$Ba(OH)_2$	H_2S	HF

are dissociation equations for several strong electrolytes. In all cases, the ions are actually hydrated:

$$NaOH \xrightarrow{H_2O} Na^+(aq) + OH^-(aq) \quad \text{2 ions in solution per formula unit}$$

$$Na_2SO_4 \xrightarrow{H_2O} 2\,Na^+(aq) + SO_4^{2-}(aq) \quad \text{3 ions in solution per formula unit}$$

$$Fe_2(SO_4)_3 \xrightarrow{H_2O} 2\,Fe^{3+}(aq) + 3\,SO_4^{2-}(aq) \quad \text{5 ions in solution per formula unit}$$

One mole of NaCl will give 1 mol of Na^+ ions and 1 mol of Cl^- ions in solution, assuming complete dissociation of the salt. One mole of $CaCl_2$ will give 1 mol of Ca^{2+} ions and 2 mol of Cl^- ions in solution:

$$\underset{1\text{ mol}}{NaCl} \xrightarrow{H_2O} \underset{1\text{ mol}}{Na^+(aq)} + \underset{1\text{ mol}}{Cl^-(aq)}$$

$$\underset{1\text{ mol}}{CaCl_2} \xrightarrow{H_2O} \underset{1\text{ mol}}{Ca^{2+}(aq)} + \underset{2\text{ mol}}{2\,Cl^-(aq)}$$

Example 15.2

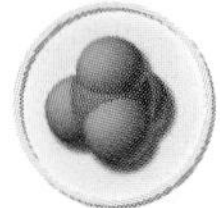

What is the molarity of each ion in a solution of (a) 2.0 M NaCl and (b) 0.40 M K_2SO_4? Assume complete dissociation.

SOLUTION

(a) According to the dissociation equation,

$$\underset{1\text{ mol}}{NaCl} \xrightarrow{H_2O} \underset{1\text{ mol}}{Na^+(aq)} + \underset{1\text{ mol}}{Cl^-(aq)}$$

the concentration of Na^+ is equal to that of NaCl: 1 mol NaCl $\longrightarrow$ 1 mol Na^+ and the concentration of Cl^- is also equal to that of NaCl. Therefore the concentrations of the ions in 2.0 M NaCl are 2.0 M Na^+ and 2.0 M Cl^-.

(b) According to the dissociation equation

$$\underset{1\text{ mol}}{K_2SO_4} \xrightarrow{H_2O} \underset{2\text{ mol}}{2\,K^+(aq)} + \underset{1\text{ mol}}{SO_4^{2-}(aq)}$$

the concentration of K^+ is twice that of K_2SO_4 and the concentration of SO_4^{2-} is equal to that of K_2SO_4. Therefore the concentrations of the ions in 0.40 M K_2SO_4 are 0.80 M K^+ and 0.40 M SO_4^{2-}.

Practice 15.4

What is the molarity of each ion in a solution of (a) 0.050 M $MgCl_2$ and (b) 0.070 M $AlCl_3$?

Colligative Properties of Electrolyte Solutions

We have learned that when 1 mol of sucrose, a nonelectrolyte, is dissolved in 1000 g of water, the solution freezes at $-1.86°C$. When 1 mol NaCl is dissolved in 1000 g of water, the freezing point of the solution is not $-1.86°C$, as might be expected, but is closer to $-3.72°C$ (-1.86×2). The reason for the lower freezing point is that 1 mol NaCl in solution produces 2 mol of particles ($2 \times 6.022 \times 10^{23}$ ions) in solution. Thus the freezing point depression produced by 1 mol NaCl is essentially equivalent to that produced by 2 mol of a nonelectrolyte. An electrolyte such as $CaCl_2$, which yields three ions in water, gives a freezing point depression of about three times that of a nonelectrolyte. These freezing point data provide additional evidence that electrolytes dissociate when dissolved in water. The other colligative properties are similarly affected by substances that yield ions in aqueous solutions.

Practice 15.5

What is the boiling point of a 1.5 *m* solution of KCl(*aq*)?

15.8 Ionization of Water

Pure water is a *very* weak electrolyte, but it does ionize slightly. Two equations commonly used to show how water ionizes are

$$\underset{\text{acid}}{H_2O} + \underset{\text{base}}{H_2O} \rightleftharpoons \underset{\text{acid}}{H_3O^+} + \underset{\text{base}}{OH^-}$$

and

$$H_2O \rightleftharpoons H^+ + OH^-$$

The first equation represents the Brønsted–Lowry concept, with water reacting as both an acid and a base, forming a hydronium ion and a hydroxide ion. The second equation is a simplified version, indicating that water ionizes to give a hydrogen and a hydroxide ion. Actually, the proton (H^+) is hydrated and exists as a hydronium ion. In either case equal molar amounts of acid and base are produced so that water is neutral, having neither H^+ nor OH^- ions in excess. The ionization of water at 25°C produces an H^+ ion concentration of 1.0×10^{-7} mol/L and an OH^- ion concentration of 1.0×10^{-7} mol/L. Square brackets, [ ], are used to indicate that the concentration is in moles per liter. Thus $[H^+]$ means the concentration of H^+ is in moles per liter. These concentrations are usually expressed as

$$[H^+] \text{ or } [H_3O^+] = 1.0 \times 10^{-7} \text{ mol/L}$$
$$[OH^-] = 1.0 \times 10^{-7} \text{ mol/L}$$

These figures mean that about two out of every billion water molecules are ionized. This amount of ionization, small as it is, is a significant factor in the behavior of water in many chemical reactions.

15.9 Introduction to pH

The acidity of an aqueous solution depends on the concentration of hydrogen or hydronium ions. The pH scale of acidity provides a simple, convenient, numerical way to state the acidity of a solution. Values on the pH scale are obtained by mathematical conversion of H^+ ion concentrations to pH by the expression

$$pH = -\log[H^+]$$

where $[H^+]$ = H^+ or H_3O^+ ion concentration in moles per liter. The **pH** is defined as the *negative* logarithm of the H^+ or H_3O^+ concentration in moles per liter:

pH

$$pH = -\log[H^+] = -\log(1 \times 10^{-7}) = -(-7) = 7$$

For example, the pH of pure water at 25°C is 7 and is said to be neutral; that is, it is neither acidic nor basic, because the concentrations of H^+ and OH^- are equal. Solutions that contain more H^+ ions than OH^- ions have pH values less than 7, and solutions that contain less H^+ ions than OH^- ions have pH values greater than 7.

$pH < 7.00$ is an acidic solution
$pH = 7.00$ is a neutral solution
$pH > 7.00$ is a basic solution

When $[H^+] = 1 \times 10^{-5}$ mol/L, pH = 5 (acidic)

When $[H^+] = 1 \times 10^{-9}$ mol/L, pH = 9 (basic)

Instead of saying that the hydrogen ion concentration in the solution is 1×10^{-5} mol/L, it's customary to say that the pH of the solution is 5. The smaller the pH value, the more acidic the solution (see Figure 15.4).

The pH scale, along with its interpretation, is given in Table 15.4, and Table 15.5 lists the pH of some common solutions. Note that a change of only one pH unit means a 10-fold increase or decrease in H^+ ion concentration. For example, a solution with a pH of 3.0 is 10 times more acidic than a solution with a pH of 4.0. A simplified method of determining pH from $[H^+]$ follows:

$[H^+] = 1 \times 10^{-5}$ ← pH = this number (5)
pH = 5
(↗ when this number is exactly 1)

$[H^+] = 2 \times 10^{-5}$ ← pH is between this number and next lower number (4 and 5)
pH = 4.7
(↗ when this number is between 1 and 10)

Calculating the pH value for H^+ ion concentrations requires the use of logarithms, which are exponents. The **logarithm** (log) of a number is simply the power to which 10 must be raised to give that number. Thus the log of 100 is 2 ($100 = 10^2$), and the log of 1000 is 3 ($1000 = 10^3$). The log of 500 is 2.70, but you can't determine this value easily without a scientific calculator.

logarithm

Help on using calculators is found in Appendix II.

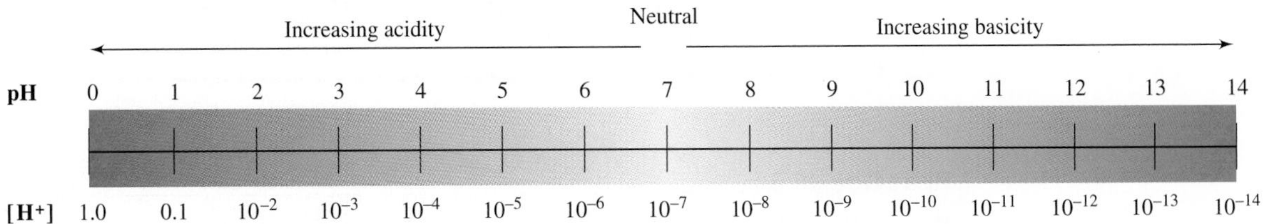

Figure 15.4
The pH scale of acidity and basicity.

Table 15.4 pH Scale for Expressing Acidity

$[H^+]$ (mol/L)	pH	
1×10^{-14}	14	↑
1×10^{-13}	13	
1×10^{-12}	12	Increasing
1×10^{-11}	11	basicity
1×10^{-10}	10	
1×10^{-9}	9	
1×10^{-8}	8	
1×10^{-7}	7	Neutral
1×10^{-6}	6	
1×10^{-5}	5	
1×10^{-4}	4	
1×10^{-3}	3	Increasing
1×10^{-2}	2	acidity
1×10^{-1}	1	
1×10^{0}	0	↓

Table 15.5 The pH of Common Solutions

Solution	pH
Gastric juice	1.0
0.1 *M* HCl	1.0
Lemon juice	2.3
Vinegar	2.8
0.1 *M* $HC_2H_3O_2$	2.9
Orange juice	3.7
Tomato juice	4.1
Coffee, black	5.0
Urine	6.0
Milk	6.6
Pure water (25°C)	7.0
Blood	7.4
Household ammonia	11.0
1 *M* NaOH	14.0

Remember: Change the sign on your calculator since pH = $-\log[H^+]$.

Let's determine the pH of a solution with $[H^+] = 2 \times 10^{-5}$ using a calculator. Enter 2×10^{-5} into your calculator and press the log key. The number $-4.69\ldots$ will be displayed. The pH is then

$$pH = -\log[H^+] = -(-4.69\ldots) = 4.7$$

Next we must determine the correct number of significant figures in the logarithm. The rules for logs are different from those we use in other math operations. The number of decimal places for a log must equal the number of significant figures in the original number. Since 2×10^{-5} has one significant figure, we should round the log to one decimal place $(4.69\ldots) = 4.7$.

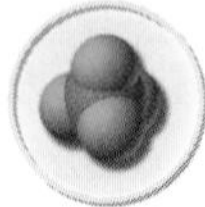

Example 15.3 What is the pH of a solution with an $[H^+]$ of (a) 1.0×10^{-11}, (b) 6.0×10^{-4}, and (c) 5.47×10^{-8}?

SOLUTION

(a) $[H^+] = 1.0 \times 10^{-11}$
(2 significant figures)

$pH = -\log(1.0 \times 10^{-11})$

$pH = 11.00$

(2 decimal places)

(b) $[H^+] = 6.0 \times 10^{-4}$
(2 significant figures)

$\log(6.0 \times 10^{-4}) = -3.22$

$pH = -\log[H^+]$

$pH = -(-3.22) = 3.22$

(2 decimal places)

(c) $[H^+] = 5.47 \times 10^{-8}$
(3 significant figures)

$\log(5.47 \times 10^{-8}) = -7.262$

$pH = -\log[H^+]$

$pH = -(-7.262) = 7.262$

(3 decimal places)

Practice 15.6

What is the pH of a solution with $[H^+]$ of (a) 3.9×10^{-12} *M*, (b) 1.3×10^{-3} *M*, and (c) 3.72×10^{-6} *M*?

CHEMISTRY IN ACTION • In the Pink—A Sign of Corrosion

As our fleet of commercial airplanes ages, scientists are looking for new ways to detect corrosion and stop it *before* it becomes a safety issue. The problem is that planes are subjected to heat, rain, and wind—all factors that can contribute to metal corrosion. How can a maintenance crew detect corrosion on a huge aircraft since many imaging techniques work on only small areas at a time? Imagine a plane that could tell the maintenance crew it needed repair.

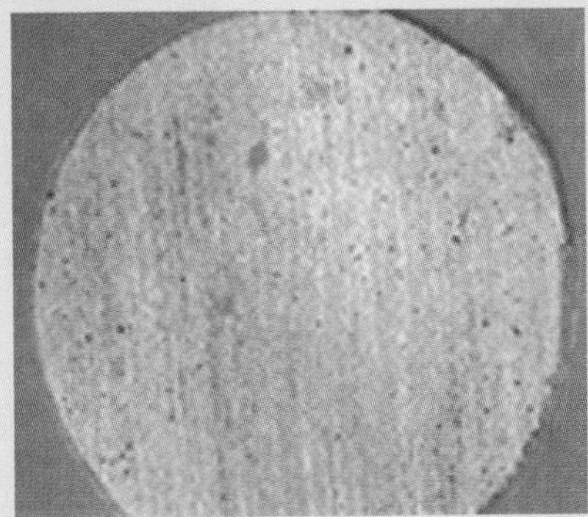
Phenolphthalein paint after eight days.

Gerald Frankel and Jim Zhang from the Ohio State University have developed a paint that detects changes in pH. Corrosion of metals (in contact with air and water) results in a chemical reaction that produces hydroxide ions, which increase pH. Frankel and Zhang made a clear acrylic coating that they mixed with phenolphthalein. This acid–base indicator turns bright fuschia when the pH increases above 8.0. The scientists can show the "paint" turns pink at corrosion sites as small as 15 μm deep.

Technicians at Wright-Patterson Air Force Base in Ohio think this could lead to a whole new way to detect corrosion. William Mullins says, "You could walk down the vehicle and see there's a pink spot." This method would be especially good at detecting corrosion concealed around rivets and where metals overlap. The only limitation appears to be that the coating must be clear.

The next time your airline maintenance crew sees pink spots, you may be in for a delay or a change of aircraft!

The measurement and control of pH is extremely important in many fields. Proper soil pH is necessary to grow certain types of plants successfully. The pH of certain foods is too acidic for some diets. Many biological processes are delicately controlled pH systems. The pH of human blood is regulated to very close tolerances through the uptake or release of H^+ by mineral ions, such as HCO_3^-, HPO_4^{2-}, and $H_2PO_4^-$, Changes in the pH of the blood by as little as 0.4 pH unit result in death.

Compounds with colors that change at particular pH values are used as indicators in acid–base reactions. For example, phenolphthalein, an organic compound, is colorless in acid solution and changes to pink at a pH of 8.3. When a solution of sodium hydroxide is added to a hydrochloric acid solution containing phenolphthalein, the change in color (from colorless to pink) indicates that all the acid is neutralized. Commercially available pH test paper contains chemical indicators. The indicator in the paper takes on different colors when wetted with solutions of different pH. Thus the pH of a solution can be estimated by placing a drop on the test paper and comparing the color of the test paper with a color chart calibrated at different pH values. Common applications of pH test indicators are the kits used to measure and adjust the pH of swimming pools, hot tubs, and saltwater aquariums. Electronic pH meters are used for making rapid and precise pH determinations (see Figure 15.5).

Figure 15.5
The pH of a substance (in this case garden soil) can be measured by using a pH meter.

15.10 Neutralization

neutralization

The reaction of an acid and a base to form a salt and water is known as **neutralization.** We've seen this reaction before, but now with our knowledge about ions and ionization, let's reexamine the process of neutralization.

Consider the reaction that occurs when solutions of sodium hydroxide and hydrochloric acid are mixed. The ions present initially are Na^+ and OH^- from the base and H^+ and Cl^- from the acid. The products, sodium chloride and water, exist as Na^+ and Cl^- ions and H_2O molecules. A chemical equation representing this reaction is

$$HCl(aq) + NaOH(aq) \longrightarrow NaCl(aq) + H_2O(l)$$

This equation, however, does not show that HCl, NaOH, and NaCl exist as ions in solution. The following total ionic equation gives a better representation of the reaction:

$$(H^+ + Cl^-) + (Na^+ + OH^-) \longrightarrow Na^+ + Cl^- + H_2O(l)$$

spectator ion

This equation shows that the Na^+ and Cl^- ions did not react. These ions are called **spectator ions** because they were present but did not take part in the reaction. The only reaction that occurred was that between the H^+ and OH^- ions. Therefore the equation for the neutralization can be written as this net ionic equation:

$$\underset{\text{acid}}{H^+(aq)} + \underset{\text{base}}{OH^-(aq)} \longrightarrow \underset{\text{water}}{H_2O(l)}$$

titration

This simple net ionic equation represents not only the reaction of sodium hydroxide and hydrochloric acid, but also the reaction of any strong acid with any water-soluble hydroxide base in an aqueous solution. The driving force of a neutralization reaction is the ability of an H^+ ion and an OH^- ion to react and form a molecule of water.

The amount of acid, base, or other species in a sample can be determined by **titration,** which measures the volume of one reagent required to react with a measured mass or volume of another reagent. Consider the titration of an acid with a base. A measured volume of acid of unknown concentration is placed in a flask, and a few drops of an indicator solution are added. Base solution of known concentration is slowly added from a buret to the acid until the indicator changes color. The indicator selected is one that changes color when the stoichiometric quantity (according to the equation) of base has been added to the acid. At this

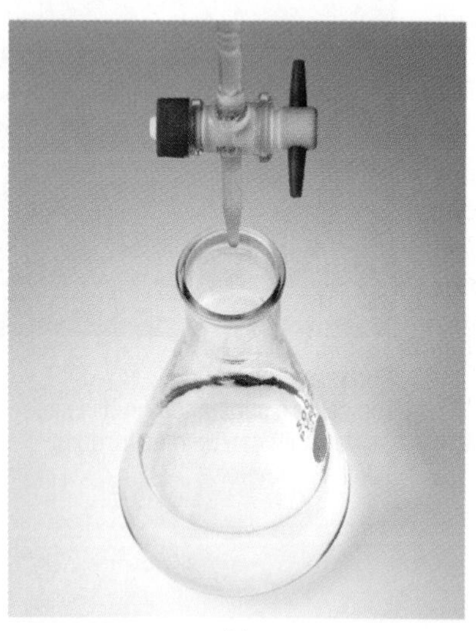
(a)

(b)

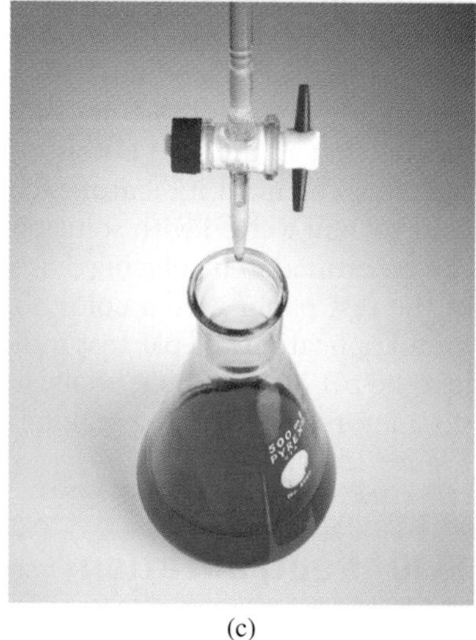
(c)

Tritation. (a) Before the titration begins, the liqiuid in the flask is colorless and clear. (b) At the endpoint a pole pink color persists in the flash. (c) After the endpoint, the color of the solution turns bright pink, indicating an excess of base.

point, known as the *end point of the titration,* the titration is complete, and the volume of base used to neutralize the acid is read from the buret. The concentration or amount of acid in solution can be calculated from the titration data and the chemical equation for the reaction. Let's look at some examples.

Example 15.4

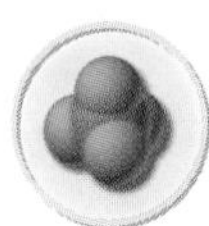

Suppose that 42.00 mL of 0.150 M NaOH solution are required to neutralize 50.00 mL of hydrochloric acid solution. What is the molarity of the acid solution?

SOLUTION

The equation for the reaction is

$$NaOH(aq) + HCl(aq) \longrightarrow NaCl(aq) + H_2O(l)$$

In this neutralization NaOH and HCl react in a 1:1 mole ratio. Therefore the moles of HCl in solution are equal to the moles of NaOH required to react with it. First we calculate the moles of NaOH used, and from this value we determine the moles of HCl:

Data: 42.00 mL of 0.150 M NaOH   50.00 mL HCl
molarity of acid = M (unknown)

Determine the moles of NaOH:

$$M = \text{mol/L} \qquad 42.00\ \text{mL} = 0.04200\ \text{L}$$

$$(0.04200\ \cancel{\text{L}})\left(\frac{0.150\ \text{mol NaOH}}{1\ \cancel{\text{L}}}\right) = 0.00630\ \text{mol NaOH}$$

Since NaOH and HCl react in a 1:1 ratio, 0.00630 mol HCl was present in the 50.00 mL of HCl solution. Therefore, the molarity of the HCl is

$$M = \frac{\text{mol}}{\text{L}} = \frac{0.00630\ \text{mol HCl}}{0.05000\ \text{L}} = 0.126\ M\ \text{HCl}$$

Example 15.5

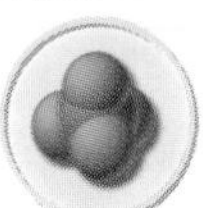

Suppose that 42.00 mL of 0.150 M NaOH solution are required to neutralize 50.00 mL of H_2SO_4 solution. What is the molarity of the acid solution?

SOLUTION

The equation for the reaction is

$$2\,NaOH(aq) + H_2SO_4(aq) \longrightarrow Na_2SO_4(aq) + 2\,H_2O(l)$$

The same amount of base (0.00630 mol NaOH) is used in this titration as in Example 15.4, but the mole ratio of acid to base in the reaction is 1:2. The moles of H_2SO_4 reacted can be calculated:

Data: 42.00 mL of 0.150 M NaOH = 0.00630 mol NaOH

$$(0.00630\ \cancel{\text{mol NaOH}})\left(\frac{1\ \text{mol}\ H_2SO_4}{2\ \cancel{\text{mol NaOH}}}\right) = 0.00315\ \text{mol}\ H_2SO_4$$

Therefore 0.00315 mol H_2SO_4 was present in 50.00 mL of H_2SO_4 solution. The molarity of the H_2SO_4 is

$$M = \frac{\text{mol}}{\text{L}} = \frac{0.00315\ \text{mol}\ H_2SO_4}{0.05000\ \text{L}} = 0.0630\ M\ H_2SO_4$$

Example 15.6 A 25.00-mL sample of H_2SO_4 solution required 14.26 mL of 0.2240 M NaOH for complete neutralization. What is the molarity of the sulfuric acid?

SOLUTION The equation for the reaction is

$$2\,\text{NaOH}(aq) + \text{H}_2\text{SO}_4(aq) \longrightarrow \text{Na}_2\text{SO}_4(aq) + 2\,\text{H}_2\text{O}(l)$$

The moles of NaOH needed are

$$\begin{aligned}\text{moles NaOH} &= (V_{\text{NaOH}})(M_{\text{NaOH}}) \\ &= (0.01426\,\text{L})\left(0.2240\,\frac{\text{mol}}{\text{L}}\right) \\ &= 0.003194\,\text{mol NaOH}\end{aligned}$$

Since the mole ratio of acid to base is $\frac{1\,\text{H}_2\text{SO}_4}{2\,\text{NaOH}}$, the moles of acid in the sample are

$$(0.003194\,\text{mol NaOH})\left(\frac{1\,\text{H}_2\text{SO}_4}{2\,\text{NaOH}}\right) = 0.001597\,\text{mol H}_2\text{SO}_4$$

Now to find the molarity of the sample, we divide the moles of acid by its original volume:

$$\frac{0.001597\,\text{mol H}_2\text{SO}_4}{0.02500\,\text{L}} = 0.06388\,M\,\text{H}_2\text{SO}_4$$

Practice 15.7

A 50.0-mL sample of HCl required 24.81 mL of 0.1250 M NaOH for neutralization. What is the molarity of the acid?

15.11 Writing Net Ionic Equations

In Section 15.10, we wrote the reaction of hydrochloric acid and sodium hydroxide in three different equations:

1. $\text{HCl}(aq) + \text{NaOH}(aq) \longrightarrow \text{NaCl}(aq) + \text{H}_2\text{O}(l)$
2. $(\text{H}^+ + \text{Cl}^-) + (\text{Na}^+ + \text{OH}^-) \longrightarrow \text{Na}^+ + \text{Cl}^- + \text{H}_2\text{O}(l)$
3. $\text{H}^+ + \text{OH}^- \longrightarrow \text{H}_2\text{O}$

formula equation
total ionic equation
net ionic equation

In the **formula equation** (1), compounds are written in their molecular, or formula, expressions. In the **total ionic equation** (2), compounds are written to show the form in which they are predominantly present: strong electrolytes as ions in solution and nonelectrolytes, weak electrolytes, precipitates, and gases in their molecular forms. In the **net ionic equation** (3), only those molecules or ions that have reacted are included in the equation; ions or molecules that do not react (the spectators) are omitted.

When balancing equations thus far, we've been concerned only with the atoms of the individual elements. Because ions are electrically charged, ionic equations often end up with a net electrical charge. A balanced equation must have the same net charge on each side, whether that charge is positive, negative, or zero. Therefore, when balancing ionic equations, we must make sure that both the same number of each kind of atom and the same net electrical charge are present on each side.

Here is a list of rules for writing ionic equations:

1. Strong electrolytes in solution are written in their ionic form.
2. Weak electrolytes are written in their molecular form.
3. Nonelectrolytes are written in their molecular form.
4. Insoluble substances, precipitates, and gases are written in their molecular forms.
5. The net ionic equation should include only substances that have undergone a chemical change. Spectator ions are omitted from the net ionic equation.
6. Equations must be balanced, both in atoms and in electrical charge.

Study the following examples. In each one, the formula equation is given. Write the total ionic equation and the net ionic equation for each.

Example 15.7

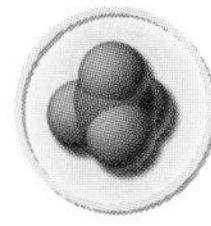

$$HNO_3(aq) + KOH(aq) \longrightarrow KNO_3(aq) + H_2O(l)$$

formula equation

SOLUTION

$$(H^+ + NO_3^-) + (K^+ + OH^-) \longrightarrow K^+ + NO_3^- + H_2O$$

total ionic equation

$$H^+ + OH^- \longrightarrow H_2O$$

net ionic equation

The HNO_3, KOH, and KNO_3 are soluble, strong electrolytes. The K^+ and NO_3^- ions are spectator ions, have not changed, and are not included in the net ionic equation. Water is a nonelectrolyte and is written in the molecular form.

Example 15.8

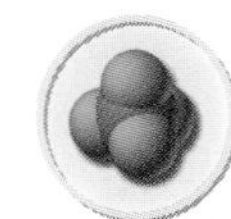

$$2\,AgNO_3(aq) + BaCl_2(aq) \longrightarrow 2\,AgCl(s) + Ba(NO_3)_2(aq)$$

formula equation

SOLUTION

$$(2\,Ag^+ + 2\,NO_3^-) + (Ba^{2+} + 2\,Cl^-) \longrightarrow 2\,AgCl(s) + Ba^{2+} + 2\,NO_3^-$$

total ionic equation

$$Ag^+ + Cl^- \longrightarrow AgCl(s)$$

net ionic equation

Although AgCl is an ionic compound, it is written in the un-ionized form on the right side of the ionic equations because most of the Ag^+ and Cl^- ions are no longer in solution but have formed a precipitate of AgCl. The Ba^{2+} and NO_3^- ions are spectator ions.

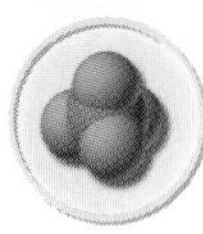

Example 15.9 $Na_2CO_3(aq) + H_2SO_4(aq) \longrightarrow Na_2SO_4(aq) + H_2O(l) + CO_2(g)$

formula equation

SOLUTION $(2\,Na^+ + CO_3^{2-}) + (2\,H^+ + SO_4^{2-}) \longrightarrow 2\,Na^+ + SO_4^{2-} + H_2O(l) + CO_2(g)$

total ionic equation

$CO_3^{2-} + 2\,H^+ \longrightarrow H_2O(l) + CO_2(g)$

net ionic equation

Carbon dioxide (CO_2) is a gas and evolves from the solution; Na^+ and SO_4^{2-} are spectator ions.

Example 15.10 $HC_2H_3O_2(aq) + NaOH(aq) \longrightarrow NaC_2H_3O_2(aq) + H_2O(l)$

formula equation

SOLUTION $HC_2H_3O_2 + (Na^+ + OH^-) \longrightarrow Na^+ + C_2H_3O_2^- + H_2O$

total ionic equation

$HC_2H_3O_2 + OH^- \longrightarrow C_2H_3O_2^- + H_2O$

net ionic equation

Acetic acid, $HC_2H_3O_2$, a weak acid, is written in the molecular form, but sodium acetate, $NaC_2H_3O_2$, a soluble salt, is written in the ionic form. The Na^+ ion is the only spectator ion in this reaction. Both sides of the net ionic equation have a -1 electrical charge.

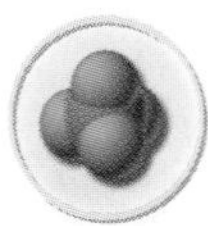

Example 15.11 $Mg(s) + 2\,HCl(aq) \longrightarrow MgCl_2(aq) + H_2(g)$

formula equation

SOLUTION $Mg + (2\,H^+ + 2\,Cl^-) \longrightarrow Mg^{2+} + 2\,Cl^- + H_2(g)$

total ionic equation

$Mg + 2\,H^+ \longrightarrow Mg^{2+} + H_2(g)$

net ionic equation

The net electrical charge on both sides of the equation is $+2$.

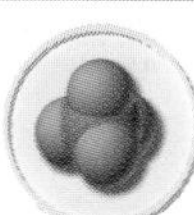

Example 15.12 $H_2SO_4(aq) + Ba(OH)_2(aq) \longrightarrow BaSO_4(s) + 2\,H_2O(l)$

formula equation

SOLUTION $(2\,H^+ + SO_4^{2-}) + (Ba^{2+} + 2\,OH^-) \longrightarrow BaSO_4(s) + 2\,H_2O(l)$

total ionic equation

$2\,H^+ + SO_4^{2-} + Ba^{2+} + 2\,OH^- \longrightarrow BaSO_4(s) + 2\,H_2O(l)$

net ionic equation

Barium sulfate ($BaSO_4$) is a highly insoluble salt. If we conduct this reaction using the conductivity apparatus described in Section 15.5, the light glows brightly at first but goes out when the reaction is complete because almost no ions are left in solution. The $BaSO_4$ precipitates out of solution, and water is a nonconductor of electricity.

Practice 15.8

Write the net ionic equation for

$3\,H_2S(aq) + 2\,Bi(NO_3)_3(aq) \longrightarrow Bi_2S_3(s) + 6\,HNO_3(aq)$

15.12 Acid Rain

Acid rain is defined as any atmospheric precipitation that is more acidic than usual. The increase in acidity might be from natural or industrial sources. Rain acidity varies throughout the world and across the United States. The pH of rain is generally lower in the eastern United States and higher in the West. Unpolluted rain has a pH of 5.6, and so is slightly acidic. This acidity results from the dissolution of carbon dioxide in the water producing carbonic acid:

$$CO_2(g) + H_2O(l) \rightleftharpoons H_2CO_3(aq) \rightleftharpoons H^+(aq) + HCO_3^-(aq)$$

The general process involves the following steps:

1. emission of nitrogen and sulfur oxides into the air
2. transportation of these oxides throughout the atmosphere
3. chemical reactions between the oxides and water, forming sulfuric acid (H_2SO_4) and nitric acid (HNO_3)
4. rain or snow, which carries the acids to the ground

The oxides may also be deposited directly on a dry surface and become acidic when normal rain falls on them.

Acid rain is not a new phenomenon. Rain was probably acidic in the early days of our planet as volcanic eruptions, fires, and decomposition of organic matter released large volumes of nitrogen and sulfur oxides into the atmosphere. Use of fossil fuels, especially since the Industrial Revolution, has made significant changes in the amounts of pollutants being released into the atmosphere. As increasing amounts of fossil fuels have been burned, more and more sulfur and nitrogen oxides have poured into the atmosphere, thus increasing the acidity of rain.

This relief work on St. Bartholomew's Church in New York shows the destructive power of acid rain. The marble is being dissolved slowly over time.

Acid rain affects a variety of factors in our environment. For example, freshwater plants and animals decline significantly when rain is acidic; large numbers of fish and plants die when acidic water from spring thaws enters the lakes. Acidic rainwater leaches aluminum from the soil into lakes, where the aluminum compounds adversely affect the gills of fish. In addition to leaching aluminum from the soil, acid rain also causes other valuable minerals, such as magnesium and calcium, to dissolve and run into lakes and streams. It can also dissolve the waxy protective coat on plant leaves, making them vulnerable to attack by bacteria and fungi.

In our cities, acid rain is responsible for extensive and continuing damage to buildings, monuments, and statues. It reduces the durability of paint and promotes the deterioration of paper, leather, and cloth. In short, we are just beginning to explore the effects of acid rain on human beings and on our food chain.

15.13 Colloids

When we add sugar to a flask of water and shake it, the sugar dissolves and forms a clear homogeneous *solution.* When we do the same experiment with very fine sand and water, the sand particles form a *suspension,* which settles when the shaking stops. When we repeat the experiment again using ordinary cornstarch, we find that the starch does not dissolve in cold water. But if the mixture is heated and stirred, the starch forms a cloudy, opalescent *dispersion.* This dispersion does not appear to be clear and homogeneous like the sugar solution, yet it is not obviously heterogeneous and does not settle like the sand suspension. In short, its properties are intermediate between those of the sugar solution and those of the sand suspension. The starch dispersion is actually a *colloid,* a name derived from the Greek *kolla,* meaning "glue," and was coined by the English scientist Thomas Graham in 1861.

colloid

As it is now used, the word **colloid** means a dispersion in which the dispersed particles are larger than the solute ions or molecules of a true solution and smaller than the particles of a mechanical suspension. The term does not imply a gluelike quality, although most glues are colloidal materials. The size of colloidal particles ranges from a lower limit of about 1 nm (10^{-7} cm) to an upper limit of about 1000 nm (10^{-4} cm). There are eight types of colloids, which are summarized in Table 15.6.

Table 15.6 Types of Colloidal Dispersions

Type	Name	Examples
Gas in liquid	foam	whipped cream, soapsuds
Gas in solid	solid foam	Styrofoam, foam rubber, pumice
Liquid in gas	liquid aerosol	fog, clouds
Liquid in liquid	emulsion	milk, vinegar in oil salad dressing, mayonnaise
Liquid in solid	solid emulsion	cheese, opals, jellies
Solid in gas	solid aerosol	smoke, dust in air
Solid in liquid	sol	india ink, gold sol
Solid in solid	solid sol	tire rubber, certain gems (e.g., rubies)

CHEMISTRY IN ACTION • Foam Cars—Wave of the Future?

Metal foams are a new class of materials that may revolutionize the car industry. Automotive makers know that one key to making cars that are fuel efficient is to decrease weight. Up till now, that weight reduction meant higher cost (in materials like titanium and aluminum) and problems in crash testing (since light vehicles don't absorb energy well).

German automotive supplier Wilhelm Karmann (whose company manufactured the Volkswagen Karmann-Ghia) developed an aluminum foam composite material with some amazing properties. Parts made of this foam composite weigh 30–50% less than an equivalent steel part and are 10 times stiffer. The material is so light it floats in water and although it costs 20–25% more than steel, it could be used for as much as 20% of a compact car. Since the surface of the new material is not smooth, it would likely be used in structural areas of the car (firewalls, roof panels, luggage compartment walls, etc.)

How is aluminum foam made? Two layers of aluminum sheet and a middle powder layer (made of titanium metal hydride and aluminum powder) are rolled together under very high pressure to make a single flat sheet. This sheet metal is then processed in traditional ways to make a variety of 3-D shapes. Then the sheet is placed in a 1148°F oven for two minutes. This quick bake allows the Al metal to melt and mix with H_2 gas (released from the titanium hydride), making foam. The sheet rises just like a cake (increasing five to seven times in thickness). When the foam cools it is a rigidly formed 3-D structure between two aluminum skins.

Not only is the new aluminum foam part much lighter than its steel counterpart; it also performs well in crash tests. One of these days you may climb into a car that is really a foam colloid—with great fuel economy and crash resistance.

The fundamental difference between a colloidal dispersion and a true solution is the size, not the nature, of the particles. The solute particles in a solution are usually single ions or molecules that may be hydrated to varying degrees. Colloidal particles are usually aggregations of ions or molecules. However, the molecules of some polymers, such as proteins, are large enough to be classified as colloidal particles when in solution. To fully appreciate the differences in relative sizes, the volumes (not just the linear dimensions) of colloidal particles and solute particles must be compared. The difference in volumes can be approximated by assuming that the particles are spheres. A large colloidal particle has a diameter of about 500 nm, whereas a fair-sized ion or molecule has a diameter of about 0.5 nm. Thus the diameter of the colloidal particle is about 1000 times that of the solute particle. Because the volumes of spheres are proportional to the cubes of their diameters, we can calculate that the volume of a colloidal particle can be up to a billion ($10^3 \times 10^3 \times 10^3 = 10^9$) times greater than that of a solution particle.

15.14 Properties of Colloids

In 1827, while observing a strongly illuminated aqueous suspension of pollen under a high-powered microscope, Robert Brown (1773–1858) noted that the pollen grains appeared to have a trembling, erratic motion. He later determined that this erratic motion is not confined to pollen but is characteristic of colloidal particles in general. This random motion of colloidal particles is called **Brownian movement.** We can readily observe such movement by confining cigarette smoke in a small transparent chamber and illuminating it with a strong beam of light at right angles to the optical axis of the microscope. The smoke particles appear as tiny randomly moving lights because the light is reflected from their surfaces. This motion is due to the continual bombardment of the smoke particles by air molecules. Since Brownian movement can be seen when colloidal particles are dispersed in either a gaseous or a liquid medium, it affords nearly direct visual proof that matter at the molecular level is moving randomly, as postulated by the kinetic-molecular theory.

Brownian movement

When an intense beam of light is passed through an ordinary solution and viewed at an angle, the beam passing through the solution is hardly visible. A beam of light, however, is clearly visible and sharply outlined when it is passed through a colloidal dispersion. This phenomenon is known as the **Tyndall effect.** The Tyndall effect, like Brownian movement, can be observed in nearly all colloidal dispersions. It occurs because the colloidal particles are large enough to scatter the rays of visible light. The ions or molecules of true solutions are too small to scatter light and therefore do not exhibit a noticeable Tyndall effect.

Tyndall effect

A beam of lights is visible in a colloid solution (left) but not in a true solution. The size of the particles in the colloid are large enough to scatter the light beam making it visible.

Another important characteristic of colloids is that the particles have relatively huge surface areas. We saw in Section 14.4 that the surface area is increased 10-fold when a 1-cm cube is divided into 1000 cubes with sides of 0.1 cm. When a 1-cm cube is divided into colloidal-size cubes measuring 10^{-6} cm, the combined surface area of all the particles becomes a million times greater than that of the original cube.

Colloidal particles become electrically charged when they adsorb ions on their surfaces. *Adsorption* should not be confused with *absorption.* Adsorption refers to the adhesion of molecules or ions to a surface, whereas absorption refers to the taking in of one material by another material. Adsorption occurs because the atoms or ions at the surface of a particle are not completely surrounded by other atoms or ions as are those in the interior. Consequently, these surface atoms or ions attract and adsorb ions or polar molecules from the dispersion medium onto the surfaces of the colloidal particles. This property is directly related to the large surface area presented by the many tiny particles.

15.15 Applications of Colloidal Properties

Activated charcoal has an enormous surface area, approximately 1 million square centimeters per gram in some samples. Hence, charcoal is very effective in selectively adsorbing the polar molecules of some poisonous gases and is therefore used in gas masks. Charcoal can be used to adsorb impurities from liquids as well as from gases, and large amounts are used to remove substances that have objectionable tastes and odors from water supplies. In sugar refineries, activated charcoal is used to adsorb colored impurities from the raw sugar solutions.

A process widely used for dust and smoke control in many urban and industrial areas was devised by an American, Frederick Cottrell (1877–1948). The Cottrell process takes advantage of the fact that the particulate matter in dust and smoke is electrically charged. Air to be cleaned of dust or smoke is passed between electrode plates charged with a high voltage. Positively charged particles are attracted to, neutralized, and thereby precipitated at the negative

Colloidal particles become electrically charged when they adsorb ions on their surface.

electrodes. Negatively charged particles are removed in the same fashion at the positive electrodes. Large Cottrell units are fitted with devices for automatic removal of precipitated material. Small units, designed for removing dust and pollen from air in the home, are now on the market. Unfortunately, Cottrell units remove only particulate matter; they cannot remove gaseous pollutants such as carbon monoxide, sulfur dioxide, and nitrogen oxides.

Thomas Graham found that a parchment membrane would allow the passage of true solutions but would prevent the passage of colloidal dispersions. Dissolved solutes can be removed from colloidal dispersions through the use of such a membrane by a process called **dialysis.** The membrane itself is called a *dialyzing membrane*. Artificial membranes are made from such materials as parchment paper, collodion, or certain kinds of cellophane. Dialysis can be demonstrated by putting a colloidal starch dispersion and some copper(II) sulfate solution in a parchment paper bag and suspending it in running water. In a few hours, the blue color of the copper(II) sulfate has disappeared, and only the starch dispersion remains in the bag.

dialysis

A life-saving application of dialysis has been the development of artificial kidneys. The blood of a patient suffering from partial kidney failure is passed through the artificial kidney machine for several hours, during which time the soluble waste products are removed by dialysis.

Chapter 15 Review

15.1 Acids and Bases

KEY TERM

Hydronium ion

- Characteristic properties of acids include:
  - They taste sour.
  - They change litmus from blue to red.
  - They react with:
    - Metals to form hydrogen gas
    - Hydroxide bases to form water and a salt
    - Carbonates to produce CO_2
- Characteristic properties of bases include:
  - They taste bitter or caustic.
  - They have a slippery feeling.
  - They change litmus from red to blue.
  - They interact with acids to form water and a salt.
- Arrhenius definition of acids and bases:
  - Acids contain excess H^+ ions in aqueous solutions.
  - Bases contain excess OH^- ions in aqueous solutions.
- Brønsted–Lowry definition of acids and bases:
  - Acids are proton donors.
  - Bases are proton acceptors.
- Lewis definition of acids and bases:
  - Acids are electron pair acceptors.
  - Bases are electron pair donors.

15.2 Reactions of Acids

- Acids react with metals above hydrogen in the activity series to form hydrogen gas and a salt.
- Acids react with bases to form water and a salt (neutralization reaction).
- Acids react with metal oxides to form water and salt.
- Acids react with carbonates to form water, a salt, and carbon dioxide.

15.3 Reactions of Bases

KEY TERM

Amphoteric

- Bases react with acids to form water and a salt (neutralization reaction).
- Some amphoteric metals react with NaOH or KOH to form hydrogen and a salt.

15.4 Salts

- Salts can be considered to be derived from the reaction of an acid and a base.
- Salts are crystalline and have high melting and boiling points.

15.5 Electrolytes and Nonelectrolytes

KEY TERMS

Electrolyte
Nonelectrolyte

- A substance whose aqueous solution conducts electricity is called an electrolyte:
  - Aqueous solution contains ions.
- A substance whose aqueous solution does not conduct electricity is called a nonelectrolyte:
  - Aqueous solution contains molecules, not ions.

15.6 Dissociation and Ionization of Electrolytes

KEY TERMS

Dissociation
Ionization

- Dissociation is the process by which the ions of a salt separate as the salt dissolves.
- Ionization is the formation of ions that occurs as a result of a chemical reaction with water.

15.7 Strong and Weak Electrolytes

KEY TERMS

Strong electrolyte
Weak electrolyte

- Strong electrolytes are 100% ionized in solution:
  - Strong acids and strong bases
- Weak electrolytes are much less ionized in solution:
  - Weak acids and bases
- Colligative properties of electrolyte solutions depend on the number of particles produced during the ionization of the electrolyte.

15.8 Ionization of Water

- Water can self-ionize to form H^+ and OH^- ions.
- Concentrations of ions in water at 25°C:
  - $[H^+] = 1.0 \times 10^{-7}$
  - $[OH^-] = 1.0 \times 10^{-7}$

15.9 Introduction to pH

KEY TERMS

pH
Logarithm

- $pH = -\log[H^+]$:
  - $pH < 7$ for acidic solution
  - $pH = 7$ in neutral solution
  - $pH > 7$ for basic solution
- The number of decimal places in a logarithm equals the number of significant figures in the original number.

15.10 Neutralization

KEY TERMS

Neutralization
Spectator ion
Titration

- The reaction of an acid and a base to form water and a salt is called neutralization:
  - The general equation for a neutralization reaction is

$$H^+(aq) + OH^-(aq) \longrightarrow H_2O(l)$$

- The quantitative study of a neutralization reaction is called a titration.

15.11 Writing Net Ionic Equations

KEY TERMS

Formula equation
Total ionic equation
Net ionic equation

- In a formula equation the compounds are written in their molecular, or formula, expressions.
- In the total ionic equation compounds are written as ions if they are strong electrolytes in solution and as molecules if they are precipitates, nonelectrolytes, or weak electrolytes in solution.
- The net ionic equation shows only the molecules or ions that have changed:
  - The spectators (nonreactors) are omitted.

15.12 Acid Rain

- Acid rain is any atmospheric precipitation that is more acidic that usual.
- Acid rain causes significant damage and destruction to our environment.

15.13 Colloids

KEY TERM

Colloid

- A colloid is a dispersion containing particles between 1 nm and 1000 nm.
- Colloid particles are usually aggregates of ions or molecules.
- The difference between a colloid and a true solution is the size of the particles.

15.14 Properties of Colloids

KEY TERMS

Brownian movement
Tyndall effect

- The random motion of particles in a colloid is called Brownion movement.
- Colloid particles cause light from an intense beam to be scattered clearly, showing the path of the light through the colloid.

15.15 Application of Colloidal Properties

KEY TERM

Dialysis

- Activated charcoal can be used to adsorb impurities from liquids and gases.
- Dialysis is based on the idea that a true solution will pass through a parchment membrane but a colloid will not.

Review Questions

All questions with blue numbers have answers in the appendix of the text.

1. Since a hydrogen ion and a proton are identical, what differences exist between the Arrhenius and Brønsted–Lowry definitions of an acid? (Table 15.1)
2. According to Figure 15.1, what type of substance must be in solution for the bulb to light?
3. Which of the following classes of compounds are electrolytes: acids, alcohols, bases, salts? (Table 15.2)
4. What two differences are apparent in the arrangement of water molecules about the hydrated ions as depicted in Figure 15.2?
5. The pH of a solution with a hydrogen ion concentration of 0.003 *M* is between what two whole numbers? (Table 15.4)
6. Which is more acidic, tomato juice or blood? (Table 15.5)
7. Use the three acid–base theories (Arrhenius, Brønsted–Lowry, and Lewis) to define an acid and a base.
8. For each acid–base theory referred to in Question 7, write an equation illustrating the neutralization of an acid with a base.
9. Write the Lewis structure for the (a) bromide ion, (b) hydroxide ion, and (c) cyanide ion. Why are these ions considered to be bases according to the Brønsted–Lowry and Lewis acid–base theories?
10. Into what three classes of compounds do electrolytes generally fall?
11. Name each compound listed in Table 15.3.
12. A solution of HCl in water conducts an electric current, but a solution of HCl in hexane does not. Explain this behavior in terms of ionization and chemical bonding.
13. How do ionic compounds exist in their crystalline structure? What occurs when they are dissolved in water?
14. An aqueous methyl alcohol, CH_3OH, solution does not conduct an electric current, but a solution of sodium hydroxide, NaOH, does. What does this information tell us about the OH group in the alcohol?
15. Why does molten NaCl conduct electricity?
16. Explain the difference between dissociation of ionic compounds and ionization of molecular compounds.
17. Distinguish between strong and weak electrolytes.
18. Explain why ions are hydrated in aqueous solutions.
19. What is the main distinction between water solutions of strong and weak electrolytes?
20. What are the relative concentrations of $H^+(aq)$ and $OH^-(aq)$ in (a) a neutral solution, (b) an acid solution, and (c) a basic solution?
21. Write the net ionic equation for the reaction of a strong acid with a water-soluble hydroxide base in an aqueous solution.
22. The solubility of HCl gas in water, a polar solvent, is much greater than its solubility in hexane, a nonpolar solvent. How can you account for this difference?
23. Pure water, containing equal concentrations of both acid and base ions, is neutral. Why?
24. Indicate the fundamental difference between a colloidal dispersion and a true solution.
25. Explain the process of dialysis, giving a practical application in society.
26. A solution with a pH of 7 is neutral. A solution with a pH less than 7 is acidic. A solution with a pH greater than 7 is basic. What do these statements mean?
27. How does acid rain form?
28. Explain the purpose of a titration.

Paired Exercises

All exercises with blue numbers have answers in the appendix of the text.

1. Identify the conjugate acid–base pairs in the following equations:
   (a) $HC_2H_3O_2 + H_2SO_4 \rightleftharpoons H_2C_2H_3O_2^+ + HSO_4^-$
   (b) The two-step ionization of sulfuric acid,
   $$H_2SO_4 + H_2O \longrightarrow H_3O^+ + HSO_4^-$$
   $$HSO_4^- + H_2O \rightleftharpoons H_3O^+ + SO_4^{2-}$$
   (c) $HClO_4 + H_2O \longrightarrow H_3O^+ + ClO_4^-$
   (d) $CH_3O^- + H_3O^+ \longrightarrow CH_3OH + H_2O$

2. Identify the conjugate acid–base pairs in the following equations:
   (a) $HCl + NH_3 \longrightarrow NH_4^+ + Cl^-$
   (b) $HCO_3^- + OH^- \rightleftharpoons CO_3^{2-} + H_2O$
   (c) $HCO_3^- + H_3O^+ \rightleftharpoons H_2CO_3 + H_2O$
   (d) $HC_2H_3O_2 + H_2O \rightleftharpoons H_3O^+ + C_2H_3O_2^-$

3. Complete and balance these equations:
(a) $Zn(s) + HCl(aq) \longrightarrow$
(b) $Al(OH)_3(s) + H_2SO_4(aq) \longrightarrow$
(c) $Na_2CO_3(aq) + HC_2H_3O_2(aq) \longrightarrow$
(d) $MgO(s) + HI(aq) \longrightarrow$
(e) $Ca(HCO_3)_2(s) + HBr(aq) \longrightarrow$
(f) $KOH(aq) + H_3PO_4(aq) \longrightarrow$

5. For each of the formula equations in Question 3, write total and net ionic equations.

7. Which of these compounds are electrolytes? Consider each substance to be mixed with water.
(a) HCl
(b) CO_2
(c) $CaCl_2$
(d) $C_{12}H_{22}O_{11}$ (sugar)
(e) C_3H_7OH (rubbing alcohol)
(f) CCl_4 (insoluble)

9. Calculate the molarity of the ions present in these salt solutions. Assume each salt to be 100% dissociated:
(a) 0.015 *M* NaCl
(b) 4.25 *M* $NaKSO_4$
(c) 0.20 *M* $CaCl_2$
(d) 22.0 g KI in 500. mL of solution

11. In Exercise 9, how many grams of each ion would be present in 100. mL of each solution?

13. Calculate the $[H^+]$ for:
(a) a solution with a pH = 8.5
(b) pure water
(c) a solution with a pH = 2.5

15. What is the molar concentration of all ions present in a solution prepared by mixing the following? (Neglect the concentration of H^+ and OH^- from water and assume that volumes of solutions are additive.)
(a) 30.0 mL of 1.0 *M* NaCl and 40.0 mL of 1.0 *M* NaCl
(b) 30.0 mL of 1.0 *M* HCl and 30.0 mL of 1.0 *M* NaOH
(c) 100.0 mL of 0.40 *M* KOH and 100.0 mL of 0.80 *M* HCl

17. Given the data for the following separate titrations, calculate the molarity of the HCl:

	mL HCl	Molarity HCl	mL NaOH	Molarity NaOH
(a)	40.13	*M*	37.70	0.728
(b)	19.00	*M*	33.66	0.306
(c)	27.25	*M*	18.00	0.555

4. Complete and balance these equations:
(a) $Fe_2O_3(s) + HBr(aq) \longrightarrow$
(b) $Al(s) + H_2SO_4(aq) \longrightarrow$
(c) $NaOH(aq) + H_2CO_3(aq) \longrightarrow$
(d) $Ba(OH)_2(s) + HClO_4(aq) \longrightarrow$
(e) $Mg(s) + HClO_4(aq) \longrightarrow$
(f) $K_2O(s) + HI(aq) \longrightarrow$

6. For each of the formula equations in Question 4, write total and net ionic equations.

8. Which of these compounds are electrolytes? Consider each substance to be mixed with water.
(a) $NaHCO_3$ (baking soda)
(b) N_2 (insoluble gas)
(c) $AgNO_3$
(d) HCOOH (formic acid)
(e) RbOH
(f) K_2CrO_4

10. Calculate the molarity of the ions present in these salt solutions. Assume each salt to be 100% dissociated:
(a) 0.75 *M* $ZnBr_2$
(b) 1.65 *M* $Al_2(SO_4)_3$
(c) 900. g $(NH_4)_2SO_4$ in 20.0 L of solution
(d) 0.0120 g $Mg(ClO_3)_2$ in 1.00 mL of solution

12. In Exercise 10, how many grams of each ion would be present in 100. mL of each solution?

14. Calculate the $[H^+]$ for:
(a) a solution with a pH = 2.4
(b) a solution with a pH = 10.0
(c) tap water with a pH = 6.4

16. What is the molar concentration of all ions present in a solution prepared by mixing the following? (Neglect the concentration of H^+ and OH^- from water and assume volumes of solutions are additive.)
(a) 100.0 mL of 2.0 *M* KCl and 100.0 mL of 1.0 *M* $CaCl_2$
(b) 35.0 mL of 0.20 *M* $Ba(OH)_2$ and 35.0 mL of 0.20 *M* H_2SO_4
(c) 1.00 L of 1.0 *M* $AgNO_3$ and 500. mL of 2.0 *M* NaCl

18. Given the data for the following separate titrations, calculate the molarity of the NaOH:

	mL HCl	Molarity HCl	mL NaOH	Molarity NaOH
(a)	37.19	0.126	31.91	*M*
(b)	48.04	0.482	24.02	*M*
(c)	13.13	1.425	39.39	*M*

19. Rewrite the following unbalanced equations, changing them into balanced net ionic equations (assume that all reactions are in water solution):
(a) $K_2SO_4(aq) + Ba(NO_3)_2(aq) \longrightarrow KNO_3(aq) + BaSO_4(s)$
(b) $CaCO_3(s) + HCl(aq) \longrightarrow CaCl_2(aq) + CO_2(g) + H_2O(l)$
(c) $Mg(s) + HC_2H_3O_2(aq) \longrightarrow Mg(C_2H_3O_2)_2(aq) + H_2(g)$

20. Rewrite the following unbalanced equations, changing them into balanced net ionic equations (assume that all reactions are in water solution):
(a) $H_2S(g) + CdCl_2(aq) \longrightarrow CdS(s) + HCl(aq)$
(b) $Zn(s) + H_2SO_4(aq) \longrightarrow ZnSO_4(aq) + H_2(g)$
(c) $AlCl_3(aq) + Na_3PO_4(aq) \longrightarrow AlPO_4(s) + NaCl(aq)$

21. For each of the given pairs, determine which solution is more acidic. All are water solutions. Explain your answer.
(a) 1 *M* HCl or 1 *M* H_2SO_4?
(b) 1 *M* HCl or 1 *M* $HC_2H_3O_2$?

22. For each of the given pairs, determine which solution is more acidic. All are water solutions. Explain your answer.
(a) 1 *M* HCl or 2 *M* HCl?
(b) 1 *M* HNO_3 or 1 *M* H_2SO_4?

23. What volume (in milliliters) of 0.245 *M* HCl will neutralize 10.0 g $Al(OH)_3$? The equation is

$$3\,HCl(aq) + Al(OH)_3(s) \longrightarrow AlCl_3(aq) + 3\,H_2O(l)$$

24. What volume (in milliliters) of 0.245 *M* HCl will neutralize 50.0 mL of 0.100 *M* $Ca(OH)_2$? The equation is

$$2\,HCl(aq) + Ca(OH)_2(aq) \longrightarrow CaCl_2(aq) + 2\,H_2O(l)$$

*25. A 0.200-g sample of impure NaOH requires 18.25 mL of 0.2406 *M* HCl for neutralization. What is the percent of NaOH in the sample?

*26. A batch of sodium hydroxide was found to contain sodium chloride as an impurity. To determine the amount of impurity, a 1.00-g sample was analyzed and found to require 49.90 mL of 0.466 *M* HCl for neutralization. What is the percent of NaCl in the sample?

*27. What volume of H_2 gas, measured at 27°C and 700. torr, can be obtained by reacting 5.00 g of zinc metal with 100. mL of 0.350 *M* HCl? The equation is

$$Zn(s) + 2\,HCl(aq) \longrightarrow ZnCl_2(aq) + H_2(g)$$

*28. What volume of H_2 gas, measured at 27°C and 700. torr, can be obtained by reacting 5.00 g of zinc metal with 200. mL of 0.350 *M* HCl? The equation is

$$Zn(s) + 2\,HCl(aq) \longrightarrow ZnCl_2(aq) + H_2(g)$$

29. Calculate the pH of solutions having these H^+ ion concentrations:
(a) 0.01 *M*
(b) 1.0 *M*
(c) 6.5×10^{-9} *M*

30. Calculate the pH of solutions having these H^+ ion concentrations:
(a) 1×10^{-7} *M*
(b) 0.50 *M*
(c) 0.00010 *M*

31. Calculate the pH of
(a) orange juice, $3.7 \times 10^{-4}\,M\,H^+$
(b) vinegar, $2.8 \times 10^{-3}\,M\,H^+$

32. Calculate the pH of
(a) black coffee, $5.0 \times 10^{-5}\,M\,H^+$
(b) limewater, $3.4 \times 10^{-11}\,M\,H^+$

33. Determine whether each of the following is a strong acid, weak acid, strong base, or weak base. Then write an equation describing the process that occurs when the substance is dissolved in water.
(a) NH_3
(b) HCl
(c) KOH
(d) $HC_2H_3O_2$

34. Determine whether each of the following is a strong acid, weak acid, strong base, or weak base. Then write an equation describing the process that occurs when the substance is dissolved in water.
(a) $H_2C_2O_4$
(b) $Ba(OH)_2$
(c) $HClO_4$
(d) HBr

Additional Exercises

All exercises with blue numbers have answers in the appendix of the text.

35. Determine whether each of the following describes a substance that is acidic, basic, or neutral.
(a) $[H^+] = 1 \times 10^{-4}$
(b) Phenolphthalein turns pink.
(c) pH = 8
(d) $[H^+] = 1 \times 10^{-9}$
(e) $[OH^-] = 1 \times 10^{-7}$
(f) Blue litmus paper turns red.

36. Draw pictures similar to those found in Figure 15.2 to show what happens to the following compounds when each is dissolved in water.
(a) $CaCl_2$ (b) KF (c) $AlBr_3$

37. What is the concentration of Ca^{2+} ions in a solution of CaI_2 having an I^- ion concentration of 0.520 *M*?

38. If 29.26 mL of 0.430 *M* HCl neutralize 20.40 mL of $Ba(OH)_2$ solution, what is the molarity of the $Ba(OH)_2$ solution? The reaction is

$$Ba(OH)_2(aq) + 2\,HCl(aq) \longrightarrow BaCl_2(aq) + 2\,H_2O(l)$$

39. A 1 *m* solution of acetic acid, $HC_2H_3O_2$, in water freezes at a lower temperature than a 1 *m* solution of ethyl alcohol, C_2H_5OH, in water. Explain.

40. At the same cost per pound, which alcohol, CH_3OH or C_2H_5OH, would be more economical to purchase as an antifreeze for your car? Why?

41. How does a hydronium ion differ from a hydrogen ion?

42. Arrange, in decreasing order of freezing points, 1 *m* aqueous solutions of HCl, $HC_2H_3O_2$, $C_{12}H_{22}O_{11}$ (sucrose), and $CaCl_2$. (List the one with the highest freezing point first.)

43. At 100°C the H^+ concentration in water is about 1×10^{-6} mol/L, about 10 times that of water at 25°C. At which of these temperatures is
(a) the pH of water the greater?
(b) the hydrogen ion (hydronium ion) concentration the higher?
(c) the water neutral?

44. What is the relative difference in H^+ concentration in solutions that differ by one pH unit?

45. What is the mole percent of a 1.00 *m* aqueous solution?

46. A sample of pure sodium carbonate with a mass of 0.452 g was dissolved in water and neutralized with 42.4 mL of hydrochloric acid. Calculate the molarity of the acid:

$$Na_2CO_3(aq) + 2\,HCl(aq) \longrightarrow 2\,NaCl(aq) + CO_2(g) + H_2O(l)$$

47. What volume (mL) of 0.1234 *M* HCl is needed to neutralize 2.00 g $Ca(OH)_2$?

48. How many grams of KOH are required to neutralize 50.00 mL of 0.240 *M* HNO_3?

49. Two drops (0.1 mL) of 1.0 *M* HCl are added to water to make 1.0 L of solution. What is the pH of this solution if the HCl is 100% ionized?

50. What volume of concentrated (18.0 *M*) sulfuric acid must be used to prepare 50.0 L of 5.00 *M* solution?

***51.** If 3.0 g NaOH are added to 500. mL of 0.10 *M* HCl, will the resulting solution be acidic or basic? Show evidence for your answer.

*52. If 380 mL of 0.35 *M* $Ba(OH)_2$ are added to 500. mL of 0.65 *M* HCl, will the mixture be acidic or basic ? Find the pH of the resulting solution.

***53.** If 50.00 mL of 0.2000 *M* HCl are titrated with 0.2000 *M* NaOH, find the pH of the solution after the following amounts of base have been added:
(a) 0.000 mL (d) 49.00 mL (f) 49.99 mL
(b) 10.00 mL (e) 49.90 mL (g) 50.00 mL
(c) 25.00 mL
Plot your answers on a graph with pH on the *y*-axis and mL NaOH on the *x*-axis.

54. Sulfuric acid reacts with NaOH:
(a) Write a balanced equation for the reaction producing Na_2SO_4.
(b) How many milliliters of 0.10 *M* NaOH are needed to react with 0.0050 mol H_2SO_4?
(c) How many grams of Na_2SO_4 will also form?

***55.** A 10.0-mL sample of HNO_3 was diluted to a volume of 100.00 mL. Then, 25 mL of that diluted solution were needed to neutralize 50.0 mL of 0.60 *M* KOH. What was the concentration of the original nitric acid?

56. The pH of a solution of a strong acid was determined to be 3. If water is then added to dilute this solution, would the pH change? Why or why not? Could enough water ever be added to raise the pH of an acid solution above 7?

Challenge Exercises

All exercises with blue numbers have answers in the appendix of the text.

***57.** An HCl solution has a pH of −0.300. What volume of water must be added to 200 mL of this solution to change the pH to −0.150?

***58.** Lactic acid (found in sour milk) has an empirical formula of $HC_3H_5O_3$. A 1.0-g sample of lactic acid required 17.0 mL of 0.65 *M* NaOH to reach the end point of a titration. What is the molecular formula for lactic acid?

Answers to Practice Exercises

15.1 (a) HCO_3^-, (b) NO_2^-, (c) $C_2H_3O_2^-$
15.2 (a) H_2SO_4, (b) NH_4^+, (c) H_2O
15.3 (a) KOH, potassium hydroxide; H_3PO_4, phosphoric acid
(b) $Mg(OH)_2$, magnesium hydroxide; HBr, hydrobromic acid
(c) LiOH, lithium hydroxide; HCl, hydrochloric acid
(d) $Fe(OH)_2$, iron(II) hydroxide; H_2CO_3, carbonic acid
15.4 (a) 0.050 *M* Mg^{2+}, 0.10 *M* Cl^-,
(b) 0.070 *M* Al^{3+}, 0.21 *M* Cl^-
15.5 approximately 101.54°C
15.6 (a) 11.41, (b) 2.89, (c) 5.429
15.7 0.0620 *M* HCl
15.8 $3\,H_2S(aq) + 2\,Bi^{3+}(aq) \longrightarrow Bi_2S_3(s) + 6\,H^+(aq)$

CHAPTER 16

Chemical Equilibrium

A coral reef is a system in ecological equilibrium with the ocean surrounding it.

Chapter Outline

Thus far, we've considered chemical change as proceeding from reactants to products. Does that mean that the change then stops? No, but often it appears to be the case at the macroscopic level. A solute dissolves until the solution becomes saturated. Once a solid remains undissolved in a container, the system appears to be at rest. The human body is a marvelous chemical factory, yet from day to day it appears to be quite the same. For example, the blood remains at a constant pH, even though all sorts of chemical reactions are taking place. Another example is a terrarium, which can be watered and sealed for long periods of time with no ill effects. Or an antacid, which absorbs excess stomach acid and does *not* change the pH of the stomach. In all of these cases, reactions are proceeding, even though visible signs of chemical change are absent. Similarly, when a system is at equilibrium, chemical reactions are dynamic at the molecular level. In this chapter, we will consider chemical systems as they approach equilibrium conditions.

16.1 Reversible Reactions

In the preceding chapters, we treated chemical reactions mainly as reactants changing to products. However, many reactions do not go to completion. Some reactions do not go to completion because they are reversible; that is, when the products are formed, they react to produce the starting reactants.

We've encountered reversible systems before. One is the vaporization of a liquid by heating and its subsequent condensation by cooling:

$$\text{liquid} + \text{heat} \longrightarrow \text{vapor}$$
$$\text{vapor} + \text{cooling} \longrightarrow \text{liquid}$$

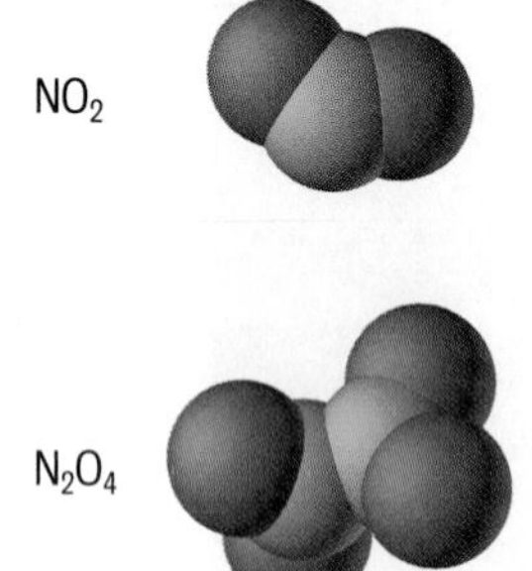

The conversion between nitrogen dioxide, NO_2, and dinitrogen tetroxide, N_2O_4, shows us visible evidence of the reversibility of a reaction. The NO_2 is a reddish-brown gas that changes with cooling to N_2O_4, a colorless gas. The reaction is reversible by heating N_2O_4:

$$2\,NO_2(g) \xrightarrow{\text{cooling}} N_2O_4(g)$$
$$N_2O_4(g) \xrightarrow{\text{heating}} 2\,NO_2(g)$$

These two reactions may be represented by a single equation with a double arrow, $\rightleftharpoons$, to indicate that the reactions are taking place in both directions at the same time:

$$2\,NO_2(g) \rightleftharpoons N_2O_4(g)$$

This reversible reaction can be demonstrated by sealing samples of NO_2 in two tubes and placing one tube in warm water and the other in ice water (see Figure 16.1).

reversible chemical reaction

A **reversible chemical reaction** is one in which the products formed react to produce the original reactants. Both the forward and reverse reactions occur simultaneously. The forward reaction is called *the reaction to the right*, and the reverse reaction is called *the reaction to the left*. A double arrow is used in the equation to indicate that the reaction is reversible.

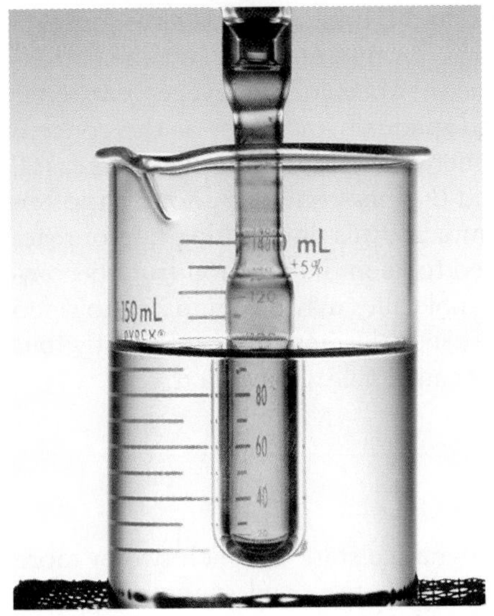

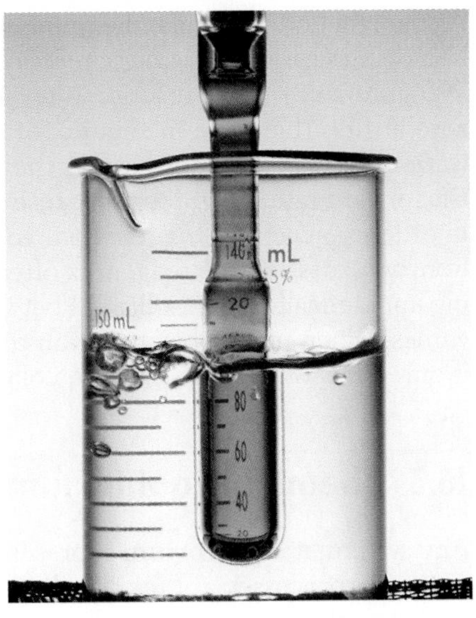

Figure 16.1
Reversible reaction of NO_2 and N_2O_4. More of the dark brown molecules are visible in the heated container on the right than in the room-temperature tube on the left.

16.2 Rates of Reaction

Every reaction has a rate, or speed, at which it proceeds. Some are fast, and some are extremely slow. The study of reaction rates and reaction mechanisms is known as **chemical kinetics.**

chemical kinetics

The rate of a reaction is variable and depends on the concentration of the reacting species, the temperature, the presence of catalysts, and the nature of the reactants. Consider the hypothetical reaction

$$A + B \longrightarrow C + D \quad \text{(forward reaction)}$$
$$C + D \longrightarrow A + B \quad \text{(reverse reaction)}$$

in which a collision between A and B is necessary for a reaction to occur. The rate at which A and B react depends on the concentration, or the number of A and B molecules present; it will be fastest, for a fixed set of conditions, when they are first mixed (as shown by the height of the red line in Figure 16.2).

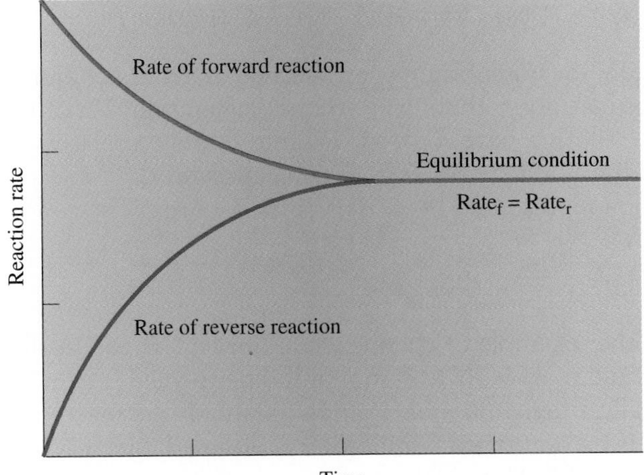

Figure 16.2
The rates of the forward and reverse reactions become equal at some point in time. The forward reaction rate (red) decreases as a result of decreasing amounts of reactants. The reverse reaction rate (blue) starts at zero and increases as the amount of product increases. When the two rates become equal (purple), a state of chemical equilibrium has been reached.

As the reaction proceeds, the number of A and B molecules available for reaction decreases, and the rate of reaction slows down (seen as the red line flattens in Figure 16.2). If the reaction is reversible, the speed of the reverse reaction is zero at first (blue line in Figure 16.2) and gradually increases as the concentrations of C and D increase. As the number of A and B molecules decreases, the forward rate slows down because A and B cannot find one another as often in order to accomplish a reaction. To counteract this diminishing rate of reaction, an excess of one reagent is often used to keep the reaction from becoming impractically slow. Collisions between molecules may be compared to video games. When many objects are on the screen, collisions occur frequently; but if only a few objects are present, collisions can usually be avoided.

16.3 Chemical Equilibrium

equilibrium

chemical equilibrium

Any system at **equilibrium** represents a dynamic state in which two or more opposing processes are taking place at the same time and at the same rate. A chemical equilibrium is a dynamic system in which two or more opposing chemical reactions are going on at the same time and at the same rate. When the rate of the forward reaction is exactly equal to the rate of the reverse reaction, a condition of **chemical equilibrium** exists (see purple line in Figure 16.2). The concentrations of the products and the reactants are not changing, and the system appears to be at a standstill because the products are reacting at the same rate at which they are being formed.

Chemical equilibrium:
rate of forward reaction = rate of reverse reaction

A saturated salt solution is in a condition of equilibrium:

$$NaCl(s) \rightleftharpoons Na^+(aq) + Cl^-(aq)$$

At equilibrium, salt crystals are continuously dissolving, and Na^+ and Cl^- ions are continuously crystallizing. Both processes are occurring at the same rate.

The ionization of weak electrolytes is another chemical equilibrium system:

$$HC_2H_3O_2(aq) + H_2O(l) \rightleftharpoons H_3O^+(aq) + C_2H_3O_2^-(aq)$$

In this reaction, the equilibrium is established in a 1 *M* solution when the forward reaction has gone about 1%—that is, when only 1% of the acetic acid molecules in solution have ionized. Therefore, only a relatively few ions are present, and the acid behaves as a weak electrolyte.

The reaction represented by

$$H_2(g) + I_2(g) \xrightleftharpoons{700\ K} 2\,HI(g)$$

provides another example of chemical equilibrium. Theoretically, 1.00 mol of hydrogen should react with 1.00 mol of iodine to yield 2.00 mol of hydrogen iodide. Actually, when 1.00 mol H_2 and 1.00 mol I_2 are reacted at 700 K, only 1.58 mol HI is present when equilibrium is attained. Since 1.58 is 79% of the theoretical yield of 2.00 mol HI, the forward reaction is only 79% complete at

CHEMISTRY IN ACTION • New Ways in Fighting Cavities and Avoiding the Drill

Dentists have understood for more than 20 years what causes cavities, but, until now, there have been only a limited number of over-the-counter products to help us avoid our dates with the drill. Bacteria in the mouth break down sugars remaining in the mouth after eating. Acids produced during this process slip through tooth enamel, dissolving minerals below the surface in a process called demineralization. Saliva works to rebuild teeth by adding calcium and phosphate back in a process called remineralization. Under ideal conditions (assuming that you brush after eating), these two processes form an equilibrium.

Unfortunately, bacteria in plaque (resulting from not brushing) shift the equilibrium toward demineralization (shown in the figure) and a cavity can begin to form. Scientists realized that fluoride encourages remineralization in teeth by replacing hydroxyl ions in nature's calcium phosphate (hydroxyapatite). The substitution changes the hydroxyapatite to fluorapatite, which is more acid resistant.

The good news is that scientists have now figured out how to shift the equilibrium between demineralization and remineralization toward remineralization. This new understanding has resulted in products that actually stop cavities and prevent the need for a filling. Scientists now know that adding calcium and phosphate to the mouth regularly speeds up remineralization of the teeth. The largest problem in doing it this way is to figure out how to dispense the calcium and phosphate through the surface enamel gradually. If a calcium ion finds a phosphate ion before this happens, a precipitate forms that can't penetrate the enamel. Researchers have developed new products such as chewing gum and sticky substances (see photo) that can be applied between the teeth. These products work better than brushing or rinsing because they last longer—a brushing or rinsing lasts only a minute.

Several types of remineralizing chewing gums are currently available, including Trident Advantage and Trident for Kids. The calcium and phosphate ions are stablilized by using a system that mimics nature. Researchers looked at how newborns get their large supply of calcium and phosphate. Eric Reynolds of the University of Melbourne in Australia says, "Nature's evolved this system in milk to carry very high levels of calcium and phosphate in a highly bioavailable form." They used casein (a milk protein) to produce microscopic particles of stabilized calcium phosphate that diffuse through the surface of teeth. When these particles were placed into the chewing gum and chewed, the results were better than hoped. In Japan, researchers at Tokyo's FAP Dental Institute have designed a toothpaste that repairs cavities smaller than 50 μm. The toothpaste grows nanocrystals of hydroxyapatite, which is treated with fluoride right on the cavity. The synthetic enamel not only repairs the tiny cavity as you brush it but also strengthens the natural enamel. Brushing can now repair cavities and prevent new ones at the same time.

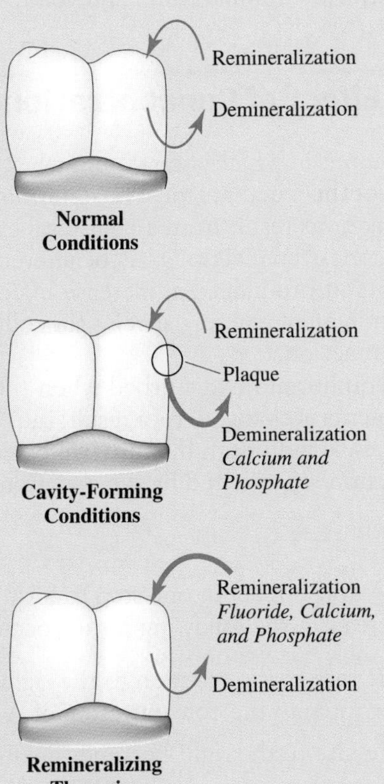

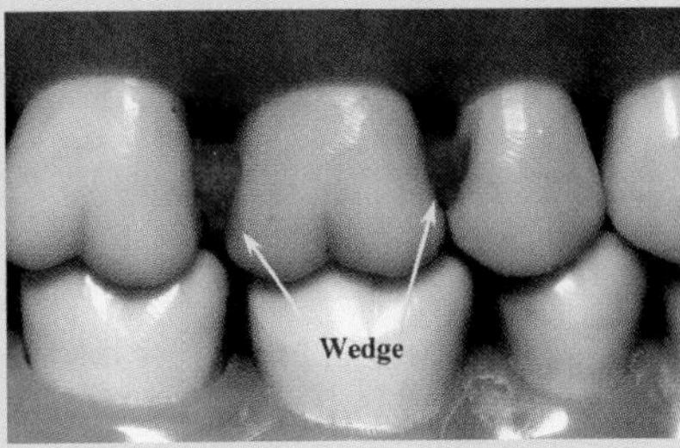

These sticky wedges dissolve slowly to release fluoride, calcium, and phosphate to teeth.

equilibrium. The equilibrium mixture will also contain 0.21 mol each of unreacted H_2 and I_2 (1.00 mol − 0.79 mol = 0.21 mol):

$$H_2(g) + I_2(g) \xrightarrow{700\text{ K}} 2\,HI(g)$$

This equation represents the condition if the reaction were 100% complete; 2.00 mol HI would be formed and no H_2 and I_2 would be left unreacted.

$$\underset{0.21\text{ mol}}{H_2(g)} + \underset{0.21\text{ mol}}{I_2(g)} \overset{700\text{ K}}{\rightleftharpoons} \underset{1.58\text{ mol}}{2\,HI(g)}$$

This equation represents the actual equilibrium attained starting with 1.00 mol each of H_2 and I_2. It shows that the forward reaction is only 79% complete.

16.4 Le Châtelier's Principle

Le Châtelier's principle

In 1888, the French chemist Henri Le Châtelier (1850–1936) set forth a simple, far-reaching generalization on the behavior of equilibrium systems. This generalization, known as **Le Châtelier's principle,** states:

> If a stress is applied to a system in equilibrium, the system will respond in such a way as to relieve that stress and restore equilibrium under a new set of conditions.

The application of Le Châtelier's principle helps us predict the effect of changing conditions in chemical reactions. We will examine the effect of changes in concentration, temperature, and volume.

16.5 Effect of Concentration on Equilibrium

The manner in which the rate of a chemical reaction depends on the concentration of the reactants must be determined experimentally. Many simple, one-step reactions result from a collision between two molecules or ions. The rate of such one-step reactions can be altered by changing the concentration of the reactants or products. An increase in concentration of the reactants provides more individual reacting species for collisions and results in an increase in the rate of reaction.

An equilibrium is disturbed when the concentration of one or more of its components is changed. As a result, the concentration of all species will change, and a new equilibrium mixture will be established. Consider the hypothetical equilibrium represented by the equation

$$A + B \rightleftharpoons C + D$$

where A and B react in one step to form C and D. When the concentration of B is increased, the following results occur:

1. The rate of the reaction to the right (forward) increases. This rate is proportional to the concentration of A times the concentration of B.
2. The rate to the right becomes greater than the rate to the left.
3. Reactants A and B are used faster than they are produced; C and D are produced faster than they are used.
4. After a period of time, rates to the right and left become equal, and the system is again in equilibrium.
5. In the new equilibrium the concentration of A is less, and the concentrations of B, C, and D are greater than in the original equilibrium.
   Conclusion: The equilibrium has shifted to the right.

Concentration	Change
$[H_2]$	?
$[I_2]$	increase
$[HI]$	?

Applying this change in concentration to the equilibrium mixture of 1.00 mol of hydrogen and 1.00 mol of iodine from Section 16.3, we find that, when an additional 0.20 mol I_2 is added, the yield of HI (based on H_2) is 85% (1.70 mol) instead of 79%. Here is how the two systems compare after the new equilibrium mixture is reached:

Original equilibrium	New equilibrium
1.00 mol H_2 + 1.00 mol I_2	1.00 mol H_2 + 1.20 mol I_2
Yield: 79% HI	Yield: 85% HI (based on H_2)
Equilibrium mixture contains:	Equilibrium mixture contains:
1.58 mol HI	1.70 mol HI
0.21 mol H_2	0.15 mol H_2
0.21 mol I_2	0.35 mol I_2

Analyzing this new system, we see that, when 0.20 mol I_2 is added, the equilibrium shifts to the right to counteract the increase in I_2 concentration. Some of the H_2 reacts with added I_2 and produces more HI, until an equilibrium mixture is established again. When I_2 is added, the concentration of I_2 increases, the concentration of H_2 decreases, and the concentration of HI increases.

Concentration	Change
$[H_2]$	decrease
$[I_2]$	increase
[HI]	increase

Practice 16.1

Use a chart like those in the margin to show what would happen to the concentrations of each substance in the system

$$H_2(g) + I_2(g) \rightleftharpoons 2\,HI(g)$$

upon adding (a) more H_2 and (b) more HI.

The equation

$$\underset{\text{pale yellow}}{Fe^{3+}(aq)} + \underset{\text{colorless}}{SCN^-(aq)} \rightleftharpoons \underset{\text{red}}{Fe(SCN)^{2+}(aq)}$$

represents an equilibrium that is used in certain analytical procedures as an indicator because of the readily visible, intense red color of the complex $Fe(SCN)^{2+}$ ion. A very dilute solution of iron(III) (Fe^{3+}) and thiocyanate (SCN^-) is light red. When the concentration of either Fe^{3+} or SCN^- is increased, the equilibrium shift to the right is observed by an increase in the intensity of the color, resulting from the formation of additional $Fe(SCN)^{2+}$.

If either Fe^{3+} or SCN^- is removed from solution, the equilibrium will shift to the left, and the solution will become lighter in color. When Ag^+ is added to the solution, a white precipitate of silver thiocyanate (AgSCN) is formed, thus removing SCN^- ion from the equilibrium:

$$Ag^+(aq) + SCN^-(aq) \rightleftharpoons AgSCN(s)$$

The system accordingly responds to counteract the change in SCN^- concentration by shifting the equilibrium to the left. This shift is evident by a decrease in the intensity of the red color due to a decreased concentration of $Fe(SCN)^{2+}$.

Now consider the effect of changing the concentrations in the equilibrium mixture of chlorine water. The equilibrium equation is

$$Cl_2(aq) + 2\,H_2O(l) \rightleftharpoons HOCl(aq) + H_3O^+(aq) + Cl^-(aq)$$

The variation in concentrations and the equilibrium shifts are tabulated in the following table. An X in the second or third column indicates that the reagent is increased or decreased. The fourth column indicates the direction of the equilibrium shift.

	Concentration		
Reagent	Increase	Decrease	Equilibrium shift
Cl_2	—	X	Left
H_2O	X	—	Right
HOCl	X	—	Left
H_3O^+	—	X	Right
Cl^-	X	—	Left

Consider the equilibrium in a 0.100 *M* acetic acid solution:

$$HC_2H_3O_2(aq) + H_2O(l) \rightleftharpoons H_3O^+(aq) + C_2H_3O_2^-(aq)$$

In this solution, the concentration of the hydronium ion (H_3O^+), which is a measure of the acidity, is 1.34×10^{-3} mol/L, corresponding to a pH of 2.87. What will happen to the acidity when 0.100 mol of sodium acetate ($NaC_2H_3O_2$) is added to 1 L of 0.100 *M* acetic acid ($HC_2H_3O_2$)? When $NaC_2H_3O_2$ dissolves, it dissociates into sodium ions (Na^+) and acetate ions ($C_2H_3O_2^-$). The acetate ion from the salt is a common ion to the acetic acid equilibrium system and increases the total acetate ion concentration in the solution. As a result the equilibrium shifts to the left, decreasing the hydronium ion concentration and lowering the acidity of the solution. Evidence of this decrease in acidity is shown by the fact that the pH of a solution that is 0.100 *M* in $HC_2H_3O_2$ and 0.100 *M* in $NaC_2H_3O_2$ is 4.74. The pH of several different solutions of $HC_2H_3O_2$ and $NaC_2H_3O_2$ is shown in the table that follows. Each time the acetate ion is increased, the pH increases, indicating a further shift in the equilibrium toward un-ionized acetic acid.

Concentration	Change
$[HC_2H_3O_2]$	increase
$[H_3O^+]$	decrease
$[C_2H_3O_2^-]$	increase

Solution	pH
1 L 0.100 *M* $HC_2H_3O_2$	2.87
1 L 0.100 *M* $HC_2H_3O_2$ + 0.100 mol $NaC_2H_3O_2$	4.74
1 L 0.100 *M* $HC_2H_3O_2$ + 0.200 mol $NaC_2H_3O_2$	5.05
1 L 0.100 *M* $HC_2H_3O_2$ + 0.300 mol $NaC_2H_3O_2$	5.23

In summary, we can say that when the concentration of a reagent on the left side of an equation is increased, the equilibrium shifts to the right. When the concentration of a reagent on the right side of an equation is increased, the equilibrium shifts to the left. In accordance with Le Châtelier's principle, the equilibrium always shifts in the direction that tends to reduce the concentration of the added reactant.

Practice 16.2

Aqueous chromate ion, CrO_4^{2-}, exists in equilibrium with aqueous dichromate ion, $Cr_2O_7^{2-}$, in an acidic solution. What effect will (a) increasing the dichromate ion and (b) adding HCl have on the equilibrium?

$$2\,CrO_4^{2-}(aq) + 2\,H^+(aq) \rightleftharpoons Cr_2O_7^{2-}(aq) + H_2O(l)$$

16.6 Effect of Volume on Equilibrium

Changes in volume significantly affect the reaction rate only when one or more of the reactants or products is a gas and the reaction is run in a closed container. In these cases the effect of decreasing the volume of the reacting gases is equivalent to increasing their concentrations. In the reaction

$$CaCO_3(s) \overset{\Delta}{\rightleftharpoons} CaO(s) + CO_2(g)$$

calcium carbonate decomposes into calcium oxide and carbon dioxide when heated about 825°C. Decreasing the volume of the container speeds up the reverse reaction and causes the equilibrium to shift to the left. Decreasing the volume increases the concentration of CO_2, the only gaseous substance in the reaction.

If the volume of the container is decreased, the pressure of the gas will increase. In a system composed entirely of gases, this decrease in the volume of the container will cause the reaction and the equilibrium to shift to the side that contains the smaller number of molecules. To clarify what's happening: When the container volume is decreased, the pressure in the container is increased. The system tries to lower this pressure by reducing the number of molecules. Let's consider an example that shows these effects.

Prior to World War I, Fritz Haber (1868–1934) invented the first major process for the fixation of nitrogen. In this process nitrogen and hydrogen are reacted together in the presence of a catalyst at moderately high temperature and pressure to produce ammonia:

$$\begin{array}{lllll} N_2(g) & + \; 3\,H_2(g) & \rightleftharpoons 2\,NH_3(g) & + & 92.5\text{ kJ} \\ \text{1 mol} & \text{3 mol} & \text{2 mol} & & \\ \text{1 volume} & \text{3 volumes} & \text{2 volumes} & & \end{array}$$

Haber received the Nobel Prize in chemistry for this process in 1918.

The left side of the equation in the Haber process represents 4 mol of gas combining to give 2 mol of gas on the right side of the equation. A decrease in the volume of the container shifts the equilibrium to the right. This decrease in volume results in a higher concentration of both reactants and products. The equilibrium shifts to the right toward fewer molecules.

Gaseous ammonia is often used to add nitrogen to the fields before planting and during early growth.

When the total number of gaseous molecules on both sides of an equation is the same, a change in volume does not cause an equilibrium shift. The following reaction is an example:

$N_2(g)$	+	$O_2(g)$	$\rightleftharpoons$	$2\ NO(g)$
1 mol		1 mol		2 mol
1 volume		1 volume		2 volumes
6.022×10^{23} molecules		6.022×10^{23} molecules		$2(6.022 \times 10^{23})$ molecules

When the volume of the container is decreased, the rate of both the forward and the reverse reactions will increase because of the higher concentrations of N_2, O_2, and NO. But the equilibrium will not shift because the number of molecules is the same on both sides of the equation and the effects on concentration are the same on both forward and reverse rates.

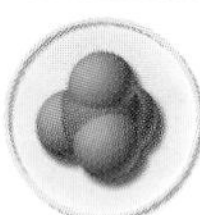

Example 16.1 What effect would a decrease in volume of the container have on the position of equilibrium in these reactions?

(a) $2\ SO_2(g) + O_2(g) \rightleftharpoons 2\ SO_3(g)$
(b) $H_2(g) + Cl_2(g) \rightleftharpoons 2\ HCl(g)$
(c) $N_2O_4(g) \rightleftharpoons 2\ NO_2(g)$

SOLUTION

(a) The equilibrium will shift to the right because the substance on the right has a smaller number of moles than those on the left.
(b) The equilibrium position will be unaffected because the moles of gases on both sides of the equation are the same.
(c) The equilibrium will shift to the left because $N_2O_4(g)$ represents the smaller number of moles.

Practice 16.3

What effect would a decrease in the container's volume have on the position of the equilibrium in these reactions?

(a) $2\ NO(g) + Cl_2(g) \rightleftharpoons 2\ NOCl(g)$
(b) $COBr_2(g) \rightleftharpoons CO(g) + Br_2(g)$

16.7 Effect of Temperature on Equilibrium

High temperatures can cause the destruction or decomposition of the reactants or products.

When the temperature of a system is raised, the rate of reaction increases because of increased kinetic energy and more frequent collisions of the reacting species. In a reversible reaction, the rate of both the forward and the reverse reactions is increased by an increase in temperature; however, the reaction that absorbs heat increases to a greater extent, and the equilibrium shifts to favor that reaction.

An increase in temperature generally increases the rate of reaction. Molecules at elevated temperatures have more kinetic energy; their collisions are thus more likely to result in a reaction.

When heat is applied to a system in equilibrium, the reaction that absorbs heat is favored. When the process, as written, is endothermic, the forward reaction is increased. When the reaction is exothermic, the reverse reaction is favored. In this sense heat may be treated as a reactant in endothermic reactions or as a product in exothermic reactions. Therefore temperature is analogous to concentration when applying Le Châtelier's principle to heat effects on a chemical reaction.

Hot coke (C) is a very reactive element. In the reaction

$$C(s) + CO_2(g) + \text{heat} \rightleftharpoons 2\,CO(g)$$

very little if any CO is formed at room temperature. At 1000°C, the equilibrium mixture contains about an equal number of moles of CO and CO_2. Since the reaction is endothermic, the equilibrium is shifted to the right at higher temperatures.

When phosphorus trichloride reacts with dry chlorine gas to form phosphorus pentachloride, the reaction is exothermic:

$$PCl_3(l) + Cl_2(g) \rightleftharpoons PCl_5(s) + 88\text{ kJ}$$

Heat must continuously be removed during the reaction to obtain a good yield of the product. According to Le Châtelier's principle, heat will cause the product, PCl_5, to decompose, re-forming PCl_3 and Cl_2. The equilibrium mixture at 200°C contains 52% PCl_5, and at 300°C it contains 3% PCl_5, verifying that heat causes the equilibrium to shift to the left.

Light sticks. The chemical reaction that produces light in these light sticks is endothermic. Placing the light stick in hot water (right) favors this reaction, producing a brighter light than when the light stick is in ice water (left).

Example 16.2

What effect would an increase in temperature have on the position of the equilibrium in these reactions?

$$4\,HCl(g) + O_2(g) \rightleftharpoons 2\,H_2O(g) + 2\,Cl_2(g) + 95.4\text{ kJ} \quad (1)$$
$$H_2(g) + Cl_2(g) \rightleftharpoons 2\,HCl(g) + 185\text{ kJ} \quad (2)$$
$$CH_4(g) + 2\,O_2(g) \rightleftharpoons CO_2(g) + 2\,H_2O(g) + 890\text{ kJ} \quad (3)$$
$$N_2O_4(g) + 58.6\text{ kJ} \rightleftharpoons 2\,NO_2(g) \quad (4)$$
$$2\,CO_2(g) + 566\text{ kJ} \rightleftharpoons 2\,CO(g) + O_2(g) \quad (5)$$
$$H_2(g) + I_2(g) + 51.9\text{ kJ} \rightleftharpoons 2\,HI(g) \quad (6)$$

SOLUTION

Reactions 1, 2, and 3 are exothermic; an increase in temperature will cause the equilibrium to shift to the left. Reactions 4, 5, and 6 are endothermic; an increase in temperature will cause the equilibrium to shift to the right.

Practice 16.4

What effect would an increase in temperature have on the position of the equilibrium in these reactions?

(a) $2\,SO_2(g) + O_2(g) \rightleftharpoons 2\,SO_3(g) + 198\text{ kJ}$

(b) $H_2(g) + CO_2(g) + 41\text{ kJ} \rightleftharpoons H_2O(g) + CO(g)$

16.8 Effect of Catalysts on Equilibrium

catalyst

A **catalyst** is a substance that influences the rate of a chemical reaction and can be recovered essentially unchanged at the end of the reaction. A catalyst does not shift the equilibrium of a reaction; it affects only the speed at which the equilibrium is reached. It does this by lowering the activation energy for the

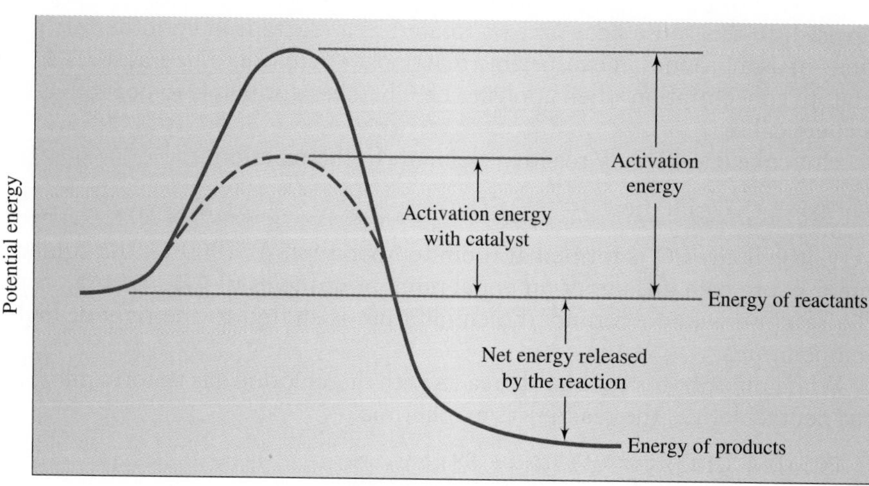

Figure 16.3
Energy diagram for an exothermic reaction. Energy is put into the reaction (activation energy) to initiate the process. In the reaction shown, all of the activation energy and the net energy are released as the reaction proceeds to products. Note that the presence of a catalyst lowers the activation energy but does not change the energies of the reactants or the products.

activation energy

reaction (see Figure 16.3). **Activation energy** is the minimum energy required for the reaction to occur. A catalyst speeds up a reaction by lowering the activation energy while not changing the energies of reactants or products. If a catalyst does not affect the equilibrium, then it follows that it must affect the rate of both the forward and the reverse reactions equally.

The reaction between phosphorus trichloride and sulfur is highly exothermic, but it's so slow that very little product, thiophosphoryl chloride, is obtained, even after prolonged heating. When a catalyst such as aluminum chloride is added, the reaction is complete in a few seconds:

$$PCl_3(l) + S(s) \xrightarrow{AlCl_3} PSCl_3(l)$$

The lab preparation of oxygen uses manganese dioxide as a catalyst to increase the rates of decomposition of both potassium chlorate and hydrogen peroxide:

$$2\,KClO_3(s) \xrightarrow[\Delta]{MnO_2} 2\,KCl(s) + 3\,O_2(g)$$

$$2\,H_2O_2(aq) \xrightarrow{MnO_2} 2\,H_2O(l) + O_2(g)$$

Catalysts are extremely important to industrial chemistry. Hundreds of chemical reactions that are otherwise too slow to be of practical value have been put to commercial use once a suitable catalyst was found. And in the area of biochemistry, catalysts are of supreme importance because nearly all chemical reactions in all forms of life are completely dependent on biochemical catalysts known as *enzymes*.

16.9 Equilibrium Constants

In a reversible chemical reaction at equilibrium, the concentrations of the reactants and products are constant. At equilibrium, the rates of the forward and reverse reactions are equal, and an equilibrium constant expression can be written relating the products to the reactants. For the general reaction

$$aA + bB \rightleftharpoons cC + dD$$

at a given temperature, the equilibrium constant expression can be written as

$$K_{eq} = \frac{[C]^c[D]^d}{[A]^a[B]^b}$$

where K_{eq}, the **equilibrium constant,** is constant at a particular temperature. The quantities in brackets are the concentrations of each substance in moles per liter. The superscript letters *a*, *b*, *c*, and *d* are the coefficients of the substances in the balanced equation. According to convention, we place the concentrations of the products (the substances on the right side of the equation as written) in the numerator and the concentrations of the reactants in the denominator.

equilibrium constant, K_{eq}

Note: The exponents are the same as the coefficients in the balanced equation.

Example 16.3

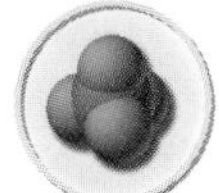

Write equilibrium constant expressions for

(a) $3\,H_2(g) + N_2(g) \rightleftharpoons 2\,NH_3(g)$

(b) $CO(g) + 2\,H_2(g) \rightleftharpoons CH_3OH(g)$

SOLUTION

(a) The only product, NH_3, has a coefficient of 2. Therefore, the numerator will be $[NH_3]^2$. Two reactants are present: H_2, with a coefficient of 3, and N_2, with a coefficient of 1. The denominator will thus be $[H_2]^3[N_2]$. The equilibrium constant expression is

$$K_{eq} = \frac{[NH_3]^2}{[H_2]^3[N_2]}$$

(b) For this equation, the numerator is $[CH_3OH]$ and the denominator is $[CO][H_2]^2$. The equilibrium constant expression is

$$K_{eq} = \frac{[CH_3OH]}{[CO][H_2]^2}$$

Practice 16.5

Write equilibrium constant expressions for

(a) $2\,N_2O_5(g) \rightleftharpoons 4\,NO_2(g) + O_2(g)$

(b) $4\,NH_3(g) + 3\,O_2(g) \rightleftharpoons 2\,N_2(g) + 6\,H_2O(g)$

The magnitude of an equilibrium constant indicates the extent to which the forward and reverse reactions take place. When K_{eq} is greater than 1, the amount of products at equilibrium is greater than the amount of reactants. When K_{eq} is less than 1, the amount of reactants at equilibrium is greater than the amount of products. A very large value for K_{eq} indicates that the forward reaction goes essentially to completion. A very small K_{eq} means that the reverse reaction goes nearly to completion and that the equilibrium is far to the left (toward the reactants). Consider the following two examples:

$$H_2(g) + I_2(g) \rightleftharpoons 2\,HI(g) \qquad K_{eq} = 54.8 \text{ at } 425°C$$

$$COCl_2(g) \rightleftharpoons CO(g) + Cl_2(g) \qquad K_{eq} = 7.6 \times 10^{-4} \text{ at } 400°C$$

In the first example, K_{eq} indicates that more product than reactant exists at equilibrium.

Units are generally not included in values of K_{eq} for reasons beyond the scope of this book.

In the second equation, K_{eq} indicates that $COCl_2$ is stable and that very little decomposition to CO and Cl_2 occurs at 400°C. The equilibrium is far to the left.

When the molar concentrations of all species in an equilibrium reaction are known, the K_{eq} can be calculated by substituting the concentrations into the equilibrium constant expression.

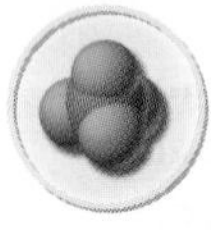

Example 16.4 Calculate K_{eq} for the following reaction based on concentrations of $PCl_5 = 0.030$ mol/L, $PCl_3 = 0.97$ mol/L, and $Cl_2 = 0.97$ mol/L at 300°C.

$$PCl_5(g) \rightleftharpoons PCl_3(g) + Cl_2(g)$$

SOLUTION First write the K_{eq} expression; then substitute the respective concentrations into this equation and solve:

$$K_{eq} = \frac{[PCl_3][Cl_2]}{[PCl_5]} = \frac{(0.97)(0.97)}{(0.030)} = 31$$

Remember: Units are not included for K_{eq}.

This K_{eq} is considered to be a fairly large value, indicating that at 300°C the decomposition of PCl_5 proceeds far to the right.

Practice 16.6

Calculate the K_{eq} for this reaction. Is the forward or the reverse reaction favored?

$$2\,NO(g) + O_2(g) \rightleftharpoons 2\,NO_2(g)$$

when $[NO] = 0.050\,M$, $[O_2] = 0.75\,M$, and $[NO_2] = 0.25\,M$.

16.10 Ion Product Constant for Water

We've seen that water ionizes to a slight degree. This ionization is represented by these equilibrium equations:

$$H_2O + H_2O \rightleftharpoons H_3O^+ + OH^- \qquad (1)$$
$$H_2O \rightleftharpoons H^+ + OH^- \qquad (2)$$

Equation 1 is the more accurate representation of the equilibrium because free protons (H^+) do not exist in water. Equation 2 is a simplified and often-used representation of the water equilibrium. The actual concentration of H^+ produced in pure water is minute and amounts to only 1.00×10^{-7} mol/L at 25°C. In pure water,

$$[H^+] = [OH^-] = 1.00 \times 10^{-7}\ \text{mol/L}$$

since both ions are produced in equal molar amounts, as shown in equation 2.

ion product constant for water, K_w

The $H_2O \rightleftharpoons H^+ + OH^-$ equilibrium exists in water and in all water solutions. A special equilibrium constant called the **ion product constant for water, K_w**, applies to this equilibrium. The constant K_w is defined as the product of the H^+ ion concentration and the OH^- ion concentration, each in moles per liter:

$$K_w = [H^+][OH^-]$$

The numerical value of K_w is 1.00×10^{-14}, since for pure water at 25°C,

$$K_w = [H^+][OH^-] = (1.00 \times 10^{-7})(1.00 \times 10^{-7}) = 1.00 \times 10^{-14}$$

Table 16.1 Relationship of H^+ and OH^- Concentrations in Water Solutions

$[H^+]$	$[OH^-]$	K_w	pH	pOH
1.00×10^{-2}	1.00×10^{-12}	1.00×10^{-14}	2.00	12.00
1.00×10^{-4}	1.00×10^{-10}	1.00×10^{-14}	4.00	10.00
2.00×10^{-6}	5.00×10^{-9}	1.00×10^{-14}	5.70	8.30
1.00×10^{-7}	1.00×10^{-7}	1.00×10^{-14}	7.00	7.00
1.00×10^{-9}	1.00×10^{-5}	1.00×10^{-14}	9.00	5.00

The value of K_w for all water solutions at 25°C is the constant 1.00×10^{-14}. It is important to realize that as the concentration of one of these ions, H^+ or OH^-, increases, the other decreases. However, the product of $[H^+]$ and $[OH^-]$ always equals 1.00×10^{-14}. This relationship can be seen in the examples shown in Table 16.1. If the concentration of one ion is known, the concentration of the other can be calculated from the K_w expression.

$$K_w = [H^+][OH^-] \qquad [H^+] = \frac{K_w}{[OH^-]} \qquad [OH^-] = \frac{K_w}{[H^+]}$$

Example 16.5

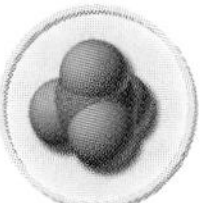

What is the concentration of (a) H^+ and (b) OH^- in a 0.001 *M* HCl solution? Remember that HCl is 100% ionized.

SOLUTION

(a) Since all the HCl is ionized, $H^+ = 0.001$ mol/L, or 1×10^{-3} mol/L:

$$\underset{}{HCl} \longrightarrow \underset{0.001\,M}{H^+} + \underset{0.001\,M}{Cl^-}$$

$$[H^+] = 1 \times 10^{-3}\ \text{mol/L}$$

(b) To calculate the $[OH^-]$ in this solution, use the following equation and substitute the values for K_w and $[H^+]$:

$$[OH^-] = \frac{K_w}{[H^+]}$$

$$[OH^-] = \frac{1.00 \times 10^{-14}}{1 \times 10^{-3}} = 1 \times 10^{-11}\ \text{mol/L}$$

Practice 16.7

Determine the $[H^+]$ and $[OH^-]$ in

(a) 5.0×10^{-5} *M* HNO_3 (b) 2.0×10^{-6} *M* KOH

Example 16.6

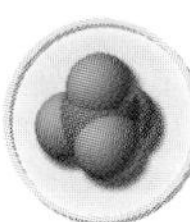

What is the pH of a 0.010 *M* NaOH solution? Assume that NaOH is 100% ionized.

SOLUTION

Since all the NaOH is ionized, $[OH^-] = 0.010$ mol/L or 1.0×10^{-2} mol/L.

$$NaOH \longrightarrow \underset{0.010\,M}{Na^+} + \underset{0.010\,M}{OH^-}$$

To find the pH of the solution, we first calculate the H^+ ion concentration. Use the following equation and substitute the values for K_w and $[OH^-]$:

$$[H^+] = \frac{K_w}{[OH^-]} = \frac{1.00 \times 10^{-14}}{1.0 \times 10^{-2}} = 1.0 \times 10^{-12}\ \text{mol/L}$$

$$pH = -\log[H^+] = -\log(1.0 \times 10^{-12}) = 12.00$$

Practice 16.8

Determine the pH for the following solutions:

(a) $5.0 \times 10^{-5}\ M\ HNO_3$ (b) $2.0 \times 10^{-6}\ M\ KOH$

Just as pH is used to express the acidity of a solution, pOH is used to express the basicity of an aqueous solution. The pOH is related to the OH^- ion concentration in the same way that the pH is related to the H^+ ion concentration:

$$pOH = -\log[OH^-]$$

Thus a solution in which $[OH^-] = 1.0 \times 10^{-2}$, as in Example 16.6, will have $pOH = 2.00$.

In pure water, where $[H^+] = 1.00 \times 10^{-7}$ and $[OH^-] = 1.00 \times 10^{-7}$, the pH is 7.0 and the pOH is 7.0. The sum of the pH and pOH is always 14.0:

$$pH + pOH = 14.00$$

In Example 16.6, the pH can also be found by first calculating the pOH from the OH^- ion concentration and then subtracting from 14.00.

$$pH = 14.00 - pOH = 14.00 - 2.00 = 12.00$$

Table 16.1 summarizes the relationship between $[H^+]$ and $[OH^-]$ in water solutions.

16.11 Ionization Constants

In addition to K_w, several other equilibrium constants are commonly used. Strong acids are essentially 100% ionized. Weak acids are only slightly ionized. Let's consider the equilibrium constant for acetic acid in solution. Because it is a weak acid, an equilibrium is established between molecular $HC_2H_3O_2$ and its ions in solution:

$$HC_2H_3O_2(aq) \rightleftharpoons H^+(aq) + C_2H_3O_2^-(aq)$$

The ionization constant expression is the concentration of the products divided by the concentration of the reactants:

$$K_a = \frac{[H^+][C_2H_3O_2^-]}{[HC_2H_3O_2]}$$

acid ionization constant, K_a

The constant is called the **acid ionization constant, K_a**, a special type of equilibrium constant. The concentration of water in the solution is large compared to other concentrations and does not change appreciably. It is therefore part of the constant K_{eq}.

At 25°C, a 0.100 M $HC_2H_3O_2$ solution is 1.34% ionized and has an $[H^+]$ of 1.34×10^{-3} mol/L. From this information we can calculate the ionization constant for acetic acid.

A 0.100 *M* solution initially contains 0.100 mol of acetic acid per liter. Of this 0.100 mol, only 1.34%, or 1.34×10^{-3} mol, is ionized, which gives an $[H^+] = 1.34 \times 10^{-3}$ mol/L. Because each molecule of acid that ionizes yields one H^+ and one $C_2H_3O_2^-$, the concentration of $C_2H_3O_2^-$ ions is also 1.34×10^{-3} mol/L. This ionization leaves $0.100 - 0.00134 = 0.099$ mol/L of un-ionized acetic acid.

Acid	Initial concentration (mol/L)	Equilibrium concentration (mol/L)
$[HC_2H_3O_2]$	0.100	0.099
$[H^+]$	0	0.00134
$[C_2H_3O_2^-]$	0	0.00134

Substituting these concentrations in the equilibrium expression, we obtain the value for K_a:

$$K_a = \frac{[H^+][C_2H_3O_2^-]}{[HC_2H_3O_2]} = \frac{(1.34 \times 10^{-3})(1.34 \times 10^{-3})}{(0.099)} = 1.8 \times 10^{-5}$$

The K_a for acetic acid, 1.8×10^{-5}, is small and indicates that the position of the equilibrium is far toward the un-ionized acetic acid. In fact, a 0.100 *M* acetic acid solution is 99% un-ionized.

Once the K_a for acetic acid is established, it can be used to describe other systems containing H^+, $C_2H_3O_2^-$, and $HC_2H_3O_2$ in equilibrium at 25°C. The ionization constants for several other weak acids are listed in Table 16.2.

Table 16.2 Ionization Constants (K_a) of Weak Acids at 25°C

Acid	Formula	K_a	Acid	Formula	K_a
Acetic	$HC_2H_3O_2$	1.8×10^{-5}	Hydrocyanic	HCN	4.0×10^{-10}
Benzoic	$HC_7H_5O_2$	6.3×10^{-5}	Hypochlorous	HClO	3.5×10^{-8}
Carbolic (phenol)	HC_6H_5O	1.3×10^{-10}	Nitrous	HNO_2	4.5×10^{-4}
Cyanic	HCNO	2.0×10^{-4}	Hydrofluoric	HF	6.5×10^{-4}
Formic	$HCHO_2$	1.8×10^{-4}			

Example 16.7

What is the $[H^+]$ in a 0.50 *M* $HC_2H_3O_2$ solution? The ionization constant, K_a, for $HC_2H_3O_2$ is 1.8×10^{-5}.

SOLUTION

To solve this problem, first write the equilibrium equation and the K_a expression:

$$HC_2H_3O_2 \rightleftharpoons H^+ + C_2H_3O_2^- \qquad K_a = \frac{[H^+][C_2H_3O_2^-]}{[HC_2H_3O_2]} = 1.8 \times 10^{-5}$$

We know that the initial concentration of $HC_2H_3O_2$ is 0.50 *M*. We also know from the ionization equation that one $C_2H_3O_2^-$ is produced for every H^+ produced; that is, the $[H^+]$ and the $[C_2H_3O_2^-]$ are equal. To solve, let $Y = [H^+]$, which also equals the $[C_2H_3O_2^-]$. The un-ionized $[HC_2H_3O_2]$ remaining will then be $0.50 - Y$, the starting concentration minus the amount that ionized:

	Initial	Equilibrium
$[H^+]$	0	Y
$[C_2H_3O_2^-]$	0	Y
$[HC_2H_3O_2]$	0.5	$0.5 - Y$

$$[H^+] = [C_2H_3O_2^-] = Y \qquad [HC_2H_3O_2] = 0.50 - Y$$

Substituting these values into the K_a expression, we obtain

$$K_a = \frac{(Y)(Y)}{0.50 - Y} = \frac{Y^2}{0.50 - Y} = 1.8 \times 10^{-5}$$

The quadratic equation is $y = \frac{-b \pm \sqrt{b^2 - 4ac}}{2a}$ **for the equation** $ay^2 + by + c = 0$

An exact solution of this equation for Y requires the use of a mathematical equation known as the *quadratic equation*. However, an approximate solution is obtained if we assume that Y is small and can be neglected compared with 0.50. Then $0.50 - Y$ will be equal to approximately 0.50. The equation now becomes

$$\frac{Y^2}{0.50} = 1.8 \times 10^{-5}$$

$$Y^2 = 0.50 \times 1.8 \times 10^{-5} = 0.90 \times 10^{-5} = 9.0 \times 10^{-6}$$

Taking the square root of both sides of the equation, we obtain

$$Y = \sqrt{9.0 \times 10^{-6}} = 3.0 \times 10^{-3}\ \text{mol/L}$$

Thus, the $[H^+]$ is approximately 3.0×10^{-3} mol/L in a 0.50 M $HC_2H_3O_2$ solution. The exact solution to this problem, using the quadratic equation, gives a value of 2.99×10^{-3} mol/L for $[H^+]$, showing that we were justified in neglecting Y compared with 0.50.

Practice 16.9

Calculate the hydrogen ion concentration in (a) 0.100 M hydrocyanic acid (HCN) solution and (b) 0.0250 M carbolic acid (HC_6H_5O) solution.

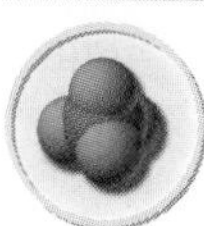

Example 16.8 Calculate the percent ionization in a 0.50 M $HC_2H_3O_2$ solution.

SOLUTION The percent ionization of a weak acid, $HA(aq) \rightleftharpoons H^+(aq) + A^-(aq)$, is found by dividing the concentration of the H^+ or A^- ions at equilibrium by the initial concentration of HA and multiplying by 100. For acetic acid,

$$\frac{\text{concentration of } [H^+] \text{ or } [C_2H_3O_2^-]}{\text{initial concentration of } [HC_2H_3O_2]} \times 100 = \text{percent ionized}$$

To solve this problem, we first need to calculate the $[H^+]$. This calculation has already been done for a 0.50 M $HC_2H_3O_2$ solution:

$$[H^+] = 3.0 \times 10^{-3}\ \text{mol/L in a } 0.50\ M \text{ solution} \quad \text{(from Example 16.7)}$$

This $[H^+]$ represents a fractional amount of the initial 0.50 M $HC_2H_3O_2$. Therefore

$$\frac{3.0 \times 10^{-3}\ \text{mol/L}}{0.50\ \text{mol/L}} \times 100 = 0.60\%\ \text{ionized}$$

A 0.50 M $HC_2H_3O_2$ solution is 0.60% ionized.

Practice 16.10

Calculate the percent ionization for
(a) 0.100 M hydrocyanic acid (HCN)
(b) 0.0250 M carbolic acid (HC_6H_5O)

16.12 Solubility Product Constant

solubility product constant, K_{sp}

The **solubility product constant, K_{sp}**, is the equilibrium constant of a slightly soluble salt. To evaluate K_{sp}, consider this example. The solubility of AgCl in water is 1.3×10^{-5} mol/L at 25°C. The equation for the equilibrium between AgCl and its ions in solution is

$$AgCl(s) \rightleftharpoons Ag^+(aq) + Cl^-(aq)$$

The equilibrium constant expression is

$$K_{eq} = \frac{[Ag^+][Cl^-]}{[AgCl(s)]}$$

The amount of solid AgCl does not affect the equilibrium system provided that some is present. In other words, the concentration of solid AgCl is constant whether 1 mg or 10 g of the salt are present. Therefore the product obtained by multiplying the two constants K_{eq} and $[AgCl(s)]$ is also a constant. This is the solubility product constant, K_{sp}:

$$K_{eq} \times [AgCl(s)] = [Ag^+][Cl^-] = K_{sp}$$

$$K_{sp} = [Ag^+][Cl^-]$$

The K_{sp} is equal to the product of the $[Ag^+]$ and the $[Cl^-]$, each in moles per liter. When 1.3×10^{-5} mol/L of AgCl dissolves, it produces 1.3×10^{-5} mol/L each of Ag^+ and Cl^-. From these concentrations, the K_{sp} can be calculated:

$$[Ag^+] = 1.3 \times 10^{-5}\ \text{mol/L} \qquad [Cl^-] = 1.3 \times 10^{-5}\ \text{mol/L}$$

$$K_{sp} = [Ag^+][Cl^-] = (1.3 \times 10^{-5})(1.3 \times 10^{-5}) = 1.7 \times 10^{-10}$$

Once the K_{sp} value for AgCl is established, it can be used to describe other systems containing Ag^+ and Cl^-.

The K_{sp} expression does not have a denominator. It consists only of the concentrations (mol/L) of the ions in solution. As in other equilibrium expressions, each of these concentrations is raised to a power that is the same number as its coefficient in the balanced equation. Here are equilibrium equations and the K_{sp} expressions for several other substances:

$$AgBr(s) \rightleftharpoons Ag^+(aq) + Br^-(aq) \qquad K_{sp} = [Ag^+][Br^-]$$

$$BaSO_4(s) \rightleftharpoons Ba^{2+}(aq) + SO_4^{2-}(aq) \qquad K_{sp} = [Ba^{2+}][SO_4^{2-}]$$

$$Ag_2CrO_4(s) \rightleftharpoons 2\,Ag^+(aq) + CrO_4^{2-}(aq) \qquad K_{sp} = [Ag^+]^2[CrO_4^{2-}]$$

$$CuS(s) \rightleftharpoons Cu^{2+}(aq) + S^{2-}(aq) \qquad K_{sp} = [Cu^{2+}][S^{2-}]$$

$$Mn(OH)_2(s) \rightleftharpoons Mn^{2+}(aq) + 2\,OH^-(aq) \qquad K_{sp} = [Mn^{2+}][OH^-]^2$$

$$Fe(OH)_3(s) \rightleftharpoons Fe^{3+}(aq) + 3\,OH^-(aq) \qquad K_{sp} = [Fe^{3+}][OH^-]^3$$

Table 16.3 lists K_{sp} values for these and several other substances.

> When the product of the molar concentration of the ions in solution (each raised to its proper power) is greater than the K_{sp} for that substance, precipitation will occur. If the ion product is less than the K_{sp} value, no precipitation will occur.

Table 16.3 Solubility Product Constants (K_{sp}) at 25°C

Compound	K_{sp}	Compound	K_{sp}
AgCl	1.7×10^{-10}	CaF_2	3.9×10^{-11}
AgBr	5.2×10^{-13}	CuS	8.5×10^{-45}
AgI	8.5×10^{-17}	$Fe(OH)_3$	6.1×10^{-38}
$AgC_2H_3O_2$	2.1×10^{-3}	PbS	3.4×10^{-28}
Ag_2CrO_4	1.9×10^{-12}	$PbSO_4$	1.3×10^{-8}
$BaCrO_4$	8.5×10^{-11}	$Mn(OH)_2$	2.0×10^{-13}
$BaSO_4$	1.5×10^{-9}		

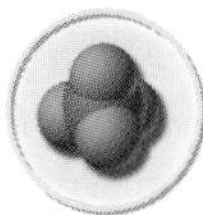

Example 16.9 Write K_{sp} expressions for AgI and PbI_2, both of which are slightly soluble salts.

SOLUTION First write the equilibrium equations:

$$AgI(s) \rightleftharpoons Ag^+(aq) + I^-(aq)$$

$$PbI_2(s) \rightleftharpoons Pb^{2+}(aq) + 2\,I^-(aq)$$

Since the concentration of the solid crystals is constant, the K_{sp} equals the product of the molar concentrations of the ions in solution. In the case of PbI_2, the $[I^-]$ must be squared:

$$K_{sp} = [Ag^+][I^-] \qquad K_{sp} = [Pb^{2+}][I^-]^2$$

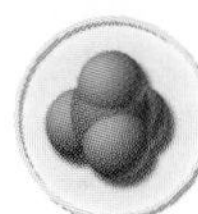

Example 16.10 The K_{sp} value for lead sulfate is 1.3×10^{-8}. Calculate the solubility of $PbSO_4$ in grams per liter.

SOLUTION First write the equilibrium equation and the K_{sp} expression:

$$PbSO_4 \rightleftharpoons Pb^{2+}(aq) + SO_4^{2-}(aq)$$

$$K_{sp} = [Pb^{2+}][SO_4^{2-}] = 1.3 \times 10^{-8}$$

Since the lead sulfate that is in solution is completely dissociated, the $[Pb^{2+}]$ or $[SO_4^{2-}]$ is equal to the solubility of $PbSO_4$ in moles per liter. Let

$$Y = [Pb^{2+}] = [SO_4^{2-}]$$

Substitute Y into the K_{sp} equation and solve:

$$[Pb^{2+}][SO_4^{2-}] = (Y)(Y) = 1.3 \times 10^{-8}$$

$$Y^2 = 1.3 \times 10^{-8}$$

$$Y = 1.1 \times 10^{-4}\ \text{mol/L}$$

The solubility of $PbSO_4$ therefore is 1.1×10^{-4} mol/L. Now convert mol/L to g/L:

1 mol of $PbSO_4$ has a mass of (207.2 g + 32.07 g + 64.00 g) 303.3 g

$$\left(\frac{1.1 \times 10^{-4}\ \cancel{\text{mol}}}{\text{L}}\right)\left(\frac{303.3\ \text{g}}{\cancel{\text{mol}}}\right) = 3.3 \times 10^{-2}\ \text{g/L}$$

The solubility of $PbSO_4$ is 3.3×10^{-2} g/L.

Practice 16.11

Write the K_{sp} expression for
(a) $Cr(OH)_3$ (b) $Cu_3(PO_4)_2$

Practice 16.12

The K_{sp} value for CuS is 8.5×10^{-45}. Calculate the solubility of CuS in grams per liter.

An ion added to a solution already containing that ion is called a *common ion*. When a common ion is added to an equilibrium solution of a weak electrolyte or a slightly soluble salt, the equilibrium shifts according to Le Châtelier's principle. For example, when silver nitrate ($AgNO_3$) is added to a saturated solution of AgCl,

$$AgCl(s) \rightleftharpoons Ag^+ + Cl^-$$

the equilibrium shifts to the left due to the increase in the $[Ag^+]$. As a result, the $[Cl^-]$ and the solubility of AgCl decreases. The AgCl and $AgNO_3$ have the common ion Ag^+. A shift in the equilibrium position upon addition of an ion already contained in the solution is known as the **common ion effect.**

common ion effect

Example 16.11

Silver nitrate is added to a saturated AgCl solution until the $[Ag^+]$ is 0.10 *M*. What will be the $[Cl^-]$ remaining in solution?

SOLUTION

This is an example of the common ion effect. The addition of $AgNO_3$ puts more Ag^+ in solution; the Ag^+ combines with Cl^- and causes the equilibrium to shift to the left, reducing the $[Cl^-]$ in solution. After the addition of Ag^+ to the mixture, the $[Ag^+]$ and $[Cl^-]$ in solution are no longer equal.

We use the K_{sp} to calculate the $[Cl^-]$ remaining in solution. The K_{sp} is constant at a particular temperature and remains the same no matter how we change the concentration of the species involved:

$$K_{sp} = [Ag^+][Cl^-] = 1.7 \times 10^{-10} \qquad [Ag^+] = 0.10\ \text{mol/L}$$

We then substitute the $[Ag^+]$ into the K_{sp} expression and calculate the $[Cl^-]$:

$$[0.10][Cl^-] = 1.7 \times 10^{-10}$$

$$[Cl^-] = \frac{1.7 \times 10^{-10}}{0.10} = 1.7 \times 10^{-9}\ \text{mol/L}$$

This calculation shows a 10,000-fold reduction of Cl^- ions in solution. It illustrates that Cl^- ions may be quantitatively removed from solution with an excess of Ag^+ ions.

Practice 16.13

Sodium sulfate (Na_2SO_4) is added to a saturated solution of $BaSO_4$ until the concentration of the sulfate ion is 2.0×10^{-2} *M*. What will be the concentration of the Ba^{2+} ions remaining in solution?

16.13 Acid–Base Properties of Salts

hydrolysis **Hydrolysis** is the term used for the general reaction in which a water molecule is split. For example, the net ionic hydrolysis reaction for a sodium acetate solution is

$$C_2H_3O_2^-(aq) + HOH(l) \rightleftharpoons HC_2H_3O_2(aq) + OH^-(aq)$$

In this reaction the water molecule is split, with the H^+ combining with $C_2H_3O_2^-$ to give the weak acid $HC_2H_3O_2$ and the OH^- going into solution, making the solution more basic.

Salts that contain an ion of a weak acid undergo hydrolysis. For example, a 0.10 *M* NaCN solution has a pH of 11.1. The hydrolysis reaction that causes this solution to be basic is

$$CN^-(aq) + HOH(l) \rightleftharpoons HCN(aq) + OH^-(aq)$$

If a salt contains the ion of a weak base, the ion produces an acidic solution by transferring a proton to water. An example is ammonium chloride, which produces the NH_4^+ and Cl^- in solution. The NH_4^+ dissociates to give an acidic solution:

$$NH_4^+(aq) + H_2O(l) \rightleftharpoons NH_3(aq) + H_3O^+(aq)$$

The ions of a salt derived from a strong acid and a strong base, such as NaCl, do not undergo hydrolysis and thus form neutral solutions. Table 16.4 lists the ionic composition of various salts and the nature of the aqueous solutions that they form.

Table 16.4 Ionic Composition of Salts and the Nature of the Aqueous Solutions They Form

Type of salt	Nature of aqueous solution	Examples
Weak base–strong acid	Acidic	NH_4Cl, NH_4NO_3
Strong base–weak acid	Basic	$NaC_2H_3O_2$, K_2CO_3
Weak base–weak acid	Depends on the salt	$NH_4C_2H_3O_2$, NH_4NO_2
Strong base–strong acid	Neutral	NaCl, KBr

Practice 16.14

Indicate whether these salts would produce an acidic, basic, or neutral aqueous solution:
(a) KCN (b) $NaNO_3$ (c) NH_4Br

16.14 Buffer Solutions: The Control of pH

The control of pH within narrow limits is critically important in many chemical applications and vitally important in many biological systems. For example, human blood must be maintained between pH 7.35 and 7.45 for the efficient transport of oxygen from the lungs to the cells. This narrow pH range is maintained by buffer systems in the blood.

buffer solution A **buffer solution** resists changes in pH when diluted or when small amounts of acid or base are added. Two common types of buffer solutions are (1) a weak acid mixed with a salt of its conjugate base and (2) a weak base mixed with a salt of its conjugate acid.

CHEMISTRY IN ACTION • Exchange of Oxygen and Carbon Dioxide in the Blood

The transport of oxygen and carbon dioxide between the lungs and tissues is a complex process that involves several reversible reactions, each of which behaves in accordance with Le Châtelier's principle.

The binding of oxygen to hemoglobin is a reversible reaction. The oxygen molecule must attach to the hemoglobin (Hb) and then later detach. The equilibrium equation for this reaction can be written:

$$Hb + O_2 \rightleftharpoons HbO_2$$

In the lungs the concentration of oxygen is high and favors the forward reaction. Oxygen quickly binds to the hemoglobin until it is saturated with oxygen.

In the tissues the concentration of oxygen is lower and in accordance with Le Châtelier's principle: The equilibrium position shifts to the left and the hemoglobin releases oxygen to the tissues. Approximately 45% of the oxygen diffuses out of the capillaries into the tissues, where it may be picked up by *myoglobin*, another carrier molecule.

Myoglobin functions as an oxygen-storage molecule, holding the oxygen until it is required in the energy-producing portions of the cell. The reaction between myoglobin (Mb) and oxygen can be written as an equilibrium reaction:

$$Mb + O_2 \rightleftharpoons MbO_2$$

The hemoglobin and myoglobin equations are very similar, so what accounts for the transfer of the oxygen from the hemoglobin to the myoglobin? Although both equilibria involve similar interactions, the affinity between oxygen and hemoglobin is different from the affinity between myoglobin and oxygen. In the tissues the position of the hemoglobin equilibrium is such that it is 55% saturated with oxygen, whereas the myoglobin is at 90% oxygen saturation. Under these conditions hemoglobin will release oxygen, while myoglobin will bind oxygen. Thus oxygen is loaded onto hemoglobin in the lungs and unloaded in the tissue's cells.

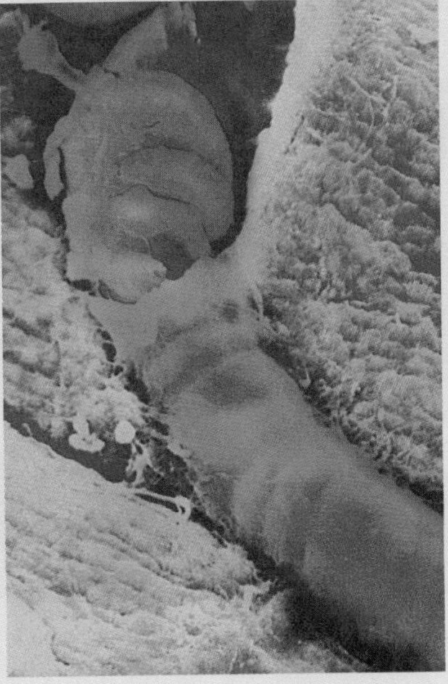

Oxygen and carbon dioxide are exchanged in the red blood cells when they are in capillaries.

Carbon dioxide produced in the cells must be removed from the tissues. Oxygen-depleted hemoglobin molecules accomplish this by becoming carriers of carbon dioxide. The carbon dioxide does not bind at the heme site as the oxygen does, but rather at one end of the protein chain. When carbon dioxide dissolves in water, some of the CO_2 reacts to release hydrogen ions:

$$CO_2 + H_2O \rightleftharpoons HCO_3^- + H^+$$

To facilitate the removal of CO_2 from the tissues, this equilibrium needs to be moved toward the right. This shift is accomplished by the removal of H^+ from the tissues by the hemoglobin molecule. The deoxygenated hemoglobin molecule can bind H^+ ions as well as CO_2. In the lungs this whole process is reversed so the CO_2 is removed from the hemoglobin and exhaled.

Molecules that are similar in structure to the oxygen molecule can become involved in competing equilibria. Hemoglobin is capable of binding with carbon monoxide (CO), nitrogen monoxide (NO), and cyanide (CN^-). The extent of the competition depends on the affinity. Since these molecules have a greater affinity for hemoglobin than oxygen, they will effectively displace oxygen from hemoglobin. For example,

$$HbO_2 + CO \rightleftharpoons HbCO + O_2$$

Since the affinity of hemoglobin for CO is 150 times stronger than its affinity for oxygen, the equilibrium position lies far to the right. This explains why CO is a poisonous substance and why oxygen is administered to victims of CO poisoning. The hemoglobin molecules can only transport oxygen if the CO is released and the oxygen shifts the equilibrium toward the left.

The action of a buffer system can be understood by considering a solution of acetic acid and sodium acetate. The weak acid ($HC_2H_3O_2$) is mostly un-ionized and is in equilibrium with its ions in solution. The sodium acetate is completely ionized:

$$HC_2H_3O_2(aq) \rightleftharpoons H^+(aq) + C_2H_3O_2^-(aq)$$

$$NaC_2H_3O_2(aq) \longrightarrow Na^+(aq) + C_2H_3O_2^-(aq)$$

Because the sodium acetate is completely ionized, the solution contains a much higher concentration of acetate ions than would be present if only acetic acid were in solution. The acetate ion represses the ionization of acetic acid and

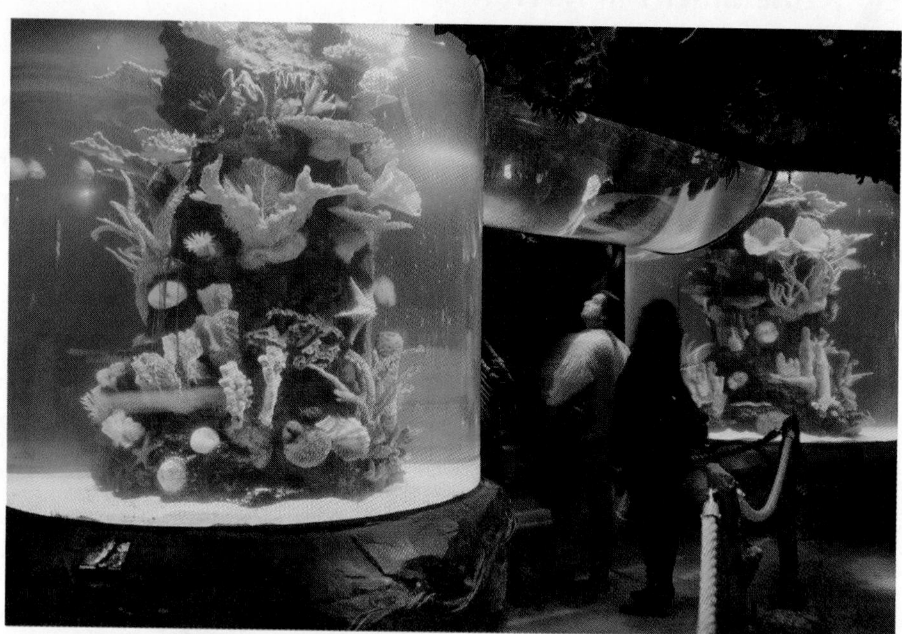

The aquariums at the MGM Grand Hotel are a buffer system.

also reacts with water, causing the solution to have a higher pH (be more basic) than an acetic acid solution (see Section 16.5). Thus, a 0.1 *M* acetic acid solution has a pH of 2.87, but a solution that is 0.1 *M* in acetic acid and 0.1 *M* in sodium acetate has a pH of 4.74. This difference in pH is the result of the common ion effect.

A buffer solution has a built-in mechanism that counteracts the effect of adding acid or base. Consider the effect of adding HCl or NaOH to an acetic acid–sodium acetate buffer. When a small amount of HCl is added, the acetate ions of the buffer combine with the H^+ ions from HCl to form un-ionized acetic acid, thus neutralizing the added acid and maintaining the approximate pH of the solution. When NaOH is added, the OH^- ions react with acetic acid to neutralize the added base and thus maintain the approximate pH. The equations for these reactions are

$$H^+(aq) + C_2H_3O_2^-(aq) \rightleftharpoons HC_2H_3O_2(aq)$$

$$OH^-(aq) + HC_2H_3O_2(aq) \rightleftharpoons H_2O(l) + C_2H_3O_2^-(aq)$$

Data comparing the changes in pH caused by adding HCl and NaOH to pure water and to an acetic acid–sodium acetate buffer solution are shown in Table 16.5. Notice that adding 0.01 mol HCl or 0.01 mol NaOH to 100 mL water changes the pH of the solution by 5 units. But adding the same amount of HCl or NaOH to a buffer solution causes a change of only 0.08 or 0.09 pH units. So buffers really protect the pH of a solution.

Buffers cannot withstand the addition of large amounts of acids or bases. They overpower the capacity of the buffer to absorb acid or base. The maximum buffering effect occurs when the weak acid or base and its conjugate are of equal molar concentrations.

The human body has a number of buffer systems. One of these, the hydrogen carbonate–carbonic acid buffer, $HCO_3^-—H_2CO_3$, maintains the blood plasma at a pH of 7.4. The phosphate system, $HPO_4^{2-}—H_2PO_4^-$, is an important buffer in the red blood cells, as well as in other places in the body.

Table 16.5 Changes in pH Caused by the Addition of HCl and NaOH

Solution	pH	Change in pH
H_2O (1000 mL)	7	—
H_2O + 0.010 mol HCl	2	5
H_2O + 0.010 mol NaOH	12	5
Buffer solution (1000 mL)		
0.10 M $HC_2H_3O_2$ + 0.10 M $NaC_2H_3O_2$	4.74	—
Buffer + 0.010 mol HCl	4.66	0.08
Buffer + 0.010 mol NaOH	4.83	0.09

Chapter 16 Review

16.1 Reversible Reactions

KEY TERM

Reversible chemical reaction

- In a reversible chemical reaction, the products formed react to produce the original reaction mixture.
- The forward reaction is called the reaction to the right.
- The reverse reaction is called the reaction to the left.

16.2 Rates of Reaction

KEY TERM

Chemical kinetics

- The rate of a reaction is variable and depends on:
  - Concentration of reacting species
  - Temperature
  - Presence of catalysts
  - Nature of the reactants

16.3 Chemical Equilibrium

KEY TERMS

Equilibrium
Chemical equilibrium

- A system at equilibrium is dynamic.
- At chemical equilibrium, the rate of the forward reaction equals the rate of the reverse reaction.

16.4 Le Châtelier's Principle

KEY TERM

Le Châtelier's principle

- If stress is applied to a system at equilibrium, the system will respond in such a way as to relive the stress and restore equilibrium under a new set of conditions.

16.5 Effect of Concentration on Equilibrium

- When the concentration of a reactant is increased, the equilibrium shifts toward the right.
- When the concentration of a product on the right side of the equation is increased, the equilibrium shifts to the left.
- If the concentration of a reactant is decreased, the equilibrium shifts toward the side where the reactant that is decreased is.

16.6 Effect of Volume on Equilibrium

- In gaseous reactions:
  - A decrease in volume causes the equilibrium position to shift toward the side of the reaction with the fewest molecules.
  - If the number of molecules is the same on both sides of a reaction, a decrease in volume has no effect on the equilibrium position.

16.7 Effect of Temperature on Equilibrium

- When heat is added to a reaction, the side of the equation that absorbs heat is favored:
  - If the reaction is endothermic, the forward reaction increases.
  - If the reaction is exothermic, the reverse reaction increases.

16.8 Effect of Catalysts on Equilibrium

KEY TERMS

Catalyst
Activation energy

- A catalyst does not shift the equilibrium of a reaction; it only affects the speed at which equilibrium is reached.
- The activation energy for a reaction is the minimum energy required for the reaction to occur:
  - A catalyst lowers the activation energy for a reaction by providing a different pathway for the reaction.

16.9 Equilibrium Constants

KEY TERM

Equilibrium constant, K_{eq}

- For the general reaction $aA + bB \rightleftharpoons cC + dD$

$$K_{eq} = \frac{[C]^c[D]^d}{[A]^a[B]^b}$$

- The magnitude of K_{eq} indicates the extent of the reaction:
  - If $K_{eq} > 1$: Amount of products is greater than reactants at equilibrium.
  - If $K_{eq} < 1$: Amount of reactants is less than products at equilibrium.

16.10 Ion Product Constant for Water

KEY TERM

Ion product constant for water, K_w

- $K_w = [H^+][OH^-]$
- $[H^+] = \dfrac{K_w}{[OH^-]}$
- $[OH^-] = \dfrac{K_w}{[H^+]}$

16.11 Ionization Constants

KEY TERM

Acid ionization constant, K_a

- The ionization constant can be used in the equilibrium expression to calculate the concentration of a reactant, the percent ionization of a substance, or the pH of a weak acid.

16.12 Solubility Product Constant

KEY TERMS

Solubility product constant, K_{sp}
Common ion effect

- The solubility product constant is used to calculate the solubility of a slightly soluble salt in water.
- The solubility product constant can also be used to determine whether or not precipitation will occur in a solution:
  - If the product of the molar concentrations of the ions in solution is greater than the K_{sp}, precipitation will occur.
- A shift in the equilibrium position upon addition of an ion already contained in the solution is known as the common ion effect.

16.13 Acid–Base Properties of Salts

KEY TERM

Hydrolysis

- Salts of weak acids and weak bases can react with water to cause hydrolysis and form solutions that are not neutral.

Type of salt	Nature of aqueous solution	Examples
Weak base–strong acid	Acidic	NH_4Cl, NH_4NO_3
Strong base–weak acid	Basic	$NaC_2H_3O_2$, K_2CO_3
Weak base–weak acid	Depends on the salt	$NH_4C_2H_3O_2$, NH_4NO_2
Strong base–strong acid	Neutral	$NaCl$, KBr

16.14 Buffer Solutions: The Control of pH

KEY TERM

Buffer solution

- A buffer solution could contain:
  - A weak acid and a salt of its conjugate base
  - A weak base and a salt of its conjugate acid
- A buffer solution resists a change in pH by neutralizing small amounts of acid or base added to it.

Review Questions

All questions with blue numbers have answers in the appendix of the text.

1. How would you expect the two tubes in Figure 16.1 to appear if both are at 25°C?
2. Is the reaction $N_2O_4 \rightleftharpoons 2\,NO_2$ exothermic or endothermic? (Figure 16.1)
3. Why is heat treated as a reactant in an endothermic process and as a product in an exothermic process?
4. At equilibrium how do the forward and reverse reaction rates compare? (Figure 16.2)
5. Why don't free protons (H^+) exist in water?
6. For each solution in Table 16.1, what is the sum of the pH plus the pOH? What would be the pOH of a solution whose pH is -1?
7. Of the acids listed in Table 16.2, which ones are stronger than acetic acid and which are weaker?
8. Tabulate the relative order of molar solubilities of $AgCl$, $AgBr$, AgI, $AgC_2H_3O_2$, $PbSO_4$, $BaSO_4$, $BaCrO_4$, and PbS. List the most soluble first. (Table 16.3)

9. Which compound in the following pairs has the greater molar solubility? (Table 16.3)
   (a) $Mn(OH)_2$ or Ag_2CrO_4
   (b) $BaCrO_4$ or Ag_2CrO_4

10. Explain how the acetic acid–sodium acetate buffer system maintains its pH when 0.010 mol of HCl is added to 1 L of the buffer solution. (Table 16.5)

11. Explain why a precipitate of NaCl forms when HCl gas is passed into a saturated aqueous solution of NaCl.

12. Why does the rate of a reaction usually increase when the concentration of one of the reactants is increased?

13. If pure hydrogen iodide, HI, is placed in a vessel at 700 K, will it decompose? Explain.

14. Why does an increase in temperature cause the rate of reaction to increase?

15. What does a catalyst do?

16. Describe how equilibrium is reached when the substances A and B are first mixed and react as

$$A + B \rightleftharpoons C + D$$

17. With dilution, aqueous solutions of acetic acid ($HC_2H_3O_2$) show increased ionization. For example, a 1.0 *M* solution of acetic acid is 0.42% ionized, whereas a 0.10 *M* solution is 1.34% ionized. Explain the behavior using the ionization equation and equilibrium principles.

18. A 1.0 *M* solution of acetic acid ionizes less and has a higher concentration of H^+ ions than a 0.10 *M* acetic acid solution. Explain this behavior. (See Question 17 for data.)

19. What would cause two separate samples of pure water to have slightly different pH values?

20. Why are the pH and pOH equal in pure water?

21. Explain why silver acetate is more soluble in nitric acid than in water. (*Hint*: Write the equilibrium equation first and then consider the effect of the acid on the acetate ion.) What would happen if hydrochloric acid were used in place of nitric acid?

22. Dissolution of sodium acetate ($NaC_2H_3O_2$) in pure water gives a basic solution. Why? (*Hint*: A small amount of $HC_2H_3O_2$ is formed.)

23. Describe why the pH of a buffer solution remains almost constant when a small amount of acid or base is added to it.

24. Describe the similarities and differences between K_a, K_b, K_w and K_{sp}.

Paired Exercises

All exercises with blue numbers have answers in the appendix of the text.

1. Express these reversible systems in equation form:
   (a) a mixture of ice and liquid water at 0°C
   (b) crystals of Na_2SO_4 in a saturated aqueous solution of Na_2SO_4

2. Express these reversible systems in equation form:
   (a) liquid water and vapor at 100°C in a pressure cooker
   (b) a closed system containing boiling sulfur dioxide, SO_2

3. Consider the following system at equilibrium:

$$4\,NH_3(g) + 3\,O_2(g) \rightleftharpoons 2\,N_2(g) + 6\,H_2O(g) + 1531\text{ kJ}$$

   (a) Is the reaction exothermic or endothermic?
   (b) If the system's state of equilibrium is disturbed by the addition of O_2, in which direction, left or right, must the reaction occur to reestablish equilibrium? After the new equilibrium has been established, how will the final molar concentrations of NH_3, O_2, N_2, and H_2O compare (increase or decrease) with their concentrations before the addition of the O_2?

4. Consider the following system at equilibrium:

$$4\,NH_3(g) + 3\,O_2(g) \rightleftharpoons 2\,N_2(g) + 6\,H_2O(g) + 1531\text{ kJ}$$

   (a) If the system's state of equilibrium is disturbed by the addition of N_2, in which direction, left or right, must the reaction occur to reestablish equilibrium? After the new equilibrium has been established, how will the final molar concentrations of NH_3, O_2, N_2, and H_2O compare (increase or decrease) with their concentrations before the addition of the N_2?
   (b) If the system's state of equilibrium is disturbed by the addition of heat, in which direction will the reaction occur, left or right, to reestablish equilibrium?

5. Consider the following system at equilibrium:

$N_2(g) + 3H_2(g) \rightleftharpoons 2NH_3(g) + 92.5\text{ kJ}$

Complete the table that follows. Indicate changes in moles by entering I, D, N, or ? in the table. (I = increase, D = decrease, N = no change, ? = insufficient information to determine.)

Change of stress imposed on the system at equilibrium	Direction of reaction, left or right, to reestablish equilibrium	Change in number of moles		
		N_2	H_2	NH_3
(a) Add N_2				
(b) Remove H_2				
(c) Decrease volume of reaction vessel				
(d) Increase temperature				

6. Consider the following system at equilibrium:

$N_2(g) + 3H_2(g) \rightleftharpoons 2NH_3(g) + 92.5\text{ kJ}$

Complete the table that follows. Indicate changes in moles by entering I, D, N, or ? in the table. (I = increase, D = decrease, N = no change, ? = insufficient information to determine.)

Change of stress imposed on the system at equilibrium	Direction of reaction, left or right, to reestablish equilibrium	Change in number of moles		
		N_2	H_2	NH_3
(a) Add NH_3				
(b) Increase volume of reaction vessel				
(c) Add catalyst				
(d) Add both H_2 and NH_3				

7. For the following equations, tell in which direction, left or right, the equilibrium will shift when these changes are made: The temperature is increased, the pressure is increased by decreasing the volume of the reaction vessel, and a catalyst is added.

(a) $3O_2(g) + 271\text{ kJ} \rightleftharpoons 2O_3(g)$
(b) $CH_4(g) + Cl_2(g) \rightleftharpoons CH_3Cl(g) + HCl(g) + 110\text{ kJ}$
(c) $2NO(g) + 2H_2(g) \rightleftharpoons N_2(g) + 2H_2O(g) + 665\text{ kJ}$

8. For the following equations, tell in which direction, left or right, the equilibrium will shift when these changes are made: The temperature is increased, the pressure is increased by decreasing the volume of the reaction vessel, and a catalyst is added.

(a) $2SO_3(g) + 197\text{ kJ} \rightleftharpoons 2SO_2(g) + O_2(g)$
(b) $4NH_3(g) + 3O_2(g) \rightleftharpoons 2N_2(g) + 6H_2O(g) + 1531\text{ kJ}$
(c) $OF_2(g) + H_2O(g) \rightleftharpoons O_2(g) + 2HF(g) + 318\text{ kJ}$

9. Utilizing Le Châtelier's principle, indicate the shift (if any) that would occur to

$C_2H_6(g) + \text{heat} \rightleftharpoons C_2H_4(g) + H_2(g)$

(a) if the concentration of hydrogen gas is decreased
(b) if the temperature is lowered
(c) if a catalyst is added

10. Utilizing Le Châtelier's principle, indicate the shift (if any) that would occur to

$C_2H_6(g) + \text{heat} \rightleftharpoons C_2H_4(g) + H_2(g)$

(a) if C_2H_6 is removed from the system
(b) if the volume of the container is increased
(c) if the temperature is raised

11. Write the equilibrium constant expression for these reactions:

(a) $2NO_2(g) + 7H_2(g) \rightleftharpoons 2NH_3(g) + 4H_2O(g)$
(b) $H_2CO_3(aq) \rightleftharpoons H^+(aq) + HCO_3^-(aq)$
(c) $2COF_2(g) \rightleftharpoons CO_2(g) + CF_4(g)$

12. Write the equilibrium constant expression for these reactions:

(a) $H_3PO_4(aq) \rightleftharpoons H^+(aq) + H_2PO_4^-(aq)$
(b) $CS_2(g) + 4H_2(g) \rightleftharpoons CH_4(g) + 2H_2S(g)$
(c) $4NO_2(g) + O_2(g) \rightleftharpoons 2N_2O_5(g)$

13. Write the solubility product expression, K_{sp}, for these substances:

(a) AgCl
(b) $PbCrO_4$
(c) $Zn(OH)_2$
(d) $Ca_3(PO_4)_2$

14. Write the solubility product expression, K_{sp}, for these substances:

(a) $MgCO_3$
(b) CaC_2O_4
(c) $Tl(OH)_3$
(d) $Pb_3(AsO_4)_2$

15. What effect will decreasing the $[H^+]$ of a solution have on (a) pH, (b) pOH, (c) $[OH^-]$, and (d) K_w?

16. What effect will increasing the $[H^+]$ of a solution have on (a) pH, (b) pOH, (c) $[OH^-]$, and (d) K_w?

17. Decide whether these salts form an acidic, basic, or neutral solution when dissolved in water:

(a) $CaBr_2$
(b) $Al(NO_3)_3$
(c) NaCN
(d) K_3PO_4

18. Decide whether these salts form an acidic, basic, or neutral solution when dissolved in water:

(a) NH_4Cl
(b) $NaC_2H_3O_2$
(c) $CuSO_4$
(d) KI

19. Write hydrolysis equations for aqueous solutions of these salts:
(a) KNO_2 (b) $Mg(C_2H_3O_2)_2$

20. Write hydrolysis equations for aqueous solutions of these salts:
(a) NH_4NO_3 (b) Na_2SO_3

21. Write hydrolysis equations for these ions:
(a) HCO_3^- (b) NH_4^+

22. Write hydrolysis equations for these ions:
(a) OCl^- (b) ClO_2^-

23. One of the important pH-regulating systems in the blood consists of a carbonic acid–sodium hydrogen carbonate buffer:
$H_2CO_3(aq) \rightleftharpoons H^+(aq) + HCO_3^-(aq)$
$NaHCO_3(aq) \longrightarrow Na^+(aq) + HCO_3^-(aq)$
Explain how this buffer resists changes in pH when excess acid, H^+, gets into the bloodstream.

24. One of the important pH-regulating systems in the blood consists of a carbonic acid–sodium hydrogen carbonate buffer:
$H_2CO_3(aq) \rightleftharpoons H^+(aq) + HCO_3^-(aq)$
$NaHCO_3(aq) \longrightarrow Na^+(aq) + HCO_3^-(aq)$
Explain how this buffer resists changes in pH when excess base, OH^-, gets into the bloodstream.

25. Calculate (a) the $[H^+]$, (b) the pH, and (c) the percent ionization of a 0.25 *M* solution of $HC_2H_3O_2$. ($K_a = 1.8 \times 10^{-5}$)

26. Calculate (a) the $[H^+]$, (b) the pH, and (c) the percent ionization of a 0.25 *M* solution of phenol, HC_6H_5O. ($K_a = 1.3 \times 10^{-10}$)

27. A 0.025 *M* solution of a weak acid, HA, is 0.45% ionized. Calculate the ionization constant, K_a, for the acid.

28. A 0.500 *M* solution of a weak acid, HA, is 0.68% ionized. Calculate the ionization constant, K_a, for the acid.

29. Calculate the percent ionization and pH of solutions of $HC_2H_3O_2$ ($K_a = 1.8 \times 10^{-5}$) having the following molarities: (a) 1.0 *M*, (b) 0.10 *M*, and (c) 0.010 *M*.

30. Calculate the percent ionization and pH of solutions of HClO ($K_a = 3.5 \times 10^{-8}$) having the following molarities: (a) 1.0 *M*, (b) 0.10 *M*, and (c) 0.010 *M*.

*31. A 0.37 *M* solution of a weak acid (HA) has a pH of 3.7. What is the K_a for this acid?

*32. A 0.23 *M* solution of a weak acid (HA) has a pH of 2.89. What is the K_a for this acid?

33. A student needs a sample of 1.0 *M* NaOH for a laboratory experiment. Calculate the $[H^+]$, $[OH^-]$, pH, and pOH of this solution.

34. A laboratory cabinet contains a stock solution of 3.0 *M* HNO_3. Calculate the $[H^+]$, $[OH^-]$, pH, and pOH of this solution.

35. Calculate the pH and the pOH of these solutions:
(a) 0.250 *M* HBr
(b) 0.333 *M* KOH
(c) 0.895 *M* $HC_2H_3O_2$ ($K_a = 1.8 \times 10^{-5}$)

36. Calculate the pH and the pOH of these solutions:
(a) 0.0010 *M* NaOH
(b) 0.125 *M* HCl
(c) 0.0250 *M* HC_6H_5O ($K_a = 1.3 \times 10^{-10}$)

37. Calculate the $[OH^-]$ in these solutions:
(a) $[H^+] = 1.0 \times 10^{-4}$ (b) $[H^+] = 2.8 \times 10^{-6}$

38. Calculate the $[OH^-]$ in these solutions:
(a) $[H^+] = 4.0 \times 10^{-9}$ (b) $[H^+] = 8.9 \times 10^{-2}$

39. Calculate the $[H^+]$ in these solutions:
(a) $[OH^-] = 6.0 \times 10^{-7}$ (b) $[OH^-] = 1 \times 10^{-8}$

40. Calculate the $[H^+]$ in these solutions:
(a) $[OH^-] = 4.5 \times 10^{-6}$ (b) $[OH^-] = 7.3 \times 10^{-4}$

41. Given the following solubility data, calculate the solubility product constant for each substance:
(a) $BaSO_4$, 3.9×10^{-5} mol/L
(b) Ag_2CrO_4, 7.8×10^{-5} mol/L
(c) $CaSO_4$, 0.67 g/L
(d) AgCl, 0.0019 g/L

42. Given the following solubility data, calculate the solubility product constant for each substance:
(a) ZnS, 3.5×10^{-12} mol/L
(b) $Pb(IO_3)_2$, 4.0×10^{-5} mol/L
(c) Ag_3PO_4, 6.73×10^{-3} g/L
(d) $Zn(OH)_2$, 2.33×10^{-4} g/L

43. Calculate the molar solubility for these substances:
(a) CaF_2, $K_{sp} = 3.9 \times 10^{-11}$
(b) $Fe(OH)_3$, $K_{sp} = 6.1 \times 10^{-38}$

44. Calculate the molar solubility for these substances:
(a) $PbSO_4$, $K_{sp} = 1.3 \times 10^{-8}$
(b) $BaCrO_4$, $K_{sp} = 8.5 \times 10^{-11}$

45. For each substance in Exercise 43, calculate the solubility in grams per 100. mL of solution.

46. For each substance in Exercise 44, calculate the solubility in grams per 100. mL of solution.

47. Solutions containing 100. mL of 0.010 M Na_2SO_4 and 100. mL of 0.001 M $Pb(NO_3)_2$ are mixed. Show by calculation whether or not a precipitate will form. Assume that the volumes are additive. (K_{sp} for $PbSO_4 = 1.3 \times 10^{-8}$)

48. Solutions containing 50.0 mL of 1.0×10^{-4} M $AgNO_3$ and 100. mL of 1.0×10^{-4} M NaCl are mixed. Show by calculation whether or not a precipitate will form. Assume the volumes are additive. (K_{sp} for $AgCl = 1.7 \times 10^{-10}$)

49. How many moles of AgBr will dissolve in 1.0 L of 0.10 M NaBr? ($K_{sp} = 5.2 \times 10^{-13}$ for AgBr)

50. How many moles of AgBr will dissolve in 1.0 L of 0.10 M $MgBr_2$? ($K_{sp} = 5.2 \times 10^{-13}$ for AgBr)

51. Calculate the $[H^+]$ and the pH of a buffer solution that is 0.20 M in $HC_2H_3O_2$ and contains sufficient sodium acetate to make the $[C_2H_3O_2^-]$ equal to 0.10 M. (K_a for $HC_2H_3O_2 = 1.8 \times 10^{-5}$)

52. Calculate the $[H^+]$ and the pH of a buffer solution that is 0.20 M in $HC_2H_3O_2$ and contains sufficient sodium acetate to make the $[C_2H_3O_2^-]$ equal to 0.20 M. (K_a for $HC_2H_3O_2 = 1.8 \times 10^{-5}$)

53. When 1.0 mL of 1.0 M HCl is added to 50. mL of 1.0 M NaCl, the $[H^+]$ changes from 1×10^{-7} M to 2.0×10^{-2} M. Calculate the initial pH and the pH change in the solution.

54. When 1.0 mL of 1.0 M HCl is added to 50. mL of a buffer solution that is 1.0 M in $HC_2H_3O_2$ and 1.0 M in $NaC_2H_3O_2$, the $[H^+]$ changes from 1.8×10^{-5} M to 1.9×10^{-5} M. Calculate the initial pH and the pH change in the solution.

Additional Exercises

All exercises with blue numbers have answers in the appendix of the text.

55. In a K_{sp} expression the concentration of solid salt is not included. Why?

56. The energy diagram in Figure 16.3 is for an exothermic reaction. How can you tell?

57. Sketch the energy diagram for an endothermic reaction using a solid line. Sketch what would happen to the energy profile if a catalyst is added to the reaction using a dotted line.

58. What is the maximum number of moles of HI that can be obtained from a reaction mixture containing 2.30 mol I_2 and 2.10 mol H_2?

59. (a) How many moles of HI are produced when 2.00 mol H_2 and 2.00 mol I_2 are reacted at 700 K? (Reaction is 79% complete.)
(b) Addition of 0.27 mol I_2 to the system increases the yield of HI to 85%. How many moles of H_2, I_2, and HI are now present?
(c) From the data in part (a), calculate K_{eq} for the reaction at 700 K.

*60. After equilibrium is reached in the reaction of 6.00 g H_2 with 200. g I_2 at 500. K, analysis shows that the flask contains 64.0 g of HI. How many moles of H_2, I_2, and HI are present in this equilibrium mixture?

61. What is the equilibrium constant of the reaction

$PCl_3(g) + Cl_2(g) \rightleftharpoons PCl_5(g)$

if a 20.-L flask contains 0.10 mol PCl_3, 1.50 mol Cl_2, and 0.22 mol PCl_5?

62. If the rate of a reaction doubles for every 10°C rise in temperature, how much faster will the reaction go at 100°C than at 30°C?

63. Would you believe that an NH_4Cl solution is acidic? Calculate the pH of a 0.30 M NH_4Cl solution.
($NH_4^+ \rightleftharpoons NH_3 + H^+$, $K_{eq} = 5.6 \times 10^{-10}$)

64. Calculate the pH of an acetic acid buffer composed of $HC_2H_3O_2$ (0.30 M) and $C_2H_3O_2^-$ (0.20 M).

65. Solutions of 50.0 mL of 0.10 M $BaCl_2$ and 50.0 mL of 0.15 M Na_2CrO_4 are mixed, forming a precipitate of $BaCrO_4$. Calculate the concentration of Ba^{2+} that remains in solution.

66. Calculate the ionization constant for the given acids. Each acid ionizes as follows: $HA \rightleftharpoons H^+ + A^-$.

Acid	Acid concentration	$[H^+]$
Hypochlorous, HOCl	0.10 M	5.9×10^{-5} mol/L
Propanoic, $HC_3H_5O_2$	0.15 M	1.4×10^{-3} mol/L
Hydrocyanic, HCN	0.20 M	8.9×10^{-6} mol/L

67. The K_{sp} of CaF_2 is 3.9×10^{-11}. Calculate (a) the molar concentrations of Ca^{2+} and F^- in a saturated solution and (b) the grams of CaF_2 that will dissolve in 500. mL of water.

68. Calculate whether or not a precipitate will form in the following mixed solutions:
(a) 100 mL of 0.010 M Na_2SO_4 and 100 mL of 0.001 M $Pb(NO_3)_2$
(b) 50.0 mL of 1.0×10^{-4} M $AgNO_3$ and 100. mL of 1.0×10^{-4} M NaCl
(c) 1.0 g $Ca(NO_3)_2$ in 150 mL H_2O and 250 mL of 0.01 M NaOH

$K_{sp}PbSO_4 = 1.3 \times 10^{-8}$

$K_{sp}AgCl = 1.7 \times 10^{-10}$

$K_{sp}Ca(OH)_2 = 1.3 \times 10^{-6}$

69. If $BaCl_2$ is added to a saturated $BaSO_4$ solution until the $[Ba^{2+}]$ is 0.050 M,
(a) what concentration of SO_4^{2-} remains in solution?
(b) how much $BaSO_4$ remains dissolved in 100. mL of the solution? ($K_{sp} = 1.5 \times 10^{-9}$ for $BaSO_4$)

70. The K_{sp} for $PbCl_2$ is 2.0×10^{-5}. Will a precipitate form when 0.050 mol $Pb(NO_3)_2$ and 0.010 mol NaCl are in 1.0 L solution? Show evidence for your answer.

71. Suppose the concentration of a solution is 0.10 M Ba^{2+} and 0.10 M Sr^{2+}. Which sulfate, $BaSO_4$ or $SrSO_4$, will precipitate first when a dilute solution of H_2SO_4 is added dropwise to the solution? Show evidence for your answer. ($K_{sp} = 1.5 \times 10^{-9}$ for $BaSO_4$ and $K_{sp} = 3.5 \times 10^{-7}$ for $SrSO_4$)

72. Calculate the K_{eq} for the reaction

$SO_2(g) + O_2(g) \rightleftharpoons SO_3(g)$

when the equilibrium concentrations of the gases at 530°C are $[SO_3] = 11.0\,M$, $[SO_2] = 4.20\,M$, and $[O_2] = 0.60 \times 10^{-3}\,M$.

73. If it takes 0.048 g BaF_2 to saturate 15.0 mL of water, what is the K_{sp} of BaF_2?

74. The K_{eq} for the formation of ammonia gas from its elements is 4.0. If the equilibrium concentrations of nitrogen gas and hydrogen gas are both 2.0 M, what is the equilibrium concentration of the ammonia gas?

75. The K_{sp} of $SrSO_4$ is 7.6×10^{-7}. Should precipitation occur when 25.0 mL of $1.0 \times 10^{-3}\,M$ $SrCl_2$ solution are mixed with 15.0 mL of $2.0 \times 10^{-3}\,M$ Na_2SO_4? Show proof.

*76. The solubility of Hg_2I_2 in H_2O is 3.04×10^{-7} g/L. The reaction $Hg_2I_2 \rightleftharpoons Hg_2^{2+} + 2\,I^-$ represents the equilibrium. Calculate the K_{sp}.

77. Under certain circumstances, when oxygen gas is heated, it can be converted into ozone according to the following reaction equation:

$3\,O_2(g) + \text{heat} \rightleftharpoons 2\,O_3(g)$

Name three different ways that you could increase the production of the ozone.

78. One day in a laboratory, some water spilled on a table. In just a few minutes the water had evaporated. Some days later, a similar amount of water spilled again. This time, the water remained on the table after 7 or 8 hours. Name three conditions that could have changed in the lab to cause this difference.

79. For the reaction $CO(g) + H_2O(g) \rightleftharpoons CO_2(g) + H_2(g)$ at a certain temperature, K_{eq} is 1. At equilibrium would you expect to find
(a) only CO and H_2
(b) mostly CO_2 and H_2
(c) about equal concentrations of CO and H_2O, compared to CO_2 and H_2
(d) mostly CO and H_2O
(e) only CO and H_2O
Explain your answer briefly.

80. Write the equilibrium constant expressions for these reactions:
(a) $3\,O_2(g) \rightleftharpoons 2\,O_3(g)$
(b) $H_2O(g) \rightleftharpoons H_2O(l)$
(c) $MgCO_3(s) \rightleftharpoons MgO(s) + CO_2(g)$
(d) $2\,Bi^{3+}(aq) + 3\,H_2S(aq) \rightleftharpoons Bi_2S_3(s) + 6\,H^+(aq)$

81. Reactants A and B are mixed, each initially at a concentration of 1.0 M. They react to produce C according to this equation:

$2\,A + B \rightleftharpoons C$

When equilibrium is established, the concentration of C is found to be 0.30 M. Calculate the value of K_{eq}.

82. At a certain temperature, K_{eq} is 2.2×10^{-3} for the reaction

$2\,ICl(g) \rightleftharpoons I_2(g) + Cl_2(g)$

Now calculate the K_{eq} value for the reaction

$I_2(g) + Cl_2(g) \rightleftharpoons 2\,ICl(g)$

83. One drop of 1 M OH^- ion is added to a 1 M solution of HNO_2. What will be the effect of this addition on the equilibrium concentration of the following?

$HNO_2(aq) \rightleftharpoons H^+(aq) + NO_2^-(aq)$

(a) $[OH^-]$
(b) $[H^+]$
(c) $[NO_2^-]$
(d) $[HNO_2]$

84. How many grams of $CaSO_4$ will dissolve in 600. mL of water? ($K_{sp} = 2.0 \times 10^{-4}$ for $CaSO_4$)

85. A student found that 0.098 g of PbF_2 was dissolved in 400. mL of saturated PbF_2. What is the K_{sp} for the lead(II) fluoride?

Challenge Exercises

All exercises with blue numbers have answers in the appendix of the text.

*86. All a snowbound skier had to eat were walnuts! He was carrying a bag holding 12 dozen nuts. With his mittened hands, he cracked open the shells. Each nut that was opened resulted in one kernel and two shell halves. When he tired of the cracking and got ready to do some eating, he discovered he had 194 total pieces (whole nuts, shell halves, and kernels). What is the K_{eq} for this reaction?

*87. At 500°C, the reaction

$SO_2(g) + NO_2(g) \rightleftharpoons NO(g) + SO_3(g)$

has $K_{eq} = 81$. What will be the equilibrium concentrations of the four gases if the two reactants begin with equal concentrations of 0.50 M?

Answers to Practice Exercises

16.1 (a)

Concentration	Change
$[H_2]$	increase
$[I_2]$	decrease
$[HI]$	increase

(b)

Concentration	Change
$[H_2]$	increase
$[I_2]$	increase
$[HI]$	increase

16.2 (a) equilibrium shifts to the left;
(b) equilibrium shifts to the right

16.3 (a) equilibrium shifts to the right;
(b) equilibrium shifts to the left

16.4 (a) equilibrium shifts to the left;
(b) equilibrium shifts to the right

16.5 (a) $K_{eq} = \dfrac{[NO_2]^4[O_2]}{[N_2O_5]^2}$; (b) $K_{eq} = \dfrac{[N_2]^2[H_2O]^6}{[NH_3]^4[O_2]^3}$

16.6 $K_{eq} = 33$; the forward reaction is favored

16.7 (a) $[H^+] = 5.0 \times 10^{-5}$ $[OH^-] = 2.0 \times 10^{-10}$
(b) $[H^+] = 5.0 \times 10^{-9}$ $[OH^-] = 2.0 \times 10^{-6}$

16.8 (a) 4.30 (b) 8.30

16.9 (a) 6.3×10^{-6} (b) 1.8×10^{-6}

16.10 (a) 6.3×10^{-3}% ionized
(b) 7.2×10^{-3}% ionized

16.11 (a) $K_{sp} = [Cr^{3+}][OH^-]^3$
(b) $K_{sp} = [Cu^{2+}]^3[PO_4^{3-}]^2$

16.12 8.8×10^{-21} g/L

16.13 7.5×10^{-8} mol/L

16.14 (a) basic
(b) neutral
(c) acidic

CHAPTER 17

Oxidation–Reduction

The Golden Gate Bridge in San Francisco requires frequent painting to prevent oxidation of the metal in the sea air.

Chapter Outline

The variety of oxidation–reduction reactions that affect us every day is amazing. Our society runs on batteries—in our calculators, laptop computers, cars, toys, radios, televisions, and more. We paint iron railings and galvanize nails to combat corrosion. We electroplate jewelry and computer chips with very thin coatings of gold or silver. We bleach our clothes using chemical reactions that involve electron transfer. We test for glucose in urine or alcohol in the breath with reactions that show vivid color changes. Plants turn energy into chemical compounds through a series of reactions called photosynthesis. These reactions all involve the transfer of electrons between substances in a chemical process called *oxidation–reduction*.

17.1 Oxidation Number

The oxidation number of an atom (sometimes called its *oxidation state*) represents the number of electrons lost, gained, or unequally shared by an atom. Oxidation numbers can be zero, positive, or negative. An oxidation number of zero means the atom has the same number of electrons assigned to it as there are in the free neutral atom. A positive oxidation number means the atom has fewer electrons assigned to it than in the neutral atom, and a negative oxidation number means the atom has more electrons assigned to it than in the neutral atom.

The oxidation number of an atom that has lost or gained electrons to form an ion is the same as the positive or negative charge of the ion. (See Table 17.1.) In the ionic compound NaCl, the oxidation numbers are clearly +1 for the Na^+ ion and −1 for the Cl^- ion. The Na^+ ion has one less electron than the neutral Na atom, and the Cl^- ion has one more electron than the neutral Cl atom. In $MgCl_2$, two electrons have transferred from the Mg atom to the two Cl atoms; the oxidation number of Mg is +2.

Table 17.1 Oxidation Numbers for Common Ions

Ion	Oxidation number
H^+	+1
Na^+	+1
K^+	+1
Li^+	+1
Ag^+	+1
Cu^{2+}	+2
Ca^{2+}	+2
Ba^{2+}	+2
Fe^{2+}	+2
Mg^{2+}	+2
Zn^{2+}	+2
Al^{3+}	+3
Fe^{3+}	+3
Cl^-	−1
Br^-	−1
F^-	−1
I^-	−1
S^{2-}	−2
O^{2-}	−2

In covalently bonded substances, where electrons are shared between two atoms, oxidation numbers are assigned by an arbitrary system based on relative electronegativities. For symmetrical covalent molecules such as H_2 and Cl_2, each atom is assigned an oxidation number of zero because the bonding pair of electrons is shared equally between two like atoms, neither of which is more electronegative than the other:

H:H   :C̈l:C̈l:

When the covalent bond is between two unlike atoms, the bonding electrons are shared unequally because the more electronegative element has a greater attraction for them. In this case the oxidation numbers are determined by assigning both electrons to the more electronegative element.

Thus in compounds with covalent bonds such as NH_3 and H_2O,

the pairs of electrons are unequally shared between the atoms and are attracted toward the more electronegative elements, N and O. This causes the N and O atoms to be relatively negative with respect to the H atoms. At the same time, it causes the H atoms to be relatively positive with respect to the N and O atoms. In H_2O, both pairs of shared electrons are assigned to the O atom, giving it two electrons more than the neutral O atom, and each H atom is assigned one electron less than the neutral H atom. Therefore, the oxidation number of the O atom is -2, and the oxidation number of each H atom is $+1$. In NH_3, the three pairs of shared electrons are assigned to the N atom, giving it three electrons more than the neutral N atom, and each H atom has one electron less than the neutral atom. Therefore, the oxidation number of the N atom is -3, and the oxidation number of each H atom is $+1$.

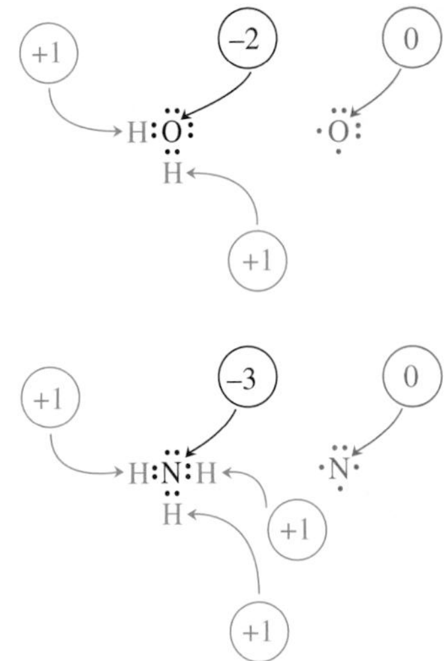

Assigning correct oxidation numbers to elements is essential for balancing oxidation–reduction equations. The **oxidation number** or **oxidation state** of an element is an integer value assigned to each element in a compound or ion that allows us to keep track of electrons associated with each atom. Oxidation numbers have a variety of uses in chemistry—from writing formulas to predicting properties of compounds and assisting in the balancing of oxidation–reduction reactions in which electrons are transferred.

Oxidation number
Oxidation state

As a starting point, the oxidation number of an uncombined element, regardless of whether it is monatomic or diatomic, is zero. Rules for assigning oxidation numbers are summarized in Table 17.2. Many elements have multiple oxidation numbers; for example, nitrogen

	N_2	N_2O	NO	N_2O_3	NO_2	N_2O_5	NO_3^-
Oxidation number	0	+1	+2	+3	+4	+5	+5

The oxidation number for Cu metal is 0 while the oxidation number for Cu^{2+} ions in the crystal is $+2$.

Use the following steps to find the oxidation number for an element within a compound:

Step 1 Write the oxidation number of each known atom below the atom in the formula.

Step 2 Multiply each oxidation number by the number of atoms of that element in the compound.

Step 3 Write an expression indicating the sum of all the oxidation numbers in the compound. *Remember*: The sum of the oxidation numbers in a compound must equal zero.

Table 17.2 Rules for Assigning Oxidation Number

1. All elements in their free state (uncombined with other elements) have an oxidation number of zero (e.g., Na, Cu, Mg, H_2, O_2, Cl_2, N_2).
2. H is $+1$, except in metal hydrides, where it is -1 (e.g., NaH, CaH_2).
3. O is -2, except in peroxides, where it is -1, and in OF_2, where it is $+2$.
4. The metallic element in an ionic compound has a positive oxidation number.
5. In covalent compounds the negative oxidation number is assigned to the most electronegative atom.
6. The algebraic sum of the oxidation numbers of the elements in a compound is zero.
7. The algebraic sum of the oxidation numbers of the elements in a polyatomic ion is equal to the charge of the ion.

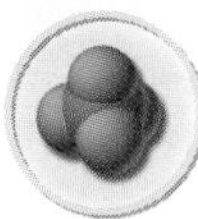

Example 17.1 Determine the oxidation number for carbon in carbon dioxide:

SOLUTION

$$CO_2$$

Step 1 -2
Step 2 $(-2)2$
Step 3 $C + (-4) = 0$
Step 4 $C = +4$ (oxidation number for carbon)

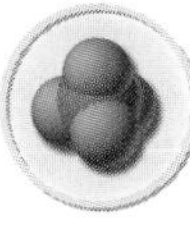

Example 17.2 Determine the oxidation number for sulfur in sulfuric acid:

SOLUTION

$$H_2SO_4$$

Step 1 $+1$ -2
Step 2 $2(+1) = +2$ $4(-2) = -8$
Step 3 $+2 + S + (-8) = 0$
Step 4 $S = +6$ (oxidation number for sulfur)

Practice 17.1

Determine the oxidation number of (a) S in Na_2SO_4, (b) As in K_3AsO_4, and (c) C in $CaCO_3$.

Oxidation numbers in a polyatomic ion (ions containing more than one atom) are determined in a similar fashion, except that in a polyatomic ion the sum of the oxidation numbers must equal the charge on the ion instead of zero.

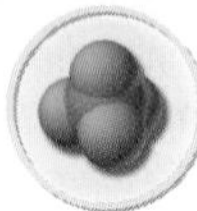

Example 17.3 Determine the oxidation number for manganese in the permanganate ion MnO_4^-:

SOLUTION

$$MnO_4^-$$

Step 1 -2
Step 2 $(-2)4$
Step 3 $Mn + (-8) = -1$ (the charge on the ion)
$Mn = +7$ (oxidation number for manganese)

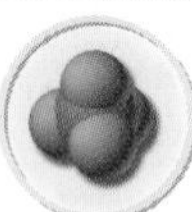

Example 17.4 Determine the oxidation number for carbon in the oxalate ion $C_2O_4^{2-}$:

SOLUTION

$$C_2O_4^{2-}$$

-2
Step 1 $(-2)4$
Step 2 $2C + (-8) = -2$ (the charge on the ion)
Step 3 $2C = +6$
$C = +3$ (oxidation number for C)

Practice 17.2

Determine the oxidation numbers of (a) N in NH_4^+, (b) Cr in $Cr_2O_7^{2-}$, and (c) P in PO_4^{3-}.

Example 17.5

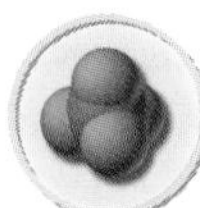

Determine the oxidation number of each element in (a) KNO_3 and (b) SO_4^{2-}.

SOLUTION

(a) Potassium is a Group 1A metal; therefore, it has an oxidation number of +1. The oxidation number of each O atom is −2 (Table 17.2, Rule 3). Using these values and the fact that the sum of the oxidation numbers of all the atoms in a compound is zero, we can determine the oxidation number of N:

$$KNO_3$$

$$+1 + N + 3(-2) = 0$$

$$N = +6 - 1 = +5$$

The oxidation numbers are K, +1; N, +5; O, −2.

(b) Because SO_4^{2-} is an ion, the sum of oxidation numbers of the S and the O atoms must be −2, the charge of the ion. The oxidation number of each O atom is −2 (Table 17.2, Rule 3). Then

$$SO_4^{2-}$$

$$S + 4(-2) = -2, \quad S - 8 = -2$$

$$S = -2 + 8 = +6$$

The oxidation numbers are S, +6; O, −2.

Practice 17.3

Determine the oxidation number of each element in these species:
(a) $BeCl_2$ (b) HClO (c) H_2O_2 (d) NH_4^+ (e) BrO_3^-

17.2 Oxidation–Reduction

oxidation–reduction
redox

Oxidation–reduction, also known as **redox,** is a chemical process in which the oxidation number of an element is changed. The process may involve the complete transfer of electrons to form ionic bonds or only a partial transfer or shift of electrons to form covalent bonds.

oxidation
reduction

Oxidation occurs whenever the oxidation number of an element increases as a result of losing electrons. Conversely, **reduction** occurs whenever the oxidation number of an element decreases as a result of gaining electrons. For example, a change in oxidation number from +2 to +3 or from −1 to 0 is oxidation; a change from +5 to +2 or from −2 to −4 is reduction (see Figure 17.1). Oxidation and reduction occur simultaneously in a chemical reaction; one cannot take place without the other.

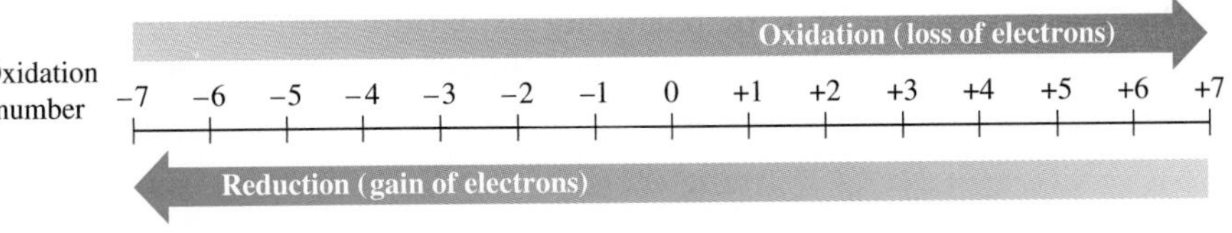

Figure 17.1
Oxidation and reduction. Oxidation results in an increase in the oxidation number, and reduction results in a decrease in the oxidation number.

Many combination, decomposition, and single-displacement reactions involve oxidation–reduction. Let's examine the combustion of hydrogen and oxygen from this point of view:

$$2\,H_2 + O_2 \longrightarrow 2\,H_2O$$

oxidizing agent

reducing agent

Both reactants, hydrogen and oxygen, are elements in the free state and have an oxidation number of zero. In the product (water), hydrogen has been oxidized to +1 and oxygen reduced to −2. The substance that causes an increase in the oxidation state of another substance is called an **oxidizing agent.** The substance that causes a decrease in the oxidation state of another substance is called a **reducing agent.** In this reaction the oxidizing agent is free oxygen, and the reducing agent is free hydrogen. In the reaction

$$Zn(s) + H_2SO_4(aq) \longrightarrow ZnSO_4(aq) + H_2(g)$$

metallic zinc is oxidized, and hydrogen ions are reduced. Zinc is the reducing agent, and hydrogen ions, the oxidizing agent. Electrons are transferred from the zinc metal to the hydrogen ions. The electron transfer is more clearly expressed as

$$Zn^0 + 2\,H^+ + SO_4^{2-} \longrightarrow Zn^{2+} + SO_4^{2-} + H_2^0$$

In the reaction between Zn and H_2SO_4 the Zn is oxidized while hydrogen is reduced.

Oxidation:	**Increase in oxidation number** **Loss of electrons**
Reduction:	**Decrease in oxidation number** **Gain of electrons**

The oxidizing agent is reduced and gains electrons. The reducing agent is oxidized and loses electrons. The transfer of electrons is characteristic of all redox reactions.

17.3 Balancing Oxidation–Reduction Equations

Many simple redox equations can be balanced readily by inspection, or by trial and error:

$$Na + Cl_2 \longrightarrow NaCl \quad \text{(unbalanced)}$$
$$2\,Na + Cl_2 \longrightarrow 2\,NaCl \quad \text{(balanced)}$$

Balancing this equation is certainly not complicated. But as we study more complex reactions and equations such as

$$P + HNO_3 + H_2O \longrightarrow NO + H_3PO_4 \quad \text{(unbalanced)}$$
$$3\,P + 5\,HNO_3 + 2\,H_2O \longrightarrow 5\,NO + 3\,H_3PO_4 \quad \text{(balanced)}$$

the trial-and-error method of balancing equations takes an unnecessarily long time.

One systematic method for balancing oxidation–reduction equations is based on the transfer of electrons between the oxidizing and reducing agents. Consider the first equation again:

The superscript 0 shows that the oxidation number is 0 for elements in their uncombined state.

$$Na^0 + Cl_2^0 \longrightarrow Na^+Cl^- \quad \text{(unbalanced)}$$

In this reaction, sodium metal loses one electron per atom when it changes to a sodium ion. At the same time chlorine gains one electron per atom. Because chlorine is diatomic, two electrons per molecule are needed to form a chloride ion from each atom. These electrons are furnished by two sodium atoms. Stepwise, the reaction may be written as two half-reactions, the oxidation half-reaction and the reduction half-reaction:

$$\begin{array}{lll} 2\,Na^0 \longrightarrow 2\,Na^+ + \cancel{2e^-} & & \text{oxidation half-reaction} \\ Cl_2^0 + \cancel{2e^-} \longrightarrow 2\,Cl^- & & \text{reduction half-reaction} \\ \hline Cl_2^0 + 2\,Na^0 \longrightarrow 2\,Na^+Cl^- & & \end{array}$$

When the two half-reactions, each containing the same number of electrons, are added together algebraically, the electrons cancel out. In this reaction there are no excess electrons; the two electrons lost by the two sodium atoms are utilized by chlorine. In all redox reactions the loss of electrons by the reducing agent must equal the gain of electrons by the oxidizing agent. Here sodium is oxidized and chlorine is reduced. Chlorine is the oxidizing agent; sodium is the reducing agent.

In the following examples, we use the change-in-oxidation-number method, a system for balancing more complicated redox equations.

Example 17.6

Balance the equation

$$Sn + HNO_3 \longrightarrow SnO_2 + NO_2 + H_2O \quad \text{(unbalanced)}$$

SOLUTION

Step 1 Assign oxidation numbers to each element to identify the elements being oxidized and those being reduced. Write the oxidation numbers below each element to avoid confusing them with ionic charge:

$$\underset{0}{Sn} + \underset{+1\ +5\ -2}{HNO_3} \longrightarrow \underset{+4\ -2}{SnO_2} + \underset{+4\ -2}{NO_2} + \underset{+1\ -2}{H_2O}$$

Note that the oxidation numbers of Sn and N have changed.

Step 2 Now write two new equations (half-reactions), using only the elements that change in oxidation number. Then add electrons to bring the equations into electrical balance. One equation represents the oxidation step; the other represents the reduction step. *Remember*: Oxidation produces electrons; reduction uses electrons.

$$Sn^0 \longrightarrow Sn^{4+} + 4\,e^- \quad \text{(oxidation)}$$

Sn^0 loses 4 electrons

$$N^{5+} + 1\,e^- \longrightarrow N^{4+} \quad \text{(reduction)}$$

N^{5+} gains 1 electron

Step 3 Multiply the two equations by the smallest whole numbers that will make the electrons lost by oxidation equal to the number of electrons gained by reduction. In this reaction the oxidation step is multiplied by 1 and the reduction step by 4. The equations become

$$Sn^0 \longrightarrow Sn^{4+} + 4\,e^- \quad \text{(oxidation)}$$

Sn^0 loses 4 electrons

$$4\,N^{5+} + 4\,e^- \longrightarrow 4\,N^{4+} \quad \text{(reduction)}$$
$4\,N^{5+}$ gains 4 electrons

We have now established the ratio of the oxidizing to the reducing agent as being four atoms of N to one atom of Sn.

Step 4 Transfer the coefficient in front of each substance in the balanced oxidation–reduction equations to the corresponding substance in the original equation. We need to use 1 Sn, 1 SnO_2, 4 HNO_3, and 4 NO_2:

$$Sn + 4\,HNO_3 \longrightarrow SnO_2 + 4\,NO_2 + H_2O \quad \text{(unbalanced)}$$

Step 5 In the usual manner, balance the remaining elements that are not oxidized or reduced to give the final balanced equation:

$$Sn + 4\,HNO_3 \longrightarrow SnO_2 + 4\,NO_2 + 2\,H_2O \quad \text{(balanced)}$$

In balancing the final elements, we must not change the ratio of the elements that were oxidized and reduced.

Finally, check to ensure that both sides of the equation have the same number of atoms of each element. The final balanced equation contains 1 atom of Sn, 4 atoms of N, 4 atoms of H, and 12 atoms of O on each side.

Because each new equation presents a slightly different problem and because proficiency in balancing equations requires practice, let's work through two more examples.

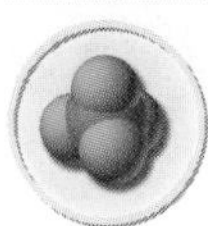

Example 17.7 Balance the equation

$$I_2 + Cl_2 + H_2O \longrightarrow HIO_3 + HCl \quad \text{(unbalanced)}$$

SOLUTION **Step 1** Assign oxidation numbers:

$$\underset{0}{I_2} + \underset{0}{Cl_2} + \underset{+1\;-2}{H_2O} \longrightarrow \underset{+1\;+5\;-2}{HIO_3} + \underset{+1\;-1}{HCl}$$

The oxidation numbers of I_2 and Cl_2 have changed, I_2 from 0 to +5, and Cl_2 from 0 to −1.

Step 2 Write the oxidation and reduction steps. Balance the number of atoms and then balance the electrical charge using electrons:

$$I_2 \longrightarrow 2\,I^{5+} + 10\,e^- \quad \text{(oxidation)}$$
I_2 loses 10 electrons ($10\,e^-$ are needed to balance the +10 charge)

$$Cl_2 + 2\,e^- \longrightarrow 2\,Cl^- \quad \text{(reduction)}$$
Cl_2 gains 2 electrons ($2\,e^-$ are needed to balance the −2 charge)

Step 3 Adjust loss and gain of electrons so that they are equal. Multiply the oxidation step by 1 and the reduction step by 5:

$$I_2 \longrightarrow 2\,I^{5+} + 10\,e^- \quad \text{(oxidation)}$$
I_2 loses 10 electrons

$$5\,Cl_2 + 10\,e^- \longrightarrow 10\,Cl^- \quad \text{(reduction)}$$
$5\,Cl_2$ gain 10 electrons

Step 4 Transfer the coefficients from the balanced redox equations into the original equation. We need to use 1 I_2, 2 HIO_3, 5 Cl_2, and 10 HCl:

$$I_2 + 5\,Cl_2 + H_2O \longrightarrow 2\,HIO_3 + 10\,HCl \quad \text{(unbalanced)}$$

Step 5 Balance the remaining elements, H and O:

$$I_2 + 5\,Cl_2 + 6\,H_2O \longrightarrow 2\,HIO_3 + 10\,HCl \quad \text{(balanced)}$$

Check: The final balanced equation contains 2 atoms of I, 10 atoms of Cl, 12 atoms of H, and 6 atoms of O on each side.

Example 17.8

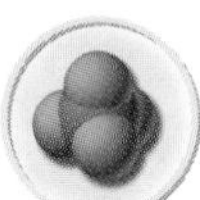

Balance the equation

$$K_2Cr_2O_7 + FeCl_2 + HCl \longrightarrow CrCl_3 + KCl + FeCl_3 + H_2O \quad \text{(unbalanced)}$$

SOLUTION

Step 1 Assign oxidation numbers (Cr and Fe have changed):

$$\underset{+1\ +6\ -2}{K_2Cr_2O_7} + \underset{+2\ -1}{FeCl_2} + \underset{+1\ -1}{HCl} \longrightarrow \underset{+3\ -1}{CrCl_3} + \underset{+1\ -1}{KCl} + \underset{+3\ -1}{FeCl_3} + \underset{+1\ -2}{H_2O}$$

Step 2 Write the oxidation and reduction steps. Balance the number of atoms and then balance the electrical charge using electrons:

$$Fe^{2+} \longrightarrow Fe^{3+} + 1\,e^- \quad \text{(oxidation)}$$

Fe^{2+} loses 1 electron

$$2\,Cr^{6+} + 6\,e^- \longrightarrow 2\,Cr^{3+} \quad \text{(reduction)}$$

$2\,Cr^{6+}$ gain 6 electrons

Step 3 Balance the loss and gain of electrons. Multiply the oxidation step by 6 and the reduction step by 1 to equalize the transfer of electrons.

$$6\,Fe^{2+} \longrightarrow 6\,Fe^{3+} + 6\,e^- \quad \text{(oxidation)}$$

$6\,Fe^{2+}$ lose 6 electrons

$$2\,Cr^{6+} + 6\,e^- \longrightarrow 2\,Cr^{3+} \quad \text{(reduction)}$$

$2\,Cr^{6+}$ gain 6 electrons

Step 4 Transfer the coefficients from the balanced redox equations into the original equation. (Note that one formula unit of $K_2Cr_2O_7$ contains two Cr atoms.) We need to use 1 $K_2Cr_2O_7$, 2 $CrCl_3$, 6 $FeCl_2$, and 6 $FeCl_3$:

$$K_2Cr_2O_7 + 6\,FeCl_2 + HCl \longrightarrow 2\,CrCl_3 + KCl + 6\,FeCl_3 + H_2O \quad \text{(unbalanced)}$$

Step 5 Balance the remaining elements in the order K, Cl, H, O:

$$K_2Cr_2O_7 + 6\,FeCl_2 + 14\,HCl \longrightarrow 2\,CrCl_3 + 2\,KCl + 6\,FeCl_3 + 7\,H_2O \quad \text{(balanced)}$$

Check: The final balanced equation contains 2 K atoms, 2 Cr atoms, 7 O atoms, 6 Fe atoms, 26 Cl atoms, and 14 H atoms on each side.

Practice 17.4

Balance these equations using the change-in-oxidation-number method:

(a) $HNO_3 + S \longrightarrow NO_2 + H_2SO_4 + H_2O$

(b) $CrCl_3 + MnO_2 + H_2O \longrightarrow MnCl_2 + H_2CrO_4$

(c) $KMnO_4 + HCl + H_2S \longrightarrow KCl + MnCl_2 + S + H_2O$

17.4 Balancing Ionic Redox Equations

The main difference between balancing ionic redox equations and molecular redox equations is in how we handle the ions. In the ionic redox equations, besides having the same number of atoms of each element on both sides of the final equation, we must also have equal net charges. In assigning oxidation numbers, we must therefore remember to consider the ionic charge.

Several methods are used to balance ionic redox equations, including, with slight modification, the oxidation-number method just shown for molecular equations. But the most popular method is probably the ion–electron method.

The ion–electron method uses ionic charges and electrons to balance ionic redox equations. Oxidation numbers are not formally used, but it is necessary to determine what is being oxidized and what is being reduced. The method is as follows:

Step 1 Write the two half-reactions that contain the elements being oxidized and reduced using the entire formula of the ion or molecule.

Step 2 Balance the elements other than oxygen and hydrogen.

Step 3 Balance oxygen and hydrogen.

Acidic solution: For reactions in acidic solution, use H^+ and H_2O to balance oxygen and hydrogen. For each oxygen needed, use one H_2O. Then add H^+ as needed to balance the hydrogen atoms.

Basic solution: For reactions in alkaline solutions, first balance as though the reaction were in an acidic solution, using Steps 1–3. Then add as many OH^- ions to each side of the equation as there are H^+ ions in the equation. Now combine the H^+ and OH^- ions into water (for example, $4\,H^+$ and $4\,OH^-$ give $4\,H_2O$). Rewrite the equation, canceling equal numbers of water molecules that appear on opposite sides of the equation.

Step 4 Add electrons (e^-) to each half-reaction to bring them into electrical balance.

Step 5 Since the loss and gain of electrons must be equal, multiply each half-reaction by the appropriate number to make the number of electrons the same in each half-reaction.

Step 6 Add the two half-reactions together, canceling electrons and any other identical substances that appear on opposite sides of the equation.

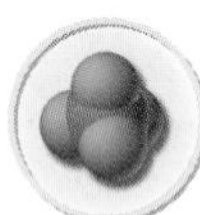

Example 17.9 Balance this equation using the ion–electron method:

$$MnO_4^- + S^{2-} \longrightarrow Mn^{2+} + S^0 \quad \text{(acidic solution)}$$

SOLUTION **Step 1** Write two half-reactions, one containing the element being oxidized and the other the element being reduced (use the entire molecule or ion):

$$S^{2-} \longrightarrow S^0 \quad \text{(oxidation)}$$

$$MnO_4^- \longrightarrow Mn^{2+} \quad \text{(reduction)}$$

Step 2 Balance elements other than oxygen and hydrogen (accomplished in Step 1: 1 S and 1 Mn on each side).

CHEMISTRY IN ACTION • Sensitive Sunglasses

Oxidation–reduction reactions are the basis for many interesting applications. Consider photochromic glass, which is used for lenses in light-sensitive glasses. These lenses, manufactured by the Corning Glass Company, can change from transmitting 85% of light to transmitting only 22% of light when exposed to bright sunlight.

Photochromic glass is composed of linked tetrahedrons of silicon and oxygen atoms jumbled in a disorderly array, with crystals of silver chloride caught between the silica tetrahedrons. When the glass is clear, the visible light passes right through the molecules. The glass absorbs ultraviolet light, however, and this energy triggers an oxidation–reduction reaction between Ag^+ and Cl^-:

$$Ag^+ + Cl^- \xrightarrow{\text{UV light}} Ag^0 + Cl^0$$

To prevent the reaction from reversing itself immediately, a few ions of Cu^+ are incorporated into the silver chloride crystal. These Cu^+ ions react with the newly formed chlorine atoms:

$$Cu^+ + Cl^0 \longrightarrow Cu^{2+} + Cl^-$$

The silver atoms move to the surface of the crystal and form small colloidal clusters of silver metal. This metallic silver absorbs visible light, making the lens appear dark (colored).

As the glass is removed from the light, the Cu^{2+} ions slowly move to the surface of the crystal, where they interact with the silver metal:

$$Cu^{2+} + Ag^0 \longrightarrow Cu^+ + Ag^+$$

The glass clears as the silver ions rejoin chloride ions in the crystals.

An oxidation–reduction reaction causes these photochromic glasses to change from light to dark in bright sunlight.

Step 3 Balance O and H. Remember the solution is acidic. The oxidation requires neither O nor H, but the reduction equation needs $4\,H_2O$ on the right and $8\,H^+$ on the left:

$$S^{2-} \longrightarrow S^0$$

$$8\,H^+ + MnO_4^- \longrightarrow Mn^{2+} + 4\,H_2O$$

Step 4 Balance each half-reaction electrically with electrons:

$$S^{2-} \longrightarrow S^0 + 2\,e^-$$

net charge = −2 on each side

$$5\,e^- + 8\,H^+ + MnO_4^- \longrightarrow Mn^{2+} + 4\,H_2O$$

net charge = +2 on each side

Step 5 Equalize loss and gain of electrons. In this case, multiply the oxidation equation by 5 and the reduction equation by 2:

$$5\,S^{2-} \longrightarrow 5\,S^0 + 10\,e^-$$

$$10\,e^- + 16\,H^+ + 2\,MnO_4^- \longrightarrow 2\,Mn^{2+} + 8\,H_2O$$

Step 6 Add the two half-reactions together, canceling the $10\,e^-$ from each side, to obtain the balanced equation:

$$5\,S^{2-} \longrightarrow 5\,S^0 + \cancel{10\,e^-}$$

$$\cancel{10\,e^-} + 16\,H^+ + 2\,MnO_4^- \longrightarrow 2\,Mn^{2+} + 8\,H_2O$$

$$16\,H^+ + 2\,MnO_4^- + 5\,S^{2-} \longrightarrow 2\,Mn^{2+} + 5\,S^0 + 8\,H_2O \quad \text{(balanced)}$$

Check: Both sides of the equation have a charge of +4 and contain the same number of atoms of each element.

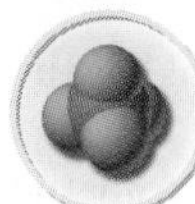

Example 17.10 Balance this equation:

$$CrO_4^{2-} + Fe(OH)_2 \longrightarrow Cr(OH)_3 + Fe(OH)_3 \quad \text{(basic solution)}$$

SOLUTION **Step 1** Write the two half-reactions:

$$Fe(OH)_2 \longrightarrow Fe(OH)_3 \quad \text{(oxidation)}$$

$$CrO_4^{2-} \longrightarrow Cr(OH)_3 \quad \text{(reduction)}$$

Step 2 Balance elements other than H and O (accomplished in Step 1).

Step 3 Remember the solution is basic. Balance O and H as though the solution were acidic. Use H_2O and H^+. To balance O and H in the oxidation equation, add 1 H_2O on the left and 1 H^+ on the right side:

$$Fe(OH)_2 + H_2O \longrightarrow Fe(OH)_3 + H^+$$

Add 1 OH^- to each side:

$$Fe(OH)_2 + H_2O + OH^- \longrightarrow Fe(OH)_3 + H^+ + OH^-$$

Combine H^+ and OH^- as H_2O and rewrite, canceling H_2O on each side:

$$Fe(OH)_2 + \cancel{H_2O} + OH^- \longrightarrow Fe(OH)_3 + \cancel{H_2O}$$

$$\boxed{Fe(OH)_2 + OH^- \longrightarrow Fe(OH)_3} \quad \text{(oxidation)}$$

To balance O and H in the reduction equation, add 1 H_2O on the right and 5 H^+ on the left:

$$CrO_4^{2-} + 5\,H^+ \longrightarrow Cr(OH)_3 + H_2O$$

Add 5 OH^- to each side:

$$CrO_4^{2-} + 5\,H^+ + 5\,OH^- \longrightarrow Cr(OH)_3 + H_2O + 5\,OH^-$$

Combine $5\,H^+ + 5\,OH^- \longrightarrow 5\,H_2O$:

$$CrO_4^{2-} + 5\,H_2O \longrightarrow Cr(OH)_3 + H_2O + 5\,OH^-$$

Rewrite, canceling 1 H_2O from each side:

$$\boxed{CrO_4^{2-} + 4\,H_2O \longrightarrow Cr(OH)_3 + 5\,OH^-} \quad \text{(reduction)}$$

Step 4 Balance each half-reaction electrically with electrons:

$$Fe(OH)_2 + OH^- \longrightarrow Fe(OH)_3 + e^- \quad \text{(balanced oxidation equation)}$$

$$CrO_4^{2-} + 4\,H_2O + 3\,e^- \longrightarrow Cr(OH)_3 + 5\,OH^- \quad \text{(balanced reduction equation)}$$

Step 5 Equalize the loss and gain of electrons. Multiply the oxidation reaction by 3:

$$3\,Fe(OH)_2 + 3\,OH^- \longrightarrow 3\,Fe(OH)_3 + 3\,e^-$$

$$CrO_4^{2-} + 4\,H_2O + 3\,e^- \longrightarrow Cr(OH)_3 + 5\,OH^-$$

Step 6 Add the two half-reactions together, canceling the $3\,e^-$ and $3\,OH^-$ from each side of the equation:

$$
\begin{array}{rcl}
3\,Fe(OH)_2 + 3\,OH^- & \longrightarrow & 3\,Fe(OH)_3 + \cancel{3e^-} \\
CrO_4^{2-} + 4\,H_2O + \cancel{3e^-} & \longrightarrow & Cr(OH)_3 + 5\,OH^- \\
\hline
CrO_4^{2-} + 3\,Fe(OH)_2 + 4\,H_2O & \longrightarrow & Cr(OH)_3 + 3\,Fe(OH)_3 + 2\,OH^-
\end{array}
\quad \text{(balanced)}
$$

Check: Each side of the equation has a charge of −2 and contains the same number of atoms of each element.

Practice 17.5

Balance these equations using the ion–electron method:

(a) $I^- + NO_2^- \longrightarrow I_2 + NO$ (acidic solution)
(b) $Cl_2 + IO_3^- \longrightarrow IO_4^- + Cl^-$ (basic solution)
(c) $AuCl_4^- + Sn^{2+} \longrightarrow Sn^{4+} + AuCl + Cl^-$

Ionic equations can also be balanced using the change-in-oxidation-number method shown in Example 17.6. To illustrate this method, let's use the equation from Example 17.10.

Example 17.11

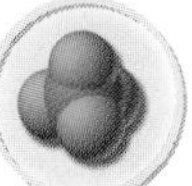

Balance this equation using the change-in-oxidation-number method:

$$CrO_4^{2-} + Fe(OH)_2 \longrightarrow Cr(OH)_3 + Fe(OH)_3 \quad \text{(basic solution)}$$

SOLUTION

Steps 1 and 2 Assign oxidation numbers and balance the charges with electrons:

$$Cr^{6+} + 3\,e^- \longrightarrow Cr^{3+} \quad \text{(reduction)}$$
Cr^{6+} gains $3\,e^-$

$$Fe^{2+} \longrightarrow Fe^{3+} + e^- \quad \text{(oxidation)}$$
Fe^{2+} loses $1\,e^-$

Step 3 Equalize the loss and gain of electrons, and then multiply the oxidation step by 3:

$$Cr^{6+} + 3\,e^- \longrightarrow Cr^{3+} \quad \text{(reduction)}$$
Cr^{6+} gains $3\,e^-$

$$3\,Fe^{2+} \longrightarrow 3\,Fe^{3+} + 3\,e^- \quad \text{(oxidation)}$$
$3\,Fe^{2+}$ loses $3\,e^-$

Step 4 Transfer coefficients back to the original equation:

$$CrO_4^{2-} + 3\,Fe(OH)_2 \longrightarrow Cr(OH)_3 + 3\,Fe(OH)_3$$

Step 5 Balance electrically. Because the solution is basic, use OH^- to balance charges. The charge on the left side is −2 and on the right side is 0. Add $2\,OH^-$ ions to the right side of the equation:

$$CrO_4^{2-} + 3\,Fe(OH)_2 \longrightarrow Cr(OH)_3 + 3\,Fe(OH)_3 + 2\,OH^-$$

Adding $4\,H_2O$ to the left side balances the equation:

$$CrO_4^{2-} + 3\,Fe(OH)_2 + 4\,H_2O \longrightarrow Cr(OH)_3 + 3\,Fe(OH)_3 + 2\,OH^- \quad \text{(balanced)}$$

Check: Each side of the equation has a charge of −2 and contains the same number of atoms of each element.

Figure 17.2
A coil of copper placed in a silver nitrate solution forms silver crystals on the wire. The pale blue of the solution indicates the presence of copper ions.

Practice 17.6

Balance these equations using the change-in-oxidation-number method:

(a) $Zn \longrightarrow Zn(OH)_4^{2-} + H_2$ (basic solution)
(b) $H_2O_2 + Sn^{2+} \longrightarrow Sn^{4+}$ (acidic solution)
(c) $Cu + Cu^{2+} \longrightarrow Cu_2O$ (basic solution)

17.5 Activity Series of Metals

Knowledge of the relative chemical reactivities of the elements helps us predict the course of many chemical reactions. For example, calcium reacts with cold water to produce hydrogen, and magnesium reacts with steam to produce hydrogen. Therefore, calcium is considered a more reactive metal than magnesium:

$$Ca(s) + 2\,H_2O(l) \longrightarrow Ca(OH)_2(aq) + H_2(g)$$

$$Mg(s) + \underset{\text{steam}}{H_2O(g)} \longrightarrow MgO(s) + H_2(g)$$

The difference in their activity is attributed to the fact that calcium loses its two valence electrons more easily than magnesium and is therefore more reactive and/or more readily oxidized than magnesium.

When a coil of copper is placed in a solution of silver nitrate ($AgNO_3$), free silver begins to plate out on the copper. (See Figure 17.2.) After the reaction has continued for some time, we can observe a blue color in the solution, indicating the presence of copper(II) ions. The equations are

$$Cu^0(s) + 2\,AgNO_3(aq) \longrightarrow 2\,Ag^0(s) + Cu(NO_3)_2(aq)$$

$$Cu^0(s) + 2\,Ag^+(aq) \longrightarrow 2\,Ag^0(s) + Cu^{2+}(aq) \quad \text{(net ionic equation)}$$

$$Cu^0(s) \longrightarrow Cu^{2+}(aq) + 2\,e^- \quad \text{(oxidation of } Cu^0\text{)}$$

$$Ag^+(aq) + e^- \longrightarrow Ag^0(s) \quad \text{(reduction of } Ag^+\text{)}$$

If a coil of silver is placed in a solution of copper(II) nitrate, $Cu(NO_3)_2$, no reaction is visible.

$$Ag^0(s) + Cu(NO_3)_2(aq) \longrightarrow \text{no reaction}$$

In the reaction between Cu and $AgNO_3$, electrons are transferred from Cu^0 atoms to Ag^+ ions in solution. Copper has a greater tendency than silver to lose electrons, so an electrochemical force is exerted upon silver ions to accept electrons from copper atoms. When an Ag^+ ion accepts an electron, it is reduced to an Ag^0 atom and is no longer soluble in solution. At the same time, Cu^0 is oxidized and goes into solution as Cu^{2+} ions. From this reaction, we can conclude that copper is more reactive than silver.

Table 17.3 Activity Series of Metals

Ease of oxidation (increasing upward)	
↑	$K \longrightarrow K^+ + e^-$
	$Ba \longrightarrow Ba^{2+} + 2\,e^-$
	$Ca \longrightarrow Ca^{2+} + 2\,e^-$
	$Na \longrightarrow Na^+ + e^-$
	$Mg \longrightarrow Mg^{2+} + 2\,e^-$
	$Al \longrightarrow Al^{3+} + 3\,e^-$
	$Zn \longrightarrow Zn^{2+} + 2\,e^-$
	$Cr \longrightarrow Cr^{3+} + 3\,e^-$
	$Fe \longrightarrow Fe^{2+} + 2\,e^-$
	$Ni \longrightarrow Ni^{2+} + 2\,e^-$
	$Sn \longrightarrow Sn^{2+} + 2\,e^-$
	$Pb \longrightarrow Pb^{2+} + 2\,e^-$
	$\mathbf{H_2 \longrightarrow 2\,H^+ + 2\,e^-}$
	$Cu \longrightarrow Cu^{2+} + 2\,e^-$
	$As \longrightarrow As^{3+} + 3\,e^-$
	$Ag \longrightarrow Ag^+ + e^-$
	$Hg \longrightarrow Hg^{2+} + 2\,e^-$
	$Au \longrightarrow Au^{3+} + 3\,e^-$

Metals such as sodium, magnesium, zinc, and iron that react with solutions of acids to liberate hydrogen are more reactive than hydrogen. Metals such as copper, silver, and mercury that do not react with solutions of acids to liberate hydrogen are less reactive than hydrogen. By studying a series of reactions such as these, we can list metals according to their chemical activity, placing the most active at the top and the least active at the bottom. This list is called the **activity series of metals**. Table 17.3 lists some of the common metals in the series. The arrangement corresponds to the ease with which the elements

activity series of metals

are oxidized or lose electrons, with the most easily oxidizable element listed first. More extensive tables are available in chemistry reference books.

The general principles governing the arrangement and use of the activity series are as follows:

1. The reactivity of the metals listed decreases from top to bottom.
2. A free metal can displace the ion of a second metal from solution, provided that the free metal is above the second metal in the activity series.
3. Free metals above hydrogen react with nonoxidizing acids in solution to liberate hydrogen gas.
4. Free metals below hydrogen do not liberate hydrogen from acids.
5. Conditions such as temperature and concentration may affect the relative position of some of these elements.

Here are two examples using the activity series of metals.

Example 17.12

Will zinc metal react with dilute sulfuric acid?

SOLUTION

From Table 17.3, we see that zinc is above hydrogen; therefore, zinc atoms will lose electrons more readily than hydrogen atoms. Hence zinc atoms will reduce hydrogen ions from the acid to form hydrogen gas and zinc ions. In fact, these reagents are commonly used for the laboratory preparation of hydrogen. The equation is

$$Zn(s) + H_2SO_4(aq) \longrightarrow ZnSO_4(aq) + H_2(g)$$

$$Zn(s) + 2\,H^+(aq) \longrightarrow Zn^{2+}(aq) + H_2(g) \qquad \text{(net ionic equation)}$$

Example 17.13

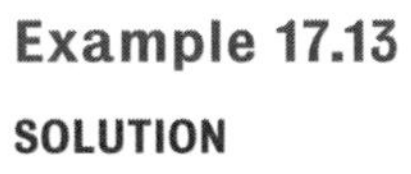

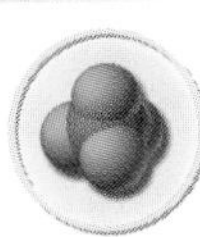

Will a reaction occur when copper metal is placed in an iron(II) sulfate solution?

SOLUTION

No, copper lies below iron in the series, loses electrons less easily than iron, and therefore will not displace iron(II) ions from solution. In fact, the reverse is true. When an iron nail is dipped into a copper(II) sulfate solution, it becomes coated with free copper. The equations are

$$Cu(s) + FeSO_4(aq) \longrightarrow \text{no reaction}$$

$$Fe(s) + CuSO_4(aq) \longrightarrow FeSO_4(aq) + Cu(s)$$

From Table 17.3, we may abstract the following pair in their relative position to each other:

$$Fe \longrightarrow Fe^{2+} + 2\,e^-$$

$$Cu \longrightarrow Cu^{2+} + 2\,e^-$$

According to Principle 2 on the use of the activity series, we can predict that free iron will react with copper(II) ions in solution to form free copper metal and iron(II) ions in solution:

$$Fe(s) + Cu^{2+}(aq) \longrightarrow Fe^{2+}(aq) + Cu(s) \qquad \text{(net ionic equation)}$$

Practice 17.7

Indicate whether these reactions will occur:

(a) Sodium metal is placed in dilute hydrochloric acid.
(b) A piece of lead is placed in magnesium nitrate solution.
(c) Mercury is placed in a solution of silver nitrate.

17.6 Electrolytic and Voltaic Cells

electrolysis
electrolytic cell

The process in which electrical energy is used to bring about chemical change is known as **electrolysis.** An **electrolytic cell** uses electrical energy to produce a chemical reaction. The use of electrical energy has many applications in industry—for example, in the production of sodium, sodium hydroxide, chlorine, fluorine, magnesium, aluminum, and pure hydrogen and oxygen, and in the purification and electroplating of metals.

What happens when an electric current is passed through a solution? Let's consider a hydrochloric acid solution in a simple electrolytic cell, as shown in Figure 17.3. The cell consists of a source of direct current (a battery) connected to two electrodes that are immersed in a solution of hydrochloric acid.

cathode
anode

The negative electrode is called the **cathode** because cations are attracted to it. The positive electrode is called the **anode** because anions are attracted to it. The cathode is attached to the negative pole and the anode to the positive pole of the battery. The battery supplies electrons to the cathode.

When the electric circuit is completed, positive hydronium ions (H_3O^+) migrate to the cathode, where they pick up electrons and evolve as hydrogen gas. At the same time the negative chloride ions (Cl^-) migrate to the anode, where they lose electrons and evolve as chlorine gas.

Reaction at the cathode:

$$H_3O^+ + 1\,e^- \longrightarrow H^0 + H_2O \quad \text{(reduction)}$$
$$H^0 + H^0 \longrightarrow H_2$$

Reaction at the anode:

$$Cl^- \longrightarrow Cl^0 + 1\,e^- \quad \text{(oxidation)}$$
$$Cl^0 + Cl^0 \longrightarrow Cl_2$$

$$2\,HCl(aq) \xrightarrow{\text{electrolysis}} H_2(g) + Cl_2(g) \quad \text{(net reaction)}$$

Note that oxidation–reduction has taken place. Chloride ions lost electrons (were oxidized) at the anode, and hydronium ions gained electrons (were reduced) at the cathode.

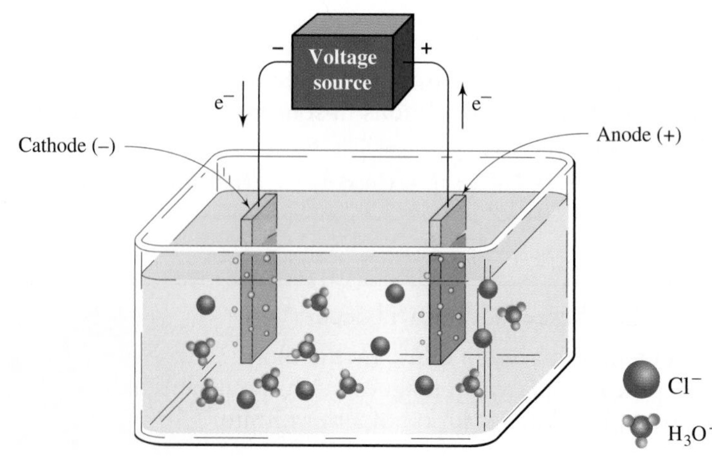

Figure 17.3
During the electrolysis of a hydrochloric acid solution, positive hydronium ions are attracted to the cathode, where they gain electrons and form hydrogen gas. Chloride ions migrate to the anode, where they lose electrons and form chlorine gas. The equation for this process is
$2\,HCl(aq) \longrightarrow H_2(g) + Cl_2(g)$.

Oxidation always occurs at the anode and reduction at the cathode.

When concentrated sodium chloride solutions (brines) are electrolyzed, the products are sodium hydroxide, hydrogen, and chlorine. The overall reaction is

$$2\,Na^+(aq) + 2\,Cl^-(aq) + 2\,H_2O(l) \xrightarrow{\text{electrolysis}} 2\,Na^+(aq) + 2\,OH^-(aq) + H_2(g) + Cl_2(g)$$

The net ionic equation is

$$2\,Cl^-(aq) + 2\,H_2O(l) \longrightarrow 2\,OH^-(aq) + H_2(g) + Cl_2(g)$$

During electrolysis, Na^+ ions move toward the cathode and Cl^- ions move toward the anode. The anode reaction is similar to that of hydrochloric acid; the chlorine is liberated:

$$2\,Cl^-(aq) \longrightarrow Cl_2(g) + 2\,e^-$$

Even though Na^+ ions are attracted by the cathode, the facts show that hydrogen is liberated there. No evidence of metallic sodium is found, but the area around the cathode tests alkaline from the accumulated OH^- ions. The reaction at the cathode is

$$2\,H_2O(l) + 2\,e^- \longrightarrow H_2(g) + 2\,OH^-(aq)$$

If electrolysis is allowed to continue until all the chloride is reacted, the solution remaining will contain only sodium hydroxide, which on evaporation yields solid NaOH. Large amounts of sodium hydroxide and chlorine are made by this process.

When molten sodium chloride (without water) is subjected to electrolysis, metallic sodium and chlorine gas are formed:

$$2\,Na^+(l) + 2\,Cl^-(l) \xrightarrow{\text{electrolysis}} 2\,Na(s) + Cl_2(g)$$

An important electrochemical application is the electroplating of metals. Electroplating is the art of covering a surface or an object with a thin adherent electrodeposited metal coating. Electroplating is done for protection of the surface of the base metal or for a purely decorative effect. The layer deposited is surprisingly thin, varying from as little as 5×10^{-5} cm to 2×10^{-3} cm, depending on the metal and the intended use. The object to be plated is set up as the cathode and is immersed in a solution containing ions of the plating metal. When an electric current passes through the solution, metal ions that migrate to the cathode are reduced, depositing on the object as the free metal. In most cases, the metal deposited on the object is replaced in the solution by using an anode of the same metal. The following equations show the chemical changes in the electroplating of nickel:

Reaction at the cathode: $Ni^{2+}(aq) + 2\,e^- \longrightarrow Ni(s)$ (Ni plated out on an object)

Reaction at the anode: $Ni(s) \longrightarrow Ni^{2+}(aq) + 2\,e^-$ (Ni replenished in solution)

Metals commonly used in commercial electroplating are copper, nickel, zinc, lead, cadmium, chromium, tin, gold, and silver.

Jewelry and eyeglasses are electroplated with rhodium to prevent tarnishing.

In the electrolytic cell shown in Figure 17.3, electrical energy from the voltage source is used to bring about nonspontaneous redox reactions. The hydrogen and chlorine produced have more potential energy than was present in the hydrochloric acid before electrolysis.

Conversely, some spontaneous redox reactions can be made to supply useful amounts of electrical energy. When a piece of zinc is put in a copper(II) sulfate solution, the zinc quickly becomes coated with metallic copper. We expect this coating to happen because zinc is above copper in the activity series; copper(II) ions are therefore reduced by zinc atoms:

$$Zn^0(s) + Cu^{2+}(aq) \longrightarrow Zn^{2+}(aq) + Cu^0(s)$$

This reaction is clearly a spontaneous redox reaction, but simply dipping a zinc rod into a copper(II) sulfate solution will not produce useful electric current. However, when we carry out this reaction in the cell shown in Figure 17.4, an electric current is produced. The cell consists of a piece of zinc immersed in a zinc sulfate solution and connected by a wire through a voltmeter to a piece of copper immersed in copper(II) sulfate solution. The two solutions are connected by a salt bridge. Such a cell produces an electric current and a potential of about 1.1 volts when both solutions are 1.0 *M* in concentration. A cell that produces electric current from a spontaneous chemical reaction is called a **voltaic cell.** A voltaic cell is also known as a *galvanic cell.*

voltaic cell

The driving force responsible for the electric current in the zinc–copper cell originates in the great tendency of zinc atoms to lose electrons relative to the tendency of copper(II) ions to gain electrons. In the cell shown in Figure 17.4, zinc atoms lose electrons and are converted to zinc ions at the zinc electrode surface; the electrons flow through the wire (external circuit) to the copper electrode. Here copper(II) ions pick up electrons and are reduced to copper atoms, which plate out on the copper electrode. Sulfate ions flow from the $CuSO_4$ solution via the salt bridge into the $ZnSO_4$ solution (internal circuit) to complete the circuit.

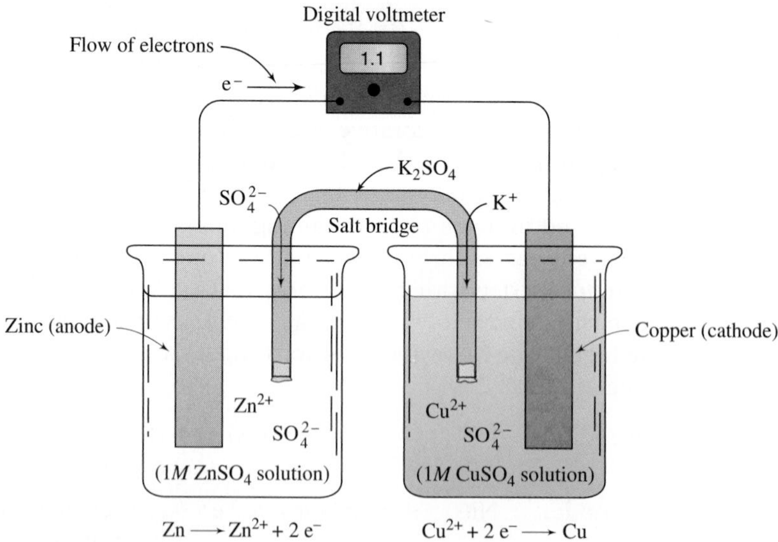

Figure 17.4
Zinc–copper voltaic cell. The cell has a potential of 1.1 volts when $ZnSO_4$ and $CuSO_4$ solutions are 1.0 M. The salt bridge provides electrical contact between the two half-cells.

The equations for the reactions of this cell are

anode	$Zn^0(s) \longrightarrow Zn^{2+}(aq) + 2\,e^-$	(oxidation)
cathode	$Cu^{2+}(aq) + 2\,e^- \longrightarrow Cu^0(s)$	(reduction)
net ionic	$Zn^0(s) + Cu^{2+}(aq) \longrightarrow Zn^{2+}(aq) + Cu^0(s)$	
overall	$Zn(s) + CuSO_4(aq) \longrightarrow ZnSO_4(aq) + Cu(s)$	

The redox reaction, the movement of electrons in the metallic or external part of the circuit, and the movement of ions in the solution or internal part of the circuit of the zinc–copper cell are very similar to the actions that occur in the electrolytic cell of Figure 17.3. The only important difference is that the reactions of the zinc–copper cell are spontaneous. This spontaneity is the crucial difference between all voltaic and electrolytic cells.

> Voltaic cells use chemical reactions to produce electrical energy. Electrolytic cells use electrical energy to produce chemical reactions.

Calculators, watches, radios, cell phones, and portable CD players are powered by small efficient voltaic cells called *dry cell batteries*. These batteries are called dry cells because they do not contain a liquid electrolyte (like the voltaic cells discussed earlier). Dry cell batteries are found in several different versions.

The *acid*-type dry cell battery contains a zinc inner case that functions as the anode. A carbon (graphite) rod runs through the center and is in contact with the zinc case at one end and a moist paste of solid MnO_2, NH_4Cl, and carbon that functions as the cathode. (See Figure 17.5a.) The cell produces about 1.5 volts. The *alkaline*-type dry cell battery is the same as the acid type except the NH_4Cl is replaced by either KOH or NaOH. These dry cells typically last longer because the zinc anode corrodes more slowly in basic conditions. A third type of dry cell is the *zinc–mercury* cell shown in Figure 17.5b. The reactions occurring in this cell are

anode	$Zn^0 + 2\,OH^- \longrightarrow ZnO + H_2O + 2\,e^-$	(oxidation)
cathode	$HgO + H_2O + 2\,e^- \longrightarrow Hg^0 + 2\,OH^-$	(reduction)
net ionic	$Zn^0 + Hg^{2+} \longrightarrow Zn^{2+} + Hg^0$	
overall	$Zn^0 + HgO \longrightarrow ZnO + Hg^0$	

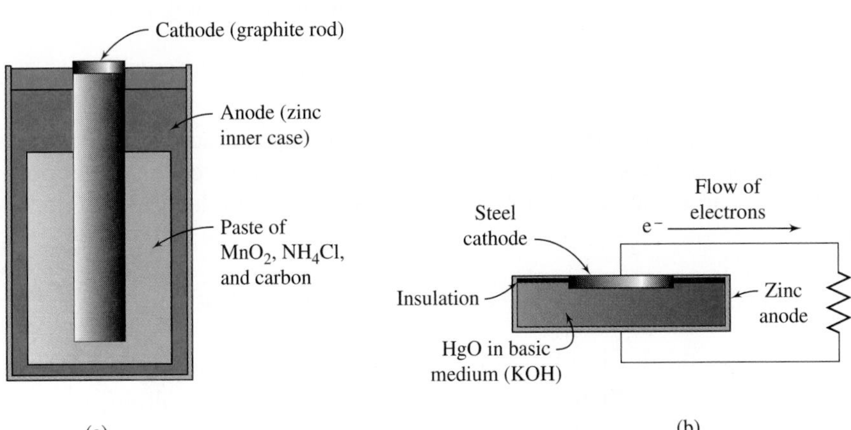

Figure 17.5
(a) A common acid-type dry cell.
(b) Diagram of an alkaline zinc–mercury cell.

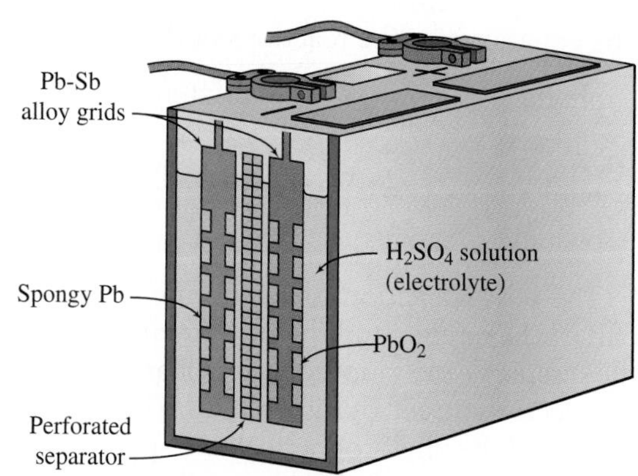

Figure 17.6
Cross-sectional diagram of a lead storage battery cell.

To offset the relatively high initial cost, this cell (1) provides current at a very steady potential of about 1.5 volts; (2) has an exceptionally long service life—that is, high energy output to weight ratio; (3) is completely self-contained; and (4) can be stored for relatively long periods of time when not in use.

An automobile storage battery is an energy reservoir. The charged battery acts as a voltaic cell and through chemical reactions furnishes electrical energy to operate the starter, lights, radio, and so on. When the engine is running, a generator or alternator produces and forces an electric current through the battery and, by electrolytic chemical action, restores it to the charged condition.

The cell unit consists of a lead plate filled with spongy lead and a lead(IV) oxide plate, both immersed in dilute sulfuric acid solution, which serves as the electrolyte (see Figure 17.6). When the cell is discharging, or acting as a voltaic cell, these reactions occur:

Pb plate (anode): $Pb^0 \longrightarrow Pb^{2+} + 2\,e^-$ (oxidation)

PbO_2 plate (cathode): $PbO_2 + 4\,H^+ + 2\,e^- \longrightarrow Pb^{2+} + 2\,H_2O$ (reduction)

Net ionic redox reaction: $Pb^0 + PbO_2 + 4\,H^+ \longrightarrow 2\,Pb^{2+} + 2\,H_2O$

Precipitation reaction on plates: $Pb^{2+}(aq) + SO_4^{2-}(aq) \longrightarrow PbSO_4(s)$

Because lead(II) sulfate is insoluble, the Pb^{2+} ions combine with SO_4^{2-} ions to form a coating of $PbSO_4$ on each plate. The overall chemical reaction of the cell is

$$Pb(s) + PbO_2(s) + 2\,H_2SO_4(aq) \xrightarrow[\text{cycle}]{\text{discharge}} 2\,PbSO_4(s) + 2\,H_2O(l)$$

The cell can be recharged by reversing the chemical reaction. This reversal is accomplished by forcing an electric current through the cell in the opposite direction. Lead sulfate and water are reconverted to lead, lead(IV) oxide, and sulfuric acid:

$$2\,PbSO_4(s) + 2\,H_2O(l) \xrightarrow[\text{cycle}]{\text{charge}} Pb(s) + PbO_2(s) + 2\,H_2SO_4(aq)$$

The electrolyte in a lead storage battery is a 38% by mass sulfuric acid solution having a density of 1.29 g/mL. As the battery is discharged, sulfuric acid is removed, thereby decreasing the density of the electrolyte solution. The state of

CHEMISTRY IN ACTION • Superbattery Uses Hungry Iron Ions

Scientists are constantly trying to make longer-lasting environmentally friendly batteries to fuel our many electronic devices. Now a new type of alkaline batteries called "super-iron" batteries have been developed by chemists at Technion-Israel Institute of Technology in Israel. A traditional cell of this type is shown on p. 455. In the new battery, the heavy manganese dioxide cathode is replaced with "super iron." (See accompanying diagram.) This special type of iron compound contains iron(VI) in compounds such as K_2FeO_4 or $BaFeO_4$. Iron typically has an oxidation state of +2 or +3. In super iron, each iron atom is missing six electrons instead of the usual 2 or 3. This allows the battery to store 1505 J more energy than other alkaline batteries. When the battery is used (or the cell discharges), the following reaction occurs:

$$2\,MFeO_4 + 3\,Zn \longrightarrow Fe_2O_3 + ZnO + 2\,MZnO_2 \quad (M = K_2 \text{ or } Ba)$$

The iron compounds used in this battery are much less expensive than the current MnO_2 compounds and the products are more environmentally friendly (Fe_2O_3 is a form of rust).

Manganese dioxide molecules in conventional batteries can only each accept 1 electron while iron(VI) compounds can absorb 3 electrons each. The super-iron compounds are highly conductive, which means the super-iron battery will work well in our high-drain-rate electronic items. The accompanying graph shows a comparison between a conventional battery and a super-iron battery (both AAA). The conventional AAA battery lasts less than half as long as a super-iron AAA battery. Just think how many more CDs you could play in that time without changing the batteries!

Super-iron battery

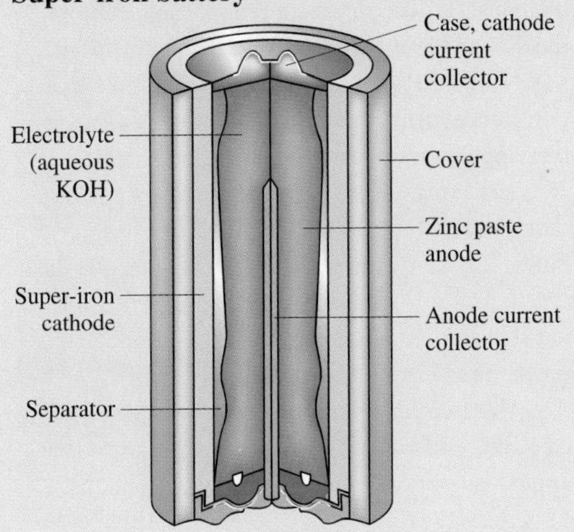

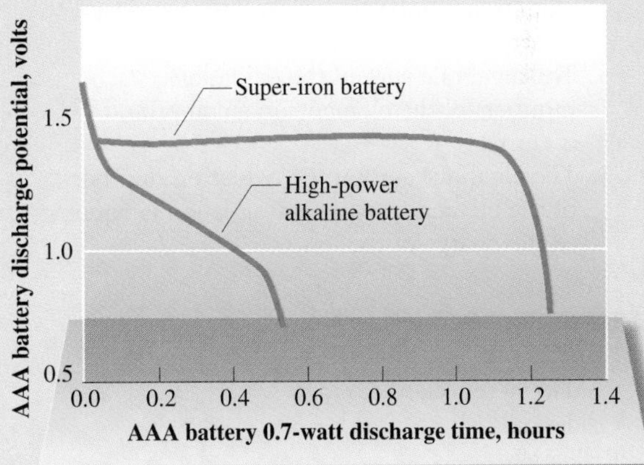

charge or discharge of the battery can be estimated by measuring the density (or specific gravity) of the electrolyte solution with a hydrometer. When the density has dropped to about 1.05 g/mL, the battery needs recharging.

In a commerical battery, each cell consists of a series of cell units of alternating lead–lead(IV) oxide plates separated and supported by wood, glass wool, or fiberglass. The energy storage capacity of a single cell is limited, and its electrical potential is only about 2 volts. Therefore a bank of six cells is connected in series to provide the 12-volt output of the usual automobile battery.

Practice 17.8

Consider the reaction

$$Sn^{2+}(aq) + Cu(s) \longrightarrow Sn(s) + Cu^{2+}(aq)$$

(a) Which metal is oxidized and which is reduced?
(b) Write the reaction occurring at the anode and at the cathode.

Chapter 17 Review

17.1 Oxidation Number

KEY TERM

Oxidation number or oxidation state

- To assign an oxidation number:

1. All elements in their free state (uncombined with other elements) have an oxidation number of zero (e.g., Na, Cu, Mg, H_2, O_2, Cl_2, N_2).
2. H is +1, except in metal hydrides, where it is −1 (e.g., NaH, CaH_2).
3. O is −2, except in peroxides, where it is −1, and in OF_2, where it is +2.
4. The metallic element in an ionic compound has a positive oxidation number.
5. In covalent compounds the negative oxidation number is assigned to the most electronegative atom.
6. The algebraic sum of the oxidation numbers of the elements in a compound is zero.
7. The algebraic sum of the oxidation numbers of the elements in a polyatomic ion is equal to the charge of the ion.

17.2 Oxidation–Reduction

KEY TERMS

Oxidation–reduction or redox
Oxidation
Reduction
Oxidizing agent
Reducing agent

- Oxidation–reduction is a chemical process in which electrons are transferred from one atom to another to change the oxidation number of the atom.
- When the oxidation number increases, oxidation occurs, resulting in the loss of electrons.
- When the oxidation number decreases, reduction occurs, resulting in the gain of electrons.

17.3 Balancing Oxidation–Reduction Equations

- Trial and error or inspection methods.
- Can be balanced by writing and balancing the half-reactions for the overall reaction.
- Can be balanced by the change-in-oxidation-number method.

17.4 Balancing Ionic Redox Equations

- To balance equations that are ionic, charge must also be balanced (in addition to atoms and ions).
- The ion–electron method for balancing equations is:

Step 1 Write the two half-reactions that contain the elements being oxidized and reduced using the entire formula of the ion or molecule.

Step 2 Balance the elements other than oxygen and hydrogen.

Step 3 Balance oxygen and hydrogen.
Acidic solution: For reactions in acidic solution, use H^+ and H_2O to balance oxygen and hydrogen. For each oxygen needed, use one H_2O. Then add H^+ as needed to balance the hydrogen atoms.
Basic solution: For reactions in alkaline solutions, first balance as though the reaction were in an acidic solution, using Steps 1–3. Then add as many OH^- ions to each side of the equation as there are H^+ ions in the equation. Now combine the H^+ and OH^- ions into water (for example, $4\,H^+$ and $4\,OH^-$ give $4\,H_2O$). Rewrite the equation, canceling equal numbers of water molecules that appear on opposite sides of the equation.

Step 4 Add electrons (e^-) to each half-reaction to bring them into electrical balance.

Step 5 Since the loss and gain of electrons must be equal, multiply each half-reaction by the appropriate number to make the number of electrons the same in each half-reaction.

Step 6 Add the two half-reactions together, canceling electrons and any other identical substances that appear on opposite sides of the equation.

17.5 Activity Series of Metals

KEY TERM

Activity series of metals

- The activity series lists metals from most to least reactive.
- A free metal can displace anything lower on the activity series.

17.6 Electrolytic and Voltaic Cells

KEY TERMS

Electrolysis
Electrolytic cell
Cathode
Anode
Voltaic cell

- Electrolysis is the process of using electricity to bring about chemical change.
- A typical electrolytic cell is shown next:

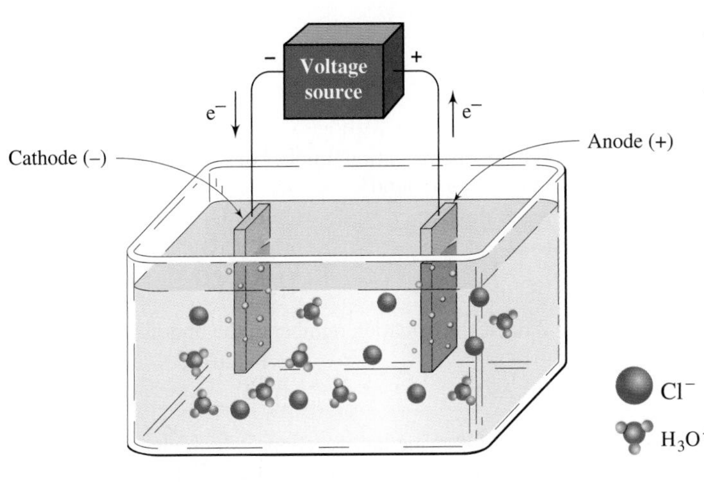

- Oxidation always occurs at the anode and reduction at the cathode.
- A cell that produces electric current from a spontaneous chemical reaction is a voltaic or galvanic cell.

Review Questions

All questions with blue numbers have answers in the appendix of the text.

1. Why do we say that the more active a metal is, the more easily it will be oxidized?
2. In the equation
   $I_2 + 5\,Cl_2 + 6\,H_2O \longrightarrow 2\,HIO_3 + 10\,HCl$
   (a) has iodine been oxidized or has it been reduced?
   (b) has chlorine been oxidized or has it been reduced? (Figure 17.1)
3. Which element of each pair is more active? (Table 17.3)
   (a) Ag or Al (b) Na or Ba (c) Ni or Cu
4. Will the following combinations react in aqueous solution? (Table 17.3)
   (a) $Zn(s) + Cu^{2+}(aq)$ (e) $Ba(s) + FeCl_2(aq)$
   (b) $Ag(s) + H^+(aq)$ (f) $Pb(s) + NaCl(aq)$
   (c) $Sn(s) + Ag^+(aq)$ (g) $Ni(s) + Hg(NO_3)_2(aq)$
   (d) $As(s) + Mg^{2+}(aq)$ (h) $Al(s) + CuSO_4(aq)$
5. Why doesn't a silver wire placed in a solution of copper(II) nitrate ($Cu(NO_3)_2$) produce a reaction?
6. The reaction between powdered aluminum and iron(III) oxide (in the thermite process) producing molten iron is very exothermic.
   (a) Write the equation for the chemical reaction that occurs.
   (b) Explain in terms of Table 17.3 why a reaction occurs.
   (c) Would you expect a reaction between powdered iron and aluminum oxide?
   (d) Would you expect a reaction between powdered aluminum and chromium(III) oxide?
7. Write equations for the chemical reaction of aluminum, chromium, gold, iron, copper, magnesium, mercury, and zinc with dilute solutions of (a) hydrochloric acid and (b) sulfuric acid. If a reaction will not occur, write "no reaction" as the product. (Table 17.3)
8. What is the difference between an oxidation number for an atom in an ionic compound and an oxidation number for an atom in a covalently bonded compound?
9. State the charge and purpose of the anode and the cathode in an electrolytic or volatic cell.
10. A $NiCl_2$ solution is placed in the apparatus shown in Figure 17.3, instead of the HCl solution shown. Write equations for the following:
    (a) the anode reaction
    (b) the cathode reaction
    (c) the net electrochemical reaction
11. What is the major distinction between the reactions occurring in Figures 17.3 and 17.4?
12. In the cell shown in Figure 17.4,
    (a) what would be the effect of removing the voltmeter and connecting the wires shown coming to the voltmeter?
    (b) what would be the effect of removing the salt bridge?
13. Why are oxidation and reduction said to be complementary processes?
14. When molten $CaBr_2$ is electrolyzed, calcium metal and bromine are produced. Write equations for the two half-reactions that occur at the electrodes. Label the anode half-reaction and the cathode half-reaction.
15. Why is direct current used instead of alternating current in the electroplating of metals?
16. What property of lead(IV) oxide and lead(II) sulfate makes it unnecessary to have salt bridges in the cells of a lead storage battery?
17. Explain why the density of the electrolyte in a lead storage battery decreases during the discharge cycle.
18. In one type of alkaline cell used to power devices such as portable radios, Hg^{2+} ions are reduced to metallic mercury when the cell is being discharged. Does this reduction occur at the anode or the cathode? Explain.
19. Differentiate between an electrolytic cell and a voltaic cell.
20. Why is a porous barrier or a salt bridge necessary in some voltaic cells?

Paired Exercises

All exercises with blue numbers have answers in the appendix of the text.

1. What is the oxidation number of the underlined element in each compound?

(a) $\underline{Na}Cl$
(b) $Fe\underline{Cl}_3$
(c) $\underline{Pb}O_2$
(d) $Na\underline{N}O_3$
(e) $H_2\underline{S}O_3$
(f) $\underline{N}H_4Cl$

3. What is the oxidation number of the underlined elements?

(a) $\underline{S}^{2-}$
(b) $\underline{N}O_2^-$
(c) $Na_2\underline{O}_2$
(d) $\underline{Bi}^{3+}$

5. In the following half-reactions, which element is changing oxidation state? Is the half-reaction an oxidation or a reduction? Supply the proper number of electrons to each side to balance each equation.

(a) $Zn^{2+} \longrightarrow Zn$
(b) $2\,Br^- \longrightarrow Br_2$
(c) $MnO_4^- + 8\,H^+ \longrightarrow Mn^{2+} + 4\,H_2O$
(d) $Ni \longrightarrow Ni^{2+}$

7. In the following unbalanced equations, identify

(a) the oxidized element and the reduced element
(b) the oxidizing agent and the reducing agent

1. $Cr + HCl \longrightarrow CrCl_3 + H_2$
2. $SO_4^{2-} + I^- + H^+ \longrightarrow H_2S + I_2 + H_2O$

9. Determine whether the following oxidation–reduction reactions are balanced correctly. If they are not, provide the correct balanced reaction.

(a) unbalanced:
$CH_4(g) + O_2(g) \longrightarrow CO_2(g) + H_2O(g)$
balanced:
$CH_4(g) + 2\,O_2(g) \longrightarrow CO_2(g) + 2\,H_2O(g)$

(b) unbalanced:
$NO^{2-}(aq) + Al(s) \longrightarrow NH_3(g) + AlO_2^-(aq)$ (basic solution)
balanced: $2\,H_2O(l) + Al(s) + NO^{2-}(aq) \longrightarrow AlO_2^-(aq) + NH_3(g) + OH^-(aq)$

(c) unbalanced:
$Mg(s) + HCl(aq) \longrightarrow Mg^{2+}(aq) + Cl^-(aq) + H_2(g)$
balanced: $Mg(s) + 2\,HCl(aq) \longrightarrow Mg^{2+}(aq) + Cl^-(aq) + H_2(g)$

(d) unbalanced: $CH_3OH(aq) + Cr_2O_7^{2-}(aq) \longrightarrow CH_2O(aq) + Cr^{3+}(aq)$ (acidic solution)
balanced:
$3\,CH_3OH(aq) + 14\,H^+(aq) + Cr_2O_7^{2-}(aq) \longrightarrow 2\,Cr^{3+}(aq) + 3\,CH_2O(aq) + 7\,H_2O(l) + 6\,H^+(aq)$

2. What is the oxidation number of the underlined element in each compound?

(a) $K\underline{Mn}O_4$
(b) $\underline{I}_2$
(c) $\underline{N}H_3$
(d) $K\underline{Cl}O_3$
(e) $K_2\underline{Cr}O_4$
(f) $K_2\underline{Cr}_2O_7$

4. What is the oxidation number of the underlined elements?

(a) $\underline{O}_2$
(b) $\underline{As}O_4^{3-}$
(c) $Fe(\underline{O}H)_3$
(d) $\underline{I}O_3^-$

6. In the following half-reactions, which element is changing oxidation state? Is the half-reaction an oxidation or a reduction? Supply the proper number of electrons to each side to balance each equation.

(a) $SO_3^{2-} + H_2O \longrightarrow SO_4^{2-} + 2\,H^+$
(b) $NO_3^- + 4\,H^+ \longrightarrow NO + 2\,H_2O$
(c) $S_2O_4^{2-} + 2\,H_2O \longrightarrow 2\,SO_3^{2-} + 4\,H^+$
(d) $Fe^{2+} \longrightarrow Fe^{3+}$

8. In the following unbalanced equations, identify

(a) the oxidized element and the reduced element
(b) the oxidizing agent and the reducing agent

1. $AsH_3 + Ag^+ + H_2O \longrightarrow H_3AsO_4 + Ag + H^+$
2. $Cl_2 + NaBr \longrightarrow NaCl + Br_2$

10. Determine whether the following oxidation–reduction reactions are balanced correctly. If they are not, provide the correct balanced reaction.

(a) unbalanced:
$MnO_2(s) + Al(s) \longrightarrow Mn(s) + Al_2O_3(s)$
balanced:
$MnO_2(s) + 2\,Al(s) \longrightarrow Mn(s) + Al_2O_3(s)$

(b) unbalanced:
$Cu(s) + Ag^+(aq) \longrightarrow Cu^{2+}(aq) + Ag(s)$
balanced:
$Cu(s) + 2\,Ag^+(aq) \longrightarrow Cu^{2+}(aq) + 2\,Ag(s)$

(c) unbalanced:
$Br^-(aq) + MnO_4^-(aq) \longrightarrow Br_2(l) + Mn^{2+}(aq)$ (acidic solution)
balanced:
$16\,H^+(aq) + 10\,Br^-(aq) + 2\,MnO_4^-(aq) \longrightarrow 5\,Br_2(l) + 2\,Mn^{2+}(aq) + 8\,H_2O(l)$

(d) unbalanced:
$MnO_4^-(aq) + S^{2-}(aq) \longrightarrow MnS(s) + S(s)$ (basic solution)
balanced: $8\,H^+(aq) + MnO_4^-(aq) + S^{2-}(aq) \longrightarrow S(s) + MnS(s) + 4\,H_2O(l)$

11. Balance these equations using the change-in-oxidation-number method:
(a) $Zn + S \longrightarrow ZnS$
(b) $AgNO_3 + Pb \longrightarrow Pb(NO_3)_2 + Ag$
(c) $Fe_2O_3 + CO \longrightarrow Fe + CO_2$
(d) $H_2S + HNO_3 \longrightarrow S + NO + H_2O$
(e) $MnO_2 + HBr \longrightarrow MnBr_2 + Br_2 + H_2O$

12. Balance these equations using the change-in-oxidation-number method.
(a) $Cl_2 + KOH \longrightarrow KCl + KClO_3 + H_2O$
(b) $Ag + HNO_3 \longrightarrow AgNO_3 + NO + H_2O$
(c) $CuO + NH_3 \longrightarrow N_2 + Cu + H_2O$
(d) $PbO_2 + Sb + NaOH \longrightarrow PbO + NaSbO_2 + H_2O$
(e) $H_2O_2 + KMnO_4 + H_2SO_4 \longrightarrow O_2 + MnSO_4 + K_2SO_4 + H_2O$

13. Balance these ionic redox equations using the ion–electron method. These reactions occur in acidic solution.
(a) $Zn + NO_3^- \longrightarrow Zn^{2+} + NH_4^+$
(b) $NO_3^- + S \longrightarrow NO_2 + SO_4^{2-}$
(c) $PH_3 + I_2 \longrightarrow H_3PO_2 + I^-$
(d) $Cu + NO_3^- \longrightarrow Cu^{2+} + NO$
*(e) $ClO_3^- + Cl^- \longrightarrow Cl_2$

14. Balance these ionic redox equations using the ion–electron method. These reactions occur in acidic solution.
(a) $ClO_3^- + I^- \longrightarrow I_2 + Cl^-$
(b) $Cr_2O_7^{2-} + Fe^{2+} \longrightarrow Cr^{3+} + Fe^{3+}$
(c) $MnO_4^- + SO_2 \longrightarrow Mn^{2+} + SO_4^{2-}$
(d) $H_3AsO_3 + MnO_4^- \longrightarrow H_3AsO_4 + Mn^{2+}$
*(e) $Cr_2O_7^{2+} + H_3AsO_3 \longrightarrow Cr^{3+} + H_3AsO_4$

15. Balance these ionic redox equations using the ion–electron method. These reactions occur in basic solution.
(a) $Cl_2 + IO_3^- \longrightarrow Cl^- + IO_4^-$
(b) $MnO_4^- + ClO_2^- \longrightarrow MnO_2 + ClO_4^-$
(c) $Se \longrightarrow Se^{2-} + SeO_3^{2-}$
*(d) $Fe_3O_4 + MnO_4^- \longrightarrow Fe_2O_3 + MnO_2$
*(e) $BrO^- + Cr(OH)_4^- \longrightarrow Br^- + CrO_4^{2-}$

16. Balance these ionic redox equations using the ion–electron method. These reactions occur in basic solution.
(a) $MnO_4^- + SO_3^{2-} \longrightarrow MnO_2 + SO_4^{2-}$
(b) $ClO_2 + SbO_2^- \longrightarrow ClO_2^- + Sb(OH)_6^-$
(c) $Al + NO_3^- \longrightarrow NH_3 + Al(OH)_4^-$
*(d) $P_4 \longrightarrow HPO_3^{2-} + PH_3$
*(e) $Al + OH^- \longrightarrow Al(OH)_4^- + H_2$

17. Balance these reactions:
(a) $IO_3^-(aq) + I^-(aq) \longrightarrow I_2(aq)$ (acid solution)
(b) $Mn^{2+}(aq) + S_2O_8^{2-}(aq) \longrightarrow MnO_4^-(aq) + SO_4^{2-}(aq)$ (acid solution)
(c) $Co(NO_2)_6^{3-}(aq) + MnO_4^-(aq) \longrightarrow Co^{2+}(aq) + Mn^{2+}(aq) + NO_3^-(aq)$ (acid solution)

18. Balance these reactions:
(a) $Mo_2O_3(s) + MnO_4^-(aq) \longrightarrow MoO_3(s) + Mn^{2+}(aq)$ (acid solution)
(b) $BrO^-(aq) + Cr(OH)_4^-(aq) \longrightarrow Br^-(aq) + CrO_4^{2-}(aq)$ (basic solution)
(c) $S_2O_3^{2-}(aq) + MnO_4^-(aq) \longrightarrow SO_4^{2-}(aq) + MnO_2(s)$ (basic solution)

Additional Exercises

All exercises with blue numbers have answers in the appendix of the text.

19. Draw a picture of an electrolytic cell made from an aqueous HBr solution.

20. The chemical reactions taking place during discharge in a lead storage battery are

$Pb + SO_4^{2-} \longrightarrow PbSO_4$

$PbO_2 + SO_4^{2-} + 4\,H^+ \longrightarrow PbSO_4 + 2\,H_2O$

(a) Complete each half-reaction by supplying electrons.
(b) Which reaction is oxidation and which is reduction?
(c) Which reaction occurs at the anode of the battery?

21. Use this unbalanced redox equation

$KMnO_4 + HCl \longrightarrow KCl + MnCl_2 + H_2 + Cl_2$

to indicate
(a) the oxidizing agent
(b) the reducing agent
(c) the number of electrons that are transferred per mole of oxidizing agent

22. Brass is an alloy of zinc and copper. When brass is in contact with salt water, it corrodes as the zinc dissolves from the alloy, leaving almost pure copper. Explain why the zinc is preferentially dissolved.

23. How many moles of NO gas will be formed by the reaction of 25.0 g of silver with nitric acid?

$Ag + HNO_3 \longrightarrow AgNO_3 + NO + H_2O$ (acid solution)

24. What volume of chlorine gas, measured at STP, must react with excess KOH to form 0.300 mol $KClO_3$?

$Cl_2 + KOH \longrightarrow KCl + KClO_3 + H_2O$ (acid solution)

25. What mass of $KMnO_4$ is needed to react with 100. mL H_2O_2 solution? ($d = 1.031$ g/mL, 9.0% H_2O_2 by mass)

$H_2O_2 + KMnO_4 + H_2SO_4 \longrightarrow O_2 + MnSO_4 + K_2SO_4 + H_2O$ (acid solution)

* **26.** What volume of 0.200 M $K_2Cr_2O_7$ will be required to oxidize 5.00 g H_3AsO_3?

$Cr_2O_7^{2-} + H_3AsO_3 \longrightarrow Cr^{3+} + H_3AsO_4$ (acid solution)

* **27.** What volume of 0.200 M $K_2Cr_2O_7$ will be required to oxidize the Fe^{2+} ion in 60.0 mL of 0.200 M $FeSO_4$ solution?

$Cr_2O_7^{2-} + Fe^{2+} \longrightarrow Cr^{3+} + Fe^{3+}$ (acid solution)

28. How many moles of H_2 can be produced from 100.0 g Al according to this reaction?

$Al + OH^- \longrightarrow Al(OH)_4^- + H_2$ (basic solution)

29. There is something incorrect about these half-reactions:
(a) $Cu^+ + e^- \longrightarrow Cu^{2+}$ (b) $Pb^{2+} + e^{2-} \longrightarrow Pb$

Identify what is wrong, and correct it.

30. Why can oxidation *never* occur without reduction?

31. The following observations were made concerning metals A, B, C, and D.
(a) When a strip of metal A is placed in a solution of B^{2+} ions, no reaction is observed.
(b) Similarly, A in a solution containing C^+ ions produces no reaction.
(c) When a strip of metal D is placed in a solution of C^+ ions, black metallic C deposits on the surface of D, and the solution tests positively for D^{2+} ions.
(d) When a piece of metallic B is placed in a solution of D^{2+} ions, metallic D appears on the surface of B and B^{2+} ions are found in the solution.

Arrange the ions—A^+, B^{2+}, C^+, and D^{2+}—in order of their ability to attract electrons. List them in order of increasing ability.

32. Tin normally has oxidation numbers of 0, +2, and +4. Which of these species can be an oxidizing agent, which can be a reducing agent, and which can be both? In each case, what product would you expect as the tin reacts?

33. Manganese is an element that can exist in numerous oxidation states. In each of these compounds, identify the oxidation number of the manganese. Which compound would you expect to be the best oxidizing agent and why?
(a) $Mn(OH)_2$ (c) MnO_2 (e) $KMnO_4$
(b) MnF_3 (d) K_2MnO_4

34. Which equations represent oxidations?
(a) $Mg \longrightarrow Mg^{2+}$ (c) $KMnO_4 \longrightarrow MnO_2$
(b) $SO_2 \longrightarrow SO_3$ (d) $Cl_2O_3 \longrightarrow Cl^-$

* **35.** In the following equation, note the reaction between manganese(IV) oxide and bromide ions:

$MnO_2 + Br^- \longrightarrow Br_2 + Mn^{2+}$

(a) Balance this redox reaction in acidic solution.
(b) How many grams of MnO_2 would be needed to produce 100.0 mL of 0.05 M Mn^{2+}?
(c) How many liters of bromine vapor at 50°C and 1.4 atm would also result?

36. Use the table shown to complete the following reactions. If no reaction occurs, write NR:
(a) $F_2 + Cl^- \longrightarrow$
(b) $Br_2 + Cl^- \longrightarrow$
(c) $I_2 + Cl^- \longrightarrow$
(d) $Br_2 + I^- \longrightarrow$

ease of reduction ↑	Activity
	F_2
	Cl_2
	Br_2
	I_2

37. Manganese metal reacts with HCl to give hydrogen gas and the Mn^{2+} ion in solution. Write a balanced equation for the reaction.

38. If zinc is allowed to react with dilute nitric acid, zinc is oxidized to the 2+ ion, while the nitrate ion can be reduced to ammonium, NH_4^+. Write a balanced equation for the reaction in acidic solution.

39. In the following equations, identify the
(a) atom or ion oxidized
(b) atom or ion reduced
(c) oxidizing agent
(d) reducing agent
(e) change in oxidation number associated with each oxidizing process
(f) change in oxidation number associated with each reducing process

1. $C_3H_8 + O_2 \longrightarrow CO_2 + H_2O$
2. $HNO_3 + H_2S \longrightarrow NO + S + H_2O$
3. $CuO + NH_3 \longrightarrow N_2 + H_2O + Cu$
4. $H_2O_2 + Na_2SO_3 \longrightarrow Na_2SO_4 + H_2O$
5. $H_2O_2 \longrightarrow H_2O + O_2$

40. In the galvanic cell shown in the diagram, a strip of silver is placed in a solution of silver nitrate, and a strip of lead is placed in a solution of lead(II) nitrate. The two beakers are connected with a salt bridge. Determine
(a) the anode
(b) the cathode
(c) where oxidation occurs
(d) where reduction occurs
(e) in which direction electrons flow through the wire
(f) in which direction ions flow through the solution

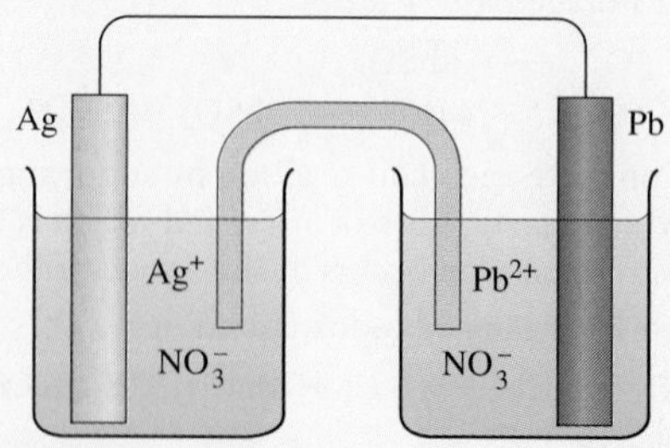

Challenge Exercises

All exercises with blue numbers have answers in the appendix of the text.

***41.** A sample of crude potassium iodide was analyzed using this reaction (not balanced):

$I^- + SO_4^{2-} \longrightarrow I_2 + H_2S$ (acid solution)

If a 4.00-g sample of crude KI produced 2.79 g of iodine, what is the percent purity of the KI?

***42.** What volume of NO gas, measured at 28°C and 744 torr, will be formed by the reaction of 0.500 mol Ag reacting with excess nitric acid?

$Ag + HNO_3 \longrightarrow AgNO_3 + NO + H_2O$ (acid solution)

Answers to Practice Exercises

17.1 (a) S = +6, (b) As = +5, (c) C = +4

17.2 (*Note:* H = +1 even though it comes second in the formula: N is a nonmetal.)
(a) N = −3, (b) Cr = +6, (c) P = +5

17.3 (a) Be = +2; Cl = −1, (b) H = +1; Cl = +1; O = −2, (c) H = +1; O = −1, (d) N = −3; H = +1, (e) Br = +5; O = −2

17.4 (a) $6\,HNO_3 + S \longrightarrow 6\,NO_2 + H_2SO_4 + 2\,H_2O$
(b) $2\,CrCl_3 + 3\,MnO_2 + 2\,H_2O \longrightarrow 3\,MnCl_2 + 2\,H_2CrO_4$
(c) $2\,KMnO_4 + 6\,HCl + 5\,H_2S \longrightarrow 2\,KCl + 2\,MnCl_2 + 5\,S + 8\,H_2O$

17.5 (a) $4\,H^+ + 2\,I^- + 2\,NO_2^- \longrightarrow I_2 + 2\,NO + 2\,H_2O$
(b) $2\,OH^- + Cl_2 + IO_3^- \longrightarrow IO_4^- + H_2O + 2\,Cl^-$
(c) $AuCl_4^- + Sn^{2+} \longrightarrow Sn^{4+} + AuCl + 3\,Cl^-$

17.6 (a) $Zn + 2\,H_2O + 2\,OH^- \longrightarrow Zn(OH)_4^{2-} + H_2$
(b) $H_2O_2 + Sn^{2+} + 2\,H^+ \longrightarrow Sn^{4+} + 2\,H_2O$
(c) $Cu + Cu^{2+} + 2\,OH^- \longrightarrow Cu_2O + H_2O$

17.7 (a) yes, (b) no, (c) no

17.8 (a) Cu is oxidized; Sn is reduced
(b) anode: $Cu(s) \longrightarrow Cu^{2+}(aq) + 2e^-$
cathode: $Sn^{2+}(aq) + 2e^- \longrightarrow Sn(s)$

PUTTING IT TOGETHER: Review for Chapters 15–17

Multiple Choice:

Choose the correct answer to each of the following.

1. When the reaction
   $Al + HCl \longrightarrow$
   is completed and balanced, this term appears in the balanced equation:
   (a) $3\,HCl$ (b) $AlCl_2$ (c) $3\,H_2$ (d) $4\,Al$

2. When the reaction
   $CaO + HNO_3 \longrightarrow$
   is completed and balanced, this term appears in the balanced equation:
   (a) H_2 (b) $2\,H_2$ (c) $2\,CaNO_3$ (d) H_2O

3. When the reaction
   $H_3PO_4 + KOH \longrightarrow$
   is completed and balanced, this term appears in the balanced equation:
   (a) H_3PO_4 (c) KPO_4
   (b) $6\,H_2O$ (d) $3\,KOH$

4. When the reaction
   $HCl + Cr_2(CO_3)_3 \longrightarrow$
   is completed and balanced, this term appears in the balanced equation:
   (a) Cr_2Cl (b) $3\,HCl$ (c) $3\,CO_2$ (d) H_2O

5. Which of these is not a salt?
   (a) $K_2Cr_2O_7$ (c) $Ca(OH)_2$
   (b) $NaHCO_3$ (d) $Na_2C_2O_4$

6. Which of these is not an acid?
   (a) H_3PO_4 (b) H_2S (c) H_2SO_4 (d) NH_3

7. Which of these is a weak electrolyte?
   (a) NH_4OH (c) K_3PO_4
   (b) $Ni(NO_3)_2$ (d) $NaBr$

8. Which of these is a nonelectrolyte?
   (a) $HC_2H_3O_2$ (c) $KMnO_4$
   (b) $MgSO_4$ (d) CCl_4

9. Which of these is a strong electrolyte?
   (a) H_2CO_3 (c) NH_4OH
   (b) HNO_3 (d) H_3BO_3

10. Which of these is a weak electrolyte?
    (a) $NaOH$ (c) $HC_2H_3O_2$
    (b) $NaCl$ (d) H_2SO_4

11. A solution has an H^+ concentration of $3.4 \times 10^{-5}\,M$. The pH is
    (a) 4.47 (c) 3.53
    (b) 5.53 (d) 5.47

12. A solution with a pH of 5.85 has an H^+ concentration of
    (a) $7.1 \times 10^{-5}\,M$ (c) $3.8 \times 10^{-4}\,M$
    (b) $7.1 \times 10^{-6}\,M$ (d) $1.4 \times 10^{-6}\,M$

13. If 16.55 mL of 0.844 M NaOHare required to titrate 10.00 mL of a hydrochloric acid solution, the molarity of the acid solution is
    (a) 0.700 M (c) 1.40 M
    (b) 0.510 M (d) 0.255 M

14. What volume of 0.462 M NaOH is required to neutralize 20.00 mL of 0.391 M HNO_3?
    (a) 23.6 mL (c) 9.03 mL
    (b) 16.9 mL (d) 11.8 mL

15. 25.00 mL of H_2SO_4 solution requires 18.92 mL of 0.1024 M NaOH for complete neutralization. The molarity of the acid is
    (a) 0.1550 M (c) 0.07750 M
    (b) 0.03875 M (d) 0.06765 M

16. Dilute hydrochloric acid is a typical acid, as shown by its
    (a) color (c) solubility
    (b) odor (d) taste

17. What is the pH of a 0.00015 M HCl solution?
    (a) 4.0 (c) between 3 and 4
    (b) 2.82 (d) no correct answer given

18. The chloride ion concentration in 300. mL of 0.10 M $AlCl_3$ is
    (a) 0.30 M (c) 0.030 M
    (b) 0.10 M (d) 0.90 M

19. The amount of $BaSO_4$ that will precipitate when 100. mL of 0.10 M $BaCl_2$ and 100. mL of 0.10 M Na_2SO_4 are mixed is
    (a) 0.010 mol (c) 23 g
    (b) 0.10 mol (d) no correct answer given

20. The freezing point of a 0.50 m NaCl aqueous solution will be about what?
    (a) $-1.86°C$ (c) $-2.79°C$
    (b) $-0.93°C$ (d) no correct answer given

21. The equation
    $HC_2H_3O_2 + H_2O \rightleftharpoons H_3O^+ + C_2H_3O_2^-$
    implies that
    (a) If you start with 1.0 mol $HC_2H_3O_2$, 1.0 mol H_3O^+ and 1.0 mol $C_2H_3O_2^-$ will be produced.
    (b) An equilibrium exists between the forward reaction and the reverse reaction.
    (c) At equilibrium, equal molar amounts of all four will exist.
    (d) The reaction proceeds all the way to the products and then reverses, going all the way back to the reactants.

22. If the reaction $A + B \rightleftharpoons C + D$ is initially at equilibrium and then more A is added, which of the following is not true?
    (a) More collisions of A and B will occur; the rate of the forward reaction will thus be increased.
    (b) The equilibrium will shift toward the right.
    (c) The moles of B will be increased.
    (d) The moles of D will be increased.

23. What will be the H^+ concentration in a 1.0 *M* HCN solution? ($K_a = 4.0 \times 10^{-10}$)
(a) 2.0×10^{-5} *M* (c) 4.0×10^{-10} *M*
(b) 1.0 *M* (d) 2.0×10^{-10} *M*

24. What is the percent ionization of HCN in Exercise 23?
(a) 100% (c) 2.0×10^{-3}%
(b) 2.0×10^{-8}% (d) 4.0×10^{-8}%

25. If $[H^+] = 1 \times 10^{-5}$ *M*, which of the following is not true?
(a) pH = 5 (c) $[OH^-] = 1 \times 10^{-5}$ *M*
(b) pOH = 9 (d) The solution is acidic.

26. If $[H^+] = 2.0 \times 10^{-4}$ *M*, then $[OH^-]$ will be
(a) 5.0×10^{-9} *M* (c) 2.0×10^{-4} *M*
(b) 3.70 *M* (d) 5.0×10^{-11} *M*

27. The solubility product of $PbCrO_4$ is 2.8×10^{-13}. The solubility of $PbCrO_4$ is
(a) 5.3×10^{-7} *M* (c) 7.8×10^{-14} *M*
(b) 2.8×10^{-13} *M* (d) 1.0 *M*

28. The solubility of AgBr is 7.2×10^{-7} *M*. The value of the solubility product is
(a) 7.2×10^{-7} (c) 5.2×10^{-48}
(b) 5.2×10^{-13} (d) 5.2×10^{-15}

29. Which of these solutions would be the best buffer solution?
(a) 0.10 *M* $HC_2H_3O_2$ + 0.10 *M* $NaC_2H_3O_2$
(b) 0.10 *M* HCl
(c) 0.10 *M* HCl + 0.10 *M* NaCl
(d) pure water

30. For the reaction $H_2(g) + I_2(g) \rightleftharpoons 2\,HI(g)$, at 700 K, $K_{eq} = 56.6$. If an equilibrium mixture at 700 K was found to contain 0.55 *M* HI and 0.21 *M* H_2, the I_2 concentration must be
(a) 0.046 *M* (c) 22 *M*
(b) 0.025 *M* (d) 0.21 *M*

31. The equilibrium constant for the reaction
$2\,A + B \rightleftharpoons 3\,C + D$ is
(a) $\dfrac{[C]^3[D]}{[A]^2[B]}$ (c) $\dfrac{[3C][D]}{[2A][B]}$
(b) $\dfrac{[2A][B]}{[3C][D]}$ (d) $\dfrac{[A]^2[B]}{[C]^3[D]}$

32. In the equilibrium represented by
$N_2(g) + O_2(g) \rightleftharpoons 2\,NO_2(g)$
as the pressure is increased, the amount of NO_2 formed
(a) increases (c) remains the same
(b) decreases (d) increases and decreases irregularly

33. Which factor will not increase the concentration of ammonia as represented by this equation?
$3\,H_2(g) + N_2(g) \rightleftharpoons 2\,NH_3(g) + 92.5\text{ kJ}$
(a) increasing the temperature
(b) increasing the concentration of N_2
(c) increasing the concentration of H_2
(d) increasing the pressure

34. If HCl(g) is added to a saturated solution of AgCl, the concentration of Ag^+ in solution
(a) increases
(b) decreases
(c) remains the same
(d) increases and decreases irregularly

35. The solubility of $CaCO_3$ at 20°C is 0.013 g/L. What is the K_{sp} for $CaCO_3$?
(a) 1.3×10^{-8} (c) 1.7×10^{-8}
(b) 1.3×10^{-4} (d) 1.7×10^{-4}

36. The K_{sp} for $BaCrO_4$ is 8.5×10^{-11}. What is the solubility of $BaCrO_4$ in grams per liter?
(a) 9.2×10^{-6} (c) 2.3×10^{-3}
(b) 0.073 (d) 8.5×10^{-11}

37. What will be the $[Ba^{2+}]$ when 0.010 mol Na_2CrO_4 is added to 1.0 L of saturated $BaCrO_4$ solution? See Exercise 36 for K_{sp}.
(a) 8.5×10^{-11} *M* (c) 9.2×10^{-6} *M*
(b) 8.5×10^{-9} *M* (d) 9.2×10^{-4} *M*

38. Which would occur if a small amount of sodium acetate crystals, $NaC_2H_3O_2$, were added to 100 mL of 0.1 *M* $HC_2H_3O_2$ at constant temperature?
(a) The number of acetate ions in the solution would decrease.
(b) The number of acetic acid molecules would decrease.
(c) The number of sodium ions in solution would decrease.
(d) The H^+ concentration in the solution would decrease.

39. If the temperature is decreased for the endothermic reaction
$A + B \rightleftharpoons C + D$
which of the following is true?
(a) The concentration of A will increase.
(b) No change will occur.
(c) The concentration of B will decrease.
(d) The concentration of D will increase.

40. In K_2SO_4, the oxidation number of sulfur is
(a) +2 (c) +6
(b) +4 (d) −2

41. In $Ba(NO_3)_2$, the oxidation number of N is
(a) +5 (c) +4
(b) −3 (d) −1

42. In the reaction
$H_2S + 4\,Br_2 + 4\,H_2O \longrightarrow H_2SO_4 + 8\,HBr$
the oxidizing agent is
(a) H_2S (c) H_2O
(b) Br_2 (d) H_2SO_4

43. In the reaction
$VO_3^- + Fe^{2+} + 4\,H^+ \longrightarrow VO^{2+} + Fe^{3+} + 2\,H_2O$
the element reduced is
(a) V (c) O
(b) Fe (d) H

Questions 44–46 pertain to the activity series below.

K Ca Mg Al Zn Fe H Cu Ag

44. Which of these pairs will not react in water solution?
(a) Zn, $CuSO_4$
(b) Cu, $Al_2(SO_4)_3$
(c) Fe, $AgNO_3$
(d) Mg, $Al_2(SO_4)_3$

45. Which element is the most easily oxidized?
(a) K
(b) Mg
(c) Zn
(d) Cu

46. Which element will reduce Cu^{2+} to Cu but will not reduce Zn^{2+} to Zn?
(a) Fe
(b) Ca
(c) Ag
(d) Mg

47. In the electrolysis of fused (molten) $CaCl_2$, the product at the negative electrode is
(a) Ca^{2+}
(b) Cl^-
(c) Cl_2
(d) Ca

48. In its reactions, a free element from Group 2A in the periodic table is most likely to
(a) be oxidized
(b) be reduced
(c) be unreactive
(d) gain electrons

49. In the partially balanced redox equation

$3\,Cu + HNO_3 \longrightarrow 3\,Cu(NO_3)_2 + 2\,NO + H_2O$

the coefficient needed to balance H_2O is
(a) 8
(b) 6
(c) 4
(d) 3

50. Which reaction does not involve oxidation–reduction?
(a) burning sodium in chlorine
(b) chemical union of Fe and S
(c) decomposition of $KClO_3$
(d) neutralization of NaOH with H_2SO_4

51. How many moles of Fe^{2+} can be oxidized to Fe^{3+} by 2.50 mol Cl_2 according to this equation?

$Fe^{2+} + Cl_2 \longrightarrow Fe^{3+} + Cl^-$

(a) 2.50 mol
(b) 5.00 mol
(c) 1.00 mol
(d) 22.4 mol

52. How many grams of sulfur can be produced in this reaction from 100. mL of 6.00 *M* HNO_3?

$HNO_3 + H_2S \longrightarrow S + NO + H_2O$

(a) 28.9 g
(b) 19.3 g
(c) 32.1 g
(d) 289 g

53. Which of these ions can be reduced by H_2?
(a) Hg^{2+}
(b) Sn^{2+}
(c) Zn^{2+}
(d) K^+

54. Which of the following is *not* true of a zinc–mercury cell?
(a) It provides current at a steady potential.
(b) It has a short service life.
(c) It is self-contained.
(d) It can be stored for long periods of time.

Balancing Oxidation–Reduction Equations

Balance each equation.

55. $P + HNO_3 \longrightarrow HPO_3 + NO + H_2O$

56. $MnSO_4 + PbO_2 + H_2SO_4 \longrightarrow HMnO_4 + PbSO_4 + H_2O$

57. $Cr_2O_7^{2-} + Cl^- \longrightarrow Cr^{3+} + Cl_2$ (acidic solution)

58. $MnO_4^- + AsO_3^{3-} \longrightarrow Mn^{2+} + AsO_4^{3-}$ (acidic solution)

59. $S^{2-} + Cl_2 \longrightarrow SO_4^{2-} + Cl^-$ (basic solution)

60. $Zn + NO_3^- \longrightarrow Zn(OH)_4^{2-} + NH_3$ (basic solution)

61. $KOH + Cl_2 \longrightarrow KCl + KClO + H_2O$ (basic solution)

62. $As + ClO_3^- \longrightarrow H_3AsO_3 + HClO$ (acidic solution)

63. $MnO_4^- + Cl^- \longrightarrow Mn^{2+} + Cl_2$ (acidic solution)

64. $H_2O_2 + Cl_2O_7 \longrightarrow ClO_2^- + O_2$ (basic solution)

Free Response Questions

Answer each of the following. Be sure to include your work and explanations in a clear, logical form.

1. You are investigating the properties of two new metallic elements found on Pluto, Bz and Yz. Bz reacts with aqueous HCl. However, the formula for the compound it forms with chlorine is $YzCl_2$. Write the balanced reaction that should occur if a galvanic cell is set up with Bz and Yz electrodes in solutions containing the metallic ions.

2. Suppose that 25 mL of an iron(II) nitrate solution are added to a beaker containing aluminum metal. Assume the reaction went to completion with no excess reagents. The solid iron produced was removed by filtration.
(a) Write a balanced redox equation for the reaction.
(b) For which solution, the initial iron(II) nitrate or the solution after the solid iron was filtered out, would the freezing point be lower? Explain your answer.

3. 50. mL of 0.10 *M* HCl solution are poured equally into two flasks.
(a) What is the pH of the HCl solution?
(b) Next, 0.050 mol Zn is added to Flask A and 0.050 mol Cu is added to Flask B. Determine the pH of each solution after approximately 20 minutes.

4. (a) Write a balanced acid–base reaction that produces Na_2S.
(b) If Na_2S is added to an aqueous solution of H_2S ($K_a = 9.1 \times 10^{-8}$), will the pH of the solution rise or fall? Explain.

5. (a) Would you expect a reaction to take place between HCN (*aq*) and $AgNO_3$ (*aq*) (K_{sp} for AgCN $= 5.97 \times 10^{-17}$)? Explain, and if a reaction occurs, write the net ionic equation.
(b) If NaCN is added to distilled water, would you expect the solution to be acidic, basic, or neutral? Explain using any chemical equations that may be appropriate.

6. For each set of beakers below, draw a picture you might expect to see when the contents of the beakers are mixed together and allowed to react.

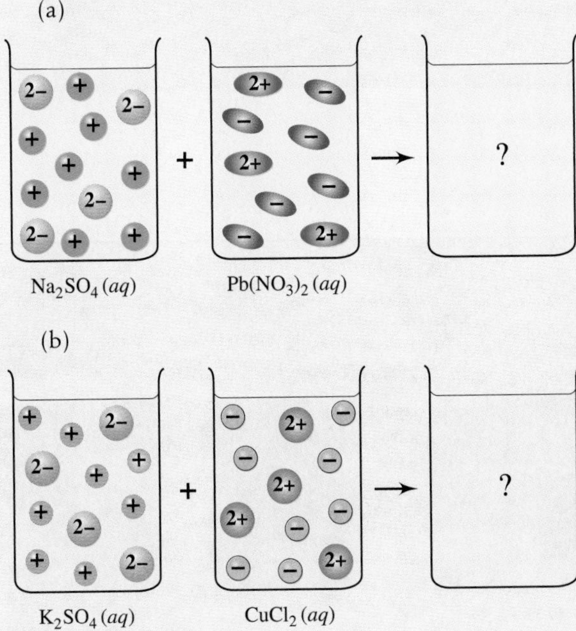

7. The picture below represents the equilibrium condition of $2\,A_3X \rightleftharpoons 2\,A_2X + A_2$.

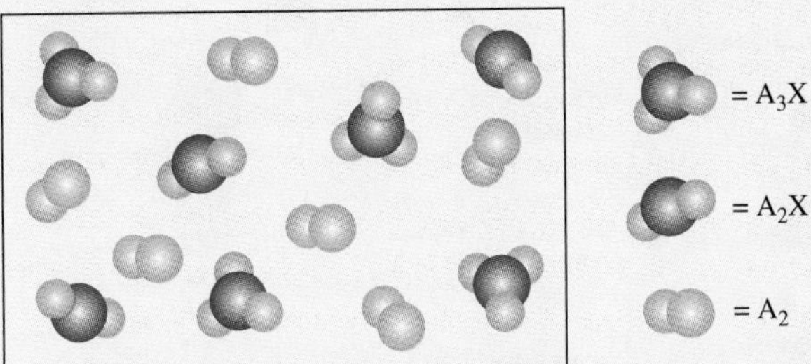

(a) What is the equilibrium constant?
(b) Does the equilibrium lie to the left or to the right?
(c) Do you think the reaction is a redox reaction? Explain your answer.

8. The picture below represents the equilibrium condition of the reaction $X_2 + 2G \rightleftharpoons X_2G_2$.

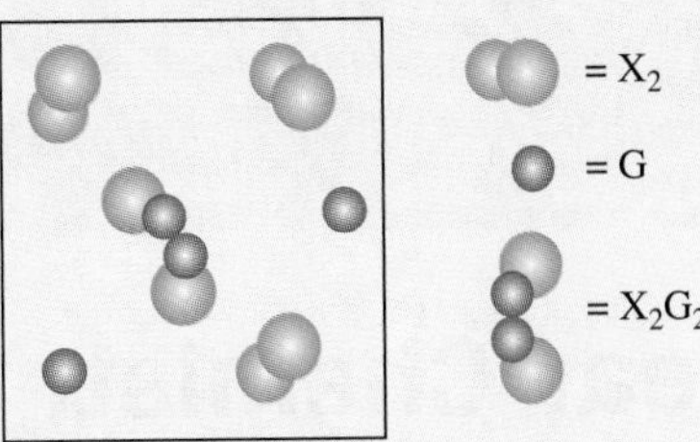

(a) What is the equilibrium constant?
(b) If the ratio of reactants to products increased when the temperature was raised, was the reaction exothermic or endothermic?
(c) Provide a logical explanation for why the equilibrium shifts to the right when the pressure is increased at constant temperature.

9. The hydroxide ion concentration of a solution is 3.4×10^{-10} *M*. Balance the following equation:

$$Fe^{2+}(aq) + MnO_4^-(aq) \longrightarrow Fe^{3+}(aq) + Mn^{2+}(aq)$$

CHAPTER 18

Nuclear Chemistry

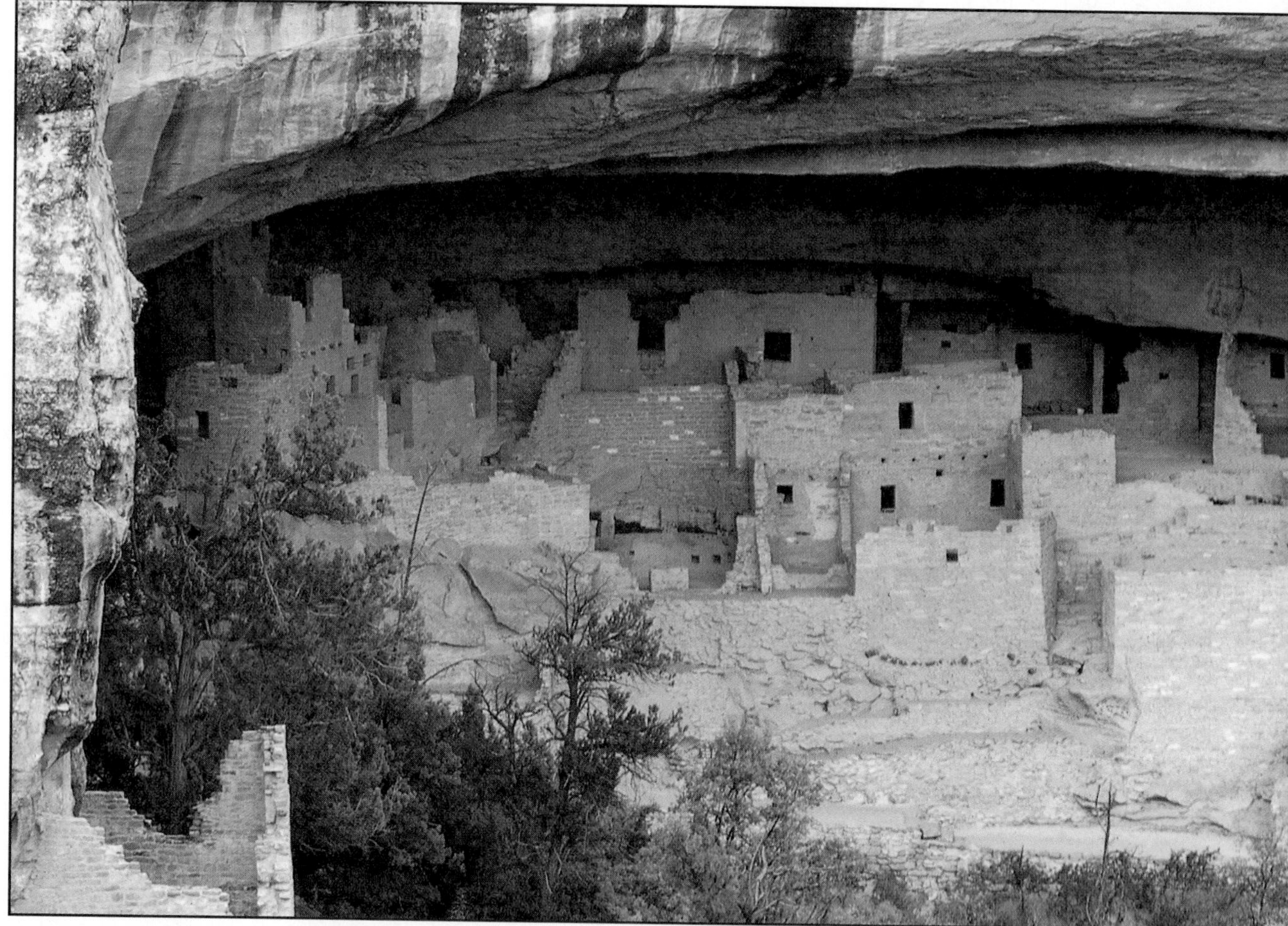

Carbon dating is used to determine the age of civilizations that lived in Cliff Palace, at Mesa Verde, Colorado.

Chapter Outline

The nucleus of the atom is a source of tremendous energy. Harnessing this energy has enabled us to fuel power stations, treat cancer, and preserve food. We use isotopes in medicine to diagnose illness and to detect minute quantities of drugs or hormones. Researchers use radioactive tracers to sequence the human genome. We also use nuclear processes to detect explosives in luggage and to establish the age of objects such as human artifacts and rocks. In this chapter we'll consider properties of atomic nuclei and their applications in our lives.

18.1 Discovery of Radioactivity

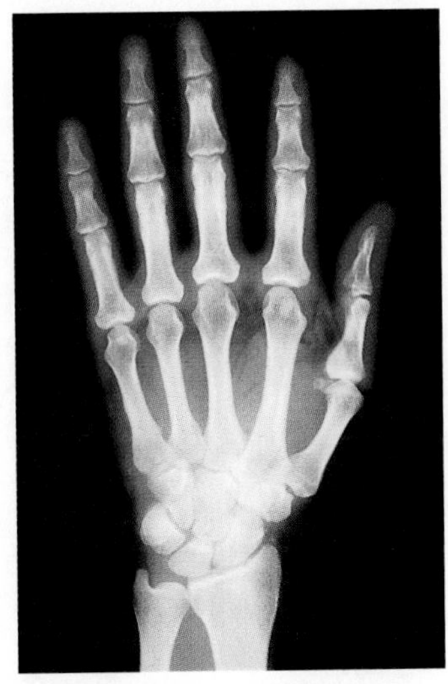

X-ray technology opened the door for a new world of diagnosis and treatment.

In 1895, Wilhelm Conrad Roentgen (1845-1923) made an important breakthrough that eventually led to the discovery of radioactivity. Roentgen discovered X-rays when he observed that a vacuum discharge tube enclosed in a thin, black cardboard box caused a nearby piece of paper coated with the salt barium platinocyanide to glow with a brilliant phosphorescence. From this and other experiments, he concluded that certain rays, which he called X-rays, were emitted from the discharge tube, penetrated the box, and caused the salt to glow. Roentgen's observations that X-rays could penetrate other bodies and affect photographic plates led to the development of X-ray photography.

Shortly after this discovery, Antoine Henri Becquerel (1852–1908) attempted to show a relationship between X-rays and the phosphorescence of uranium salts. In one experiment he wrapped a photographic plate in black paper, placed a sample of uranium salt on it, and exposed it to sunlight. The developed photographic plate showed that rays emitted from the salt had penetrated the paper. When Becquerel attempted to repeat the experiment, the sunlight was intermittent, so he placed the entire setup in a drawer. Several days later he developed the photographic plate, expecting to find it only slightly affected. To his amazement, he found an intense image on the plate. He repeated the experiment in total darkness and obtained the same results, proving that the uranium salt emitted rays that affected the photographic plate without its being exposed to sunlight. Thus did the discovery of radioactivity come about, a combination of numerous experiments by the finest minds of the day—and serendipity. Becquerel later showed that the rays coming from uranium are able to ionize air and are also capable of penetrating thin sheets of metal.

The name *radioactivity* was coined two years later (in 1898) by Marie Curie. **Radioactivity** is the spontaneous emission of particles and/or rays from the nucleus of an atom. Elements having this property are said to be radioactive.

radioactivity

In 1898, Marie Sklodowska Curie (1867–1934) and her husband Pierre Curie (1859–1906) turned their research interests to radioactivity. In a short time, the Curies discovered two new elements, polonium and radium, both of which are radioactive. To confirm their work on radium, they processed 1 ton of pitchblende residue ore to obtain 0.1 g of pure radium chloride, which they used to make further studies on the properties of radium and to determine its atomic mass.

In 1899, Ernest Rutherford began to investigate the nature of the rays emitted from uranium. He found two particles, which he called *alpha* and *beta particles*. Soon he realized that uranium, while emitting these particles, was changing into another element. By 1912, over 30 radioactive isotopes were known, and many more are known today. The *gamma ray*, a third type of emission from radioactive materials similar to an X-ray, was discovered by Paul Villard (1860–1934) in 1900. Rutherford's description of the nuclear atom led scientists to attribute the phenomenon of radioactivity to reactions taking place in the nuclei of atoms.

The symbolism and notation we described for isotopes in Chapter 5 are also very useful in nuclear chemistry:

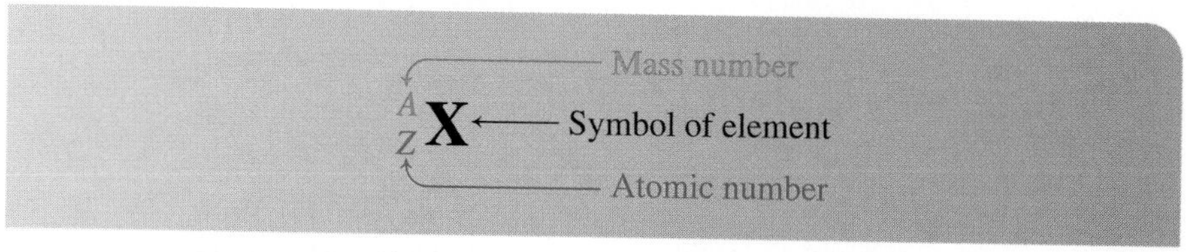

For example, $^{238}_{92}U$ represents a uranium isotope with an atomic number of 92 and a mass number of 238. This isotope is also designated as U-238, or uranium-238, and contains 92 protons and 146 neutrons. The protons and neutrons collectively are known as **nucleons**. The mass number is the total number of nucleons in the nucleus. Table 18.1 shows the isotopic notations for several particles associated with nuclear chemistry.

nucleon

When we speak of isotopes, we mean atoms of the same element with different masses, such as $^{16}_{8}O$, $^{17}_{8}O$, $^{18}_{8}O$. In nuclear chemistry we use the term **nuclide** to mean any isotope of any atom. Thus, $^{16}_{8}O$ and $^{235}_{92}U$ are referred to as nuclides. Nuclides that spontaneously emit radiation are referred to as *radionuclides*.

nuclide

Table 18.1 Isotopic Notation for Several Particles (and Small Isotopes) Associated with Nuclear Chemistry

Particle	Symbol	Z Atomic number	A Mass number
Neutron	$^{1}_{0}n$	0	1
Proton	$^{1}_{1}H$	1	1
Beta particle (electron)	$^{0}_{-1}e$	−1	0
Positron (positive electron)	$^{0}_{+1}e$	1	0
Alpha particle (helium nucleus)	$^{4}_{2}He$	2	4
Deuteron (heavy hydrogen nucleus)	$^{2}_{1}H$	1	2

18.2 Natural Radioactivity

radioactive decay

Radioactive elements continuously undergo **radioactive decay,** or disintegration, to form different elements. The chemical properties of an element are associated with its electronic structure, but radioactivity is a property of the nucleus. Therefore, neither ordinary changes of temperature and pressure nor the chemical or physical state of an element has any effect on its radioactivity.

The principal emissions from the nuclei of radionuclides are known as alpha particles, beta particles, and gamma rays. Upon losing an alpha or beta particle, the radioactive element changes into a different element. We will explain this process in detail later.

Each radioactive nuclide disintegrates at a specific and constant rate, which is expressed in units of half-life. The **half-life** ($t_{1/2}$) is the time required for one-half of a specific amount of a radioactive nuclide to disintegrate. The half-lives of the elements range from a fraction of a second to billions of years. To illustrate, suppose we start with 1.0 g of $^{226}_{88}Ra$ ($t_{1/2}$ = 1620 years):

half-life

Element	$t_{1/2}$
$^{238}_{92}U$	4.5×10^9 years
$^{226}_{88}Ra$	1620 years
$^{15}_{6}C$	2.4 seconds

$$1.0 \text{ g } ^{226}_{88}\text{Ra} \xrightarrow[\text{1620 years}]{t_{1/2}} 0.50 \text{ g } ^{226}_{88}\text{Ra} \xrightarrow[\text{1620 years}]{t_{1/2}} 0.25 \text{ g } ^{226}_{88}\text{Ra}$$

The half-lives of the various radioisotopes of the same element differ dramatically. Half-lives for certain isotopes of radium, carbon, and uranium are listed in Table 18.2.

Table 18.2 Half-Lives for Radium, Carbon, and Uranium Isotopes

Isotope	Half-life	Isotope	Half-life
Ra-223	11.7 days	C-14	5730 years
Ra-224	3.64 days	C-15	2.4 seconds
Ra-225	14.8 days	U-235	7.1×10^8 years
Ra-226	1620 years	U-238	4.5×10^9 years
Ra-228	6.7 years		

Example 18.1

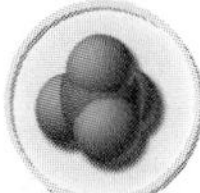

The half-life of $^{131}_{53}I$ is 8 days. How much $^{131}_{53}I$ from a 32-g sample remains after five half-lives?

SOLUTION

Using the following graph, we can find the number of grams of ^{131}I remaining after one half-life:

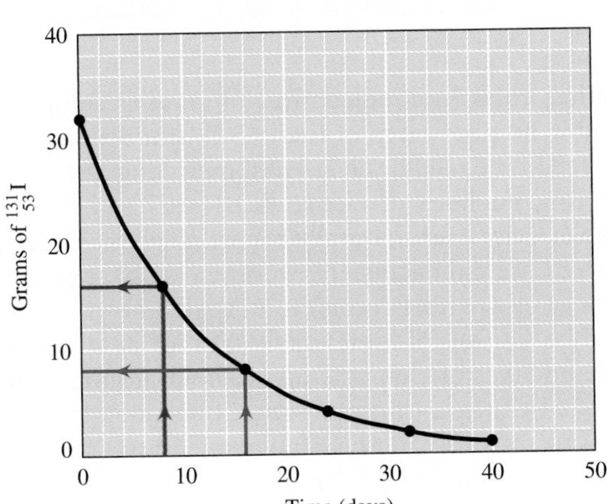

Trace a perpendicular line from 8 days on the x-axis to the line on the graph.

Now trace a horizontal line from this point on the plotted line to the y-axis and read the corresponding grams of ^{131}I. Continue this process for each half-life, adding 8 days to the previous value on the x-axis:

Half-lives	0	1	2	3	4	5
Number of days		8	16	24	32	40
Amount remaining	32 g	16 g	8 g	4 g	2 g	1 g

Starting with 32 g, 1 g ^{131}I remains after five half-lives (40 days).

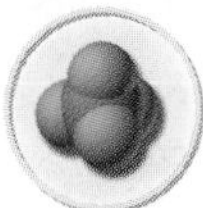

Example 18.2 In how many half-lives will 10.0 g of a radioactive nuclide decay to less than 10% of its original value?

SOLUTION We know that 10% of the original amount is 1.0 g. After the first half-life, half the original material remains and half has decayed (5.00 g). After the second half-life, one-fourth of the original material remains (i.e., one-half of 5.00 g). This progression continues, reducing the quantity remaining by half for each half-life that passes.

Half-lives	0	1	2	3	4
Percent remaining	100%	50%	25%	12.5%	6.25%
Amount remaining	10.0 g	5.00 g	2.50 g	1.25 g	0.625 g

Therefore the amount remaining will be less than 10% sometime between the third and the fourth half-lives.

Practice 18.1

The half-life of $^{14}_{6}C$ is 5730 years. How much $^{14}_{6}C$ will remain after six half-lives in a sample that initially contains 25.0 g?

Nuclides are said to be either *stable* (nonradioactive) or *unstable* (radioactive). Elements that have atomic numbers greater than 83 (bismuth) are naturally radioactive, although some of the nuclides have extremely long half-lives. Some of the naturally occurring nuclides of elements 81, 82, and 83 are radioactive, and some are stable. Only a few naturally occurring elements that have atomic numbers less than 81 are radioactive. However, no stable isotopes of element 43 (technetium) or of element 61 (promethium) are known.

Radioactivity is believed to be a result of an unstable ratio of neutrons to protons in the nucleus. Stable nuclides of elements up to about atomic number 20 generally have about a 1:1 neutron-to-proton ratio. In elements above number 20, the neutron-to-proton ratio in the stable nuclides gradually increases to about 1.5:1 in element number 83 (bismuth). When the neutron-to-proton ratio is too high or too low, alpha, beta, or other particles are emitted to achieve a more stable nucleus.

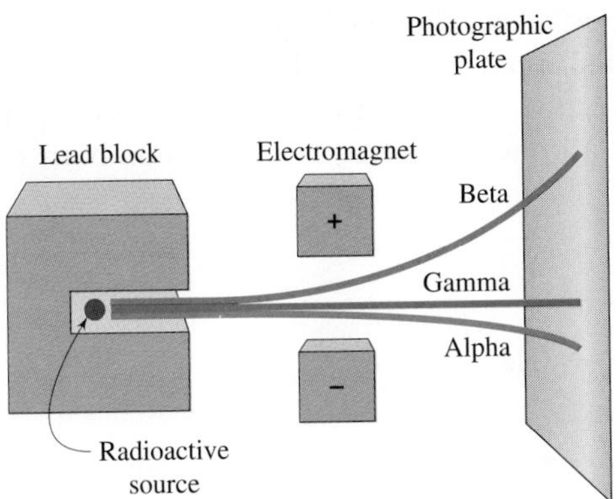

Figure 18.1
The effect of an electromagnetic field on alpha particles, beta particles, and gamma rays. Lighter beta particles are deflected considerably more than alpha particles. Alpha and beta particles are deflected in opposite directions. Gamma radiation is not affected by the electromagnetic field.

18.3 Alpha Particles, Beta Particles, and Gamma Rays

The classical experiment proving that alpha and beta particles are oppositely charged was performed by Marie Curie (see Figure 18.1). She placed a radioactive source in a hole in a lead block and positioned two poles of a strong electromagnet so that the radiations that were given off passed between them. The paths of three different kinds of radiation were detected by means of a photographic plate placed some distance beyond the electromagnet. The lighter beta particles were strongly deflected toward the positive pole of the electromagnet; the heavier alpha particles were less strongly deflected and in the opposite direction. The uncharged gamma rays were not affected by the electromagnet and struck the photographic plates after traveling along a path straight out of the lead block.

Alpha Particles

alpha particle

An **alpha particle** (α) consists of two protons and two neutrons, has a mass of about 4 amu, and has a charge of +2. It is a helium nucleus that is usually given one of the following symbols: α or ${}^{4}_{2}\text{He}$. When an alpha particle is emitted from the nucleus, a different element is formed. The atomic number of the new element is 2 less, and the mass is 4 amu less, than that of the starting element.

> Loss of an alpha particle from the nucleus results in
> loss of 4 in the mass number (A)
> loss of 2 in the atomic number (Z)

For example, when ${}^{238}_{92}\text{U}$ loses an alpha particle, ${}^{234}_{90}\text{Th}$ is formed, because two neutrons and two protons are lost from the uranium nucleus. This disintegration may be written as a nuclear equation:

$${}^{238}_{92}\text{U} \longrightarrow {}^{234}_{90}\text{Th} + \alpha \quad \text{or} \quad {}^{238}_{92}\text{U} \longrightarrow {}^{234}_{90}\text{Th} + {}^{4}_{2}\text{He}$$

For the loss of an alpha particle from ${}^{226}_{88}\text{Ra}$, the equation is

$${}^{226}_{88}\text{Ra} \longrightarrow {}^{222}_{86}\text{Rn} + {}^{4}_{2}\text{He} \quad \text{or} \quad {}^{226}_{88}\text{Ra} \longrightarrow {}^{222}_{86}\text{Rn} + \alpha$$

A nuclear equation, like a chemical equation, consists of reactants and products and must be balanced. To have a balanced nuclear equation, the sum of the mass numbers (superscripts) on both sides of the equation must be equal and the sum of the atomic numbers (subscripts) on both sides of the equation must be equal:

sum of mass numbers equals 226

$$^{226}_{88}\text{Ra} \longrightarrow {}^{222}_{86}\text{Rn} + {}^{4}_{2}\text{He}$$

sum of atomic numbers equals 88

sum of mass numbers is 230

$$^{230}_{90}\text{Th} \longrightarrow {}^{226}_{88}? + {}^{4}_{2}\text{He}$$

sum of atomic numbers is 90

What new nuclide will be formed when $^{230}_{90}\text{Th}$ loses an alpha particle? The new nuclide will have a mass of 226 amu and will contain 88 protons, so its atomic number is 88. Locate the corresponding element on the periodic table—in this case, $^{226}_{88}\text{Ra}$ or radium-226.

Beta Particles

beta particle

The **beta particle** (β) is identical in mass and charge to an electron; its charge is -1. Both a beta particle and a proton are produced by the decomposition of a neutron:

$$^{1}_{0}\text{n} \longrightarrow {}^{1}_{0}\text{p} + {}^{0}_{-1}\text{e}$$

The beta particle leaves, and the proton remains in the nucleus. When an atom loses a beta particle from its nucleus, a different element is formed that has essentially the same mass but an atomic number that is 1 greater than that of the starting element. The beta particle is written as β or $^{0}_{-1}\text{e}$.

> Loss of a beta particle from the nucleus results in
> no change in the mass number (A)
> increase of 1 in the atomic number (Z)

Examples of equations in which a beta particle is lost are

$$^{234}_{90}\text{Th} \longrightarrow {}^{234}_{91}\text{Pa} + \beta$$

$$^{234}_{91}\text{Pa} \longrightarrow {}^{234}_{92}\text{U} + {}^{0}_{-1}\text{e}$$

$$^{210}_{82}\text{Pb} \longrightarrow {}^{210}_{83}\text{Bi} + \beta$$

Gamma Rays

gamma ray

Gamma rays (γ) are photons of energy. A gamma ray is similar to an X-ray but is more energetic. It has no electrical charge and no measurable mass. Gamma rays are released from the nucleus in many radioactive changes along with either alpha or beta particles. Gamma radiation does not result in a change of atomic number or the mass of an element.

> Loss of a gamma ray from the nucleus results in
> no change in mass number (A) or atomic number (Z)

Example 18.3

(a) Write an equation for the loss of an alpha particle from the nuclide $^{194}_{78}Pt$.
(b) What nuclide is formed when $^{228}_{88}Ra$ loses a beta particle from its nucleus?

SOLUTION

(a) Loss of an alpha particle, $^{4}_{2}He$, results in a decrease of 4 in the mass number and a decrease of 2 in the atomic number:

Mass of new nuclide:	$A - 4$	or	$194 - 4 = 190$
Atomic number of new nuclide:	$Z - 2$	or	$78 - 2 = 76$

Looking up element number 76 on the periodic table, we find it to be osmium, Os. The equation then is

$$^{194}_{78}Pt \longrightarrow {}^{190}_{76}Os + {}^{4}_{2}He$$

(b) The loss of a beta particle from a $^{228}_{88}Ra$ nucleus means a gain of 1 in the atomic number with no essential change in mass. The new nuclide will have an atomic number of $(Z + 1)$, or 89, which is actinium, Ac:

$$^{228}_{88}Ra \longrightarrow {}^{228}_{89}Ac + {}^{0}_{-1}e$$

Example 18.4

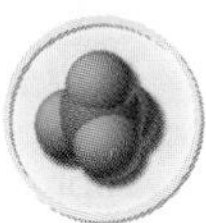

What nuclide will be formed when $^{214}_{82}Pb$ successively emits two beta particles, then one alpha particle from its nucleus? Write successive equations showing these changes.

SOLUTION

The changes brought about in the three steps outlined are as follows:

β loss:	Increase of 1 in the atomic number; no change in mass
β loss:	Increase of 1 in the atomic number; no change in mass
α loss:	Decrease of 2 in the atomic number; decrease of 4 in the mass

The equations are

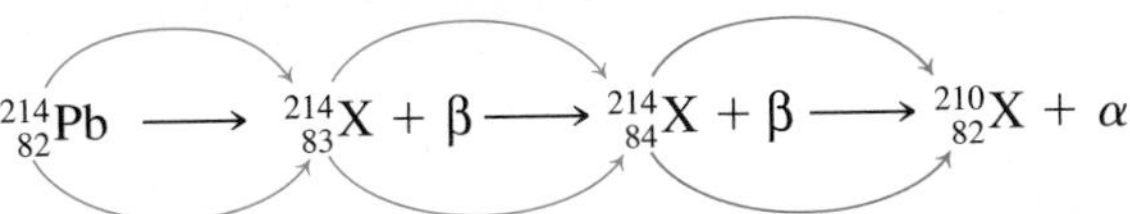

$$^{214}_{82}Pb \longrightarrow {}^{214}_{83}X + \beta \longrightarrow {}^{214}_{84}X + \beta \longrightarrow {}^{210}_{82}X + \alpha$$

where X stands for the new nuclide formed. Looking up each of these elements by their atomic numbers, we rewrite the equations

$$^{214}_{82}Pb \underset{\beta}{\longrightarrow} {}^{214}_{83}Bi \underset{\beta}{\longrightarrow} {}^{214}_{84}Po \underset{\alpha}{\longrightarrow} {}^{210}_{82}Pb$$

Practice 18.2

What nuclide will be formed when $^{222}_{86}Rn$ emits an alpha particle?

The ability of radioactive rays to pass through various objects is in proportion to the speed at which they leave the nucleus. Gamma rays travel at the velocity of light (186,000 miles per second) and are capable of penetrating several inches of lead. The velocities of beta particles are variable, the fastest being about nine-tenths the velocity of light. Alpha particles have velocities less than

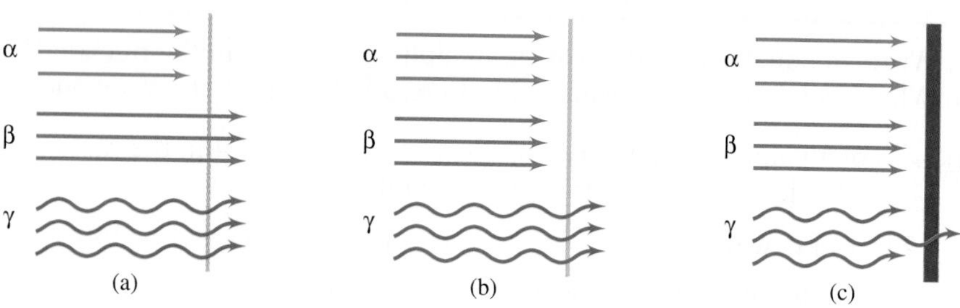

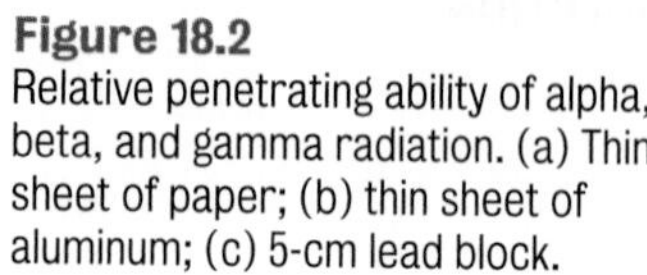
Figure 18.2
Relative penetrating ability of alpha, beta, and gamma radiation. (a) Thin sheet of paper; (b) thin sheet of aluminum; (c) 5-cm lead block.

one-tenth the velocity of light. Figure 18.2 illustrates the relative penetrating power of these rays. A few sheets of paper will stop alpha particles; a thin sheet of aluminum will stop both alpha and beta particles; and a 5-cm block of lead will reduce, but not completely stop, gamma radiation. In fact, it is difficult to stop all gamma radiation. Table 18.3 summarizes the properties of alpha, beta, and gamma radiation.

Table 18.3 Characteristics of Nuclear Radiation

Radiation	Symbol	Mass (amu)	Electrical charge	Velocity	Composition	Ionizing power
Alpha	α, ${}^{4}_{2}\text{He}$	4	+2	Variable, less than 10% the speed of light	He nucleus	High
Beta	β, ${}^{0}_{-1}\text{e}$	$\frac{1}{1837}$	−1	Variable, up to 90% the speed of light	Identical to an electron	Moderate
Gamma	γ	0	0	Speed of light	Photons or electromagnetic waves of energy	Almost none

18.4 Radioactive Disintegration Series

These three series begin with the elements uranium, thorium, and actinium.

The naturally occurring radioactive elements with a higher atomic number than lead (Pb) fall into three orderly disintegration series. Each series proceeds from one element to the next by the loss of either an alpha or a beta particle, finally ending in a nonradioactive nuclide. The uranium series starts with ${}^{238}_{92}\text{U}$ and ends with ${}^{206}_{82}\text{Pb}$. The thorium series starts with ${}^{232}_{90}\text{Th}$ and ends with ${}^{208}_{82}\text{Pb}$. The actinium series starts with ${}^{235}_{92}\text{U}$ and ends with ${}^{207}_{82}\text{Pb}$. A fourth series, the neptunium series, starts with the synthetic element ${}^{241}_{94}\text{Pu}$ and ends with the stable bismuth nuclide ${}^{209}_{83}\text{Bi}$. The uranium series is shown in Figure 18.3. Gamma radiation, which accompanies alpha and beta radiation, is not shown.

By using these series and the half-lives of their members, scientists have been able to approximate the age of certain geologic deposits. This approximation is done by comparing the amount of ${}^{238}_{92}\text{U}$ with the amount of ${}^{206}_{82}\text{Pb}$ and other nuclides in the series that are present in a particular geologic formation. Rocks found in Canada and Finland have been calculated to be about 3.0×10^{9} (3 billion) years old. Some meteorites have been determined to be 4.5×10^{9} years old.

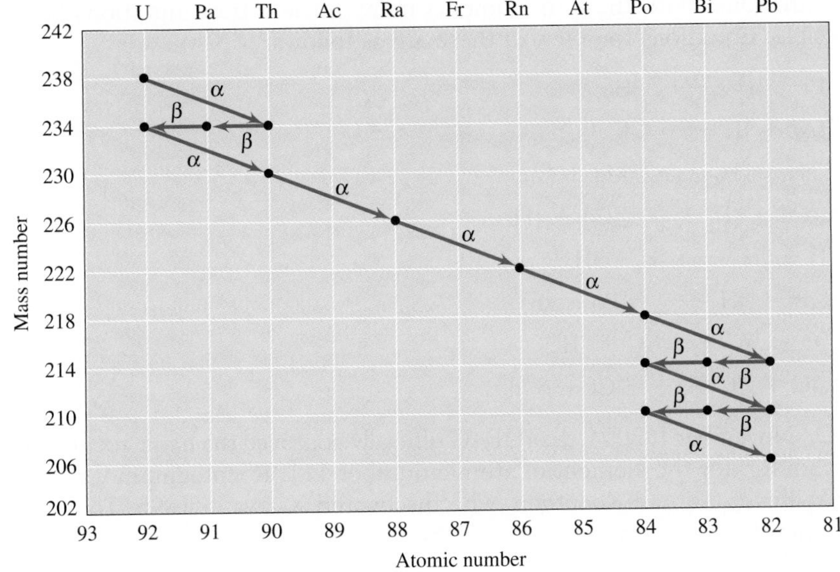

Figure 18.3
The uranium disintegration series. $^{238}_{92}U$ decays by a series of alpha (α) and beta (β) emissions to the stable nuclide $^{206}_{82}Pb$.

Practice 18.3

What nuclides are formed when $^{238}_{92}U$ undergoes the following decays?

(a) an alpha particle and a beta particle
(b) three alpha particles and two beta particles

18.5 Transmutation of Elements

transmutation

Transmutation is the conversion of one element into another by either natural or artificial means. Transmutation occurs spontaneously in natural radioactive disintegrations. Alchemists tried for centuries to convert lead and mercury into gold by artificial means, but transmutation by artificial means was not achieved until 1919, when Ernest Rutherford succeeded in bombarding the nuclei of nitrogen atoms with alpha particles and produced oxygen nuclides and protons. The nuclear equation for this transmutation can be written as

$$^{14}_{7}N + \alpha \longrightarrow {}^{17}_{8}O + {}^{1}_{1}H \quad \text{or} \quad {}^{14}_{7}N + {}^{4}_{2}He \longrightarrow {}^{17}_{8}O + {}^{1}_{1}H$$

It is believed that the alpha particle enters the nitrogen nucleus, forming $^{18}_{9}F$ as an intermediate, which then decomposes into the products.

Rutherford's experiments opened the door to nuclear transmutations of all kinds. Atoms were bombarded by alpha particles, neutrons, protons, deuterons ($^{2}_{1}H$), electrons, and so forth. Massive instruments were developed for accelerating these particles to very high speeds and energies to aid their penetration of the nucleus. The famous cyclotron was developed by E. O. Lawrence (1901–1958) at the University of California; later instruments include the Van de Graaf electrostatic generator, the betatron, and the electron and proton

Electrons and positrons are accelerated along the tunnels at the Stanford Linear Accelerator Center.

synchrotrons. With these instruments many nuclear transmutations became possible. Equations for a few of these are as follows:

$$^{7}_{3}Li + ^{1}_{1}H \longrightarrow 2\,^{4}_{2}He$$
$$^{40}_{18}Ar + ^{1}_{1}H \longrightarrow ^{40}_{19}K + ^{1}_{0}n$$
$$^{23}_{11}Na + ^{1}_{1}H \longrightarrow ^{23}_{12}Mg + ^{1}_{0}n$$
$$^{114}_{48}Cd + ^{2}_{1}H \longrightarrow ^{115}_{48}Cd + ^{1}_{1}H$$
$$^{2}_{1}H + ^{2}_{1}H \longrightarrow ^{3}_{1}H + ^{1}_{1}H$$
$$^{209}_{83}Bi + ^{2}_{1}H \longrightarrow ^{210}_{84}Po + ^{1}_{0}n$$
$$^{16}_{8}O + ^{1}_{0}n \longrightarrow ^{13}_{6}C + ^{4}_{2}He$$
$$^{238}_{92}U + ^{12}_{6}C \longrightarrow ^{244}_{98}Cf + 6\,^{1}_{0}n$$

On November 1, 2004, the IUPAC officially approved the name roentgenium, symbol Rg, for the element of atomic number 111. Roentgenium was named for Wilhelm Conrad Roentgen, who discovered X-rays in 1895. The nuclear equation for its formation is

$$^{209}_{83}Bi + ^{64}_{28}Ni \longrightarrow ^{272}_{111}Rg + ^{1}_{0}n$$

18.6 Artificial Radioactivity

Irene Joliot-Curie (daughter of Pierre and Marie Curie) and her husband Frederic Joliot-Curie observed that when aluminum-27 is bombarded with alpha particles, neutrons and positrons (positive electrons) are emitted as part of the products. When the source of alpha particles is removed, neutrons cease to be produced, but positrons continue to be emitted. This observation suggested that the neutrons and positrons come from two separate reactions. It also indicated that a product of the first reaction is radioactive. After further investigation, they discovered that, when aluminum-27 is bombarded with alpha particles, phosphorus-30 and neutrons are produced. Phosphorus-30 is radioactive, has a half-life of 2.5 minutes, and decays to silicon-30 with the emission of a positron. The equations for these reactions are

$$^{27}_{13}Al + ^{4}_{2}He \longrightarrow ^{30}_{15}P + ^{1}_{0}n$$
$$^{30}_{15}P \longrightarrow ^{30}_{14}Si + ^{0}_{+1}e$$

artificial radioactivity
induced radioactivity

The radioactivity of nuclides produced in this manner is known as **artificial radioactivity** or **induced radioactivity.** Artificial radionuclides behave like natural radioactive elements in two ways: They disintegrate in a definite fashion and they have a specific half-life. The Joliot-Curies received the Nobel Prize in chemistry in 1935 for the discovery of artificial, or induced, radioactivity.

18.7 Measurement of Radioactivity

Radiation from radioactive sources is so energetic that it is called *ionizing radiation.* When it strikes an atom or a molecule, one or more electrons are knocked off, and an ion is created. The Geiger counter, an instrument commonly used to detect and measure radioactivity, depends on this property. The instrument consists of a Geiger-Müller detecting tube and a counting device.

The detector tube is a pair of oppositely charged electrodes in an argon gas-filled chamber fitted with a thin window. When radiation, such as a beta particle, passes through the window into the tube, some argon is ionized, and a momentary pulse of current (discharge) flows between the electrodes. These current pulses are electronically amplified in the counter and appear as signals in the form of audible clicks, flashing lights, meter deflections, or numerical readouts (Figure 18.4).

Figure 18.4
Geiger–Müller survey meter.

The amount of radiation that an individual encounters can be measured by a film badge. This badge contains a piece of photographic film in a light-proof holder and is worn in areas where radiation might be encountered. The silver grains in the film will darken when exposed to radiation. The badges are processed after a predetermined time interval to determine the amount of radiation the wearer has been exposed to.

A scintillation counter is used to measure radioactivity for biomedical applications. A scintillator contains molecules that emit light when exposed to ionizing radiation. A light-sensitive detector counts the flashes and converts them to a numerical readout.

curie (Ci)

The *curie* is the unit used to express the amount of radioactivity produced by an element. One **curie (Ci)** is defined as the quantity of radioactive material giving 3.7×10^{10} disintegrations per second. The basis for this figure is pure radium, which has an activity of 1 Ci/g. Because the curie is such a large quantity, the millicurie and microcurie, representing one-thousandth and one-millionth of a curie, respectively, are more practical and more commonly used.

roentgen (R)

rad (radiation absorbed dose)

The curie only measures radioactivity emitted by a radionuclide. Different units are required to measure exposure to radiation. The **roentgen (R)** quantifies exposure to gamma or X-rays; 1 roentgen is defined as the amount of radiation required to produce 2.1×10^9 ions/cm^3 of dry air. The **rad (*r*adiation *a*bsorbed *d*ose)** is defined as the amount of radiation that provides 0.01 J of energy per kilogram of matter. The amount of radiation absorbed will change depending on the type of matter. The roentgen and the rad are numerically similar; 1 roentgen of gamma radiation provides 0.92 rad in bone tissue.

rem (roentgen equivalent to man)

Neither the rad nor the roentgen indicates the biological damage caused by radiation. One rad of alpha particles has the ability to cause 10 times more damage than 1 rad of gamma rays or beta particles. Another unit, **rem (*r*oentgen *e*quivalent to *m*an)**, takes into account the degree of biological effect caused by the type of radiation exposure; 1 rem is equal to the dose in rads multiplied by a factor specific to the form of radiation. The factor is 10 for alpha particles and 1 for both beta particles and gamma rays. Units of radiation are summarized in Table 18.4.

Table 18.4 Radiation Units

Unit	Measure	Equivalent
curie (Ci)	rate of decay of a radioactive substance	1 Ci = 3.7×10^{10} disintegrations/sec
roentgen (R)	exposure based on the quantity of ionization produced in air	1 R = 2.1×10^9 ions/cm^3
rad	absorbed dose of radiation	1 rad = 0.01 J/kg matter
rem	radiation dose equivalent	1 rem = 1 rad × factor
gray (Gy) (SI unit)	energy absorbed by tissue	1 Gy = 1 J/kg tissue (1 Gy = 100 rad)

18.8 Nuclear Fission

nuclear fission

In **nuclear fission** a heavy nuclide splits into two or more intermediate-sized fragments when struck in a particular way by a neutron. The fragments are called *fission products.* As the atom splits, it releases energy and two or three neutrons, each of which can cause another nuclear fission. The first instance of nuclear fission was reported in January 1939 by the German scientists Otto Hahn (1879–1968) and Fritz Strassmann (1902–1980). Detecting isotopes of barium, krypton, cerium, and lanthanum after bombarding uranium with neutrons led scientists to believe that the uranium nucleus had been split.

Characteristics of nuclear fission are as follows:

1. Upon absorption of a neutron, a heavy nuclide splits into two or more smaller nuclides (fission products).
2. The mass of the nuclides formed ranges from about 70 to 160 amu.
3. Two or more neutrons are produced from the fission of each atom.
4. Large quantities of energy are produced as a result of the conversion of a small amount of mass into energy.
5. Most nuclides produced are radioactive and continue to decay until they reach a stable nucleus.

One process by which this fission takes place is illustrated in Figure 18.5. When a heavy nucleus captures a neutron, the energy increase may be sufficient to cause deformation of the nucleus until the mass finally splits into two fragments, releasing energy and usually two or more neutrons.

In a typical fission reaction, a $^{235}_{92}U$ nucleus captures a neutron and forms unstable $^{236}_{92}U$. This $^{236}_{92}U$ nucleus undergoes fission, quickly disintegrating into two fragments, such as $^{139}_{56}Ba$ and $^{94}_{36}Kr$, and three neutrons. The three neutrons in turn may be captured by three other $^{235}_{92}U$ atoms, each of which undergoes

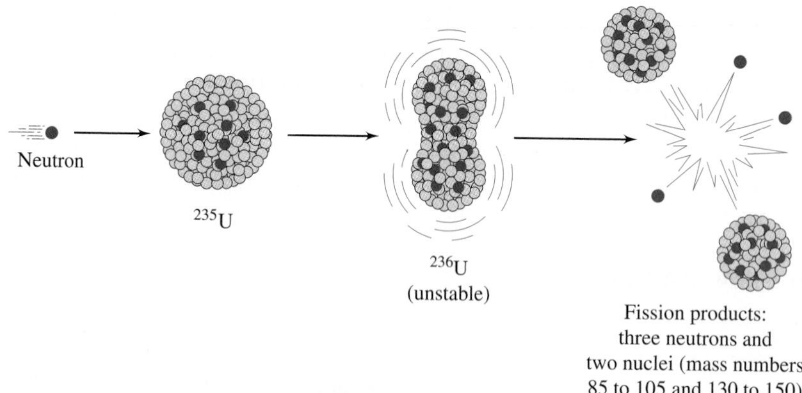

Figure 18.5
The fission process. When a neutron is captured by a heavy nucleus, the nucleus becomes more unstable. The more energetic nucleus begins to deform, resulting in fission. Two nuclear fragments and three neutrons are produced by this fission process.

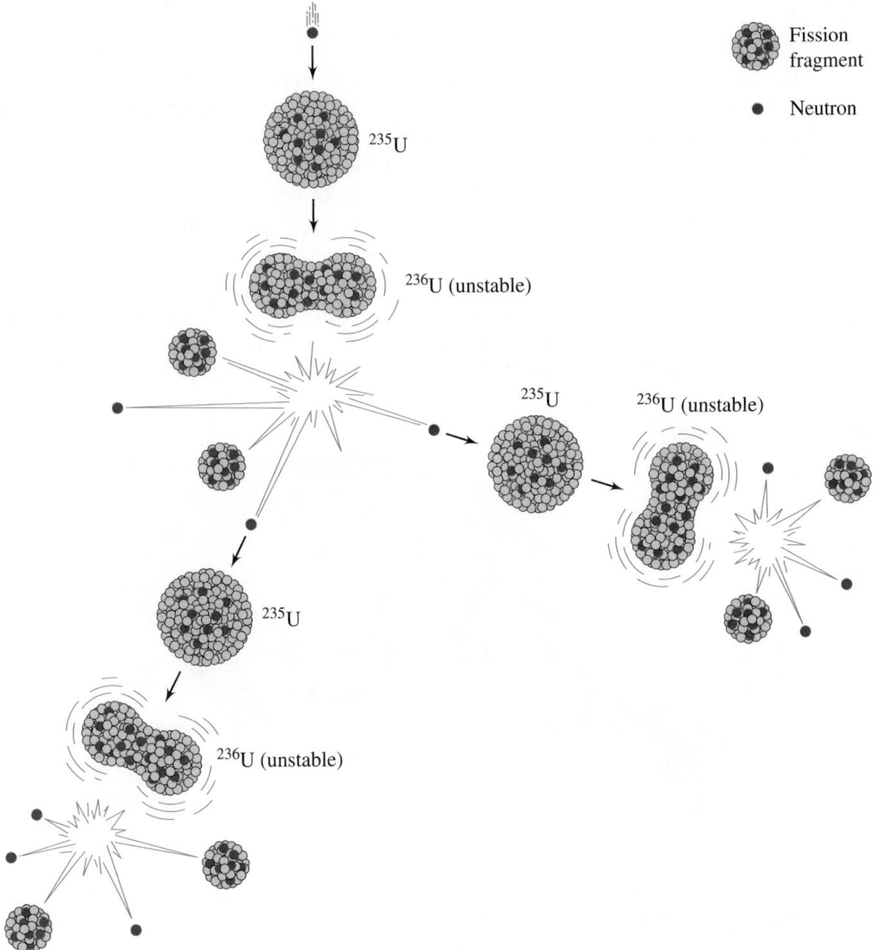

Figure 18.6
Fission and chain reaction of $^{235}_{92}U$. Each fission produces two major fission fragments and three neutrons, which may be captured by other $^{235}_{92}U$ nuclei, continuing the chain reaction.

fission, producing nine neutrons, and so on. A reaction of this kind, in which the products cause the reaction to continue or magnify, is known as a **chain reaction.** For a chain reaction to continue, enough fissionable material must be present so that each atomic fission causes, on average, at least one additional fission. The minimum quantity of an element needed to support a self-sustaining chain reaction is called the **critical mass.** Since energy is released in each atomic fission, chain reactions provide a steady supply of energy. (A chain reaction is illustrated in Figure 18.6.) Two of the many possible ways in which $^{235}_{92}U$ may fission are shown by these nuclear equations:

chain reaction

critical mass

$$^{235}_{92}U + ^{1}_{0}n \longrightarrow ^{139}_{56}Ba + ^{94}_{36}Kr + 3\,^{1}_{0}n$$

$$^{235}_{92}U + ^{1}_{0}n \longrightarrow ^{144}_{54}Xe + ^{90}_{38}Sr + 2\,^{1}_{0}n$$

CHEMISTRY IN ACTION • Does Your Food Glow in the Dark?

The Food and Drug Administration (FDA) has approved irradiation to delay ripening or kill microbes and insects in wheat, potatoes, fresh fruits, and poultry as well as in spices. But the only regular use of irradiation on food in the United States has been on spices.

Now the FDA has approved irradiation of red meat, and new legislation allows the labels about irradiated food to be in much smaller type. The general public has not yet accepted the use of radiation to reduce microbes and help preserve food. Why? Many people believe that irradiating food makes it radioactive. This is not true; in fact, the most that irradiation of food does is to produce compounds similar to those created by cooking and also to reduce the vitamin content of some food.

How does irradiation of food work? The radioactive source currently used is cobalt-60. It is contained in a concrete cell (see diagram) with 6-foot-thick walls. Inside the cell is a pool of water with racks of thin Co-60 rods suspended above. When the rods are not being used, they are submerged in the water, which absorbs the gamma radiation. When food to be irradiated moves into the cells, the Co-60 rods are lifted from the water, and the boxes of food move among them on a conveyor being irradiated from all sides.

Scientists who have investigated the process say that food irradiation is safe. The Center for Disease Control and Prevention in Atlanta estimates that food-borne illness causes as many as 9000 deaths per year. Irradiation provides one way to reduce microbial contamination in food. It does not solve the problem of careless handling of food by processors, and long-term studies on humans have not yet been concluded regarding irradiated food supplies. Once again, we face the issue of balancing benefit and risk.

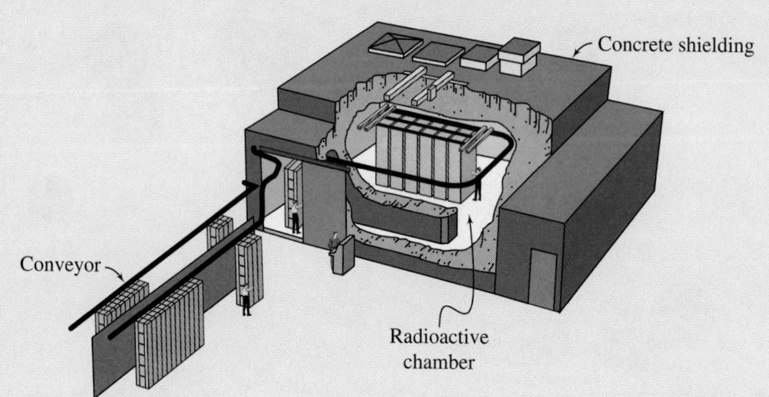

18.9 Nuclear Power

Nearly all electricity for commercial use is produced by machines consisting of a turbine linked by a drive shaft to an electrical generator. The energy required to run the turbine can be supplied by falling water, as in hydroelectric power plants, or by steam, as in thermal power plants.

The world's demands for energy, largely from fossil fuels, is heavy. At the present rates of consumption, the estimated world supply of fossil fuels is sufficient for only a few centuries. Although the United States has large coal and oil shale deposits, it currently imports over 40% of its oil supply. We clearly need to develop alternative energy sources. At present, uranium is the most productive alternative energy source, and about 17% of the electrical energy used in the United States is generated from power plants using uranium fuel.

A nuclear power plant is a thermal power plant in which heat is produced by a nuclear reactor instead of by combustion of fossil fuel. The major components of a nuclear reactor are

1. an arrangement of nuclear fuel, called the reactor core
2. a control system, which regulates the rate of fission and thereby the rate of heat generation
3. a cooling system, which removes the heat from the reactor and also keeps the core at the proper temperature

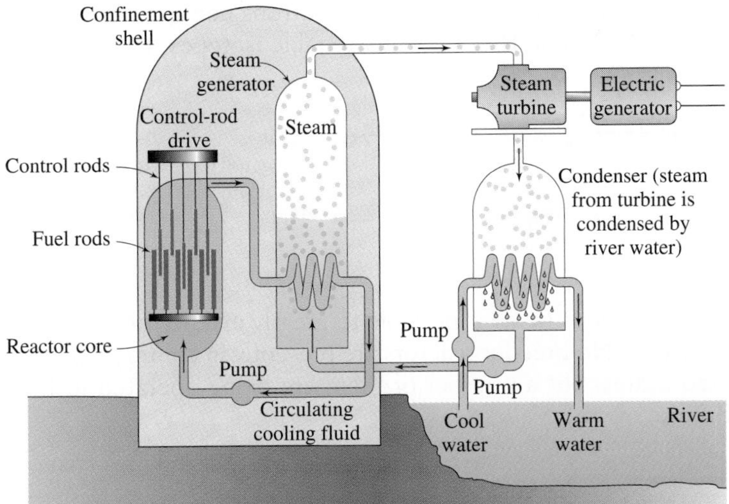

Figure 18.7
Diagram of a nuclear power plant. Heat produced by the reactor core is carried by a cooling fluid to a steam generator, which turns a turbine that produces electricity.

One type of reactor uses metal slugs containing uranium enriched from the normal 0.7% U-235 to about 3% U-235. The self-sustaining fission reaction is moderated, or controlled, by adjustable control rods containing substances that slow down and capture some of the neutrons produced. Ordinary water, heavy water, and molten sodium are typical coolants used. Energy obtained from nuclear reactions in the form of heat is used in the production of steam to drive turbines for generating electricity. (See Figure 18.7.)

The potential dangers of nuclear power were tragically demonstrated by the accidents at Three Mile Island, Pennsylvania (1979), and Chernobyl in the former U.S.S.R. (1986). Both accidents resulted from the loss of coolant to the reactor core. The reactors at Three Mile Island were covered by concrete containment buildings and therefore released a relatively small amount of radioactive material into the atmosphere. But because the Soviet Union did not require containment structures on nuclear power plants, the Chernobyl accident resulted in 31 deaths and the resettlement of 135,000 people. The release of large quantities of I-131, Cs-134, and Cs-137 appears to be causing long-term health problems in that exposed population.

Another problem with nuclear power plants is their radioactive waste production. The United States has debated where and how to permanently store nuclear wastes. In July 2002, an underground site was authorized at Yucca Mountain, Nevada, about 90 miles north of Las Vegas. The spent nuclear fuel will remain radioactive for thousands of years.

In the United States, reactors designed for commercial power production use uranium oxide, U_3O_8, that is enriched with the relatively scarce fissionable U-235 isotope. Because the supply of U-235 is limited, a new type of reactor known as the *breeder reactor* has been developed. Breeder reactors produce additional fissionable material at the same time that the fission reaction is occurring. In a breeder reactor, excess neutrons convert nonfissionable isotopes, such as U-238 or Th-232, to fissionable isotopes, Pu-239 or U-233, as shown below:

$$^{238}_{92}U + ^{1}_{0}n \longrightarrow ^{239}_{92}U \xrightarrow[\beta]{} ^{239}_{93}Np \xrightarrow[\beta]{} ^{239}_{94}Pu$$

$$^{232}_{90}Th + ^{1}_{0}n \longrightarrow ^{233}_{90}Th \xrightarrow[\beta]{} ^{233}_{91}Pa \xrightarrow[\beta]{} ^{233}_{92}U$$

These transmutations make it possible to greatly extend the supply of fuel for nuclear reactors. No breeder reactors are presently in commercial operation in the United States, but a number of them are being operated in Europe and Great Britain.

18.10 The Atomic Bomb

The atomic bomb is a fission bomb; it operates on the principle of a very fast chain reaction that releases a tremendous amount of energy. An atomic bomb and a nuclear reactor both depend on self-sustaining nuclear fission chain reactions. The essential difference is that in a bomb the fission is "wild," or uncontrolled, whereas in a nuclear reactor the fission is moderated and carefully controlled. A minimum critical mass of fissionable material is needed for a bomb; otherwise a major explosion will not occur. When a quantity smaller than the critical mass is used, too many neutrons formed in the fission step escape without combining with another nucleus, and a chain reaction does not occur. Therefore the fissionable material of an atomic bomb must be stored as two or more subcritical masses and brought together to form the critical mass at the desired time of explosion. The temperature developed in an atomic bomb is believed to be about 10 million degrees Celsius.

The nuclides used in atomic bombs are U-235 and Pu-239. Uranium deposits contain about 0.7% of the U-235 isotope, the remainder being U-238. Uranium-238 does not undergo fission except with very high-energy neutrons. It was discovered, however, that U-238 captures a low-energy neutron without undergoing fission and that the product, U-239, changes to Pu-239 (plutonium) by a beta-decay process. Plutonium-239 readily undergoes fission upon capture of a neutron and is therefore useful for nuclear weapons. The equations for the nuclear transformations are

$$^{238}_{92}U + ^{1}_{0}n \longrightarrow ^{239}_{92}U \xrightarrow[\beta]{} ^{239}_{93}Np \xrightarrow[\beta]{} ^{239}_{94}Pu$$

The hazards of an atomic bomb explosion include not only shock waves from the explosive pressure and tremendous heat, but also intense radiation in the form of alpha particles, beta particles, gamma rays, and ultraviolet rays. Gamma rays and X-rays can penetrate deeply into the body, causing burns, sterilization, and gene mutation, which can adversely affect future generations. Both radioactive fission products and unfissioned material are present after the explosion. If the bomb explodes near the ground, many tons of dust are lift-

The mushroom cloud is a signature of uncontrolled fission in an atomic bomb.

ed into the air. Radioactive material adhering to this dust, known as *fallout,* is spread by air currents over wide areas of the land and constitutes a lingering source of radiation hazard.

Today nuclear war is probably the most awesome threat facing civilization. Only two rather primitive fission-type atom bombs were used to destroy the Japanese cities of Hiroshima and Nagasaki and bring World War II to an early end. The threat of nuclear war is increased by the fact that the number of nations possessing nuclear weapons is steadily increasing.

18.11 Nuclear Fusion

nuclear fusion

The process of uniting the nuclei of two light elements to form one heavier nucleus is known as **nuclear fusion.** Such reactions can be used for producing energy, because the masses of the two nuclei that fuse into a single nucleus are greater than the mass of the nucleus formed by their fusion. The mass differential is liberated in the form of energy. Fusion reactions are responsible for the tremendous energy output of the sun. Thus, aside from relatively small amounts from nuclear fission and radioactivity, fusion reactions are the ultimate source of our energy, even the energy from fossil fuels. They are also responsible for the devastating power of the thermonuclear, or hydrogen, bomb.

Fusion reactions require temperatures on the order of tens of millions of degrees for initiation. Such temperatures are present in the sun but have been produced only momentarily on Earth. For example, the hydrogen, or fusion, bomb is triggered by the temperature of an exploding fission bomb. Two typical fusion reactions are

$$\underset{\text{tritium}}{{}^{3}_{1}\text{H}} + \underset{\text{deuterium}}{{}^{2}_{1}\text{H}} \longrightarrow {}^{4}_{2}\text{He} + {}^{1}_{0}\text{n} + \text{energy}$$

$$\underset{\substack{3.0150\\ \text{amu}}}{{}^{3}_{1}\text{H}} + \underset{\substack{1.0079\\ \text{amu}}}{{}^{1}_{1}\text{H}} \longrightarrow \underset{\substack{4.0026\\ \text{amu}}}{{}^{4}_{2}\text{He}} + \text{energy}$$

The total mass of the reactants in the second equation is 4.0229 amu, which is 0.0203 amu greater than the mass of the product. This difference in mass is manifested in the great amount of energy liberated.

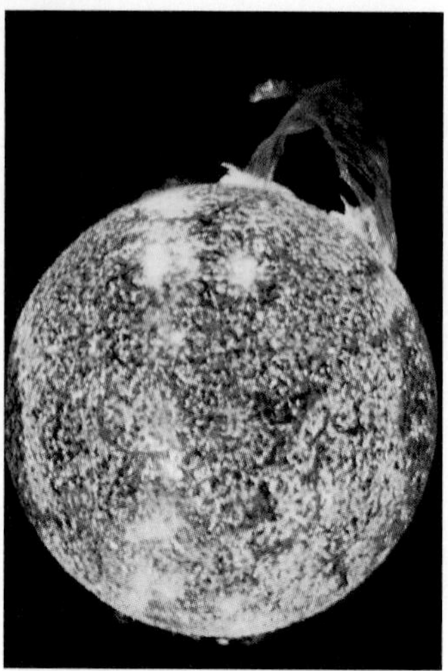

Solar flares such as these are indications of fusion reactions occurring at temperatures of millions of degrees.

A great deal of research in the United States and in other countries, especially the former Soviet Union, has focused on controlled nuclear fusion reactions. The goal of controlled nuclear fusion has not yet been attained, although the required ignition temperature has been reached in several devices. Evidence to date leads us to believe that we can develop a practical fusion power reactor. Fusion power, if we can develop it, will be far superior to fission power for the following reasons:

1. Virtually infinite amounts of energy are possible from fusion. Uranium supplies for fission power are limited, but heavy hydrogen, or deuterium (the most likely fusion fuel), is abundant. It is estimated that the deuterium in a cubic mile of seawater used as fusion fuel could provide more energy than the petroleum reserves of the entire world.
2. From an environmental viewpoint, fusion power is much "cleaner" than fission power because fusion reactions (in contrast to uranium and plutonium fission reactions) do not produce large amounts of long-lived and dangerously radioactive isotopes.

18.12 Mass–Energy Relationship in Nuclear Reactions

Large amounts of energy are released in nuclear reactions; thus significant amounts of mass are converted to energy. We stated earlier that the amount of mass converted to energy in chemical changes is insignificant compared to the amount of energy released in a nuclear reaction. In fission reactions, about 0.1% of the mass is converted into energy. In fusion reactions as much as 0.5% of the mass may be changed into energy. The Einstein equation, $E = mc^2$, can be used to calculate the energy liberated, or available, when the mass loss is known. For example, in the fusion reaction

$$\underset{7.016\text{ g}}{{}^{7}_{3}\text{Li}} + \underset{1.008\text{ g}}{{}^{1}_{1}\text{H}} \longrightarrow \underset{4.003\text{ g}}{{}^{4}_{2}\text{He}} + \underset{4.003\text{ g}}{{}^{4}_{2}\text{He}} + \text{energy}$$

the mass difference between the reactants and products (8.024 g − 8.006 g) is 0.018 g. The energy equivalent to this amount of mass is 1.62×10^{12} J. By comparison, this is more than 4 million times greater than the 3.9×10^{5} J of energy obtained from the complete combustion of 12.01 g (1 mol) of carbon.

The mass of a nucleus is actually less than the sum of the masses of the protons and neutrons that make up that nucleus. The difference between the mass of the protons and the neutrons in a nucleus and the mass of the nucleus is known as the **mass defect.** The energy equivalent to this difference in mass is known as the **nuclear binding energy.** This energy is the amount that would be required to break a nucleus into its individual protons and neutrons. The higher the binding energy, the more stable the nucleus. Elements of intermediate atomic masses have high binding energies. For example, iron (element number 26) has a very high binding energy and therefore a very stable nucleus. Just as electrons attain less energetic and more stable arrangements through ordinary chemical reactions, neutrons and protons attain less energetic and more stable arrangements through nuclear fission or fusion reactions. Thus, when uranium undergoes fission, the products have less mass (and greater binding energy) than the original uranium. In like manner, when hydrogen and

mass defect

nuclear binding energy

lithium fuse to form helium, the helium has less mass (and greater binding energy) than the hydrogen and lithium. It is this conversion of mass to energy that accounts for the very large amounts of energy associated with both fission and fusion reactions.

Example 18.5

Calculate the mass defect and the nuclear binding energy for an α particle (helium nucleus).

Data:
$$\begin{aligned} \text{proton mass} &= 1.0073\ \text{g/mol} \\ \text{neutron mass} &= 1.0087\ \text{g/mol} \\ \alpha\ \text{mass} &= 4.0015\ \text{g/mol} \\ 1.0\ \text{g} &= 9.0 \times 10^{13}\ \text{J} \end{aligned}$$

SOLUTION

First calculate the sum of the individual parts of an α particle (2 protons +2 neutrons).

$$\begin{aligned} &\text{2 protons:} \quad 2 \times 1.0073\ \text{g/mol} = 2.0146\ \text{g/mol} \\ &\text{2 neutrons:} \quad 2 \times 1.0087\ \text{g/mol} = \underline{2.0174\ \text{g/mol}} \\ &\qquad\qquad\qquad\qquad\qquad\qquad\quad 4.0320\ \text{g/mol} \end{aligned}$$

The mass defect is the difference between the mass of the α particle and its component parts:

$$\begin{aligned} &4.0320\ \text{g/mol} \\ &\underline{4.0015\ \text{g/mol}} \\ &0.0305\ \text{g/mol} \quad \text{(mass defect)} \end{aligned}$$

To find the nuclear binding energy, convert the mass defect to its energy equivalent:

$$(0.0305\ \text{g/mol})(9.0 \times 10^{13}\ \text{J/g}) = 2.7 \times 10^{12}\ \text{J/mol}\ (6.5 \times 10^{11}\ \text{cal})$$

18.13 Transuranium Elements

transuranium element

The elements that follow uranium on the periodic table and that have atomic numbers greater than 92 are known as the **transuranium elements.** They are synthetic radioactive elements; none of them occur naturally.

The first transuranium element, number 93, was discovered in 1939 by Edwin M. McMillan (1907–1991) at the University of California while he was investigating the fission of uranium. He named it neptunium for the planet Neptune. In 1941, element 94, plutonium, was identified as a beta-decay product of neptunium:

$$^{238}_{93}\text{Np} \longrightarrow {}^{238}_{94}\text{Pu} + {}^{\ 0}_{-1}\text{e}$$

$$^{239}_{93}\text{Np} \longrightarrow {}^{239}_{94}\text{Pu} + {}^{\ 0}_{-1}\text{e}$$

Plutonium is one of the most important fissionable elements known today.

Since 1964, the discoveries of eight new transuranium elements, numbers 104–111, have been announced. These elements have been produced in minute quantities by high-energy particle accelerators.

18.14 Biological Effects of Radiation

ionizing radiation

Radiation with energy to dislocate bonding electrons and create ions when passing through matter is classified as **ionizing radiation.** Alpha particles, beta particles, gamma rays, and X-rays fall into this classification. Ionizing radiation can damage or kill living cells and can be particularly devastating when it strikes the cell nuclei and affects molecules involved in cell reproduction. The effects of radiation on living organisms fall into these general categories: (1) acute or short-term effects, (2) long-term effects, and (3) genetic effects.

Acute Radiation Damage

High levels of radiation, especially from gamma rays or X-rays, produce nausea, vomiting, and diarrhea. The effect has been likened to a sunburn throughout the body. If the dosage is high enough, death will occur in a few days. The damaging effects of radiation appear to be centered in the nuclei of the cells, and cells that are undergoing rapid cell division are most susceptible to damage. It is for this reason that cancers are often treated with gamma radiation from a Co-60 source. Cancerous cells multiply rapidly and are destroyed by a level of radiation that does not seriously damage normal cells.

Long-Term Radiation Damage

Protracted exposure to low levels of any form of ionizing radiation can weaken an organism and lead to the onset of malignant tumors, even after fairly long time delays. The largest exposure to synthetic sources of radiation is from X-rays. Evidence suggests that the lives of early workers in radioactivity and X-ray technology may have been shortened by long-term radiation damage.

Strontium-90 isotopes are present in the fallout from atmospheric testing of nuclear weapons. Strontium is in the same periodic-table group as calcium, and its chemical behavior is similar to that of calcium. Hence when foods contaminated with Sr-90 are eaten, Sr-90 ions are laid down in the bone tissue along with ordinary calcium ions. Strontium-90 is a beta emitter with a half-life of 28 years. Blood cells manufactured in bone marrow are affected by the radiation from Sr-90. Hence there is concern that Sr-90 accumulation in the environment may cause an increase in the incidence of leukemia and bone cancers.

Genetic Effects

The information needed to create an individual of a particular species, be it a bacterial cell or a human being, is contained within the nucleus of a cell. This genetic information is encoded in the structure of DNA (deoxyribonucleic acid) molecules, which make up genes. The DNA molecules form precise duplicates of themselves when cells divide, thus passing genetic information from one generation to the next. Radiation can damage DNA molecules. If the damage is not severe enough to prevent the individual from reproducing, a mutation may result. Most mutation-induced traits are undesirable. Unfortunately, if the bearer of the altered genes survives to reproduce, these traits are passed along to succeeding generations. In other words, the genetic effects of increased radiation exposure are found in future generations, not only in the present generation.

Because radioactive rays are hazardous to health and living tissue, special precautions must be taken in designing laboratories and nuclear reactors, in disposing of waste materials, and in monitoring the radiation exposure of people working in this field.

CHEMISTRY IN ACTION • A Window into Living Organisms

Imagine viewing a living process as it is occurring. This was the dream of many scientists in the past as they tried to extract this knowledge from dead tissue. Today, because of innovations in nuclear chemistry, this dream is a common, everyday occurrence.

Compounds containing a radionuclide are described as being *labeled,* or *tagged.* These compounds undergo their normal chemical reactions, but their location can be detected because of their radioactivity. When such compounds are given to a plant or an animal, the movement of the nuclide can be traced through the organism by the use of a Geiger counter or other detecting device.

In an early use of the tracer technique, the pathway by which CO_2 becomes fixed into carbohydrate ($C_6H_{12}O_6$) during photosynthesis was determined. The net equation for photosynthesis is

$$6\,CO_2 + 6\,H_2O \longrightarrow C_6H_{12}O_6 + 6\,O_2$$

Radioactive $^{14}CO_2$ was injected into a colony of green algae, and the algae were then placed in the dark and killed at selected time intervals. When the radioactive compounds were separated by paper chromatography and analyzed, the results elucidated a series of light-independent photosynthetic reactions.

Biological research using tracer techniques have determined

1. the rate of phosphate uptake by plants, using radiophosphorus
2. the flow of nutrients in the digestive tract using radioactive barium compounds.
3. the accumulation of iodine in the thyroid gland, using radioactive iodine
4. the absorption of iron by the hemoglobin of the blood, using radioactive iron

In chemistry, uses for tracers are unlimited. The study of reaction mechanisms, the measurement of the rates of chemical reactions, and the determination of physical constants are just a few of the areas of application.

A radioactive tracer is injected into this patient and absorbed by the brain. The PET scanner detects photons emitted by the tracer and produces an image used in medical diagnosis and research.

Radioactive tracers are commonly used in medical diagnosis. The radionuclide must be effective at a low concentration and have a short half-life to reduce the possibility of damage to the patient.

Radioactive iodine (I-131) is used to determine thyroid function, where the body concentrates iodine. In this process a small amount of radioactive potassium or sodium iodide is ingested. A detector is focused on the thyroid gland and measures the amount of iodine in the gland. This picture is then compared to that of a normal thyroid to detect any differences.

Doctors examine the heart's pumping performance and check for evidence of obstruction in coronary arteries by *nuclear scanning.* The radionuclide Tl-201, when injected into the bloodstream, lodges in healthy heart muscle. Thallium-201 emits gamma radiation, which is detected by a special imaging device called a *scintillation camera.* The data obtained are simultaneously translated into pictures by a computer. With this technique doctors can observe whether heart tissue has died after a heart attack and whether blood is flowing freely through the coronary passages.

One of the most recent applications of nuclear chemistry is the use of positron emission tomography (PET) in the measurement of dynamic processes in the body, such as oxygen use or blood flow. In this application, a compound is made that contains a positron-emitting nuclide such as C-11, O-15, or N-13. The compound is injected into the body, and the patient is placed in an instrument that detects the positron emission. A computer produces a three-dimensional image of the area.

PET scans have been used to locate the areas of the brain involved in epileptic seizures. Glucose tagged with C-11 is injected, and an image of the brain is produced. Since the brain uses glucose almost exclusively for energy, diseased areas that use glucose at a rate different than normal tissue can then be identified.

Chapter 18 Review

18.1 Discovery of Radioactivity

KEY TERMS

Radioactivity
Nucleon
Nuclide

- Radioactivity is the spontaneous emission of particles and energy from the nucleus of an atom.
- Protons and neutrons are known as nucleons.
- In nuclear chemistry isotopes are also known as nuclides.

18.2 Natural Radioactivity

KEY TERMS

Radioactivite decay
Half-life

- Radioactive elements undergo radioactive decay to form different elements.
- Radioactivity is not affected by changes in temperature, pressure, or the state of the element.
- Principal emissions:
  - Alpha particles
  - Beta particles
  - Gamma rays
- The half-life of a nuclide is the time required for $^1/_2$ of a specific amount of the nuclide to disintegrate.

18.3 Alpha Particles, Beta Particles, and Gamma Rays

KEY TERMS

Alpha particle
Beta particle
Gamma ray

- Alpha particles:
  - Consist of 2 protons and 2 neutrons with mass of 4 amu and charge of +2
  - Loss of an alpha particle from the nucleus results in:
    - Loss of 4 in mass number
    - Loss of 2 in atomic number
  - High ionizing power, low penetration
- Beta particles:
  - Same as an electron
  - Loss of a beta particle from the nucleus results in:
    - No change in mass number
    - Increase of 1 in atomic number
  - Moderate ionizing power and penetration
- Gamma rays:
  - Photons of energy
  - Loss of a gamma ray from the nucleus results in:
    - No change in mass number (A)
    - No change in atomic number (Z)
  - Almost no ionizing power, high penetration

18.4 Radioactive Disintegration Series

- As element undergo disintegration, they eventually become stable after the loss of a series of particles and energy.
- The disintegration series can be used to determine the age of geologic deposits.

18.5 Transmutation of Elements

KEY TERM

Transmutation

- Transmutation of an element can occur spontaneously or artificially.

18.6 Artificial Radioactivity

KEY TERM

Artificial radioactivity or induced radioactivity

- Artificial radionuclides behave like natural radioactive elements in two ways:
  - They disintegrate in a definite fashion.
  - They have a specific half-life.

18.7 Measurement of Radioactivity

KEY TERMS

Curie (Ci)
Roentgen (R)
Rad (radiation absorbed dose)
Rem (roentgen equivalent to man)

- Radioactivity can be measured in several ways:
  - Geiger–Müller counter
  - Film badge
  - Scintillation counter
- Units for measuring radiation:
  - Gray
  - Curie
  - Roentgen
  - Rad
  - Rem

18.8 Nuclear Fission

KEY TERMS

Nuclear fission
Chain reaction
Critical mass

- Characteristics of fission:
  - A heavy nuclide splits into two or more smaller nuclides.
  - The mass of nuclides formed is 70 to 160 amu.
  - Two or more neutrons are produced per fission.
  - Large quantities of energy are released.
  - Most nuclides produced are radioactive and continue to decay until they reach stability.

18.9 Nuclear Power

- A nuclear power plant is a thermal plant with a nuclear fission source of energy.
- Major components of a nuclear reactor:
  - Reactor core
  - Control system
  - Cooling system

18.10 The Atomic Bomb

- Out-of-control fission reaction

18.11 Nuclear Fusion

KEY TERM

Nuclear fusion

- Nuclear fusion is the process of uniting 2 light nuclei to form 1 heavier nucleus.
- Nuclear fusion requires extreme temperatures for ignition.

18.12 Mass–Energy Relationship in Nuclear Reactions

KEY TERM

Mass defect
Nuclear binding energy

- The mass defect is the difference between the actual mass of an atom and the calculated mass of the protons and neutrons in the nucleus of that atom.
- The nuclear binding energy is the energy equivalent to the mass defect. $E = mc^2$ is used to determine the energy liberated in a nuclear reaction.

18.13 Transuranium Elements

KEY TERM

Transuranium elements

- Transuranium elements are the elements that follow uranium on the periodic table.

18.14 Biological Effects of Radiation

KEY TERM

Ionizing radiation

- Effects of radiation fall into 3 general categories:
  - Acute or short-term
  - Long-term
  - Genetic

Review Questions

All questions with blue numbers have answers in the appendix of the text.

1. To afford protection from radiation injury, which kind of radiation requires (a) the most shielding? (b) the least shielding?
2. Why is an alpha particle deflected less than a beta particle in passing through an electromagnetic field?
3. Name three pairs of nuclides that might be obtained by fissioning U-235 atoms.
4. Identify these people and their associations with the early history of radioactivity:
   (a) Antoine Henri Becquerel
   (b) Marie and Pierre Curie
   (c) Wilhelm Roentgen
   (d) Ernest Rutherford
   (e) Otto Hahn and Fritz Strassmann
5. Why is the radioactivity of an element unaffected by the usual factors that affect the rate of chemical reactions, such as ordinary changes of temperature and concentration?
6. Distinguish between the terms *isotope* and *nuclide*.
7. The half-life of Pu-244 is 76 million years. If Earth's age is about 5 billion years, discuss the feasibility of finding this nuclide as a naturally occurring nuclide.
8. Tell how alpha, beta, and gamma radiation are distinguished from the standpoint of
   (a) charge
   (b) relative mass
   (c) nature of particle or ray
   (d) relative penetrating power
9. Distinguish between natural and artificial radioactivity.
10. What is a radioactive disintegration series?
11. Briefly discuss the transmutation of elements.
12. Stable Pb-208 is formed from Th-232 in the thorium disintegration series by successive α, β, β, α, α, α, α, β, β, α particle emissions. Write the symbol (including mass and atomic number) for each nuclide formed in this series.
13. The nuclide Np-237 loses a total of seven alpha particles and four beta particles. What nuclide remains after these losses?
14. Bismuth-211 decays by alpha emission to give a nuclide that in turn decays by beta emission to yield a stable nuclide. Show these two steps with nuclear equations.
15. What were Otto Hahn and Fritz Strassmann's contribution to nuclear physics?
16. What is a breeder reactor? Explain how it accomplishes the "breeding."
17. What is the essential difference between the nuclear reactions in a nuclear reactor and those in an atomic bomb?
18. Why must a certain minimum amount of fissionable material be present before a self-supporting chain reaction can occur?
19. What are mass defect and nuclear binding energy?
20. Explain why radioactive rays are classified as ionizing radiation.

21. Give a brief description of the biological hazards associated with radioactivity.

22. Strontium-90 has been found to occur in radioactive fallout. Why is there so much concern about this radionuclide being found in cow's milk? (Half-life of Sr-90 is 28 years.)

23. What is a radioactive tracer? How is it used?

24. Describe the radiocarbon method for dating archaeological artifacts.

25. How might radioactivity be used to locate a leak in an underground pipe?

26. Anthropologists have found bones whose age suggests that the human line may have emerged in Africa as much as 4 million years ago. If wood or charcoal were found with such bones, would C-14 dating be useful in dating the bones? Explain.

27. What are the disadvantages of nuclear power?

28. What are the hazards associated with an atomic bomb explosion?

29. What types of elements undergo fission? How about fusion?

30. Using Figure 18.7, describe how nuclear power is generated.

Paired Exercises

All exercises with blue numbers have answers in the appendix of the text.

1. Indicate the number of protons, neutrons, and nucleons in these nuclei:
(a) $^{35}_{17}Cl$ (b) $^{226}_{88}Ra$

2. Indicate the number of protons, neutrons, and nucleons in these nuclei:
(a) $^{235}_{92}U$ (b) $^{82}_{35}Br$

3. How are the mass and the atomic number of a nucleus affected by the loss of an alpha particle?

4. How are the mass and the atomic number of a nucleus affected by the loss of a beta particle?

5. Write nuclear equations for the alpha decay of
(a) $^{218}_{85}At$ (b) $^{221}_{87}Fr$

6. Write nuclear equations for the alpha decay of
(a) $^{192}_{78}Pt$ (b) $^{210}_{84}Po$

7. Write nuclear equations for the beta decay of
(a) $^{14}_{6}C$ (b) $^{137}_{55}Cs$

8. Write nuclear equations for the beta decay of
(a) $^{239}_{93}Np$ (b) $^{90}_{38}Sr$

9. Determine the type of emission or emissions (alpha, beta, or gamma) that occurred in the following transitions:
(a) $^{226}_{88}Ra$ to $^{222}_{86}Rn$
(b) $^{222}_{88}Ra$ to $^{222}_{87}Fr$ to $^{222}_{87}Fr$
(c) $^{238}_{92}U$ to $^{238}_{93}Np$

10. Determine the type of emission or emissions (alpha, beta, or gamma) that occurred in the following transitions:
(a) $^{210}_{82}Pb$ to $^{210}_{82}Pb$
(b) $^{234}_{91}Pa$ to $^{230}_{89}Ac$ to $^{230}_{90}Th$
(c) $^{234}_{90}Th$ to $^{230}_{88}Ra$ to $^{230}_{88}Ra$

11. Write a nuclear equation for the conversion of $^{13}_{6}C$ to $^{14}_{6}C$.

12. Write a nuclear equation for the conversion of $^{30}_{15}P$ to $^{30}_{14}Si$.

13. Complete and balance these nuclear equations by supplying the missing particles:
(a) $^{66}_{29}Cu \longrightarrow ^{66}_{30}Zn +$ ______
(b) $^{0}_{-1}e +$ ______ $\longrightarrow ^{7}_{3}Li$
(c) $^{27}_{13}Al + ^{4}_{2}He \longrightarrow ^{30}_{14}Si +$ ______
(d) $^{85}_{37}Rb +$ ______ $\longrightarrow ^{82}_{35}Br + ^{4}_{2}He$

14. Complete and balance these nuclear equations by supplying the missing particles:
(a) $^{27}_{13}Al + ^{4}_{2}He \longrightarrow ^{30}_{15}P +$ ______
(b) $^{27}_{14}Si \longrightarrow ^{0}_{+1}e +$ ______
(c) ______ $+ ^{2}_{1}H \longrightarrow ^{13}_{7}N + ^{1}_{0}n$
(d) ______ $\longrightarrow ^{82}_{36}Kr + ^{0}_{-1}e$

15. Strontium-90 has a half-life of 28 years. If a 1.00-mg sample was stored for 112 years, what mass of Sr-90 would remain?

16. Strontium-90 has a half-life of 28 years. If a sample was tested in 1980 and found to be emitting 240 counts/ min, in what year would the same sample be found to be emitting 30 counts/min? How much of the original Sr-90 would be left?

17. By what series of emissions does $^{230}_{90}Th$ disintegrate to $^{218}_{84}Po$?

18. By what series of emissions does $^{222}_{88}Ra$ disintegrate to $^{210}_{83}Bi$?

*19. Consider the fission reaction

$$^{235}_{92}U + ^{1}_{0}n \longrightarrow ^{94}_{38}Sr + ^{139}_{54}Xe + 3\,^{1}_{0}n + \text{energy}$$

Calculate the following using this mass data (1.0 g is equivalent to 9.0×10^{13} J):

U-235 = 235.0439 amu	Sr-94 = 93.9154 amu
Xe-139 = 138.9179 amu	n = 1.0087 amu

(a) the energy released in joules for a single event (one uranium atom splitting)
(b) the energy released in joules per mole of uranium splitting
(c) the percentage of mass lost in the reaction

20. Consider the fusion reaction

$$^{1}_{1}H + ^{2}_{1}H \longrightarrow ^{3}_{2}He + \text{energy}$$

Calculate the following using this mass data (1.0 g is equivalent to 9.0×10^{13} J):

$^{1}_{1}H$ = 1.00794 amu
$^{2}_{1}H$ = 2.01410 amu
$^{3}_{2}H$ = 3.01603 amu

(a) the energy released in joules per mole of He-3 formed
(b) the percentage of mass lost in the reaction

Additional Exercises

All exercises with blue numbers have answers in the appendix of the text.

21. The Th-232 disintegration series starts with $^{232}_{90}Th$ and emits the following rays successively: α, β, β, α, α, α, β, α, β, α. The series ends with the stable $^{208}_{82}Pb$. Write the formula for each nuclide in the series.

22. The half-life of Ra-224 is 3.64 days. How long will it take for 7/8 of an 8.0-g sample to disappear?

23. When the nuclide $^{249}_{98}Cf$ was bombarded with $^{15}_{7}N$, four neutrons and a new transuranium element were formed. Write the nuclear equation for this transmutation.

24. What percent of the mass of the nuclide $^{226}_{88}Ra$ is neutrons? electrons?

25. If radium costs $50,000 a gram, how much will 0.0100 g of $^{226}RaCl_2$ cost if the price is based only on the radium content?

26. An archaeological specimen was analyzed and found to be emitting only 25% as much C-14 radiation per gram of carbon as newly cut wood. How old is this specimen?

27. Barium-141 is a beta emitter. What is the half-life if a 16.0-g sample of the nuclide decays to 0.500 g in 90 minutes?

*28. Calculate (a) the mass defect and (b) the binding energy of $^{7}_{3}Li$ using the mass data:

$^{7}_{3}Li$ = 7.0160 g	n = 1.0087 g
p = 1.0073 g	e^- = 0.00055 g

1.0 g ≡ 9.0×10^{13} J (from $E = mc^2$)

*29. In the disintegration series $^{235}_{92}U \longrightarrow ^{207}_{82}Pb$, how many alpha and beta particles are emitted?

30. List three devices used for radiation detection and explain their operation.

31. The half-life of I-123 is 13 hours. If 10 mg of I-123 are administered to a patient, how much I-123 remains after 3 days and 6 hours?

32. Clearly distinguish between fission and fusion. Give an example of each.

33. Starting with 1 g of a radioactive isotope whose half-life is 10 days, sketch a graph showing the pattern of decay for that material. On the *x*-axis, plot time (you may want to simply show multiples of the half-life), and on the *y*-axis, plot mass of material remaining. Then after completing the graph, explain why a sample never really gets to the point where *all* of its radioactivity is considered to be gone.

34. Identify each missing product (name the element and give its atomic number and mass number) by balancing the following nuclear equations:

(a) $^{235}U + ^{1}_{0}n \longrightarrow ^{143}Xe + 3\,^{1}_{0}n +$ ______
(b) $^{235}U + ^{1}_{0}n \longrightarrow ^{102}Y + 3\,^{1}_{0}n +$ ______
(c) $^{14}N + ^{1}_{0}n \longrightarrow ^{1}H +$ ______

35. Consider these reactions:

(a) $H_2O(l) \longrightarrow H_2O(g)$
(b) $2\,H_2(g) + O_2(g) \longrightarrow 2\,H_2O(g)$
(c) $^{2}_{1}H + ^{2}_{1}H \longrightarrow ^{3}_{1}H + ^{1}_{1}H$

The following energy values belong to one of these equations:

energy$_1$	115.6 kcal released
energy$_2$	10.5 kcal absorbed
energy$_3$	7.5×10^7 kcal released

Match the equation to the energy value and briefly explain your choices.

36. When $^{235}_{92}U$ is struck by a neutron, the unstable isotope $^{236}_{92}U$ results. When that daughter isotope undergoes fission, there are numerous possible products. If strontium-90 and three neutrons are the results of one such fission, what is the other product?

37. Write balanced nuclear equations for

(a) beta emission by $^{29}_{12}Mg$
(b) alpha emission by $^{150}_{60}Nd$
(c) positron emission by $^{72}_{33}As$

38. How much of a sample of cesium-137 ($t_{1/2}$ = 30 years) must have been present originally if, after 270 years, 15.0 g remain?

39. Suppose that the existence of element 114 were confirmed and reported. What element would this new substance fall beneath on the periodic table? Would it be a metal? What typical ion might you expect it to form in solution?

40. Cobalt-60 has a half-life of 5.26 years. If 1.00 g of ^{60}Co were allowed to decay, how many grams would be left after
(a) one half-life?
(b) two half-lives?
(c) four half-lives?
(d) ten half-lives?

41. Write balanced equations to show these changes:
(a) alpha emission by boron-11
(b) beta emission by strontium-88
(c) neutron absorption by silver-107
(d) proton emission by potassium-41
(e) electron absorption by antimony-116

42. The $^{14}_{6}C$ content of an ancient piece of wood was found to be one-sixteenth of that in living trees. How many years old is this piece of wood if the half-life of carbon-14 is 5730 years?

43. The curie is equal to 3.7×10^{10} disintegration/sec, and the becquerel is equivalent to just 1 disintegration/sec. Suppose a hospital has a 150-g radioactive source with an activity of 1.24 Ci. What is its activity in becquerels?

Challenge Exercises

All exercises with blue numbers have answers in the appendix of the text.

*44. Rubidium-87, a beta emitter, is the product of positron emission. Identify
(a) the product of rubidium-87 decay
(b) the precursor of rubidium-87

*45. Potassium-42 is used to locate brain tumors. Its half-life is 12.5 hours. Starting with 15.4 mg, what fraction will remain after 100 hours? If it was necessary to have at least 1 μg for a particular procedure, could you hold the original sample for 200 hours before using it?

Answers to Practice Exercises

18.1 0.391 g

18.2 $^{218}_{84}Po$

18.3 (a) $^{234}_{91}Pa$
(b) $^{226}_{88}Ra$

PUTTING IT TOGETHER: Review for Chapter 18

Multiple Choice:

Choose the correct answer to each of the following:

1. If $^{238}_{92}U$ loses an alpha particle, the resulting nuclide is
(a) $^{237}_{92}U$ (b) $^{234}_{90}Th$ (c) $^{238}_{93}Np$ (d) $^{236}_{90}Th$

2. If $^{210}_{82}Pb$ loses a beta particle, the resulting nuclide is
(a) $^{209}_{83}Bi$ (b) $^{210}_{81}Ti$ (c) $^{206}_{80}Hg$ (d) $^{210}_{83}Bi$

3. In the equation $^{209}_{83}Bi + ? \longrightarrow ^{210}_{84}Po + ^{1}_{0}n$, the missing bombarding particle is
(a) $^{2}_{1}H$ (b) $^{1}_{0}n$ (c) $^{4}_{2}He$ (d) $^{0}_{-1}e$

4. Which of the following is not a characteristic of nuclear fission?
(a) Upon absorption of a proton, a heavy nucleus splits into two or more smaller nuclei.
(b) Two or more neutrons are produced from the fission of each atom.
(c) Large quantities of energy are produced.
(d) Most nuclei formed are radioactive.

5. The half-life of Sn-121 is 10 days. If you started with 40 g of this isotope, how much would you have left 30 days later?
(a) 10 g (b) none (c) 15 g (d) 5 g

6. $^{241}_{94}Pu$ successively emits $\beta, \alpha, \alpha, \beta, \alpha, \alpha$. At that point, the nuclide has become
(a) $^{225}_{94}Pu$ (b) $^{225}_{88}Ra$ (c) $^{207}_{84}Po$ (d) $^{219}_{84}Po$

7. The roentgen is the unit of radiation that measures
(a) an absorbed dose of radiation
(b) exposure to gamma or X-rays
(c) the dose from a different type of radiation
(d) the rate of decay of a radioactive substance

8. The radioactivity ray with the greatest penetrating ability is
(a) alpha (c) gamma
(b) beta (d) proton

9. In a nuclear reaction,
(a) mass is lost
(b) mass is gained
(c) mass is converted into energy
(d) energy is converted into mass

10. As the temperature of a radionuclide increases, its half-life
(a) increases
(b) decreases
(c) remains the same
(d) fluctuates

11. The nuclide that has the longest half-life is
(a) $^{238}_{92}U$ (b) $^{210}_{82}Pb$ (c) $^{234}_{90}Th$ (d) $^{222}_{88}Ra$

12. Which of the following is not a unit of radiation?
(a) curie (b) roentgen (c) rod (d) rem

13. When $^{235}_{92}U$ is bombarded by a neutron, the atom can fission into
(a) $^{124}_{53}I + ^{109}_{47}Ag + 2\,^{1}_{0}n$
(b) $^{124}_{50}Sn + ^{110}_{42}Mo + 2\,^{1}_{0}n$
(c) $^{134}_{56}Ba + ^{128}_{36}Xe + 2\,^{1}_{0}n$
(d) $^{90}_{38}Sr + ^{143}_{58}Ce + 2\,^{1}_{0}n$

14. In the nuclear equation
$^{45}_{21}Sc + ^{1}_{0}n \longrightarrow X + ^{1}_{1}H$
the nuclide X that is formed is
(a) $^{45}_{22}Ti$ (b) $^{45}_{20}Ca$ (c) $^{46}_{22}Ti$ (d) $^{45}_{20}K$

15. What type of radiation is a very energetic form of photon?
(a) alpha
(b) beta
(c) gamma
(d) positron

16. When $^{239}_{92}U$ decays to $^{239}_{93}Np$, what particle is emitted?
(a) positron (c) alpha particle
(b) neutron (d) beta particle

Free Response Questions

1. The nuclide $^{223}_{87}Fr$ emits three radioactive particles, losing 8 mass units and 3 atomic number units. Propose a radioactive decay series and write the symbol for the resulting nuclide.

2. How many half-lives are required to change 96 g of a sample of a radioactive isotope to 1.5 g over approximately 24 days? What is the approximate half-life of this isotope?

3. (a) What type of process is $^{6}_{3}Li + ^{1}_{0}n \longrightarrow ^{3}_{1}H + ^{4}_{2}He$?
(b) How is this process different from radioactive decay?
(c) Could nuclear fission be classified as the same type of process?

4. Calculate the nuclear binding energy of $^{56}_{26}Fe$.
Mass data:
$^{56}_{26}Fe$ = 55.9349 g/mol
n = 1.0087 g/mol e^- = 0.00055 g/mol
p = 1.0073 g/mol   1.0 g = 9.0×10^{13} J

CHAPTER 19

Organic Chemistry: Saturated Hydrocarbons

Organic molecules are studied with advanced technology. Lasers are used to initiate organic reactions to study the details of organic reaction mechanisms.

Chapter Outline

Many substances throughout nature contain silicon or carbon within their molecular structures. Silicon is the staple of the geologist: It combines with oxygen in a variety of ways to produce silica and a family of compounds known as the silicates. These compounds form the chemical foundation of most types of sand, rocks, and soil, which are essential materials of the construction industry.

In the living world, carbon, in combination with hydrogen, oxygen, nitrogen, and sulfur, forms the basis for millions of organic compounds. Carbon compounds provide us with energy sources in the form of hydrocarbons and their derivatives that allow us to heat and light our homes, drive our automobiles to work, and fly off to Hawaii for vacation. Small substitutions in these carbon molecules can produce chlorofluorocarbons, the compounds used in plastics and refrigerants. An understanding of these molecules and their effect upon our global environment is vital in the continuing search to find ways to maintain our lifestyles while preserving the planet.

19.1 Organic Chemistry: History and Scope

During the late 18th and the early 19th centuries, chemists were baffled by the fact that compounds obtained from animal and vegetable sources defied the established rules for inorganic compounds—namely, that compound formation is due to a simple attraction between positively and negatively charged elements. In their experience with inorganic chemistry, groups of two or three elements formed only one, or at most a few, different compounds. However, they observed that one group—carbon, hydrogen, oxygen, and nitrogen—gave rise to a large number of different compounds that often were remarkably stable.

Because no organic compounds had been synthesized from inorganic substances and because there was no other explanation for the complexities of organic compounds, chemists believed that organic compounds were formed by some "vital force." The **vital force theory** held that organic substances could originate only from living material. In 1828, a German chemist Friedrich Wöhler (1800–1882) did a simple experiment that eventually proved to be the death blow to this theory. In attempting to prepare ammonium cyanate (NH_4CNO) by heating cyanic acid ($HCNO$) and ammonia, Wöhler obtained a white crystalline substance that he identified as urea. Wöhler knew that urea must be an authentic organic substance because it is a product of metabolism that can be isolated from urine. Although Wöhler's discovery was not immediately and generally recognized, the vital force theory was overthrown by this simple observation that an organic compound had been made from nonliving materials.

vital force theory

The formula for urea is

$$H_2N-\overset{\overset{\displaystyle O}{\|}}{C}-NH_2$$

After the work of Wöhler, it was apparent that no vital force other than skill and knowledge was needed to make organic chemicals in the laboratory and that inorganic as well as organic substances could be used as raw materials. Today, **organic chemistry** designates the branch of chemistry that deals with carbon compounds, but does not imply that these compounds must originate from some form of life. A few special kinds of carbon compounds (e.g., carbon oxides, metal carbides, and carbonates) are often excluded from the organic classification because their chemistry is more conveniently related to that of inorganic substances.

organic chemistry

Organic molecules form the products used in cosmetics and perfumes.

The field of organic chemistry is vast, for it includes not only the composition of all living organisms, but also that of a great many other materials that we use daily. Examples of organic materials are nutrients (fats, proteins, carbohydrates), fuels, fabrics (cotton, wool, rayon, nylon), wood and paper products, paints and varnishes, plastics, dyes, soaps and detergents, cosmetics, medicinals, rubber products, and explosives.

What makes carbon special and different from the other elements in the periodic table? Carbon has the unique ability to bond to itself in long chains and rings of varying size. The greater the number of carbon atoms in a molecule, the more ways there are to link these atoms in different arrangements. This flexibility in the arrangement of atoms produces compounds with the same chemical composition and different structures. There is no theoretical limit on the number of organic compounds that can exist. In addition to this unique bonding property, carbon forms strong covalent bonds with a variety of elements, especially hydrogen, nitrogen, oxygen, sulfur, phosphorus, and the halogens. These are the elements most commonly found in organic compounds.

19.2 The Carbon Atom: Bonding and Shape

The carbon atom is central to all organic compounds. The atomic number of carbon is 6, and its electron structure is $1s^22s^22p^2$. Two stable isotopes of carbon exist, ^{12}C and ^{13}C. In addition, there are several radioactive isotopes; ^{14}C is the most widely known of these because of its use in radiocarbon dating. With four electrons in its outer shell, carbon has oxidation numbers ranging from $+4$ to -4, and it forms predominantly covalent bonds. Carbon occurs as the free element in diamond, graphite, coal, coke, carbon black, charcoal, lampblack, and buckminsterfullerene.

A carbon atom usually forms four covalent bonds; each bond results from two atoms sharing a pair of electrons. (See Section 11.5.) The number of electron pairs that two atoms share determines whether the bond is single or multiple. In a single bond, only one pair of electrons is shared by the atoms. If both atoms have the same electronegativity (see Section 11.6), the bond is classified as nonpolar. In this type of bond there is no separation of positive and negative charge between the atoms.

Carbon also forms multiple bonds by sharing two or three pairs of electrons between two atoms. The double bond formed by sharing two electron pairs is stronger than a single bond, but not twice as strong. It is also shorter than a single bond. Similarly, the triple bond formed by sharing three electron pairs is stronger and shorter than a double bond. An organic compound is classified as **saturated** if it contains only single bonds and as **unsaturated** if the molecules possess one or more multiple carbon–carbon bonds.

saturated compound

unsaturated compound

Lewis structures are useful in representing the bonding between atoms in a molecule, but these representations tell us little about the geometry of the molecules. A number of bonding theories predict the shape of molecules; one common theory is the valence shell electron pair repulsion (VSEPR) theory. (See Section 11.11) This is a fairly simple, yet accurate method for determining the shape of a molecule.

VSEPR theory states that electron pairs repel each other because they have like charges. The electron pairs will therefore try to spread out as far as possible around an atom. In addition, unshared pairs of electrons occupy more space than shared electron pairs.

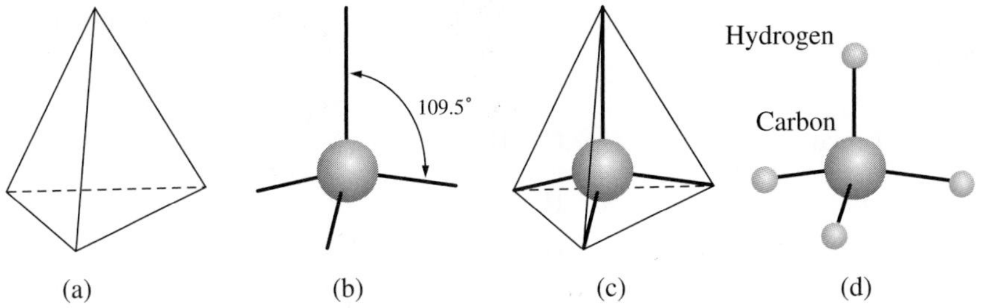

Figure 19.1
Tetrahedral structure of carbon: (a) a regular tetrahedron; (b) a carbon atom with tetrahedral bonds; (c) a carbon atom within a regular tetrahedron; (d) a methane molecule, CH_4.

What does VSEPR theory tell us about the shapes of carbon-containing compounds? Consider the simplest organic molecule, methane, CH_4. It contains one carbon atom with four single bonds to hydrogen atoms. The four shared electron pairs must be placed as far apart as possible in three dimensions. This results in the hydrogen atoms forming the corners of a tetrahedron with the carbon atom in the center. (See Figure 19.1.) The angle between these tetrahedral bonds is 109.5°, which is the least strained of the possible angles.

Double or triple bonds have a significant effect on the shape of the molecule. Because of increased repulsion between the pairs, additional electron pairs that are in close proximity take up more space than those in a single bond. Consider the Lewis structure for C_2H_4. (See Figure 19.2a.) Each carbon atom has three separate regions for shared electrons. To place them as far apart as possible requires placing each atom at the corner of a triangle. (See Figure 19.2b.) The bond angles around the carbon atoms are 120°.

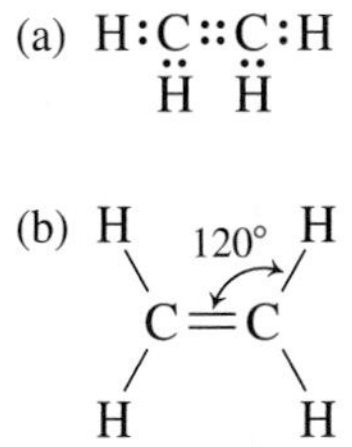

(a) H:C::C:H
      H H

(b) H   120°   H
      \   /
      C=C
      /   \
    H   H

Figure 19.2
(a) Lewis structure for C_2H_4.
(b) Shape of a molecule with a carbon–carbon double bond. The hydrogens and carbon form the vertices of a triangle. The bond angles around the carbon are 120°.

In a triple bond, the carbon has only two regions for shared electrons. To be placed as far apart as possible, a linear arrangement is required. The bond angle is 180°:

H:C:::C:H   H—C≡C—H (180°)

We will study more about the carbon–carbon double and triple bonds in Chapter 20.

Practice 19.1

Estimate the bond angles around the starred carbon (C*) for the following:

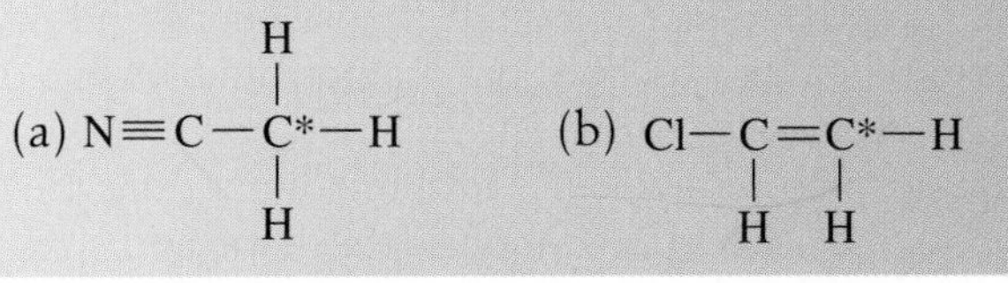

(a) N≡C—C*—H, with H above and H below C*

(b) Cl—C=C*—H, with H below C and H below C*

19.3 Organic Formulas and Molecular Models

Formulas of organic molecules are represented differently from those of inorganic compounds. A formula gives information about the composition of a compound. Inorganic compounds contain relatively few groups of atoms and are frequently represented by *empirical formulas*, which give only the simplest ratio of the atoms in a molecule. Larger molecules are often represented by a *molecular formula*, which gives more information. The molecular formula gives the actual number of atoms in a molecule. For inorganic compounds the empirical and molecular formulas are often the same.

In a *structural formula*, the arrangement of the atoms within the molecule is clearly shown. Thus,

C_2H_6 is shown as

$$\begin{array}{ccccccc} & & H & & H & & \\ & & | & & | & & \\ H & - & C & - & C & - & H \\ & & | & & | & & \\ & & H & & H & & \end{array}$$

and C_3H_8 is shown as

$$\begin{array}{ccccccccc} & & H & & H & & H & & \\ & & | & & | & & | & & \\ H & - & C & - & C & - & C & - & H \\ & & | & & | & & | & & \\ & & H & & H & & H & & \end{array}$$

condensed structural formulas

line structures

Organic chemists often shorten these structural formulas into a another type called **condensed structural formulas.** In these formulas, each carbon is grouped with the hydrogens bonded to it and then written as a formula—for example, CH_3CH_3 or $CH_3CH_2CH_3$. The diversity of methods for writing organic formulas is illustrated in Table 19.1.

An even simpler representation of organic molecules is given by **line structures.** For hydrocarbons these structures consist of zigzag (diagonal) lines in which a carbon atom is located at each junction where two lines meet and change direction. This method can be used for any compound but is generally used for cyclic and large complicated molecules. To explain the system used, let's consider the compound, C_7H_{16}. The condensed structure is

$$\overset{1}{C}H_3\overset{2}{C}H_2\overset{3}{C}H_2\overset{4}{C}H_2\overset{5}{C}H_2\overset{6}{C}H_2\overset{7}{C}H_3$$

The line structure is

2 4 6
1 3 5 7

The zigzag line represents the chain of seven carbon atoms. At each junction where the zigzag line changes direction, C-2 to C-6, there is a carbon atom. At the ends of the carbon chain, C-1 and C-7, there is also a carbon atom. The lines represent carbon to carbon bonds (C—C). No carbon–hydrogen bonds (C—H) are shown; they are assumed to complete the four bonds that carbon forms. So the line structure of C_7H_{16} has three C—H bonds on carbons l and 7, and two C—H bonds on carbons 2 to 6. When there is a carbon—carbon double bond or triple bond in the structure, two or three lines, respectively, are used.

Today chemists can accurately picture molecules in three dimensions. Several structural models show both relative three-dimensional shape and size—that is, a picture of the actual molecule. The two most common are the ball-and-stick model (Figure 19.3e) and the spacefilling model (Figure 19.3f). In the ball-and-stick model, different atoms are represented by different colored balls and bonds are represented by sticks. The commonly used colors are C, gray or black; H, light blue; O, red; S, yellow; N, blue; P, orange; F, green; Cl, light orange; Br, dark orange; I, tan. The spacefilling model gives a more accurate picture of the actual molecule but is not as clear in showing chemical bonds between the atoms. Figure 19.3 summarizes the different formulas and molecular models for C_3H_8. In the Lewis structure, the structural formula, and the line structure, it is important to remember that the bond angle is not the way it appears in the drawing. Rather, it is 109.5° because the molecule is actually three-dimensional.

$$\begin{array}{ccccccc} & & H & & H & & H \\ & & \cdot\cdot & & \cdot\cdot & & \cdot\cdot \\ H & : & C & : & C & : & C & : & H \\ & & \cdot\cdot & & \cdot\cdot & & \cdot\cdot \\ & & H & & H & & H \end{array}$$

(a)

$$\begin{array}{ccccccccc} & & H & & H & & H & & \\ & & | & & | & & | & & \\ H & - & C & - & C & - & C & - & H \\ & & | & & | & & | & & \\ & & H & & H & & H & & \end{array}$$

(b)

$CH_3CH_2CH_3$

(c)

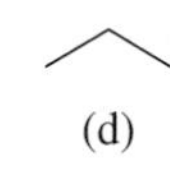

(d)

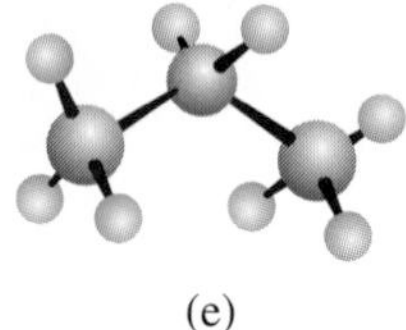

(e)

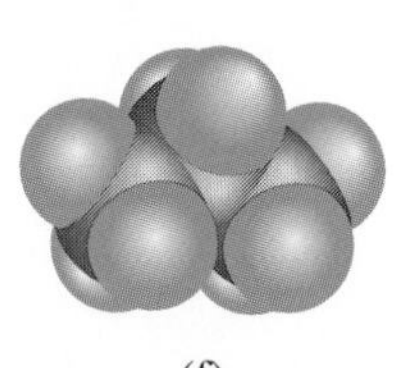

(f)

Figure 19.3
Models of a three-carbon, saturated compound, C_3H_8 (propane): (a) Lewis structure; (b) structural formula; (c) condensed structural formula; (d) line structure; (e) ball-and-stick model; (f) space-filling model.

Table 19.1 Classes of Organic Compounds

Class of compound	General formula*	IUPAC name**, ***	Molecular formula	Condensed structural formula	Structural formula
Alkane	RH	Eth*ane* (Ethane)	C_2H_6	CH_3CH_3	H H | | H—C—C—H | | H H
Alkene	R—CH=CH_2	Eth*ene* (Ethylene)	C_2H_4	$H_2C{=}CH_2$	H H C=C H H
Alkyne	R—C≡C—H	Eth*yne* (Acetylene)	C_2H_2	HC≡CH	H—C≡C—H
Alkyl halide	RX	Chloroethane (Ethyl chloride)	C_2H_5Cl	CH_3CH_2Cl	H H | | H—C—C—Cl | | H H
Alcohol	ROH	Ethan*ol* (Ethyl alcohol)	C_2H_6O	CH_3CH_2OH	H H | | H—C—C—OH | | H H
Ether	R—O—R	Methoxymethane (Dimethyl ether)	C_2H_6O	CH_3OCH_3	H H | | H—C—O—C—H | | H H
Aldehyde	R—C=O | H	Ethan*al* (Acetaldehyde)	C_2H_4O	CH_3CHO	H | H—C—C—H | ‖ H O
Ketone	R—C—R ‖ O	Propan*one* (Dimethyl ketone)	C_3H_6O	CH_3COCH_3	H H | | H—C—C—C—H | ‖ | H O H
Carboxylic acid	R—C—OH ‖ O	Ethan*oic acid* (Acetic acid)	$C_2H_4O_2$	CH_3COOH	H | H—C—C—OH | ‖ H O
Ester	R—C—OR ‖ O	Methyl ethan*oate* (Methyl acetate)	$C_3H_6O_2$	CH_3COOCH_3	H H | | H—C—C—O—C—H | ‖ | H O H
Amide	R—C—NH_2 ‖ O	Ethan*amide* (Acetamide)	C_2H_5ON	CH_3CONH_2	H H | / H—C—C—N | ‖ \\ H O H
Amine	R—CH_2—NH_2	Aminoethane (Ethylamine)	C_2H_7N	$CH_3CH_2NH_2$	H H | | H—C—C—N—H | | | H H H

* The letter R is used to indicate any of the many possible alkyl groups. ** Class name ending in italic. *** Common name in parentheses.

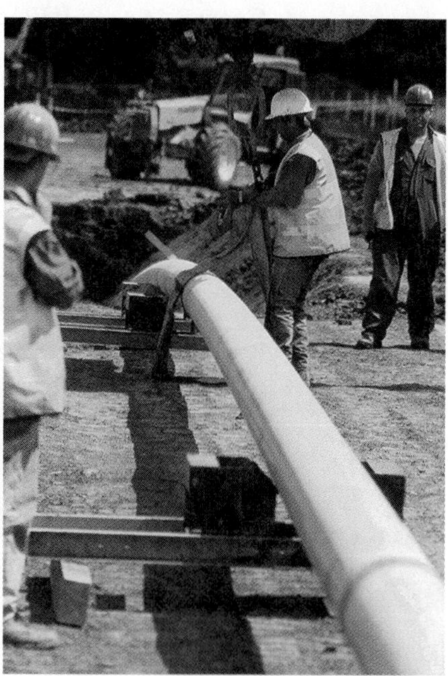

Fossil fuels are mined from many sources: an oil exploration rig in Alaska, a coal miner in Pennsylvania, and a natural gas operation in Washington (left to right).

19.4 Classifying Organic Compounds

functional group

It is impossible for anyone to study the properties of each of the millions of organic compounds. Organic compounds with similar structures are grouped into classes as shown in Table 19.1. The members of each class of compounds contain a characteristic atom or group of atoms called a **functional group**, shown as the red portion of the structural formula in Table 19.1. Organic compounds from different classes may have the same molecular formula, but completely different chemical and physical properties.

19.5 Hydrocarbons

hydrocarbon

aliphatic

Hydrocarbons are compounds that are composed entirely of carbon and hydrogen atoms bonded to each other by covalent bonds. Hydrocarbons are classified into two major categories, aliphatic and aromatic. The term *aromatic* refers to compounds that contain benzene rings. All hydrocarbons that are not aromatic are often described as **aliphatic** (from the Greek word *aleiphar*, meaning "fat"). The aliphatic hydrocarbons include the alkanes, alkenes, alkynes, and cycloalkanes. (See Figure 19.4.)

Coal tar-containing shampoos are used to treat dandruff

Fossil fuels—natural gas, petroleum, and coal—are the principal sources of hydrocarbons. Natural gas is primarily methane with small amounts of ethane, propane, and butane. Petroleum is a mixture of hydrocarbons from which gasoline, kerosene, fuel oil, lubricating oil, paraffin wax, and petrolatum (themselves mixtures of hydrocarbons) are separated. Coal tar, a volatile by-product of making coke from coal for use in the steel industry, is the source of many valuable chemicals, including the aromatic hydrocarbons benzene, toluene, and naphthalene.

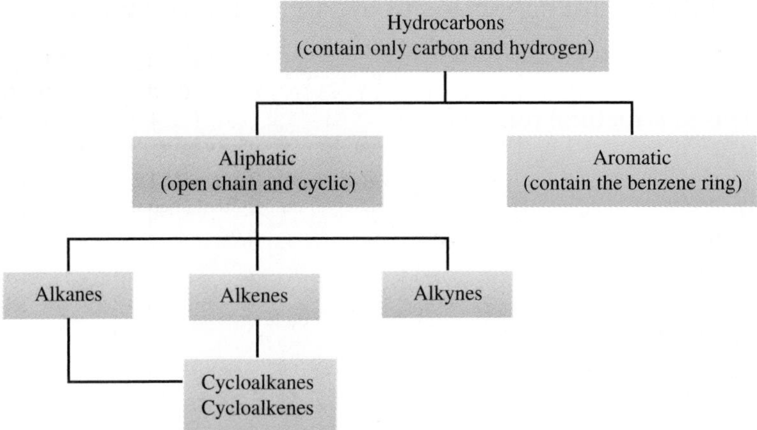

Figure 19.4
Classes of hydrocarbons.

The fossil fuels provide a rich resource of hydrocarbons for human society. In the past, these resources have been used primarily as a source of heat (via combustion). We now see that extensive combustion can have severe environmental consequences (e.g., air pollution and global warming). Fossil fuels also serve as the raw materials for much of today's chemical industry. One theme that runs through the study of organic chemistry is the synthetic relationship between compounds; for example, acids are often formed from alcohols. Fossil fuels are the starting materials for many of these synthetic sequences. And, in the long run, the fossil fuels may well prove to be more valuable to us as a source of organic chemicals than as a source of heat.

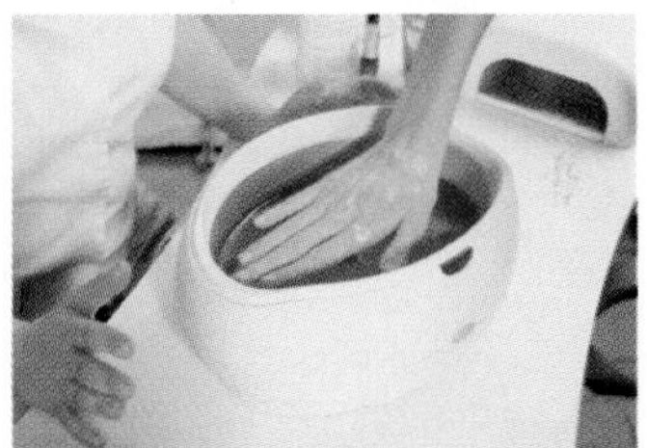

A hot paraffin wax dip can be used to relieve arthritis pain.

19.6 Saturated Hydrocarbons: Alkanes

The **alkanes**, also known as *paraffins* or *saturated hydrocarbons*, are straight- or branched-chain hydrocarbons with only single covalent bonds between the carbon atoms. We shall study the alkanes in some detail because many other classes of organic compounds can be considered as derivatives of these substances. For example, it is necessary to learn the names of the first 10 members of the alkane series because these names are used as a basis for naming other classes of compounds.

alkane

$CH_3CH_2CH_2CH_2CH_3$
straight-carbon chain

$$\begin{array}{c} CH_3 \\ | \\ CH_3CHCH_2CH_3 \end{array}$$
branched-chain of carbon atoms

Methane, CH_4, is the first member of the alkane series. Alkanes with two-, three-, and four-carbon atoms are ethane, propane, and butane, respectively. The names of the first four alkanes are of common or trivial origin and must be memorized, but the names beginning with the fifth member, pentane, are derived from Greek numbers and are relatively easy to recall. The names and formulas of the first 10 members of the series are given in Table 19.2.

Successive compounds in the alkane series differ from each other in composition by one carbon and two hydrogen atoms. When each member of a series differs from the next member by a CH_2 group, the series is called a **homologous series**. The members of a homologous series are similar in structure, but have a regular difference in formula. All common classes of organic compounds exist in homologous series. Each homologous series can be represented by a general formula. For all open-chain alkanes, the general

homologous series

C_nH_{2n+2}
general formula for open-chain alkanes

Table 19.2 Names, Formulas, and Physical Properties of Straight-Chain Alkanes

Name	Molecular formula C_nH_{2n+2}	Condensed structural formula	Boiling point (°C)	Melting point (°C)
Methane	CH_4	CH_4	−161	−183
Ethane	C_2H_6	CH_3CH_3	−89	−172
Propane	C_3H_8	$CH_3CH_2CH_3$	−42	−187
Butane	C_4H_{10}	$CH_3CH_2CH_2CH_3$	−0.6	−135
Pentane	C_5H_{12}	$CH_3CH_2CH_2CH_2CH_3$	36	−130
Hexane	C_6H_{14}	$CH_3CH_2CH_2CH_2CH_2CH_3$	69	−95
Heptane	C_7H_{16}	$CH_3CH_2CH_2CH_2CH_2CH_2CH_3$	98	−90
Octane	C_8H_{18}	$CH_3CH_2CH_2CH_2CH_2CH_2CH_2CH_3$	125	−57
Nonane	C_9H_{20}	$CH_3CH_2CH_2CH_2CH_2CH_2CH_2CH_2CH_3$	151	−54
Decane	$C_{10}H_{22}$	$CH_3CH_2CH_2CH_2CH_2CH_2CH_2CH_2CH_2CH_3$	174	−30

formula is C_nH_{2n+2}, where n corresponds to the number of carbon atoms in the molecule. The molecular formula of any specific alkane is easily determined from this general formula. Thus, for pentane, $n = 5$ and $2n + 2 = 12$, so the formula is C_5H_{12}. For hexadecane, the 16-carbon alkane, the formula is $C_{16}H_{34}$.

Practice 19.2

(a) Write molecular formulas for the alkanes that contain 12, 14, and 20 carbons.
(b) Write the condensed structural formulas for the straight-chain alkanes that contain 12 and 14 carbons.

19.7 Carbon Bonding in Alkanes

A carbon atom is capable of forming single covalent bonds with one, two, three, or four other atoms. To understand this remarkable bonding ability, we must look at the electron structure of carbon. The valence electrons of carbon in their ground state are $2s^22p_x^12p_y^1$. All of carbon's valence electrons can be shared to make a total of four bonds. When a carbon atom is bonded to other atoms by single bonds (e.g., to four hydrogen atoms in CH_4), it would appear at first that there should be two different types of bonds: bonds involving the 2*s* electrons and bonds involving the 2*p* electrons of the carbon atom. However, this is not the case. All four carbon–hydrogen bonds are identical.

If the carbon atom is to form four equivalent bonds, its electrons in the 2*s* and 2*p* orbitals must rearrange to four equivalent orbitals. To form the four equivalent orbitals, imagine that a 2*s* electron is promoted to a 2*p* orbital, giving carbon an outer shell electron structure of $2s^12p_x^12p_y^12p_z^1$. The 2*s* orbital and the three 2*p* orbitals then hybridize to form four equivalent hybrid orbitals, which are designated sp^3 orbitals. The orbitals formed (sp^3) are neither *s* orbitals nor *p* orbitals, but are instead a hybrid of those orbitals, having one-fourth *s* orbital character and three-fourths *p* orbital character. This process is illustrated in Figure 19.5. It is these sp^3 orbitals that are directed toward the corners of a regular tetrahedron. (See Figure 19.6.)

$2p$ ↿ ↿ —
$2s$ ⇅

A $2s$ electron is promoted to a $2p$ orbital →

$2p$ ↿ ↿ ↿
$2s$ ↿

Hybridization →

sp^3 ↿ ↿ ↿ ↿

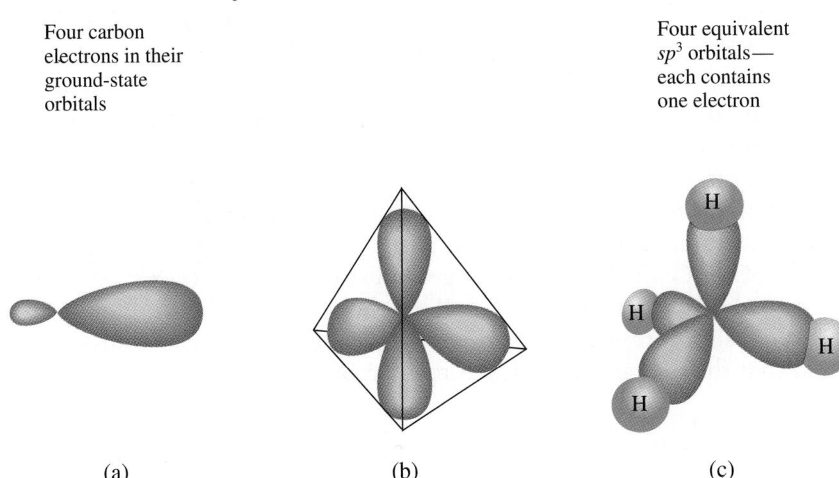

Figure 19.5
Schematic hybridization of $2s^22p_x^12p_y^1\,2p_z^0$ orbitals of carbon to form four sp^3 electron orbitals.

Figure 19.6
The tetrahedral nature of sp^3 orbitals: (a) a single sp^3-hybridized orbital; (b) four sp^3-hybridized orbitals in tetrahedral arrangement; (c) sp^3 and s orbitals overlapping to form C—H bonds in methane.

A single bond is formed when one of the sp^3 orbitals overlaps an orbital of another atom. Thus, each C—H bond in methane is the result of the overlapping of a carbon sp^3 orbital and a hydrogen s orbital, Figure 19.6c. Once the bond is formed, the pair of bonding electrons constituting it are said to be in a molecular orbital. In a similar way, a C—C single bond results from the overlap of sp^3 orbitals between two carbon atoms. This type of bond (C—H and C—C) is called a sigma (σ) bond. A **sigma bond** exists if the electron cloud formed by the pair of bonding electrons lies on a straight line drawn between the nuclei of the bonded atoms.

sigma bond

19.8 Isomerism

The properties of an organic substance are dependent on its molecular structure. The majority of organic compounds are made from relatively few elements: carbon, hydrogen, oxygen, nitrogen, and the halogens. The valence bonds or points of attachment may be represented in structural formulas by a corresponding number of dashes attached to each atom:

—C— (with bonds above and below)   H—   —O—   —N— (with bond below)   Cl—   Br—   I—   F—

Hence, carbon has four bonds to each atom, nitrogen three bonds, oxygen two bonds, and hydrogen and the halogens one bond to each atom.

In an alkane, each carbon atom is joined to four other atoms by four single covalent bonds. These bonds are separated by angles of 109.5°. (The angles correspond to those formed by lines drawn from the center of a regular tetrahedron to its corners.) Alkane molecules contain only carbon–carbon and carbon–hydrogen bonds and are essentially nonpolar. Alkane molecules are nonpolar because (1) carbon–carbon bonds are nonpolar, since they are between like atoms; (2) carbon–hydrogen bonds are only slightly polar because the difference in electronegativity between carbon and hydrogen atoms is small; and (3) the bonds in an alkane are symmetrically directed toward the corners of a tetrahedron. In virtue of their low polarity, alkane molecules have very little intermolecular attraction and therefore have relatively low boiling points compared with other organic compounds of similar molar mass.

Without the use of models or perspective drawings, the three-dimensional character of atoms and molecules is difficult to portray accurately. However, concepts of structure can be conveyed to some extent by structural formulas.

To write the correct structural formula for propane, C_3H_8, we must determine how to place each atom in the molecule. An alkane contains only single bonds. Therefore, each carbon atom must be bonded to four other atoms by either C—C or C—H bonds. Hydrogen must be bonded to only one carbon atom by a C—H bond, because C—H—C bonds do not occur and an H—H bond would simply represent a hydrogen molecule. Applying this information, we find that the only possible structure for propane is

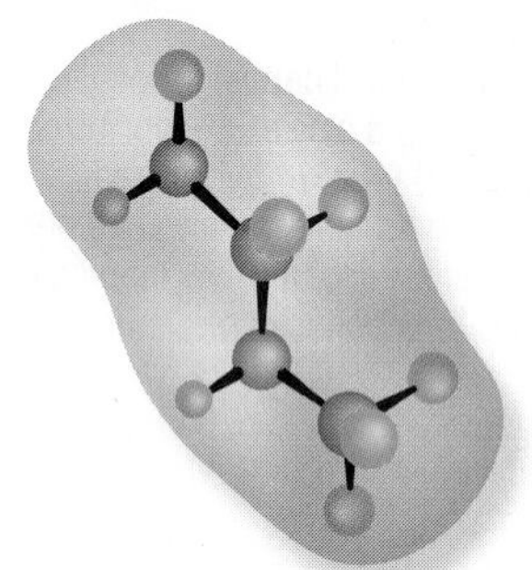

```
    H   H   H
    |   |   |
H — C — C — C — H
    |   |   |
    H   H   H
```

propane

However, it is possible to write two structural formulas corresponding to the molecular formula C_4H_{10}:

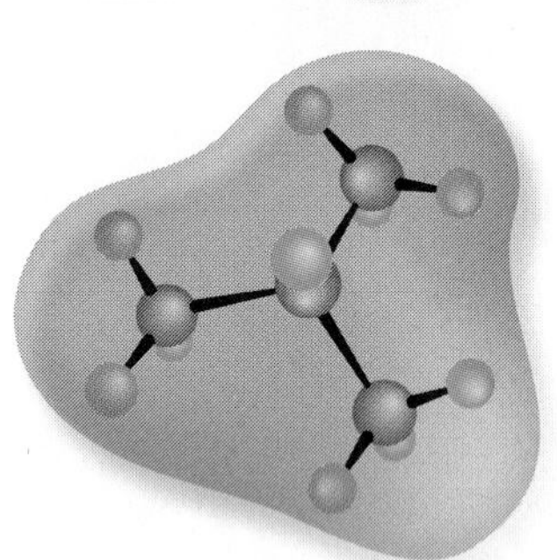

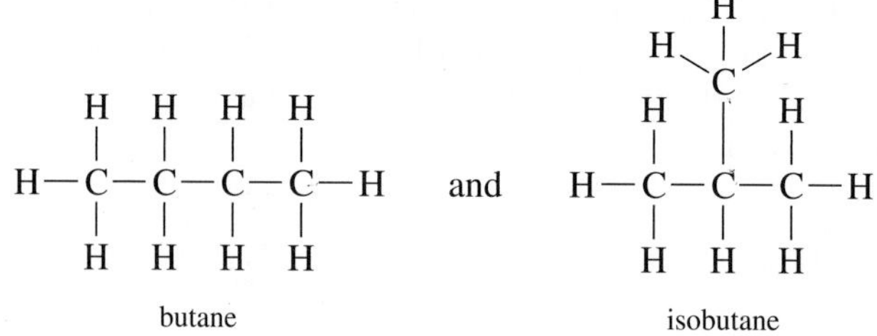

butane and isobutane

Butane and isobutane ball-and-stick models within transparent space-filling models.

Two C_4H_{10} compounds with the structural formulas shown here actually exist. The butane with the unbranched carbon chain is called *butane* or *normal butane* (abbreviated *n*-butane); it boils at −0.6°C and melts at −135°C. The common name for the branched-chain butane is *isobutane*; it boils at −11.7°C and melts at −159.5°C. These differences in physical properties are sufficient to establish that the two compounds are different substances even though they have the same molecular formula. Models illustrating the structural arrangement of the atoms in methane, ethane, propane, butane, and isobutane are shown in Figure 19.7.

"Isobutane" is a common name; a more systematic nomenclature will be introduced in Section 19.9.

isomerism
isomer

This phenomenon of two or more compounds having the same molecular formula but different structural arrangements of their atoms is called **isomerism**. The various individual compounds are called **isomers**. For example, there are two isomers of butane, C_4H_{10}. Isomerism is very common among organic compounds and is another reason for the large number of known compounds. There are 3 isomers of pentane, 5 isomers of hexane, 9 isomers of heptane, 18 isomers of octane, 35 isomers of nonane, and 75 isomers of decane. The phenomenon of isomerism is a compelling reason for the use of structural formulas.

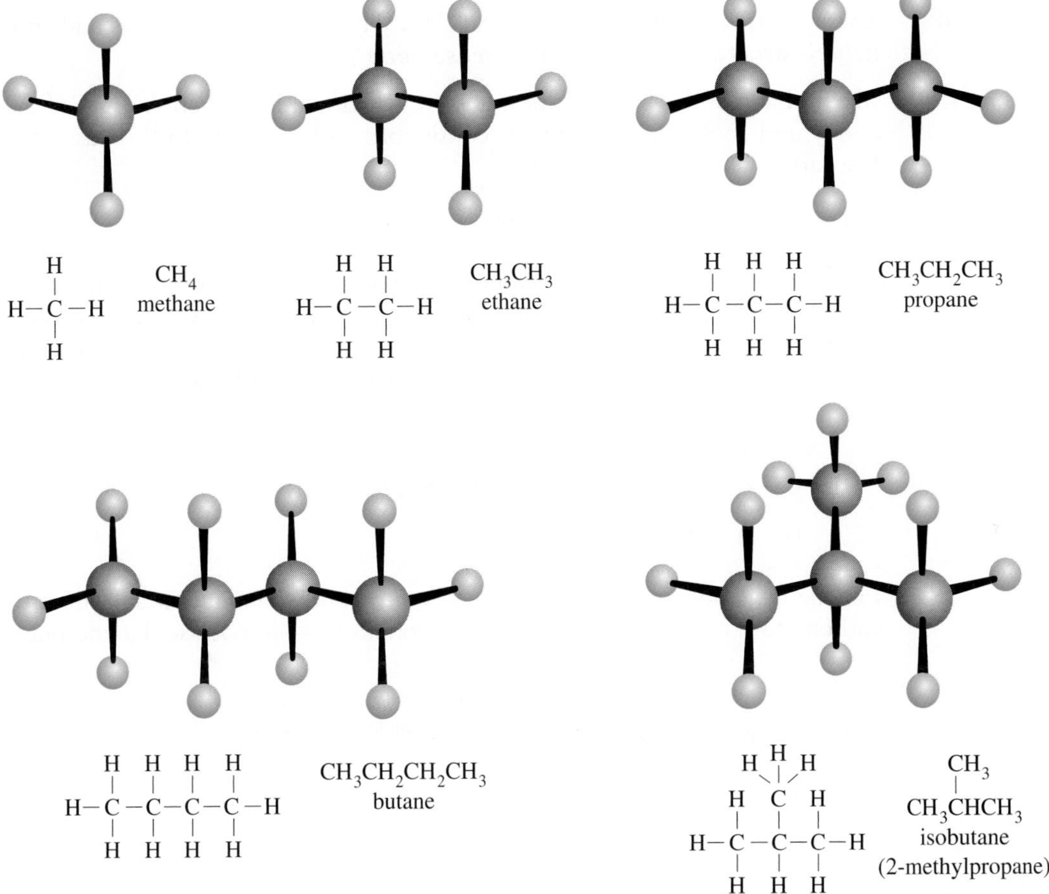

Figure 19.7
Ball-and-stick models illustrating structural formulas of methane, ethane, propane, butane, and isobutane.

> Isomers are compounds that have the same molecular formula, but different structural formulas.

To save time and space in writing, condensed structural formulas are often used. In the condensed structural formulas, atoms and groups that are attached to a carbon atom are generally written to the right of that carbon atom.

Let's interpret the condensed structural formula for propane:

$$\overset{1}{C}H_3-\overset{2}{C}H_2-\overset{3}{C}H_3 \quad \text{or} \quad CH_3CH_2CH_3$$

Carbon 1 has three hydrogen atoms attached to it and is bonded to carbon 2, which has two hydrogen atoms on it and is bonded to carbon 3, which has three hydrogen atoms bonded to it.

When structures are written as in butane, $CH_3CH_2CH_2CH_3$, the bonds (long dashes) are often not shown. But when a group is shown above or below the main carbon chain, as in isobutane (2-methylpropane), a vertical dash must be used to indicate the point of attachment of the group to the main carbon chain.

Example 19.1 There are three isomers of pentane, C_5H_{12}. Write structural formulas and condensed structural formulas for these isomers.

SOLUTION In a problem of this kind, it is best to start by writing the carbon skeleton of the compound containing the longest continuous carbon chain. In this case, it is five carbon atoms (from pentane):

```
C—C—C—C—C
```

Now complete the structure by attaching hydrogen atoms around each carbon atom so that each carbon atom has four bonds attached to it. The carbon atoms at each end of the chain need three hydrogen atoms. The three inner carbon atoms each need two hydrogen atoms to give them four bonds:

```
   H  H  H  H  H
   |  |  |  |  |
H—C—C—C—C—C—H      CH3CH2CH2CH2CH3
   |  |  |  |  |
   H  H  H  H  H
```

For the next isomer, start by writing a four-carbon chain and attach the fifth carbon atom to either of the middle carbon atoms—do not use the end ones:

```
      C            C
      |            |
C—C—C—C      C—C—C—C
```

Both of these structures represent the same compound.

Now add the 12 hydrogen atoms to complete the structure:

```
           H
        H\ | /H
           C
   H  H    |  H                 CH3
   |  |    |  |                  |
H—C—C—C—C—H         CH3CH2CHCH3      or      CH3CH2CH(CH3)2
   |  |  |  |
   H  H  H  H
```

This is the only four-chain compound that can be written.

For the third isomer, start with a three-carbon chain, attach the other two carbon atoms to the central carbon atom, and complete the structure by adding the 12 hydrogen atoms:

```
                    H
                 H\ | /H
                    C
   C             H  |  H           CH3
   |             |  |  |            |
C—C—C        H—C—C—C—H        CH3CCH3      or      C(CH3)4
   |             |  |  |            |
   C             H  |  H           CH3
                    C
                 H/ | \H
                    H
```

No more isomers of C_5H_{12} can be written.

Practice 19.3

Write structural and condensed formulas for the isomers of hexane, C_6H_{14}.

19.9 Naming Organic Compounds

In the early development of organic chemistry, each new compound was given a name, usually by the person who had isolated or synthesized it. Names were not systematic, but often conveyed some information—usually about the origin of the substance. Wood alcohol (methanol), for example, was so named because it was obtained by destructive distillation or pyrolysis of wood. Methane, formed during underwater decomposition of vegetable matter in marshes, was originally called marsh gas. A single compound was often known by several names. For example, the active ingredient in alcoholic beverages has been called alcohol, ethyl alcohol, methyl carbinol, grain alcohol, spirit, and ethanol.

A meeting in Geneva in 1892 initiated the development of an international system for naming compounds. In its present form, the method recommended by the International Union of Pure and Applied Chemistry is systematic, generally unambiguous, and internationally accepted. It is called the **IUPAC System**. Despite the existence of the official IUPAC System, a great many well-established common, or trivial, names and abbreviations (e.g., TNT and DDT) are used because of their brevity or convenience. It is thus necessary to have a knowledge of both the IUPAC System and many common names.

IUPAC System

In order to name organic compounds systematically, you must be able to recognize certain common alkyl groups. **Alkyl groups** have the general formula C_nH_{2n+1} (one less hydrogen atom than the corresponding alkane). The missing H atom may have been detached from any carbon in the alkane. The name of the group is formed from the name of the corresponding alkane by simply dropping *-ane* and substituting a *-yl* ending. The names and formulas of selected alkyl groups up to and including four carbon atoms are given in Table 19.3. The letter R is often used in formulas to mean any of the many possible alkyl groups:

alkyl group

$R = C_nH_{2n+1}$ Any alkyl group

Three prefixes commonly indicate structural information. They are *iso-*, *sec-* (for secondary) and *tert-* or *t-* (for tertiary). (See Table 19.3.) *Iso-* indicates that a compound or alkyl group has the following structure at one end of the carbon chain:

$$\begin{array}{c} CH_3 \\ | \\ CH_3-CH- \end{array}$$

A carbon atom bonded to only one other carbon is a primary carbon atom (1°). A carbon atom bonded to two other carbons is a secondary (*sec-*) carbon atom (2°). Thus, the name *sec-*butyl indicates that this group contains four carbons (butyl) and the carbon that connects this group to other atoms is bonded to two other carbon atoms. A carbon atom bonded to three other carbons is a tertiary carbon atom (3°). So the name *tert-*butyl indicates that this group contains four carbons (butyl) and the carbon that connects this group to other atoms is bonded to three other carbons (see Table 19.3). 1°, 2°, and 3° carbon atoms are identified in the following example:

$$\begin{array}{ccccccccc} & & 1°\rightarrow CH_3 & & & & & & \\ & & | & & & & & & \\ CH_3 & - & CH & - & CH_2 & - & CH_2 & - & CH_3 \\ \uparrow & & \uparrow & & \uparrow & & \uparrow & & \uparrow \\ 1° & & 3° & & 2° & & 2° & & 1° \end{array}$$

Practice 19.4

Application to Biochemistry The amino acids leucine and isoleucine, each contain relatively long carbon chains:

$$\begin{array}{ccc} & CH_3 & \\ & | & \\ & CH-CH_3 & \\ & | & \\ & CH_2 & \\ & | & \\ NH_2- & CH & -COOH \end{array} \qquad \begin{array}{ccc} & CH_3 & \\ & | & \\ & CH_2 & \\ & | & \\ & CH-CH_3 & \\ & | & \\ NH_2- & CH & -COOH \end{array}$$

leucine isoleucine

Identify the 1°, 2° and 3° carbons along each chain (marked in blue)

According to the IUPAC System, a few relatively simple rules are all that are needed to name a great many alkanes. In later sections, these rules will be extended to cover other classes of compounds, but advanced texts or references must be consulted for the complete system.

IUPAC Rules for Naming Alkanes

1. Select the longest continuous chain of carbon atoms as the parent compound, and consider all alkyl groups attached to it as branch chains that have replaced hydrogen atoms of the parent hydrocarbon. If two chains of equal length are present, use the chain that has the larger number of substituents attached to it as the parent compound. The name of the alkane consists of the name of the parent compound prefixed by the names of the branch-chain alkyl groups attached to it.
2. Number the carbon atoms in the parent carbon chain from one end to the other, starting from the end closest to the first carbon atom that has a branch chain:

$$\begin{array}{cccccc} & & CH_3 & & & \\ & & | & & & \\ CH_3 & CH_2 & CH & CH_2 & CH_2 & CH_3 \\ 1 & 2 & 3 & 4 & 5 & 6 \end{array} \quad (\textit{not}\ \begin{array}{cccccc} & & CH_3 & & & \\ & & | & & & \\ CH_3 & CH_2 & CH & CH_2 & CH_2 & CH_3 \\ 6 & 5 & 4 & 3 & 2 & 1 \end{array})$$

3. Name each branch-chain alkyl group and designate its position on the parent carbon chain by a number (e.g., 3-methyl means a methyl group attached to carbon 3).
4. When the same alkyl-group branch chain occurs more than once, indicate this by a prefix (*di-*, *tri-*, *tetra-*, etc.) written in front of the alkyl-group name (e.g., *dimethyl* indicates two methyl groups). The numbers indicating the positions of these alkyl groups are separated by a comma, followed by a hyphen, and placed in front of the name (e.g., 2,3-dimethyl).

Adding numerical prefixes (*di-*, *tri-*, *tetra-*, etc.) to the alkyl name does not change the alphabetical order. For example, ethyl is still named before dimethyl. The prefixes *sec-* and *tert-* are not used, but the prefix *iso-* is used in alphabetical naming. For example, isopropyl is named before methyl.

5. When several different alkyl groups are attached to the parent compound, list them in alphabetical order (e.g., *ethyl* before *methyl* as in 3-ethyl-4-methyloctane).

Table 19.3 Names and Formulas of Selected Alkyl Groups

Formula	Name	Formula	Name
CH_3-	methyl		
CH_3CH_2-	ethyl	$\underset{\displaystyle CH_3CH-}{\overset{\displaystyle CH_3}{\vert}}$	isopropyl
$CH_3CH_2CH_2-$	propyl		
$CH_3CH_2CH_2CH_2-$	butyl	$\underset{\displaystyle CH_3CHCH_2-}{\overset{\displaystyle CH_3}{\vert}}$	isobutyl
$CH_3(CH_2)_3CH_2-$	pentyl		
$CH_3(CH_2)_4CH_2-$	hexyl	$\underset{\displaystyle CH_3CH_2CH-}{\overset{\displaystyle CH_3}{\vert}}$	*sec*-butyl (secondary butyl)
$CH_3(CH_2)_5CH_2-$	heptyl		
$CH_3(CH_2)_6CH_2-$	octyl		
$CH_3(CH_2)_7CH_2-$	nonyl	$CH_3\underset{\underset{\displaystyle CH_3}{\vert}}{\overset{\overset{\displaystyle CH_3}{\vert}}{C}}-$	*tert*-butyl or *t*-butyl (tertiary butyl)
$CH_3(CH_2)_8CH_2-$	decyl		

The compound shown here is commonly called isopentane:

$$\overset{4}{CH_3}-\overset{3}{CH_2}-\underset{\underset{\displaystyle CH_3}{\vert}}{\overset{2}{CH}}-\overset{1}{CH_3}$$

2-methylbutane
(isopentane)

Let's name it by using the IUPAC System.

Explanation The longest continuous carbon chain contains four carbon atoms. Therefore, the parent compound is butane and the compound is named as a butane (Rule 1). The methyl group (CH_3-) attached to carbon 2 is named as a prefix to butane, the "2-" indicating the point of attachment of the methyl group on the butane chain (Rule 3).

How would we write the structural formula for 2-methylpentane? An analysis of its name gives us this information:

1. The parent compound, pentane, contains five carbons. Write and number the five-carbon skeleton of pentane (at this point, numbering can be either from right to left or from left to right):

$$\overset{5}{C}-\overset{4}{C}-\overset{3}{C}-\overset{2}{C}-\overset{1}{C}$$

2. Put a methyl group on carbon 2 ("2-methyl" in the name gives this information):

$$\overset{5}{C}-\overset{4}{C}-\overset{3}{C}-\underset{\underset{\displaystyle CH_3}{\vert}}{\overset{2}{C}}-\overset{1}{C}$$

3. Add hydrogens to give each carbon four bonds. The structural formula is

```
CH3—CH2—CH2—CH—CH3
             |
            CH3
```

2-methylpentane

Should this compound be called 4-methylpentane? No, the IUPAC System specifically states that the parent carbon chain is numbered starting from the end nearest to the side or branch chain.

It is important to understand that it is the longest continuous sequence of carbon atoms and alkyl groups that determines the name of a compound, and not the way the sequence is written. Each of the following formulas represents 2-methylpentane:

```
 1    2    3    4    5                                CH3
CH3—CH—CH2—CH2—CH3              5    4    3    |2   1
     |                         CH3—CH2—CH2—CH—CH3
    CH3

 5          1
CH3        CH3
 |         2|
CH2—CH2—CH
 4   3     |
          CH3

    CH3
 1  2|   3
CH3—CH—CH2
         |
        CH2—CH3
         4    5

     CH3
     2|        4
      CH      CH2
 1   /   \    /   \
CH3       CH2      CH3
           3        5

                   5
    CH3           CH3
     |             |
CH3—CH—CH2—CH2
 1   2   3   4
```

The following formulas and names demonstrate other aspects of the official nomenclature system:

```
 4    3    2    1
CH3—CH—CH—CH3
     |    |
    CH3  CH3
```

2,3-dimethylbutane

Explanation The name of this compound is 2,3-dimethylbutane. The longest carbon chain is four, indicating butane; "dimethyl" indicates two methyl groups; "2,3-" means that one CH_3 is on carbon 2 and one is on carbon 3 (Rule 4).

```
               CH3
 4    3        |2   1
CH3—CH2—C—CH3
               |
              CH3
```

2,2-dimethylbutane

Explanation In 2,2-dimethylbutane, both methyl groups are on the same carbon atom; both numbers are required (Rule 4).

```
                         CH3
 1    2    3             |4   5    6
CH3—CH—CH2—CH—CH2—CH3
     |
    CH3
```

2,4-dimethylhexane

Explanation In 2,4-dimethylhexane, the molecule is numbered from left to right (Rule 2).

```
       3     4     5     6
CH3—CH—CH2—CH2—CH3
      2|
     CH2
      1|
     CH3
```

3-methylhexane

Explanation In 3-methylhexane, there are six carbons in the longest continuous chain (Rule 1).

```
                    CH3—CH—CH3
 8      7    6    5      |4   3      2      1
CH3—CH2CH2CH2—C——CH—CH—CH3
                         |     |        |
                       CH3  Cl     CH3
```

3-chloro-4-isopropyl-2,4-dimethyloctane

Cl

line structure

Explanation In this compound, the longest carbon chain is eight. The groups that are attached or substituted for H on the octane chain are named in alphabetical order (Rule 5).

Example 19.2

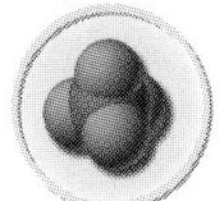

Write the formulas for
(a) 3-ethylpentane (b) 2,2,4-trimethylpentane

SOLUTION

(a) The name pentane indicates a five-carbon chain:

```
1   2   3   4   5
C—C—C—C—C
```

Write five connected carbon atoms and number them.

An ethyl group is written as CH_3CH_2—. Attach this group to carbon 3:

```
1   2   3   4   5
C—C—C—C—C
        |
      CH2CH3
```

Now add hydrogen atoms to give each carbon atom four bonds. Carbons 1 and 5 each need three H atoms; carbons 2 and 4 each need two H atoms; and carbon 3 needs one H atom. The formula is complete:

```
CH3CH2CHCH2CH3
         |
       CH2CH3
```

(b) Pentane indicates a five-carbon chain:

```
1   2   3   4   5
C—C—C—C—C
```

Write five connected carbon atoms and number them.

There are three methyl groups (CH_3—) in the compound (*trimethyl*), two attached to carbon 2 and one attached to carbon 4. Attach these three methyl groups to their respective carbon atoms:

```
    CH3      CH3
 1  2|   3  4|   5
 C—C—C—C—C
    |
    CH3
```

Now add H atoms to give each carbon atom four bonds. Carbons 1 and 5 each need three H atoms; carbon 2 does not need any H atoms; carbon 3 needs two H atoms; and carbon 4 needs one H atom. The formula is complete:

```
     CH3  CH3
     |     |
 CH3CCH2CHCH3
     |
     CH3
```

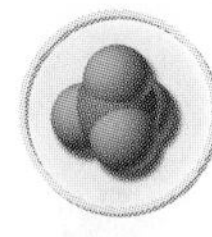

Example 19.3 Name the following compounds:

```
                          CH3
                          |
(a) CH3CH2CH2CH2CHCH3

                     CH2CH3
                     |
(b) CH3CH2CH2CHCH2CHCH3
                           |
                           CH2CH3
```

SOLUTION (a) The longest continuous carbon chain contains six carbon atoms (Rule 1). Thus, the parent name of the compound is hexane. Number the carbon chain from right to left so that the methyl group attached to carbon 2 is given the lowest possible number (Rule 2). With a methyl group on carbon 2, the name of the compound is 2-methylhexane (Rule 3).

(b) The longest continuous carbon chain contains eight carbon atoms (Rule 1). Number the chain from right to left (Rule 2):

```
                 C—C
 8   7   6   |5  4   3
 C—C—C—C—C—C—C
                     2|   1
                      C—C
```

Thus, the parent name is octane. As the chain is numbered, there is a methyl group on carbon 3 and an ethyl group on carbon 5, so the name of the compound is 5-ethyl-3-methyloctane. Note that ethyl is named before methyl (alphabetical order, Rule 5).

Practice 19.5

Write the formula for (a) 2-methylhexane, (b) 3,4-dimethylheptane, (c) 2-chloro-3-ethylpentane.

Practice 19.6

Write the line structures and name the following compounds:

$$\text{(a)}\ \begin{array}{c} \quad CH_3CHCH_3 \\ \quad | \\ CH_3CHCH_2CHCH_2CHCH_3 \\ |\qquad\qquad\quad | \\ CH_3 \qquad\quad CH_3 \end{array} \qquad \text{(b)}\ \begin{array}{c} \qquad\qquad CH_2CH_3 \\ \qquad\qquad | \\ CH_3CH_2CHCH_2CHCH_3 \\ | \\ CH_3 \end{array}$$

19.10 Introduction to the Reactions of Carbon

Spacefilling model of amoxicillin, ($C_{16}H_{19}N_3O_5S$) a potent antibiotic.

One reason that organic chemistry is so important in our modern world is the versatility of carbon. Many diverse reactions are possible with this element. One of the most important carbon reactions is oxidation reduction. *When carbon atoms are oxidized, they often form additional bonds to oxygen.* In the following example, carbon in methane is oxidized to carbon dioxide:

$$\underset{\text{methane}}{\begin{array}{c} H \\ | \\ H-C-H \\ | \\ H \end{array}} + 2\,O_2 \longrightarrow \underset{\text{carbon dioxide}}{O{=}C{=}O} + 2\,H_2O$$

When carbon atoms are reduced, they often form additional bonds to hydrogen. During this reaction, the carbon may decrease the number of bonds to oxygen and increase the number of bonds to hydrogen. The following reaction shows the reduction of a carbon compound:

$$\underset{\text{carbon dioxide}}{O{=}C{=}O} + 3\,H_2 \longrightarrow \underset{\text{methanol}}{\begin{array}{c} H \\ | \\ H-C-OH \\ | \\ H \end{array}} + H_2O$$

Each class of organic compounds can undergo important oxidation–reduction reactions.

Plants reduce carbon dioxide to make carbohydrates.

Although countless different carbon-containing molecules exist, there are relatively few types of organic reactions, and only a few carbon atoms, at most, are involved in any common reaction. These changes can be usefully categorized by counting the number of atoms bonded to a reactive carbon. Carbon can be bonded to a maximum of four atoms.

substitution reaction

1. In a **substitution reaction** one atom or group of atoms is exchanged by another atom or group of atoms; for example,

$$\text{H}-\underset{\text{H}}{\overset{\text{H}}{\underset{|}{\overset{|}{\text{C}}}}}-\text{H} + \text{Br}_2 \longrightarrow \text{H}-\underset{\text{H}}{\overset{\text{H}}{\underset{|}{\overset{|}{\text{C}}}}}-\text{Br} + \text{HBr}$$

methane, colorless gas → bromomethane, colorless liquid

elimination reaction

2. An **elimination reaction** is a reaction in which a single reactant is split into two or more products, and one of the products is eliminated; for example,

$$\text{H}-\underset{\text{H}}{\overset{\text{H}}{\underset{|}{\overset{|}{\text{C}}}}}-\underset{\text{H}}{\overset{\text{H}}{\underset{|}{\overset{|}{\text{C}}}}}-\text{Br} \longrightarrow \underset{\text{H}}{\overset{\text{H}}{\underset{|}{\overset{|}{\text{C}}}}}=\underset{\text{H}}{\overset{\text{H}}{\underset{|}{\overset{|}{\text{C}}}}} + \text{HBr}$$

bromoethane, colorless liquid → ethene, colorless gas

Elimination reactions form multiple bonds.

addition reaction

3. Two reactants adding together to form a single product is called an **addition reaction**. An addition reaction can be thought of as the reverse of an elimination reaction; for example,

$$\underset{\text{H}}{\overset{\text{H}}{\underset{|}{\overset{|}{\text{C}}}}}=\underset{\text{H}}{\overset{\text{H}}{\underset{|}{\overset{|}{\text{C}}}}} + \text{HBr} \longrightarrow \text{H}-\underset{\text{H}}{\overset{\text{H}}{\underset{|}{\overset{|}{\text{C}}}}}-\underset{\text{H}}{\overset{\text{H}}{\underset{|}{\overset{|}{\text{C}}}}}-\text{Br}$$

ethene, colorless gas → bromoethane, colorless liquid

Practice 19.7

The following reactions form ethanol:

(a) $CH_2 = CH_2 + H_2O \xrightarrow{H^+} CH_3CH_2OH$

(b) $CH_3CH_2I + H_2O \longrightarrow CH_3CH_2OH + HI$

Categorize each reaction as substituion, elimination, or addition.

19.11 Reactions of Alkanes

One type of alkane reaction has inspired people to explore equatorial jungles, endure the heat and sandstorms of the deserts of Africa and the Middle East, mush across the frozen Arctic, and drill holes—some more than 30,000 feet deep—on land and on the ocean floor! These strenuous and expensive activities have been undertaken because alkanes, as well as other hydrocarbons, undergo combustion with oxygen with the evolution of large amounts of heat energy. Methane, for example, reacts with oxygen as follows:

$$CH_4(g) + 2\,O_2(g) \longrightarrow CO_2(g) + 2\,H_2O(g) + 802.5\text{ kJ }(191.8\text{ kcal})$$

The combustion of hydrocarbons is spectacular but can be a huge problem, as shown in this oil-well fire in Wyoming.

When carbon dioxide is formed, the alkane has undergone *complete* oxidation. The resulting thermal energy can be converted to mechanical and electrical energy. Combustion reactions overshadow all other reactions of alkanes in economic importance. But combustion reactions of hydrocarbons are not usually of great interest to organic chemists because carbon dioxide and water are the only chemical products of complete combustion.

Aside from their combustibility, alkanes are relatively sluggish and limited in reactivity. But with proper activation, such as with high temperature or catalysts, alkanes can be made to react in a variety of ways. Some important noncombustion reactions of alkanes include the following:

1. **Halogenation** (a *substitution* reaction). A halogen is substituted for a hydrogen atom in halogenation. When a specific halogen such as chlorine is used, the reaction is called chlorination; RH is an alkane (alkyl group + H atom) that reacts with halogens in this manner:

   Example: $RH + X_2 \longrightarrow RX + HX$ (X = Cl or Br)

   $$CH_3CH_3 + Cl_2 \longrightarrow CH_3CH_2Cl + HCl$$

   This reaction yields alkyl halides, RX, which are useful as intermediates for the manufacture of other substances.

2. **Dehydrogenation** (an *elimination* reaction). Hydrogen is lost from an alkane during dehydrogenation:

   Example: $\underset{\text{alkane}}{C_nH_{2n+2}} \xrightarrow{700-900^\circ C} \underset{\text{alkene}}{C_nH_{2n}} + H_2$

   $$\underset{\text{propane}}{CH_3CH_2CH_3} \xrightarrow{\Delta} \underset{\text{propene}}{CH_3CH=CH_2} + H_2$$

   This reaction yields alkenes, which, like alkyl halides, are useful chemical intermediates. Hydrogen is a valuable by-product.

3. **Cracking** (breaking up large molecules to form smaller ones):

Example: $\underset{\text{alkane}}{C_{16}H_{34}} \xrightarrow[\text{catalyst}]{\Delta} \underset{\text{alkane}}{C_8H_{18}} + \underset{\text{alkene}}{C_8H_{16}}$ (one set of many possible products)

4. **Isomerization** (rearrangement of molecular structures):

Example: $CH_3CH_2CH_2CH_2CH_3 \xrightarrow[\Delta,\ \text{pressure}]{\text{Catalyst}} CH_3CH_2CHCH_3$ (with CH_3 attached to the CH)

Halogenation is used extensively in the manufacture of petrochemicals (chemicals derived from petroleum and used for purposes other than fuels). The other three reactions—dehydrogenation, cracking, and isomerization—singly or in combination, are of great importance in the production of motor fuels and petrochemicals.

A well-known reaction of methane and chlorine is shown by the equation

$$CH_4 + Cl_2 \longrightarrow \underset{\substack{\text{chloromethane}\\ \text{(methyl chloride)}}}{CH_3Cl} + HCl$$

The reaction of methane and chlorine gives a mixture of mono-, di-, tri-, and tetra-substituted chloromethanes:

$$CH_4 \xrightarrow{Cl_2} CH_3Cl \xrightarrow{Cl_2} CH_2Cl_2 \xrightarrow{Cl_2} CHCl_3 \xrightarrow{Cl_2} CCl_4 + 4\,HCl$$

However, if an excess of chlorine is used, the reaction can be controlled to give all tetrachloromethane (carbon tetrachloride). On the other hand, if a large ratio of methane to chlorine is used, the product will be predominantly chloromethane (methyl chloride). Table 19.4 lists the formulas and names for all the chloromethanes. The names for the other halogen-substituted methanes follow the same pattern as for the chloromethanes; for example, CH_3Br is bromomethane, or methyl bromide, and CHI_3 is triiodomethane, or iodoform.

Table 19.4 Chlorination Products of Methane

Formula	IUPAC name	Common name
CH_3Cl	Chloromethane	Methyl chloride
CH_2Cl_2	Dichloromethane	Methylene chloride
$CHCl_3$	Trichloromethane	Chloroform
CCl_4	Tetrachloromethane	Carbon tetrachloride

Chloromethane Dichloromethane Trichloromethane Tetrachloromethane

Chloromethane is a monosubstitution product of methane. The term **monosubstitution** refers to the fact that one hydrogen atom in an organic molecule is substituted by another atom or by a group of atoms. In hydrocarbons, for example, when we substitute one chlorine atom for a hydrogen atom, the new compound is a monosubstitution (monochlorosubstitution) product. In a like manner we can have *di-*, *tri-*, *tetra-*, and so on, substitution products.

monosubstitution

This kind of chlorination (or bromination) is general with alkanes. There are nine different chlorination products of ethane. See if you can write the structural formulas for all of them.

When propane is chlorinated, two isomeric monosubstitution products are obtained because a hydrogen atom can be replaced on either the first or second carbon:

$$CH_3CH_2CH_3 + Cl_2 \xrightarrow[25°C]{light} \underset{\text{1-chloropropane}}{CH_3CH_2CH_2Cl} + \underset{\text{2-chloropropane}}{CH_3CHClCH_3} + HCl$$

The letter X is commonly used to indicate a halogen atom. The formula RX indicates a halogen atom attached to an alkyl group and represents the class of compounds known as the **alkyl halides**. When R is CH_3, then CH_3X can be CH_3F, CH_3Cl, CH_3Br, or CH_3I.

alkyl halide

Alkyl halides are named systematically in the same general way as alkanes. Halogen atoms are identified as *fluoro-*, *chloro-*, *bromo-*, or *iodo-* and are named as substituents like branch-chain alkyl groups. Study these examples:

Names for halogen atoms in organic compounds.

F	fluoro
Cl	chloro
Br	bromo
I	iodo

$$\underset{\text{2-chlorobutane}}{CH_3{-}CHCl{-}CH_2{-}CH_3} \qquad \underset{\text{2-bromo-1-chloropropane}}{CH_2Cl{-}CHBr{-}CH_3}$$

$$\underset{\text{2-chloro-3-methylpentane}}{CH_3{-}CH_2{-}\underset{\displaystyle CH_3}{\underset{|}{CH}}{-}CHCl{-}CH_3} \qquad \underset{\text{1,1-dichloropropane}}{CH_3{-}CH_2{-}CHCl_2}$$

Example 19.4

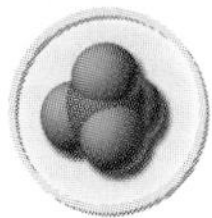

How many monochlorosubstitution products can be obtained from pentane?

SOLUTION

First write the formula for pentane:

$$\overset{5}{CH_3}\overset{4}{CH_2}\overset{3}{CH_2}\overset{2}{CH_2}\overset{1}{CH_3}$$

Now rewrite the formula five times, substituting a Cl atom for an H atom on each C atom:

I	$CH_3CH_2CH_2CH_2\overset{1}{CH_2}Cl$	Cl on carbon 1
II	$CH_3CH_2CH_2\overset{2}{CH}ClCH_3$	Cl on carbon 2
III	$CH_3CH_2\overset{3}{CH}ClCH_2CH_3$	Cl on carbon 3
IV	$CH_3\overset{4}{CH}ClCH_2CH_2CH_3$	Cl on carbon 4
V	$\overset{5}{CH_2}ClCH_2CH_2CH_2CH_3$	Cl on carbon 5

Compounds I and V are identical. By numbering compound V from left to right, we find that both compounds (I and V) are 1-chloropentane. Compounds II and IV are identical. By numbering compound IV from left to right, we find that both compounds (II and IV) are 2-chloropentane. Thus, there are three monochlorosubstitution products of pentane: 1-chloropentane, 2-chloropentane, and 3–chloropentane.

Practice 19.8

How many dichlorosubstitution isomers can be written for hexane?

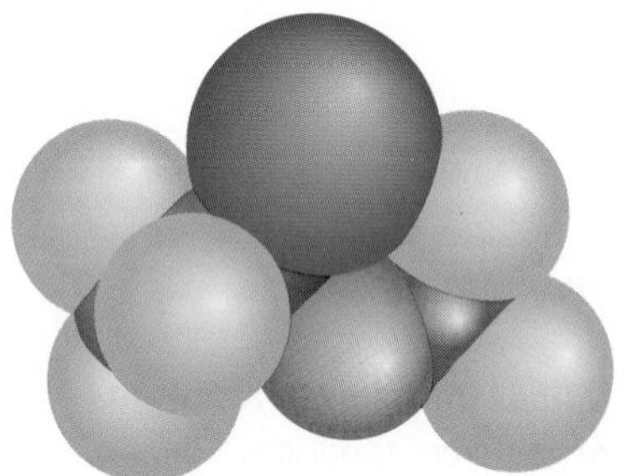

Spacefilling model of isoflurane.

You may be familiar with Teflon® nonstick surfaces. This material is fluorocarbon polymer. The fluorine atoms cause the surface to be both unreactive and slick. Alkyl halides also have many uses in medicine. Inhalational general anesthetics are mostly carbon compounds that contain chlorine, fluorine, or bromine. Adding halogens improves the anesthetic by stabilizing and limiting metabolism of the compound. Several examples of common inhalational anesthetics are as follows (isoflurane and desflurane also contain an oxygen and are classified as ethers):

$CF_3CHClBr$	$CF_3CHCl—O—CHF_2$	$CF_3CHF—O—CHF_2$
halothane	isoflurane	desflurane

Medical implants such as pacemakers and infusion ports are often coated with alkyl halide polymers. The coating minimizes both bacterial attachment (and the possibility of infection) and blood clotting.

Free-Radical Mechanism for Halogenation of Alkanes

mechanism The **mechanism** of a reaction is the path it takes for the reactants to go to products. As an example, let's consider the mechanism for the chlorination of methane.

free radical A **free radical** is an atom (or a group of atoms) that has an unpaired electron, for example:

$$:\ddot{\underset{\cdot\cdot}{Cl}}:\ddot{\underset{\cdot\cdot}{Cl}}: \xrightarrow[\text{light}]{\text{UV}} 2:\ddot{\underset{\cdot\cdot}{Cl}}\cdot$$ (usually written as Cl· , a chlorine free radical)

A free-radical mechanism is a three-step reaction: (1) initiation, (2) propagation, (3) termination. In initiation, the initial free radical is formed; in propagation, more free radicals are formed, the reaction perpetuates itself, and the product is formed; in termination, the reaction is usually stopped by free radicals combining to form stable compounds.

Initiation: A chlorine molecule reacts with ultraviolet light to form chlorine free radicals. One electron from the single bond in the chlorine molecule goes with each chlorine atom.

$$:\ddot{\underset{\cdot\cdot}{Cl}}:\ddot{\underset{\cdot\cdot}{Cl}}: \xrightarrow[\text{light}]{\text{UV}} 2\,Cl\cdot$$

Propagation: A chlorine free radical then reacts with a molecule of methane to form a methyl free radical and hydrogen chloride.

$$Cl\cdot + CH_4 \longrightarrow \underset{\text{methyl free radical}}{\cdot CH_3} + HCl$$

The methyl free radical then reacts with a chlorine molecule to produce chloromethane (methyl chloride) and a chlorine free radical.

$$\cdot CH_3 + Cl_2 \longrightarrow \underset{\text{chloromethane}}{CH_3Cl} + Cl\cdot$$

Termination:

$$Cl\cdot + Cl\cdot \longrightarrow Cl_2$$

$$Cl\cdot + \cdot CH_3 \longrightarrow CH_3Cl$$

$$\cdot CH_3 + \cdot CH_3 \longrightarrow CH_3CH_3$$

The overall reaction is

$$CH_4 + Cl_2 \xrightarrow{\text{UV light}} CH_3Cl + HCl$$

In the monochlorination of a hydrocarbon with three or more carbon atoms, there is the possibility of forming more than one carbon free radical. For example, propane can form two different free radicals, a propyl free radical and an isopropyl free radical.

$$\underset{\text{propyl free radical}}{CH_3CH_2CH_2\cdot} \qquad \underset{\text{isopropyl free radical}}{CH_3\dot{C}HCH_3}$$

In general there can be three types of carbon free radicals: primary (1°), secondary (2°), and tertiary (3°). In the primary free radical, the unpaired electron is on a carbon atom that is bonded to one other carbon atom; in a secondary free radical, the unpaired electron is on a carbon atom that is bonded to two carbon atoms; in a tertiary free radical, the unpaired electron is on a carbon atom that is bonded to three carbon atoms.

```
                                          H
                                          |
                                        H—C—H
  H  H         H  H  H               H    |    H
  |  |         |  |  |               |    |    |
H—C—C•       H—C—C—C—H             H—C————C————C—H
  |  |         |  •  |               |    •    |
  H  H         H     H               H         H

                                         CH3
                                          |
CH3CH2•       CH3CHCH3                 CH3CCH3
                 •                        •
  1°             2°                       3°
```

Experimentally, we find that a tertiary free radical is more stable and more easily formed than a secondary free radical, which is more stable and more easily formed than a primary free radical.

$$3^\circ > 2^\circ > 1^\circ > CH_3$$

Getting back to the chlorination of propane, two products can be formed: 1-chloropropane and 2-chloropropane. However, we find experimentally that 2-chloropropane is the major product of the reaction because the secondary free radical is easier to form than the primary free radical.

$$2\ CH_3CH_2CH_3 + Cl_2 \xrightarrow{\text{UV light}} \underset{\text{major product}}{CH_3CHClCH_3} + \underset{\text{minor product}}{CH_3CH_2CH_2Cl} + 2\ HCl$$

There is a lot more to be learned about halogenation, but it will have to be left to a more intensive course in organic chemistry. For example, the choice of halogen reactant can change the mixture of product isomers. Monochlorination of propane yields 55% 2-chloropropane and 45% 1-chloropropane, whereas the monobromination of propane yields almost 100% 2-bromopropane.

Practice 19.9

Write an equation for each step in the free-radical mechanism forming 2-chloropropane.

19.12 Sources of Alkanes

The two main sources of alkanes are natural gas and petroleum. Natural gas is formed by the anaerobic decay of plants and animals. The composition of natural gas varies in different locations. Its main component is methane (80–95%), the balance being varying amounts of other hydrocarbons, hydrogen, nitrogen, carbon monoxide, carbon dioxide, and in some locations, hydrogen sulfide. Economically significant amounts of methane are now obtained by the decomposition of sewage, garbage, and other organic waste products.

Petroleum, also called *crude oil*, is a viscous black liquid consisting of a mixture of hydrocarbons with smaller amounts of nitrogen and sulfur-containing organic compounds. Petroleum is formed by the decomposition of plants and animals over millions of years. The composition of petroleum varies widely from one locality to another. Crude oil is refined into such useful products as gasoline, kerosene, diesel fuel, jet fuel, lubricating oil, heating oil, paraffin wax, petroleum jelly (petrolatum), tars, and asphalt.

At the rate that natural gas and petroleum are being used, these sources of hydrocarbons are destined to be in short supply and virtually exhausted in the not-too-distant future. Alternative sources of fuels must be developed.

Petroleum is converted to the starting materials for many chemicals at a refinery.

19.13 Gasoline: A Major Petroleum Product

A large fraction of petroleum is burned as fuel. (See Figure 19.8.) Historically, fuels were used for heating and cooking; today, most fuels provide power for transportation. Because different engines require different fuels, there are many different fuels. However, befitting the importance of the automobile in transportation, gasoline is the fuel produced in the largest quantity.

Gasoline, aside from the additives put into it, consists primarily of hydrocarbons. Gasoline, as it is distilled from crude oil, causes "knocking" when burned in high-compression automobile engines. Knocking, caused by a too-rapid combustion or detonation of the air–gasoline mixture, is a severe problem in high-compression engines. The knock resistance of gasolines, a quality that varies widely, is usually expressed in terms of *octane number*, or *octane rating*.

Because of its highly branched chain structure, isooctane (2,2,4-trimethylpentane), is a motor fuel that is resistant to knocking. Mixtures of isooctane and heptane, a straight-chain alkane that knocks badly, have been used as standards to establish octane ratings of gasolines. Isooctane is arbitrarily assigned an octane number of 100 and heptane an octane number of 0. To determine the octane rating,

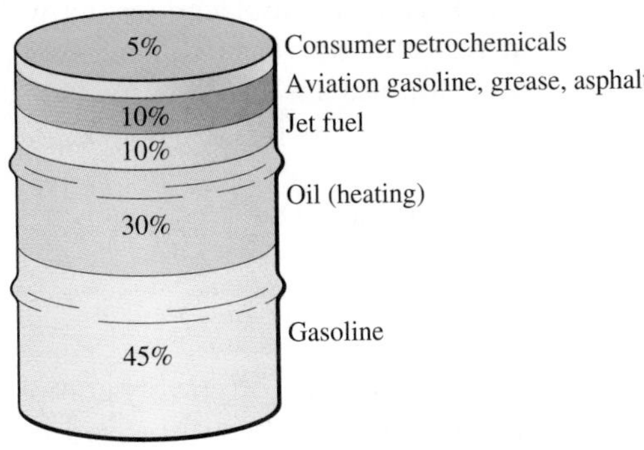

Figure 19.8
Uses of petroleum.

Unleaded gasolines are formulated to different octane ratings so they can be used in different engines.

a gasoline is compared with mixtures of isooctane and heptane in a test engine. The octane number of the gasoline corresponds to the percentage of isooctane present in the isooctane–heptane mixture that matches the knocking characteristics of the gasoline being tested. Thus, a 90-octane gasoline has knocking characteristics matching those of a mixture of 90% isooctane and 10% heptane.

When first used to establish octane numbers, isooctane was the most knock-resistant substance available. However, because technological advances have resulted in engines with greater power and compression ratios, higher quality fuels were subsequently necessary. Fuels containing more highly branched hydrocarbons, unsaturated hydrocarbons, or aromatic hydrocarbons burn more smoothly than isooctane and have a higher octane rating than 100.

Practice 19.10

Which of the following molecules would be predicted to yield a higher octane rating? (Remember that more highly branched alkanes commonly yield higher octane numbers.) Draw each structure before reaching your conclusion.

(a) 3-isopropyl-2,4-dimethylpentane
(b) 3,4-diethylhexane

In recent years, manufacturers have "reformulated" gasoline to solve environmental problems as well as to boost the octane rating. Tetraethyl lead, $(C_2H_5)_4Pb$, a gasoline additive that very effectively increased octane ratings, has been eliminated from gasoline because of environmental dangers from lead. Compounds that contain oxygen have replaced tetraethyl lead. These additives increase the octane ratings and decrease the hydrocarbon pollution from automobile exhaust. Two examples are as follows:

$$CH_3-\underset{\displaystyle CH_3}{\overset{\displaystyle CH_3}{\underset{|}{\overset{|}{C}}}}-O-CH_3 \qquad CH_3CH_2OH$$

methyl-*tert*-butyl ether (MTBE) — ethanol

Unfortunately, environmental issues are often very complex. The use of MTBE is a case in point. While this additive has improved air quality, MTBE has leaked from underground gasoline storage tanks and polluted groundwater.

Thus, scientists are now looking for new, environmentally benign additives to improve gasoline combustion.

Although scientists have worked long and hard to improve gasoline, to make a better fuel and to decrease air pollution, it is an unavoidable fact that even the cleanest burning gasoline increases atmospheric carbon dioxide, which can contribute to the greenhouse effect. Furthermore, combustion uses hydrocarbons in a nonrecyclable way. Burning gasoline pollutes the air and consumes a valuable resource. As petroleum chemists look to the future, they recommend (1) development of nonpolluting energy sources and (2) shifting the major use of petroleum from fuel to the manufacture of other organic compounds.

A new fuel mixture on the market in the United States is E85–85% ethanol and 15% gasoline. This mixture decreases the use of a nonrenewable resource, petroleum, because it only contains 15% gasoline. The ethanol can be produced from renewable sources such as corn or sugar cane. Also, this mixture is the cleanest-burning fuel now available. Given these substantial advantages, it is unfortunate that only specially adapted cars can burn this fuel (auto manufacturers label these as "flexible fuel vehicles"). Experts predict that E85 may be the fuel of choice until more exotic alternates such as H_2 can be used.

19.14 Cycloalkanes

cycloalkane

Cyclic, or closed-chain, alkanes also exist. These substances, called **cycloalkanes**, *cycloparaffins*, or *naphthenes*, have the general formula C_nH_{2n}. Their names are formed by adding the prefix *cyclo-* to the name of the open-chain alkane with the same number of carbon atoms. For example, the six-carbon cycloalkane is called cyclohexane. Note that this series of compounds has two fewer hydrogen atoms than the open-chain alkanes. The bonds for the two missing hydrogen atoms are accounted for by an additional carbon–carbon bond in the cyclic ring of carbon atoms. Structures for the four smallest cycloalkanes are shown in Figure 19.9.

Guayule and sunflower are renewable sources of hydrocarbons

With the exception that follows for cyclopropane and cyclobutane, cycloalkanes are generally similar to open-chain alkanes in both physical properties and chemical reactivity. Cycloalkanes are saturated hydrocarbons; they contain only single bonds between carbon atoms.

The reactivity of cyclopropane, and to a lesser degree that of cyclobutane, is greater than that of other alkanes. This greater reactivity exists because the carbon–carbon bond angles in these substances deviate substantially from the normal tetrahedral angle. The carbon atoms form a triangle in cyclopropane,

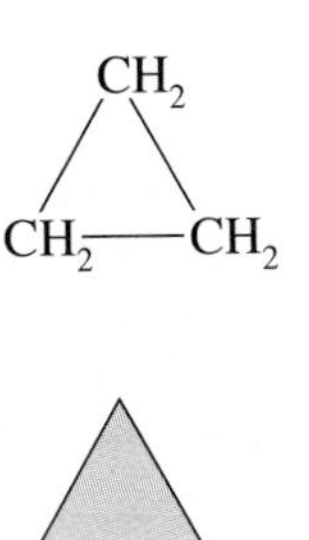

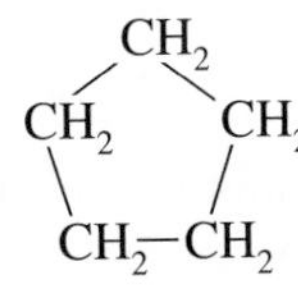

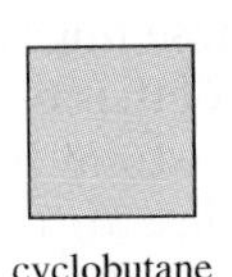

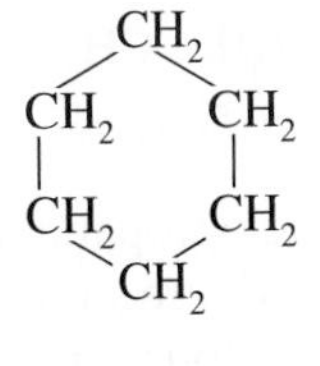

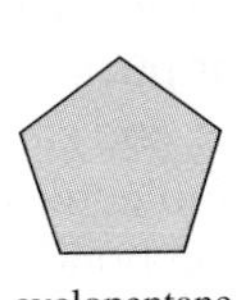

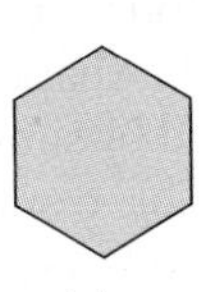

cyclopropane C_3H_6 | cyclobutane C_4H_8 | cyclopentane C_5H_{10} | cyclohexane C_6H_{12}

Figure 19.9
Cycloalkanes. In the line representations, each corner of the diagram represents a CH_2 group.

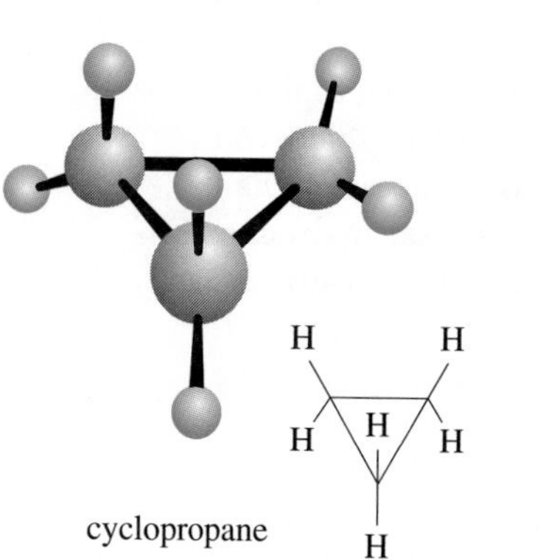

hexane $CH_3CH_2CH_2CH_2CH_2CH_3$

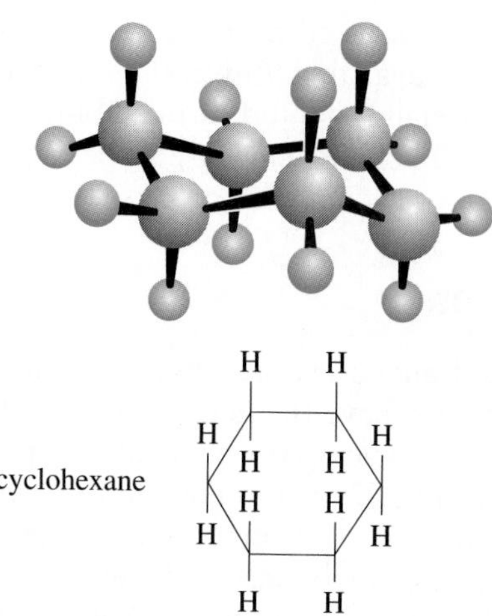

Figure 19.10
Ball-and-stick models illustrating cyclopropane, hexane, and cyclohexane. In cyclopropane, all the carbon atoms are in one plane. The angle between the carbon atoms is 60°, not the usual 109.5°; therefore, the cyclopropane ring is strained. In cyclohexane, the carbon–carbon bonds are not strained. This is because the molecule is puckered (as shown in the chair conformation) with carbon–carbon bond angles about 109.5°, as found in hexane.

and in cyclobutane, they approximate a square. Cyclopropane molecules therefore have carbon–carbon bond angles of 60°, and in cyclobutane, the bond angles are about 90°. In the open-chain alkanes and in larger cycloalkanes, the carbon atoms are in a three-dimensional zigzag pattern in space and have normal (tetrahedral) bond angles of about 109.5°.

Bromine adds to cyclopropane readily and to cyclobutane to some extent. In this reaction, the ring breaks and an open-chain dibromopropane is formed:

$$\underset{\text{cyclopropane}}{\overset{CH_2}{CH_2{-}CH_2}} + Br_2 \longrightarrow \underset{\text{1,3-dibromopropane}}{BrCH_2CH_2CH_2Br}$$

Cyclopropane and cyclobutane react in this way because their carbon–carbon bonds are strained and therefore weakened. Cycloalkanes, whose rings have more than four carbon atoms, do not react in the same way, because their molecules take the shape of nonplanar puckered rings. These rings can be considered to be formed by simply joining the end carbon atoms of the corresponding normal alkanes. The resulting cyclic molecules are nearly strain free, with carbon atoms arranged in space so that the bond angles are close to 109.5° (Figure 19.10).

conformation

The **conformation** of a molecule is its three-dimensional shape in space. Molecular models show that cyclohexane can assume two distinct nonplanar

(a) (b)

Figure 19.11
Conformations of cyclohexane: (a) chair conformation; (b) boat conformation. Axial hydrogens are shown in blue in the chair conformation.

conformations. One form is shaped like a chair, while the other is shaped like a boat. (See Figure 19.11.) In the chair form, the hydrogen atoms are separated as effectively as possible, so this is the more stable conformation. Six of the hydrogens in "chair" cyclohexane lie approximately in the same plane as the carbon ring. These are called **equatorial hydrogens**. The other six hydrogen atoms are approximately at right angles above or below the plane of the ring. These are called **axial hydrogens**. Substituent groups are usually found in equatorial positions where they are farthest from hydrogens and other groups. Five- and six-membered rings are a common occurrence in organic chemistry and biochemistry. These conformations and isomers are evident in carbohydrates as well as in nucleic acids. For simplicity, we draw cyclic pentanes and hexanes as flat pentagon and hexagon rings, respectively. (See Figure 19.10.)

equatorial hydrogen

axial hydrogen

Cyclopropane is a useful general anesthetic and, along with certain other cycloalkanes, is used as an intermediate in some chemical syntheses. The high reactivity of cyclopropane requires great care in its use as an anesthetic because it is an extreme fire and explosion hazard. The cyclopentane and cyclohexane ring structures are present in many naturally occurring molecules such as prostaglandins, steroids (e.g., cholesterol and sex hormones), and some vitamins.

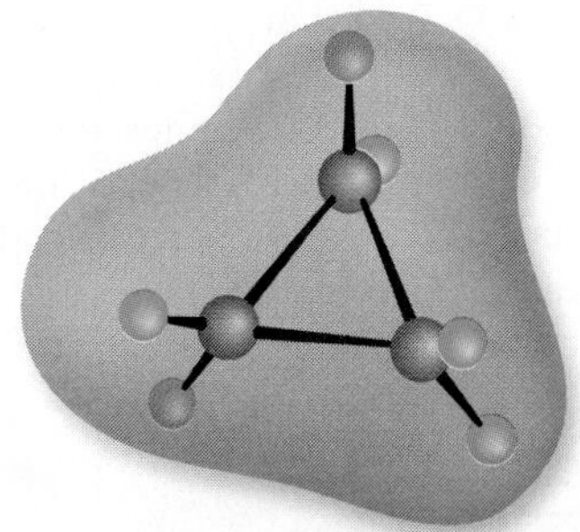

Ball-and-stick model of cyclopropane enclosed in the spacefilling model.

Practice 19.11

Draw the structures for the two cycloalkanes that have the molecular formula C_4H_8.

Naming Cycloalkanes

Substituted cycloalkanes are named by identifying and numbering the location of the substituent(s) on the carbon ring, followed by the name of the parent cycloalkane. For example, for a monosubstituted cycloalkane, it does not matter on which carbon atom the substituted group is placed. The structural formulas

CH_3 and CH_3

$C_6H_{11}CH_3$ $C_6H_{11}CH_3$

are both named methylcyclohexane. However, when a cycloalkane has two or more substituted groups, the ring is numbered (clockwise or counterclockwise) to give the location of the substituted groups the smallest numbers.

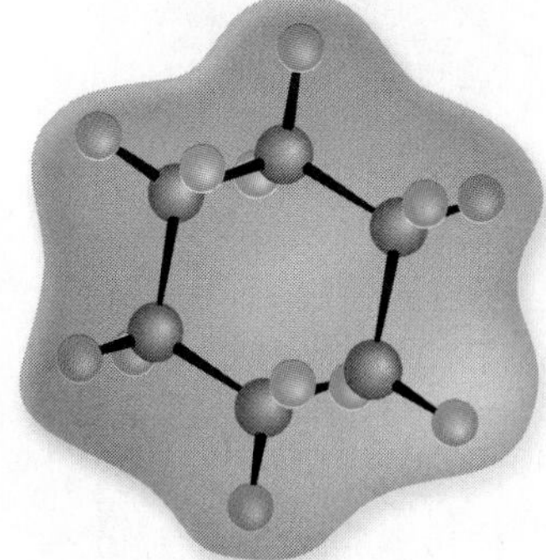

Ball-and-stick model of cyclohexane (chair form) enclosed in the spacefilling model.

CHEMISTRY IN ACTION • A Flammable Planet

Titan, Saturn's largest moon, is overflowing with hydrocarbons. On Titan, a truly alien moon, it rains methane and ethane—the same compounds burned in natural gas stoves. Under the right conditions, Titan could ignite. What strange conditions exist on such a flammable moon?

We have learned much about Titan since the *Cassini* space probe arrived at the Saturn system in 2004. Sophisticated radar has peered beneath Titan's orange-brown atmosphere to map the surface. Other instruments have recorded the environmental chemistry and remarkable weather cycle. Finally, in 2005, a smaller probe was released from *Cassini* and parachuted to touch down on this alien world (see www.esa.int for a movie made during the descent of this spacecraft to Titan's surface).

Hydrocarbons participate in Titan's weather, much like water does on Earth. Methane evaporates and moves many miles into the atmosphere. At these very high altitudes, methane condenses and forms raindrops. The raindrops fall slowly back toward the planet's surface; in low gravity, each moves more like a snowflake than a raindrop. Picture countless raindrops slowly drifting, and you have an idea of a typical day on Titan. Because the process just described is slow and continuous, areas of Titan's atmosphere can become supersaturated with methane. Then, a small change will cause the extra methane to fall out of the atmosphere—a storm. On Earth, we might call it a cloudburst. Suddenly, huge quantities of methane pour toward the surface, picking up speed as the liquid approaches the surface. Rivers of alkanes flow into large lakes. (The accompanying photo above) covers about 190 miles × 90 miles, and the river in the lower left is about 62 miles long.) Scientists expect that these downpours cause surface erosion; hence, the surface of Titan is similar to the landscape of the southwestern United States. Gullies and washes that are dry most of the time are suddenly filled to overflowing by a cloudburst of methane. The landing probe photographed such a landscape (see below).

On Earth a weather cycle such as this would create a huge hazard—one mistaken spark and the whole atmosphere would explode. So why isn't Titan in flames? Just one chemical is missing from this scenario. Alkane combustion (as chemistry students learn) requires oxygen. Titan, like Earth before photosynthesis, is devoid of molecular oxygen. Without this reactant, Titan remains a flammable moon—but with no flames.

All the other rules for naming alkanes also apply. Study these examples for dimethylcycloalkane, $C_6H_{10}(CH_3)_2$:

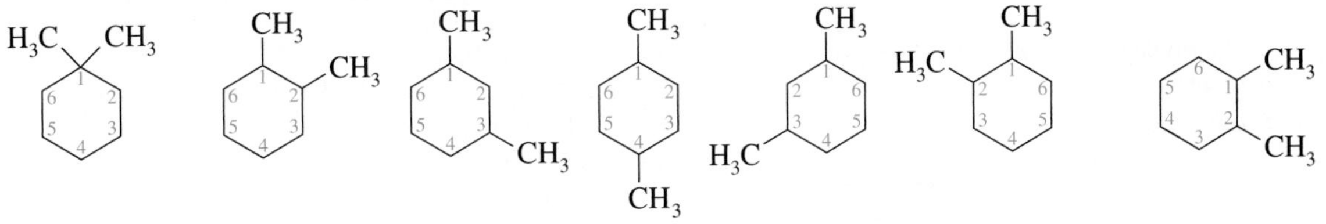

A B C D E F G

A is 1,1-dimethylcyclohexane
B is 1,2-dimethylcyclohexane
C is 1,3-dimethylcyclohexane
D is 1,4-dimethylcyclohexane
E is 1,3-dimethylcyclohexane (numbered counterclockwise)
F is 1,2-dimethylcyclohexane (numbered counterclockwise)
G is 1,2-dimethylcyclohexane (same as B and F)

Note that B, F, and G are the same compound, and C and E are the same compound. Therefore, there are four isomers of dimethylcyclohexane.

Practice 19.12

Write structural formulas for

(a) 1,4-diethylcyclohexane
(b) 1-ethyl-3-methylcyclohexane
(c) 1,2,4-trimethylcyclopentane

Chapter 19 Review

19.1 Organic Chemistry: History and Scope

KEY TERMS

Organic chemistry
Vital force theory

- Organic chemistry is the chemistry of carbon compounds.
- Organic compounds were originally thought to be formed only by living organisms (vital force theory).
- Versatile carbon chemistry allows a vast array of different compounds.

19.2 The Carbon Atom: Bonding and Shape

KEY TERMS

Saturated compound
Unsaturated compound

- A carbon atom usually has four covalent bonds.
- In a saturated compound, each carbon atom is bonded to four other atoms by single bonds with bond angles of about 109.5°.
- An unsaturated compound contains double and/or triple bonds between carbon atoms.

19.3 Organic Formulas and Molecular Models

KEY TERMS

Condensed structural formulas
Line structures

- Organic molecules are three-dimensional, with shapes best represented by ball-and-stick or spacefilling models.
- In contrast to inorganic compounds, organic compounds are often written in structural or condensed structural formulas.
- Line structures simplify structural formulas and are often used for molecules that are complex or that contain rings.

19.4 Classifying Organic Compounds

KEY TERM

Functional group

- Organic molecules are classified based on functional groups.
- Functional groups represent different classes of compounds.

19.5 Hydrocarbons

KEY TERMS

Aliphatic
Hydrocarbon

- Compounds that contain only hydrogen and carbon are hydrocarbons.
- Compounds that contain benzene rings are aromatic, while all other organic compounds are classed as aliphatic.

19.6 Saturated Hydrocarbons: Alkanes

KEY TERMS

Alkane
Homologous series

- Alkanes are saturated hydrocarbons.
- By successively adding CH_2 groups to methane, a series of different alkanes is generated–an homologous series.
- The general formula for the homologous series of open-chain alkanes is C_nH_{2n+2}

19.7 Carbon Bonding in Alkanes

KEY TERM

Sigma bond

- In alkanes, carbon hybridizes its valence shell electron orbitals to form four sp^3 hybrids.
- Sigma bonds occur when two atoms share two electrons along a line drawn between the two nuclei.

19.8 Isomerism

KEY TERMS

Isomerism
Isomers

- Isomerism is the phenomenon that two or more different compounds have the same molecular formula but different structures.
- Molecules with different structural formulas that have the same molecular formula are isomers of each other.
- Alkane isomers differ with respect to the length of the longest carbon chain and/or the number, type, and position of branch chains.

19.9 Naming Organic Compounds

KEY TERMS

IUPAC system
Alkyl group

- Alkyl groups have one less hydrogen than the corresponding alkanes. Their formulas are C_nH_{2n+1}.
- Common names are often based on an organic compound's origin
- The IUPAC system derives a name using a set of rules based on an organic compound's structural formula.
- Most alkanes can be named in the IUPAC system using five rules.
  - The longest continuous carbon chain (parent chain) is given the name of the corresponding alkane.
  - The longest continuous carbon chain is numbered starting at the end that is closest to the first branch.
  - Each branch is given its alkyl name and number location on the longest continuous chain.
  - Prefixes are added to the alkyl name when there is more than one of the same branch (*di-*, *tri-*, *tetra-*, etc.).
  - Branches are listed alphabetically in front of the parent name.

19.10 Introduction to the Reactions of Carbon

KEY TERMS

Substitution reaction
Elimination reaction
Addition reaction

- There are three common classes of reactions for organic compounds.
- A substitution reaction occurs when an atom bonded to carbon is replaced by a different atom or group of atoms.
- An elimination reaction is a reaction in which a single reactant is split into two products and one of the products is eliminated.
- In an addition reaction two reactants add together to form a single product.

19.11 Reactions of Alkanes

KEY TERMS

Monosubstitution
Alkyl halide
Mechanism
Free radical

- Modern societies depend on the energy derived from alkane combustion reactions.
- Common noncombustion alkane reactions include halogenation, dehydrogenation, cracking, and isomerization.
- Substituting one halogen atom for a hydrogen atom in an alkane is an example of a monosubstitution reaction.
- Alkyl halides are formed when alkanes react with halogens.
- The mechanism for a reaction is the path it takes to go from reactants to products.
- A free radical is an atom or group of atoms that has an unpaired electron.
- Alkane halogenation goes by a multistep, free-radical mechanism.

19.12 Sources of Alkanes

- The two main sources of alkanes are natural gas and petroleum.

19.13 Gasoline: A Major Petroleum Product

- Gasoline is the most common petroleum product.
- Gasoline additives can ease environmental problems and decrease engine "knocking." The quality of gasoline is rated by its octane number.

19.14 Cycloalkanes

KEY TERMS

Cycloalkane
Conformation
Equatorial hydrogen
Axial hydrogen

- Cycloalkanes are cyclic structures with the general formula C_nH_{2n}.
- Cycloalkanes have well-defined three-dimensional shapes-conformations.
- The conformations of cyclohexane hold the molecule's hydrogen atoms in either the plane of the ring (equatorial hydrogens) or roughly perpendicular to the ring (axial hydrogens).
- Cycloalkanes are named according to the number of carbon atoms in the ring and are prefixed with the term *cyclo-*; for example, cyclohexane has six carbon atoms in the ring. With one ring substituent the carbons in the ring need not be numbered; with two or more substituents, the ring is numbered, clockwise or counterclockwise, so as to give the substituents the lowest numbers.

Review Questions

All questions with blue numbers have answers in the appendix of the text.

1. What are the main reasons for the large number of organic compounds?
2. Why is it believed that a carbon atom must form hybrid electron orbitals when it bonds to hydrogen atoms to form methane?
3. Write the names and formulas for the first 10 normal alkanes.
4. State the number of sigma bonds in a molecule of
   (a) ethane (b) butane (c) 2-methylpropane
5. What are the major advantages to using alkyl halides as inhalation anesthetics?
6. What is the difference between the meaning of the prefixes *iso*-, *sec*-, and *tert*-?
7. Differentiate between
   (a) a substitution reaction versus an elimination reaction
   (b) an addition reaction versus an elimination reaction
8. What is a major advantage of gasoline with a higher octane rating?
9. List two advantages of E85 as an alternative to gasoline.
10. What would be a major advantage of using a plant such a guayule or sunflower as a source of combustible hydrocarbons?

Paired Exercises

All exercises with blue numbers have answers in the appendix of the text.

1. Write the Lewis structures for
   (a) CCl_4 (b) C_2Cl_6 (c) $CH_3CH_2CH_3$

2. Write the Lewis structures for
   (a) methane (b) propane (c) pentane

3. Which of these formulas represent isomers?

(a) $CH_3CH_2CH_2CH_3$

(b) $CH_3CH_2CH_2CH_2CH_3$

(c)
```
CH3CHCH3
   |
 CH3CH2
```

(d) $CH_3CH_2CH_2CH_2CH_2CH_3$

(e)
```
CH3     CH3
   \   /
   CH—CH
   /   \
CH3     CH3
```

(f)
```
CH2—CH2
 |    |
CH2  CH2
  \  /
   CH2
```

(g)
```
CH3CHCH2CH2CH3
    |
   CH3
```

(h)
```
CH2
 |\
 | CHCH2CH3
 |/
CH2
```

(i)
```
CH3CH2
   |
 CH2CH3
```

4. Which of these formulas represent the same compound?

(a)
```
CH3CHCH2CHCH3
    |     |
   CH3   CH3
```

(b)
```
CH3
 |
CH2CHCH2CHCH3
    |     |
   CH3   CH3
```

(c)
```
              CH3
               |
CH3CHCH2CH2CHCH3
    |
   CH3
```

(d)
```
     CH3
      |
CH3CHCHCH2CH2
    |      |
   CH3    CH3
```

(e)
```
   CH3
    |
 CH3CHCH2
       |
CH3CH2CHCH3
```

(f)
```
  CH3
   |
CH3CH
   |
  CH2CHCH3
      |
    CH2CH3
```

5. How many methyl groups are in each formula of Exercise 3?

6. How many methyl groups are in each formula of Exercise 4?

7. Write the condensed structural formulas for the nine isomers of heptane.

8. Write the condensed structural formulas for the five isomers of hexane.

9. Draw structural formulas for all the isomers of
(a) CH_2Cl_2 (b) C_3H_7Br
(c) $C_3H_6Cl_2$ (d) $C_4H_8Cl_2$

10. Draw structural formulas for all the isomers of
(a) CH_3Br (b) C_2H_5Cl
(c) C_4H_9I (d) C_3H_6BrCl

11. How many carbons are in the parent chains for each of the following?

```
                CH3
                |
                CH—CH3
                |
(a) CH3—CH—CH2—CH3

          CH3  CH3
             \ /
              CH
              |
(b) CH3—CH2—C—CH2—CH3
              |
              CH3

              CH3
              |
(c)    CH3—C—CH3
              |
    CH3—CH2—C—CH2—CH3
              |
     CH3—CH—CH2
          |
          CH3
```

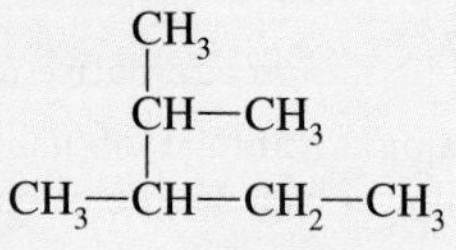

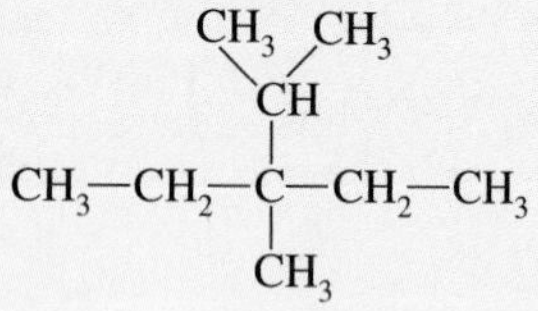

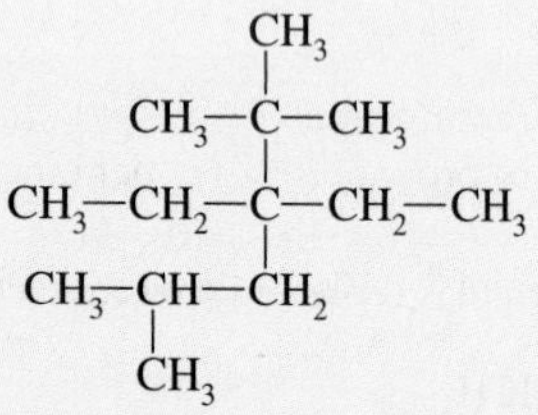

12. How many carbons are in the parent chains for each of the following?

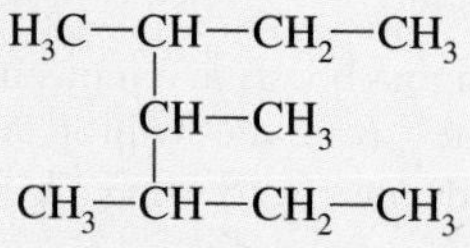

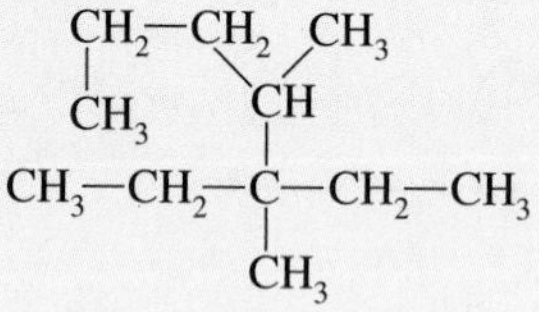

```
       H3C—CH—CH2—CH3
            |
            CH—CH3
            |
(a) CH3—CH—CH2—CH3

          CH2—CH2  CH3
          |        \ /
          CH3       CH
                    |
(b) CH3—CH2—C—CH2—CH3
              |
              CH3

               CH3
               |
(c)       CH3—C—CH3
               |
     CH3—CH2—CH—CH3
```

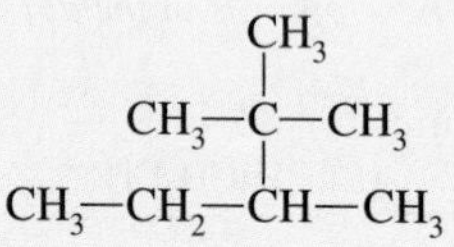

13. Give IUPAC names for the following:
(a) $CH_3CH_2CH_2Cl$
(b) $CH_3CHClCH_3$
(c) $(CH_3)_3CCl$
(d)

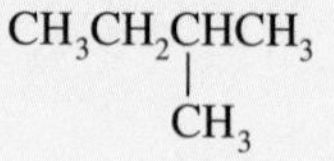

```
CH3CH2CHCH3
        |
        CH3
```

(e)

```
          CH3CHCH3
             |
CH3CH2CH2CHCH3
```

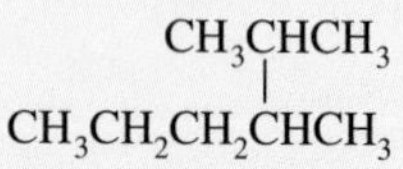

14. Give the IUPAC name for each of the following compounds:
(a) CH_3CH_2Cl
(b) $(CH_3)_2CHCH_2Cl$
(c) $CH_3CHClCH_2CH_3$
(d)

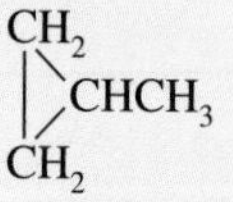

```
CH2
| \
|  CHCH3
| /
CH2
```

(e) $(CH_3)_2CHCH_2CH(CH_3)_2$

15. Circle all tertiary carbons in the structures of Exercise 11.

16. Circle all tertiary carbons in the structures of Exercise 12.

17. Draw structural formulas for the following compounds:
(a) 2,4-dimethylpentane
(b) 2,2-dimethylpentane
(c) 3-isopropyloctane
(d) 5,6-diethyl-2,7-dimethyl-5-propylnonane
(e) 2-ethyl-1,3-dimethylcyclohexane

18. Draw structural formulas for the following compounds:
(a) 4-ethyl-2-methylhexane
(b) 4-tert-butylheptane
(c) 4-ethyl-7-isopropyl-2,4,8-trimethyldecane
(d) 3-ethyl-2,2-dimethyloctane
(e) 1,3-diethylcyclohexane

19. The following names are incorrect:
(a) 3-methylbutane
(b) 2-ethylbutane
(c) 2-dimethylpentane
(d) 1,4-dimethylcyclopentane
Explain why the name is wrong and give the correct name.

20. The following names are incorrect:
(a) 3-methyl-5-ethyloctane
(b) 3,5,5-triethylhexane
(c) 4,4-dimethyl-3-ethylheptane
(d) 1,6-dimethylcyclohexane
Explain why the name is wrong and give the correct name.

21. Draw the structures for the 10 dichlorosubstituted isomers, $C_5H_{10}Cl_2$, of 2-methylbutane.

22. The structure for hexane is $CH_3CH_2CH_2CH_2CH_2CH_3$
Draw the structural formulas for all the monochlorohexanes, $C_6H_{13}Cl$, that have the same linear carbon structure as hexane.

23. Complete the equations for (a) the monochlorination, and (b) the complete combustion, of butane.
(a) $CH_3CH_2CH_2CH_3 + Cl_2 \xrightarrow{hv}$
(b) $CH_3CH_2CH_2CH_3 + O_2 \xrightarrow{\Delta}$

24. Complete the equations for (a) the monobromination and (b) the complete combustion of propane.
(a) $CH_3CH_2CH_3 + Br_2 \xrightarrow{hv}$
(b) $CH_3CH_2CH_3 + O_2 \xrightarrow{\Delta}$

25. Write names for

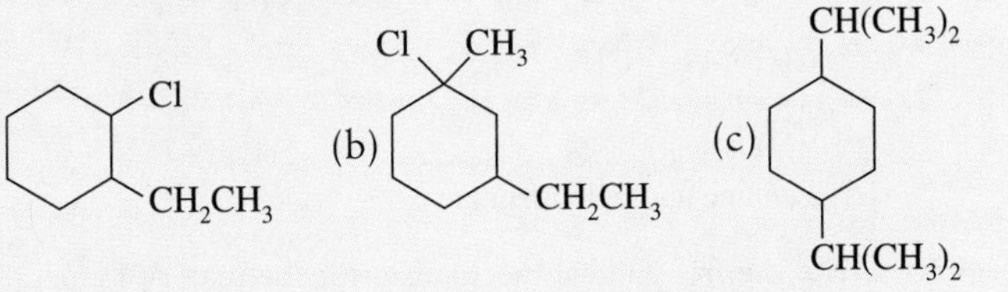

(a) cyclohexane ring with Cl and CH_2CH_3 on adjacent carbons (b) cyclohexane ring with Cl and CH_3 on one carbon and CH_2CH_3 (c) cyclohexane ring with $CH(CH_3)_2$ at positions 1 and 4

26. Write structural formulas for
(a) 1,4-di(*tert*-butyl)cyclohexane
(b) 1-methyl-3-propylcyclohexane
(c) 1,3-dichloro-2,5 diethylcyclohexane

27. Draw condensed structural formulas for all isomers of pentane that contain tertiary carbons.

28. Draw condensed structural formulas for all isomers of hexane that contain quaternary carbons.

29. In each dehydrogenation reaction, give the condensed structural formula and name for the alkane reactant:

(a) 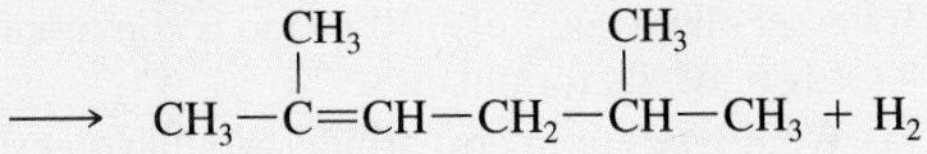 $\longrightarrow CH_3-C(CH_3){=}CH-CH_2-CH(CH_3)-CH_3 + H_2$

(b) 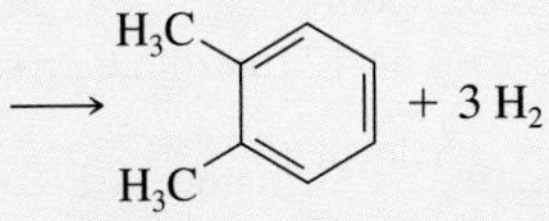 $\longrightarrow$ (benzene ring with two adjacent H_3C groups) $+ 3\,H_2$

30. In each halogenation reaction, give the condensed structural formula and name for the alkane reactant:

(a) 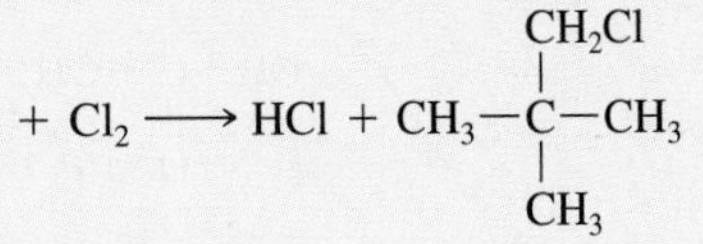 $+ Cl_2 \longrightarrow HCl + CH_3-C(CH_2Cl)(CH_3)-CH_3$

(b) 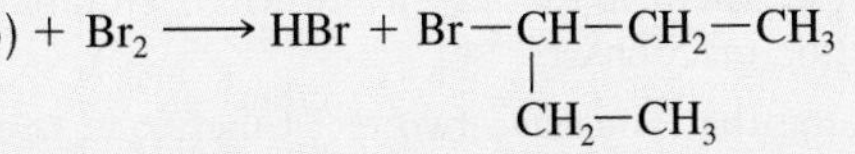 $+ Br_2 \longrightarrow HBr + Br-CH(CH_2-CH_3)-CH_2-CH_3$

31. Give structural formulas and names for all isomers formed by monochlorination of
(a) 2-methylpentane
(b) methylcyclohexane

32. Give structural formulas and names for all isomers formed by dibromination of
(a) cyclopentane
(b) 2,2-dimethylbutane

Additional Exercises

All exercises with blue numbers have answers in the appendix of the text.

33. Explain why cyclohexane forms either a chair or boat conformation and not a flat ring.

34. The name of the compound of formula $C_{11}H_{24}$ is undecane. What is the formula for dodecane, the next higher homologue in the alkane series?

35. The newer, more environmentally safe substitutes for chlorofluorocarbons (CFCs) are termed hydrochlorofluorocarbons (HCFCs). Write an equation for the monochlorination of 1,2-difluoroethane to produce an HCFC.

36. High-efficiency cars use gasoline at about the rate of 60 miles per gallon. One gallon of gasoline contains about 19 moles of octane. Assuming complete combustion of the octane, how many moles of carbon dioxide will be exhausted to travel 60 miles? How many liters of carbon dioxide will be exhausted (at 20.°C and 1.0 atm pressure)?

37. Write the structures and names for all the cycloalkanes that have the molecular formula C_5H_{10}.

38. The following reactions are important in biochemistry:

(a) $$\mathrm{HO{-}\overset{O}{\overset{\|}{C}}{-}\underset{CH_2OH}{\underset{|}{C}}H(H_2PO_4)} \longrightarrow \mathrm{HO{-}\overset{O}{\overset{\|}{C}}{-}\underset{CH_2}{\underset{\|}{C}}(H_2PO_4)} + \mathrm{H_2O}$$

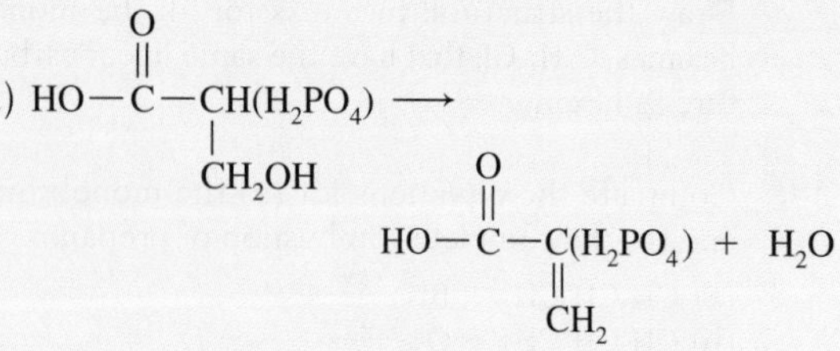

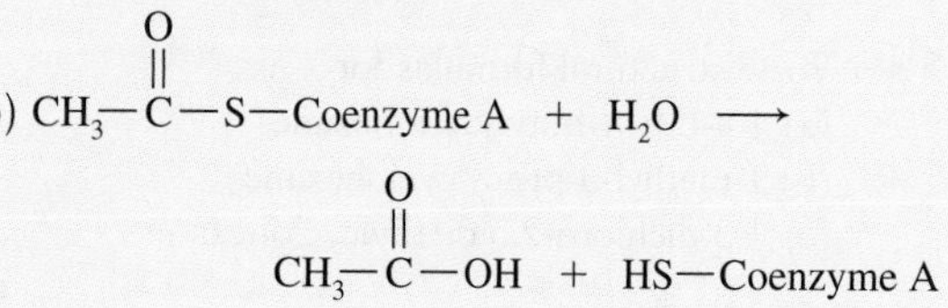

(b) $$\mathrm{CH_3{-}\overset{O}{\overset{\|}{C}}{-}S{-}Coenzyme\ A} + \mathrm{H_2O} \longrightarrow \mathrm{CH_3{-}\overset{O}{\overset{\|}{C}}{-}OH} + \mathrm{HS{-}Coenzyme\ A}$$

(c) $$\mathrm{HO{-}\overset{O}{\overset{\|}{C}}{-}CH{=}\underset{\underset{\underset{O}{\|}}{C-OH}}{\underset{|}{C}}CH_2{-}\overset{O}{\overset{\|}{C}}{-}OH} + \mathrm{H_2O} \xrightarrow{H^+} \mathrm{HO{-}\overset{O}{\overset{\|}{C}}{-}\underset{OH}{\underset{|}{C}}H{-}\underset{\underset{\underset{O}{\|}}{C-OH}}{\underset{|}{C}}HCH_2{-}\overset{O}{\overset{\|}{C}}{-}OH}$$

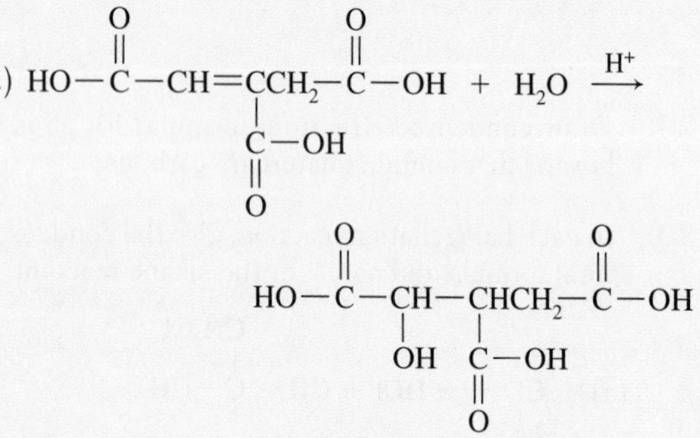

Classify them as either substitution, elimination, or addition reactions.

39. Assume that you have two test tubes, one containing hexane and the other containing 3-methylheptane. Can you tell which one is in which tube by testing their solubility in water? Why or why not?

40. Gasoline is a mixture of hydrocarbons. The components must be blended so that the gasoline has the correct volatility, and gasoline must have enough highly volatile compounds so that sufficient vapors are available to burn when the spark plug fires. In general, the lower the boiling point of a substance, the higher is the volatility. Place the following compounds in their order of increasing volatility: butane, propane, and hexane.

41. In each of the following, state whether the compounds given are (i) isomers of one another, (ii) the same compound, or (iii) neither isomers nor the same compound:

(a) $CH_3CH_2CHBrCH_2Br$ and $CH_2BrCH_2CH_2CH_2Br$

(b) 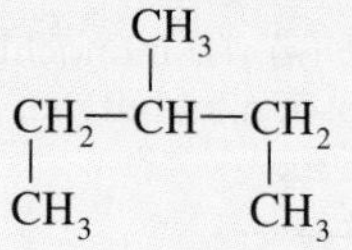 $$\mathrm{\underset{CH_3}{\underset{|}{C}}H_2{-}\overset{CH_3}{\overset{|}{C}}H{-}\underset{CH_3}{\underset{|}{C}}H_2}$$ and 2-methylhexane

(c) 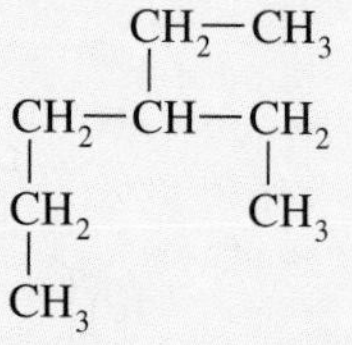 $$\mathrm{\underset{\underset{\underset{CH_3}{|}}{CH_2}}{\underset{|}{C}}H_2{-}\overset{CH_2-CH_3}{\overset{|}{C}}H{-}\underset{CH_3}{\underset{|}{C}}H_2}$$ and $CH_3(CH_2)_5CH_3$

(d) Pentane and cylcopentane

42. Write the free radical mechanism for the major product formed in the monochlorination of (a) chloromethane, (b) 2-methylpropane, (c) cyclopentane, (d) methylcyclohexane.

43. In the monochlorination of methane, what technique(s) would you use to minimize the formation of di-, tri-, and tetracloromethane.

44. Freon-12 is nontoxic, nonflammable, and noncorrosive. It has a boiling point of $-30°C$ and is commonly used as a refrigerant. Its formula is CCl_2F_2.

(a) What types of hybrid orbitals would you expect to be present in this compound?

(b) Would you expect Freon-12 to exhibit structural isomerism? Why or why not?

(c) Give an appropriate IUPAC name for Freon-12.

(d) Based on Table 19.1, to which class of compounds does Freon-12 belong?

45. Pheromones are chemical communicators. When an ant is disturbed, it can release undecane, $C_{11}H_{24}$, and tridecane, $C_{13}H_{28}$, to send an alarm signal to other ants.

(a) Write the structural formulas for undecane and tridecane.

(b) Are either of these compounds alkanes? How can you tell?

46. Write the structure for, and comment on, the name 1,3,5-trimethylcyclopentane.

Challenge Exercise

47. A hydrocarbon sample of formula C_4H_{10} is brominated and four different monobromo compounds of formula C_4H_9Br are isolated. Is the original sample a pure compound or a mixture of compounds? Explain your answer by showing the structural isomers where needed.

Answers to Practice Exercises

19.1 (a) 109.5° (b) 120°

19.2 (a) $C_{12}H_{26}$, $C_{14}H_{30}$, $C_{20}H_{42}$

(b) $CH_3CH_2CH_2CH_2CH_2CH_2CH_2CH_2CH_2CH_2CH_2CH_3$

$CH_3CH_2CH_2CH_2CH_2CH_2CH_2CH_2CH_2CH_2CH_2CH_2CH_2CH_3$

19.3

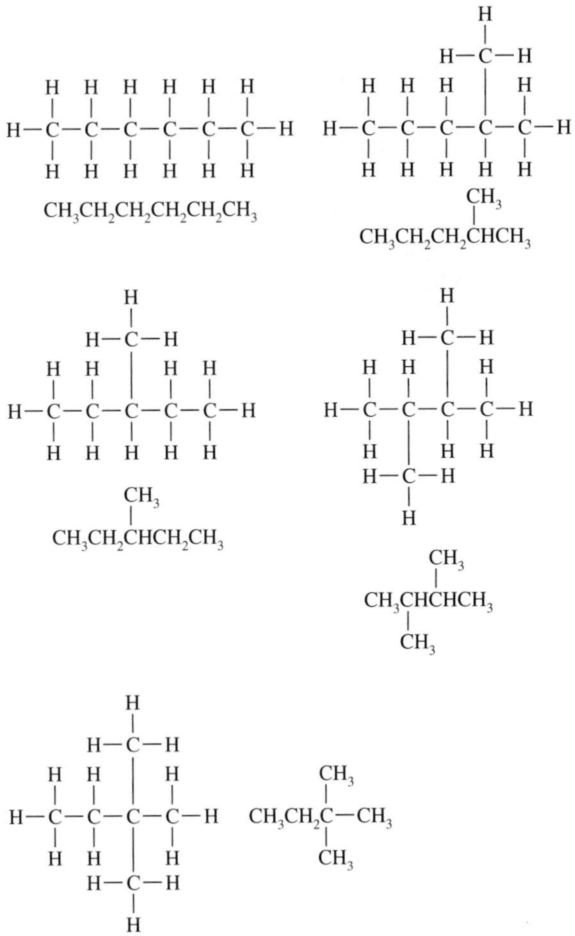

```
    H  H  H  H  H  H
    |  |  |  |  |  |
 H—C—C—C—C—C—C—H
    |  |  |  |  |  |
    H  H  H  H  H  H
```

$CH_3CH_2CH_2CH_2CH_2CH_3$

```
                 H
                 |
              H—C—H
    H  H  H      |   H
    |  |  |      |   |
 H—C—C—C—C—C—H
    |  |  |  |   |
    H  H  H  H   H
```

$CH_3CH_2CH_2CH(CH_3)CH_3$ (drawn as $CH_3CH_2CH_2\overset{CH_3}{\overset{|}{C}}HCH_3$)

```
          H
          |
       H—C—H
    H  H  |  H  H
    |  |  |  |  |
 H—C—C—C—C—C—H
    |  |  |  |  |
    H  H  H  H  H
```

$CH_3CH_2CH(CH_3)CH_2CH_3$

```
             H
             |
          H—C—H
    H  H  |  H
    |  |  |  |
 H—C—C—C—C—H
    |  |  |  |
    H  |  H  H
    H—C—H
       |
       H
```

$CH_3CH(CH_3)CH(CH_3)CH_3$

```
          H
          |
       H—C—H
    H  H  |  H
    |  |  |  |
 H—C—C—C—C—H
    |  |  |  |
    H  H  |  H
       H—C—H
          |
          H
```

$CH_3CH_2C(CH_3)_2CH_3$

19.4

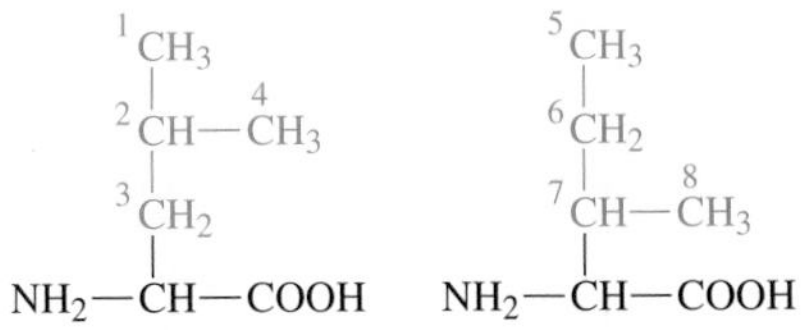

leucine: $^{1}CH_3$—^{2}CH($^{4}CH_3$)—$^{3}CH_2$—$CH(NH_2)$—COOH

isoleucine: $^{5}CH_3$—$^{6}CH_2$—^{7}CH($^{8}CH_3$)—$CH(NH_2)$—COOH

Primary carbons: 1, 4, 5, 8

Secondary carbons: 3, 6

Tertiary carbons: 2, 7

19.5

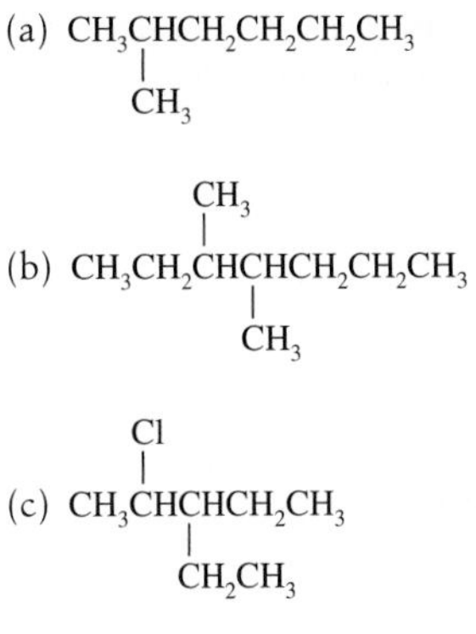

(a) $CH_3CH(CH_3)CH_2CH_2CH_2CH_3$

(b) $CH_3CH_2CH(CH_3)CH(CH_3)CH_2CH_2CH_3$

(c) $CH_3CH(Cl)CH(CH_2CH_3)CH_2CH_3$

19.6 (a) 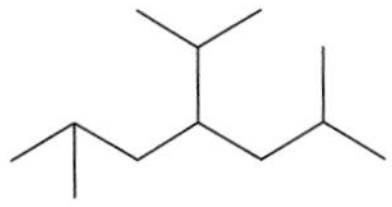 4-isopropyl-2,6-dimethylheptane

(b) 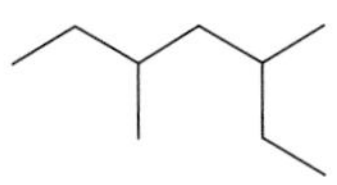 3,5-dimethylheptane

19.7 (a) addition

(b) substitution

19.8 There are 12 dichloro products of hexane

19.9 Initiation:

$$:\ddot{Cl}:\ddot{Cl}: \xrightarrow[\text{light}]{\text{UV}} 2\ Cl\cdot$$

Propagation:

$$Cl\cdot + CH_3CH_2CH_3 \longrightarrow CH_3\dot{C}HCH_3 + HCl$$

$$CH_3\dot{C}HCH_3 + Cl_2 \longrightarrow CH_3CHClCH_3 + Cl\cdot$$

Termination:

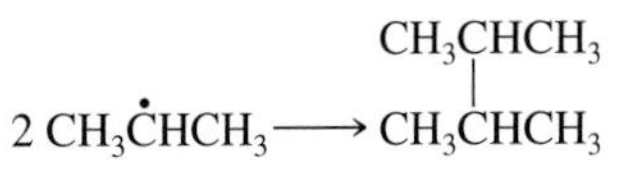

$$2\ CH_3\dot{C}HCH_3 \longrightarrow \begin{array}{c} CH_3CHCH_3 \\ | \\ CH_3CHCH_3 \end{array}$$

and/or

$$CH_3\dot{C}HCH_3 + Cl\cdot \longrightarrow CH_3CHClCH_3$$

and/or

$$Cl\cdot + Cl\cdot \longrightarrow Cl_2$$

19.10 (a) $CH_3CH(CH_3)CH(CH(CH_3)CH_3)CH(CH_3)CH_3$

$$\begin{array}{c} CH_3 \\ | \\ CH_3-CH \\ | \\ CH_3CHCHCHCH_3 \\ |\quad\quad | \\ CH_3\quad CH_3 \end{array}$$

more branching, predict higher octane rating

(b)

$$\begin{array}{c} CH_2CH_3 \\ | \\ CH_3CH_2CHCHCH_2CH_3 \\ | \\ CH_2CH_3 \end{array}$$

19.11 Cyclobutane (square), C_4H_8; methylcyclopropane (triangle—CH_3), C_4H_8

19.12 (a) Cyclohexane ring with CH_2CH_3 at positions 1 and 4

(b) Cyclohexane ring with CH_2CH_3 and CH_3 at positions 1 and 3

(c) Cyclopentane ring with three CH_3 groups (positions 1, 2, 4)

CHAPTER 20

Unsaturated Hydrocarbons

Perfumes and colognes interact with chemicals in our skin to produce unique fragrances.

Chapter Outline

Think of the many images a particular fragrance or smell can evoke. A favorite perfume triggers memories of a romantic evening, while the aroma of a light-bodied red wine reminds us of a favorite restaurant. The compounds responsible for such evocative odors are the essential oils in plants. Many of these molecules are classified as unsaturated hydrocarbons.

Unsaturated hydrocarbons enhance our lives in many ways. The polymers, manufactured from unsaturated molecules, would be difficult to do without. There are large numbers of these; you are probably familiar with polyethylene plastic bags and bottles and polystyrene Styrofoam cups, and plastic wrap. The molecules associated with the essential oils in plants often contain multiple bonds beween carbon atoms. These oils are widely used in cosmetics, medicines, flavorings, and perfumes.

Hydrocarbons also form rings of carbon atoms. These ring molecules are the basis for many consumer products such as detergents, insecticides, and dyes. Such molecules are known as aromatic carbon compounds and are also found in living organisms.

20.1 Bonding in Unsaturated Hydrocarbons

The unsaturated hydrocarbons consist of three families of homologous compounds that contain multiple bonds between carbon atoms. In each family, every compound contains fewer hydrogens than the alkane with the corresponding number of carbons. Compounds in the first family, known as the **alkenes**, contain carbon–carbon double bonds. Those in the second family, known as the **alkynes**, contain carbon–carbon triple bonds, and those in the third family, known as **aromatic compounds**, contain benzene rings.

alkene

alkyne

aromatic compound

The double bonds in alkenes are different from the sp^3 hybrid bonds found in the alkanes. The hybridization of the carbons that form the double bond in alkenes may be visualized in the following way: One of the $2s$ electrons of carbon is promoted to a $2p$ orbital to form the four half-filled orbitals, $2s^1 2p_x^1 2p_y^1 2p_z^1$. Three of these orbitals ($2s^1 2p_x^1 2p_y^1$) hybridize, thereby forming three equivalent orbitals designated as sp^2. Thus, the four orbitals available for bonding are three sp^2 orbitals and one p orbital. This process is illustrated in Figure 20.1.

The three sp^2 hybrid orbitals form angles of 120° with each other and lie in a single plane. The remaining $2p$ orbital is oriented perpendicular to this plane, with one lobe above and one lobe below the plane. (See Figure 20.2.) In the formation of a double bond, an sp^2 orbital of one carbon atom overlaps an identical sp^2 orbital of another carbon to form a sigma bond. At the same time the two perpendicular p orbitals (one on each carbon atom) overlap to form a **pi (π) bond** between the two carbon atoms. This pi bond

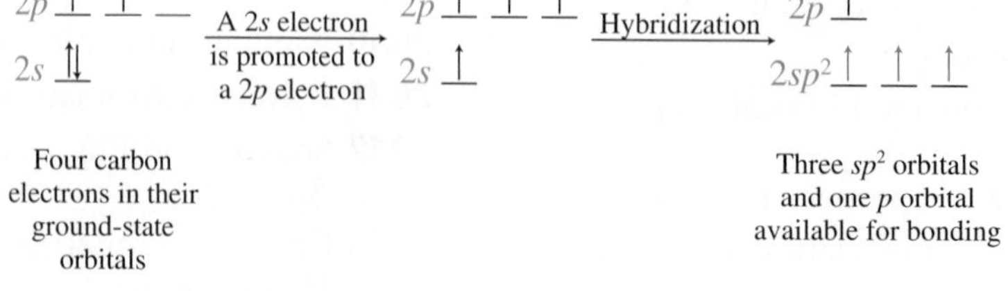

pi bond

$2p$ ↿ ↿ __ A $2s$ electron is promoted to a $2p$ electron → $2p$ ↿ ↿ ↿ Hybridization → $2p$ ↿

$2s$ ↿⇂ $2s$ ↿ $2sp^2$ ↿ ↿ ↿

Four carbon electrons in their ground-state orbitals

Three sp^2 orbitals and one p orbital available for bonding

Figure 20.1
Schematic hybridization of $2s^2 2p_x^1 2p_y^1$ orbitals of carbon to form three sp^2 electron orbitals and one p electron orbital.

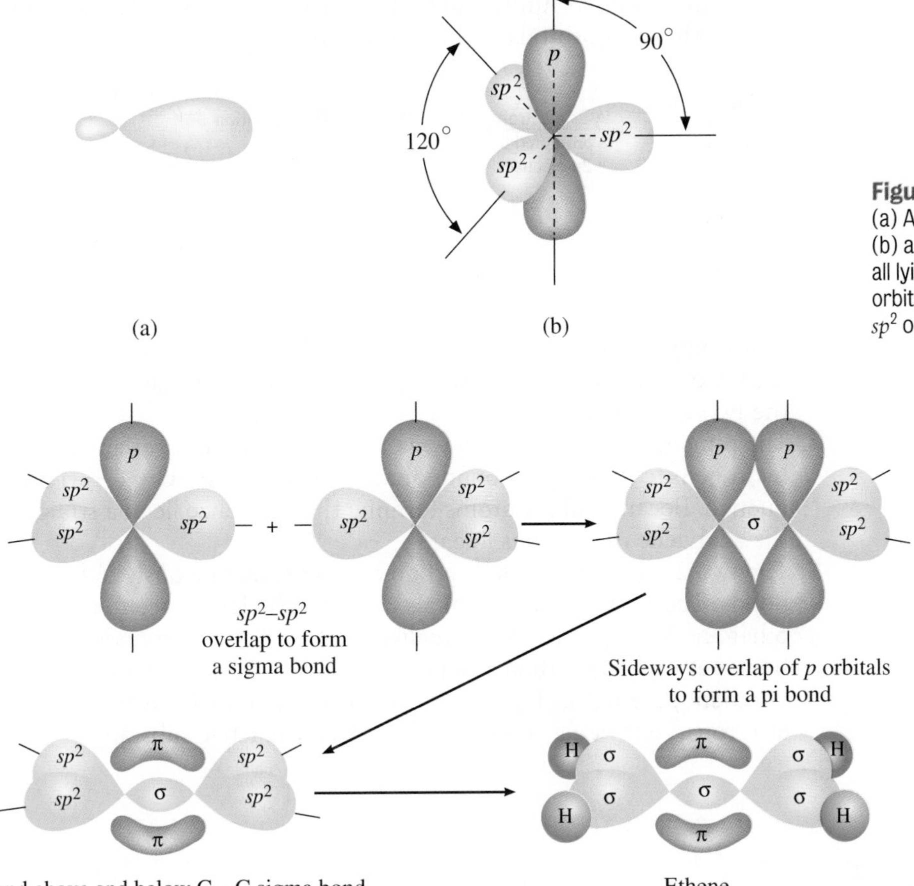

Figure 20.2
(a) A single sp^2 electron orbital and (b) a side view of three sp^2 orbitals all lying in the same plane with a p orbital perpendicular to the three sp^2 orbitals.

Figure 20.3
Pi (π) and sigma (σ) bonding in ethene.

consists of two electron clouds: one above and one below the sigma bond. (See Figure 20.3.) The $CH_2{=}CH_2$ molecule is completed as the remaining sp^2 orbitals (two on each carbon atom) overlap hydrogen s orbitals to form sigma bonds between the carbon and hydrogen atoms. Thus, there are five sigma bonds and one pi bond in an ethene molecule.

In the formula commonly used to represent ethene, $CH_2{=}CH_2$, no distinction is made between the sigma bond and the pi bond in the carbon–carbon double bond. However, these bonds are actually very different from each other. The sigma bond is formed by the overlap of sp^2 orbitals; the pi bond is formed by the overlap of p orbitals. The sigma bond electron cloud is distributed above a line joining the carbon nuclei, but the pi bond electron cloud is distributed above and below the sigma bond region. (See Figure 20.3.) The carbon–carbon pi bond is much weaker and, as a consequence, much more reactive than the carbon–carbon sigma bond.

The formation of a triple bond between carbon atoms, as in acetylene, $CH{\equiv}CH$, may be visualized as follows:

1. A carbon atom $2s$ electron is promoted to a $2p$ orbital ($2s^12p_x^12p_y^12p_z^1$).
2. The $2s$ orbital hybridizes with one of the $2p$ orbitals to form two equivalent orbitals known as sp orbitals. These two hybrid orbitals lie on a straight line that passes through the center of the carbon atom.

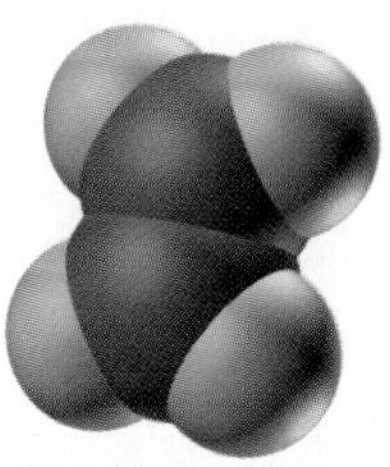

Spacefilling model of ethene.

The remaining two unhybridized $2p$ orbitals are oriented at right angles to these sp orbitals and to each other.

$2p$ ↑ ↑ —	A $2s$ electron →	$2p$ ↑ ↑ ↑	Hybridization →	$2p$ ↑ ↑ —
$2s$ ⇅	is promoted to a $2p$ electron	$2s$ ↑		$2sp$ ↑ ↑
Four carbon atoms in their ground-state orbitals				Two sp orbitals and $2p$ orbitals available for bonding

3. In forming carbon–carbon bonds, one carbon sp orbital overlaps an identical sp orbital on another carbon atom to establish a sigma bond between the two carbon atoms.
4. The remaining sp orbitals (one on each carbon atom) overlap s orbitals on hydrogens to form sigma bonds and establish the H—C—C—H bond sequence. Because all the atoms forming this sequence lie in a straight line, the acetylene molecule is linear.
5. The two $2p$ orbitals on each carbon overlap simultaneously to form two pi bonds. These two pi bond orbitals occupy sufficient space to overlap each other and form a continuous tubelike electron cloud surrounding the sigma bond between the carbon atoms (Figure 20.4). These pi bond electrons (as in ethene) are not as tightly held by the carbon nuclei as the sigma bond electrons. Acetylene, consequently, is a very reactive substance.

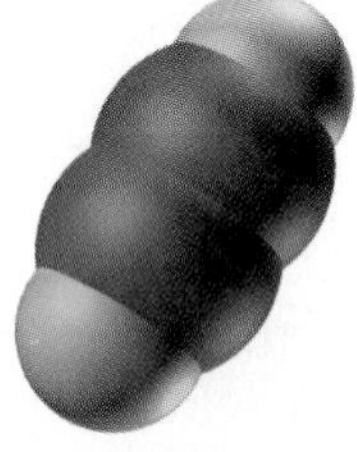
Spacefilling model of acetylene (ethyne).

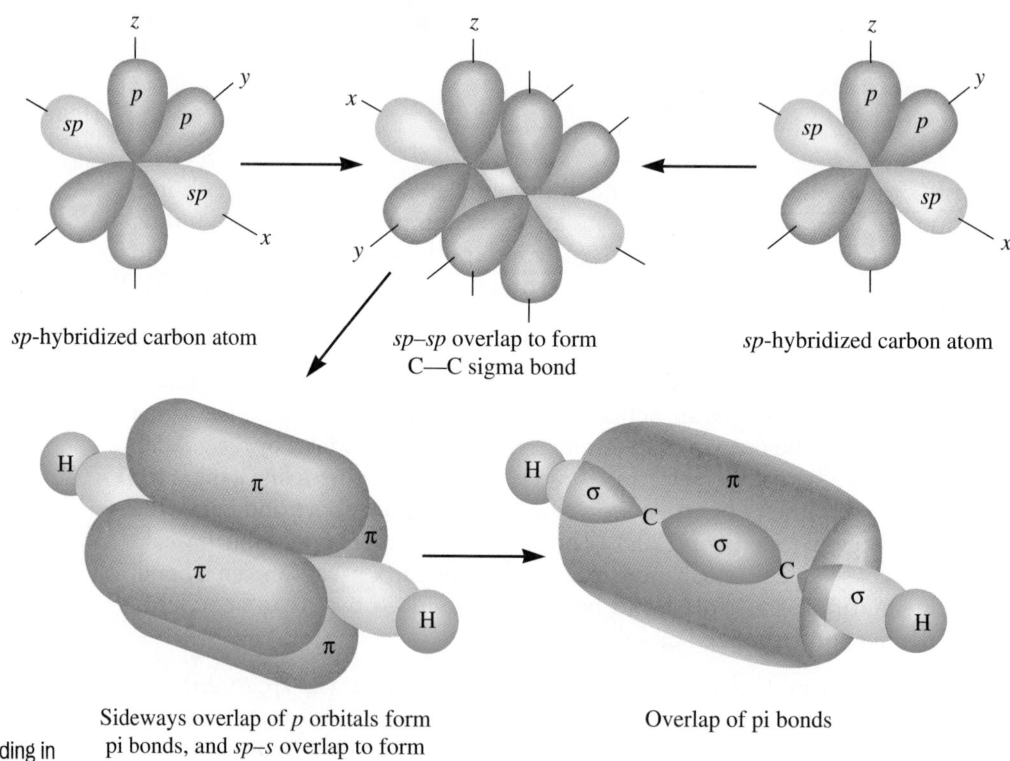

Figure 20.4
Pi (π) and sigma (σ) bonding in acetylene.

Practice 20.1

Identify the hybrid orbitals around each carbon in the following compounds:

(a) $CH_3-CH=\overset{\overset{\displaystyle CH_3}{|}}{C}-CH_3$

(b) $CH\equiv C-CH_2CH_3$

20.2 Nomenclature of Alkenes

The names of alkenes are derived from the names of corresponding alkanes.

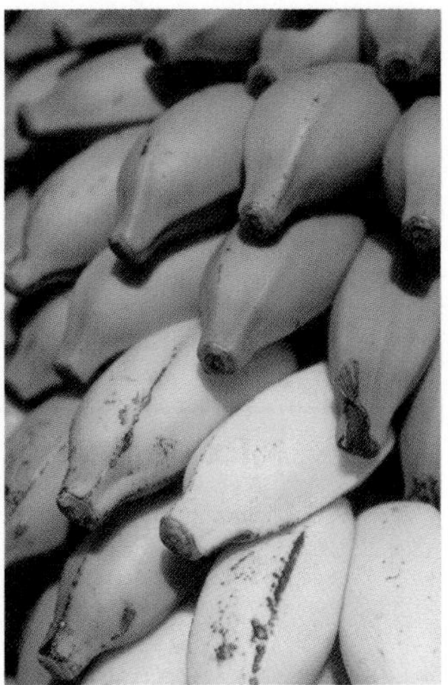

Ethene acts as a plant hormone, signaling fruit to ripen.

IUPAC Rules for Naming Alkenes

1. Select the longest continuous carbon–carbon chain that contains the double bond.
2. Name this parent compound as you would an alkane, but change the *-ane* ending to *-ene*; for example, propane is changed to propene.

$$\underset{\text{propane}}{CH_3CH_2CH_3} \qquad \underset{\text{propene}}{CH_3CH=CH_2}$$

3. Number the carbon chain of the parent compound starting with the end nearer to the double bond. Use the smaller of the two numbers on the double-bonded carbon atoms to indicate the position of the double bond. Place this number in front of the alkene name; for example, 2-butene means that the carbon–carbon double bond is between carbons 2 and 3.
4. Branch chains and other groups are treated as in naming alkanes, by numbering and assigning them to the carbon atom to which they are bonded.

Study the following examples of named alkenes:

General formula for alkenes is C_nH_{2n}

$$\underset{\text{1-butene}}{\overset{4}{C}H_3\overset{3}{C}H_2\overset{2}{C}H=\overset{1}{C}H_2} \qquad \underset{\text{2-butene}}{\overset{1}{C}H_3\overset{2}{C}H=\overset{3}{C}H\overset{4}{C}H_3}$$

$$\underset{\text{3-methyl-1-butene}}{\overset{4}{C}H_3\overset{\overset{\displaystyle CH_3}{|}}{\overset{3}{C}}H\overset{2}{C}H=\overset{1}{C}H_2} \qquad \underset{\text{3-propyl-1-hexene}}{\overset{6}{C}H_3\overset{5}{C}H_2\overset{4}{C}H_2\overset{\overset{\displaystyle CH_2CH_2CH_3}{|}}{\overset{3}{C}}H\overset{2}{C}H=\overset{1}{C}H_2}$$

To write a structural formula from a systematic name, the naming process is reversed. For example, how would we write the structural formula for 4-methyl-2-pentene? The name indicates

1. Five carbons in the longest chain
2. A double bond between carbons 2 and 3
3. A methyl group on carbon 4

Write five carbon atoms in a row. Place a double bond between carbons 2 and 3, and place a methyl group on carbon 4:

$$\overset{1}{C}-\overset{2}{C}=\overset{3}{C}-\underset{\displaystyle CH_3}{\overset{4}{C}}-\overset{5}{C}$$

carbon skeleton

Now add hydrogen atoms to give each carbon atom four bonds. Carbons 1 and 5 each need three H atoms; carbons 2, 3, and 4 each need one H atom. The complete formula is

$$CH_3CH{=}CH\underset{\displaystyle CH_3}{CH}CH_3$$

4-methyl-2-pentene

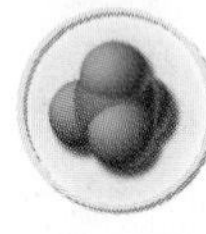

Example 20.1 Write structural formulas for (a) 2-pentene and (b) 7-methyl-2-octene.

SOLUTION (a) The stem *pent-* indicates a five-carbon chain; the suffix *-ene* indicates a carbon–carbon double bond; the number 2 locates the double bond between carbons 2 and 3. Write five carbon atoms in a row and place a double bond between carbons 2 and 3:

$$\overset{1}{C}-\overset{2}{C}=\overset{3}{C}-\overset{4}{C}-\overset{5}{C}$$

Add hydrogen atoms to give each carbon atom four bonds. Carbons 1 and 5 each need three H atoms; carbons 2 and 3 each need one H atom; carbon 4 needs two H atoms. The complete formula is

$$CH_3CH{=}CHCH_2CH_3$$

2-pentene

(b) Octene, like octane, indicates an eight-carbon chain. The chain contains a double bond between carbons 2 and 3 and a methyl group on carbon 7. Write eight carbon atoms in a row, place a double bond between carbons 2 and 3, and place a methyl group on carbon 7:

$$\overset{1}{C}-\overset{2}{C}=\overset{3}{C}-\overset{4}{C}-\overset{5}{C}-\overset{6}{C}-\underset{\displaystyle CH_3}{\overset{7}{C}}-\overset{8}{C}$$

Now add hydrogen atoms to give each carbon atom four bonds. The complete formula is

$$CH_3CH{=}CHCH_2CH_2CH_2\underset{\displaystyle CH_3}{CH}CH_3$$

7-methyl-2-octene

line structure

Example 20.2

Name this compound:

$$CH_3CH_2CCH_2CH_2CH_3$$
$$\|$$
$$CH_2$$

SOLUTION

The longest carbon chain contains six carbons. However, since the compound is an alkene, we must include the double bond in the chain. The longest carbon chain containing the double bond has five carbons. Therefore, the compound is named as a pentene (Rules 1 and 2):

$$CH_3CH_2\overset{2}{C}\overset{3}{C}H_2\overset{4}{C}H_2\overset{5}{C}H_3$$
$$\|$$
$$\overset{1}{C}H_2$$

Attached to carbon 2 is an ethyl group. The name is 2-ethyl-1-pentene.

Practice 20.2

Write structural formulas for (a) 3-hexene, (b) 4-ethyl-2-heptene, (c) 3,4-dimethyl-2-pentene, and (d) 2-cyclohexyl-3-hexene.

Practice 20.3

Name these compounds:

(a) $CH_3C{=}CHCH_3$ with CH_2CH_3 attached to the second carbon

(b) $CH_3C{=}CHCHCH_3$ with CH_3 attached to the second carbon and CH_3 attached to the fourth carbon

Practice 20.4 Application to Biochemistry

(a) The human body accumulates cholesterol from the diet *and* from synthesis in the liver. This synthesis starts with a close relative of the following compound:

$$CH_2$$
$$\|$$
$$CH_3CCH_2CH_3$$

Name this compound.

(b) Cholesterol biosynthesis then forms an isomer similar to the following:

$$CH_3$$
$$|$$
$$CH_3C{=}CHCH_3$$

Name this compound.

Plant pigments contain many double bonds.

The common name for a compound is often more widely used than the IUPAC name. For example, "ethylene" is the common name for ethene. Ethylene is used to make many important products such as the plastic polyethylene and the major component of automobile antifreeze, ethylene glycol.

$$CH_2CH_2$$
$$|\quad|$$
$$OH\ OH$$

ethylene glycol

20.3 Geometric Isomerism in Alkenes

Only two dichloroethanes are known: 1,1-dichloroethane, $CHCl_2CH_3$, and 1,2-dichloroethane, CH_2ClCH_2Cl. But surprisingly there are three dichloroethenes—namely, 1,1-dichloroethene, $CCl_2{=}CH_2$, and *two* isomers of 1,2-dichloroethene, $CHCl{=}CHCl$. There is only one 1,2-dichloroethane because carbon atoms rotate freely about a single bond. Thus, the structural formulas I and II that follow represent the same compound. The chlorine atoms are simply shown in different conformations in the two formulas due to rotation of the CH_2Cl group about the carbon–carbon single bond:

$$\begin{array}{ccc} \begin{array}{ccccc} & Cl & & Cl & \\ & | & & | & \\ H- & C & \circlearrowleft & C & -H \\ & | & & | & \\ & H & & H & \end{array} & \qquad & \begin{array}{ccccc} & Cl & & H & \\ & | & & | & \\ H- & C & - & C & -H \\ & | & & | & \\ & H & & Cl & \end{array} \\ \text{I} & & \text{II} \end{array}$$

1,2-dichloroethane

Compounds containing a carbon–carbon double bond (pi bond) have restricted rotation about that double bond. Rotation cannot occur unless the pi bond breaks—a difficult process. Restricted rotation in a molecule gives rise to a type of isomerism known as *geometric isomerism*. Isomers that differ from each other only in the geometry of their molecules and not in the order of their atoms are known as **geometric isomers**. They are also called **cis-trans isomers**. Two isomers of 1,2-dichloroethene exist because of geometric isomerism.

geometric isomer

cis-trans isomer

For a further explanation, let's look at the geometry of an ethene molecule. This molecule is planar, or flat, with all six atoms lying in a single plane as in a rectangle:

$$\begin{array}{ccccc} H & & & & H \\ & \diagdown & & \diagup & \\ & & C{=}C & & \\ & \diagup & & \diagdown & \\ H & & & & H \end{array}$$

Since the hydrogen atoms are identical, only one structural arrangement is possible for ethene. But if one hydrogen atom on each carbon atom is replaced by chlorine, for example, two different geometric isomers are possible:

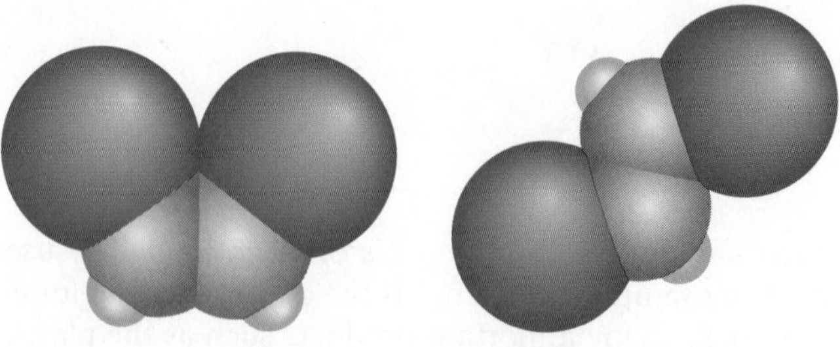

cis-1,2-dichloroethene *trans*-1,2-dichloroethene

```
Cl      Cl               H       Cl
  \    /                  \     /
   C=C          and        C=C
  /    \                  /     \
H       H               Cl       H
```

cis-1,2-dichloroethene (bp = 60.1°C) *trans*-1,2-dichloroethene (bp = 48.4°C)

Both of these isomers are known and have been isolated. The fact that they have different boiling points, as well as other different physical properties, is proof that they are not the same compound. Their spacefilling models look very different. Note that, in naming these geometric isomers, the prefix *cis-* is used to designate the isomer with the substituent groups (chlorine atoms) on the same side of the double bond, and the prefix *trans-* designates the isomer with the substituent groups on opposite sides of the double bond.

Molecules of *cis-* and *trans-*1,2-dichloroethene are not superimposable. That is, we cannot pick up one molecule and place it over the other in such a way that all the atoms in each molecule occupy the same relative positions in space. Nonsuperimposability is a general test for isomerism that all kinds of isomers must meet.

> An alkene shows cis–trans isomerism when each carbon atom of the double bond has two different kinds of groups attached to it:

```
a     a          a     b
 \   /            \   /
  C=C              C=C
 /   \            /   \
b     b          b     a
```

cis isomer *trans* isomer

An alkene does not show *cis–trans* isomerism if one carbon atom of the double bond has two identical groups attached to it:

```
H     H  \                      /  H     CH3
 \   /    \                    /    \   /
  C=C      } <-- Two groups --> {    C=C
 /   \    /      the same      \    /   \
H     H  /                      \  H     H
```

Thus, there are no geometric isomers of ethene or propene.

Four structural isomers of butene, C_4H_8, are known. Two of these, 1-butene and 2-methylpropene, do not show geometric isomerism.

```
CH3CH2      H  \                      /  CH3     H
      \    /    \                    /      \   /
       C=C       } <-- Two groups --> {      C=C
      /    \    /      the same      \      /   \
     H      H  /                      \  CH3     H
```

1-butene 2-methylpropene

The other two butenes are the cis–trans isomers shown here:

```
CH3   CH3             CH3    H
   \ /                   \  /
   C=C        and        C=C
   / \                   /  \
  H   H                 H    CH3
```

cis-2-butene *trans*-2-butene

```
H3C    CH2
    \\ //
     C
     |
  (cyclohexene ring)
     |
    CH3
```

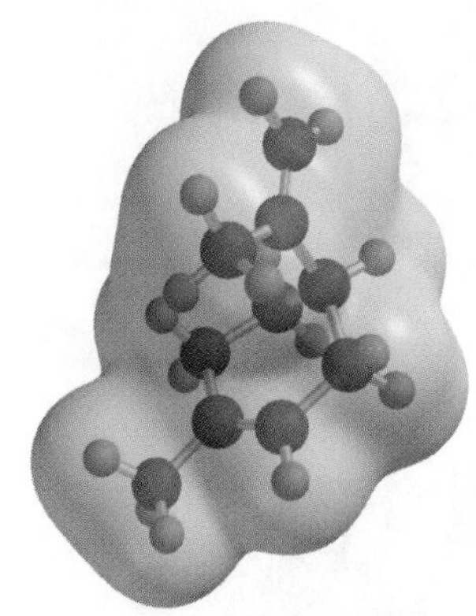

cis-1,2-dichloroethene

Aroma therapy uses many unsaturated plant compounds. For example, a clean, citrus odor is provided by limonene (oil of lemon).

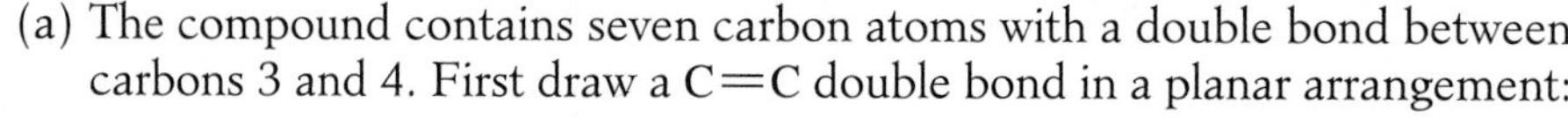

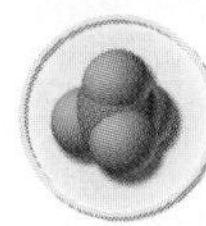

Example 20.3 Draw structures for (a) *trans*-3-heptene and (b) *cis*-5-chloro-2-hexene.

SOLUTION (a) The compound contains seven carbon atoms with a double bond between carbons 3 and 4. First draw a C=C double bond in a planar arrangement:

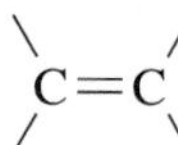

In the trans positions attach a two-carbon chain to one carbon atom and a three-carbon chain to the other carbon atom to give a continuous chain of seven carbon atoms:

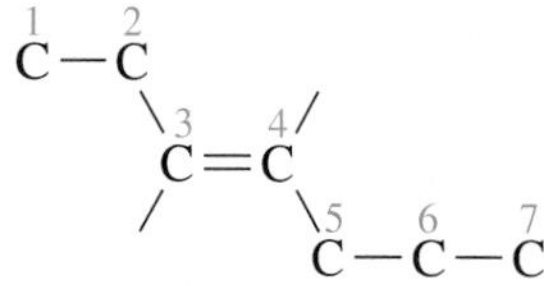

Now attach hydrogen atoms to give each carbon atom four bonds. Carbons 3 and 4 need only one hydrogen atom apiece; these two hydrogen atoms are also trans to each other. The structure is

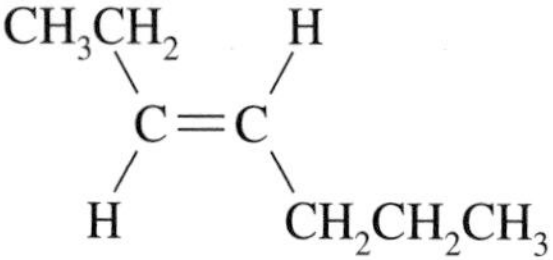

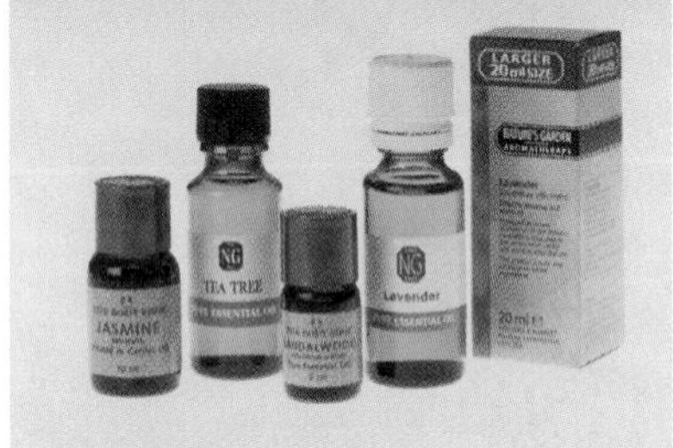

Essential oils are unsaturated organic compounds derived from plants. These oils have been used medicinally and in perfumes since antiquity.

(b) The compound contains six carbons, a double bond between carbons 2 and 3, and a Cl atom on carbon 5. First, draw the straight-chain compound:

$$\overset{1}{C}H_3\overset{2}{C}H=\overset{3}{C}H\overset{4}{C}H_2\overset{5}{C}HCl\overset{6}{C}H_3$$

Each carbon of the double bond has two different groups attached to it, a requirement of cis–trans isomerism. Now draw the C=C double bond in a planar arrangement:

```
\2  3/
 C=C
/    \
```

In the cis position, place the CH_3 group on carbon 2 and the $CH_2CHClCH_3$ group on carbon 3:

```
H3C       CH2CHClCH3
   \     /
    C=C
   /   \
```

Add H atoms to give each carbon atom four bonds. The structure is

```
H3C       CH2CHClCH3
   \     /
    C=C
   /   \
  H     H
```

Example 20.4

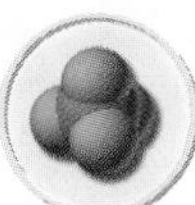

Draw structural formulas and names for all the isomers of pentene, C_5H_{10}. Identify all geometric isomers.

SOLUTION

Start by drawing the isomers of the five-carbon chain, placing the C=C in all possible positions. (There are two possible isomers: 1-pentene and 2-pentene.) Then proceed to the four-carbon chains with a methyl group branch chain. Locate the C=C and the CH_3 group in all possible positions. No three-carbon chain pentenes are possible. Check for duplications from the names of the compounds. The structures are:

$CH_2{=}CHCH_2CH_2CH_3$ — 1-pentene

$CH_3CH{=}CHCH_2CH_3$ — 2-pentene

```
   CH3
   |
CH2=CCH2CH3
```
2-methyl-1-butene

```
   CH3
   |
CH3C=CHCH3
```
2-methyl-2-butene

```
       CH3
       |
CH2=CHCHCH3
```
3-methyl-1-butene

Of these five compounds, only 2-pentene can have cis–trans isomers. The others have two identical groups on one carbon atom of the double bond. Draw the cis–trans isomers. The result should look like this:

```
H       H
 \     /
  C=C
 /     \
CH3     CH2CH3
```
cis-2-pentene

```
H       CH2CH3
 \     /
  C=C
 /     \
CH3     H
```
trans-2-pentene

Example 20.5

Is the compound below the cis or the trans isomer?

```
1
CH3      CH3
  \2   3/
   C=C
  /    \4  5
H       CH2CH3
```

SOLUTION

In branched-chain alkenes, the cis–trans designation is ordinarily given to the structure containing the longest continuous carbon chain that includes the carbon–carbon double bond. In this case, the longest chain is the 2-pentene, in which the methyl group on carbon 2 and the ethyl group on carbon 3 are trans to each other. Thus, the name of the compound shown is *trans*-3-methyl-2-pentene. The cis isomer is

```
CH3     CH2CH3
  \    /
   C=C
  /    \
H       CH3
```

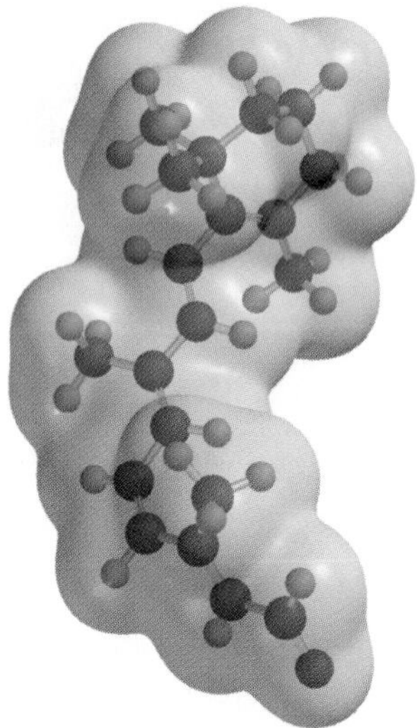

11-*cis*-retinal

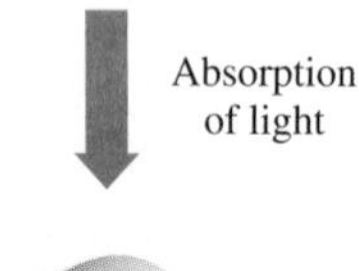

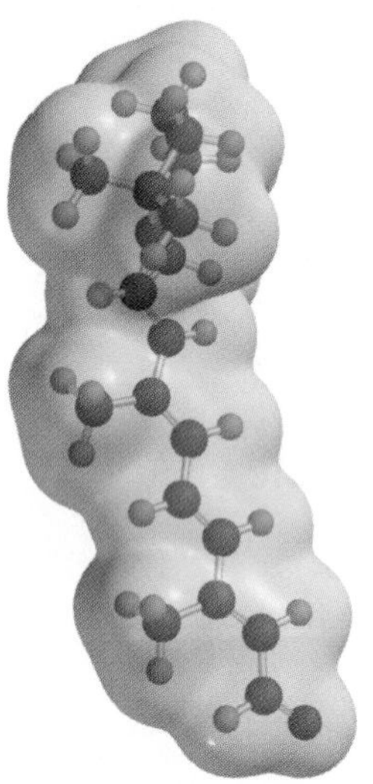

all-*trans*-retinal

Geometric isomerism allows our eyes to sense (see) light. Can you see the shape change (geometric isomerization) when our visual pigment, 11-*cis*-retinal absorbs light?

Practice 20.5

Draw structures (a) *cis*-3-methyl-2-pentene and (b) *trans*-2-bromo-2-pentene.

Practice 20.6

Determine whether geometric isomers exist for the following compounds: (a) 1-chloro-3-methyl-2-pentene (b) 3-hexene (c) 2,3-dimethyl-2-pentene Draw structures for the cis and trans isomers.

Practice 20.7

Identify each of the following as the cis or trans isomer:

```
(a)     H         CH3       (b) CH3CH2         CH2CH3
         \       /                    \       /
          C = C                        C = C
         /       \                    /       \
   CH3CH2         Cl               CH3         Br
```

Practice 20.8 Application to Biochemistry

Biochemical processes are very specific. (a) Only the trans isomer of the following two compounds can be used in the synthesis of fatty acids:

```
        H         H                CH3CH2CH2      H
         \       /                          \    /
          C = C                              C = C
         /       \                          /     \
CH3CH2CH2         C—SACP                   H       C—SACP
                  ||                               ||
                  O                                O

            I                                  II
```

(ACP means acyl carrier protein)

Which isomer, I or II, will be used in fatty acid synthesis?

(b) Only the *trans* isomer of the following two compounds is used for energy by the mitochondria:

```
  H        COOH       H        H
   \      /            \      /
    C = C               C = C
   /      \            /      \
COOH       H        COOH      COOH

     I                   II
```

Which isomer is used for energy, I or II?

Living cells are sensitive to, and respond to, specific isomers. For example, the eye responds to light only when a specific *cis* isomer is converted to its trans form. The molecule involved, retinal, is derived from vitamin A and β-carotene (found in vegetables such as carrots). One of the double bonds in retinal is converted to the *cis* form. The structure is as follows:

```
CH3   CH3  H      CH3    H   H       H
   \  /    |      |      |   |       |   H
    C      C      C      C = C       C   |
   / \   /   \\  /  \\  /      \   // \  |
CH2    C       C      C          C     C = O
|      ||      |      |          |
CH2    C       H      H          CH3
   \  / \
   CH2   CH3
```

11-*cis*-retinal — *cis* double bond

The 11-*cis*-retinal is combined with a protein (opsin) to form the visual pigment, rhodopsin. When light is absorbed, the cis double bond changes to a *trans* double bond:

all-*trans*-retinal — *trans* double bond

The isomerization of 11-*cis*-retinal to all-*trans*-retinal starts a cascade of events that our brains interpret as "seeing."

Many compounds have more than one carbon–carbon double bond. Compounds with two (*di*) double bonds are called *dienes*; those with three (*tri*) double bonds are *trienes*, and so forth. The location of the double bonds are numbered like in alkenes. One notable example is 2-methyl-1,3-butadiene, the component of natural rubber.

Study the names and structures of the following compounds:

$$\overset{1}{C}H_2{=}\overset{2}{C}H{-}\overset{3}{C}H{=}\overset{4}{C}H_2$$

1,3-butadiene

Position of the double bonds — Four carbon atoms — Inserted to ease pronunciation — Two double bonds

$$CH_2{=}CH{-}CH_2{-}CH{=}CH_2$$

1,4-pentadiene

$$CH_2{=}\underset{}{\overset{CH_3}{\overset{|}{C}}}{-}CH{=}CH_2$$

2-methyl-1,3-butadiene (isoprene)

Dienes are components of both natural and synthetic rubbers, essential ingredients in tire manufacturing.

20.4 Cycloalkenes

cycloalkene

As the name implies, **cycloalkenes** are cyclic compounds that contain a carbon–carbon double bond in the ring. The two most common cycloalkenes are cyclopentene and cyclohexene. The double bond may be placed between any two carbon atoms. Note the simplicity of the line structures.

The general formula for cyclohexenes is C_nH_{2n-2}

cyclopentene (C_5H_8)

cyclohexene (C_6H_{10})

In cycloalkenes the carbons of the double bond are assigned numbers 1 and 2. Thus, the positions of the double bond need not be indicated in the name of the compound. Other substituents on the ring are named in the usual manner, and their positions on the ring are indicated with the smallest possible numbers.

CHEMISTRY IN ACTION • An Ancient Aromatherapy

Frankincense and *myrrh*—the words evoke images of musty tombs, incense-filled temples, and ancient kings. Frankincense and myrrh have been literally worth their weight in gold. They are produced in the Arabian peninsula and the Horn of Africa; in ancient times, Arabian princes became rich by shipping the spices as far as Europe to the west and China to the east. Why are frankincense and myrrh prized so richly? The answer to this question lies with the evocative nature of these spices.

Frankincense and myrrh come from the sap of small trees or bushes of the botanical family Burseraceae. When the tree's bark is cut, sap exudes and forms drops of dried resin, "tears." These tears are collected and sorted by hand; high-quality frankincense sells for about $100 per pound today. The hardened resin carries with it the smell of the plant, and it is the smell that carries the evocative mood.

These resins contain essential oils, large unsaturated organic compounds from plants. Airborne oil molecules bind to our olfactory receptors, and we recognize the smell of individual plants. Recognition can be very specific; thus, we might be able to distinguish a tangerine from a mandarin orange or a clementine based on smell alone. Yet smell alone doesn't account for the value of frankincense and myrrh.

A flexible light panel sold in commercial production.

Frankincense and myrrh are prized because their aroma elicits specific emotions. We are all familiar with evocative smells,—the smell of a turkey roasting at Thanksgiving or that "new car smell." In the aftermath of the World Trade Center disaster, people remembered a smell that was described by a journalist covering the scene (Stefan Fatsis) as "the scent of devastation." Yet smells unconnected with specific experiences also can change our emotional state. The aromas of frankincense and myrrh are examples.

Such aromas serve a therapeutic use in aromatherapy. Over the past 20 years scientists have rigorously studied aromatherapy. Animal studies show that aromas can have both stimulant and sedative effects. Since the mid-1990s, carefully designed human studies have been undertaken. The National Cancer Institute reports that essential oils can relieve depression and create a better emotional environment for cancer patients under some conditions. In the near future, scientists hope to better define aromatherapy and improve its utility in clinical settings.

Frankincense and myrrh have provided "aromatherapy" for thousands of years. Emotions they elicit have been variously described as "calming," "sweetly sad," or "meditative." These aroma–bearing resins may be worth more than gold. While gold can buy a palace, frankincense and myrrh bring the sense of living richly.

In the following examples, note that a Cl or a CH_3 has replaced one H atom in the molecule:

3-chlorocyclohexene	1-methylcyclopentene	1, 3-dimethylcyclohexene
(C_6H_9Cl)	($C_5H_7CH_3$)	[$C_6H_8(CH_3)_2$]

The ring is numbered either clockwise or counterclockwise starting with the carbon–carbon double bond so that the substituted group(s) have the smallest possible numbers.

Cycloalkenes can have more than one double bond. For example,

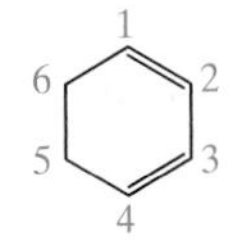

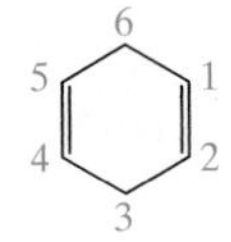

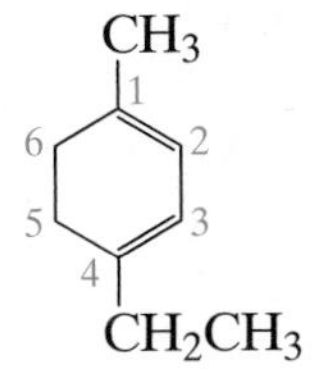

1,3-cyclohexadiene 1,4-cyclohexadiene 4-ethyl-1-methyl-1,3-cyclohexadiene

The names of these compounds must include the prefix *cyclo-*, the location of the double bonds in the ring, and the number of double bonds, such as *diene* or *triene*.

Practice 20.9

Name the following compounds:

(a) 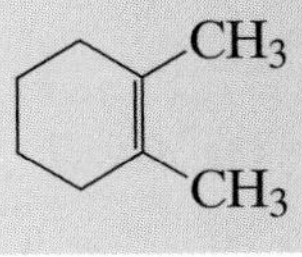 (b) 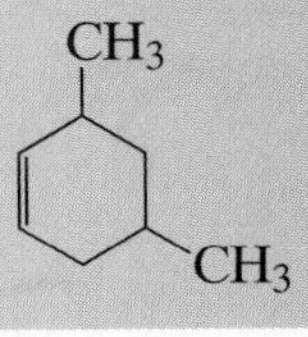 (c)

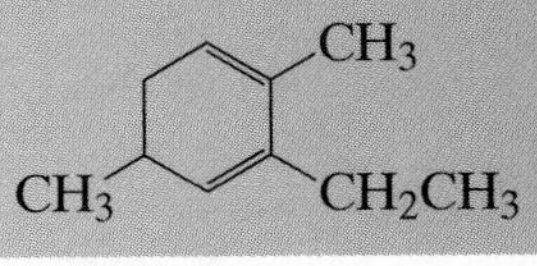

No cis–trans designation is necessary for cycloalkenes that contain up to seven carbon atoms in the ring. Cyclooctene has been shown to exist in both cis and trans forms.

20.5 Preparation and Physical Properties of Alkenes

Common preparation methods for alkenes start with saturated organic molecules; each carbon in these molecules is bonded to four other atoms. Then, to form double bonds, atoms must be removed. The product contains double-bonded carbons that are connected to only three atoms each. Alkene synthesis commonly means "getting rid" of some atoms—an elimination reaction. Watch for evidence of elimination in the following two examples of alkene preparation.

Cracking

Ethene can be produced by the cracking of petroleum. **Cracking**, or pyrolysis, is the process in which saturated hydrocarbons are heated to very high temperatures in the presence of a catalyst (usually silica–alumina):

cracking

$$\text{alkane } (C_nH_{2n+2}) \xrightarrow[\text{catalyst}]{\Delta} \text{mixture of alkenes} + \text{alkanes} + \text{hydrogen gas}$$

$$2\ CH_3CH_2CH_3 \xrightarrow{\sim 500°C} CH_3CH{=}CH_2 + CH_2{=}CH_2 + CH_4 + H_2$$

This results in large molecules breaking into smaller ones, with the elimination of hydrogen, to form alkenes and small hydrocarbons like methane and ethane. Unfortunately, cracking always results in mixtures of products and is therefore not used often in the laboratory.

Dehydration of Alcohols

Dehydration involves the elimination of a molecule of water from a reactant molecule. Dehydration reactions are very common in organic chemistry as well as in biochemistry.

dehydration

To produce an alkene by dehydration, an alcohol is heated in the presence of concentrated sulfuric acid:

Dehydration is an elimination reaction

$$CH_3\overset{H}{\underset{H}{C}}-\overset{H}{\underset{OH}{C}}CH_3 \xrightarrow[\Delta]{\text{conc. } H_2SO_4} CH_3\overset{H}{C}=\overset{H}{C}CH_3 + H_2O$$

The alkene is formed as a result of elimination of H and OH (shown in blue) from adjacent carbon atoms.

Practice 20.10 Application to Biochemistry

Living cells use common organic reactions such as alcohol dehydration. In the mitochondria, citric acid is dehydrated to form an alkene-containing compound, aconitic acid:

$$HO-\overset{CH_2-COOH}{\underset{CH_2-COOH}{C}}-COOH \longrightarrow \text{Aconitic acid}$$

citric acid

Draw the *cis* and *trans* structures of this acid.

Physical Properties

Alkenes have physical properties very similar to the corresponding alkanes. This is not surprising, since the difference between an alkane and an alkene is simply two hydrogen atoms.

General formula for alkanes	C_nH_{2n+2}
General formula for alkenes	C_nH_{2n}
General formula for alkynes and cycloalkenes	C_nH_{2n-2}

Because alkenes have slightly smaller molar masses, their boiling points are slightly lower than the corresponding alkanes. The smaller alkenes (to 5 carbons) are gases at room temperature. As the chain lengthens (5–17 carbons), the alkenes are liquid, and above 17 carbons they are solid. The alkenes are nonpolar, like the other hydrocarbons, and so are insoluble in water, but soluble in organic solvents. The densities of most alkenes are much less than water. Table 20.1 shows the properties of some alkenes. Notice that isomers (C_4H_8) have similar

Table 20.1 Physical Properties of Alkenes

Molecular formula	Structural formula	IUPAC name	Density (g/mL)	Melting point (°C)	Boiling point (°C)
C_2H_4	$CH_2=CH_2$	Ethene	—	−169	−104
C_3H_6	$CH_3CH=CH_2$	Propene	—	−185	−48
C_4H_8	$CH_3CH_2CH=CH_2$	1-Butene	0.595	−185	−6
C_4H_8	$(CH_3)_2C=CH_2$	2-Methylpropene	0.594	−14	−7
C_5H_{10}	$CH_3(CH_2)_2CH=CH_2$	1-Pentene	0.641	−138	30

boiling points, although they differ significantly in melting points. Melting-point differences occur because the isomers have different shapes and therefore fit into their crystalline structures in significantly different ways.

20.6 Chemical Properties of Alkenes

What type of reaction might be expected for an alkene (or an alkyne)? Note that ethene ($CH_2{=}CH_2$) has only three atoms bonded to each carbon; acetylene ($CH{\equiv}CH$) has only two atoms bonded to each carbon. Both alkenes and alkynes have fewer than the maximum of four atoms bonded per carbon. These molecules are more reactive than the corresponding alkanes and readily undergo addition reactions.

Alkene chemistry is especially important in the biochemistry of fats.

Addition

Addition at the carbon–carbon double bond is the most common reaction of alkenes. Hydrogen, halogens (Cl_2 or Br_2), hydrogen halides, sulfuric acid, and water are some of the reagents that can be added to unsaturated hydrocarbons. Ethene, for example, reacts in the presence of a platinum catalyst in this fashion:

The double bond is broken, and unsaturated alkene molecules become saturated by an addition reaction.

$$\underset{\text{ethene}}{CH_2{=}CH_2} + H_2 \xrightarrow[\text{1 atm}]{\text{Pt, 25°C}} \underset{\text{ethane}}{CH_3{-}CH_3}$$

$$\underset{\text{colorless}}{CH_2{=}CH_2} + \underset{\text{reddish brown}}{Br{-}Br} \longrightarrow \underset{\substack{\text{1,2-dibromoethane}\\ \text{colorless}}}{CH_2Br{-}CH_2Br} \quad (Cl_2 \text{ reacts similarly})$$

The disappearance of the reddish-brown bromine color provides visible evidence of reaction. Other reactions of ethene include the following:

$$CH_2{=}CH_2 + HCl \longrightarrow \underset{\substack{\text{chloroethane}\\ \text{(ethyl chloride)}}}{CH_3CH_2Cl}$$

$$CH_2{=}CH_2 + \underset{\text{sulfuric acid}}{HOSO_3H\,(conc.)} \longrightarrow \underset{\text{ethyl hydrogen sulfate}}{CH_3CH_2OSO_3H}$$

$$CH_2{=}CH_2 + HOH \xrightarrow{H^+} \underset{\substack{\text{ethanol}\\ \text{(ethyl alcohol)}}}{CH_3CH_2OH}$$

Bromine has the characteristic orange color (left), but when added to an alkene the color disappears (right).

The H^+ indicates that the reaction is carried out under acidic conditions.

Unsaturated alkenes are readily converted to saturated molecules by addition. Also, note that addition is the reverse of elimination. For example, by the addition of water, an alkene can be converted to an alcohol, and by the elimination of water, an alcohol can be converted to an alkene.

The preceding examples dealt with ethene, but reactions of this kind can be made to occur on almost any molecule that contains a carbon–carbon double bond. If a symmetrical molecule such as Cl_2 is added to propene, only one product, 1,2-dichloropropane, is formed:

$$CH_2{=}CH{-}CH_3 + Cl_2 \longrightarrow \underset{\text{1,2-dichloropropane}}{CH_2Cl{-}CHCl{-}CH_3}$$

But if an unsymmetrical molecule such as HCl is added to propene, two products are theoretically possible, depending upon which carbon atom adds the hydrogen. The two possible products are 1-chloropropane and 2-chloropropane. Experimentally, we find that 2-chloropropane is formed almost exclusively:

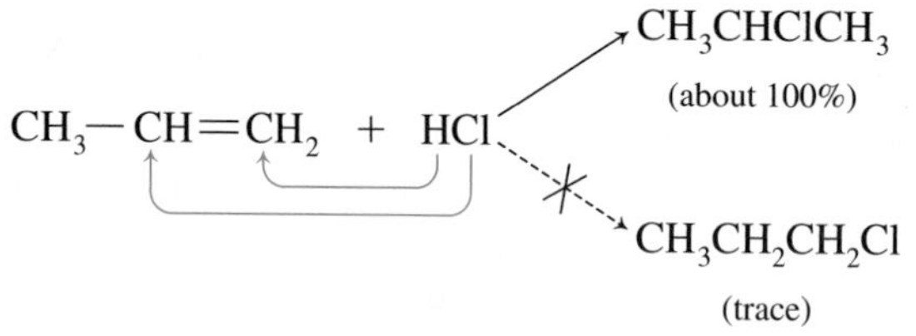

$$CH_3{-}CH{=}CH_2 + HCl \longrightarrow \underset{\text{(about 100\%)}}{CH_3CHClCH_3}$$

$$CH_3{-}CH{=}CH_2 + HCl \not\longrightarrow \underset{\text{(trace)}}{CH_3CH_2CH_2Cl}$$

A single product is obtained because the reaction follows specific steps. The sum of these steps is termed a *reaction mechanism*. The reaction between propene and HCl proceeds by the following addition mechanism:

1. A proton (H^+) from HCl bonds to carbon 1 of propene by utilizing the pi bond electrons. The intermediate formed is a positively charged alkyl group, or carbocation. The positive charge is localized on carbon 2 of this carbocation. Their reaction is diagrammed as follows:

$$\overset{1}{C}H_2{=}\overset{2}{C}H{-}\overset{3}{C}H_3 + HCl \longrightarrow \underset{\text{isopropyl carbocation}}{CH_3{-}\overset{+}{C}H{-}CH_3} + Cl^-$$

2. The chloride ion then adds to the positively charged carbon atom to form a molecule of 2-chloropropane:

$$CH_3{-}\overset{+}{C}H{-}CH_3 + Cl^- \longrightarrow \underset{\text{2-chloropropane}}{CH_3{-}CHCl{-}CH_3}$$

carbocation

An ion in which a carbon atom has a positive charge is known as a **carbocation**. There are four types of carbocations: methyl, primary (1°), secondary (2°), and tertiary (3°). Examples of these four types are as follows:

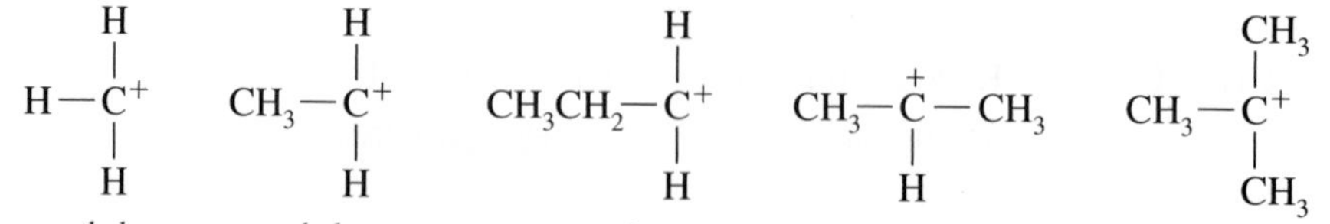

H_3C^+	$CH_3{-}CH_2^+$	$CH_3CH_2{-}CH_2^+$	$CH_3{-}\overset{+}{C}H{-}CH_3$	$(CH_3)_3C^+$
methyl carbocation	ethyl carbocation (primary)	propyl carbocation (primary)	isopropyl carbocation (secondary)	*t*-butyl carbocation (tertiary)

A carbon atom is designated as primary if it is bonded to one carbon atom, secondary if it is bonded to two carbon atoms, and tertiary if it is bonded to three carbon atoms. Thus, in a primary carbocation, the positive carbon atom is bonded to only one carbon atom. In a secondary carbocation, the positive carbon atom is bonded to two carbon atoms. In a tertiary carbocation, the positive carbon atom is bonded to three carbon atoms. There are no quaternary carbocations.

The order of stability of carbocations and hence the ease with which they are formed is tertiary > secondary > primary:

Stability of carbocations: $3° > 2° > 1° > \overset{+}{C}H_3$

The addition also depends on the electronegativities of the elements in the adding molecule. In hydrogen chloride, Cl is more electronegative than H, resulting in $H^{\delta+}Cl^{\delta-}$ molecules. Thus, in the reaction of propene and HCl, a proton adds to propene to give the more stable isopropyl carbocation (secondary) as an intermediate in preference to n-propyl carbocation (primary).

In the middle of the 19th century, a Russian chemist, V. Markovnikov, observed reactions of this kind, and in 1869 he formulated a useful generalization now known as **Markovnikov's rule**. This rule in essence states:

Markovnikov's rule

> When an unsymmetrical molecule such as HX(HCl) adds to a carbon–carbon double bond, the hydrogen from HX goes to the carbon atom that has the greater number of hydrogen atoms.

As you can see, the addition of HCl to propene, discussed previously, follows Markovnikov's rule. The addition of HI to 2-methylpropene is another example illustrating this rule:

$$CH_3-\overset{\overset{\displaystyle CH_3}{|}}{C}=CH_2 + HI \longrightarrow CH_3-\underset{\underset{\displaystyle I}{|}}{\overset{\overset{\displaystyle CH_3}{|}}{C}}-CH_3$$

2-iodo-2-methylpropane
(*tert*-butyl iodide)

In this reaction, the $=CH_2$ is the alkene carbon with the most H atoms (two); it thus adds the H^+, followed by the addition of I^- to the other alkene carbon atom.

General rules of this kind help us predict the products of reactions; however, exceptions are known for most such rules.

Example 20.6

Write formulas for the organic products formed when 2-methyl-1-butene reacts with (a) H_2, Pt/25°C; (b) Cl_2; (c) HCl; and (d) H_2O, H^+.

SOLUTION

First write the formula for 2-methyl-1-butene:

$$\overset{1}{C}H_2=\overset{\overset{\displaystyle CH_3}{|}}{\overset{2}{C}}-\overset{3}{C}H_2-\overset{4}{C}H_3$$

(a) The double bond is broken when a hydrogen molecule adds. One H atom adds to each carbon atom of the double bond. Platinum, Pt, is a necessary catalyst in this reaction. The product is

$$CH_3\overset{\overset{\displaystyle CH_3}{|}}{C}HCH_2CH_3$$

2-methylbutane

(b) The Cl_2 molecule adds to the carbons of the double bond. One Cl atom adds to each carbon atom of the double bond. The product is

$$\begin{array}{l} \quad\;\; CH_3 \\ \quad\;\;\, | \\ CH_2CCH_2CH_3 \\ \;|\quad\;\, | \\ Cl\;\; Cl \end{array}$$

1,2-dichloro-2-methylbutane

(c) HCl adds to the double bond according to Markovnikov's rule. The H^+ goes to carbon 1 (the more stable 3° carbocation is formed as an intermediate product), and the Cl^- goes to carbon 2. The product is

$$\begin{array}{l} \quad\;\; CH_3 \\ \quad\;\;\, | \\ CH_3CCH_2CH_3 \\ \quad\;\;\, | \\ \quad\;\; Cl \end{array}$$

2-chloro-2-methylbutane

(d) The net result of this reaction is a molecule of water added across the double bond. The H adds to carbon 1 (the carbon with the greater number of hydrogen atoms), and the OH adds to carbon 2 (the carbon of the double bond with the lesser number of hydrogen atoms). The product is

$$\begin{array}{l} \quad\;\; CH_3 \\ \quad\;\;\, | \\ CH_3CCH_2CH_3 \\ \quad\;\;\, | \\ \quad\;\; OH \end{array}$$

2-methyl-2-butanol
(an alcohol)

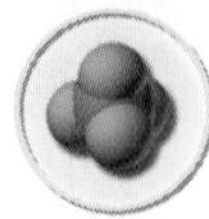

Example 20.7 Write equations for the addition of HCl to (a) 1-pentene and (b) 2-pentene.

SOLUTION (a) In the case of 1-pentene, $CH_3CH_2CH_2CH{=}CH_2$, the proton from HCl adds to carbon 1 to give the more stable secondary carbocation, followed by the addition of Cl^- to give the product 2-chloropentane. The addition is directly in accordance with Markovnikov's rule:

$$CH_3CH_2CH_2CH{=}CH_2 \xrightarrow{HCl} \underset{2^\circ \text{ carbocation}}{CH_3CH_2CH_2\overset{+}{C}HCH_3} \xrightarrow{Cl^-} \underset{\text{2-chloropentane}}{CH_3CH_2CH_2CHClCH_3}$$

(b) In 2-pentene, $CH_3CH_2CH{=}CHCH_3$, each carbon of the double bond has one H atom, and the addition of a proton to either one forms a secondary carbocation. After the addition of Cl^-, the result is two isomeric products that are formed in almost equal quantities:

$$CH_3CH_2CH{=}CHCH_3 \xrightarrow{HCl} CH_3CH_2CH_2\overset{+}{C}HCH_3 \xrightarrow{Cl^-} \underset{\text{2-chloropentane}}{CH_3CH_2CH_2CHClCH_3}$$

$$CH_3CH_2CH{=}CHCH_3 \xrightarrow{HCl} CH_3CH_2\overset{+}{C}HCH_2CH_3 \xrightarrow{Cl^-} \underset{\text{3-chloropentane}}{CH_3CH_2CHClCH_2CH_3}$$

Practice 20.11

Write formulas for the organic products formed when 3-methyl-2-pentene reacts with (a) H_2/Pt; (b) Br_2; (c) HCl; and (d) H_2O, H^+.

Practice 20.12

Write equations for (a) addition of water to 1-methylcyclopentene and (b) addition of HI to 2-methyl-2-butene.

Practice 20.13 Application to Biochemistry

The nutritionally essential amino acid, threonine, can be made by hydration of the following alkene:

$$\begin{array}{c} CH_2 \\ \| \\ CH \\ | \\ NH_2-CH-COOH \end{array}$$

The hydration follows Markovnikov's rule. Show the structure of the alcohol formed; that is, show the structure of threonine.

Oxidation

Another typical reaction of alkenes is oxidation at the double bond. For example, when shaken with a cold, dilute solution of potassium permanganate, $KMnO_4$, an alkene is converted to a glycol (glycols are dihydroxy alcohols). Ethene reacts in this manner:

$$\underset{\substack{\text{ethene} \\ \text{(ethylene)}}}{CH_2{=}CH_2} + \underset{\text{(purple)}}{KMnO_4(aq)} + H_2O \longrightarrow \underset{\substack{\text{1,2 ethanediol} \\ \text{(ethylene glycol)}}}{\begin{array}{cc} CH_2- & CH_2 \\ | & | \\ OH & OH \end{array}} + \underset{\text{(brown)}}{MnO_2} + KOH$$

The *Baeyer test* makes use of this reaction to detect or confirm the presence of double (or triple) bonds in hydrocarbons. Evidence of reaction (positive Baeyer test) is the disappearance of the purple color of permanganate ions. The Baeyer test is not specific for detecting unsaturation in hydrocarbons because other classes of compounds may also give a positive Baeyer test.

Carbon–carbon double bonds are found in many different kinds of molecules. Most of these substances react with potassium permanganate and undergo somewhat similar reactions with other oxidizing agents, including oxygen in the air, and especially, with ozone. Such reactions are frequently troublesome. For example, premature aging and cracking of automobile tires in smoggy atmospheres occur because ozone attacks the double bonds in rubber molecules. Cooking oils and fats sometimes develop disagreeable odors and flavors because the oxygen of the air reacts with the double bonds present in these materials. Potato chips, because of their large surface area, are particularly subject to flavor damage caused by oxidation of the unsaturated cooking oils that they contain.

Antioxidants like vitamin E protect living cells from similar damage.

20.7 Alkynes: Nomenclature and Preparation

Nomenclature

IUPAC Rules for Naming Alkynes

The rules for naming alkynes are the same as those for alkenes, but the ending *-yne* is used to indicate the presence of a triple bond. Table 20.2 lists names and formulas for some common alkynes.

Practice 20.14

Write formulas for

(a) 3-methyl-1-butyne
(b) 4-ethyl-4-methyl-2-hexyne
(c) cyclohexylethyne

Preparation

Although triple bonds are very reactive, it is relatively easy to synthesize alkynes. Acetylene can be prepared inexpensively from calcium carbide and water:

$$CaC_2 + 2\,H_2O \longrightarrow HC{\equiv}CH + Ca(OH)_2$$

Acetylene is also prepared by the cracking of methane in an electric arc:

$$2\,CH_4 \xrightarrow{1500^\circ C} HC{\equiv}CH + 3\,H_2$$

Table 20.2 Nomenclature for Some Common Alkynes

The general formula for alkynes is C_nH_{2n-2}.

Molecular formula	Structural formula	IUPAC name
C_2H_2	$H{-}C{\equiv}C{-}H$	Ethyne*
C_3H_4	$CH_3{-}C{\equiv}C{-}H$	Propyne
C_4H_6	$CH_3CH_2{-}C{\equiv}C{-}H$	1-Butyne
C_4H_6	$CH_3{-}C{\equiv}C{-}CH_3$	2-Butyne

*Ethyne is commonly known as acetylene.

20.8 Physical and Chemical Properties of Alkynes

Physical Properties

Alkynes are only rarely found in living organisms.

Acetylene is a colorless gas, with little odor when pure. The disagreeable odor we associate with it is the result of impurities (usually PH_3). Acetylene is partially water soluble (one volume of acetylene gas dissolves in one volume of water) and is a gas at normal temperature and pressure (bp = −84°C).

As a liquid, acetylene is very sensitive and may decompose violently (explode), either spontaneously or from a slight shock. The reaction is as follows:

$$HC{\equiv}CH \longrightarrow H_2 + 2\,C + 227\text{ kJ } (54.3\text{ kcal})$$

To eliminate the danger of explosions, acetylene is dissolved under pressure in acetone and is packed in cylinders that contain a porous inert material.

Chemical Properties

Acetylene is used mainly (1) as fuel for oxyacetylene cutting and welding torches and (2) as an intermediate in the manufacture of other substances. Both uses are dependent upon the great reactivity of acetylene. Acetylene and oxygen mixtures produce flame temperatures of about 2,800°C. Acetylene readily undergoes addition reactions rather similar to those of ethene. It reacts with chlorine and bromine, and decolorizes a permanganate solution (Baeyer's test). Either one or two molecules of bromine or chlorine can be added:

$$HC{\equiv}CH + Br_2 \longrightarrow \underset{\text{1,2-dibromoethene}}{CHBr{=}CHBr}$$

or

$$HC{\equiv}CH + 2\,Br_2 \longrightarrow \underset{\text{1,1,2,2-tetrabromoethane}}{CHBr_2{-}CHBr_2}$$

It is apparent that either unsaturated or saturated compounds can be obtained as addition products of acetylene. Often, unsaturated compounds capable of undergoing further reactions are made from acetylene. For example, vinyl chloride, which is used to make the plastic polyvinyl chloride (PVC), can be made by simple addition of HCl to acetylene:

$$CH_2{\equiv}CH + HCl \longrightarrow \underset{\substack{\text{chloroethene}\\ \text{(vinyl chloride)}}}{CH_2{=}CHCl}$$

(*Note*: The common name for the $CH_2{=}CH{-}$ group is *vinyl*.) If the reaction is not properly controlled, another HCl adds to the chloroethene, in accordance with Markovinkov's rule.

$$CH_2{=}CHCl + HCl \longrightarrow \underset{\text{1,1-dichloroethane}}{CH_3CHCl_2}$$

Addition reactions are common for all alkynes.

Hydrogen chloride reacts with other alkynes in a similar fashion to form substituted alkenes. The addition follows Markovnikov's rule. Alkynes can react with 1 or 2 mol of HCl. Consider the reaction of propyne with HCl:

$$CH_3C{\equiv}CH + HCl \longrightarrow \underset{\text{2-chloropropene}}{CH_3CCl{=}CH_2}$$

$$CH_3CCl{=}CH_2 + HCl \longrightarrow \underset{\text{2,2-dichloropropane}}{CH_3CCl_2CH_3}$$

Superabsorbants used in disposable diapers increase the amount of liquid the diaper will absorb without leaking.

Certain reactions are unique to alkynes. Alkynes are capable of reacting at times when alkenes will not. Acetylene, with certain catalysts, reacts with HCN to form $CH_2{=}CHCN$ (acrylonitrile). This chemical is used industrially to manufacture Orlon, a polymer commonly found in clothing. It is also used to form the superabsorbants, which are capable of retaining up to 2,000 times their mass of water. These superabsorbants are used in disposable diapers as well as in soil additives to retain water.

The reactions of other alkynes resemble those of acetylene. Although many other alkynes are known, acetylene is by far the most important industrially.

Practice 20.15

Write formulas for the products formed by the reaction of HCl with (a) 1-butyne and (b) 2-pentyne to form saturated compounds.

20.9 Aromatic Hydrocarbons: Structure

Benzene and all substances with structures and chemical properties that resemble benzene are classified as *aromatic compounds*. The word *aromatic* originally referred to the rather pleasant odor possessed by many of these substances, but this meaning has been dropped. Benzene, the parent substance of the aromatic hydrocarbons, was first isolated by Michael Faraday in 1825; its correct molecular formula, C_6H_6, was established a few years later. The establishment of a reasonable structural formula that would account for the properties of benzene was a difficult problem for chemists in the mid-19th century.

Finally, in 1865, August Kekulé proposed that the carbon atoms in a benzene molecule are arranged in a six-membered ring with one hydrogen atom bonded to each carbon atom and with three carbon–carbon double bonds, as shown in the following structure:

```
          H
          |
 H     ,C=      H
   \C/    \C/
    ||     |
    C      C
 H/   \C=/   \H
       |
       H
```

Kekulé soon realized that there should be two dibromobenzenes, based on double- and single-bond positions relative to the two bromine atoms:

```
          Br                              Br
          |                               |
 H     ,C=      Br              H      =C      Br
   \C/    \C/            and      \C/    \C/
    ||     |                       |      ||
    C      C                       C      C
 H/   \C=/   \H                 H/  \=C/    \H
       |                              |
       H                              H
```

Since only one dibromobenzene (with bromine atoms on adjacent carbons) could be produced, Kekulé suggested that the double bonds are in rapid oscillation within the molecule. He therefore proposed representing the structure of benzene in this fashion:

```
          H                               H
          |                               |
 H     ,C=      H               H      =C      H
   \C/    \C/                     \C/    \C/
    ||     |          ⇌            |      ||
    C      C                       C      C
 H/   \C=/   \H                 H/  \=C/    \H
       |                              |
       H                              H
```

Kekulé's concepts are a landmark in the history of chemistry. They are the basis of the best representation of the benzene molecule devised in the 19th century, and they mark the beginning of our understanding of structure in aromatic compounds.

Kekulé's formulas have one serious shortcoming: They represent benzene and related substances as highly unsaturated compounds. Yet benzene does not react like a typical alkene; it does not decolorize bromine solutions rapidly, nor does it destroy the purple color of permanganate ions (Baeyer's test). Instead, benzene behaves chemically like a typical alkane. Its reactions are usually the substitution type, wherein a hydrogen atom is replaced by some other group; for example,

$$C_6H_6 + Cl_2 \xrightarrow{Fe} C_6H_5Cl + HCl$$

This problem was not fully resolved until the technique of X-ray diffraction, developed in the years following 1912, permitted chemists to determine the actual distances between the nuclei of carbon atoms in molecules.

The center-to-center distances between carbon atoms in different kinds of hydrocarbon molecules are

Ethane (single bond)	0.154 nm
Ethene (double bond)	0.134 nm
Benzene	0.139 nm

Because only one carbon–carbon distance (bond length) is found in benzene, it is apparent that alternating single and double bonds do not exist in the benzene molecule.

Modern theory accounts for the structure of the benzene molecule in this way: The orbital hybridization of the carbon atoms is sp^2. (See the structure of ethene in Section 20.1.) A planar hexagonal ring is formed by the overlapping of two sp^2 orbitals on each of six carbon atoms. The other sp^2 orbital on each carbon atom overlaps an s orbital of a hydrogen atom, bonding the carbon to the hydrogen by a sigma bond. The remaining six p orbitals, one on each carbon atom, overlap each other and form doughnut-shaped pi electron clouds above and below the plane of the ring.*(See Figure 20.5.) The electrons composing these clouds are not attached to particular carbon atoms, but are delocalized and associated with the entire molecule. This electronic structure imparts unusual stability to benzene and is responsible for many of the characteristic properties of aromatic compounds. Because of this stable electronic structure, benzene does not readily undergo addition or elimination reactions, but it does undergo substitution reactions.

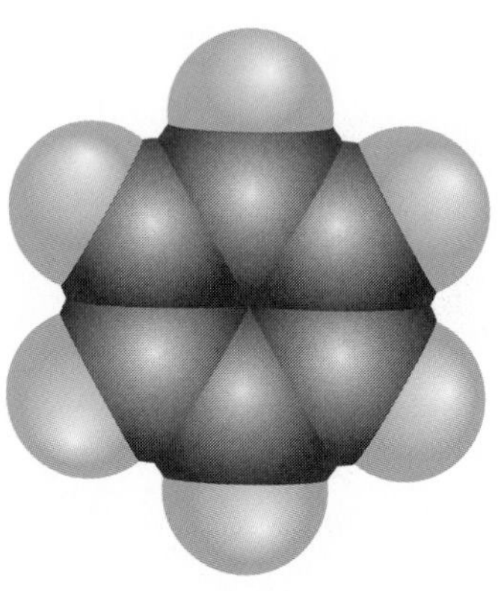

Spacefilling model of benzene.

For convenience, present-day chemists usually write the structure of benzene as one or the other of these abbreviated forms:

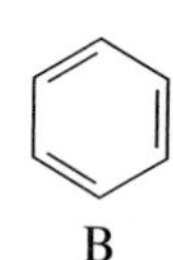

A B C

In all of these representations, it is understood that there is a carbon atom and a hydrogen atom at each corner of the hexagon. The classical Kekulé structures are represented by formulas A and B. However, neither of these Kekulé structures actually exists, although in some circumstances they are still used. The real benzene molecule is a hybrid of these structures and is commonly represented by formula C. The circle indicates the special nature of the benzene pi bonds, as shown in Figure 20.5. We will use the hexagon with the circle to represent a benzene ring.

Hexagons are used to represent the structural formulas of benzene derivatives—that is, substances in which one or more hydrogen atoms in the ring have been replaced by other atoms or groups. Chlorobenzene, for example, is written in this fashion:

Cl

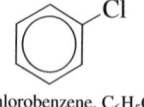

chlorobenzene, C_6H_5Cl

This notation indicates that the chlorine atom has replaced a hydrogen atom and is bonded directly to a carbon atom in the ring. Thus, the correct formula for chlorobenzene is C_6H_5Cl, not C_6H_6Cl.

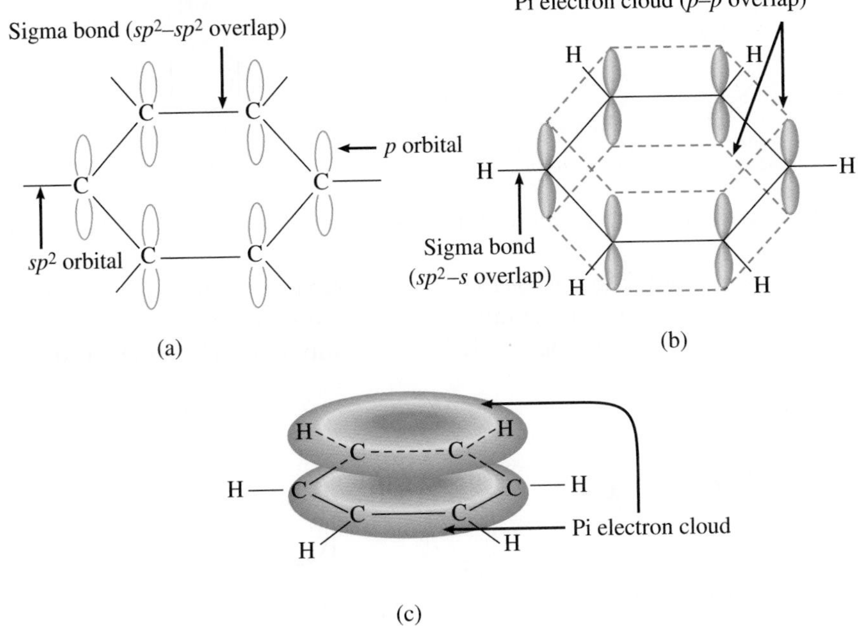

Figure 20.5
Bonding in a benzene molecule: (a) sp^2–sp^2 orbital overlap to form the carbon ring structure; (b) carbon–hydrogen bonds formed by sp^2–s orbital overlap and overlapping of p orbitals; (c) pi electron clouds above and below the plane of the carbon ring.

*The detailed pi molecular orbital theory of the benzene structure is beyond the scope of this book.

20.10 Naming Aromatic Compounds

Substituted benzenes are the most common benzene derivatives because substitution is the most common reaction type for benzene. A substituted benzene is derived by replacing one or more hydrogen atoms of benzene by another atom or group of atoms. Thus, a monosubstituted benzene has the formula C_6H_5G, where G is the group replacing a hydrogen atom.

Monosubstituted Benzenes

Substituted benzenes are found in biochemical pigments, hormones, and nerve transmitters.

Some monosubstituted benzenes are named by adding the name of the substituent group as a prefix to the word *benzene*. The name is written as one word. Note that the position of the substituent is not important here, as all the positions in the hexagon are equivalent. Examples include the following:

O_2N — nitrobenzene

CH_2CH_3 — ethylbenzene

Cl — chlorobenzene

Br — bromobenzene

Certain monosubstituted benzenes have special names:

CH_3 — toluene (methylbenzene)

OH — phenol (hydroxybenzene)

$CH{=}CH_2$ — styrene (vinylbenzene)

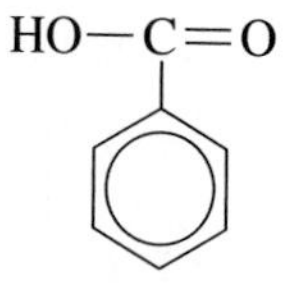

HO—C=O H—C=O NH_2

benzoic acid
(benzene carboxylic acid)

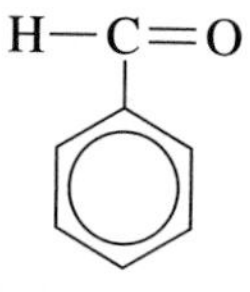

benzaldehyde
(benzene carboxaldehyde)

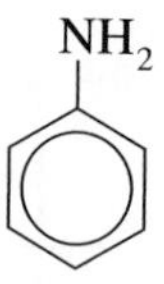

aniline
(aminobenzene)

These are used as parent names for further substituted compounds and should be memorized.

The C_6H_5— group is known as the phenyl group (pronounced *fen-il*), and the name *phenyl* is used to name compounds that cannot easily be named as benzene derivatives. For example, the following compounds are named as derivatives of alkanes:

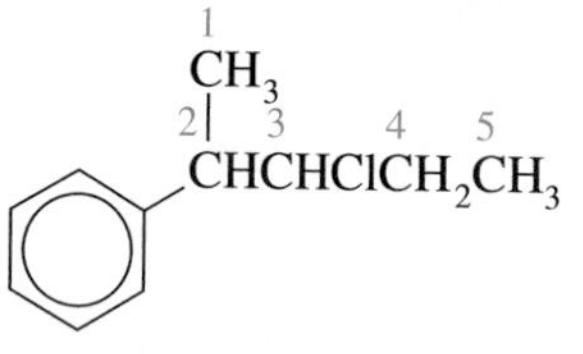

1 CH_3
2 3 4 5
$CHCHClCH_2CH_3$

CH_2

3-chloro-2-phenylpentane

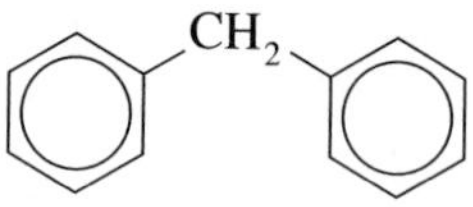

diphenylmethane

Practice 20.16 Application to Biochemistry

There are 20 common amino acids. They differ from one another at the side chain marked in blue in the following example:

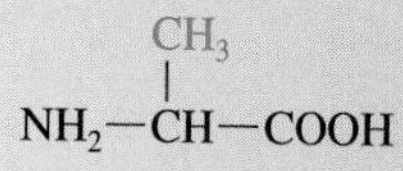

CH_3
|
NH_2—CH—COOH

alanine

Based on what you know about the use of the name *phenyl*, propose a structure for the amino acid, phenylalanine.

Disubstituted Benzenes

When two substituent groups replace two hydrogen atoms in a benzene molecule, three different isomeric compounds are possible. The prefixes *ortho-*, *meta-* and *para-* (abbreviated *o-*, *m-*, and *p-*) are used to name disubstituted benzenes. In ortho disubstituted compounds, the two substituents are located on adjacent carbon atoms of the benzene ring (the 1,2-positions). In meta disubstituted compounds, the two substituents are one carbon apart (the 1,3-position). In para disubstituted compounds, the two substituents are located on opposite points of the hexagon (the 1,4-position). Thus, with respect to a group G on the benzene ring, there are two ortho positions, two meta positions, and one para position.

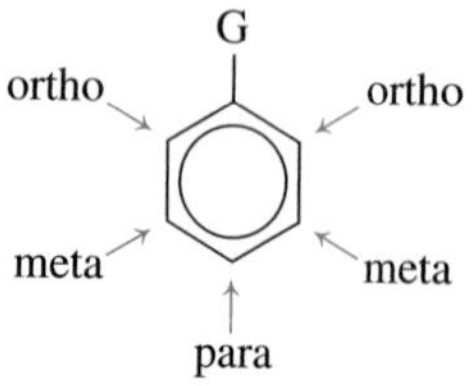

G
ortho ortho
meta meta
para

Consider the dichlorobenzenes, $C_6H_4Cl_2$. Note that the three isomers have different physical properties, indicating that they are truly different substances:

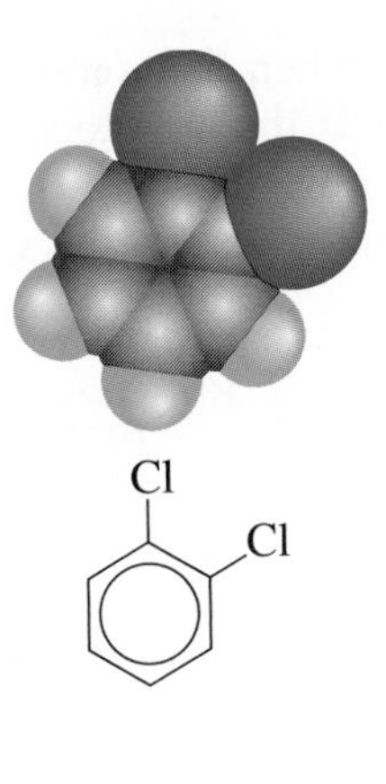

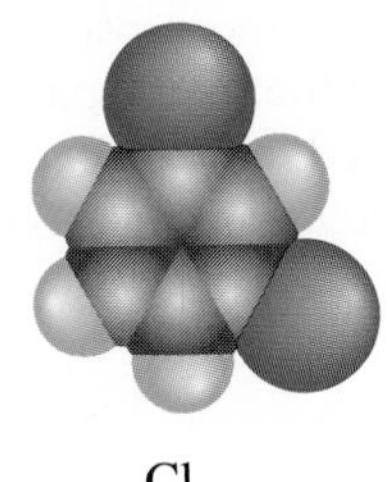

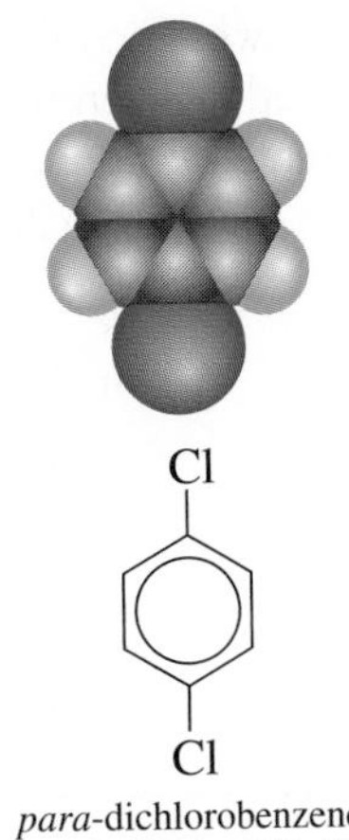

ortho-dichlorobenzene (mp –17.2°C, bp 180.4°C)

meta-dichlorobenzene (mp –24.8°C, bp 172°C)

para-dichlorobenzene (mp 53.1°C, bp 174.4°C)

Many sunscreens include p-aminobenzoic acid (PABA).

When the two substituents are different and neither is part of a compound with a special name, the names of the two substituents are given in alphabetical order, followed by the word *benzene*, as in the following examples:

o-bromochlorobenzene

m-ethylnitrobenzene

The dimethylbenzenes have the special name *xylene*:

ortho-xylene

meta-xylene

para-xylene

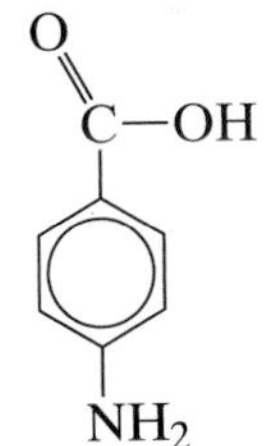

When one of the substituents corresponds to a monosubstituted benzene that has a special name, the disubstituted compound is named as a derivative of that parent compound. In the following examples, the parent compounds are phenol, aniline, and toluene:

o-nitrophenol

p-bromoaniline

m-nitrotoluene

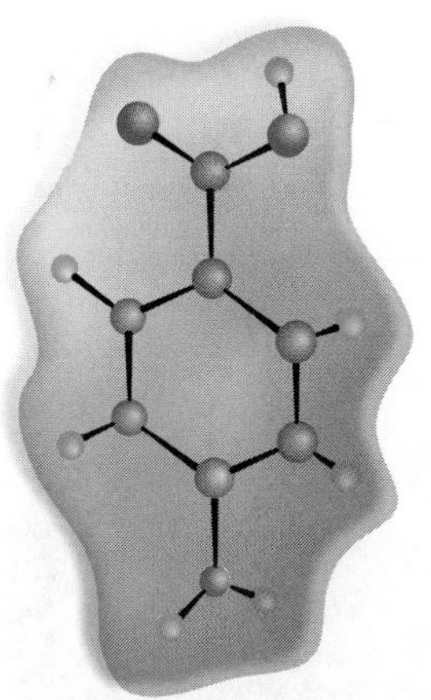

p-aminobenzoic acid

The numbering system as described under polysubstituted benzenes is also used to name disubstituted benzenes.

Polysubstituted Benzenes

When there are more than two substituents on a benzene ring, the carbon atoms in the ring are numbered starting at one of the substituted groups. Numbering may be either clockwise or counterclockwise, but must be done in the direction that gives the lowest possible numbers to the substituent groups. When the compound is named as a derivative of one of the special parent compounds, the substituent of the parent compound is considered to be on carbon 1 of the ring (e.g., the CH_3 group is on carbon 1 in 2,4,6-trinitrotoluene). The following examples illustrate this system:

1,3,5-trinitrobenzene

1,2,4-tribromobenzene (not 1,4,6-)

2,4,6-trinitrotoluene (TNT)

5-bromo-2-chlorophenol

Example 20.8 Write formulas and names for all the possible isomers of (a) chloronitrobenzene, $C_6H_4Cl(NO_2)$, and (b) tribromobenzene, $C_6H_3Br_3$.

SOLUTION

(a) The name and formula indicate a chloro group (Cl) and a nitro group (NO_2) attached to a benzene ring. There are six positions in which to place these two groups. They can be ortho, meta, or para to each other:

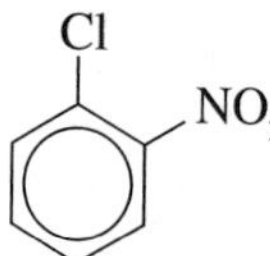

o-chloronitrobenzene

m-chloronitrobenzene

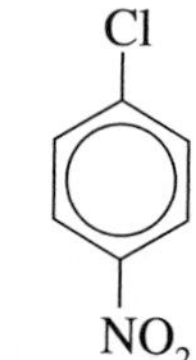

p-chloronitrobenzene

(b) For tribromobenzene, start by placing the three bromo groups in the 1-, 2-, and 3-positions; then the 1-, 2-, and 4-positions; and so on until all possible isomers are formed:

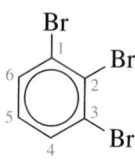

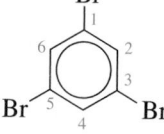

1,2,3-tribromobenzene

1,2,4-tribromobenzene

1,3,5-tribromobenzene

The name of each isomer will allow you to check that the formulas have not been duplicated.

There are only three isomers of tribromobenzene. If one erroneously writes the 1,2,5- compound, a further check will show that, by numbering the rings as indicated, it is in reality the 1,2,4- isomer:

1,2,5-tribromobenzene (incorrect name)

1,2,4-tribromobenzene (correct name)

The dramatic collapse of this building is from the explosive TNT (common name for 2,4,6-trinitrotoluene).

Practice 20.17

Write formulas and names for all possible isomers of chlorophenol.

Practice 20.18

Name the following:

(a) COOH, Br

(b) NH_2, F

(c) O, ‖, C—H, Cl, NO_2

Practice 20.19 Application to Biochemistry

Substituted benzenes are important parts of some amino acids and hormones. We focus only on the biologically important substituted benzenes. Name the following:

(a) OH, OH

found in the hormone adrenalin

(b)

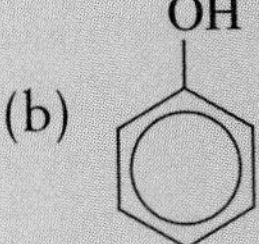

found in the amino acid tyrosine

(c) OH, I, I

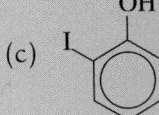

found in the thyroid hormones

20.11 Polycyclic Aromatic Compounds

There are many other aromatic ring systems besides benzene. Their structures consist of two or more rings in which two carbon atoms are common to two rings. These compounds are known as **polycyclic or fused aromatic ring systems**. Three of the most common hydrocarbons in this category are naphthalene, anthracene, and phenanthrene. One hydrogen is attached to each carbon atom, except at the carbons common to two rings.

polycyclic or fused aromatic ring system

naphthalene, $C_{10}H_8$

anthracene, $C_{14}H_{10}$

phenanthrene, $C_{14}H_{10}$

All three of these substances can be obtained from coal tar. Naphthalene is known as moth balls and has been used as a moth repellant for many years.

A number of the polycyclic aromatic hydrocarbons (and benzene) have been shown to be carcinogenic (cancer producing). Formulas for some of the more notable ones, found in coal tar, tar from cigarette smoke, and soot in urban environments, are as follows:

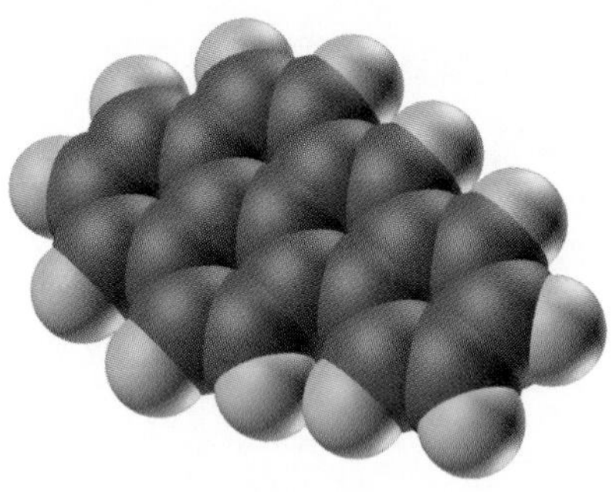

Spacefilling model of 3,4-benzpyrene.

1,2-benzanthracene

1,2,5,6-dibenzanthracene

3,4-benzpyrene

These compounds are processed in living cells and become potent carcinogens. For example, the lung cells absorb polycyclic aromatic hydrocarbons from inhaled cigarette smoke. The aromatic rings are chemically modified, and the compounds become active carcinogens.

20.12 Sources and Physical Properties of Aromatic Hydrocarbons

When coal is heated to high temperatures (450–1200°C) in the absence of air to produce coke (C), coal gas and a complex mixture of condensable substances called coal tar are driven off:

$$\text{Coal} \xrightarrow{\Delta} \text{Coke} + \text{Coal gas} + \text{Coal tar}$$

The aromatic hydrocarbons, such as benzene, toluene, xylene, naphthalene, and anthracene, were first obtained in significant quantities from coal tar. Since coal tar itself is a by-product of the manufacture of coke, the total amount of aromatics that can be obtained from this source is limited. The demand for aromatic hydrocarbons, which are used in the production of a vast number of materials such as drugs, dyes, detergents, explosives, insecticides, plastics, and synthetic rubber, became too great to be obtained from coal tar alone. Processes were devised to make aromatic hydrocarbons from the relatively inexpensive alkanes found in petroleum. Currently, about one-third of our benzene supply and the greater portion of our toluene and xylene supplies are obtained from petroleum.

Aromatic hydrocarbons are essentially nonpolar substances, insoluble in water but soluble in many organic solvents. They are liquids or solids and usually have densities less than that of water. Aromatic hydrocarbons burn readily, usually with smoky yellow flames as a result of incomplete carbon combustion. Some are blended with other hydrocarbons to make good motor fuels with excellent antiknock properties.

20.13 Chemical Properties of Aromatic Hydrocarbons

Substitution

The most characteristic reactions of aromatic hydrocarbons involve the substitution of some group for a hydrogen on one of the ring carbons. Recall that a substitution reaction does not change the number of atoms bonded to each carbon. For example, each carbon in benzene will start and end a substitution reaction bonded to three other atoms.

The following examples depict typical aromatic substitution reactions, and in each of these reactions, a functional group is substituted for a hydrogen atom:

1. Halogenation (chlorination or bromination) When benzene reacts with chlorine or bromine in the presence of a catalyst such as iron(III) chloride or iron(III) bromide, a Cl or a Br atom replaces an H atom to form the products. The result is

$$C_6H_6 + X_2 \xrightarrow{FeX_3} C_6H_5X + HX$$

benzene; chlorine or bromine; chlorobenzene or bromobenzene

2. Nitration When benzene reacts with a mixture of concentrated nitric acid and concentrated sulfuric acid at about 50°C, nitrobenzene is formed. In this reaction a nitro group, $—NO_2$, is substituted for an H atom of benzene:

$$C_6H_6 + HO—NO_2 \xrightarrow{H_2SO_4} C_6H_5NO_2 + H_2O$$

benzene; nitric acid; nitrobenzene

3. Alkylation (Friedel–Crafts reaction) There are many variations of the Friedel–Crafts reaction. In this type the alkyl group from an alkyl halide (RX), in the presence of $AlCl_3$ catalyst, substitutes for an H atom on the benzene ring:

$$C_6H_6 + CH_3CH_2Cl \xrightarrow{AlCl_3} C_6H_5CH_2CH_3 + HCl$$

benzene; chloroethane; ethylbenzene

From about 1860 onward, especially in Germany, a great variety of useful substances such as dyes, explosives, and drugs were synthesized from aromatic hydrocarbons by reactions of the types just described. These early syntheses were developed by trial-and-error methods. A good picture of the reaction mechanism, or step-by-step sequence of intermediate stages in the overall reaction, was not obtained until about 1940.

We now recognize that aromatic substitution reactions usually proceed by a mechanism called *electrophilic substitution*. Three steps are involved:

Step 1 An electrophile (electron-seeking group) is formed.
Step 2 The electrophile is attached to the benzene ring, forming a positively charged carbocation intermediate.
Step 3 A hydrogen ion is lost from the carbocation to form the product. This reaction mechanism is illustrated in the chlorination of benzene catalyzed by iron(III) chloride:

Step 1 $FeCl_3 + Cl_2 \longrightarrow FeCl_4^- + Cl^+$

iron(III) chloride — chloronium ion (electrophile)

Step 2 benzene $+ Cl^+ \longrightarrow$ [carbocation with H and Cl]

a carbocation

Step 3 carbocation $+ FeCl_4^- \longrightarrow$ chlorobenzene $+ FeCl_3 + HCl$

chlorobenzene

In Step 1, the electrophile (chloronium ion) is formed. In Step 2, the chloronium ion adds to benzene to form an intermediate carbocation, which loses a proton (H^+) in Step 3 to form the products, C_6H_5Cl and HCl. The catalyst, $FeCl_3$, is regenerated in Step 3. Nitration and alkylation also proceed by an electrophilic mechanism.

This same mechanism is used by living organisms when aromatic rings gain or lose substituents. For example, the thyroxines (thyroid gland hormones) contain aromatic rings that are iodinated by following an electrophilic substitution mechanism. Iodination is a key step in producing these potent hormones. In general, scientists find that most of life's reactions follow mechanisms that have been elucidated in the organic chemist's laboratory.

Practice 20.20 Application to Biochemistry

In the thyroid gland, a part of the amino acid tyrosine is iodinated to begin thyroid hormone formation:

OH, R (phenol ring) $\longrightarrow$ OH, I, R (iodinated ring)

a part of tyrosine

(*Note*: R represents the remainder of each biochemical.)

Show Step 2 of the electrophilic substitution mechanism for this reaction. (Assume that iodine reacts in the same way as chlorine.)

Side-Chain Oxidation

Carbon chains attached to an aromatic ring are fairly easy to oxidize. Reagents most commonly used to accomplish this in the laboratory are $KMnO_4$ or $K_2Cr_2O_7$ + H_2SO_4. No matter how long the side chain is, the carbon atom attached to the aromatic ring is oxidized to a carboxylic acid group, —COOH. For example, toluene, ethylbenzene, and propylbenzene are all oxidized to benzoic acid. For this oxidation to occur, there must be at least one H atom on the carbon bonded to the aromatic ring:

$$C_6H_5CH_3 \xrightarrow[\Delta]{KMnO_4/H_2O} C_6H_5COOH$$

toluene → benzoic acid

$$C_6H_5CH_2CH_3 \xrightarrow[\Delta]{K_2Cr_2O_7/H_2SO_4} C_6H_5COOH + CO_2$$

ethylbenzene

$$C_6H_5CH_2CH_2CH_3 \xrightarrow[\Delta]{K_2Cr_2O_7/H_2SO_4} C_6H_5COOH + CH_3COOH$$

propylbenzene

Chapter 20 Review

20.1 Bonding in Unsaturated Hydrocarbons

KEY TERMS

Alkene
Alkyne
Aromatic compound
Pi bond

- Alkenes are hydrocarbons that contain carbon-carbon double bonds.
- Alkynes are hydrocarbons that contain carbon-carbon triple bonds.
- Aromatic compounds are hydrocarbons that contain benzene rings.
  - The four carbon electrons are hybridized to three sp^2 and one p orbitals.
- A pi (π) bond is formed by the sideways overlap of two p orbitals and consists of two electron clouds: one above and one below the sigma bond.
- All unsaturated hydrocarbons have pi bonds.
  - Alkenes and aromatic compounds have double bonds composed of one sigma and one pi bond each.
  - Alkynes have triple bonds composed of one sigma bond and two pi bonds each.

20.2 Nomenclature of Alkenes

- Naming alkenes follows a similar process to that used for alkanes.
  - The longest continuous chain (parent chain) *that contains the double bond* is given the name of the corresponding alkane modified by changing the *-ane* ending to *-ene*.
  - The longest continuous chain is numbered starting at the end that is closest to the double bond.
  - Use the smaller of the two numbers on the double-bonded carbons to indicate the double-bond position.
  - Each branch is given its alkyl name and number location on the longest continuous chain.
  - Prefixes (such as *di-*, *tri-*, *tetra-*) are added to the alkyl name when there is more than one of the same branch.
  - Branches are listed alphabetically in front of the parent name.
- Ethylene is a common name that is often used for ethene.

20.3 Geometric Isomerism in Alkenes

KEY TERM

Geometric or *cis–trans* isomers

- Geometric isomers differ in their molecular geometry but not in the bonded order of their atoms.
- The restricted rotation around double bonds allows alkenes to have geometric isomerism.
- Alkene geometric isomerism *is not* present if two identical groups are found on one of the double-bonded carbon atoms.
- Alkene geometric isomerism *is* present if two different groups are present on each of the double-bonded carbon atoms. The isomers are distinguished by picturing a boundary that parallels the double bond and runs through the double-bonded carbon atoms.
  - *Cis*-isomers have atoms or groups on the same side of the boundary. (Alternatively, the parent chain is all on one side of the boundary.)
  - *Trans*-isomers have atoms or groups on opposite sides of the boundary. (Alternatively, the parent chain crosses from one side of the boundary to the other.)

20.4 Cycloalkenes

KEY TERM

Cycloalkene

- Cycloalkenes are cyclic compounds that contain a double bond in the ring.
- The double-bonded carbons are always numbered 1 and 2.

20.5 Preparation and Physical Properties of Alkenes

KEY TERMS

Cracking
Dehydration

- A cracking process breaks saturated hydrocarbons into smaller pieces by heating the molecules to high temperatures in the presence of a catalyst.
- Dehydration removes water from alcohols to form alkenes.
- Alkenes have physical properties that are very similar to those of alkanes.

20.6 Chemical Properties of Alkenes

KEY TERMS

Carbocation
Markovnikov's rule

- The most common alkene reaction is the addition of two new groups to the double-bonded carbons.
- The alkene addition reaction follows a series of steps (a reaction mechanism) that involves a carbocation.
- A carbocation is a positively–charged carbon ion.
- Markovnikov's rule states that when a hydrogen-containing unsymmetric molecule (HX) adds to a double bond, the hydrogen will add to the carbon with the greatest number of hydrogens.
- Markovnikov's rule can be explained based on the stability of carbocations.
- Another typical alkene reaction is oxidation; this reaction impacts many biological materials.

20.7 Alkynes: Nomenclature and Preparation

- The rules for naming alkynes are the same as for alkenes, except the suffix *-yne* is used to indicate the presence of a triple bond.
- Acetylene (ethyne) is commonly prepared by (a) reacting calcium carbide with water or (b) cracking methane in an electric arc.

20.8 Physical and Chemical Properties of Alkynes

- Acetylene is a colorless, odorless gas (at room temperature and pressure) that is partially water soluble (one volume of acetylene gas dissolves in an equal volume of water).
- Acetylene reacts with oxygen to yield a very hot flame—a property used in oxyacetylene cutting and welding.
- Acetylene, like other alkynes, undergoes addition reactions.

20.9 Aromatic Hyrocarbons: Structure

- The term *aromatic hydrocarbon* refers to a special arrangement of double bonds in a cyclic compound, not to a pleasant-smelling compound.
- In benzene, the alternating pi bonds overlap to create a circular pi electron cloud above and below the benzene ring.

20.10 Naming Aromatic Compounds

- Common monosubstituted benzenes can be named by adding the name of the substituent as a prefix to the word *benzene*.
- Special names for some monosubstituted benzenes remain common.
- When a parent chain has a benzene group attached, this substituent is known as a phenyl group.
- Disubstituted benzenes use a common nomenclature that tells the relative position of the two substituents.
  - The prefix *ortho-* means the two substituents are located on adjacent carbons in the benzene ring.
  - The prefix *meta-* means the two substituents are bound to benzene carbons with one intervening carbon.
  - The prefix *para-* means the two substituents are located on opposite points of the benzene hexagon.
- Polysubstituted benzenes have substituent locations that are numbered.

20.11 Polycyclic Aromatic Compounds

KEY TERMS

Polycyclic or fused aromatic ring system

- Polycyclic or fused aromatic ring systems contain two or more benzene rings (*polycyclic*) in which two carbon atoms are common to two rings (*fused*).

20.12 Sources and Physical Properties of Aromatic Hydrocarbons

- The two main sources of aromatic hydrocarbons are coal tar and petroleum.
- As with other hydrocarbons, aromatics are nonpolar and water insoluble with densities less than water.

20.13 Chemical Properties of Aromatic Hydrocarbons

- The most characteristic aromatic hydrocarbon reaction is electrophilic substitution.
- Electrophilic substitution involves attack by a positively charged group (the electrophile) on a benzene carbon with subsequent loss of a hydrogen.
- Alkyl side chains on aromatic rings are easily oxidized; alkyl benzenes are all oxidized to benzoic acid no matter how long the side chain.
- Common aromatic reactions include halogenation, nitration and alkylation (Friedel-Crafts reaction).

Review Questions

All questions with blue numbers have answers in the appendix of the text.

1. The double bond in ethene, C_2H_4, is made up of a sigma bond and a pi bond. Explain how the pi bond differs from the sigma bond.
2. Why is it possible to obtain cis and trans isomers of 1,2-dichloroethene but not of 1,2-dichloroethane?
3. Why do many rubber products deteriorate rapidly in smog-ridden areas?
4. Explain the two different types of explosion hazards present when acetylene is being handled.
5. Why is an elimination reaction often considered to be the reverse of an addition reaction?
6. In terms of historical events, why did the major source of aromatic hydrocarbons shift from coal tar to petroleum during the 10-year period 1935–1945?
7. Explain how the reactions of benzene provide evidence that its structure does not include double bonds like those found in alkenes.
8. How is the retinal structure changed when light is absorbed by the visual pigment rhodopsin? p. 548–549
9. What is the primary health danger associated with polycyclic aromatic hydrocarbons?
10. Cyclohexane can be found in either a "boat" or "chair" conformation. Why doesn't benzene also assume a "boat" or "chair" shape?
11. How does the carbocation reaction mechanism help to explain Markovnikov's rule?
12. Why do you think the term "cracking" was chosen to describe pyrolysis of hydrocarbons?

Paired Exercises

All exercises with blue numbers have answers in the appendix of the text.

1. Draw Lewis structures to represent the following molecules:
(a) ethane (b) ethene (c) ethyne

2. Draw Lewis structures to represent the following molecules:
(a) propane (b) propene (c) propyne

3. There are 11 possible isomeric iodobutenes, C_4H_7I, including cis–trans isomers.
(a) Write the structural formula for each isomer.
(b) Name each isomer and include the prefix *cis-* or *trans-* where apppropriate.

4. Four isomeric chloropropenes, C_3H_5Cl, are possible.
(a) Draw and name the structural formula for each isomer.
(b) There is another compound with this same molecular formula. What is its structure and name?

5. Draw structural formulas for the following:
(a) 2,5-dimethyl-3-hexene
(b) *cis*-4-methyl-2-pentene
(c) 3-pentene-1-yne
(d) *trans*-3-hexene
(e) 3-methyl-1-pentyne
(f) 3-methyl-2-phenylhexane

6. Draw structural formulas for the following:
(a) 3-ethyl-3-methyl-1-pentene
(b) *cis*-1,2-diphenylethene
(c) 3-phenyl-1-butyne
(d) cyclopentene
(e) 1-methylcyclohexene
(f) 3-isopropylcyclopentene

7. The following names are incorrect:
(a) 3-methyl-3-butene
(b) *cis*-3-pentene
(c) *cis*-2-methyl-2-pentene
State why each name is wrong and give the correct name.

8. The following names are incorrect:
(a) 3-ethyl-3-butene
(b) 2-chlorocyclopentene
(c) *trans*-4-hexene
State why each name is wrong and give the correct name.

9. Name the following compounds:

(a) $CH_3CH_2(CH_3)C=C(H)CH_2CH_3$ — CH_3CH_2 and H on one carbon; CH_3 and CH_2CH_3 on the other

(b) $C_6H_5CH_2C\equiv CH$ (benzene ring with $CH_2C\equiv CH$)

(c) $CH_3CHBrCHBrC\equiv CCH_3$

10. Name the following compounds:

(a) H and CH_3 on one carbon of C=C; H and $CHCH_2CH_3$ (with CH_3 branch) on the other

(b) $(CH_3)_2C=C(CH_3)_2$

(c) $CH_3CH_2CHCH=CH_2$ with $CH(CH_3)_2$ substituent (H_3C–CH–CH_3) on the third carbon

11. Write the structural formulas and IUPAC names for all the hexynes.

12. Write the structural formulas and IUPAC names for all the pentynes.

13. Which of the following molecules have structural formulas that permit cis–trans isomers to exist?
(a) $(CH_3)_2C=CHCH_3$
(b) $CH_3CH=CHCl$
(c) $CCl_2=CBr_2$

14. Which of the following molecules have structural formulas that permit cis–trans isomers to exist?
(a) $CH_2=CHCl$
(b) $CH_3CH_2C\equiv CCH_3$
(c) $CH_2ClCH=CHCH_2Cl$

15. Complete the following equations:

(a) $CH_3CH_2CH_2CH=CH_2 + Br_2 \longrightarrow$

(b) $CH_3CH_2C(CH_3)=CHCH_3 + HI \longrightarrow$

(c) $CH_3CH_2CH=CH_2 + H_2O \xrightarrow{H^+}$

(d) $C_6H_5CH=CH_2 + H_2 \xrightarrow[\text{1 atm}]{\text{Pt, 25°C}}$

(e) $CH_3CH=CHCH_3 + KMnO_4 \xrightarrow[\text{cold}]{H_2O}$

16. Complete the following equations:

(a) $CH_3CH_2CH_2CH=CH_2 + H_2O \xrightarrow{H^+}$

(b) $CH_3CH_2CH=CHCH_3 + HBr \longrightarrow$

(c) $CH_2=CHCl + Br_2 \longrightarrow$

(d) $C_6H_5CH=CH_2 + HCl \longrightarrow$

(e) $CH_2=CHCH_2CH_3 + KMnO_4 \xrightarrow[\text{cold}]{H_2O}$

17. The following names are incorrect:
(a) 1-ethyne
(b) *cis*-2-butyne
(c) 2-propyne
Draw each structure and give its correct name.

18. The following names are incorrect:
(a) 3-butyne
(b) 1-propyne
(c) *trans*-2-pentyne
Draw each structure and give its correct name.

19. Complete the following equations:

(a) $CH_3C\equiv CCH_3 + Br_2$ (1 mol) $\longrightarrow$

(b) The two step reaction

$CH\equiv CH + HCl \longrightarrow$ First product $\xrightarrow{HCl}$

(c) $CH_3CH_2C\equiv CH + H_2$ (1 mol) $\xrightarrow[\text{25°C, 1 atm}]{\text{Pt}}$

20. Complete the following equations:

(a) $CH_3C\equiv CH + H_2$ (1 mol) $\xrightarrow[\text{25°C, 1 atm}]{\text{Pt}}$

(b) $CH_3C\equiv CCH_3 + Br_2$ (2 mol) $\longrightarrow$

(c) The two step reaction

$CH_3C\equiv CH + HCl \longrightarrow$ First product $\xrightarrow{HCl}$

21. Write the formula and name for the product when cyclohexene reacts with
(a) Br_2
(b) HI
(c) H_2O, H^+
(d) $KMnO_4(aq)$ (cold)

22. Write the formula and name for the product when cyclopentene reacts with
(a) Cl_2
(b) HBr
(c) H_2, Pt
(d) H_2O, H^+

23. Name and draw the condensed structural formula for the alkene reactant in each of the following reactions:

(a) $+ \ HBr \longrightarrow CH_3-\overset{\overset{\large CH_3}{|}}{\underset{\underset{\large Br}{|}}{C}}-CH_2CH_3$

(b) $+ \ H_2O \longrightarrow CH_3-\overset{\overset{\large CH_3}{|}}{\underset{\underset{\large CH_3}{|}}{C}}-\overset{\overset{\large OH}{|}}{CH}-CH_3$

(c) $+ \ H_2 \longrightarrow CH_3-\underset{\underset{\large CH_3}{|}}{CH}-CH_3$

(d) $+ \ Cl_2 \longrightarrow$ (cyclohexane ring bearing CH_3, Cl, Cl, H_3C on adjacent carbons)

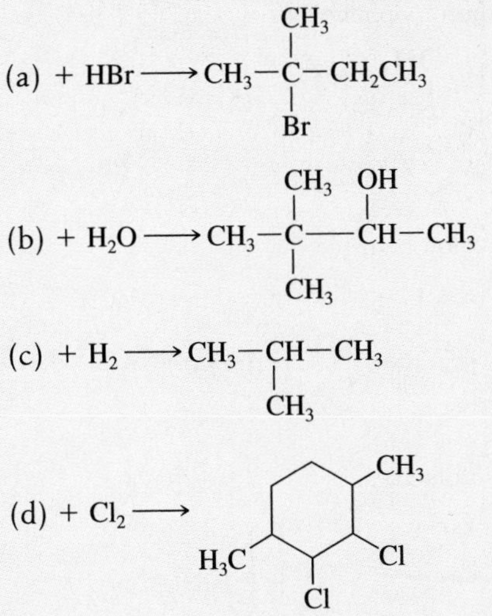

24. Name and draw the condensed structural formula for the alkyne reactant in each of the following reactions:

(a) $+ \ 2Br_2 \longrightarrow CH_3-\overset{\overset{\large Br}{|}}{\underset{\underset{\large Br}{|}}{C}}-\overset{\overset{\large Br}{|}}{\underset{\underset{\large Br}{|}}{C}}-\overset{\overset{\large CH_2CH_3}{|}}{CH_2}$

(b) $+ \ 2HCl \longrightarrow CH_3-\overset{\overset{\large CH_3}{|}}{\underset{\underset{\large CH_3}{|}}{C}}-\overset{\overset{\large Cl}{|}}{\underset{\underset{\large Cl}{|}}{C}}-CH_3$

(c) $+ \ H_2 \longrightarrow CH_2{=}CH-CH_2\overset{\overset{\large CH_3}{|}}{\underset{\underset{\large CH_3}{|}}{C}}-CH_3$

(d) $+ \ HBr \longrightarrow H_3C-CH{=}\overset{\overset{\large Br}{|}}{C}-CH_3$

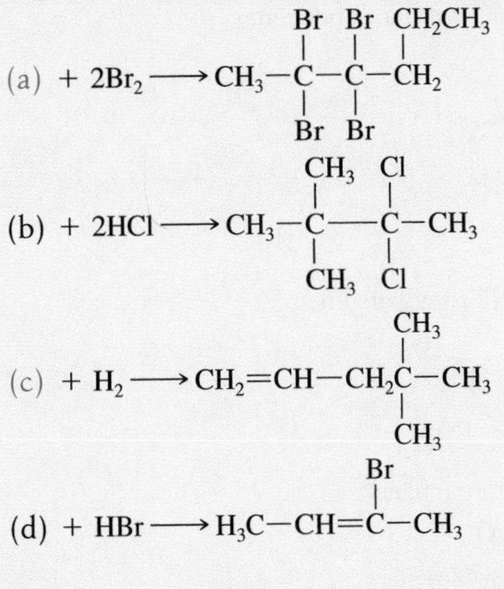

25. Write equations to show how 2-butyne can be converted into
(a) 2,3-dibromobutane
(b) 2,2-dibromobutane
(c) 2,2,3,3-tetrabromobutane

26. Write equations to show how 2-pentyne can be converted into
(a) 2,3-dichloropentane
(b) 2,2-dichloropentane
(c) 2,2,3,3-tetrachloropentane

27. Write structural formulas for
(a) benzene
(b) aniline
(c) benzoic acid
(d) naphthalene

28. Write structural formulas for
(a) toluene
(b) ethylbenzene
(c) phenol
(d) anthracene

29. Write structural formulas for
(a) 1,3,5-tribromobenzene
(b) *o*-bromochlorobenzene
(c) *tert*-butylbenzene
(d) *p*-xylene

30. Write structural formulas for
(a) 1,3-dichloro-5-nitrobenzene
(b) *m*-dinitrobenzene
(c) 1,1-diphenylethane
(d) styrene

31. Write structural formulas and names for all the isomers of
(a) dichlorobromobenzene, $C_6H_3Cl_2Br$
(b) the toluene derivatives of formula C_9H_{12}

32. Write structural formulas and names for all the isomers of
(a) trichlorobenzene, $C_6H_3Cl_3$
(b) the benzene derivatives of formula C_8H_{10}

33. Write the structures and names for all the isomers that can be written by substituting an additional chlorine atom in *o*-chlorobromobenzene.

34. Write the structures and names for all the isomers that can be written by substituting a third chlorine atom in *o*-dichlorobenzene.

35. Name the following compounds:

(a) benzene ring with CH_2CH_3 and Cl (para)

(b) benzene ring with $CH_2CH_2CH_3$

(c) benzene ring with NH_2 and NO_2 (meta)

(d) benzene ring with OH and Br (para)

(e) CH bonded to three benzene rings

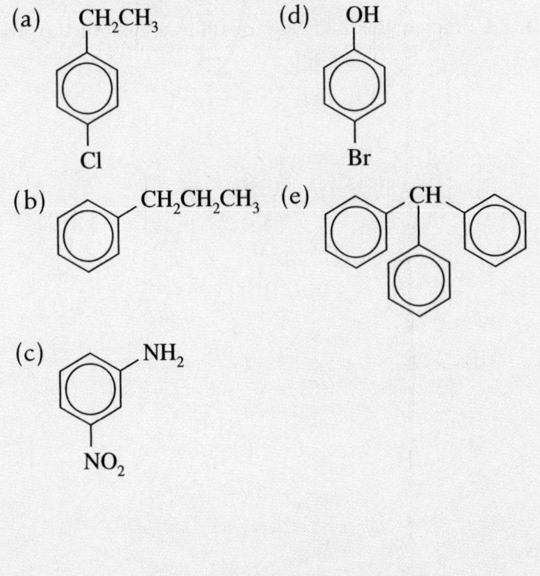

36. Name the following compounds:

(a) benzene ring with $CH{=}CH_2$

(b)

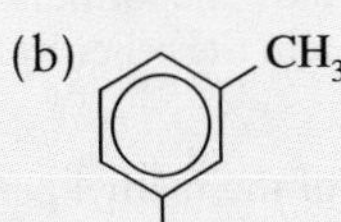

(c) benzene ring with COOH, Br, Br (3,5)

(d) benzene ring with $\underset{\underset{\large CH_3}{|}}{CH}-CH_3$

(e)

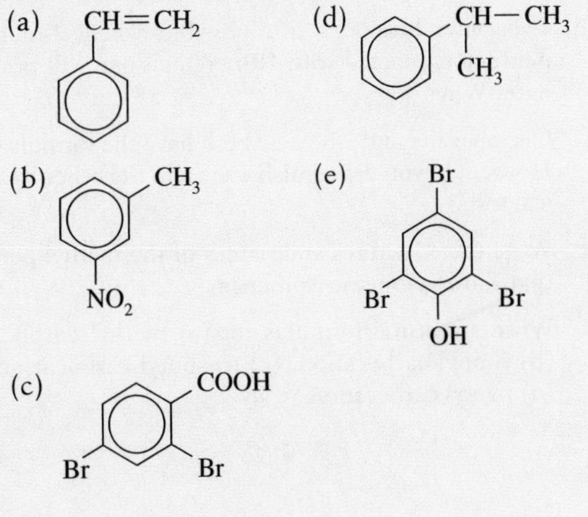

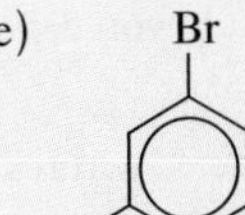

37. Each of the following structures is associated with an incorrect name. Give the correct name.
(a) 2-chloro-p-diiodobenzene

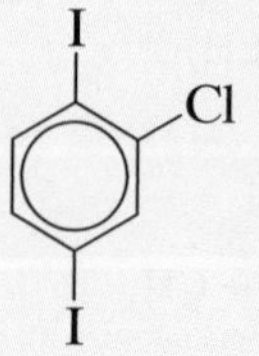

I, Cl, I

(b) 2,3-dibromobenzene

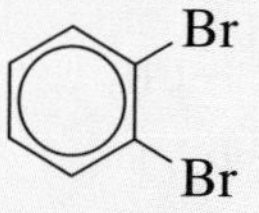

Br, Br

(c) o-chloroanaline

NH_2, Cl

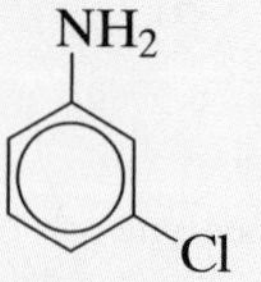

38. Each of the following structures is associated with an incorrect name. Give the correct name.
(a) 5,6-dibromophenol

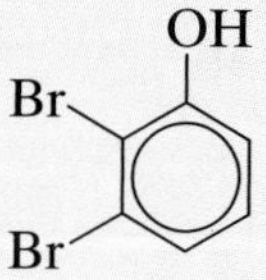

OH, Br, Br

(b) *m*-trichlorobenzene

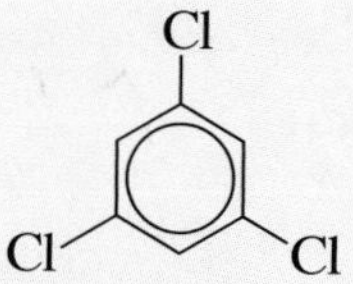

Cl, Cl, Cl

(c) *p*-iodotoluene

CH_3, I

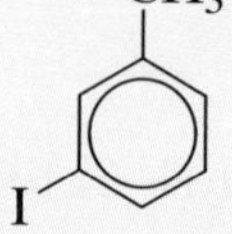

39. Complete the following equations and name the organic products:

(a) benzene + Br_2 $\xrightarrow{FeBr_3}$

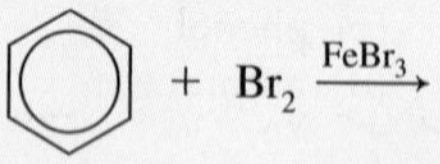

(b) CH_3–C$_6$H$_4$–CH_3 + HNO_3 $\xrightarrow{H_2SO_4}$

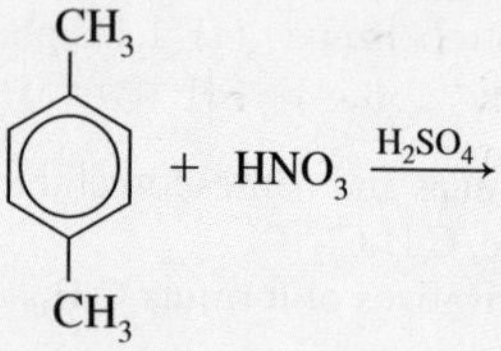

40. Complete the following equations and name the organic products:

(a) benzene + CH_3CHCH_3 (Cl on C-2) $\xrightarrow{AlCl_3}$

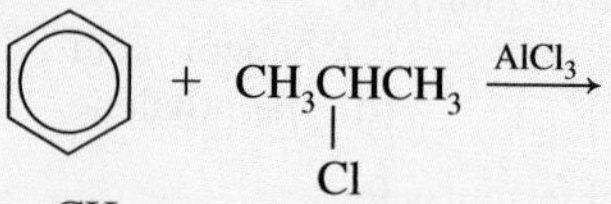

(b) CH_3–C$_6$H$_5$ + $KMnO_4$ $\xrightarrow{H_2O}$

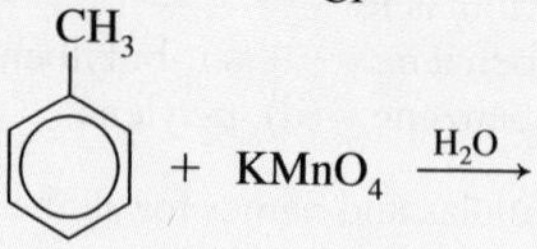

Additional Exercises

All exercises with blue numbers have answers in the appendix of the text.

41. Two alkyl bromides are possible when 2-methyl-1-pentene is reacted with HBr. Which one will predominate? Why?

42. Cyclohexane and 2-hexene both have the formula C_6H_{12}. How could you distinguish one from the other by chemical tests?

43. Write the structures and names of the methyl pentenes that show geometric isomerism.

44. Write structural formulas for (a) methyl carbocation, (b) propyl carbocation, (c) *tert*-butyl carbocation, and (d) pentyl carbocation.

45. Complete the following equations and name the products:

(a) cyclohexene + Br_2 ⟶

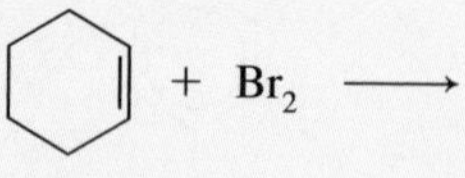

(b) cyclohexene + HCl ⟶

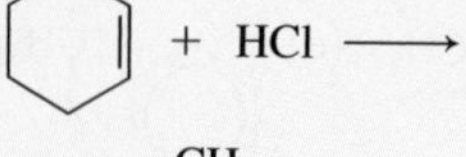

(c) 1-methylcyclohexene (CH_3) + HCl ⟶

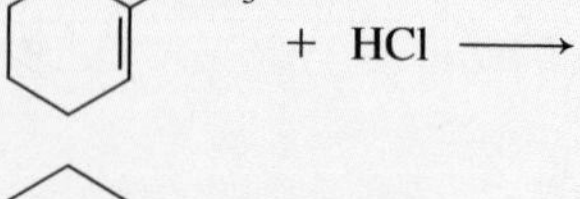

(d) 4-methylcyclohexene (CH_3) + HCl ⟶

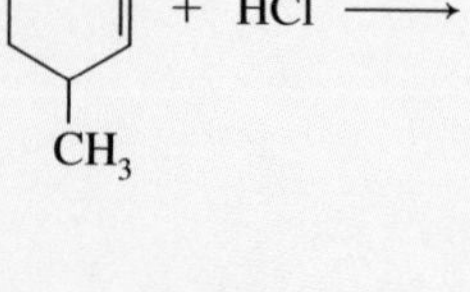

46. Describe the reaction mechanism by which benzene is brominated in the presence of $FeBr_3$. Show the equations.

47. Another method for making alkenes is by dehydrohalogenation (removal of HX) from an alkyl halide. What alkenes will be formed when HCl is removed from (a) 2-chlorobutane, (b) 1-chloropentane, and (c) chlorocyclohexane?

48. An aliphatic hydrocarbon, C_4H_8, reacts with Br_2, but no HBr is formed. Will there be a color change? Explain. Give possible structures for the original molecule.

49. Heptane and 1-heptene are both colorless liquids that boil in the range of 90–100°C. Describe a simple chemical test that would distinguish between the two liquids.

50. Describe the bonding in ethane, ethene, and ethyne. Are they different or the same? Explain.

51. An unknown compound has the molecular formula C_5H_{10}. This compound does not show addition reactions with hydrogen or water. Write one possible structure for this compound. Draw four other structures that have the formula C_5H_{10}.

52. Draw structures for all benzene compounds with the formula C_8H_9Cl. (There are 14 isomers.)

53. Identify the hybrid orbitals that surround each carbon in (a) 3,3-dimethyl-1-pentene, (b) propyne, and (c) m-xylene.

54. Name and draw the condensed structural formulas for all alkenes with the molecular formula (a) C_4H_8 and (b) C_5H_8.

55. Which Kekulé type structure for naphthalene, (a) or (b), is correct? Why is the other structure incorrect?

(a) 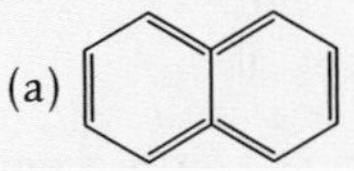 (b) 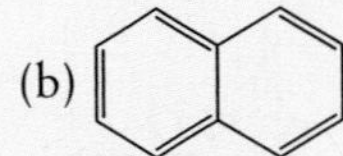

56. Why does 2,3-dimethyl-2-pentene not show geometric isomerism?

57. Classify each of the following compounds as a member of the alkane, alkene, alkyne or cycloalkene series:

(a) C_9H_{16} (b) C_7H_{14} (c) $C_{24}H_{50}$ (d) C_6H_{10}

Challenge Exercise

All exercises with blue numbers have answers in the appendix of the text.

* 58. You are given three colorless liquid samples that are known to be benzene, 1-hexene, and 1-hexyne. Using simple chemical tests, devise a scheme to identify these samples.

59. Starting with propyne and any other reactants, show by equations the synthesis of:
(a) 2-bromo-2-chloropropane
(b) 1,2-dibromo-2-chloropropane
(c) 1,1,2,2-tetrachloropropane

Answers to Practice Exercises

20.1 (a) $\overset{①}{C}H_3-\overset{②}{C}H=\overset{③}{C}(\overset{⑤}{C}H_3)-\overset{④}{C}H_3$ (b) $\overset{①}{C}H\equiv\overset{②}{C}-\overset{③}{C}H_2-\overset{④}{C}H_3$

(a) sp^2 hybrids: 2, 3; sp^3 hybrids: 1, 4, 5
(b) sp hybrids: 1, 2; sp^3 hybrids: 3, 4

20.2 (a) $CH_3CH_2CH=CHCH_2CH_3$
(b) $CH_3CH=CHCH(CH_2CH_3)CH_2CH_2CH_3$
(c) $CH_3CH=C(CH_3)CH(CH_3)CH_3$
(d) $CH_3CH(C_6H_{11})CH=CHCH_2CH_3$ (cyclohexyl on C-2)

20.3 (a) 3-methyl-2-pentene (b) 2,4-dimethyl-2-pentene

20.4 (a) 2-methyl-1-butene (b) 2-methyl-2-butene

20.5 (a) $(CH_3CH_2)(CH_3)C=C(CH_3)(H)$ — CH_3CH_2 and CH_3 on one carbon; CH_3 (upper) and H on the other
(b) $(Br)(CH_3)C=C(CH_2CH_3)(H)$ — Br and CH_3 on one carbon; CH_2CH_3 (upper) and H on the other

20.6 (a) Yes
$(CH_2Cl)(H)C=C(CH_3)(CH_2CH_3)$ — trans
$(CH_2Cl)(H)C=C(CH_2CH_3)(CH_3)$ — cis

(b) Yes
$(CH_3CH_2)(H)C=C(CH_2CH_3)(H)$ — cis
$(H)(CH_3CH_2)C=C(CH_2CH_3)(H)$ — trans

(c) No geometric isomers

20.7 (a) trans (b) cis

20.8 (a) Isomer II will be used in fatty acid synthesis.
(b) Isomer I will be used for energy.

20.9 (a) 1,2-dimethylcyclohexene
(b) 3,5-dimethylcyclohexene
(c) 3-ethyl-2,5-dimethyl-1,3-cyclohexadiene

20.10

$$\mathrm{(H)(COOH)C{=}C(HOOC)(CH_2COOH)}$$
cis

$$\mathrm{(HOOC)(H)C{=}C(HOOC)(CH_2COOH)}$$
trans

20.11 (a) $CH_3CH_2CHCH_2CH_3$ with CH_3 on the central C

(b) $CH_3CHCCH_2CH_3$ with CH_3 on the third C, and Br on the second and third C

(c) $CH_3CH_2CCH_2CH_3$ with CH_3 and Cl on the central C

(d) $CH_3CH_2CCH_2CH_3$ with CH_3 and OH on the central C

20.12 (a) 1-methylcyclopentene $+ \; H_2O \xrightarrow{H^+}$ 1-methylcyclopentanol (cyclopentane bearing CH_3 and OH on the same carbon)

(b) $\mathrm{(H)(CH_3)C{=}C(CH_3)(CH_3)} + HI \longrightarrow CH_3CH_2CCH_3$ with CH_3 and I on the third C

20.13

$$\mathrm{NH_2{-}CH{-}COOH}$$
with $HC{-}OH$ on the CH, which in turn bears CH_3

20.14 (a) $CH_3CHC{\equiv}CH$ with CH_3 on the second C

(b) $CH_3C{\equiv}C{-}C{-}CH_2CH_3$ with CH_3 and CH_2CH_3 on the fourth C

(c) cyclohexyl$-C{\equiv}CH$

20.15 (a) $CH_3CH_2CCl_2CH_3$
(b) $CH_3CH_2CCl_2CH_2CH_3 + CH_3CCl_2CH_2CH_2CH_3$

20.16

$$\mathrm{NH_2{-}CH{-}COOH}$$
with CH_2–phenyl ($C_6H_5CH_2$) on the CH

20.17 phenol rings bearing OH and Cl:

o-chlorophenol *m*-chlorophenol *p*-chlorophenol

20.18

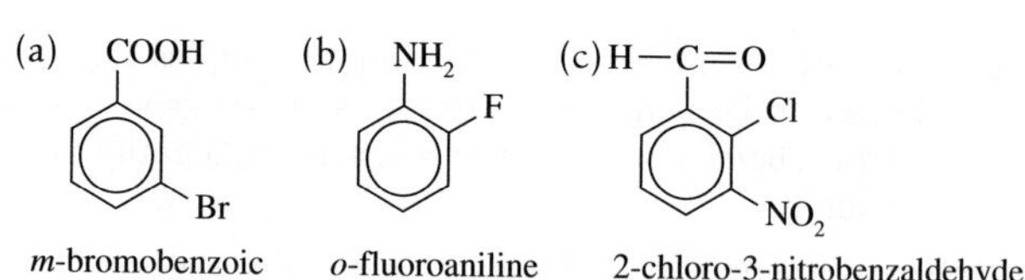

m-bromobenzoic acid *o*-fluoroaniline 2-chloro-3-nitrobenzaldehyde

20.19 (a) *o*-hydroxyphenol or *o*-dihydroxybenzene
(b) phenol (c) 2,6-diiodophenol

20.20 4-R-phenol (OH, R para) $+ \; I^+ \longrightarrow$ [ring carbocation bearing OH, with H and I on the same ring carbon, R para to OH, positive charge delocalized (+)]

CHAPTER 21

Polymers: Macromolecules

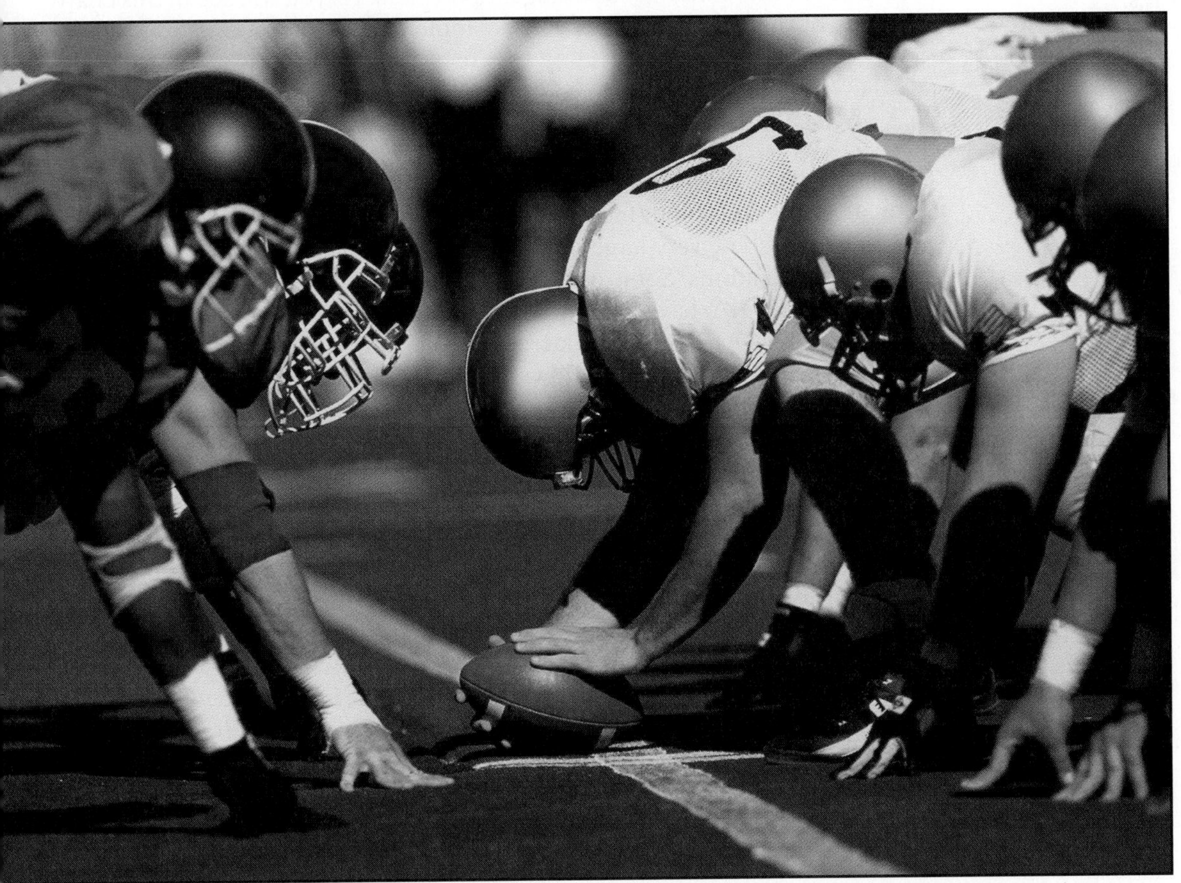

The helmets protecting the football players' heads are made of plastic, as are many other parts of their uniforms.

Chapter Outline

This concept car is made almost entirely of plastics.

What is a "mer"? The terms *polymer* and *monomer* are part of our everyday speech, but we are usually familiar only with the prefixes, *poly* meaning "many" and *mono* meaning "one." Although "mer" sounds like something from a child's cartoon show, scientists use this small three-letter root to convey an important concept. It is derived from the Greek *meros*, meaning "part." So, a monomer is a "one part" and a polymer is a "many part."

Look around you and observe the objects that could be described in these terms. A brick building could be described as a polymer, with each brick representing a monomer. A string of pearls is an elegant polymer, with each pearl a monomer. Starting with simple chemical monomers, chemists have learned to construct both utilitarian and elegant polymers. As you will see in this chapter, many of the polymers are of great commercial importance.

21.1 Macromolecules

Thus far, we have dealt mainly with rather small organic molecules that contain up to 50 atoms and some (fats) that contain up to about 150 atoms. But there exist in nature some very large molecules (macromolecules) that contain tens of thousands of atoms. Some of these, such as starch, glycogen, cellulose, proteins, silk, and DNA, have molar masses in the millions and are central to many of our life processes. Synthetic macromolecules touch every phase of our lives. Today, it is hard to imagine a world without polymers. Textiles for clothing, carpeting, draperies, shoes, toys, automobile parts, construction materials, synthetic rubber, chemical equipment, medical supplies, cooking utensils, synthetic leather, recreational equipment—the list could go on and on—all these and a host of others that we consider to be essential in our daily lives are wholly or partly synthetic polymers. Until about 1930, most of these polymers were unknown. The vast majority of these polymeric materials are based on petroleum. Because petroleum is a nonreplaceable resource, our dependence on polymers is another good reason for not squandering the limited world supply of petroleum.

Polyethylene is an example of a synthetic polymer. Ethylene, derived from petroleum, is made to react with itself to form polyethylene (or polythene). Polyethylene is a long-chain hydrocarbon made from many ethylene units:

$$\text{Many units } CH_2{=}CH_2 \longrightarrow -\underbrace{CH_2CH_2}_{\text{Ethylene unit}}-\underbrace{CH_2CH_2}_{\text{Ethylene unit}}-\underbrace{CH_2CH_2}_{\text{Ethylene unit}}-\underbrace{CH_2CH_2}_{\text{Ethylene unit}}-$$

$$\underset{\text{ethylene}}{n\,CH_2{=}CH_2} \xrightarrow{\text{catalyst}} \underset{\text{polyethylene}}{-\!(CH_2CH_2)_n\!-}$$

A typical polyethylene molecule contains anywhere from 2,500 to 25,000 ethylene units joined in a continuous chain.

The process of forming very large, high-molar-mass molecules from smaller units is called **polymerization**. The large molecule, or unit, is called the **polymer** and the small unit, the **monomer**. The term *polymer* is derived from the Greek word *polumerēs*, meaning "having many parts." Ethylene is a

polymerization
polymer
monomer

monomer, and polyethylene is a polymer. Because of their large size, polymers are often called *macromolecules*. Another commonly used term is *plastics*. The word *plastic* means to be capable of being molded, or pliable. Although not all polymers are pliable and capable of being remolded, *plastics* has gained general acceptance and has come to mean any of a variety of polymeric substances.

21.2 Synthetic Polymers

Some of the early commercial polymers were merely modifications of naturally occurring substances. One chemically modified natural polymer, nitrated cellulose, was made and sold as Celluloid late in the 19th century. But the first commercially successful, fully synthetic polymer, Bakelite, was made from phenol and formaldehyde by Leo Baekeland in 1909. This was the beginning of the modern plastics industry. Chemists began to create many synthetic polymers in the late 1920s. Since then, ever-increasing numbers of synthetic macromolecular materials have transformed the modern world. Even greater numbers of polymers have been made and discarded for technical or economic reasons. Polymers are used extensively in nearly every industry. For example, the electronics industry uses "plastics" in applications, ranging from microchip production to fabrication of heat and impact-resistant cases for the assembled products. In the auto industry, huge amounts of polymers are used as body, engine, transmission, and electrical system components, as well as for tires. Vast quantities of polymers with varied and sometimes highly specialized characteristics are used for packaging; for example, packaging for frozen foods must withstand subfreezing temperatures as well as those experienced in either conventional or microwave ovens.

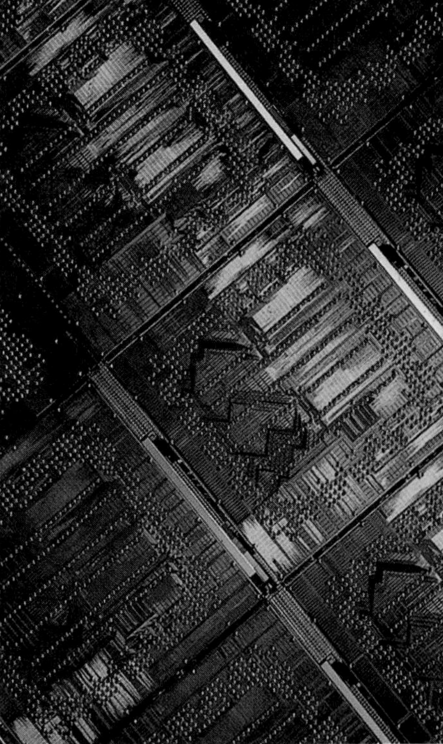

Synthetic polymers are used in the production of microchips.

Although there is a great variety of synthetic polymers, based on their properties and uses, they can be classified into the following general groups:

1. Rubberlike materials or elastomers
2. Flexible films
3. Synthetic textiles and fibers
4. Resins (or plastics) for casting, molding, and extruding
5. Coating resins for dip-, spray-, or solvent-dispersed applications
6. Miscellaneous (e.g., hydraulic fluids, foamed insulation, ion-exchange resins)

21.3 Polymer Types

Organic chemistry has had perhaps the greatest impact on our lives via the polymers (plastics) that we use. Although these polymers often have complex structures, the underlying chemistry of polymerization is relatively simple. Two classes of organic reactions are routinely used, the addition reaction and the substitution reaction. An **addition polymer** is a polymer that is produced by successive *addition reactions*. Polyethylene is an example of an addition polymer. A **condensation polymer** is a polymer that is formed when monomers combine and split out water or some other simple substance—a *substitution reaction*. Nylon is a condensation polymer.

addition polymer

condensation polymer

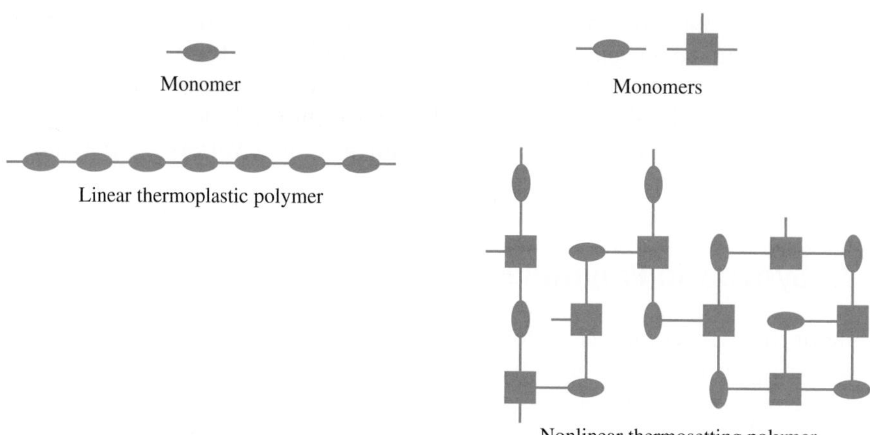

Figure 21.1
Diagrams of thermoplastic and thermosetting polymer structures.

thermoplastic polymers
thermosetting polymers

Polymers are either thermoplastic or thermosetting. Those that soften on reheating are **thermoplastic polymers**; those that set to an infusible solid and do not soften on reheating are **thermosetting polymers**. Thermoplastic polymers are formed when monomer molecules join end to end in a linear chain with little or no cross-linking between the chains. Thermosetting polymers are macromolecules in which the polymeric chains are cross-linked to form a network structure. The structures of thermoplastic and thermosetting polymers are illustrated in Figure 21.1.

21.4 Addition Polymerization

Addition polymerization starts with monomers that contain C=C double bonds. When these double bonds react, each alkene carbon bonds to another monomer:

$$\mathrm{C{=}C} + \mathrm{C{=}C} \longrightarrow -\mathrm{C}-\mathrm{C}-\mathrm{C}-\mathrm{C}-$$

This reaction is typical of many alkene addition reactions. (See Chapter 20.)

Ethylene polymerizes to form polyethylene according to the reaction shown in Section 21.1. Polyethylene is one of the most important and also the most widely used polymer on the market today. It is a tough, inert, but flexible thermoplastic material. Over 1.1×10^{10} kg of polyethylene is produced annually in the United States alone. Polyethylene is made into hundreds of different articles, many of which we use every day. Low-density polyethylene is stretchable and is used in objects such as food wrap and plastic squeeze bottles; high-density polyethylene is used in items that require more strength, such as plastic grocery bags and cases for electronic components.

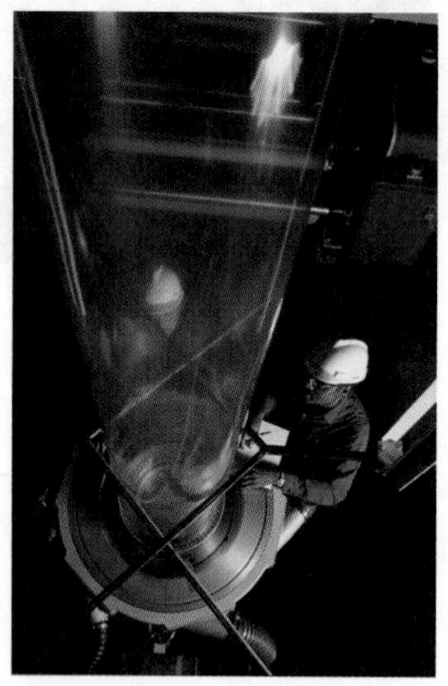

Polyethylene film is produced by blowing a "plastic bubble."

The double bond is the key structural feature involved in the polymerization of ethylene. Ethylene derivatives, in which one or more hydrogen atoms have been replaced by other atoms or groups, can also be polymerized. This is often called *vinyl polymerization*. Many of our commercial synthetic polymers are made from such modified ethylene monomers. The names, structures, and uses of some of these polymers are given in Table 21.1.

Table 21.1 Polymers Derived from Modified Ethylene Monomers

Monomer	Polymer	Identification code for most commonly recycled polymers	Uses
$CH_2{=}CH_2$ ethylene	$\text{-}(CH_2{-}CH_2)_n\text{-}$ polyethylene	4 LDPE (Low-Density Polyethylene)	Packing material, molded articles, plastic films, garbage bags, flexible bottles, containers, and toys
		2 HDPE (High-Density Polyethylene)	Plastic grocery bags; bottles for milk, juice, water, and laundry products; medical products
$CH_2{=}CH{-}CH_3$ propylene	$\text{-}(CH_2{-}CH(CH_3))_n\text{-}$ polypropylene	5 PP	Textile fibers, molded articles, automotive parts, lightweight ropes, and autoclavable biological equipment
$CH_2{=}C(CH_3)_2$ isobutylene	$\text{-}(CH_2{-}C(CH_3)_2)_n\text{-}$ polyisobutylene		Pressure-sensitive adhesives, butyl rubber (contains some isoprene as copolymer)
$CH_2{=}CH{-}Cl$ vinyl chloride	$\text{-}(CH_2{-}CH(Cl))_n\text{-}$ polyvinyl chloride (PVC)	3 V	Garden hoses, pipes, molded articles, floor tile, electrical insulation, medical tubing, blood bags, and vinyl leather
$CH_2{=}CCl_2$ vinylidene chloride	$\text{-}(CH_2{-}CCl_2)_n\text{-}$ Saran		Food packaging, textile fibers, pipes, and tubing (contains some vinyl chloride as copolymer)
$CH_2{=}CH{-}CN$ acrylonitrile	$\text{-}(CH_2{-}CH(CN))_n\text{-}$ Orlon, Acrilan		Textile fibers
$CF_2{=}CF_2$ tetrafluoroethylene	$\text{-}(CF_2{-}CF_2)_n\text{-}$ Teflon		Gaskets, valves, insulation, heat-resistant and chemical-resistant coatings, linings for pots and pans
$CH_2{=}CH{-}C_6H_5$ styrene	$\text{-}(CH_2{-}CH(C_6H_5))_n\text{-}$ polystyrene	6 PS	Molded articles, Styrofoam, insulation, toys, disposable food containers
$CH_2{=}CH{-}OC({=}O){-}CH_3$ vinyl acetate	$\text{-}(CH_2{-}CH(OC({=}O){-}CH_3))_n\text{-}$ polyvinyl acetate		Adhesives, paint, and varnish
$CH_2{=}C(CH_3){-}C({=}O){-}O{-}CH_3$ methylmethacrylate	$\text{-}(CH_2{-}C(CH_3)(C({=}O){-}O{-}CH_3))_n\text{-}$ Lucite, Plexiglas (acrylic resins)		Contact lenses, clear sheets for windows and optical uses, molded articles, and automobile finishes

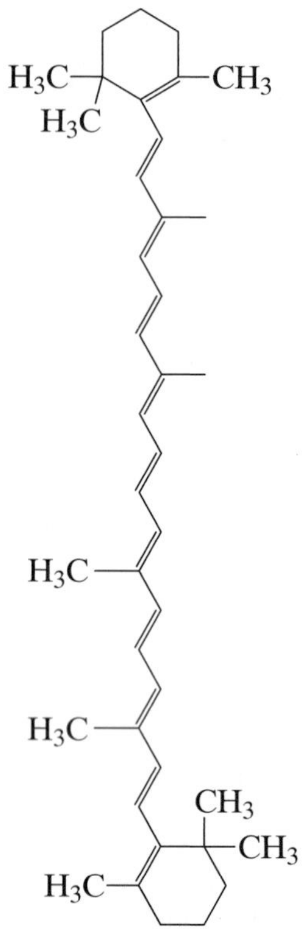

Many important biochemicals are formed by addition polymerization. Beta-carotene, a major source of dietary vitamin A, is a good example.

Free radicals catalyze or initiate many addition polymerizations. Organic peroxides (ROOR) are frequently used for this purpose. The reaction proceeds as shown in a typical free-radical mechanism:

Step 1 **Free-radical formation.** The peroxide splits into free radicals:

$$\text{RO:OR} \longrightarrow 2\,\text{RO}\cdot$$

Step 2 **Propagation of polymeric chain.** The initial free radical adds to ethylene to form a new free radical. The chain continues to elongate (polymerize) as long as free radicals continue to add to ethylene:

$$\text{RO}\cdot + \text{CH}_2\text{=CH}_2 \longrightarrow \text{ROCH}_2\text{CH}_2\cdot$$

$$\text{ROCH}_2\text{CH}_2\cdot + \text{CH}_2\text{=CH}_2 \longrightarrow \text{ROCH}_2\text{CH}_2\text{CH}_2\text{CH}_2\cdot$$

$$\text{ROCH}_2\text{CH}_2\text{CH}_2\text{CH}_2\cdot + \text{CH}_2\text{=CH}_2 \longrightarrow \text{RO(CH}_2\text{CH}_2)_n\cdot$$

Step 3 **Termination.** Polymerization stops when the free radicals are used up. This occurs when free radicals combine to form a stable compound:

$$\text{RO(CH}_2\text{CH}_2)_n\cdot + \cdot\text{OR} \longrightarrow \text{RO(CH}_2\text{CH}_2)_n\text{OR}$$

$$\text{RO(CH}_2\text{CH}_2)_n\cdot + \cdot(\text{CH}_2\text{CH}_2)_n\text{OR} \longrightarrow \text{RO(CH}_2\text{CH}_2)_n(\text{CH}_2\text{CH}_2)_n\text{OR}$$

Addition (or vinyl) polymerization of ethylene and its substituted derivatives yields saturated polymers—that is, polymer chains without carbon–carbon double bonds. The pi bond is eliminated when a free radical adds to an ethylene molecule. One electron of the pi bond pairs with the unpaired electron of the free radical, thus bonding the radical to the ethylene unit. The other pi bond electron remains unpaired, generating a new and larger free radical. This new free radical then adds another ethylene molecule, continuing the building of the polymeric chain. This process is illustrated by the following electron-dot diagram:

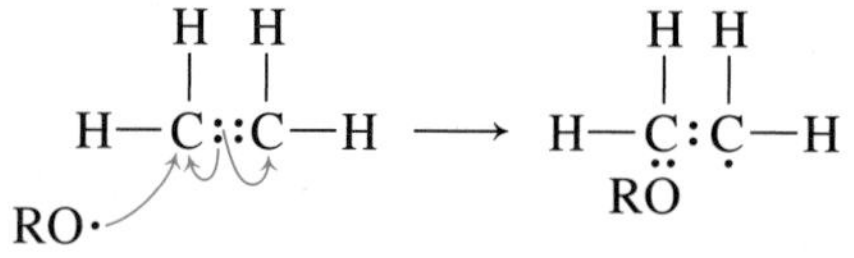

$$\text{RO}\cdot + \text{H}-\overset{\text{H}}{\underset{}{\text{C}}}::\overset{\text{H}}{\text{C}}-\text{H} \longrightarrow \text{H}-\overset{\text{H}}{\underset{\text{RO}}{\text{C}}}:\overset{\text{H}}{\underset{\cdot}{\text{C}}}-\text{H} \quad (\text{ROCH}_2\text{CH}_2\cdot)$$

Practice 21.1

Complete this free-radical addition equation:

$$\text{ROCH}_2\text{CH}_2\cdot + \text{CH}_2\text{=CHCH}_3 \longrightarrow$$

21.5 Addition Polymers: Use and Reuse (Recycling)

Over 50% of all polymers produced and used in the United States are made from ethylene and related compounds. These polymers are inexpensive to produce and can be shaped and molded into almost any form. Their very utility makes disposal a severe problem.

Plastic is ground up during the recycling process before it is used to make new articles such as these play structures.

For millions of years, microorganisms have broken down large molecules to smaller ones, starting the biosynthetic process over again. Unfortunately, the C—C single bonds of common addition polymers, like polyethylene, are different from those of most natural polymers and cannot be metabolized by many microorganisms. Consequently, while paper and cardboard in our landfills are very slowly disintegrated by microorganisms, most of the plastic waste is not.

Recycling seems to be the best solution to the problem of disposing of these long-lived addition polymers. Reuse of polyethylene (or any other of these plastics in relatively pure form) is fairly easy—simply melt and re-form into the desired product. The big problems of recycling lie with separating the different types of plastic; some plastics are not compatible with others and must be separated before reuse. The symbols on plastic goods (see Table 21.1) are one step toward plastic identification and separation. Each plastic container used in commerce carries a number that identifies its constituent polymer. It is possible (although difficult) to manually separate containers based on this code. Current research has focused on machine-separation techniques that depend on properties of plastics such as density. We recycle a relatively small amount of all plastics (about 40% of all soft-drink bottles), compared with about 55% of aluminum; we clearly could recycle much more.

21.6 Butadiene Polymers

A diene is a compound that contains two carbon–carbon double bonds. Another type of addition polymer is based on the compound 1,3-butadiene or its derivatives:

$$\overset{1}{C}H_2{=}\overset{2}{C}H{-}\overset{3}{C}H{=}\overset{4}{C}H_2$$

1,3-butadiene

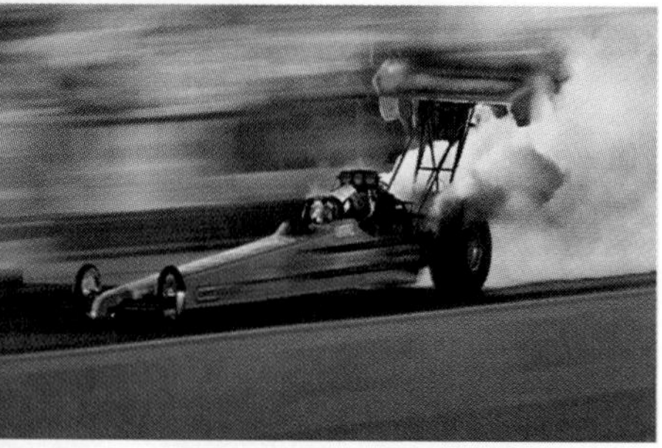

The tires of these cars are made of rubber that has been vulcanized. This process improves the temperature range for use and resistance to abrasion.

Natural rubber is a polymer of isoprene (2-methyl-1,3-butadiene). Many synthetic elastomers or rubberlike materials are polymers of isoprene or of butadiene. Unlike the saturated ethylene polymers, these polymers are unsaturated; that is, they have double bonds in their polymeric structures:

$$n\ CH_2{=}\underset{\displaystyle}{\overset{\displaystyle CH_3}{\overset{|}{C}}}{-}CH{=}CH_2 \longrightarrow {+}CH_2{-}\overset{\displaystyle CH_3}{\overset{|}{C}}{=}CH{-}CH_2{+}_n$$

isoprene (2-methyl-1,3-butadiene) → rubber polymer chain (polyisoprene)

$$n\ CH_2{=}CH{-}CH{=}CH_2 \longrightarrow {+}CH_2{-}CH{=}CH{-}CH_2{+}_n$$

1,3-butadiene → butadiene polymer chain

$$n\ CH_2{=}\overset{\displaystyle Cl}{\overset{|}{C}}{-}CH{=}CH_2 \longrightarrow {+}CH_2{-}\overset{\displaystyle Cl}{\overset{|}{C}}{=}CH{-}CH_2{+}_n$$

2-chloro-1,3-butadiene (chloroprene) → neoprene polymer chain (polychloroprene)

In this kind of polymerization, the free radical adds to the butadiene monomer at carbon 1 of the carbon–carbon double bond. At the same time, a double bond is formed between carbons 2 and 3, and a new free radical is formed at carbon 4. This process is illustrated in the following diagram:

$$RO\cdot + H{-}\overset{\displaystyle H}{\overset{|1}{C}}{::}\overset{\displaystyle H}{\overset{|2}{C}}{:}\overset{\displaystyle H}{\overset{|3}{C}}{::}\overset{\displaystyle H}{\overset{|4}{C}}{-}H \longrightarrow RO{:}\underset{\displaystyle H}{\underset{|}{\overset{\displaystyle H}{\overset{|}{C}}}}{:}\overset{\displaystyle H}{\overset{|}{C}}{::}\overset{\displaystyle H}{\overset{|}{C}}{:}\underset{\displaystyle H}{\underset{|}{\overset{\displaystyle H}{\overset{|}{C}}}}\cdot$$

free radical; 1,3-butadiene → free-radical chain lengthened by four carbon atoms

The process continues by adding butadiene molecules until the polymeric chain is terminated.

One of the outstanding synthetic rubbers (styrene–butadiene rubber, SBR) is made from two monomers, styrene and 1,3-butadiene. These substances form a **copolymer**—that is, a polymer containing two different kinds of monomer units. Styrene and butadiene do not necessarily have to combine in a 1 : 1 ratio. In the actual manufacture of SBR polymers, about 3 mol of butadiene is used per mole of styrene. Thus, the butadiene and styrene units are intermixed, but in a ratio of about 3:1 and the polymer is represented in the following way:

copolymer

$$-CH_2CH{=}CHCH_2{-}|{-}\underset{\displaystyle C_6H_5}{CH_2CH}{-}|{-}CH_2CH{=}CHCH_2{-}|{-}CH_2CH{=}CHCH_2{-}$$

styrene unit; butadiene unit

segment of styrene–butadiene rubber (SBR)

Practice 21.2

Show one unit of a copolymer made from the monomers vinyl chloride and 1,1-dichloroethene.

The presence of double bonds at intervals along the chains of rubber and rubberlike synthetic polymers designed for use in tires is almost a necessity and, at the same time, a disadvantage. On the positive side, double bonds make vulcanization possible. On the negative side, double bonds afford sites where ozone, present especially in smoggy atmospheres, can attack the rubber, causing "age hardening" and cracking. Vulcanization extends the useful temperature range of rubber products and imparts greater abrasion resistance to them. The vulcanization process is usually accomplished by heating raw rubber with sulfur and other auxiliary agents. It consists of introducing sulfur atoms that connect or cross-link the long strands of polymeric chains. Vulcanization was devised through trial-and-error experimentation by the American inventor Charles Goodyear in 1839, long before the chemistry of the process was understood. Goodyear's patent on "Improvement in India Rubber" was issued on June 15, 1844. In the segment of vulcanized rubber shown here, the chains of polymerized isoprene are cross-linked by sulfur–sulfur bonds, giving the polymer more strength and elasticity:

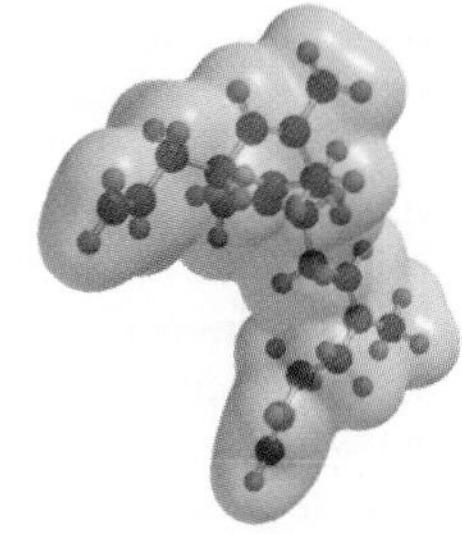

A segment of all-cis natural rubber: cis isomers give rubber molecules a "kinked" shape.

$$
\begin{array}{c}
\quad CH_3 \qquad\qquad\quad CH_3 \qquad\qquad\quad CH_3 \\
-CH_2C{=}CHCHCH_2C{=}CHCH_2CHC{=}CHCH_2- \\
\qquad S \qquad\qquad\qquad\qquad S \\
\qquad | \qquad\qquad\qquad\qquad | \\
\qquad S \qquad\qquad\qquad\qquad S \\
-CH_2C{=}CHCHCH_2C{=}CHCH_2CHC{=}CHCH_2- \\
\quad CH_3 \qquad\qquad\quad CH_3 \qquad\qquad\quad CH_3
\end{array}
$$

segment of vulcanized rubber

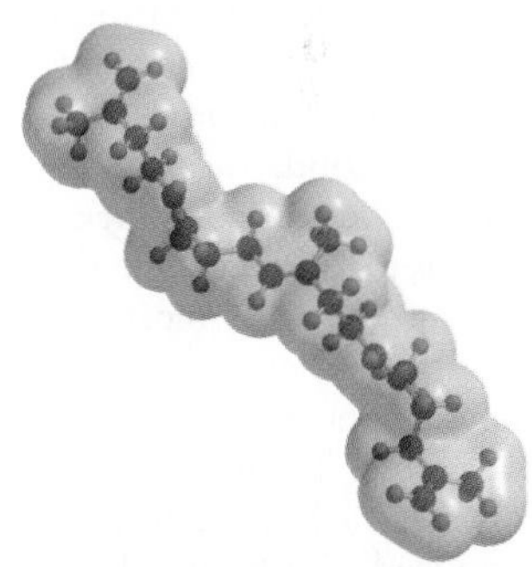

A segment of all-trans gutta-percha: trans isomers give gutta-percha molecules a linear shape.

21.7 Geometric Isomerism in Polymers

The recurring double bonds in isoprene and butadiene polymers make it possible to have polymers with specific spatial orientation as a result of cis–trans isomerism. Recall from Section 20.3 that two carbon atoms joined by a double bond are not free to rotate and thus give rise to cis–trans isomerism. An isoprene polymer can have all-cis, all-trans, or a random distribution of cis and trans configurations about the double bonds.

Natural rubber is *cis*-polyisoprene with an all-cis configuration about the carbon–carbon double bonds. Gutta-percha, also obtained from plants, is a *trans*-polyisoprene with an all-trans configuration. Although these two polymers have the same composition, their properties are radically different. The cis natural rubber is a soft, elastic material, whereas the trans gutta-percha is a tough, nonelastic, hornlike substance. Natural rubber has many varied uses.

CHEMISTRY IN ACTION • Adjusting Polyethylene to Bear the Load

Bone failure is often a symptom of aging: Osteoporosis is very common, and joint failure is a major concern. Take, for example, the hip joint. The upper leg bone's rounded end fits into a socket on the hip; it is where the bones meet (the bearing surface) that failure is common. Disease states such as rheumatoid arthritis as well as simply extended use can overwhelm our body's ability to compensate. Discomfort, pain, and a loss of mobility result, and a new hip joint is needed. Scientists have been working for the past 50 years to perfect artificial hip joints.

Finding the best bearing surface has proven difficult. The bearing surface must meet diverse and exacting criteria. First, the bearing surface is exposed to about seven times the body's weight in force, so it must be tough. Second, the bearing surface must provide some "shock absorber" character, so it must be flexible. Finally, and perhaps most importantly, the surface needs to withstand many millions of eccentric motions; it must be durable. Today most hip replacement joints have a bearing surface composed of a hard ball (a stainless steel alloy) and a soft cup. Much engineering has gone into choosing the best soft cup material, and it may be surprising to learn that the most commonly chosen material is not a high-tech material but simply polyethylene.

We tend to take common plastics for granted, yet polyethylene, like other plastics, offers a very special engineering advantage: Its physical properties can be adjusted to fit the application. Low-density polyethylene (LDPE; the polymer chains are randomly packed) provides the toughness needed for the bearing surface, but it is too stretchable. Think of a clear plastic bread wrapper to visualize LDPE characteristics. You can stretch it, poke your finger into it, and yet it tears only with great difficulty. A more rigid polyethylene can be formed by forcing the polymer chains to align (high-density polyethylene, HDPE; milk containers are often made of HDPE). Although HDPE is rigid enough for the bearing surface of artificial joints, it is too brittle and can crack. The aligned chains give some crystallinity to the plastic, and, as with most crystals, HDPE has a tendency to crack.

To solve the cracking problem, very long polyethylene chains were synthesized. These extra-long chains didn't align quite as well as HDPE (lower crystallinity), but the extra-long chains provided the stiffness needed for artificial joints. This plastic (called ultra-high-molecular-weight polyethylene, UHMWPE) has been the plastic most commonly used in hip replacement joints since about 1970.

Over time a new problem has become apparent–artificial joints slowly lose contact with the bones. This failure commonly results from bone loss, osteolysis, and requires a new hip replacement about every 10 to 15 years. The root cause of the osteolysis seems to be microscopic particles of plastic that break away from the hip joint during normal daily activities. After these particles build to a certain concentration around the joint, the body's white cells start an immune reaction. In the process, bone cells around the artificial joint are killed (osteolysis) and the artificial joint no longer adheres to the bone.

So, again, plastics chemists were asked to make a polyethylene with new properties. Scientists reasoned that if the polymers were bonded together (cross-linked), wear would be inhibited. Radiation used to sterilize the artificial joint was also used to displace hydrogen from some of the polyethylene carbons; these carbons then bonded to each other, covalently connecting one chain to another. The lab tests indicate that this polyethylene is about ten times more wear resistant than HMWPE. Clinical trials have been in progress for several years, but the artificial joints must be in place for 10–20 years before the efficacy of this change can be judged. Meanwhile, polyethylene shows the true nature of a plastic: Its physical properties are adjustable to meet exacting conditions.

Some uses of gutta-percha include electrical insulation, dentistry, and golf balls.

$$-CH_2\diagdown\quad CH_2-CH_2\diagdown\quad CH_2-CH_2\diagdown\quad CH_2-$$
$$C=C\qquad C=C\qquad C=C$$
$$CH_3\quad H\quad CH_3\quad H\quad CH_3\quad H$$

all-*cis* configuration of natural rubber

$$-CH_2\diagdown\quad H\quad CH_3\diagdown\quad CH_2-CH_2\diagdown\quad H$$
$$C=C\qquad C=C\qquad C=C$$
$$CH_3\quad CH_2-CH_2\quad H\quad CH_3\quad CH_2-$$

all-*trans* configuration of gutta-percha

Chicle is another natural substance containing polyisoprenes. It is obtained by concentrating the latex from the sapodilla tree, which contains about 5% *cis*-polyisoprene and 12% *trans*-polyisoprene. The chief use of chicle is in chewing gum.

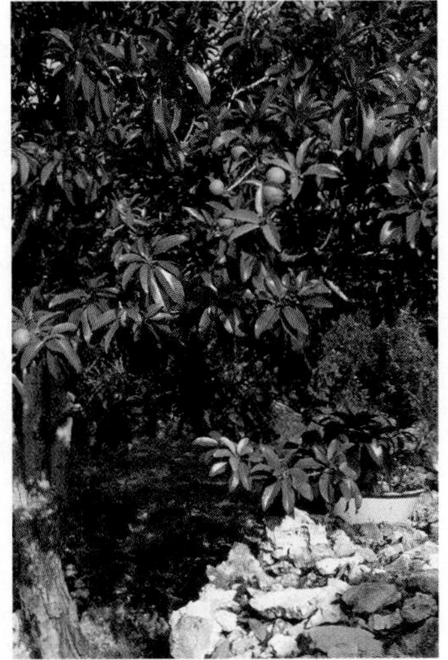

Natural chicle (for chewing gum) is obtained from the sapodilla tree found in North and South America.

Practice 21.3 Application to Biochemistry

Natural rubber is a member of a class of biochemicals called terpenes. Many plant fragrances and pigments are also terpenes. Two examples follow:

Carvone structure: ring ketone (O) bearing CH_3 on ring double bond (b); ring substituent $CH_3-C(=CH_2)-$ with double bond (a).

carvone, oil of spearmint

Geraniol structure: CH_3 (d) $-C{=}CH-CH_2-OH$; $CH_2-CH_2-CH{=}C(CH_3)-CH_3$ with double bond (c).

geraniol, oil of geranium

For each double bond, (a) through (d), decide whether (1) it is a cis isomer or (2) it is a trans isomer or (3) it does not show geometric isomerism.

Only random or nonstereospecific polymers are obtained by free-radical polymerization. Synthetic polyisoprenes made by free-radical polymerization are much inferior to natural rubber, since they contain both the cis and the trans isomers. But in the 1950s, Karl Ziegler (1898–1973) of Germany and Giulio Natta (1903–1979) of Italy developed catalysts [e.g., $(C_2H_5)_3Al/TiCl_4$] that allowed polymerization to proceed by an ionic mechanism, producing stereochemically controlled polymers. Ziegler–Natta catalysts made possible the synthesis of polyisoprene with an all-cis configuration and with properties fully comparable to those of natural rubber. This material is known by the odd, but logical name *synthetic natural rubber*. In 1963, Natta and Ziegler were jointly awarded the Nobel prize for their work on stereochemically controlled polymerization reactions.

Other types of polymers will be discussed in later chapters.

Practice 21.4

Random free-radical polymerization of isoprene produces a polymer with a mixture of cis and trans double bonds. Draw four units of a polymer that contains both the cis and trans forms.

Chapter 21 Review

21.1 Macromolecules

KEY TERMS

Polymerization
Polymer
Monomer

- Macromolecules commonly contain thousands of atoms.
- A monomer is a small bonded group of atoms that forms polymers by reacting with other monomers.
- A polymer is a macromolecule made up of many monomers.
- Polymerization is the chemical reaction that bonds monomers together to make a polymer.

21.2 Synthetic Polymers

- The first commercially successful synthetic polymer (Bakelite) was made in 1909.
- Today, nearly every commercial product employs some synthetic polymers.

21.3 Polymer Types

KEY TERMS

Addition polymer
Condensation polymer
Thermoplastic polymers
Thermosetting polymers

- An addition polymer (e.g., polyethylene) uses an addition reaction to connect monomers together.
- A condensation polymer (e.g., nylon) is formed when monomers bond together with a loss of water (a substitution reaction).
- Thermoplastic polymers soften and can be remolded every time they are heated.
- Thermosetting plastics set to an infusible solid upon heating and do not soften upon reheating.
- Thermoplastic polymers have few cross-links between polymer chains, while thermosetting polymers are cross-linked many times to form a network of interconnected polymer chains.

21.4 Addition Polymerization

- To make synthetic addition polymers, modified ethene (ethylene) molecules are commonly used.
- The double bond in one monomer adds to the double bond in an adjacent monomer to form these polymers.
- Addition polymerization is a free-radical chain reaction.

21.5 Addition Polymers: Use and Reuse (Recycling)

- Most synthetic addition polymers are not biodegradable.
- Recycling addition polymers seems to be the best means of decreasing plastic waste and reusing valuable organic molecules.
- Currently the biggest problem with recycling plastics may be sorting one polymer from another.

21.6 Butadiene Polymers

KEY TERMS

Copolymer

- A butadiene polymer is an addition polymer made from 1,3-butadiene or its derivatives.
- Many elastomers (rubber-like polymers) are butadiene polymers.
- A copolymer is made up of two different kinds of monomers.
- A copolymer's properties depend on the ratio of the two different monomers in the mixture.
- The double bonds in butadiene polymers allow vulcanization but also cause these polymers to be sensitive to attack by ozone.
  - Vulcanization uses sulfur to cross-link butadiene polymers in order to impart greater abrasion resistance.
  - Ozone (often from smog) reacts with butadiene polymers to cause "age hardening" and cracking.

21.7 Geometric Isomerism in Polymers

- Polymers with double bonds commonly have geometric isomerism.
- Natural rubber is an elastomer of all-cis polyisoprene while gutta-percha is tough nonelastic material made from all-trans polyisoprene.

Rewiew Questions

All questions with blue numbers have answers in the appendix of the text.

1. How does condensation polymerization differ from addition polymerization?
2. What property distinguishes a thermoplastic polymer from a thermosetting polymer?
3. Why must at least some monomer molecules be trifunctional to form a thermosetting polymer?
4. How are rubber molecules modified by vulcanization?
5. Strictly speaking, what does the word "plastic" mean? Why does this word apply to many organic polymers?
6. In addition polymerization, how does the termination step differ from the propagation step?
7. What is perhaps the biggest problem associated with recycling plastics?
8. Why does polyethylene not fit the definition of a copolymer?

Paired Exercises

All exercises with blue numbers have answers in the appendix of the text.

1. How many ethylene units are in a polyethylene molecule that has a molar mass of approximately 25,000?

2. Calculate the molar mass of a polystyrene molecule consisting of 3,000 monomer units.

3. Write formulas showing the structure of
(a) polyvinyl chloride
(b) poly-2-phenyl-2-butene
(c) polypropylene
(d) Lucite

4. Write structural formulas for the following polymers:
(a) Saran
(b) Orlon
(c) Teflon
(d) polystyrene

5. Draw the structure of the free radical that is formed when the first two propylene units combine during propagation of the polypropylene chain.

6. Draw the structure of the free radical that is formed when the first two ethylene units combine during propagation of the polyethylene chain.

7. Write a structural formula, showing the polymer that can be formed from
(a) propylene
(b) 2-methylpropene
(c) 2-butene

8. Write a structural formula showing the polymer that can be formed from
(a) ethylene
(b) chloroethene
(c) 1-butene

9. Write structures showing two possible ways in which vinyl chloride can polymerize to form polyvinyl chloride. Show four units in each structure.

10. Write structures showing two possible ways in which acrylonitrile can polymerize to form an "acrylic" fiber (Orlon, Acrilan). Show four units in each structure.

11. Natural rubber is the all-cis polymer of isoprene. Write the structure of the polymer showing at least three isoprene units.

12. Gutta-percha is the all-trans polymer of isoprene. Write the structure of the polymer showing at least three isoprene units.

13. Is the useful life of natural rubber shortened in smoggy atmospheres? Briefly explain your answer.

14. Is the useful life of styrene–butadiene rubber (SBR) shortened in smoggy atmospheres? Briefly explain your answer.

15. Write the chemical structures for the monomers of
(a) natural rubber
(b) synthetic (SBR) rubber

16. Write the chemical structures for the monomers of
(a) synthetic natural rubber
(b) neoprene rubber

17 Nitrile rubber (Buna N) is a copolymer of two parts of 1,3-butadiene and one part of acrylonitrile ($CH_2 = CHCN$). Write a structure for this synthetic rubber, showing at least two units of the polymer.

18. Ziegler–Natta catalysts can orient the polymerization of propylene to form isotactic polypropylene—that is, polypropylene with all the methyl groups on the same side of the long carbon chain. Write the structure for (a) isotactic polypropylene and (b) another possible geometric form of polypropylene.

19. Using Table 21.1, what polymer is symbolized by ♲6 PS?
Draw the structure of the monomer used to make this polymer.

20. Using Table 21.1, what polymer is symbolized by ♲3 V?
Draw the structure of the monomer used to make this polymer.

21. Draw the representative structure for the unsaturated polymer that is widely used and is also a copolymer.

22. Draw the representative structure for the polymer that is most widely used in the market today.

23. The following monomers undergo free-radical addition polymerization: (a) propene (propylene); (b) isoprene. In each case, how many double bonds per monomer unit remain *after* polymerization?

24. The following monomers undergo free-radical addition polymerization: (a) 1,3-butadiene; (b) tetrafluoroethylene. In each case, how many double bonds per monomer unit remain *after* polymerization?

Additional Exercises

All exercises with blue numbers have answers in the appendix of the text.

25. Of polyethylene, polypropylene, and polystyrene, which polymer contains the highest mass percent of carbon? Calculate the percentage.

26. Draw three units of the polymer that can be made from

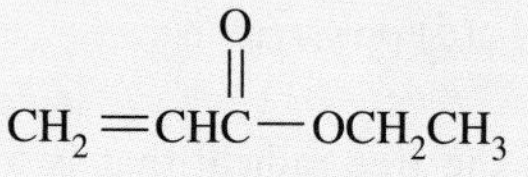

$$CH_2{=}CH{-}\overset{\overset{\displaystyle O}{\|}}{C}{-}OCH_2CH_3$$

27. Superglue contains the following monomer:

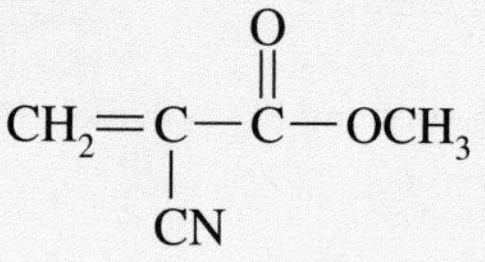

$$CH_2{=}\underset{\underset{\displaystyle CN}{|}}{C}{-}\overset{\overset{\displaystyle O}{\|}}{C}{-}OCH_3$$

cyanoacrylate ester

When this monomer is spread between two surfaces, it rapidly polymerizes to give a very strong bond. Write the structure of the polymer formed.

28. (a) Write the structure for a polymer that can be formed from 2,3-dimethyl-1,3-butadiene. (b) Can 2,3-dimethyl-1,3-butadiene form cis and trans isomers? Explain.

29. Would you expect it to be easier to recycle a thermosetting polymer or a thermoplastic polymer? Why?

30. A copolymer formed from ethylene and tetrafluoroethylene can be used to make wire insulation. Write the structural formulas for the monomers. Write a portion of the copolymer that you would expect from an addition polymerization reaction.

31. How many isoprene units are in a natural rubber that has a molar mass of 250,000 g/mol? Show your calculation.

32. The polymer used in hard contact lenses is polymethyl methacrylate. The structures of the monomer methyl methacrylate and a section of the polymer are shown here:

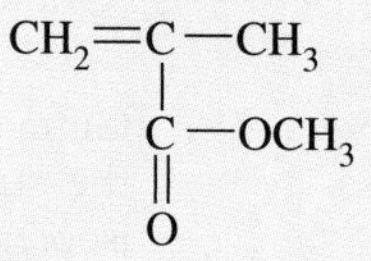

$$CH_2{=}\underset{\underset{\displaystyle \underset{\underset{\displaystyle O}{\|}}{C}{-}OCH_3}{|}}{C}{-}CH_3$$

methyl methacrylate

$$-CH_2{-}\overset{\overset{\displaystyle CH_3}{|}}{\underset{\underset{\displaystyle \underset{\underset{\displaystyle O}{\|}}{C}{-}OCH_3}{|}}{C}}{-}CH_2{-}\overset{\overset{\displaystyle CH_3}{|}}{\underset{\underset{\displaystyle \underset{\underset{\displaystyle O}{\|}}{C}{-}OCH_3}{|}}{C}}{-}CH_2{-}\overset{\overset{\displaystyle CH_3}{|}}{\underset{\underset{\displaystyle \underset{\underset{\displaystyle O}{\|}}{C}{-}OCH_3}{|}}{C}}{-}$$

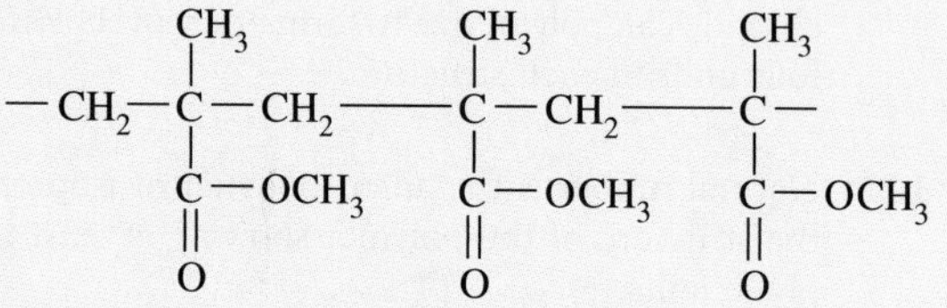

polymer

Would you expect this polymer to result from an addition reaction or a condensation reaction? Why?

Challenge Exercise

33. A self-proclaimed "plastics expert" tells you that recycling HDPE is difficult because this plastic must be broken down to monomers and then polymerized again. Do you agree or disagree? Explain.

Answers to Practice Exercises

21.1 $ROCH_2CH_2CH_2\underset{\underset{\displaystyle CH_3}{|}}{CH}\cdot$

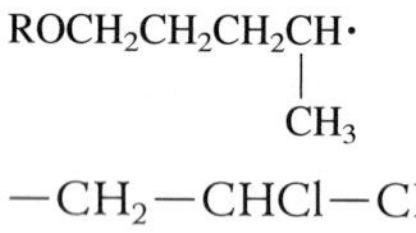

21.2 $-CH_2{-}CHCl{-}CH_2{-}CCl_2{-}$ or $-CH_2{-}CHCl{-}CCl_2{-}CH_2{-}$

21.3 (a) no geometric isomerism; (b) cis isomer; (c) no geometric isomerism; (d) trans isomer

21.4

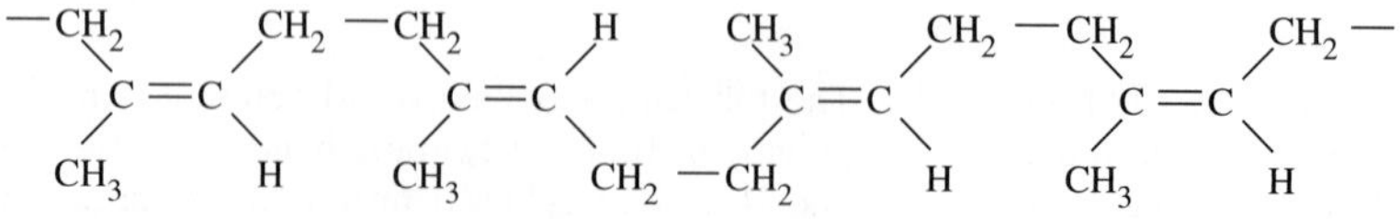

```
—CH2        CH2 —CH2        H      CH3        CH2 —CH2        CH2 —
    \      /        \      /          \      /        \      /
     C = C           C = C             C = C           C = C
    /      \        /      \          /      \        /      \
 CH3        H    CH3        CH2 —CH2          H    CH3        H
      cis              trans              trans             cis
```

PUTTING IT TOGETHER: Review for Chapters 19–21

Multiple Choice:

Choose the correct answer to each of the following.

1. The element present in every organic compound is
(a) hydrogen (b) oxygen (c) carbon (d) nitrogen

2. Compounds containing only carbon and hydrogen are known as
(a) methane (c) isomers
(b) hydrocarbons (d) carbohydrates

3. The simplest hydrocarbon compound is
(a) CH_2 (b) CH_3 (c) CH_4 (d) C_2H_4

4. The carbon–hydrogen sigma bond in methane is made by the overlap of
(a) an *s* orbital and an sp^3 orbital
(b) two sp^3 orbitals
(c) two *s* orbitals
(d) electron orbitals from two carbon atoms

5. In the alkane homologous series, the formula of each member differs from the preceding member by
(a) CH_3 (c) CH_2
(b) CH (d) no correct answer given

6. The general formula for a saturated open chain hydrocarbon is
(a) C_nH_{2n} (b) C_nH_n (c) C_nH_{2n+1} (d) C_nH_{2n+2}

7. The general formula for a saturated cycloalkane is
(a) C_nH_{2n} (b) C_nH_n (c) C_nH_{2n+1} (d) C_nH_{2n+2}

8. The general formula for an alkene is
(a) C_nH_{2n} (b) C_nH_n (c) C_nH_{2n+1} (d) C_nH_{2n+2}

9. The name for the alkane C_5H_{12} is
(a) propane (b) butane (c) nonane (d) pentane

10. Compounds made up of the same number and kinds of atoms, but differing in their molecular structure are known as
(a) isotopes (c) isomers
(b) homologs (d) hydrocarbons

11. The products of the complete combustion of hydrocarbons are
(a) carbon and hydrogen
(b) carbon monoxide and water
(c) carbon dioxide and water
(d) no correct answer given

12. When pentane is chlorinated, the number of possible monochlorosubstitution products is
(a) one (b) three (c) five (d) six

13. When hexane is chlorinated, the number of possible monochlorosubstitution products is
(a) one (b) three (c) five (d) six

14. The IUPAC name for $CH_3CH_2CH_2CHClCH_3$ is
(a) 4-chloropentane (c) 2-chloropentane
(b) 4-chlorohexane (d) 2-chloropropane

15. How many methyl and ethyl groups are in this formula?

```
            CH3
            |
CH3CH2CHCHCH2CH3
              |
              CH2CH3
```

(a) four (b) five (c) six (d) seven

16. The molecular formula for chlorocyclohexane is
(a) $C_6H_{11}Cl$ (c) $C_6H_{13}Cl$
(b) $C_6H_{12}Cl$ (d) no correct formula given

17. Which formula is iodoform?
(a) CH_3I (b) CH_2I_2 (c) CHI_3 (d) CI_4

18. Which compound can be both a cycloalkane and a cycloalkene?
(a) C_3H_6 (c) C_5H_8
(b) C_4H_8 (d) no correct answer given

19. What is the correct name for

```
   CH2CH3     CH2CH3
   |          |
CH3CH—CH2—CHCH3
```

(a) 4-ethyl-2-methylhexane
(b) 3,5-dimethylheptane
(c) 2-ethyl-4-methylhexane
(d) 2,4-diethylpentane

20. A pi bond is formed from
(a) two sp^3 electron orbitals
(b) two sp^2 electron orbitals
(c) an sp^3 and an sp^2 electron orbital
(d) two *p* orbitals

21. A carbon–carbon triple bond consists of
(a) a sigma bond and two pi bonds
(b) three pi bonds
(c) three sigma bonds
(d) two sigma bonds and one pi bond

22. The compound of formula C_6H_{12} cannot be
(a) an alkene (c) an alkyne
(b) a cycloalkane (d) no correct answer given

23. How many hydrogen atoms does the open-chain compound C_8H_{10} need to become a saturated hydrocarbon?
(a) 2 (b) 4 (c) 6 (d) 8

24. Which formula represents an alkyne hydrocarbon?
(a) C_nH_{2n} (b) C_nH_n (c) C_nH_{2n-2} (d) C_nH_{2n-4}

25. Which formula represents a cycloalkene hydrocarbon?
(a) C_nH_n (b) C_nH_{2n} (c) C_nH_{2n-2} (d) C_nH_{2n-4}

26. Which statement is *not* true? The compound C_6H_{10} can have in its structure
(a) two carbon–carbon double bonds
(b) one carbon–carbon triple bond
(c) one cyclic ring and one carbon–carbon double bond
(d) one cyclic ring and two carbon–carbon double bonds

27. The name of this compound is

$$CH_3CH{=}CHCHCH_3$$ with CH_3 on the fourth carbon

(a) 4-methyl-2-pentene (c) isohexene
(b) 2-methyl-3-pentene (d) no correct name given

28. The name of this compound is

$$CH_3C{\equiv}C{-}CHCH_3$$ with CH_2CH_3 on the fourth carbon

(a) 3-methyl-4-hexyne (c) 4-methyl-2-hexyne
(b) 4-ethyl-2-pentyne (d) 4-methyl-2-hexene

29. Which is an incorrect name for the compounds shown?

(a) $CH_3CH_2CH_2Cl$
1-chloropropane

(b) $CH_3CH_2CH{=}CHCH_3$
2-pentene

(c)
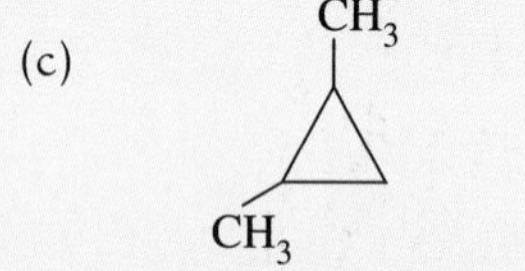

1,3-dimethylcyclopropane

(d) 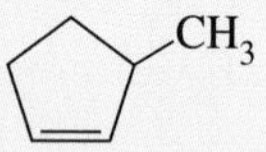 CH_3
3-methylcyclopentene

30. The formula $CH_3CH_2CH_2^+$ is a

(a) primary carbocation (c) tertiary carbocation
(b) secondary carbocation (d) quaternary carbocation

31. Which of the following compounds has a structure that can have cis–trans isomers?

(a) $CH_3CH{=}CHCl$

(b) $CH_3CH_2C{\equiv}CCH_3$

(c) $(CH_3)_2C{=}CHCl$

(d) $CH_3CH_2C{=}C{-}CH_2Cl$ with CH_3 and CH_2Cl on the double-bond carbons

32. Which of the following compounds is *cis*-2,3-dichloro-2-butene?

(a) 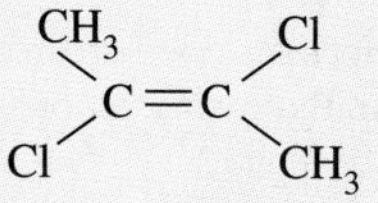 CH_3, Cl / C=C / Cl, CH_3 (c) CH_3, CH_2Cl / C=C / Cl, H

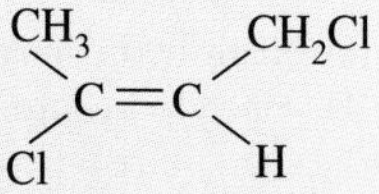

(b) 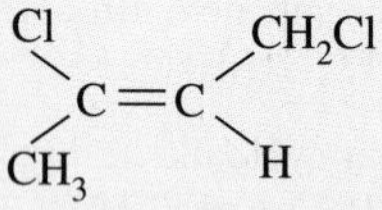 Cl, CH_2Cl / C=C / CH_3, H (d) 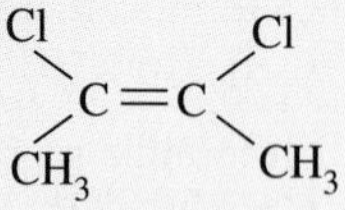 Cl, Cl / C=C / CH_3, CH_3

33. Which of the following compounds is aniline?

(a) 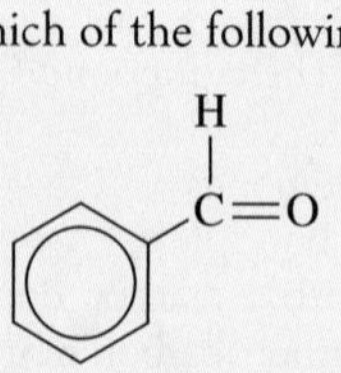 benzene ring with H–C=O (c) benzene ring with NO_2

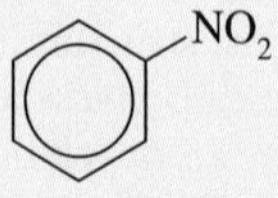

(b) 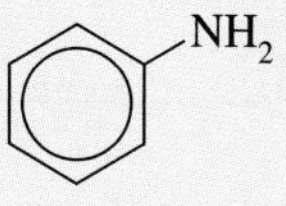 benzene ring with NH_2 (d) 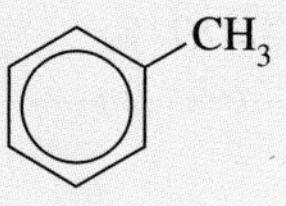 benzene ring with CH_3

34. Which of the following compounds is 2-chloro-4-nitrotoluene?

(a) 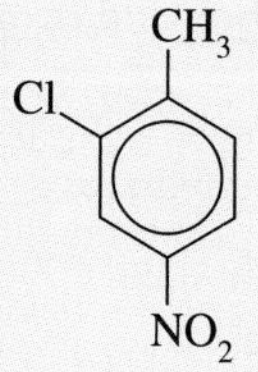 CH_3, Cl, NO_2 (c) 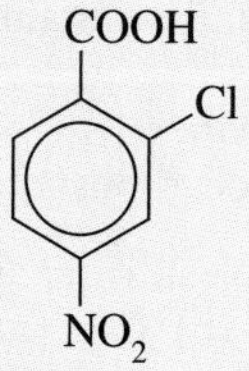 COOH, Cl, NO_2

(b) 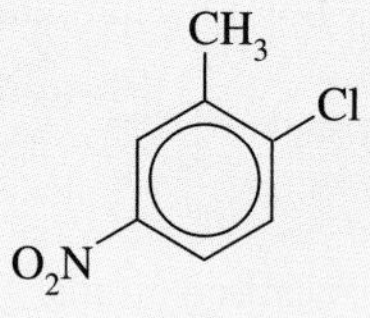 CH_3, Cl, O_2N (d) 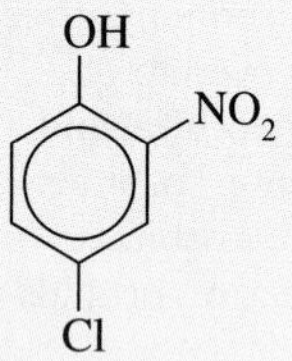 OH, NO_2, Cl

35. Which of the following compounds is *m*-xylene?

(a)
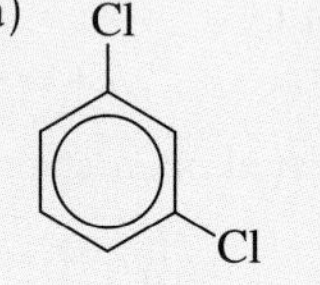

(c) 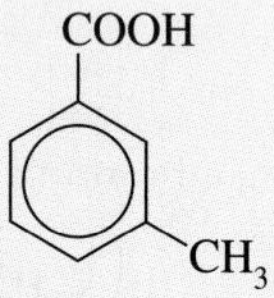 COOH, CH_3

(b)
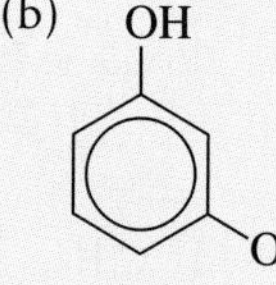

(d) 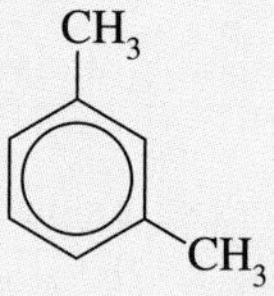 CH_3, CH_3

36. Which of the following compounds is benzaldehyde?

(a)
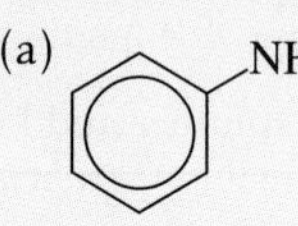

(c) 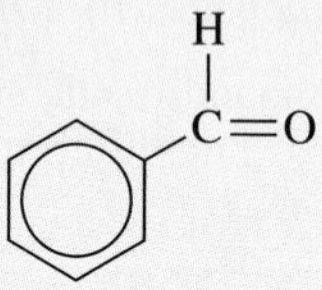 H–C=O

(b)
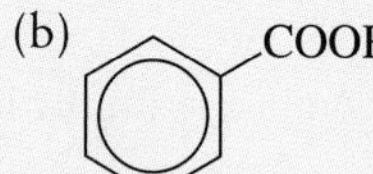

(d) 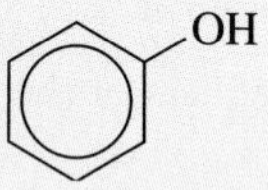 OH

37. Which of the following compounds is *p*-dihydroxybenzene?

(a) 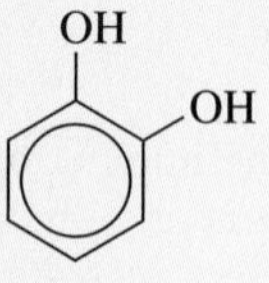 OH, OH (c) 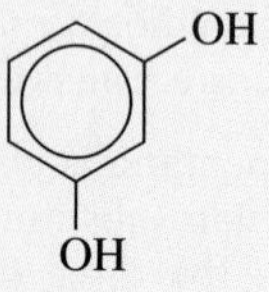 OH, OH

(b)
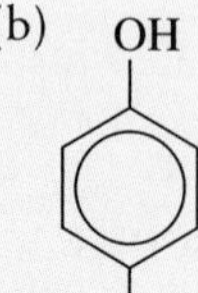

(d) 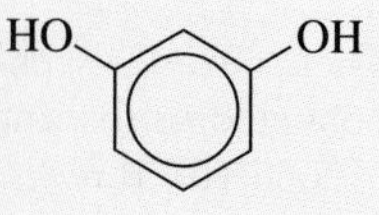 HO, OH

38. Which of the following compounds is naphthalene?

(a) 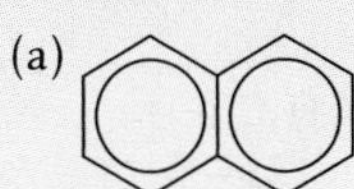

(b) 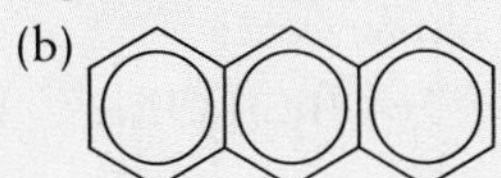

(c) 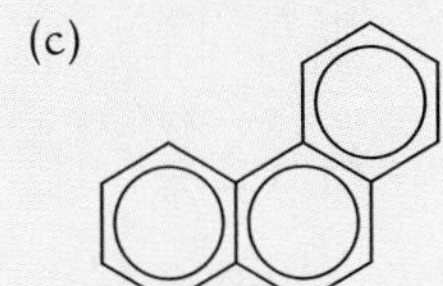

(d) 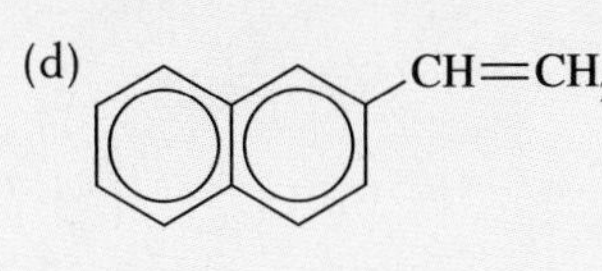

39. How many benzene isomers can be written for the formula C_8H_9Br?
(a) 6 (b) 10 (c) 12 (d) 14

Write the products for the reactions given in Questions 40–46.

40. $(CH_3)_2C{=}CHCH_3 + H_2O \xrightarrow{H^+}$

(a) $(CH_3)_2CH_2CHCH_3$ (OH on CH)

(b) $(CH_3)_2CCH_2CH_3$ (OH on C)

(c) $(CH_3)_2C{-}CHCH_3$ (OH on each carbon: OH OH)

(d) $(CH_3)_2CH{-}O{-}CH_2CH_3$

41. 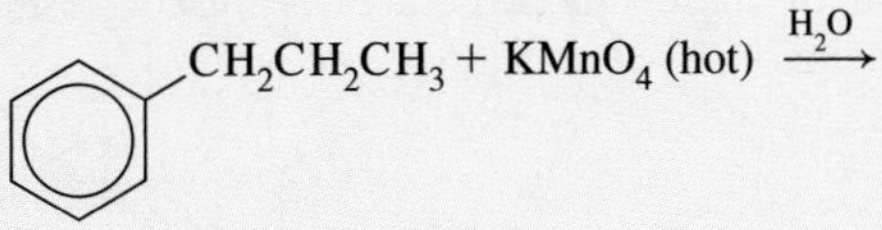 (benzene ring)–$CH_2CH_2CH_3 + KMnO_4$ (hot) $\xrightarrow{H_2O}$

(a) 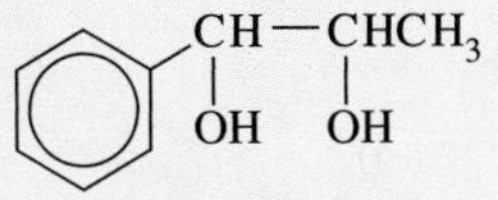 (benzene ring)–$CH{-}CHCH_3$ (OH OH)

(b) 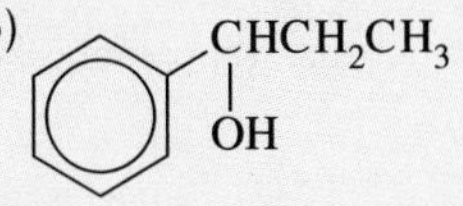 (benzene ring)–$CHCH_2CH_3$ (OH)

(c) 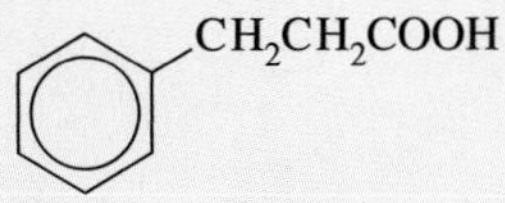 (benzene ring)–CH_2CH_2COOH

(d) 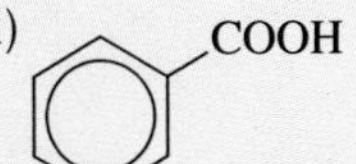 (benzene ring)–$COOH$

42. $CH_3CH_2CH{=}CHCH_2CH_3 + HCl \longrightarrow$
(a) $CH_3CH_2CH_2CH_2CH_2CH_3$
(b) $CH_3CH_2CHClCHClCH_2CH_3$
(c) $CH_3CH_2CH_2CHClCH_2CH_3$
(d) no correct answer given

43. $CH_3C{\equiv}CH + 2\ HBr \longrightarrow$
(a) $CH_3CBr_2CH_3$ (c) $CH_3CHBrCHBr_2$
(b) $CH_3CH_2CHBr_2$ (d) $CH_2BrCH_2CH_2Br$

44. $CH_3\overset{CH_3}{\overset{|}{C}}{=}CHCH_3 + Br_2 \longrightarrow$

(a) $CH_3\overset{CH_3}{\overset{|}{C}}BrCHBrCH_3$ (c) $CH_3\overset{CH_3}{\overset{|}{C}}{=}CHCH_2Br$

(b) $CH_3\overset{CH_3}{\overset{|}{C}}HCBr_2CH_3$ (d) $CH_3\overset{CH_3}{\overset{|}{C}}BrCH_2CH_3$

45. 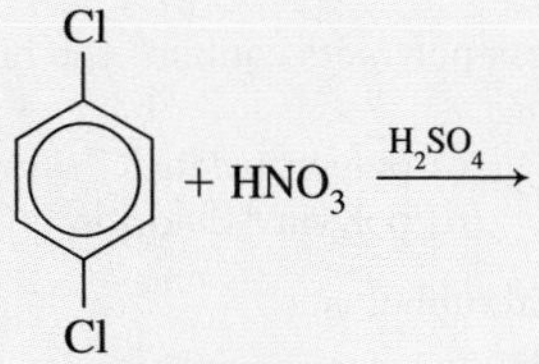 (benzene ring with Cl at both para positions) $+ HNO_3 \xrightarrow{H_2SO_4}$

(a) 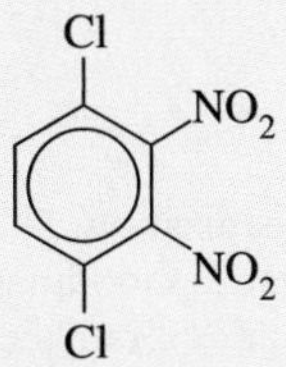 (benzene ring: Cl, NO_2, NO_2, Cl)

(b) 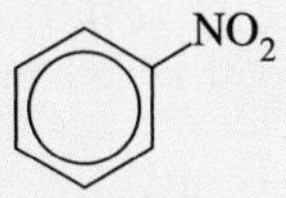 (benzene ring)–NO_2

(c) 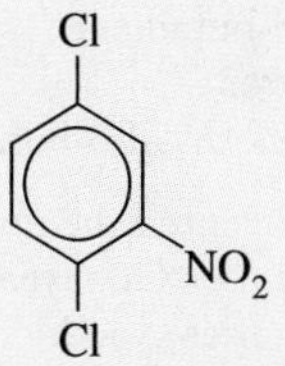 (benzene ring: Cl, NO_2, Cl)

(d) 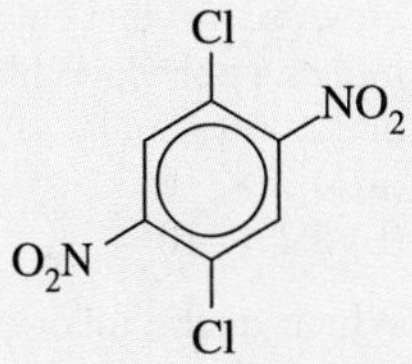 (benzene ring: Cl, NO_2, O_2N, Cl)

46. 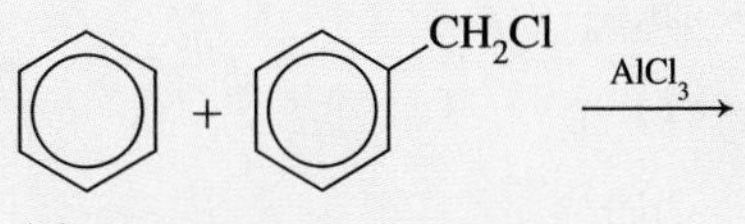 (benzene ring) + (benzene ring)–CH_2Cl $\xrightarrow{AlCl_3}$

(a) 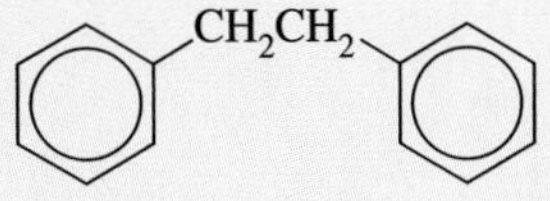 (benzene ring)–CH_2CH_2–(benzene ring)

(b) 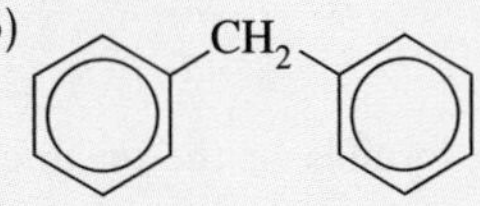 (benzene ring)–CH_2–(benzene ring)

(c) 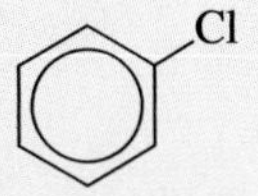 (benzene ring)–Cl

(d) no correct answer given

47. Macromolecules are made from small units called
(a) telemers (c) monomers
(b) polymers (d) dimers

48. Which of the following is not a natural macromolecule?
(a) cellulose (b) protein (c) orlon (d) glycogen

49. Polymers such as polyethylene, polystyrene, and polyvinyl chloride are known as
(a) condensation polymers
(b) substitution polymers
(c) addition polymers
(d) transition polymers

50. Which of the following is a free radical?
(a) CH_3^+ (c) $CH_2{=}CH^-$
(b)

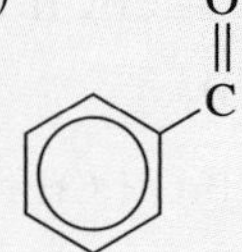

(d) $CH_3CH_2^-$

51. Which monomer of these polymers contains the highest mass percent of carbon?
(a) polyethylene (c) polypropylene
(b) polystyrene (d) polyvinyl chloride

52. The monomer of natural rubber is
(a) 2-chloro-1,3-butadiene
(b) 1,3-butadiene
(c) styrene
(d) 2-methyl-1,3-butadiene

53. Who invented the process of vulcanization?
(a) Charles Goodyear (c) Charles Goodman
(b) Charles Goodrich (d) Charles Granville

54. How many vinyl chloride units ($CH_2{=}CHCl$) are in 1.0 mol of a polyvinyl chloride polymer that has a molar mass of 32,000 g/mol?
(a) 4.2×10^2 (c) 6.2×10^2
(b) 5.1×10^2 (d) 2.0×10^6

55. Which of the following formulas is the polymer Teflon?
(a) $-(CF_2-CF_2)-_n$
(b) $-(CH_2-CCl_2)-_n$
(c) $-(CH_2-\underset{\displaystyle CN}{\underset{|}{CH}})-_n$
(d) $-(CH_2-\underset{\displaystyle CH_3}{\underset{|}{CH}})-_n$

56. Which of the following formulas is the polymer Saran?
(a) $-(CF_2CF_2)-_n$
(b) $-(CH_2CCl_2)-_n$
(c) $-(CH_2-\underset{\displaystyle CN}{\underset{|}{CH}})-_n$
(d) $-(CH_2\underset{\displaystyle CH_3}{\underset{|}{CH}})-_n$

57. Which of the following formulas is the polymer polypropylene?
(a) $-(CH_2CH_2CH_2)-_n$ (c) $-(CH_2CH)-_n$ (with a benzene ring on CH)
(b) $-(CH_2CCl_2)-_n$ (d) $-(CH_2\underset{\displaystyle CH_3}{\underset{|}{CH}})-_n$

58. Which of the following formulas is a diene?
(a) $CH_3CH{=}CHCH_3$ (c) C_6H_5–$CH{=}CH_2$ (benzene ring with CH=CH$_2$)
(b) $CH_2{=}C{=}CH_2$ (d) no correct answer given

59. The formula for the polymer that can be formed from 2-butene is
(a) $-(CH_2CH_2CH_2CH_2)-_n$ (c) $-(\overset{\displaystyle CH_3}{\overset{|}{CH}}-\overset{\displaystyle CH_3}{\overset{|}{CH}})-_n$
(b) $-(CH_2CH{=}CHCH_2)-_n$ (d) $-(CH_2-\overset{\displaystyle CH_2CH_3}{\overset{|}{CH}})-_n$

60. A polymer that is made from two different monomer units is known as a
(a) double polymer (c) dipolymer
(b) copolymer (d) bipolymer

61. Which of these line structures represents 2-methyl-2-heptene?
(a) (c)
(b) (d)

62. Which of these line structures represents 2-methyl-1-pentanol?
(a) OH (c) OH
(b) OH (d)

Free Response Questions

1. Why are alkanes unable to undergo addition reactions?

2. What general type of polymer would be most useful for making the handles for pots that can be used on a stove?

3. What is the difference in name, functional group, and reactivity (common type of reaction) between the following three molecules?

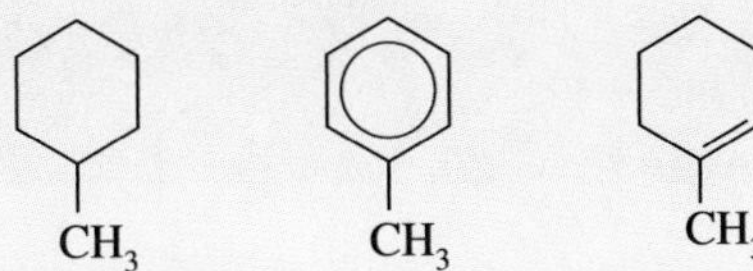

4. A bottle on the shelf in your lab is missing part of its label. What is left of the label indicates the formula for the compound is a six-carbon hydrocarbon. You are pretty sure it is one of these compounds:

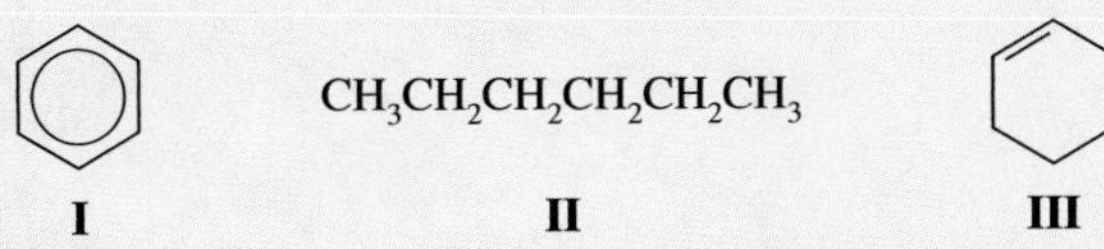

Among the other lab chemicals readily available is bromine. Suggest what test(s) you could perform that would give quick visual results to determine the identity of the compound.

5. Can alkynes have geometric isomers? Why or why not? Are there any groups of alkanes that could have geometric isomers?

6. Each of the following pairs of compounds have similar molar masses:

(a) benzene and cyclohexane
(b) hexane and 1-hexyne
(c) cyclohexene and cyclohexane

Would you prefer to use physical properties (e.g., boiling point or solubility) or chemical properties to distinguish between the compounds in each pair? Explain and give examples where appropriate.

7. Plexiglass is the trademarked name for a polymer often used as a glass substitute partly because of its transparent qualities. From the structure of Plexiglass shown here, do you think Plexiglass is an addition or a condensation polymer and what is(are) the monomer(s)?

$$\left(-\underset{CO_2CH_3}{\overset{CH_3}{C}}-CH_2-\underset{CO_2CH_3}{\overset{CH_3}{C}}-CH_2-\underset{CO_2CH_3}{\overset{CH_3}{C}}-CH_2-\underset{CO_2CH_3}{\overset{CH_3}{C}}-CH_2-\underset{CO_2CH_3}{\overset{CH_3}{C}}-CH_2- \right)_n$$

8. Draw and name the structures for compounds I–III obtained in the following reaction sequence:

$$CH_3CH_3 \xrightarrow[\text{light}]{Cl_2} \text{I} \xrightarrow[AlCl_3]{\text{benzene}} \text{II} \xrightarrow[H_2SO_4]{K_2Cr_2O_7} \text{III}$$

CHAPTER 22

Alcohols, Ethers, Phenols, and Thiols

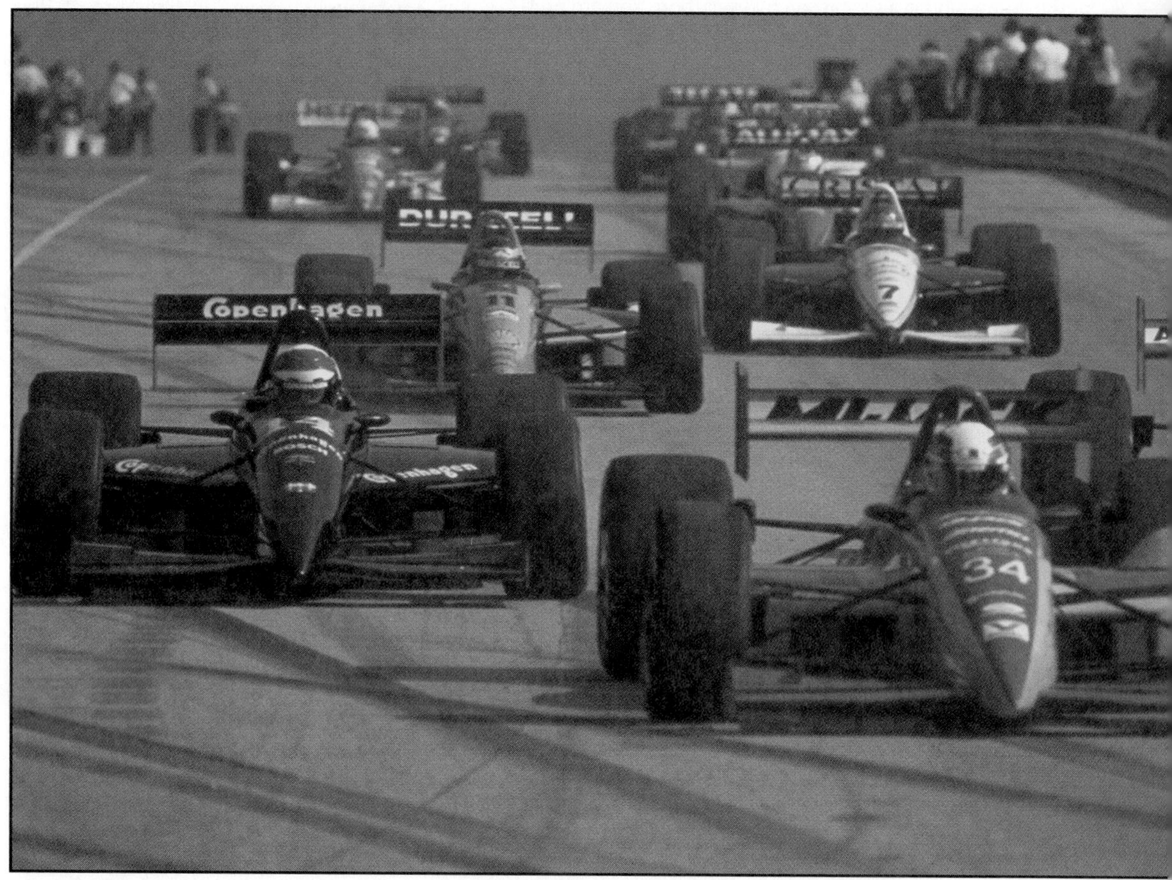

These Indy cars have just refueled with methanol during the annual race at Elkhart Lake, Wisconsin.

Chapter Outline

We tend to play as strenuously as we work. Our "vacations" often involve driving to areas where we can indulge in vigorous physical activities like hiking, cycling, skiing, or swimming. After a hard day "torturing" our bodies, we relax and commiserate over our sore muscles. During these activities, organic molecules that contain an —OH group play a significant role. Ethylene glycol acts as a coolant in the radiator of our car, sugar and other carbohydrates provide the biochemical energy for our physical activities, ethyl alcohol is a component in any alcoholic beverage we might consume, and phenolic compounds are an active ingredient in the muscle rubs and analgesics that relieve our sore muscles.

Just what changes does the addition of an —OH group produce in the physical and chemical properties of an organic molecule? In this chapter, we begin to examine the effect of various functional groups on organic molecules.

22.1 Functional Groups

Organic compounds were originally obtained from plants and animals, which even today are still the direct sources of many important chemicals. As a case in point, millions of tons of sucrose (table sugar) are obtained from sugar cane and sugar beet juices each year. However, as our knowledge of chemistry has increased, we have been able to synthesize many naturally occurring compounds, often at far less cost than the natural products. Of even greater significance than the cheaper manufacture of natural substances has been the synthesis of new substances that are totally unlike any natural product. These syntheses have been greatly aided by the realization that organic chemicals can be divided into a relatively small number of classes and studied on the basis of similar chemical properties. (See Table 19.1.) The various classes of compounds are identified by the presence of certain characteristic groups, called functional groups. For example, if a hydroxyl group (—OH) is substituted for a hydrogen atom in an alkane molecule, the resulting compound is an **alcohol**. Thus alcohols are a class of compounds in which the functional group is the hydroxyl group.

alcohol

Through the chemical reactions of functional groups, it is possible to create or synthesize new substances. The synthesis of new and possibly useful compounds and the more economical synthesis of known compounds are two main concerns of modern organic chemistry. Most chemicals used today do not occur in nature, but are synthesized from naturally occurring materials. The chemical and physical properties of an organic compound depend on (1) the kinds and number of functional groups present and (2) the shape and size of the molecule.

The structures of alcohols, ethers, and phenols may be derived from the structure of water by replacing the hydrogen atoms of water with alkyl groups (R) or aromatic rings:

H–O–H (water) R–O–H (alcohol) R–O–R (ether) C_6H_5–O–H (phenol)

The R— groups in ethers can be the same or different and can be alkyl groups or aromatic rings.

22.2 Classification of Alcohols

Structurally, an alcohol is derived from an aliphatic hydrocarbon by the replacement of at least one hydrogen atom with a hydroxyl group (—OH). Alcohols are represented by the general formula ROH, with methanol (CH_3OH) being the first member of the homologous series. (R represents an alkyl or substituted alkyl group.) Models illustrating the structural arrangements of the atoms in methanol and ethanol are shown in Figure 22.1.

primary alcohol
secondary alcohol
tertiary alcohol

Alcohols are classified as **primary** (1°), **secondary** (2°), or **tertiary** (3°), depending on whether the carbon atom to which the —OH group is attached is directly bonded to one, two, or three other carbon atoms, respectively. (The terms *secondary* and *tertiary* are used as they were for alkyl groups in Chapter 19.) Generalized formulas for 1°, 2°, and 3° alcohols are as follows:

$$\begin{array}{ccc} \begin{array}{c} H \\ | \\ R-C-OH \\ | \\ H \end{array} & \begin{array}{c} R \\ | \\ R-C-OH \\ | \\ H \end{array} & \begin{array}{c} R \\ | \\ R-C-OH \\ | \\ R \end{array} \\ \text{primary alcohol} & \text{secondary alcohol} & \text{tertiary alcohol} \end{array}$$

Formulas of specific examples of these classes of alcohols are shown in Table 22.1. Methanol (CH_3OH) is grouped with the primary alcohols.

Molecular structures with more than one —OH group attached to a single carbon atom are generally not stable. But an alcohol molecule can contain two or more —OH groups if each —OH is attached to a different carbon atom. Accordingly, alcohols are also classified as monohydroxy, dihydroxy, trihydroxy, etc., on the basis of the number of hydroxyl groups per molecule.

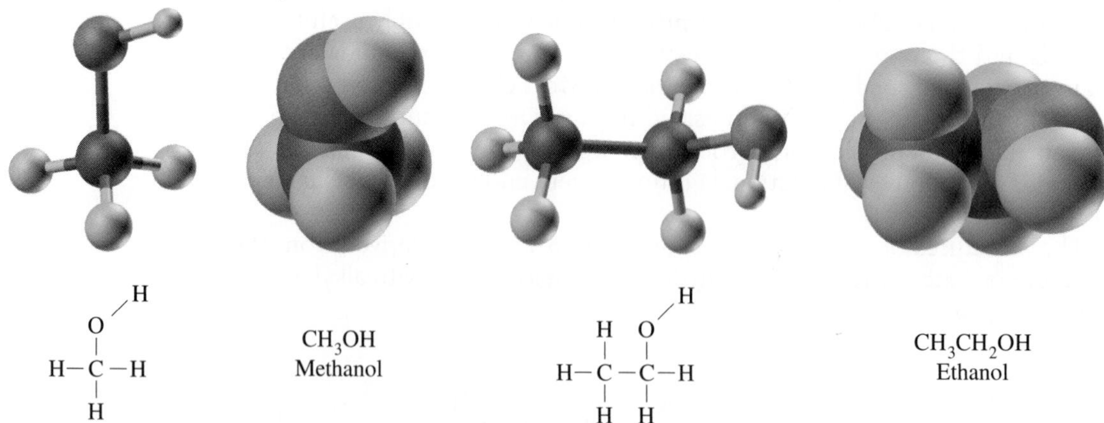

Figure 22.1
Ball-and-stick and spacefilling models illustrating structural formulas of methanol and ethanol.

Table 22.1 Names and Classification of Alcohols

Class	Formula	IUPAC name	Common name*	Boiling point (°C)
Primary	CH_3OH	Methanol	Methyl alcohol	65.0
Primary	CH_3CH_2OH	Ethanol	Ethyl alcohol	78.5
Primary	$CH_3CH_2CH_2OH$	1-Propanol	*n*-Propyl alcohol	97.4
Primary	$CH_3CH_2CH_2CH_2OH$	1-Butanol	*n*-Butyl alcohol	118
Primary	$CH_3(CH_2)_3CH_2OH$	1-Pentanol	*n*-Pentyl alcohol (*n*-amyl)	138
Primary	$CH_3CH(CH_3)CH_2OH$	2-Methyl-1-propanol	Isobutyl alcohol	108
Primary	$C_6H_5{-}CH_2OH$	Phenylmethanol	Benzyl alcohol	205.0
Secondary	$CH_3CH(OH)CH_3$	2-Propanol	Isopropyl alcohol	82.5
Secondary	$CH_3CH_2CH(OH)CH_3$	2-Butanol	*sec*-Butyl alcohol	91.5
Tertiary	$CH_3{-}C(CH_3)_2{-}OH$	2-Methyl-2-propanol	*t*-Butyl alcohol	82.9
Dihydroxy	$HOCH_2CH_2OH$	1,2-Ethanediol	Ethylene glycol	197
Trihydroxy	$HOCH_2CH(OH)CH_2OH$	1,2,3-Propanetriol	Glycerol or glycerine	290

*The abbreviations *n*, *sec*, and *t* stand for normal, secondary, and tertiary, respectively. $C_6H_5CH_2{-}$ is a benzyl group.

Polyhydroxy alcohols and *polyols* are general terms for alcohols that have more than one —OH group per molecule. Polyhydroxy compounds are very important molecules in living cells, as they include the carbohydrate class of biochemicals.

polyhydroxy alcohol

An alcohol such as 2-butanol can be written in a single-line formula by enclosing the —OH group in parentheses and placing it after the carbon to which it is bonded. For example, the following two formulas represent the same compound:

$$CH_3CH_2\underset{\displaystyle OH}{\underset{|}{C}}HCH_3 \qquad CH_3CH_2CH(OH)CH_3$$

Practice 22.1 Application to Biochemistry

Blood sugar (glucose) contains five alcohol groups. Using the structure of glucose shown here, label each alcohol group as 1°, 2° or 3°.

$$O{=}CHCH(OH)CH(OH)CH(OH)CH(OH)CH_2OH$$

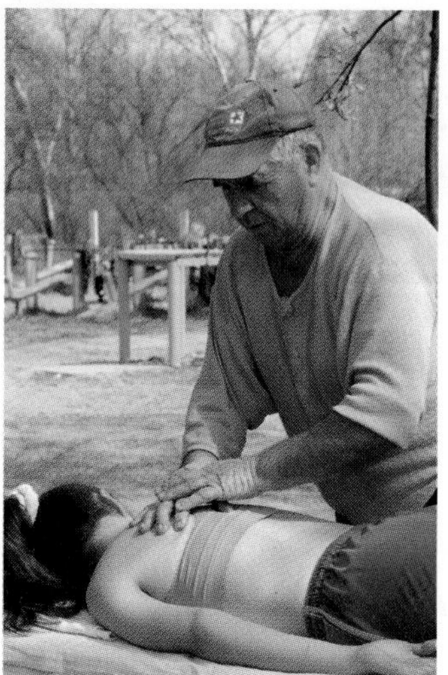

Athletic trainers use 2-propanol (isopropanol) as a muscle rub.

22.3 Naming Alcohols

If you know how to name alkanes, it is easy to name alcohols by the IUPAC System:

IUPAC Rules for Naming Alcohols

1. Select the longest continuous chain of carbon atoms containing the hydroxyl group.
2. Number the carbon atoms in this chain so that the one bearing the —OH group has the lowest possible number.
3. Form the parent alcohol name by replacing the final *-e* of the corresponding alkane name by *-ol*. When isomers are possible, locate the position of the —OH by placing the number (hyphenated) of the carbon atom to which the —OH is bonded immediately before the parent alcohol name.
4. Name each alkyl side chain (or other group), and designate its position by number.

Now we will use the IUPAC rules just given to name the alcohol $CH_3CH_2CH_2CH_2OH$:

Rule 1 The longest carbon chain containing the —OH group has four carbons.

Rule 2 Number the carbon atoms, giving the number 1 to the carbon bonded to the —OH:

$$\overset{4}{C}-\overset{3}{C}-\overset{2}{C}-\overset{1}{C}-OH$$

Rule 3 The name of the four-carbon alkane is butane. Replace the final *e* in butane with *ol*, forming the name *butanol*. Since the —OH is on carbon 1, place a *1-* before butanol to give the complete alcohol name 1-butanol.

Rule 4 No groups of atoms other than hydrogen are attached to the butanol chain, so the name of this alcohol is 1-butanol.

Study how the following examples and those shown in Table 22.1 are named with the use of the IUPAC System:

$$\overset{3}{C}H_3-\overset{2}{C}H_2-\overset{1}{C}H_2OH$$
1-propanol

$$\overset{1}{C}H_3-\underset{\displaystyle OH}{\overset{2}{C}H}-\overset{3}{C}H_3$$
2-propanol

cyclohexanol (cyclohexane ring with OH)

1,3-cyclohexanediol (cyclohexane ring numbered 1–6, with OH on C1 and C3)

$$\overset{4}{C}H_3-\underset{\displaystyle CH_3}{\overset{3}{C}H}-\overset{2}{C}H_2-\overset{1}{C}H_2OH$$
3-methyl-1-butanol

$$HO\overset{2}{C}H_2-\overset{1}{C}H_2OH$$
1,2-ethanediol

Example 22.1

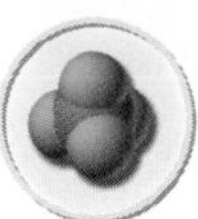

Name the following alcohol by the IUPAC method:

```
CH3CH2CHCH2CHCH3
      |     |
     CH3    OH
```

SOLUTION

Rule 1 The longest continuous carbon chain containing the —OH group has six carbon atoms.

Rule 2 This carbon chain is numbered from right to left, such that the —OH group has the lowest possible number:

```
6   5   4   3   2   1
C — C — C — C — C — C
        |       |
       CH3      OH
```

In this case, the —OH is on carbon 2.

Rule 3 The name of the six-carbon alkane is hexane. Replace the final *e* in hexane by *ol*, to form the name *hexanol*. Since the —OH is on carbon 2, place a 2- before hexanol to give the parent alcohol name 2-hexanol.

Rule 4 A methyl group (—CH_3) is located on carbon 4. Therefore, the full name of the compound is 4-methyl-2-hexanol.

Example 22.2

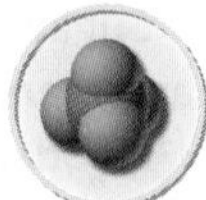

Write the structural formula of 3,3-dimethyl-2-hexanol.

SOLUTION

Rule 1 The "2-hexanol" refers to a six-carbon chain with an —OH group on carbon 2. Write six carbons in a row and place an —*OH* on carbon 2:

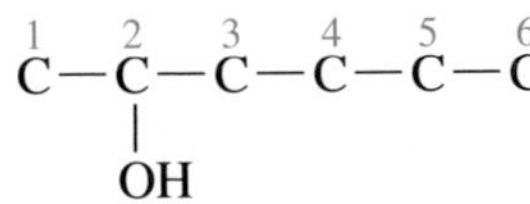

Rule 2 Place the two methyl groups ("3,3-dimethyl") on carbon 3:

```
           CH3
1    2     |3   4   5   6
C —  C  —  C — C — C — C
     |     |
     OH   CH3
```

Rule 3 Finally, add H atoms to give each carbon atom four bonds:

```
          CH3
          |
CH3CH  —  C — CH2CH2CH3
    |     |
    OH   CH3
```

3,3-dimethyl-2-hexanol

Practice 22.2

Write the IUPAC name for each of the following:

(a) $CH_3CH_2CHCH_2CH_2CH_3$ with CH_2OH attached to the third carbon (C-3)

(b) $CH_3CH_2CHCH_2CCH_3$ with OH attached to the third carbon (C-3), and Br and CH_3 attached to the fifth carbon (C-5)

Practice 22.3

Write the structural formula for each of the following:
(a) 3-methylcyclohexanol (b) 4-ethyl-2-methyl-3-heptanol

Practice 22.4 Application to Biochemistry

Glycerol ($HOCH_2CH(OH)CH_2OH$) is a key component of most fats. Give the IUPAC name for glycerol.

Several of the alcohols are generally known by their common names, so it may be necessary to know more than one name for a given alcohol. The common name is usually formed from the name of the alkyl group that is attached to the —OH group, followed by the word *alcohol*. (See the examples in Table 22.1.)

22.4 Physical Properties of Alcohols

Sugars are easily dissolved and transported through the bloodstream, because they contain many hydroxyl groups.

The physical properties of alcohols are related to those of both water and alkane hydrocarbons. This is easily understandable if we recall certain facts about water and the alkanes. Water molecules are quite polar. The properties of water, such as its high boiling point and its ability to dissolve many polar substances, are due largely to the polarity of its molecules. Alkane molecules possess almost no polarity. The properties of the alkanes—for example, their relatively low boiling points and inability to dissolve water and other polar substances—reflect this lack of polarity. An alcohol molecule is made up of a waterlike hydroxyl group joined to a hydrocarbonlike alkyl group:

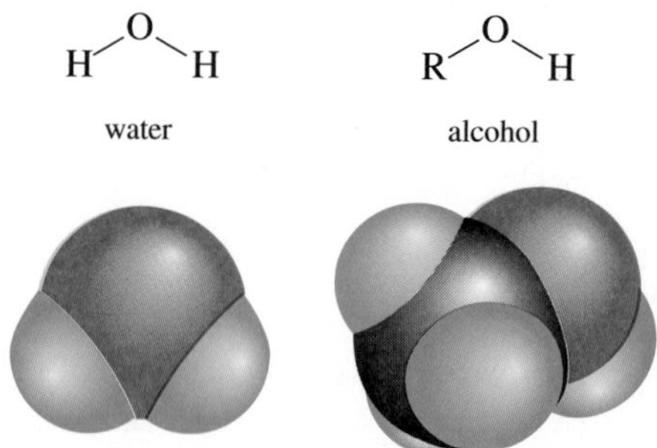

One striking property of alcohols is their relatively high boiling points. The simplest alcohol, methanol, boils at 65°C. But methane, the simplest hydrocarbon, boils at −162°C. The boiling points of the normal alcohols increase in a regular fashion with increasing number of carbon atoms. Branched-chain alcohols have lower boiling points than the corresponding straight-chain alcohols. (See Table 22.1.)

Alcohols containing up to three carbon atoms are infinitely soluble in water. With one exception (2-methyl-2-propanol), alcohols with four or more carbon atoms have limited solubility in water. In contrast, all hydrocarbons are essentially insoluble in water.

The hydroxyl group on the alcohol molecule is responsible for both the water solubility and the relatively high boiling points of the low-molar-mass alcohols. Hydrogen bonding (Section 13.8) between water and alcohol molecules accounts for the solubility, and hydrogen bonding between alcohol molecules accounts for their high boiling points. The following diagrams show the hydrogen bonding in water–alcohol and alcohol–alcohol clearly:

```
      O         H····         O          R
····H   H····O              R    H····O
              H                        H
              :                        :
····H         O            R           O
     O····H     R            O····H      R
····H                   ····H

     water–alcohol          alcohol–alcohol
```

As the length of the hydrocarbon chain increases, the polar effect of the hydroxyl group becomes relatively less important. Alcohols with 5–11 carbons are oily liquids of slight water solubility, and in physical behavior they resemble the corresponding alkane hydrocarbons. However, the effect of the —OH group is still noticeable in that their boiling points are higher than those of alkanes with similar molar masses. (See Table 22.2.) Alcohols containing 12 or more carbons are waxlike solids that resemble solid alkanes in physical appearance.

In alcohols with two or more hydroxyl groups, the effect of the hydroxyl groups on intermolecular attractive forces is, as we might suspect, even more striking than in monohydroxy alcohols. Ethanol, CH_3CH_2OH, boils at 78°C, but the boiling point of 1,2-ethanediol, $CH_2(OH)CH_2OH$, is 197°C. Comparison of the boiling points of ethanol, 1-propanol, and 1,2-ethanediol shows that increased

Table 22.2 Comparison of Boiling Points of Alkanes and Monohydroxy Alcohols

Name	Boiling point (°C)	Name	Boiling point (°C)
Pentane	36	1-Pentanol	138
Hexane	69	1-Hexanol	157
Octane	126	1-Octanol	195
Decane	174	1-Decanol	233

Spacefilling models of 1-propanol (left) and 1-octanol (right).

molar mass does not account for the high boiling point of 1,2-ethanediol. (See Table 22.3.) The higher boiling point is primarily a result of additional hydrogen bonding due to the two —OH groups in the 1,2-ethanediol molecule.

Table 22.3 Comparison of the Boiling Points of Ethanol, 1,2-Ethanediol, and 1-Propanol

Name	Formula	Molar mass	Boiling point (°C)
Ethanol	CH_3CH_2OH	46	78
1,2-Ethanediol	$CH_2(OH)CH_2OH$	62	197
1-Propanol	$CH_3CH_2CH_2OH$	60	97

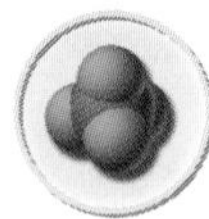

Example 22.3 Glucose is one of the most important carbohydrates in biochemistry. It has six carbons and five alcohol groups (molar mass = 180.2 g). How would you predict the water solubility of glucose to differ from that of 1-hexanol?

SOLUTION Each polar alcohol group attracts water molecules and increases the solubility of organic compounds in water. Because glucose contains five —OH groups, whereas hexanol contains only one, glucose should dissolve to a much greater extent than 1-hexanol. In fact, only about 0.6 g of hexanol will dissolve in 100 g of water (at 20°C). In contrast, about 95 g of glucose will dissolve in 100 g of water. The high water solubility of glucose is important, because this molecule is transported in a water solution to the body's cells.

Practice 22.5

List the following alcohols in order of increasing solubility in water:
(a) 1-hexanol (b) 1-pentanol (c) 1,3-propanediol (d) 1-butanol

22.5 Chemical Properties of Alcohols

An alcohol contains the hydroxyl functional group (—OH). Chemists have chosen the term *functional group* to indicate that this group brings a "function" to an organic molecule. Just as adding new electronic chips to a calculator adds new functions, attaching a hydroxyl group to an alkyl chain allows a molecule to function in new ways. As we have seen, the —OH group interacts with water so that when a molecule adds a hydroxyl group, its solubility in water increases. In general, the "functions" of a hydroxyl group help determine the properties of an alcohol.

Acidic and Basic Properties

Aliphatic alcohols are similar to water in their acidic/basic properties. If an alcohol is mixed with a strong acid, it will accept a proton (act as a Brønsted–Lowry base) to form a protonated alcohol or **oxonium ion**:

oxonium ion

$$CH_3{-}\ddot{\underset{..}{O}}H + H_2SO_4 \longrightarrow CH_3{-}\overset{+}{\ddot{O}}(H)_2 + HSO_4^-$$

Alcohols also can act as Brønsted–Lowry acids. Methanol and ethanol have approximately the same acid strength as water, while the larger alcohols are weaker acids than water, reflecting the properties of the longer alkenelike carbon chains. Both water and alcohols react with alkali metals to release hydrogen gas and an anion:

$$2\ H_2O + 2\ Na \longrightarrow \underset{\text{sodium hydroxide}}{2\ Na^+\ {}^-OH} + H_2(g)$$

$$2\ CH_3CH_2OH + 2\ Na \longrightarrow \underset{\text{sodium ethoxide}}{2\ Na^+\ {}^-OCH_2CH_3} + H_2(g)$$

The resulting anion in the alcohol reaction is known as an **alkoxide ion** (RO^-). Alkoxides are strong bases (stronger than hydroxide), and so they are used in organic chemistry when a strong base is required in a nonaqueous solution.

alkoxide ion

The order of reactivity of alcohols with sodium or potassium is primary > secondary > tertiary. Alcohols do not react with sodium as vigorously as water. Reactivity decreases with increasing molar mass, since the —OH group becomes a relatively smaller, less significant part of the molecule.

As you will see in later chapters, the —OH group is also part of the carboxylic acid functional group ($-\overset{O}{\overset{\|}{C}}-OH$). The —OH behaves much differently in the carboxylic acid group than it does in an alcohol.

Oxidation

We will consider only a few of the many reactions that alcohols are known to undergo. One important reaction is oxidation. We saw in Chapter 17 that the oxidation number of an element increases as a result of oxidation. Carbon can exist in several oxidation states, ranging from −4 to +4. In the −4 oxidation state, such as in methane, the carbon atom is considered to be completely reduced. In carbon dioxide, the carbon atom is completely oxidized; that is, it is in its highest oxidation state (+4). In many cases, oxidation reactions in organic chemistry and biochemistry can be considered in a simple manner without the use of oxidation numbers. Oxidation is the loss of hydrogen or the gain of bonds to oxygen by the organic reactant. Table 22.4 illustrates the progression of oxidation states for various compounds containing one carbon atom.

Table 22.4 Oxidation States of Carbon in One-Carbon Compounds

	Compound	Number of C—O bonds	Oxidation state
CH_4	Methane	0	−4
CH_3OH	Methanol	1	−2
$H_2C{=}O$	Methanal (formaldehyde)	2	0
$H\overset{O}{\overset{\|}{C}}-OH$	Methanoic acid (formic acid)	3	+2
$O{=}C{=}O$	Carbon dioxide	4	+4

Carbon atoms exist in progressively higher stages of oxidation in different functional-group compounds:

$$\text{Alkanes} \longrightarrow \text{Alcohols} \longrightarrow \left\{\begin{matrix}\text{Aldehydes}\\ \text{Ketones}\end{matrix}\right\} \longrightarrow \text{Carboxylic acids} \longrightarrow \text{Carbon dioxide}$$

Increasing oxidation state ⟶

The synthesis of alcohols enables us to make many compounds that are more oxidized. In general, it is not practical to convert alkanes directly to alcohols, but once the —OH functional group has been attached, further oxidations are possible. The hydroxyl group gives an organic compound the capability ("function") of forming an aldehyde, ketone, or carboxylic acid. Industrial chemists use this hydroxyl-group function often: Compounds as diverse as plastics, antibiotics, and fertilizers are synthesized starting with alcohols. The following equations represent generalized oxidation reactions in which the oxidizing agent is represented by [O]:

$$\underset{\text{primary alcohol}}{R{-}\overset{\displaystyle H}{\overset{|}{C}}H{-}OH} \xrightarrow{[O]} \underset{\text{aldehyde}}{R{-}\overset{\displaystyle O}{\overset{\|}{C}}{-}H} + H_2O \xrightarrow{[O]} \underset{\text{carboxylic acid}}{R{-}\overset{\displaystyle O}{\overset{\|}{C}}{-}OH}$$

$$\underset{\text{secondary alcohol}}{R{-}\underset{\displaystyle OH}{\underset{|}{\overset{\displaystyle H}{\overset{|}{C}}}}{-}R} \xrightarrow{[O]} \underset{\text{ketone}}{R{-}\overset{\displaystyle O}{\overset{\|}{C}}{-}R} + H_2O$$

$$\underset{\text{tertiary alcohol}}{R{-}\underset{\displaystyle OH}{\underset{|}{\overset{\displaystyle R}{\overset{|}{C}}}}{-}R} \xrightarrow{[O]} \text{No reaction}$$

Tertiary alcohols do not have a hydrogen on the —OH carbon and so cannot react with oxidizing agents, except by drastic procedures such as combustion. Both primary and secondary alcohols are oxidized as shown in the equations by the loss of the blue hydrogen atoms.

Alcohols in biochemistry react in a similar manner. For example, after a meal that includes an alcoholic beverage, our liver oxidizes the ethanol. Think about how ethanol might react given that ethanol is a primary alcohol.

First, ethanol is converted to an aldehyde:

$$\underset{\text{ethanol}}{CH_3{-}CH_2OH} \xrightarrow{[O]} \underset{\text{acetaldehyde (ethanal)}}{CH_3{-}\overset{\displaystyle O}{\overset{\|}{C}}{-}H}$$

The color change in the ethanol–dichromate reaction (from orange to green at arrow) is used in alcohol breath analyzers.

Unfortunately, the aldehyde is toxic and often damages the liver. Excess alcohol consumption can cause cirrhosis of the liver. Second, the aldehyde is converted to a carboxylic acid:

$$\underset{\text{acetaldehyde}}{CH_3-\overset{\overset{\large O}{\|}}{C}H} \xrightarrow{[O]} \underset{\text{acetic acid}}{CH_3-\overset{\overset{\large O}{\|}}{C}-OH}$$

Acetic acid can be used as a source of biochemical energy. Thus, excess consumption of alcohol can result in obesity.

Some common oxidizing agents used for specific reactions are potassium permanganate, $KMnO_4$, in an alkaline solution, potassium dichromate, $K_2Cr_2O_7$, in an acid solution, or oxygen in the air. A complete equation for an alcohol oxidation is fairly complex. Since our main interest is in the changes that occur in the functional groups, we can convey this information in abbreviated form:

$$\underset{\text{ethanol}}{CH_3CH_2OH} \xrightarrow[\Delta]{K_2Cr_2O_7/H_2SO_4} \underset{\text{ethanal}}{CH_3\overset{\overset{\large O}{\|}}{C}-H} + H_2O$$

Visible evidence that a reaction occurred is a change in color of $Cr_2O_7^{2-}$. When reduced in an acid medium, orange-colored $Cr_2O_7^{2-}$ changes to green (Cr^{3+}).

Although the abbreviated equation lacks some of the details, it does show the overall reaction involving the organic compounds. Additional information is provided by notations above and below the arrow, which indicate that this reaction is carried out in heated potassium dichromate–sulfuric acid solution. Abbreviated equations of this kind will be used frequently in the remainder of this book.

Example 22.4

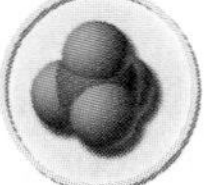

What are the products when (a) 1-propanol, (b) 2-propanol, and (c) cyclohexanol are oxidized with $K_2Cr_2O_7/H_2SO_4$?

SOLUTION

(a) The formula for 1-propanol is $CH_3CH_2CH_2OH$. Since it is a primary alcohol, it can be oxidized to an aldehyde or a carboxylic acid. The oxidation occurs at the carbon bonded to the —OH group, and this carbon atom becomes an aldehyde or a carboxylic acid. The rest of the molecule remains the same:

$$\underset{\text{1-propanol}}{CH_3CH_2CH_2OH} \xrightarrow{[O]} \underset{\text{propanal}}{CH_3CH_2\overset{\overset{\large H}{|}}{C}=O}$$

$$\underset{\text{1-propanol}}{CH_3CH_2CH_2OH} \xrightarrow{[O]} \underset{\text{propanoic acid}}{CH_3CH_2COOH}$$

(b) 2-Propanol is a secondary alcohol and is oxidized to a ketone. Ketones resist further oxidation. The oxidation occurs at the carbon bonded to the —OH group:

$$\underset{\text{2-propanol}}{CH_3\underset{\underset{\large OH}{|}}{C}HCH_3} \xrightarrow{[O]} \underset{\text{propanone (acetone)}}{CH_3\underset{\underset{\large O}{\|}}{C}CH_3}$$

CHEMISTRY IN ACTION • Ethanol: A Simple Molecule—A Complex Impact

There is a Russian proverb that states, "Drink a glass of schnapps after your soup, and you steal a ruble from the doctor." Is there truth underlying this old saying?

Ethanol is a potent drug with a long recorded history. These records clearly document the adverse effects of alcohol overconsumption. Ethanol is poisonous when too much is consumed. Continued excessive use can lead to dependence, irreversible liver damage, and cognitive dysfunction. As with most potent drugs, an overdose of ethanol produces severe side effects. However, unlike most drugs that receive extensive clinical trials before being made available to the public, only now is ethanol receiving careful study. Biochemists are measuring the impact of ethanol when consumed in moderate amounts on a regular basis (one drink per day for women or two drinks per day for men). This is analogous to measuring the impact of a *therapeutic* drug dose.

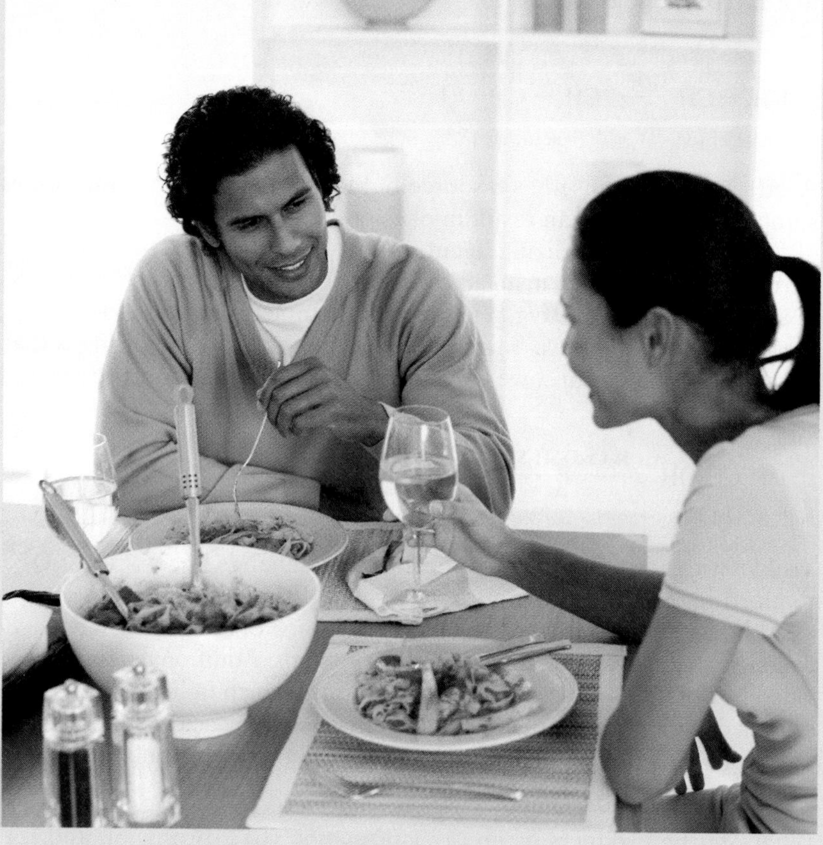

A popular misconception is that strong drugs must be complex molecules. In contrast, ethanol is a very simple organic molecule, just two saturated carbons with one hydroxyl functional group. We might predict that such a simple molecule would have simple physiological effects; however, we would be wrong. Ethanol impacts almost every cell in some way. Several important examples follow.

Ethanol and the Brain: Ethanol is known as a depressant. Under this blanket term is hidden a complex interaction between ethanol and the brain. This alcohol specifically enhances inhibitory neurotransmitters and limits the effects of excitatory neurotransmitters. Ethanol also modulates the interaction between various brain regions. The short-term effect of moderate ethanol consumption is often described as "relaxation," although sleep patterns are disrupted. Longer-term, continued moderate consumption may improve the retention of cognitive abilities.

Ethanol and the Cardiovascular System: Epidemiological studies have consistently shown that moderate ethanol consumption decreases the risk of heart attack and stroke. These findings are especially significant because cardiovascular diseases account for more deaths among Americans than any other group of diseases. Biochemists are unsure of the root causes of ethanol's impact but suggest that this alcohol may (1) reduce inappropriate blood clotting and (2) lessen damage to blood vessels caused by lipid plaques.

Ethanol and the Liver: Ethanol has a clearly destructive impact on the liver. This organ is responsible for removing foreign molecules (including drugs) from the bloodstream. As the liver metabolizes ethanol, by-products cause harm. Liver inflammation (alcohol hepatitis) and cirrhosis (liver scarring) increase when the amount of alcohol consumed increases.

Ethanol and Cancer: Epidemiological studies have shown a connection between ethanol consumption and some cancers. The strongest link seems to be with cancers of the airway and digestive tract. As the cancer is not well understood, biochemists are unsure as to how ethanol might cause a cancer.

Thus, many tissues are impacted in some way by ethanol—a potent drug. There are both health benefits and health risks to ethanol consumption. In societies such as the United States where cardiovascular disease is widespread, moderate alcohol consumption decreases overall mortality by about 18%. Perhaps the old Russian proverb quoted at the beginning should read, "Drink a glass of schnapps, *but no more,* after your soup, and you steal a ruble from the doctor."

(c) Cyclohexanol is also a secondary alcohol and is oxidized to cyclohexanone, a ketone:

$$\text{cyclohexanol} \xrightarrow{[O]} \text{cyclohexanone}$$

cyclohexanol cyclohexanone

Practice 22.6

Write the structure for the products of the oxidation of the following alcohols with $K_2Cr_2O_7$ and H_2SO_4, and indicate what evidence you would see that a reaction took place:
(a) 1-butanol (b) 2-methyl-2-butanol (c) cyclopentanol

Practice 22.7 Application to Biochemistry

Many cells produce energy via the citric acid cycle. One citric acid cycle reaction oxidizes the following acid:

$$\begin{array}{l} HO-CH-COOH \\ \quad\;\; | \\ \quad\;\; CH_2-COOH \end{array}$$

malic acid

Write the structure of the oxidized product.

Dehydration

The term *dehydration* implies the elimination of water. Alcohols can be dehydrated to form alkenes (Section 20.5) or ethers. One of the more effective dehydrating agents is sulfuric acid. Whether an ether or an alkene is formed depends on the ratio of alcohol to sulfuric acid, the reaction temperature, and the type of alcohol (1°, 2°, or 3°).

Alkenes: Intramolecular Dehydration The formation of alkenes from alcohols requires a relatively high temperature. Water is removed from within a *single* alcohol molecule, and a carbon–carbon double bond forms as shown here:

$$\begin{array}{c} H\;\;\;H \\ |\;\;\;\;| \\ H-C-C-H \\ |\;\;\;\;| \\ H\;\;\;OH \end{array} \xrightarrow[180°C]{96\%\ H_2SO_4} CH_2{=}CH_2 + H_2O$$

(only possible alkene product)

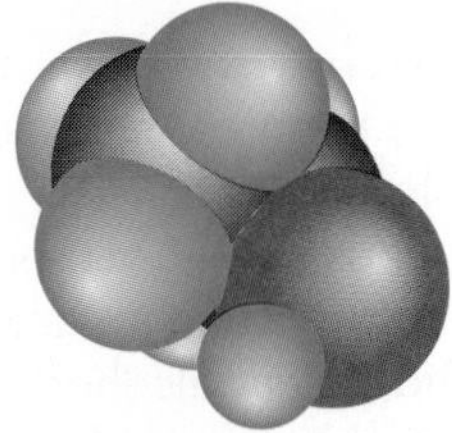

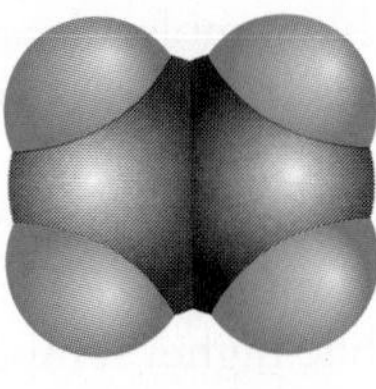

For many alcohols, there is more than one way to remove water. Therefore, the double bond can be located in different positions. The major product in such cases is the alkene in which the carbon–carbon double bond has the greatest number of alkyl substituents on it (or the least number of hydrogens). For example, 2-butanol can lose water either from carbon 1 and carbon 2 or from carbon 2 and carbon 3:

$$\underset{\text{2-butanol}}{\underset{\displaystyle\quad\ \ |\atop \displaystyle\quad\ \ OH}{CH_3CH_2CHCH_3}} \xrightarrow[100°C]{60\%\ H_2SO_4} \underset{\substack{\text{2-butene}\\ \text{(major product)}}}{CH_3CH{=}CHCH_3} + \underset{\text{1-butene}}{CH_3CH_2CH{=}CH_2} + H_2O$$

To predict the major product in an intramolecular dehydration, follow these steps: (1) Remove H and OH from adjacent carbons forming a carbon–carbon double bond; (2) if there are choices (multiple hydrogen-containing neighboring carbon atoms adjacent to the —OH carbon), remove the hydrogen from the carbon with the least number of hydrogens. This procedure is known as **Saytzeff's rule**:

Saytzeff's rule

Saytzeff's rule

During intramolecular dehydration, if there is a choice of positions for the carbon–carbon double bond, the preferred location is the one that generally gives the more highly substituted alkene—that is, the alkene with the most alkyl groups attached to the double-bond carbons.

Ethers: Intermolecular Dehydration A dehydration reaction can take place between two alcohol molecules to produce an ether. However, the dehydration to make an ether is useful only for primary alcohols, since secondary and tertiary alcohols predominantly yield alkenes. The reaction is

$$\begin{matrix} CH_3CH_2O\boxed{H} \\ CH_3CH_2\boxed{OH} \end{matrix} \xrightarrow[140°C]{96\%\ H_2SO_4} \underset{\text{diethyl ether}}{CH_3CH_2OCH_2CH_3} + H_2O$$

condensation reaction

A reaction in which *two* molecules are combined by removing a small molecule is known as a **condensation reaction**. There are many examples of condensation reactions in the formation of biochemical molecules.

The type of dehydration that occurs depends on the temperature and the number of reactant molecules. Lower temperatures and *two* alcohol molecules produce ethers, while higher temperatures and a *single* alcohol produce alkenes.

Dehydration reactions are often used in industry to make relatively expensive ethers from lower priced alcohols. Alkenes are less expensive than alcohols and so are not produced industrially by this method.

Esterification (Conversion of Alcohols to Esters)

An alcohol can react with a carboxylic acid to form an ester and water. The reaction is represented as follows:

$$\underset{\text{carboxylic acid}}{R{-}\overset{\overset{\displaystyle O}{\|}}{C}{-}OH} + \underset{\text{alcohol}}{H{-}OR'} \xrightleftharpoons{H^+} \underset{\text{ester}}{R\overset{\overset{\displaystyle O}{\|}}{C}{-}OR'} + HOH$$

$$\underset{\text{acetic acid}}{CH_3\overset{\overset{\displaystyle O}{\|}}{C}{-}OH} + \underset{\text{ethanol}}{HOCH_2CH_3} \xrightleftharpoons{H^+} \underset{\text{ethyl acetate}}{CH_3\overset{\overset{\displaystyle O}{\|}}{C}{-}OCH_2CH_3} + HOH$$

Esterification is an important reaction of alcohols and is discussed in greater detail in Chapter 24.

Utility of the Hydroxyl Functional Group

The hydroxyl group is a particularly important functional group; it introduces a myriad of possible reactions leading to a variety of other valuable organic compounds. (See Figure 22.2.) An organic chemist might term the hydroxyl group as a valuable intermediate, a "gateway" functional group.

The hydroxyl group has a special place in biochemical reactions as well. Commonly, a hydroxyl is created to "open the gateway" for other reactions. For example, fats are degraded in reactions that are fundamentally the same as those employed by industrial organic chemists. First, a portion of the fat is converted to an alkene:

$$\underset{\substack{\text{part of a fat}\\ \text{(W and Z abbreviate the}\\ \text{rest of the fat molecule)}}}{W{-}CH_2{-}CH_2{-}Z} \longrightarrow \underset{\text{an alkene}}{W{-}CH{=}CH{-}Z}$$

Then, the alkene is hydrated to form an alcohol:

$$W{-}CH{=}CH{-}Z + H_2O \xrightarrow{H^+} \underset{\text{an alcohol}}{W{-}\overset{\overset{\displaystyle OH}{|}}{C}H{-}CH_2{-}Z}$$

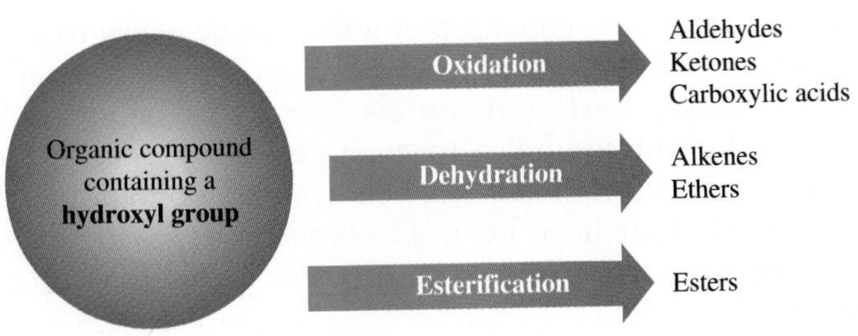

Figure 22.2
Hydroxyl group reactions.

Oxidation follows, converting the alcohol into a ketone:

$$\mathrm{W{-}\overset{\overset{\displaystyle OH}{|}}{C}H{-}CH_2{-}Z \xrightarrow{[O]} W{-}\overset{\overset{\displaystyle O}{\|}}{C}{-}CH_2{-}Z}$$

a ketone

Forming the ketone allows fats to be broken down into smaller pieces as degradation continues.

Serious physiological consequences result when cells cannot degrade fats. About 10% of the infants who die from SIDS (sudden infant death syndrome) are unable to degrade fats. Therefore, their bodies cannot depend on fat for energy. It is theorized that the infants' lack of energy leads to sudden death.

22.6 Common Alcohols

Three general methods for making alcohols are

1. *Hydrolysis of an ester:*

$$\underset{\text{ester}}{\mathrm{R\overset{\overset{\displaystyle O}{\|}}{C}{-}OR'}} + \mathrm{HOH} \xrightarrow[\Delta]{H^+} \underset{\text{carboxylic acid}}{\mathrm{R\overset{\overset{\displaystyle O}{\|}}{C}{-}OH}} + \underset{\text{alcohol}}{\mathrm{R'OH}}$$

   Hydrolysis is a reaction of water with another species in which the water molecule is split. (See Section 16.13.) The hydrolysis of an ester is the reverse reaction of esterification. A carboxylic acid and an alcohol are formed as products. The reaction can be conducted in an acid or an alkaline medium.

2. *Alkaline hydrolysis of an alkyl halide* (1° and 2° alcohols only):

$$\underset{\text{alkyl halide}}{\mathrm{RX}} + \mathrm{NaOH}(aq) \longrightarrow \underset{\text{alcohol}}{\mathrm{ROH}} + \mathrm{NaX}$$

$$\mathrm{CH_3CH_2Cl} + \mathrm{NaOH}(aq) \longrightarrow \mathrm{CH_3CH_2OH} + \mathrm{NaCl}$$

3. *Catalytic reduction of aldehydes and ketones* to produce primary and secondary alcohols. These reactions are discussed in Chapter 23.

In theory, the preceding general methods provide a way to make almost any desired alcohol, but they may not be practical for a specific alcohol because the necessary starting material—ester, alkyl halide, aldehyde, or ketone—cannot be obtained at a reasonable cost. Hence, for economic reasons, most of the widely used alcohols are made on an industrial scale by special methods that have been developed for specific alcohols. The preparation and properties of several of these alcohols are described in the paragraphs that follow.

Methanol

When wood is heated to a high temperature in an atmosphere lacking oxygen, methanol (wood alcohol) and other products are formed and driven off. The process is called *destructive distillation*, and until about 1925, nearly all methanol was obtained in this way. In the early 1920s, the synthesis of methanol by high-pressure catalytic hydrogenation of carbon monoxide was developed in Germany. The reaction is

$$CO + 2\,H_2 \xrightarrow[300\text{–}400^\circ C,\ 200\ atm]{ZnO\text{–}Cr_2O_3} CH_3OH$$

The most economical nonpetroleum source of carbon monoxide for making methanol is coal. In addition to coal, burnable materials such as wood, agricultural wastes, and sewage sludge also are potential sources of methanol.

Methanol is a volatile (bp 65°C), highly flammable liquid. It is poisonous and capable of causing blindness or death if taken internally. Exposure to methanol vapors for even short periods of time is dangerous. Despite this danger, over 5.1×10^9 kg is manufactured annually. Methanol is used for

1. Conversion to formaldehyde (methanal), primarily for use in the manufacture of polymers
2. Manufacture of other chemicals, especially various kinds of esters
3. Denaturing ethyl alcohol (rendering it unfit as a beverage)
4. An industrial solvent

Ethanol

Ethanol is without doubt the earliest and most widely known alcohol. It has been known by a variety of other names, such as ethyl alcohol, "alcohol," grain alcohol, and spirit. Huge quantities of this substance are prepared by fermentation. Starch and sugar are the raw materials. Starch is first converted to sugar by enzyme- or acid-catalyzed hydrolysis. (An enzyme is a biological catalyst, as discussed in Chapter 30.) The conversion of simple sugars to ethanol is accomplished by yeast:

$$\underset{\text{glucose}}{C_6H_{12}O_6} \xrightarrow[H_2O]{\text{yeast}} \underset{\text{ethanol}}{2\,CH_3CH_2OH} + 2\,CO_2$$

Pure ethanol (100%) is highly hygroscopic. It takes up water rapidly, to a stable concentration of 95.6% ethanol.

For legal use in beverages, ethanol is made by fermentation, but a large part of the alcohol for industrial uses (5.9×10^8 kg annually) is made by the acid-catalyzed addition of water to ethylene (made from petroleum):

$$CH_2{=}CH_2 + H_2O \xrightarrow{H^+} CH_3CH_2OH$$

Some of the economically significant uses of ethanol include the following:

1. An intermediate in the manufacture of other chemicals such as acetaldehyde, acetic acid, ethyl acetate, and diethyl ether
2. A solvent for many organic substances

3. A compounding ingredient for pharmaceuticals, perfumes, flavorings, and so on
4. An essential ingredient in alcoholic beverages

Ethanol acts physiologically as a food, as a drug, and as a poison. It is a food in the limited sense that the body is able to metabolize it to carbon dioxide and water with the production of energy. As a drug, ethanol is often mistakenly considered to be a stimulant, but it is in fact a depressant. In moderate quantities, ethanol causes drowsiness and depresses brain functions, so that activities requiring skill and judgment (such as automobile driving) are impaired. In larger quantities, ethanol causes nausea, vomiting, impaired perception, and incoordination. If a very large amount is consumed, unconsciousness and ultimately death may occur.

Authorities maintain that the effects of ethanol on automobile drivers are a factor in about half of all fatal traffic accidents in the United States. When you realize that traffic accidents are responsible for many thousands of deaths each year in the United States, the gravity of the problem cannot be overstated.

Heavy taxes are imposed on alcohol in beverages. A gallon of pure alcohol costs only a few dollars to produce, but in a distilled beverage it bears a U.S. federal tax of about 27 dollars.

Ethanol for industrial use is often denatured (rendered unfit for drinking) and thus is not taxed. Denaturing is done by adding small amounts of methanol and other denaturants that are extremely difficult to remove. Denaturing is required by the federal government to protect the beverage-alcohol tax source. Special tax-free use permits are issued to scientific and industrial users who require pure ethanol for nonbeverage uses.

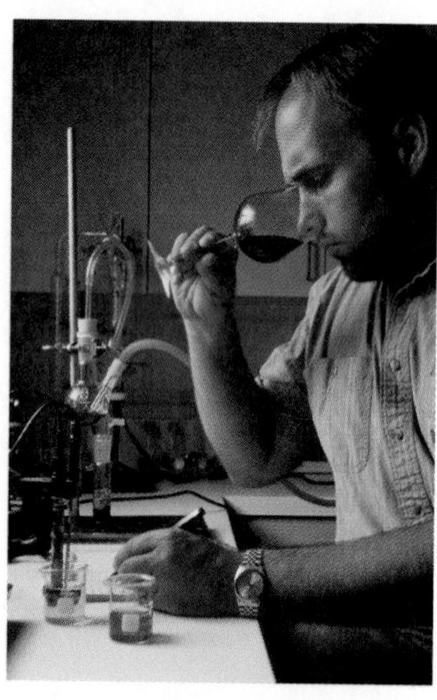

Before bottling the wine, a vintner checks alcohol content, color, and sediment.

2-Propanol (Isopropyl Alcohol)

2-Propanol (isopropyl alcohol) is made from propene derived from petroleum. This synthesis is analogous to that used for making ethanol from ethylene:

$$\underset{\text{propene}}{CH_3CH{=}CH_2} + H_2O \xrightarrow{H^+} \underset{\text{2-propanol}}{CH_3\underset{\displaystyle OH}{\underset{|}{C}}HCH_3}$$

Note that 2-propanol, not 1-propanol, is produced. This is because, in the first step of the reaction, an H^+ adds to carbon 1 of propene according to Markovnikov's rule (Section 20.6). The —OH group then ends up on carbon 2 to give the final product.

2-Propanol is a relatively low-cost alcohol that is manufactured in large quantities, about 6.3×10^8 kg annually. It is not a potable alcohol, and merely breathing large quantities of the vapor may cause dizziness, headache, nausea, vomiting, mental depression, and coma. Isopropyl alcohol is used (1) to manufacture other chemicals (especially acetone), (2) as an industrial solvent, and (3) as the principal ingredient in rubbing-alcohol formulations.

Ethylene Glycol (1,2-Ethanediol)

Ethylene glycol is the simplest alcohol, containing two —OH groups. Like most other relatively cheap, low-molar-mass alcohols, it is commercially derived from petroleum. One industrial synthesis is from ethylene via ethylene oxide (oxirane):

$$\underset{\text{ethylene}}{2\,CH_2{=}CH_2} + O_2 \xrightarrow[200\text{–}300^\circ C]{\text{Ag catalyst}} \underset{\substack{\text{ethylene oxide}\\ \text{(oxirane)}}}{2\,CH_2\overset{O}{—}CH_2}$$

$$CH_2\overset{O}{—}CH_2 + H_2O \xrightarrow{H^+} \underset{\substack{\text{1,2-ethanediol}\\ \text{(ethylene glycol)}}}{HOCH_2CH_2OH}$$

This alcohol is commonly referred to as ethylene glycol in commercial products. Major uses of ethylene glycol are (1) in the preparation of the synthetic polyester fiber Dacron and film Mylar, (2) as a major ingredient in "permanent-type" antifreeze for cooling systems, (3) as a solvent in the paint and plastics industries, and (4) in the formulations of printing ink and ink for ballpoint pens.

Its low molar mass, complete water solubility, low freezing point, and high boiling point make ethylene glycol a nearly ideal antifreeze. A 58%-by-mass aqueous solution of this alcohol freezes at $-48°C$. Its high boiling point and high heat of vaporization prevent it from being boiled away and permit higher, and therefore more efficient, engine-operating temperatures than are possible with water alone. The U.S. production of ethylene glycol amounts to about 2.4×10^9 kg annually. Ethylene glycol is extremely toxic when ingested.

Glycerol (1,2,3-Propanetriol)

Glycerol, also known as *glycerine* or 1,2,3-propanetriol, is an important trihydroxy alcohol. Glycerol is a syrupy liquid with a sweet, warm taste. It is about six-tenths as sweet as cane sugar. It is obtained as a by-product of the processing of animal and vegetable fats to make soap and other products and is also synthesized commercially from propene. The major uses of glycerol are (1) as a raw material in the manufacture of polymers and explosives, (2) as an emollient in cosmetics, (3) as a humectant in tobacco products, and (4) as a sweetener. Each use is directly related to the three —OH groups on glycerol.

The —OH groups provide sites through which the glycerol unit bonds to other molecules to form a polymer. The explosive nitroglycerine, or glyceryltrinitrate, is made by reacting the —OH groups with nitric acid:

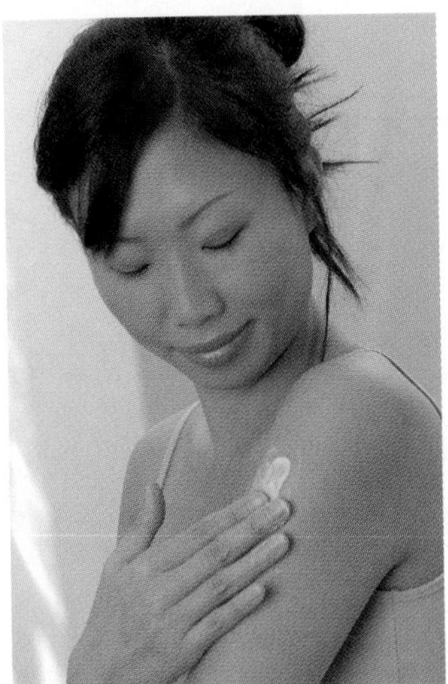

Glycerol is emulsified with other ingredients to make skin moisturizers.

$$\underset{\text{glycerol}}{\begin{array}{l}CH_2OH\\ |\\ CHOH\\ |\\ CH_2OH\end{array}} + \underset{\text{nitric acid}}{3\,HONO_2} \longrightarrow \underset{\substack{\text{glyceryltrinitrate}\\ \text{(nitroglycerine)}}}{\begin{array}{l}CH_2ONO_2\\ |\\ CHONO_2\\ |\\ CH_2ONO_2\end{array}} + 3\,H_2O$$

The three polar —OH groups on the glycerol molecule are able to hold water molecules by hydrogen bonding. Consequently, glycerol is a hygroscopic substance; that is, it has the ability to take up water vapor from the air. It is therefore used as a skin moisturizer in cosmetic preparations. Glycerol is also used as an additive in tobacco products; by taking up moisture from the air, it prevents the tobacco from becoming excessively dry and crumbly.

22.7 Phenols

phenol

The term **phenol** (pronounced *fee-nol*) is used for the class of compounds that have a hydroxy group attached to an aromatic ring. The parent compound is also called phenol, C_6H_5OH.

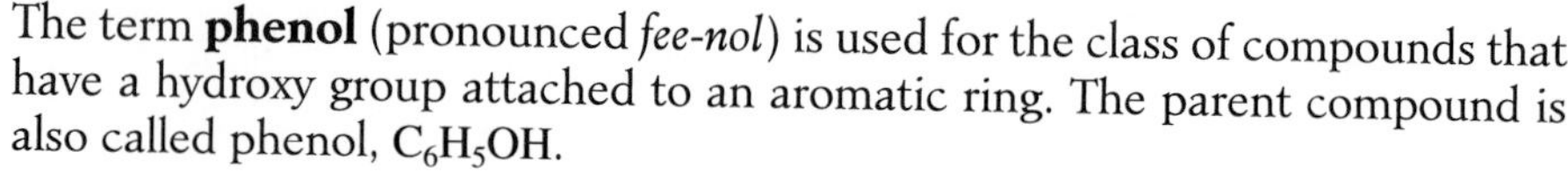

Space-filling model of phenol.

Naming Phenols

Many phenols are named as derivatives of the parent compound, via the general methods for naming aromatic compounds. The following compounds are examples:

phenol | *m*-bromophenol | *p*-aminophenol | 2,4,6-trinitrophenol (picric acid)

Common Phenols

Many natural substances have phenolic groups in their structures. The rest of this section consists of formulas for several phenolic compounds, accompanied by brief descriptions of some of their properties and uses:

catechol (*o*-dihydroxybenzene) | resorcinol (*m*-dihydroxybenzene) | hydroquinone (*p*-dihydroxybenzene)

The *ortho-*, *meta-*, and *para*-dihydroxybenzenes have the special names *catechol*, *resorcinol*, and *hydroquinone*, respectively. The catechol structure occurs in many natural substances, and hydroquinone, a manufactured product, is commonly used as a photographic reducer and developer.

vanillin | eugenol | thymol

Vanillin is the principal odorous component of the vanilla bean. It is one of the most widely used flavorings and is also used for masking undesirable odors in many products, such as paints.

Eugenol is the essence of oil of cloves. It is used in the manufacture of synthetic vanillin. Thymol occurs in the oil of thyme. It has a pleasant odor and flavor and is used as an antiseptic in preparations such as mouthwashes. Thymol is the starting material for the synthesis of menthol, the main constituent of oil of peppermint. Thymol is a widely used flavoring and pharmaceutical.

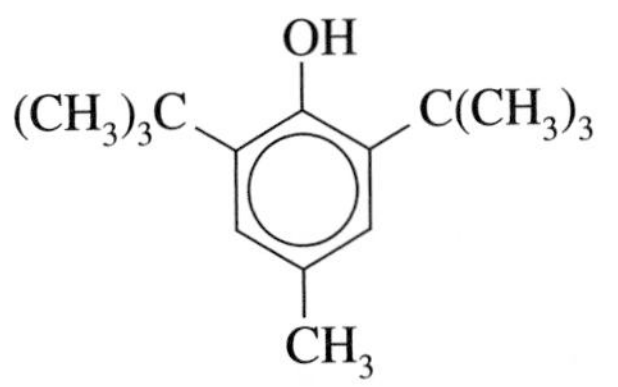

2,6-di-*t*-butyl-4-methylphenol (butylated hydroxytoluene, BHT)

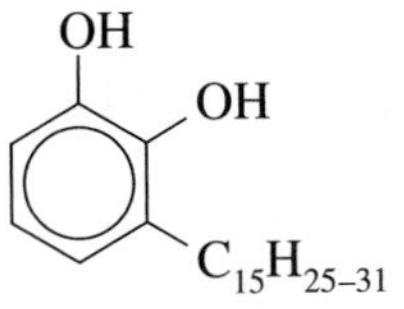

urushiols ($C_{15}H_{25-31}$ side chain)

Butylated hydroxytoluene (BHT) is used in small amounts as an antioxidant preservative for food, synthetic rubber, vegetable oils, soap, and some plastics.

The active irritants in poison ivy and poison oak are called urushiols. They are catechol derivatives with an unbranched 15-carbon side chain in position 3 on the phenol ring.

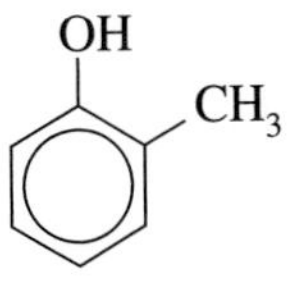

o-cresol

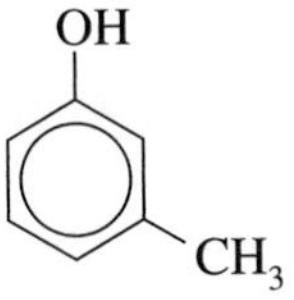

m-cresol

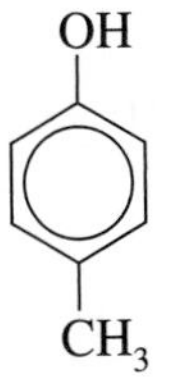

p-cresol

The *ortho*-, *meta*-, and *para*-methylphenols are present in coal tar and are known as cresols. They are all useful disinfectants.

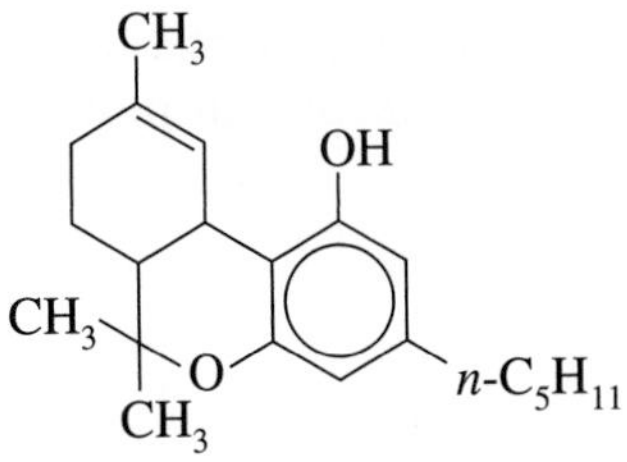

tetrahydrocannabinol (from marijuana)

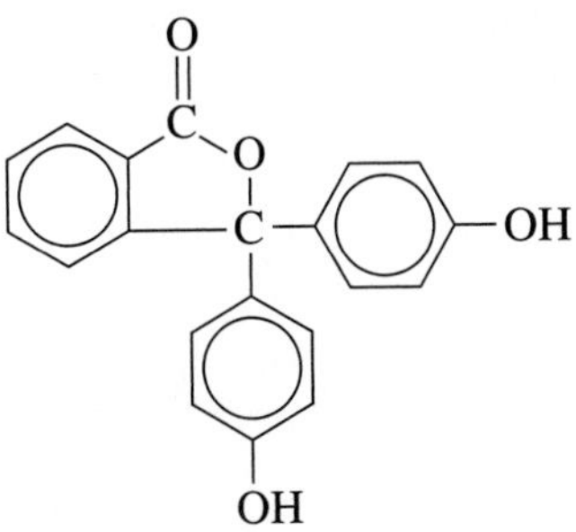

phenolphthalein

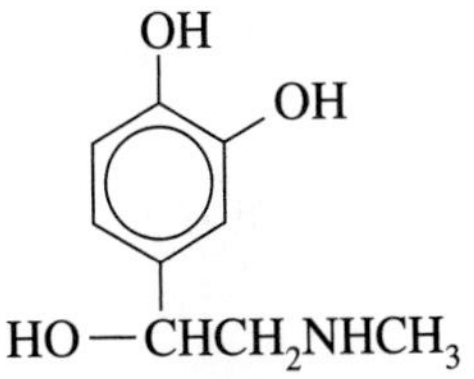

$HO{-}CHCH_2NHCH_3$

adrenalin (epinephrine)

The principal active component of marijuana is tetrahydrocannabinol. It is obtained from the dried leaves and flowering tops of the hemp plant and has been used since antiquity for its physiological effects. The common acid–base indicator phenolphthalein is a phenol derivative. Epinephrine (adrenalin) is secreted by the adrenal gland in response to stress, fear, anger, or other heightened emotional states. It stimulates the conversion of glycogen to glucose in the body. Phenol is the starting material for the manufacture of aspirin, one of the most widely used drugs for self-medication.

22.8 Properties of Phenols

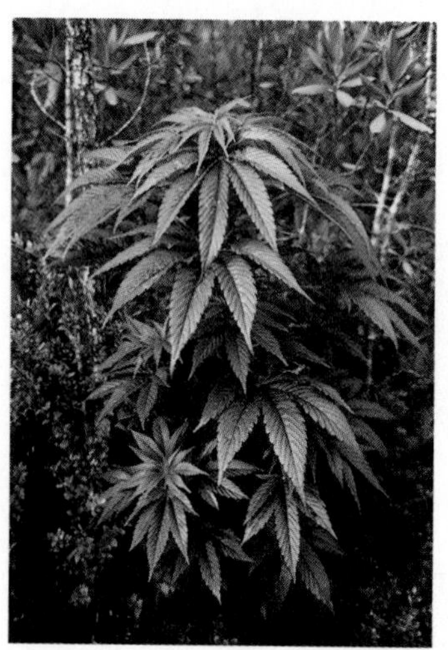

Marijuana is a member of the hemp family. It contains the physiologically active molecule tetrahydrocannabinol.

In the pure state, phenol is a colorless crystalline solid with a melting point of about 41°C and a characteristic odor. Phenol is highly poisonous. The ingestion of even small amounts of it may cause nausea, vomiting, circulatory collapse, and death from respiratory failure.

Phenol is a weak acid; it is more acidic than alcohols and water, but less acidic than acetic and carbonic acids. The pH values are as follows: 0.1 *M* acetic acid, 2.87; water, 7.0; 0.1 *M* phenol, 5.5. Thus, phenol reacts with sodium hydroxide solution to form a salt, but does not react with sodium hydrogen carbonate. The salt formed is called sodium phenoxide or sodium phenolate. Sodium hydroxide does not remove a hydrogen atom from an alcohol, because alcohols are weaker acids than water:

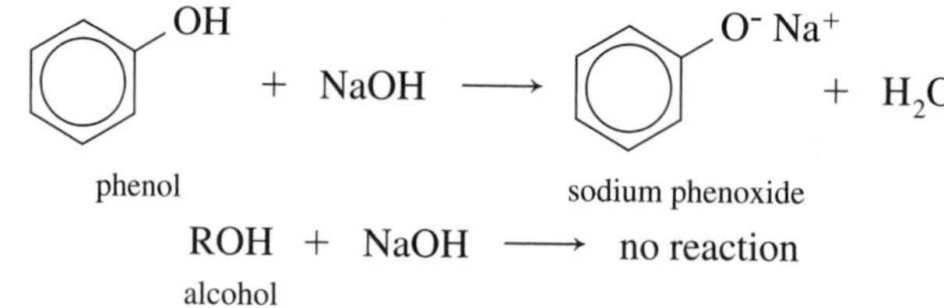

$$\underset{\text{alcohol}}{ROH} + NaOH \longrightarrow \text{no reaction}$$

In general, the phenols are toxic to microorganisms. They are widely used as antiseptics and disinfectants. Phenol was the first compound to be used extensively as an operating room disinfectant. Joseph Lister (1827–1912) first used phenol for this purpose in 1867. The antiseptic power of phenols is increased by substituting alkyl groups (up to six carbons) in the benzene ring. For example, 4-hexylresorcinol is used as an antiseptic in numerous pharmaceuticals.

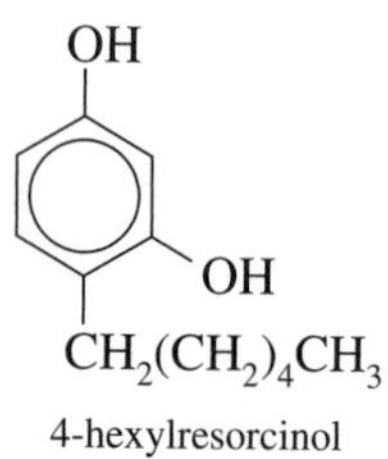

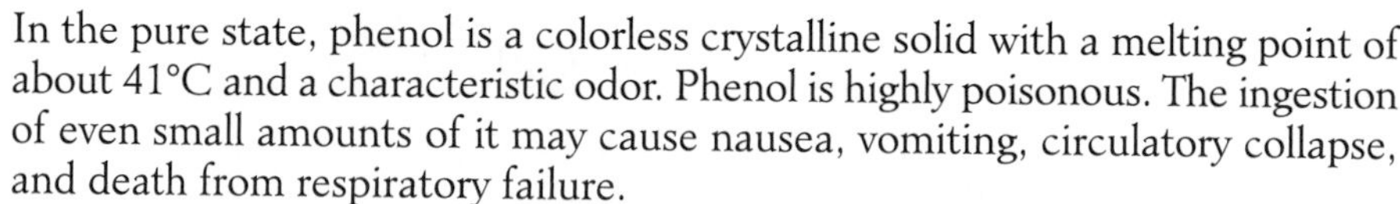

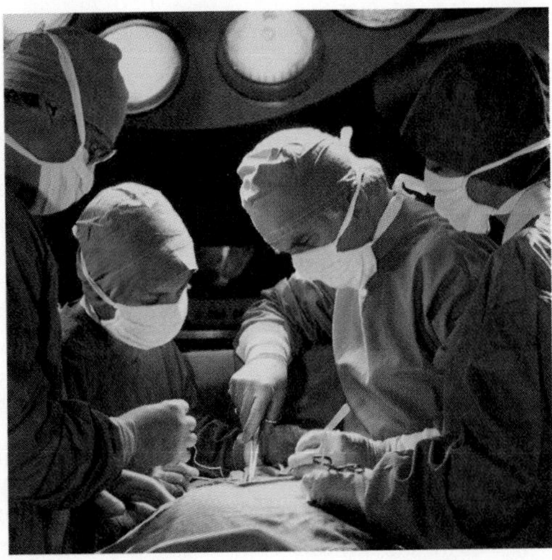

Phenols were among the first antiseptics to be used in operating rooms to prevent the spread of bacteria. Today, other antiseptics, such as iodine solutions and germicidal soaps, have replaced phenols.

22.9 Production of Phenol

Phenol is obtained from coal tar. In addition, several commercial methods are used to produce phenol synthetically. The most economical of these methods starts with benzene and propene, which react to form cumene. Cumene is then oxidized by air to cumene hydroperoxide, which is treated with dilute sulfuric acid to obtain phenol and acetone. The economic feasibility of the process is due to the fact that two important commercial products are produced. The equations for the reactions are as follows:

$$C_6H_6 \text{ (benzene)} + CH_3CH{=}CH_2 \text{ (propene)} \xrightarrow{H_2SO_4} C_6H_5{-}CH(CH_3)_2 \text{ (cumene (isopropylbenzene))}$$

$$\xrightarrow{O_2} C_6H_5{-}C(CH_3)_2{-}O{-}O{-}H \text{ (cumene hydroperoxide)} \xrightarrow[H_2SO_4]{\text{dilute}} C_6H_5OH \text{ (phenol)} + CH_3\overset{\overset{\displaystyle O}{\|}}{C}CH_3 \text{ (acetone)}$$

Over 1.9×10^9 kg of synthetic phenol is produced annually in the United States. The chief use of phenol is for the manufacture of phenol–formaldehyde resins and plastics. (See Section 23.6.)

22.10 Ethers

ether

Ethers have the general formula ROR′. The groups R and R′ can be derived from saturated, unsaturated, or aromatic hydrocarbons, and for a given ether, R and R′ may be alike or different. Cyclic ethers are formed by joining the ends of a single hydrocarbon chain through an oxygen atom to form a ring structure. Table 22.5 shows structural formulas and names for some of the different kinds of ethers.

Spacefilling model of dimethyl ether.

Naming Ethers

Individual ethers, like alcohols, may be known by several names. The ether having the formula $CH_3CH_2-O-CH_2CH_3$ and formerly widely used as an anesthetic is called diethyl ether, ethyl ether, ethoxyethane, or simply ether. Common names of ethers are formed from the names of the groups attached to the oxygen atom, followed by the word *ether* (the IUPAC allows this method for naming simple ethers):

$$\underset{\text{methyl}}{CH_3}-\underset{\text{ether}}{\boxed{O}}-\underset{\text{methyl}}{CH_3} \qquad \underset{\text{methyl}}{CH_3}-\underset{\text{ether}}{\boxed{O}}-\underset{\text{ethyl}}{CH_2CH_3}$$

dimethyl ether — ethyl methyl ether

Table 22.5 Names and Structural Formulas of Ethers

Name*	Formula	Boiling point (°C)
Dimethyl ether (methoxymethane)	CH_3-O-CH_3	−25
Ethyl methyl ether (methoxyethane)	$CH_3CH_2-O-CH_3$	8
Diethyl ether (ethoxyethane)	$CH_3CH_2-O-CH_2CH_3$	35
Ethyl isopropyl ether (2-ethoxypropane)	$CH_3CH_2-O-\underset{\displaystyle CH_3}{\underset{\vert}{C}}HCH_3$	54
Divinyl ether (ethenyloxyethene)	$CH_2=CH-O-CH=CH_2$	39
Anisole, methyl phenyl ether (methoxybenzene)	$C_6H_5-OCH_3$	154
Diphenyl ether (phenoxybenzene)	$C_6H_5-O-C_6H_5$	259
Ethylene oxide (oxirane)	CH_2-CH_2 ring closed by O	11
Tetrahydrofuran (THF) (1,4-epoxybutane)	CH_2-CH_2, CH_2 CH_2 ring closed by O	66

*The IUPAC names are in parentheses.

To name ethers by the IUPAC System, we need to learn the name of another group, the alkoxy group (RO—). An alkoxy group consists of an alkyl or aryl group and an oxygen atom. It is named by dropping the *-yl* of the alkyl name and adding the suffix *-oxy*. For example,

CH_3O- is called methoxy (meth + oxy)

CH_3CH_2O- is called ethoxy (eth + oxy)

C_6H_5O- is called phenoxy (phen + oxy)

In the IUPAC System, ethers are named as alkoxy (RO—) derivatives of the alkane corresponding to the longest carbon–carbon chain in the molecule.

IUPAC Rules for Naming Ethers

1. Select the longest continuous carbon chain and label it with the name of the corresponding alkane.
2. Change the *-yl* ending of the other hydrocarbon group to *-oxy* to obtain the alkoxy group name. For example, CH_3O— is called *methoxy*.
3. Combine the two names from Steps 1 and 2, giving the alkoxy name and its position on the longest carbon chain first, to form the ether name.

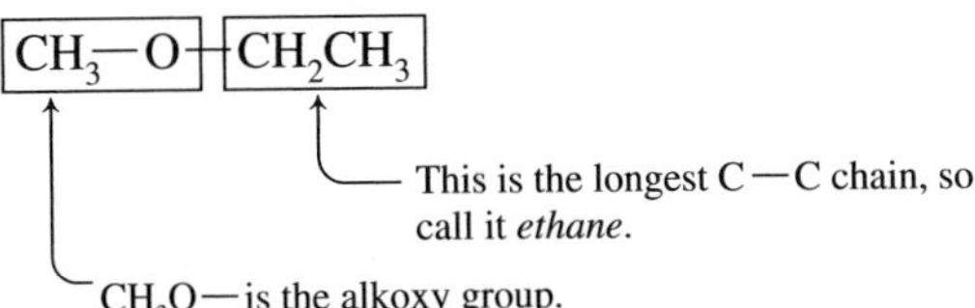
CH_3—O—CH_2CH_3
This is the longest C—C chain, so call it *ethane*.
CH_3O—is the alkoxy group.

Modify the name for the CH_3O— group from *methyl* to *methoxy*, and combine this term with *ethane* to obtain the name of the ether, methoxyethane. Thus,

CH_3CH_2—O—CH_2CH_3 is ethoxyethane

$CH_3CH_2CH_2$—O—$CH_2CH_2CH_2CH_3$ is 1-propoxybutane

$CH_3CH(CH_3)$—O—$CH(CH_3)CH_2CH(CH_3)CH_3$ is 2-isopropoxy-4-methylpentane

Additional examples are found in Table 22.5.

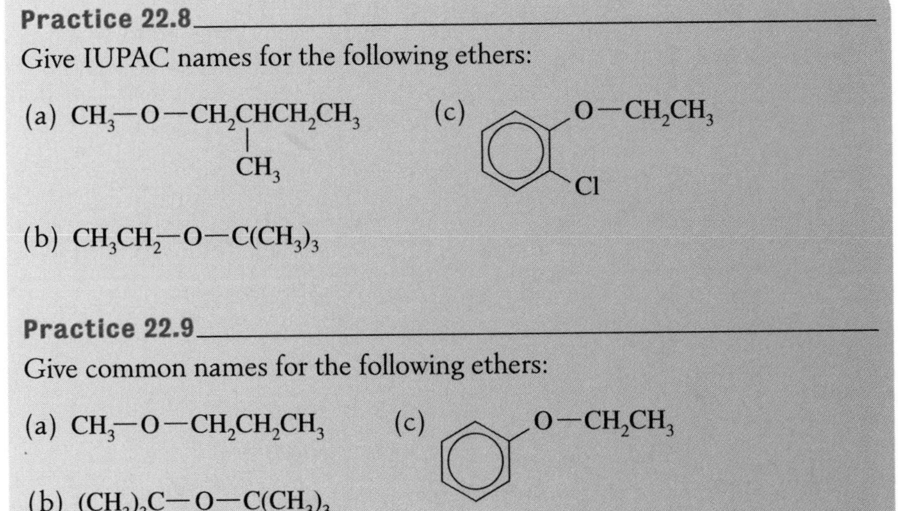

Practice 22.8

Give IUPAC names for the following ethers:

(a) CH_3—O—$CH_2CH(CH_3)CH_2CH_3$

(b) CH_3CH_2—O—$C(CH_3)_3$

(c) O—CH_2CH_3 and Cl on benzene ring (ortho)

Practice 22.9

Give common names for the following ethers:

(a) CH_3—O—$CH_2CH_2CH_3$

(b) $(CH_3)_3C$—O—$C(CH_3)_3$

(c) C_6H_5—O—CH_2CH_3

22.11 Structures and Properties of Ethers

An oxygen atom linking two carbon atoms is the key structural feature of an ether molecule. This oxygen atom causes ether molecules to have a bent shape somewhat like that of water and alcohol molecules:

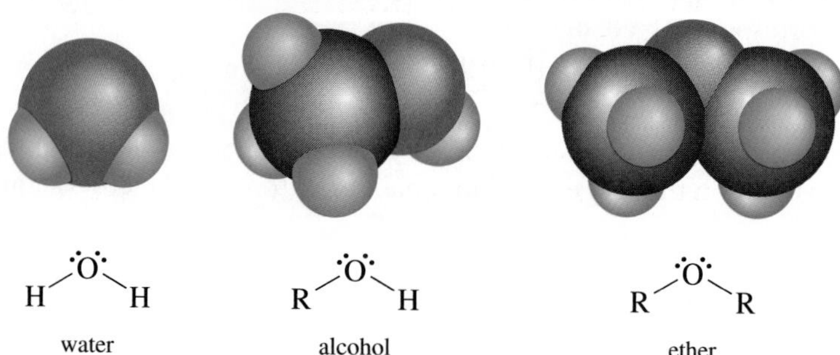

Ethers are somewhat more polar than alkanes, because alkanes lack the oxygen atom with its exposed, nonbonded electrons. But ethers are much less polar than alcohols, since no hydrogen is attached to the oxygen atom in an ether. The solubility and boiling point (vapor pressure) characteristics of ethers are related to the C—O—C structure. Alkanes have virtually no solubility in water or acid, but about 7.5 g of diethyl ether will dissolve in 100 g of water at 20°C. Diethyl ether also dissolves in sulfuric acid. Hydrogen bonding between ether and water molecules and between ether and acid molecules is responsible for this solubility:

O—H
H ←—— Hydrogen bond ——→ H—O—S(=O)(=O)—O—H
CH_3CH_2—O—CH_2CH_3 CH_3CH_2—O—CH_2CH_3
ether ··· water ether ··· acid

Because no —OH group is present, hydrogen bonding does not occur between ether molecules. This lack of hydrogen bonding can be seen by comparing the boiling points of a hydrocarbon, an ether, and an alcohol of similar molar mass, as in Table 22.6. The boiling point of the ether is somewhat above that of the hydrocarbon, but much lower than that of the more polar alcohol.

Table 22.6 Boiling Points of Ethers, Alkanes, and Alcohols

Name	Formula	Molar mass	Boiling point (°C)
Dimethyl ether	CH_3OCH_3	46	−24
Propane	$CH_3CH_2CH_3$	44	−42
Ethanol	CH_3CH_2OH	46	78
Ethyl methyl ether	$CH_3OCH_2CH_3$	60	8
Butane	$CH_3CH_2CH_2CH_3$	58	−0.6
1-Propanol	$CH_3CH_2CH_2OH$	60	97
2-Propanol	$CH_3CH(OH)CH_3$	60	83

Ethers—especially diethyl ether—are exceptionally good solvents for organic compounds. Many polar compounds, including water, acids, alcohols, and other oxygenated organic compounds, dissolve, at least to some extent, in ethers. This solubility is a result of intermolecular attractions between the slightly polar ether molecules and the molecules of the other polar substance. Nonpolar compounds such as hydrocarbons and alkyl halides also dissolve in ethers. These substances dissolve because the ether molecules are not very polar and therefore are not strongly attracted either to one another or to the other kinds of molecules. Thus, ether molecules are able to intermingle freely with the molecules of a nonpolar substance and form a solution by simple mixing.

In sum, ethers are polar enough to dissolve some polar substances, but their polarity is so slight that they act as nonpolar solvents for a great many nonpolar substances.

Ethers have little chemical reactivity, but because a great many organic substances dissolve readily in ethers, they are often used as solvents in laboratory and manufacturing operations. Their use can be dangerous, since low-molar-mass ethers are volatile and their highly flammable vapors form explosive mixtures with air. Another hazard of ethers is that, despite their generally low chemical reactivity, oxygen of the air slowly reacts with them to form unstable peroxides that are subject to explosive decomposition:

$$CH_3CH_2{-}O{-}CH_2CH_3 + O_2 \longrightarrow CH_3\underset{\displaystyle O{-}O{-}H}{\underset{|}{C}}H{-}O{-}CH_2CH_3$$

diethyl ether hydroperoxide

22.12 Preparation of Ethers

We have seen that ethers can be made by intermolecular dehydration of alcohols by heating in the presence of an acid. (See Section 22.5.) Ethers are also made from alkyl halides and sodium alkoxides or sodium phenoxides via a substitution reaction known as the Williamson synthesis:

Organic chemistry is replete with reactions named after scientists who discovered them.

$$\underset{\text{alkyl halide}}{RX} + \underset{\text{alkoxide}}{R'ONa} \longrightarrow \underset{\text{ether}}{ROR'} + NaX$$

The alkyl halide, RX, may be a methyl or primary alkyl group, but not a secondary or tertiary alkyl group or an aryl group. This reaction requires a strong base such as sodium or potassium alkoxide. The alkoxide, R′ONa, may be methyl, primary, secondary, tertiary, or an aryl group.

Alkyl halides are especially capable of substitution reactions. Here, the halide is replaced by an alkoxide or phenoxide to form an ether. This synthesis is generally useful in the preparation of mixed ethers (where $R \neq R'$) and aromatic ethers:

$$\underset{\text{sodium ethoxide}}{CH_3CH_2ONa} + \underset{\text{bromomethane}}{CH_3Br} \longrightarrow \underset{\text{ethyl methyl ether}}{CH_3CH_2{-}O{-}CH_3} + NaBr$$

$$\underset{\text{sodium phenoxide}}{C_6H_5ONa} + \underset{\text{bromomethane}}{CH_3Br} \longrightarrow \underset{\text{methoxybenzene (anisole)}}{C_6H_5OCH_3} + NaBr$$

Practice 22.10

Write equations for the preparation of (a) 1-propoxybutane, (b) methyl phenyl ether, and (c) benzyl ethyl ether, starting with an alcohol, a phenol, and an alkyl halide.

22.13 Thiols

Sulfur and oxygen are found next to each other in the same family on the periodic table. This proximity indicates some similarity in the formulas of their compounds. Organic compounds that contain the —SH group are analogs of alcohols. The —SH-containing compounds are known as **thiols**, or **mercaptans**.

thiol
mercaptan

Thiols are named by the same system as alcohols, except that the suffix *-thiol* is used in place of *-ol*:

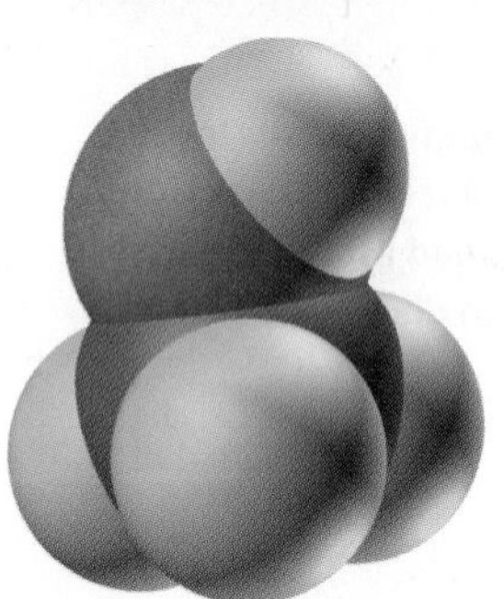

Spacefilling models of methanethiol (top) and methanol (bottom). Notice the difference in size between the sulfur atom and the oxygen atom.

CH_3SH
methanethiol
(methyl mercaptan)

$$CH_3CH_2\underset{\displaystyle SH}{\underset{|}{C}}HCH_3$$
2-butanethiol
(*sec*-butyl mercaptan)

Thiols have a higher molar mass than corresponding alcohols, but boil at lower temperatures (ethanol, 78°C; ethanethiol, 36°C). The reason for this discrepancy lies in the fact that alcohols form hydrogen bonds while thiols do not.

The major important properties of thiols are summarized as follows:

1. Foul odors: Some of these compounds smell so awful that companies make special labels to warn consumers. The odor given off by a frightened skunk has thiols as the active ingredient. Natural gas is odorized to be detectable by adding small amounts of methanethiol.
2. Oxidation to disulfides:

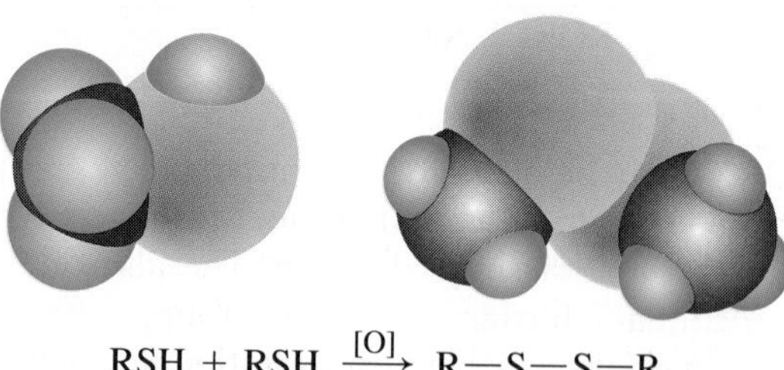

$$RSH + RSH \xrightarrow{[O]} R{-}S{-}S{-}R$$

$$CH_3SH + CH_3SH \xrightarrow{[O]} CH_3{-}S{-}S{-}CH_3$$

This reaction can be accomplished with many oxidizing agents. The sulfur is being oxidized just as the carbon was being oxidized in alcohols.

Thiols serve important functions in living systems. The disulfide structure often binds proteins into biologically useful three-dimensional shapes. (See Chapter 29.) The thiol is also a key part of a molecule (coenzyme A) that plays a central role in the metabolism of carbohydrates, lipids, and proteins. (See Chapter 34.)

Chapter 22 Review

22.1 Functional Groups

KEY TERM

Alcohol

- An alcohol is an organic compound that contains the hydroxyl functional group.
  - Functional groups are especially reactive groups of atoms on organic molecules.
- Alcohols, ethers, and phenols are structurally related to water.

H–O–H R–O–H R–O–R O–H

22.2 Classification of Alcohols

KEY TERMS

Primary alcohol
Secondary alcohol
Tertiary alcohol
Polyhydroxy alcohol

- A hydroxyl-group carbon of a primary alcohol is bonded to one other carbon (methanol is grouped with the 1° alcohols).
- For a secondary alcohol, two carbons are bonded to the hydroxyl-group carbon; three carbons are bonded to the hydroxyl-group carbon for a tertiary alcohol.
- Polyhydroxy alcohols (or polyols) have more than one hydroxyl group per molecule.

22.3 Naming Alcohols

- Naming alcohols follows a similar process to that used for alkanes.
  - The longest continuous carbon chain (parent chain) *that contains the hydroxyl group* is given the name of the corresponding alkane modified by replacing the final *-e* with *-ol*.
  - The parent chain is numbered starting at the end that is closest to the hydroxyl group.
  - When isomers are possible, locate the position of the hydroxyl group by placing the number (hyphenated) of the carbon to which the hydroxyl group is bonded immediately before the parent alcohol name.
  - Each branch is given its alkyl name and number location on the parent chain.
- Several alcohols are generally known by their common names.
  - Some examples are methyl alcohol, ethyl alcohol, and isopropyl alcohol.
- Several common polyhydroxy alcohols are known by common names, including ethylene glycol (1,2-ethanediol) and glycerol (1,2,3-propanetriol).

22.4 Physical Properties of Alcohols

- Because of the hydroxyl group, alcohols have higher boiling points than corresponding alkanes.
- Alcohols of three carbons or less are infinitely soluble in water in contrast to essentially water-insoluble alkanes.
- Alcohols become more like hydrocarbons as the alkyl chain is made longer; alcohols of more than five carbons are only slightly water soluble.
- Hydrogen bonding between water and alcohol molecules accounts for their solubility; hydrogen bonding between alcohol molecules accounts for their relatively high boiling points.

22.5 Chemical Properties of Alcohols

KEY TERMS

Oxonium ion
Alkoxide ion
Saytzeff's rule
Condensation reaction

- Aliphatic alcohols, like water, can be protonated or deprotonated.
  - An oxonium ion forms when an alcohol is protonated.
  - An alkoxide ion is formed when an alcohol is deprotonated.
- Alcohols can be oxidized to form many different organic compounds.
  - Primary alcohols oxidize to sequentially form aldehydes and then carboxylic acids.
  - Secondary alcohols oxidize to form ketones.
  - Tertiary alcohols do not oxidize with common oxidizing agents.
- Alcohols lose water (dehydrate) in the presence of a strong acid.
  - 1°, 2°, and 3° alcohols will undergo an intramolecular dehydration to form alkenes.
    - Saytzeff's rule states that when there is a choice of positions for the carbon–carbon double bond, the more highly substituted alkene is preferred.
  - Also, 1° alcohols can be easily dehydrated to form ethers, an intermolecular dehydration.
    - A reaction in which two molecules (e.g., two 1° alcohols) are combined by removing a small molecule (e.g., water) is known as a condensation reaction.
- Alcohols react with carboxylic acids to form esters, an esterification reaction.

22.6 Common Alcohols

- There are three general methods for making alcohols:
  - Hydrolysis of an ester.
  - Alkaline hydrolysis of an alkyl halide.
  - Catalytic reduction of aldehydes and ketones.
- Methanol is commonly used as an industrial solvent and in the manufacture of other chemicals such as formaldehyde (used to make polymers).
  - Most methanol is synthesized from carbon monoxide by high-pressure catalytic hydrogenation.

- Ethanol is the most widely used alcohol.
  - Alcoholic beverages contain ethanol, a physiological depressant.
    - Beverage ethanol is formed by fermentation.
  - Ethanol is an important organic solvent and is used as a compounding ingredient for pharmaceuticals, perfumes, flavorings, and so on.
  - Ethanol is an intermediate in the manufacture of chemicals such as acetaldehyde, acetic acid, ethyl acetate, and diethyl ether.
  - Industrial ethanol is commonly formed by adding water to ethene.
- 2-Propanol (isopropyl alcohol) is used as an industrial intermediate and solvent as well as the principal ingredient in rubbing-alcohol formulations.
  - 2-Propanol is formed by adding water to propene.
- Ethylene glycol (1,2-ethanediol) has several important uses: in preparation of the polyester fiber Dacron and the film Mylar, as a major ingredient in cooling system antifreezes, as a solvent in paints and plastics, in formulations for printing ink and ink for ballpoint pens.
  - One common industrial synthesis produces ethylene glycol from ethylene (ethene).
- Glycerol (1,2,3-propanetriol) is used as an cosmetic emollient, as a sweetener, and as a raw material to make polymers and explosives (e.g., nitroglycerine).

22.7 Phenols

KEY TERM

Phenol

- A phenol contains a hydroxyl group bound to an aromatic ring.
- Phenols are named as aromatic compounds.
- Many natural products are phenols. These compounds have a variety of physiological effects.

22.8 Properties of Phenols

- Phenol is a weak acid, stronger than alcohol or water but weaker than acetic or carbonic acids.
- Phenol is toxic to microorganisms and was the first compound to be extensively used as an operating room disinfectant.

22.9 Production of Phenol

- Phenol is produced in an economical way from benzene and propene. The intermediate, cumene, is oxidized and gives rise to phenol and another important commercial product, acetone.

22.10 Ethers

KEY TERM

Ether

- Ethers have the general formula, ROR′.
- Simple ether names combine the names of the alkyl groups attached to the ether O followed by *ether*.
- IUPAC naming follows some standard steps.
  - Select the longest continuous carbon chain and give it the corresponding alkane name.
  - Name the other alkyl group by changing the suffix from *-yl* to *-oxy*. For example, *methyl* is changed to *methoxy*.
  - Combine the alkoxy name with the alkane name, giving the alkoxy name and its position on the alkane chain first, to form the ether name. For example: $CH_3—O—CH_2CH_3$, methoxyethane.

22.11 Structures and Properties of Ethers

- The oxygen in an ether gives the ether a bent shape.
- Because of the oxygen, ethers are more polar than alkanes.
- The ether oxygen hydrogen-bonds to water molecules; this accounts for low-molar-mass ether's significant water solubility.
- Ether's intermediate polarity allows it to serve as a solvent for both polar and nonpolar compounds.

22.12 Preparation of Ethers

- Ethers can be synthesized by intermolecular dehydration of alcohols.
- Ethers are also formed by reacting alkyl halides with sodium alkoxides or sodium phenoxides (Williamson synthesis).

22.13 Thiols

KEY TERMS

Thiol
Mercaptan

- Thiols or mercaptans are organic compounds that contain the —SH group.
- Naming thiols is analogous to naming alcohols except that the suffix *-thiol* is used in place of the suffix *-ol*.
- Unlike alcohols, thiols have foul odors and do not hydrogen-bond.
- Unlike alcohols, two thiols can be oxidized and link together via a disulfide bond.

Review Questions

All questions with blue numbers have answers in the appendix of the text.

1. Write the structural formulas for, and give an example of,
   (a) an alkyl halide
   (b) a phenol
   (c) an ether
   (d) an aldehyde
   (e) a ketone
   (f) a carboxylic acid
   (g) an ester
   (h) a thiol
2. Although it is possible to make alkenes from alcohols, alkenes are seldom, if ever, made in this way on an industrial scale. Why not?
3. Isopropyl alcohol is usually used in rubbing alcohol formulations. Why is this alcohol used in preference to normal propyl alcohol?

4. What classes of compounds can be formed by the oxidation of primary alcohols? Cite examples.
5. Why is 1,2-ethanediol (ethylene glycol) superior to methanol as an antifreeze for automobile radiators?
6. Briefly explain why glucose (blood sugar) is soluble (and therefore easily transported) in the bloodstream.
7. Briefly outline the physiological effects of (a) methanol and (b) ethanol.
8. Write equations for the cumene hydroperoxide synthesis of phenol and acetone.
9. List three common phenols and their commercial uses.
10. What two hazards may be present when one is working with low-molar-mass ethers?
11. Explain, in terms of molecular structure, why ethanol, CH_3CH_2OH (molar mass = 46), is a liquid at room temperature and dimethyl ether, CH_3OCH_3 (molar mass = 46), is a gas.
12. What is one possible cause of sudden infant death syndrome (SIDS)? Relate this cause to the chemistry of alcohols.

Paired Exercises

All exercises with blue numbers have answers in the appendix of the text.

1. Write structural formulas for the following:
   (a) methanol
   (b) 3-methyl-1-hexanol
   (c) 1,2-propanediol
   (d) 1-phenylethanol
   (e) 2,3-butanediol
   (f) 2-propanethiol

2. Write structural formulas for the following:
   (a) 2-butanol
   (b) 2-methyl-2-butanol
   (c) 2-propanol
   (d) cyclopentanol
   (e) 1-pentanethiol
   (f) 4-ethyl-3-hexanol

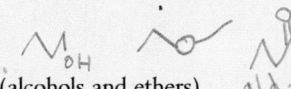

3. There are eight open-chain isomeric alcohols that have the formula $C_5H_{11}OH$. Write the structural formula and the IUPAC name for each of these alcohols.

4. Write structures for all the isomers (alcohols and ethers) with the formula
   (a) C_3H_8O (b) $C_4H_{10}O$

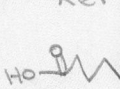

5. Which of the isomers in Exercise 3 are
   (a) primary alcohols?
   (b) secondary alcohols?
   (c) tertiary alcohols?

6. Which of the isomers in Exercise 4 are
   (a) primary alcohols?
   (b) secondary alcohols?
   (c) tertiary alcohols?

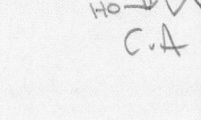

7. Name the following compounds:
   (a) $CH_3CH_2CH_2CH_2OH$
   (b) $CH_3CH(OH)CH_3$
   (c) C_6H_5–CH_2CHCH_2OH with CH_3 on the CH (benzene ring attached to CH_2)
   (d) H_2C—CH_2 bridged by O (epoxide ring)
   (e) CH_3CCH_3 with CH_2CH_3 above and OH below the central C
   (f) cyclohexane ring bearing OH and, on the adjacent carbon, CH_3
   (g) $HOCH_2CHCHCH_2OH$ with CH_3 on each of the two central carbons (one above, one below)

8. Name the following compounds:
   (a) CH_3CH_2OH
   (b) C_6H_5–CH_2CH_2OH (benzene ring)
   (c) $CH_3CCH_2CH_3$ with CH_2CH_3 above and OH below the second C
   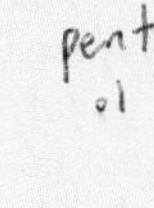
   (d) cyclopentane ring bearing OH and CH_3 on the same carbon
   (e) $CH_3CH_2CH(OH)CH_2CH_3$
   (f) CH_3CHCH_2 with HO on the CH and OH on the CH_2
   (g) $CH_3CH_2CHCH_2CHCH_3$ with CH_2CH_3 below the third carbon and OH above the fifth carbon

9. Write the formula and the name of the chief product when the following alcohols are dehydrated to alkenes:

(a) $CH_3CH(CH_3)CH(OH)CH_3$ — $CH_3CHCHCH_3$ with CH_3 on C-2 and OH on C-3

(b) $CH_3CH(OH)CH_2CH_2CH_3$

(c) 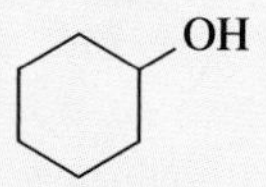 cyclohexanol (cyclohexane ring bearing OH)

10. Write the formula and the name of the chief product when the following alcohols are dehydrated to alkenes:

(a) 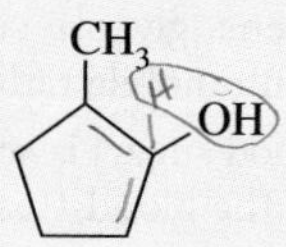 cyclopentane ring bearing CH_3 and OH on the same carbon

(b) $CH_3CH_2CH(OH)CH_3$

(c) $CH_3CH(CH_3)CH(OH)CH_2CH_3$ — $CH_3CHCHCH_2CH_3$ with CH_3 on C-2 and OH on C-3

11. Write the formula and the name of the alcohol that, when dehydrated, gives rise to the following alkene only:

(a) $CH_3CH_2CH_2CH{=}CHCH_2CH_3$

(b) 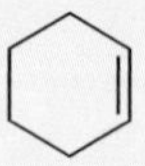 cyclohexene

(c) 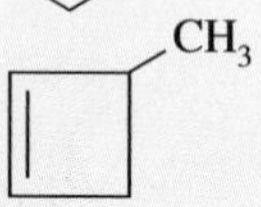 cyclobutene ring bearing CH_3

12. Write the formula and the name of the alcohol that, when dehydrated, gives rise to the following alkene only:

(a) 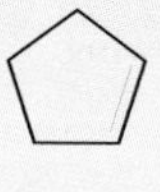 cyclopentene (b) 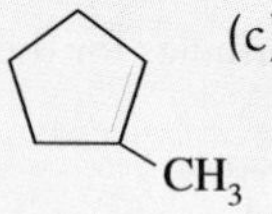 cyclopentene ring bearing CH_3 (c) $CH_3CH(CH_3)CH{=}C(CH_3)CH_3$ — $CH_3CHCH{=}CCH_3$ with CH_3 on C-2 and C-4

13. 3-Ethyl-1-hexanol is a primary alcohol. Write the formulas of two different organic compounds that can be obtained by oxidizing this alcohol.

14. 3-Methyl-2-pentanol is a secondary alcohol. Write the formula of one organic compound that can be obtained by oxidizing this alcohol.

15. Write the equation for the preparation of alcohols by reacting each of the following alkenes with sulfuric acid and water:
(a) propene
(b) 1-butene
(c) 2-pentene

16. Write the equation for the preparation of alcohols by reacting each of the following alkenes with sulfuric acid and water:
(a) 2-butene
(b) 1-pentene
(c) 2-methyl-2-butene

17. Alcohols can be made by reacting alkyl halides with sodium hydroxide as follows:

$$RX + NaOH \longrightarrow ROH + NaX$$

Give the names and formulas of the alkyl bromides (RBr) needed to prepare the following alcohols by this method:
(a) 2-propanol
(b) cyclohexanol
(c) 3-methyl-1-butanol

18. Alcohols can be made by reacting alkyl halides with sodium hydroxide as follows:

$$RX + NaOH \longrightarrow ROH + NaX$$

Give the names and formulas of the alcohols produced from the following alkyl bromides by this method:
(a) 2-bromobutane
(b) 2-bromo-3-ethylpentane
(c) bromocyclopentane

19. Complete the following equations and name the principal organic product(s) formed in each case:

(a) $2\,CH_3CH_2OH \xrightarrow[140°C]{96\%\ H_2SO_4}$

(b) $CH_3CH_2CH_2OH \xrightarrow[180°C]{96\%\ H_2SO_4}$

(c) $CH_3CH(OH)CH_2CH_3 \xrightarrow[\Delta]{K_2Cr_2O_7/H_2SO_4}$

(d) $CH_3CH_2\overset{O}{\overset{\|}{C}}{-}OCH_2CH_3 + H_2O \xrightarrow{H^+}$

20. Complete the following equations, giving only the major organic products:

(a) $CH_3CH_2OH + Na \longrightarrow$

(b) $CH_3CH_2CH_2CH_2OH \xrightarrow[H_2SO_4]{K_2Cr_2O_7}$

(c) 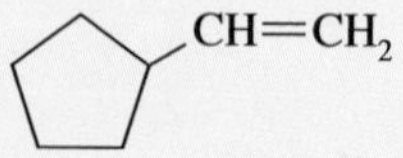 cyclopentyl$-CH{=}CH_2 \xrightarrow{H_2O/H_2SO_4}$

(d) $CH_3CH_2\overset{O}{\overset{\|}{C}}{-}OCH_3 + NaOH \longrightarrow$

21. Write structural formulas for each of the following:
(a) *o*-methylphenol
(b) *m*-dihydroxybenzene
(c) 4-hydroxy-3-methoxybenzaldehyde (vanillin)

22. Write structural formulas for each of the following:
(a) *p*-nitrophenol
(b) 2,6-dimethylphenol
(c) *o*-dihydroxybenzene

23. Give the structural formula and name for the alcohol that produces the product shown:

(a) 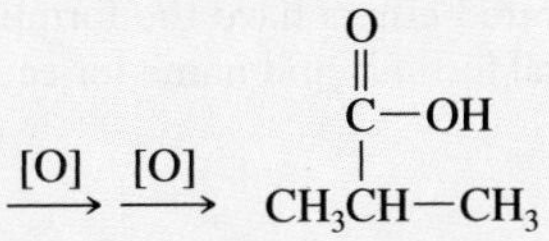

(b) 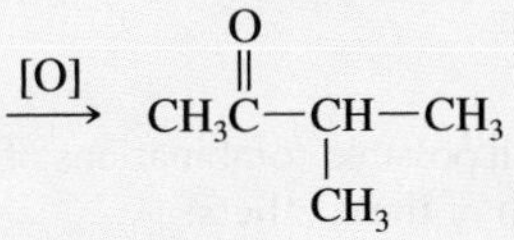

(c) 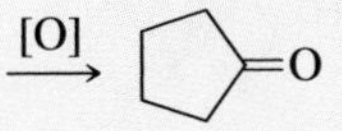

24. Give the structural formula and name for the alcohol that produces the product shown:

(a) 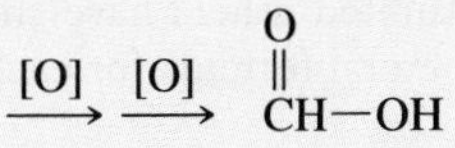

(b) 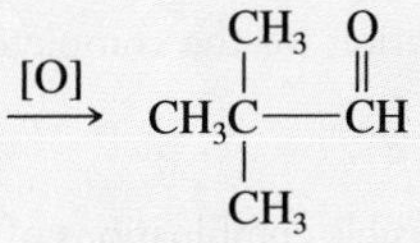

(c) 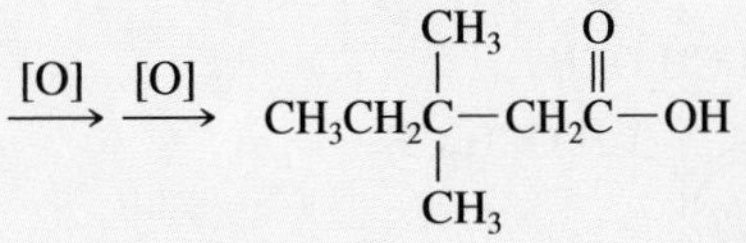

25. Name the alcohol used to make the following esters:

(a) 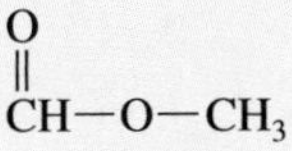

(b) 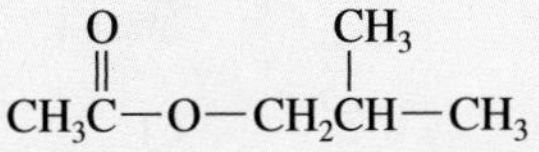

(c) 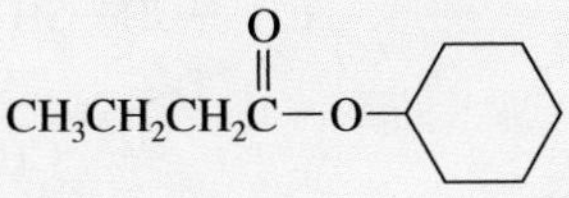

26. Name the alcohol used to make the following esters:

(a) 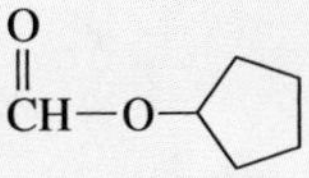

(b) 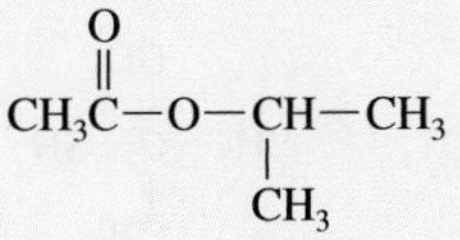

(c) 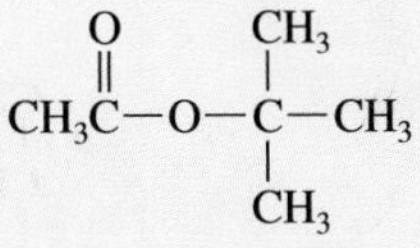

27. Name the following compounds:

(a) 

(b) 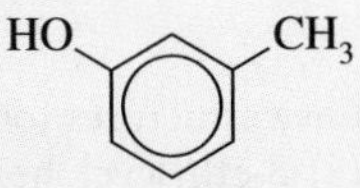

(c) 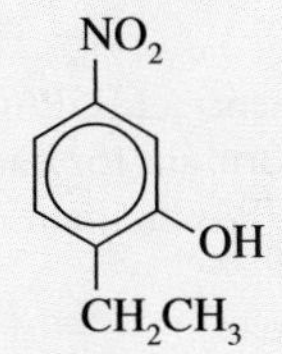

(d) 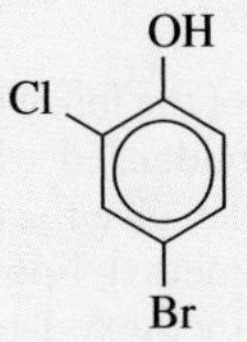

28. Name the following compounds:

(a) 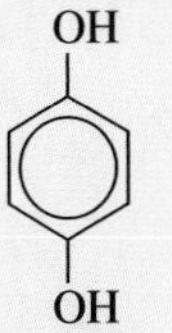

(b) 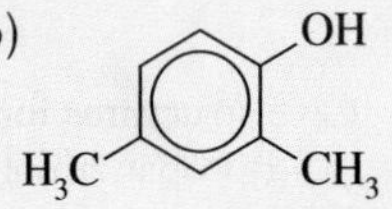

(c) 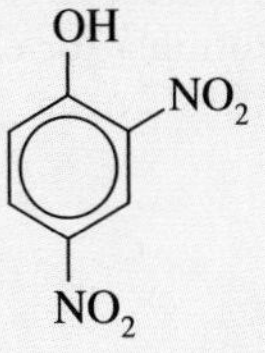

(d) 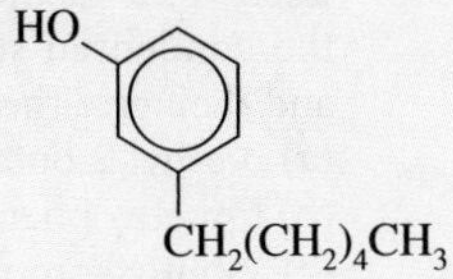

29. Arrange the following substances in order of increasing solubility in water:
(a) $CH_3CH_2OCH_2CH_2CH_3$
(b) $CH_3CH(OH)CH_2CH_2CH_3$
(c) $CH_3CH_2CH_2CH_2CH_3$
(d) $CH_3CH(OH)CH(OH)CH_2CH_3$

30. Arrange the following substances in order of decreasing solubility in water:
(a) $CH_3CH(OH)CH(OH)CH_2OH$
(b) $CH_3CH_2OCH_2CH_3$
(c) $CH_3CH_2CH_2CH_2OH$
(d) $CH_3CH(OH)CH_2CH_2OH$

31. Fourteen isomeric saturated ethers have the formula $C_6H_{14}O$. Write the structural formula for each of these ethers.

32. Six isomeric saturated ethers have the formula $C_5H_{12}O$. Write the structural formula and name for each of these ethers.

33. Write the balanced equation for the complete combustion of 2-butanol.

34. Write the balanced equation for the complete combustion of diethyl ether.

35. Write formulas for all possible combinations of RONa and RCl for making each of these ethers:
(a) $CH_3CH_2OCH_3$
(b) C_6H_5–$CH_2OCH_2CH_3$ (benzene ring bearing $CH_2OCH_2CH_3$)
(c) C_6H_5–$O-CH_2CH_3$ (benzene ring bearing $O-CH_2CH_3$)

36. Write formulas for all possible combinations of RONa and RCl for making each of these ethers:
(a) $CH_3CH_2CH_2OCH_2CH_2CH_3$
(b) CH_3–$\underset{|}{\overset{|}{HC}}OCH_2CH_2CH_3$ with CH_3 above and CH_3 below the CH (i.e., $(CH_3)_2CHOCH_2CH_2CH_3$)
(c) benzene ring–O–cyclohexane ring

37. Give IUPAC names for the following thiols:
(a) $CH_3-CH_2-C(SH)(CH_3)-CH_3$ (SH above, CH_3 below the central C)
(b) cyclohexane ring bearing SH
(c) $CH_3-C(CH_3)_2-SH$ (CH_3 above and below the C)
(d) $CH_3-CH_2-CH(SH)-CH(CH_3)-CH_3$, drawn as $CH_3-CH_2-CH-SH$ with $CH-CH_3$ above the CH and CH_3 above that

38. Give IUPAC names for the following thiols:
(a) cyclopentane ring bearing SH and an adjacent CH_3
(b) $CH_3-CH_2-C(SH)(CH_3)-CH_3$ with CH_3-CH_2 below the first CH_2: $CH_3-CH(CH_2CH_3)-C(SH)(CH_3)-CH_3$, drawn as $CH_3-CH-C-CH_3$ with SH above the C, CH_3 below the C, and CH_3-CH_2 below the CH
(c) $CH_3-CH(CH_3)-SH$
(d) $CH_3-CH_2-CH(SH)-$ with $CH-CH_3$ and CH_3 above, and CH_3-C-CH_3 with CH_3 below, drawn as:
CH_3 / $CH-CH_3$ / $CH_3-CH_2-CH-SH$ / CH_3-C-CH_3 / CH_3

39. Give structures for the following:
(a) 2-propanethiol
(b) 3-isopropyl-1-hexanethiol
(c) 2-methylcyclopentanethiol

40. Give structures for the following:
(a) 3-ethylcyclohexanethiol
(b) 3-pentanethiol
(c) 2,2-dimethyl-1-butanethiol

41. Each of the following is an incorrect IUPAC name. Draw the condensed structural formula for each compound and then give the correct IUPAC name.
(a) 3-ethyl-2-butanol
(b) 5-methylphenol
(c) 2-butoxypropane
(d) 1-propanethiol

42. Each of the following is an incorrect IUPAC name. Draw the condensed structural formula for each compound and then give the correct IUPAC name.
(a) 3-methyl-4-pentanethiol
(b) 1-methoxy-1-methylpropane
(c) 1,2-ethandiol
(d) 2-methyl-3-ethylphenol

Additional Exercises

All exercises with blue numbers have answers in the appendix of the text.

43. Phenol and its derivatives have been used as antiseptics since the middle 1800s. One common present-day antiseptic is related to phenol:

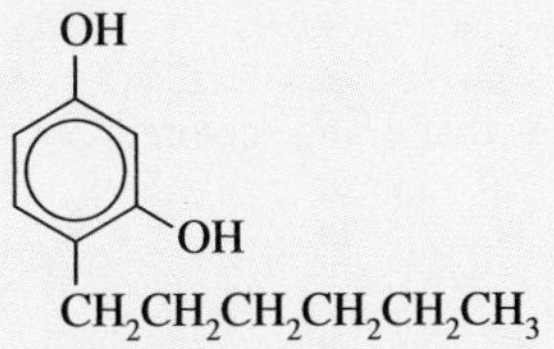

(structure: benzene ring with OH at position 1, OH at position 2, and $CH_2CH_2CH_2CH_2CH_2CH_3$ at position 4)

Name this compound.

44. Cyclic glycols show cis–trans isomerism because of restricted rotation about the carbon–carbon bonds in the ring. Draw and label the structures for *cis*- and *trans*-1,2-cyclopentanediol.

45. Arrange these three compounds in order of increasing acidity:

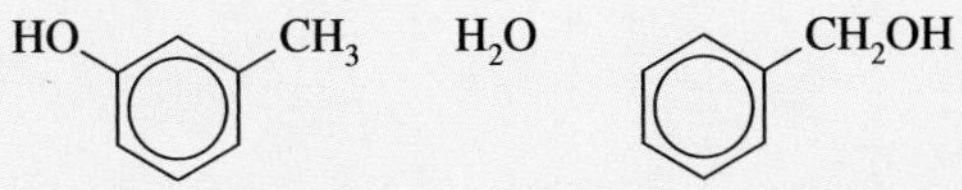

HO–C_6H_4–CH_3 (meta) $\quad H_2O \quad C_6H_5CH_2OH$

46. A deficiency in the production of dopamine by the brain is believed to be the cause of Parkinson's disease. Dopamine is part of a class of compounds called catecholamines. Draw the structure of the common phenol derivative that is a part of the dopamine molecule.

47. Methyl alcohol is toxic partly because it is metabolized to formaldehyde ($O{=}CH_2$) after ingestion. What type of reaction converts methanol to formaldehyde?

48. Write equations to show how the following transformations can be accomplished (some conversions may require more than one step, and some reactions studied in previous chapters may be needed):

(a) $CH_3CH(OH)CH_3 \longrightarrow CH_3C({=}O)CH_3$

(b) $CH_3CH_2CH_2CH{=}CH_2 \longrightarrow CH_3CH_2CH_2CH(OH)CH_3$

(c) $CH_3CH_2OH \longrightarrow CH_3CH_2O^-Na^+$

(d) $CH_3CH_2CH{=}CH_2 \longrightarrow CH_3CH_2C({=}O)CH_3$

(e) $CH_3CH_2CH_2CH_2OH \longrightarrow CH_3CH_2CH(Cl)CH_3$

(f) $CH_3CH_2CH_2Cl \longrightarrow CH_3CH_2C(H){=}O$

49. Starting with *p*-methylphenol and ethane, show the equations for the synthesis of *p*-ethoxytoluene.

50. Give a simple chemical test that will distinguish between the compounds in each of the following pairs:
(a) ethanol and dimethyl ether
(b) 1-pentanol and 1-pentene
(c) *p*-methylphenol and methoxybenzene

51. In what important ways do phenols differ from alcohols?

52. Draw all the isomeric compounds that contain a benzene ring and have the molecular formula $C_8H_{10}O$.

53. Which of the following compounds would you expect to react with sodium hydroxide?

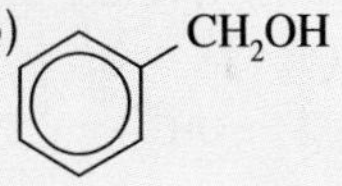

(a) C_6H_5OH (b) $C_6H_5CH_2OH$

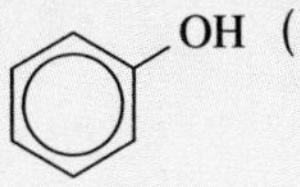

(c) $C_6H_5OC_6H_5$

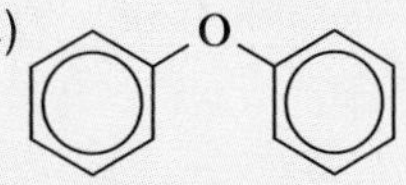

Write equations for those which react.

54. Arrange these compounds in order of *increasing* boiling points:
1-pentanol 1,2-pentanediol 1-octanol
Give reasons for your answer.

Challenge Exercise

All exercises with blue numbers have answers in the appendix of the text.

55. (a) When 2-methyl-1-butanol is dehydrated in an acid medium to an alkene, it yelds mainly 2-methyl-2-butene rather than 2-methyl-1-butene. This indicates that the dehydration to an alkene is at least a two-step reaction. Suggest a mechanism to explain the reaction. (*Hint:* A primary carbocation is formed initially.)

(b) When this product (2-methyl-2-butene) is again hydrated, 2-methyl-2-butanol is formed. Explain why this hydration does not yield the original reactant, 2-methyl-1-butanol.

Answers to Practice Exercises

22.1 O=CHCH(OH)CH(OH)CH(OH)CH(OH)CH$_2$OH

(carbons marked, left to right: 2°, 2°, 2°, 2°, 1°)

22.2 (a) 2-ethyl-1-pentanol
(b) 5-bromo-5-methyl-3-hexanol

22.3 (a) 3-methylcyclohexanol (cyclohexane ring with OH and CH$_3$)
(b) CH$_3$CH(CH$_3$)CH(OH)CH(CH$_2$CH$_3$)CH$_2$CH$_2$CH$_3$

22.4 1,2,3-propanetriol

22.5 1-hexanol, 1-pentanol, 1-butanol, 1,3-propanediol

22.6 (a) CH$_3$CH$_2$CH$_2$CH$_2$OH $\xrightarrow{[O]}$ CH$_3$CH$_2$CH$_2$C(=O)—H

CH$_3$CH$_2$CH$_2$CH$_2$OH $\xrightarrow{[O]}$ CH$_3$CH$_2$CH$_2$C(=O)—OH

(b) No reaction (2-methyl-2-butanol is a tertiary alcohol)

(c) cyclopentanone (cyclopentane ring with =O)

Evidence of reaction is that orange colored $Cr_2O_7^{2-}$ turned green (Cr^{3+}).

22.7 O=C—COOH
 |
CH$_2$—COOH

22.8 (a) 1-methoxy-2-methylbutane
(b) 2-ethoxy-2-methylpropane
(c) *o*-chloroethoxybenzene

22.9 (a) methyl propyl ether
(b) di-*tert*-butyl ether
(c) ethyl phenyl ether

22.10 (a) 2 CH$_3$CH$_2$CH$_2$OH + 2 Na ⟶ 2 CH$_3$CH$_2$CH$_2$ONa + H$_2$

CH$_3$CH$_2$CH$_2$ONa + CH$_3$CH$_2$CH$_2$CH$_2$Br ⟶ CH$_3$CH$_2$CH$_2$—O—CH$_2$CH$_2$CH$_2$CH$_3$ + NaBr

OR

(a) 2 CH$_3$CH$_2$CH$_2$CH$_2$OH + 2 Na ⟶ 2 CH$_3$CH$_2$CH$_2$CH$_2$ONa + H$_2$

CH$_3$CH$_2$CH$_2$CH$_2$ONa + CH$_3$CH$_2$CH$_2$Br ⟶ CH$_3$CH$_2$CH$_2$CH$_2$—O—CH$_2$CH$_2$CH$_3$ + NaBr

(b) 2 C$_6$H$_5$OH + 2 Na ⟶ 2 C$_6$H$_5$ONa + H$_2$

C$_6$H$_5$ONa + CH$_3$Br ⟶ C$_6$H$_5$OCH$_3$ + NaBr

(c) 2 C$_6$H$_5$CH$_2$OH + 2 Na ⟶ 2 C$_6$H$_5$CH$_2$ONa + H$_2$

C$_6$H$_5$CH$_2$ONa + CH$_3$CH$_2$Br ⟶ C$_6$H$_5$CH$_2$—O—CH$_2$CH$_3$ + NaBr

OR

(d) 2 CH$_3$CH$_2$OH + 2 Na ⟶ 2 CH$_3$CH$_2$ONa + H$_2$

CH$_3$CH$_2$ONa + C$_6$H$_5$CH$_2$Br ⟶ C$_6$H$_5$CH$_2$—O—CH$_2$CH$_3$ + NaBr

CHAPTER 23

Aldehydes and Ketones

How do you wish to pay for your purchase: cash, check, or plastic? Credit cards and many other plastic products are made with molecules known as aldehydes.

Chapter Outline

Many organic molecules contain a carbon atom double-bonded to an oxygen atom. This grouping of atoms is particularly reactive and is present in both aldehydes and ketones. Formaldehyde is by far the most common aldehyde molecule. It is used commercially as a preservative for animal specimens and in the formation of formaldehyde polymers. Examples of these polymers are found in Melmac plastic dishes, in Formica tabletops, and in the adhesives used in the manufacture of plywood and fiberboard. Other aldehydes are found in nature as spices.

Acetone is the simplest and most common ketone in our lives. It is used in nail polish remover, paints, varnishes, and resins. Acetone is also produced in the body during lipid metabolism. The presence of acetone in a urine sample or on the breath is a positive indicator of diabetes. An understanding of aldehydes and ketones forms the basis for a discussion of many important organic and biochemical reactions.

23.1 Structures of Aldehydes and Ketones

carbonyl group
aldehyde
ketone

The aldehydes and ketones are closely related classes of compounds. Their structures contain the **carbonyl group**, $\gt C{=}O$, a carbon–oxygen double bond. **Aldehydes** have at least one hydrogen atom bonded to the carbonyl group, whereas **ketones** have only alkyl or aryl (aromatic, denoted Ar) groups bonded to the carbonyl group:

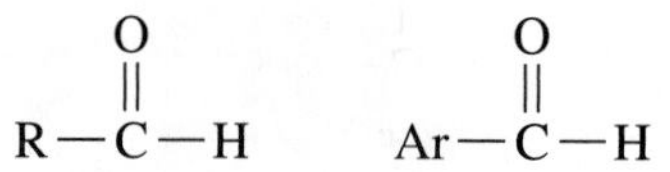

$$\underset{}{R-\overset{\overset{\displaystyle O}{\|}}{C}-H} \qquad Ar-\overset{\overset{\displaystyle O}{\|}}{C}-H$$

aldehydes

$$R-\overset{\overset{\displaystyle O}{\|}}{C}-R \qquad R-\overset{\overset{\displaystyle O}{\|}}{C}-Ar \qquad Ar-\overset{\overset{\displaystyle O}{\|}}{C}-Ar$$

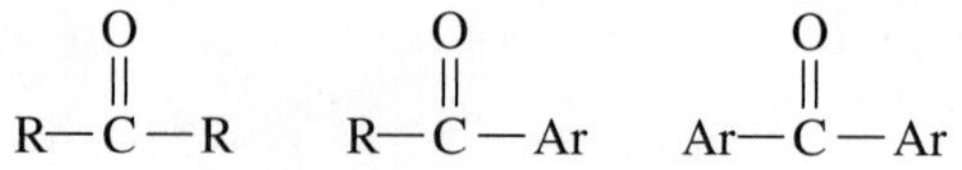

ketones

In a linear expression, the aldehyde group is often written as CHO. For example,

$$CH_3CHO \quad \text{is equivalent to} \quad CH_3\overset{\overset{\displaystyle O}{\|}}{C}-H$$

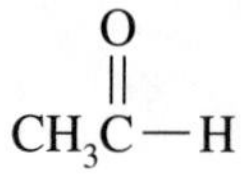

In the linear expression of a ketone, the carbonyl group is written as CO. For example,

$$CH_3COCH_3 \quad \text{is equivalent to} \quad CH_3\overset{\overset{\displaystyle O}{\|}}{C}CH_3$$

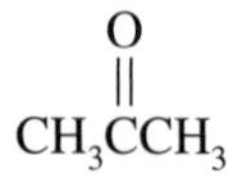

The general formula for the saturated homologous series of aldehydes and ketones is $C_nH_{2n}O$.

23.2 Naming Aldehydes and Ketones

Aldehydes

IUPAC Rules for Naming Aldehydes

1. To establish the parent name, select the longest continuous chain of carbon atoms that contains the aldehyde group.
2. The carbons of the parent chain are numbered starting with the aldehyde group. Since the aldehyde group is at the beginning (or end) of a chain, it is understood to be number 1.
3. Form the parent aldehyde name by dropping the *-e* from the corresponding alkane name and adding the suffix *-al.*
4. Other groups attached to the parent chain are named and numbered as we have done before.

The first member of the homologous series, $H_2C{=}O$, is methanal. The name *methanal* is derived from the hydrocarbon methane, which contains one carbon atom. The second member of the series is ethanal, the third member is propanal, and so on. The following diagrams illustrate how the name is formed for the first two members of the series:

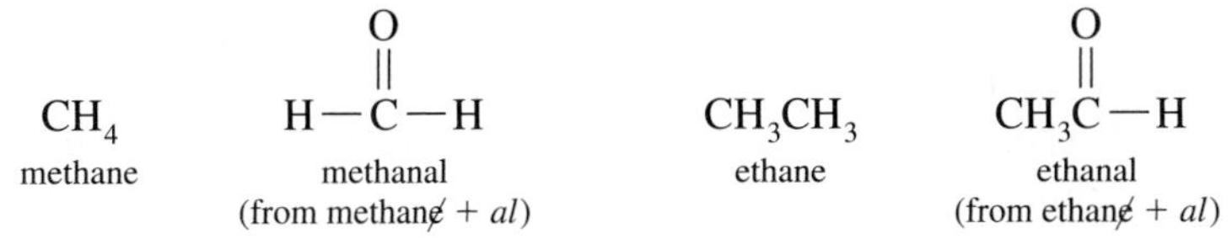

$$CH_4$$
methane

$$H{-}\overset{\overset{\displaystyle O}{\|}}{C}{-}H$$
methanal
(from methan~~e~~ + *al*)

$$CH_3CH_3$$
ethane

$$CH_3\overset{\overset{\displaystyle O}{\|}}{C}{-}H$$
ethanal
(from ethan~~e~~ + *al*)

The longest carbon chain containing the aldehyde group is the parent compound. Other groups attached to this chain are numbered and named as before, as the following structural formula shows:

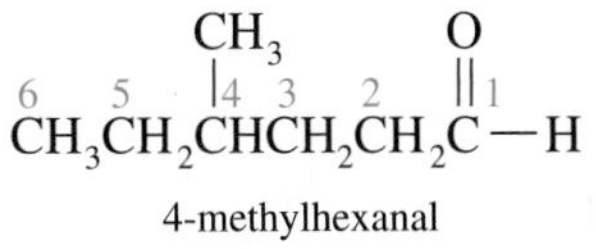

$$\overset{6}{C}H_3\overset{5}{C}H_2\overset{\overset{\displaystyle CH_3}{|}}{\underset{4}{C}}H\overset{3}{C}H_2\overset{2}{C}H_2\overset{\overset{\displaystyle O}{\|}}{\underset{1}{C}}{-}H$$
4-methylhexanal

The naming of aldehydes is also illustrated in Table 23.1. The common names for some aldehydes are widely used. Common names for the aliphatic aldehydes are derived from the common names of the carboxylic acids. (See Table 24.1.) The *-ic acid* or *-oic acid* ending of the acid name is dropped and is replaced with the suffix *-aldehyde*. Thus, the name of the one-carbon acid, formic acid, becomes formaldehyde for the one-carbon aldehyde, and the name for the two-carbon acid, acetic acid, becomes acetaldehyde for the two-carbon aldehyde:

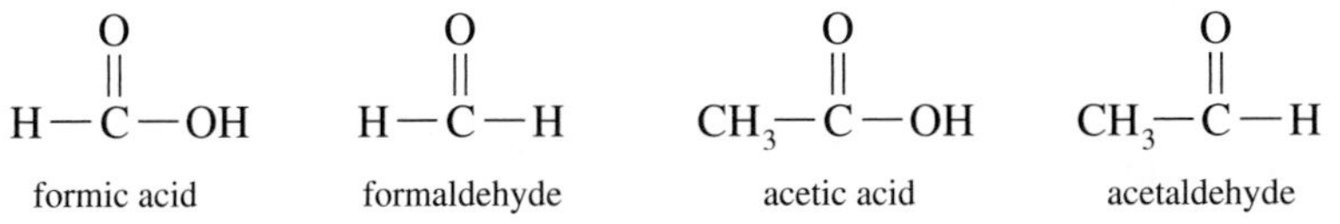

$$H{-}\overset{\overset{\displaystyle O}{\|}}{C}{-}OH$$
formic acid

$$H{-}\overset{\overset{\displaystyle O}{\|}}{C}{-}H$$
formaldehyde

$$CH_3{-}\overset{\overset{\displaystyle O}{\|}}{C}{-}OH$$
acetic acid

$$CH_3{-}\overset{\overset{\displaystyle O}{\|}}{C}{-}H$$
acetaldehyde

Table 23.1 IUPAC and Common Names of Selected Aldehydes

Formula	IUPAC name	Commonly used names
$H{-}C({=}O){-}H$	Methanal	Formaldehyde
$CH_3C({=}O){-}H$	Ethanal	Acetaldehyde
$CH_3CH_2C({=}O){-}H$	Propanal	Propionaldehyde
$CH_3CH_2CH_2C({=}O){-}H$	Butanal	Butyraldehyde
$CH_3CH(CH_3)C({=}O){-}H$	2-Methylpropanal	Isobutyraldehyde

Aromatic aldehydes contain an aldehyde group bonded to an aromatic ring. These aldehydes are named after the corresponding carboxylic acids. Thus, the name benzaldehyde is derived from benzoic acid, and the name *p*-tolualdehyde is derived from *p*-toluic acid:

$C_6H_5{-}C({=}O){-}H$ benzaldehyde

$C_6H_5{-}C({=}O){-}OH$ benzoic acid

$CH_3{-}C_6H_4{-}C({=}O){-}H$ *p*-tolualdehyde

$CH_3{-}C_6H_4{-}C({=}O){-}OH$ *p*-toluic acid

Aldehydes are used in the manufacture of building materials.

In dialdehydes, the suffix *-dial* is added to the corresponding hydrocarbon name. For example,

$$H{-}C({=}O)CH_2CH_2C({=}O){-}H$$

is named *butanedial.*

Ketones

IUPAC Rules for Naming Ketones

1. To establish the parent name, select the longest continuous chain of carbon atoms that contains the ketone group.
2. Form the parent name by dropping the *-e* from the corresponding alkane name and adding the suffix *-one*.
3. If the chain is longer than four carbons, it is numbered so that the carbonyl group has the smallest number possible; this number is prefixed to the parent name of the ketone.
4. Other groups attached to the parent chain are named and numbered as we have done before.

The following examples of the rules for naming ketones are illustrative:

$$CH_3-\overset{O}{\overset{\|}{C}}-CH_3$$

propanone

$$\overset{5}{C}H_3\overset{4}{C}H_2\overset{3}{C}H_2-\overset{O}{\overset{\|}{\underset{2}{C}}}-\overset{1}{C}H_3$$

2-pentanone

$$\overset{1}{C}H_3\overset{2}{C}H_2-\overset{O}{\overset{\|}{\underset{3}{C}}}-\underset{CH_3}{\underset{|}{\overset{4}{C}H}}\overset{5}{C}H_2\overset{6}{C}H_3$$

4-methyl-3-hexanone

Note that in 4-methyl-3-hexanone, the carbon chain is numbered from left to right to give the carbonyl group the lowest possible number.

An alternative non-IUPAC method that is often used to name simple ketones lists the names of the alkyl or aromatic groups attached to the carbonyl carbon together with the word *ketone*. Thus, butanone ($CH_3COCH_2CH_3$) is methyl ethyl ketone:

$$\underset{\text{methyl}}{\underset{\uparrow}{CH_3}}-\underset{\text{ketone}}{\underset{\uparrow}{\overset{O}{\overset{\|}{C}}}}-\underset{\text{ethyl}}{\underset{\uparrow}{CH_2CH_3}}$$

Two of the most widely used ketones have special common names: Propanone is called acetone, and butanone is known as methyl ethyl ketone, or MEK.

Aromatic ketones are named in a fashion similar to that for aliphatic ketones and often have special names as well:

$$C_6H_5-\overset{O}{\overset{\|}{C}}-CH_3$$

1-phenylethanone (IUPAC)
(methyl phenyl ketone)
(acetophenone)

$$C_6H_5-\overset{O}{\overset{\|}{C}}-CH_2CH_3$$

1-phenyl-1-propanone (IUPAC)
(ethyl phenyl ketone)
(propiophenone)

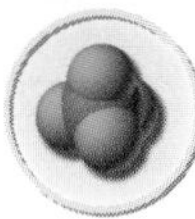

Example 23.1 Write both the formulas and the names for the straight-chain five- and six-carbon aldehydes.

SOLUTION The IUPAC names are based on the five- and six-carbon alkanes. Drop the *-e* of the alkane name and add the suffix *-al*. Thus, pentane, C_5, becomes pentanal and hexane, C_6, becomes hexanal. (The aldehyde group does not need to be numbered; it is understood to be on carbon 1.) The common names (in parentheses) are derived from valeric acid and caproic acid, respectively:

$$CH_3CH_2CH_2CH_2\overset{\overset{\displaystyle O}{\|}}{C}-H \qquad CH_3CH_2CH_2CH_2CH_2\overset{\overset{\displaystyle O}{\|}}{C}-H$$

pentanal (valeraldehyde) hexanal (caproaldehyde)

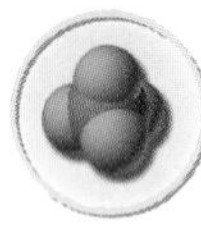

Example 23.2 Give two names for each of the following ketones:

(a) $CH_3CH_2\overset{\overset{\displaystyle O}{\|}}{C}CH_2\overset{\overset{\displaystyle CH_3}{|}}{C}HCH_3$ (b) $CH_3CH_2CH_2\overset{\overset{\displaystyle O}{\|}}{C}-C_6H_5$

SOLUTION (a) The parent carbon chain that contains the carbonyl group has six carbons. Number this chain from the end nearer to the carbonyl group. The ketone group is on carbon 3, and a methyl group is on carbon 5. The six-carbon alkane is hexane. Drop the *-e* from hexane and add *-one* to give the parent name, hexanone. Prefix the name hexanone with a 3- to locate the ketone group and with 5-methyl- to locate the methyl group. The name is 5-methyl-3-hexanone. The common name is ethyl isobutyl ketone, since the C=O has an ethyl group and an isobutyl group bonded to it.

(b) The longest aliphatic chain has four carbons. The parent ketone name is *butanone*, derived by dropping the *-e* of butane and adding *-one*. The butanone has a phenyl group attached to carbon 1. The IUPAC name is therefore 1-phenyl-1-butanone. The common name for this compound is phenyl *n*-propyl ketone, since the C=O group has a phenyl and an *n*-propyl group bonded to it.

Practice 23.1

Write structures for the following carbonyl compounds:

(a) 4-bromo-5-hydroxyhexanal
(b) phenylethanal
(c) 3-buten-2-one
(d) diphenylmethanone (diphenyl ketone)
(e) 2,2,4,4-tetramethyl-3-pentanone
(f) 4-cyclohexylbutanal

Practice 23.2

Name each compound, using the IUPAC System:

(a) cyclohexanone structure (ring)=O

(c) $CH_3CHCH_2\overset{O}{\overset{\|}{C}}-H$ with CH_3 on the second carbon

(b) $ClCH_2CH_2\overset{O}{\overset{\|}{C}}-CH_3$

(d) $CH=CH-\overset{O}{\overset{\|}{C}}-H$ with a benzene ring on the first CH

Practice 23.3 Application to Biochemistry

Honey is especially sweet because of the sugar, fructose. Give the IUPAC name for fructose, whose structure is

$$CH_2(OH)\overset{O}{\overset{\|}{C}}CH(OH)CH(OH)CH(OH)CH_2OH$$

fructose

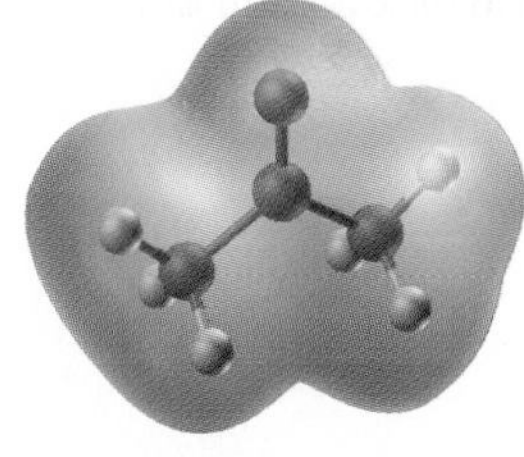

Spacefilling model of acetone showing e^- density. Acetone's electron density surface clearly shows the polarity of the carbonyl. Extra electron density (shown in red) centers at the oxygen atom.

23.3 Bonding and Physical Properties

The carbon atom of the carbonyl group is sp^2-hybridized and is joined to three other atoms by sigma bonds. The fourth bond is made by overlapping *p* electrons of carbon and oxygen to form a pi bond between the carbon and oxygen atoms. Since the carbon has a double bond, the possibility of adding another bonded atom (an addition reaction) exists for carbonyl carbon atoms.

Because the oxygen atom is considerably more electronegative than carbon, the carbonyl group is polar, with the electrons shifted toward the oxygen atom. This makes the oxygen atom partially negative (δ^-) and leaves the carbon atom partially positive (δ^+):

The hydrogen in a hydrogen bond must also be covalently bonded to a nitrogen, an oxygen, or a fluorine.

$$\overset{\delta^+}{C}=\overset{\delta^-}{\ddot{O}}:$$

polarity

H-bonds

no H-bonds

Many of the chemical reactions of aldehydes and ketones are due to this polarity.

Unlike alcohols, aldehydes and ketones cannot hydrogen-bond to themselves, because no hydrogen atom is attached to the oxygen atom of the carbonyl group. Aldehydes and ketones, therefore, have lower boiling points than alcohols of comparable molar mass (Table 23.2).

Low-molar-mass aldehydes and ketones are soluble in water, but for five or more carbons, the solubility decreases markedly. Ketones are highly efficient organic solvents.

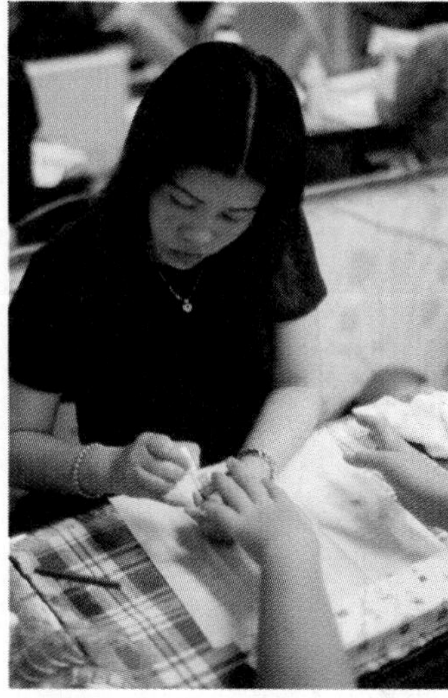

Both acetone and MEK (butanone) are highly effective solvents and are used to remove nail polish.

Table 23.2 Boiling Points of Selected Aldehydes and Ketones and Corresponding Alcohols

Name	Molar mass	Boiling point (°C)
1-Propanol	60	97
Propanal	58	49
Propanone	58	56
1-Butanol	74	118
Butanal	72	76
Butanone	72	80
1-Pentanol	88	138
Pentanal	86	103
2-Pentanone	86	102

The lower-molar-mass aldehydes have a penetrating, disagreeable odor and are partially responsible for the taste of some rancid and stale foods. As the molar mass increases, the odor of both aldehydes and ketones—especially the aromatic ones—becomes more fragrant. Some are even used in flavorings and perfumes. A few of these and other selected aldehydes and ketones are shown in Figure 23.1.

$C_6H_5-\overset{O}{\overset{\|}{C}}-H$

benzaldehyde
(oil of bitter almonds)

$C_6H_5-CH{=}CH-\overset{O}{\overset{\|}{C}}-H$

cinnamaldehyde
(oil of cinnamon)

CH_3, O, $CH_3-C{=}CH_2$

carvone
(chief component of spearmint oil)

$$\begin{array}{l} \quad\quad\quad\quad\quad\quad CH_3 \\ CH_2CH_2CH_2CH_2CH_2CHCH_2 \\ | \quad\quad\quad\quad\quad\quad\quad\quad\quad\quad C{=}O \\ CH_2CH_2CH_2CH_2CH_2CH_2CH_2 \end{array}$$

muscone
(gland of male musk deer, used in perfume)

$$\begin{array}{l} CHCH_2CH_2CH_2CH_2CH_2CH_2CH_2 \\ \| \quad\quad\quad\quad\quad\quad\quad\quad\quad\quad\quad C{=}O \\ CHCH_2CH_2CH_2CH_2CH_2CH_2CH_2 \end{array}$$

civetone
(secretion of the civet cat, used in perfume)

H_3C CH_3, CH_3, O

camphor
(from the camphor tree)

CH_2OH, $C{=}O$, OH, H_3C, O, H_3C, O

cortisone
(hormone; regulation of carbohydrate and protein metabolism; used to reduce inflammation)

Figure 23.1
Selected naturally occurring aldehydes and ketones.

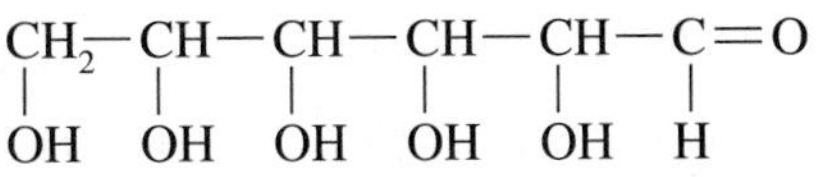

$$\underset{\text{OH}}{\underset{|}{\text{CH}_2}}-\underset{\text{OH}}{\underset{|}{\text{CH}}}-\underset{\text{OH}}{\underset{|}{\text{CH}}}-\underset{\text{OH}}{\underset{|}{\text{CH}}}-\underset{\text{OH}}{\underset{|}{\text{CH}}}-\underset{\text{H}}{\underset{|}{\text{C}}}=\text{O}$$

glucose
(sugar)

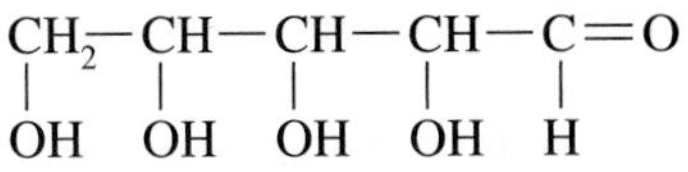

$$\underset{\text{OH}}{\underset{|}{\text{CH}_2}}-\underset{\text{OH}}{\underset{|}{\text{CH}}}-\underset{\text{OH}}{\underset{|}{\text{CH}}}-\underset{\text{OH}}{\underset{|}{\text{CH}}}-\underset{\text{H}}{\underset{|}{\text{C}}}=\text{O}$$

ribose
(sugar)

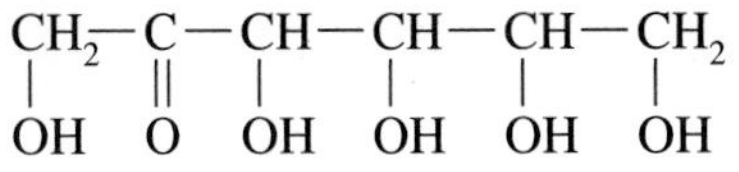

$$\underset{\text{OH}}{\underset{|}{\text{CH}_2}}-\underset{\text{O}}{\underset{\|}{\text{C}}}-\underset{\text{OH}}{\underset{|}{\text{CH}}}-\underset{\text{OH}}{\underset{|}{\text{CH}}}-\underset{\text{OH}}{\underset{|}{\text{CH}}}-\underset{\text{OH}}{\underset{|}{\text{CH}_2}}$$

fructose
(sugar)

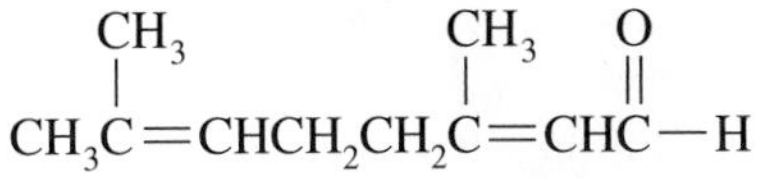

$$\text{CH}_3\overset{\text{CH}_3}{\overset{|}{\text{C}}}=\text{CHCH}_2\text{CH}_2\overset{\text{CH}_3}{\overset{|}{\text{C}}}=\text{CH}\overset{\text{O}}{\overset{\|}{\text{C}}}-\text{H}$$

citral
(oil of lemon)

$$\text{(naphthoquinone ring with CH}_3\text{)}-\text{CH}_2\text{CH}=\underset{\text{CH}_3}{\underset{|}{\text{C}}}-\left[\text{CH}_2\text{CH}_2\text{CH}_2\overset{\text{CH}_3}{\overset{|}{\text{CH}}}\right]_3-\text{CH}_3$$

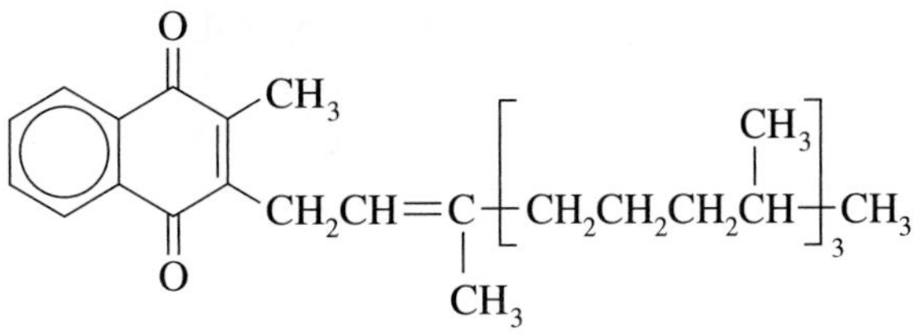

vitamin K_1
(antihemorrhagic vitamin)

Figure 23.1 (cont.)

23.4 Chemical Properties of Aldehydes and Ketones

The carbonyl group ($\rangle C{=}O$) is the functional group of both aldehydes and ketones. Associated with every functional group are characteristic reactions. Aldehydes undergo both oxidation and reduction reactions; ketones undergo reduction reactions.

The carbon atom in the carbonyl group also has a double bond. What kind of reactivity does this suggest for aldehydes and ketones? This question can be answered by reviewing the chemistry of the alkenes in Chapter 20. Like alkenes, the aldehydes and ketones have a carbon atom that is bonded to three other atoms—one less than the usual maximum of four. Such a carbon atom can easily bond to one more atom. Alkenes readily undergo addition reactions. This suggests that the aldehydes and ketones also undergo addition reactions. In fact, addition is the characteristic reaction of aldehydes and ketones.

An addition reaction for aldehydes and ketones can be envisioned as

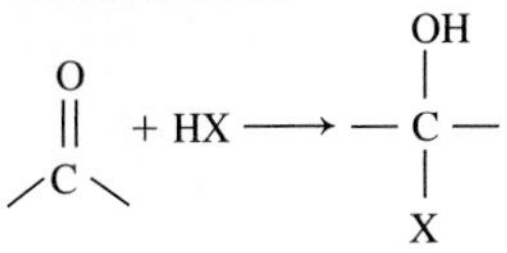

$$\overset{\text{O}}{\overset{\|}{\diagup\text{C}\diagdown}} + \text{HX} \longrightarrow -\underset{\text{X}}{\underset{|}{\overset{\text{OH}}{\overset{|}{\text{C}}}}}-$$

Oxidation

Aldehydes are easily oxidized to carboxylic acids by a variety of oxidizing agents, including (under some conditions) oxygen of the air. Oxidation is the reaction in which aldehydes differ most from ketones. In fact, aldehydes and ketones may be separated into classes by their relative susceptibilities to oxidation. Aldehydes are easily oxidized to carboxylic acids by $K_2Cr_2O_7 + H_2SO_4$ and by mild oxidizing agents such as Ag^+ and Cu^{2+} ions; ketones are unaffected by such reagents. Ketones can be oxidized under drastic conditions—for example, by treatment with hot potassium permanganate solution. However, under these conditions, carbon–carbon bonds are broken, and a variety of products are formed. Equations for the oxidation of aldehydes by dichromate are

$$3\,\text{R}\overset{\text{O}}{\overset{\|}{\text{C}}}-\text{H} + \text{Cr}_2\text{O}_7^{2-} + 8\,\text{H}^+ \longrightarrow \underset{\text{carboxylic acid}}{3\,\text{R}\overset{\text{O}}{\overset{\|}{\text{C}}}-\text{OH}} + 2\,\text{Cr}^{3+} + 4\,\text{H}_2\text{O}$$

$$3\,CH_3\overset{\overset{\displaystyle O}{\|}}{C}-H + Cr_2O_7^{2-} + 8\,H^+ \longrightarrow 3\,CH_3\overset{\overset{\displaystyle O}{\|}}{C}-OH + 2\,Cr^{3+} + 4\,H_2O$$

acetaldehyde (ethanal) acetic acid

Tollens test

The **Tollens test** (silver-mirror test) for aldehydes is based on the ability of silver ions to oxidize aldehydes. The Ag^+ ions are thereby reduced to metallic silver. In practice, a little of the suspected aldehyde is added to a solution of silver nitrate and ammonia in a clean test tube. The appearance of a silver mirror on the inner wall of the tube is a positive test for the aldehyde group. The abbreviated equation is

Tollens test: Silver metal is deposited on the inside of this test tube as aldehydes are oxidized.

$$R\overset{\overset{\displaystyle O}{\|}}{C}-H + 2\,Ag^+ \xrightarrow[H_2O]{NH_3} R\overset{\overset{\displaystyle O}{\|}}{C}-O^-NH_4^+ + 2\,Ag(s) \quad \text{(general reaction)}$$

$$CH_3\overset{\overset{\displaystyle O}{\|}}{C}-H + 2\,Ag^+ \xrightarrow[H_2O]{NH_3} CH_3\overset{\overset{\displaystyle O}{\|}}{C}-O^-\,NH_4^+ + 2\,Ag(s)$$

Fehling and the Benedict tests

Fehling and Benedict solutions contain Cu^{2+} ions in an alkaline medium. In the **Fehling and the Benedict tests**, the aldehyde group is oxidized to an acid by Cu^{2+} ions. The blue Cu^{2+} ions are reduced and form brick-red copper(I) oxide (Cu_2O), which precipitates during the reaction. The tests can be used for detecting carbohydrates that have an available aldehyde group. The abbreviated equation is

$$R\overset{\overset{\displaystyle O}{\|}}{C}-H + \underset{\text{(blue)}}{2\,Cu^{2+}} \xrightarrow[H_2O]{NaOH} R\overset{\overset{\displaystyle O}{\|}}{C}-O^-Na^+ + \underset{\text{(brick-red)}}{Cu_2O(s)}$$

Because most ketones do not give a positive test with Tollens, Fehling, or Benedict solutions, these tests are used to distinguish between aldehydes and ketones.

$$R-\overset{\overset{\displaystyle O}{\|}}{C}-R + Ag^+ \xrightarrow[H_2O]{NH_3} \text{No reaction}$$

$$R-\overset{\overset{\displaystyle O}{\|}}{C}-R + Cu^{2+} \xrightarrow[H_2O]{OH^-} \text{No reaction}$$

Aldehydes and ketones are often involved in industrial syntheses, because oxidation is a relatively simple reaction that can start with inexpensive reactants. Petroleum, natural gas, and the oxygen in the air are inexpensive and abundant materials; many aldehydes and ketones can be prepared from these starting materials. In turn, the aldehydes and ketones are useful for making even more valuable products. For example, simple hydrocarbons are oxidized to form acetaldehyde, which can be oxidized by air to acetic acid, the most widely used carboxylic acid. Air oxidation of 2-propanol yields acetone, an important commercial solvent. Methanol is air oxidized to formaldehyde, which is the precursor for a wide variety of plastics and glues.

The oxidation of aldehydes is an important reaction in biochemistry. When our cells "burn" carbohydrates, they take advantage of the aldehyde reactivity. The aldehyde group is oxidized to a carboxylic acid and is eventually converted to carbon dioxide, which is then exhaled. This stepwise oxidation provides some of the energy necessary to sustain life.

Practice 23.4

Write the structures of the products (or indicate "no reaction") for the following:

(a) pentanal in the Tollens test
(b) 4-methyl-2-hexanone in the Tollens test
(c) 3-pentanone in the Fehling test
(d) 4-methylhexanal in the Fehling test

Reduction

Aldehydes and ketones are easily reduced to alcohols, either by elemental hydrogen in the presence of a catalyst or by chemical reducing agents such as lithium aluminum hydride ($LiAlH_4$) or sodium borohydride ($NaBH_4$). Aldehydes yield primary alcohols, whereas ketones yield secondary alcohols:

$$\mathrm{R{-}\overset{\overset{\Large O}{\|}}{C}{-}H} \xrightarrow[\Delta]{H_2/Ni} \underset{1°\ \text{alcohol}}{\mathrm{RCH_2OH}} \qquad \text{(general reaction)}$$

$$\mathrm{R{-}\overset{\overset{\Large O}{\|}}{C}{-}R} \xrightarrow[\Delta]{H_2/Ni} \underset{2°\ \text{alcohol}}{\mathrm{R{-}\overset{\overset{\large OH}{|}}{CH}{-}R}} \qquad \text{(general reaction)}$$

$$\mathrm{CH_3\overset{\overset{\Large O}{\|}}{C}{-}H} \xrightarrow[\Delta]{H_2/Ni} \mathrm{CH_3CH_2OH}$$

$$\mathrm{CH_3\overset{\overset{\Large O}{\|}}{C}CH_3} \xrightarrow[\Delta]{H_2/Ni} \mathrm{CH_3\overset{\overset{\large OH}{|}}{C}HCH_3}$$

In biochemistry, aldehyde/ketone reduction is commonly reversible. For example, hardworking muscles generate an excess of pyruvic acid, which is then reduced to an alcohol (lactic acid) before being secreted into the bloodstream:

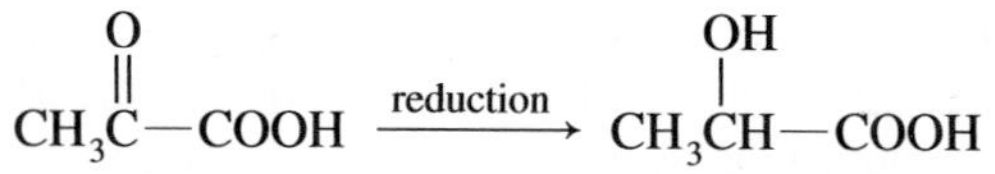

$$\underset{\text{pyruvic acid}}{CH_3\overset{\overset{\displaystyle O}{\|}}{C}-COOH} \xrightarrow{\text{reduction}} \underset{\text{lactic acid}}{CH_3\overset{\overset{\displaystyle OH}{|}}{C}H-COOH}$$

The product, lactic acid, contributes to the burning sensation of hardworking muscles and is removed from the blood by the liver. The first reaction of lactic acid in the liver is a conversion *back* to the ketone:

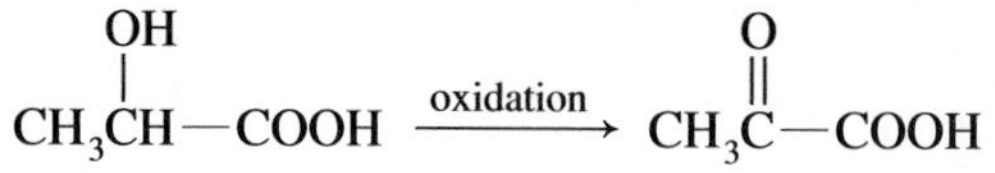

$$\underset{\text{lactic acid}}{CH_3\overset{\overset{\displaystyle OH}{|}}{C}H-COOH} \xrightarrow{\text{oxidation}} \underset{\text{pyruvic acid}}{CH_3\overset{\overset{\displaystyle O}{\|}}{C}-COOH}$$

This reversible reaction completes a cycle: The ketone is reduced to lactic acid, which is then oxidized to the ketone once again. Of what advantage is a cycle? This is a common question in biochemistry. In this case, the oxidation of lactic acid by the liver accelerates glucose production—glucose that the muscle needs to do more work.

Practice 23.5 Application to Biochemistry

Human starvation leads to a condition called "ketosis." A specific ketone is formed by the liver and released into the bloodstream:

$$CH_3-\overset{\overset{\displaystyle O}{\|}}{C}-CH_2COOH$$

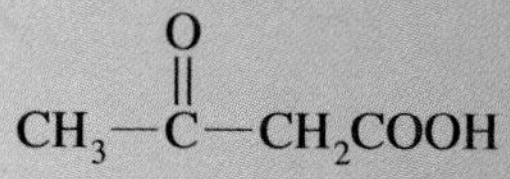

At the same time the liver reduces some of this ketone. The product alcohol also is released into the blood. Show the structure of the alcohol.

Addition Reactions

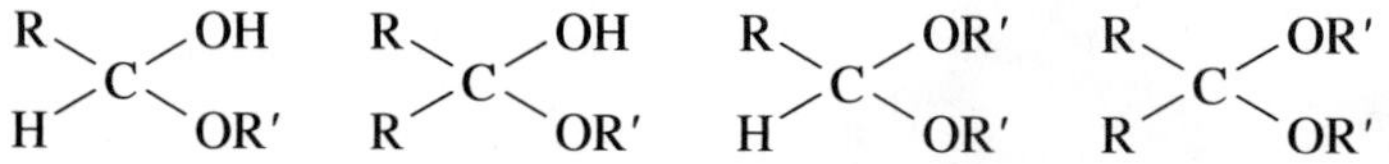

hemiacetal
hemiketal
acetal
ketal

The alkoxy group in these compounds forms an ether linkage.

Addition of Alcohols Compounds derived from aldehydes and ketones that contain an alkoxy and a hydroxy group on the same carbon atom are known as **hemiacetals** and **hemiketals**. In a like manner, compounds that have two alkoxy groups on the same carbon atom are known as **acetals** and **ketals**. The following diagram illustrates all four compounds:

$$\underset{\text{hemiacetal}}{\begin{matrix} R & & OH \\ & C & \\ H & & OR' \end{matrix}} \qquad \underset{\text{hemiketal}}{\begin{matrix} R & & OH \\ & C & \\ R & & OR' \end{matrix}} \qquad \underset{\text{acetal}}{\begin{matrix} R & & OR' \\ & C & \\ H & & OR' \end{matrix}} \qquad \underset{\text{ketal}}{\begin{matrix} R & & OR' \\ & C & \\ R & & OR' \end{matrix}}$$

Most open-chain hemiacetals and hemiketals are so unstable that they cannot be isolated. On the other hand, acetals and ketals are stable in alkaline solutions, but are unstable in acid solutions, in which they are hydrolyzed back to the original aldehyde or ketone.

In the reactions illustrated next, we show only aldehydes in the equations, but keep in mind that ketones behave in a similar fashion, although they are not as reactive.

Aldehydes react with alcohols in the presence of a trace of acid to form hemiacetals:

$$\underset{\text{propanal}}{CH_3CH_2\overset{\overset{\displaystyle O}{\|}}{C}-H} + \underset{\text{methanol}}{CH_3OH} \overset{H^+}{\rightleftharpoons} CH_3CH_2\overset{\overset{\displaystyle OH}{|}}{\underset{\underset{\displaystyle OCH_3}{|}}{CH}}$$

1-methoxy-1-propanol
(propionaldehyde methyl hemiacetal)

In the presence of excess alcohol and a strong acid such as dry HCl, aldehydes or hemiacetals react with a second molecule of the alcohol to give an acetal:

$$CH_3CH_2\overset{\overset{\displaystyle OH}{|}}{\underset{\underset{\displaystyle OCH_3}{|}}{CH}} + CH_3OH \overset{\text{dry HCl}}{\rightleftharpoons} CH_3CH_2\overset{\overset{\displaystyle OCH_3}{|}}{\underset{\underset{\displaystyle OCH_3}{|}}{CH}} + H_2O$$

1,1-dimethoxypropane
(propionaldehyde dimethyl acetal)

The vanilla bean is a source of an aldehyde, vanillin, that gives the vanilla flavor.

A hemiacetal has both an alcohol and an ether group attached to the aldehyde carbon. An acetal has two ether groups attached to the aldehyde carbon.

If the alcohol and carbonyl groups are within the same molecule, the result is the formation of a cyclic hemiacetal (or hemiketal):

$$\begin{array}{c} CH_2-O-H \quad\quad \\ / \quad\quad\quad\quad O \\ CH_2 \quad\quad\quad C\!\!\swarrow\!\!\!\!// \quad \\ \backslash \quad\quad / \quad \backslash H \\ CH_2-CH_2 \end{array} \overset{H^+}{\rightleftharpoons} \begin{array}{c} CH_2-O \quad\quad\quad \\ / \quad\quad\quad \backslash \quad OH \\ CH_2 \quad\quad\quad C \quad\quad \\ \backslash \quad\quad / \quad \backslash H \\ CH_2-CH_2 \end{array}$$

5-hydroxypentanal — stable hemiacetal

This *intramolecular* cyclization is particularly significant in carbohydrate chemistry during the study of monosaccharides (Chapter 27).

Practice 23.6

Write structures and IUPAC names for the hemiacetal and acetal, and the hemiketal and ketal, formed by reacting (a) ethanal and ethanol and (b) butanone and methanol.

Practice 23.7 Application to Biochemistry

The following three sugars are cyclic hemiacetals:

(a) 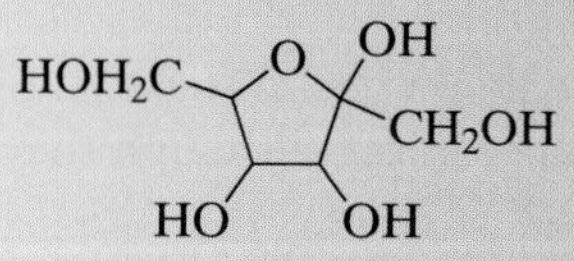 Fructose

(b) 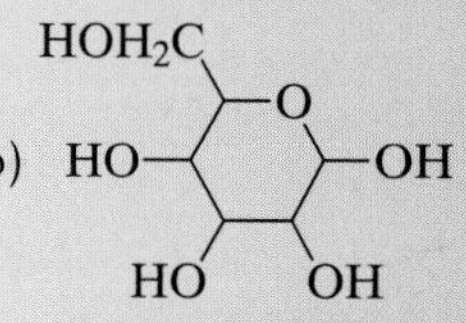 Glucose

(c) 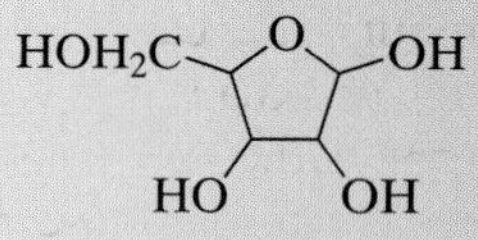 Ribose

Circle the hemiacetal carbon in each.

Addition of Hydrogen Cyanide The addition of hydrogen cyanide, HCN, to aldehydes and ketones forms a class of compounds known as cyanohydrins.

cyanohydrin **Cyanohydrins** have a cyano (—CN) group and a hydroxyl group on the same carbon atom. This reaction is typical of carbonyl addition reactions. The reaction takes place in a basic medium.

$$\underset{\text{acetaldehyde}}{CH_3\overset{O}{\overset{\|}{C}}-H} + HCN \xrightarrow{OH^-} \underset{\text{acetaldehyde cyanohydrin}}{CH_3\overset{OH}{\overset{|}{C}}HCN}$$

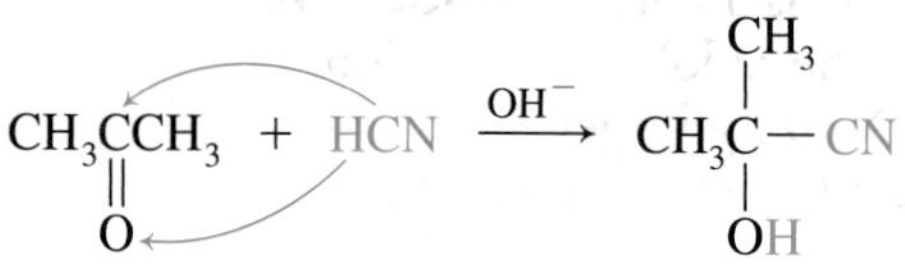

$$\underset{\text{acetone}}{CH_3\underset{O}{\underset{\|}{C}}CH_3} + HCN \xrightarrow{OH^-} \underset{\text{acetone cyanohydrin}}{CH_3\overset{CH_3}{\overset{|}{\underset{OH}{\underset{|}{C}}}}-CN}$$

In the cyanohydrin reaction, the more positive H atom of HCN adds to the oxygen of the carbonyl group, and the —CN group adds to the carbon atom of the carbonyl group. In the aldehyde addition, the length of the carbon chain is increased by one carbon. The ketone addition product also contains an additional carbon atom.

Cyanohydrins are useful intermediates for the synthesis of several important compounds. For example, the hydrolyses of cyanohydrins produce α-hydroxy acids:

$$CH_3\underset{OH}{\underset{|}{C}}H-CN + H_2O \xrightarrow{H^+} \underset{\substack{\text{lactic acid}\\ \text{(an }\alpha\text{-hydroxy acid)}}}{CH_3\underset{OH}{\underset{|}{C}}HCOOH} + NH_4^+$$

Acetaldehyde can also be converted into other important biochemical compounds, such as the amino acid alanine:

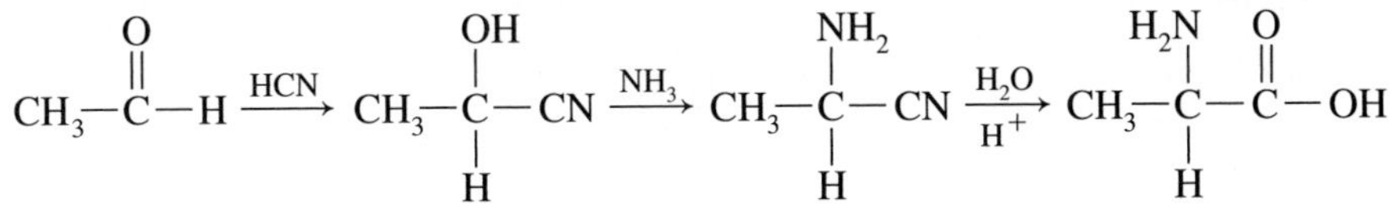

$$CH_3-\overset{\overset{\displaystyle O}{\|}}{C}-H \xrightarrow{HCN} CH_3-\underset{\underset{\displaystyle H}{|}}{\overset{\overset{\displaystyle OH}{|}}{C}}-CN \xrightarrow{NH_3} CH_3-\underset{\underset{\displaystyle H}{|}}{\overset{\overset{\displaystyle NH_2}{|}}{C}}-CN \xrightarrow[H^+]{H_2O} CH_3-\underset{\underset{\displaystyle H}{|}}{\overset{\overset{\displaystyle H_2N}{|}}{C}}-\overset{\overset{\displaystyle O}{\|}}{C}-OH$$

alanine

Some commercial reactions also involve the use of cyanohydrins. Acetone cyanohydrin can be converted to methyl methacrylate when refluxed with methanol and a strong acid. The methyl methacrylate can then be polymerized to Lucite or Plexiglas, both transparent plastics. (See Table 21.1.)

Aldol Condensation (Self-addition) In a carbonyl compound, the carbon atoms are labeled using the Greek alphabet: alpha (α), beta (β), gamma (γ), delta (δ), and so on, according to their positions with respect to the carbonyl group. The α-carbon is adjacent to the carbonyl carbon, the β-carbon is next, the γ-carbon is third, and so forth. The hydrogens attached to the α-carbon atom are therefore called α-hydrogens, and so on, as shown here:

The last carbon in the chain is known as the omega (ω) carbon.

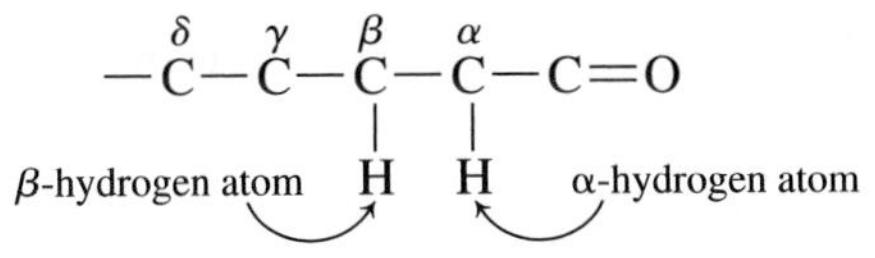

$$\overset{\delta}{-C}-\overset{\gamma}{C}-\underset{\underset{\displaystyle H}{|}}{\overset{\beta}{C}}-\underset{\underset{\displaystyle H}{|}}{\overset{\alpha}{C}}-C=O$$

β-hydrogen atom → H H ← α-hydrogen atom

The hydrogen atoms attached to the α-carbon atom have the unique ability to be more easily released as protons than other hydrogens within the molecule.

Remember that in a condensation reaction two smaller molecules combine to form a larger molecule.

An aldehyde or ketone that contains α-hydrogens may add to itself or to another α-hydrogen containing aldehyde or ketone. The product of this reaction contains both a carbonyl group and an alcohol group within the same molecule. The reaction is known as an **aldol condensation** and is carried out in a dilute basic medium. The aldol condensation is similar to the other carbonyl addition reactions. An α-hydrogen adds to the carbonyl oxygen, and the remainder of the molecule adds to the carbonyl carbon. The reaction may be depicted as follows:

aldol condensation

$$CH_3-\overset{\overset{\displaystyle O}{\|}}{C}-H \;+\; H-\underset{\underset{\displaystyle H}{|}}{\overset{\overset{\displaystyle (H)}{|}}{C}}-\overset{\overset{\displaystyle O}{\|}}{C}-H \xrightarrow{\text{dilute NaOH}} CH_3\overset{\overset{\displaystyle OH}{|}}{C}HCH_2\overset{\overset{\displaystyle O}{\|}}{C}-H$$

αH

aldol
(3-hydroxybutanal)

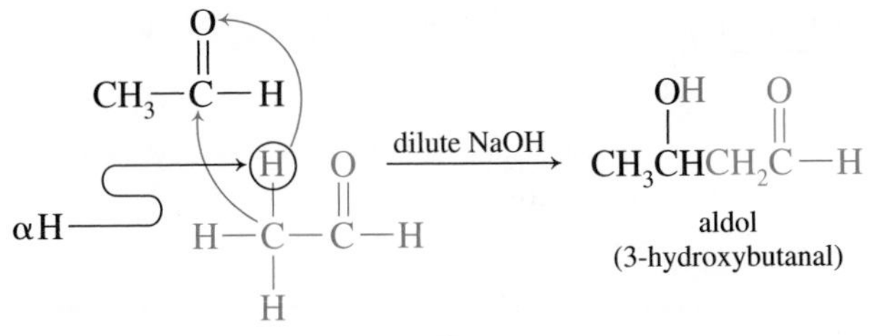

In this reaction, an alpha hydrogen first transfers from one molecule to the oxygen of the other molecule. This breaks the C=O pi bond, leaving intermediates in which a carbon atom of each molecule has three bonds:

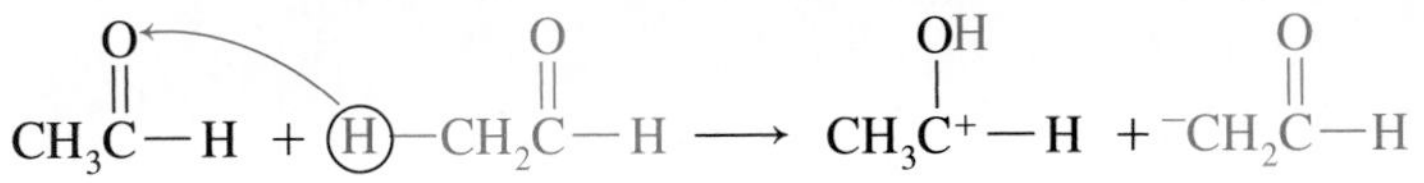

$$CH_3\overset{\overset{\displaystyle O}{\|}}{C}-H + (H)-CH_2\overset{\overset{\displaystyle O}{\|}}{C}-H \longrightarrow CH_3\overset{\overset{\displaystyle OH}{|}}{C^+}-H + {}^-CH_2\overset{\overset{\displaystyle O}{\|}}{C}-H$$

For a molecule to undergo an aldol condensation reaction, the molecule must have at least one H atom on the carbon alpha to the carbonyl group.

The two intermediates then bond to each other, forming the product:

$$CH_3\overset{OH}{\overset{|}{C}}^{+}-H + {}^{-}CH_2\overset{O}{\overset{\|}{C}}-H \longrightarrow CH_3\overset{OH}{\overset{|}{C}}HCH_2\overset{O}{\overset{\|}{C}}-H$$

Acetone also undergoes the aldol condensation:

$$\underset{\text{acetone}}{CH_3\overset{O}{\overset{\|}{C}}CH_3} + \underset{\text{acetone}}{(H)-CH_2\overset{O}{\overset{\|}{C}}CH_3} \xrightarrow[\text{NaOH}]{\text{dilute}} \underset{\substack{\text{diacetone alcohol}\\ \text{(4-hydroxy-4-methyl-2-pentanone)}}}{CH_3\overset{OH}{\overset{|}{\underset{CH_3}{\underset{|}{C}}}}-CH_2\overset{O}{\overset{\|}{C}}CH_3}$$

Example 23.3 Write the equation for the aldol condensation of propanal.

SOLUTION First write the structure for propanal and locate the α-hydrogen atoms:

$$CH_3\underset{\underset{(H)\longleftarrow \alpha H}{|}}{C}H\overset{H}{\overset{|}{C}}=O$$

Now write two propanal molecules and transfer an α-hydrogen from one molecule to the oxygen of the second molecule. After the pi bond breaks, the two carbon atoms that are bonded to only three other atoms bond to each other to form the product:

$$\begin{matrix} CH_3CH_2\overset{O}{\overset{\|}{C}}-H \\ CH_3\overset{(H)}{\overset{|}{C}}H\overset{O}{\overset{\|}{C}}-H \end{matrix} \xrightarrow[\text{NaOH}]{\text{dilute}} \begin{matrix} CH_3CH_2\overset{OH}{\overset{|}{\underset{H}{\underset{|}{C}}}}{}^{+} \\ {}^{-}\underset{CH_3}{\underset{|}{C}}H\overset{O}{\overset{\|}{C}}-H \end{matrix} \longrightarrow \underset{\text{3-hydroxy-2-methylpentanal}}{CH_3CH_2\overset{OH}{\overset{|}{C}}H\underset{CH_3}{\underset{|}{C}}H\overset{O}{\overset{\|}{C}}-H}$$

Practice 23.8

(a) Write the equation for the aldol condensation of butanal.
(b) Write the equation for the aldol condensation of 3-pentanone.

CHEMISTRY IN ACTION • Formaldehyde—A "Sticky" Subject

Formaldehyde is a very simple organic molecule, the simplest carbonyl-containing molecule. As such, it is widely used in the synthesis of more complex compounds. About 4.3 million metric tons per year are produced in the United States each year. Formaldehyde is the fifth most common chemical product in the United States.

Formaldehyde is one of the best chemicals for making industrial adhesives—a sticky subject. In fact, the largest single use of formaldehyde is to make adhesives for particle board and plywood. Other adhesives have been tried over the past 50 years but the formaldehyde-based chemicals continue to offer the best result. They are cheap, can be spread on or mixed with wood easily, and form a very tight bond after heating. In general, once the bonds form, they can be exposed to a variety of environmental changes without weakening. The latter part of the 20th century saw a marked increase in formaldehyde-based adhesive production as particle board continues to replace lumber in one application after another.

The latter part of the 20th century also saw an increased concern for and awareness of the environment. Environmental chemists discovered that formaldehyde is toxic even at low concentrations. Furthermore, since formaldehyde is a gas, it can spread easily. Finally, one of the most common formaldehyde adhesives (one made by adding urea) was found to release formaldehyde gas, over time. So not only does formaldehyde form excellent adhesives but it also creates a sticky situation for our environment.

The wood products industry has responded by modifying the formaldehyde mixture to significantly reduce formaldehyde emissions, but in the long term (over the next five years) this is not enough. Recently, chemists have reported a feasible solution to this sticky problem—modified soy protein. A renewable resource, soy protein with an organic additive yields an adhesive as strong as the formaldehyde resin at about the same price.

Particle board manufacturing.

The history of formaldehyde-based adhesives aptly summarizes the iterative nature of the scientific enterprise. With persistence (by "sticking to it"), chemists developed an important product, then continued to study the product, found problems, and discovered a potential solution.

Practice 23.9 Application to Biochemistry

One step in the biochemical process that produces blood sugar (glucose) depends on an aldol condensation to connect an aldehyde to a ketone:

$$\begin{matrix} W \\ | \\ C{=}O \\ | \\ {}^{*}CH_2OH \end{matrix} \quad + \quad \begin{matrix} O \\ \| \\ CH \\ | \\ CHOH \\ | \\ W' \end{matrix} \longrightarrow$$

(W and W′ represent the remainder of the molecules.)

Write the structure of the product, given that the hydrogen on the starred carbon (*C) reacts with the aldehyde.

Both the aldol condensation and the addition of hydrogen cyanide form new C—C bonds. These reactions are used in industry to build larger molecules from smaller precursors. For example, nifedipine (Procardia, Adalat) is an important heart medication (part of a class of drugs known as calcium-channel blockers) used to treat various forms of angina. Although nifedipine is a complex molecule, it can be synthesized from the simpler compound, *o*-nitrobenzaldehyde (which contains a reactive aldehyde group):

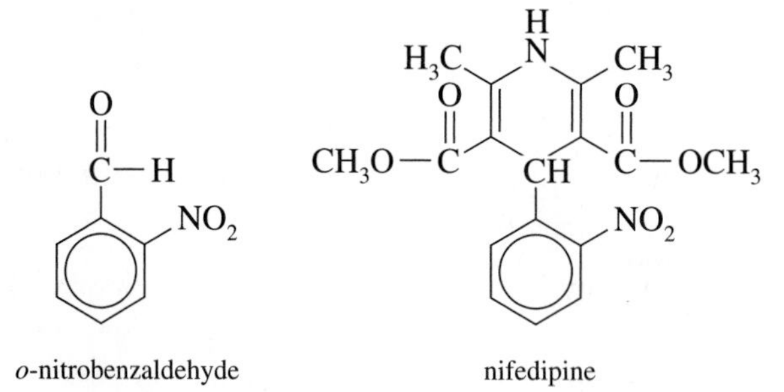

o-nitrobenzaldehyde

nifedipine

By using an aldol condensation, the drug is synthesized in one step. The carbonyl functional group provides the reactivity needed by pharmaceutical chemists to create a complex molecule with important biological activity.

Aldol condensations are common in biochemistry. For example, the strength of collagen depends on aldol condensation reactions. Collagen is the most abundant protein in humans. (See Chapter 29.) It is found (1) in the bone matrix around which the mineral calcium is deposited, (2) in the material that gives skin its tear-resistant quality, and (3) in tendons for strength. After the collagen molecule is formed, aldehyde groups are added along its length. Then, when collagen molecules are layered next to each other, aldol condensations occur naturally.

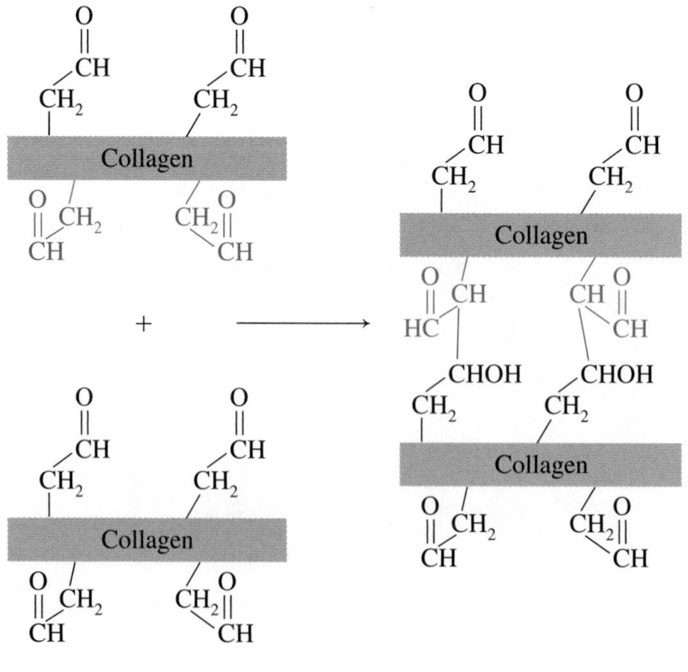

The cross-linking bonds form a strong network. Without cross-links, collagen fibers are weak. Genetic diseases that limit cross-linking commonly result in malformed bones, hyperextendable joints, and an early death from disrupted blood vessels. Of course, too much cross-linking can also pose problems. A loss of joint flexibility for older adults has been attributed partly to increased collagen cross-linking.

23.5 Common Aldehydes and Ketones

Numerous methods have been devised for making aldehydes and ketones. The oxidation of alcohols is a very general method. Special methods are often used for the commercial production of individual aldehydes and ketones.

Formaldehyde (Methanal) This aldehyde is made from methanol by reaction with oxygen (air) in the presence of a silver or copper catalyst:

$$2\,CH_3OH + O_2 \xrightarrow[400°C]{\text{Ag or Cu}} \underset{\substack{\text{formaldehyde}\\ \text{(methanal)}}}{2\,H_2C{=}O} + 2\,H_2O$$

Formaldehyde is a poisonous, irritating gas that is highly soluble in water. It is marketed as a 37% aqueous solution called *formalin*, which also contains 10–15% methanol to keep formaldehyde from polymerizing. By far the largest use of formaldehyde is in the manufacture of polymers. About 1.33×10^9 kg of formaldehyde is manufactured annually in the United States.

Formaldehyde vapors are intensely irritating to the mucous membranes. Ingestion may cause severe abdominal pains, leading to coma and death.

It is of interest that formaldehyde may have had a significant role in chemical evolution. Formaldehyde is believed to have been a component of the primitive atmosphere of the earth. Scientists theorize that the reactivity of this single-carbon aldehyde enabled it to form more complex organic molecules that were precursors of the still more complicated substances that today are essential components of every living organism.

Formaldehyde readily polymerizes to form a linear solid polymer, paraformaldehyde $\mathrm{+CH_2O+_{\it n}}$, which releases formaldehyde when heated. Formaldehyde has been used to preserve biological specimens, in fumigation, and in a number of other polymers.

Acetaldehyde (Ethanal) Acetaldehyde is a volatile liquid (bp 21°C) with a pungent, irritating odor. It has a general narcotic action and, in large doses, may cause respiratory paralysis. Its principal use is as an intermediate in the manufacture of other chemicals, such as acetic acid and 1-butanol. Acetic acid, for example, is made by air oxidation of acetaldehyde:

$$2\,CH_3\overset{\overset{\displaystyle O}{\|}}{C}{-}H + O_2 \xrightarrow[\Delta]{Mn^{2+}} 2\,CH_3\overset{\overset{\displaystyle O}{\|}}{C}{-}OH$$

Acetaldehyde undergoes reactions in which three or four molecules combine or polymerize to form the cyclic compounds paraldehyde and metaldehyde:

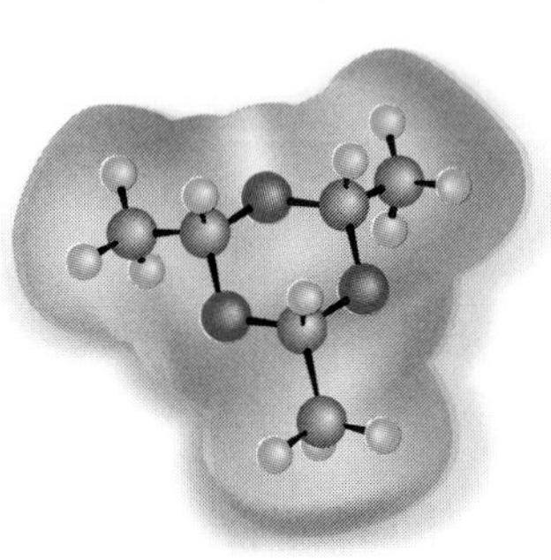

Spacefilling model of paraldehyde.

$$3\,CH_3-\overset{\overset{\displaystyle O}{\|}}{C}-H \xrightarrow[\Delta]{H^+} \text{paraldehyde}$$

paraldehyde
(bp 125°C)

$$4\,CH_3\overset{\overset{\displaystyle O}{\|}}{C}-H \xrightarrow[-20^\circ C]{Ca(NO_3)_2 + HBr} \text{metaldehyde}$$

metaldehyde
(mp 246°C)

Paraldehyde is a controlled substance and has been used as a sedative. Metaldehyde is very attractive and highly poisonous to slugs and snails. For this reason, it is an active ingredient in some pesticides that are sold for lawn and garden use. Metaldehyde is also used as a solid fuel.

Acetone and Methyl Ethyl Ketone Ketones are widely used organic solvents. Acetone, in particular, is used in very large quantities for this purpose. The U.S. production of acetone is about 1.26×10^8 kg annually. Acetone is used as a solvent in the manufacture of drugs, chemicals, and explosives; for the removal of paints, varnishes, and fingernail polish; and as a solvent in the plastics industry. Methyl ethyl ketone (MEK) is also widely used as a solvent, especially for lacquers. Both acetone and MEK are made by oxidation (dehydrogenation) of secondary alcohols:

$$\underset{\text{2-propanol}}{CH_3\underset{\underset{\displaystyle OH}{|}}{C}HCH_3} \xrightarrow[250\text{–}300^\circ C]{Cu} \underset{\text{acetone (propanone)}}{CH_3\underset{\underset{\displaystyle O}{\|}}{C}CH_3} + H_2$$

$$\underset{\text{2-butanol}}{CH_3CH_2\underset{\underset{\displaystyle OH}{|}}{C}HCH_3} \xrightarrow[250\text{–}300^\circ C]{Cu} \underset{\text{methyl ethyl ketone (2-butanone)}}{CH_3CH_2\underset{\underset{\displaystyle O}{\|}}{C}CH_3} + H_2$$

Acetone is also a coproduct in the manufacture of phenol. (See Section 22.9.) Acetone is formed in the human body as a by-product of lipid metabolism. Normal concentrations of acetone in the body are less than 1 mg/100 mL of blood. In patients with diabetes mellitus, the concentration of acetone may rise, and it is then excreted in the urine, where it can be easily detected. Sometimes the odor of acetone can be detected on the breath of these patients.

23.6 Condensation Polymers

Condensation polymers are formed by reactions between functional groups on adjacent monomers. As a rule, a smaller molecule, usually water, is eliminated in the reaction. Each monomer loses a small group of atoms and gains a larger group, a substitution reaction that allows the polymer to continue growing. Condensation polymerization uses functional groups that favor substitution reactions. For the polymer to continue to grow, each monomer must undergo at least two substitution reactions and must thus be at least bifunctional. If cross-linking is to occur, there must be more than two functional groups on some monomer molecules.

Many different condensation polymers have been synthesized. Important classes include the polyesters, polyamides, phenol–formaldehyde polymers, and polyurethanes. Most biochemical polymers are the condensation type. We will consider only phenol–formaldehyde polymers at this time.

Phenol–Formaldehyde Polymers

As noted earlier, a phenol–formaldehyde condensation polymer (Bakelite) was first marketed over 90 years ago. Polymers of this type are still widely used, especially in electrical equipment, because of their insulating and fire-resistant properties. Polymers made from phenol are known as *phenolics*.

Each phenol molecule can react with formaldehyde to lose an H atom from the para position and from each ortho position (indicated by arrows):

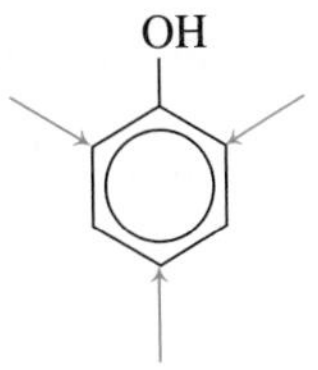

OH

Each formaldehyde molecule reacts with two phenol molecules to eliminate water:

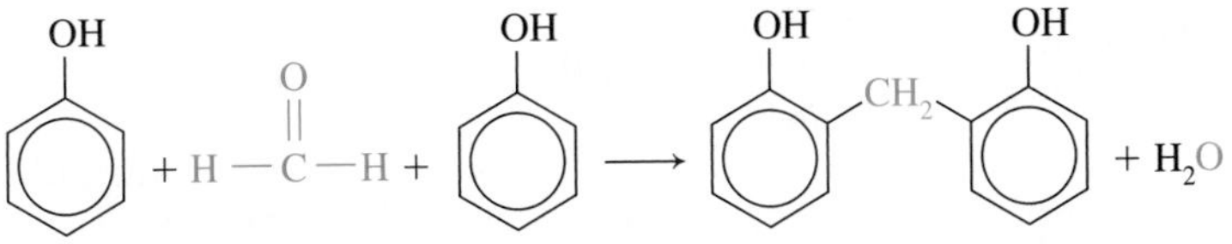

OH + H—C(=O)—H + OH → OH CH$_2$ OH + H_2O

Similar reactions can occur at the other two reactive sites on each phenol molecule, leading to the formation of the polymer. This polymer is thermosetting because it has an extensively cross-linked network structure. The following is a typical section of that structure:

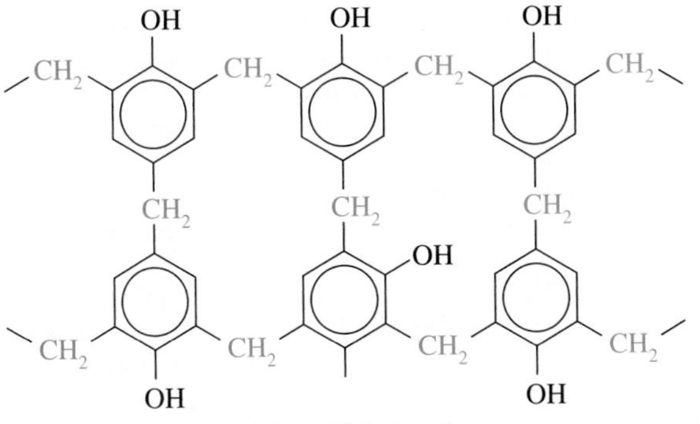

OH OH OH
CH$_2$ CH$_2$ CH$_2$ CH$_2$
CH$_2$ CH$_2$ CH$_2$
OH
CH$_2$ CH$_2$ CH$_2$ CH$_2$
OH OH

phenol–formaldehyde polymer

Chapter 23 Review

23.1 Structures of Aldehydes and Ketones

KEY TERMS

Carbonyl group
Aldehyde
Ketone

- Both aldehydes and ketones contain the carbonyl group, $\rangle C{=}O$, a carbon–oxygen double bond.
- Aldehydes have at least one hydrogen atom bonded to the carbonyl carbon.
- Ketones have only alkyl or aryl groups bonded to the carbonyl carbon.

23.2 Naming Aldehydes and Ketones

- The IUPAC naming of aldehydes follows a similar process to that used for other organic molecules.
  - The longest continuous carbon chain (parent chain) *that contains the carbonyl* is given the name of the corresponding alkane modified by dropping the last *-e* and adding *-al*.
  - The parent chain is numbered from the aldehyde end—the carbonyl group is understood to be number 1.
  - Other groups attached to the parent chain are named and numbered in the usual IUPAC fashion.
- The IUPAC naming of ketones is the same as that for aldehydes except for the following:
  - The longest continuous carbon chain (parent chain) *that contains the carbonyl* is given the name of the corresponding alkane modified by dropping the last *-e* and adding *-one*.
  - If the parent chain is longer than four carbons, it is numbered so that the carbonyl group has the smallest number possible; this number is prefixed to the parent ketone name.

23.3 Bonding and Physical Properties

- The carbonyl group in aldehydes and ketones is very polar, with the oxygen pulling electrons from the carbon.
- Unlike alcohols, aldehydes and ketones do not hydrogen-bond to themselves. They have lower boiling points than alcohols of comparable molar mass.
- Aldehydes and ketones with four or fewer carbons are at least partly water soluble.
- Lower-molar-mass aldehydes have a penetrating, disagreeable odor, while higher-molar-mass aldehydes and ketones are more fragrant, some being used in flavorings and perfumes.

23.4 Chemical Properties of Aldehydes and Ketones

KEY TERMS

Tollens test
Fehling and the Benedict tests
Hemiacetal
Hemiketal
Acetal
Ketal
Cyanohydrin
Aldol condensation

- Aldehydes are easily oxidized to carboxylic acids. Under similar conditions, ketones are unreactive.
- In the Tollens test, Ag^+ is reduced to form metallic silver as the aldehyde is oxidized.
- In the Fehling and Benedict tests, Cu^{2+} is reduced to form Cu_2O; a reddish-brown precipitate. Aldehydes are oxidized; simple ketones do not react.
- Both aldehydes and ketones can be reduced—aldehydes yield primary alcohols, while ketones yield secondary alcohols.
- A hemiacetal is formed when an alcohol adds across an aldehyde carbonyl double bond. If an alcohol adds to a ketone, a hemiketal is produced.
  - The carbon in a hemiacetal or hemiketal is bonded to both an ether oxygen and an alcohol oxygen.
- With an excess of alcohol, two alcohols can react with a carbonyl to form an acetal (from an aldehyde) and a ketal (from a ketone).
  - The carbon in an acetal or ketal is bonded to two ether oxygens.
- Intramolecular hemiacetal/hemiketal formation is significant in carbohydrate chemistry.
- Cyanohydrins are formed by the addition of HCN to a carbonyl group.
  - The cynaohydrin carbon is bonded to an hydroxyl group and a cyano (—CN) group.
- In an aldol condensation, two carbonyl-containing molecules (either aldehydes or ketones) connect together.
  - One aldehyde or ketone adds across the carbonyl group of the second aldhyde or ketone.
  - The alpha hydrogen from one molecule adds to the carbonyl oxygen of the second molecule, while the remainder of the first molecule, adds to the carbonyl carbon.

23.5 Common Aldehydes and Ketones

- Formaldehyde (methanal) is produced from methanol by air oxidation and is used primarily in the manufacture of polymers.
- Acetaldehyde (ethanal) is used to produce acetic acid and 1-butanol.
- Acetone and methyl ethyl ketone (MEK) are formed by oxidation from the secondary alcohols 2-propanol and 2-butanol, respectively.
  - These ketones are excellent organic solvents and are found in many commercial products such as paints, lacquers, varnishes, and fingernail polish remover.

23.6 Condensation Polymers

- Condensation polymers form when monomers combine and eliminate a small-molecule product such as water.
- Common examples of condensation polymers are the phenol-formaldehyde polymers, also known as phenolics. These thermosetting polymers are widely used in electrical equipment because of their insulating and fire-resistant properties.

Review Questions

All questions with blue numbers have answers in the appendix of the text.

1. Write generalized structures for
   (a) an aldehyde
   (b) a ketone
   (c) a dialdehyde
   (d) a hemiacetal
   (e) a hemiketal
   (f) an acetal
   (g) a ketal
   (h) a cyanohydrin
2. Explain, in terms of structure, why aldehydes and ketones have lower boiling points than alcohols of similar molar masses.
3. Write structural formulas for propanal and propanone. Judging from these formulas, do you think that aldehydes and ketones are isomeric with each other? Show evidence and substantiate your answer by testing with a four-carbon aldehyde and a four-carbon ketone.
4. How are collagen fibers affected by the aldol condensation reaction?
5. Why are high-molar-mass aldehydes and ketones not very soluble in water?
6. What type of orbital bonds make up the carbonyl group?
7. Describe a positive test for aldehydes that uses
   (a) Benedict solution
   (b) Tollens solution
8. Based on the structure of a phenol–formaldehyde polymer, why is this material rigid and not stretchable or flexible like, for example, polyethylene? (See also Chapter 21.)
9. Why is formaldehyde commonly used to make particle board and plywood?
10. What is MEK and why is it used in lacquers (and paints)?
11. Why do we not need to identify the location of the keto group in propanone and butanone?

Paired Exercises

All exercises with blue numbers have answers in the appendix of the text.

1. Give the IUPAC name for each of these aldehydes (unless otherwise indicated):

   (a) $H_2C{=}O$ (give both the IUPAC name and the common name)

   (b) $CH_3CH(CH_3)CH_2C(=O)-H$

   (c) $H-C(=O)CH_2CH_2C(=O)-H$

   (d) $C_6H_5-CH{=}CHC(=O)-H$ (benzene ring attached)

   (e) $CH_3(H)C{=}C(H)C(=O)-H$

2. Give the IUPAC name for each of these aldehydes (unless otherwise indicated):

   (a) $CH_3C(=O)-H$ (give both the IUPAC name and the common name)

   (b) $CH_3CH_2CH_2C(=O)-H$

   (c) $C_6H_5-C(=O)-H$ (benzene ring attached)

   (d) benzene ring bearing $C(=O)-H$, with Cl on the adjacent (ortho) carbon and $CH(CH_3)_2$ ($H_3C-CH-CH_3$) at the para position

   (e) $CH_3CH(OH)CH_2C(=O)-H$

3. Give the IUPAC name for each of these ketones (unless otherwise indicated):

(a) CH_3COCH_3 (three names)

(b) $C_6H_5\overset{O}{\overset{\|}{C}}{-}CH_2CH_3$ (two names)

(c) cyclopentanone ($C_5H_8{=}O$ ring)

(d) $CH_3\overset{CH_3}{\overset{|}{\underset{\underset{OH}{|}}{C}}}CH_2\overset{O}{\overset{\|}{C}}CH_3$

4. Give the IUPAC name for each of these ketones (unless otherwise indicated):

(a) $CH_3CH_2COCH_3$ (two names)

(b) $CH_3\overset{O}{\overset{\|}{C}}{-}\overset{CH_3}{\overset{|}{\underset{\underset{CH_3}{|}}{C}}}CH_3$ (two names)

(c) $CH_3\overset{O}{\overset{\|}{C}}CH_2CH_2\overset{O}{\overset{\|}{C}}CH_3$

(d) $C_6H_5CH_2\overset{O}{\overset{\|}{C}}CH_3$ (two names)

5. Write the structural formulas for
(a) 1,3-dichloropropanone
(b) 3-butenal
(c) 4-phenyl-3-hexanone
(d) hexanal
(e) 3-ethyl-2-pentanone

6. Write the structural formulas for
(a) 3-hydroxypropanal
(b) 4-methyl-3-hexanone
(c) cyclohexanone
(d) 2,4,6-trichloroheptanal
(e) 3-pentenal

7. Each of the following IUPAC names is incorrect. Draw the condensed structural formula for each molecule and give it the correct IUPAC name.
(a) 1-pentanal
(b) 2,2,3-trimethyl-4-hexanone
(c) 1-ethyl-3-pentanone

8. Each of the following IUPAC names is incorrect. Draw the condensed structural formula for each molecule and give it the correct IUPAC name.
(a) 2,2-dichloro-6-heptanone
(b) 3-ethyl-1-pentanal
(c) 1-hexanone

9. Which compound in each of the following pairs has the higher boiling point? (Try to answer without consulting tables.)
(a) 2-hexanone or 2,5-hexanedione
(b) hexane or hexanal
(c) 2-pentanone or 2-pentanol
(d) propanone or butanone

10. Which compound in each of the following pairs has the higher boiling point? (Try to answer without consulting tables.)
(a) pentane or pentanal
(b) benzaldehyde or benzyl alcohol
(c) 2-hexanone or butanone
(d) 1-butanol or butanal

11. Which compound in each of the following pairs has the higher aqueous solubility? (Try to answer without consulting tables.)
(a) 2-hexanone or 2,5-hexanedione
(b) propane or propanal
(c) heptanal or ethanal

12. Which compound in each of the following pairs has the higher aqueous solubility? (Try to answer without consulting tables.)
(a) acetaldehyde or ethane
(b) 2-pentanone or 2,4-pentanediol
(c) propanal or hexanal

13. Write equations to show how the following compounds are oxidized by (1) $K_2Cr_2O_7 + H_2SO_4$ and (2) air + Cu or Ag + heat:
(a) 3-pentanol (b) 3-ethyl-1-hexanol

14. Write equations to show how the following compounds are oxidized by (1) $K_2Cr_2O_7 + H_2SO_4$ and (2) air + Cu or Ag + heat:
(a) 1-propanol (b) 2,3-dimethyl-2-butanol

15. Give the structural formula for the organic product from each of the following reactions:

(a) $CH_3CH_2\overset{O}{\overset{\|}{C}}H + K_2Cr_2O_7 + H_2SO_4 \longrightarrow$

(b) $H\overset{O}{\overset{\|}{C}}\underset{\underset{CH_3}{|}}{C}HCH_3 + 2Cu^{2+} \xrightarrow[H_2O]{OH^-}$

(c) $CH_3\underset{\underset{\underset{H}{C=O}}{|}}{C}HCH_3 + 2Ag^+ \xrightarrow[H_2O]{NH_3}$

16. Give the structural formula for the organic product from each of the following reactions:

(a) $C_6H_{11}{-}\overset{O}{\overset{\|}{C}}H + 2Cu^{2+} \xrightarrow[H_2O]{OH^-}$ (cyclohexyl)

(b) $\overset{O}{\overset{\|}{C}}HCH_2CH_2\overset{O}{\overset{\|}{C}}H + 4Ag^+ \xrightarrow[H_2O]{NH_3}$

(c) $CH_3CH_2CH_2\overset{O}{\overset{\|}{C}}H + K_2Cr_2O_7 + H_2SO_4 \longrightarrow$

17. Give the structural formula and name for the organic reactant that yields the product shown:

(a) $\xrightarrow[H_2O]{NH_3} CH_3CH_2\overset{O}{\overset{\|}{C}}O^- \ NH_4^+ + 2Ag$

(b) $\xrightarrow[\Delta]{H_2/Ni} CH_3\overset{OH}{\overset{|}{C}}HCH_2CH_3$

(c) $\xrightarrow[H_2O]{NaOH} CH_3CH_2CH_2\overset{O}{\overset{\|}{C}}O^-Na^+ + Cu_2O$

18. Give the structural formula and name for the organic reactant that yields the product shown:

(a) $\xrightarrow[H_2O]{NaOH} H\overset{O}{\overset{\|}{C}}O^-Na^+ + 2Cu_2O$

(b) $\longrightarrow CH_3CH_2\overset{O}{\overset{\|}{C}}OH + 2Cr^{3+} + 4H_2O$

(c) $\xrightarrow[\Delta]{H_2/Ni} CH_3CH_2CH_2OH$

19. (a) What functional group is present in a compound that gives a positive Tollens test?
(b) What is the visible evidence for a positive Tollens test?
(c) Write an equation showing the reaction involved in a positive Tollens test.

20. (a) What functional group is present in a compound that gives a positive Fehling test?
(b) What is the visible evidence for a positive Fehling test?
(c) Write an equation showing the reaction involved in a positive Fehling test.

21. Give the products for the reaction of the following compounds with Tollens reagent:
(a) butanal
(b) benzaldehyde
(c) methyl ethyl ketone

22. Give the products for the reaction of the following compounds with Fehling reagent:
(a) propanal
(b) acetone
(c) 3-methylpentanal

23. Give the structural formula and name for the organic reactant that yields the product shown:

(a) $\xrightarrow[\Delta]{H_2/Ni}$ (cyclohexane ring)—OH

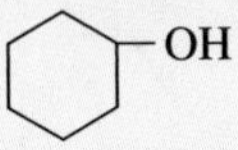

(b) $\xrightarrow[\Delta]{H_2/Ni} CH_3\underset{CH_3}{\underset{|}{C}}H\overset{CH_3}{\overset{|}{C}}H\underset{OH}{\underset{|}{C}}HCH_3$

(c) $\xrightarrow[\Delta]{H_2/Ni} CH_3\underset{CH_3}{\underset{|}{C}}HCH_2CH_2OH$

24. Give the structural formula and name for the organic reactant that yields the product shown:

(a) $\xrightarrow[\Delta]{H_2/Ni} CH_2\overset{OH}{\overset{|}{C}}H_2\overset{CH_3}{\overset{|}{C}}HCH_3$

(b) $\xrightarrow[\Delta]{H_2/Ni}$ (cyclopentane ring)—OH

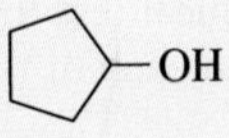

(c) $\xrightarrow[\Delta]{H_2/Ni} CH_3\overset{CH_3}{\underset{CH_3}{\overset{|}{\underset{|}{C}}}}CH_2OH$

25. Write equations showing the aldol condensation for the following compounds:
(a) butanal (b) phenylethanal

26. Write equations showing the aldol condensation for the following compounds:
(a) 3-pentanone (b) propanal

27. Complete the following equations:

(a) $CH_3\underset{O}{\underset{\|}{C}}CH_3 + \underset{OH}{\underset{|}{C}}H_2\underset{OH}{\underset{|}{C}}H_2 \xrightleftharpoons{\text{dry HCl}}$

(b) $CH_3CH_2\overset{O}{\overset{\|}{C}}{-}H + CH_3CH_2OH \xrightleftharpoons{H^+}$

(c) $CH_3CH(CH_3)CH_2CH(OCH_3)_2 \xrightarrow[H^+]{H_2O}$

28. Complete the following equations:

(a) $CH_3CH_2\overset{O}{\overset{\|}{C}}{-}H + CH_3CH_2CH_2OH \xrightleftharpoons{\text{dry HCl}}$

(b) 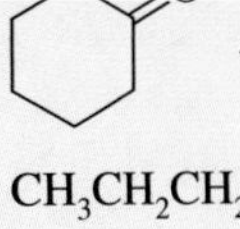 (cyclohexanone, ring C=O) $+ CH_3OH \xrightleftharpoons{H^+}$

(c) $CH_3CH_2CH_2CH(OCH_3)_2 \xrightarrow[H^+]{H_2O}$

29. Write equations for the following sequence of reactions:
(a) propanone + HCN $\longrightarrow$
(b) product of part (a) + $H_2O \longrightarrow$
(c) product of part (b) + acetaldehyde + dry HCl $\longrightarrow$

30. Write equations for the following sequence of reactions:
(a) benzaldehyde + HCN $\longrightarrow$
(b) product of part (a) + $H_2O \longrightarrow$
(c) product of part (b) + $K_2Cr_2O_7 + H_2SO_4 \longrightarrow$

Additional Exercises

All exercises with blue numbers have answers in the appendix of the text

31. 4-hydroxybutanal can form an intramolecular cyclic hemiacetal. What is the structure of the hemiacetal?

32. The following cyanohydrin is used in a multistep synthesis to make the artificial fiber Orlon (polyacrylonitrile):

$$CH_3\underset{\large OH}{CH}C{\equiv}N$$

From what carbonyl-containing compound is this cyanohydrin synthesized? (Give both the structural formula and name.)

33. Write the structure for each aldol condensation product that is possible when a mixture of ethanal and propanal is reacted with dilute NaOH.

34. The following molecules ("ketone bodies") are used by the human body as an emergency energy supply, primarily for the muscles:

$$\underset{\text{compound I}}{CH_3\overset{OH}{CH}CH_2COOH} \qquad \underset{\text{compound II}}{CH_3\overset{O}{\overset{\|}{C}}CH_2COOH}$$

These two molecules can be interconverted in a single biochemical reaction. What chemical change takes place when compound I is converted to compound II?

35. Give a simple visible chemical test that will distinguish between the compounds in each of the following pairs:

(a) $CH_3CH_2\overset{O}{\overset{\|}{C}}{-}H$ and $CH_3\overset{O}{\overset{\|}{C}}CH_3$

(b) $CH_3CH_2\overset{O}{\overset{\|}{C}}{-}H$ and $CH_2{=}CH\overset{O}{\overset{\|}{C}}{-}H$

(c) $C_6H_5CH_2CH_2OH$ and $C_6H_5\underset{OH}{CH}CH_3$

36. Write equations to show how you could prepare lactic acid, $CH_3CH(OH)COOH$, from acetaldehyde through a cyanohydrin intermediate.

37. The millipede carries the following cyanohydrin as a chemical defense weapon:

$$C_6H_5\overset{OH}{CH}C{\equiv}N$$

When attacked, the millipede reverses the reaction that formed this cyanohydrin, and deadly hydrogen cyanide gas is released. One millipede is said to be able to release enough poisonous gas to kill a small mouse. Write the reaction that releases hydrogen cyanide and name the aldehyde product.

38. Pyruvic acid is formed during muscle exertion. If the muscles are working strenuously, this acid is converted to lactic acid. The structures of the two acids are as follows:

$$\underset{\text{pyruvic acid}}{CH_3\overset{O}{\overset{\|}{C}}COOH} \qquad \underset{\text{lactic acid}}{CH_3\overset{OH}{CH}COOH}$$

What chemical change takes place as pyruvic acid is converted to lactic acid?

39. Ketones are prepared by oxidation of secondary alcohols. Name the alcohol that should be used to prepare
(a) 3-pentanone
(b) methyl ethyl ketone
(c) 4-phenyl-2-butanone

40. Write structures for all isomeric aldehydes and ketones with the molecular formula $C_6H_{12}O$.

41. Write structures for all the benzaldehyde isomers with the molecular formula $C_9H_{10}O$.

42. (a) Starting with ethanal, write equations showing the synthesis of 1-butanol.
(b) Choose your starting material from methanal, ethanal, propanal, and butanal, and write equations showing the synthesis of 1,3-propanediol.

43. Write a structure showing four units of a linear polymer made from acetaldehyde.

44. Using *p*-cresol in place of phenol to form a phenol–formaldehyde polymer results in a thermoplastic rather than a thermosetting polymer. Explain why this occurs.

$$HO{-}C_6H_4{-}CH_3$$

p-cresol

Challenge Exercise

45. During the biological synthesis of glucose, two 3-carbon compounds are connected via a new carbon–carbon single bond to form a 6-carbon straight-chain molecule (a sugar). The reactants in a simplified version are shown as follows:

$$\underset{\text{OH OH}}{CH_2CHCH}(=O) + \underset{\text{OH OH}}{CH_2CCH_2}(=O) \longrightarrow$$

(a) Name both reactants.
(b) Use your knowledge of the aldol condensation reaction to write the structure of the product.

Answers to Practice Exercises

23.1 (a) $CH_3CH(OH)CH(Br)CH_2CH_2C(=O)-H$ (c) $CH_3C(=O)CH=CH_2$

(b) $C_6H_5-CH_2C(=O)-H$ (d) $C_6H_5-C(=O)-C_6H_5$

(e) $CH_3C(CH_3)_2-C(=O)-C(CH_3)_2CH_3$ (f) $C_6H_{11}-CH_2CH_2CH_2C(=O)-H$ (cyclohexyl)

23.2 (a) cyclohexanone (c) 3-methylbutanal
(b) 4-chloro-2-butanone (d) 3-phenylpropenal

23.3 1,3,4,5,6-pentahydroxy-2-hexanone

23.4 (a) $CH_3CH_2CH_2CH_2C(=O)-O^-NH_4^+ + Ag(s)$

(b) No reaction

(c) No reaction

(d) $CH_3CH_2CH(CH_3)CH_2CH_2C(=O)-O^-Na^+ + Cu_2O(s)$

23.5 $CH_3-CH(OH)-CH_2-COOH$

23.6 (a) $CH_3CH(OH)(OCH_2CH_3)$ 1-ethoxyethanol $CH_3CH(OCH_2CH_3)(OCH_2CH_3)$ 1,1-diethoxyethane

(b) $CH_3CH_2C(OH)(OCH_3)CH_3$ 2-methoxy-2-butanol $CH_3CH_2C(OCH_3)_2CH_3$ 2,2-dimethoxybutane

23.7 (a) Five-membered ring (ring O) bearing HOH_2C, OH, CH_2OH; ring substituents HO, OH — Fructose

(b) Six-membered ring (ring O) bearing HOH_2C, HO, OH; ring substituents HO, OH — Glucose

(c) Five-membered ring (ring O) bearing HOH_2C, OH; ring substituents HO, OH — Ribose

23.8 (a) $2\ CH_3CH_2CH_2C(=O)-H \xrightarrow{\text{dilute } OH^-} CH_3CH_2CH_2C(OH)(H)-CH(CH_2CH_3)C(=O)-H$

(b) $2\ CH_3CH_2C(=O)CH_2CH_3 \xrightarrow{\text{dilute } OH^-} CH_3CH_2C(OH)(CH_2CH_3)-CH(CH_3)C(=O)CH_2CH_3$

23.9 W–C=O–CHOH–CHOH–CHOH–W′

$$\begin{array}{c} W \\ | \\ C{=}O \\ | \\ CHOH \\ | \\ CHOH \\ | \\ CHOH \\ | \\ W' \end{array}$$

PUTTING IT TOGETHER: Review for Chapters 22–23

Multiple Choice:

Choose the correct answer to each of the following.

1. Which of the following is not a primary alcohol?

(a) CH_3CHCH_2OH (with CH_3 on C-2)
(c) C_6H_5–CH_2OH (benzene ring)
(b) cyclopentane–OH
(d) $CH_3CH_2CH_2CH_2OH$

2. Which of the following formulas is a tertiary alcohol?

(a) CH_3CHCH_3 (OH on C-2)
(c) CH_2CHCH_2 (OH OH OH)
(b) $CH_3CH_2CHCH_3$ (OH on C-2)
(d) cyclohexane with CH_3 and OH on the same carbon

3. Arrange the following compounds in order of increasing solubility in water:

(A) C_5H_{12}
(C) CH_3CHCH_2OH (OH on C-2)
(B) $CH_3CH_2OCH_2CH_3$
(D) $CH_3CH_2CH_2CH_2OH$

(a) BDCA (b) DCAB (c) BDAC (d) ABDC

4. Arrange the following alcohols in order of increasing boiling points:

(A) CH_3CH_2OH
(C) $CH_3CH(OH)CH_3$
(B) $CH_3CH_2CH_2OH$
(D) $HOCH_2CH_2OH$

(a) ABCD (b) ABDC (c) ACBD (d) DCBA

5. Which of the following formulas is isoamyl alcohol?

(a) CH_3CHCH_2OH (with CH_3 on C-2)
(c) C_6H_5–CH_2OH (benzene ring)
(b) $CH_3CH_2CH_2CH_2OH$
(d) $CH_3CHCH_2CH_2OH$ (with CH_3 on C-2)

6. Give the IUPAC name for

$$CH_3C(CH_3)_2\text{—}CH(OH)CH_3$$

(a) 1,1,1-trimethyl-2-propanol
(b) 1-*tert*-butylethanol
(c) 3,3-dimethyl-2-butanol
(d) 2,2-dimethyl-3-butanol

7. Which of the following formulas is 3-methylcyclohexanol?

(a) cyclohexane with OH on C-1 and CH_3 on C-3
(c) cyclohexane with OH on C-1 and CH_3 on C-2
(b) cyclohexane with OH on C-1 and CH_3 on C-4
(d) no correct formula given

8. Give the correct IUPAC name for

$$CH_3CH(CH_3)\text{—}C(OH)(CH_2CH_3)\text{—}CH_2CH_2CH_3$$

(a) 2-methyl-3-ethyl-3-hexanol
(b) 4-ethyl-5-methyl-4-hexanol
(c) 3-ethyl-2-methyl-3-hexanol
(d) 4-isopropyl-4-hexanol

9. Methanol has all of the following properties *except*

(a) colorless liquid
(b) poisonous
(c) boils below 100°C
(d) does not hydrogen-bond

10. The alcohol present in alcoholic beverages is

(a) methyl alcohol
(c) isopropyl alcohol
(b) ethyl alcohol
(d) ethylene glycol

11. What is the oxidation number of the tertiary carbon atom in this compound?

$$CH_3CH_2C(CH_3)_2\text{—}OH$$

(a) −1 (b) +1 (c) 0 (d) +2

12. What is the product of this chemical reaction?

$$CH_3CH_2OH + Na \longrightarrow$$

(a) CH_3CH_2ONa
(b) $CH_3CH_2OCH_2CH_3$
(c) $CH_3CH_2CH_2CH_3$
(d) no reaction

13. Oxidation of 2-pentanol yields
(a) pentanal (c) pentanoic acid
(b) 2-pentanone (d) 3-pentanone

14. Give the structure of the major alkene formed by the dehydration of

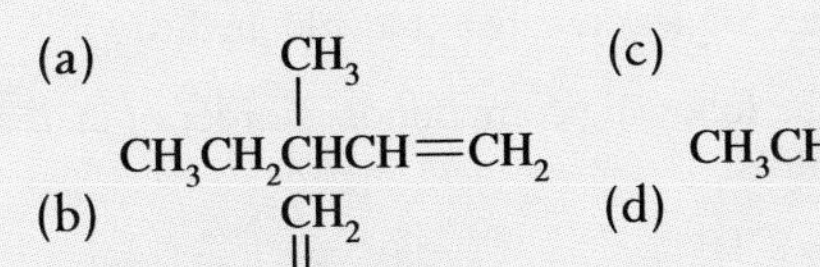

(a) $CH_3CH_2CH(CH_3)CH{=}CH_2$
(b) $CH_3CH_2C({=}CH_2)CH_2CH_3$
(c) $CH_3CH_2C(CH_3){=}CHCH_3$
(d) $CH_3CH{=}C(CH_3)CH{=}CH_2$

15. Which of the following compounds is ethylene glycol?
(a) $HOCH_2OCH_2OH$
(b) $HOCH_2CH_2OH$
(c) CH_2CH_2 (with O bridging the two carbons)
(d) $HOCH{=}CHOH$

16. What is the name for $CH_3CH_2{-}O{-}CH_2CH_2CH_3$?
(a) propoxyethane
(b) 1-propoxyethane
(c) ethyl propyl ether
(d) diethyl oxide

17. Which of the following formulas is methyl-*tert*-butyl ether (MTBE)?

(a) $CH_3C(CH_3)_2{-}O{-}C(CH_3)_2CH_3$
(b) $CH_3{-}O{-}C(CH_3)_2CH_3$
(c) $CH_3OC(CH_3)_2CH_2CH_3$
(d) $CH_3{-}O{-}CH(CH_3)CH_2CH_3$

18. Give the name for

$CH_3CH(CH_3){-}O{-}CH_2CH_3$

(a) 2-methyl diethyl ether
(b) ethyl isopropyl ether
(c) 2-pentyl ether
(d) ethoxy propyl ether

19. Which of the following is the structure of the ether formed by the dehydration of 2-methyl-1-propanol?

(a) $CH_3CH(CH_3){-}O{-}CH(CH_3)CH_3$
(b) $CH_3CH_2CH_2{-}O{-}CH_2CH_2CH_3$
(c) $CH_3CH(CH_3)CH_2{-}O{-}CH_2CH(CH_3)CH_3$
(d) $CH_3C(CH_3){=}CH{-}O{-}CH{=}C(CH_3)CH_3$

20. How many aliphatic isomeric ethers are possible for $C_5H_{12}O$?
(a) 4 (b) 6 (c) 8 (d) 9

21. Which of the following compounds is phenol?

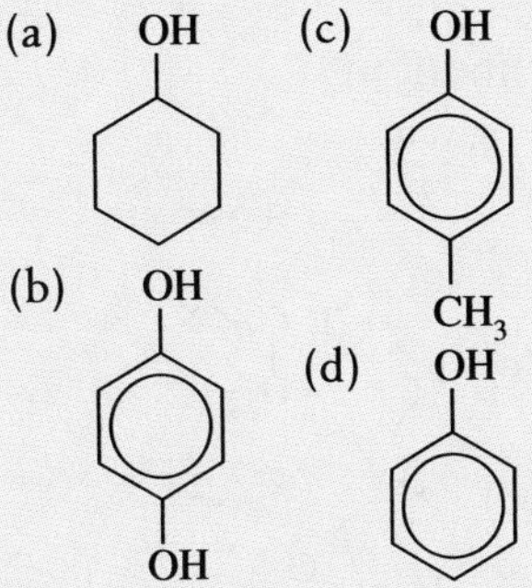

22. Which of the following compounds is catechol?

(a) OH, CH_3 (ortho-substituted benzene ring)
(b) OH, OH (ortho-substituted benzene ring)
(c) OCH_3 (benzene ring)
(d) OH, OH (meta-substituted benzene ring)

23. Give the name for

(benzene ring bearing OH, Br and CH_3)

(a) *m*-methyl-*p*-bromophenol
(b) 4-bromo-5-methylphenol
(c) 4-bromo-3-methylphenol
(d) 3-bromo-2-methylphenol

In Problems 24–28, which product is formed in the following chemical reactions?

24. $CH_3CH_2\underset{OH}{\overset{CH_3}{C}}HCH_3 \xrightarrow[180°C]{96\%\ H_2SO_4}$ alkene

(a) $CH_3\overset{CH_3}{C}{=}CHCH_3$

(b) $CH_3CH_2CH{=}\overset{CH_3}{C}CH_3$

(c) $CH_3CH_2\overset{CH_3}{C}{=}CHCH_3$

(d) $CH_3CH_2\overset{CH_2}{\overset{\|}{C}}CH_2CH_3$

25. $CH_3CH_2CH_2CH{=}CH_2 + H_2O \xrightarrow{H^+}$

(a) $CH_3CH_2CH_2CH(OH)CH_3$
(b) $CH_3CH_2CH_2CH_2CH_2OH$
(c) $CH_3CH_2CH_2CH(OH)CH_2OH$
(d) no correct answer given

26. —OH $\xrightarrow[H^+]{Cr_2O_7^{2-}}$

(a)

(b) =O

(c) —O—

(d) O

27. $CH_3CH_2Br + CH_3\overset{CH_3}{C}HCH_2ONa \longrightarrow$

(a) $CH_3\overset{CH_3}{C}HCH_2{-}O{-}CH_2\overset{CH_3}{C}HCH_3$
(b) $CH_3CH_2{-}O{-}CH_2CH_2CH_3$
(c) $CH_3CH_2{-}O{-}\overset{CH_3}{C}HCH_3$
(d) $CH_3CH_2{-}O{-}CH_2\overset{CH_3}{C}HCH_3$

28. Br $+ CH_3CH_2ONa \longrightarrow$

(a) OCH_2CH_3

(b) Br, CH_2CH_3

(c) CH_2CH_3

(d) no reaction

29. The carbonyl group, $>C{=}O$, contains
(a) two sigma bonds
(b) two pi bonds
(c) a sigma bond and a pi bond
(d) an sp^3 electron orbital

30. Which statement is not true?
(a) Aliphatic aldehydes and ketones are isomeric.
(b) Aliphatic aldehydes and ethers are isomeric.
(c) Aromatic aldehydes contain a benzene ring.
(d) Ketones are less reactive than aldehydes.

31. Arrange the following compounds in order of increasing boiling points:

(A) $CH_3CH_2\overset{H}{C}{=}O$ (C) $CH_3CH_2CH_2CH_2OH$
(B) $CH_3CH_2CH_2OH$ (D) $CH_3\underset{O}{\underset{\|}{C}}CH_2CH_3$

(a) ADBC (b) BCDA (c) ACBD (d) DCBA

32. Which of these compounds has the lowest vapor pressure?
(a) ethanal (b) butanal (c) pentanal (d) propanal

33. Which of these compounds has the highest boiling point?
(a) ethanal (b) butanal (c) pentanal (d) propanal

34. Give the name for

$CH_3CH_2CH_2CH_2\overset{H}{C}{=}O$

(a) 1-pentanol (c) pentanal
(b) propanal (d) pentanaldehyde

35. Give the name for

$C_6H_5{-}\overset{CH_3CH_2}{C}H\underset{CH_3}{C}H\overset{H}{C}{=}O$

(a) 3-ethyl-2-methyl-3-phenylpropanal
(b) 4-methyl-3-phenylpentanal
(c) 2-methyl-3-phenylhexanal
(d) no correct name given

36. Give the name for

$CH_3CH_2\underset{O}{\underset{\|}{C}}\overset{CH_3}{C}HCH_2CH_3$

(a) 4-methyl-3-hexanone (c) 3-methyl-4-hexanone
(b) ethyl isobutyl ketone (d) butyl ethyl ketone

37. Which of the following compounds is benzaldehyde?

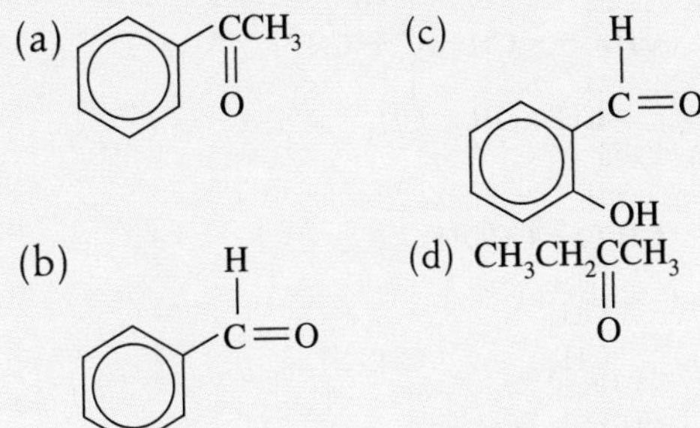

38. A 37% aqueous solution of formaldehyde is called
(a) acetone (b) aldol (c) formalin (d) methanal

39. What is the correct name for this compound?

$$\overset{\text{H}}{\text{O}=\text{C}}\text{CH}_2\text{CH}_2\text{CH}_2\overset{\text{H}}{\text{C}}=\text{O}$$

(a) 1,5-pentanedial (c) ethanedial
(b) 1,5-propanedial (d) pentanedial

40. What is the correct name for this compound?

$$CH_3CH_2CH_2\underset{\|}{\overset{}{C}}CH_2CH_3 \quad (C=O)$$

(a) 4-hexanone (c) 3-hexanone
(b) butyl ethyl ketone (d) dipropyl ketone

41. What is the correct name for this compound?

$CH_3\underset{CH_3}{CH}CH_2\underset{O}{\overset{\|}{C}}CH_2$–(benzene ring)

(a) benzyl *sec*-butyl ketone
(b) 4-methyl-1-phenyl-2-pentanone
(c) 2-methyl-5-phenyl-4-pentanone
(d) no correct name given

42. Choose the *incorrect* name for

(benzene ring)–$\underset{O}{\overset{\|}{C}}-CH_3$

(a) methyl phenyl ketone
(b) benzophenone
(c) 1-phenylethanone
(d) acetophenone

43. Which of the following compounds is a cyanohydrin?

(a) CH_3CH_2CN (c) CH_3-O-CH_2CN
(b) $CH_3\underset{OH}{CH}CN$ (d)

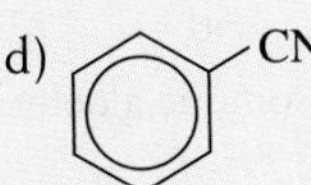

44. Choose the compounds that, when oxidized, will yield

$$CH_3CH_2CH_2CH_2\overset{H}{C}=O$$

(a) 1-pentanol (c) 3-pentanol
(b) 2-pentanol (d) 1-propanol

45. What is the evidence that a reaction occurred when an aldehyde is oxidized by $Cr_2O_7^{2-} + H^+$?
(a) orange-colored solution formed
(b) silver mirror formed
(c) brick red Cu_2O formed
(d) green solution formed

46. Reduction of an aldehyde produces a
(a) ketone (c) secondary alcohol
(b) primary alcohol (d) hydrocarbon

47. What is the product of this reaction?

$$CH_3CH_2\overset{H}{C}=O + HCN \longrightarrow$$

(a) $CH_3\underset{OH}{CH}CN$ (c) $CH_3CH_2\underset{OH}{CH}CN$
(b) $CH_3CH_2-O-CH_2CN$ (d) $CH_3\underset{OH}{CH}CH_2CN$

48. Which of the following compounds will give a positive Tollens test?

(a) $CH_3\underset{OH}{CH}\overset{H}{C}=O$ (c) $CH_3\overset{OH}{C}=O$
(b) (cyclopentanone ring)=O (d) $CH_3\underset{O}{\overset{\|}{C}}CH_2CH_3$

In Problems 49 and 50, use these formulas:

(A) $CH_3CH_2\overset{OH}{CH}-OCH_2CH_3$
(B) $(CH_3CH_2)_2\overset{OCH_2CH_3}{C}-CH_2CH_3$
(C) $CH_3\overset{OCH_2CH_3}{\underset{CH_3}{C}}-OCH_2CH_3$
(D) $(CH_3CH_2)_2\overset{OH}{C}-OCH_2CH_3$

49. Which formula represents a hemiacetal?
(a) A (b) B (c) C (d) D

50. Which formula represents a ketal?
(a) A (b) B (c) C (d) D

51. Which of the following compounds will *not* undergo an aldol condensation?

(a) $H_2C{=}O$ (c) $CH_3CH_2C(=O)CH_3$

(b) $CH_3C(H){=}O$ (d) $C_6H_5CH_2C(H){=}O$

52. Which of the following compounds will be isolated from this reaction?

$$CH_3C(=O)CH_3 \xrightarrow[H^+]{Cr_2O_7^{2-}}$$

(a) CH_3CH_2COOH (c) $CH_3C(=O)COOH$

(b) $CH_3C(=O)CH_3$ (d) $CH_3CH(OH)CH_3$

53. Ketones can be prepared by the oxidation of
(a) primary alcohols
(b) secondary alcohols
(c) tertiary alcohols
(d) cannot be prepared by oxidation reactions

54. What is the product of this aldol condensation reaction?

$$CH_3CH(CH_3)C(H){=}O \xrightarrow[NaOH]{dilute}$$

(a) $CH_3CH(CH_3){-}COONa$ (c) $CH_3CH(CH_3){-}CH(OH){-}C(CH_3)_2{-}C(H){=}O$

(b) $CH_3CH(CH_3)CH(CH_3)CH(OH)CH_3$ (d) no reaction

55. What is the product of this aldol condensation reaction?

$$CH_3C(CH_3)_2{-}C(H){=}O \xrightarrow[NaOH]{dilute}$$

(a) $CH_3C(CH_3)_2{-}CH(OH){-}C(CH_3)_2{-}CH_3$

(b) $CH_3C(CH_3)_2{-}COOH$

(c) $CH_3C(CH_3)_2{-}CH(OH){-}C(CH_3)_2{-}C(H){=}O$

(d) no reaction

56. The polymer of phenol and formaldehyde is known as
(a) paraformaldehyde (c) paraldehyde
(b) Bakelite (d) meta

57. When hydrolyzed, cyanohydrins form
(a) α-keto acids (c) glycols
(b) α-hydroxy acids (d) α-hydroxy aldehydes

58. What is the product of this reaction?

$$\text{cyclohexanone } (=O) + 2\,CH_3OH \xrightarrow[HCl]{dry}$$

(a) OH, OCH$_3$ (c) OCH$_3$, OCH$_3$

(b) OCH$_3$, =O, OCH$_3$ (d) OCH$_3$

59. The substance that precipitates in a positive Benedict test is
(a) Ag (b) Ag_2O (c) CuO (d) Cu_2O

60. Which of the following compounds will give both a positive Tollens and Benedict test?

(a) CH_3CH_2OH (c) $CH_3CH_2C(=O)CH_3$

(b) 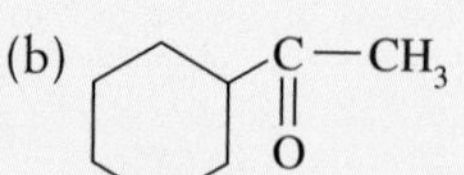

(d) 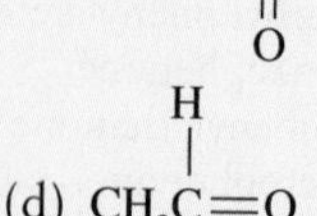 $CH_3C(H){=}O$

61. The oxidation of an aldehyde will yield a
(a) primary alcohol (c) ketone
(b) carboxylic acid (d) secondary alcohol

62. How many isomeric aldehydes and ketones are possible for $C_5H_{10}O$?
(a) 5 (b) 6 (c) 7 (d) 9

Free Response Questions

Answer each of the following. Be sure to include your work and explanations in a clear, logical form.

1. An open-chain compound has the formula $C_6H_{12}O$. Chemical tests show that there is no alkene or aromatic group in the compound. How do you identify whether the compound is an alcohol, an ether, an aldehyde, or a ketone? You are allowed to perform only one chemical test.

2. Why are aldehydes and ketones easily able to undergo addition reactions, but alcohols, ethers, and thiols are not?

3. A compound has the formula
$CH_3OCH_2CH(OH)CH_2CH_2COCH(CH_3)CH_2CHO$
What functional group(s) is (are) present? Draw the compound showing the relevant functional group(s).

4. Why is it possible to have a straight-chain triol, triether, trithiol, or trione, but it is not possible to have a straight-chain trialdehyde? Draw a structure for each type of compound.

5. Which of the following would you expect to be most soluble in water? least soluble in water? Explain.

   methoxyethane, butene, 1-propanol, propanal

6. Draw and name three structural isomers of the following compound, each containing a *different* functional group.

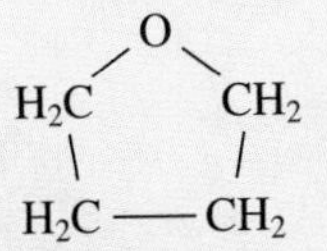

   O, H_2C, CH_2, H_2C—CH_2

7. Draw the organic products I–III for the following reaction sequence:

   OH (cyclohexanol) $\xrightarrow[H_2SO_4]{K_2Cr_2O_7}$ I $\xrightarrow{HCN}$ II $\xrightarrow[H^+]{H_2O}$ III

8. Fill in the missing reagents or organic structures, I–III for the following reactions:

   I —(aldol condensation)→ IV: $CH_3CH_2CH_2CH(OH)CH(CH_2CH_3)CHO$

   I —(II)→ $CH_3CH_2CH_2CH_2OH$ $\xrightarrow{Na}$ $\xrightarrow{CH_3CH_2Br}$ III

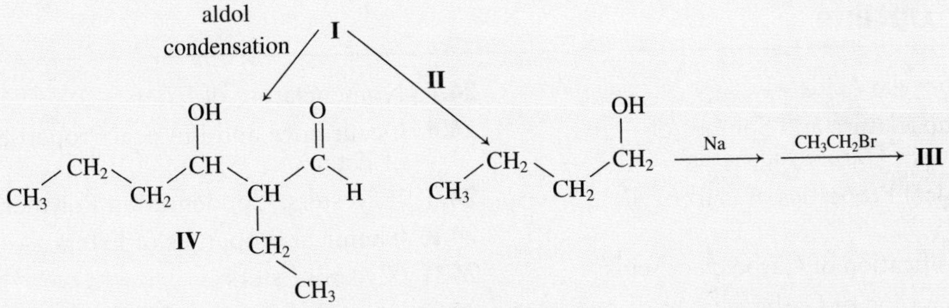

   Give the IUPAC names for compounds I, III, and IV.

9. Which of the compounds that follow can easily be identified from the others by taking a boiling point? Explain briefly. For those which cannot be easily identified by boiling point, indicate how you may distinguish among them.

   2-hexanone, 1-hexanol, hexanal, dipropyl ether, 1-pentanethiol

10. Suggest at least two visual chemical tests that could be used to distinguish between the members of the following pairs of compounds:
    (a) phenol and 1-hexanol
    (b) phenol and hexanal
    (c) 1-hexanol and 3-hexanone

11. Acetals and ketals are very commonly used as protecting groups in organic synthesis (converting the carbonyl into a functional group that is unreactive to the conditions used to react with another functional group). Draw the structure of the *acetal* produced when the following compound is reacted with ethylene glycol in the presence of an acid catalyst:

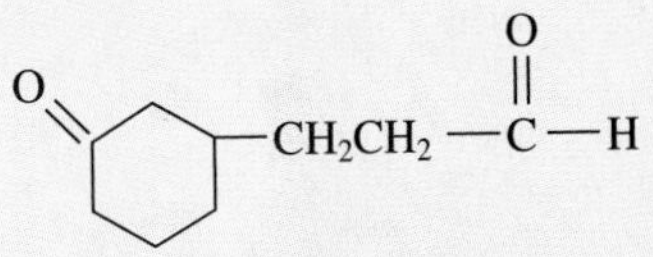

    O=(cyclohexanone ring)—CH_2CH_2—C(=O)—H

CHAPTER 24

Carboxylic Acids and Esters

All citrus fruits contain a variety of carboxylic acids.

Chapter Outline

Whenever we eat food with a sour or tart taste, it is very likely that at least one carboxylic acid is present in that food. For example, citric acid is present in lemons, acetic acid is present in vinegar, and lactic acid is present in sour milk. Carboxylic acids are also important compounds in biochemistry. Citric acid is found in our blood, and lactic acid is produced in our muscles during the breakdown of glucose. In living systems, these acids are most often found in the form of salts or acid derivatives. In the world around us, we use carboxylic acid salts as preservatives, especially in cheeses and breads, and to treat skin irritations like diaper rash and athlete's foot.

Carboxylic acid derivatives known as esters are responsible for the sweet and pleasant odors and tastes of our food. We frequently use these compounds as artificial flavors in foods in place of more expensive natural extracts. In biochemistry, esterlike molecules act as the energy carriers in many cells. As you will see, the properties of carboxylic acids are quite distinct from those of aldehydes and alcohols.

24.1 Carboxylic Acids

The functional group of the carboxylic acids is called a **carboxyl group** and is represented in the following ways:

carboxyl group

$$-\overset{\overset{\displaystyle O}{\|}}{C}-OH \quad \text{or} \quad -COOH \quad \text{or} \quad -CO_2H$$

Carboxylic acids can be either aliphatic or aromatic:

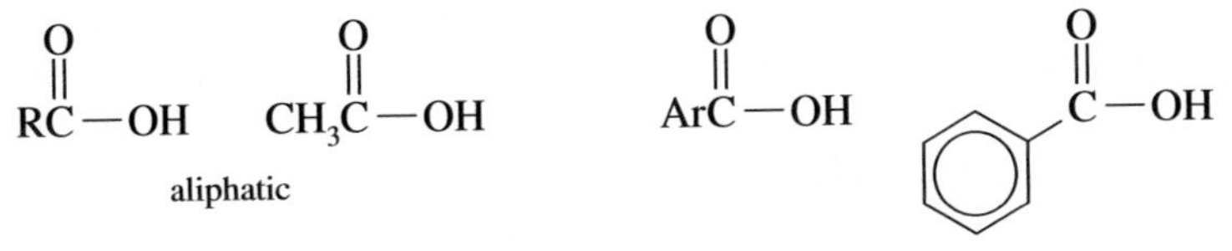

$$R\overset{\overset{\displaystyle O}{\|}}{C}-OH \qquad CH_3\overset{\overset{\displaystyle O}{\|}}{C}-OH$$

aliphatic

$$Ar\overset{\overset{\displaystyle O}{\|}}{C}-OH \qquad C_6H_5-\overset{\overset{\displaystyle O}{\|}}{C}-OH$$

aromatic

"R" is often used to symbolize an aliphatic group; "Ar" symbolizes an aromatic group.

24.2 Nomenclature and Sources of Aliphatic Carboxylic Acids

Aliphatic carboxylic acids form a homologous series. The carboxyl group is always at the beginning of a carbon chain, and the carbon atom in this group is understood to be carbon number 1 when the compound is named.

IUPAC Rules for Naming Carboxylic Acids

1. To establish the parent name, identify the longest carbon chain that includes the carboxyl group.
2. Drop the final *e* from the name of the hydrocarbon with the same length carbon chain.
3. Add the suffix *-oic acid*.

Thus the names corresponding to the one-, two-, and three-carbon acids are methanoic acid, ethanoic acid, and propanoic acid. These names are derived from methane, ethane, and propane:

CH_4	methane	$HCOOH$	methanoic acid
CH_3CH_3	ethane	CH_3COOH	ethanoic acid
$CH_3CH_2CH_3$	propane	CH_3CH_2COOH	propanoic acid

Other groups bonded to the parent chain are numbered and named as we have done previously. For example, we have

$$\overset{5}{C}H_3\overset{4}{C}H_2\underset{\displaystyle CH_3}{\overset{3}{C}H}\overset{2}{C}H_2\overset{1}{C}OOH$$

3-methylpentanoic acid

The sting of an ant bite is caused by formic acid.

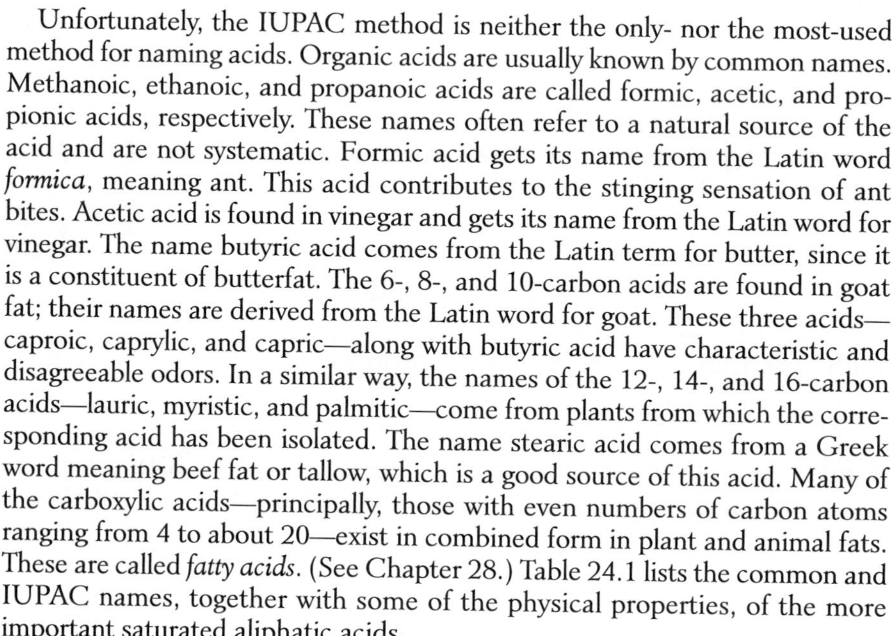

Unfortunately, the IUPAC method is neither the only- nor the most-used method for naming acids. Organic acids are usually known by common names. Methanoic, ethanoic, and propanoic acids are called formic, acetic, and propionic acids, respectively. These names often refer to a natural source of the acid and are not systematic. Formic acid gets its name from the Latin word *formica*, meaning ant. This acid contributes to the stinging sensation of ant bites. Acetic acid is found in vinegar and gets its name from the Latin word for vinegar. The name butyric acid comes from the Latin term for butter, since it is a constituent of butterfat. The 6-, 8-, and 10-carbon acids are found in goat fat; their names are derived from the Latin word for goat. These three acids—caproic, caprylic, and capric—along with butyric acid have characteristic and disagreeable odors. In a similar way, the names of the 12-, 14-, and 16-carbon acids—lauric, myristic, and palmitic—come from plants from which the corresponding acid has been isolated. The name stearic acid comes from a Greek word meaning beef fat or tallow, which is a good source of this acid. Many of the carboxylic acids—principally, those with even numbers of carbon atoms ranging from 4 to about 20—exist in combined form in plant and animal fats. These are called *fatty acids*. (See Chapter 28.) Table 24.1 lists the common and IUPAC names, together with some of the physical properties, of the more important saturated aliphatic acids.

Common and IUPAC nomenclature systems should not be intermixed.

Another nomenclature method uses letters of the Greek alphabet (α, β, γ, δ, . . .) to name certain acid derivatives, especially hydroxy, amino, and halogen acids. When Greek letters are used, the carbon atoms, beginning with the one adjacent to the carboxyl group, are labeled α, β, γ, δ, ... When numbers are used (the IUPAC System), the numbers begin with the carbon in the $-COOH$ group.

$$\underset{5}{\overset{\delta}{C}}-\underset{4}{\overset{\gamma}{C}}-\underset{3}{\overset{\beta}{C}}-\underset{2}{\overset{\alpha}{C}}-\underset{1}{\overset{O}{\overset{\|}{C}}}-OH$$

	$CH_3CH_2\underset{\displaystyle OH}{C}HCOOH$	$CH_3\underset{\displaystyle NH_2}{C}HCOOH$	CH_2ClCH_2COOH
Common name:	α-hydroxybutyric acid	α-aminopropionic acid	β-chloropropionic acid
IUPAC name:	2-hydroxybutanoic acid	2-aminopropanoic acid	3-chloropropanoic acid

Table 24.1 Names, Formulas, and Physical Properties of Saturated Aliphatic Carboxylic Acids

Common name (IUPAC name)	Formula	Melting point (°C)	Boiling point (°C)	Solubility in water*
Formic acid (methanoic acid)	$HCOOH$	8.4	100.8	∞
Acetic acid (ethanoic acid)	CH_3COOH	16.6	118	∞
Propionic acid (propanoic acid)	CH_3CH_2COOH	−21.5	141.4	∞
Butyric acid (butanoic acid)	$CH_3(CH_2)_2COOH$	−6	164	∞
Valeric acid (pentanoic acid)	$CH_3(CH_2)_3COOH$	−34.5	186.4	3.3
Caproic acid (hexanoic acid)	$CH_3(CH_2)_4COOH$	−3.4	205	1.1
Caprylic acid (octanoic acid)	$CH_3(CH_2)_6COOH$	16.3	239	0.1
Capric acid (decanoic acid)	$CH_3(CH_2)_8COOH$	31.4	269	Insoluble
Lauric acid (dodecanoic acid)	$CH_3(CH_2)_{10}COOH$	44.1	225**	Insoluble
Myristic acid (tetradecanoic acid)	$CH_3(CH_2)_{12}COOH$	54.2	251**	Insoluble
Palmitic acid (hexadecanoic acid)	$CH_3(CH_2)_{14}COOH$	63	272**	Insoluble
Stearic acid (octadecanoic acid)	$CH_3(CH_2)_{16}COOH$	69.6	287**	Insoluble
Arachidic acid (eicosanoic acid)	$CH_3(CH_2)_{18}COOH$	77	298**	Insoluble

*Grams of acid per 100 g of water.
**Boiling point is given at 100 mm Hg pressure instead of atmospheric pressure, because thermal decomposition occurs before this acid reaches its boiling point at atmospheric pressure.

Example 24.1

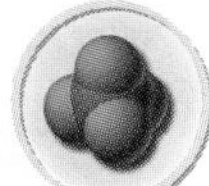

Write formulas for the following:
(a) 3-chloropentanoic acid (b) γ-hydroxybutyric acid (c) phenylacetic acid

SOLUTION

(a) This is an IUPAC name, as indicated by the use of the position number. *Pentanoic* indicates a five-carbon acid. Substituted on carbon 3 is a chlorine atom. Write five carbon atoms in a row. Make carbon 1 a carboxyl group, place a Cl on carbon 3, and add hydrogens to give each carbon four bonds. The formula is

$CH_3CH_2CHClCH_2COOH$

(b) This is a common name, as indicated by the Greek letter. *Butyric* indicates a four-carbon acid. The γ-position is three carbons removed from the carboxyl group. Therefore, the formula is

(c) *Acetic acid* (a common name) is the familiar two-carbon acid. There is only one place to substitute the phenyl group and still call the compound an acid—that is, at the CH_3 group. Substitute a phenyl group for one of the three H atoms to give the formula:

$$C_6H_5-CH_2COOH$$

Practice 24.1

Write formulas for (a) 2-methylpropanoic acid, (b) β-chlorocaproic acid, and (c) cyclohexanecarboxylic acid.

Practice 24.2 Application to Biochemistry

Uncontrolled diabetics have high blood levels of the following carboxylic acid:

$$CH_3-\underset{|}{\overset{OH}{CH}}-CH_2-\overset{O}{\overset{\|}{C}}-OH$$

Give the IUPAC name.

24.3 Physical Properties of Carboxylic Acids

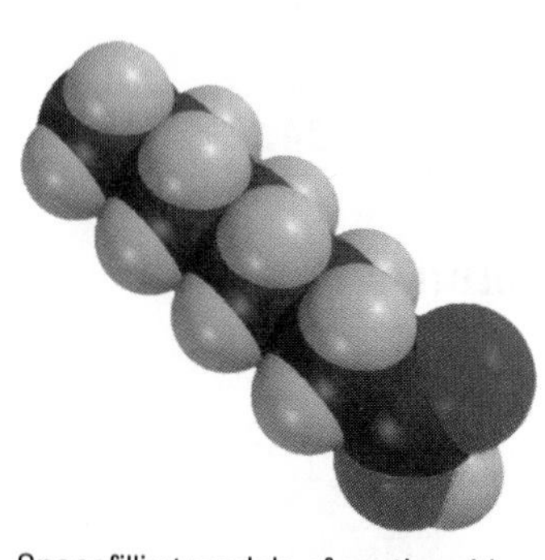

Spacefilling models of acetic acid (top) and octanoic acid (bottom).

Each aliphatic carboxylic acid molecule is polar and consists of a carboxyl group and a hydrocarbon group (R). These two unlike parts have great bearing on the physical, as well as chemical, behavior of the molecule as a whole. The first four acids, formic through butyric, are completely soluble (miscible) in water (Table 24.1). Beginning with pentanoic acid (valeric acid), the water solubility falls sharply and is only about 0.1 g of acid per 100 g of water for octanoic acid (caprylic acid). Acids of this series with more than eight carbons are virtually insoluble in water. The water-solubility characteristics of the first four acids are determined by the highly soluble polar carboxyl group. Thereafter, the water insolubility of the nonpolar hydrocarbon chain is dominant.

The polarity due to the carboxyl group is evident in the boiling-point data. Formic acid (HCOOH) boils at about 101°C. Carbon dioxide, a nonpolar substance of similar molar mass, remains in the gaseous state until it is cooled to −78°C. In like manner, the boiling point of acetic acid (molar mass 60 g/mol) is 118°C, whereas nonpolar butane (molar mass 58 g/mol) boils at −0.6°C. The comparatively high boiling points for carboxylic acids are due to intermolecular attractions resulting from hydrogen bonding. In fact, molar-mass determinations on gaseous acetic acid (near its boiling point) show a value of about 120 g/mol, indicating that two molecules are joined together to form a *dimer* $(CH_3COOH)_2$:

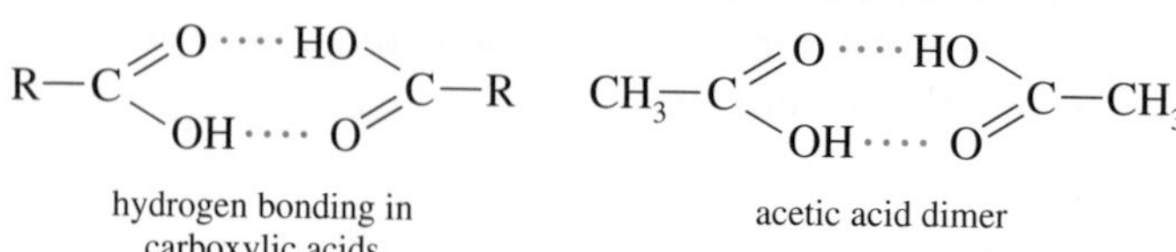

hydrogen bonding in carboxylic acids
acetic acid dimer

Saturated monocarboxylic acids that have fewer than 10 carbon atoms are liquids at room temperature, whereas those with more than 10 carbon atoms are waxlike solids.

Carboxylic acids and phenols, like mineral acids such as HCl, ionize in water to produce hydronium ions and anions (Chapter 15). Carboxylic acids are generally weak acids; that is, they are only slightly ionized in water. Phenols are, in general, even weaker acids than carboxylic acids. For example, the ionization constant for acetic acid is 1.8×10^{-5} (1.3% ionized for a 0.10 *M* solution), and that for phenol is 1.3×10^{-10} (0.00036% ionized for a 0.10 *M* solution). Equations illustrating these ionizations are as follows:

$$\underset{\text{hydrogen chloride}}{HCl} + H_2O \longrightarrow \underset{\text{hydronium ion}}{H_3O^+} + \underset{\text{chloride ion}}{Cl^-}$$

$$\underset{\text{acetic acid}}{CH_3\overset{\overset{O}{\|}}{C}-OH} + H_2O \rightleftharpoons \underset{\text{hydronium ion}}{H_3O^+} + \underset{\text{acetate ion}}{CH_3\overset{\overset{O}{\|}}{C}-O^-}$$

$$\underset{\text{phenol}}{C_6H_5OH} + H_2O \rightleftharpoons \underset{\text{hydronium ion}}{H_3O^+} + \underset{\text{phenoxide ion}}{C_6H_5O^-}$$

A weak monopratic acid (HA) ionizes according to the following equilibrium:

$$HA \rightleftharpoons H^+ + A^-$$

The acid equilibrium constant expression is

$$K_a = \frac{[H^+][A^-]}{[HA]}$$

Carboxylic acids are common in biochemistry. When these molecules ionize, both the carboxylate anion and the H^+ can have significant effects. For example, if the liver releases too many carboxylic acids, the blood becomes too acidic. This happens during uncontrolled diabetes, and the resulting ketoacidosis can be deadly.

Under biological conditions, most carboxylic acids form anions. In turn, living cells respond to changes in ionic charge. A good example starts with a positively charged molecule, dopamine. When dopamine enters the brain, it eases the effects of Parkinson's disease, a chronic disease of the central nervous system. Unfortunately, dopamine is not able to cross the blood–brain barrier. So, scientists added a carboxylic acid group to dopamine, creating Dopa. Dopa contains both a negative and a positive charge when it enters the blood. The charges neutralize each other, and Dopa thus passes through the blood–brain barrier. Once into the brain, Dopa is converted to dopamine, which alleviates the symptoms of Parkinson's disease.

HO, HO, NH_2

dopamine ($C_8H_{11}NO_2$)

HO, HO, NH_2, COOH

dopa ($C_9H_{11}NO_4$)

CHEMISTRY IN ACTION • A Sour Success

About 10,000 years ago, a sour-tasting solution was first fermented from fruit. For many people, "sour tasting" might have put an end to this story. However, by 5000 B.C.E., the Babylonians were commercially producing this sour liquid for use as both a beverage and a food preservative. This liquid was recognized to have many applications. Egyptians started to use it as an antiseptic around 3000 B.C.E. The great physician Hippocrates prescribed a mixture of honey and this sour solution for respiratory problems. Roman legionnaires drank it diluted—perhaps the first example of an "energy drink." This sour liquid was also recognized as an excellent solvent. It is said that Cleopatra dissolved a pearl in this solvent and then drank the result during a dinner she would claim "cost a fortune." Today this sour liquid still has a surprisingly wide range of successful applications. These successes derive from the properties of one specific carboxylic acid—acetic acid. The sour solution that has found so many uses for over 7,000 years is common vinegar.

Vinegar is derived from the French *vin*, meaning "wine," and *aigre*, meaning "sour". Many different grains and fruits can be fermented to vinegar and, depending on the source, vinegar may contain from 4% to 7% acetic acid. Because carboxylic acids are weak, this solution is not acidic enough to be especially dangerous, but it is acidic enough to be very useful.

Vinegar is one of the oldest cooking ingredients in human history. Acetic acid liberates protons that are recorded by our taste buds as sour. So vinegar adds a little tang to a variety of recipes. Fermentation of different fruits or grains produces vinegars of subtly different flavors. Several common vinegars include: balsamic vinegar, made from a specific Italian grape and aged in wooden casks, is sweetly–sour; cider vinegar made from apples, has a crisp flavor; malt vinegar, made from fermented grain, carries a heartier taste. Thai cooking commonly uses coconut vinegar, sugarcane vinegar can be found in Philippine cooking, and Chinese dishes may be served with rice wine vinegar.

Vinegar is a mild and inexpensive antimicrobial. When vinegar is used to preserve foods, the process is called "pickling." Vinegar's acetic acid adds a sour taste *and* kills microorganisms. Foods ranging from meats to fruits and vegetables can be pickled. Although vinegar pickling helps preserve foods, long-term preservation requires a combination of processes, for example, pickling *and* cold storage or canning.

As an antiseptic, vinegar is one of the few household chemicals that effectively kills germs and is relatively nontoxic to humans. Thus, a wash solution of vinegar and hydrogen peroxide effectively and safely cleans bacterial contamination from fresh fruits and vegetables.

Photo of decanter-type containers of various vinegars

Vinegar is a likely component of any household cleaner. From coffee pots to rug stains to tarnished metal, vinegar can clean, too. The key to its success is, again, the acetic acid. Protons react with many stains; the product becomes water soluble and can be washed away.

Vinegar—this sour solution—has been used successfully since antiquity. For some very good reasons, vinegar remains a staple commodity today and, perhaps, far into the future.

24.4 Classification of Carboxylic Acids

Thus far, our discussion has dealt mainly with a single type of acid—that is, saturated monocarboxylic acids. But various other kinds of carboxylic acids are known. We discuss some of the more important ones here.

Unsaturated Carboxylic Acids

An unsaturated acid contains one or more carbon–carbon double bonds. The first member of the homologous series of unsaturated carboxylic acids that contains one carbon–carbon double bond is acrylic acid, $CH_2{=}CHCOOH$. The IUPAC name for $CH_2{=}CHCOOH$ is propenoic acid. Derivatives of acrylic acid are used to manufacture a class of synthetic polymers known as the acrylates. (See Chapter 21.) These polymers are widely used as textiles and in paints and lacquers. Unsaturated carboxylic acids undergo the reactions of both an unsaturated hydrocarbon and a carboxylic acid.

Even one carbon–carbon double bond in the molecule exerts a major influence on the physical and chemical properties of an acid. The effect of a double bond can be seen by comparing the two 18-carbon acids, stearic and oleic. Stearic acid, $CH_3(CH_2)_{16}COOH$, a solid that melts at 70°C, shows only the reactions of a carboxylic acid. On the other hand, oleic acid,

$$CH_3(CH_2)_7CH{=}CH(CH_2)_7COOH \text{ (mp 16°C)}$$

with one double bond, is a liquid at room temperature and shows the reactions of an unsaturated hydrocarbon as well as those of a carboxylic acid.

Aromatic Carboxylic Acids

In an aromatic carboxylic acid, the carbon of the carboxyl group (—COOH) is bonded directly to a carbon in an aromatic ring. The parent compound of this series is benzoic acid, and other common examples are the three isomeric toluic acids:

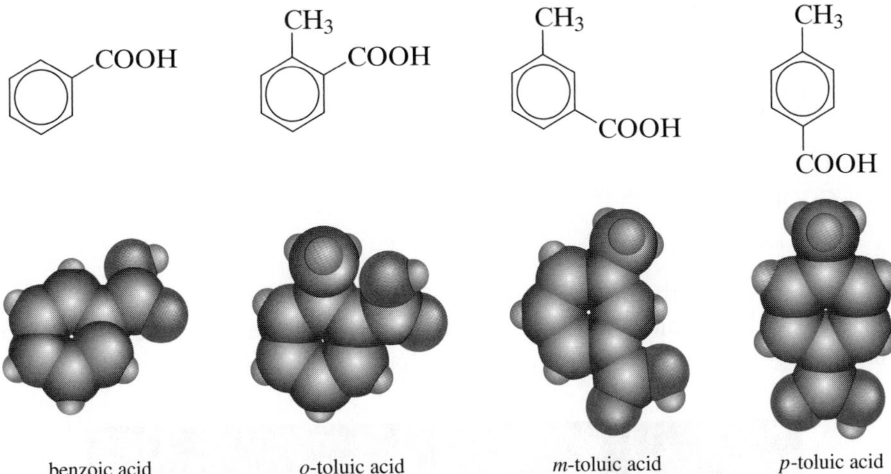

benzoic acid | *o*-toluic acid | *m*-toluic acid | *p*-toluic acid

Dicarboxylic Acids

Acids of both the aliphatic and aromatic series that contain two or more carboxyl groups are known. Those with two —COOH groups are called dicarboxylic acids. The simplest member of the aliphatic series is oxalic acid. The next member in the homologous series is malonic acid. Several dicarboxylic acids and their names are listed in Table 24.2.

The IUPAC names for dicarboxylic acids are formed by modifying the corresponding hydrocarbon names to end in *-dioic acid*. Thus, the two-carbon acid is ethanedioic acid (derived from ethane). However, the common names for dicarboxylic acids are frequently used.

Oxalic acid is found in various plants, including spinach, cabbage, and rhubarb. Among its many uses are bleaching straw and leather and removing rust and ink stains. Although oxalic acid is poisonous, the amounts present in the previously mentioned vegetables are usually not harmful.

Table 24.2 Names and Formulas of Selected Dicarboxylic Acids

Common name*	IUPAC name	Formula
Oxalic acid	Ethanedioic acid	HOOCCOOH
Malonic acid	Propanedioic acid	$HOOCCH_2COOH$
Succinic acid	Butanedioic acid	$HOOC(CH_2)_2COOH$
Glutaric acid	Pentanedioic acid	$HOOC(CH_2)_3COOH$
Adipic acid	Hexanedioic acid	$HOOC(CH_2)_4COOH$
Pimelic acid	Heptanedioic acid	$HOOC(CH_2)_5COOH$
Fumaric acid	*trans*-2-Butenedioic acid	HOOCCH=CHCOOH
Maleic acid	*cis*-2-Butenedioic acid	HOOCCH=CHCOOH

*A mnemonic for remembering the common names of the saturated dicarboxylic acid uses the first letter of the acid name:
Oh My Such Good Apple Pie

Malonic acid is made synthetically, but was originally prepared from malic acid, which is commonly found in apples and many fruit juices. Malonic acid is one of the major compounds used in the manufacture of the class of drugs known as barbiturates. When heated above their melting points, malonic acid and substituted malonic acids lose carbon dioxide to give monocarboxylic acids. Thus, malonic acid yields acetic acid when strongly heated:

$$\begin{array}{c}\boxed{COO}H \\ | \\ CH_2 \\ | \\ COOH \\ \text{malonic acid}\end{array} \xrightarrow{150^\circ C} \underset{\text{acetic acid}}{CH_3COOH} + CO_2(g)$$

Malonic acid is the biological precursor for the synthesis of fatty acids. In living cells, when malonic acid loses carbon dioxide, the acetic acid units are linked together and begin forming long-chain fatty acids.

This vendor is selling fresh spinach, which contains oxalic acid in addition to important vitamins.

Succinic acid has been known since the 16th century, when it was obtained as a distillation product of amber. Succinic, fumaric, and citric acids are among the important acids in the energy-producing metabolic pathway known as the citric acid cycle. (See Chapter 34.) Citric acid is a tricarboxylic acid that is widely distributed in plant and animal tissue, especially in citrus fruits. (Lemon juice contains 5–8%.) The formula for citric acid is

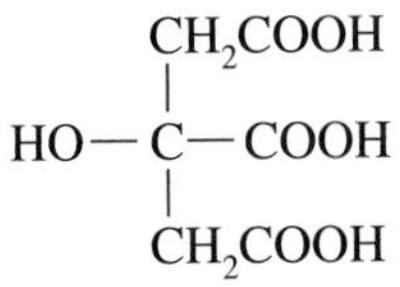

$$\begin{array}{c} CH_2COOH \\ | \\ HO-C-COOH \\ | \\ CH_2COOH \end{array}$$

citric acid

When succinic acid is heated, it loses water, forming succinic anhydride, an acid anhydride.

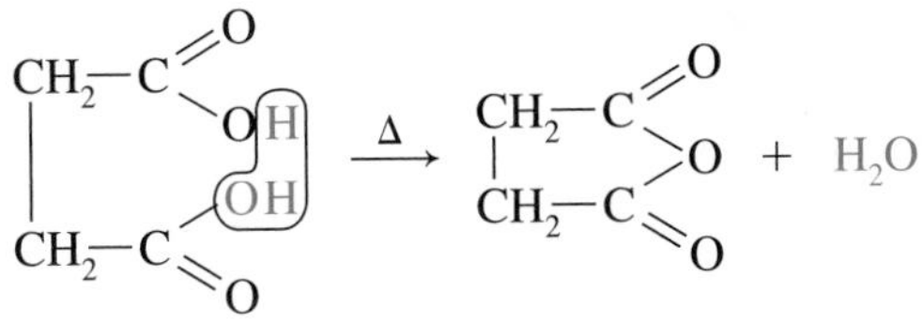

$$\begin{array}{l} CH_2-C(=O)-OH \\ | \\ CH_2-C(=O)-OH \end{array} \xrightarrow{\Delta} \begin{array}{l} CH_2-C(=O) \\ | \qquad\quad > O \\ CH_2-C(=O) \end{array} + H_2O$$

succinic acid — succinic anhydride

Glutaric acid behaves similarly, forming glutaric anhydride, a six-membered ring.

Adipic acid is the most important commercial dicarboxylic acid. It is made from benzene by converting it first to cyclohexene and then, by oxidation, to adipic acid. About 8.2×10^8 kg of adipic acid is produced annually in the United States. Most of the adipic acid is used to produce nylon (Chapter 25). Adipic acid is also used in polyurethane foams, plasticizers, and lubricating-oil additives.

Aromatic dicarboxylic acids contain two carboxyl groups attached directly to an aromatic ring.

Examples are the three isomeric phthalic acids, $C_6H_4(COOH)_2$:

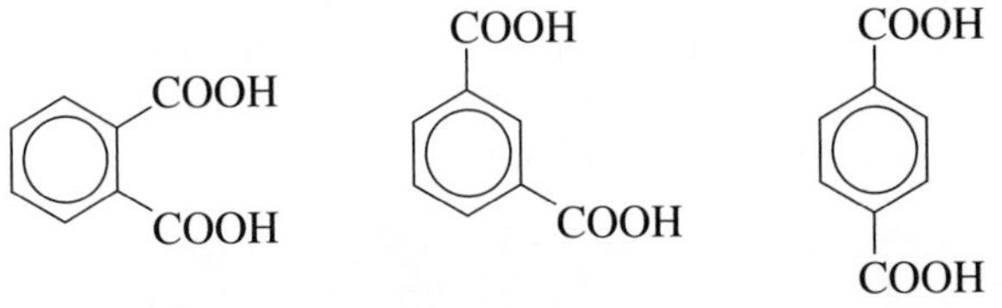

o-phthalic acid (phthalic acid) — *m*-phthalic acid (isophthalic acid) — *p*-phthalic acid (terephthalic acid)

Dicarboxylic acids are *bifunctional*; that is, they have two sites at which reactions can occur. Therefore, they are often used as monomers in the preparation of synthetic polymers such as Dacron polyester (Section 24.9).

Hydroxy Acids

Lactic acid, found in sour milk, sauerkraut, and dill pickles, has the functional groups of both a carboxylic acid and an alcohol. Lactic acid is the end product when our muscles use glucose for energy in the absence of oxygen, a process called *glycolysis*. (See Chapter 34.) Salicylic acid is both a carboxylic acid and a phenol. It is of special interest because a family of useful drugs—the

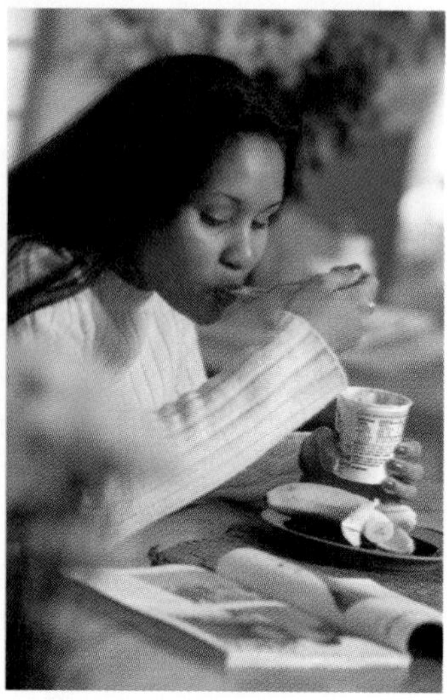

Yogurt has a "sour tang" that comes from carboxylic acids.

salicylates—are derivatives of this acid. The salicylates include aspirin and function as *analgesics* (pain relievers) and as *antipyretics* (fever reducers). The structural formulas of several hydroxy acids are as follows:

$$\underset{\displaystyle \text{OH}}{\text{CH}_3\underset{|}{\text{C}}\text{HCOOH}}$$

lactic acid
(2-hydroxypropanoic acid)

$$\begin{array}{c} \text{COOH} \\ | \\ \text{H}-\text{C}-\text{OH} \\ | \\ \text{CH}_2 \\ | \\ \text{COOH} \end{array}$$

malic acid
(2-hydroxybutanedioic acid)

$$C_6H_4(OH)COOH$$

salicylic acid
(*o*-hydroxybenzoic acid)

$$\begin{array}{l} \text{HO}-\text{CHCOOH} \\ \quad\quad\ | \\ \text{HO}-\text{CHCOOH} \end{array}$$

tartaric acid
(2,3-dihydroxybutanedioic acid)

Amino Acids

Naturally occuring amino acids have this general formula, with the amino group in the alpha position:

$$\begin{array}{c} \quad\quad\quad\ \ \text{O} \\ \quad\quad\quad\ \ \| \\ \text{R}-\text{CH}-\text{C}-\text{OH} \\ | \quad\quad\quad\quad \\ \text{NH}_2 \quad\quad\quad \end{array}$$

Each amino acid molecule has a carboxyl group that acts as an acid and an amino group that acts as a base. About 20 biologically important amino acids, each with a different group represented by R, are found in nature. (In amino acids, R does not always represent an alkyl group.) The immensely complicated protein molecules, found in every form of life, are built from amino acids. Some protein molecules contain more than 10,000 amino acid units. Amino acids and proteins are discussed in more detail in Chapter 29.

Practice 24.3

Write a structural formula for the smallest molecule of each of the following acids:
(a) a carboxylic acid, (b) an α-amino acid, (c) an α-hydroxy acid,
(d) a β-bromo acid, (e) a dicarboxylic acid, (f) an aromatic carboxylic acid

Practice 24.4 Application to Biochemistry

The food additive monosodium glutamate (MSG) is derived from 2-aminopentanedioic acid. Write the structural formula for this acid.

24.5 Preparation of Carboxylic Acids

Many different methods of preparing carboxylic acids are known. We consider only a few examples.

Oxidation of an Aldehyde or a Primary Alcohol This general method is used to convert an aldehyde or primary alcohol to the corresponding carboxylic acid:

$$RC(=O)-H \xrightarrow{[O]} RCOOH$$

$$RCH_2OH \xrightarrow{[O]} RCOOH$$

Butyric acid (butanoic acid) can be obtained by oxidizing either 1-butanol or butanal with potassium dichromate in the presence of sulfuric acid:

$$CH_3CH_2CH_2CH_2OH \xrightarrow[H_2SO_4,\ \Delta]{Cr_2O_7^{2-}} CH_3CH_2CH_2COOH$$

1-butanol   butanoic acid

$$CH_3CH_2CH_2CH(=O) \xrightarrow[H_2SO_4,\ \Delta]{Cr_2O_7^{2-}} CH_3CH_2CH_2COOH$$

butanal   butanoic acid

Aromatic acids may be prepared by the same general method. For example, benzoic acid is obtained by oxidizing benzyl alcohol:

$$C_6H_5CH_2OH \xrightarrow[H_2SO_4,\ \Delta]{Cr_2O_7^{2-}} C_6H_5COOH$$

benzyl alcohol   benzoic acid

Carboxylic acids can also be obtained by the hydrolysis or saponification of esters. (See Section 24.10.)

Practice 24.5

Name the compound formed when (a) 1-hexanol and (b) 2-hexanol are oxidized with $K_2Cr_2O_7/H_2SO_4$.

Oxidation of Alkyl Groups Attached to Aromatic Rings When reacted with a strong oxidizing agent (alkaline permanganate solution or potassium dichromate and sulfuric acid), alkyl groups bonded to aromatic rings are oxidized to carboxyl groups. Regardless of the size or length of the alkyl group, the carbon atom adjacent to the ring remains bonded to the ring and is oxidized to a carboxyl group. The remainder of the alkyl group goes either to carbon dioxide or to a salt of a carboxylic acid. Thus, sodium benzoate is obtained when toluene, ethylbenzene, or propylbenzene is heated with alkaline permanganate solution:

$$C_6H_5CH_3 \xrightarrow[NaOH,\ \Delta]{NaMnO_4} C_6H_5C(=O){-}O^- Na^+$$

toluene — sodium benzoate

$$C_6H_5CH_2CH_3 \xrightarrow[NaOH,\ \Delta]{NaMnO_4} C_6H_5C(=O){-}O^- Na^+ + CO_2(g)$$

ethylbenzene — sodium benzoate

$$C_6H_5CH_2CH_2CH_3 \xrightarrow[NaOH,\ \Delta]{NaMnO_4} C_6H_5C(=O){-}O^- Na^+ + CH_3COO^- Na^+$$

propylbenzene — sodium benzoate — sodium acetate

Since the reaction is conducted in an alkaline medium, a salt of the carboxylic acid (sodium benzoate) is formed instead of the free acid. To obtain the free carboxylic acid, the reaction mixture is acidified with a strong mineral acid (HCl or H_2SO_4) in a second step:

$$C_6H_5COONa + H^+ \longrightarrow C_6H_5COOH + Na^+$$

sodium benzoate — benzoic acid

Practice 24.6

What aromatic hydrocarbon of formula C_9H_{12}, when oxidized with $NaMnO_4/NaOH$, will give (a) benzoic acid and (b) *o*-phthalic acid?

Hydrolysis of Nitriles Nitriles, RCN, which can be prepared by adding HCN to aldehydes and ketones (Section 23.4) or by reacting alkyl halides with KCN, can be hydrolyzed to carboxylic acids:

$$\underset{\text{alkyl halide}}{RX} + KCN \longrightarrow \underset{\text{a nitrile}}{RCN} + KX$$

$$RCN + 2\,H_2O \xrightarrow{H^+} RCOOH + NH_4^+$$

$$CH_3CN + 2\,H_2O \xrightarrow{H^+} CH_3COOH + NH_4^+$$

Practice 24.7

Ethanal is reacted with HCN. Write equations for the nitrile addition product and the subsequent hydrolysis to the acid. Name the acid.

The biological importance of carboxylic acids has resulted in numerous biochemical syntheses. Most living cells form carboxylic acids by the oxidation of aldehydes—a common reaction in organic chemistry. In contrast, the vitamin biotin provides a unique synthesis. This vitamin traps and then transfers carbon dioxide to an acceptor molecule to form a carboxylic acid. Malonic acid (see Section 24.4) is formed using biotin, and the synthesis of fatty acids (see Section 24.2) depends on biotin. In fact, biotin is essential for human good health.

24.6 Chemical Properties of Carboxylic Acids

The structural formula of the carboxyl group consists of a carbonyl group and a hydroxyl group in combination. From such a combination, carboxylic acids might well be expected to have reactions characteristic of both aldehydes and alcohols. But such is not the case; carboxylic acids have their own unique properties.

Hybrids are not uncommon in our world. The nectarines available in our markets are large, sweet, firm-fleshed fruits. The shopper may not be aware that the nectarine is a hybrid created from a peach and a plum, because this hybrid does not have the flavor and texture characteristics of either parent fruit. In the same way, the carboxyl group is a new combination of atoms—a new functional group with properties different from those of both the carbonyl and hydroxyl groups.

The carboxyl group is the most acidic of any organic functional group discussed so far. The —OH group acts as a proton donor, even more effectively than phenol.

The carboxyl group commonly is involved in substitution-type reactions. The C=O retains its double-bond character, while the —OH is replaced by another group or atom. Thus, carboxylic acids react with many substances to produce derivatives. In these reactions, the hydroxyl group is replaced by a halogen (—Cl), an acyloxy group (—OOCR), an alkoxy group (—OR), or an amino group (—NH_2). The general reactions are summarized here:

Substitution reactions require an exchange of atoms or groups of atoms: A–X + Y→A–Y + X

$$\mathrm{R{-}\overset{\overset{\displaystyle O}{\|}}{C}{-}OH} \xrightarrow{-\mathrm{Cl}} \mathrm{R{-}\overset{\overset{\displaystyle O}{\|}}{C}{-}Cl} \quad \text{(acyl halide)}$$

$$\mathrm{R{-}\overset{\overset{\displaystyle O}{\|}}{C}{-}OH} \xrightarrow{-\mathrm{OOCR'}} \mathrm{R{-}\overset{\overset{\displaystyle O}{\|}}{C}{-}O{-}\overset{\overset{\displaystyle O}{\|}}{C}{-}R'} \quad \text{(acid anhydride)}$$

$$\mathrm{R{-}\overset{\overset{\displaystyle O}{\|}}{C}{-}OH} \xrightarrow{-\mathrm{OR'}} \mathrm{R{-}\overset{\overset{\displaystyle O}{\|}}{C}{-}OR'} \quad \text{(ester)}$$

$$\mathrm{R{-}\overset{\overset{\displaystyle O}{\|}}{C}{-}OH} \xrightarrow{-\mathrm{NH_2}} \mathrm{R{-}\overset{\overset{\displaystyle O}{\|}}{C}{-}NH_2} \quad \text{(amide)}$$

Acid–Base Reactions Because of their ability to form hydrogen ions in solution, acids in general have the following properties:

1. Sour taste
2. Change blue litmus to red and affect other suitable indicators
3. Form water solutions with pH values of less than 7
4. Undergo neutralization reactions with bases to form water and a salt

All of the foregoing general properties of an acid are readily seen in low-molar-mass carboxylic acids such as acetic acid. However, these general properties are greatly influenced by the size of the hydrocarbon chain attached to the carboxyl group. In stearic acid, for example, taste, effect on indicators, and pH are not detectable, because the long hydrocarbon chain makes the acid insoluble in water. But stearic acid reacts with a base to form water and a salt. With sodium hydroxide, the equation for the reaction is

$$\underset{\text{stearic acid}}{C_{17}H_{35}COOH} + NaOH \longrightarrow \underset{\text{sodium stearate}}{C_{17}H_{35}COONa} + H_2O$$

The salts formed from this neutralization reaction have properties different from those of the acids. Sodium and potassium salts are soluble in water and dissociate completely in solution. These properties assist in the separation of carboxylic acids from other nonpolar compounds. For example, when a base like NaOH is added to a reaction mixture containing a carboxylic acid, the carboxylic acid reacts, forming its sodium salt and water. The salt dissolves in water, while the remaining nonpolar molecules stay in the organic layer. Once the layers are separated, some mineral acid like HCl can be added to the carboxylic acid salt to re-form and to recover the acid.

Carboxylic acids generally react with sodium hydrogen carbonate to release carbon dioxide. This reaction can be used to distinguish a carboxylic acid from a phenol (also a weak acid). Phenols do not react with sodium hydrogen carbonate, although they will be neutralized by a strong base.

Acid Chloride Formation Thionyl chloride ($SOCl_2$) reacts with carboxylic acids to form acid chlorides. In this substitution reaction, a chlorine atom replaces an —OH group.

$$\underset{\text{acid}}{R\overset{O}{\overset{\|}{C}}-OH} + \underset{\text{thionyl chloride}}{SOCl_2} \longrightarrow \underset{\text{acid chloride}}{R\overset{O}{\overset{\|}{C}}-Cl} + SO_2 + HCl$$

$$CH_3\overset{O}{\overset{\|}{C}}-OH + SOCl_2 \longrightarrow \underset{\text{acetyl chloride}}{CH_3\overset{O}{\overset{\|}{C}}-Cl} + SO_2 + HCl$$

Acid chlorides undergo substitution reactions.

Acid chlorides are extremely reactive substances. They must be kept away from moisture, or they will hydrolyze back to the carboxylic acid:

$$R\overset{O}{\overset{\|}{C}}-Cl + H_2O \longrightarrow R\overset{O}{\overset{\|}{C}}-OH + HCl$$

Acid chlorides are more reactive than acids and can be used in place of acids to prepare esters and amides:

$$CH_3\overset{O}{\overset{\|}{C}}-Cl + CH_3OH \longrightarrow \underset{\text{methyl acetate}}{CH_3\overset{O}{\overset{\|}{C}}-OCH_3} + HCl$$

$$CH_3\overset{\overset{O}{\|}}{C}-Cl + 2\,NH_3 \longrightarrow \underset{\text{acetamide}}{CH_3\overset{\overset{O}{\|}}{C}-NH_2} + NH_4Cl$$

Acid Anhydride Formation Inorganic anhydrides are formed by the elimination of a molecule of water from an acid or a base:

$$H_2SO_3 \longrightarrow SO_2 + H_2O$$

$$Ba(OH)_2 \xrightarrow{\Delta} BaO + H_2O$$

An organic anhydride is formed by the elimination of a molecule of water from two molecules of a carboxylic acid:

$$R-\overset{\overset{O}{\|}}{C}-OH + HO-\overset{\overset{O}{\|}}{C}-R' \longrightarrow R-\overset{\overset{O}{\|}}{C}-O-\overset{\overset{O}{\|}}{C}-R' + H_2O$$

The most commonly used organic anhydride is acetic anhydride. It can be prepared by the reaction of acetyl chloride with sodium acetate:

$$CH_3\overset{\overset{O}{\|}}{C}-Cl + Na^{+}\,{}^{-}O-\overset{\overset{O}{\|}}{C}-CH_3 \longrightarrow \underset{\text{acetic anhydride}}{CH_3\overset{\overset{O}{\|}}{C}-O-\overset{\overset{O}{\|}}{C}-CH_3} + NaCl$$

Acid anhydrides are highly reactive and can be used to synthesize amides and esters, although not as often as the acid chlorides in organic syntheses. In living cells, acid anhydrides commonly activate carboxylic acids for further reaction.

The "energy currency" for living cells is the acid anhydride. (See Chapter 33.)

ester

Ester Formation An **ester** is an organic compound formed by the reaction of an acid with an alcohol or a phenol; water is also a product in this reaction:

$$\underset{\text{carboxylic acid}}{R\overset{\overset{O}{\|}}{C}-\boxed{OH}} + \underset{\text{alcohol}}{R'O\boxed{H}} \overset{H^+}{\rightleftharpoons} \underset{\text{ester}}{R\overset{\overset{O}{\|}}{C}-OR'} + H_2O$$

(R can be H, alkyl, or Ar, but R′ cannot be H)

$$\underset{\text{formic acid}}{H-\overset{\overset{O}{\|}}{C}-OH} + \underset{\text{ethanol}}{CH_3CH_2OH} \overset{H^+}{\rightleftharpoons} \underset{\text{ethyl formate}}{H-\overset{\overset{O}{\|}}{C}-OCH_2CH_3} + H_2O$$

At first glance, this looks like the familiar acid–base neutralization reaction. But that is not the case, because the alcohol does not yield OH^- ions, and the ester, unlike a salt, is a molecular, not an ionic, substance. The forward reaction of an acid and an alcohol is called *esterification;* the reverse reaction of an ester with water is called *hydrolysis.* Chemists frequently modify reaction conditions to favor the formation of either esters or their component parts, alcohols and acids.

Esterification is one of the most important reactions of carboxylic acids. Many biologically significant substances are esters.

24.7 Nomenclature of Esters

Butyric acid smells like rancid butter.

The general formula for an ester is RCOOR′, where R may be a hydrogen, an alkyl group, or an aryl group and R′ may be an alkyl group or an aryl group, but *not* a hydrogen. Esters are found throughout nature. The ester linkage is particularly important in the study of fats and oils, both of which are esters. Esters of phosphoric acid are also of vital importance to life.

Esters are alcohol derivatives of carboxylic acids. They are named in much the same way as salts. The alcohol part is named first, followed by the name of the acid modified to end in *-ate*. The *-ic* ending of the organic acid name is replaced by the ending *-ate*. Thus, in the IUPAC System, *ethanoic acid* becomes *ethanoate*. In the common names, *acetic acid* becomes *acetate*. To name an ester, it is necessary to recognize the portion of the ester molecule that comes from the acid and the portion that comes from the alcohol. In the general formula for an ester, the RC=O comes from the acid, and the R′O comes from the alcohol:

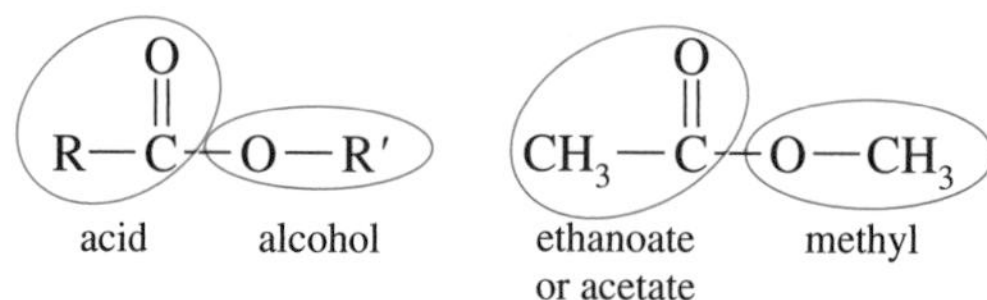

Ester name: methyl ethanoate or methyl acetate

Ethyl butyrate smells like pineapple.

The R′ in R′O is named first, followed by the name of the acid, modified by replacing *-ic acid* with *-ate*. The ester derived from ethyl alcohol and acetic acid is called ethyl acetate or ethyl ethanoate. Consider the ester formed from CH_3CH_2COOH and CH_3OH:

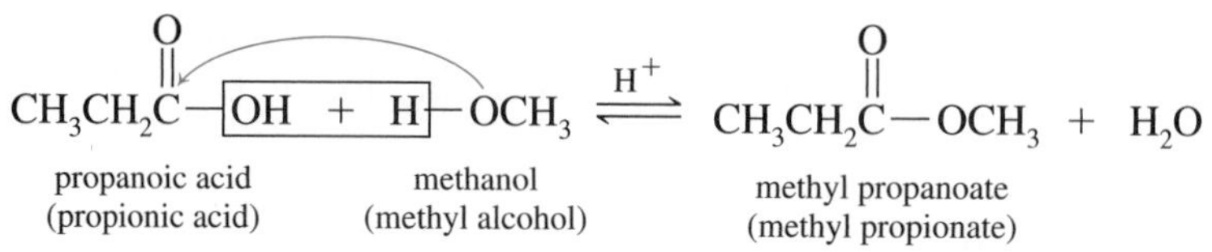

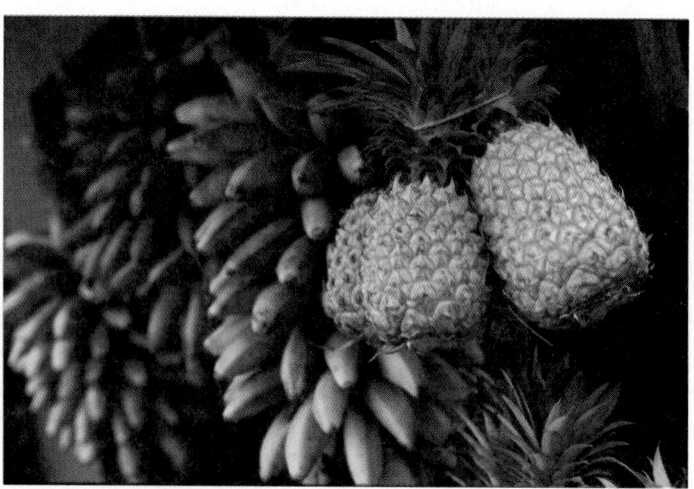
The odor of pineapple is due to the ester ethyl butyrate, and the odor of ripe bananas is due to isoamyl acetate.

Table 24.3 Formulas, Names, and Odors of Selected Esters

Formula	IUPAC name	Common name	Odor or flavor
$CH_3C(=O)-OCH_2CH_2CH(CH_3)CH_3$	Isopentyl ethanoate	Isoamyl acetate	Banana, pear
$CH_3CH_2CH_2C(=O)-OCH_2CH_3$	Ethyl butanoate	Ethyl butyrate	Pineapple
$HC(=O)-OCH_2CH(CH_3)CH_3$	Isobutyl methanoate	Isobutyl formate	Raspberry
$CH_3C(=O)-OCH_2(CH_2)_6CH_3$	Octyl ethanoate	Octyl acetate	Orange
$C_6H_4(OH)C(=O)-OCH_3$	Methyl 2-hydroxybenzoate	Methyl salicylate	Wintergreen

Esters of aromatic acids are named in the same general way as those of aliphatic acids. For example, the ester of benzoic acid and isopropyl alcohol is

$$C_6H_5C(=O)-O-CH(CH_3)_2$$

isopropyl benzoate

Formulas and names for additional esters are given in Table 24.3.

Example 24.2

Name the following esters:

(a) $H-C(=O)-OCH_2CH_2CH_3$

(b) $C_6H_5C(=O)-OCH_2CH_3$

(c) $O=C(-OCH_2CH_3)-CH_2-C(=O)-OCH_2CH_3$

SOLUTION

(a) First, identify the acid and alcohol components. The acid contains one carbon and is formic acid. The alcohol is propyl alcohol.

$$\underbrace{H-\overset{\overset{\Large O}{\|}}{C}}_{\text{formic acid}}-\underbrace{OCH_2CH_2CH_3}_{\text{propyl alcohol}}$$

Change the *-ic* ending of the acid to *-ate*, making the name formate or methanoate. The name of the ester, then, is propyl formate or propyl methanoate.

(b) The acid is benzoic acid; the alcohol is ethyl alcohol. Using the same procedure as in part (a), we obtain the name of the ester: ethyl benzoate.

(c) The acid is the three-carbon dicarboxylic acid, malonic acid. The alcohol is ethyl alcohol. Both acid groups are in the ester form. The name, therefore, is diethyl malonate.

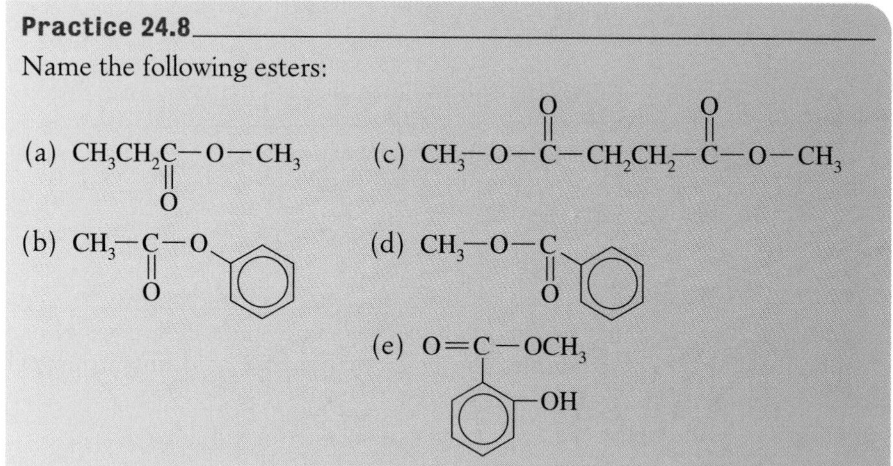

24.8 Occurrence and Physical Properties of Esters

Since many acids and many alcohols are known, the number of esters that are theoretically possible is very large. In fact, both natural and synthesized esters exist in almost endless variety. Simple esters derived from monocarboxylic acids and monohydroxy alcohols are colorless, generally nonpolar liquids or solids. The low polarity of ester molecules is substantiated by the fact that both their water solubility and boiling points are lower than those of either acids or alcohols of similar molar masses.

Low- and intermediate-molar-mass esters (from both acids and alcohols up to about 10 carbons) are liquids with characteristic (usually fragrant or fruity) odors. The distinctive odor and flavor of many fruits are caused by one or more of these esters. The difference in properties between an acid and its esters is remarkable. For example, in contrast to the extremely unpleasant odor of butyric acid, ethyl butyrate has the pleasant odor of pineapple and methyl butyrate the odor of rum. Esters are used in flavoring and scenting agents. (See Table 24.3.) They are generally good solvents for organic substances, and those with relatively low molar masses are volatile. Therefore, esters such as ethyl acetate, butyl acetate, and isoamyl acetate are used extensively in paints, varnishes, and lacquers.

High-molar-mass esters (formed from acids and alcohols of 16 or more carbons) are waxes and are obtained from various plants. They are used in furniture wax and automobile wax preparations; for example, carnauba wax contains esters of 24- and 28-carbon fatty acids and 32- and 34-carbon alcohols. Polyesters with very high molar masses, such as Dacron, are widely used in the textile industries.

24.9 Polyesters: Condensation Polymers

Condensation polymers are formed by substitution reactions between neighboring monomers. The polyesters are joined by ester linkages between carboxylic acid and alcohol groups; the macromolecule formed may be linear or cross-linked. A linear polyester is obtained from the bifunctional monomers terephthalic acid and ethylene glycol. Esterification occurs between the alcohol and acid groups on both ends of both monomers, forming long-chain macromolecules:

Polyesters give fabrics durability and wrinkle resistance.

$$\text{diacid} \quad + \quad \text{diol} \longrightarrow \quad \text{polyester}$$

$$\mathrm{HOOC(CH_2)_xCOOH + HO(CH_2)_yOH \longrightarrow \left[\overset{O}{\overset{\|}{C}}(CH_2)_x\overset{O}{\overset{\|}{C}}-O(CH_2)_yO \right]_n}$$

$$\mathrm{HOOC-C_6H_4-COOH} \quad \mathrm{HOCH_2CH_2OH} \quad \text{(monomers)} \longrightarrow$$

terephthalic acid ethylene glycol

$$\mathrm{-OCH_2CH_2O\left[-\overset{O}{\overset{\|}{C}}-C_6H_4-\overset{O}{\overset{\|}{C}}-OCH_2CH_2O- \right]_n \overset{O}{\overset{\|}{C}}-C_6H_4-\overset{O}{\overset{\|}{C}}-} \quad \text{(polymer)}$$

This polymer may be drawn into fibers or formed into transparent films of great strength. Dacron and Terylene synthetic textiles and Mylar films are made from this polyester, as is the familiar 2-L soft-drink bottle. The symbol on objects made of this material is shown in the margin. In actual practice, the dimethyl ester of terephthalic acid is used as the monomer, and the molecule split out is methyl alcohol instead of water.

When trifunctional acids or alcohols are used as monomers, cross-linked thermosetting polyesters are obtained. (See Chapter 21.) One common example is the reaction of glycerol and *o*-phthalic acid. The polymer formed is one of a group of polymers known as alkyd resins. Glycerol has three functional —OH groups, and phthalic acid has two functional —COOH groups:

$$\mathrm{HOCH_2\overset{OH}{\overset{|}{C}}HCH_2OH} \qquad \mathrm{C_6H_4(COOH)_2}$$

glycerol *o*-phthalic acid

A cross-linked macromolecular structure is formed that, with modifications, has proved to be one of the most outstanding materials used in the coatings industry.

24.10 Chemical Properties of Esters

The most important reaction of esters is *hydrolysis*—the splitting of molecules through the addition of water. The majority of organic and biochemical substances react only very slowly, if at all, with water. In order to increase the rate of these reactions, a catalyst is required. In the laboratory, chemists often employ an acid or base as a catalyst for hydrolysis. In living systems, enzymes act as catalysts.

Acid Hydrolysis

The hydrolysis of an ester involves reaction with water to form a carboxylic acid and an alcohol. The hydrolysis is catalyzed by strong acids (H_2SO_4 and HCl) or by certain enzymes:

$$\underset{\text{ester}}{RC(=O)—OR'} + H_2O \xrightarrow[\text{or enzyme}]{H^+} \underset{\text{acid}}{RC(=O)—OH} + \underset{\text{alcohol}}{R'OH}$$

$$\underset{\text{methyl propanoate}}{CH_3CH_2C(=O)—OCH_3} + H_2O \xrightarrow{H^+} \underset{\text{propanoic acid}}{CH_3CH_2COOH} + \underset{\text{methanol}}{CH_3OH}$$

$$\underset{\text{methyl salicylate}}{C_6H_4(OH)—C(=O)—OCH_3} + H_2O \xrightarrow{H^+} \underset{\text{salicylic acid}}{C_6H_4(OH)COOH} + \underset{\text{methanol}}{CH_3OH}$$

Alkaline Hydrolysis (Saponification)

saponification

Saponification is the hydrolysis of an ester by a strong base (NaOH or KOH) to produce an alcohol and a salt (or soap if the salt formed is from a high-molar-mass acid):

$$\underset{\text{ester}}{RC(=O)—OR'} + NaOH \xrightarrow[\Delta]{H_2O} \underset{\text{salt}}{RC(=O)—O^-Na^+} + \underset{\text{alcohol}}{R'OH}$$

$$\underset{\text{ethyl stearate}}{CH_3(CH_2)_{16}C(=O)—OCH_2CH_3} + NaOH \xrightarrow[\Delta]{H_2O} \underset{\text{sodium stearate}}{CH_3(CH_2)_{16}C(=O)ONa} + CH_3CH_2OH$$

The carboxylic acid may be obtained by reacting the salt with a strong acid:

$$CH_3(CH_2)_{16}COONa + HCl \longrightarrow \underset{\text{stearic acid}}{CH_3(CH_2)_{16}COOH} + NaCl$$

Notice that in saponification, the base is a reactant and not a catalyst.

Practice 24.9

Write equations showing (a) the acid hydrolysis and (b) the alkaline hydrolysis of isobutyl pentanoate.

24.11 Glycerol Esters

fat, oil
triacylglycerol
triglyceride

Fats and **oils** are esters of glycerol and predominantly long-chain fatty acids. Fats and oils are also called **triacylglycerols** or **triglycerides**, since each molecule is derived from one molecule of glycerol and three molecules of fatty acid:

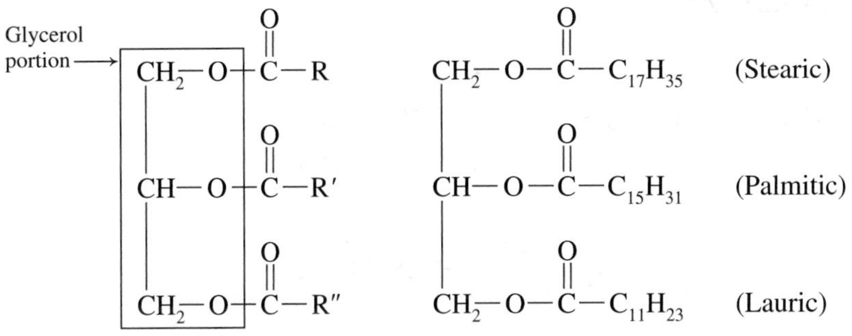

Glycerol portion →

$$\begin{array}{l} CH_2-O-\overset{O}{\overset{\|}{C}}-R \\ | \\ CH-O-\overset{O}{\overset{\|}{C}}-R' \\ | \\ CH_2-O-\overset{O}{\overset{\|}{C}}-R'' \end{array}$$

General formula for a triacylglycerol

$$\begin{array}{ll} CH_2-O-\overset{O}{\overset{\|}{C}}-C_{17}H_{35} & \text{(Stearic)} \\ | & \\ CH-O-\overset{O}{\overset{\|}{C}}-C_{15}H_{31} & \text{(Palmitic)} \\ | & \\ CH_2-O-\overset{O}{\overset{\|}{C}}-C_{11}H_{23} & \text{(Lauric)} \end{array}$$

Typical triacylglycerol containing three different fatty acids

The structural formulas of triacylglycerol molecules differ because

1. The length of the fatty acid chain varies from 4 to 20 carbons, but the number of carbon atoms in the chain is nearly always even.
2. Each fatty acid may be saturated or unsaturated and may contain one, two, or three carbon–carbon double bonds.
3. A triacylglycerol may, and frequently does, contain three different fatty acids.

The most abundant saturated fatty acids in fats and oils are lauric, myristic, palmitic, and stearic acids (Table 24.4). The most abundant unsaturated acids in fats and oils contain 18 carbon atoms and have one, two, or three carbon–carbon double bonds. In all of these naturally occurring unsaturated acids, the configuration about the double bond is cis. Their formulas are

$$CH_3(CH_2)_7CH{=}CH(CH_2)_7COOH$$

oleic acid

$$CH_3(CH_2)_4CH{=}CHCH_2CH{=}CH(CH_2)_7COOH$$

linoleic acid

$$CH_3CH_2CH{=}CHCH_2CH{=}CHCH_2CH{=}CH(CH_2)_7COOH$$

linolenic acid

The major physical difference between fats and oils is that fats are solid and oils are liquid at room temperature. (See Section 28.2.) Since the glycerol part of the structure is the same for a fat and an oil, the difference must be due to

Table 24.4 Fatty Acid Composition of Selected Fats and Oils

	Fatty acid (%)				
Fat or oil	Myristic acid	Palmitic acid	Stearic acid	Oleic acid	Linoleic acid
Animal fat					
Butter*	7–10	23–26	10–13	30–40	4–5
Lard	1–2	28–30	12–18	41–48	6–7
Tallow	3–6	24–32	14–32	35–48	2–4
Vegetable oil					
Olive	0–1	5–15	1–4	49–84	4–12
Peanut	—	6–9	2–6	50–70	13–26
Corn	0–2	7–11	3–4	43–49	34–42
Cottonseed	0–2	19–24	1–2	23–33	40–48
Soybean	0–2	6–10	2–4	21–29	50–59
Linseed**	—	4–7	2–5	9–38	3–43

*Butyric acid, 3–4%
**Linolenic acid, 25–58%

the fatty acid end of the molecule. Fats contain a larger proportion of saturated fatty acids, whereas oils contain greater amounts of unsaturated fatty acids. The term *polyunsaturated* has been popularized in recent years; it means that each molecule of fat in a particular product contains several double bonds.

Fats and oils are obtained from natural sources. In general, fats come from animal sources and oils from vegetable sources. Thus, lard is obtained from hogs, tallow from cattle and sheep. Olive, cottonseed, corn, peanut, soybean, canola, linseed, and other oils are obtained from the fruit or seed of their respective vegetable sources. Table 24.4 shows the major constituents of several fats and oils.

The caloric value per unit mass for triacylglycerols is about 9 kcal/g, while proteins and carbohydrates yield about 4 kcal/g.

Triacylglycerols are the principal form in which energy is stored in the body. The caloric value per unit of mass is over twice as great as that for carbohydrates and proteins. As a source of energy, triacylglycerols can be completely replaced by either carbohydrates or proteins. However, some minimum amount of fat is needed in the diet, because fat supplies the nutritionally essential unsaturated fatty acids (linoleic, linolenic, and arachidonic) required by the body. (See Section 28.2.)

Hydrogenation of Glycerides The addition of hydrogen is a characteristic reaction of the carbon–carbon pi bonds. Industrially, low-cost vegetable oils are partially hydrogenated to obtain solid fats that are useful as shortening in baking or in making margarine. In this process, hydrogen gas is bubbled through hot oil that contains a finely dispersed nickel catalyst. The hydrogen adds to the carbon–carbon double bonds of the oil to saturate the double bonds and form fats:

$$H_2 + \underset{\text{in oil or fat}}{-CH{=}CH-} \xrightarrow{Ni} -CH_2-CH_2-$$

In practice, only some of the double bonds are allowed to become saturated. The degree of hydrogenation can be controlled to obtain a product of any desired degree of saturation. The products resulting from the partial hydrogenation of oils are marketed as solid shortening (Crisco, Spry, etc.) and are used for cooking and baking. Oils and fats are also partially hydrogenated to improve their keeping qualities. Rancidity in fats and oils results from air oxidation at points of unsaturation, producing low-molar-mass aldehydes and acids of disagreeable odor and flavor.

Hydrogenation can also change the double-bond geometric isomers. (See Chapter 20.) Naturally occurring unsaturated oils commonly contain cis double bonds. During hydrogenation, some of these double bonds convert to the trans isomer. Humans have difficulty metabolizing these "unnatural" fats and oils. Recent evidence indicates that "trans fats" increase the risk of atherosclerosis (heart disease).

Hydrogenolysis Triacylglycerols can be split and reduced in a reaction called **hydrogenolysis** (splitting by hydrogen). Hydrogenolysis requires higher temperatures and pressures and a different catalyst (copper chromite) than does the hydrogenation of double bonds. Each triacylglycerol molecule yields a molecule of glycerol and three primary alcohol molecules. The hydrogenolysis of glyceryl trilaurate is represented as follows:

hydrogenolysis

$$\begin{array}{l} CH_2-O-\overset{O}{\overset{\|}{C}}-C_{11}H_{23} \\ | \\ CH-O-\overset{O}{\overset{\|}{C}}-C_{11}H_{23} \\ | \\ CH_2-O-\overset{O}{\overset{\|}{C}}-C_{11}H_{23} \end{array} + 6\,H_2 \xrightarrow[\Delta,\ \text{pressure}]{\text{copper chromite}} 3\,CH_3(CH_2)_{10}CH_2OH + \begin{array}{l} CH_2OH \\ | \\ CHOH \\ | \\ CH_2OH \end{array}$$

glyceryl trilaurate; lauryl alcohol (1-dodecanol); glycerol

Long-chain primary alcohols obtained by this reaction are important, since they are used to manufacture other products, especially synthetic detergents. (See Section 24.12.)

Hydrolysis Triacylglycerols can be hydrolyzed, yielding fatty acids and glycerol. The hydrolysis is catalyzed by digestive enzymes at room temperatures and by mineral acids at high temperatures:

$$\begin{array}{l} CH_2-O-\overset{O}{\overset{\|}{C}}-R \\ | \\ CH-O-\overset{O}{\overset{\|}{C}}-R' \\ | \\ CH_2-O-\overset{O}{\overset{\|}{C}}-R'' \end{array} + 3\,H_2O \xrightarrow[\text{enzymes}]{H^+\text{ or}} \begin{array}{l} RCOOH \\ R'COOH \\ R''COOH \end{array} + \begin{array}{l} CH_2OH \\ | \\ CHOH \\ | \\ CH_2OH \end{array}$$

triacylglycerol; fatty acids (3 molecules); glycerol

The enzyme-catalyzed reaction occurs in digestive reactions and in biological degradation (or metabolic) processes. The acid-catalyzed reaction is employed in the commercial preparation of fatty acids and glycerol.

Saponification The saponification of a fat or oil involves the alkaline hydrolysis of a triester. The products formed are glycerol and the alkali metal salts of fatty acids, which are called soaps. As a specific example, glyceryl tripalmitate reacts with sodium hydroxide to produce sodium palmitate and glycerol:

$$\begin{array}{l} CH_2-O-\overset{\overset{\displaystyle O}{\|}}{C}-C_{15}H_{31} \\ | \\ CH-O-\overset{\overset{\displaystyle O}{\|}}{C}-C_{15}H_{31} \\ | \\ CH_2-O-\overset{\overset{\displaystyle O}{\|}}{C}-C_{15}H_{31} \end{array} + 3\ NaOH \xrightarrow{\Delta} \underset{\substack{\text{sodium palmitate} \\ \text{(a soap)}}}{3\ C_{15}H_{31}COONa} + \underset{\text{glycerol}}{\begin{array}{l} CH_2OH \\ | \\ CHOH \\ | \\ CH_2OH \end{array}}$$

glyceryl tripalmitate

Glycerol, fatty acids, and soaps are valuable articles of commerce, and the processing of fats and oils to obtain these products is a major industry.

Practice 24.10

(a) Write the structure for a triacylglycerol that has one unit each of myristic, palmitic, and oleic acids.
(b) How many isomers of this compound can exist?

24.12 Soaps and Synthetic Detergents

In the broadest sense possible, a detergent is simply a cleansing agent. Soap has been used as a cleansing agent for at least 2000 years and is thereby classified as a detergent under this definition. However, beginning about 1930, a number of new cleansing agents that were superior in many respects to ordinary soap began to appear on the market. Because they were both synthetic organic products and detergents, they were called **synthetic detergents**, or syndets. A soap is distinguished from a synthetic detergent on the basis of chemical composition and not on the basis of function or usage.

synthetic detergent

Soaps

In former times, soapmaking was a crude operation. Surplus fats were boiled with wood ashes or with some other alkaline material. Today, soap is made in large manufacturing plants under controlled conditions. Salts of long-chain fatty acids are called **soaps**. However, only the sodium and potassium salts of carboxylic acids that contain 12–18 carbon atoms are of great value as soaps, because of their abundance in fats. The reaction proceeds as follows:

soap

$$\text{fat or oil} + NaOH \longrightarrow \text{soap} + \text{glycerol}$$

To understand how a soap works as a cleansing agent, let's consider sodium palmitate, $CH_3(CH_2)_{14}COONa$, as an example of a typical soap. In water, this substance exists as sodium ions, Na^+, and palmitate ions, $CH_3(CH_2)_{14}COO^-$. The sodium ion is an ordinary hydrated metal ion. The cleansing property, then, must be centered in the palmitate ion. The palmitate ion contains both a **hydrophilic** (water-loving) and a **hydrophobic** (water-fearing) group. The hydrophilic end is the polar, negatively charged carboxylate group. The hydrophobic end is the long hydrocarbon group. The hydrocarbon group is soluble in oils and greases, but not in water. The hydrophilic carboxylate group is soluble in water.

hydrophilic
hydrophobic

The cleansing action of a soap is explained in this fashion: When the soap comes in contact with grease on a soiled surface, the hydrocarbon end of the soap dissolves in the grease, leaving the negatively charged carboxylate end exposed on the grease surface. Because the negatively charged carboxylate groups are strongly attracted by water, small droplets are formed, and the grease is literally lifted or floated away from the soiled object. (See Figure 24.1.)

The cleansing property of a soap is due to its ability to act as an emulsifying agent between water and water-insoluble greases and oils. The grease–soap emulsion is stable, because the oil droplets repel each other due to the negatively charged carboxyl groups on their surfaces. Some insoluble particulate matter is carried away with the grease; the remainder is wetted and mechanically washed away in the water. Synthetic detergents function in a similar way.

Ordinary soap is a good cleansing agent in soft water, but it is not satisfactory in hard water because insoluble calcium, magnesium, and iron(III) salts are formed. Palmitate ions, for example, are precipitated by calcium ions:

$$\underset{\text{}}{Ca^{2+}(aq)} + 2\,\underset{\text{palmitate ion}}{CH_3(CH_2)_{14}COO^-(aq)} \longrightarrow \underset{\text{calcium palmitate}}{[CH_3(CH_2)_{14}COO]_2Ca(s)}$$

These precipitates are sticky substances and are responsible for "bathtub ring" and the sticky feel of hair after it is shampooed with soap in hard water.

The cleansing action of soap is of little interest to this baby, who is enjoying the water!

Soaps are ineffective in acidic solutions, because water-insoluble molecular fatty acids are formed:

$$CH_3(CH_2)_{14}COO^- + H^+ \longrightarrow \underset{\text{palmitic acid molecule}}{CH_3(CH_2)_{14}COOH}$$

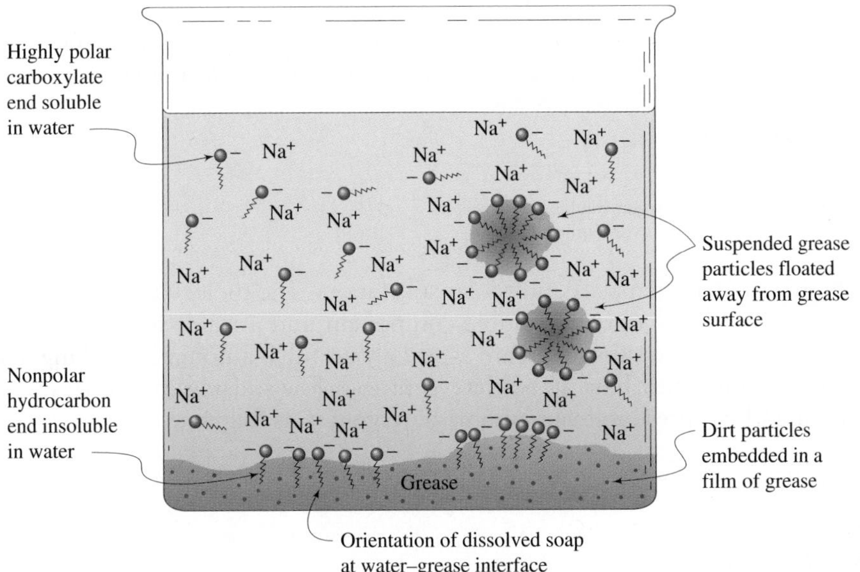

Figure 24.1
Cleansing action of soap: Dirt particles are embedded in a surface film of grease. The hydrocarbon ends of negative soap ions dissolve in the grease film, leaving exposed carboxylate groups. These carboxylate groups are attracted to water, and small droplets of grease-bearing dirt are formed and floated away from the surface.

Synthetic Detergents

Once it was recognized that the insoluble hydrocarbon radical joined to a highly polar group was the key to the detergent action of soaps, chemists set out to make new substances that would have similar properties. About 1930, synthetic detergents (syndets) began to replace soaps, and now about 4 lb of syndets are sold for each pound of soap.

Although hundreds of substances with detergent properties are known, an idea of their general nature can be obtained from a consideration of sodium lauryl sulfate and sodium *p*-dodecylbenzene sulfonate:

$$CH_3(CH_2)_{10}CH_2OSO_3^-Na^+$$

sodium lauryl sulfate

$$CH_3(CH_2)_{10}CH_2\text{—}C_6H_4\text{—}SO_3^-Na^+$$

sodium *p*-dodecylbenzene sulfonate

Sodium lauryl sulfate and sodium *p*-dodecylbenzene sulfonate act in water in much the same way as sodium palmitate does. Like the palmitate ion, the negative lauryl sulfate ion has a long hydrocarbon chain that is soluble in grease and a sulfate group that is attracted to water:

$$\underbrace{CH_3CH_2CH_2CH_2CH_2CH_2CH_2CH_2CH_2CH_2CH_2CH_2}_{\text{nonpolar hydrophobic end, grease soluble}}\text{—}\underbrace{OSO_3^-}_{\text{polar hydrophilic end, water soluble}}$$

The one great advantage these synthetic detergents have over soap is that their calcium, magnesium, and iron(III) salts, as well as their sodium salts, are soluble in water. Therefore, they are nearly as effective in hard water as in soft water.

The foregoing are anionic detergents, because they contain long-chain negatively charged ions. Other detergents, both cationic and nonionic, have been developed for special purposes. A cationic detergent has a long hydrocarbon chain and a positive charge. A representative structure is as follows:

$$\underbrace{CH_3(CH_2)_{14}CH_2}_{\text{grease soluble, hydrophobic}}\text{—}\underbrace{\overset{+}{N}(CH_3)_3}_{\text{water soluble, hydrophilic}}$$

Nonionic detergents are molecular substances. The molecule of a nonionic detergent contains a grease-soluble component and a water-soluble component. Some of these substances are especially useful in automatic washing machines because they have good detergent, but low sudsing, properties. The structure of a representative nonionic detergent is as follows:

$$\underbrace{CH_3(CH_2)_{10}CH_2}_{\text{grease soluble, hydrophobic}}\text{—}O\text{—}\underbrace{(CH_2CH_2O)_7\text{—}CH_2CH_2OH}_{\text{water soluble, hydrophilic}}$$

Biodegradability

Organic substances that are readily decomposed by microorganisms in the environment are said to be **biodegradable**. All naturally occurring organic substances are eventually converted to simple inorganic molecules and ions such as CO_2, H_2O, N_2, Cl^-, and SO_4^{2-}. Most of these conversions are catalyzed by enzymes produced by microorganisms. These enzymes attack only certain specific molecular configurations that are found in substances occurring in nature.

biodegradable

A number of years ago a serious environmental pollution problem arose in connection with synthetic detergents. Some of the early syndets, which contained highly branched chain hydrocarbons, had no counterparts in nature. Therefore, enzymes capable of degrading them did not exist, and the detergents were essentially nonbiodegradable and broke down very, very slowly. As a result, these syndets accumulated in water supplies, where they caused severe pollution problems due to excessive foaming and other undesirable effects.

Detergent manufacturers, acting on the recommendations of chemists and biologists, changed from a branched-chain alkyl benzene to a straight-chain alkyl benzene raw material. Detergents that contain the straight-chain alkyl groups are biodegradable. The structural difference is as follows:

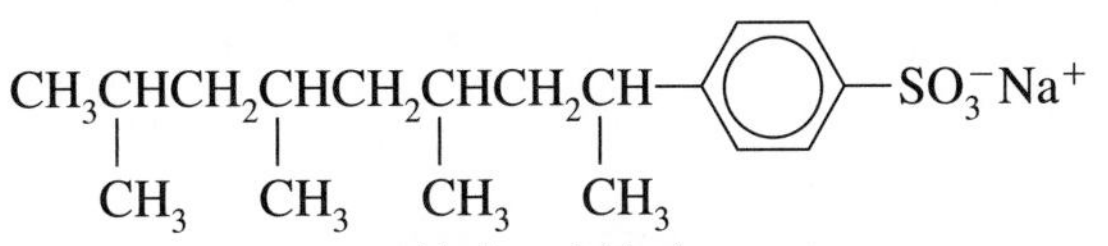

$$CH_3\underset{\displaystyle CH_3}{\underset{|}{C}}HCH_2\underset{\displaystyle CH_3}{\underset{|}{C}}HCH_2\underset{\displaystyle CH_3}{\underset{|}{C}}HCH_2\underset{\displaystyle CH_3}{\underset{|}{C}}H-C_6H_4-SO_3^-Na^+$$

a nonbiodegradable detergent

$$CH_3CH_2CH_2CH_2CH_2CH_2CH_2CH_2CH_2CH_2CH_2CH_2-C_6H_4-SO_3^-Na^+$$

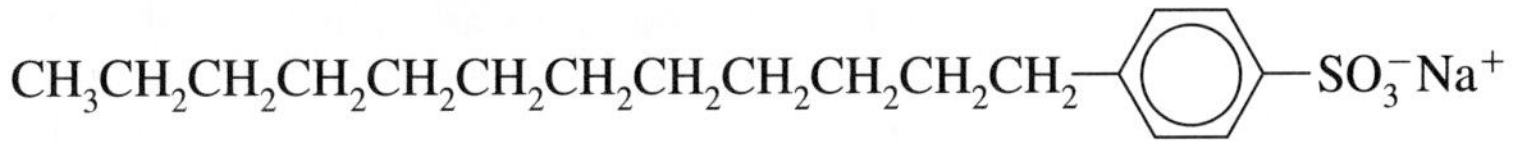

a biodegradable detergent

24.13 Esters and Anhydrides of Phosphoric Acid

Phosphoric acid has a Lewis structure similar to that of a carboxylic acid:

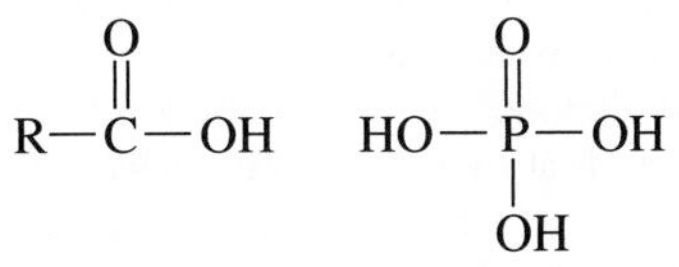

$$R-\overset{\displaystyle O}{\overset{\|}{C}}-OH \qquad HO-\underset{\displaystyle OH}{\underset{|}{\overset{\displaystyle O}{\overset{\|}{P}}}}-OH$$

carboxylic acid phosphoric acid

In both molecules, an —OH is attached to an element that is double-bonded to an oxygen. In fact, phosphoric acid has three such —OH groups. This similarity in structure permits phosphoric acid to behave as a carboxylic acid in reaction with an alcohol:

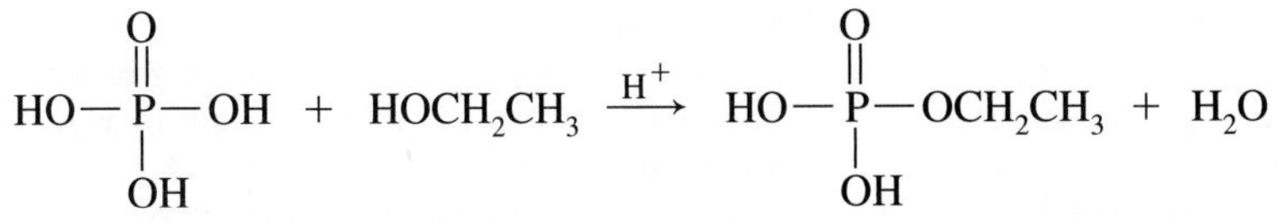

$$HO-\underset{\displaystyle OH}{\underset{|}{\overset{\displaystyle O}{\overset{\|}{P}}}}-OH + HOCH_2CH_3 \xrightarrow{H^+} HO-\underset{\displaystyle OH}{\underset{|}{\overset{\displaystyle O}{\overset{\|}{P}}}}-OCH_2CH_3 + H_2O$$

phosphoric acid ethanol monoethyl phosphate

phosphate ester The product of the esterification reaction is called a **phosphate ester**. This phosphate ester still has two —OH groups that can form additional esters. The result of one further esterification reaction is a molecule with two phosphate ester linkages. The diester then can contain two different alcohol groups in the same molecule. This structure is common in biochemistry—specifically in nucleic acids (Chapter 31) and phospholipids (Chapter 28).

Pyrophosphoric acid is an acid anhydride formed by splitting out a molecule of water between two molecules of phosphoric acid:

$$\underset{\text{phosphoric acid}}{\text{HO}-\overset{\overset{\displaystyle \text{O}}{\|}}{\underset{\underset{\displaystyle \text{OH}}{|}}{\text{P}}}-\text{OH}} + \underset{\text{phosphoric acid}}{\text{HO}-\overset{\overset{\displaystyle \text{O}}{\|}}{\underset{\underset{\displaystyle \text{OH}}{|}}{\text{P}}}-\text{OH}} \longrightarrow \underset{\text{pyrophosphoric acid}}{\text{HO}-\overset{\overset{\displaystyle \text{O}}{\|}}{\underset{\underset{\displaystyle \text{OH}}{|}}{\text{P}}}-\text{O}-\overset{\overset{\displaystyle \text{O}}{\|}}{\underset{\underset{\displaystyle \text{OH}}{|}}{\text{P}}}-\text{OH}} + \text{H}_2\text{O}$$

In general, both phosphate esters and phosphoric acid anhydrides have significant biological importance. The phosphate esters "tag" many biochemicals, labeling them for specific biological purposes. The phosphoric acid anhydrides serve as a temporary store of metabolic energy. (See Chapter 33.)

Chapter 24 Review

24.1 Carboxylic Acids

KEY TERM

Carboxyl group

- Carboxylic acids contain the carboxyl group, which can be represented as follows:

$$-\overset{\overset{\displaystyle \text{O}}{\|}}{\text{C}}-\text{OH} \quad \text{or} \quad -\text{COOH} \quad \text{or} \quad -\text{CO}_2\text{H}$$

24.2 Nomenclature and Sources of Aliphatic Carboxylic Acids

- The IUPAC naming of carboxylic acids follows a similar process to that used for other organic molecules.
  - The longest continuous carbon chain (parent chain) *that contains the carboxyl* is given the name of the corresponding alkane modified by dropping the last *-e* and adding *-oic acid*.
  - The parent chain is numbered from the carboxyl end; that is, the carboxyl group is understood to be number 1.
  - Other groups attached to the parent chain are named and numbered in the usual IUPAC fashion.
- Many carboxylic acids are known by common names.
  - Methanoic, ethanoic and propanoic acids are more commonly known as formic, acetic, and propionic acids, respectively.
  - Each common name generally refers to a natural source for the carboxylic acid.
  - Substituent positions are sometimes designated with small Greek letters rather than the numbers; each parent chain carbon is given a Greek letter starting with the carbon *adjacent to* the carboxyl group (given the letter alpha, α).

24.3 Physical Properties of Carboxylic Acids

- The polar carboxyl group causes the small carboxylic acids to be water soluble (carboxylic acids larger than about five carbons are essentially water insoluble).
- Carboxyl groups can also hydrogen-bond to each other at two places.
  - This strong noncovalent bonding causes high boiling points for carboxylic acids.
  - Smaller saturated monocarboxylic acids (less than 10 carbons) are liquids at room temperature, while larger carboxylic acids (greater than 10 carbons) are solids.
- Carboxylic acids are weak acids; the carboxylic acid is in equilibrium with its conjugate base.

24.4 Classification of Carboxylic Acids

- Some important carboxylic acids differ from the saturated monocarboxylic acids in the following ways:
  - Unsaturated carboxylic acids also contain at least one multiple carbon–carbon bond.
  - Common aromatic carboxylic acids contain a benzene ring.
  - Dicarboxylic acids have two carboxyl groups.
  - Hydroxy acids contain a hydroxy group as well as a carboxyl group.
  - Amino acids combine an amine ($-NH_2$) on the same molecule with the carboxyl group.

24.5 Preparation of Carboxylic Acids

- Strong oxidizing agents such as potassium dichromate in sulfuric acid will convert aldehydes and primary alcohols to carboxylic acids.
- Alkyl chains on benzene rings can be oxidized by strong oxidizing agents to form benzoic acid.
- Nitriles, RCN, will react with water to yield carboxylic acids.

24.6 Chemical Properties of Carboxylic Acids

KEY TERM

Ester

- Because the carboxyl group is a weak acid, carboxylic acids have a sour taste, affect pH indicators, form water solutions with a pH of less than 7, and undergo neutralization reactions.
- Carboxylic acids react with bases to form salts.
  - Sodium and potassium salts of carboxylic acids are water soluble.
  - This property is used to distinguish carboxylic acids from other organic compounds.
- Many common carboxylic acid reactions involve substitution of the —OH group for another atom or group of atoms.
  - Thionyl chloride ($SOCl_2$) reacts with carboxylic acids to form acid chlorides—a chlorine atom substitutes for the —OH.
  - When one carboxylic acid substitutes for the —OH on another carboxylic acid, an acid anhydride is formed.
    - *Anhydride* refers to the loss of water when the two carboxylic acids combine.
  - An ester is formed when an alcohol or phenol substitutes for the carboxylic acid —OH, an *esterification*.
    - Unlike a common acid–base neutralization reaction, esterification forms a covalent compound (the ester), not a salt.
    - *Hydrolysis* reverses *esterification*: Water in an acid medium reacts to break the ester into its component parts, the carboxylic acid and the alcohol.

24.7 Nomenclature of Esters

- Esters are named as alcohol derivatives of carboxylic acids.
  - The alcohol part is named first by naming it as an alkyl group.
  - The carboxylic acid is named second by dropping the *-ic acid* and adding the suffix *-ate*.

24.8 Occurrence and Physical Properties of Esters

- Low- to intermediate-molar-mass esters are liquids with characteristically fruity or fragrant aromas.
- Esters such as ethyl acetate, butyl acetate, and isoamyl acetate are good solvents and commonly used in paints, varnishes, and lacquers.

24.9 Polyesters: Condensation Polymers

- Condensation polymers are produced when monomers combine with the loss of a small molecule. Polyesters are formed when alcohol monomers react with carboxylic acid monomers and water is lost.
- Polyesters have great utility; they are used, for example, in fabrics (Dacron, Terylene), plastic films (Mylar), and the familiar 2-L soft drink bottles.

24.10 Chemical Properties of Esters

KEY TERM

Saponification

- The most important reaction of esters is *hydrolysis*—water is used to split esters into alcohols and carboxylic acids.
  - Adding a strong acid (e.g., HCl, H_2SO_4) will speed up hydrolysis—acid hydrolysis.
  - When a strong base is used to speed up hydrolysis, the process is called alkaline hydrolysis or saponification.
    - Alkaline hydrolysis yields a carboxylic acid salt.
    - This reaction is called saponification because a soap is formed when the carboxylic acid salt has a long alkyl chain.

24.11 Glycerol Esters

KEY TERMS

Fat, oil
Triacylglycerol
Triglyceride
Hydrogenolysis

- Fats and oils are esters of glycerol and predominantly long-chain carboxylic acids (fatty acids).
- A triacylglycerol or triglyceride is a glycerol that has been esterified with three fatty acids, generally three *different* fatty acids.
  - Fatty acids generally have long carbon chains (4 to about 20 carbon atoms) and almost always contain an even number of carbons.
  - Fatty acids may be either saturated or unsaturated, with from one to four double bonds.
  - Fatty acid double bonds are generally cis isomers.
- Fats are solids at room temperature while oils are liquids.
- Triacylglycerols are the principal energy-storage molecules in the human body.
- Hydrogenation of unsaturated fatty acids adds hydrogen to the carbon–carbon double bonds.
  - Unsaturated fatty acids yield saturated fatty acids.
  - Low-cost vegetable oils form fats.
  - Partial hydrogenation selectively changes the physical properties of triacylglycerols.
- Hydrogenolysis breaks a triacylglycerol into glycerol and three long-chain primary alcohols using hydrogen gas.
- Triacylglycerols can be hydrolyzed in acid to form fatty acids (and glycerol) or base to form soaps (and glycerol).

24.12 Soaps and Synthetic Detergents

KEY TERMS

Synthetic detergent
Soap
Hydrophilic
Hydrophobic
Biodegradable

- A synthetic detergent is an organic molecule that has been synthesized in a chemical factory and acts as a cleansing agent.
- Soaps are salts of long-chain fatty acids.
  - The alkali metal salts of fatty acids are the most effective soaps.
  - The fatty acid anion combines properties described as hydrophilic (water-loving) and hydrophobic (water-fearing).
  - A soap acts as a wetting agent for grease.
    - The hydrophobic alkyl chains of the soap dissolve in the grease.
    - The hydrophilic anions of the soap coat the surface of a grease particle, allowing it to be washed away in water.
- Synthetic detergents also combine hydrophilic and hydrophobic properties.
  - Long alkyl chains contribute the hydrophobic property.
  - Anionic detergents gain a hydrophilic character from acid anions such as sulfates or sulfonates.
  - Cationic detergents use positively charged nitrogens for hydrophilic character.
  - Nonionic detergents have hydrophilic properties because of functional groups like ethers or alcohols.
- Most detergents produced today are biodegradable.
- Biodegradable refers to organic substances that are readily decomposed by microorganisms in the environment.

24.13 Esters and Anhydrides of Phosphoric Acid

KEY TERM

Phosphate ester

- Like a carboxylic acid, phosphoric acid can form esters and anhydrides.
  - A phosphate ester combines a phosphoric acid with an alcohol.
- These derivatives of phosphoric acid are very important in biochemistry.
  - Phosphate esters label biochemicals for specific metabolic purposes.
  - Phosphoric acid anhydrides serve as a temporary store for metabolic energy.

Review Questions

All questions with blue numbers have answers in the appendix of the text.

1. Using a specific compound in each case, write structural formulas for the following:
   (a) an aliphatic carboxylic acid
   (b) an aromatic carboxylic acid
   (c) an α-hydroxy acid
   (d) an α-amino acid
   (e) a β-chloro acid
   (f) a dicarboxylic acid
   (g) an unsaturated carboxylic acid
   (h) an ester
   (i) a nitrile
   (j) a sodium salt of a carboxylic acid
   (k) an acid halide
   (l) a triacylglycerol
   (m) a soap
2. Which of the following would have the more objectionable odor? Briefly explain.
   (a) a 1% solution of butyric acid (C_3H_7COOH) or
   (b) a 1% solution of sodium butyrate (C_3H_7COONa)
3. Which has the greater solubility in water?
   (a) methyl propanoate or propanoic acid
   (b) sodium palmitate or palmitic acid
   (c) sodium stearate or barium stearate
   (d) phenol or sodium phenoxide
4. Explain why carboxylic acids have a much higher boiling point than alkanes.
5. Explain the difference between the following:
   (a) a fat and an oil
   (b) a soap and a syndet
   (c) hydrolysis and saponification
6. What taste is associated with acids? Name several foods that taste like acids.
7. Explain the cleansing action of detergents.
8. Cite the principal advantages that synthetic detergents (syndets) have over soaps.
9. With what family of drugs is aspirin associated?
10. Why is a polyester defined as a condensation polymer?
11. What is an important use for phosphoric acid anhydrides in metabolism?
12. Name the carboxylic acid (both the common name and the IUPAC name) that is found in vinegar.
13. List three important and common uses for vinegar.
14. Referring to Table 24.1, note that formic acid is infinitely water soluble while capric acid is almost completely insoluble. Based on a comparison of structures, explain the large difference in water solubility.

Paired Exercises

All exercises with blue numbers have answers in the appendix of the text.

1. Name the following compounds:

(a) $CH_3(CH_2)_4COOH$ (IUPAC name)

(b) $HOOCCH{=}CHCOOH$ (common name for the trans isomer)

(c) benzene ring with COOH and COOH on adjacent carbons (common name)

(d) $CH_3(CH_2)_{16}COOH$ (common name)

(e) $CH_3CH_2COO^-Na^+$ (IUPAC name)

2. Name the following compounds:

(a) benzene ring with COOH and Cl on adjacent carbons (IUPAC name)

(b) $CH_3(CH_2)_7CH{=}CH(CH_2)_7COOH$ (common name)

(c) benzene ring with COOH and CH_3 in meta positions (common name)

(d) $CH_3CH_2CH(OH)COOH$ (IUPAC name)

(e) $CH_3CH_2COO^-NH_4^+$ (IUPAC name)

3. Write structures for the following compounds:
(a) hexanoic acid
(b) malonic acid
(c) sodium benzoate
(d) *o*-toluic acid
(e) stearic acid
(f) 2-chloropropanoic acid
(g) potassium butyrate

4. Write structures for the following compounds:
(a) oxalic acid
(b) pentanoic acid
(c) *o*-phthalic acid
(d) linolenic acid
(e) sodium *p*-aminobenzoate
(f) ammonium propanoate
(g) β-hydroxybutyric acid

5. Assume that you have 0.01 *M* solutions of the following substances:

(a) NH_3 (b) HCl (c) NaCl (d) NaOH (e) CH_3COOH (f) benzene ring–OH

Arrange them in order of increasing pH (list the most acidic solution first).

6. Assume that you have 0.04 *M* solutions of the following substances:

(a) NH_3 (b) KOH (c) HBr (d) HCOOH (e) KBr (f) benzene ring–OH

Arrange them in order of decreasing pH (list the most basic solution first).

7. Choose the compound with the higher boiling point between each of the following pairs:
(a) succinic acid vs. butanoic acid
(b) ethyl acetate vs. butanoic acid
(c) 1-propanol vs. acetic acid
(d) butanoic acid vs. 1-pentanol

8. Choose the compound with the higher boiling point between each of the following pairs:
(a) propanone vs. ethanoic acid
(b) propanoic acid vs. 2-butanol
(c) methylmethanoate vs. ethanoic acid
(d) propanoic acid vs. malonic acid

9. Give IUPAC and common names for the following compounds:

(a) $CH_2{=}CHC(=O){-}OCH_3$

(b) $CH_3CH_2CH_2C(=O){-}OCH_2CH_3$

(c) benzene ring bearing $C(=O){-}O{-}C_6H_5$ and, on the adjacent carbon, OH

10. Give IUPAC and common names for the following compounds:

(a) $HC(=O){-}OCH_3$

(b) benzene ring–$C(=O){-}OCH_2CH_2CH_3$

(c) $CH_3CH_2C(=O){-}OCH_2CH_3$

11. Write the structural formulas for the following:
(a) methyl formate (c) ethyl benzoate
(b) butyl butanoate

12. Write the structural formulas for the following:
(a) propyl acetate (c) ethyl hexanoate
(b) methyl benzoate

13. Write the structural formula and name of the principal organic product for the following reactions:

(a) $CH_3(CH_2)_7CH{=}CH(CH_2)_7COOH + H_2 \xrightarrow{Ni}$

(b) $C_6H_5\overset{O}{\overset{\|}{C}}{-}H \xrightarrow[H_2SO_4]{Na_2Cr_2O_7}$

(c) $CH_3CH_2COOH + NaOH \longrightarrow$

(d) $CH_3(CH_2)_3CH_2\overset{O}{\overset{\|}{C}}{-}OCH_2CH_2CH_3 + NaOH \xrightarrow{\Delta}$

(e) $C_6H_5CH_2CH_2CH_3 \xrightarrow[\Delta]{NaMnO_4/NaOH}$

14. Write the structural formula and name of the principal organic product for the following reactions:

(a) $C_6H_4(CH_2CH_3)(CH_3)$ (meta) $\xrightarrow[\Delta]{NaMnO_4/NaOH}$

(b) $HOOCCH{=}CHCOOH + H_2 \xrightarrow{Ni}$

(c) $CH_3(CH_2)_7CH{=}CH(CH_2)_7COOH + NaOH \longrightarrow$

(d) $CH_3CH_2CH_2\overset{O}{\overset{\|}{C}}{-}H \xrightarrow[H_2SO_4]{Na_2Cr_2O_7}$

(e) $C_6H_5\overset{O}{\overset{\|}{C}}{-}O{-}C_6H_5 + NaOH \xrightarrow{\Delta}$

15. In each of the following, are the two compounds the same or different? If the two compounds are different, name each.

(a) $CH_3CH_2\overset{O}{\overset{\|}{C}}OCH_3$ vs. $CH_3\overset{O}{\overset{\|}{C}}OCH_2CH_3$

(b) $H\overset{O}{\overset{\|}{C}}OCH_3$ vs. $CH_3O\overset{O}{\overset{\|}{C}}H$

(c) $CH_3\underset{CH_3}{\underset{|}{C}}H\overset{O}{\overset{\|}{C}}OCH_3$ vs. $CH_3\overset{CH_3}{\overset{|}{C}}HO\overset{O}{\overset{\|}{C}}CH_3$

16. In each of the following, are the two compounds the same or different? If the two compounds are different, name each.

(a) $C_6H_5{-}\overset{O}{\overset{\|}{C}}OCH_3$ vs. $CH_3\overset{O}{\overset{\|}{C}}O{-}C_6H_5$

(b) $CH_3\overset{O}{\overset{\|}{C}}OCH_3$ vs. $CH_3O\overset{O}{\overset{\|}{C}}CH_2CH_3$

(c) $CH_3\overset{O}{\overset{\|}{C}}OCH_2CH_2CH_3$ vs. $CH_3CH_2CH_2O\overset{O}{\overset{\|}{C}}CH_3$

17. The following compounds have incorrect IUPAC names. Write the condensed structural formula and give the correct IUPAC name for each compound.
(a) ethyl formate
(b) 2,2-dichloro-4-butanoic acid
(c) methyl 2-methylpropionate

18. The following compounds have incorrect IUPAC names. Write the condensed structural formula and give the correct IUPAC name for each compound.
(a) 2-methyl-l-pentanoic acid
(b) ethyl acetate
(c) 4-chlorobutyric acid

19. Write the structural formula of the ester that, when hydrolyzed, would yield the following:
(a) methanol and acetic acid
(b) ethanol and formic acid
(c) 2-propanol and benzoic acid

20. Write the structural formula of the ester that, when hydrolyzed, would yield the following:
(a) methanol and propanoic acid
(b) 1-octanol and acetic acid
(c) ethanol and butanoic acid

21. Write structural formulas for the reactants that will yield the following esters:
(a) methyl palmitate
(b) phenyl propionate
(c) dimethyl succinate

22. Write structural formulas for the reactants that will yield the following esters:
(a) isopropyl formate
(b) diethyl adipate
(c) benzyl benzoate

23. Write structural formulas for the organic products of the following reactions:

(a) $C_6H_5C(=O)Cl + H_2O \longrightarrow$

(b) $CH_2{=}CHCOOH + Br_2 \longrightarrow$

(c) $C_6H_5CH_2C{\equiv}N + H_2O \xrightarrow{H^+}$

(d) $CH_3CH_2OH + CH_3\overset{O}{\overset{\|}{C}}{-}Cl \longrightarrow$

(e) $CH_3\overset{O}{\overset{\|}{C}}{-}Cl + NH_3 \longrightarrow$

24. Write structural formulas for the organic products of the following reactions:

(a) $C_6H_5C(=O)Cl + NH_3 \longrightarrow$

(b) $CH_3CH_2COOH + SOCl_2 \longrightarrow$

(c) $CH_3CH_2\overset{O}{\overset{\|}{C}}{-}Cl + CH_3OH \longrightarrow$

(d) $CH_3CH_2CH_2C{\equiv}N + H_2O \xrightarrow{H^+}$

(e) $C_6H_5CH_2CH_2COOH + SOCl_2 \longrightarrow$

25. What simple tests can be used to distinguish between the following pairs of compounds?
(a) benzoic acid and sodium benzoate
(b) maleic acid and malonic acid

26. What simple tests can be used to distinguish between the following pairs of compounds?
(a) benzoic acid and ethyl benzoate
(b) succinic acid and fumaric acid

27. Write structural formulas and names for the organic products of the following reactions:

(a) $CH_3CH(COOH)_2 \xrightarrow{150^\circ C}$

(b) $C_6H_5\overset{O}{\overset{\|}{C}}{-}OCH_2CH_3 + H_2O \xrightarrow[\Delta]{H^+}$

(c) $CH_3CH_2COOH + C_6H_5OH \xrightarrow{H^+}$

28. Write structural formulas and names for the organic products of the following reactions:

(a) $HOOC{-}CH_2{-}CH_2{-}CH_2{-}COOH \xrightarrow{\Delta}$
glutaric acid

(b) $HOOC{-}CH_2{-}COOH + 2\,CH_3CH_2OH \xrightarrow{H^+}$

(c) $C_6H_5O{-}\overset{O}{\overset{\|}{C}}CH_3 + H_2O \xrightarrow[\Delta]{H^+}$

29. Trans isomers of naturally occurring fatty acids can act as metabolic inhibitors. Draw the structural formulas of *cis,cis,cis*-linolenic acid and its all-trans isomer.

30. The geometric configuration of naturally occurring unsaturated 18-carbon acids is all cis. Draw structural formulas for the following:
(a) *cis*-oleic acid (b) *cis,cis*-linoleic acid

31. Predict the products for each of the following hydrolysis reactions:

(a) $HC(=O)OCH_3 + NaOH \xrightarrow[\Delta]{H_2O}$

(b) $CH_3C(=O)OC(=O)CH_3 + H_2O \xrightarrow{H^+}$

(c) $CH_3C(=O)OCH_3 + KOH \xrightarrow[\Delta]{H_2O}$

32. Predict the products for each of the following hydrolysis reactions:

(a) $CH_3C(=O)Cl + H_2O \longrightarrow$

(b) $HC(=O)OCH_2CH_3 + KOH \xrightarrow[\Delta]{H_2O}$

(c) $CH_3CH(CH_3)OC(=O)CH_3 + NaOH \xrightarrow[\Delta]{H_2O}$

33. Would $CH_3(CH_2)_{12}COOH$ or $CH_3(CH_2)_{12}COONa$ be the more useful cleansing agent in soft water? Explain.

34. Would $CH_3(CH_2)_{11}OSO_3Na$ (sodium lauryl sulfate) or $CH_3CH_2CH_2OSO_3Na$ (sodium propyl sulfate) be the more effective detergent in hard water? Explain.

35. Which one of the following substances is a good detergent in water? Is this substance a nonionic, anionic, or cationic detergent?

(a) $C_{16}H_{33}N(CH_3)_3^+Cl^-$
hexadecyltrimethyl ammonium chloride

(b) $C_{16}H_{34}$
hexadecane

(c) $C_{15}H_{31}COOH$
palmitic acid (hexadecanoic acid)

(d) $C_{15}H_{31}C(=O){-}OC_{16}H_{33}$
cetyl palmitate (hexadecyl hexadecanoate)

36. Which one of the following substances is a good detergent in water? Is this substance a nonionic, anionic, or cationic detergent?

(a) $CH_3(CH_2)_{11}O{-}C(=O)(CH_2)_{14}CH_3$

(b) $HO{-}C(=O)(CH_2)_{14}CH_3$

(c) $CH_3(CH_2)_{10}CH_2O(CH_2CH_2O)_7CH_2CH_2OH$

(d) 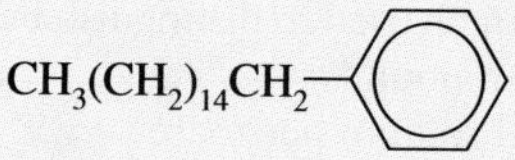 $CH_3(CH_2)_{14}CH_2{-}C_6H_5$

37. Show the products of the reaction of 1 mol of phosphoric acid with the following:
(a) 1 mol of ethanol to make a phosphate monoester
(b) 1 mol of ethanol followed by 1 mol of 1-propanol to make a phosphate diester

38. Show the products of the reaction of 1 mol of phosphoric acid with the following:
(a) 3 mol of methanol to make a phosphate triester
(b) 1 mol of methanol followed by 1 mol of 2-butanol to make a phosphate diester

Additional Exercises

All exercises with blue numbers have answers in the appendix of the text

39. The FDA allows the addition of up to 0.1% by weight of sodium benzoate to baked goods to retard spoilage. How many grams of sodium benzoate can be included in 1 lb of dinner rolls (454 g = 1 lb)? How many moles?

40. Aspirin is synthesized on an industrial scale by combining the following reactants:

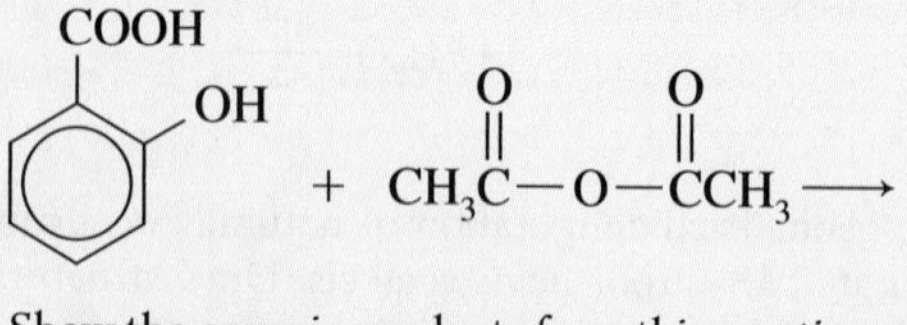

$$C_6H_4(COOH)(OH) + CH_3C(=O){-}O{-}C(=O)CH_3 \longrightarrow$$

Show the organic products from this reaction, and name both reactants and products.

41. Upon hydrolysis, an ester of formula $C_6H_{12}O_2$ yields an acid A and an alcohol B. When B is oxidized, it yields a product identical to A. What is the structure of the ester? Explain your answer.

42. The gas, phosgene, with structure

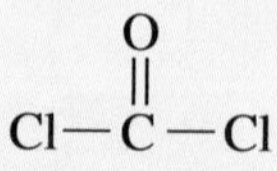

$$Cl{-}C(=O){-}Cl$$

was used in chemical warfare during World War I. Today this chemical is used extensively to form amide linkages during industrial syntheses. Identify the functional groups on this molecule and name the acid from which phosgene is derived.

43. Most plant oils are high in unsaturated fatty acids; the average soybean triacylglycerol has two linoleic acid esters and one oleic acid ester. However, several plant oils are considered less healthy because they contain higher amounts of saturated fatty acids. For example, an average triacylglycerol from palm oil contains two palmitic acid esters and one oleic acid ester. Give structural formulas for both the soybean and the palm triacylglycerols.

44. Write the structural formula of a triacylglycerol that contains one unit each of lauric acid, palmitic acid, and oleic acid. How many other triacylglycerols, each containing all three of these acids, are possible?

45. Choose one of the triacylglycerols from Exercise 44. Write the names and formulas of all products expected when this triacylglycerol is
(a) reacted with water at high temperature and pressure in the presence of mineral acid
(b) reacted with hydrogen at relatively high pressure and temperature in the presence of a copper chromite catalyst
(c) boiled with potassium hydroxide solution
(d) reacted with hydrogen in the presence of Ni

46. Write a balanced chemical equation for the reaction of KOH with glyceryl tristearate. Discuss any changes in solubility that might occur as a result of this reaction.

47. Waxes can be formed from the reaction between a fatty acid and a long-chain alcohol. A major component of beeswax is a combination of a 16-carbon acid and a 30-carbon alcohol. Write the condensed formula for this component of beeswax.

48. Compound A, C_7H_8, reacted with $KMnO_4$ to form compound B, $C_7H_6O_2$. Compound B reacted with methanol to yield a sweet-smelling liquid, compound C. Compound A did not react with Br_2 in CCl_4. Write structural formulas of A, B, and C.

49. A good lipstick is uniform in color and sticks to the lips once it is applied. The stickiness of a lipstick is controlled by the addition of a compound such as isopropyl myristate. Write the structural formula of isopropyl myristate.

50. An ester A of formula $C_{10}H_{12}O_2$ contains a benzene ring. When hydrolyzed, A gives an alcohol B, C_7H_8O, and an acid C, $C_3H_6O_2$. What are the structures of A, B, and C?

51. You are given two unlabeled bottles, one of which contains butanoic acid and the other ethyl butanoate. Describe a simple test that you could do to determine which compound is in which bottle.

52. If 1.00 kg of triolein (glycerol trioleate) is converted to tristearin (glycerol tristearate) by hydrogenation:
(a) How many liters of hydrogen (at STP) are required?
(b) What is the mass of the tristearin that is produced?

53. Consider this statement: "When methyl propanoate is hydrolyzed, formic acid and 1-propanol are formed." If this statement is true, write a balanced chemical reaction for it. If it is false, explain why.

54. Write structural formulas for all the esters that have a molecular formula of $C_5H_{10}O_2$.

55. (a) Calculate the mass percent of oxygen in Dacron.
(b) The number of units in Dacron polyester varies from 80 to 130. Calculate the molar mass of a Dacron that consists of 105 units.

56. (a) Write the structure of a polyester that is made from the monomers *m*-phthalic acid and 1,3-propanediol.
(b) Write the structures for the monomers of

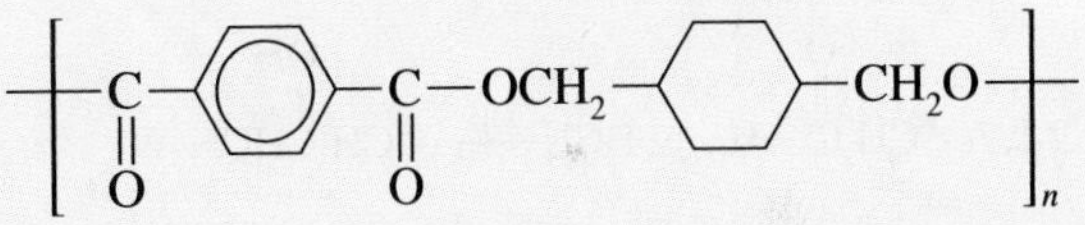

$$\left[-\overset{\displaystyle O}{\overset{\|}{C}}-C_6H_4-\overset{\displaystyle O}{\overset{\|}{C}}-OCH_2-C_6H_{10}-CH_2O- \right]_n$$

This polyester is known as Kodel and produces a fabric that has good crease resistance.

57. Alkyd polyesters are made from glycerol and phthalic acid. Would this kind of polymer more likely be thermosetting or thermoplastic? Explain.

Challenge Exercises

58. Starting with ethyl alcohol as the only source of organic material and using any other reagents you desire, write equations to show the synthesis of the following:
(a) acetic acid
(b) ethyl acetate
(c) β-hydroxybutyric acid

59. Starting with bromoethane as the only organic compound, and using any other reactants, show equations for the synthesis of the following:
(a) ethyl propanoate
(b) 1-butanol

60. Starting with cyclohexanol and any other reactants, show equations for the synthesis of cyclohexane carboxylic acid.

Answers to Practice Exercises

24.1 (a) $CH_3\underset{\displaystyle|}{\overset{\displaystyle CH_3}{C}}HCOOH$ (with CH_3 on the second carbon)

(b) $CH_3CH_2CH_2\overset{Cl}{C}HCH_2COOH$ (Cl on the fourth carbon from the left)

(c) cyclohexane ring with COOH substituent

24.2 3-hydroxybutanoic acid

24.3 (a) $H\overset{O}{\overset{\|}{C}}{-}OH$

(b) NH_2CH_2COOH

(c) $HOCH_2COOH$

(d) CH_2BrCH_2COOH

(e) $HO{-}\overset{O}{\overset{\|}{C}}{-}\overset{O}{\overset{\|}{C}}{-}OH$

(f) benzene ring–COOH

24.4

$$\begin{array}{l} CH_2COOH \\ | \\ CH_2 \\ | \\ H_2NCHCOOH \end{array}$$

24.5 (a) caproic acid or hexanoic acid (b) 2-hexanone

24.6 (a) benzene ring–$CH_2CH_2CH_3$ (b) benzene ring with CH_3 and CH_2CH_3 on adjacent carbons

24.7

$$CH_3\overset{O}{\overset{\|}{C}}{-}H + HCN \xrightarrow{OH^-} CH_3\overset{OH}{\overset{|}{C}}HCN$$

$$CH_3\overset{OH}{\overset{|}{C}}HCN + H_2O \xrightarrow{H^+} CH_3\underset{OH}{\underset{|}{C}}HCOOH + NH_4^+$$

lactic acid, 2-hydroxypropanoic acid

24.8 (a) methyl propanoate (b) phenyl ethanoate (phenyl acetate) (c) dimethyl succinate (d) methyl benzoate (e) methyl salicylate (wintergreen)

24.9

(a)
$$CH_3CH_2CH_2CH_2\overset{O}{\overset{\|}{C}}{-}OCH_2\overset{CH_3}{\overset{|}{C}}HCH_3 + H_2O \xrightarrow{H^+}$$

$$CH_3CH_2CH_2CH_2\overset{O}{\overset{\|}{C}}{-}OH + CH_3\overset{CH_3}{\overset{|}{C}}HCH_2OH$$

(b)
$$CH_3CH_2CH_2CH_2\overset{O}{\overset{\|}{C}}{-}OCH_2\overset{CH_3}{\overset{|}{C}}HCH_3 + NaOH \xrightarrow[\Delta]{H_2O}$$

$$CH_3CH_2CH_2CH_2\overset{O}{\overset{\|}{C}}{-}O^-Na^+ + CH_3\overset{CH_3}{\overset{|}{C}}HCH_2OH$$

24.10 (a)

$$\begin{array}{l} CH_2{-}O{-}\overset{O}{\overset{\|}{C}}(CH_2)_7CH{=}CH(CH_2)_7CH_3 \\ | \\ CH{-}O{-}\overset{O}{\overset{\|}{C}}(CH_2)_{12}CH_3 \\ | \\ CH_2{-}O{-}\overset{O}{\overset{\|}{C}}(CH_2)_{14}CH_3 \end{array}$$

The sequence of the three esters in the triacylglycerol may vary.

(b) Three isomers can exist.

CHAPTER 25

Amides and Amines: Organic Nitrogen Compounds

Nylon is one of the materials used to give these colorful sails their strength and durability.

Chapter Outline

Many organic compounds contain nitrogen. The amines and amides are the two major classes of nitrogen-containing compounds. Amines isolated from plants form a group of compounds called alkaloids. Thousands of alkaloids have been isolated. Many of these compounds exhibit physiological activity. Examples of common alkaloid compounds include quinine, used in the treatment of malaria; strychnine, a poison; morphine, a narcotic; and caffeine, a stimulant. Many other drugs are also nitrogen-containing compounds.

Amides are nitrogen derivatives of carboxylic acids. These compounds are found as polymers, both commercially, as in nylon, and biologically, as in proteins. An understanding of the chemistry of organic nitrogen compounds is the cornerstone of genetics and is essential to unlocking the chemical secrets of living organisms.

25.1 Amides: Nomenclature and Physical Properties

Carboxylic acids react with ammonia to form ammonium salts:

$$\underset{\text{carboxylic acid}}{\overset{\displaystyle O}{\overset{\|}{RC}}\text{—OH}} + \underset{\text{ammonia}}{NH_3} \longrightarrow \underset{\text{ammonium salt}}{\overset{\displaystyle O}{\overset{\|}{RC}}\text{—}O^- NH_4^+}$$

$$\underset{\text{acetic acid}}{CH_3\overset{\displaystyle O}{\overset{\|}{C}}\text{—OH}} + NH_3 \longrightarrow \underset{\text{ammonium acetate}}{CH_3\overset{\displaystyle O}{\overset{\|}{C}}\text{—}O^- NH_4^+}$$

Ammonium salts of carboxylic acids are ionic substances. Ammonium acetate, for example, is ionized and exists as ammonium ions and acetate ions, both in the crystalline form and when dissolved in water.

When heated, ammonium salts of carboxylic acids lose a molecule of water and are converted to *amides*:

$$\underset{\text{ammonium salt}}{R\text{—}\overset{\displaystyle O}{\overset{\|}{C}}\text{—}O^- NH_4^+} \xrightarrow{\Delta} \underset{\text{amide}}{\overset{\displaystyle O}{\overset{\|}{RC}}\text{—}NH_2} + H_2O$$

$$\underset{\text{ammonium acetate}}{CH_3\overset{\displaystyle O}{\overset{\|}{C}}\text{—}O^- NH_4^+} \xrightarrow{\Delta} \underset{\substack{\text{ethanamide}\\ \text{(acetamide)}}}{CH_3\overset{\displaystyle O}{\overset{\|}{C}}\text{—}NH_2} + H_2O$$

Other methods of making amides are starting with acyl halides or esters.

amide **Amides** are neutral (nonbasic) molecular substances; they exist as molecules (not ions) both in the crystalline form and when dissolved in water. An amide contains the following characteristic structures:

$$R\text{—}\overset{\displaystyle O}{\overset{\|}{C}}\text{—}NH_2 \qquad R\text{—}\overset{\displaystyle O}{\overset{\|}{C}}\text{—}NHR' \qquad R\text{—}\overset{\displaystyle O}{\overset{\|}{C}}\text{—}NR'R''$$

amide structures

In amides, the carbon atom of a carbonyl group is bonded directly to a nitrogen atom of an $-NH_2$, $-NHR$, or $-NR_2$ group. The amide structure occurs in numerous substances, including proteins and some synthetic polymers, such as nylon.

Following are the rules for forming both the IUPAC and the common names for amides:

IUPAC Rules for Naming Amides

1. To establish the parent name, identify the longest continuous carbon chain that includes the amide group.
2. Drop the *-oic acid* ending from the name of the carboxylic acid containing the same length carbon chain as the amide.
3. Add the suffix *-amide*.
4. When the nitrogen atom of the amide is bonded to an alkyl or other group, the group is named as a prefix to the amide name proceded by the letter *N*, such as *N*-methyl.

Using the IUPAC rules,

methanoic acid becomes methanamide

$HCOOH$ → $HC(=O)-NH_2$

ethanoic acid becomes ethanamide

CH_3COOH → $CH_3C(=O)-NH_2$

Spacefilling model of butanamide

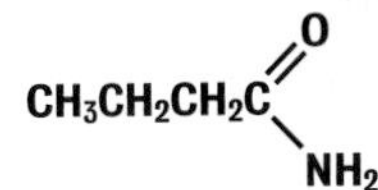

In a like manner, the common names for amides are formed from the common names of the corresponding carboxylic acids, by dropping the *-ic* or *-oic acid* ending and adding the suffix *-amide*. Thus,

formic acid becomes formamide

$HCOOH$ → $HC(=O)-NH_2$

butyric acid becomes butyramide

$CH_3CH_2CH_2COOH$ → $CH_3CH_2CH_2C(=O)-NH_2$

benzoic acid becomes benzamide

C_6H_5COOH → $C_6H_5C(=O)-NH_2$

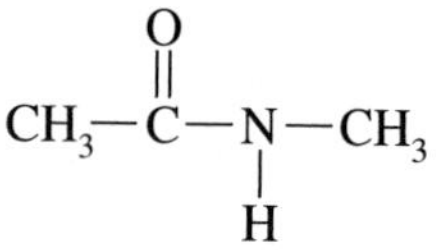

$$CH_3-\overset{\overset{\large O}{\|}}{C}-\underset{\underset{\large H}{|}}{N}-CH_3$$ is *N*-methylacetamide (*N*-methylethanamide)

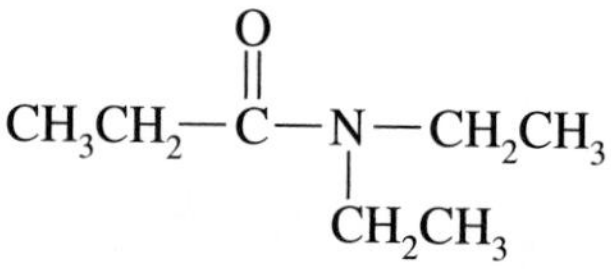

$$CH_3CH_2-\overset{\overset{\large O}{\|}}{C}-\underset{\underset{\large CH_2CH_3}{|}}{N}-CH_2CH_3$$ is *N,N*-diethylpropionamide (*N,N*-diethylpropanamide)

Formulas and names of selected amides are shown in Table 25.1.

Table 25.1 Formulas and Names of Selected Amides

Formula	IUPAC name	Common name	Boiling point (°C)
$H\overset{\overset{O}{\|}}{C}-NH_2$	Methanamide	Formamide	210
$CH_3\overset{\overset{O}{\|}}{C}-NH_2$	Ethanamide	Acetamide	222
$CH_3CH_2\overset{\overset{O}{\|}}{C}-NH_2$	Propanamide	Propionamide	222
$CH_3\underset{\underset{CH_3}{\|}}{C}H\overset{\overset{O}{\|}}{C}-NH_2$	2-Methylpropanamide	Isobutyramide	—
$CH_3\overset{\overset{O}{\|}}{C}-\overset{\overset{H}{\|}}{N}-C_6H_5$	*N*-Phenylethanamide	Acetanilide	304
$CH_3\overset{\overset{O}{\|}}{C}-\underset{\underset{H}{\|}}{N}-C_6H_4-OH$	*N*-*p*-Hydroxyphenylethanamide	Acetaminophen (Tylenol)	—
$H_2N-\overset{\overset{O}{\|}}{C}-C_6H_5$	Benzamide	Benzamide	288
$CH_3NH-\overset{\overset{O}{\|}}{C}-C_6H_5$	*N*-Methylbenzamide	*N*-Methylbenzamide	—

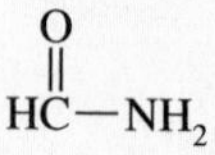

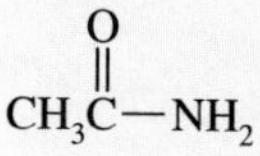

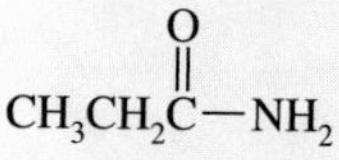

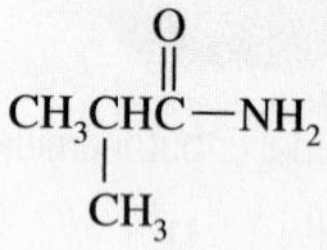

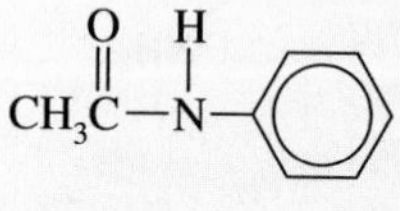

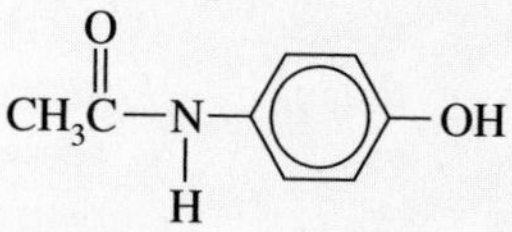

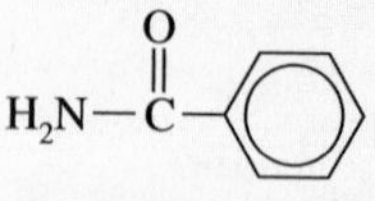

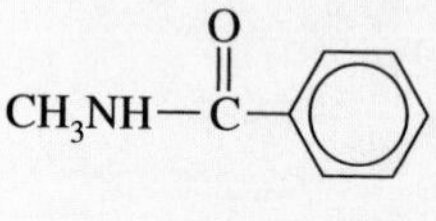

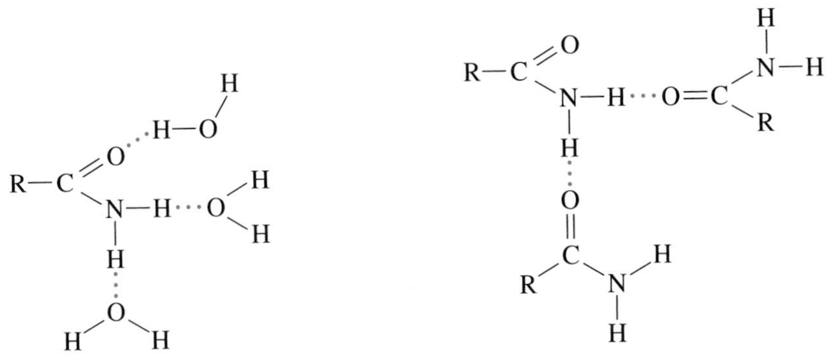

(a) Hydrogen bonding between amides and water molecules

(b) Intermolecular hydrogen bonding

Figure 25.1
Hydrogen bonding in amides.

Practice 25.1

Write formulas for

(a) *N*-isopropylmethanamide
(b) *N*-ethylbutanamide
(c) *N*-methyl-2-methylpropanamide
(d) *N,N*-diethylhexanamide
(e) *p*-aminobenzamide

Except for formamide, a liquid, all other unsubstituted amides are solids at room temperature. Many are odorless and colorless. Low-molar-mass amides are soluble in water, but solubility decreases quickly as molar mass increases. The amide functional group is polar, and nitrogen is capable of hydrogen bonding. The solubility of these molecules and their exceptionally high melting points and boiling points are the result of this polarity and hydrogen bonding between molecules, shown in Figure 25.1.

25.2 Chemical Properties of Amides

Hydrolysis is one of the more important reactions of amides. This type of reaction is analogous to the hydrolysis of carboxylic acid esters. The amide is cleaved into two parts, the carboxylic acid portion and the nitrogen-containing portion. As in ester hydrolysis, this reaction requires the presence of a strong acid or a strong base for it to occur in the laboratory. Amide hydrolysis is accomplished in living systems during the degradation of proteins by enzymatic reactions under much milder conditions (Chapter 30). Hydrolysis of an unsubstituted amide in an acid solution produces a carboxylic acid and an ammonium salt:

$$CH_3CH_2CH_2\overset{\overset{\displaystyle O}{\|}}{C}-NH_2 + H_2O + HCl \xrightarrow{\Delta} CH_3CH_2CH_2\overset{\overset{\displaystyle O}{\|}}{C}-OH + NH_4Cl$$

Basic hydrolysis results in the production of ammonia and the salt of a carboxylic acid:

$$CH_3-\overset{\overset{\displaystyle O}{\|}}{C}-NH_2 + NaOH \xrightarrow{\Delta} CH_3\overset{\overset{\displaystyle O}{\|}}{C}-O^- Na^+ + NH_3(g)$$

Example 25.1 Show the products of (a) the acid hydrolysis and (b) the basic hydrolysis of

$$C_6H_5-\overset{O}{\overset{\|}{C}}-NH_2$$

SOLUTION (a) In acid hydrolysis, the C—N bond is cleaved, and the carboxylic acid is formed. The $-NH_2$ group is converted into an ammonium ion:

$$C_6H_5-\overset{O}{\overset{\|}{C}}-NH_2 + H_2O + H^+ \longrightarrow C_6H_5-\overset{O}{\overset{\|}{C}}-OH + NH_4^+$$

(b) In basic solution, the C—N bond is also cleaved, but since the solution is basic, the salt of the carboxylic acid is formed, along with ammonia:

$$C_6H_5-\overset{O}{\overset{\|}{C}}-NH_2 + NaOH \longrightarrow C_6H_5-\overset{O}{\overset{\|}{C}}-O^-Na^+ + NH_3$$

Practice 25.2

Give the products of the acidic and basic hydrolysis of

(a) $CH_3CH_2\overset{O}{\overset{\|}{C}}-NH_2$ (b) $CH_3\overset{O}{\overset{\|}{C}}-N(CH_3)_2$

Practice 25.3 Application to Biochemistry

Many digestive enzymes are produced by the pancreas and then activated when secreted into the small intestine. Thus, for our own safety, these enzymes become functional only when our body needs them to digest food. The activation process is hydrolysis of specific amide bonds. For example, chymotrypsin becomes active when the amide bond between amino acid segments #15 and #16 is hydrolyzed. Show the product of this reaction:

$$\overset{\#14}{CH}-\overset{O}{\overset{\|}{C}}-NH-\overset{\#15}{CH}-\overset{O}{\overset{\|}{C}}-NH-\overset{\#16}{CH}-\overset{O}{\overset{\|}{C}}-NH-\overset{\#17}{CH} + H_2O \longrightarrow$$

the rest of the long chymotrypsin polymer

(# represents a complex side chain for each amino acid segment)

25.3 Polyamides: Condensation Polymers

Nylon fibers are spun as the polymer forms.

Polyamides are condensation polymers. They can be formed from two monomers, a dicarboxylic acid and a diamine, or from a single monomer having a carboxylic acid and an amine in the same molecule. Proteins are biological polyamides. (See Practice 25.3 and Chapter 29.) Perhaps the most famous of the synthetic polyamides is Nylon-66.

Although there are several nylons, Nylon-66 is one of the best known and the first commercially successful polyamide. This polymer was so named because it is made from two 6-carbon monomers, adipic acid, $HOOC(CH_2)_4COOH$, and 1,6-hexanediamine, $H_2N(CH_2)_6NH_2$. The polymer chains of polyamides contain recurring amide linkages. The amide linkage can be made by reacting a carboxylic acid group with an amine group:

$$\underset{\text{carboxylic acid group}}{R-\overset{O}{\overset{\|}{C}}-OH} + \underset{\text{amine group}}{H\overset{H}{\overset{|}{N}}-CH_2-R'} \xrightarrow{\Delta} R-\underset{\text{amide linkage}}{\overset{O}{\overset{\|}{C}}-NH}-CH_2-R' + H_2O$$

The repeating structural unit of the Nylon-66 chain consists of one adipic acid unit and one 1,6-hexanediamine unit:

$$\underset{\text{adipic acid}}{HOOC-(CH_2)_4-COOH} \qquad \underset{\text{1,6-hexanediamine (hexamethylenediamine; 1,6-diaminohexane)}}{H_2N-(CH_2)_6-NH_2}$$

$$-NH(CH_2)_6-NH\left[\overset{O}{\overset{\|}{C}}(CH_2)_4-\overset{O}{\overset{\|}{C}}-NH(CH_2)_6-NH\right]_n\overset{O}{\overset{\|}{C}}(CH_2)_4-\overset{O}{\overset{\|}{C}}-$$

segment of Nylon-66 polyamide

Nylon was developed as a synthetic fiber for stockings and other wearing apparel and introduced to the public at the New York World's Fair in 1939. Today Nylon is used to make fibers for clothing and carpeting, filaments for fishing lines and ropes, and bristles for brushes, as well as tents, shower curtains, umbrellas, surgical sutures, tire cords, sports equipment, artificial human hair, synthetic animal fur, and molded objects such as gears and bearings. For the latter application, no lubrication is required, because nylon surfaces are inherently slippery.

25.4 Urea

The body disposes of nitrogen by the formation of a diamide known as urea:

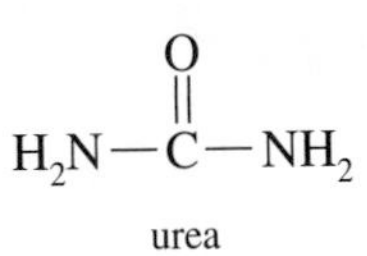

$$H_2N-\overset{O}{\overset{\|}{C}}-NH_2$$

urea

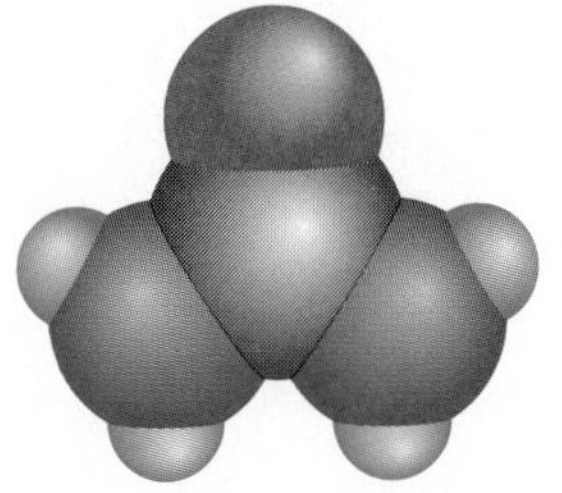

Crops are often fertilized with urea.

Urea is a white solid that melts at 133°C. It is soluble in water and therefore is excreted from the body in the urine. The normal adult excretes about 30 g of urea daily.

Urea is a common commercial product as well. It is widely used in fertilizers to add nitrogen to the soil or as a starting material in the production of plastics and barbiturates. Here's an example:

$$\underset{\text{urea}}{O{=}C(NH_2)_2} + \underset{\text{diethylmalonate}}{CH_2(C(=O)OCH_2CH_3)_2} \longrightarrow \underset{\text{barbituric acid}}{O{=}C(NH{-}C(=O))_2CH_2} + 2\,CH_3CH_2OH$$

The various barbiturate drugs have certain organic groups substituted for the hydrogen atoms on the CH_2. When two ethyl groups replace these hydrogen atoms, the compound is known as *barbital*, which is a controlled substance (depressant) and a hypnotic. The unsubstituted barbituric acid has no hypnotic properties. There are many barbiturate drugs.

25.5 Amines: Nomenclature and Physical Properties

amine

An **amine** is a substituted ammonia molecule with basic properties and has the general formula RNH_2, R_2NH, or R_3N, where R is an alkyl or an aryl group. Amines are classified as primary (1°), secondary (2°), or tertiary (3°), depending on the number of hydrocarbon groups attached to the nitrogen atom. Some examples include the following:

$H{-}NH_2$ — ammonia

$H_2N{-}CH_3$ — methylamine (1° amine)

$(CH_3)(H)N{-}CH_2CH_3$ — methylethylamine (2° amine)

$CH_3CH_2{-}N(CH_2CH_3)_2$ — triethylamine (3° amine)

$C_6H_5NH_2$ — aniline (1° amine)

$C_6H_{11}NH_2$ — cyclohexylamine (1° amine)

Practice 25.4

Identify these compounds as 1°, 2°, or 3° amines:

(a) $CH_3CH_2CH_2NH_2$

(b) $CH_3C(CH_3)_2{-}NH_2$

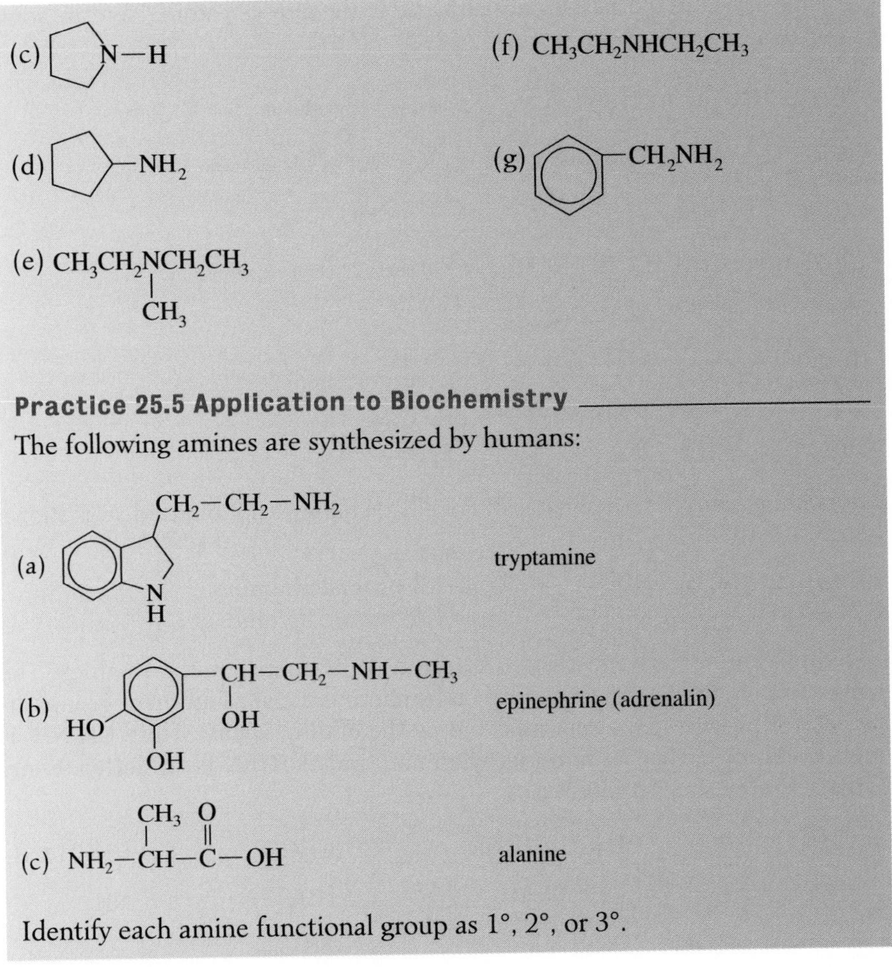

Naming Amines

The IUPAC System names simple aliphatic amines by using the name of the alkane (omitting the final *-e*) and adding the ending *-amine*. The NH_2 and side chains are located by number on the parent carbon chain, as shown here:

$CH_3CH_2NH_2$
ethanamine

$CH_3CH_2CH_2NH_2$
1-propanamine

CH_3CHCH_3 (with NH_2 on C-2)
2-propanamine

$CH_3C(CH_3)_2$—NH_2
2-methyl-2-propanamine

(cyclohexane ring)—NH_2
cyclohexanamine

Secondary and tertiary amines are named as N-substituted primary amines by using the longest continuous carbon chain as the parent name. A capital N in the name indicates the alkyl groups attached to the nitrogen atom. Here are some examples:

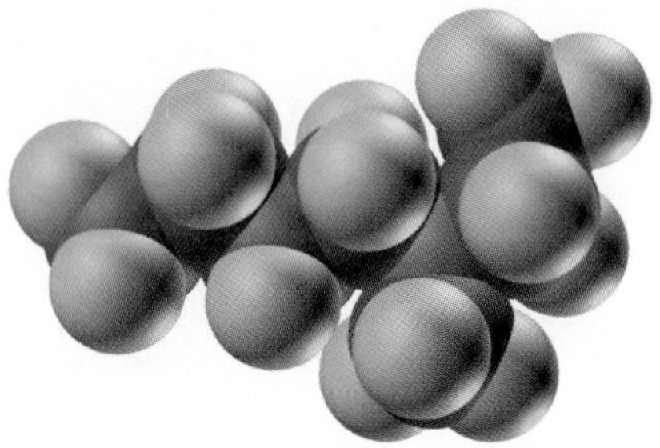

Spacefilling model of *N*-ethyl-*N*-methyl-1-butanamine

$CH_3{-}NH{-}CH_2CH_2CH_3$ *N*-methyl-1-propanamine

$CH_3{-}N(CH_2CH_3){-}CH_2CH_2CH_2CH_3$ *N*-ethyl-*N*-methyl-1-butanamine

$(CH_3)_2N{-}CH_2CH_2CH(CH_2CH_3)CH_2CH_3$ *N,N*-dimethyl-3-ethyl-1-pentanamine

$C_5H_9{-}N(CH_3)_2$ (cyclopentyl) *N,N*-dimethylcyclopentanamine

For naming diamines, the final *-e* of the alkane name is not omitted, as is shown here:

$H_2N{-}CH_2CH_2CH_2{-}NH_2$ 1,3-propanediamine
$H_2N{-}(CH_2)_6{-}NH_2$ 1,6-hexanediamine

Simple amines are most often referred to by their common names. The names for aliphatic amines are formed by naming the alkyl group or groups attached to the nitrogen atom, followed by the ending *-amine*. Thus, CH_3NH_2 is methylamine, $(CH_3)_2NH$ is dimethylamine, and $(CH_3)_3N$ is trimethylamine. Other examples are as follows:

$CH_3CH_2NH_2$ ethylamine

$CH_3CH_2CH_2NH_2$ propylamine

$CH_3CH(NH_2)CH_3$ isopropyl amine

$CH_3CH(NH_2)CH_2CH_3$ *sec*-butylamine

$CH_3C(CH_3)_2{-}NH_2$ *tert*-butylamine

$CH_3N(CH_3)CH_2CH_3$ ethyldimethylamine

The most important aromatic amine is aniline ($C_6H_5NH_2$). Derivatives are named as substituted anilines. To identify a substituted aniline in which the substituent group is attached to the nitrogen atom, an *N*- is placed before the group name to indicate that the substituent is bonded to the nitrogen atom and not to a carbon atom in the ring. For example, the following compounds are called aniline, *N*-methylaniline, *N,N*-dimethylaniline, and *N*-ethyl-*N*-methylaniline:

$C_6H_5NH_2$ aniline

$C_6H_5NHCH_3$ *N*-methylaniline

$C_6H_5N(CH_3)_2$ *N,N*-dimethylaniline

$C_6H_5N(CH_2CH_3)(CH_3)$ *N*-ethyl-*N*-methylaniline

When a group is substituted for a hydrogen atom in the ring, the resulting ring-substituted aniline is named as we have previously done with aromatic compounds. The monomethyl ring-substituted anilines are known as toluidines. Study the names for the following substituted anilines:

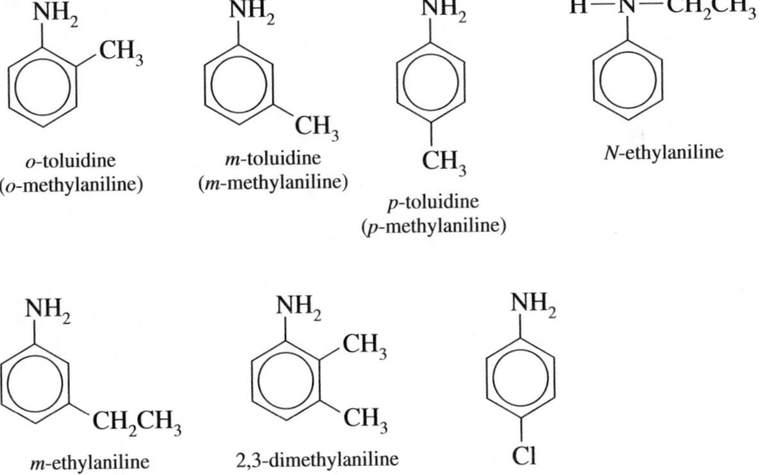

p-chloroaniline

The smell of fertilizers or decaying plant matter is characteristic of nitrogen-containing compounds, found, for example, in a pile of mulch.

Physiologically, aniline is a toxic substance. It is easily absorbed through the skin and affects both the blood and the nervous system. Aniline reduces the oxygen-carrying capacity of the blood by converting hemoglobin to methemoglobin. Methemoglobin is the oxidized form of hemoglobin in which the iron has gone from a +2 to a +3 oxidation state.

Name the following two compounds:

(a) $CH_3CH(CH_3)CH_2—NH_2$ (b) benzene ring with COOH at top and NH_2 at the para position

SOLUTION

(a) The alkyl group attached to NH_2 is an isobutyl group. Thus, the common name is isobutylamine. The longest chain containing the NH_2 has three carbons. Therefore, the parent carbon chain is propane, and the IUPAC name is 2-methyl-1-propanamine.

(b) The parent compound on which the name is based is benzoic acid. With an amino group in the para position, the name is *p*-aminobenzoic acid. The acronym for *p*-aminobenzoic acid is PABA. Esters of PABA are some of the most effective ultraviolet screening agents and are used in suntanning lotions. PABA is a component of folic acid, an essential vitamin for humans.

Practice 25.6

Name the following compounds:

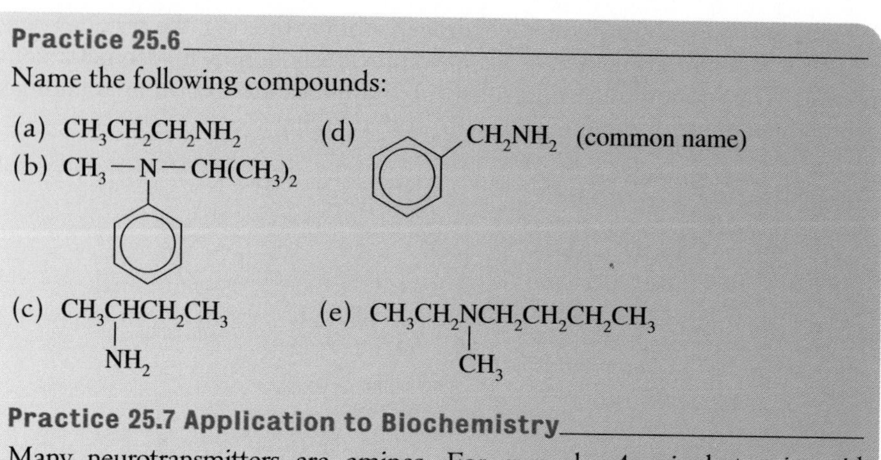

Practice 25.7 Application to Biochemistry

Many neurotransmitters are amines. For example, 4-aminobutanoic acid (gamma-aminobutyric acid, GABA) is a neurotransmitter that is found in about 30% of the brain's nerve cells. Draw the structure of this amino acid.

heterocyclic compound

Ring compounds in which the atoms in the ring are not all alike are known as **heterocyclic compounds**. The most common heteroatoms are oxygen, nitrogen, and sulfur. A number of the nitrogen-containing heterocyclic compounds are present in naturally occurring biological substances such as DNA, which controls heredity. The structural formulas of several nitrogen-containing heterocyclics are as follows:

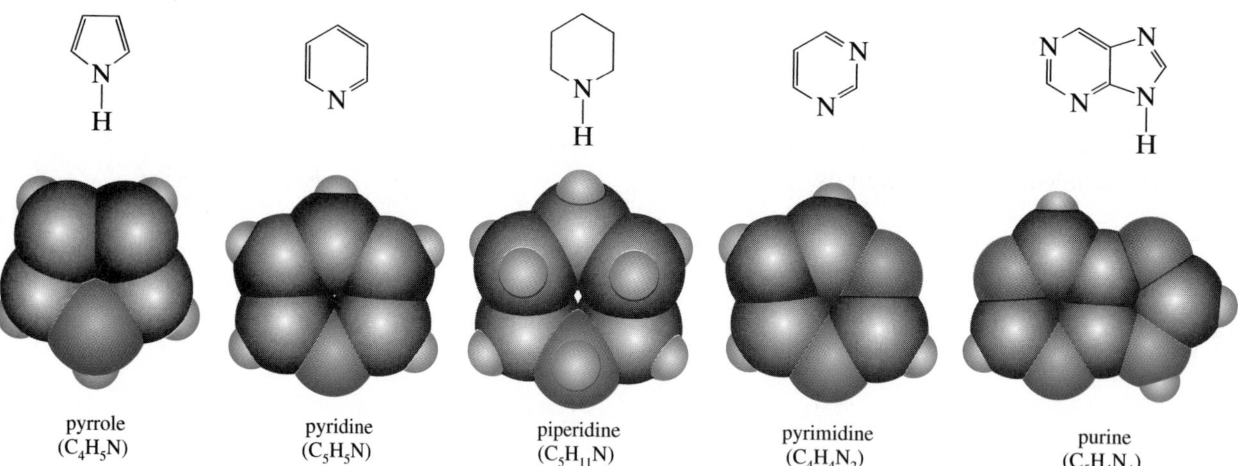

pyrrole (C_4H_5N) pyridine (C_5H_5N) piperidine ($C_5H_{11}N$) pyrimidine ($C_4H_4N_2$) purine ($C_5H_4N_4$)

Amines are capable of hydrogen bonding with water. As a result, the aliphatic amines with up to six carbons are quite soluble in water. Methylamine and ethylamine are flammable gases with a strong ammoniacal odor. Trimethylamine has a "fishy" odor. Higher-molar-mass amines have obnoxious odors. The foul odors arising from dead fish and decaying flesh are due to amines released by bacterial decomposition. Two of these compounds are diamines, 1,4-butanediamine and 1,5-pentanediamine. Each compound contains two amino groups:

$H_2NCH_2CH_2CH_2CH_2NH_2$
1,4-butanediamine
(putrescine)

$H_2NCH_2CH_2CH_2CH_2CH_2NH_2$
1,5-pentanediamine
(cadaverine)

Simple aromatic amines are all liquids or solids. They are colorless or almost colorless when freshly prepared, but become dark brown or red when exposed to air and light.

25.6 Preparation of Amines

Alkylation of Ammonia and Amines The substitution of alkyl groups for hydrogen atoms of ammonia can be done by reacting ammonia with alkyl halides. Thus, in successive reactions, a primary, a secondary, and a tertiary amine can be formed:

$$NH_3 \xrightarrow{CH_3Br} \underset{\substack{\text{methylamine} \\ (1°)}}{CH_3NH_2} \xrightarrow{CH_3Br} \underset{\substack{\text{dimethylamine} \\ (2°)}}{(CH_3)_2NH} \xrightarrow{CH_3Br} \underset{\substack{\text{trimethylamine} \\ (3°)}}{(CH_3)_3N}$$

Tertiary amines can be further alkylated so that four organic groups bond to the nitrogen atom. In compounds that have four organic groups bonded to a nitrogen atom, the nitrogen atom is positively charged and the compounds are called **quaternary ammonium salts**. The positive charge is on the nitrogen atom, as in ammonium salts. Here's an example:

quaternary ammonium salt

$$CH_3-\ddot{N}(CH_3)-CH_3 + CH_3Br \longrightarrow CH_3-N^+(CH_3)_2-CH_3\,Br^-$$

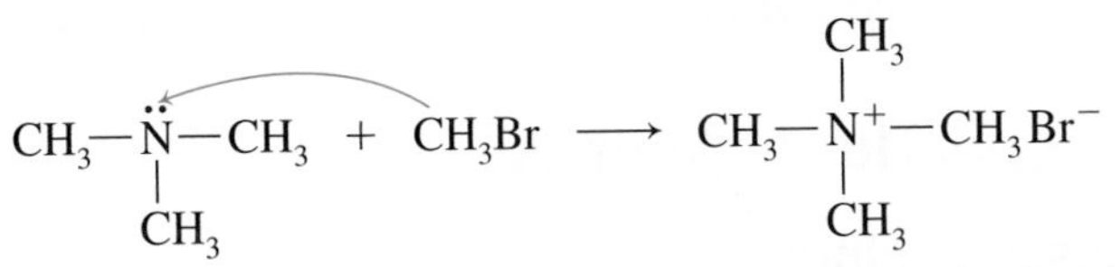

tetramethylammonium bromide

Quaternary ammonium salts are well known in biologically active compounds and in many popular medicinals. For example, acetylcholine, an active neurotransmitter in the brain, is a quaternary ammonium salt. Choline is an important component of many biological membranes.

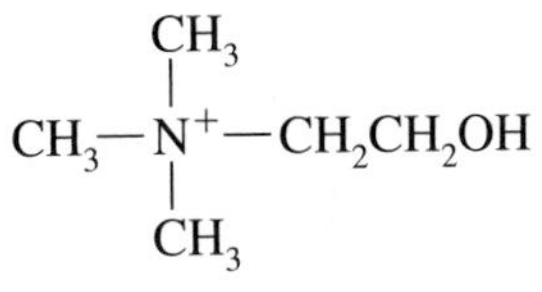

$$CH_3-N^+(CH_3)_2-CH_2CH_2OH \qquad CH_3-N^+(CH_3)_2-CH_2CH_2O\overset{\displaystyle O}{\overset{\|}{C}}CH_3$$

choline 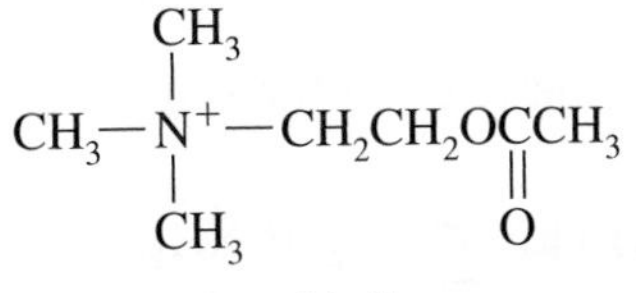 — acetylcholine

The quaternary ammonium salt thiamine hydrochloride is marketed as vitamin B_1. Many well-known fabric softening agents used in laundering clothes are quaternary ammonium salts.

Reduction of Amides and Nitriles Amides can be reduced with lithium aluminum hydride to give amines. For example, acetamide can be reduced to ethylamine; and when *N,N*-diethylacetamide is reduced, triethylamine is formed:

$$CH_3\overset{\overset{\displaystyle O}{\|}}{C}-NH_2 \xrightarrow{LiAlH_4} \underset{\text{ethylamine}}{CH_3CH_2NH_2}$$

$$CH_3\overset{\overset{\displaystyle O}{\|}}{C}-N(CH_2CH_3)_2 \xrightarrow{LiAlH_4} \underset{\text{triethylamine}}{(CH_3CH_2)_3N}$$

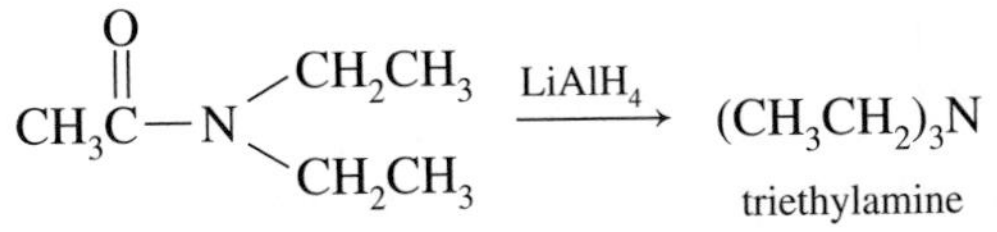

Nitriles, RCN, are also reducible to amines, using hydrogen and a metal catalyst:

$$CH_3CH_2C{\equiv}N \xrightarrow{H_2/Ni} CH_3CH_2CH_2NH_2$$

propionitrile → propylamine

Reduction of Aromatic Nitro Compounds Aniline, the most widely used aromatic amine, is made by reducing nitrobenzene. The nitro group can be reduced by several reagents; Fe and HCl, or Sn and HCl, are commonly used:

$$C_6H_5NO_2 \xrightarrow[HCl]{Sn} C_6H_5NH_3^+Cl^- \xrightarrow{NaOH} C_6H_5NH_2$$

nitrobenzene → phenylammonium chloride (anilinium chloride) → aniline

Practice 25.8

Write the structure of an amide or nitrile that, when reduced, forms
(a) ethylmethylamine
(b) 2-methylpropanamine
(c) 3-ethyl-2-methylpentanamine

25.7 Chemical Properties of Amines

Alkaline Properties of Amines

In many respects, amines resemble ammonia in their reactions. Thus, amines are bases and, like ammonia, produce OH^- ions in water:

A Brønsted–Lowry base is a proton (H^+) acceptor.

$$\ddot{N}H_3 + HOH \rightleftharpoons NH_4^+ + OH^-$$

ammonia molecule; ammonium ion; hydroxide ion

Methylamine and aniline react in the same manner:

$$CH_3\ddot{N}H_2 + HOH \rightleftharpoons CH_3\overset{+}{N}H_3 + OH^-$$

methylamine molecule; methylammonium ion; hydroxide ion

$$C_6H_5\ddot{N}H_2 + HOH \rightleftharpoons C_6H_5\overset{+}{N}H_3 + OH^-$$

aniline molecule; anilinium ion; hydroxide ion

The ions formed are substituted ammonium ions. They are named by replacing the *amine* ending by *ammonium* and, for the aromatic amines, by replacing the *aniline* name by *anilinium*.

$$(CH_3)CH_3\overset{+}{N}H_2 \qquad o\text{-}CH_3C_6H_4\overset{+}{N}H_3 \qquad CH_3\overset{+}{N}H_2C_6H_5$$

dimethylammonium ion; *o*-methylanilinium ion; *N*-methylanilinium ion

Like ammonia, amines are weak bases. Methylamine is a slightly stronger base than ammonia, and aniline is considerably weaker than ammonia. The pH values for 0.1 *M* solutions are methylamine, 11.8; ammonia, 11.1; and aniline, 8.8.

Amines are much more basic than amides, and our bodies recognize this difference. The liver converts ammonia, which is basic and toxic, into the safer diamide, urea (see Chapter 35, Section 35.9):

$$2\,NH_3 + H_2CO_3 \longrightarrow NH_2{-}\overset{\overset{\displaystyle O}{\|}}{C}{-}NH_2 + 2\,H_2O$$

(a partial summary of the urea cycle)

Urea does not react with water to form the hydroxide ion. Thus, we normally excrete nitrogen without making the blood or urine too basic. In contrast, blood ammonia levels increase during liver cirrhosis, and the increased blood pH that results can be life threatening.

Because amine groups form substituted ammonium ions under physiological conditions, they provide the positive charge for biological molecules. For example, neurotransmitters are often positively charged. The structures of two such transmitter compounds, dopamine and serotonin, are shown here:

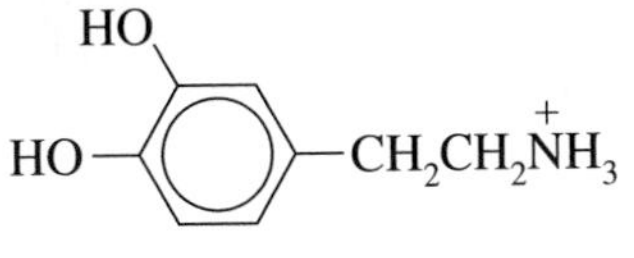

(3,4-dihydroxyphenyl)–$CH_2CH_2\overset{+}{N}H_3$

dopamine

(5-hydroxyindol-3-yl, N–H)–$CH_2CH_2\overset{+}{N}H_3$

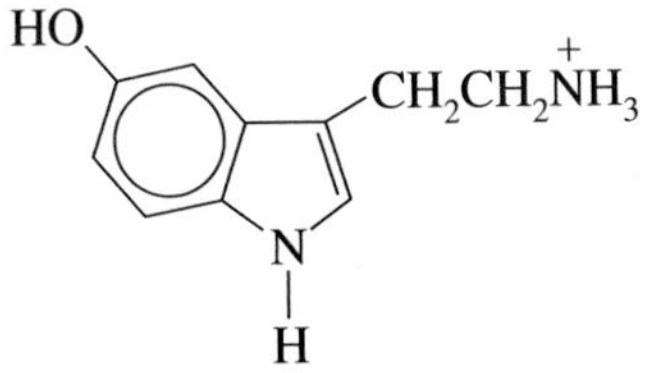

serotonin

Salt Formation

An amine reacts with a strong acid to form a salt; for example, methylamine and hydrogen chloride react in this fashion:

$$CH_3NH_2(g) + HCl(g) \longrightarrow CH_3\overset{+}{N}H_3Cl^-$$

methylamine molecule	hydrogen chloride molecule	methylammonium chloride (salt)

Methylammonium chloride is made up of methylammonium ions, $CH_3N^+H_3$, and chloride ions, Cl^-. It is a white crystalline salt that in physical appearance resembles ammonium chloride very closely.

Aniline reacts in a similar manner, forming anilinium chloride:

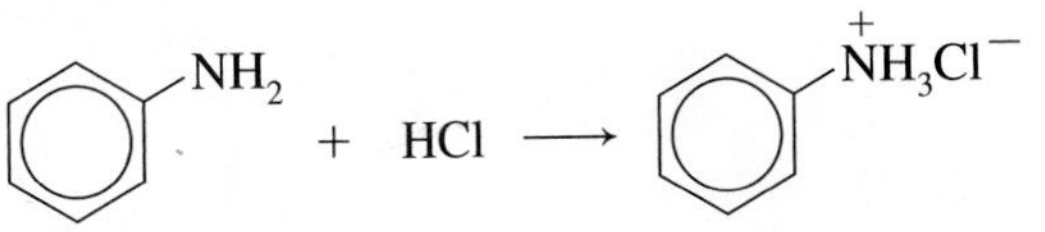

$$C_6H_5NH_2 + HCl \longrightarrow C_6H_5\overset{+}{N}H_3Cl^-$$

anilinium chloride
(aniline hydrochloride salt)

Many amines or amino compounds are more stable in the form of the hydrochloride salt. When the free amine is wanted, the HCl is neutralized to liberate the free amine. Thus,

$$\underset{\text{an amine hydrochloride salt}}{R\overset{+}{N}H_3\,Cl^- \text{ (or } RNH_2 \cdot HCl)} + NaOH \longrightarrow \underset{\text{free amine}}{RNH_2} + NaCl + H_2O$$

Formation of Amides

Primary and secondary amines react with acid chlorides to form substituted amides, as shown here:

$$CH_3\overset{O}{\overset{\|}{C}}-Cl + 2\,(CH_3CH_2)_2NH \longrightarrow \underset{N,N\text{-diethylacetamide}}{CH_3\overset{O}{\overset{\|}{C}}-N(CH_2CH_3)_2} + (CH_3CH_2)_2\overset{+}{N}H_2Cl^-$$

As seen from the previous example, amide formation creates larger molecules from smaller precursors. Building larger, more complex molecules is often a goal in both industrial chemistry and biochemistry. Amide formation plays an important role in this process. The barbiturates provide a simple example from pharmaceutical chemistry. Amide bonds form the central six-membered ring that is characteristic of barbiturates (the amide bonds are shown in red):

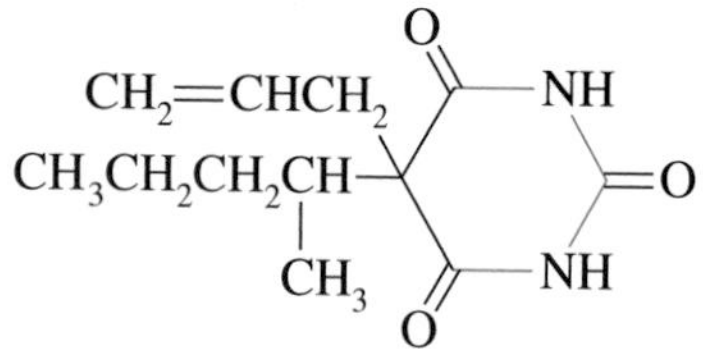

CH$_2$=CHCH$_2$ / CH$_3$CH$_2$CH$_2$CH(CH$_3$)— attached to a six-membered ring with C=O, NH, C=O, NH, C=O

seconal

Living cells also use amide linkages to create proteins. In proteins, small reactants (amino acids) are connected via amide bonds. It is not uncommon for several hundred amino acids to be linked together to form a single protein molecule.

Practice 25.9 Application to Biochemistry

The pyrimidine heterocycles are a very important part of a cell's genetic information (nucleic acids). The pyrimidine ring is closed by amide formation. Show the products of the following reaction:

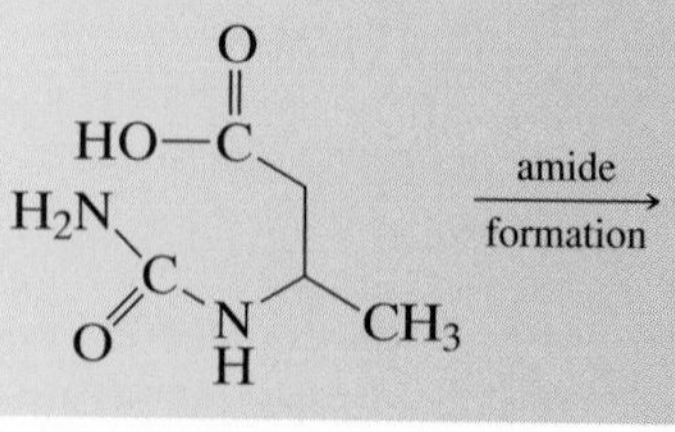

HO—C(=O)—CH$_2$—CH(CH$_3$)—NH—C(=O)—NH$_2$ $\xrightarrow[\text{formation}]{\text{amide}}$

25.8 Sources and Uses of Selected Amines

Nitrogen compounds are found throughout the plant and animal kingdoms. Amines, substituted amines, and amides occur in every living cell. Many of these compounds have important physiological effects. Several examples of well-known nitrogen compounds follow.

Many antibacterial agents contain nitrogen. Common examples include the synthetic sulfa drugs and penicillin-related antibiotics, which are synthesized by molds:

sulfanilamide

ampicillin

Many drinks, coffee for example, contain caffeine, an alkaloid.

Amines and amides are part of the structures of the B-complex vitamins. Examples are thiamine (vitamin B_1) and nicotinamide (niacin):

thiamine (vitamin B_1)

nicotinamide (niacin)

caffeine

A wide variety of nitrogen-containing compounds affect both the central and peripheral nervous system. Local anesthetics like the common drug procaine contain amines:

procaine hydrochloride (novocaine)

Basic compounds that are derived from plants and show physiological activity are known as **alkaloids**. These substances are usually amines. Nicotine, an alkaloid derived from tobacco leaves, acts to stimulate the nervous system:

alkaloid

nicotine

The opium alkaloids are often called opiates and include both compounds derived from the opium poppy and synthetic compounds that have morphinelike activity (sleep-inducing and analgesic properties). These drugs are classified as narcotics because they produce physical addiction, and they are strictly regulated by federal law. An example is methadone:

methadone

Another common narcotic alkaloid, cocaine, is obtained from the leaves of the coca plant:

cocaine

Amphetamines, as the name implies, are amine-containing compounds. These drugs act to stimulate the central nervous system by mimicking the action of compounds such as epinephrine, which is produced naturally by the body. They are used to treat depression, narcolepsy, and obesity. Use of amphetamines produces a feeling of well-being, loss of fatigue, and increasing alertness. The most widely abused amphetamine is methamphetamine, commonly called "speed." The structures of epinephrine and methamphetamine are shown here:

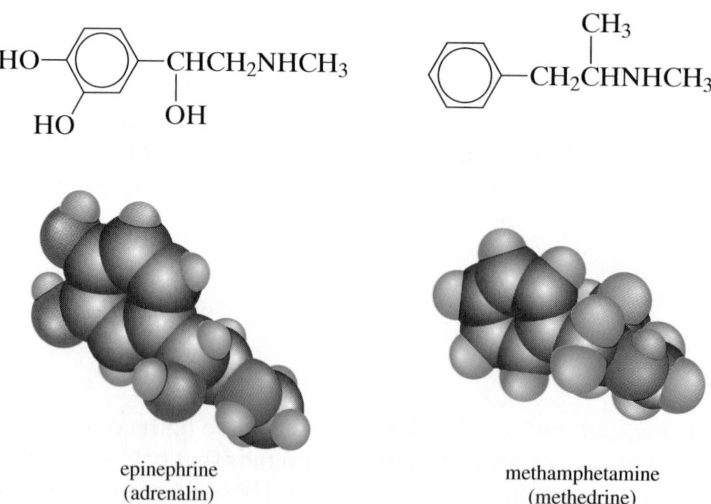

epinephrine (adrenalin)

methamphetamine (methedrine)

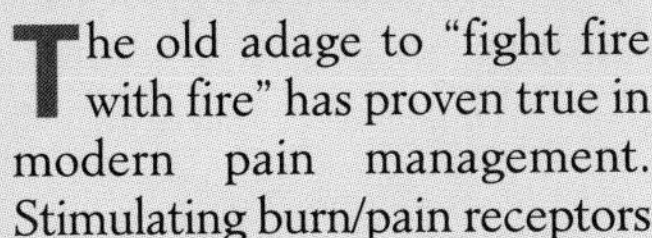

CHEMISTRY IN ACTION • CIA – The Hottest Pain Killer

The old adage to "fight fire with fire" has proven true in modern pain management. Stimulating burn/pain receptors can help relieve pain sensations (neuralgia) in conditions as diverse as toothache, shingles, and arthritis. The chemical used to alleviate pain comes from the lowly (but hot) chili pepper and is called capsaicin.

Photo of red chilis.

Capsaicin, like many neuroactive substances, contains nitrogen. The nitrogen atom is found in an amide as shown in the structures to the right.

It is the amide plus the polar groups on the benzene ring that give capsaicin its painkilling ability and also its hot flavor. Capsaicin is one of the hottest-tasting chemicals known; the hottest pepper (the habañero) is ranked at 350,000 on the Scoville scale of pepper pungency, while capsaicin scores about 15 million on the same scale. The hotness or pungency of a chemical partly correlates with its ability to block pain. Other molecules with less pungent flavors—eugenol from cloves, piperine from black pepper, and zingerone from ginger—have less painkilling potency.

Capsaicin has a dual impact on burn/pain receptors. First, the chemical elicits a burning sensation. The capsaicin then lingers at the receptor and blocks further pain signals. This second effect provides pain relief. This also means that capsaicin must be applied where the pain originates. For example, capsaicin cream is effective in relieving shingles pain when applied directly to the skin inflammation; capsaicin cream relieves arthritis pain best when the arthritic joint is close to the skin.

Another limitation to capsaicin treatment is that the initial burning sensation may be too severe. For serious pain, higher concentrations of capsaicin are sometimes combined with a local anesthetic. Scientists are now creating analogs of capsaicin, searching for a drug that does not burn but does block further pain signals.

Barbiturates are synthetic drugs classified as sedatives. They are prepared from urea and substituted malonic acid, and they contain amide groups. Barbiturates act to depress the activity of brain cells. For this reason, they are often called "downers"; they are one of the more widely abused drugs. An example is pentobarbital:

pentobarbital
(Nembutal)

Many common tranquilizers contain amines and amides. These drugs are used to modify psychotic behavior without inducing sleep, or to reduce anxiety or restlessness. Psychotic behavior is treated with strong tranquilizers such as

Thorazine. The pressure and anxiety of daily life are often relieved by the use of milder tranquilizers such as diazepam (Valium):

diazepam
(Valium)

Other amine-containing drugs are being developed as scientists learn more about the central nervous system. Fluoxetine (Prozac) is a widely used and very effective antidepressant:

$F_3C-C_6H_4-O-CHCH_2CH_2NHCH_3$ (with C_6H_5 on the CH)

fluoxetine (Prozac)

Recently, Prozac has been found useful in treating other central nervous system disorders.

Chapter 25 Review

25.1 Amides: Nomenclature and Physical Properties

KEY TERM

Amide

- Carboxylic acids react with ammonia to form salts.
- When ammonium salts of carboxylic acids are heated, amides are formed.
- Amides are neutral, nonionic compounds that contain the following characteristic structure:

$$-\overset{\displaystyle O}{\overset{\|}{C}}-\overset{|}{N}-$$

- The IUPAC naming of amides follows a similar process to that used for other organic molecules.
  - The longest continuous carbon chain (parent chain) that contains the amide is given the IUPAC name of the corresponding carboxylic acid modified by dropping *-oic acid* and adding *-amide*.
- Common naming follows the same procedure, dropping either *-ic acid* or *-oic acid* and adding *-amide*.
- If the nitrogen is bonded to an alkyl or aryl group, the group is named as a prefix preceded by the letter *N*.
- Unsubstituted amides are solid at room temperature except for formamide.
- Strong intermolecular hydrogen bonding accounts for the amides' high melting and boiling points as well as the water solubility of low-molar-mass amides.

25.2 Chemical Properties of Amides

- Amide hydrolysis uses water to split an amide into a carboxylic acid portion and a nitrogen-containing portion.
  - Hydrolysis of an unsubstituted amide in acid yields a carboxylic acid and an ammonium ion.
  - Hydrolysis of an unsubstituted amide in base yields a carboxylic acid salt and ammonia.

25.3 Polyamides: Condensation Polymers

- Polyamides form when a carboxylic acid from one monomer reacts with an amine from a second monomer, forming an amide bond and splitting out water.
  - One of the most common commercial polyamides is Nylon-66, formed from a six-carbon dicarboxylic acid (adipic acid) and a six-carbon diamine (1,6-hexanediamine).

25.4 Urea

- Urea is a simple diamide of the structure

$$H_2N-\overset{\overset{\displaystyle O}{\|}}{C}-NH_2$$

- Animals excrete nitrogen in the form of urea.
- Urea is also a valuable fertilizer and industrial chemical.

25.5 Amines: Nomenclature and Physical Properties

KEY TERMS

Amine
Heterocyclic compound

- An amine is a substituted ammonia molecule bonded to at least one alkyl or aryl group (R).
  - Primary (1°) amines have the general formula RNH_2, secondary (2°) amines have the general formula R_2NH, and tertiary (3°) amines have the general formula R_3N.
- The IUPAC System names simple amines by dropping the *-e* from the alkane name and adding the ending *-amine*.
  - Secondary and tertiary amines are named as *N*-substituted primary amines by using the longest carbon chain as the parent.
  - A capital *N* indicates the additional alkyl groups found in secondary and tertiary amines.
- Many simple amines are referred to by their common names.
- The most important aromatic amine is aniline,

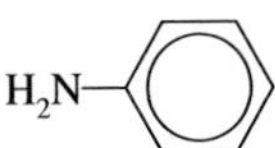

- Heterocyclic compounds are ring structures where the atoms forming the ring are not all alike.
  - Many nitrogen-containing heterocyclic compounds are present in naturally occurring biochemicals.
- Since amines can hydrogen-bond with water, aliphatic amines containing up to six carbons are quite water soluble.
- In general, amines have sharp or obnoxious odors.

25.6 Preparation of Amines

KEY TERM

Quaternary ammonium salt

- Alkylation of ammonia (with alkyl halides) will successively form primary, secondary, and tertiary amines.
- Quaternary ammonium salts are formed when tertiary amines are alkylated so that four organic groups are bonded to the nitrogen.
  - Many well-known biologically active compounds are quaternary ammonium salts.
- Reduction of several different nitrogen-containing compounds will form amines.
  - $LiAlH_4$ will reduce amides to form amines.
  - Nitriles (RCN) are reducible to amines with hydrogen gas and a metal cartalyst.
  - Nitrobenzene can be reduced to aniline using Sn and HCl or Fe and HCl.

25.7 Chemical Properties of Amines

- Like ammonia, amines are bases.
  - Amines accept a proton from water, forming substituted ammonium ions.
  - When the nitrogen is protonated, the *-amine* ending is changed to *-ammonium* and the *-aniline* is changed to *-anilinium*.
  - Amines are only weak bases like ammonia.
- Amines react with strong acids to form salts.
- Primary and secondary amines react with acid chlorides to form amides.
- Forming amides commonly creates more complex products, an important synthetic process in both industry and living cells.
  - Proteins are polymers of amino acids linked by amide bonds.

25.8 Sources and Uses of Selected Amines

KEY TERM

Alkaloid

- Many antibacterial agents contain nitrogen.
- Amines (and amides) are important components of the B-complex vitamins.
- Basic compounds that are derived from plants and show physiological activity are known as alkaloids.
- Many alkaloids are amines.
  - Opium alkaloids have sleep-inducing and analgesic properties.
  - Amphetamines stimulate the central nervous system.
  - Barbiturates are sedatives.
- Many other common drugs, including tranquilizers and antidepressants, are amines.

Review Questions

All questions with blue numbers have answers in the appendix of the text.

1. Write the structural formula and name of (a) a typical amine and (b) a typical amide. Explain how the difference in functional groups affects the chemical properties of amines and amides.
2. Contrast the physical properties of amides with those of amines.
3. Explain why amines have approximately the same water solubility as alcohols of similar molar mass.

4. Explain why unsubstituted amides have a much higher melting point than esters of similar molar mass.

5. (a) What is a heterocyclic compound?
(b) How many heterocyclic rings are present in
(i) purine, (ii) ampicillin, (iii) methadone, and
(iv) nicotine?

6. Indicate the functional groups present in
(a) procaine hydrochloride (c) nicotinamide
(b) cocaine (d) methamphetamine

7. Explain why it is advantageous for the body to excrete urea rather than ammonia.

8. Explain what is meant by "condensation polymer." Illustrate your answer by showing the first reaction between adipic acid and 1,6-hexanediamine to start formation of Nylon-66.

9. What is the meaning of the term *quaternary* in quaternary ammonium salts?

10. Why do tertiary amines not react with acid chlorides to form amides?

Paired Exercises

All exercises with blue numbers have answers in the appendix of the text.

1. Draw structural formulas for the following:
(a) *p*-methylaniline
(b) 2-butanamine
(c) *N*-methyl-*p*-bromobenzamide

2. Draw structural formulas for the following:
(a) 4-chloro-2-pentanamine
(b) *N,N*-diethylbenzamide
(c) *m*-chloroaniline

3. Name the following compounds:

(a) $CH_3C(=O){-}NH_2$ (common name)

(b) $CH_3CH(OH)CH_2CH(NH_2)CH_3$ (IUPAC name)

(c) benzene ring bearing $C(=O){-}NHCH_3$ and, meta to it, CH_3 (IUPAC name)

(d) $CH_3{-}N(CH_2CH_3){-}CH_2C(CH_3)_2CH_2CH_3$ (IUPAC name)

4. Name the following compounds:

(a) $CH_3CH(CH_3)CH_2CH_2NH_2$ (IUPAC name)

(b) $HC(=O){-}NH_2$ (common name)

(c) $CH_3CH_2C(=O){-}NH{-}C_6H_5$ (IUPAC name)

(d) $CH_3CH(CH_3){-}N(CH_2CH_3){-}CH(CH_3)CH_3$ (common and IUPAC names)

5. Arrange the following set of compounds in order of increasing solubility in water:

$CH_3CH_2C(=O){-}NHCH_3$

$CH_3CH_2C(=O){-}N(CH_3)_2$

$CH_3CH_2C(=O){-}NH_2$

6. Arrange the following set of compounds in order of increasing solubility in water:

$CH_3C(=O){-}NH_2$

$CH_3(CH_2)_4C(=O){-}NH_2$

$C_6H_5C(=O){-}NH_2$

7. Acetamide is soluble in water. Show the possible hydrogen bonding that helps explain this property.

8. Urea is soluble in water. Show the possible hydrogen bonding that helps explain this property.

9. Classify the following compounds as an acid, a base, or neither:

(a) CH_3CH_2OH

(b) $CH_3CH_2NH_2$

(c) 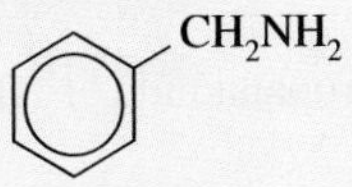 $C_6H_5CH_2NH_2$ (benzene ring with CH_2NH_2)

(d) 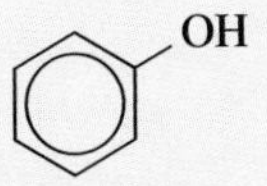 C_6H_5OH (benzene ring with OH)

(e) $CH_3CH_2\overset{O}{\overset{\|}{C}}—NH_2$

(f) $CH_3CH_2\overset{O}{\overset{\|}{C}}—OH$

10. Classify the following compounds as an acid, a base, or neither:

(a) $CH_3CH_2OCH_2CH_3$

(b) $CH_3CH_2NHCH_2CH_3$

(c) 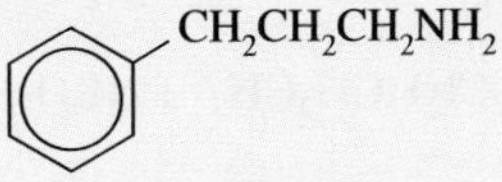 benzene ring with $CH_2CH_2CH_2NH_2$

(d) 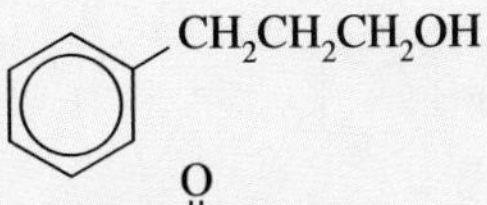 benzene ring with $CH_2CH_2CH_2OH$

(e) 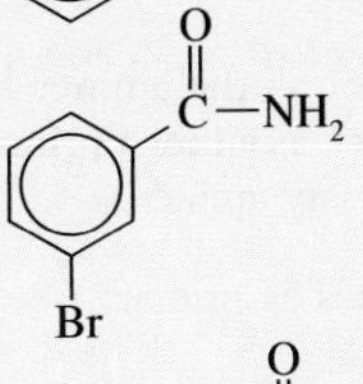 benzene ring with $\overset{O}{\overset{\|}{C}}—NH_2$ and Br

(f) 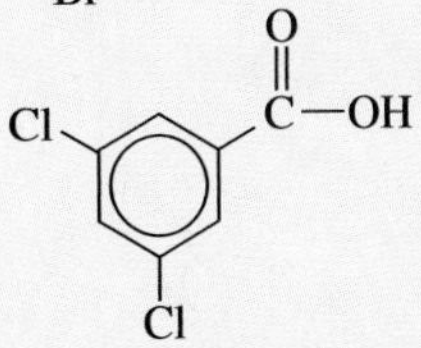 benzene ring with $\overset{O}{\overset{\|}{C}}—OH$, Cl, and Cl

11. Give the structural formula and name for the organic product of the following reaction:

(a) $CH_3\overset{O}{\overset{\|}{C}}NHCH_3 \xrightarrow{LiAlH_4}$

(b) $N{\equiv}C\overset{CH_3}{\overset{|}{C}}HCH_3 \xrightarrow{H_2/Pt}$

(c) $CH_3\underset{CH_3}{\underset{|}{C}}HNH_2 + CH_3Br\ (1\ mole) \longrightarrow$

(d) 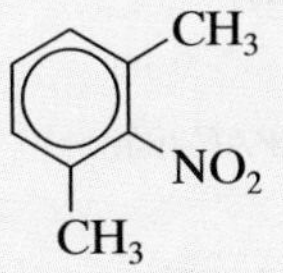 benzene ring with CH_3, NO_2, CH_3 $\xrightarrow[HCl]{Sn} \xrightarrow{NaOH}$

12. Give the structural formula and name for the organic product of the following reaction:

(a) 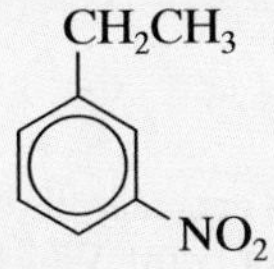 benzene ring with CH_2CH_3 and NO_2 $\xrightarrow[HCl]{Sn} \xrightarrow{NaOH}$

(b) $H\overset{O}{\overset{\|}{C}}NH_2 \xrightarrow{LiAlH_4}$

(c) $CH_3CH_2\overset{CH_3}{\overset{|}{C}}HC{\equiv}N \xrightarrow{H_2/Pt}$

(d) $CH_3Br\ (2\ moles) + NH_2CH_2CH_3 \longrightarrow$

13. Predict the organic products for the following reactions:

(a) $CH_3NH—\overset{O}{\overset{\|}{C}}CH_3 + H_2O + H^+ \xrightarrow{\Delta}$

(b) Acetic acid + isopropylamine $\xrightarrow{\Delta}$

(c) 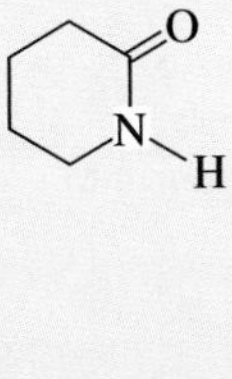 six-membered lactam ring (C=O, N–H) + NaOH $\xrightarrow{\Delta}$

14. Predict the organic products for the following reactions:

(a) Acetic acid + diethylamine $\xrightarrow{\Delta}$

(b) 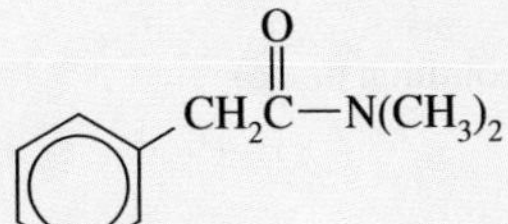 benzene ring with $CH_2\overset{O}{\overset{\|}{C}}—N(CH_3)_2$ $+ H_2O + H^+ \xrightarrow{\Delta}$

(c) $CH_3\underset{CH_3}{\underset{|}{C}}HCH_2\overset{O}{\overset{\|}{C}}—OH + CH_3NH_2 \xrightarrow{\Delta}$

15. Predict the reactants that yield the following product(s):

(a) $\xrightarrow{\Delta}$ *N*-methylbutanamide + H_2O

(b) $\xrightarrow{\Delta}$ $CH_3CH_2\overset{\overset{O}{\|}}{C}NHCH_2CH_3$ + H_2O

(c) $\xrightarrow{\Delta}$ $H\overset{\overset{O}{\|}}{C}OH$ + (cyclohexyl)–$\overset{+}{N}H_3\overset{-}{Cl}$

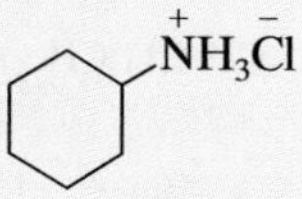

17. Draw structural formulas for all the amines having the formula $C_4H_{11}N$. Classify each as either a primary amine, a secondary amine, or a tertiary amine.

19. Classify the following amines as primary, secondary, or tertiary:

(a) $CH_3CH_2CH_2NH_2$

(b) 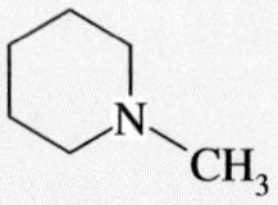 (piperidine ring) N–CH_3

(c) 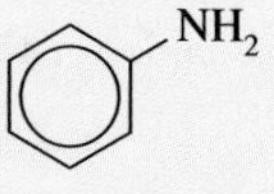 (benzene ring)–NH_2

(d) $H_2NCH_2CH_2CH_2NH_2$

21. Low-molar-mass aliphatic amines generally have odors suggestive of ammonia or stale fish (or both). Which of the following solutions would have the stronger odor? Explain.
(a) a 1% trimethylamine solution in 1.0 *M* sulfuric acid
(b) a 1% trimethylamine solution in 1.0 *M* NaOH

23. Name the following compounds:

(a) CH_3NHCH_3 (common name)

(b) (benzene ring with CH_2CH_3 and NH_2 on adjacent carbons) (IUPAC name)

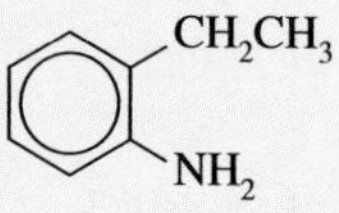

(c) (cyclohexane ring)–NH_2 (IUPAC name)

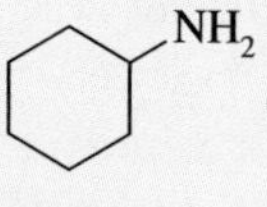

(d) $(C_2H_5)_4N^+I^-$ (common name)

(e) (benzene ring with NH_2 and NO_2, meta) (IUPAC name)

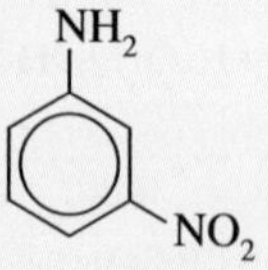

(f) $(CH_3CH_2)_2N$–(cyclohexane ring) (common name)

16. Predict the reactants that yield the following product(s):

(a) $\xrightarrow{\Delta}$ $CH_3\overset{\overset{O}{\|}}{C}O^-\overset{+}{Na}$ + $CH_3CH_2NH_2$

(b) $\xrightarrow{\Delta}$ *N,N*-dimethylpropanamide + H_2O

(c) $\xrightarrow{\Delta}$ NH_4Cl + $CH_3\overset{\overset{O}{\|}}{C}OH$

18. Draw structural formulas for all the amines having the formula C_3H_9N. Classify each as either a primary amine, a secondary amine, or a tertiary amine.

20. Classify the following amines as primary, secondary, or tertiary:

(a) CH_3NHCH_3

(b) $CH_3CH_2N(CH_3)_2$

(c) (cyclohexane ring)–NH–(benzene ring)

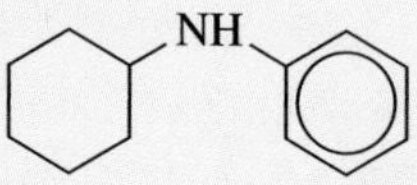

(d) $CH_3-\underset{\underset{CH_2CH_3}{|}}{N}-CH_2CH_2CH_2CH_3$

22. Low-molar-mass aliphatic amines generally have odors suggestive of ammonia or stale fish (or both). Which of the solutions that follow would have the stronger odor? Explain.
(a) a 1% isopropylamine solution in 1.0 *M* hydrochloric acid
(b) a 1% isopropylamine solution in 1.0 *M* KOH

24. Name the following compounds:

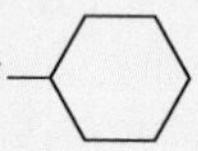

(a) 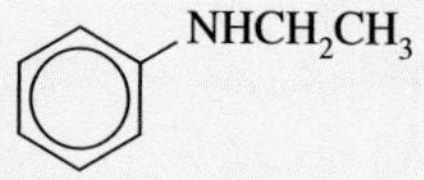 (benzene ring)–$NHCH_2CH_3$ (IUPAC name)

(b) $CH_3CH_2\underset{\underset{NH_2}{|}}{C}HCH_3$ (IUPAC name)

(c) 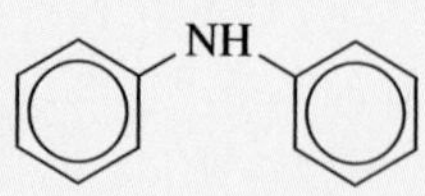 (benzene ring)–NH–(benzene ring) (IUPAC name)

(d) $CH_3CH_2NH_3^+Br^-$ (common name)

(e) 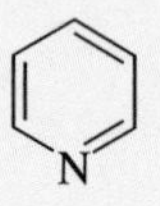 (pyridine ring, N) (common name, which is also IUPAC name)

(f) $CH_3\overset{\overset{O}{\|}}{C}-NHCH_2CH_3$ (common name)

25. Draw structural formulas for the following compounds:
(a) *N*-methylethanamine
(b) aniline
(c) 1,4-butanediamine
(d) ethylisopropylmethylamine
(e) pyridine
(f) triethylammonium chloride

26. Draw structural formulas for the following compounds:
(a) tributylamine
(b) *N*-methylanilinium chloride
(c) ethylammonium chloride
(d) 2-amino-1-pentanol
(e) 3,3-dimethyl-*N*-phenyl-2-aminohexane
(f) 1,2-ethanediamine

27. Write equations to show how each conversion may be accomplished (some conversions may require more than one step):

(a) $CH_3CH_2CH_2Br \longrightarrow CH_3CH_2CH_2NH_2$

(b) $CH_3CH_2CH_2Br \longrightarrow CH_3CH_2CH_2CH_2NH_2$

(c) 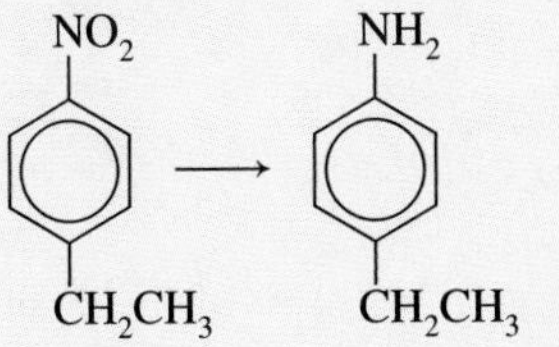 [benzene ring with NO_2 and para CH_2CH_3] $\longrightarrow$ [benzene ring with NH_2 and para CH_2CH_3]

(d) $CH_3CH_2CH_2NH_2 \longrightarrow CH_3\overset{O}{\overset{\|}{C}}NHCH_2CH_2CH_3$

28. Show structural formulas for the organic products from the following reactions:

(a) $CH_3CH_2\overset{O}{\overset{\|}{C}}{-}O^-\ NH_4^+ \xrightarrow{\Delta}$

(b) $CH_3CH_2CH_2C{\equiv}N + H_2 \xrightarrow{Ni}$

(c) 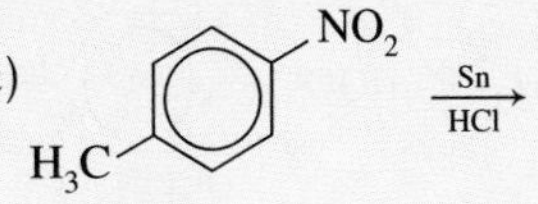 [benzene ring with NO_2 and H_3C substituents] $\xrightarrow[HCl]{Sn}$

(d) 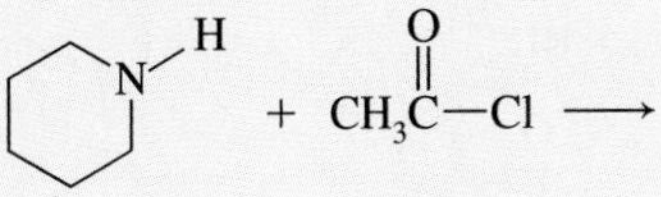 [piperidine ring, N–H] $+\ CH_3\overset{O}{\overset{\|}{C}}{-}Cl \longrightarrow$

29. Draw structures for each of the following and give correct IUPAC names:
(a) 2-ethyl-2-butamine
(b) triethylamine
(c) *N*-propylethanamine

30. Draw structures for each of the following and give correct IUPAC names:
(a) dipropylamine
(b) *N*-propylmethanamine
(c) 2-ethyl-4-pentanamine

Additional Exercises

All exercises with blue numbers have answers in the appendix of the text

31. Propylamine, ethylmethylamine, and trimethylamine have the same molecular formula, C_3H_9N. Explain why the boiling point of trimethylamine is considerably lower than the boiling points of the other two compounds.

32. List three classes of drugs that are also amines, and indicate the major physiological effects of each.

33. When you experience a sudden fright, your blood epinephrine (adrenalin) level rises quickly to 1.0×10^{-7} *M*. Given a blood volume of 5.0 L, how many grams of epinephrine are in circulation when you are in this condition?

34. Show the structure known as an amide linkage. Indicate some important classes of biochemicals that contain this linkage.

35. After cleaning or packing fish, workers often use lemon juice to clean their hands. What is the purpose of the lemon juice? Explain.

36. Cephalosporins are antibiotics that are often effective against penicillin-resistant bacteria.

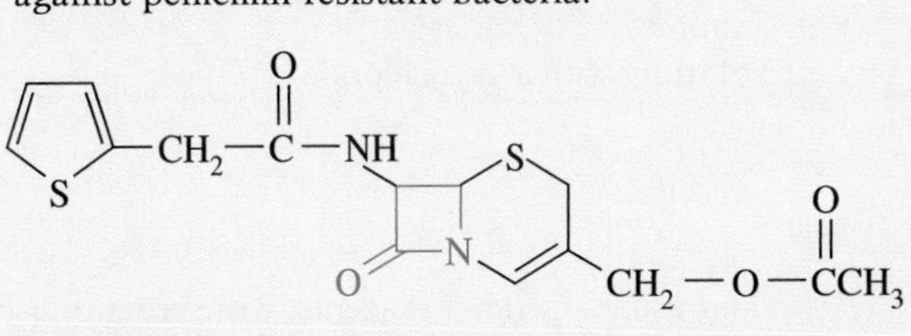

[Structure: thiophene ring–CH_2–C(=O)–NH– attached to a β-lactam fused to a sulfur-containing six-membered ring bearing $CH_2{-}O{-}\overset{O}{\overset{\|}{C}}CH_3$]

(The blue ring is critical to the cephalosporin's antibiotic activity.) Identify the amide and ester groups.

37. Write a chemical reaction in which aniline acts as a base.

38. Putrescine, a product of decay of organic matter, is necessary for the growth of cells and occurs in animal tissues. Draw the structure and identify the type of nitrogen bonds that are present.

39. Why are many drugs given as an ammonium salt?

40. Aspartame is 200 times sweeter than sucrose. Circle any amine or amide in this structure:

$$H_2NCH(CH_2COOH)C(=O)-NHCH(CH_2C_6H_5)C(=O)-OCH_3$$

41. The polymer Kevlar is many times stronger than steel and is often used to make bulletproof vests. This polymer is formed by linking terephthalic acid and *p*-aminoaniline via amide bonds. Show the structure of the polymer (two units).

$$H_2N-C_6H_4-NH_2 \qquad HOOC-C_6H_4-COOH$$

42. (a) Write structures for the monomers used to make the polymer shown here:

$$-C(=O)-(CH_2)_8-C(=O)-NH-(CH_2)_6-NH-C(=O)-(CH_2)_8-C(=O)-NH-(CH_2)_6-NH-$$

(b) The polymer Quiana is spun into a soft silky fabric. The structure of the polymer is as follows:

$$-C(=O)-(CH_2)_6-C(=O)-NH-C_6H_{10}-CH_2-C_6H_{10}-NH-C(=O)-(CH_2)_6-C(=O)-NH-C_6H_{10}-CH_2-C_6H_{10}-NH-$$

Draw the structures of the two monomers used to make Quiana.

Challenge Exercise

43. (a) Starting with ethanol and using any other reagents, write reactions for the synthesis of 1,4-butanediamine.

(b) Starting with acetyl chloride,

$$CH_3C(=O)-Cl$$

and using any other reagents, write reactions for the synthesis of a primary, a secondary, and a tertiary amine.

Answers to Practice Exercises

25.1

(a) $HC(=O)-NHCH(CH_3)CH_3$

(b) $CH_3CH_2CH_2C(=O)-NHCH_2CH_3$

(c) $CH_3CH(CH_3)C(=O)-NHCH_3$

(d) $CH_3CH_2CH_2CH_2CH_2C(=O)-N(CH_2CH_3)-CH_2CH_3$

(e) $H_2N-C_6H_4-C(=O)-NH_2$

25.2

(a) acid hydrolysis $CH_3CH_2C(=O)-OH + NH_4^+$

basic hydrolysis $CH_3CH_2C(=O)-O^- + NH_3$

(b) acid hydrolysis $CH_3C(=O)-OH + \overset{+}{N}H_2(CH_3)_2$

basic hydrolysis $CH_3C(=O)-O^- + NH(CH_3)_2$

25.3

$\sim CH(\#14)-C(=O)-NH-CH(\#15)-C(=O)-OH + NH_2-CH(\#16)-C(=O)-NH-CH(\#17)\sim$

25.4 (a) 1° (b) 1° (c) 2° (d) 1° (e) 3° (f) 2° (g) 1°

25.5 (a) tryptamine, 1°, 2°
(b) epinephrine, 2°
(c) alanine, 1°

25.6 (a) propanamine (propylamine)
(b) *N*-isopropyl-*N*-methylaniline
(c) 2-butanamine (sec-butylamine)
(d) benzylamine
(e) *N*-ethyl-*N*-methylbutanamine

25.7 $NH_2CH_2CH_2CH_2COOH$

25.8

(a) $CH_3C(=O)-NHCH_3$ or $HC(=O)-NHCH_2CH_3$

(b) $CH_3CH(CH_3)C\equiv N$ or $CH_3CH(CH_3)C(=O)-NH_2$

(c) $CH_3CH_2CH(CH_2CH_3)CH(CH_3)C\equiv N$ or $CH_3CH_2CH(CH_2CH_3)CH(CH_3)C(=O)-NH_2$

25.9

Ring: $H-N-C(=O)-CH_2-CH(COOH)-N(H)-C(=O)-$ (six-membered ring) $+ H_2O$

PUTTING IT TOGETHER: Review for Chapters 24–25

Multiple Choice:

Choose the correct answer to each of the following.

1. The functional group for carboxylic acids is

(a) —COOH (c) —C=O

(b) >CHOH (d) 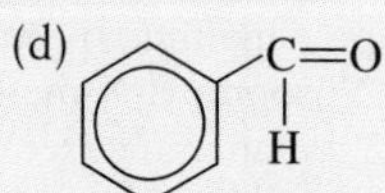

2. The IUPAC name for $CH_3CH_2CH_2COOH$ is

(a) butyric acid (c) butanoic acid

(b) butonic acid (d) propanoic acid

3. The common name for the six-carbon dioic acid is

(a) oxalic acid (c) adipic acid

(b) succinic acid (d) malonic acid

4. How many carbon atoms are in the compound diethylmalonate?

(a) 4 (b) 5 (c) 7 (d) 8

5. What is the name for this compound?

COONa

CH_2CH_3

(a) *m*-ethylbenzoic acid

(b) sodium *m*-ethylbenzoate

(c) sodium *m*-toluate

(d) sodium *m*-methyltoluate

6. The name for $CH_3CHBrCH_2COOH$ is

(a) α-bromobutyric acid (c) γ-bromobutyric acid

(b) β-bromobutyric acid (d) 2-bromobutanoic acid

7. The formula for stearic acid is

(a) $CH_3(CH_2)_{10}COOH$ (c) $CH_3(CH_2)_{14}COOH$

(b) $CH_3(CH_2)_{12}COOH$ (d) $CH_3(CH_2)_{16}COOH$

8. Which compound can show cis–trans isomerism?

(a) succinic acid (c) adipic acid

(b) fumaric acid (d) *m*-toluic acid

9. Which compound is oxalic acid?

(a) COOH, COOH

(b) COOH, COOH

(c) COOH—COOH

(d) COOH—CH_2—COOH

10. Which of these acids is the least soluble in water?

(a) HCOOH (c) $CH_3CH_2CH_2COOH$

(b) CH_3COOH (d) $CH_3(CH_2)_6COOH$

11. Which class of compounds generally has pleasant odors?

(a) carboxylic acids (c) amides

(b) acyl halides (d) esters

12. Which of these compounds does *not* have intermolecular hydrogen bonding?

(a) $CH_3C(=O)—OCH_3$ (c) $CH_3C(=O)—NHCH_3$

(b) CH_3COOH (d) CH_3CH_2OH

13. Which compound, when oxidized, yields terphthalic acid?

(a) 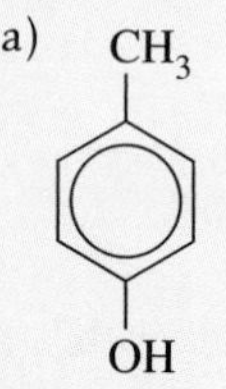

(c) CH_3, CH_2CH_3

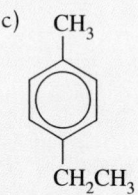

(b) 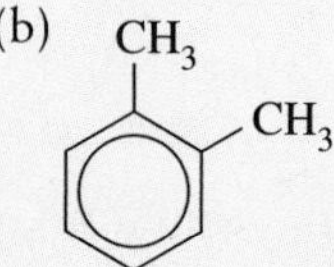

(d) 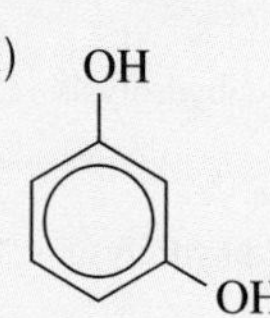

14. What product is formed in this reaction?

$$CH_3CH_2COOH + CH_3CH_2OH \xrightarrow{H^+}$$

(a) $CH_3C(=O)—OCH_2CH_2CH_3$

(b) $CH_3CH_2C(=O)—OCH_2CH_3$

(c) $CH_3CH_2CH_3—O—CH_2CH_3$

(d) $CH_3CH_2C(=O)—O—C(=O)CH_2CH_3$

15. What products are formed when this compound is hydrolyzed in an alkaline solution?

$$CH_3CH_2C(=O)—OCH_2CH_3$$

(a) $CH_3CH_2COOH + CH_3CH_2OH$

(b) $CH_3COOH + CH_3CH_2CH_2OH$

(c) $CH_3CH_2COONa + CH_3CH_2OH$

(d) $CH_3CH_2COONa + CH_3CH_2ONa$

16. What is the name for this compound?

$$CH_3\overset{\overset{\displaystyle O}{\|}}{C}-O-\overset{\overset{\displaystyle O}{\|}}{C}CH_3$$

(a) dimethylcarbonyl ether (c) diacetyl ether
(b) acetic anhydride (d) no correct name given

17. Which compound will be formed in this reaction?

(benzene ring with $\overset{\overset{\displaystyle O}{\|}}{C}-OH$ and OH) $+\ CH_3\overset{\overset{\displaystyle O}{\|}}{C}-OH \xrightarrow{H^+}$

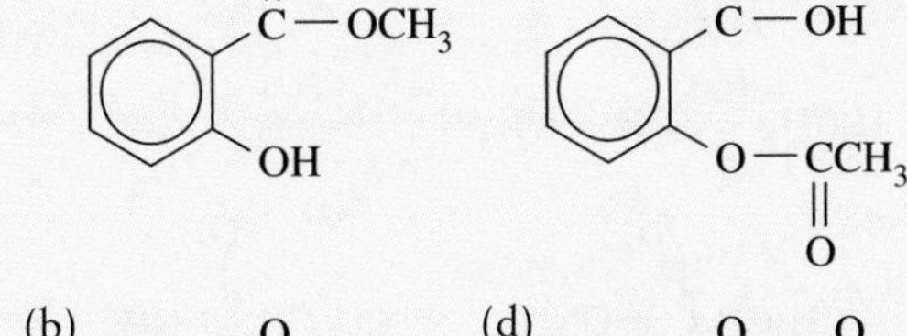

(b) (benzene ring with $\overset{\overset{\displaystyle O}{\|}}{C}-OH$ and OCH_3)

(d) (benzene ring with $\overset{\overset{\displaystyle O}{\|}}{C}-O\overset{\overset{\displaystyle O}{\|}}{C}CH_3$ and OH)

18. Which ester has the odor of pineapple?
(a) octyl acetate (c) ethyl butyrate
(b) isoamyl acetate (d) methyl salicylate

19. Which compound is aspirin?
(a) (benzene ring with COOH and $O\overset{}{C}CH_3$, $\|$ O)
(c) (benzene ring with $\overset{\overset{\displaystyle O}{\|}}{C}-OH$ and O–phenyl)
(b) (benzene ring with $\overset{\overset{\displaystyle O}{\|}}{C}-OCH_3$ and OH)
(d) (benzene ring with $\overset{\overset{\displaystyle O}{\|}}{C}-OCH_2CH_3$ and OH)

20. Dacron polyester is a copolymer of
(a) oxalic acid and ethylene glycol
(b) terphthalic acid and ethylene glycol
(c) malonic acid and ethylene glycol
(d) adipic acid and 1,6-hexanediamine

21. Which formula is linoleic acid?
(a) $CH_3(CH_2)_{16}COOH$
(b) $CH_3(CH_2)_4CH{=}CH(CH_2)_7COOH$
(c) $CH_3(CH_2)_7CH{=}CH(CH_2)_7COOH$
(d) $CH_3(CH_2)_4CH{=}CHCH_2CH{=}CH(CH_2)_7COOH$

22. If glyceryl tripalmitate is saponified with NaOH, the products are
(a) glycerol and tripalmitate
(b) glycerol and $CH_3(CH_2)_{16}COONa$
(c) glycerol and $CH_3(CH_2)_{14}COONa$
(d) glycerol and $CH_3(CH_2)_{14}CH_2ONa$

23. Which of these compounds is a soap?
(a) $CH_3CH_2CH_2COONa$
(b) $CH_3(CH_2)_{14}CH_2ONa$
(c) $CH_3(CH_2)_{14}COONa$
(d) $[CH_3(CH_2)_{14}COO]_2Ca$

Consider formulas **A**, **B**, **C**, *and* **D** *for Questions 24–26.*

(A) $CH_3(CH_2)_{12}CH_2OSO_3Na$
(B) $CH_3(CH_2)_{14}CH_2N(CH_3)_3Cl$
(C) $CH_3(CH_2)_{10}CH_2-O-(CH_2CH_2O)_7-CH_2CH_2OH$

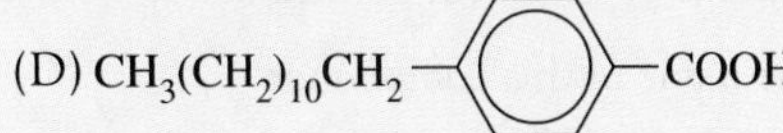

24. Which formula is a cationic detergent?
(a) A (b) B (c) C (d) D

25. Which formula is an anionic detergent?
(a) A (b) B (c) C (d) D

26. Which formula is a nonionic detergent?
(a) A (b) B (c) C (d) D

27. How many ester linkages can be made from one molecule of phosphoric acid?
(a) 1 (b) 2 (c) 3 (d) 4

28. Which formula is an acyl halide?

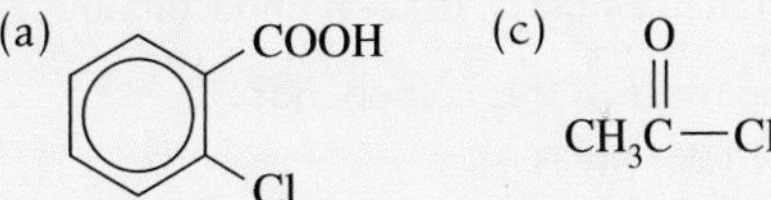

(c) $CH_3\overset{\overset{\displaystyle O}{\|}}{C}-Cl$

(b) $CH_3CHClCOOH$

(d) $CH_3\overset{\overset{\displaystyle O}{\|}}{C}-CH_2Cl$

29. The amide structure in a compound is
(a) neutral (c) basic
(b) acidic (d) both acidic and basic

30. In an amide, the nitrogen atom is bonded to
(a) an alkyl group (c) a carbonyl group
(b) a benzene ring (d) a secondary carbon atom

31. The formula for 2-methylbutanamide is

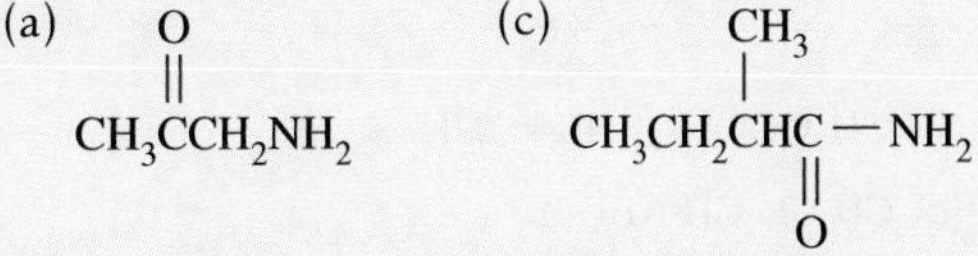

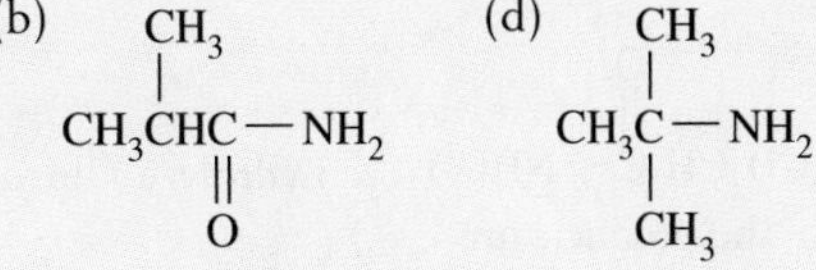

32. What is the name of this compound?

$C_6H_5-C(=O)-N(CH_3)-CH_2CH_3$

(a) *N*-isopropylbenzamide
(b) ethylmethylbenzamide
(c) *N*-ethyl-*N*-methylbenzoamide
(d) *N*-ethyl-*N*-methylbenzamide

33. What is the name of this compound?

$C_6H_5-NH-C(=O)CH_3$

(a) aniline ethanoate (c) acetanilide
(b) methanilide (d) benzanilide

34. What is the name of this compound?

$CH_3CH(CH_2CH_3)CH_2CH_2C(=O)-NH_2$

(a) 4-ethylpentanamide (c) 4-methylhexanamide
(b) isohexanamide (d) 3-methylcaproamide

35. What is the IUPAC name for this compound:

$H_2NCH_2CH_2CH_2CH_2CH_2CH_2CH_2C(=O)-NH_2$

(a) 7-aminooctanamide (c) 1,8-aminooctanamide
(b) 1,8-octanediamine (d) 8-aminooctanamide

36. What is the name of this compound?

$CH_3CH(CH_3)CH_2C(=O)-N(CH_3)_2$

(a) trimethylbutyramide
(b) *N,N*-dimethyl-3-methylbutanamide
(c) *N,N*-dimethyl-2-methylbutanamide
(d) no correct answer given

37. When $CH_3CH_2C(=O)-NH_2$ is hydrolyzed in an acid solution, the products are
(a) $CH_3CH_2COOH + NH_4^+$
(b) $CH_3COOH + HC(=O)-NH_2$
(c) $CH_3CH_2CH_2NH_2$
(d) $CH_3CH_2COOH + NH_4OH$

38. When $CH_3CH_2C(=O)-NHCH_3$ is hydrolyzed in a basic solution, the products are
(a) $CH_3CH_2CH(OH)NHCH_3$
(b) $CH_3CH_2COOH + HC(=O)-NHCH_3$
(c) $CH_3CH_2COONa + CH_3NH_2$
(d) $CH_3CH_2C(=O)-H + CH_3NH_2$

What are the product(s) for the reactions shown in Questions 39 and 40?

39. $CH_3CH_2C(=O)-OH + NH_3 \longrightarrow$
(a) $CH_3CH_2C(OH)-ONH_2$ (c) $CH_3CH_2CH(OH)NH_2$
(b) $CH_3CH_2C(=O)-ONH_4$ (d) $CH_3CH(NH_2)COOH$

40. $CH_3CH_2C(=O)-Cl + NH_3 \longrightarrow$
(a) $CH_3CH_2C(=O)-NH_2 + NH_4Cl$
(b) $CH_3CH_2CH(OCl)NH_2$
(c) $CH_3CH_2CH_2NH_2 + NH_4Cl$
(d) $CH_3CH_2C(OH)ClNH_2$

41. Which of the following compounds is the most acidic?
(a) $CH_3C(=O)-H$ (c) $CH_3C(=O)-NH_2$
(b) CH_3COOH (d) $CH_3CH_2NH_2$

42. Which of the following compounds is a secondary amine?
(a) $CH_3CH_2NH_2$ (c) $C_6H_5-CH_2NH_2$
(b) $CH_3CH_2NHCH_3$ (d) $CH_3CH_2C(=O)-N(CH_3)_2$

43. Which class of amine is this compound?

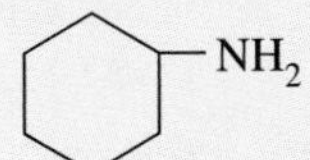

(a) primary (c) tertiary
(b) secondary (d) quaternary

44. Which compound is *not* capable of hydrogen bonding with water?

(a) CH_3NHCH_3
(b) $(CH_3CH_2)_2\overset{+}{N}(CH_3)CH_3$
(c) pyrrolidine ring with N–H
(d) $CH_3CH_2NH_2$

45. The formula for urea is

(a) $CH_3C(=O)—NHCH_3$
(b) $H_2NCH_2CH_2NH_2$
(c)
(d) $H_2N—C(=O)—NH_2$

46. The formula for anilinium bromide is

(a) benzene ring–NHBr
(b) benzene ring with NH_2 and Br (meta)
(c) benzene ring with Br and NH_2 (ortho)
(d) benzene ring–$\overset{+}{N}H_3Br^-$

47. Which formula is a quaternary ammonium ion?

(a) $[CH_3NH_3]^+$ (c) $[(CH_3)_3NH]^+$
(b) $[CH_3NH_2CH_3]^+$ (d) $[(CH_3)_4N]^+$

What is the product of the reactions shown in Questions 48–50?

48. $CH_3CH_2CN \xrightarrow{H_2/Ni}$

(a) $CH_3CH_2NH_2$
(b) $CH_3CH_2CH_2NH_2$
(c) $CH_3CH_2C(=O)—NH_2$
(d) $CH_3CH_2CH_2NHCH_2CH_2CH_3$

49. $CH_3CH_2Cl + CH_3NH_2 \longrightarrow$

(a) $CH_3CH_2NHCH_3$
(b) $CH_3C(CH_3)(NH_2)Cl$
(c) $CH_3CHClNHCH_3$
(d) $CH_3N(CH_3)CH_3$

50. benzene ring–$C(=O)—NH_2 \xrightarrow{LiAlH_4}$

(a) benzene ring–CN
(b) benzene ring–NH_2
(c) benzene ring–CH_2NH_2
(d) benzene ring–$C(=O)—NH—C(=O)$–benzene ring

51. Which set of reagents will convert nitrobenzene to aniline?

(a) $AlCl_3 + NH_3$ (c) Sn + HCl/NaOH
(b) $NaOH + NH_3$ (d) $FeCl_3$

52. Which of the following compounds is nicotine?

(a) benzene ring–$CH_2CH(NHCH_3)CH_3$
(b) HO, HO-substituted benzene ring–$CH(OH)CH_2NHCH_3$
(c) pyridine ring–pyrrolidine ring with N–CH_3
(d) benzene ring–$C(=O)—NHCH_3$

53. Which statement does *not* apply to Nylon-66?

(a) It is a copolymer.
(b) It is a condensation polymer.
(c) It is a thermosetting polymer.
(d) It contains carbon, hydrogen, oxygen, and nitrogen.

54. Which formula is the repeating structure for Nylon-66?

(a) $-[\overset{O}{\overset{\|}{C}}-(CH_2)_4NH(CH_2)_6NH]_n-$

(b) $-[\overset{O}{\overset{\|}{C}}-(CH_2)_4-\overset{O}{\overset{\|}{C}}-NH(CH_2)_6-NH]_n-$

(c) $-[\overset{O}{\overset{\|}{C}}-(CH_2)_6\overset{O}{\overset{\|}{C}}-NH(CH_2)_6-NH]_n-$

(d) $-[\overset{O}{\overset{\|}{C}}-(CH_2)_6-NH]_n-$

55. How many isomers (amines) can be written that have a molecular formula of $C_4H_{11}N$?
(a) 2 (b) 4 (c) 6 (d) 8

Free Response Questions

1. Which of the fatty acids that follow is most likely synthetically produced? Explain.

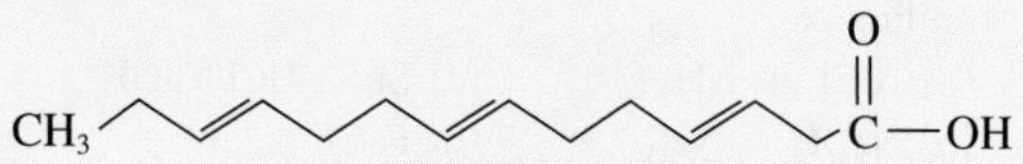

CH_3 ... $\overset{O}{\overset{\|}{C}}-OH$

I (trans isomer)

CH_3 ... $\overset{O}{\overset{\|}{C}}-OH$

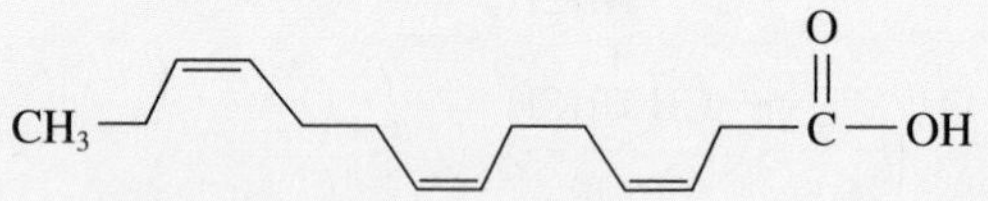

II (cis isomer)

2. In the Krebs cycle, succinic acid is converted to fumaric acid, which in turn is converted to malic acid. Draw the reaction sequence and indicate what type of reaction is occurring in each of these two processes. In the body, enzymes facilitate these reactions, but we can also carry them out synthetically in the lab. Suggest how the second step could be done in the lab.

3. What are the formulas and names of the products expected when phenyl benzoate is saponified by using NaOH? Would the products be different if a weaker base were used?

4. Which of the following compounds contain(s) a carboxylic acid functional group?

$CH_3CH_2CH_2\underset{OH}{\underset{|}{C}}H\overset{O}{\overset{\|}{C}}-H$

$CH_3CH_2CH_2CH_2COOH$

$HOCH_2\overset{O}{\overset{\|}{C}}CH_2CH_2CH_2\overset{O}{\overset{\|}{C}}-OH$

5. Would you expect phosphate monoesters to be more or less soluble in water than carboxylic esters? Explain. Which would you expect to have a higher melting and boiling point? (Assume R is such that the molar masses are similar.)

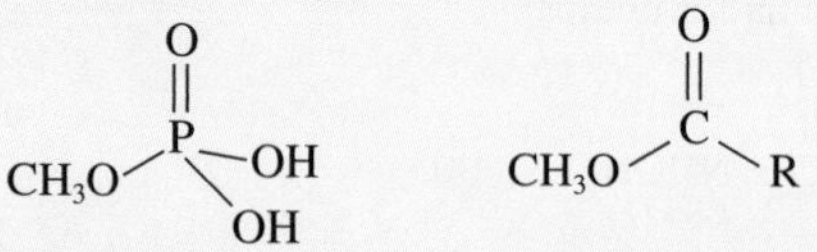

$CH_3O-\overset{O}{\overset{\|}{P}}(-OH)-OH$ $\qquad$ $CH_3O-\overset{O}{\overset{\|}{C}}-R$

6. Draw the organic compounds I–III in the following reaction sequence also indicate what type of reaction is occurring in each step:

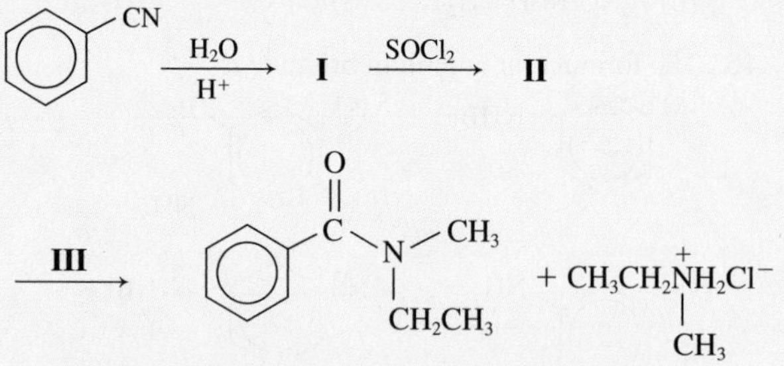

$C_6H_5-CN \xrightarrow[H^+]{H_2O}$ I $\xrightarrow{SOCl_2}$ II

III $\longrightarrow C_6H_5-\overset{O}{\overset{\|}{C}}-N(CH_3)(CH_2CH_3) + CH_3CH_2\overset{+}{N}H_2(CH_3)Cl^-$

7. Although amides can be prepared by the reaction of a carboxylic acid with ammonia or an amine, the acid halide (RCOCl) is more commonly used in place of the carboxylic acid. Suggest two advantages of using the acid halide over the carboxylic acid.

8. Draw structures I–III for the following reaction sequence:

I (a five carbon carboxylic acid with a double bond between the γ and δ carbons) $\xrightarrow{H_2O/H^+}$ II

$\xrightarrow[heat]{H^+}$ III (product of a condensation reaction)

CHAPTER 26

Stereoisomerism

The mirror image of this children's ballet class is not superimposable on the class.

Chapter Outline

Many of us grew up hearing such comments as "Can't you tell your left from your right?" Have you ever watched a small child try to differentiate between a right and left shoe? Not surprisingly, the distinction between right and left is difficult. After all, our bodies are reasonably symmetrical. For example, both hands are made up of the same components (four fingers, a thumb, and a palm) ordered in the same way (from thumb through little finger). Yet, there is a difference if we try to put a left-handed glove on our right hand or a right shoe on our left foot.

Molecules possess similar, subtle structural differences, which can have a major impact on their chemical reactivity. For example, although there are two forms of blood sugar, related as closely as our left and right hands, only one of these structures can be used by our bodies for energy. Stereoisomerism is a subject that attempts to define these subtle differences in molecular structure.

26.1 Review of Isomerism

Isomerism is the phenomenon of two or more compounds having the same number and kinds of atoms. (See Section 19.8.) There are two types of isomerism. In the first type, known as structural isomerism, the difference between isomers is due to different structural arrangements of the atoms that form the molecules. For example, butane and isobutane, ethanol and dimethyl ether, and 1-chloropropane and 2-chloropropane are structural isomers:

$CH_3CH_2CH_2CH_3$ butane	CH_3CH_2OH ethanol	$CH_3CH_2CH_2Cl$ 1-chloropropane
CH_3CHCH_3 (with CH_3 attached to the middle carbon) 2-methylpropane (isobutane)	CH_3OCH_3 methoxymethane (dimethyl ether)	$CH_3CHClCH_3$ 2-chloropropane

stereoisomerism
stereoisomer

In the second type of isomerism, the isomers have the same structural formulas, but differ in the spatial arrangement of the atoms. This type of isomerism is known as **stereoisomerism**. Thus, compounds that have the same structural formulas but differ in their spatial arrangement are called **stereoisomers**. There are two types of stereoisomers: cis–trans or geometric isomers, which we have already considered, and optical isomers, the subject of this chapter. One outstanding feature of optical isomers is that they have the ability to rotate the plane of plane-polarized light.

26.2 Plane-Polarized Light

plane-polarized light

Plane-polarized light is light that is vibrating in only one plane. Ordinary (unpolarized) light consists of electromagnetic waves vibrating in all directions (planes) perpendicular to the direction in which it is traveling. When ordinary light passes through a polarizer, it emerges vibrating in only one plane and is called plane-polarized light (Figure 26.1).

Polarizers are made from calcite or tourmaline crystals or from a Polaroid filter, which is a transparent plastic that contains properly oriented embedded

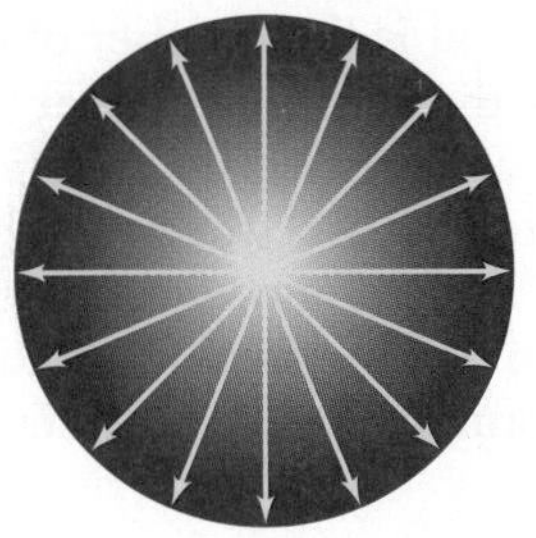
(a) Ordinary (unpolarized) light

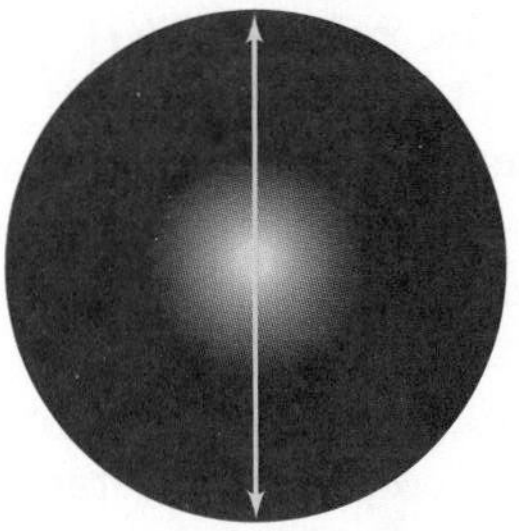
(b) Plane-polarized light

Figure 26.1
(a) Diagram of ordinary light vibrating in all possible directions (planes) and (b) diagram of plane-polarized light vibrating in a single plane. The beam of light is coming toward the reader.

crystals. Two Polaroid filters with parallel axes allow the passage of plane-polarized light. But when one filter is placed so that its axis is at right angles to that of the other filter, the passage of light is blocked and the filters appear black (Figure 26.2).

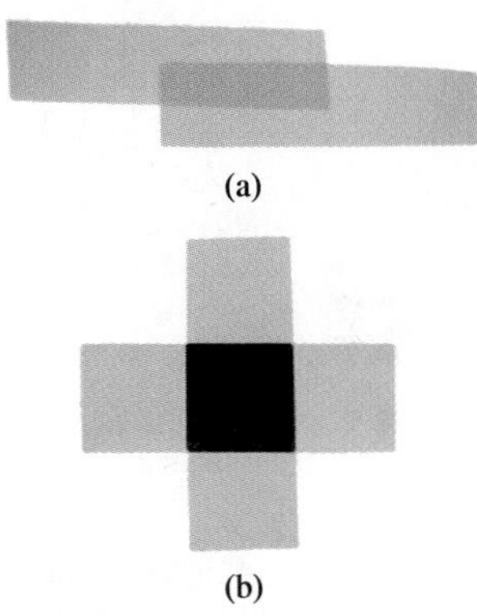

Figure 26.2
Two Polaroid filters (top) with axes parallel and (bottom) with axes at right angles. In (a), light passes through both filters and emerges polarized. In (b), the polarized light that emerges from one filter is blocked and does not pass through the second filter, which is at right angles to the first. With no light emerging, the filters appear black.

Specific Rotation

The rotation of plane-polarized light is quantitatively measured with a polarimeter. The essential features of this instrument are (1) a light source (usually a sodium lamp), (2) a polarizer, (3) a sample tube, (4) an analyzer (which is another matched polarizer), and (5) a calibrated scale (360°) for measuring the number of degrees the plane of polarized light is rotated. The calibrated scale is attached to the analyzer. (See Figure 26.3.) When the sample tube contains a solution of an optically inactive material, the axes of the polarizer and the analyzer are parallel, and the scale is at zero degrees; the light passing through is at maximum intensity. When a solution of an optically active substance is placed in the sample tube, the plane in which the polarized light is vibrating is rotated through an angle (α). The analyzer is then rotated to the position where the emerging light is at maximum intensity. The number of degrees and the direction of rotation by the solution are then read from the scale as the observed rotation.

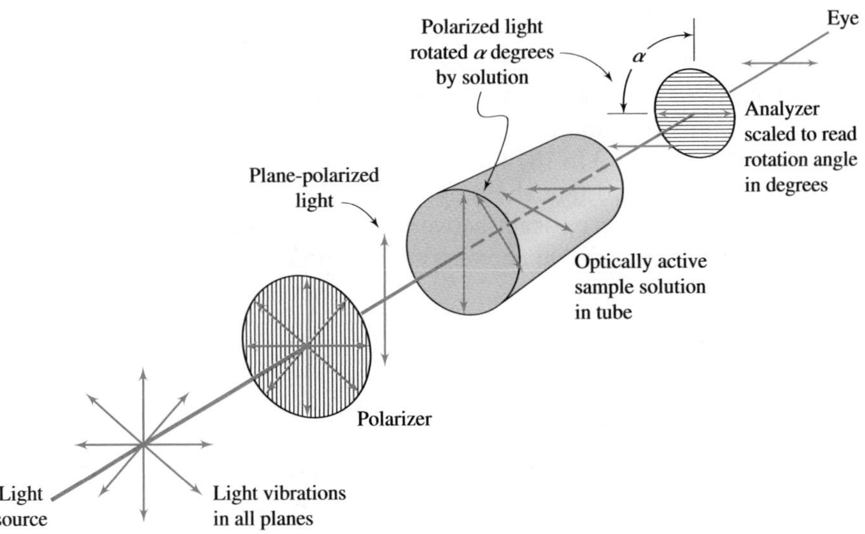

Figure 26.3
Schematic diagram of a polarimeter. This instrument measures the angle α through which an optically active substance rotates the plane of polarized light.

specific rotation

The **specific rotation**, $[\alpha]$, of a compound is the number of degrees that polarized light would be rotated by passing through 1 decimeter (dm) of a solution of the substance at a concentration of 1 g/mL. The specific rotation of an optically active substance varies with temperature, wavelength of light, and solvent used. The wavelength usually used is 589.3 nm, which is the D-line of sodium obtained from a sodium vapor lamp. The specific rotation is designated by the symbol $[\alpha]$. The following formula is used to calculate specific rotation from polarimeter data:

$$[\alpha] = \frac{\text{Observed rotation in degrees}}{\left(\begin{array}{c}\text{Length of}\\ \text{sample tube in decimeters}\end{array}\right)\left(\begin{array}{c}\text{Sample concentration in}\\ \text{grams per milliliter}\end{array}\right)}$$

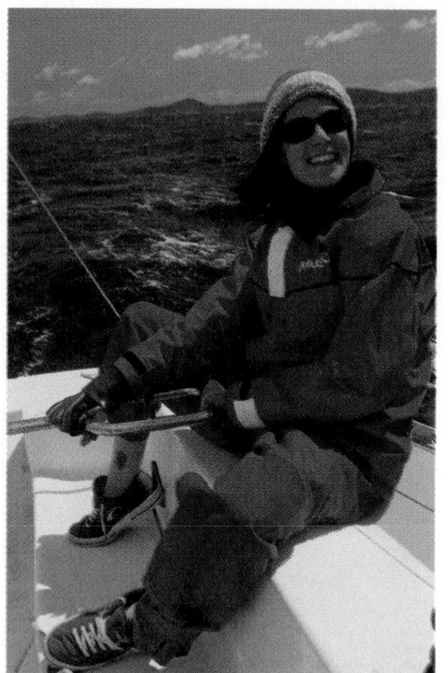

Polarized lenses in sunglasses reduce glare from the sun.

26.3 Optical Activity

optically active
dextrorotatory
levorotatory
asymmetric carbon atom

Many naturally occurring substances are able to rotate the plane of polarized light. Because of this ability to rotate polarized light, such substances are said to be **optically active**. When plane-polarized light passes through an optically active substance, the plane of the polarized light is rotated. If the rotation is to the right (clockwise), the substance is said to be **dextrorotatory**; if the rotation is to the left (counterclockwise), the substance is said to be **levorotatory**.

Some minerals, notably quartz, rotate the plane of polarized light. (In fact, optical activity was discovered in minerals.) However, when such mineral crystals are melted, the optical activity disappears. This means that the optical activity of these crystals must be due to an ordered arrangement within the crystals.

In 1848, Louis Pasteur (1822–1895) observed that sodium ammonium tartrate, a salt of tartaric acid, exists as a mixture of two kinds of crystals. Pasteur carefully hand separated the two kinds of crystals. Investigating their properties, he found that solutions made from either kind of crystal would rotate the plane of polarized light, but in opposite directions. Since this optical activity was present in a solution, it could not be caused by a specific arrangement within a crystal. Instead, Pasteur concluded that this difference in optical activity occurred because of a difference in molecular symmetry.

Quartz is optically active.

The tetrahedral arrangement of single bonds around a carbon atom makes asymmetry (lack of symmetry) possible in organic molecules. When four different atoms or functional groups are bonded to a carbon atom, the molecule formed is asymmetric, and the carbon atom is called an **asymmetric carbon atom**. (See Figure 26.4.) In 1874, J. H. van't Hoff (1852–1911) and J. A. Le Bel (1847–1930) concluded that the presence of at least one asymmetric carbon atom in a molecule of an optically active substance is a key factor for optical activity. The first Nobel Prize in chemistry was awarded to van't Hoff in 1901.

D
A B
C

○ = Carbon atom

A, B, C, D = Four different atoms or groups of atoms

Figure 26.4
Three-dimensional representation of an asymmetric carbon atom with four different groups bonded to it. The carbon atom is a sphere. Bonds to A and B project from the sphere toward the observer. Bonds to C and D project from the sphere away from the observer.

Molecules of optically active substances must have at least one center of dissymmetry. Although there are optically active compounds that do not contain asymmetric carbon atoms (their center of dissymmetry is due to some other structural feature), most optically active organic substances contain one or more asymmetric carbon atoms.

Your right and your left hands are mirror images of each other; that is, your left hand is a mirror reflection of your right hand, and vice versa. Furthermore, your hands are not superimposable on each other. (See Figure 26.5.) **Superimposable** means that, when we lay one object upon another, all parts of both objects coincide exactly.

A molecule that is not superimposable on its mirror image is said to be **chiral**. The word *chiral* comes from the Greek word *cheir*, meaning hand. Chiral molecules have the property of "handedness"; that is, they are related to each other in the same manner as the right and left hands. An asymmetric carbon atom is also called a **chiral carbon atom** or chiral center. Molecules or objects that are superimposable on each other are **achiral**. A molecule cannot be chiral if it has a plane of symmetry. For example, if you can pass a plane through a molecule (or an object) in such a way that one-half of the molecule is the mirror image of the other half, the molecule is not chiral. Some chiral and achiral objects are shown in Figure 26.6.

Figure 26.5
The left hand is the same as the mirror image of the right hand. Right and left hands are not superimposable; hence, they are chiral.

superimposable
chiral
chiral carbon atom
achiral

(a) chiral

(b) achiral

Figure 26.6
Can you explain why these objects are chiral or achiral?

26.4 Fischer Projection Formulas

Molecules of a compound that contain one chiral carbon atom occur in two optically active isomeric forms. This is because the four different groups bonded to the chiral carbon atom can be oriented in space in two different configurations. It is important to understand how we represent such isomers on paper.

Let's consider the spatial arrangement of a lactic acid molecule, $CH_3CH(OH)COOH$, that contains one chiral carbon atom:

$$\begin{array}{c} {}^{1}COOH \\ | \\ H-{}^{2}C^{*}-OH \\ | \\ {}^{3}CH_3 \end{array} \qquad C^{*} = \text{chiral carbon atom}$$

lactic acid

CHEMISTRY IN ACTION • The Shape of Our Senses

Mirror-image stereoisomers (enantiomers) have exactly the same molecular weight, exactly the same functional groups, exactly the same alkyl and aryl substituents, exactly the same melting points and boiling points, and almost exactly the same chemical reactions. We might expect enantiomers to be indistinguishable in biochemical systems—but we would be very wrong. Because most biological molecules are recognized by shape, mirror images are often treated very differently.

Mirror-image recognition depends on sensing whether groups are on one side or the other of a fixed reference. For example, if our hands are palms down on a table, the right hand has the thumb on the left while the left hand has the thumb on the right—we can see the difference. Living systems differentiate between enantiomers in a similar way. For example, the enantiomers of the amino acid asparagine bind to human taste buds differently. As shown below, the two molecules have a mirror-image arrangement and bind with groups on opposite sides—no matter how these molecules rotate they can never be superimposed.

So, we recognize D-asparagine as sweet while L-asparagine is tasteless. A more important commercial product, the artificial sweetener aspartame (N-L-α-aspartyl-L-phenylalanine), has a very sweet taste while its enantiomer (N-D-α-aspartyl-D-phenylalanine) is bitter.

The human nose is an even more discriminating sensor of molecular shape. With eyes closed a person can sniff the difference between an orange ((+)-limonene) and a lemon ((−)-limonene).

Sometimes only one enantiomer produces an aroma. The smell of jasmine is recognizable when the molecule is at a concentration of 0.003 ppm; its enantiomer has no odor. Other times, one isomer produces a much stronger odor than the other. The musk aroma ("warm, sensual, natural") used in perfumes comes from an isomer that can be smelled at concentrations as low as 0.06 ppm. Its enantiomer is only detectable at 0.2 ppm. The characteristic aroma of a white wine such as a Gwürztraminer is produced by a molecule that can be sensed in concentrations as low as 0.01 picograms per liter of air while its enantiomer must be present in concentrations greater than 1 million picograms per liter. Molecular shape greatly impacts a molecule's aroma and taste.

What is apparent for taste and smell holds true throughout all biochemistry: Hormones and neurotransmitters are recognized by shape; metabolism only works on molecules of the right shapes. A mirror-image difference in shape can change everything.

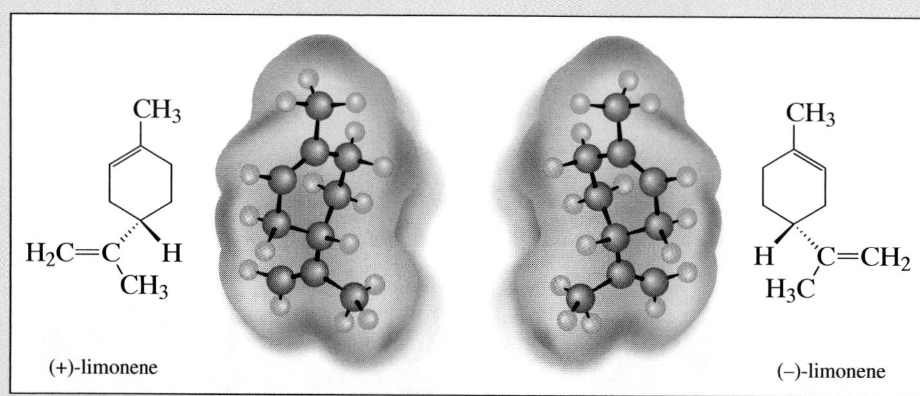

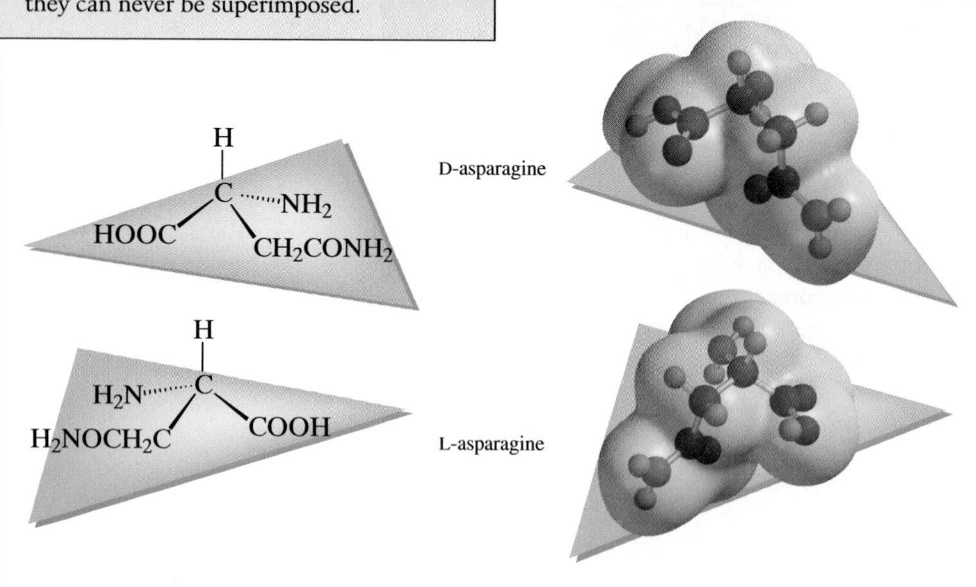

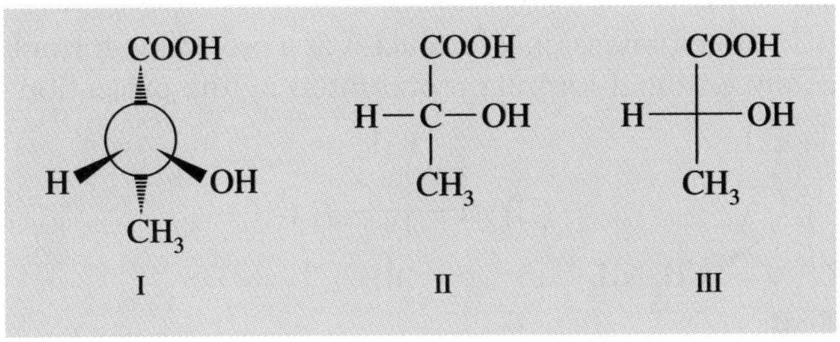

COOH H OH CH_3 I | COOH H—C—OH CH_3 II | COOH H OH CH_3 III

Figure 26.7
Methods of representing three-dimensional formulas of a compound that contains one chiral carbon atom. All three structures represent the same molecule.

Three-dimensional models are the best means of representing such a molecule, but by adopting certain conventions and by using imagination, we can formulate the images on paper. The geometrical arrangement of the four groups about the chiral carbon (carbon 2) is the key to the stereoisomerism of lactic acid. The four bonds attached to carbon 2 are separated by angles of about 109.5°. Diagram I in Figure 26.7 is a three-dimensional representation of lactic acid in which the chiral carbon atom is represented as a sphere, with its center in the plane of the paper. The —H and —OH groups are projected forward from the paper (toward the observer), and the —COOH and $-CH_3$ groups are projected back from the paper (away from the observer).

For convenience, simpler diagrams such as II and III are used for writing structures of chiral compounds. These are much easier and faster to draw. In II, it is understood that the groups (—H and —OH) attached to the horizontal bonds are coming out of the plane of the paper toward the observer and that the groups attached to the vertical bonds are projected back from the paper. The molecule represented by formula III is made by drawing a cross and attaching the four groups in their respective positions, as in formula II. The chiral carbon atom is understood to be located where the lines cross. Formulas II and III are called **Fischer projection formulas**.

Fischer projection formula

It is important to be careful when comparing projection formulas. Two rules apply: (1) Projection formulas must **not** be turned 90°; (2) projection formulas must **not** be lifted or flipped out of the plane of the paper. Projection formulas may, however, be turned 180° in the plane of the paper without changing the spatial arrangement of the molecule. Consider the following projection formulas:

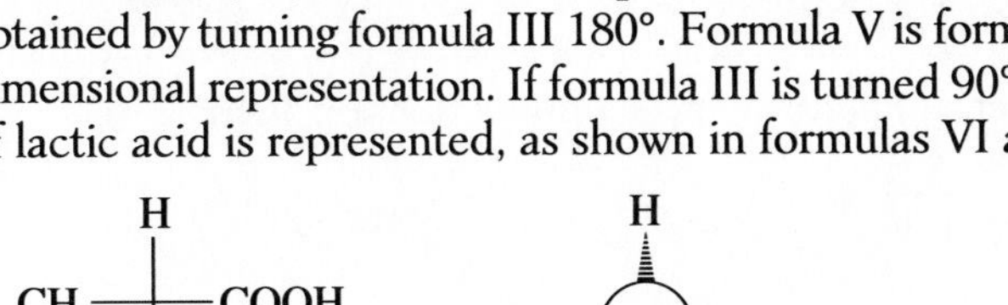

COOH H OH CH_3 III | CH_3 HO H COOH IV | CH_3 HO H COOH V

Formulas I, II, III, IV, and V represent the same molecule. Formula IV was obtained by turning formula III 180°. Formula V is formula IV drawn in a three-dimensional representation. If formula III is turned 90°, the other stereoisomer of lactic acid is represented, as shown in formulas VI and VII:

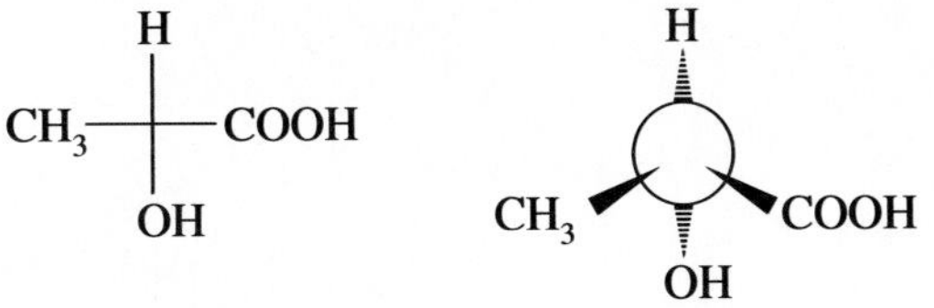

H CH_3 COOH OH VI | H CH_3 COOH OH VII

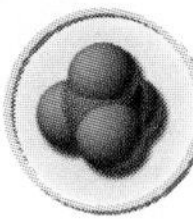

Example 26.1 (a) Redraw the three-dimensional formula A as a projection formula. (b) Draw the three-dimensional formula represented by the projection formula B.

(A) CH_3 (top), H (left), CH_2NH_2 (right), Br (bottom)

(B) CH_3 (top), H— (left), $—CH_2CH_3$ (right), OH (bottom)

$CH_3CHBrCH_2NH_2$ $CH_3CH(OH)CH_2CH_3$

SOLUTION (a) Draw a vertical and a horizontal line crossing each other. Place the CH_3 at the top and the Br at the bottom of the vertical line. Place the H at the left and the CH_2NH_2 at the right of the horizontal line to complete the following projection formula:

CH_3 (top), H— (left), $—CH_2NH_2$ (right), Br (bottom) (projection formula of A)

(b) Draw a small circle to represent the chiral carbon atom (carbon 2) that is located where the two lines cross in the projection formula. Draw a short line extending from the top and from the bottom of the circle. Place the CH_3 on the top and the OH on the bottom at the end of these lines. Now draw two short lines from within the circle coming toward your left and right arms. Place the H at the end of the left line and the CH_2CH_3 at the end of the right line. The finished formula should look like this:

CH_3 (top), H (left), CH_2CH_3 (right), OH (bottom) (three-dimensional formula of B)

Practice 26.1

Redraw this projection formula as a three-dimensional formula:

Br (top), Cl— (left), $—CH_3$ (right), F (bottom)

Practice 26.2

Redraw this three-dimensional formula as a projection formula:

COOH (top), H (left), NH_2 (right), CH_2CH_3 (bottom)

26.5 Enantiomers

The formulas developed in Section 26.4 dealt primarily with a single kind of lactic acid molecule. But two stereoisomers of lactic acid are known, one that rotates the plane of polarized light to the right and one that rotates it to the left. These two forms of lactic acid are shown in Figure 26.8. If we examine these two structural formulas carefully, we can see that they are mirror images of each other. The reflection of either molecule in a mirror corresponds to the structure of the other molecule. Even though the two molecules have the same molecular formula and the same four groups attached to the central carbon atom, they are not superimposable. Therefore, the two molecules are not identical, but are isomers. One molecule rotates the plane of polarized light to the left and is termed *levorotatory;* the other molecule rotates it to the right and is *dextrorotatory.* A plus (+) or a minus (−) sign written in parentheses and placed in front of a name or formula indicates the direction of rotation of polarized light and becomes part of the name of the compound. Plus (+) indicates rotation to the right and minus (−) to the left. Using projection formulas, we write the two lactic acids as shown here:

$$\begin{array}{c} \text{COOH} \\ | \\ \text{H}-\text{C}-\text{OH} \\ | \\ \text{CH}_3 \\ (-)\text{-lactic acid} \end{array} \qquad \begin{array}{c} \text{COOH} \\ | \\ \text{HO}-\text{C}-\text{H} \\ | \\ \text{CH}_3 \\ (+)\text{-lactic acid} \end{array}$$

Mirror Mirror

COOH, H, OH, CH_3 — (−)-lactic acid

COOH, HO, H, CH_3 — (+)-lactic acid

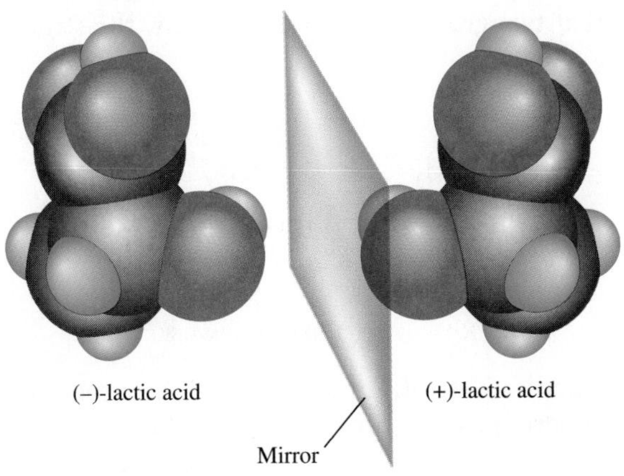

Figure 26.8
Mirror-image isomers of lactic acid. Each isomer is the mirror reflection of the other. (−)-Lactic acid rotates plane-polarized light to the left, and (+)-lactic acid rotates plane-polarized light to the right.

Originally, it was not known which lactic acid structure was the (+) or the (−) compound. However, we now know that they are as shown. Chiral molecules that are mirror images of each other are stereoisomers and are called **enantiomers**. The word *enantiomer* comes from the Greek word *enantios,* which means opposite.

enantiomer

Many, but not all, molecules that contain a chiral carbon are chiral. To decide whether a molecule is chiral and has an enantiomer, make models of the molecule and of its mirror image, and see if they are superimposable. This is the ultimate test, but instead of making models every time, first examine the formula to see if it has a chiral carbon atom. If a chiral carbon atom is found, draw a cross and attach the four groups on the chiral carbon to the four ends of the cross. The chiral carbon is understood to be located where the lines cross. Remember that a chiral carbon atom has four different groups attached to it. Let's test the compounds 2-butanol and 2-chloropropane:

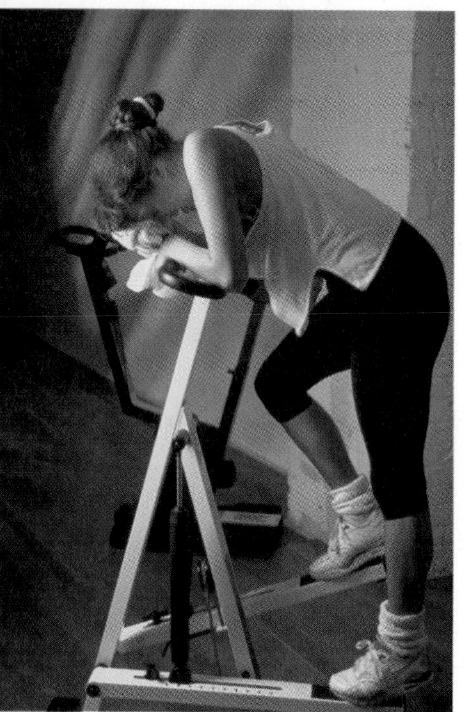

Overuse of muscle tissue results in the buildup of (+)-lactic acid.

$CH_3CH_2CHCH_3$ (with OH on carbon 2) — 2-butanol

CH_3CHCH_3 (with Cl on carbon 2) — 2-chloropropane

In 2-butanol, carbon 2 is chiral. The four groups attached to carbon 2 are H, OH, CH_3, and CH_2CH_3. Draw the structure and its mirror image:

Mirror

VIII: CH_3 (top), H (left), OH (right), CH_2–CH_3 (bottom)

IX: CH_3 (top), HO (left), H (right), CH_2–CH_3 (bottom)

X (IX turned 180°): CH_3–CH_2 (top), H (left), OH (right), CH_3 (bottom)

VIII and IX: enantiomers

Turning structure IX 180° in the plane of the paper allows H and OH to coincide with their position in VIII, but CH_3 and CH_2CH_3 do not coincide. Therefore, we conclude that the mirror-image structures VIII and IX are enantiomers, since they are not superimposable.

In 2-chloropropane, the four groups attached to carbon 2 are H, Cl, CH_3, and CH_3. Note that two groups are the same. Draw the structure and its mirror image:

Mirror

XI: CH_3 (top), H (left), Cl (right), CH_3 (bottom)

XII: CH_3 (top), Cl (left), H (right), CH_3 (bottom)

XIII (XII turned 180°): CH_3 (top), H (left), Cl (right), CH_3 (bottom)

When we turn structure XII 180° in the plane of the paper, the two structures XI and XIII are superimposable, proving that 2-chloropropane, which does not have a chiral carbon, does not exist in enantiomeric forms.

Example 26.2

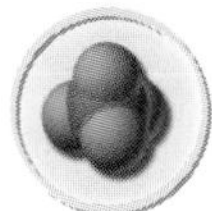

Draw mirror-image isomers for any of the following compounds that can exist as enantiomers:

(a) $CH_3CH_2CH_2OH$
(b) $CH_3CH_2CHClCH_2CH_3$
(c) $CH_3CH_2CHClCH_2CH_2CH_3$

SOLUTION

First check each formula for chiral carbon atoms:

(a) No chiral carbon atoms; each carbon has at least two groups that are the same.
(b) No chiral carbon atoms; carbon 3 has H, Cl, and two CH_3CH_2 groups.
(c) Carbon 3 is chiral; the four groups on carbon 3 are H, Cl, CH_3CH_2, and $CH_3CH_2CH_2$. Draw mirror images:

Mirror

```
          H                                  H
          |                                  |
CH3CH2—C—CH2CH2CH3     ¦     CH3CH2CH2—C—CH2CH3
          |                                  |
          Cl                                 Cl
```

Practice 26.3

Draw the mirror-image isomers for the following compounds that can exist as enantiomers:

(a) $NH_2CH_2CHBrCH_2OH$
(b) $NH_2CH_2CHBrCH_2NH_2$
(c) $NH_2CH_2CH_2CH_2OH$

Practice 26.4

Draw structural formulas for the pentyl alcohols, $C_5H_{11}OH$, that will show optical activity.

Practice 26.5 Application to Biochemistry

Two important compounds in our body's energy metabolism are citric acid and isocitric acid:

```
       COOH                 COOH
        |                    |
    H—C—H               HO—C—H
        |                    |
   HO—C—COOH             H—C—COOH
        |                    |
    H—C—H                H—C—H
        |                    |
       COOH                 COOH
```

citric acid isocitric acid

Star any carbons that are chiral, and draw an enantiomer for any compound that is chiral.

The relationship between enantiomers is such that if we change the position of any two groups attached to a chiral carbon atom on a compound that contains only one chiral carbon atom, we obtain the structure of its enantiomer. If we make a second change, the structure of the original isomer is obtained again.

In both cases shown here, (+)-lactic acid is formed by interchanging the positions of two groups on (−)-lactic acid:

(+)-lactic acid: CH_3 (top), H (left), OH (right), COOH (bottom) ← change position of CH_3 and COOH ← (−)-lactic acid: COOH (top), H (left), OH (right), CH_3 (bottom) → change position of H and OH → (+)-lactic acid: COOH (top), HO (left), H (right), CH_3 (bottom)

To test whether two projection formulas are the same structure or are enantiomers, (1) turn one structure 180° in the plane of the paper and compare to see if they are superimposable or (2) make successive group interchanges until the formulas are identical. If an odd number of interchanges are made, the two original formulas represent enantiomers; if an even number of interchanges are made, the two formulas represent the same compound. The following two examples illustrate this method:

XIV: CH_3 (top), Br (left), H (right), Cl (bottom)   XV: H (top), Br (left), Cl (right), CH_3 (bottom)

XV: H (top), Br (left), Cl (right), CH_3 (bottom) → interchange CH_3 and Cl → H (top), Br (left), CH_3 (right), Cl (bottom) → interchange CH_3 and H → XIV: CH_3 (top), Br (left), H (right), Cl (bottom)

Are structures XIV and XV the same compound? Two interchanges were needed to make structure XV identical to structure XIV. Therefore, structures XIV and XV represent the same compound.

In the following example, do structures XVI and XVII represent the same compound?

XVI: CH_3 (top), HO (left), H (right), CH_2CH_3 (bottom)   XVII: OH (top), CH_3CH_2 (left), CH_3 (right), H (bottom)

XVII: OH (top), CH_3CH_2 (left), CH_3 (right), H (bottom) → interchange CH_3CH_2 and OH → CH_2CH_3 (top), HO (left), CH_3 (right), H (bottom) → interchange CH_3 and H → CH_2CH_3 (top), HO (left), H (right), CH_3 (bottom) → interchange CH_3 and CH_3CH_2 → XVI: CH_3 (top), HO (left), H (right), CH_2CH_3 (bottom)

Three interchanges were needed to make structure XVII identical to structure XVI. Therefore, structures XVI and XVII do not represent the same compound; they are enantiomers.

Enantiomers ordinarily have the same chemical properties, and other than optical rotation, they also have the same physical properties. (See Table 26.1.) They rotate plane-polarized light the same number of degrees, but in opposite directions.

Enantiomers usually differ in their biochemical properties. In fact, most living cells are able to use only one isomer of a mirror-image pair. As an example, (+)-glucose ("blood sugar") is used in the body for metabolic energy, whereas (−)-glucose is not. Enantiomers are truly different molecules and are treated as such by most organisms. The key factors of enantiomers and optical isomerism can be summarized as follows:

1. A carbon atom that has four different groups bonded to it is called an asymmetric or a chiral carbon atom.
2. A compound with one chiral carbon atom can exist in two stereoisomeric forms called enantiomers.
3. Enantiomers are nonsuperimposable mirror-image isomers.
4. Enantiomers are optically active; that is, they rotate plane-polarized light.
5. One isomer of an enantiomeric pair rotates polarized light to the left (counterclockwise). The other isomer rotates polarized light to the right (clockwise). The degree of rotation is the same but in opposite directions.
6. Rotation of polarized light to the right is indicated by (+), placed in front of the name of the compound, and rotation to the left is indicated by (−); for example, (+)-lactic acid and (−)-lactic acid.

26.6 Racemic Mixtures

A mixture containing equal amounts of a pair of enantiomers is known as a **racemic mixture**. Such a mixture is optically inactive and shows no rotation of polarized light when tested in a polarimeter. Each enantiomer rotates the plane of polarized light by the same amount, but in opposite directions. Thus, the rotation by each isomer is canceled. The (±) symbol is often used to designate racemic mixtures. For example, a racemic mixture of lactic acid is written as (±)-lactic acid because this mixture contains equal molar amounts of (+)-lactic acid and (−)-lactic acid.

racemic mixture

Racemic mixtures are usually obtained in laboratory syntheses of compounds in which a chiral carbon atom is formed. Thus, catalytic reduction of pyruvic acid (an achiral compound) to lactic acid produces a racemic mixture containing equal amounts of (+)- and (−)-lactic acid:

$$\underset{\text{pyruvic acid}}{CH_3\overset{}{\underset{\underset{O}{\|}}{C}}COOH} + H_2 \xrightarrow{Ni} \underset{(\pm)\text{-lactic acid}}{CH_3\underset{\underset{OH}{|}}{C}HCOOH}$$

As a general rule, in the biological synthesis of optically active compounds, only one of the isomers is produced. For example, (+)-lactic acid is produced by reactions occurring in muscle tissue, and (−)-lactic acid is produced by lactic acid bacteria in the souring of milk. These stereospecific reactions occur because biochemical syntheses are enzyme catalyzed. The preferential production of one isomer over another is often due to the configuration (shape) of the specific enzyme involved. Returning to the hand analogy, if the "right-handed" enantiomer is produced, then the enzyme responsible for the product can be likened to a right-handed glove.

The mirror-image isomers (enantiomers) of a racemic mixture are alike in all ordinary physical properties, except in their action on polarized light. It is possible to separate or resolve racemic mixtures into their optically active components. In fact, Pasteur's original work with sodium ammonium tartrate involved such a separation. But a general consideration of the methods involved in such separations is beyond the scope of our present discussion.

Many pharmaceuticals are synthesized as racemic mixtures, since organic synthetic processes are often not stereospecific. Unfortunately, often only one enantiomer has biological activity because most biomolecules are almost always stereospecific. Thus, the contents of a racemic mixture drug may be one-half bioinactive. Furthermore, the inactive enantiomer may cause adverse effects. The most famous example of drug-related problems with racemic mixtures involved the drug thalidomide:

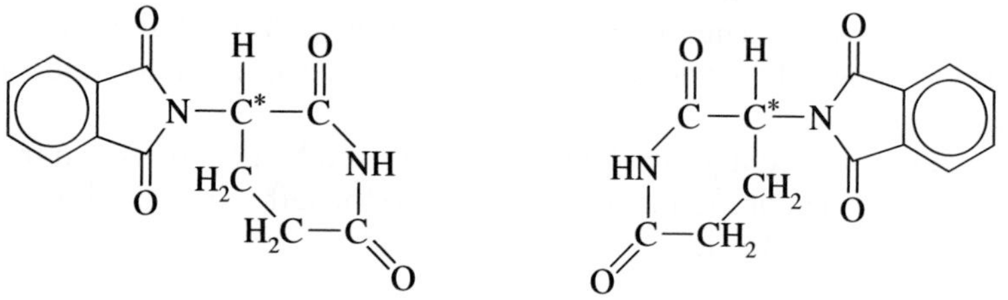

thalidomide racemic mixture
(the chiral carbon is marked with a star, *)

One isomer is effective at suppressing some immune responses, while the other isomer causes birth defects. During the period that the racemic mixture was on sale in Europe (from about 1957 to 1962), about 10,000 babies were born with serious malformations.

Drug companies and governmental agencies have learned from the thalidomide disaster. New syntheses that yield only one isomer, coupled with techniques for separating enantiomeric pairs, have improved the feasibility of selling only pure stereoisomeric drugs. The U.S. Food and Drug Administration requires that "Specifications of the final product [drug] should assure identity, strength, quality, and purity from a stereochemical viewpoint." More and more drugs are being produced as stereochemically pure isomers. The formulas for several chiral drugs are shown in Figure 26.9.

Practice 26.6 Application to Biochemistry

The chirality of drugs is very important because the reactions of biomolecules are generally stereospecific. Star the chiral carbons in the drug formulas in Figure 26.9.

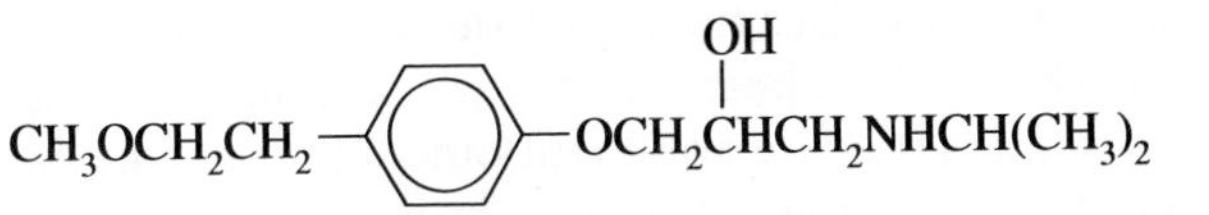

$CH_3OCH_2CH_2$–C$_6$H$_4$–$OCH_2CH(OH)CH_2NHCH(CH_3)_2$

metoprolol (Lopressor®)
treats high blood pressure

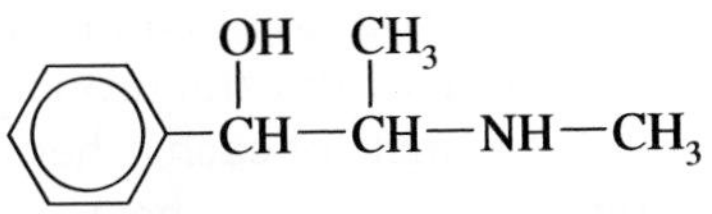

C$_6$H$_5$–CH(OH)–CH(CH_3)–NH–CH_3

pseudoephedrine (Sudafed®)
treats nasal congestion

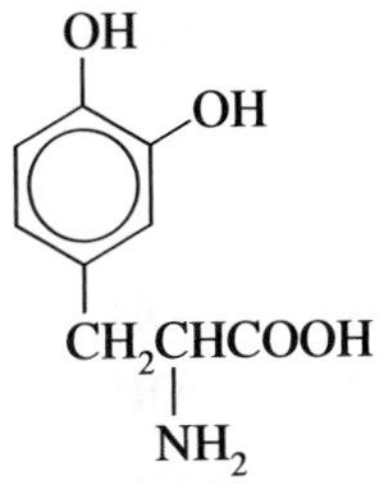

3,4-(OH)$_2$C$_6$H$_3$–$CH_2CH(NH_2)COOH$

3-(3,4-dihydroxyphenyl)-alanine (L-Dopa)
treats Parkinson's disease

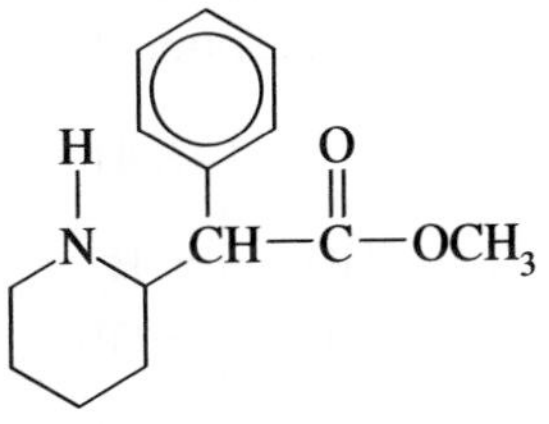

methylphenidate (Ritalin®)
treats attention deficit disorder

Figure 26.9
Some examples of common chiral drugs.

26.7 Diastereomers and Meso Compounds

The enantiomers discussed in the preceding sections are stereoisomers. That is, they differ only in the spatial arrangement of the atoms and groups within the molecule. The number of stereoisomers increases as the number of chiral carbon atoms increases. The maximum number of stereoisomers for a given compound is obtained by the formula 2^n, where n is the number of chiral carbon atoms.

2^n = Maximum number of stereoisomers for a given chiral compound
n = Number of chiral carbon atoms in a molecule

As we have seen, there are two ($2^1 = 2$) stereoisomers of lactic acid. But for a substance with two nonidentical chiral carbon atoms, such as 2-bromo-3-chlorobutane ($CH_3CHBrCHClCH_3$), four stereoisomers are possible ($2^2 = 4$). These four possible stereoisomers are written as projection formulas in the following way (carbons 2 and 3 are chiral):

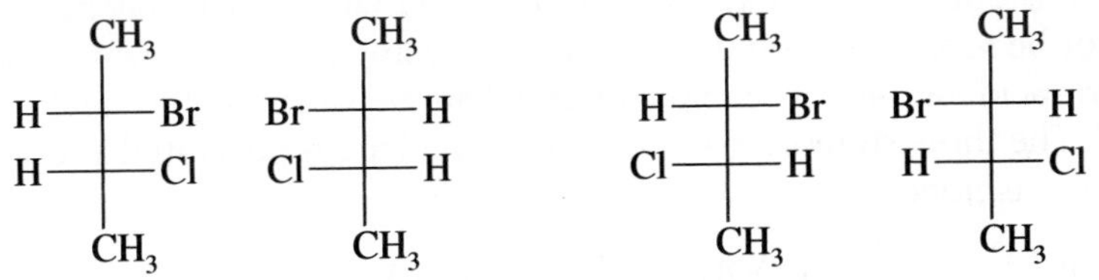

XVIII	XIX	XX	XXI
CH_3	CH_3	CH_3	CH_3
H–C–Br	Br–C–H	H–C–Br	Br–C–H
H–C–Cl	Cl–C–H	Cl–C–H	H–C–Cl
CH_3	CH_3	CH_3	CH_3

XVIII and XIX: enantiomers; XX and XXI: enantiomers

Remember that, for comparison, projection formulas may be turned 180° in the plane of the paper, but they cannot be lifted (flipped) out of the plane. Formulas XVIII and XIX, and formulas XX and XXI, represent two pairs of nonsuperimposable mirror-image isomers and are therefore two pairs of enantiomers.

All four compounds are optically active. But the properties of XVIII and XIX differ from the properties of XX and XXI because they are not mirror-image isomers of each other. Stereoisomers that are not enantiomers (not mirror images of each other) are called **diastereomers**. There are four different pairs of diastereomers of 2-bromo-3-chlorobutane: They are XVIII and XX, XVIII and XXI, XIX and XX, and XIX and XXI.

diastereomer

The 2^n formula indicates that four stereoisomers of tartaric acid are possible. The projection formulas of these four possible stereoisomers are written in this way (carbons 2 and 3 are chiral):

```
      COOH            COOH              COOH            COOH
       |               |                 |               |
 HO ---+--- H     H ---+--- OH      H ---+--- OH   HO ---+--- H
  H ---+--- OH   HO ---+--- H       H ---+--- OH   HO ---+--- H
       |               |                 |               |
      COOH            COOH              COOH            COOH

      XXII            XXIII             XXIV            XXV
```

Formulas XXII and XXIII represent nonsuperimposable mirror-image isomers and are thus enantiomers. Formulas XXIV and XXV are also mirror images. But by turning XXV 180°, we see that it is exactly superimposable on XXIV. Therefore, XXIV and XXV represent the same compound, and only *three* stereoisomers of tartaric acid actually exist. Compound XXIV is achiral and does not rotate polarized light. A plane of symmetry can be passed between carbons 2 and 3 so that the top and bottom halves of the molecule are mirror images:

```
      COOH
       |
   H — C — OH
---------------  ← Plane of symmetry
   H — C — OH
       |
      COOH
```

Thus, the molecule is internally compensated. The rotation of polarized light in one direction by half of the molecule is exactly compensated by an opposite rotation by the other half. Stereoisomers that contain chiral carbon atoms and are superimposable on their own mirror images are called **meso compounds**, or **meso structures**. All meso compounds are optically inactive.

meso compound
meso structure

The term *meso* comes from the Greek word *mesos,* meaning middle. It was first used by Pasteur to name a kind of tartaric acid that was optically inactive and could not be separated into different forms by any means. Pasteur called it *meso*-tartaric acid, because it seemed intermediate between the (+)- and (−)-tartaric acid. The three stereoisomers of tartaric acid are represented and designated in this fashion:

```
      COOH               COOH               COOH
       |                  |                  |
 HO — C — H          H — C — OH         H — C — OH
       |                  |                  |
  H — C — OH        HO — C — H          H — C — OH
       |                  |                  |
      COOH               COOH               COOH

 (−)-tartaric acid   (+)-tartaric acid   meso-tartaric acid
```

The physical properties of tartaric acid stereoisomers are given in Table 26.1. Note that the properties of (+)-tartaric acid and (−)-tartaric acid are identical, except for opposite rotation of polarized light. However, *meso*-tartaric acid has properties that are entirely different from those of the other isomers. But most surprising is the fact that the racemic mixture, though composed of equal parts of the (+) and (−) enantiomers, differs from them in specific gravity, melting point, and solubility. Why, for example, is the melting point of the racemic mixture higher than that of any of the other forms? The melting point of any substance is largely dependent on the attractive forces holding the ions or molecules together. The melting point of the racemic mixture is higher than that of either enantiomer. Therefore, we can conclude that the attraction between molecules of the (+) and (−) enantiomers in the racemic mixture is greater than the attraction between molecules of the (+) and (+) or the (−) and (−) enantiomers.

Table 26.1 Properties of Tartaric Acid, HOOCCH(OH)CH(OH)COOH

Name	Specific gravity	Melting point (°C)	Solubility (g/100 g H_2O)	Specific rotation [α]
(+)-Tartaric acid	1.760	170	$147^{20°C}$	+12°
(−)-Tartaric acid	1.760	170	$147^{20°C}$	−12°
(±)-Tartaric acid (racemic mixture)	1.687	206	$20.6^{20°C}$	0°
meso-Tartaric acid	1.666	140	$125^{15°C}$	0°

Example 26.3

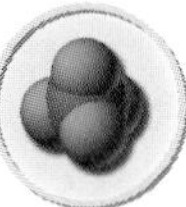

How many stereoisomers can exist for the following compounds?

(a) $CH_3CHBrCHBrCH_2CH_3$ (b) $CH_2BrCHClCHClCH_2Br$

Write their structures and label any pairs of enantiomers and meso compounds. Point out any diastereomers.

SOLUTION

(a) Carbons 2 and 3 are chiral, so there can be a maximum of four stereoisomers ($2^2 = 4$). Write structures around the chiral carbons:

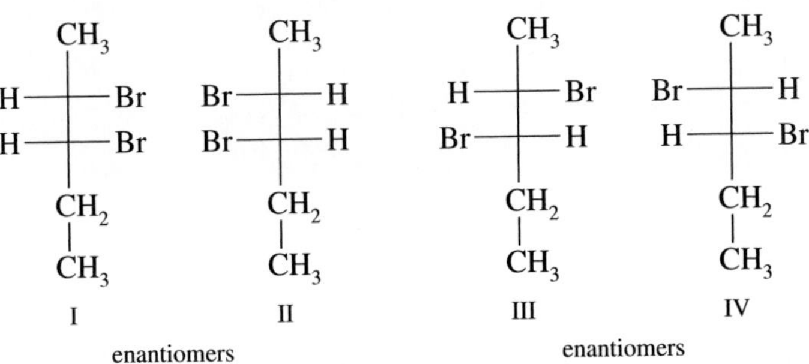

There are four stereoisomers: two pairs of enantiomers (I and II, and III and IV) and no meso compounds. Structures I and III, I and IV, II and III, and II and IV are diastereomers.

(b) Carbons 2 and 3 are chiral, so there can be a maximum of four stereoisomers. Write structures around the chiral carbons:

V	VI	VII	VIII
CH_2Br / H—C—Cl / H—C—Cl / CH_2Br	CH_2Br / Cl—C—H / Cl—C—H / CH_2Br	CH_2Br / Cl—C—H / H—C—Cl / CH_2Br	CH_2Br / H—C—Cl / Cl—C—H / CH_2Br
meso compound		enantiomers	

There are three stereoisomers: one pair of enantiomers (VII and VIII) and one meso compound (V). Structures V and VI represent the meso compound because there is a plane of symmetry between carbons 2 and 3, and turning VI 180° makes it superimposable on V. Structures V and VII and V and VIII are diastereomers.

Practice 26.7

Write all stereoisomer structures and label any pairs of enantiomers and meso compounds for the following compound (also, point out any diastereomers):

$HOOCCHClCH_2CH_2CHClCOOH$

Practice 26.8 Application to Biochemistry

D-2-Deoxyribose is a sugar that is very important in the structure of nucleic acids, the cell's genetic information:

O
‖
C—H
|
H—C—H
|
H—C—OH
|
H—C—OH
|
CH_2OH

Write all the stereoisomers of D-2-deoxyribose. Label any pairs of enantiomers.

Chapter 26 Review

26.1 Review of Isomerism

KEY TERMS

Stereoisomerism
Stereoisomer

- Isomerism is the phenomenon of two or more compounds having the same number and kind of atoms.
- In stereoisomerism the isomers have the same structural formula but differ in the spatial arrangement of atoms.

- Stereoisomers have the same structural formula but differ in their spatial arrangement.

26.2 Plane-Polarized Light

KEY TERMS

Plane-polarized light
Specific rotation

- Plane-polarized light is light vibrating in only one plane.
- A polarimeter uses two polarizers to measure the rotation of plane-polarized light caused by a solution that contains an optically active compound.
- The specific rotation is the number of degrees that polarized light is rotated by passing through 1 decimeter (dm) of a solution whose concentration is 1 g/mL.

26.3 Optical Activity

KEY TERMS

Optically active
Dextrorotatory
Levorotatory
Asymmetric carbon atom
Superimposable
Chiral
Chiral carbon atom
Achiral

- Compounds that are able to rotate polarized light are said to be optically active.
- If the plane-polarized light is rotated in a clockwise manner as it passes through a solution, the solute is dextrorotatory.
- If the plane-polarized light is rotated in a counterclockwise manner as it passes through a solution, the solute is levorotatory.
- Optical activity is commonly associated with asymmetric carbon atoms.
  - An asymmetric carbon atom is bonded to four different groups.
- A compound with an asymmetric carbon atom is not superimposable on its mirror image.
  - Superimposable means that, when we lay one object upon another, all parts of both objects coincide exactly.
- A molecule that is not superimposable on its mirror image is said to be chiral.
  - Thus, an asymmetric carbon atom can also be described as a chiral carbon atom.
  - Molecules that are superimposable on their mirror images are described as achiral.

26.4 Fischer Projection Formulas

KEY TERM

Fischer projection formula

- A molecule with one chiral carbon atom can be in two optically active isomeric forms.
- Projection formulas depict a three-dimensional molecule as a flat, two-dimensional drawing.
  - In a projection formula, vertical and horizontal lines carry different information concerning a three-dimensional molecule.
  - Vertical lines symbolize the bonds around a tetrahedral carbon atom that project away from the observer.
  - Horizontal lines symbolize the bonds around a tetrahedral carbon atom that project toward the observer.
- When testing projection formulas for superimposability, it is important to adhere to the following:
  - Projection formulas must not be turned 90°.
  - Projection formulas must not be lifted or flipped out of the plane of the paper.
  - Projection formulas may be turned 180° in the plane of the paper.

26.5 Enantiomers

KEY TERM

Enantiomer

- Chiral molecules that are mirror images of each other are stereoisomers and are called enantiomers.
- In the projection formula, when any two groups around a chiral carbon are interchanged, the mirror image is formed

26.6 Racemic Mixture

KEY TERM

Racemic mixture

- A mixture containing equal amounts of a pair of enantiomers is known as a racemic mixture.
- A racemic mixture is usually obtained in a laboratory synthesis, while only one isomer of the enantiomeric pair is produced in a biological synthesis.
- To improve safety, more and more pharmaceuticals are being marketed as single isomers rather than racemic mixtures.

26.7 Diastereomers and Meso Compounds

KEY TERMS

Diastereomer
Meso compound
Meso structure

- The maximum number of stereoisomers for a given chiral compound is equal to 2^n where n equals the number of chiral carbon atoms in the molecule.
- Stereoisomers that are not mirror images (enantiomers) are called diastereomers.
- Stereoisomers that contain chiral carbon atoms and are superimposable on their mirror images are called meso compounds or meso structures.
- Meso compounds are not optically active and are achiral compounds.

Review Questions

All questions with blue numbers have answers in the appendix of the text.

1. What is a chiral carbon atom? Draw structural formulas of three different compounds that contain one chiral carbon atom, and mark the chiral carbon in each with an asterisk.
2. How can you tell when the axes of two Polaroid filters are parallel? How can you tell when one filter has been rotated by 90°?
3. What is a necessary and sufficient condition for a compound to show enantiomerism?
4. Differentiate between an enantiomer and a diastereomer.
5. Differentiate between a diastereomer and a meso compound.
6. The following are the physical properties for (+)-2-methyl-1-butanol: specific rotation, +5.76°; bp, 129°C; density, 0.819 g/mL. What are these same properties for (−)-2-methyl-1-butanol?
7. Why are single-isomer drugs commonly more effective than racemic mixtures?
8. Differentiate between a meso compound and a racemic mixture.
9. If (+)-fructose has a specific rotation of +92°, what will be the specific rotation for its enantiomer?
10. If (+)-glucose has a specific rotation of +53°, what will be the specific rotation for the glucose racemic mixture?

Paired Exercises

All exercises with blue numbers have answers in the appendix of the text.

1. Explain why it is not possible to separate enantiomers by ordinary chemical and physical means.

2. Explain why it is usually possible to separate diastereomers by ordinary chemical and physical means.

3. Which of these objects is chiral?
(a) your ear
(b) a pair of pliers
(c) a coiled spring
(d) the letter b

4. Which of these objects is chiral?
(a) a wood screw
(b) the letter o
(c) the letter g
(d) this textbook

5. How many chiral carbon atoms are present in each of the following compounds?

```
       H  Cl
       |  |
(a) Cl—C—C—Br
       |  |
       H  H
```

(b) $CH_3CH_2CH_2CHClCH_3$

```
(c) H—C=O
       |
    H—C—OH
       |
    H—C—OH
       |
    H—C—OH
       |
       H
```

```
         H  H  O
         |  |  ‖
(d) HO—C—C—C—OH
         |  |
         H  NH2
```

6. How many chiral carbon atoms are present in each of the following compounds?

```
      Br Cl
       |  |
(a) H—C—C—H
       |  |
       H  H
```

```
              OH
              |
(b) CH3CH2CHCH2CH3
```

```
(c)     H
        |
     H—C—OH
        |
        C=O
        |
    HO—C—H
        |
     H—C—OH
        |
     H—C—OH
        |
     H—C—OH
        |
        H
```

(d) $CH_3CH_2CHBrCHClCH_3$

7. Write the formulas and decide which of the following compounds will show optical activity:
(a) 1-chloropentane
(b) 3-chloropentane
(c) 2-chloro-2-methylpentane
(d) 4-chloro-2-methylpentane

8. Write the formulas and decide which of the following compounds will show optical activity:
(a) 2-chloropentane
(b) 1-chloro-2-methylpentane
(c) 3-chloro-2-methylpentane
(d) 3-chloro-3-methylpentane

9. Glucose, $C_6H_{12}O_6$, has four chiral carbon atoms. How many stereoisomers of glucose are theoretically possible?

10. Fructose, $C_6H_{12}O_6$, has three chiral carbon atoms. How many stereoisomers of fructose are theoretically possible?

11. Do structures A and B represent enantiomers or the same compound? Justify your answer.

(A)
Cl
Br—C—H
F

(B)
Br
H—C—Cl
F

12. Do structures A and B represent enantiomers or the same compound? Justify your answer.

(A)
H
Br—C—F
CH_3

(B)
CH_3
Br—C—H
F

13. Which of these projection formulas represent (−)-lactic acid and which represent (+)-lactic acid?

COOH
H—OH
CH_3
(−)-lactic acid

COOH
HO—H
CH_3
(+)-lactic acid

(a)
CH_3
HO—H
COOH

(b)
OH
H—COOH
CH_3

(c)
COOH
H—CH_3
OH

(d)
CH_3
H—OH
COOH

(e)
COOH
CH_3—H
OH

(f)
H
CH_3—OH
COOH

14. Which of these projection formulas represent the naturally occurring amino acid (+)-alanine and which represent (−)-alanine?

COOH
H_2N—H
CH_3
(+)-alanine

COOH
H—NH_2
CH_3
(−)-alanine

(a)
CH_3
H_2N—H
COOH

(b)
NH_2
H—COOH
CH_3

(c)
COOH
H—CH_3
NH_2

(d)
CH_3
H—NH_2
COOH

(e)
COOH
CH_3—H
NH_2

(f)
H
CH_3—NH_2
COOH

15. Draw all the diastereomers of

CH_3
H—Cl
H_2N—H
COOH

16. Draw all the diastereomers of

CH_3
H—NH_2
H—OH
COOH

17. Draw all the enantiomers of

CH_2OH
H—OH
Br—H
H—OH
CH_3

18. Draw all the enantiomers of

CH_2OH
H—Cl
Br—H
Cl—H
COOH

19. Draw projection formulas for all possible stereoisomers of the following compounds:
(a) 1,2-dibromopropane
(b) 2-butanol
(c) 3-chlorohexane
Label pairs of enantiomers and meso compounds.

21. Draw projection formulas for all the stereoisomers of 1,2,3-trihydroxybutane. Point out enantiomers, meso compounds, and diastereomers, where present.

23. Write structures for the four stereoisomers of 2-hydroxy-3-pentene.

25. (a) Draw the nine structural isomers of $C_4H_8Cl_2$.
(b) Identify which structures represent chiral molecules, and draw all possible pairs of enantiomers and meso compounds, if any.

27. (+)-2-Bromopentane is further brominated to give dibromopentanes, $C_5H_{10}Br_2$. Write structures for all the possible isomers formed, and indicate which of these isomers will be optically active. [Remember that (+)-2-bromopentane is optically active.]

29. In the chlorination of propane, 1-chloropropane and 2-chloropropane are obtained as products. After separation by distillation, neither product rotates the plane of polarized light. Explain these results.

31. Which, if any, of the following are meso compounds?

(a)
COOH
H—OH
HO—COOH
H

(b)
CH_3
H—Cl
CH_3—Cl
H

(c)
CH_3
H—Cl
Br—CH_3
H

(d)
CH_3
H—Cl
H—Br
H—Cl
CH_3

33. Some substituted cycloalkanes are chiral. Draw the structures and enantiomers for any of the following that are chiral:

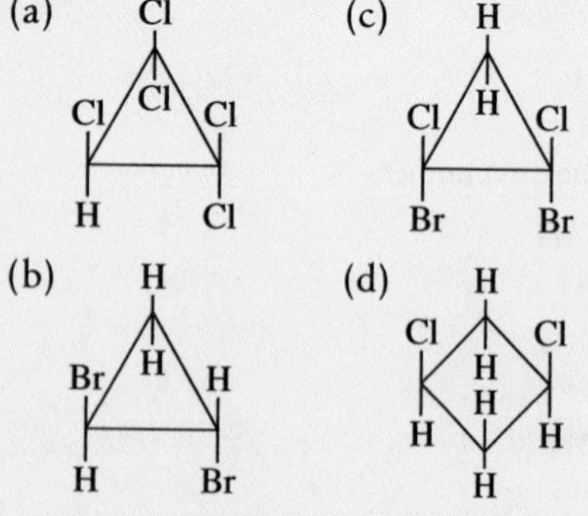

20. Draw projection formulas for all possible stereoisomers of the following compounds:
(a) 2,3-dichlorobutane
(b) 2,4-dibromopentane
(c) 3-hexanol
Label pairs of enantiomers and meso compounds.

22. Draw projection formulas for all the stereoisomers of 3,4-dichloro-2-methylpentane. Point out enantiomers, meso compounds, and diastereomers, where present.

24. Write structures for the four stereoisomers of 2-chloro-3-hexene.

26. (a) Draw the four structural isomers of $C_3H_6Br_2$.
(b) Identify which structures represent chiral molecules, and draw all possible pairs of enantiomers and meso compounds, if any.

28. (+)-2-Chlorobutane is further chlorinated to give dichlorobutanes, $C_4H_8Cl_2$. Write structures for all the possible isomers formed, and indicate which of these isomers will be optically active. [Remember that (+)-2-chlorobutane is optically active.]

30. In the chlorination of butane, 1-chlorobutane and 2-chlorobutane are obtained as products. After separation by distillation, neither product rotates the plane of polarized light. Explain these results.

32. Which, if any, of the following are meso compounds?

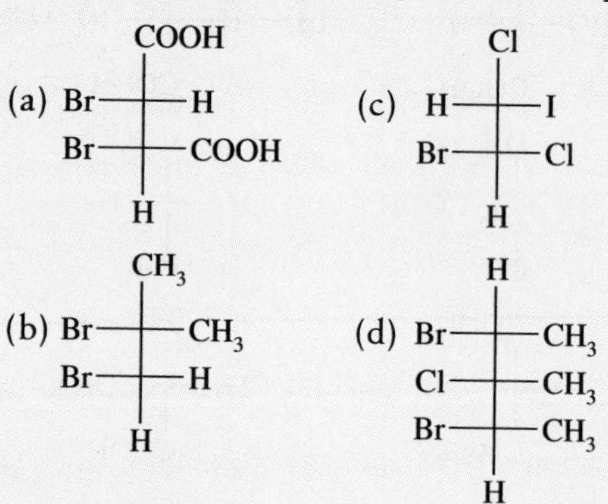

34. Some substituted cycloalkanes are chiral. Draw the structures and enantiomers for any of the following that are chiral:

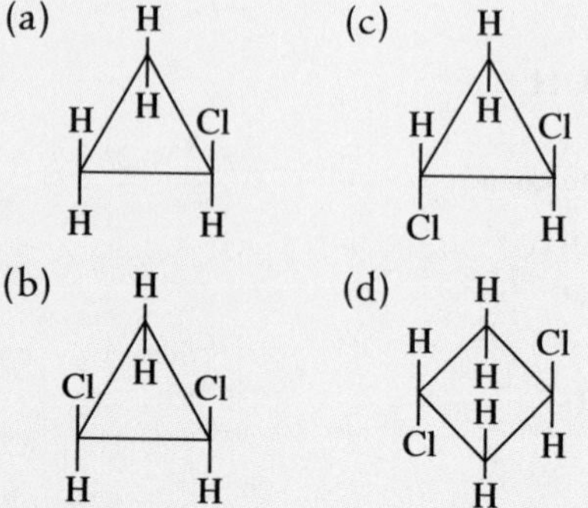

35. The following compounds are related as (i) enantiomers, (ii) diastereomers, or (iii) non-isomers. Choose the correct relationship for each pair.

(a) CH_2OH / —NH_2 / —OH / CH_2OH and CH_2OH / —OH / —NH_2 / CH_2OH

(b) CH_2OH / —NH_2 / —OH / CH_2OH and CH_2OH / H_2N— / H_2N— / CH_2OH

(c) CH_2OH / —NH_2 / HO— / —NH_2 / CH_3 and CH_2OH / —NH_2 / HO— / H_2N— / CH_3

36. The following compounds are related as (i) enantiomers, (ii) diastereomers, or (iii) non-isomers. Choose the correct relationship for each pair.

(a) O=CH / —OH / —OH / CH_2OH and CH_2OH / HO— / —OH / CH_2OH

(b) O=CH / — / —OH / HO— / CH_2OH and O=CH / — / HO— / —OH / CH_2OH

(c) O=CH / —OH / —OH / HO— / CH_3 and O=CH / —OH / HO— / HO— / CH_3

Additional Exercises

All exercises with blue numbers have answers in the appendix of the text

37. Suppose a carbon atom is located at the center of a square with four different groups attached to the corners in a planar arrangement. Would the compound rotate polarized light? Explain.

38. Ibuprofen is a common pain reliever found in medications such as Motrin and Advil. It is sold as a racemic mixture. The structure of ibuprofen is as follows:

$(CH_3)_2CHCH_2$—C_6H_4—$CH(CH_3)COOH$

(a) Identify the chiral carbon atom(s) in ibuprofen.
(b) Using projection formulas, draw the enantiomers of this drug.

39. (a) A chiral substance was identified as a primary alcohol of the formula $C_5H_{12}O$. What is its structure?
(b) The compound $C_6H_{14}O_3$ has three primary alcohol groups and is chiral. What is its structure?

40. The following dextrorotatory isomer of carvone gives caraway seeds their distinctive smell:

(structure: ring with CH_3, =O, H, H, H, H, H and $C(H)=CH_2$ substituent)

(a) Identify any chiral carbon atoms in this molecule.
(b) The enantiomer of this molecule is responsible for the refreshing smell of spearmint. How does the "spearmint" molecule differ from the "caraway seed" molecule?

41. Ephedrine is a potent drug originally found in a Chinese herbal medicine. It acts as a bronchial dilator and is used to treat asthma. The structure of ephedrine is

C_6H_5—$CH(OH)CH(CH_3)NHCH_3$

How many stereoisomers are possible for this drug? Briefly explain your reasoning.

42. One significant property of most amino acids is their chirality. Identify the chiral carbon in the amino acid alanine shown here:

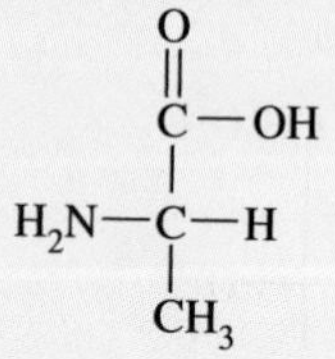

```
        O
        ||
        C—OH
        |
H2N—C—H
        |
        CH3
```

Draw the mirror image of this compound.

43. When biological interaction of a drug involves optical isomers, drug design becomes complicated because often only one isomer will have the desired property. For example, levomethorphan is an addictive opiate and dextromethorphan is a nonaddictive cough suppressant. What symbols would you use to differentiate between the two? Can you tell which one rotates light to the right? Explain why or why not.

44. Write structures for all stereoisomers, and label the enantiomers and meso compounds, for the following compounds:
(a) 2-bromo-3-chlorobutane
(b) 2,3,4-trichloro-1-pentanol

45. Draw all possible meso structures for alkanes that have the following molecular formulas:
(a) C_8H_{18}
(b) C_9H_{20}
(c) $C_{10}H_{22}$

46. Draw structures for all optically active alcohols having the formula $C_6H_{14}O$. (Draw only one enantiomer of each structure.)

47. Almost all natural amino acids share a common structure: Each molecule has a chiral carbon at the same position. Find and circle this chiral carbon atom. Briefly describe how the arrangement of groups around this chiral carbon is similar for all three examples that follow:

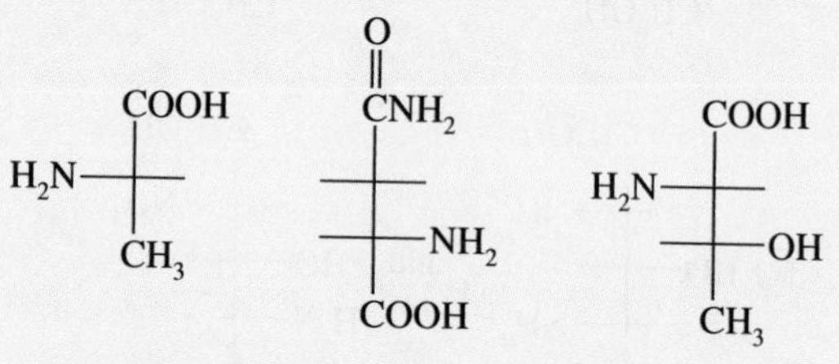

```
                              O
                              ||
      COOH                   CNH2              COOH
H2N——+——                ——+——          H2N——+——
      CH3                     +——NH2            +——OH
                             COOH               CH3
```

Challenge Exercise

48. What is the structure of a substance of formula $C_3H_8O_2$ that is (a) chiral and contains two —OH groups; (b) chiral and contains one —OH group; (c) achiral and contains two —OH groups? (Only one —OH group can be bonded to a carbon atom.)

Answers to Practice Exercises

26.1

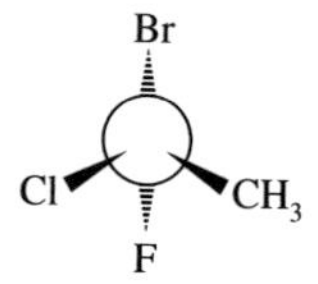

```
        Br
   Cl ◯ CH3
        F
```

26.2

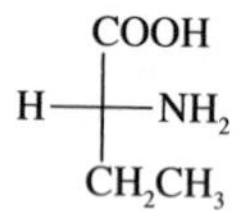

```
    COOH
H——+——NH2
    CH2CH3
```

26.3 (a)

```
            H                 ← Mirror      H
NH2CH2—C—CH2OH        |      HOCH2—C—CH2NH2
            Br                              Br
```

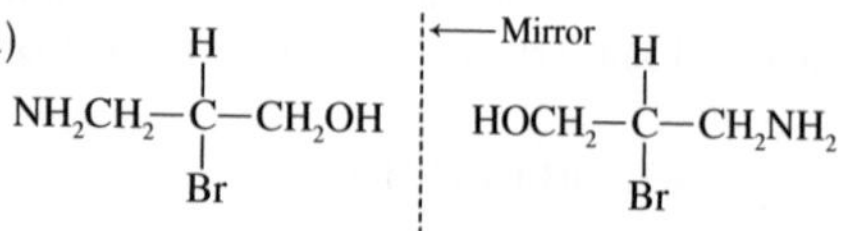

(b), (c) have no chiral carbons

26.4

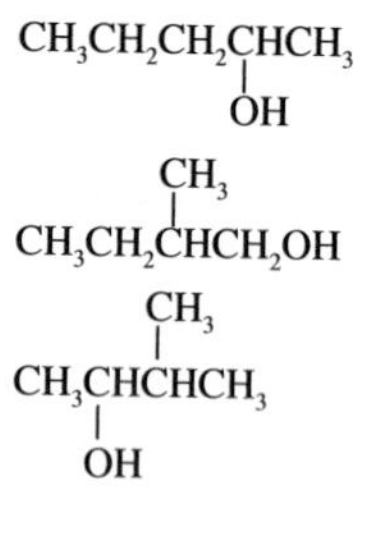

```
CH3CH2CH2CHCH3
             |
             OH

        CH3
        |
CH3CH2CHCH2OH

    CH3
    |
CH3CHCHCH3
      |
      OH
```

26.5

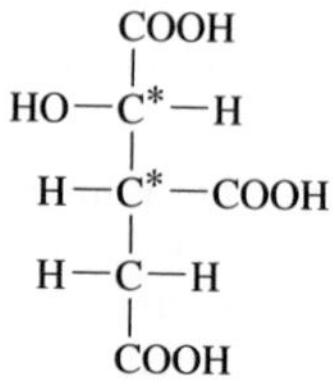

```
       COOH                         COOH
HO—C*—H                    H—C*—OH
 H—C*—COOH            HOOC—C*—H
 H—C—H                      H—C—H
       COOH                         COOH
   isocitric acid               enantiomer
```

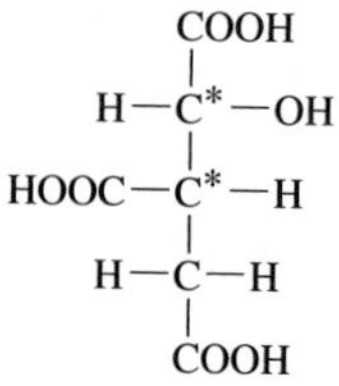

26.6

$CH_3OCH_2CH_2$–(benzene ring)–$OCH_2{}^*CH(OH)$–CH_2–NH–$CH(CH_3)$–CH_3

(benzene ring with OH, OH)–$CH_2{}^*CH(NH_2)COOH$

(benzene ring)–$^*CH(OH)$–$^*CH(CH_3)$–NH–CH_3

(piperidine ring, N–H)*–$^*CH$(benzene ring)–C(=O)–OCH_3

26.7

I	II	III	IV
COOH	COOH	COOH	COOH
H–C–Cl	Cl–C–H	H–C–Cl	Cl–C–H
H–C–H	H–C–H	H–C–H	H–C–H
H–C–H	H–C–H	H–C–H	H–C–H
H–C–Cl	Cl–C–H	Cl–C–H	H–C–Cl
COOH	COOH	COOH	COOH

I and II: meso compound; III and IV: enantiomers

Structures I and III, I and IV, II and III, and II and IV are diasteromers.

26.8

I	II	III	IV
O=C–H	O=C–H	O=C–H	O=C–H
H–C–H	H–C–H	H–C–H	H–C–H
H–C–OH	HO–C–H	HO–C–H	H–C–OH
H–C–OH	HO–C–H	H–C–OH	HO–C–H
CH_2OH	CH_2OH	CH_2OH	CH_2OH

Enantiomers are structures I and II, and III and IV.

CHAPTER 27

Carbohydrates

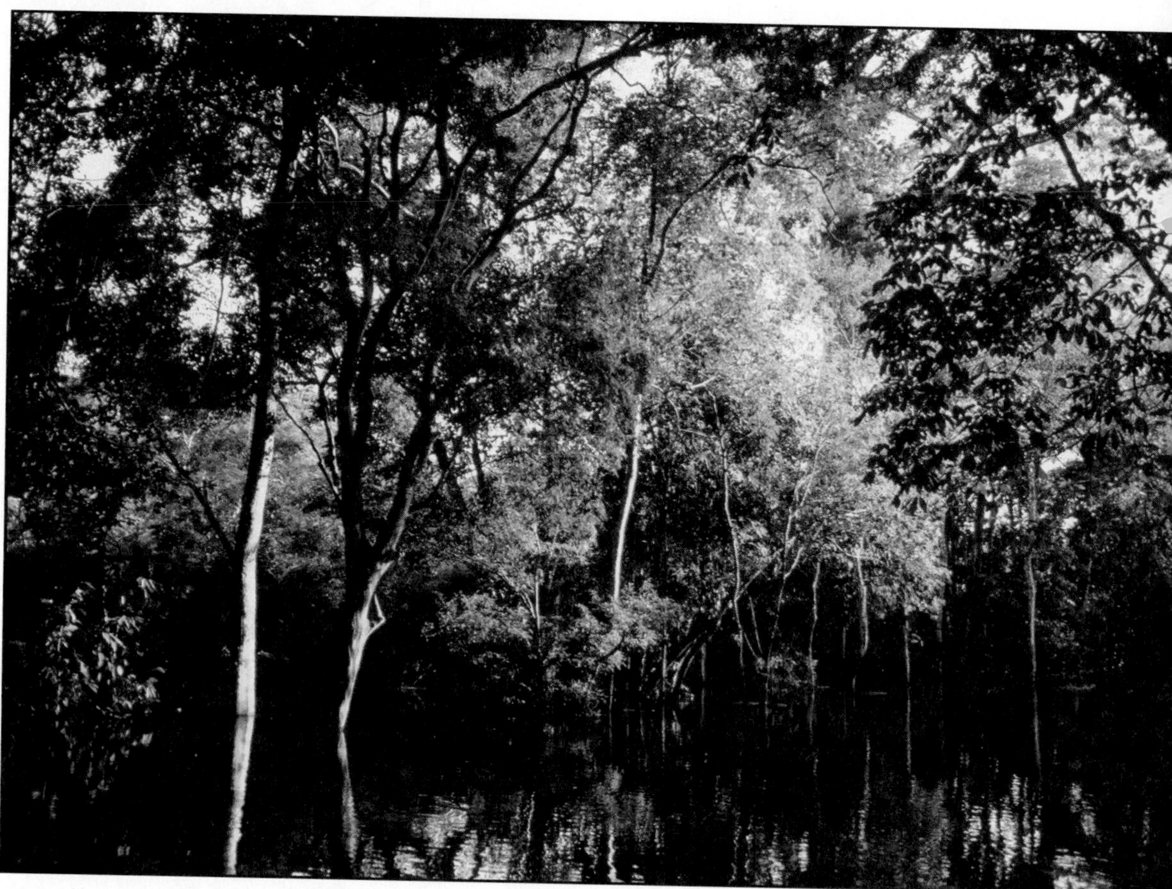

Green plants turn H_2O, CO_2, and sunlight into carbohydrates.

Chapter Outline

What is the most abundant organic chemical in the world? The answer is not petroleum products, plastics, or drugs. Rather, it is cellulose. An amazing 10 billion tons of cellulose are formed daily in the biosphere. Aggregates of cellulose allow the California redwoods to stretch hundreds of feet toward the sky and make a Brazil nut a "hard nut to crack." Products as diverse as the paper in this book and cotton in clothing are derived from cellulose. So, perhaps it is not surprising that this carbohydrate is the most widespread organic chemical in the world.

Carbohydrates are molecules of exceptional utility. They provide basic diets for many of us (starch and sugar), roofs over our heads, and clothes for our bodies (cellulose). They also thicken our ice cream, stick postage stamps to our letters, and provide biodegradable plastic trash sacks. From relatively simple components (carbon, hydrogen, and oxygen), nature has created one of the premier classes of biochemicals.

27.1 Carbohydrates: A First Class of Biochemicals

Carbohydrates are among the most widespread and important biochemicals. Most of the matter in plants, except water, consists of these substances. Carbohydrates are one of the three principal classes of energy-yielding nutrients; the other two are fats and proteins. Because of their widespread distribution and their role in many vital metabolic processes such as photosynthesis, carbohydrates have been scrutinized by scientists for over 150 years.

The name *carbohydrates* was given to this class of compounds many years ago by French scientists, who called them *hydrates de carbone* because their empirical formulas approximated $(C \cdot H_2O)_n$. It was found later that not all substances classified as carbohydrates conform to this formula (e.g., rhamnose, $C_6H_{12}O_5$, and deoxyribose, $C_5H_{10}O_4$). It seems clear that carbohydrates are not simply hydrated carbon; they are complex substances that contain from three to many thousands of carbon atoms. **Carbohydrates** are generally defined as polyhydroxy aldehydes or polyhydroxy ketones or substances that yield these compounds when hydrolyzed.

carbohydrate

The simplest carbohydrates are glyceraldehyde and dihydroxyacetone:

```
                  Carbonyl            H
                  carbon              |
H—C=O                            H—C—OH
  |                                   |
H—C—OH                               C=O
  |                                   |
H—C—OH                           H—C—OH
  |                                   |
  H                                   H
```

glyceraldehyde, $C_3H_6O_3$ dihydroxyacetone, $C_3H_6O_3$

These substances are "polyhydroxy" because each molecule has more than one hydroxyl group. Glyceraldehyde contains a carbonyl carbon in a terminal position and is therefore an aldehyde. The internal carbonyl of dihydroxyacetone identifies it as a ketone. Much of the chemistry and biochemistry of

When sugar (sucrose) and sulfuric acid (left) are combined, water is removed from the sugar, leaving a black column of carbon (right).

carbohydrates can be understood from a basic knowledge of the chemistry of the hydroxyl and carbonyl functional groups. (See Chapters 22 and 23 for review.)

27.2 Classification of Carbohydrates

monosaccharide

disaccharide

A carbohydrate is classified as a monosaccharide, a disaccharide, an oligosaccharide, or a polysaccharide, depending on the number of monosaccharide units linked to form the molecule. A **monosaccharide** is a carbohydrate that cannot be hydrolyzed to simpler carbohydrate units. The monosaccharide is the basic carbohydrate unit of cellular metabolism. A **disaccharide** yields two monosaccharides—either alike or different—when hydrolyzed:

$$\text{disaccharide} + \text{water} \xrightarrow{\text{H}^+ \text{ or enzymes}} 2 \text{ monosaccharides}$$

Disaccharides are often used by plants or animals to transport monosaccharides from one cell to another. The monosaccharides and disaccharides generally have names ending in *-ose*—for example, glucose, sucrose, and lactose. These water-soluble carbohydrates, which have a characteristically sweet taste, are also called *sugars*.

oligosaccharide

An **oligosaccharide** has two to six monosaccharide units linked together. *Oligo* comes from the Greek word *oligos*, which means small or few. Free oligosaccharides that contain more than two monosaccharide units are rarely found in nature.

polysaccharide

A **polysaccharide** is a macromolecular substance that can be hydrolyzed to yield many monosaccharide units:

$$\text{polysaccharide} + \text{water} \xrightarrow{\text{H}^+ \text{ or enzymes}} \text{many monosaccharide units}$$

Polysaccharides are important as structural supports, particularly in plants, and also serve as a storage depot for monosaccharides, which cells use for energy.

Carbohydrates can also be classified in other ways. A monosaccharide might be described with respect to several of these categories:

Sugar cane is a major source for sucrose, or table sugar.

1. As a triose, tetrose, pentose, hexose, or heptose:

Trioses	$C_3H_6O_3$	Hexoses	$C_6H_{12}O_6$
Tetroses	$C_4H_8O_4$	Heptoses	$C_7H_{14}O_7$
Pentoses	$C_5H_{10}O_5$		

   Theoretically, a monosaccharide can have any number of carbons greater than three, but only monosaccharides of three to seven carbons are commonly found in the biosphere.
2. As an aldose or ketose, depending on whether an aldehyde group (—CHO) or keto group (>C=O) is present. For ketoses, the >C=O is normally located on carbon 2.
3. As a D or L isomer, depending on the spatial orientation of the —H and —OH groups attached to the carbon atom adjacent to the terminal primary alcohol group. When the —OH is written to the right of this carbon in the projection formula (see Section 26.4), the D isomer is represented. When this —OH is written to the left, the L isomer is represented. The reference compounds for this classification are the trioses D-glyceraldehyde and L-glyceraldehyde, whose formulas follow. Also shown are two aldohexoses (D- and L-glucose) and a ketohexose (D-fructose).

```
                      H—C=O               H—C=O
                        |                   |
D-Configuration —→    H—C—OH           HO—C—H ←— L-Configuration
                        |                   |
Terminal 1° ROH ——→    CH2OH              CH2OH ←—— Terminal 1° ROH

                 D-glyceraldehyde     L-glyceraldehyde
```

```
                      H—C=O              H—C=O                                CH2OH
                        |                  |                                    |
                     HO—C—H             H—C—OH                                  C=O
                        |                  |                                    |
                      H—C—OH           HO—C—H                               HO—C—H
                        |                  |                                    |
                     HO—C—H             H—C—OH                               H—C—OH
                        |                  |                                    |
L-Configuration —→   HO—C—H             H—C—OH ←— D-Configuration —→         H—C—OH
                        |                  |                                    |
Terminal 1° ROH ——→    CH2OH              CH2OH ←—— Terminal 1° ROH ——→        CH2OH

                     L-glucose           D-glucose                          D-fructose
```

   The letters D and L do not in any way refer to the direction of optical rotation of a carbohydrate. The D and L forms of any specific compound are enantiomers (e.g., D- and L-glucose).

Remember, enantiomers are mirror-image isomers.

4. As a (+) or (−) isomer, depending on whether the monosaccharide rotates the plane of polarized light to the right (+) or to the left (−). (See Section 26.5.)
5. As a furanose or a pyranose, depending on whether the cyclic structure of the carbohydrate is related to that of the five- or six-membered heterocyclic ring compound furan or pyran (a heterocyclic ring contains more than one kind of atom in the ring):

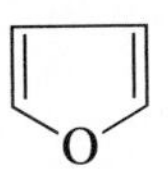

furan, C_4H_4O
(five-membered ring with oxygen in the ring)

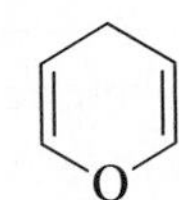

pyran, C_5H_6O
(six-membered ring with oxygen in the ring)

6. As having an alpha (α) or beta (β) configuration, based on the orientation of the —H and —OH groups about a specific chiral carbon in the cyclic form of the monosaccharide (Section 27.6).

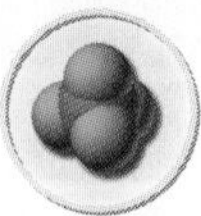

Example 27.1 Write projection formulas for (a) an L-aldotriose, (b) a D-ketotetrose, and (c) a D-aldopentose.

SOLUTION (a) Triose indicates a three-carbon carbohydrate; aldo indicates that the compound is an aldehyde; L- indicates that the —OH on carbon 2 (adjacent to the terminal CH_2OH) is on the left. The aldehyde group is carbon 1:

```
   H—¹C=O
      |
  HO—²C—H    ←—— L-configuration
      |
     ³CH₂OH
```

an L-aldotriose

(b) Tetrose indicates a four-carbon carbohydrate; keto indicates a ketone group (on carbon 2); D- indicates that the —OH on carbon 3 (adjacent to the terminal CH_2OH) is on the right. Carbons 1 and 4 have the configuration of primary alcohols:

```
     ¹CH₂OH
      |
     ²C=O
      |
   H—³C—OH   ←—— D-configuration
      |
     ⁴CH₂OH
```

a D-ketotetrose

(c) Pentose indicates a five-carbon carbohydrate; aldo indicates an aldehyde group (on carbon 1); D- indicates that the —OH on carbon 4 (adjacent to the terminal CH_2OH) is on the right. The orientation of the —OH groups on carbons 2 and 3 is not specified here and therefore can be written in either direction for this problem:

```
   H—¹C=O
      |
   H—²C—OH
      |
   H—³C—OH
      |
   H—⁴C—OH   ←—— D-configuration
      |
     ⁵CH₂OH
```

a D-aldopentose

Practice 27.1

Write the projection formula for an L-ketopentose.

Practice 27.2

In a projection formula for a D-aldotriose, is the —OH of the secondary alcohol carbon written on the right or the left side?

27.3 Importance of Carbohydrates

As we stated earlier, carbohydrates are the most abundant organic chemical—they must be molecules of exceptional utility. But what makes them so special? Why are they so important in our daily lives?

If a carbohydrate carbon has less than 4 bonds to oxygen, it can be oxidized further.

Carbohydrates are very effective energy-yielding nutrients. In general, a molecule can provide biological energy if it (1) contains carbon, (2) is relatively reactive (contains some polar bonds), and (3) contains carbons that have less than four bonds to oxygen. Carbohydrates fit this description very well. The average carbohydrate contains about 40% carbon by mass. Carbohydrates are relatively reactive because they contain many polar groups (e.g., hydroxyl groups). Finally, each carbohydrate may be oxidized further.

Carbohydrates can serve as very effective building materials. These materials need to be water-soluble and strong but not brittle. As you will see in Section 27.14, large, water-insoluble polymers (polysaccharides) are formed from simple carbohydrates. Cellulose polymers can form durable fibers by hydrogen bonding. These fibers provide building materials for strong, relatively rigid wood construction; for flexible, comfortable cotton clothing; and for flexible yet smooth paper writing material. Cellulose also forms at least part of the cell wall for all higher plants.

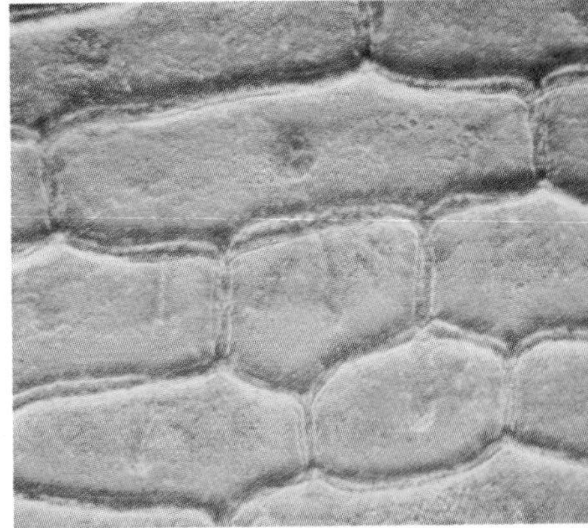

Carbohydrates are important building blocks in the formation of cell walls in higher plants. Shown here are the cell walls of an onion.

Carbohydrates are important water-soluble molecules. Many smaller ones, especially the mono- and disaccharides, are very water soluble. These carbohydrates are often mixed into foods to give them a sweet taste. Water solubility allows these molecules to pass easily from living cell to living cell.

27.4 Monosaccharides

Although a great many monosaccharides have been synthesized, only a very few appear to be of much biological significance. One pentose monosaccharide (ribose) and its deoxy derivative are essential components of ribonucleic acid (RNA) and of deoxyribonucleic acid (DNA). (See Chapter 31.) However, the hexose monosaccharides are the most important carbohydrate sources of cellular energy. Three hexoses—glucose, galactose, and fructose—are of major significance in nutrition. All three have the same molecular formula, $C_6H_{12}O_6$, and thus deliver the same amount of cellular energy. They differ in structure, but are biologically interconvertible. Glucose plays a central role in carbohydrate energy utilization. Other carbohydrates are usually converted to glucose before cellular utilization. The structure of glucose is considered in detail in Section 27.5.

Glucose

Glucose is the most important of the monosaccharides. It is an aldohexose and is found in the free state in plant and animal tissue. Glucose is also known as *dextrose* or *grape sugar*. It is a component of the disaccharides sucrose, maltose, and lactose and is the monomer of the polysaccharides starch, cellulose, and glycogen. Among the common sugars, glucose is of intermediate sweetness. (See Section 27.12.)

Glucose is called dextrose because it is dextrorotatory and grape sugar because it occurs in grapes.

Glucose is the key sugar of the body and is carried by the bloodstream to all body parts. The concentration of glucose in the blood is normally 80–100 mg per

100 mL of blood. Because glucose is the most abundant carbohydrate in the blood, it is commonly called *blood sugar*. Glucose requires no digestion and may therefore be given intravenously to patients who cannot take food by mouth. Glucose is found in the urine of those who have diabetes mellitus (sugar diabetes). The condition in which glucose is excreted in the urine is called glycosuria.

Galactose

Galactose is also an aldohexose and occurs, along with glucose, in lactose and in many oligo- and polysaccharides such as pectin, gums, and mucilages. Galactose is an isomer of glucose, differing only in the spatial arrangement of the —H and —OH groups around carbon 4. (See Section 27.5.) Galactose is synthesized in the mammary glands to make the lactose of milk. It is also a constituent of glycolipids and glycoproteins in many cell membranes, such as those in nervous tissue. Galactose is less than half as sweet as glucose.

The inability of infants to metabolize galactose is an inherited condition called galactosemia. In galactosemia, the galactose concentration increases markedly in the blood and also appears in the urine. Galactosemia causes vomiting, diarrhea, enlargement of the liver, and often mental retardation. If not recognized within a few days after birth, it can lead to death. If diagnosis is made early and lactose is excluded from the diet, the symptoms disappear and normal growth may be resumed.

Fructose

Fructose, also known as *levulose*, is a ketohexose that occurs in fruit juices, honey, and, along with glucose, as a constituent of sucrose. Fructose is the major constituent of the polysaccharide inulin, a starchlike substance present in many plants such as dahlia tubers, chicory roots, and Jerusalem artichokes. Fructose is the sweetest of all the common sugars, being about twice as sweet as glucose. This accounts for the sweetness of high-fructose corn syrup and honey. The enzyme invertase, present in bees, splits sucrose into glucose and fructose. Fructose is metabolized directly, but is also readily converted to glucose in the liver.

27.5 Structure of Glucose and Other Aldoses

In one of the classic feats of research in organic chemistry, Emil Hermann Fischer (1852–1919), working in Germany, established the structural configuration of glucose along with that of many other sugars. For his work, he received the Nobel prize in chemistry in 1902. Fischer devised projection formulas that relate the structure of a sugar to one or the other of the two enantiomeric forms of glyceraldehyde. These projection formulas represent three-dimensional stereoisomers (see Chapter 26) in a two-dimensional plane. (Remember that stereoisomers cannot be interconverted without breaking and reforming covalent bonds.) Each carbohydrate isomer has a different shape and thus reacts differently in biological systems.

A projection formula is commonly written in either of two ways, with or without the chiral carbon atoms.

In Fischer projection formulas, the molecule is represented with the aldehyde (or ketone) group at the top. The —H and —OH groups attached to interior carbons are written to the right or to the left as they would appear when projected toward the observer. The two glyceraldehydes are represented as follows:

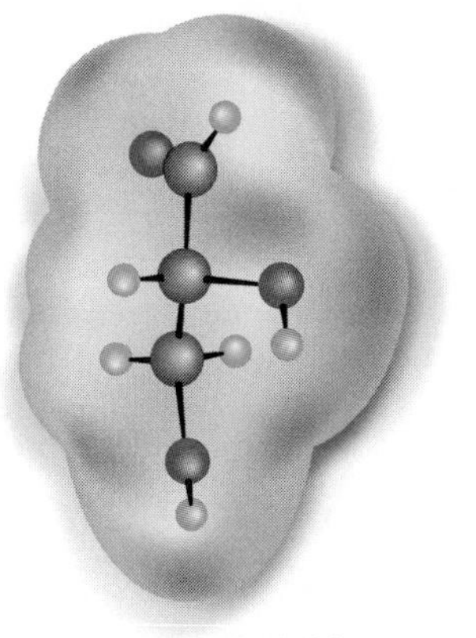

```
    O=C—H
      ⋮
 H         OH
      ⋮
    CH2OH
```

```
O=C—H          O=C—H
  |              |
H—C—OH         H—+—OH
  |              |
 CH2OH         CH2OH
```

Ball-and-stick/ spacefilling model | 3-dimensional representation | Fischer projection formulas

D-glyceraldehyde

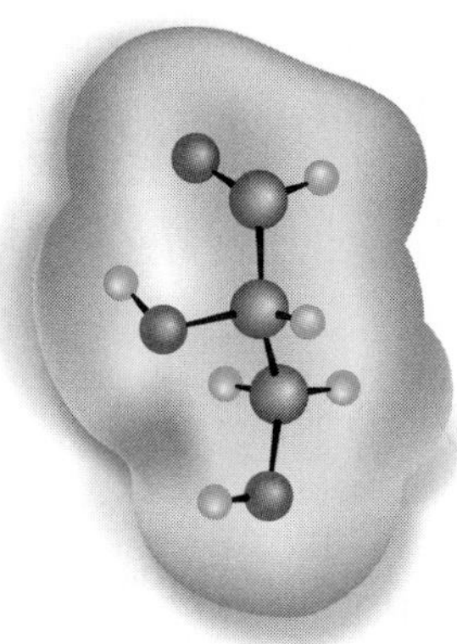

```
    O=C—H
      ⋮
 HO        H
      ⋮
    CH2OH
```

```
 O=C—H           O=C—H
   |               |
HO—C—H          HO—+—H
   |               |
  CH2OH          CH2OH
```

Ball-and-stick/ spacefilling model | 3-dimensional representation | Fischer projection formulas

L-glyceraldehyde

In the three-dimensional molecules represented by these formulas, the carbon 2 atoms are in the plane of the paper. The —H and —OH groups project forward (toward the observer); the —CHO and —CH_2OH groups project backward (away from the observer). Any two monosaccharides that differ only in the configuration around a single carbon atom are called **epimers**. Thus, D- and L-glyceraldehyde are epimers.

epimer

Fischer recognized that there are two enantiomeric forms of glucose. To these forms, he assigned the following structures and names:

```
 H—C=O       H—C=O          H—C=O       H—C=O
   |           |              |           |
 H—C—OH      H—+—OH        HO—C—H     HO—+—H
   |           |              |           |
HO—C—H      HO—+—H          H—C—OH      H—+—OH
   |           |              |           |
 H—C—OH      H—+—OH        HO—C—H     HO—+—H
   |           |              |           |
 H—C—OH      H—+—OH        HO—C—H     HO—+—H
   |           |              |           |
  CH2OH       CH2OH          CH2OH       CH2OH
```

D-glucose | L-glucose

Enantiomers appear to have similar structures, but they differ at every chiral carbon.

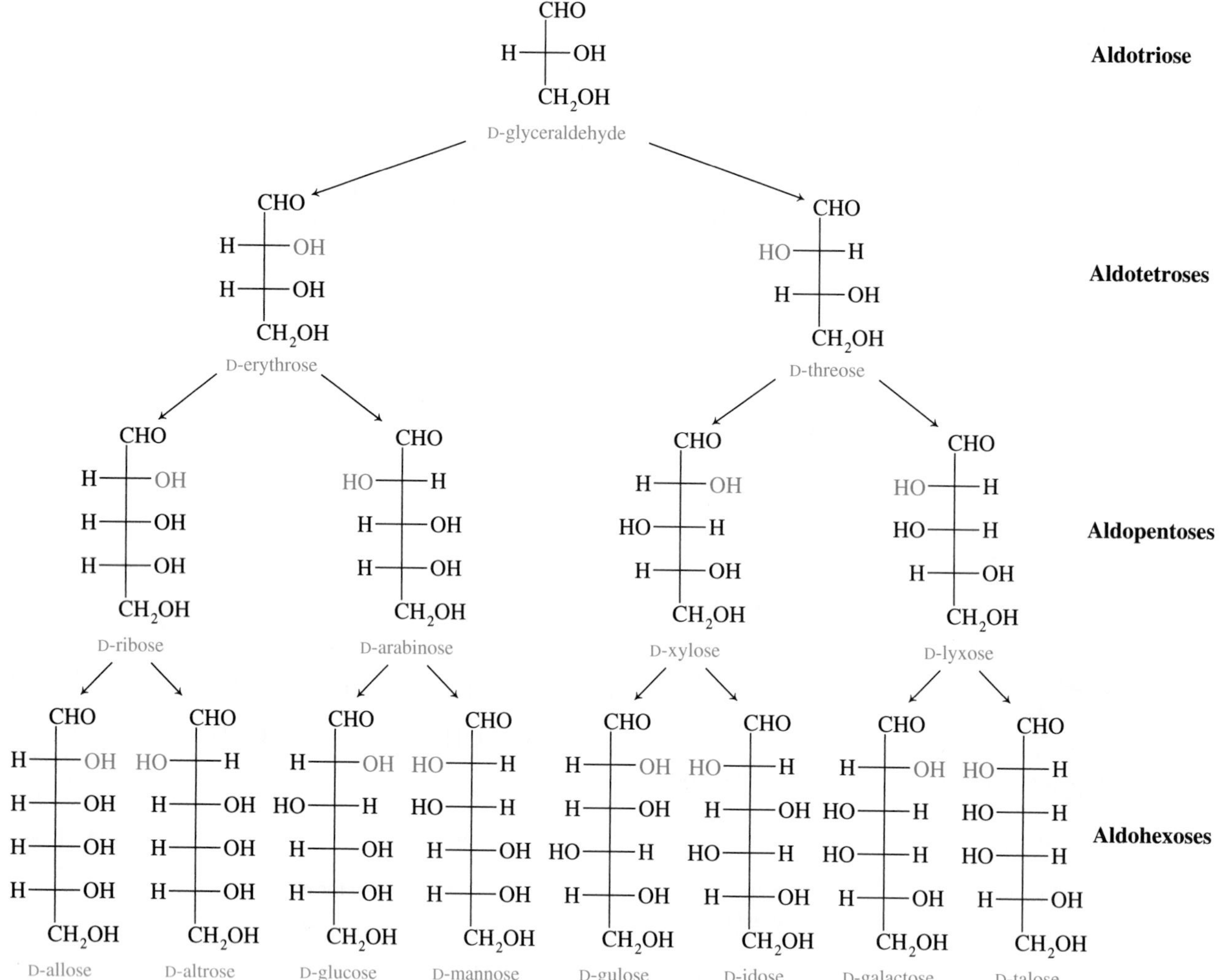

Figure 27.1
Configurations of the D-family of aldoses. The hydroxyl group on the new chiral carbon atom, added in going from triose to tetrose to pentose to hexose, is shown in red.

The structure called D-glucose is so named because the —H and —OH on carbon 5 are in the same configuration as the —H and —OH on carbon 2 in D-glyceraldehyde. The configuration of the —H and —OH on carbon 5 in L-glucose corresponds to the —H and —OH on carbon 2 in L-glyceraldehyde.

Fischer recognized that 16 different aldohexoses, 8 with the D configuration and 8 with the L configuration, were possible. This follows our formula 2^n for optical isomers. (See Section 26.7.) Glucose has four chiral carbon atoms and should have 16 stereoisomers (2^4). The configurations of the D-aldose family are shown in Figure 27.1. In this family, new chiral carbon atoms are formed as we go from triose to tetrose to pentose to hexose. Each time a new chiral carbon is added, a pair of epimers is formed that differ only in the structure around carbon 2. The sequence continues until eight D-aldohexoses are created. A similar series starting with L-glyceraldehyde is known, making a total of 16 aldohexoses.

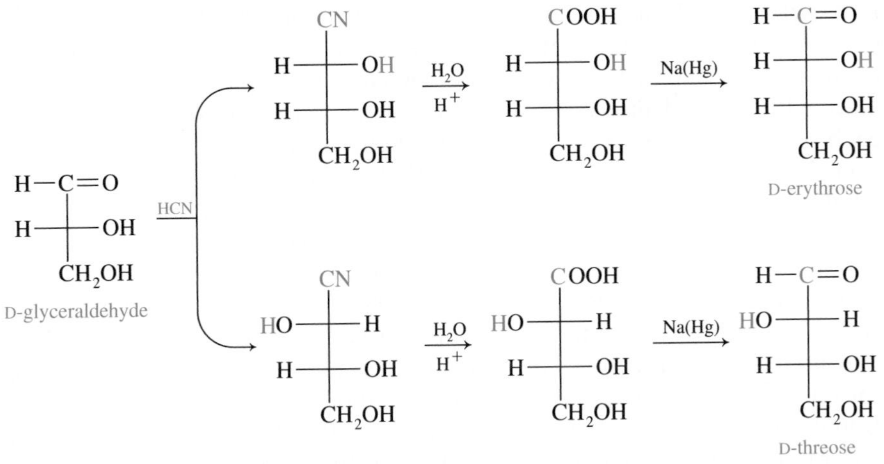

H—C=O
H—C—OH
CH$_2$OH
D-glyceraldehyde

HCN →

CN | H—C—OH | H—C—OH | CH$_2$OH $\xrightarrow[H^+]{H_2O}$ COOH | H—C—OH | H—C—OH | CH$_2$OH $\xrightarrow{Na(Hg)}$ H—C=O | H—C—OH | H—C—OH | CH$_2$OH (D-erythrose)

CN | HO—C—H | H—C—OH | CH$_2$OH $\xrightarrow[H^+]{H_2O}$ COOH | HO—C—H | H—C—OH | CH$_2$OH $\xrightarrow{Na(Hg)}$ H—C=O | HO—C—H | H—C—OH | CH$_2$OH (D-threose)

The aldehyde carbon of D-glyceraldehyde undergoes an addition reaction during Kiliani–Fischer synthesis.

Figure 27.2
An example of the Kiliani–Fischer synthesis in which two aldotetrose molecules are formed from an aldotriose molecule.

Practice 27.3

Refer to Figure 27.1 and draw the structure for

(a) the enantiomer of D-idose
(b) the enantiomer of D-xylose

The 16 aldohexoses have all been synthesized, but only D-glucose and D-galactose appear to be of considerable biological importance. Since the metabolism of most living organisms revolves about D-glucose, our discussion will focus on this substance.

The laboratory conversion of one aldose into another aldose containing one more carbon atom is known as the *Kiliani–Fischer synthesis*. This synthesis makes use of the aldehyde's ability to bond to an additional group. (See Section 23.5.) Heinrich Kiliani (1855–1945) discovered that the addition of HCN to an aldose formed a cyanohydrin. Emil Fischer subsequently published a method for converting the cyanohydrin nitrile group (—CN) into an aldehyde. The Kiliani–Fischer synthesis involves (1) the addition of HCN to form a cyanohydrin; (2) hydrolysis of the —CN group to —COOH; and (3) reduction with sodium amalgam, Na(Hg), to form the aldehyde. As an example, the formation of D-erythrose and D-threose from D-glyceraldehyde is shown in Figure 27.2.

Practice 27.4

Write the structures of the two aldoheptoses formed when D-glucose is subjected to the Kiliani–Fischer synthesis.

27.6 Cyclic Structure of Glucose; Mutarotation

Straight open-chain D-glucose is so reactive that almost all molecules quickly rearrange their bonds to form two new structures. These two forms are diastereomers that differ with respect to their rotation of polarized light. One form, labeled α-D-glucopyranose, has a specific rotation, [α], of +112°; the other, labeled β-D-glucopyranose, has a specific rotation of +18.7°. An interesting phenomenon occurs when these two forms of glucose are put into separate solutions and allowed to stand for several hours. The specific rotation of each solution changes to +52.7°. This phenomenon is known as mutarotation. An explanation of mutarotation is that D-glucose exists in solution as an equilibrium mixture of two cyclic forms and the open-chain form. (See Figure 27.3.) The two cyclic molecules are optical isomers, differing only in the orientation of the —H and —OH groups about carbon 1. When dissolved, some α-D-glucopyranose molecules are transformed into β-D-glucopyranose, and vice versa, until an equilibrium is reached between the α and β forms. (Note that no other chiral centers in D-glucose are altered when this sugar is dissolved.) The equilibrium solution contains about 36% α molecules and 64% β molecules, with a trace of open-chain molecules. When two cyclic isomers differ only in their stereo arrangement about the carbon involved in mutarotation, they are called **anomers**. For example, α- and β-D-glucopyranose are anomers. (See Figure 27.3.) **Mutarotation** is the process by which anomers are interconverted.

anomer
mutarotation

The cyclic forms of D-glucose may be represented by either Fischer projection formulas or by Haworth perspective formulas. These structures are shown in Figure 27.3. In the cyclic Fischer projection formulas of the D-aldoses, the α form has the —OH on carbon 1 written to the right; in the β form, the —OH on carbon 1 is on the left. The Haworth formula represents the molecule as a flat hexagon with the —H and —OH groups above and below the plane of the hexagon. In the α form, the —OH on carbon 1 is written below the plane; in the β form, the —OH on carbon 1 is above the plane. In converting the projection formula

The aldehyde carbon of the open-chain D-glucose undergoes an addition reaction when the cyclic structure of D-glucose forms.

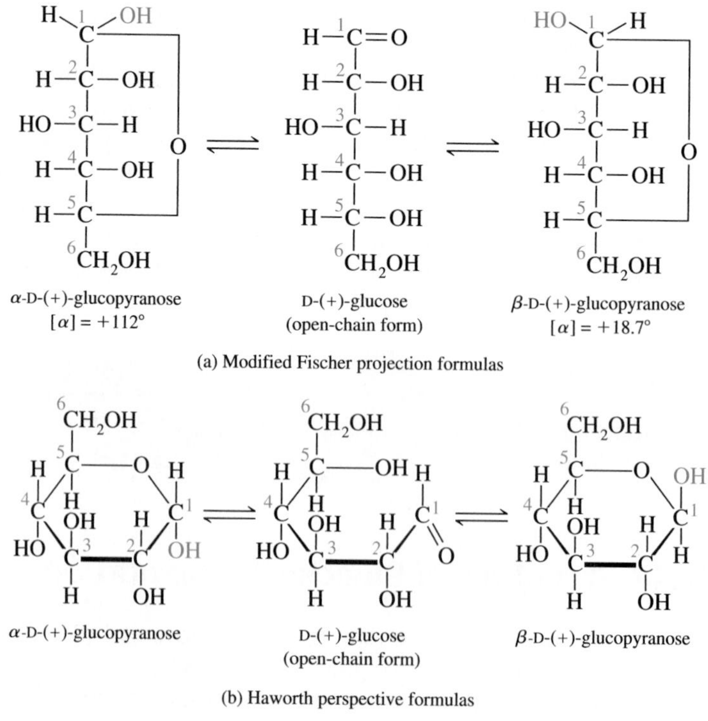

α-D-(+)-glucopyranose [α] = +112° ⇌ D-(+)-glucose (open-chain form) ⇌ β-D-(+)-glucopyranose [α] = +18.7°

(a) Modified Fischer projection formulas

α-D-(+)-glucopyranose ⇌ D-(+)-glucose (open-chain form) ⇌ β-D-(+)-glucopyranose

(b) Haworth perspective formulas

Figure 27.3
Mutarotation of D-glucose.

of a D-aldohexose to the Haworth formula, the —OH groups on carbons 2, 3, and 4 are written below the plane if they project to the right and above the plane if they project to the left. Carbon 6 is written above the plane.

Haworth formulas are sometimes shown in abbreviated schematic form. For example, α-D-(+)-glucopyranose is shown as follows:

The four open bonds are understood to have hydrogen atoms on them.

α-D-(+)-glucopyranose
(abbreviated form)

Although both the Fischer projection formula and the Haworth formula provide useful representations of carbohydrate molecules, it is important to understand that these structures only approximate the true molecular shapes. We know, for example, that the pyranose ring is not flat, but, rather, can assume either a chair or boat conformation like the cycloalkanes. (See Section 19.14.) Most naturally occurring monosaccharides are found in the chair form as shown in Figure 27.4 for α-D-glucopyranose. Even this three-dimensional structure does not truly capture how a sugar molecule must appear. For, unlike this representation, we know that atoms move as close together as possible when they form molecules. Perhaps the most accurate representation of a sugar molecule is the spacefilling model. A spacefilling model of α-D-glucopyranose is shown in Figure 27.4. At best, any two-dimensional representation is a compromise in portraying the three-dimensional configuration of such molecules. Structural models are much more effective, especially if constructed by the student.

The two cyclic forms of D-glucose differ only in the relative positions of the —H and —OH groups attached to carbon 1. Yet, this seemingly minor structural difference has important biochemical consequences because the physical shape

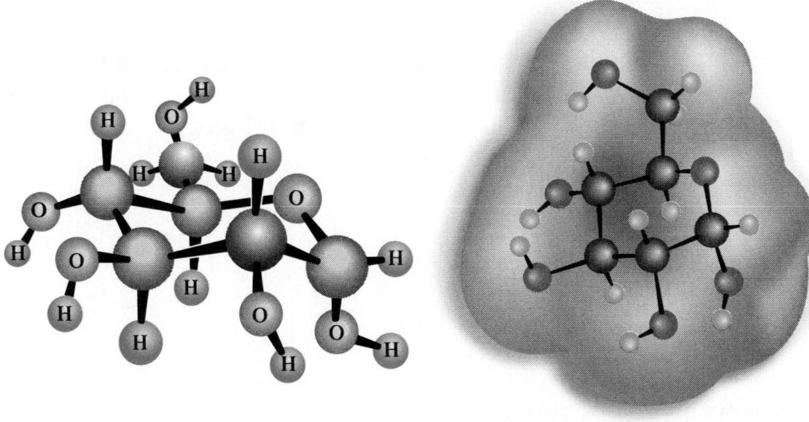

Figure 27.4
Three-dimensional representations of the chair form of α-D-glucopyranose.

of a molecule often determines its biological use. For example, the fundamental structural difference between starch and cellulose is that starch is a polymer of α-D-glucopyranose, whereas cellulose is a polymer of β-D-glucopyranose. As a consequence, starch is easily digested by humans, but we are totally unable to digest cellulose.

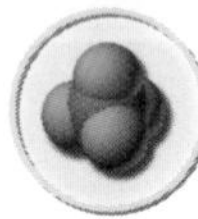

Example 27.2 Write the pyranose Haworth perspective formulas and names for the two anomers of D-mannose.

SOLUTION First, write the open-chain Fischer projection formula for D-mannose (you must memorize this structure or know where to find it in the text) and number the carbons from top (the aldehyde group) to bottom (the primary alcohol group):

H—C=O (1)
HO—(2)—H
HO—(3)—H
H—(4)—OH
H—(5)—OH
(6) CH_2OH

Next, draw the structure of the Haworth pyranose ring. Number the carbons from the right-hand point of the hexagon clockwise around the cyclic form, placing the CH_2OH (C6) group in the up position on the ring. Then, refer to the open-chain Fischer projection formula. All the hydroxyl groups on the right of the open chain should be written *down* in the Haworth formula, and all the hydroxyl groups on the left should be written *up*. Since this rule only applies to chiral centers, we can ignore the hydroxyl on carbon 6:

The carbon involved in mutarotation, carbon 1, can have either of two configurations, the α anomer when the hydroxyl is pointed down or the β anomer when pointed up. The last step is to add the hydroxyl group at carbon 1 and name the anomers:

α-D-mannopyranose

β-D-mannopyranose

Practice 27.5

Draw the Haworth formulas for (a) α-D-galactopyranose and (b) β-D-glucopyranose.

27.7 Hemiacetals and Acetals

In Chapter 23, we studied the reactions of aldehydes (and ketones) to form hemiacetals and acetals. The hemiacetal structure consists of an ether linkage and an alcohol linkage on the same carbon atom (shown in red), whereas the acetal structure has two ether linkages to the same carbon atom:

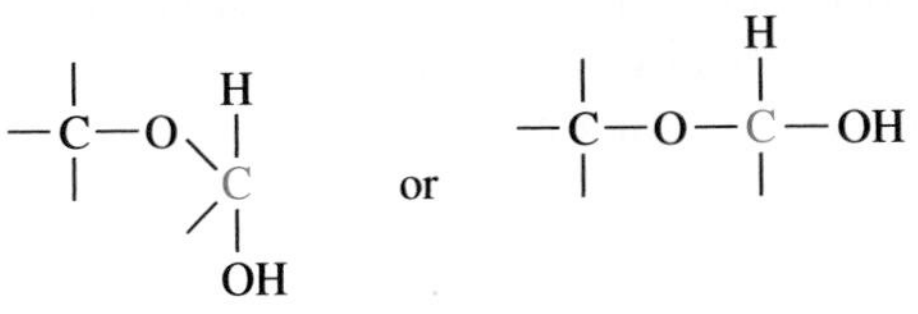

Cyclic structures of monosaccharides are intramolecular hemiacetals. Five- or six-membered rings are especially stable:

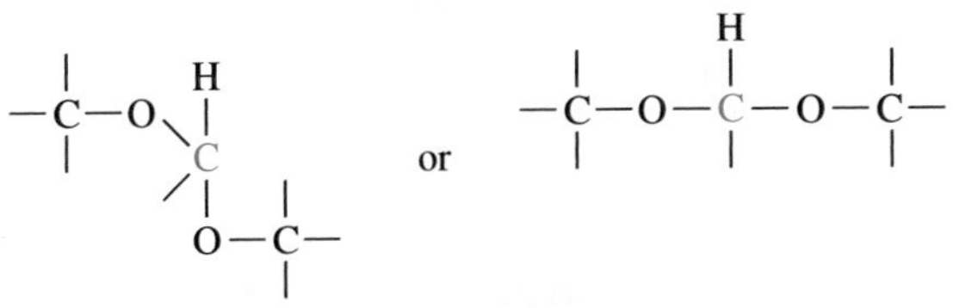

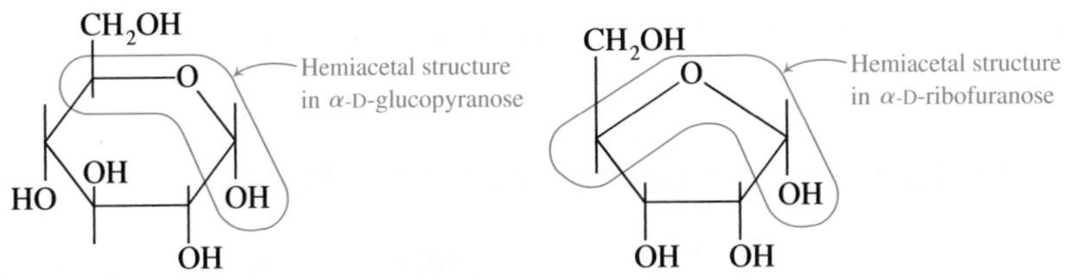

However, in an aqueous solution, the ring often opens and the hemiacetal momentarily reverts to the open-chain aldehyde. When the open chain closes, it forms either the α or the β anomer. Mutarotation results from this opening and closing of the hemiacetal ring. (See Figure 27.3.)

When an alcohol, ROH, reacts with another alcohol, R′OH, to split out H_2O, the product formed can be an ether, ROR′. Carbohydrates are alcohols and behave accordingly. When a monosaccharide hemiacetal reacts with an alcohol, the product is an acetal. In carbohydrate terminology, this acetal structure is called a **glycoside** (derived from the Greek word *glykys*, meaning sweet). In the case of glucose, it would be a glucoside; if galactose, a galactoside; and so on.

glycoside

For example, an α-glycoside is shown as follows:

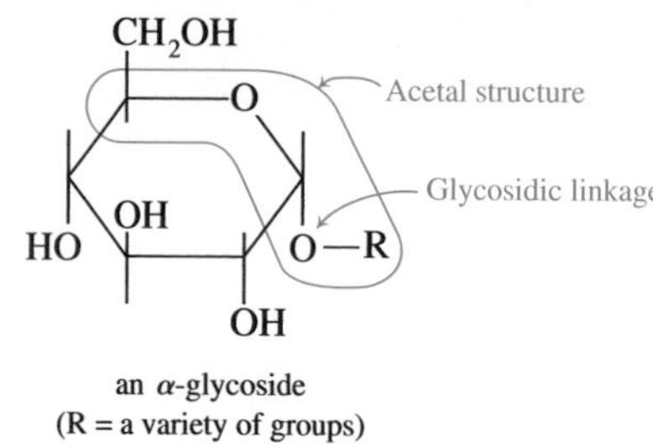

an α-glycoside
(R = a variety of groups)

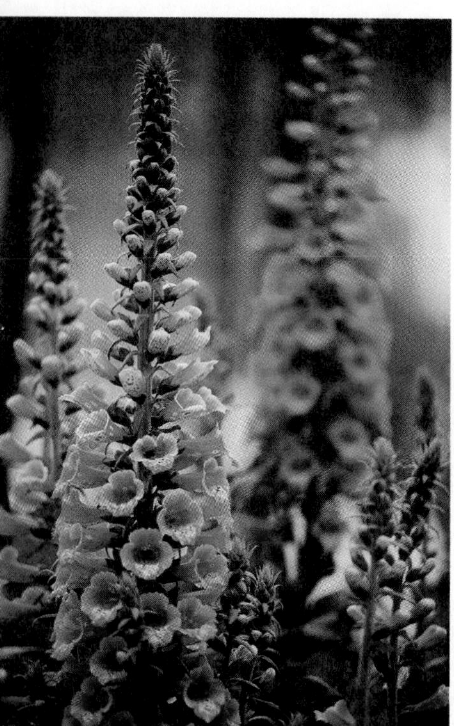

Digitalis, a well-known heart stimulant, is produced in nature by the foxglove plant.

A glycoside differs significantly from a monosaccharide with respect to chemical reactivity.

When α-D-glucopyranose is heated with methyl alcohol and a small quantity of hydrogen chloride is added, two optically active isomers are formed—methyl α-D-glucopyranoside and methyl β-D-glucopyranoside:

α-D-glucopyranose $\underset{HCl}{\overset{CH_3OH}{\rightleftharpoons}}$ methyl α-D-glucopyranoside (mp 165°C, [α] = + 158°) + methyl β-D-glucopyranoside (mp 107°C, [α] = − 33°)

Unlike D-glucose, the two glycoside products no longer undergo mutarotation. They do not form open-chain compounds in aqueous solution. Acetals tend to be more stable and less reactive than hemiacetals.

The glycosidic linkage occurs in a wide variety of natural substances. All carbohydrates other than monosaccharides are glycosides. Heart stimulants such as digitalis and ouabain are known as heart glycosides. Several antibiotics such as streptomycin and erythromycin are also glycosides.

27.8 Structures of Galactose and Fructose

Galactose, like glucose, is an aldohexose and differs structurally from glucose only in the configuration of the —H and —OH group on carbon 4:

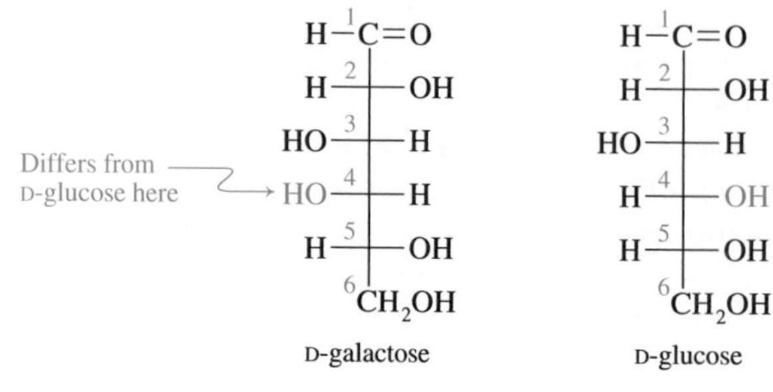

D-galactose D-glucose

Galactose, like glucose, also exists primarily in two cyclic pyranose forms that have hemiacetal structures and undergo mutarotation:

To identify anomers in the Haworth formula, focus on carbon 1 for aldoses and carbon 2 for ketoses. If the —OH on this carbon is written below the plane of the ring, the molecule is an α anomer; if the —OH is written above the plane of the ring, the molecule is a β anomer.

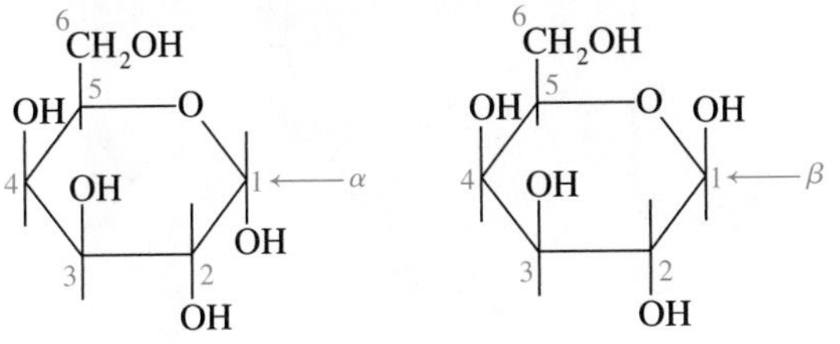

α-D-galactopyranose β-D-galactopyranose

Fructose is a ketohexose. The open-chain form may be represented in a Fischer projection formula:

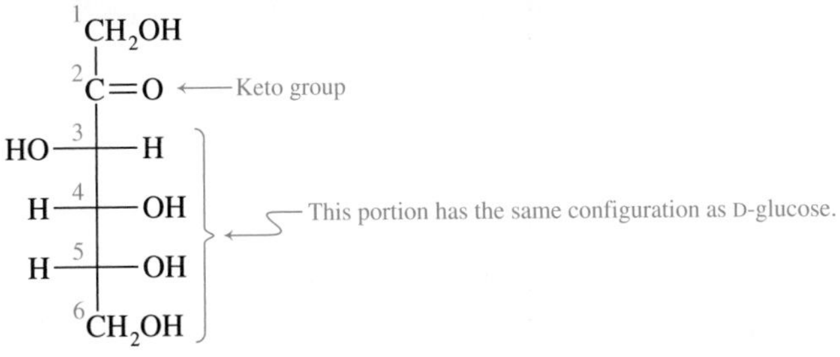

Keto group

This portion has the same configuration as D-glucose.

D-fructose

Like glucose and galactose, fructose exists in both cyclic and open-chain forms. One common cyclic structure is a five-membered furanose ring in the β configuration:

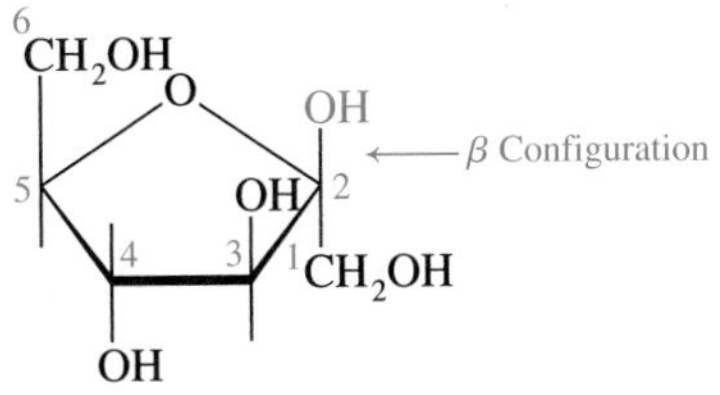

β Configuration

β-D-fructofuranose

27.9 Pentoses

An open-chain aldopentose has three chiral carbon atoms. Therefore, eight (2^3) isomeric aldopentoses are possible. The four possible D-pentoses are shown in Figure 27.1. Arabinose and xylose occur in some plants as polysaccharides called pentosans. D-Ribose and its derivative, D-2-deoxyribose, are the most interesting pentoses because of their relationship to nucleic acids and the genetic code (Chapter 31). Note the difference between the two names, D-ribose and D-2-deoxyribose. In the latter name, the 2-deoxy means that oxygen is missing from the D-ribose molecule at carbon 2. Check the formulas that follow to verify this difference.

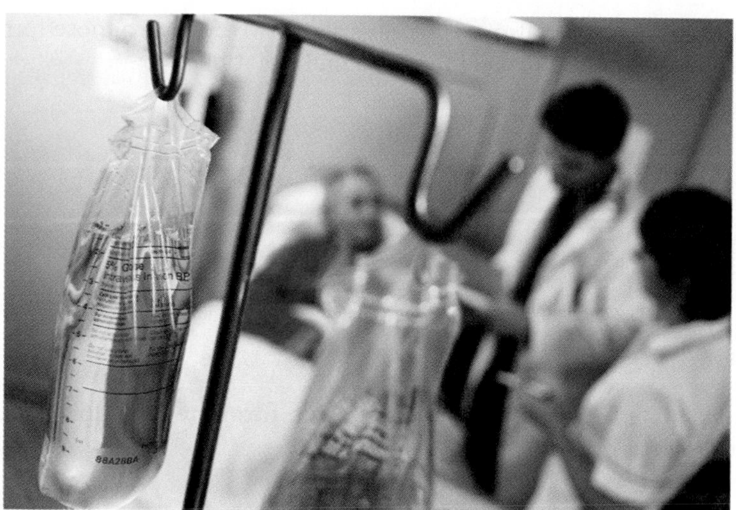

Glucose is often given intravenously to patients suffering from major trauma.

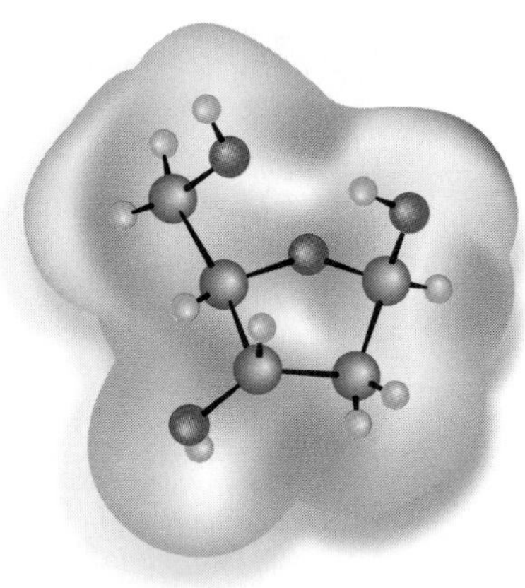

Spacefilling model of β-D-2-deoxyribofuranose. Do you see the carbon *not* bonded to an oxygen?

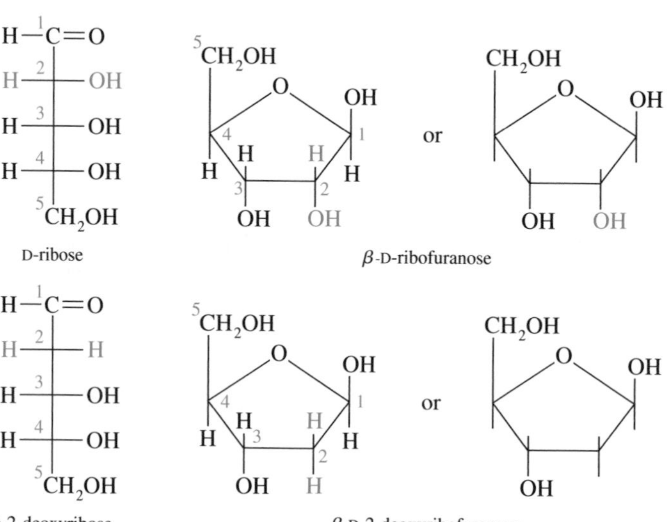

The ketose that is closely related to D-ribose is named D-ribulose. (Ketose names are often derived from the corresponding aldose name by changing the suffix from *-ose* to *-ulose*.) This ketose is an intermediate that allows cells to make many other monosaccharides. In photosynthetic organisms, D-ribulose is used to capture carbon dioxide and thus make new carbohydrates. The structure of D-ribulose is as follows:

CH_2OH
C=O
H—C—OH
H—C—OH
CH_2OH

D-ribulose

Practice 27.6

Write Haworth formulas for the following:

(a) α-D-fructofuranose
(b) α-D-ribofuranose

27.10 Disaccharides

Disaccharides are carbohydrates composed of two monosaccharide residues united by a glycosidic linkage. The two important disaccharides found in the free state in nature are sucrose and lactose ($C_{12}H_{22}O_{11}$). Sucrose, commonly known as *table sugar*, exists throughout the plant kingdom. Sugar cane contains 15–20% sucrose, and sugar beets 10–17%. Maple syrup and sorghum are also good sources of sucrose. Lactose, also known as *milk sugar*, is found free in nature mainly in the milk of mammals. Human milk contains about 6.7% lactose and cow milk about 4.5% lactose.

Milk sugar, or lactose, is found mainly in the milk of mammals.

Unlike sucrose and lactose, several other important disaccharides are derived directly from polysaccharides by hydrolysis. (See Section 27.14.) For example, maltose, isomaltose, and cellobiose are formed when specific polysaccharides are hydrolyzed. Of this group, maltose is the most common and is found as a constituent of sprouting grain.

Upon hydrolysis, disaccharides yield two monosaccharide molecules. The hydrolysis is catalyzed by hydrogen ions (acids), usually at elevated temperatures, or by certain enzymes that act effectively at room or body temperatures. A different enzyme is required for the hydrolysis of the three disaccharides:

$$\text{sucrose} + \text{water} \xrightarrow[\text{sucrase}]{\text{H}^+ \text{ or}} \text{glucose} + \text{fructose}$$

$$\text{lactose} + \text{water} \xrightarrow[\text{lactase}]{\text{H}^+ \text{ or}} \text{galactose} + \text{glucose}$$

$$\text{maltose} + \text{water} \xrightarrow[\text{maltase}]{\text{H}^+ \text{ or}} \text{glucose} + \text{glucose}$$

The enzyme lactase is present in the small intestine of infants and allows them to easily digest lactose from their milk diet. Unfortunately, as people mature their intestines often stop producing the lactase enzyme, and they lose the ability to digest lactose. Instead, this sugar is metabolized by common bacteria that live in the large intestine. The gas and intestinal discomfort that results is termed milk or lactose intolerance and is a condition that afflicts many adults. Carefully note the small structural differences between lactose and the other common disaccharides in the next section. Small differences in molecular shape often determine what our bodies do with specific molecules.

27.11 Structures and Properties of Disaccharides

Disaccharides contain an acetal structure (glycosidic linkage), and some also contain a hemiacetal structure. The acetal structure in maltose may be considered as being derived from two glucose molecules, by the elimination of a molecule of water between the —OH group on carbon 1 of one glucose unit and the —OH group of carbon 4 on the other glucose unit. This is an α-1,4-glycosidic linkage, since the glucose units have the α configuration and are joined at carbons 1 and 4. In a more systematic nomenclature, this form of maltose is known as α-D-glucopyranosyl-(1,4)-α-D-glucopyranose. If the structure of glucose is known, this name provides a complete description for drawing the maltose formula:

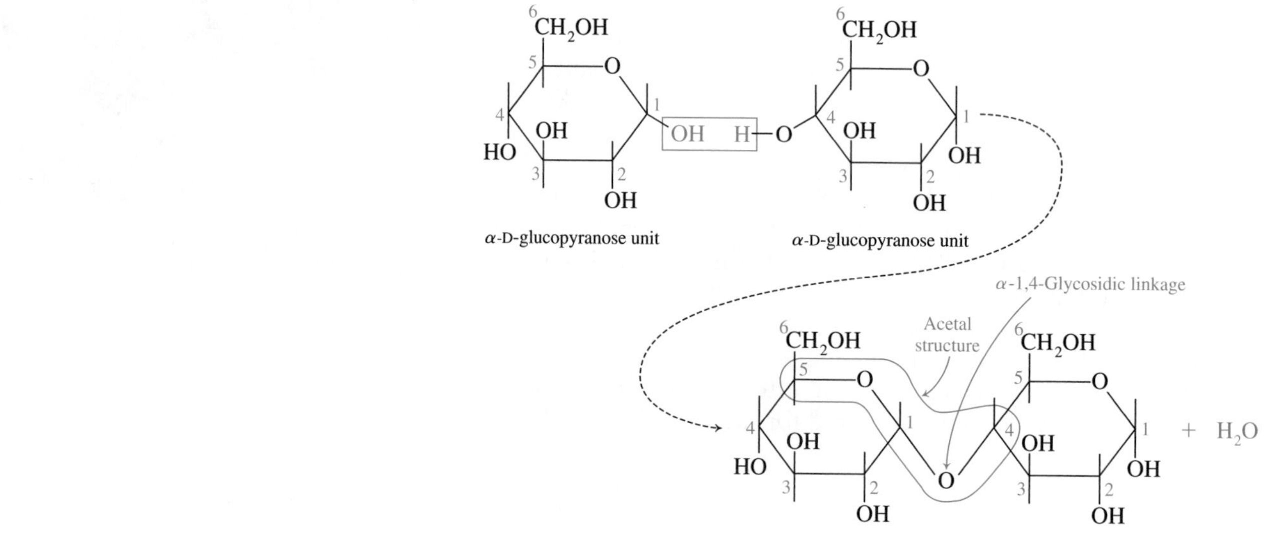

α-D-glucopyranose unit　　α-D-glucopyranose unit

α-1,4-Glycosidic linkage

Acetal structure

$+ H_2O$

maltose
α-D-glucopyranosyl-(1,4)-α-D-glucopyranose

Lactose consists of a β-D-galactopyranose unit linked to an α-D-glucopyranose unit. These are joined by a β-1,4-glycosidic linkage from carbon 1 on galactose to carbon 4 on glucose. The more systematic name for lactose is β-D-galactopyranosyl-(1,4)-α-D-glucopyranose.

Two conventions are used for writing Haworth formulas connecting two or more monosaccharide units through a glycoside bond. Using the first convention for lactose, for example, we find that the glycosidic bond is shown in a bent fashion to provide stereochemical information:

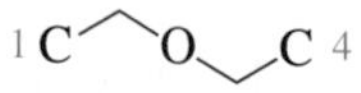

1 C—O—C 4 (bent structure)

The bond from carbon 1 in galactose is the β configuration pointing up and is bonded through oxygen to carbon 4 of glucose, whose bond to oxygen points down.

In the second convention, the two monosaccharides are shown in a "stacked" position. In lactose, the glycosidic bond from carbon 1 of galactose in the

β position is pointed up and is directly bonded through oxygen to carbon 4 of glucose, whose bond to oxygen is pointed down. These two conventions are illustrated as follows:

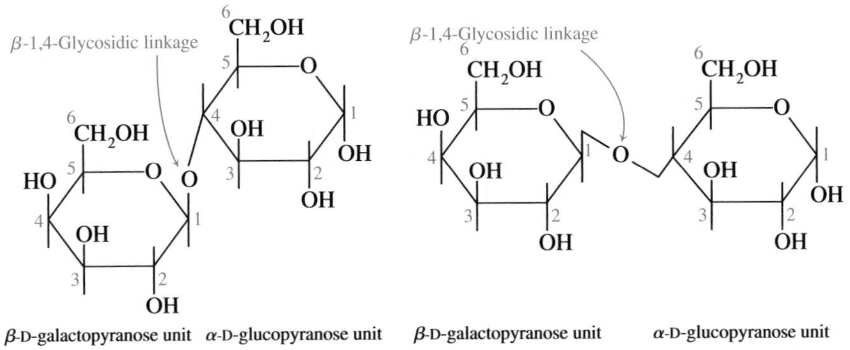

β-1,4-Glycosidic linkage

β-D-galactopyranose unit α-D-glucopyranose unit

β-D-galactopyranose unit α-D-glucopyranose unit

lactose

β-D-galactopyranosyl-(1,4)-α-D-glucopyranose

Sucrose consists of an α-D-glucopyranose unit and a β-D-fructofuranose unit. These monosaccharides are joined by an oxygen bridge from carbon 1 on glucose to carbon 2 on fructose—that is, by an α-1,2-glycosidic linkage:

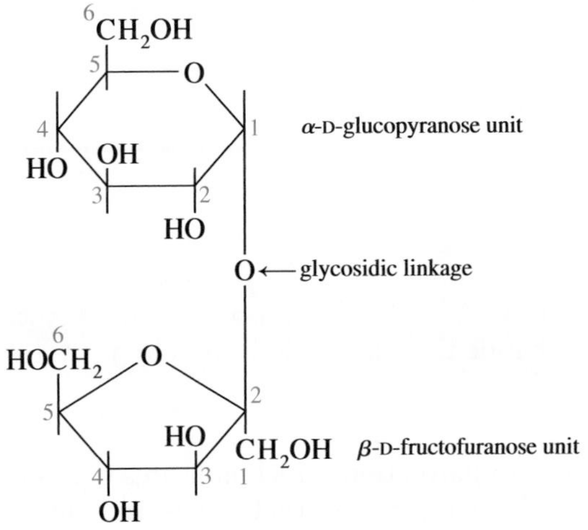

α-D-glucopyranose unit

glycosidic linkage

β-D-fructofuranose unit

α-D-glucopyranosyl-(1,2)-β-D-fructofuranose

In this perspective formula, fructose is bonded to glucose through the β-2 position of fructose to the α-1 position of glucose.

Example 27.3

Write the Haworth formula for isomaltose, which is α-D-glucopyranosyl-(1,6)-α-D-glucopyranose.

SOLUTION

Recognize that this disaccharide is composed of two α-D-glucopyranose units linked between carbon 6 of one unit and carbon 1 of the other. First, write the Haworth formula for the monosaccharides, and number their carbons:

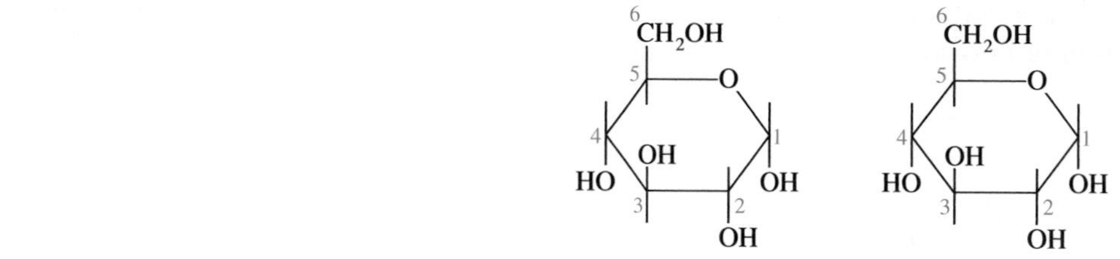

The two α-D-glucopyranose units must be linked in such a way that the stereochemistry at carbon 1 is preserved. (Carbon 6 is not a chiral center.) One correct way to write the isomaltose structure is as follows:

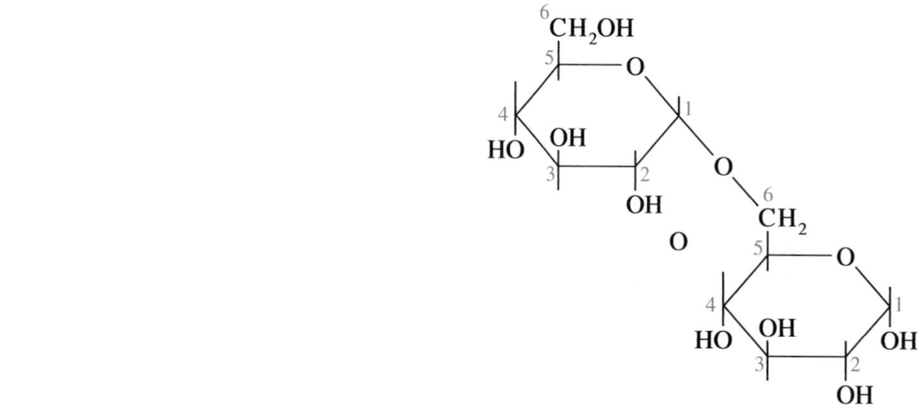

isomaltose (α 1,6-glycosidic linkage)

Practice 27.7

Write the structure for cellobiose, β-D-glucopyranosyl-(1,4)-β-D-glucopyranose. Cellobiose is a disaccharide that can be derived from plants.

Lactose and maltose both show mutarotation, which indicates that one of the monosaccharide units has a hemiacetal ring that opens and closes to interchange anomers. Sucrose has no hemiacetal structure and hence does not mutarotate.

The three disaccharides sucrose, lactose, and maltose have physical properties associated with large polar molecules. All three are crystalline solids and are quite soluble in water; the solubility of sucrose amounts to 200 g per 100 g of water at 0°C. Hydrogen bonding between the polar —OH groups on the sugar molecules and the water molecules is a major factor in this high solubility. These sugars are not easily melted. In fact, lactose is the only one with a clearly defined melting point (201.6°C). Sucrose and maltose begin to decompose when heated to 186°C and 103°C, respectively. When sucrose is heated to melting, it darkens and undergoes partial decomposition. The resulting mixture is known as caramel, or burnt sugar, and is used as coloring, as well as a flavoring agent in foods.

27.12 Sweeteners and Diet

Some sugar substitutes can be used in baking.

Carbohydrates have long been valued for their ability to sweeten foods. Fructose is the sweetest of the common sugars, although sucrose (table sugar) is the most commonly used sweetener. (A scale of relative sweetness is given in Table 27.1.) Astonishingly large amounts of sucrose are produced from sugar beets and cane: World production is on the order of 90 million tons annually. There are no essential chemical differences between cane and beet sugar. In the United States, approximately 20–30% of the average caloric intake is sucrose (about 150 g/day per person). Low price and sweet taste are the major reasons for high sucrose consumption. Note that sucrose is only 58% as sweet as fructose (Table 27.1). However, because sucrose is inexpensive to produce and is amenable to a variety of food processing techniques, approximately 40–60% of all sweeteners are sucrose.

Sucrose has a tendency to crystallize from concentrated solutions or syrups. Therefore, in commercial food preparations (e.g., candies, jellies, and canned fruits) the sucrose is often hydrolyzed.

$$\text{sucrose} + H_2O \xrightarrow{H^+} \text{glucose} + \text{fructose}$$

The resulting mixture of glucose and fructose, usually in solution, is called **invert sugar**. Invert sugar has less tendency to crystallize than sucrose, and it has greater sweetening power than an equivalent amount of sucrose. The nutritive value of the sucrose is not affected in any way by the conversion to invert sugar, because the same hydrolysis reaction occurs in normal digestion.

invert sugar

High-fructose syrups, derived from cornstarch, are used as sweeteners in products such as soft drinks. Starch is a polymer of glucose, contains no fructose, and is not sweet. However, through biotechnology, a method was developed to convert starch to a very sweet syrup of high fructose content. The starch polymers are hydrolyzed to glucose; then, part of the glucose is enzymatically converted to fructose, yielding the syrup. High-fructose corn syrups are used because they are more economical sweetening agents than either cane or beet sugar.

Unfortunately, high sugar consumption presents health problems. For many people, sucrose is a source of too many calories. Oral bacteria also find

Table 27.1 Relative Sweetness of Sugars and Sugar Substitutes (Based on fructose = 100)

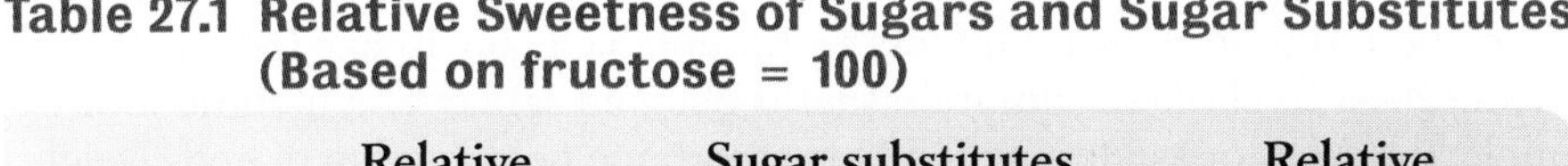

Sugars	Relative sweetness	Sugar substitutes (common brand names)	Relative sweetness
Fructose	100	Sucralose (Splenda®)	3.5×10^4
Invert sugar	75	Saccharin (Sweet 'N Low®)	1.7×10^4
Sucrose	58	Acesulfame potassium (Sunette®, Sweet One®)	1.2×10^4
Glucose	43	Aspartame (Equal®, NutraSweet®)	1.2×10^4
Maltose	19		
Galactose	19		
Lactose	9.2		

sucrose easy to metabolize, increasing the incidence of dental caries. Finally, because the monosaccharides, fructose and glucose, are quickly absorbed from the small intestine, sugar consumption leads to a rapid increase in blood sugar. Such a sharp rise can be dangerous for people with impaired carbohydrate metabolism—for example, those who have diabetes mellitus.

Scientists have searched diligently for sugar substitutes and successfully have identified both natural and artificial sweeteners. Several artificial sweeteners (some are shown here) are currently approved as additives in U.S. foods (other countries may have different guidelines):

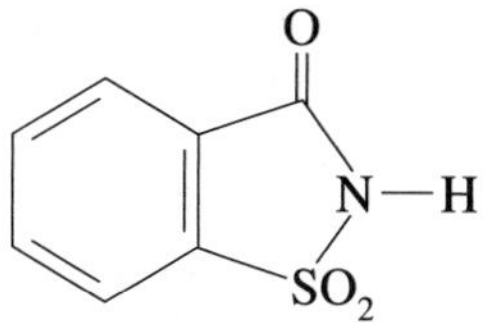

saccharin

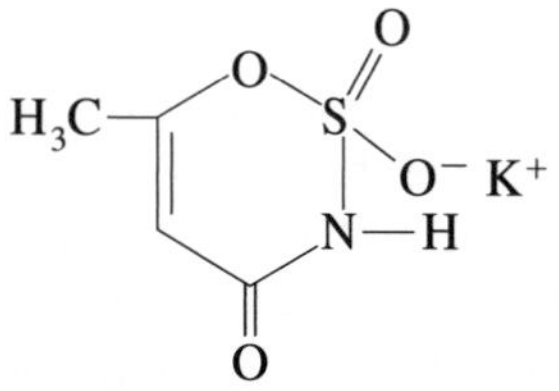

acesulfame potassium

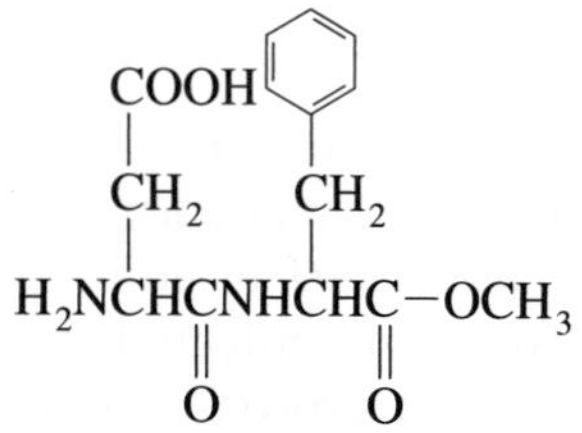

aspartame

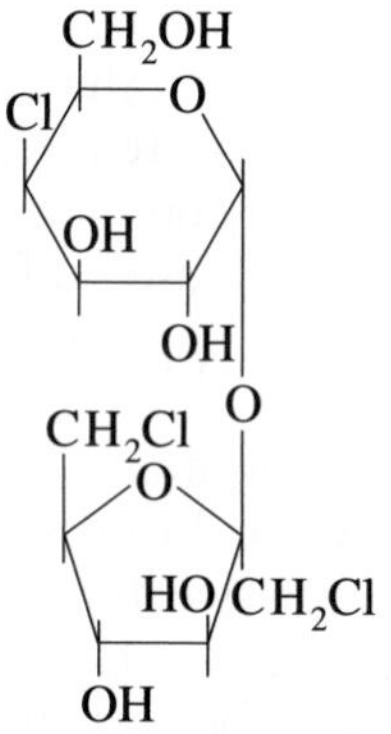

sucralose

Notice the similarities between sucralose and sucrose.

Saccharin, which was discovered in 1879, was the first purely artificial, noncarbohydrate sweetener to be developed. Much sweeter than the natural sugars saccharin has a distinctive aftertaste. Sugar substitutes need to be as sweet as possible, so that only small amounts have to be added to foods. Note that in all cases, these chemicals are much sweeter than fructose. (See Table 27.1.) Each chemical may have slightly different uses; for example, aspartame intensifies fruit flavors, while acesulfame potassium can be used for baking.

The preceding group of sweeteners has been shown to be safe in many animal studies. Yet, safety concerns are difficult to answer completely. The U.S. Food and Drug Administration sets an Acceptable Daily Intake (ADI) for each chemical—a level of consumption that is deemed safe over a lifetime. For example, the ADI for aspartame is 50 mg aspartame per kilogram of body weight. This corresponds to about 15 cans of aspartame-containing soft drink per day over the average adult lifespan.

The difficulties of health-risk assessment are illustrated by the history of sodium cyclamate, whose structure is as follows:

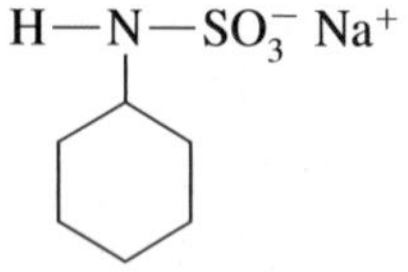

sodium cyclamate

Sodium cyclamate was discovered in the 1930s and was approved for food use in the 1960s. However, animal studies indicated some cancer risk from cyclamate, and the sweetener was banned in 1970. Since that time, many studies have rebutted the earlier findings. Cyclamate probably does not pose a great health risk, yet, it is no longer needed because of the large number of other non-nutritive sweeteners.

Partly because of continued health concerns, natural substitutes for sucrose are also being developed. Again, the aim is to find a chemical that sweetens without the calories of sucrose or the risk of dental caries. Several products that may gain wide approval in the United States as food additives include (1) nonnutritive sweeteners derived from citrus fruits (the dihydrochalcones, 300–2000 times sweeter than sucrose); (2) a nonnutritive extract from licorice root (glycyrrhizin, 50–100 times sweeter than sucrose); (3) an extract from a South American plant leaf (stevioside, 300 times sweeter than sucrose); and (4) a mixture of sweet-tasting proteins from an African fruit (thaumatin, 2000–3000 times sweeter than sucrose).

27.13 Redox Reactions of Monosaccharides

Oxidation

The aldehyde groups in monosaccharides can be oxidized to monocarboxylic acids by mild oxidizing agents such as bromine water. The carboxylic acid group is formed at carbon 1. The name of the resulting acid is formed by changing the *-ose* ending to *-onic acid*. Glucose yields gluconic acid, galactose yields galactonic acid, and so on.

Carbohydrate oxidation increases the number of bonds between carbon and oxygen.

$$\begin{array}{rcl} & \text{H}-\text{C}=\text{O} & \\ \text{H} & - & \text{OH} \\ \text{HO} & - & \text{H} \\ \text{H} & - & \text{OH} \\ \text{H} & - & \text{OH} \\ & \text{CH}_2\text{OH} & \end{array} + Br_2 + H_2O \longrightarrow \begin{array}{rcl} & \text{COOH} & \\ \text{H} & - & \text{OH} \\ \text{HO} & - & \text{H} \\ \text{H} & - & \text{OH} \\ \text{H} & - & \text{OH} \\ & \text{CH}_2\text{OH} & \end{array} + 2\,HBr$$

D-glucose D-gluconic acid

Dilute nitric acid, a vigorous oxidizing agent, oxidizes both carbon 1 and carbon 6 of aldohexoses to form dicarboxylic acids. The resulting acid is named by changing the *-ose* sugar suffix to *-aric acid*. Glucose yields glucaric acid (saccharic acid) and galactose yields galactaric acid (mucic acid):

Notice that galactaric acid is a meso compound: It is superimposable on its mirror image.

$$\begin{array}{rcl} & \text{H}-\text{C}=\text{O} & \\ \text{H} & - & \text{OH} \\ \text{HO} & - & \text{H} \\ \text{H} & - & \text{OH} \\ \text{H} & - & \text{OH} \\ & \text{CH}_2\text{OH} & \end{array} \xrightarrow{\text{warm dilute HNO}_3} \begin{array}{rcl} & \text{COOH} & \\ \text{H} & - & \text{OH} \\ \text{HO} & - & \text{H} \\ \text{H} & - & \text{OH} \\ \text{H} & - & \text{OH} \\ & \text{COOH} & \end{array} \qquad \begin{array}{rcl} & \text{COOH} & \\ \text{H} & - & \text{OH} \\ \text{HO} & - & \text{H} \\ \text{HO} & - & \text{H} \\ \text{H} & - & \text{OH} \\ & \text{COOH} & \end{array}$$

D-glucose glucaric acid (saccharic acid) galactaric acid (mucic acid)

Reduction

Monosaccharides can be reduced to their corresponding polyhydroxy alcohols by reducing agents such as H_2/Pt or sodium amalgam, Na(Hg). For example, glucose yields sorbitol (glucitol), galactose yields galactitol (dulcitol), and mannose yields mannitol; all of these are hexahydric alcohols (containing six —OH groups).

Carbohydrate reduction decreases the number of bonds between carbon and oxygen.

$$\begin{array}{ccc} \text{H—C=O} & & \text{CH}_2\text{OH} \\ \text{H—C—OH} & & \text{H—C—OH} \\ \text{HO—C—H} & \xrightarrow{H_2/Pt} & \text{HO—C—H} \\ \text{H—C—OH} & & \text{H—C—OH} \\ \text{H—C—OH} & & \text{H—C—OH} \\ \text{CH}_2\text{OH} & & \text{CH}_2\text{OH} \\ \text{D-glucose} & & \text{D-glucitol (sorbitol)} \end{array}$$

Hexahydric alcohols have properties resembling those of glycerol (Section 22.7). Because of their affinity for water, they are used as moisturizing agents in food and cosmetics. Sorbitol, galactitol, and mannitol occur naturally in a variety of plants.

reducing sugar

Redox Tests for Carbohydrates

Sorbitol occurs naturally in plants and has an affinity for water. It is an ingredient in many moisturizers and lotions.

Under prescribed conditions, some sugars reduce silver ions to free silver, and copper(II) ions to copper(I) ions. Such sugars are called **reducing sugars**. This reducing ability, which is useful in classifying sugars and, in certain clinical tests, is dependent on the presence of (1) aldehydes, (2) α-hydroxyketone groups (—CH_2COCH_2OH) such as in fructose, or (3) hemiacetal structures in cyclic molecules such as maltose. These groups are easily oxidized to carboxylic acid (or carboxylate ion) groups; the metal ions are thereby reduced ($Ag^+ \rightarrow Ag^0$; $Cu^{2+} \rightarrow Cu^+$). Several different reagents, including Tollens, Fehling, Benedict, and Barfoed, are used to detect reducing sugars. (See Section 23.5.) The Benedict, Fehling, and Barfoed tests depend on the formation of copper(I) oxide precipitate to indicate a positive reaction (see Figure 27.5):

$$\underset{\text{aldehyde group}}{\text{R}\overset{\text{O}}{\overset{\|}{\text{C}}}\text{—H}} + \underset{\text{(blue)}}{2\,\text{Cu}^{2+}} + 5\,\text{OH}^- \longrightarrow \underset{\text{carboxylate ion group}}{\text{R}\overset{\text{O}}{\overset{\|}{\text{C}}}\text{—O}^-} + \underset{\text{copper(I) oxide (brick red)}}{\text{Cu}_2\text{O}(s)} + 3\,\text{H}_2\text{O}$$

Barfoed reagent contains Cu^{2+} ions in the presence of acetic acid. It is used to distinguish reducing monosaccharides from reducing disaccharides. Under the same reaction conditions, the reagent is reduced more rapidly by monosaccharides.

Glucose and galactose contain aldehyde groups; fructose contains an α-hydroxyketone group. Therefore, all three of these monosaccharides are reducing sugars.

A carbohydrate molecule need not have a free aldehyde or α-hydroxyketone group to be a reducing sugar. A hemiacetal structure, as shown here, also reacts:

Hemiacetal structure

glucose   maltose

Maltose and the cyclic form of glucose are examples of molecules with the hemiacetal structure.

Under mildly alkaline conditions, the rings open at the points indicated by the arrows to form aldehyde groups:

Ring opens here

$\xrightarrow[H_2O]{OH^-}$

glucose (ring structure)   glucose (open-chain structure)

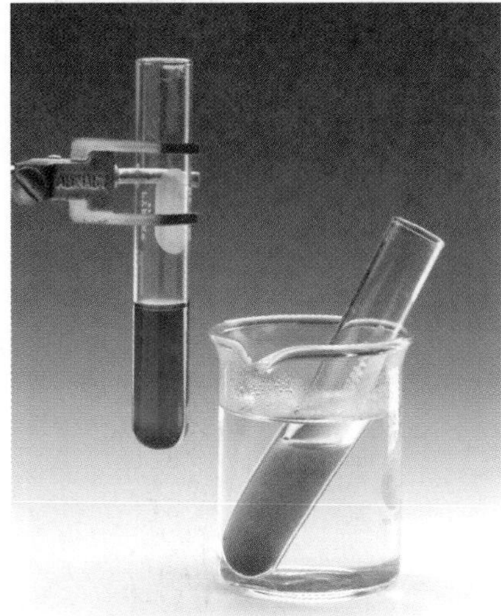

Figure 27.5
Benedict test for reducing sugars. The test tube on the left contains Benedict reagent. The test tube on the right shows the results of a positive Benedict test with glucose [formation of a brick-red precipitate of copper(I) oxide].

Among the disaccharides, lactose and maltose have hemiacetal structures and are therefore reducing sugars. Sucrose is not a reducing sugar, because it does not have the hemiacetal structure.

Many clinical tests monitor glucose as a reducing sugar. For example, Benedict and Fehling reagents are used to detect the presence of glucose in urine. Initially, the reagents are deep blue in color. A positive test is indicated by a color change to greenish yellow, yellowish orange, or brick red, corresponding to an increasing glucose (reducing sugar) concentration. These tests are used to estimate the amount of glucose in the urine of diabetics, in order to adjust the amount of insulin needed for proper glucose utilization.

Alternatively, a clinical test (glucose oxidase test) uses an enzyme-catalyzed oxidation of glucose to test for urine sugar. The inclusion of an enzyme ensures a reaction that is specific for the glucose structure, allowing a more selective test for glucose in the urine.

Color reactions such as those mentioned in the preceding paragraphs provide the basis for many home glucose monitoring kits. These kits allow diabetics and others to measure blood glucose levels easily and on a regular basis. Perhaps the biggest disadvantage to the kits now in use is the need to draw a drop of blood from a finger. However, such a disadvantage may be soon solved. Glucose tests have become so sensitive that they can detect glucose levels in the liquid between the skin cells (the interstitial fluid); these tests no longer require blood. Also, an implant is in development that may provide "real-time" continuous glucose monitoring.

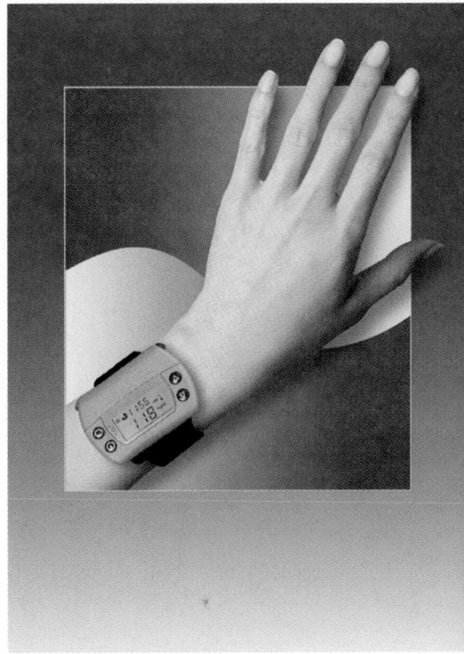

A non-invasive glucose monitor provides a continuous measure of blood glucose levels.

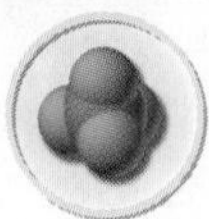

Example 27.4 Two samples labeled A and B are known to be D-threose and D-erythrose. Water solutions of each sample are optically active. However, when each solution was warmed with nitric acid, the solution from sample A became optically inactive, while that from sample B was still optically active. Determine which sample (A or B) contains D-threose and which sample contains D-erythrose.

SOLUTION In this problem, we need to examine the structures of D-threose and D-erythrose, write equations for the reaction with nitric acid, and examine the products to see why one is optically active and the other optically inactive. Start by writing the formulas for D-threose and D-erythrose:

```
   H—C=O          H—C=O
HO——+——H        H——+——OH
 H——+——OH       H——+——OH
   CH2OH          CH2OH
 D-threose      D-erythrose
```

Oxidation of these tetroses with HNO_3 yields dicarboxylic acids:

```
   H—C=O                  COOH    |    H—C=O                COOH
HO——+——H    warm      HO——+——H    |  H——+——OH    warm    H——+——OH
 H——+——OH  ----->      H——+——OH   |  H——+——OH   ----->   H——+——OH
            HNO3                  |              HNO3
   CH2OH                  COOH    |    CH2OH                COOH
 D-threose                 I      |  D-erythrose             II
```

Product I is a chiral molecule and is optically active. Product II is a meso compound and is optically inactive. Therefore, sample A is D-erythrose, since oxidation yields the meso acid. Sample B, then, must be D-threose.

Practice 27.8

A disaccharide yields no copper(I) oxide when reacted with Benedict reagent. This carbohydrate is composed of two α-D-galactopyranose units. Identify the carbon from each monosaccharide involved in the acetal linkage.

Practice 27.9

Write the structure of the product formed when D-galactose is reduced with H_2/Pt. Is this compound optically active?

27.14 Polysaccharides Derived from Glucose

Although many naturally occurring polysaccharides are known, three—starch, cellulose, and glycogen—are of major importance. All three, when hydrolyzed, yield D-glucose as the only product, according to this approximate general equation:

$$\underset{\text{polysaccharide molecule (approximate formula)}}{(C_6H_{10}O_5)_n + n\,H_2O} \longrightarrow \underset{\text{D-glucose}}{n\,C_6H_{12}O_6}$$

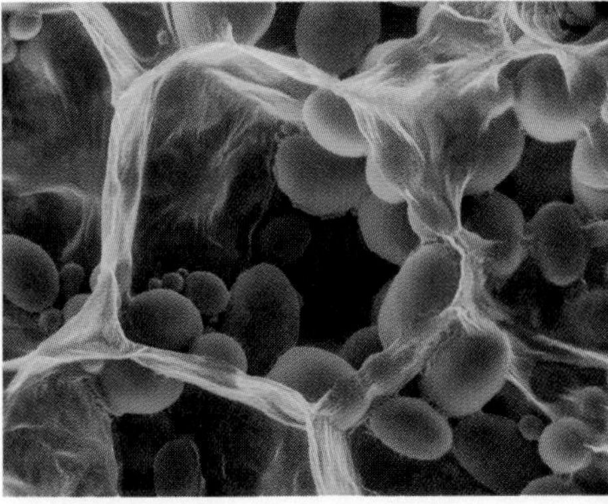

Figure 27.6
Scanning electron micrograph of starch granules in potato tuber cells.

The hydrolysis reaction establishes that all three polysaccharides are polymers made up of glucose monosaccharide units. It also means that the differences in properties among the three polysaccharides must be due to differences in the structures or sizes of these molecules (or both).

Many years of research were required to determine the detailed structures for polysaccharide molecules. Consideration of all this work is beyond the scope of our discussion, so what follows is an abbreviated summary of the results.

Starch

Starch ($(C_6H_{10}O_5)_n$)is found in plants, mainly in the seeds, roots, or tubers. (See Figure 27.6.) Corn, wheat, potatoes, rice, and cassava are the chief sources of dietary starch. The two main components of starch are amylose and amylopectin. Amylose molecules are unbranched chains composed of about 25–1300 α-D-glucose units joined by α-1,4-glycosidic linkages, as shown in Figure 27.7(a). The stereochemistry of the α anomer causes amylose to coil into a helical conformation. Partial hydrolysis of this linear polymer yields the disaccharide maltose.

Amylopectin is a branched-chain polysaccharide with much larger molecules than those of amylose. Amylopectin molecules consist, on average, of several thousand α-D-glucose units with molar masses ranging up to 1 million or more. The main chain contains glucose units connected by α-1,4-glycosidic linkages. Branch chains are linked to the main chain through α-1,6-glycosidic linkages about every 25 glucose units, as shown in Figure 27.7(c). This molecule has a characteristic treelike structure because of its many branch chains. Partial hydrolysis of amylopectin yields both maltose and the related disaccharide isomaltose, α-D-glucopyranosyl-(1,6)-α-D-glucopyranose.

Despite the presence of many polar —OH groups, starch molecules are insoluble in cold water, apparently because of their very large size. Starch readily forms colloidal dispersions in hot water. Such starch "solutions" form an intense blue-black color in the presence of free iodine. Hence, a starch solution can be used to detect free iodine, or a dilute iodine solution can be used to detect starch.

Starch is readily converted to glucose by heating with water and a little acid (e.g., hydrochloric or sulfuric acid). It is also readily hydrolyzed at room temperature by certain digestive enzymes. The hydrolysis of starch to glucose is shown in the following equation:

$$\text{starch} \xrightarrow[\text{or salivary and pancreatic amylase}]{\text{acid} + \Delta} \text{dextrins} + \text{maltose} + \text{isomaltose} \xrightarrow[\text{or maltase and other intestinal enzymes}]{\text{acid} + \Delta} \text{D-glucose}$$

The hydrolysis of starch can be followed qualitatively by periodically testing samples from a mixture of starch and saliva with a very dilute iodine solution. The change of color sequence is blue–black → blue → purple → pink → colorless, as the starch molecules are broken down into smaller and smaller fragments.

Fermentation of grain starch is an important source of ethanol. Ethanol is fast becoming an important renewable alternate energy source for fuel to operate automobiles and power plants, replacing fossil fuels.

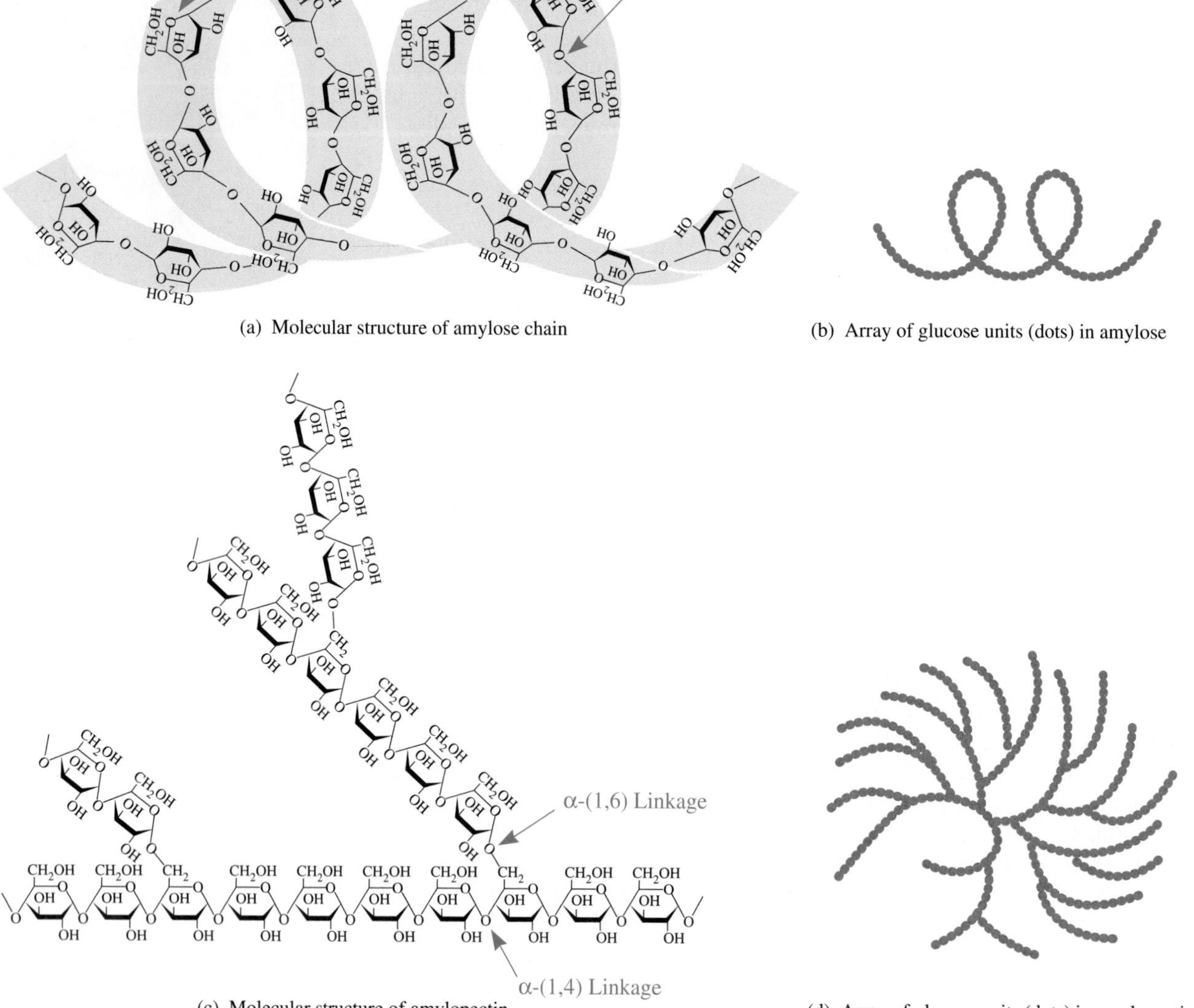

Figure 27.7
Representations of amylose and amylopectin.

Hydrolysis is a key chemical reaction in the digestion of starchy foods. If these foods are well chewed, salivary amylase normally decreases the starch polymer chain length from on the order of a thousand glucose units to about eight per chain. In the small intestine, pancreatic amylase continues digestion to form maltose. Enzymes in the small intestine membranes complete the conversion of starch to glucose, which is then absorbed into the bloodstream.

Starch is the most important energy-storage carbohydrate of the plant kingdom. In turn, humans and other animals consume huge quantities of starch. This polymer is an important food source, because it has the appropriate structure to be readily broken down to D-glucose. The carbons from starch provide much of our daily energy needs as they are oxidized to carbon dioxide.

CHEMISTRY IN ACTION • Screening for Cancer: Finding a Needle in a Haystack

Cancer is a hidden disease—it is difficult to diagnose, especially in its early stages, when treatment is most effective. So scientists continue to look for better screening procedures.

Screening tests should be comprehensive, accurate, and relatively noninvasive. Blood tests often fit these criteria. Because the blood travels to all tissues in the body, it carries molecules that give information about all tissues—truly a wealth of information. The problem lies with finding the right information. Among the billions of cells releasing molecules to the bloodstream, cancer screening tests must choose molecules specific for the cancer—they must "find a needle in a haystack."

A cancer cell, like other cells, carries specific coded information on its surface (antigens). On a molecular level, this information is commonly built into cell surface carbohydrates. These complex polysaccharides contain many different sugars linked together in many different ways, much different from simple polysaccharides like glycogen or cellulose. These carbohydrates are bonded to either membrane proteins or membrane lipids and are released into the surrounding fluids. Ideally, this release can provide advance warning that a hidden cancer is growing. Furthermore, knowing the identity of the carbohydrate can point toward the tissue in which the cancer may be hiding.

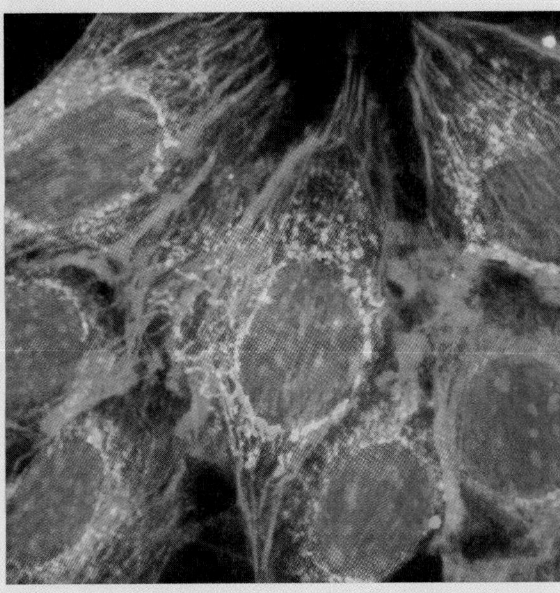
Fluorescent-stained cancer cells

Today, clinicians use several carbohydrate antigens (CA): CA 15-3 for breast cancer, CA 19-9 for pancreatic cancer, and CA 125 for ovarian cancer. Unfortunately, these screening tests need to be almost *absolutely* specific, and they are not. For example, the screening test for CA 19-9 has good specificity; it is 98.5% specific for pancreatic cancer. However, because the incidence of pancreatic cancer is only about 0.015% of the general population, CA 19-9 screening would result in about 350 false positives for every pancreatic cancer diagnosed—not a very accurate diagnostic tool. Scientists are now seeking to combine CA screening with other blood test results to improve specificity.

In the meantime, these carbohydrate antigens are proving very useful at tracking the course of the disease. In general, if the CA level is high initially, the prognosis is poorer. Also, when surgery is used to remove a tumor, if the CA level does not drop, there may be additional unremoved tumors. Finally, successful chemotherapy will cause a progressive decrease in the CA levels, and treatment progress can be easily measured.

Glycogen

Glycogen $(C_6H_{10}O_5)_n$ is the energy-storage carbohydrate of the animal kingdom. It is formed by polymerization of glucose and is stored in the liver and in muscle tissues. Structurally, it is very similar to the amylopectin fraction of starch, except that it is more highly branched. The α-1,6-glycosidic linkages occur on one of every 12–18 glucose units.

Cellulose

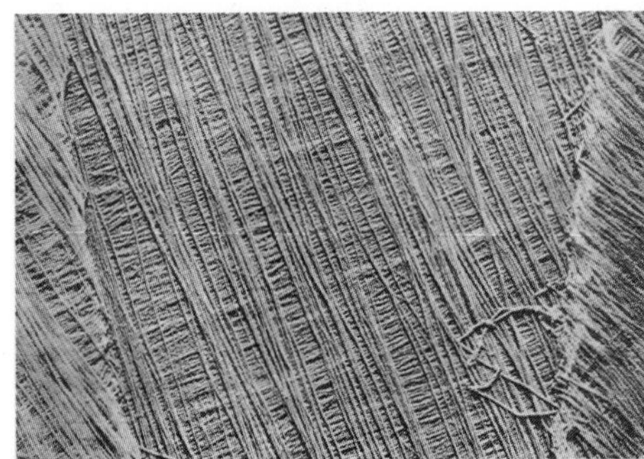
Figure 27.8
Scanning electron micrograph of cellulose fibers in a plant cell wall.

Cellulose $(C_6H_{10}O_5)_n$ is the most abundant organic substance found in nature. It is the chief structural component of plants and wood (Figure 27.8). Cotton fibers are almost pure cellulose; wood, after removal of moisture, consists of about 50% cellulose. Cellulose is an important substance in the textile and paper industries.

Cellulose, like starch and glycogen, is a polymer of glucose. But cellulose differs from starch and glycogen in that the glucose units are joined by

β-1,4-glycosidic linkages, instead of α-1,4-glycosidic linkages. The stereochemistry of the β anomer allows the polymer to form an extended chain that can hydrogen-bond to adjacent cellulose molecules. The large number of hydrogen bonds so formed partially accounts for the strength of the resulting plant cell walls. The cellulose structure is illustrated in Figure 27.9.

When partial hydrolysis of cellulose occurs, it produces the disaccharide cellobiose, β-D-glucopyranosyl-(1,4)-β-D-glucopyranose. However, cellulose has greater resistance to hydrolysis than either starch or glycogen. It is not appreciably hydrolyzed when boiled in a 1% sulfuric acid solution. It does not show a color reaction with iodine. Humans cannot digest cellulose, because they have no enzymes capable of catalyzing its hydrolysis. Fortunately, some microorganisms found in soil and in the digestive tracts of certain animals produce enzymes that do catalyze the breakdown of cellulose. The presence of these microorganisms explains why cows and other herbivorous animals thrive on grass, and why termites thrive on wood.

The —OH groups of starch and cellulose can be reacted without destruction of the macromolecular structures. For example, nitric acid converts an —OH group to a nitrate group in this fashion:

$$\underset{\substack{\text{hydroxyl group}\\\text{on cellulose}\\\text{molecule}}}{\text{cellulose}-\boxed{\text{OH}} + \boxed{\text{H}}}\underset{\substack{\text{nitric}\\\text{acid}}}{\text{ONO}_2} \longrightarrow \underset{\substack{\text{nitrate group}\\\text{on cellulose}\\\text{molecule}}}{\text{cellulose}-\text{ONO}_2} + \underset{\text{water}}{\text{H}_2\text{O}}$$

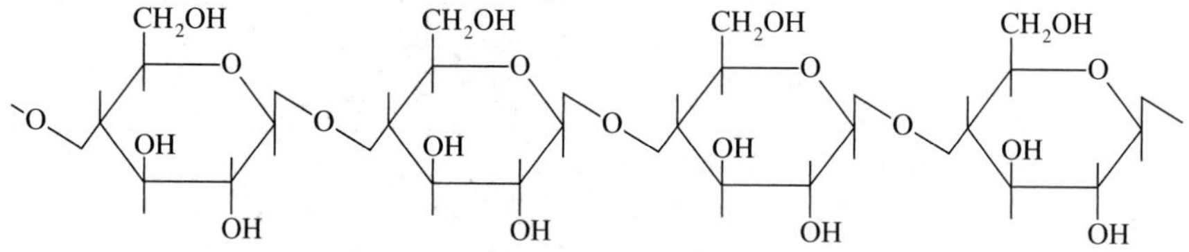

(a) Haworth formula

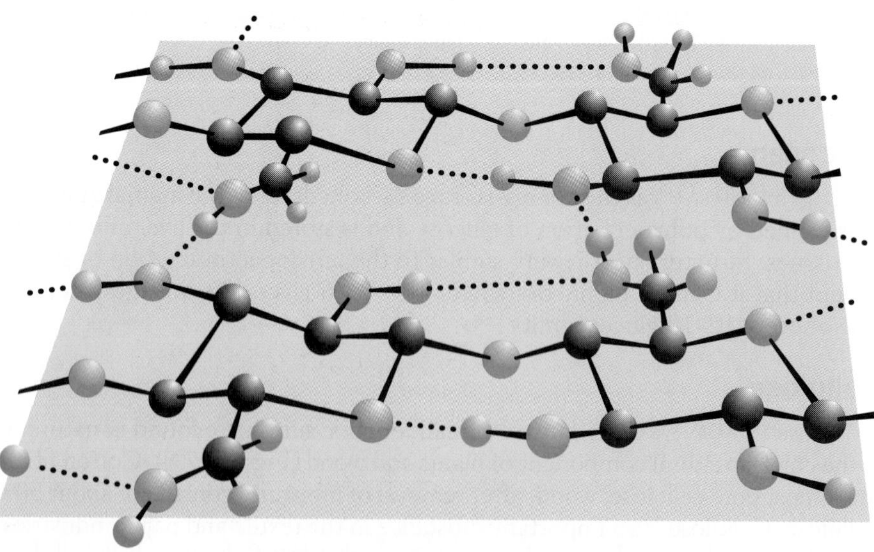

(b) Three-dimensional representation

Figure 27.9
Two representations of cellulose. To simplify the three-dimensional drawing, only C—H bonds to C-6 of each glucose are shown. Note the hydrogen bonding that links the extended cellulose polymers into cellulose fibers.

If only a portion of the —OH groups on the cellulose molecules are nitrated, a plastic nitrocellulose material known as celluloid or pyroxylin is obtained. This material has been used to make such diverse articles as billiard balls, celluloid shirt collars, and photographic film. By nitration of nearly all —OH groups, a powerful high explosive is obtained. This highly nitrated cellulose, or "guncotton," is the basic ingredient in modern "smokeless" gunpowder.

Another modified cellulose, cellulose acetate, is made by esterification of —OH groups with acetic acid (acetic anhydride). About two-thirds of the —OH groups are esterified:

$$\underset{\text{hydroxyl group on cellulose molecule}}{\text{cellulose}-\text{OH}} + \underset{\text{acetic anhydride}}{CH_3-\overset{O}{\overset{\|}{C}}-O-\overset{O}{\overset{\|}{C}}-CH_3} \longrightarrow \underset{\text{acetate group on cellulose molecule}}{\text{cellulose}-O-\overset{O}{\overset{\|}{C}}-CH_3} + CH_3COOH$$

Cellulose acetate, unlike the dangerously flammable cellulose nitrate, can be made to burn only with difficulty. For this reason, cellulose acetate has displaced cellulose nitrate in almost all kinds of photographic films. The textile known as acetate rayon is made from cellulose acetate. Cellulose acetate is also used as a clear, transparent packaging film. In another process, cellulose reacts with carbon disulfide in the presence of sodium hydroxide to form a soluble cellulose derivative called cellulose xanthate, from which cellulose can be regenerated. Viscose rayon textiles and cellophane packaging materials are made of regenerated cellulose prepared by this process.

27.15 Complex Polysaccharides

Very complex polysaccharides are found in animals. These carbohydrates are linked together by glycosidic bonds at various positions on the monosaccharides. Thus a sugar may be bonded to its neighbor from carbon 1 to carbon 3, or carbon 1 to carbon 4, or carbon 1 to carbon 6, and so on. Because of these numerous linkages, the polysaccharides often have a complex, branching structure. Also, it is not uncommon for these molecules to contain many different kinds of monosaccharides. These complex carbohydrates serve a variety of functions in animals.

Glycosaminoglycans (mucopolysaccharides) make up part of the connective tissue. These polymers act as shock absorbers between bones at joints, and they keep our bones from rubbing against each other as we move. Osteoarthritis, the most common form of arthritis, is the result of the loss of glycosaminoglycans at joints.

Even more complicated polysaccharides are found on the surfaces of almost all cells. These carbohydrates serve as "labels," or antigens, allowing organisms to distinguish their own cells from invading bacteria, for example. Antigen recognition illustrates a very important biochemical principle: *Molecular shape carries information that guides the reactions of life.*

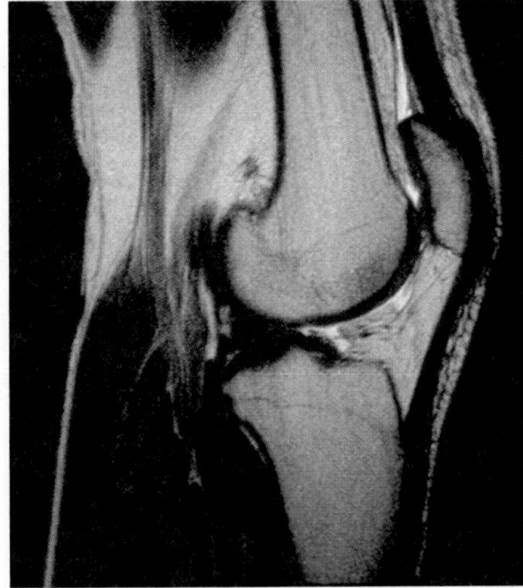

The connective tissue inside the knee contains mucopolysaccharides.

In humans, the polysaccharides on the surface of the red blood cells give rise to different blood types, often classified by the ABO system. Cells of different blood types have surface polysaccharides with different structures (see Figure 27.10). Cells carrying one carbohydrate structure are commonly not tolerated by an individual of another blood type. For example, if type A blood is transfused into someone with type O blood cells, the new cells will be attacked by the immune system. Red cell destruction can lead to serious injury or death.

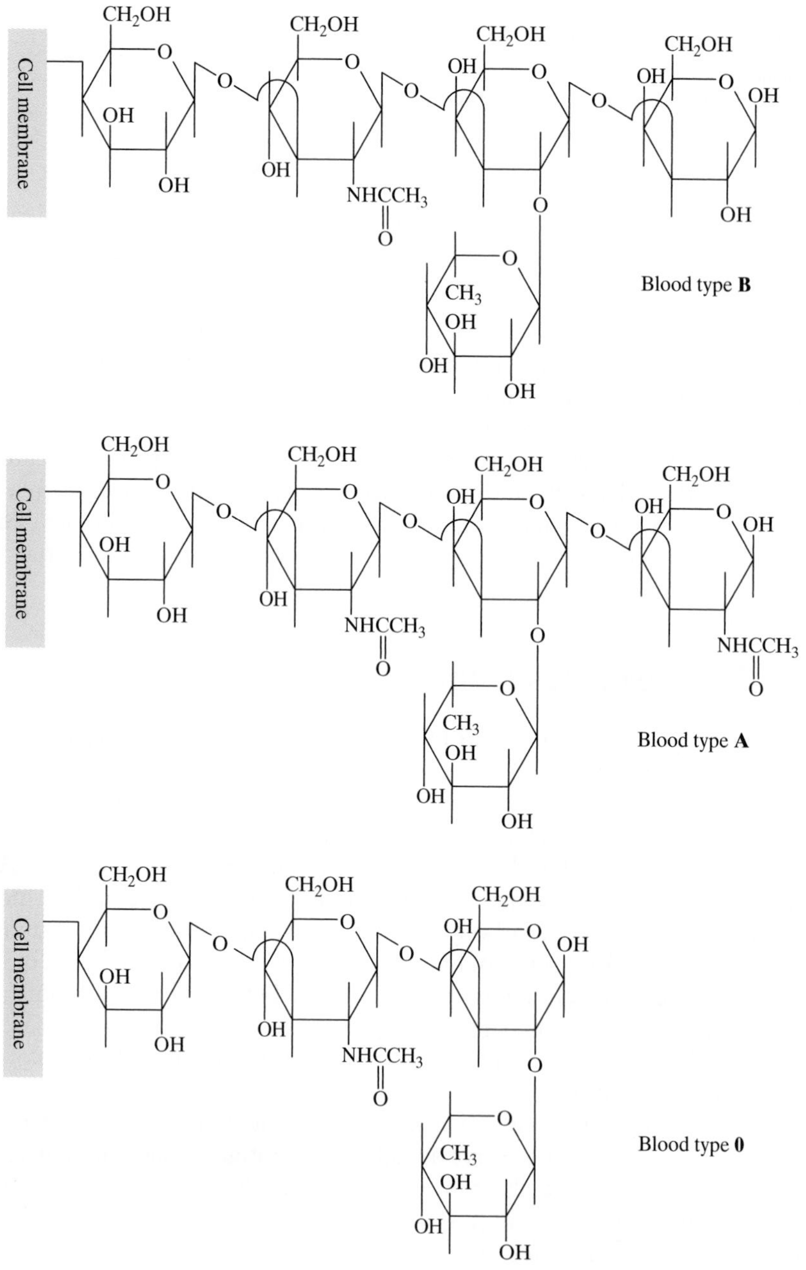

Cell membrane

Blood type **B**

Cell membrane

Blood type **A**

Cell membrane

Blood type **0**

Figure 27.10
Common blood group substances (antigens) that serve to identify red blood cells in the ABO blood group system.

Chapter 27 Review

27.1 Carbohydrates: A First Class of Biochemicals

KEY TERM

Carbohydrate

- Carbohydrates are among the most important classes of biochemicals.
- Carbohydrates are generally defined as polyhydroxy aldehydes or polyhydroxy ketones or substances that yield these compounds when hydrolyzed.

27.2 Classification of Carbohydrates

KEY TERMS

Monosaccharide
Disaccharide
Oligosaccharide
Polysaccharide

- Carbohydrates are classified based on the number of monosaccharide units linked to form the molecule.
  - A monosaccharide is a carbohydrate that cannot be hydrolyzed to simpler carbohydrate units.
  - A disaccharide yields two monosaccharides when hydrolyzed.
  - An oligosaccharide is composed of two to six monosaccharide units.
  - A polysaccharide is a macromolecule that can be hydrolyzed to yield many monosaccharide units.
- A monosaccharide is commonly classified in the following ways:
  - Based on the number of carbons in the molecule, a common monosaccharide will be classified as a triose (three C), a tetrose (four C), a pentose (five C), a hexose (six C) or a heptose (seven C).
  - Based on the type of carbonyl, a monosaccharide will be classified as either an aldose (aldehyde-containing monosaccharide) or a ketose (ketone-containing monosaccharide).
  - Based on the configuration around the chiral carbon closest to the terminal primary alcohol, a monosaccharide is classified as a D-monosaccharide if the —OH is on the right in the Fischer projection formula or an L-monosaccharide if the —OH is on the left.
  - Based on the optical rotation, a monosaccharide will be classified as either a (+) isomer if the monosaccharide rotates plane-polarized light clockwise (to the right) or a (−) isomer if the monosaccharide rotates plane-polarized light counterclockwise (to the left).
  - Based on the size of the ring structure, a monosaccharide is classified as a furanose if the monosaccharide forms a five-membered ring or a pyranose if the monosaccharide forms a six-membered ring.
  - Based on the configuration around the hemiacetal/hemiketal carbon, a monosaccharide will be classified as either alpha (α) if the —OH is on the right in the Fischer projection formula for D-monosaccharides or beta (β) if the —OH is on the left in the Fischer projection formula for D-monosaccharides.

27.3 Importance of Carbohydrates

- Carbohydrates are very effective energy-yielding nutrients.
- Carbohydrates can serve as very effective building materials.
- Many carbohydrates are important water-soluble molecules.

27.4 Monosaccharides

- Of all the many possible monosaccharides, only a few are of much biological significance.
- The pentose ribose and its deoxy derivitive deoxyribose are found in ribonucleic acid (RNA) and deoxyribonucleic acid (DNA) respectively.
- Three hexoses—glucose, galactose, and fructose—are of major significance in nutrition.
  - Glucose, an aldose, is the most important of the monosaccharides. It is also known as *blood sugar* because this sugar is transported through the bloodstream and is directly metabolized by all human cells.
  - Galactose, an aldose, differs from glucose at only one chiral carbon. Galactose is combined with glucose in the disaccharide lactose, or milk sugar.
  - Fructose, the only common ketose, is found with glucose in the disaccharide sucrose, or table sugar.

27.5 Structure of Glucose and Other Aldoses

KEY TERM

Epimer

- Emil Fischer won the Nobel Prize in part for establishing the three-dimensional structure of glucose and other sugars.
- The Fischer projection formula is commonly used with carbohydrates: vertical bonds represent bonds going away from the observer in three-dimensions; horizontal bonds represent bonds going toward the observer in three-dimensions.
- Any two monosaccharides that differ in configuration around a single carbon atom are called epimers.
- There are two enantiomeric forms of glucose, D-glucose and L-glucose.
- The Kiliani-Fischer synthesis in which HCN adds to an aldehyde to form a cyanohydrin can be used to convert trioses to tetroses to pentoses to hexoses and so on.

27.6 Cyclic Structure of Glucose; Mutarotation

KEY TERMS

Anomer
Mutarotation

- Glucose is seldom found in its straight-chain form because it undergoes a reaction common for most monosaccharides—glucose reacts to form ring structures.
- When glucose cyclizes, the aldehyde carbon becomes chiral and two new isomers exist for glucose.
- When two cyclic isomers differ only in their stereo arrangement at this new chiral carbon, they are called anomers.
  - The alpha (α) anomer has the new chiral carbon's —OH on the right in the Fischer projection formula.
  - The beta (β) anomer has the new chiral carbon's —OH on the left in the Fischer projection formula.
- Mutarotation is the process by which anomers are interconverted.
- A Haworth projection formula shows the glucose ring structures as hexagons.
  - If an —OH is shown to the right in the Fischer projection formula, it is written below the plane of the hexagon in the Haworth projection formula.
  - If an —OH is shown to the left in the Fischer projection formula, it is written above the plane of the hexagon in the Haworth projection formula.
  - The primary alcohol (C-6) of D-glucose is written above the plane of the hexagon in the Haworth projection formula.

27.7 Hemiacetals and Acetals

KEY TERM

Glycoside

- A hemiacetal structure consists of an ether linkage and an alcohol linkage on the same carbon atom, whereas an acetal structure has two ether linkages to the same carbon atom.
- Cyclic structures of monosaccharides are intramolecular hemiacetals.
- When a monosaccharide hemiacetal reacts with an alcohol, the product is an acetal; this product is a glycoside in carbohydrate terminology.

27.8 Structures of Galactose and Fructose

- Galactose is an aldohexose that differs from glucose only in the configuration at carbon 4.
- Fructose is a ketohexose and commonly exists in a furanose ring structure.

27.9 Pentoses

- D-ribose is an aldopentose that occurs in nucleic acids (RNA).
- D-2-deoxyribose occurs in nucleic acids (DNA) and is like D-ribose except that there is no oxygen on carbon 2 of D-2-deoxyribose.
- D-ribulose is a common ketopentose.

27.10 Disaccharides

- Upon hydrolysis, disaccharides yield two monosaccharides.
- Both sucrose and lactose are disaccharides that are found free in nature.
  - Sucrose (table sugar) is composed of glucose and fructose.
  - Lactose (milk sugar) is composed of glucose and galactose.
- Some disaccharides are formed as polysaccharides are degraded.
- Maltose is a common disaccharide composed of two glucose molecules; it is formed as polysaccharides are degraded in sprouting grain.

27.11 Structures and Properties of Disaccharides

- Disaccharides contain an acetal structure (glycosidic linkage), and some also contain a hemiacetal structure.
- The maltose structure can be described in a systematic way as α-D-glucopyranosyl-(1,4)-α-D-glucopyranose.
- Lactose consists of a β-D-galactopyranose and an α-D-glucopyranose joined by a β-1,4-glycosidic linkage (β-D-galactopyranosyl-(1,4)-α-D-glucopyranose).
- The sucrose structure can be described in a systematic way as α-D-glucopyranosyl-(1,2)-β-D-fructofuranose.
- Only disaccharides with hemiacetals can mutarotate; lactose and maltose show mutarotation, but sucrose does not.

27.12 Sweetners and Diet

KEY TERM

Invert sugar

- Carbohydrates are used to sweeten foods.
  - Sucrose is the most common sweetener, but it tends to crystallize from concentrated solutions or syrups.
  - Invert sugar is made by hydrolyzing sucrose to form a mixture of glucose and fructose.
  - Invert sugar has less tendency to crystallize and has greater sweetening power because fructose is a very sweet monosaccharide.
- Fructose is an important sweetener in high-fructose corn syrups—produced by hydrolyzing starch to glucose and then converting some of the glucose to fructose.
- Nonnutritive sweeteners have become an important component of many diets.
- Foods can be sweetened with only tiny amounts of these sweeteners as they are all very much sweeter than fructose.
- Each sweetener has its own advantages; different sweeteners serve different purposes.

27.13 Redox Reactions of Monosaccharides

KEY TERM

Reducing sugar

- The aldehyde groups of monosaccharides can be oxidized to monocarboxylic acids by mild oxidizing agents.

- Mild oxidizing agents include Br_2, Cu^{2+}, and Ag^+.
- The resulting acids are named by changing the *-ose* sugar suffix to *-onic acid*.
- Dilute nitric acid oxidizes both the aldehyde group and the primary alcohol of aldoses to form dicarboxylic acids.
  - The resulting acids are named by changing the *-ose* sugar suffix to *-aric acid*.
- Monosaccharides can be reduced by agents such as H_2/Pt to form polyhydroxy alcohols.
- Reducing sugars will reduce copper(II) ions to copper(I) ions or silver ions to silver metal.
  - In the Tollens test, reducing sugars convert silver ions to silver metal that coats the glass to form a mirror.
  - In the Benedict, Fehling, and Barfoed tests, a blue Cu^{2+} solution is converted to a brick-red copper(I) oxide precipitate.
  - Reducing sugars contain (1) aldehyde groups, (2) α-hydroxy ketone groups or (3) hemiacetal groups.
- Many clinical tests monitor glucose as a reducing sugar.

27.14 Polysaccharides Derived from Glucose

- Starch is the most important energy-storage polysaccharide in plants.
- Starch is a mixture of two polysaccharides, amylose and amylopectin.
  - Amylose is a linear polymer of α-D-glucose units joined by α-1,4-glycosidic linkages.
  - Amylopectin is a branched polymer of long chains of α-D-glucose joined by α-1,4-glycosidic linkages. The chains are connected at branch points that are α-1,6-glycosidic linkages.
- Glycogen is the energy-storage polysaccharide of the animal kingdom.
- Glycogen has a structure like amylopectin, except that glycogen has more branch points.
- Cellulose is major component of most plant cell walls.
- Most living organisms cannot digest cellulose.
- Cellulose is a linear polymer of β-D-glucose joined by β-1,4-glycosidic linkages.

27.15 Complex Polysaccharides

- Glycosaminoglycans (mucopolysaccharides) are complex carbohydrates that are part of animal connective tissue.
  - These carbohydrates act as shock absorbers between bones at joints.
- Even more complex carbohydrates are found on the surface of cells and act as "labels," or antigens, to identify the cells.
  - The ABO system is used to type red blood cells based on their surface carbohydrates.

Review Questions

All questions with blue numbers have answers in the appendix of the text.

1. Why is the oxidation state of a carbohydrate carbon important in metabolism?
2. What is the significance of the notations D and L in the name of a carbohydrate?
3. What is the significance of the notations (+) and (−) in the name of a carbohydrate?
4. What is galactosemia and what are its effects on humans?
5. Which of the D-aldohexoses in Figure 27.1 are epimers?
6. Explain how a carbohydrate with a pyranose structure differs from one with a furanose structure.
7. Explain how α-D-glucopyranose differs from β-D-glucopyranose.
8. Are the cyclic forms of monosaccharides hemiacetals or glycosides?
9. Explain the phenomenon of mutarotation.
10. What is (are) the major sources of each of these carbohydrates?
    (a) sucrose (b) lactose (c) maltose
11. Consider the eight aldohexoses given in Figure 27.1 and answer the following:
    (a) The eight aldohexoses are oxidized by nitric acid to dicarboxylic acids. Which of these give meso (optically inactive) dicarboxylic acids?
    (b) Write the structures and names for the enantiomers of D-altrose and D-idose.
12. Explain why invert sugar is sweeter than sucrose.
13. What are the structural differences between amylose and cellulose?
14. What visual difference would you expect between a dilute glucose solution and a concentrated glucose solution in the Benedict test?
15. What are the two main components of starch? How are they alike and how do they differ?
16. What change in the digestive system leads to lactose intolerance?
17. What are glycosaminoglycans?
18. Using Figure 27.10, describe the difference between the blood type A molecule and blood type B molecule.
19. What is the main purpose of complex carbohydrates that are located on the cell surface?

Paired Exercises

All exercises with blue numbers have answers in the appendix of the text.

1. Dihydroxyacetone is the simplest ketose.
(a) Write a Fischer projection formula for this ketose. Are there any chiral carbon atoms in this molecule?
(b) Write the Fischer projection formula of the product obtained by reacting dihydroxyacetone with hydrogen in the presence of platinum.

2. Glyceraldehyde is the simplest aldose.
(a) Write Fischer projection formulas, and identify the D and L forms, for this aldose.
(b) Write the Fischer projection formula of the product obtained by reacting D-glyceraldehyde with hydrogen in the presence of platinum.

3. Give the Fischer structure of an epimer of D-mannose.

4. Give the Fischer structure of an epimer of D-galactose.

5. Write Fischer structures and names for the enantiomers of D-galactose, D-mannose, and D-ribose.

6. Write Fischer structures and names for the enantiomers of D-glucose, D-fructose, and D-2-deoxyribose.

7. Write the cyclic structures for α-D-glucopyranose, β-D-galactopyranose, and α-D-mannopyranose.

8. Write the cyclic structures of β-D-glucopyranose, α-D-galactopyranose, and β-D-mannopyranose.

9. Briefly explain why starch is easily biodegradable in human digestion.

10. Briefly explain why cellulose is difficult to biodegrade in human digestion.

11. Starting with the proper D-tetrose, show the steps for the synthesis of D-glucose by the Kiliani–Fischer synthesis.

12. Starting with the proper D-triose, show the steps for the synthesis of D-ribose by the Kiliani–Fischer synthesis.

13. Is D-2-deoxymannose the same as D-2-deoxyglucose? Briefly explain.

14. Is D-2-deoxygalactose the same as D-2-deoxyglucose? Briefly explain.

15. What is the monosaccharide composition of the following?
(a) sucrose
(b) glycogen
(c) amylose
(d) maltose

16. What is the monosaccharide composition of the following?
(a) lactose
(b) amylopectin
(c) cellulose
(d) sucrose

17. How does the structure of cellobiose differ from that of isomaltose?

18. How does the structure of maltose differ from that of isomaltose?

19. Of the disaccharides sucrose and lactose, which show mutarotation? Explain why.

20. Of the disaccharides maltose and isomaltose, which show mutarotation? Explain why.

21. Draw structural formulas for cellobiose and isomaltose. Point out the portion of the structure that is responsible for making these reducing sugars.

22. Draw structural formulas for maltose and lactose. Point out the portion of the structure that is responsible for making these reducing sugars.

23. Give systematic names for isomaltose and cellobiose.

24. Give systematic names for maltose and lactose.

25. Describe the principal structural differences and similarities between the members of the following pairs:
(a) D-glucose and D-fructose
(b) maltose and sucrose
(c) cellulose and glycogen

26. Describe the principal structural differences and similarities between the members of the following pairs:
(a) D-ribose and D-2-dexoyribose
(b) amylose and amylopectin
(c) lactose and isomaltose

27. Write the structural formula and the name of the dicarboxylic acid that is formed when D-galactose is oxidized with nitric acid.

28. Write the structural formula and the name of the dicarboxylic acid that is formed when D-mannose is oxidized with nitric acid.

29. Write the formulas for the four L-ketohexoses. Indicate which pair (or pairs) of the L-ketohexoses are epimers.

30. Write the formulas for the four L-aldopentoses. Indicate which pair (or pairs) of the L-aldopentoses are epimers.

31. Draw the Haworth formulas for
(a) β-D-glucopyranosyl-(1,4)-α-D-galactopyranose
(b) β-D-galactopyranosyl-(1,6)-β-D-mannopyranose

32. Draw the Haworth formulas for
(a) β-D-mannopyranosyl-(1,4)-β-D-galactopyranose
(b) β-D-galactopyranosyl-(1,6)-α-D-glucopyranose

33. All the disaccharides formed upon partial digestion of amylose have the same structure. Draw that structure.

34. All the disaccharides formed upon partial digestion of cellulose have the same structure. Draw that structure.

35. The disaccharides formed upon partial digestion of glycogen have two different structures. Draw both structures.

36. The disaccharides formed upon partial digestion of amylopectin have two different structures. Draw both structures.

37. Name the following.

(a)
```
      O
      ‖
      CH
  H —|— OH
 HO —|— H
 HO —|— H
  H —|— OH
     CH2OH
```

(b) CH₂OH, O, OH, OH, OH, OH

(c) CH₂OH, O, OH, OH OH

38. Name the following.

(a)
```
      O
      ‖
      CH
  H —|— OH
  H —|— OH
  H —|— OH
     CH2OH
```

(b) CH₂OH, O, OH, OH, OH, OH

(c) CH₂OH, O, CH₂OH, HO, OH, OH

39. Give both the common and systematic name for the following:

CH₂OH, CH₂OH, HO, O, O, OH, OH, OH, OH, OH

40. Give both the common and systematic name for the following:

CH₂OH, CH₂OH, O, O, OH, O, OH, OH, OH, OH, OH

Additional Exercises

All exercises with blue numbers have answers in the appendix of the text.

41. Why is glucose sometimes called "blood sugar"?

42. Cite two advantages of aspartame, as a sweetener, over sucrose.

43. What changes must take place to convert cornstarch into the sweetener high-fructose corn syrup?

44. A thickener commonly used in ice cream, alginic acid, is a seaweed polymer composed of the following monomer:

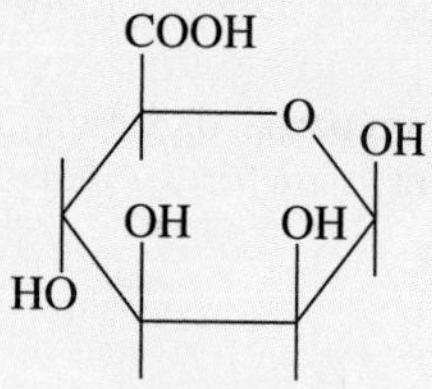

COOH
O
OH
OH OH
HO

(a) From what monosaccharide might this monomer be derived?

(b) Alginic acid is made by linking these monomers together via β-(1,4)-glycosidic bonds. Draw a short section (three units) of this polymer.

45. A reddish color is obtained when compound A (a disaccharide) is reacted with Benedict solution. Is this compound more likely to be maltose or sucrose? Briefly explain.

46. Refer to this compound to answer the questions that follow:

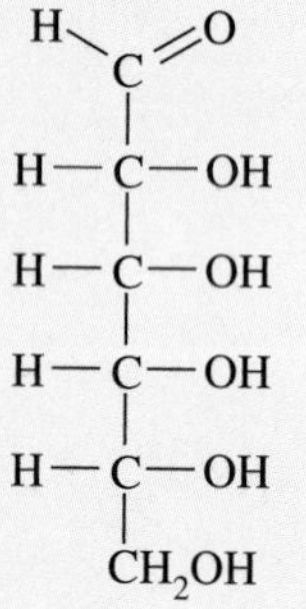

H—C=O
H—C—OH
H—C—OH
H—C—OH
H—C—OH
CH_2OH

(a) How many chiral carbons are present in this compound as it is written? Label them.

(b) Draw the α-anomer pyranose ring form of this compound.

(c) How many chiral carbons are present in the structure you drew in part (b)? Label them.

47. If the compound shown below rotates light plane-polarized 25° to the right, draw the structure of a closely related compound that rotates light 25° to the left. Are the two compounds epimers of one another? Why or why not?

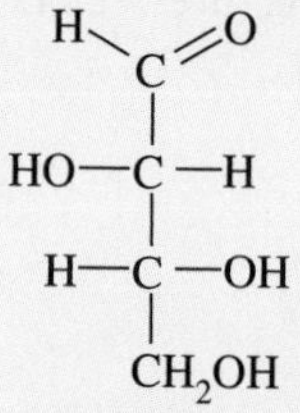

H—C=O
HO—C—H
H—C—OH
CH_2OH

48. Draw the structure of a nonreducing disaccharide composed of two molecules of α-D-galactopyranose.

49. Why is cellulose considered "fiber" in your diet but starch is not? In your answer, refer specifically to the structures of cellulose, amylose, and amylopectin.

50. When lactose is metabolized, the first step separates it into glucose and galactose. Galactose must be changed into glucose before it can be further metabolized in the body.

(a) What changes have to occur in the structure of galactose to make it into glucose?

(b) Which of these terms best describes the relationship of glucose to galactose: disastereomers, enantiomers, or epimers?

51. One of your classmates has told you that the optical rotation of D-glucose and D-mannose are equal and opposite because they are epimers of one another. Should you believe your classmate? Why or why not?

Challenge Exercise

52. Candy makers who want an especially sweet treat may add some lemon juice (an acid) to their boiling sucrose solution before letting the candy harden.

(a) Draw the structure of sucrose.

(b) Write the chemical reaction that is aided (catalyzed) by the acid in lemon juice.

(c) How will this chemical reaction make the candy sweeter?

Answers to Practice Exercises

27.1

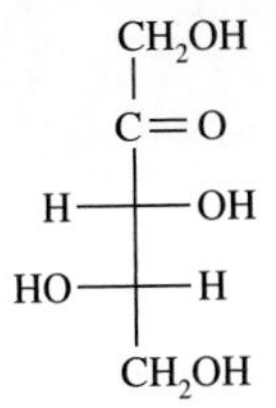

CH_2OH
C=O
H—OH
HO—H
CH_2OH

The orientation of the —OH on carbon 3 is not specified and can be written in either direction.

27.2 The right side

27.3 (a)

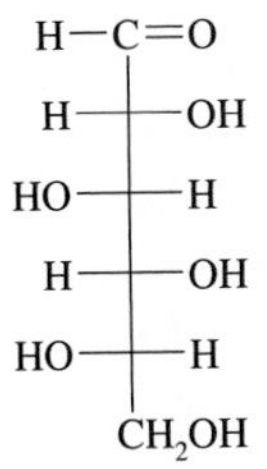

H—C=O
H—OH
HO—H
H—OH
HO—H
CH_2OH

(b)

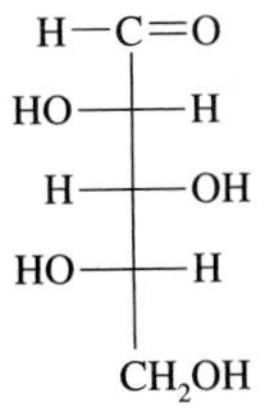

H—C=O
HO—H
H—OH
HO—H
CH_2OH

27.4

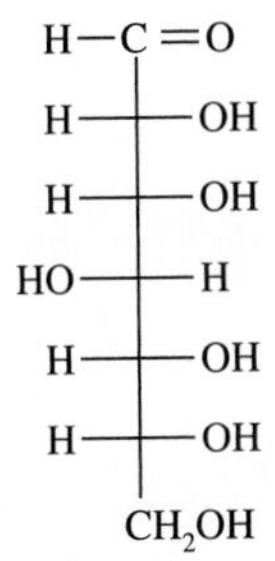

H—C=O
H—OH
H—OH
HO—H
H—OH
H—OH
CH_2OH

and

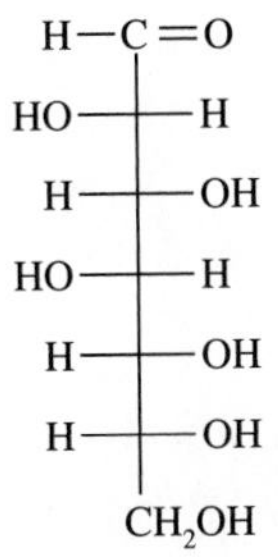

H—C=O
HO—H
H—OH
HO—H
H—OH
H—OH
CH_2OH

27.5 (a) $^{6}CH_2OH$, HO, O, OH, OH, OH; 1, 2, 3, 4, 5

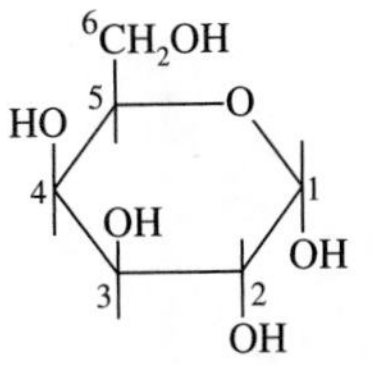

(b) $^{6}CH_2OH$, O, OH, OH, HO, OH; 1, 2, 3, 4, 5

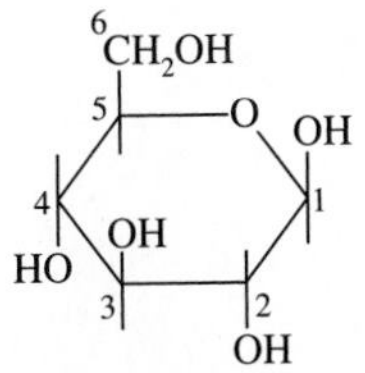

27.6 (a) CH_2OH, CH_2OH, O, HO, OH, OH

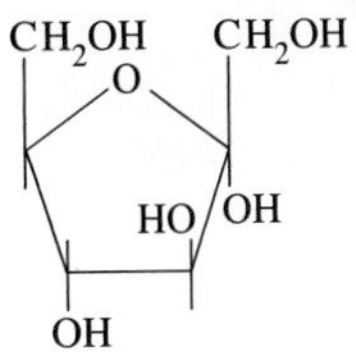

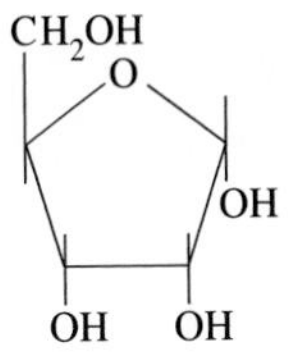

(b) CH_2OH, O, OH, OH, OH

27.7 $^{6}CH_2OH$, O, OH, HO, OH, O, $^{6}CH_2OH$, O, OH, OH, OH, OH; 1, 2, 3, 4, 5

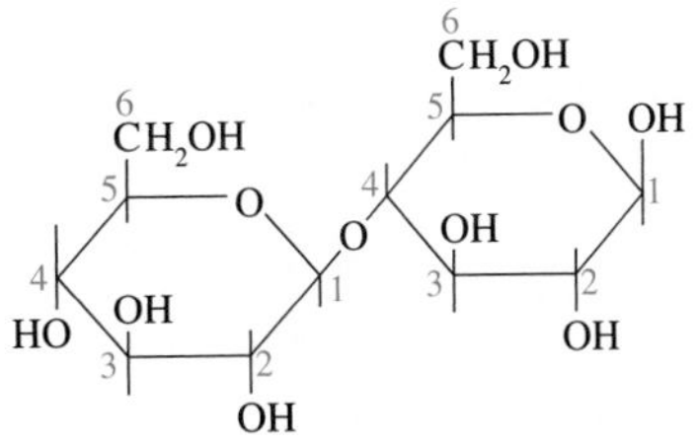

27.8 Carbon 1 from each α-D-galactopyranose

27.9

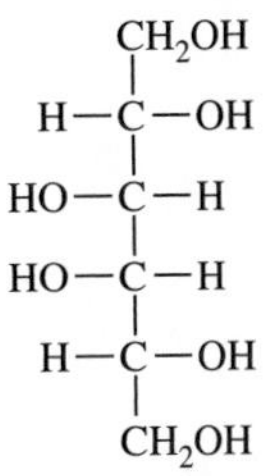

CH_2OH
H—C—OH
HO—C—H
HO—C—H
H—C—OH
CH_2OH

This compound is not optically active, it is a meso compound.

CHAPTER 28

Lipids

Polar bears have a large reserve of lipids.

Chapter Outline

The behavior of lipids in water—the fact that they are insoluble—is a key to their importance in nature. For example, an oil slick can spread for many square miles on the surface of the ocean partly because oil and water don't mix. This same principle enables cells to surround themselves with a thin film of lipid, the cell membrane. We protect a fine wood floor with wax, another lipid, because we can depend on this material to adhere to the floor and not dissolve in water. A lipid's stickiness and water insolubility also create diseases such as atherosclerosis, where arteries become partially clogged by cholesterol-containing lipids. Lipids bring both benefits and problems—these molecules are truly a mixed blessing.

28.1 Lipids: Hydrophobic Molecules

Lipids are water-insoluble, oily, or greasy biochemical compounds that can be extracted from cells by nonpolar solvents such as ether, chloroform, or benzene. Unlike carbohydrates, lipids share no common chemical structure. **lipid**

What makes a molecule such as a lipid insoluble in water? To answer this question, we must establish two important principles about water solutions: A compound dissolves in water (1) if the water molecules bond well to the potential solute and (2) if the water molecules can still move relatively freely around the dissolved compound. For example, salt (sodium chloride) dissolves because it forms ions to which water molecules can bond. Also, water molecules are free to move around these small ions. Sugar (sucrose) dissolves because it forms hydrogen bonds with water and because it is still a relatively small molecule.

Lipid structures and solubilities differ from both salts and carbohydrates. Lipid molecules are big enough to substantially affect the free movement of water molecules. In addition, lipids cannot hydrogen-bond to the extent that carbohydrates can, nor do they form the large number of positive and negative charges found in a salt solution. What makes a lipid water insoluble? Lipids are *large* and relatively *nonpolar* molecules and thus are water insoluble.

Many carbohydrates and salts are said to be *hydrophilic* ("water loving"). In contrast, lipids are said to be *hydrophobic* ("water fearing").

Consider fatty acids, which are common components of lipids. As shown in Table 28.1, when the number of atoms in a fatty acid molecule increases, the water solubility of the fatty acid decreases dramatically. Water molecules can easily maneuver around smaller compounds like butyric acid, which is infinitely soluble in water. However, these same water molecules run into a huge barrier when they encounter the 18-carbon chain of stearic acid, which is essentially insoluble in water.

The hydrophobic nature of lipids contributes significantly to the biological functions of these molecules. Their water insolubility allows lipids to serve as barriers to aqueous solutions. This property, as we shall see later, is of great importance when lipids form cellular membranes.

Table 28.1 Some Naturally Occurring Fatty Acids

Fatty acid	Number of C atoms	Formula	Solubility (g/100 g water)	Melting point (°C)
Saturated acids				
Butyric acid	4	$CH_3CH_2CH_2COOH$	∞	−7.9
Caproic acid	6	$CH_3(CH_2)_4COOH$	1.08	−3.4
Caprylic acid	8	$CH_3(CH_2)_6COOH$	0.07	17
Capric acid	10	$CH_3(CH_2)_8COOH$	0.015	31
Lauric acid	12	$CH_3(CH_2)_{10}COOH$	insoluble	44
Myristic acid	14	$CH_3(CH_2)_{12}COOH$	insoluble	59
Palmitic acid	16	$CH_3(CH_2)_{14}COOH$	insoluble	63
Stearic acid	18	$CH_3(CH_2)_{16}COOH$	insoluble	70
Arachidic acid	20	$CH_3(CH_2)_{18}COOH$	insoluble	76
Unsaturated acids*				
Palmitoleic acid	16	$CH_3(CH_2)_5CH{=}CH(CH_2)_7COOH$	—	0.5
Oleic acid	18	$CH_3(CH_2)_7CH{=}CH(CH_2)_7COOH$	—	4
Linoleic acid	18	$CH_3(CH_2)_4CH{=}CHCH_2CH{=}CH(CH_2)_7COOH$	—	−12
Linolenic acid	18	$CH_3CH_2CH{=}CHCH_2CH{=}CHCH_2CH{=}CH(CH_2)_7COOH$	—	−11
Arachidonic acid	20	$CH_3(CH_2)_4(CH{=}CHCH_2)_4CH_2CH_2COOH$	—	−50

*Omega (ω) is the last letter in the Greek alphabet. Correspondingly, the last carbon atom in a carbon chain of a compound is often referred to as the omega carbon. In reference to unsaturated carboxylic acids, omega plus a number (e.g., ω-3) indicates the location of the first carbon–carbon double bond, counting from the omega carbon.

The omega designation of these acids are palmitoleic (ω-7); oleic (ω-9); linoleic (ω-6); linolenic (ω-3); arachidonic (ω-6).

28.2 Classification of Lipids

An ester is formed by splitting out a molecule of water between an alcohol and an acid.

Lipids are hydrophobic molecules; their structures are relatively large and nonpolar. Yet, within this broad description, lipid structures vary markedly. The following classification scheme recognizes important structural differences:

1. **Simple lipids**
   (a) *Fats and oils:* esters of fatty acids and glycerol.
   (b) *Waxes:* esters of high-molar-mass fatty acids and high-molar-mass alcohols.
2. **Compound lipids**
   (a) *Phospholipids:* substances that yield glycerol, phosphoric acid, fatty acids, and a nitrogen-containing base upon hydrolysis.
   (b) *Sphingolipids:* substances that yield an unsaturated amino alcohol (sphingosine), a long-chain fatty acid, and either a carbohydrate or phosphate and a nitrogen base upon hydrolysis.
   (c) *Glycolipids:* substances that yield sphingosine, a fatty acid, and a carbohydrate upon hydrolysis.
3. **Steroids**
   Substances that possess the steroid nucleus, which is a 17-carbon structure consisting of four fused carbocyclic rings. Cholesterol and several hormones are in this class.
4. **Miscellaneous lipids**
   Substances that do not fit into the preceding classifications; these include the fat-soluble vitamins A, D, E, and K, and lipoproteins.

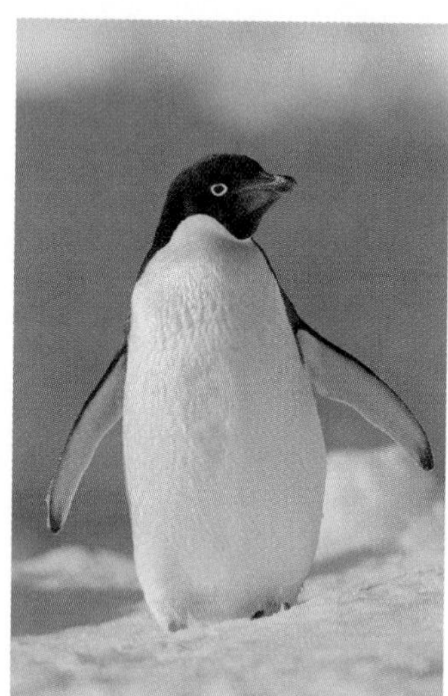

A layer of fat under the penguin's skin and stiff feathers serve as insulation against the extreme Antarctic cold winter mean temperature (−57°C).

The most abundant lipids are the fats and oils. These substances constitute one of the three important classes of foods. In the discussion that follows, we examine fats and oils as they pertain to biochemistry. A more complete consideration of the properties and composition of various fats and oils is given in Section 24.10.

28.3 Simple Lipids

Fatty Acids

Fatty acids, which form part of most lipids, are carboxylic acids with long, hydrophobic carbon chains. The formulas for some of the most common fatty acids are shown in Table 28.1. All these fatty acids are straight-chain compounds with an even number of carbon atoms. Five of the fatty acids in this table—palmitoleic, oleic, linoleic, linolenic, and arachidonic—are unsaturated, having carbon–carbon double bonds in their structures. Animal and higher plant cells produce lipids in which palmitic, oleic, linoleic, and stearic acids predominate. Over one-half of plant and animal fatty acids are unsaturated, plant lipids tending to be more unsaturated than their animal counterparts.

Double bonds impart some special characteristics to the unsaturated fatty acids. Remember that the presence of double bonds raises the possibility of geometric isomerism (Section 20.3). Unsaturated fatty acids may be either cis or trans isomers. To illustrate the effect of these double bonds on fatty acid structure, the two fatty acids in Figure 28.1 are portrayed in a simplified manner, with each of the many $-CH_2-$ groups as an apex at the intersection between two single bonds. Note that the trans isomer is almost a linear molecule, while the double bond in the cis isomer introduces a kink into the fatty acid structure. Unsaturated fatty acids found in nature are almost always cis isomers. These kinked fatty acids cannot stack closely together and hence do not solidify easily. As shown in Table 28.1, unsaturated fatty acids have lower melting points than saturated fatty acids of a similar size. Cooking oils are liquids at room temperature because a high percentage of their fatty acids are unsaturated. In like manner, biological membranes are very fluid because of the presence of fatty acid cis isomers. (See Section 28.6.)

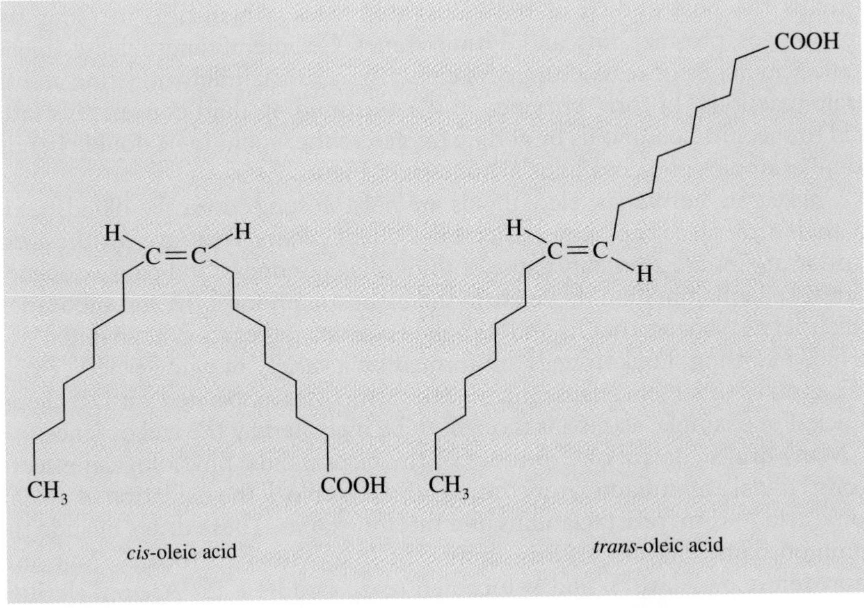

Figure 28.1
Cis- and *trans*-oleic acids. Notice the bent structure of the cis isomer.

Practice 28.1

Draw the cis and trans isomers for palmitoleic acid having the formula $CH_3(CH_2)_5CH{=}CH(CH_2)_7COOH$.

Polyunsaturated means the compound has more than one double bond.

Scientists continue to learn more about the nutritional importance of unsaturated fatty acids. If unsaturated fatty acids are converted from cis to trans, they lose much of their special dietary importance. For example, trans isomers are formed when margarine is produced by partially hydrogenating vegetable oils. Many health organizations suggest the use of softer and less hydrogenated fats where feasible.

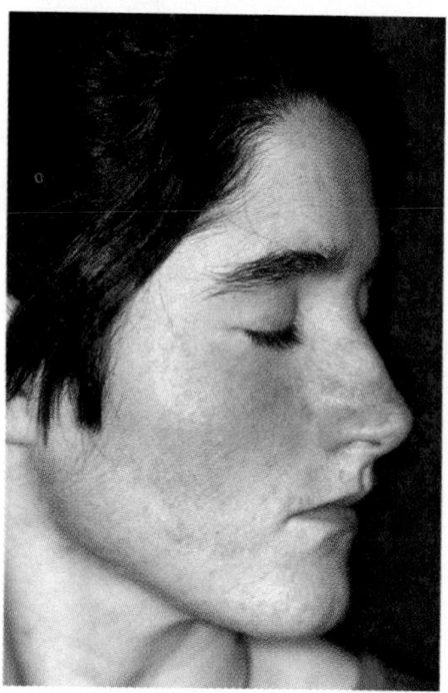

Prostaglandins are synthesized at the site in response to allergic reactions such as the dermatitis shown here.

Three common polyunsaturated fatty acids are linoleic, linolenic, and arachidonic acids. These fatty acids represent two nutrient classes that are metabolized differently. The difference lies with how close the double bonds are to the terminal methyl group. Linolenic acid has a double bond on the third carbon from the end of the carbon chain [the omega (ω) position]. Thus, linolenic acid is an ω-3 fatty acid. In contrast, linoleic and arachidonic acids have a double bond on the sixth carbon from the end of the chain; they are ω-6 fatty acids. Good health requires a diet that supplies both classes of polyunsaturated fatty acids.

The three unsaturated fatty acids—linoleic, linolenic, and arachidonic—are essential for animal nutrition and must be present in the diet. Diets lacking these fatty acids lead to impaired growth and reproduction and skin disorders such as eczema and dermatitis. A dermatitis disorder can be attributed to an unsaturated fatty acid deficiency if the symptoms clear up when that fatty acid is supplied in the diet.

Certain fatty acids, as well as other lipids, are biochemical precursors of several classes of hormones. The well-known steroid hormones are synthesized from cholesterol; they will be discussed later in this chapter. Arachidonic acid and, to a lesser extent, linolenic acid are also used by the body to make hormonelike substances. The biochemicals derived from arachidonic acid are collectively termed eicosanoids, using a derivative of the Greek word for 20 (*eikosi*) to indicate that these compounds have 20 carbon atoms. Prostaglandins are perhaps the best known of the eicosanoid class, which also includes the leukotrienes, prostacyclins, and thromboxanes. Cell membranes release arachidonic acid in response to a variety of circumstances, including infection and allergic reactions. In turn, enzymes in the surrounding fluid convert this fatty acid to specific eicosanoids by adding oxygen to the arachidonic double bonds. Some examples of eicosanoids are shown in Figure 28.2.

Unlike true hormones, eicosanoids are not transported via the bloodstream to their site of action, but rather take effect where they are synthesized. Prostaglandins are a primary cause of the swelling, redness, and pain associated with tissue inflammation. Platelets in the bloodstream form the thromboxanes, which act as vasoconstrictors and stimulate platelet aggregation as an initial step in blood clotting. Leukotrienes are formed by a variety of white blood cells as well as other tissues and cause many of the symptoms associated with an allergy attack. For example, asthma is thought to be mediated by the leukotrienes.

Many drugs control one or more of the eicosanoids' physiological effects. Nonsteroidal anti-inflammatory drugs (NSAIDs) block the oxidation of arachidonic acid to form prostaglandins and thromboxanes. These drugs include the common pain relievers aspirin, ibuprofen [e.g., Advil®; Motrin®, Nuprin®, naproxen (e.g., Aleve®), and ketoprofen (e.g., Orudix KT®, Actron®)]. Such

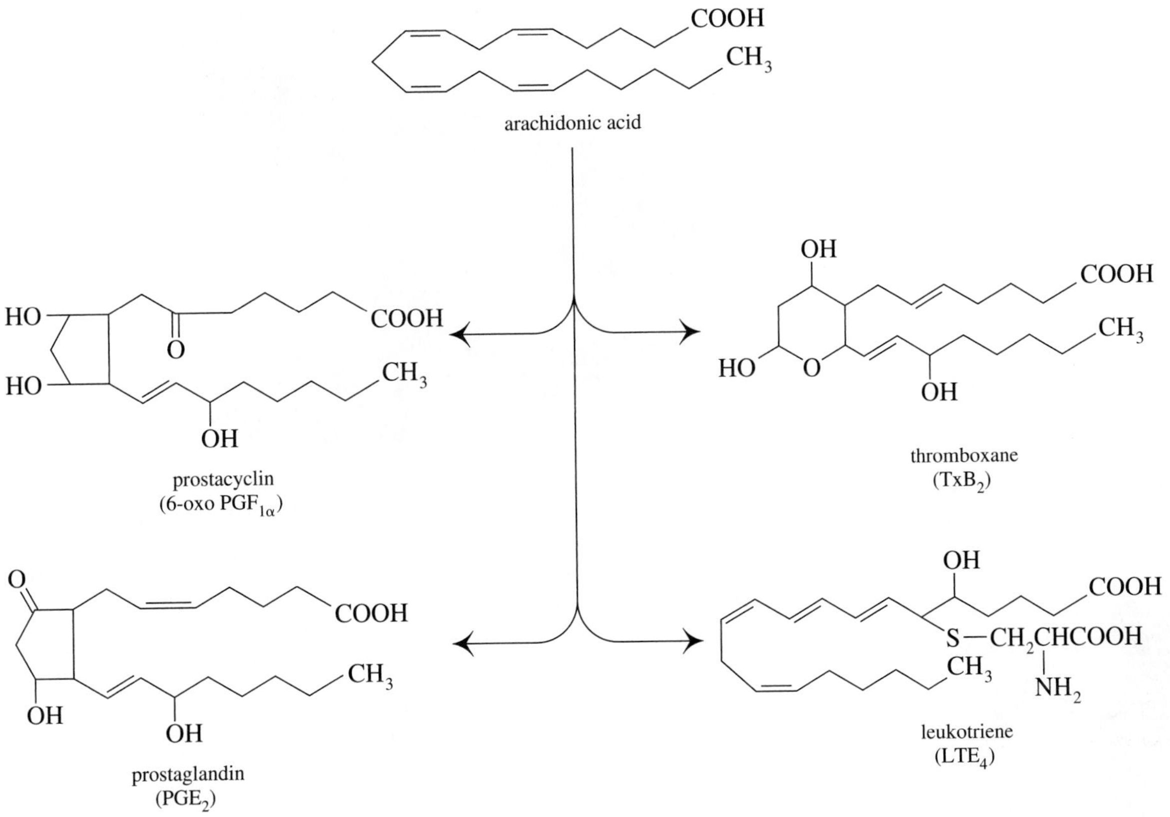

Figure 28.2
Several examples of eicosanoids. Each of these molecules is derived from arachidonic acid.

drugs are effective at controlling pain, fever, and inflammation. Unfortunately, they also cause acid damage to the stomach. About 20% of patients on long-term NSAID treatment (e.g., arthritis sufferers) develop stomach ulcers. Thus, a second group of drugs, the COX-2 inhibitors, has been designed to inhibit prostaglandin syntheses for pain control, while only minimally affecting the stomach. A third group of drugs specifically controls leukotriene synthesis, which blocks one of the triggers responsible for asthma attacks. Aspirin has been shown to prevent heart attacks and strokes, probably by blocking thromboxane synthesis. Cortisone blocks the release of arachidonic acid and therefore stops inflammation. The number of drugs devoted to eicosanoid control emphasizes the importance of these "local hormones" to life.

Fats and Oils

Chemically, fats and oils are esters of glycerol and the higher molar-mass fatty acids. They have the following general formula:

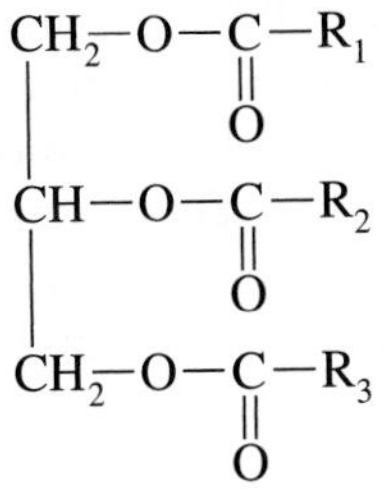

$$\begin{array}{l} CH_2-O-\overset{\displaystyle O}{\overset{\|}{C}}-R_1 \\ \,| \\ CH-O-\overset{\displaystyle O}{\overset{\|}{C}}-R_2 \\ \,| \\ CH_2-O-\overset{\displaystyle O}{\overset{\|}{C}}-R_3 \end{array}$$

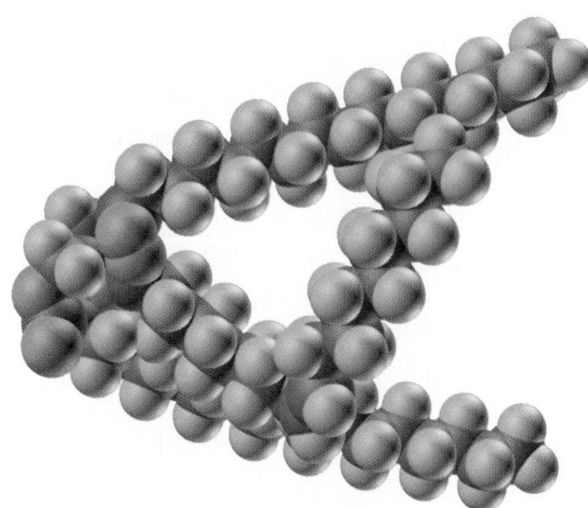

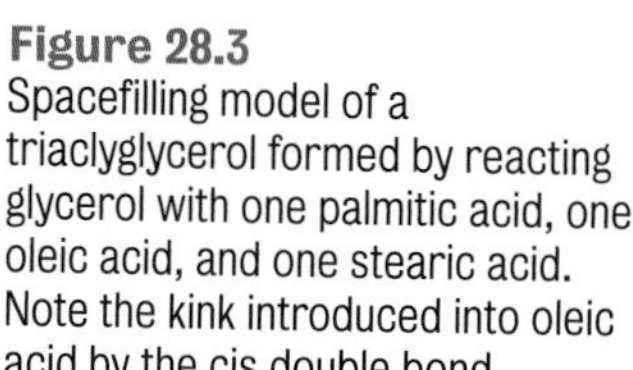

Figure 28.3
Spacefilling model of a triaclyglycerol formed by reacting glycerol with one palmitic acid, one oleic acid, and one stearic acid. Note the kink introduced into oleic acid by the cis double bond.

Cultures that use fish as a dietary staple have a low level of heart disease.

The "R"'s can be either long-chain saturated or unsaturated hydrocarbon groups. Figure 28.3 is a spacefilling model of a typical fat.

Fats may be considered to be triesters formed from the trihydroxy alcohol glycerol and three molecules of fatty acids:

$$\begin{array}{lcl}
CH_2{-}O{-}\boxed{H + H{-}O}{-}\underset{\displaystyle O}{\underset{\|}{C}}{-}R_1 & & CH_2{-}O{-}\underset{\displaystyle O}{\underset{\|}{C}}{-}R_1 \\
| & & | \\
CH{-}O{-}\boxed{H + H{-}O}{-}\underset{\displaystyle O}{\underset{\|}{C}}{-}R_2 & \longrightarrow & CH{-}O{-}\underset{\displaystyle O}{\underset{\|}{C}}{-}R_2 + 3\,H_2O \\
| & & | \\
CH_2{-}O{-}\boxed{H + H{-}O}{-}\underset{\displaystyle O}{\underset{\|}{C}}{-}R_3 & & CH_2{-}O{-}\underset{\displaystyle O}{\underset{\|}{C}}{-}R_3 \\
\text{glycerol} \quad\quad \text{fatty acids} & & \text{a triacylglycerol (triglyceride)}
\end{array}$$

Most of the fatty acids in these esters have 14–18 carbons. Because there are three ester groups per glycerol, the molecules are called triacylglycerols, or triglycerides (an older name that is still commonly used). The three R groups are usually different.

Practice 28.2

Write the balanced equation (showing structures of reactants and products) for the reaction between a molecule of glycerol, a molecule of stearic acid, and two palmitic acid molecules to form a triacylglycerol.

Fats and oils fit the general description of a lipid. They are large molecules, averaging more than 50 carbon atoms per molecule and have many nonpolar, uncharged groups. Triacylglycerols are insoluble in water.

Nutritionists recommend that less than 30% of the daily dietary caloric intake be derived from fats.

Fats are an important food source for humans and normally account for about 25–50% of our caloric intake. When oxidized to carbon dioxide and water, fats supply about 40 kJ of energy per gram (9.5 kcal/g), which is more than twice the amount obtained from carbohydrates or proteins. This is partly because triacylglycerol carbons are more reduced than those of most other foods. As shown in the following structures, a typical carbohydrate carbon has one bond to oxygen while a typical fat carbon has no bonds to oxygen.

CHEMISTRY IN ACTION • When Is a Balanced Diet Balanced? A Case for Omega-3 Fatty Acids

We are told by nutritionists that having the right amount of each nutrient in a diet leads to a balance. Usually this means the dietary intake balances the metabolic need. However, sometimes it is relative proportion of nutrients that must be in balance. This seems to be the case for the essential fatty acids.

The essential fatty acids include linoleic acid, linolenic acid, and, sometimes, arachidonic acid. Each of these fatty acids is polyunsaturated. They are essential in metabolism to make prostaglandins, leukotrienes, thromboxanes, and prostacyclins—the human body's "local hormone" set; they are essential in diets because human metabolism is incapable of synthesizing the double bonds in the last half of the molecules (at carbons beyond C-10).

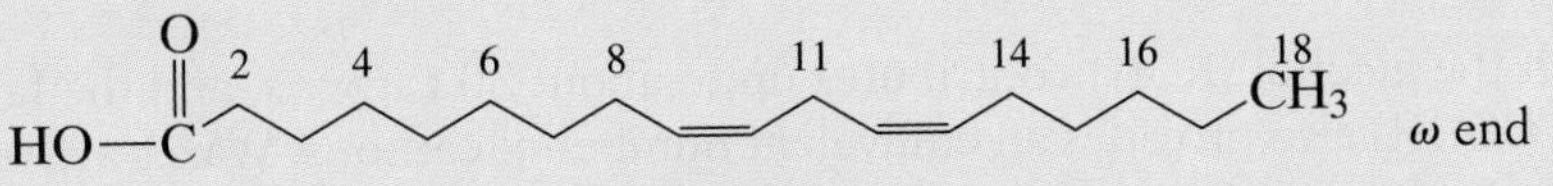

linoleic acid

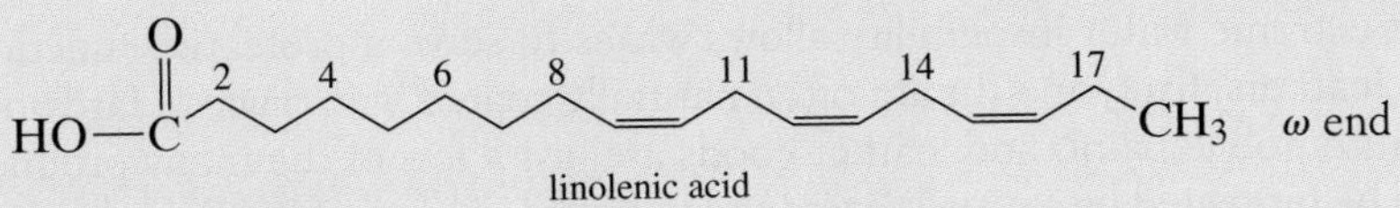

linolenic acid

To emphasize the importance of the last half of the fatty acid structure, the methyl group is given a special designation—the omega (ω) end (using the last letter in the Greek alphabet). The terminal double bond is numbered from the omega end to highlight its position in the last half of the fatty acid chain. Thus, linoleic acid is an omega-6 acid while linolenic acid is an omega-3 acid.

Essential fatty acid consumption has changed along with the typical Western diet over the last century. One hundred years ago most of our essential fatty acids were obtained from meat and vegetables; we consumed about equal amounts of omega-6 and omega-3 fatty acids. To decrease saturated fat intake, our diet now includes more plant oils. Since most plant oils are higher in omega-6 than omega-3 fatty acids, a side effect of this diet is an increase in the proportion of omega-6 to omega-3 fatty acids; currently the ratio is about 16:1.

Much biochemical evidence shows that when this ratio changes, so does health. Although all essential fatty acids allow a "local hormone" set to be made, not all sets have the same impact. Omega-3 fatty acids tend to give a "local hormone" set that is anti-inflammatory, antiarrhythmic, and antithrombotic. Omega-6 fatty acids yield a "local hormone" set with opposite effects. Thus, biochemical studies support increased omega-3 fatty acid consumption to lower the risk of heart arrhythmias, cardiovascular disease, and hypertension. Some studies also show benefits for treatment of rheumatoid arthritis and cancer.

So, should we eat foods that contain a high level of omega-3 fatty acids (oily fish such as salmon, mackerel, sardines and herring; plant oils such as canola, flaxseed, soybean, and walnut)?

Epidemiological studies give mixed results, as might be expected for such a complex situation; some studies show a definite benefit from a high omega-3 diet and some show no benefit. In general, nutritionists recommend a conservative change in our diet: the Canadian government has published recommendations that the omega-6:omega-3 ratio be decreased to 6:1, while the American Heart Association recommends switching to high omega-3 plant oils and having two oily fish meals per week.

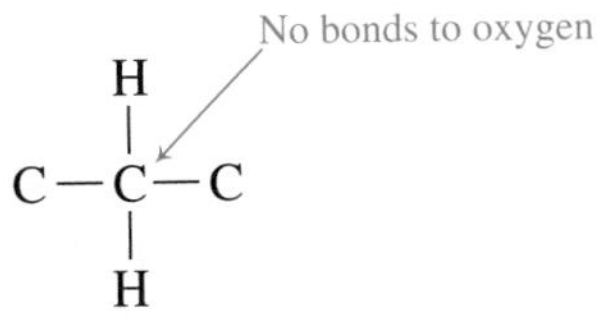

No bonds to oxygen

in a fat

One bond to oxygen

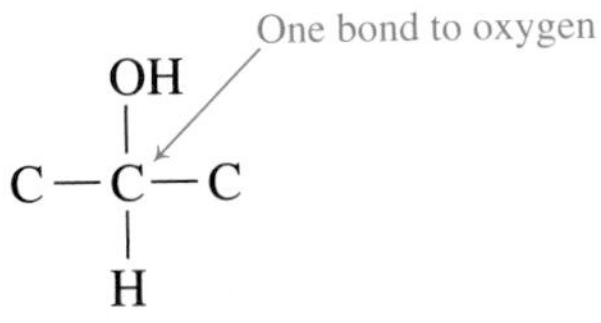

in a carbohydrate

In addition, the average fat contains about 75% carbon by mass, whereas the average carbohydrate contains only about 40% carbon. Fats are indeed a rich source of biochemical energy.

In a calorie-conscious society, fatty foods can be "too much of a good thing." Thus, scientists have developed fat substitutes. In general, fat substitutes need to give a creamy texture and taste to foods without the energy content of fats. A cellulose-based substitute was introduced in the 1960s. Since that time, a form of processed protein (microparticulated) with a creamy texture has also been used as a fat substitute. Both carbohydrate- and protein-based fat substitutes are naturally lower in calories than fats. Yet, neither cooks like a fat. More recently, a sucrose derivative, olestra, has been used as a fat substitute. Olestra is a polyester of long-chain fatty acids and sucrose. It tastes and cooks like a fat. But, because the fatty acids are esterified to sucrose and not to glycerol, olestra is indigestable and passes through the digestive tract without contributing calories to the diet. Olestra has been approved for use in snack foods; however medical studies continue on potential long-term impacts of this fat substitute.

Waxes

wax

Waxes are esters of high-molar-mass fatty acids and high-molar-mass alcohols. They have the general formula

The name, wax, derives from the old English word, "weax," which means "material from a honeycomb."

$$R'-\overset{\overset{\displaystyle O}{\|}}{C}-O-R$$

in which the alcohol (ROH) contributes up to about 30 carbons, and the fatty acid (R′COOH) also provides an equivalent number of carbons. Waxes are very large molecules with almost no polar groups. They represent one of the most hydrophobic classes of lipids.

Their extreme water insolubility allows waxes to serve a protective function. Leaves, feathers, fruit, and fur are often naturally coated with a wax. Hardwood floors, automobile paint, and leather goods are just a few of the many products that can be protected by a wax. Waxes tend to be the hardest of the lipids because their carbon chains are long and have very few double bonds. As with fats and oils, the size of the wax molecule and the number of double bonds contained in its carbon chains determine whether the wax will be a liquid or a solid.

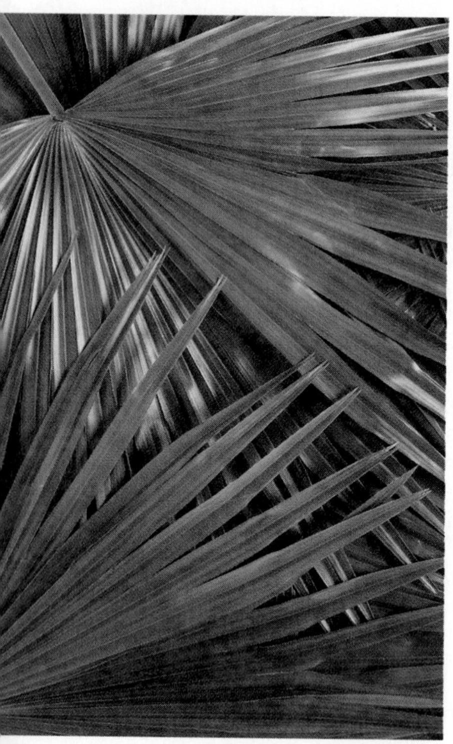

The "shine" on these leaves is due to a thick, protective wax coating.

Practice 28.3

Write the formula for a wax formed from stearic acid and 1-octacosanol [$CH_3(CH_2)_{26}CH_2OH$].

28.4 Compound Lipids

Phospholipids

phospholipid

The **phospholipids** are a group of compounds that yield one or more fatty acid molecules, a phosphate group, and usually a nitrogenous base upon hydrolysis. The phosphate group and the nitrogenous base, which are found at one end of the phospholipid molecule, often have negative and positive charges. Consequently, in contrast to the triacylglycerols, phospholipids have a hydrophilic end that interacts with water:

$$\left.\begin{array}{l} CH_2-O-\boxed{\text{Fatty acid}} \\ | \\ CH-O-\boxed{\text{Fatty acid}} \\ | \\ CH_2-O-\boxed{\text{Fatty acid}} \end{array}\right\} \text{All hydrophobic}$$

a triacylglycerol

$$\begin{array}{l} \left.\begin{array}{l} CH_2-O-\boxed{\text{Fatty acid}} \\ | \\ CH-O-\boxed{\text{Fatty acid}} \end{array}\right\} \text{Hydrophobic} \\ \quad | \\ \left. CH_2-O-\boxed{\text{Phosphate} + \text{Nitrogen base}} \right\} \text{Hydrophilic} \end{array}$$

a phospholipid

As will be seen later in this chapter, a lipid with both hydrophobic and hydrophilic character is needed to make membranes. It is not surprising that phospholipids are one of the most important membrane components.

Phospholipids are also involved in the metabolism of other lipids and nonlipids. Although they are produced to some extent by almost all cells, most of the phospholipids that enter the bloodstream are formed in the liver. Descriptions of representative phospholipids follow.

Phosphatidic Acids Phosphatidic acids are glyceryl esters of fatty acids and phosphoric acid. The phosphatidic acids are important intermediates in the synthesis of triacylglycerols and other phospholipids. Their structure is as follows:

$$\begin{array}{l} \left.\begin{array}{l} CH_2-O-\overset{\displaystyle O}{\overset{\|}{C}}-R_1 \\ | \\ CH-O-\overset{\displaystyle O}{\overset{\|}{C}}-R_2 \end{array}\right\} \text{Hydrophobic} \\ \quad | \\ \left. CH_2-O-\underset{\displaystyle O^-}{\underset{|}{\overset{\displaystyle O}{\overset{\|}{P}}}}-O^- \right\} \text{Hydrophilic} \end{array}$$

a phosphatidic acid

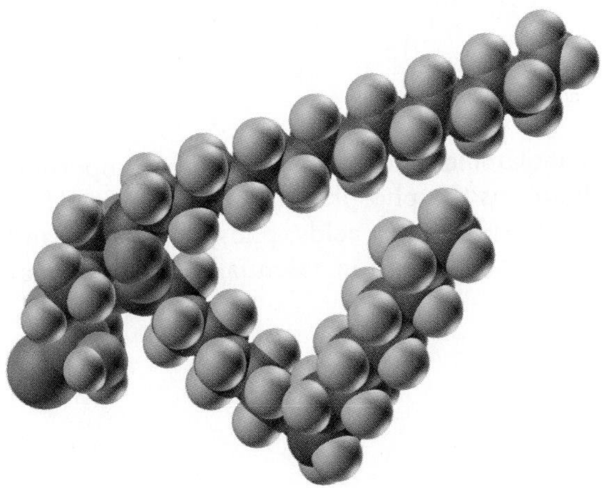

Figure 28.4
Spacefilling model of a phosphatidic acid that is esterified to palmitic acid at the top glycerol carbon, to oleic acid at the middle carbon, and to a phosphate group at the bottom glycerol carbon. Note the kink introduced into the oleic acid by the cis double bond.

Figure 28.4 is a spacefilling model of a typical phosphatidic acid. As with all common phospholipids, the fatty acid chains are large relative to the rest of this molecule. Other phospholipids are formed from a phosphatidic acid when specific nitrogen-containing compounds are linked to the phosphate group by an ester bond. Three commonly used nitrogen compounds are choline, ethanolamine, and L-serine:

$$\underset{\text{choline}}{HOCH_2CH_2\overset{CH_3}{\overset{|}{N^+}}\underset{\underset{CH_3}{|}}{}{-}CH_3} \qquad \underset{\text{ethanolamine}}{HOCH_2CH_2\overset{+}{N}H_3} \qquad \underset{\text{L-serine}}{HOCH_2\underset{\underset{\overset{+}{N}H_3}{|}}{C}HCOO^-}$$

Because other phospholipids are structurally related to phosphatidic acids, their names are also closely related.

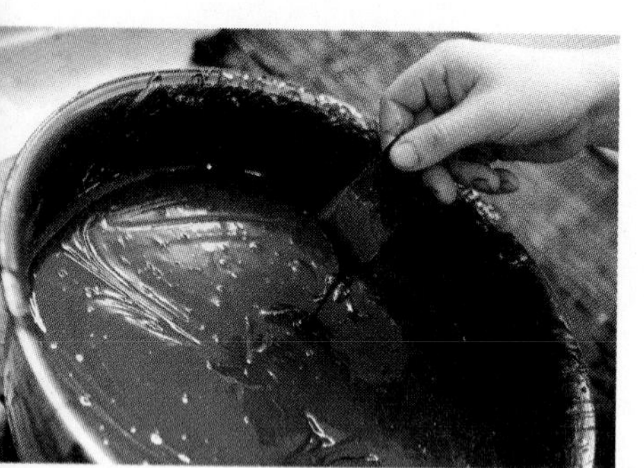

The creamy texture of this chocolate is partly due to phosphatidyl choline.

Phosphatidyl Cholines (Lecithins) Phosphatidyl cholines (lecithins) are glyceryl esters of fatty acids, phosphoric acid, and choline. The synonym *lecithin* is an older term that is still used, particularly in commercial products that contain phosphatidyl choline. Phosphatidyl cholines are synthesized in the liver and are present in considerable amounts in nerve tissue and brain substance. Most commercial phosphatidyl choline is obtained from soybean oil and contains palmitic, stearic, palmitoleic, oleic, linoleic, linolenic, and arachidonic acids. Phosphatidyl choline is an edible and digestible emulsifying agent used extensively in the food industry. For example, chocolate and margarine are generally emulsified with phosphatidyl choline. Phosphatidyl choline is also used as an emulsifier in many pharmaceutical preparations.

The single most important biological function for phosphatidyl choline is as a membrane component. This phospholipid makes up between 10 and 20% of many membranes. Its structure is as follows:

$$\begin{array}{l} CH_2{-}O{-}\underset{\underset{O}{\|}}{C}{-}R_1 \\ | \\ CH{-}O{-}\underset{\underset{O}{\|}}{C}{-}R_2 \\ | \\ CH_2{-}O{-}\overset{\overset{O}{\|}}{\underset{\underset{O^-}{|}}{P}}{-}\overbrace{O{-}CH_2CH_2\overset{\overset{CH_3}{|}}{\underset{\underset{CH_3}{|}}{N^+}}{-}CH_3}^{\text{choline}} \end{array} \quad \begin{array}{l} \left.\right\} \text{Hydrophobic} \\ \\ \left.\right\} \text{Hydrophilic} \end{array}$$

a phosphatidyl choline (lecithin) molecule

Phosphatidyl Ethanolamines (Cephalins) Another important constituent of biological membranes is the phosphatidyl ethanolamines (cephalins). These lipids are glyceryl esters of fatty acids, phosphoric acid, and ethanolamine ($HOCH_2CH_2NH_2$). They are found in essentially all living organisms, with the following structure:

$$\begin{array}{l} CH_2{-}O{-}\underset{\underset{O}{\|}}{C}{-}R_1 \\ | \\ CH{-}O{-}\underset{\underset{O}{\|}}{C}{-}R_2 \\ | \\ CH_2{-}O{-}\overset{\overset{O}{\|}}{\underset{\underset{O^-}{|}}{P}}{-}\overbrace{OCH_2CH_2NH_3^+}^{\text{ethanolamine}} \end{array} \quad \begin{array}{l} \left.\right\} \text{Hydrophobic} \\ \\ \left.\right\} \text{Hydrophilic} \end{array}$$

a phosphatidyl ethanolamine (cephalin) molecule

Example 28.1

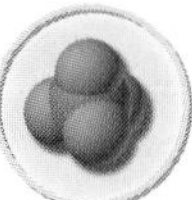

Write the formula for a phosphatidyl ethanolamine that contains two palmitic acid groups.

SOLUTION

Phosphatidyl ethanolamine is a phospholipid and so contains glycerol, phosphate, fatty acids, and a nitrogen base. First, write the structure for glycerol. Then add two palmitic acid units, which are linked by ester bonds to the top two carbons of glycerol:

$$\begin{array}{l} CH_2OH \\ | \\ CHOH \\ | \\ CH_2OH \end{array} \qquad \begin{array}{l} CH_2{-}O{-}\overset{\displaystyle O}{\overset{\|}{C}}(CH_2)_{14}CH_3 \\ | \\ CH{-}O{-}\overset{\displaystyle O}{\overset{\|}{C}}(CH_2)_{14}CH_3 \\ | \\ CH_2OH \end{array}$$

Next, connect a phosphate group via an ester bond to the bottom glycerol carbon in order to form a phosphatidic acid. Finally, link the ethanolamine to the phosphate group to yield phosphatidyl ethanolamine:

$$\begin{array}{l} CH_2{-}O{-}\overset{\displaystyle O}{\overset{\|}{C}}(CH_2)_{14}CH_3 \\ | \\ CH{-}O{-}\overset{\displaystyle O}{\overset{\|}{C}}(CH_2)_{14}CH_3 \\ | \\ CH_2{-}O{-}\overset{\displaystyle O}{\overset{\|}{P}}{-}O^- \\ \qquad\quad | \\ \qquad\quad O^- \end{array} \qquad \begin{array}{l} CH_2{-}O{-}\overset{\displaystyle O}{\overset{\|}{C}}(CH_2)_{14}CH_3 \\ | \\ CH{-}O{-}\overset{\displaystyle O}{\overset{\|}{C}}(CH_2)_{14}CH_3 \\ | \\ CH_2{-}O{-}\overset{\displaystyle O}{\overset{\|}{P}}{-}OCH_2CH_2\overset{+}{N}H_3 \\ \qquad\quad | \\ \qquad\quad O^- \end{array}$$

Practice 28.4

Write the structure of a phosphatidyl choline that contains palmitic acid and stearic acid. How many structures are possible for this compound?

Sphingolipids

sphingolipid

Sphingolipids are compounds that, when hydrolyzed, yield a hydrophilic group (either phosphate and choline or a carbohydrate), a long-chain fatty acid (18–26 carbons), and sphingosine (an unsaturated amino alcohol). When drawn as follows, the similarities between sphingosine and glycerol esterified to one fatty acid become recognizable:

$$\begin{array}{l} OH \\ | \\ CH{-}CH{=}CH(CH_2)_{12}CH_3 \\ | \\ CH{-}\overset{*}{N}H_2 \\ | \\ CH_2{-}\overset{*}{O}H \end{array} \qquad \begin{array}{l} CH_2{-}O{-}\overset{\displaystyle O}{\overset{\|}{C}}(CH_2)_nCH_3 \\ | \\ CH{-}\overset{*}{O}H \\ | \\ CH_2{-}\overset{*}{O}H \end{array}$$

sphingosine — glycerol esterified with one fatty acid

The starred atoms on sphingosine react further to make sphingolipids, just as the starred atoms on the glycerol compound react further to give triacylglycerols or phospholipids.

Sphingolipids are common membrane components because they have both hydrophobic and hydrophilic character. For example, sphingomyelins are found in the myelin sheath membranes that surround nerves:

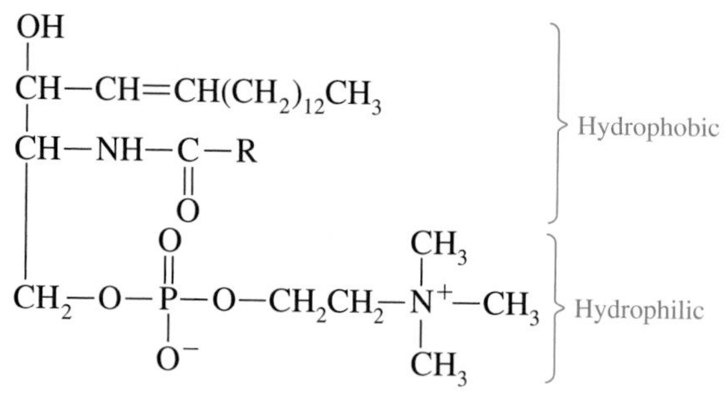

```
OH
|
CH—CH=CH(CH2)12CH3                                  } Hydrophobic
|
CH—NH—C—R
|      ||
|      O
|         O                 CH3
|         ||                |
CH2—O—P—O—CH2CH2—N+—CH3     } Hydrophilic
          |                 |
          O−                CH3
```

a sphingomyelin

Notice the hydrophobic and hydrophilic parts of this molecule. Sphingomyelins can also be classified as phospholipids.

> **Practice 28.5**
>
> Write the structure of a sphingomyelin that contains stearic acid.

Glycolipids

glycolipid Sphingolipids that contain a carbohydrate group are also known as **glycolipids**. The two most important classes of glycolipids are cerebrosides and gangliosides. These substances are found mainly in cell membranes of nerve and brain tissue. A cerebroside may contain either D-galactose or D-glucose. The following formula of a galactocerebroside shows the typical structure of cerebrosides:

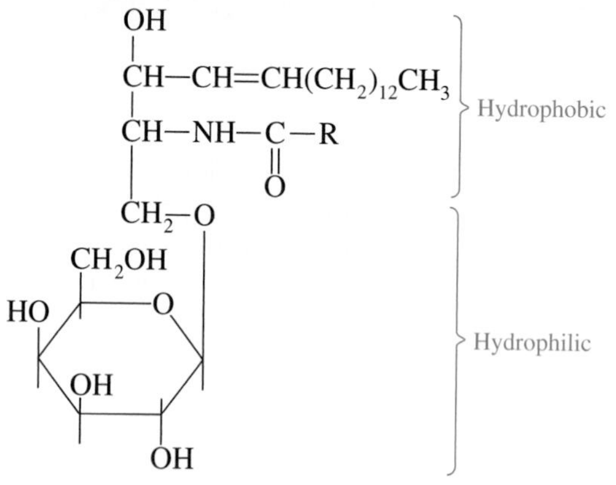

```
OH
|
CH—CH=CH(CH2)12CH3                    } Hydrophobic
|
CH—NH—C—R
|      ||
|      O
CH2—O
CH2OH
HO                                    } Hydrophilic
OH
OH
```

a β-galactocerebroside

Gangliosides resemble cerebrosides in structure, but contain complex oligosaccharides instead of simple monosaccharides.

28.5 Steroids

steroids

Steroids are compounds that have the steroid nucleus, which consists of four fused carbocyclic rings. This nucleus contains 17 carbon atoms in one five-membered and three six-membered rings:

steroid ring nucleus

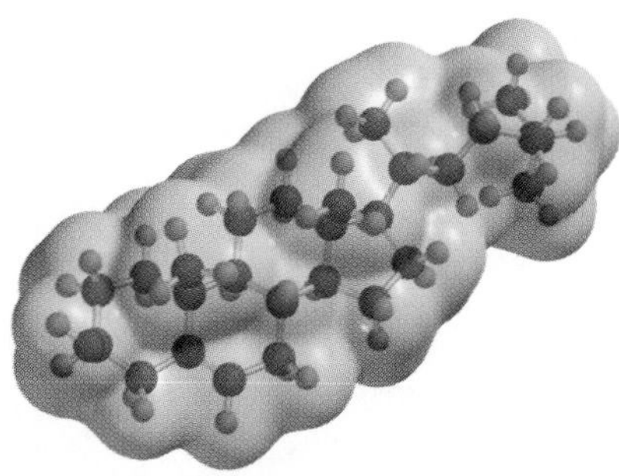

Spacefilling model of cholesterol. Compare with the bond-line drawing of cholesterol in Figure 28.5.

Modifications of this nucleus in the various steroid compounds include, for example, added side chains, hydroxyl groups, carbonyl groups, and ring double bonds.

Steroids are closely related in structure, but are highly diverse in function. Examples of steroids and steroid-containing materials are (1) cholesterol, the most abundant steroid in the body, which is widely distributed in all cells and serves as a major membrane component; (2) bile salts, which aid in the digestion of fats; (3) ergosterol, a yeast steroid, which is converted to vitamin D by ultraviolet radiation; (4) digitalis and related substances called cardiac glycosides, which are potent heart drugs; (5) the adrenal cortex hormones, which are involved in metabolism; and (6) the sex hormones, which control sexual characteristics and reproduction. The formulas for several steroids are given in Figure 28.5.

Cholesterol is the parent compound from which the steroid hormones are synthesized. As we will see, small changes in steroid structure can lead to large changes in hormonal action. Cholesterol is first converted to progesterone, a compound that helps control the menstrual cycle and pregnancy. This hormone is, in turn, the parent compound from which testosterone and the adrenal corticosteroids are produced. Notice that the long side chain on carbon 17 in cholesterol (Figure 28.5) is smaller in progesterone and is eliminated when testosterone is formed. Interestingly, testosterone is the precursor for the female sex hormones such as estradiol. These sex hormones are produced by the gonads—that is, either the male testes or the female ovaries. The small structural differences between testosterone and estradiol trigger vastly different physiological responses.

Practice 28.6

Identify the functional groups in (a) progesterone, (b) cholic acid, and (c) estradiol.

Cholesterol is also used to build cell membranes, many of which contain about 25% by mass of this steroid. In fact, often there is as much cholesterol as there is phospholipid, sphingolipid, or glycolipid. These latter three lipid classes cause the membrane to be more oily; cholesterol solidifies the membrane.

This different behavior arises from an important structural difference. Cholesterol's four fused rings make it a rigid molecule, while other membrane lipids are more flexible. When biological membranes are synthesized, their cholesterol level is adjusted to achieve an appropriate balance between a solid and a liquid consistency.

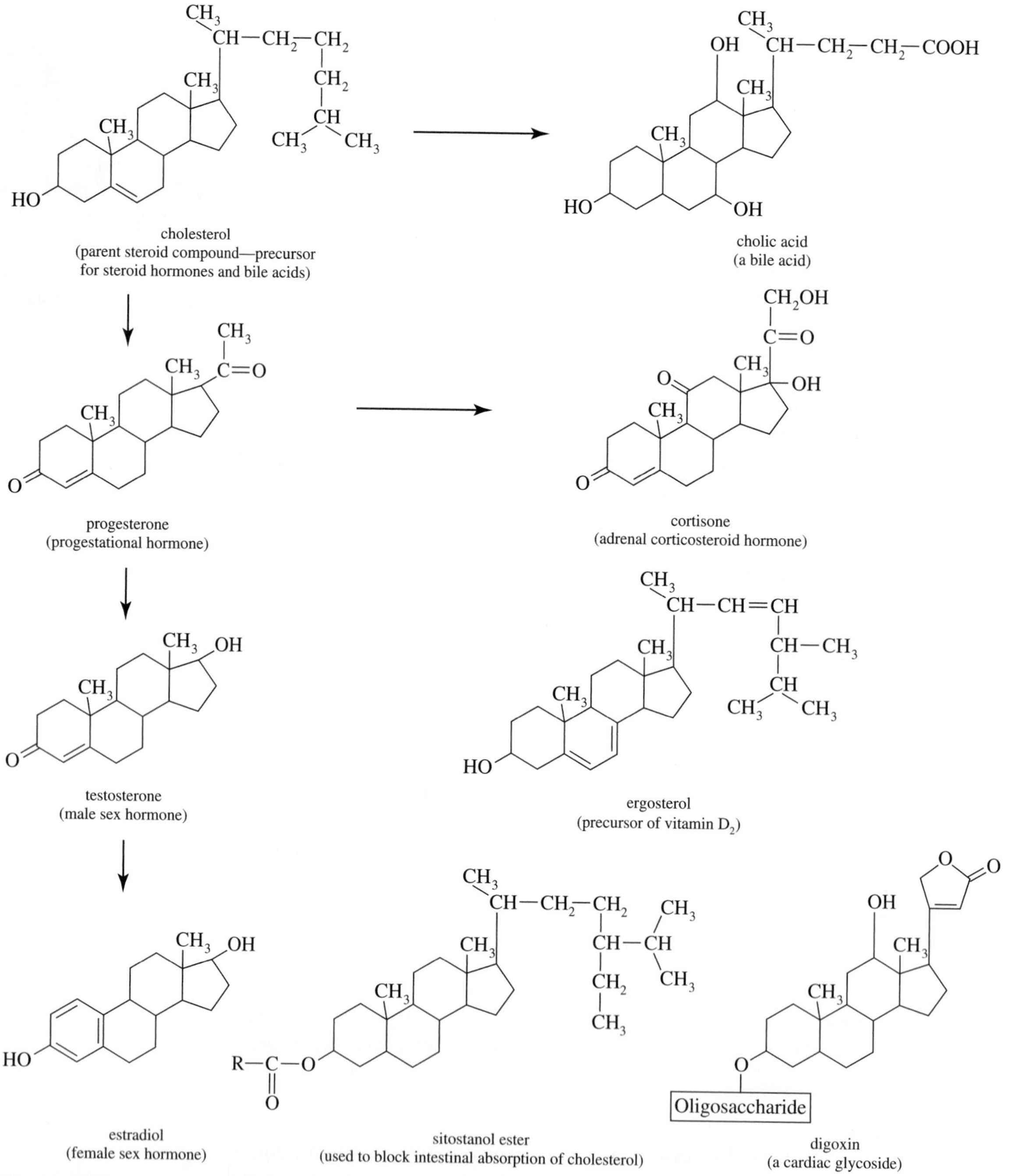

cholesterol
(parent steroid compound—precursor for steroid hormones and bile acids)

cholic acid
(a bile acid)

progesterone
(progestational hormone)

cortisone
(adrenal corticosteroid hormone)

testosterone
(male sex hormone)

ergosterol
(precursor of vitamin D_2)

estradiol
(female sex hormone)

sitostanol ester
(used to block intestinal absorption of cholesterol)

Oligosaccharide

digoxin
(a cardiac glycoside)

Figure 28.5
Structures of selected steroids. Arrows show the biosynthetic relationships between steroids derived from cholesterol. Ergosterol and digoxin are from plant sources.

28.6 Hydrophobic Lipids and Biology

Olive oil and vinegar. Olive oil is a hydrophobic lipid. If allowed to stand, the olive oil molecules will separate from the vinegar.

The hydrophobic nature of lipids has many important biological consequences. The water insolubility of lipids results in (1) lipid aggregation that causes atherosclerosis, and (2) lipid aggregation that forms biological membranes. When a lipid is surrounded by water, it is in a hostile environment. The lipid molecules aggregate to minimize their contact with water. This explains why olive oil (a mixture of triacylglycerols) separates from vinegar if it is allowed to stand. This same process of separation occurs continuously in biological solutions.

Let's look at this process on a molecular level. The hydrophilic part of lipid molecules is attracted to water and forms an interface with it, but the hydrophobic part distances itself from water molecules. Depending on the general shape of the lipid molecules, different-shaped aggregates form. Smaller lipids like fatty acids will come together to make micelles. If the fatty acid is shown as

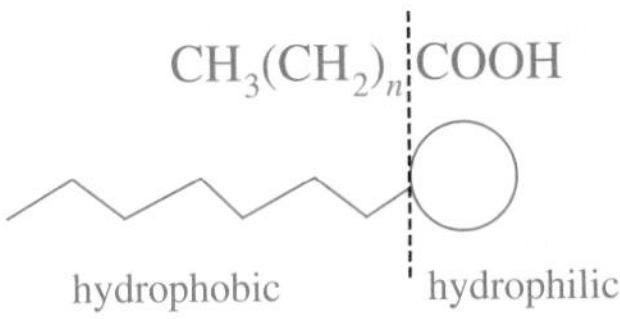

then a micelle can be visualized as in Figure 28.6. Notice that the hydrophilic carboxyl groups coat the aggregate and protect the hydrophobic alkyl groups from water.

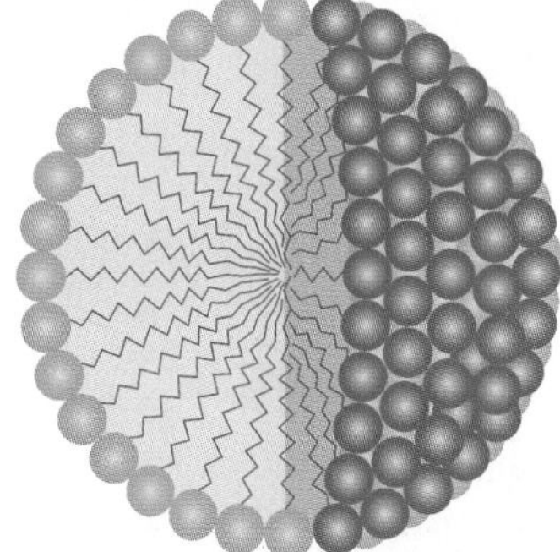

Figure 28.6
Cutaway of a micelle. Note that the hydrophobic chains extend to the center of this aggregate.

More complex lipids such as phospholipids and sphingolipids are shaped differently than fatty acids and will aggregate differently in water solutions. These lipids have two hydrophobic alkyl groups and can be represented as follows:

Having two hydrophobic chains, these lipids are not as thin as fatty acids. They don't aggregate to form micelles in aqueous solution, but rather form liposomes. We see in Figure 28.7 that a simple liposome is bounded by two layers of lipid. The hydrophobic alkyl chains are covered by hydrophilic groups on both the liposome's inside and outside. Thus, unlike micelles, liposomes have a water core.

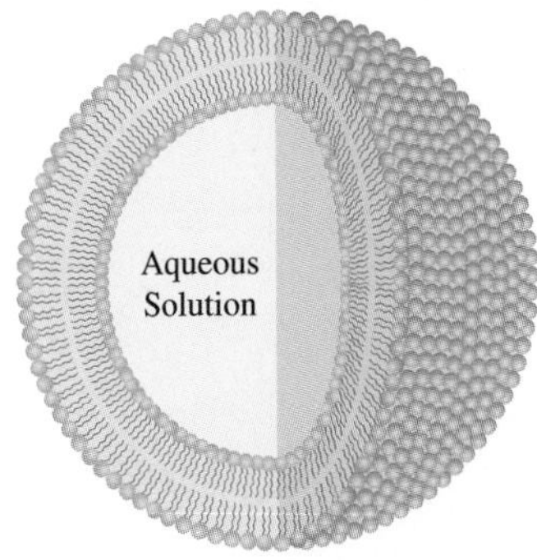

Figure 28.7
A cutaway of a single liposome. Note the bilayer shell and the water solution core.

Pharmaceutical companies are actively developing liposomes for use in drug delivery systems that consist of specially prepared liposomes carrying drug molecules inside. If the liposome surface is modified appropriately, then the drug will be absorbed only by a specific organ or tissue. The simplest modification is to place positive charges on the liposomes' exterior; since most cells have a negatively-charged cell membrane, the liposomes are attracted to, and merge with, the cells.

Today's medicines use liposomes to package especially toxic drugs for serious chemotherapies. Deadly systemic fungal infections can be treated with an antifungal (amphotericin B) wrapped in a liposome coat. Cancer chemotherapy drugs such as daunorubicin are also infused within a liposome package. Currently, researchers are examining ways of using liposomes to transfer DNA in order to correct deadly genetic diseases such as cystic fibrosis.

Atherosclerosis

Atherosclerosis is a metabolic disease that leads to deposits of cholesterol and other lipids on the inner walls of the arteries. The name *athere* is Greek for "mush" and nicely describes the appearance of these fatty deposits (called *plaque*). As plaque accumulates, the arterial passages become progressively narrower. The walls of the arteries also lose their elasticity and their ability to expand to accommodate the volume of blood pumped by the heart. Blood pressure increases as the heart works harder to pump sufficient blood through the narrowed passages, which may eventually lead to a heart attack. The accumulation of plaque also causes the inner walls to have a rough rather than a normal smooth surface, which is a condition that may lead to coronary thrombosis (heart attack due to blood clots).

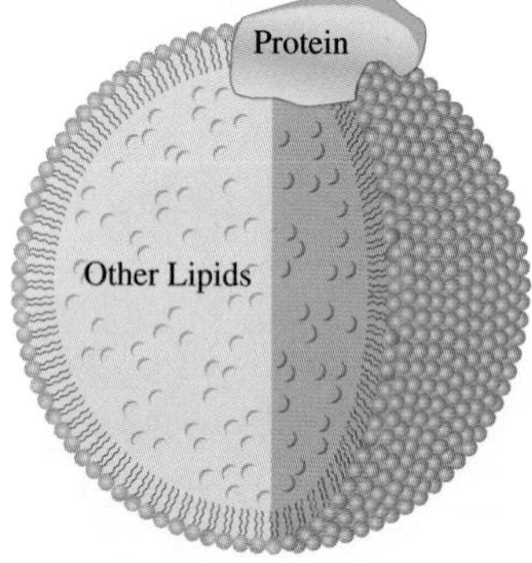

Figure 28.8
A cutaway of VLDL. Note the single-layer shell and the lipid core.

Plaque formation begins because of a lipid's natural tendency to aggregate. In fact, atherosclerosis might be described as a disease partly caused by unwanted lipid aggregation. Scientists trace plaque formation as follows: (1) Lipids (including cholesterol) are trapped in the arterial wall and become oxidized; (2) white cells (macrophages) scavenge these oxidized lipids and become bloated with excess cholesterol, forming "foam cells"; (3) foam cells collect on the arterial lining together with cell debris, narrowing and hardening the blood vessel and thereby producing atherosclerosis.

Improper transport of cholesterol through the blood contributes to atherosclerosis. Cholesterol (and other lipids) must be packaged for transport because lipids aggregate in the aqueous bloodstream. The liver packages dietary lipid into aggregates known as *very-low-density lipoproteins* (VLDL). A cutaway of a VLDL is shown in Figure 28.8. The VLDL surface is hydrophilic and contains a single layer of lipids (and some protein), while the interior provides space for the more hydrophobic triacylglycerols and cholesterol. The VLDLs travel through the bloodstream delivering triacylglycerols to fat cells (the adipose tissue). (See Figure 28.9.) The VLDL packages are converted into smaller

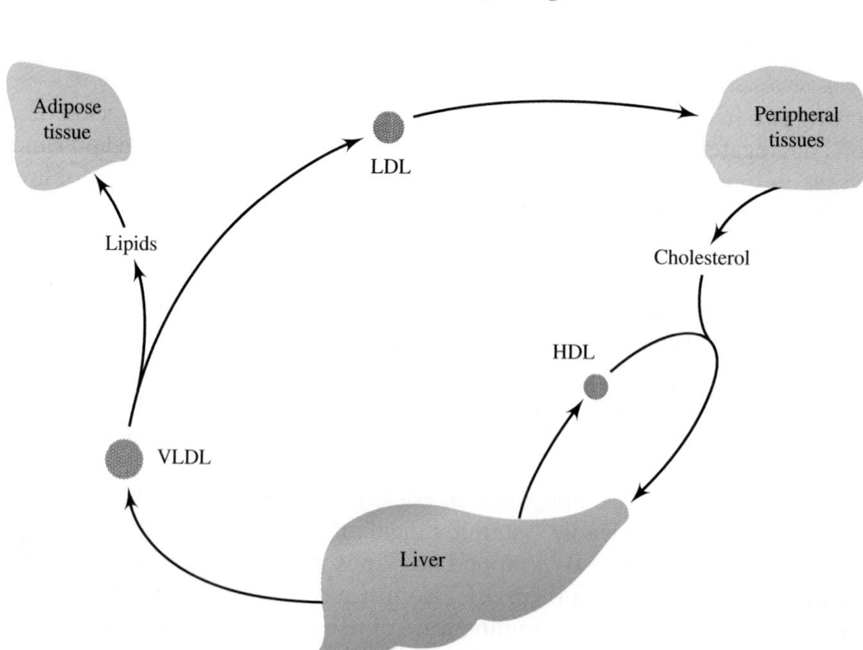

Figure 28.9
The lipid distribution system through the bloodstream.

low-density lipoproteins (LDL), which deliver cholesterol to peripheral tissues. Proper circulation of cholesterol also depends on a third lipoprotein, the *high-density lipoprotein* (HDL). The HDL acts as a cholesterol scavenger by collecting cholesterol and returning it to the liver, essentially opposing the action of LDL.

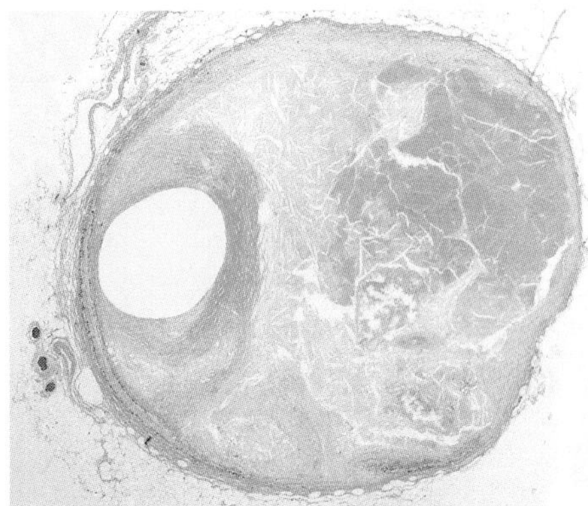
This clogged artery shows plaque formation on the artery wall.

People with high-plasma LDL concentrations are prone to atherosclerosis, even though they may be on low-cholesterol diets. In contrast, large amounts of plasma HDL appear to prevent plaque formation, and HDL levels can be increased by strenuous exercise, weight loss, and administration of estradiol.

However, the most common drug treatments for preventing atherosclerosis still involve the reduction of serum cholesterol levels by (1) decreasing triglyceride production (often associated with increasing HDL levels); (2) inhibiting metabolic synthesis of cholesterol; or (3) increasing excretion of the cholesterol derivatives, the bile acids (or a combination of all three steps). Note that each of these drug treatments assumes a low-cholesterol diet, and so the treatment focuses on lowering the body's cholesterol production and storage. The fibrate drugs (e.g., Lopid® and Tricor®) directly lower triglyceride levels and may increase HDL. The statins (e.g., Mevacor,® Lipitor,® and Zocor®) block metabolic production of cholesterol. These drugs are especially effective at lowering serum cholesterol, since up to 80% of the total body cholesterol is metabolically produced. Resin drugs (e.g., Colestid® and Questran®) bind the bile acids so they will be excreted. This can lead to a net decrease in body cholesterol levels because bile acids are synthesized from cholesterol. Much the same effect is caused by soluble dietary fiber. Thus, diets with good sources of soluble fiber (e.g., oat and barley meal) can significantly decrease serum cholesterol.

Another route to lower serum cholesterol is to decrease intestinal absorption of cholesterol. Certain plant sterols have such an effect. Recently, scientists have modified plant sterols and included them in special margarines. Sitostanol ester (see Figure 28.5) is included in Benecol®, while sitosterol is included in Take Control®. When these margarines are eaten daily (1- to 2-g servings), they have been shown to decrease serum cholesterol levels by about 10–20%.

Biological Membranes

Biological membranes are thin, semipermeable cellular barriers. The general function of these barriers is to exclude dangerous chemicals from the cell while allowing nutrients to enter. Membranes also confine special molecules to specific sections of the cell. Because almost all the dangerous chemicals, nutrients, and special molecules are water soluble, the membranes can act as effective barriers only if they impede the movement of hydrophilic (water-soluble) molecules.

To act as such a barrier, a membrane must have some special properties. To exclude water and water solutes, the bulk of a membrane must be hydrophobic. But a membrane necessarily touches water both inside and outside the cell. Therefore, the surface of a membrane must be hydrophilic. Thus, a membrane can be visualized as being layered much like a piece of laminated plywood.

The hydrophobic interior provides the barrier while the hydrophilic exterior interacts with the aqueous environment:

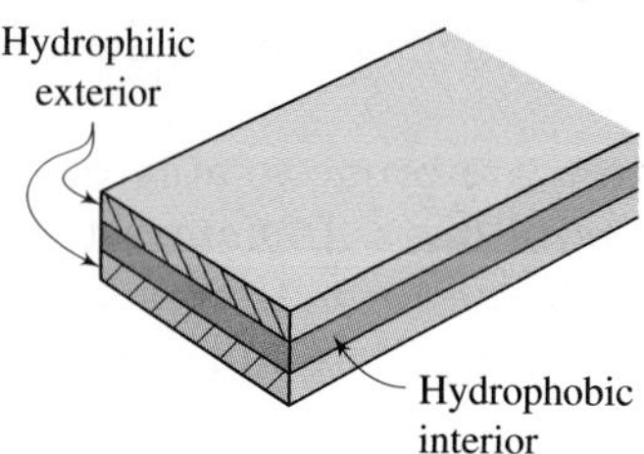

Plywood model of a simple membrane

The cell uses lipids to give the membrane its hydrophobic nature. In fact, by selecting the right lipids, both the hydrophobic and hydrophilic portions of a membrane can be assembled. There are several classes of membrane lipids. The most important of these are the phospholipids and sphingolipids. Remember that these lipids have both a hydrophobic and a hydrophilic section.

lipid bilayer

Membrane lipids naturally aggregate to form lipid bilayers. A **lipid bilayer** is composed of two adjoining layers of lipid molecules aligned so that their hydrophobic portions form the bilayer interior while their hydrophilic portions form the bilayer exterior. A lipid bilayer has the necessary properties of a membrane—a hydrophobic barrier and a hydrophilic surface:

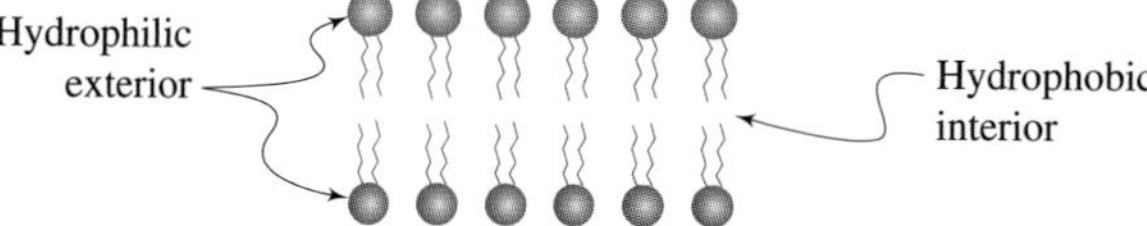

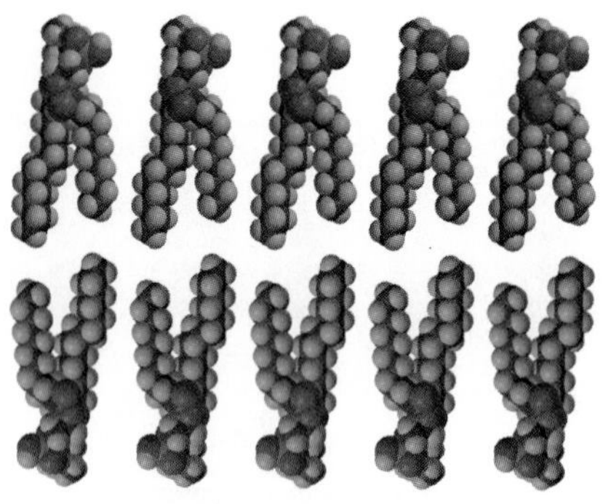

A spacefilling model of a phosphatidyl choline lipid bilayer.

Lipids give this barrier an oily or fluid appearance. The more unsaturated fatty acids in the membrane, the more fluid it will be. Other lipids, most importantly cholesterol, cause the membrane to be less fluid.

Membrane fluidity can have significant effects on cell function. It is thought that general anesthetics (e.g., ether, halothane) are effective partly because they dissolve in membranes, altering the fluidity of the lipid bilayer. During severe cirrhosis of the liver, red cells are forced to take abnormally large amounts of cholesterol into their membranes, causing these membranes to be less fluid. The red cells become more rigid, have greater difficulty passing through narrow capillaries, and are destroyed more easily.

It is truly amazing that lipid bilayers form spontaneously, simply because of the ability of lipids to aggregate. With no complicated planning, a barrier forms with both a hydrophobic interior and a hydrophilic exterior. Such a barrier provides the basis of the biological membrane. In fact, scientists have proposed that once lipidlike molecules were formed on primal earth, lipid bilayers must have spontaneously aggregated. These bilayers may have provided the boundaries needed as primitive cells first developed.

All known cells in today's world need a membrane that is more complicated than a simple lipid bilayer. A membrane must function as more than just a barrier. Tasks such as passing molecules from one side of a bilayer to the other are an essential part of life. Yet many molecules are hydrophilic and have

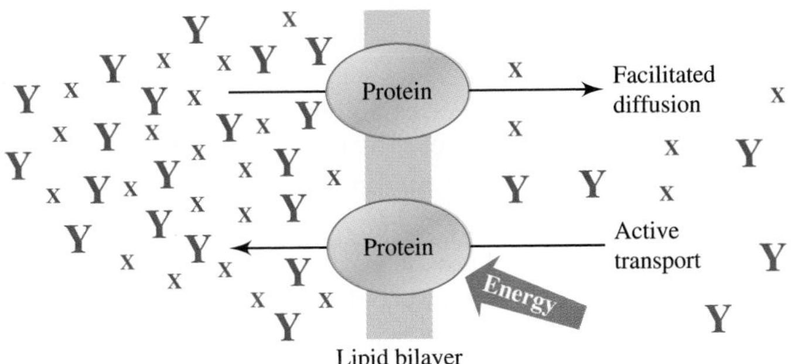

Figure 28.10
How proteins aid membrane transport: Molecules or ions (symbolized by Y and x) can move from high concentration to low concentration without energy (facilitated diffusion), but movement in the reverse direction requires energy (active transport).

difficulty traversing a lipid bilayer. We can imagine the cell being faced with a dilemma: How can it selectively allow some hydrophilic molecules to cross the lipid bilayer while excluding others?

Proteins in the fluid bilayer solve this dilemma. They allow specific molecular transport through the hydrophobic interior. (For a general discussion of proteins, see Chapter 29.) They recognize specific molecules on the exterior of a cell membrane and shuttle these molecules into the cell. This process can be as simple as providing a tunnel through the membrane for selected nutrients. If the protein helps (facilitates) transport without using energy, the process is called **facilitated diffusion**. Some transport requires energy, such as when molecules are moved from areas of low concentration to areas of high concentration (the opposite direction from that of diffusion). This energy-requiring transport is termed **active transport**. (See Figure 28.10.)

facilitated diffusion

active transport

Thus, a complete cellular membrane must have both lipid and protein. A typical membrane includes about 60% protein, 25% phospholipid, 10% cholesterol, and 5% sphingolipid. The fluid lipid bilayer is studded with many solid proteins. The proteins form a random pattern on the outer surface of the oily lipid. This general membrane is called the **fluid-mosaic model**. (See Figure 28.11.) An *intrinsic membrane protein* is one in which the bulk of the protein is inside the lipid bilayer; a protein that is mainly on the surface is an *extrinsic membrane protein*.

fluid-mosaic model

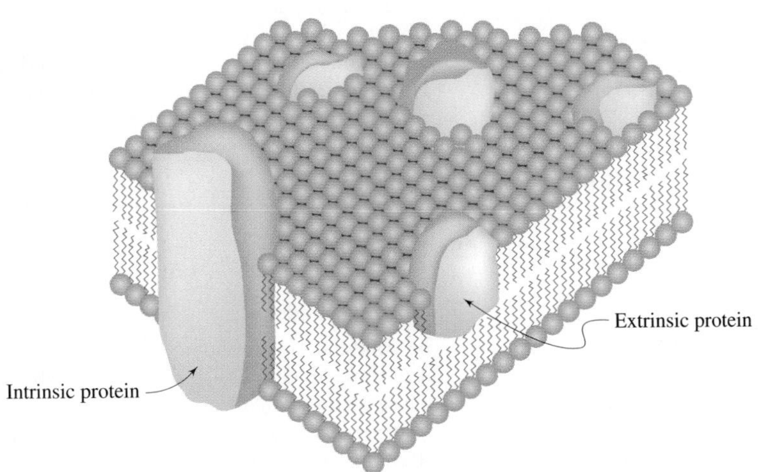

Figure 28.11
The fluid-mosaic model of a membrane.

Chapter 28 Review

28.1 Lipids: Hydrophobic Molecules

KEY TERM

Lipid

- Lipids are water-insoluble, oily, or greasy biochemicals.
- Lipids are described as hydrophobic (water fearing).

28.2 Classification of Lipids

- All lipids are relatively large and nonpolar, yet lipid structures may differ markedly.
- Simple lipids are esters of fatty acids and alcohols.
  - Fats and oils are esters of fatty acids and glycerol.
  - Waxes are esters of fatty acids and high-molar-mass alcohols.
- Compound lipids are composed of fatty acid esters and other components as well.
  - Phospholipids yield a phosphate and a nitrogen-containing base as well as glycerol and fatty acids upon hydrolysis.
  - Sphingolipids yield an unsaturated amino alcohol (sphingosine) and a carbohydrate or phosphate and nitrogen base in addition to a fatty acid upon hydrolysis.
  - Glycolipids yield sphingosine and a carbohydrate as well as a fatty acid upon hydrolysis.
- Steroids posses the steroid nucleus—a 17-carbon structure consisting of four fused carbocyclic rings.
- Miscellaneous lipids are substances that don't fit the preceding classifications and are large and nonpolar.
  - The fat-soluble vitamins (A, D, E, K) are miscellaneous lipids.
  - Lipoproteins are miscellaneous lipids.

28.3 Simple Lipids

KEY TERM

Wax

- Fatty acids, which form part of most lipids, are carboxylic acids with long, hydrophobic carbon chains.
  - Common fatty acids are straight-chain compounds with an even number of carbons.
  - Unsaturated (double bond–containing) fatty acids are commonly cis isomers.
- Unsaturation lowers the melting point of fatty acids.
- Some unsaturated fatty acids are dietary essentials (linoleic, linolenic, and arachidonic acids).
- The unsaturation in arachidonic acid allows it to serve as a precursor for a variety of intercellular signaling compounds, the eicosanoids.
  - The eicosanoids are involved in processes such as inflammation, platelet aggregation, and smooth muscle constriction.
- Fats and oils are esters of glycerol and long carbon chain fatty acids.
- Waxes are esters of fatty acids and a high-molar-mass alcohol (up to about 30 carbons long).
- Waxes are extremely water insoluble and serve a protective function on leaves, feathers, fruit, and fur.

28.4 Compound Lipids

KEY TERMS

Phospholipid
Sphingolipid
Glycolipid

- The phospholipids are a group of compounds that yield one or more fatty acid molecules, a phosphate group, and a nitrogenous base upon hydrolysis.
- Phospholipids are made starting with phosphatidic acids, glyceryl esters of fatty acids, and phosphoric acid.
- To complete the phospholipid structure, the phosphate in phosphatidic acid is esterified to a small, nitrogen-containing compound, either ethanolamine, choline, or L-serine.
  - Phosphatidyl choline (lecithin) is the phospholipid containing choline and is an important constituent of biological membranes.
  - Phosphatidyl ethanolamine (cephalin) contains ethanolamine and is also found in biological membranes.
- Phospholipids combine hydrophobic and hydrophilic properties, essential features of membrane lipids.
- Sphingolipids are compounds that, when hydrolyzed, yield a hydrophilic group (either phosphate and choline or a carbohydrate), a fatty acid, and sphingosine (an unsaturated amino alcohol).
- Sphingolipids are important membrane components; as with other membrane lipids, they combine hydrophilic and hydrophobic properties.
- Glycolipids are sphingolipids that contain carbohydrates.

28.5 Steroids

KEY TERM

Steroids

- Steroids are compounds that have the steroid nucleus.
  - The steroid nucleus is composed of 17 carbons in four fused carbocyclic rings.
- Cholesterol is the most common steroid and serves as (1) an important membrane component and (2) the precursor for other steroid hormones.

28.6 Hydrophobic Lipids and Biology

KEY TERMS

Lipid bilayer
Facilitated diffusion
Active transport
Fluid-mosaic model

- The combined hydrophobic and hydrophilic character of some lipids causes them to aggregate into specific structures when placed in water.
  - Fatty acids aggregate into micelles, oily balls with the fatty acid alkyl chains on the interior and the carboxyl groups on the surface.

  - Compound lipids aggregate into liposomes, small vesicles with an aqueous interior and a lipid bilayer surface.
- Atherosclerosis is a metabolic disease in which fatty material (plaque) is deposited on the inner walls of arteries.
  - Arteries lose elasticity and blood pressure rises.
  - Arteries narrow and the inner walls roughen, making a heart attack due to blood clots more likely.
- Improper transport of cholesterol through the blood system contributes to atherosclerosis.
  - Lipoproteins (specific protein-lipid aggregates) transport lipids through the blood.
    - Very-low-density lipoproteins (VLDL) carry lipids from the liver to peripheral tissues.
    - As triacylglycerols are removed by the adipose tissue, the VLDL are converted to low-density lipoproteins (LDL) that deliver cholesterol to the peripheral tissues.
    - High-density lipoproteins (HDL) scavenge cholesterol and return it to the liver.
  - High LDL levels predispose persons to atherosclerosis, while high HDL levels decrease the chances of contracting this metabolic disease.
- A lipid bilayer is composed of two adjoining layers of compound lipid molecules so that their hydrophobic portions form the bilayer interior while their hydrophilic portions form the bilayer exterior.
- Proteins aid passage of molecules through a lipid bilayer.
  - Facilitated diffusion is transport across a lipid bilayer with the aid of a protein but without the need for energy input.
  - Active transport is aided by a protein and also requires energy.
- The fluid-mosaic model pictures a biological membrane as a fluid lipid bilayer with many imbedded proteins.

Review Questions

All questions with blue numbers have answers in the appendix of the text.

1. Why are the lipids, which are dissimilar substances, classified as a group?
2. Briefly explain why caproic acid is more water soluble than stearic acid.
3. Briefly explain why arachidonic acid is of special biological importance.
4. What are the three essential fatty acids? What are the consequences of their being absent from the diet?
5. List two reasons why fats contain more biochemical energy than carbohydrates.
6. How is aspirin thought to relieve inflammation?
7. Wax serves a protective function on many types of leaves. How does it do this?
8. What two properties must a membrane lipid possess?
9. In what organ in the body are phospholipids mainly produced?
10. List the four classes of eicosanoids.
11. What are some health benefits of omega-3 fatty acids?
12. What two structural features allow sphingomyelin to serve as a membrane lipid?
13. What is atherosclerosis? In general, how is it produced and what are its symptoms?
14. Briefly describe the body's cholesterol distribution system.
15. How does a diet that contains a large amount of fish possibly decrease the risk of a heart attack?
16. Why is HDL a potential aid in controlling serum cholesterol levels?
17. Why can a lipid bilayer be described as a barrier?
18. What advantages does a liposome provide as a vehicle for drug delivery?
19. Distinguish active transport from facilitated diffusion.
20. What common structural feature do all steroids possess? Write the structural formulas for two steroids.
21. Why are the statin drugs so effective at reducing serum cholesterol levels?

Paired Exercises

All exercises with blue numbers have answers in the appendix of the text.

1. Write the structure of a monoacylglycerol and a diacylglycerol, using linoleic acid.
2. Write the structure of a monoacylglycerol and a diacylglycerol, using linolenic acid.
3. Write the structural formula of a triacylglycerol that contains one unit each of palmitic, stearic, and oleic acids. How many other triacylglycerols are possible that use one unit of each of these acids?
4. Write the structural formula of a triacylglycerol that contains two units of palmitic acid and one unit of oleic acid. How many other triacylglycerols are possible that use this same set of fatty acids?

5. Would a triacylglycerol that contains three units of lauric acid be more hydrophobic than a triacylglycerol that contains three units of stearic acid? Explain.

7. Draw formulas for the products of a hydrolysis reaction involving a triacylglycerol that contains palmitic acid, oleic acid, and linoleic acid.

9. Write the structure of a wax that contains a 26-carbon, straight-chain primary alcohol and stearic acid.

11. Write the structure of the most hydrophobic monoacylglycerol choosing from the following components: palmitic acid, linoleic acid, arachidic acid.

13. Give the structure of a phospholipid that is formed from glycerol, phosphoric acid, palmitic acid, and ethanolamine.

15. Given the following components, draw the structure of a sphingolipid: sphingosine, oleic acid, phosphoric acid, choline.

17. Given the following components, draw the structure of a glycolipid: sphingosine, palmitic acid, D-glucose.

19. Sketch the portion of a micelle that forms from palmitic acid. Show the structural formula for one palmitic acid in your sketch.

21. How does LDL differ in both structure and function from VLDL?

23. In what ways is sphingosine similar to a glycerol molecule that has been esterified to one fatty acid?

25. If sodium ions are moved from a solution (0.1 *M*) on one side of a membrane to a solution (0.001 *M*) on the other side, is this process facilitated diffusion or active transport? Briefly explain.

27. Differentiate between the function of a thromboxane and a prostaglandin.

29. Name the fatty acids that were used to make the following lipids:

(a)

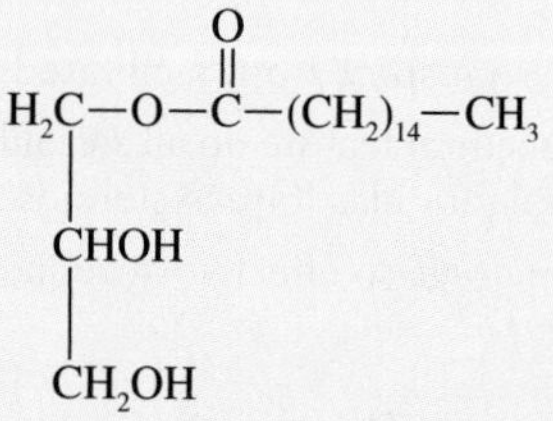

$$H_2C-O-\overset{\overset{O}{\|}}{C}-(CH_2)_{14}-CH_3$$
$$|$$
$$CHOH$$
$$|$$
$$CH_2OH$$

(b)

$$H_2C-O-\overset{\overset{O}{\|}}{C}-(CH_2)_7-\overset{H}{C}=\overset{H}{C}-(CH_2)_7CH_3$$
$$|$$
$$HC-O-\overset{\overset{O}{\|}}{C}-(CH_2)_{16}CH_3$$
$$|$$
$$H_2C-O-\overset{\overset{O}{\|}}{\underset{\underset{O^-}{|}}{P}}-O-CH_2-CH_2-\overset{\overset{CH_3}{|}}{\underset{\underset{CH_3}{|}}{N^+}}-CH_3$$

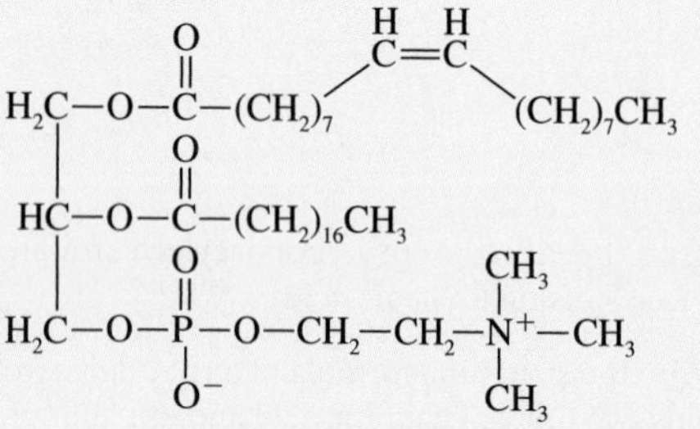

6. Would a triacylglycerol that contains three units of stearic acid be more hydrophobic than a triacylglycerol that contains three units of myristic acid? Explain.

8. Draw formulas for the products of a hydrolysis reaction involving a triacylglycerol that contains oleic acid, stearic acid, and palmitic acid.

10. Write the structure of a wax that contains a 30-carbon, straight-chain primary alcohol and palmitic acid.

12. Write the structure of the most hydrophobic monoacylglycerol choosing from the following components: oleic acid, palmitic acid, myrisitic acid.

14. Give the structure of a phospholipid that is formed from glycerol, phosphoric acid, stearic acid, and choline.

16. Given the following components, draw the structure of a sphingolipid: sphingosine, stearic acid, phosphoric acid, ethanolamine.

18. Given the following components, draw the structure of a glycolipid: sphingosine, oleic acid, D-galactose.

20. Sketch the portion of a micelle that forms from myristic acid. Show the structural formula for one myristic acid in your sketch.

22. How does HDL differ in both structure and function from LDL?

24. In what ways is a diacylglycerol similar to a sphingosine that has been linked to a fatty acid by an amide bond?

26. If phosphate ions are moved from a solution (0.1 *M*) on one side of a membrane to a solution (0.5 *M*) on the other side, is this process facilitated diffusion or active transport? Briefly explain.

28. Differentiate between the function of a thromboxane and a leukotriene.

30. Name the fatty acids that were used to make the following lipids:

(a)

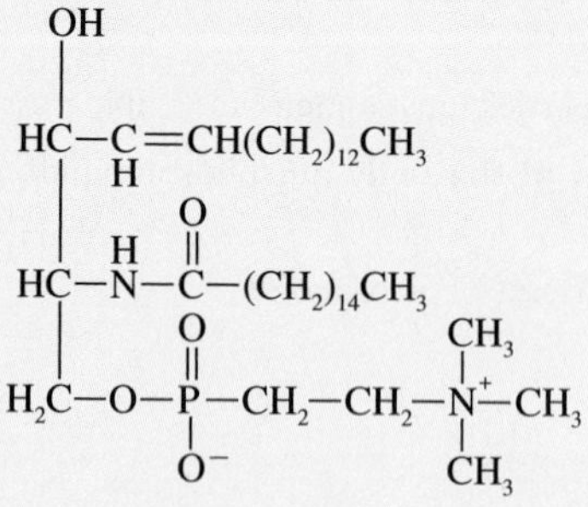

$$\overset{\overset{OH}{|}}{HC}-\underset{H}{C}=CH(CH_2)_{12}CH_3$$
$$|$$
$$HC-\overset{H}{N}-\overset{\overset{O}{\|}}{C}-(CH_2)_{14}CH_3$$
$$|$$
$$H_2C-O-\overset{\overset{O}{\|}}{\underset{\underset{O^-}{|}}{P}}-CH_2-CH_2-\overset{\overset{CH_3}{|}}{\underset{\underset{CH_3}{|}}{N^+}}-CH_3$$

(b)

$$H_2C-O-\overset{\overset{O}{\|}}{C}-(CH_2)_{16}-CH_3$$
$$|$$
$$CHOH$$
$$|$$
$$CH_2O-\overset{\overset{O}{\|}}{C}-(CH_2)_7-\overset{H}{C}=\overset{H}{C}-CH_2-\overset{H}{C}=\overset{H}{C}-(CH_2)_4CH_3$$

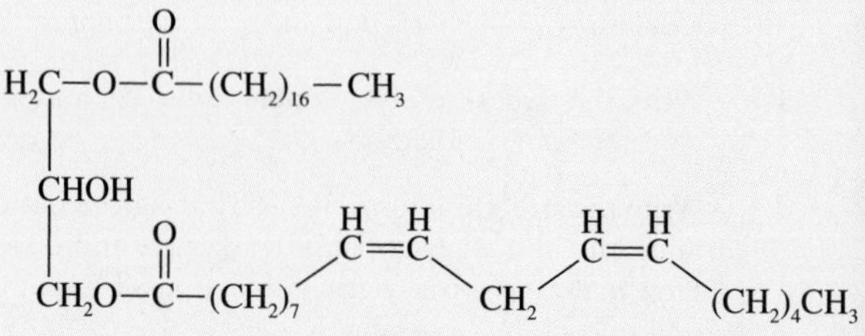

31. Name the lipid class for each of the following molecules:

(a) 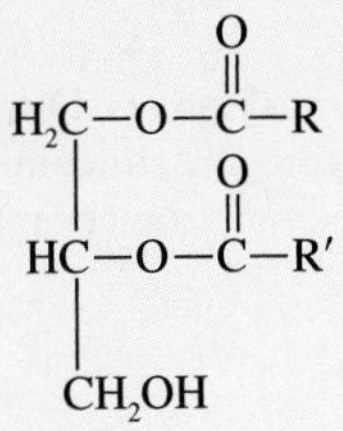 $H_2C{-}O{-}\overset{O}{\overset{\|}{C}}{-}R$
$HC{-}O{-}\overset{O}{\overset{\|}{C}}{-}R'$
CH_2OH

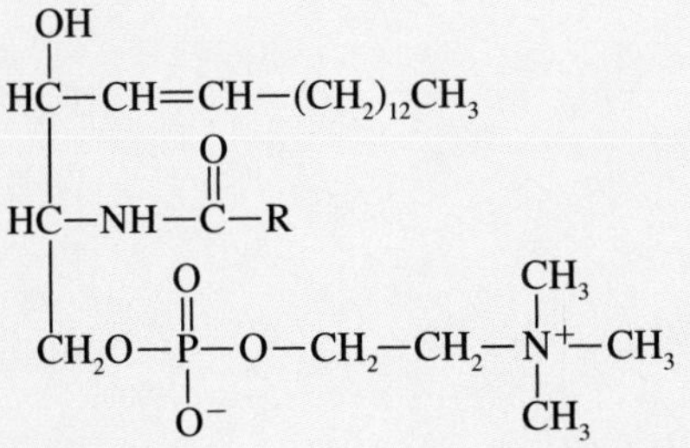

(b) $\overset{OH}{HC}{-}CH{=}CH{-}(CH_2)_{12}CH_3$
$HC{-}NH{-}\overset{O}{\overset{\|}{C}}{-}R$
$CH_2O{-}\overset{O}{\overset{\|}{P}}(O^-){-}O{-}CH_2{-}CH_2{-}N^+(CH_3)_2{-}CH_3$

(c) $H_2C{-}O{-}\overset{O}{\overset{\|}{C}}{-}R$
$HC{-}O{-}\overset{O}{\overset{\|}{C}}{-}R'$
$CH_2O{-}\overset{O}{\overset{\|}{P}}(O^-){-}O{-}CH_2{-}\overset{NH_3^+}{CH}{-}\overset{O}{\overset{\|}{C}}{-}O^-$

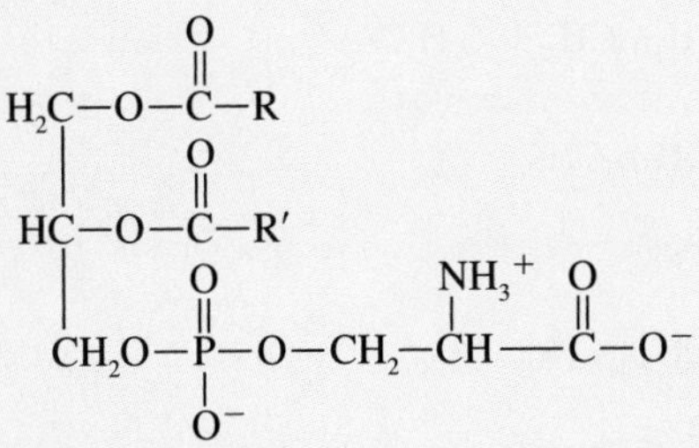

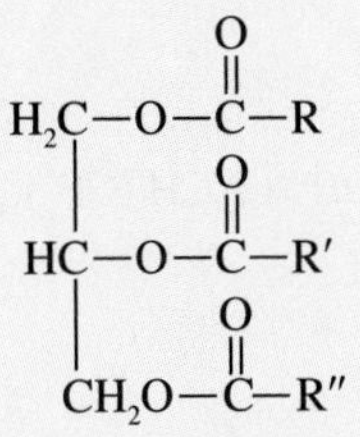

32. Name the lipid class for each of the following molecules:

(a) 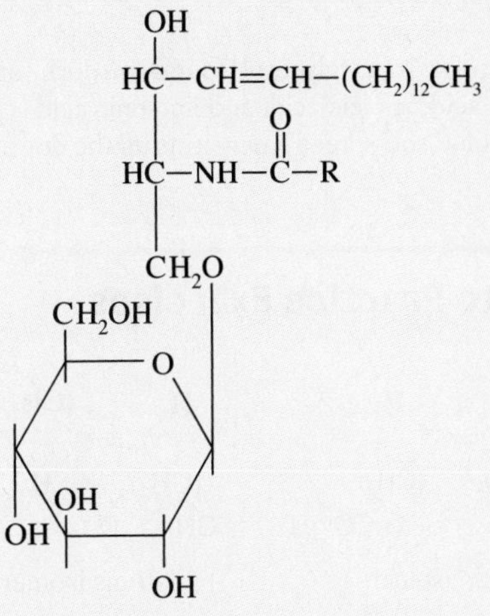 $\overset{OH}{HC}{-}CH{=}CH{-}(CH_2)_{12}CH_3$
$HC{-}NH{-}\overset{O}{\overset{\|}{C}}{-}R$
CH_2O — (sugar ring with CH_2OH, O, OH, OH, OH)

(b) $H_2C{-}O{-}\overset{O}{\overset{\|}{C}}{-}R$
$HC{-}O{-}\overset{O}{\overset{\|}{C}}{-}R'$
$CH_2O{-}\overset{O}{\overset{\|}{C}}{-}R''$

(c) $H_2C{-}O{-}\overset{O}{\overset{\|}{C}}{-}R$
$HC{-}O{-}\overset{O}{\overset{\|}{C}}{-}R'$
$CH_2O{-}\overset{O}{\overset{\|}{P}}(O^-){-}O{-}CH_2{-}CH_2{-}NH_3^+$

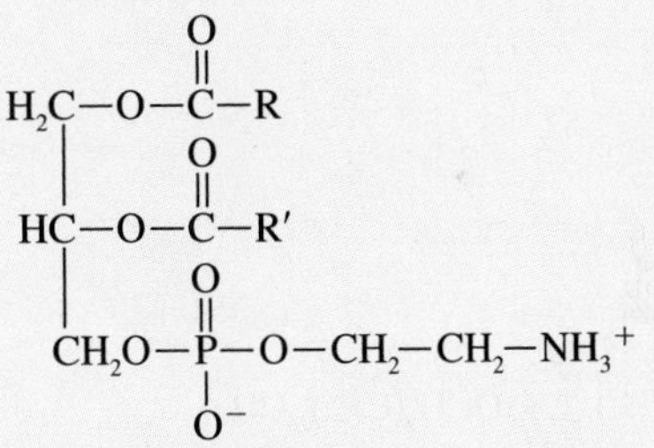

Additional Exercises

All exercises with blue numbers have answers in the appendix of the text.

33. What biochemical effect of ibuprofen decreases inflammation, redness, and swelling?

34. How does olestra differ structurally from natural fat? Explain why olestra passes through the body without being digested.

35. (a) Draw structural formulas for the three essential fatty acids.
(b) What are the consequences of these fatty acids being absent from the diet?

36. Suppose a simple meal is composed of 10 g fat, 17 g carbohydrate, and 19 g protein. The total energy from this meal equals approximately 234 Calories (kcal). If the amount of fat intake is increased to 15 g, what is the energy content now?

37. Beeswax is a mixture of long-chain hydrocarbons and esters. The esters consist of straight-chain alcohols with 24–36 carbon atoms esterified with straight-chain carboxylic acids having up to 36 carbon atoms. Write the structure for triacontanyl hexadecanoate, a C_{30} alcohol and a C_{16} acid.

38. Write the structure for the 18-carbon, ω-9 fatty acid. Give another name for this compound.

Challenge Exercise

39. A particular triacylglycerol found in corn oil contains palmitic acid, linoleic acid, and linolenic acid. During production of a "soft-spread" margarine *all* the double bonds in this molecule are changed to trans isomers. Draw the original triacylglycerol and the final product structures. Be careful to show the correct double-bond isomers in each case.

Answers to Practice Exercises

28.1

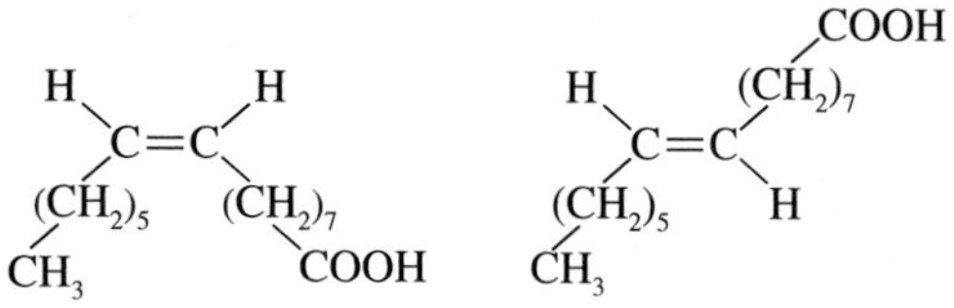

cis isomer: H(H)C=C(H) with substituents $(CH_2)_5CH_3$ and $(CH_2)_7COOH$ on the same side.

trans isomer: H and $(CH_2)_7COOH$ on one carbon; $(CH_2)_5CH_3$ and H on the other, on opposite sides.

cis isomer trans isomer

28.2

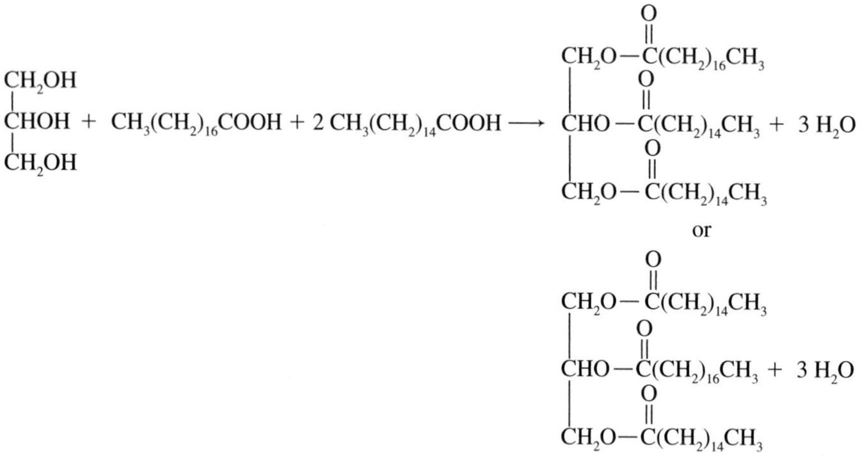

$$\begin{array}{l} CH_2OH \\ CHOH \\ CH_2OH \end{array} + CH_3(CH_2)_{16}COOH + 2\ CH_3(CH_2)_{14}COOH \longrightarrow \begin{array}{l} CH_2O-C(=O)(CH_2)_{16}CH_3 \\ CHO-C(=O)(CH_2)_{14}CH_3 \\ CH_2O-C(=O)(CH_2)_{14}CH_3 \end{array} + 3\ H_2O$$

or

$$\begin{array}{l} CH_2O-C(=O)(CH_2)_{14}CH_3 \\ CHO-C(=O)(CH_2)_{16}CH_3 \\ CH_2O-C(=O)(CH_2)_{14}CH_3 \end{array} + 3\ H_2O$$

28.3 $CH_3(CH_2)_{16}C(=O)-OCH_2(CH_2)_{26}CH_3$

28.4 Two possible structures:

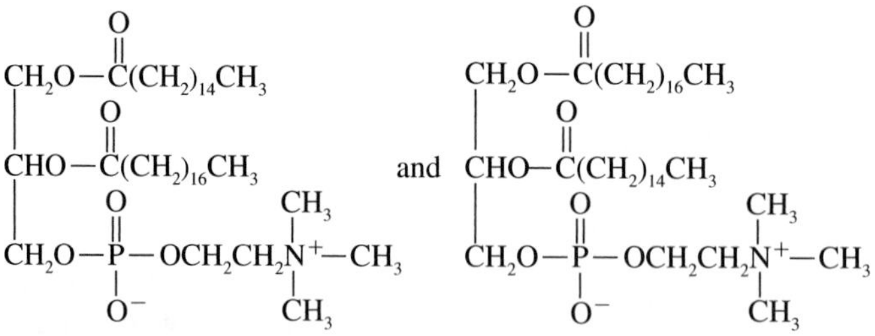

$$\begin{array}{l} CH_2O-C(=O)(CH_2)_{14}CH_3 \\ CHO-C(=O)(CH_2)_{16}CH_3 \\ CH_2O-P(=O)(O^-)-OCH_2CH_2N^+(CH_3)_2-CH_3 \end{array}$$

and

$$\begin{array}{l} CH_2O-C(=O)(CH_2)_{16}CH_3 \\ CHO-C(=O)(CH_2)_{14}CH_3 \\ CH_2O-P(=O)(O^-)-OCH_2CH_2N^+(CH_3)_2-CH_3 \end{array}$$

28.5

$$\begin{array}{l} CH(OH)-CH=CH(CH_2)_{12}CH_3 \\ CH-NH-C(=O)(CH_2)_{16}CH_3 \\ CH_2-O-P(=O)(O^-)-OCH_2CH_2-N^+(CH_3)_2-CH_3 \end{array}$$

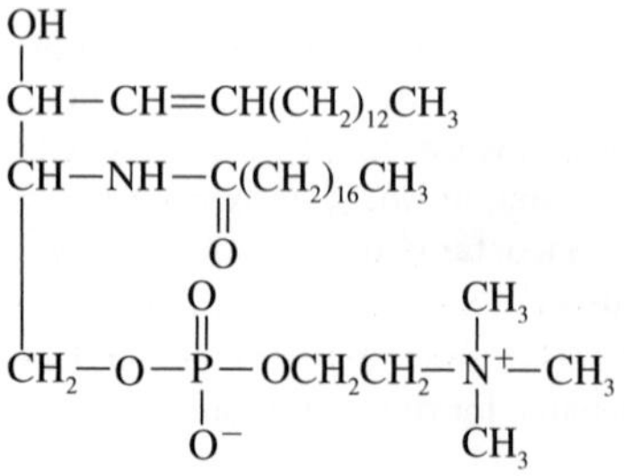

28.6
(a) Progesterone: two ketone functional groups and one C=C double bond
(b) Cholic acid: three alcohol functional groups, one carboxylic acid functional group
(c) Estradiol: one alcohol functional group, one phenol functional group

PUTTING IT TOGETHER: Review for Chapters 26–28

Multiple Choice:

Choose the correct answer to each of the following.

1. The instrument that measures the rotation of plane-polarized light is a
   (a) polariscope (c) polarimeter
   (b) polaroid (d) polar light meter

2. Cis-trans isomers that do not have a chiral carbon atom
   (a) will rotate polarized light
   (b) are not stereoisomers
   (c) will not be enantiomers
   (d) are optically active

3. A compound that rotates plane-polarized light is said to be
   (a) optically active (c) polarized
   (b) optically reactive (d) achiral

4. An object that cannot be superimposed on its mirror image is said to be
   (a) achiral (b) chiral (c) meso (d) dextrorotatory

5. A substance in solution that rotates plane-polarized light to the left is
   (a) levorotatory (c) achiral
   (b) dextrorotatory (d) left handed

6. Which of these items is chiral?
   (a) a square box (c) a glove
   (b) CH_3Cl (d) a bowling pin

7. How many chiral carbon atoms are in this structure?

$$\underset{}{CH_3}-\underset{\displaystyle OH}{\underset{|}{CH}}-\underset{\displaystyle Br}{\underset{|}{CH}}-\underset{\displaystyle Cl}{\underset{|}{CH}}-CH_3$$

   (a) 3 (b) 4 (c) 5 (d) 6

8. Compound X has an optical rotation of $-12.4°$. What is the rotation of its enantiomer?
   (a) $+24.8°$ (b) $+12.4°$ (c) $+6.2°$ (d) $-12.4°$

9. Which physical property will be different for the isomers in a pair of enantiomers?
   (a) boiling point (c) melting point
   (b) optical rotation (d) density

10. What is the maximum number of stereoisomers for glucose, which has the following structure?

$$\underset{\displaystyle OH}{\underset{|}{CH_2}}-\underset{\displaystyle OH}{\underset{|}{CH}}-\underset{\displaystyle OH}{\underset{|}{CH}}-\underset{\displaystyle OH}{\underset{|}{CH}}-\underset{\displaystyle OH}{\underset{|}{CH}}-\underset{\displaystyle H}{\underset{|}{C}}=O$$

   (a) 4 (b) 8 (c) 16 (d) 32

11. Which statement is *not* true about a meso compound?
   (a) A meso compound has a plane of symmetry in the molecule.
   (b) A meso compound does not rotate the plane of polarized light.
   (c) A meso compound is superimposable on its mirror image.
   (d) A meso compound is a racemic mixture.

12. How many meso forms does the following compound have?

$$CH_3-\underset{\displaystyle OH}{\underset{|}{CH}}-\underset{\displaystyle Br}{\underset{|}{CH}}-\underset{\displaystyle OH}{\underset{|}{CH}}-CH_3$$

   (a) 1 (b) 2 (c) 3 (d) 4

13. The compound shown in Question 12 has how many pairs of enantiomers?
   (a) 0 (b) 1 (c) 2 (d) 3

14. Which of the following structures is *not* a mirror image of the following compound?

$$CH_3\overset{\displaystyle H}{\overset{|}{\underset{\displaystyle OH}{\underset{|}{C}}}}-CH_2CH_3$$

   (a) $CH_3CH_2-\overset{\displaystyle H}{\overset{|}{\underset{\displaystyle OH}{\underset{|}{C}}}}-CH_3$
   (b) $CH_3-\overset{\displaystyle OH}{\overset{|}{\underset{\displaystyle H}{\underset{|}{C}}}}-CH_2CH_3$
   (c) $CH_3CH_2-\overset{\displaystyle OH}{\overset{|}{\underset{\displaystyle H}{\underset{|}{C}}}}-CH_3$
   (d) $H-\overset{\displaystyle OH}{\overset{|}{\underset{\displaystyle CH_2CH_3}{\underset{|}{C}}}}-CH_3$

15. What do these two structures represent?

$$H\overset{\displaystyle COOH}{\underset{\displaystyle CH_3}{+}}OH \quad \text{and} \quad CH_3\overset{\displaystyle COOH}{\underset{\displaystyle OH}{+}}H$$

   (a) a pair of enantiomers
   (b) a racemic mixture
   (c) a pair of diastereomers
   (d) the same compound

16. Three of these projection structures are the same compound and the other one is their enantiomer:

   (a) $Br\overset{\displaystyle CH_3}{\underset{\displaystyle H}{+}}Cl$
   (b) 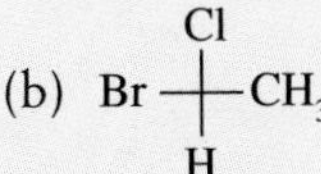 $Br\overset{\displaystyle Cl}{\underset{\displaystyle H}{+}}CH_3$
   (c) $H\overset{\displaystyle Cl}{\underset{\displaystyle CH_3}{+}}Br$
   (d) $Cl\overset{\displaystyle CH_3}{\underset{\displaystyle H}{+}}Br$

   Which structure is the enantiomer?

17. A solution of a racemic mixture shows no optical rotation because
   (a) the two enantiomers contain two chiral carbon atoms
   (b) it contains equal amounts of the (+) and (−) forms of the two enantiomers
   (c) the two enantiomers cannot be separated
   (d) the two enantiomers have the same boiling and melting points

18. Explain why, in the following reaction, a chiral carbon atom is formed, but the product shows no optical activity:

$CH_3CH_2CH{=}CH_2 + HCl \longrightarrow CH_3CH_2CHClCH_3$

(a) The reaction follows Markovnikov's rule.
(b) The product contains a mixture of 50% dextro- and 50% levorotatory stereoisomers.
(c) No optical activity can result from a chemical reaction.
(d) A meso compound is formed.

19. Which of the following two sugars are related as enantiomers?
(a) D-glucose and L-mannose
(b) D-galactose and D-mannose
(c) L-glucose and L-mannose
(d) L-galactose and D-galactose

20. Which of the following two sugars are related as epimers?
(a) D-glucose and L-mannose
(b) D-galactose and D-mannose
(c) L-glucose and L-mannose
(d) L-galactose and D-galactose

21. α-D-Glucopyranose is related to β-D-glucopyranose as
(a) a pair of epimers
(b) a pair of anomers
(c) a pair of diastereomers
(d) no correct answer given

22. D-Ribose is *not* classified as
(a) a tetrose (c) an aldose
(b) a monosaccharide (d) a pentose

23. Sucrose is classified as a
(a) monosaccharide
(b) trisaccharide
(c) reducing sugar
(d) nonreducing sugar

24. Which of the following is an example of a ketose?
(a) D-glyceraldehyde
(b) α-D-glucofuranose
(c) L-galactose
(d) dihydroxyacetone

25. What is the name of the following carbohydrate?

CH_2OH
OH O OH
OH
OH

(a) α-D-glucopyranose
(b) lactose
(c) β-D-galactopyranose
(d) β-D-glucopyranose

26. What is the name of the mirror image of the following sugar?

CH_2OH
$C{=}O$
$H{-}C{-}OH$
$HO{-}C{-}H$
$HO{-}C{-}H$
CH_2OH

(a) D-fructose (c) sucrose
(b) D-ketose (d) L-fructose

27. The ketone carbon in the sugar structure of Question 26 has an oxidation number of
(a) −2 (c) +3
(b) +2 (d) −1

28. What is the name of the following carbohydrate?

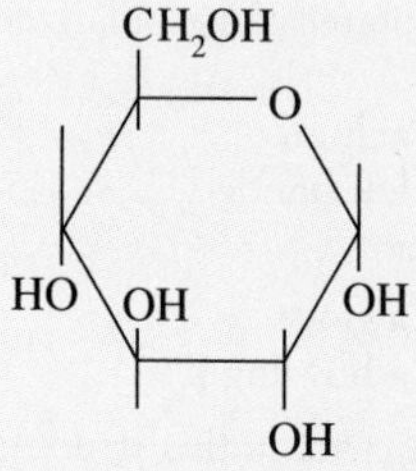

(a) β-D-glucopyranose (c) α-D-galactopyranose
(b) α-D-mannopyranose (d) no correct answer given

29. The conversion of β-D-mannopyranose to α-D-mannopyranose is called
(a) mutarotation (c) reduction
(b) oxidation (d) epimerization

30. Lactose is also known as
(a) levulose (c) grape sugar
(b) invert sugar (d) milk sugar

31. In a Benedict test, which of the following functional groups does *not* react?
(a) hemiacetal (c) acetal
(b) hemiketal (d) aldehyde

32. In a Benedict test, D-glucose is converted to
(a) D-glucitol (c) glucaric acid
(b) D-gluconic acid (d) saccharic acid

33. Amylopectin can be described as
(a) a linear oligosaccharide
(b) a branching oligosaccharide
(c) a component of glycogen
(d) a component of starch

34. When maltose is reacted in the Benedict test,
(a) Cu_2O is produced
(b) the hemiacetal is reduced
(c) no change is observed
(d) the acetal is cleaved

35. Cellulose can be described as a polymer of
(a) α-D-glucopyranose (c) β-D-glucofuranose
(b) α-D-glucofuranose (d) no correct answer given

36. Invert sugar is formed from
(a) α-D-glucopyranosyl-(1,4)-α-D-glucopyranose
(b) β-D-glucopyranosyl-(1,4)-β-D-glucopyranose
(c) α-D-glucopyranosyl-(1,2)-β-D-fructofuranose
(d) β-D-galactopyranosyl-(1,4)-α-D-glucopyranose

37. The FDA sets dietary levels for sugar substitutes. These levels are called a person's
(a) recommended dietary allowance
(b) acceptable daily intake
(c) acceptable dietary consumption
(d) recommended daily consumption

38. Breaking the acetal bond of the following disaccharide produces what two compounds?

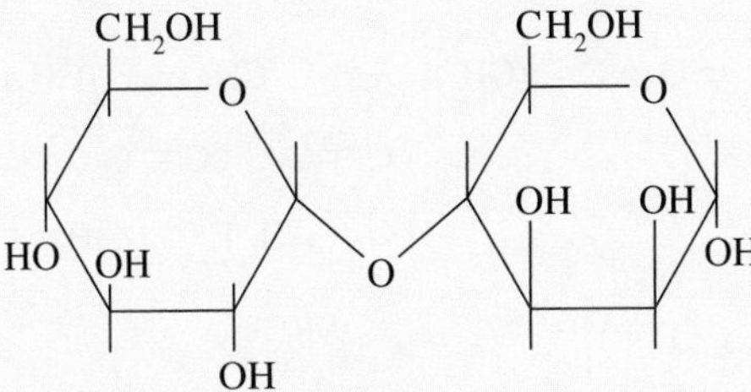

(a) D-galactose and D-mannose
(b) D-mannose and D-glucose
(c) D-galactose and D-glucose
(d) D-glucose and D-ribose

39. Dilute nitric acid converts D-galactose into
(a) glucaric acid (c) sorbitol
(b) D-galactonic acid (d) galactaric acid

40. When D-glucose reacts with H_2/Pt, the aldehyde carbon changes from an oxidation number of +1 to an oxidation number of
(a) 0 (b) +3 (c) −1 (d) −2

41. Which of the following lipids contain a total of three ester functional groups?
(a) phospholipid (c) wax
(b) triacylglycerol (d) sphingolipid

42. Stearic acid should have a higher melting point than oleic acid because
(a) stearic acid has no carbon–carbon double bonds, whereas oleic acid has one double bond
(b) oleic acid is only 14 carbons long, whereas stearic acid contains 18 carbons
(c) oleic acid has a trans double bond, whereas stearic acid has a cis double bond
(d) stearic acid has two double bonds, whereas oleic acid has only one

43. Which of the following is an example of an ω-3 fatty acid?
(a) oleic acid (c) linolenic acid
(b) linoleic acid (d) no correct answer given

44. Aspirin blocks the oxidation of which fatty acid?
(a) arachidonic acid
(b) stearic acid
(c) oleic acid
(d) arachidic acid

45. The only common lipid class that contains a carbohydrate is a
(a) phospholipid (c) triacylglycerol
(b) wax (d) glycolipid

46. An unsaturated fatty acid contains some carbon atoms with an oxidation number of
(a) 0 (b) −1 (c) +4 (d) +1

47. Commercial products refer to which lipid class as lecithin?
(a) phosphatidyl serine
(b) cholesterol
(c) phosphatidyl choline
(d) phosphatidyl ethanolamine

48. Which of the following compounds is *not* an eicosanoid:
(a) prostaglandin (c) leukotriene
(b) cephalin (d) prostacyclin

49. The lipoprotein that acts as a "cholesterol scavenger" is abbreviated
(a) HDL (b) VLDL (c) LDL (d) MLDL

50. The fluid-mosaic model of a biological membrane has
(a) a hydrophilic interior
(b) a hydrophobic exterior
(c) no cholesterol in the lipid bilayer
(d) a hydrophobic interior

51. A transport process that requires an energy input is called
(a) facilitated diffusion (c) passive transport
(b) active diffusion (d) active transport

52. Which of the following classes of lipids contains an amide bond?
(a) triacylglycerol (c) phospholipid
(b) sphingolipid (d) wax

53. Which of the following classes of lipids is the most hydrophobic?
(a) phospholipid (c) wax
(b) sphingolipid (d) glycolipid

54. Bile salts are derived from
(a) arachidonic acid (c) cholesterol
(b) lecithin (d) prostaglandin

55. Because of their shape, fatty acids tend to form one of the following in aqueous solution:
(a) liposomes (c) HDLs
(b) micelles (d) LDLs

56. Which of the following classes of lipids contains four ester bonds?
(a) triacylglycerol (c) sphingolipid
(b) phospholipid (d) glycolipid

Free Response Questions

Answer each of the following. Be sure to include your work and explanation in a clear, logical form.

1. (a) How are the molecules shown classified? Are they enantiomers or epimers of each other? Explain.
   (b) Based on your knowledge of the anomers of carbohydrates, which molecule has an α-OH and which has a β-OH group?

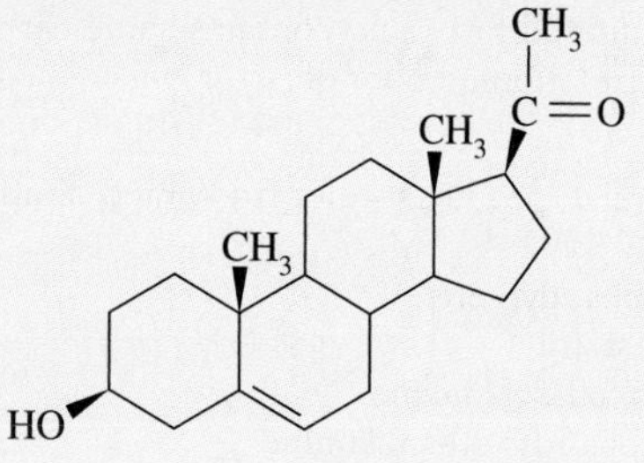

pregnenolone

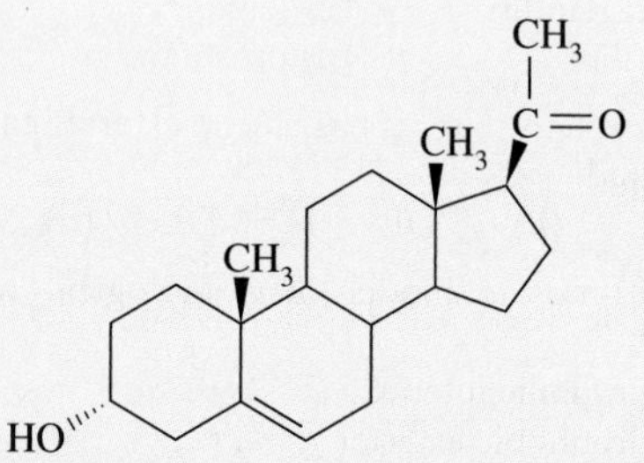

2. The partial hydrolysis of a glucose polymer (cellulose, glycogen, amylose, or amylopectin) yielded the following disaccharide only:

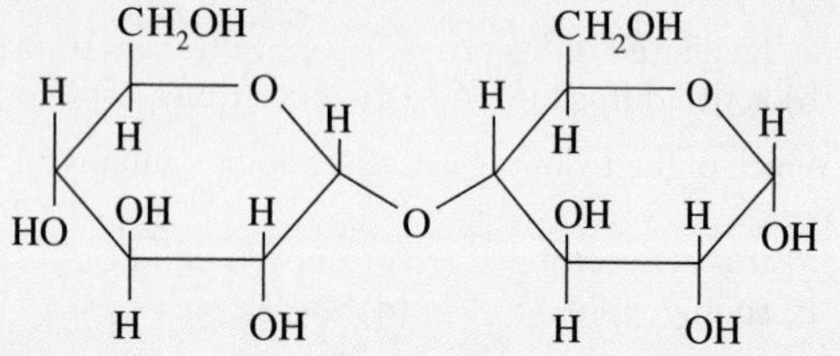

   Give the systematic name for the disaccharide, state which polysaccharide was hydrolyzed, and indicate whether a dilute solution of sulfuric acid would have accomplished this hydrolysis.

3. L-Rhamnose is a common 6-deoxy sugar. The Haworth structure is as follows:

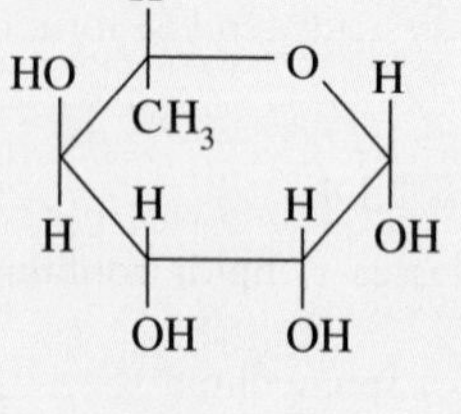

   (a) What is the maximum possible number of diastereomers of this sugar?
   (b) Is D-rhamnose an enantiomer, a diastereomer, or an anomer of L-rhamnose?
   (c) Draw the disaccharide α-D-glucopyranosyl-(1,4)-α-L-rhamnopyranose.
   (d) Is the disaccharide you drew a reducing sugar?

4. Are the following triacylglycerols structural isomers, optical isomers, geometric isomers, or identical molecules? Explain.

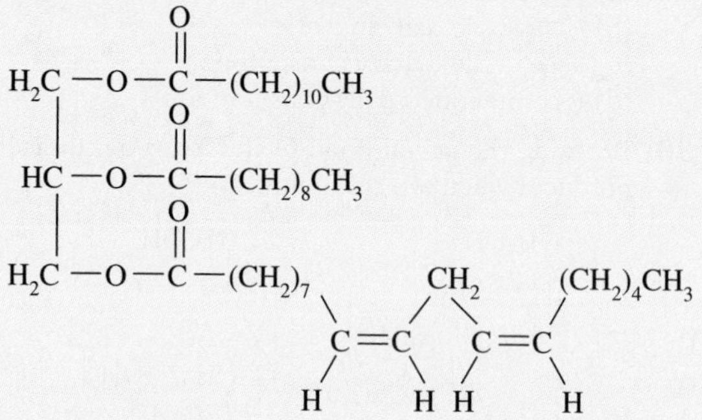

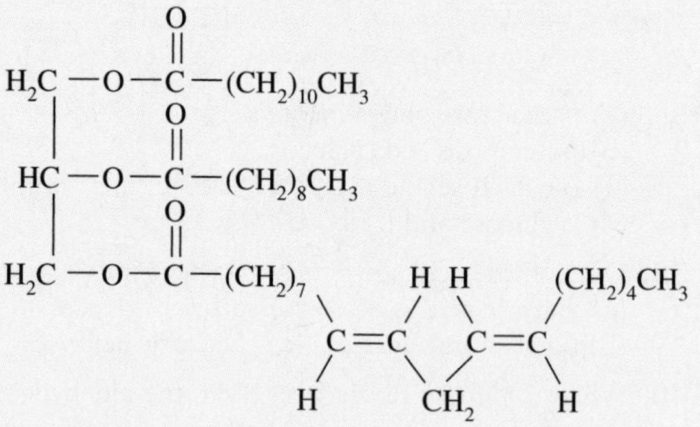

5. (a) Does the following lipid contain a glycerol or a sphingosine structure?

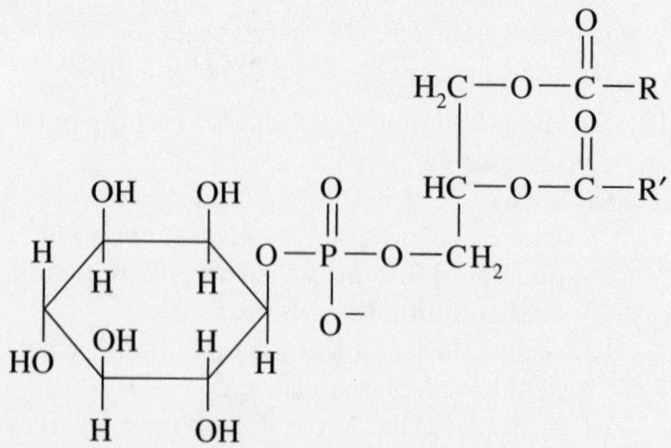

   (b) Does the substituent on the phosphate group represent an aldohexose? Explain.

6. In the following sphingomyelin, give the name of the fatty acid that is present, the functional group that resulted from its bonding to sphingosine, and the type of reaction that most likely occurred between sphingosine and the fatty acid:

OH
|
$HC-CH=CH(CH_2)_{12}CH_3$
|　　　O
|　　　‖
$HC-NH-C-CH_2(CH_2)_{17}CH_3$
|　　O
|　　‖　　　　　　　　+
$H_2C-O-P-O-CH_2CH_2N(CH_3)_3$
　　　　|
　　　　O^-

7. Name the fatty acids used in the following triacylglycerol, and indicate which one could be classified as an ω-6 fatty acid:

　　　　O
　　　　‖
$H_2C-O-C-(CH_2)_{12}CH_3$
|　　　O
|　　　‖
$HC-O-C-(CH_2)_7CH=CHCH_2CH=CH(CH_2)_4CH_3$
|　　　O
|　　　‖
$H_2C-O-C-(CH_2)_7CH=CH(CH_2)_5CH_3$

How many optical isomers are possible for this lipid?

8. If the ketone in L-psicose is reduced to an alcohol, explain why two molecules could be expected to form, what their relationship to each other is, and whether either or both will have optical activity. The structure of L-psicose is as follows:

CH_2OH
|
$C=O$
$HO-C-H$
$HO-C-H$ ⟶
$HO-C-H$
CH_2OH

L-psicose

9. The following biological compounds exhibit different solubility properties when water is the solvent:

(polymer of glucose units in pyranose rings: ring substituents CH_2OH, H, O, H, OH, H, H, OH; units joined through O; repeat unit in parentheses, subscript n)

(*n* is very large)

(a)

(furanose ring: $HOCH_2$, O, CH_2OH, H, OH, H, OH, H, OH)

(b)

　　　　　　　　　　　　　　O
　　　　　　　　　　　　　　‖
$CH_3(CH_2)_{18}CH=CH(CH_2)_{10}C-O(CH_2)_{35}CH_3$

(c)

Classify each compound by type—for example, as a lipid. (Be specific if possible.) Note whether they are soluble, partially soluble, or insoluble in water, and explain why.

10. In the following glycolipid, what functional group results from the connection of the carbohydrate to the sphingosine structure?

OH
|
$HC-CH=CH(CH_2)_{12}CH_3$
|　　　O
|　　　‖
$HC-NH-C-(CH_2)_{16}CH_3$
|
H_2C-O

(pyranose ring attached through O: CH_2OH, O, HO, H, OH, H, H, H, H, OH)

Name the carbohydrate used, and indicate how many chiral carbons are in the entire lipid.

CHAPTER 29

Amino Acids, Polypeptides, and Proteins

The spider has three sets of spinnerets that produce protein-containing fluids that harden as they are drawn out to form silk threads.

Chapter Outline

Proteins are present in every living cell. Their very name, derived from the Greek word *proteios*, which means "holding first place," signifies the importance of these substances. Think of the startling properties of these molecules. Spider-web protein is many times stronger than the toughest steel; hair, feathers, and hooves are all made from one related group of proteins; another protein provides the glass-clear lens material found in the eyes and needed for vision. Very small quantities (milligram amounts) of certain proteins missing from the blood can signify that a person's metabolic processes are out of control. Juvenile-onset diabetes mellitus results from a lack of the insulin protein. Dwarfism can arise when the growth hormone protein is lacking. A special "antifreeze" blood protein allows Antarctic fish to survive at body temperatures below freezing.

This list could go on and on, but what is perhaps most amazing is that the great variety of proteins are made from the same, relatively small, group of amino acids. By using various amounts of these amino acids in different sequences, nature creates biochemical compounds that are essential to the many functions needed to sustain life.

Another aspect of proteins is their importance in our nourishment. Proteins are one of the three major classes of foods. The other two, carbohydrates and fats, are needed for energy; proteins are needed for growth and maintenance of body tissue. Some common foods with high (over 10%) protein content are fish, beans, nuts, cheese, eggs, poultry, and meat. These foods tend to be scarce and relatively expensive. Therefore, proteins are the class of foods that is least available to the undernourished people of the world. Hence, the question of how to secure an adequate supply of high-quality protein for an ever-increasing population is one of the world's more critical problems. (See Chapter 32.)

29.1 The Structure–Function Connection

Proteins function as structural materials and as enzymes (catalysts) that regulate the countless chemical reactions taking place in every living organism, including the reactions involved in the decomposition and synthesis of proteins.

All proteins are polymeric substances that yield amino acids on hydrolysis. Those which yield only amino acids when hydrolyzed are classified as **simple proteins**; those which yield amino acids and one or more additional products are classified as **conjugated proteins**. There are approximately 200 different known amino acids in nature. Some are found in only one particular species of plant or animal, others in only a few life-forms. But 20 of these amino acids are found in almost all proteins. Furthermore, these same 20 amino acids are used by all forms of life in the synthesis of proteins.

simple protein
conjugated protein

All proteins contain carbon, hydrogen, oxygen, and nitrogen. Some proteins contain additional elements, usually sulfur, phosphorus, iron, copper, or zinc. The significant presence of nitrogen in all proteins sets them apart from carbohydrates and lipids. The average nitrogen content of proteins is about 16%.

Proteins are highly specific in their functions. The amino acid units in a given protein molecule are arranged in a definite sequence. An amazing fact about proteins is that, in some cases, if just one of the hundreds or thousands of amino acid units is missing or out of place, the biological function of that protein is seriously damaged or destroyed.

> The sequence of amino acids in a protein establishes the function of that protein.

This relationship between structure and function contrasts sharply with that for other classes of biochemicals. For example, carbohydrates provide cellular energy because they contain one particular type of atom, reduced carbon, that is readily oxidizable. This important function does not depend directly on the sequence in which the atoms are arranged. On the other hand, an appropriate sequence of amino acids produces a protein strong enough to form a horse's hoof; a different sequence produces a protein capable of absorbing oxygen in the lungs and releasing it to needy cells; and yet another sequence produces a hormone capable of directing carbohydrate metabolism for an entire organism. Full understanding of the function of a protein requires an understanding of its structure.

Spacefilling model of α-amino butyric acid.

29.2 The Nature of Amino Acids

Each amino acid has at least two functional groups: an amino group ($-NH_2$) and a carboxyl group ($-COOH$). The amino acids found in proteins are called alpha (α) amino acids because the amino group is attached to the first or α-carbon atom adjacent to the carboxyl group. The beta (β) position is the next adjacent carbon, the gamma (γ) position the next, and so on. The following formula represents an α-amino acid:

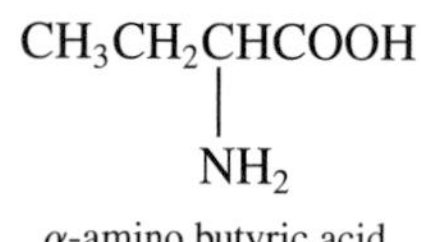

α-amino butyric acid

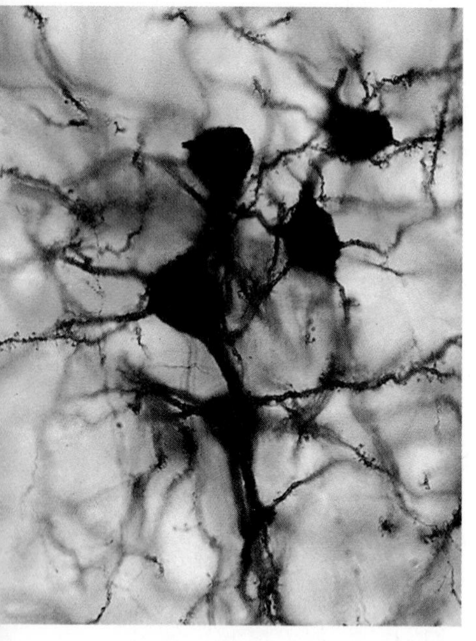

Cerebrum nerve synapse. Neurotransmitters, many of which are amino acids, cross these synapses to send messages between cells.

Amino acids as a whole are represented by this general formula:

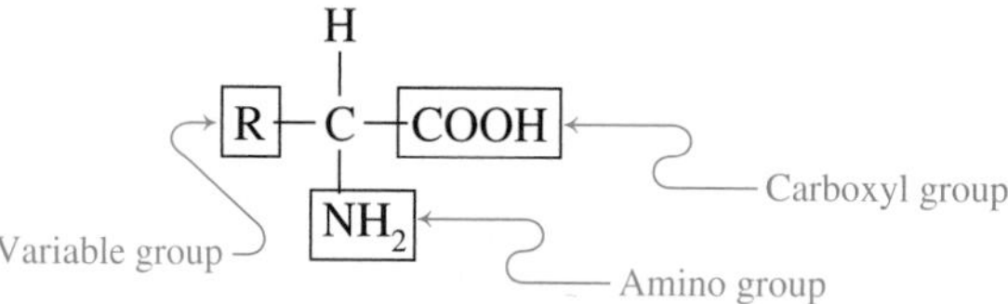

The portion of the molecule designated R is commonly referred to as the *amino acid side chain*. It is not restricted to alkyl groups and may contain (a) open-chain, cyclic, or aromatic hydrocarbon groups; (b) additional amino or carboxyl groups; (c) hydroxyl groups; or (d) sulfur-containing groups.

Amino acids are divided into three groups: neutral, acidic, and basic. They are classified as

1. *Neutral* when their molecules have the same number of amino and carboxyl groups
2. *Acidic* when their molecules have more carboxyl groups than amino groups
3. *Basic* when their molecules have more amino groups than carboxyl groups

The names, formulas, and abbreviations of the common amino acids are given in Table 29.1 on pages 836 and 837. Two of the these (aspartic acid and glutamic acid) are classified as acidic, three (lysine, arginine, and histidine) as basic, and the remainder as neutral amino acids.

Perhaps the most important role played by amino acids is as the building blocks for proteins. However, selected amino acids also have physiological importance on their own. Many neurotransmitters are amino acids or their derivatives. Glycine and glutamic acids function as chemical messengers between nerve cells in some organisms. Tyrosine is converted to the very important neurotransmitter dopamine. A deficiency of this amino acid derivative causes Parkinson's disease, which can be relieved by another compound formed from tyrosine, L-dopa. Tyrosine is also the parent compound for the "flight-or-fight" hormone epinephrine (adrenalin) and the metabolic hormone thyroxine. Histidine, which is converted in the body to histamine, is yet another amino acid with an important physiological role. Histamine causes the stomach lining to secrete HCl, but is probably best known for causing many of the symptoms associated with tissue inflammation and colds; this is the reason antihistamines are such popular and important over-the-counter medications.

Antihistamines lessen the symptoms of colds and allergies by reducing tissue inflammation.

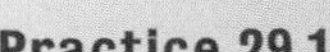

Practice 29.1

Write the structures for the simplest α-amino acid and β-amino acid.

29.3 Essential Amino Acids

During digestion, protein is broken down into its constituent amino acids, which supply much of the body's need for amino acids. (See Chapter 32.) Ten of the amino acids are **essential amino acids** (see Table 29.2) because they are essential to the normal functioning of the human body. Since the body is not capable of synthesizing them, they must be supplied in our diets if we are to enjoy normal health. Some animals require other amino acids in addition to those listed for humans.

essential amino acids

On a nutritional basis, proteins are classified as *complete* or *incomplete*. A complete protein supplies all the essential amino acids; an incomplete protein is deficient in one or more essential amino acids. Many proteins, especially those from vegetable sources, are incomplete. For example, protein from corn (maize) is deficient in lysine. The nutritional quality of such vegetable proteins can be greatly improved by supplementing them with the essential amino acids that are lacking, if these can be synthesized at reasonable costs. Lysine, methionine, and tryptophan are now used to enrich human food and livestock feed as a way to extend the world's limited supply of high-quality food protein. In still another approach to the problem of obtaining more high-quality protein, plant breeders have developed maize varieties with greatly improved lysine content. Genetic engineering may hold the key to further significant improvements in plant protein quality.

Vegetarians must carefully choose combinations of foods that include all essential amino acids.

Practice 29.2

Which of the following are essential amino acids for humans?
Gly, Ala, Leu, Lys, Tyr, Trp, His

Adding essential amino acids to livestock feed is a way to extend high-quality protein.

Table 29.1 Common Amino Acids Derived from Proteins

Name	Abbreviation	Formula
Alanine	Ala	$CH_3CH(NH_2)COOH$
Arginine*	Arg	$NH_2-C(=NH)-NH-CH_2CH_2CH_2CH(NH_2)COOH$
Asparagine	Asn	$NH_2C(=O)-CH_2CH(NH_2)COOH$
Aspartic acid	Asp	$HOOCCH_2CH(NH_2)COOH$
Cysteine	Cys	$HSCH_2CH(NH_2)COOH$
Glutamic acid	Glu	$HOOCCH_2CH_2CH(NH_2)COOH$
Glutamine	Gln	$NH_2C(=O)CH_2CH_2CH(NH_2)COOH$
Glycine	Gly	$HCH(NH_2)COOH$
Histidine*	His	imidazole ring (N=CH, HC=C, N–H) $-CH_2CH(NH_2)COOH$
Isoleucine*	Ile	$CH_3CH_2CH(CH_3)-CH(NH_2)COOH$
Leucine*	Leu	$(CH_3)_2CHCH_2-CH(NH_2)COOH$
Lysine*	Lys	$NH_2CH_2CH_2CH_2CH_2CH(NH_2)COOH$
Methionine*	Met	$CH_3SCH_2CH_2CH(NH_2)COOH$

*Amino acids essential in human nutrition

Table 29.1 Continued

Name	Abbreviation	Formula
Phenylalanine*	Phe	C_6H_5—$CH_2CH(NH_2)COOH$
Proline	Pro	pyrrolidine ring (N—H) with COOH
Serine	Ser	$HOCH_2CH(NH_2)COOH$
Threonine*	Thr	$CH_3CH(OH)—CH(NH_2)COOH$
Tryptophan*	Trp	indole (N—H, C=CH)—$CH_2CH(NH_2)COOH$
Tyrosine	Tyr	HO—C_6H_4—$CH_2CH(NH_2)COOH$
Valine*	Val	$(CH_3)_2CHCH(NH_2)COOH$

*Amino acids essential in human nutrition

Table 29.2 Essential Amino Acids for Humans

Arginine
Histidine
Isoleucine
Leucine
Lysine
Methionine
Phenylalanine
Threonine
Tryptophan
Valine

29.4 D-Amino Acids and L-Amino Acids

All amino acids, except glycine, have at least one asymmetric carbon atom. For example, two stereoisomers of alanine are possible:

Refer to Chapter 26 to review stereoisomerism.

$$\begin{array}{ccc} & COOH & \\ & | & \\ H— & C & —NH_2 \\ & | & \\ & CH_3 & \end{array} \qquad \begin{array}{ccc} & COOH & \\ & | & \\ H_2N— & C & —H \\ & | & \\ & CH_3 & \end{array}$$

D-(−)-alanine   L-(+)-alanine

Fischer projection formulas illustrate the D and L configurations of amino acids in the same way they illustrate the configurations of D- and L- glyceraldehyde (Section 27.2). The —COOH group is written at the top of the projection formula, and the D configuration is indicated by writing the alpha —NH_2 to the right of carbon 2. The L configuration is indicated by writing the alpha $-NH_2$ to the left of carbon 2. Although some D-amino acids occur in nature, only L-amino acids occur in proteins. The (+) and (−) signs in the name indicate the direction of rotation of plane-polarized light by the amino acid. Some amino acids have relatively complex structures that make the use of the projection formula difficult. Thus, unless stereochemical information is explicitly considered, amino acids will be shown using a condensed, structural formula.

29.5 Amphoterism

A Brønsted–Lowry acid is a proton donor; a Brønsted–Lowry base is a proton acceptor.

Amino acids are *amphoteric* (or *amphiprotic*); that is, they can react either as an acid or as a base. For example, with a strong base such as sodium hydroxide, alanine reacts as an acid:

$$\underset{\text{alanine}}{CH_3\underset{\underset{NH_2}{|}}{C}HCOOH} + NaOH \longrightarrow \underset{\text{sodium alanate}}{CH_3\underset{\underset{NH_2}{|}}{C}HCOO^- Na^+} + H_2O$$

With a strong acid such as HCl, alanine reacts as a base:

$$\underset{\text{alanine}}{CH_3\underset{\underset{NH_2}{|}}{C}HCOOH} + HCl \longrightarrow \underset{\text{alanyl ammonium chloride}}{CH_3\underset{\underset{NH_3^+Cl^-}{|}}{C}HCOOH}$$

zwitterion

Even in neutral biological solutions, amino acids do not actually exist in the molecular form shown in the preceding equations. Instead, they exist mainly as dipolar ions called **zwitterions**. Again, using alanine as an example, the proton on the carboxyl group transfers to the amino group, forming a zwitterion by an acid–base reaction within the molecule:

A dipolar ion is formed by the internal acid–base reaction of amino acids.

$$\underset{\text{alanine molecule}}{CH_3\underset{\underset{H_2N:}{|}}{C}HCOO\textcircled{H}} \longrightarrow \underset{\text{alanine zwitterion}}{CH_3\underset{\underset{NH_3^+}{|}}{C}HCOO^-}$$

On an ionic basis, the reaction of alanine with OH^- and H^+ is

$$\underset{\text{alanine zwitterion}}{CH_3\underset{\underset{NH_3^+}{|}}{C}HCOO^-} + OH^- \longrightarrow \underset{\text{alanate anion}}{CH_3\underset{\underset{NH_2}{|}}{C}HCOO^-} + H_2O \qquad (1)$$

$$\underset{\text{alanine zwitterion}}{CH_3\underset{\underset{NH_3^+}{|}}{C}HCOO^-} + H^+ \longrightarrow \underset{\text{alanyl ammonium cation}}{CH_3\underset{\underset{NH_3^+}{|}}{C}HCOOH} \qquad (2)$$

Most other amino acids behave like alanine. Together with protein molecules that contain —COOH and $—NH_2$ groups, they help to buffer or stabilize the pH of the blood at about 7.4. The pH is maintained close to 7.4 because any excess acid or base in the blood is neutralized by reactions such as equations (1) and (2).

When an amino acid in solution has equal positive and negative charges, it is electrically neutral and does not migrate toward either the positive or negative electrode when placed in an electrolytic cell. This is shown in formula I:

$$\underset{\substack{\text{cation form}\\ \text{II}}}{\underset{NH_3^+}{\underset{|}{RCHCOOH}}} \underset{H^+}{\overset{OH^-}{\rightleftharpoons}} \underset{\substack{\text{zwitterion form}\\ \text{I}}}{\underset{NH_3^+}{\underset{|}{RCHCOO^-}}} \underset{H^+}{\overset{OH^-}{\rightleftharpoons}} \underset{\substack{\text{anion form}\\ \text{III}}}{\underset{NH_2}{\underset{|}{RCHCOO^-}}} \quad (3)$$

Electrophoresis is a common technique used by genetic engineers.

The pH at which there is no migration toward either electrode is called the **isoelectric point**. (See Table 29.3.) If acid (H^+) is added to an amino acid at its isoelectric point, the equilibrium is shifted toward formula II, and the cation formed migrates toward the negative electrode. When base (OH^-) is added, the anion formed (formula III) migrates toward the positive electrode. Differences in isoelectric points are important in isolating and purifying amino acids and proteins, since their rates and directions of migration can be controlled in an electrolytic cell by adjusting the pH. This method of separation is called *electrophoresis*.

isoelectric point

Table 29.3 Isoelectric Points of Selected Amino Acids

Amino acid	pH at isoelectric point
Arginine	10.8
Lysine	9.7
Alanine	6.0
Glycine	6.0
Serine	5.7
Glutamic acid	3.2
Aspartic acid	2.9

Amino acids are classified as basic, neutral, or acidic, depending on whether the ratio of $—NH_2$ to —COOH groups in the molecules is greater than 1 : 1, equal to 1 : 1, or less than 1 : 1, respectively. Furthermore, this ratio differs from 1 : 1 only if the amino acid side chain (R—) contains an additional amino or carboxyl group. For example, if the side chain contains a carboxyl group, the amino acid is considered acidic. Thus, the R— group determines whether an amino acid is classified as basic, neutral, or acidic.

Isoelectric points are found at pH values ranging from 7.8 to 10.8 for basic amino acids, 4.8 to 6.3 for neutral amino acids, and 2.8 to 3.3 for acidic amino acids. It is logical that a molecule such as glutamic acid, with one amino group and two carboxyl groups, would be classified as acidic and that its isoelectric point would be at a pH lower than 7.0. It might also seem that the isoelectric point of an amino acid classified as neutral, such as alanine, with one amino and one carboxyl group, would have an isoelectric point of 7.0. However, the isoelectric point of alanine is 6.0, not 7.0. This is because the carboxyl group and the amino group are not equally ionized. The carboxyl group of alanine ionizes to a greater degree as an acid than the amino group ionizes as a base.

Example 29.1

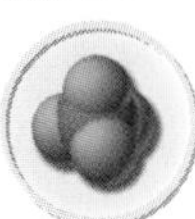

Draw the structure of L-serine in a strongly acidic solution.

SOLUTION

First, draw the structure for L-serine as the molecule would exist in a neutral solution. The α-amino and carboxylic acid groups form a zwitterion:

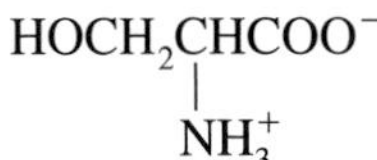

$$\underset{NH_3^+}{\underset{|}{HOCH_2CHCOO^-}}$$

When the solution is made acidic (i.e., the concentration of hydrogen ions is increased), the amine group is unaffected because it is already protonated. However, the carboxylate anion bonds to a hydrogen ion, resulting in an L-serine structure with a net positive charge:

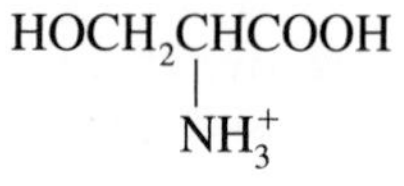

$$\underset{\displaystyle NH_3^+}{HOCH_2\overset{}{C}HCOOH}$$

Practice 29.3

Draw the structure of L-valine in a strongly basic solution.

29.6 Formation of Polypeptides

Proteins are polyamides consisting of amino acid units joined through amide structures. If we react two glycine molecules, so that a molecule of water is eliminated, we form a compound containing the amide structure, also called the **peptide linkage**, or peptide bond. The elimination of water occurs between the carboxyl group of one amino acid and the α-amino group of a second amino acid. (See Section 25.1.) The product formed from two glycine molecules is called glycylglycine (abbreviated Gly-Gly). Because it contains two amino acid units, it is called a dipeptide:

peptide linkage

An amide is formed by reaction between a carboxylic acid and an amine.

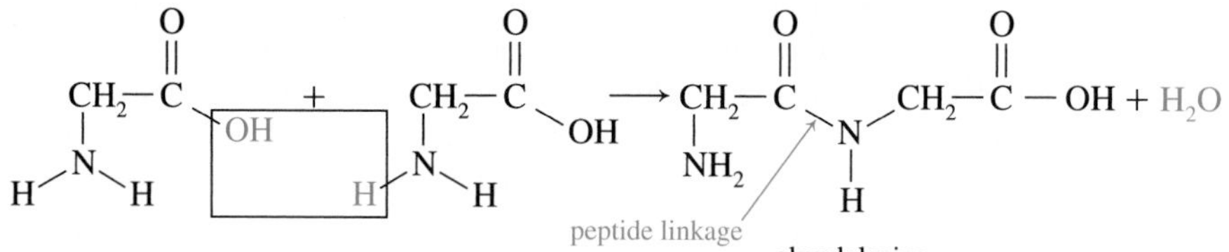

$$H_2N{-}CH_2{-}C(=O){-}OH + H{-}NH{-}CH_2{-}C(=O){-}OH \longrightarrow H_2N{-}CH_2{-}C(=O){-}NH{-}CH_2{-}C(=O){-}OH + H_2O$$

peptide linkage

glycylglycine (Gly-Gly)

If three amino acid residues are included in a molecule, it is a tripeptide; if four, a tetrapeptide; if five, a pentapeptide; and so on. Peptides containing up to about 40–50 amino acid units in a chain are called **polypeptides**. The units making up the peptide are amino acids, minus the elements of water, and are referred to as *amino acid residues* or, simply, residues. Still longer chains of amino acids are known as proteins.

polypeptides

When amino acids form a polypeptide chain, a carboxyl group and an α-amino group are involved in each peptide bond (amide bond). While these groups are joined in peptide bonds, they cannot ionize as acids or bases. Consequently, the properties of a polypeptide/protein are determined to a large extent by the side chains of the amino acid residues.

In linear peptides, one end of the chain has a free amino group and the other end a free carboxyl group. The amino-group end is called the *N-terminal residue* and the other end the *C-terminal residue:*

1 2 3 4 5 6 7
Ala-Pro-Tyr-Met-Gly-Lys-Gly

N-terminal residue (Ala) — C-terminal residue (Gly)

The sequence of amino acids in a chain is numbered starting with the N-terminal residue, which is written on the left. The C-terminal residue is written on the right. Any segment of the sequence that is not specifically known is placed in parentheses. Thus, in the preceding heptapeptide, if the order of tyrosine and methionine were not known, the structure would be written as

Ala-Pro-(Met, Tyr)-Gly-Lys-Gly

Peptides are named as acyl derivatives of the C-terminal amino acid, with the C-terminal unit keeping its complete name. The *-ine* ending of all but the C-terminal amino acid is changed to *-yl*, and these are listed in the order in which they appear, starting with the N-terminal amino acid:

$$\underbrace{CH_3CH(NH_2)C(=O)}_{\text{alanyl}}-\underbrace{NHCH(CH_2C_6H_4OH)C(=O)}_{\text{tyrosyl}}-\underbrace{NHCH_2COOH}_{\text{glycine}}$$

Ala-Tyr-Gly

Hence, Ala-Tyr-Gly is called alanyltyrosylglycine. The name of Arg-Glu-His-Ala is arginylglutamylhistidylalanine.

Alanine and glycine can form two different dipeptides, Gly-Ala and Ala-Gly, using each amino acid only once:

$$\underset{\text{glycylalanine (Gly-Ala)}}{CH_2(NH_2)-C(=O)-N(H)-CH(CH_3)-C(=O)-OH} \quad \text{and} \quad \underset{\text{alanylglycine (Ala-Gly)}}{CH_3CH(NH_2)-C(=O)-N(H)-CH_2-C(=O)-OH}$$

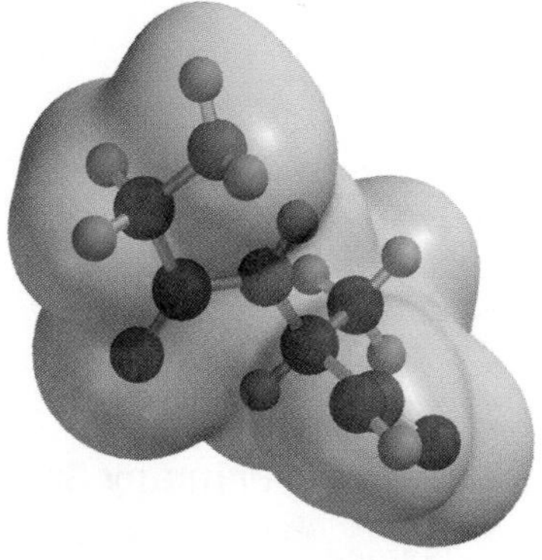

Spacefilling model of glycylalanine: Can you see the amide bond?

If three different amino acids react—for example, glycine, alanine, and threonine—six tripeptides, in which each amino acid appears only once, are possible:

Gly-Ala-Thr	Ala-Thr-Gly	Thr-Ala-Gly
Gly-Thr-Ala	Ala-Gly-Thr	Thr-Gly-Ala

The number of possible peptides rises very rapidly as the number of amino acid units increases. For example, there are 120 ($1 \times 2 \times 3 \times 4 \times 5 = 120$) different ways to combine five different amino acids to form a pentapeptide, using each amino acid only once in each molecule. If the same constraints are applied to 15 different amino acids, the number of possible combinations is greater than 1 trillion (10^{12})! Since a protein molecule can contain several hundred amino acid units, with individual amino acids occurring several times, the number of possible combinations from 20 amino acids is simply beyond imagination.

There are a number of small, naturally occurring polypeptides with significant biochemical functions. (Over 30 different peptides are known at present.)

In general, these substances serve as hormones or nerve transmitters. Their functions range from controlling pain and pleasure responses in the brain to controlling smooth muscle contraction or kidney fluid excretion rates. (See Table 29.4.)

The amino acid sequence and chain length give a polypeptide its biological effectiveness and specificity. For example, recent research has shown that the effects of opiates (opium derivatives) on the brain are also exhibited by two naturally occurring pentapeptides: Leu-enkephalin (Tyr-Gly-Gly-Phe-Leu) and Met-enkephalin (Tyr-Gly-Gly-Phe-Met). These two pentapeptides are natural painkillers. Alterations of the amino acid sequence—which alter the side-chain characteristics—cause drastic changes in the analgesic effects of Leu-enkephalin and Met-enkephalin. The substitution of L-alanine for either of the glycine residues in these compounds (simply changing one side-chain group from —H to $-CH_3$) causes approximately a thousand-fold decrease in their effectiveness as painkillers! The substitution of L-tyrosine for L-phenylalanine causes a comparable loss of activity. Even the substitution of D-tyrosine for the L-tyrosine residue causes a considerable loss in the analgesic effectiveness of the pentapeptides.

It is clearly evident that a particular sequence of amino acid residues is essential for proper polypeptide function. This sequence aligns the side-chain characteristics (large or small; polar or nonpolar; acidic, basic, or neutral) in the proper positions for a specific polypeptide function.

Oxytocin and vasopressin are similar nonapeptides, differing at only two positions in their primary structure. (See Table 29.4.) Yet their biological functions differ dramatically. Oxytocin controls uterine contractions during labor in childbirth and also causes contraction of the smooth muscles of the mammary glands, resulting in milk excretion. Vasopressin in high concentration raises the blood pressure and has been used to treat surgical shock. Vasopressin is also an antidiuretic, regulating the excretion of fluid by the

Table 29.4 Primary Structures and Functions of Some Biological Polypeptides*

Name	Primary structure	General biological function
Substance P	Arg-Pro-Lys-Pro-Gln-Gln-Phe-Phe-Gly-Leu-Met-NH_2	Is a pain-producing agent
Bradykinin	Arg-Pro-Pro-Gly-Phe-Ser-Pro-Phe-Arg	Affects tissue inflammation and blood pressure
Angiotensin II	Asp-Arg-Val-Tyr-Val-His-Pro-Phe	Maintains water balance and blood pressure
Leu-enkephalin Met-enkephalin	Tyr-Gly-Gly-Phe-Leu Tyr-Gly-Gly-Phe-Met	Relieves pain, produces sense of well-being
Vasopressin	S—S (linking Cys-1 and Cys-6) Cys-Tyr-Phe-Gln-Asn-Cys-Pro-Arg-Gly-NH_2	Increases blood pressure, decreases kidney water excretion
Oxytocin	S—S (linking Cys-1 and Cys-6) Cys-Tyr-Ile-Gln-Asn-Cys-Pro-Leu-Gly-NH_2	Initiates childbirth labor, causes mammary gland milk release, affects kidney excretion of water and sodium

*Where $-NH_2$ is indicated at the end of the sequence, the C-terminal amino acid has an amide structure rather than a free —COOH.

Table 29.4 Continued

Name	Primary structure	General biological function
Thyrotropin-releasing hormone	Glu-His-Pro	Stimulates release of hormones from the pituitary
Orexin	33 amino acids long	Causes wakefulness
Neuropeptide FF	Phe-Leu-Phe-Gln-Pro-Gln-Arg-Phe-NH_2	Modulates pain sensations
Neurotensin	Glu-Leu-Tyr-Glu-Asn-Lys-Pro-Arg-Arg-Pro-Tyr-Ile-Leu	Involved in brain memory functions
Neuropeptide Y	36 amino acids long	Stimulates eating
Endomorphin	Tyr-Pro-Trp-Phe-NH_2	Acts as a morphine-like analgesic

kidneys. The absence of vasopressin leads to diabetes insipidus. This condition is characterized by the excretion of up to 30 L of urine per day, but it can be controlled by the administration of vasopressin or any of its derivatives.

The isolation and synthesis of oxytocin and vasopressin was accomplished by Vincent du Vigneaud (1901–1978) and coworkers at Cornell University. Du Vigneaud was awarded the Nobel prize in chemistry in 1955 for this work. Synthetic oxytocin is indistinguishable from the natural material. It is available commercially and is used to induce labor in the late stages of pregnancy.

Example 29.2

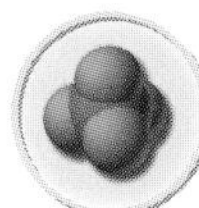

Write the structure of the tripeptide Ser-Gly-Ala.

SOLUTION

First, write the structures of the three amino acids in this tripeptide:

$$\underset{\text{serine (Ser)}}{\mathrm{HOCH_2\underset{\displaystyle NH_2}{\underset{|}{C}}HCOOH}} \qquad \underset{\text{glycine (Gly)}}{\mathrm{\underset{\displaystyle NH_2}{\underset{|}{C}}H_2COOH}} \qquad \underset{\text{alanine (Ala)}}{\mathrm{CH_3\underset{\displaystyle NH_2}{\underset{|}{C}}HCOOH}}$$

By convention, the amino acid residue written at the left end of the tripeptide has a free amino group, while the residue at the right end has a free carboxylic acid group. Now split out a water molecule in two places: between the carboxyl group of serine and the amino group of glycine and between the carboxyl group of glycine and the amino group of alanine. When the amino acids are connected by peptide linkages, the following structure results:

$$\mathrm{HOCH_2\underset{\displaystyle NH_2}{\underset{|}{C}}H\overset{\displaystyle O}{\overset{\|}{C}}-NHCH_2\overset{\displaystyle O}{\overset{\|}{C}}-NH\underset{\displaystyle CH_3}{\underset{|}{C}}HCOOH}$$

Practice 29.4

Write the structure of the pentapeptide Gly-Leu-Asp-Ser-Cys.

29.7 Protein Structure

By 1940, a great deal of information concerning proteins had been assembled. Their elemental composition was known, and they had been carefully classified according to their solubility in various solvents. Proteins were known to be polymers of amino acids, and the different amino acids had, for the most part, been isolated and identified. Protein molecules were known to be very large, with molar masses ranging from several thousand to several million.

Knowledge of protein structure would answer many chemical and biological questions. But for a while, the task of determining the actual structure of molecules of such colossal size appeared to be next to impossible. Then Linus Pauling, at the California Institute of Technology, attacked the problem by taking a new approach. Using X-ray diffraction techniques, Pauling and his collaborators painstakingly determined the bond angles and dimensions of amino acids and of dipeptides and tripeptides. After building accurate scale models of the dipeptides and tripeptides, they determined how they could fit together into likely polypeptide configurations. In 1951, on the basis of this work, Pauling and Robert Corey (1897–1971) proposed that two different conformations—the *α-helix* and the *β-pleated sheet*—were the most probable stable polypeptide chain configurations of protein molecules. These two macromolecular structures are illustrated in Figure 29.1. Within a short time, it was established that many proteins do have structures corresponding to those predicted by Pauling and Corey. This work was a very great achievement. Pauling received the 1954 Nobel prize in chemistry for his work on protein structure. Pauling's and Corey's work inspired another great biochemical breakthrough: the concept of the double-helix structure for deoxyribonucleic acid (DNA; see Section 31.6).

Proteins are very large molecules. But just how many amino acid units must be present for a substance to be a protein? There is no universally agreed-upon answer to this question. Some authorities state that a protein must have a molar mass of at least 6000 or contain about 50 amino acid residues. Smaller amino acid polymers, containing from 5 to 50 amino acid residues, are classified as polypeptides and are not proteins. In reality, there is no clearly defined lower limit to the molecular size of proteins. The distinction is made to emphasize (1) that proteins usually serve structural or enzymatic functions, while polypeptides usually serve hormone-related functions, and (2) that the three-dimensional conformation of proteins is directly related to function, while the relationship is not so definitive with polypeptides.

In general, for a protein molecule to serve a specific biological function, it must have a closely defined overall shape. Chemists typically describe large proteins on several levels: (1) a primary structure, (2) a secondary structure, (3) a tertiary structure, and, for the most complex proteins, (4) a quaternary structure.

primary structure

The **primary structure** of a protein is established by the number, kind, and sequence of amino acid residues composing the polypeptide chain or chains making up the molecule. The primary structure determines the alignment of side-chain characteristics, which, in turn, determines the three-dimensional shape into which the protein folds. In this sense, the amino acid sequence is of primary importance in establishing protein shape.

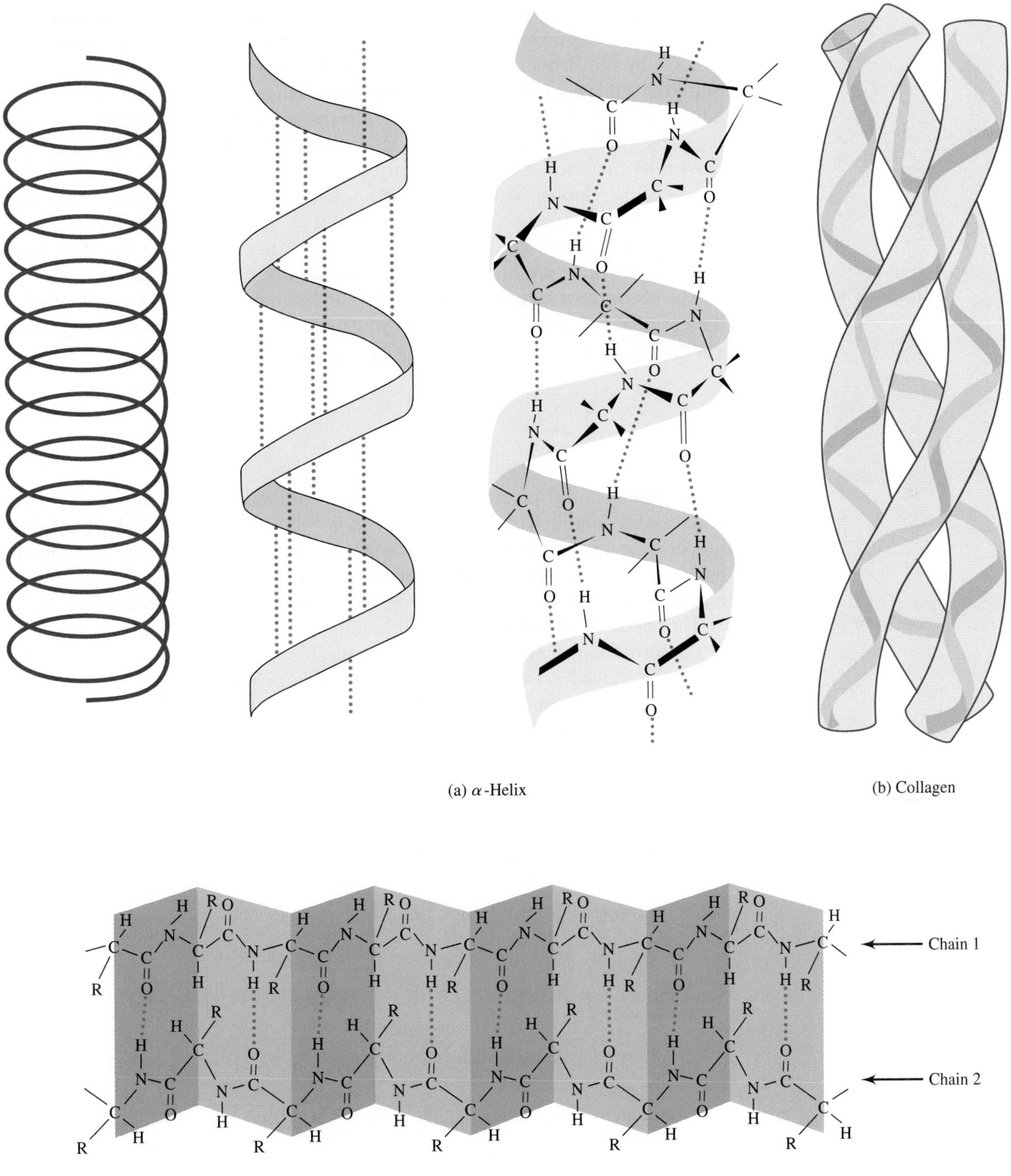

(a) α-Helix

(b) Collagen

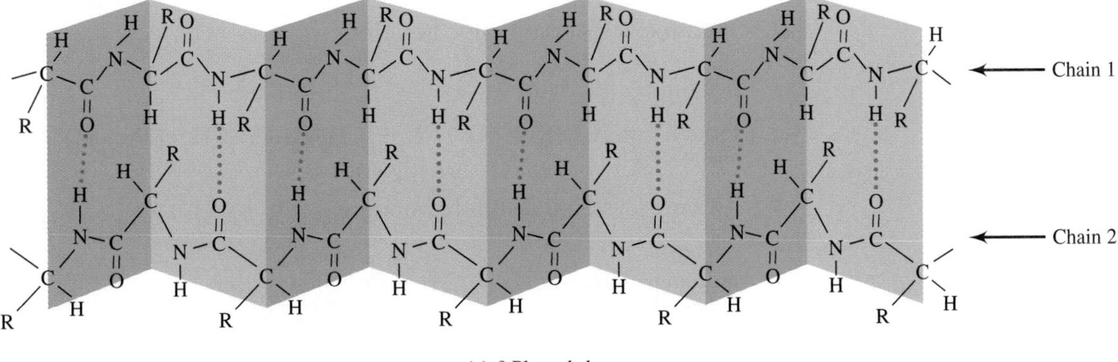

(c) β-Pleated sheet

Figure 29.1
Three common protein secondary structures.

Determining the sequence of the amino acids in even one protein molecule was a formidable task. The amino acid sequence of beef insulin was announced in 1955 by the British biochemist Frederick Sanger. This accomplishment required several years of effort by a team under Sanger's direction. He was awarded the 1958 Nobel prize in chemistry for this work. Insulin is a hormone that regulates the blood-sugar level. A deficiency of insulin leads to the condition known as diabetes. Human insulin consists of 51 amino acid units in two polypeptide chains. The two chains are connected by disulfide linkages (—S—S—) of two cysteine residues at two different sites. The primary structure is shown in Figure 29.2. Insulin from other animals, differs slightly by one, two, or three amino acid residues in chain A.

secondary structure

The **secondary structure** of proteins can be characterized as a regular three-dimensional structure held together by hydrogen bonding between the oxygen of the $>C{=}O$ and the hydrogen of the $H{-}N<$ groups in the polypeptide chains:

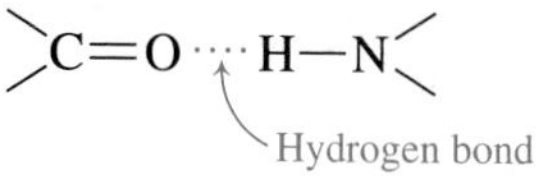

$$>C{=}O \cdots H{-}N<$$

Hydrogen bond

The α-helical and β-pleated-sheet structures of Pauling and Corey are two examples of secondary structure. As shown in Figure 29.1, essentially every peptide bond is involved in at least one hydrogen bond in these structures. Proteins having α-helical or β-pleated-sheet secondary structures are strongly held in particular conformations by virtue of the large number of hydrogen bonds.

tertiary structure

The **tertiary structure** of a protein refers to the distinctive and characteristic conformation, or shape, of a protein molecule. This overall three-dimensional conformation is held together by a variety of interactions between amino acid side chains. These interactions include (1) hydrogen bonding, (2) ionic bonding, and (3) disulfide bonding. Here are some examples:

1. Glutamic acid–tyrosine hydrogen bonding

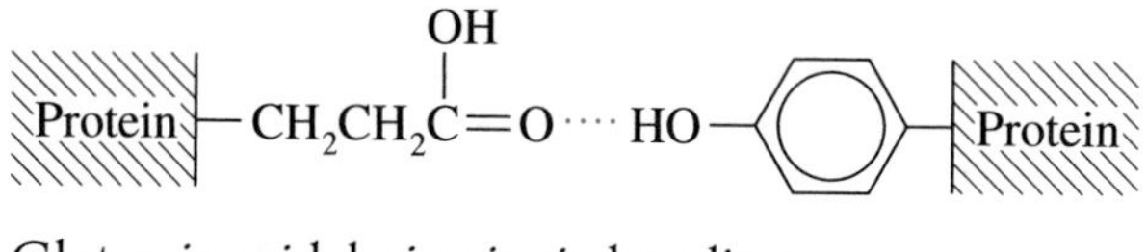

$$\text{Protein}-CH_2CH_2\overset{\overset{OH}{|}}{C}{=}O \cdots HO-C_6H_4-\text{Protein}$$

2. Glutamic acid–lysine ionic bonding

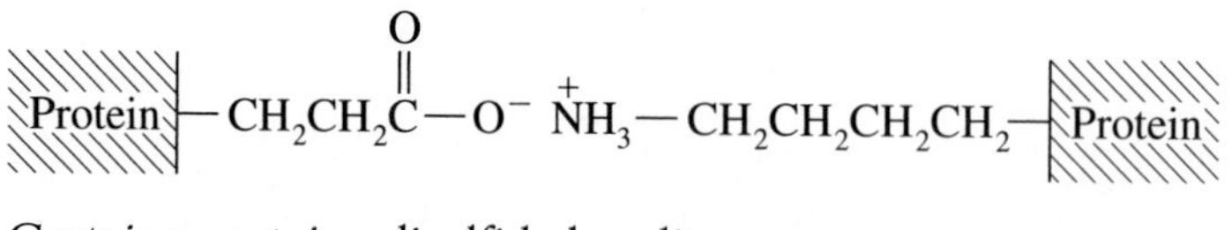

$$\text{Protein}-CH_2CH_2\overset{\overset{O}{\|}}{C}-O^- \; \overset{+}{N}H_3-CH_2CH_2CH_2CH_2-\text{Protein}$$

3. Cysteine–cysteine disulfide bonding

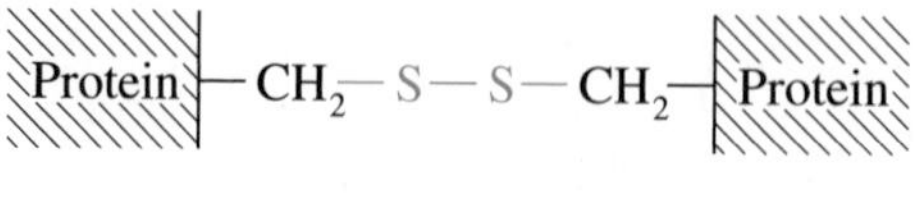

$$\text{Protein}-CH_2-S-S-CH_2-\text{Protein}$$

The tertiary structure depends on the number and locations of these interactions, variables that are fixed when the primary structure is synthesized. Thus, the tertiary structure depends on the primary structure. For example, there are three locations in the insulin molecule (Figure 29.2) where the primary sequence permits disulfide bonding. This specificity has an obvious bearing on the shape of insulin.

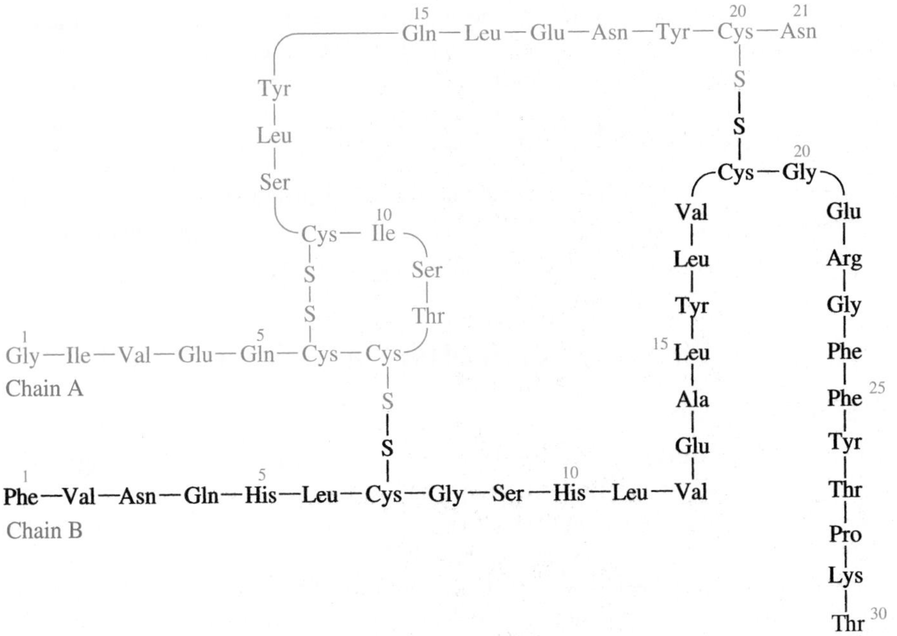

Chain A: Gly—Ile—Val—Glu—Gln—Cys—Cys—Thr—Ser—Ile—Cys—Ser—Leu—Tyr—Gln—Leu—Glu—Asn—Tyr—Cys—Asn (residues 1, 5, 10, 15, 20, 21)

Chain B: Phe—Val—Asn—Gln—His—Leu—Cys—Gly—Ser—His—Leu—Val—Glu—Ala—Leu—Tyr—Leu—Val—Cys—Gly—Glu—Arg—Gly—Phe—Phe—Tyr—Thr—Pro—Lys—Thr (residues 1, 5, 10, 15, 20, 25, 30)

Disulfide bonds (S—S): Chain A Cys 6 – Chain A Cys 11; Chain A Cys 7 – Chain B Cys 7; Chain A Cys 20 – Chain B Cys 19

Figure 29.2 Amino acid sequence of human insulin.

Hair is especially rich in disulfide bonds. These can be broken by certain reducing agents and restored by an oxidizing agent. This fact is the key to "cold" permanent waving of hair. Some of the disulfide bonds are broken by applying a reducing agent to the hair. The hair is then styled with the desired curls or waves. These are then permanently set by using an oxidizing agent to reestablish the disulfide bonds at different points. The two reactions involved are as follows:

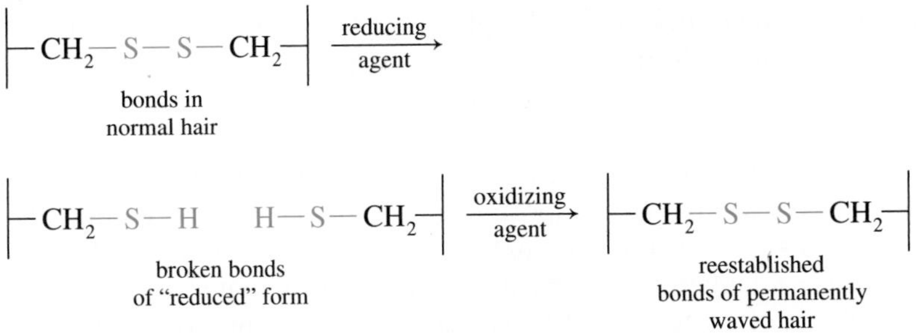

$$\vdash CH_2-S-S-CH_2 \dashv \xrightarrow[\text{agent}]{\text{reducing}}$$

bonds in normal hair

$$\vdash CH_2-S-H \quad H-S-CH_2 \dashv \xrightarrow[\text{agent}]{\text{oxidizing}} \vdash CH_2-S-S-CH_2 \dashv$$

broken bonds of "reduced" form — reestablished bonds of permanently waved hair

In a permanent wave, disulfide bonds in the hair are broken with reducing agents, set to a curling pattern, and then restored by an oxidizing agent.

If we examine Table 29.1 closely, we can conclude that most amino acids have side chains that form neither hydrogen, ionic, nor disulfide bonds. What, then, do the majority of amino acids do to hold together the tertiary structure of a protein? This is an important question and leads to an equally important answer. The uncharged relatively nonpolar amino acids form the center, or core, of most proteins. These amino acids have side chains that do not bond very well to water; they are like saturated hydrocarbons or lipids and are hydrophobic. When a protein is synthesized, the uncharged, nonpolar amino acids turn inward toward each other, excluding water and forming the core of the protein structure.

quaternary structure

A fourth type of structure, called a **quaternary structure**, is found in some complex proteins. These proteins are made of two or more smaller protein subunits (polypeptide chains). Nonprotein components may also be present.

The quaternary structure refers to the shape of the entire complex molecule and is determined by the way in which the subunits are held together by *noncovalent* bonds—that is, by hydrogen bonding, ionic bonding, and so on.

Practice 29.5

Show a tertiary structure for the following pentapeptide *after* a disulfide bond forms: Ala-Cys-Gly-Gly-Cys. (Use the amino acid structures given in Table 29.1.)

29.8 Some Examples of Proteins and Their Structures

Fibrous Proteins

fibrous protein

Collagen **derives from the Greek word *kolla*, meaning "glue."**

The **fibrous proteins** are an important class of proteins that contain highly developed secondary structures. As their name suggests, fibrous proteins have a "fiberlike," or elongated, shape. Because secondary structures provide strength, these proteins tend to function in support roles. Several such proteins are shown in Figure 29.3. The α-keratins depend on the α-helix for strength and are found in such diverse structures as hair, feathers, horns, and hooves. The silk protein, fibroin, is folded into a β-pleated-sheet secondary structure.

The most abundant protein in the animal kingdom, collagen, is a fibrous protein with a unique secondary structure (Figure 29.3). Collagen forms the bone matrix around which the calcium phosphate mineral crystallizes. Ligaments, tendons, and skin are composed of a large proportion of collagen. The structure of this protein allows it to provide a strong framework for each of these tissues. Collagen is a long, slender protein whose three strands are wrapped one around another as a rope would be woven, and, like a rope, the finished product is much stronger than a single strand. These triple helices are in turn stacked like cordwood to form collagen fibers. The resulting tough material can be aligned to form tendons or deposited as a network to form a bone or skin structure. Next time you take a stride, think about the force exerted on your Achilles tendon and how your being able to function depends on the strength of collagen.

Collagen takes on an added importance for artificial biological implants. In fact, collagen is an almost ideal biomaterial: It is strong, is stable, and does not generally cause a dangerous allergic reaction. Artificial materials such as plastics or metals are often coated with collagen before implantation. Collagen can be used directly in artificial skin, in bone graft substitutes, and as vascular or tissue sealant.

Feathers contain a fibrous protein.

Globular Proteins

globular protein

The **globular proteins** share a similar compact shape. Although it may not be apparent at first, this roughly spherical structure allows each globular protein its own complex tertiary structure. Each unique structure is associated with a particular function in the biological world. Some proteins function to bind specific biochemicals. This function is important for biological transport (e.g., oxygen transport by hemoglobin) and for processes such as the immune response (e.g., immunoglobulins). Other globular proteins bind biochemicals and then catalyze a chemical reaction (e.g., enzymes). A catalyst is required for essentially all metabolic reactions.

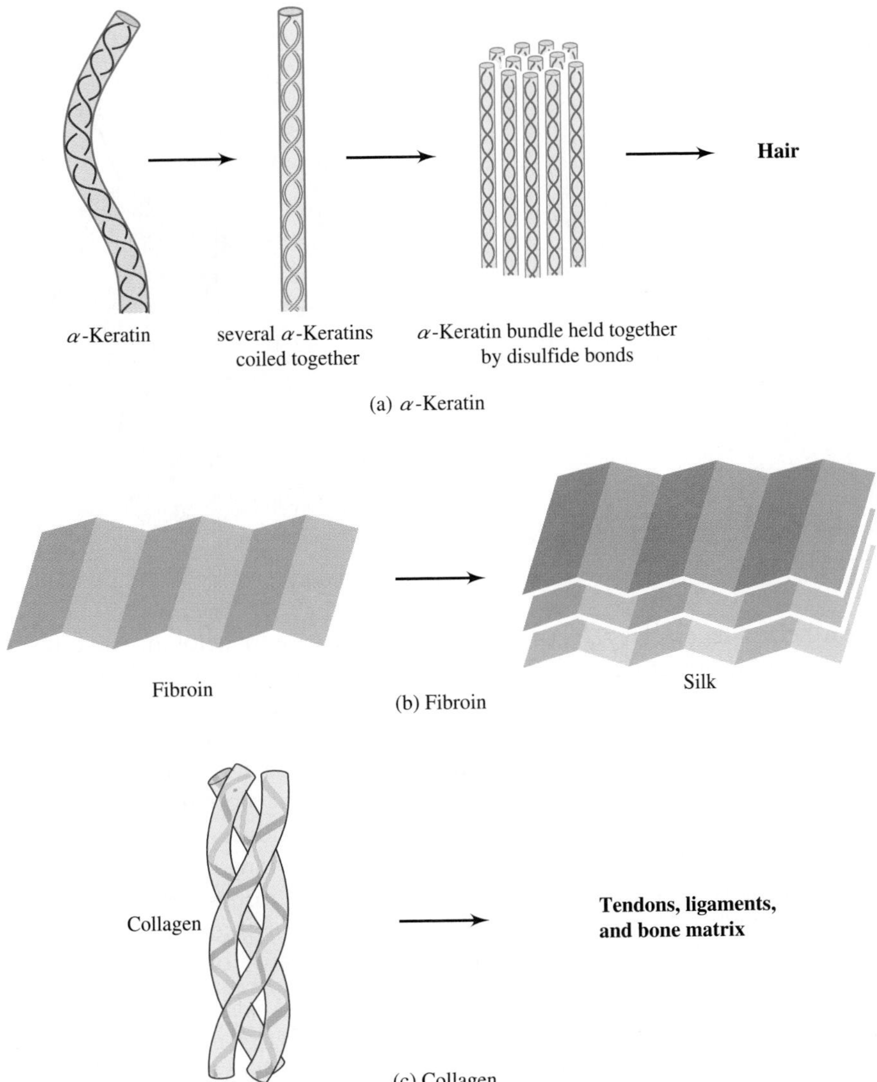

Figure 29.3
Three examples of fibrous proteins.

We will consider several protein structures in detail. These proteins contain hundreds to thousands of atoms and look very complicated at first glance. However, scientists have learned to look for much simpler, major structural features. These features are often related to a protein's function. To help visualize the various molecular relationships, we will use drawing techniques that emphasize major structural features.

Myoglobin—A Binding Protein Myoglobin is one of the simplest of the globular proteins (153 amino acid residues with a molar mass of about 17,000 g/mol). It functions as a *binding protein*, carrying oxygen for muscle tissue. The spacefilling model in Figure 29.4a shows myoglobin to be a globular protein. To better visualize the details of myoglobin's structure, we must strip away the amino acid side chains and focus only on the covalent chain linking the amino acids, as seen in Figure 29.4b. This simplified view shows myoglobin with a folded-sausage structure, where each section of the sausage represents a segment of an α-helix.

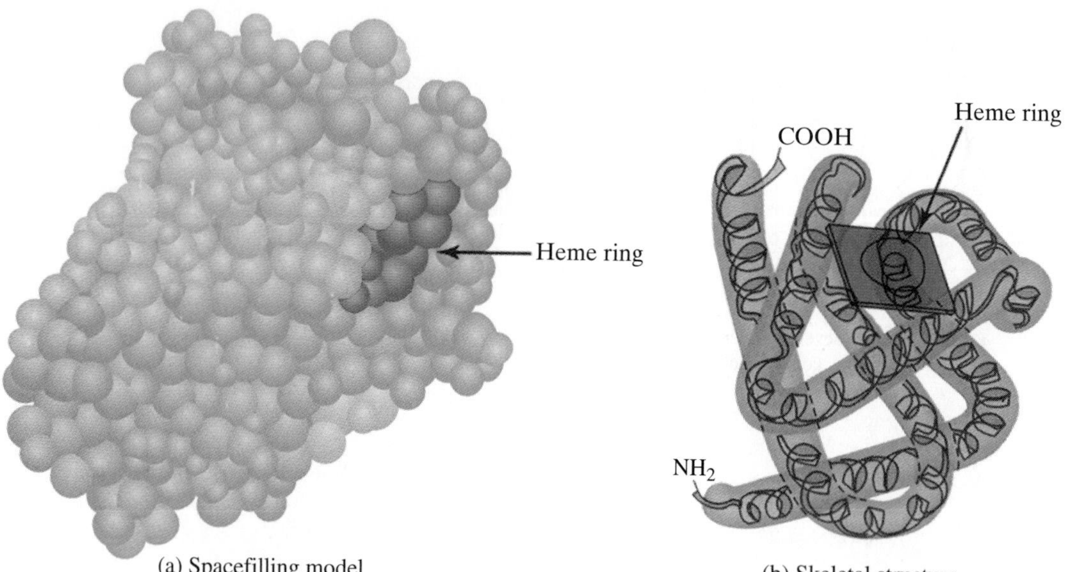

Figure 29.4
Representations of myoglobin.

The α-helical secondary structure provides stability for myoglobin. The folds of the myoglobin structure are held firmly in place by hydrogen bonds and other interactions between amino acid side chains. Within the center of this structure is an organically bound iron atom (in the heme group, shown in Figure 29.4b by a rectangular solid) that allows myoglobin to store and transport oxygen. Because of the heme group, myoglobin is a *conjugated* protein—that is, a protein which contains groups other than amino acid residues.

Jaundice is a disease caused by the improper processing of the heme group. As the heme is degraded, bilirubin forms. If the liver does not continuously excrete this yellow pigment, bilirubin builds up in body tissues. Jaundice is characterized by a yellowing of the skin and may be indicative of liver malfunction.

Carboxypeptidase A—An Enzyme Carboxypeptidase A is a small enzyme (307 amino acid residues with a molar mass of 34,500 g/mol) that catalyzes protein digestion in the small intestine. The spacefilling model shows carboxypeptidase A to be a globular protein (Figure 29.5a). The major structural features of carboxypeptidase A are illustated with a "ribbon structure" (Figure 29.5b). As we saw in myoglobin, this structure shows the position of the protein chain and leaves out the location of the amino acid side chains. An α-helix is represented by a curling line,

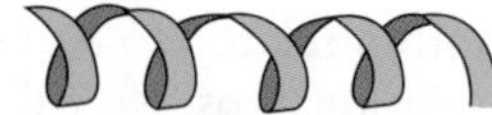

while each strand of β-pleated sheet is a broader line with an arrowhead at its end, to show which way the protein chain is running (from the amino end toward the carboxyl terminus):

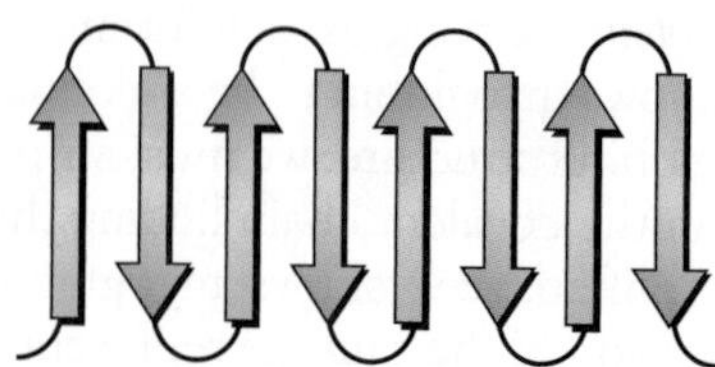

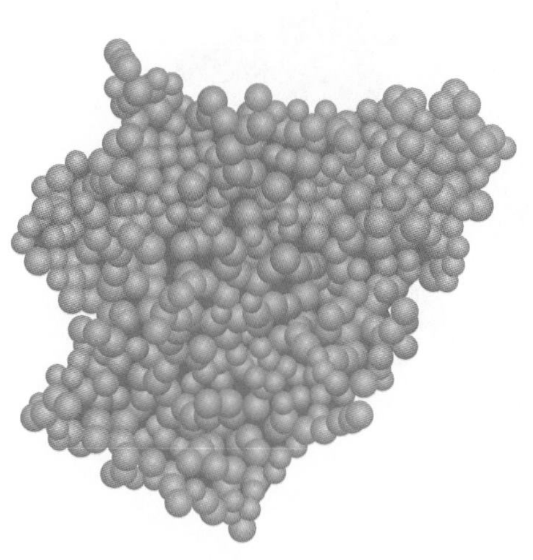

(a) Spacefilling model

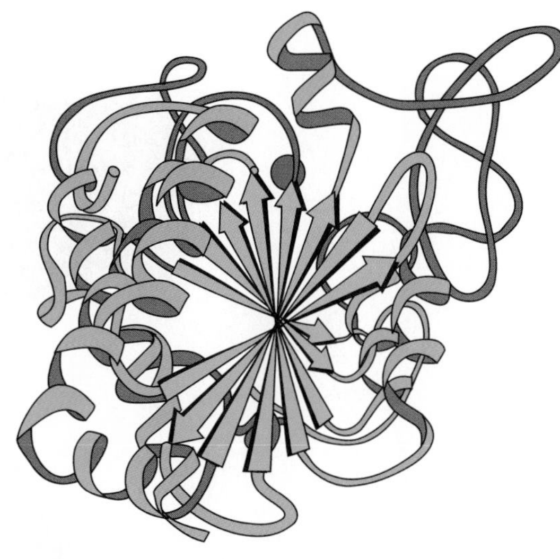

(b) Ribbon structure

Figure 29.5
Representations of carboxypeptidase A.

A β-pleated sheet provides a stable core for carboxypeptidase A. This sheet is often called a fan structure in order to describe the appearance of the twisted β-pleated sheet. A set of α-helices surrounds carboxypeptidase A's β-pleated sheet and further strengthens the protein structure.

Just as various animal species have common bone structures, proteins share common sketal structures. Some proteins use the α-helix like myoglobin (see Figure 29.4); spiral tubes are twisted around to support the bulk of these proteins. In other proteins, the β-pleated sheet is configured to provide a rigid protein framework. Some proteins have a fan of β-pleated sheet (like carboxypeptidase A), while, for other molecules, the β-pleated sheet has wrapped around on itself to form a barrel. The β-pleated sheets often act as the core of a protein, with the α-helices on the outside.

Larger proteins are commonly folded into a number of compact globular units. These units are about the size of myoglobin or carboxypeptidase A and are called **protein domains**. In this way, a larger protein is like several myoglobin-sized proteins connected together by peptide bonds. Each protein domain has its own secondary structure that usually performs a discrete task in a protein's overall function.

protein domain

Immunoglobulin G—A Binding Protein Immunoglobulins (antibodies) are large binding proteins (for immunoglobulin G, about 1320 amino acid residues with a molar mass of about 145,000 g/mol) that serve a critical biological function. These proteins bind molecules that are foreign to the body, as a defense against disease. Figure 29.6 shows immunoglobulins to be complex molecules. They are composed of more than one polypeptide chain and have a quaternary structure, as well as a primary, a secondary, and a tertiary structure. Each polypeptide chain has several protein domains, and β-pleated sheets form a stable core within each protein domain.

Diagrams of an immunoglobulin are often used to describe the binding function of these proteins (Figure 29.7). A successful immunological response requires that our bodies be prepared with a protein whose binding site is closely complementary to each of millions of different invading molecules (antigens).

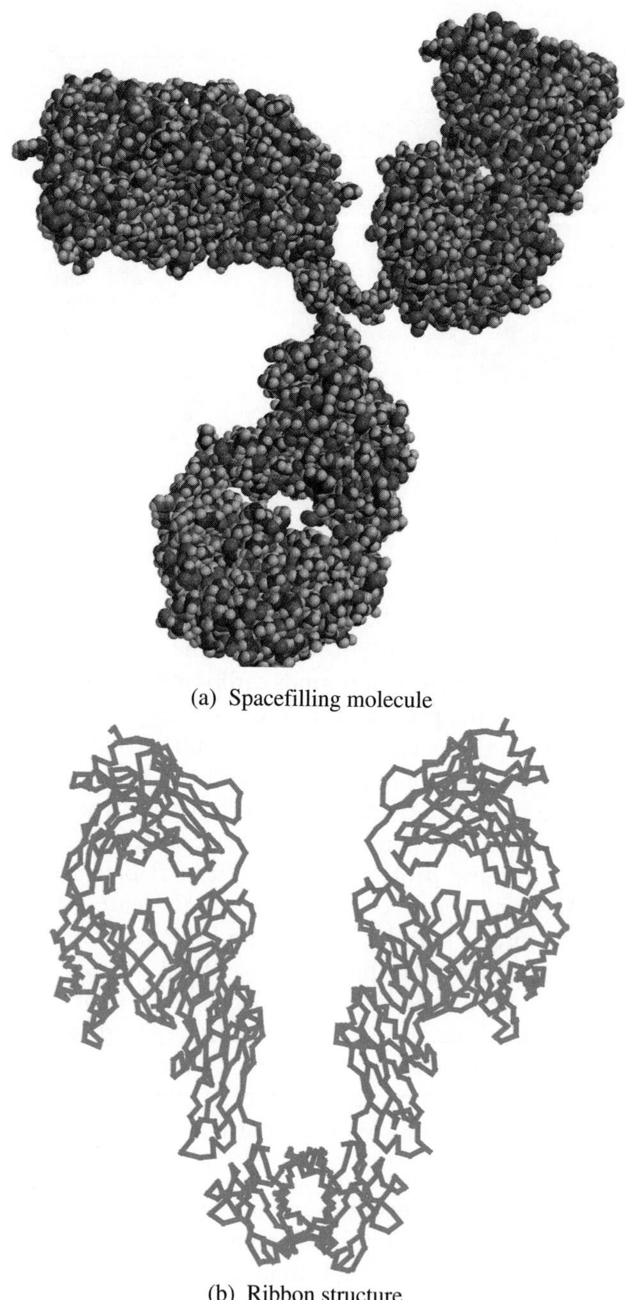

(a) Spacefilling molecule

(b) Ribbon structure

Figure 29.6
Representations of an immunoglobulin G protein.

The way in which our body accomplishes this feat appears to start with the shape of a typical immunoglobulin G (Figure 29.7). Antigen binding regions are located at the ends of two arms, hinged together so that the distance between binding sites varies with the size of the invading particle. The immunoglobulin protein chains are endowed with special hypervariable regions; every different immunoglobulin has a distinct amino acid sequence in its antigen binding sites and therefore has sites of unique size and shape. The

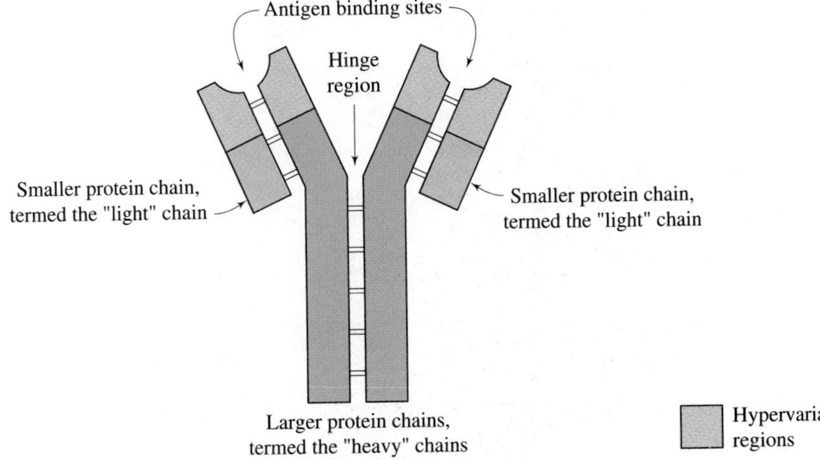

Figure 29.7
Diagrammatic shape of a typical immunoglobulin G protein.

body ensures that there will be immunoglobulins for many different antigens by producing millions of proteins with different hypervariable regions. Note that two different protein chains make up the antigen binding site. Since each chain has a hypervariable region that can be changed independently, the body has the ability to produce upwards of 10^{10} (10 billion) immunoglobulins with different binding sites—more than enough to handle exposure to invading particles.

Hemoglobin—A Binding Protein Quaternary structure is important in proteins that are involved in the control of metabolic processes. For example, hemoglobin (574 amino acid residues with a total molar mass of 64,500 g/mol), the oxygen transport protein of the blood, is composed of four subunits (Figure 29.8). Each subunit resembles myoglobin in that a set of helical segments surrounds the oxygen-binding heme group. The important structural difference is that hemoglobin has a quaternary structure. This leads to an equally important functional difference: Whereas myoglobin always binds and releases oxygen in the same way, hemoglobin changes its oxygen-binding characteristics according to the amount of available oxygen. When hemoglobin binds oxygen on one subunit, the protein's conformation changes to facilitate the binding of three additional oxygen molecules. (See Figure 29.9.) Conversely, the loss of one oxygen from hemoglobin facilitates the release of oxygen from the other sites. The oxygen binding is said to be *cooperative*. Cooperativity implies a shared action: When one O_2 binds, three more are encouraged to bind. When one O_2 leaves, three more are encouraged to leave.

Even minor alterations in primary structure may have drastic effects on the three-dimensional structural function of a protein. Sickle-cell anemia is a graphic example of this fact. This crippling genetic disease is due to red blood cells assuming a sickle shape. Sickled cells have impaired vitality and function. Sickle-cell anemia has been traced to a small change in the primary structure, or amino acid residue sequence, of hemoglobin. Normal hemoglobin molecules contain four polypeptide subunits—two identical α-polypeptide chains and two identical β-polypeptide chains. Sickle-cell and normal β-polypeptide chains of hemoglobin each contain 146 amino acid residues and differ only

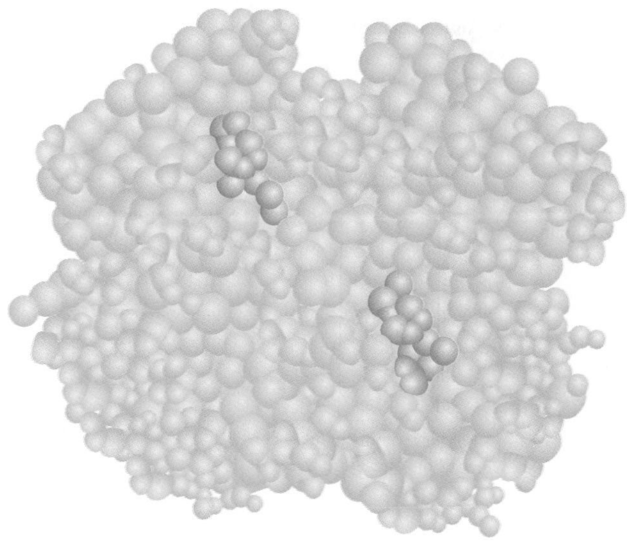

(a) Spacefilling model

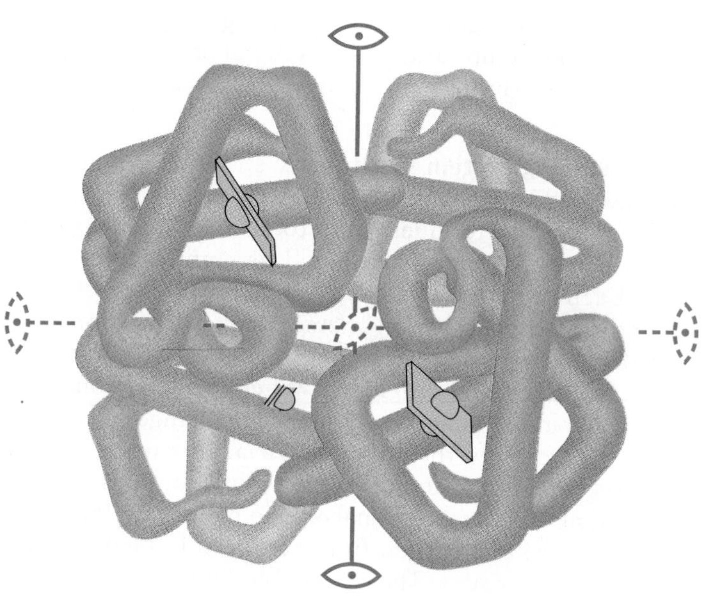

(b) Skeletal structure

Figure 29.8
Representations of hemoglobin. In the spacefilling model, just the edges of the heme groups (shown in blue) are visible in the two closest subunits.

in the residue at the sixth position. In the sickle-cell hemoglobin chain, a glutamic acid residue has been replaced by a valine residue at the sixth position. This change is sufficient, under some circumstances, to cause hemoglobin to aggregate into long filaments—a different quaternary structure. Large amounts of hemoglobin are present in red blood cells, and the changed quaternary structure causes sickling of the cell and greatly diminishes its vitality. (See Figure 29.10.) This cell affliction has led to the premature deaths of many affected individuals.

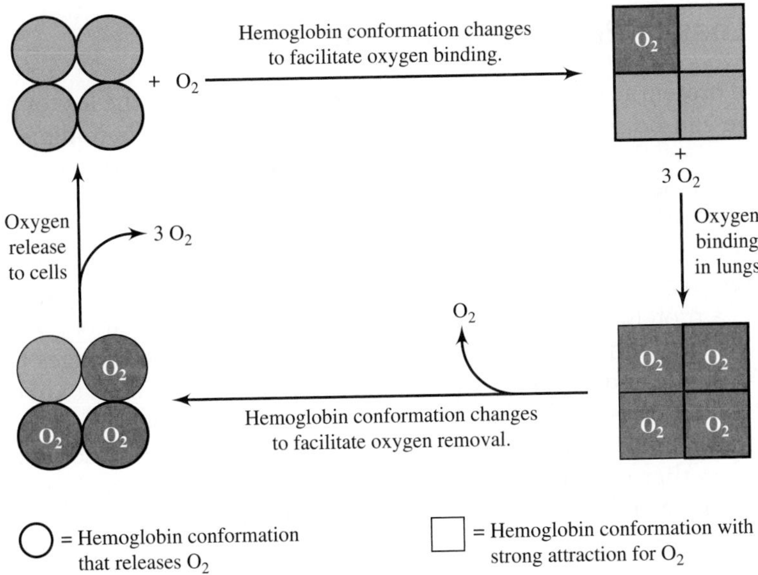

Figure 29.9
The oxygen–hemoglobin binding process. The circles and the squares represent two different conformations of the hemoglobin molecule. Oxygen binding or release causes hemoglobin to change its conformation.

Practice 29.6

(a) Circle each alpha-helix in the following protein:

cytochrome b–a mitochondrial protein (notice the heme group shown as a ball-and-stick model in the lower left-hand part of the cytochrome)

(b) Estimate how many domains are present in the following protein:

calmodulin—a calcium-binding protein

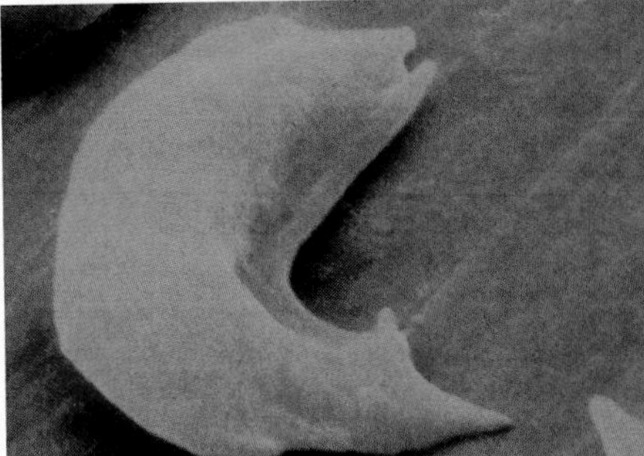

Figure 29.10
Scanning electron micrograph of sickled red blood cell (top) and normal red blood cell (bottom).

29.9 Loss of Protein Structure

denaturation

Because protein structure is so important to life's functions, the loss of protein structure can be crucial. If a protein loses only its natural three-dimensional conformation, the process is referred to as **denaturation**. In contrast, the hydrolysis of peptide bonds ultimately converts proteins into their constitutent amino acids. Denaturation often precedes hydrolysis.

Denaturation involves the alteration or disruption of the secondary, tertiary, or quaternary—but not primary—structure of proteins. (See Figure 29.11.) Because a protein's function depends on its natural conformation, biological activity is lost with denaturation. This process may involve changes ranging from the subtle and reversible alterations caused by a slight shift in pH to the extreme alterations involved in tanning a skin to form leather.

Environmental changes may easily disrupt natural protein structure, which is held together predominantly by noncovalent, relatively weak bonds. As gentle an act as pouring a protein solution can cause denaturation. Purified proteins must often be stored under ice-cold conditions because room temperature denatures them. It is not surprising that a wide variety of chemical and physical agents can also denature proteins, including strong acids and strong bases, salts (especially those of heavy metals), certain specific reagents such as tannic acid and picric acid, alcohol and other organic solvents, detergents, mechanical action such as whipping, high temperature, and ultraviolet radiation. Denatured proteins are generally less soluble than natural proteins and often coagulate or precipitate from solution. Cooks have taken advantage of this for many years. When egg white, which is a concentrated solution of egg albumin protein, is stirred vigorously (as with an egg beater), an unsweetened meringue forms; the albumin denatures and coagulates. A cooked egg solidifies partially because egg proteins, including albumin, are denatured by heat. (See Figure 29.12.)

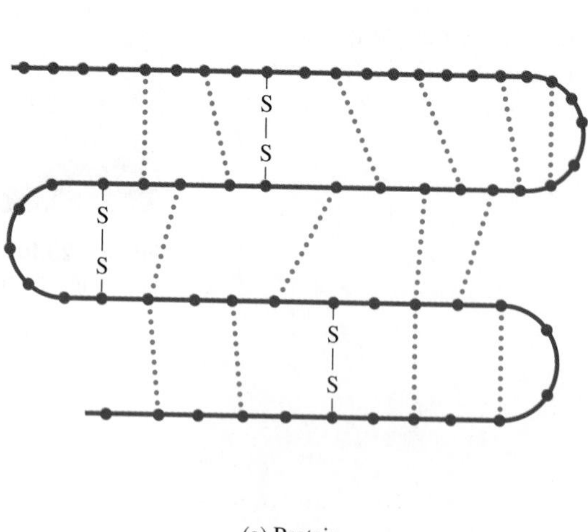

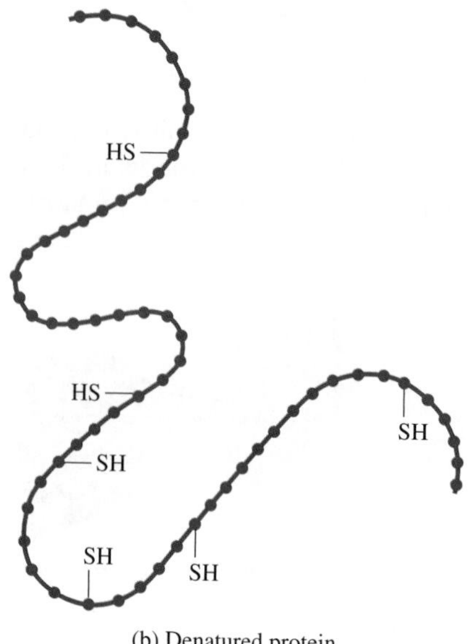

Figure 29.11
Structural change when a protein molecule is denatured: The relatively weak hydrogen, electrostatic, and disulfide bonds are broken, resulting in a change of structure and properties (...denotes the noncovalent bonds that stabilize a protein's natural conformation).

Figure 29.12
Left: Raw egg white is concentrated protein. Middle: In a cooked egg, the proteins are denatured by heat. Right: If egg whites are beaten, the protein denatures and coagulates.

In clinical laboratories, the analysis of blood serum for small molecules such as glucose and uric acid is hampered by the presence of serum protein. This problem is resolved by first treating the serum with an acid to denature and precipitate the protein. The precipitate is removed, and the protein-free liquid is then analyzed.

Loss of protein structure also occurs with hydrolysis of the peptide bonds to produce free amino acids. This chemical reaction destroys the protein's primary structure. Proteins can be hydrolyzed by boiling in a solution containing a strong acid such as HCl or a strong base such as NaOH. At ordinary temperatures, proteins can be hydrolyzed with the use of enzymes. (See Chapter 30.) These molecules, called proteolytic enzymes, are themselves proteins that function to catalyze or speed up the hydrolysis reaction. The essential reaction of hydrolysis is the breaking of a peptide linkage and the addition of the elements of water:

Peptide linkage in protein chain

$$\sim\!\underset{}{\overset{\large H}{\overset{|}{N}}}-\underset{\underset{\large H}{|}}{\overset{\large R}{\overset{|}{C}}}-\overset{\large O}{\overset{\|}{C}}-\underset{\underset{\large H}{|}}{N}-\underset{\underset{\large R}{|}}{\overset{\large H}{\overset{|}{C}}}-\underset{\underset{\large O}{\|}}{C}\!\sim + H_2O \longrightarrow \sim\!\overset{\large H}{\overset{|}{N}}-\underset{\underset{\large H}{|}}{\overset{\large R}{\overset{|}{C}}}-\overset{\large O}{\overset{\|}{C}}-OH + H_2N-\underset{\underset{\large R}{|}}{\overset{\large H}{\overset{|}{C}}}-\underset{\underset{\large O}{\|}}{C}\!\sim$$

(ww = the rest of the protein chain)

Any molecule with one or more peptide bonds, from the smallest dipeptide to the largest protein molecule, can be hydrolyzed. During hydrolysis, proteins are broken down into smaller and smaller fragments until all component amino acids are liberated.

Dietary protein must have its primary structure completely destroyed before it can provide nutrition for the body. Thus, digestion involves both denaturation and hydrolysis. Stomach acid causes most proteins to denature. Then, proteolytic enzymes in the stomach and the small intestine hydrolyze the proteins to smaller and smaller fragments, until the free amino acids are formed. They can then be absorbed through the intestinal membranes into the bloodstream.

CHEMISTRY IN ACTION • Proteins Gone Wrong

One of the basic tenets of biochemistry is that a correctly functioning protein requires a correct three–dimensional structure. Normally, loss of this correct structure means loss of function. Occasionally, loss of the correct three–dimensional shape means disaster. The following short sequence of events illustrates how important the correct protein three–dimensional structure is to life.

The prion is a common protein existing on the membrane of many cells. It is a small protein with not a very unusual structure. (In the accompanying picture, the alpha helical portions are marked in red while beta pleated sheet is shown in blue.)

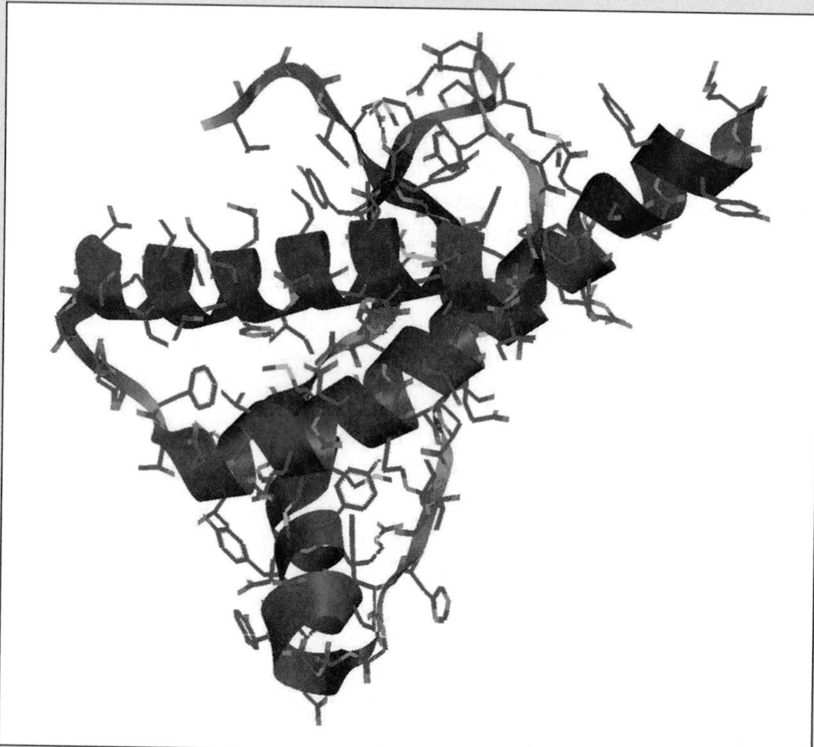

The high percentage of alpha helix (about 40% of the total structure) stabilizes the protein structure. The prion's function is not known for certain; perhaps it is important in metabolizing copper, an essential nutrient; perhaps it helps one cell recognize another. Mice have been raised without prions ("knockout" mice) and seem to live a fairly normal existence. Whatever the prions' function, they are known to be most numerous in the central nervous system.

Once in a great many times a prion will fold into a different three–dimensional shape. The reason for this misfolding is unknown. The prion amino acid sequence is unchanged, yet the three–dimensional structure folds incorrectly. The structure now has about 30% alpha helix and 45% beta pleated sheet—a firmly bonded protein. Normally a misfolded protein is destroyed. However, a misfolded prion is impervious to the cell's attempts to destroy it. Disaster follows.

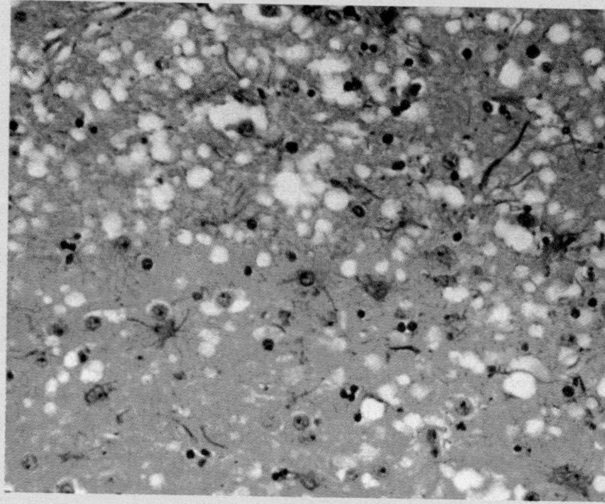

Prion disease gives brain tissue a spongy appearance (spongiform encephalopathy)

When an incorrectly shaped prion contacts a normal prion, the normal prion's shape is converted. Like water turning to ice, the normal prion's structure "freezes " into the incorrect structure. Then, the misshapen prions aggregate and kill the surrounding nerve tissue.

One misshapen prion can infect a the entire central nervous system. The disease progresses from insanity to death. Many different animal species can be infected: in cattle, it is called mad cow disease; in sheep, it is called scrapie. Humans have a one in a million chance of contracting this disease. This is the first disease known to be transmitted without the aid of a virus or bacteria, and scientists are uncertain about treatment regimes. However, suggested treatments focus on forcing the prion back to its normal three–dimensional shape. Correct protein shape is crucial to life.

29.10 Tests for Proteins and Amino Acids

Many tests have been devised to detect and distinguish among amino acids, peptides, and proteins. Some examples are described here.

Xanthoproteic Test Proteins containing a benzene ring—for example, the amino acids phenylalanine, tryptophan, and tyrosine—react with concentrated nitric acid to give yellow products. Nitric acid on skin produces a positive xanthoproteic test, as skin proteins are modified by this reaction.

Biuret Test A violet color is produced when dilute copper(II) sulfate is added to an alkaline solution of a peptide or protein. At least two peptide bonds must be present, as the color changes only when peptide bonds can surround the Cu^{2+} ion. Thus, amino acids and dipeptides do not give a positive biuret test.

Lowry Assay The Lowry assay is used to detect proteins. A dark violet-blue color is produced by a combination of the biuret reaction and the reduction of molybdates/tungstates by tyrosine and tryptophan amino acids. This combination of two reactions makes the Lowry assay much more sensitive than the biuret test.

Bradford Assay The Bradford assay is the most sensitive common test for proteins. A dark blue color develops as the protein binds to a specific dye, Coomassie Brilliant Blue.

Ninhydrin Test Triketohydrindene hydrate, generally known as *ninhydrin*, is an extremely sensitive reagent for amino acids:

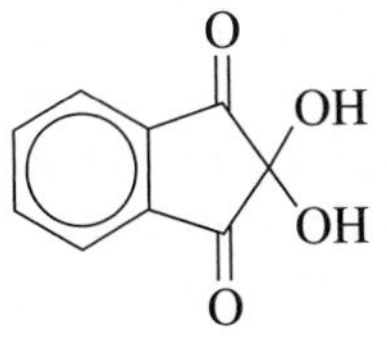

ninhydrin

All amino acids, except proline and hydroxyproline, give a blue solution with ninhydrin. Proline and hydroxyproline produce a yellow solution. Less than $1 \mu g$ ($10^{-6} g$) of an amino acid can be detected with ninhydrin.

Chromatographic Separation

Complex mixtures of amino acids are readily separated by thin-layer, paper, or column chromatography. In chromatographic methods, the components of a mixture are separated by means of differences in their distributions between two phases. Separation depends on the relative tendencies of the components to remain in one phase or the other. In *thin-layer chromatography* (TLC), for example, a liquid and a solid phase are used. The procedure is as follows: A tiny drop of a solution containing a mixture of amino acids (obtained by hydrolyzing a protein) is spotted on a strip (or sheet) coated with a thin layer of dried alumina or some other adsorbant. After the spot has dried, the bottom edge of the strip is placed in a suitable solvent. The solvent ascends the strip (by diffusion), carrying the different amino acids upward at different rates. When the solvent front nears the top, the strip is removed from the solvent

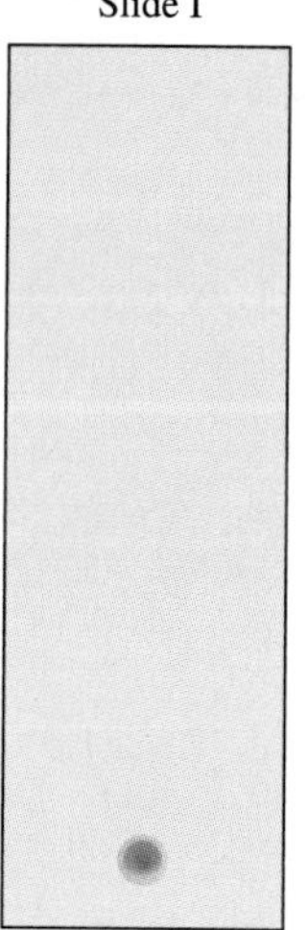

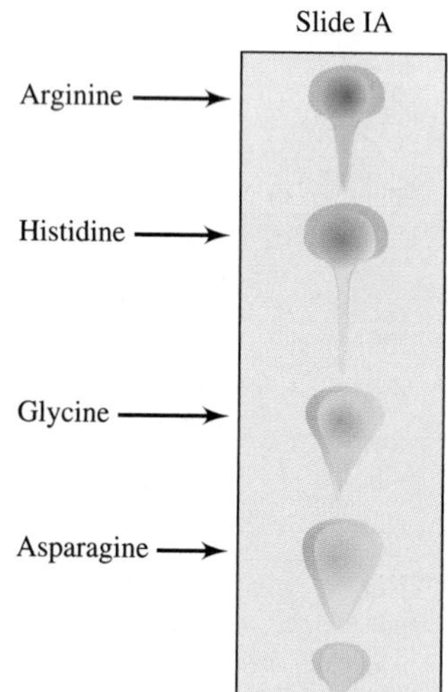

Figure 29.13
Chromatogram showing separation of selected amino acids. On slide I, spotted on the chromatographic strip is an amino acid mixture containing arginine, histidine, glycine, and asparagine. On slide IA is the developed chromatogram showing the separated amino acids after treatment with ninhydrin. The solvent is a 250 : 60 : 250 volume ratio of 1-butanol–acetic acid–water.

and dried. The locations of the different amino acids are established by spraying the strip with ninhydrin solution and noting where colored spots appear. The pattern of colored spots is called a chromatogram. The identities of the amino acids in an unknown mixture can be established by comparing the chromatogram of the mixture with a chromatogram produced by known amino acids. A typical chromatogram of amino acids is shown in Figure 29.13.

Since proteins tend to denature during thin-layer chromatography, they are often separated by *column chromatography.* In this technique, a solution containing a mixture of proteins (liquid phase) is allowed to percolate through a column packed with beads of a suitable polymer (solid phase). Separation of the mixture is accomplished as the proteins partition between the solid and liquid phases. The separated proteins move at different rates and are collected as they leave the column. Unlike the conditions under which amino acid chromatography takes place, mild conditions must be maintained to limit protein denaturation. In the past, it was not uncommon for protein column chromatography to take from 6 to 12 hours. Recent developments in the field of high-performance liquid chromatography (HPLC) have shortened this time to minutes.

Electrophoresis

Perhaps the most precise separation technique available to the protein chemist is electrophoresis. This technique separates proteins on the basis of a difference in size and a difference in charge. Electrophoresis is based on the principle that charged particles move in an electric field. Since different proteins contain differing numbers of positively and negatively charged amino acids, proteins will separate in an electric field. (See Figure 29.14.) Sometimes proteins with an overall positive charge are separated from proteins with an overall negative charge (Figure 29.14a). Most often, however, conditions are chosen so that all proteins have either a net positive or a net negative charge. Proteins with a greater charge are attracted more strongly to an electrode and move faster during electrophoresis. (See Figure 29.14b.) The faster moving proteins separate from the slower moving proteins.

Protein electrophoresis commonly is carried out in a gel matrix. This gel may be a natural polymer like starch or agarose, or it may be a purely synthetic polymer like polyacrylamide. No matter which gel is used, the proteins are forced to move through a sievelike material. They are acted upon by friction; the larger the protein, the more friction there is, and the slower the protein will tend to move. Thus, the speed at which a protein moves depends on both charge *and* size during standard electrophoresis. Tens of proteins may be separated precisely using one gel electrophoresis.

This technique has proved not only valuable, but also versatile. Several modifications of the standard procedure are used routinely:

1. *SDS (sodium dodecyl sulfate) electrophoresis:* In this procedure, a negatively charged detergent (SDS) is added to the protein solution and acts to mask differences in charge among proteins. The electrophoretic separation depends primarily on size differences. SDS electrophoresis is often used to determine the approximate molar mass of proteins.
2. *Isoelectric focusing:* During this type of electrophoresis, proteins are forced to move through a pH gradient. The proteins stop moving when they reach their *isoelectric point* pH, because, at this pH, the protein carries no net

At beginning of electrophoresis

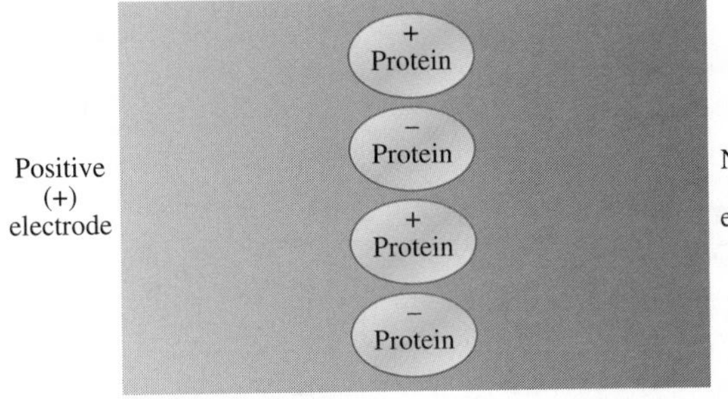

At a later time

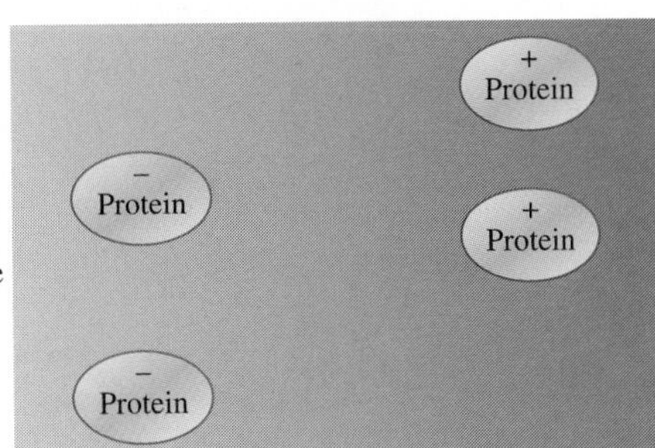

(a) Separating negatively charged proteins from positively charged proteins

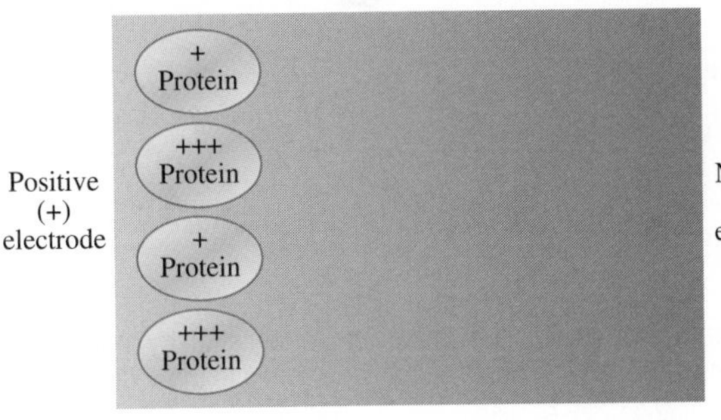

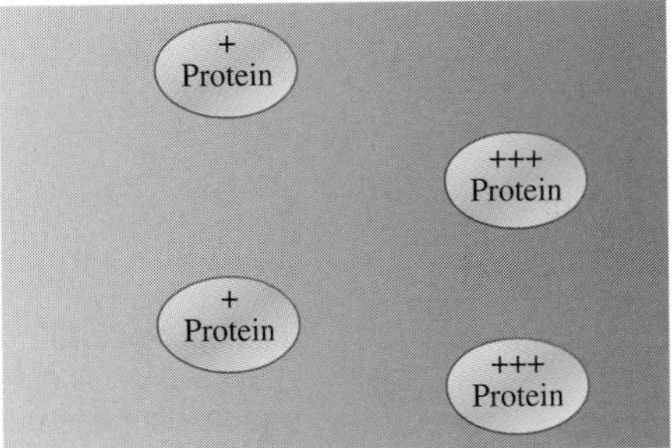

(b) Separating the more positively charged proteins from the less positively charged ones

Figure 29.14
Two forms of electrophoresis.

charge.(See Section 29.5.) The electrophoresis is completed once all the proteins have been "focused" at their isoelectric points. This technique separates proteins on the basis of the number and kinds of titratable amino acid side chains that they contain.

Today, two-dimensional electrophoresis is not uncommon. This procedure multiplies the benefits of an electrophoretic separation by running one separation in the y-direction, followed by a second electrophoresis in the x-direction, using the same gel slab. Figure 29.15 shows a separation of all proteins from a bacterium (*E. coli*). The proteins were first separated by molar mass, using SDS electrophoresis in the y-direction. Each band was then separated in the x-direction by using isoelectric focusing, a technique that separates proteins based on their charge. Over 1000 proteins have been separated on this one gel!

Electrophoresis can help a medical diagnosis that depends on protein identification. If this identification distinguishes proteins by their size and charge, electrophoresis is an ideal technique. For example, genetic diseases that affect hemoglobin can be diagnosed via isoelectric focusing. These diseases (e.g., sickle-cell anemia, the α- and β-thalassemias) may change only one amino acid in hemoglobin, yet they can be extremely serious. The thalassemias lead to an early death as a consequence of severe anemia.

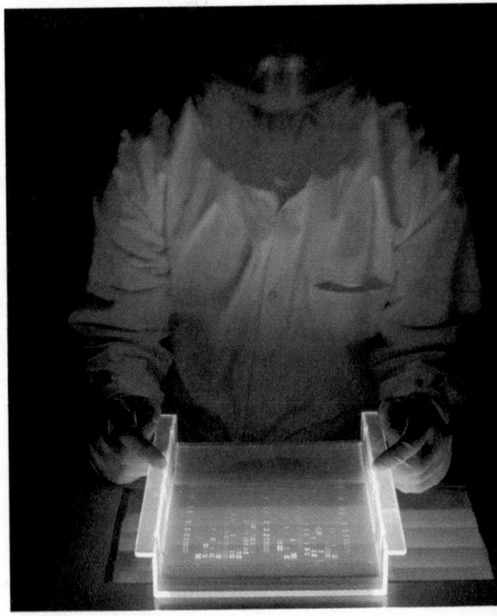

Electrophoresis is used in most biotech labs.

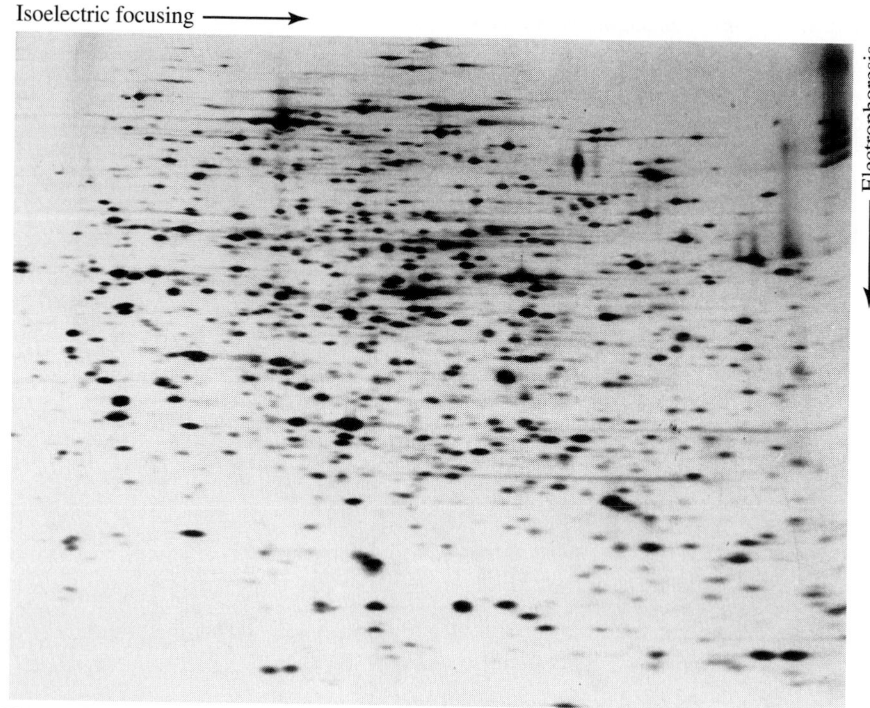

Figure 29.15
Two-dimensional electrophoresis of proteins from the *E. coli* bacterium. Separation by molar mass has been used in the y-direction and separation by charge has been used in the x-direction.

Often, these hemoglobinopathies are characterized by changes in hemoglobin's size and charge. Standard electrophoresis can distinguish these changes; however, isoelectric focusing provides added sensitivity that is especially important when small amounts of blood are used (e.g., with a newborn). For adults, these screening techniques have been especially useful as a genetic counseling tool.

29.11 Determination of the Primary Structure of Polypeptides

Sanger's Reagent The pioneering work of Frederick Sanger gave us the first complete primary structure of a protein, insulin, in 1955. He used specific enzymes to hydrolyze insulin into smaller peptides and amino acids, then separated and identified the hydrolytic products by various chemical reactions.

Sanger's reagent, 2-4-dinitrofluorobenzene (DNFB), reacts with the α-amino group of the N-terminal amino acid of a polypeptide chain. The carbon–nitrogen bond between the amino acid and the benzene ring of Sanger's reagent is more resistant to hydrolysis than are the remaining peptide linkages. Thus, when the substituted polypeptide is hydrolyzed, the terminal amino acid remains with the dinitrobenzene group and can be isolated and identified. The remaining peptide chain is hydrolyzed to free amino acids in the process. This method marked an important step in determining the amino acid sequence in a protein.

The Sanger's reagent reaction is illustrated in the following equations:

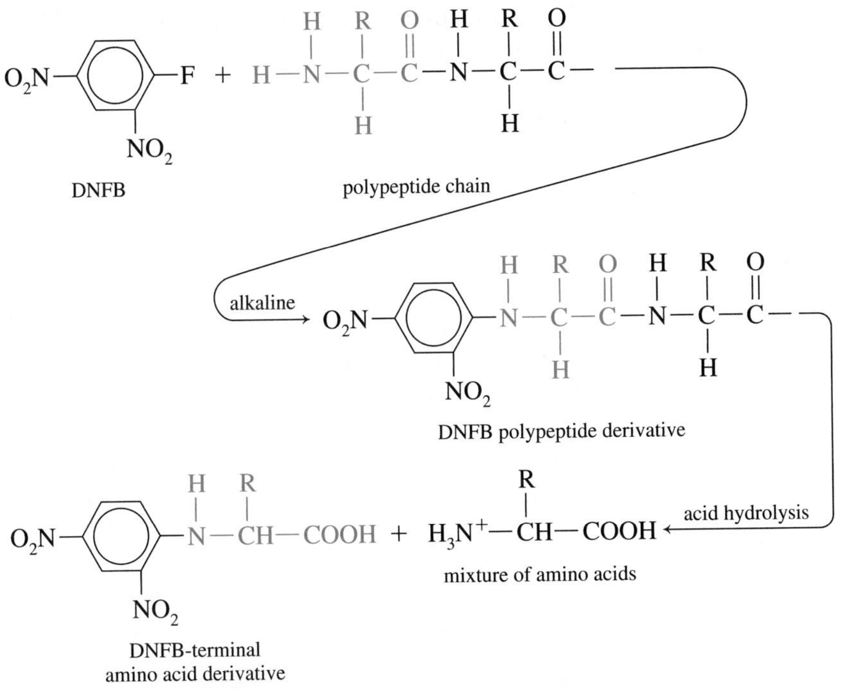

$$O_2N\text{-}C_6H_3(NO_2)\text{-}F + H\text{-}N(H)\text{-}CH(R)\text{-}C(=O)\text{-}N(H)\text{-}CH(R)\text{-}C(=O)\text{-}$$

DNFB polypeptide chain

alkaline →

$$O_2N\text{-}C_6H_3(NO_2)\text{-}N(H)\text{-}CH(R)\text{-}C(=O)\text{-}N(H)\text{-}CH(R)\text{-}C(=O)\text{-}$$

DNFB polypeptide derivative

acid hydrolysis →

$$O_2N\text{-}C_6H_3(NO_2)\text{-}N(H)\text{-}CH(R)\text{-}COOH + H_3N^+\text{-}CH(R)\text{-}COOH$$

DNFB-terminal amino acid derivative

mixture of amino acids

From the DNFB hydrolysis, Sanger learned which amino acids are present in insulin. By less drastic hydrolysis, he split the insulin molecule into peptide fragments consisting of two, three, four, or more amino acid residues. After analyzing vast numbers of fragments via the N-terminal method, he pieced them together in the proper sequence by combining fragments with overlapping structures at their ends, finally elucidating the entire insulin structure. As an example, consider the overlap that occurs between the hexapeptide and heptapeptide shown here:

Gly-Glu-Arg-	Gly-Phe-Phe		Hexapeptide
	Gly-Phe-Phe-	Tyr-Thr-Pro-Lys	Heptapeptide
Gly-Glu-Arg-	Gly-Phe-Phe-	Tyr-Thr-Pro-Lys	Decapeptide

The three residues, Gly-Phe-Phe, at the end of the hexapeptide match the three residues at the beginning of the heptatpeptide. By using these three residues in common, the structure of the decapeptide shown, which occurs in chain B of insulin (residues 20–29), is determined. (See Figure 29.2.)

Edman Degradation The Edman degradation method has been developed to split off amino acids one at a time from the N-terminal end of a polypeptide chain. In this procedure, the reagent, phenylisothiocyanate, is first added to the N-terminal amino group. The N-terminal amino acid–phenylisothiocyanate addition product is then cleaved from the polypeptide chain. The resultant substituted phenyl-thiohydantoin is isolated and identified. The shortened polypeptide chain is then ready to undergo another Edman degradation.

By repeating this set of reactions, chemists determine the amino acid sequence of a polypeptide directly. A machine, the protein sequenator, which carries out protein sequencing by Edman degradation automatically, is now available.

The Edman degradation process occurs as follows:

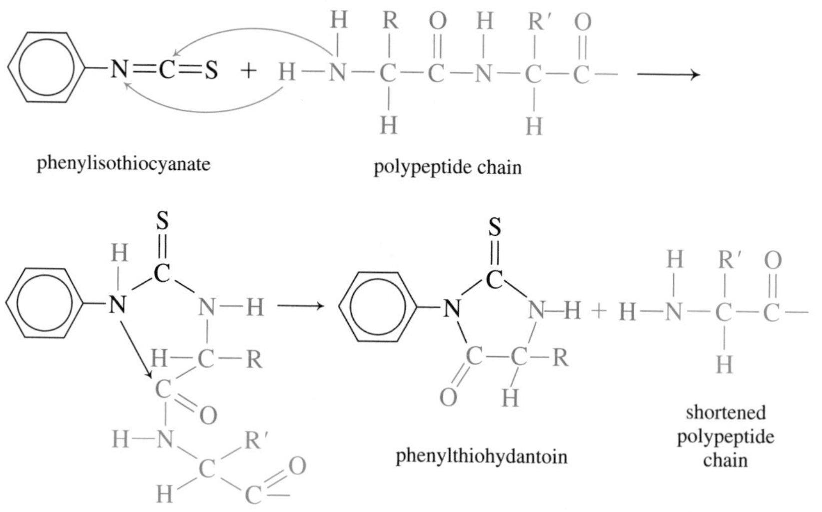

phenylisothiocyanate

polypeptide chain

phenylthiohydantoin

shortened polypeptide chain

Practice 29.7

What is the amino acid sequence of a nonapeptide (nine amino acids) based on the following information:

(1) Three rounds of Edman degradation yield sequentially Ala, then Gly, and then Lys;

(2) Hydrolysis yields three fragments: a tripeptide, Gly-Gly-Gly; a dipeptide, Ala-Gly; and a pentapeptide, Lys-His-Ala-Thr-Gly.

29.12 Synthesis of Peptides and Proteins

Since each amino acid has two functional groups, it is not difficult to form dipeptides or even fairly large polypeptide molecules. As mentioned in Section 29.6, glycine and alanine react to form two different dipeptides: glycylalanine and alanylglycine. If threonine, alanine, and glycine react, six tripeptides, each made up of three different amino acid units, are obtained. By reacting a mixture of several amino acids, polypeptides are produced. This clarifies the fact that even a small protein like insulin cannot be produced simply by reacting a mixture of the required amino acids. When such a large polypeptide is synthesized, it is formed by joining amino acids, one by one, in the proper sequence. Within cells, the proper sequence is maintained by careful genetic control. (See Section 31.10.)

When a polypeptide chain of known primary structure is synthesized *in vitro* (synthesis in "glass" without the aid of living tissue), the process must be controlled so that only one particular amino acid is added at each stage. Remarkable progress has been made in developing the necessary techniques.

Amino acids are used in which either the amino or carboxylic acid group has been inactivated or blocked with a suitable *blocking agent*. The blocked amino or carboxylic acid group is reactivated by removing the blocking agent in the next stage of the synthesis. For example, if the tripeptide Thr-Ala-Gly is to be made, amino-blocked threonine is reacted with carboxyl-blocked alanine to make the doubly blocked Thr-Ala dipeptide. The carboxyl-blocked alanine end of the dipeptide can then be reactivated and reacted with carboxyl-blocked glycine to make the blocked Thr-Ala-Gly tripeptide. The free tripeptide can then be obtained by removing the blocking agents from both ends. Using [B] to designate the blocking agents, this synthesis is represented schematically as follows:

$$\boxed{B}\,H_2N\underset{\underset{\underset{CH_3}{|}}{H-\overset{|}{C}-OH}}{\overset{}{CH}}COOH + H_2N\underset{\underset{CH_3}{|}}{CH}COOH\,\boxed{B} \longrightarrow \boxed{B}\,H_2N\underset{\underset{\underset{CH_3}{|}}{H-\overset{|}{C}-OH}}{CH}\overset{\overset{O}{||}}{C}-NH\underset{\underset{CH_3}{|}}{CH}COOH\,\boxed{B}$$

amino-blocked threonine; carboxyl-blocked alanine; double-blocked Thr-Ala dipeptide

$$\boxed{B}\,H_2N\underset{\underset{\underset{CH_3}{|}}{H-\overset{|}{C}-OH}}{CH}\overset{\overset{O}{||}}{C}-NH\underset{\underset{CH_3}{|}}{CH}COOH + H_2NCH_2COOH\,\boxed{B} \longrightarrow$$

amino-blocked Thr-Ala dipeptide; carboxyl-blocked glycine

$$\boxed{B}\,H_2N\underset{\underset{\underset{CH_3}{|}}{H-\overset{|}{C}-OH}}{CH}\overset{\overset{O}{||}}{C}-NH\underset{\underset{CH_3}{|}}{CH}\overset{\overset{O}{||}}{C}-NHCH_2COOH\,\boxed{B} \longrightarrow \text{Thr-Ala-Gly tripeptide}$$

double-blocked Thr-Ala-Gly tripeptide

Longer polypeptide chains of specified amino acid sequence could be made by this general technique, but the procedure was very tedious. The synthesis of insulin required an effort equivalent to one person working steadily for several years.

Fortunately, it is not necessary to prepare an entire synthetic protein polypeptide chain by starting at one end and adding amino acids one at a time. Previously prepared shorter polypeptide chains of known structure can be joined to form a long polypeptide chain. This method was used to make synthetic insulin. The insulin molecule, shown in Figure 29.2, consists of two polypeptide chains bonded together by two disulfide bonds. A third disulfide bond, between two amino acids in one chain, makes a small loop in that chain. Once the two chains had been assembled from fragments, a seemingly formidable problem remained: how to form the disulfide bonds in the correct positions. As it turned out, it was necessary only to bring the two chains together, with the cysteine side chains in the reduced condition, and to treat them with an oxidizing agent. Disulfide bonds formed at the right places, and biologically active insulin molecules were obtained.

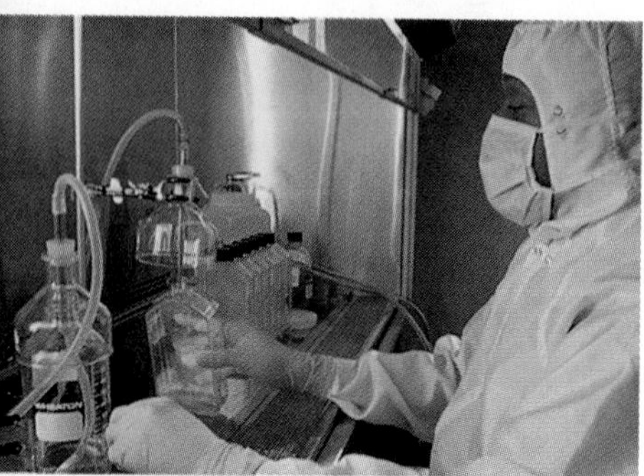

Production of a cloned protein. The technician is holding a container with cultured cells in a nutrient medium.

In the mid-1960s, a machine capable of automatically synthesizing polypeptide chains of known amino acid sequence was designed by R. Bruce Merrifield of Rockefeller University. The starting amino acid is bonded to a plastic surface (polystyrene bead) in the reaction chamber of the apparatus. Various reagents needed for building a chain of predetermined structures are automatically delivered to the reaction chamber in a programmed sequence. Twelve reagents and about 100 operations are needed to lengthen the chain by a single amino acid residue. But the machine is capable of adding residues to the chain at the rate of six a day. Such a machine makes it possible to synthesize complex molecules like insulin in a few days. The first large polypeptide (pancreatic ribonuclease), containing 124 amino acid residues, was synthesized by this method in 1969. More recently, a human growth hormone (HGH) containing 188 amino acid residues was synthesized with the use of this technique.

Over the past 30 years, techniques have been developed that use biological systems to create and produce new proteins. These procedures start with genetic material that codes for the protein of interest. Then, this material is properly modified and introduced into rapidly growing cells that are treated to overproduce the foreign protein. The result can be a harvest of much-needed protein, such as the human insulin used by diabetics or the human growth hormone used to treat dwarfism. These procedures constitute a part of genetic engineering and will be discussed in more detail in Chapter 31. The advantages of the new techniques become evident when we compare the rate of biological protein synthesis with that achieved by the machine designed by Merrifield. If a bacterium such as *E. coli* is used to produce a human protein such as growth hormone, one amino acid residue can be added every 0.01 second; a growth hormone molecule can be produced every 10–20 seconds. In contrast, the same growth hormone would take days to produce in the chemistry laboratory.

Chapter 29 Review

29.1 The Structure–Function Connection

KEY TERMS

Simple protein
Conjugated protein

- Proteins are polymers of amino acids.
- Simple proteins yield only amino acids upon hydrolysis.
- Conjugated proteins yield amino acids and additional products upon hydrolysis.
- The sequence of amino acids in a protein determines its function.

29.2 The Nature of Amino Acids

- Amino acids found in proteins are called alpha (α) amino acids because the amine group is bonded to the first, or α-carbon, adjacent to the carboxyl group.
- Common amino acids are represented by the following structure,

$$\begin{array}{c} \text{H} \\ | \\ \text{R}-\text{C}-\text{COOH} \\ | \\ \text{NH}_2 \end{array}$$

where the R group is referred to as the amino acid side chain.
- The amino acid side chain may have many different functional groups and many different structures.
- Depending on the amino acid side chain functional groups, amino acids may be
  - *neutral*, when they have the same number of carboxyl and amino groups,

  - *acidic*, when they have more carboxyl groups than amino groups, or
  - *basic*, when they have more amino groups than carboxyl groups.
- Amino acids are used to make neurotransmitters and hormones in addition to the amino acids' main function as building blocks of proteins.

29.3 Essential Amino Acids

KEY TERM

Essential amino acids

- Ten of the common amino acids are essential amino acids because they are needed for good health and cannot be synthesized in the human body.
- Dietary protein is classified as either complete or incomplete.
  - Complete dietary protein provides all the essential amino acids.
  - Incomplete dietary protein does not provide all the essential amino acids.

29.4 D-Amino Acids and L-Amino Acids

- The α-carbon is chiral for all the common amino acids but one (glycine).
- When shown in a Fischer projection formula with the carboxyl group at the top and the R group at the bottom:
  - L-amino acids have the amine group on the left side of the α-carbon.
  - D-amino acids have the amine group on the right side of the α-carbon.
- Only L-amino acids are found in proteins.

29.5 Amphoterism

KEY TERMS

Zwitterion
Isoelectric point

- Amino acids are amphoteric (or amphiprotic) because they can react either as an acid or as a base.
- In neutral solutions, amino acids form dipolar ions called zwitterions.
- When amino acids have a net charge, they will migrate toward either the positive or negative electrode in an electrolytic cell.
  - An amino acid's charge depends on pH.
  - The pH at which an amino acid has no overall charge (and will not migrate in an electrolytic cell) is called the isoelectric point.
  - Basic amino acids have isoelectric points from 7.8 to 10.8, neutral amino acids have isoelectric points from 4.8 to 6.3, and acidic amino acids have isoelectric points from 2.8 to 3.3.

29.6 Formation of Polypeptides

KEY TERMS

Peptide linkage
Polypeptides

- An amide bond is formed when a carboxyl group on one amino acid reacts with the α-amine group of a second amino acid to eliminate water.
  - This amide bond links amino acids together and is termed a peptide linkage or peptide bond.
- Two amino acids linked together form a dipeptide, three amino acids linked together form a tripeptide, and so on.
- Polymers containing up to about 50 amino acids are called polypeptides.
- Linear peptides have one end with a free amine group (the N-terminal residue) and one end with a free carboxyl group (the C-terminal residue).
  - The sequence of amino acids is numbered starting with the N-terminal residue, which is written on the left.
  - Peptides are named as acyl derivatives of the C-terminal residue.
    - The C-terminal residue keeps its complete name.
    - The rest of the amino acid names are changed by substituting a *-yl* ending in place of the *-ine* suffix and are listed in sequence order starting with the N-terminal residue.
- A variety of small, naturally occurring peptides function as hormones and nerve transmitters.

29.7 Protein Structure

KEY TERMS

Primary structure
Secondary structure
Tertiary structure
Quaternary structure

- Commonly, the term *protein* refers to a peptide that contains 50 or more amino acids.
- The primary structure of a protein is the amino acid sequence of that protein.
- The secondary structure of a protein can be characterized as a regular, three-dimensional structure held together by hydrogen bonds between the components of the peptide linkages (the $\rangle C{=}O$ and the $H{-}N\langle$).
  - The α-helix and β-pleated sheet are two very common examples of secondary structures.
  - Secondary structures provide strength and stability for proteins.
- The tertiary structure of a protein refers to the distinctive and characteristic conformation, or shape, of a protein molecule.
  - The tertiary structure is held together by interactions between amino acid side chains.
  - These interactions include (1) hydrogen bonding, (2) ionic bonding, and (3) disulfide bonding.
- A quaternary structure is found in proteins with more than one peptide chain and refers to the relative three-dimensional arrangement of these peptides with respect to each other.

29.8 Some Examples of Proteins and Their Structures

KEY TERMS

Fibrous protein
Globular protein
Protein domain

- Fibrous proteins have highly developed secondary structures and often function in support roles.
  - The silk protein, fibroin, is composed primarily of β-pleated sheet.
  - The α-keratins make up hair, feathers, horns, and hooves and have a high percentage of α-helix.
  - Collagen is the most abundant protein in animals and is found in most connective tissue; collagen has a unique secondary structure.
- Globular proteins have a more complex tertiary structure and a compact shape.
- Globular proteins provide more complex biochemical functions than do fibrous proteins.
  - Myoglobin is an oxygen-binding protein in muscle; myoglobin contains a heme group and is a conjugated protein.
  - Carboxypeptidase A is a digestive enzyme found in the small intestine; carboxypeptidase A structure is stabilized by a core of β-pleated sheet surrounded by a set of α-helices.
- Larger proteins are often composed of several globular sections about the size of myoglobin or carboxypeptidase A—called protein domains.
  - Immunoglobulin G is a binding protein that binds to different invading molecules (an antibody); immunoglobulin G has more than one protein chain (a quaternary structure) and several protein domains.
  - Hemoglobin is the red blood cell protein that binds and transports molecular oxygen; hemoglobin is composed of four peptide chains, and its quaternary structure changes when it binds molecular oxygen.

29.9 Loss of Protein Structure

KEY TERM

Denaturation

- If a protein loses only its natural three-dimensional conformation, the process is referred to as denaturation.
- Environmental changes (such as adding strong acids or bases, adding salts, or adding organic solvents), mechanical action (such as whipping), increasing the temperature, or ultraviolet radiation, can cause denaturation.
- Protein hydrolysis or digestion destroys a protein's primary structure and simultaneously denatures the protein.

29.10 Tests for Proteins and Amino Acids

- Many colorimetric tests are available for amino acids and proteins.
  - The xanthoproteic test reacts aromatic amino acids with nitric acid to form a yellow color.
  - The biuret test forms a violet color when copper(II) ion reacts with peptide bonds.
  - In the ninhydrin test, ninhydrin (triketohydrindene) reacts with almost all amino acids to give a blue color (proline and hydroxyproline react to give a yellow color).
- Mixtures of amino acids and/or proteins are readily separated by chromatography.
- Chromatography separates mixtures based on the relative tendencies of components to partition between a mobile phase and a stationary phase.
- In thin-layer chromatography, a liquid mobile phase travels through a thin stationary phase made of alumina or some other absorbant.
- Since proteins tend to denature during thin-layer chromatography, they are often separated by column chromatography—a liquid phase containing a mixture of proteins is separated as it percolates through a column packed with beads of a suitable polymer (stationary phase).
- One of the most precise protein separation techniques is electrophoresis, wherein proteins move in response to an electric field.
- Since protein electrophoresis is commonly carried out in a gell matrix, proteins separate based both on charge and size.
  - SDS (sodium dodecyl sulfate) electrophoresis uses a detergent (SDS) to mask differences in protein charges so that separation depends primarily on size differences.
  - In isoelectric focusing, proteins are electrophoresed through a pH gradient; the proteins move until they reach their isoelectric point.
- Today, it is not uncommon to use two-dimensional electrophoresis—separation in the x-direction is followed by a different separation in the y-direction.

29.11 Determination of the Primary Structure of Polypeptides

- Determining the primary structure of a protein requires tagging amino acids based on their location in the primary structure.
  - Sanger's reagent, 2,4-dinitrofluorobenzene, reacts with the α-amino group of the N-terminal residue to tag that specific amino acid with a yellow color.
  - The Edman degradation uses phenylisothiocyanate to sequentially remove the N-terminal amino acids from protein.

29.12 Synthesis of Peptides and Proteins

- Laboratory synthesis of peptides with specific primary structures requires a complex series of chemical reactions; specific amino and carboxylic acid groups must be blocked during specific reactions.
- Today the most efficient method to make a specific protein is to use genetic engineering; new genetic information is introduced into a microorganism, which then rapidly synthesizes a specific, new protein.

Review Questions

All questions with blue numbers have answers in the appendix of the text.

1. Why are the common amino acids in proteins called α-amino acids?
2. What elements are present in amino acids and proteins?
3. Why are proteins from some foods of greater nutritional value than others?
4. Write the names of the amino acids that are essential to humans.
5. Why are amino acids amphoteric? Why are they optically active?
6. What can you say about the number of positive and negative charges on a protein molecule at its isoelectric point?
7. Explain what is meant by (a) the primary structure, (b) the secondary structure, (c) the tertiary structure, and (d) the quaternary structure of a protein.
8. What special role does the sulfur-containing amino acid cysteine have in protein structure?
9. How do myoglobin and hemoglobin differ in structure and function?
10. Differentiate between an α-helix and a β-pleated sheet structure of a protein.
11. Explain how the hydrolysis of a protein differs from its denaturation.
12. Which amino acids give a positive xanthoproteic test?
13. What is the visible evidence observed in a positive reaction for the following tests?
    (a) xanthoproteic test
    (b) biuret test
    (c) ninhydrin test
    (d) Lowry assay
    (e) Bradford assay
14. Briefly describe the separation of proteins by column chromatography.
15. (a) What is thin-layer chromatography?
    (b) Describe how amino acids are separated with this technique.
    (c) What reagent is commonly used to locate the amino acids in the chromatogram?
16. How does SDS electrophoresis differ from standard electrophoresis?
17. Both a prion and a bacteria can cause disease. In general, how do they differ?

Paired Exercises

All exercises with blue numbers have answers in the appendix of the text.

1. Write the structural formulas for the D and L forms of alanine:

   $CH_3CHCOOH$ (with NH_2 on the central carbon)

   Which form is found in proteins?

2. Write the structural formulas for the D and L forms of serine:

   $HOCH_2CHCOOH$ (with NH_2 on the CH carbon)

   Which form is found in proteins?

3. A 100.0-g sample of a food product is analyzed and found to contain 6.0 g of nitrogen. If protein contains an average of 16% nitrogen, what percentage of the food is protein?

4. A 250.0-g sample of hamburger is analyzed and found to contain 5.2 g of nitrogen. If protein contains an average of 16% nitrogen, how many grams of protein are contained in the hamburger?

5. Write the structural formula representing the following amino acid at its isoelectric point:

   $CH_3CH{-}CHCOOH$ (with OH on the first CH and NH_2 on the second CH)

   threonine

6. Write the structural formula representing the following amino acid at its isoelectric point:

   $NH_2CCH_2CHCOOH$ (with =O on the first C and NH_2 on the CH)

   asparagine

7. For the amino acid

$$C_6H_5-CH_2CH(NH_2)COOH$$

phenylalanine

write
(a) the zwitterion formula
(b) the formula in 0.1 M H_2SO_4
(c) the formula in 0.1 M NaOH

8. For the amino acid

$$\text{indol-3-yl}-CH_2CH(NH_2)COOH$$

tryptophan

write:
(a) the zwitterion formula
(b) the formula in 0.1 M H_2SO_4
(c) the formula in 0.1 M NaOH

9. Write ionic equations to show how alanine, with structure

$$CH_3CH(NH_2)COOH$$

acts as a buffer toward
(a) H^+ (b) OH^-

10. Write ionic equations to show how leucine, with structure

$$(CH_3)_2CHCH_2CH(NH_2)COOH$$

acts as a buffer toward
(a) H^+ (b) OH^-

11. Write the structural formula representing the following amino acid at its isoelectric point:

$$CH_3SCH_2CH_2CH(NH_2)COOH$$

methionine

12. Write the structural formula representing the following amino acid at its isoelectric point:

$$(CH_3)_2CHCH(NH_2)COOH$$

valine

13. Write the full structural formula of the two dipeptides containing

$HOCH_2CH(NH_2)COOH$ (serine) and $CH_3CH(NH_2)COOH$ (alanine)

Indicate the location of the peptide bonds.

14. Write the full structural formula of the two dipeptides containing

$CH_2(NH_2)COOH$ (glycine) and $CH_3CH(OH)CH(NH_2)COOH$ (threonine)

Indicate the location of the peptide bonds.

15. Write a structure of the common amino acid that has a 2° amine. Use this amino acid to make a dipeptide.

16. Write a structure of the common amino acid that has a thiol. Use this amino acid to make a disulfide bond.

17. Use the amino acids

$CH_2(NH_2)COOH$ (glycine) $HOCH_2CH(NH_2)COOH$ (serine) $CH_3CH(NH_2)COOH$ (alanine)

to write structures for
(a) glycylglycine
(b) alanylglycylserine
(c) glycylserylglycine

18. Use the amino acids

$CH_2(NH_2)COOH$ (glycine) $HOCH_2CH(NH_2)COOH$ (serine) $CH_3CH(NH_2)COOH$ (alanine)

to write structures for
(a) alanylalanine
(b) serylglycylglycine
(c) serylglycylalanine

19. Using amino acid abbreviations, write all the possible tripeptides that contain one unit each of glycine, phenylalanine, and leucine.

20. Using amino acid abbreviations, write all the possible tripeptides that contain one unit each of tyrosine, aspartic acid, and alanine.

21. The following stucture is part of a protein:

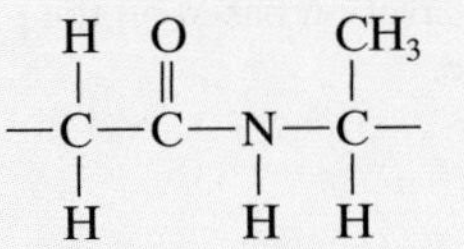

```
 H  O     CH3
 |  ||    |
—C—C—N—C—
 |      |  |
 H      H  H
```

Circle the atoms that will form bonds when this protein folds into a beta-pleated sheet.

22. The following stucture is part of a protein:

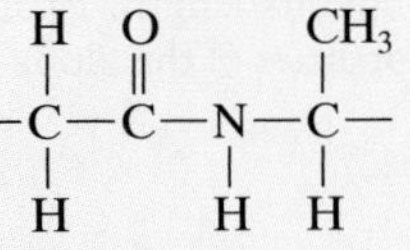

```
 H  O     CH3
 |  ||    |
—C—C—N—C—
 |      |  |
 H      H  H
```

Circle the atoms that will form bonds when this protein folds into an alpha helix.

23. At pH = 7, draw the bond that might form between two serine side chains ($HOCH_2$—) to hold together a tertiary protein structure.

24. At pH = 7, draw the bond that might form between a lysine side chain ($NH_2CH_2CH_2CH_2CH_2$—) and an aspartic acid side chain ($HOOCCH_2$—) to hold together a tertiary protein structure.

25. Would the compound

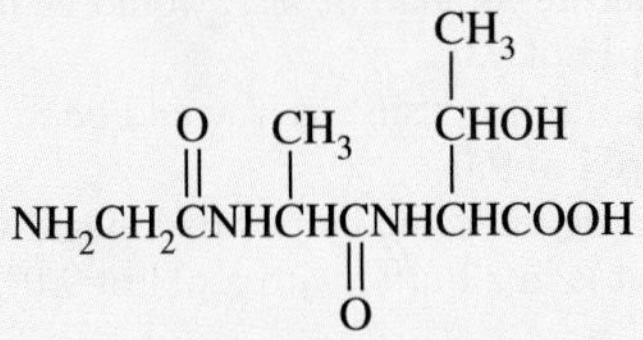

```
                    CH3
                    |
     O  CH3         CHOH
     ||  |          |
NH2CH2CNHCHCNHCHCOOH
             ||
             O
```

react with
(a) copper(II) sulfate in the biuret test?
(b) concentrated HNO_3 to give a positive xanthoproteic test?
(c) ninhydrin?

26. Would the compound

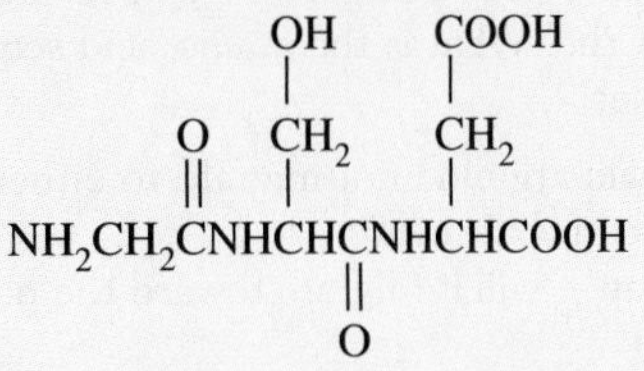

```
             OH        COOH
             |         |
     O  CH2       CH2
     ||  |         |
NH2CH2CNHCHCNHCHCOOH
             ||
             O
```

react with
(a) copper(II) sulfate in the biuret test?
(b) concentrated HNO_3 to give a positive xanthoproteic test?
(c) ninhydrin?

27. Write the products for the complete hydrolysis of

```
          C6H5 (benzene ring)
           |         COOH
           |         |
  CH3     CH2      CH2
  |        |         |
NH2CHCNHCHCNHCHCOOH
      ||       ||
      O        O
```

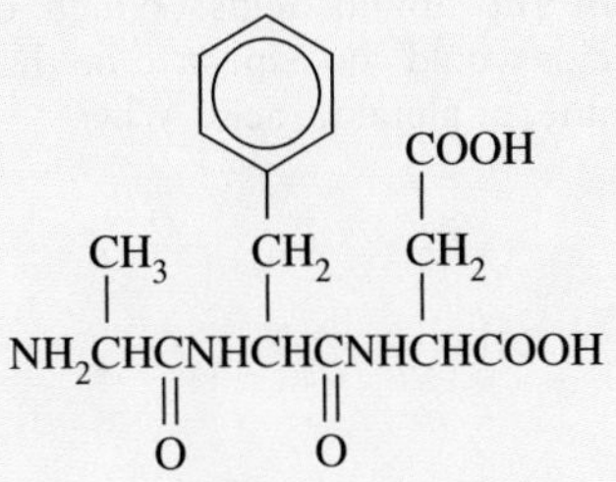

28. Write the products for the complete hydrolysis of

```
                              OH
           HOOC                |
            |          C6H4 (benzene ring)
            CH2                |
            |                  |
            CH2                |
  CH3       |                 CH2
  |         |                  |
NH2CHCNHCHCNHCHCOOH
      ||       ||
      O        O
```

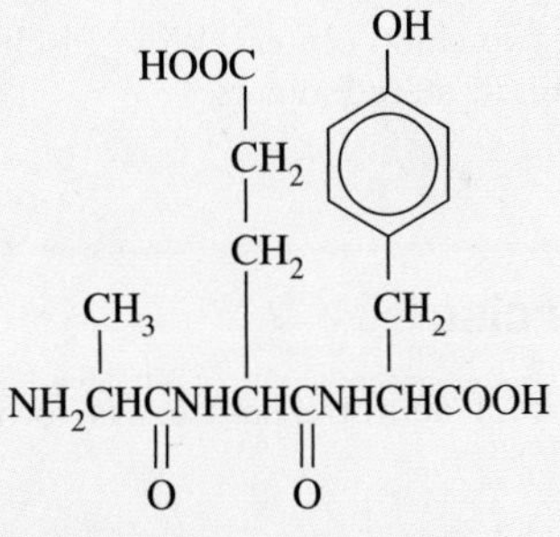

29. Human cytochrome c contains 0.43% iron by mass. If each cytochrome c contains one iron atom, what is the molar mass of cytochrome c?

30. Human hemoglobin contains 0.33% iron by mass. If each hemoglobin contains four iron atoms, what is the molar mass of hemoglobin?

31. What is the amino acid sequence of a heptapeptide that, upon hydrolysis, yields the tripeptides Gly-Phe-Leu, Phe-Ala-Gly, and Leu-Ala-Tyr? The heptapeptide contains one residue each of Gly, Leu, and Tyr and two residues each of Ala and Phe.

32. What is the amino acid sequence of a heptapeptide that, upon hydrolysis, yields the tripeptides Phe-Gly-Tyr, Phe-Ala-Ala, and Ala-Leu-Phe? The heptapeptide contains one residue each of Gly, Leu, and Tyr and two residues each of Ala and Phe.

33. A newly discovered protein has about 95% beta–pleated sheet structure. Would you predict that this protein will function as an enzyme or as a structural support protein? Briefly explain.

34. A newly discovered protein is globular in shape with beta–pleated sheet at its core. Would you predict that this protein will function as a binding protein or as a structural support protein? Briefly explain.

35. A newly discovered protein has a molar mass of 452,000 g/mole. Would you predict that this protein will have more than one domain? Briefly explain.

36. A newly discovered protein is reported to have two domains. Would you predict the molar mass for this protein to be in the 40,000–60,000 g/mole range or the 100,000–150,000 g/mole range? Briefly explain.

37. A human hair (composed of alpha-keratins) can be carefully stretched to about double its length. Explain this observation based on the properties of the alpha–keratin 2° structure.

38. Silk (composed of fibroin) is very flexible but not easily stretched. Explain this observation based on the properties of fibroin's 2° structure.

Additional Exercises

All exercises with blue numbers have answers in the appendix of the text.

39. A nonapeptide is obtained from blood plasma by treatment with the enzyme trypsin. Analysis shows that both terminal amino acids are arginine. Total hydrolysis of this peptide yields Gly, Ser, 2 Arg, 2 Phe, 3 Pro. Partial hydrolysis gives Phe-Ser, Phe-Arg, Arg-Pro, Pro-Pro, Pro-Gly-Phe, Ser-Pro-Phe. What is the amino acid sequence of the nonapeptide?

40. (a) At what pH will arginine not migrate to either electrode in an electrolytic cell?
   (b) In what pH range will it migrate toward the positive electrode?

41. Is it possible for a protein to have a primary structure if it does not have a secondary structure? Is it possible for a protein to have a secondary structure if it does not have a primary structure? Explain.

42. Ribonuclease is a protein that is slightly smaller than myoglobin. How many protein domains would you predict for the structure of ribonuclease? Explain your answer.

43. Threonine has two chiral centers. Write Fischer projection formulas for its stereoisomers.

44. How are the hypervariable regions in immunoglobulin chains important to this protein's function?

45. The isoelectric point for alanine is 6.00 and for lysine is 9.74.
   (a) Write the structure for alanine as it would be in a buffer with a pH of 9.0.
   (b) Write the structure for lysine as it would be in a buffer with a pH of 9.0.
   (c) Would the charge on lysine be positive, negative, or neutral when it is in a buffer with a pH of 9.0?

46. How many dipeptides containing glycine can be written using the 20 amino acids from Table 29.1?

47. The structures of vasopressin and oxytocin are given in Table 29.3. Which one would you expect to have the higher isoelectric point? Why?

48. Thin-layer chromatography is used to separate amino acids from one another. The separation is based on the relative polarity of the amino acids. Which of the following amino acids would you expect to be the most polar, leucine, alanine, or glutamic acid? Why?

Challenge Exercise

49. A whole family of neurotransmitters is formed from L-tyrosine:

L-tyrosine ($NH_2-CH(CH_2C_6H_4OH)-COOH$) → L-dopa ($NH_2-CH(CH_2C_6H_3(OH)_2)-COOH$) → dopamine ($NH_2-CH_2-CH_2C_6H_3(OH)_2$) → norepinephrine ($NH_2-CH_2-CHOH-C_6H_3(OH)_2$) → epinephrine ($CH_3-NH-CH_2-CHOH-C_6H_3(OH)_2$)

   (a) L-tyrosine is often incorporated into proteins. Show the dipeptide that is formed by connecting two L-tyrosines together.
   (b) Parkinson's disease patients often receive a racemic mixture of dopa. Draw the two isomers in this mixture.
   (c) Explain briefly why "dopamine" does not have an "L-" prefix.
   (d) Of norepinephrine and epinephrine, which is a primary amine? Which is a secondary amine?

Answers to Practice Exercises

29.1 α-Amino acid: $CH_2(NH_2)COOH$

β-Amino acid: $CH_2(NH_2)CH_2COOH$

29.2 Leu (leucine), Lys (lysine), Trp (tryptophan), His (histidia)

29.3 $(CH_3)_2CHCH(NH_2)COO^-$

29.4 $CH_2(NH_2)C(=O)-NHCH(CH_2CH(CH_3)_2)C(=O)-NHCH(CH_2COOH)C(=O)-NHCH(CH_2OH)C(=O)-NHCH(CH_2SH)COOH$

29.5 $CH_3CH(NH_2)C(=O)-NHCH(CH_2-S)C(=O)-NHCH_2C(=O)-NH-CH_2-C(=O)-NH-CH(COOH)-CH_2-S$ (the two S atoms joined by a disulfide bond: $CH_2-S-S-CH_2$)

The tertiary structure must contain a cyclic structure joined by the disulfide bond.

29.6 (a)

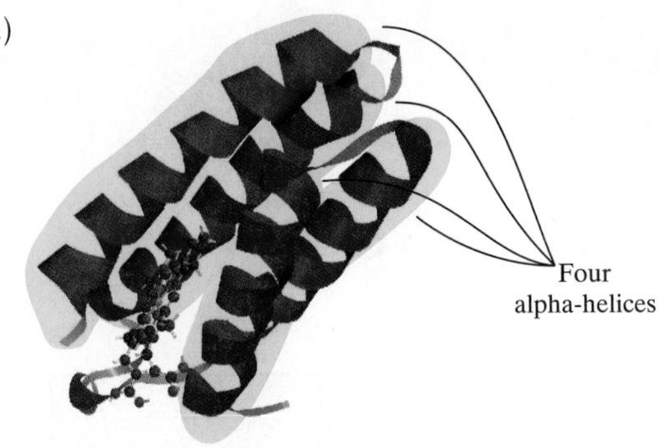

(b) Two domains (counting only the two end lobes) or three domains (counting the two end lobes and the connecting piece of protein).

29.7

From Edman degradation	Ala-Gly-Lys
Dipeptide	Ala-Gly
Pentapeptide	Lys-His-Ala-Thr-Gly
Tripeptide	Gly-Gly-Gly
Nonapeptide	Ala-Gly-Lys-His-Ala-Thr-Gly-Gly-Gly

CHAPTER 30

Enzymes

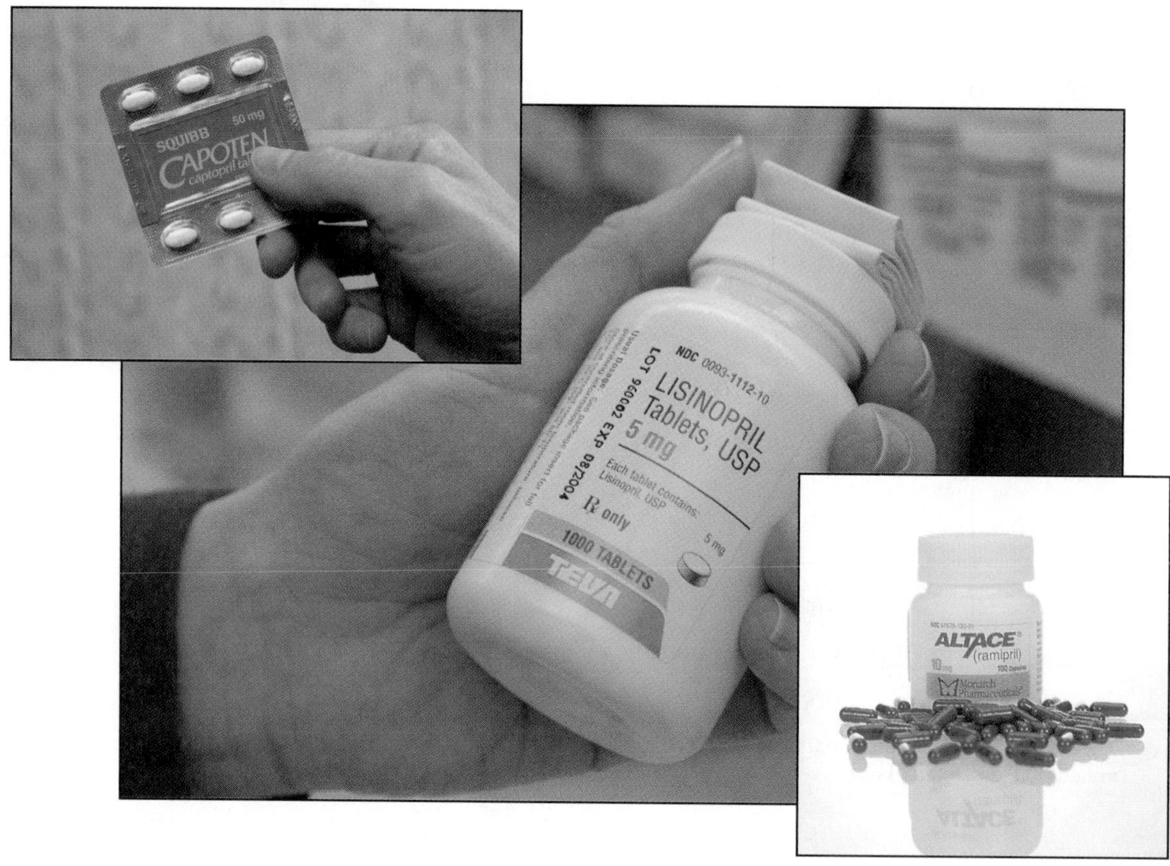

Like many drugs, ACE inhibitors target an enzyme and so cause a very selective metabolic change.

Chapter Outline

Perhaps you have seen laundry detergents advertised as "containing enzymes." Enzymes are added to detergents to clean especially hard-to-remove spots. In living organisms, enzymes also take on the "tough" metabolic tasks—in fact, life cannot continue in the absence of enzymes.

Enzymes are important because they can accelerate a chemical reaction by 1 million to 100 million times. Imagine what would happen if some of our tasks were accelerated by that amount! A two-hour daily homework assignment would be completed in about one-thousandth of a second. A flight from Los Angeles to New York would take about one-hundredth of a second. The building of Hoover Dam, a monumental task requiring five years of earth moving and complex steel and concrete work, would occur in about 30 seconds.

Scientists now understand the general characteristics of enzymes. And, as we will see in this chapter, these molecules achieve almost miraculous results by following some very basic chemical principles.

30.1 Molecular Accelerators

Enzymes are proteins that catalyze biochemical reactions. Enzymes catalyze nearly all the myriad reactions that occur in living cells. Uncatalyzed reactions that require hours of boiling in the presence of a strong acid or a strong base in the laboratory can occur in a fraction of a second in the presence of the proper enzyme at room temperature and nearly neutral pH. The catalytic functions of enzymes are directly dependent on their three-dimensional structures. It was believed until quite recently that all biological catalysts were proteins. However, research carried out since about 1980 has shown that certain ribonucleic acids (RNAs, Chapter 31), called ribozymes, also catalyze some biochemical reactions.

enzyme

Louis Pasteur was one of the first scientists to study enzyme-catalyzed reactions. He believed that living yeasts or bacteria were required for these reactions, which he called *fermentations*—that is, the conversion of glucose to alcohol by yeast. In 1897, Eduard Büchner (1860–1917) made a cell-free filtrate that contained enzymes prepared by grinding yeast cells with very fine sand. The enzymes in this filtrate converted glucose to alcohol, thus proving that the presence of living cells was not required for enzyme activity. For this work, Büchner received the Nobel prize in chemistry in 1907.

Enzymes are essential to life. In the absence of enzymes, the critical biochemical reactions occur too slowly for life to be maintained. The typical biochemical reaction occurs more than a million times faster when catalyzed by an enzyme. For example, we know that the reduced carbons of carbohydrates can react with oxygen to produce carbon dioxide and energy. Yet, the sucrose in the sugar bowl at home never reacts significantly with the oxygen in the air. These sucrose molecules must overcome an energy barrier (activation energy) before reaction can occur. (See Figure 30.1.) In the sugar bowl, the energy barrier is too large. But in the biological cell, enzymes lower the activation energy and enable sucrose to react rapidly enough to provide the energy needed for life processes.

Each organism contains thousands of enzymes. Some are simple proteins consisting only of amino acid units. Others are conjugated and consist of a protein part, or **apoenzyme**, and a nonprotein part, or **coenzyme**. Both parts are essential, and a functioning enzyme that consists of both the protein and nonprotein parts is called a **holoenzyme**:

apoenzyme
coenzyme
holoenzyme

Apoenzyme + Coenzyme = Holoenzyme

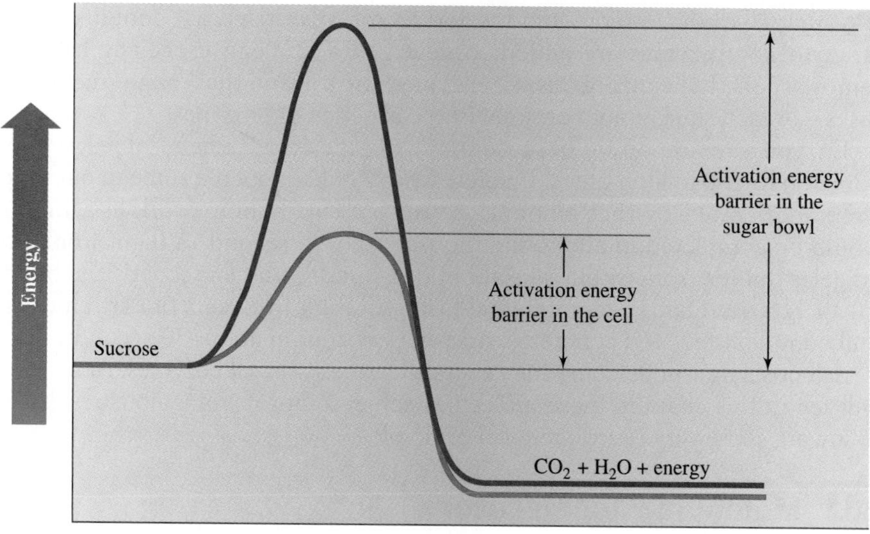

Figure 30.1
A typical reaction-energy profile: The lower activation energy in the cell is due to the catalytic effect of enzymes.

Often the coenzyme is derived from a vitamin, and one coenzyme may be associated with many different enzymes.

For some enzymes, an inorganic component such as a metal ion (e.g., Ca^{2+}, Mg^{2+}, or Zn^{2+}) is required. This inorganic component is an *activator*. From the standpoint of function, an activator is analogous to a coenzyme, but inorganic components are not called coenzymes.

Another remarkable property of enzymes is their specificity of reaction—that is, a certain enzyme catalyzes the reaction of a specific type of substance. For example, the enzyme *maltase* catalyzes the reaction of maltose and water to form glucose. Maltase has no effect on the other two common disaccharides, sucrose and lactose. Each of these sugars requires a specific enzyme—sucrase to hydrolyze sucrose, lactase to hydrolyze lactose. Such reactions are indicated by the following equations:

$$\underset{\text{maltose}}{C_{12}H_{22}O_{11}} + H_2O \xrightarrow{\text{maltase}} \underset{\text{glucose}}{C_6H_{12}O_6} + \underset{\text{glucose}}{C_6H_{12}O_6}$$

$$\underset{\text{sucrose}}{C_{12}H_{22}O_{11}} + H_2O \xrightarrow{\text{sucrase}} \underset{\text{glucose}}{C_6H_{12}O_6} + \underset{\text{fructose}}{C_6H_{12}O_6}$$

$$\underset{\text{lactose}}{C_{12}H_{22}O_{11}} + H_2O \xrightarrow{\text{lactase}} \underset{\text{glucose}}{C_6H_{12}O_6} + \underset{\text{galactose}}{C_6H_{12}O_6}$$

The substance acted on by an enzyme is called the *substrate*. Sucrose is the substrate of the enzyme sucrase. Enzymes have been named by adding the suffix *-ase* to the root of the substrate name. Note the derivations of maltase, sucrase, and lactase from maltose, sucrose, and lactose. Many enzymes, especially digestive enzymes, have common names such as pepsin, rennin, trypsin, and so on. The names have no systematic significance.

In the International Union of Biochemistry (IUB) System, enzymes are assigned to one of six classes, the names of which clearly describe the nature of the reaction they catalyze. The six main classes of enzymes are as follows:

1. *Oxidoreductases:* Enzymes that catalyze the oxidation–reduction between two substrates

2. *Transferases:* Enzymes that catalyze the transfer of a functional group between two substrates
3. *Hydrolases:* Enzymes that catalyze the hydrolysis of esters, carbohydrates, and proteins (polypeptides)
4. *Lyases:* Enzymes that catalyze the removal of groups from substrates by mechanisms other than hydrolysis
5. *Isomerases:* Enzymes that catalyze the interconversion of stereoisomers and structural isomers
6. *Ligases:* Enzymes that catalyze the linking of two compounds by breaking a phosphate anhydride bond in adenosine triphosphate (ATP, Chapter 31)

Because the systematic name is usually long and often complex, working or practical names are used for enzymes. For example, adenosine triphosphate creatine phosphotransferase is called creatine kinase, and acetylcholine acylhydrolase is called acetylcholine esterase.

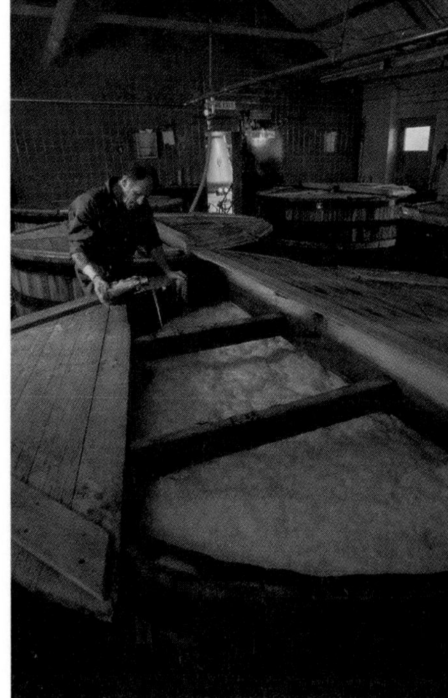

A worker adds yeast to a vat in a whiskey distillery.

30.2 Rates of Chemical Reactions

Enzymes catalyze biochemical reactions and thus increase the rate of these chemical reactions—but how does the process take place? To answer this question, we must first consider some general properties of a chemical reaction.

Every chemical reaction starts with at least one reactant and finishes with a minimum of one product. As the reaction proceeds, the reactant concentration decreases and the product concentration increases. We often plot these changes as a function of time, as shown in Figure 30.2 for the hypothetical conversion of reactant A into product B:

$$A \rightarrow B$$

A reaction rate is defined as a change in concentration with time—the rate at which the reactants of a chemical reaction disappear and the products form.

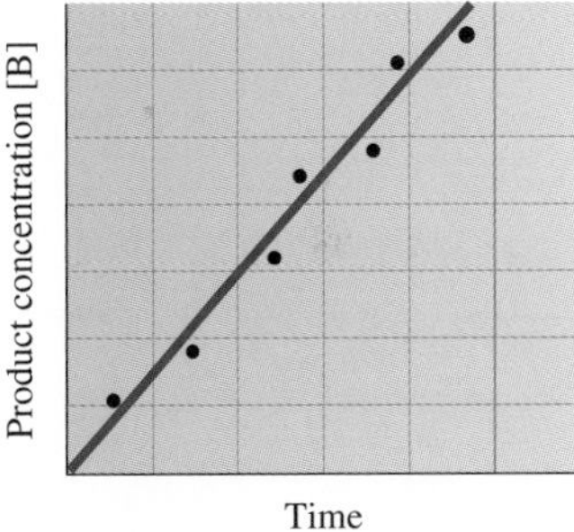

Figure 30.2
The change in product concentration [B] as a function of time. The reaction rate is determined by measuring the slope of this line.

Practice 30.1

Calculate the reaction rate for each of the following:

(a)

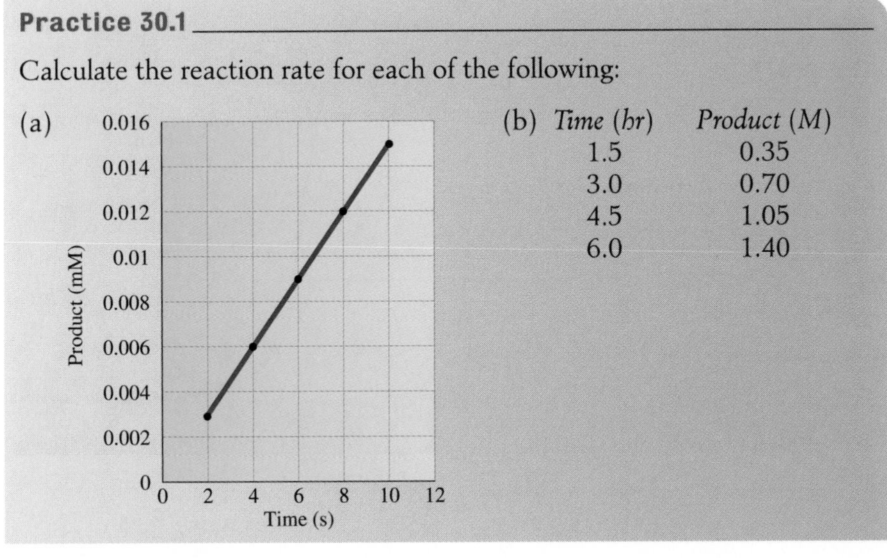

(b)

Time (hr)	*Product (M)*
1.5	0.35
3.0	0.70
4.5	1.05
6.0	1.40

transition state

The reactant must pass through a high-energy **transition state** to be converted into a product. This transition state is an unstable structure with characteristics of both the reactant and the product. The energy necessary to move a reactant to the transition state is termed the activation energy. The larger this energy barrier is, the slower the reaction rate will be.

For example, carbon dioxide can be reacted with water to yield the hydrogen carbonate ion:

$$CO_2 + H_2O \longrightarrow HCO_3^- + H^+$$

However, as shown in Figure 30.3, energy is required to form the transition state. Once the transition state is created, the remainder of the process proceeds easily. As with almost all chemical reactions, reaching the transition state is difficult and limits the rate at which reactants are converted to products.

There are three common ways to increase a reaction rate:

1. *Increasing the reactant concentration:* When the reactant concentrations are made larger, the number of reactant molecules with the necessary activation energy also increases. For simple reactions in the absence of a catalyst, the reaction rate increases with reactant concentration.
2. *Increasing the reaction temperature:* An increase in temperature generally means that each reactant molecule becomes more energetic. A larger fraction of the reactants have the activation energy necessary to be converted to products, and the reaction rate increases.
3. *Adding a catalyst:* A catalyst lowers the activation energy by allowing a new, lower energy transition state. Since this new process has a lower activation energy, more reactants have the energy to become products. Hence, the reaction rate increases.

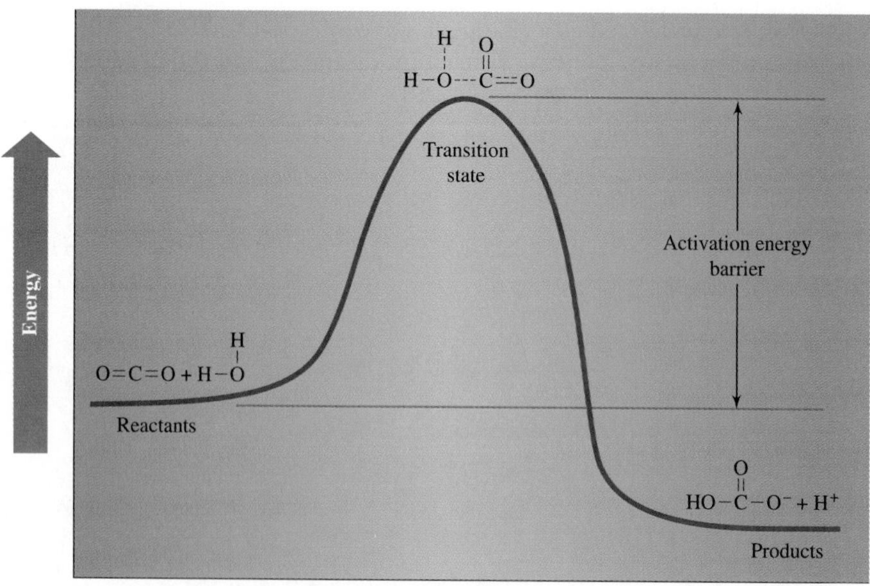

Figure 30.3
An energy profile for the reaction between water and carbon dioxide. The dashed lines in the transition-state formula indicate bonds being formed and broken.

Biological systems can rarely change reactant concentration or temperature upon demand. Thus, to alter reaction rates, the system developed a set of superb catalysts—the enzymes.

Practice 30.2

For the carbon dioxide + water reaction shown in Figure 30.3, predict which set of conditions will yield a faster reaction rate. Briefly explain.

(a) CO_2 pressure = 100 torr, $T = 37°C$, Activation energy = 31 kcal/mol *vs.* CO_2 pressure = 100 torr, $T = 37°C$, Activation energy = 26 kcal/mol

(b) CO_2 pressure = 100 torr, $T = 37°C$, Activation energy = 31 kcal/mol *vs.* CO_2 pressure = 130 torr, $T = 37°C$, Activation energy = 31 kcal/mol

30.3 Enzyme Kinetics

In 1913, two German researchers, Leonor Michaelis (1875–1949) and Maud Menten (1879–1960), measured enzyme-catalyzed reaction rates as a function of substrate (reactant) concentration. They observed that most enzyme-catalyzed reactions show an increasing rate with increasing substrate concentration, *but* only to a specific maximum velocity, V_{max}. A graph like that in Figure 30.4 is often called a Michaelis–Menten plot.

Scientists have learned much about the nature of enzymes by studying Michaelis–Menten plots. First, because the rate approaches a maximum, we can conclude that enzymes have a limited catalytic ability. Once these enzymes are operating at a maximum, a further increase in substrate (reactant) concentration does not change the reaction rate. By analogy, an enzyme

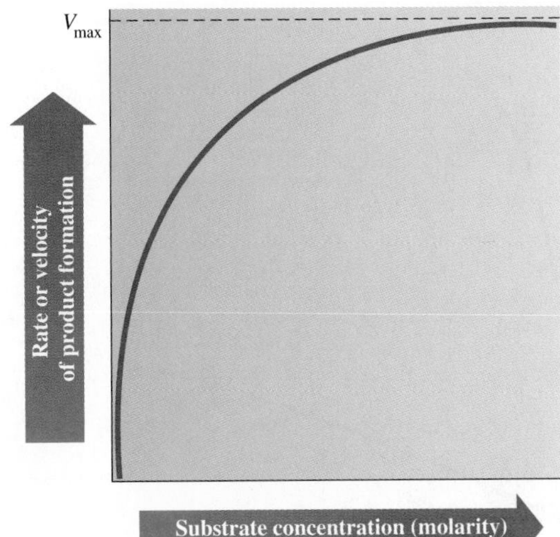

Figure 30.4
A Michaelis–Menten plot showing the rate of an enzyme-catalyzed reaction as a function of substrate concentration. The lower left portion of the graph marks the approximate area where an enzyme responds best to concentration changes.

shuttles reactants to the transition state (and on to products) like an usher seating patrons at a theater. No matter how many people arrive at the entrance, a conscientious usher can only lead the ticket holders to their places at a set pace. Similarly, each specific enzyme has a maximum catalytic rate that is fixed.

Scientists believe that specific enzymes are tailored to fit specific metabolic needs. Some enzymes work well at low substrate concentrations, whereas others require much higher concentrations before they operate efficiently. Some enzymes catalyze reactions very quickly, while others fit a biochemical need by reacting much more slowly. Several examples will illustrate these concepts.

Glucose metabolism is partly controlled by two mammalian enzymes that work effectively in different concentration ranges: *Hexokinase* has a strong attraction for glucose, while *glucokinase* has a much weaker attraction. (See Figure 30.5.) Thus, hexokinase has first priority to use the available glucose, because the enzyme reacts at lower glucose concentrations. Only if there is excess glucose will the glucokinase enzyme start to convert glucose into glycogen.

turnover number

An enzyme's catalytic speed is also matched to an organism's metabolic needs. This catalytic speed is commonly measured as a **turnover number**—the number of molecules an enzyme can react or "turn over" in a given time span. For example, the enzyme catalase has a large turnover number, 10,000,000/s. This enzyme protects us from the toxin hydrogen peroxide and effectively destroys 10 million hydrogen peroxide molecules per second. In contrast, the enzyme chymotrypsin has a much smaller turnover number of about 0.2/s, digesting two protein peptide bonds per 10 seconds as a meal is passed slowly through the small intestine. Since the digestive process is slow, chymotrypsin is not required to have the catalytic speed of catalase. In general, an enzyme's turnover number matches the speed required for a biological process.

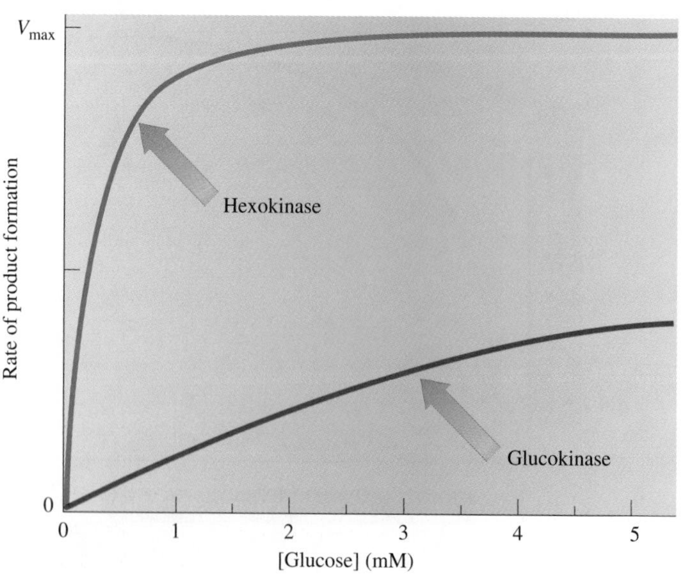

Figure 30.5
Michaelis–Menten plots for two glucose metabolic enzymes.

Example 30.1

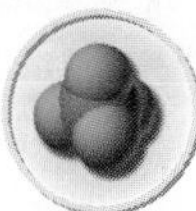

Two enzymes react with the same substrate under identical conditions. Enzyme A is found to have a turnover number of 1500/s, whereas enzyme B shows a turnover number of 500/s. Compare the catalytic efficiency of the two enzymes.

SOLUTION

The turnover number is a measure of catalytic efficiency. Under identical conditions, the larger the turnover number, the more efficient an enzyme is as a catalyst. Because enzyme A has a larger turnover number than enzyme B, enzyme A is the more efficient catalyst.

Practice 30.3

(a) Under optimal conditions, the digestive enzyme pepsin has a turnover number of about 30/min, while a second digestive enzyme, trypsin, has a turnover number of 12/min. Both of these enzymes digest proteins. Which would you judge to be the more efficient? Briefly explain.

(b) Calculate a turnover number from the following information:
(i) 6300 epinephrine (adreniline) molecules are destroyed in ten minutes;
(ii) 580 glucose molecules are oxidized in two minutes.

30.4 Industrial-Strength Enzymes

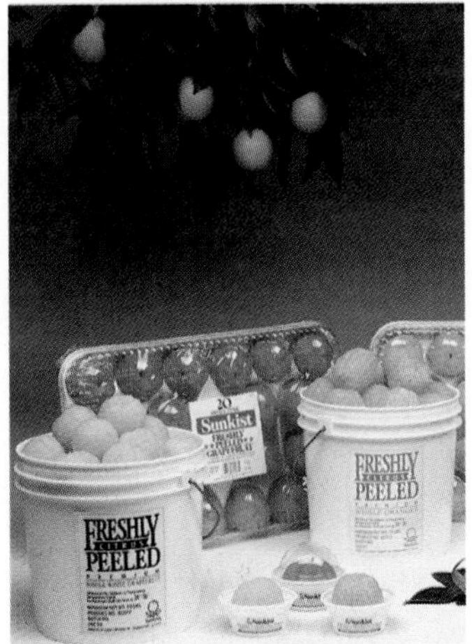

The enzyme pectinase can effectively peel an orange, leaving the fruit in perfect condition.

Not only are enzymes important in biology, they are increasingly important in industry. Enzymes offer two major advantages to manufacturing processes and in commercial products: First, enzymes cause very large increases in reaction rates even at room temperature; second, enzymes are relatively specific and can be used to target selected reactants. Perhaps their biggest disadvantage is their relative scarcity, and therefore higher cost, compared with traditional chemical treatments. Recent developments in biotechnology offer supplies of less expensive enzymes through genetic engineering.

The following sections focus on common industrial enzymes. Many of these enzymes have a digestive or "breakdown" function. They are hydrolases, but are known by common names: **proteases** (proteolytic enzymes) break down proteins; **lipases** digest lipids; cellulases, amylases, lactases, and pectinases break down the carbohydrates, cellulose, amylose, lactose, and pectin, respectively.

proteases
lipases

Food Processing

Enzymes have long been used in food processing. The citrus industry has perfected a process to remove peel from oranges or grapefruits by using the enzyme pectinase. The pectinase penetrates the peel in a vacuum infusion process. There it dissolves the albedo (the white stringy material) that attaches the peel to the fruit. When the fruit is removed from the solution, the skin can be peeled easily by machine or hand. The industry markets prepeeled citrus to hospitals, airlines, and restaurants.

The enzyme lactase is used on an industrial scale to convert lactose to glucose and galactose. This change gives people who are lactose intolerant the ability to consume milk products. Among the great number of different molecules in milk, lactase reacts only with lactose. Since the glucose and galactose formed are sweeter than lactose and differ in solubility, lactase treatment

Customer is looking at a pair of denim jeans that have been softened by "biostoning" with a cellulase enzyme.

produces milk with different properties. Ice cream made from lactase-treated milk, for example, tends to be creamier.

About 25% of all industrial enzymes are used to convert cornstarch into syrups that are equivalent in sweetness and in calories to ordinary table sugar. More than 10 million metric tons of such syrups are produced annually. The process uses three enzymes: (1) α-amylase catalyzes the liquefaction of starch to dextrins; (2) a glucoamylase catalyzes the breakdown of dextrins to glucose; (3) glucose isomerase converts glucose to fructose. The process is as follows:

$$\text{Starch} \xrightarrow{\alpha\text{-amylase}} \text{Dextrins}$$

$$\text{Dextrins} \xrightarrow{\text{glucoamylase}} \text{Glucose}$$

$$\text{Glucose} \xrightarrow[\text{isomerase}]{\text{glucose}} \text{Fructose}$$

The final product is a high-fructose syrup that is equivalent in sweetness to sucrose. One of these syrups, sold commercially since 1968, contains by dry weight about 42% fructose, 50% glucose, and 8% other carbohydrates.

Technology

Industrial enzymes offer solutions to environmental pollution problems for some industries. Cotton has traditionally been "scoured" with the strong base sodium hydroxide to prepare it for dyeing. Unfortunately, wastewater from the process can be harmful to the environment. Recently, a pectinase has successfully replaced this toxic chemical. The enzyme operates under milder conditions and achieves the same "scoured" result.

A number of industries use cellulases. Denim manufacturers use it to give a soft appearance to this cotton fabric, "stonewashing" or "biostoning" the cloth. In this process, the cellulases digest some of the cellulose on the surface of the cotton. Paper producers use cellulases to complete the breakdown of wood chips to paper pulp. The enzyme treatment forms a better pulp, and paper can be produced with smaller amounts of bleach, again easing environmental pollution.

Consumer Goods

Industrial enzymes are used most often as detergent additives. These enzymes clean under relatively mild conditions and can remove spots not accessible to soaps. Proteases digest the most difficult clothing stains (e.g., grass, blood, and sweat, which contain proteins). Lipases team up with soaps to remove greasy stains. Finally, amylases are often added to digest starchy residues from foods such as oatmeal, chocolate, and mashed potatoes. As a side benefit, detergents that are effective in mild conditions also save on water-heating bills.

Medical Uses

Enzymes are used in medicine primarily because of their specificity. For example, several enzymes are used to dissolve blood clots in patients with such diseases as lung embolism (clot in the lung), stroke (clot in the brain), and heart attack (clot in the heart). These proteases are specific for the blood protein plasminogen, which is converted to plasmin and degrades the blood clot. The proteases are called plasminogen activators; treatment can dissolve a clot in 30–60 minutes.

Practice 30.4

Briefly explain why cellulases are not used to "biostone" polyester fabrics.

30.5 Enzyme Active Site

Catalysis takes place on a small portion of the enzyme structure called the enzyme active site. Often this is a crevice or pocket on the enzyme that represents only 1–5% of the total surface area. Figure 30.6 shows a spacefilling model of the enzyme hexokinase, which catalyzes a first step in the breakdown of glucose to provide metabolic energy. Notice that the active site is located in a crevice. Glucose enters the site and is bound. The enzyme then must change shape before the reaction takes place. Thus, although catalysis occurs at the small active site, the entire three-dimensional structure of the enzyme is important.

By examining values such as the turnover number and the effective substrate concentration range, scientists have gained a basic understanding of what takes place at an enzyme active site. To function effectively, an enzyme must attract and bind the substrate. Once the substrate is bound, a chemical reaction is catalyzed. This two-step process is described by the following general sequence: Enzyme (E) and substrate (S) combine to form an enzyme–substrate intermediate (E–S). The intermediate decomposes to give the product (P) and regenerate the enzyme:

$$E + S \xrightarrow[\text{binding}]{} E\text{–}S \xrightarrow[\text{catalysis}]{} E + P$$

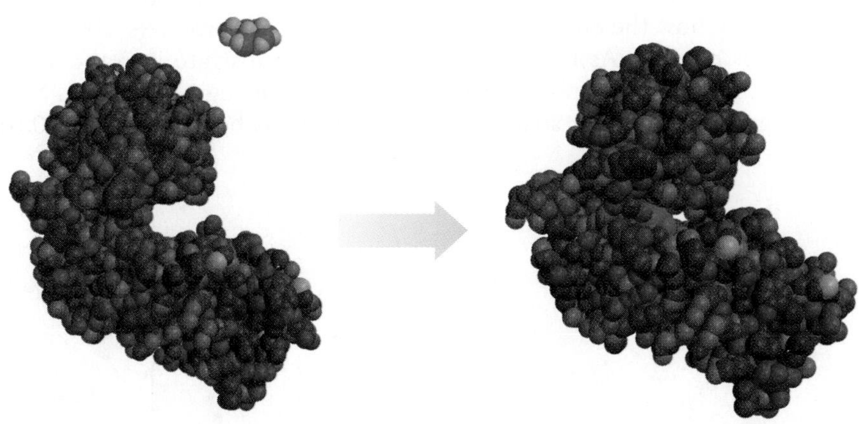

(a) Hexokinase *before* binding glucose (b) Hexokinase *after* binding glucose

Figure 30.6
Spacefilling models showing (a) the enzyme hexokinase and its substrate, glucose (in pink), before binding and (b) after binding. Note the top half of the hexokinase closes down on top of the glucose as predicted by the induced-fit model.

For the hydrolysis of maltose by the enzyme maltase, the sequence is

$$\underset{\text{E}}{\text{Maltase}} + \underset{\text{S}}{\text{Maltose}} \xrightarrow[\text{binding}]{} \underset{\text{E–S}}{\text{Maltase–Maltose}}$$

$$\underset{\text{E–S}}{\text{Maltase–Maltose}} + H_2O \xrightarrow[\text{catalysis}]{} \underset{\text{E}}{\text{Maltase}} + \underset{\text{P}}{2\ \text{Glucose}}$$

Each different enzyme has its own unique active site whose shape determines, in part, which substrates can bind. Enzymes are said to be stereospecific—that is, each enzyme catalyzes reactions for only a limited number of different reactant structures. For example, maltase binds to maltose (two glucose units linked by an α-1,4-glycosidic bond), but not to lactose (galactose coupled to glucose). In fact, the enzyme even distinguishes maltose from cellobiose (glucose coupled to glucose by a β-1,4-linkage), in which only the glycosidic linkage is different. Enzyme stereospecificity is a very important means by which the cell controls its biochemistry.

lock-and-key hypothesis

Enzyme specificity is partially due to a complementary relationship between the active site and substrate structures. This **lock-and-key hypothesis** envisions the substrate as a key that fits into the appropriate active site: the lock. Although the hypothesis describes a fundamental property of enzyme–substrate binding, scientists have known for some time that enzyme active sites are not rigid, as a lock would be. Instead, the active site bends to a certain degree when the appropriate substrate binds. The **induced-fit model** proposes that the active site adjusts its structure in order to prepare the substrate–enzyme complex for catalysis. Enzyme stereospecificity is thus explained in terms of an active site having a somewhat flexible shape. This shape is rigid enough to exclude very dissimilar substrates (lock-and-key hypothesis), but flexible enough to accommodate (induced-fit model) and allow catalysis of appropriate substrates. (See Figure 30.7.)

induced-fit model

The fact that an enzyme is flexible helps explain enzyme-binding specificity and also how the enzyme converts reactants into products. *An enzyme is a dynamic catalyst.* As the enzyme attracts the substrate into the active site, the enzyme's shape and the reactant's shape both begin to change. Note the change in hexokinase structure when glucose is bound (Figure 30.6).

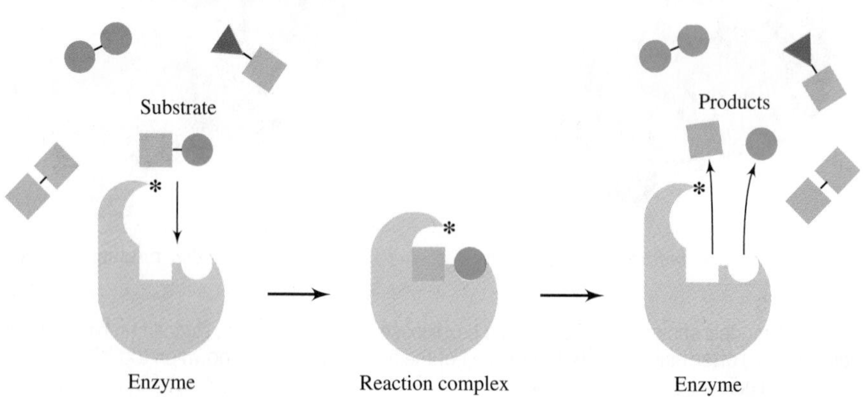

Figure 30.7
Enzyme–substrate interaction illustrating both the lock-and-key hypothesis and the induced-fit model. The correct substrate (■–●) fits the active site (lock-and-key hypothesis). This substrate also causes an enzyme conformation change that positions a catalytic group (*) to cleave the appropriate bond (induced-fit model).

The alteration of enzyme shape aids the transformation of the reactant into the product. Figure 30.7 depicts an enzyme where a shape change leads to catalysis. This only exemplifies the numerous alterations that enzymes bring about (e.g., bonds broken or formed, charges moved, and new molecular substituents added or removed).

An actual metabolic reaction will illustrate some of the basic features of enzyme catalysis. For example, let's take the enzyme hexokinase, which catalyzes the transformation of glucose to glucose-6-phosphate:

$$\text{glucose} \xrightarrow[\text{Mg}^{2+},\ \text{ATP}]{\text{hexokinase}} \text{glucose-6-phosphate}$$

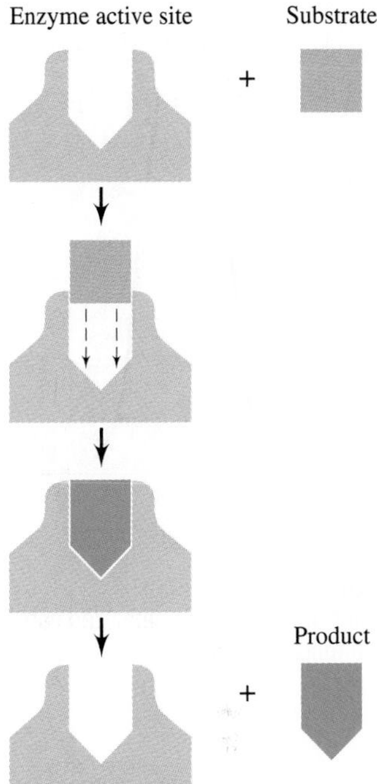

Figure 30.8
Strain hypothesis: The substrate is being forced toward the product shape by enzyme binding.

Like many enzymes, hexokinase requires a second reactant. In this case, adenosine-5′-triphosphate (ATP) supplies the phosphate group that is transferred to the glucose. ATP is a very important molecule within the cell, because it often transfers energy as well as phosphate groups from enzyme to enzyme. (See Chapter 31.) Additionally, the metal ion, Mg^{2+}, is required as an activator in this reaction. With all of these components present, the enzyme can start the catalytic process.

When glucose binds, the hexokinase shape changes to bring the ATP close to the carbon 6 of the glucose, thus forcing the transfer of phosphate to this specific carbon. Hexokinase speeds this reaction in several ways. First, the enzyme acts to bring the reactants close together, a process termed **proximity catalysis**. Second, hexokinase positions the reactants so the proper bonds will break and form. (The enzyme ensures that a phosphate is added to carbon 6 of glucose and not to one of the other carbons.) This is often termed the **productive binding hypothesis** because reactants are bound or oriented in such a way that products result. Hexokinase is a successful catalyst—glucose reacts with ATP 10 billion (10^{10}) times faster than it would in the absence of this enzyme. Increases in reaction rate of this magnitude are essential to life. In fact, without enzymes, cellular reactions are too slow to keep cells alive.

proximity catalysis

productive binding hypothesis

Biotechnologists make use of our understanding of enzyme catalysis to design (with nature's help) completely new enzymes. Antibodies (immunoglobulin proteins) are produced that bind tightly to a molecule, as in the transition state. When these antibodies bind reactant molecules, the strong attractive forces "strain" the reactants, as illustrated in Figure 30.8. The reactant molecule is impelled to change shape to fit the binding site. Catalysis occurs, and these antibodies act as enzymes! Scientists have termed this mode of catalysis the **strain hypothesis**, and it is thought to be important in many natural enzymes as well as these "antibody enzymes." In the not-too-distant future, such artificial enzymes may serve important industrial and medical applications.

strain hypothesis

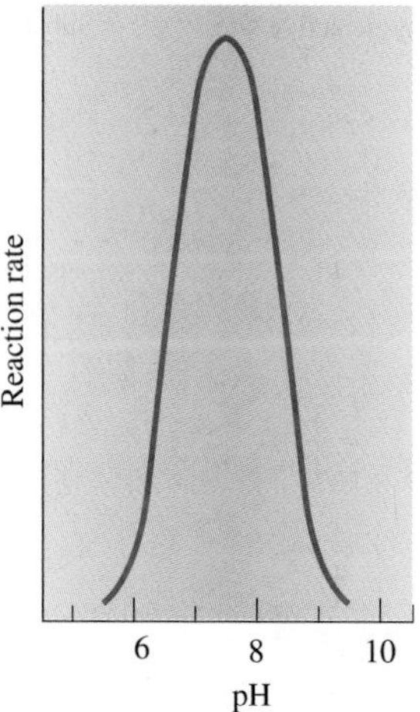

Figure 30.9
A plot of the enzyme-catalyzed rate as a function of pH.

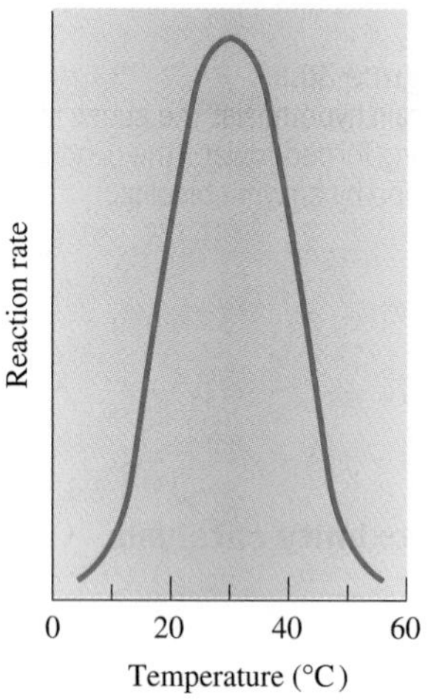

Figure 30.10
A plot of the temperature dependence of an enzyme-catalyzed reaction.

30.6 Temperature and pH Effects on Enzyme Catalysis

Essentially, any change that affects protein structure also affects an enzyme's catalytic function. If an enzyme is denatured, its activity will be lost. Thus, strong acids and bases, organic solvents, mechanical action, and high temperature are examples of treatments that decrease an enzyme-catalyzed rate of reaction.

Even slight changes in the pH can have profound effects on enzyme catalysis. Remember that some of the amino acids that make up enzymes have side chains whose charge depends on pH. For example, side chains with carboxylic acid functional groups may be either neutral or negatively charged; those with amino groups may be either neutral or positively charged. Thus, as the pH changes, the charges on an enzyme, or more specifically at the active site of an enzyme, also change. Enzyme catalysis is affected. For each enzyme, there is an optimal pH; shifts to more acidic or more basic conditions decrease the enzyme activity, as shown in Figure 30.9.

Because enzymes are so sensitive to small pH changes, our bodies have developed elaborate mechanisms to control the amount of acid and base present in cellular fluids. The kidneys and lungs share the responsibility for maintaining blood acid and base levels. If the blood pH shifts outside of the narrow range 6.8–7.8, death often results. The body's metabolic processes can no longer maintain life, partly because of a pH-induced change in enzyme activities.

Body temperature is also carefully controlled, partly because enzyme activities are particularly temperature sensitive. A typical plot of temperature versus enzyme activity is shown in Figure 30.10. At very low temperatures, few reactant molecules have enough energy to overcome the activation energy barrier, and the reaction occurs slowly. As the temperature increases, more reactants have the necessary activation energy, and the rate increases. But a point is reached where the high temperature causes enzyme denaturation, and the rate of reaction starts to decrease. Hence, both high and low temperatures cause slow reaction rates, and, as with pH, an optimal temperature exists for each enzyme. Not surprisingly, the optimal temperature for many human enzymes is approximately body temperature (about 37°C).

Not all enzymes operate optimally around the physiological temperature (37°C) and pH (7). In fact, chemists have long sought enzymes that have optimal activity at higher temperatures and more acidic or basic pHs, conditions that favor industrial reactions. Such enzymes can be harvested from microorganisms. It has been estimated that fewer than 5% of all fungal and bacterial species have been discovered. Thus scientists travel the world to find new microorganisms (and new enzymes). Especially valuable species have been found in hot springs, deep ocean vents, and tropical rain forests.

Table 30.1 gives some characteristics for bacterial proteases commonly used with detergents. Each protease has different optimum conditions. Both Alcalase® and Savinase® with temperature optima at about 55°C, are ideal for warm-water washes. However, Savinase® has a better pH optimum for detergent powders, which commonly have a pH around 9–10. In contrast, Alcalase® has a pH optimum of about 8, which is better adapted for liquid soaps. Esperase® is frequently used in industrial or institutional laundries.

Table 30.1 Some Examples of Enzymes Used with Detergents

Enzyme name (the most common trademarked name)	Optimum pH	Optimum temperature	Common use
Alcalase®	7–9	40–60°C	Liquid soaps
Savinase®	8–11	40–60°C	Powder detergents
Esperase®	10–12	50–70°C	Industrial or institutional laundries

This enzyme remains active at temperatures up to 70°C and a pH of 12. In each case, knowing the conditions needed for washing enabled scientists to find the right enzyme.

30.7 Enzyme Regulation

Since enzymes are vital to life, enzyme catalysis is under careful cellular control. Living cells use a variety of mechanisms to change the rate of substrate conversion to product. Sometimes a new group of atoms is covalently joined to the enzyme in a process called *covalent modification.* In other cases, another molecule is noncovalently bound to the enzyme. The protein structural change that results causes a decrease in enzyme activity, **enzyme inhibition**, or an increase in activity, **enzyme activation**.

enzyme inhibition
enzyme activation

Hexokinase is an example of an enzyme that is under cellular control. Because this enzyme begins the breakdown of glucose to yield cellular energy, the hexokinase-catalyzed reaction is often not needed when cellular energy levels are high. Control occurs in the following way: (1) If glucose is no longer needed for cellular energy, the glucose-6-phosphate concentration increases; (2) this compound binds to the hexokinase enzyme and inhibits the enzyme activity. Because glucose-6-phosphate is a product of the reaction, this form of enzyme control is called **product inhibition**. Hexokinase responds to the overall cellular energy state and does not use more glucose than needed.

product inhibition

Feedback inhibition and **feedforward activation** are two other common forms of enzyme control. To understand these mechanisms, visualize the various cellular processes as assembly lines, with a different enzyme at each step from raw material to finished product. Feedback inhibition affects enzymes at the beginning of the molecular assembly line. A final product acts as "feedback" and inhibits an enzyme from using too many molecular starting materials. In contrast, feedforward activation often controls enzymes at the end of the molecular assembly line. If there is an excess of starting materials, these molecules will "feedforward" and activate enzymes, which, in turn, cause the whole process to move faster. As with other control mechanisms, both feedback inhibition and feedforward activation serve to coordinate enzyme processes within the cell.

feedback inhibition
feedforward activation

A variety of drug therapies make use of enzyme control to selectively affect target cells. Careful drug design can create a molecule that binds to only one type of enzyme. By binding, the drug blocks normal catalysis and causes enzyme inhibition.

CHEMISTRY IN ACTION • Aspirin and Stroke: A Very Potent Enzyme Inhibitor

Pharmaceutical companies are in the business of drug design. This very successful endeavor takes advantage of enzyme specificity: a carefully designed drug will selectively target only one specific enzyme. Specificity is key, and the pharmaceutical industry invests about $1 billion per new drug to achieve the desired result.

So it is noteworthy when nature provides a very effective drug for little expense. Aspirin is such a drug. It has been used for centuries as part of willow bark tea to reduce fever and ease aches and pains. Willow bark contains the active ingredient, salicylic acid, that is improved by making the acetic acid ester acetylsalicylic acid, or aspirin. This medicine was first widely available about 1900.

Aspirin potently inhibits one particular enzyme, cyclooxygenase. This enzyme has an active site in a protein tunnel. A positively charged amino acid is at the tunnel's mouth, and a reactive hydroxyl is just inside. Aspirin's structure

[Structure: benzene ring bearing $-C(=O)-O^-$ and $-O-C(=O)-CH_3$ on adjacent carbons]

is ideal for blocking this active site. First, aspirin's negatively charged carboxylate is attracted to the positive charge at the tunnel entrance. Then aspirin's aromatic ring squeezes into and blocks the tunnel. Finally, aspirin's acetyl group esterifies the reactive alcohol so that even after the salicylic acid leaves, the active site remains inhibited.

A good example of a specific aspirin effect is aspirin's protection against heart attack and stroke. Protection is achieved at very low aspirin concentrations. Aspirin's target is the blood platelet cyclooxygenase. Cyclooxygenase helps platelets aggregate during clot formation, so, when aspirin blocks the cyclooxygenase, fewer clots are formed. Furthermore, because aspirin's inhibition is permanent, a small daily dose will have a cumulative effect. One baby aspirin a day (81 mg of the drug) decreases the risk of heart attack and stroke by about 20%. And patients can be treated for years with little or no side effects.

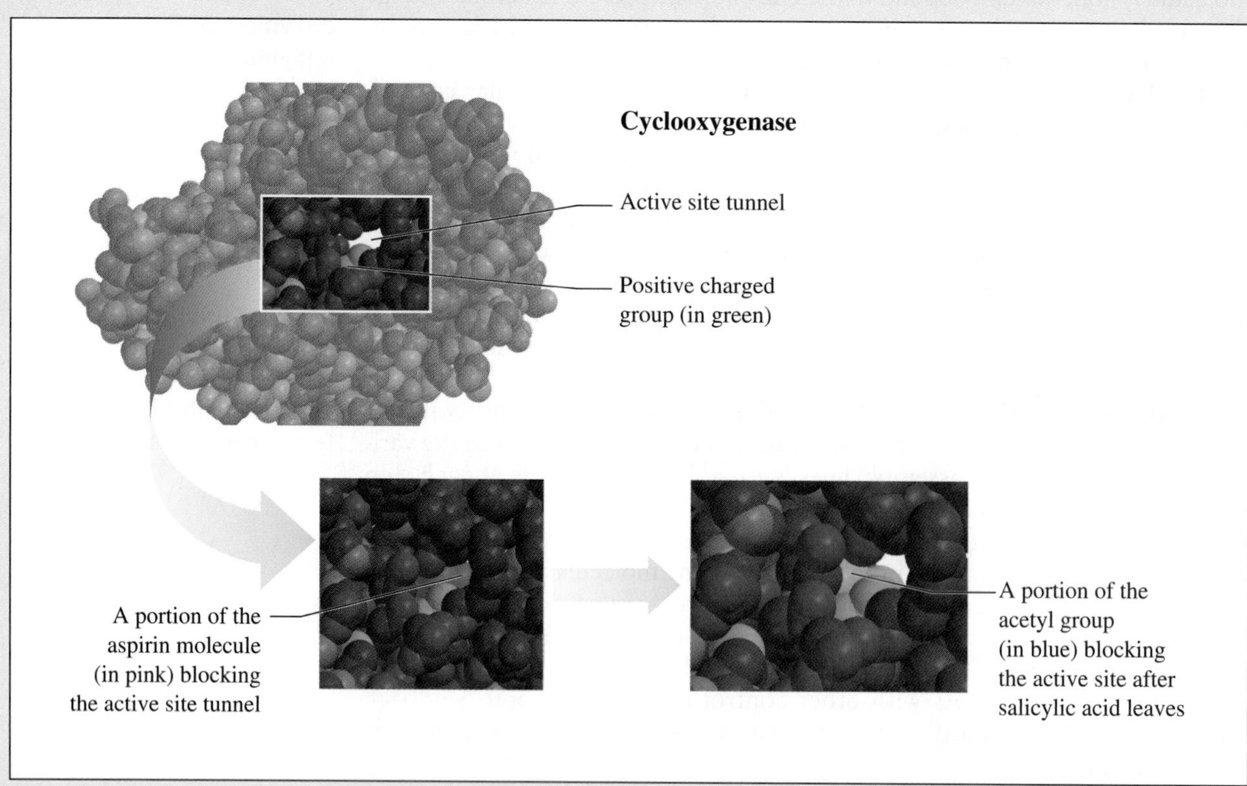

A revolution in mental health treatments has followed the introduction of drugs to inhibit specific brain processes. For example, Prozac® and related drugs specifically inhibit reuptake of serotonin, a neurotransmitter. More serotonin remains to activate neurons that, in turn, alleviate mental depression. Other neurotransmitters are also amenable to treatment by specific enzyme-inhibitor drugs.

The statin drugs (e.g., lovastatin, pravastatin, simvastatin, fluvastatin, atorvastatin) cause a marked decrease in blood cholesterol levels. These drugs specifically inhibit a key enzyme in the cholesterol synthesis pathway. Although the human body contains thousands of different enzymes, the statins specifically impact one.

These examples of drug therapy illustrate a very important principle: Because biological processes depend on enzymes, enzyme control often has a major impact on life.

Chapter 30 Review

30.1 Molecular Accelerators

KEY TERMS

Enzyme
Apoenzyme
Coenzyme
Holoenzyme

- Enzymes are proteins that catalyze biochemical reactions.
- Some ribonucleic acids (called ribozymes) also catalyze biochemical reactions.
- Some enzymes are conjugated proteins; the protein part is the apoenzyme, and the nonprotein part is the coenzyme. As a whole, the conjugated protein is termed a holoenzyme.
- Enzymes accelerate reactions by millions of times and can specifically react one molecule within a large mixture of different molecules.
- Although there is a systematic nomenclature for enzymes (from the International Union of Biochemistry), enzymes are often designated with common names.

30.2 Rates of Chemical Reactions

KEY TERM

Transition state

- An enzyme is commonly studied by measuring a reaction rate, the change in concentration of reactants or products with time.
- A reactant increases in energy in the process of becoming a product.
- When the reactant reaches the highest energy point during reaction, it reaches the transition state—an unstable structure with characteristics of both the reactant and the product.
- There are three common ways to increase a reaction rate:
  - Increase the reactant(s) concentration.
  - Increase the reaction temperature.
  - Add a catalyst.

30.3 Enzyme Kinetics

KEY TERM

Turnover number

- A Michaelis–Menten plot shows the rate of an enzyme-catalyzed reaction on the y-axis versus the substrate (reactant) concentration on the x-axis.
- Enzymes have a fixed, limited catalytic capability.
- The turnover number, the number of substrates an enzyme can react in a given time span, is a measure of an enzyme's catalytic ability.

30.4 Industrial-Strength Enzymes

KEY TERMS

Proteases
Lipases

- Enzymes are becoming more important in industry because of the following:
  - They are selective and target specific molecules.
  - They are excellent catalysts, so reactions will proceed quickly even at room temperature.
- Many common industrial enzymes fall into the two following classes:
  - Proteases, enzymes that break down proteins.
  - Lipases that digest lipids.
- Some examples of industrial enzyme use include the following:
  - Food processing.
  - Environmentally-benign cotton processing.
  - Detergent production.
  - Medical treatments.

30.5 Enzyme Active Site

KEY TERMS

Lock-and-key hypothesis
Induced-fit model
Proximity catalysis
Productive binding hypothesis
Strain hypothesis

- Catalysis takes place on a small portion of the enzyme surface called the active site.
- An enzyme must bind the substrate at the active site before reaction occurs.
- Enzyme specificity is partly due to a complementary relationship between the structure of the active site and the structure of the substrate; this is the lock-and-key hypothesis.
- Also, the active site structure changes slightly in response to substrate binding.
  - This is called the induced-fit model.
  - The slight change in active site structure optimizes catalytic activity.
- Most enzymes use three common means of catalysis:
  - Proximity catalysis occurs when reactants are drawn together at the enzyme active site.
  - The productive binding hypothesis proposes that as reactants bind to the enzyme active site, they align for optimum reaction.
  - The strain hypothesis states that the binding forces that draw a reactant to the enzyme active site also force the reactant to change into the transition state.

30.6 Temperature and pH Effects on Enzyme Catalysis

- Essentially anything that affects an enzyme's structure also affects an enzyme's activity.
  - Protein denaturing agents cause a loss of enzyme activity.
- Each enzyme has an optimum pH for catalytic activity.
  - Most enzymes in humans have an optimum pH around 7.
- Each enzyme has an optimum temperature for catalytic activity.
  - Most enzymes in humans have an optimum temperature around 37°C.

30.7 Enzyme Regulation

KEY TERMS

Enzyme inhibition
Enzyme activation
Product inhibition
Feedback inhibition
Feedforward activation

- Since life (and metabolism) can only occur when enzymes are active, enzyme catalysis is under careful cellular control.
  - Enzyme inhibition means a decrease in enzyme activity.
  - Enzyme activation means an increase in enzyme activity.
- Product inhibition occurs when the product of an enzyme-catalyzed reaction causes a decrease in enzyme activity.
- When an initial enzyme-catalyzed step is inhibited by a product of subsequent metabolism, feedback inhibition is said to apply.
- When the product of an initial enzyme-catalyzed step activates a subsequent enzyme-catalyzed step, feedforward activation applies.

Review Questions

All questions with blue numbers have answers in the appendix of the text.

1. What is activation energy and how is this energy affected by enzymes?
2. What is the general role of enzymes in the body?
3. Distinguish between a coenzyme and an apoenzyme.
4. Give the names of enzymes that catalyze the hydrolysis of (a) sucrose, (b) lactose, and (c) maltose.
5. What are the six general classes of enzymes?
6. A catalyst increases the rate of a chemical reaction. List two other means of increasing reaction rates.
7. Differentiate between the lock-and-key hypothesis and the induced-fit model for enzymes.
8. List three ways that substrate binding to the active site helps the reactants convert to products.
9. How does an enzyme inhibitor differ from an enzyme substrate?
10. Describe the process and function of the enzyme pectinase, which is used in food processing.
11. Describe the function of proteases in detergents.
12. How do specific proteases aid stroke victims?
13. How does an enzyme activator differ from an enzyme substrate?
14. Briefly describe the medical effect of the enzyme inhibitors known as the *statin* drugs.

Paired Exercises

All exercises with blue numbers have answers in the appendix of the text.

1. An enzyme reacts 0.005 *M* substrate every 3.5 min. What is the reaction rate in units of *M* per second?

2. An enzyme reacts 0.02 *M* of substrate every 8 min. What is the reaction rate in units of *M* per second?

3. A lactase enzyme breaks down 0.03 *M* lactose every 5 minutes. What is the reaction rate in
(a) M/min?
(b) mM/s?

4. A sucrase enzyme breaks down 0.15 *M* sucrose every 37 minutes. What is the reaction rate in
(a) M/min?
(b) μM/min?

5. From the following graph, calculate the reaction rate in M/min units:

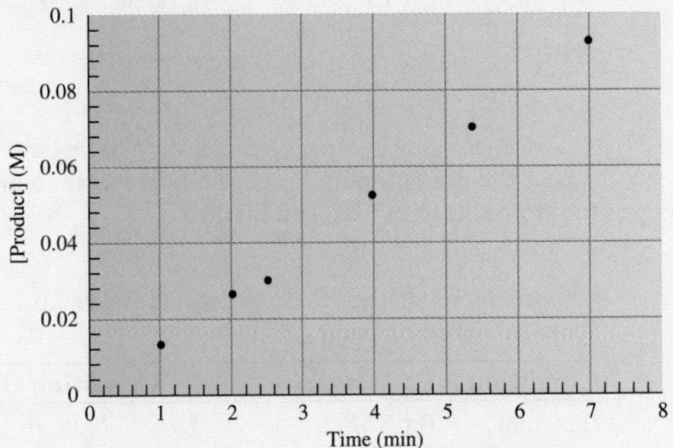

6. From the following graph, calculate the reaction rate in M/min units:

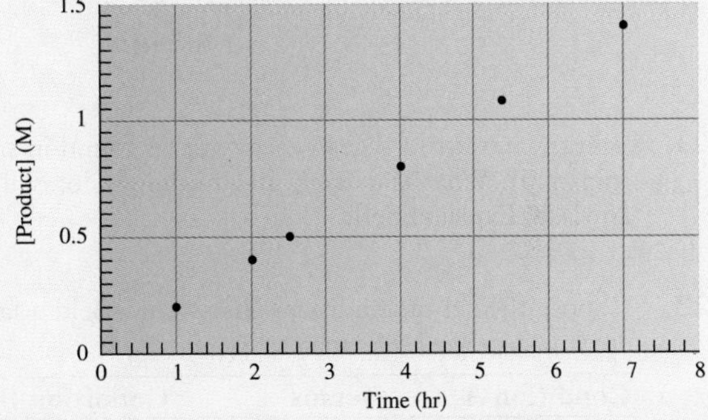

7. Determine the reaction rate for Exercise #5 in M/s units.

8. Determine the reaction rate for Exercise #6 in mM/min units.

9. Calculate the turnover number for an enzyme, given that 2470 substrate molecules react in 360s.

10. Calculate the turnover number for an enzyme, given that 5.2×10^4 substrate molecules react in 500s.

11. Lysozyme, an enzyme that cleaves the cell wall of many bacteria, has a turnover number of 0.5/s. How many reactants can be converted to products by lysozyme in 1 min?

12. Pepsin, a digestive enzyme found in the stomach, has a turnover number of 1.2/s for a specific protein substrate. How many proteins can be digested by three pepsin molecules in 5 min?

13. Two enzymes are being studied; the first enzyme uses 1-butanol, $C_4H_{10}O$, as a substrate, while the second enzyme uses 2-methyl-2-propanol, $C_4H_{10}O$, as a substrate. Based on the lock-and-key hypothesis, how might the shapes for the enzyme active sites differ?

14. Two enzymes are being studied; the first enzyme uses butanoic acid, $C_4H_8O_2$, as a substrate, while the second enzyme uses acetic acid, $C_2H_4O_2$, as a substrate. Based on the lock-and-key hypothesis, how might the shapes for the enzyme active sites differ?

15. Why does an enzyme-catalyzed reaction rate decrease at lower temperatures?

16. Why does an enzyme-catalyzed reaction rate decrease at higher temperatures?

17. If the V_{max} for an enzyme is decreased, would you predict activation or inhibition? Briefly explain.

18. If the turnover number for an enzyme is increased, would you predict activation or inhibition? Briefly explain.

19. Based on the following graph, which enzyme has a greater attraction for substrate? Briefly explain.

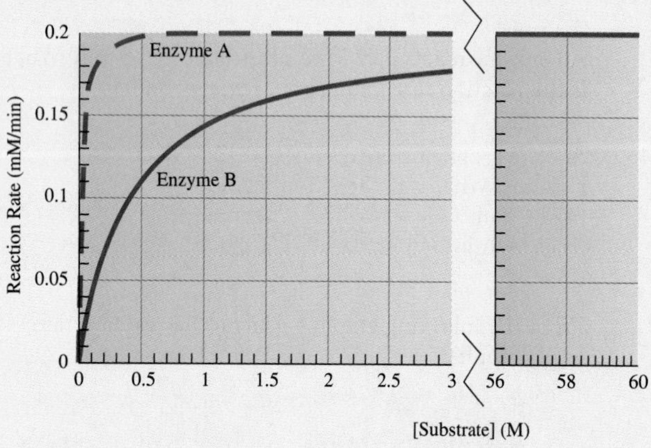

20. Based on the following graph, which enzyme has a greater maximum velocity V_{max}? Briefly explain.

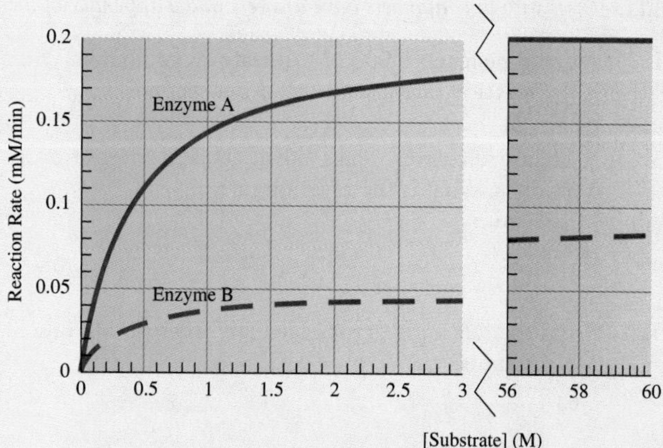

21. Salivary amylase, a digestive enzyme, is found in saliva (pH = 6). What is a likely pH optimum for salivary amylase? Explain briefly.

22. The stomach has a $[H_3O^+] = 0.01\ M$. What would you expect the pH optimum for pepsin (a protease found in the stomach) to be? Explain briefly.

23. Choose the set of conditions that would yield a faster enzyme-catalyzed reaction:

Condition A	versus	Condition B
[reactant]	= 0.013 *M*	0.013 *M*
temp.	= 68°C	55°C
activation energy	= 38 kcal/mol	38 kcal/mol

Explain briefly.

24. Choose the set of conditions that would yield a faster enzyme-catalyzed reaction:

Condition A	versus	Condition B
[reactant]	= 0.05 *M*	0.05 *M*
temp.	= 68°C	68°C
activation energy	= 38 kcal/mol	35 kcal/mol

Explain briefly.

25. A scientist studies two enzymes that catalyze the same reaction. Enzyme A has a turnover number of 225/s, while enzyme B has a turnover number of 120/s. The scientist concludes that enzyme B is more effective than enzyme A. Do you agree? Briefly explain.

26. A scientist studies two enzymes that catalyze the same reaction. Enzyme A has a turnover number of 0.05/s, while enzyme B has a turnover number of 9.8×10^{-1}/s. The scientist concludes that enzyme B is more effective than enzyme A. Do you agree? Briefly explain.

27. The amino acid glutamine is produced by a metabolic pathway in the liver. As the concentration of glutamine goes up, the metabolic pathway slows down. Is this control feedforward activation or feedback inhibition? Briefly explain.

28. The amino acid glutamine is used by a liver metabolic pathway to make urea. As the concentration of glutamine goes up, the metabolic pathway speeds up. Is this control feedforward activation or feedback inhibition? Briefly explain.

29. Briefly explain why cellulase is used in paper manufacturing.

30. Briefly explain why proteases are used to treat stroke.

31. Briefly describe the function of a lipase in laundry detergent.

32. Briefly describe the function of an amylase in laundry detergent.

33. Briefly explain why lactase-treated milk can help lactose-intolerant individuals.

34. Briefly explain why sugar solutions are sweeter after being treated with sucrase.

Additional Exercises

All exercises with blue numbers have answers in the appendix of the text.

35. Chymotrypsin has a turnover number for a glycine- containing substrate of 0.05/s and a turnover number for an L-tyrosine–containing substrate of 200/s. For which substrate is chymotrypsin a more efficient catalyst? Briefly explain.

36. Feedback inhibition is an important form of enzyme regulation. Based on what you know about this control mechanism, how might a different process, "feedback activation," cause regulatory problems for a cell?

37. Explain the meaning of V_{max} by drawing a Michaelis–Menten plot.

38. The enzyme lactase is used in the food industry to modify milk products. What chemical change does lactase catalyze?

Challenge Exercise

39. A "learned professor" tells you that a protein must be absolutely rigid to be a good catalyst. Do you agree or disagree? Explain briefly.

Answers to Practice Exercises

30.1 (a)

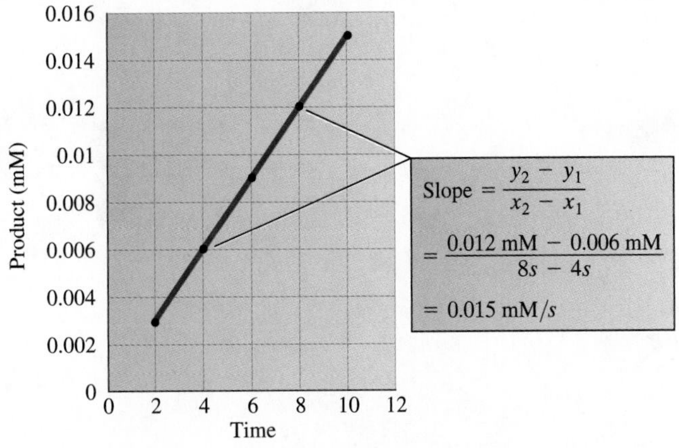

Product (mM): 0, 0.002, 0.004, 0.006, 0.008, 0.01, 0.012, 0.014, 0.016

Time: 0, 2, 4, 6, 8, 10, 12

$$\text{Slope} = \frac{y_2 - y_1}{x_2 - x_1} = \frac{0.012\ \text{mM} - 0.006\ \text{mM}}{8s - 4s} = 0.015\ \text{mM}/s$$

(b) $$\text{rate} = \frac{[\text{product}]_2 - [\text{product}]_1}{\text{time}_2 - \text{time}_1} = \frac{0.70\ \text{M} - 0.35\ \text{M}}{3.0\ \text{hr} - 1.5\ \text{hr}} = 0.23\ \text{M/hr}.$$

30.2 (a) The condition set with the lower activation energy (26 kcal/mol) yields the faster reaction rate because the energy barrier to reaction is smaller. (b) The condition set with the larger CO_2 pressure (130 torr) yields the faster reaction rate because there are more reactant molecules.

30.3 (a) Under these conditions, pepsin is more efficient, because it has the larger turnover number, converting 30 reactant molecules to products per minute.
(b) (i) 630/min;
(ii) 290/min.

30.4 You would not use cellulase to "biostone" polyester fabric, because cellulase is an enzyme that acts specifically on cellulose. Since polyesters are not made of cellulose, the enzyme would have no effect on them.

CHAPTER 31

Nucleic Acids and Heredity

The science of genetics began when Gregor Mendel studied pea plants in the mid-nineteenth century.

Chapter Outline

The plight of Doctor Frankenstein's monster touches a chord in all of us. A scientist has given this monster life, but cannot control his creation. The monster is "unnatural" and the story ends tragically.

The advent of genetic engineering has raised a similar specter. Genetic engineers work with the molecules that code life—the nucleic acids. By changing the code, they can produce new life-forms. Already, bacteria have been altered to make needed human proteins. Recently, both cows and goats have been genetically engineered to produce a human protein in their milk.

The scientists involved in these initial programs have followed careful protocols and have produced valuable medicines. However, the day may come when we can decide whether humans should be made smarter or stronger via genetic engineering. How this decision will be made and what it will be are the topics of heated discussions and much controversy; the potential ramifications are enormous—and not only in the scientific community. The power to even consider such decisions and to possibly open "Pandora's box" in ways that were once only the stuff of science fiction is the result of our understanding of the biochemistry of nucleic acids.

31.1 Molecules of Heredity—A Link

The question of how hereditary material duplicates itself was one of the most baffling problems of biology for many years. Generations of biologists attempted in vain to solve this problem and to answer the question "Why are the offspring of a species undeniably of that species?" Many thought the chemical basis for heredity lay in the structure of the proteins. But no one was able to provide evidence showing how protein could reproduce itself. The answer to the heredity question was finally found in the structure of the nucleic acids.

The unit structure of all living things is the cell. Suspended in the nuclei of cells are chromosomes, which consist largely of proteins and nucleic acids. A simple protein bonded to a nucleic acid is called a **nucleoprotein**. **Nucleic acids** are polymers of nucleotides and contain either the sugar *deoxyribose* or the sugar *ribose*. Accordingly, they are called deoxyribonucleic acid (DNA) and ribonucleic acid (RNA), respectively. Although many of us think of DNA as a recent discovery, it was actually discovered in 1869 by the Swiss physiologist Friedrich Miescher (1844–1895), who extracted it from the nuclei of cells.

nucleoprotein
nucleic acid

31.2 Bases and Nucleosides

Nucleic acids are complex chemicals that combine several different classes of smaller molecules. As with many complex structures, it is easier to understand the whole by first studying its component parts. We begin our examination of nucleic acids by learning about a critical part of these molecules, two classes of heterocyclic bases called the *purines* and the *pyrimidines*:

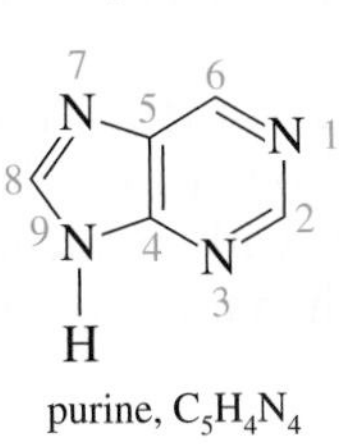

purine, $C_5H_4N_4$ 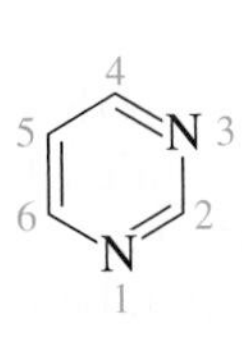 pyrimidine, $C_4H_4N_2$

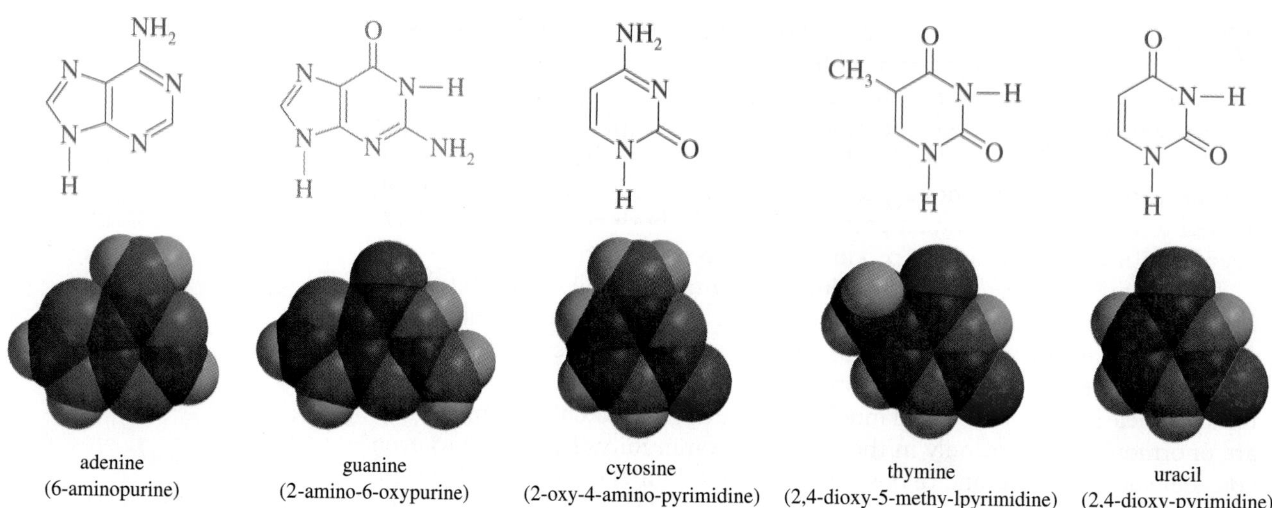

Figure 31.1
Purine and pyrimidine bases found in living matter.

The parent compounds are related in structure: The pyrimidine is a six-membered heterocyclic ring, while the purine contains both a five- and six-membered ring. The nitrogen atoms cause these compounds to be known as *heterocycles* (the rings are made up of more than just carbon atoms) and also as bases. Like the ammonia nitrogen, heterocycles react with hydrogen ions to make a solution more basic.

Five major bases are commonly found in nucleic acids—two purine bases (adenine and guanine) and three pyrimidine bases (cytosine, thymine, and uracil). Figure 31.1 gives one stable form for each compound. These bases must be available to the cell in order to reproduce genetic information (DNA). The importance of the bases is underscored by their use in cancer chemotherapy. Specific "modified" bases are used to kill fast-growing cancer cells. Perhaps the most common anticancer drug in this category is 5-fluorouracil:

5-fluorouracil

Note that 5-fluorouracil is similar to thymine: It acts to inhibit the enzyme that catalyzes the formation of thymine. Cancer cells require a rapid synthesis of DNA. Without sufficient thymine to form DNA, cancer cells die.

The natural bases differ one from another in their ring substituents. Each base has a lowermost nitrogen, which is bonded to a hydrogen as well as two carbons. This specific —NH shares chemical similarities with an alcohol (—OH) group. Just as two sugars molecules can be linked when an alcohol of

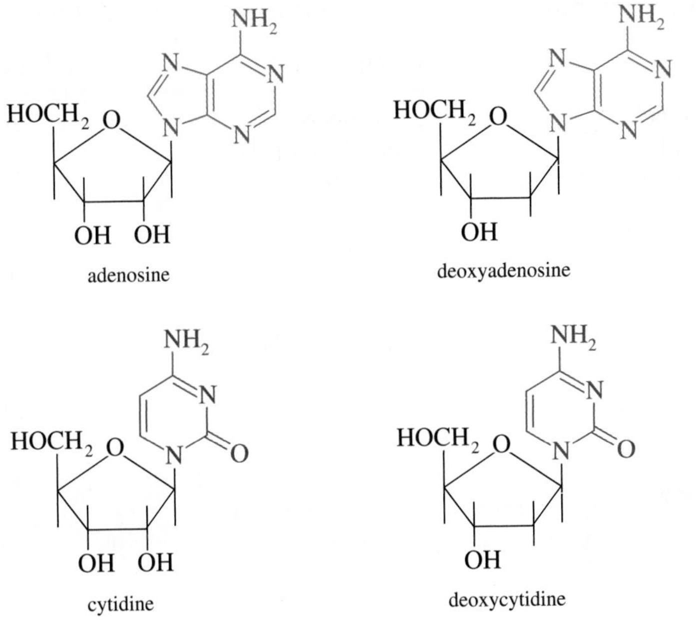

NH$_2$ N N HOCH$_2$ O N N OH OH adenosine

NH$_2$ N N HOCH$_2$ O N N OH deoxyadenosine

NH$_2$ N HOCH$_2$ O N O OH OH cytidine

NH$_2$ N HOCH$_2$ O N O OH deoxycytidine

Figure 31.2 Structures of typical ribonucleosides and deoxyribonucleosides.

one monosaccharide reacts with a second monosaccharide (see Chapter 27), so a purine or pyrimidine molecule can be bonded to a sugar molecule by a reaction with the —NH group.

nucleoside

A **nucleoside** is formed when either a purine or pyrimidine base is linked to a sugar molecule, usually D-ribose or D-2′-deoxyribose:

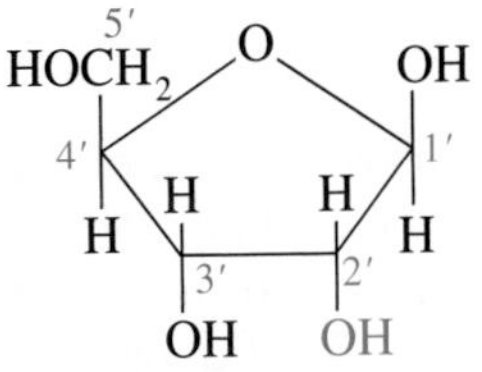

5′ HOCH$_2$ O OH 4′ 1′ H H H H 3′ 2′ OH OH D-ribose

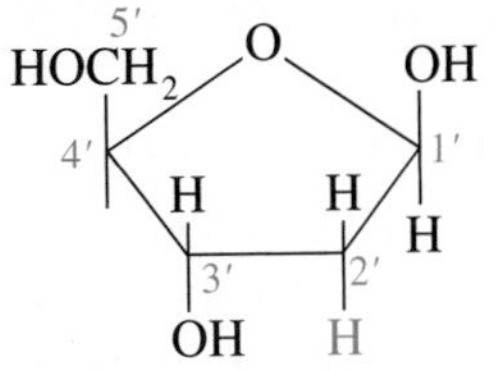

5′ HOCH$_2$ O OH 4′ 1′ H H H 3′ 2′ OH H D-2′-deoxyribose

The base and sugar are bonded together between carbon 1′ of the sugar and either the purine nitrogen at position 9 or the pyrimidine nitrogen at position 1 by splitting out a molecule of water. Typical structures of nucleosides are shown in Figure 31.2. A prime is added to the position number to differentiate the sugar numbering system from the purine or pyrimidine numbering system.

The name of each nucleoside emphasizes the importance of the base to the chemistry of the molecule. Thus, adenine and D-ribose react to yield adenosine, whereas cytosine and D-2′-deoxyribose yield deoxycytidine. The root of the nucleoside name derives from the purine or pyrimidine name. The compositions of the common ribonucleosides and the deoxyribonucleosides are given in Table 31.1.

Table 31.1 Composition of Ribonucleosides and Deoxyribonucleosides

Name	Composition	Abbreviation
Adenosine	Adenine–ribose	A
Deoxyadenosine	Adenine–deoxyribose	dA
Guanosine	Guanine–ribose	G
Deoxyguanosine	Guanine–deoxyribose	dG
Cytidine	Cytosine–ribose	C
Deoxycytidine	Cytosine–deoxyribose	dC
Thymidine	Thymine–ribose	T
Deoxythymidine	Thymine–deoxyribose	dT
Uridine	Uracil–ribose	U
Deoxyuridine	Uracil–deoxyribose	dU

Practice 31.1

Draw the structures of (a) guanosine and (b) deoxycytidine.

31.3 Nucleotides: Phosphate Esters

nucleotide

A more complex set of biological molecules is formed by linking phosphate groups to nucleosides. Phosphate esters of nucleosides are termed **nucleotides**. These molecules consist of a purine or a pyrimidine base linked to a sugar, which in turn is bonded to at least one phosphate group:

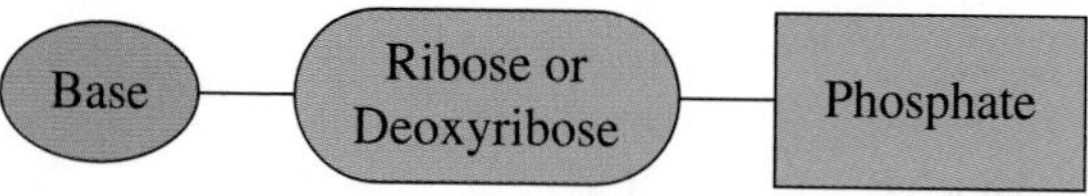

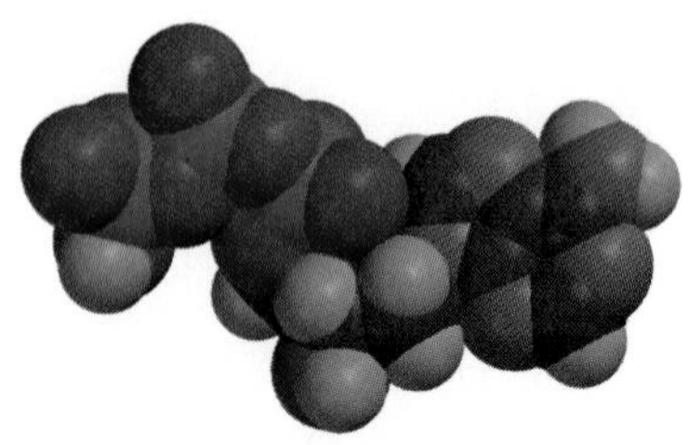

Spacefilling model of ATP.

The ester may be a monophosphate, a diphosphate, or a triphosphate. When two or more phosphates are linked together, a high-energy phosphate anhydride bond is formed. (See Section 31.4.) The ester linkage may be to the hydroxyl group of position 2′, 3′, or 5′ of ribose or to position 3′ or 5′ of deoxyribose. Examples of nucleotide structures are shown in Figure 31.3.

Nucleotide abbreviations start with the corresponding nucleoside abbreviation. (See Table 31.1.) The letters MP (monophosphate) can be added to any of these to designate the corresponding nucleotide. Thus, GMP is guanosine

adenosine-5′-monophosphate (AMP)

deoxyadenosine-5′-monophosphate (dAMP)

Figure 31.3
Examples of nucleotides.

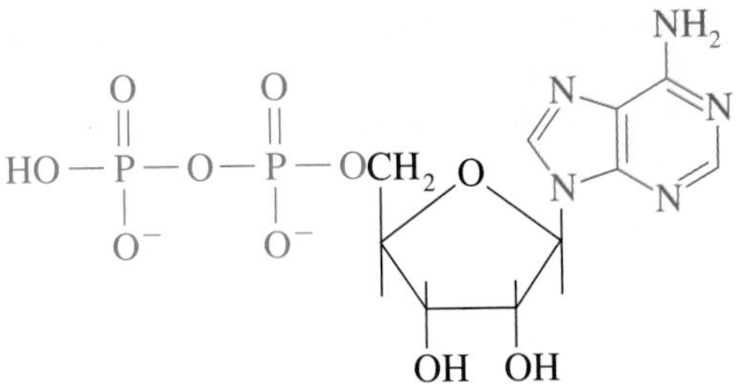

adenosine-5′-diphosphate
(ADP)

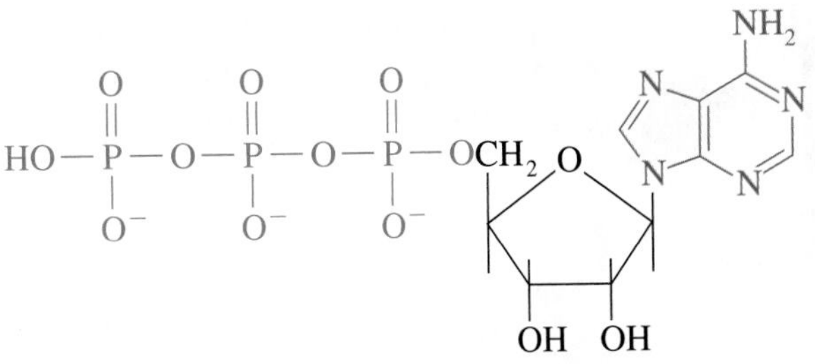

adenosine-5′-triphosphate
(ATP)

Figure 31.4
Structures of ADP and ATP.

monophosphate. A lowercase "d" is placed in front of GMP if the nucleotide contains the deoxyribose sugar (dGMP). When the letters such as AMP or GMP are given, it is generally understood that the phosphate group is attached to position 5′ of the ribose unit (5′-AMP). If attachment is elsewhere, it will be designated, for example, as 3′-AMP.

Two other important adenosine phosphate esters are adenosine diphosphate (ADP) and adenosine triphosphate (ATP). Note that the letters DP are used for diphosphate and TP for triphosphate. In these molecules, the phosphate groups are linked together. The structures are similar to AMP, except that they contain two and three phosphate residues, respectively. (See Figure 31.4.) All the nucleosides form mono-, di-, and triphosphate nucleotides.

Practice 31.2

Name and give abbrevations for the following structures:

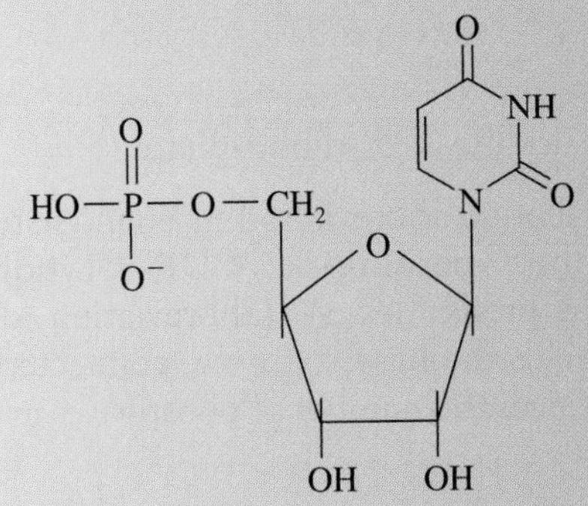

(a)

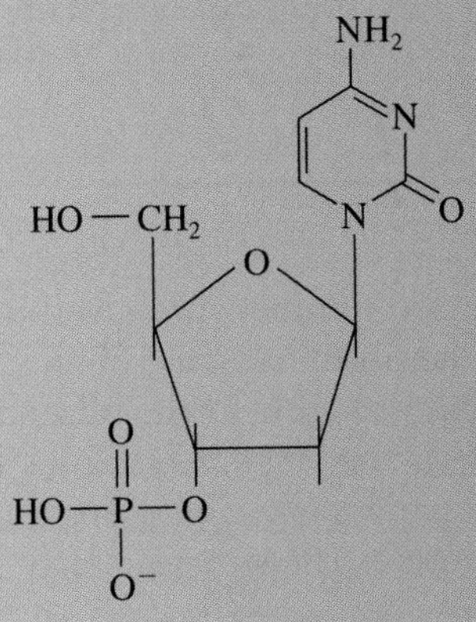

(b)

31.4 High-Energy Nucleotides

Nucleotides have a central role in the energy transfers in many metabolic processes. ATP and ADP are especially important in these processes, since these two nucleotides store and release energy to the cells and tissues. The source of energy is the foods we eat, particularly carbohydrates and fats. Energy is released as the carbons from these foods are oxidized. (See Chapter 33.)

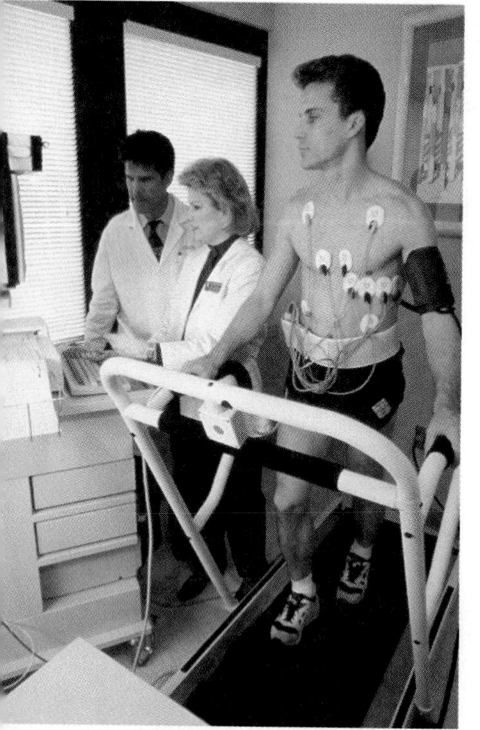

Muscle movements, including heartbeats, depend on energy from ATP.

Part of this energy is used to maintain body temperature, and part is stored in the phosphate anhydride bonds of such molecules as ADP and ATP. Because a relatively large amount of energy is stored in these bonds, they are known as high-energy phosphate anhydride bonds:

$$HO-\overset{\overset{O}{\|}}{\underset{\underset{O^-}{|}}{P}}\sim O-\overset{\overset{O}{\|}}{\underset{\underset{O^-}{|}}{P}}\sim O-\overset{\overset{O}{\|}}{\underset{\underset{O^-}{|}}{P}}-O-\text{Adenosine} \qquad HO-\overset{\overset{O}{\|}}{\underset{\underset{O^-}{|}}{P}}\sim O-\overset{\overset{O}{\|}}{\underset{\underset{O^-}{|}}{P}}-O-\text{Adenosine}$$

High-energy phosphate anhydride bonds (ATP) High-energy anhydride bond (ADP)

Energy is released during the hydrolysis of high-energy phosphate anhydride bonds in ADP and ATP. In the hydrolysis, ATP forms ADP and inorganic phosphate (P_i) yielding about 35 kJ of energy per mole of ATP:

$$ATP + H_2O \underset{\text{energy storage}}{\overset{\text{energy utilization}}{\rightleftharpoons}} ADP + P_i + \sim 35\ kJ$$

The hydrolysis reaction is reversible, with ADP being converted to ATP by still higher energy molecules. In this manner, energy is supplied to the cells from ATP, and energy is stored by the synthesis of ATP. Processes such as muscle movement, nerve sensations, vision, and even the maintenance of our heartbeats are all dependent on energy from ATP.

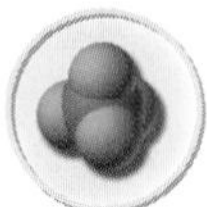

Example 31.1 Draw the structures of (a) CDP and (b) uridine-3′,5′-diphosphate.

SOLUTION (a) When no numbers are given in a nucleotide abbreviation, it is understood that the phosphates are connected at the 5′-position (i.e., CDP = cytidine-5′-diphosphate). In the absence of the prefix "deoxy-" (abbreviation "d-"), ribose is the sugar. Cytidine ("C") signals the base, cytosine, connected to the sugar. Finally, diphosphate ("DP") gives the number of phosphates in the nucleotide. The structure is as follows:

$$HO-\overset{\overset{O}{\|}}{\underset{\underset{O^-}{|}}{P}}-O-\overset{\overset{O}{\|}}{\underset{\underset{O^-}{|}}{P}}-O-CH_2 \text{ (ribose, 2'-OH, 3'-OH; N-linked cytosine with } NH_2 \text{ and C=O)}$$

(b) The full name provides all the information needed to draw the nucleotide structure. In the absence of the prefix "deoxy," the sugar is ribose. "Uridine" signals the base uracil, connected to the sugar. "-3′,5′-diphosphate" gives the number and locations of the phosphates. The structure is as follows:

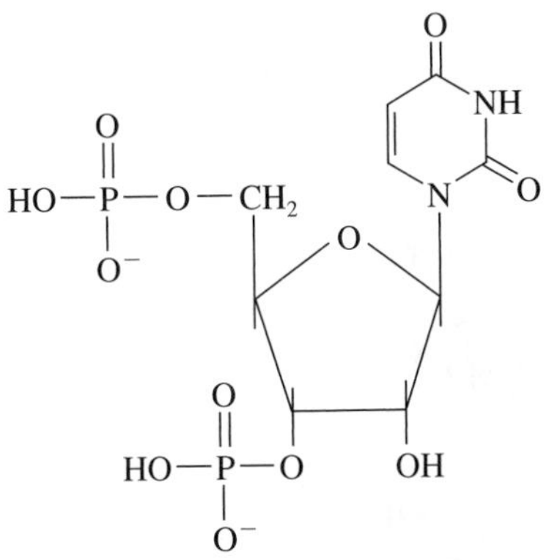

Practice 31.3

Draw structures for the following: (a) guanosine-5′-diphosphate; (b) guanosine-3′,5′-diphosphate.

31.5 Polynucleotides; Nucleic Acids

Starting with two nucleotides, a dinucleotide is formed by splitting out a molecule of water between the —OH of the phosphate group of one nucleotide and the —OH on carbon 3′ of the ribose or deoxyribose of the other nucleotide. Then, another and another nucleotide can be added in the same manner, until a polynucleotide chain is formed. Each nucleotide is linked to its neighbors by phosphate ester bonds. (See Section 24.13.)

Two series of polynucleotide chains are known, one containing D-ribose and the other D-2′-deoxyribose. One polymeric chain consists of the monomers AMP, GMP, CMP, and UMP and is known as a *polyribonucleotide*. The other chain contains the monomers dAMP, dGMP, dCMP, and dTMP and is known as a *polydeoxyribonucleotide*. The two chains are as follows:

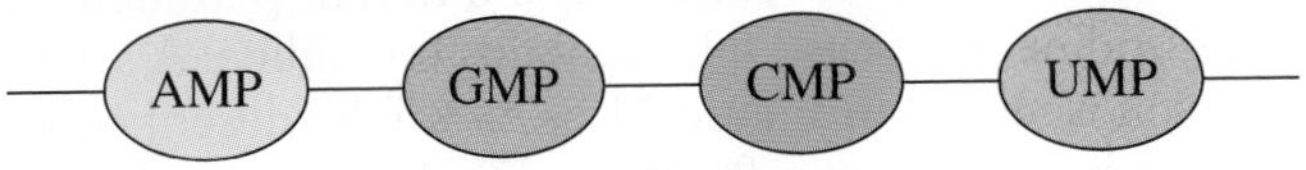

—AMP—GMP—CMP—UMP—

polyribonucleotide (RNA)

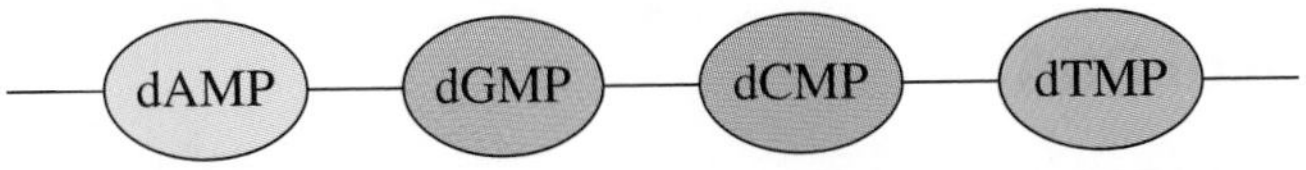

—dAMP—dGMP—dCMP—dTMP—

polydeoxyribonucleotide (DNA)

ribonucleic acid (RNA)

The nucleic acids DNA and RNA are polynucleotides. **Ribonucleic acid (RNA)** is a polynucleotide that, upon hydrolysis, yields D-ribose, phosphoric acid, and the four purine and pyrimidine bases adenine, guanine, cytosine,

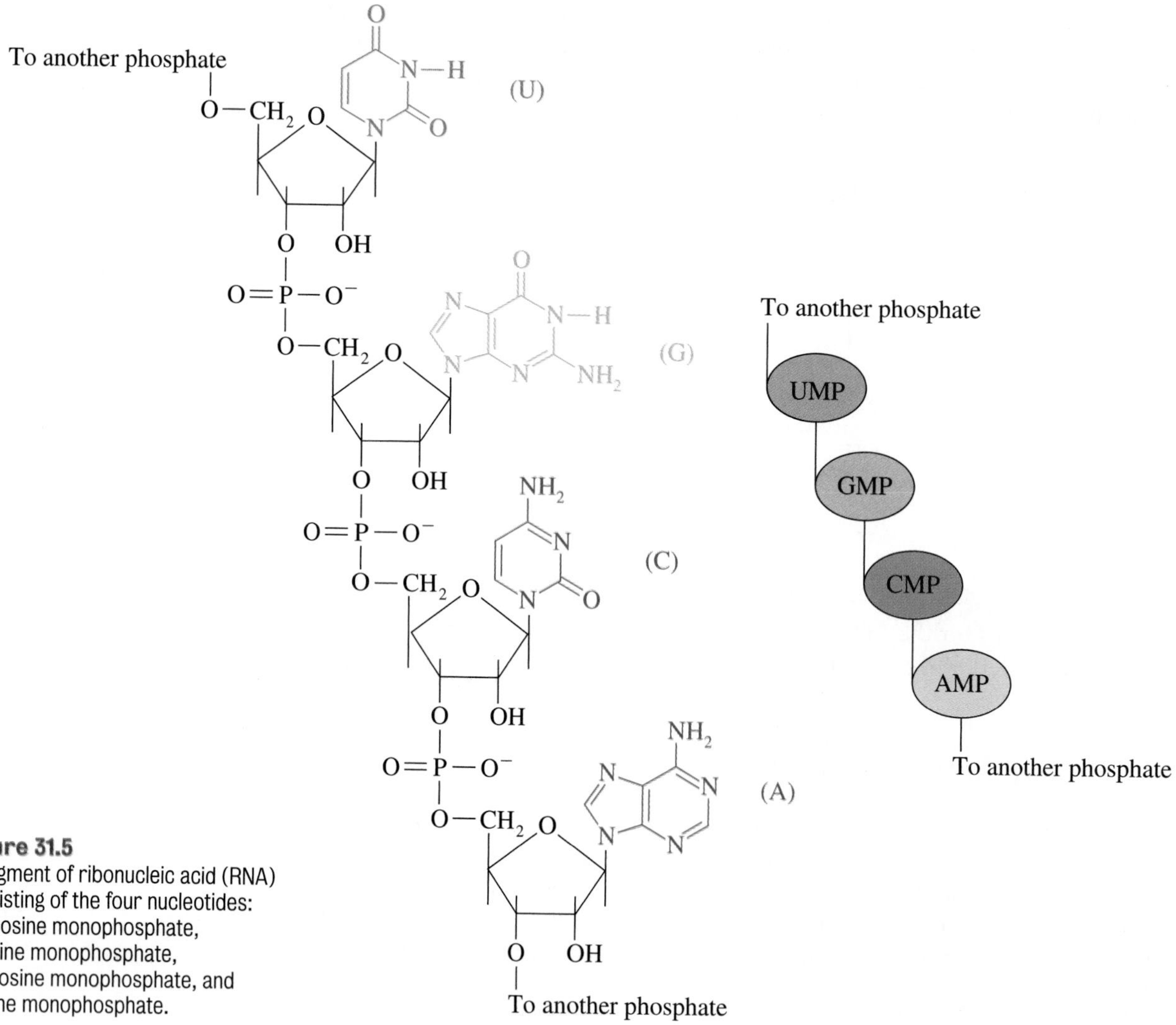

Figure 31.5
A segment of ribonucleic acid (RNA) consisting of the four nucleotides: adenosine monophosphate, cytidine monophosphate, guanosine monophosphate, and uridine monophosphate.

deoxyribonucleic acid (DNA)

and uracil. **Deoxyribonucleic acid (DNA)** is a polynucleotide that yields D-2′-deoxyribose, phosphoric acid, and the four bases adenine, guanine, cytosine, and thymine. Note that RNA and DNA contain one different pyrimidine nucleotide: RNA contains uridine, whereas DNA contains thymidine. A segment of a ribonucleic acid chain is shown in Figure 31.5. As will be described later, RNA and DNA also commonly differ in function: DNA serves as the storehouse for genetic information; RNA aids in expressing genetic characteristics.

31.6 Structure of DNA

Deoxyribonucleic acid (DNA) is a polymeric substance made up of the four nucleotides dAMP, dGMP, dCMP, and dTMP. The size of the DNA polymer varies with the complexity of the organism; more complex organisms tend to have larger DNAs. For example, simple bacteria like *Escherichia coli* have about 8 million nucleotides in their DNA, while human DNA contains up to 500 million

Table 31.2 Relative Amounts of Purines and Pyrimidines in Samples of DNA

Source	Adenine	Thymine	Ratio A/T	Guanine	Cytosine	Ratio G/C
Beef thymus	29.0	28.5	**1.02**	21.2	21.2	**1.00**
Beef liver	28.8	29.0	**0.99**	21.0	21.1	**1.00**
Beef sperm	28.7	27.2	**1.06**	22.2	22.0	**1.01**
Human thymus	30.9	29.4	**1.05**	19.9	19.8	**1.00**
Human liver	30.3	30.3	**1.00**	19.5	19.9	**0.98**
Human sperm	30.9	31.6	**0.98**	19.1	18.4	**1.04**
Hen red blood cells	28.8	29.2	**0.99**	20.5	21.5	**0.96**
Herring sperm	27.8	27.5	**1.01**	22.2	22.6	**0.98**
Wheat germ	26.5	27.0	**0.98**	23.5	23.0	**1.02**
Yeast	31.7	32.6	**0.97**	18.3	17.4	**1.05**
Vaccinia virus	29.5	29.9	**0.99**	20.6	20.0	**1.03**
Bacteriophage T_2	32.5	32.6	**1.00**	18.2	18.6	**0.98**

nucleotides. The order in which these nucleotides occur varies in different DNA molecules, and it is within this order that the genetic information in a cell is stored.

Scientists have tried to understand DNA structure by determining the nucleotide composition of this molecule from many different sources. For a long time, it was thought that the four nucleotides occurred in equal amounts in DNA. However, more refined analyses showed that the amounts of purine and pyrimidine bases vary in different DNA molecules. Surprisingly, careful consideration of these data also showed that the ratios of adenine to thymine and guanine to cytosine are always essentially 1 : 1. This observation served as an important key to unraveling the structure of DNA. Analyses of DNA from several species are shown in Table 31.2.

A second important clue to the special configuration and structure of DNA came from X-ray diffraction studies. Most significant was the work of Maurice Wilkins (b. 1916) of Kings College of London. Wilkins's X-ray pictures implied that the nucleotide bases were stacked one on top of another like a stack of saucers. From his work, as well as that of others, the American biologist James D. Watson (b. 1928) and British physicist Francis H.C. Crick

James Watson and Francis Crick with their DNA model.

(1916–2004), working at Cambridge University, designed and built a scale model of a DNA molecule. In 1953, Watson and Crick announced their now famous double-stranded helical structure for DNA. This was a milestone in the history of biology, and in 1962 Watson, Crick, and Wilkins were awarded the Nobel Prize in medicine and physiology for their studies of DNA.

The structure of DNA, according to Watson and Crick, consists of two polymeric strands of nucleotides in the form of a double helix, with both nucleotide strands coiled around the same axis. (See Figure 31.6.) Along each strand are alternate phosphate and deoxyribose units, with one of the four bases adenine, guanine, cytosine, or thymine attached to deoxyribose as a side group. The double helix is held together by hydrogen bonds extending from the base on one strand of the double helix to a complementary base on the other strand. The four bases are flat in their ring structures. The structure of DNA has been likened to a ladder that is twisted into a double helix, with the rungs of the ladder kept perpendicular to the twisted railings. The phosphate and deoxyribose units alternate along the two railings of the ladder, and two nitrogen bases form each rung of the ladder.

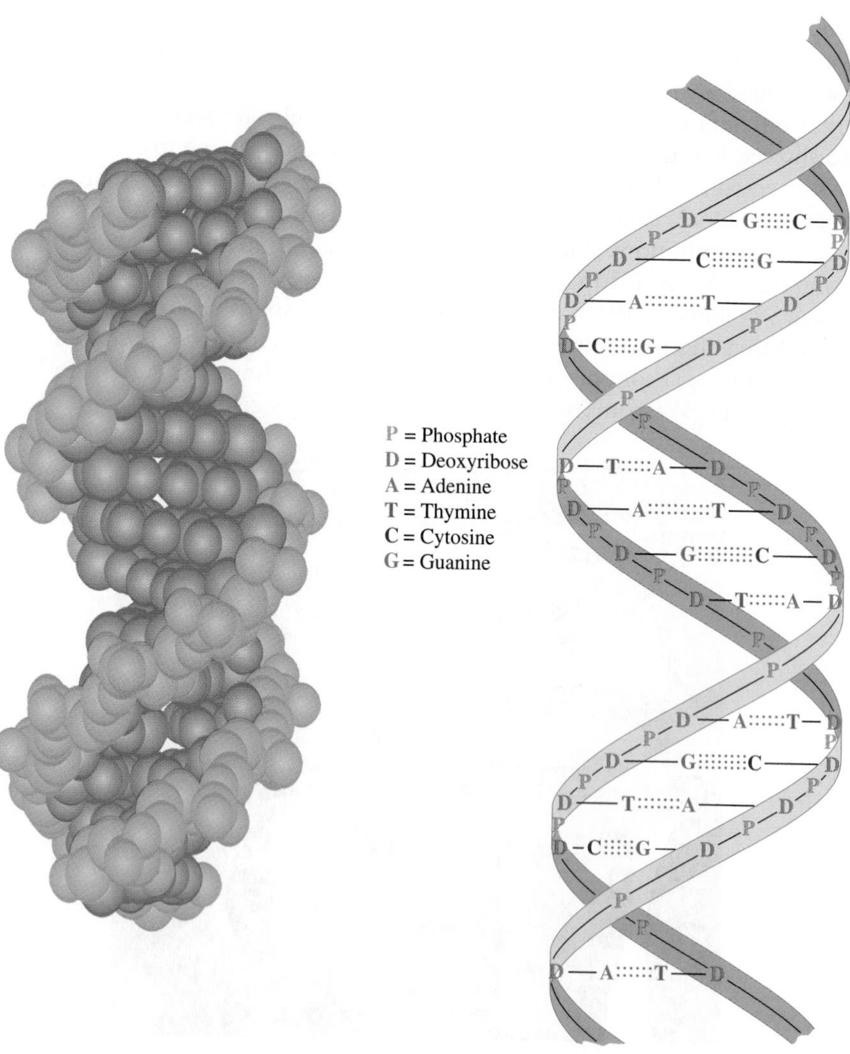

Figure 31.6
Right: Double-stranded helical structure of DNA (....denotes a hydrogen bond between adjoining bases).
Left: Spacefilling model of DNA. The sugar-phospate backbone is colored yellow; the bases are blue.

In the Watson–Crick model of DNA, the two polynucleotide strands fit together best when a purine base is adjacent to a pyrimidine base. Although this allows for four possible base pairings, namely, A–T, A–C, G–C, and G–T, only the adenine–thymine and guanine–cytosine base pairs can effectively hydrogen-bond together. Under normal conditions, an adenine on one polynucleotide strand is paired with a thymine on the other strand; a guanine is paired with a cytosine. (See Figure 31.7.) This pairing results in A:T and G:C ratios of 1:1, as substantiated by the data in Table 31.2. The hydrogen bonding of complementary base pairs is shown in Figure 31.8. Note that if the sequence of one strand is known, the sequence of the other strand can be inferred. The two DNA polymers are said to be *complementary* to each other. As will be discussed in Section 31.7, the cell is able to chemically "read" one strand in order to synthesize its complementary partner.

The double helix is an important part of DNA structure. Still, it does not explain how the very large DNA can be packed into a cell or the even smaller cell nucleus. For example, a human DNA molecule can be extended to almost 10 cm in length and yet is contained in a nucleus with a diameter about a hundred thousand times smaller. To begin the necessary packing, the DNA is looped around small aggregates of positively charged histone proteins. Hundreds of these aggregates are associated with each DNA molecule, so that the DNA is foreshortened and has the appearance of a string of pearls. Further condensation is achieved by wrapping this structure into a tight coil called a solenoid, as shown in Figure 31.9. Finally, the solenoid nucleoprotein complex is wound around a protein scaffold within the nucleus.

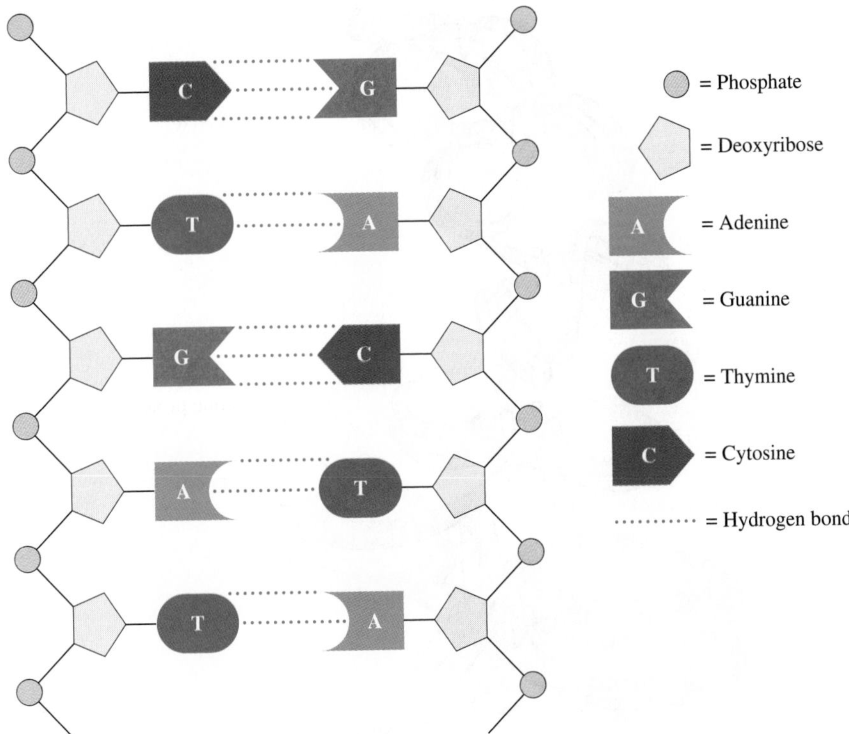

Figure 31.7
DNA segment showing phosphate, deoxyribose, and complementary base pairings held together by hydrogen bonds.

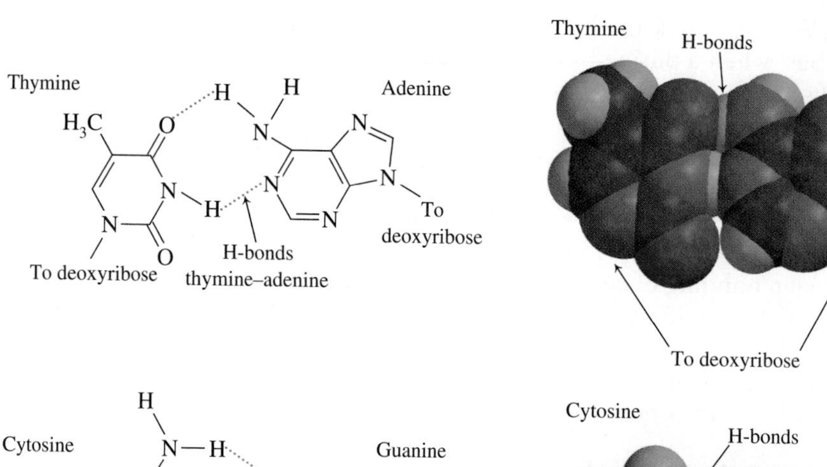

Figure 31.8
Hydrogen bonding between the complementary bases thymine and adenine (T∷A) and cytosine and guanine (C⋮⋮G). Note that one pair of bases has two hydrogen bonds and the other pair has three hydrogen bonds.

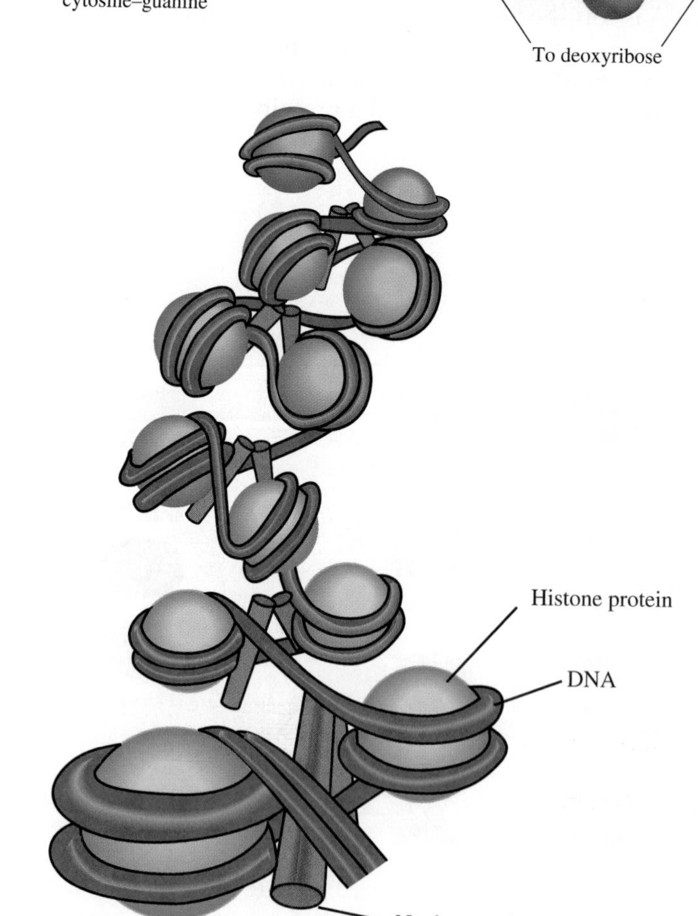

Figure 31.9
Condensed form of DNA. After the DNA polymer wraps twice around the histone protein aggregates (shown as round balls), it coils around a central axis to form a solenoid structure.

By following this complete procedure, human DNA can be reduced in length to about 10 mm in condensed chromosomes. Like a high-density computer disk, DNA takes up only a little space relative to the large amount of genetic information it contains.

31.7 DNA Replication

The foundations of our present concepts of heredity and evolution were laid within the span of a decade. Charles Robert Darwin (1809–1882), in *The Origin of Species* (1859), presented evidence supporting the concept of organic evolution and his theory of natural selection. Gregor Johann Mendel (1822–1884) discovered the basic laws of heredity in 1866, and Friedrich Miescher discovered nucleic acid in 1869. Although Darwin's views were widely discussed and generally accepted by biologists within a few years, Mendel's and Miescher's work went unnoticed for many years.

Mendel's laws were rediscovered about 1900 and led to our present understanding of heredity and the science of genetics. Interest in nucleic acids lagged until nearly the 1950s, when chemical and X-ray data led to Watson and Crick's idea that DNA exists in a double helix and has the possible copying mechanism for genetic material.

Heredity is the process by which the physical and mental characteristics of parents are transferred to their offspring. For this to occur, it is necessary for the material responsible for genetic transfer to be able to make exact copies of itself. The polymeric DNA molecule is the chemical basis for heredity. The genetic information needed for transmitting a species' characteristics is coded along the polymeric chain. Although the chain is made from only four different nucleotides, the information content of DNA resides in the sequence of these nucleotides.

heredity

The **genome** is the sum of all hereditary material contained in a cell. Within the eucaryotic genome are chromosomes, which are long, threadlike bodies composed of nucleic acids and proteins that contain the fundamental units of heredity, called *genes*. A **gene** is a segment of the DNA chain that controls the formation of a molecule of RNA. In turn, many RNAs determine the amino acid sequences for specific polypeptides or proteins. Usually, one gene directs the synthesis of only one polypeptide or protein molecule. The cell has the capability of producing a multitude of different proteins, because each DNA molecule contains a large number of different genes.

genome

gene

For life to continue relatively unchanged, genetic information must be reproduced exactly each time a cell divides. **Replication**, as the name implies, is the biological process for duplicating the DNA molecule. The DNA structure of Watson and Crick holds the key to replication; because of the complementary nature of DNA's nitrogen bases, adenine bonds only to thymine and guanine only to cytosine. Nucleotides with complementary bases can hydrogen-bond to each single strand of DNA and hence be incorporated into a new DNA double helix. Every double-stranded DNA molecule that is produced contains one template strand and one newly formed complementary strand.

replication

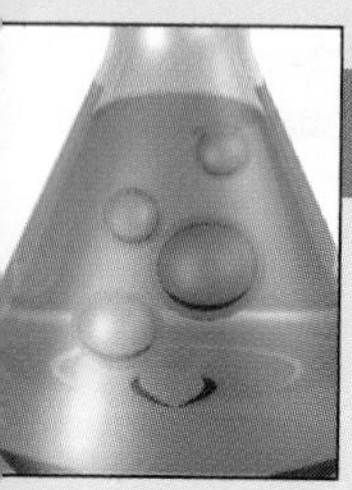

CHEMISTRY IN ACTION • The Double Helix Revisited

By James Watson

The man who launched the Human Genome Project celebrates its success

None of us so privileged few who first saw the double helix in the spring of 1953 ever contemplated that we might in our lifetime see it completely decoded. All our dreams at the time centered on the next big objective—finding how the four letters of the DNA alphabet (A, T, G and C) spell out the linear sequences of amino acids in the synthesis of proteins, the main actors in the drama of cellular life. As it turns out, the essence of the genetic code and of the molecular machinery that reads it was solidly established by 1966, only 13 years after Francis Crick and I discovered the double helix.

Then the creative juices of science turned to how to read the messages of DNA. To our surprise, Frederick Sanger at Cambridge University and Walter Gilbert at Harvard, working independently, needed less than a decade to develop powerful methods for determining the order of DNA letters. At roughly the same time, Herbert Boyer and Stanley Cohen devised elegantly simple procedures for cutting and rejoining DNA molecules to produce "recombinant DNA."

Then voices of doom proclaimed that these procedures would create life forms as threatening to our existence as nuclear weapons. Such false alarms held us back only a few years, however. By 1980 the immense powers of recombinant DNA were let loose for the public good. Soon they were to change irreversibly the faces of biology and medicine and bring modern biotechnology into existence.

It was my desire to help speed up human genetics that drove me in 1986 to become an early partisan of the Human Genome Project, whose ultimate objective was to sequence the roughly 3 billion DNA letters that comprise our genetic code. Though many young hotshots argued that the time for the project had not yet arrived, those of us a generation older were seeing at too close hand our parents and spouses falling victim to diseases of genetic predisposition. And virtually all of us knew couples rearing children whose future was clouded by a bad throw of the genetic dice. So the National Academy of Sciences assembled an expert committee that reported one year later that the human instruction book could be established within 15 years—if we assembled the appropriate scientific leadership and give it $3 billion to spend wisely over those 15 years.

Our first report emphasized that we should begin by sequencing the relatively tiny genomes (1 million to 13 million letters) of bacteria and yeast and then move on to the 100 million letter-size genomes of worms and flies. We were confident that by the time we were done, sequencing technology would cost less than 50¢ a letter, and that by then, we would be ready to tackle the human genome. We were also confident that genomics would pay scientific and medical dividends long before the final letters of the human genome were in place.

So when I went up to Capitol Hill in May 1987 with the N.A.S. report in hand, I promised that long before the genome project was completed we would have cloned many of the key genes predisposing humans to Alzheimer's or to cancers of the breast and colon, all diseases known to run in families. Happily, time has seen our science so move.

Congress accepted this message much faster than many of my fellow molecular biologists, soon appropriating moneys that let the genome project start with a bang. In October 1988, I went to Washington to direct the National Institutes of Health's major role in the effort. From the start, I worked to ensure that the project was an international one, supported by all the major countries of the developed world. That way no one nation or private body would be perceived as controlling the human genome. We also wanted all the data placed on the Internet so that they would be available, free of cost, all over the world.

Today those involved in the international consortium take pride in posting their new DNA sequences on the Web within 24 hours of assembly. Twelve years ago, no one could have imagined that nearly 500 million base pairs of assembled DNA would be posted in just one month.

Watson, with the world's most famous molecule.

With the project essentially complete three years ahead of schedule, we must note the change of heart of its early enemies. Instead of wanting us shut down, as they did as late as 1991, they now beg us to move quickly on the genomes of the mouse, the rat and the dog. Equally important, we note that no other big science project, save possibly the Manhattan Project, has been carried out with such zeal for the common good. In sharing their sequences so freely and so quickly, members of our genome community have little time left to promote their scientific reputations.

By contrast, the large infusions of private capital over the past two years support companies that aim to find and patent key DNA sequences before they become publicly available. Not surprisingly, the leaders of these companies have implied that those of us who started the project were no longer needed. To our vast relief, the publicly supported effort received not less but more money. Our backers want to ensure that all the essential features of the human genome are available without cost to all the people of the world. The events of the past few weeks have shown that those who work for the public good do not necessarily fall behind those driven by personal gain.

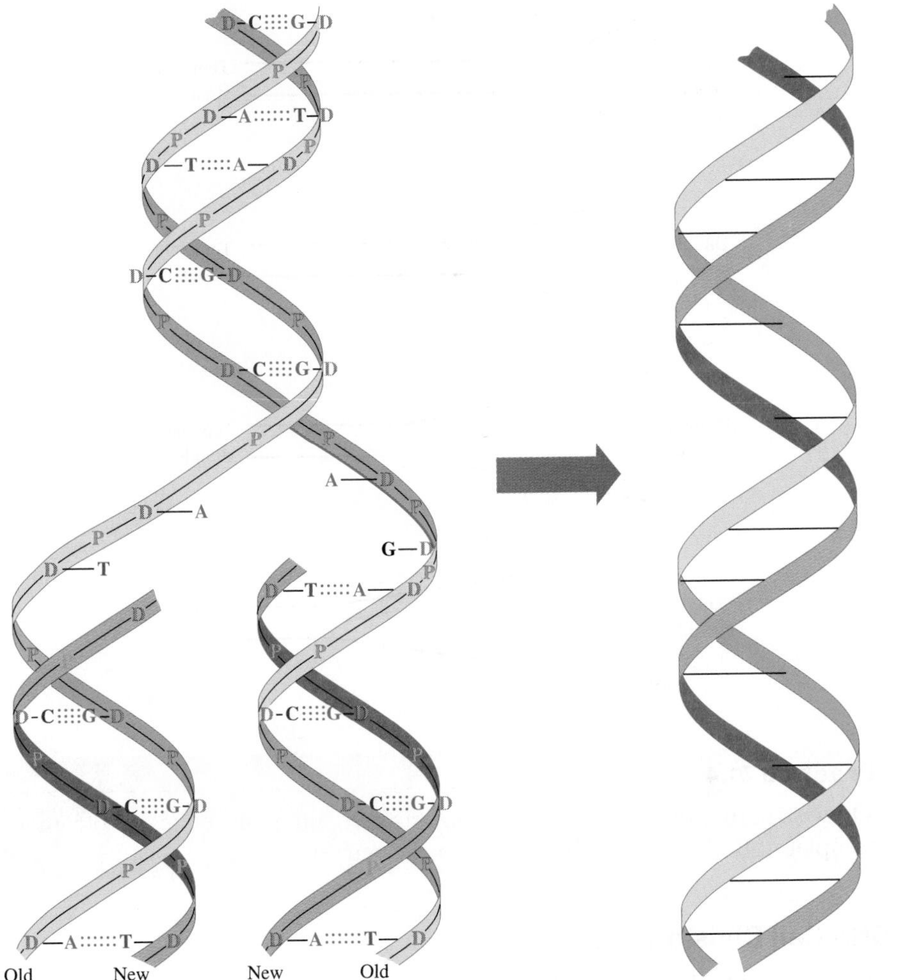

Figure 31.10
Semiconservative replication of DNA: The two helices unwind, separating at the hydrogen bonds. Each strand then serves as a template, recombining with the proper nucleotides and forming two new double-stranded helices. The newly synthesized DNA strands are shown in red.

This form of DNA synthesis, known as *semiconservative* replication, is illustrated in Figure 31.10.

Replication is one of the most complicated enzyme-catalyzed processes in life. Enzymes are required to unwind the DNA before replication and to repackage the DNA after synthesis. The two template strands are copied differently: One daughter strand grows directly toward the point at which the templates are unwinding, while the other daughter strand is synthesized away from this point. In this way, the same enzyme-catalyzed reaction can be used to synthesize both strands. However, the one daughter strand that has been created in small fragments must be connected before replication is complete. So, while one DNA strand is formed by continuous synthesis, the other new strand is formed by a repetition of fragment synthesis followed by a coupling reaction. (See Figure 31.11.)

Other processes "proofread" the new polymers to check for errors. Amazingly, replication is so carefully coordinated that a mistake is passed on to the new strands only once in about a billion times. Of course, this important fact means that new cells retain the genetic characteristics of their parents.

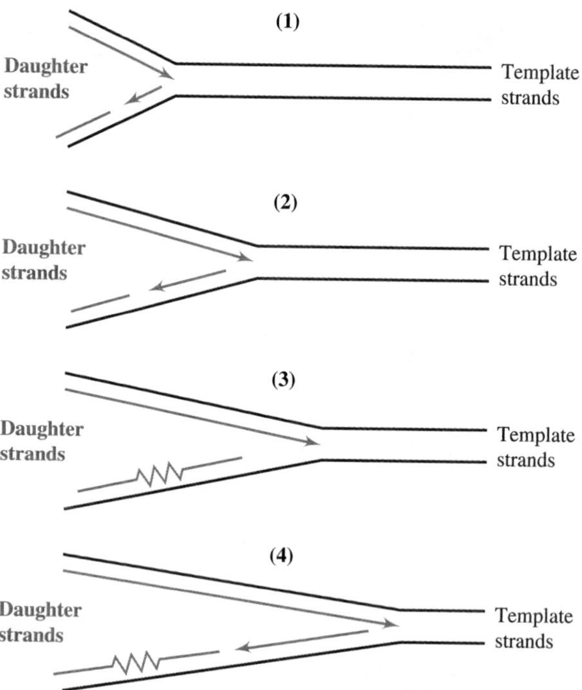

Figure 31.11
The basic replication process. Arrows indicate the direction of DNA synthesis. Between (1) and (2), the template DNA strands unwind, and some synthesis occurs for both daughter strands. Moving to (3), the DNA fragments are connected on one daughter strand (-ᐱᐱᐱ-), while DNA synthesis continues on the other strand and the template strands unwind further. DNA synthesis again takes place for both daughter strands in (4).

Practice 31.4

Draw structure of a small, "new" DNA strand formed in replication from a template DNA strand with the nucleotide sequence GG.

DNA Cell Division and Cloning

The DNA content of cells doubles just before the cell divides, and one-half of the DNA goes to each daughter cell. After cell division is completed, each daughter cell contains DNA and the full genetic code that was present in the original cell. This process of ordinary cell division is known as **mitosis** and occurs in all the cells of the body except the reproductive cells.

mitosis

As we have indicated before, DNA is an integral part of the chromosomes. Each species carries a specific number of chromosomes in the nucleus of each of its cells. The number of chromosomes varies with the species. Humans have 23 pairs, or 46 chromosomes. The fruit fly has 4 pairs, or 8 chromosomes. Each chromosome contains DNA molecules. Mitosis produces cells with the same chromosomal content as the parent cell. However, in the sexual reproductive cycle, cell division occurs by a different process, known as meiosis.

In sexual reproduction, two cells—the sperm cell from the male and the egg cell (or ovum) from the female—unite to form the cell of the new individual. If reproduction took place with mitotic cells, the normal chromosome content would double when two cells united. However, in **meiosis**, the cell splits in such a way as to reduce the number of chromosomes to one-half of the number normally present (23 in humans). The sperm cell carries half of the chromosomes from its original cell, and the egg cell also carries half of the chromosomes from its original cell. When the sperm and the egg cells unite

meiosis

CHEMISTRY IN ACTION • Fingerprints in the Genes

Just as for a set of fingerprints, no two individuals' DNA sequences are exactly the same. More than 100 years ago, detectives first started to use fingerprints from crime scenes to identify criminals. Today, forensic chemists use biochemistry to unlock the DNA sequence in order to identify possible criminals.

Because the human genome contains approximately 3 billion base pairs, it is not practical for forensic chemists to examine the entire genome sequence. Instead, they identify randomly selected pieces of DNA. If enough DNA is available (e.g., from a 1-cm-diameter blood stain), this genetic material can be broken into pieces and separated on a gel by electrophoresis. The gel is then treated with radioactive molecules that bind to specific sites and serve as tags on the DNA fragments. These tags form a pattern on the gel known as a DNA "fingerprint."

A second technique uses the enzyme DNA polymerase to replicate the DNA from a very small sample. (Only several nanograms of DNA are needed.) The replication process multiplies the amount of DNA available, until there are enough DNA pieces to separate and visualize in gel electrophoresis. Again, a pattern is formed that can be treated like a fingerprint.

Unfortunately, DNA fingerprinting uses only a small segment of the total genome. Although each person has a unique total genome, there is an extremely small chance that two people will have the same DNA sequence for a particular segment. So scientists must estimate the possibility of a match by chance, and courtroom decisions often hinge on complex statistical reasoning.

Still, DNA fingerprinting, a technique available only since 1990, continues to make courtroom history. Because most body fluids contain cells that carry DNA, DNA "fingerprints" can often be found at crime scenes, and both "DNA" convictions and exonerations have occurred because of this. Many experts believe that DNA fingerprinting will become more and more widely used in court proceedings in the future.

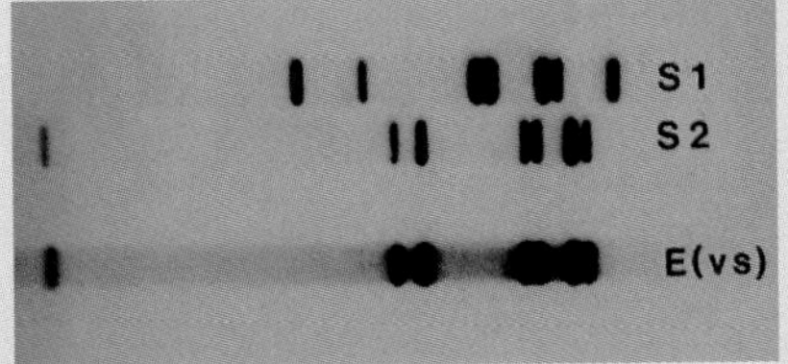

DNA fingerprint matching. Note that Suspect S2 matches evidence blood sample E(vs).

during fertilization, the cell once again contains the correct number of chromosomes and the hereditary characteristics of the species. Thus, the offspring derives half its genetic characteristics from the father and half from the mother.

Although sexual reproduction allows for genetic diversity, there are times when the reproduction of an exact copy might be useful. For example, to replicate a sheep that grows especially fine wool or a cow that gives especially good milk could be quite beneficial. Recently, such copies, called clones, have been produced from adult animals (sheep and cattle). Molecular biologists first transplant an animal cell nucleus into an egg cell. Once the egg has started to divide, it is implanted into a female animal, and gestation proceeds. Each cloned offspring carries a DNA sequence matched to the DNA from the original adult animal donor. The first cloned animal was a sheep named Dolly, who was produced in Scotland in 1996. Cloning is not a routine procedure; there are still many technical difficulties. Yet, the scientific problems are being solved. It will be much harder to solve the societal and ethical implications of routine cloning.

"Dolly," the first sheep cloned from adult cells.

DNA, Cancer, and Chemotherapy

Scientists understood a key principle about tissue growth once they realized that cells must replicate DNA in order to reproduce. For example, fast-growing cancer cells must make DNA much faster than normal cells. Thus, many common cancer chemotherapies are designed to interfere with DNA replication. Some drugs (e.g., topotecan, originally derived from the campotheca tree in Japan) block the unwinding of DNA that is necessary before replication can proceed. Some flat molecules, such as cisplatin (Platinol-AQ®) and adriamycin, slip in between the flat DNA bases and disrupt replication. Cisplatin's structure is as follows:

Cl Cl
Pt
H_2N NH_2

cisplatin

Still other drugs (e.g., gemcitabine, or Gemzar®) are unnatural nucleosides that cause an abortive DNA replication. Note that the sugar ring of gemcitabine has been altered so that it no longer forms a 3′ phosphate ester to the next nucleotide in a growing DNA polymer:

NH_2
N
CH_2OH
O N O
F
OH F

gemcitabine

Of course, the downside to these treatments is that normal cells are affected to a certain extent as well. Fast-growing cells, such as those in bone marrow, are almost always affected adversely by cancer chemotherapies. Thus, scientists continue to search for more specific and better treatments for cancer.

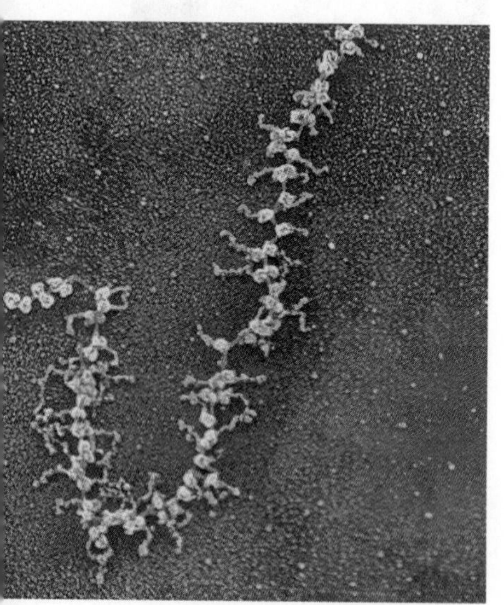

These ribosomes are translating mRNA strands to produce proteins.

31.8 RNA: Genetic Transcription

One of the main functions of DNA is to direct the synthesis of ribonucleic acids (RNAs). RNA differs from DNA in the following ways: (1) It consists of a single polymeric strand of nucleotides rather than a double helix; (2) it contains the pentose D-ribose instead of D-2′-deoxyribose; (3) it contains the pyrimidine base uracil instead of thymine; and (4) some types of RNA have a significant number of modified bases in addition to the common four. RNA also differs functionally from DNA: Whereas DNA serves as the storehouse of genetic information, RNA is used to process this information into proteins. Three types of RNA are needed to produce proteins: ribosomal RNA (rRNA), messenger RNA (mRNA), and transfer RNA (tRNA).

More than 80% of the cellular RNA is ribosomal RNA. It is found in the ribosomes, where it is associated with protein in proportions of about 60–65% protein to 30–35% rRNA. Ribosomes are the sites for protein synthesis.

Messenger RNA carries genetic information from DNA to the ribosomes. It is a template made from DNA, and it carries the code that directs the synthesis of proteins. The size of mRNA varies according to the length of the polypeptide chain it will encode.

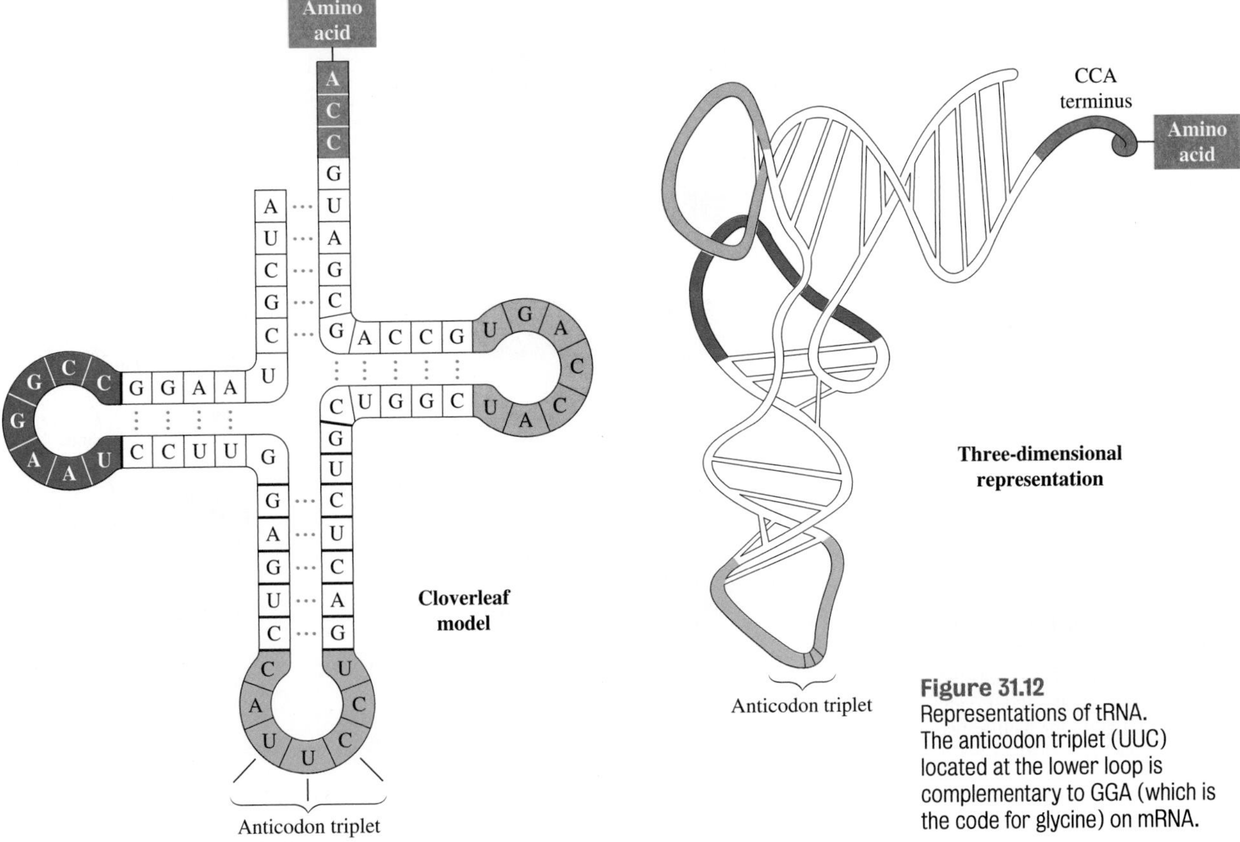

Figure 31.12
Representations of tRNA. The anticodon triplet (UUC) located at the lower loop is complementary to GGA (which is the code for glycine) on mRNA.

The primary function of tRNA is to bring amino acids to the ribosomes for incorporation into protein molecules. Consequently, at least one tRNA exists for each of the 20 amino acids required for proteins. Transfer RNA molecules have a number of structural features in common. The end of the chain of all tRNA molecules, to which is attached the amino acid to be transferred to a protein chain, terminates with the three nucleotides C, C, and A. The primary structure of tRNA allows extensive folding of the molecule, such that complementary bases are hydrogen-bonded to each other to form a structure that appears like a cloverleaf. The cloverleaf model of tRNA has an anticodon loop consisting of seven unpaired nucleotides. Three of these nucleotides make up an anticodon. (See Figure 31.12.) The anticodon is complementary to, and hydrogen-bonds with, three bases on an mRNA. The other two loops in the cloverleaf structure enable the tRNA to bind to the ribosome and other specific enzymes during protein synthesis. (See Section 31.11.)

transcription

The making of RNA from DNA is called **transcription**. The verb *transcribe* literally means *to copy*, often into a different format. When the nucleotide sequence of one strand of DNA is transcribed into a single strand of RNA, genetic information is copied from DNA to RNA. This transcription occurs in a complementary fashion and depends on hydrogen-bonded pairing between appropriate bases. (See Figure 31.13.) For example, where a DNA segment has a guanine base, RNA has a cytosine base. Cytosine is transcribed to guanine,

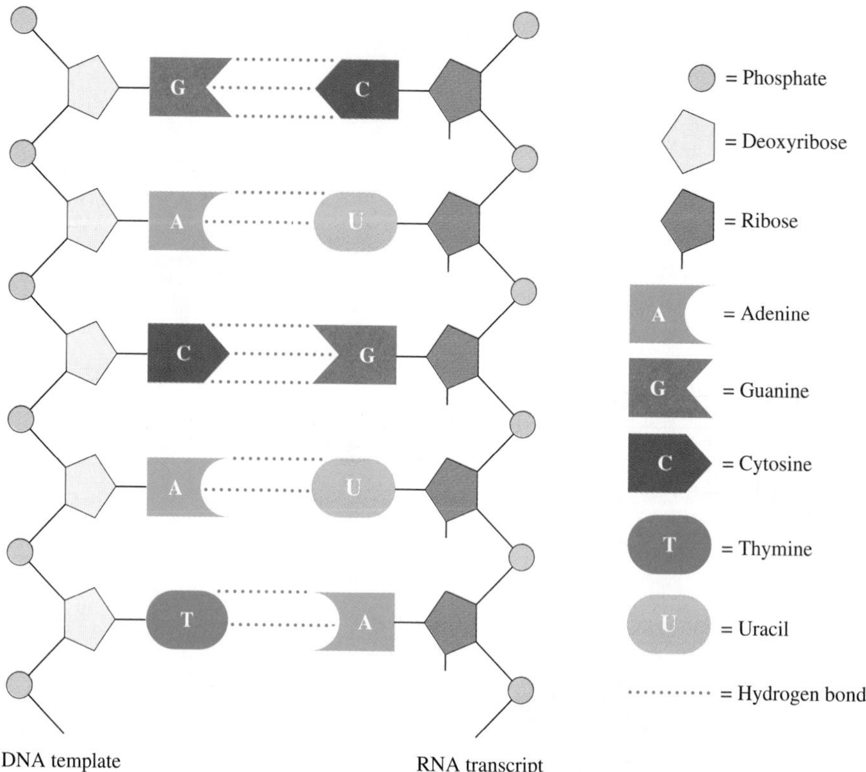

Figure 31.13
Transcription of RNA from DNA. The sugar in RNA is ribose. The complementary base of adenine is uracil. After transcription is complete, the new RNA separates from its DNA template and travels to another location for further use.

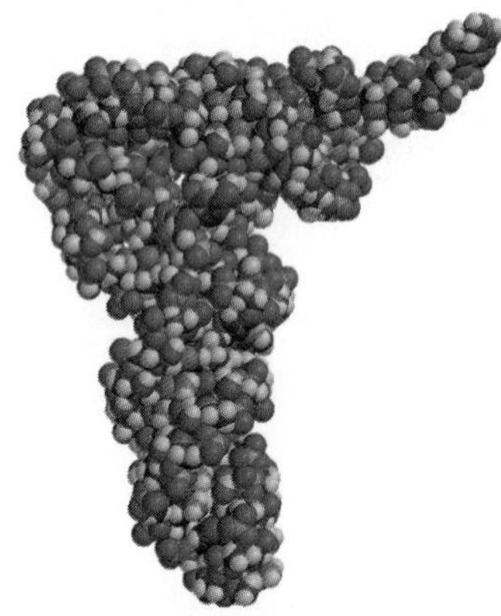

Spacefilling model of tRNA.

thymine to adenine, and adenine to uracil (the thymine-like base that is found in RNA).

Because transcription is the initial step in the expression of genetic information, this process is under stringent cellular control. Only a small fraction of the total information stored in DNA is used at any one time. In procaryotic cells (see Section 33.5), related genes are often located together so that they can be transcribed in concert. The control of eucaryotic gene expression is more complex.

Following transcription, the RNA molecules produced from DNA are often modified before they are put to use. Ribosomal RNA and tRNA molecules form as larger precursors and are then trimmed to the correct size. After transcription, many of the bases in tRNA are modified by methylation, saturation of a double bond, or isomerization of the ribose-base linkage. This posttranscriptional modification or processing changes the information content of the RNA.

Eucaryotic mRNA undergoes considerable alteration before the message that it carries is ready to guide protein synthesis. These changes may include the elimination of portions of the mRNA molecule, the splicing together of two or more mRNA molecules, or the addition of new bases to one end of the mRNA. During this processing, the message encoded in the mRNA molecule is apparently refined so that it can be correctly read by the ribosome.

Figure 31.14 summarizes the processes by which genetic information is transferred within a cell. These steps culminate in the conversion of a coded nucleotide message into a protein amino acid sequence.

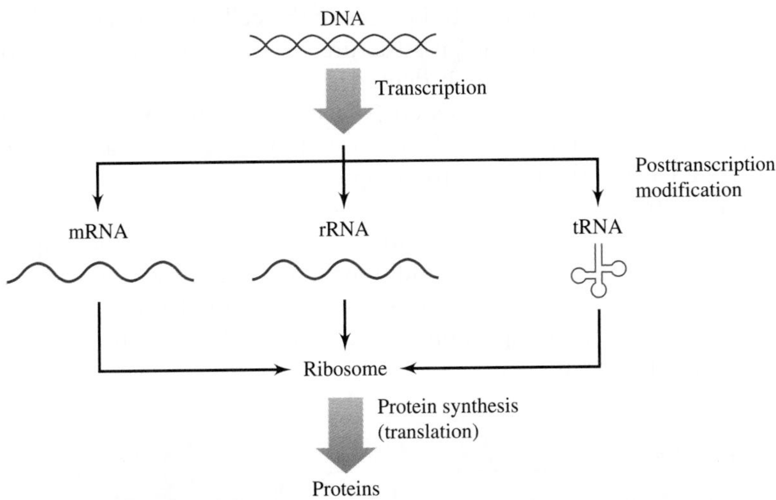

Figure 31.14
A flow diagram representing the processing of cellular genetic information.

RNA and Cancer

An improved understanding of cancer has increased our general knowledge of cell control processes. Cell controls often start at the transcription level, which is the beginning of gene expression, but they also affect replication and translation. Cancer results from a loss of cell control. Given the complexity of cellular control, perhaps it is not surprising that cancer is so widespread.

The importance of transcription control is emphasized by the discovery of the **oncogenes**. These genes are present in cancerous or malignant cells and code for proteins that control cell growth. To the surprise of many investigators, it was found that the same oncogenes are present in many normal mammalian cells. The question arose, "Why do some cells become cancerous and some remain normal?" The answer is that cancerous cells lose control of oncogene transcription. Control is lost when the DNA structure is altered by, for example, a viral infection or a chemical carcinogen. The oncogenes are then transcribed too often, cell growth is uncontrolled—and a cancer develops.

oncogenes

Equally surprising to investigators was the discovery of genes that actively block cancer development, the **tumor-suppressor genes**. These genes code for proteins that allow cell growth *only* if the cells are correctly functioning. For example, some check the DNA for damage before replication reproduces potential mistakes. Studies of rare hereditary cancers have identified over 20 tumor-suppressor genes. When these genes are not active, a cancer can develop. Just one of these tumor-suppressor genes, p53, is inactivated in over 50% of all cancers. On the other hand, when a tumor-suppressor gene is active *and* recognizes cellular problems, it triggers commands that lead to cell destruction. Potential cancer cells are automatically eliminated. This self-destruction is called **apoptosis**, and it maintains the organism's health at the expense of the individual cells.

tumor-suppressor genes

apoptosis

The existence of apoptosis also came as a surprise to investigators. This type of cell death follows a coordinated sequence and so is called *programmed cell death*. Apoptosis often starts with the release of a protein (cytochrome c) from the mitochondria. Then, this protein activates specific digestive enzymes—

caspases—that break apart selected pieces of the cellular machinery. Proteins that hold together the intracellular structure are the first to be broken down. Then, an enzyme is activated that degrades the DNA. The cell disintegrates from the inside out. Apoptosis is essential in development, because some cells must die to make room for others. One reason that cancer cells are so insidious is that they have lost the triggers for apoptosis.

New cancer treatments are using our knowledge of cell control systems. Clinical trials are currently testing the feasibility of reintroducing tumor-suppressor genes (e.g., p53) into cancer cells. Additionally, trials of drugs known to induce terminal cell differentiation and apoptosis are in progress. These new treatments are more selective in their impact on cancer cells and therefore have milder side effects on the cancer patients.

Practice 31.5

You are told that a newly discovered oncogene causes apoptosis. Do you agree? Briefly explain.

31.9 The Genetic Code

For a long time after the structure of DNA was elucidated, scientists struggled with the problem of how the information stored in DNA could specify the synthesis of so many different proteins. Since the backbone of the DNA molecule contains a regular structure of repeated and identical phosphate and deoxyribose units, the key to the code had to lie with the four bases: adenine, guanine, cytosine, and thymine.

The code, using only the four nucleotides A, G, C, and T, must at least be capable of coding for the 20 amino acids that occur in proteins. If each nucleotide codes 1 amino acid, only 4 can be represented. If the code uses two nucleotides to specify an amino acid, 16 (4×4) combinations are possible—still not enough. But, using three nucleotides, we get 64 ($4 \times 4 \times 4$) possible combinations—which is more than enough to specify the 20 common amino acids in proteins. We now know that each code word requires a sequence of three nucleotides. The code is therefore a triplet. Each triplet of three nucleotides is called a **codon**, and, in general, each codon specifies 1 amino acid. Thus, to describe a protein containing 200 amino acid units, a gene containing at least 200 codons, or 600 nucleotides, is required.

codon

In the sequence of biological events, the code from a gene in DNA is first transcribed to a coded RNA, which, in turn, directs the synthesis of a protein. The 64 possible codons for mRNA are given in Table 31.3. In this table, a three-letter sequence (first nucleotide–second nucleotide–third nucleotide) specifies a particular amino acid. For example, the codon CAC (cytosine–adenine–cytosine) is the code for the amino acid histidine (His). Note that 3 codons in the table, marked TC, do not encode any amino acids. These are called *nonsense* or *termination codons*. They act as signals to indicate where the synthesis of a protein molecule is to end. The other 61 codons identify 20 amino acids. Methionine and tryptophan have only 1 codon each. For the other amino acids, the code is redundant; that is, each amino acid is specified by at least 2, and sometimes by as many as 6, codons.

Table 31.3 The Genetic Code for Messenger RNA

First nucleotide	Second nucleotide	Third nucleotide and amino acid coded: U	C	A	G
U	U	Phe	Phe	Leu	Leu
	C	Ser	Ser	Ser	Ser
	A	Tyr	Tyr	TC*	TC*
	G	Cys	Cys	TC*	Trp
C	U	Leu	Leu	Leu	Leu
	C	Pro	Pro	Pro	Pro
	A	His	His	Gln	Gln
	G	Arg	Arg	Arg	Arg
A	U	Ile	Ile	Ile	Met
	C	Thr	Thr	Thr	Thr
	A	Asn	Asn	Lys	Lys
	G	Ser	Ser	Arg	Arg
G	U	Val	Val	Val	Val
	C	Ala	Ala	Ala	Ala
	A	Asp	Asp	Glu	Glu
	G	Gly	Gly	Gly	Gly

*Termination or nonsense codon

This table shows the sequence of nucleotides in the triplet codons of messenger RNA that specify a given amino acid. For example, UUU or UUC is the codon for Phe, UCU is the codon for Ser, and CAU or CAC is the codon for His.

Scientists believe that the genetic code is a universal code for all living organisms (with a few exceptions); that is, the same nucleotide triplet specifies a given amino acid, regardless of whether that amino acid is synthesized by a bacterial cell, a pine tree, or a human being.

31.10 Genes and Medicine

Two facts have been known for many years: (1) Most human traits can be traced to the genetic makeup of the individual, and (2) some diseases (e.g., sickle-cell anemia and Tay-Sach's disease) occur mainly within certain ethnic groups, clearly indicating a genetic connection for these maladies. These two facts motivated scientists to learn more about the human genome. By the early 1980s, techniques were available to "read" genetic material—that is, to determine the sequence of bases in the strings of DNA-comprising genes.

A complete sequencing or "mapping" of the human genome was desirable but daunting—given that this genome contains about 3 billion base pairs. The time required to sequence or decode the entire genome in a single laboratory was estimated to be about a thousand years! However, after much discussion and correspondence among many of the world's leading scientists, the Human Genome Project was organized in 1988.

The first five years of the project were used to locate genetic markers—"benchmarks" on the different human chromosome pairs. As these genetic markers have been identified, the genes associated with a variety of diseases have been located. Genes have been found that trigger muscular dystrophy, Huntington's disease, cystic fibrosis, and some breast cancers, to name a few.

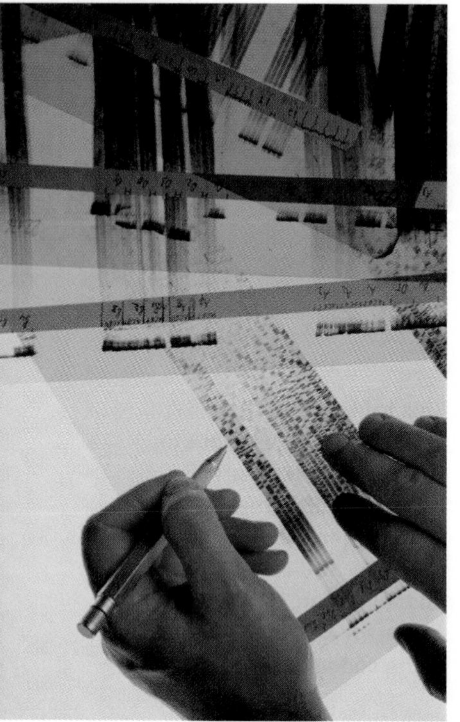

Scientist studies a gene map as part of the Human Genome Project.

Recent research has turned to sequencing the chromosomes. This involves breaking a selected segment of the genome into pieces, sequencing each piece, and then matching overlapping sequences—like fitting a giant jigsaw puzzle together. Computer programs for fitting the pieces together have facilitated the project. The first billion bases in the human genome were sequenced by November 1999. The complete mapping of the human genome (more than 90%) was announced in June 2000.

Enormous benefits from the Human Genome Project will arise, because the DNA sequence codes for all life processes. Genetic counseling will be more accurate and birth defects more predictable. The same Human Genome Project techniques applied to microbes, plants, and animals will aid agriculture and industry. One primary benefit is that the genetic basis of inheritable diseases will be better understood, and this understanding will eventually be used to correct genetic errors through *gene therapy*.

For example, one chromosome that is already sequenced (chromosome 19) has been found to contain the genetic defect for nine related diseases, including Huntington's disease. Knowing the DNA sequence error and where to locate this error gives gene therapists a target to work on. Now, scientists are working on ways to delete the error and add a corrected DNA sequence. The techniques used in gene therapy are still in the development stages, yet clinical trials are already being conducted because of this therapy's potential benefits. These early trials have mostly been unsuccessful, but scientists have high hopes for gene therapy in the immediate future.

31.11 Biosynthesis of Proteins

translation

The biosynthesis of proteins is extremely complex, and what follows is only a cursory description of the overall process. The production of a polypeptide by means of an mRNA template is called **translation**. This term is used because it literally means a change from one language to another. The genetic code is translated into the primary structure of a polypeptide or a protein.

The biosynthesis of proteins begins when messenger RNA leaves the cellular nucleus and travels to the cytoplasm. Each mRNA is then bound to five or more ribosomes, the bodies responsible for protein synthesis. Amino acids must also be transferred to the ribosomes. To accomplish this step, cellular energy in the form of ATP is used to couple amino acids to tRNAs, in order to form aminoacyl–tRNA complexes that can bind to the ribosomes:

$$\mathrm{R{-}CH(NH_2){-}C({=}O){-}OH} + \mathrm{tRNA} + \mathrm{ATP} \xrightarrow[\text{synthetase}]{\text{aminoacyl–tRNA}}$$

$$\underset{\text{aminoacyl–tRNA}}{\mathrm{R{-}CH(NH_2){-}C({=}O){-}tRNA}} + \mathrm{AMP} + \mathrm{HO{-}P({=}O)(OH){-}O{-}P({=}O)(OH){-}OH}$$

A different specific enzyme is utilized for binding each of the 20 amino acids to a corresponding tRNA. Although only 20 amino acids are involved, there are about 60 different tRNA molecules in the cells. Thus, the function of tRNA is to bring amino acids to the ribosome synthesis site.

Initiation The next step in the process is the initiation of polypeptide synthesis. Two codons, AUG and GUG, signal the start of protein synthesis. (AUG is the more common.) As shown in Table 31.3, these codons also code for the incorporation of the amino acids methionine and valine. Most mRNAs have more than one AUG and GUG, and the ribosome must choose at which codon to begin. It appears that the ribosome uses information in addition to the AUG or GUG triplet to choose the correct starting codon for protein synthesis.

Once the correct starting point has been identified, a special initiator tRNA binds to the ribosome. This specific tRNA carries an *N*-formyl methionine in procaryotic cells (e.g., bacterial cells; see Section 33.5):

$$\begin{array}{c} \mathrm{HC{=}O} \\ | \\ \mathrm{NH} \quad \mathrm{O} \\ | \quad\;\; \| \\ \mathrm{CH_3{-}S{-}CH_2CH_2CH{-}C{-}tRNA} \end{array}$$

N-formyl methionine–tRNA

Because of the attached formyl group (CHO), a second amino acid will react only with the carboxyl group of methionine, the correct direction for protein synthesis. In eucaryotic cells, the initiator tRNA carries a methionine group.

Elongation The next stage involves the elongation or growth of the peptide chain, which is assembled one amino acid at a time. After the initiator tRNA attaches to the mRNA codon, the elongation of the polypeptide chain involves the following steps (see Figure 31.15):

1. The next aminoacyl–tRNA enters the ribosome and attaches to mRNA through the hydrogen bonding of the tRNA anticodon to the mRNA codon.
2. The peptide bond between the two amino acids forms by transferring the amino acid from the initial aminoacyl–tRNA to the incoming aminoacyl–tRNA. In this step, which is catalyzed by the enzyme peptidyl transferase, the carboxyl group of the first amino acid separates from its tRNA and forms a peptide bond with the free amino group of the incoming aminoacyl–tRNA.
3. The tRNA carrying the peptide chain (now known as a peptidyl–tRNA) moves over in the ribosome, the free tRNA is ejected, and the next aminoacyl–tRNA enters the ribosome.

The peptide chain is transferred to the incoming amino acid, and the sequence is repeated over and over as the ribosome moves over to the next codon on the mRNA chain. Each time the ribosome moves to the next mRNA codon, the entire peptide chain is transferred to the incoming amino acid.

The initiation and elongation of polypeptide chains requires cellular energy. The primary source of this energy is the nucleotide guanosine-5′-triphosphate (GTP). At various steps in the growth of the protein chain, a high-energy GTP phosphate anhydride bond is hydrolyzed, yielding GDP and a phosphate group. The energy that is released drives protein synthesis. This reaction is analogous to that involving ATP. (See Section 31.4.)

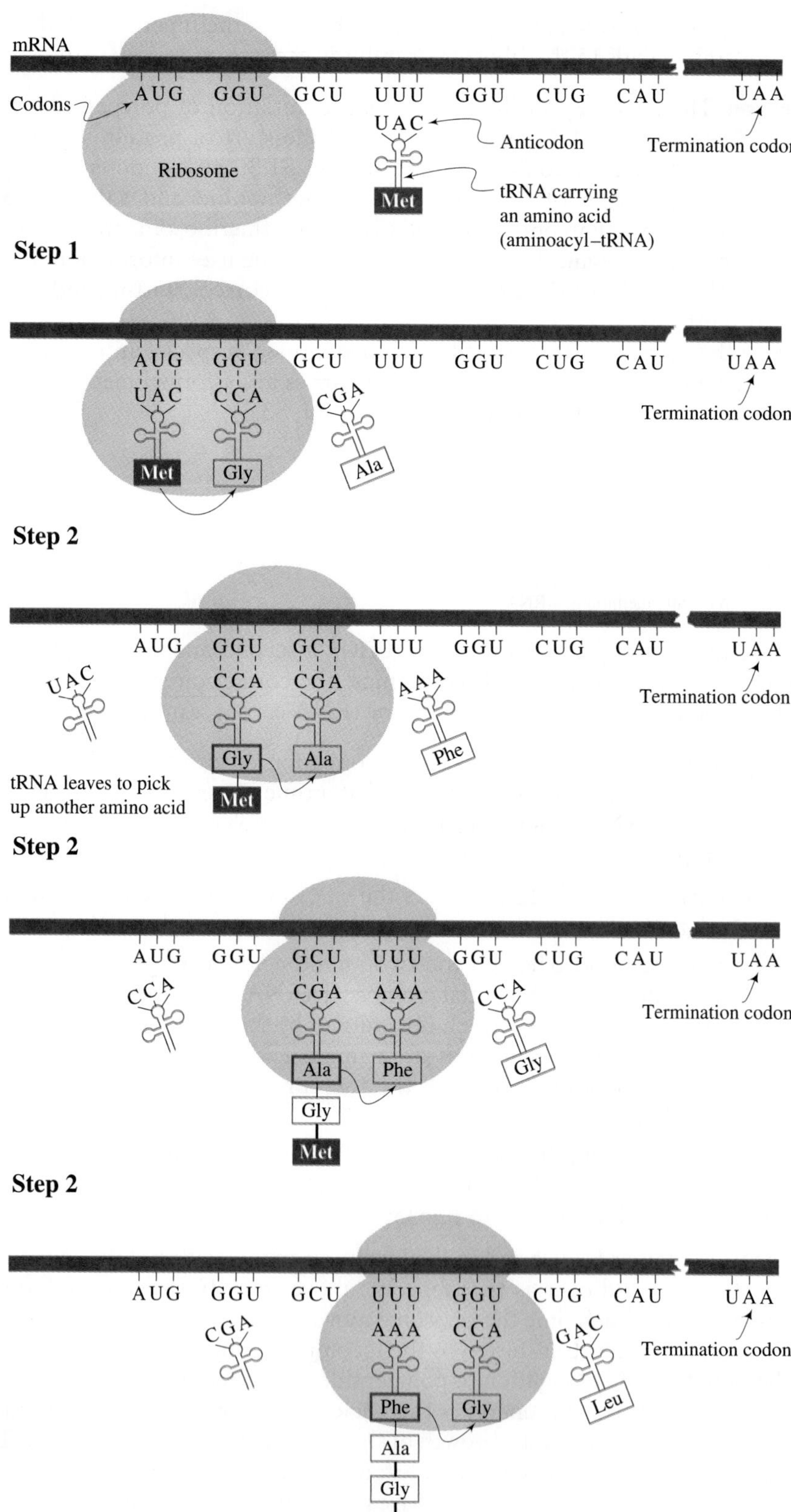

Figure 31.15
Biosynthesis of proteins.
Step 1, initiation: mRNA enters and complexes with the ribosomes. tRNA carrying an amino acid (aminoacyl–tRNA) enters the ribosome and attaches to mRNA at its complementary codon.
Step 2, elongation: Another aminoacyl–tRNA enters the ribosome and attaches to the next codon on mRNA. The peptide chain elongates when a peptide bond is formed by the transfer of the peptide chain from the initial (Met) to the incoming amino acid. This sequence is repeated until the protein chain of amino acids is complete. Step 3, termination (not shown in this figure): The sequence ends when a termination codon moves onto the ribosome.

Termination The polypeptide chain is terminated when a "nonsense" or termination codon appears. In normal cells, no tRNAs have complementary anticodons to the termination codons. Because there is no new aminoacyl–tRNA to bind to the ribosome, the peptidyl–tRNA is hydrolyzed, and the free polypeptide (protein) is released. All of these amazing, coordinated steps are accomplished at a high rate of speed—about 1 minute for a 146-amino-acid chain of human hemoglobin and 10–20 seconds for a 300–500 amino acid chain in the bacterium *Escherichia coli (E. coli)*. This mechanism of protein synthesis is illustrated in Figure 31.15.

31.12 Changing the Genome: Mutations and Genetic Engineering

From time to time, a new trait appears in an individual that is not present in either its parents or its ancestors. These traits, which are generally the result of genetic or chromosomal changes, are called **mutations**. Some mutations are beneficial, but most are harmful. Because mutations are genetic, they can be passed on to the next and future generations. **mutation**

Mutations occur spontaneously or are caused by chemical agents or various types of radiation, such as X rays, cosmic rays, and ultraviolet rays. The agent that causes the mutation is called a **mutagen**. Exposure to mutagens can produce changes in the DNA of the sperm or ovum. The likelihood of such changes is increased by the intensity and length of exposure to the mutagen. Mutations may then show up as birth defects in the next generation. Common types of genetic alterations include the substitution of one purine or pyrimidine for another during DNA replication. Such a substitution changes the genetic code and causes misinformation to be transcribed from the DNA. A mutagen can also alter genetic material by causing a chromosome or chromosome fragment to be added or removed. **mutagen**

Practice 31.6

A mutation changes an mRNA sequence from

-GGU UGG AUC-

to

-GGU UAG AUC-

How will this change the protein product?

In recent years, laboratory techniques for controlling genetic change have been developed. These techniques, collectively known as genetic engineering, are responsible for considerable progress in medicine and biology. Genetic engineers are now able to insert specific genes into the genome of a host cell and thus program it to produce new and different proteins.

Genetic engineering is a reality because of several basic advances in nucleic acid biochemistry. First, scientists are now able to isolate, identify, and then synthesize multiple copies of specific genetic messages. A key to this process is the *DNA polymerase chain reaction*. DNA polymerase is an enzyme that replicates DNA when supplied a starting fragment (the primer) and a DNA strand to copy (the template). With the appropriate primer supplied, the polymerase can be induced to synthesize a specific gene, even in the presence of a wide variety of other genetic material. What makes this reaction especially valuable is that the polymerase uses the newly synthesized DNA as a template for further replication. Thus, after 20 synthetic cycles, almost a million copies of the original gene can be produced. Like a radioactive chain reaction, the polymerase chain reaction leads to an explosion in the numbers of copied genes.

A second important process in genetic engineering is the insertion of genetic material into a "foreign" genome. Special enzymes, the *restriction endonucleases*, provide a key step during gene insertion. These enzymes split double-stranded DNA at very specific locations:

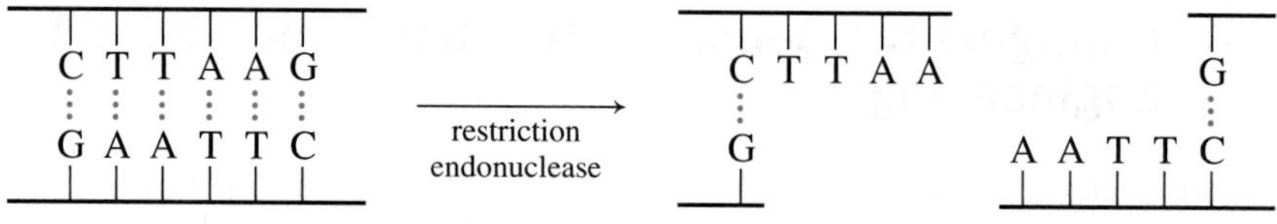

Often the newly formed break has what are termed "sticky ends"; that is, one end of the break "sticks" to the end of another break via hydrogen bonding between complementary bases. When both a gene and a foreign genome are processed in this way, they bond together. Then, with the aid of other enzymes (ligases), the gene can be covalently bonded into place:

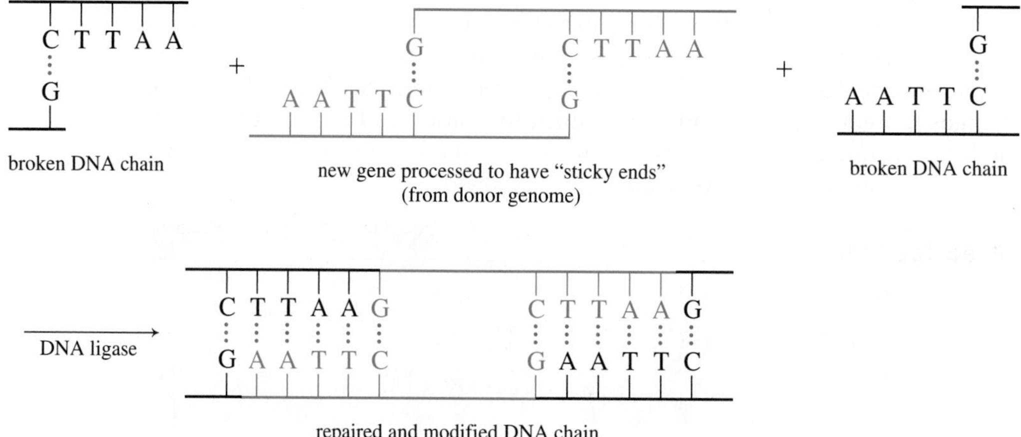

The modified and repaired foreign genome now contains a new gene. The result of this process is a form of **recombinant DNA**, a DNA whose genes have been rearranged to contain new or different hereditary information.

recombinant DNA

Agriculture is greatly affected by genetic engineering. Planners estimate that the amount of arable land worldwide will not change much, but the population will continue to increase. Genetic engineering can improve land use by, for example, increasing yield and decreasing crop losses due to insects.

Increased yields can be achieved in a variety of ways. Presently, much effort is going into aquaculture; fish are genetically engineered to produce larger amounts of the fish growth hormone. These fish grow much faster than wild types (coho salmon up to 37 times faster, catfish up to 60% faster). Since fish

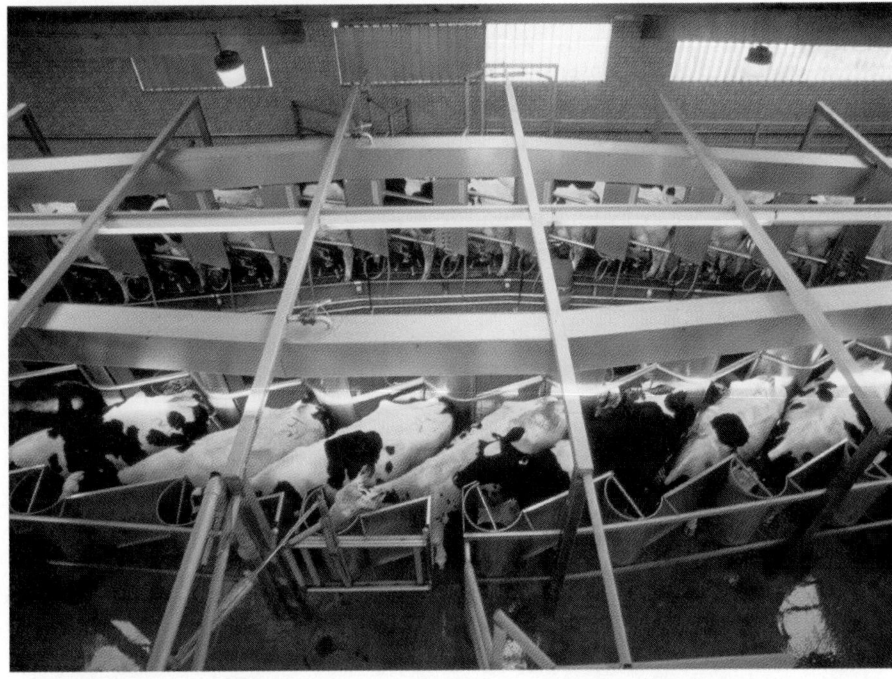

Bovine growth hormone injected into cows causes a 10% increase in milk production.

continue to grow throughout their life span, this faster growth rate should result in bigger fish and more protein for the world's population. A modification of the procedure has been used to increase milk production in cows. Bovine growth hormone is made by genetically modified microorganisms. This hormone is then injected into the cows, causing a 10% increase in milk production.

Efforts to make plants more resistant to diseases, insect pests, and other hazards constitute another promising avenue of research. One approach to this goal is to modify the plant so that it makes its own insecticide. For example, the most damaging insect to potatoes in the United States is the potato beetle. A bacterial gene that codes for a toxin that kills this pest has been successfully inserted into potato plants. A related technique is used to modify plants so that they are not affected by specific herbicides. A genetically altered soybean that tolerates the common herbicide glyphosate (Roundup®) is now commercially available.

Genetic engineering is also being used to change the quality of the final products. For example, slow-ripening tomatoes are now widely available. Because they ripen slowly, the tomatoes can be picked later in the growing season and should thus be better tasting when purchased by the consumer. An important oil-producing plant, rapeseed (canola), has been genetically altered to produce up to 40% lauric acid, an important component of many detergents. In the near future, this same plant may be producing another oil that is used to make nylon.

The techniques used in genetic engineering are becoming routine. It is clear that many benefits can be drawn from this technology. It is equally clear that there may be unknown negative consequences of these procedures. Genetic engineering remains controversial because it is difficult to foresee the future consequences of such a major innovation.

These genetically engineered tomatoes are disease resistant and ripen slowly, so that they can be picked later in the growing season.

CHEMISTRY IN ACTION • RNAi: A Muffled Message

RNA carries the message of life. Whereas DNA stores genetic information, RNA transmits that information, giving direction to metabolism. If the wrong message is delivered, serious consequences can result. Viral infections commonly take advantage of RNA message delivery—by introducing their own RNA, they change the messages and the cell makes viral particles. Cells become cancerous by following misleading messages. Inheritable diseases arise when cells carry the wrong message. With all of these bad possibilities, it is surprising to learn that cells can also cancel faulty messages. The 2006 Nobel Prize in physiology or medicine was given to the two scientists, Andrew Fire and Craig Mello, who discovered the cellular process used to "muffle the message."

When people hear a message and they understand it, they say that "it makes sense." In a similar way, messenger RNA has a sequence that "makes sense," and the message is used to make a protein. In contrast to messenger RNA, a newly discovered form of RNA is called "antisense RNA." This is not nonsense (making no sense) but antisense—the RNA sequence follows the message but the bases are complementary. Because the bases are complementary, the antisense RNA can hydrogen-bond to the messenger RNA, just as the strands of DNA can hydrogen-bond together. When an antisense RNA is present in a cell, it specifically selects a message and forms double-stranded RNA.

One of the earliest discovered examples of RNA interference; the message for purple color is destroyed to create patterns in these petunias.

Double-stranded RNA is very rare in a cell; it is recognized by the cellular machinery, and protective action follows. The double-stranded RNA is cut up into small pieces. In fact, from this time on, *any* RNA that hydrogen-bonds to the pieces is also cut up. Instead of the message being transmitted, now the RNA is degraded—the message has been muffled. The technical term for this process is RNA interference, or RNAi.

Scientists believe that antisense RNA is a cell defense mechanism. For example, a viral message might be "permanently" blocked by RNA interference. Since human cells are not very good at forming antisense RNA to block wrong messages, medical research is stepping in to help. New treatments for serious viral infections such as hepatitis and HIV are possible by introducing artificial antisense RNA. Other major therapies are on the horizon. Scientists are attempting to kill cancer by incorporating antisense RNA to silence the oncogene messages. Even genetic diseases such as diabetes may eventually be curable by muffling the wrong cell messages. Looking to the future, antisense RNA seems to make good sense.

Chapter 31 Review

31.1 Molecules of Heredity—A Link

KEY TERMS

Nucleoprotein
Nucleic acid

- The basis for heredity is found in the structure of nucleic acids.
- Chromosomes consist largely of protein bonded to nucleic acid, a nucleoprotein.
- Nucleic acids are polymers of nucleotides that contain either the sugar deoxyribose (deoxyribonucleic acid, DNA) or the sugar ribose (ribonucleic acid, RNA).

31.2 Bases and Nucleosides

KEY TERM

Nucleoside

- Two classes of heterocyclic bases are found in nucleic acids:
  - Purines
  - Pyrimidines
- There are two common purine bases and three common pyrimidine bases.
- A nucleoside is formed when either a purine or a pyrimidine base is linked to a sugar molecule, either D-ribose or D-2′-deoxyribose.
- In a nucleoside, the base is bonded to the C-1′ of the sugar.
- The root of the nucleoside name is derived from the name of the base; the prefix *deoxy-* is added if the nucleoside contains deoxyribose.

31.3 Nucleotides: Phosphate Esters

KEY TERM

Nucleotide

- Nucleotides are phosphate esters of nucleosides.
- The ester may be a monophosphate (MP), a diphosphate (DP), or a triphosphate (TP).
- Nucleotide abbreviations start with the corresponding nucleoside abbreviation, include the ester position on the sugar (2′, 3′, or 5′), and conclude with the number of phosphates (ether MP, DP or TP).

Name	Composition	Abbreviation
Adenosine	Adenine–ribose	A
Deoxyadenosine	Adenine–deoxyribose	dA
Guanosine	Guanine–ribose	G
Deoxyguanosine	Guanine–deoxyribose	dG
Cytidine	Cytosine–ribose	C
Deoxycytidine	Cytosine–deoxyribose	dC
Thymidine	Thymine–ribose	T
Deoxythymidine	Thymine–deoxyribose	dT
Uridine	Uracil–ribose	U
Deoxyuridine	Uracil–deoxyribose	dU

31.4 High-Energy Nucleotides

- ATP and ADP are especially important in metabolic energy transfers.
- An anhydride bond linking two phosphates together is a high-energy bond.
- About 35 kJ of energy is released during hydroysis of one anhydride bond.

31.5 Polynucleotides; Nucleic Acids

KEY TERMS

Ribonucleic acid (RNA)
Deoxyribonucleic acid (DNA)

- Polynucleotides contain nucleotides linked together by phosphate esters.
- Ribonucleic acid (RNA) is a polynucleotide that, upon hydrolysis, yields ribose, phosphoric acid, and the four purine and pyrimidine bases adenine, guanine, cytosine, and uracil.
- Deoxyribonucleic acid (DNA) is a polynucleotide that, upon hydrolysis, yields deoxyribose, phosphoric acid, and the four purine and pyrimidine bases adenine, guanine, cytosine, and thymine.

31.6 Structure of DNA

- Genetic information is stored in the sequential order of nucleotides in DNA.
- Although the nucleotide composition may vary among different DNAs, the ratios of A to T and G to C are always essentially 1:1.
- According to the structure proposed by Watson and Crick, the structure of DNA consists of two polymeric strands in the form of a double helix with both nucleotide strands coiled around the same axis.
- The two DNA polynucleotide strands fit together best when a purine base is adjacent to a pyrimidine base.
    - Adenine hydrogen bonds best to thymine, and guanine hydrogen bonds best to cytosine; these are complementary base pairs.
    - Once the sequence of one DNA strand is known, the sequence of the other strand can be inferred; these strands are said to be complementary to each other.
- The DNA must be folded into more compact structures to fit into the nucleus.

31.7 DNA Replication

KEY TERMS

Heredity
Genome
Gene
Replication
Mitosis
Meiosis

- Heredity is the process by which the physical and mental characteristics of parents are transferred to their offspring.
- The genome is the sum of all hereditary material contained in a cell.
- A gene is a segment of the DNA chain that controls the formation of a molecule of RNA.
- Replication is the biological process for duplication of the DNA molecule.
    - Replication creates two new DNA molecules from one original.
    - Every newly synthesized double-stranded DNA contains one strand from the original; this form of synthesis is said to be *semiconservative*.
    - The two newly formed strands are synthesized by slightly different processes; one synthesis is continuous while the other is discontinuous.
- Mitosis occurs during normal cell division when each daughter cell gains the full genetic code that was present in the parent.
- In contrast, meiosis occurs when the cell splits in such a way as to reduce the number of chromosomes to one-half the normal amount.
    - Meiosis prepares the way for sexual reproduction when two cells unite and bring the chromosome amount back to normal.
- An exact copy of an individual can be produced by cloning.
- Because cancer cells are rapidly dividing, DNA replication is much more active in cancers than normal tissues.
    - Many cancer drugs impact DNA replication.

31.8 RNA: Genetic Transcription

KEY TERMS

Transcription
Oncogenes
Tumor-suppressor genes
Apoptosis

- One of the main functions of DNA is to direct synthesis of RNA.
    - RNA differs from DNA in four important ways:
        - RNA is single stranded instead of double stranded.
        - RNA contains D-ribose instead of D-2′-deoxyribose.
        - RNA contains uracil instead of thymine.

- RNA contains a significant number of modified bases.
- There are three types of RNA used to make protein.
  - Ribosomal RNA (rRNA) comprises 60–65% of the ribosome, the site for protein synthesis.
  - Messenger RNA (mRNA) carries the genetic information from the DNA to the ribosome.
  - Transfer RNA (tRNA) brings the amino acids to the ribosomes for protein synthesis.
    - There must be at least one tRNA for each of the 20 amino acids required to make proteins.
    - A specific amino acid is covalently attached to the end of each tRNA.
    - The anticodon loop in each tRNA contains three bases that are complementary to a three-base sequence on the mRNA.
- Transcription is the making of RNA from DNA.
  - Transcription occurs in a complementary fashion: Guanine transcribes to cytosine, adenine to uracil, thymine to adenine, and cytosine to guanine.
  - Only a small fraction of the total DNA information is transcribed at any one time.
  - Following transcription, RNA molecules are often modified before they are put to use.
- Cell control often starts at transcription.
- Oncogenes cause the cell to lose control and become cancerous.
- Tumor-suppressor genes actively block cancer development; tumor-suppressor genes will trigger commands that lead to cell destruction if normal cellular control fails.
- Apoptosis is cell instigated self-destruction.
  - This processs maintains the organism's health at the expense of individual cells.
  - Cellular control signals that the cellular machinery should be degraded; the cell disintegrates from the inside out.

31.9 The Genetic Code

KEY TERM

Codon

- Each code word in DNA is a three-nucleotide sequence and is called a codon.
- Almost every codon signals a specific amino acid.
- Three codons are nonsense or termination codons and signal the end of protein synthesis.

31.10 Genes and Medicine

- Scientists have successfully determined the DNA sequence for the human genome.
- Information from the human genome is now being used to better understand and treat inherited diseases.

31.11 Biosynthesis of Proteins

KEY TERM

Translation

- Translation is the production of a polypeptide by means of an mRNA.
- Each amino acid needed for protein synthesis is covalently linked to a tRNA.
- Protein synthesis occurs on the ribosome and is initiated by a special tRNA (carrying either *N*-formyl methionine or methionine).
- Elongation or growth of the polypeptide chain takes place one amino acid at a time.
  - An aminoacyl-tRNA attaches to the mRNA by hydrogen bonds between the tRNA anticodon and the mRNA codon.
  - A new peptide bond between the two amino acids forms by transferring the amino acids from the initial aminoacyl-tRNA to the incoming aminoacyl-tRNA.
  - The tRNA carrying the peptide chain (now known as a peptidyl-tRNA) moves over in the ribosome, the free tRNA is ejected, and the next aminoacyl-tRNA enters the ribosome.
- Protein synthesis is terminated when a nonsense, or termination, codon appears.

31.12 Changing the Genome: Mutations and Genetic Engineering

KEY TERMS

Mutation
Mutagen
Recombinant DNA

- Mutations are traits that are generally the result of genetic or chromosomal changes.
- A mutagen is an agent that causes a mutation.
- Genetic engineering is a process for introducing controlled genetic change.
  - DNA polymerase can be used to make many copies of a single gene.
  - Restriction endonucleases specifically cut a DNA molecule, forming a break to which a foreign gene can be connected.
- The modified and repaired genome now contains a new gene and is called recombinant DNA.

Review Questions

All questions with blue numbers have answers in the appendix of the text.

1. Write the names and structural formulas of the five nitrogen bases found in nucleotides.

2. What is the difference between a nucleoside and a nucleotide?

3. What are the principal structural differences between DNA and RNA?
4. What is the major function of ATP in the body?
5. Briefly describe the structure of DNA as proposed by Watson and Crick.
6. What is meant by the term *complementary bases*?
7. What is the genetic code?
8. Why are at least three nucleotides needed for one unit of the genetic code?
9. Starting with DNA, briefly outline the biosynthesis of proteins.
10. In protein synthesis, how does initiation differ from elongation?
11. In protein synthesis, how does termination differ from elongation?
12. How does a codon differ from an anticodon?
13. Explain the role of *N*-formyl methionine in procaryotic protein synthesis.
14. What is a mutation?
15. What is meant by a "DNA fingerprint"?
16. Why will gene therapy benefit from the Human Genome Project?
17. What is an oncogene?
18. Why is apoptosis of potential importance in curing cancer?
19. In general, how does a tumor-suppressor gene block cancer formation?

Paired Exercises

All exercises with blue numbers have answers in the appendix of the text.

1. Identify the compounds represented by the following letters: A, AMP, dADP, UTP.
2. Identify the compounds represented by the following letters: G, GMP, dGDP, CTP.
3. Write structural formulas for the substances represented by
   (a) A
   (b) AMP
   (c) CDP
   (d) dGMP
4. Write structural formulas for the substances represented by
   (a) U
   (b) UMP
   (c) CTP
   (d) dTMP
5. Draw the structures of
   (a) cytidine-2′,5′-diphosphate
   (b) deoxythymidine-3′-monophosphate
6. Draw the structures of
   (a) deoxyguanosine-3′,5′-diphosphate
   (b) uridine-2′,5′-diphosphate
7. Draw the structure of dTDP. Use a solid line to circle the ester functional group.
8. Draw the structure of ADP. Use a solid line to circle the ester functional group.
9. Using the bases C, T, and A, draw the structure of a three-nucleotide segment of single-stranded DNA.
10. Using the bases G, U, and A, draw the structure of a three-nucleotide segment of RNA.
11. Use structural formulas to show the hydrogen bonding between adenine and uracil.
12. Use structural formulas to show the hydrogen bonding between guanine and cytosine.
13. What is the role of RNA in the genetic process?
14. What is the role of DNA in the genetic process?
15. Differentiate between replication and transcription.
16. Differentiate between transcription and translation.
17. How does the function of tRNA differ from that of mRNA?
18. How does the function of mRNA differ from that of rRNA?
19. There are 146 amino acid residues in the β-polypeptide chain of hemoglobin. How many nucleotides are needed to code for the amino acids in this chain?
20. There are 573 amino acid residues in the phosphoglycerate kinase enzyme. How many nucleotides are needed to code for the amino acids in this chain?
21. A segment of a DNA strand consists of TCAATACCCGCG.
   (a) What is the nucleotide order in the complementary mRNA?
   (b) What is the anticodon order in the tRNA?
   (c) What is the sequence of amino acids coded by the DNA?
22. A segment of a DNA strand consists of GCTTAGACCTGA.
   (a) What is the nucleotide order in the complementary mRNA?
   (b) What is the anticodon order in the tRNA?
   (c) What is the sequence of amino acids coded by the DNA?

23. Describe the bond (e.g., the atoms involved, the functional groups that react) that forms during transcription to link nucleotides together.

24. Describe the bond (e.g., the atoms involved, the functional groups which react) that forms during translation to link amino acids together.

25. Briefly describe the events that occur during translation termination.

26. Briefly describe the events that occur during translation initiation.

27. What is the tRNA anticodon if the codon in mRNA is as follows?
(a) GUC (c) UUU
(b) AGG (d) CCA

28. What is the tRNA anticodon if the codon in mRNA is as follows?
(a) CGC (c) GAU
(b) ACA (d) UUC

29. Mark where the following DNA segment must be cleaved by an endonuclease in order to form equivalent "sticky ends" on the two strands of DNA (each "sticky end" has the same four-base sequence):

-AGTACATGTTT-
-TCATGTACAAA-

30. Mark where the following DNA segment must be cleaved by an endonuclease in order to form equivalent "sticky ends" on the two strands of DNA (each "sticky end" has the same four-base sequence):

-GGGGATCTTCG-
-CCCCTAGAAGC-

31. A mutagen causes one purine base to substitute for another in DNA. If the original base was adenine, what is the structure of the new base?

32. A mutagen causes one pyrimidine base to substitute for another in DNA. If the original base was thymine, what is the structure of the new base?

33. A mutation changes the mRNA sequence from

-CUU CCC ACU-
to
-UUU CCC ACU-.

How will this mutation change the protein product?

34. A mutation changes the mRNA sequence from

-GUG CAA AAA-
to
-GUG UAA AAA-.

How will this mutation change the protein product?

35. What is the meaning of *recombinant* in *recombinant DNA*?

36. What is the meaning of *nonsense* in *nonsense codons*?

Additional Exercises

All exercises with blue numbers have answers in the appendix of the text.

37. Explain why the ratio of thymine to adenine in DNA is 1 : 1, but the ratio of thymine to guanine is not necessarily 1 : 1.

38. Occasionally, thymine isomerizes to form the following structure:

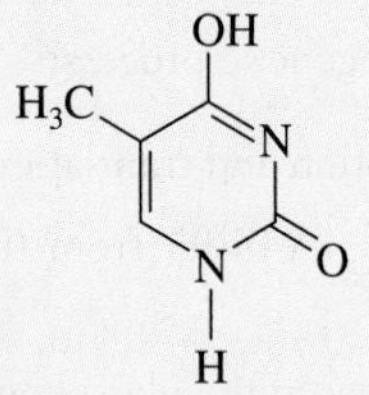

OH, H_3C, N, N, O, H

How might this change H-bonded pairing in DNA?

39. Mutations in DNA can occur when one base is substituted for another or when an extra base is inserted into the DNA strand. Explain why an insertion mutation is apt to be more harmful than a substitution mutation.

40. The sequence of bases in a segment of mRNA is UUUCAUAAG. Answer the following questions:
(a) What amino acids would be coded for by this segment of mRNA?
(b) What is the sequence of DNA that would produce the given segment of mRNA?

41. Describe the use of DNA polymerase in DNA fingerprinting.

42. What is the Human Genome Project?

Challenge Exercise

43. As shown in Table 31.2, DNA has a 1 : 1 ratio of guanine to cytosine. Would you expect RNA to have the same ratio? Explain.

Answers to Practice Exercises

31.1

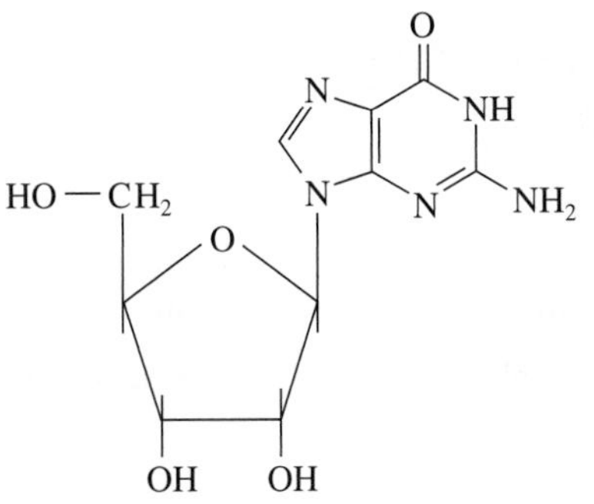

(a)

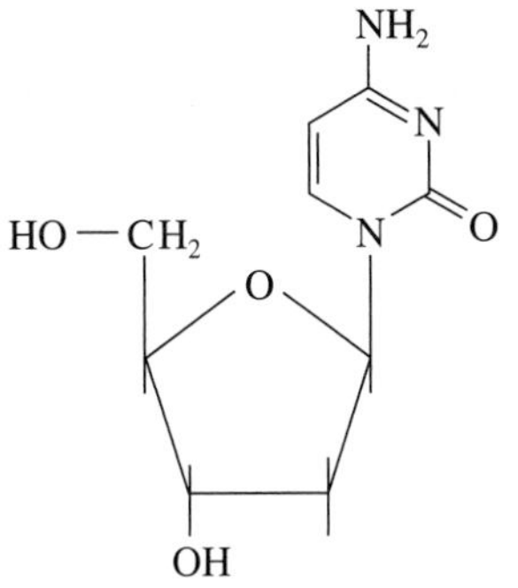

(b)

31.2 (a) uridine-5′-monophosphate, UMP
(b) cytidine-3′-monophosphate, 3′-CMP

31.3

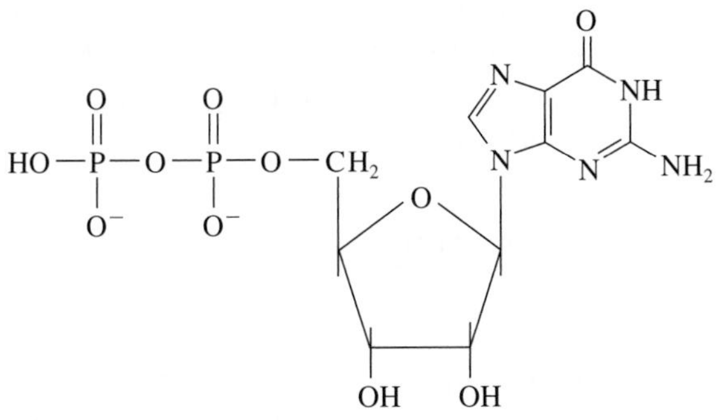

(a)

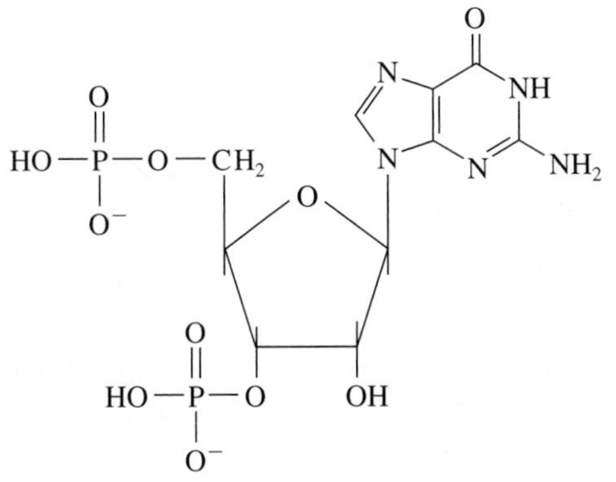

(b)

31.4

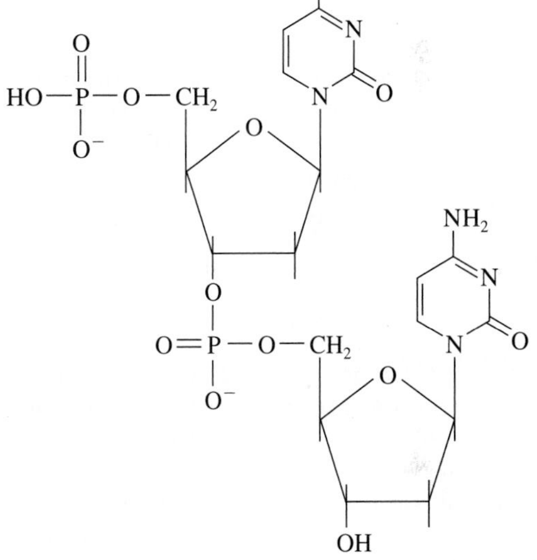

31.5 Disagree. An oncogene causes cancer, an uncontrolled growth of cells. If apoptosis happened, the cells would die instead of growing.

31.6 The original mRNA codes for a protein of the sequence, -Gly Trp Ile- After mutation, the codon UGG becomes UAG, a termination codon. The protein product will be shorter because translation terminates too soon.

PUTTING IT TOGETHER: Review for Chapters 29–31

Multiple Choice:

Choose the correct answer to each of the following.

1. Which of the following describes this structure?

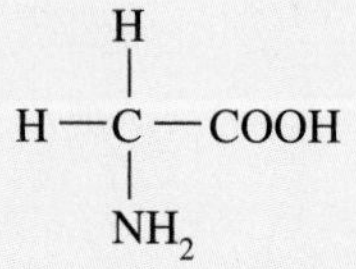

$$\mathrm{H-\overset{\overset{\large H}{|}}{\underset{\underset{\large NH_2}{|}}{C}}-COOH}$$

(a) an α-amino acid
(b) an amino acid with one chiral center
(c) an α-hydroxy acid
(d) a β-hydroxy acid

2. How is this structure classified?

$$\mathrm{CH_3-\overset{\overset{\large H}{|}}{\underset{\underset{\large NH_2}{|}}{C}}-COOH}$$

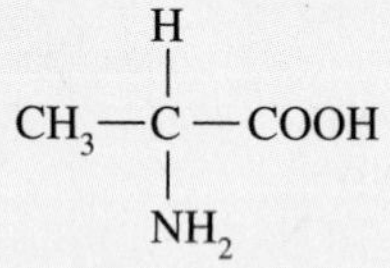

(a) as a basic amino acid
(b) as an acidic hydroxy acid
(c) as a neutral amino acid
(d) no correct answer given

3. How is this structure classified?

$$\mathrm{HOOC-CH_2-\overset{\overset{\large H}{|}}{\underset{\underset{\large NH_2}{|}}{C}}-COOH}$$

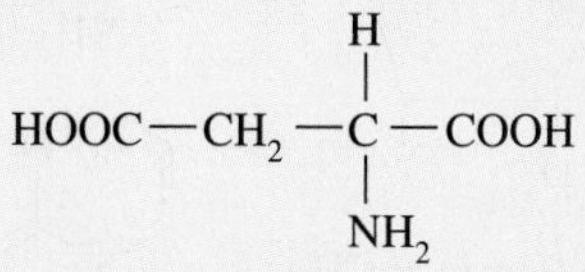

(a) as a basic hydroxy acid
(b) as an acidic amino acid
(c) as a basic amino acid
(d) as an acidic hydroxy acid

4. A complete protein provides
(a) all essential enzymes
(b) all essential nutrients
(c) all essential amino acids
(d) all neutral amino acids

5. What does this structure represent?

$$\mathrm{H-\overset{\overset{\large COOH}{|}}{\underset{\underset{\large CH_3}{|}}{C}}-NH_2}$$

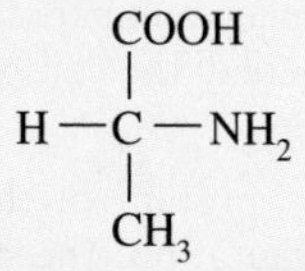

(a) a biologically common acid
(b) a D-amino acid
(c) an L-amino acid amino acid
(d) a β-amino acid

6. What does this structure represent?

$$\mathrm{H-\overset{\overset{\large COOH}{|}}{\underset{\underset{\large H}{|}}{C}}-NH_2}$$

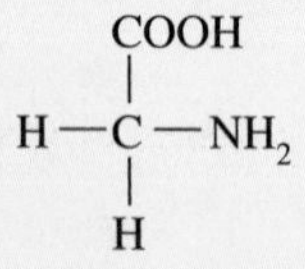

(a) a D-amino acid
(b) an L-amino acid
(c) an α-amino acid
(d) no correct answer given

7. Under strongly basic conditions, the molecule

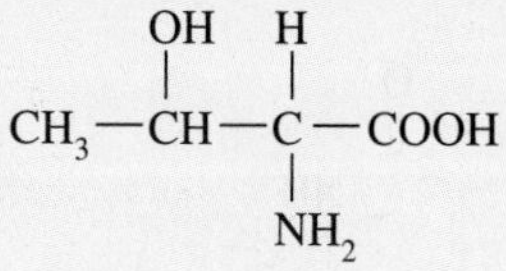

$$\mathrm{CH_3-\overset{\overset{\large OH}{|}}{CH}-\overset{\overset{\large H}{|}}{\underset{\underset{\large NH_2}{|}}{C}}-COOH}$$

will form an ion of
(a) 0 charge
(b) −2 charge
(c) +1 charge
(d) −1 charge

8. In an electrophoresis cell, to which electrode(s) will the molecule

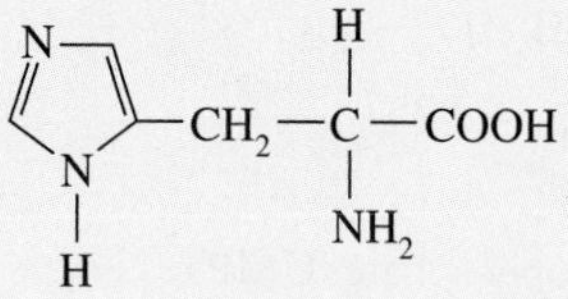

(imidazole ring: N, N–H) $\mathrm{-CH_2-\overset{\overset{\large H}{|}}{\underset{\underset{\large NH_2}{|}}{C}}-COOH}$

migrate at its isoelectric point?
(a) the positive electrode
(b) neither electrode
(c) the negative electrode
(d) both electrodes

9. Under strongly acidic conditions, which electrode(s) does this molecule

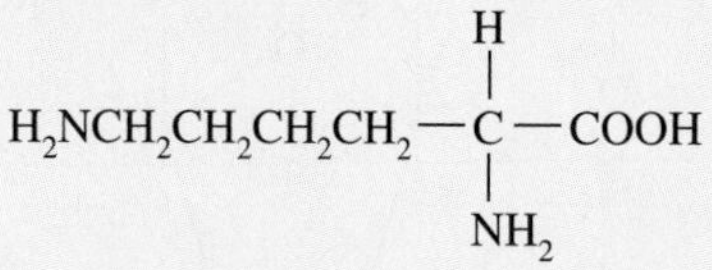

$$\mathrm{H_2NCH_2CH_2CH_2CH_2-\overset{\overset{\large H}{|}}{\underset{\underset{\large NH_2}{|}}{C}}-COOH}$$

migrate toward?
(a) the positive electrode
(b) neither electrode
(c) the negative electrode
(d) both electrodes

10. Using Table 29.1, choose the correct abbreviation for the amino acids in the following structure:

$$\mathrm{\underset{\underset{\large NH_2}{|}}{CH_2}-\overset{\overset{\large O}{\|}}{C}-\underset{\underset{\large H}{|}}{N}-\underset{\underset{\large CH_3}{|}}{CH}-\overset{\overset{\large O}{\|}}{C}-\underset{\underset{\large H}{|}}{N}-\underset{\underset{\large CH_2OH}{|}}{CH}-\overset{\overset{\large O}{\|}}{C}-OH}$$

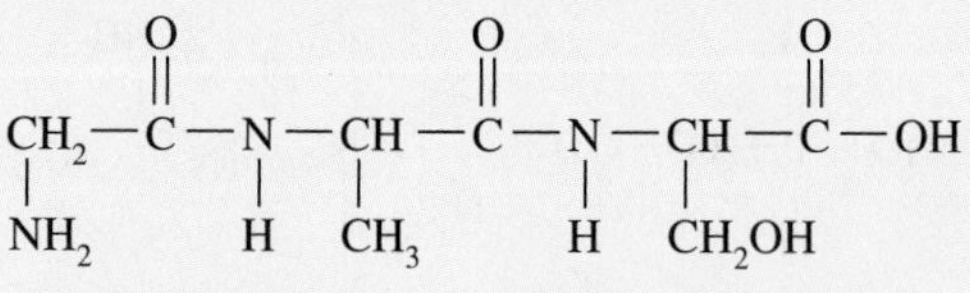

(a) gly-ala-ser
(b) ala-thr-gly
(c) gly-thr-ser-asp
(d) ser-cys-ala

11. Using Table 29.1, choose the correct abbreviation for the amino acids in the following structure:

$$\mathrm{\underset{\underset{\large NH_2}{|}}{CH_2}-\overset{\overset{\large O}{\|}}{C}-\underset{\underset{\large H}{|}}{N}-\underset{\underset{\underset{\large COOH}{|}}{CH_2}}{\underset{|}{CH}}-\overset{\overset{\large O}{\|}}{C}-\underset{\underset{\large H}{|}}{N}-\underset{\underset{\underset{\large CH_3}{|}}{CHOH}}{\underset{|}{CH}}-\overset{\overset{\large O}{\|}}{C}-\underset{\underset{\large H}{|}}{N}-\underset{\underset{\large CH_2SH}{|}}{CH}-\overset{\overset{\large O}{\|}}{C}-OH}$$

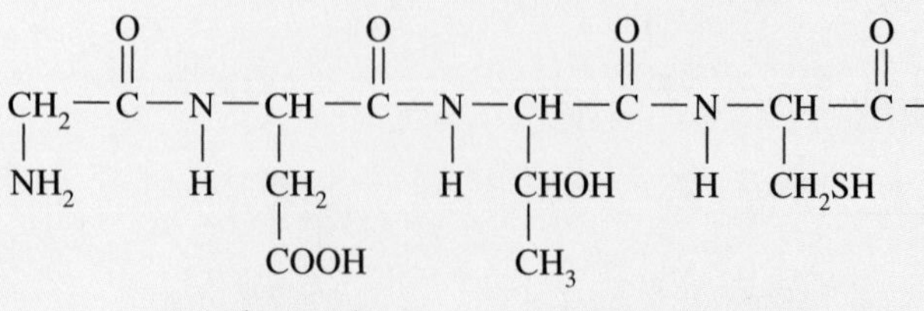

(a) gly-asp-thr-cys
(b) gly-ala-ser-glu
(c) asp-ala-ser-cys
(d) ser-cys-ala-pro

12. The structure in Question 10 is an example of
(a) a secondary structure (c) a primary structure
(b) a tertiary structure (d) a quaternary structure

13. The structure in Question 11 is an example of
(a) a tripeptide (c) a pentapeptide
(b) a dipeptide (d) no correct answer given

14. When a polypeptide is formed from seven amino acid molecules, how many molecules of water are formed?
(a) seven (c) six
(b) five (d) none

15. Which of the following is an example of a pain-producing polypeptide?
(a) bradykinin (c) oxytocin
(b) substance P (d) leu-enkephalin

16. Which of the following polypeptides is used to induce labor in childbirth?
(a) bradykinin (c) oxytocin
(b) substance P (d) leu-enkephalin

17. Which of the following protein structures is held together by hydrogen bonds between the amide "N—H" and the amide "C=O"?
(a) primary structure (c) secondary structure
(b) tertiary structure (d) quaternary structure

18. A high percentage of the secondary structure, the β-pleated sheet, is found in
(a) myoglobin (c) hemoglobin
(b) collagen (d) fibroin

19. A high percentage of α-helix is found in
(a) carboxypeptidase A (c) α-keratin
(b) fibroin (d) no correct answer given

20. When two cysteine amino acid side chains form a bond, it is called
(a) an ionic bond (c) a hydrophobic bond
(b) a hydrogen bond (d) a disulfide bond

21. Bonds between amino acid side chains commonly hold together
(a) tertiary structures (c) primary structures
(b) secondary structures (d) delta structures

22. Myoglobin is an example of
(a) a fibrous protein
(b) a β-pleated sheet containing protein
(c) a conjugated protein
(d) an enzyme

23. Hemoglobin is said to be "cooperative" because
(a) four O_2's bind, each successive binding being more difficult
(b) four O_2's bind, each successive binding being easier
(c) one O_2 binds and hemoglobin works with other proteins
(d) eight O_2's bind, each successive binding being easier

24. Which of the following proteins is *not* classified as a binding protein?
(a) myoglobin (c) immunoglobulin G
(b) hemoglobin (d) carboxypeptidase A

25. In electrophoresis, when proteins are separated by moving through a pH gradient, the process is called
(a) SDS electrophoresis
(b) isoelectric focusing
(c) thin-layer chromatography
(d) high-performance liquid chromatography

26. Denaturation of proteins does *not* cause a loss of
(a) primary structure (c) tertiary structure
(b) quaternary structure (d) secondary structure

27. Enzymes increase the rate of a reaction by
(a) decreasing the reaction temperature
(b) decreasing the activation energy
(c) increasing the activation energy
(d) increasing the reactant concentration

28. An enzyme that is also a conjugated protein can be called
(a) a holoenzyme (c) a coenzyme
(b) an activator enzyme (d) an apoenzyme

29. The enzymes that digest proteins are classed as
(a) lyases (c) ligases
(b) hydrolases (d) transferases

30. When an enzyme lowers the activation energy of the transition state,
(a) the reaction rate also decreases
(b) the substrate concentration increases
(c) the product concentration also decreases
(d) no correct answer given

31. If enzyme reaction results in a large turnover number, then
(a) the reaction rate is slow
(b) the reaction temperature is low
(c) the enzyme is an efficient catalyst
(d) the substrate concentration is low

32. In commercial products, proteases are *not* used to
(a) soften denim (c) dissolve blood clots
(b) remove clothing stains (d) no correct answer given

33. Statin drugs that lower blood cholesterol levels can be classified as
(a) proteases (c) enzyme inhibitors
(b) enzyme denaturants (d) hydrolases

34. Industrial detergents depend upon enzymes that are stable at
(a) low pH and high temperatures
(b) low pH and low temperatures
(c) high pH and low temperatures
(d) high pH and high temperatures

35. Nucleic acids
(a) do not contain deoxyribose
(b) do not contain uracil
(c) are polymers
(d) are also called nucleotides

36. What is the name of this compound?

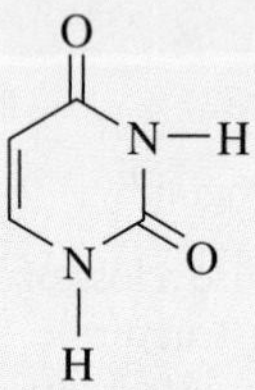

(a) adenine (c) thymine
(b) uracil (d) cytosine

37. What is the name of this compound?

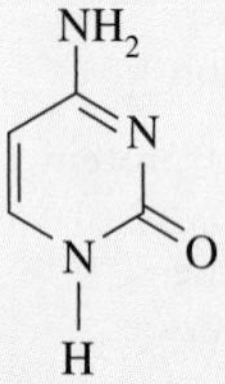

(a) cytidine (c) thymine
(b) adenine (d) no correct answer given

38. Thymidine 5′-monophosphate is an example of a
(a) nucleoside (c) nucleotide
(b) purine base (d) pyrimidine base

39. Which of the following is the structure for deoxyguanosine?

(a)

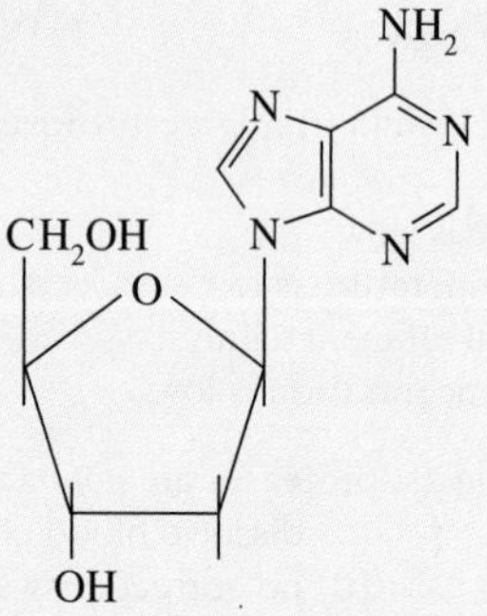

(b)

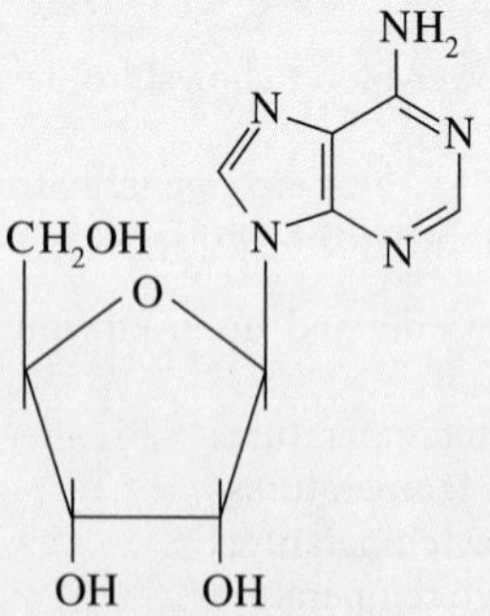

(c)

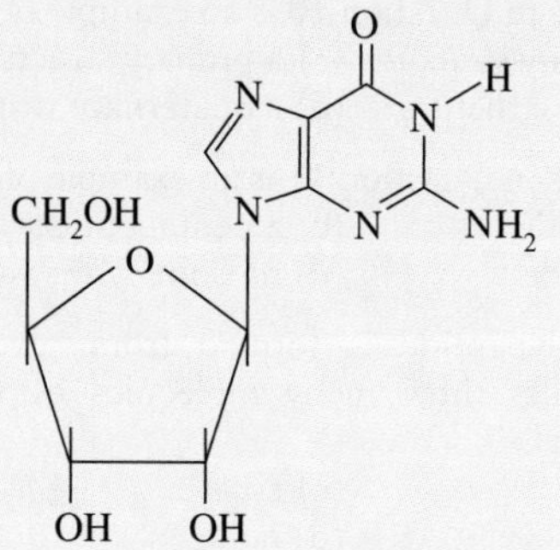

(d)

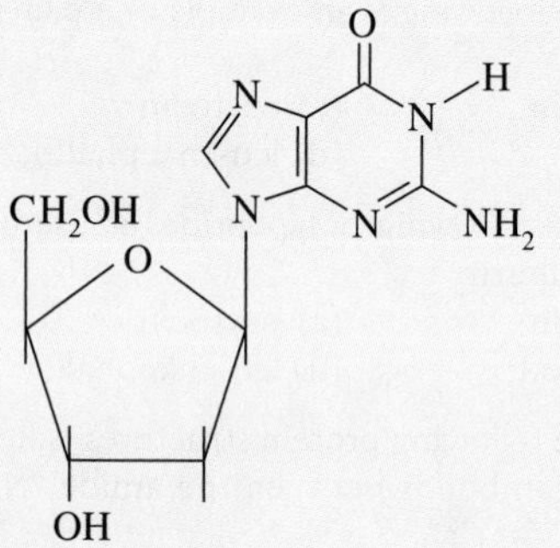

40. Which of the following is the abbreviation for this compound?

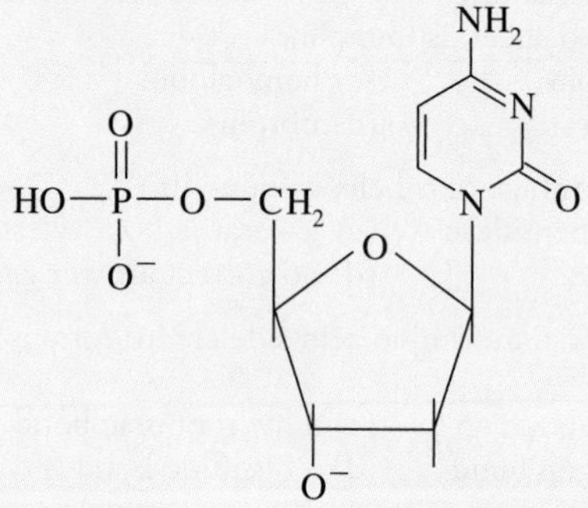

(a) UMP (c) CMP
(b) dCMP (d) dTMP

41. What is the name of this compound?

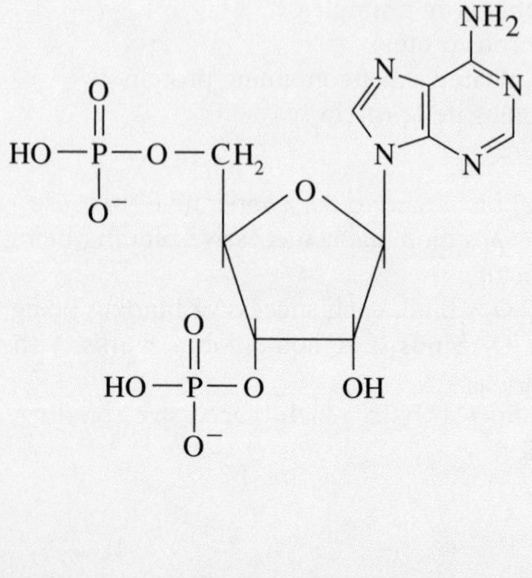

(a) adenosine-5′-diphosphate
(b) adenosine-4′,5′-diphosphate
(c) adenosine-3′,5′-diphosphate
(d) adenidine-3′,5′-diphosphate

42. ATP is a high-energy
(a) nucleoside (c) nucleotide
(b) nucleic acid (d) purine base

43. Which of the following is *not* a complementary base pair?
(a) A-G (c) A-U
(b) A-T (d) C-G

44. Semiconservative replication means
(a) RNA is synthesized from DNA
(b) each new DNA has one new strand and one template strand
(c) some DNA is lost during each replication
(d) no correct answer given

45. Which of the following scientists discovered the basic laws of heredity?
(a) Darwin (c) Watson and Crick
(b) Mendel (d) Miescher

46. The ribonucleic acid that carries an amino acid to a ribosome is
(a) a transfer RNA (c) a ribosomal RNA
(b) a messenger RNA (d) a nuclear RNA

47. When transcription takes place,
(a) a U codes for a T (c) a T codes for a U
(b) an A codes for a T (d) an A codes for a U

48. The biosynthesis of proteins is termed
(a) replication (c) initiation
(b) translation (d) transcription

49. A key role in genetic engineering is played by
(a) restriction endonucleases
(b) 3′ to 5′ exonucleases
(c) retardant exonucleases
(d) 5′ to 3′ exonucleases

50. A nonsense codon is an example of a
(a) mutation (c) termination signal
(b) mutagen (d) a codon that nature has not used

Free Response Questions

1. Ingesting of ethylene glycol, $HOCH_2CH_2OH$ (commonly found in antifreeze), can lead to death. The ethylene glycol undergoes the following enzyme-catalyzed reactions to produce oxalic acid, which is actually the lethal material:

$$\underset{\text{ethylene glycol}}{HOCH_2CH_2OH} \xrightarrow{\text{Step 1}} HOCH_2\overset{\overset{O}{\|}}{C}-H \rightarrow\rightarrow\rightarrow \underset{\text{oxalic acid}}{HOOCCOOH}$$

What type of enzyme catalyzes the first step of the process?

2. Photosynthesis involves the conversion of CO_2 and H_2O in the presence of sunlight to carbohydrates (primarily glucose) and oxygen. The following transformations are part of the process occurring during the dark reactions of photosynthesis:

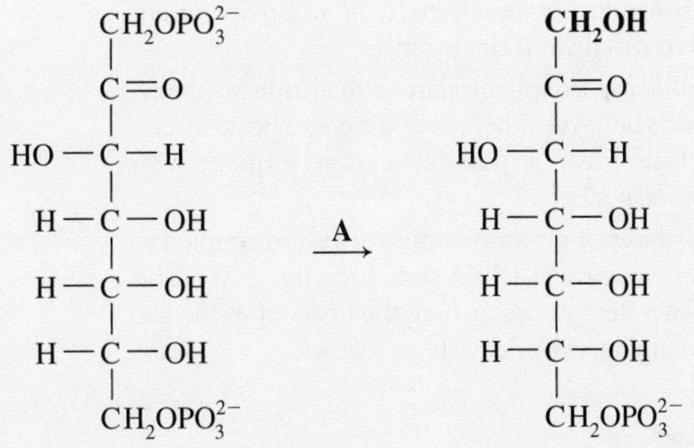

```
      CH2OPO3(2-)                  CH2OH
      |                            |
      C=O                          C=O
      |                            |
  HO—C—H                       HO—C—H
      |                            |
   H—C—OH          A            H—C—OH
      |          ——→               |
   H—C—OH                       H—C—OH
      |                            |
   H—C—OH                       H—C—OH
      |                            |
      CH2OPO3(2-)                  CH2OPO3(2-)
sedoheptulose-1,7-bisphosphate   sedoheptulose-7-phosphate
```

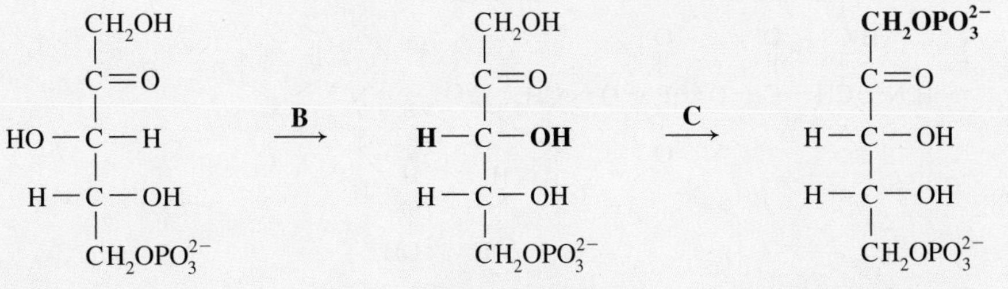

```
      CH2OH                    CH2OH                    CH2OPO3(2-)
      |                        |                        |
      C=O                      C=O                      C=O
      |           B            |           C            |
  HO—C—H        ——→        H—C—OH        ——→        H—C—OH
      |                        |                        |
   H—C—OH                   H—C—OH                   H—C—OH
      |                        |                        |
      CH2OPO3(2-)              CH2OPO3(2-)              CH2OPO3(2-)
xylulose-5-phosphate      ribulose-5-phosphate     ribulose-1,5-bisphosphate
```

What type of enzyme is needed to catalyze each of these reactions?

3. Name and draw the peptide resulting from the codon sequence UCAAUGCCGUAG.
4. As the name implies, globular proteins tend to have a roundish, globular shape. What types of side chains might you expect to find on the amino acid residues that face the outside of a globular protein if it is primarily found in an aqueous environment in the body?
5. If a peptide were synthesized according to the following codon, how many carboxylate groups would be present?

   AUCGAUCAGGCAGAGUAA
6. Carboxypeptidase A is a digestive enzyme that cleaves amino acids from a peptide chain from the C-terminal end. If the following peptide were being digested, what type of reaction would take place if carboxypeptidase A were to act upon it?

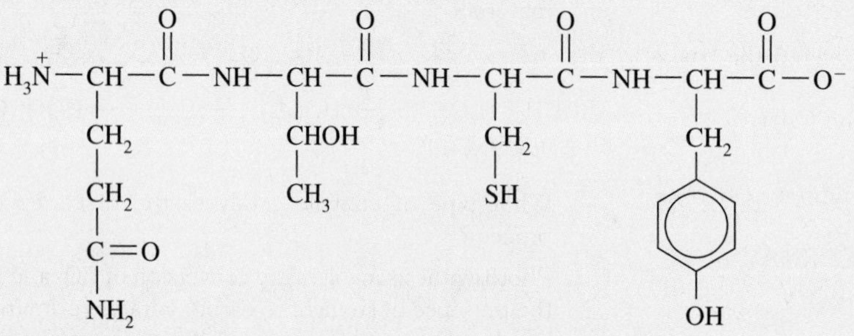

$$H_3\overset{+}{N}-\underset{\displaystyle CH_2-CH_2-C(=O)-NH_2}{CH}-\overset{O}{\overset{\|}{C}}-NH-\underset{\displaystyle CHOH-CH_3}{CH}-\overset{O}{\overset{\|}{C}}-NH-\underset{\displaystyle CH_2-SH}{CH}-\overset{O}{\overset{\|}{C}}-NH-\underset{\displaystyle CH_2-C_6H_4-OH}{CH}-\overset{O}{\overset{\|}{C}}-O^-$$

   Draw and name the residue that would be the first product of the reaction.
7. Suggest a reason that digestive enzymes which work in the stomach would not be efficient in an environment similar to that of saliva in the mouth.
8. Is the dipeptide aspartylphenylalanine the same as phenylalanylaspartic acid? Will either (or both) give a positive xanthoproteic test? Give a possible codon sequence for aspartylphenylalanine.
9. In the biosynthesis of proteins, amino acids are coupled to tRNAs to form aminoacyl–tRNA complexes using ATP. This is actually a two-step process, where the first step is the formation of an aminoacyl–adenylate as follows:

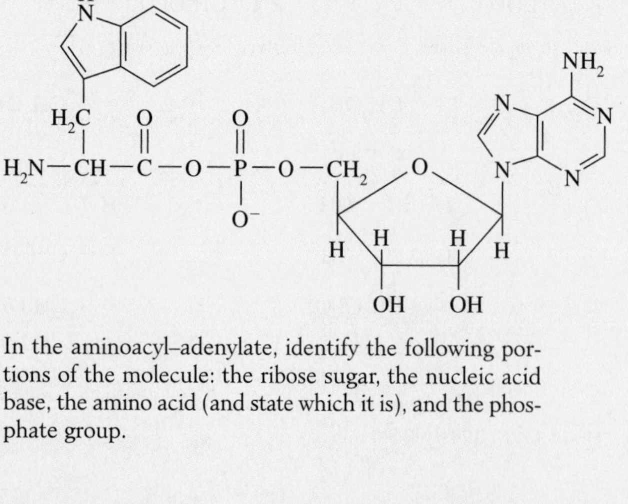

H N
NH_2
H_2C O O N N
$H_2N-CH-C-O-P-O-CH_2$ O N N
O^-
H H H H
OH OH

   In the aminoacyl–adenylate, identify the following portions of the molecule: the ribose sugar, the nucleic acid base, the amino acid (and state which it is), and the phosphate group.

CHAPTER 32

Nutrition

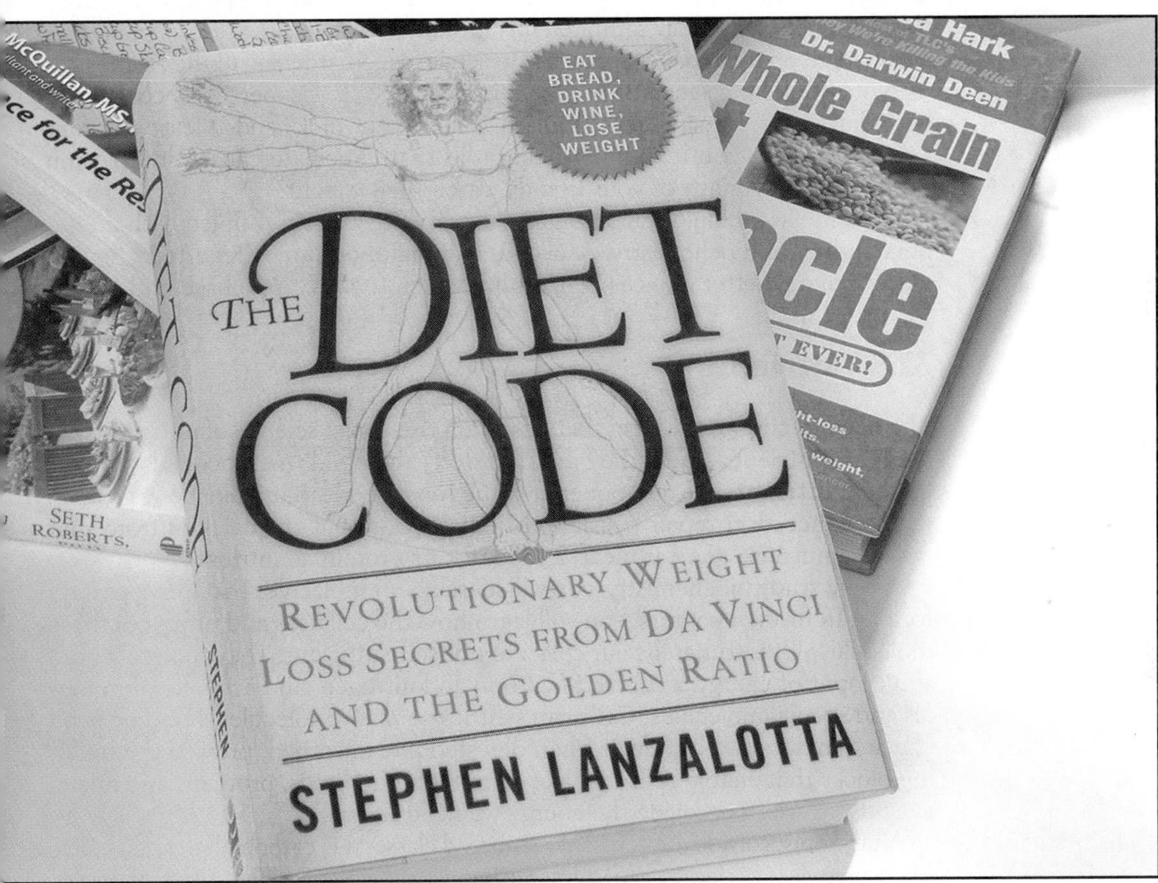

Nutrition is a common concern that is often a topic in popular publications.

Chapter Outline

Jack Sprat could eat no fat,
his wife could eat no lean,
yet twixt the two of them,
they licked the platter clean.

—MOTHER GOOSE

For hundreds of years, we have known that diet is important to good health and that dietary needs may vary from one individual to another. Yet only recently have we been able to understand nutritional needs on a molecular level. In the nursery rhyme, Jack Sprat is clearly on a low-fat diet. Today, we can point to evidence that fat intake (especially saturated fat) is related to heart disease. Fat molecules accumulate on the blood vessel walls, which leads to hardening of the arteries.

Nutrition is a science of practical importance. Every person is concerned with diet. As children, we often heard "Finish your milk" and "Please eat your vegetables." Do you remember being told that chocolate causes complexion problems? How many of your friends watch their weight?

An understanding of nutrition and digestion is closely coupled with an understanding of biochemistry. As we learn more about how diet affects health, we can make better choices concerning which foods to purchase and to eat.

32.1 Nutrients

The need to choose the right diet has given rise to the science of nutrition, the study of nutrients—how they are digested, absorbed, metabolized, and excreted. **Nutrients** are components of the food we eat that make body growth, maintenance, and repair possible. Milk is a food, but the calcium from that milk is a nutrient. We eat meat, a food, for its protein, a nutrient.

nutrient

As you study biochemistry, you will learn about molecules that are necessary for life. In future chapters, you will learn how the cell uses and produces these molecules. Unfortunately, our cells are not self-sufficient. They require a constant input of energy, a source of carbon and nitrogen, and a variety of minerals and special molecules. Cells can synthesize many molecules, but they need starting materials. Nutrients, when digested and absorbed, provide the building blocks that enable cells to make carbohydrates, lipids, proteins, and nucleic acids, as well as provide the energy we need to live.

Nutritionists divide nutrients into six broad classes: (1) carbohydrates, (2) lipids, (3) proteins, (4) vitamins, (5) minerals, and (6) water. The first three classes—carbohydrates, lipids, and proteins—are the major sources of the building materials, replacement parts, and energy needs of the cells. These nutrients are used in relatively large amounts and are known as **macronutrients**. The next three classes of nutrients—vitamins, minerals, and water—have functions other than being sources of energy or building materials. Vitamins are components of some enzyme systems that are vital to the cells. Minerals are required to maintain specific concentrations of certain inorganic ions in the cellular and extracellular fluids. They are also utilized in a variety of other ways—for example, calcium and phosphorus in bones and teeth, and iron in hemoglobin. Needed in relatively small amounts, vitamins and minerals are termed **micronutrients**. Water is absolutely essential to every diet, because most biochemical reactions occur in aqueous solution.

macronutrient

micronutrient

Nutrients supply the needs of the body's many different cells. For example, red blood cells must produce hemoglobin and thus need iron. Muscle cells produce muscle fibers and need much protein. As they do their jobs, cells wear out and must be replaced. Red blood cells are renewed, on average, every six weeks, whereas the cells lining the digestive tract must be replaced every three days.

The six classes of nutrients provide the basis for a healthy diet. This chapter will briefly describe (1) the relationship between diet and nutrition, (2) some characteristics of each nutrient category, and (3) the processes by which some nutrients are digested.

32.2 Diet

Diet is the food and drink that we consume. The foods we consume determine the nutrients available to our bodies. Thus, our health depends directly on our diet.

diet

Unfortunately, some dietary choices are not clear. Controversy surrounds many food selections. Are "fast foods" unhealthy? Should we avoid food additives? What are "natural" foods? To eat intelligently, we must understand the role of the food nutrients in maintaining good health.

Nutritionists have the difficult problem of deciding the kinds of nutrients that should be in a diet. They often establish dietary need by correlating physical well-being with nutrient consumption. James Lind (1716–1794), a physician in the British Navy during the 18th century, was one of the first to use this approach. In a study of scurvy, a disease that afflicted sailors on long voyages, he placed seamen who suffered from scurvy on various diets, some of which contained citrus fruits. By observing changes in the conditions of the seamen, Lind was able to conclude that citrus fruits provide a nutrient that prevents scurvy. This and later work eventually led to the requirement of limes and lemons in the diets of the British Merchant Marine (1765) and the British Navy (1795) (hence the origin of the name "limey" for British sailor). Scurvy is now recognized as a deficiency disease caused by a lack of vitamin C, ascorbic acid.

As scientists like Lind continued nutrition research, they found that there were no universal minimum quantities of nutrients needed for good health; these quantities varied from person to person. Also, whereas one quantity might be needed to avoid deficiency symptoms, another quantity might be needed to ensure the best health. Furthermore, some nutrients are healthful at lower levels but become toxic at higher levels. In recognition of this complexity, today's nutritionists refer to **Dietary Reference Intakes (DRI)**. DRIs have been carefully defined and take into account the many different types of nutrient data needed to judge a healthful diet. The DRI takes the place of the older **recommended dietary allowance (RDA)**. The recommended dietary allowance is defined as the daily dietary intake that is sufficient to meet the requirements of nearly all individuals in a specific age and gender group. RDAs are still established for nutrients that are relatively well understood and form the basis for many of the DRIs. Where an RDA cannot be determined, the Adequate Intake (AI) is given. The AI is an experimentally determined, average requirement. Table 32.1 shows the DRIs for a variety of important nutrients.

dietary reference intake (DRI)

recommended dietary allowance (RDA)

Practice 32.1

From Table 32.1,
(a) What is a young woman's (age 19–30) iron DRI in grams?
(b) Which vitamin for adults has the smallest DRI?

Table 32.1 Selected Daily Dietary Reference Intakes (DRIs)[a]

Life Stage Group	Age (yrs)	Water (L/d)	Carbo-hydrate (g/d)	Total Fiber (g/d)	Fat (g/d)	Protein (g/d)	Fat-soluble vitamins: Vitamin A (μg RE/d)[b]	Vitamin D (μg/d)[c]	Vitamin E (mg α-TE/d)[d]	Vitamin K (μg/d)
Infants	0.0–0.5	0.7	60	ND	31	9.1	400*	5*	4*	2.0*
	0.5–1.0	0.8	95	ND	30	11.0	500*	5*	5*	2.5*
Children	1–3	1.3	130	19	ND	13	300	5*	6	30*
	4–8	1.7	130	25	ND	19	400	5*	7	55*
Males	4–13	2.4	130	31	ND	34	600	5*	11	60*
	14–18	3.3	130	38	ND	52	900	5*	15	75*
	19–30	3.7	130	38	ND	56	900	5*	15	120*
	31–50	3.7	130	38	ND	56	900	5*	15	120*
	51–70	3.7	130	30	ND	56	900	10*	15	120*
	>70	3.7	130	30	ND	56	900	15*	15	120*
Females[e]	9–13	2.1	130	26	ND	34	600	5*	11	60*
	14–18	2.3	130	26	ND	46	700	5*	15	75*
	19–30	3.7	130	25	ND	46	700	5*	15	90*
	31–50	3.7	130	25	ND	46	700	5*	15	90*
	51–70	3.7	130	21	ND	46	700	10*	15	90*
	>70	3.7	130	21	ND	46	700	15*	15	90*

	Water-soluble vitamins: Vitamin C (mg/d)	Thiamin (mg/d)	Riboflavin (mg/d)	Niacin (mg/d)	Vitamin B_6 (mg/d)	Folate (μg/d)	Vitamin B_{12} (μg/d)	Minerals: Ca (mg/d)	P (mg/d)	Mg (mg/d)	Fe (mg/d)	Zn (mg/d)	I (μg/d)	Se (μg/d)
Infants	40*	0.2*	0.3*	2*	0.1*	65*	0.4*	210*	100*	30*	0.27*	2*	110*	15*
	50*	0.3*	0.4*	4*	0.3*	80*	0.5*	270*	275*	75*	11	3	130*	20*
Children	15	0.5	0.5	6	0.5	150	0.9	500*	460	80*	7	3	90	20
	25	0.6	0.6	8	0.6	200	1.2	800*	500	130	10	5	90	30
Males	45	0.9	0.9	12	1.0	300	1.8	1300*	1250	240	8	8	120	40
	75	1.2	1.3	16	1.3	400	2.4	1300*	1250	410	11	11	150	55
	90	1.2	1.3	16	1.3	400	2.4	1000*	700	400	8	11	150	55
	90	1.2	1.3	16	1.3	400	2.4	1000*	700	420	8	11	150	55
	90	1.2	1.3	16	1.7	400	2.4	1200*	700	420	8	11	150	55
	90	1.2	1.3	16	1.7	400	2.4	1200*	700	420	8	11	150	55
Females	45	0.9	0.9	12	1.0	300	1.8	1300*	1250	240	8	8	120	40
	65	1.0	1.0	14	1.2	400	2.4	1300*	1250	360	15	9	150	55
	75	1.1	1.1	14	1.3	400	2.4	1000*	700	310	18	8	150	55
	75	1.1	1.1	14	1.3	400	2.4	1000*	700	320	18	8	150	55
	75	1.1	1.1	14	1.5	400	2.4	1200*	700	320	8	8	150	55
	75	1.1	1.1	14	1.5	400	2.4	1200*	700	320	8	8	150	55

Reference: Food and Nutrition Board, Institute of Medicine, National Academy of Sciences, 2003.

[a] Most of the DRIs given in this table are recommended dietary allowances (RDAs). Where a value has an asterisk (*), the value is an Adequate Intake (AI).

[b] Retinol equivalents; 1 retinol equivalent = 1 μg retinol or 6 μg β-carotene.

[c] As cholecalciferol; 10 μg cholecalciferol = 400 IU of vitamin D.

[d] α-Tocopherol equivalents; 1 mg α-tocopherol = 1 α-TE.

[e] Pregnant and lactating females have special DRIs.

ND = non-determinable.

32.3 Energy in the Diet

An important component of every diet is the **Estimated Energy Requirement (EER)**. The EER is the average dietary energy intake that is predicted to maintain energy balance for good health. It derives primarily from the energy-containing nutrients: the carbohydrates, lipids, and proteins. These molecules are a rich source of reduced carbons. As we will see in Chapter 33, almost all of the energy for life is derived from reactions in which oxidation of carbon compounds occurs in the cells.

Estimated Energy Requirement (EER)

The dietary energy allowance varies with activity, body size, age, and sex. Thus, a 65-kg male office worker requires a diet that furnishes 2600 kcal/day, whereas a man of similar weight working as a carpenter requires about 2900 kcal/day. Women generally use less energy than men, and energy use decreases with age. Nutritionists express energy in units of kilocalories (kcal) or its equivalent, the large Calorie (Cal). Remember, there are 4.184 kJ per kilocalorie (or Calorie).

The balance between energy needs and the energy allowance is of vital importance. Calorie deficiency leads to a condition called *marasmus*, which affects many of the world's poor, particularly children. Marasmus is a wasting disease due to starvation. People suffering from this disease have limited diets that often consist of bulky, carbohydrate-containing foods—for example, sweet potatoes (32% carbohydrate, 1% fat, 2% protein). These foods are not energy rich (about 160 kcal for a medium-size sweet potato). Children, with their small stomachs, have difficulty consuming enough to satisfy their energy requirements. It is estimated that 15–20% of the people in underdeveloped countries suffer from malnutrition due to insufficient calories.

Food for Thought: For the first time in history, a large proportion of the world's population, mainly in developed countries, has an overabundance of food.

In contrast, many people in developed countries consume far more calories than they need for health and well-being. Food is abundant, much of it rich in energy. For example, a fast-food lunch might consist of a hamburger (560 kcal), french fries (220 kcal), and a chocolate shake (380 kcal), which would supply nearly one-half of the daily energy allowance for an adult male. Excess accumulated calories lead to a condition called *obesity*, which is characterized by an overabundance of fatty tissue and by many attendant health problems.

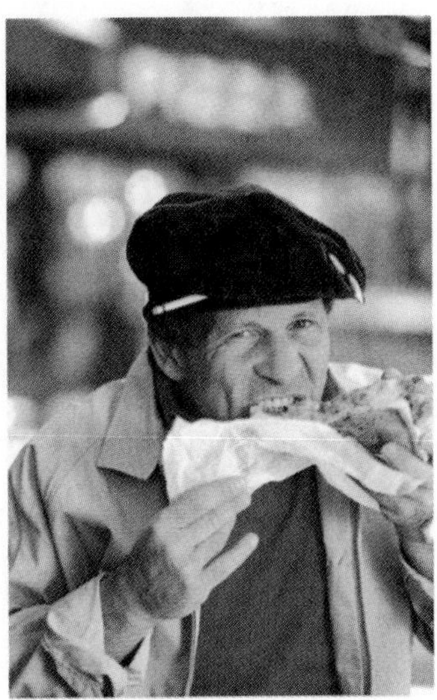

Fast foods make up a large portion of the daily diet for many people.

Each pound of body fat contains about 3500 kcal of energy, or between 1 and 2 days' total EER for energy. Because fat contains so much energy, it is normally accumulated only slowly. Unfortunately, for the same reason, fat is also very difficult to lose. To lose 1 lb of fat, a 65-kg man would have to swim nonstop for about 10 hours, play tennis continuously for about 8 hours, or run for 4 hours. Exercise alone is generally not sufficient to cure obesity; a change in diet is also necessary.

Thus, many people in the affluent, developed countries find it necessary to choose a restricted-calorie diet to lose weight. The difficulties associated with selecting a restricted-calorie diet have led to the creation of numerous fad diets that fail to provide sound nutrition. Nutritionists counsel that successful and nutritionally sound dieting can be accomplished by following these simple guidelines: (1) Reduce calorie intake by only a moderate amount (a 500-kcal/day reduction causes a loss of about 1 lb of fat per week), and be prepared to continue the diet for a long time; (2) select foods carefully, so that the diet contains adequate amounts of all nutrients. The primary function of any diet is to maintain good health.

32.4 Carbohydrates in the Diet

Food for Thought: Carbohydrates are the most versatile of energy nutrients. Humans have more different metabolic options for carbohydrates than for any other energy nutrient.

Carbohydrates are the major nutrient in most human diets. (See Figure 32.1.) They are justly described as macronutrients, because they are a major source of usable, reduced carbon atoms and, therefore, of dietary energy. About half of the average daily calorie requirement derives from carbohydrates. These molecules are perhaps the most easily metabolized (see Chapter 34) of the energy-supplying nutrients and can be used for energy under both aerobic and anaerobic conditions. Additionally, carbons from carbohydrates are used to build other cellular molecules—amino acids and nucleic acids, as well as other carbohydrates. Excess dietary carbohydrate is most often converted to fat.

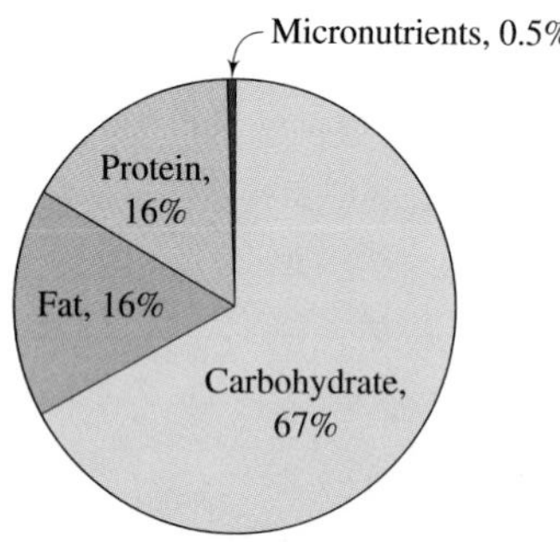

Figure 32.1
Relative masses of nutrients in a typical diet.

Dietary carbohydrates are primarily the polysaccharides starch and cellulose, the disaccharides lactose and sucrose, and the monosaccharides glucose and fructose. Seeds are the most common source of starch, grains being about 70% starch by mass, while dried peas and beans contain about 40% starch. A second major source of starch is tuber and root crops, such as potatoes, yams, and cassava. The disaccharide lactose is an important component of milk, and sucrose is usually consumed as refined sugar (derived from sugar beets or sugar cane). The monosaccharides are often found in fruits.

The polysaccharides, also called *complex carbohydrates*, are difficult to digest because of their complex structures. (See Section 27.15.) Starch is digested slowly, enabling the body to control the distribution of this energy nutrient. Cellulose is not digested by humans; however, it is a major source of *dietary fiber*. As cellulose passes through the digestive tract, it absorbs water and provides dietary bulk. This bulk acts to prevent constipation and diverticulosis, a weakening of the intestinal walls. Many nutritionists recommend a daily fiber intake of 25–30 g, supplied by such foods as whole wheat bread (1.8 g/slice) and bran cereals (7.5 g/0.5 cup).

For the first time an DRI for carbohydrates has been set; it is 130 g or about 45–60% of the total calories per day. In the American diet, most of these calories are derived from starch, although in recent years an increasing amount has come from sucrose.

The high sucrose content of many modern diets is due primarily to the large amounts of sucrose in commercially prepared foods. In 1909, an American consumed an average of about 84 lb of sucrose annually in prepared foods and beverages. By 2006, an average American was consuming about 140 lbs. of added sugars (both sucrose and other caloric sweeteners). This large increase was caused mainly by (1) increased consumption of prepared foods and (2) an increased percentage of sucrose added to these foods by the manufacturers in attempts to gain larger shares of the market. (It is well known that many people, especially children, have a preference for sweet foods.) Although the consumption of other sweeteners, such as high-fructose corn syrup and aspartame, has risen, the consumption of sucrose remains high.

The increase in sucrose consumption troubles nutritionists for several reasons. First, sucrose is a prime factor in the incidence of dental caries. Because it is readily used by oral bacteria, sucrose promotes growth of the microorganisms that cause tooth decay. Second, ingested sucrose is rapidly hydrolyzed to monosaccharides, which are promptly absorbed from the intestine, leading to a rapid increase in blood-sugar levels. Wide variations in the blood-sugar level can stress the body's hormonal system. Finally, sucrose is said to provide "empty calories." This means that sucrose supplies metabolic energy (calories), but lacks other nutrients. Nutritionists usually recommend starch over sucrose as a major source of dietary carbohydrate.

32.5 Fats in the Diet

Fat is a more concentrated source of dietary energy than carbohydrate. Not only does fat contain more carbons per unit mass, but the carbons in fats are more reduced. As an energy source, fats provide about 9 kcal/g, whereas carbohydrates and proteins provide only about 4 kcal/g. As we will see in Chapter 35, energy can be obtained from fat only when oxygen is present; fat metabolism is strictly aerobic in humans. The carbons from fats can be used to synthesize amino acids, nucleic acids, and other fats, but our bodies cannot achieve a net synthesis of carbohydrate from fat. Thus, fats tend not to be as versatile a nutrient as carbohydrates.

Fats contribute much of the dietary energy of many foods. For example, french fried potatoes contain about 18% fat by weight, yet the fat provides about 40% of the calories in this food. A cup of whole milk contains 170 kcal, while a cup of skim (nonfat) milk contains 80 kcal. Many nutritionists counsel that the best way to reduce calorie intake is to eat foods containing less fat. Although fats are macronutrients, it is recommended that fat intake not exceed 25–30% of the daily energy allowance.

In a diet, both the kind and the amount of fat are important. Fatty acids from meat and dairy products are relatively saturated, whereas those from plant sources are generally more unsaturated. Because there is a probable link between high consumption of saturated fats and atherosclerosis, the U.S. Department of Agriculture and other agencies concerned with nutrition recommend that American diets contain about equal portions of polyunsaturated and saturated fatty acids. Polyunsaturated fats in the diet can be increased by (1) cooking with vegetable oils such as corn or canola oil and (2) using soft margarine, which usually contains more unsaturated fats than hard, or stick, margarine or butter.

Food for Thought: Of the three essential fatty acids, linolenic acid has the added benefit of being an ω-3 fatty acid. (See Chapter 28.)

Polyunsaturated fats generally also contain the essential fatty acids. Essential fatty acids provide a source of both omega-6 and omega-3 fatty acids. The most common essential fatty acids are linoleic acid, linolenic acid, and arachidonic acid.

$$CH_3(CH_2)_4CH{=}CHCH_2CH{=}CH(CH_2)_7CO_2H$$

linoleic acid (omega-6)

$$CH_3CH_2CH{=}CHCH_2CH{=}CHCH_2CH{=}CH(CH_2)_7CO_2H$$

linolenic acid (omega-3)

$$CH_3(CH_2)_4CH{=}CHCH_2CH{=}CHCH_2CH{=}CHCH_2CH{=}CH(CH_2)_3CO_2H$$

arachidonic acid (omega-6)

These unsaturated fatty acids have been shown to relieve the deleterious physiological changes that result from a totally fat-free diet, such as poor growth, skin lesions, kidney damage, and impaired fertility. One essential biochemical function of these fatty acids is as a precursor for prostaglandin synthesis.

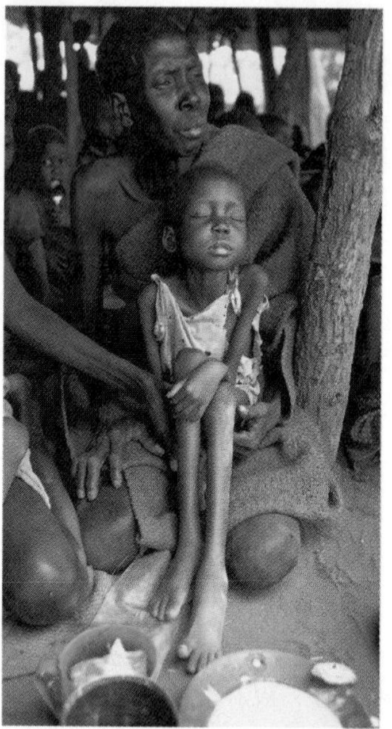

Kwashiorkor (protein deficiency) is a serious problem for children in many poverty-ridden areas of the world.

32.6 Proteins in the Diet

The third macronutrient is protein, with an average energy yield of about 4.2 kcal/g. This energy is derived primarily from reduced carbons. However, recall that proteins also contain about 16% nitrogen; in fact, protein is the primary source of nitrogen in our diets. As discussed in Chapter 35, perhaps the most important function of dietary protein is to provide for the synthesis of nitrogen-containing molecules such as nucleic acids, enzymes and other proteins, nerve transmitters, and many hormones. Because these materials are critical for human growth, protein malnutrition, or *kwashiorkor,* is a particularly serious problem. A specific DRI has been long established for dietary protein. (See Table 32.1.)

Kwashiorkor can occur even when one's calorie intake is sufficient; thus, it is an especially insidious form of malnutrition. In many poverty-ridden areas of the world, the only reliable source of protein for children is mother's milk. After weaning, the children eat a protein-poor grain diet. Although the caloric intake is sufficient, these children show the stunted growth, poor resistance to disease, and general body wasting that are characteristic of kwashiorkor.

Food for Thought: Kwashiorkor, like many nutritional deficiencies, can occur even with more than enough food to eat.

Proteins are obtained from animal sources, such as meat, milk, cheese, and eggs, and from plant sources, such as cereals, nuts, and legumes (peas, beans, and soybeans). Animal proteins have nutritive values that are generally superior to those of vegetable proteins, in that they supply all of the 20 amino acids that the body uses. In contrast, a single vegetable source often lacks several amino acids. This deficiency is critical, because humans cannot synthesize a group of amino acids called the essential amino acids (Table 32.2). These 10 *essential amino acids* must be obtained from the diet.

complete protein

Nutritionists often speak of a dietary source (or sources) of complete protein. A **complete protein** is a protein that supplies all of the essential amino acids. Animal products are, in general, sources of complete proteins. However, by either choice or necessity, animal protein is seldom consumed by a significant fraction of the world's population. Besides the moral, ethical, and religious reasons given for limiting consumption of animal protein, these proteins are relatively expensive sources of amino acids. Thus, most people in underdeveloped countries subsist primarily on vegetable protein. A constant danger inherent in a vegetarian diet is that the source of vegetable protein may be deficient in several of the essential amino acids; that is, the protein may be incomplete. For this reason, nutritionists recommend that several sources of

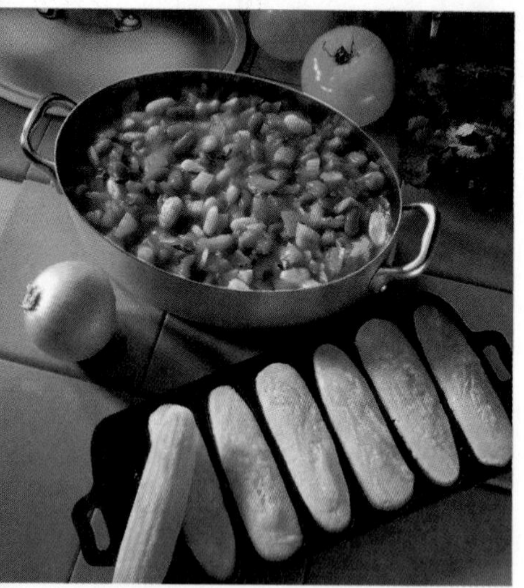

In a vegetarian diet, it is important to include several sources of protein, such as beans and rice, to produce a complete protein.

Table 32.2 Common Dietary Amino Acids

Essential		Nonessential	
Arginine*	Methionine	Alanine	Glutamine
Histidine*	Phenylalanine	Asparagine	Glycine
Isoleucine	Threonine	Aspartic acid	Proline
Leucine	Tryptophan	Cysteine	Serine
Lysine	Valine	Glutamic acid	Tyrosine

*Essential for growing children

vegetable protein be included with each meal in a vegetarian diet. As an example, soybeans, which are rich in lysine, might supplement wheat, which is lysine deficient.

32.7 Vitamins

Vitamins are a group of naturally occurring organic compounds that are essential for good nutrition and that must be supplied in the diet. Whereas the energy-supplying nutrients are digested and metabolized extensively, vitamins are often used after only minimal modification. Some of the vitamins necessary for humans are listed in Table 32.3. Note that vitamins are often classified according to their solubility: those which are fat soluble and those which are water soluble. The structural formulas of several vitamins are shown in Figure 32.2.

vitamin

Table 32.3 Some of the Most Important Vitamins

Vitamin	Important dietary sources	Some deficiency symptoms
FAT SOLUBLE		
Vitamin A (Retinol)	Green and yellow vegetables, butter, eggs, nuts, cheese, fish liver oil	Poor teeth and gums, night blindness
Vitamin D (Ergocalciferol, D_2; cholecalciferol, D_3)	Egg yolk, milk, fish liver oils; formed from provitamin in the skin when exposed to sunlight	Rickets (low blood-calcium level, soft bones, distorted skeletal structure)
Vitamin E (α-Tocopherol)	Meat, egg yolk, wheat germ oil, green vegetables; widely distributed in foods	Not definitely known in humans
Vitamin K (Phylloquinone, K_1; menaquinone, K_2)	Eggs, liver, green vegetables; produced in the intestines by bacterial reactions	Blood is slow to clot (antihemorrhagic vitamin)
WATER SOLUBLE		
Vitamin B_1 (Thiamin)	Meat, whole-grain cereals, liver, yeast, nuts	Beriberi (nervous system disorders, heart disease, fatigue)
Vitamin B_2 (Riboflavin)	Meat, cheese, eggs, fish, liver	Sores on the tongue and lips, bloodshot eyes, anemia
Vitamin B_6 (Pyridoxine)	Cereals, liver, meat, fresh vegetables	Skin disorders (dermatitis)
Vitamin B_{12} (Cyanocobalamin)	Meat, eggs, liver, milk, fish	Pernicious anemia
Vitamin C (Ascorbic acid)	Citrus fruits, tomatoes, green vegetables	Scurvy (bleeding gums, loose teeth, swollen joints, slow healing of wounds, weight loss)
Niacin (Nicotinic acid and amide)	Meat, yeast, whole wheat	Pellagra (dermatitis, diarrhea, mental disorders)
Biotin (Vitamin H)	Liver, yeast, egg yolk	Skin disorders (dermatitis)
Folic acid	Liver, wheat germ, yeast, green leaves	Macrocytic anemia, gastrointestinal disorders

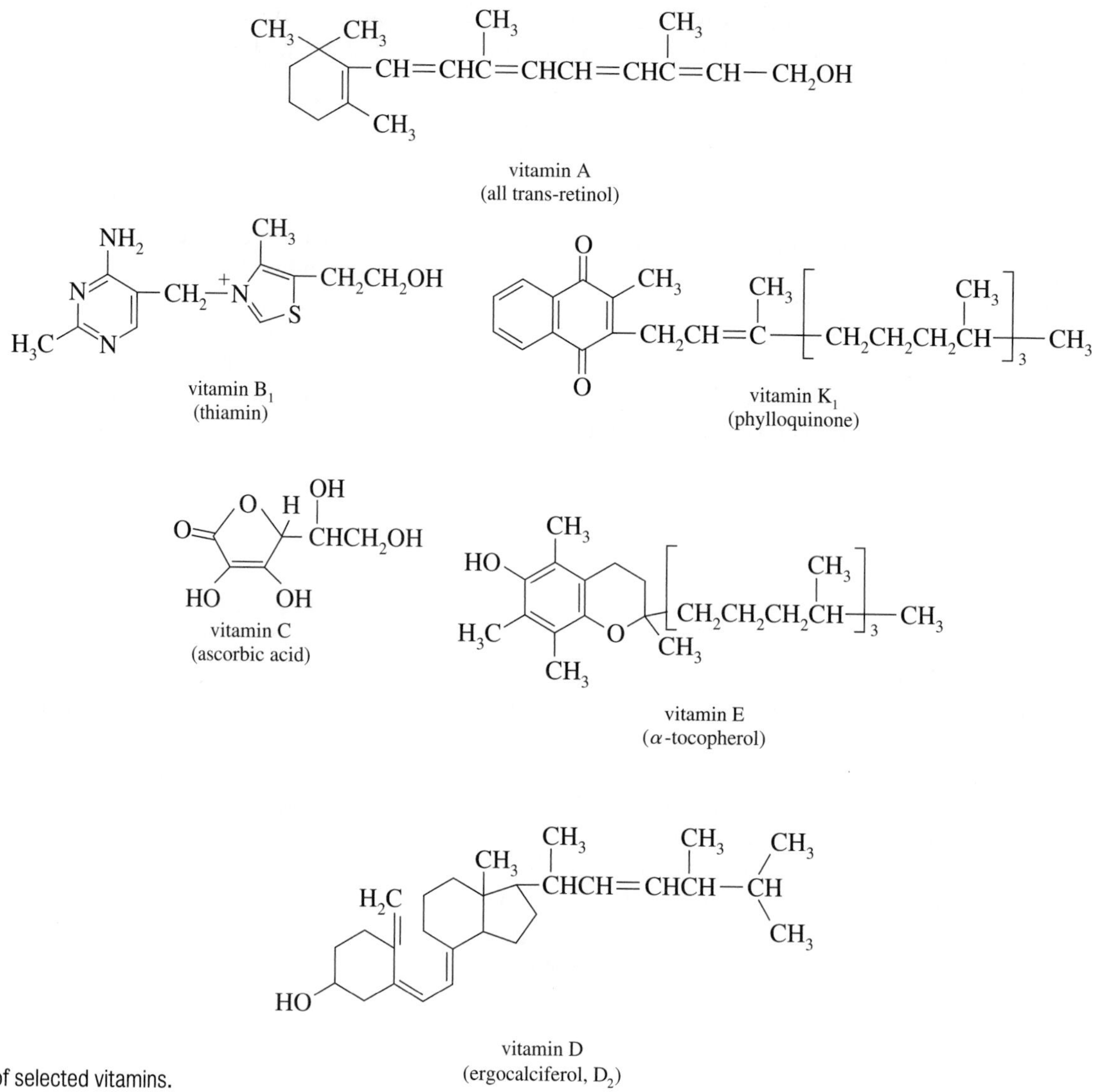

vitamin A
(all trans-retinol)

vitamin B_1
(thiamin)

vitamin K_1
(phylloquinone)

vitamin C
(ascorbic acid)

vitamin E
(α-tocopherol)

vitamin D
(ergocalciferol, D_2)

Figure 32.2
The structure of selected vitamins.

A prolonged lack of vitamins in the diet leads to vitamin deficiency diseases such as beriberi, pellagra, pernicious anemia, rickets, and scurvy. Left uncorrected, a vitamin deficiency ultimately results in death. Even when supplementary amounts of vitamins are provided, impaired growth due to a vitamin deficiency may be irreversible. For example, it is difficult to correct the distorted bone structures resulting from a childhood lack of vitamin D. Thus, it is especially important that children receive sufficient vitamins for proper growth and development.

Vitamins are required in only small amounts (see the RDAs in Table 32.1) and are classed as micronutrients. For example, a typical diet might include 250 g of carbohydrate (a macronutrient) and only 2 mg of vitamin B_6. However, the biochemistry of life cannot continue without vitamins. Each vitamin serves at least one specific purpose for an organism. The water-soluble compounds are generally involved in cellular metabolism of the energy-supplying nutrients. For example, thiamin is required to achieve a maximum energy yield from carbohydrates, and pyridoxine is of central importance in protein metabolism. (See Chapters 34 and 35, respectively.) Niacin and riboflavin are key components in almost all cellular redox reactions. The fat-soluble vitamins often serve very specialized functions. Vitamin D acts as a regulator of calcium metabolism. One function of vitamin A is to furnish the pigment that makes vision possible, while vitamin K enables blood clotting to occur normally.

Because some vitamin functions are not well understood, miraculous properties have been ascribed to these substances, such as vitamins C and E. In the absence of conclusive scientific studies, it is difficult to judge the merits of some of these claims.

Although vitamins are required only in small quantities, their natural availability is also low. Nutritionists caution that a diet should be balanced to include adequate sources of vitamins. As you can see from Table 32.3, the dietary sources of vitamins are varied. In general, fruits, vegetables, and meats are rich sources of the water-soluble vitamins; and eggs, milk products, and liver are good sources of the fat-soluble vitamins.

Food for Thought: There is no difference between "artificial vitamins" and "natural vitamins." Since vitamins are well-characterized organic chemicals, chemists can synthesize vitamins of a potency equal to that of the naturally synthesized vitamins.

A seemingly balanced diet may be deficient in vitamins, due to losses incurred in food processing, storing, and cooking. As much as 50–60% of the water-soluble vitamins in vegetables can be lost during cooking. Vitamin C can be destroyed by exposure to air. Removal of the outside hull from grains drastically decreases their B vitamin content. Thus, when polished rice became a dietary staple in the Orient, beriberi (the thiamin-deficiency disease) grew to epidemic proportions.

32.8 Minerals

A number of inorganic elements are needed for good health. Like vitamins, minerals are classified as micronutrients. The body does not synthesize minerals. Those which must be ingested in relatively large amounts—the *major elements*—include sodium, potassium, calcium, magnesium, chloride, and phosphate (phosphorus). Elements in a second group, the *trace elements*, are required in much smaller amounts. As scientific studies of the body's mineral requirements proceed, the list of needed trace elements continually lengthens. A recent compilation of these trace elements is given in Table 32.4.

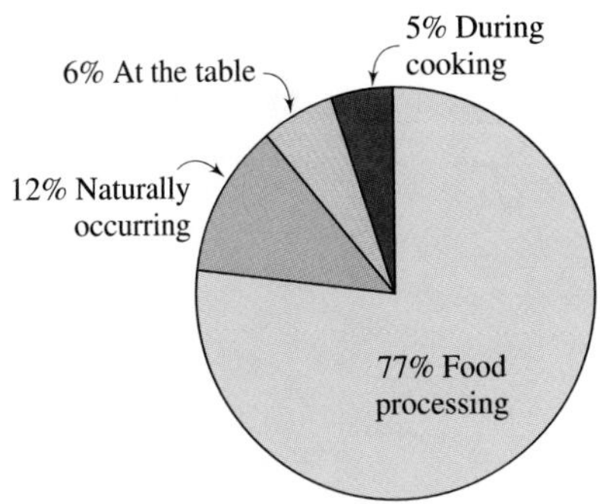

Relative amount of dietary sodium in the American diet.

Mineral nutrients differ from organic nutrients in that the body, in general, uses minerals in the ionic form in which they are absorbed. Although these elements are required for good health, they can also be toxic if ingested in quantities that are too large. Nutritionists have established RDAs for some of these minerals and warn against excess intake.

The major minerals—sodium, potassium, and chloride—are responsible for maintaining the appropriate salt levels in body fluids. Many enzyme reactions require an optimal salt concentration of 0.1–0.3 *M*. In addition, individual elements serve specific functions; nerve transmission, for example, requires a supply of extracellular sodium and intracellular potassium. Although there is no RDA for these two elements, 2–4 g/day represents an average NaCl consumption. But because high levels of NaCl can contribute to high blood pressure, many nutritionists advise using salt in moderation.

Table 32.4 Required Trace Elements

Element	Function	Human deficiency signs
Fluorine	Structure of teeth, possibly of bones; possible growth effect	Increased incidence of dental caries; possibly risk factor for osteoporosis
Silicon	Calcification; possible function in connective tissue	Not known
Vanadium	Not known	Not known
Chromium	Efficient use of insulin	Relative insulin resistance, impaired glucose tolerance, elevated serum lipids
Manganese	Mucopolysaccharide metabolism, superoxide dismutase enzyme	Not known
Iron	Oxygen and electron transport, hemoglobin replenishment	Anemia
Cobalt	Part of vitamin B_{12}	Only as vitamin B_{12} deficiency
Nickel	Interaction with iron absorption	Not known
Copper	Oxidative enzymes; interaction with iron; cross-linking of elastin connective protein	Anemia, changes of ossification; possibly elevated serum cholesterol
Zinc	Numerous enzymes involved in energy metabolism and in transcription and translation	Growth depression, sexual immaturity, skin lesions, depression of immunocompetence, change in acuity of taste
Arsenic	Not known	Not known
Selenium	Glutathione peroxidase; interaction with heavy metals	Endemic heart problems conditioned by selenium deficiency
Molybdenum	Xanthine, aldehyde, and sulfide oxidase enzymes	Not known
Iodine	Constituent of thyroid hormones	Goiter, depression of thyroid function, cretinism

Calcium and magnesium serve many roles in the body, including being required by some enzymes. Fully 90% of all body calcium, and a significant percentage of the magnesium, are found in the bones and teeth. Calcium also is needed for nerve transmission and blood clotting.

Trace elements are similar to vitamins in that (1) they are required in small amounts and (2) food contains only minute quantities of them. Usually, a normal diet contains adequate quantities of all trace elements. The exact function of many trace elements remains unknown.

The most notable exception to the preceding generalizations is iron. This element is part of hemoglobin, the oxygen-binding protein of the blood. (Iron is also a critical component of some enzymes.) Relatively large amounts of iron are required for hemoglobin replenishment. Unfortunately, many foods are not rich enough in iron to provide the necessary DRI. This is especially true for the higher iron DRI for women. Therefore, nutritionists sometimes recommend a daily iron dietary supplement.

Practice 32.2

From Table 32.1, why is the largest DRI for calcium linked to the largest DRI for vitamin D?

32.9 Water

Water is the solvent of life. As such, it carries nutrients to the cells, allows biochemical reactions to proceed, and carries waste from the cells. Water makes up approximately two–thirds of the total body mass.

Our bodies are constantly losing water via urine (700–1400 mL/day), feces (150 mL/day), sweat (500–900 mL/day), and expired air (400 mL/day). If dehydration is to be prevented, this water output must be offset by water intake. In normal diets, liquids provide about 1200–1500 mL of water per day. The remaining increment is derived from food (700–1000 mL/day) and water formed by biochemical reactions (metabolic water; 200–300 mL/day). Water losses vary and depend on the intake and the activity of the individual.

A proper water balance is maintained through the control of water intake and water excretion. The kidney is the center for control of water output. Water intake is controlled by the thirst sensation. We feel thirsty when one or both of the two following events occur: (1) As the body's water stores are depleted, the salt concentration rises. Specific brain cells monitor this salinity and initiate the thirst feeling. (2) As water is drawn away from the salivary glands, the mouth feels dry; this event also leads to the sensation of thirst.

Water is a universal nutrient. It has many functions for keeping our bodies healthy.

Food for Thought: Water is an important nutrient that is often overlooked; many Americans drink too little water each day.

32.10 Nutrition Content Labeling

Nearly all foods in our supermarkets have been processed to some degree, and loss of nutrients is a problem with some kinds of food processing. In general, the more extensive the processing, the greater is the loss. Processing can be quite extensive for such items as luncheon meats, TV dinners, and breakfast cereals.

In an effort to aid the consumer, nutritionists have established procedures for monitoring the nutritional values of processed foods. These values are reported on a package panel termed *Nutrition Facts*. (See Figure 32.3.) Much can be learned about food nutrients by examining such panels. First, let us make some general observations: (1) Many people's daily energy requirement centers around a 2000-kcal (Cal) or a 2500-kcal (Cal) diet; the Nutrition Facts calculations are based on these approximations. (2) Fats are subdivided to recognize the important health difference between saturated and unsaturated fatty acids as well as the danger of trans fats; cholesterol is given its own category. (3) Carbohydrates are categorized as sugars and fiber, with the remainder being complex carbohydrates such as starch. (4) Total protein is always listed. (5) These nutritional facts are based on a standard serving size (set by the federal government).

Figure 32.3 compares the Nutrition Facts for two breakfast cereals. Notice that the serving size is the same (about 50 g), although the volume of one cereal is smaller. While both cereals are primarily carbohydrates, the "granola" cereal contains about 25% (12 g out of 48 g) sugar. In contrast, the "shredded wheat" cereal contains only complex carbohydrates, with a slightly higher percent daily value of fiber. Both cereals are relatively low in fat, but the "granola" cereal contains more. Since common grains contain

NUTRITION FACTS
Serving Size 1 cup (50g)
Servings Per Container about 7

Amount Per Serving	1 Cup Cereal	Cereal with 1 Cup Vitamins A&D Skim Milk
Calories	180	270
Calories from Fat	15	15
		% Daily Value*
Total Fat 1.5g†	2%	2%
Saturated Fat 0g	0%	0%
Trans Fat 0g		
Cholesterol 0mg	0%	1%
Sodium 0mg	0%	5%
Total Carbohydrate 41g	14%	18%
Dietary Fiber 6g	26%	26%
Sugars 0g		
Protein 5g		
Vitamin A	**	10%
Vitamin C	**	4%
Calcium	2%	30%
Iron	8%	10%
Vitamin D	**	25%
Thiamin	6%	10%
Riboflavin	4%	20%
Niacin	15%	15%
Phosphorus	20%	45%
Magnesium	15%	20%

†Amount in 1 Cup Cereal. One cup skim milk contributes an additional 90 calories, less than 5mg cholesterol, 125mg sodium, 12g total carbohydrate (12g sugars), and 8g protein.
*Percent Daily Values are based on a 2,000 calorie diet. Your daily values may be higher or lower depending on your calorie needs.
**Contains less than 2% of the daily value of these nutrients.

	Calories:	2000	2500
Total Fat	Less than	65g	80g
Sat Fat	Less than	20g	25g
Cholesterol	Less than	300mg	300mg
Sodium	Less than	2,400mg	2,400mg
Total Carbohydrate		300g	375g
Dietary Fiber		25g	30g

(a) "Shredded wheat" cereal

NUTRITION FACTS
Serving Size ½ cup (48g)
Servings Per Container about 9

Amount Per Serving	
Calories 210	
Calories from Fat 70	
	% Daily Value*
Total Fat 8g	12%
Saturated Fat 3.5g	17%
Trans Fat 0g	
Cholesterol 0mg	0%
Sodium 15mg	1%
Potassium 210mg	6%
Total Carbohydrate 33g	11%
Other Carbohydrate 18g	
Dietary Fiber 4g	16%
Sugars 12g	
Protein 5g	
Vitamin A	0%
Vitamin C	0%
Calcium	4%
Iron	6%
Thiamin	10%
Phosphorus	10%
Magnesium	10%
Copper	15%

*Percent Daily Values are based on a 2,000 calorie diet. Your daily values may be higher or lower depending on your calorie needs.

	Calories:	2000	2500
Total Fat	Less than	65g	80g
Sat Fat	Less than	20g	25g
Cholesterol	Less than	300mg	300mg
Sodium	Less than	2,400mg	2,400mg
Total Carbohydrate		300g	375g
Dietary Fiber		25g	30g
Potassium		3,500mg	3,500mg

Calories per gram:
Fat 9 • Carbohydrate 4 • Protein 4

(b) "Granola" cereal

Figure 32.3
Two examples of Nutrition Facts panels.

about the same percentage protein, both cereals contain 5 g of protein. The "shredded wheat" cereal manufacturer has opted to include nutritional information after adding 1 cup of skim milk. An additional food item such as milk can significantly change nutrient levels. Both cereals list vitamins and minerals; the list may vary, depending on what is included in the food item. Finally, the recommended nutrient amounts for a healthful daily diet are given at the bottom of these Nutrition Facts panels (based on either 2000-Cal or 2500-Cal diets).

The Nutrition Facts panel gives us the amount of nutrients suggested for a 2000-kcal or a 2500-kcal diet. The percent daily values are based on these suggestions. For example, the percent total carbohydrate in 48 g of granola cereal is readily calculated for a 2000-kcal diet (Figure 32.3) thus:

$$\frac{33 \text{ g carbohydrate}}{300 \text{ g carbohydrate}} \times 100 = 11\%$$

The nutrient amounts given at the bottom of the Nutrition Facts panel have been chosen with these important guidelines in mind: (1) Carbohydrate should be the bulk of a diet; (2) saturated fat should be a small proportion of the total fat intake; (3) consumption of both sodium and cholesterol should be minimized. We can estimate our nutrient intake and also learn more about nutrition from the Nutrition Facts panel.

Practice 32.3

Referring to the "shredded wheat" Nutrition Facts in Figure 32.3a, if 6 g of dietary fiber supplies 26% of the percent daily value, how many grams of dietary fiber should be consumed per day?

32.11 Food Additives

Various chemicals are often added to foods during processing. In fact, more than 3000 of these *food additives* have been given the "generally recognized as safe" (GRAS) rating by the U.S. Food and Drug Administration.

The purpose of food additives varies. Some additives enhance the nutritional value (e.g., when a food is vitamin enriched). Other additives serve as preservatives. For example, sodium benzoate is used to inhibit bacterial growth, and BHA (butylated hydroxyanisole) and BHT (butylated hydroxytoluene) are used as antioxidants. Still other additives—for example, emulsifiers, thickeners, anticaking agents, flavors, flavor enhancers, nonsugar sweeteners, and colors—improve the appearance and flavor of a food. (See Table 32.5.)

There has been, and continues to be, a great deal of controversy concerning the use of food additives. Due to the nature and complexity of the subject, this debate will no doubt continue. Controversy centers around the problem of balancing the benefits derived from additives against the risk to consumers. The use of at least some additives is necessary to prepare many foods. Discovering whether a risk exists and assessing the degree of risk requires long, difficult, and expensive research. As a case in point, salting and smoking were used in curing meats for centuries before there was any knowledge that these

Table 32.5 Some Common Food Additives

Food additive	Purpose
Sodium benzoate Calcium lactate Sorbic acid	Antimicrobials (prevent food spoilage)
BHA (Butylated hydroxyanisole) BHT (Butylated hydroxytoluene) EDTA (Ethylenediaminetetraacetic acid)	Antioxidants (prevent changes in color and flavor)
Calcium silicate Silicon dioxide Sodium silicoaluminate	Anticaking agents (keep powders and salt free flowing)
Carrageenan Lecithin	Emulsifiers (aid even distribution of suspended particles)
Pectin Propylene glycol	Thickeners (impart body and texture)
MSG (Monosodium glutamate) Hydrolyzed vegetable protein	Flavor enhancers (supplement or modify taste)

processes might involve a hazard. Research has shown that both processes involve risks to at least some consumers: Salt aggravates certain cardiovascular conditions, and smoke produces carcinogens in the meat. Yet in the eyes of many consumers, these risks are outweighed by the benefits to be had from salted and smoked meats. On the other hand, few consumers indeed would accept a compound known to be highly carcinogenic as an additive in their ice cream, even though that compound could improve the flavor of the ice cream remarkably!

Owing to the technical nature of the task, consumers must rely largely on the judgment and integrity of the professional people who have been assigned the responsibility of protecting our food supply. Consumers should also inform themselves as fully as possible concerning the nature, purpose, and possible hazards of the additives that are used or proposed for use in our foods and, as responsible citizens, make sure that our government provides adequate support to the professionals charged with protecting our food.

32.12 A Balanced Diet

To summarize the preceding sections, the macronutrients (carbohydrates, fats, and proteins) constitute the majority of our diets. These nutrients supply the molecules we need for energy, growth, and maintenance. A second group of nutrients—vitamins, minerals, and water—is not used for energy, but is nevertheless essential to our existence. Vitamins provide organic molecules that cannot be made in the body, and minerals provide the inorganic ions needed for life. Water is the solvent in which most of the chemical reactions essential to life occur. Vitamins and minerals are required only in small amounts, from a few micrograms to a few milligrams per day, but water must be consumed in large quantities, 2–3 L/day.

Figure 32.4
The most recent U.S. government food pyramid, MyPyramid emphasizes the need for a personal approach to a balanced diet. The recommended proportions for each food group are given by the colored bands: orange, grains; green, vegetables; red, fruits; yellow, oils; blue, milk products; purple, meats and beans.

For health and well-being, each of six groups of nutrients must be in our diet. With the variety of foods available, how can we make sure that our diets contain enough of all the needed nutrients? How can we make sure that our diet is balanced?

To answer these questions, nutritionists have divided foods into six groups: (1) milk products, (2) vegetables, (3) fruits, (4) grains, (5) meats and beans, and (6) oils. Each group is a good source of one or more nutrients. To obtain a balanced mixture of carbohydrate, fat, vitamins, and minerals, a diet should contain food from several classes. Figure 32.4 is a pictorial guide to a balanced diet.

A balanced diet must include several food groups, even though each food may contain many nutrients. For example, compare the three breakfasts in Figure 32.5, each of which supplies about 600 kcal of food energy. By choosing just doughnuts (cereal food group) and coffee, the consumer would gain some nutrients, but only in small quantities. A breakfast of cold cereal (cereal group) and milk (milk group), toast (cereal group) with margarine (fats) and jelly (sweets), and coffee includes more food groups. Still more food groups are found in a breakfast of orange juice (fruit group), a fried egg (meat group), pancakes (cereal group) with margarine (fats) and syrup (sweets), milk (milk group), and coffee. Many other breakfasts could be chosen instead, yet an important generalization can be drawn from these three examples: The nutritional value of a meal improves if at least several food groups are included. The third breakfast in this illustration is the most balanced. Although the consumer often does not know the nutrient content of a specific food, overall nutrition can be ensured by choosing a balanced diet.

After the food composing a balanced diet is eaten, it must be digested, absorbed, and transported in order for the proper nutrients to reach the cells. Eating puts food into the alimentary canal, which includes the mouth, esophagus, stomach, small intestine, and large intestine. In a sense, this canal is an extension of our external environment; that is, until food is broken down and processed, it cannot enter our internal environment to reach the cells. After digestion and absorption, the nutrients are transported to the cells by the blood and lymph systems. The liver plays a vital role in controlling nutrient levels in the blood and in neutralizing toxic substances.

Figure 32.5 A comparison of nutritional values for three breakfasts.

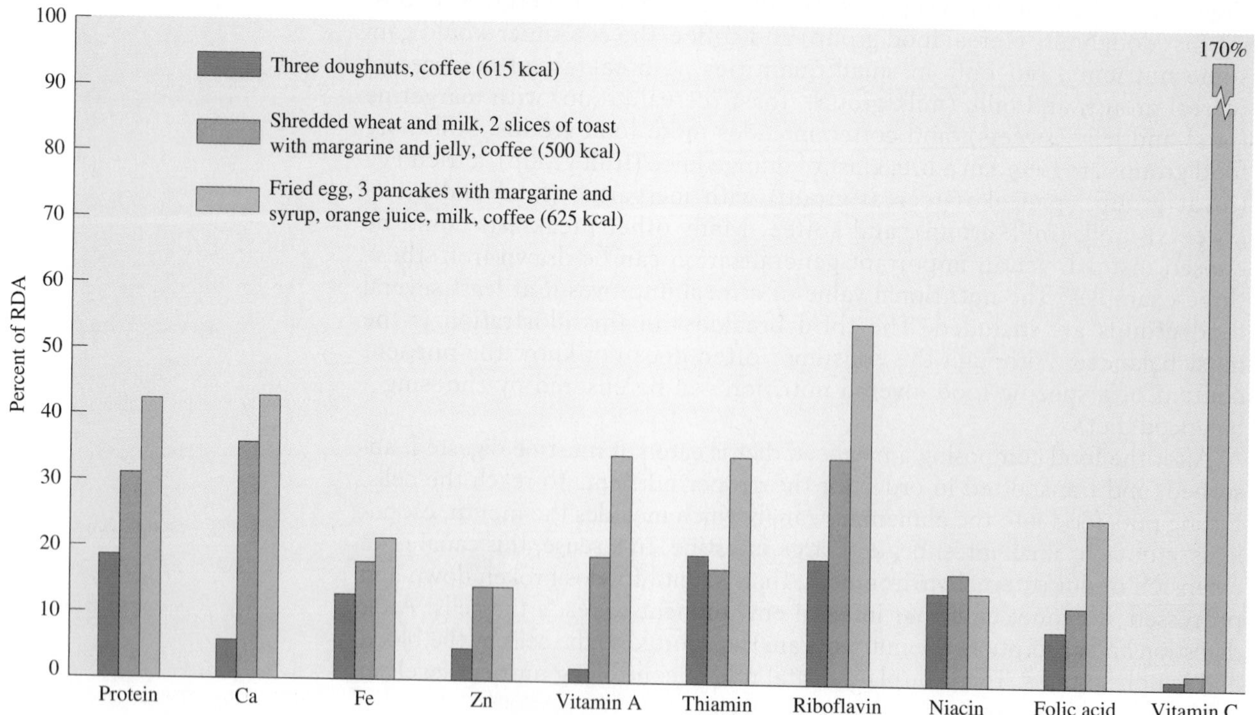

32.13 Human Digestion

The human digestive tract is shown in Figure 32.6. Although food is broken up mechanically by chewing and by a churning action in the stomach, digestion is a chemical process. **Digestion** is a series of enzyme-catalyzed reactions by which large molecules are hydrolyzed to molecules small enough to be absorbed through the intestinal membranes. Foods are digested to smaller molecules according to this general scheme:

digestion

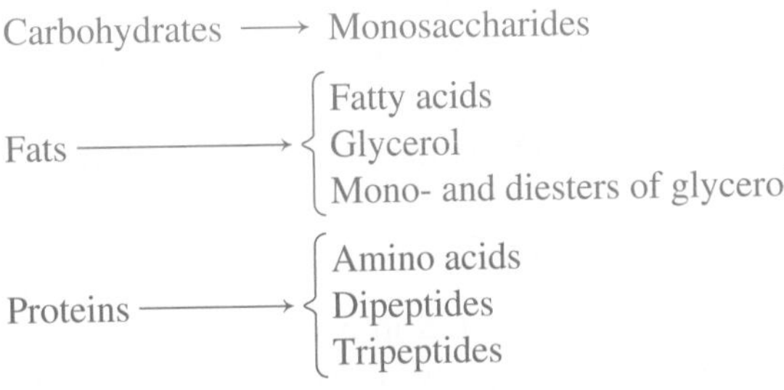

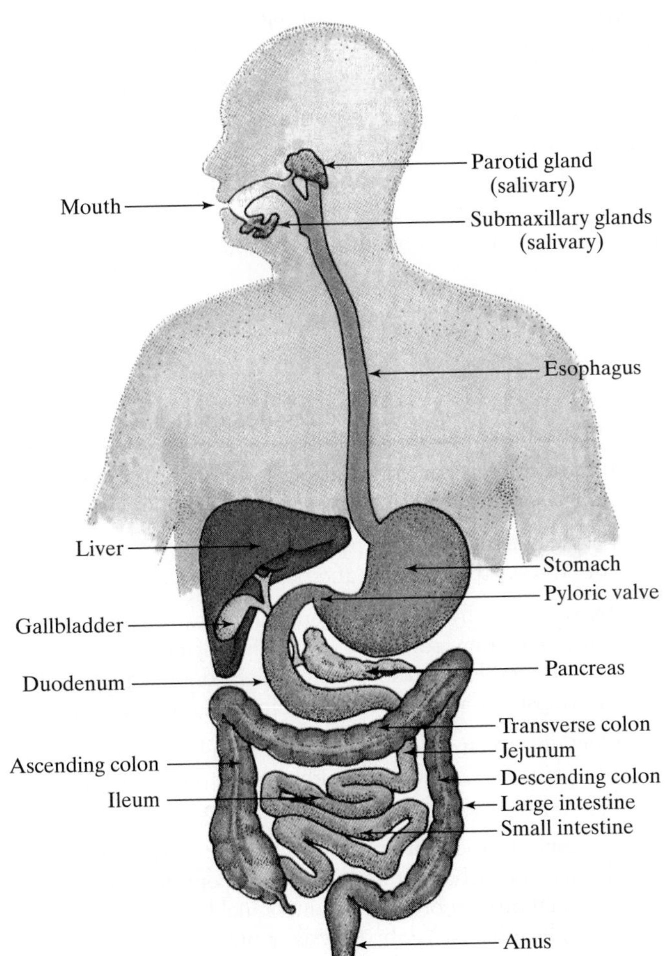

Figure 32.6
The human digestive tract.

Food passes through the human digestive tract (Figure 32.6) in this sequence:

Mouth ⟶ esophagus ⟶ stomach ⟶ small intestine (duodenum, jejunum, and ileum) ⟶ large intestine

Five principal digestive juices (or fluids) enter the digestive tract at various points:

1. Saliva from three pairs of salivary glands in the mouth
2. Gastric juice from glands in the walls of the stomach
3. Pancreatic juice, secreted by the pancreas and entering the duodenum through the pancreatic duct
4. Bile, secreted by the liver and entering the duodenum via a duct from the gallbladder
5. Intestinal juice from intestinal mucosal cells

Detailed accounts of the various stages of digestion are to be found in biochemistry and physiology textbooks.

The main functions and principal enzymes found in each of these fluids are summarized in Table 32.6. The important digestive enzymes occur in gastric, pancreatic, and intestinal juices. An outline of the digestive process follows.

Salivary Digestion Food is chewed (masticated) and mixed with saliva in the mouth, and the hydrolysis of starch begins. The composition of saliva depends on many factors: age, diet, condition of teeth, time of day, and so on. Normal saliva is about 99.5% water. Saliva also contains mucin (a glycoprotein); a number of mineral ions, such as K^+, Ca^{2+}, Cl^-, PO_4^{3-}, SCN^-; and one enzyme, salivary amylase (ptyalin). The pH of saliva ranges from slightly acidic to slightly basic, with the optimal pH about 6.6–6.8.

Mucin acts as a lubricant and facilitates the chewing and swallowing of food.

Table 32.6 Digestive Fluids

Fluid (volume produced daily)	Source	Principal enzymes or function
Saliva (1000–1500 mL)	Salivary glands	Lubricant, aids chewing and swallowing; also, contains salivary amylase (ptyalin), which begins the digestion of starch
Gastric juice (2000–3000 mL)	Glands in stomach wall	Pepsin, gastric lipase; pepsin catalyzes the partial hydrolysis of proteins to proteoses and peptones in the stomach
Pancreatic juice (500–800 mL)	Pancreas	Trypsinogen, chymotrypsinogen, and procarboxypeptidase (after secretion, converted to trypsin, chymotrypsin, and carboxypeptidase, respectively, which continue protein digestion); amylopsin (α-amylase, a starch-digestive enzyme), steapsin (a lipase)
Bile (500–1000 mL)	Liver	Contains no enzymes, but does contain bile salts, which aid digestion by emulsifying lipids; serves to excrete cholesterol and bile pigments derived from hemoglobin
Intestinal juice	Intestinal mucosal cells	Contains a variety of finishing enzymes: sucrase, maltase, and lactase for carbohydrates; aminopolypeptidase and dipeptidase for final protein breakdown; intestinal lipase; nucleases and phosphatase for hydrolysis of nucleic acids

The enzyme salivary amylase catalyzes the hydrolysis of starch to maltose:

$$\text{Starch} + \text{Water} \xrightarrow{\text{salivary amylase}} \text{Maltose}$$

Salivary amylase is inactivated at a pH of 4.0, so it has very little time to act before the food reaches the highly acidic stomach juices.

Saliva is secreted continuously, but the rate of secretion is greatly increased by the sight and odor, or even the thought, of many foods. The mouth-watering effect of the sight or thought of pickles is familiar to many of us. This is an example of a *conditioned reflex*.

Gastric Digestion When food is swallowed, it passes through the esophagus to the stomach. In the stomach, mechanical action continues; food particles are reduced in size and mixed with gastric juices until a material of liquid consistency, known as chyme, is obtained.

Gastric juice is a clear, pale yellow, acidic fluid having a pH of about 1.5–2.5. It contains hydrochloric acid; the mineral ions Na^+, K^+, Cl^-; some phosphates; and the digestive enzymes pepsin and lipase. The flow of gastric juice is accelerated by conditioned reflexes and by the presence of food in the stomach. The secretion of the hormone *gastrin* is triggered by food entering the stomach. This hormone, produced by the gastric glands, is absorbed into the bloodstream and returned to the stomach wall, where it stimulates the secretion of additional gastric juice. Control of gastric secretion by this hormone is but one of many chemical control systems that exist in the body.

The chief digestive function of the stomach is the partial digestion of protein. The principal enzyme of gastric juice is pepsin, which digests protein. The enzyme is secreted in an inactive form called pepsinogen, which is activated by hydrochloric acid to pepsin. Pepsin catalyzes the hydrolysis of proteins to fragments called proteoses and peptones, which are still fairly large molecules. Pepsin splits the peptide bonds adjacent to only a few amino acid residues, particularly tyrosine and phenylalanine:

$$\text{Protein} + \text{Water} \xrightarrow{\text{pepsin}} \text{Proteoses} + \text{Peptones}$$

The second enzyme in the stomach, gastric lipase, is a fat-digesting enzyme. Its action in the stomach is slight, because the acidity is too high for lipase activity. Food can be retained in the stomach for as long as six hours. It then passes through the pyloric valve into the duodenum.

Intestinal Digestion The next section of the digestive tract, the small intestine, is where most of the digestion occurs. The stomach contents are first made alkaline by secretions from the pancreatic and bile ducts. The pH of the pancreatic juice is 7.5–8.0, and the pH of bile is 7.1–7.7. The shift in pH is necessary because the enzymes of the pancreatic and intestinal juices are active only in an alkaline medium. Enzymes that digest all three kinds of food—carbohydrates, fats, and proteins—are secreted by the pancreas. Pancreatic secretion is stimulated by hormones secreted into the bloodstream by the duodenum and the jejunum.

The enzymes occurring in the small intestine include pancreatic amylases (diastase) that hydrolyze most of the starch to maltose, and carbohydrases (α-amylase, maltase, sucrase, and lactase) that complete the hydrolysis of disaccharides to monosaccharides. The proteolytic enzymes *trypsin* and *chymotrypsin* attack proteins, proteoses, and peptones, hydrolyzing them to dipeptides. Then the peptidases—carboxypeptidase, aminopeptidase, and dipeptidase—complete the hydrolysis of proteins to amino acids. Pancreatic lipases catalyze the hydrolysis of almost all fats. Fats are split into fatty acids, glycerol, and mono- and diesters of glycerol by these enzymes.

The liver is another important organ in the digestive system. A fluid known as bile is produced by the liver and stored in the gallbladder, a small organ located on the surface of the liver. When food enters the duodenum, the gallbladder contracts, and the bile enters the duodenum through a duct that is also used by the pancreatic juice. In addition to water, the major constituents of bile are bile acids (as salts), bile pigments, inorganic salts, and cholesterol. The bile acids are steroid monocarboxylic acids, two of which are shown here:

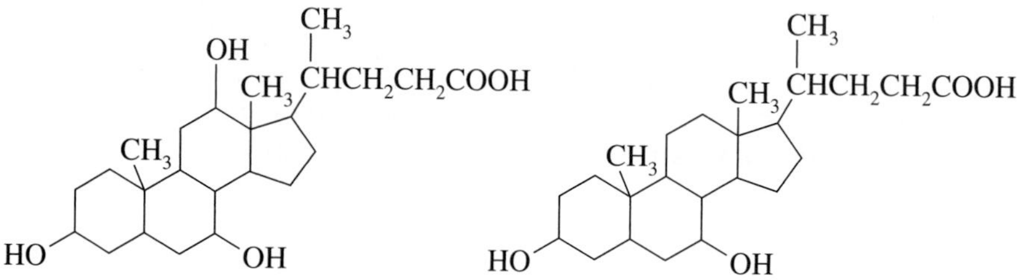

cholic acid

chenodeoxycholic acid

The bile acids are synthesized in the liver from cholesterol, which is also synthesized in the liver. The presence of bile in the intestine is important for the digestion and absorption of fats. When released into the duodenum, the bile acids emulsify the fats, allowing them to be hydrolyzed by the pancreatic lipases. About 90% of the bile salts are reabsorbed in the lower part of the small intestine and are transported back to the liver and used again.

Most of the digested food is absorbed from the small intestine. Undigested and indigestible material passes from the small intestine to the large intestine, where it is retained for varying periods of time before final elimination as feces. Additional chemical breakdown, sometimes with the production of considerable amounts of gases, is brought about by bacteria (or rather, by bacterial enzymes) in the large intestine. For a healthy person, this additional breakdown is not important from the standpoint of nutrition, because nutrients are not absorbed from the large intestine. However, large amounts of water, partly from digestive juices, are absorbed from the large intestine, so that the contents become more solid before elimination as feces.

Practice 32.4

The complete digestion of starch yields which monosaccharide?

32.14 Absorption

For digested food to be utilized in the body, it must pass from the intestine into the blood and lymph systems. The process by which digested foods pass through the membrane linings of the small intestine and enter the blood and lymph is called **absorption**. Absorption is complicated, and we shall describe it only briefly here.

absorption

After the food you have eaten is digested, your body must absorb billions upon billions of nutrient molecules and ions into the bloodstream. The absorption system is in the membranes of the small intestine, which, upon microscopic inspection, are seen to be wrinkled into hundreds of folds. These folds are covered with thousands of small projections called *villi*. Each projection is itself covered by many minute folds called *microvilli*. The small intestine's wrinkles, folds, and projections increase the surface area available for absorption. The average small intestine is about 4 m (13.3 ft) long and is estimated to contain about 8360 m^2 (90,000 ft^2) of absorbing surface.

The inner surface of the small intestine is composed of mucosal cells that produce many enzymes needed to complete the digestive process, such as disaccharidases, aminopeptidases, and dipeptidases. As digestion is completed, the resulting nutrient molecules are absorbed by the mucosal cells and transferred to the blood and lymph systems. Water-soluble nutrients, such as monosaccharides, glycerol, short-chain fatty acids, amino acids, and minerals, enter directly into the bloodstream. Fat-soluble nutrients, such as long-chain fatty acids and monoacylglycerols, first enter the lymph fluid and then enter the bloodstream where the two fluids come together.

An important factor in the absorption process is that the membranes of the intestine are selectively permeable; that is, they prevent the passage of most large molecules, but allow the passage of smaller molecules. For example, polysaccharides, disaccharides, and proteins are not ordinarily absorbed, but, generally, monosaccharides and amino acids are.

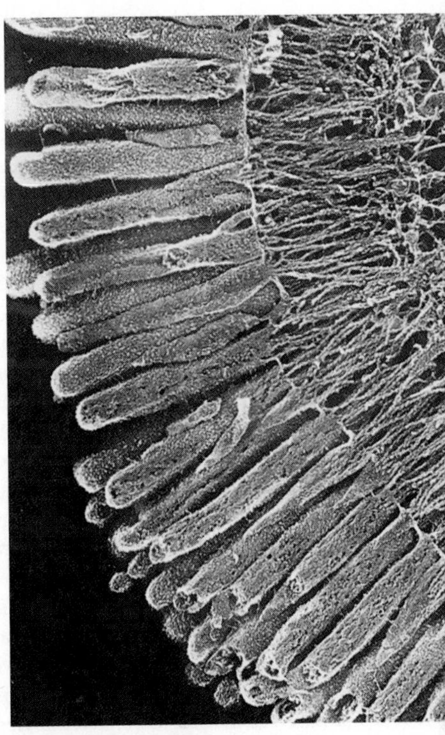

Absorption of digested food occurs in the small intestine through the villi and microvilli shown here.

32.15 Liver Function

The liver is the largest organ in the body, and it performs several vital functions. Two of these functions are the regulation of the concentrations of organic nutrients in the blood and the removal of toxic substances from the blood.

The concentration of blood sugar (glucose) is controlled and maintained by processes that occur in the liver. After absorption, excess glucose and other monosaccharides are removed from the blood and converted to glycogen in the liver. The liver is the principal storage organ for glycogen. As glucose is used in other cells, the stored liver glycogen is gradually hydrolyzed to maintain the appropriate blood-glucose concentration. Liver function is under sensitive hormonal control, as are most vital body functions. This regulation will be discussed further in Chapter 34.

A second major function of the liver is the detoxification of harmful and potentially harmful substances. This function apparently developed as higher vertebrates appeared in the evolutionary time scale. The liver is able to deal with most of the toxic molecules that occur in nature. For example, ethanol is

CHEMISTRY IN ACTION • Indigestion

It is in the digestive tract that we most closely interact with our environment. Foodstuffs are sorted to be used or discarded; chemicals are interconverted. The alimentary canal must act as a barrier, defend against invading organisms, and cause chemical changes that allow for nutrient absorption. When this process is not successful, trouble develops. This can be because our digestive system is not up to the challenge or has not adjusted to a change in diet. Sometimes, our diet remains the same, but the digestive system changes. Following are descriptions of three common problems caused by a misfit between the diet and the digestive tract.

Lactose intolerance is a common malady, afflicting approximately 55 million people in the United States. In such people, lactose is not broken into monosaccharides and therefore is not absorbed. Instead, lactose feeds the bacterial flora in the large intestine, leading to gas, diarrhea, bloating, and, perhaps, cramps. The disease represents a case of our body changing while our diet remains constant. At birth, mammals are capable of digesting the lactose from milk via the enzyme lactase. By the time most humans are in their teenage years, however, the amount of lactase enzyme they have is only 5% of that of an infant. Our bodies, then, adapt to a diet that, over the years, has lacked milk. Lactose intolerance can be treated simply by decreasing milk product consumption or by consuming milk products that have been treated to decrease their lactose content (e.g., lactase-treated milk or yogurt).

Irritable bowel syndrome may arise because our digestive tracts have not evolved to match our eating habits. This disease is commonly started by an allergic reaction to gluten, a protein found in many grains. An allergy attack can cause great misery; irritable bowel syndrome affects about 30 million people in the United States alone. Scientists postulate that this allergy was not a problem for primitive humans, who were hunter–gatherers; a diet of meats and nuts would not contain gluten. When the human diet began to depend on cultivated crops, especially wheat and its relatives, our intake of gluten increased. People suffering from irritable bowel syndrome can control their problems by moving to a diet that does not contain gluten, a simple change in behavior that, unfortunately, is difficult to stick to. Foods must be chosen carefully, and when starch is used, a corn, potato, or tapioca base should be substituted for wheat.

Diverticulitis is another digestive problem that arises as our digestive system begins to wear out. In this case, the intestinal walls weaken over time. Roughly half of all people over 70 suffer from this digestive change. As the intestinal walls weaken, small pockets form that can trap partly digested food. In turn, this trapped food can become infected, leading to diverticuloses. A change in diet can prevent these attacks of infection. By ensuring that our diet contains a high fiber component, the digesting food retains water and remains relatively soft as it passes through the alimentary canal. Because it is soft, the food will be less likely to enter these small intestinal pockets and lead to infection.

The digestive tract is similar to other bodily systems. We try to avoid too much sun so that our skin is not damaged; we choose the right type and intensity of exercise so that our muscles will remain healthy. If we understand the limitations and capabilities of the digestive system, our food will be digested much more smoothly.

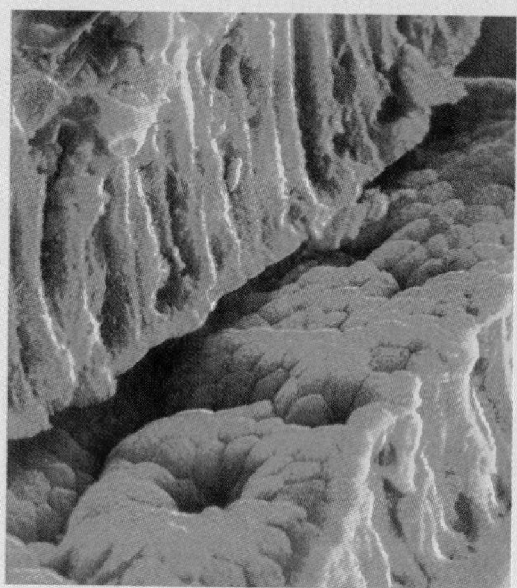

The absorbing surface of the small intestine.

oxidized in the liver, and nitrogenous metabolic waste products are converted to urea for excretion.

Organic chemists have learned to synthesize new substances that have no counterparts in the biological world. These potentially toxic substances are used as industrial chemicals, insecticides, drugs, and food additives. When they are ingested, even in small amounts, the body is faced with the difficult challenge of metabolizing or destroying substances that are unlike any found in nature.

The liver meets this challenge and deals with most of these foreign molecules through an oxidation system located in the endoplasmic reticulum. Bound to these intracellular membranes are enzymes that catalyze reactions between oxygen and the foreign molecules. As the latter molecules are oxidized, they become more polar and water soluble. Finally, the oxidation products of the potential toxins are excreted in the urine or bile fluid.

For example, most automobile antifreeze solutions contain ethylene glycol, a toxic substance. Even though this compound does not occur naturally, the liver can metabolize ethylene glycol. When small amounts are ingested, the following chemical changes occur:

```
 OH OH                      OH O                     OH O
 |  |                       |  ||                    |  ||
H—C—C—H   —oxidation→    H—C—C—H   —oxidation→    H—C—C—O⁻   —oxidation→
 |  |                       |                        |
 H  H                       H                        H

ethylene                  glyoxal                  glycolate
 glycol
```

```
   O  O                          O                 O
   ||  ||                        ||                ||
H—C—C—O⁻   —oxidation→      H—C—O⁻    +    HO—C—O⁻

 glyoxalate                  formate          bicarbonate
```

Thus, ethylene glycol is converted by oxidation to two more polar (charged) acid anions that are easily eliminated from the body.

Unfortunately, oxidation is not effective with some compounds. Halogenated hydrocarbons, which are particularly inert to oxidation, accumulate in fatty tissue or in the liver itself. Examples of halogenated hydrocarbons are carbon tetrachloride, hexachlorobenzene, DDT, dioxins, and polychlorinated biphenyls (PCBs). Some compounds become more toxic after oxidation. For example, polycyclic hydrocarbons (which can be formed when food is barbecued) become carcinogenic upon partial oxidation. Methanol becomes particularly toxic because it is converted by oxidation to formaldehyde. One of the most serious dangers of environmental pollution lies in the introduction of compounds that the liver cannot detoxify.

In a sense, the liver is the final guardian along the pathway by which nutrients pass to the cells. This pathway starts with a balanced diet. Once foods are digested and absorbed, the liver adjusts nutrient levels and removes potential toxins. The blood can then provide nutrients for the cellular biochemistry that constitutes life.

Chapter 32 Review

32.1 Nutrients

KEY TERMS

Nutrient
Macronutrient
Micronutrient

- Nutrients are components of the food we eat that make body growth, maintenance, and repair possible.
- Nutrients, when digested and absorbed, provide the building blocks and energy for life.
- Carbohydrates, lipids, and proteins are classified as macronutrients because they are used in relatively large amounts by cells.
- Vitamins and minerals are used in relatively small amounts and are classed as micronutrients.
- Water is a special nutrient in that it serves as the solvent for life.

32.2 Diet

KEY TERMS

Diet
Dietary Reference Intake (DRI)
Recommended Dietary Allowance (RDA)

- A diet is the food and drink that we consume.
- There is no universal minimum quantity of each nutrient needed by every person.

- The Dietary Reference Intake (DRI) takes into account the many different types of nutrient data needed to judge a healthful diet. It is based generally on either a Recommended Dietary Allowance or an Adequate Intake.
- The Recommended Dietary Allowance (RDA) is defined as the daily dietary intake that is sufficient to meet the requirements of nearly all individuals in a specific age and gender group.
- The Adequate Intake (AI) is an experimentally determined, average requirement.

32.3 Energy in the Diet

KEY TERM

Estimated Energy Requirement (EER)

- The Estimated Energy Requirement (EER) is the average dietary intake that is required to maintain energy balance for good health.
- Nutrients that are a rich source of reduced carbons (carbohydrates, lipids, and proteins) provide most of the EER.
- Nutritionists use the large Calorie (Cal) to measure dietary energy; one Cal is equal to one kcal.
- Marasmus occurs when a diet is energy deficient.
- Overweight and obesity result from an excess of dietary calories.

32.4 Carbohydrates in the Diet

- Carbohydrates are the major nutrient in most human diets based on mass.
- Complex carbohydrates are slowly digested if they are digested at all.
  - Starch releases glucose slowly into the blood.
  - Cellulose is not digested but instead provides dietary fiber.
- Consumption of large amounts of simple sugars such as sucrose can cause serious health problems.

32.5 Fats in the Diet

- Fat is a more concentrated source of dietary energy than carbohydrate.
- Fats that contain polyunsaturated fatty acids are more healthful than those that contain saturated fatty acids.
- The essential fatty acids provide a source of omega-3 and omega-6 fatty acids; these fatty acids are required for good health. The essential fatty acids are linoleic, linolenic and arachidonic acids.

32.6 Proteins in the Diet

KEY TERM

Complete protein

- Protein is an energy-providing nutrient that is also the primary source of nitrogen in the diet.
- Kwashiorkor, protein malnutrition, is an especially serious problem in underdeveloped countries.
- Proteins also serve as a source of amino acids that humans cannot synthesize, the essential amino acids.
- A complete protein provides all the essential amino acids.

32.7 Vitamins

KEY TERM

Vitamin

- Vitamins are a group of naturally occurring organic compounds that are essential for good nutrition and that must be supplied in the diet.
- Vitamins can be classified according to their solubility: those that are fat soluble and those that are water soluble.
- A prolonged lack of vitamins in the diet leads to vitamin deficiency diseases. It is especially important that children receive sufficient vitamins for proper growth and development.
- Vitamins are classed as micronutrients.

32.8 Minerals

- Minerals are micronutrients that come from inorganic sources.
  - Major elements include sodium, potassium, calcium, magnesium, chloride, and phosphorus and are needed in relatively larger amounts.
  - Trace elements are needed in relatively small amounts and include metal cations derived from elements such as iron, zinc, and copper.

32.9 Water

- Water is the solvent of life. It constitutes about two-thirds of the human body mass.
- A proper water balance requires that water intake compensate for a variety of water loses.

32.10 Nutrition Content Labeling

- Many food packages contain a Nutrition Facts label.
- Nutrition Facts report the nutrients contained in each food product.
  - Values reference a 2000 or 2500 Cal/ day diet.
  - Serving size is chosen based on U.S. federal guidelines.
  - The % Daily Value compares the nutrient amount in a serving size of each food with the DRI for that nutrient.

32.11 Food Additives

- Food additives are chemicals that are added to foods during processing.
- Over 3000 additives are given the "generally recognized as safe" (GRAS) rating by the U.S. Food and Drug Administration.
- Food additives serve a variety of functions: antimicrobials, antioxidants, anticaking agents, emulsifiers, thickeners, flavor enhancers.

32.12 A Balanced Diet

- A diet that supplies all the necessary nutrients is said to be "balanced."

- Nutritionists recommend that a healthful diet contain a mixture of foods from six groups: (1) milk products; (2) vegetables; (3) fruits; (4) grains; (5) meats and beans; (6) oils.

32.13 Human Digestion

KEY TERM

Digestion

- Digestion is a series of enzyme-catalyzed reactions by which large molecules are hydrolyzed to molecules small enough to be absorbed through the intestinal membranes.
- There are five principal digestive juices (or fluids) that enter the digestive tract at various points:
  - Saliva in the mouth
  - Gastric juice in the stomach
  - Pancreatic juice in the duodenum of the small intestine
  - Bile in the duodenum of the small intestine
  - Intestinal juice in the small intestine
- Salivary digestion hydrolyzes starch to maltose, lubricates food, and breaks large food pieces into smaller ones.
- Gastric digestion hydrolyzes proteins to proteoses and peptones, and it reduces the food pieces to even smaller sizes.
- Intestinal digestion is the most extensive digestion; enzymes digest carbohydrates, fats, and proteins.
- Bile salts are made by the liver and aid in fat digestion.

32.14 Absorption

KEY TERM

Absorption

- Absorption is the process by which digested foods pass through the membrane linings of the small intestine and enter the blood and lymph.

32.15 Liver Function

- Two important functions for the liver are the regulation of organic nutrient concentrations in the blood and the removal of toxic substances from the blood.
- Many molecules that are foreign to the body are first oxidized in the liver before they are eliminated.
  - Oxidation makes molecules more polar and/or charged and, thus, more water soluble.
  - Unfortunately, some foreign molecules become more toxic upon oxidation.

Review Questions

All questions with blue numbers have answers in the appendix of the text.

1. How does a food differ from a nutrient?
2. What is meant by the term *energy allowance*?
3. How does marasmus differ from kwashiorkor?
4. What is meant by the statement that candy provides "empty calories?"
5. What structural features do the essential fatty acids have in common?
6. What is meant by the term *essential fatty acid*?
7. List the essential amino acids.
8. How do animal proteins differ nutritionally from vegetable proteins?
9. Which vitamins are water soluble? Which are fat soluble?
10. List three functions that can be attributed to vitamins.
11. Distinguish the major elements from the trace elements.
12. List two major biological functions of calcium.
13. What percentage of the average person's water consumption comes from solid foods? How much from liquids?
14. List five common categories of food additives.
15. What are the five principal digestive juices?
16. Which federal agency is responsible for regulating the use of food additives?
17. What enzymes are in each of the digestive juices?
18. What is chyme?
19. What is the digestive function of the liver?
20. How do the intestinal mucosal cells aid digestion?
21. How does the liver metabolize toxic compounds?
22. What is the meaning of "GRAS"?
23. Describe the common digestive disorder lactose intolerance.
24. What component of the diet is responsible for irritable bowel syndrome?
25. How do vitamins differ from minerals?
26. Which food group should be the largest proportion of the diet? Which food group should be the smallest proportion of the diet?

Paired Exercises

All exercises with blue numbers have answers in the appendix of the text.

1. A sirloin steak (85 g) has the following nutritional composition:

Protein	23	g
Carbohydrates	0	
Fats	19	g
Sodium	52	mg
Potassium	297	mg
Magnesium	23	mg
Iron	2.5	mg
Zinc	4.7	mg
Calcium	9.0	mg
Vitamin A	15	μg
Vitamin C	0	
Thiamin	92	μg
Riboflavin	218	μg
Niacin	3.2	mg
Vitamin B_6	330	μg
Folic acid	7	μg
Vitamin B_{12}	2	μg

(a) What is the mass of macronutrients contained in this food?
(b) What is the mass of minerals?
(c) What is the mass of vitamins?

2. A battered and fried chicken breast (280 g) has the following nutritional composition:

Protein	70	g
Carbohydrates	25	g
Fats	37	g
Sodium	770	mg
Potassium	564	mg
Magnesium	68	mg
Iron	3.5	mg
Zinc	2.7	mg
Calcium	56	mg
Vitamin A	56.5	μg
Vitamin C	0	
Thiamin	322	μg
Riboflavin	408	μg
Niacin	29.5	mg
Vitamin B_6	1.2	mg
Folic acid	16	μg
Vitamin B_{12}	0.82	μg

(a) What is the mass of macronutrients contained in this food?
(b) What is the mass of minerals?
(c) What is the mass of vitamins?

3. Fat yields about 9 kcal/g.
(a) From Exercise 1, find how much energy is available from fat in sirloin steak.
(b) If the total energy content is 270 kcal, what percentage of energy is derived from the fat?

4. Fat yields about 9 kcal/g.
(a) From Exercise 2, find how much energy is available from fat in chicken breast.
(b) If the total energy content is 728 kcal, what percentage of energy is derived from the fat?

5. For 14- to 18-year-old women, the DRI for calcium is 1.3 g. What percentage of the DRI is supplied by the sirloin steak described in Exercise 1?

6. For women 50 years or older, the DRI for calcium is 1.2 g. What percentage of the DRI is supplied by the chicken breast described in Exercise 2?

7. The DRI for vitamin B_{12} is 2.0 μg for adults. What percentage of the DRI is supplied by the sirloin steak described in Exercise 1?

8. The DRI for vitamin B_{12} is 2.0 μg for adults. What percentage of the DRI is supplied by the chicken breast described in Exercise 2?

9. List three classes of nutrients that are considered macronutrients.

10. List two classes of nutrients that are considered micronutrients.

11. Rank order the vitamins in steak from the one that is in the largest amount to the one in the smallest amount (using the information in Exercise 1).

12. Rank order the minerals in chicken breast (using the information in Exercise 2) from the one that is in the largest amount to the one in the smallest amount.

13. List three important reasons to chew your food well as an aid to digestion.

14. How does the stomach aid digestion of proteins?

15. List three classes of nutrients that do not commonly supply energy for the cell.

16. List three classes of nutrients that commonly supply energy for the cell.

17. Why is starch important in the diet?

18. Why is cellulose important in the diet?

19. Is saliva slightly acidic, slightly basic, or neutral?

20. Is pancreatic juice slightly acidic, slightly basic, or neutral?

21. In what parts of the digestive system are proteins digested?

22. In what parts of the digestive system are carbohydrates digested?

23. Pancreatic amylase is a digestive enzyme found in the small intestine. What nutrient class does this enzyme digest?

24. Pepsin is a digestive enzyme found in the stomach. What nutrient class does this enzyme digest?

25. According to the food pyramid guide, which food group should supply the majority of our diet?

26. According to the food pyramid guide, which food group should be the second most abundant in our diet?

27. Sailors were afflicted with scurvy on long voyages when fresh fruits were not available in their diets. What vitamin were these sailors missing?

28. When rice was first "polished" to remove the brown outer coat, a major outbreak of beriberi occurred. What vitamin was lost in the rice "polishing"?

Additional Exercises

All exercises with blue numbers have answers in the appendix of the text.

29. Milk is a food. List four nutrients that can be obtained from milk.

30. You are told that a new diet will cause you to lose 9 kg (20 lb) of fat in 1 week. Is this reasonable? Explain briefly.

31. In what ways are the trace elements similar to vitamins? How do they differ?

32. Explain why a tablespoon of butter (a fatty food) approximately doubles the calorie content of a medium-size baked potato (a carbohydrate food).

33. Why must the food of higher animals be digested before it can be utilized?

34. Galactosemia is an inherited condition in some babies. It is a deficiency in the enzyme that catalyzes the conversion of galactose to glucose. Symptoms are lack of appetite, weight loss, diarrhea, and jaundice. What staple of a baby's diet must be changed to correct this problem?

35. One large egg contains 270 mg of cholesterol. How does this compare with the daily recommended amount of cholesterol? (See the Nutrition Facts panel in Figure 32.3.)

Challenge Exercise

36. Given the Nutrition Facts panel in the adjacent column from cream-of-chicken soup,
   (a) What percentage of total energy is provided by the fat?
   (b) How many grams of protein will be provided by two servings of this food?
   (c) For a person on a reduced-calorie diet (1000 Cal/day), what percentage of the total energy is supplied by one serving?
   (d) What percentage of the total carbohydrates is fiber?

NUTRITION FACTS Serv. Size 1/2 cup (125g) Servings about 2 1/2	
Amount Per Serving	
Calories 100	Fat Cal 50
	% Daily Value*
Total Fat 5g	8%
Saturated fat 2g	9%
Cholesterol 10mg	3%
Sodium 860mg	36%
Total carbohydrate 10g	3%
Fiber 2g	9%
Sugars 1g	
Protein 3g	
Vitamin A 10% • Vitamin C 0%	
Calcium 0% • Iron 2%	
*Percent Daily Values are based on a 2,000 calorie diet.	

Answers to Practice Exercises

32.1 (a) A young woman's iron

$$\text{DRI} = (15\text{ mg})\left(\frac{1\text{ g}}{1000\text{ mg}}\right) = 1.5 \times 10^{-2}\text{g}$$

(b) Vitamin B_{12} has the smallest DRI.

32.2 Vitamin D acts as a regulator for calcium metabolism, and so the Vitamin D DRI is high when the calcium DRI is also high.

32.3 Fiber daily value: $\dfrac{6\text{ g} \times 100\%}{26\%} = X\text{ g}$

Fiber daily value = 20 g

32.4 Glucose

CHAPTER 33

Bioenergetics

Finding adequate sources of energy is a constant challenge for all living organisms, including this bear.

Chapter Outline

If you travel through the midwestern United States on a warm night in early summer, you may be lucky enough to see small yellow lights dancing through the fields. This beautiful sight is evidence that the fireflies, or lightning bugs, have returned. Their flickering light does not arise from a small fire or electrical discharge, but rather from a chemical reaction. Bioenergetics are at work there.

The monarch butterfly is a striking insect with a wingspan of several inches. As it flies, it is buffeted by the wind, first in one direction and then in another. Clearly, the monarch is light and not a powerful flier. Yet, each year (in one leg of a complex, multigeneration migration), these small creatures fly from southern Canada to winter in central Mexico, a distance of over 2000 miles. Chemical reactions in the insects' cells make energy available for this arduous trip. Bioenergetics are at work again.

While you have been reading these paragraphs, fully 20% of the oxygen you breathed has been used by your brain. Chemical processes have reacted this oxygen with blood glucose, yielding the energy needed to comprehend your reading. This is another example of bioenergetics—the chemical processes directly related to energy needs and uses in life.

33.1 Energy Changes in Living Organisms

bioenergetics

One of the basic requirements for life is a source of energy. **Bioenergetics** is the study of the transformation, distribution, and utilization of energy by living organisms. Bioenergetics includes the radiant energy of sunlight used in photosynthesis, electrical energy of nerve impulses, mechanical energy of muscle contractions, heat energy liberated by chemical reactions within cells, and potential energy stored in "energy-rich" chemical bonds. The major source of biological energy is the chemical reactions occurring inside cells.

The bioenergetics of a cell can be compared to the energetics of a manufacturing plant. (See Figure 33.1.) First, energy is delivered to the cell or plant.

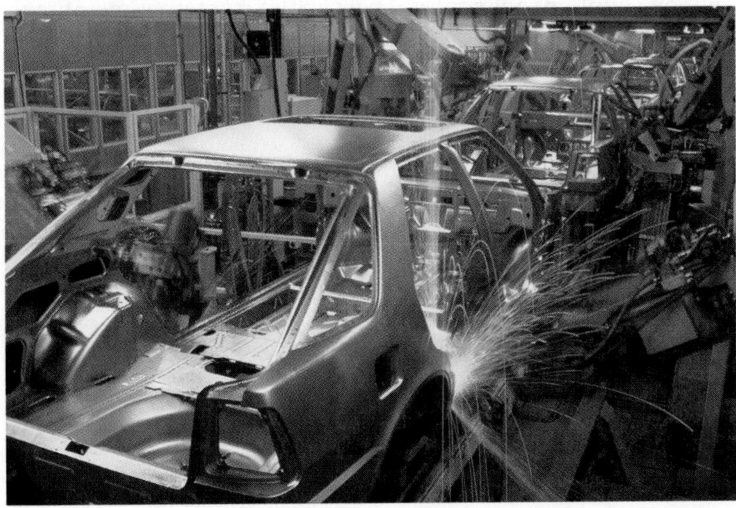

Figure 33.1
A typical manufacturing plant converts electrical energy to mechanical energy and uses this energy to change raw materials to finished products, which is illustrated here as robots weld the frame of a car on an assembly line.

Table 33.1 Delivery, Conversion, and Use of Energy in Energetics

Step	Manufacturing plant energetics	Animal cell bioenergetics	Plant cell bioenergetics
1. Energy delivery	Energy is delivered as electricity, coal, oil, gas, etc.	Energy is delivered as reduced carbon atoms.	Energy is delivered as sunlight.
2. Energy conversion	Electricity, coal, oil, gas, etc., is converted to mechanical energy.	Energy in reduced carbon atoms is converted to high-energy phosphate bonds.	Sunlight is converted to chemical energy.
3. Work done	Mechanical energy converts raw materials into products.	High-energy phosphate bonds are used to do the work of the cell.	High-energy phosphate bonds are used to do the work of the cell.

Second, this energy is commonly converted to a more usable form. Third, work is done. (See Table 33.1.)

33.2 Metabolism and Cell Structure

The sum of all chemical reactions that occur within a living organism is defined as **metabolism**. Many hundreds of different chemical reactions occur in a typical cell. To make sense of this myriad of reactions, biochemists subdivide metabolism into two contrasting categories: *anabolism* and *catabolism*. **Anabolism** is the process by which simple substances are synthesized (built up) into complex substances. **Catabolism** is the process by which complex substances are broken down into simpler substances. Anabolic reactions usually involve carbon reduction and consume cellular energy, whereas catabolic reactions usually involve carbon oxidation and produce energy for the cell.

metabolism **anabolism** **catabolism**

Cells segregate many of their metabolic reactions into specific, subcellular locations. The simple **procaryotes**—cells without internal membrane-bound bodies—have a minimum amount of spatial organization. (See Figure 33.2.) The anabolic processes of DNA and RNA synthesis in these cells are localized in the nuclear material, whereas most other metabolic reactions are spread throughout the cytoplasm.

procaryote

In contrast, metabolic reactions in the cells of higher plants and animals are often segregated into specialized compartments. These cells, the **eucaryotes**, contain internal, membrane-bound bodies called **organelles**. (See Figure 33.2.) It is within the organelles that many specific metabolic processes occur.

eucaryote **organelle**

In the eucaryotic cell, most of the DNA and RNA syntheses are localized in the nucleus. Anabolism of proteins takes place in the ribosomes, whereas that of carbohydrates and lipids occurs primarily in the cytoplasm.

There are a variety of specialized catabolic organelles within a eucaryotic cell. The lysosome contains the cell's digestive enzymes, and the peroxisome

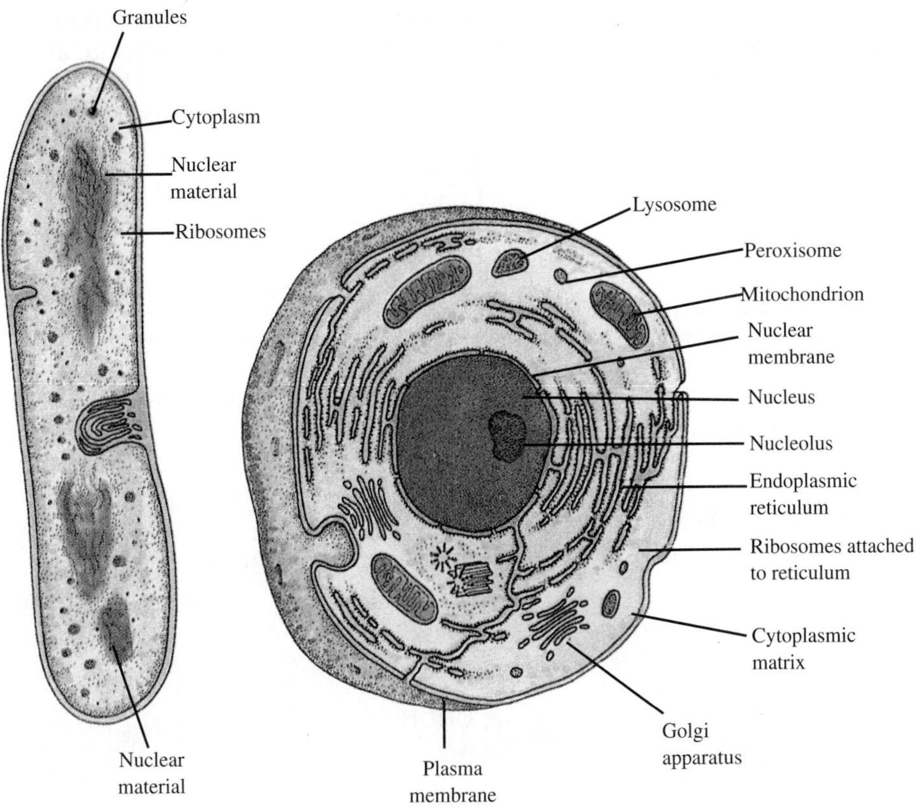

Figure 33.2
A procaryotic and a eucaryotic cell. The procaryote lacks much of the organized structure and the numerous organelles found in the eucaryote.

is the site of oxidative reactions that form hydrogen peroxide. Perhaps the most important catabolic organelle is the mitochondrion (plural, mitochondria). This membrane-bound body provides most of the energy for a typical cell. The energy is released by catabolic processes, which oxidize carbon-containing nutrients. Mitochondria consume most of the O_2 that is inhaled and produce most of the CO_2 that is exhaled by the lungs.

33.3 Biological Oxidation–Reduction: Energy Delivery

The ultimate source of biological energy on earth is sunlight. Plants capture light energy and transform it to chemical energy by a process called *photosynthesis*. (See Section 33.7.) This chemical energy is stored in the form of reduced carbon atoms in carbohydrate molecules. It is important to understand that the energy contained in carbohydrates, lipids, and proteins originally came from sunlight.

Humans, as well as other animals, draw most of their energy from foodstuffs that contain reduced carbons. (See Chapter 32.) For example, after we eat a meal, nutrients such as carbohydrates and fats are metabolized and transported to our cells. The carbons in these compounds are in a reduced state and thus contain stored energy. Typical structures are as follows:

```
     H                 OH
     |                 |
   —C—               —C—                O=C=O
     |                 |
     H                 H
```

typical fatty acid carbon (oxidation number = −2)

typical carbohydrate carbon (oxidation number = 0)

carbon dioxide carbon (oxidation number = +4)

Through the cell's metabolism, these reduced carbons are oxidized, step by step, and are eventually converted to carbon dioxide.

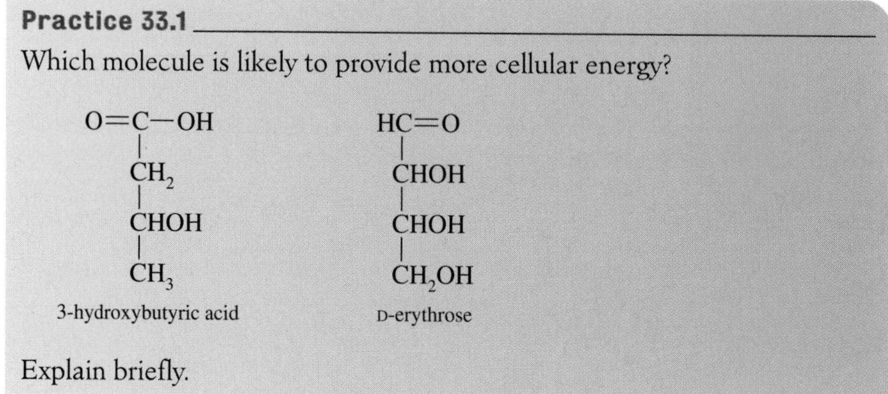

Practice 33.1

Which molecule is likely to provide more cellular energy?

```
O=C—OH            HC=O
  |                 |
  CH2               CHOH
  |                 |
  CHOH              CHOH
  |                 |
  CH3               CH2OH
```

3-hydroxybutyric acid

D-erythrose

Explain briefly.

Figure 33.3 is an energy diagram summarizing the energy flow through metabolism. Such a diagram will be used several times in this chapter. The black arrows trace progress from one chemical to the next. Because

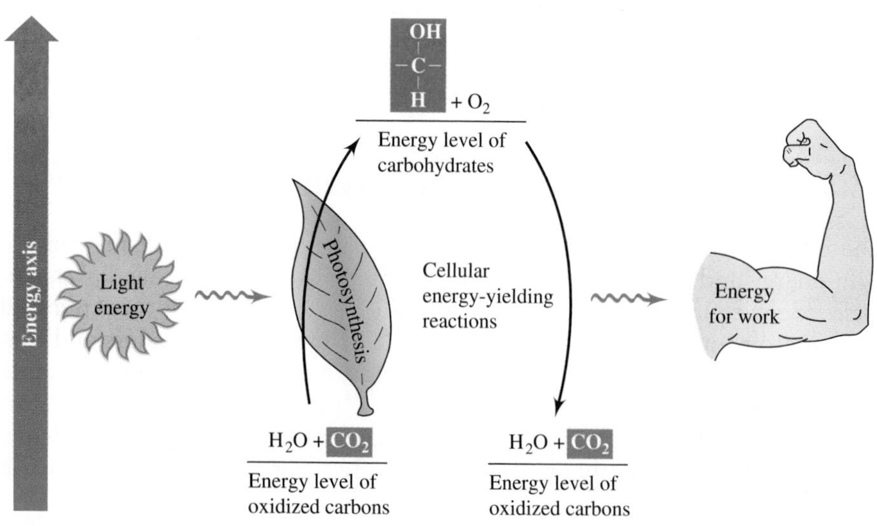

Figure 33.3
Energy flow through metabolism, using the important energy nutrients, carbohydrates. Reduction of carbon stores energy, while oxidation of carbon releases energy.

each chemical has its own special energy level, it is presented at a particular height with respect to the energy axis (the higher the level, the more energy the carbon atoms contain). When energy levels change, energy must be released or absorbed, as shown by the red arrows. Thus, in Figure 33.3, photosynthesis causes carbons to move to a higher energy level, as they are reduced to carbohydrate carbons. The carbohydrate carbons are then oxidized to a lower energy level with the release of energy to do work.

In eucaryotic cells, specific organelles are present that specialize in redox reactions. The *mitochondria* (Figure 33.4), often called the "powerhouses" of the cell, are the sites for most of the catabolic redox reactions. *Chloroplasts* (Figure 33.4) are organelles found in higher plants and contain an electron-transport system that is responsible for the anabolic redox reactions in photosynthesis. (See Section 33.7.)

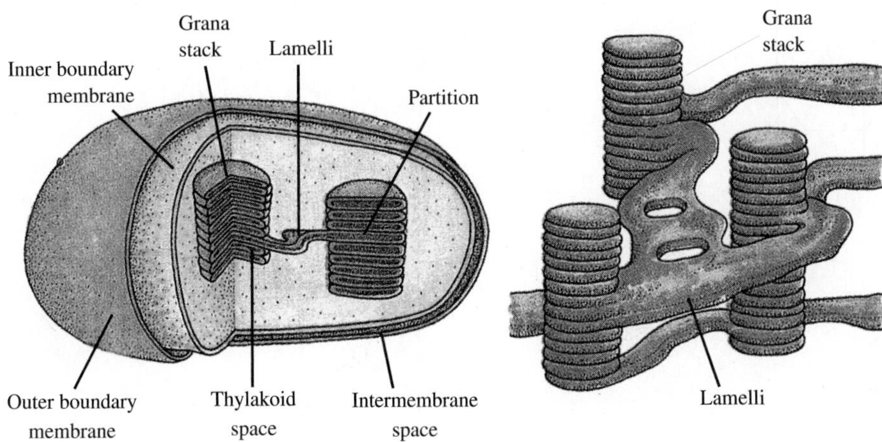

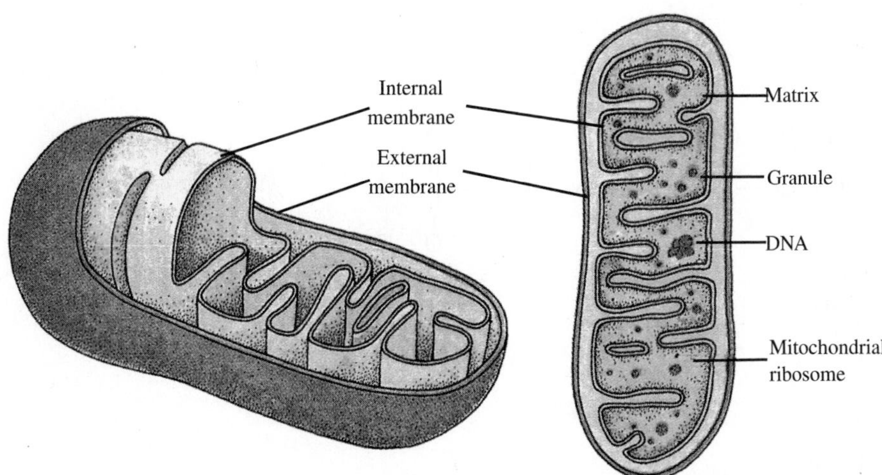

Figure 33.4
(a) A chloroplast and (b) a mitochondrion.

Recall from Section 30.1 that a coenzyme is an organic compound that is used and reused to help an enzyme-catalyzed reaction.

To move electrons from one place to another (often outside of the mitochondrion or chloroplast), the cell uses a set of redox coenzymes. The redox coenzymes act as temporary storage places for electrons. The three most common redox coenzymes, shown in Figure 33.5, are nicotinamide adenine dinucleotide, NAD^+; nicotinamide adenine dinucleotide phosphate, $NADP^+$; and flavin adenine dinucleotide, FAD. Humans synthesize NAD^+ and $NADP^+$ from the vitamin niacin, while FAD is made from the vitamin riboflavin. In each case, the vitamin provides the reaction center of the coenzyme.

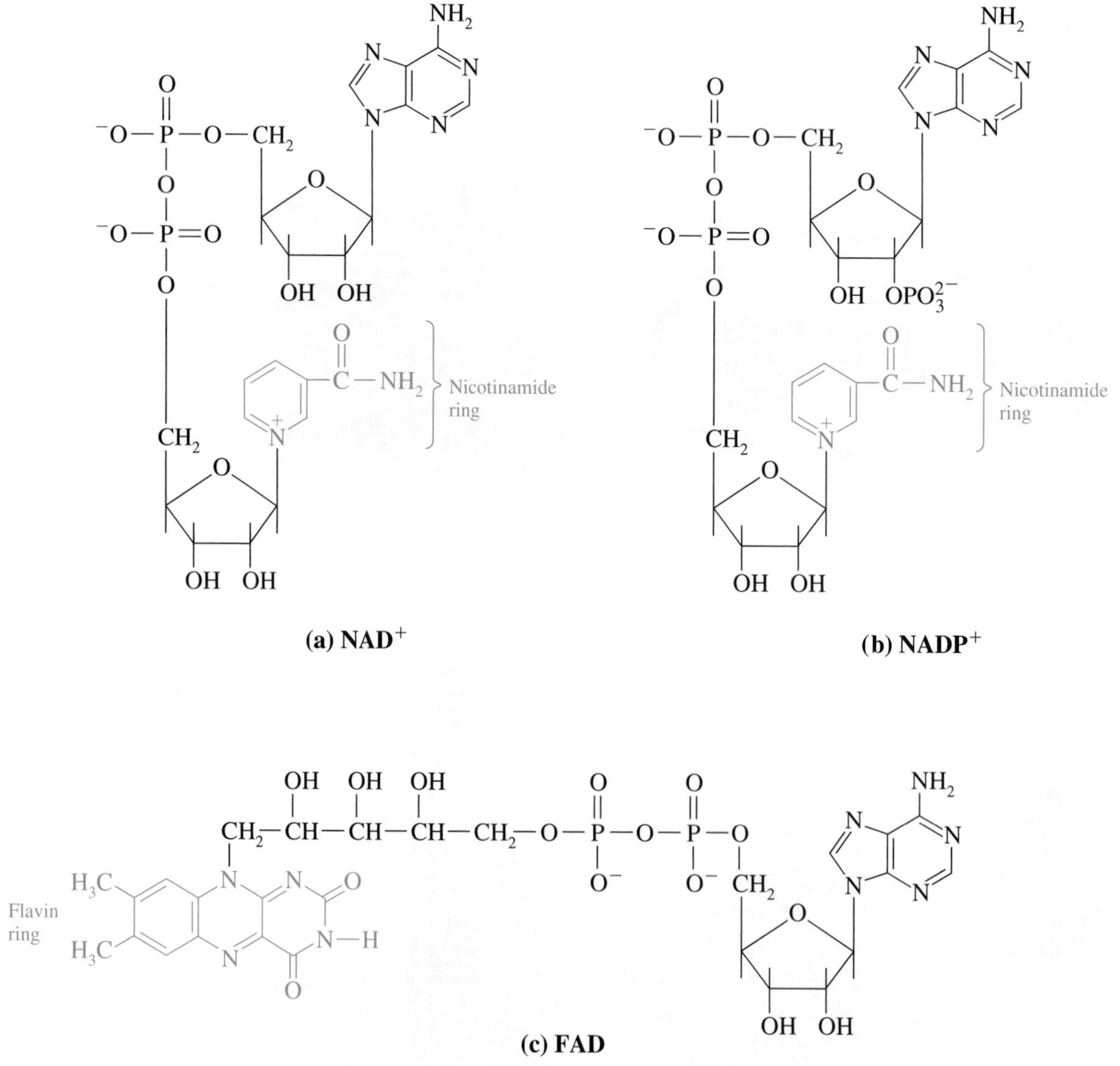

(a) NAD^+ (b) $NADP^+$ (c) FAD

Figure 33.5
Structures of the redox coenzymes (a) nicotinamide adenine dinucleotide (NAD^+), (b) nicotinamide adenine dinucleotide phosphate ($NADP^+$), and (c) flavin adenine dinucleotide (FAD).

The addition or removal of electrons occurs in only one portion of each of these complex molecules. The nicotinamide ring is the reactive component within NAD^+ or $NADP^+$:

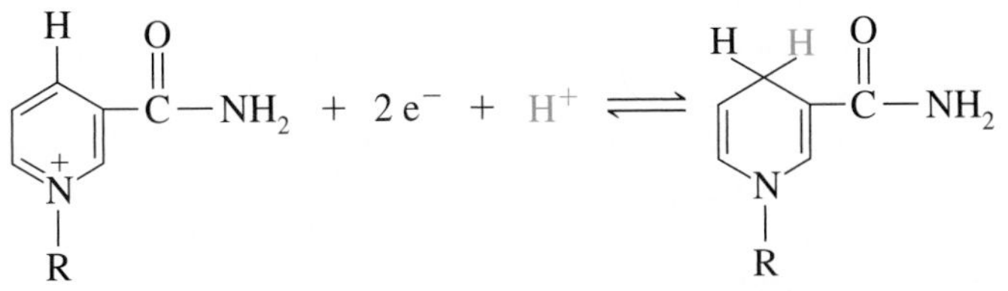

$$\text{NAD}^+ \text{ or NADP}^+ \text{ (oxidized form)} + 2\,e^- + H^+ \rightleftharpoons \text{NADH or NADPH (reduced form)}$$

NAD^+ or $NADP^+$
(oxidized form)

NADH or NADPH
(reduced form)

(R represents the remainder of each molecule.)

For FAD, the flavin ring is the reactive component:

$$\text{FAD (oxidized form)} + 2\,e^- + 2\,H^+ \rightleftharpoons \text{FADH}_2 \text{ (reduced form)}$$

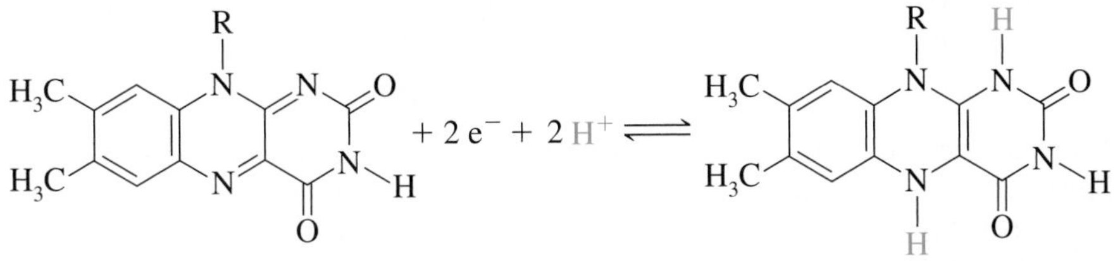

FAD
(oxidized form)

$FADH_2$
(reduced form)

(R represents the remainder of each molecule.)

A very important function of these redox coenzymes is to carry electrons to the mitochondrial electron-transport system. As the coenzymes are oxidized, molecular oxygen is reduced:

$$\underset{\text{reduced coenzyme}}{2\ FADH_2} + O_2 \xrightarrow[\text{electron transport}]{\text{mitochondrial}} \underset{\text{oxidized coenzyme}}{2\ FAD} + 2\ H_2O$$

$$\underset{\text{reduced coenzyme}}{2\ NADH} + 2\ H^+ + O_2 \xrightarrow[\text{electron transport}]{\text{mitochondrial}} \underset{\text{oxidized coenzyme}}{2\ NAD^+} + 2\ H_2O$$

With the movement of electrons, energy is released. In fact, over 85% of a typical cell's energy is derived from this redox process. However, the released energy is not used immediately by the cell, but is instead stored, usually in high-energy phosphate bonds such as those in ATP.

Remember, the mitochondria are the powerhouses of the cell!

Practice 33.2

What is wrong with the following equation for biological redox?

$$CH_3{-}OH + FADH_2 \rightarrow CH_2{=}O + FAD$$

Explain briefly.

33.4 Molecular Oxygen and Metabolism

Although a discussion of metabolism justifiably focuses on carbon, molecular oxygen also plays a critical role in energy production. This diatomic molecule acts as the final receptacle for electrons in the mitochondrial electron-transport system. Aerobic metabolism (i.e., metabolism in the presence of molecular oxygen) is the best way to produce energy for most cells. However, like a very potent drug, oxygen can also be dangerous to life.

Mitochondria have become experts at handling this dangerous chemical. The chemical reactions inside these organelles are specifically designed to carry out a four-electron redox reaction with diatomic oxygen:

$$O_2 + 4e^- + 4H^+ \longrightarrow 2H_2O$$

Water is a *very* safe product.

Other reduced products of O_2 are dangerous. They are known as reactive oxygen species (ROS), and they can react with and destroy many vital cell molecules. Sometimes a two-electron redox reaction occurs and makes hydrogen peroxide:

$$O_2 + 2e^- + 2H^+ \longrightarrow H_2O_2$$

More dangerous ROS are formed if redox reactions involve only one electron. *Superoxide* is formed when molecular oxygen is reduced with one electron, O_2^-. The most dangerous ROS is the hydroxyl radical, a neutral OH that can react with almost anything in the cell.

Although ROS form in small amounts wherever aerobic metabolism occurs, large bursts of ROS contribute to the destructiveness of many diseases. Inflammation is associated with ROS toxicity. When skin is exposed to UV wavelengths in sunlight, ROS production can lead to cancer. Cell destruction in stroke and heart attack is closely associated with ROS formation. A blood clot first limits the oxygen supply (ischemia), weakening cells. Then, when the blood supply is restored (reperfusion), the cells can't handle the aerobic environment, and they experience a destructive burst of ROS.

Since cells have to live with the danger of ROS at all times, they have developed some defense mechanisms. Perhaps the two most important protective tools are specific enzymes. Most cells carry an enzyme, superoxide dismutase, that destroys two superoxides by making only one hydrogen peroxide:

$$2O_2^- + 2H^+ \longrightarrow H_2O_2 + O_2$$

Then, a second enzyme, catalase, can convert the hydrogen peroxide into water:

$$2H_2O_2 \longrightarrow O_2 + 2H_2O$$

These enzymes are effective under normal circumstances. However, oxidative damage can easily get out of control.

33.5 High-Energy Phosphate Bonds

Cells need an energy delivery system. Most cellular energy is produced in the mitochondria, but this energy must be transported throughout the cell. Such a delivery system must carry relatively large amounts of energy and be easily accessible to cellular reactions. Molecules that contain high-energy phosphate bonds meet this need.

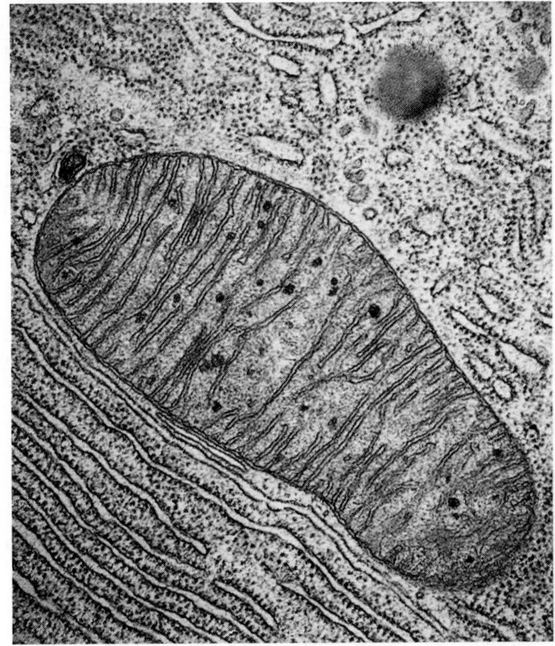

Photomicrograph of a mitochondrion.

The most common high-energy phosphate bond within the cell is the phosphate anhydride bond (or phosphoanhydride bond) (see also Section 31.4):

$$\underset{\text{phosphate}}{HO{-}\overset{O^-}{\underset{O}{\overset{|}{\underset{||}{P}}}}{-}OH} + \underset{\text{phosphate}}{HO{-}\overset{O^-}{\underset{O}{\overset{|}{\underset{||}{P}}}}{-}OR} + 35\ \text{kJ} \rightleftharpoons \underset{\text{diphosphate anhydride}}{HO{-}\overset{O^-}{\underset{O}{\overset{|}{\underset{||}{P}}}}{\sim}O{-}\overset{O^-}{\underset{O}{\overset{|}{\underset{||}{P}}}}{-}O{-}R} + H_2O$$

Phosphate anhydride bond

(R represents the remainder of each molecule.)

A relatively large amount of energy is required to bond the two negatively charged phosphate groups. The repulsion between these phosphates causes the phosphate anhydride bond to behave somewhat like a coiled spring. When the bond is broken, the phosphates separate rapidly, and energy is released.

The phosphate anhydride bond is an important component of the nucleotide triphosphates, the most important of which is adenosine triphosphate (ATP):

$$\underbrace{HO{-}\overset{O^-}{\underset{O}{\overset{|}{\underset{||}{P}}}}{\sim}O{-}\overset{O^-}{\underset{O}{\overset{|}{\underset{||}{P}}}}{\sim}O{-}\overset{O^-}{\underset{O}{\overset{|}{\underset{||}{P}}}}}_{\text{triphosphate}}{-}OCH_2\text{–ribose (OH, OH)–adenine }(NH_2)$$

triphosphate

adenine

ribose

adenosine triphosphate (ATP)

Adenosine triphosphate plays an important role in all cells, from the simplest to the most complex. It functions by storing and transporting the energy in its high-energy phosphate bonds to the places in the cell where energy is needed. ATP is the common intermediary in energy metabolism.

The cell realizes several advantages by storing energy in ATP. First, the stored energy is easily accessible to the cell; it is readily released by a simple hydrolysis reaction yielding adenosine diphosphate (ADP) and an inorganic phosphate ion (P_i):

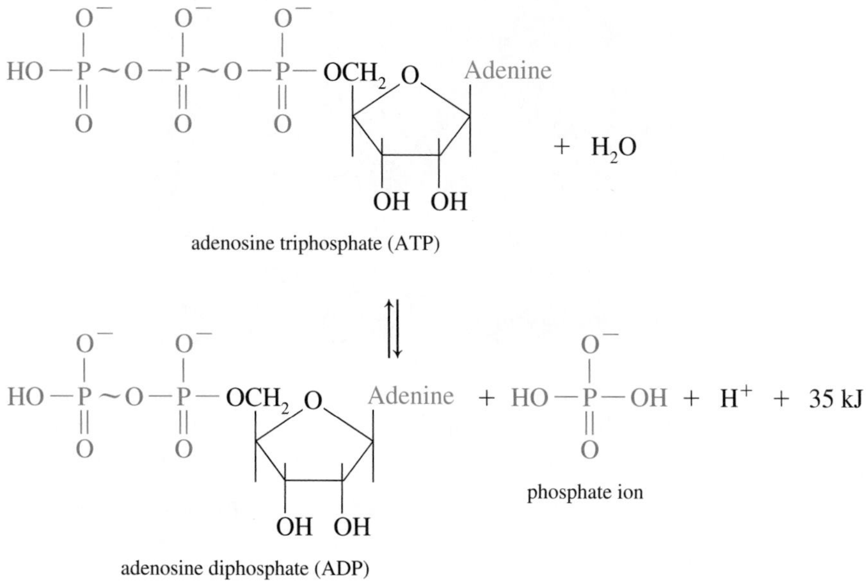

$$HO-\overset{O^-}{\underset{O}{\overset{|}{\underset{||}{P}}}}\sim O-\overset{O^-}{\underset{O}{\overset{|}{\underset{||}{P}}}}\sim O-\overset{O^-}{\underset{O}{\overset{|}{\underset{||}{P}}}}-OCH_2 \text{ (ribose ring: O, Adenine; OH OH)} + H_2O$$

adenosine triphosphate (ATP)

$$\rightleftharpoons$$

$$HO-\overset{O^-}{\underset{O}{\overset{|}{\underset{||}{P}}}}\sim O-\overset{O^-}{\underset{O}{\overset{|}{\underset{||}{P}}}}-OCH_2 \text{ (ribose ring: O, Adenine; OH OH)} + HO-\overset{O^-}{\underset{O}{\overset{|}{\underset{||}{P}}}}-OH + H^+ + 35\ \text{kJ}$$

adenosine diphosphate (ADP) phosphate ion

Second, ATP serves as the common energy currency for the cell. Energy from catabolism of many different kinds of molecules is stored in ATP. For example, the energy obtained from oxidation of carbohydrates, lipids, and proteins is stored in ATP. This process is analogous to an economic system that values all goods and services in terms of a common currency such as the dollar. Buying and selling within the system is thus greatly simplified. In the cell, energy utilization is greatly simplified by converting stored energy to ATP, the common energy currency.

33.6 Phosphorylation: Energy Conversion

Thus far, we have considered two forms for chemical storage of biological energy: reduced carbon atoms and high-energy phosphate bonds. It is vital that the cell be able to convert one form of stored energy to the other.

Figure 33.6 summarizes biological energy conversion. In the first process, a variety of cellular reactions oxidize the carbons of nutrients, converting these molecules from a high energy level to a lower energy state. The energy released (red arrow) is used to create high-energy phosphate bonds. Finally, when a cellular reaction needs energy, or work must be accomplished, molecules with high-energy phosphate bonds are recycled to low-energy molecules plus inorganic phosphate.

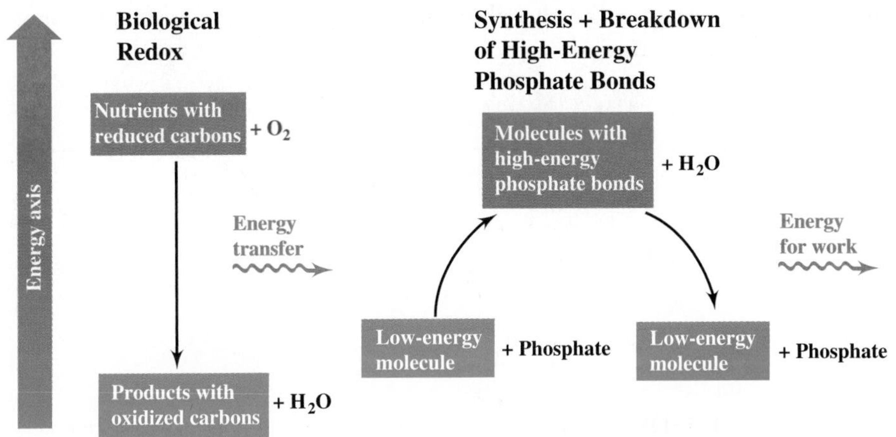

Figure 33.6
Energy flow from nutrients with reduced carbons (energy-yielding nutrients) to high-energy phosphate bonds that are used to do work.

Energy is stored in phosphate anhydride bonds through two biological processes: *substrate-level phosphorylation* and *oxidative phosphorylation*. **Substrate-level phosphorylation** is the process whereby energy derived from oxidation is used to form high-energy phosphate bonds on various biochemical molecules (substrates) (Figure 33.7). Such a reaction proceeds as follows:

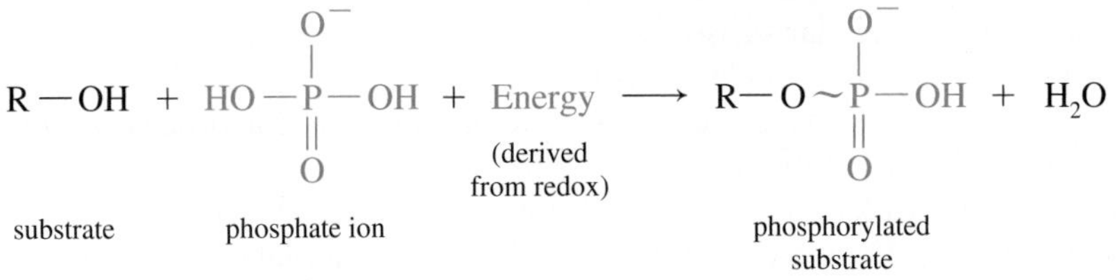

substrate-level phosphorylation

$$\underset{\text{substrate}}{R{-}OH} + \underset{\text{phosphate ion}}{HO{-}\overset{O^-}{\overset{|}{P}}(=O){-}OH} + \underset{\text{(derived from redox)}}{\text{Energy}} \longrightarrow \underset{\text{phosphorylated substrate}}{R{-}O{\sim}\overset{O^-}{\overset{|}{P}}(=O){-}OH} + H_2O$$

(R represents the remainder of the substrate molecule.)

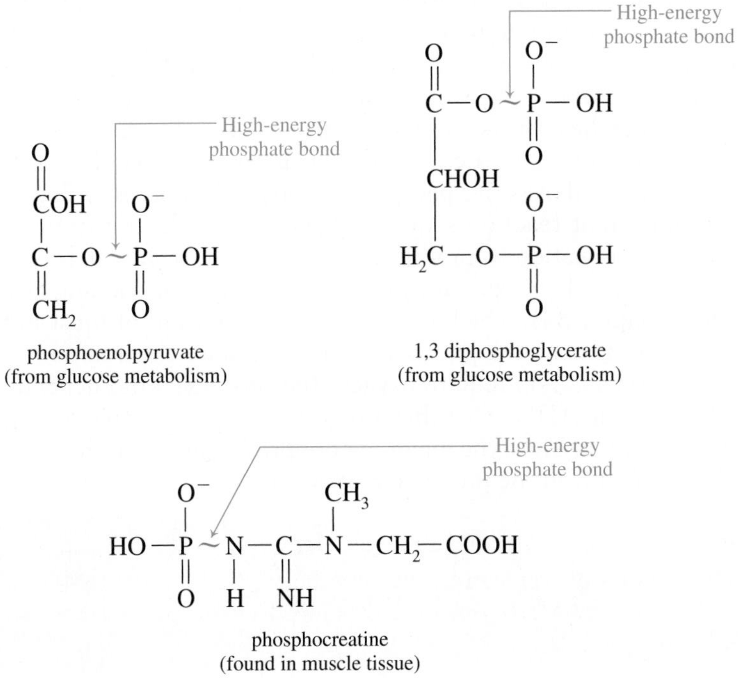

$$CH_2{=}C(COOH){-}O{\sim}P(O^-)(=O){-}OH$$

(High-energy phosphate bond marked ~)

phosphoenolpyruvate
(from glucose metabolism)

$$HO{-}P(O^-)(=O){-}O{-}CH_2{-}CHOH{-}C(=O){-}O{\sim}P(O^-)(=O){-}OH$$

(High-energy phosphate bond marked ~)

1,3 diphosphoglycerate
(from glucose metabolism)

$$HO{-}P(O^-)(=O){\sim}N(H){-}C(=NH){-}N(CH_3){-}CH_2{-}COOH$$

(High-energy phosphate bond marked ~)

phosphocreatine
(found in muscle tissue)

Figure 33.7
Some biological molecules that contain high-energy phosphate bonds (phosphorylated substrates).

In a succeeding reaction, the phosphorylated substrate often transfers the phosphate to ADP and forms ATP:

$$\underset{\text{phosphorylated substrate}}{\mathrm{R{-}O{\sim}P(=O)(O^-){-}OH}} + \underset{\text{adenosine diphosphate (ADP)}}{\mathrm{HO{-}P(=O)(O^-){\sim}O{-}P(=O)(O^-){-}OCH_2}\text{(ribose, OH OH)}\text{—Adenine}} + H^+ \longrightarrow$$

$$\underset{\text{substrate}}{\mathrm{R{-}OH}} + \underset{\text{adenosine triphosphate (ATP)}}{\mathrm{HO{-}P(=O)(O^-){\sim}O{-}P(=O)(O^-){\sim}O{-}P(=O)(O^-){-}OCH_2}\text{(ribose, OH OH)}\text{—Adenine}}$$

Fireflies convert ATP energy to light.

This process is called substrate-level phosphorylation, because ADP gains a phosphate from a cellular substrate.

Substrate-level phosphorylation is found most commonly in the catabolism of carbohydrates—that is, glycolysis. (See Chapter 34.) This process accounts for only a small amount of a resting cell's total energy production and does not require oxygen. Under anaerobic conditions (e.g., when the bloodstream cannot deliver enough O_2 to hardworking muscles), substrate-level phosphorylation may be the cell's principal means of forming ATP.

oxidative phosphorylation

Oxidative phosphorylation is a process that directly uses energy from redox reactions to form ATP. This process occurs in the mitochondria and depends on the mitochondrial electron-transport system. The enzyme-catalyzed oxidation and phosphorylation reactions are coupled in such a way that energy released by the oxidation of a coenzyme is used to form ATP. The overall process is shown in Figure 33.8. Note that, for each $FADH_2$ oxidized, two ATPs are formed; for each NADH oxidized, three ATPs are produced. This combination of mitochondrial electron transport and oxidative phosphorylation produces most cellular ATP.

Several important reaction sequences depend on electron transport and oxidative phosphorylation to produce ATP. Cells can derive energy from fats (Chapter 35) only when oxidative phosphorylation is functioning. The citric acid cycle (Chapter 34), which completes the oxidation of most nutrients, forms ATP by using oxidative phosphorylation. Because electron transport and oxidative phosphorylation require oxygen, the processes that depend on this means of producing ATP are aerobic. Oxygen must be available during oxidative phosphorylation. Thus, the major energy-producing reaction sequences in the cell function only in the presence of oxygen.

Practice 33.3

How many moles of ATP can be produced from 0.25 mol of ($NADH + H^+$) and 0.45 mol of $FADH_2$ during mitochondrial electron transport and oxidative phosphorylation?

CHEMISTRY IN ACTION • Brown Fat

Obesity is endemic in many industrialized countries. Numerous studies have documented the serious health risks associated with gaining weight. Yet many people find it very difficult to stop overeating and adding extra pounds of fat. The average adult American is now classified as overweight; a middle-aged male is about 30 pounds heavier, and a female is about 25 pounds heavier, than a comparable person from 1960. Thus, it may be surprising that some scientists are suggesting that we need to put on some fat to avoid obesity—not just any fat, of course, but *brown fat*.

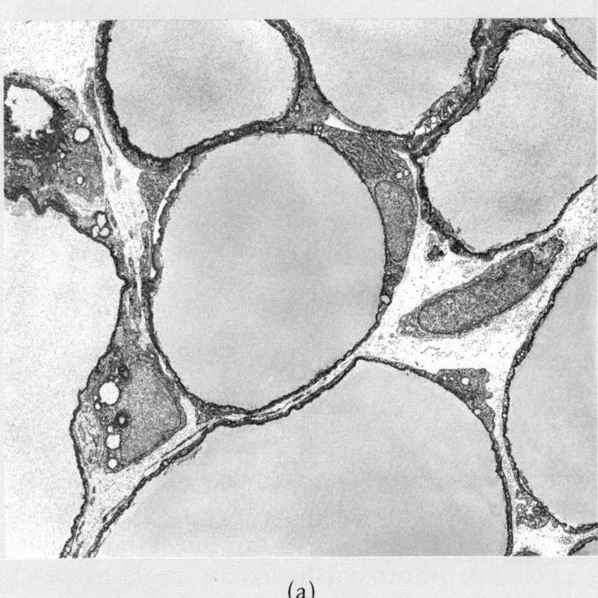

(a)

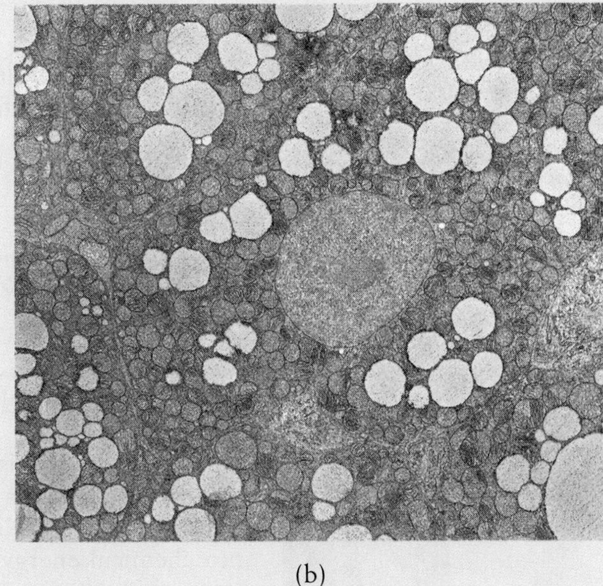

(b)

Microscopic photographs of (a) white fat and (b) brown fat.

Brown fat is found in hibernating animals, animals that are adapted to cold climates, and newborn infants. Whereas standard adipose tissue (white fat) is a storage depot for triacylglycerols and has a low metabolism, brown fat is metabolically very active and contains many mitochondria. Mitochondria are the "powerhouses" of the cell, using molecular oxygen to provide most of the cellular ATP. However, even active brown fat cells don't need much energy. Instead, the mitochondria in brown fat serve a different purpose: O_2 is used in a normal way but little ATP is produced. Since the energy is not trapped, it is given off as heat. Brown fat cells serve as small metabolic heaters.

Evidence shows that hibernating animals use brown fat cells to warm up as they awake. Animals that are cold-adapted stay warm with brown fat tissue. Babies use brown fat cells to maintain body temperature before they are old enough to shiver. For humans, brown fat serves only a temporary function; as the human body ages, brown fat cells disappear. There is no evidence for active brown fat in adult humans. This raises an important question: Why are scientists focusing attention on an absent tissue (brown fat) to solve a very present problem, obesity?

For some years it has been known that adult animals that retain brown fat (e.g., mice) use these cells to adjust overall metabolic rates. Brown fat mitochondria literally burn up excess calories, and the mice can eat copiously without gaining weight. Recently, progenitor brown fat cells have been discovered in adult human fat tissue. Furthermore, these cells can be triggered to mature into actively metabolizing brown fat. If these new tissues work as they do in the animal models, people might become thin on demand. Obesity will be a curable disease like smallpox, polio, and so on. Drug treatment could adjust the human metabolic rate to fit eating habits and activity levels. Health problems associated with obesity might become a thing of the past.

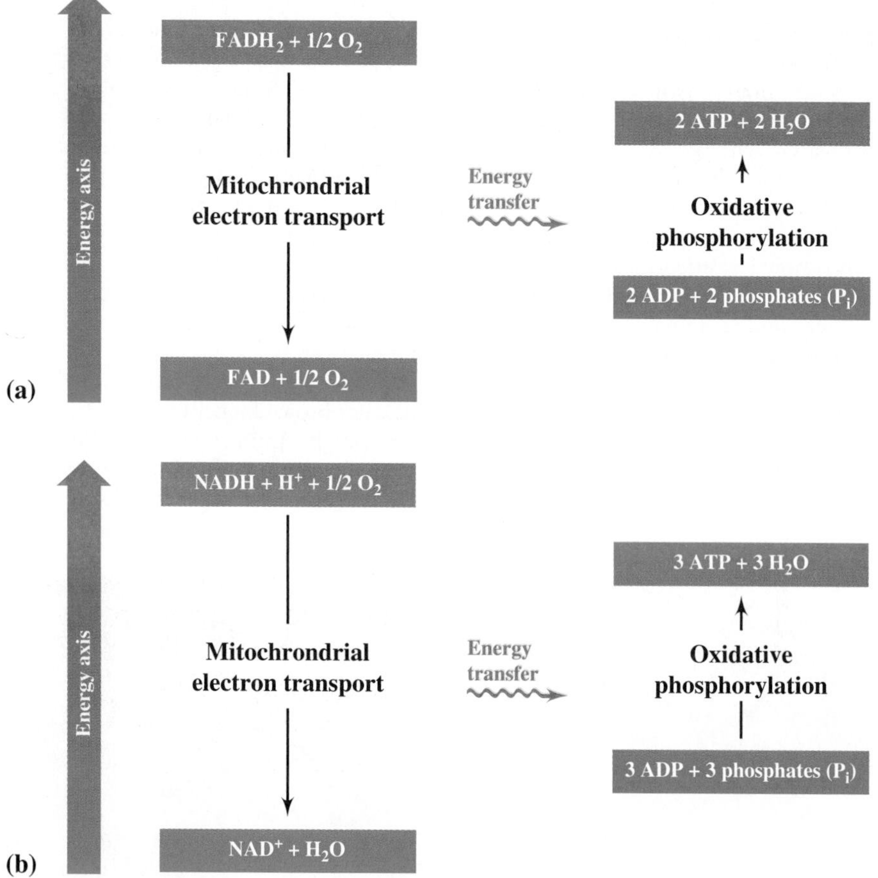

Figure 33.8
Mitochondrial electron transport and oxidative phosphorylation. (a) $FADH_2$ oxidation releases less energy than (b) NADH oxidation.

33.7 Photosynthesis

photosynthesis

Green algae shows spiral chloroplasts within the cytoplasm in the cells.

Light from the sun is the original source of nearly all energy for biological systems. Many kinds of cells can transform chemical energy to a form useful for doing work. However, there are also cells that can transform sunlight into chemical energy. Such cells use **photosynthesis**, a process by which energy from the sun is converted to chemical energy that is stored in chemical bonds.

Photosynthesis is performed by a wide variety of organisms, both eucaryotic and procaryotic. Besides the higher plants, photosynthetic eucaryotes include multicellular green, brown, and red algae, and unicellular organisms such as euglena. Photosynthetic procaryotes include the green and purple bacteria and the blue–green algae. Although the photosynthetic importance of higher plants is usually emphasized, it has been estimated that more than half of the world's photosynthesis is carried out by unicellular organisms.

Photosynthesis in higher plants is a complex series of reactions in which carbohydrates are synthesized from atmospheric carbon dioxide and water:

$$6\ CO_2 + 6\ H_2O + 2820\ \text{kJ} \longrightarrow \underset{\text{glucose}}{C_6H_{12}O_6} + 6\ O_2$$

Sunlight provides the large energy requirement for this process. An important side benefit of photosynthesis is the generation of oxygen, which is crucial to all aerobic metabolism.

The 1961 Nobel prize in chemistry was awarded to the American chemist Melvin Calvin (1911–1992), of the University of California at Berkeley, for his work on photosynthesis. Calvin and his coworkers used radioactive carbon tracer techniques to discover the details of the complicated sequence of chemical reactions that occur in photosynthesis.

Photosynthesis traps light energy by reducing carbons. For eucaryotes the necessary electron-transfer reactions are segregated in the chloroplast. (See Figure 33.4a.) Like the mitochondrion, the chloroplast contains an electron-transport system within its internal membranes. Unlike the mitochondrial system, which oxidizes coenzymes to liberate energy, the chloroplast electron-transport system reduces coenzymes with an input of energy:

$$\text{NADP}^+ + \text{H}_2\text{O} + \text{Energy} \xrightarrow[\text{electron transport}]{\text{chloroplast}} \text{NADPH} + \text{H}^+ + \tfrac{1}{2}\text{O}_2$$

The chloroplasts capture light energy and place it in chemical storage.

The photosynthetic mechanism is complex, but it can be divided into two general components: the *dark reactions* and the *light reactions*. The dark reactions produce glucose from carbon dioxide, reduced coenzymes, and ATP. No light is needed, and, in nature, these reactions continue during the night.

The light reactions of photosynthesis form the ATP and NADPH needed to produce glucose. The mechanism for capturing light energy is unique to the photosynthetic process. Although much research has been devoted to this topic, not all of the details are clear. In general, light is absorbed by colored compounds (pigments) located in the chloroplasts. The most abundant of these pigments is chlorophyll. Once the light energy is absorbed, it is transferred to specific molecules (special chlorophylls) that lose electrons. These energized electrons travel through the chloroplast electron-transport system, as shown in Figure 33.9. Two events follow in quick succession: First, the electrons lost by these special chlorophylls are moved to higher energy levels until they can reduce molecules of the coenzyme NADP^+; second, the special chlorophylls that lost electrons now regain them. Water is the electron donor, giving up electrons and producing oxygen gas (and hydrogen ions) in the process.

This greenhouse creates ideal conditions for photosynthesis.

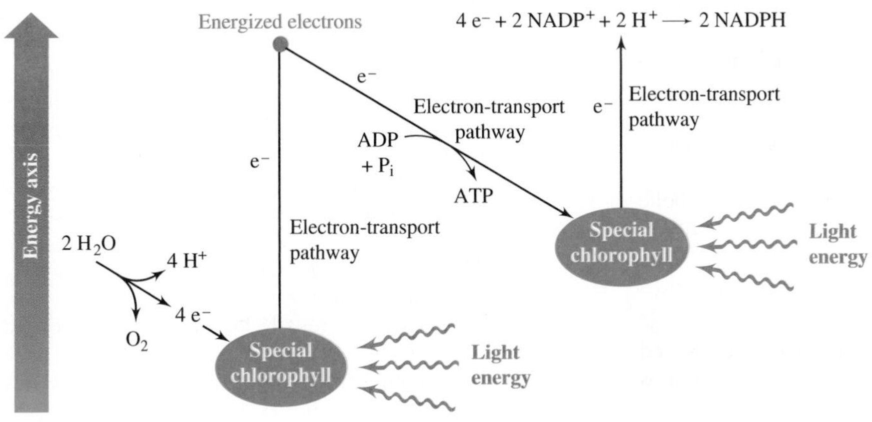

Figure 33.9
The movement of electrons from water to NADP^+ in the photosynthetic electron-transport pathway. Note that light energy causes the electrons to become more energetic, so that they can reduce NADP^+.

The overall redox reaction moves four electrons from two water molecules to produce two molecules of NADPH:

$$2\ H_2O + 2\ NADP^+ \xrightarrow{\text{light}} 2\ NADPH + O_2 + 2\ H^+$$

Figure 33.9 shows how photosynthesis uses light energy to force electrons to higher energy levels. As noted in the preceding paragraph, these energetic electrons are used to reduce $NADP^+$. But also notice that, when the electron loses energy in the middle of this electron-transport process, the released energy is used to make ATP. Thus, the light reactions of photosynthesis supply both the NADPH and ATP needed to make glucose.

Chapter 33 Review

33.1 Energy Changes in Living Organisms

KEY TERM

Bioenergetics

- Life requires energy.
- Bioenergetics is the study of the transformation, distribution, and utilization of energy by living organisms.

33.2 Metabolism and Cell Structure

KEY TERMS

Metabolism
Anabolism
Catabolism
Procaryote
Eucaryote
Organelle

- Metabolism is the sum of all chemical reactions that occur within a living organism.
- Metabolism is divided into two parts:
  - Anabolism is the process by which simple substances are synthesized into complex substances.
  - Catabolism is the process by which complex substances are broken down into simpler substances.
- Anabolism usually involves carbon reduction while consuming cellular energy; catabolism reverses this process.
- Living cells spatially segregate metabolism.
  - Procaryotes are cells without internal membrane-bound bodies and have a minimum amount of spatial organization.
  - Eucaryotes are cells that contain internal, membrane-bound bodies called organelles.
- The mitochondrion is the most important catabolic organelle in eukaryotes and provides the majority of energy for most cells.

33.3 Biological Oxidation–Reduction: Energy Delivery

- Plants capture light energy during photosynthesis and this energy is stored in the form of reduced carbons in carbohydrates.
- Animals draw most of their energy from foodstuffs containing reduced carbons.
- Chloroplasts are organelles in which photosynthesis occurs.
- Mitochondria are the cell's "powerhouses"—organelles where most of the catabolic redox reactions occur.
- Redox coenzymes carry electrons from one place to another inside the cell.
- Nicotinamide adenine dinucleotide (NAD^+) and nicotinamide adenine dinucleotide phosphate ($NADP^+$) are reduced as follows:

$$NAD^+ \text{ or } NADP^+ + 2e^- + H^+ \rightleftharpoons NADH \text{ or } NADPH$$

- Flavin adenine dinucleotide (FAD) is reduced as follows:

$$FAD + 2e^- + 2H^+ \rightleftharpoons FADH_2$$

- A very important function of these redox coenzymes is to reduce molecular oxygen in the mitochondrial electron transport system.

33.4 Molecular Oxygen and Metabolism

- Aerobic metabolism is the most efficient metabolic energy production and it requires molecular oxygen.
  - Molecular oxygen is completely reduced during mitochondrial electron transport.

$$O_2 + 4e^- + 4H^+ \longrightarrow 2H_2O$$

- Other reduced products of molecular oxygen are dangerous and are called reactive oxygen species (ROS).
  - A two-electron reduction of molecular oxygen yields the toxic chemical hydrogen peroxide, H_2O_2.
  - A one-electron reduction of molecular oxygen gives the dangerous superoxide O_2^{-}.
  - The most dangerous ROS is the neutral hydroxyl radical OH.
- Cells contain two common enzymes to protect themselves from ROS.

- Superoxide dismutase uses two superoxides to make one hydrogen peroxide.
- Catalase converts hydrogen peroxide to water.

33.5 High-Energy Phosphate Bonds

- The most common high-energy phosphate bond in the cell is the phosphate anhydride (or phosphoanhydride bond).
- Phosphate anhydride bonds are important components of nucleotide triphosphates, the most common of which is adenosine triphosphate (ATP).
- ATP is the common energy currency of the cell:
  - The phosphate anhydride stored energy is easily accessible via hydrolysis reactions.
  - Almost all catabolism stores energy in ATP before work is done.

33.6 Phosphorylation: Energy Conversion

KEY TERMS

Substrate-level phosphorylation
Oxidative phosphorylation

- There are two common forms of chemical energy storage in cells: reduced carbon atoms in molecules such as carbohydrates and phosphate anhydride bonds in molecules like ATP.
- Substrate-level phosphorylation uses energy derived from oxidation to form high-energy phosphate bonds on substrates, biochemicals in metabolic pathways.
  - In a succeeding reaction, the phosphate is reacted with ADP to form ATP.
- Oxidative phosphorylation directly uses energy from redox reactions to form ATP.
  - Oxidative phosphorylation occurs in the mitochondria.

33.7 Photosynthesis

KEY TERM

Photosynthesis

- Photosynthesis is a process by which energy from the sun is converted to chemical energy that is stored in chemical bonds.
- Photosynthesis produces both carbohydrates and molecular oxygen.
- Photosynthetic light reactions use sunlight to make reduced coenzymes like NADPH.
- Photosynthetic dark reactions use the reduced coenzymes to convert carbon dioxide to glucose.
- An electron transport pathway is used for photosynthesis in chloroplasts just as one is used for oxidative phosphorylation in the mitochondria.

Review Questions

All questions with blue numbers have answers in the appendix of the text.

1. Contrast the type of energy delivered to plants with that delivered to animals.
2. Describe a common energy conversion found in both plants and animals.
3. Why are fats and carbohydrates good sources of cellular energy?
4. What are the common (and dangerous) reactive oxygen species?
5. Describe general problems with oxygen metabolism that are caused by a stroke.
6. Give a general structure for the most common high-energy phosphate bond found in the cell. With what compound is it generally associated?
7. Why is ATP known as the "common energy currency" of the cell?
8. What is an oxidation–reduction coenzyme?
9. Draw the ring structure portion of NAD^+ that becomes reduced during metabolism.
10. How does oxidative phosphorylation differ from substrate-level phosphorylation?
11. In what part of the eucaryotic cell does oxidative phosphorylation occur?
12. Compare the structural similarities between choroplasts and mitochondria.
13. What role do chloroplast pigments serve in photosynthesis?
14. Give the overall reaction for photosynthesis in higher plants.
15. What is the physiological function for brown fat?

Paired Exercises

All exercises with blue numbers have answers in the appendix of the text.

1. Which molecule delivers more energy in biological redox?

   CH_3OH or CO_2

   Explain briefly.

2. Which molecule delivers more energy in biological redox?

   $CH_3CH{=}O$ (acetaldehyde, $CH_3\overset{O}{\overset{\|}{C}}H$) or $CH_3\overset{O}{\overset{\|}{C}}-OH$

   Explain briefly.

3. Which biochemical contains more metabolic energy, 1 mole of glucose or 1 mole of a six-carbon saturated fatty acid? Explain briefly.

4. Which biochemical contains more metabolic energy, a 10-carbon saturated fatty acid or two ribose molecules? Explain briefly.

5. How does the typical eucaryotic cell protect itself from superoxide?

6. How does the typical eucaryotic cell protect itself from hydrogen peroxide?

7. Give an example of a neutral ROS.

8. Give an example of a negatively charged ROS.

9. What is wrong with the following equation for biological redox?

$$NADH + FAD \rightarrow NAD^+ + FADH_2$$

Explain briefly.

10. What is wrong with the following equation for biological redox?

$$NAD^+ + FAD \rightarrow NADH + H^+ + FADH_2$$

Explain briefly.

11. Many cells convert glucose to carbon dioxide. Is this an anabolic or a catabolic process? Explain briefly.

12. Many cells convert acetate to long-chain fatty acids. Is this an anabolic or a catabolic process? Explain briefly.

13. What is wrong with this statement? The chloroplasts in this eucaryotic cell are photosynthesizing, a catabolic process.

14. What is wrong with this statement? The mitochondria in this eucaryotic cell are undergoing oxidative phosphorylation, an anabolic process.

15. What is wrong with this statement? The chloroplasts in this procaryotic cell are undergoing photosynthesis.

16. What is wrong with this statement? The chloroplasts in this eucaryotic cell are undergoing oxidative phosphorylation.

17. Explain why the chemical changes in the mitochondria are said to be catabolic.

18. Explain why the chemical changes in the chloroplasts are said to be anabolic.

19. List three important characteristics of an anabolic process.

20. List three important characteristics of a catabolic process.

21. Name the ring component of NAD^+ that is the reaction center.

22. Name the ring component of FAD that is the reaction center.

23. You discover a new metabolic pathway that uses 3 moles of NAD^+ and yields 3 moles of NADH (among other products). Would you classify this pathway as catabolic or anabolic? Explain briefly.

24. You discover a new metabolic pathway that uses 2 moles of $FADH_2$ and yields 2 moles of FAD (among other products). Would you classify this pathway as catabolic or anabolic? Explain briefly.

25. In mitochondrial electron transport, how many moles of oxygen gas are reduced to water by 2.38 mol of $FADH_2$?

26. In mitochondrial electron transport, how many moles of oxygen gas are reduced to water by 0.67 mol of NADH?

27. In mitochondrial chemical processes, how many moles of electrons will reduce 11.75 mol of NAD^+ to NADH?

28. In mitochondrial chemical processes, how many moles of electrons will reduce 0.092 mol of FAD to $FADH_2$?

29. How many high-energy phosphate anhydride bonds are contained in the following formula?

```
                            O
                            ‖
                  CH2—O—P—O⁻
  O       O        |        |
  ‖       ‖        |        O⁻
⁻O—P—O—P—O—CH
   |       |       |        O⁻      O
   O⁻      O⁻      |        |       ‖
                  CH2—O—P—O—P—O⁻
                            ‖       |
                            O       O⁻
```

30. How many high-energy phosphate anhydride bonds are contained in the following formula?

```
                                     O
                                     ‖
                           CH2—O—P—O⁻
   O       O       O        |        |
   ‖       ‖       ‖        |        O⁻
⁻O—P—O—P—O—P—O—CH
   |       |       |        |        O⁻
  ⁻O       O⁻      O⁻       |        |
                           CH2—O—P—O⁻
                                     ‖
                                     O
```

31. The equation for conversion of ADP to ATP shows an energy requirement of 35 kJ. How much energy would be required to convert 0.55 mol of ADP to ATP?

32. The equation for conversion of ATP to ADP shows an energy release of 35 kJ. How much energy would be released from 1.65 mol of ATP?

33. The following compound is formed during glucose catabolism:

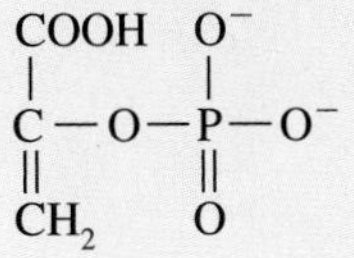

```
COOH  O⁻
 |    |
 C—O—P—O⁻
 ||   ||
 CH2  O
```

How many ATPs can be formed from one molecule of this compound during substrate-level phosphorylation? Explain.

34. The following compound is formed during glucose catabolism:

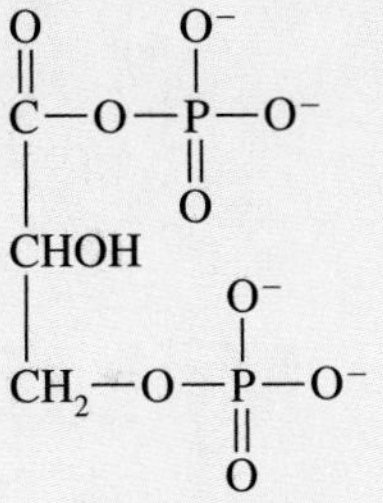

```
O      O⁻
||     |
C—O—P—O⁻
|      ||
|      O
CHOH
|      O⁻
|      |
CH2—O—P—O⁻
       ||
       O
```

How many ATPs can be formed from one molecule of this compound during substrate-level phosphorylation? Explain.

35. How many ATPs would be formed from 4 NADH and 2 $FADH_2$, using mitochondrial electron transport and oxidative phosphorylation?

36. How many ATPs would be formed from 2 NADH and 3 $FADH_2$, using mitochondrial electron transport and oxidative phosphorylation?

37. What vitamin is required to make NAD^+?

38. What vitamin is required to make FAD?

39. The structures of the common redox coenzymes are given in Figure 33.5. How does $NADP^+$ differ from NAD^+?

40. The structures of the common redox coenzymes are given in Figure 33.5. What nucleic acid base is found in all three redox coenzymes?

Additional Exercises

All exercises with blue numbers have answers in the appendix of the text.

41. One of your student colleagues has just told you that photosynthesis is a process by which carbon atoms are oxidized. Do you agree? Explain briefly.

42. Based on the function of the mitochondria, explain why these organelles are composed of about 90% membrane by mass.

43. Could a higher plant cell survive with chloroplasts but no mitochondria? Explain.

44. What structural similarities, if any, exist between ATP and NAD^+?

45. Describe the major differences between the photosynthetic dark and light reactions.

Challenge Exercise

46. A new metabolic pathway is discovered that converts pyruvic acid,

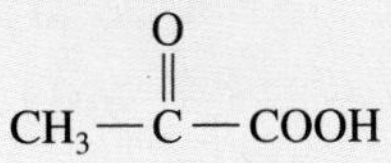

```
       O
       ||
CH3—C—COOH
```

to glyceraldehyde:

```
          OH   O
          |    ||
HOCH2—CH—CH
```

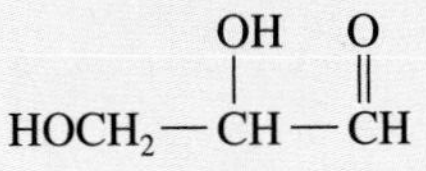

(a) This pathway uses a nicotinamide-containing coenzyme. Do you expect NADH to be a reactant or a product? Explain briefly.
(b) Would you expect ATP to be a reactant or a product? Explain briefly.

Answers to Practice Exercises

33.1 3-Hydroxybutyric acid is likely to provide more energy, because the carbons are more reduced than those in D-erythrose.

33.2 The oxidized form (FAD) should react with CH_3OH to yield the reduced form ($FADH_2$) and $H_2C{=}O$.

33.3 1.65 mol of ATP

CHAPTER 34

Carbohydrate Metabolism

Quick bursts of energetic movement depend upon carbohydrate metabolism.

Chapter Outline

Have you ever stopped to think about why carbohydrates are a staple of our world? Carbohydrates are everywhere. Our diets consist mainly of carbohydrates, many building materials in our homes are carbohydrates, and most of the plant kingdom consists of carbohydrates. But *why* is this class of molecules so prevalent? Biochemists might answer this question in a number of ways, but one of the most important answers is in the air around us. Carbohydrates can be made from carbon dioxide, water, and sunlight through *photosynthesis*, so the constituents for carbohydrate metabolism are incredibly abundant.

Of course, a solid carbohydrate like sugar, which is sweet to the taste, has little in common with gaseous carbon dioxide, which is tasteless and can be hazardous in high concentrations. But a series of chemical reactions can convert carbon dioxide and water to sugar. Photosynthesis is an example of carbohydrate metabolism, and it illustrates the fact that metabolism can achieve remarkable conversions. Our world depends on such metabolic processes as photosynthesis.

34.1 Metabolic Pathways

metabolic pathway

Biochemists define a **metabolic pathway** as a series of biochemical reactions that serve a specific purpose. Every cell contains thousands of reactions comprising many metabolic pathways (Figure 34.1). A select group of pathways provides cellular energy using carbohydrates. Carbohydrates are rich sources of energy for most organisms. As we will see in this chapter, a number of different metabolic pathways (and a large number of chemical reactions) are required to obtain energy.

Sunlight provides the energy that is stored in carbohydrates:

$$6\ CO_2 + 6\ H_2O + \underset{\substack{\text{from}\\ \text{sunlight}}}{2820\ \text{kJ}} \xrightarrow[\text{chlorophyll}]{\text{enzymes}} C_6H_{12}O_6 + 6\ O_2 \qquad (1)$$

Equation (1) summarizes the production of glucose (a carbohydrate) and oxygen by the endothermic process called *photosynthesis*. Note that, in the overall transformation, carbon atoms from carbon dioxide are reduced and oxygen atoms from water are oxidized. The light energy is used to cause a net movement of electrons from water to carbon dioxide. As the carbons become more reduced, more energy is stored. Photosynthesis also supplies free oxygen to the atmosphere and is an example of an anabolic pathway.

The equation

$$C_6H_{12}O_6 + 6\ O_2 \xrightarrow{\text{enzymes}} 6\ CO_2 + 6\ H_2O + 2820\ \text{kJ} \qquad (2)$$

represents the oxidation of glucose and corresponds to the reversal of the overall photosynthesis reaction. In this reversal reaction, electrons move from carbohydrate carbons to oxygen. Energy stored in the reduced carbon atoms in glucose is released and can then be used by the cell to do work. This is an example of a catabolic process.

Glucose can be burned in oxygen in the laboratory to produce carbon dioxide, water, and heat. But in the living cell, the oxidation of glucose does not proceed directly to carbon dioxide and water. Instead, like photosynthesis, the overall process proceeds by a series of enzyme-catalyzed intermediate reactions. These intermediate steps channel some of the liberated energy into uses other than heat production. Specifically, a portion of the energy is stored in the chemical bonds of ATP.

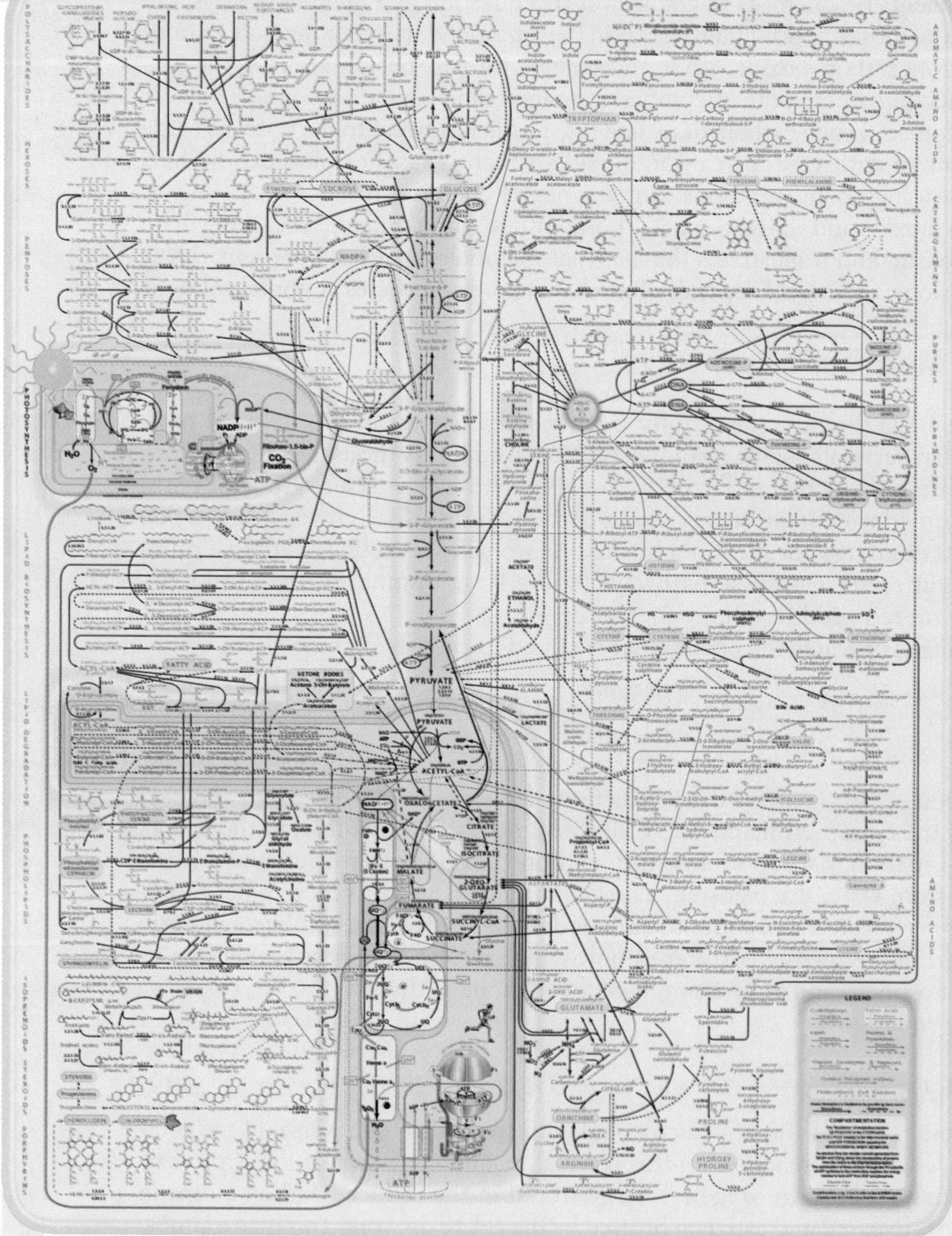

Figure 34.1
A complete map of all metabolic pathways in a typical cell. Each line/arrow signifies a metabolic pathway/reaction. The map is shown to demonstrate what a complex metabolic system is occurring in a typical cell.

Practice 34.1

Is the following chemical process catabolic or anabolic? Explain briefly.

$$\begin{array}{l} \mathrm{COOH} \\ | \\ \mathrm{HOCH} \\ | \\ \mathrm{HC{-}COOH} \\ | \\ \mathrm{CH_2} \\ | \\ \mathrm{COOH} \end{array} + \mathrm{NAD^+} \longrightarrow \begin{array}{l} \mathrm{COOH} \\ | \\ \mathrm{C{=}O} \\ | \\ \mathrm{CH_2} \\ | \\ \mathrm{CH_2} \\ | \\ \mathrm{COOH} \end{array} + \mathrm{NADH} + \mathrm{H^+} + \mathrm{CO_2}$$

34.2 Exercise and Energy Metabolism

Working or playing, studying or watching TV—everything we do requires metabolic energy. Sometimes we need the energy quickly, and sometimes it can be delivered more slowly. Metabolism is a complex interplay: Chemical reactions within cells lead to chemical transport between cells.

Carbohydrate catabolism is designed to release energy relatively quickly, so this form of catabolism is activated during strenuous muscular exercise. When a muscle contracts, energy is consumed. Muscle contraction, as with almost all mechanical work, uses ATP, but just like ready cash, ATP is in short supply. Muscle tissue can contract for no more than several seconds before the supply of high-energy phosphate bonds is depleted.

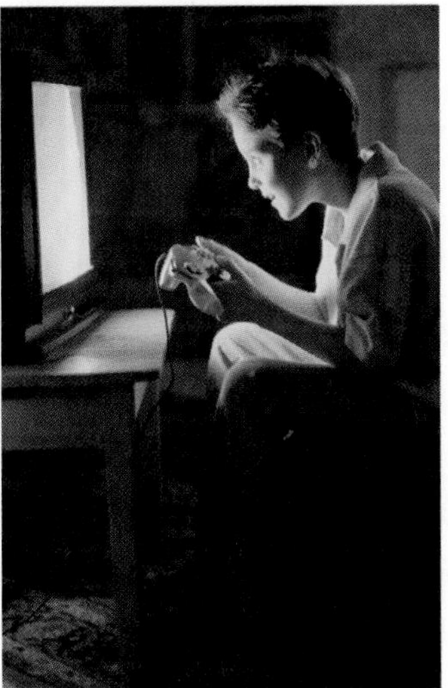

Even relaxing and playing a video game requires metabolic energy.

After the initial contraction, the muscle cells look for other energy sources. Muscle glycogen is the next available source. This polymer breaks down to glucose, which is oxidized to replenish the ATP supply. Because glucose oxidation is a complex process, muscle contraction must proceed at a slower rate. Unfortunately, even this energy supply is useful only for about two minutes of work: Muscles rapidly deplete their glycogen stores and build up lactic acid.

If that were the end of the story, our work schedule would always be very short. In fact, certain animals respond in just this way. For example, an alligator can run quickly and thrash its tail powerfully while capturing a prey, but only for several minutes. The alligator's exertion must be followed by a rest of several hours, as the muscles recuperate their energy stores. Humans (and other mammals) are not so limited, because their livers replenish the muscle glucose supply and remove lactic acid. Muscle contractions can continue with help from the liver, but the rate of contraction is slowed further as muscle cells wait on glucose transport.

34.3 The Bloodstream: A Metabolic Connection

The bloodstream transports chemicals from one cell to another. Nutrients (e.g., glucose, amino acids, and fatty acids) and oxygen are delivered; metabolic products (e.g., lactate) and carbon dioxide are removed.

Human muscle cells, like many other cells, must have an adequate oxygen supply. About 5 g of oxygen (4 L at 25° C and 1 atm) is required for every minute of strenuous activity by a young adult. The necessary oxygen is supplied by arterial blood (using the oxygen-transport protein, hemoglobin, Section 29.8), and waste carbon dioxide is removed by venous blood.

Second, hydrogen ions are formed as the muscles work. Two products of cellular metabolism—carbonic acid and lactic acid—are the primary sources of this acidity:

$$CO_2 + H_2O \rightleftharpoons \underset{\text{carbonic acid}}{H_2CO_3} \rightleftharpoons H^+ + HCO_3^-$$

$$\underset{\text{lactic acid}}{CH_3CH(OH)COOH} \rightleftharpoons H^+ + \underset{\text{lactate}}{CH_3CH(OH)COO^-}$$

The lungs restore the blood-gas balance: Oxygen is inhaled and carbon dioxide is exhaled. The blood becomes less acidic and hemoglobin binds more oxygen. The overall relationship between the muscles and lungs can be summarized in the following way:

Energy use at the muscle	Gas balance restored at the lungs
O_2 used	O_2 inhaled
CO_2 formed	CO_2 exhaled
Acidity increases	Acidity decreases
More O_2 released from hemoglobin	More O_2 binds to hemoglobin

The liver maintains many of the nutrient levels in the blood. For example, the liver sets normal blood-glucose levels (70–100 mg/100 mL of blood for a healthy person before eating) by either releasing glucose to the blood or removing glucose from the blood. When blood concentrations of lactate increase, the lactate is converted back to glucose in the liver. The relationship between the muscles and the liver can be summarized as follows:

Energy use at the muscle	Nutrient balance restored at the liver
Glucose consumed	Glucose released
Lactate produced	Lactate converted to glucose
Acidity increases	Acidity decreases

Figure 34.2 relates the blood levels of glucose and lactate to cellular metabolic processes within the muscle and liver. As muscles or other tissues use more glucose, the liver increases glucose output to the blood. After a meal, the liver removes excess glucose from the blood.

When blood glucose is in excess, it is converted to glycogen in the liver and in muscle tissue. Glycogen is a storage polysaccharide; it quickly hydrolyzes to replace depleted glucose supplies in the blood. The synthesis of glycogen from glucose is called **glycogenesis**; the hydrolysis, or breakdown, of glycogen to glucose is known as **glycogenolysis** (see Figure 34.2).

glycogenesis
glycogenolysis

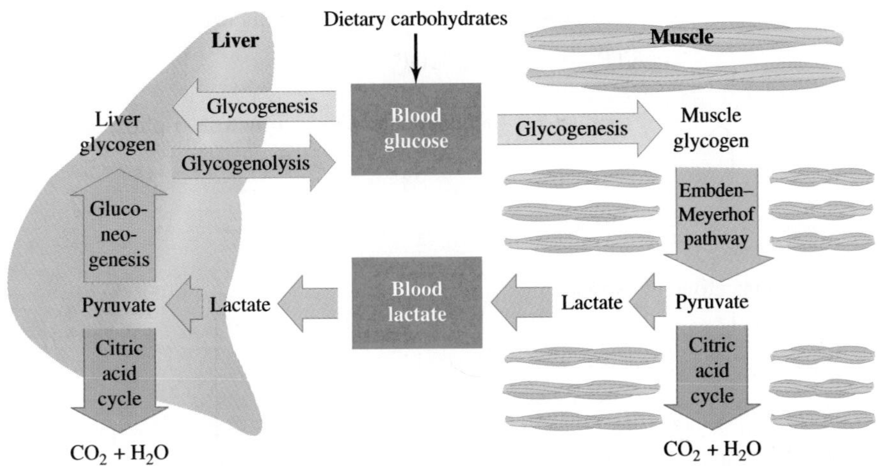

Figure 34.2
Overview of carbohydrate metabolism.

Glucose supplies cellular energy via several metabolic pathways. Under anaerobic (oxygen not required) conditions, a general sequence of reactions known as the *Embden–Meyerhof pathway* oxidizes glucose. This reaction sequence is especially important for humans (and other animals) during strenuous activity, when muscle tissue generates much lactate. Part of the lactate produced is sent to the liver and is converted to glucose in the *gluconeogenesis* pathway and then to glycogen. The rest is converted to carbon dioxide and water via the *citric acid cycle*, an aerobic (free or respiratory oxygen required) sequence. The *citric acid cycle* is also known as the Krebs cycle or the tricarboxylic acid cycle.

Figure 34.2 gives a general overview of carbohydrate metabolism. The entire process requires specific enzymes at each stage of the reaction and is controlled by chemical regulatory compounds. The net result of these reactions is a conversion of carbohydrate to CO_2 and H_2O.

The importance of nutrient transport between cells is illustrated by the glycogen-storage diseases and by diabetes mellitus (see Section 34.9). These diseases affect the body's ability to shuttle glucose between cells and make use of it when it arrives.

The earliest understood glycogen-storage disease is von Gierke disease. Patients with this disease lack the ability to export glucose from the liver. They suffer from a liver that is engorged with glycogen, and they exhibit severely low blood glucose. The importance of the liver in maintaining blood glucose levels is emphasized by von Gierke disease.

Patients with McArdle disease can no longer break down muscle glycogen to provide glucose for the Embden–Meyerhof pathway (see Figure 34.3). These patients' muscles are healthy when resting, but during light exercise they often cramp painfully. The muscles can obtain glucose from the blood, but when they are stressed, blood glucose is not sufficient and muscle glycogen is not available. By default, energy sources other than glucose are used by these muscles. Thus, they produce only very low levels of lactate. McArdle disease emphasizes the importance of glycogen as a storage form of glucose.

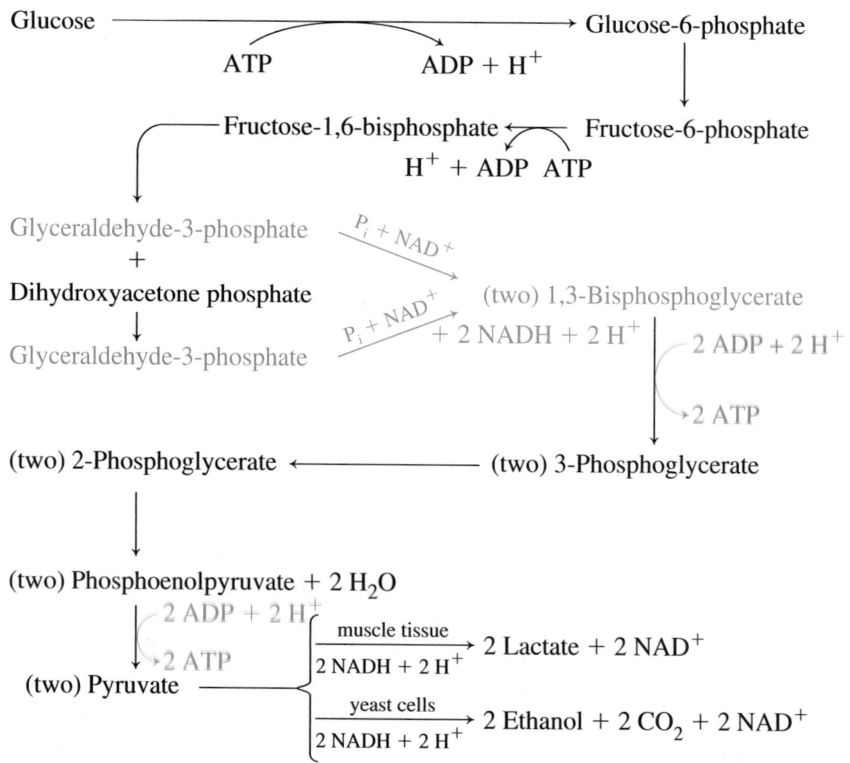

Figure 34.3
Ten-step conversion of glucose to pyruvate via the Embden–Meyerhof pathway (an anaerobic sequence). Glycolysis proceeds further, forming lactate. The formation of ATP is denoted in green. The only oxidation–reduction reaction in this pathway is shown in red.

Practice 34.2

Enumerate several important fates for pyruvate as it is formed from lactate in the liver.

34.4 Anaerobic Sequence

In the absence of oxygen, glucose in living cells is converted to a variety of end products, including lactic acid (in muscle) and alcohol (in yeast). A similar conversion occurs in many different cells. At least a dozen reactions, many different enzymes, ATP, and inorganic phosphate (P_i) are required. Such a sequence of reactions from a particular reactant to end products is called a metabolic pathway.

Embden–Meyerhof pathway

Carbon oxidation commonly provides energy for life. (See Section 27.3.)

The anaerobic conversion of glucose to pyruvate is known as the **Embden–Meyerhof pathway**. The sequence is a catabolic one in which glucose is oxidatively degraded:

$$\underset{\substack{\text{D-glucose} \\ \text{(sum of oxidation numbers} \\ \text{for six carbons = 0)}}}{C_6H_{12}O_6} \xrightarrow[\text{pathway}]{\text{Embden–Meyerhof}} \underset{\substack{\text{pyruvate} \\ \text{(sum of oxidation numbers} \\ \text{for six carbons = +4)}}}{2\,CH_3\overset{\overset{\displaystyle O}{\|}}{C}{-}COO^-}$$

In glucose, the sum of the oxidation numbers of the six carbon atoms is zero. After processing one molecule of glucose to two molecules of pyruvate by the pathway, the sum of the oxidation numbers of the six carbon atoms is +4. This makes it evident that an overall oxidation of carbon must occur in the Embden–Meyerhof pathway.

As with most catabolic processes, the Embden–Meyerhof pathway produces energy for the cell. Carbon atoms are oxidized, and energy is released and stored in the form of ATP. The pathway uses the process of substrate-level phosphorylation (see Section 33.6): The energy released from carbon oxidation is used to form a high-energy substrate–phosphate bond, which in turn is used to form ATP. It is interesting to note that there is only one oxidation–reduction reaction in the Embden–Meyerhof pathway: The conversion of glyceraldehyde-3-phosphate to 1,3-bisphosphoglycerate (marked in red in Figure 34.3):

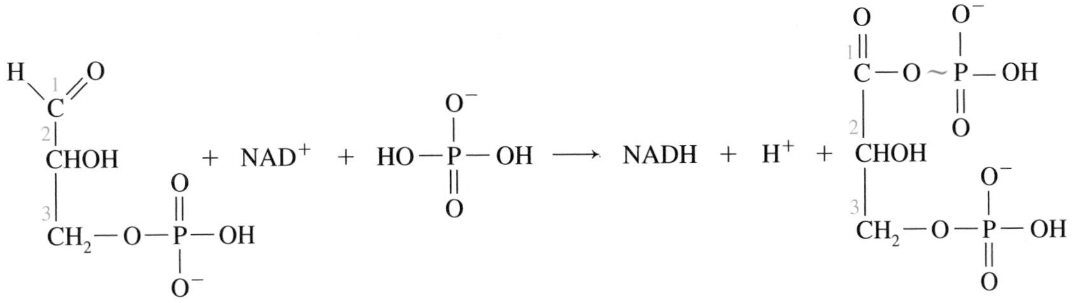

$$\text{glyceraldehyde-3-phosphate} + NAD^+ + HO-PO(O^-)-OH \longrightarrow NADH + H^+ + \text{1,3 bisphosphoglycerate}$$

H–C(=O) (C1) / CHOH (C2) / $CH_2-O-P(=O)(O^-)-OH$ (C3)

glyceraldehyde-3-phosphate

C(=O)–O~P(=O)(O⁻)–OH (C1) / CHOH (C2) / $CH_2-O-P(=O)(O^-)-OH$ (C3)

1,3 bisphosphoglycerate

Carbon 1 is oxidized from the aldehyde to the carboxylate oxidation state. Simultaneously, the oxidation–reduction coenzyme NAD^+ is reduced. This single oxidation–reduction supplies most of the energy generated by the Embden–Meyerhof pathway, which is used to make these two different high-energy phosphate bonds:

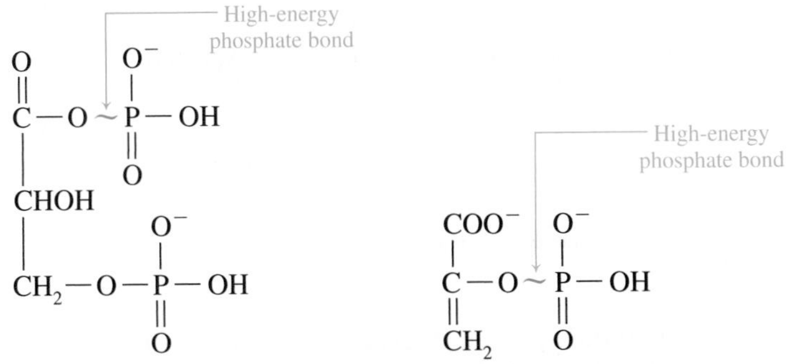

High-energy phosphate bond

C(=O)–O~P(=O)(O⁻)–OH / CHOH / $CH_2-O-P(=O)(O^-)-OH$

1,3-bisphosphoglycerate

High-energy phosphate bond

COO^- / C–O~P(=O)(O⁻)–OH / $=CH_2$

phosphoenolpyruvate

The two high-energy bonds are found on intermediate compounds or pathway substrates. Substrate-level phosphorylation is complete when these substrates transfer their high-energy phosphate bonds to ADP, forming ATP (the ATP formation is shown in green in Figure 34.3):

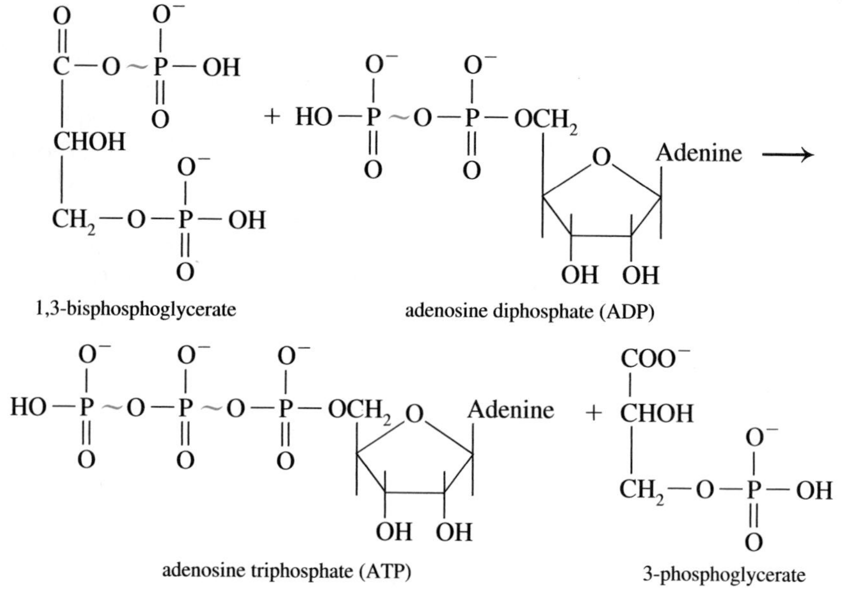

1,3-bisphosphoglycerate + adenosine diphosphate (ADP) → adenosine triphosphate (ATP) + 3-phosphoglycerate

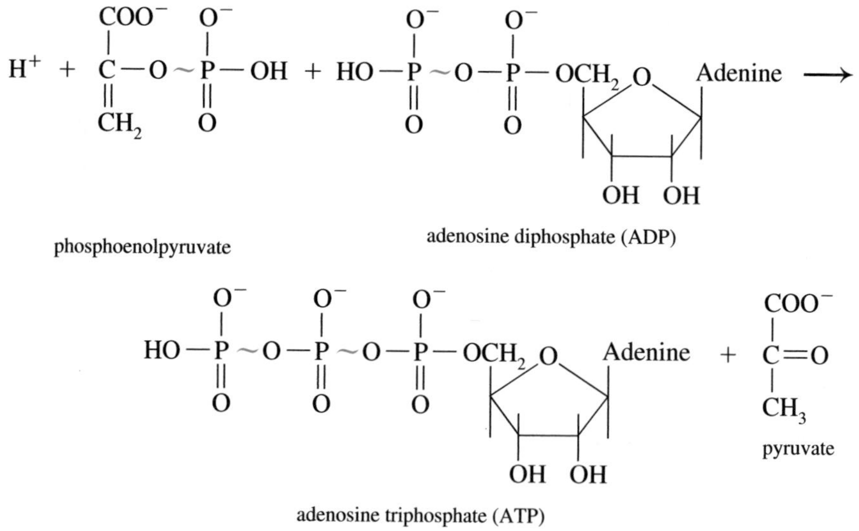

H^+ + phosphoenolpyruvate + adenosine diphosphate (ADP) → adenosine triphosphate (ATP) + pyruvate

It is important to note that the Embden–Meyerhof pathway oxidizes carbon and produces ATP in the absence of molecular oxygen. The pathway provides for *anaerobic* energy production. For this reaction sequence to continue, the coenzyme NADH must be recycled back to NAD^+; that is, an additional oxidation–reduction reaction is needed to remove electrons from NADH. In the absence of oxygen, pyruvate is used as an electron acceptor. In human muscle cells, pyruvate is reduced directly to lactate:

$$\underset{\text{pyruvate}}{CH_3\overset{\overset{\large O}{\|}}{C}COO^-} + NADH + H^+ \rightleftharpoons \underset{\text{lactate}}{CH_3\underset{\underset{\large OH}{|}}{C}HCOO^-} + NAD^+$$

In yeast cells, pyruvate is converted to acetaldehyde, which is then reduced to ethanol:

$$\underset{\text{pyruvate}}{CH_3\underset{\underset{\large O}{\|}}{C}COO^-} + H^+ \xrightarrow[\text{cells}]{\text{in yeast}} \underset{\text{acetaldehyde}}{CH_3\overset{\overset{\large O}{\|}}{C}-H} + CO_2$$

$$\underset{\text{acetaldehyde}}{CH_3\overset{\overset{\large O}{\|}}{C}-H} + NADH + H^+ \rightleftharpoons \underset{\text{ethanol}}{CH_3CH_2OH} + NAD^+$$

In each case, NADH is reoxidized to NAD^+, and a carbon atom from pyruvate is reduced. When lactate is the final product of anaerobic glucose catabolism, the pathway is termed **glycolysis**. As the Embden–Meyerhof pathway produces equal amounts of pyruvate and NADH, there is just enough pyruvate to recycle all the NADH. This is a good example of an important general characteristic of metabolism: Chemical reactions in the cell are precisely balanced, so there is never a large surplus or a large deficit of any metabolic product. If such a situation does occur, the cell may die.

glycolysis

What glycolysis does for the cell can be summarized with the following net chemical equation:

$$C_6H_{12}O_6 + 2\ ADP + 2\ P_i \longrightarrow 2\ CH_3CH(OH)COO^- + 2\ ATP + 150\ kJ$$

This anaerobic process is not very efficient, because only 2 mol of ATP is formed per mole of glucose. In fact, if the energy stored in the two ATPs is added to the energy released as heat during the Embden–Meyerhof pathway, a total of only about 220 kJ/mol is found to have been removed from glucose. Compare this result with the complete oxidation of glucose:

$$C_6H_{12}O_6 + 6\ O_2 \longrightarrow 6\ CO_2 + 6\ H_2O + 2820\ kJ$$

We see that the Embden–Meyerhof pathway releases less than one-tenth of the total energy available in glucose. Thus, lactate must still contain much energy in its reduced carbon atoms.

Practice 34.3

If 1.7 mol of glucose is processed via the Embden–Meyerhof pathway, (a) how many moles of pyruvate are produced? (b) how many moles of ATP are produced?

34.5 Citric Acid Cycle (Aerobic Sequence)

As discussed previously, only a small fraction of the energy that is potentially available from glucose is liberated during the anaerobic conversion to lactate (glycolysis). Consequently, lactate remains valuable to the cells. The lactate formed may be (1) circulated back to the liver and converted to glycogen at the expense of some ATP, or (2) converted back to pyruvate in order to enter the citric acid cycle:

$$\underset{\text{lactate}}{CH_3\overset{\overset{\displaystyle OH}{|}}{C}HCOO^-} + NAD^+ \rightleftharpoons \underset{\text{pyruvate}}{CH_3\overset{\overset{\displaystyle O}{\|}}{C}COO^-} + NADH + H^+$$

Pyruvate is the link between the anaerobic sequence (Embden–Meyerhof pathway) and the aerobic sequence (citric acid cycle). Pyruvate itself does not enter into the citric acid cycle. It is first converted to acetyl coenzyme A (acetyl-CoA), a complex substance that, like ATP, is of great importance in metabolism:

$$CH_3\overset{\overset{\displaystyle O}{\|}}{C}-COO^- + \underset{\text{coenzyme A}}{CoASH} + NAD^+ \longrightarrow \underset{\text{acetyl-CoA}}{CH_3\overset{\overset{\displaystyle O}{\|}}{C}-SCoA} + NADH + CO_2$$

This important reaction depends on the availability of several vitamins. In addition to niacin (needed for the synthesis of NAD^+), riboflavin and thiamine are also required. This is one of the few metabolic reactions that require thiamine, but because the reaction is key to obtaining large amounts of energy from carbohydrate, thiamine is a vitamin of major importance. Also, our bodies need the vitamin pantothenic acid to synthesize coenzyme A.

Acetyl coenzyme A consists of an acetyl group bonded to a coenzyme A group. Coenzyme A contains the following units: adenine, ribose-3-phosphate, diphosphate, pantothenic acid, and thioethanolamine. Coenzyme A is abbreviated as CoASH or CoA. Acetyl coenzyme A is abbreviated as acetyl-CoA or acetyl-SCoA. The following is the simplified structure of acetyl-CoA:

Acetyl — Thioethanolamine — Pantothenic acid — Diphosphate — Ribose-3-phosphate — Adenine

acetyl-CoA

The acetyl group is the group that is actually oxidized in the citric acid cycle. This group is attached to the large carrier molecule as a thioester—that is, by an ester linkage in which oxygen is replaced by sulfur:

$$CH_3-\overset{\overset{\displaystyle O}{\|}}{C}-S-CoA$$

acetyl group (CH_3-C); thioester linkage (S); carrier group (CoA)

Not only is acetyl-CoA a key component of carbohydrate metabolism, but also (as we will see in Chapter 35), it serves to tie the metabolism of fats and that of certain amino acids to the citric acid cycle.

The **citric acid cycle** was elucidated by Hans A. Krebs (1900–1981), a British biochemist; thus, it is also called the Krebs cycle. For his studies in intermediary metabolism, Krebs shared the 1953 Nobel Prize in medicine and physiology with Fritz A. Lipmann (1899–1986), an American biochemist who discovered coenzyme A.

citric acid cycle

Krebs showed the citric acid cycle to be a series of eight reactions in which the acetyl group of acetyl-CoA is oxidized to carbon dioxide and water. As these reactions take place, many reduced coenzymes (both NADH and $FADH_2$) are formed. These reduced coenzymes then pass electrons through the electron-transport system and, in the process, produce ATP. The citric acid cycle occurs in the mitochondria, which are the primary sites for the production of cellular energy.

The sequence of reactions involved in the citric acid cycle is shown in Figure 34.4. It is important to note that this cycle produces little usable cellular energy directly (only one GTP, or guanosine triphosphate, convertible to ATP). However, many ATPs are produced from two other processes: electron transport and oxidative phosphorylation. (See Section 33.6.) Electron transport recycles the large number of reduced coenzymes formed by the citric acid cycle. As a consequence, oxidative phosphorylation produces ATP from ADP and phosphate. Because electron transport requires oxygen and the citric acid cycle depends upon electron transport, the citric acid cycle is an aerobic process.

All three processes—the citric acid cycle (plus the initial conversion of pyruvate to acetyl-CoA), electron transport, and oxidative phosphorylation—team up to produce energy for the cell. All three take place in the mitochondria. Overall, these three processes result in the oxidation of 1 pyruvate ion to 3 carbon dioxide molecules (and the formation of a maximum of 15 ATP molecules):

$$CH_3\overset{\overset{\displaystyle O}{\|}}{C}-COO^- \longrightarrow 3\,CO_2$$

A large quantity of energy (1260 kJ) becomes available to the cell when these three carbon atoms are fully oxidized. Most of this energy is used to reduce coenzymes:

$$4\,NAD^+ + FAD + 10\,H^+ + 10\,e^- \longrightarrow 4\,NADH + 4\,H^+ + FADH_2$$

These coenzymes, in turn, yield a total of 14 ATP molecules via mitochondrial electron transport and oxidative phosphorylation. Each mole of ATP stores approximately 35 kJ of energy. If we include the single GTP (convertible to ATP) that is formed in the citric acid cycle, the cell obtains about 462 kJ from each mole of pyruvate that is oxidized. By using an aerobic process, the cell produces 30 ATP molecules from the two lactates after gaining only 2 ATPs via the anaerobic pathway—that is, glycolysis. The presence of oxygen yields a large energy bonus for the cell.

Energy summation:
Glucose

glycolysis	2 ATP
2 Lactate	
citric acid cycle, e^- transport, oxidative phosphorylation	+30 ATP
6 CO_2	32 ATP

Myocardial infarction and stroke are two injuries that are especially serious because they deprive rapidly metabolizing tissue of oxygen. When the heart muscle loses at least part of its normal blood supply, myocardial infarction results; a similar occurrence in the brain results in a stroke. In both cases, a lack of blood circulation means that at least part of the tissues lose their normal oxygen supply. Suddenly, these cells lose most of their capacity to generate ATP. Therefore, the anaerobic process of glycolysis cannot support continued cell viability. Permanent tissue damage results from only minutes of oxygen deprivation.

Practice 34.4

Using the citric acid cycle, calculate how many moles of (a) NADH and (b) $FADH_2$ are produced from 7.8 mol of acetyl-CoA.

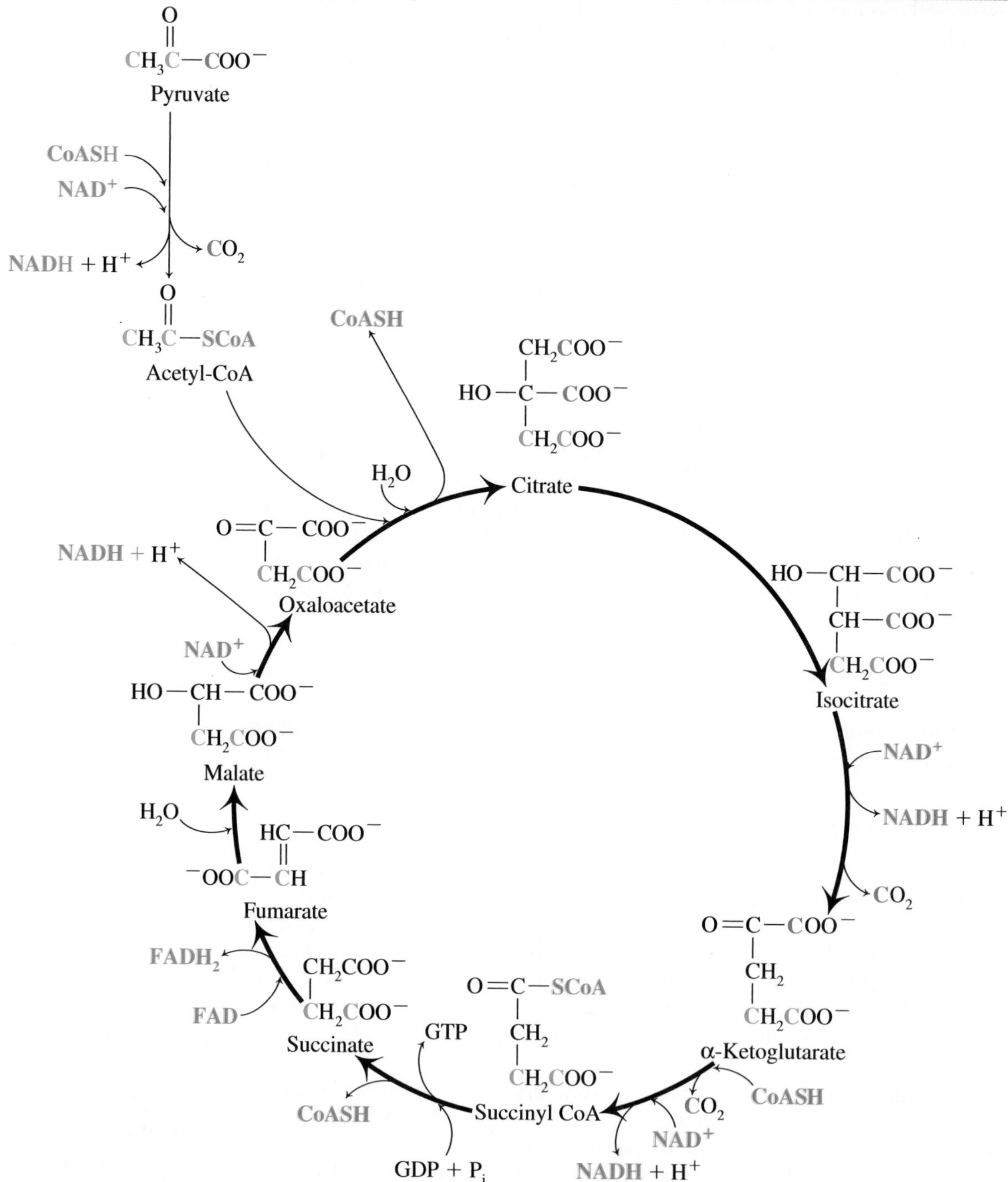

Figure 34.4
The citric acid cycle (Krebs cycle). During one cycle, (1) the carbons marked in blue enter the cycle, and (2) the carbons marked in red are lost as CO_2. Coenzymes are marked in green. The cycle is started by the reaction of acetyl-CoA with oxaloacetate to form citrate.

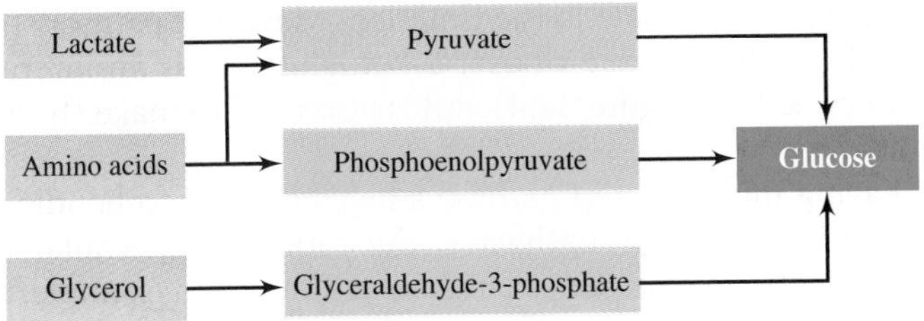

Figure 34.5
An overview of gluconeogenesis. All transformations except lactate to pyruvate require a series of reactions.

34.6 Gluconeogenesis

A continuous supply of glucose is needed by the body, especially for the brain and the nervous system. But the amount of glucose and glycogen present in the body will last for only about four hours at normal metabolic rates. Thus, there is a metabolic need for a pathway that produces glucose from noncarbohydrate sources. The formation of glucose from noncarbohydrate sources is called **gluconeogenesis**.

gluconeogenesis

Most of the glucose formed during gluconeogenesis comes from lactate, certain amino acids, and the glycerol of fats. In each case, these molecules are converted to Embden–Meyerhof pathway intermediates. (Figure 34.3.) As shown previously, lactate is converted to pyruvate. Amino acids first undergo *deamination* (loss of amino groups) and then are converted to pyruvate or phosphoenolpyruvate. Glycerol is converted to glyceraldehyde-3-phosphate. (See Figure 34.5.) Most of the steps in the Embden–Meyerhof pathway are reversed to transform these pathway intermediates into glucose.

Gluconeogenesis takes place primarily in the liver and also in the kidneys. These organs have the enzymes that catalyze reversal of the Embden–Meyerhof pathway. Because of this capability, the liver is the organ that is primarily responsible for maintaining normal blood-sugar levels.

Long-distance runners use gluconeogenesis to provide a source of glucose when more readily available supplies are exhausted.

34.7 Overview of Complex Metabolic Pathways

When we examine a metabolic pathway such as the Embden–Meyerhof pathway (Figure 34.3) or the citric acid cycle (Figure 34.4), the sheer complexity can be overwhelming. We are tempted to ask why a cell uses so many reactions to achieve its goal. Biochemists have asked this question for many years. It appears that a physiological design that includes a number of reactions per pathway achieves several vital objectives for the cell.

First, many of the chemicals that form in the middle of a pathway, *pathway intermediates*, are also used in other metabolic processes within the cell:

$$A \rightleftharpoons B \rightleftharpoons C \rightleftharpoons D \qquad \text{Pathway I}$$
(B, C = pathway intermediates)

$$B \rightleftharpoons X \rightleftharpoons Y \rightleftharpoons Z \qquad \text{Pathway II}$$

$$C \rightleftharpoons I \rightleftharpoons J \rightleftharpoons K \qquad \text{Pathway III}$$

These intermediates are like interchangeable machine parts. Once such parts are made, they can be used in more than one machine. For example, glucose-

6-phosphate is an intermediate in the Embden–Meyerhof pathway (see Figure 34.3) and is also used to make ribose; α-ketoglutarate is an intermediate in the citric acid cycle (see Figure 34.4) and is also used to make the amino acid L-glutamic acid.

Second, having multiple-step pathways helps the cell to handle metabolic energy efficiently. For many pathways, the total energy available is much greater than the cell can handle in a single reaction. As an example, a one-step complete oxidation of glucose yields enough energy to cook a cell, figuratively speaking. To avoid such disasters, the cell extracts only a little energy from glucose at each chemical reaction (Figure 34.6). The quantity of energy released in each step is small enough to be handled by the cell. Thus, for a cell to extract the maximum energy, a metabolic pathway must have a number of steps.

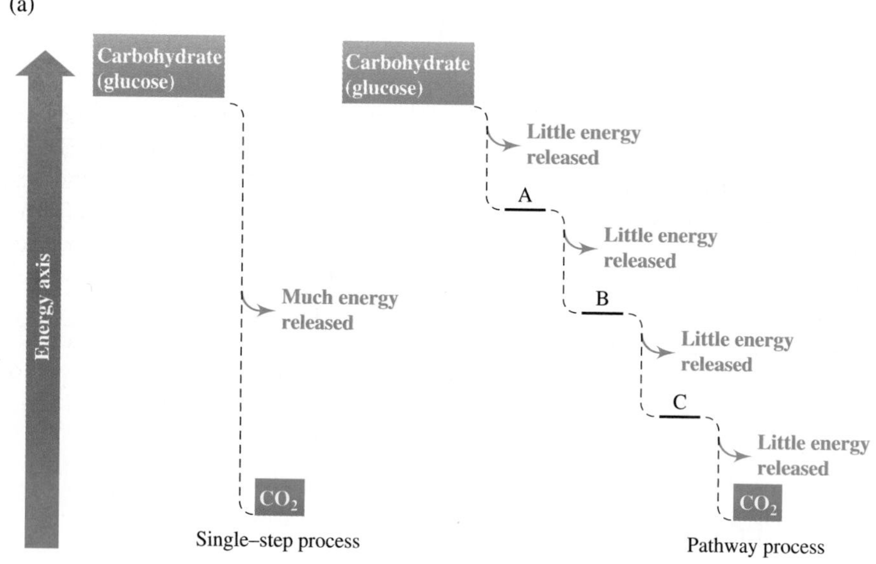

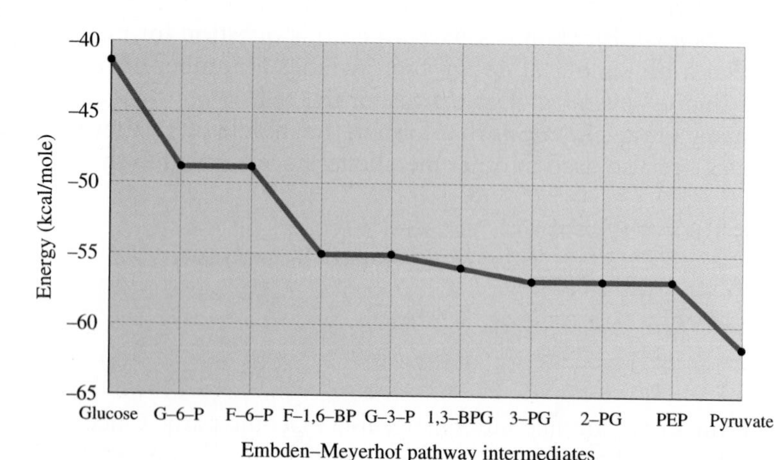

Figure 34.6
Metabolic energy changes.
(a) A single-step oxidation process compared with a multiple-step process. In the pathway, A, B, and C represent hypothetical pathway intermediates. (b) The actual energy change for each reaction in the Embden–Meyerhof path (G-6-P=glucose-6-phosphate, F-6-P=fructose-6-phosphate, F-1,6-BP=fructose-1,6-bisphosphate, G-3-P=glyceraldehyde-3-phosphate, 1,3-BPG=1,3-bisphosphoglycerate, 3-PG=3-phopshoglycerate, 2-PG=2-phosphoglycerate, PEP=phosphoenolpyruvate).

Metabolic reaction sequences seem needlessly complex because we see only a small part of the total cellular function. Although we study the processes separately, it is important to remember that they are all interrelated.

34.8 Hormones

hormones

Hormones are chemical substances that act as control agents in the body, often regulating metabolic pathways. Hormones help to adjust physiological processes such as digestion, metabolism, growth, and reproduction. For example, the concentration of glucose in the blood is maintained within definite limits by hormonal action. Hormones are secreted by the endocrine, or ductless, glands directly into the bloodstream and are transported to various parts of the body where they exert specific control functions. The endocrine glands include the thyroid, parathyroid, pancreas, adrenal, pituitary, ovaries, testes, placenta, and certain portions of the gastrointestinal tract. A hormone produced by one species is usually active in some other species. For example, the insulin used to treat diabetes mellitus in humans can be obtained from the pancreas of animals slaughtered in meatpacking plants.

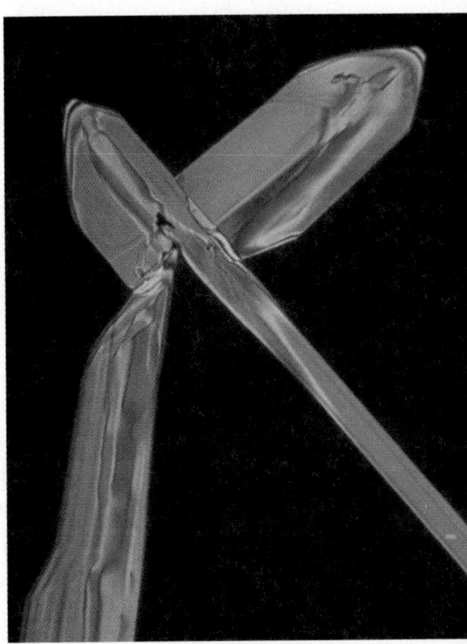

A crystal of thyroxine. This hormone is produced in the thyroid gland and controls the metabolic rate.

Hormones are often called the chemical messengers of the body. They do not fit into any single chemical structural classification. Many are proteins or polypeptides, some are steroids, and others are phenol or amino acid derivatives; examples are given in Figure 34.7. Because a lack of any hormone often produces serious physiological disorders, many hormones are produced synthetically or are extracted from their natural sources and made available for medical use.

Like vitamins, hormones are generally needed in only minute amounts. Concentrations range from 10^{-6} M to 10^{-12} M. Unlike vitamins, which must be supplied in the diet, the necessary hormones are produced in the body of a healthy person. A number of hormones and their functions are listed in Table 34.1.

I, I, I, I, HO, O, $CH_2CHCOOH$, NH_2

thyroxine

Cys-Tyr-Ile-Gln-Asn-Cys-Pro-Leu-Gly-NH_2 (S—S bridge between the two Cys)

oxytocin

1, 10, 15
His-Ser-Gln-Gly-Thr-Phe-Thr-Ser-Asp-Tyr-Ser-Lys-Tyr-Leu-Asp-Ser-Arg-Arg-
20, 29
Ala-Gln-Asp-Phe-Val-Gln-Tyr-Leu-Met-Asn-Thr

glucagon

CH_3, OH, CH_3, O

testosterone

CH_3, OH, HO

estradiol (estrogen)

OH, OH, $HOCHCH_2NHCH_3$

epinephrine (adrenalin)

Figure 34.7
Structure of selected hormones. Thyroxine is produced in the thyroid gland; oxytocin is a polypeptide produced in the posterior lobe of the pituitary gland; glucagon is a polypeptide produced in the pancreas; testosterone and estradiol are steroid hormones produced in the testes and the ovaries, respectively; epinephrine is produced in the adrenal glands.

Table 34.1 Selected Hormones and Their Functions

Hormone	Source	Principal functions
Insulin	Pancreas	Controls blood-glucose level and storage of glycogen
Glucagon	Pancreas	Stimulates conversion of glycogen to glucose; raises blood-glucose level
Oxytocin	Pituitary gland	Stimulates contraction of the uterine muscles and secretion of milk by the mammary glands
Vasopressin	Pituitary gland	Controls water excretion by the kidneys; stimulates constriction of the blood vessels
Growth hormone	Pituitary gland	Stimulates growth
Adrenocorticotrophic hormone (ACTH)	Pituitary gland	Stimulates the adrenal cortex, which, in turn, releases several steroid hormones
Prolactin	Pituitary gland	Stimulates milk production by mammary glands after birth of a baby
Epinephrine (adrenalin)	Adrenal glands	Stimulates rise in blood pressure, accelerates the heartbeat, decreases secretion of insulin, and increases blood glucose
Cortisone	Adrenal glands	Helps control carbohydrate metabolism, salt and water balance, formation and storage of glycogen
Thyroxine and liothyronine (triiodothyronine)	Thyroid gland	Increases the metabolic rate of carbohydrates and proteins
Calcitonin	Thyroid gland	Prevents the rise of calcium in the blood above the required level
Parathyroid hormone	Parathyroid gland	Regulates the metabolism of calcium and phosphate in the body
Gastrin	Stomach	Stimulates secretion of gastric juice
Secretin	Duodenum	Stimulates secretion of pancreatic juice
Estrogen	Ovaries	Stimulates development and maintenance of female sexual characteristics
Progesterone	Ovaries	Stimulates female sexual characteristics and maintains pregnancy
Testosterone	Testes	Stimulates development and maintenance of male sexual characteristics

Practice 34.5

Name the hormone in Figure 34.7 that can be described as an amino acid.

34.9 Blood Glucose and Hormones

An adequate blood-glucose level must be maintained to ensure good health. To achieve this goal, hormones regulate and coordinate metabolism in specific organs. The hormones control selected enzymes, which, in turn, regulate the rates of reaction in the appropriate metabolic pathways. Let's examine some of the physiological mechanisms that maintain proper blood-glucose levels.

Glucose concentrations average about 70–100 mg/100 mL of blood under normal fasting conditions—that is, when no nourishment has been taken for several hours. Most people are in a normal fasting condition before eating breakfast. After the ingestion of carbohydrates, the glucose concentration rises above the normal level, and a condition of *hyperglycemia* exists. If the concentration of glucose rises still further, the renal threshold for glucose is eventually reached. The *renal threshold* is the concentration of a substance in the blood above which the kidneys begin to excrete that substance into the urine. The renal threshold for glucose is about 140–170 mg/100 mL of blood. Glucose excreted by the kidneys can be detected in the urine by a test for reducing sugars (e.g., the Benedict test; see Section 27.13). When the glucose concentration of the blood is below the normal fasting level, hypoglycemia exists. (See Figure 34.8.)

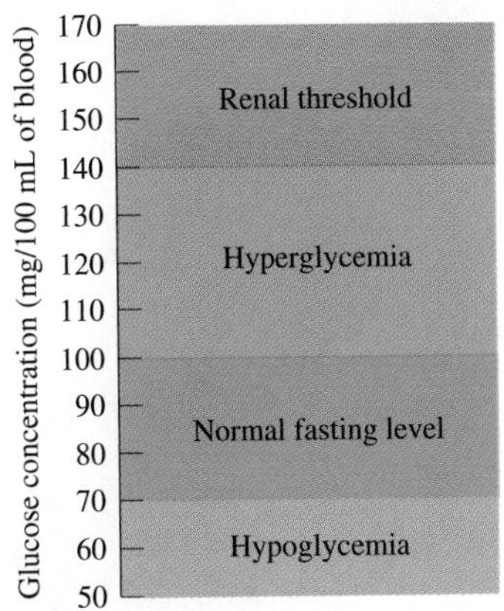

Figure 34.8
Conditions related to the concentration of glucose in the blood.

Glucose concentration in the blood is under the control of various hormones. These hormones act as checks on one another and establish an equilibrium condition called *homeostasis*—that is, self-regulated equilibrium. Three hormones—insulin, epinephrine (adrenalin), and glucagon—are of special significance in maintaining the glucose concentration within the proper limits. Insulin, secreted by the islets of Langerhans in the pancreas, reduces blood-glucose levels by increasing the rate of glycogen formation. Epinephrine from the adrenal glands and glucagon from the pancreas increase the rate of glycogen breakdown (glycogenolysis) and thereby increase blood-glucose levels. These opposing effects are summarized as follows:

$$\text{Blood glucose} \underset{\text{glycogenolysis}}{\overset{\text{glycogenesis}}{\rightleftharpoons}} \text{Glycogen}$$

Glycogenesis: Stimulated by insulin
Glycogenolysis: Stimulated by epinephrine or glucagon

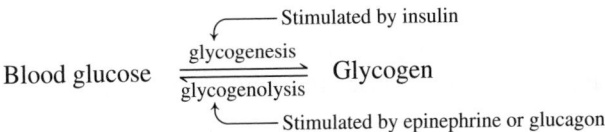

During the digestion of a meal rich in carbohydrates, the blood-glucose level of a healthy person rises into the hyperglycemic range. This stimulates insulin secretion, and the excess glucose is converted to glycogen, thereby returning the glucose level to normal. A large amount of ingested carbohydrates can overstimulate insulin production and thereby produce a condition of mild hypoglycemia. This in turn triggers the secretion of additional epinephrine and glucagon, and the blood-glucose levels are again restored to normal. The body is able to maintain the normal fasting level of blood glucose for long periods without food by drawing on liver glycogen, muscle glycogen, and, finally, body fat as glucose replacement sources. Thus, in a normal person, neither hyperglycemia nor mild hypoglycemia has serious consequences, since the body is able to correct these conditions. However, either condition, if not corrected, can have very serious consequences. Since the brain is heavily dependent on blood glucose for energy, hypoglycemia affects the brain and the central nervous system. Mild hypoglycemia can result in impaired vision, dizziness, and fainting spells. Severe hypoglycemia produces convulsions and unconsciousness; prolonged, it may result in permanent brain damage and death.

Hyperglycemia can be induced by fear or anger, because the rate of epinephrine secretion is increased under emotional stress. Glycogen hydrolysis is thereby speeded up and glucose-concentration levels rise sharply. This whole sequence readies the individual for the strenuous effort of either fighting or fleeing as the situation demands.

Glucose monitors can help diabetics adjust insulin levels. For example, CGMS® System Gold is used by physicians to track continuous blood sugar patterns in people with diabetes.

Type 1 diabetes mellitus (also known as juvenile-onset or insulin-dependent diabetes) is a serious metabolic disorder characterized by hyperglycemia, glycosuria (glucose in the urine), frequent urination, thirst, weakness, and loss of weight. Prior to 1921, diabetes often resulted in death. In that year, Frederick Banting and Charles Best, working at the University of Toronto, discovered insulin and devised methods for extracting the hormone from animal pancreases. For his work on insulin, Banting, with J. J. MacLeod, received the Nobel Prize in medicine and physiology in 1923. Insulin is highly effective in controlling diabetes. It must be given by injection, because, like any other protein, it would be hydrolyzed to amino acids in the gastrointestinal tract.

People with mild or borderline diabetes may show normal fasting blood-glucose levels, but they are unable to produce sufficient insulin for prompt control of ingested carbohydrates. As a result, their blood glucose rises to an abnormally high level and does not return to normal for a long period of time. Such a person has a decreased tolerance for glucose, which may be diagnosed by a glucose-tolerance test. After fasting for at least 12 hours, blood and urine specimens are taken to establish a reference level. The person then drinks a solution containing 100 g of glucose (the standard amount for adults). Blood and urine specimens are collected at 0.5-, 1-, 2-, and 3-hour intervals and tested for glucose content. In a normal situation, the blood-glucose level returns to normal after about 3 hours. Individuals with mild diabetes show a slower drop in glucose levels, but the glucose level in a severe diabetic remains high for the entire 3 hours. Responses to a glucose tolerance test are shown in Figure 34.9.

In 1978, the synthesis of insulin, identical in structure to that made by the human pancreas, was announced. In 1982, this insulin became the first genetically engineered human protein to receive FDA approval for sale. Today, many diabetics choose to use human insulin rather than the animal insulin obtained as a by-product of meatpacking.

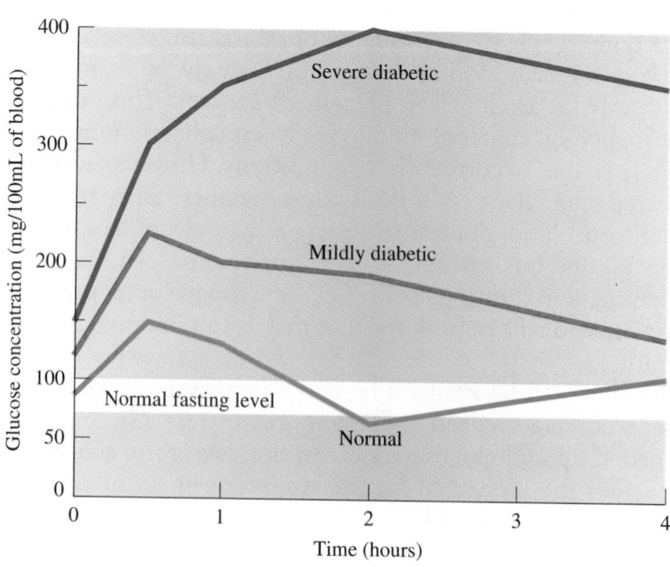

Figure 34.9
Typical responses to a glucose-tolerance test.

CHEMISTRY IN ACTION • Eating Too Much? Don't Listen to Your Stomach, Listen To Your Fat Cells!

When people say "listen to your stomach," they often mean "eat only when you are hungry." If we eat only when we are hungry, we may stay thinner. However, most obesity is intractable to this simple solution. For example, it has little impact on metabolic syndrome, a complicated problem associated with obesity, high blood pressure, and insulin resistance that is fast becoming a new American epidemic. Metabolic syndrome is metabolism out of balance. Although a hormone dysfunction is suspected, the normal culprits like insulin are not directly implicated. Instead, a new set of chemical signals appear to be involved.

These people are probably suffering from metabolic syndrome.

Some important observations pointed to an unusual source of hormonal signals. Danger is associated with body fat in particular locations. Fat around the waist (giving an "apple body shape") leads to more problems than fat in the lower extremities (giving a "pear body shape"). A specific excess of visceral fat (fat in the abdominal cavity or around internal organs) correlates with the symptoms of metabolic syndrome. Visceral fat cells influence overall body metabolism like endocrine glands.

In the past, fat cells were defined as simple depots for the storage of triacylglycerols. Micrographs of these cells show that more than 90% of their volume is occupied by a large fat droplet. Scientists assumed that fat cells were metabolically simple. Wrong!! Over the last 10 years, it has been shown that fat cells are the source of a group of polypeptides with strong hormone-like activities, the adipocytokines. Instead of simply storing excess metabolic energy as triacylglycerols, fat cells help coordinate how energy is used by all tissues.

Fat cells take charge in the following ways:

1. When fat cells are filled with triacylglycerol, they send a signal to the brain that suppresses appetite and increases the overall metabolic rate. The signal is the adipocytokine, leptin.
2. Fat cells alter the function of insulin, a key hormone in energy metabolism. When fat cells are filled with triacylglycerol, they limit the insulin impact by secreting the adipocytokine resistin. Alternatively, when fat cells are lean and short of triacylglycerol, they release adiponectin to make tissues more sensitive to the impact of insulin.

Scientists are far from understanding all the actions of the adipocytokines (they may not even have discovered them all). However, they do conclude that adipocytokines control food intake and metabolism based on the amount of stored fat. Metabolic syndrome may result from not listening (at least figuratively) to our fat cells.

Chapter 34 Review

34.1 Metabolic Pathways

KEY TERM

Metabolic pathway

- A metabolic pathway is a series of biochemical reactions that serve a specific purpose.
- Photosynthesis is an example of an anabolic pathway; it uses sunlight to make glucose from carbon dioxide.
- Glucose releases metabolic energy as it is converted back to carbon dioxide by catabolic pathways.

34.2 Exercise and Energy Metabolism

- All life's processes require metabolic energy.
- Metabolic energy for exercise is supplied via multiple routes:
  - ATP provides an immediately available energy source but is in short supply.

- The next available energy source is glycogen.
- In humans, the liver converts lactic acid back to glucose, allowing extended periods of muscle exertion.

34.3 The Bloodstream: A Metabolic Connection

KEY TERMS

Glycogenesis
Glycogenolysis

- The bloodstream carries nutrients and oxygen to tissues and removes waste molecules, hydrogen ions, and carbon dioxide.
- Lungs restore the gas balance of the bloodstream by replacing carbon dioxide with oxygen.
- The liver maintains the blood-glucose level by converting lactate to glucose and releasing glucose into the blood.
- Glycogen is the storage form for glucose.
  - Glycogenesis is the synthesis of glycogen from glucose.
  - Glycogenolysis is the catabolic pathway that converts glycogen to glucose.
- Glucose suppies metabolic energy via several pathways:
  - In the absence of molecular oxygen, the Embden–Meyerhof pathway oxidizes glucose to pyruvate and lactate.
  - In the presence of O_2, pyruvate is converted to carbon dioxide by the citric acid cycle.
- Glycogen-storage diseases illustrate the importance of coordination between various tissues during glucose metabolism.
  - Von Gierke disease limits the liver's ability to export glucose: blood-glucose levels are low and the liver has excess glycogen.
  - McArdle disease limits the muscle's ability to break down glycogen: The muscles are healthy when resting but are not able to do severe exercise.

34.4 Anaerobic Sequence

KEY TERMS

Embden–Meyerhof pathway
Glycolysis

- The Embden–Meyerhof pathway is the anaerobic conversion of glucose to pyruvate.
  - This metabolic pathway has ten reactions.
  - It is catabolic since glucose is oxidized.
- A single oxidation–reduction reaction supplies most of the energy derived from the Embden–Meyerhof pathway.
- Compounds containing high-energy phosphate bonds, 1,3-bisphosphoglycerate and phosphoenolpyruvate, are used to make ATP in this pathway.
- The Embden–Meyerhof pathway is anaerobic because no molecular oxygen is used in this process.
- Glycolysis produces lactate from glucose by combining the Embden–Meyerhof pathway with a reaction that reduces pyruvate to lactate.
  - Yeast converts pyruvate to ethanol following the Embden–Meyerhof pathway.
- The Embden–Meyerhof pathway is inefficient, releasing only about 10% of the energy available from the glucose molecule.

34.5 Citric Acid Cycle (Aerobic Sequence)

KEY TERM

Citric acid cycle

- Pyruvate is used by the citric acid cycle after being converted to acetyl-CoA.
  - Coenzyme A is a complex molecule that forms a thioester with acetate.
- The citric acid cycle is a cyclic metabolic pathway that converts the acetate carbons of acetyl-CoA to carbon dioxide.
  - This pathway produces many reduced coenzymes (NADH and $FADH_2$).
  - The citric acid cycle is found in the mitochondria.
  - By combining electron transport and oxidative phosphorylation with the citric acid cycle, the mitochondria produce many ATPs.
- Glycolysis yields two ATPs, while these mitochondrial processes yield thirty more ATPs as two lactates are converted to carbon dioxide.

34.6 Gluconeogenesis

KEY TERM

Gluconeogenesis

- Gluconeogenesis is the formation of glucose from noncarbohydrate precursors.
- Gluconeogenesis commonly starts from lactate, amino acids, or glycerol.
- The liver and the kidneys are responsible for most gluconeogenesis in humans.
- Gluconeogenesis uses the reverse of many of the Embden–Meyerhof reactions.

34.7 Overview of Complex Metabolic Pathways

- Multiple-step metabolic pathways are advantageous to life:
  - Pathway intermediates may be used in more than one metabolic process.
  - Cellular energy use is more effective when only small amounts of energy are released in each step.

34.8 Hormones

KEY TERM

Hormones

- Hormones are chemical substances that act as control agents in the body, often regulating metabolic pathways.
- Hormones are secreted by the endocrine glands and act as chemical messengers.
- Hormones are effective at very small concentrations, $10^{-6}\ M$ to $10^{-12}\ M$.

34.9 Blood Glucose and Hormones

- Normal fasting blood-glucose levels range from about 70 to 100 mg/100 mL blood.

- After eating, blood-glucose concentrations rise and a condition of hyperglycemia exists.
- If the blood-glucose concentration increases further, the renal threshold is reached and glucose is passed into the urine.
- When the blood glucose is below fasting levels, hypoglycemia pertains.
- The hormones insulin, glucagon, and epinephrine are primarily responsible for maintaining a self-regulated equilibrium (homeostasis) for blood glucose.
  - Insulin, secreted by the pancreas, lowers blood-glucose levels and stimulates glycogen synthesis.
  - Epinephrine (from the adrenal glands) and glucagon (from the pancreas) increase the rate of glycogen breakdown and increase blood-glucose levels.
- Type 1 diabetes (juvenile-onset diabetes, insulin-dependent diabetes) is characterized by insufficient amounts of insulin.

Review Questions

All questions with blue numbers have answers in the appendix of the text.

1. Why are carbohydrates considered to be energy-storage molecules?
2. List two general ways in which glucose catabolism differs from photosynthesis.
3. How much energy would be released if 3 mol of glucose were burned to form carbon dioxide and water?
4. Why are enzymes important in metabolism?
5. What form of chemical energy is used first for muscle contraction? What is used next?
6. How is the liver important in muscle contraction?
7. What is the purpose of the final reactions in anaerobic glucose metabolism?
8. How many high-energy phosphate bonds are directly formed in the citric acid cycle? List and identify the reduced coenzymes that are produced in this cycle.
9. What is acetyl-CoA and what is its function in metabolism?
10. What are the general functions of hormones and where are they produced?
11. (a) What is the range of glucose concentration in blood under normal fasting conditions?
    (b) What blood-glucose concentrations are considered to be hyperglycemic? Hypoglycemic?
12. What is meant by the renal threshold?
13. Why is gluconeogenesis defined as an anaerobic process?
14. Your friend tells you that the citric acid cycle doesn't involve carbon oxidation. Do you agree or disagree? Briefly explain.

Paired Exercises

All exercises with blue numbers have answers in the appendix of the text.

1. Draw the structure of the carbohydrate that is produced during glycogenolysis.
2. Draw the structure of the carbohydrate that is produced during glycogenesis.
3. Would you define glycogenesis as anabolic or catabolic? Explain briefly.
4. Would you define glycogenolysis as anabolic or catabolic? Explain briefly.
5. In hyperventilation too much carbon dioxide is exhaled. How does this affect blood acidity?
6. A drug-induced coma leads to very shallow breathing. How might this affect blood acidity?
7. Is the ATP production that is associated with the citric acid cycle substrate-level phosphorylation or oxidative phosphorylation? Explain briefly.
8. Is the ATP production that is associated with the Embden–Meyerhof pathway substrate-level phosphorylation or oxidative phosphorylation? Explain briefly.
9. Circle the high-energy phosphate bond(s) in the following structure:

```
O     O
||    ||
C—O—P—OH
|     |
|     O⁻
CHOH
|      O
|      ||
CH₂—O—P—OH
       |
       O⁻
```

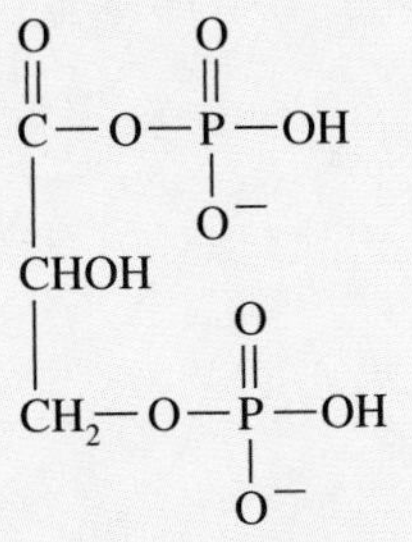

10. Circle the high-energy phosphate bond(s) in the following structure:

```
O
||
C—O⁻  O
|     ||
C—O—P—OH
||    |
CH₂   O⁻
```

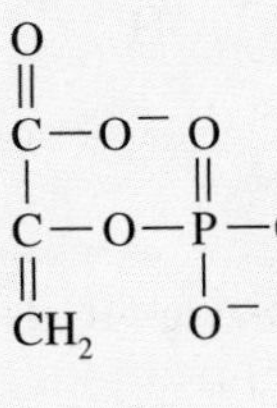

11. Explain why the Embden–Meyerhof pathway is considered to be anaerobic, even though the oxidation of glucose occurs there.

12. Explain why the citric acid cycle is considered to be aerobic, even though no molecular oxygen (O_2) is used in this cycle.

13. How many moles of lactate will be formed from 2.8 mol of glucose through glycolysis?

14. How many moles of ATP will be formed from 0.85 mol of glucose through glycolysis?

15. What are the end products of the anaerobic catabolism of glucose in yeast cells?

16. What are the end products of the anaerobic catabolism of glucose in muscle tissue?

17. Is the Embden–Meyerhof pathway catabolic or anabolic? Explain briefly.

18. Is the gluconeogenesis pathway catabolic or anabolic? Explain briefly.

19. Fill in the coenzyme reactant and product of the following reaction:

$$\begin{array}{l} CH_2COO^- \\ | \\ CH_2COO^- \end{array} + \_\_ \longrightarrow \begin{array}{l} HC{-}COO^- \\ \| \\ CH{-}COO^- \end{array} + \_\_$$

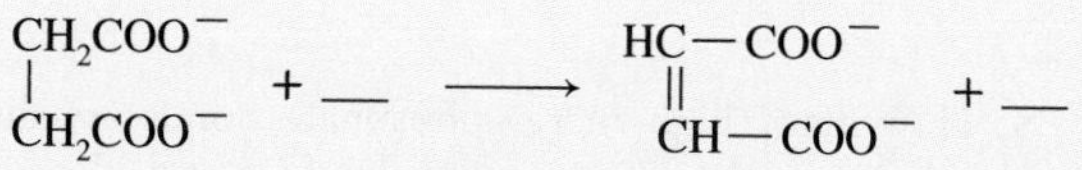

20. Fill in the coenzyme reactant and product of the following reaction:

$$\begin{array}{l} HO{-}CHCOO^- \\ \quad\;\; | \\ \quad CH_2COO^- \end{array} + \_\_ \longrightarrow \begin{array}{l} O{=}C{-}COO^- \\ \quad\; | \\ \quad CH_2{-}COO^- \end{array} + \_\_ + H^+$$

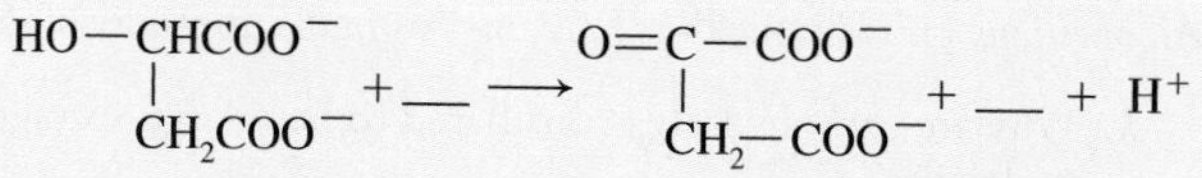

21. Why are electron transport and oxidative phosphorylation needed when the cell uses the citric acid cycle to produce energy?

22. Why are electron transport and oxidative phosphorylation not needed when the cell uses the glycolysis pathway to produce energy?

23. How many moles of ATP can be formed from 0.6 mol of acetyl CoA, using the citric acid cycle, electron transport, and oxidative phosphorylation?

24. How many moles of ATP can be formed from 0.38 mol of acetyl CoA, using the citric acid cycle, electron transport, and oxidative phosphorylation?

25. In what way(s) are hormones like vitamins?

26. In what way(s) do hormones differ from vitamins?

27. Exercise and a low-carbohydrate diet can cause a blood-glucose concentration of 65 mg/100 mL of blood. Is this hyperglycemic or hypoglycemic? Explain briefly.

28. After a meal, a diabetic can have a blood-glucose concentration of 350 mg/100 mL of blood. Is this hyperglycemic or hypoglycemic? Explain briefly.

29. Briefly describe several common fates for pyruvate produced through the Embden–Meyerhof pathway in muscle tissue.

30. Briefly describe several common fates for glucose produced through gluconeogenesis in the liver.

31. Choose and give the name for the steroid in the following hormones:

32. Choose and give the name for the polypeptide found in the hormones in Exercise 31.

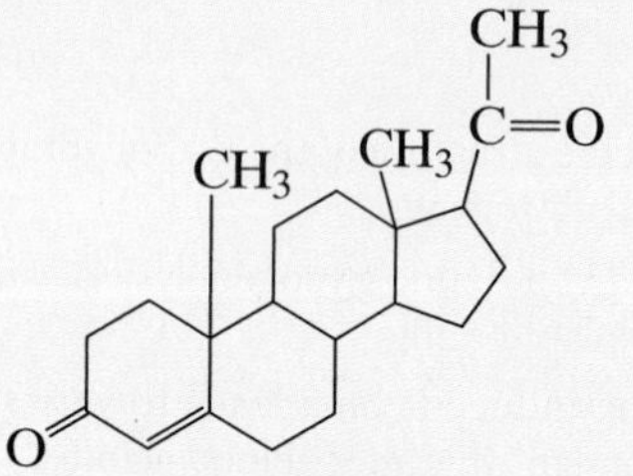

CH_3, $C{=}O$, CH_3, CH_3, O

Progesterone

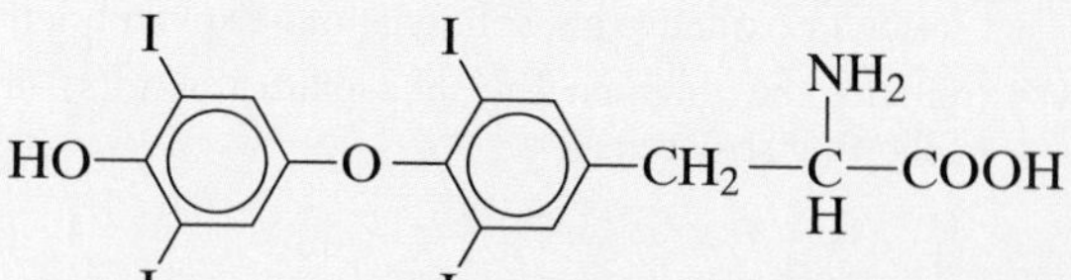

I, I, NH_2, HO, O, $CH_2{-}C{-}COOH$, H, I, I

thyroxine

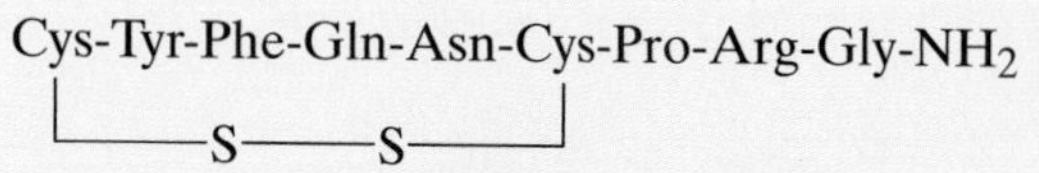

Cys-Tyr-Phe-Gln-Asn-Cys-Pro-Arg-Gly-NH_2

S——S

vasopressin

Additional Exercises

All exercises with blue numbers have answers in the appendix of the text.

33. Why was the citric acid cycle not involved in the metabolism of life-forms before the evolution of photosynthesis?

34. How does the function of hormones in the body differ from that of enzymes?

35. Why is the hormone insulin not effective when taken orally?

36. Predict what might happen to blood-glucose concentrations if a large overdose of insulin is taken by accident.

37. Why is epinephrine sometimes called the emergency or crisis hormone?

38. Write the structure of lactate. Would you predict that this molecule could provide cellular energy? Explain briefly.

39. In our bodies, what would be the most efficient way of using the energy produced: a multireaction system or a single-step reaction? Explain.

40. How does the use of pyruvate differ in aerobic catabolism and anaerobic catabolism?

41. What are the common structural features in NAD, FAD, ADP, ATP, and acetyl-CoA?

42. In animal metabolism, glucose is converted to carbon dioxide. Is this overall process oxidation or reduction? Explain.

43. Define the following terms: (a) glycolysis, (b) gluconeogenesis, (c) glycogenesis, (d) glycogenolysis.

44. Write the net reaction for the formation of two pyruvate from glucose via the Embden–Meyerhof pathway. (Your answer should include NAD^+ and ADP.)

45. Write structural formulas for (a) lactate, (b) pyruvate, (c) glucose-6-phosphate, (d) fructose-6-phosphate, (e) fructose-1,6-bisphosphate, and (f) adenosine triphosphate.

Challenge Exercise

46. Glucose can be catabolized under anaerobic or aerobic conditions.
    (a) Calculate the change in the average carbon oxidation number and the number of ATPs formed per carbon under anaerobic conditions. (Glucose is converted to two lactic acids.) What happens to net ATP production if the liver recycles the two lactic acids back to glucose? (Gluconeogenesis requires four ATP and two GTP to make glucose.)
    (b) Calculate the change in the average carbon oxidation number and the number of ATPs formed per carbon under aerobic conditions. (Glucose is converted to six carbon dioxides.) Can the body recycle the carbon dioxide? What happens to this carbon dioxide?

Answers to Practice Exercises

34.1 The process is catabolic because (a) a six-carbon compound is converted to a five-carbon compound and CO_2, and (b) the top two carbons of the reactant are oxidized.

34.2 As shown in Figure 34.2, pyruvate can be oxidized to carbon dioxide in the citric acid cycle or pyruvate can be converted back to glycogen via gluconeogenesis.

34.3 (a) 3.4 mol pyruvate (b) 3.4 mol ATP

34.4 (a) 23.4 mol NADH (b) 7.8 mol $FADH_2$

34.5 Thyroxine is the hormone in Figure 34.7 that has the general structure of an amino acid.

CHAPTER 35

Metabolism of Lipids and Proteins

Whales have an insulating layer of up to 30 cm in some species. Whale oil, extracted from this blubber, is used to make soap, leather dressing, lubricants, and hydrogenated fats. Whales provide Eskimos with food, clothing, illumination, and cooking oil.

Chapter Outline

One of the world's most pervasive nutritional problems is protein deficiency, known as kwashiorkor. It especially afflicts children and, when untreated, has a mortality rate between 30 and 90%. These young people suffer from growth retardation, anemia, and liver damage and often appear bloated because of excess water absorption.

In more affluent societies, nutritional problems are often associated with a high intake of saturated fat—stroke and heart disease are closely correlated with lipid intake.

These disparate nutritional problems point toward an important similarity between protein and lipid metabolism. Our biochemical processes require specific amino acids (proteins) and lipids. No matter how much food is available for our diet, we must also be concerned with meeting requirements for selected nutrients.

35.1 Metabolic Energy Sources

The ability to produce cellular energy is a vital characteristic of every cell's metabolism. As we have seen, carbohydrates are one of the major sources of cellular energy. The other two are lipids and proteins.

Of all the lipids, fatty acids are the most commonly used for cellular energy. Each fatty acid contains a long chain of reduced carbon atoms that can be oxidized to yield energy. Palmitic acid is one example:

$$CH_3CH_2CH_2CH_2CH_2CH_2CH_2CH_2CH_2CH_2CH_2CH_2CH_2CH_2CH_2COOH$$

palmitic acid

The average oxidation number of the carbon atoms in fatty acids is about -2, compared with 0 in carbohydrates. Thus, when catabolized (oxidized), fatty acids yield more energy per carbon atom than do carbohydrates.

Proteins (amino acids) are also a source of reduced carbon atoms that can be catabolized to provide cellular energy. In addition, amino acids provide the major pool of usable nitrogen for cells. Proteins and amino acids also perform diverse functions, some of which will be considered later in this chapter.

35.2 Fatty Acid Oxidation (Beta Oxidation)

Fats are the most energy-rich class of nutrients. Most of the energy from fats is derived from their constituent fatty acids. Palmitic acid derived from fat yields 39.1 kJ (9.35 kcal) per gram when burned to form carbon dioxide and water. By contrast, glucose yields only 15.6 kJ (3.73 kcal) per gram. Of course, fats are not actually burned in the body simply to produce heat. They are broken down in a series of enzyme-catalyzed reactions that also produce useful potential chemical energy in the form of ATP. In complete biochemical oxidation, the carbon and hydrogen of a fat ultimately are combined with oxygen (from respiration) to form carbon dioxide and water.

In 1904, Franz Knoop, a German biochemist, demonstrated that the catabolism of fatty acids involves a process whereby their carbon chains are shortened, two carbon atoms at a time. Knoop knew that animals do not metabolize benzene groups to carbon dioxide and water. Instead, the benzene nucleus remains attached to at least one carbon atom and is eliminated in the urine as a derivative of either benzoic acid or phenylacetic acid:

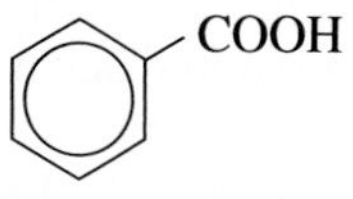

$$C_6H_5-COOH \qquad C_6H_5-CH_2COOH$$

benzoic acid 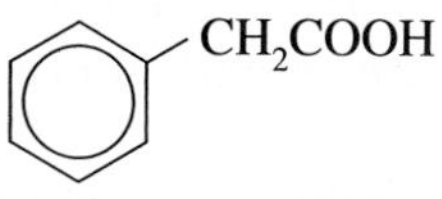 phenylacetic acid

Accordingly, Knoop prepared a homologous series of straight-chain fatty acids, with a phenyl group at one end and a carboxyl group at the other. He then fed these benzene-tagged acids to test animals. Phenylaceturic acid was identified in the urine of the animals that ate acids with an even number of carbon atoms; hippuric acid was present in the urine of the animals that consumed acids with an odd number of carbon atoms:

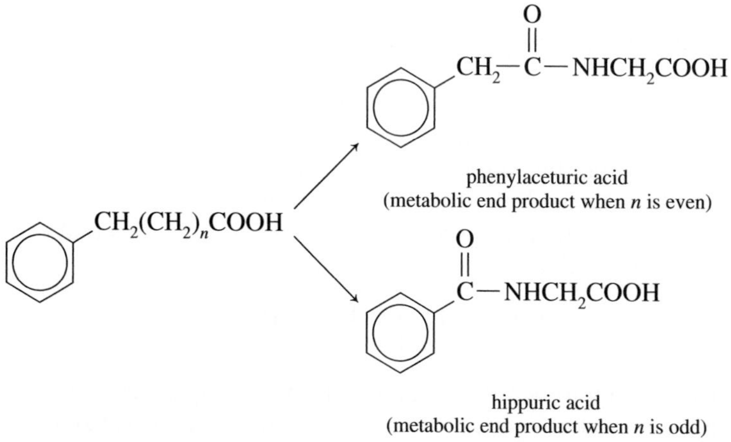

$$C_6H_5-CH_2(CH_2)_nCOOH \longrightarrow C_6H_5-CH_2-\overset{O}{\overset{\|}{C}}-NHCH_2COOH$$

phenylaceturic acid
(metabolic end product when n is even)

$$C_6H_5-CH_2(CH_2)_nCOOH \longrightarrow C_6H_5-\overset{O}{\overset{\|}{C}}-NHCH_2COOH$$

hippuric acid
(metabolic end product when n is odd)

These results indicated a metabolic pathway for fatty acids in which the carbon chain is shortened by two carbon atoms at each stage.

Knoop's experiments were remarkable for their time. They involved the use of tagged molecules and served as prototypes for modern research that utilizes isotopes to tag molecules.

Knoop postulated that the carbon chain of a fatty acid is shortened by successive removals of acetic acid units. The process involves the oxidation of the β-carbon atom and cleavage of the chain between the α and β carbons. A six-carbon fatty acid would produce three molecules of acetic acid thus:

This C is oxidized.

First reaction sequence

$$CH_3CH_2CH_2\overset{\beta}{C}H_2\overset{\alpha}{C}H_2COOH \longrightarrow CH_3CH_2CH_2COOH + CH_3COOH$$

Chain is cleaved here.

caproic acid — butyric acid — acetic acid

This C is oxidized.

Second reaction sequence

$$CH_3\overset{\beta}{C}H_2\overset{\alpha}{C}H_2COOH \longrightarrow CH_3COOH + CH_3COOH$$

Chain is cleaved here.

The general validity of Knoop's theory of β-carbon oxidation has been confirmed. However, the detailed pathway for fatty acid oxidation was not established until about 50 years after his original work. Like those of the Embden–Meyerhof and citric acid pathways, the sequence of reactions involved is another fundamental metabolic pathway. **Beta oxidation**, or the *two-carbon chop*, is accomplished in a series of reactions whereby the first two carbon atoms of the fatty acid chain become the acetyl group in a molecule of acetyl-CoA.

beta oxidation

In many ways, a thioester is like the more common ester that is formed from an alcohol and a carboxylic acid.

The catabolism proceeds with a fatty acid reacting with coenzyme A (CoASH) to form an activated thioester. The energy needed for this step of the catabolism is obtained from ATP:

Step 1 **Activation (formation of thioester with CoA):**

$$\underset{\text{fatty acid}}{RCH_2CH_2CH_2\overset{\overset{\displaystyle O}{\|}}{C}{-}OH} + CoASH + ATP \longrightarrow \underset{\text{CoA thioester of a fatty acid}}{RCH_2CH_2CH_2\overset{\overset{\displaystyle O}{\|}}{C}{-}SCoA} + AMP + \underset{\substack{\text{inorganic}\\\text{phosphate}}}{2\,P_i}$$

Next, the activated thioester undergoes four more steps in the reaction sequence, involving *oxidation*, *hydration*, *oxidation*, and *cleavage*, to produce acetyl-CoA and an activated thioester shortened by two carbon atoms. The cleavage reaction requires an additional molecule of CoA. The steps are as follows:

Hydration of a double bond means adding the components of water to the double-bonded carbons.

Step 2 **Oxidation [dehydrogenation at carbons 2 and 3 (α- and β-carbons)]:**

$$RCH_2CH_2CH_2\overset{\overset{\displaystyle O}{\|}}{C}{-}SCoA + FAD \longrightarrow RCH_2CH{=}CH\overset{\overset{\displaystyle O}{\|}}{C}{-}SCoA + FADH_2$$

Step 3 **Hydration (conversion to a secondary alcohol):**

$$RCH_2CH{=}CH\overset{\overset{\displaystyle O}{\|}}{C}{-}SCoA + H_2O \longrightarrow RCH_2\overset{\overset{\displaystyle OH}{|}}{C}HCH_2\overset{\overset{\displaystyle O}{\|}}{C}{-}SCoA$$

Step 4 **Oxidation [dehydrogenation of carbon 3 (β-carbon) to a keto group]:**

$$RCH_2\overset{\overset{\displaystyle OH}{|}}{C}HCH_2\overset{\overset{\displaystyle O}{\|}}{C}{-}SCoA + NAD^+ \longrightarrow RCH_2\overset{\overset{\displaystyle O}{\|}}{C}CH_2\overset{\overset{\displaystyle O}{\|}}{C}{-}SCoA + NADH + H^+$$

Step 5 **Carbon-chain cleavage (reaction with CoA to produce acetyl-CoA and an activated thioester of a fatty acid shortened by two carbons):**

$$RCH_2\overset{\overset{\displaystyle O}{\|}}{C}CH_2\overset{\overset{\displaystyle O}{\|}}{C}{-}SCoA + CoASH \longrightarrow RCH_2\overset{\overset{\displaystyle O}{\|}}{C}{-}SCoA + \underset{\text{acetyl-CoA}}{CH_3\overset{\overset{\displaystyle O}{\|}}{C}{-}SCoA}$$

The shortened-chain thioester repeats the reaction sequence of oxidation, hydration, oxidation, and cleavage to shorten the carbon chain further and produce another molecule of acetyl-CoA. Thus, for example, eight molecules of acetyl-CoA can be produced from one molecule of palmitic acid.

As in the metabolic pathways for glucose, each reaction in the fatty acid oxidation pathway is enzyme catalyzed. No ATP is directly produced during fatty acid catabolism. Instead, ATP forms when the reduced coenzymes, $FADH_2$ and NADH, are oxidized by the mitochondrial electron-transport system in concert with oxidative phosphorylation. Fatty acid oxidation is aerobic because the products, $FADH_2$ and NADH, can be reoxidized only when oxygen is present.

The ruby-throated hummingbird stores enough fat during the summer to supply most of the energy needed to migrate from the southeastern United States across the Gulf of Mexico to the Yucatan peninsula.

In general, fatty acid catabolism yields more energy than can be derived from the breakdown of glucose. (See Figure 35.1.) For example, the reduced coenzymes derived from stearic acid (18 carbons) yield 139 ATPs via electron transport and oxidative phosphorylation. Nine additional ATPs can be obtained from the nine GTPs formed in the citric acid cycle, while 1 ATP is used to start the beta oxidation process. Thus, the 18 carbons of stearic acid yield a total of 147 ATPs, or about 8.2 ATPs per carbon atom. In contrast, glucose yields between 36 and 38 ATPs, as its six carbons are completely oxidized to carbon dioxide. About 6 ATPs per carbon atom are gained from glucose, compared with about 8 ATPs per carbon atom from a fatty acid. Because fatty acid carbons are, in general, more reduced than glucose carbons, fatty acids are a more potent source of energy and yield more ATP molecules during metabolism.

Not surprisingly, the energy-storage molecule of choice in the human body is the fatty acid. On the average, a 70-kg male adult carries about 15 kg of fat (as triacylglycerols), but only about 0.22 kg of carbohydrate (as glycogen). Fat is such an effective way to store energy that obese people could exist for about one year without food. Unfortunately, fat is not the best energy-supply molecule for all tissues. For example, the brain normally derives all of its energy needs from glucose, using about 60% of all glucose metabolized by an adult at rest. Thus, fatty acids are not a universal energy source, although they are our most concentrated supply of energy.

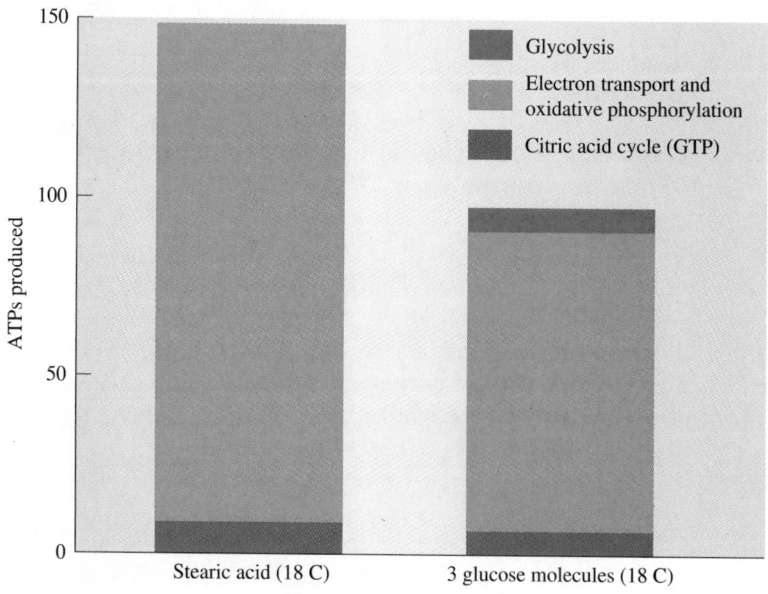

Figure 35.1
A comparison of the ATPs produced from 18 carbons of one stearic acid molecule and 18 carbons of three glucose molecules.

The high energy stored in fats means that diets with a high fat content may have too much energy. To decrease calories, and because fats are not a universal energy source, many people seek to limit their dietary fat intake. Most nutritionists agree that a diet should be limited to no more than 30% calories from fat. Fats—especially saturated fats—are known to increase the risk of heart attack, stroke, and some forms of cancer. Therefore, more and more consumers are shopping for foods with low fat content. However, two important points should be made. First, as shown in Chapter 34 and Section 35.5, carbohydrates and proteins also can be rich energy sources, so eating less fat may not be enough to achieve a low-calorie diet. Second, as shown in Section 32.5, some fatty acids are essential for good health, so eating too little fat can actually cause health problems. In general, nutritionists advise against trying for a "zero-fat" diet.

The essential fatty acids are linoleic, linolenic, and arachidonic acids.

Practice 35.1

Write the structure of palmitic acid.
(a) Star the carbon that first reacts with coenzyme-A in beta oxidation.
(b) Circle the beta carbon.
(c) Draw a line through each carbon–carbon bond that is broken when palmitic acid is converted to eight molecules of acetyl-CoA.

35.3 Fat Storage and Utilization

Fats (triacylglycerols) are stored primarily in adipose tissue, which is widely distributed in the body. Fat tends to accumulate under the skin (subcutaneous fat), in the abdominal region, and around some internal organs, especially the kidneys. Fat deposited around internal organs acts as a shock absorber, or cushion. Subcutaneous fat acts as an insulating blanket. It is developed to an extreme degree in mammals such as seals, walruses, and whales, which live in cold water.

Fat is the major reserve of potential energy. It is metabolized continuously. Stored fat does not remain in the body unchanged: There is a rapid exchange between the triacylglycerols of the plasma lipoproteins and the triacylglycerols in the adipose tissue. The plasma lipoprotein–bound triacylglycerols are broken down by the enzyme lipoprotein lipase that is found on the walls of all capillaries, and the resulting free fatty acids are transported into the adipose cells. The fatty acids are then converted back to triacylglycerols. When the body needs energy from fat, adipose cell enzymes hydrolyze triacylglycerols, and the fatty acids are exported to other body tissues. This vital process is under careful hormonal control. For example, a part of the "fight or flight" response caused by the hormone epinephrine (adrenalin) is an increased fatty acid output from the adipose tissue. Conversely, when there is more energy available in the diet than the body needs, the excess energy is used to make body fat. Continued eating of more food than the body can use results in obesity.

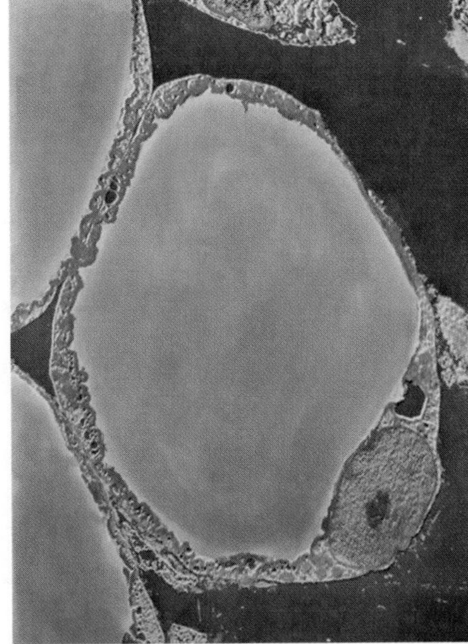

A microscopic photo of a fat (adipose) cell. This cell contains a large fat droplet (shown in yellow). The smaller body is the nucleus. All the fat cell's metabolism is forced into a thin layer adjacent to the membrane.

35.4 Biosynthesis of Fatty Acids (Lipogenesis)

The biosynthesis of fatty acids from acetyl-CoA is called **lipogenesis**. Acetyl-CoA can be obtained from the catabolism of carbohydrates, fats, or proteins. After they are synthesized, fatty acids combine with glycerol to form triacylglycerols, which are stored in adipose tissue. Consequently, lipogenesis is the pathway by which all three of the major classes of nutrients are ultimately converted to body fat.

lipogenesis

Is lipogenesis just the reverse of fatty acid oxidation (beta oxidation)? By analogy with carbohydrate metabolism (see Chapter 34), we might expect this to be the case. However, fatty acid biosynthesis is not simply a reversal of fatty acid oxidation. The following are the major differences between the two pathways:

1. Fatty acid catabolism occurs in the mitochondria, but fatty acid anabolism (lipogenesis) occurs in the cytoplasm.
2. Lipogenesis requires a set of enzymes that are different from the enzymes used in the catabolism of fats.
3. In the anabolic pathway (lipogenesis), the growing fatty acid chain bonds to a special acyl carrier protein, ACP—SH, which acts as a "handle" to transfer the growing chain from one enzyme to another through the series of enzyme-catalyzed reactions in the pathway. Coenzyme A is the carrier in fatty acid catabolism.
4. A preliminary set of reactions, involving malonyl-CoA, occurs for each two-carbon addition cycle in the synthesis. "Malonyl-" refers to a three-carbon group and has no counterpart in the catabolic pathway. Malonyl-CoA is synthesized from acetyl-CoA and carbon dioxide in the presence of the enzyme acetyl-CoA carboxylase, ATP, and the vitamin biotin:

See Table 24.2: Dicarboxylic acids.

$$\underset{\text{acetyl-CoA}}{CH_3\overset{O}{\overset{\|}{C}}-SCoA} + CO_2 \xrightarrow[\text{acetyl-CoA carboxylase}]{\text{ATP, biotin}} \underset{\text{malonyl-CoA}}{HO\overset{O}{\overset{\|}{C}}CH_2\overset{O}{\overset{\|}{C}}-SCoA}$$

The biosynthesis of a fatty acid occurs by the addition of successive two-carbon increments to a lengthening chain starting with acetyl-CoA. Each incremental addition follows this five-step pathway or reaction sequence:

Step 1 Acetyl-CoA and malonyl-CoA bond to separate acyl carrier proteins:

$$\underset{\text{acetyl-CoA}}{CH_3\overset{O}{\overset{\|}{C}}-SCoA} + \underset{\text{malonyl-CoA}}{HO\overset{O}{\overset{\|}{C}}CH_2\overset{O}{\overset{\|}{C}}-SCoA} + 2\ \underset{\text{acyl protein carrier}}{ACP-SH} \longrightarrow$$

$$\underset{\text{acetyl-ACP}}{CH_3\overset{O}{\overset{\|}{C}}-SACP} + \underset{\text{malonyl-ACP}}{HO\overset{O}{\overset{\|}{C}}CH_2\overset{O}{\overset{\|}{C}}-SACP} + 2\ CoASH$$

Step 2 Acetyl-ACP and malonyl-ACP condense, with loss of carbon dioxide (decarboxylation):

$$\underset{\text{acetyl-ACP}}{CH_3\overset{O}{\overset{\|}{C}}-SACP} + \underset{\text{malonyl-ACP}}{HO\overset{O}{\overset{\|}{C}}CH_2\overset{O}{\overset{\|}{C}}-SACP} \longrightarrow$$

$$\underset{\text{acetoacetyl-ACP}}{CH_3\overset{O}{\overset{\|}{C}}CH_2\overset{O}{\overset{\|}{C}}-SACP} + CO_2 + ACP-SH$$

The three steps that follow are approximate reversals of three steps in fatty acid beta oxidation (Section 35.2):

Step 3 Reduction [hydrogenation of carbon 3 (β-keto group)]:

$$CH_3\overset{O}{\overset{\|}{C}}CH_2\overset{O}{\overset{\|}{C}}{-}SACP + NADPH + H^+ \longrightarrow CH_3\overset{OH}{\overset{|}{C}H}CH_2\overset{O}{\overset{\|}{C}}{-}SACP + NADP^+$$

β-hydroxybutyryl-ACP

Step 4 Dehydration (formation of a double bond between carbons 2 and 3):

$$CH_3\overset{OH}{\overset{|}{C}H}CH_2\overset{O}{\overset{\|}{C}}{-}SACP \longrightarrow CH_3CH{=}CH\overset{O}{\overset{\|}{C}}{-}SACP + H_2O$$

crotonyl-ACP

Step 5 Reduction (hydrogenation of carbons 2 and 3):

$$CH_3CH{=}CH\overset{O}{\overset{\|}{C}}{-}SACP + NADPH + H^+ \longrightarrow CH_3CH_2CH_2\overset{O}{\overset{\|}{C}}{-}SACP + NADP^+$$

butyryl-ACP

This completes the first cycle of the synthesis; the chain has been lengthened by 2 carbon atoms. The biosynthesis of longer-chain fatty acids proceeds by a series of such cycles, each lengthening the carbon chain by an increment of 2 carbon atoms. The next cycle begins with the reaction of butyryl-ACP and malonyl-ACP, leading to a 6-carbon chain, and so on. This synthesis commonly produces palmitic acid (16 carbons) as its end product. The synthesis of palmitic acid from acetyl-CoA and malonyl-CoA requires cycling through the series of steps seven times. The net equation for the formation of palmitic acid is

$$\underset{\text{acetyl-CoA}}{CH_3\overset{O}{\overset{\|}{C}}{-}SCoA} + 7\ \underset{\text{malonyl-CoA}}{HO\overset{O}{\overset{\|}{C}}CH_2\overset{O}{\overset{\|}{C}}{-}SCoA} + 14\ NADPH + 14\ H^+ \longrightarrow$$

$$\underset{\text{palmitic acid}}{CH_3(CH_2)_{14}COOH} + 7\ CO_2 + 6\ H_2O + 8\ CoASH + 14\ NADP^+$$

Nearly all naturally occurring fatty acids have even numbers of carbon atoms. A sound reason for this fact is that both the catabolism and the synthesis proceed by two-carbon increments.

In conclusion, it should be noted that the metabolism of fats has some features in common with that of carbohydrates. The acetyl-CoA produced in the catabolism of both carbohydrates and fatty acids can be used as a raw material for making other substances and as an energy source. When acetyl-CoA is oxidized via the citric acid cycle, more potential energy can be trapped in ATP. The ATP in turn serves as the source of energy needed for the production of other substances, including the synthesis of carbohydrates and fats.

CHEMISTRY IN ACTION • Ketone Bodies: Retooling for a Different Metabolism

The human body regularly functions with a standard set of metabolic tools. Carbohydrates or lipids or proteins yield energy via aerobic metabolism. Yet not all energy foodstuffs are equivalent; carbohydrates are special. They are the only foodstuff that can be used anaerobically and are the preferred energy source for the brain. So the human body has an emergency procedure to limit carbohydrate use—a major retooling of the metabolic machinery.

When humans are low in carbohydrates, most cells shift to a metabolism that is based on fatty acids. This leaves the remaining glucose for the brain and some anaerobic work by the muscles. As the liver moves to a fatty acid metabolism, it starts to produce a group of fatty acid derivatives called ketone bodies.

$$CH_3\overset{\overset{\large O}{\|}}{C}CH_2\overset{\overset{\large O}{\|}}{C}-OH$$

acetoacetic acid

$$CH_3\overset{\overset{\large OH}{|}}{C}HCH_2\overset{\overset{\large O}{\|}}{C}-OH$$

β-hydroxybutyric acid

$$CH_3\overset{\overset{\large O}{\|}}{C}CH_3$$

acetone

These molecules are water soluble (like glucose) but are metabolized aerobically (like fat). As blood-glucose levels drop, blood ketone body concentrations rise. Meanwhile, the brain retools its metabolism to generate energy from ketone bodies instead of glucose. By week 4 of a starvation diet, the brain is operating by using 70% ketone bodies and 30% glucose for energy. The new metabolic machinery creates a condition called *ketosis*. Historically, ketosis has been found when people are starving or suffering from uncontrolled diabetes. Today, ketosis may result from "low-carbohydrate" or "high-fat/high-protein" weight-loss programs that rigidly restrict carbohydrate intake.

It is clear that ketosis means a general retooling of metabolic machinery. However, specific changes at the cellular level are not well understood. For example, scientists have no clear explanation for the fact that many epileptics experience fewer seizures on a ketotic diet.

So it is with diets that depend on low carbohydrates and high fat/high protein. Their impact is not understood. These diets achieve a greater weight loss in comparison with the standard low-fat diet according to the most current studies. Yet there is no clear explanation. Greater weight loss may come from the extra water loss during ketosis, a decreased feeling of hunger during ketosis, or some unknown metabolic change. Scientists need more time to unravel all the changes that follow such a major retooling of the metabolic machinery.

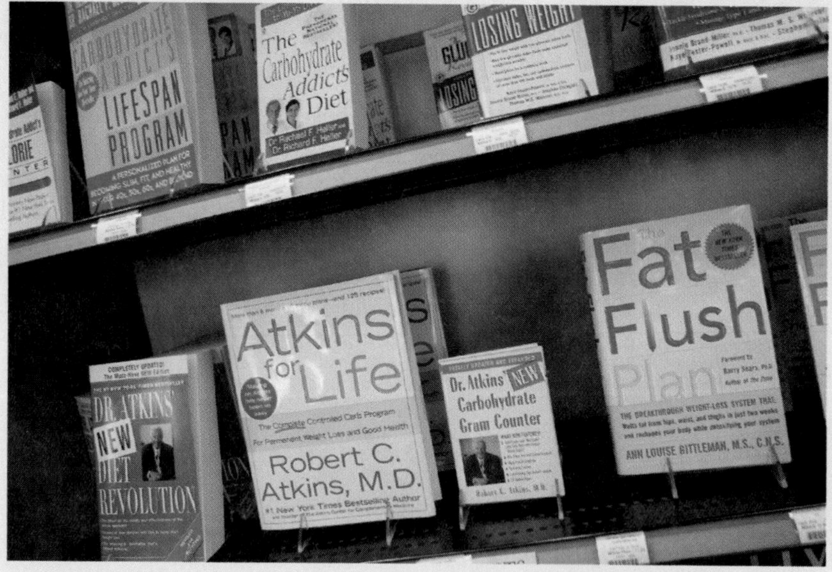

Many of today's popular diets rigidly restrict dietary carbohydrates.

35.5 Amino Acid Metabolism

Amino acids serve an important and unique role in cellular metabolism; they are the building blocks of proteins and also provide most of the nitrogen for other nitrogen-containing compounds.

Amino acid metabolism differs markedly from the biochemistry of carbohydrates or fatty acids. Amino acids always contain nitrogen, and the chemistry of this element presents unique problems for the cell. In addition, there is no structure common to all the carbon skeletons of amino acids. The carbohydrates can share a common metabolic pathway, as can fatty acids, because they share common structures. But the carbon skeletons of amino acids vary widely, and the cell must use a different metabolic pathway for almost every amino acid. Thus, the metabolism of the carbon structures of amino acids is more complex. In the sections that follow, we present a brief overview of amino acid metabolism, together with a more detailed examination of nitrogen metabolism.

35.6 Metabolic Nitrogen Fixation

Nitrogen is an important component of many biochemicals. In addition to being a component of amino acids, nitrogen is found in proteins, nucleic acids, hemoglobin, and many vitamins. Every cell must have a continuous supply of nitrogen. This might seem easy to obtain, because the atmosphere contains about 78% free (elemental) nitrogen. Unfortunately, most cells cannot use elemental nitrogen.

Elemental nitrogen on earth exists as N_2 molecules, the two atoms being bonded together by a strong and stable triple bond. On an industrial scale, high temperature (400–500°C), high pressure (several hundred atmospheres), and a catalyst are required to react nitrogen gas with the reducing agent, hydrogen, to form ammonia (the Haber process). In the biosphere, only a few procaryotes, including the *Azobacter*, *Clostridium pasteurianum*, and *Rhizobium* species, have the metabolic machinery necessary to use the abundant atmospheric nitrogen. By converting elemental nitrogen to compounds, these organisms make nitrogen available to the rest of the biological world.

The conversion of diatomic nitrogen to a biochemically useful form is termed **nitrogen fixation**. The process by which nitrogen is circulated and recirculated from the atmosphere through living organisms and back to the atmosphere is known as the **nitrogen cycle**.

nitrogen fixation

nitrogen cycle

Nitrogen compounds are required by all living organisms. Despite the fact that the atmosphere is about 78% nitrogen, all animals, as well as the higher plants, are unable to utilize free nitrogen. Higher plants require inorganic nitrogen compounds, and animals must have nitrogen in the form of organic compounds.

Atmospheric nitrogen is fixed—that is, converted into chemical compounds that are useful to higher forms of life—by three general routes:

1. *Bacterial action* Certain bacteria are capable of converting N_2 into nitrates. Most of these bacteria live in the soil in association with legumes (e.g., peas, beans, clover). Some free-living soil bacteria and the blue-green algae, which live in water, are also capable of fixing nitrogen. Of the procaryotes that are able to catalyze nitrogen fixation, *Rhizobium* bacteria deserve special attention. These bacteria flourish on the roots of legumes, in a symbiotic relationship. The bacteria have degenerated essentially into nitrogen-fixing machines that are maintained by the plants in the root nodules. The legumes even provide a special hemoglobin protein to assist the bacteria in their task.
2. *High temperature* The high temperature of lightning flashes causes substantial amounts of nitrogen oxide in the atmosphere. This NO is dissolved in rainwater and eventually is converted to nitrate ions in the soil. The combustion of fuels also provides temperatures high enough to form NO

Legume root nodules containing nitrogen-fixing bacteria.

in the atmosphere. The total amount of nitrogen fixed by combustion is relatively insignificant on a worldwide basis. However, NO produced by combustion, especially in automobile engines, is a serious air pollution problem in some areas.

3. *Chemical fixation* Chemical processes have been devised for making nitrogen compounds directly from atmospheric nitrogen. By far the most important of these is the Haber process for making ammonia from nitrogen and hydrogen. This process is the major means of production for the millions of tons of nitrogen fertilizers that are produced synthetically each year.

The nitrogen cycle is diagrammed in Figure 35.2. The cycle begins in the atmosphere, with the fixation of nitrogen by any one of the three routes. In the soil, nitrogen or nitrogen compounds are converted to nitrates, taken up by higher plants, and converted to organic compounds. The plants eventually die or are eaten by animals. During the life of the animal, part of the nitrogen from plants in its diet is returned to the soil in the form of fecal and urinary excreta. Eventually, after death, both plants and animals decompose by bacterial action. Part of the nitrogen from plant and animal tissues returns to the atmosphere as free nitrogen, and part is retained in the soil. The cycle thus continues.

Of the common elements needed for life, nitrogen is the most difficult to obtain. For animals, nitrogen comes primarily from protein sources; protein is a very important dietary component. Examination of the nitrogen cycle reveals that new, usable nitrogen is obtained directly from plants, which in turn gain much of their nitrogen from bacterial sources; about 60% of all newly fixed nitrogen comes from bacteria. Thus, biological nitrogen supplies depend on a small number of nitrogen-fixing microorganisms.

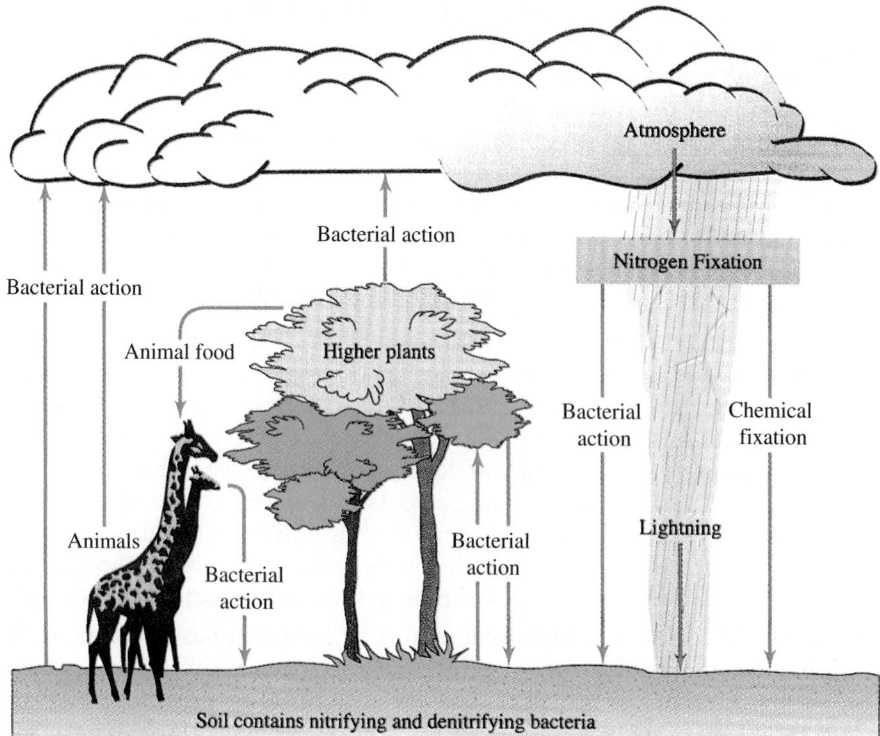

Figure 35.2
The nitrogen cycle. Arrows indicate the movement of nitrogen through the cycle.

Currently, in many university and industrial laboratories, scientists are searching for ways to increase bacterial nitrogen fixation via genetic engineering. The immediate focus is on the legumes, nature's "champion" nitrogen-enriching plants. The nitrogen-fixing symbiosis between these higher plants and bacteria is complex, requiring many bacterial genes as well as many plant genes. For example, the bacteria can fix nitrogen under anaerobic conditions (without O_2) partly because the plants produce a protein (leghemoglobin) that "sponges up" oxygen. Genetic engineering of this complex system is a daunting goal. However, scientists have already identified legume strains that produce extra root nodules and therefore contain genes that do a better job of nitrogen fixation. An intermediate goal is to start symbiosis between nonleguminous plants and nitrogen-fixing bacteria. Finally, as a long-term goal, plant biochemists hope to be able to introduce the nitrogen-fixing genes directly into higher plant genomes. Success will mean that grains and other commercially important crops can be grown in poor soils, using little or no nitrogen fertilizer. This achievement would vastly increase the world's food production capability.

Higher plants use nitrogen compounds primarily to produce proteins, which, in turn, enter the animal food chain. Both plant and animal proteins are important human nutrients.

Practice 35.2

Identify the following as either (i) bacterial action nitrogen fixation, (ii) high-temperature nitrogen fixation, or (iii) chemical nitrogen fixation:

(a) A brown haze of nitrogen oxides from automobile exhaust forms over Los Angeles.

(b) A fertilizer company makes ammonia by compressing and heating nitrogen and hydrogen gas together.

35.7 Amino Acids and Metabolic Nitrogen Balance

Protein is digested and absorbed to provide the amino acid dietary requirements. (See Chapter 32.) Once absorbed, an amino acid can be

1. incorporated into a protein, or
2. used to synthesize other nitrogenous compounds such as nucleic acids, or
3. deaminated to a keto acid, which can either be used to synthesize other compounds or be oxidized to carbon dioxide and water to provide energy.

amino acid pool

Absorbed amino acids enter the **amino acid pool**—the total supply of amino acids available for use throughout the body. The amount of amino acids in the pool is maintained in balance with other cellular nitrogen pools. (See Figure 35.3.)

One particularly important nitrogen pool is composed of the proteins in all the body's tissues. Amino acids continually move back and forth between the amino acid pool and the tissue proteins. In other words, our body proteins are constantly being broken down and resynthesized. The rate of turnover varies with different proteins. In the liver and other active organs, protein molecules may have a half-life of less than one week, whereas the half-life of some muscle proteins is about six months.

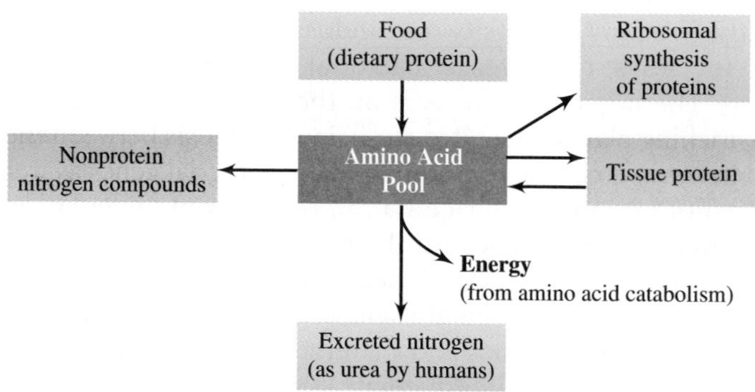

Figure 35.3
Major biological nitrogen pools related to the central amino acid pool.

nitrogen balance

In a healthy, well-nourished adult, the amount of nitrogen excreted is equal to the amount of nitrogen ingested. The nitrogen pools within the body remain constant (see Figure 35.3), and such a person is said to be in **nitrogen balance**.

positive nitrogen balance

For a growing child, the amount of nitrogen consumed is more than that excreted. Dietary protein feeds into the amino acid pool, which is used to make many new nitrogen-containing molecules. The contents of the nitrogen pools increase, and the child is said to be in **positive nitrogen balance**.

negative nitrogen balance

A fasting or starving person excretes more nitrogen than is ingested; such a person is said to be in **negative nitrogen balance**. Tissue protein breaks down to supply more amino acids to the amino acid pool. (See Figure 35.3.) These amino acids are used as an energy source, and nitrogen is excreted. The body's nitrogen pools are depleted during negative nitrogen balance.

Most of the amino acid pool is derived originally from dietary protein, digested to yield amino acids. (See Section 32.13.) In contrast, only a very small proportion of the amino acid pool comes directly from dietary amino acids. Recently, it has been suggested that some diets be supplemented by additional amino acids. Extra amino acid intake reportedly aids muscle building and counters mental depression. Most studies, however, show that the protein in a balanced diet is sufficient to provide all needed amino acids. In fact, sometimes added amino acids are harmful. The food additive monosodium glutamate (MSG; the sodium salt of the amino acid glutamic acid) causes allergic reactions in some people. However, more research is certainly needed on the effect of ingesting extra amino acids.

Nitrogen balance depends on the transition between nitrogen pools. If these pathways are disturbed, severe health problems can develop. For example, people suffering from phenylketonuria (PKU) are unable to catabolize phenylalanine. When people with PKU ingest food protein containing this amino acid, it builds up in their body's amino acid pool and eventually causes toxic side reactions. Babies who suffer from PKU develop extensive mental retardation. Fortunately, these symptoms can be avoided by removing most of the phenylalanine from the diet. Products that contain large amounts of phenylalanine, such as the artificial sweetener aspartame (Nutrasweet®), carry a warning label for people with the PKU genetic defect.

As shown in Figure 35.3, amino acids are central to many different metabolic processes. They are the raw material for protein synthesis (see Section 31.11) and can also be used to make other nitrogen-containing molecules. In addition, amino acids can supply metabolic energy if the amino acid nitrogen is first removed (or transferred to another molecule).

35.8 Amino Acids and Nitrogen Transfer

Amino acids are important in metabolism as carriers of usable nitrogen. If an amino acid is not directly incorporated into tissue proteins, its nitrogen may be incorporated into various molecules, such as other amino acids, nucleic acid bases, the heme of hemoglobin, and some lipids. In general, when an amino acid is used for some purpose other than protein synthesis, the amino acid carbon skeleton is separated from the amino acid nitrogen.

A process called transamination is responsible for most of the nitrogen transfer to and from amino acids. **Transamination** is the transfer of an amino group from an α-amino acid to an α-keto acid:

transamination

$$\underset{\alpha\text{-amino acid}}{\mathrm{RCH(NH_2)-COOH}} + \underset{\alpha\text{-keto acid}}{\mathrm{R'C(=O)-COOH}} \rightleftharpoons \mathrm{RC(=O)-COOH} + \mathrm{R'CH(NH_2)-COOH}$$

Transamination involves many different molecules, with each different transamination requiring a different enzyme (transaminase). For example, one enzyme (glutamic-pyruvic transaminase) catalyzes the conversion of pyruvic acid to L-alanine:

$$\underset{\text{pyruvic acid}}{\mathrm{CH_3C(=O)-COOH}} + \underset{\text{L-glutamic acid}}{\mathrm{HOOC-CH_2-CH_2-CH(NH_2)-COOH}} \rightleftharpoons \underset{\text{L-alanine}}{\mathrm{CH_3CH(NH_2)-COOH}} + \underset{\alpha\text{-ketoglutaric acid}}{\mathrm{HOOC-CH_2-CH_2-C(=O)-COOH}}$$

A different enzyme catalyzes the production of L-aspartic acid from oxaloacetic acid:

$$\underset{\text{oxaloacetic acid}}{\mathrm{HOOC-CH_2-C(=O)-COOH}} + \underset{\text{L-glutamic acid}}{\mathrm{HOOC-CH_2-CH_2-CH(NH_2)COOH}} \rightleftharpoons \underset{\text{L-aspartic acid}}{\mathrm{HOOC-CH_2-CH(NH_2)COOH}} + \underset{\alpha\text{-ketoglutaric acid}}{\mathrm{HOOC-CH_2-CH_2-C(=O)-COOH}}$$

Note that, in both of these reactions, L-glutamic acid is converted to α-ketoglutaric acid. In fact, most transaminations use L-glutamic acid. Thus, this amino acid plays a central role in cellular nitrogen transfer.

Transamination is the first step in the conversion of the carbon skeletons of amino acids to energy-storage compounds. Amino acids that are used to produce glucose are termed **glucogenic amino acids**. Most amino acids are glucogenic (see Table 35.1), but some amino acids are converted to acetyl-CoA.

glucogenic amino acid

Table 35.1 Classification of Amino Acids as Sources of Energy-Storage Molecules

Glucogenic	Ketogenic and glucogenic	Ketogenic
Alanine	Isoleucine	Leucine
Arginine	Phenylalanine	Lysine
Aspartic acid	Tryptophan	
Asparagine	Tyrosine	
Cysteine		
Glutamic acid		
Glutamine		
Glycine		
Histidine		
Methionine		
Proline		
Serine		
Threonine		
Valine		

ketogenic amino acid

Fed to starving animals, these amino acids cause an increase in the rate of ketone body formation and are therefore called **ketogenic amino acids**. Only leucine and lysine are completely ketogenic, but a few amino acids can be converted to either glucose or acetyl-CoA, and are both ketogenic and glucogenic.

Relatively recently, high-protein/low-carbohydrate diets have been touted as a means of losing weight quickly. This diet forces protein to be used for energy. In the liver, glucogenic and ketogenic amino acids are converted to glucose and ketone bodies, respectively. These molecules are transported to other cells via the bloodstream. The high-ketone body concentration can lead to a condition called *ketosis*. (See the "Chemistry in Action" feature "Ketone Bodies: Retooling for a Different Metabolism.") This same situation is experienced during fasting, starvation, or diabetes. Weight can certainly be lost by a high-protein/low-carbohydrate diet. But unfortunately much of this lost weight is water as the kidneys attempt to remove ketone bodies from the blood. The body is forced into an emergency situation through this unbalanced diet.

The amino acid pool of Figure 35.3 can now be described in more detail. At the center of this pool is L-glutamic acid, and other amino acids can either add or remove nitrogen from this central compound. L-Glutamic acid can also accept a second nitrogen atom to form L-glutamine:

An amide is formed when ammonia or an amine reacts with a carboxylic acid.

$$NH_3 + \underset{\text{L-glutamic acid}}{HOOC{-}CH_2{-}CH_2{-}CH(NH_2)COOH} \rightleftharpoons \underset{\text{L-glutamine}}{H_2N{-}CO{-}CH_2{-}CH_2{-}CH(NH_2)COOH} + H_2O$$

Although this reaction can be considered as the simple addition of ammonia to yield an amide, the actual cellular reactions are more complex. The product, L-glutamine, facilitates biological nitrogen transfer, the amide nitrogen being transferable in a number of cellular reactions.

It is worthwhile to examine the synthesis of L-glutamine in more detail. Ammonia is a base and therefore is toxic to the cell. When ammonia forms an amide bond to L-glutamic acid, the nitrogen becomes less basic and also nontoxic. Thus L-glutamine serves as a safe package for transporting nitrogen. In the human body, L-glutamine is the major compound for transferring nitrogen from one cell to another via the bloodstream.

Fish excrete ammonia through their gills, where this toxic molecule is rapidly diluted.

35.9 Nitrogen Excretion and the Urea Cycle

Nitrogen—unlike carbon, hydrogen, and oxygen—is often conserved for reuse by the cell. Nitrogen excretion does occur, however, when an excess of this element exists or when the carbon skeletons of nitrogen-containing compounds are needed for other purposes. Two examples from human nutrition arise (1) when we consume more protein than is needed (an excess of nitrogen-containing molecules) or (2) when we experience starvation (protein is destroyed to provide energy). Under normal conditions, adult humans excrete 6–18 g of nitrogen per day.

The nitrogen elimination process poses a major problem for the cell. The simplest excretion product is ammonia, but ammonia is basic and therefore toxic. Fish can excrete ammonia through their gills because ammonia is soluble and is swept away by water passing through the gills. Land animals and birds excrete nitrogen in less toxic forms. Birds and reptiles excrete nitrogen as the white solid uric acid, a derivative of the purine bases. Mammals excrete the water-soluble compound urea. Both of these compounds contain a high percentage of nitrogen in a nontoxic form.

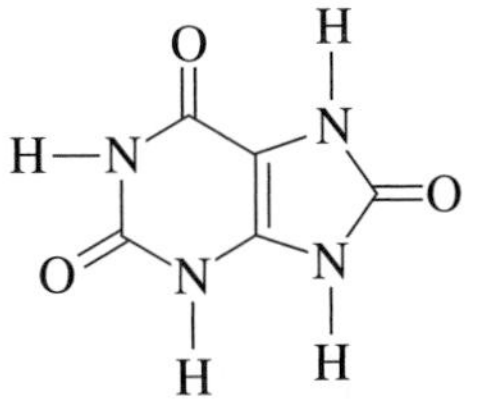

uric acid

$$H_2N-\overset{O}{\overset{\|}{C}}-NH_2$$

urea

Urea synthesis in mammals follows a pathway called the **urea cycle** (Figure 35.4), which takes place in the liver. The urea cycle uses ATP energy to make urea from ammonia, bicarbonate, and L-aspartic acid. First, ammonium ion is produced from L-glutamic acid in an oxidation–reduction reaction:

urea cycle

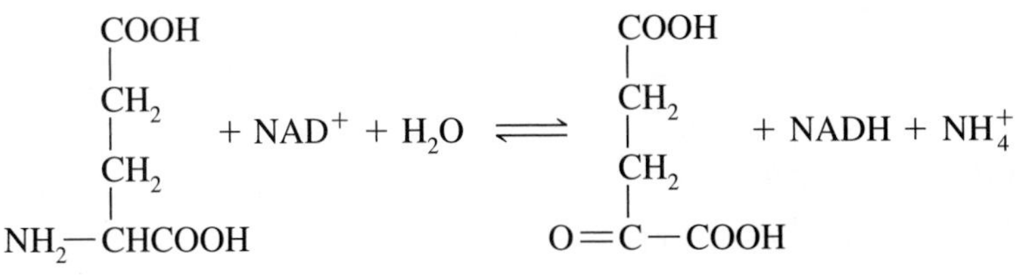

$$HOOC-CH_2-CH_2-CH(NH_2)-COOH + NAD^+ + H_2O \rightleftharpoons HOOC-CH_2-CH_2-C(=O)-COOH + NADH + NH_4^+$$

L-glutamic acid $\qquad$ α-ketoglutaric acid

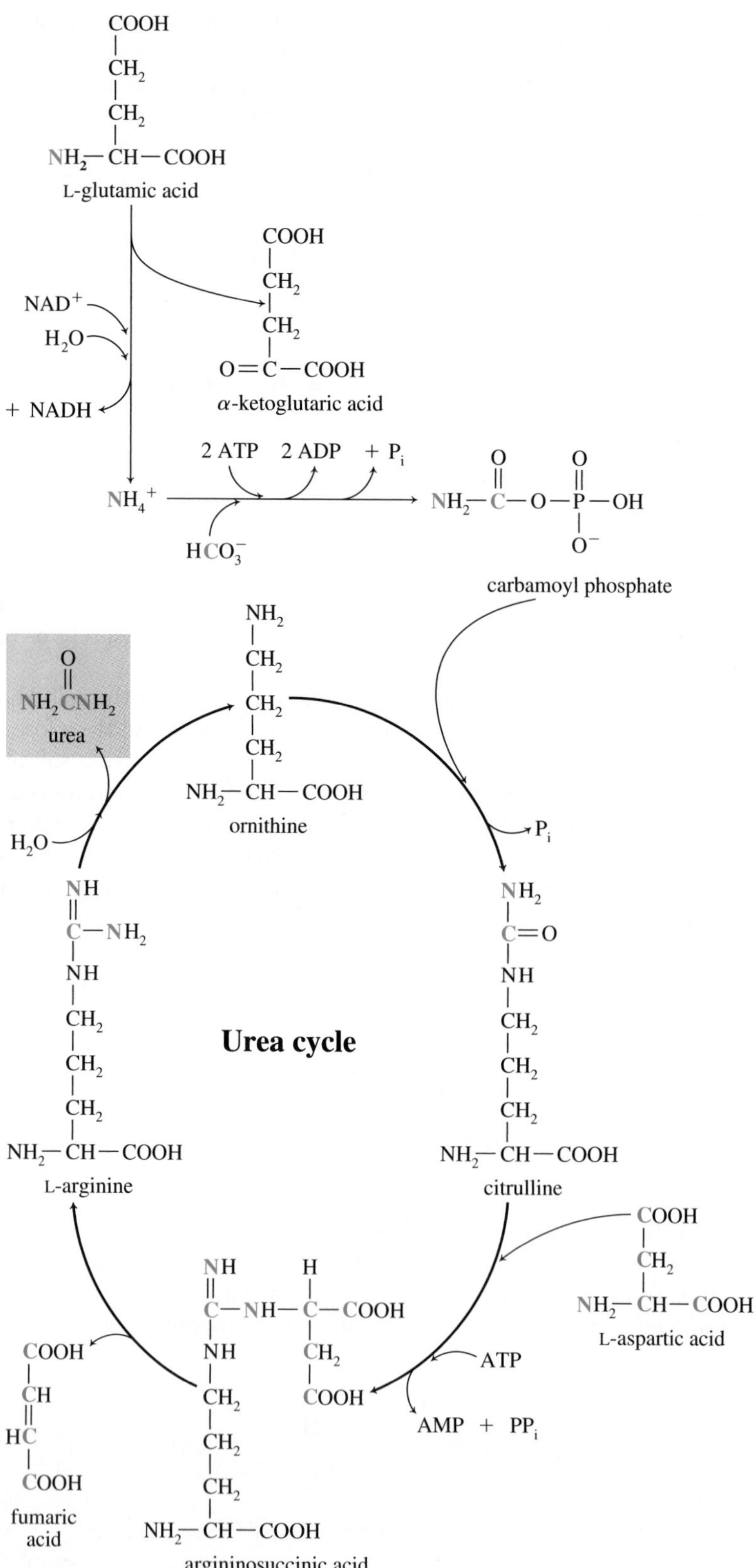

Figure 35.4
Urea cycle. The cycle begins with the reaction between carbamoyl phosphate and ornithine.

Then, the ammonium ion quickly reacts with the hydrogen carbonate ion and ATP to form carbamoyl phosphate:

$$NH_4^+ + HCO_3^- + 2\,ATP \longrightarrow \underset{\text{carbamoyl phosphate}}{NH_2{-}\overset{\overset{\displaystyle O}{\|}}{C}{-}O{-}\overset{\overset{\displaystyle O^-}{|}}{\underset{\underset{\displaystyle O}{\|}}{P}}{-}OH} + 2\,ADP + P_i$$

Finally, carbamoyl phosphate enters the urea cycle. The overall reaction of the urea cycle is

$$\underset{\text{carbamoyl phosphate}}{NH_2{-}\overset{\overset{\displaystyle O}{\|}}{C}{-}O{-}\overset{\overset{\displaystyle O^-}{|}}{\underset{\underset{\displaystyle O}{\|}}{P}}{-}OH} + \underset{\text{L-aspartic acid}}{NH_2{-}\overset{\overset{\displaystyle COOH}{|}}{\overset{\displaystyle CH_2}{\overset{|}{\underset{\underset{\displaystyle H}{|}}{C}}}}{-}COOH} + ATP + H_2O \xrightarrow{\text{urea cycle}}$$

$$\underset{\text{urea}}{NH_2{-}\overset{\overset{\displaystyle O}{\|}}{C}{-}NH_2} + \underset{\text{fumaric acid}}{\begin{matrix} COOH \\ | \\ CH \\ \| \\ HC \\ | \\ COOH \end{matrix}} + AMP + P_i + PP_i$$

Like the intermediate compounds of the citric acid cycle, the urea cycle intermediates do not appear in the overall reaction. Note that the urea cycle intermediates are α-amino acids. These amino acids are rarely used in protein synthesis. Their primary role is in the formation of urea.

Also, it is important to recognize that the cell must expend energy, in the form of ATP, to produce urea. The formation of a nontoxic nitrogen excretion product is essential. In fact, one of the major problems of liver cirrhosis, caused by alcoholism, is the impairment of the urea cycle. As liver function is impaired, more nitrogen is excreted as toxic ammonia.

Finally, let's look at the sources of the nitrogen that is excreted as urea. One nitrogen atom in each urea molecule comes from L-glutamic acid. The other nitrogen atom comes from L-aspartic acid, which may have gained its nitrogen from L-glutamic acid by transamination. Thus, the amino acid that is central to nitrogen-transfer reactions is also the major contributor to nitrogen excretion.

35.10 Acetyl-CoA, a Central Molecule in Metabolism

Thinking back through metabolism, we can identify some especially important compounds—molecules that are central to this part of biochemistry. For example, glucose is the central compound in carbohydrate metabolism; glutamic acid is central to amino acid metabolism. One compound is at the hub of all common metabolic processes: acetyl-CoA, which is central to the metabolism of fats, proteins, and carbohydrates (Figure 35.5). This molecule is a critical intermediate in the processes that form and break down both fats and amino acids. In addition, essentially all compounds that enter the citric acid cycle must first be catabolized to acetyl-CoA. In this section, we will examine (1) the characteristics of acetyl-CoA that make it a central metabolic compound and (2) the advantages of a centralized metabolism for the cell.

Recall that acetyl-CoA consists of a small two-carbon unit (an acetyl group) bonded by a thioester linkage to a large organic coenzyme molecule, coenzyme A:

$$\underbrace{CH_3-\overset{\overset{\large O}{\|}}{C}}_{\text{acetyl group}}-\underset{\uparrow\ \text{thioester linkage}}{S}-\boxed{\text{CoA}}$$

This structure makes for an almost ideal central metabolic molecule with the following major advantages:

1. *Potential use in a wide variety of syntheses.* The small size and simple structure of the two-carbon acetyl fragment enable this molecule to be used to build a variety of diverse structures. Complex molecules with very different shapes and functions, such as long-chain fatty acids, amino acids, and steroid hormones, are synthesized from acetyl-CoA.
2. *Special reactivity of bonds.* The thioester causes both carbons in the acetyl fragment to be specially reactive. These carbons are "primed" to form new bonds as the acetyl-CoA enters various metabolic pathways.
3. *Structure that is recognizable by a wide variety of enzymes.* Coenzyme A acts as a kind of "handle" for the various enzymes that catalyze reactions of the acetyl group. Because many enzymes bind tightly to CoA, the acetyl group of acetyl-CoA is involved in a great number of diverse reactions.

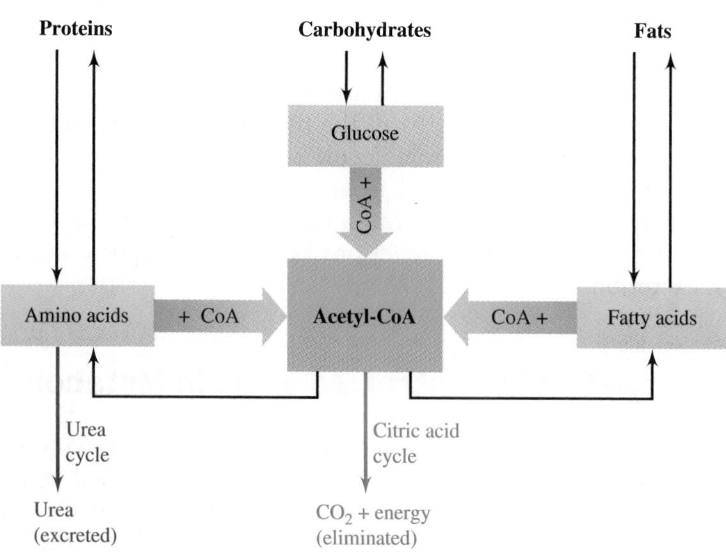

Figure 35.5
Simplified diagram showing acetyl-CoA is at the hub of protein, carbohydrate, and fat metabolism.

Acetyl-CoA is comparable to ATP as a central metabolic molecule. Remember that ATP has (1) potential uses in a wide variety of syntheses, (2) a special reactivity in its phosphate anhydride bonds, and (3) a structure that is recognizable and used by a wide variety of enzymes. For these important reasons, ATP serves as the "common energy currency" for the cell. In analogous fashion, acetyl-CoA might be termed the "common carbon currency" for the cell.

A consideration of these central metabolic compounds raises an important question: "How does centralization aid the cell?" We have seen numerous examples of centralization in biochemistry. Not only are there important central metabolites, such as ATP and acetyl-CoA, but there are pathways, such as the citric acid cycle, that centralize the cellular metabolic machinery. A general answer to this question, then, is that centralization improves the efficiency of metabolism and ensures that biochemistry is under careful metabolic control.

Greater efficiency results from central metabolic pathways that handle a variety of different nutrients. For example, the citric acid cycle completes carbon oxidation for all energy-supplying nutrients that are first converted to the central metabolite acetyl-CoA.

Greater control results from the dependence of many "feeder" pathways on a single central process. Control of a central path like the citric acid cycle will affect the metabolism of a variety of nutrients in a coordinated way. As scientists have learned more about metabolism, it has become clear that centralization is an important attribute of the chemistry of life.

Chapter 35 Review

35.1 Metabolic Energy Sources

- Three common biochemical energy sources are carbohydrates, lipids, and proteins.
- Fatty acids are the most commonly metabolized lipid.
- Fatty acids are more reduced (less oxidized) than carbohydrates, consequently, fatty acids are richer in energy than carbohydrates.

35.2 Fatty Acid Oxidation (Beta Oxidation)

KEY TERM

Beta oxidation

- Early research by Franz Knoop, a German biochemist, indicated that fatty acid catabolism shortens the fatty acid chain successively by two carbon atoms at a time.
- Knoop's theory has proven correct—a fatty acid becomes two carbons shorter each time it passes through the fatty acid oxidation pathway.
- This catabolic pathway is called fatty acid oxidation because the fatty acid carbon 3 (numbering from the carboxylic acid end) is oxidized to form a new carboxylic acid.
  - This oxidation is also known as beta oxidation because carbon 3 is two carbons removed from the terminal carboxylic acid (at the beta position).
- Beta oxidation, or the two-carbon chop, is a series of reactions whereby the first two carbon atoms of the fatty acid chain become the acetyl group in a molecule of acetyl-CoA.
- Beta oxidation repeats a set of reactions (oxidation, hydration, oxidation, cleavage) until the whole fatty acid chain has been converted into acetyl-CoA molecules.
- In general, complete fatty acid oxidation to carbon dioxide yields more energy than can be derived from glucose.
- Fatty acids are the energy-storage molecules of choice in the human body.

35.3 Fat Storage and Utilization

- Fats (triacylglycerols) are stored in the adipose tissue.
- Adipose tissue not only stores fat but also serves as insulation against cold weather and a shock absorber to protect internal organs.
- Fatty acids are transported to and from the adipose tissue via the blood.

35.4 Biosynthesis of Fatty Acids (Lipogenesis)

KEY TERM

Lipogenesis

- The biosynthesis of fatty acids from acetyl-CoA is called lipogenesis.

- Lipogenesis is not simply a reverse of beta oxidation.
  - Lipogenesis occurs in the cytoplasm, while beta oxidation is found in the mitochondria.
  - Lipogenesis is catalyzed by a different set of enzymes than those used in beta oxidation.
  - A special "acyl carrier protein" is used in lipogenesis instead of the coenzyme A that is used in beta oxidation.
  - A unique set of preliminary reactions forms malonyl-CoA in lipogenesis. (These reactions are not found in beta oxidation.)
- Lipogenesis lengthens a fatty acid successively two carbons at a time.

35.5 Amino Acid Metabolism

- Amino acids serve as building blocks for protein synthesis and also provide nitrogen for other nitrogen-containing biochemicals.
- Amino acid metabolism is very different from that for carbohydrates or fatty acids.
  - Amino acids contain nitrogen, which requires special metabolism.
  - Each different amino acid has a unique carbon skeleton and requires its own metabolic pathway.

35.6 Metabolic Nitrogen Fixation

KEY TERMS

Nitrogen fixation
Nitrogen cycle

- Although the atmosphere contains about 78% elemental nitrogen, most cells cannot use this as a source of metabolic nitrogen.
- Nitrogen fixation is the conversion of diatomic nitrogen to a biochemically usable form.
- The nitrogen cycle describes the process by which nitrogen is circulated and recirculated from the atmosphere through living organisms and back to the atmosphere.
- Nitrogen is fixed by three general routes:
  - Certain bacteria are capable of fixing nitrogen. Of special note, the *Rhizobium* bacteria are maintained in a symbiotic relationship in legume root nodules.
  - The high temperature of lightning flashes oxidizes diatomic nitrogen, making it available to plants.
  - A major means of chemical fertilizer production reacts hydrogen gas with nitrogen gas (the Haber process) to produce ammonia.
- Of the common elements needed for life, nitrogen is the most difficult to obtain; the availability of most biological nitrogen depends on a relatively small number of nitrogen-fixing bacteria.

35.7 Amino Acids and Metabolic Nitrogen Balance

KEY TERMS

Amino acid pool
Nitrogen balance
Positive nitrogen balance
Negative nitrogen balance

- Protein is digested and absorbed to provide the amino acid dietary requirements.
- Absorbed amino acids enter the amino acid pool—the total supply of amino acids available for use throughout the body.
- The amino acid pool supplies other nitrogen pools such as the tissue proteins.
- When the amount of excreted nitrogen equals the amount of nitrogen ingested, nitrogen balance is achieved.
- A growing child ingests more nitrogen than is excreted; the child is in a positive nitrogen balance.
- A fasting or starving person ingests less nitrogen than is excreted and is in a negative nitrogen balance.

35.8 Amino Acids and Nitrogen Transfer

KEY TERMS

Transamination
Glucogenic amino acid
Ketogenic amino acid

- In general, when an amino acid is used for some purpose other than protein synthesis, nitrogen is removed from the amino acid.
- Transamination is the transfer of an amino group from an alpha amino acid to an alpha keto acid:

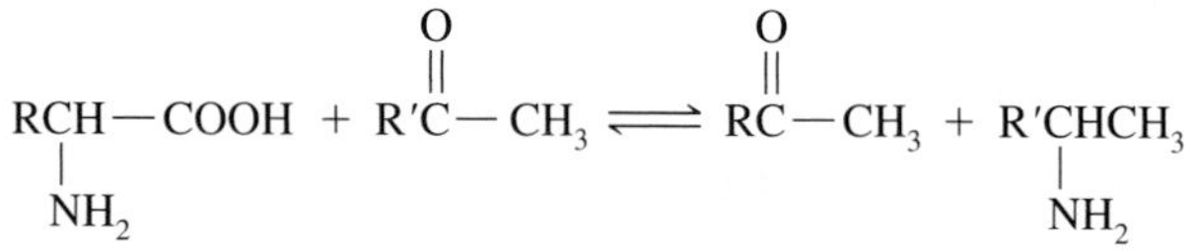

$$\underset{\displaystyle NH_2}{RCH}-COOH + R'\overset{\displaystyle O}{\overset{\|}{C}}-CH_3 \rightleftharpoons R\overset{\displaystyle O}{\overset{\|}{C}}-CH_3 + R'\underset{\displaystyle NH_2}{CH}CH_3$$

- Glucogenic amino acids can be used to produce glucose.
- Some amino acids can be directly converted to acetyl-CoA and can cause an increase in ketone body concentrations; these amino acids are ketogenic amino acids.
- Some amino acids are both glucogenic and ketogenic.

35.9 Nitrogen Excretion and the Urea Cycle

KEY TERM

Urea cycle

- Nitrogen excretion occurs when there is an excess of nitrogen.
- Since the simplest nitrogen excretion product, ammonia, is toxic, humans must eliminate nitrogen in another form, urea:

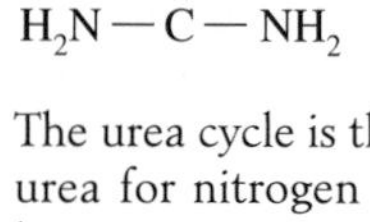

$$H_2N-\overset{\displaystyle O}{\overset{\|}{C}}-NH_2$$

- The urea cycle is the metabolic pathway that creates urea for nitrogen excretion and takes place in the liver.
- ATP energy is used to synthesize urea and excrete nitrogen in a nontoxic form.

35.10 Acetyl-CoA, a Central Molecule in Metabolism

- Acetyl-CoA is at the hub of most common metabolic processes.
- Acetyl-CoA is central for three important reasons:
  - The small size and simple structure of the acetyl group enables this two carbon fragment to be used in many biochemical syntheses.
  - The thioester causes both carbons of the acetyl fragment to be especially reactive.
  - Coenzyme A is bound by many enzymes, so acetyl-CoA can be involved in many enzyme-catalyzed reactions.

Review Questions

All questions with blue numbers have answers in the appendix of the text.

1. What major characteristic of a fatty acid allows it to serve as an energy-storage molecule?
2. Briefly describe Knoop's experiment on fatty acid oxidation and catabolism.
3. What is meant by the terms *beta oxidation* and *beta cleavage* in relation to fatty acid catabolism?
4. How are ketone bodies important in energy metabolism?
5. Define *ketosis*.
6. Aside from being a food reserve, what are the two principal functions of body fat?
7. Is fatty acid synthesis simply the reverse of beta oxidation? Explain briefly.
8. Briefly describe the nitrogen cycle in nature.
9. In general terms, how is the citric acid cycle involved in obtaining energy from fats?
10. Briefly describe why soybeans are a crop that enriches the soil.
11. Briefly describe why L-glutamic acid is considered to be the central amino acid of the amino acid pool.
12. In general, what are the metabolic fates of amino acids?
13. Why is L-glutamine a better nitrogen-transport molecule than ammonia?
14. What nitrogen compound is excreted by (a) fish and (b) birds?
15. Give the structures of urea and uric acid.
16. How many ATPs are used to produce urea in the urea cycle?
17. List three reasons that acetyl-CoA makes a good central intermediate in metabolism.
18. Briefly explain why a growing child is in a positive nitrogen balance.
19. From what important energy nutrients are ketone bodies derived? Into what important catabolic pathway do ketone bodies go?
20. In general, what metabolic changes occur when a person is placed on a very low carbohydrate diet?

Paired Exercises

All exercises with blue numbers have answers in the appendix of the text.

1. Circle the two carbons that are found in acetyl-CoA after one pass through the beta oxidation pathway by butyric acid, $CH_3CH_2CH_2COOH$.
2. Circle the two carbons that remain after two passes through the beta oxidation pathway by caproic acid, $CH_3CH_2CH_2CH_2CH_2COOH$.
3. Write the structure and mark the β-carbon in palmitic acid.
4. Write the structure and mark the β-carbon in myristic acid.
5. How many molecules of acetyl-CoA can be formed from lauric acid via the beta oxidation pathway?
6. How many molecules of acetyl-CoA can be formed from palmitic acid via the beta oxidation pathway?
7. One $FADH_2$ is produced for each pass through the beta oxidation pathway. How many of these reduced coenzymes will be formed from lauric acid?
8. One NADH is produced for each pass through the beta oxidation pathway. How many of these reduced coenzymes will be formed from palmitic acid?
9. Each $FADH_2$ can yield two ATPs, and each NADH can yield three ATPs. Given that one NADH and one $FADH_2$ is made for each pass through the beta oxidation pathway, how many ATPs will be formed from myristic acid?
10. Each $FADH_2$ can yield two ATPs, and each NADH can yield three ATPs. Given that one NADH and one $FADH_2$ is made for each pass through the beta oxidation pathway, how many ATPs will be formed from lauric acid?
11. How is the acyl carrier protein (ACP) similar to coenzyme A (CoA)?
12. How is the acyl carrier protein (ACP) different from coenzyme A (CoA)?

13. How might a low-protein diet cause a negative nitrogen balance?

14. How might a rapid growth spurt cause a positive nitrogen balance?

15. A diet high in L-leucine was observed to cause an increase in blood levels of

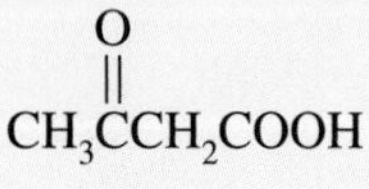

$$CH_3\overset{\overset{\displaystyle O}{\|}}{C}CH_2COOH$$

acetoacetic acid

Is L-leucine a glucogenic or ketogenic amino acid? Explain briefly.

16. A diet high in L-aspartic acid was observed to cause no change in blood levels of

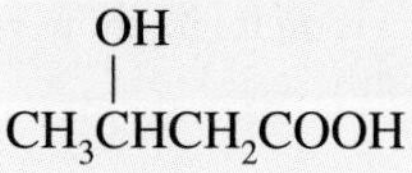

$$CH_3\overset{\overset{\displaystyle OH}{|}}{C}HCH_2COOH$$

β-hydroxybutyric acid

Is L-aspartic acid a glucogenic or ketogenic amino acid? Explain briefly.

17. Write the structural formulas of the compounds produced by transamination of the following amino acids:
(a) L-alanine (b) L-serine

18. Write the structural formulas of the compounds produced by transamination of the following amino acids:
(a) L-aspartic acid (b) L-phenylalanine

19. What portion of the carbamoyl phosphate molecule, with structure

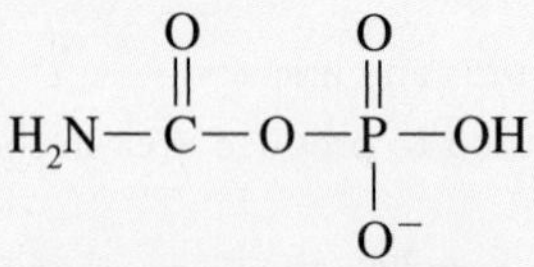

$$H_2N-\overset{\overset{\displaystyle O}{\|}}{C}-O-\underset{\underset{\displaystyle O^-}{|}}{\overset{\overset{\displaystyle O}{\|}}{P}}-OH$$

is incorporated into urea?

20. What portion of the L-aspartic acid molecule, with structure

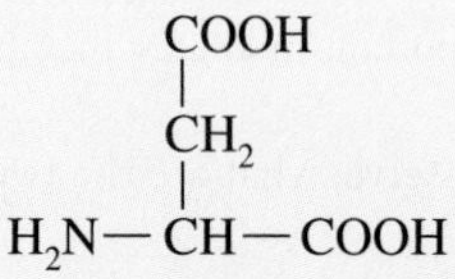

$$H_2N-\overset{\overset{\displaystyle COOH}{|}}{\overset{\displaystyle CH_2}{\underset{}{|}}}\!\!\!\!\!\!\!\!\!\quad CH-COOH$$

is incorporated into urea?

21. Given 1.6 moles of L-glutamic acid and the reactions associated with the urea cycle: How many moles of urea can be formed (all other reactants are readily available)? How many moles of ATP are used?

22. Given 0.84 mol of L-aspartic acid and the reactions associated with the urea cycle: How many moles of urea can be formed (all other reactants are readily available)? How many moles of L-glutamic acid will be used as well?

23. Dietary amino acids are used in transamination to make L-glutamic acid. Is this process nitrogen fixation? Briefly explain.

24. Bacteria convert ammonia to nitrate, which is then taken up by plants and used to make proteins. Is this process nitrogen fixation? Briefly explain.

Additional Exercises

All exercises with blue numbers have answers in the appendix of the text.

25. How is it possible to accumulate fatty tissue (or adipose tissue) even though very little fat is included in the diet?

26. Give the name and structure of a ketone body that does not contain a ketone functional group.

27. Beta oxidation is used by the cell to produce energy, yet no ATPs are formed in this pathway. Briefly explain this seeming contradiction.

28. Starting with acetyl-CoA and malonyl-CoA, write the condensed equation for the lipogenesis of myristic acid, $CH_3(CH_2)_{12}COOH$.

29. Starting with L-glutamic acid and

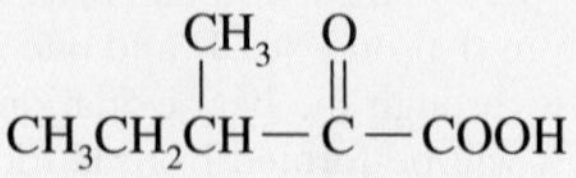

$$CH_3CH_2\overset{\overset{\displaystyle CH_3}{|}}{C}H-\overset{\overset{\displaystyle O}{\|}}{C}-COOH$$

write an equation showing the transamination to isoleucine.

30. You have been invited to a lecture entitled "The Role of Malonyl CoA." Would you expect the lecture to be about anabolism of fats, the catabolism of fats, or both? Explain briefly.

31. In Step 2 of beta oxidation of a fatty acid, dehydrogenation occurs. Is the fatty acid oxidized or reduced? Explain briefly.

32. The energy content of fats is 9 kcal/g. Calculate the energy content of 1.0 mol of palmitic acid.

33. A balanced diet has carbohydrate as the single largest component. A friend suggests that simply increasing fat intake will cause ketoses. Do you agree? Explain briefly.

Challenge Exercise

34. Based on your knowledge of catabolic pathways, why is the ATP yield from a six-carbon fatty acid greater than the ATP yield from a six-carbon hexose (glucose)?

Answers to Practice Exercises

35.1 (a), (b), (c)

$$CH_3{-}CH_2 \mid CH_2{-}CH_2 \mid CH_2{-}CH_2 \mid CH_2{-}CH_2 \mid CH_2{-}CH_2 \mid CH_2{-}CH_2 \mid CH_2{-}\text{(}CH_2\text{)} \mid CH_2{-}\overset{\overset{\displaystyle O}{\|}}{C^*}{-}OH$$

35.2 (a) High-temperature nitrogen fixation.

(b) Chemical nitrogen fixation.

PUTTING IT TOGETHER: Review for Chapters 32–35

Multiple Choice:

Choose the correct answer for each of the following:

1. Milk is a good example of a
   (a) macronutrient (c) mineral
   (b) food (d) carbohydrate
2. Which of the following is a micronutrient?
   (a) glucose (c) iron
   (b) stearic acid (d) water
3. *RDA* stands for
   (a) recommended daily allowance
   (b) recommended dietary allowance
   (c) required daily allowance
   (d) required dietary allowance
4. Amylose, a component of starch, is an example of a
   (a) micronutrient (c) vitamin
   (b) mineral (d) macronutrient
5. Starvation leads to a condition called
   (a) kwashiorkor (c) marasmus
   (b) scurvy (d) no correct answer given
6. Which of the following is *not* an example of a macronutrient?
   (a) water (c) fructose
   (b) triacylglycerol (d) soy protein
7. Each pound of fat is equivalent to about
   (a) 3500 kcal (c) 4.184 kJ
   (b) 500 kcal (d) 65 kg
8. An example of a complex carbohydrate is
   (a) blood sugar (c) cellulose
   (b) lactose (d) sucrose
9. Which of the following is an example of an ω-3 fatty acid?
   (a) linoleic acid (c) arachidonic acid
   (b) linolenic acid (d) oleic acid
10. Vitamin C is known as
    (a) a fat-soluble vitamin (c) ascorbic acid
    (b) citric acid (d) pantothenic acid
11. A person suffering from beriberi has a deficiency of
    (a) vitamin B_1 (c) vitamin B_2
    (b) vitamin B_{12} (d) niacin
12. Potassium is an example of a
    (a) trace element (c) major mineral
    (b) macronutrient (d) no correct answer given
13. A deficiency of iodine commonly causes
    (a) anemia (c) dental caries
    (b) goiter (d) impaired glucose tolerance
14. How many grams of calcium are supplied if a serving of cereal (plus milk) gives 30% of a calcium RDA (1200 mg)?
    (a) 0.36 g (c) 360 g
    (b) 3.6×10^5 g (d) 0.00036 g
15. If 0.9 mg of zinc is 6% of the daily value, the RDA for zinc is
    (a) 54 mg (c) 1.5 mg
    (b) 15 mg (d) 0.15 g
16. Many food additives have
    (a) an RDA
    (b) a GARS rating
    (c) a "generally recognized as sufficient" rating
    (d) no correct answer given
17. Ultimately, carbohydrates are digested to
    (a) disaccharides (c) lactic acid
    (b) monosaccharides (d) carbon dioxide
18. A very important enzyme in salivary digestion is
    (a) pepsin (c) amylase
    (b) lipase (d) chymotrypsin
19. Bile is important for
    (a) fat digestion (c) protein digestion
    (b) protein absorption (d) carbohydrate digestion
20. Intestinal microvilli
    (a) secrete bile (c) synthesize bile
    (b) absorb nutrients (d) no correct answer given
21. Blood-glucose concentration is maintained by the
    (a) liver (c) kidneys
    (b) gallbladder (d) gastric juices
22. Hemoglobin binds to carbon dioxide to form
    (a) a heme–carboxyl ion (c) a bicarbonate ion
    (b) a carbonate ion (d) a carbamino ion
23. The formula for carbonic acid is
    (a) H_3CO_4 (c) H_2CO_3
    (b) HCO_3^- (d) H_2CO_4
24. Two major waste products of cellular energetics are
    (a) carbonic acid and pyruvic acid
    (b) carbonic acid and lactic acid
    (c) carbolic acid and lactic acid
    (d) carbolic acid and pyruvic acid
25. Which carbon in the following compound provides the most energy upon oxidation?

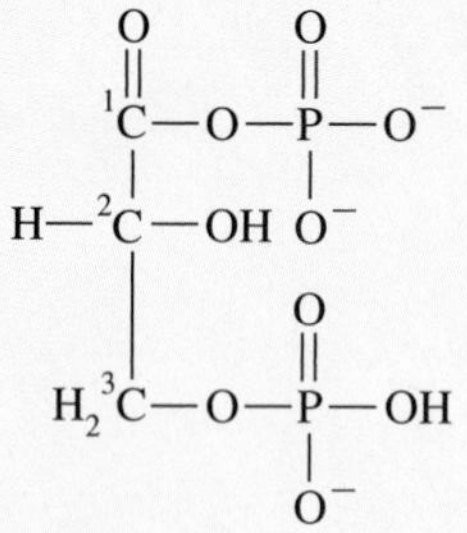

    O O
    ‖ ‖
    ^{1}C—O—P—O$^-$
    | |
    H—^{2}C—OH O$^-$
    |
    | O
    | ‖
    H_2^{3}C—O—P—OH
    |
    O$^-$

    (a) carbon 1 (c) carbon 3
    (b) carbon 2 (d) no correct answer given
26. The name of FAD is
    (a) flavin adenine dinucleotide
    (b) flavin arginine dinucleotide
    (c) flavodoxin adenine dinucleotide
    (d) flavodoxin arginine dinucleotide

27. NADH is the
(a) oxidized form of nicotinamide adenine dinucleotide
(b) oxidized form of nicotinamide adenine dinucleotide phosphate
(c) reduced form of nitcotinamide adenine dinucleotide
(d) reduced form of nicotinamide adenine dinucleotide phosphate

28. NAD^+ is derived from the vitamin
(a) niacin (c) pantothenic acid
(b) thiamine (d) riboflavin

29. Which reaction is *not* part of the mitochondrial electron transport system?
(a) $2\,NADH + 2\,H^+ + O_2 \longrightarrow 2\,NAD^+ + 2\,H_2O$
(b) $2\,NAD^+ + 2\,H_2O \longrightarrow 2\,NADH + 2\,H^+ + O_2$
(c) $2\,FADH_2 + O_2 \longrightarrow 2\,FAD + 2\,H_2O$
(d) $4\,NADH + 4\,H^+ + 2\,O_2 \longrightarrow 4\,NAD^+ + 4\,H_2O$

30. ATP is composed of a triphosphate, an adenine, and
(a) a deoxyribose (c) a glucose
(b) a fructose (d) a ribose

31. How many high-energy phosphate bonds does the following structure have?

```
                                        O
                              CH3       ‖
                                  \   /   \  N — H
     O        O                    ‖        |
     ‖        ‖                    |        |
HO — P — O — P — O — CH2            \  N  /  ‖ O
     |        |          \   O        |
     O−       O−          \ / \       |
                           |    |
                            \__/
                             |
                             OH
```

(a) two (c) one
(b) four (d) three

32. A process that directly uses redox reaction energy to form ATP is called
(a) mitochondrial electron transport
(b) chloroplast electron transport
(c) oxidative phosphorylation
(d) reductive phosphorylation

33. Photosynthesis uses 6 CO_2 to make
(a) one ribose-5-phosphate
(b) glucose
(c) two 3-phosphoglycerates
(d) fructose

34. The chloroplast electron transport produces
(a) NADH (c) $NADP^+$
(b) $FADH_2$ (d) no correct answer given

35. In photosynthesis, light energy is first absorbed by
(a) $NADP^+$ (c) ATP
(b) chlorophyll (d) glucose

36. Which of the following is an anabolic pathway?
(a) the Embden–Meyerhof pathway
(b) glycogenesis
(c) glycogenolysis
(d) glycolysis

37. How many high-energy phosphate bonds does the following compound (1,3-bisphosphoglycerate) have?

```
      O        O
      ‖        ‖
      C — O — P — OH
      |        |
  H — C — OH  O−
      |        O
      |        ‖
    H2C — O — P — OH
               |
               O−
```

(a) one (c) three
(b) two (d) four

38. The Embden–Meyerhof pathway converts glucose to two pyruvates. The sum of the oxidation numbers of carbon changes from zero in glucose to what value in two pyruvates?
(a) +6 (c) −2
(b) +4 (d) +2

39. In order to cycle NADH back to NAD^+, muscle tissue converts pyruvate to
(a) acetaldehyde (c) ethanol
(b) glyceraldehyde (d) lactate

40. Glycolysis is termed anaerobic because
(a) molecular oxygen is required
(b) only small amounts of ATP are formed
(c) molecular oxygen is not required
(d) NADH is produced

41. The synthesis of coenzyme A requires
(a) pantothenic acid (c) vitamin D
(b) lactic acid (d) nicotinic acid

42. The citric acid cycle takes place in the
(a) cytoplasm (c) chloroplasts
(b) nuclei (d) mitochondria

43. The citric acid cycle was elucidated by
(a) Fritz Lipmann (c) Hans Krebs
(b) Franz Meyerhof (d) Melvin Calvin

44. In the citric acid cycle, the acetyl group of acetyl CoA is
(a) reduced to two carbon dioxides
(b) oxidized to two carbon dioxides
(c) oxidized to two NADHs
(d) reduced to two NADHs

45. The citric acid cycle is aerobic because
(a) O_2 is used in the cycle
(b) the cycle depends on chloroplast electron transport
(c) the cycle depends on mitochondrial electron transport
(d) the cycle uses O_2 to convert pyruvate to acetyl CoA

46. How many NADHs are produced from 13.2 mol of acetyl CoA by the citric acid cycle?
(a) 6.6 mol (c) 39.6 mol
(b) 13.2 mol (d) 26.4 mol

47. Which of the following compounds *cannot* be used to make glucose in gluconeogenesis?
(a) glycerol (c) lactate
(b) acetyl CoA (d) amino acids

48. Which of the following is *not* an advantage to a multistep process (compared with a single-step process)?
(a) Energy can be handled efficiently.
(b) Fewer enzymes are needed.
(c) Intermediates can be used in several pathways.
(d) The energy release per step is small.

49. The glucose renal threshold is reached when
(a) a condition of hyperglycemia occurs
(b) blood-glucose levels are too low
(c) insulin levels are too high
(d) a fasting state occurs

50. In studying beta oxidation, which of the following scientists designed prototypes of modern equipment that tags molecules with radioactive isotopes?
(a) Franz Meyerhof (c) Franz Knoop
(b) Melvin Calvin (d) no correct answer given

51. How many ATPs are produced in beta oxidation from one stearic acid molecule?
(a) 2 (c) 139
(b) 0 (d) 98

52. The anabolic pathway for fatty acids is called
(a) "two-carbon chop" (c) glycolysis
(b) lipogenesis (d) glycogenesis

53. The vitamin biotin is very important in reactions that
(a) add carbon dioxide to a molecule
(b) cause hydration of a double bond
(c) form a thioester
(d) reduce $NADP^+$

54. Which of the following is *not* a method for fixing nitrogen?
(a) lightning (c) transamination
(b) Haber process (d) soil bacteria

55. Acetyl CoA is to the citric acid cycle as which of the following compounds is to the urea cycle?
(a) glutamic cycle (c) carbamoyl phosphate
(b) fumaric acid (d) glutamine

56. When a person takes in more nitrogen than is excreted, he or she is said to be in
(a) negative nitrogen balance
(b) nitrogen balance
(c) positive nitrogen balance
(d) no correct answer given

57. Transamination is the first step in the conversion of glucogenic amino acids to
(a) ketogenic amino acids (c) glycogen
(b) fats (d) essential amino acids

58. Fish commonly excrete nitrogen in the form of
(a) urea (c) uric acid
(b) ammonia (d) glutamine

59. How many nitrogen atoms are in one molecule of urea?
(a) one (c) three
(b) two (d) four

60. The urea cycle occurs in the
(a) liver (c) pancreas
(b) small intestine (d) skeletal muscles

Free Response Questions

1. Is the vitamin

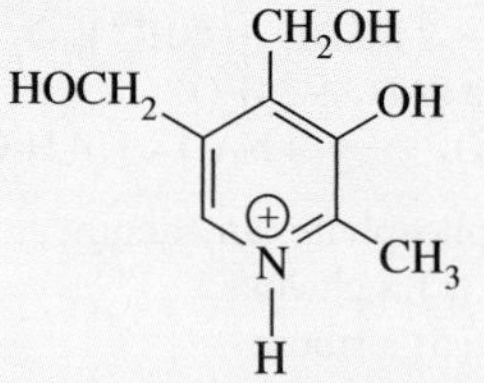

CH_2OH, $HOCH_2$, OH, $N^{\oplus}$, CH_3, H

likely to be water soluble or fat soluble? Explain.

2. What are the end products of the digestion of polysaccharides, fats and oils, and proteins before they enter the metabolic pathways? Into what common metabolic intermediate are most of these digestive end products often converted?

3. Are the two nutrients shown macronutrients or micronutrients?

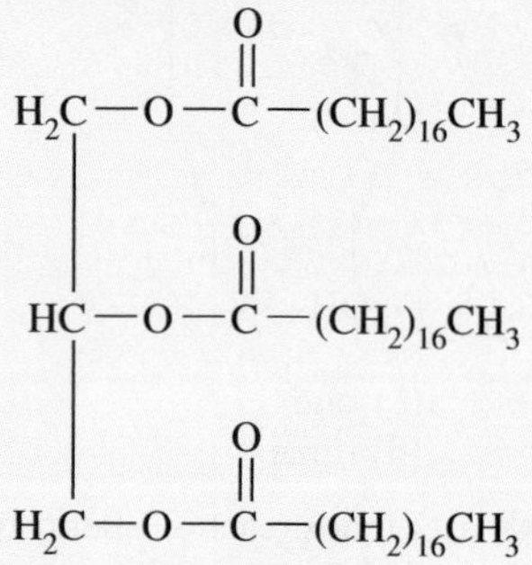

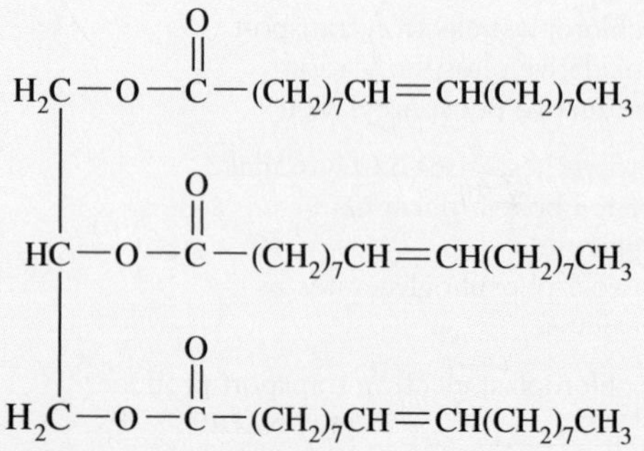

$$H_2C{-}O{-}\overset{O}{\overset{\|}{C}}{-}(CH_2)_{16}CH_3$$
$$HC{-}O{-}\overset{O}{\overset{\|}{C}}{-}(CH_2)_{16}CH_3$$
$$H_2C{-}O{-}\overset{O}{\overset{\|}{C}}{-}(CH_2)_{16}CH_3$$

$$H_2C{-}O{-}\overset{O}{\overset{\|}{C}}{-}(CH_2)_7CH{=}CH(CH_2)_7CH_3$$
$$HC{-}O{-}\overset{O}{\overset{\|}{C}}{-}(CH_2)_7CH{=}CH(CH_2)_7CH_3$$
$$H_2C{-}O{-}\overset{O}{\overset{\|}{C}}{-}(CH_2)_7CH{=}CH(CH_2)_7CH_3$$

What class of nutrients do they fall under? Draw the products of digestion (hydrolysis) of these two nutrients, and indicate which is more likely to provide more cellular energy. Explain briefly.

4. The three macronutrients can be converted into glucose. For example, carbohydrates are digested via salivary and pancreatic juices to give monosaccharides, including glucose directly. *Briefly* outline how proteins and fats or oils can also lead to the production of glucose in humans, and indicate the name of the pathway that includes both processes.

5. Complete the following reaction using nicotinamide adenine dinucleotide as the redox coenzyme:

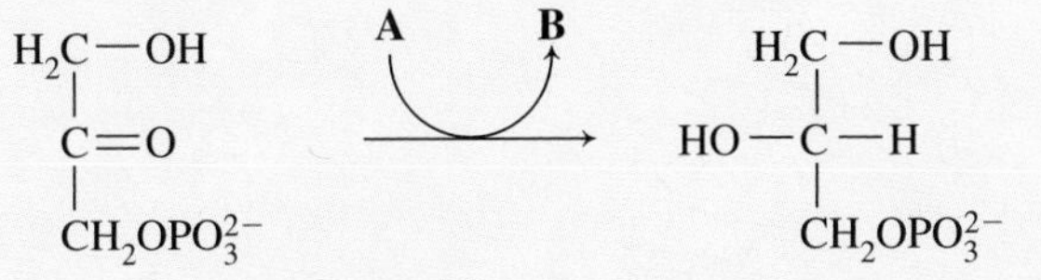

$$\begin{array}{c} H_2C-OH \\ | \\ C=O \\ | \\ CH_2OPO_3^{2-} \end{array} \xrightarrow{A \curvearrowright B} \begin{array}{c} H_2C-OH \\ | \\ HO-C-H \\ | \\ CH_2OPO_3^{2-} \end{array}$$

Is the coenzyme being reduced or oxidized in this process? Identify coenzymes A and B. Would this coenzyme be shunted to the electron-transport chain to produce ATP? Explain.

6. The complete oxidation of glucose ($C_6H_{12}O_6$) to carbon dioxide and water can be broken down into three main processes: glycolysis, pyruvate oxidation, and the citric acid cycle. The glycolysis of 1 mol of glucose in the skeletal muscle yields 2 mol NADH, 2 mol ATP, and 2 mol of pyruvate. The oxidation of 1 mol of pyruvate yields 1 mol NADH and 1 mol acetyl-CoA. The citric acid cycle produces 3 mol NADH, 1 mol $FADH_2$, and 1 mol ATP (actually GTP, but it can be converted to ATP) for every mole of acetyl-CoA entering the cycle. Before a race, if an athlete eats 45 g of glucose powder, how many moles of ATP would be produced, assuming that no ATP is used to transport the glucose to the skeletal muscle and that all the glucose and coenzymes are oxidized?

7. One mole of which saturated fatty acid was fully catabolized via beta oxidation if the acetyl-CoA from the catabolism produced 96 moles of ATP when used in the citric acid cycle? The citric acid cycle produces 3 moles NADH, 1 mol $FADH_2$, and 1 mol ATP for every mole of acetyl-CoA entering the cycle. Explain briefly.

8. Fill in the missing information in the following metabolic summary:

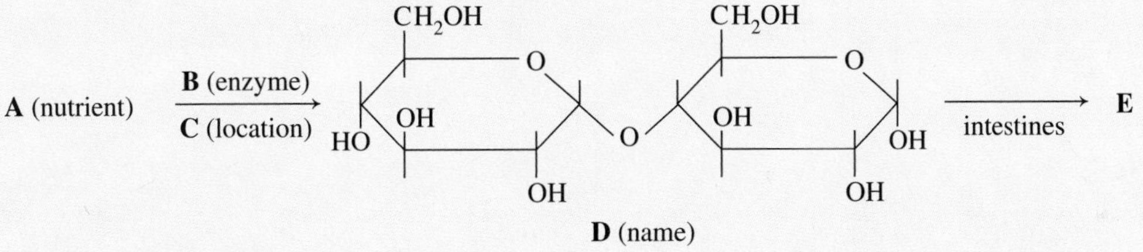

A (nutrient) $\xrightarrow[\textbf{C}\text{ (location)}]{\textbf{B}\text{ (enzyme)}}$ [disaccharide structure: two pyranose rings, each with CH_2OH, OH, OH groups, joined by O; HO at left, OH at right] $\xrightarrow[\text{intestines}]{}$ **E**

D (name)

E $\xrightarrow{\textbf{F}\text{ (pathway)}}$ pyruvate $\longrightarrow$ acetyl-CoA $\xrightarrow{\textbf{G}\text{ (cycle)}}$ CO_2

9. A possible pathway for the nitrogen ingested could be represented as follows:

glu-phe-asp-asn-val-tyr-trp-ala-val-phe $\xrightarrow{A}$ glu-phe, asp-asn-val-tyr, trp-ala-val-phe $\xrightarrow{B}$ glu, phe, asp, asn, val, tyr, trp, ala, val, phe $\xrightarrow{C}$ urea

Indicate in which part of the body where each of the processes (A, B, C) shown takes place.

APPENDIX I

Mathematical Review

Multiplication Multiplication is a process of adding any given number or quantity to itself a certain number of times. Thus, 4 times 2 means 4 added two times, or 2 added together four times, to give the product 8. Various ways of expressing multiplication are

$$ab \quad a \times b \quad a \cdot b \quad a(b) \quad (a)(b)$$

Each of these expressions means a times b, or a multiplied by b, or b times a.

When $a = 16$ and $b = 24$, we have $16 \times 24 = 384$.

The expression $°F = (1.8 \times °C) + 32$ means that we are to multiply 1.8 times the Celsius degrees and add 32 to the product. When °C equals 50,

$$°F = (1.8 \times 50) + 32 = 90 + 32 = 122°F$$

The result of multiplying two or more numbers together is known as the *product.*

Division The word *division* has several meanings. As a mathematical expression, it is the process of finding how many times one number or quantity is contained in another. Various ways of expressing division are

$$a \div b \quad \frac{a}{b} \quad a/b$$

Each of these expressions means a divided by b.

When $a = 15$ and $b = 3$, $\frac{15}{3} = 5$.

The number above the line is called the *numerator;* the number below the line is the *denominator.* Both the horizontal and the slanted (/) division signs also mean "per." For example, in the expression for density, we determine the mass per unit volume:

$$\text{density} = \text{mass/volume} = \frac{\text{mass}}{\text{volume}} = \text{g/mL}$$

The diagonal line still refers to a division of grams by the number of milliliters occupied by that mass. The result of dividing one number into another is called the *quotient.*

Fractions and Decimals A fraction is an expression of division, showing that the numerator is divided by the denominator. A *proper fraction* is one in which the numerator is smaller than the denominator. In an *improper fraction,* the numerator is the larger number. A decimal or a decimal fraction is a proper fraction in which the

Proper fraction		Decimal fraction		Proper fraction
$\frac{1}{8}$	=	0.125	=	$\frac{125}{1000}$
$\frac{1}{10}$	=	0.1	=	$\frac{1}{10}$
$\frac{3}{4}$	=	0.75	=	$\frac{75}{100}$
$\frac{1}{100}$	=	0.01	=	$\frac{1}{100}$
$\frac{1}{4}$	=	0.25	=	$\frac{25}{100}$

denominator is some power of 10. The decimal fraction is determined by carrying out the division of the proper fraction. Examples of proper fractions and their decimal fraction equivalents are shown in the accompanying table.

Addition of Numbers with Decimals To add numbers with decimals, we use the same procedure as that used when adding whole numbers, but we always line up the decimal points in the same column. For example, add 8.21 + 143.1 + 0.325:

$$\begin{array}{r} 8.21 \\ +143.1 \\ +\ \ \ 0.325 \\ \hline 151.635 \end{array}$$

When adding numbers that express units of measurement, we must be certain that the numbers added together all have the same units. For example, what is the total length of three pieces of glass tubing: 10.0 cm, 125 mm, and 8.4 cm? If we simply add the numbers, we obtain a value of 143.4, but we are not certain what the unit of measurement is. To add these lengths correctly, first change 125 mm to 12.5 cm. Now all the lengths are expressed in the same units and can be added:

$$\begin{array}{r} 10.0\text{ cm} \\ 12.5\text{ cm} \\ 8.4\text{ cm} \\ \hline 30.9\text{ cm} \end{array}$$

Subtraction of Numbers with Decimals To subtract numbers containing decimals, we use the same procedure as for subtracting whole numbers, but we always line up the decimal points in the same column. For example, subtract 20.60 from 182.49:

$$\begin{array}{r} 182.49 \\ -\ \ 20.60 \\ \hline 161.89 \end{array}$$

When subtracting numbers that are measurements, be certain that the measurements are in the same units. For example, subtract 22 cm from 0.62 m. First change m to cm, then do the subtraction.

$$(0.62\text{ m})\left(\frac{100\text{ cm}}{\text{m}}\right) = 62\text{ cm} \qquad \begin{array}{r} 62\text{ cm} \\ -22\text{ cm} \\ \hline 40.\text{ cm} \end{array}$$

Multiplication of Numbers with Decimals To multiply two or more numbers together that contain decimals, we first multiply as if they were whole numbers. Then, to locate the decimal point in the product, we add together the number of digits to the right of the decimal in all the numbers multiplied together. The product should have this same number of digits to the right of the decimal point.

Multiply $2.05 \times 2.05 = 4.2025$ (total of four digits to the right of the decimal). Here are more examples:

If a number is a measurement, the answer must be adjusted to the correct number of significant figures.

$14.25 \times 6.01 \times 0.75 = 64.231875$ (six digits to the right of the decimal)

$39.26 \times 60 = 2355.60$ (two digits to the right of the decimal)

Division of Numbers with Decimals To divide numbers containing decimals, we first relocate the decimal points of the numerator and denominator by moving them to the right as many places as needed to make the denominator a whole number. (Move the decimal of both the numerator and the denominator the same amount and in the same direction.) For example,

$$\frac{136.94}{4.1} = \frac{1369.4}{41}$$

The decimal point adjustment in this example is equivalent to multiplying both numerator and denominator by 10. Now we carry out the division normally, locating the decimal point immediately above its position in the dividend:

$$\begin{array}{r}33.4\\41\overline{)1269.4}\end{array} \qquad \frac{0.441}{26.25} = \frac{44.1}{2625} = \begin{array}{r}0.0168\\2625\overline{)44.1000}\end{array}$$

These examples are guides to the principles used in performing the various mathematical operations illustrated. Every student of chemistry should learn to use a calculator for solving mathematical problems (see Appendix II). The use of a calculator will save many hours of doing tedious calculations. After solving a problem, the student should check for errors and evaluate the answer to see if it is logical and consistent with the data given.

Algebraic Equations Many mathematical problems that are encountered in chemistry fall into the following algebraic forms. Solutions to these problems are simplified by first isolating the desired term on one side of the equation. This rearrangement is accomplished by treating both sides of the equation in an identical manner until the desired term is isolated.

(a) $a = \dfrac{b}{c}$

To solve for b, multiply both sides of the equation by c:

$$a \times c = \frac{b}{\cancel{c}} \times \cancel{c}$$

$$b = a \times c$$

To solve for c, multiply both sides of the equation by $\dfrac{c}{a}$:

$$\cancel{a} \times \frac{c}{\cancel{a}} = \frac{b}{\cancel{c}} \times \frac{\cancel{c}}{a}$$

$$c = \frac{b}{a}$$

(b) $\dfrac{a}{b} = \dfrac{c}{d}$

To solve for a, multiply both sides of the equation by b:

$$\frac{a}{\cancel{b}} \times \cancel{b} = \frac{c}{d} \times b$$

$$a = \frac{c \times b}{d}$$

To solve for b, multiply both sides of the equation by $\dfrac{b \times d}{c}$:

$$\frac{a}{\cancel{b}} \times \frac{\cancel{b} \times d}{c} = \frac{\cancel{c}}{\cancel{d}} \times \frac{b \times \cancel{d}}{\cancel{c}}$$

$$\frac{a \times d}{c} = b$$

(c) $a \times b = c \times d$

To solve for a, divide both sides of the equation by b:

$$\frac{a \times \cancel{b}}{\cancel{b}} = \frac{c \times d}{b}$$

$$a = \frac{c \times d}{b}$$

(d) $\frac{b - c}{a} = d$

To solve for b, first multiply both sides of the equation by a:

$$\frac{\cancel{a}(b - c)}{\cancel{a}} = d \times a$$

$$b - c = d \times a$$

Then add c to both sides of the equation:

$$b - \cancel{c} + \cancel{c} = d \times a + c$$

$$b = (d \times a) + c$$

When $a = 1.8, c = 32$, and $d = 35$,

$$b = (35 \times 1.8) + 32 = 63 + 32 = 95$$

Expression of Large and Small Numbers In scientific measurement and calculations, we often encounter very large and very small numbers—for example, 0.00000384 and 602,000,000,000,000,000,000,000. These numbers are troublesome to write and awkward to work with, especially in calculations. A convenient method of expressing these large and small numbers in a simplified form is by means of exponents, or powers, of 10. This method of expressing numbers is known as **scientific, or exponential, notation.**

scientific, or exponential notation

exponent

An **exponent** is a number written as a superscript following another number. Exponents are often called *powers* of numbers. The term *power* indicates how many times the number is used as a factor. In the number 10^2, 2 is the exponent, and the number means 10 squared, or 10 to the second power, or $10 \times 10 = 100$. Three other examples are

$$3^2 = 3 \times 3 = 9$$

$$3^4 = 3 \times 3 \times 3 \times 3 = 81$$

$$10^3 = 10 \times 10 \times 10 = 1000$$

For ease of handling, large and small numbers are expressed in powers of 10. Powers of 10 are used because multiplying or dividing by 10 coincides with moving the decimal point in a number by one place. Thus, a number multiplied by 10^1 would move the decimal point one place to the right; 10^2, two places to the right; 10^{-2}, two places to the left. To express a number in powers of 10, we move the decimal point in the original number to a new position, placing it so that the number is a value between 1 and 10. This new decimal number is multiplied by 10 raised to the proper power. For example, to write the number 42,389 in exponential form, the decimal point is placed between the 4 and the 2 (4.2389), and the number is multiplied by 10^4; thus, the number is 4.2389×10^4:

$$\underset{4321}{42{,}389} = 4.2389 \times 10^4$$

The exponent of 10 (4) tells us the number of places that the decimal point has been moved from its original position. If the decimal point is moved to the left, the exponent is a positive number; if it is moved to the right, the exponent is a negative number. To express the number 0.00248 in exponential notation (as a power of 10), the decimal point is moved three places to the right; the exponent of 10 is −3, and the number is 2.48×10^{-3}.

$$0.00248 = 2.48 \times 10^{-3}$$

1 2 3

Study the following examples of changing a number to scientific notation.

$$1237 = 1.237 \times 10^3$$
$$988 = 9.88 \times 10^2$$
$$147.2 = 1.472 \times 10^2$$
$$2{,}200{,}000 = 2.2 \times 10^6$$
$$0.0123 = 1.23 \times 10^{-2}$$
$$0.00005 = 5 \times 10^{-5}$$
$$0.000368 = 3.68 \times 10^{-4}$$

Exponents in multiplication and division The use of powers of 10 in multiplication and division greatly simplifies locating the decimal point in the answer. In multiplication, first change all numbers to powers of 10, then multiply the numerical portion in the usual manner, and finally add the exponents of 10 algebraically, expressing them as a power of 10 in the product. **In multiplication, the exponents (powers of 10) are added algebraically.**

$$10^2 \times 10^3 = 10^{(2+3)} = 10^5$$
$$10^2 \times 10^2 \times 10^{-1} = 10^{(2+2-1)} = 10^3$$

Multiply:	$(40{,}000)(4200)$
Change to powers of 10:	$(4 \times 10^4)(4.2 \times 10^3)$
Rearrange:	$(4 \times 4.2)(10^4 \times 10^3)$
	$16.8 \times 10^{(4+3)}$
	16.8×10^7 or 1.68×10^8 (Answer)

Multiply: $(380)(0.00020)$
$(3.80 \times 10^2)(2.0 \times 10^{-4})$
$(3.80 \times 2.0)(10^2 \times 10^{-4})$
$7.6 \times 10^{(2-4)}$
7.6×10^{-2} or 0.076 (Answer)

Multiply: $(125)(284)(0.150)$
$(1.25 \times 10^2)(2.84 \times 10^2)(1.50 \times 10^{-1})$
$(1.25)(2.84)(1.50)(10^2 \times 10^2 \times 10^{-1})$
$5.325 \times 10^{(2+2-1)}$
5.33×10^3 (Answer) (3 significant figures)

In division, after changing the numbers to powers of 10, move the 10 and its exponent from the denominator to the numerator, changing the sign of the exponent. Carry out the division in the usual manner, and evaluate the power of 10. Change the sign(s) of the exponent(s) of 10 in the denominator, and move the 10 and its exponent(s) to the numerator. Then add all the exponents of 10 together. For example,

$$\frac{10^5}{10^3} = 10^5 \times 10^{-3} = 10^{(5-3)} = 10^2$$

$$\frac{10^3 \times 10^4}{10^{-2}} = 10^3 \times 10^4 \times 10^2 = 10^{(3+4+2)} = 10^9$$

Significant Figures in Calculations The result of a calculation based on experimental measurements cannot be more precise than the measurement that has the greatest uncertainty. (See Section 2.4 for additional discussion.)

Addition and subtraction The result of an addition or subtraction should contain no more digits to the right of the decimal point than are contained in the quantity that has the least number of digits to the right of the decimal point.

Perform the operation indicated and then round off the number to the proper number of significant figures:

(a)		(b)	
142.8 g		93.45 mL	
18.843 g		−18.0 mL	
36.42 g		75.45 mL	
198.063 g		75.5 mL	(Answer)
198.1 g	(Answer)		

(a) The answer contains only one digit after the decimal point since 142.8 contains only one digit after the decimal point.

(b) The answer contains only one digit after the decimal point since 18.0 contains one digit after the decimal point.

Multiplication and division In calculations involving multiplication or division, the answer should contain the same number of significant figures as the measurement that has the least number of significant figures. In multiplication or division, the position of the decimal point has nothing to do with the number of significant figures in the answer. Study the following examples:

	Round off to
$(2.05)(2.05) = 4.2025$	4.20
$(18.48)(5.2) = 96.096$	96
$(0.0126)(0.020) = 0.000252$ or	
$(1.26 \times 10^{-2})(2.0 \times 10^{-2}) = 2.520 \times 10^{-4}$	2.5×10^{-4}
$\frac{1369.4}{41} = 33.4$	33
$\frac{2268}{4.20} = 540.$	540.

Dimensional Analysis Many problems of chemistry can be readily solved by dimensional analysis using the factor-label or conversion-factor method. Dimensional analysis involves the use of proper units of dimensions for all factors that are multiplied, divided, added, or subtracted in setting up and solving a problem. Dimensions are physical quantities such as length, mass, and time, which are expressed in such units as centimeters, grams, and seconds, respectively. In solving a problem, we treat these units mathematically just as though they were numbers, which gives us an answer that contains the correct dimensional units.

A measurement or quantity given in one kind of unit can be converted to any other kind of unit having the same dimension. To convert from one kind of unit to another, the original quantity or measurement is multiplied or divided by a conversion factor. The key to success lies in choosing the correct conversion factor. This general method of calculation is illustrated in the following examples.

Suppose we want to change 24 ft to inches. We need to multiply 24 ft by a conversion factor containing feet and inches. Two such conversion factors can be written relating inches to feet:

$$\frac{12 \text{ in.}}{1 \text{ ft}} \quad \text{or} \quad \frac{1 \text{ ft}}{12 \text{ in.}}$$

We choose the factor that will mathematically cancel feet and leave the answer in inches. Note that the units are treated in the same way we treat numbers, multiplying or dividing as required. Two possibilities then arise to change 24 ft to inches:

$$(24\ \not{ft})\left(\frac{12\ \text{in.}}{1\ \not{ft}}\right) \quad \text{or} \quad (24\ \text{ft})\left(\frac{1\ \text{ft}}{12\ \text{in.}}\right)$$

In the first case (the correct method), feet in the numerator and the denominator cancel, giving us an answer of 288 in. In the second case, the units of the answer are $\text{ft}^2/\text{in.}$, the answer being $2.0\ \text{ft}^2/\text{in.}$ In the first case, the answer is reasonable because it is expressed in units having the proper dimensions. That is, the dimension of length expressed in feet has been converted to length in inches according to the mathematical expression

$$\not{ft} \times \frac{\text{in.}}{\not{ft}} = \text{in.}$$

In the second case, the answer is not reasonable because the units ($\text{ft}^2/\text{in.}$) do not correspond to units of length. The answer is therefore incorrect. The units are the guiding factor for the proper conversion.

The reason we can multiply 24 ft times 12 in./ft and not change the value of the measurement is that the conversion factor is derived from two equivalent quantities. Therefore, the conversion factor 12 in./ft is equal to unity. When you multiply any factor by 1, it does not change the value:

$$12\ \text{in.} = 1\ \text{ft} \quad \text{and} \quad \frac{12\ \text{in.}}{1\ \text{ft}} = 1$$

Convert 16 kg to milligrams. In this problem it is best to proceed in this fashion:

$$\text{kg} \longrightarrow \text{g} \longrightarrow \text{mg}$$

The possible conversion factors are

$$\frac{1000\ \text{g}}{1\ \text{kg}} \quad \text{or} \quad \frac{1\ \text{kg}}{1000\ \text{g}} \qquad \frac{1000\ \text{mg}}{1\ \text{g}} \quad \text{or} \quad \frac{1\ \text{g}}{1000\ \text{mg}}$$

We use the conversion factor that leaves the proper unit at each step for the next conversion. The calculation is

$$(16\ \not{kg})\left(\frac{1000\ \not{g}}{1\ \not{kg}}\right)\left(\frac{1000\ \text{mg}}{1\ \not{g}}\right) = 1.6 \times 10^7\ \text{mg}$$

Regardless of application, the basis of dimensional analysis is the use of conversion factors to organize a series of steps in the quest for a specific quantity with a specific unit.

Graphical Representation of Data A graph is often the most convenient way to present or display a set of data. Various kinds of graphs have been devised, but the most common type uses a set of horizontal and vertical coordinates to show the relationship of two variables. It is called an *x–y* graph because the data of one variable are represented on the horizontal or *x*-axis (abscissa) and the data of the other variable are represented on the vertical or *y*-axis (ordinate). See Figure I.1.

As a specific example of a simple graph, let us graph the relationship between Celsius and Fahrenheit temperature scales. Assume that initially we have only the information in the table next to Figure I.2.

On a set of horizontal and vertical coordinates (graph paper), scale off at least 100 Celsius degrees on the *x*-axis and at least 212 Fahrenheit degrees on the *y*-axis. Locate and mark the three points corresponding to the three temperatures given and draw a line connecting these points (see Figure I.2).

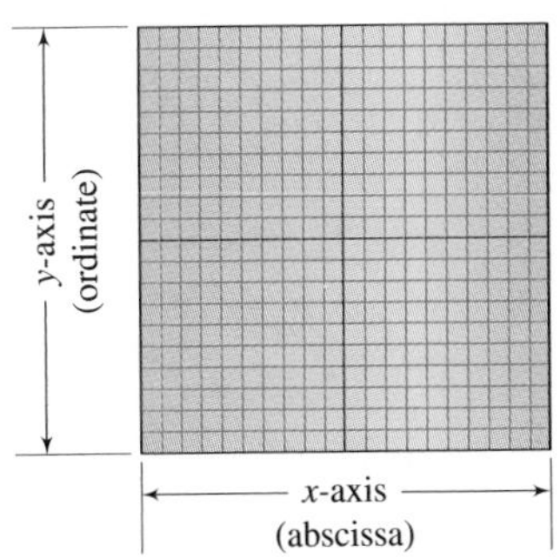

Figure I.1

°C	°F
0	32
50	122
100	212

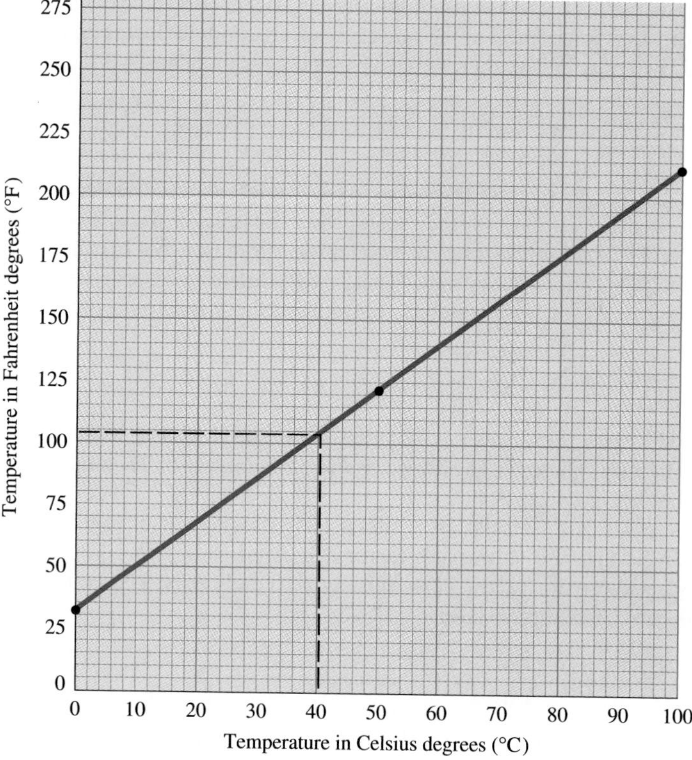

Figure I.2

Here is how a point is located on the graph: Using the (50°C, 122°F) data, trace a vertical line up from 50°C on the *x*-axis and a horizontal line across from 122°F on the *y*-axis and mark the point where the two lines intersect. This process is called *plotting*. The other two points are plotted on the graph in the same way. (*Note:* The number of degrees per scale division was chosen to give a graph of convenient size. In this case, there are 5 Fahrenheit degrees per scale division and 2 Celsius degrees per scale division.)

The graph in Figure I.2 shows that the relationship between Celsius and Fahrenheit temperature is that of a straight line. The Fahrenheit temperature corresponding to any given Celsius temperature between 0° and 100° can be determined from the graph. For example, to find the Fahrenheit temperature corresponding to 40°C, trace a perpendicular line from 40°C on the *x*-axis to the line plotted on the graph. Now trace a horizontal line from this point on the plotted line to the *y*-axis and read the corresponding Fahrenheit temperature (104°F). See the dashed lines in Figure I.2. In turn, the Celsius temperature corresponding to any Fahrenheit temperature between 32° and 212° can be determined from the graph by tracing a horizontal line from the Fahrenheit temperature to the plotted line and reading the corresponding temperature on the Celsius scale directly below the point of intersection.

The mathematical relationship of Fahrenheit and Celsius temperatures is expressed by the equation $°F = (1.8 \times °C) + 32$. Figure I.2 is a graph of this equation. Because the graph is a straight line, it can be extended indefinitely at either end. Any desired Celsius temperature can be plotted against the corresponding Fahrenheit temperature by extending the scales along both axes as necessary.

Figure I.3 is a graph showing the solubility of potassium chlorate in water at various temperatures. The solubility curve on this graph was plotted from the data in the table next to the graph.

In contrast to the Celsius–Fahrenheit temperature relationship, there is no simple mathematical equation that describes the exact relationship between temperature and the

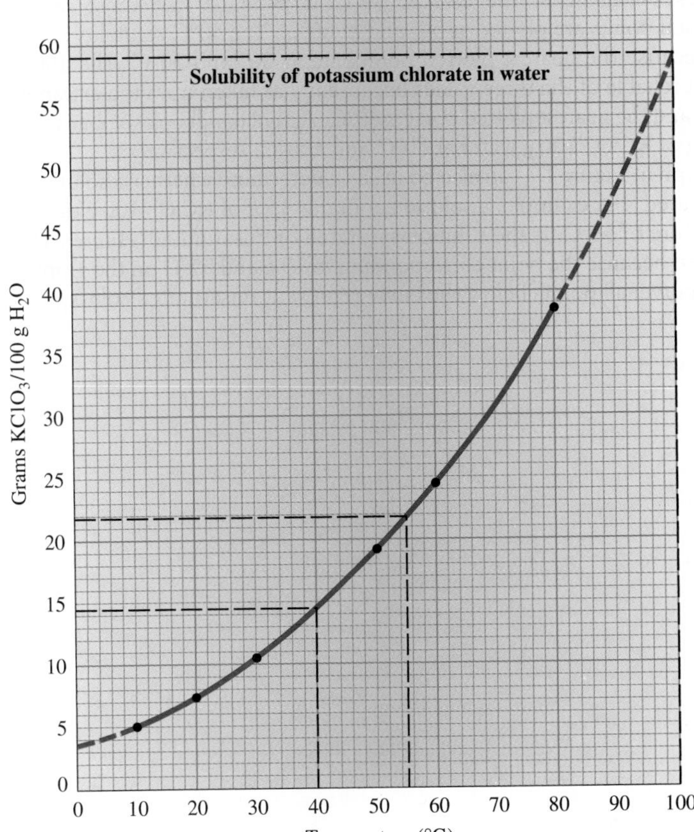

Temperature (°C)	Solubility (g $KClO_3$/ 100 g water)
10	5.0
20	7.4
30	10.5
50	19.3
60	24.5
80	38.5

Figure I.3

solubility of potassium chlorate. The graph in Figure I.3 was constructed from experimentally determined solubilities at the six temperatures shown. These experimentally determined solubilities are all located on the smooth curve traced by the unbroken-line portion of the graph. We are therefore confident that the unbroken line represents a very good approximation of the solubility data for potassium chlorate over the temperature range from 10° to 80°C. All points on the plotted curve represent the composition of saturated solutions. Any point below the curve represents an unsaturated solution.

The dashed-line portions of the curve are *extrapolations;* that is, they extend the curve above and below the temperature range actually covered by the plotted solubility data. Curves such as this one are often extrapolated a short distance beyond the range of the known data, although the extrapolated portions may not be highly accurate. Extrapolation is justified only in the absence of more reliable information.

The graph in Figure I.3 can be used with confidence to obtain the solubility of $KClO_3$ at any temperature between 10° and 80°C, but the solubilities between 0° and 10°C and between 80° and 100°C are less reliable. For example, what is the solubility of $KClO_3$ at 55°C, at 40°C, and at 100°C?

First draw a perpendicular line from each temperature to the plotted solubility curve. Now trace a horizontal line to the solubility axis from each point on the curve and read the corresponding solubilities. The values that we read from the graph are

40°C	14.2 g $KClO_3$/100 g water
55°C	22.0 g $KClO_3$/100 g water
100°C	59 g $KClO_3$/100 g water

Of these solubilities, the one at 55°C is probably the most reliable because experimental points are plotted at 50° and at 60°C. The 40°C solubility value is a bit less reliable because the nearest plotted points are at 30° and 50°C. The 100°C solubility is the least reliable of the three values because it was taken from the extrapolated part of the curve, and the nearest plotted point is 80°C. Actual handbook solubility values are 14.0 and 57.0 g of $KClO_3$/100 g water at 40°C and 100°C, respectively.

The graph in Figure I.3 can also be used to determine whether a solution is saturated or unsaturated. For example, a solution contains 15 g of $KClO_3$/100 g of water and is at a temperature of 55°C. Is the solution saturated or unsaturated? *Answer:* The solution is unsaturated because the point corresponding to 15 g and 55°C on the graph is below the solubility curve; all points below the curve represent unsaturated solutions.

APPENDIX II

Using a Scientific Calculator

A calculator is useful for most calculations in this book. You should obtain a scientific calculator, that is, one that has at least the following function keys on its keyboard.

Addition [+]	Second function [F], [INV], [Shift]
Subtraction [−]	Change sign [+/−]
Multiplication [×]	Exponential number [Exp]
Division [÷]	Logarithm [Log]
Equals [=]	Antilogarithm [10^x]

Not all calculators use the same symbolism for these function keys, nor do all calculators work in the same way. The following discussion may not pertain to your particular calculator. Refer to your instruction manual for variations from the function symbols shown above and for the use of other function keys.

Some keys have two functions, upper and lower. In order to use the upper (second) function, the second function key [F] must be pressed in order to activate the desired upper function after entering the number.

The second function key may have a different designation on your calculator.

The display area of the calculator shows the numbers entered and often shows more digits in the answer than should be used. Therefore, the final answer should be rounded to reflect the proper number of significant figures of the calculations.

Addition and Subtraction To add numbers using your calculator,

1. Enter the first number to be added followed by the plus key [+].
2. Enter the second number to be added followed by the plus key [+].
3. Repeat Step 2 for each additional number to be added, except the last number.
4. After the last number is entered, press the equal key [=]. You should now have the answer in the display area.
5. When a number is to be subtracted, use the minus key [−] instead of the plus key.

As an example, to add 16.0 + 1.223 + 8.45, enter 16.0 followed by the [+] key; then enter 1.223 followed by the [+] key; then enter 8.45 followed by the [=] key. The display shows 25.673, which is rounded to the answer 25.7.

Examples of Addition and Subtraction

Calculation	Enter in sequence	Display	Rounded answer
a. 12.0 + 16.2 + 122.3	12.0 [+] 16.2 [+] 122.3 [=]	150.5	150.5
b. 132 − 62 + 141	132 [−] 62 [+] 141 [=]	211	211
c. 46.23 + 13.2	46.23 [+] 13.2 [=]	59.43	59.4
d. 129.06 + 49.1 − 18.3	129.06 [+] 49.1 [−] 18.3 [=]	159.86	159.9

Multiplication To multiply numbers using your calculator,

1. Enter the first number to be multiplied followed by the multiplication key [×].
2. Enter the second number to be multiplied followed by the multiplication key [×].
3. Repeat Step 2 for all other numbers to be multiplied except the last number.
4. Enter the last number to be multiplied followed by the equal key [=]. You now have the answer in the display area.

Round off to the proper number of significant figures.

As an example, to calculate (3.25)(4.184)(22.2), enter 3.25 followed by the [×] key; then enter 4.184 followed by the [×] key; then enter 22.2 followed by the [=] key. The display shows 301.8756, which is rounded to the answer 302.

Examples of Multiplication

Calculation	Enter in sequence	Display	Rounded answer
a. 12 × 14 × 18	12 [×] 14 [×] 18 [=]	3024	3.0×10^3
b. 122 × 3.4 × 60.	122 [×] 3.4 [×] 60. [=]	24888	2.5×10^4
c. 0.522 × 49.4 × 6.33	0.522 [×] 49.4 [×] 6.33 [=]	163.23044	163

Division To divide numbers using your calculator,

1. Enter the numerator followed by the division key [÷].
2. Enter the denominator followed by the equal key to give the answer.
3. If there is more than one denominator, enter each denominator followed by the division key except for the last number, which is followed by the equal key.

As an example, to calculate $\frac{126}{12}$, enter 126 followed by the [÷] key; then enter 12 followed by the [=] key. The display shows 10.5, which is rounded to the answer 11.

Examples of Division

Calculation	Enter in sequence	Display	Rounded answer
a. $\frac{142}{25}$	142 [÷] 25 [=]	5.68	5.7
b. $\frac{0.422}{5.00}$	0.422 [÷] 5.00 [=]	0.0844	0.0844
c. $\frac{124}{0.022 \times 3.00}$	124 [÷] 0.022 [÷] 3.00 [=]	1878.7878	1.9×10^3

Exponents In scientific measurements and calculations, we often encounter very large and very small numbers. To express these large and small numbers conveniently, we use exponents, or powers, of 10. A number in exponential form is treated like any other number; that is, it can be added, subtracted, multiplied, or divided.

To enter an exponential number into your calculator, first enter the nonexponential part of the number and then press the exponent key [Exp], followed by the exponent. For example, to enter 4.9×10^3, enter 4.94, then press [Exp], and then press 3. When the exponent of 10 is a negative number, press the Change of Sign key [+/−] after entering the exponent. For example, to enter 4.94×10^{-3}, enter in sequence 4.94 [Exp] 3 [+/−]. In most calculators, the exponent will appear in the display a couple of spaces after the nonexponent part of the number—for example, 4.94 03 or 4.94 −03.

Examples Using Exponential Numbers

Calculation	Enter in sequence	Display	Rounded answer
a. $(4.94 \times 10^3)(21.4)$	4.94 [Exp] 3 [×] 21.4 [=]	105716	1.06×10^5
b. $(1.42 \times 10^4)(2.88 \times 10^{-5})$	1.42 [Exp] 4 [×] 2.88 [Exp] 5 [+/−] [=]	0.40896	0.409
c. $\frac{8.22 \times 10^{-5}}{5.00 \times 10^7}$	8.22 [Exp] 5 [+/−] [÷] 5.00 [Exp] 7 [=]	1.644 −12	1.64×10^{-12}

Logarithms The logarithm of a number is the power (exponent) to which some base number must be raised to give the original number. The most commonly used base number is 10. The base number that we use is 10. For example, the log of 100 is 2.0 ($\log 100 = 10^{2.0}$). The log of 200 is 2.3 ($\log 200 = 10^{2.3}$). Logarithms are used in chemistry to calculate the pH of an aqueous acidic solution. The answer (log) should contain the same number of significant figures to the right of the decimal as is in the original number. Thus, log 100 = 2.0, but log 100. is 2.000.

The log key on most calculators is a function key. To determine the log using your calculator, enter the number and then press the log function key. For example, to determine the log of 125, enter 125 and then the [Log] key. The answer is 2.097.

Examples Using Logarithms
Determine the log of the following:

	Enter in sequence	Display	Rounded answer
a. 42	42 [Log]	1.6232492	1.62
b. 1.62×10^5	1.62 [Exp] 5 [Log]	5.209515	5.210
c. 6.4×10^{-6}	6.4 [Exp] 6 [+/−] [Log]	−5.19382	−5.19

Antilogarithms (Inverse Logarithms) An antilogarithm is the number from which the logarithm has been calculated. It is calculated using the [10^x] key on your calculator. For example, to determine the antilogarithm of 2.891, enter 2.891 into your calculator and then press the second function key followed by the [10^x] key: 2.891 [F] [10^x].

The display shows 778.03655, which rounds to the answer 778.

Examples Using Antilogarithms
Determine the antilogarithm of the following:

	Enter in sequence	Display	Rounded answer
a. 1.628	1.628 [F] [10^x]	42.461956	42.5
b. 7.086	7.086 [F] [10^x]	12189896	1.22×10^7
c. −6.33	6.33 [+/−] [F] [10^x]	4.6773514 −07	4.7×10^{-7}

Additional Practice Problems*

Problem	Display	Rounded answer
1. $143.5 + 14.02 + 1.202$	158.722	158.7
2. $72.06 - 26.92 - 49.66$	−4.52	−4.52
3. $2.168 + 4.288 - 1.62$	4.836	4.84
4. $(12.3)(22.8)(1.235)$	346.3434	346
5. $(2.42 \times 10^6)(6.08 \times 10^{-4})(0.623)$	916.65728	917
6. $\frac{(46.0)(82.3)}{19.2}$	197.17708	197
7. $\frac{0.0298}{243}$	1.2263374 −04	1.23×10^{-4}
8. $\frac{(5.4)(298)(760)}{(273)(1042)}$	4.2992554	4.3
9. $(6.22 \times 10^6)(1.45 \times 10^3)(9.00)$	8.1171 10	8.12×10^{10}
10. $\frac{(1.49 \times 10^6)(1.88 \times 10^6)}{6.02 \times 10^{23}}$	4.6531561 −12	4.65×10^{-12}
11. $\log 245$	2.389166	2.389
12. $\log 6.5 \times 10^{-6}$	−5.1870866	−5.19
13. $\log 24 \times \log 34$	2.1137644	2.11
14. antilog 6.34	2187761.6	2.2×10^6
15. antilog −6.34	4.5708818 −07	4.6×10^{-7}

*Only the problem, the display, and the rounded answer are given.

APPENDIX III

Units of Measurement

Physical Constants

Constant	Symbol	Value
Atomic mass unit	amu	1.6606×10^{-27} kg
Avogadro's number	N	6.022×10^{23} particles/mol
Gas constant	R (at STP)	0.08205 L atm/K mol
Mass of an electron	m_e	9.109×10^{-28} g
		5.486×10^{-4} amu
Mass of a neutron	m_n	1.675×10^{-27} kg
		1.00866 amu
Mass of a proton	m_p	1.673×10^{-27} kg
		1.00728 amu
Speed of light	c	2.997925×10^{8} m/s

SI Units and Conversion Factors

Length

SI unit: meter (m)

1 meter	= 1.0936 yards
	= 100 centimeters
	= 1000 millimeters
1 centimeter	= 0.3937 inch
1 inch	= 2.54 centimeters (exactly)
1 kilometer	= 0.62137 mile
1 mile	= 5280 feet
	= 1.609 kilometers
1 angstrom	= 10^{-10} meter

Mass

SI unit: kilogram (kg)

1 kilogram	= 1000 grams
	= 2.20 pounds
1 gram	= 1000 milligrams
1 pound	= 453.59 grams
	= 0.45359 kilogram
	= 16 ounces
1 ton	= 2000 pounds
	= 907.185 kilograms
1 ounce	= 28.3 grams
1 atomic mass unit	= 1.6606×10^{-27} kilogram

Volume

SI unit: cubic meter (m^3)

1 liter	= 10^{-3} m^3
	= 1 dm^3
	= 1.0567 quarts
	= 1000 milliliters
1 gallon	= 4 quarts
	= 8 pints
	= 3.785 liters
1 quart	= 32 fluid ounces
	= 0.946 liter
1 fluid ounce	= 29.6 milliliters

Temperature

SI unit: kelvin (K)

0 K = −273.15°C

= −459.67°F

K = °C + 273.15

$$°C = \frac{°F - 32}{1.8}$$

°F = 1.8(°C) + 32

$$°C = \frac{5}{9}(°F - 32)$$

Energy

SI unit: joule (J)

1 joule	= 1 kg m^2/s^2
	= 0.23901 calorie
1 calorie	= 4.184 joules

Pressure

SI unit: pascal (Pa)

1 pascal	= 1 kg/m s^2
1 atmosphere	= 101.325 kilopascals
	= 760 torr (mm Hg)
	= 14.70 pounds per square inch (psi)

APPENDIX IV

Vapor Pressure of Water at Various Temperatures

Temperature (°C)	Vapor pressure (torr)	Temperature (°C)	Vapor pressure (torr)
0	4.6	26	25.2
5	6.5	27	26.7
10	9.2	28	28.3
15	12.8	29	30.0
16	13.6	30	31.8
17	14.5	40	55.3
18	15.5	50	92.5
19	16.5	60	149.4
20	17.5	70	233.7
21	18.6	80	355.1
22	19.8	90	525.8
23	21.2	100	760.0
24	22.4	110	1074.6
25	23.8		

APPENDIX V

Solubility Table

	F^-	Cl^-	Br^-	I^-	O^{2-}	S^{2-}	OH^-	NO_3^-	CO_3^{2-}	SO_4^{2-}	$C_2H_3O_2^-$
H^+	*aq*	*aq*	*aq*	*aq*	*aq*	*sl.aq*	*aq*	*aq*	*sl.aq*	*aq*	*aq*
Na^+	*aq*	*aq*	*aq*	*aq*	*aq*	*aq*	*aq*	*aq*	*aq*	*aq*	*aq*
K^+	*aq*	*aq*	*aq*	*aq*	*aq*	*aq*	*aq*	*aq*	*aq*	*aq*	*aq*
NH_4^+	*aq*	*aq*	*aq*	*aq*	—	*aq*	*aq*	*aq*	*aq*	*aq*	*aq*
Ag^+	*aq*	*I*	*I*	*I*	*I*	*I*	—	*aq*	*I*	*I*	*I*
Mg^{2+}	*I*	*aq*	*aq*	*aq*	*I*	*d*	*I*	*aq*	*I*	*aq*	*aq*
Ca^{2+}	*I*	*aq*	*aq*	*aq*	*I*	*d*	*I*	*aq*	*I*	*I*	*aq*
Ba^{2+}	*I*	*aq*	*aq*	*aq*	*sl.aq*	*d*	*sl.aq*	*aq*	*I*	*I*	*aq*
Fe^{2+}	*sl.aq*	*aq*	*aq*	*aq*	*I*	*I*	*I*	*aq*	*sl.aq*	*aq*	*aq*
Fe^{3+}	*I*	*aq*	*aq*	—	*I*	*I*	*I*	*aq*	*I*	*aq*	*I*
Co^{2+}	*aq*	*aq*	*aq*	*aq*	*I*	*I*	*I*	*aq*	*I*	*aq*	*aq*
Ni^{2+}	*sl.aq*	*aq*	*aq*	*aq*	*I*	*I*	*I*	*aq*	*I*	*aq*	*aq*
Cu^{2+}	*sl.aq*	*aq*	*aq*	—	*I*	*I*	*I*	*aq*	*I*	*aq*	*aq*
Zn^{2+}	*sl.aq*	*aq*	*aq*	*aq*	*I*	*I*	*I*	*aq*	*I*	*aq*	*aq*
Hg^{2+}	*d*	*aq*	*I*	*I*	*I*	*I*	*I*	*aq*	*I*	*d*	*aq*
Cd^{2+}	*sl.aq*	*aq*	*aq*	*aq*	*I*	*I*	*I*	*aq*	*I*	*aq*	*aq*
Sn^{2+}	*aq*	*aq*	*aq*	*sl.aq*	*I*	*I*	*I*	*aq*	*I*	*aq*	*aq*
Pb^{2+}	*I*	*I*	*I*	*I*	*I*	*I*	*I*	*aq*	*I*	*I*	*aq*
Mn^{2+}	*sl.aq*	*aq*	*aq*	*aq*	*I*	*I*	*I*	*aq*	*I*	*aq*	*aq*
Al^{3+}	*I*	*aq*	*aq*	*aq*	*I*	*d*	*I*	*aq*	—	*aq*	*aq*

Key: *aq* = soluble in water
sl.aq = slightly soluble in water
I = insoluble in water (less than 1 g/100 g H_2O)
d = decomposes in water

APPENDIX VI

Answers to Selected Review Questions and Exercises

Chapter 1

Review Questions

3. Six

8. Mercury and water.

9. Air.

15. (a) sugar, a compound and (c) gold, an element

Exercises

2. Two states are present; solid and liquid.

4. The maple leaf represents a heterogeneous mixture.

6. (a) homogeneous
(b) homogeneous
(c) heterogeneous
(d) heterogeneous

Chapter 2

Review Questions

2. 7.6 cm

6. 0.789 g/mL < ice < 0.91 g/mL

8. $d = m/V$ specific gravity $= \dfrac{d_{\text{substance}}}{d_{\text{water}}}$

Exercises

2. (a) 1000 meters = 1 kilometer
(c) 0.000001 liter = 1 microliter
(e) 0.001 liter = 1 milliliter

4. (a) mg
(c) m
(e) Å

6. (a) not significant.
(c) not significant.
(e) significant

8. (a) 40.0 (3 sig fig)
(b) 0.081 (2 sig fig)
(c) 129,042 (6 sig fig)
(d) 4.090×10^{-3} (4 sig fig)

10. (a) 8.87
(b) 21.3
(c) 130 (1.30×10^2)
(d) 2.00×10^6

12. (a) 4.56×10^{-2}
(b) 4.0822×10^3
(c) 4.030×10^1
(d) 1.2×10^7
(e) 2.49×10^{-4}

14. (a) 28.1
(c) 4.0×10^1

16. (a) $\frac{1}{4}$
(b) $\frac{5}{8}$
(c) $1\frac{2}{3}$ or $\frac{5}{3}$
(d) $\frac{8}{9}$

18. (a) 1.0×10^2
(b) 4.6 mL
(c) 22

20. (a) 4.5×10^8 Å
(c) 8.0×10^6 mm
(e) 6.5×10^5 mg
(g) 468 mL

22. (a) 117 ft
(c) 7.4×10^4 mm^3
(e) 75.7 L

24. 12 mi/hr

26. 8.33 grains

28. 79 days

30. 2.54×10^{-3} lb

32. 5.94×10^3 cm/s

34. $4500

36. 3.0×10^5 straws

38. 160 L

40. 4×10^5 m^2

42. 6 gal

44. 113°F Summer!

46. (a) 90°F
(b) −22.6°C
(c) 546 K
(d) −300 F

48. −11.4°C = 11.4°F

50. −297°F

52. 3.12 g/mL

54. 1.28 g/mL

56. 3.40×10^2 g

58. A graduated cylinder would be the best choice for adding 100 mL of solvent to a reaction. While the volumetric flask is also labeled 100 mL, volumetric flasks are typically used for doing dilutions. The other three pieces of glassware could also be used, but they hold smaller volumes, so it would take a longer time to measure out 100 mL. Also, because you would have to repeat the measurement many times using the other glassware, there is a greater chance for error.

60. 26 mL

62. 7.0 lb

64. Yes, 116.5 L additional solution

66. $-15°C > 4.5°F$

68. 20.9 lb

70. 5.1×10^3 L

72. 16.4 cm^3

74. 0.965 g/mL

76. ethyl alcohol, because it has the lower density

78. 54.3 mL

82. 76.9 g

Chapter 3

Review Questions

2. (a) Ag (e) Fe
(c) H (g) Mg

4. The symbol of an element represents the element itself.

7. sodium, neon
fluorine, helium
nickel, calcium
zinc, chlorine

10. 86 metals, 7 metalloids, 18 nonmetals

12. 1 metal 0 metalloids 5 nonmetals

14. (a) iodine (b) bromine

19. H_2–hydrogen, Cl_2–chlorine
N_2–nitrogen, Br_2–bromine
O_2–oxygen, I_2–iodine
F_2–fluorine

Exercises

2. (c) H_2 (g) ClF
(f) NO

4. (a) magnesium, bromine
(c) hydrogen, nitrogen, oxygen
(e) aluminum, phosphorus, oxygen

6. (a) $AlBr_3$ (c) $PbCrO_4$

8. (a) 1 atom Al, 3 atoms Br
(c) 12 atoms C, 22 atoms H, 11 atoms O

10. (a) 2 atoms (e) 17 atoms
(c) 9 atoms

12. (a) 2 atoms H (e) 8 atoms H
(c) 12 atoms H

14. (a) mixture (e) mixture
(c) pure substance

16. (c) compound (d) element

18. (a) compound (c) mixture
(b) compound

20. (a) compound (c) mixture

22. No. The only common liquid elements (at room temperature) are mercury and bromine.

24. 72% solids

25. A physical change is reversible. Therefore, boil the salt–water solution. The water will evaporate and leave the salt behind.

29. (a) 1 carbon atom and 1 oxygen atom; total number of atoms = 2
(c) 1 hydrogen atom, 1 nitrogen atom, and 3 oxygen atoms; total number of atoms = 5
(e) 1 calcium atom, 2 nitrogen atoms, and 6 oxygen atoms; total number of atoms = 9

32. 40 atoms H

34. (a) magnesium, manganese, molybdenum, mendelevium, mercury, meitnerium
(c) sodium, potassium, iron, silver, tin, antimony

36. 420 atoms

39. (a) NaCl (g) $C_6H_{12}O_6$
(c) K_2O (i) $Cr(NO_3)_3$
(e) K_3PO_4

Chapter 4

Review Questions

2. liquid

4. A new substance is always formed during a chemical change, but never formed during physical changes.

6. (a) $118.0°C + 273 = 391.0$ K
(b) $(118.0°C)\ 1.8 + 32 = 244.4$ °F

8. Water vapor.

Exercises

2. (a) physical (e) physical
(b) physical (f) physical
(c) chemical (g) chemical
(d) physical (h) chemical

4. The copper wire, like the platinum wire, changed to a glowing red color when heated. Upon cooling, a new substance, black copper(II) oxide, had appeared.

6. Reactant: water
Products: hydrogen, oxygen

8. (a) physical (e) chemical
(c) chemical

10. (a) potential energy (e) potential energy
(c) kinetic energy

12. the transformation of kinetic energy to thermal energy

14. (a) + (b) − (c) − (d) + (e) −

16. 2.2×10^3 J

18. 5.58×10^3 J

20. 5.03×10^{-2} J/g °C

22. 5°C

26. sp. ht. = 1.02 J/g °C

28. 3.0×10^2 cal

31. 44 g coal

33. 654°C

40. 8 g fat

43. A chemical change has occurred. Hydrogen molecules and oxygen molecules have combined to form water molecules.

Chapter 5

Review Questions

1. (a) copper, 29 (e) zinc, 30
(c) phosphorus, 15

5. Isotopic notation $^{A}_{Z}E$
Z represents the atomic number
A represents the mass number

Exercises

2. The formula for hydrogen peroxide is H_2O_2. There are two atoms of oxygen for every two atoms of hydrogen. The molar mass of oxygen is 16.00 g, and the molar mass of hydrogen is 1.01 g. For hydrogen peroxide, the total mass of hydrogen is 2.016 g and the total mass of oxygen is 32.00 g, for a ratio of hydrogen to oxygen of approximately 2 : 32. or 1 : 16. Therefore, there is 1 gram of hydrogen for every 16 grams of oxygen.

4. (a) The nucleus of the atom contains most of the mass.
(c) The atom is mostly empty space.

6. The nucleus of an atom contains nearly all of its mass.

8. Electrons:
Dalton—Electrons are not part of his model.
Thomson—Electrons are scattered throughout the positive mass of matter in the atom.
Rutherford—Electrons are located out in space away from the central positive mass.
Positive matter:
Dalton—No positive matter in his model.
Thomson—Positive matter is distributed throughout the atom.
Rutherford—Positive matter is concentrated in a small central nucleus.

10. Yes. The mass of the isotope $^{12}_{6}C$, 12, is an exact number. The mass of other isotopes are not exact numbers.

12. Three isotopes of hydrogen have the same number of protons and electrons, but differ in the number of neutrons.

14. All five isotopes have nuclei that contain 30 protons and 30 electrons. The numbers of neutrons are:

Isotope mass number	Neutrons
64	34
66	36
67	37
68	38
70	40

16. (a) $^{109}_{47}Ag$ (c) $^{57}_{26}Fe$

18. (a) $^{25}_{12}Mg$ (c) $^{122}_{50}Sn$

20. (a) 35 (c) 45

22. 24.31 amu

24. 6.716 amu

26. $1.0 \times 10^5 : 1.0$

28. (a) These two atoms are isotopes.

29. 1.9×10^8 enlargement

31. ^{210}Bi has the largest number of neutrons (127).

33. 131 amu

40. (a) 0.02554% (c) 0.02741%

42. The electron region is the area around the nucleus where electrons are most likely to be located.

Chapter 6

Review Questions

1. (a) $NaClO_3$ (e) $Zn(HCO_3)_2$
(c) $Sn(C_2H_3O_2)_2$

3. (a) HBrO hypobromous acid
$HBrO_2$ bromous acid
$HBrO_3$ bromic acid
$HBrO_4$ perbromic acid

5. Chromium(III) compounds
(a) $Cr(OH)_3$ (g) $CrPO_4$
(c) $Cr(NO_2)_3$ (i) Cr_2O_3
(e) $Cr_2(CO_3)_3$

7. (a) Metals are located in groups 1A (except for hydrogen), 2A, 1B–8B and atomic numbers 13, 31, 49, 50, 81, 82, 83, lanthanides and actinides.
(c) The transition metals are located in groups 1B–8B in the center of the periodic table. The lanthanides and actinides are located below the main body of the periodic table.

Exercises

2. (a) BaO (c) $AlCl_3$ (d) $BeBr_2$ (f) Mg_3P_2

4.

Cl^-	HSO_4^-
Br^-	HSO_3^-
F^-	CrO_4^{2-}
I^-	CO_3^{2-}
CN^-	HCO_3^-
O^{2-}	$C_2H_3O_2^-$
OH^-	ClO_3^-
S^{2-}	MnO_4^-
SO_4^{2-}	$C_2O_4^{2-}$

6. (a) calcium hydroxide
(c) sulfur
(d) sodium hydrogen carbonate
(f) potassium carbonate

8.

Ion	SO_4^{2-}	OH^-	AsO_4^{3-}	$C_2H_3O_2^-$	CrO_4^{2-}
NH_4^+	$(NH_4)_2SO_4$	NH_4OH	$(NH_4)_3AsO_4$	$NH_4C_2H_3O_2$	$(NH_4)_2CrO_4$
Ca^{2+}	$CaSO_4$	$Ca(OH)_2$	$Ca_3(AsO_4)_2$	$Ca(C_2H_3O_2)_2$	$CaCrO_4$
Fe^{3+}	$Fe_2(SO_4)_3$	$Fe(OH)_3$	$FeAsO_4$	$Fe(C_2H_3O_2)_3$	$Fe_2(CrO_4)_3$
Ag^+	Ag_2SO_4	$AgOH$	Ag_3AsO_4	$AgC_2H_3O_2$	Ag_2CrO_4
Cu^{2+}	$CuSO_4$	$Cu(OH)_2$	$Cu_3(AsO_4)_2$	$Cu(C_2H_3O_2)_2$	$CuCrO_4$

10. NH_4^+ compounds: ammonium sulfate, ammonium hydroxide, ammonium arsenate, ammonium acetate, ammonium chromate.

Ca^{2+} compounds: calcium sulfate, calcium hydroxide, calcium arsenate, calcium acetate, calcium chromate.

Fe^{3+} compounds: iron(III) sulfate, iron(III) hydroxide, iron(III) arsenate, iron(III) acetate, iron(III) chromate.

Ag^+ compounds: silver sulfate, silver hydroxide, silver arsenate, silver acetate, silver chromate.

Cu^{2+} compounds: copper(II) sulfate, copper(II) hydroxide, copper(II) arsenate, copper(II) acetate, copper(II) chromate.

12. (a) sodium nitrate, $NaNO_3$
(c) barium hydroxide, $Ba(OH)_2$
(e) silver carbonate, Ag_2CO_3
(g) potassium nitrite, KNO_2

14. (a) potassium oxide (c) calcium iodide (e) sodium phosphate (g) zinc nitrate

16. (a) $SnBr_4$ (c) $Fe_2(CO_3)_3$ (d) $Hg(NO_2)_2$ (f) $Fe(C_2H_3O_2)_2$

18. (a) $HC_2H_3O_2$ (c) H_2S (d) H_3BO_3 (f) $HClO$

20. (a) phosphoric acid (c) iodic acid (d) hydrochloric acid (f) nitric acid (g) hydroiodic acid

22. (a) Na_2CrO_4 (c) $Ni(C_2H_3O_2)_2$ (e) $Pb(NO_3)_2$ (g) $Mn(OH)_2$ (i) $NaClO$ (k) $Cr_2(SO_3)_3$ (m) $Na_2C_2O_4$

24. (a) calcium hydrogen sulfate
(c) tin(II) nitrite
(e) potassium hydrogen carbonate
(f) bismuth(III) arsenate
(h) ammonium monohydrogen phosphate
(j) potassium permanganate

28. (a) sulfate (e) hydroxide (c) nitrate

30. (a) K_2SO_4 (e) KOH (c) KNO_3

32. Formula: KCl Name: potassium chloride

34. (a) $AgNO_3 + NaCl \longrightarrow AgCl + NaNO_3$
(c) $KOH + H_2SO_4 \longrightarrow K_2SO_4 + H_2O$

37. $Li_3Fe(CN)_6$
$AlFe(CN)_6$
$Zn_3[Fe(CN)_6]_2$

Chapter 7

Review Questions

2. A mole of gold has a higher mass than a mole of potassium.

4. A mole of gold atoms contains more electrons than a mole of potassium atoms.

7. 6.022×10^{23}

9. (a) 6.022×10^{23} (c) 1.204×10^{24} (e) 32.00

10. 6.022×10^{23} molecules in one molar mass of H_2SO_4.
4.215×10^{24} atoms in one molar mass of H_2SO_4.

Exercises

2. Molar masses

(a) $NaOH$	40.00
(c) Cr_2O_3	152.0
(e) $Mg(HCO_3)_2$	146.3
(g) $C_6H_{12}O_6$	180.2
(i) $BaCl_2 \cdot 2\,H_2O$	244.2

4. (a) 0.675 mol NaOH (c) 7.18×10^{-3} mol $MgCl_2$ (e) 2.03×10^{-2} mol Na_2SO_4

6. (a) 0.0417 g H_2SO_4 (b) 11 g CCl_4 (c) 0.122 g Ti (d) 8.0×10^{-7} g S

8. (a) 1.05×10^{24} molecules Cl_2
(b) 1.6×10^{23} molecules C_2H_6O
(c) 1.64×10^{23} molecules CO_2
(d) 3.75×10^{24} molecules CH_4

10. (a) 1.3×10^2 atoms N_2O_5
(b) 6.02×10^{24} atoms Au
(c) 2.67×10^{24} atoms BF_3
(d) 3.85×10^{22} atoms U

12. (a) 3.271×10^{-22} g Au
(b) 3.952×10^{-22} g U
(c) 2.828×10^{-23} g NH_3
(d) 1.795×10^{-22} g $C_6H_4(NH_2)_2$

14. (a) 0.886 mol S
(b) 42.8 mol NaCl
(c) 1.05×10^{24} atoms Mg
(d) 9.47 mol Br_2

16. One mole of ammonia contains
(a) 6.022×10^{23} molecules NH_3
(b) 6.022×10^{23} N atoms
(c) 1.807×10^{24} H atoms
(d) 2.409×10^{24} atoms

18. (a) 6.0×10^{24} atoms O
(b) 5.46×10^{24} atoms O
(c) 5.0×10^{18} atoms O

20. (a) 1.27 g Cl (c) 2.74 g H
(b) 9.25×10^{-2} g H

22. (a) 47.98% Zn
52.02% Cl
(c) 12.26% Mg
31.24% P
56.48% O
(e) 23.09% Fe
17.37% N
59.53% O

24. (a) KCl 47.55% Cl
(b) $BaCl_2$ 34.05% Cl
(c) $SiCl_4$ 83.46% Cl
(d) LiCl 83.63% Cl
highest % Cl is LiCl
lowest % Cl is $BaCl_2$

26. 24.2% C
4.04% H
71.72% Cl

28. (a) $KClO_3$ lower % Cl
(b) $KHSO_4$ higher % S
(c) Na_2CrO_4 lower % Cr

30. (a) KBr (c) $AgNO_3$

32. Empirical formulas
(a) CuCl (e) $BaCr_2O_7$
(c) Cr_2S_3

34. V_2O_5

36. $HgCl_2$

38. The molecular formula is $C_6H_{12}O_6$.

40. The molecular formula is $C_4H_8O_2$

42. 40.0% C, 6.73% H, 53.3% O. Empirical formula is CH_2O, molecular formula is $C_6H_{12}O_6$.

44. $Al_2(SiO_3)_3$

45. 8.43×10^{23} atoms P

47. 10.8 g/molar mass

49. 5.54×10^{19} m

50. 9.9×10^{13} dollars/person

52. (a) 10.3 cm^3 (b) 2.18 cm

54. 41.58 g Fe_2S_3

56. 1.0 ton of iron ore contains 2×10^4 g Fe

58. (a) 76.98% Hg (c) 17.27% N

62. 42.10% C
6.480% H
51.42% O

64. 4.77 g O in 8.50 g $Al_2(SO_4)_3$

66. The empirical formula is $C_4H_4CaO_6$

68. 1.910×10^{16} years

70. 1.529×10^{-7} g $C_3H_8O_3$

72. $Co_3Mo_2C_{11}$

74. The empirical formula is C_5H_5N.

76. (a) CH_2O (e) $C_6H_2Cl_2O$
(c) CH_2O

Chapter 8

Review Questions

1. The purpose of balancing chemical equations is to conform to the Law of Conservation of Mass.

5. The charts are one way of keeping track of the number of atoms of each element on the reactant side of a chemical equation and on the product side of an equation. The top row in a chart gives the number and types of atoms on the reactant side and the bottom row gives the number and types of atoms on the product side of a chemical equation.

6. The symbols indicate whether a substance is a solid, a liquid, a gas, or is in an aqueous solution. A solid is indicated by (*s*), a liquid by (*l*), a gas by (*g*), and a aqueous solution by (*aq*).

8. A combustion reaction is an exothermic process (usually burning) done in the presence of oxygen.

Exercises

2. (a) endothermic (e) exothermic
(c) endothermic

4. (a) $H_2 + Br_2 \longrightarrow 2\,HBr$ combination
(c) $Ba(ClO_3)_2 \longrightarrow BaCl_2 + 3\,O_2$ decomposition
(e) $2\,H_2O_2 \longrightarrow 2\,H_2O + O_2$ decomposition

6. A metal and a nonmetal can react to form a salt; also an acid plus a base.

8. (a) $2\,SO_2 + O_2 \longrightarrow 2\,SO_3$
(c) $2\,Na + 2\,H_2O \longrightarrow 2\,NaOH + H_2$
(e) $Bi_2S_3 + 6\,HCl \longrightarrow 2\,BiCl_3 + 3\,H_2S$
(g) $2\,LiAlH_4 \xrightarrow{\Delta} 2\,LiH + 2\,Al + 3\,H_2$
(i) $2\,K_3PO_4 + 3\,BaCl_2 \longrightarrow 6\,KCl + Ba_3(PO_4)_2$

10. (a) $2\,Cu + S \xrightarrow{\Delta} Cu_2S$
(c) $2\,Ag_2O \xrightarrow{\Delta} 4\,Ag + O_2$
(e) $Ni_3(PO_4)_2 + 3\,H_2SO_4 \longrightarrow 3\,NiSO_4 + 2\,H_3PO_4$
(g) $3\,AgNO_3 + AlCl_3 \longrightarrow 3\,AgCl + Al(NO_3)_3$

12. (a) $CuSO_4(aq) + 2\,KOH(aq) \longrightarrow Cu(OH)_2(s) + K_2SO_4(aq)$
(c) $3\,NaHCO_3(s) + H_3PO_4(aq) \longrightarrow Na_3PO_4(aq) + 3\,H_2O(l) + 3\,CO_2(g)$

14. (a) $Cu(s) + FeCl_3(aq) \longrightarrow$ no reaction
(c) $2\,Al(s) + 6\,HBr(aq) \longrightarrow 3\,H_2(g) + 2\,AlBr_3(aq)$

16. (a) $SO_2 + H_2O \longrightarrow H_2SO_3$
(c) $Ca + 2\,H_2O \longrightarrow Ca(OH)_2 + H_2$

18. (a) $C + O_2 \longrightarrow CO_2$
(c) $CuBr_2 + Cl_2 \longrightarrow CuCl_2 + Br_2$
(e) $2\,NaNO_3 \xrightarrow{\Delta} 2\,NaNO_2 + O_2$

20. (a) 2 mol of Na react with 1 mol of Cl_2 to produce 2 mol of NaCl and release 822 kJ of energy. Exothermic
(b) 1 mol of PCl_5 absorbs 92.9 kJ of energy to produce 1 mol of PCl_3 and 1 mol of Cl_2. Endothermic

22. (a) $2\,Al + 3\,I_2 \longrightarrow 2\,AlI_3 + \text{heat}$
(b) $4\,CuO + CH_4 + \text{heat} \longrightarrow 4\,Cu + CO_2 + 2\,H_2O$

24. (a) single displacement, $Ni(s) + Pb(NO_3)_2(aq) \longrightarrow Pb(s) + Ni(NO_3)_2(aq)$
(c) decomposition, $2\,HgO(s) \longrightarrow 2\,Hg(l) + O_2(g)$

25. (a) change in color and texture of the bread
(b) change in texture of the white and the yolk
(c) the flame (combustion), change in matchhead, odor

26. $P_4O_{10} + 12\,HClO_4 \longrightarrow 6\,Cl_2O_7 + 4\,H_3PO_4$
58 atoms O on each side

32. (a) $4\,K + O_2 \longrightarrow 2\,K_2O$
(c) $CO_2 + H_2O \longrightarrow H_2CO_3$

33. (a) $2\,HgO \xrightarrow{\Delta} 2\,Hg + O_2$
(b) $2\,NaClO_3 \xrightarrow{\Delta} 2\,NaCl + 3\,O_2$
(c) $MgCO_3 \xrightarrow{\Delta} MgO + CO_2$

34. (a) $Zn + H_2SO_4 \longrightarrow H_2 + ZnSO_4$
(c) $Mg + 2\,AgNO_3 \longrightarrow Mg(NO_3)_2 + 2\,Ag$

35. (a) $ZnCl_2 + 2\,KOH \longrightarrow Zn(OH)_2 + 2\,KCl$
(c) $3\,Ca(OH)_2 + 2\,H_3PO_4 \longrightarrow 6\,H_2O + Ca_3(PO_4)_2$
(e) $Ba(OH)_2 + 2\,HNO_3 \longrightarrow 2\,H_2O + Ba(NO_3)_2$

38. (a) $2\,C_2H_6 + 7\,O_2 \longrightarrow 4\,CO_2 + 6\,H_2O$
(c) $C_7H_{16} + 11\,O_2 \longrightarrow 7\,CO_2 + 8\,H_2O$

Chapter 9

Review Questions

2. In order to convert grams to moles, the molar mass of the compound under consideration needs to be determined.

4. (a) correct (c) correct (e) correct

6. You can calculate the percent yield of a chemical reaction by dividing the actual yield by the theoretical yield and multiplying by one hundred.

Exercises

2. (a) 25.0 mol $NaHCO_3$ (c) 16 mol CO_2

4. (a) 1.31 g $NiSO_4$ (e) 18 g K_2CrO_4
(c) 373 g Bi_2S_3

6. HCl

8. (a) $\dfrac{3\text{ mol }CaCl_2}{1\text{ mol }Ca_3(PO_4)_2}$ (d) $\dfrac{1\text{ mol }Ca_3(PO_4)_2}{2\text{ mol }H_3PO_4}$
(e) $\dfrac{3\text{ mol }CaCl_2}{2\text{ mol }H_3PO_4}$ (f) $\dfrac{2\text{ mol }H_3PO_4}{6\text{ mol }HCl}$

10. 2.80 mol Cl_2

12. (a) 8.33 mol H_2O (b) 0.800 mol $Al(OH)_3$

14. 19.7 g $Zn_3(PO_4)_2$

16. 117 g H_2O, 271 g Fe

18. (a) 0.500 mol Fe_2O_3 (e) 0.871 mol O_2
(c) 6.20 mol SO_2

20. (a)

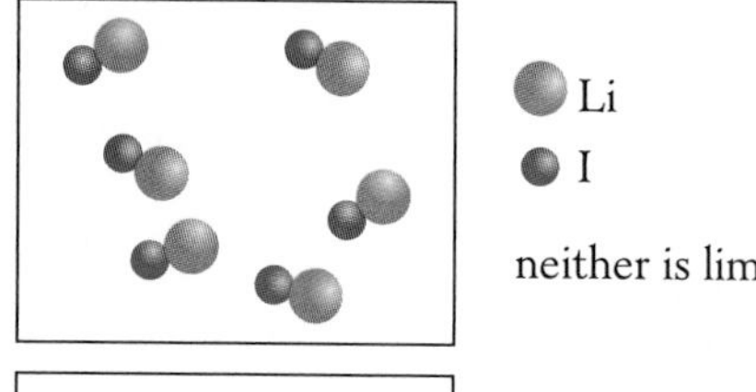

Li
I
neither is limiting

(b)

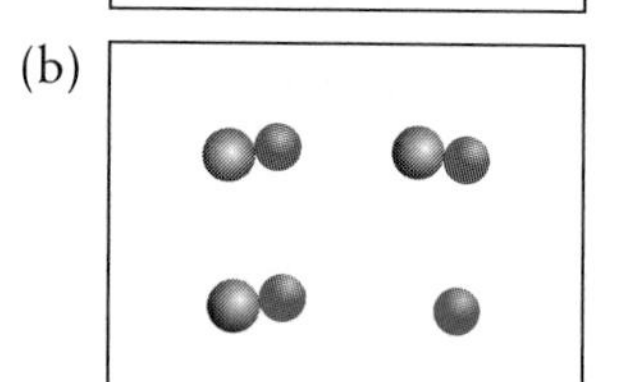

Ag
Cl
Ag is limiting

22. (a) Oxygen is the limiting reactant.
Nitrogen Oxygen

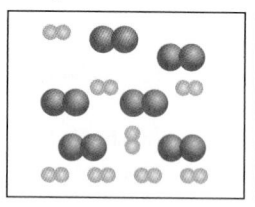

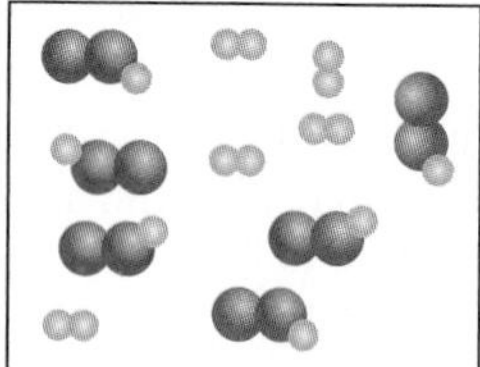

Before After

(b) Water is the limiting reactant.

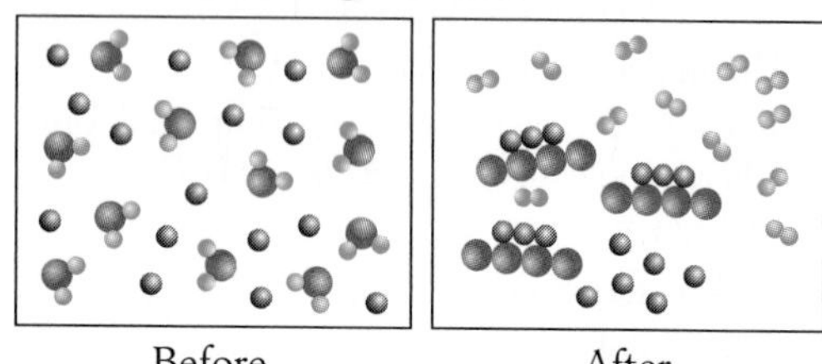

Before After

24. (a) H_2S is the limiting reactant and $Bi(NO_3)_3$ is in excess.
(b) H_2O is the limiting reactant and Fe is in excess.

26. (a) 0.313 mol H_2O.
(b) 0.521 mol H_2O.
(c) 10 mol CO_2; C_3H_6 will be left over.

28. Zinc

30. 95.0% yield of Cu

32. 77.8% CaC_2

36. (a) 0.19 mol KO_2 (b) 28 g O_2/hr

38. 65g O_2

40. (a) 0.16 mol $MnCl_2$. (e) 11 g $KMnO_4$
(c) 82% yield

42. beaker weighs 24.835 g

44. 22.5% KCl; 77.5% KNO_3

48. (a) 3.2×10^2 g C_2H_5OH
(b) 1.10×10^3 g $C_6H_{12}O_6$

50. 3.7×10^2 kg Li_2O

52. 223.0 g H_2SO_4; 72.90% yield

54. 67.2% $KClO_3$

Chapter 10

Review Questions

1. An electron orbital is a region in space around the nucleus of an atom where an electron is most probably found.

3. The valence shell is the outermost energy level of an atom.

8. 1*s*, 2*s*, 2*p*, 3*s*, 3*p*, 4*s*, 3*d*, 4*p*

9. *s*–2 electrons per shell
p–6 electrons per shell after the first energy level
d–10 electrons per shell after the second energy level

13. 3 is the third energy level
d indicates an energy sublevel
7 indicates the number of electrons in the *d* sublevel

15. Elements in the *s*-block all have one or two electrons in their outermost energy level. These valence electrons are located in an *s*-orbital.

18. The greatest number of elements in any period is 32. The 6th period has this number of elements.

20. Pairs of elements which are out of sequence with respect to atomic masses are: Ar and K; Co and Ni; Te and I; Th and Pa; U and Np; Pu and Am; Sg and Bh; Hs and Mt.

Exercises

2. (a) F 9 protons
(b) Ag 47 protons
(c) Br 35 protons
(d) Sb 51 protons

4. (a) Cl $1s^22s^22p^63s^23p^5$
(c) Li $1s^22s^1$
(e) I $1s^22s^22p^63s^23p^64s^23d^{10}4p^65s^24d^{10}5p^5$

6. Bohr said that a number of orbits were available for electrons, each corresponding to an energy level. When an electron falls from a higher energy orbit to a lower orbit, energy is given off as a specific wavelength of light. Only those energies in the visible range are seen in the hydrogen spectrum. Each line corresponds to a change from one orbit to another.

8. 32 electrons in the fourth energy level

10. (a) (14p 14n) $2e^-8e^-4e^-$ $^{28}_{14}Si$

(c) (18p 22n) $2e^-8e^-8e^-$ $^{40}_{18}Ar$

(e) (15p 16n) $2e^-8e^-5e^-$ $^{31}_{15}P$

12. (a) Li $1s^22s^1$
(c) Zn $1s^22s^22p^63s^23p^64s^23d^{10}$
(e) K $1s^22s^22p^63s^23p^64s^1$

14. (a) Sc (c) Sn

16.

Atomic number	Electron structure
(a) 9	$[He]2s^22p^5$
(c) 31	$[Ar]4s^23d^{10}4p^1$
(d) 39	$[Kr]5s^24d^1$
(f) 10	$[He]2s^22p^6$

18. (a) Phosphorus (P)
↑↓ | ↑↓ | ↑↓ ↑↓ ↑↓ | ↑↓ | ↑ ↑ ↑

(c) Calcium (Ca)
↑↓ | ↑↓ | ↑↓ ↑↓ ↑↓ | ↑↓ | ↑↓ ↑↓ ↑↓ | ↑↓

(e) Potassium (K)
↑↓ | ↑↓ | ↑↓ ↑↓ ↑↓ | ↑↓ | ↑↓ ↑↓ ↑↓ | ↑

20. (a) Cl ↑↓ | ↑↓ | ↑↓ ↑↓ ↑↓ | ↑↓ | ↑↓ ↑↓ ↑

(c) Ni ↑↓ | ↑↓ | ↑↓ ↑↓ ↑↓ | ↑↓ | ↑↓ ↑ ↑
↑↓ | ↑↓ ↑↓ ↑↓ ↑ ↑

(e) Ba ↑↓ | ↑↓ | ↑↓ ↑↓ ↑↓ | ↑↓ | ↑↓ ↑↓ ↑↓
↑↓ | ↑↓ ↑↓ ↑↓ ↑↓ ↑↓ | ↑↓ ↑↓ ↑↓
↑↓ | ↑↓ ↑↓ ↑↓ ↑↓ ↑↓ | ↑↓ ↑↓ ↑↓ | ↑↓

22. (a) (13p 14n) $2e^-8e^-3e^-$ $^{27}_{13}Al$

(b) (22p 26n) $2e^-8e^-10e^-2e^-$ $^{48}_{22}Ti$

24. The last electron in potassium is located in the fourth energy level because the 4*s* orbital is at a lower energy level than the 3*d* orbital. Also the properties of potassium are similar to the other elements in Group IA.

26. Noble gases each have filled *s* and *p* orbitals in the outermost energy level.

28. The elements in a group have the same number of outer energy level electrons. They are located vertically on the periodic table.

30. Valence shell electrons
(a) 5 (c) 6 (e) 3

32. All of these elements have an s^2d^{10} electron configuration in their outermost energy levels.

34. (a) and (f) (e) and (h)

36. 7, 33 since they are in the same periodic group.

38. (a) nonmetal, I (c) metal, Mo

40. Period 4, Group 3B

42. Group 3A contains 3 valence electrons. Group 3B contains 2 electrons in the outermost level and 1 electron in an inner *d* orbital. Group A elements are representative, while Group B elements are transition elements.

44. (a) valence energy level 2, 1 valence electron
(c) valence energy level 3, 4 valence electrons
(e) valence energy level 2, 2 valence electrons

46. (a) 7A, Halogens (e) 8A, Noble Gases
(c) 1A, Alkali Metals

48. (a) Any orbital can hold a maximum of two electrons.
(c) The third principal energy level can hold two electrons in 3*s*, six electrons in 3*p*, and ten electrons in 3*d* for a total of eighteen electrons.
(e) An *f* sublevel can hold a maximum of fourteen electrons.

52. (a) Ne (c) F
(b) Ge (d) N

54. Transition elements are found in Groups 1B–8B, lanthanides and actinides.

56. Elements number 8, 16, 34, 52, 84 all have 6 electrons in their outer shell.

58. (a) sublevel *p* (c) sublevel *f*
(b) sublevel *d*

60. If 36 is a noble gas, 35 would be in periodic Group 7A and 37 would be in periodic Group 1A.

62. (a) $[Rn]7s^25f^{14}6d^{10}7p^5$
(c) F, Cl, Br, I, At

Chapter 11

Review Questions

1. smallest Cl, Mg, Na, K, Rb largest.

3. When a third electron is removed from beryllium, it must come from a very stable electron structure corresponding to that of the noble gas, helium. In addition, the third electron must be removed from a +2 beryllium ion, which increases the difficulty of removing it.

5. The first ionization energy decreases from top to bottom because the outermost electrons in the successive noble gases are farther away from the nucleus and are more shielded by additional inner electron energy levels.

7. The first electron removed from a sodium atom is the one valence electron, which is shielded from most of its nuclear charge by the electrons of lower levels. To remove a second electron from the sodium ion requires breaking into the noble gas structure. This requires much more energy than that necessary to remove the first electron, because the Na^+ is already positive.

8. (a) K > Na (e) Zr > Ti
(c) O > F

10. Atomic size increases down the column, since each successive element has an additional energy level that contains electrons located farther from the nucleus.

12. Metals are less electronegative than nonmetals. Therefore, metals lose electrons more easily than nonmetals. So, metals will transfer electrons to nonmetals, leaving the metals with a positive charge and the nonmetals with a negative charge.

14. A polar bond is a bond between two atoms with very different electronegativities. A polar molecule is a dipole due to unequal electrical charge between bonded atoms resulting from unequal sharing of electrons.

16. A Lewis structure is a visual representation of the arrangement of atoms and electrons in a molecule or an ion. It shows how the atoms in a molecule are bonded together.

18. The dots in a Lewis structure represent nonbonding pairs of electrons. The lines represent bonding pairs of electrons.

22. Valence electrons are the electrons found in the outermost energy level of an atom.

24. An aluminum ion has a +3 charge because it has lost 3 electrons in acquiring a noble-gas electron structure.

Exercises

2. (a) Fe^{2+} (e) Rubidium ion.
(c) Chloride ion.

4.

	+	−		+	−
(a)	H	Cl	(e)	Mg	H
(c)	C	Cl			

6. (a) covalent (c) covalent

8. (a) $F + 1\,e^- \longrightarrow F^-$
(b) $Ca \longrightarrow Ca^{2+} + 2\,e^-$

10. $Ca\!: + \cdot\ddot{O}\!: \longrightarrow CaO$
$Na\cdot + \cdot\ddot{Br}\!: \longrightarrow NaBr$

12. Si(4) N(5) P(5) O(6) Cl(7)

14. (a) Chloride ion, none
(b) Nitrogen atom, gain 3 e^- or lose 5 e^-
(c) Potassium atom, lose 1 e^-

16. (a) Nonpolar covalent compound; O_2.
(c) Polar covalent compound; H_2O.

18. (a) SbH_3, Sb_2O_3 (c) HCl, Cl_2O_7

20. $BeBr_2$, beryllium bromide
$MgBr_2$, magnesium bromide
$SrBr_2$, strontium bromide
$BaBr_2$, barium bromide
$RaBr_2$, radium bromide

22. (a) Ga: (b) $[Ga]^{3+}$ (c) $[Ca]^{2+}$

24. (a) covalent (c) covalent
(b) ionic (d) covalent

26. (a) covalent (c) ionic
(b) covalent

28. (a) :Ö::Ö: (b) :Br:Br: (c) :I:I:

30. (a) :S:H with H below S (c) H:N:H with H below N

32. (a) $[:\ddot{I}:]^-$ (e) $[:\ddot{O}:N::O:$ with :Ö: below N$]^-$
(c) $[:\ddot{O}:C::\ddot{O}$ with :Ö: below C$]^{2-}$

34. (a) F_2, nonpolar (c) NH_3, polar
(b) CO_2, nonpolar

36. (a) 3 electron pairs, trigonal planar
(b) 4 electron pairs, tetrahedral
(c) 4 electron pairs, tetrahedral

38. (a) tetrahedral (c) tetrahedral
(b) pyramidal

40. (a) tetrahedral (b) bent (c) bent

42. potassium

43. N_2H_4 14 e^- :N:N: with H H above and H H below; HN_3 16 e^- H—N=N=N

46. (a) Hg (c) N (e) Au

47. (a) Zn (c) N

51. $SnBr_2$, $GeBr_2$

54. This structure shown is incorrect since the bond is ionic. It should be represented as:
$[Na]^+[:\ddot{O}:]^{2-}[Na]^+$

59. (a) 105° (c) 109.5°

63. Empirical formula is SO_3
:Ö: bonded to S; :Ö—S=Ö:

Chapter 12

Review Questions

2. The pressure of a gas is the force that gas particles exert on the walls of a container. It depends on the temperature, the number of molecules of the gas, and the volume of the container.

4. The major components of dry air are nitrogen and oxygen.

6. The molecules of H_2 at 100°C are moving faster. Temperature is a measure of average kinetic energy. At higher temperatures, the molecules will have more kinetic energy.

7. 1 atm corresponds to 4 L.

13. The order of increasing molecular velocities is the order of decreasing molar masses.

increasing molecular velocity →
Rn, F_2, N_2, CH_4, He, H_2
decreasing molar mass →

At the same temperature the kinetic energies of the gases are the same and equal to $1/2\ mv^2$. For the kinetic energies to be the same, the velocities must increase as the molar masses decrease.

15. Gases are described by the following parameters:
(a) pressure (c) temperature
(b) volume (d) number of moles

18. Charles' law: $V_1/T_1 = V_2/T_2$,
ideal gas equation: $PV = nRT$
Rearrange the ideal gas equation to: $V/T = nR/P$
If you have an equal number of moles of two gases at the same pressure the right side of the rearranged ideal gas equation will be the same for both. You can set V/T for the first gas equal to V/T for the second gas (Charles' law) because the right side of both equations will cancel.

21. Equal volumes of H_2 and O_2 at the same T and P:
(a) have equal number of molecules (Avogadro's law)
(c) moles O_2 = moles H_2
(e) rate H_2 = 4 times the rate of O_2 (Graham's Law of Effusion)

24. We refer gases to STP because some reference point is needed to relate volume to moles. A temperature and pressure must be specified to determine the moles of gas in a given volume, and 0°C and 760 torr are convenient reference points.

27. Heating a mole of N_2 gas at constant pressure has the following effects:
(a) Density will decrease. Heating the gas at constant pressure will increase its volume. The mass does not change, so the increased volume results in a lower density.
(c) Average kinetic energy of the molecules increases. This is a basic assumption of the Kinetic Molecular Theory.
(e) Number of N_2 molecules remains unchanged. Heating does not alter the number of molecules present, except if extremely high temperatures were attained. Then, the N_2 molecules might dissociate into N atoms resulting in fewer N_2 molecules.

Exercises

2. (a) 715 torr (c) 95.3 kPa
(b) 953 bar
4. (a) 0.082 atm (c) 0.296 atm
6. (a) 241 mL (b) 577 mL
8. 3.3 atm
10. (a) 6.17 L (b) 8.35 L
12. 7.8×10^2 mL
14. 33.4 L
16. −150°C
18. 681 torr
20. 1.450×10^3 torr
22. 1.19 L C_3H_8
24. 0.156 mol N_2
26. (a) 67.0 L SO_3 (c) 7.96 L Cl_2
(b) 170 L C_2H_6
28. 1.78 g C_3H_6
30. 2.69×10^{22} molecules CH_4
32. 2.1 g CH_4
34. (a) 9.91 g/L Rn (c) 2.50 g/L C_4H_8
(b) 1.96 g/L CO_2
36. (a) 3.165 g/L Cl_2 (b) 1.46 g/L Cl_2
38. 19 L Kr
40. 72.8 L
42. 0.13 mol N_2
44. (a) 1.65×10^4 mL O_2 (b) 0.0201 mol H_2O_2
46. (a) 36 L O_2 (c) 12 L H_2O
(b) 210 g CO_2
48. 1.12×10^3 L CO_2
51. (a) the pressure will be cut in half
(c) the pressure will be cut in half
54. 22.4 L
56. SF_6 has the greatest density.
57. (a) CH_3
(b) C_2H_6
(c) H—C(H)(H)—C(H)(H)—H

60. 2.3×10^7 K
63. 327°C
65. 65 atm 211 K (−62°C)
67. 3.55 atm
70. 1.03×10^4 mm (33.8 ft)
71. 330 mol O_2
73. 43.2 g/mol (molar mass)
75. (a) 8.4 L H_2 (c) 8.48 g/L CO_2
77. 279 L H_2 at STP
78. Helium effuses 2.64 times faster than nitrogen.
80. C_4H_8

Chapter 13

Review Questions

2. H_2S, H_2Se, and H_2Te are gases at 0°C
4. Liquids take the shape of the container they are in. Gases also exhibit this property.
6. 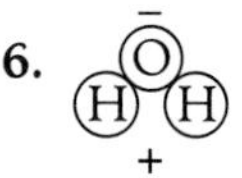

10. about 70°C,
13. In Figure 13.1, it would be case (b) in which the atmosphere would reach saturation. The vapor pressure of water is the same in both (a) and (b), but since (a) is an open container the vapor escapes into the atmosphere and doesn't reach saturation.
15. The vapor pressure observed in (c) would remain unchanged. The presence of more water in (b) does not change the magnitude of the vapor pressure of the water. The temperature of the water determines the magnitude of the vapor pressure.
17. (a) At a pressure of 500 torr, water boils at 88°C.
(b) The normal boiling point of ethyl alcohol is 78°C.
(c) At a pressure of 0.50 atm (380 torr), ethyl ether boils at 16°C.
20. For water, to have its maximum density, the temperature must be 4°C, and the pressure sufficient to keep it liquid. $d = 1.0$ g/mL
30. (a) mercury, acetic acid, water, toluene, benzene, carbon tetrachloride, methyl alcohol, bromine
(b) Highest boiling point is mercury; lowest is bromine.
33. Vapor pressure varies with temperature. The temperature at which the vapor pressure of a liquid equals the prevailing pressure is the boiling point of the liquid.
36. Ammonia would have a higher vapor pressure than SO_2 at −40°C because it has a lower boiling point (NH_3 is more volatile than SO_2).
40. 34.6°C, the boiling point of ethyl ether. (See Table 13.2)

Exercises

2. SO_2, SO_3, N_2O_5
4. [KOH, K_2O] [$Ba(OH)_2$, BaO] [$Ca(OH)_2$CaO]
6. (a) $Li_2O + H_2O \longrightarrow 2\ LiOH$
   (c) $Ba + 2\ H_2O \longrightarrow Ba(OH)_2 + H_2$
   (e) $SO_3 + H_2O \longrightarrow H_2SO_4$
8. (a) magnesium ammonium phosphate hexahydrate
   (b) iron(II) sulfate heptahydrate
   (c) tin(IV) chloride pentahydrate
10. (a) Distilled water has been vaporized by boiling and recondensed.
    (b) Natural waters are generally not pure, but contain dissoved minerals and suspended matter and can even contain harmful bacteria.
12. (a) HF will hydrogen bond; hydrogen is bonded to fluorine
    (c) H_2O_2 will hydrogen bond; hydrogen is bonded to oxygen.
    (e) H_2 will not hydrogen bond; hydrogen is not bonded to fluorine, oxygen, or nitrogen.
14. 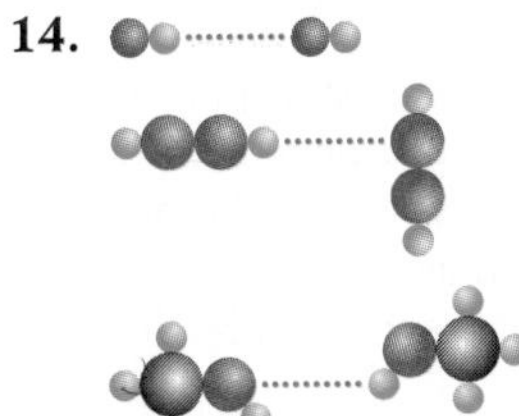
16. Water forming beaded droplets is an example of cohesive forces. The water molecules have stronger attractive forces for other water molecules then they do for the surface.
18. 0.262 mol $FeI_2 \cdot 4\ H_2O$
20. 1.05 mol H_2O
22. 48.67% H_2O
24. $FePO_4 \cdot 4\ H_2O$
26. 5.5×10^4 J
28. 42 g of steam needed; not enough steam (35 g)
30. The system will be at 0°C. It will be a mixture of ice and water.
32. (a) 18.0 g H_2O (c) 18.0 g H_2O
    (b) 36.0 g H_2O
34. Steam molecules will cause a more severe burn. Steam molecules contain more energy at 100°C than water molecules at 100°C due to the energy absorbed during the vaporization state (heat of vaporization).
36. When one leaves the swimming pool, water starts to evaporate from the skin of the body. Part of the energy needed for evaporation is absorbed from the skin, resulting in the cool feeling.
38. During phase changes (ice melting to liquid water or liquid water evaporating to steam), all the heat energy is used to cause the phase change. Once the phase change is complete, the heat energy is once again used to increase the temperature of the substance.
40. 75°C at 270 torr pressure
42. $MgSO_4 \cdot 7\ H_2O$ $Na_2HPO_4 \cdot 12\ H_2O$
44. chlorine
46. When organic pollutants in water are oxidized by dissolved oxygen, there may not be sufficient dissolved oxygen to sustain marine life.
48. Na_2 zeolite(s) + $Mg^{2+}(aq) \longrightarrow$ Mg zeolite(s) + $2\ Na^+(aq)$
51. 6.78×10^5 J
52. 40.7 kJ/mol
54. 2.30×10^6 J
56. 40.2 g H_2O
58. 1.00 mole of water vapor at STP has a volume of 22.4 L (gas)
60. 3.9×10^3 J
63. 2.71×10^4 J

Chapter 14

Review Questions

2. From Table 14.2, approximately 4.5 g of NaF would be soluble in 100 g of water at 50°C.
3. From Figure 14.4, solubilities in water at 25°C are:
   (a) KCl 35g/100 g H_2O
   (b) $KClO_3$ 9g/100 g H_2O
   (c) KNO_3 39g/100 g H_2O
6. KNO_3
8. 6×10^2 cm^2
9. The solution would be unsaturated at 40°C and 50°C.
12. The solution level in the thistle tube will fall.
16. The ions or molecules of a dissolved solute do not settle out because the individual particles are so small that the force of molecular collisions is large compared with the force of gravity.
24. For a given mass of solute, the smaller the particles, the faster the dissolution of the solute. This is due to the smaller particles having a greater surface area exposed to the dissolving action of the solvent.
28. The solution contains 16 moles of HNO_3 per liter of solution.
30. The champagne would spray out of the bottle.
34. A lettuce leaf immersed in salad dressing containing salt and vinegar will become limp and wilted as a result of osmosis.

36. (a) 1 *M* NaOH 1 L
(b) 0.6 *M* $Ba(OH)_2$ 0.83 L
(c) 2 *M* KOH 0.50 L
(d) 1.5 *M* $Ca(OH)_2$ 0.33 L

42. The molarity of a 5 molal solution is less than 5 *M*.

Exercises

2. Reasonably soluble: (c) $CaCl_2$
Insoluble: (a) PbI_2 (e) $BaSO_4$

4. (a) 16.7% (c) 39% K_2CrO_4

6. 250. g solution

8. (a) 9.0 g $BaCl_2$ (b) 66 g solvent

10. 33.6% NaCl

12. 22% C_6H_{14}

14. (a) 4.0 *M* (c) 1.97 *M* $C_6H_{12}O_6$

16. (a) 1.1 mol HNO_3 (c) 0.088 mol LiBr
(b) 7.5×10^{-3} mol $NaClO_3$

18. (a) 2.1×10^3 g H_2SO_4 (c) 1.2 g $Fe_2(SO_4)_3$
(b) 6.52 g $KMnO_4$

20. (a) 3.4×10^3 mL (b) 1.88×10^3 mL

22. (a) 1.2 *M* (c) 0.333 *M*
(b) 0.070 *M*

24. (a) 25 mL 18 *M* H_2SO_4 (b) 5.0 mL 15 *M* NH_3

26. (a) 1.2 *M* HCl
(b) 4.21×10^{-3} *M* HCl

28. (a) 3.6 mol Na_2SO_4 (e) 38.25 mL H_2SO_4
(c) 0.625 mol NaOH

30. (a) 1.88×10^{-3} mol H_2O (e) 1.8 L CO_2
(c) 90.2 mL $HC_2H_3O_2$

32. (a) 5.5 *m* $C_6H_{12}O_6$ (b) 9.9×10^{-4} *m* I_2

34. (a) 10.74 *m* (c) 105.50°C
(b) −20.0°C

36. 163 g/mol

43. 60. g 10% NaOH solution

45. (a) 160 g sugar (c) 0.516 *m*
(b) 0.47 *M*

47. 16.2 *m* HCl

49. 455 g solution

51. (a) 4.5 g NaCl
(b) 450. mL H_2O must evaporate.

53. 210. mL solution

55. 6.72 *M* HNO_3

59. 1.20 g $Mg(OH)_2$ is the more effective acid neutralizer.

61. 6.2 *m* H_2SO_4
5.0 *M* H_2SO_4

62. (a) 2.9 *m* (b) 101.5°C

64. (a) 8.04×10^3 g $C_2H_6O_2$ (c) −4.0°F
(b) 7.24×10^3 mL $C_2H_6O_2$

65. 47% $NaHCO_3$

66. (a) 7.7 L H_2O must be added
(b) 0.0178 mol
(c) 0.0015 mol H_2SO_4 in each mL

69. 2.84 g $Ba(OH)_2$ is formed.

70. (a) 0.011 mol Li_2CO_3
(c) 3.2×10^2 mL solution

Chapter 15

Review Questions

2. An electrolyte must be present in the solution for the bulb to glow.

4. First, the orientation of the polar water molecules about the Na^+ and Cl^- ions is different. Second, more water molecules will fit around Cl^-, since it is larger than the Na^+ ion.

6. tomato juice

8. Arrhenius: $HCl + NaOH \longrightarrow NaCl + H_2O$
Brønsted–Lowry: $HCl + KCN \longrightarrow HCN + KCl$
Lewis: $AlCl_3 + NaCl \longrightarrow AlCl_4^- + Na^+$

10. acids, bases, salts

13. In their crystalline structure, salts exist as positive and negative ions in definite geometric arrangement to each other, held together by the attraction of the opposite charges. When dissolved in water, the salt dissociates as the ions are pulled away from each other by the polar water molecules.

15. Molten NaCl conducts electricity because the ions are free to move. In the solid state, however, the ions are immobile and do not conduct electricity.

18. Ions are hydrated in solution because there is an electrical attraction between the charged ions and the polar water molecules.

20. (a) $[H^+] = [OH^-]$ (c) $[OH^-] > [H^+]$
(b) $[H^+] > [OH^-]$

21. The net ionic equation for an acid–base reaction in aqueous solutions is: $H^+ + OH^- \longrightarrow H_2O$.

24. The fundamental difference is in the size of the particles. The particles in true solutions are less than 1 nm in size. The particles in colloids range from 1–1000 nm in size.

Exercises

2. Conjugate acid–base pairs:
(a) $HCl - Cl^-$; $NH_4^+ - NH_3$
(c) $H_3O^+ - H_2O$; $H_2CO_3 - HCO_3^-$

4. (a) $Fe_2O_3(s) + 6\,HBr(aq) \longrightarrow 2\,FeBr_3(aq) + 3\,H_2O(l)$
(c) $2\,NaOH(aq) + H_2CO_3(aq) \longrightarrow Na_2CO_3(aq) + 2\,H_2O(l)$
(e) $Mg(s) + 2\,HClO_4(aq) \longrightarrow Mg(ClO_4)_2(aq) + H_2(g)$

6. (a) $Fe_2O_3 + (6\ H^+ + 6\ Br^-) \longrightarrow (2\ Fe^{3+} + 6\ Br^-) + 3\ H_2O$

$Fe_2O_3 + 6\ H^+ \longrightarrow 2\ Fe^{3+} + 3\ H_2O$

(c) $(2\ Na^+ + 2\ OH^-) + H_2CO_3 \longrightarrow (2\ Na^+ + CO_3^{2-}) + 2\ H_2O$

$2\ OH^- + H_2CO_3 \longrightarrow CO_3^{2-} + 2\ H_2O$

(e) $Mg + (2\ H^+ + 2\ ClO_4^-) \longrightarrow (Mg^{2+} + 2\ ClO_4^-) + H_2$

$Mg + 2\ H^+ \longrightarrow Mg^{2+} + H_2$

8. (a) $NaHCO_3$—salt (e) RbOH—base
(c) $AgNO_3$—salt

10. (a) 0.75 M Zn^{2+}, 1.5 M Br^-
(b) 4.95 M SO_4^{2-}, 3.30 M Al^{3+}
(c) 0.680 M NH_4^+, 0.340 M SO_4^{2-}
(d) 0.0628 M Mg^{2+}, 0.126 M ClO_3^-

12. (a) 4.9 g Zn^{2+}, 12 g Br^-
(b) 8.90 g Al^{3+}, 47.6 g SO_4^{2-}
(c) 1.23 g NH_4^+, 3.28 g SO_4^{2-}
(d) 0.153 g Mg^{2+}, 1.05 g ClO_3^-

14. (a) $[H^+] = 3.98 \times 10^{-3}$
(b) $[H^+] = 1.0 \times 10^{-10}$
(c) $[H^+] = 3.98 \times 10^{-7}$

16. (a) $[K^+] = 1.0\ M$, $[Ca^{2+}] = 0.5\ M$, $[Cl^-] = 2.0\ M$
(b) No ions are present in the solution.
(c) 0.67 M Na^+, 0.67 M NO_3^-

18. (a) 0.147 M NaOH
(b) 0.964 M NaOH
(c) 0.4750 M NaOH

20. (a) $H_2S(g) + Cd^{2+}(aq) \longrightarrow CdS(s) + 2\ H^+(aq)$
(b) $Zn(s) + 2H^+(aq) \longrightarrow Zn^{2+}(aq) + H_2(g)$
(c) $Al^{3+}(aq) + PO_4^{3-}(aq) \longrightarrow AlPO_4(s)$

22. (a) 2 M HCl
(b) 1 M H_2SO_4

24. 40.8 mL of 0.245 M HCl

26. 7.0% NaCl in the sample

28. 0.936 L H_2

30. (a) pH = 7.0 (c) pH = 4.00
(b) pH = 0.30

32. (a) pH = 4.30 (b) pH = 10.47

34. (a) weak acid (c) strong acid
(b) strong base (d) strong acid

35. (a) acidic (e) neutral
(c) basic

37. 0.260 M Ca^{2+}

38. 0.309 M $Ba(OH)_2$

42. Freezing point depression is directly related to the concentration of particles in the solution.

highest freezing point	$C_{12}H_{22}O_{11}$	>	$HC_2H_3O_2$	>	HCl	>	$CaCl_2$	lowest freezing point
	1 mol		1 + mol		2 mol		3 mol	

(particles in solution)

44. As the pH changes by 1 unit, the concentration of H^+ in solution changes by a factor of 10. For example, the pH of 0.10 M HCl is 1.00, while the pH of 0.0100 M HCl is 2.00.

45. 1.77% solute, 99.22% H_2O

46. 0.201 M HCl

48. 0.673 g KOH

50. 13.9 L of 18.0 M H_2SO_4

52. pH = 1.1, acidic

54. (a) $2\ NaOH(aq) + H_2SO_4(aq) \longrightarrow Na_2SO_4(aq) + 2\ H_2O(l)$
(b) 1.0×10^2 mL NaOH
(c) 0.71 g Na_2SO_4

Chapter 16

Review Questions

2. The reaction is endothermic because the increased temperature increases the concentration of product (NO_2) present at equilibrium.

4. At equilibrium, the rate of the forward reaction equals the rate of the reverse reaction.

6. The sum of the pH and the pOH is 14. A solution whose pH is –1 would have a pOH of 15.

8. The order of molar solubilities is $AgC_2H_3O_2$, $PbSO_4$, $BaSO_4$, AgCl, $BaCrO_4$, AgBr, AgI, PbS.

9. (a) Ag_2CrO_4 (b) Ag_2CrO_4

12. The rate increases because the number of collisions increases due to the added reactant.

15. A catalyst speeds up the rate of a reaction by lowering the activation energy. A catalyst is not used up in the reaction.

18. The statement does not contradict Le Chatetier's Principle. The previous question deals with the case of dilution. If pure acetic acid is added to a dilute solution, the reaction will shift to the right, producing more ions in accordance with Le Chatelier's Principle. But the concentration of the un-ionized acetic acid will increase faster than the concentration of the ions, thus yielding a smaller percent ionization.

20. In pure water, H^+ and OH^- are produced in equal quantities by the ionization of the water molecules, $H_2O \rightleftharpoons H^+ + OH^-$. Since pH = −log[$H^-$], and pOH = −log[$OH^-$], they will always be identical for pure water. At 25°C, they each have the value of 7, but at higher temperatures, the degree of ionization is greater, so the pH and pOH would both be less than 7, but still equal.

Exercises

2. (a) $H_2O(l) \xrightleftharpoons{100°C} H_2O(g)$
(b) $SO_2(l) \rightleftharpoons SO_2(g)$

4. (a) [NH_3], [O_2], and [N_2], will be increased. [H_2O], will be decreased. Reaction shifts left.
(b) The addition of heat will shift the reaction to the left.

6.
(a) left I I D (add NH_3)
(b) left I I D (increase volume)
(c) no change N N N (add catalyst)
(d) ? ? I I (add H_2 and N_2)

8.

Reaction	Increase temperature	Increase pressure	Add catalyst
(a)	right	left	no change
(b)	left	left	no change
(c)	left	left	no change

10. Equilibrium shift
(a) left (b) right (c) right

12. (a) $K_{eq} = \dfrac{[H^+][H_2PO_4^-]}{[H_3PO_4]}$ (c) $K_{eq} = \dfrac{[N_2O_5]^2}{[NO_2]^4[O_2]}$

(b) $K_{eq} = \dfrac{[CH_4][H_2S]^2}{[CS_2][H_2]^4}$

14. (a) $K_{sp} = [Mg^{2+}][CO_3^{2-}]$ (c) $K_{sp} = [Tl^{3+}][OH^-]^3$
(b) $K_{sp} = [Ca^{2+}][C_2O_4^{2-}]$ (d) $K_{sp} = [Pb^{2+}]^3[AsO_4^{3-}]^2$

16. If H^+ is increased,
(a) pH is decreased
(b) pOH is increased
(c) OH^- is decreased
(d) K_W, remains unchanged

18. (a) NH_4Cl acidic (c) $CuSO_4$ acidic

20. (a) $NH_4^+(aq) + H_2O(l) \rightleftharpoons H_3O^+(aq) + NH_3(aq)$
(b) $SO_3^{2-}(aq) + H_2O(l) \rightleftharpoons OH^-(aq) + HSO_3^-(aq)$

22. (a) $OCl^-(aq) + H_2O(l) \rightleftharpoons OH^-(aq) + HOCl(aq)$
(b) $ClO_2^-(aq) + H_2O(l) \rightleftharpoons OH^-(aq) + HClO_2(aq)$

24. When excess base gets into the bloodstream, it reacts with H^+ to form water. Then H_2CO_3 ionizes to replace H^+, thus maintaining the approximate pH of the solution.

26. (a) $K_a = 5.7 \times 10^{-6} M$ (c) 2.3×10^{-3}% ionized
(b) pH = 5.24

28. $K_a = 2.3 \times 10^{-5}$

30. (a) pH = 3.72 (c) pH = 4.72
(b) pH = 4.23

32. $K_a = 7.3 \times 10^{-6}$

34. [H^+] = 3.0 *M*; pH = −0.47; pOH = 14.47; [OH^-] = 3.3×10^{-15}

36. (a) pOH = 3.00; pH = 11.0
(b) pH = 0.903; pOH = 13.1
(c) pH = 5.74; pOH = 8.3

38. (a) [OH^-] = 2.5×10^{-6}
(b) [OH^-] = 1.1×10^{-13}

40. (a) [H^+] = 2.2×10^{-9}
(b) [H^+] = 1.4×10^{-11}

42. (a) 1.2×10^{-23} ZnS
(c) 1.81×10^{-18} Ag_3PO_4

44. (a) 1.1×10^{-4} *M*
(b) 9.2×10^{-6} *M*

46. (a) 3.3×10^{-3} g $PbSO_4$
(b) 2.3×10^{-4} g $BaCrO_4$

48. Precipitation occurs.

50. 2.6×10^{-12} mol AgBr will dissolve.

52. pH = 4.74

54. Change in pH = 4.74 − 4.72 = 0.02 units in the buffered solution. Initial pH = 4.74.

58. 4.2 mol HI

60. 0.500 mol HI, 2.73 mol H_2, 0.538 mol I_2

62. 128 times faster

64. pH = 4.6

66.
Hypochlorous acid: $K_a = 3.5 \times 10^{-5}$
Propanoic acid: $K_a = 1.3 \times 10^{-5}$
Hydrocyanic acid: $K_a = 4.0 \times 10^{-10}$

68. (a) Precipitation occurs.
(b) Precipitation occurs.
(c) No precipitation occurs.

70. No precipitate of $PbCl_2$

72. $K_{eq} = 1.1 \times 10^4$

74. 8.0 *M* NH_3

76. $K_{sp} = 4.00 \times 10^{-28}$

78. (a) The temperature could have been cooler.
(b) The humidity in the air could have been higher.
(c) The air pressure could have been greater.

80. (a) $K_{eq} = \dfrac{[O_3]^2}{[O_2]^3}$ (c) $K_{eq} = [CO_2]$

(b) $K_{eq} = \dfrac{[H_2O(l)]}{[H_2O(g)]}$ (d) $K_{eq} = \dfrac{[H^+]^6}{[Bi^{3+}]^2[H_2S]^3}$

82. $K_{eq} = 450$

84. 1.1 g $CaSO_4$

Chapter 17

Review Questions

2. (a) I_2 has been oxidized.
(b) Cl_2 has been reduced.

4. If the free element is higher on the list than the ion with which it is paired, the reaction occurs.
(a) Yes. $Zn(s) + Cu^{2+}(aq) \longrightarrow Zn^{2+}(aq) + Cu(s)$
(c) Yes. $Sn(s) + 2\,Ag^{+}(aq) \longrightarrow Sn^{2+}(aq) + 2\,Ag(s)$
(e) Yes. $Ba(s) + FeCl_2(aq) \longrightarrow BaCl_2(aq) + Fe(s)$
(g) Yes. $Ni(s) + Hg(NO_3)_2(aq) \longrightarrow Ni(NO_3)_2(aq) + Hg(l)$

6. (a) $2\,Al + Fe_2O_3 \longrightarrow Al_2O_3 + 2\,Fe + \text{heat}$
(c) No. Fe is less reactive than Al.

10. (a) Oxidation occurs at the anode.
$2\,Cl^{-}(aq) \longrightarrow Cl_2(g) + 2\,e^{-}$
(b) Reduction occurs at the cathode.
$Ni^{2+}(aq) + 2\,e^{-} \longrightarrow Ni(s)$
(c) The net chemical reaction is
$Ni^{2+}(aq) + 2\,Cl^{-}(aq) \xrightarrow[\text{energy}]{\text{electrical}} Ni(s) + Cl_2(g)$

12. (a) It would not be possible to monitor the voltage produced, but the reactions in the cell would still occur.
(b) If the salt bridge were removed, the reaction would stop.

14. $Ca^{2+} + 2\,e^{-} \longrightarrow Ca$ cathode reaction, reduction
$2\,Br^{-} \longrightarrow Br_2 + 2\,e^{-}$ anode reaction, oxidation

16. Since lead dioxide and lead(II) sulfate are insoluble, it is unnecessary to have salt bridges in the cells of a lead storage battery.

18. Reduction occurs at the cathode.

20. A salt bridge permits movement of ions in the cell. This keeps the solution neutral with respect to the charged particles (ions) in the solution.

Exercises

2. (a) $K\underline{Mn}O_4$ +7
(c) $\underline{N}H_3$ −3
(e) $K_2\underline{Cr}O_4$ +6

4. (a) $\underline{O_2}$ 0
(c) $Fe(\underline{O}H)_3$ −2

6. (a) $SO_3^{2-} + H_2O \longrightarrow SO_4^{2-} + 2\,H^{+} + 2\,e^{-}$
S is oxidized
(c) $S_2O_4^{2-} + 2\,H_2O \longrightarrow 2\,SO_3^{2-} + 4\,H^{+} + 2\,e^{-}$
S is oxidized

8. Equation (1):
(a) As is oxidized, Ag^{+} is reduced.
(b) Ag^{+} is the oxiding agent, AsH_3 the reducing agent.
Equation (2):
(a) Br is oxidized, Cl is reduced.
(b) Cl_2 is the oxidizing agent, NaBr the reducing agent.

10. (a) incorrectly balanced
$3\,MnO_2(s) + 4\,Al(s) \longrightarrow 3\,Mn(s) + 2\,Al_2O_3(s)$
(c) correctly balanced

12. (a) $3\,Cl_2 + 6\,KOH \longrightarrow KClO_3 + 5\,KCl + 3\,H_2O$
(c) $3\,CuO + 2\,NH_3 \longrightarrow N_2 + 3\,Cu + 3\,H_2O$
(e) $5\,H_2O_2 + 2\,KMnO_4 + 3\,H_2SO_4 \longrightarrow 5\,O_2 + 2\,MnSO_4 + K_2SO_4 + 8\,H_2O$

14. (a) $6\,H^{+} + ClO_3^{-} + 6\,I^{-} \longrightarrow 3\,I_2 + Cl^{-} + 3\,H_2O$
(c) $2\,H_2O + 2\,MnO_4^{-} + 5\,SO_2 \longrightarrow 4\,H^{+} + 2\,Mn^{2+} + 5\,SO_4^{2-}$
(e) $8\,H^{+} + Cr_2O_7^{2-} + 3\,H_3AsO_3 \longrightarrow 2\,Cr^{3+} + 3\,H_3AsO_4 + 4\,H_2O$

16. (a) $H_2O + 2\,MnO_4^{-} + 3\,SO_3^{2-} \longrightarrow 2\,MnO_2 + 3\,SO_4^{2-} + 2\,OH^{-}$
(c) $8\,Al + 3\,NO_3^{-} + 18\,H_2O + 5\,OH^{-} \longrightarrow 3\,NH_3 + 8\,Al(OH)_4^{-}$
(e) $2\,Al + 6\,H_2O + 2\,OH^{-} \longrightarrow 2\,Al(OH)_4^{-} + 3\,H_2$

18. (a) $5\,Mo_2O_3 + 6\,MnO_4^{-} + 18\,H^{+} \longrightarrow 10\,MoO_3 + 6\,Mn^{2+} + 9\,H_2O$
(b) $3\,BrO^{-} + 2\,Cr(OH)_4^{-} + 2\,OH^{-} \longrightarrow 3\,Br^{-} + 2\,CrO_4^{2-} + 5\,H_2O$
(c) $3\,S_2O_3^{2-} + 8\,MnO_4^{-} + H_2O \longrightarrow 6\,SO_4^{2-} + 8\,MnO_2 + 2\,OH^{-}$

20. (a) $Pb + SO_4^{2-} \longrightarrow PbSO_4 + 2\,e^{-}$
$PbO_2 + SO_4^{2-} + 4\,H^{+} + 2\,e^{-} \longrightarrow PbSO_4 + 2\,H_2O$
(b) The first reaction is oxidation (Pb^{0} is oxidized to Pb^{2+}).
The second reaction is reduction (Pb^{4+} is reduced to Pb^{2+}).
(c) The first reaction (oxidation) occurs at the anode of the battery.

22. Zinc is a more reactive metal than copper, so when corrosion occurs, the zinc preferentially reacts. Zinc is above hydrogen in the Activity Series of Metals; copper is below hydrogen.

24. 20.2 L Cl_2

26. 66.2 mL of 0.200 *M* $K_2Cr_2O_7$ solution

28. 5.560 mol H_2

30. The electrons lost by the species undergoing oxidation must be gained (or attracted) by another species, which then undergoes reduction.

32. Sn^{4+} can only be an oxidizing agent.
Sn^{0} can only be a reducing agent.
Sn^{2+} can be both oxidizing and reducing agents.

34. Equations (a) and (b) represent oxidations.

36. (a) $F_2 + 2\,Cl^{-} \longrightarrow 2\,F^{-} + Cl_2$
(b) $Br_2 + Cl^{-} \longrightarrow$ NR
(c) $I_2 + Cl^{-} \longrightarrow$ NR
(d) $Br_2 + 2\,I^{-} \longrightarrow 2\,Br^{-} + I_2$

38. $4\,Zn + NO_3^{-} + 10\,H^{+} \longrightarrow 4\,Zn^{2+} + NH_4^{+} + 3\,H_2O$

40. (a) Pb is the anode.
(c) Pb (anode)
(e) Electrons flow from the lead through the wire to the silver.

Chapter 18

Review Questions

2. Alpha particles are much heavier.

4. Contributions to the early history of radioactivity include
(a) Henri Becquerel—discovered radioactivity.
(b) Marie and Pierre Curie—discovered polonium and radium.
(c) Wilhelm Roentgen—discovered X rays and developed the technique for producing them.
(d) Ernest Rutherford—discovered alpha and beta particles, established the link between radioactivity and transmutation, and produced the first successful man-made transmutation.
(e) Otto Hahn and Fritz Strassmann—were first to produce nuclear fission.

6. *Isotope* is used with reference to atoms of the same element that contain different masses. *Nuclide* refers to any isotope of any atom.

10. A radioactive disintegration series is a series of alpha and beta emissions from a radioactive nuclide until it reaches a stable nuclide.

12. $^{232}_{90}Th \xrightarrow{-\alpha} {}^{228}_{88}Ra \xrightarrow{-\beta} {}^{228}_{89}Ac \xrightarrow{-\beta} {}^{228}_{90}Th \xrightarrow{-\alpha}$

$^{224}_{88}Ra \xrightarrow{-\alpha} {}^{220}_{86}Rn \xrightarrow{-\alpha} {}^{216}_{84}Po \xrightarrow{-\alpha}$

$^{212}_{82}Pb \xrightarrow{-\beta} {}^{212}_{83}Bi \xrightarrow{-\beta} {}^{212}_{84}Po \xrightarrow{-\alpha} {}^{208}_{82}Pb$

14. $^{211}_{83}Bi \longrightarrow {}^{4}_{2}He + {}^{207}_{81}Ti$ $\qquad$ $^{207}_{81}Ti \longrightarrow {}^{0}_{-1}e + {}^{207}_{82}Pb$

17. The fission reaction in a nuclear reactor and in an atomic bomb are essentially the same. The difference is that the fissioning is "wild" or uncontrolled in the bomb. In a nuclear reactor, the fissioning rate is controlled by means of moderators, such as graphite, to slow the neutrons and control rods of cadmium or boron to absorb some of the neutrons.

20. When radioactive rays pass through normal matter, they cause that matter to become ionized (usually by knocking out electrons). Therefore, the radioactive rays are classified as ionizing radiation.

23. A radioactive "tracer" is a radioactive material, whose presence is traced by a Geiger counter or some other detecting device. Tracers are often injected into the human body, animals, and plants to determine chemical pathways, rates of circulation, etc.

26. The half-life of carbon-14 is 5730 years.

$$(4 \times 10^6 \text{ years})\left(\frac{1 \text{ half-life}}{5730 \text{ years}}\right) = 7 \times 10^2 \text{ half-lives}$$

700 half-lives would pass in 4 million years. Not enough C-14 would remain to allow detection with any degree of reliability. C-14 dating would not prove useful in this case.

30. In a nuclear power plant, a controlled nuclear fission reaction provides heat energy that is used to produce steam. The steam turns a turbine that generates electricity.

Exercises

2.

	Protons	Neutrons	Nucleons
(a) $^{235}_{92}U$	92	143	235
(b) $^{82}_{35}Br$	35	47	82

4. Its atomic number increases by one, and its mass number remains unchanged.

6. Equations for alpha decays:
(a) $^{192}_{78}Pt \longrightarrow {}^{4}_{2}He + {}^{188}_{76}Os$
(b) $^{210}_{84}Po \longrightarrow {}^{4}_{2}He + {}^{206}_{82}Pb$

8. (a) $^{239}_{93}Np \longrightarrow {}^{0}_{-1}e + {}^{239}_{94}Pu$
(b) $^{90}_{38}Sr \longrightarrow {}^{0}_{-1}e + {}^{90}_{39}Y$

10. (a) gamma-emission
(b) alpha-emission then beta-emission
(c) alpha-emission then gamma-emission

12. $^{30}_{15}P \longrightarrow {}^{30}_{14}Si + {}^{0}_{+1}e$

14. (a) $^{27}_{13}Al + {}^{4}_{2}He \longrightarrow {}^{30}_{15}P + {}^{1}_{0}n$
(c) $^{12}_{6}C + {}^{2}_{1}He \longrightarrow {}^{13}_{7}He + {}^{1}_{0}n$

16. The year 2064; 1/8 of Sr-90 remaining

18. $^{226}_{88}Ra$, alpha-emission; $^{222}_{86}Rn$, alpha-emission; $^{218}_{84}Po$, alpha-emission; $^{214}_{82}Pb$, beta-emission; $^{214}_{83}Bi$, beta-emission; $^{214}_{84}Po$, alpha-emission; $^{210}_{82}Pb$, beta-emission; $^{210}_{83}Bi$

20. (a) 5.4×10^{11} J/mol
(b) 0.199% mass loss

22. 10.9 days

24. 61.59% neutrons by mass. 0.021% electrons by mass

26. 11,460 years old

28. (a) 0.0424 g/mol
(b) 3.8×10^{12} J/mol

31. 0.16 mg remaining

34. (a) $^{235}_{92}U + {}^{1}_{0}n \longrightarrow {}^{143}_{54}Xe + 3\,{}^{1}_{0}n + {}^{90}_{38}Sr$
(b) $^{235}_{92}U + {}^{1}_{0}n \longrightarrow {}^{102}_{39}Y + 3\,{}^{1}_{0}n + {}^{131}_{53}I$
(c) $^{14}_{7}N + {}^{1}_{0}n \longrightarrow {}^{1}_{1}H + {}^{14}_{6}C$

36. $^{236}_{92}U \longrightarrow {}^{90}_{38}Sr + 3\,{}^{1}_{0}n + {}^{143}_{54}Xe$

38. 7680 g

40. (a) 0.500 g left
(b) 0.250 g left
(c) 0.0625 g
(d) 9.77×10^{-4} g

42. 2.29×10^4 years

Chapter 19

Review Questions

4. (a) A molecule of ethane, C_2H_6, contains seven sigma bonds.
(b) A molecule of butane, C_4H_{10}, contains thirteen sigma bonds.
(c) A molecule of 2-methylpropane also contains thirteen sigma bonds.

7. (a) A substitution reaction allows an exchange of atoms or groups of atoms between reactants while in an elimination reaction a single reactant is split into two products.
(b) Two reactants combine together in addition reaction to give one product, while one reactant is split into two products in an elimination reaction.

9. E85 is a gasoline that contains 85% ethanol and 15% petroleum. This mixture reduces the use of petroleum, a nonrenewable resource. E85 is the cleanest burning gasoline now available.

Exercises

2. Lewis structures:

```
              H
              ..
(a) CH4    H:C:H
              ..
              H

              H H H
              .. .. ..
(b) C3H8   H:C:C:C:H
              .. .. ..
              H H H

              H H H H H
              .. .. .. .. ..
(c) C5H12  H:C:C:C:C:C:H
              .. .. .. .. ..
              H H H H H
```

4. Formulas (b), (e), and (f) are identical. The others are different.

6. The formulas in Exercise 4 contain the following numbers of methyl groups:
(a) 4 (b) 4 (c) 4 (d) 4 (e) 4 (f) 4

8.

```
hexane  CH3CH2CH2CH2CH2CH3        CH3CH2CH2CHCH3
                                           |
                                           CH3

                         CH3           CH3
                         |             |
CH3CH2CHCH2CH3      CH3CH2CCH3     CH3CHCHCH3
      |                  |             |
      CH3                CH3           CH3
```

10. (a) CH_3Br, one: CH_3Br (b) C_2H_5Cl, one: C_2H_5Cl
(c) C_4H_9I, four: $CH_3CH_2CH_2CH_2I$ $CH_3CH_2CHICH_3$ $(CH_3)_2CHCH_2I$ $(CH_3)_3CI$
(d) C_3H_6BrCl, five: $CH_3CH_2CHBrCl$ $CH_3CHClCH_2Br$ $CH_3CHBrCH_2Cl$ $CH_2ClCH_2CH_2Br$ $CH_3CBrClCH_3$

12. Carbon atoms in parent chain
(a) seven (b) seven (c) five.

14. IUPAC names
(a) chloroethane
(b) 1-chloro-2-methylpropane
(c) 2-chlorobutane
(d) methylcyclopropane
(e) 2,4-dimethylpentane

16. Tertiary carbon atoms (circled)

```
         H3C—(CH)—CH2—CH3
              |
             (CH)—CH3
              |
(a)      CH3—(CH)—CH2—CH3

         CH2—CH2  CH3
         |       \/
         CH3     (CH)
                  |
(b)      CH3—CH2—C—CH2—CH3
                  |
                  CH3

                  CH3
                  |
             CH3—C—CH3
                  |
(c)      CH3—CH2—(CH)—CH3
```

```
                        CH3
                        |
18. (a)  CH3CH2CHCH2CHCH3
                |
                CH2CH3

    (b)  CH3CH2CH2CHCH2CH2CH3
                      |
                      C(CH3)3

                                  CH3
                                  |
             CH3    CH2CH3      CHCH3
             |      |           |
    (c)  CH3CHCH2CCH2CH2CHCHCH2CH3
                    |           |
                    CH3         CH3

             H3C  CH2CH3
              |   |
    (d)  CH3—C—CHCH2CH2CH2CH2CH3
              |
              CH3

    (e)  cyclohexane ring bearing CH2CH3 and CH2CH3
```

20. (a) 3-methyl-5-ethyloctane

```
CH3CH2CHCH2CHCH2CH2CH3
      |      |
      CH3    CH2CH3
```

Ethyl should be named before methyl (alphabetical order): The numbering is correct. The correct name is 5-ethyl-3-methyloctane.

(b) 3,5,5-triethylhexane $CH_3CH_2CHCH_2C(CH_2CH_3)_2CH_3$ (with CH_2CH_3 on C-3)

The name is not based on the longest carbon chain (7 carbons). The correct name is 3,5-diethyl-3-methylheptane.

(c) 4,4-dimethyl-3-ethylheptane

$CH_3CH_2CH(CH_2CH_3)—C(CH_3)_2CH_2CH_2CH_3$

Ethyl should be named before dimethyl (alphabetical order). The correct name is 3-ethyl-4,4-dimethylheptane.

22. $CH_3CH_2CH_2CH_2CH_2CH_2Cl$
$CH_3CH_2CH_2CH_2CHClCH_3$
$CH_3CH_2CH_2CHClCH_2CH_3$

24. (a) $CH_3CH_2CH_3 + Br_2 \xrightarrow{hv} CH_3CH_2CH_2Br + CH_3CHBrCH_3 + HBr$

(b) $CH_3CH_2CH_3 + 5\ O_2 \xrightarrow{\Delta} 3\ CO_2 + 4\ H_2O$

26. (a) 1,4-di-tert-butylcyclohexane: $(CH_3)_3C$–cyclohexane–$C(CH_3)_3$ (b) cyclohexane with CH_3 and $CH_2CH_2CH_3$ substituents

(c) cyclohexane with Cl, CH_2CH_3, Cl, CH_3CH_2 substituents

28. One isomer: $CH_3C(CH_3)_2CH_2CH_3$

30. (a) $CH_3C(CH_3)_2CH_3$, 2,2-dimethylpropane
(b) $CH_3CH_2CH_2CH_2CH_3$, pentane

32. (a) Three isomers:

1,2-dibromocyclopentane 1,3-dibromocyclopentane

1,1-dibromocyclopentane

(b) Seven isomers:

$CHBr_2—C(CH_3)_2—CH_2—CH_3$
1,1-dibromo-2,2-dimethylbutane

$H_3C—C(CH_3)_2—CH_2—CHBr_2$
1,1-dibromo-3,3-dimethylbutane

$CH_2Br—C(CH_3)_2—CHBr—CH_3$
1,3-dibromo-2,2-dimethylbutane

$H_3C—C(CH_3)_2—CHBr—CH_2Br$
1,2-dibromo-3,3-dimethylbutane

$H_3C—C(CH_3)_2—CBr_2—CH_3$
3,3-dibromo-2,2-dimethylbutane

$CH_2Br—C(CH_2Br)(CH_3)—CH_2—CH_3$
1-bromo-2-bromomethyl-2-methylbutane

$CH_2Br—C(CH_3)_2—CH_2—CH_2Br$
1,4-dibromo-2,2-dimethylbutane

34. The formula for dodecane is $C_{12}H_{26}$

35. $FCH_2CH_2F + Cl_2 \longrightarrow FCH_2CH_2FCl + HCl$

38. (a) elimination
(b) substitution
(c) addition

40. Propane is the most volatile.
Volatility: propane > butane > hexane

41. (a) The compounds are isomers.
(b) The compounds are not the same and are not isomers.
(c) The compounds are not the same and are not isomers.
(d) The compounds are not the same and are not isomers.

43. Using a high mole ratio of methane to chlorine will allow a chlorine free radical to react with a methane molecule rather than a chloromethane molecule and minimize the formation of di-, tri-, and tetrachloromethane.

Chapter 20

Review Questions

2. Cl(H)C=C(Cl)H — cis-1,2-dichloroethene; H(Cl)C=C(Cl)H — trans-1,2-dichloroethene; H–CHCl–CClH–H — 1,2-dichloroethane

We are able to get *cis-trans* isomers of ethene because there is no rotation around the carbon–carbon double bond, so the two structures shown are not the same. With 1,2-dichloroethane, there is free rotation of the carbon–carbon single bond. In the structure shown, exchanging the chlorine on the right with either hydrogen atom on that carbon appears to make a different isomer, but rotation makes any of the three positions equivalent.

7. Benzene does not undergo the typical reactions of an alkene. Benzene does not decolorize bromine rapidly, and it does not destroy the purple color of permanganate ions. The reactions of benzene are more like those of alkanes.

9. Polycyclic aromatic hydrocarbons are potent carcinogens.

10. Cyclohexane carbons form bond angles of about 109° *in a tetrahedron*; this causes the ring to be either a "boat" or a "chair" shape. Benzene carbons form bond angles of 120° *in a plane*. Therefore, the benzene molecule must be planar.

11. According to the mechanism, the addition of an unsymmetrical molecule such as HX adds to a carbon–carbon double bond. In the first step the H adds to the carbon of the carbon–carbon double bond that has the most hydrogen atoms on it, according to Markovnikoff's Rule. The second step completes the addition by adding the more negative element, X, of the HX.

Exercises

2. (a) propane H:C:C:C:H (each C with H above and below) (b) propene H:C:C::C:H (c) propyne H:C:C:::C:H

4. (a) C_3H_5Cl

$CH_3(H)C=C(H)Cl$ — *trans*-1-chloropropene; $CH_3(H)C=C(Cl)H$ — *cis*-1-chloropropene; $CH_3CCl{=}CH_2$ — 2-chloropropene; $CH_2ClCH{=}CH_2$ — 3-chloropropene

(b) chlorocyclopropane (cyclopropane ring with H and Cl on one carbon)

6. (a) $CH_2{=}CHC(CH_2CH_3)(CH_3)CH_2CH_3$

(b) H(C₆H₅)C=C(C₆H₅)H (cis)

(c) $CH{\equiv}CCH(C_6H_5)CH_3$

(d) cyclopentene

(e) 1-methylcyclohexene (CH_3 on the double-bond carbon)

(f) $CH(CH_3)_2$ on cyclopentene

8. (a) $CH_2{=}CHCH(CH_2CH_3)CH_3$ Longest chain contains five carbon atoms. Correct name: 3-methyl-1-pentene

(b) cyclohexene with Cl on the double-bond carbon. The C=C bond in cyclohexene is numbered so that substituted groups have the smallest numbers. Correct name: 1-chlorocyclohexene

(c) $CH_3(H)C=C(H)CH_2CH_2CH_3$ Numbering was started from the wrong end of the structure. Correct name: *trans*-2-hexene

10. (a) cis-4-methyl-2-hexene (b) 2,3-dimethyl-2-butene (c) 3-isopropyl-1-pentene

12. all the pentynes, C_5H_8

1-pentyne $CH_3CH_2CH_2C{\equiv}CH$; 2-pentyne $CH_3CH_2C{\equiv}CCH_3$; 3-methyl-1-butyne $CH_3CH(CH_3)C{\equiv}CH$

14. Only structure (c) will show *cis-trans* isomers: $CH_2ClCH{=}CHCH_2Cl$

16. (a) $CH_3CH_2CH_2CH{=}CH_2 + H_2O \xrightarrow{H^+} CH_3CH_2CH_2CH(OH)CH_3$

(b) $CH_3CH_2CH{=}CHCH_3 + HBr \longrightarrow CH_3CH_2CHBrCH_2CH_3 + CH_3CH_2CH_2CHBrCH_3$

(c) $CH_2{=}CHCl + Br_2 \longrightarrow CH_2BrCHClBr$

(d) $C_6H_5CH{=}CH_2 + HCl \longrightarrow C_6H_5CHClCH_3$

(e) $CH_2{=}CHCH_2CH_3 + KMnO_4 \xrightarrow[cold]{H_2O} CH_2(OH)CH(OH)CH_2CH_3$

18. (a) 1-butyne ↓ $CH{\equiv}CCH_2CH_3$ (b) propyne ↓ $CH{\equiv}CCH_3$ (c) 2-pentyne ↓ $CH_3C{\equiv}CCH_2CH_3$

20. (a) $CH_3C{\equiv}CH + H_2(1\ mol) \xrightarrow[1\ atm]{Pt,\ 25°C} CH_3CH{=}CH_2$

(b) $CH_3C{\equiv}CCH_3 + Br_2(2\ mol) \longrightarrow CH_3CBr_2CBr_2CH_3$

(c) two-step reaction:
$CH_3C{\equiv}CH + HCl \longrightarrow CH_3CCl{=}CH_2 \xrightarrow{HCl} CH_3CCl_2CH_3$

22. When cyclopentene, reacts with

(a) Cl_2 the product is 1,2-dichlorocyclopentane

(b) HBr the product is bromocyclopentane

(c) H_2, Pt the product is cyclopentane

(d) H_2O, H^+ the product is cyclopentanol

24. (a) $CH_3C{\equiv}CCH_2CH_2CH_3$ 2-hexyne
(b) $CH_3C(CH_3)_2C{\equiv}CH$ 3,3-dimethyl-1-butyne
(c) $HC{\equiv}CCH_2C(CH_3)_3$ 4,4-dimethyl-1-pentyne
(d) $CH_3C{\equiv}CCH_3$ 2-butyne

26. (a) $CH_3C{\equiv}CCH_2CH_3 + Cl_2(1\ mole) \longrightarrow CH_3CCl{=}CClCH_2CH_3$

$CH_3CCl{=}CClCH_2CH_3 + H_2 \xrightarrow[1\ atm]{25°C} CH_3CHClCHClCH_2CH_3$

(b) $CH_3C{\equiv}CCH_2CH_3 + HCl \longrightarrow CH_3CCl{=}CHCH_2CH_3$

$CH_3CCl{=}CHCH_2CH_3 + HCl \longrightarrow CH_3CCl_2CH_2CH_2CH_3$

Can also yield $CH_3CH_2CCl_2CH_2CH_3$ at the same time.

(c) $CH_3C{\equiv}CCH_2CH_3 + 2\ Cl_2 \longrightarrow CH_3CCl_2CCl_2CH_2CH_3$

28. (a) CH_3 (c) OH

(b) CH_2CH_3 (d)

30. (a) Cl, Cl, NO_2 (c) CH_3, CH

(b) NO_2, NO_2 (d) $CH{=}CH_2$

32. (a) trichlorobenzenes

1, 2, 3-trichlorobenzene 1, 2, 4-trichlorobenzene 1, 3, 5-trichlorobenzene

(b) the benzene derivatives of formula C_8H_{10}:

o-xylene or 1, 2-dimethylbenzene m-xylene or 1, 3-dimethylbenzene

p-xylene or 1, 4-dimethylbenzene ethylbenzene

34. The isomers that can be written substituting a third chlorine atom on *o*-dichlorobenzene are:

1, 2, 3-trichlorobenzene 1, 2, 4-trichlorobenzene

36. (a) styrene
(b) *m*-nitrotoluene
(c) 2,4-dibromobenzoic acid
(d) isopropylbenzene
(e) 2,4,6-tribromophenol

38. (a) 2,3-dibromophenol
(b) 1,3,5-trichlorobenzene
(c) *m*-iodotoluene

40. (a) $C_6H_6 + CH_3CHClCH_3 \xrightarrow{AlCl_3} C_6H_5CH(CH_3)CH_3 + HCl$

isopropylbenzene

(b) $C_6H_5CH_3 + KMnO_4 \xrightarrow{H_2O} C_6H_5COOH$

benzoic acid

44. (a) $\overset{+}{C}H_3$
(b) $CH_3CH_2\overset{+}{C}H_2$
(c) $CH_3C^+(CH_3)_2$
(d) $CH_3CH_2CH_2CH_2\overset{+}{C}H_2$

47. (a) $CH_3CHClCH_2CH_3 \xrightarrow{-HCl} CH_3CH{=}CHCH_3 + CH_2{=}CHCH_2CH_3$

(b) $CH_2ClCH_2CH_2CH_2CH_3 \xrightarrow{-HCl} CH_2{=}CHCH_2CH_2CH_3$

(c) 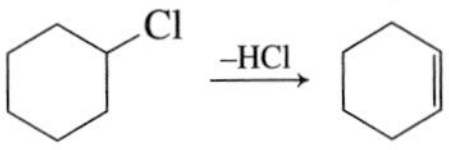 chlorocyclohexane $\xrightarrow{-HCl}$ cyclohexene

54. (a) Four isomers:

$CH_2{=}CHCH_2CH_3$ 1-butene

$CH_3CH{=}CHCH_3$ 2-butene (cis and trans)

$CH_2{=}C(CH_3)_2$ 2-methyl-1-propene

(b) One isomer

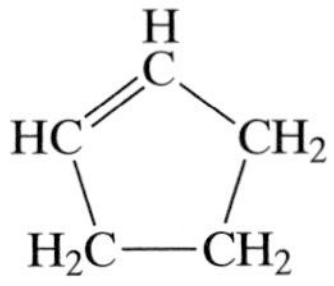

cyclopentene

56. Carbon-2 has two methyl groups on it. The configuration of two of the same groups on a carbon of a carbon–carbon double bond does not show cis-trans isomerism.

57. (a) alkyne or cycloalkene
(b) alkene
(c) alkane
(d) alkyne or cycloalkene

Chapter 21

Review Questions

1. An addition polymer is produced by the successive addition of a single monomer molecule. A condensation polymer is one that is formed from monomer molecules in a reaction that splits out water or some other small molecule. Condensation polymerization usually involves two different monomers.

2. Those polymers which soften on reheating are thermoplastic polymers; those which set to an infusible solid and do not soften on reheating are thermosetting polymers.

Exercises

2. polystyrene: $\left[CH_2{-}CH(C_6H_5)\right]_n$ C_8H_8

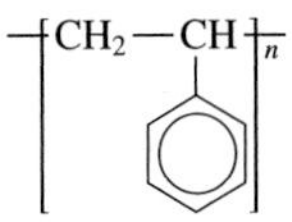

molar mass of 1 unit:
$8(12.01\ g/mol) + 8(1.008\ g/mol) = 104.1\ g/mol$

molar mass: $(3000\ units)\left(\frac{104.1\ g/mol}{unit}\right)$

$= 3 \times 10^5\ g/mol =$ molar mass

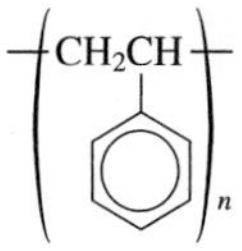

4. (a) Saran $\left[CH_2CCl_2\right]_n$
(b) Orlon $\left[CH_2CH(CN)\right]_n$
(c) Teflon $\left[CF_2CF_2\right]_n$
(d) Polystyrene $\left[CH_2CH(C_6H_5)\right]_n$

6. polyethylene free radical (2 units) starting with RO· free radical $ROCH_2CH_2CH_2CH_2\cdot$

8. (a) ethylene $\left[CH_2{-}CH_2\right]_n$
(b) chloroethene $\left[CH_2CHCl\right]$
(c) 1-butene $\left[CH_2CH(CH_2CH_3)\right]_n$

10. Two possible ways in which acrylonitrile can polymerize to form Orlon are:

$$\left[CH_2CH(CN)CH_2CH(CN)CH_2CH(CN)CH_2CH(CN)\right]_n$$

$$\left[CH_2CH(CN){-}CH(CN)CH_2CH_2CH(CN){-}CH(CN)CH_2\right]_n$$

12. Gutta percha (all trans)

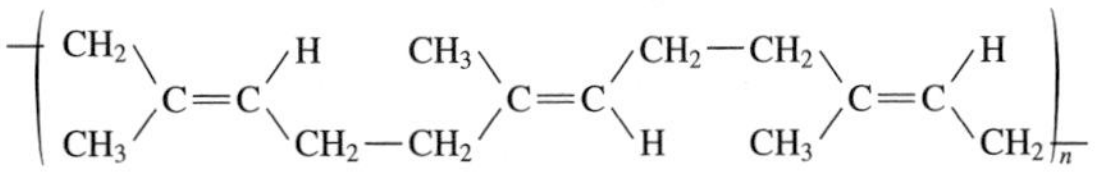

$$\left[CH_2(CH_3)C{=}CH{-}CH_2{-}CH_2(CH_3)C{=}CH{-}CH_2{-}CH_2(CH_3)C{=}CH{-}CH_2\right]_n$$

14. Yes, styrene—butadiene rubber contains carbon–carbon double bonds and is attacked by the ozone in smog, causing "age hardening" and cracking.

16. (a) synthetic natural rubber: $CH_2{=}C(CH_3)CH{=}CH_2$
(b) neoprene rubber: $CH_2{=}CCl{-}CH{=}CH_2$

18. (a) Isotactic polypropylene

$$\left[CH_2CH(CH_3)CH_2CH(CH_3)CH_2CH(CH_3)CH_2CH(CH_3)\right]_n$$

(b) Another form of polypropylene

$$\left[CH_2CH(CH_3)CH_2CH(CH_3)CH_2CH(CH_3)CH_2CH(CH_3)\right]_n$$

(other structures are possible)

20. The symbol, ♲3 V, signifies polyvinyl chloride. The monomer used to make this plastic is vinyl chloride, $CH_2{=}CHCl$.

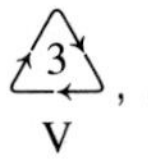

22. $\left[CH_2{-}CH_2\right]_n$

24. (a) one double bond $-CH_2CH{=}CHCH_2-$ (b) none

25. Polystyrene has the highest mass percent carbon.

$\mathrm{+CH_2-CH(C_6H_5)+_n}$ $\quad C_8H_8$ molar mass = 104.1 g/mol

$$\frac{96.08\text{ g C}}{104.1\text{ g}} \times 100\% = 92.30\%\text{ C}$$

Chapter 22

Review Questions

2. Alkenes are almost never made from alcohols because the alcohols are almost always the higher value material. This is because recovering alkenes from hydrocarbon sources in an oil refinery (primarily catalytic cracking) is a relatively cheap process.

4. Oxidation of primary alcohols yields aldehydes. Further oxidation yields carboxylic acids. Examples are:

$$CH_3CH_2OH + [O] \longrightarrow CH_3C(=O)-H + H_2O$$

$$CH_3C(=O)-H + [O] \longrightarrow CH_3C(=O)-OH$$

10. Low molar mass ethers present two hazards.
(a) They are very volatile and their highly flammable vapors form explosive mixtures with air.
(b) They also slowly react with oxygen in the air to form unstable explosive peroxides.

11. Ethanol (molar mass = 46.07) is a liquid at room temperature because it has a significant amount of hydrogen bonding between molecules and thus has a much higher boiling point than would be predicted from molar mass alone. Dimethyl ether (molar mass = 46.07) is not capable of hydrogen bonding to itself, so it has low attraction between molecules, making it a gas at room temperature.

Exercises

2. (a) $CH_3CH(OH)CH_2CH_3$
(b) $CH_3C(CH_3)(OH)CH_2CH_3$
(c) $CH_3CH(OH)CH_3$
(d) cyclopentane ring–OH
(e) $CH_3CH_2CH_2CH_2CH_2SH$
(f) $CH_3CH_2CH(OH)CH(CH_2CH_3)CH_2CH_3$

4. (a) C_3H_8O

Alcohols	*Ethers*
$CH_3CH_2CH_2OH$	$CH_3OCH_2CH_3$
$CH_3CH(OH)CH_3$	

(b) $C_4H_{10}O$

Alcohols	*Ethers*
$CH_3CH_2CH_2CH_2OH$	$CH_3OCH_2CH_2CH_3$
$CH_3CH_2CH(OH)CH_3$	$CH_3OCH(CH_3)_2$
$(CH_3)_2CHCH_2OH$	$CH_3CH_2OCH_2CH_3$
$(CH_3)_3C-OH$	

6. The primary alcohols in Exercise 4 are:
$CH_3CH_2CH_2OH$ $\quad CH_3CH_2CH_2CH_2OH$
$(CH_3)_2CHCH_2OH$

The secondary alcohols are:
$CH_3CH(OH)CH_3$ $\quad CH_3CH_2CH(OH)CH_3$

The tertiary alcohol is $(CH_3)_3C-OH$

8. (a) ethanol (ethyl alcohol)
(b) 2-phenylethanol
(c) 3-methyl-3-pentanol
(d) 1-methylcyclopentanol
(e) 3-pentanol
(f) 1,2-propanediol
(g) 4-ethyl-2-hexanol

10. (a) cyclopentene ring with CH_3 on the double-bond carbon — 1-methylcyclopentene
(b) $CH_3CH{=}CHCH_3$ 2-butene
(c) $CH_3C(CH_3){=}CHCH_2CH_3$ 2-methyl-2-pentene

12. (a) cyclopentane ring–OH — cyclopentanol
(b) cyclopentane ring with CH_3 and OH on the same carbon — 1-methylcyclopentanol
(c) $CH_3CH(CH_3)CH(OH)CH(CH_3)CH_3$ 2,4-dimethyl-3-pentanol

14. Oxidation of a secondary alcohol:

$$CH_3CH(OH)CH(CH_3)CH_2CH_3 \longrightarrow CH_3C(=O)CH(CH_3)CH_2CH_3$$

ketone

16. (a) $CH_3CH{=}CHCH_3 + H_2O \xrightarrow{H^+} CH_3CH_2CH(OH)CH_3$

(b) $CH_3CH_2CH_2CH{=}CH_2 + H_2O \xrightarrow{H^+} CH_3CH_2CH_2CH(OH)CH_3$

(c) $CH_3C(CH_3){=}CHCH_3 + H_2O \xrightarrow{H^+} CH_3C(OH)(CH_3)CH_2CH_3$

18. (a) $CH_3CH(OH)CH_2CH_3$ 2-butanol

(b) $CH_3CH(OH)CH(CH_2CH_3)CH_2CH_3$ 3-ethyl-2-pentanol

(c) 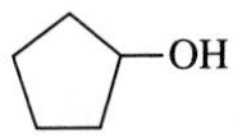 cyclopentyl–OH cyclopentanol

20. (a) $2\ CH_3CH_2OH + 2\ Na \longrightarrow 2\ CH_3CH_2O^-Na^+ + H_2$

(b) $CH_3CH_2CH_2CH_2OH \xrightarrow[H_2SO_4]{K_2Cr_2O_7} CH_3CH_2CH_2C(=O){-}H$

(c) 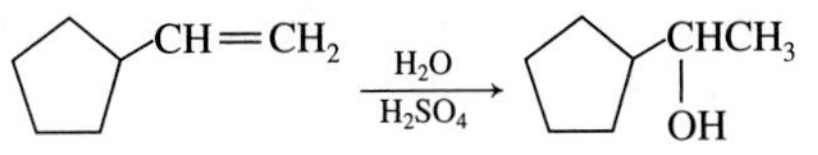 cyclopentyl–$CH{=}CH_2 \xrightarrow[H_2SO_4]{H_2O}$ cyclopentyl–$CH(OH)CH_3$

(d) $CH_3CH_2C(=O){-}OCH_3 + NaOH \longrightarrow$

$CH_3CH_2C(=O){-}O^-Na^+ + CH_3OH$

22. (a) 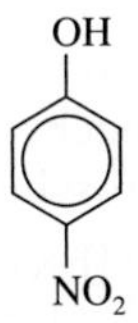 4-nitrophenol (OH and NO_2 para on benzene ring)

(b) 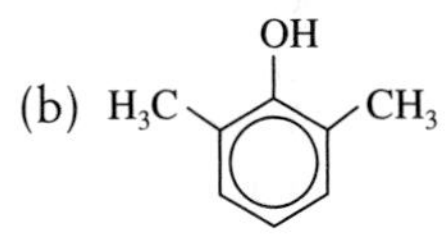 2,6-dimethylphenol (H_3C, OH, CH_3 on benzene ring)

(c) 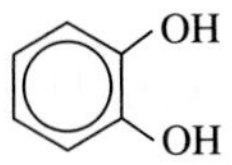 1,2-dihydroxybenzene (two OH ortho on benzene ring)

24. (a) Methanol, $CH_3{-}OH$

(b) 2,2-dimethyl-1-propanol

$CH_3{-}C(CH_3)_2{-}CH_2{-}OH$

(c) 3,3-dimethyl-1-pentanol

$CH_3{-}CH_2{-}C(CH_3)_2{-}CH_2CH_2{-}OH$

26. (a) cyclopentanol
(b) 2-propanol
(c) 2-methyl-2-propanol

28. (a) *p*-dihydroxybenzene (hydroquinone)
(b) 2,4-dimethylphenol
(c) 2,4-dinitrophenol
(d) *m*-hexylphenol

30. Order of decreasing solubility in water: [a (highest), d, c, b (lowest)]
(a) $CH_3CH(OH)CH(OH)CH_2OH$
(d) $CH_3CH(OH)CH_2CH_2OH$
(c) $CH_3CH_2CH_2CH_2OH$
(b) $CH_3CH_2OCH_2CH_3$

32. The six isomeric ethers of $C_5H_{12}O$ are:

$CH_3OCH_2CH_2CH_2CH_3$
methyl *n*-butyl ether
(1-methoxybutane)

$CH_3OCH(CH_3)CH_2CH_3$
methyl sec-butyl ether
(2-methoxybutane)

$CH_3OCH_2CH(CH_3)CH_3$
methyl isobutyl ether
(1-methoxy-2-methylpropane)

$CH_3OC(CH_3)_2CH_3$
methyl *t*-butyl ether
(2-methoxy-2-methylpropane)

$CH_3CH_2OCH_2CH_2CH_3$
ethyl propyl ether
(1-ethoxypropane)

$CH_3CH_2OCH(CH_3)CH_3$
ethyl isopropyl ether
(2-ethoxypropane)

34. $CH_3CH_2OCH_2CH_3 + 6\ O_2 \xrightarrow{\Delta} 4\ CO_2 + 5\ H_2O$

36. These are possible combinations of reactants to make the following ethers by the Williamson syntheses:
(a) $CH_3CH_2CH_2OCH_2CH_2CH_3$
$CH_3CH_2CH_2ONa + CH_3CH_2CH_2Cl$

(b) $HC(CH_3)_2OCH_2CH_2CH_3$

$HC(CH_3)_2ONa + CH_3CH_2CH_2Cl$ or $CH_3CH_2CH_2ONa + CH_3CH(Cl)CH_3$

(c) 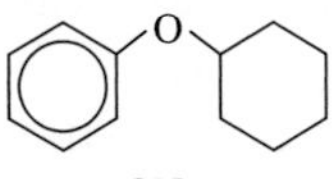 phenyl–O–cyclohexyl

phenyl–ONa + cyclohexyl–Br

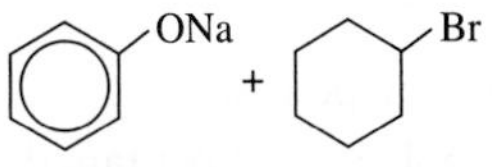

Cannot use a 2° alkyl halide in the Williamson synthesis.

38. IUPAC names
(a) 2-methylcyclopentanethiol
(b) 2,3-dimethyl-2-pentanethiol
(c) 2-propanethiol
(d) 3-ethyl-2,2,4-trimethyl-3-pentanethiol

40. (a) SH (on cyclohexane ring), CH_2CH_3

(b) $CH_3CH_2CHCH_2CH_3$ with SH on C-3

(c) $CH_3CH_2CCH_2SH$ with CH_3 above and CH_3 below C-3

42. Structures and correct names
(a) $CH_3CH_2CHCHCH_3$ (CH_3 on C-3, SH on C-2) 3-methyl-2-pentanethiol
(b) $CH_3CH_2CH{-}OCH_3$ (CH_3 below) 2-methoxybutane
(c) $HOCH_2CH_2OH$ 1,2-ethanediol
(d) OH, CH_3, CH_2CH_3 (on benzene ring) 3-ethyl-2-methylphenol

43. The name is 4-hexylresorcinol or 4-hexyl-1,3-dihydroxybenzene.

44.

OH OH — cis-1,2-cyclopentanediol OH, OH — trans-1,2-cyclopentanediol

46. The common phenolic structure in the catecholamines is catechol, *o*-dihydroxybenzene:

OH, OH (on benzene ring) catechol

47. Methyl alcohol is converted to formaldehyde by an oxidation reaction.

53. Only compound (a), phenol, will react with NaOH.

$$C_6H_5OH + NAOH \longrightarrow C_6H_5ONa + H_2O$$

54. Order of increasing boiling points.
1-pentanol < 1-octanol < 1,2-pentanediol
138°C 194°C 210°C

All three compounds are alcohols. 1-pentanol has the lowest molar mass and hence the lowest boiling point. 1-octanol has a higher molar mass and therefore a higher boiling point than 1-pentanol. 1,2-pentanediol has two —OH groups and therefore forms more hydrogen bonds than the other two alcohols, which causes its higher boiling point.

Chapter 23

Review Questions

2. Aldehydes and ketones have lower boiling points than alcohols of similar molar mass because they do not form hydrogen bonds, since they contain no —OH groups.

7. (a) When an aldehyde reacts with Benedict solution, the blue color of copper(II) ion disappears and a red precipitate of Cu_2O forms.
(b) When an aldehyde reacts with Tollens solution, the silver ion in solution forms a silver metal mirror on the inside of the glass container used in the test.

10. MEK is an abbreviation for methyl ethyl ketone. Its main use is as a solvent, especially for lacquers and paints.

11. A ketone group, $>C{=}O$, cannot be located at the end of a carbon–carbon chain. Consequently its only possible location in both propanone and butanone is on C—2 of these ketones; therefore, its location need not be numbered.

Exercises

2. (a) ethanal, acetaldehyde
(b) butanal
(c) benzaldehyde
(d) 2-chloro-5-isopropylbenzaldehyde
(e) 3-hydroxybutanal

4. (a) 2-butanone, methyl ethyl ketone (MEK)
(b) 3,3-dimethylbutanone, *t*-butyl methyl ketone
(c) 2,5-hexanedione
(d) 1-phenyl-2-propanone, benzyl methyl ketone

6. (a) $HOCH_2CH_2C({=}O){-}H$

(b) $CH_3CH_2C({=}O)CHCH_2CH_3$ with CH_3 on C-4

(c) cyclohexanone (ring with =O)

(d) $CH_3CHCH_2CHCH_2CHC({=}O){-}H$ with Cl, Cl, Cl

(e) $CH_3CH{=}CHCH_2C({=}O){-}H$

8. Structures and correct names

(a) $CH_3CCl_2CH_2CH_2CH_2C(=O)CH_3$

6,6-dichloro-2-heptanone

(b) $CH_3CH_2CH(CH_2CH_3)CH_2C(=O){-}H$ 3-ethylpentanal

(c) $CH_3CH_2CH_2CH_2CH_2C(=O){-}H$

hexanal

10. Higher boiling point

(a) pentanal
(b) benzyl alcohol
(c) 2-hexanone
(d) 1-butanol

12. Higher aqueous solubility

(a) acetaldehyde
(b) 2,4-pentanediol
(c) propanal

14. (a) 1-propanol:

$$CH_3CH_2CH_2OH \xrightarrow[H_2SO_4]{K_2Cr_2O_7} CH_3CH_2C(=O){-}H \text{ or } CH_3CH_2COOH$$

$$CH_3CH_2CH_2OH + O_2 \xrightarrow[\Delta]{Cu} CH_3CH_2C(=O){-}H$$

(b) 2,3-dimethyl-2-butanol:

$$CH_3CH(CH_3)C(OH)(CH_3)CH_3 + O_2 \xrightarrow[\Delta]{Ag} \text{No reaction (3° alcohol) with either oxidizing agent}$$

16. (a) $C_6H_{11}{-}C(=O){-}O^-$ (cyclohexyl)

(b) $^-O{-}C(=O){-}CH_2CH_2{-}C(=O){-}O^-$

(c) $CH_3{-}CH_2CH_2{-}C(=O){-}OH$

18. (a) formaldehyde (methanal) $CH_2{=}O$

(b) propanal $CH_3{-}CH_2{-}CH{=}O$

(c) propanal $CH_3{-}CH_2{-}CH{=}O$

20. (a) An aldehyde group, $-C(H){=}O$ must be present to give a positive Fehling test.

(b) The visible evidence for a positive Fehling test is the formation of brick-red Cu_2O, which precipitates during the reaction.

(c) $$CH_3C(H){=}O + 2\,Cu^{2+} \xrightarrow[H_2O]{NaOH} CH_3COONa + Cu_2O(s) \text{ (brick red)}$$

22. (a) propanal $\longrightarrow$ $CH_3CH_2COONa + Cu_2O(s)$

(b) acetone $\longrightarrow$ no reaction

(c) 3-methylpentanal $\longrightarrow$ $CH_3CH_2CH(CH_3)CH_2COONa + Cu_2O(s)$

24. (a) 3-methylbutanal $HC(=O){-}CH_2{-}CH(CH_3){-}CH_3$

(b) cyclopentanone (cyclopentane ring with =O)

(c) 2,2-dimethylpropanal $CH_3{-}C(CH_3)_2{-}CH{=}O$

26. (a) 3-pentanone:

$$2\,CH_3CH_2C(=O)CH_2CH_3 \xrightarrow[NaOH]{dilute} CH_3CH_2C(OH)(CH_2CH_3){-}CH(CH_3){-}C(=O)CH_2CH_3$$

(b) propanal:

$$2\,CH_3CH_2C(H){=}O \xrightarrow[NaOH]{dilute} CH_3CH_2CH(OH)CH(CH_3)C(H){=}O$$

28. (a) $$CH_3CH_2C(H){=}O + CH_3CH_2CH_2OH \overset{dry\ HCl}{\rightleftharpoons} CH_3CH_2C(OH)(H){-}OCH_2CH_2CH_3 + H_2O$$

(b) cyclohexanone $+ CH_3OH \overset{H^+}{\rightleftharpoons}$ 1-methoxycyclohexanol (ring bearing OH and OCH_3)

(c) $$CH_3CH_2CH_2CH(OCH_3)_2 \underset{H^+}{\overset{H_2O}{\rightleftharpoons}} CH_3CH_2CH_2C(H){=}O + 2\,CH_3OH$$

30. (a) C_6H_5CHO (H, C=O on benzene ring) + HCN $\xrightarrow{OH^-}$ $C_6H_5CH(OH)$—CN

(b) $C_6H_5CH(OH)$—CN + H_2O $\xrightarrow{H^+}$ $C_6H_5CH(OH)COOH$ (OH, CHCOOH)

(c) $C_6H_5CH(OH)COOH$ (OH, CHCOOH) $\xrightarrow[H_2SO_4]{K_2Cr_2O_7}$ $C_6H_5C(=O)$—COOH

31. Cyclic hemiacetal: five-membered ring with O, and H and OH on the carbon adjacent to O.

33. Four aldol condensation products from a mixture of ethanal (E) and propanal (P) are possible: EE PP EP and PE.

EE is $CH_3CH(OH)CH_2C(=O)$—H PP is $CH_3CH_2CH(OH)CH(CH_3)C(=O)$—H

EP is $CH_3CH(OH)CH(CH_3)C(=O)$—H PE is $CH_3CH_2CH(OH)CH_2C(=O)$—H

38. Pyruvic acid is changed to lactic acid by a reduction reaction.

39. Alcohols that should be oxidized to give the ketones listed.

(a) 3-pentanol

(b) 2-butanol

(c) 4-phenyl-2-butanol

Chapter 24

Review Questions

4. Unlike alkanes, carboxylic acids hydrogen bond tightly to each other. Thus, they have much higher boiling points than alkanes.

9. Aspirin acts in the body as an antipyretic, an analgesic, and as an anti-inflammatory agent.

11. Phosphoric acid anhydrides are storage molecules for metabolic energy for biological reactions.

14. Both compounds have a hydrophilic structural component. Capric acid contains a nine carbon hydrocarbon chain which decreases its solubility in water. Formic acid does not have a hydrocarbon chain.

Exercises

2. (a) *o*-chlorobenzoic acid (d) 2-hydroxybutanoic acid
(b) oleic acid (e) ammonium propanoate
(c) *m*-toluic acid

4. (a) COOH—COOH
(b) $CH_3CH_2CH_2CH_2COOH$
(c) benzene ring with two adjacent COOH groups
(d) $CH_3CH_2CH{=}CHCH_2CH{=}CHCH_2CH{=}CH(CH_2)_7COOH$
(e) COONa on benzene ring, NH_2 para (g) $CH_3CH(OH)CH_2COOH$
(f) $CH_3CH_2COONH_4$

6. Decreasing pH means increasing acidity:

KOH < NH_3 < KBr < C_6H_5OH (phenol) < HCOOH < HBr

8. (a) ethanoic acid (c) ethanoic acid
(b) propanoic acid (d) malonic acid

10. (a) methyl methanoate methyl formate
(b) propyl benzoate propyl benzoate
(c) ethyl propanoate ethyl propionate

12. (a) $CH_3C(=O)$—$OCH_2CH_2CH_3$ (b) $C_6H_5C(=O)$—OCH_3
(c) $CH_3CH_2CH_2CH_2CH_2C(=O)$—$OCH_2CH_3$

14. (a) benzene ring with COONa groups in meta positions disodium-*m*-phthalate
(b) $HOOCCH_2CH_2COOH$ succinic acid
(c) $CH_3(CH_2)_7CH{=}CH(CH_2)_7COONa$ + H_2O
sodium oleate
(d) $CH_3CH_2CH_2COOH$ butanoic acid
(e) C_6H_5COONa + C_6H_5ONa
sodium benzoate sodium phenolate

16. (a) Compounds are different: methyl benzoate and phenyl acetate
(b) Compounds are different: methyl acetate and methyl methanoate
(c) Compounds are the same

18. (a) $CH_3CH_2CH_2CH(CH_3)COOH$ 2-methylpentanoic acid

(b) $CH_3C(=O){-}OCH_2CH_3$ ethyl ethanoate

(c) $CH_2ClCH_2CH_2COOH$ 4-chlorobutanoic acid

20. (a) methanol and propanoic acid: $CH_3CH_2C(=O){-}OCH_3$

(b) 1-octanol and acetic acid: $CH_3C(=O){-}OCH_2(CH_2)_6CH_3$

(c) ethanol and butanoic acid: $CH_3CH_2CH_2C(=O){-}OCH_2CH_3$

22. (a) isopropyl formate: $CH_3CH(OH)CH_3 + HCOOH$

(b) diethyl adipate: $CH_3CH_2OH + HOOC(CH_2)_4COOH$

(c) benzyl benzoate: $C_6H_5CH_2OH + C_6H_5COOH$

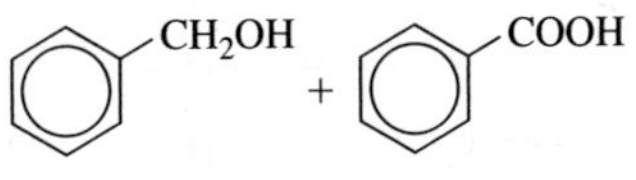

24. (a) $C_6H_5C(=O){-}Cl + NH_3 \longrightarrow C_6H_5C(=O){-}NH_2$

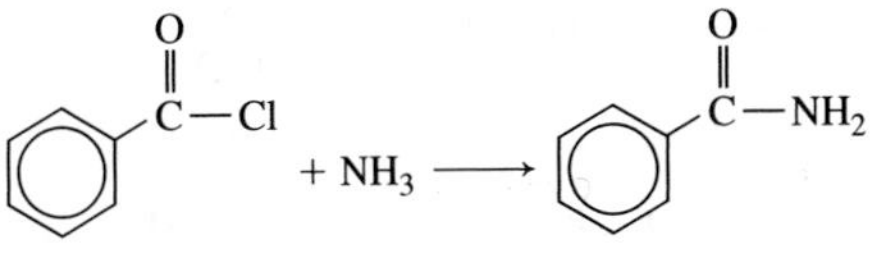

(b) $CH_3CH_2COOH + SOCl_2 \longrightarrow CH_3CH_2C(=O){-}Cl$

(c) $CH_3CH_2C(=O){-}Cl + CH_3OH \longrightarrow CH_3CH_2C(=O){-}OCH_3$

(d) $CH_3CH_2CH_2C{\equiv}N + H_2O \xrightarrow{H^+} CH_3CH_2CH_2COOH$

(e) $C_6H_5CH_2CH_2COOH + SOCl_2 \longrightarrow C_6H_5CH_2CH_2C(=O){-}Cl$

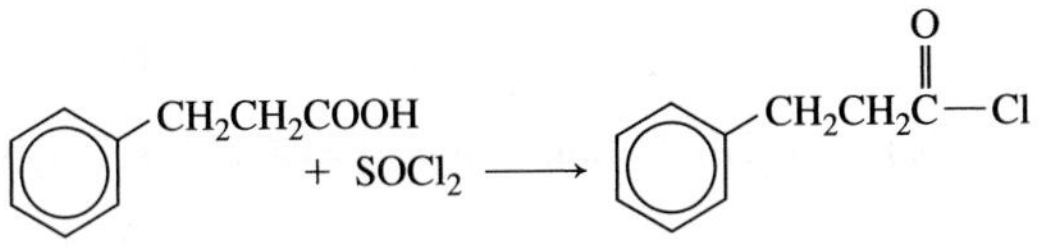

26. (a) Benzoic acid and ethyl benzoate: benzoic acid is an odorless solid; ethyl benzoate is a fragrant liquid.
(b) Succinic acid and fumaric acid: fumaric acid has a carbon–carbon double bond and will readily add and decolorize bromine; succinic acid will not decolorize bromine.

28. (a) Cyclic anhydride ring: $H_2C{-}C(=O){-}O{-}C(=O){-}CH_2{-}CH_2$ glutaric anhydride

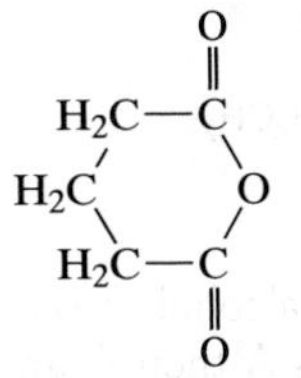

(b) $CH_2(COOCH_2CH_3)_2$ diethyl malonate

(c) C_6H_5OH + CH_3COOH
phenol acetic acid

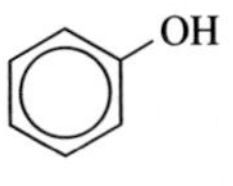

30. (a) cis-oleic acid: $CH_3(CH_2)_7CH{=}CH(CH_2)_7COOH$ (cis, H and H on the same side)

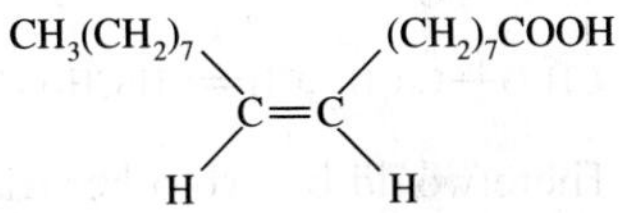

(b) cis,cis-linoleic acid: $CH_3(CH_2)_4CH{=}CHCH_2CH{=}CH(CH_2)_7COOH$ (both double bonds cis)

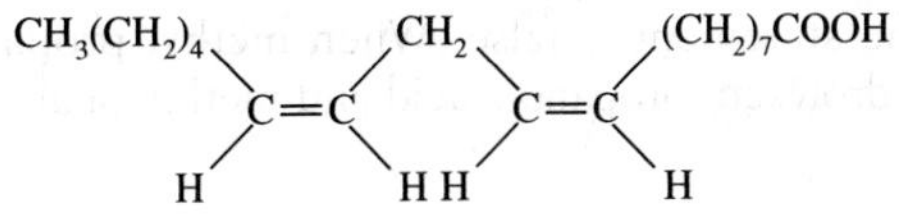

32. (a) $CH_3{-}C(=O){-}OH + HCl$

(b) $HC(=O){-}O^- K^+ + HOCH_2CH_3$

(c) $CH_3{-}CH(OH)CH_3 + Na^{+}\,{}^{-}O{-}C(=O){-}CH_3$

34. Sodium lauryl sulfate would be more effective as a detergent in hard water than sodium propyl sulfate because the hydrocarbon chain is only three carbons long in the latter, not long enough to dissolve grease well.

36. Only (c), $CH_3(CH_2)_{10}CH_2O(CH_2CH_2O)_7CH_2CH_2OH$, would be a good detergent in water. It is nonionic.

38. (a) $CH_3O{-}P(=O)(OCH_3){-}OCH_3 + 3\,H_2O$

(b) $CH_3O{-}P(=O)(OH){-}OCH(CH_3)CH_2CH_3$

41. The ester is propyl propanoate

$$CH_3CH_2\overset{\overset{\displaystyle O}{\|}}{C}{-}OCH_2CH_2CH_3$$

Compound A is an acid; B is an alcohol. If B is oxidized to an acid which is the same as A then A and B both have the same carbon structure, three carbon atoms each. The acid A must be propanoic acid, CH_3CH_2COOH. The alcohol can be 1-propanol or 2-propanol. Only 1-propanol can be oxidized to propanoic acid which is the same as compound A.

44.

$$\begin{array}{l} CH_2O{-}\overset{O}{\overset{\|}{C}}(CH_2)_{10}CH_3 \\ |\\ CHO{-}\overset{O}{\overset{\|}{C}}(CH_2)_{14}CH_3 \\ |\\ CH_2O{-}\overset{O}{\overset{\|}{C}}(CH_2)_7CH{=}CH(CH_2)_7CH_3 \end{array}$$

There would be two other triacylglycerols containing all three of these acids. Each of the three acids can be at the middle position.

53. The statement is false. When methyl propanoate is hydrolyzed, propanoic acid and methanol are formed.

Chapter 25

Review Questions

2. Amides: Unsubstituted amides (except formamide) are solids at room temperature. Many are odorless and colorless. Low molar-mass amides are water soluble. Solubility in water decreases as the molar mass increases. Amides are neutral compounds. The NH_2 group is capable of hydrogen bonding. Amines: Low molar-mass amines are flammable gases with an ammonia-like odor. Aliphatic amines up to six carbon atoms are water soluble. Many amines have a "fishy" odor and many have very foul odors. Aromatic amines occur as liquids and solids. Soluble aliphatic amines give basic solutions. Aromatic amines are less soluble in water and less basic than aliphatic amines. The NH_2 group is capable of hydrogen bonding.

7. Ammonia is a toxic, basic, water soluble compound which can increase the pH of the blood and the urine and would be painful to pass through bodily tissues. However, ammonia is converted in the liver to the neutral diamide, urea, which is water soluble and is excreted in the urine.

9. The nitrogen in a compound that has four groups bonded to it is positively charged and is called a quaternary ammonium nitrogen. The compound is called a quaternary ammonium salt.

Exercises

2. (a) $CH_3\underset{\underset{\displaystyle NH_2}{|}}{C}HCH_2CHClCH_3$

(b) $C_6H_5{-}\overset{}{\underset{\underset{\displaystyle O}{\|}}{C}}{-}N(CH_2CH_3)_2$

(c) NH_2 on benzene ring with Cl in the meta position (3-chloroaniline)

4. (a) 1-amino-3-methylbutane
(b) formamide
(c) *N*-phenylpropanamide
(d) ethyldiisopropylamine
N-ethyl-*N*-isopropyl-2-propanamine

6. Increasing solubility in water

$$C_6H_5\overset{\overset{\displaystyle O}{\|}}{C}{-}NH_2 < CH_3(CH_2)_4\overset{\overset{\displaystyle O}{\|}}{C}{-}NH_2 < CH_3\overset{\overset{\displaystyle O}{\|}}{C}{-}NH_2$$

8. Several possibilities:

$$O{=}C\begin{array}{l} \diagup N(H){-}H\cdots\ddot{O}(H){-}H \\ \diagdown N(H){-}H\cdots\ddot{O}(H){-}H \end{array} \qquad O{=}C\begin{array}{l} \diagup NH_2\cdots\ddot{O}(H){-}H \\ \diagdown NH_2\cdots\ddot{O}(H){-}H \end{array}$$

10. (a) neither acid nor base
(b) base
(c) base
(d) neither acid nor base
(e) neither acid nor base
(f) acid

12. (a) 3-ethylaniline $C_6H_4(CH_2CH_3)(NH_2)$

(b) methanamine $CH_3{-}NH_2$

(c) 2-methylbutanamine $CH_3CH_2\overset{\overset{\displaystyle CH_3}{|}}{C}HCH_2NH_2$

(d) N,N-dimethylethanamine $CH_3{-}\overset{\overset{\displaystyle CH_3}{|}}{N}{-}CH_2CH_3$

14. Organic products

(a) $CH_3\overset{\overset{\displaystyle O}{\|}}{C}{-}N(CH_2CH_3)_2$

(b) $C_6H_5CH_2COOH + CH_3\overset{+}{N}H_2CH_3$

(c) $CH_3CH(CH_3)CH_2C(=O)-NHCH_3$

16. (a) $CH_3C(=O)-NHCH_2CH_3 + NaOH$

(b) $CH_3CH_2C(=O)-OH + CH_3NHCH_3$

(c) $CH_3C(=O)-NH_2 + HCl + H_2O$

18. Structures of amines with formula C_3H_9N:

$CH_3CH_2CH_2NH_2$	$CH_3CH(NH_2)CH_3$	$CH_3CH_2NHCH_3$	$CH_3N(CH_3)CH_3$
1°	1°	2°	3°

20. (a) secondary (b) tertiary (c) secondary (d) tertiary

22. The isopropylamine solution in 1.0 *M* KOH would have the more objectionable odor because it would be in the form of the free amine, while in the acid solution the amine would form a salt that will have little or no odor.

24. (a) N-ethylaniline
(b) 2-aminobutane
(c) diphenylamine
(d) ethylammonium bromide
(e) pyridine
(f) N-ethylacetamide

26. (a) $CH_3CH_2CH_2CH_2N(CH_2CH_2CH_2CH_3)CH_2CH_2CH_2CH_3$

(b) $C_6H_5\overset{+}{N}H_2(CH_3)Cl^-$

(c) $CH_3CH_2NH_3^+Cl^-$

(d) $CH_3CH_2CH_2CH(NH_2)CH_2OH$

(e) $CH_3CH_2CH_2C(CH_3)_2-CH(CH_3)-NH-C_6H_5$

(f) $H_2NCH_2CH_2NH_2$

28. (a) $CH_3CH_2C(=O)-NH_2$ (b) $CH_3CH_2CH_2CH_2NH_2$

(c) $H_3C-C_6H_4-NH_2$ (para) (d) $C_5H_{10}N-C(=O)-CH_3$ (N-acetylpiperidine)

30. (a) $(CH_3CH_2CH_2)_2NH$ N-propylpropanamine
(b) $CH_3CH_2CH_2NHCH_3$ N-methylpropanamine
(c) $CH_3CH(NH_2)CH_2CH(CH_2CH_3)CH_3$ 4-methyl-2-hexanamine

32. *Amphetamines:* stimulate the central nervous system; used to treat depression, narcolepsy, and obesity
Tranquilizers: used to modify psychotic behavior and relieve pressure and anxiety
Antibacterial agents: used as antibiotics

36. (thiophene)$-CH_2-C(=O)-NH-$(β-lactam/dihydrothiazine ring system)$-CH_2O-C(=O)CH_3$; labels: amide (side-chain C(=O)–NH), amide (β-lactam ring), ester (CH_2O-CCH_3)

Chapter 26

Review Questions

4. Enantiomers are nonsuperimposable mirror-image isomers. Diastereomers are stereoisomers that are not enantiomers (not mirror-image isomers).

8. A meso compound contains chiral carbons but is not optically active. A racemic mixture contains two optically active compounds that exactly cancel out each other's optical activity (the mixture is not optically active).

9. The enantiomer will have a specific rotation of −92°.

Exercises

2. Diastereoisomers do not have identical physical properties, so the differences form a basis for chemical or physical separation. Differences of boiling point, freezing point, and solubilities are most commonly used.

4. (a) a wood screw, (c) the letter g, (d) this textbook

6. Number of chiral carbon atoms
(a) 0 (b) 0 (c) 3 (d) 2

8. (a), (b), (c) will show optical activity.

10. Fructose, which has three chiral carbon atoms, will have eight possible stereoisomers. This can be determined from $2^n = 2^3 = 8$.

12. The two projection formulas (A) and (B) are the same compound, for it takes two changes to make (B) identical to (A).

(A): Br—C(H)(CH$_3$)—F (H above, CH$_3$ below)
(B): Br—C(CH$_3$)(F)—H (CH$_3$ above, F below)
1st change in (B) (H and CH$_3$): Br—C(H)(F)—CH$_3$ (H above, F below)
2nd change in (B) (F and CH$_3$): Br—C(H)(CH$_3$)—F (H above, CH$_3$ below)

14. (+)-alanine: COOH top; H$_2$N left, H right; CH$_3$ bottom
(−)-alanine: COOH top; H left, NH$_2$ right; CH$_3$ bottom

(b) NH$_2$ top; H left, COOH right; CH$_3$ bottom
(a) CH$_3$ top; H$_2$N left, H right; COOH bottom
(c) COOH top; H left, CH$_3$ right; NH$_2$ bottom
(e) COOH top; CH$_3$ left, H right; NH$_2$ bottom
(d) CH$_3$ top; H left, NH$_2$ right; COOH bottom
(f) H top; CH$_3$ left, NH$_2$ right; COOH bottom

16.
CH$_3$ / H$_2$N—H / H—OH / COOH
CH$_3$ / H—NH$_2$ / HO—H / COOH

18.
CH$_2$OH / Cl—H / H—Br / H—Cl / COOH

20. (a) 2,3-dichlorobutane

CH$_3$ / H—Cl / Cl—H / CH$_3$ and CH$_3$ / Cl—H / H—Cl / CH$_3$: enantiomers
CH$_3$ / H—Cl / H—Cl / CH$_3$: meso compound

(b) 2,4-dibromopentane

CH$_3$ / H—Br / H—H / Br—H / CH$_3$ and CH$_3$ / Br—H / H—H / H—Br / CH$_3$: enantiomers
CH$_3$ / H—Br / H—H / H—Br / CH$_3$: meso compound

(c) 3-hexanol

CH$_2$CH$_3$ / H—OH / CH$_2$CH$_2$CH$_3$ and CH$_3$CH$_2$ / HO—H / CH$_3$CH$_2$CH$_2$
enantiomers

22. All the stereoisomers of 3,4-dichloro-2-methylpentane:

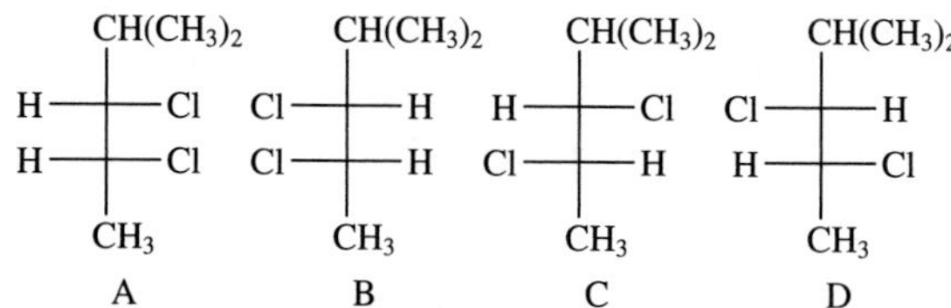

Compounds A and B and C and D are pairs of enantiomers.There are no meso compounds. Pairs of diastereomers are A and C, A and D, B and C, and B and D.

24. The four stereoisomers of 2-chloro-3-hexene:

CH$_3$ / H—C—Cl / H—C / ‖ / H—C / CH$_2$CH$_3$ — cis
CH$_3$ / Cl—C—H / H—C / ‖ / H—C / CH$_2$CH$_3$ — cis
CH$_3$ / H—C—Cl / H—C / ‖ / C—H / CH$_2$CH$_3$ — trans
CH$_3$ / Cl—C—H / H—C / ‖ / C—H / CH$_2$CH$_3$ — trans

The two cis compounds are enantiomers and the two trans compounds are enantiomers.

26. (a) $CH_3CH_2CHBr_2$ (i) $CH_3CHBrCH_2Br$ (ii)
$CH_3CBr_2CH_3$ (iii) $CH_2BrCH_2CH_2Br$ (iv)

(b) (ii) is chiral
CH$_2$Br / H—Br / CH$_3$ and CH$_2$Br / Br—H / CH$_3$ enantiomers

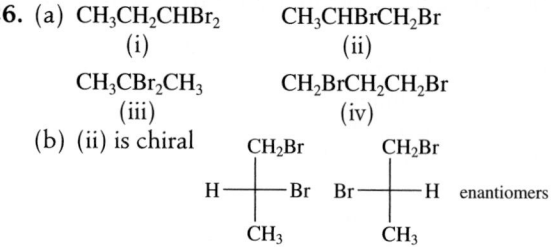

(i), (iii), and (iv) are achiral; there are no meso compounds.

28. Assume (+)-2-chlorobutane is CH$_3$—CH$_2$—C(H)(Cl)—CH$_3$ (CH$_3$ / CH$_2$ / H—C—Cl / CH$_3$)

All possible isomers formed when (+)-2-chlorobutane is further chlorinated to dichlorobutane are:

A: CH$_3$ / CH$_2$ / H—C—Cl / CH$_2$Cl
B: CH$_2$Cl / CH$_2$ / H—C—Cl / CH$_3$
C: CH$_3$ / CH$_2$ / Cl—C—Cl / CH$_3$
D: CH$_3$ / H—C—Cl / H—C—Cl / CH$_3$
E: CH$_3$ / Cl—C—H / H—C—Cl / CH$_3$

Compounds A, B, and E would be optically active: C does not have a chiral carbon atom; D is a meso compound.

30. If 1-chlorobutane and 2-chlorobutane were obtained by chlorinating butane, and then distilled, they would be separated into the two fractions, because their boiling points are different. 1-Chlorobutane has no chiral carbon, so would not be optically active. 2-Chlorobutane would exist as a racemic mixture (equal quantities of enantiomers) because substitution of Cl for H on carbon-2 gives equal amounts of the two enantiomers. Distillation would not separate the enantiomers because their boiling points are identical. The optical rotation of the two enantiomers of the 2-chlorobutane fraction would exactly cancel, and thus would not show optical activity.

32. Compound (d) is meso.

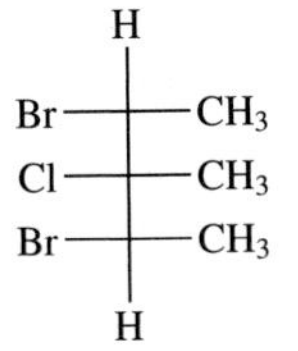

H
Br — CH₃
Cl — CH₃
Br — CH₃
H

34. (c) is chiral.

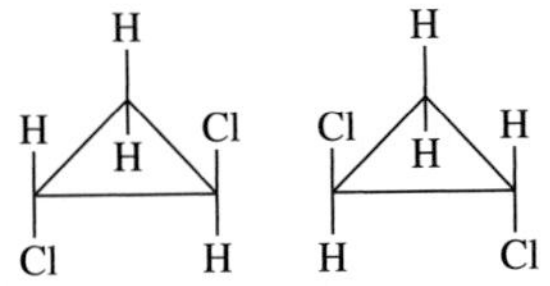

enantiomers

36. (a) diastereomers (b) enantiomers (c) diastereomers.

38. $(CH_3)_2CHCH_2$–C₆H₄–C(CH₃)(H)–COOH and HOOC–C(CH₃)(H)–C₆H₄–$CH_2CH(CH_3)_2$ (chiral carbon)

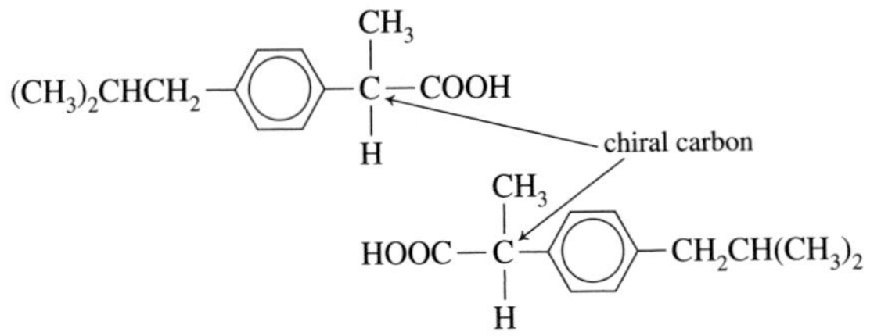

44. Stereoisomer structures

(a) 2-bromo-3-chlorobutane

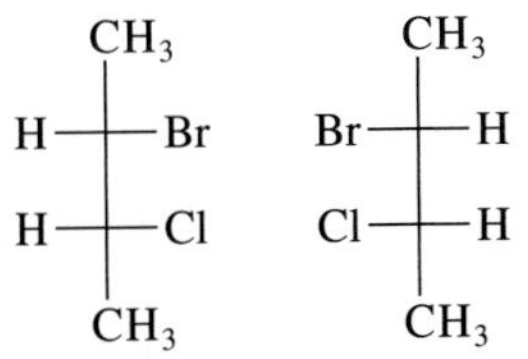

CH_3	CH_3
H — Br	Br — H
H — Cl	Cl — H
CH_3	CH_3

enantiomers

CH_3	CH_3
H — Br	Br — H
Cl — H	H — Cl
CH_3	CH_3

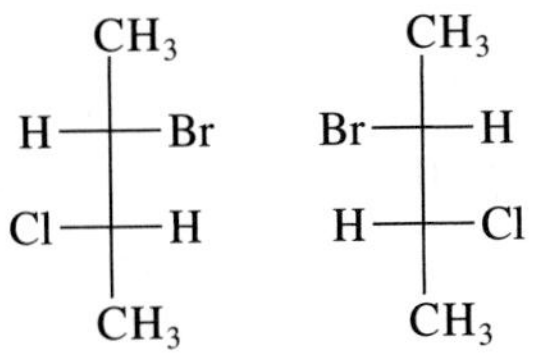

enantiomers

(b) 2,3,4-trichloro-1-pentanol

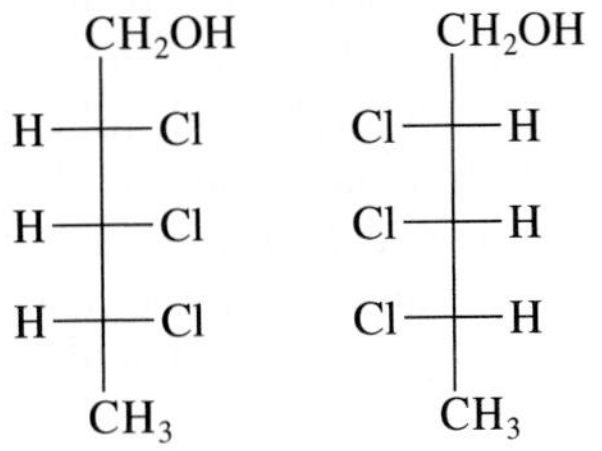

CH_2OH	CH_2OH
H — Cl	Cl — H
H — Cl	Cl — H
H — Cl	Cl — H
CH_3	CH_3

enantiomers

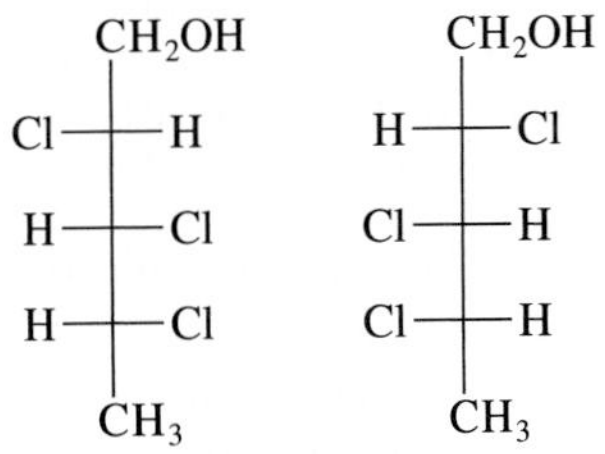

CH_2OH	CH_2OH
Cl — H	H — Cl
H — Cl	Cl — H
H — Cl	Cl — H
CH_3	CH_3

enantiomers

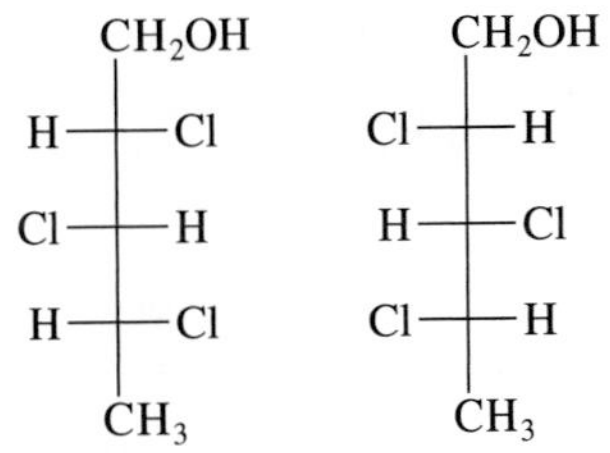

CH_2OH	CH_2OH
H — Cl	Cl — H
Cl — H	H — Cl
H — Cl	Cl — H
CH_3	CH_3

enantiomers

CH_2OH	CH_2OH
H — Cl	Cl — H
H — Cl	Cl — H
Cl — H	H — Cl
CH_3	CH_3

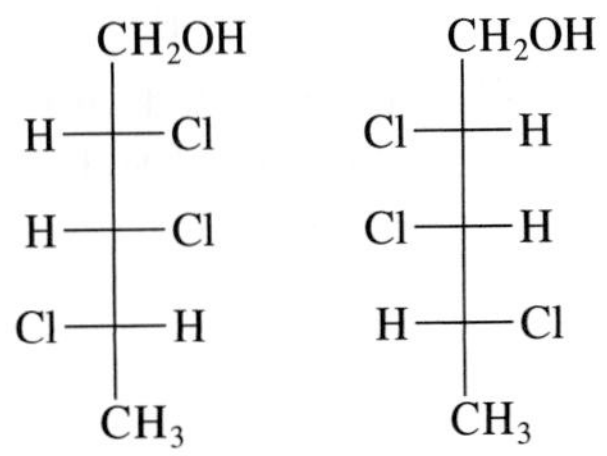

enantiomers

There are no meso compounds.

47.

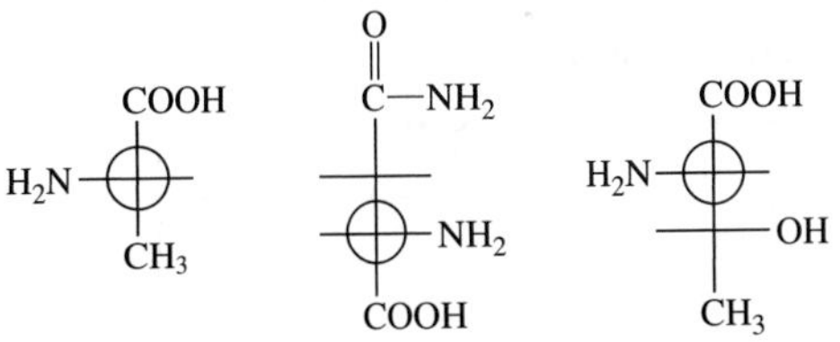

COOH; H_2N; CH_3 — $C(=O)-NH_2$; NH_2; COOH — COOH; H_2N; OH; CH_3

The similarity is that the chiral carbon in each compound is bonded to an NH_2, a COOH, and an H group.

Chapter 27

Review Questions

2. The notations D and L in the name of a carbohydrate specify the configuration on the last chiral carbon atom (from C-1) in the Fischer projection formula.

If the —OH is written to the right of that carbon the compound is a D-carbohydrate. If the —OH is written to the left, it is an L-carbohydrate. For example, D-glyceraldehyde and L-glyceraldehyde differ only at the chiral C—OH; D-glyceraldehyde has the —OH on the right while L-glyceraldehyde has the —OH on the left.

3. The notations (+) and (−) in the name of a carbohydrate specify whether the compound rotates the plane of polarized light to the right (+) or to the left (−).

7. α-D-glucose and β-D-glucose differ in the configuration at carbon-1 in the cyclic structure. In the open chain structure, carbon-1 is an aldehyde group and is not chiral. In the cyclic structure that carbon contains a hemiacetal structure and is chiral. When the ring forms, carbon-1 can have two configurations leading to the two structures called α and β-D-glucose.

12. Invert sugar is sweeter than sucrose because it is a 50-50 mixture of fructose and glucose. Glucose is somewhat less sweet than sucrose, but fructose is much sweeter, so the mixture is sweeter.

18. Blood types A and B differ in one place. In the fourth pyran ring at C—2, blood type B has an OH group while blood type A has an amide group.

Exercises

2. (a) D-glyceraldehyde: H—C=O / H—C—OH / CH_2OH; L-glyceraldehyde: H—C=O / HO—C—H / CH_2OH

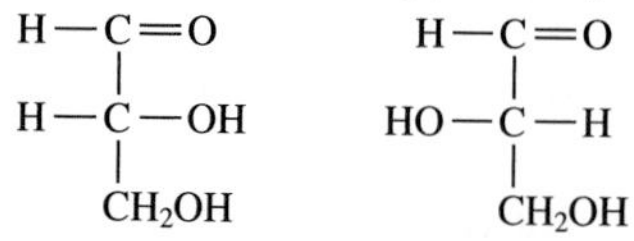

(b) If D-glyceraldehyde is reacted with hydrogen in the presence of a platinum catalyst, the product will be glycerol.

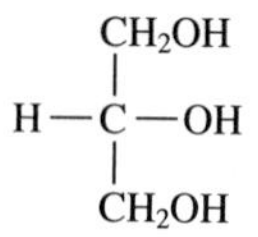

CH_2OH / H—C—OH / CH_2OH

4. The epimer could differ at carbon 2 *or* carbon 3 *or* carbon 4 *or* carbon 5. One possible answer is:

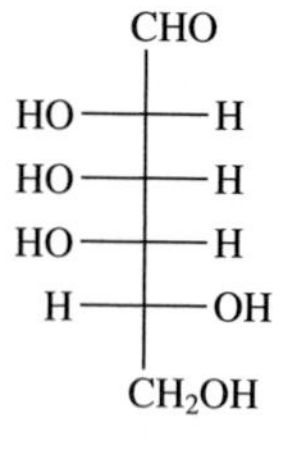

CHO / HO—H / HO—H / HO—H / H—OH / CH_2OH

6. 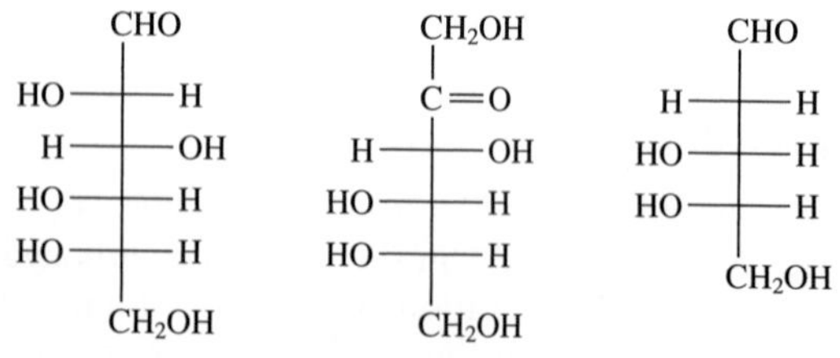 L-glucose: CHO / HO—H / H—OH / HO—H / HO—H / CH_2OH

L-fructose: CH_2OH / C=O / H—OH / HO—H / HO—H / CH_2OH

L-2-deoxyribose: CHO / H—H / HO—H / HO—H / CH_2OH

8. Either the Fischer projection formulas or Haworth formulas are satisfactory.

β-D-glucopyranose: HO—C—H / H—C—OH / HO—C—H / H—C—OH / H—C (ring O to C-1) / CH_2OH

α-D-galactopyranose: H—C—OH / H—C—OH / HO—C—H / HO—C—H / H—C (ring O to C-1) / CH_2OH

β-D-mannopyranose: HO—C—H / HO—C—H / HO—C—H / H—C—OH / H—C (ring O to C-1) / CH_2OH

Haworth formulas:

β-D-glucopyranose: CH_2OH, H, O, OH, H, OH, H, HO, H, H, OH

α-D-galactopyranose: CH_2OH, HO, O, H, H, OH, H, H, OH, H, OH

β-D-mannopyranose: CH_2OH, H, O, OH, H, OH, HO, HO, H, H, H

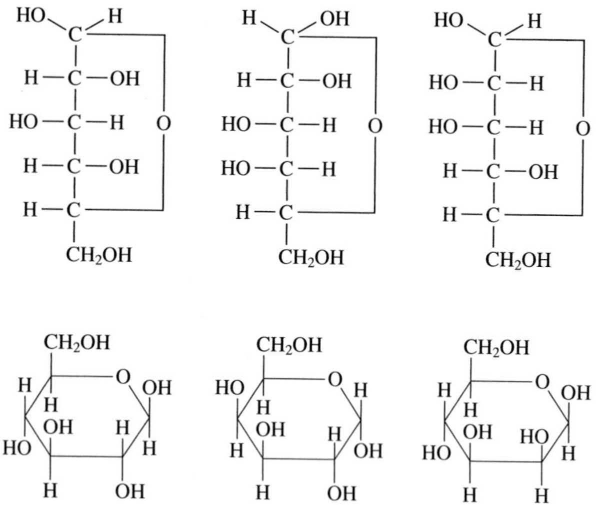

10. The glucose units in cellulose are connected by β-1,4-glycosidic linkages. The human digestive system does not have the enzymes to catalyze the hydrolysis of cellulose.

12. The Kiliani–Fischer synthesis of D-ribose starts with the proper D-triose, D-glyceraldehyde.

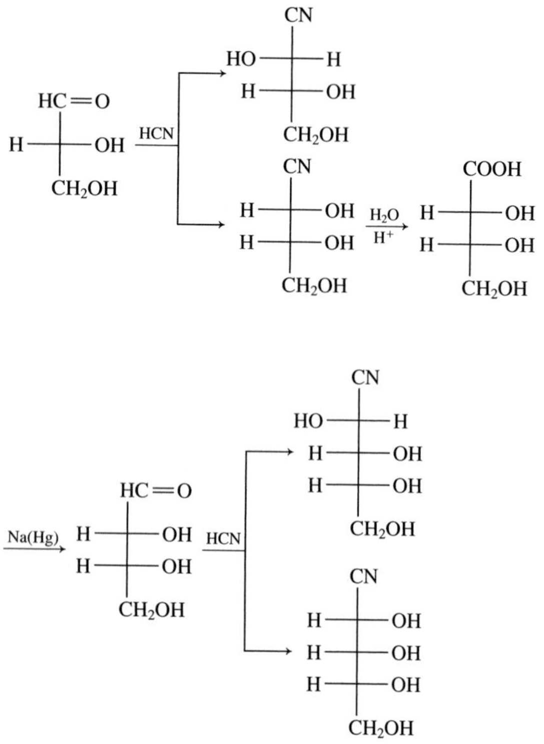

HC=O / H—OH / CH_2OH —HCN→ (CN / HO—H / H—OH / CH_2OH) and (CN / H—OH / H—OH / CH_2OH) —H_2O, H^+→ COOH / H—OH / H—OH / CH_2OH

—Na(Hg)→ HC=O / H—OH / H—OH / CH_2OH —HCN→ (CN / HO—H / H—OH / H—OH / CH_2OH) and (CN / H—OH / H—OH / H—OH / CH_2OH)

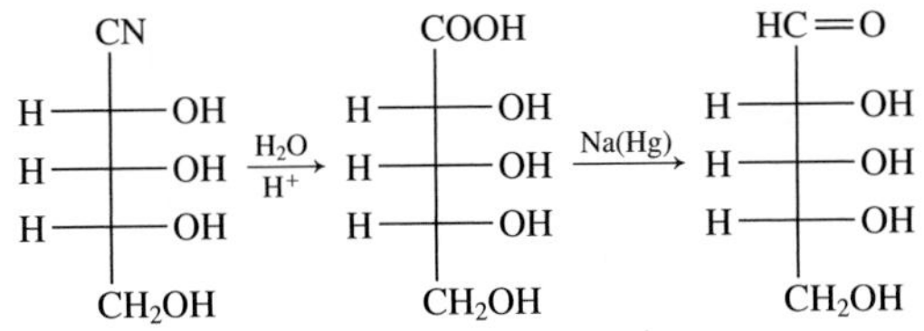

D-ribose

14. No, D-2-deoxygalactose is not the same as D-2-deoxyglucose. D-Galactose differs from D-glucose at carbon 4, so replacement of the carbon 2 OH with an H does not make these two sugars identical.

16. The monosaccharide composition of:
   (a) lactose: one glucose and one galactose unit
   (b) amylopectin: many glucose units
   (c) cellulose: many glucose units
   (d) sucrose: one glucose and one fructose unit

18. Both maltose and isomaltose are disaccharides composed of two glucose units. The glucose units in maltose are linked by an α-1,4-glycosidic bond while the glucose units of isomaltose are linked by an α-1,6-glycosidic bond.

20. Both maltose and isomaltose will show mutarotation. Both disaccharides contain a hemiacetal structure that will open, allowing mutarotation.

22.

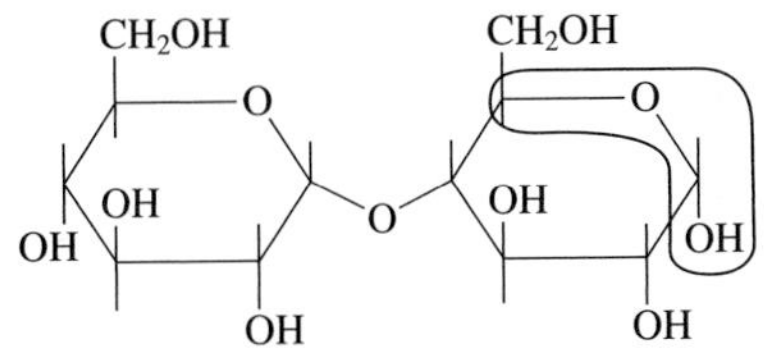

maltose

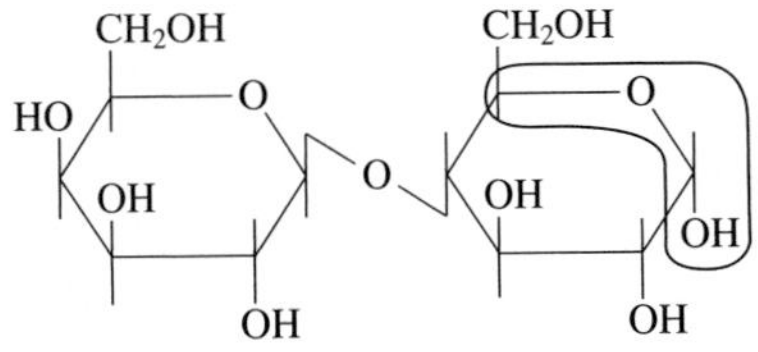

lactose

The circled hemiacetal structures allow these two disaccharides to be reducing sugars.

24. The systematic name for maltose is α-D-glucopyranosyl-(1,4)-α-D-glucopyranose. The systematic name for lactose is β-D-galactopyranosyl-(1,4)-α-D-glucopyranose.

26. (a) D-ribose and D-2-deoxyribose. The D-2-deoxyribose has no OH group on the number 2 carbon, only 2 hydrogen atoms. Both are five-carbon sugars.
   (b) Amylose and amylopectin. Amylose is a straight-chain polysaccharide; amylopectin has branched chains and more monomer units per molecule. Both are large polysaccharides composed of α-D-glucose units.
   (c) Lactose and isomaltose. These sugars are disaccharides. Lactose is composed of one galactose unit and one glucose unit while isomaltose is composed of two glucose units. The monosaccharide units in lactose are linked by a β-1,4-glycosidic bond whereas the units in isomaltose are linked by an α-1,6-glycosidic bond. Both disaccharides also contain hemiacetal structures.

28.

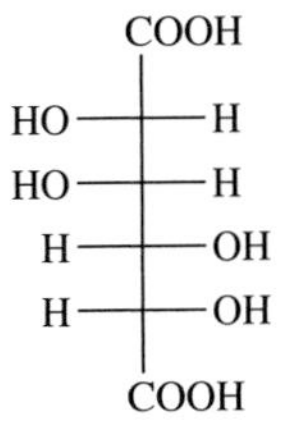

mannaric acid

30.

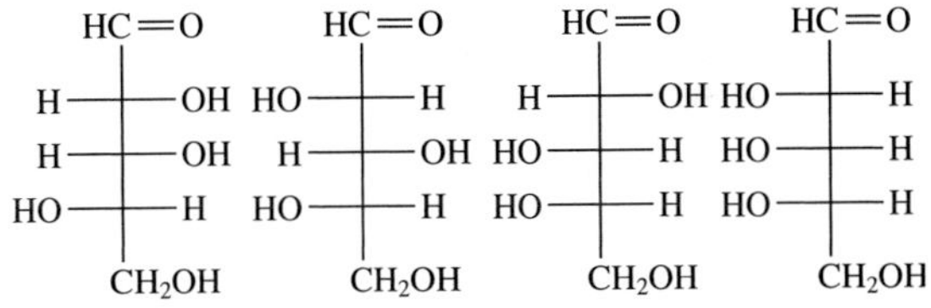

A B C D

Epimers are A and B, A and C, B and D, C and D.

32. (a) β-D-mannopyranosyl-(1,4)-β-D-galactopyranose

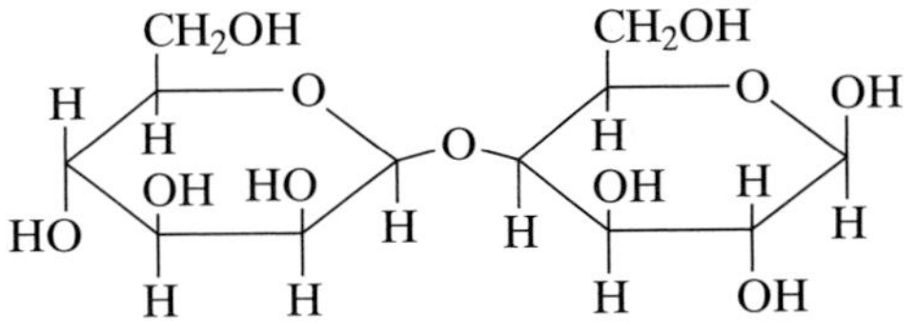

   (b) β-D-galactopyranosyl-(1,6)-α-D-glucopyranose

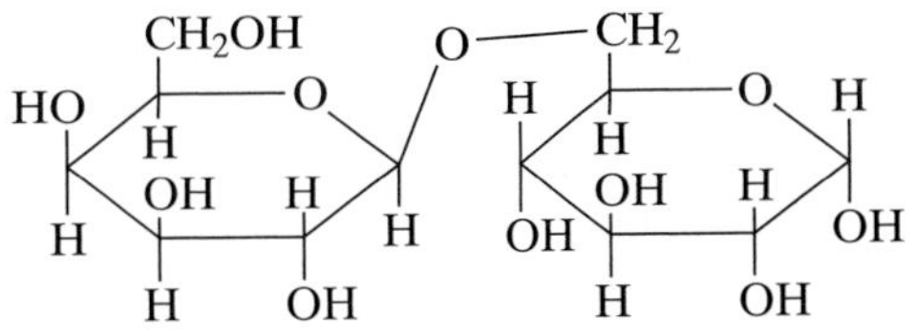

34.

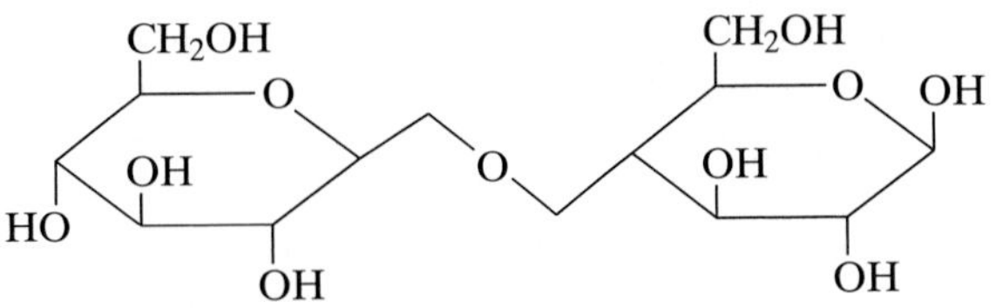

cellobiose

36.

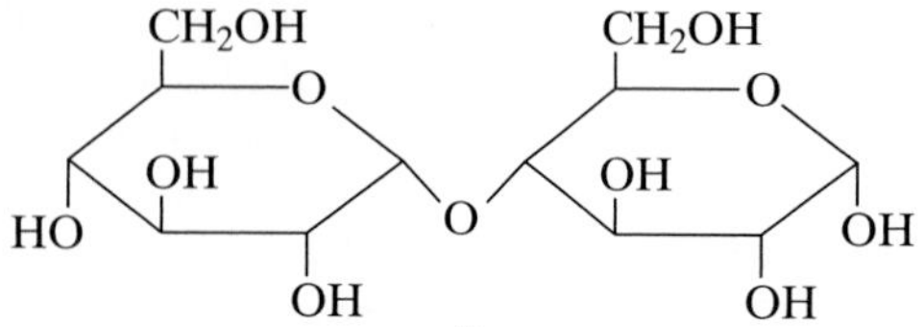

maltose

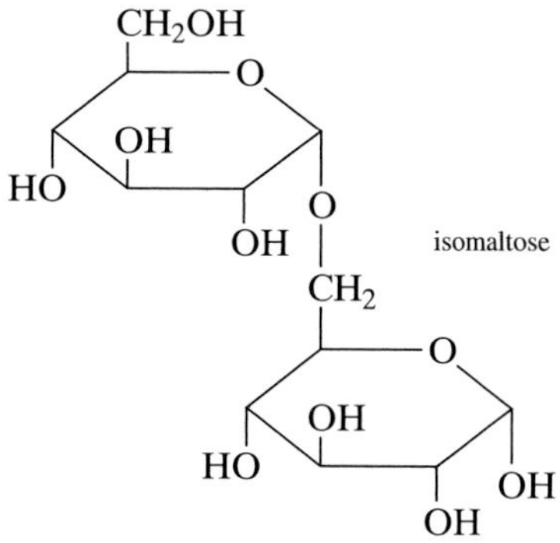

isomaltose

38. (a) D-ribose (b) β-D-glucopyranose (c) α-D-fructofuranose

40. maltose, α-D-glucopyranosyl-(1,4)-α-D-glucopyranose

41. Glucose is called blood sugar because it is the most abundant carbohydrate in the blood and is carried by the bloodstream to all parts of the body.

48. A nonreducing disaccharide composed of two molecules of α-D-galactopyranose can have no hemiacetal structures. Thus, the hemiacetal structure of one α-D-galactopyranose must be used to form the glycosidic link to the hemiacetal structure of the other α-D-galactopyranose unit.

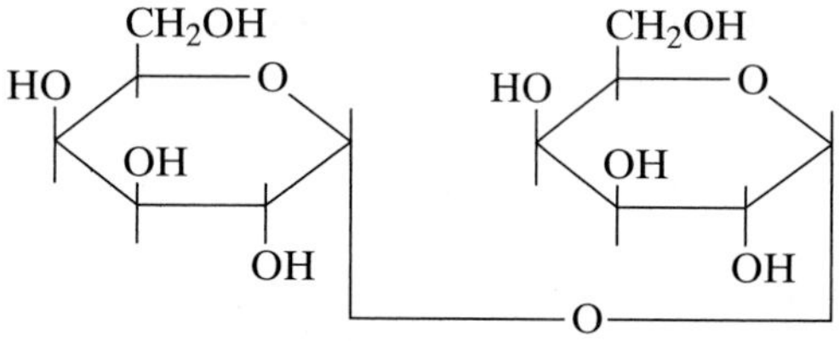

51. No, the classmate should not be believed. Although D-glucoase and D-mannose are related as epimers, it is pairs of enantiomers which have equal and opposite optical rotation.

Chapter 28

Review Questions

2. Although caproic acid, $CH_3(CH_2)_4COOH$, has the same number of polar bonds as stearic acid, $CH_3(CH_2)_{16}COOH$, caproic acid has a shorter, nonpolar hydrocarbon chain and, therefore, is more water soluble.

6. Aspirin relieves inflammation by blocking the conversion of arachidonic acid to prostaglandins.

11. The omega-3 fatty acids can substitute for arachidonic acid and block its action as a trigger for heart attack and stroke.

16. HDL is a cholesterol scavenger, picking up this steroid in the serum and returning it to the liver.

Exercises

2.

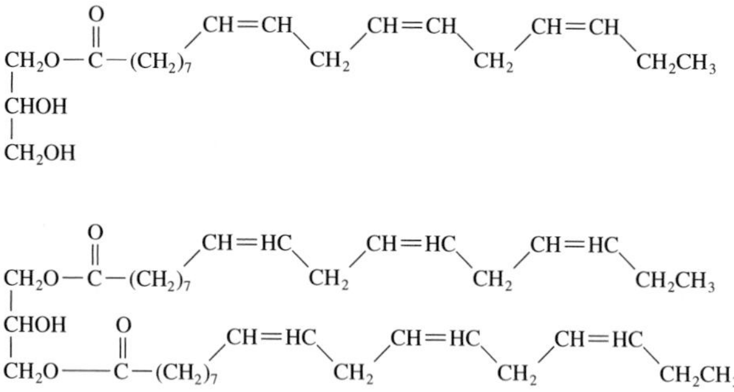

(Linolenic acid can be located at any two of the three hydroxyls on glycerol.)

4.

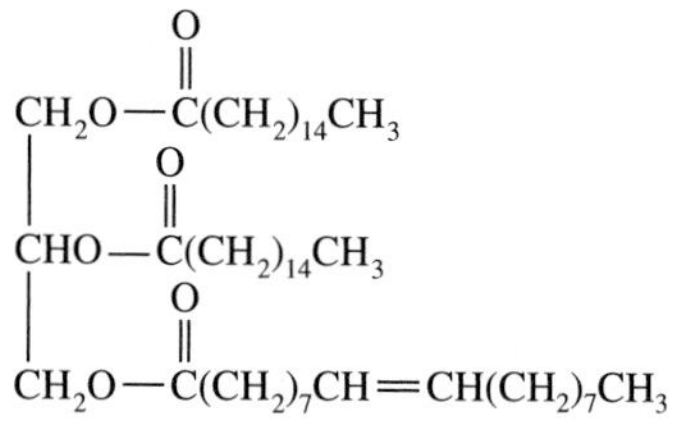

palmitic acid

palmitic acid

oleic acid

There is one other possible triacylglycerol with the same components.

6. Yes, a triacylglycerol that contains three units of stearic acid would be more hydrophobic than a triacylglycerol that contains three units of myristic acid. Larger molecules tend to be more hydrophobic than smaller molecules, and a triacylglycerol with three stearic acid units is larger than a triacylglycerol with three myristic acid units.

8.

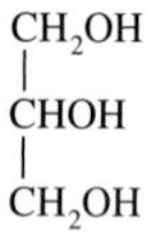

glycerol

$CH_3(CH_2)_7CH{=}CH(CH_2)_7COOH$ oleic acid

$CH_3(CH_2)_{14}COOH$ palmitic acid

$CH_3(CH_2)_{16}COOH$ stearic acid

10. $CH_3{-}(CH_2)_{28}{-}CH_2{-}O{-}\overset{O}{\overset{\|}{C}}{-}(CH_2)_{14}{-}CH_3$

12.

$$\begin{array}{l} CH_2O{-}\overset{O}{\overset{\|}{C}}{-}(CH_2)_7{-}CH{=}HC{-}(CH_2)_7{-}CH_3 \\ | \\ CHOH \\ | \\ CH_2OH \end{array}$$

(The oleic acid may be located at any one of the three hydroxyls on glycerol.)

14.

$$\begin{array}{l} CH_2O{-}\overset{O}{\overset{\|}{C}}(CH_2)_{16}CH_3 \\ | \\ CHO{-}\overset{O}{\overset{\|}{C}}(CH_2)_{16}CH_3 \\ | \\ CH_2O{-}\underset{O^-}{\underset{|}{\overset{O}{\overset{\|}{P}}}}{-}OCH_2CH_2\underset{CH_3}{\underset{|}{\overset{CH_3}{\overset{|}{N^+}}}}{-}CH_3 \end{array}$$

16.

$$\begin{array}{l} OH \\ | \\ CHCH{=}CH(CH_2)_{12}CH_3 \\ | \\ CHNH{-}\overset{O}{\overset{\|}{C}}(CH_2)_{16}CH_3 \\ | \\ CH_2O{-}\underset{O^-}{\underset{|}{\overset{O}{\overset{\|}{P}}}}{-}OCH_2CH_2\overset{+}{N}H_3 \end{array}$$

18.

$$\begin{array}{l} OH \\ | \\ CHCH{=}CH(CH_2)_{12}CH_3 \\ | \\ CHNH{-}\overset{O}{\overset{\|}{C}}(CH_2)_7CH{=}CH(CH_2)_7CH_3 \\ | \\ CH_2O{-}\text{(galactose ring: } CH_2OH,\ OH,\ OH,\ OH\text{)} \end{array}$$

or

$$\begin{array}{l} OH \\ | \\ CHCH{=}CH(CH_2)_{12}CH_3 \\ | \\ CHNH{-}\overset{O}{\overset{\|}{C}}(CH_2)_7CH{=}CH(CH_2)_7CH_3 \\ | \\ CH_2O{-}\text{(galactose ring: } CH_2OH,\ OH,\ OH,\ OH\text{)} \end{array}$$

20.

$HO{-}\overset{O}{\overset{\|}{C}}(CH_2)_{12}CH_3$

hydrophilic exterior (—COOH)

hydrophobic interior ($-(CH_2)_{12}CH_3$)

22. HDL (high density lipoprotein) differs from LDL (low density lipoprotein) in that
(a) LDL has a lower density than HDL
(b) LDL delivers cholesterol to peripheral tissues whereas HDL scavenges cholesterol and returns it to the liver.

24. (a) Both compounds have two long hydrophobic chains.
(b) For both compounds, the primary alcohol can react further with either acids (to form esters) or sugars (to form acetals).

26. Phosphate ion will move from a region of low concentration to a region of high concentration as it moves from a 0.1 *M* solution across a membrane to 0.5 *M* solution. This process requires energy and can be accomplished by active transport.

28. Thromboxanes act as vasoconstrictors and stimulate platelet aggregation. Leukotrienes have been associated with many of the symptoms of an allergy attack (e.g., an asthma attack).

30. (a) palmitic acid (b) stearic acid and linoleic acid.

32. (a) glycolipid
(b) triacylglycerol (triglyceride)
(c) phospholipid (phosphatidyl ethanolamine).

36. This meal is changed by increasing the fat content from 10 grams to 15 grams, an increase of 5 grams. Since each gram of fat yields an average of 9.5 Cal an increase of 5 grams equates to an increase of 47.5 Cal. Thus the meal will now contain 282 Cal.

38. $CH_3(CH_2)_7CH{=}CH(CH_2)_7COOH$
oleic acid or 9-octadeceneoic acid

Chapter 29

Review Questions

1. The amino acids of proteins are called alpha amino acids because the amine group is always attached to the alpha carbon atom, that is, the carbon atom next to the carboxyl group, COOH.

$$R{-}\underset{H}{\underset{|}{\overset{NH_2}{\overset{|}{C}}}}{-}COOH$$

(C = alpha carbon atom)

6. At its isoelectric point, a protein molecule must have an equal number of positive and negative charges.

11. Hydrolysis breaks the peptide bonds, thus disrupting the primary structure of the protein. Denaturation involves alteration or disruption of the secondary, tertiary, or quaternary structure but not the primary structure of proteins.

14. Protein column chromatography uses a column packed with polymer beads (solid phase) through which a protein solution (liquid phase) is passed. Proteins separate based on differences in how they interact with the solid phase. The proteins move through the column at different rates and can be collected separately.

Exercises

2. D-serine: COOH / H—C—NH$_2$ / CH$_2$OH; L-serine (form commonly found in proteins): COOH / H$_2$N—C—H / CH$_2$OH

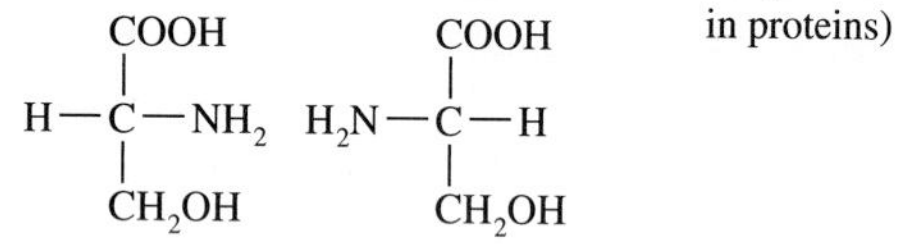

4. 33 g

6. $NH_2C(=O)—CH_2—CH(NH_3^+)—COO^-$

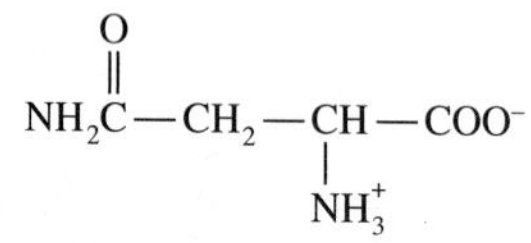

8. For tryptophan:
(a) Zwitterion formula: indole ring (N—H) with —$CH_2CH(NH_3^+)COO^-$
(b) formula in 0.1 *M* H_2SO_4: indole ring (N—H) with —$CH_2CH(NH_3^+)COOH$
(c) formula in 0.1 *M* NaOH: indole ring (N—H) with —$CH_2CH(NH_2)COO^-$

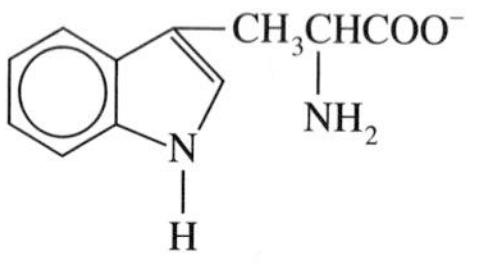

10. Ionic equations showing how leucine acts as a buffer toward:

(a) H^+

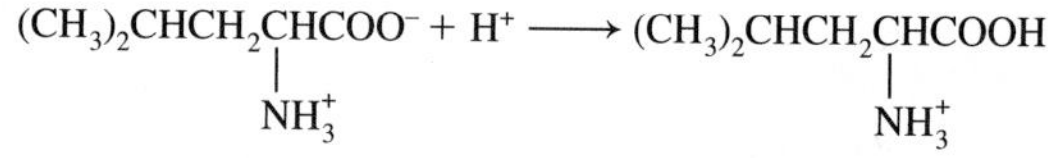

$(CH_3)_2CHCH_2CH(NH_3^+)COO^- + H^+ \longrightarrow (CH_3)_2CHCH_2CH(NH_3^+)COOH$

(b) OH^-

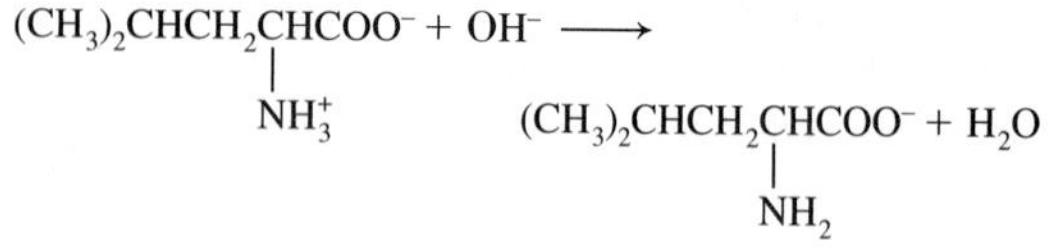

$(CH_3)_2CHCH_2CH(NH_3^+)COO^- + OH^- \longrightarrow (CH_3)_2CHCH_2CH(NH_2)COO^- + H_2O$

12. $(CH_3)_2CHCH(NH_3^+)COO^-$

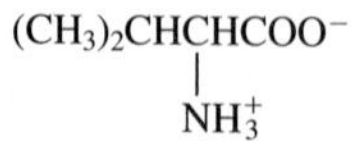

14. Gly-Thr: $NH_2CH_2C(=O)$—$NHCH(CHOHCH_3)COOH$ (peptide bond between C=O and NH)

Thr-Gly: $NH_2CH(CHOHCH_3)C(=O)$—$NHCH_2COOH$ (peptide bond between C=O and NH)

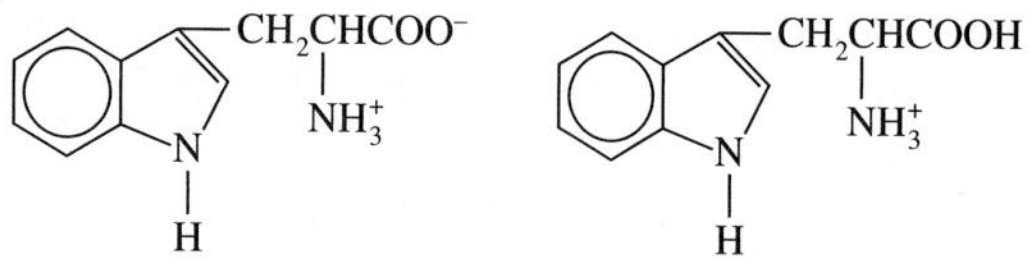

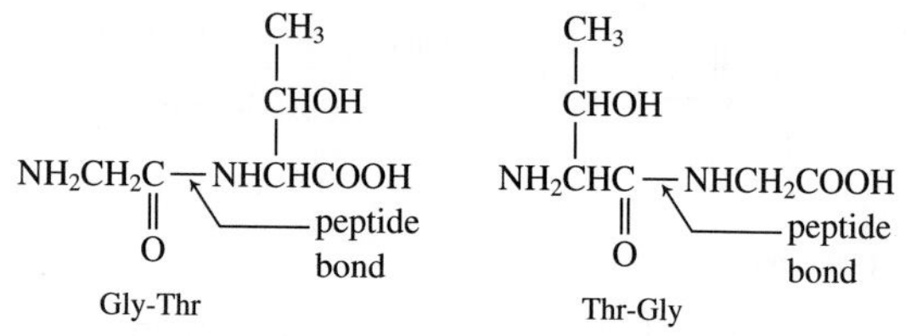

16. $H_2N—CH(CH_2SH)—COOH$

$H_2N—CH(COOH)—CH_2—S—S—CH_2—CH(COOH)—NH_2$

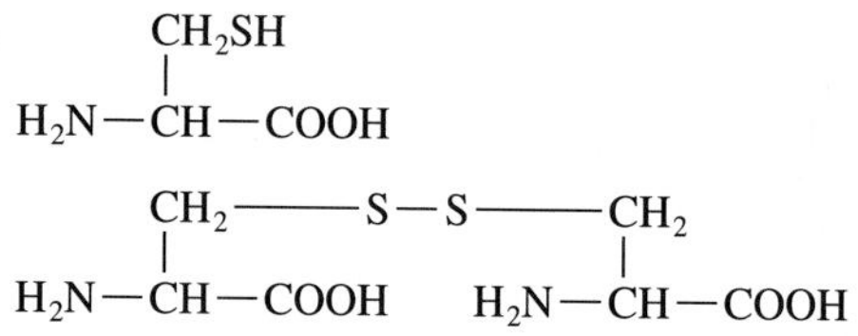

18. (a) $NH_2CH(CH_3)C(=O)—NHCH(CH_3)COOH$

(b) $NH_2CH(CH_2OH)C(=O)—NHCH_2C(=O)—NHCH_2COOH$

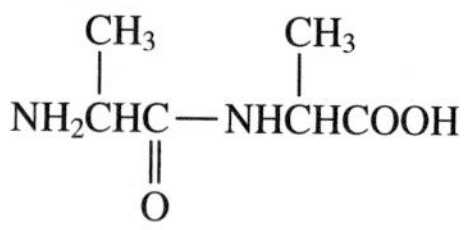

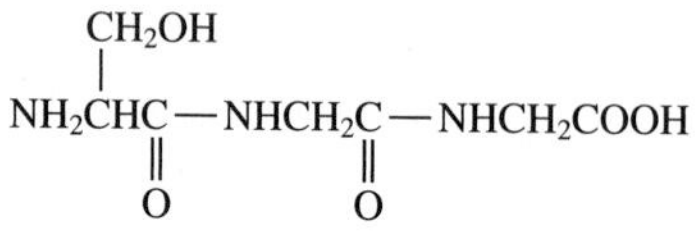

(c) $NH_2CH(CH_2OH)C(=O)—NHCH_2C(=O)—NHCH(CH_3)COOH$

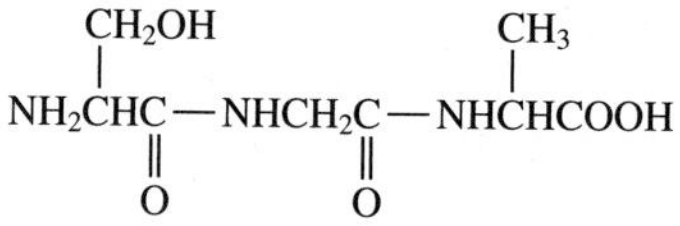

20. Tyr-Asp-Ala, Tyr-Ala-Asp, Asp-Tyr-Ala, Asp-Ala-Tyr, Ala-Tyr-Asp, Ala-Asp-Tyr

22. —C(H)(H)—C(=O)—N(H)—C(CH$_3$)(H)— (the O and the N—H are highlighted)

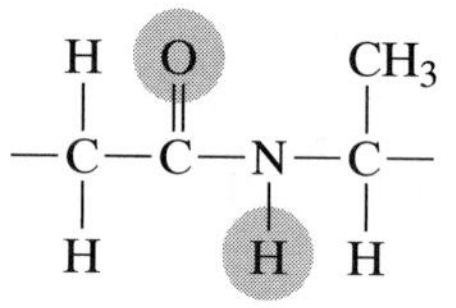

24. $—CH_2COO^- \; {}^+NH_3CH_2CH_2CH_2CH_2—$ (ionic bond)

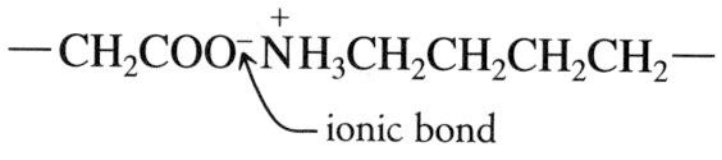

26. The tripeptide, Gly-Ser-Asp, will
(a) react with $CuSO_4$ (biuret test) to give a violet color. The tripeptide has the required two peptide bonds.
(b) not react to give a positive xanthoproteic test because there are no benzene rings in this tripeptide.
(c) react with ninhydrin to give a blue solution.

28. $NH_2CH(CH_3)COOH$, $NH_2CH(CH_2CH_2COOH)COOH$, $NH_2CH(CH_2C_6H_4OH)COOH$,

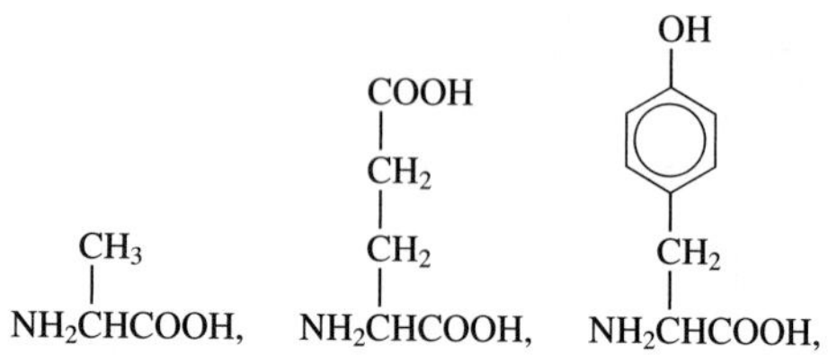

30. $6.8 \times 10^4 \frac{\text{g}}{\text{mole}}$ (molar mass of hemoglobin)

32. Phe-Ala-Ala-Leu-Phe-Gly-Tyr

34. This newly discovered protein is probably not a structural-support protein because it is globular in shape and has secondary structure (beta pleated sheet) at its core. Thus, it is more likely to be a binding protein.

36. A domain is a compact piece of protein structure of about 20,000 g/mole. The newly discovered protein with two domains is more likely to have a molar mass between 40,000 and 60,000 g/mole.

38. The silk protein, fibroin, has a high percentage of the secondary structure, beta-pleated sheet. This secondary structure is like a sheet of paper in that it is flexible but not stretchable. Thus, fibroin if not easily stretched.

39. The amino acid sequence of the nonapeptide is:
Arg-Pro-Pro-Gly-Phe-Ser-Pro-Phe-Arg

43. The stereoisomers of threonine:

```
   COOH           COOH            COOH           COOH
    |              |               |              |
H—C—NH2     H2N—C—H       H—C—NH2     H2N—C—H
    |              |               |              |
H—C—OH      HO—C—H        HO—C—H        H—C—OH
    |              |               |              |
   CH3            CH3             CH3            CH3
```

46. thirty-nine dipeptides

Chapter 30

Review Questions

1. Activation energy is the energy barrier to chemical reaction and is measured as the difference between the reactant(s) energy level and the transition state energy level. Enzymes lower the activation energy barrier and, thus, increase biochemical reaction rates.

5. The six general classes of enzymes are:
(a) oxidoreductases, (b) transferases, (c) hydrolases, (d) lyases, (e) isomerases, and (f) ligases.

14. The statin drugs inhibit a key enzyme in cholesterol biosynthesis and, thus, lower blood cholesterol levels.

Exercises

2. 4×10^{-7} M/s

4. (a) 0.0041 M/min
(b) 4100 μM/min

6. Approximately 0.003 M/min

8. Approximately 3 mM/min

10. 104/sec

12. 1×10^3 reactants converted to products by three pepsin molecules

14. The first enzyme would have a narrow active site which is long enough to fit the four carbon carboxylic acid, butanoic acid ($CH_3CH_2CH_2COOH$). The second enzyme would have a narrow active site which would be shorter because it need only fit the two carbon carboxylic acid, acetic acid (CH_3COOH).

16. An enzyme-catalyzed rate decreases at high temperatures because the enzyme loses its natural shape (denatures). As an enzyme denatures it ceases to be an effective catalyst.

18. If the turnover number for an enzyme is increased, the enzyme can convert more reactants to products per unit time; the enzyme is a better catalyst. Any process that increases an enzyme's catalytic abilities is termed activation.

20. Enzyme A: the reaction rate goes up much more gradually for Enzyme B because of the lower attraction of this enzyme for substrate.

22. Enzymes tend to have pH optima to fit their environment. Since pepsin is found in the stomach ($[H_3O^+] = 0.001$ M, pH = 3), this enzyme's pH optimum is predicted to be about 3.

24. Condition B has a lower activation energy. Since the barrier to reaction is smaller, the reaction rate will be faster.

26. Yes. Enzyme B is more effective than enzyme A based on a comparison of turnover numbers (9.8×10^{-1}/s for enzyme B vs. 0.05/s for enzyme A).

28. Glutamine is a starting material (reactant) for the liver metabolic pathway that produces urea. Thus, glutamine control of this pathway must be "feedforward." Since an increase in glutamine concentration causes the metabolic pathway to speed up, this control must be feedforward activation.

30. Blood clots in the brain are the major cause of strokes. Specific proteases are used to dissolve these blood clots to alleviate the cause of strokes.

32. Amylase is present in laundry detergents to digest starchy residues.

34. Sucrase converts sucrose to a mixture of glucose and fructose. Since fructose is sweeter than sucrose, a sweeter solution is created.

35. The turnover number measures the number of substance molecules converted to product by one enzyme molecule under optimum conditions. Chymotrypsin can convert one glycine-containing substance to product every 20 seconds while the enzyme can convert 200 L-tyrosine-containing substrates to products each second. Chymotrypsin is a much more efficient catalyst for the L-tyrosine-containing substrate.

Chapter 31

Review Questions

3. There are three structural differences between DNA and RNA.
   (a) In RNA the sugar molecule is always ribose. In DNA the sugar molecule is always deoxyribose.
   (b) Both molecules use a mixture of four nitrogen bases. Both use cytosine, adenine, and guanine. In DNA the fourth base is thymine. In RNA; the fourth base is uracil.
   (c) DNA exists as a double helix whereas RNA is a single strand of nucleotides.
4. The major function of ATP in the body is to store chemical energy, and to release it when called upon to carry out many of the complex reactions that are essential to most of our life processes.
6. Complementary bases are the pairs that hydrogen bond ("fit") to each other between the two helixes of DNA.
12. A codon is a triplet of three nucleotides, and each codon specifies one amino acid. The cloverleaf model of transfer RNA has an anticodon loop consisting of seven unpaired nucleotides. Three of these make up the anticodon, which is complementary to, and hydrogen-bonds to the codon on mRNA.
14. From time to time a new trait appears in an individual that is not present in either parents or ancestors. These traits, which are generally the result of genetic or chromosomal changes, are called mutations.

Exercises

2. (a) G, guanosine
   (b) GMP, guanosine-5′-monophosphate
   (c) dGDP, deoxyguanosine-5′-diphosphate
   (d) CTP, cytidine-5′-triphosphate
4. (a) U

O, N—H, $HOCH_2$, O, N, O, OH, OH

(b) UMP

O, N—H, O, HO—P—OCH_2, O, N, O, O^-, OH, OH

(c) CTP

NH_2, N, O, O, O, HO—P—O—P—O—P—OCH_2, O, N, O, O^-, O^-, O^-, OH, OH

(d) dTMP

O, H_3C, N—H, O, HO—P—OCH_2, O, N, O, O^-, OH

6. (a)

O, N, NH, N, N, NH_2, O, O, HO—P—O—CH_2, O^-, O, HO—P—O, O^-

(b)

O, NH, N, O, O, O, O^-—P—O—CH_2, OH, OH, HO, O—P—O^-, O

8.

NH_2, N, N, N, N, O, O, O, HO—P—O—P—O—CH_2, O^-, O^-, HO, OH

10. There are several possible sequences for the three-nucleotide single-stranded RNA. One possible structure follows:

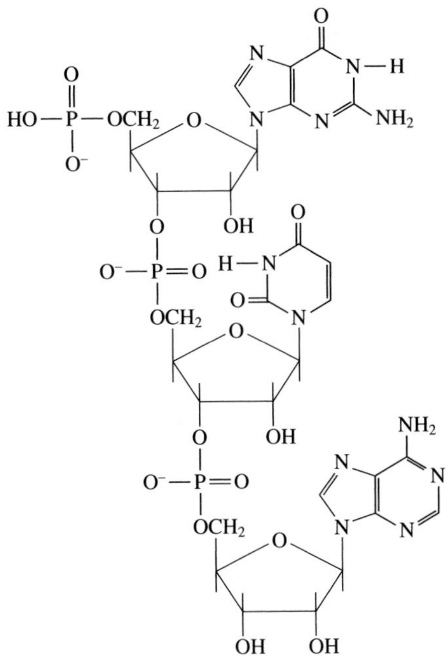

12. 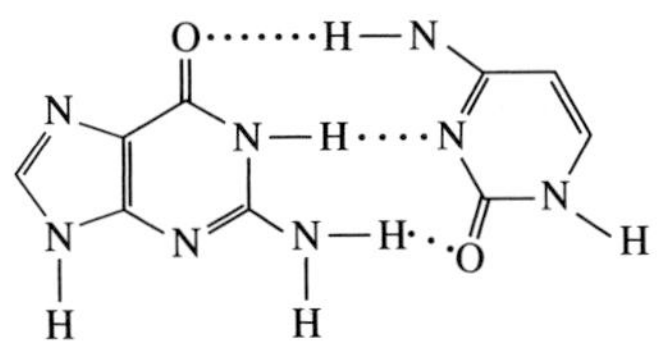 (dotted lines are hydrogen bonds)

14. DNA is considered to be the genetic substance of life, because it contains the sequence of bases that carries the code for genetic characteristics.

16. Transcription is the process of making RNA using a DNA template, whereas translation is a process for making protein using an mRNA template.

18. Both mRNA and rRNA can be found in the ribosomes. rRNA serves as part of the ribosome structure, whereas mRNA serves as a template for protein synthesis.

20. $3 \times 573 = 1719$

22. (a) CGAAUCUGGACU
(b) GCUUAGACCUGA
(c) Arg-Ile-Trp-Thr

24. Translation makes a polymer of amino acids by forming amide bonds to connect the amino acids to each other. The amide bond combines an amine with a carboxylic acid.

26. Translation initiation occurs when the ribosome reaches a special AUG or GUG codon along the mRNA. Since there is commonly more than one AUG or GUG codon, the ribosome must use other information to choose the special AUG or GUG. This codon is the starting point for protein synthesis and is bound by either a special tRNA carrying N-formyl methionine (in procaryotes) or a tRNA carrying methionine (in eucaryotes).

28. (a) GCG (c) CUA
(b) UGU (d) AAG

30. -GGGGATC↓TTCG-
-CCC↑CTAGAAGC-

32.

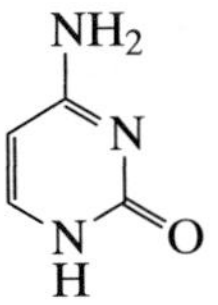

34. The original mRNA sequence codes for the amino acids ... Val-Gln-Lys ... The mutation changes the second codon to a termination or nonsense codon UAA. The protein will end at this point and may cause a major disruption of the protein function.

36. "Nonsense" means the codon does not code for an amino acid. Instead, this codon causes translation to stop. (It is also known as a termination codon.)

39. A substitution mutation can occur and will change only one codon when one base is substituted for another. An insertion mutation will shift the sequence position of all bases following the insertion by one base. Each base will then take the position of its neighbor and every codon following an insertion will be changed.

Chapter 32

Review Questions

5. The essential fatty acids have 18 or more carbon atoms per molecule and are polyunsaturated.

19. The digestive function of the liver is the production of bile, one of the important digestive juices. The bile is stored in the gallbladder.

22. GRAS means Generally Recognized As Safe.

26. The grains (carbohydrates) should be the largest proportion of the diet while the oils (fats) should be the smallest proportion.

Exercises

2. (a) 132 g (b) 1464 mg (c) 31.5 mg

4. (a) 3×10^2 kcal (b) 40%

6. 4.7%

8. 41%

10. The two micronutrient classes are minerals and vitamins.

12. Based on the nutrition information in Exercise 2, the ranking of minerals in a fried chicken breast is as follows: Na (770 mg), K (564 mg), Mg (68 mg), Ca (56 mg), Fe (3.5 mg), Zn (2.7 mg).
14. When food enters the stomach it triggers the release of gastric juice which contains, among other things, hydrochloric acid and the digestive enzymes pepsin and lipase. The main digestive function of the stomach is the partial digestion of proteins aided by the enzyme pepsin.
16. The three classes of nutrients that commonly supply energy for the cells are proteins, carbohydrates, and lipids.
18. Cellulose is important as dietary fiber. Although it is not digested, cellulose absorbs water and provides dietary bulk which helps maintain a healthy digestive tract.
20. Pancreatic juice is slightly basic.
22. Carbohydrates are digested in the mouth and small intestine.
24. Pepsin digests proteins.
26. Vegetables should be the second most abundant food group in our diet.
28. Vitamin B_1 (thiamin) is the vitamin that is lost in beri-beri.

Chapter 33

Review Questions

3. Fats and carbohydrates are good sources of cellular energy because they contain many reduced carbon atoms.
4. The dangerous reactive oxygen species are hydrogen peroxide, H_2O_2, superoxide, O_2^-, and the neutral hydroxyl radical, OH.
11. Oxidative phosphorylation takes place in the mitochondria.

Exercises

2. The average oxidation state of carbon in ethanal in −2. The average oxidation state of carbon in acetic acid is 0. The carbons in ethanal, being in a lower oxidation state, will deliver more energy in biological redox reactions.
4. The ten carbon, saturated fatty acid contains as many carbons as two ribose molecules but the fatty acid carbons are more reduced. Being more extensively reduced, they will release more energy upon complete oxidation.
6. The typical eucaryotic cell contains the enzyme catalase that removes hydrogen peroxide by converting it to oxygen and water.
8. Superoxide is a negatively charged ROS.
10. Both oxidized forms of the coenzymes (NAD^+ and FAD) are on the same side of the equation. For a reaction to occur one reduced coenzyme (NADH or $FADH_2$) must be paired with one of the oxidized coenzymes.
12. The conversion of acetate to long-chain fatty acids is anabolic because:
    (a) a smaller molecule (acetate) is converted to a larger molecule (long-chain fatty acids)
    (b) the carbons from acetate become reduced (on the average) as they are converted to long-chain fatty acids
14. Oxidative phosphorylation creates ATP and is a catabolic process, not an anabolic one.
16. Chloroplasts carry out photosynthesis, not oxidative phosphorylation.
18. The chemical changes in the chloroplast are said to be anabolic because:
    (a) smaller carbon dioxide molecules are converted to larger glucose molecules
    (b) as this transformation takes place, the carbons become progressively more reduced
    (c) this process requires an input of energy (from sunlight)
20. Three important characteristics of a catabolic process are (1) complex substances are broken down into simpler substances; (2) carbons are often oxidized; (3) cellular energy is often produced.
22. The flavin ring is the reactive center of FAD.
24. Since the conversion of $FADH_2$ to FAD is oxidation, some pathway metabolites are being reduced. A pathway that reduces carbon compounds is probably anabolic.
26. 0.34 mol O_2
28. 0.18 mol e^-
30. two high energy phosphate bonds
32. 58 kJ
34. Since the phosphorylated substrate 1,3-diphosphoglycerate contains only one high energy phosphate bond, substrate-level phosphorylation can produce only one ATP.
36. 12 ATPs
38. The vitamin, riboflavin, is required to make FAD.
40. Adenine is present in all three redox coenzymes.

Chapter 34

Review Questions

6. The liver replenishes muscle glucose (via the blood) by (a) releasing glucose from liver glycogen stores and (b) by converting lactic acid back into glucose.
10. Hormones are chemical substances that act as control or regulatory agents in the body. Hormones are secreted by the endocrine, or ductless glands directly into the bloodstream and are transported to various parts of the body to exert specific control functions.

13. In gluconeogenesis, glucose is synthesized from noncarbohydrate sources: lactate, amino acids, and glycerol. These three substances are converted to Emden–Meyerhof pathway intermediates, which are then converted to glucose. Since the Embden–Meyerhof pathway is anaerobic, gluconeogenesis is considered to be an anaerobic process.

Exercises

2. Glycogenesis is the process by which glucose is used to produce glycogen.

CH_2OH, O, OH, O ← α-(1,6)-bond, OH, CH_2, CH_2OH, O, O, OH, O, OH, O, OH, OH ← α-(1,4)-bond

Glycogen

4. Glycogenolysis is catabolic because it starts with larger precursors (glycogen molecules) and forms smaller products (glucose molecules).

6. Very shallow breathing increases the blood concentration of carbon dioxide. More dissolved carbon dioxide will increase the concentration of carbonic acid and acidify the blood.

8. The Embden-Meyerhof pathway uses substrate-level phosphorylation to produce ATP. That is, following carbon oxidation a phosphate group is transferred from a substrate to ADP, forming ATP.

10. O=C(—O⁻)—C(=CH₂)—O～P(=O)(—OH)—O⁻ (high energy phosphate bond)

12. Although no molecular oxygen (O_2) is used in the citric acid cycle, reduced coenzymes, NADH and $FADH_2$ are produced. These reduced coenzymes must be oxidized by electron transport so that the citric acid cycle can continue. And, it is electron transport that uses molecular oxygen. Thus, the citric acid cycle depends on the presence of molecular oxygen and is considered to be aerobic.

14. 1.7 mol ATP

16. The end product of the anaerobic catabolism of glucose in muscle tissue is lactate.

18. The gluconeogenesis pathway is considered to be anabolic because (a) smaller, noncarbohydrate precursors such as lactate are converted to the larger product, glucose, and (b) as this conversion takes place, the carbons become progressively more reduced.

20. HO—CH(COOH)—CH_2COOH + <u>NAD^+</u> ⟶ O=C(—COOH)—CH_2—COOH + <u>NADH</u> + H^+

22. The glycolysis pathway uses substrate-level phosphorylation to produce ATP and thus does not need electron transport and oxidative phosphorylation.

24. (0.38 moles acetyl CoA)(10 ATP/acetyl CoA) = 3.8 moles of ATP.

26. Unlike vitamins, which must be supplied in the diet, hormones are produced in the body.

28. The normal blood glucose concentration is 70–100 mg/100 mL of blood under fasting conditions. Hyperglycemia occurs when the blood glucose concentration rises above this range. Thus, 350 mg glucose/100 mL blood is hyperglycemic.

30. Glucose produced through liver gluconeogenesis may be exported into the blood stream to be used by other tissues (e.g., muscles) or the glucose may be stored in liver glycogen.

32. The polypeptide hormone is vasopressin.

33. The citric acid cycle depends on molecular oxygen. Large amounts of free oxygen were probably not available on earth until photosynthetic organisms had evolved. (Oxygen is a product of photosynthesis.)

34. Hormones function as chemical messengers. They are chemical substances that act as control or regulatory agents in the body. Enzymes are catalysts for specific reactions, allowing these reactions to occur faster and under milder conditions than would otherwise be possible.

38. $CH_3CH(OH)COO^-$

lactate

All three carbons of lactate are more reduced than the carbon in CO_2. Thus, lactate can provide more cellular energy upon further oxidation of the carbons.

Chapter 35

Review Questions

3. In the oxidation of fatty acids, the carbon atom beta to the carboxyl group is oxidized forming a β-keto acid which is then cleaved between the α and β carbon atoms leaving a new fatty acid that is two carbons shorter than the original fatty acid.

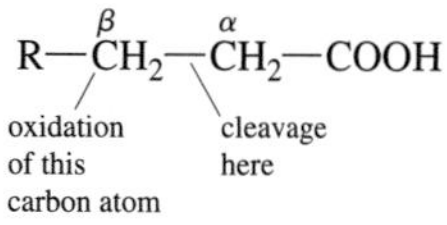

$$R-\overset{\beta}{C}H_2-\overset{\alpha}{C}H_2-COOH$$

oxidation of this carbon atom (β); cleavage here (between β and α)

9. In the enzymatic oxidation of fatty acids from fats, the fatty acids are oxidized, ultimately forming Acetyl-CoA. Acetyl-CoA enters the citric acid cycle to produce energy.

13. The nitrogen in L-glutamine is less basic than ammonia and, thus, is less toxic to biological organisms.

16. three ATPs

18. For a growing child the amount of nitrogen consumed is more than that excreted. Consequently the amount of amino acids from dietary protein increases in the nitrogen pools and the child is said to be in positive nitrogen balance.

Exercises

2. (CH_3CH_2)$CH_2CH_2CH_2COOH$; the two carbons circled remain.

4. $CH_3(CH_2)_{10}{}^*CH_2CH_2COOH$

6. 8 acetyl-CoA

8. 7 NADH

10. 25 ATP formed from lauric acid

12. (a) ACP is a protein whereas CoA is a coenzyme
(b) ACP is used in fatty acid synthesis whereas CoA is used in the β-oxidation pathway.

14. A rapid growth spurt commonly means the body is using ingested amino acids to build new proteins. Thus, less nitrogen will be excreted and the nitrogen balance should become more positive.

16. L-aspartic acid is not a ketogenic amino acid because a diet high in L-aspartic acid caused no increase in blood levels of the ketone body β-hydroxybutyric acid. (Other ketone bodies would not increase their blood levels also.) Because L-aspartic acid is not a ketogenic amino acid, it must be a glucogenic amino acid.

18. (a) $HOOCCH_2\overset{O}{\overset{\|}{C}}COOH$

(b) $C_6H_5-CH_2\overset{O}{\overset{\|}{C}}COOH$

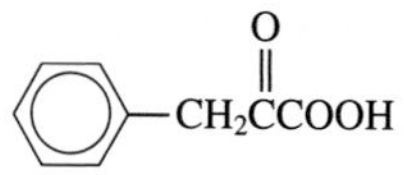

20.

$$\begin{array}{l} \quad\;\; COOH \\ \quad\;\; | \\ \quad\;\; CH_2 \\ \quad\;\; | \\ (H_2N)-CH-COOH \end{array}$$

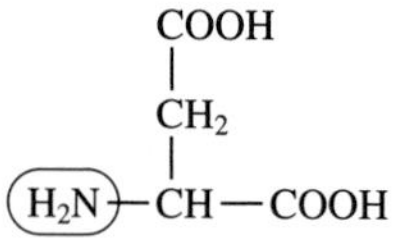

22. One mole of urea is produced for every mole of L-aspartic acid used. 0.84 moles of aspartic acid will yield 0.84 moles of urea. One mole of L-glutamic acid is used for every mole of L-aspartic acid. 0.84 moles of aspartic acid uses 0.84 moles of glutamic acid.

24. Although this is part of the nitrogen cycle, it is not nitrogen fixation. The bacteria are starting with ammonia rather than elemental nitration as is required in nitrogen fixation.

25. It is possible to accumulate body fat despite a low fat diet because the body is capable of synthesizing fats from nonfat starting materials. Fatty acids are synthesized from Acetyl-CoA (lipogenesis) which then combine with glycerol to form triacylglycerols (fats).

26. β-hydroxybutyric acid:

$$\begin{array}{l} CH_3CHCH_2COOH \\ \quad\;\;\; | \\ \quad\;\; OH \end{array}$$

32. $$(1.0 \text{ mol palmitic acid})\left(\frac{256.4 \text{ g}}{\text{mol palmitic acid}}\right) = 2.6 \times 10^2 \text{ g}$$

$$\left(\frac{9 \text{ kcal}}{\text{g}}\right)(2.6 \times 10^2 \text{ g}) = 2 \times 10^3 \text{ kcal}$$

APPENDIX VII

Answers to Putting It Together Review Exercises

Chapters 1–4

Multiple Choice: **1.** d **2.** a **3.** d **4.** b **5.** d **6.** c **7.** a **8.** d **9.** a **10.** b **11.** a **12.** d **13.** b **14.** c **15.** a **16.** d **17.** c **18.** b **19.** d **20.** c **21.** c **22.** a **23.** b **24.** c **25.** c **26.** a **27.** d **28.** d **29.** c **30.** a **31.** b **32.** a **33.** c **34.** c **35.** c **36.** d **37.** a **38.** a **39.** d **40.** d **41.** c **42.** c **43.** b **44.** b **45.** c **46.** b **47.** b

Free Response:

1. $(1.5\text{ m})\left(\frac{100\text{ cm}}{1\text{ m}}\right)\left(\frac{1\text{ in.}}{2.54\text{ cm}}\right)\left(\frac{1\text{ ft}}{12\text{ in.}}\right) = 4.9\text{ ft}$

$(4\text{ m})\left(\frac{100\text{ cm}}{1\text{ m}}\right)\left(\frac{1\text{ in.}}{2.54\text{ cm}}\right)\left(\frac{1\text{ ft}}{12\text{ in.}}\right) = 13\text{ ft}$

$(27°\text{C} \times 1.8) + 32 = 81°\text{F}$

2. Jane needs to time how long it took from starting to heat to when the butter is just melted. From this information, she can determine how much heat the pot and butter absorbed. Jane can look up the specific heat of copper. Jane should weigh the pot and measure the temperature of the pot and the temperature at which the butter just melted. This should allow Jane to calculate how much heat the pot absorbed. Then she simply has to subtract the heat the pot absorbed from the heat the stove put out to find out how much heat the butter absorbed.

3. $CaCO_3 \longrightarrow CaO + CO_2$

75 g   42 g   X

$X = 75\text{ g} - 42\text{ g} = 33\text{ g}$

44 g CO_2 occupies 24 dm^3

Therefore, 33 g CO_2 occupies $(33\text{ g})\left(\frac{24\text{ dm}^3}{44\text{ g}}\right)\left(\frac{1\text{ L}}{1\text{ dm}^3}\right) = 18\text{ L}$

4. (a), (b), and (c) Picture (2) best represents a homogeneous mixture. Pictures (1) and (3) show heterogeneous mixtures, and picture (4) does not show a mixture, as only one species is present.

Picture (1) likely shows a compound, as one of the components of the mixture is made up of more than one type of "ball." Picture (2) shows a component with more than one part, but the parts seem identical, and therefore it could be representing a diatomic molecule.

5. (a) Picture (3) because fluorine gas exists as a diatomic molecule.

(b) Other elements that exist as diatomic molecules are oxygen, nitrogen, chlorine, hydrogen, bromine, and iodine.

(c) Picture (2) could represent SO_3 gas.

6. (a) Tim's bowl should require less energy. Both bowls hold the same volume, but since snow is less dense than a solid block of ice, the mass of water in Tim's bowl is less than the mass of water in Sue's bowl. (Both bowls contain ice at 12°F.)

(b) $\dfrac{12°F - 32}{1.8} = -11°C$

Temperature change: $-11°C$ to $25°C = 36°C$

(c) temperature change: $-11°C$ to $0°C = 11°C$

specific heat of ice $= 2.059\ J/g°C$

$$\text{vol. of } H_2O = 1\ qt = (0.946\ L)\left(\frac{1000\ mL}{L}\right) = 946\ mL$$

$$\text{mass of ice} = \text{mass of water} = (946\ \cancel{mL})\left(\frac{1\ g}{1\ \cancel{mL}}\right) = 946\ g$$

$$\text{heat required} = (m)(\text{sp. ht.})(\Delta t)$$

$$= (946\ g)\left(\frac{2.059\ J}{g°C}\right)(11°C)\left(\frac{1\ kJ}{1000\ J}\right) = 21\ kJ$$

(d) Physical changes

7. (a) Let x = RDA of iron

60% of x = 11 mg Fe

$$x = \frac{11\ mg\ Fe \times 100\ \%}{60\ \%} = 18\ mg\ Fe$$

(b) density of iron $= 7.86\ g/mL$

$$V = \frac{m}{d} = (11\ mg\ Fe)\left(\frac{1\ g}{1000\ mg}\right)\left(\frac{1\ mL}{7.86\ g}\right) = 1.4 \times 10^{-3}\ mL\ Fe$$

8. (a) $Ca_3(PO_4)_2$: molar mass $= 3(40.08) + 2(30.97) + 8(16.00) = 310.3$

$$\%Ca \text{ in } Ca_3(PO_4)_2 = \frac{3(40.08)}{310.3}(100) = 38.7\%$$

Let x = mg $Ca_3(PO_4)_2$

38.7% of x = 162 mg Ca

$$x = \frac{(162\ mg)(100)}{38.7} = 419\ mg\ Ca_3(PO_4)_2$$

(b) $Ca_3(PO_4)_2$ is a compound.

(c) Convert 120 mL to cups

$$(120\ mL)\left(\frac{1\ L}{1000\ mL}\right)\left(\frac{1.059\ qt}{1\ L}\right)\left(\frac{4\ cups}{1\ qt}\right) = 0.51\ cup$$

13% of x = 0.51 cup

$$x = \frac{(0.51\ cup)(100)}{13} = 3.9\ cups$$

9. If Alfred inspects the bottles carefully, he should be able to see whether the contents are solid (silver) or liquid (mercury). Alternatively, since mercury is more dense than silver, the bottle of mercury should weigh more than the bottle of silver (the question indicated that both bottles were of similar size and both were full). Density is mass/volume.

10. (a) Container holds a mixture of sulfur and oxygen.

(b) No. If the container were sealed, the total mass would remain the same whether a reaction took place or not. The mass of the reactants must equal the mass of the products.

(c) No. Density is mass/volume. The volume is the container volume, which does not change. Since the total mass remains constant even if a reaction has taken place, the density of the container, including its contents, remains constant. The density of each individual component within the container may have changed, but the total density of the container is constant.

Chapters 5–6

Multiple Choice: **1.** b **2.** d **3.** b **4.** d **5.** b **6.** b **7.** a **8.** b **9.** d **10.** c **11.** b **12.** d **13.** a **14.** c **15.** d

Names and Formulas: The following are correct: 1, 2, 4, 5, 6, 7, 9, 11, 12, 15, 16, 17, 18, 19, 21, 22, 25, 28, 30, 32, 33, 34, 36, 37, 38, 40.

Free Response:

1. (a) An ion is a charged atom or group of atoms. The charge can be either positive or negative.

(b) Electrons have negligible mass compared with the mass of protons and neurons. The only difference between Ca and Ca^{2+} is two electrons. The mass of those two electrons is insignificant compared with the mass of the protons and neutrons present (and whose numbers do not change).

2. (a) Let x = abundance of heavier isotope.

$$303.9303(x) + 300.9326(1 - x) = 303.001$$

$$303.9303x - 300.9326x = 303.001 - 300.9326$$

$$2.9977x = 2.068$$

$$x = 0.6899$$

$$1 - x = 0.3101$$

% abundance of heavier isotope = 68.99%

% abundance of lighter isotope = 31.01%

(b) $^{304}_{120}Wz$, $^{301}_{120}Wz$

(c) mass number − atomic number = 303 − 120 = 183 neutrons

3. Cl_2O_7 Cl: $17p \times 2 = 34$ protons

O: $8p \times 7 = 56$ protons

90 protons in Cl_2O_7

Since the molecule is electrically neutral, the number of electrons is equal to the number of protons, so Cl_2O_7 has 90 electrons. The number of neutrons cannot be precisely determined unless it is known which isotopes of Cl and O are in this particular molecule.

4. Phosphate has a −3 charge; therefore, the formula for the ionic compound is $M_3(PO_4)_2$.

P has 15 protons; therefore, $M_3(PO_4)_2$ has 30 phosphorus protons.

$$3\text{ (number of protons in M)} = \frac{30 \times 6}{5} = 36 \text{ protons in 3 M}$$

$$\text{number of protons in M} = \frac{36}{3} = 12 \text{ protons}$$

from the periodic table, M is Mg.

5. (a) Iron can form cations with different charges (e.g., Fe^{2+} or Fe^{3+}). The Roman numeral indicating which cation of iron is involved is missing. This name cannot be fixed unless the particular cation of iron is specified.

(b) $K_2Cr_2O_7$. Potassium is generally involved in ionic compounds. The naming system used was for covalent compounds. The name should be potassium dichromate. (Dichromate is the name of the $Cr_2O_7^{2-}$ anion.)

(c) Sulfur and oxygen are both nonmetals and form a covalent compound. The number of each atom involved needs to be specified for covalent compounds. There are two common oxides of sulfur—SO_2 and SO_3. Both elements are non-metales, so the names should be sulfur dioxide and sulfur trioxide, respectively.

6. No. Each compound, SO_2 and SO_3, has a definite composition of sulfur and oxygen by mass. The law of multiple proportions says that two elements may combine in different ratios to form more than one compound.

7. (a) Electrons are not in the nucleus.
(b) When an atom becomes an anion, its size increases.
(c) An ion of Ca (Ca^{2+}) and an atom of Ar have the same number of electrons.

8. (a) $12 \text{ amu} \times 7.18 = 86.16 \text{ amu}$
(b) The atom is most likely Rb or Sr. Other remote possibilities are Kr or Y.
(c) Because of the possible presence of isotopes, the atom cannot be positively identified. The periodic table gives average masses.
(d) M forms a +1 cation and is most likely in group 1A. The unknown atom is most likely $^{86}_{37}Rb$.

9. The presence of isotopes contradicts Dalton's theory that all atoms of the same element are identical. Also, the discovery of protons, neutrons, and electrons suggests that there are particles smaller than the atom and that the atom is not indivisible.
Thomson proposed a model of an atom with no clearly defined nucleus.
Rutherford passed alpha particles through gold foil and inspected the angles at which the alpha particles were deflected. From his results, he proposed the idea of an atom having a small dense nucleus.

Chapters 7-9

Multiple Choice: **1.** a **2.** a **3.** d **4.** d **5.** a **6.** b **7.** a **8.** b **9.** b **10.** d **11.** a **12.** d **13.** b **14.** c **15.** c **16.** d **17.** d **18.** b **19.** b **20.** a **21.** b **22.** c **23.** d **24.** b **25.** b **26.** a **27.** d **28.** c **29.** c **30.** b **31.** c **32.** c **33.** b **34.** d **35.** b **36.** d **37.** c **38.** b **39.** c **40.** d **41.** a **42.** c **43.** c **44.** b **45.** b **46.** a **47.** d **48.** b **49.** b **50.** a

Free Response:

1. (a) $104 \text{ g } O_2 = (104 \text{ g } O_2)\left(\frac{1 \text{ mol}}{32.00 \text{ g}}\right) = 3.25 \text{ mol } O_2$

$$\underset{}{X} + \underset{3.25 \text{ mol}}{O_2} \longrightarrow \underset{2 \text{ mol}}{CO_2} + \underset{2.5 \text{ mol}}{H_2O} \quad \text{(multiply moles by 4)}$$

$$4X + 13\,O_2 \longrightarrow 8\,CO_2 + 10\,H_2O$$

Oxygen is balanced. By inspection, X must have 8/4 C atoms and 20/4 H atoms (2 C and 5 H).
Empirical formula is C_2H_5.

(b) Additional information needed is the molar mass of X.

2. (a)

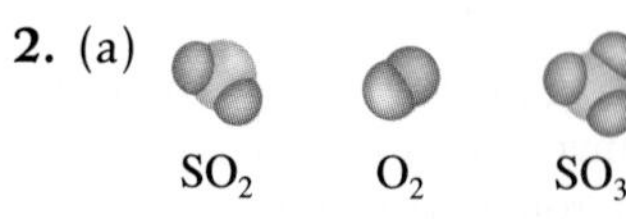

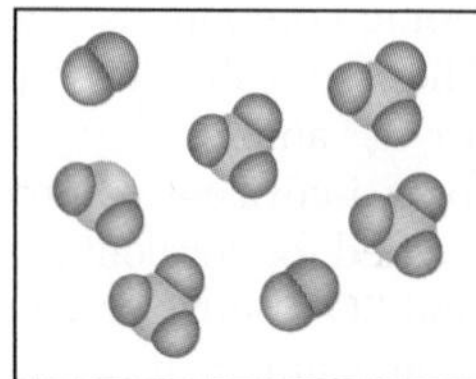

(b) $25\ g\ SO_2\left(\frac{1\ mol}{64.07\ g}\right) = 0.39\ mol\ SO_2$

$5\ g\ O_2\left(\frac{1\ mol}{32.00\ g}\right) = 0.16\ mol\ O_2$

$\text{mol ratio} = \frac{0.39}{0.16} = \frac{2.4\ mol\ SO_2}{1\ mol\ O_2}$

O_2 is the limiting reagent

(c) False. The percentages given are not mass percentages. The percent composition of S in SO_2 is $(32/64) \times 100 = 50.\%$ S. The percent composition of S in SO_3 is $(32/80) \times 100 = 40.\%$ S.

3. (a) $\%O = 100 - (63.16 + 8.77) = 28.07\%\ O$

Start with 100 g compound Z

C: $(63.16\ g)\left(\frac{1\ mol}{12.01\ g}\right) = 5.259\ mol$ $\qquad \frac{5.259\ mol}{1.75\ mol} = 2.998$

H: $(8.77\ g)\left(\frac{1\ mol}{1.008\ g}\right) = 8.70\ mol$ $\qquad \frac{8.70\ mol}{1.754\ mol} = 4.96$

O: $(28.07\ g)\left(\frac{1\ mol}{16.00\ g}\right) = 1.754\ mol$ $\qquad \frac{1.754\ mol}{1.754\ mol} = 1.000$

The ratio of C:H:O is 3:5:1.
The empirical formula is C_3H_5O.
molar mass = 114; mass of empirical formula is 57
Therefore, the molecular formula is $C_6H_{10}O_2$.

(b) $2\,C_6H_{10}O_2 + 15\,O_2 \longrightarrow 12\,CO_2 + 10\,H_2O$

4. (a) Compound A must have a lower activation energy than compound B because B requires heat to overcome the activation energy for the reaction.

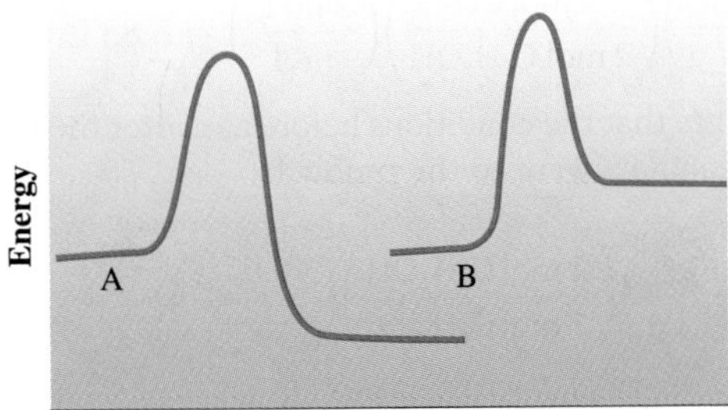

(b) (i) $2\,NaHCO_3 \xrightarrow{\Delta} Na_2CO_3 + H_2O + CO_2$

Decomposition of 0.500 mol $NaHCO_3$ requires 85.5 kJ of heat.
If 24.0 g CO_2 is produced, then

$(24.0\ g\ CO_2)\left(\frac{1\ mol\ CO_2}{44.01\ g}\right)\left(\frac{1\ mol\ H_2O}{1\ mol\ CO_2}\right)\left(\frac{18.02\ g}{1\ mol\ H_2O}\right) = 9.83\ g\ H_2O$

produced

0.500 mol $NaHCO_3$ produces

$(0.500\ mol\ NaHCO_3)\left(\frac{1\ mol\ CO_2}{2\ mol\ NaHCO_3}\right)\left(\frac{44.01\ g\ CO_2}{1\ mol\ CO_2}\right) = 11.0\ g\ CO_2$

producing 11.0 g CO_2 required 85.5 kJ

producing 24.0 g CO_2 requires $\left(\frac{24.0\ g}{11.0\ g}\right)(85.5\ kJ) = 187\ kJ$

(ii) $NaHCO_3$ could be compound B. Since heat was absorbed for the decomposition of $NaHCO_3$, the reaction was endothermic. Decomposition of A was exothermic.

5. (a) Double-displacement reaction

(b) $2\,NH_4OH(aq) + CoSO_4(aq) \longrightarrow (NH_4)_2SO_4(aq) + Co(OH)_2(s)$

(c) 8.09 g is 25% yield

Therefore, 100% yield $= (8.09\,g\,(NH_4)_2SO_4)\left(\frac{100\%}{25\%}\right) = 32.4\,g\,(NH_4)_2SO_4$ (theoretical yield)

(d) molar mass of $(NH_4)_2SO_4 = 132.2\,g/mol$

theoretical moles $(NH_4)_2SO_4 = (32.4\,g)\left(\frac{1}{132.2\,g/mol}\right) = 0.245\,mol\,(NH_4)_2SO_4$

Calculate the moles of $(NH_4)_2SO_4$ produced from 38.0 g of each reactant.

$$(38.0\,g\,NH_4OH)\left(\frac{1\,mol}{35.05\,g}\right)\left(\frac{1\,mol\,(NH_4)_2SO_4}{2\,mol\,NH_4OH}\right) = 0.542\,mol\,(NH_4)_2SO_4$$

$$(38.0\,g\,CoSO_4)\left(\frac{1\,mol}{155.0\,g}\right)\left(\frac{1\,mol\,(NH_4)_2SO_4}{1\,mol\,CoSO_4}\right) = 0.254\,mol\,(NH_4)_2SO_4$$

Limiting reactant is $CoSO_4$; NH_4OH is in excess

6. (a) $C_6H_{12}O_6 \longrightarrow 2\,C_2H_5OH + 2\,CO_2(g)$

Calculate the grams of $C_6H_{12}O_6$ that produced 11.2 g C_2H_5OH.

$$(11.2\,g\,C_2H_5OH)\left(\frac{1\,mol}{46.07\,g}\right)\left(\frac{1\,mol\,C_6H_{12}O_6}{2\,mol\,C_2H_5OH}\right)\left(\frac{180.1\,g}{1\,mol}\right) = 21.9\,g\,C_6H_{12}O_6$$

$25.0\,g - 21.9\,g = 3.1\,g\,C_6H_{12}O_6$ left unreacted

Volume of CO_2 produced:

$$(11.2\,g\,C_2H_5OH)\left(\frac{1\,mol}{46.07\,g}\right)\left(\frac{2\,mol\,CO_2}{2\,mol\,C_2H_5OH}\right)\left(\frac{24.0\,L}{mol}\right) = 5.83\,L$$

The assumptions made are that the conditions before and after the reaction are the same and that all reactants went to the products.

(b) theoretical yield

$$= (25.0\,g\,C_6H_{12}O_6)\left(\frac{1\,mol}{180.1\,g}\right)\left(\frac{2\,mol\,C_2H_5OH}{1\,mol\,C_6H_{12}O_6}\right)\left(\frac{46.07\,g}{mol}\right)$$

$= 12.8\,g\,C_2H_5OH$

$$\%\ yield = \left(\frac{11.2\,g}{12.8\,g}\right)(100) = 87.5\%$$

(c) decomposition reaction

7. (a) double decomposition (precipitation)

(b) lead(II) iodide (PbI_2)

(c) $Pb(NO_3)_2(aq) + 2\,KI(aq) \longrightarrow 2\,KNO_3(aq) + PbI_2(s)$

If $Pb(NO_3)_2$ is limiting, the theoretical yield is

$$(25\,g\,Pb(NO_3)_2)\left(\frac{1\,mol}{331.2\,g}\right)\left(\frac{1\,mol\,PbI_2}{1\,mol\,Pb(NO_3)_2}\right)\left(\frac{461.0\,g}{mol}\right) = 35\,g\,PbI_2$$

If KI is limiting, the theoretical yield is

$$(25\,g\,KI)\left(\frac{1\,mol}{166.0\,g}\right)\left(\frac{1\,mol\,PbI_2}{2\,mol\,KI}\right)\left(\frac{461.0\,g}{mol}\right) = 35\,g\,PbI_2$$

$$\text{percent yield} = \left(\frac{7.66\,g}{35\,g}\right)(100) = 22\%$$

8. (a) Balance the equation

$$2\,XNO_3 + CaCl_2 \longrightarrow 2\,XCl + Ca(NO_3)_2$$

$$(30.8\text{ g } CaCl_2)\left(\frac{1\text{ mol}}{111.0\text{ g}}\right)\left(\frac{2\text{ mol XCl}}{1\text{ mol } CaCl_2}\right) = 0.555\text{ mol XCl}$$

$$\text{Therefore, molar mass of XCl} = \frac{79.6\text{ g}}{0.555\text{ mol}} = 143\text{ g/mol}$$

mass of (X + Cl) = mass of XCl
mass of X = 143 − 35.45 = 107.6
X = Ag (from periodic table)

(b) No. Ag is below H in the activity series.

9. (a) $2\,H_2O_2 \longrightarrow 2\,H_2O + O_2$
There must have been eight H_2O_2 molecules and four O_2 molecules in the flask at the start of the reaction.
(b) The reaction is exothermic.
(c) Decomposition reaction
(d) The empirical formula is OH.

Chapters 10–11

Multiple Choice: **1.** c **2.** a **3.** b **4.** a **5.** a **6.** b **7.** b **8.** b **9.** c **10.** a **11.** d **12.** d **13.** a **14.** b **15.** c **16.** c **17.** b **18.** c **19.** d **20.** d **21.** d **22.** a **23.** c **24.** c **25.** a **26.** b **27.** d **28.** a **29.** b **30.** b **31.** a **32.** b **33.** c **34.** c **35.** d **36.** a **37.** a **38.** c **39.** d **40.** b **41.** c **42.** c **43.** c **44.** a

Free Response:

1. The compound will be ionic because there is a very large difference in electronegativity between elements in Group 2A and those in Group 7A of the Periodic Table. The Lewis structure is

   $[M]^{2+}$ $[:\ddot{\underset{..}{X}}:]^-$ $[:\ddot{\underset{..}{X}}:]^-$

2. Having an even atomic number has no bearing on electrons being paired. An even atomic number means only that there is an even number of electrons. For example, carbon is atomic number six, and it has two unpaired *p* electrons: $1s^2 2s^2 2p_x^1 2p_y^1$.
3. False. The noble gases do not have any unpaired electrons. Their valence shell electron structure is ns^2np^6 (except He).
4. The outermost electron in potassium is farther away from the nucleus than the outermost electrons in calcium, so the first ionization energy of potassium is lower than that of calcium. However, once potassium loses one electron, it achieves a noble gas electron configuration, and therefore taking a second electron away requires considerably more energy. For calcium, the second electron is still in the outermost shell and does not require as much energy to remove it.
5. The ionization energy is the energy required to *remove* an electron. A chlorine atom forms a chloride ion by *gaining* an electron to achieve a noble gas configuration.
6. The anion is Cl^-; therefore, the cation is K^+ and the noble gas is Ar. K^+ has the smallest radius, while Cl^- will have the largest. K loses an electron, and therefore, in K^+, the remaining electrons are pulled in even closer. Cl was originally larger than Ar, and gaining an electron means that, since the nuclear charge is exceeded by the number of electrons, the radius will increase relative to a Cl atom.
7. The structure shown in the question implies covalent bonds between Al and F, since the lines represent shared electrons. Solid AlF_3 is an ionic compound and therefore probably exists as an Al^{3+} ion and three F^- ions. Only valence electrons are shown in Lewis structures.

8. Carbon has four valence electrons; it needs four electrons to form a noble gas electron structure. By sharing four electrons, a carbon atom can form four covalent bonds.
9. NCl_3 is pyramidal. The presence of three pairs of electrons and a lone pair of electrons around the central atom (N) gives the molecule a tetrahedral structure and a pyramidal shape. BF_3 has three pairs of electrons and no lone pairs of electrons around the central atom (B), so both the structure and the shape of the molecule are trigonal planar.
10. The atom is Br ($35e^-$), which should form a slightly polar covalent bond with sulfur. The Lewis structure of Br is $:\ddot{\underset{..}{Br}}\cdot$.

Chapters 12–14

Multiple Choice: **1.** b **2.** a **3.** b **4.** c **5.** a **6.** d **7.** d **8.** b **9.** a **10.** b **11.** c **12.** a **13.** c **14.** c **15.** d **16.** a **17.** c **18.** a **19.** a **20.** d **21.** b **22.** c **23.** a **24.** a **25.** d **26.** d **27.** a **28.** b **29.** c **30.** a **31.** a **32.** c **33.** c **34.** c **35.** a **36.** b **37.** b **38.** c **39.** d **40.** a **41.** c **42.** d **43.** b **44.** c **45.** c **46.** c **47.** a **48.** c **49.** d **50.** b **51.** a **52.** c **53.** d **54.** b **55.** b **56.** a **57.** d **58.** c **59.** b **60.** b

Free Response:

1. 10.0% (m/v) has 10.0 g KCl per 100. mL of solution:

   Therefore, KCl solution contains

   $$\left(\frac{10.0\text{ g KCl}}{100.\text{ mL}}\right)(215\text{ mL})\left(\frac{1\text{ mol}}{74.55\text{ g}}\right) = 0.288\text{ mol KCl}$$

   NaCl solution contains $\left(\frac{1.10\text{ mol NaCl}}{\text{L}}\right)\left(\frac{1\text{ L}}{1000\text{ mL}}\right)(224\text{ mL}) = 0.246\text{ mol NaCl}$

   The KCl solution has more particles in solution and will have the higher boiling point.

2. $$\text{Mass of } CO_2 \text{ in solution} = (\text{molar mass})(\text{moles}) = (\text{molar mass})\left(\frac{PV}{RT}\right)$$

   $$= \left(\frac{44.01\text{ g }CO_2}{\text{mol}}\right)\left(\frac{1\text{ atm} \times 1.40\text{ L}}{\frac{0.08206\text{ L atm}}{\text{mol K}} \times 298\text{ K}}\right)$$

   $$= 2.52\text{ g }CO_2$$

   $$\text{mass of soft drink} = (345\text{ mL})\left(\frac{0.965\text{ g}}{\text{mL}}\right) = 333\text{ g}$$

   $$\text{ppm of }CO_2 = \left(\frac{2.52\text{ g}}{333\text{ g} + 2.52\text{ g}}\right)(10^6) = 7.51 \times 10^3\text{ ppm}$$

3. 10% KOH m/v solution contains 10 g KOH in 100 mL solution.

   10% KOH by mass solution contains 10 g KOH + 90 g H_2O.

   The 10% by mass solution is the more concentrated solution and therefore would require less volume to neutralize the HCl.

4. (a) $\dfrac{0.355\text{ mol}}{0.755\text{ L}} = 0.470\text{ M}$

   (b) The lower pathway represents the evaporation of water; only a phase change occurs; no new substances are formed. The upper path represents the decomposition of water. The middle path is the ionization of water.

5. Zack went to Ely, Gaye went to the Dead Sea, and Lamont was in Honolulu. Zack's b.p. was lowered, so he was in a region of lower atmospheric pressure (on a mountain). Lamont was basically at sea level, so his b.p. was about normal. Since Gaye's boiling point was raised, she was at a place of higher atmospheric pressure and therefore was possibly in a location below sea level. The Dead Sea is below sea level.

6. The particles in solids and liquids are close together (held together by intermolecular attractions), and an increase in pressure is unable to move them significantly closer to each other. In a gas, the space between molecules is significant, and an increase in pressure is often accompanied by a decrease in volume.

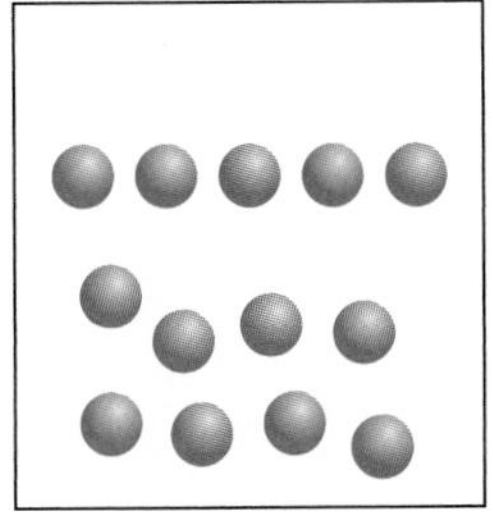
Liquid

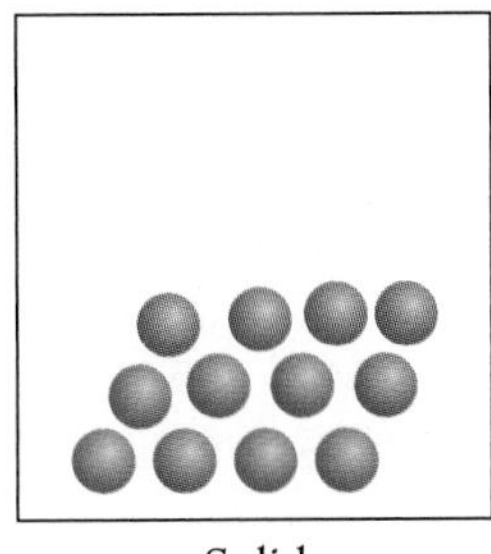
Solid

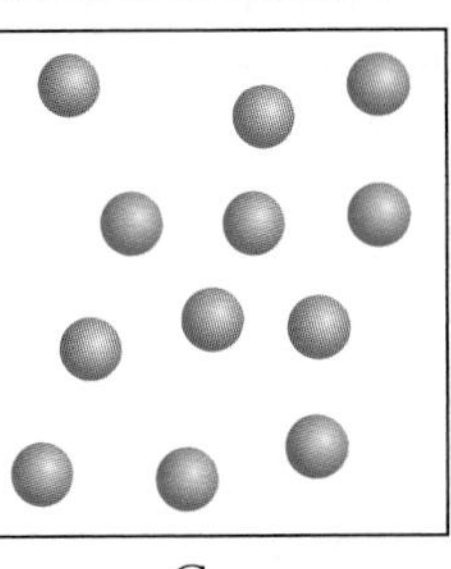
Gas

7. (a) The CO_2 balloon will be heaviest, followed by the Ar balloon. The H_2 balloon would be the lightest. Gases at the same temperature, pressure, and volume contain the same number of moles. All balloons will contain the same number of moles of gas molecules, so when moles are converted to mass, the order from heaviest to lightest is CO_2, Ar, H_2.

(b) Molar mass: O_2, 32.00; N_2, 28.02; Ne, 20.18

Using equal masses of gas, we find that the balloon containing O_2 will have the lowest number of moles of gas. Since pressure is directly proportional to moles, the balloon containing O_2 will have the lowest pressure.

8. Ray probably expected to get 0.050 moles $Cu(NO_3)_2$, which is

$$(0.050\ \text{mol Cu(NO}_3)_2)\left(\frac{187.6\ \text{g}}{\text{mol}}\right) = 9.4\ \text{g Cu(NO}_3)_2$$

The fact that he got 14.775 g meant that the solid blue crystals were very likely a hydrate containing water of crystallization.

9. For most reactions to occur, molecules or ions need to collide. In the solid phase, the particles are immobile and therefore do not collide. In solution or in the gas phase, particles are more mobile and can collide to facilitate a chemical reaction.

10. $\Delta t_b = 81.48°\text{C} - 80.1°\text{C} = 1.38°\text{C}$

$$K_b = 2.53\frac{°\text{C kg solvent}}{\text{mol solute}}$$

$$\Delta t_b = mK_b$$

$$1.38°\text{C} = m\left(2.53\frac{°\text{C kg solvent}}{\text{mol solute}}\right)$$

$$0.545\frac{\text{mol solute}}{\text{kg solvent}} = m$$

Now we convert molality to molarity.

$$\left(\frac{5.36\ \text{g solute}}{76.8\ \text{g benzene}}\right)\left(\frac{1000\ \text{g benzene}}{1\ \text{kg benzene}}\right)\left(\frac{1\ \text{kg benzene}}{0.545\ \text{mol solute}}\right) = 128.\ \text{g/mol}$$

Chapters 15–17

Multiple Choice: **1.** c **2.** d **3.** d **4.** c **5.** c **6.** d **7.** a **8.** d **9.** b **10.** c **11.** a **12.** d **13.** c **14.** b **15.** b **16.** d **17.** c **18.** a **19.** a **20.** a **21.** b **22.** c **23.** a **24.** c **25.** c **26.** d **27.** a **28.** b **29.** a **30.** b **31.** a **32.** c **33.** a **34.** b **35.** c **36.** c **37.** b **38.** d **39.** a **40.** c **41.** a **42.** b **43.** a **44.** b **45.** a **46.** a **47.** d **48.** a **49.** c **50.** d **51.** b **52.** a **53.** a **54.** b

Balanced Equations:

55. $3\,P + 5\,HNO_3 \longrightarrow 3\,HPO_3 + 5\,NO + H_2O$

56. $2\,MnSO_4 + 5\,PbO_2 + 3\,H_2SO_4 \longrightarrow 2\,HMnO_4 + 5\,PbSO_4 + 2\,H_2O$

57. $Cr_2O_7^{2-} + 14\,H^+ + 6\,Cl^- \longrightarrow 2\,Cr^{3+} + 7\,H_2O + 3\,Cl_2$

58. $2\,MnO_4^- + 5\,AsO_3^{3-} + 6\,H^+ \longrightarrow 2\,Mn^{2+} + 5\,AsO_4^{3-} + 3\,H_2O$

59. $S^{2-} + 4\,Cl_2 + 8\,OH^- \longrightarrow SO_4^{2-} + 8\,Cl^- + 4\,H_2O$

60. $4\,Zn + NO_3^- + 6\,H_2O + 7\,OH^- \longrightarrow 4\,Zn(OH)_4^{2-} + NH_3$

61. $2\,KOH + Cl_2 \longrightarrow KCl + KClO + H_2O$

62. $4\,As + 3\,ClO_3^- + 6\,H_2O + 3\,H^+ \longrightarrow 4\,H_3AsO_3 + 3\,HClO$

63. $2\,MnO_4^- + 10\,Cl^- + 16\,H^+ \longrightarrow 2\,Mn^{2+} + 5\,Cl_2 + 8\,H_2O$

64. $Cl_2O_7 + 4\,H_2O_2 + 2\,OH^- \longrightarrow 2\,ClO_2^- + 4\,O_2 + 5\,H_2O$

Free Response:

1. $2\,Bz + 3\,Yz^{2+} \longrightarrow 2\,Bz^{3+} + 3\,Yz$
Bz is above Yz in the activity series.

2. (a) $2\,Al(s) + 3\,Fe(NO_3)_2(aq) \longrightarrow 2\,Al(NO_3)_3(aq) + 3\,Fe(s)$
(b) The initial solution of $Fe(NO_3)_2$ will have the lower freezing point. It has more particles in solution than the product.

3. (a) $pH = -\log[H^+] = -\log[0.10] = 1.00$
(b) $\text{mol HCl} = 0.050\text{ L} \times \dfrac{0.10\text{ mol}}{\text{L}} = 0.0050\text{ mol HCl}$

Flask A
$Zn(s) + 2\,HCl(aq) \longrightarrow ZnCl_2(aq) + H_2(g)$
HCl is the limiting reactant, so no HCl will remain in the product.
$pH = 7.0$

Flask B
No reaction occurs in flask B, so the pH does not change.
$pH = 1.00$

4. (a) $2\,NaOH(aq) + H_2S(aq) \longrightarrow Na_2S(aq) + 2\,H_2O(l)$
(b) $H_2S \rightleftharpoons 2H^+ + S^{2-}$ (aqueous solution)
$Na_2S \longrightarrow 2\,Na^+ + S^{2-}$ (aqueous solution)
The addition of S^{2-} to a solution of H_2S will shift the equilibrium to the left, reducing the $[H^+]$ and thereby increasing the pH (more basic).

5. (a) Yes. The K_{eq} of AgCN indicates that it is slightly soluble in water, so a precipitate will form.
Net ionic equation: $Ag^+(aq) + CN^-(aq) \rightleftharpoons AgCN(s)$
(b) NaCN is a salt of a weak acid and a strong base and will hydrolyze in water.
$CN^-(aq) + H_2O(l) \rightleftharpoons HCN(aq) + OH^-(aq)$
The solution will be basic due to increased OH^- concentration.

6. (a)

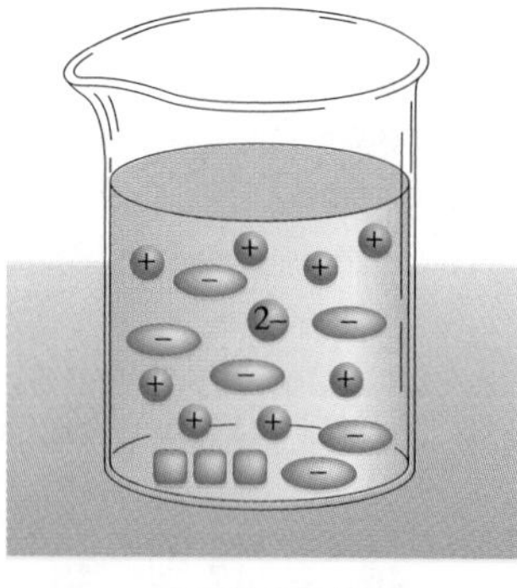

■ = $PbSO_4(s)$

(b)

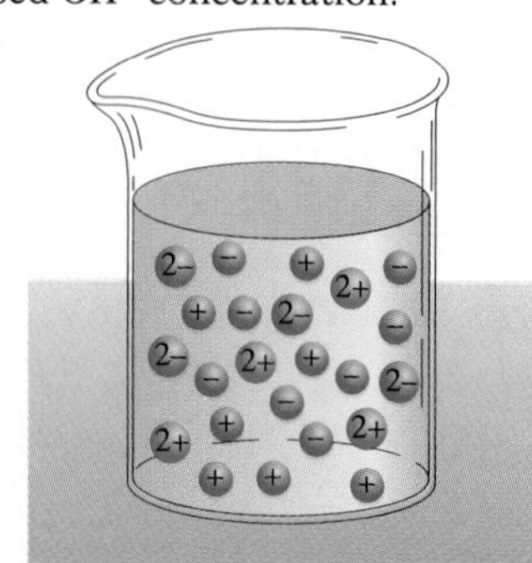

No reaction—contents are merely mixed.

7. (a) $2\,A_3X \rightleftharpoons 2\,A_2X + A_2$

$$K_{eq} = \frac{(A_2X)^2(A_2)}{(A_3X)^2} = \frac{(3)^2(6)}{(4)^2} = 3.375$$

(b) The equilibrium lies to the right. $K_{eq} > 1$

(c) Yes, it is a redox reaction because the oxidation state of A has changed. The oxidation state of A in A_2 must be 0, but the oxidation state of A in A_3X is not 0.

8. (a) $X_2 + 2\,G \rightleftharpoons X_2G_2$

$$K_{eq} = \frac{(X_2G_2)}{(X_2)(G)^2} = \frac{(1)}{(3)(2)^2} = 8.33 \times 10^{-2}$$

(b) Exothermic. An increase in the amount of reactants means that the equilibrium shifted to the left.

(c) An increase in pressure will cause an equilibrium to shift by reducing the number of moles of gas in the equilibrium. If the equilibrium shifts to the right, there must be fewer moles of gas in the product than in the reactants.

9. The pH of the solution is 4.5. (Acid medium)

$$5\,Fe^{2+} + MnO_4^- + 8\,H^+ \longrightarrow 5\,Fe^{3+} + Mn^{2+} + 4\,H_2O$$

Chapter 18

Multiple Choice: **1.** b **2.** d **3.** a **4.** a **5.** d **6.** b **7.** b **8.** c **9.** c **10.** c **11.** a **12.** c **13.** b **14.** b **15.** c **16.** d

Free Response:

1. The nuclide lost two alpha particles and one beta particle in any order. One series could be ${}^{223}_{87}Fr \xrightarrow{-\alpha} {}^{219}_{85}At \xrightarrow{-\beta} {}^{219}_{86}Rn \xrightarrow{-\alpha} {}^{215}_{84}Po$. ${}^{215}_{84}Po$ is the resulting nuclide.

2. $96\text{ g} \longrightarrow 48\text{ g} \longrightarrow 24\text{ g} \longrightarrow 12\text{ g} \longrightarrow 6\text{ g} \longrightarrow 3\text{ g} \longrightarrow 1.5\text{ g}$

Each change involves $\frac{1}{2}$ life. Six (6) half-lives are required to go from 96 g to 1.5 g.

The half-life is four days delays.

3. (a) The process is transmutation.

(b) Transmutation produces a different element by bombarding an element with a small particle such as hydrogen, deuterium, or a neutron. Radioactive decay is the formation of a different element by the loss of an alpha or beta particle from the nucleus of a radioactive element.

(c) No. Nuclear fission involves the bombardment of a heavy element with a neutron, causing the element to split (fission) into two or more lighter elements and produce two or more neutrons. A large amount of energy also is formed in a nuclear fission.

4. ${}^{56}_{26}Fe$ is made up of 26 protons, 30 neutrons, and 26 electrons.

Calculated mass:

26 protons	26 (1.0073 g/mol) =	26.1898 g/mol
30 neutrons	30 (1.0087 g/mol) =	30.2610 g/mol
26 electrons	26 (0.00055 g/mol) =	0.0143 g/mol
calculated mass		56.4651 g/mol

mass defect = calculated mass – actual mass

= 56.4651 g/mol – 55.9349 g/mol = 0.5302 g/mol

$$\text{Nuclear binding energy} = \left(\frac{0.5302\text{ g}}{\text{mol}}\right)\left(\frac{9.0 \times 10^{13}\text{ J}}{\text{g}}\right) = 4.8 \times 10^{13}\text{ J/mol}$$

Chapters 19–21

Multiple Choice: **1.** c **2.** b **3.** c **4.** a **5.** c **6.** d **7.** a **8.** a **9.** d **10.** c **11.** c **12.** b **13.** b **14.** c **15.** d **16.** a **17.** c **18.** d **19.** b **20.** d **21.** a **22.** c **23.** d **24.** c **25.** c **26.** d **27.** a **28.** c **29.** c **30.** a **31.** a **32.** d **33.** b **34.** a **35.** d **36.** c **37.** b **38.** a **39.** d **40.** b **41.** d **42.** c **43.** a **44.** a **45.** c **46.** b **47.** c **48.** c **49.** c **50.** b **51.** b **52.** d **53.** a **54.** b **55.** a **56.** c **57.** d **58.** b **59.** c **60.** b **61.** c **62.** b

Free Response:

1. Alkane carbon atoms are already bonded to four different atoms and thus cannot add any more atoms to their structure.
2. Since pot handles are going to be subject to heating, a thermosetting plastic is preferable over a thermoplastic polymer. Thermosetting plastics tend to be significantly crosslinked and therefore are better plastics for making pot handles. Thermosetting plastics tend to be condensation polymers.
3. From left to right, the names are methylcyclohexane, toluene, 1-methylcyclohexene. The functional groups are alkane, aromatic ring, alkene. The common type of reactions these functional groups undergo are substitution reactions for the alkane and aromatic compounds, and addition reactions for the alkene.
4. Each of these compounds will react with bromine but if the unknown is either of compounds I and II, additional reagents/conditions will be needed. In all situations the disappearance of the bromine color will indicate a reaction (positive result) and therefore the identity of the unknown. Add bromine to the unknown. A positive result indicates compound III is the unknown. If a negative result, try adding $FeBr_3$. A positive result indicates compound I. Two negative results indicate compound II. This can be verified by adding bromine in the presence of light or high heat to compound III and looking for a positive result.
5. Alkynes cannot have geometric isomers because they have only one group attached to the alkyne carbons and they are linear compounds around the alkyne bond. Cycloalkanes with substituents on different carbons can have geometric isomers.
6. Because each pair of compounds are hydrocarbons with approximately the same molar mass and relatively similarly shaped, predicting a significant difference in boiling point or solubility would be difficult. Their intermolecular attractions are similar and no molecule is significantly more polar. In actuality, their boiling points are very similar. All would be expected to be insoluble in water, again because they have similar types of intermolecular forces. However, their chemical properties are markedly different, for example:
   (a) Reaction with $Br_2/FeBr_3$.—benzene will react, cyclohexane will not.
   (b) Reaction with $KMnO_4$—1-hexyne will react, hexane will not.
   (c) Reaction with $KMnO_4$—cyclohexene will react, cyclohexane will not unless subjected to harsh conditions.
7. Because there is only one monomer involved, Plexiglass is likely an addition polymer. The structure of the monomer therefore probably has a double bond (alkene) which is used in the addition reaction.

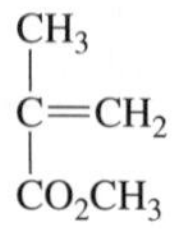

$$\begin{array}{c} CH_3 \\ | \\ C{=}CH_2 \\ | \\ CO_2CH_3 \end{array}$$

8. $CH_3CH_3 \xrightarrow[\text{light}]{Cl_2} CH_3CH_2Cl \xrightarrow[AlCl_3]{\text{benzene}} C_6H_5CH_2CH_3 \xrightarrow[H_2SO_4]{K_2Cr_2O_7} C_6H_5COOH$

chloroethane ethylbenzene benzoic acid

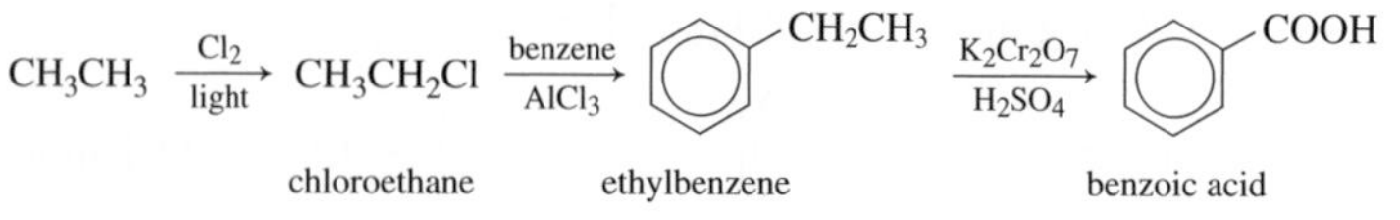

Chapters 22–23

Multiple Choice: **1.** b **2.** d **3.** d **4.** c **5.** d **6.** c **7.** a **8.** c **9.** d **10.** b **11.** b **12.** a **13.** b **14.** c **15.** b **16.** c **17.** b **18.** b **19.** c **20.** b **21.** d **22.** b **23.** c **24.** c **25.** a **26.** b **27.** d **28.** a **29.** c **30.** b **31.** a **32.** c **33.** c **34.** c **35.** d **36.** a **37.** b **38.** c **39.** d **40.** c **41.** b **42.** b **43.** b **44.** a **45.** d **46.** b **47.** c **48.** a **49.** a **50.** c **51.** a **52.** b **53.** b **54.** c **55.** d **56.** b **57.** b **58.** c **59.** d **60.** d **61.** b **62.** c

Free Response:

1. Based on the formula, and the other conditions given, the compound cannot be an alcohol or an ether. Therefore it must be either an aldehyde or a ketone. Try oxidizing a sample of the compound with dichromate. If a color change from orange to green is obtained, the compound underwent oxidation and is an aldehyde. If there is no color change, the compound did not oxidize and is a ketone.
2. Aldehydes and ketones have a carbon that is attached to only three different atoms and therefore involve a pi bond. Addition reactions generally involve the breaking of a pi bond and the formation of a sigma bond, adding a new atom to each of the original participants in the pi bond.
3. The compound contains an ether, an alcohol, a ketone, and an aldehyde (from left to right).

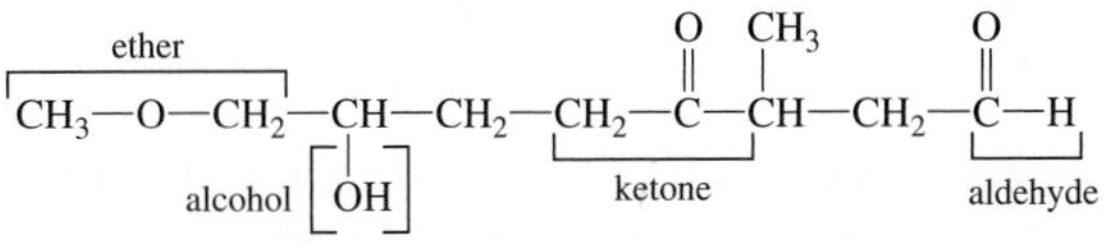

$$\underbrace{CH_3{-}O{-}CH_2}_{\text{ether}}{-}\underset{\underset{\text{alcohol}}{OH}}{CH}{-}CH_2{-}CH_2{-}\underbrace{\overset{O}{\overset{\|}{C}}{-}\overset{CH_3}{\overset{|}{CH}}}_{\text{ketone}}{-}CH_2{-}\underbrace{\overset{O}{\overset{\|}{C}}{-}H}_{\text{aldehyde}}$$

4. An aldehyde must be at the end of a carbon chain since a hydrogen must be attached to the carbonyl carbon. None of the other groups listed have this restriction. Examples are illustrated below.

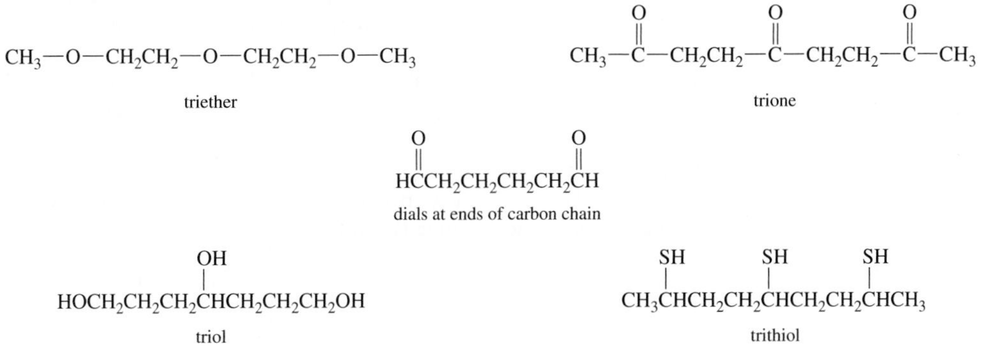

$$CH_3{-}O{-}CH_2CH_2{-}O{-}CH_2CH_2{-}O{-}CH_3$$
triether

$$CH_3{-}\overset{O}{\overset{\|}{C}}{-}CH_2CH_2{-}\overset{O}{\overset{\|}{C}}{-}CH_2CH_2{-}\overset{O}{\overset{\|}{C}}{-}CH_3$$
trione

$$H\overset{O}{\overset{\|}{C}}CH_2CH_2CH_2CH_2\overset{O}{\overset{\|}{C}}H$$
dials at ends of carbon chain

$$HOCH_2CH_2CH_2\overset{OH}{\overset{|}{C}}HCH_2CH_2CH_2OH$$
triol

$$CH_3\overset{SH}{\overset{|}{C}}HCH_2CH_2\overset{SH}{\overset{|}{C}}HCH_2CH_2\overset{SH}{\overset{|}{C}}HCH_3$$
trithiol

5. The most soluble should be the one that is able to participate most effectively in hydrogen bonding with water, that is, 1-propanol. The least soluble compound should be the least polar compound, in this case. butene. Butene has no polar groups and is unable to effectively associate with water.
6.

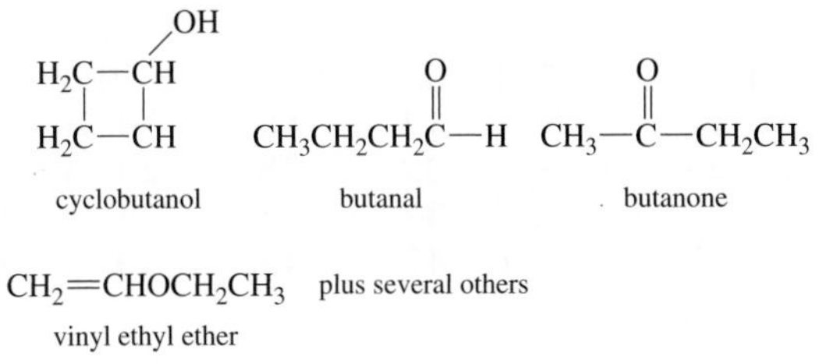

$$\begin{matrix} H_2C{-}CH{-}OH \\ | \quad\ \ | \\ H_2C{-}CH \end{matrix}$$
cyclobutanol

$$CH_3CH_2CH_2\overset{O}{\overset{\|}{C}}{-}H$$
butanal

$$CH_3{-}\overset{O}{\overset{\|}{C}}{-}CH_2CH_3$$
butanone

$CH_2{=}CHOCH_2CH_3$ plus several others
vinyl ethyl ether

7. Cyclohexyl–CH–OH $\xrightarrow[H_2SO_4]{K_2Cr_2O_7}$ cyclohexyl–C=O (I) $\xrightarrow{HCN}$ cyclohexyl–C(OH)(CN) (II) $\xrightarrow[H^+]{H_2O}$ cyclohexyl–C(OH)(COOH) (III)

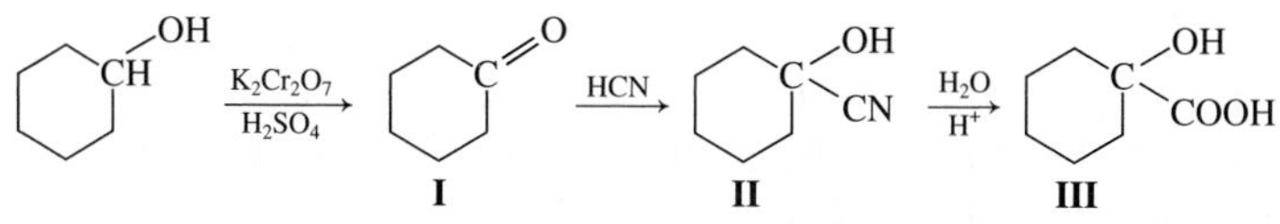

8. $CH_3CH_2CH_2\overset{O}{\overset{\|}{C}}-H$ butanal **I**

aldol condensation (I) → $CH_3CH_2CH_2\overset{OH}{\overset{|}{C}}H\underset{CH_2CH_3}{\underset{|}{C}}H\overset{O}{\overset{\|}{C}}-H$ **IV**
2-ethyl-3-hydroxyhexanal

$LiAlH_4$ or $NaBH_4$ (I) → **II** $CH_3CH_2CH_2\overset{OH}{\overset{|}{C}}H_2$ $\xrightarrow{Na}$ $\xrightarrow{CH_3CH_2Br}$ $CH_3CH_2CH_2CH_2OCH_2CH_3$
ethoxybutane or butyl ethyl ether **III**

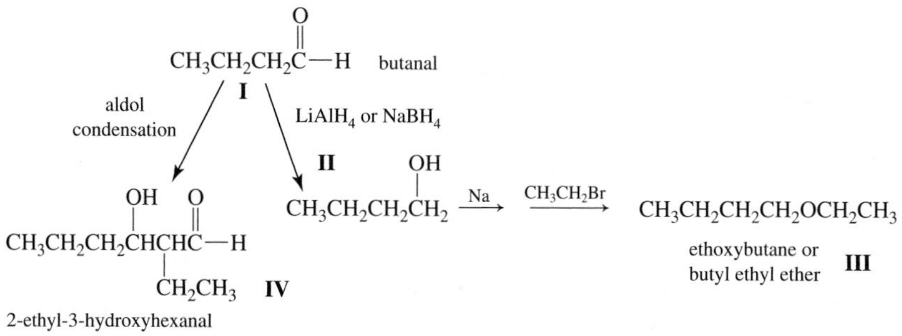

9. Aldehydes and ketones of the same molar mass are difficult to distinguish from each other based on boiling points so they will need to be identified based on a chemical test, such as a Tollens test, which will give a positive result (silver mirror) for the aldehyde. The rest of the compounds should boil in the following order from lowest boiling point to highest:

dipropyl ether < 1-pentanethiol < 1-hexanol
91°C 126°C 155°C

These boiling points would be based on the strengths of the intermolecular attractions. The alcohol has hydrogen-bonding, while the thiol is more polar than the symmetrical ether. Note that the thiol may be difficult to distinguish from the aldehyde and ketone in terms of boiling point also, but can probably be fairly easily identified based on odor.

10. (a) phenol and 1-hexanol—treat with NaOH and see which becomes miscible with water (phenol); or treat with $K_2Cr_2O_7/H_2SO_4$ (or $KMnO_4$ in alkaline solution) and look for a color change (hexanol)
(b) phenol and hexanal—treat with NaOH and see which becomes miscible with water (phenol); or perform the Tollens test and look for a positive result (hexanal)
(c) 1-hexanol and 3-hexanone—treat with Na and see which bubbles (reacts with hexanol to produce H_2); or treat with $K_2Cr_2O_7/H_2SO_4$ (or $KMnO_4$ in alkaline solution) and look for a color change (hexanol)

11. O=C(cyclohexanone)–$CH_2CH_2-\overset{O}{\overset{\|}{C}}-H$ $\xrightarrow[H^+]{HOCH_2CH_2OH}$ O=C(cyclohexanone)–CH_2CH_2–C(H) with cyclic acetal O–H_2C–CH_2–O

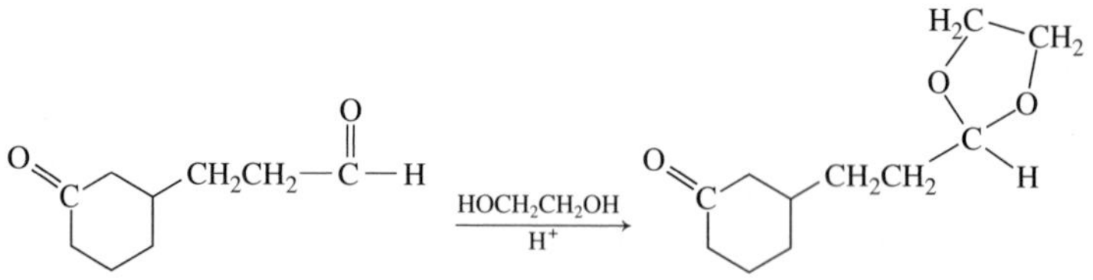

Chapters 24–25

Multiple Choice: **1.** a **2.** c **3.** c **4.** c **5.** b **6.** b **7.** d **8.** b **9.** c **10.** d
11. d **12.** a **13.** c **14.** b **15.** c **16.** b **17.** c **18.** c **19.** a **20.** b
21. d **22.** c **23.** c **24.** b **25.** a **26.** c **27.** c **28.** c **29.** a **30.** c
31. c **32.** d **33.** c **34.** c **35.** d **36.** b **37.** a **38.** c **39.** b **40.** a
41. b **42.** b **43.** a **44.** b **45.** d **46.** d **47.** d **48.** b **49.** a **50.** c
51. c **52.** c **53.** c **54.** b **55.** d

Free Response:

1. Fatty acid I is most likely synthetically produced since most naturally occurring unsaturated fatty acids contain cis-double-bonds, not trans-double-bonds.
2. Succinic acid undergoes an elimination reaction (dehydrogenation) to produce fumaric acid. Fumaric acid is hydrated (an addition reaction) to produce malic acid.

$$HOOC{-}CH_2{-}CH_2{-}COOH \longrightarrow HOOC{-}CH{=}CH{-}COOH \xrightarrow[H_2SO_4]{H_2O} HOOC{-}\underset{OH}{\underset{|}{CH}}{-}CH_2{-}COOH$$

succinic acid; fumaric acid (trans); malic acid

3. phenyl benzoate $\xrightarrow{NaOH}$ sodium benzoate ($C_6H_5C(=O)O^-\,{}^+Na$) + sodium phenolate ($Na^+\,{}^-OC_6H_5$)

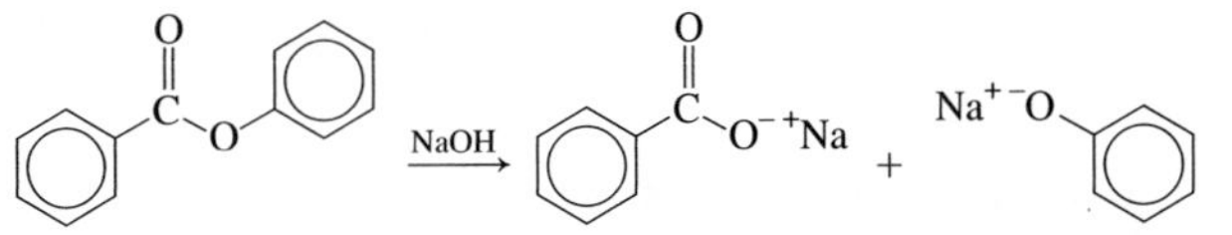

Yes, one of the organic products would be different if a weaker base were used. The carboxylate salt would still be obtained but the alcohol would most likely be obtained as the neutral alcohol and not as the alkoxide, i.e., phenol would be the product. Phenol is not as acidic as carboxylic acids and will not be deprotonated by weak bases.

4. $CH_3CH_2CH_2CH_2COOH$ $\quad HOCH_2\overset{O}{\overset{\|}{C}}CH_2CH_2CH_2\overset{O}{\overset{\|}{C}}OH$

5. Phosphate monoesters should be more soluble in water than carboxylic esters because they can participate in hydrogen bonding with water much more effectively. Although both have lone pairs of electrons on their oxygens that can hydrogen bond to the hydrogen atoms of water, the carboxylic ester has no hydrogen atoms that can participate in hydrogen bonding while the phosphate monoester has two hydrogen bond donors. The melting and boiling point of the phosphate ester should be higher since it can participate in hydrogen bonding when in a pure sample. The carboxylic ester will not exhibit hydrogen bonding as a pure sample.

6. $C_6H_5CN \xrightarrow[H^+]{H_2O}$ (hydrolysis) $C_6H_5C(=O)OH$ (I) $\xrightarrow{SOCl_2}$ (substitution) $C_6H_5C(=O)Cl$ (II) $\xrightarrow[\text{III}]{CH_3NHCH_2CH_3}$ (condensation) $C_6H_5C(=O)N(CH_3)CH_2CH_3$

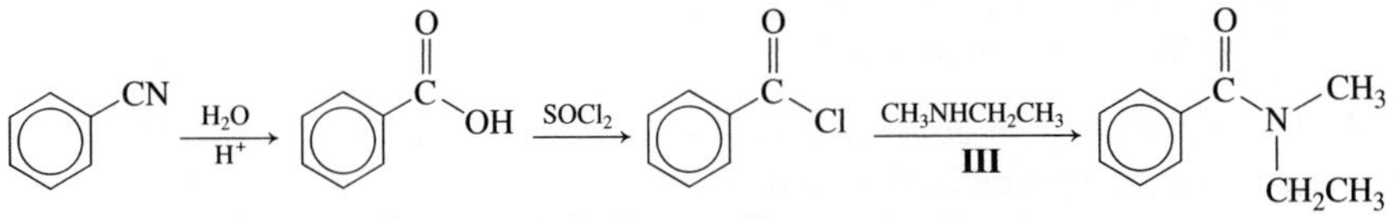

7. Acid halides are much more reactive than carboxylic acids. Also, the initial reaction between a carboxylic acid and ammonia (or an amine) is an acid–base reaction forming a salt, not the desired amide. In order for the amide to form, the initially produced salt must be strongly heated and not all compounds or other functional groups are always stable to high heat conditions.

8. $\underset{\delta}{CH_2}{=}\underset{\gamma}{CH}{-}\overset{\beta}{CH_2}{-}\underset{\alpha}{CH_2}{-}C(=O)OH$ (I) $\xrightarrow[H^+]{H_2O}$ $CH_3{-}\underset{OH}{\underset{|}{CH}}{-}CH_2{-}CH_2{-}C(=O)OH$ (II) $\xrightarrow[\text{heat}]{H^+}$ cyclic ester (a lactone) III $+ H_2O$

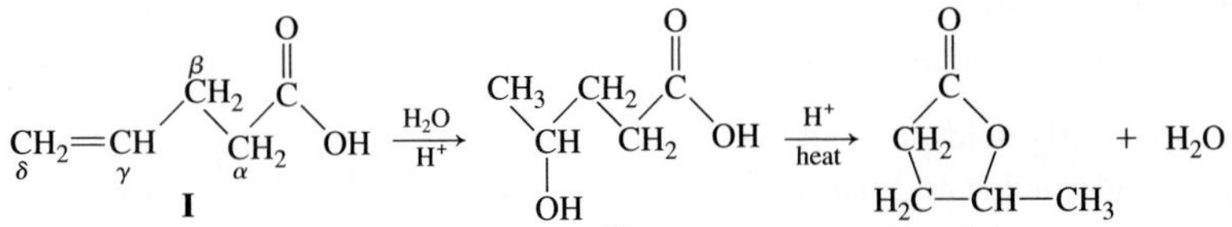

(III: five-membered ring C(=O)–CH_2–H_2C–$CH(CH_3)$–O)

Chapters 26–28

Multiple Choice: **1.** c **2.** c **3.** a **4.** b **5.** a **6.** c **7.** a **8.** b **9.** b **10.** c **11.** d **12.** b **13.** b **14.** d **15.** d **16.** a **17.** b **18.** b **19.** d **20.** c **21.** b **22.** a **23.** d **24.** d **25.** c **26.** a **27.** b **28.** d **29.** a **30.** d **31.** c **32.** b **33.** d **34.** a **35.** d **36.** c **37.** b **38.** b **39.** d **40.** c **41.** b **42.** a **43.** c **44.** a **45.** d **46.** b **47.** c **48.** b **49.** a **50.** d **51.** d **52.** b **53.** c **54.** c **55.** b **56.** b

Free Response:

1. (a) The molecules are steroids. They are epimers of each other since only one (the OH) of the several chiral carbons has the opposite stereochemistry.
(b) In carbohydrate chemistry, the (β-anomer is the structure with the OH being up, therefore the upper structure has a β-OH while the lower structure has the α-OH.

2. The disaccharide is α-D-glucopyranosyl-(1,4)-α-D-glucopyranose (maltose). Of the four glucose polymers, only amylose produces only maltose as the only disaccharide upon partial hydrolysis. Amylose is fairly easily hydrolyzed (unlike cellulose) so a dilute sulfuric acid solution should accomplish the partial hydrolysis.

3. (a) There are 5 chiral carbons in this molecule and therefore there are 2^5 or 32 possible diastereomers.
(b) D-rhamnose is the enantiomer of L-rhamnose.
(c) α-D-glucopyranosyl-(1,4)-α-L-rhamnopyranose

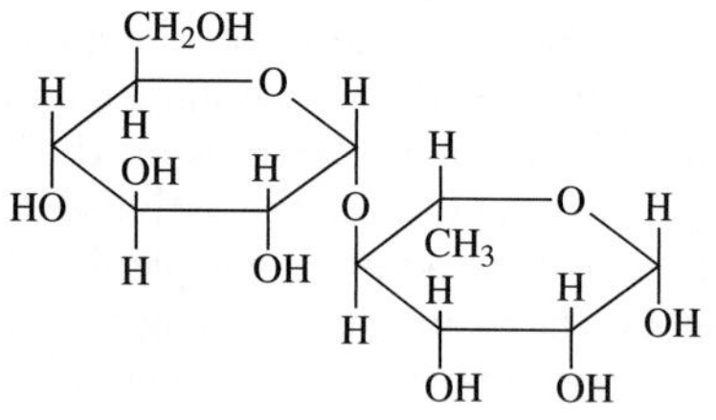

(d) Yes, this disaccharide should be a reducing sugar. The L-rhamnose ring will open to form an aldehyde.

4. The molecules are geometric isomers of each other. The only difference is the cis/trans orientation of the double bonds in the lower fatty acid chain.

5. (a) The lipid is made up from a triglycerol structure.
(b) The substituent on the phosphate group does not represent an aldohexose as there is no oxygen in the ring. It is not a pyranose ring, and there is no hemiacetal or acetal group present.

6. The fatty acid has a total of 20 carbons and is therefore arachidic acid. The sphingosine backbone has an amine group and therefore a condensation reaction between the amine of sphingosine and the carboxylic acid group of the fatty acid should result in the amide functional group shown in the molecule.

7. The three fatty acids involved are (from the top down) myristic acid, linoleic acid (an ω-6 fatty acid) and palmitoleic acid. The only chiral carbon in this lipid is the middle carbon of the glycerol unit, and therefore there are only two possible optical isomers.

8.

```
        CH2OH              CH2OH               CH2OH
         |                  |                   |
        C=O          HO ----+---- H       H ----+---- OH
         |                  |                   |
  HO ----+---- H     HO ----+---- H      HO ----+---- H
         |       --->       |          +        |
  HO ----+---- H     HO ----+---- H      HO ----+---- H
         |                  |                   |
  HO ----+---- H     HO ----+---- H      HO ----+---- H
         |                  |                   |
        CH2OH              CH2OH               CH2OH
     L-psicose               A                   B
```

In the reduction a new chiral carbon is formed and therefore since the reaction is not selective, a mix of the two products shown above will be formed. These two compounds are diastereomers, or more specifically, epimers of each other, differing only at the second carbon. The molecule on the left (A) will not have optical activity as it is a meso compound, while the molecule on the right (B) should exhibit optical activity.

9. Compound A is a glucose polymer, specifically cellulose (has β-1,4-linkages). Although it has many sites for hydrogen bonding, they are usually occupied with hydrogen bonding to other cellulose strands; due to that and to its large size, cellulose is insoluble in water.

Compound B is a monosaccharide, specifically a ketohexose. It is small and has many sites for hydrogen bonding with water and is thus expected to be soluble in water.

Compound C is a large molar-mass ester and therefore could be classified as lipid, specifically a wax. The only significant polarity in the molecule, the ester group, is not sufficient to compensate for the large nonpolar portions of the molecule in terms of water solubility and is therefore insoluble in water.

10. The functional group connecting the carbohydrate and the sphingosine structure is an acetal (ether). The carbohydrate used is β-D-galactose. There are 7 chiral carbons in the molecule (as shown by the *).

```
          OH
       *  |
          HC—CH=CH(CH2)12CH3
          |           O
          |           ||
       *  HC—NH—C—(CH2)16CH3
          |
         H2C—O
        CH2OH  |
   HO  |*  —O  |
     *  H       |
       OH  H   |*
   H           H
       |*  |*
       H   OH
```

Chapters 29–31

Multiple Choice: **1.** a **2.** c **3.** b **4.** c **5.** b **6.** c **7.** d **8.** b **9.** c **10.** a **11.** a **12.** c **13.** d **14.** c **15.** b **16.** c **17.** c **18.** d **19.** c **20.** d **21.** a **22.** c **23.** b **24.** d **25.** b **26.** a **27.** b **28.** a **29.** b **30.** d **31.** c **32.** a **33.** c **34.** d **35.** c **36.** b **37.** d **38.** c **39.** d **40.** b **41.** c **42.** c **43.** a **44.** b **45.** b **46.** a **47.** d **48.** b **49.** a **50.** c

Free Response:

1. In the first step, ethylene glycol is being oxidized to an aldehyde. An oxidoreductase enzyme is involved (specifically alcohol dehydrogenase).

2. Enzyme A can be classified as a hydrolase or as a transferase (depending on what happened to the phosphate group).

   Enzyme B can be classified as an isomerase (epimerase).

   Enzyme C can be classified as a transferase (phosphotransferase).

3.

```
         O  H        O
    +    ||  |        ||              O
H3N—CH—C—N—CH—C—N—CH—C//
       |          |        /   \      \
      CH2        CH2   CH2     CH2     O−
       |          |        \  /
      OH         CH2        CH2
                  |
                  S
                  |
                 CH3
```

ser-met-pro or serylmethionylproline-termination

4. Polar, acidic or basic amino acids are more likely to be found on the outside as these side chains are the ones that will interact with water. Hydrophobic (nonpolar) side chains will not have favorable interactions with water and would be more likely to be folded towards the inside of a globular protein.

5. The peptide coded by the nucleotide sequence is Ile-Asp-Gln-Ala-Glu-termination. There are two acidic amino acids (Asp and Glu) that have carboxylate groups (COO^-), and the C-terminal end has a carboxylate group. Therefore, there are three carboxylate groups present on this pentapeptide.

6. A hydrolysis reaction takes place, and the tyrosine group at the C-terminal end of the tetrapeptide gets cleaved.

```
             O
    +        ||
H3N—CH—C—O−
       |
      CH2
       |
   (benzene ring)
       |
      OH
```

tyrosine

7. The stomach is very acidic, and so the optimum pH for the stomach enzymes is likely to be in the acidic pH range. Altering the pH range for an enzyme often affects its turnover rate if not actually causing its denaturation. The pH range in the stomach is 1.5–2.5, and that of the saliva is 6.6–6.8.

8. The dipeptides are not the same as peptides, which are always named from the N-terminal to the C-terminal end. In phe-asp there is an N-terminal phe group and a C-terminal asp group while in asp-phe there is an N-terminal asp group and a C-terminal phe group. So asp-phe is not the same as phe-asp. Both peptides will give a positive xanthoproteic test as they contain phe which has an aromatic side group. Possible codon sequences for asp-phe are GAUUUU, GAUUUC, GACUUU or GACUUC.

9.

tryptophan — phosphate — ribose sugar — nucleic acid base (adenine)

$H_2N—CH(CH_2\text{-indolyl})—C(=O)—O—P(=O)(O^-)—O—CH_2$—ribose (H, H, H, H; OH, OH)—adenine (NH_2)

H
N
NH_2
phosphate
H_2C O O
N N
$H_2N—CH—C—O—P—O—CH_2$
N N
O
tryptophan
O^- H H nucleic acid base
H H (adenine)
OH OH
ribose sugar

Chapters 32-35

Multiple Choice: **1.** b **2.** c **3.** c **4.** d **5.** c **6.** a **7.** a **8.** c **9.** b **10.** c
11. a **12.** c **13.** b **14.** a **15.** b **16.** d **17.** b **18.** c **19.** a **20.** b
21. a **22.** d **23.** c **24.** b **25.** b **26.** a **27.** c **28.** a **29.** b **30.** d
31. c **32.** c **33.** b **34.** d **35.** b **36.** b **37.** a **38.** b **39.** d **40.** c
41. a **42.** d **43.** c **44.** b **45.** c **46.** c **47.** b **48.** b **49.** a **50.** c
51. b **52.** b **53.** a **54.** c **55.** c **56.** c **57.** c **58.** b **59.** b **60.** a

Free Response:

1. The vitamin is water-soluble. The relatively small size of the molecule and the presence of three polar, hydrophilic OH groups as well as an NH group should enable this molecule to hydrogen bond with water fairly effectively. (The vitamin is pyridoxine or vitamin B_6.)
2. Polysaccharides are digested (hydrolyzed) into monosaccharides; fats and oils hydrolyze to give fatty acids and glycerol; while proteins produce amino acids. Most of these hydrolysis (digestive) products may be converted into the metabolic intermediate acetyl-CoA.
3. $H_2C—OH$ / $HC—OH$ / $H_2C—OH$ (glycerol) $+ 3\ CH_3(CH_2)_{16}COOH$ (stearic acid)

   $H_2C—OH$ / $HC—OH$ / $H_2C—OH$ (glycerol) $+ 3\ CH_3(CH_2)_7CH{=}CH(CH_2)_7COOH$ (oleic acid)

   The molecules represent triacylglycerols, which are lipids. Lipids are macronutrients. The above molecules specifically represent a fat (glycerol and stearic acid products) and an oil (glycerol and oleic acid products). The fat provides a little more cellular energy than the oil because its carbons are in a more reduced state. Both have the same number of carbon atoms, but the unsaturated carbons in the oil are in a more oxidized state than the saturated carbons in the fat.
4. Proteins are digested mainly in the stomach and the intestines (stomach and pancreatic juices) to give amino acids. These amino acids go into the amino acid pool where the glucogenic amino acids can first undergo transamination, then conversion to pyruvate or phosphoenolpyruvate with the subsequent transformation into glucose.

   Fats and oils are hydrolysed mainly in the intestines to produce fatty acids and glycerol. Glycerol is converted to glyceraldehyde-3-phosphate that is then converted into glucose.

   In both cases, glucose is being formed from noncarbohydrate sources via gluconeogenesis.

5. $H_2C{-}OH$ / $C{=}O$ / $CH_2OPO_3^-$ $\xrightarrow{\text{NADH} \to \text{NAD}^+}$ $H_2C{-}OH$ / $HO{-}C{-}H$ / $CH_2OPO_3^-$

Since the reaction shown involves the reduction of a ketone to an alcohol, the coenzyme must be oxidized and therefore goes from the reduced state (NADH) to the oxidized state (NAD^+). The NAD^+ is in an oxidized state and therefore does not have the "extra" electrons to be used in the electron transport chain. Only the reduced forms of the coenzymes enter the electron transport chain to produce ATP.

6. I mol glucose $\longrightarrow$ 2 mol pyruvate (2 NADH)

1 mol pyruvate $\longrightarrow$ acetyl-CoA (1 NADH)

1 mol acetyl-CoA $\longrightarrow$ cycle (3 NADH, 1 $FADH_2$, 1 ATP)

Therefore, a total of 10 NADH, 2 $FADH_2$, and 2 ATPs are produced per mole of glucose oxidized in the skeletal muscle. Each NADH provides 3 ATPs when oxidized and each $FADH_2$ provides 2 ATPs. The total ATPs produced per mole of glucose is therefore 36 moles. The athlete ate 45 g glucose or 0.25 mol of glucose and therefore should produce 9.0 moles of ATP upon complete oxidation.

7. Fatty acid catabolism (breakdown) occurs two carbon atoms at a time. Each set of two carbons cleaved from the fatty acid, produces a molecule of acetyl CoA. Each mole of acetyl CoA produces 12 moles of ATP from the citric acid cycle. Therefore if 96 moles of ATP are produced, 8 moles of acetyl CoA entered the citric acid cycle. The fatty acid had 16 carbons and must have been palmitic acid.

8.

carbohydrate $\xrightarrow[\text{mouth or saliva}]{\alpha\text{-amylase}}$ maltose $\xrightarrow{\text{intestines}}$ glucose

glucose $\xrightarrow[\text{pathway}]{\text{Embden-Meyerhof}}$ pyruvate $\longrightarrow$ acetyl-CoA $\xrightarrow[\text{cycle}]{\text{citric acid}}$ CO_2

9.

glu-phe-asp-asn-val--tyr-trp-ala-val-phe- $\xrightarrow{\text{stomach}}$ glu-phe -asp-asn-val-tyr- trp-ala-val-phe $\xrightarrow{\text{intestines}}$ glu, phe, asp, asn, val, tyr, trp, ala, val, phe $\xrightarrow{\text{liver}}$ urea

GLOSSARY

A

absolute zero −273°C, the zero point on the Kelvin (absolute) temperature scale. *See also* Kelvin scale. [2.8, 12.6]

absorption The process by which digested foods pass through the membrane linings of the small intestine and enter the blood and lymph systems. [32.14]

acetal or ketal A compound that has two alkoxy groups on the same carbon atom. [23.4]

achiral Molecules or other objects that are superimposable on one another. [26.3]

acid (1) A substance that produces H^+ (H_3O^+) when dissolved in water. (2) A proton donor. (3) An electron-pair acceptor. A substance that bonds to an electron pair. [15.1]

acid anhydride A nonmetal oxide that reacts with water to form an acid. [13.8]

acid ionization constant (K_a) The equilibrium constant for the ionization of a weak acid in water. [16.11]

activation energy The amount of energy needed to start a chemical reaction. [8.5, 16.8]

active transport The movement of molecules from areas of low concentration to areas of high concentration requiring energy. [28.6]

activity series of metals A listing of metallic elements in descending order of reactivity. [17.5]

actual yield The amount of product actually produced in a chemical reaction (compared with the theoretical yield). [9.6]

addition polymer A polymer that is produced by successive *addition reactions*. [21.3]

addition reaction In organic chemistry, a reaction in which two substances join together to produce one substance. [19.10]

aerobic metabolism Chemical reactions that require molecular oxygen (O_2) and that are found in living organisms. This metabolism is the best way to produce energy for most cells. [33.4, 34.5]

alcohol An organic compound consisting of an —OH group bonded to a carbon atom in a nonaromatic hydrocarbon group; alcohols are classified as primary (1°), secondary (2°), or tertiary (3°), depending on whether the carbon atom to which the —OH group is attached is bonded to one, two, or three other carbon atoms, respectively. [22.1]

aldehyde An organic compound that contains the —CHO group. The general formula is RCHO. [23.1]

aldol condensation A reaction in which an aldehyde or ketone that contains α-hydrogen adds to itself or to another α-hydrogen containing aldehyde or ketone. [23.4]

aliphatic From the Greek *alciphar*, meaning "fat," a hydrocarbon that is not aromatic, including alkanes, alkenes, alkynes and cycloalkanes. [19.5]

alkali metal An element (except H) from Group IA of the periodic table. [10.4]

alkaline earth metal An element from Group IIA of the periodic table. [10.4]

alkaloid A basic compound derived from plants that shows physiological activity. [25.8]

alkane (saturated hydrocarbon) A compound composed of carbon and hydrogen, having only single bonds between the atoms; also known as paraffin hydrocarbon. *See also* alkene and alkyne. [19.6]

alkene (unsaturated hydrocarbon) A hydrocarbon whose molecules have at least one carbon–carbon double bond. [20.1]

alkoxide ion The anion that results when an alcohol hydroxyl group loses a proton. [22.5]

alkyl group An organic group derived from an alkane by removal of one H atom. The general formula is C_nH_{2n+1} (e.g., CH_3, methyl). Alkyl groups are generally indicated by the letter R. [19.9]

alkyl halide A class of compounds in which a halogen atom is attached to an alkyl group. [19.11]

alkyne (unsaturated hydrocarbon) A hydrocarbon whose molecules have at least one carbon–carbon triple bond. [20.1]

allotrope A substance existing in two or more molecular or crystalline forms (example: graphite and diamond are two allotropic forms of carbon). [12.17]

alpha particle (α) A particle emitted from a nucleus of an atom during radioactive decay; it consists of two protons and two neutrons with a mass of about 4 amu and a charge of +2; it is considered to be a doubly charged helium atom. [18.3]

amide Neutral (nonbasic) molecular substance formed by the reaction of a carboxylic acid with ammonia or an amine. [25.1]

amine A substituted ammonia molecule with basic properties and the general formula RNH_2, R_2NH, or R_3N. [25.5]

amino acid An organic compound containing two functional groups—an amino group (NH_2) and a carboxyl group (COOH). Amino acids are the building blocks for proteins. [29.2, 35.7]

amino acid pool The total supply of amino acids available for use throughout the body. [35.7]

amorphous A solid without shape or form. [1.7]

amphoteric (substance) A substance having properties of both an acid and a base. [15.3]

anabolism The metabolic process by which simple substances are synthesized (built up) into complex substances. [33.2]

anaerobic metabolism Chemical reactions that do not depend on molecular oxygen (O_2) and that are found in living organisms. [34.4]

anion A negatively charged ion. *See also* ion. [3.9, 5.5, 6.2]

anode The electrode where oxidation occurs in an electrochemical reaction. [17.6]

anomers Cyclic sugars that differ only in their stereo arrangement about the carbon involved in mutarotation. [27.6]

apoenzyme The protein part of an enzyme that is a conjugated protein. Often the apoenzyme is not catalytically competent [30.1]

apoptosis The process of self-destruction in which cells are automatically eliminated. [31.8]

aqueous solution A water solution. [16.3]

aromatic compound An organic compound whose molecules contain a benzene ring or that has properties resembling benzene. [20.1]

artificial radioactivity Radioactivity produced in nuclides during some types of transmutations. Artificial radioactive nuclides behave like natural radioactive elements in two ways: They disintegrate in a definite fashion, and they have a specific half-life. Sometimes called *induced radioactivity*. [18.6]

asymmetric carbon atom A carbon atom bonded to four different atoms or groups of atoms. [26.3]

1 atmosphere The standard atmospheric pressure; that is, the pressure exerted by a column of mercury 760 mm high at a temperature of 0°C. *See also* atmospheric pressure. [12.3]

atmospheric pressure The pressure experienced by objects on Earth as a result of the layer of air surrounding our planet. A pressure of 1 atmosphere (1 atm) is the pressure that will support a column of mercury 760 mm high at 0°C. [12.3]

atom The smallest particle of an element that can enter into a chemical reaction. [3.1]

atomic mass The average relative mass of the isotopes of an element referred to the atomic mass of carbon-12. [5.9]

atomic mass number (A) The sum of protons and neutrons in the nucleus of the atom. [5.11]

atomic mass unit (amu) A unit of mass equal to one-twelfth the mass of a carbon-12 atom. [5.11]

atomic number (Z) The number of protons in the nucleus of an atom of a given element. *See also* isotopic notation. [5.7]

atomic theory The theory that substances are composed of atoms, and that chemical reactions are explained by the properties and the interactions of these atoms. [5.2, Ch. 10]

Avogadro's law Equal volumes of different gases at the same temperature and pressure contain equal numbers of molecules. [12.11]

Avogadro's number 6.022×10^{23}; the number of formula units in 1 mole. [7.1, 9.1]

axial hydrogen Hydrogens whose bonds lie approximately perpendicular to the plane of a cycloalkane ring. [19.14]

B

balanced equation A chemical equation having the same number of each kind of atom and the same electrical charge on each side of the equation. [8.2]

barometer A device used to measure atmospheric pressure. [12.3]

base A substance whose properties are due to the liberation of hydroxide (OH^-) ions into a water solution. [15.1]

basic anhydride A metal oxide that reacts with water to form a base. [13.8]

beta oxidation The *two-carbon chop* which is accomplished in a series of reactions whereby the first two carbon atoms of the fatty acid chain become the acetyl group in a molecule of acetyl-CoA. [35.2]

beta particle (β) A particle identical in charge (-1) and mass to an electron. [18.3]

binary compound A compound composed of two different elements. [6.4]

biochemistry The branch of chemistry concerned with chemical reactions occurring in living organisms. [27.1]

biodegradable Refers to organic substances that are readily decomposed by microorganisms in the environment. [24.12]

bioenergetics The study of the transformation, distribution, and utilization of energy by living organisms. [33.1]

boiling point The temperature at which the vapor pressure of a liquid is equal to the pressure above the liquid. It is called the normal boiling point when the pressure is 1 atmosphere. [13.5]

bond length The distance between two nuclei that are joined by a chemical bond. [13.10]

Boyle's law At constant temperature, the volume of a fixed mass of gas is inversely proportional to the pressure (PV = constant). [12.5]

Brownian movement The random motion of colloidal particles. [15.14]

buffer solution A solution that resists changes in pH when diluted or when small amounts of a strong acid or strong base are added. [16.14]

C

calorie (cal) A commonly used unit of heat energy; 1 calorie is a quantity of heat energy that will raise the temperature of 1 g of water 1°C (e.g., from 14.5 to 15.5°C). Also, 4.184 joules = 1 calorie exactly. *See also* joule. [4.6]

capillary action The spontaneous rising of a liquid in a narrow tube, which results from the cohesive forces within the liquid and the adhesive forces between the liquid and the walls of the container. [13.4]

carbocation An ion in which a carbon atom has a positive charge. [20.6]

carbohydrate A polyhydroxy aldehyde or polyhydroxy ketone, or a compound that upon hydrolysis yields a polyhydroxy aldehyde or ketone; sugars, starch, and cellulose are examples. [27.1]

carbonyl group The structure $>C{=}O$. [23.1]

carboxyl group The functional group of carboxylic acids:

$$-\overset{\overset{\displaystyle O}{\|}}{C}-OH$$ [24.1]

carboxylic acid An organic compound having a carboxyl group. [24.1]

catabolism The metabolic process by which complex substances are broken down into simpler substances. [33.2]

catalyst A substance that influences the rate of a reaction and can be recovered essentially unchanged at the end of the reaction. [16.8]

cathode The electrode where reduction occurs in an electrochemical reaction. [17.6]

cation A positively charged ion. *See also* ion. [3.9, 5.5]

Celsius temperature scale (°C) The temperature scale on which water freezes at 0°C and boils at 100°C at 1 atm pressure. [2.8]

chain reaction A self-sustaining nuclear or chemical reaction in which the products cause the reaction to continue or to increase in magnitude. [18.8]

Charles' law At constant pressure, the volume of a fixed mass of any gas is directly proportional to the absolute temperature (V/T = constant). [12.6]

chemical bond The attractive force that holds atoms together in a compound. [Ch. 11]

chemical change A change producing products that differ in composition from the original substances. [4.3]

chemical equation A shorthand expression showing the reactants and the products of a chemical change (for example, $2\,H_2O = 2\,H_2 + O_2$). [4.3, 8.1]

chemical equilibrium The state in which the rate of the forward reaction equals the rate of the reverse reaction in a chemical change. [16.3]

chemical family *See* groups or families of elements.

chemical formula A shorthand method for showing the composition of a compound, using symbols of the elements. [3.9]

chemical kinetics The study of reaction rates and reaction mechanisms. [16.2]

chemical properties The ability of a substance to form new substances either by reaction with other substances or by decomposition. [4.1]

chemistry The science of the composition, structure, properties, and reactions of matter, especially of atomic and molecular systems. [1.2]

chiral A molecule that is not superimposable on its mirror image. [26.3]

chiral carbon atom An asymmetric carbon atom or chiral center. [26.3]

chlorofluorocarbons (CFCs) A group of compounds made of carbon, chlorine, and fluorine. [12.17]

citric acid cycle A series of eight reactions in which the acetyl group of acetyl-CoA is oxidized to carbon dioxide and water, and where many reduced coenzymes are formed. [34.5]

codon A triplet code of three nucleotides in the genetic code. [31.9]

coenzyme The non-protein part of an enzyme that is a conjugated protein. [30.1]

colligative properties Properties of a solution that depend on the number of solute particles in solution and not on the nature of the solute (examples: vapor-pressure lowering, freezing-point depression, boiling-point elevation). [14.7]

colloid A dispersion in which the dispersed particles are larger than the solute ions or molecules of a true solution and smaller than the particles of a mechanical suspension. [15.13]

combination reaction A direct union or combination of two substances to produce one new substance. [8.4]

combustion A chemical reaction in which heat and light are given off; generally, the process of burning or uniting a substance with oxygen. [19.11]

common-ion effect The shift of a chemical equilibrium caused by the addition of an ion common to the ions in the equilibrium. [16.12]

common names Arbitrary names that are not based on the chemical composition of compounds (examples: quicksilver for mercury, laughing gas for nitrous oxide). [6.1]

complete protein A protein that supplies all of the essential amino acids. [32.6]

compound A distinct substance composed of two or more elements combined in a definite proportion by mass. [3.8]

concentrated solution A solution containing a relatively large amount of dissolved solute. [14.6]

concentration of a solution A quantitative expression of the amount of dissolved solute in a certain quantity of solvent. [14.2]

condensation The process by which molecules in the gaseous state return to the liquid state. [13.3]

condensation polymer A polymer that is formed when monomers combine and split out water or some other simple substance, a *substitution reaction*. [21.3]

condensation reaction A reaction in which two molecules are combined by removing a small molecule. [22.5]

condensed structural formula A formula in which the arrangement of atoms is shown and each carbon is shown grouped with the hydrogens bonded to it. Thus, the structural formula $\left(\begin{array}{c}\ \ \ H\ \ \ H\\ \ \ \ |\ \ \ \ |\\ H-C-C-H\\ \ \ \ |\ \ \ \ |\\ \ \ \ H\ \ \ H\end{array}\right)$ becomes the condensed structural formula (CH_3CH_3). [19.3]

conformation The three-dimensional shape of a molecule in space. [19.14]

conjugate acid–base Two molecules or ions whose formulas differ by one H^+. (The acid is the species with the H^+, and the base is the species without the H^+.) [15.1]

conjugated protein A protein that is made up of amino acids and one or more additional components. [29.1]

copolymer A polymer containing two different kinds of monomer units. [21.6]

covalent bond A chemical bond formed between two atoms by sharing a pair of electrons. [11.5]

cracking, or pyrolysis The process in which saturated hydrocarbons are broken down by heating to very high temperatures in the presence of a catalyst. [20.5]

critical mass The minimum quantity of mass required to support a self-sustaining chain reaction. [18.8]

curie (Ci) A unit of radioactivity indicating the rate of decay of a radioactive substance: 1 Ci = 3.7×10^{10} disintegrations per second. [18.7]

cyanohydrin A class of compounds with a cyano (—CN) group and a hydroxyl group on the same carbon atom. [23.4]

cycloalkane A closed-chain (cyclic) alkane substance and a series of compounds with two fewer hydrogen atoms than the open-chain alkanes. [19.14]

cycloalkene A cyclic compound that contains a carbon–carbon double bond in the ring. [20.4]

D

Dalton's atomic theory The first modern atomic theory to state that elements are composed of minute individual particles called *atoms*. [5.2]

Dalton's law of partial pressures The total pressure of a mixture of gases is the sum of the partial pressures exerted by each of the gases in the mixture. [12.10]

decomposition reaction A breaking down, or decomposition, of one substance into two or more different substances. [8.4]

dehydration The elimination of a molecule of water from a reactant molecule. [20.5]

denaturation The process in which a protein loses only its natural three-dimensional conformation. [29.9]

density The mass of an object divided by its volume. [2.19]

Deoxyribonucleic acid (DNA) A high molar-mass polymer of nucleotides, present in all living matter, that contains the genetic code that transmits hereditary characteristics. DNA nucleotides contain deoxribose. [31.5]

dextrorotatory Rotation of plane-polarized light to the right by an optically active substance. [26.3]

dialysis The process by which a parchment membrane allows the passage of true solutions, but prevents the passage of colloidal dispersions. [15.16]

diastereomers Stereoisomers that are not mirror images of each other. [26.7]

diatomic molecules The molecules of elements that always contain two atoms. Seven elements occur as diatomic molecules: H_2, N_2, O_2, F_2, Cl_2, Br_2, and I_2. [3.17]

diet The food and drink that a person consumes. [32.2]

dietary reference intake (DRI) Takes into account the many different types of nutrient data needed to judge a healthful diet and is based generally on either a Recommended Dietary Allowance or an Adequate Intake. [32.2]

diffusion The property by which gases and liquids mix spontaneously because of the random motion of their particles. [12.2]

digestion A series of enzyme-catalyzed reactions by which large molecules are hydrolyzed to molecules small enough to be absorbed through the intestinal membranes. [32.13]

dilute solution A solution containing a relatively small amount of dissolved solute. [14.6]

dipeptide Two α-amino acids joined by a peptide linkage. [29.6]

dipole A molecule that is electrically asymmetrical, causing it to be oppositely charged at two points. [11.6]

disaccharide A carbohydrate that yields two monosaccharide units when hydrolyzed. [27.2]

dissociation The process by which a salt separates into individual ions when dissolved in water. [15.6]

double bond A covalent bond in which two pairs of electrons are shared. [11.5]

double-displacement reaction A reaction of two compounds to produce two different compounds by exchanging the components of the reacting compounds. [8.4]

ductile A property of metals; can be drawn to wires. [3.4]

E

effusion The process by which gas molecules pass through a tiny orifice from a region of high pressure to a region of lower pressure. [12.2]

Einstein's mass–energy equation $E = mc^2$: the relationship between mass and energy. [18.12]

electrolysis The process whereby electrical energy is used to bring about a chemical change. [17.6]

electrolyte A substance whose aqueous solution conducts electricity. [15.5]

electrolytic cell An electrolysis apparatus in which electrical energy from an outside source is used to produce a chemical change. [17.6]

electron A subatomic particle that exists outside the nucleus and carries a negative electrical charge. [5.6]

electron configuration The orbital arrangement of electrons in an atom. [10.5]

electron-dot structure *See* Lewis structure.

electronegativity The relative attraction that an atom has for a pair of shared electrons in a covalent bond. [11.6]

electron shell *See* principal energy levels of electrons.

element A basic building block of matter that cannot be broken down into simpler substances by ordinary chemical changes; in 1994, there were 111 known elements. [3.1]

elimination reaction A reaction in which a single reactant is split into two products, and one of the products is eliminated. [19.10]

Embden-Meyerhof pathway The anaerobic conversion of glucose to pyruvate. [34.4]

empirical formula A chemical formula that gives the smallest whole-number ratio of atoms in a compound—that is, the relative number of atoms of each element in the compound; also known as the simplest formula. [7.4]

enantiomers Chiral molecules that are mirror images of each other and are stereoisomers. [26.5]

endothermic reaction A chemical reaction that absorbs heat. [8.5]

energy The capacity of matter to do work. [4.5]

energy levels of electrons Areas in which electrons are located at various distances from the nucleus. [10.4]

energy sublevels The *s, p, d,* and *f* orbitals within a principal energy level occupied by electrons in an atom. [10.4]

enzyme A protein that catalyzes a biochemical reaction. [30.1]

enzyme activation A protein structural change that causes an increase in enzyme activity. [30.7]

enzyme inhibition A protein structural change that causes a decrease in enzyme activity. [30.7]

epimer Any two monosaccharides that differ only in the configuration around a single carbon atom. [27.5]

equatorial hydrogen Hydrogens whose bonds lie approximately in the plane of a cycloalkane ring. [19.14]

equilibrium A dynamic state in which two or more opposing processes are taking place at the same time and at the same rate. [16.3]

equilibrium constant (K_{eq}) A value representing the unchanging concentrations of the reactants and the products in a chemical reaction at equilibrium. [16.9]

essential amino acid An amino acid that is not synthesized by the body and therefore must be supplied in the diet. [29.3]

ester An organic compound derived from a carboxylic acid and an alcohol. The general formula is

$$R-\underset{\underset{O}{\|}}{C}-OR' \qquad [24.6]$$

estimated energy requirement (EER) The average dietary intake that is required to maintain energy balance for good health. [32.3]

ether An organic compound having two hydrocarbon groups attached to an oxygen atom. The general formula is $R-O-R'$. [22.10]

eucaryote A cell that contains internal, membrane-bound bodies called "organelles." [33.2]

evaporation The escape of molecules from the liquid state to the gas or vapor state. [13.2]

exothermic reaction A chemical reaction in which heat is released as a product. [8.5]

F

facilitated diffusion The process in which a protein helps (facilitates) transport without using energy. [28.6]

Fahrenheit temperature scale (°F) The temperature scale on which water freezes at 32°F and boils at 212°F at 1 atm pressure. [2.8]

fats and oils Esters of fatty acids and glycerol. *See also* triacylglycerol. [24.11]

fatty acids Long-chain carboxylic acids present in lipids (fats and oils). [28.3]

feedback inhibition A common form of enzyme control in which the final product acts as "feedback" and inhibits an enzyme from using too many molecular starting materials. [30.7]

feedforward activation A form of enzyme control at the end of the molecular assembly-line: if there is an excess of starting materials, these molecules will "feedforward" and activate enzymes, causing the whole process to move faster. [30.7]

Fehling and Benedict test A test in which the aldehyde group is oxidized to an acid by Cu^{2+} ions, used for detecting carbohydrates that have an available aldehyde group. [23.4]

fibrous protein An important class of proteins that contain highly developed secondary structures, with a "fiber-like" or elongated shape. [29.8]

fluid-mosaic model A model of a biological membrane containing a lipid bilayer studded with membrane proteins. [28.6]

formula equation A chemical equation in which all the reactants and products are written in their molecular, or normal, formula expression; also called a molecular equation. [15.12]

formula unit The atom or molecule indicated by the formula of the substance under consideration (examples: Mg, O_2, H_2O). [7.1]

free radical A neutral atom or group of atoms having one or more unpaired electrons. [12.17, 19.11]

freezing or melting point The temperature at which the solid and liquid states of a substance are in equilibrium. [13.6]

frequency A measurement of the number of waves that pass a particular point per second. [10.2]

functional group An atom or group of atoms that characterizes a class of organic compounds. For example, —COOH is the functional group of carboxylic acids. [19.4]

G

galvanic cell *See* voltaic cell.

gamma ray (γ) High-energy photons emitted by radioactive nuclei; they have no electrical charge and no measurable mass. [18.3]

gas A state of matter that has no shape or definite volume so that the substance completely fills its container. [1.7]

Gay-Lussac's law At constant volume, the pressure of a fixed mass of gas is directly proportional to the absolute temperature ($P/T =$ constant). [12.7]

Gay-Lussac's law of combining volumes (of gases) When measured at the same temperature and pressure, the ratios of the volumes of reacting gases are small whole numbers. [12.11]

genes Basic units of heredity that consist primarily of DNA and proteins and occur in the chromosomes. [31.7]

genome The sum of all hereditary material contained in a cell. [31.7]

geometric isomers Isomers that differ from one another only in the geometry of their molecules and not in the order of their atoms. [20.3]

globular protein A protein with a roughly spherical shape and a complex tertiary structure. [29.8]

glucogenic amino acid An amino acid that is used to produce glucose. [35.8]

gluconeogenesis The formation of glucose from noncarbohydrate sources. [34.6]

glycogenesis The synthesis of glycogen from glucose. [34.3]

glycogenolysis The hydrolysis, or breakdown, of glycogen to glucose. [34.3]

glycolipid A sphingolipid that contains a carbohydrate group, including cerebrosides and gangliosides, substances found mainly in cell membranes of nerve and brain tissue. [28.4]

glycolysis The metabolic pathway formed when lactate is the final product of anaerobic glucose catabolism. [34.4]

glycoside From the Greek word *glykys*, meaning "sweet," the acetal structure that is produced when a monosaccharide hemiacetal reacts with an alcohol. [27.7]

Graham's law of effusion The rates of effusion of two gases at the same temperature and pressure are inversely proportional to the square roots of their densities or molar masses. [12.2]

greenhouse effect The effect of CO_2 and other greenhouse gasses to warm the atmosphere by absorbing energy near the surface of the earth, which can lead to global warming. [8.6]

ground state The lowest available energy level within an atom. [10.3]

groups or families (of elements) Vertical groups of elements in the periodic table (IA, IIA, and so on). Families of elements that have similar outer-orbital electron structures. [10.6]

H

half-life The time required for one-half of a specific amount of a radioactive nuclide to disintegrate; half-lives of the elements range from a fraction of a second to billions of years. [18.2]

halogenation The substitution of a halogen atom for a hydrogen atom in an organic compound. [19.11]

halogens Group VIIA of the periodic table; consists of the elements fluorine, chlorine, bromine, iodine, and astatine. [10.4]

heat A form of energy associated with the motion of small particles of matter. [2.8]

heat of fusion The energy required to change 1 gram of a solid into a liquid at its melting point. [13.7]

heat of reaction The quantity of heat produced by a chemical reaction. [8.5]

heat of vaporization The amount of heat required to change 1 gram of a liquid to a vapor at its normal boiling point. [13.7]

hemiacetal or hemiketal A compound derived from aldehydes and ketones that contains an alkoxy and a hydroxy group on the same carbon atom. [23.4]

heterocyclic compound A ring compound in which the atoms in the ring are not all alike. [25.5]

heterogeneous Matter without a uniform composition—having two or more components or phases. [1.8]

holoenzyme The combination of the protein and non-protein parts of an enzyme that is a conjugated protein. The holoenzyme is normally an active catalyst. [30.1]

homogeneous Matter that has uniform properties throughout. [1.8]

homologous series A series of compounds in which the members differ from one another by a regular increment. For example, each successive member of the alkane series of hydrocarbons differs by a CH_2 group. [19.6]

hormones The chemical substances that act as control agents in the body, often regulating metabolic pathways. [34.8]

hydrate A solid that contains water molecules as a part of its crystalline structure. [13.13]

hydrocarbon A compound composed entirely of carbon and hydrogen. [8.5, 19.5]

hydrogen bond The intermolecular force acting between molecules that contain hydrogen covalently bonded to the highly electronegative elements, F, O, and N. [13.8]

hydrogenolysis A reaction that breaks a triacylglycerol into glycerol and three long chain primary alcohols using hydrogen gas. [24.11]

hydrolysis Chemical reaction with water in which the water molecule is split into H^+ and OH^-. [16.13]

hydronium ion The result of a proton combining with a polar water molecule to form a hydrated hydrogen ion, H_3O^+. [15.1]

hydrophilic Water-loving (dissolves in water). [24.12]

hydrophobic Water-fearing (does not dissolve in water). [24.12]

hypothesis A tentative explanation of certain facts to provide a basis for further experimentation. [1.5]

I

ideal gas A gas that behaves precisely according to the Kinetic Molecular Theory; also called a perfect gas. [12.2]

ideal-gas equation $PV = nRT$; that is, the volume of a gas varies *directly* with the number of gas molecules and the absolute temperature and *inversely* with the pressure. [12.14]

immiscible Incapable of mixing; immiscible liquids do not form a solution with one another. [14.2]

induced-fit model A model of enzyme-substrate binding in which the active site adjusts its structure in order to prepare the enzyme-substrate complex for catalysis. [30.5]

induced radioactivity *See* artificial radioactivity.

invert sugar The mixture of glucose and fructose, usually in solution, resulting from hydrolysis of sucrose. [27.12]

ion A positively or negatively charged atom or group of atoms. *See also* cation, anion. [3.8, 5.5]

ionic bond The chemical bond between a positively charged ion and a negatively charged ion. [11.3]

ionic compound A compound that is composed of ions (e.g., Na^+Cl^-) [3.8, 6.3]

ionization The formation of ions, which occurs as the result of a chemical reaction of certain substances with water. [15.6]

ionization energy The energy required to remove an electron from an atom, an ion, or a molecule. [11.1]

ionizing radiation Radiation with enough energy to dislocate bonding electrons and create ions when passing through matter. [18.4]

ion product constant for water (K_w) An equilibrium constant defined as the product of the H^+ ion concentration and the OH^- ion concentration, each in moles per liter. $K_w = [H^+][OH^-] = 1 \times 10^{-14}$ at 25°C. [16.10]

isoelectric point The pH at which a molecule will migrate toward neither electrode in an applied electric field. [29.5]

isomerism The phenomenon of two or more compounds having the same molecular formula but different molecular structures. [19.8]

isomers Compounds having identical molecular formulas, but different structural formulas. [19.8]

isotope An atom of an element that has the same atomic number, but a different atomic mass. Since their atomic numbers are identical, isotopes vary only in the number of neutrons in the nucleus. [5.10]

isotopic notation Notation for an isotope of an element where the subscript is the atomic number, the superscript is the mass number, and they are attached on the left of the symbol for the element. (For example, hydrogen-1 is notated as ${}^1_1H$.) *See also* atomic number, mass number. [5.10]

IUPAC International Union of Pure and Applied Chemistry, which devised (in 1921) and continually upgrades the system of nomenclature for inorganic and organic compounds. [6.1, 19.9]

J

joule (J) The SI unit of energy. *See also* calorie. [4.6]

K

Kelvin temperature scale (K) Absolute temperature scale starting at absolute zero, the lowest temperature possible. Freezing and boiling points of water on this scale are 273 K and 373 K, respectively, at 1 atm pressure. *See also* absolute zero. [2.8, 12.6]

ketogenic amino acid An amino acid that causes an increase in the rate of ketone body formation. [35.8]

ketone An organic compound that contains a carbonyl group between two other carbon atoms. The general formula is $R_2C = O$. [23.1]

kilocalorie (kcal) 1000 cal; the kilocalorie is also known as the nutritional or large Calorie, used for measuring the energy produced by food. [4.6]

kilogram (kg) The standard unit of mass in the metric system; 1 kilogram equals 2.205 pounds. [2.9]

kilojoule (kJ) 1000 J. [4.6]

kinetic energy (KE) The energy that matter possesses due to its motion; $KE = 1/2\, mv^2$. [4.5]

kinetic-molecular theory (KMT) A group of assumptions used to explain the behavior and properties of gases. [12.2]

L

law A statement of the occurrence of natural phenomena that occur with unvarying uniformity under the same conditions. [1.5]

law of conservation of energy Energy can be neither created nor destroyed, but it can be transformed from one form to another. [4.8]

law of conservation of mass No change is observed in the total mass of the substances involved in a chemical reaction; that is, the mass of the products equals the mass of the reactants. [4.4]

law of definite composition A compound always contains two or more elements in a definite proportion by mass. [5.3]

law of multiple proportions Atoms of two or more elements may combine in different ratios to produce more than one compound. [5.3]

Le Chatelier's principle If a stress is applied to a system in equilibrium, the system will respond in such a way as to relieve that stress and restore equilibrium under a new set of conditions. [16.4]

levorotatory Rotation of plane-polarized light to the left by an optically active substance. [26.3]

Lewis structure A method of indicating the covalent bonds between atoms in a molecule or an ion such that a pair of electrons (:) represents the valence electrons forming the covalent bond. [11.2]

limiting reactant A reactant that limits the amount of product formed because it is present in insufficient amount compared with the other reactants. [9.6]

linear structure In the VSEPR model, an arrangement where the pairs of electrons are arranged 180° apart for maximum separation. [11.11]

line structures of organic compounds Those structures are even simpler than the condensed structural formula. For hydrocarbons, these structures consist of zig zag (diagonal) lines in which a carbon is located at each junction where two lines meet and change direction. [19.3]

line spectrum Colored lines generated when light emitted by a gas is passed through a spectroscope. Each element possesses a unique set of line spectra. [10.3]

lipase An enzyme that digests lipids. [30.4]

lipid bilayer A basic structure composed of two adjoining layers of lipid molecules aligned so that their hydrophobic portions form the bilayer interior while their hydrophilic portions form the bilayer exterior. [28.6]

lipids Organic compounds found in living organisms that are water insoluble, but soluble in such fat solvents as diethyl ether, benzene, and carbon tetrachloride; examples are fats, oils, and steroids. [28.1]

lipogenesis The biosynthesis of fatty acids from acetyl-CoA. [35.4]

liquid A state of matter in which the particles move about freely while the substance retains a definite volume; thus, liquids flow and take the shape of their containers. [1.7]

liter (L) A unit of volume commonly used in chemistry; 1 L = 1000 mL; the volume of a kilogram of water at 4°C. [2.10]

lock-and-key hypothesis A hypothesis describing a fundamental property of enzyme-substrate binding, which envisions the substrate as a key that fits into the appropriate active site, the lock. [30.5]

logarithm (log) The power to which 10 must be raised to give a certain number. The log of 100 is 2.0. [15.9]

M

macromolecule *See* polymer.

macronutrients Carbohydrates, lipids, and proteins—the major sources of building materials, replacement parts, and energy needs of the cells that are used in relatively large amounts. [32.1]

malleable A property of metals; can be rolled or hammered into sheets. [3.4]

Markovnikov's rule Formulated in the mid-19th century by Russian chemist V. Markovnikov: when an unsymmetrical molecule such as HX (e.g., HCl) adds to a carbon–carbon double bond, the hydrogen from HX goes to the carbon atom that has the greater number of hydrogen atoms. [20.6]

mass The quantity or amount of matter that an object possesses. [2.7]

mass defect The difference between the actual mass of an atom of an isotope and the calculated mass of the protons and neutrons in the nucleus of that atom. [18.12]

mass number The sum of the protons and neutrons in the nucleus of a given isotope of an atom. *See also* isotopic notation. [5.10]

mass percent solution The grams of solute in 100 g of a solution. [14.6]

matter Anything that has mass and occupies space. [1.6]

mechanism The path taken by a chemical reaction to go from reactants to products. [19.11]

meiosis The process of cell division to form a sperm cell and an egg cell in which each cell formed contains half of the chromosomes found in the normal single cell. [31.7]

melting or freezing point *See* freezing or melting point.

meniscus The shape of the surface of a liquid when placed in a glass cylinder. It can be concave or convex. [13.4]

mercaptan The —SH-containing compound, also known as thiol. [22.13]

meso compound or meso structure Stereoisomers that contain chiral carbon atoms and are superimposable on their own mirror images. [26.7]

metabolic pathway A series of biochemical reactions that serve a specific purpose. [34.1]

metabolism The sum of all chemical reactions that occur within a living organism. [33.2]

metal An element that is solid at room temperature and whose properties include luster, ductility, malleability, and good conductivity of heat and electricity; metals tend to lose their valence electrons and become positive ions. [3.4]

metalloid An element having properties that are intermediate between those of metals and nonmetals (for example, silicon); these elements are useful in electronics. [3.4]

meter (m) The standard unit of length in the SI and metric systems; 1 meter equals 39.37 inches. [2.5]

metric system A decimal system of measurements. *See also* SI. [2.5]

micronutrients Vitamins and minerals—nutrients that are needed in relatively small amounts. [32.1]

miscible Capable of mixing and forming a solution. [14.2]

mitosis Ordinary cell division in which a DNA molecule is duplicated by uncoiling to single strands and then reassembling with complementary nucleotides. Each new cell contains the normal number of chromosomes. [31.7]

mixture Matter containing two or more substances, which can be present in variable amounts; mixtures can be homogeneous (sugar water) or heterogeneous (sand and water). [1.8]

molality (*m*) An expression of the number of moles of solute dissolved in 1000 g of solvent. [14.7]

molarity (*M*) The number of moles of solute per liter of solution. [14.6]

molar mass The mass in grams of Avogadro's number of atoms or molecules. The sum of all the atoms in an element, compound or ion. The mass in grams of a mole of any formula unit. It is also known as the molecular weight. [7.1, 9.1]

molar solution A solution containing 1 mole of solute per liter of solution. [14.6]

molar volume (of a gas) The volume of 1 mol of a gas at STP equals 22.4 L/mol. [12.12]

mole The amount of a substance containing the same number of formula units (6.022×10^{23}) as there are in exactly 12 g of ^{12}C. One mole is equal to the molar mass in grams of any substance. [7.1]

mole ratio A ratio between the number of moles of any two species involved in a chemical reaction; the mole ratio is used as a conversion factor in stoichiometric calculations. [9.2]

molecular formula The total number of atoms of each element present in one molecule of a compound; also known as the true formula. *See also* empirical formula. [7.4]

molecule The smallest uncharged individual unit of a compound formed by the union of two or more atoms. [3.8]

monomer The small unit or units that undergo polymerization to form a polymer. [21.1]

monosaccharide A carbohydrate that cannot be hydrolyzed to simpler carbohydrate units (for example, simple sugars like glucose or fructose). [27.2]

monosubstitution When one atom in an organic molecule is substituted by another atom or by a group of atoms. [19.11]

mutagen A chemical agent or type of radiation such as x-rays, cosmic rays, and ultraviolet rays that causes mutations. [31.12]

mutarotation The process by which anomers are interconverted. [27.6]

mutation A new trait that appears in an individual that is not present in either its parents or ancestors, and generally is the result of genetic or chromosomal changes. [31.12]

N

negative nitrogen balance When the amount of nitrogen excreted from a body exceeds that ingested, as in a fasting or starving person. [35.7]

net ionic equation A chemical equation that includes only those molecules and ions that have changed in the chemical reaction. [15.12]

neutralization The reaction of an acid and a base to form a salt plus water. [15.10]

neutron A subatomic particle that is electrically neutral and is found in the nucleus of an atom. [5.6]

nitrogen balance When the amount of nitrogen excreted from the body is equal to the amount ingested, the person is said to be in "nitrogen balance." [35.7]

nitrogen cycle The process by which nitrogen is circulated and recirculated from the atmosphere through living organisms and back to the atmosphere. [35.6]

nitrogen fixation The conversion of diatomic nitrogen to a biochemically useful form. [35.6]

noble gases A family of elements in the periodic table—helium, neon, argon, krypton, xenon, and radon—that contain a particularly stable electron structure. [10.4]

nonelectrolyte A substance whose aqueous solutions do not conduct electricity. [15.5]

nonmetal An element that has properties the opposite of metals: lack of luster, relatively low melting point and density, and generally poor conduction of heat and electricity. Nonmetals may or may not be solid at room temperature (examples: carbon, bromine, nitrogen); many are gases. They are located mainly in the upper right-hand corner of the periodic table. [3.8]

nonpolar covalent bond A covalent bond between two atoms with the same electronegativity value; thus, the electrons are shared equally between the two atoms. [11.6]

normal boiling point The temperature at which the vapor pressure of a liquid equals 1 atm or 760 torr pressure. [13.5]

nuclear binding energy The energy equivalent to the mass defect; that is, the amount of energy required to break a nucleus into its individual protons and neutrons. [18.12]

nuclear fission The splitting of a heavy nuclide into two or more intermediate-sized fragments when struck in a particular way by a neutron. As the atom is split, it releases energy and two or three more neutrons that can then cause another nuclear fission. [18.8]

nuclear fusion The uniting of two light elements to form one heavier nucleus, which is accompanied by the release of energy. [18.11]

nucleic acids Complex organic acids essential to life and found in the nucleus of living cells. They consist of thousands of units called nucleotides. Includes DNA and RNA. [31.1]

nucleon A collective term for the neutrons and protons in the nucleus of an atom. [18.1]

nucleoprotein A simple protein bonded to a nucleic acid. [31.1]

nucleoside A structure formed when either a purine or pyrimidine base is linked to a sugar molecule, usually D-ribose or D-2′-deoxyribose. [31.2]

nucleotide The building-block unit for nucleic acids. A phosphate group, a sugar residue, and a nitrogenous organic base are bonded together to form a nucleotide. [31.3]

nucleus The central part of an atom that contains all its protons and neutrons. The nucleus is very dense and has a positive electrical charge. [5.8]

nuclide A general term for any isotope of any atom. [18.1]

nutrient Component of food that makes body growth, maintenance, and repair possible. [32.1]

O

oils *See* fats and oils.

oligosaccharide A carbohydrate that has two to six monosaccharide units linked together. [27.2]

oncogene A gene present in cancerous or malignant cells and codes for proteins that control cell growth; they are present in many normal mammalian cells. [31.8]

one atmosphere The standard atmospheric pressure: that is, the pressure exerted by a column of mercury 760 mm high at a temperature of 0°C. [12.3]

optical activity The ability to rotate a plane of polarized light. [26.3]

orbital A cloudlike region around the nucleus where electrons are located. Orbitals are considered to be energy sublevels (*s, p, d, f*) within the principal energy levels. *See also* principal energy levels. [10.3, 10.4]

orbital diagram A way of showing the arrangement of electrons in an atom, where boxes with small arrows indicating the electrons represent orbitals. [10.5]

organelle An internal membrane-bound body found in eucaryotic cells. [33.2]

organic chemistry The branch of chemistry that deals with carbon compounds, but does not imply that these compounds must originate from some form of life. *See also* vital-force theory. [19.1]

osmosis The diffusion of water, either pure or from a dilute solution, through a semipermeable membrane into a solution of higher concentration. [14.8]

oxidation An increase in the oxidation number of an atom as a result of losing electrons. [17.2]

oxidation number A small number representing the state of oxidation of an atom. For an ion, it is the positive or negative charge on the ion; for covalently bonded atoms, it is a positive or negative number assigned to each atom based on differences in electronegativity; in free elements, it is zero. [17.1]

oxidation–reduction A chemical reaction wherein electrons are transferred from one element to another; also known as redox. [17.2]

oxidation state *See* oxidation number. [17.1]

oxidative phosphorylation A process that directly uses energy from redox reactions to form ATP. [33.6]

oxidizing agent A substance that causes an increase in the oxidation state of another substance. The oxidizing agent is reduced during the course of the reaction. [17.2]

oxonium ion A protonated alcohol formed when an alcohol is mixed with a strong acid. [22.5]

ozone layer A high concentration of ozone located in the stratosphere that shields the earth from damaging ultraviolet radiation. [12.15]

P

partial pressure The pressure exerted independently by each gas in a mixture of gases. [12.10]

parts per million (ppm) A measurement of the concentration of dilute solutions now commonly used by chemists in place of mass percent. [14.6]

Pauli exclusion principle An atomic orbital can hold a maximum of two electrons, which must have opposite spins. [10.4]

peptide linkage The amide bond in a protein molecule; bonds one amino acid to another. [29.6]

percent composition of a compound The mass percent represented by each element in a compound. [7.3]

percent yield The ratio of the actual yield to the theoretical yield multiplied by 100. [9.6]

perfect gas A gas that behaves precisely according to theory; also called an *ideal gas*. [12.2]

period of elements The horizontal groupings (rows) of elements in the periodic table. [3.5]

periodic table An arrangement of the elements according to their atomic numbers. The table consists of horizontal rows, or periods, and vertical columns, or families, of elements. Each period ends with a noble gas. [3.5]

pH A method of expressing the H^+ concentration (acidity) of a solution; $pH = -\log[H^+]$, $pH = 7$ is a neutral solution, $pH < 7$ is acidic, and $pH > 7$ is basic. [15.9]

phase A homogeneous part of a system separated from other parts by a physical boundary. [1.8]

phenol The class of compounds that have a hydroxy group attached to an aromatic ring. [22.7]

phosphate ester A compound formed by reacting an alcohol with phosphoric acid. [24.13]

phospholipid A group of compounds that yield one or more fatty acid molecules, a phosphate group, and usually a nitrogenous base upon hydrolysis. [28.4]

photon Theoretically, a tiny packet of energy that streams with others of its kind to produce a beam of light. [10.2]

photosynthesis The process by which green plants utilize light energy to synthesize carbohydrates. [33.7]

physical change A change in form (such as size, shape, or physical state) without a change in composition. [4.2]

physical properties Inherent physical characteristics of a substance that can be determined without altering its composition: color, taste, odor, state of matter, density, melting point, boiling point. [4.1]

physical states of matter Solids, liquids, and gases. [1.7]

physiological saline solution A solution of 0.90% sodium chloride that is isotonic (has the same osmotic pressure) with blood plasma. [14.8]

pi bond A double bond formed by two perpendicular *p* orbitals (one on each carbon atom) that overlap, with two electron clouds, one above and one below the sigma bond. [20.1]

plane-polarized light Light that vibrates in only one plane. [26.2]

pOH A method of expressing the basicity of a solution. $pOH = -\log[OH^-]$. $pOH = 7$ is a neutral solution, $pOH < 7$ is basic, and $pOH > 7$ is acidic. [16.11]

polar covalent bond A covalent bond between two atoms with differing electronegativity values, resulting in unequal sharing of bonding electrons. [11.5]

polyatomic ion An ion composed of more than one atom. [6.5]

polycyclic or fused aromatic ring system A compound consisting of two or more rings in which two carbon atoms are common to two rings. [20.11]

polyhydroxy alcohol An alcohol that has more than one —OH group. [22.2]

polymer (macromolecule) A natural or synthetic giant molecule formed from smaller molecules (monomers). [21.1]

polymerization The process of forming large, high-molar-mass molecules from smaller units. [21.1]

polypeptide A peptide chain containing up to 50 amino acid units. [29.6]

polysaccharide A carbohydrate that can be hydrolyzed to many monosaccharide units; cellulose, starch, and glycogen are examples. [27.2]

positive nitrogen balance When the amount of nitrogen consumed is more than that excreted from a body, as in a growing child. [35.7]

positron A particle with a +1 charge having the mass of an electron (a positive electron). [18.1]

potential energy (PE) Stored energy, or the energy of an object due to its relative position. [4.5]

pressure Force per unit area; expressed in many units, such as mm Hg, atm, lb/in.^{2}, torr, and pascal. [12.3]

primary alcohol An alcohol in which the carbon atom bonded to the —OH group is bonded to only one other carbon atom. [22.2]

primary structure A large protein structure established by the number, kind, and sequence of amino acid units composing the polypeptide chain or chains making up the molecule. This protein structure determines the alignment of side-chain characteristics which, in turn, determines the three-dimensional shape into which the protein folds. [29.7]

principal energy levels of electrons Existing within the atom, these energy levels contain orbitals within which electrons are found. *See also* orbital, electron. [10.4]

procaryote Cells without internal membrane-bound bodies. [33.2]

product A chemical substance produced from reactants by a chemical change. [4.3]

product inhibition A form of enzyme control in which the product of an enzyme-catalyzed reaction inhibits further enzyme catalysis. [30.7]

productive binding hypothesis As molecules bind to an enzyme active site they are oriented to favor reaction. [30.5]

projection formula A style of diagram used for representing three-dimensional structures of chiral compounds in two dimensions. [26.4]

properties The characteristics, or traits, of substances that give them their unique identities. Properties are classified as physical or chemical. [4.1]

protease An enzyme that breaks down proteins. [30.4]

protein A polymer consisting mainly of α-amino acids linked together; occurs in all animal and vegetable matter. [29.1]

protein domain A compact globular unit about the size of myoglobin or carboxypeptidase A. [29.8]

proton A subatomic particle found in the nucleus of the atom that carries a positive electrical charge and a mass of about 1 amu. An H^+ ion is a proton. [5.6]

proximity catalysis A process in which an enzyme acts to bring reactants close together. [30.5]

Q

quanta Small discrete increments of energy. From the theory proposed by physicist Max Planck that energy is emitted in energy *quanta* rather than a continuous stream. [10.3]

quantum mechanics or wave mechanics The modern theory of atomic structure based on the wave properties of matter. [10.3]

quaternary ammonium salt A compound in which four organic groups bond to the nitrogen atom. [25.6]

quaternary structure A protein structure, found in some complex proteins, made of two or more smaller protein subunits (polypeptide chains). The shape of the entire complex molecule is determined by the way in which the subunits are held together by *noncovalent* bonds. [29.7]

R

racemic mixture A mixture containing equal amounts of a pair of enantiomers. [26.6]

rad (radiation absorbed dose) A unit of absorbed radiation indicating the energy absorbed from any ionizing radiation; 1 rad = 0.01 J of energy absorbed per kilogram of matter. [18.7]

radioactive decay The process by which a radioactive element emits particles or rays and is transformed into another element. [18.2]

radioactive disintegration series The spontaneous decay of a certain radioactive nuclide by emission of alpha and beta particles from the nucleus, finally stopping at a stable isotope of lead or bismuth. [18.4]

radioactivity The spontaneous emission of radiation from the nucleus of an atom. [18.1]

rate of reaction The rate at which the reactants of a chemical reaction disappear and the products form. [16.2]

reactant A chemical substance entering into a reaction. [4.3]

recombinant DNA DNA whose genes have been rearranged to contain new or different hereditary information. [31.12]

recommended dietary allowance (RDA) An average value of a nutrient that has been shown to maintain health for large groups of people. [32.2]

redox *See* oxidation–reduction. [17.2]

reducing agent A substance that causes a decrease in the oxidation state of another substance; the reducing agent is oxidized during the course of a reaction. [17.2]

reducing sugar A sugar that reduces silver ions to free silver, and copper(II) ions to copper(I) ions under prescribed conditions. [27.13]

reduction A decrease in the oxidation number of an element as a result of gaining electrons. [17.2]

rem (roentgen equivalent to man) A unit of radiation-dose equivalent taking into account that the energy absorbed from different sources does not produce the same degree of biological effect. [18.7]

replication The biological process for duplicating the DNA molecule. [31.7]

representative element An element in one of the A groups in the periodic table. [10.6]

resonance structure A molecule or ion that has multiple Lewis structures. *See also* Lewis structure. [11.8]

reversible chemical reaction A chemical reaction in which the products formed react to produce the original reactants. A double arrow is used to indicate that a reaction is reversible. [16.1]

Ribonucleic acid (RNA) A high-molar-mass polymer of nucleotides present in all living matter. Its main function is to direct the synthesis of proteins. RNA nucleotides contain ribose. [31.5]

roentgen (R) A unit of exposure of gamma radiation based on the quantity of ionization produced in air. [18.7]

rounding off numbers The process by which the value of the last digit retained is determined after dropping nonsignificant digits. [2.3]

S

salts Ionic compounds of cations and anions. [Ch. 6, 15.4]

saponification The hydrolysis of an ester by a strong base (NaOH or KOH) to produce an alcohol and a salt (or a soap if the salt formed is from a high-molar-mass acid). [24.10]

saturated compound An organic compound containing only single bonds. [19.2]

saturated hydrocarbon A hydrocarbon that has only single bonds between carbon atoms; classified as *alkanes*. [19.2]

saturated solution A solution containing dissolved solute in equilibrium with undissolved solute. [14.3]

Saytzeff's rule During intermolecular dehydration, if there is a choice of positions for the carbon–carbon double bond, the preferred location is the one that generally gives the more highly substituted alkene—that is, the alkene with the most alkyl groups attached to the double-bond carbons. [22.5]

scientific laws Simple statements of natural phenomena to which no exceptions are known under the given conditions. [1.5]

scientific method A method of solving problems by observation; recording and evaluating data of an experiment; formulating hypotheses and theories to explain the behavior of nature; and devising additional experiments to test the hypotheses and theories to see if they are correct. [1.5]

scientific notation A convenient way of expressing large and small numbers using powers of 10. To write a number as a power of 10, move the decimal point in the original number so that it is located after the first nonzero digit, and follow the new number by a multiplication sign and 10 with an exponent (called its *power*) that is the number of places the decimal point was moved. Example: $2468 = 2.468 \times 10^3$. [2.4]

secondary alcohol An alcohol in which the carbon atom bonded to the —OH group is bonded to two other carbon atoms. [22.2]

secondary structure A protein structure characterized as a regular, three-dimensional shape held together by hydrogen bonding. [29.7]

semipermeable membrane A membrane that allows the passage of water (solvent) molecules through it in either direction but prevents the passage of larger solute molecules or ions. [14.8]

SI An agreed-upon standard system of measurements used by scientists around the world (*Système Internationale*). *See also* metric system. [2.5]

sigma bond A bond in which the electron cloud formed by the pair of bonding electrons lies on a straight line drawn between the nuclei of the bonded atom. [19.7]

significant figures The number of digits that are known plus one estimated digit are considered significant in a measured quantity; also called significant digits. [2.3, Appendix I]

simple protein A protein that yields only amino acids when hydrolyzed. [29.1]

simplest formula *See* empirical formula.

single bond A covalent bond in which one pair of electrons is shared between two atoms. [11.5, 19.2]

single-displacement reaction A reaction of an element and a compound that yields a different element and a different compound. [8.4]

soap A salt of a long-carbon-chain fatty acid. [24.12]

solid A state of matter having a definite shape and a definite volume, whose particles cohere rigidly to one another, so that a solid can be independent of its container. [1.7]

solubility An amount of solute that will dissolve in a specific amount of solvent under stated conditions. [14.2]

solubility product constant (K_{sp}) The equilibrium constant for the solubility of a slightly soluble salt. [16.12]

solute The substance that is dissolved—or the least abundant component—in a solution. [14.1]

solution A system in which one or more substances are homogeneously mixed or dissolved in another substance. [14.1]

solvent The dissolving agent or the most abundant component in a solution. [14.1]

specific gravity The ratio of the density of one substance to the density of another substance taken as a standard. Water is usually the standard for liquids and solids; air, for gases. [2.9]

specific heat The quantity of heat required to change the temperature of 1 g of any substance by 1°C. [4.6]

specific rotation The number of degrees that polarized light is rotated by passing through 1 decimeter (dm) of a solution of the substance at a concentration of 1 g/mL. [26.2]

spectator ion An ion in solution that does not undergo chemical change during a chemical reaction. [15.10]

speed (of a wave) A measurement of how fast a wave travels through space. [10.2]

sphingolipid A compound that, when hydrolyzed, yields a hydrophilic group (either phosphate and chlorine or a carbohydrate), a long-chain fatty acid (18–26 carbons) and sphingosine (an unsaturated amino alcohol). [28.4]

spin A property of an electron that describes its appearance of spinning on an axis like a globe; the electron can spin in only two directions, and, to occupy the same orbital, two electrons must spin in opposite directions. *See also* orbital. [10.4]

standard boiling point *See* normal boiling point.

standard conditions *See* STP.

standard temperature and pressure *See* STP.

steroid A compound that has the steroid nucleus, which consists of four fused carbocyclic rings. [28.5]

stereoisomers Compounds that have the same structural formulas but differ in the spatial arrangement of the atoms. [26.1]

stereoisomerism A type of isomerism in which the isomers have the same structural formulas but differ in the spatial arrangement of atoms. [26.1]

Stock (nomenclature) System A system that uses Roman numerals to name elements that form more than one type of cation. (For example: Fe^{2+}, iron(II); Fe^{3+}, iron(III).) [6.4]

stoichiometry The area of chemistry that deals with the quantitative relationships among reactants and products in a chemical reaction. [9.2]

STP (standard temperature and pressure) 0°C (273 K) and 1 atm (760 torr); also known as standard conditions. [12.8]

strain hypothesis The mode of catalysis in which a reactant molecule is impelled to change shape to fit the binding site. [30.5]

strong electrolyte An electrolyte that is essentially 100% ionized in aqueous solution. [15.7]

subatomic particles Particles found within the atom, mainly protons, neutrons, and electrons. [5.6]

sublimation The process of going directly from the solid state to the vapor state without becoming a liquid. [13.2]

subscript Number that appears partially below the line and to the right of a symbol of an element (example: H_2SO_4). [3.9]

substance Matter that is homogeneous and has a definite, fixed composition; substances occur in two forms—as elements and as compounds. [1.8]

substitution reaction A reaction in which one atom in a molecule is exchanged by another atom or group of atoms. [19.10]

substrate In biochemical reactions, the substrate is the unit acted upon by an enzyme. [30.1]

substrate-level phosphorylation The process whereby energy derived from oxidation is used to form high-energy phosphate bonds on various biochemical molecules. [33.6]

superimposable When one object is placed on another, all parts of both objects coincide exactly. [26.3]

supersaturated solution A solution containing more solute than needed for a saturated solution at a particular temperature. Supersaturated solutions tend to be unstable; jarring the container or dropping in a "seed" crystal will cause crystallization of the excess solute. [14.3]

surface tension The resistance of a liquid to an increase in its surface area. [13.4]

symbol In chemistry, an abbreviation for the name of an element. [3.4]

synthetic detergent (syndet) A cleansing agent, including both synthetic organic products and detergents. [24.12]

system A body of matter under consideration. [1.8]

T

temperature A measure of the intensity of heat, or of how hot or cold a system is; the SI unit is the kelvin (K). [2.8]

tertiary alcohol An alcohol in which the carbon atom bonded to the —OH group is bonded to three other carbon atoms. [22.2]

tertiary structure A protein structure referring to the distinctive and characteristic conformation, or shape, of a protein molecule: an overall three-dimensional formation held together by a variety of interactions between amino acid side chains. [29.7]

tetrahedral structure An arrangement of the VSEPR model where four pairs of electrons are placed 109.5° degrees apart to form a tetrahedron. [11.11, 19.2]

theoretical yield The maximum amount of product that can be produced according to a balanced equation. [9.6]

theory An explanation of the general principles of certain phenomena with considerable evidence to support it; a well-established hypothesis. [1.5]

thermoplastic polymers Polymers that soften on reheating. [21.3]

thermosetting polymers Polymers that set to an infusible solid and do not soften on reheating. [21.3]

thiol The —SH-containing compound, also known as mercaptan. [22.13]

Thomson model of the atom Thomson asserted that atoms are not indivisible, but are composed of smaller parts; they contain both positively and negatively charged particles—protons as well as electrons. [5.6]

titration The process of measuring the volume of one reagent required to react with a measured mass or volume of another reagent. [15.10]

Tollens test The silver-mirror test for aldehydes, based on the ability of silver ions to oxidize aldehydes. [23.4]

torr A unit of pressure (1 torr = 1 mm Hg). [12.3]

total ionic equation An equation that shows compounds in the form in which they actually exist. Strong electrolytes are written as ions in solution, whereas nonelectrolytes, weak electrolytes, precipitates, and gases are written in the un-ionized form. [15.12]

transamination The transfer of an amino group from an α-amino acid to an α-keto acid. [35.8]

transcription The process of forming RNA from DNA. [31.8]

transition elements The metallic elements characterized by increasing numbers of *d* and *f* electrons. These elements are located in Groups 1B–8B and in Group VIII of the periodic table. [10.6]

transition state An unstable structure through which a reactant must pass to be converted into a product with characteristics of both reactant and product. [30.2]

translation The production of a polypeptide using an mRNA template. [31.11]

transmutation The conversion of one element into another element. [18.5]

transuranium element An element that has an atomic number higher than that of uranium (>92). [18.13]

triacylglycerol (triglyceride) An ester of glycerol and three molecules of fatty acids. [24.11]

trigonal planar An arrangement of atoms in the VSEPR model where the three pairs of electrons are placed 120° apart on a flat plane. [11.11]

triple bond A covalent bond in which three pairs of electrons are shared between two atoms. [11.5, 19.2]

tumor-suppressor genes Genes that trigger commands that lead to cell destruction if normal cellular control fails. [31.8]

turnover number The number of molecules an enzyme can react or "turn over" in a given time span. [30.3]

Tyndall effect An intense beam of light is clearly visible when passed through a colloidal dispersion, but is not visible when passed through a true solution. [15.14]

U

unsaturated compound An organic compound in which the molecules possess one or more multiple carbon–carbon bonds. [19.2]

unsaturated hydrocarbon A hydrocarbon whose molecules contain one or more double or triple bonds between two carbon atoms; classified as *alkenes*, *alkynes*, and *aromatic* compounds. [14.4, Ch. 20]

unsaturated solution A solution containing less solute per unit volume than its corresponding saturated solution. [14.6]

urea cycle A cyclic metabolic pathway that creates urea for nitrogen excretion and that takes place in the liver. [35.9]

V

valence electron An electron in the outermost energy level of an atom; these electrons are the ones involved in bonding atoms together to form compounds. [10.5]

valence shell electron pair repulsion (VSEPR) theory A theory for predicting the shape of covalent compounds based on the repulsion of bonded and nonbonded pairs of valence electrons. [11.11]

vapor pressure The pressure exerted by a vapor in equilibrium with its liquid. [13.3]

vaporization *See* evaporation. [13.2]

vapor-pressure curve A graph generated by plotting the temperature of a liquid on the *x*-axis and its vapor pressure on the *y*-axis. Any point on the curve represents an equilibrium between the vapor and liquid. [13.5]

vital-force theory A theory that held that organic substances could originate only from some form of living material. The theory was overthrown early in the 19th century. [19.1]

vitamin A group of naturally occurring organic compounds that are essential for good nutrition and must be supplied in the diet. [32.7]

volatile (substance) A substance that evaporates readily; a liquid with a high vapor pressure and a low boiling point. [13.3]

voltaic cell A cell that produces electric current from a spontaneous chemical reaction. [17.6]

volume The amount of space occupied by matter; measured in SI units by cubic meters (m^3), but also commonly in liters and milliliters. [2.7]

volume percent (solution) The volume of solute in 100 mL of solution. [14.6]

W

water of crystallization Water molecules that are part of a crystalline structure, as in a hydrate; also called water of hydration. [13.9]

water of hydration *See* water of crystallization.

wavelength The distance between consecutive peaks and troughs in a wave; symbolized by the Greek letter lambda. [10.2]

wax Esters of high-molar-mass fatty acids and high-molar-mass alcohols. [28.3]

weak electrolyte A substance that is ionized to a small extent in aqueous solution. [15.7]

weight A measure of the earth's gravitational attraction for a body (object). [2.7]

word equation A statement in words, in equation form, of the substances involved in a chemical reaction. [8.2]

Z

zwitterion Dipolar ion, a common form in which amino acids exist. [29.5]

PHOTO CREDITS

Chapter 1
Page 1: Yuriko Nakao/Reuters//Corbis Images. Page 3: J. David Andrews/Masterfile. Page 4: Michael Rubner. Page 5 (top): Bob Krist/Corbis Images. Page 5 (bottom): Elena Korenbaum/iStockphoto. Page 7: Charles D. Winters/Photo Researchers, Inc. Page 9 (left): Richard Megna/Fundamental Photographs. Page 9 (right): Richard Megna/Fundamental Photographs. Page 10 (top): Paul Silverman/Fundamental Photographs. Page 10 (left): Rita Amaya. Page 10 (right): Rita Amaya. Page 12: Richard Megna/Fundamental Photographs. Page 13 (left): Richard Megna/Fundamental Photographs. Page 13 (right): Mark A. Schneider/Visuals Unlimited.

Chapter 2
Page 14: Photodisc Blue/Getty Images, Inc. Page 16 (left): SKA Archives. Page 16 (right): David M. Phillips/Visuals Unlimited. Page 23: ExLibris. Page 29: JPL/ NASA. Page 30 (left): Rita Amaya. Page 30 (right): Rita Amaya. Page 33: Official contribution of the National Institute of Standards and Technology, not subject to copyright in the United States. Sissy Riley, Information Services division/NIST (National Inst of Standards and Technology). Page 36: Poterfield/Chickoring/Photo Researchers, Inc.

Chapter 3
Page 47: John McAnulty/Corbis Images. Page 47 (right): Jack Plekan/Fundamental Photographs. Page 48: Photodisc/Getty Images, Inc. Page 49 (left): Richard Megna/ Fundamental Photographs. Page 49 (right): Richard Megna/ Fundamental Photographs. Page 50 (top): NOAO/ AURA/NSF/Photo Researchers, Inc. Page 50 (center left): NASA Earth Observatory. Page 50 (bottom): Photodisc/ Getty Images, Inc. Page 51 (top): Tom Pantages. Page 51 (bottom): Andrew Lambert Photography/Photo Researchers, Inc. Page 52: PaulSilverman/Fundamental Photographs. Page 53: Photodisc/Getty Images, Inc. Page 55: The Andy Warhol Foundation/Art Resource. Page 58 (top): Courtesy DaimlerChrysler Corporation. Page 58 (bottom left): Christoph Ermel/iStockphoto. Page 58 (bottom center): Andrew Lambert Photography/Photo Researchers, Inc. Page 58 (bottom right): Charles D. Winters/Photo Researchers, Inc.

Chapter 4
Page 65: Charles Rotkin/Corbis Images. Page 67: Corbis Digital Stock. Page 68 (left): Tom Pantages. Page 68 (center): Tom Pantages. Page 68 (right): Tom Pantages. Page 69: Charles D. Winters/Photo Researchers, Inc. Page 71 (top): Derrick Ditchburn/Visuals Unlimited. Page 71 (bottom): Yuri Gripas/Reuters/Landov LLC. Page 72: Brian Yaruin/The Image Works. Page 74 (left): Jaques Jangoux/Visuals Unlimited. Page 75: National Magazines/Retna.

Chapter 5
Page 84: Byron Aughenbaugh/AFP/Getty Images. Page 88: Richard Megna/Fundamental Photographs. Page 89: Courtesy of IBM Corporation. Page 89 (left): Tom Pantages. Page 89 (center): Richard Megna/Fundamental Photographs. Page 89 (right): Larry Stepanowicz/Fundamental Photographs. Page 94: PhotoDisc, Inc. Page 95: Roger Weller. Page 96: Lester Lefkowitz/Corbis Images.

Chapter 6
Page 103: Antonio Rosario/The Image Bank/Getty Images. Page 106 (top left): The Image Works. Page 106 (top right): Bettmann/Corbis Images. Page 106 (center left): Bettmann/Corbis Images. Page 106 (right): Hulton/Getty Images, Inc. Page 106 (bottom): Science Source/Photo Researchers, Inc. Page 109: Ken Karp. Page 117: Ken Karp. Page 127: Stockbyte Gold/AFP/Getty Images. Page 128: PhotoDisc, Inc./Getty Images.

Chapter 7
Page 129 (left): Royalty-Free/Corbis Images. Page 129 (center): Copyright John Wiley & Sons, Inc. Page 129 (right): Tom Pantages. Page 133: Tom Pantages. Page 139: Copyright John Wiley & Sons, Inc. Page 144: Tony Freeman/PhotoEdit.

Chapter 8
Page 150: Ken Karp. Page 157: Mark Harmel/Photo Researchers. Page 159: Charles D. Winters/Photo Researchers. Page 160: Richard Megna/Fundamental Photographs. Page 160: Charles D. Winters/Photo Researchers. Page 162: Richard Megna/Fundamental Photographs. Page 163: Richard Megna/Fundamental Photographs. Page 164 (top left): Michael Watson. Page 164 (top right): Michael Watson. Page 164 (bottom left): Michael Watson. Page 165 (left): Hannelie Coetzee/Masterfile. Page 165 (right): Paul Damien/National Geographic Society. Page 166: Richard Megna/Fundamental Photographs. Page 167 (top): Phil Degginger/Alamy Limited. Page 167 (bottom): Tom Pantages.

Chapter 9
Page 176: Photodisc Blue/Getty Images, Inc. Page 177: Richard Megna/Fundamental Photographs. Page 186: Michael Knight/iStockphoto.

Chapter 10
Page 203: Carl & Ann Purcell/Corbis Images. Page 204: Tony Freeman/PhotoEdit. Page 205 (left): Courtesy of NJ Marshall, University of Queensland. Page 205 (right): Courtesy of NJ Marshall, University of Queensland. Page 210: Clayton Hansen/iStockphoto.

Chapter 11
Page 224: Adam Woolfit/Corbis Images. Page 232: Tom Pantages. Page 239: Tom Pantages. Page 243: Courtesy IBM Almaden Research Center. Page 243: Bharat Bhushan, Ohio State University. Page 248: Tom McCarthy. Page 255: Dwayne Newton/PhotoEdit.

Chapter 12
Page 265: SUPERSTOCK. Page 266: Tom Pantages. Page 277 (left): Rita Amaya. Page 277 (center): Rita Amaya. Page 277 (right): Rita Amaya. Page 284: Jean-Paul Chassenet/Photo Researchers. Page 287: Tom Pantages. Page 297 (top): Wesley Bocxe/Photo Researchers. Page 297 (bottom): NASA Earth Observatory. Page 307: S.R. Maglione/ Photo Researchers. Page 311: Charles D. Winters /Photo Researchers.

Chapter 13
Page 312 (right): Nature's Images/Photo Researchers. Page 312: Yoav levy/Phototake. Page 312: Tom Pantages. Page 320: John Pinkston/Laura Stern//USGS. Page 321 (left): Richard Megna/Fundamental Photographs. Page 321 (right): Richard Megna/Fundamental Photographs. Page 324: Richard Megna/Fundamental Photographs. Page 327: Dana Bartekoske/iStockphoto. Page 328: Mitch Kaufman/ Courtesy of Southern California Edison.

Chapter 14
Page 336: Corbis Digital Stock. Page 338 (top): Frederic Mikulec/Advanced Materials, 2002, 14, No. 1 Jan. 4, (Fig. 1). By permission of Wiley-VCH. Page 338: Richard Megna/Fundamental Photographs. Page 339: Kip Peticolas/Fundamental Photographs. Page 342: Ex Libris. Page 343: Richard Megna/Fundamental Photographs. Page 356: David Young-Woff/PhotoEdit. Page 357: Craig Newbauer/Peter Arnold, Inc. Page 359: Christine Balderas/ iStockphoto. Page 360 (left): Dennis Kunkel/Phototake. Page 360 (center): Dennis Kunkel/Phototake. Page 360 (right): Dennis Kunkel/Phototake.

Chapter 15
Page 373: Burke/Triolo Productions/Jupiter Images Corp. Page 379: Prof. P. Motta/Dept. of Anatomy/University "La Sapienza", Rome/Photo Researchers. Page 380: Charles C. Place/The Image Bank/Getty Images. Page 381 (left): Tom Pantages. Page 381 (center): Tom Pantages. Page 381 (right): Tom Pantages. Page 381 (bottom): Leonard Lessin/Peter Arnold, Inc. Page 389 (top): Courtesy of Wright Patterson Air Force Base. Page 390 (left): Richard Megna/Fundamental Photographs. Page 390 (center): Richard Megna/Fundamental Photographs. Page 390 (right): Fundamental Photographs. Page 395: John Kaprielian/Photo Researchers. Page 397: Courtesy of Karmann, Inc. Page 398: Kip Peticolas/Fundamental Photographs.

Chapter 16
Page 406: Jeff Hunter/The Image Bank/Getty Images. Page 407 (left): Richard Megna/Fundamental Photographs. Page 407 (right): Richard Megna/Fundamental Photographs. Page 409: Courtesy Frederick C. Eichmiller, ADA Paffenberger Research Center. Page 413: Grant Heilman Photography. Page 415: Richard Megna/Fundamental Photographs. Page 427: Charles D. Wiinters/Photo Researchers. Page 428: Mark E. Gibson/Corbis Images.

Chapter 17
Page 437: Roger Ressmeyer/Corbis Images. Page 439: Tom Pantages. Page 442: Tom Pantages. Page 447 (left): Courtesy Transitional Optical, Inc. Page 447 (right): Courtesy Transitions Optical, Inc. Page 450: Peticolas/Megna/Fundamental Photographs. Page 453: James L. Amos/Corbis Images. Page 456: Charles D. Winters/Photo Researchers.

Chapter 18
Page 468: John Gerlach/Visuals Unlimited. Page 469: Eric Schrempp/Photo Researchers. Page 478: Courtesy Stanford Linear Accelerator Center. Page 479: Rennie Van Munchow/Phototake. Page 485: SKA Archive. Page 486: Visuals Unlimited. Page 489: Hank Morgan/Photo Researchers.

Chapter 19
Page 496: Science VU/NASA/ARC/Visuals Unlimited. Page 498: Pixtal/Age Fotostock America, Inc. Page 502 (left): Ken Graham/Age Fotostock America, Inc. Page 502 (center): Larry Mangino/The Image Works. Page 502 (right): Chris Knapton/Photo Researchers. Page 503: BE&W agencja fotograficzna Sp. z o.o./Alamy. Page 517: Lowell Georgia/Photo Researchers. Page 523: J.L. Bohin/Photo Researchers. Page 524: John Gress/Reuters/Corbis. Page 525 (top): Clarence A. Rechenthin. Courtesy of USDA NRCS Texas State Office. Page 525 (bottom): Robert H. Mohlenbrock. USDA SCS. 1991. Southern wetland Flora: Technical Center, Fort Worth, TX. Courtesy of USDA NRCS Wetland Science Institute. Page 528 (bottom left): NASA/JPL/Space Science Institute. Page 528 (top): NASA/JPL. Page 528 (bottom right): NASA/JPL/ESA/University of Arizona.

Chapter 20
Page 537: Robb Kendrick/Aurora Photos. Page 541: Butch Martin/Alamy. Page 543: Nic Miller/Alamy. Page 546: Steve Horrell/Photo Researchers, Inc. Page 549: Jeff Smtih/Image State. Page 550: Diccon Alexander/Photo Researchers, Inc. Page 553: Wiley Image Archive. Page 560: Andy Cox/Stone/Getty Images. Page 565: Kevin Dodge/Masterfile. Page 566: ¸AP/Wide World Photos.

Chapter 21
Page 579: Jerry Miievoi/Age Fotostock America, Inc. Page 580: AFP/Getty Images. Page 581: Courtesy of International Business Machines Corporation. Page 582: PhotoDisc, Inc./Getty Images. Page 585 (left): Richard T. Nowitz/Corbis Images. Page 585 (right): Tony Freeman/PhotoEdit. Page 586: Peter Walton//Index Stock. Page 588: Tek Image/Photo Researchers, Inc.

Chapter 22
Page 589: geophotos/Alamy. Page 598: Alvis Uptitis/The Image Bank/Getty Images. Page 602: Oleksiv Maksymenko/

Alamy. Page 608 (left): Courtesy of Jim Maynard and Jerry Jacobsen/University of Wisconsin Department of Chemistry. Page 608 (right): Courtesy of Jim Maynard and Jerry Jacobsen/University of Wisconsin Department of Chemistry. Page 610: Stockdisc/Age Fotostock. Page 616: Billy Hustace/Stone/Getty Images. Page 617: Pierre Arsenault/Masterfile. Page 620: Phil Schermeister/Stone/Getty Images. Page 621: Mauritius GMBH/Phototake.

Chapter 23
Page 635: John Mahar/Stock Boston. Page 638: Kurt Wittman/Omni-Photo Communications. Page 642: Robert Brenner/PhotoEdit. Page 644: Andrew Lambert Photography/Science Photo Library/Photo Researchers, Inc. Page 647: Alamy Images. Page 651: ITAR-TASS/Yuri Mashkov/NewsCom.

Chapter 24
Page 668: Alain Altair/The Image Bank/Getty Images. Page 670: Anthony Bannister/Corbis Images. Page 674: Croydon Studio/Age Fotostock. Page 676: Kindra Clineff//Index Stock. Page 678: Steven Dunwell//Index Stock. Page 684: Alamy Images. Page 687: PhotoDisc, Inc./Getty Images. Page 693: Thierry Dosogne/The Image Bank/Getty Images.

Chapter 25
Page 705: Paolo Curto/The Image Bank/Getty Images. Page 711 (top): John Madere/Corbis Images. Page 711 (bottom): Lloyd Sutton/Masterfile. Page 715: PhotoDisc, Inc./Getty Images. Page 721: Getty Images News and Sport Services. Page 723: Du Boisberranger Jean/Age Fotostock.

Chapter 26
Page 737: Paul Morris/Corbis Images. Page 740 (top): Rino Gropuzzo/Marka/Age Fotostock. Page 740 (bottom): J.A. Kraulis/Masterfile. Page 741 (top): Julie Kranhold/Ex Libris. Page 741 (bottom left): PhotoDisc/Getty Images. Page 741 (bottom left): Maximilian Weinzierl/Alamy. Page 741 (bottom left): PhotoDisc/Getty Images. Page 746: Mark Burnside/Stone/Getty Images.

Chapter 27
Page 762: Will & Deni McIntyre/Stone/Getty Images. Page 764 (left): Kristen Brochmann/Fundamental Photographs. Page 764 (right): Kristen Brochmann/Fundamental Photographs. Page 765: Jeremy Horner/Corbis Images. Page 767: Lester V. Bergman/Corbis Images. Page 776: Royalty-Free/Corbis Images. Page 778: Brand X Pictures/Age Fotostock. Page 779: Jaume Gual/Age Fotostock America, Inc. Page 783: PhotoDisc/Fotosearch, LLC. Page 786: Ron Chapple/Taxi/Getty Images. Page 787 (top): Richard Megna/Fundamental Photographs. Page 787 (bottom): ¸2003 Cygnus, Inc. Page 789: Pascal Goetgheluck/Photo Researchers. Page 791 (top): Dr. Gopal Murti/Photo Researchers, Inc. Page 791 (bottom): Biophoto Associates/Photo Researchers. Page 793: Susan Leavines/Photo Researchers, Inc.

Chapter 28
Page 802: Daniel J. Cox/Stone/Getty Images. Page 804: Ingo Arndt/Minden Pictures/Getty Images. Page 806: ISM/Phototake. Page 808: David Hiser/Stone/Getty Images. Page 809: Michael Alberstat/Masterfile. Page 810: PhotoDisc, Inc./Getty Images. Page 812: Elleringmann/laif/Aurora Photos. Page 817: Cathy Melloan/PhotoEdit. Page 819: ¸imagingbody.com.

Chapter 29
Page 832: Joseph De Sciose/Aurora Photos. Page 834: Manfred Kage/Peter Arnold, Inc. Page 835: Color Day Production/The Image Bank/Getty Images. Page 836: Alamy Images. Page 847: Cheryl Maeder/Corbis Images. Page 848: Brand X Pictures/Getty Images. Page 855 (top right): Stanley Fleger/Visuals Unlimited. Page 855 (bottom right): Stanley Flegler/Visuals Unlimited. Page 857 (left): Julie Kranhold/Exlibris. Page 857 (center): Julie Kranhold/Exlibris. Page 857 (right): Julie Kranhold/Exlibris. Page 858: Science Source/Photo Researchers, Inc. Page 861: S. Fraser/Photo Researchers, Inc. Page 862: Courtesy of Patrick O'Farrell/UCSF. Page 866: CNRI/Phototake.

Chapter 30
Page 874 (top left): Will & Deni McIntyre/Science Photo Library/Photo Researchers, Inc. Page 874 (center): Daniel Acker/Bloomberg News/Landov. Page 874 (bottom right): King Pharmaceuticals/Bloomberg News/Landov LLC. Page 877: Woodfin Camp & Associates. Page 881: Courtesy of Hunt, Rook, and Scheibel. Page 882: David Young-Wolff/PhotoEdit.

Chapter 31
Page 894: Robert P. Carr/Bruce Coleman, Inc. Page 900: Fotopic/Omni-Photo Communications. Page 903: A. Barrington Brown/Photo Researchers. Page 908: Frank Boellman/(c) Science Photo Library. Page 911 (top): Leonard Lessin/Peter Arnold, Inc. Page 911 (bottom): ¸AP/Wide World Photos. Page 912: Dr. Elena Kiseleva/Science Photo Library/Photo Researchers, Inc. Page 918: Charles Thatcher/Stone/Getty Images. Page 923 (top): Peter Menzel/Stock Boston. Page 923 (bottom): Gary Wagner/Stock Boston. Page 924 (left): Courtesy Dr. Richard Jorgensen, Dept. of Plant Sciences, University of Arizona. Page 924 (right): Courtesy Dr. Richard Jorgensen, Dept. of Plant Sciences, University of Arizona.

Chapter 32
Page 935: Peter Tobia/Philadelphia Inquirer/KRT/NewsCom. Page 939: Comstock Images. Page 942 (top): Betty Press/Woodfin Camp & Associates. Page 942 (bottom): Triolo Burke/Foodpix/Jupiter Images. Page 947: Mark Downey/Stock Boston. Page 951: Courtesy USDA. Page 952: Courtesy Julie Kranhold. Page 957: Fawcett/Hirokawa/Heuser/Science Source/Photo Researchers. Page 958: Dennis Kunkel/Phototake.

Chapter 33
Page 964: Taxi/Getty Images. Page 965: Chad Ehlers//Photo Network. Page 973: K.R. Porter/Science Source/Photo Researchers. Page 976: Keith Kent/Photo Researchers, Inc. Page 977 (left): S. Cinti "The Adipose Organ" Kurtis , Milan

1999. Page 977 (right): S. Cinti "The Adipose Organ" Kurtis, Milan 1999. Page 978: Ed Reschke/Peter Arnold, Inc. Page 979: Kevin Morris/Stone/Getty Images.

Chapter 34

Page 984: Stephen Wilkes/The Image Bank/Getty Images. Page 986: Dr. Donald Nicholson. Page 987: Kwame Zikomo/SUPERSTOCK. Page 997: Peter Southwick/Stock Boston. Page 999: Dennis Kunkel/Phototake. Page 1002: Courtesy of Medtronic MiniMed. Page 1003: Dennis MacDonald/Age Fotostock America, Inc.

Chapter 35

Page 1008: Tom Brakefield/Bruce Coleman, Inc. Page 1012: William Leaman/Alamy. Page 1013: CNRI/Phototake. Page 1016: Chris Hondros/Getty/NewsCom. Page 1018: CNRI/Visuals Unlimited. Page 1023: Stockbyte/Age Fotostock America, Inc.

INDEX

Q

R

Names, Formulas, and Charges of Common Ions

Positive Ions (Cations)

Charge	Name	Formula
1+	Ammonium	NH_4^+
	Copper(I) (Cuprous)	Cu^+
	Hydrogen	H^+
	Potassium	K^+
	Silver	Ag^+
	Sodium	Na^+
2+	Barium	Ba^{2+}
	Cadmium	Cd^{2+}
	Calcium	Ca^{2+}
	Cobalt(II)	Co^{2+}
	Copper(II) (Cupric)	Cu^{2+}
	Iron(II) (Ferrous)	Fe^{2+}
	Lead(II)	Pb^{2+}
	Magnesium	Mg^{2+}
	Manganese(II)	Mn^{2+}
	Mercury(II) (Mercuric)	Hg^{2+}
	Nickel(II)	Ni^{2+}
	Tin(II) (Stannous)	Sn^{2+}
	Zinc	Zn^{2+}
3+	Aluminum	Al^{3+}
	Antimony(III)	Sb^{3+}
	Arsenic(III)	As^{3+}
	Bismuth(III)	Bi^{3+}
	Chromium(III)	Cr^{3+}
	Iron(III) (Ferric)	Fe^{3+}
	Titanium(III) (Titanous)	Ti^{3+}
4+	Manganese(IV)	Mn^{4+}
	Tin(IV) (Stannic)	Sn^{4+}
	Titanium(IV) (Titanic)	Ti^{4+}
5+	Antimony(V)	Sb^{5+}
	Arsenic(V)	As^{5+}

Negative Ions (Anions)

Charge	Name	Formula
1−	Acetate	$C_2H_3O_2^-$
	Bromate	BrO_3^-
	Bromide	Br^-
	Chlorate	ClO_3^-
	Chloride	Cl^-
	Chlorite	ClO_2^-
	Cyanide	CN^-
	Fluoride	F^-
	Hydride	H^-
	Hydrogen carbonate (Bicarbonate)	HCO_3^-
	Hydrogen sulfate (Bisulfate)	HSO_4^-
	Hydrogen sulfite (Bisulfite)	HSO_3^-
	Hydroxide	OH^-
	Hypochlorite	ClO^-
	Iodate	IO_3^-
	Iodide	I^-
	Nitrate	NO_3^-
	Nitrite	NO_2^-
	Perchlorate	ClO_4^-
	Permanganate	MnO_4^-
	Thiocyanate	SCN^-
2−	Carbonate	CO_3^{2-}
	Chromate	CrO_4^{2-}
	Dichromate	$Cr_2O_7^{2-}$
	Oxalate	$C_2O_4^{2-}$
	Oxide	O^{2-}
	Peroxide	O_2^{2-}
	Silicate	SiO_3^{2-}
	Sulfate	SO_4^{2-}
	Sulfide	S^{2-}
	Sulfite	SO_3^{2-}
3−	Arsenate	AsO_4^{3-}
	Borate	BO_3^{3-}
	Phosphate	PO_4^{3-}
	Phosphide	P^{3-}
	Phosphite	PO_3^{3-}